The Quadratic Formula

The roots of $ax^2 + bx + c = 0$ are

$$x = \frac{-b \pm \sqrt{b^2 - 4ac}}{2a}$$

Factorials

$$n! = n \times (n - 1) \times \ldots \times 3 \times 2 \times 1, \qquad \binom{n}{k} = \frac{n!}{(k!(n - k)!}$$

The Binomial Theorem

$$(a + b)^n = a^n + \binom{n}{1} a^{n-1}b + \binom{n}{2} a^{n-2}b^2 + \ldots + \binom{n}{k} a^{n-k}b^k + \ldots + \binom{n}{n-1} ab^{n-1} + b^n$$

Trigonometric and Hyperbolic Identities

$$\cos^2 \theta + \sin^2 \theta = 1, \qquad 1 + \tan^2 \theta = \sec^2 \theta, \qquad \cosh^2 x - \sinh^2 x = 1$$

$$\cos^2 \theta = \frac{1 + \cos(2\theta)}{2}, \qquad \sin^2 \theta = \frac{1 - \cos(2\theta)}{2}$$

$$\cos^2 \frac{\theta}{2} = \frac{1 + \cos \theta}{2}, \qquad \sin^2 \frac{\theta}{2} = \frac{1 - \cos \theta}{2}$$

$$\sin(\theta + \phi) = \cos \theta \sin \phi + \sin \theta \cos \phi, \qquad \cos(\theta + \phi) = \cos \theta \cos \phi - \sin \theta \sin \phi$$

$$\sin 2\theta = 2 \cos \theta \sin \theta$$

$$\cos 2\theta = \cos^2 \theta - \sin^2 \theta = 2 \cos^2 \theta - 1 = 1 - 2 \sin^2 \theta$$

$$\sin(-\theta) = -\sin \theta \qquad \cos(-\theta) = \cos \theta$$

$$\sin\left(\frac{\pi}{2} - \theta\right) = \cos \theta \qquad \cos\left(\frac{\pi}{2} - \theta\right) = \sin \theta \qquad \tan\left(\frac{\pi}{2} - \theta\right) = \cot \theta$$

$$\sin(\pi - \theta) = \sin \theta \qquad \cos(\pi - \theta) = -\cos \theta \qquad \tan(\pi - \theta) = -\tan \theta$$

$$\sin(\pi + \theta) = -\sin \theta \qquad \cos(\pi + \theta) = -\cos \theta \qquad \tan(\pi + \theta) = \tan \theta$$

(Continued inside back cover)

Calculus *with Analytic Geometry*

Calculus with *Analytic Geometry*

Joe Repka

University of Toronto

Wm. C. Brown Publishers

Dubuque, Iowa • Melbourne, Australia • Oxford, England

Book Team

Editor *Paula Christy-Heighton*
Developmental Editor *Theresa Grutz*
Publishing Services Specialist—Production *Julie Avery Kennedy*

 Wm. C. Brown Publishers
A Division of Wm. C. Brown Communications, Inc.

Vice President and General Manager *Beverly Kolz*
Vice President, Director of Sales and Marketing *Virginia S. Moffat*
Marketing Manager *Julie Keck*
Advertising Manager *Janelle Keeffer*
Director of Production *Colleen A. Yonda*
Publishing Services Manager *Karen J. Slaght*

 Wm. C. Brown Communications, Inc.

President and Chief Executive Officer *G. Franklin Lewis*
Corporate Vice President, President of WCB Manufacturing *Roger Meyer*
Vice President and Chief Financial Officer *Robert Chesterman*

Cover image © Tony Stone Worldwide/Stephen Johnson
   The cover shows a computer-generated image of part of the Mandelbrot set.
This set is defined in terms of a very simple formula, but it produces pictures of
astounding variety and complexity. Each part of the picture reveals additional
structure at higher and higher magnifications.
   The Mandelbrot set is a key element in the study of Fractal Geometry, which
is part of a larger field known as Chaos Theory. Many structures that occur in
nature are too complicated to predict in detail but have overall behavior that can
usefully be discussed in terms of Chaos Theory.
   This approach is giving new insight into a wide range of problems, ranging
from weather patterns to the mysteries of the brain.

Cover and interior design by York Production Services, Inc.

Copyediting, permissions, and production by York Production Services, Inc.

Illustrations by Debby Cohen

The credits section for this book begins on page 1321 and is considered an
extension of the copyright page.

. . . for Debby,

*whose patience and help
and love make practically
anything possible . . .*

Brief Contents

Calculus with Analytic Geometry presents all chapters.
Single Variable Calculus presents Chapters 1 through 13.

1   Preliminaries   *1*

2   Limits and Continuity   *62*

3   Derivatives   *128*

4   Applications of Derivatives   *227*

5   Integration   *346*

6   Applications of Integration   *413*

7   Logarithms and Exponential Functions   *509*

8   Techniques of Integration   *560*

9   Taylor's Formula   *614*

10   Series   *637*

11   Conics   *725*

12   Polar Coordinates   *755*

13   Curves in the Plane   *796*

14   Vector Geometry   *824*

15   Curves and Surfaces in Space   *871*

16   Partial Derivatives   *928*

17   Multiple Integrals   *1053*

18   Vector Calculus   *1141*

19   Differential Equations   *1211*

Detailed Contents

Calculus with Analytic Geometry presents all chapters.
Single Variable Calculus presents Chapters 1 through 13.

1 *Preliminaries* 1

Introduction		**1**
1.1	**The Real Numbers**	**2**
	Scientific Notation	3
1.2	**Inequalities**	**4**
	Intervals	6
1.3	**Polynomials and Rational Functions**	**7**
	Factoring	8
	Completing the Square and the Quadratic Formula	9
	The Factor Theorem	11
	Factorials and the Binomial Theorem	14
	Rational Functions	16
1.4	**Functions and Graphs**	**17**
	Graphing Functions	19
1.5	**Geometry**	**22**
	Similar Triangles	24
	Straight Lines	25
	Graphing Inequalities	29
	Areas	30
1.6	**Graphing Conics**	**32**
	Parabolas	33
	Ellipses	36
	Hyperbolas	37
1.7	**Trigonometric Functions**	**43**
	Radian Measure	45
	Special Values of Trigonometric Functions	46
	Law of Cosines, Addition Formulas	50
	Polar Coordinates	53
1.8	**Logarithms and Exponentials**	**56**
	Logarithms	56
Chapter Summary		**59**
Review Exercises		**60**

2 Limits and Continuity 62

Introduction		**62**
2.1	**Introduction to Limits**	**63**
2.2	**Limits of Polynomials and Rational Functions**	**70**
2.3	**Working with Limits**	**80**
	Limits of Noninteger Powers of x	80
	Limits of Composite Functions	84
	The Squeeze Theorem	86
	The Meaning of Limits	90
2.4	**One-Sided and Infinite Limits and Limits at Infinity**	**92**
	Infinite Limits	97
	Limits at Infinity	100
2.5	**Continuity**	**105**
2.6	**Limits of Trigonometric Functions**	**114**
2.7	**The Precise Definition of Limits**	**119**
Chapter Summary		**125**
Review Exercises		**126**

3 Derivatives 128

Introduction		**128**
3.1	**Introduction to Differentiation**	**129**
3.2	**Differentiation Rules; Derivatives of Polynomials and Rational Functions**	**140**
3.3	**Maximum/Minimum Problems**	**154**
3.4	**Derivatives of Trigonometric Functions**	**166**
3.5	**Derivatives of Logarithms, Exponentials, and Hyperbolic Functions**	**174**
	The Exponential Function	177
	Hyperbolic Functions	179
3.6	**The Chain Rule**	**185**
3.7	**Implicit Differentiation**	**195**
	Implicit Differentiation and Maximum/Minimum Problems	198
	Logarithmic Differentiation	200
3.8	**Derivatives of Inverse Trigonometric and Inverse Hyperbolic Functions**	**202**
	Existence of Inverse Functions	212
	Inverse Hyperbolic Functions	214
3.9	**Second and Higher-Order Derivatives**	**218**
Chapter Summary		**224**
Review Exercises		**225**

4 Applications of Derivatives 227

Introduction		**227**
4.1	**Related Rates**	**228**
4.2	**Mean Value Theorem, Increasing and Decreasing Functions**	**236**
	Rolle's Theorem and the Mean Value Theorem	240

4.3	**Critical Points and Extrema**	**248**
4.4	**The Second Derivative Test and Classifying Extrema**	**254**
4.5	**Curve Sketching and Asymptotes**	**263**
	Asymptotes	268
4.6	**Concavity and Inflection Points; Applications to Curve Sketching**	**276**
	Inflection Points	281
	Cusps	285
4.7	**Applied Maximum/Minimum Problems**	**290**
4.8	**Antiderivatives**	**304**
4.9	**Velocity and Acceleration**	**314**
4.10	**Linear Approximation and Differentials**	**319**
	Differentials	322
4.11	**Newton's Method**	**325**
4.12	**l'Hôpital's Rule**	**331**
4.13	**Marginal Cost, Revenue, and Profit**	**339**
	Chapter Summary	**342**
	Review Exercises	**342**

5 *Integration* 346

	Introduction	**346**
5.1	**The Area Under a Curve**	**347**
5.2	**Summation Notation**	**353**
5.3	**Riemann Sums**	**362**
	The Precise Definition of the Integral	370
5.4	**Properties of Integrals**	**373**
	Integrating Functions with Removable or Jump Discontinuities	379
5.5	**Indefinite Integrals and the Fundamental Theorem of Calculus**	**383**
	Derivative of an Indefinite Integral	385
5.6	**Definite Integrals**	**392**
5.7	**Integration by Substitution**	**398**
	Definite Integrals and Substitution	406
	Chapter Summary	**410**
	Review Exercises	**410**

6 *Applications of Integration* 413

	Introduction	**413**
6.1	**Areas**	**414**
6.2	**Setting Up Integrals; Limits of Integration**	**425**
	Functions with Specified Rate of Change	429
6.3	**Volumes of Solids of Revolution by Slices**	**432**
6.4	**Volumes by Cylindrical Shells**	**439**
	Slices vs Shells	443
6.5	**Average Values**	**446**
6.6	**Work**	**456**
	Gravitational Force and Work	459

6.7   Moments and Centroids   462
6.8   Fluid Pressure and Force   476
      Buoyancy   480
6.9   Arclength; Areas of Surfaces of Revolution   484
      Surface Areas of Surfaces of Revolution   487
6.10 Numerical Integration; The Trapezoidal Rule and
      Simpson's Rule   493
      The Trapezoidal Rule   494
      Simpson's Rule   498
Chapter Summary   505
Review Exercises   506

7 Logarithms and Exponential Functions   509

Introduction   509
7.1   Logarithms   510
7.2   Exponentials and Powers   514
      Powers   516
      Limits of Powers   518
      Logarithms to Other Bases   519
7.3   Exponential Growth and Decay   522
7.4   Examples of Exponential Growth and Decay   529
7.5   Logistic Growth   545
7.6   Differential Equations   551
      Separable Equations   551
      Linear Differential Equations with Constant Coefficients   553
Chapter Summary   558
Review Exercises   558

8 Techniques of Integration   560

Introduction   560
8.1   Integration by Parts   562
8.2   Trigonometric Integrals   569
      Integrals Involving Powers of $\tan x$ and $\sec x$ 573
8.3   Trigonometric Substitutions   578
8.4   Completing the Square in Quadratic Expressions
      in Integrands   585
8.5   Partial Fractions   589
8.6   Rationalizing Substitutions   598
8.7   Additional Applications   601
8.8   Improper Integrals   606
Chapter Summary   611
Review Exercises   612

9 Taylor's Formula   614

Introduction   614
9.1   Taylor's Formula   615

9.2   Taylor's Formula with Remainder; Approximations   **623**
Chapter Summary   **635**
Review Exercises   **635**

10 *Series* *637*

Introduction   **637**
10.1   Sequences   **638**
      The Precise Definition of Limits for Sequences   646
10.2   Geometric Sums and Series and Applications   **648**
      Interest, Mortgages, and Annuities   651
      Geometric Series   654
10.3   Convergence of Series   **658**
      The Harmonic Series   661
10.4   The Integral Test   **666**
      Approximating the Value of a Series   670
10.5   The Comparison Test   **675**
10.6   Alternating Series and Absolute Convergence   **680**
      Absolute Convergence   684
10.7   The Ratio Test and the nth Root Test   **687**
      The nth Root Test   690
10.8   Taylor Series   **695**
      Convergence of Taylor Series to the Original Function   701
10.9   Power Series   **704**
10.10   Derivatives and Integrals of Power Series   **711**
Chapter Summary   **722**
Review Exercises   **722**

11 *Conics* *725*

Introduction   **725**
11.1   Identifying and Graphing Conics   **726**
11.2   Geometry of Conics   **732**
11.3   Rotated Coordinates   **741**
Chapter Summary   **752**
Review Exercises   **753**

12 *Polar Coordinates* *755*

Introduction   **755**
12.1   Polar Coordinates   **756**
12.2   Graphs in Polar Coordinates   **760**
      Slopes of Tangent Lines   768
12.3   Area in Polar Coordinates   **771**
12.4   Conics in Polar Coordinates   **779**
      Circles and Lines   783
12.5   Parametric Equations in Rectangular and Polar Coordinates   **787**
Chapter Summary   **794**
Review Exercises   **794**

13 *Curves in the Plane*   796

Introduction   **796**
13.1   Parametric Equations   **797**
13.2   Vectors and Motion   **803**
         Vectors   804
         Velocity   806
         Arclength   809
13.3   Projectiles   **815**
Chapter Summary   **822**
Review Exercises   **822**

14 *Vector Geometry*   *824*

Introduction   **824**
14.1   Coordinates in Three-Dimensional Space   **825**
         Graphs in Three Dimensions   826
14.2   Vectors, Dot Product, and Cross Product   **830**
         Dot Product and Cross Product   833
14.3   Applications of Vectors   **842**
         Displacements   842
         Velocity   844
         Force and Work   848
         Cross Products   851
14.4   Lines in Three-Dimensional Space   **854**
14.5   Planes in Three-Dimensional Space   **862**
Chapter Summary   **869**
Review Exercises   **869**

15 *Curves and Surfaces in Space*   *871*

Introduction   **871**
15.1   Curves in Space   **872**
         Velocity Vectors   873
15.2   Unit Tangent and Normal Vectors and Arclength   **880**
         Normal Vectors   882
         Arclength   886
         Different Parametrizations   887
         Parametrization by Arclength   888
15.3   Acceleration and Curvature   **892**
         Curvature   894
         Radius of Curvature, Osculating Circle   897
15.4   Quadric Surfaces   **901**
15.5   Cylindrical and Spherical Coordinates   **907**
         Spherical Coordinates   909
15.6   Vectors and Coordinates in n-Dimensional Space   **913**
         Coordinates in n-Dimensional Space   915
         Curves in $\mathbb{R}^n$ 916

15.7   Kepler's Laws of Planetary Motion   920
Chapter Summary   926
Review Exercises   926

16 *Partial Derivatives*   928

Introduction   928
16.1   Functions of Several Variables   929
    Level Curves   932
    Level Surfaces   935
16.2   Limits and Continuity   938
    Functions of Three or More Variables   944
    Continuity   947
16.3   Partial Derivatives   951
    Tangent Planes   955
    Higher-Order Partial Derivatives   958
16.4   Differentials   962
16.5   The Chain Rule   974
    Implicit Differentiation   984
16.6   Directional Derivatives and Gradients   989
    Gradients   994
    Three or More Variables   996
    Gradients and Level Curves   1000
16.7   Tangent Planes   1003
16.8   Maximum/Minimum Problems   1009
    Applied Maximum/Minimum Problems   1018
16.9   The Second Derivative Test   1023
16.10   Lagrange Multipliers   1030
    Lagrange Multipliers with Two Constraints   1036
16.11   Linear Regression and Least Squares Approximation   1040
Chapter Summary   1050
Review Exercises   1050

17 *Multiple Integrals*   1053

Introduction   1053
17.1   Double Integrals   1054
17.2   Setting Up Double Integrals   1064
17.3   Areas and Volumes   1076
17.4   Double Integrals in Polar Coordinates   1086
17.5   Surface Areas   1095
17.6   Triple Integrals   1102
17.7   Moments and Centroids   1109
    Moments of Inertia and Rotational Motion   1113
17.8   Cylindrical and Spherical Coordinates   1120
    Spherical Coordinates   1122
17.9   Change of Variables   1126
Chapter Summary   1137
Review Exercises   1138

18 *Vector Calculus   1141*

Introduction   **1141**
18.1   **Line Integrals**   **1142**
      Line Integrals and Work   1147
18.2   **Path Independence**   **1151**
18.3   **Green's Theorem**   **1164**
      Fluid Flows   1170
18.4   **Surface Integrals**   **1175**
      Surface Integrals and Moments   1178
18.5   **The Divergence Theorem**   **1181**
      Normal Vectors   1183
18.6   **Stokes' Theorem**   **1191**
18.7   **Applications**   **1200**
      Rotation and Curl   1200
      Divergence and Flux Density (the Continuity Equation)   1201
      Gauss' Law   1203
      Gravitational Force and Spheres   1205
Chapter Summary   **1207**
Review Exercises   **1208**

19 *Differential Equations   1211*

Introduction   **1211**
19.1   **Homogeneous Linear Differential Equations with
      Constant Coefficients**   **1212**
19.2   **Complex Numbers and Linear Homogeneous Equations**   **1221**
19.3   **Undetermined Coefficients**   **1228**
19.4   **Vibrations**   **1236**
19.5   **First-Order Differential Equations**   **1247**
      Separable Equations   1247
      Homogeneous First-Order Equations   1251
      Linear First-Order Equations   1253
19.6   **Variation of Parameters**   **1256**
19.7   **Series Solutions**   **1260**
Chapter Summary   **1266**
Review Exercises   **1267**

Appendix   **1269**
Answer Section   **1271**
Index   **1307**

Preface

One of the main difficulties facing the teacher of calculus is the wide range of background preparation and abilities of the students, compounded by the diversity of their interests and goals. To make the subject comprehensible to the weaker students without boring the stronger ones is a difficult juggling act.

This text addresses the dilemma by presenting the material in an intuitive and reasonable way. The style is friendly without compromising accuracy and detail; the material is very accessible but has enough depth to challenge even the best students. Fine points and pitfalls are discussed in remarks in the text and are often elaborated in some of the later exercises at the end of a section; tricky computational steps and other minor difficulties are explained by notes in the margin without interrupting the exposition.

The presentation is enlivened by large numbers of examples drawn from many areas in the physical, life, and social sciences, and from business and finance. Most sections end with a "Point to Ponder," a discussion of some idea usually considered to be beyond the scope of an introductory text. Some involve technical subtleties, some involve extensions and generalizations, and others offer a glimpse into more advanced topics or other disciplines.

Many results are given complete proofs, but some proofs are omitted, especially when the level of complexity or difficulty limits their pedagogical value. In other cases, a proof is given for a special case, in order to convey the ideas without bogging down. Theoretical concepts are often discussed through examples and exercises rather than in dense technical language. Omissions are clearly acknowledged and proofs are not finessed.

Problem-solving strategies are discussed and conveniently summarized, but students are discouraged from memorizing cookbook approaches. Considerable attention is given to interpreting results, to warning the reader about complications and potential errors, and to considering shortcomings and limitations. An important part of the experience of learning calculus ought to be learning to analyze problems and to think about them in quantitative terms.

Problem sets are divided into a section of fairly routine and mechanical drill questions followed by a section of more varied exercises. These are not necessarily particularly difficult, but they are more interesting and usually require more thought. Students are encouraged not only to acquire a certain level of basic calculus skills, but also to develop simultaneously an ability to go beyond the rote and apply their skills effectively and critically.

A number of applications are given more sophisticated treatment than usual. For instance, Newton's method inspires a simple introduction to numerical analysis at a nontechnical level. Another example is a brief discussion of projectile motion using only single-variable techniques (simply by breaking the motion into horizontal and vertical components, without ever mentioning the word "vector"). This is done long before vector calculus and curves are available (though of course it makes a second appearance at that stage). In Section 7.5 there is a discussion of logistic growth, which may have more practical use than exponential growth, especially for students in the life and social sciences. This topic is not found in most other calculus texts.

Care has been taken wherever possible to avoid having later sections depend on earlier ones. Students do not usually read a calculus book through; they tend to "flip around," looking for what they need. As much as possible, the discussion in each section is self-contained. Though many of the examples in calculus relate to physics, there is an effort to keep physics intuition out of the mainstream of the exposition, to avoid alienating the students who have no physics background.

Graphics Calculators and Computers

One of the most exciting and challenging developments in mathematical education is the recent availability of graphics calculators and computers. Initially heralded as the breakthrough that would revolutionize the teaching of calculus, they have proved a little more difficult than expected to apply intelligently and effectively.

Some mathematics teachers have experienced considerable pressure to jump onto the technology bandwagon without adequate thought and preparation. Enthusiastic early claims have been pounced upon by university administrators who typically know very little about mathematics but are dazzled by anything that offers a solution to what they like to regard as "the calculus problem."

The situation is perhaps typified by "curve-sketching." Students in a traditional calculus course spend a lot of time sketching graphs. Computers and graphics calculators can now do this quickly and accurately, so why not just give the student a machine and save all that time and frustration?

The answer involves a question: Why do we have students sketch graphs? In fact, it is not particularly because we want to be able to look at the shape of the curve. Drawing graphs is an excellent way to practice most of the techniques and concepts of differential calculus. Working out a large number of examples consolidates these difficult ideas and helps the student acquire the intuition that is essential for the sophisticated use of calculus for max/min and other problems.

Eliminating this process is not necessarily a good thing. On the other hand, it is certainly possible to use the technology in concert with the traditional approach, in ways that enhance both.

A major downside to calculators and computers is the considerable amount of time required for students to learn how to use them, time *not* spent learning calculus. Another problem with in-course use of calculators and computers is that there are parts of the subject to which they are well suited and other parts where finding an appropriate use is much more problematical. But if a course is designed around expensive equipment that has been purchased by the students or by the institution, it is hard to justify letting it sit idle for substantial parts of the term. On the other hand, nothing turns students off faster than contrived examples that have been concocted just for the sake of having something.

Perhaps the single most important thing about calculators and computers is that they catch the students' attention and interest. *Anything* that does that can't be all bad. This text takes advantage of this effect by including graphics calculator or computer examples in most sections. These examples are designed to show how calculators can enhance our capabilities and in particular to illustrate some types of difficult problems that become tractable with a little help from our electronic friends.

Each example includes complete keystrokes for four of the most popular brands of graphics calculators, so that even a novice can quickly follow along and see how it works. Many include suggestions for modifications and related questions, to draw the reader into more creative use of the equipment. Even the instructor who has no computing experience can easily use the text examples to inject a little pizzazz into the classroom. The rest of the text does not depend on these examples; instructors can use them as much or as little as they want. This feature provides the perfect entry for the instructor who is interested in integrating calculators or computers into a course but who is not able to undertake the enormous task of structuring a course around them.

A few examples use a personal computer for tasks that are not possible on the graphics calculator.

Most sections end with a few exercises to be done with graphics calculators or computers. They are offered without much guidance, so in this form they are probably for the stronger students, preferably ones with some computing experience. Of course, an instructor can always include some classroom discussion or practice to help students approach these exercises.

The text will be accompanied by *The Calculator View*, by John Paulling, a more detailed supplement on the use of graphics calculators.

Calculator Example Keystrokes

The graphics calculator keystrokes are color-coded to match the keys on the calculators. In some cases light green is used to distinguish explanatory symbols from actual keystrokes. For instance, if the required key is a function key F5, and this appears with an on-screen menu item GRAPH, the keystroke F5 will be followed by (GRAPH) in parentheses and colored green. The hope is to make the code more comprehensible.

The keystrokes prepared on the Texas Instruments TI-85 will mostly work on the TI-81, either verbatim or with minor adjustments, but there are some exceptions.

Similar remarks apply to the SHARP EL-9300 and the EL-9200. In some places, the minus key − (i.e., "subtract") has to be replaced by (−) (i.e., "negative") at the beginning of expressions. Also for parametric and polar plots it may be necessary to rearrange the order of the nine RANGE parameters.

The CASIO fx-6300G is an inexpensive smaller version of the fx-7700GB, and they are quite similar, but there are some differences. A few keys have different colors on the two versions, so the printed color will not always be right for everybody. A few keys are actually different, and we have attempted to point this out. There are also a few things that the fx-7700GB can do that the fx-6300G cannot.

The Hewlett Packard HP48SX is in some ways the most sophisticated of the four types used here, but by the same token it is harder to learn and use. Programs that use the full power of the HP48SX can be nearly impenetrable to the novice, so I have opted for key sequences that are perhaps a little naive but easier to unwind. One of the main purposes of the graphics calculator examples is to make it easy for readers to get started with calculators, and I think it makes sense to sacrifice some efficiency and elegance in order to have keystrokes that can easily be modified to do related examples.

A final word of caution: All of these calculators have various settings that the user can change. If you get unexpected results or errors, it is quite likely because something has been changed from the default values. Clearing the previous example may solve the problem, and if not you should try resetting the calculator.

Organization of the Text

The organization of the text follows the standard calculus outline.

The trigonometric functions are differentiated in Chapter 3 and the derivatives of the logarithm and exponential are simply stated, so that these functions can be used in examples and exercises. Calculus courses increasingly introduce the transcendental functions early, finding that the difficulties are less serious than they used to be and the advantages are considerable.

For one thing, students are not particularly intimidated by these functions any more; they all recognize them from the buttons on their calculators. Beginning early allows the student to acquire a working familiarity with them. It also opens the door to many more interesting examples and applications, and makes various theoretical topics, such as the Chain Rule, much more interesting.

Chapters 4 and 6 give applications of differentiation and integration; there are a wide range of topics and innovative examples. Most instructors will likely include the first few sections and pick and choose among the more specialized topics. Section 4.12 covers l'Hôpital's rule, a topic that is often left until much later.

As much as possible, the chapters are written to be independent, to allow the greatest flexibility in the order of presentation. Chapter 7, on Logarithms and Exponential Functions, and Chapter 8, Techniques of Integration, are completely interchangeable, and Chapter 9 (Taylor's Formula) can be interchanged with most of Chapter 10 (Series). Section 10.8 (Taylor Series) brings the two streams together.

Chapter 7 contains an extremely brief discussion of differential equations, which are also the subject of Chapter 19.

Chapters 1 through 13 are available as *Single Variable Calculus,* while *Calculus with Analytic Geometry* contains Chapters 1 through 19.

The chapters on multivariable calculus do not assume a knowledge of linear algebra. They introduce the necessary vector techniques and determinants as they are needed.

Supplements

The Students' Solutions Manual, containing complete solutions to the odd-numbered exercises, was prepared by the author himself. The solutions are in the same style and use the same techniques as the text. They are prepared to a very high standard of accuracy and completeness.

Graphing calculator manuals, Derive® project manuals, and Mathematica® project manuals are also available for your students.

The Instructor's Solutions Manual and Instructor's Resource Manual have also been prepared by the author to follow the same approach and style found in the text.

Please contact your Wm. C. Brown Publishers representative for information on content and availability of the various supplements that support this text.

Reviewers and Acknowledgments

During its development this text has been class tested by many people; for their thoughtfulness and their suggestions, and the tact with which they have made them, I wish to thank: Stefan Bilaniuk, Trent University; Harriet Botta, York University; Julian Edward, M. P. Heble, Abe Igelfeld, Jerzy Jarosz, Val Jurdjevic, Anthony Lam, Philip Leah, Bong Lian, Dave Masson, Kunio Murasugi, D. W. Pravica, Felix Recio, Dave Roberts, Elizabeth Rowlinson, Andrew Schwartz, Luis Seco, Dipak Sen, Alexander Sobolev, Catherine Sulem, Ti Wang, all of the University of Toronto; and David Wehlau, Royal Military College.

Many others have made specific contributions or have added to my understanding of calculus and how to teach it. I want to thank: Ed Barbeau; Peter Botta; Ian Graham; George Giordano, Ryerson Polytechnical University; Barbara Japp, Clinical Physics Department, Princess Margaret Hospital; Emile Leblanc; Pierre Milman; Y. Park; Paul Selick; and John Vrolyk, George Brown College.

Among the many students who have pointed out errors in preliminary versions or made suggestions for improvements, I especially want to thank Lu Anne Burnham, Kong Wah Cora Chan, Nina Doan, David Hou, Tamarah Kagan, Shawn Langer, Wendy W. K. Lee, Dimitri Loukanidis, Ed MacDougall, Elena Poulos, Sharon M. L. Seow, Heng Sun, Erin Truscott, Cheryl Wein, and Stephanie Wiesenthal.

Various versions of the text have been thoroughly reviewed by an extensive team of experts. This is lonely work, but it is much appreciated. The reviewers were:

Dan Anderson
University of Iowa

Fred Brauer
University of Wisconsin–Madison

Mike Davidson
Cabrillo College

Kenneth Davis
Albion College

Edward Donley
Indiana University of Pennsylvania

Douglas Frank
Indian University–Penn

Dewey Furness
Ricks College

Stuart Goldenberg
California Polytechnic State
    University

Raymond Patrick Guzman
Pasadena City College

Gary S. Itzkowitz Ph.D.
Glassboro State College

Andrew Karantinos
The University of South Dakota

Judy Kasabian
El Camino College

John Krajewski
University of Wisconsin–Eau Claire

Robert F. Lax
Louisiana State University

Michael Mays
West Virginia University

Jim McKinney
California State Polytechnic
    University

Wesley J. Orser
Clark College

Dr. John Paulling
Nicholls State University

Susan Pfeifer
Butler County Community College

Claudia Pinter-Lucke
California State Polytechnic
   University

Benjamin F. Plybon
Miami University

Dr. Janice Rech
University of Nebraska at Omaha

Wayne Roberts
MacAlester College

William J. Roberts
Plymouth State College

Walter S. Sizer
Moorhead State University

Bruce Stephan
Webb Institute of Naval
   Architecture

J. R. Vanstone
University of Toronto

Gary L. Walls
University of Southern Mississippi

John L. Wulff
California State University at
   Sacramento

Anne Young
Loyola College

I owe an immeasurable debt to the magnificent staff in the University of Toronto Mathematics Department: Marie Bachtis, Pat Broughton, Ida Bulat, Nadia Cavaliere, Chibeck Graham, Beverley Leslie, Karin Smith, Haleh Vaez, and Yu Yuet-Wah.

The preparation of the graphics calculator examples in the text was made possible through the generosity of Texas Instruments, Hewlett Packard, CASIO, and the special help of Mike Brellisford at SHARP Electronics of Canada. I am indebted to "Uncle Dave" Roughley, who prepared the appropriate graphics output. I also want to thank "DesignCAD Keith" Campbell of American Small Business Computers for providing CAD images, and Giovanna Peel of the Ontario College of Art for all her help.

The people at York Graphic Services have done a wonderful job of putting together the manuscript with all its bells and whistles. I am especially grateful to Mary Jo Gregory, who patiently answered all my questions and more than once got a sore ear doing it.

It has been a pleasure for me to work with Wm. C. Brown Publishers; I want particularly to thank Earl McPeek, Theresa Grutz, Beth Kundert, the late Ed Jaffe, Shane Sheehan, and Wes Rafuse.

Finally I wish to thank Chandru Kriplani, without whom this book would never have been written.

Preliminaries

1

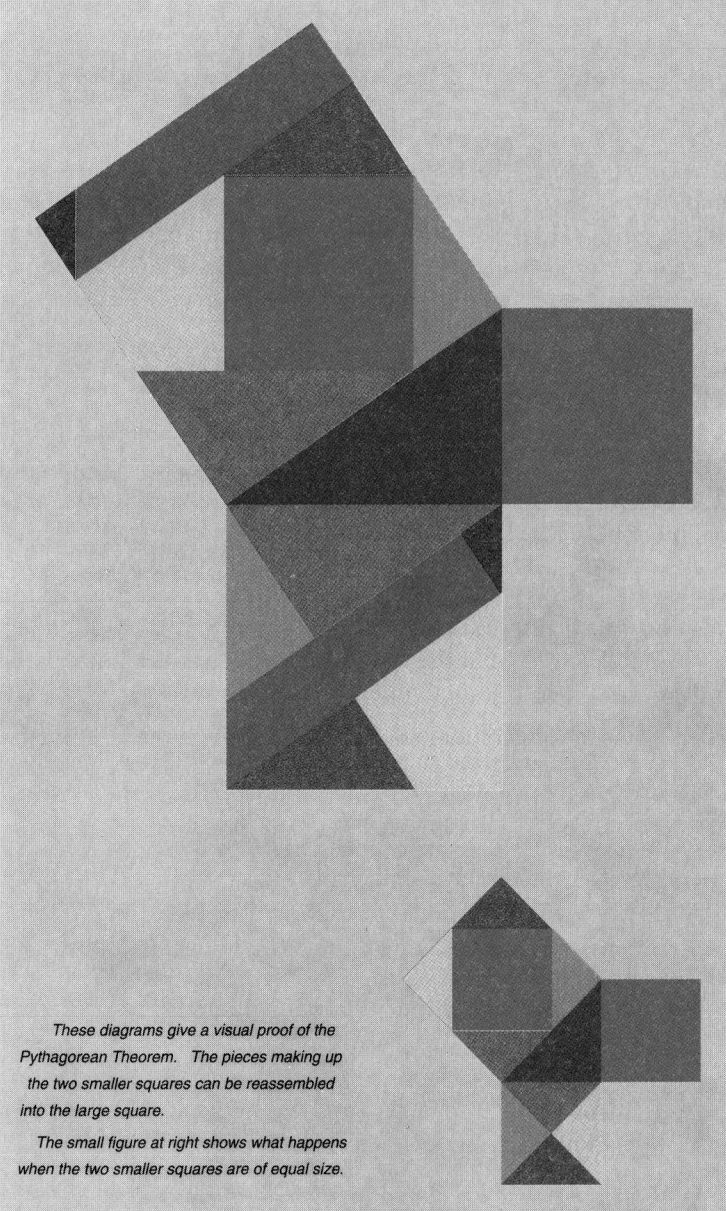

These diagrams give a visual proof of the Pythagorean Theorem. The pieces making up the two smaller squares can be reassembled into the large square.

The small figure at right shows what happens when the two smaller squares are of equal size.

INTRODUCTION

In this chapter we will discuss various concepts and formulas that will be needed later on. Many of them will be familiar, and it may not be necessary for you to spend much time reviewing them. On the other hand, it would probably be wise to read fairly carefully any sections that are new to you (a quick look at the summary at the end of the chapter should help you to identify unfamiliar topics). As you read later sections of the book, you will be able to use this chapter as a reference. Additional discussion and exercises are available in the Student's Solutions Manual.

Section 1.1 discusses the real numbers, which are fundamental to calculus and to much of the rest of mathematics. Section 1.2 is about inequalities, and Section 1.3 is about polynomials and rational functions (which simply means functions that are quotients of two polynomials). Section 1.4 discusses the concept of a function and the related idea of drawing a graph. Section 1.5 recalls some facts from geometry, principally the Pythagorean Theorem, facts about similar triangles, areas of simple regions, and volumes of simple geometric shapes. The remainder of the chapter is devoted to several specific types of graphs and functions. Section 1.6 is about conics, with particular emphasis on drawing their graphs, which is what will be most important for studying calculus. Section 1.7 is about trigonometric functions. One important item there is the idea of "radian measure" for angles; if you are not familiar with radians, you should be sure to read Section 1.7. Finally, Section 1.8 is about logarithms and exponentials.

1.1 The Real Numbers

RATIONAL NUMBERS
IRRATIONAL NUMBERS
SCIENTIFIC NOTATION

We are all familiar with the **natural numbers**, $\{1, 2, 3, 4, \ldots\}$. For many purposes it is necessary to consider negative numbers too, and this leads to the **integers** $\{\ldots, -4, -3, -2, -1, 0, 1, 2, 3, 4, \ldots\}$.

The most obvious drawback to the integers is that the quotient of two integers may not be an integer (e.g., $\frac{5}{9}$ is not an integer). This suggests considering all the numbers that are quotients of integers. These numbers are called **rational numbers**; the name is related to the word "ratio." Every rational number can be written in the form $r = \frac{a}{b}$, where a and b are integers. Of course the same rational number can be written as the quotient of many different pairs of integers; for instance,

$$\frac{-5}{9} = \frac{5}{-9} = \frac{-10}{18} = \frac{15}{-27}, \quad \text{etc.}$$

The rational number $r = \frac{a}{b}$ is said to be **in lowest terms** if a and b have no common divisors except 1 and -1. In the example above, $\frac{-5}{9}$ and $\frac{5}{-9}$ are in lowest terms (the only integers that divide ± 5 and ± 9 are 1 and -1), and the other two are not.

Rational numbers are very convenient and have the property that we can add, subtract, multiply, and divide them (except by zero) and the results will still be rational numbers.

On the other hand, the number π is not a rational number, though this is difficult to show, and as a result if the radius r of a circle is a rational number, then the area of the circle is *not* a rational number, since it equals πr^2 (see Exercise 5).

The square roots of some rational numbers are also rational numbers: $\sqrt{\frac{9}{4}} = \frac{3}{2}$, $\sqrt{\frac{1}{16}} = \frac{1}{4}$. However, there are many rational numbers whose square roots are not rational.

EXAMPLE 1   The number $\sqrt{2}$ is not a rational number.

PROOF   Suppose there was some rational number $\frac{a}{b}$ so that $\left(\frac{a}{b}\right)^2 = 2$. We can assume that $\frac{a}{b}$ is in lowest terms because otherwise we could divide out the common factor in the numerator and denominator. Multiplying both sides of the equation by b^2, we find

$$a^2 = 2b^2.$$

Looking at this equation carefully, we see that the right side is an even integer, that is, divisible by 2. The same must be true of the left side, and the only way this could happen is for a to be divisible by 2 (after all if a is odd, then a^2 is odd too).

But if a is divisible by 2, then a^2 is divisible by 4, and this means that $2b^2$ is divisible by 4, so b^2 must be divisible by 2. The only way this can happen is for b to be divisible by 2.

We have just shown that a and b must both be divisible by 2, but we assumed at the beginning that they had no common divisors. This contradiction shows that our assumption that there was a rational number whose square is 2 is impossible, which proves that $\sqrt{2}$ is not rational.

This fact was first recognized by the ancient Greeks.

If we want to be able to find square roots, at least of nonnegative numbers, we have to expand the number system. The real numbers provide the solution to this and many related problems, and they are the most appropriate set of numbers for the study of calculus. The real numbers $\mathbb{R}$ are frequently represented by drawing a horizontal straight line, called the **real line**. We mark a point on it and label it 0 (zero); it is often in the middle of the line, but not always. We mark other points evenly spaced to the right and left of 0 and label them as the integers. To the right of 0 we place 1, 2, 3, and so on, in that order, and to the left of 0 we place -1, then -2, then -3, and so on. See Figure 1.1.1.

Figure 1.1.1

Between the integers we can mark any fractions we wish; for instance, $\frac{7}{2}$ is halfway between 3 and 4. It is less easy to mark an **irrational number** like π or $\sqrt{2}$; to do that, we use the decimal expansion of these numbers.

Every real number has a decimal expansion. For instance, $\pi = 3.141\,592\,65\ldots$, and $\sqrt{2} = 1.414\,213\,56\ldots$. The dots are very important. They tell us that the expansion does not end where we have stopped writing (in fact if the decimal expansion did stop, then the number would have to be a rational number).

We cannot write out the whole expansion of a number like π or $\sqrt{2}$, since it goes on forever, but when we write down some moderate number of digits (in the two examples above we have written nine digits), this gives us a pretty good *approximation* to the number in question.

Scientific Notation

For numbers that are very large or very small, it is often convenient to write them in **scientific notation**. This means writing a number as the product of a number between 1 and 10 and a power of 10, that is, as $a \times 10^n$, where $1 \leq a < 10$ and n is an integer.

The obvious advantage of this notation is that it eliminates having to write a large number of zeroes ($1,540,000,000$ can be written more neatly as 1.54×10^9, and certainly $0.000\,000\,000\,0037$ is harder to read than 3.7×10^{-12}).

The other advantage is that the number of significant digits is clear from the way the number is written. For instance, the meaning of a measurement written as 2000 is unclear, but 2×10^3, 2.0×10^3, and 2.00×10^3 (which in a sense all equal 2000) make it quite clear what is intended.

Exercises 1.1

1. Show that $\sqrt{8}$ is not a rational number.

*2. (i) Show that $\sqrt{3}$ is not a rational number.

(ii) Show that $\sqrt{6}$ is not a rational number.

*3. Explain the remark made in the text that a decimal expansion that terminates must represent a rational number.

*4. Show that the number whose decimal expansion is

19.680 314 314 314 314 . . . (with the three digits 314 repeated again and again indefinitely) is a rational number (by writing it as the quotient of two integers).

*5. If a is a rational number and b is an irrational number, show that (i) $a + b$ is irrational, (ii) ab is irrational unless $a = 0$.

1.2 Inequalities

ABSOLUTE VALUE
TRIANGLE INEQUALITY
INTERVAL
CLOSED/OPEN INTERVAL
UNION
INTERSECTION

Figure 1.2.1

One of the special properties of the real numbers is that there is a concept of "order," meaning that given two distinct real numbers, one of them is always less than the other. We use the conventional symbols: $a < b$ means a is less than b, $a > b$ means a is greater than b (which is the same thing as $b < a$), $a \leq b$ means a is less than or equal to b, and $a \geq b$ means a is greater than or equal to b. Each of these expressions is called an **inequality**. We sometimes refer to the ones with $<$ and $>$ as **strict inequalities**.

In terms of the picture of the real number line, to say that a number a is less than b means that a is to the left of b on the line (see Figure 1.2.1). We also say that a is "negative" if $a < 0$ and that a is "positive" if $a > 0$.

There are a number of properties of inequalities that we will be using.

Summary 1.2.1

Properties of Inequalities

(i) If $a < b$ and $b < c$, then $a < c$.   (Transitivity Property)

(ii) If $a \leq b$ and $b \leq a$, then $a = b$.

(iii) If $a < b$ and $c \leq d$, then $a + c < b + d$.

(iv) If $a < b$ and $c > 0$, then $ac < bc$.

(v) If $a < b$ and $c < 0$, then $ac > bc$.

(vi) If $0 < a < b$, then $\frac{1}{a} > \frac{1}{b}$; if $a < b < 0$, then $\frac{1}{a} > \frac{1}{b}$.

(vii) $a^2 \geq 0$, for any $a \in \mathbb{R}$.

Properties (v) and (vi) say that an inequality is *reversed* if it is multiplied by a negative number or if both sides have the same sign and we take the reciprocal of both sides.

Property (vii) follows from Properties (iv) and (v).

Using Properties (iv) and (v) we also see that $ab > 0$ if a and b are both positive or both negative and $ab < 0$ if one of a and b is positive and the other is negative. In particular, the only way that the product ab can equal zero is for one (or both) of a and b to equal zero.

An important concept is the **absolute value** of a number a, written $|a|$. It is the "size" of the number and is *always nonnegative*. In symbols we write

$$|a| = \begin{cases} a, & \text{if } a \geq 0, \\ -a, & \text{if } a < 0. \end{cases}$$

> Note that if a is negative, then $|a| = -a$ (since $-a$ is positive ...).

So $|3| = 3$, $|-3| = 3$, and $|0| = 0$. It is less clear what to do with $|\sqrt{12} - 4|$, but if we notice that $4^2 = 16 > 12$, we see that $4 > \sqrt{12}$. If we subtract 4 from both sides of this last inequality, we see that $0 > \sqrt{12} - 4$, so $\sqrt{12} - 4$ is negative and $|\sqrt{12} - 4| = -(\sqrt{12} - 4) = 4 - \sqrt{12}$.

We will often encounter inequalities involving absolute values. One of the most common types is $|x| \leq c$. If $x \geq 0$, we know that $|x| = x$, so the inequality means $x \leq c$. If $x \leq 0$, we know that $|x| = -x$, and the inequality means $-x \leq c$ or $x \geq -c$. Combining these two, we find that

$$|x| \leq c \quad \text{means exactly the same thing as} \quad -c \leq x \leq c.$$

The set of all numbers x that satisfy this condition can be drawn on the line; they fill a segment between $-c$ and c. See Figure 1.2.2.

There is an extremely important inequality involving absolute values. It is called the triangle inequality.

—•——|——•—
$-c$ 0 c

Figure 1.2.2

THEOREM 1.2.2

> **The Triangle Inequality**
>
> If a and b are any real numbers, then
>
> $$|a + b| \leq |a| + |b|.$$

PROOF Since a equals either $|a|$ or $-|a|$, $-|a| \leq a \leq |a|$. Similarly, $-|b| \leq b \leq |b|$. Adding these two inequalities, we find that

$$-|a| - |b| \leq a + b \leq |a| + |b|$$

or

$$-(|a| + |b|) \leq a + b \leq |a| + |b|.$$

This is exactly the same as saying $|a + b| \leq |a| + |b|$, which is the result we wanted to show. ◢

> Recall $|x| \leq c$ means $-c \leq x \leq c$; here x is $a + b$ and c is $|a| + |b|$.

There is also a second version of the triangle inequality.

THEOREM 1.2.3

> **The Triangle Inequality (Second Version)**
>
> If a and b are real numbers, then
>
> $$|a - b| \geq |a| - |b|.$$

PROOF Applying the first version of the triangle inequality to $a = (a - b) + b$, we find $|(a - b) + b| \leq |a - b| + |b|$, or $|a| \leq |a - b| + |b|$. Rearranging gives the desired result. ◢

$\overbrace{\qquad\qquad}^{|a-b|}$

—|————|—
a b

Figure 1.2.3

The absolute value can also describe distances. If a and b are real numbers, then $|a - b|$ is the **distance** between them. See Figure 1.2.3.

Figure 1.2.4

Figure 1.2.5

Figure 1.2.6

Figure 1.2.7

Figure 1.2.8

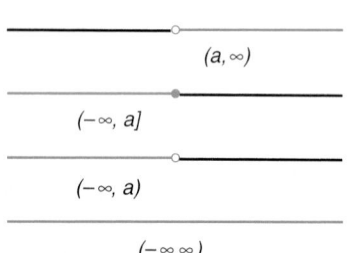

Figure 1.2.9

EXAMPLE 1   Consider the set described as

$$\{x : |x - 2| \le 3\},$$

which we read as "the set of all x for which $|x - 2| \le 3$."

Now $|x - 2| \le 3$ means $-3 \le x - 2 \le 3$; adding 2 to these inequalities gives $-1 \le x \le 5$, so the set consists of all x's that satisfy $-1 \le x \le 5$, that is, all x's lying between -1 and 5. See Figure 1.2.4.   ◢

Using the concept of distance frequently makes it easier to work with inequalities.

Intervals

If $a < b$, then the set of all points that lie between a and b is called an **interval**. The points a and b are called the **endpoints** of the interval, and it is important to specify whether or not the endpoints belong to the set.

An interval with its endpoints included is called a **closed interval**; one with the endpoints not included is called an **open interval**. We draw closed intervals with dots at the ends to indicate that the endpoints are included, and open intervals with "open" dots to show that those points are excluded. Figure 1.2.5 shows the closed interval from 1 to $2\frac{1}{2}$, and Figure 1.2.6 shows the open interval from 4 to 7.

We write $[a, b]$ to represent the closed interval between a and b and (a, b) for the open interval between a and b. In symbols,

$$[a, b] = \{x : a \le x \le b\}, \qquad (a, b) = \{x : a < x < b\}.$$

Occasionally it is necessary to consider **half-open intervals**, meaning ones that contain one endpoint but not the other. Adapting the above notation for closed and open intervals, we write

$$[a, b) = \{x : a \le x < b\}, \qquad (a, b] = \{x : a < x \le b\}.$$

These half-open intervals are shown in Figure 1.2.7.

The other type of interval that is often used is called an **infinite interval**. We use the notation $[a, \infty)$ to refer to the "interval" shown in Figure 1.2.8, whose left endpoint is a and which extends out all the way along the real line. The symbol ∞ is *not* a real number. We call it "infinity," but it is extremely important to keep in mind that it is not a number. It is only a convenient way to summarize the idea of this type of set. We also use symbols like $(-\infty, 3)$, $(-\infty, -5]$ and $(-\infty, \infty)$ (which means the entire real line $\mathbb{R}$). See Figure 1.2.9.

$$[a, \infty) = \{x : a \le x\} \qquad \qquad (a, \infty) = \{x : a < x\}$$
$$(-\infty, a] = \{x : x \le a\} \qquad \qquad (-\infty, a) = \{x : x < a\}$$
$$(-\infty, \infty) = \text{ the real line} = \mathbb{R}.$$

EXAMPLE 2   Consider the set

$$\{x : |x + 2| > 2\}.$$

We write $|x + 2| = |x - (-2)|$, which is the distance from x to the point -2.

The set in question is the set of all x's whose distance from -2 is greater than 2. From Figure 1.2.10 we see that the points whose distance from -2 is greater than 2 are all the points to the right of 0 and all the points to the left of -4.

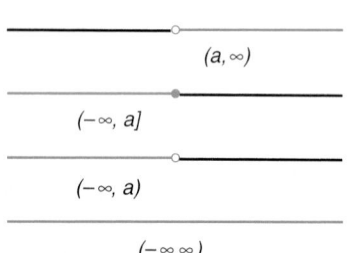

Figure 1.2.10

In Example 1 we saw that $|x - 2| \le 3$ is the same as $-3 \le x - 2 \le 3$. But $|x + 2| > 2$ is *not* the same as $-2 > x + 2 > 2$.... If we want to write it out using inequalities, we have to say that $x + 2 > 2$ *or* $x + 2 < -2$.

We could write this out in symbols as follows:

$$\{x : |x + 2| > 2\} = (-\infty, -4) \cup (0, \infty).$$

We have rewritten the set as the union of two infinite intervals.

REMARK 1.2.4

We recall some useful notation. We write $A \cup B$ for the **union** of A and B, that is, the set of all points that belong to either or both of the sets. Similarly, we write $A \cap B$ for the **intersection** of A and B, that is, the set of all points that belong to both sets. In symbols,

$$A \cup B = \{x : x \in A \quad \text{or} \quad x \in B\}, \qquad A \cap B = \{x : x \in A \quad \text{and} \quad x \in B\}.$$

Exercises 1.2

Without using a calculator, write an expression for each of the following quantities without using any absolute value signs.

1. $\left|1 - \frac{5}{12} - \frac{3}{7} - \frac{1}{6}\right|$
2. $\left|1 + \frac{5}{12} - \sqrt{2}\right|$
3. $\left|\frac{5}{12} + \frac{3}{7} + \frac{1}{6} - 1\right|$
4. $\left|(1 + \sqrt{2})(1 - \sqrt{2})\right|$
5. $\left|\sqrt{3} - 2\right|$
6. $\left|3 - 2\sqrt{2}\right|$
7. $\left|3 - \sqrt{2} - \sqrt{3}\right|$
8. $\left|5 - \sqrt{5} - \sqrt{8}\right|$
9. $\left|x^2 + 3\right|$
10. $\left|x^2 + 2x + 2\right|$

Express each of the following sets as an interval or a union of intervals.

11. $\{x : |x - 1| \le 4\}$
12. $\{x : |x + 3| < 1\}$
13. $\{x : |3x - 6| < 2\}$
14. $\{x : |4x + 1| \le 2\}$
15. $\{x : |2x - 1| > 1\}$
16. $\{x : |3x + 5| \ge 4\}$
17. $\{x : |x - 3| \le 4 \text{ and } |x - 1| \ge 1\}$
18. $\{x : |2x + 1| > 4 \text{ and } |x + 3| \ge 1\}$
19. $\{x : |1 - x^2| \ge 2\}$
20. $\{x : |x| \ge 3 \text{ and } |x + 3| \le 2\}$

1.3 Polynomials and Rational Functions

COEFFICIENT
CONSTANT TERM
QUADRATIC/LINEAR POLYNOMIAL
COMPLETING THE SQUARE
QUADRATIC FORMULA
RATIONAL ROOT THEOREM
BINOMIAL THEOREM
FACTORIALS

DEFINITION 1.3.1

A **polynomial** is an expression $p(x)$ of the form

$$p(x) = a_n x^n + a_{n-1} x^{n-1} + a_{n-2} x^{n-2} + \ldots + a_2 x^2 + a_1 x + a_0,$$

where n is a nonnegative integer and each a_i is a real number. The numbers $a_n, a_{n-1}, \ldots, a_2, a_1, a_0$ are called the **coefficients** of the polynomial.

The highest power of x that occurs in the polynomial (which would be n in this polynomial, provided $a_n \ne 0$) is called the **degree** of the polynomial.

Notice that the **subscript** i on the coefficient a_i is just a label; it tells us which power of x is multiplied by a_i. We do not bother to write $a_0 x^0$, but just a_0 for the last term, which is called the **constant term**.

EXAMPLE 1 In the polynomial $3x^4 - 2x^3 + 5x^2 - 2x + 7$, the coefficients are $a_4 = 3$, $a_3 = -2$, $a_2 = 5$, $a_1 = -2$, and $a_0 = 7$. The degree of the polynomial is 4.

EXAMPLE 2 The formula $p(x) = 3x^2 - 5x + 2$ can be worked out for any specific value of x. For instance, if $x = 2$, then $p(2) = 3(2)^2 - 5(2) + 2 = 12 - 10 + 2 = 4$. On the other hand, if $x = 1$, then $p(1) = 3(1)^2 - 5(1) + 2 = 3 - 5 + 2 = 0$.

It is very easy to evaluate $p(0) = 3(0)^2 - 5(0) + 2 = 0 - 0 + 2 = 2$.

The same idea can be used with any polynomial. We call this process **evaluating** the polynomial $p(x)$ at the point x, and call the resulting number the **value of the polynomial** p at the point x.

Since x can *vary* and take on different values, we call it the **variable**.

Factoring

It is often important to be able to write a complicated polynomial as the product of two or more simpler polynomials. For instance, it is very easy to check what is known as the **difference of squares** formula:

$$a^2 - b^2 = (a + b)(a - b). \tag{1.3.2}$$

This formula makes sense when a and b are numbers (you can check it with $a = 5$ and $b = 3$, for instance), but it also works if a and b are polynomials.

EXAMPLE 3

(i) $49 - 9 = (7 + 3)(7 - 3) = (10)(4)$;   (ii) $x^2 - 16 = (x + 4)(x - 4)$;

(iii) $(x + 1)^2 - 1 = (x + 2)x$.

The process of breaking up a polynomial $p(x)$ into a product of simpler factors is called **factoring** the polynomial. The simpler polynomials whose product is $p(x)$ are called the **factors** of $p(x)$. For instance, when we write $4x^2 - 25 = (2x + 5)(2x - 5)$, the polynomials $2x + 5$ and $2x - 5$ are the factors of $4x^2 - 25$.

Suppose we want to factor the polynomial $p(x) = x^2 - x - 6$. This is a **quadratic polynomial**, meaning one whose degree is 2. In this case it is possible to guess a factorization, namely, $x^2 - x - 6 = (x - 3)(x + 2)$.

This kind of guesswork approach can often be used.

EXAMPLE 4 Factor each of the following quadratic polynomials: (i) $x^2 + 2x + 1$; (ii) $x^2 + 2x - 8$; (iii) $2x^2 - 6x - 20$; (iv) $x^2 - 2$; (v) $x^2 + 1$.

Solution

(i) We just look for a factorization $(x - a)(x - b) = x^2 - (a + b) + ab$. The first thing to do is to try to find *integer* values for a and b. In this case the only possible pairs of integers a and b whose product is 1 are 1 and 1 or -1 and -1. Since $a + b = -2$, the obvious choice is $a = b = -1$, and the factorization is $x^2 + 2x + 1 = (x + 1)(x + 1) = (x + 1)^2$.

(ii) $x^2 + 2x - 8$. The only pairs of integers whose product is -8 are 1 and -8, or

-1 and 8, or 2 and -4, or -2 and 4. If the sum must also be -2, the only possibility is 2 and -4. The factorization is $x^2 + 2x - 8 = (x - 2)(x + 4)$.

(iii) We write $2x^2 - 6x - 20 = 2(x^2 - 3x - 10)$, and factor $x^2 - 3x - 10 = (x - 5)(x + 2)$, so $2x^2 - 6x - 20 = 2(x^2 - 3x - 10) = 2(x - 5)(x + 2)$.

(iv) $x^2 - 2$. Here we try looking for a factorization with integers but cannot find one.

Suppose that we knew a factorization $x^2 - 2 = (x - a)(x - b)$. If we let $x = a$, then the right side of this equation would be $(a - a)(a - b) = 0(a - b) = 0$. This suggests that we try to find a (or b) by looking for values of x that make $p(x) = 0$.

To solve $p(x) = 0$, we rewrite $x^2 - 2 = 0$ as $x^2 = 2$, which tells us that x should equal $\sqrt{2}$ (or $-\sqrt{2}$). We try letting a and b be $\sqrt{2}$ and $-\sqrt{2}$: $(x - \sqrt{2})(x + \sqrt{2}) = x^2 - 2$.

(v) $x^2 + 1$. If it did factor as $(x - a)(x - b)$, then the polynomial $x^2 + 1$ would have to be zero when we let x equal a or b. But we know that $x^2 \geq 0$ for any value of x, so $x^2 + 1 \geq 1$ for any value of x. This means it can never be zero, so it is impossible to find a and b. The polynomial does not factor.

In hindsight, $x^2 - 2$ is a difference of squares, $x^2 - (\sqrt{2})^2$, so we might have guessed the factorization that way too.

Some quadratic polynomials can be factored by just guessing integers a and b so that the polynomial equals $(x - a)(x - b)$. If the coefficient of x^2 in the original polynomial is not 1, then it is necessary to factor it out first.

Some polynomials cannot be factored with integer values for a and b. It may still be possible to guess a factorization, but if not, the next thing to do is look for values of x for which $p(x) = 0$. If we can find such a value, then it will serve as a in a factorization $p(x) = (x - a)(x - b)$.

We also saw in Example 4 that there are some polynomials that cannot be factored at all.

Completing the Square and the Quadratic Formula

We have already encountered some examples in which a quadratic polynomial factors as the square of a polynomial of degree 1 (polynomials of degree 1 are often called **linear polynomials**).

EXAMPLE 5   If the polynomial $x^2 - 4x + 12$ is a square, $(x - a)^2 = x^2 - 2a + a^2$, say, then its middle coefficient -4 would have to equal $-2a$, so $a = -2$. Since the constant term is *not* equal to $(-2)^2 = 4$, we see that the polynomial is not a square.

On the other hand, we can write it as $x^2 - 4x + 12 = (x^2 - 4x + 4) + 8$. Now the polynomial in parentheses is a square. We can write $x^2 - 4x + 12 = (x^2 - 4x + 4) + 8 = (x - 2)^2 + 8$. This process is called **completing the square**, for obvious reasons.

With the polynomial written as $(x - 2)^2 + 8$, we see immediately that it cannot be factored. After all, it is 8 plus a square, so for any $x \in \mathbb{R}$ it is always greater than or equal to 8, so it can *never* be zero. We can never find a and b so that the polynomial factors as $(x - a)(x - b)$.

Summary 1.3.3

Completing the Square

If $p(x) = x^2 + bx + c$, we can write $p(x)$ as the sum of the square of a linear polynomial and a number, as follows:

$$p(x) = x^2 + bx + c = x^2 + bx + \left(\frac{b}{2}\right)^2 - \left(\frac{b}{2}\right)^2 + c = \left(x + \frac{b}{2}\right)^2 + c - \left(\frac{b^2}{4}\right).$$

If $q(x) = ax^2 + bx + c$, we write $q(x) = a\left(x^2 + \frac{b}{a}x + \frac{c}{a}\right)$ and then complete the square of the polynomial inside the parentheses (for this to work, it is necessary that

$a \neq 0$). We find that

$$q(x) = a\left(x^2 + \frac{b}{a}x + \frac{c}{a}\right) = a\left(x^2 + \frac{b}{a}x + \left(\frac{b}{2a}\right)^2 - \left(\frac{b}{2a}\right)^2 + \frac{c}{a}\right)$$

$$= a\left(\left(x + \frac{b}{2a}\right)^2 + \frac{c}{a} - \left(\frac{b}{2a}\right)^2\right) = a\left(x + \frac{b}{2a}\right)^2 + c - \left(\frac{b^2}{4a}\right).$$

One advantage of completing the square in a quadratic polynomial is that it makes it easier to solve.

EXAMPLE 6 Find all solutions of $x^2 - 4x + 1 = 0$.

Solution We begin by completing the square on the left side, which gives us that $x^2 - 4x + 1 = (x^2 - 4x + 4) - 4 + 1 = (x - 2)^2 - 3$.

The original equation becomes $(x - 2)^2 - 3 = 0$, or $(x - 2)^2 = 3$. So $x - 2$ must be something whose square is 3; there are exactly two such numbers, $\sqrt{3}$ and $-\sqrt{3}$. The possibilities are $x - 2 = \sqrt{3}$ and $x - 2 = -\sqrt{3}$. The solutions of the equation are

$$x = 2 + \sqrt{3} \quad \text{and} \quad x = 2 - \sqrt{3}.$$

A common abbreviation for this is to write $x = 2 \pm \sqrt{3}$.

When we write the square root sign, it always means the **nonnegative square root**. This has the slightly surprising consequence that if $a < 0$ then $\sqrt{a^2} \neq a$; since $\sqrt{a^2} \geq 0$, we see that $\sqrt{a^2} = |a| = -a$.

The basic approach we have used in this example can be used to solve any quadratic equation. To find the solutions of $ax^2 + bx + c = 0$, we complete the square and find that the left side of the equation equals $a\left(x + \frac{b}{2a}\right)^2 + c - \left(\frac{b^2}{4a}\right)$. From this we can rewrite the original equation as $\left(x + \frac{b}{2a}\right)^2 = \frac{1}{a}\left(\left(\frac{b^2}{4a}\right) - c\right) = \frac{b^2}{4a^2} - \frac{c}{a}$.

To solve this, we should take square roots of both sides; this is possible *only if* the right side is nonnegative. In this case we find that $x + \frac{b}{2a} = \pm\sqrt{\frac{b^2}{4a^2} - \frac{c}{a}}$.

This amounts to $x = -\frac{b}{2a} \pm \sqrt{\frac{b^2}{4a^2} - \frac{c}{a}}$, and putting it over a common denominator, we find

$$x = \frac{-b \pm 2a\sqrt{\frac{b^2}{4a^2} - \frac{c}{a}}}{2a} = \frac{-b \pm \sqrt{4a^2\frac{b^2}{4a^2} - 4ac}}{2a} = \frac{-b \pm \sqrt{b^2 - 4ac}}{2a}.$$

This result is tremendously useful.

THEOREM 1.3.4

The Quadratic Formula

Consider the equation $ax^2 + bx + c = 0$, where $a \neq 0$. It has no solutions if $b^2 - 4ac < 0$, exactly one solution if $b^2 - 4ac = 0$, and two solutions if $b^2 - 4ac > 0$.

If $b^2 - 4ac \geq 0$, the solutions are given by the formula

$$x = \frac{-b \pm \sqrt{b^2 - 4ac}}{2a}.$$

The quantity $D = b^2 - 4ac$, which tells us whether or not there are solutions, is called the **discriminant** of the polynomial $ax^2 + bx + c$. Notice that when the discriminant is zero, the square root is zero, and this is why there is only one solution.

EXAMPLE 7(a)   Find all solutions of the following equations: (i) $x^2 + x - 2 = 0$; (ii) $x^2 + x - 1 = 0$; (iii) $x^2 + x = -1$.

Solution

(i) The discriminant is $D = 1^2 - 4(1)(-2) = 9$ and the solutions are $x = \frac{-1 \pm \sqrt{1^2 - 4(1)(-2)}}{2(1)} = \frac{-1 \pm \sqrt{9}}{2} = \frac{-1 \pm 3}{2}$, that is, $\frac{2}{2}$ and $\frac{-4}{2}$, that is, 1 and -2. We might have guessed this if we tried factoring $x^2 + x - 2 = (x-1)(x+2)$.

(ii) Here $D = 1^2 - 4(1)(-1) = 5$, and the solutions are $x = \frac{-1 \pm \sqrt{5}}{2}$, that is, $\frac{-1 + \sqrt{5}}{2}$ and $\frac{-1 - \sqrt{5}}{2}$.

(iii) First we rewrite $x^2 + x = -1$ as $x^2 + x + 1 = 0$. The discriminant is $D = 1 - 4 = -3$, which is negative. This means that there are no solutions.

EXAMPLE 7(b)   Describe (iv) $\{x : x^2 + 2x - 3 < 0\}$; (v) $\{x : x^2 - 4x + 2 \geq 0\}$.

Solution

(iv) We find $x^2 + 2x - 3 = x^2 + 2x + 1 - 1 - 3 = (x+1)^2 - 4$. The condition $x^2 + 2x - 3 < 0$ amounts to $(x+1)^2 - 4 < 0$, or $(x+1)^2 < 4$. Taking square roots gives $|x+1| < 2$, that is, $|x - (-1)| < 2$, so the set consists of all points whose distance from -1 is less than 2. It is the interval $(-3, 1)$.

(v) Here $x^2 - 4x + 2 = (x-2)^2 - 2$, so the condition is $(x-2)^2 - 2 \geq 0$, that is, $(x-2)^2 \geq 2$. This amounts to $|x-2| \geq \sqrt{2}$, which means $x \geq 2 + \sqrt{2}$ or $x \leq 2 - \sqrt{2}$. The set is $(-\infty, 2 - \sqrt{2}] \cup [2 + \sqrt{2}, \infty)$.

The Factor Theorem

When we studied quadratic polynomials, we found that if $p(a) = 0$, then $p(x)$ could be factored, one of the factors being $(x - a)$. From this it was easy to find the other factor. If $p(x)$ is any polynomial, not necessarily of degree 2, suppose we know that a is a root of $p(x)$, that is, that $p(a) = 0$. Our experience with quadratic polynomials suggests that $p(x)$ can be factored, with one of the factors being $(x - a)$.

THEOREM 1.3.5

> **The Factor Theorem**
>
> Suppose $p(x)$ is a polynomial, and suppose a is a **root** of $p(x)$, that is, a real number satisfying $p(a) = 0$. Then there is a polynomial $q(x)$ so that
> $$p(x) = (x - a)\,q(x).$$

The other factor $q(x)$ can be found quite easily, using long division.

EXAMPLE 8 If $p(x) = x^4 - 2x^3 + 4x^2 - 5x - 6$, we see that $p(2) = 2^4 - 2(2^3) + 4(2^2) - 5(2) - 6 = 16 - 16 + 16 - 10 - 6 = 0$. The Factor Theorem says that we should be able to find a polynomial $q(x)$ so that $p(x) = (x - 2)\,q(x)$. To find $q(x)$, we divide both sides of this equation by $(x - 2)$ and find $q(x) = \frac{p(x)}{x-2}$. We have to divide, and this suggests long division. In fact, this is easier than it sounds.

> At each stage, put in the term that makes the highest powers of x match up.

$$
\begin{array}{r}
x^3 \qquad\quad + 4x + 3 \\
(x-2)\overline{\smash{\big)}\,x^4 - 2x^3 + 4x^2 - 5x - 6} \\
\underline{x^4 - 2x^3 \qquad\qquad\qquad} \\
0 \; + 4x^2 - 5x - 6 \\
\underline{4x^2 - 8x \qquad} \\
3x - 6 \\
\underline{3x - 6} \\
0
\end{array}
$$

The quotient is $x^3 + 4x + 3$, and it is very easy to check that

$$(x - 2)(x^3 + 4x + 3) = x^4 - 2x^3 + 4x^2 - 5x - 6,$$

as desired.

EXAMPLE 9 Factor the polynomial $p(x) = x^2 + 6x + 6$.

Solution The obvious attempts to factor $p(x)$ by inspection do not work.

We can use the Quadratic Formula to find a root a, and the Factor Theorem assures us that $p(x)$ has $(x - a)$ as a factor.

The roots of $p(x)$ are $\frac{-6 \pm \sqrt{36 - 24}}{2} = -3 \pm \frac{\sqrt{12}}{2} = -3 \pm \frac{2\sqrt{3}}{2} = -3 \pm \sqrt{3}$, so $\left(x - (-3 + \sqrt{3})\right)$ and $\left(x - (-3 - \sqrt{3})\right)$ are both factors of $p(x)$, and their product $\left(x - (-3 + \sqrt{3})\right)\left(x - (-3 - \sqrt{3})\right) = (x + 3 - \sqrt{3})(x + 3 + \sqrt{3})$ must divide $p(x)$. We try multiplying out the product:

$$(x + 3 - \sqrt{3})(x + 3 + \sqrt{3}) = x^2 + (3 - \sqrt{3} + 3 + \sqrt{3})x + (3 - \sqrt{3})(3 + \sqrt{3})$$
$$= x^2 + 6x + (9 - 3) = x^2 + 6x + 6.$$

We see that the result is exactly equal to the original polynomial $p(x)$, so we have finished factoring $p(x)$:

$$p(x) = (x + 3 - \sqrt{3})(x + 3 + \sqrt{3}).$$

This procedure will always work for quadratic polynomials, provided there are roots. If $p(x) = ax^2 + bx + c$ has roots r_1 and r_2, then it factors as follows:

$$p(x) = ax^2 + bx + c = a(x - r_1)(x - r_2). \tag{1.3.6}$$

Although it can be extremely difficult to factor a polynomial, especially if its degree is large, there are two useful facts that can sometimes help us guess a root.

Suppose we know how to factor a polynomial $p(x) = k(x - r_1)(x - r_2) \times \ldots \times (x - r_n)$. Then the constant term of $p(x)$ is just $\pm k$ times the product of all the roots: $(-1)^n k r_1 \times \ldots \times r_n$. If k and all the roots happen to be integers, then this means that each root must divide the constant term of $p(x)$.

In fact, something similar is true about any integer root of $p(x)$, even if some of the other roots are not integers, provided every coefficient of $p(x)$ is an integer. There is also a related statement about rational roots.

THEOREM 1.3.7

> **The Rational Root Theorem**
>
> Suppose $p(x) = a_n x^n + a_{n-1} x^{n-1} + \ldots + a_1 x + a_0$ is a polynomial whose coefficients $a_0, a_1, \ldots, a_n$ are all *integers*.
>
> (i) If r is a root of $p(x)$ that is an integer, then a_0 is an integer multiple of r, that is, $a_0 = rm$, for some integer m. If $r \neq 0$, this just means r divides a_0.
>
> (ii) If $r = \frac{p}{q}$ is a root of $p(x)$ that is a rational number, then provided that $\frac{p}{q}$ is written in *lowest terms*, the constant term a_0 of $p(x)$ must be an integer multiple of the numerator p and the leading coefficient a_n must be an integer multiple of the denominator q.

The value of this theorem is that it limits the number of integers or rational numbers we have to consider when looking for a root.

EXAMPLE 10

(i) If $p(x) = x^3 - 5x^2 - 2x + 10$, we note that its coefficients are all integers, so any integer root must divide the constant term 10. This means that the only possible integer roots are $\pm 1, \pm 2, \pm 5, \pm 10$. We check and find that $x = 5$ is in fact a root. We saved time by not bothering to check $\pm 3, \pm 4$.

By long division we can easily find that $p(x) = x^3 - 5x^2 - 2x + 10 = (x - 5)(x^2 - 2)$. From this we see that the other two roots are $x = \pm \sqrt{2}$, which are not integers.

(ii) Consider the polynomial $q(x) = \left(x - \frac{1}{3}\right)(x - 3) = x^2 - \frac{10}{3}x + 1$. Notice that $x = 3$ is a root which is also an integer, but it does *not* divide the constant term 1. This is not a contradiction because the coefficients of $q(x)$ are not all integers (the middle one is $\frac{10}{3}$).

If the coefficients are not integers, the theorem simply does not apply.

(iii) Let $R(x) = x^4 + 2x^3 - x^2 - 6x - 6$. Since the coefficients of $R(x)$ are all integers, any integer root must divide -6. The only possibilities are $\pm 1, \pm 2, \pm 3, \pm 6$.

We evaluate $R(x)$ at each of these values of x and find that $R(x)$ is not zero at any of them. We conclude that $R(x)$ has no integer roots. (In fact, it can be written as $(x^2 - 3)(x^2 + 2x + 2)$; the second factor has no roots, so the roots of $R(x)$ are $x = \pm \sqrt{3}$.)

In part (iii) of this example we were able to check eight possible roots and then use the theorem to conclude that there are no integer roots. Without the theorem we might easily have spent our time checking larger and larger integers, reluctant to give up too soon.

The theorem allows us to save time by not checking unnecessary integers and tells us when there is no point in checking further.

Factorials and the Binomial Theorem

Frequently we have to take expressions and square them, cube them, or raise them to other powers. The Binomial Theorem gives a formula for the nth power of a sum $(a + b)$, where n is a positive integer. The quantities a and b can be numbers or polynomials or even more complicated expressions.

First we need some notation. We often need to use the product of the first n positive integers, $(1)(2)(3) \times \ldots \times (n-2)(n-1)(n)$. It is convenient to have a notation for this product; we write $(1)(2)(3) \times \ldots \times (n-2)(n-1)(n) = n!$ and read it "n **factorial**."

So, for instance, $3! = (1)(2)(3) = 6$, $5! = (1)(2)(3)(4)(5) = 120$, and so on. Factorials have the useful property that $n! = n(n-1)!$.

It would be convenient to be able to use this relation even when $n = 1$. To be able to do that, we need to say that $0! = 1$. This seems a bit peculiar at first, but in fact it is the sensible way to make the definition.

DEFINITION 1.3.8

> **Factorials**
>
> If n is a positive integer, we define $n!$ to be the product of the first n positive integers:
>
> $$n! = (1)(2)(3) \times \ldots \times (n-2)(n-1)(n).$$
>
> If $n = 0$, we define $0! = 1$.
> The symbol $n!$ is read "n **factorial**."

Using factorials, we can define another very useful symbol. It can be interpreted as "the number of different ways of choosing k elements from a set of n elements." We will not be concerned with this interpretation here but simply make the definition.

DEFINITION 1.3.9

> If n and k are integers with $0 \leq k \leq n$, then the symbol $\binom{n}{k}$ is defined as follows:
>
> $$\binom{n}{k} = \frac{n!}{k!(n-k)!}.$$
>
> We read $\binom{n}{k}$ as "n choose k."

EXAMPLE 11 Evaluate: (i) $\binom{3}{2}$; (ii) $\binom{4}{3}$; (iii) $\binom{5}{2}$; (iv) $\binom{6}{0}$.

Solution

(i) By the definition, $\binom{3}{2} = \frac{3!}{2!(3-2)!} = \frac{6}{2!1!} = \frac{6}{2} = 3$.

(ii) $\binom{4}{3} = \frac{4!}{3!(4-3)!} = \frac{24}{3!1!} = \frac{24}{6} = 4$.

(iii) $\binom{5}{2} = \frac{5!}{2!(5-2)!} = \frac{120}{2!3!} = \frac{120}{12} = 10.$

(iv) $\binom{6}{0} = \frac{6!}{0!(6-0)!} = \frac{720}{(1)6!} = \frac{720}{720} = 1.$

EXAMPLE 12 Suppose n is any positive integer. Then it is easy to evaluate

$$\binom{n}{0} = \frac{n!}{0!\,n!} = 1, \qquad \binom{n}{1} = \frac{n!}{1!\,(n-1)!} = \frac{n!}{(n-1)!} = n,$$

$$\binom{n}{n-1} = \frac{n!}{(n-1)!\,(n-(n-1))!} = \frac{n!}{(n-1)!\,1!} = n, \qquad \binom{n}{n} = \frac{n!}{n!\,0!} = 1.$$

With this notation it is now possible to state the Binomial Theorem.

THEOREM 1.3.10

The Binomial Theorem

If n is a positive integer, then

$$(a+b)^n = a^n + \binom{n}{1}a^{n-1}b + \binom{n}{2}a^{n-2}b^2 + \ldots + \binom{n}{n-2}a^2b^{n-2}$$
$$+ \binom{n}{n-1}ab^{n-1} + b^n$$
$$= a^n + \frac{n!}{1!(n-1)!}a^{n-1}b + \frac{n!}{2!(n-2)!}a^{n-2}b^2 + \ldots$$
$$+ \frac{n!}{(n-2)!2!}a^2b^{n-2} + \frac{n!}{(n-1)!1!}ab^{n-1} + b^n$$
$$= a^n + na^{n-1}b + \frac{n!}{2!(n-2)!}a^{n-2}b^2 + \ldots + \frac{n!}{(n-2)!2!}a^2b^{n-2}$$
$$+ nab^{n-1} + b^n.$$

(We have used Example 12 to simplify some of the expressions here).

Notice that each term in the sum contains the product of a power of a and a power of b; the two exponents always add up to n. Even the first and last terms can be seen this way if we remember that $b^0 = 1$ and $a^0 = 1$.

The coefficient of $a^{n-k}b^k$ is $\binom{n}{k}$.

EXAMPLE 13

(i) $(x+5)^2 = x^2 + \binom{2}{1}x(5) + 5^2 = x^2 + 2(5)x + 25 = x^2 + 10x + 25.$

(ii) $(x-7)^3 = x^3 + \binom{3}{1}x^2(-7) + \binom{3}{2}x(-7)^2 + (-7)^3$
$= x^3 + 3x^2(-7) + 3x(49) - 343 = x^3 - 21x^2 + 147x - 343.$

(iii) $(x^2 - 5x)^4 = (x^2)^4 + \binom{4}{1}(x^2)^3(-5x) + \binom{4}{2}(x^2)^2(-5x)^2 + \binom{4}{3}(x^2)(-5x)^3 +$
$(-5x)^4 = x^8 + 4x^6(-5x) + \frac{4!}{2!2!}x^4(25x^2) + 4x^2(-125x^3) + 625x^4 =$
$x^8 - 20x^7 + 150x^6 - 500x^5 + 625x^4.$

The Binomial Theorem with $n = 2, 3$ gives results about squares and cubes that we should remember:

$$(a+b)^2 = a^2 + 2ab + b^2, \qquad (a+b)^3 = a^3 + 3a^2b + 3ab^2 + b^3.$$

Rational Functions

Finally we discuss rational functions. Just as we constructed the rational numbers because we could not divide two integers, we can divide one polynomial by another and call the result a rational function.

So for instance $R(x) = \frac{x^2+x-5}{x-2}$ is a rational function. If we specify a value of x, then we can calculate what $R(x)$ is. There is one complication, however. The quotient does not make sense if the denominator is zero, because division by zero is not allowed. With this particular example, that means we cannot let $x = 2$.

DEFINITION 1.3.11

> A **rational function** is an expression of the form $R(x) = \frac{p(x)}{q(x)}$, where $p(x)$ and $q(x)$ are both polynomials.
>
> The rational function $R(x)$ is not defined at any x for which $q(x) = 0$, that is, where the denominator is zero.

EXAMPLE 14

(i) The rational function $R(x) = \frac{x^3+2x^2+3x+5}{x^2-x-6}$ is not defined when the denominator is zero. Since $x^2 - x - 6 = (x-3)(x+2)$, we see that the denominator is zero at $x = 3$ and $x = -2$, and nowhere else. The rational function $R(x)$ makes sense for all other values of x.

(ii) The rational function $Q(x) = \frac{-7}{x^2+4}$ is defined for *every* value of x because the denominator is never zero (it is 4 plus a square, so it can never be zero).

(iii) Consider the rational function $S(x) = \frac{x^2-9}{x-3}$. It does not make sense when $x = 3$ but can be evaluated at every other real number.

However, if we look at it carefully, we might recognize that the numerator is a difference of squares. If we write $\frac{x^2-9}{x-3} = \frac{(x-3)(x+3)}{x-3}$, it is very tempting to cancel out the $x - 3$ from top and bottom. This suggests that $S(x) = x + 3$.

This is true, *for every x except x = 3*. However, when $x = 3$, $S(x)$ is not defined and $x + 3$ is defined (it equals 6). The cancellation is correct when $x \neq 3$, but canceling amounts to *dividing* both numerator and denominator by $x - 3$. When $x = 3$, this would mean dividing by zero, which is not permissible.

When $x \neq 3$, we can cancel, and we find that for every such x, $S(x) = x + 3$. When $x = 3$, $S(x)$ does not make sense, but $x + 3$ does make sense. They are *not* the same. ◢

Exercises 1.3

Factor the polynomial in each of the following questions.

1. $x^2 + x - 12$

2. $x^2 - x - 12$

3. $x^2 + 3x - 4$

4. $x^2 - 2x - 15$

5. $2x^2 + 4x + 2$

6. $4x^2 - 12x + 9$

7. $x^2 - 4$

8. $x^2 - 6$

9. $(x - 7)^2 - 7$

10. $(x + 2)^2 - 16$

Find the roots (if any) of each of the following polynomials.

11. $x^2 + 3x - 2$

12. $x^2 + 3x + 2$

13. $x^2 - 4x - 1$

14. $x^2 + 2x + 2$

15. $x^2 - x + 1$

16. $9x^2 + 12x + 4$

17. $x^2 - 6x$

18. $3x^2 + 4x + 1$

19. $5x^2 - 4x + \frac{1}{2}$

20. $4x^3 + 8x^2 + 3x$

Describe each of the following sets in terms of intervals.

21. $\{x : x^2 - 2x < 0\}$

22. $\{x : x^2 + 2x \le 8\}$

23. $\{x : x^2 + 4x + 3 \ge 0\}$

24. $\{x : x^2 < 6x\}$

25. $\{x : x^2 \ge 16\}$

26. $\{x : x^2 - 4x + 5 \ge 0\}$

27. $\{x : 2x^2 - 8x + 4 \ge 0\}$

28. $\{x : x^2 + 6x + 10 \le 0\}$

29. $\{x : x^2 - 8x + 19 \ge 0\}$

30. $\{x : 0 \le x^2 + 4x \le 12\}$

For each of the following polynomials $p(x)$, find a small integer a for which $p(a) = 0$, and then find $q(x)$ for which $p(x) = (x - a) q(x)$.

31. $p(x) = x^3 + 3x^2 + 4x + 2$

32. $p(x) = x^4 - 2x + 1$

33. $p(x) = x^3 - 3x^2 + 3x - 2$

34. $p(x) = 2x^4 + x^3 - 2x - 1$

35. $p(x) = x^3 - 1$

36. $p(x) = x^4 - 1$

37. $p(x) = x^3 - 27$

38. $p(x) = x^4 - 16$

39. $p(x) = x^4 - 5x^3 + 3x^2 - 11x$

40. $p(x) = x^8 - 1$

Expand each of the following.

41. $(x + 4)^3$

42. $(x - 3)^4$

43. $(x + 1)^5$

44. $(x - 1)^6$

45. $(x^2 - 2)^3$

46. $\left(x - \frac{1}{2}\right)^4$

47. $(1 - 2)^9$

48. $(3x - 2)^4$

49. $(1 - 4x)^3$

50. $(3\sqrt{2} - 2)^4$

1.4 Functions and Graphs

DOMAIN
RANGE
COORDINATES
AXES

In Section 1.3 we noted that a polynomial $p(x)$ could be "evaluated" at any particular value of x. For example, if $p(x) = x^3 - 2x + 7$, then $p(1) = 1^3 - 2(1) + 7 = 6$, $p(0) = 7$, and $p(-2) = (-2)^3 - 2(-2) + 7 = 3$. This is an example of a **function**.

The idea is that a function f takes a number a and gives us a result, which is often written $f(a)$. It will do this for various values of a. The result $f(a)$ is often called the **value of the function $f(x)$ at $x = a$**.

For instance if $f(x)$ is the polynomial $2x^2 + 3$, we can easily find its value at any point: $f(0) = 3, f(3) = 2(3^2) + 3 = 21$, and $f(-3) = 2(-3)^2 + 3 = 21$. These last two values are equal, and of course we realize that the same will be true of the values of this $f(x)$ at $x = a$ and $x = -a$, for any real number a.

Rational functions, as the name implies, are also functions. For instance, $R(x) = \frac{x^3 + x - 4}{x^2 - 1}$ can easily be evaluated. One difference, however, is that there are some values of x at which the function cannot be evaluated (in this case the function does not make any sense at $x = 1$ or at $x = -1$). This is not especially serious. A function $f(x)$ makes sense for some x's and not for others; this is frequently described by saying the function is *defined* for some x's and *not defined* for some others.

There are examples of functions that are not defined for large sets of x's. For instance, the square root function $f(x) = \sqrt{x}$ is defined only when $x \ge 0$.

The set of points at which a function is defined is called the **domain** of the function.

EXAMPLE 1

(i) The domain of any polynomial is the set of all real numbers (because the formula for a polynomial can be evaluated at any number $x = a$).

Very often the domain of a function will be clear from the context, but sometimes it will be less clear, and sometimes it will be necessary to specify the domain very carefully or deliberately change it.

For instance, we could define a function for $x \in [1, 3) \cup (5, 7)$ by $F(x) = \frac{x^5 - 2}{x - 3}$. The rational function $\frac{x^5 - 2}{x - 3}$ makes sense for every x except $x = 3$, but we have declared that the domain of $F(x)$ is exactly $[1, 3) \cup (5, 7)$.

We say that the domain of $F(x)$ is $[1, 3) \cup (5, 7)$, but the **natural domain** of the rational function $\frac{x^5 - 2}{x - 3}$ is $(-\infty, 3) \cup (3, \infty)$. The natural domain is the set of numbers for which a formula makes sense, even if we specify some (smaller) set as the domain. This complication will not arise often.

Figure 1.4.1

(ii) The domain of a rational function $R(x) = \frac{p(x)}{q(x)}$ is the set of all x's for which $q(x) \neq 0$. It is convenient to write this in symbols as

$$\text{domain}\big(R(x)\big) = \{x : q(x) \neq 0\}.$$

(iii) The domain of the square root function is the set of all nonnegative numbers; if $f(x) = \sqrt{x}$, then $\text{domain}\big(f(x)\big) = \{x : x \geq 0\}$.

(iv) What is the domain of the function $g(x) = \sqrt{1 - 2x}$?

Solution We know that $\sqrt{1 - 2x}$ makes sense only if $1 - 2x \geq 0$, that is, if $1 \geq 2x$, that is, $\frac{1}{2} \geq x$. We see that $\text{domain}\big(g(x)\big) = \{x : x \leq \frac{1}{2}\}$.

(v) What is the domain of the function $h(x) = \sqrt{x/(x^2 - 1)}$?

Solution What we need to know is when the quantity inside the square root sign is nonnegative. First we notice that it does not even make sense if $x = \pm 1$, since the denominator is zero there. Second, a quotient is positive when both its numerator and denominator are positive or both are negative. The denominator is positive when $x^2 - 1 > 0$, that is, $x^2 > 1$, that is, $x > 1$ or $x < -1$. The denominator is negative when $-1 < x < 1$.

The numerator is positive when $x > 0$ and negative when $x < 0$, so the numerator and denominator are both positive when $x > 1$, and both negative when $-1 < x < 0$. The quotient inside the square root sign is positive when $x \in (-1, 0) \cup (1, \infty)$. It is zero when $x = 0$.

Combining these remarks, we see that the function $f(x)$ is defined when $x \in (-1, 0] \cup (1, \infty)$, so $\text{domain}\big(h(x)\big) = (-1, 0] \cup (1, \infty)$. This set is illustrated in Figure 1.4.1. ◢

It is also necessary to discuss the set of values of a function, meaning the set of all numbers that are values $f(x)$ for some $x \in \text{domain}(f)$. This set is called the **range** of the function $f(x)$. In symbols we write

$$\text{range}(f) = \{f(x) : x \in \text{domain}(f)\} = \{y : y = f(x), \text{ for some } x \in \text{domain}(f)\}.$$

EXAMPLE 2

(i) What is the range of the polynomial $p(x) = x^2 + 2$?

Solution Since x^2 is always nonnegative, we see that $x^2 + 2 \geq 2$ for every x, and the range of $p(x)$ is contained in $[2, \infty)$.

The question is whether or not *every* element of $[2, \infty)$ is in the range. Suppose $r \geq 2$; let us try to find x for which $p(x) = r$. This means $x^2 + 2 = r$, or $x^2 = r - 2$. The solutions will be $x = \pm\sqrt{r - 2}$, provided this makes sense, that is, provided that the quantity $r - 2$ inside the square root sign is nonnegative. This condition, $r - 2 \geq 0$, is the same as $r \geq 2$, which is exactly what we have assumed is true.

This shows that $\text{range}\big(p(x)\big) = [2, \infty)$.

(ii) What is the range of $q(x) = 3\sqrt{x} - 7$?

Solution We know that $\sqrt{x} \geq 0$ for every x in the domain of $\sqrt{x}$. This shows that $3\sqrt{x} \geq 0$, so $q(x) = 3\sqrt{x} - 7 \geq -7$, which makes us suspect that the range of $q(x)$ is $[-7, \infty)$.

We still have to show that every element of $[-7, \infty)$ is in the range; suppose $r \in [-7, \infty)$ and let us try to find x so that $q(x) = r$. In other words, we have to solve $3\sqrt{x} - 7 = r$, or $3\sqrt{x} = r + 7$, or $\sqrt{x} = \frac{r+7}{3}$. It is easy to find what x should be: $x = \left(\frac{r+7}{3}\right)^2$. The subtlety here is that with this value of x, the square root is $\sqrt{x} = \sqrt{\left((r+7)/3\right)^2} = |\frac{r+7}{3}|$, so we have to verify that this equals $\frac{r+7}{3}$. This amounts to checking that $\frac{r+7}{3} \geq 0$, and this is the same thing as $r + 7 \geq 0$, or $r \geq -7$. Once again, this is exactly what we have assumed (based on an intelligent guess at the answer).

We have shown that the range of $q(x)$ is $[-7, \infty)$.

When we consider a polynomial, $p(x) = 5x^4 - 7x^3 + 2x^2 - 1$, for instance, it is reasonably easy to evaluate it at any point x. The same is true of rational functions. In a way this is misleading. Many functions cannot be described in a way that makes them easy to calculate.

For one type of example, consider the polynomial $p(x) = x^7 + 3x^5 + 2x^3 + x + a$. It turns out that for any particular value of the constant a, there is exactly *one* value of x for which $p(x) = 0$, that is, for which $x^7 + 3x^5 + 2x^3 + x = -a$. This is not obvious (we will learn why it is true in Chapter 4), but accepting that it is true, let us define a function $f(x)$ as follows: For any real number a, $f(a)$ is that number x for which $x^7 + 3x^5 + 2x^3 + x = -a$.

This function $f(x)$ makes perfectly good sense, but it is not possible to write down a formula for it in terms of elementary functions (this is not to say that it is merely difficult, but rather that it cannot be done at all).

There are many other examples of functions that arise in quite natural ways but for which there is no nice formula. Part of the usefulness of calculus is that it provides means for dealing with some of these functions.

Another type of function we will need to use is one that is defined by one formula for some values of x and another for other values of x. For instance, consider the function

$$f(x) = \begin{cases} 2x + 7\sqrt{2}, & \text{if } x \leq 3, \\ \frac{1}{x^2-4}, & \text{if } x > 3. \end{cases}$$

This is a perfectly good function. The formula tells us quite clearly how to evaluate the function at every real number.

It is very important to become accustomed to the idea that not all functions are polynomials and rational functions.

Graphing Functions

We recall how to describe points in the plane by their coordinates. We start with a horizontal line, which we call the **x-axis**, and a vertical line, which we call the **y-axis**. The point at which the axes cross is called the **origin**. The position of any point P in the plane can be described by giving an ordered pair of numbers (x, y). To find the corresponding point, we start at the origin, move x units along the x-axis, interpreting a positive number x to mean a movement to the right of the origin and a negative x to mean a movement to the left. Then we move y units in the vertical direction, interpreting positive values of y to refer to positions above the x-axis and negative values to refer to positions below the axis. See Figure 1.4.2.

Several points are shown in Figure 1.4.3, marked with their coordinates. The horizontal coordinate, usually denoted x, is called the **x-coordinate** or the **abscissa** of the point, and the vertical coordinate, usually denoted y, is called the **y-coordinate** or

Figure 1.4.2

Figure 1.4.3

René Descartes

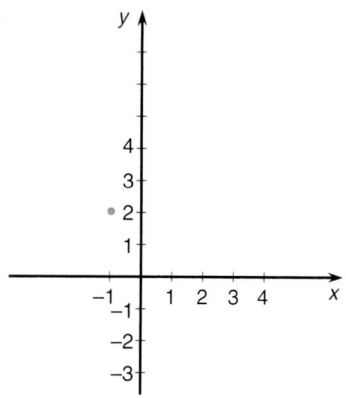

Figure 1.4.4

the **ordinate**. The ordered pair (x, y) are sometimes called the **Cartesian coordinates** of the point P, after the French philosopher and mathematician René Descartes (1596–1650).

Suppose $f(x)$ is a function. For each x in its domain, there is a value $f(x)$, and if we think of this value as the y-coordinate of a point, then we can draw the corresponding point on a graph. In other words, letting $y = f(x)$, we can plot the point (x, y).

Consider the function $f(x) = x^2 - 2x - 1$. Corresponding to each x we have the value $y = x^2 - 2x - 1$, and this is easy to evaluate for any particular x. For instance, if $x = -1$, we find that $y = (-1)^2 - 2(-1) - 1 = 1 + 2 - 1 = 2$, and we plot the point $(-1, 2)$. See Figure 1.4.4.

In a similar way we can find other points, such as $(0, -1)$, $(1, -2)$, $(2, -1)$, $(3, 2)$, $(4, 7)$, $(\frac{1}{2}, -\frac{7}{4})$, $(-1, 2)$, and so on. See Figure 1.4.5.

In Figure 1.4.6 we have plotted all the points with x moving from -1 to 4 in steps of $\frac{1}{2}$, and we begin to see a pattern emerging.

With the help of a graphics calculator it is possible to sketch the graph of a function quickly and accurately. For instance, the keystrokes below show how to use various calculators to sketch the graph of the function $y = x^4 - 2x^2 + 1$ for $x \in [-2, 2]$.

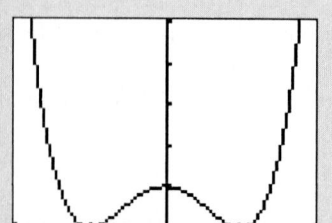

TI-85	SHARP EL9300
GRAPH F1 $(y(x) =)$ F1 $(x)\hat{}4 - 2$ F1 $(x)\hat{}2 + 1$ M2 (RANGE) $(-)2$ ▼ 2 ▼ ▼ 0 ▼ 5 F5 (GRAPH)	⌁ x/θ/T a^b 4 ▶ -2 x/θ/T $x^2 + 1$ RANGE -2 ENTER 2 ENTER 1 ENTER 0 ENTER 5 ENTER 1 ENTER ⌁

CASIO f_x-7700GB/f_x-6300G	HP 48SX
Cls EXE Range -2 EXE 2 EXE 1 EXE 0 EXE 5 EXE 1 EXE EXE EXE EXE Graph X x^y 4 $- 2$ X x^y 2 $+ 1$ EXE	⬛ PLOT PLOTR ERASE ATTN ' αY ⬛ αX y^x 4 $- 2 \times \alpha X$ y^x 2 $+ 1$ ↵ PLOT ⬛ DRAW ERASE 2 $+/-$ ENTER 2 XRNG 1 $+/-$ ENTER 5 YRNG DRAW

If you have a graphics calculator, you can use it to sketch the graphs of the functions we have been discussing. However, it is important to realize that it is not always easy to determine the overall shape of a graph. For instance, you could try editing any of the above sequences of keystrokes to graph the function $y = x^4 - 5x^3 + x$ for $x \in [-2, 2]$. You could then try the same function with $x \in [-3, 3]$ and with $-50 \le y \le 200$. Then you could try it with $x \in [-3, 6]$ and $-70 \le y \le 200$. The first two pictures will give seriously misleading impressions about the shape of the graph. . .

Figure 1.4.5

Figure 1.4.6

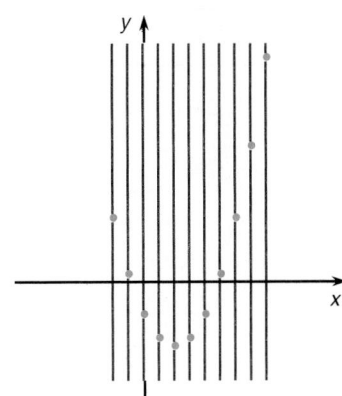

Figure 1.4.7

Since $f(x) = x^2 - 2x - 1$ is a polynomial, its domain is the set of all real numbers. There is exactly one point on the graph on each vertical line in the plane. See Figure 1.4.7.

If we could plot all the points $(x, f(x))$, we would obtain a curve that goes right across the plane from one side to the other. If we plot a reasonably large number of points, we can "fill in" the curve and get a fairly good picture of it (or at least of some part of it). This picture is called the **graph** of the function $f(x)$, or the graph of $y = f(x)$. In this particular case the curve is a "parabola," shown in Figure 1.4.8.

If we tried to graph a rational function like $R(x) = \frac{1}{x^2-1}$, for instance, we would realize that its domain is the set of all real numbers *except* ± 1. There can be no point on the graph with x-coordinate 1 or -1. On the other hand, for every other value of x, that is, for every x in the domain of f, there is exactly one point on the graph with that x as its x-coordinate. The graph of $y = \frac{1}{x^2-1}$ is shown in Figure 1.4.9.

The domain of $g(x) = \sqrt{x}$ is $\{x : x \geq 0\}$. There is a point $\left(x, g(x)\right)$ on the graph of $y = g(x)$ for each $x \geq 0$ (see Figure 1.4.10). Negative values of x are not in the domain of $g(x)$; in the picture there are no points on the graph lying to the left of the y-axis.

When we draw the graph of a function, there can never be more than one point on the graph on any vertical line. This observation shows that a circle cannot be the graph of a function, since there are vertical lines that intersect the circle in two points. See Figure 1.4.11.

Figure 1.4.8

Figure 1.4.9

Figure 1.4.10

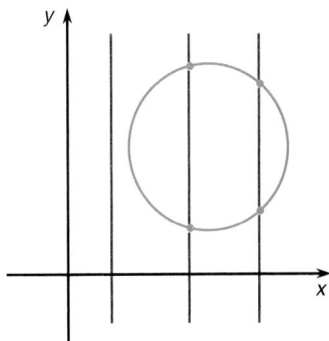

Figure 1.4.11

Exercises 1.4

Specify the domain of each of the following functions.

1. $f(x) = x^2 - 3x + 5$

2. $f(x) = x^5 - 4x^3 + 7x^2 - 4$

3. $f(x) = \frac{x^3 - 2x + 8}{x + 3}$

4. $f(x) = \frac{3x^4 - 2x + 1}{x^2 - 16}$

5. $f(x) = 3x - 5\sqrt{x}$

6. $f(x) = 1 - 2\sqrt{3 - 4x}$

7. $f(x) = x^3 - \frac{1}{\sqrt{x + 1}}$

8. $f(x) = \sqrt{x(x - 4)}$

9. $f(x) = \frac{1}{1 - \sqrt{x}}$

10. $f(x) = \frac{1}{1 - \sqrt{x^2 + 4x + 1}}$

Specify the range of each of the following functions.

11. $f(x) = x^2 + 5$

12. $f(x) = x^4 + 1$

13. $f(x) = (x - 2)^2 + 3$

14. $f(x) = x^2 - 6x + 5$

15. $f(x) = 3x^2 - 5x + 2$

16. $f(x) = 4x^2 - 2x + 1$

17. $f(x) = \sqrt{x - 6}$

18. $f(x) = 3 - 7\sqrt{2 - x}$

19. $f(x) = \frac{1}{1 + \sqrt{x}}$

20. $f(x) = x + 2\sqrt{x}$

Plot six points (or more) on the graph of each of the following functions, and then fill in the approximate shape of the rest of the graph.

21. $y = x^2 - 3$

22. $y = x^2 + 2$

23. $y = (x - 1)^2 - 3$

24. $y = (x + 2)^2 - 1$

25. $y = x^2 + 4x - 2$

26. $y = x^2 + 6x$

27. $y = 2x^2 + 1$

28. $y = 2x^2 + 4x + 3$

29. $y = 2x^2 + 3x - 8$

30. $y = \frac{1}{2}x^2 + 2x + 3$

1.5 Geometry

PYTHAGOREAN THEOREM
HYPOTENUSE
SIMILAR TRIANGLES
EQUATIONS OF LINES
INTERSECTING LINES
AREA

We begin with the Pythagorean Theorem, which is named after the ancient Greek mathematician Pythagoras (570–497 B.C.). Suppose $\triangle ABC$ is a **right-angled triangle**, with its right angle at B (i.e., $\angle ABC = 90°$). The side opposite the right angle (in this case the side AC) is called the **hypotenuse**. See Figure 1.5.1.

THEOREM 1.5.1

The Pythagorean Theorem

In a right-angled triangle the square of the length of the hypotenuse equals the sum of the squares of the lengths of the other two sides.

Consider a right-angled triangle $\triangle ABC$ with a right angle at B. If we write $|AB|$ for the length of the segment AB, and similarly for the other segments, the theorem can be stated as follows:

$$|AC|^2 = |AB|^2 + |BC|^2.$$

EXAMPLE 1

(i) Consider a right-angled triangle in which both sides adjacent to the right angle have length 1. Applying the Pythagorean Theorem, with $|AB| = |BC| = 1$, we find that $|AC|^2 = 1 + 1 = 2$, so $|AC| = \sqrt{2}$. The length of the hypotenuse is $\sqrt{2}$.

(ii) If the hypotenuse of a right-angled triangle has length 5 and one of the other sides has length 4, what is the length of the remaining side?

Solution   If we assume that $|AC| = 5$ and $|AB| = 4$, then the theorem tells us that $5^2 = 4^2 + |BC|^2$, which means that $|BC|^2 = 25 - 16 = 9$, from which it follows that the length of the third side $|BC|$ is 3.

Figure 1.5.1

Figure 1.5.2

Figure 1.5.3

Figure 1.5.4

Figure 1.5.5

Figure 1.5.6

EXAMPLE 2   Suppose a person walks 100 meters due north and then turns and walks 75 m due east. How far will she be from her starting point?

Solution   We begin by drawing a picture like the one in Figure 1.5.2, showing the starting point S, the point N which is 100 m due north of S, and the final point F which is 75 m due east of N. Then $\triangle SNF$ is a right-angled triangle with a right angle at N.

What we want to know is the distance between the starting point and the final point, which means the length of the segment SF. From the Pythagorean Theorem we see that

$$|SF|^2 = |SN|^2 + |NF|^2 = 100^2 + 75^2 = 15{,}625.$$

The length is the square root: $|SF| = \sqrt{15{,}625} = 125$. The person in the question is now 125 meters away from her starting point.

We can also use the Pythagorean Theorem to calculate distances in the plane.

EXAMPLE 3   How far is the point $P = (-2, 5)$ from the origin $O = (0, 0)$?

Solution   We begin by drawing a picture, showing the points P and O and a vertical line from P down to the x-axis. The point where this line meets the x-axis is the point $S = (-2, 0)$. See Figure 1.5.3.

The triangle PSO is a right-angled triangle with right angle at S. The distance $|OP|$ is the length of the hypotenuse.

The Pythagorean Theorem says that $|OP|^2 = |OS|^2 + |SP|^2 = 4 + 25 = 29$. The distance from P to O is $|OP| = \sqrt{29} \approx 5.385$.

The same procedure can be used to find the distance from any point to the origin, or the distance between any two points.

EXAMPLE 4

(i) Find the distance from $P = (x, y)$ to the origin $O = (0, 0)$.

(ii) Find the distance between $P = (x, y)$ and $Q = (a, b)$.

Solution

(i) As in the previous example, we draw the point S on the x-axis with the same x-coordinate as P, that is, $S = (x, 0)$. See Figure 1.5.4. Then $|OP|^2 = |OS|^2 + |SP|^2 = x^2 + y^2$, so $|OP| = \sqrt{x^2 + y^2}$.

(ii) In Figure 1.5.5 we draw a right-angled triangle whose hypotenuse is PQ and with one vertical side and one horizontal side. The length of the horizontal side is the horizontal distance between P and Q, which is $|x - a|$. Similarly the length of the vertical side is $|y - b|$. The Pythagorean Theorem says that the distance between $P = (x, y)$ and $Q = (a, b)$ is

$$|PQ| = \sqrt{|PS|^2 + |SQ|^2} = \sqrt{(x - a)^2 + (y - b)^2}. \qquad (1.5.2)$$

This formula for the distance between two points is extremely important and useful.

The distances between points in the plane satisfy the triangle inequality (see Theorem 1.2.2). What it says in this context is that the length of any side of a triangle is less than or equal to the sum of the lengths of the other two sides. See Figure 1.5.6.

THEOREM 1.5.3

> **The Triangle Inequality**
>
> If P, Q, and R are three points in the plane, then
>
> $$|PQ| \leq |PR| + |QR|.$$

Similar Triangles

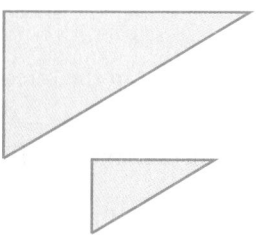

Figure 1.5.7

Two triangles $\triangle ABC$ and $\triangle DEF$ are **similar** if their respective angles are equal, that is, if $\angle ABC = \angle DEF$ and $\angle BCA = \angle EFD$ and $\angle CAB = \angle FDE$. In fact, since the angles of any triangle always add up to a total of $180°$, it is enough to check that any *two* pairs of angles are equal, since the third pair will then automatically have to be equal too.

In Figure 1.5.7 we see two right-angled triangles (notice that this means that one pair of angles, the right angles, are already equal). In fact the triangles are similar. (The angle at the extreme right end of each triangle is $30°$, so these two angles are equal. The remaining angle in each triangle is $180 - 90 - 30 = 60°$.)

The triangles are clearly not the same size; the hypotenuse of one is half as long as the hypotenuse of the other. Because the triangles are similar, that is, have the same angles, we can think of one as a "scaled-down" version of the other. In this particular case it is scaled down by a factor of $\frac{1}{2}$, and we expect that each side of the small triangle should be half as long as the corresponding side of the larger one.

The same idea will work for any similar triangles. The sides of one will all be the same multiple of the "corresponding" sides of the other. It is not immediately clear what is meant by "corresponding," but one good way to say it is to label sides by saying which angle they are *opposite*. So the hypotenuse of a right-angled triangle is the side opposite the right angle. In the example illustrated above, we can speak of the side opposite the $30°$ angle and the side opposite the $60°$ angle. By **corresponding sides** in two similar triangles we mean sides that are opposite equal angles.

THEOREM 1.5.4

> Suppose $\triangle ABC$ and $\triangle DEF$ are similar, with $\angle ABC = \angle DEF$ and $\angle BCA = \angle EFD$ (and so $\angle CAB = \angle FDE$). Then
>
> $$\frac{|AB|}{|DE|} = \frac{|AC|}{|DF|} = \frac{|BC|}{|EF|}.$$
>
> In words we say that the ratios of the lengths of each pair of corresponding sides are all the same.

Figure 1.5.8

EXAMPLE 5   Consider a person standing near a streetlight, as shown in Figure 1.5.8. If the person is 6 ft tall, standing 20 ft from the base of the streetlight, and the shadow is 8 ft long, how high is the streetlight above the ground?

Solution   We draw the picture again, labeling the light A, the person's head C, the end of the shadow B, the person's feet D, and the base of the lamppost O. See Figure 1.5.9.

The large triangle $\triangle ABO$ is similar to the smaller triangle $\triangle CBD$. This is because they have the same angle at B and each has a right angle (at O and D, respectively). We know the person is 6 ft tall, so $|DC| = 6$, and we want to find $|OA|$, the height of the streetlight.

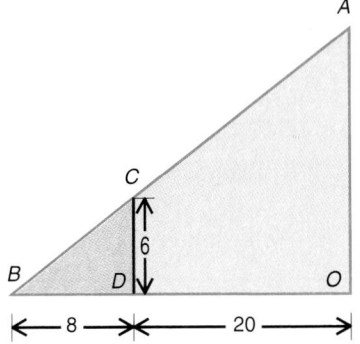

Figure 1.5.9

Now $\frac{|OA|}{|DC|} = \frac{|BO|}{|BD|}$, since these are two pairs of corresponding sides (OA and DC are both opposite to the angle at B, and BO and BD are opposite to angles $\angle BAO$ and $\angle BCD$ in their respective triangles).

We know that the length of the shadow is $|BD| = 8$, so the only thing left to find is $|BO| = |BD| + |DO| = 8 + 20 = 28$.

Putting all this together, we find that

$$\frac{|OA|}{6} = \frac{28}{8}, \qquad \text{so} \qquad |OA| = \frac{(6)(28)}{8} = 21.$$

The streetlight is 21 ft high.

Straight Lines

Consider three points P, Q, and R on a straight line ℓ. We draw some horizontal and vertical lines as shown in Figure 1.5.10 and label the points A, B, and C. There are three similar triangles in this diagram, $\triangle PQA$, $\triangle PRB$, and $\triangle QRC$. From 1.5.4 we conclude that $\frac{|QA|}{|RB|} = \frac{|PA|}{|PB|}$ and $\frac{|RB|}{|RC|} = \frac{|PB|}{|QC|}$, so

$$\frac{|QA|}{|PA|} = \frac{|RB|}{|PB|} = \frac{|RC|}{|QC|}.$$

Each of these ratios is obtained by taking two of the points on the line and taking the quotient of the vertical distance between them by the horizontal distance between them. Since the three points we started with could have been any three points on the line, we see that this ratio will be the same for any pair of points.

If the coordinates of P are (x, y) and the coordinates of Q are (u, v), then the vertical distance between them is $|y - v|$ and the horizontal distance is $|x - u|$. The ratio mentioned above is $\frac{|y-v|}{|x-u|}$, assuming that $x \neq u$.

If, for instance, this ratio equals 1, it means that the horizontal distance and vertical distance between P and Q are the same, which means that the line through the slopes at an angle of $45°$. The line could slope up or down; because of the absolute value signs, the ratio would be the same.

This suggests that it might be better to consider the ratio $\frac{y-v}{x-u}$. If the line slopes upward, then the numerator and denominator will both have the same sign, while if the line slopes downward, they will have opposite signs (check this by looking at Figure 1.5.11).

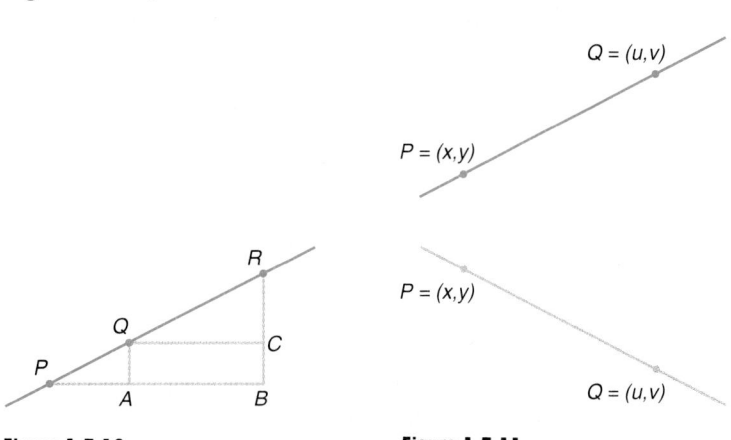

Figure 1.5.10   **Figure 1.5.11**

DEFINITION 1.5.5

If $P = (x, y)$ and $Q = (u, v)$ are two distinct points on the line ℓ, then provided $x \neq u$, the ratio

$$m = \frac{y - v}{x - u}$$

is called the **slope** of the line.
If $x = u$, the line ℓ must be a vertical line, and we say its slope is *undefined*.

REMARK 1.5.6

(i) If the slope of ℓ is $m = 0$, then $y = v$, that is, P and Q have the same y-coordinate, and ℓ is a horizontal line.

(ii) We have already noticed that $m > 0$ if ℓ slopes upward to the right, and $m < 0$ if ℓ slopes downward to the right. See Figure 1.5.12.

(iii) If $m = \frac{y-v}{x-u}$ is a large positive number, then $y - v$ is large in comparison to $x - u$, which means that the line slopes *steeply* upward. If m is a small positive number, it means that ℓ slopes gradually upward. If m is negative but close to zero, then ℓ slopes gradually downward, and if m is a negative number whose absolute value is large, then ℓ slopes steeply downward. See Figure 1.5.13.

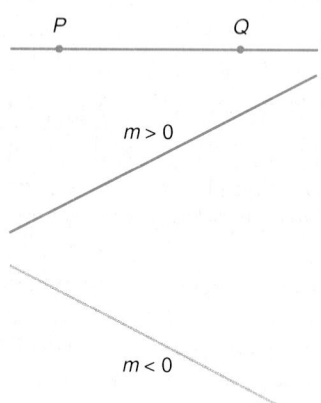

Figure 1.5.12

Now suppose ℓ is a line containing the point $Q = (x_0, y_0)$ and whose slope is m (in particular, this means the line is not vertical, since the slope is not defined for a vertical line). Let $P = (x, y)$ be any other point on the line ℓ (see Figure 1.5.14). Then we know that $\frac{y - y_0}{x - x_0} = m$. Multiplying by $x - x_0$, we find

$$y - y_0 = m(x - x_0), \qquad \text{or} \qquad y = mx + (y_0 - mx_0). \tag{1.5.7}$$

This is called the **point-slope form** of the equation of the line ℓ, because to write it, we need to know a point on the line and its slope. The points $P = (x, y)$ that satisfy it are exactly the points on the line.

If we let $b = y_0 - mx_0$, Equation 1.5.7 can be written as

$$y = mx + b. \tag{1.5.8}$$

This is called the **slope-intercept form** of the equation of the line ℓ. It involves the slope m, but we need to discover the significance of the constant b. If we let $x = 0$, we find from Equation 1.5.8 that $y = b$. So the point $(0, b)$ is on the line; b is the y-coordinate of the point at which ℓ crosses the y-axis (see Figure 1.5.15). This y-coordinate b is called the **y-intercept** of ℓ, which explains the name.

Figure 1.5.13

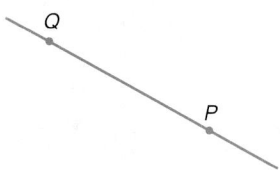

Figure 1.5.14

EXAMPLE 6   Find the equation of each of the following lines: (i) the line containing the point $(3, 1)$ and having slope $m = -2$; (ii) the line containing the points $(1, 2)$ and $(5, -2)$; (iii) the line whose slope is $m = 3$ and whose y-intercept is $b = -2$.

Solution

(i) The point-slope form is $y - 1 = -2(x - 3)$, that is, $y = 1 - 2x + 6$, or $y = 7 - 2x$. It can also be written as $y + 2x = 7$, or $y + 2x - 7 = 0$, or $2x + y - 7 = 0$, or $2x + y = 7$.

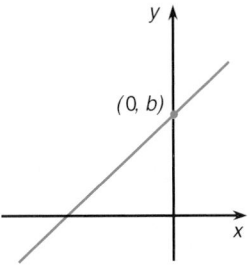

Figure 1.5.15

(ii) We need to find the slope. Using the two points we are given, we find that $m = \frac{2-(-2)}{1-5} = \frac{4}{-4} = -1$. Then from the point-slope form, the equation of the line is $y - 2 = -(x - 1)$, or $y + x - 3 = 0$.

(iii) The point-intercept form gives $y = 3x + (-2)$, or $y = 3x - 2$, which can also be written as $y - 3x + 2 = 0$.

We might observe that we have been able to write the equation of each of these lines in the form $ax + by + c = 0$, where a, b, and c are real numbers.

If a and b are not both zero, the equation

$$ax + by + c = 0 \qquad\qquad (1.5.9)$$

is called the **general form of the equation of a line**. Any straight line has an equation in this form, and any equation in this form is the equation of a straight line.

REMARK 1.5.10

(i) The line $ax + by + c = 0$ is vertical if $b = 0$, and it is horizontal if $a = 0$. Notice that if $b \neq 0$, it is possible to rearrange and divide by b to obtain $y = -\frac{ax+c}{b}$. This expresses y as a function of x. If $b = 0$, the line is vertical, and it is not possible to express y as a function of x (recall that the graph of a function can intersect any vertical line in at most one point).

(ii) If $b \neq 0$, the line $ax + by + c = 0$ is not vertical. We have seen that the equation can be written as $y = -\frac{ax+c}{b}$, or $y = -\frac{a}{b}x - \frac{c}{b}$. This is the slope-intercept form of the equation; in particular, for the line defined by Equation 1.5.9, if $b \neq 0$,

$$y\text{-intercept} = -\frac{c}{b}, \qquad \text{slope} = -\frac{a}{b}. \qquad (1.5.11)$$

While we are doing this, we might also like to find the **x-intercept**. A point (x, y) is on the x-axis if $y = 0$, so if the point $(x, 0)$ is on the line $ax + by + c = 0$, we see that $ax + b(0) + c = 0$, or $ax + c = 0$. If $a \neq 0$, this shows that the x-intercept is $x = -\frac{c}{a}$.

(iii) If $ax + by + c = 0$ is the equation of the line ℓ, then the equation $2ax + 2by + 2c = 0$ is *also* an equation of ℓ. So is the equation we get by multiplying by *any* nonzero number.

(iv) Suppose two lines ℓ_1 and ℓ_2 have the same slope. There are two possibilities: Either they are the same line or they never meet, that is, they have no points in common. In this latter case we say they are **parallel lines**. See Figure 1.5.16.

Figure 1.5.16

EXAMPLE 7 Find the point of intersection of the lines $3x + 2y + 1 = 0$ and $2x - y - 4 = 0$.

Solution If (x, y) is the point of intersection, it lies on both lines. This means it satisfies both equations:

$$3x + 2y + 1 = 0 \qquad \text{and} \qquad 2x - y - 4 = 0.$$

We use one of the equations to solve for one of the variables x and y in terms of the other. For instance, the first equation can be rearranged into $2y = -3x - 1$, so $y = -\frac{1}{2}(3x + 1)$. Then we "substitute" this expression for y into the second equation: $2x - y - 4 = 0$ becomes $2x + \frac{1}{2}(3x + 1) - 4 = 0$, or $2x + \frac{3}{2}x + \frac{1}{2} - 4 = 0$. Combining terms, we find that $\frac{7}{2}x - \frac{7}{2} = 0$, or $\frac{7}{2}x = \frac{7}{2}$, or $x = 1$.

Next we take this value and substitute it into either equation to find the corresponding value of y. For instance, the first equation $3x + 2y + 1 = 0$ becomes $3(1) + 2y + 1 = 0$, or $2y + 4 = 0$, or $2y = -4$, so $y = -2$.

The point of intersection is $(1, -2)$.

You should check that substituting $x = 1$ into the second equation also gives $y = -2$.

The technique we used was to solve for one variable in terms of the other, using one of the equations, and then substitute the resulting expression into the other equation. This gives an equation with only one variable, so it is easily solved. This process is called "eliminating" one variable; in the solution above, we eliminated y to get an equation involving only x.

We could just as well have eliminated x and solved for y. Another thing to notice is that if we had used the second equation, $2x - y - 4 = 0$, we would have found $2x - 4 = y$, or $y = 2x - 4$. This is simpler because it does not involve any fractions. The same answer will result no matter which variable is eliminated, but sometimes one choice will give easier calculations than another. Specifically, if the coefficient of either x or y is 1 or -1 in either equation, then using that equation allows us to solve for that variable without dividing.

EXAMPLE 8   Find the point of intersection of the two lines $2x - 3y + 1 = 0$ and $-4x + 6y - 3 = 0$.

Solution   We use the first equation to find $2x = 3y - 1$, so $x = \frac{1}{2}(3y - 1)$. Substituting this into the second equation, we find

$$0 = -4x + 6y - 3 = -4\left(\frac{1}{2}\right)(3y - 1) + 6y - 3$$
$$= -2(3y - 1) + 6y - 3 = -6y + 2 + 6y - 3 = -1.$$

We have just found that $0 = -1$. This is clearly impossible. The reason is that in effect we *assumed* that there was a point (x, y) on both lines. These two lines are parallel (check their slopes), so there is no point of intersection. See Figure 1.5.17.

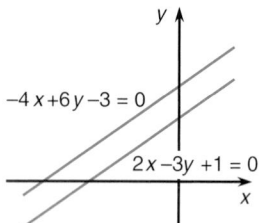

Figure 1.5.17

Finally consider two lines that are perpendicular. Let us assume that neither of them is horizontal (or vertical). In Figure 1.5.18 we have labeled the point of intersection I and have marked a point P on one of the lines. Also shown is a point Q on the other line, with the same y-coordinate as P. Finally we mark the point R between P and Q that is directly above (or below) I.

The right-angled triangles $\triangle QIP$ and $\triangle QRI$ share the angle $\angle PQI = \angle IQR$. Because of this, they are similar triangles.

In the same way, $\triangle QIP$ and $\triangle IRP$ are similar because they share the angle $\angle IPQ = \angle RPI$. Combining these facts, we see that $\triangle QRI$ is similar to $\triangle IRP$. Comparing corresponding sides, we get

$$\frac{|RP|}{|RI|} = \frac{|RI|}{|QR|}.$$

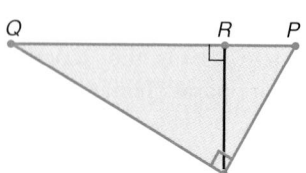

Figure 1.5.18

The right side of this equation is the absolute value of the slope of the line IQ. The left side is the reciprocal of $\frac{|RI|}{|RP|}$, which is the absolute value of the slope of IP. So the absolute values of the slopes of the two perpendicular lines are each other's reciprocals. Since one of the lines must slope up and the other down, we see that one must have positive slope and the other must have negative slope. This means that their slopes must be the *negatives* of each other's reciprocals.

In fact the reverse is also true: If the slopes are negative reciprocals, then the lines are perpendicular.

Neither of these statements applies when one line is horizontal and the other is vertical, since the slope of a vertical line is not defined.

THEOREM 1.5.12

> Suppose ℓ_1 and ℓ_2 are two lines with nonzero slopes m_1 and m_2, respectively. To say that the lines are perpendicular is equivalent to the condition
>
> $$m_1 = -\frac{1}{m_2}.$$

Graphing Inequalities

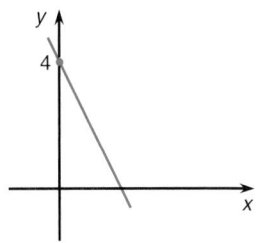

Figure 1.5.19

EXAMPLE 9 Describe each of the following sets of points: (i) $\{(x, y) : y + 2x \leq 4\}$; (ii) $\{(x, y) : 3x - y + 2 < 0\}$.

Solution

(i) We rewrite the condition $y + 2x \leq 4$ as $y \leq -2x + 4$. The corresponding equation $y = -2x + 4$ is the point-intercept form of the equation of a line, sketched in Figure 1.5.19.

    The condition $y \leq -2x + 4$ means that for any particular value of x, we are asking that y should be less than or equal to the y-coordinate for the corresponding point on the line. The required points are exactly those lying on or below the line. See Figure 1.5.20.

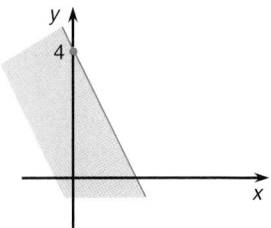

Figure 1.5.20

(ii) Again we rewrite the condition $3x - y + 2 < 0$ as $3x + 2 < y$, or $y > 3x + 2$. We sketch the corresponding line $y = 3x + 2$ and observe that $y > 3x + 2$ means that y is strictly greater than the y-coordinate of the point on that line. The set consists of all points that lie above the line $y = 3x + 2$ but not on it. See Figure 1.5.21.

Figure 1.5.21

EXAMPLE 10 Find a formula to describe the set of all points that lie inside the circle of radius 2 centered at the point $(3, -2)$.

Solution The set consists of all points whose distance from the point $(3, -2)$ is less than 2. The distance from the point (x, y) to the point $(3, -2)$ is $\sqrt{(x - 3)^2 + (y + 2)^2}$, so we need $\sqrt{(x - 3)^2 + (y + 2)^2} < 2$. Squaring both sides, we get $(x - 3)^2 + (y + 2)^2 < 4$, or $x^2 - 6x + y^2 + 4y + 13 < 4$, or $x^2 - 6x + y^2 + 4y < -9$. See Figure 1.5.22.

    Notice that the condition for the points that lie *on* the circle is the *equation* $x^2 - 6x + y^2 + 4y = -9$.

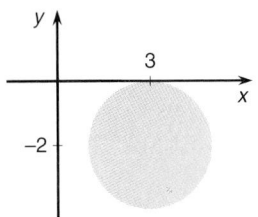

Figure 1.5.22

EXAMPLE 11 Describe the set $\{(x, y) : x + y > 1 \text{ and } x^2 + y^2 \leq 9\}$.

Solution The first condition, $x + y > 1$, can be written as $y > -x + 1$, so it describes the points that lie above the line $y = -x + 1$, the region illustrated in Figure 1.5.23.

    The second condition can be rewritten as $\sqrt{x^2 + y^2} \leq 3$. This inequality describes the points whose distance from the origin is less than or equal to 3. This is the set of all points lying inside or on the circle $x^2 + y^2 = 9$, the circle of radius 3 about the origin.

    The required set is the set of all points that lie inside or on the circle of radius 3 about the origin and above the line $y = -x + 1$.

Figure 1.5.23

Figure 1.5.24

Figure 1.5.25

Figure 1.5.26

As with parallelograms, the "base" and the "height" may not always refer to a horizontal side at the bottom and a vertical distance, but the height is always measured in a direction perpendicular to the base.

Figure 1.5.27(i)

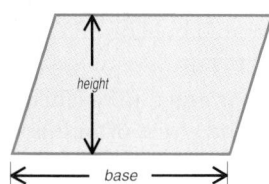

Figure 1.5.27(ii)

This set is illustrated in Figure 1.5.24. We use a solid line for the part of the circle that is included in the set and a dashed line for the line segment, which is not.

Areas

The **area of a rectangle** is the product of its length and width. The area of a circle of radius r is $A = \pi r^2$. See Figure 1.5.25.

The **area of a triangle** is a little more difficult. We first consider the area of a parallelogram, as shown in Figure 1.5.26. The parallelogram is drawn with one side (the "base") horizontal. If we cut off the right side of this parallelogram and move that piece over to the left, we can reassemble it into a rectangle, as shown in Figure 1.5.27(i). This rectangle has the same base as the original parallelogram, and the height of the rectangle is what we might call the "height" of the parallelogram, the *perpendicular distance* from the base to the opposite side. See Figure 1.5.27(ii).

The **area of the parallelogram** equals the area of the rectangle we have just constructed, which equals the product of the base of the parallelogram and the height of the parallelogram.

Strictly speaking, this argument does not always work. If we consider the long thin parallelogram shown in Figure 1.5.28, we cannot simply cut off the right side and reassemble into a rectangle. However, even in this case the area does equal the base times the height.

We drew our parallelogram with its base horizontal, but of course not every parallelogram is in this position. If we want to find the area of a tilted parallelogram, we must decide to call one of the sides its base and interpret the height accordingly. In Figure 1.5.29 the "base" is vertical, and the "height" is horizontal. This is mildly surprising but causes no real difficulty.

Now we return to the area of a triangle. We can add another triangle with the same area and end up with a parallelogram (see Figure 1.5.30). The area of the parallelogram is twice the area of the original triangle, and it is also the base times the height. Because of the way we constructed the parallelogram, its base is the base of the original triangle, and its height is the "height" of the triangle, meaning the perpendicular distance from the base to the opposite vertex. So

$$2 \times (\textit{Area of triangle}) = \textit{base} \times \textit{height},$$

so

$$\textit{Area of triangle} = \frac{1}{2}\textit{base} \times \textit{height}.$$

Figure 1.5.28

Figure 1.5.29

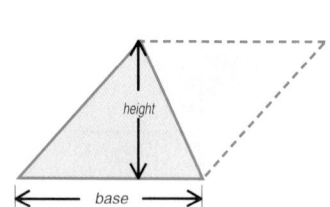

Figure 1.5.30

Summary 1.5.13   **Areas**

$$Area\ of\ rectangle = length \times width$$
$$Area\ of\ circle = \pi r^2, \qquad (r = radius)$$
$$\left(Circumference\ of\ circle = 2\pi r, \qquad (r = radius)\right)$$
$$Area\ of\ parallelogram = base \times height$$
$$Area\ of\ triangle = \frac{1}{2} base \times height$$

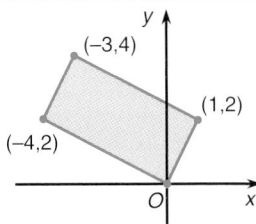

Figure 1.5.31

EXAMPLE 12   Find the area of each of the following figures: (i) the rectangle with vertices $(0,0)$, $(1,2)$, $(-4,2)$, and $(-3,4)$; (ii) the parallelogram with vertices $(1,3)$, $(1,5)$, $(3,-1)$ and $(3,1)$; (iii) the triangle with vertices $(-3,4)$, $(5,-3)$, and $(5,-2)$.

Solution

(i) Since we are told that the figure is a rectangle, all that is needed to find the area is to multiply the lengths of two adjacent sides. The length of the side joining $(0,0)$ and $(1,2)$ is $\sqrt{(1-0)^2 + (2-0)^2} = \sqrt{5}$. The length of the side joining $(0,0)$ and $(-4,2)$ is $\sqrt{(-4-0)^2 + (2-0)^2} = 2\sqrt{5}$. The area is the product of these two lengths, or $A = \sqrt{5} \times 2\sqrt{5} = 2(5) = 10$. See Figure 1.5.31.

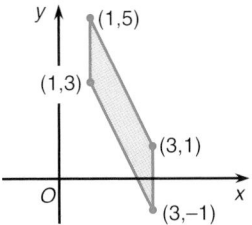

Figure 1.5.32

(ii) The side joining $(1,3)$ and $(1,5)$ is vertical, so it will be convenient to use it as the base. Its length is 2. The "height" is the *horizontal* distance from this vertical side to the other side (through $(3,-1)$ and $(3,1)$), which we see from Figure 1.5.32 is 2. The area is $A = base \times height = 2 \times 2 = 4$.

We chose the vertical side as the base to make the "height" easy to measure. If the base was a slanting side, it would be more difficult to measure the height, a distance that must be measured in the direction *perpendicular* to the base.

(iii) The side through $(5,-3)$ and $(5,-2)$ is vertical. We use it as the base; its length is 1 and the height is the distance from the vertical line $x = 5$ to the vertical line through the other vertex, that is, $x = -3$. The distance between these two lines is $5 - (-3) = 8$, and the area of the triangle is $\frac{1}{2} base \times height = \frac{1}{2}(1 \times 8) = 4$. See Figure 1.5.33.

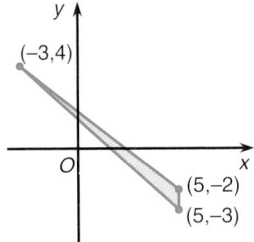

Figure 1.5.33

We summarize some useful formulas about the volumes of various solid regions as follows.

Figure 1.5.34

Summary 1.5.14   **Volume Formulas (see Figure 1.5.34)**

(i) The volume of a rectangular box is the product of its three dimensions: *length* $\times$ *width* $\times$ *height*.

(ii) The volume of a ball of radius r is $\frac{4}{3}\pi r^3$.

(iii) The volume of a cylinder of height h and with cross-sectional radius r is (*cross-sectional area* $\times$ *height*) $= \pi r^2 h$.

(iv) The volume of a cone of height h and radius r is $\frac{1}{3}\pi r^2 h$.

Exercises 1.5

Find each of the following.

1. The length of the hypotenuse of a right-angled triangle whose other two sides have lengths 2 and 24.

2. The length of the third side of a right-angled triangle whose hypotenuse has length 15 and another side of length 12.

3. The distance between the points $(1, -2)$ and $(4, 2)$.

4. The distance between the points $(5, 3)$ and $(-7, 4)$.

5. The distance from the point $(6, -8)$ to the origin.

6. The distance from the point $(3x - 1, 4 - x)$ to the origin.

7. The area of the triangle whose vertices are $(0, 0)$, $(0, 4)$, and $(2, 3)$.

8. The area of the triangle whose vertices are $(1, 3)$, $(1, 1)$, and $(-5, 4)$.

9. The area of the triangle whose vertices are $(-2, 1)$, $(2, 3)$, and $(6, 3)$.

10. The area of the triangle whose vertices are $(1, 1)$, $(2, 2)$, and $(5, 5)$.

Find the equation of each of the following straight lines.

11. The line through $(1, 2)$ and $(4, 5)$.

12. The line through $(-2, 4)$ and $(2, 2)$.

13. The line through $(3, 0)$ and $(-5, 4)$.

14. The line through $(-1, 3)$ and $(-1, 4)$.

15. The line through $(4, -6)$ with slope $m = 2$.

16. The line through $(-2, 7)$ with slope $m = -3$.

17. The line with slope $m = -1$, y-intercept $b = 3$.

18. The line with slope $m = 4$, y-intercept $b = -5$.

19. The line with slope $m = \frac{1}{3}$, x-intercept $d = -2$.

20. The line with slope $m = -\sqrt{2}$, x-intercept $d = 4$.

Find each of the following.

21. The slope of the line $3x + 2y - 4 = 0$.

22. The slope of the line $4x - y + 7 = 0$.

23. The equation of the line parallel to $x + 2y = 3$, through the point $(3, 4)$.

24. The equation of the line through $(-1, 3)$, parallel to $2x - y + 5 = 0$.

25. The y-intercept of the line $2x - 3y + 7 = 0$.

26. The y-intercept of the line $4x - 2y + 8 = 0$.

27. The x-intercept of the line $2x + y - 7 = 0$.

28. The x-intercept of the line $y - x = 3$.

29. The intersection of $x + 3y = 9$ and $3x - y - 7 = 0$.

30. The intersection of $2x - 4y - 6 = 0$ and $x - 2y = 5$.

1.6 Graphing Conics

PARABOLA
TRANSLATION OF AXES
ELLIPSE
HYPERBOLA
ASYMPTOTES

In this section we consider some important kinds of equations, with particular emphasis on drawing their graphs. The graphs include ellipses, hyperbolas, and parabolas. Each of them can be obtained by looking at the points of intersection of a flat plane with a double-ended cone (see Figure 1.6.1). For this reason these graphs are called **conics**.

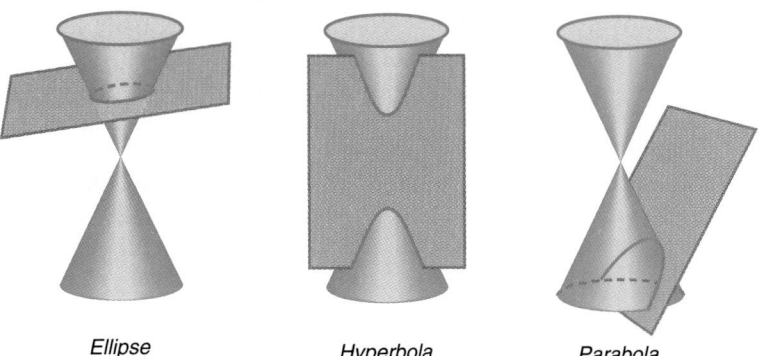

Ellipse   *Hyperbola*   *Parabola*

Figure 1.6.1

The equations we will be considering involve **polynomials in *two* variables**, x and y.

We will deal only with **quadratic polynomials**, which means polynomials whose degree is 2. Quadratic polynomials must contain a term of degree 2, that is, a term with x^2, y^2, or xy, and may also contain terms with x, y, and/or constants.

In fact, we will usually consider only quadratic polynomials that do not contain an xy-term, that is, polynomials of the form $ax^2 + by^2 + cx + dy + q$.

Parabolas

Consider the equation $y = x^2$. Every point on the graph of $y = x^2$ satisfies $y \geq 0$, so the graph never goes below the x-axis.

We also notice that the value of y will be the same for x and $-x$. We say that the graph will be "symmetrical" about the y-axis, meaning that it will be the same to the left of the y-axis as it is to the right of the axis.

We can plot a few points and begin to draw the graph. Its shape is the familiar **parabola** (see Figure 1.6.2).

If we consider instead the equation $y = ax^2$, the graph will also be a parabola. If $a < 0$, then the parabola will open downward, while if $a > 0$, the parabola will open upward. The larger $|a|$ is, the more rapidly the points on the parabola will move away from the origin, that is, the longer and narrower the parabola will be.

Similar remarks apply to equations of the form $x = ay^2$. Here the graph will be a parabola that is symmetrical about the x-axis, opening to the right if $a > 0$ and to the left if $a < 0$. See Figure 1.6.3. In each case, the origin is called the **vertex** of the parabola.

Figure 1.6.2

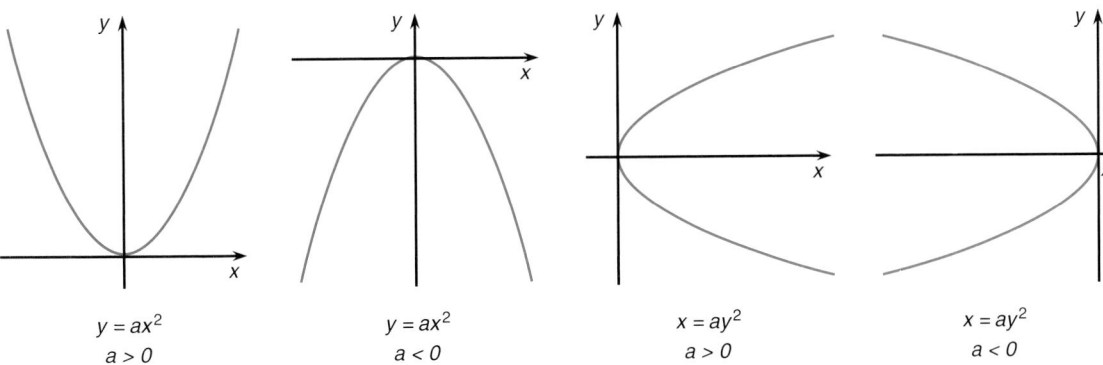

$y = ax^2$ $y = ax^2$ $x = ay^2$ $x = ay^2$
$a > 0$ $a < 0$ $a > 0$ $a < 0$

Figure 1.6.3

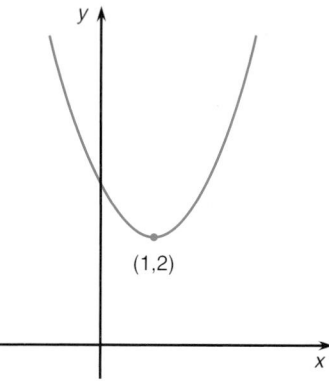

Figure 1.6.4

EXAMPLE 1   Now consider the equation $y = x^2 - 2x + 3$. To see what its graph is, we complete the square: $y = x^2 - 2x + 3 = (x - 1)^2 + 2$. The quantity $(x - 1)$ measures the horizontal position starting at the value $x = 1$ instead of $x = 0$. This suggests that the equation $y = (x - 1)^2$ is a parabola, except that its lowest point (its vertex) will be at $x = 1$.

   The equation we are considering is obtained from this by adding 2. Adding 2 to the y-coordinate raises the whole graph up by two units. The graph of $y = x^2 - 2x + 3$ is a parabola that opens upward and whose vertex is the point $(1, 2)$. See Figure 1.6.4.

EXAMPLE 2   Consider the equation $y = ax^2 + bx + c$, with $a \neq 0$. By completing the square on the right side we can rewrite it as $y = a\left(x + \frac{b}{2a}\right)^2 + \left(c - \frac{b^2}{4a}\right)$, that is, $y - \left(c - \frac{b^2}{4a}\right) = a\left(x + \frac{b}{2a}\right)^2$. If we let $h = -\frac{b}{2a}$ and $k = c - \frac{b^2}{4a}$, then the equation becomes

$$y - k = a(x - h)^2.$$

   We recognize this as the equation of a parabola. It opens upward if $a > 0$ and downward if $a < 0$. Its vertex is at the point (h, k). See Figure 1.6.5.

   The approach used in this example suggests a useful idea. Suppose h and k are real numbers; we can plot the point $O' = (h, k)$. The quantities $x' = x - h$ and $y' = y - k$ measure the horizontal and vertical position of a point, *starting at the point* O' instead of starting at the origin.

   We can draw "axes" through $O' = (h, k)$ and think of x' and y' as coordinates *relative to these axes*. The point O' is the origin for these axes and coordinates. See Figure 1.6.6.

   As we saw in Example 1, an equation may be more easily recognized if it is expressed in terms of the coordinates x' and y'. We wrote $y = x^2 - 2x + 3$ as $y - 2 = x^2 - 2x + 1$. With $y' = y - 2$ and $x' = x - 1$ it becomes $y' = x'^2$, which is a parabola whose vertex is at $O' = (1, 2)$. In Figure 1.6.7 it is easy to draw this parabola if we mark the x'-axis and the y'-axis. After we have drawn the graph, we can ignore these axes and return to the original x-axis and y-axis.

   This process is called "translating the axes."

Figure 1.6.5

Figure 1.6.6

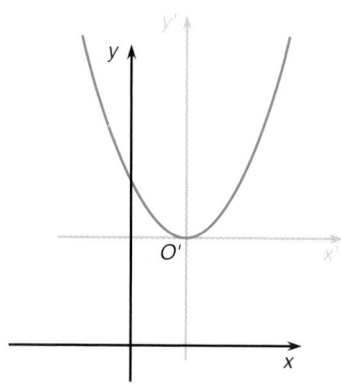

Figure 1.6.7

Summary 1.6.1

Translation of Axes

Suppose h and k are real numbers. Let $O' = (h, k)$, and let $x' = x - h$, $y' = y - k$. Draw the x'-axis and the y'-axis through O'.

To graph all the points (x, y) whose coordinates satisfy some equation $F(x, y) = 0$, it is equivalent to express $F(x, y)$ as some function of x' and y', that is,

$$F(x, y) = G(x', y') = G(x - h, y - k),$$

and graph all points (x', y') satisfying $G(x', y') = 0$; these points (x', y') must be plotted relative to the x'-axis and the y'-axis.

It is easy to find the function G. Observe that $x' = x - h$, so $x = x' + h$, and $y' = y - k$, so $y = y' + k$. Substitute these expressions into the equation above to find

$$G(x', y') = F(x, y) = F(x' + h, y' + k).$$

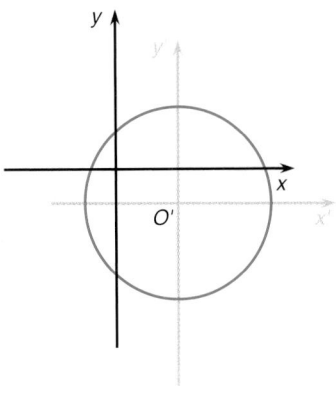

Figure 1.6.8

EXAMPLE 3 Consider the equation $x^2 - 4x + y^2 + 2y - 4 = 0$. We complete squares in both x and y, so $(x^2 - 4x) + (y^2 + 2y) - 4 = 0$ becomes $(x - 2)^2 - 4 + (y + 1)^2 - 1 - 4 = 0$, or $(x - 2)^2 + (y + 1)^2 - 9 = 0$.

This suggests that we try letting $x' = x - 2$ and $y' = y + 1$. Using the notation of the preceding summary, this means $h = 2$ and $k = -1$. It is easy to write the original equation in terms of x' and y': the "translated" equation is $x'^2 + y'^2 - 9 = 0$, or $x'^2 + y'^2 = 9$. The left-hand side is the square of the distance from (x', y') to the new origin O'. The equation $x'^2 + y'^2 = 9$ says that the distance from (x', y') to the origin O' is $\sqrt{9} = 3$. The set of all points satisfying this condition is the circle with center $O' = (2, -1)$ and radius 3. It is easily graphed (see Figure 1.6.8). ◢

Translating the equation made it much easier to recognize that its graph was a circle.

REMARK 1.6.2

Whenever a quadratic polynomial contains both an x^2-term and an x-term, we can complete the square and translate coordinates to get a polynomial with an x'^2-term but no x'-term. Similarly, a polynomial with a y^2-term and a y-term can be converted into one with a y'^2-term but no y'-term by completing the square in y.

If the equation does not have an x^2-term, it is possible to combine the x-term and the constant term into x', resulting in an equation without a constant term. A similar procedure can be used to remove the constant term from an equation with no y^2-term.

In each case the effect of these changes in terms of the graph is to translate the origin, which can also be seen as moving the center of the conic.

In Example 3 we noticed that if $A > 0$, the graph of all points satisfying an equation $x^2 + y^2 = A$ is the graph of the circle centered at the origin with radius $\sqrt{A}$. We have already remarked that a circle cannot be the graph of a function, since there can be two values of y for the same x, but we can draw its graph, plotting all the points that satisfy the equation. The parabola $y = ax^2 + bx + c$ expresses y as a function of x, but for many conics, y will not be a function of x. In this situation we say that the equation gives a **relation** between x and y.

Ellipses

Consider the equation $x^2 + 4y^2 = 4$. It is not the equation of a circle (such as $x^2 + y^2 = R^2$), but it is similar. Let us rewrite it as $x^2 + (2y)^2 = 4$. If we let $y' = 2y$, the equation reads $x^2 + y'^2 = 4$. So (x, y') lies on the circle of radius $\sqrt{4} = 2$ with center at the origin; the problem is to interpret this in terms of the original variables x and y.

Changing from y to $y' = 2y$ amounts to changing the scale on the vertical axis, stretching everything by a factor of 2 in the vertical direction. We have a circle after stretching by a factor of 2 in the vertical direction, so before stretching, the figure must have been a "flattened out circle." It helps to get a more precise idea of the shape of the original figure by finding its x- and y-intercepts.

When $x = 0$, we find $0^2 + 4y^2 = 4$, so $y^2 = 1$ and $y = \pm 1$. When $y = 0$, we find $x^2 + 4(0^2) = 4$, so $x = \pm 2$. We plot $(0, \pm 1)$ and $(\pm 2, 0)$ and then fill in the rest of the graph, plotting additional points first if greater accuracy is required. See Figure 1.6.9.

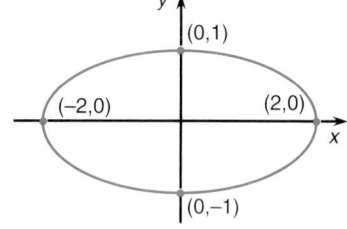

Figure 1.6.9

The graph of $x^2 + 4y^2 = 4$ is an **ellipse**. The equations $2x^2 + 8y^2 = 10$, $x^2 + \frac{y^2}{9} = 16$, etc., are also ellipses. If we multiply one of these equations by a constant, the graph will not be changed, so there are many equations that represent the same ellipse. To remove this ambiguity, we frequently multiply or divide the equation by a constant so that the right-hand side becomes 1. It is customary to write the equation in the form

$$\frac{x^2}{a^2} + \frac{y^2}{b^2} = 1 \qquad \text{or} \qquad \frac{x^2}{b^2} + \frac{y^2}{a^2} = 1, \qquad (1.6.3)$$

with $a > b > 0$.

The graph of an equation of this form is an ellipse whose center is at the origin $(0,0)$. For the first form, the intercepts are $(\pm a, 0)$ and $(0, \pm b)$. For the second form, the intercepts are $(\pm b, 0)$ and $(0, \pm a)$.

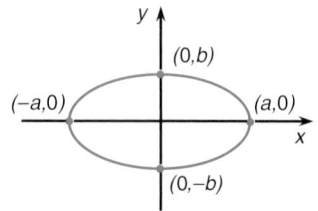

Figure 1.6.10(i)

The distinction between the two forms involves whether the larger number a is in the denominator of the x^2-term or of the y^2-term. We see that for the first form, the x-intercepts $(\pm a, 0)$ are farther apart than the y-intercepts $(0, \pm b)$. This means that the corresponding ellipse is wider than it is tall (see Figure 1.6.10(i)). In the second type of equation the situation is reversed, and the ellipse is taller than it is wide (see Figure 1.6.10(ii)). It is to help distinguish these two cases that we always choose $a > b$.

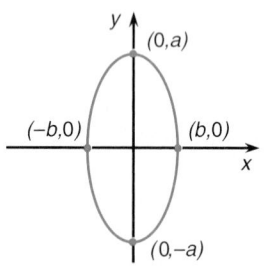

Figure 1.6.10(ii)

EXAMPLE 4 Sketch the graph of the equation $4x^2 - 16x + 9y^2 + 18y - 11 = 0$.

Solution We begin by completing the square. The equation becomes $4(x^2 - 4x) + 9(y^2 + 2y) - 11 = 0$, or $4(x^2 - 4x + 4) - 16 + 9(y^2 + 2y + 1) - 9 - 11 = 0$, or $4(x - 2)^2 + 9(y + 1)^2 - 36 = 0$. We let $x' = x - 2$, $y' = y + 1$, and translate the axes to the new origin, $O' = (2, -1)$.

The equation now reads $4x'^2 + 9y'^2 = 36$. Following the model of Equa-

Figure 1.6.11

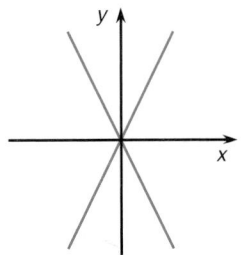

Figure 1.6.12

To illustrate this, suppose $x = 20$. Then $y^2 = 4x^2 - 4 = 4(20)^2 - 4 = 1596$. Taking the square root with a calculator, we see that $y \approx \pm 39.949\,97$. This is very close to what we get from $y^2 = 4x^2 = 1600$, from which $y = \pm 40$.

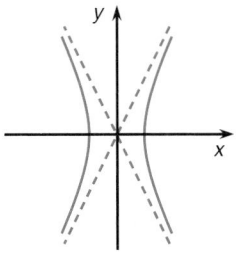

Figure 1.6.13

tions 1.6.3, we divide by 36 to obtain

$$\frac{x'^2}{9} + \frac{y'^2}{4} = 1.$$

This is the equation of an ellipse whose center is at the point O' and whose x'- and y'-intercepts are $(\pm 3, 0)$ and $(0, \pm 2)$, respectively.

It is easily graphed (see Figure 1.6.11). Note that the intercepts are described here in terms of x'- and y'-coordinates. Finally, we return to x and y coordinates and see that the graph is an ellipse whose center is at $(2, -1)$. The longest segment through the center from one side of the ellipse to the other goes from $(-1, -1)$ to $(5, -1)$ (i.e., the two x'-intercepts). The shortest such segment goes from $(2, -3)$ to $(2, 1)$.

These segments are called the **major axis** and **minor axis**, respectively. If the equation of an ellipse is written in either of the forms in Equations 1.6.3, then the major and minor axes go between the x- and y-intercepts. The major axis is the longer one (between $(a, 0)$ and $(-a, 0)$ for the first form of the equation and between $(0, -a)$ and $(0, a)$ for the second type).

The numbers a and b are half the lengths of the major and minor axes; they are sometimes called the **semi-major** and **semi-minor axes**, respectively.

If $a = b$, then the graph is a circle instead of an ellipse, and it does not make sense to talk about major and minor axes.

Hyperbolas

Consider the equation $4x^2 - y^2 = 4$. When $y = 0$, we see that $4x^2 = 4$, so $x = \pm 1$ and the x-intercepts are $(1, 0)$ and $(-1, 0)$. If we let $x = 0$, the equation becomes $-y^2 = 4$. This has no solutions, since y^2 is never negative, so there are no y-intercepts.

Next we ask what will happen if x is a large number. The equation can be written as $y^2 = 4x^2 - 4$, and if x is very large, the -4 on the right side will be so much smaller than the $4x^2$ that the equation is very nearly the same as $y^2 = 4x^2$.

So as x moves away from 0, the graph lies very close to $y^2 = 4x^2$, or $y = \pm 2x$. The graphs of $y = 2x$ and $y = -2x$ are straight lines through the origin, with slopes 2 and -2, respectively. They are sketched in Figure 1.6.12.

The equation $y^2 = 4x^2 - 4$ has no solution unless the right side is nonnegative. This amounts to $4x^2 \geq 4$, or $x^2 \geq 1$, or $|x| \geq 1$. The graph has no points with $-1 < x < 1$.

Finally we remark that if a point (x, y) satisfies $4x^2 - y^2 = 4$, then so does $(x, -y)$, and so do $(-x, y)$ and $(-x, -y)$. This means that the graph is "symmetrical" about both axes, which means that "reflecting" it in either axis does not change the picture.

From all this we can guess the shape of the graph. We know its intercepts, we know there are no points with $-1 < x < 1$, and we know that as x becomes large, the graph approaches one or other of the lines $y = 2x$ and $y = -2x$. Combined with the symmetry, these facts suggest the shape shown in Figure 1.6.13. This conic is called a **hyperbola**. The lines $y = \pm 2x$ are called its **asymptotes**. The graph consists of two "pieces," which are called its **branches**.

As with the equation of an ellipse, it is convenient to divide the equation of a hyperbola by a suitable number. We choose the number so that the constant term of the equation is 1. So, for instance, the above equation $4x^2 - y^2 = 4$ becomes

$$x^2 - \frac{y^2}{4} = 1.$$

Summary 1.6.4

Figure 1.6.14(i)

Figure 1.6.14(ii)

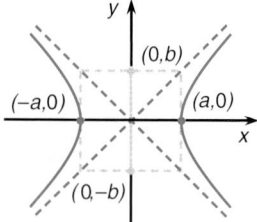

Figure 1.6.15

General Form of the Hyperbola

The equations

$$\frac{x^2}{a^2} - \frac{y^2}{b^2} = 1 \qquad \text{and} \qquad \frac{y^2}{a^2} - \frac{x^2}{b^2} = 1$$

are the equations of hyperbolas.

In the first case the x-intercepts are $(\pm a, 0)$, there are no points on the graph with $-a < x < a$, and the asymptotes are found by realizing that if x is large, then the equation is almost the same as $\frac{x^2}{a^2} = \frac{y^2}{b^2}$, or $\frac{x}{a} = \pm\frac{y}{b}$, or $y = \frac{b}{a}x$ and $y = -\frac{b}{a}x$. These are lines through the origin with slopes $\frac{b}{a}$ and $-\frac{b}{a}$. See Figure 1.6.14(i).

In the second case the roles of x and y are reversed. In this case there are no x-intercepts and no points on the graph with $-a < y < a$. The y-intercepts are $(0, \pm a)$, and the asymptotes are found by approximating the equation for large values of x by $\frac{y^2}{a^2} = \frac{x^2}{b^2}$, so they are the lines $\frac{y}{a} = \pm\frac{x}{b}$, or $y = \pm\frac{a}{b}x$. See Figure 1.6.14(ii).

Notice that we decide which denominator is a^2 and which is b^2 on the basis of the signs of the terms; a^2 is the denominator of the positive term (or, more precisely, the term with the same sign as the constant term on the right side). Unlike what happened with ellipses, it is *not* a question of the relative sizes of a and b.

In each case there are two intercepts (either two x-intercepts or two y-intercepts). The line segment joining them is called the **transverse axis**. Its length is $2a$. We also draw the segment that is perpendicular to the transverse axis, centered at the origin and of length $2b$; it is called the **conjugate axis**. (It is in order to be able to say that the transverse axis has length $2a$ that we changed around the order of the numbers a and b when we wrote the second form of the equation.) See Figure 1.6.15.

So for $\frac{x^2}{a^2} - \frac{y^2}{b^2} = 1$, the transverse axis joins $(-a, 0)$ and $(a, 0)$; the conjugate axis joins $(0, -b)$ and $(0, b)$. In the other case, $\frac{y^2}{a^2} - \frac{x^2}{b^2} = 1$, the transverse axis joins $(0, -a)$ and $(0, a)$, while the conjugate axis joins $(-b, 0)$ and $(b, 0)$.

If we fill in the rectangle passing through the ends of the transverse and conjugate axes (as in Figure 1.6.15), then its corners are on the asymptotes. This makes it easy to draw the asymptotes.

Rather than memorizing which formula represents which kind of hyperbola, it is preferable to realize that it is quite easy to distinguish the two cases simply by determining whether there are x-intercepts or y-intercepts. It is all a matter of whether the x^2-term or the y^2-term has the same sign as the constant term.

We are now able to sketch the graph of any quadratic equation that does not contain any xy-term. The idea is to inspect the equation to decide whether it is a parabola, an ellipse, a circle, or a hyperbola. We first complete squares if necessary and translate the axes accordingly.

Summary 1.6.5

Graphing Conics

Consider the quadratic equation $ax^2 + by^2 + cx + dy + q = 0$, which has no xy-term. To draw its graph, use the following steps.

(i) Complete squares if necessary; if the equation contains both an x^2-term and an x-term, that is, if $a \neq 0$ and $c \neq 0$, then complete the square in x. Similarly, if $b \neq 0$ and $d \neq 0$, complete the square in y.

(ii) Translate axes if necessary. If you found it necessary to complete squares in Step (i), and the result of completing the square in x is a term involving $x - h$, then let $x' = x - h$. If completing the square in y gives a term involving $y - k$, let $y' = y - k$.

  If there is no x^2-term, then there must be a y^2-term. In this situation, combine the x-term and the constant term as $c(x - h)$, for some constants c and h, and let $x' = x - h$. Similarly, if there is no y^2-term, combine the y-term and the constant term as $d(y - k)$ and let $y' = y - k$.

(iii) If the axes have been translated, plot the "new origin" $O' = (h, k)$. (If you changed x to $x' = x - h$ but did not have to translate y, then $O' = (h, 0)$, that is, let $k = 0$. Similarly, if it was not necessary to translate x to x', then let $h = 0$.)

(iv) Express the equation in terms of x' and y' if translations were necessary. Inspect the equation to see whether it is the equation of a parabola, an ellipse, a circle, or a hyperbola. Then find the intercepts. For a hyperbola, find the asymptotes and mark them on the picture. Sketch the graph.

(v) Occasionally we encounter a quadratic equation that is not a parabola, an ellipse, a circle, or a hyperbola. This kind of graph is called a **degenerate conic**. There are several possibilities, but we describe just two of them. In an equation like $\frac{x^2}{a^2} + \frac{y^2}{b^2} = -1$, the left side is the sum of two squares, so it is never negative, and we see that no point (x, y) can satisfy the equation. There are *no points* on the graph.

  Another possibility is an equation of the form $\frac{x^2}{a^2} - \frac{y^2}{b^2} = 0$. This is like a hyperbola except that the constant term is zero. Rearranging, we find that $\frac{x^2}{a^2} = \frac{y^2}{b^2}$, or $\frac{x}{a} = \pm \frac{y}{b}$. The graph is the two straight lines $y = \frac{b}{a}x$ and $y = -\frac{b}{a}x$. This is what results if a hyperbola "degenerates" into its two asymptotes.

EXAMPLE 5 Sketch the graph of the equation $9x^2 - 18x - 4y^2 + 32y = 19$.

Solution We complete the squares:

$$9x^2 - 18x - 4y^2 + 32y - 19 = 9(x^2 - 2x) - 4(y^2 - 8y) - 19$$
$$= 9(x^2 - 2x + 1) - 9 - 4(y^2 - 8y + 16) + 64 - 19$$
$$= 9(x - 1)^2 - 4(y - 4)^2 + 36.$$

If we let $h = 1$, $k = 4$, $x' = x - 1$, $y' = y - 4$, then the original equation becomes $9x'^2 - 4y'^2 = -36$, or

$$\frac{y'^2}{9} - \frac{x'^2}{4} = 1.$$

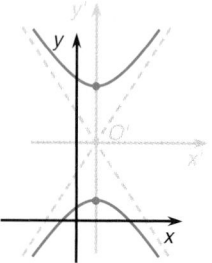

Figure 1.6.16

This is the equation of a hyperbola. Since the y'^2-term is positive, it is a hyperbola with transverse axis along the y'-axis and conjugate axis along the x'-axis. The center of the hyperbola is $O' = (h, k) = (1, 4)$. The intercepts on the y'-axis are $y' = \pm 3$. These points, where $x' = 0$ and $y' = \pm 3$, are the points where $x = x' + 1 = 0 + 1 = 1$ and $y = y' + 4 = 4 \pm 3 = 7, 1$, which are the points $(1, 1)$ and $(1, 7)$.

The conjugate axis has length $2\sqrt{4} = 4$; it runs between $(x', y') = (\pm 2, 0)$, that is, from $(x, y) = (-1, 4)$ to $(3, 4)$.

The asymptotes are the lines $\frac{y'^2}{9} = \frac{x'^2}{4}$, or $y' = \pm \frac{3}{2} x'$. They are easily sketched relative to the x'- and y'-axes.

Then it is comparatively easy to sketch in the shape of the hyperbola, making the graph symmetrical about both the x'- and y'-axes. See Figure 1.6.16. ◢

Even without completing the squares, we might have noticed that the x^2-term and the y^2-term in the original equation had opposite signs, which suggests that the graph should be a hyperbola. We cannot be absolutely certain about this because it might have turned out to be a degenerate conic—a pair of straight lines. It is also important to observe that, even assuming that the graph is a hyperbola and just looking at the original equation, we could not easily tell which way the hyperbola opens (i.e., up/down or left/right). The constant term on the right side of the original equation has the same sign as the x^2-term, but after completing squares we have a constant with the same sign as the y'^2-term, and it is *this* that decides the type of the hyperbola.

> Generally speaking, our approach to graphing conics has been to begin by completing squares and translating axes in order to simplify the equation. Then it is possible to recognize the type of conic from the simplified form of the equation. Rather than memorizing the various possibilities, we have concentrated on identifying the curve by finding intercepts, asymptotes, if any, and values of x or y for which there are no points on the curve. The calculations involved are simple and can frequently be done in your head. They use nothing more complicated than the fact that a sum of squares can never be negative.

We have observed that an ellipse is not the graph of a function, because some vertical lines intersect it twice. The same is true for hyperbolas and circles and some parabolas.

On the other hand, if we consider the ellipse $12x^2 + y^2 = 12$, it is tempting to write $y^2 = 12 - 12x^2$ and take the square root to solve for y:

$$y = \sqrt{12 - 12x^2}.$$

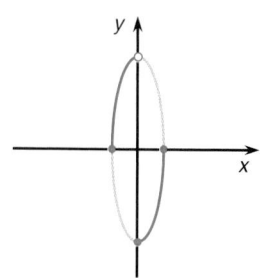

Figure 1.6.17

This formula does express y as a function of x. Since the square root is always nonnegative, the graph of this function $y = \sqrt{12 - 12x^2}$ is the *upper half* of the ellipse. See Figure 1.6.17.

We could also have tried letting $y = -\sqrt{12 - 12x^2}$, and the graph of this function is the lower half of the ellipse.

We could even have combined them in various ways, for instance,

$$y = f(x) = \begin{cases} \sqrt{12 - 12x^2}, & \text{if } x < 0, \\ -\sqrt{12 - 12x^2}, & \text{if } x \geq 0. \end{cases}$$

The graph of this function is the upper part of the ellipse some of the time and the lower part the rest of the time (see Figure 1.6.18). Clearly it is possible to do this sort of thing in many different and complicated ways.

Figure 1.6.18

Suppose we want to graph the ellipse $\frac{x^2}{4} + \frac{y^2}{9} = 1$ using a graphics calculator. Because the ellipse is not the graph of a function, we have to graph it in two parts.

We begin by solving for y: $\frac{y^2}{9} = 1 - \frac{x^2}{4}$, so $y^2 = 9 - \frac{9x^2}{4}$. This means $y = \pm\sqrt{9 - \frac{9x^2}{4}} = \pm 3\sqrt{1 - \frac{x^2}{4}}$. We get the ellipse by graphing these *two* functions.

The other important thing to note is that the ellipse consists of points (x, y) with $-2 \leq x \leq 2$. Now we can graph the ellipse.

TI-85	SHARP EL9300
GRAPH F1 $(y(x) =) 3 \sqrt{(1 - \text{F1}(x)x^2}$ $\div 4)$ ▼ $(-)3\sqrt{(1 - \text{F1}(x)x^2 \div 4)}$ M2(RANGE) $(-)$ 2 ▼ 2 ▼ ▼ $(-)$ 3 ▼ 3 F5(GRAPH)	⊬ 3 $\sqrt{1 - \text{x/}\theta\text{/T } x^2 \div 4}$ 2ndF ▼ $-3\sqrt{1 - \text{x/}\theta\text{/T } x^2 \div 4}$ RANGE -2 ENTER 2 ENTER 1 ENTER -3 ENTER 3 ENTER 1 ENTER ⊬
CASIO f_x-7700GB/f_x-6300G	HP 48SX
Cls EXE Range -2 EXE 2 EXE 1 EXE -3 EXE 3 EXE 1 EXE EXE EXE EXE Graph $3\sqrt{(1 - X x^y 2 \div 4)}$ EXE Graph $-3\sqrt{(1 - X x^y 2 \div 4)}$ EXE	↰ PLOT PLOTR ERASE ATTN ′ $\alpha Y = 3 \times \sqrt{()1 - \alpha X y^x 2 \div 4}$ ↱ PLOT ↰ DRAW ERASE 2 +/- ENTER 2 XRNG 3 +/- ENTER 3 YRNG DRAW ATTN ′$\alpha Y = -3$ $\times \sqrt{()1 - \alpha X y^x 2 \div 4}$ ↱ PLOT ↰ DRAW DRAW

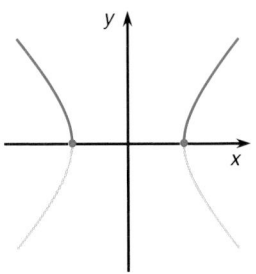

Be sure to take note of the vertical and horizontal scales when you identify the major and minor axes.

Now you might like to try graphing some other ellipses, such as $4x^2 + 3y^2 = 1$, and then some hyperbolas, such as $x^2 - y^2 = 1$, $4y^2 - 9x^2 = 16$. Try including asymptotes.

Many of the calculus techniques we will be learning are intended to be used on functions. There are times when we want to apply some of them to something like an ellipse but, to be able to do this it is first necessary to express part of the ellipse (say the upper or lower half) as the graph of a function and apply calculus to that function. Afterward, it may be possible to combine our conclusions about all the parts and draw some conclusion about the whole ellipse.

The same sort of thing will work for hyperbolas and parabolas. For instance, the hyperbola $\frac{x^2}{a^2} - \frac{y^2}{b^2} = 1$ can be divided into the graphs of two functions, $y = b\sqrt{x^2/a^2 - 1}$ and $y = -b\sqrt{x^2/a^2 - 1}$ (the top and bottom halves of the hyperbola; the top one is illustrated in Figure 1.6.19). Each of these functions has as its domain the set of all x satisfying $|x| \geq a$.

Figure 1.6.19

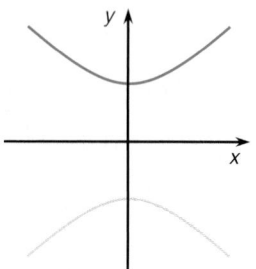

Figure 1.6.20

The hyperbola $\frac{x^2}{b^2} - \frac{y^2}{a^2} = -1$ can also be divided into the graphs of two functions, $y = a\sqrt{1 + x^2/b^2}$ and $y = -a\sqrt{1 + x^2/a^2}$ (the top and bottom branches of the hyperbola). Each of these functions has $\mathbb{R}$ as its domain. The top branch is illustrated in Figure 1.6.20.

Exercises 1.6

Identify each of the following conics, specifying its center, its intercepts, its axes, and its asymptotes, to the extent that they are applicable.

1. $16x^2 + 9y^2 = 144$
2. $4x^2 + 25y^2 = 100$
3. $x^2 - 4y^2 = 4$
4. $4y^2 - x^2 = 4$
5. $y^2 + 4x = 8$
6. $x^2 + y^2 = 4$
7. $x^2 + 4x + 9y^2 - 5 = 0$
8. $x^2 + y^2 - 6x + 2y + 9 = 0$
9. $x^2 + 2x - 4y - 5 = 0$
10. $x^2 + 4x + 4y^2 - 8y + 8 = 0$
11. $2x^2 + 8y^2 = 1$
12. $x^2 - y^2 = 1$
13. $12x^2 + 3y^2 = 27$
14. $\frac{4}{5}y^2 - 5x^2 = \frac{1}{5}$
15. $20x - 3y^2 = 2$
16. $5x^2 = 1 - 5y^2$
17. $1 - x^2 + 4x = 3y^2$
18. $9x^2 - 16y^2 = -1$
19. $16x + 25y^2 = 20$
20. $4x^2 - 24x - 9y^2 + 18y + 27 = 0$

Sketch the graph of each of the following conics, paying special attention to axes, intercepts, and asymptotes, whenever it makes sense to speak of them.

21. $4y + 16x^2 = 1$
22. $4x^2 + 25y^2 = 25$
23. $y^2 - x^2 - 1 = 0$
24. $y^2 = 1 - x^2$
25. $x^2 - 4y^2 = -2$
26. $x + 2y + y^2 + 3 = 0$
27. $4x^2 + 9y^2 - 7 = 0$
28. $x^2 = y - x$
29. $x^2 + y^2 + x + y - \frac{9}{2} = 0$
30. $4x^2 + 8x + 9y^2 + 5 = 0$

Find the equation of each of the following conics (the equations are all assumed to have no xy-term).

31. The circle with center $(-1, 3)$ and radius 2.
32. The ellipse with center $(0, 0)$, semi-minor axis $b = 3$, and horizontal major axis of length 8.
33. The hyperbola with asymptotes $y = \pm 2x$ and containing $(2, 0)$.
34. The parabola with vertex $(3, -5)$, opening downward, and containing $(4, -7)$.
35. The ellipse with center $(5, 2)$ containing $(3, 2)$ and $(5, 3)$.
36. The parabola opening to the right, with vertex $(-1, 2)$ and containing the origin.
37. The circle whose center is on the x-axis and that contains $(2, 3)$ and $(-6, 3)$.
38. The hyperbola whose center is $(-1, -2)$, whose asymptotes have slopes ± 3, and that contains $(-1, 0)$.
39. The ellipse with vertical major axis that is twice the length of the minor axis, which has center at $(2, -4)$ and contains the origin.
40. The hyperbola with transverse and conjugate axis both of length 2, transverse axis vertical, having its center at $(5, -2)$.

1.7 Trigonometric Functions

IDENTITY
RADIANS
LAW OF COSINES
ADDITION FORMULAS
POLAR COORDINATES

Figure 1.7.1

Figure 1.7.2

Figure 1.7.3

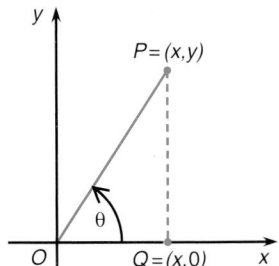

Figure 1.7.4

Consider the angle θ shown in Figure 1.7.1. It is drawn at the origin in the plane, and we have also drawn a circle of radius r, with its center at the origin. For convenience we have used the positive x-axis as one of the sides of the angle, and we label P, the point where the other side of the angle intersects the circle. We draw a vertical line through P and label Q, the point where this line intersects the x-axis. So the angle $\angle POQ$ equals θ.

The radius of the circle is $|OP| = r$. What we want to know is the lengths $|OQ|$ and $|PQ|$. If we draw the same sort of picture, with the same angle θ but with a larger circle, then r will be different, and so will the corresponding distances, $|OB|$ and $|AB|$ (see Figure 1.7.2).

However, if we draw both pictures together, as in Figure 1.7.3, then we see that $\triangle AOB$ is similar to $\triangle POQ$, which means that the ratios of the lengths of corresponding sides are equal. In particular,

$$\frac{|OQ|}{|OP|} = \frac{|OB|}{|OA|}, \qquad \frac{|PQ|}{|OP|} = \frac{|AB|}{|OA|}, \qquad \frac{|PQ|}{|OQ|} = \frac{|AB|}{|OB|}.$$

No matter how large or small the circle, if the angle at the center is θ, then these ratios will be the same. We need a way to express these ratios in terms of the angle θ.

The distance $|OP|$ is always r, the radius of the circle. If we write the coordinates of P as $P = (x, y)$, then $Q = (x, 0)$. In Figure 1.7.4, both x and y are positive, so $|OQ| = x$, and $|PQ| = y$. The ratios mentioned above can all be expressed in terms of x, y, and r.

These observations explain why the following definitions are made.

DEFINITION 1.7.1

For each angle θ we define the **trigonometric functions** $\cos\theta$, $\sin\theta$, $\tan\theta$, $\sec\theta$, $\csc\theta$, and $\cot\theta$ (abbreviations for *cosine, sine, tangent, secant, cosecant,* and *cotangent*, respectively).

Draw the positive x-axis, and draw another half-line starting at the origin that makes an angle of θ with the positive x-axis (see Figure 1.7.5). If θ is a positive angle, measure it counterclockwise from the positive x-axis; if it is negative, measure it clockwise from the positive x-axis. Now let $P = (x, y)$ be any point (other than the origin) on this second half-line. Let $r = |OP| = \sqrt{x^2 + y^2}$, and define

$$\cos\theta = \frac{x}{r}, \qquad \sin\theta = \frac{y}{r}.$$

If $x \neq 0$, we also define

$$\tan\theta = \frac{y}{x}, \qquad \sec\theta = \frac{r}{x}.$$

If $y \neq 0$, we also define

$$\cot\theta = \frac{x}{y}, \qquad \csc\theta = \frac{r}{y}.$$

Figure 1.7.5

Each of these formulas defines a function of θ; they are known as the trigonometric functions.

In fact, all the trigonometric functions can be expressed in terms of the first two, $\sin\theta$ and $\cos\theta$. For instance,

$$\tan\theta = \frac{y}{x} = \frac{\frac{y}{r}}{\frac{x}{r}} = \frac{\sin\theta}{\cos\theta}, \qquad \cot\theta = \frac{\cos\theta}{\sin\theta},$$

$$\sec\theta = \frac{1}{\cos\theta}, \qquad\qquad \csc\theta = \frac{1}{\sin\theta}.$$

If we compare the angles θ and $\theta + 360°$, we see that when the line OP is drawn it will be the same for both angles. This means that $\sin(\theta + 360°) = \sin(\theta)$, and the same for all the other trigonometric functions (see Figure 1.7.6). The function $\sin\theta$ (or $\cos\theta$, etc.) "repeats itself after $360°$"; this is frequently abbreviated by saying that $\sin\theta$ (and $\cos\theta$, etc.) are **periodic functions**.

In the diagrams we have drawn so far, the coordinates (x, y) of P are both positive. However, it is easy enough to find an angle for which x or y (or both) will be negative (see Figure 1.7.7). In this situation, some of the trigonometric functions will be negative.

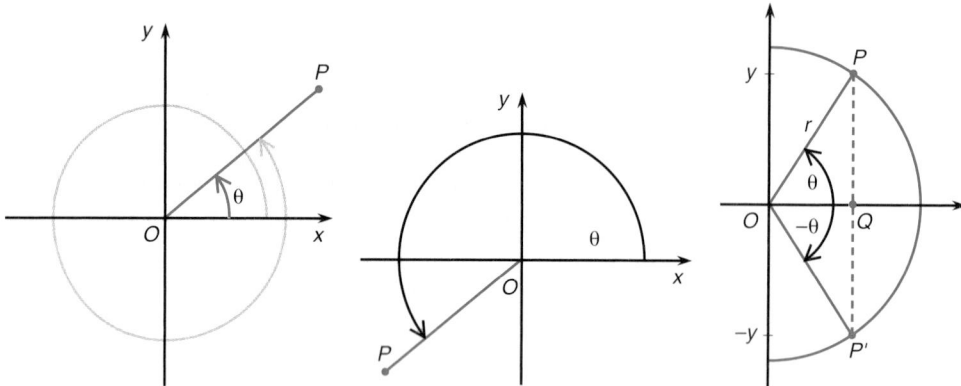

Figure 1.7.6   **Figure 1.7.7**   **Figure 1.7.8**

Figure 1.7.8 illustrates an angle θ and its negative $-\theta$, with the corresponding points P and P'. If $P = (x, y)$, then $P' = (x, -y)$, and the point $Q = (x, 0)$ is the same for both of them. The distances $|OP|$ and $|OP'|$ are the same, as are the distances $|PQ|$ and $|P'Q|$, so $\cos(-\theta) = \cos(\theta)$. However, $\sin(-\theta) = \frac{-y}{r} = -\frac{y}{r} = -\sin\theta$.

Notice that the domain of $\sin\theta$ or $\cos\theta$ is all real numbers θ. For the other functions it is necessary to exclude some points from the domain. For $\tan\theta$, for instance, it is necessary to exclude any angle for which $x = 0$. This means $\theta = 90°$, $\theta = -90°$, and any number that can be obtained by adding a multiple of $360°$ to either of these angles (such as $90 + 360 = 450°$ or $90 - 2 \times 360 = -630°$, etc).

Figure 1.7.9 is like the diagram that was used to define $\cos\theta$ and $\sin\theta$. Recalling that with $P = (x, y)$ and $r = \sqrt{x^2 + y^2}$, we have $\cos\theta = \frac{x}{r}$ and $\sin\theta = \frac{y}{r}$, we can calculate

$$\cos^2\theta + \sin^2\theta = \left(\frac{x}{r}\right)^2 + \left(\frac{y}{r}\right)^2 = \frac{x^2 + y^2}{r^2} = \frac{x^2 + y^2}{x^2 + y^2} = 1.$$

This is a fundamental **identity** (the word "identity" means an equation that holds for every possible value of the variable).

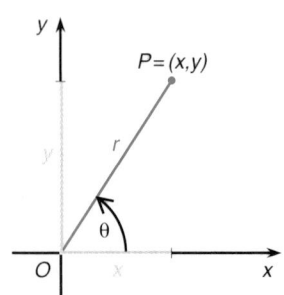

Figure 1.7.9

THEOREM 1.7.2

For every value of θ,
$$\sin^2\theta + \cos^2\theta = 1.$$

If θ is any angle for which $\cos\theta \neq 0$, we can divide this identity by $\cos^2\theta$ and obtain

$$\frac{\sin^2\theta}{\cos^2\theta} + \frac{\cos^2\theta}{\cos^2\theta} = \frac{1}{\cos^2\theta},$$

that is,

$$\tan^2\theta + 1 = \sec^2\theta \qquad \text{or} \qquad \tan^2\theta = \sec^2\theta - 1. \qquad (1.7.3)$$

Dividing the identity in Theorem 1.7.2 by $\sin^2\theta$ gives another identity, valid when $\sin\theta \neq 0$:

$$1 + \cot^2\theta = \csc^2\theta \qquad \text{or} \qquad \cot^2\theta = \csc^2\theta - 1. \qquad (1.7.4)$$

Radian Measure

We are accustomed to measuring angles in degrees. This unit for angles can be traced back to an ancient Babylonian system of numbers. It is quite familiar to us, but apart from that there is very little reason to expect, for instance, that a right angle should happen to be $90°$. To put it the other way around, why should $\frac{1}{90}$ of a right angle be chosen as the unit of measurement for angles?

It is possible to choose any units whatever to measure angles, but it turns out that there is one choice that is particularly convenient for doing calculus. The unfortunate thing is that it is fairly inconvenient for almost everything else. However, we will see in Chapter 3 that it makes calculus formulas very much simpler, and for this reason it is the choice we will adopt.

Consider the circle of radius $r = 1$, centered at the origin, as shown in Figure 1.7.10. This circle is called the **unit circle**; its circumference is $2\pi r = 2\pi$. Now if we are given an angle θ, let us measure the angle starting at the positive x-axis and rotating through an angle θ about the origin, rotating counterclockwise if the angle is positive, clockwise if it is negative. If $S = (1, 0)$ is the point where the circle intersects the positive x-axis, we can picture a point moving around the circle, starting at S, and moving through an angle θ about the origin. As before, let P be the point where it stops.

To measure the angle θ, we need some way of saying how far the point has moved in going from S to P. One fairly natural way to do this would be to measure the distance it has traveled along the circumference of the circle, with the understanding that a counterclockwise movement is measured with a positive distance and a clockwise movement with a negative number. This is **radian measure**.

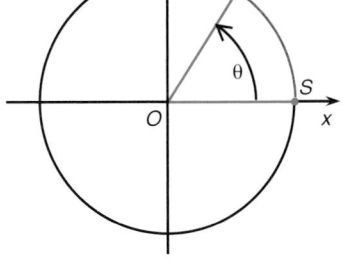

Figure 1.7.10

Figure 1.7.11

EXAMPLE 1

(i) An angle of $360°$ is the angle that goes all the way around the circle exactly once. The point on the circle goes around exactly once, so the distance it travels is the circumference of the circle, which is 2π (see Figure 1.7.11).

(ii) For an angle of $-45°$, the point moves *clockwise* $\frac{1}{8}$ of the way around the circle (since $\frac{45}{360} = \frac{1}{8}$), so the distance it travels is $\frac{1}{8}2\pi = \frac{\pi}{4}$. However, since the

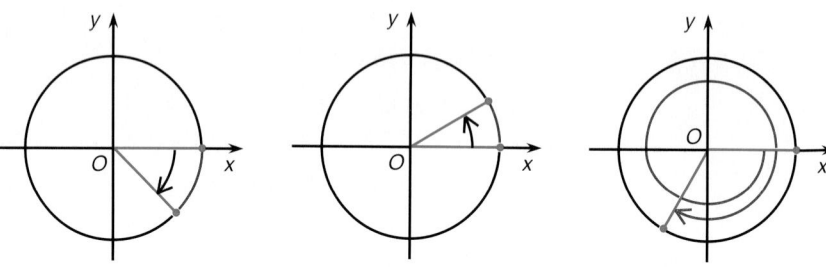

Figure 1.7.12   **Figure 1.7.13**   **Figure 1.7.14**

angle is in the negative (clockwise) direction, its radian measure is $-\frac{\pi}{4}$. See Figure 1.7.12.

(iii) An angle of $30°$ is $\frac{30}{360} = \frac{1}{12}$ of a circle, so its radian measure is $\frac{1}{12}(2\pi) = \frac{\pi}{6}$. See Figure 1.7.13.

(iv) For an angle of $-480°$ the radian measure is $\frac{-480}{360}(2\pi) = -\frac{8}{3}\pi = -\frac{8\pi}{3} \approx -8.377\,58$. See Figure 1.7.14.

Most calculators can operate either in "degree mode" or in "radian mode." It is important to develop the habit of putting your calculator into radian mode to avoid confusion.

It is very easy to convert degrees to radians and vice versa. Specifically, an angle of $\theta°$ is equal to $\frac{\theta}{360}(2\pi) = \frac{\theta}{180}\pi$ radians. In particular, an angle of $1°$ is $\frac{1}{360}(2\pi) = \frac{\pi}{180} \approx 0.017\,4533$ radians. The process can be reversed: An angle of t radians equals $\frac{t}{2\pi}\,360° = \frac{180t}{\pi}°$. In particular, 1 radian is $\frac{180(1)}{\pi} \approx 57.295\,78°$.

Summary 1.7.5   **Radian/Degree Conversions**

$$\theta° = \frac{\theta}{360}\,2\pi = \frac{\theta}{180}\pi \text{ radians}, \qquad t \text{ radians} = \frac{t}{2\pi}\,360° = \frac{180t}{\pi}°;$$

$$1° = \frac{\pi}{180} \approx 0.017\,4533 \text{ radians}, \qquad 1 \text{ radian} = \frac{180}{\pi}° \approx 57.295\,78°.$$

From now on, we will measure angles in radians unless otherwise specified.

Special Values of Trigonometric Functions

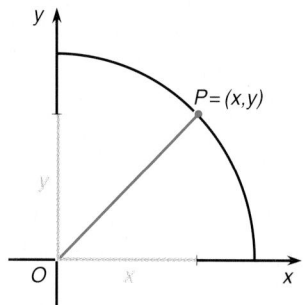

Figure 1.7.15

There are some angles for which it is quite easy to find the sin and the cos, and we should become familiar with these values.

For instance, consider the angle $\frac{\pi}{4}$ (which we used to call $45°$). In Figure 1.7.15 we observe that for the point $P = (x, y)$ the numbers x and y are equal (P lies on the line through the origin that rises at an angle of $45°$, which is the line $y = x$). On the other hand, P lies on the "unit circle," the circle of radius 1 centered at the origin, whose equation is $x^2 + y^2 = 1$. If $x = y$ and $x^2 + y^2 = 1$, then $x^2 + x^2 = 1$, or $2x^2 = 1$, so $x^2 = \frac{1}{2}$ and $x = \pm\frac{1}{\sqrt{2}}$. From the picture it is clear that it is the positive

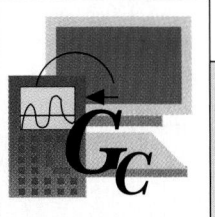

With a graphics calculator it is possible to sketch the graphs of any of the trigonometric functions. It is also possible to compare functions that can be constructed from them.

The keystrokes below show how to plot the functions $y = \sin x$ and $y = \sin(2x)$ for $x \in [-2\pi, 2\pi]$. The idea is that as x runs from -2π to 2π, $2x$ will run from -4π to 4π, so $\sin(2\pi)$ will go up and down twice as fast as $\sin x$. Be sure your calculator is in radian mode.

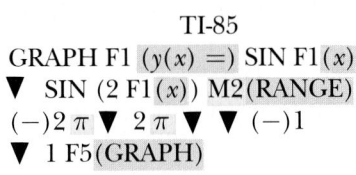

TI-85	SHARP EL9300
GRAPH F1 $(y(x) =)$ SIN F1(x)	⟳ sin x/θ/T 2ndF ▼ sin 2 x/θ/T
▼ SIN (2 F1(x)) M2(RANGE)	RANGE -2 π ENTER 2 π ENTER
$(-)2$ π ▼ 2 π ▼ ▼ $(-)1$	1 ENTER -1.5 ENTER 1.5 ENTER 1
▼ 1 F5(GRAPH)	ENTER ⟳

CASIO f_x-7700GB/f_x-6300G	HP 48SX
Cls EXE Range -2 π EXE 2 π EXE	⬑ PLOT PLOTR ERASE ATTN ′
1 EXE -1 EXE 1 EXE 1 EXE EXE	αY ▤ SIN αX ↱ PLOT ⬑
EXE EXE Graph sin X EXE Graph	DRAW ERASE 2 π +/− × ENTER 2
sin(2X) EXE	π × XRNG 1 +/− ENTER 1 YRNG
	DRAW ATTN ′ αY ▤ SIN 2 × αX
	↱ PLOT ⬑ DRAW DRAW

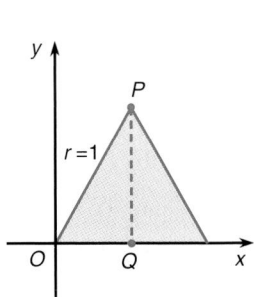

You might like to predict the behavior of $y = \sin(3x)$ before plotting it with $y = \sin x$. How will the graphs of $y = \cos x$ and $y = 2\cos(5x)$ compare? Plot $y = \sin(\frac{\pi}{2} - x)$; plot $y = \sin^2 x + \cos^2 x$. Try sketching $y = \frac{\sin 5x}{\sin(x/2)}$, for $x \in [-\pi, \pi]$.

value we want, so $x = y = \frac{1}{\sqrt{2}}$. From this we see that $\cos\frac{\pi}{4} = \frac{x}{r} = \left(\frac{1}{\sqrt{2}}\right)/1 = \frac{1}{\sqrt{2}}$ and $\sin\frac{\pi}{4} = \frac{y}{r} = \frac{x}{r} = \cos\frac{\pi}{4} = \frac{1}{\sqrt{2}}$.

For another example, let us consider an angle of $\frac{\pi}{3}$ (i.e., $60°$). Since the angles in a triangle add up to $180°$, a triangle whose sides are 1 unit long and whose angles are all equal must have angles of $60°$. This means that our angle of $\frac{\pi}{3}$ forms one corner of the equilateral triangle shown in Figure 1.7.16. The sloping edge on the left side of the triangle is a radius of the unit circle, with length $r = 1$. The length $|OQ|$ is exactly half the bottom side of the triangle; since the sides of the triangle all have the same length, 1, the length $|OQ|$ is half as much, or $\frac{1}{2}$. We find that $\cos\frac{\pi}{3} = \frac{|OQ|}{r} = \frac{1}{2}$.

To find $\sin\frac{\pi}{3}$, we find the height $|QP|$, using the Pythagorean Theorem (Theorem 1.5.1). The right-angled triangle $\triangle POQ$ has base $\frac{1}{2}$ and hypotenuse of length 1, so the length of the vertical side is $\sqrt{1^2 - \left(\frac{1}{2}\right)^2} = \sqrt{1 - \frac{1}{4}} = \sqrt{\frac{3}{4}} = \frac{\sqrt{3}}{2}$. This shows that $\sin\frac{\pi}{3} = \frac{\sqrt{3}}{2}$.

Figure 1.7.16

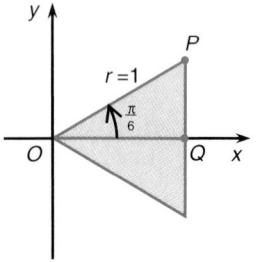

Figure 1.7.17

A similar argument can be applied to the angle $\frac{\pi}{6}$. This time the equilateral triangle has one side making an angle of $\frac{\pi}{6}$ with the positive x-axis and another side at an angle of $-\frac{\pi}{6}$ (see Figure 1.7.17). The horizontal segment OQ has length $\frac{\sqrt{3}}{2}$, and the vertical segment PQ has length $\frac{1}{2}$, so $\cos\frac{\pi}{6} = \frac{\sqrt{3}}{2}$ and $\sin\frac{\pi}{6} = \frac{1}{2}$.

Of course it is easy to see that $\cos(0) = 1$, $\sin(0) = 0$, $\cos\frac{\pi}{2} = 0$, $\sin\frac{\pi}{2} = 1$, $\cos\left(-\frac{\pi}{2}\right) = 0$, $\sin\left(-\frac{\pi}{2}\right) = -1$, $\cos\pi = -1$, and $\sin\pi = 0$.

Summary 1.7.6

Special Values of Trigonometric Functions

$$\cos 0 = 1, \qquad \sin 0 = 0, \qquad \cos\pi = -1, \qquad \sin\pi = 0,$$
$$\cos\frac{\pi}{2} = 0, \qquad \sin\frac{\pi}{2} = 1, \qquad \cos\frac{\pi}{4} = \frac{1}{\sqrt{2}}, \qquad \sin\frac{\pi}{4} = \frac{1}{\sqrt{2}},$$
$$\cos\frac{\pi}{3} = \frac{1}{2}, \qquad \sin\frac{\pi}{3} = \frac{\sqrt{3}}{2}, \qquad \cos\frac{\pi}{6} = \frac{\sqrt{3}}{2}, \qquad \sin\frac{\pi}{6} = \frac{1}{2}.$$

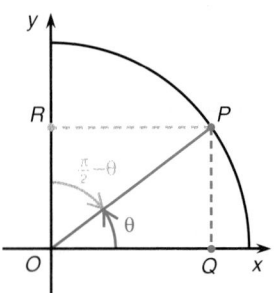

Figure 1.7.18

Now consider two angles whose sum is a right angle $\frac{\pi}{2}$. We can write them as θ and $\frac{\pi}{2} - \theta$. In Figure 1.7.18 the angle θ is drawn in the usual way, so that $\angle POQ = \theta$, and we see that $\angle POR = \frac{\pi}{2} - \theta$.

If $P = (x, y)$ is on the unit circle, then $r = 1$, so $\cos\theta = x$ and $\sin\theta = y$. Now let us consider the angle $\angle POR$. If we change our point of view and measure the angle starting from the positive y-axis and rotating *clockwise* to the line OP, then we can find $\cos\left(\frac{\pi}{2} - \theta\right)$ and $\sin\left(\frac{\pi}{2} - \theta\right)$ by looking at the triangle $\triangle POR$.

We started measuring the angle at the y-axis, so the cosine is the coordinate of P along the y-axis divided by the radius, that is, the cosine is $\frac{y}{r} = y$. The sine is the other coordinate divided by r, or $\frac{x}{r} = x$. We conclude the following:

THEOREM 1.7.7

For any angle θ we have

$$\cos\left(\frac{\pi}{2} - \theta\right) = \sin\theta \qquad \text{and} \qquad \sin\left(\frac{\pi}{2} - \theta\right) = \cos\theta.$$

If $P = (x, y)$ is the point on the unit circle determined by the angle θ as shown in Figure 1.7.19, then the corresponding point for the angle $-\theta$ is $P' = (x, -y)$. So $\cos(-\theta) = \frac{x}{r} = x = \cos(\theta)$ and $\sin(-\theta) = \frac{-y}{r} = -\frac{y}{r} = -y = -\sin(\theta)$.

From these formulas it is easy to see what happens to the other trigonometric functions if θ is replaced by $-\theta$.

THEOREM 1.7.8

The following formulas hold for every θ for which they make sense:

$$\cos(-\theta) = \cos(\theta), \qquad \sin(-\theta) = -\sin(\theta), \qquad \tan(-\theta) = -\tan(\theta),$$
$$\sec(-\theta) = \sec(\theta), \qquad \csc(-\theta) = -\csc(\theta), \qquad \cot(-\theta) = -\cot(\theta).$$

These facts and the values of sin and cos that we found allow us to find the values of trigonometric functions at many angles.

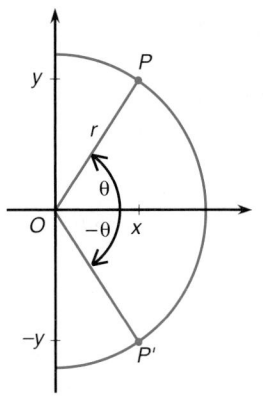

Figure 1.7.19

EXAMPLE 2   Find (i) $\sin\left(\frac{3\pi}{4}\right)$; (ii) $\cos\left(-\frac{2\pi}{3}\right)$; (iii) $\tan\left(\frac{7\pi}{6}\right)$.

Solution

(i) Since $\frac{3\pi}{4} + \left(-\frac{\pi}{4}\right) = \frac{\pi}{2}$, we know that $\sin\left(\frac{3\pi}{4}\right) = \cos\left(-\frac{\pi}{4}\right) = \cos\left(\frac{\pi}{4}\right) = \frac{1}{\sqrt{2}}$.

(ii) $\cos\left(-\frac{2\pi}{3}\right) = \cos\left(\frac{2\pi}{3}\right) = \cos\left(\frac{\pi}{2} - \left(-\frac{\pi}{6}\right)\right) = \sin\left(\left(-\frac{\pi}{6}\right)\right) = -\sin\left(\left(\frac{\pi}{6}\right)\right) = -\frac{1}{2}$.

(iii) To evaluate tan, it is simpler to begin with sin and cos: $\cos\frac{7\pi}{6} = \cos\left(\frac{\pi}{2} - \left(-\frac{2\pi}{3}\right)\right) = \sin\left(-\frac{2\pi}{3}\right) = -\sin\left(\frac{2\pi}{3}\right) = -\sin\left(\frac{\pi}{2} - \left(-\frac{\pi}{6}\right)\right) = -\cos\left(-\frac{\pi}{6}\right) = -\cos\left(\frac{\pi}{6}\right) = -\frac{\sqrt{3}}{2}$.

Similarly, $\sin\frac{7\pi}{6} = \sin\left(\frac{\pi}{2} - \left(-\frac{2\pi}{3}\right)\right) = \cos\left(-\frac{2\pi}{3}\right) = \cos\left(\frac{2\pi}{3}\right) = \cos\left(\frac{\pi}{2} - \left(-\frac{\pi}{6}\right)\right) = \sin\left(-\frac{\pi}{6}\right) = -\sin\left(\frac{\pi}{6}\right) = -\frac{1}{2}$.

From these two facts we see that

$$\tan\frac{7\pi}{6} = \frac{\sin(7\pi/6)}{\cos(7\pi/6)} = \frac{-1/2}{-\sqrt{3}/2} = \frac{1}{\sqrt{3}} \approx 0.577\,35.$$

Now that we can evaluate the trigonometric functions at many points, it is possible to plot a number of points and sketch their graphs, as in Figure 1.7.20.

$y = \cos x$

$y = \sin x$

$y = \tan x$

$y = \sec x$

$y = \csc x$

$y = \cot x$

Figure 1.7.20

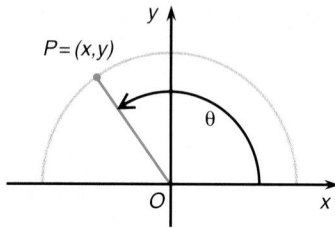

Figure 1.7.21

In Figure 1.7.21 we have drawn the point $P = (x, y)$, and labeled the angle θ between the positive x-axis and the line OP. If we let $r = \sqrt{x^2 + y^2}$, then P is on the circle of radius r centered at the origin. We know from Definition 1.7.1 that $\cos\theta = \frac{x}{r}$ and $\sin\theta = \frac{y}{r}$. These two equations can be rearranged to give

$$x = r\cos\theta \qquad \text{and} \qquad y = r\sin\theta. \tag{1.7.9}$$

We can find the coordinates of P if we know the length $|OP|$ and the angle θ.

Law of Cosines, Addition Formulas

The Pythagorean Theorem relates the squares of the lengths of the sides of a right-angled triangle. It is possible to find a relation among the lengths of the sides of any triangle, but it will depend on the angles as well as the lengths of the sides.

THEOREM 1.7.10

> **The Law of Cosines**
>
> Consider a triangle whose sides have lengths a, b, and c. Suppose the side of length c is opposite to an angle θ. Then
>
> $$c^2 = a^2 + b^2 - 2ab\cos\theta.$$

Notice that if $\theta = \frac{\pi}{2}$, then the triangle has a right angle, and since $\cos\frac{\pi}{2} = 0$, the Law of Cosines reduces to the Pythagorean Theorem.

PROOF   In Figure 1.7.22 we place the angle θ at the origin, the side of length a along the x-axis from O to $(a, 0)$, and the side of length b from the origin to the point $P = (x, y)$.

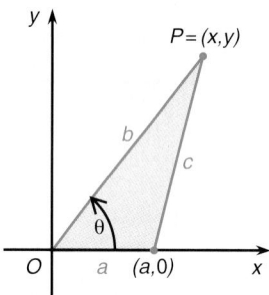

Figure 1.7.22

Then $b = \sqrt{x^2 + y^2}$, and by Equations 1.7.9 we see that $x = b\cos\theta$ and $y = b\sin\theta$. The side whose length is c lies between the points $P = (x, y)$ and $(a, 0)$, so $c = \sqrt{(x - a)^2 + (y - 0)^2}$.

From this we find $c^2 = (x - a)^2 + y^2 = x^2 - 2ax + a^2 + y^2 = (x^2 + y^2) + a^2 - 2ax$, and since $x^2 + y^2 = b^2$ and $x = b\cos\theta$, this becomes $c^2 = b^2 + a^2 - 2ab\cos\theta$, which is the desired formula. ◢

A natural and important question to ask is this: If we know the values of cos and sin for two angles θ and ϕ (ϕ is the Greek letter phi, which is sometimes pronounced "fee" and sometimes pronounced to rhyme with "eye"), how can we find $\cos(\theta + \phi)$?

It is actually slightly easier to find $\cos(\theta - \phi)$ first. Since $\cos(\theta - \phi) = \cos(\phi - \theta)$, we can assume $\theta > \phi$ (because otherwise we could interchange them, without

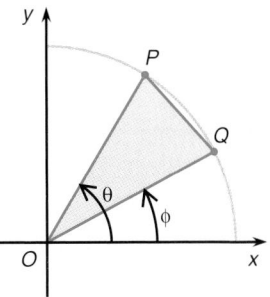

Figure 1.7.23

affecting the result). In Figure 1.7.23 we have drawn both angles θ and ϕ starting at the positive x-axis, and we have plotted the unit circle. Using Equations 1.7.9, we can find the coordinates of the points P and Q where the second arm of each angle intersects the circle. Since the radius of the circle is $r = 1$, we find that $P = (\cos\theta, \sin\theta)$ and $Q = (\cos\phi, \sin\phi)$.

Now we apply the Law of Cosines to the triangle $\triangle POQ$. The two sides OP and OQ both have length 1, and the length of the third side PQ satisfies

$$|PQ|^2 = c^2 = 1^2 + 1^2 - 2(1)(1)\cos(\theta - \phi) = 2 - 2\cos(\theta - \phi).$$

On the other hand, since $P = (\cos\theta, \sin\theta)$ and $Q = (\cos\phi, \sin\phi)$, we see that

$$\begin{aligned}
|PQ|^2 &= (\cos\theta - \cos\phi)^2 + (\sin\theta - \sin\phi)^2 \\
&= \cos^2\theta - 2\cos\theta\cos\phi + \cos^2\phi + \sin^2\theta - 2\sin\theta\sin\phi + \sin^2\phi \\
&= \cos^2\theta + \sin^2\theta + \cos^2\phi + \sin^2\phi - 2(\cos\theta\cos\phi + \sin\theta\sin\phi) \\
&= 2 - 2(\cos\theta\cos\phi + \sin\theta\sin\phi).
\end{aligned}$$

Equating these two expressions for $|PQ|^2$, we see that

$$\cos(\theta - \phi) = \cos\theta\cos\phi + \sin\theta\sin\phi.$$

If we replace ϕ in this equation with $-\phi$, then the left side becomes $\cos(\theta + \phi)$. Since $\cos(-\phi) = \cos\phi$ and $\sin(-\phi) = -\sin\phi$, the right side becomes $\cos\theta\cos\phi - \sin\theta\sin\phi$.

We could use a similar argument to find $\sin(\theta + \phi)$, but it is easier to reason as follows:

$$\begin{aligned}
\sin(\theta + \phi) &= \cos\left(\frac{\pi}{2} - (\theta + \phi)\right) = \cos\left(\left(\frac{\pi}{2} - \theta\right) - \phi\right) \\
&= \cos\left(\frac{\pi}{2} - \theta\right)\cos\phi + \sin\left(\frac{\pi}{2} - \theta\right)\sin\phi \\
&= \sin\theta\cos\phi + \cos\theta\sin\phi.
\end{aligned}$$

Again we can replace ϕ with $-\phi$ to find the formula for $\sin(\theta - \phi)$.

THEOREM 1.7.11

Addition Formulas

$$\begin{aligned}
\cos(\theta + \phi) &= \cos\theta\cos\phi - \sin\theta\sin\phi, \\
\sin(\theta + \phi) &= \sin\theta\cos\phi + \cos\theta\sin\phi, \\
\cos(\theta - \phi) &= \cos\theta\cos\phi + \sin\theta\sin\phi, \\
\sin(\theta - \phi) &= \sin\theta\cos\phi - \cos\theta\sin\phi.
\end{aligned}$$

If we let $\phi = \theta$ in these formulas, we can find formulas for $\cos(2\theta)$ and $\sin(2\theta)$.

THEOREM 1.7.12

Double-Angle Formulas

$$\cos(2\theta) = \cos^2\theta - \sin^2\theta, \qquad \sin(2\theta) = 2\cos\theta\sin\theta.$$

If we take the double-angle formula $\cos(2\theta) = \cos^2\theta - \sin^2\theta$ and substitute into it the identity $\sin^2\theta = 1 - \cos^2\theta$, we find $\cos(2\theta) = \cos^2\theta - (1 - \cos^2\theta) = 2\cos^2\theta - 1$. This can be rearranged as $2\cos^2\theta = 1 + \cos(2\theta)$, or $\cos^2\theta = \frac{1+\cos(2\theta)}{2}$.

We can use this to find that $\sin^2 \theta = 1 - \cos^2 \theta = 1 - \frac{1+\cos(2\theta)}{2} = \frac{1-\cos(2\theta)}{2}$.

These formulas can also be used to find $\cos^2\left(\frac{\theta}{2}\right)$ and $\sin^2\left(\frac{\theta}{2}\right)$ in terms of $\cos\theta$.

THEOREM 1.7.13

> **Half-Angle Formulas**
>
> $$\cos^2 \theta = \frac{1 + \cos(2\theta)}{2}, \qquad \sin^2 \theta = \frac{1 - \cos(2\theta)}{2}.$$

These formulas can be used to help us evaluate trigonometric functions at many additional points.

EXAMPLE 3 Find (i) $\sin\frac{\pi}{12}$; (ii) $\cos\frac{\pi}{8}$; (iii) $\cos\frac{11\pi}{12}$.

Solution

(i) From the half-angle formula we know that $\sin^2 \frac{\pi}{12} = \frac{1}{2}\left(1 - \cos\left(2\left(\frac{\pi}{12}\right)\right)\right) = \frac{1}{2}\left(1 - \cos\frac{\pi}{6}\right) = \frac{1}{2}\left(1 - \frac{\sqrt{3}}{2}\right) = \frac{1}{2} - \frac{\sqrt{3}}{4} \approx 0.066\,987$. What we want is $\sin\frac{\pi}{12}$, which is either the square root of this number or the negative of the square root. Since $\frac{\pi}{12}$ is between 0 and the right angle $\frac{\pi}{2}$, we see that its sin must be positive, so $\sin\frac{\pi}{12} = \sqrt{\frac{1}{2} - \frac{\sqrt{3}}{4}} \approx 0.258\,819$.

(ii) The half-angle formula gives $\cos^2 \frac{\pi}{8} = \frac{1}{2}\left(1 + \cos\left(2\left(\frac{\pi}{8}\right)\right)\right) = \frac{1}{2}\left(1 + \cos\frac{\pi}{4}\right) = \frac{1}{2}\left(1 + \frac{1}{\sqrt{2}}\right) = \frac{1}{2} + \frac{1}{2\sqrt{2}} \approx 0.853\,5534$. Since $\cos\frac{\pi}{8}$ is positive, $\cos\frac{\pi}{8} = \sqrt{\frac{1}{2} + \frac{1}{2\sqrt{2}}} \approx 0.923\,8795$.

(iii) Here $\cos^2 \frac{11\pi}{12} = \frac{1}{2}\left(1 + \cos\frac{11\pi}{6}\right) = \frac{1}{2}\left(1 + \cos\left(2\pi - \frac{\pi}{6}\right)\right) = \frac{1}{2}\left(1 + \cos\left(-\frac{\pi}{6}\right)\right)$. We have used the fact that cos is periodic, having the same value at $2\pi + \theta$ as it has at θ.

Since $\cos\left(-\frac{\pi}{6}\right) = \cos\frac{\pi}{6} = \frac{\sqrt{3}}{2}$, we find $\cos^2 \frac{11\pi}{12} = \frac{1}{2}\left(1 + \frac{\sqrt{3}}{2}\right) = \frac{1}{2} + \frac{\sqrt{3}}{4}$. Since $\frac{\pi}{2} < \frac{11\pi}{12} < \pi$, we see that if we start at the positive x-axis and rotate through an angle of $\frac{11\pi}{12}$, the resulting point will lie above the *negative x*-axis, so its x-coordinate will be negative. So $\cos\frac{11\pi}{12}$ is negative, and $\cos\frac{11\pi}{12} = -\sqrt{\frac{1}{2} + \sqrt{3}/4} \approx -0.965\,9258$. ◢

$cos\ \theta$

Figure 1.7.24(i)

$sin\ \theta$

Figure 1.7.24(ii)

The half-angle formulas (Theorem 1.7.13) permit us to find $\cos^2 \frac{\theta}{2}$ and $\sin^2 \frac{\theta}{2}$. To find $\cos\frac{\theta}{2}$ or $\sin\frac{\theta}{2}$, we must take square roots, but it is also necessary to determine the correct *sign* of the answer. Figures 1.7.24(i) and (ii) show the regions in which cos and sin are positive and negative.

If we add the formulas for $\cos(\theta + \phi)$ and $\cos(\theta - \phi)$, we find that

$$\cos(\theta + \phi) + \cos(\theta - \phi) = \cos\theta\cos\phi - \sin\theta\sin\phi + \cos\theta\cos\phi + \sin\theta\sin\phi$$
$$= 2\cos\theta\cos\phi.$$

Turning this around, we find that $\cos\theta\cos\phi = \frac{1}{2}\left(\cos(\theta + \phi) + \cos(\theta - \phi)\right)$.

From the difference of the two addition formulas we find $\sin\theta\sin\phi = \frac{1}{2}\big(\cos(\theta - \phi) - \cos(\theta + \phi)\big)$. Working with the addition formulas for sin gives an additional identity.

THEOREM 1.7.14

$$\cos\theta\cos\phi = \frac{1}{2}\big(\cos(\theta + \phi) + \cos(\theta - \phi)\big),$$

$$\sin\theta\sin\phi = \frac{1}{2}\big(\cos(\theta - \phi) - \cos(\theta + \phi)\big),$$

$$\sin\theta\cos\phi = \frac{1}{2}\big(\sin(\theta + \phi) + \sin(\theta - \phi)\big).$$

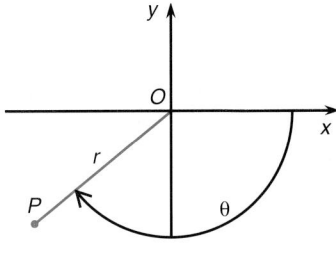

Figure 1.7.25

Polar Coordinates

Representing a point P in the plane by its coordinates (x, y) is quite familiar to us, but there is another way in which the same thing can be accomplished. Consider the point P shown in Figure 1.7.25, and join it to the origin O. The angle θ between the positive x-axis and the segment OP determines the "direction" in which the segment OP is "pointing." If we also know how far P is from the origin, then we will have determined the position of P exactly.

DEFINITION 1.7.15

Consider a point P in the plane. If the segment OP makes an angle θ with the positive x-axis, and if we let $r = |OP|$ be the distance from the origin to P, then the pair of numbers (r, θ) is called the **polar coordinates** of the point P.

The number r is sometimes called the **radius** or the **modulus** of P, and θ is called its **angle**, or its **polar angle**, or its **argument**.

There are several things to notice about this definition. One is that for any point P there are many possible choices for the angle θ. For instance, we could measure the angle θ in either direction from the positive x-axis. We could also measure an extra complete rotation about the origin, which would add 2π to θ.

So the angle θ is not completely determined. We can add 2π or 4π or -2π or *any* integer multiple of 2π to θ, and the resulting angle will still determine the same direction in the plane. See Figure 1.7.26.

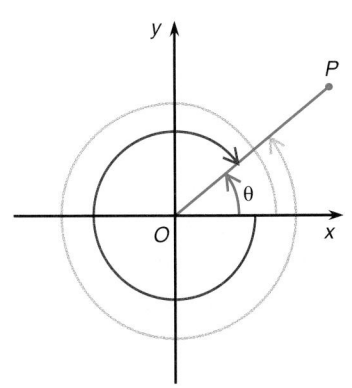

Figure 1.7.26

For instance, for the point $(0, 2)$ on the y-axis we could naturally choose the angle $\theta = \frac{\pi}{2}$. But $\frac{\pi}{2} + 2\pi = \frac{5\pi}{2}$ would be just as good. For that matter, so would $\frac{\pi}{2} - 2\pi = -\frac{3\pi}{2}$, or $\frac{9\pi}{2}$, etc. In fact if k is any integer, we could choose the angle to be $\frac{\pi}{2} + 2k\pi$.

In this situation we often write $\theta = \frac{\pi}{2} + 2k\pi$, to help remind ourselves that there are many possible values of θ.

Another complication arises with the point O, the origin. Its radius is $r = 0$, but *any* angle θ can be used. If we move a distance of zero in *any* direction, the point at which we will end up is the origin.

One obvious question is how can we go back and forth between polar coordinates (r, θ) and Cartesian coordinates (x, y). In other words, if we know the polar coordinates (r, θ) of a point P, how do we find its Cartesian (or "rectangular") coordinates (x, y), and vice versa?

In fact we have already answered one of these questions in Equations 1.7.9, where we saw that

$$x = r \cos \theta \qquad \text{and} \qquad y = r \sin \theta.$$

In the opposite direction it is a little more complicated. If P is the point whose rectangular coordinates are (x, y), then we know its modulus is $r = \sqrt{x^2 + y^2}$. If θ is its argument, then we know that $x = r \cos \theta$ and $y = r \sin \theta$. If $x \neq 0$, we divide these two equations and find that $\frac{y}{x} = \frac{\sin \theta}{\cos \theta} = \tan \theta$. This tells us that θ is an angle whose tangent equals $\frac{y}{x}$, which is often expressed by saying that $\theta = \arctan(\frac{y}{x})$.

The complication is that this does not quite determine what θ is. For instance, if $P = (-1, 1)$, then $\frac{y}{x} = -1$. If we look for angles whose tangent equals -1, then there are many possibilities, but among them are $\theta = -\frac{\pi}{4}$ and $\theta = \frac{3\pi}{4}$. These two angles point in *opposite* directions from the origin.

This reflects the fact that the quotient of the y-coordinate by the x-coordinate is the same for $P = (x, y)$ and for $-P = (-x, -y)$; in both cases it equals $\frac{y}{x}$. However, by looking at the picture we can easily tell which is the right choice (in the case of $P = (-1, 1)$, one choice of the angle lies between 0 and π, so a suitable choice is $\frac{3\pi}{4}$); see Figure 1.7.27.

The polar coordinates of a point $P = (x, y)$ are (r, θ), where

$$r = \sqrt{x^2 + y^2} \qquad \text{and} \qquad \tan \theta = \frac{y}{x}.$$

If an angle θ can be found for which $\tan \theta = \frac{y}{x}$, then it is necessary to sketch a picture and decide whether the argument is this θ or $\theta + \pi$ (i.e., the opposite angle). If θ is one correct choice for the argument, then $\theta + 2k\pi$ is also a correct choice, for any integer k.

If $x = 0$, that is, P is on the vertical axis, then $\frac{y}{x}$ is undefined, but in this case, $\theta = \frac{\pi}{2} + 2k\pi$ if $y > 0$, and $\theta = -\frac{\pi}{2} + 2k\pi$ if $y < 0$.

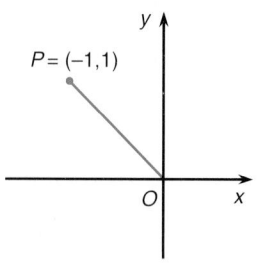

$P = (-1, 1)$

Figure 1.7.27

EXAMPLE 4   (i) Find the rectangular coordinates of the point whose polar coordinates are $\left(2\sqrt{2}, -\frac{3\pi}{4}\right)$. Find the polar coordinates of the points (ii) $P = (-3, -\sqrt{3})$; (iii) $Q = (5, -2)$; (iv) $S = (0, -3)$.

Solution

(i) From Equations 1.7.9 we see that $x = r \cos \theta = 2\sqrt{2} \cos\left(-\frac{3\pi}{4}\right) = 2\sqrt{2} \cos\left(\frac{3\pi}{4}\right)$
$= 2\sqrt{2} \sin\left(\frac{\pi}{2} - \frac{3\pi}{4}\right) = 2\sqrt{2} \sin\left(-\frac{\pi}{4}\right) = -2\sqrt{2} \sin\frac{\pi}{4} = -2\sqrt{2}\frac{1}{\sqrt{2}} = -2$.
Similarly, $y = r \sin \theta = 2\sqrt{2} \sin\left(-\frac{3\pi}{4}\right) = -2\sqrt{2} \sin\left(\frac{3\pi}{4}\right) =$
$-2\sqrt{2} \cos\left(\frac{\pi}{2} - \frac{3\pi}{4}\right) = -2\sqrt{2} \cos\left(-\frac{\pi}{4}\right) = -2\sqrt{2} \cos\frac{\pi}{4} = -2\sqrt{2}\frac{1}{\sqrt{2}} = -2$.
The rectangular coordinates of the point in question are $(-2, -2)$.

(ii) We find that $r = \sqrt{(-3)^2 + (-\sqrt{3})^2} = \sqrt{9 + 3} = \sqrt{12} = 2\sqrt{3}$.
We also know that $\tan \theta = \frac{-\sqrt{3}}{-3} = \frac{1}{\sqrt{3}}$. If we think through the angles for which we know the values of the trigonometric functions, we know that $\sin\frac{\pi}{6} = \frac{1}{2}$ and $\cos\frac{\pi}{6} = \frac{\sqrt{3}}{2}$, so $\tan\frac{\pi}{6} = \frac{1}{2}/\frac{\sqrt{3}}{2} = \frac{1}{\sqrt{3}}$. We expect that either $\frac{\pi}{6}$ or $\frac{\pi}{6} + \pi = \frac{7\pi}{6}$ will be the correct argument.

Figure 1.7.28

Figure 1.7.29

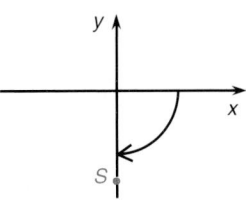

Figure 1.7.30

A quick look at Figure 1.7.28 will convince us that the angle is $\frac{7\pi}{6}$. The polar coordinates of the point $P = (-3, -\sqrt{3})$ are $\left(2\sqrt{3}, \frac{7\pi}{6} + 2k\pi\right)$, where k can be any integer.

(iii) For $Q = (5, -2)$ we find $r = \sqrt{5^2 + (-2)^2} = \sqrt{29}$ and $\tan\theta = -\frac{2}{5}$. None of the angles whose tangents we can work out easily has this as its tangent, so we resort to a calculator. Being careful to set it to *radian* mode, we find that the arctan (or "inverse tangent") of $-\frac{2}{5}$ is $\arctan\left(-\frac{2}{5}\right) \approx -0.380\,506$.

From Figure 1.7.29 we expect the argument to be a negative number between $-\pi$ and 0, which this is (recall that $\pi \approx 3.141\,59$). The angle we have found is the correct one. (If it were not, then we would simply add π to it.)

The polar coordinates of Q are $(\sqrt{29}, -0.380\,506 + 2k\pi)$, where k can be any integer and the argument is given as an approximate value.

(iv) For $S = (0, -3)$ the modulus is $r = 3$, but we cannot find $\frac{y}{x}$ because we would have to divide by zero. Fortunately, we see that S is on the vertical axis *below* the origin, so $\theta = -\frac{\pi}{2} + 2k\pi$. See Figure 1.7.30. ◢

Exercises 1.7

Find the following values of trigonometric functions, without using a calculator except to check your work.

1. $\sin\left(-\frac{\pi}{2}\right)$

2. $\cos\left(\frac{9\pi}{2}\right)$

3. $\cos\left(\frac{3\pi}{4}\right)$

4. $\sin\left(-\frac{4\pi}{3}\right)$

5. $\tan\left(-\frac{5\pi}{3}\right)$

6. $\sec\left(\frac{2\pi}{3}\right)$

7. $\csc\left(\frac{5\pi}{4}\right)$

8. $\cot\left(\frac{5\pi}{6}\right)$

9. $\sin\left(-\frac{3\pi}{4}\right)$

10. $\tan\left(\frac{7\pi}{2}\right)$

11. $\sin\frac{\pi}{8}$

12. $\cos\frac{\pi}{12}$

13. $\sin\frac{5\pi}{12}$

14. $\cos\frac{7\pi}{8}$

15. $\cos\frac{11\pi}{12}$

16. $\sin\frac{17\pi}{8}$

17. $\tan\frac{11\pi}{8}$

18. $\sec\frac{13\pi}{12}$

19. $\tan\frac{15\pi}{8}$

20. $\csc\frac{3\pi}{12}$

Find the rectangular coordinates of each of the points whose polar coordinates are given below.

21. $(3, \pi)$

22. $\left(\frac{1}{2}, 4\pi\right)$

23. $\left(2, \frac{\pi}{2}\right)$

24. $\left(5, -\frac{\pi}{2}\right)$

25. $\left(1, -\frac{\pi}{3}\right)$

26. $\left(2, \frac{7\pi}{2}\right)$

27. $\left(4, -\frac{5\pi}{4}\right)$

28. $\left(8, \frac{11\pi}{4}\right)$

29. $\left(2, \frac{5\pi}{6}\right)$

30. $\left(2, \frac{3\pi}{8}\right)$

Find the polar coordinates of each of the points whose Cartesian coordinates are given below.

31. $(3, 0)$

32. $(-2, 0)$

33. $(0, -2)$

34. $\left(0, \frac{1}{2}\right)$

35. $(2, -2)$

36. $(4, 4\sqrt{3})$

37. $(-4, -4)$

38. $(0, 0)$

39. $\left(\frac{1}{4}, -\frac{1}{4\sqrt{3}}\right)$

40. $(\cos x, \sin x)$

1.8 Logarithms and Exponentials

EXPONENT
LOGARITHM
BASE
COMMON LOGARITHMS
NATURAL LOGARITHMS

Taking powers is a familiar concept: If n is any positive integer and x is any real number, x^n means the product of x with itself n times, that is, $x \times x \times \ldots \times x$, with a total of n factors. The number n is called the **exponent**.

The product $x^m x^n$ is the product of two products, one containing m factors all equal to x and the other containing n such factors. So $x^m x^n$ is the product of $m + n$ factors, all equal to x, so it is x^{m+n}. If x is any real number and m and n are positive integers,

$$x^m x^n = x^{m+n}. \tag{1.8.1}$$

Another useful property of powers is illustrated by looking at $(x^2)^3$, the product of three copies of x^2. Since each of them is the product of two copies of x, the whole thing is the product of six copies of x, or x^6. In exactly the same way, $(x^m)^n$ is the product of mn copies of x. In other words, for any real x and any positive integers m and n,

$$(x^m)^n = x^{mn}. \tag{1.8.2}$$

Property 1.8.1 suggests a way to extend the definition to integer exponents. If we want all integer powers to satisfy the same property, then we should expect that $x^0 x^n = x^{0+n} = x^n$. For this to be true for all values of x, x^0 must equal 1.

Moreover, $x^{-n} x^n = x^{-n+n} = x^0 = 1$, so $x^{-n} = \frac{1}{x^n}$. This allows us to define x^n for any integer n and any real number x, except that when $x = 0$, it only makes sense to consider x^n for $n > 0$.

$$x^0 = 1 \qquad \text{and} \qquad x^{-n} = \frac{1}{x^n}. \tag{1.8.3}$$

Next it is possible to consider powers x^r, where the exponent r is a rational number. We expect Property 1.8.1 to hold for them too.

For instance, with $r = \frac{1}{2}$ we expect that $x^r x^r = x^{2r}$, that is, $x^{1/2} x^{1/2} = x^1$, which means $(x^{1/2})^2 = x$. This says that $x^{1/2}$ is something whose square equals x. This is not possible if $x < 0$, and if $x \geq 0$, it means that $x^{1/2} = \sqrt{x}$. Similarly, $x^{1/3}$ is the cube root of x, $x^{1/4}$ is the fourth root of x, and so on.

If Property 1.8.2 holds for rational exponents, then $x^{m/n} = x^{(1/n)m} = (x^{1/n})^m$, the mth power of the nth root of x. This makes sense whenever $x > 0$ and $\frac{m}{n}$ is a rational number (if $\frac{m}{n} < 0$, we can assume that $n > 0$ and $m < 0$, so it makes sense to consider $x^{1/n}$ and $(x^{1/n})^m$).

It is also possible to define what is meant by x^s even if s is not a rational number, provided that $x > 0$. For instance, knowing that $\sqrt{2}$ is not a rational number, we could ask what is meant by $x^{\sqrt{2}}$. Roughly speaking, the idea is to find rational numbers r that are closer and closer to $\sqrt{2}$ and hope that x^r will get close to some number w as r gets close to $\sqrt{2}$. It turns out that this is exactly what happens, so we can let $x^{\sqrt{2}} = w$. For any real number s we can find rational numbers r that get closer and closer to s; the numbers x^r will then get closer and closer to some number, which is what we call x^s. The details of this approach will be discussed more carefully in Chapter 7.

Logarithms

Suppose $a > 0$ and we want to find a real number s so that $10^s = a$. For instance, if $a = 1000$, then $s = 3$; if $a = \frac{1}{10}$, then $s = -1$; if $a = \sqrt{10}$, then $s = \frac{1}{2}$. This number s is called the **logarithm** of a, and it is often written $s = \log a$. It is determined by the property that $10^{\log a} = a$. Another way of putting it is that $\log(10^s) = s$; the logarithm of a positive number is the exponent to which we have to raise 10 in order to get that number:

$$10^{\log a} = a \qquad \text{and} \qquad \log(10^s) = s. \tag{1.8.4}$$

Property 1.8.1 tells us that

$$10^{\log(ac)} = ac = (10^{\log a})(10^{\log c}) = 10^{\log a + \log c}.$$

Matching exponents at the left and right ends of this equation, we find that $\log(ac) = \log a + \log c$, an extremely important property of logarithms. In much the same way,

$$10^{\log(a^s)} = a^s = (a)^s = (10^{\log a})^s = 10^{s \log a},$$

so $\log(a^s) = s \log a$. In particular, with $s = -1$ we see that $\log(a^{-1}) = -\log a$, that is, $\log \frac{1}{a} = -\log a$. From this we find that $\log \frac{a}{c} = \log \left(a \times \frac{1}{c}\right) = \log a + \log \frac{1}{c} = \log a - \log c$. These properties of log are summarized below. For any $a, c > 0$,

$$\begin{aligned} \log(ac) &= \log a + \log c, & \log(a^s) &= s \log a, \\ \log \frac{1}{a} &= -\log a, & \log \frac{a}{c} &= \log a - \log c. \end{aligned} \tag{1.8.5}$$

There is no particular reason why we should choose powers of 10 when we define logarithms. It would have been just as good to use 2, say, and look for a number s so that $2^s = a$. This number s is called the logarithm to the base 2 of a. It is the exponent to which 2 must be raised to get the number a.

We could have used any positive number $b \neq 1$ in the same way; if $b^s = a$, we say that s is the logarithm to the base b of a, and we write $s = \log_b(a)$. There are properties analogous to Equations 1.8.4:

$$b^{\log_b a} = a \qquad \text{and} \qquad \log_b(b^s) = s. \tag{1.8.6}$$

There are also properties analogous to Equations 1.8.5 for logarithms to the base b:

$$\begin{aligned} \log_b(ac) &= \log_b a + \log_b c, & \log_b(a^s) &= s \log_b a, \\ \log_b \frac{1}{a} &= -\log_b a, & \log_b \frac{a}{c} &= \log_b a - \log_b c. \end{aligned} \tag{1.8.7}$$

In particular, when we write $\log a$, we mean $\log_{10} a$. Because logarithms to the base 10 are used so often, they are often called **common logarithms**.

EXAMPLE 1   Find (i) $\log \frac{100}{\sqrt{10}}$; (ii) $\log_2(16)$; (iii) $\log_3 \frac{1}{27}$.

Solution

(i) By Properties 1.8.5, $\log \frac{100}{\sqrt{10}} = \log 100 - \log(\sqrt{10}) = 2 - \log(10^{1/2}) = 2 - \frac{1}{2} = \frac{3}{2}$.

(ii) $\log_2(16) = \log_2(2^4) = 4$.

(iii) $\log_3 \frac{1}{27} = -\log_3 27 = -\log_3(3^3) = -3$.

There is one other base that is used quite frequently for logarithms. It is like radian measure for trigonometric functions in the sense that it is particularly well suited for use in calculus, even though it seems fairly inconvenient for many other purposes. Because logarithms to this base are so well adapted for calculus, they are called **natural logarithms**, and we write $\ln x$ for the natural logarithm of x. The base for natural logarithms is the number $e \approx 2.718\,281\,83$, that is, $\ln x = \log_e x$. So $\ln a$ is defined for any $a > 0$, and

$$e^{\ln a} = a, \qquad \ln(e^s) = s, \qquad e^{-\ln a} = \frac{1}{a}. \tag{1.8.8}$$

Because natural logarithms are logarithms to the base e, Properties 1.8.7 hold for them too:

$$\ln(ac) = \ln a + \ln c, \qquad \ln(a^s) = s \ln a,$$
$$\ln \frac{1}{a} = -\ln a, \qquad \ln \frac{a}{c} = \ln a - \ln c. \tag{1.8.9}$$

The domain of the function $y = \ln x$ is the set of all positive real numbers x; its graph is sketched in Figure 1.8.1. Notice that $\ln 1 = 0$ and that $\ln x > 0$ when $x > 1$ and $\ln x < 0$ when $0 < x < 1$; $\ln x$ does not make sense if $x \leq 0$.

The domain of $y = e^x$ is the set of all real numbers x; the graph of $y = e^x$ is sketched in Figure 1.8.2. Notice that $e^x > 0$ for every $x \in \mathbb{R}$. The function $y = e^x$ is often called the **exponential function**. It grows very rapidly as x increases. For instance, $e^1 = e \approx 2.718\,281\,83$, $e^{10} \approx 22,026.4658$, $e^{20} \approx 485,165,195.4$, $e^{30} \approx 1.068\,647 \times 10^{13}$, $e^{100} \approx 2.688 \times 10^{43}$.

Figure 1.8.1   **Figure 1.8.2**

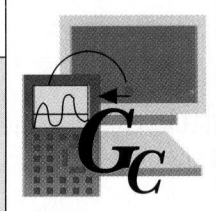

The fact that $\ln(2x) = \ln(2) + \ln(x)$ suggests that the graph of $y = \ln(2x)$ should be the graph of $y = \ln x$ "shifted up" by the constant $\ln(2)$.

We can use a graphics calculator to help visualize this fact.

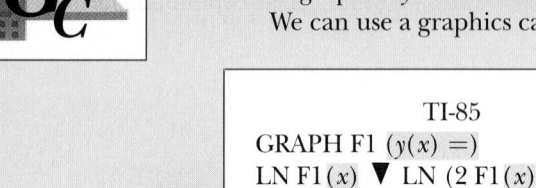

TI-85	SHARP EL9300
GRAPH F1 $(y(x) =)$ LN F1(x) ▼ LN (2 F1(x)) M2(RANGE) 0.1 ▼ 2 ▼ ▼ (−)3 ▼ 1.5 F5(GRAPH)	⤴ ln x/θ/T 2ndF ▼ ln 2 x/θ/T RANGE .1 ENTER 2 ENTER 1 ENTER −3 ENTER 1.5 ENTER 1 ENTER ⤴

CASIO f_x-7700GB/f_x-6300G	HP 48SX
Cls EXE Range .1 EXE 2 EXE 1 EXE −3 EXE 1.5 EXE 1 EXE EXE EXE EXE Graph ln X EXE Graph ln(2X) EXE	⤴ PLOT PLOTR ERASE ATTN ' αY = LN αX ↱ PLOT ⤴ DRAW ERASE .1 ENTER 2 XRNG 3 +/− ENTER 1.5 YRNG DRAW ATTN ' αY = LN 2 × αX ↱ PLOT ⤴ DRAW DRAW

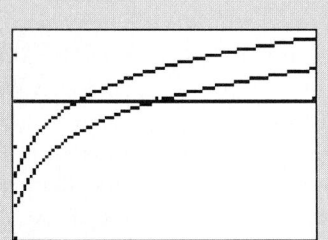

Try graphing $y = \ln(x)$ and $y = \ln(2x) - \ln(2)$ together; also try $y = \ln(x)$ and $y = \ln(x^3)$.

Exercises 1.8

Suppose a and b are two positive real numbers, and let $A = \log a$, $B = \log b$. Express each of the following in terms of A and B.

1. $\log ab$

2. $\log a^2 b^3$

3. $\log \frac{a^2}{b}$

4. $\log a^{-7} b$

5. $\log \sqrt{a}$

6. $\log \sqrt{ab}$

7. $\log a^{1/3}$

8. $\log b^{\pi}$

9. $\log a^b$

10. $\log b^{a+3}$

Evaluate each of the following.

11. $\log_2(32)$

12. $\log_5\left(\frac{1}{125}\right)$

13. $\log_4(2)$

14. $\log_9\left(\frac{1}{3}\right)$

15. $\log_{27}\left(\frac{1}{3}\right)$

16. $\log_3(81)$

17. $\log_{1/2}(8)$

18. $\log_{1/3}(27)$

19. $\log_{1/4}\left(\frac{1}{2}\right)$

20. $\log_{e^{-2}}(e)$

Chapter Summary

§1.1
natural numbers
rational numbers in lowest terms
real line
irrational numbers
scientific notation

§1.2
inequality
strict inequality
properties of inequalities
absolute value
triangle inequality (1.2.2, 1.2.3)
distance
interval
endpoint
closed interval
open interval
half-open interval
infinite interval
union, $A \cup B$
intersection, $A \cap B$

§1.3
polynomial
coefficient
degree
subscript
constant term
evaluating
value of a polynomial
variable
factoring
factor

difference of squares
quadratic polynomial
linear polynomial
completing the square
discriminant
Quadratic Formula (1.3.4)
nonnegative square root
root
Factor Theorem (1.3.5)
Rational Root Theorem (1.3.7)
factorials
Binomial Theorem (1.3.10)
rational function

§1.4
function
value of a function
domain
natural domain
range
origin
x-axis
y-axis
x-coordinate (abscissa)
y-coordinate (ordinate)
Cartesian coordinates
graph

§1.5
right-angled triangle
hypotenuse
Pythagorean Theorem (1.5.1)
similar triangles (1.5.4)
corresponding sides

slope
point–slope form (1.5.7)
slope–intercept form (1.5.8)
general form of the equation of a line (1.5.9)
x-intercept
y-intercept
parallel lines
area of rectangle
area of circle
area of parallelogram
area of triangle

§1.6
conics
polynomial in two variables
quadratic polynomials
parabola
vertex (of a parabola)
translation of axes
ellipse
relation
semi-major and semi-minor axes
major and minor axes
intercepts
hyperbola
asymptotes
transverse axis
conjugate axis
branches

§1.7
trigonometric functions
unit circle
periodic functions

radian measure	polar coordinates (1.7.15)	logarithm
radian/degree conversions	radius	base
identity	polar angle	common logarithms
$\cos^2 \theta + \sin^2 \theta = 1$ (1.7.2)	modulus	natural logarithms
Law of Cosines (1.7.10)	argument	$e \approx 2.718\,281\,83$
addition formulas (1.7.11)	**§1.8**	$\log_b(ac) = \log_b a + \log_b c$
double-angle formulas (1.7.12)	exponent	exponential function
half-angle formulas (1.7.13)		

Review Exercises

Find an equivalent way of expressing each of the following without using absolute value signs.

1. $0 < |x + 3| \le 2$
2. $1 \le |x - 1| < 2$
3. $1 < \left|\frac{1}{x}\right| < 4$
4. $3 \le \frac{2}{|x+2|}$
5. $\{x : 1 < |x + 6| < 8\}$
6. $\{x : |x + 1| \ge 3\}$
7. $\{x : |1 - x^2| \le \frac{1}{2}\}$
8. $\{x : 1 < |x^2 - 4| < 2\}$
9. $\{x : |x^3 - 4x^2 + x + 1| \le -2\}$
10. $\{x : \frac{1}{2} \le \frac{1}{|x+7|} \le \frac{1}{4}\}$

Factor each of the following polynomials as much as possible.

11. $x^2 + 7x - 60$
12. $x^2 + 7x + 12$
13. $x^2 + x - 1$
14. $x^2 + 3x - 40$
15. $x^3 - 2x^2 - 35x$
16. $x^4 + 3x^2 + 2$
17. $x^4 - 2x^2 - 15$
18. $x^3 + x^2 + x + 1$
19. $x^5 - x^3$
20. $x^6 + 4x^4$

Find the coefficient of x^3 in each of the following expressions.

21. $(x + 3)^4$
22. $(x - 2)^5$
23. $x^2(x + 6)^2$
24. $(x + 4)^{10}$
25. $x^2(x - 4)(x + 4)$
26. $\frac{1}{x}\left(32 - (x + 2)^5\right)$
27. $x^2\left((1 + x)^2 - (1 - x)^2\right)$
28. $x(x^2 + 1)^3$
29. $(x^3 - 2)^4$
30. $(x^2 - 7)^{11}$

Find each of the following.

31. The domain of $\frac{1}{x^2 - 3}$.
32. The domain of $\frac{x}{x^2 - 9}$.
33. The domain of $\sqrt{\frac{1}{1-x}}$.
34. The domain of $x \ln x$.
35. The range of $x^2 - 6x + 2$.
36. The range of $1 - x + 2x^2$.
37. The distance between the points $(2, 3)$ and $(1, -4)$.
38. The distance between the points $(1, -5)$ and $(2, 0)$.
39. The distance from the y-axis to the point $(-4, 6)$.
40. The distance from the x-axis to the point $(3, -5)$.
41. The length of the hypotenuse in a right-angled triangle whose other sides have lengths 3 and 7.
42. The length of the third side in a right-angled triangle with hypotenuse of length 4 and one other side of length 3.
43. The area of the triangle whose vertices are $(0, 0)$, $(4, 2)$, and $(3, 0)$.
44. The area of the triangle whose vertices are $(1, 1)$, $(3, -3)$, and $(3, 2)$.
45. The area of the triangle whose vertices are $(-1, 3)$, $(2, 5)$, and $(3, 3)$.
46. The area of the parallelogram whose vertices are $(1, 4)$, $(-1, 4)$, $(7, -1)$, and $(5, -1)$.
47. The equation of the straight line containing $(2, 5)$ and $(3, 4)$.
48. The equation of the line through $(1, -3)$ and with slope -3.
49. The equation of the line through $(0, -3)$ and with x-intercept -2.
50. The equation of the line through $(0, -3)$ and perpendicular to $x + 3y = 7$.
51. The equation of the ellipse whose center is $(0, 0)$ and that contains $(0, 2)$ and $(-3, 0)$.
52. The equation of the ellipse whose center is $(2, -1)$ and that has vertical major axis of length 4 and minor axis of length 2.
53. The equation of the circle whose center is $(5, 4)$ and whose radius is 3.
54. The equation of the parabola that opens upward, whose vertex is $(2, 3)$, and that contains the point $(3, 4)$.

55. The equation of the hyperbola with asymptotes $y = \pm 2x$ that contains $(0, 3)$.

56. The equation of the hyperbola with asymptotes $2x = \pm(3y + 6)$ and containing $(1, -2)$.

57. The points of intersection of the line $y = x + 3$ and the ellipse $x^2 + 4y^2 = 8$.

58. The points of intersection of the circle $x^2 + y^2 = 4$ and the ellipse $x^2 + 2y^2 = 4$.

59. The center of the ellipse $x^2 + 4x + 9y^2 - 18y + 5 = 0$.

60. The semi-major axis of the ellipse $3x^2 - 6x + 4y^2 + 3y = 0$.

Identify each of the following conics (state what type it is, its center, and, where applicable, find major and minor axes, in which direction it opens, and asymptotes).

61. $x^2 - 4y = 8$

62. $x^2 + 4x + y^2 - 6y = 0$

63. $x^2 - y^2 = -4$

64. $4x^2 - 9y^2 = -36$

65. $x^2 + 4y^2 - 12y = 8$

66. $2x^2 - 4x + 3y^2 + 6y = 4$

67. $16x^2 - 9y^2 = -1$

68. $4x - 3y^2 = -1$

69. $x^2 + 4x + 3y^2 + 3 = 0$

70. $x^2 - y^2 + 2y = 1$

Evaluate each of the following without using a calculator.

71. $\cos\left(-\frac{5\pi}{4}\right)$

72. $\sin\frac{7\pi}{6}$

73. $\cos\frac{5\pi}{8}$

74. $\sec\left(-\frac{5\pi}{12}\right)$

75. $\tan\frac{5\pi}{3}$

76. $\sec^2\frac{5\pi}{3}$

77. $\sin\left(\frac{\pi}{4} + \frac{2\pi}{3}\right)$

78. $\cos 3\pi$

79. $\cot\frac{4\pi}{3}$

80. $2\cos\frac{\pi}{8}\sin\frac{\pi}{8}$

Convert each of the following rectangular coordinates (x, y) into polar coordinates (r, θ) or vice versa, using a calculator if necessary to get approximate values.

81. $(x, y) = (-4, -4)$

82. $(x, y) = (5, -2)$

83. $(x, y) = (3, -6)$

84. $(x, y) = (-2, 2\sqrt{3})$

85. $(x, y) = (0, -8)$

86. $(x, y) = (-5, 0)$

87. $(r, \theta) = (1, -\pi)$

88. $(r, \theta) = \left(2, \frac{\pi}{4}\right)$

89. $(r, \theta) = \left(7, \frac{5\pi}{4}\right)$

90. $(r, \theta) = \left(\frac{1}{2}, -\frac{7\pi}{6}\right)$

Assume $a > 0$, $b > 0$. Express each of the following expressions in terms of a, b, $\ln a$, $\ln b$, and any necessary constants (for constants such as $\ln 10$, for instance, do not evaluate them on your calculator, but simply write them down in this form. For example, for a question that asks you to rewrite $\ln(10^s)$, write $\ln(10^s) = s\ln 10$).

91. $\ln(10a^2 b)$

92. $\ln(a^{2s})$

93. $\log a$

94. $\log_2(a^4)$

95. $\ln\frac{a}{2b}$

96. $\ln(e^5)$

97. $e^{-3\ln a}$

98. $\ln(a^b)$

99. $\ln(2a \times 3^b)$

100. $\ln\left(\frac{ab}{ba}\right)$

Limits and Continuity

2

Aristotle declared that heavy objects fall faster than light ones, in proportion to their weight, and for centuries the world believed him.

Finally Galileo showed otherwise, reputedly by dropping a cannonball and a musket ball from the tower of Pisa. He is often regarded as the father of experimental science (Aristotle purported to have conducted similar experiments, but presumably never did).

Galileo's work lead to Newton's study of gravitation, which in turn led to the development of calculus.

INTRODUCTION

This chapter is devoted to the ideas of limits and continuity. These are subtle concepts that are at the foundation of the study of calculus.

Section 2.1 serves to introduce the notion of limits by considering how it arises in various contexts. Section 2.2 shows how to work with limits of polynomials and rational functions; if you are not sufficiently familiar with polynomials and rational functions, you might like to review Section 1.3 before proceeding. In Section 2.3 we learn techniques for dealing with many different types of limits and the rules that govern their use. Section 2.4 considers some specialized types of limits: one-sided limits, infinite limits, and limits at infinity. Having learned to work with limits, in Section 2.5 we use them to discuss the idea of continuity. In Section 2.6 we study limits as they apply to the trigonometric functions; for this it is very important to be familiar with the concept of radian measure, which is discussed in Section 1.7. Finally, in Section 2.7 there is a discussion of the mathematically precise definition of limits. This is a difficult topic, which is why we have concentrated on using and understanding limits before considering it.

2.1 Introduction to Limits

TANGENT LINE
SLOPE
AREA

The fundamental idea on which calculus is based is the concept of a limit. We are drawn to this concept in the course of attempting to solve several different types of problems, some of which will be described in this section.

EXAMPLE 1   Suppose we look at a sketch of the parabola $y = x^2 - x + 2$ and note that the point $(1, 2)$ is on the graph (see Figure 2.1.1). We might want to find the equation of the tangent line to the parabola at that point. Obviously the tangent goes through the point $(1, 2)$, but to find its equation, we need to know either another point that it goes through or else its slope.

The way we usually find the slope of a line is to find two points on it and use the formula

$$Slope = \frac{rise}{run} = \frac{y_2 - y_1}{x_2 - x_1},$$

where (x_1, y_1) and (x_2, y_2) are the two points (see Figure 2.1.2). The difficulty with a tangent line is that we usually know only one point on it, namely, the point of tangency.

We could try to get around this problem by picking a second point on the graph and drawing the secant—the line through this point and $(1, 2)$. If the second point were chosen to be fairly close to $(1, 2)$, then the secant should be fairly close to the tangent line. Figure 2.1.3 illustrates a few such secants.

As the second point gets closer and closer to $(1, 2)$, without actually reaching it, the secant gets closer and closer to the tangent. More important, its slope gets closer and closer to the slope of the tangent, which is what we want to find.

Suppose the second point is (x, y). Because we are choosing it to be a point on the graph, it must satisfy the equation of the parabola. In other words, $y = x^2 - x + 2$, and we can write the point (x, y) as $(x, x^2 - x + 2)$ (see Figure 2.1.4).

Figure 2.1.1

Figure 2.1.2

Figure 2.1.3

Figure 2.1.4

Figure 2.1.5

Figure 2.1.6

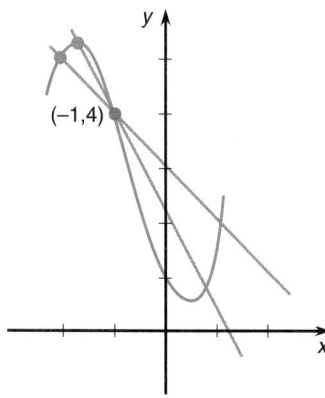

Figure 2.1.7

We can calculate the slope of the secant using the formula given above:

$$Slope = \frac{y_2 - y_1}{x_2 - x_1} = \frac{(x^2 - x + 2) - 2}{x - 1} = \frac{x^2 - x}{x - 1} = \frac{x(x-1)}{x-1} = x.$$

For this last step we divided by $x - 1$, so we have to know that $x - 1$ is not equal to zero, but it is all right because we are choosing the second point to be different from $(1, 2)$, so $x \neq 1$. As the second point approaches $(1, 2)$, the value of x must approach 1, so the slope of the secant, which is x, also approaches 1. This suggests that the slope of the tangent must be 1. See Figure 2.1.5.

Knowing this and the fact that the tangent passes through $(1, 2)$, we can find its equation using the point-slope form of the equation of a line (cf. Equations 1.5.7):

$$y - 2 = (1)(x - 1), \qquad \text{or} \qquad y - 2 = x - 1,$$

or $x - y + 1 = 0$.

Finding the slope of a tangent line is one of the most important ideas in calculus, and while the example we have just done is misleadingly simple in some respects, it does illustrate the general approach. In particular, it shows how we can approximate the tangent by secants and let them get closer and closer to the tangent. The slopes of the secants should then approach the slope of the tangent.

Unfortunately we have been vague about what we mean by letting secants "approach" the tangent and by their slopes "approaching" the slope of the tangent. The value they approach is called the limit, and the process of approaching a limit is the subject of this chapter. This introductory section will use several examples to explore some of the ways in which limits arise and how they are used.

If $f(x)$ is a function, we can consider what happens to the value of $f(x)$ as x approaches some fixed value a. If the values of $f(x)$ approach some value L, as shown in the example in Figure 2.1.6, we call L the limit of $f(x)$ as x approaches a, and we write

$$L = \lim_{x \to a} f(x).$$

EXAMPLE 2   Find the equation of the line that is tangent to the curve $y = x^3 + 2x^2 - 2x + 1$ at the point $(-1, 4)$.

Solution   If (x, y) is a point on the curve, we know that it satisfies the equation, so we can write the point as $(x, x^3 + 2x^2 - 2x + 1)$ (see Figure 2.1.7). We calculate the slope of the secant through this point and the point $(-1, 4)$, using the formula given above, and find it equals

$$\frac{(x^3 + 2x^2 - 2x + 1) - 4}{x - (-1)} = \frac{x^3 + 2x^2 - 2x - 3}{x + 1}.$$

It is easy to see that the numerator factors: we find that $x^3 + 2x^2 - 2x - 3 = (x + 1)(x^2 + x - 3)$, (verify this), so the slope is

$$\frac{(x + 1)(x^2 + x - 3)}{x + 1} = x^2 + x - 3.$$

Once again we must be careful not to divide by zero, so this formula for the slope of the secant makes sense only when $x \neq -1$. What we want is to let the point $(x, y) = (x, x^3 + 2x^2 - 2x + 1)$ approach $(-1, 4)$ and see what happens to the slope.

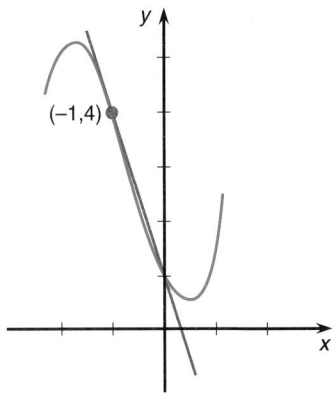

Figure 2.1.8

In other words, we want to consider

$$\lim_{x \to -1} (x^2 + x - 3).$$

Now as x approaches -1, it seems reasonable to expect that $x^2 + x - 3$ will approach $(-1)^2 + (-1) - 3 = -3$. In other words,

$$\lim_{x \to -1} (x^2 + x - 3) = -3.$$

This limit of the slopes of the secants will be the slope of the tangent line. See Figure 2.1.8.

Now we can find the equation of the tangent line, since we know that it passes through $(-1, 4)$ and has slope equal to -3. The equation is

$$y - 4 = (-3)\big(x - (-1)\big) \qquad \text{or} \qquad y - 4 = -3(x + 1) \qquad \text{or}$$
$$3x + y - 1 = 0.$$

If a vehicle travels at a constant speed for 15 minutes, say, and covers a distance of 10 miles, then we can easily find the speed by dividing the distance by the time. In this case we find that the speed is $\frac{10\,miles}{\frac{1}{4}\,hour} = 40$ miles/hour. If the vehicle is not traveling at a constant speed, then this same procedure can be used to find the *average* speed. But there is the question of what we mean by the speed of the vehicle at a particular moment, which is sometimes called the "instantaneous speed." This is a very familiar notion; after all, it is exactly what the speedometer in a car tells us.

If a car is traveling at 40 mph, we cannot conclude that it will have to travel 40 miles in the next hour. We could say that if the car continued at exactly the same speed for an hour, then it would go 40 miles; but in practice this is both unlikely and awkward to use.

Slightly better would be to say that if it continued for another 15 minutes at the same speed, then it would go 10 miles. But carrying this idea further, we could say that in one minute it would go $\frac{2}{3}$ mile or that in one second it would go $\frac{2}{3}\frac{1}{60} = \frac{1}{90}$ mile.

The advantage is that for a car moving at 40 mph, there isn't much need to worry about the speed changing in 1 second. And if it does, or if we want better accuracy, we could consider $\frac{1}{10}$ second, $\frac{1}{100}$ second, etc.

What we are doing is finding the average speed over a brief period of time. The shorter the period of time, the better we will be able to approximate the instantaneous speed. To eliminate the question of just how short the time should be, we could take the limit as it approaches zero. This limit is exactly what we mean by the instantaneous speed. In fact, it was to deal with this question that calculus was originally invented.

Suppose an object held at a height A feet above the ground is released and allowed to fall, as shown in Figure 2.1.9. The height of the object t seconds after its release will be approximately

$$h(t) = A - 16t^2,$$

where this height is measured in feet. Notice that when $t = 0$, this formula says that the height is A feet, which means that the object has not begun to fall, and that as t increases, the height will decrease. Of course the object will eventually hit the ground and stop moving, and from that point on, this formula will not apply.

Figure 2.1.9

EXAMPLE 3   Suppose an object falls from a height of 64 ft above the ground. How fast will it be falling one second after its release? When will it hit the ground? What will be its speed of impact?

Solution   We use the above formula with $A = 64$. If t is slightly greater than 1, then between 1 sec and t sec the object will fall a distance of $h(1) - h(t) = \left(64 - 16(1)^2\right) - (64 - 16t^2) = 64 - 16 - 64 + 16t^2 = 16(t^2 - 1)$ ft.

The average speed of the object between 1 sec and t sec will be

$$\frac{distance}{time} = \frac{16(t^2 - 1)}{t - 1} = \frac{16(t+1)(t-1)}{t-1} = 16(t+1) \text{ ft/sec.}$$

> Notice that since the object is falling, we expect $h(1) > h(t)$, so we have taken $h(1) - h(t)$ in order to get a positive distance.

This calculation involves dividing by $t - 1$, so it cannot be done when $t = 1$, but we can find the instantaneous speed at the moment when $t = 1$ by taking a limit. The speed of the falling object 1 sec after its release will be the limit as t approaches 1 of the average speed over the brief interval from 1 to t:

$$Speed \text{ } after \text{ } 1 \text{ } sec = \lim_{t \to 1} 16(t+1) = 16(1+1) = 32 \text{ ft/sec.}$$

To find when the object hits the ground we remark that what this means is just that the height above the ground should be zero, that is, that $h(t) = 0$. This says $0 = h(t) = 64 - 16t^2$, or $64 = 16t^2$, so $t^2 = \frac{64}{16} = 4$ and $t = \sqrt{4} = 2$. The object will hit the ground 2 seconds after its release.

To find the speed with which it hits the ground, we can pretend that it just continues falling after 2 seconds and do the same sort of calculation that we just completed.

The average speed between 2 sec and t sec is $\frac{h(2) - h(t)}{t - 2} = \frac{(64 - 16(2)^2) - (64 - 16t^2)}{t - 2} = \frac{(64 - 64) - (64 - 16t^2)}{t - 2} = \frac{16t^2 - 64}{t - 2}$. We can factor the numerator as a difference of squares to get $\frac{(4t+8)(4t-8)}{t-2} = \frac{(4t+8)4(t-2)}{t-2} = 4(4t + 8) = 16t + 32$.

The instantaneous speed with which the falling object hits the ground is

$$\lim_{t \to 2} (16t + 32) = 16(2) + 32 = 64 \text{ ft/sec.}$$

The use of limits to find instantaneous speed is actually quite similar to the way we used them to find the slope of a tangent line to a curve, even though the two problems look somewhat different. We now discuss an entirely different way in which limits arise.

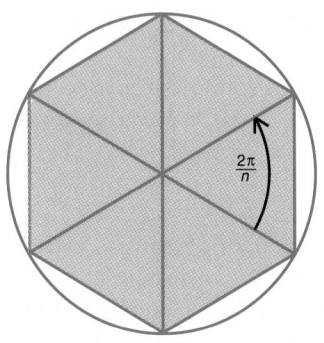

Figure 2.1.10(i)

EXAMPLE 4   Suppose we want to figure out the area inside a circle. For simplicity, assume the radius of the circle is 1. The familiar formula that the area of a circle is πr^2 if the radius is r tells us that in this case the area is π, but how would we figure this out starting from scratch, or for that matter, how would we figure out what π is?

One way to go about it would be to approximate the region inside the circle by regions whose areas are easier to calculate. Figure 2.1.10(i) shows how the region inside the circle can be approximated by n congruent triangles. To figure out the angle in each of these triangles at the vertex which is the center of the circle, we note that n of them exactly make up the angle that goes all around the circle. Since this angle measures 2π radians, each of the n equal pieces must measure $\frac{2\pi}{n}$ radians.

The area inside the circle is approximated by the total area of the n triangles, which is n times the area of one of them. Consider the triangle that lies along the

Figure 2.1.10(ii)

Figure 2.1.10(iii)

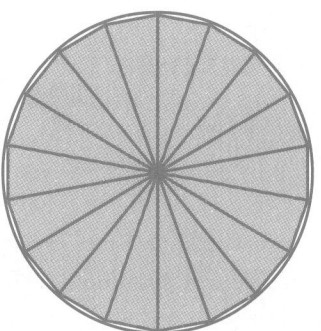

Figure 2.1.11

positive x-axis, as shown in Figure 2.1.10(ii), and look at the upper half of it, the smaller triangle illustrated in Figure 2.1.10(iii). To find its area, we need to know its base and its height. The angle inside this little triangle at the center of the circle is half the angle in the bigger triangle, or $\frac{2\pi}{2n} = \frac{\pi}{n}$. By the definitions of the trigonometric functions (cf. Definition 1.7.1), the base of the triangle is $\cos\frac{\pi}{n}$ and its height is $\sin\frac{\pi}{n}$. The area of the little triangle is therefore $\frac{1}{2} base \times height = \frac{1}{2}\cos\left(\frac{\pi}{n}\right)\sin\left(\frac{\pi}{n}\right)$.

The bigger triangle then has area twice as large, or $\cos\left(\frac{\pi}{n}\right)\sin\left(\frac{\pi}{n}\right)$. The n congruent triangles that approximate the region inside the circle have total area $n\cos\left(\frac{\pi}{n}\right)\sin\left(\frac{\pi}{n}\right)$. This approximates the area of the circle.

To get better and better approximations, we should let n get larger and larger (see Figure 2.1.11). We can describe this by saying that n should "approach infinity," which we write as $n \to \infty$. Then we can consider better and better approximations and take their limit as $n \to \infty$; we find that the area inside the circle is

$$Area = \lim_{n\to\infty} n\cos\left(\frac{\pi}{n}\right)\sin\left(\frac{\pi}{n}\right).$$

Again we have expressed something in terms of a limit. This limit is not easy to evaluate, but the point is that we have been able to use it to give a precise meaning to the difficult idea of the area of a circle.

The last example brought us face to face with the possibility of encountering perfectly reasonable limits that are difficult to evaluate. Let us look at another example.

EXAMPLE 5

Consider the following limit:

$$\lim_{x\to 0} \frac{\sin x}{x}.$$

As usual, we are measuring x in radians. There is no obvious way to go about evaluating this limit, but we could try to get an idea of what is going on by working out what we get for some values of x that are close to 0.

Using a calculator, being sure to set it to measure angles in radians, we can evaluate $\frac{\sin x}{x}$ for some small values of x:

x	$\frac{\sin x}{x}$
1.0	0.84147099
0.5	0.95885108
-0.5	0.95885108
0.2	0.99334665
-0.2	0.99334665
0.1	0.99833417
-0.1	0.99833417
0.01	0.99998333
-0.01	0.99998333
0.001	0.99999983
-0.001	0.99999983
0.0001	0.999999998

These values certainly make it appear as though $\frac{\sin x}{x}$ approaches the limit 1 as $x \to 0$. This is true, but it is difficult to prove. In passing, it is worth noticing that it does not work if angles are measured in degrees.

It is interesting to plot the function $y = \frac{\sin x}{x}$, which was discussed in Example 5. It is very important to make sure that your calculator is in RADIAN mode.

TI-85	SHARP EL9300
GRAPH F1$(y(x) =)$ SIN F1$(x) \div$ F1(x) M2(RANGE) $(-) \pi$ ▼ π ▼ ▼ $(-)$ 1.5 ▼ 1.5 F5(GRAPH)	⌁ sin x/θ/T $\div$ x/θ/T RANGE $-\pi$ ENTER π ENTER 1 ENTER -1.5 ENTER 1.5 ENTER 1 ENTER ⌁
CASIO f_x-7700GB/f_x-6300G	HP 48SX
Cls EXE Range $- \pi$ EXE π EXE 1 EXE -1.5 EXE 1.5 EXE 1 EXE EXE EXE EXE Graph sin X $\div$ X EXE	↰ PLOT PLOTR ERASE ATTN′ αY = SIN αX ▶ $\div \alpha X$ ↱ PLOT ↰ DRAW ERASE π +/− ENTER π XRNG 1.5 +/− ENTER 1.5 YRNG DRAW

What happens at the point $x = 0$, where both numerator and denominator of $\frac{\sin x}{x}$ are equal to zero?

Finally, we look at two examples to show how things can go wrong with limits.

EXAMPLE 6   What happens to the function $f(x) = \frac{x}{|x|}$ as $x \to 0$?

Solution   Notice that the function $f(x)$ is not defined when $x = 0$. When $x > 0$, we know that $|x| = x$, so $f(x) = \frac{x}{|x|} = \frac{x}{x} = 1$.

On the other hand, when $x < 0$, it is easy to see that $|x| = -x$; after all, if $x < 0$, then $-x > 0$. (For example, when $x = -3$, we see that $|x| = |-3| = 3$, and this is the same thing as $-(-3)$.) So when $x < 0, f(x) = \frac{x}{|x|} = \frac{x}{-x} = -1$. The graph of $f(x)$ is sketched in Figure 2.1.12.

Now as $x \to 0$, it depends whether x is negative or positive. If x approaches 0 from the left, that is, through negative values of x, then $f(x)$ will equal -1 all the time and so certainly approaches -1 as a limit. On the other hand, if x approaches 0 from the right, through positive values of x, then $f(x)$ can only approach the value 1 as its limit.

There is no single value that we can call the limit in this case, so we have to say that the limit does not exist. Actually, in this particular example we could talk about the "limit from the left side" or the "limit from the right side," both of which do exist, but it is possible to make up examples with even worse behavior in which even the limits from the left and right do not exist.

Figure 2.1.12

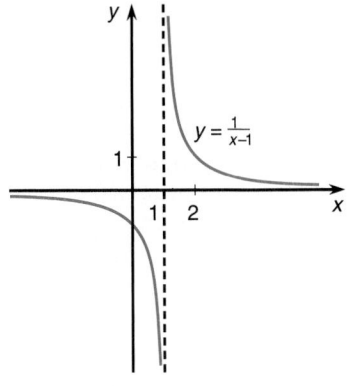

Figure 2.1.13

The fact that it may be difficult to tell whether or not a limit exists is one of the things that make them difficult to work with.

EXAMPLE 7   What happens to the function $g(x) = \frac{1}{x-1}$ as $x \to 1$?

Solution   The function $g(x)$ is not defined when $x = 1$. If x is a number slightly bigger than 1, then $x - 1$ is a small positive number, and $g(x) = \frac{1}{x-1}$ is a large positive number. The closer x gets to 1, the smaller $x - 1$ gets, and the larger $g(x) = \frac{1}{x-1}$ gets.

In fact, $g(x)$ doesn't have any limit L as $x \to 1$, because $g(x)$ gets bigger than any L when x gets sufficiently close to 1. To make matters worse, if x approaches 1 from the left, $x - 1$ is negative, and so is $g(x) = \frac{1}{x-1}$, and as x gets close to 1, $g(x)$ becomes larger and larger in the negative direction. See Figure 2.1.13.

The next section will be devoted to learning to work with limits.

Exercises 2.1

I

Find the slope of the tangent line to the graph of each of the given functions $y = f(x)$ at the specified point.

1. $f(x) = x^2$, at the point $(2, 4)$
2. $f(x) = x^2 + 2x + 1$, at the point $(-2, 1)$
3. $f(x) = 3x^2 - 4x + 1$, at the point $(1, 0)$
4. $f(x) = 2x^2 - 12x + 6$, at the point $(3, -12)$
5. $f(x) = x^3 + 2x - 3$, at the point $(-1, -6)$
6. $f(x) = x^3 - 2x^2 - 3x - 3$, at the point $(0, -3)$
7. $f(x) = x^3 - 8$, at the point $(2, 0)$
8. $f(x) = x^4 + x^2 + 1$, at the point $(-1, 3)$
9. $f(x) = x^4$, at the point $(0, 0)$
10. $f(x) = 3x + 1$, at the point $(1, 4)$

Find the equation of the tangent line to the graph of each of the given functions $y = f(x)$ at the specified point.

11. $f(x) = x^2 - 2$, at the point $(1, -1)$
12. $f(x) = x^2 - 2x + 1$, at the point $(1, 0)$
13. $f(x) = x^2 + 3x + 5$, at the point $(-1, 3)$
14. $f(x) = 3x^2 + 5x - 6$, at the point $(-2, -4)$
15. $f(x) = 5x^2 + x$, at the point $(-2, 18)$
16. $f(x) = x^3$, at the point $(2, 8)$
17. $f(x) = x^3 - x - 2$, at the point $(-2, -8)$
18. $f(x) = x^3 + x^2 + x + 1$, at the point $(0, 1)$
19. $f(x) = x^4$, at the point $(-2, 16)$
20. $f(x) = x^4 + 2x^2 + 4$, at the point $(1, 7)$

II

21. Find the equation of the tangent line to the parabola $y = x^2$ at the point (a, a^2).
22. If an object is released from a height of 144 ft above the ground, how long will it take for it to reach the ground, and what will be its speed of impact?
23. If an object is released and allowed to fall to the ground, from what height was it dropped if it takes 4 sec to reach the earth?
24. Suppose an object falls from a height A ft above the ground. How long will it take to hit the ground?
*25. Suppose an object falls from a height A ft above the ground. At what speed will it be falling at the moment t sec after its release?
*26. If an object is released and allowed to fall to the ground, from what height was it dropped if it hits the earth with an impact speed of 100 ft/sec?
27. Suppose an apple falls from the top of a tree and lands on Newton's head at a speed of 16 ft/sec. If Newton is sitting with his head 4 ft above ground level, how high is the tree?
*28. Suppose an object is dropped to the ground from a height of A ft. If its speed of impact is S ft/sec, find A, the height from which it was dropped.
*29. Suppose an object is thrown straight up into the air. It can be shown that its height in feet above the ground after t sec will be given by an expression of the form

$$h(t) = -16t^2 + vt + A.$$

Obviously, A ft is the height from which the object is thrown (since $h(0) = A$).

(i) Show that $|v|$ is the initial speed with which the object is thrown, that is, the instantaneous speed when $t = 0$.

(ii) What does it mean when v is positive, negative, or zero?

(iii) At what moment (i.e., at what value of t) will the object reach the highest point in its flight, and what is the greatest height it will reach? (*Hint:* Complete the square.)

(iv) If a ball is thrown upward from a height of 4 ft above the ground at a speed of 20 ft/sec, when will it hit the ground?

(v) If someone on a balcony throws a ball upward from a height of 28 ft above the ground and finds that it hits the ground $1\frac{3}{4}$ sec later, how fast was it thrown?

*30. In this question we discuss an alternative method for solving part (iii) of the previous question. Knowing that the height above the ground of the object is $h(t) = -16t^2 + vt + A$, find its instantaneous speed at the moment when $t = u$, where u is some fixed but unspecified number. Then to find when the object reaches the highest point in its flight, observe that this will happen at the moment when the speed is exactly equal to 0, when the object has stopped rising but not yet begun to fall.

31. For a circle of radius 1 we found that the area can be expressed as the limit

$$\lim_{n \to \infty} n \cos\left(\frac{\pi}{n}\right) \sin\left(\frac{\pi}{n}\right).$$

Use a calculator or computer to evaluate this function for various values of n (e.g., $n = 4, 8, 20, 50, 100, \ldots$). What is the first value of n for which the answer is correct to four decimal places? What is the first value of n for which your calculator gives the same answer with n and $n + 1$? How large does n have to be so that your calculator gives the same answer as its built-in value of π, which equals the area of the circle? (*Hint:* Be sure you are using *radians*.)

32. Use a calculator or computer to estimate the area of a circle of radius 3. How large does n have to be to get an answer that is correct to four decimal places?

33. Estimate to four decimal places the area of the upper semicircle of radius 4. How large does n have to be to get an answer that is correct to four decimal places?

34. Estimate to within four decimal places the area of the part of the circle of radius 4 that lies below the line $y = x$ and above the line $y = -x$. How large does n have to be to get an answer that is correct to four decimal places?

35. Use a graphics calculator or computer to graph the function $y = \frac{\sin x}{x}$ for $-2 \leq x \leq 2$. Then graph it for $-0.1 \leq x \leq 0.1$ and for $-0.01 \leq x \leq 0.01$. Compare with Example 5.

36. How do the graphs in the previous question change if you set your machine to measure angles in degrees?

37. (i) In Example 3 we discussed a falling object whose height is $h(t) = 64 - 16t^2$. Graph this function and by magnifying the graph estimate how far the object falls in the last 1 sec, the last 0.1 sec, and the last 0.01 sec.

(ii) Graph the function $d(t)$ that tells how far the object falls between t and $t + 0.1$. Graph the function $D(t)$ that tells how far the object falls between t and $t + 0.01$. Compare these graphs with the result of Example 3.

2.2 Limits of Polynomials and Rational Functions

LIMITS OF SUMS
LIMITS OF PRODUCTS
LIMITS OF QUOTIENTS

To define the notion of limit with precision is surprisingly tricky. We won't attempt it until Section 2.7. Fortunately, in many situations it is possible to work with limits using only our intuitive sense of what they mean.

In this section we explore the basic rules that apply to limit calculations. In particular, we learn to evaluate limits of polynomials and rational functions (cf. Definitions 1.3.1 and 1.3.11).

Suppose $f(x)$ is a function whose domain may be the entire real line $\mathbb{R}$ or perhaps only some subset of it. Suppose $a \in \mathbb{R}$ is fixed. We can consider what happens to $f(x)$ when x "approaches" a. This can make sense even if a is not in the domain of f, but we should assume $f(x)$ is defined at every *other* point in an interval (r, s) containing a. This means $f(x)$ is defined on $(r, a) \cup (a, s)$, which is sometimes called a "punctured interval." Of course $f(x)$ might also be defined at $x = a$.

Figure 2.2.1

Figure 2.2.2

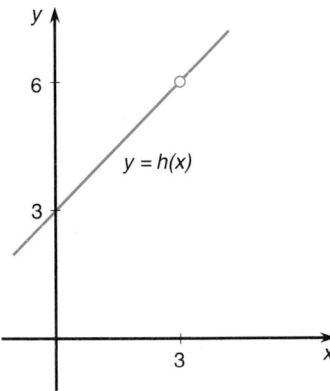

Figure 2.2.3

If there is some number L so that $f(x)$ approaches L when x approaches a (see Figure 2.2.1), we say that L is the **limit** of $f(x)$ as x approaches a or that $f(x)$ tends to L as $x \to a$, and we write

$$\lim_{x \to a} f(x) = L.$$

We have seen, in Examples 2.1.6 and 2.1.7, that there are cases in which no limit exists. When we write $\lim_{x \to a} f(x) = L$, we mean both that the limit exists and that it equals L.

EXAMPLE 1

(i) Let $g(x) = \sqrt{x}$ and consider the behavior of $g(x)$ as x approaches a, where a is some positive number. From Figure 2.2.2 it certainly looks as if $y = g(x)$ approaches $\sqrt{a}$ when x approaches a, and if you were to try with your calculator, you would find that for values of x near a, the value of $g(x) = \sqrt{x}$ would be close to $\sqrt{a}$. This leads us to suspect that

$$\lim_{x \to a} g(x) = \lim_{x \to a} \sqrt{x} = \sqrt{a},$$

which is in fact correct.

(ii) Next we consider the behavior of $h(x) = \frac{x^2 - 9}{x - 3}$ as x approaches 3. Notice that the number 3 is not in the domain of h, because $h(x) = \frac{x^2 - 9}{x - 3}$ makes sense only when $x \neq 3$. So we consider what happens to $h(x)$ when x approaches 3 without ever equaling 3.

Now $h(x) = \frac{x^2 - 9}{x - 3} = \frac{(x+3)(x-3)}{x - 3} = x + 3$, whenever $x \neq 3$. So $h(x)$ is the function whose domain is all real numbers except 3 and that equals $x + 3$ for all such numbers x. We sketch the graph of $h(x)$ in Figure 2.2.3 and leave a little empty circle at the point $(3, 6)$ to indicate that this point is left out. It certainly does appear that as x approaches 3, the point $(x, h(x))$ on the graph approaches the missing point or, to put it another way, that

$$\lim_{x \to 3} h(x) = 6.$$

There are two types of limits that are particularly easy to calculate. One we have already encountered in Example 2.1.1; there we noticed that

$$\lim_{x \to 1} x = 1.$$

This simply says that as x approaches 1, the value of x approaches 1, which is difficult to dispute. There is no reason to restrict ourselves to the number 1; 2 or $-\sqrt{3}$ would do just as well, or in fact any fixed number a.

PROPOSITION 2.2.1

If a is any constant, then

$$\lim_{x \to a} x = a.$$

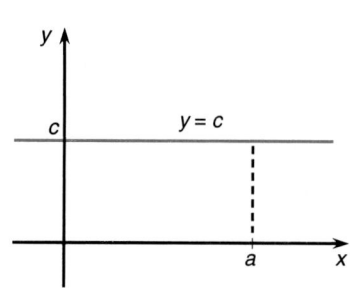

Figure 2.2.4   **Figure 2.2.5**

This result is illustrated in Figure 2.2.4, which shows the graph of $y = x$.

The other very easy limit is even easier. If $f(x)$ is the constant function $f(x) = c$, for some fixed number c, then no matter what x is, $f(x)$ always equals c (see Figure 2.2.5). So if x approaches a, $f(x)$ "approaches" c—in fact it equals c all the time. In symbols this says the following.

PROPOSITION 2.2.2

> If a and c are any constants, then
>
> $$\lim_{x \to a} c = c.$$

Now suppose we want to find $\lim_{x \to a} x^2$. If we write $x^2 = x \times x$, and realize that as $x \to a$ the value of x approaches the limit a, then we should expect that the product $x \times x$ should approach $a \times a$, the product of the limits.

A similar line of thought should convince us that as $x \to a$, the function $p(x) = x^2 + x$ should approach $a^2 + a$, the sum of the limits of x^2 and x.

This sort of thing ought to work not just for these particular functions x and x^2, but for any functions that have limits as $x \to a$. If $f(x)$ and $g(x)$ are two such functions, then their sum $f(x) + g(x)$ ought to have a limit, and the limit ought to be the sum of the limits of $f(x)$ and $g(x)$. Also the product $f(x)g(x)$ ought to have a limit, and the limit should be the product of the limits of $f(x)$ and $g(x)$. If one of the functions is a constant function, say $g(x) = c$, then Proposition 2.2.2 tells us its limit is just c, so the limit of $cf(x)$ is just c times the limit of $f(x)$.

We record these important observations as a theorem.

THEOREM 2.2.3

> Suppose c is a constant, $f(x)$ and $g(x)$ are functions, and a is a number so that
>
> $$\lim_{x \to a} f(x) = L \quad \text{and} \quad \lim_{x \to a} g(x) = M.$$
>
> (Recall that this means both limits exist and they have the specified values.)
> Then
>
> (i) $\lim_{x \to a} \big(f(x) + g(x)\big) = L + M,$
> (ii) $\lim_{x \to a} f(x)g(x) = LM,$
> (iii) $\lim_{x \to a} cf(x) = cL.$
>
> (i.e., the limits exist and have the given values).

Combining this theorem with the previous two propositions allows us to calculate many limits.

EXAMPLE 2

(i) Evaluate $\lim_{x \to 2} (x^2 + 3x)$.

> **Solution**   (i) We regard $x^2 + 3x$ as the sum of two functions, x^2 and $3x$. Theorem 2.2.3 says that its limit will be the sum of the limits of these two functions, and it also tells us how to evaluate those limits. After all, $x^2 = (x)(x)$, and by Proposition 2.2.1, x approaches the limit 2 as $x \to 2$. So by part (ii) of Theorem 2.2.3, $\lim_{x \to 2} x^2 = (\lim_{x \to 2} x)(\lim_{x \to 2} x) = (2)(2) = 4$.
>
> We can also use part (iii) of Theorem 2.2.3 to find the limit of $3x$, by writing it as the product of the constant function 3 and the function x, so $\lim_{x \to 2} 3x = 3(\lim_{x \to 2} x) = 3(2) = 6$.
>
> Finally we use these results and part (i) of Theorem 2.2.3 to see that $\lim_{x \to 2} x^2 + 3x = (\lim_{x \to 2} x^2) + (\lim_{x \to 2} 3x) = (4) + (6) = 10$. Here we used Propositions 2.2.1 and 2.2.2 to evaluate the limits of x and 3, respectively.

(ii) Evaluate $\lim_{x \to -4} (2x^3 - 5x^2 + 2x + 7)$.

> **Solution**   By repeated use of part (i) of Theorem 2.2.3 we can express this limit as the sum of the limits of $2x^3$, $-5x^2$, $2x$, and the constant function 7. By part (i) and repeated use of part (ii) we can express each of these as a product of the limits of some constant function and the function x, the appropriate number of times. So, for instance, we can write $2x^3 = 2(x)(x)(x)$ and find that its limit as $x \to -4$ is $2(-4)(-4)(-4) = 2(-4)^3 = -128$.
>
> Doing this for the other three and combining, we find that

$$\lim_{x \to -4} (2x^3 - 5x^2 + 2x + 7) = \left(\lim_{x \to -4} 2x^3 \right) + \left(\lim_{x \to -4} -5x^2 \right) + \left(\lim_{x \to -4} 2x \right)$$
$$+ \left(\lim_{x \to -4} 7 \right)$$
$$= -128 + (-5)(-4)(-4) + 2(-4) + (7)$$
$$= -128 - 80 - 8 + 7 = -209.$$

REMARK 2.2.4

> In doing the last example, we realized that Theorem 2.2.3 applies not just to sums and products of two functions, but to sums and products of any (finite) number of functions.

In Example 2 we found the limit of $2x^3$ as $x \to -4$. In fact, we could do the same thing for any power of x, as x approaches any number a, and of course we would find that

$$\lim_{x \to a} x^n = a^n.$$

Now suppose we consider a polynomial such as $p(x) = 3x^7 - 2x^4 + 9x + 2$, and let $x \to a$. Then each of the powers of x in $p(x)$ tends to the corresponding power of a, and because of Theorem 2.2.3 the whole polynomial $3x^7 - 2x^4 + 9x + 2$ tends to $3a^7 - 2a^4 + 9a + 2$, which equals $p(a)$. In other words,

$$\lim_{x \to a} p(x) = p(a),$$

or the limit of the polynomial $p(x)$ as $x \to a$ is just the value $p(a)$.

This calculation was done for a specific polynomial, $3x^7 - 2x^4 + 9x + 2$, but exactly the same argument works for any polynomial. If $p(x) = c_n x^n + c_{n-1}x^{n-1} + \ldots + c_2 x^2 + c_1 x + c_0$ and we let $x \to a$, then each power of x will approach the corresponding power of a, and $p(x)$ will approach $c_n a^n + c_{n-1}a^{n-1} + \ldots + c_2 a^2 + c_1 a + c_0 = p(a)$. We have proved a very important and useful result.

THEOREM 2.2.5

(i) If n is any positive integer, then
$$\lim_{x \to a} x^n = a^n.$$

(ii) If $p(x)$ is any polynomial, then
$$\lim_{x \to a} p(x) = p(a).$$

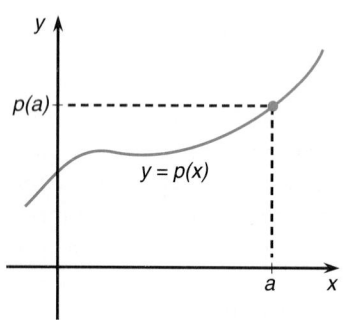

Figure 2.2.6

Notice that what this result says is that as $x \to a$, the polynomial $p(x)$ approaches the "correct" value, the value that $p(x)$ has when x actually equals a. See Figure 2.2.6.

EXAMPLE 3 Evaluate the following limits: (i) $\lim_{x \to -1} (x^3 - 2x^2 + 4x + 8)$, (ii) $\lim_{x \to 4} (x^4 - 4x^3 + x - 4)$, (iii) $\lim_{x \to \sqrt{2}} (x^4 - 2x^3 + 6x^2 + x - 5)$.

Solution

(i) We merely apply the theorem to the polynomial $p(x) = x^3 - 2x^2 + 4x + 8$ and find that $\lim_{x \to -1} (x^3 - 2x^2 + 4x + 8) = p(-1) = (-1)^3 - 2(-1)^2 + 4(-1) + 8 = -1 - 2 - 4 + 8 = 1$.

(ii) $\lim_{x \to 4} (x^4 - 4x^3 + x - 4) = 4^4 - 4(4^3) + 4 - 4 = 256 - 4(64) + 4 - 4 = 256 - 256 + 4 - 4 = 0$.

(iii) $\lim_{x \to \sqrt{2}} (x^4 - 2x^3 + 6x^2 + x - 5) = (\sqrt{2})^4 - 2(\sqrt{2})^3 + 6(\sqrt{2})^2 + \sqrt{2} - 5 = 2^2 - 2(2\sqrt{2}) + 6(2) + \sqrt{2} - 5 = 4 - 4(\sqrt{2}) + 12 + \sqrt{2} - 5 = 11 - 3\sqrt{2} \approx 6.757$.

In simplifying this last limit, wherever possible we combined square roots of 2 (using the fact that $\sqrt{2}\sqrt{2} = 2$) and then collected first the terms that no longer contained any square roots and second the terms that did contain $\sqrt{2}$. Only then did we use a calculator to approximate the final answer.

Since Theorem 2.2.3 says that limits "preserve" sums and products, it is reasonable to ask what happens with quotients. The same sort of reasoning that led us to expect the limit of a sum to be the sum of the limits and the limit of a product to be the product of the limits should also apply to quotients, but there is a complication because we have to avoid dividing by zero.

If $\lim_{x \to a} g(x) = M$, then as $x \to a$, $\frac{1}{g(x)}$ ought to approach $\frac{1}{M}$, provided $M \neq 0$. If we write

$$\frac{f(x)}{g(x)} = f(x) \frac{1}{g(x)},$$

then we can apply Theorem 2.2.3(ii) about limits of products to see that the quotient of $f(x)$ and $g(x)$ should have as its limit the quotient of the limits of $f(x)$ and $g(x)$.

THEOREM 2.2.6

Suppose $f(x)$ and $g(x)$ are functions and a is a number satisfying

$$\lim_{x \to a} f(x) = L \quad \text{and} \quad \lim_{x \to a} g(x) = M.$$

Suppose too that $M \neq 0$. Then

(i) $\lim_{x \to a} \frac{1}{g(x)} = \frac{1}{M}$,

(ii) $\lim_{x \to a} \frac{f(x)}{g(x)} = \frac{L}{M}$.

(iii) In particular, if $f(x) = \frac{p(x)}{q(x)}$ is a rational function, then

$$\lim_{x \to a} f(x) = \frac{p(a)}{q(a)} = f(a),$$

whenever $f(a)$ is defined (i.e., whenever $q(a) \neq 0$).

EXAMPLE 4 Find (i) $\lim_{x \to 3} \frac{x^2 + 1}{x + 1}$; (ii) $\lim_{x \to -1} \frac{x + 1}{x^2 + 1}$.

Solution

(i) As $x \to 3$, we know that the numerator approaches $3^2 + 1 = 10$ and the denominator approaches $3 + 1 = 4$, so the quotient approaches $\frac{10}{4} = \frac{5}{2}$.

(ii) The denominator approaches the limit $(-1)^2 + 1 = 2$, and the numerator approaches the limit $(-1) + 1 = 0$, so $\lim_{x \to -1} \frac{x + 1}{x^2 + 1} = \frac{0}{2} = 0$. ◢

Example 4 illustrates that the limit of the numerator does not have to be nonzero for the theorem to work; the only condition is that the limit of the denominator is nonzero.

We should also consider what goes wrong when the limit of the denominator does equal zero. In fact, we have already seen one such problem in Example 2.1.7, where we considered the behavior as $x \to 1$ of $g(x) = \frac{1}{x-1}$. There we realized that as x approaches 1 from the right side, $g(x)$ becomes very large and does not approach any limit. As x approaches 1 from the left side, $g(x)$ becomes very large in the negative direction and does not approach any limit.

This illustrates that when the limit of the denominator is zero, it may be that the quotient does not have a limit. It can also happen that the quotient does have a limit, as the following example demonstrates.

EXAMPLE 5 Find (i) $\lim_{x \to -3} \frac{2x^2 + 6x}{x + 3}$; (ii) $\lim_{x \to 5} \frac{x^2 - 10x + 25}{x - 5}$.

Solution

(i) $\lim_{x \to -3} \frac{2x^2 + 6x}{x + 3} = \lim_{x \to -3} \frac{2x(x + 3)}{x + 3} = \lim_{x \to -3} 2x = -6$.

(ii) $\lim_{x \to 5} \frac{x^2 - 10x + 25}{x - 5} = \lim_{x \to 5} \frac{(x - 5)^2}{x - 5} = \lim_{x \to 5} (x - 5) = 0$. ◢

We see that when the limit of the denominator is zero, the limit of the quotient is unpredictable. It might not exist, as in Example 2.1.7, but it also might exist, as in the above example, and may even equal zero.

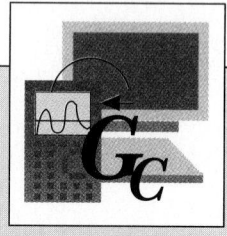

With a graphics calculator it is possible to sketch the graph of the function $y = \frac{1}{x-1}$ from Example 2.1.7, to see why it has no limit as $x \to 1$.

TI-85	SHARP EL9300
GRAPH F1 $(y(x) =)$ 1 ÷ (F1 (x) −1) ▊M2(RANGE) 0 ▼ 2 ▼ ▼ (−) 20 ▼ 20 F5 (GRAPH)	⌁ 1 ÷ (x/θ/T − 1) RANGE 0 ENTER 2 ENTER 1 ENTER −20 ENTER 20 ENTER 1 ENTER ⌁

CASIO f_x-7700 GB/f_x-6300 G	HP 48SX
Cls EXE Range 0 EXE 2 EXE 1 EXE −20 EXE 20 EXE 1 EXE EXE EXE EXE Graph 1 ÷ (▊ − 1 EXE	▊ ▊ PLOTR ERASE ATTN′ αY ▊ 1 ÷ ▊αX − 1 ↱ ▊▊ ▊ DRAW ERASE 0 ENTER 2 XRNG 20 +/− ENTER 20 YRNG DRAW

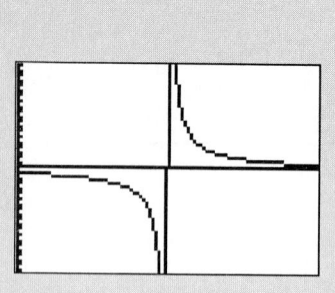

You might like to try varying the vertical scale in your plot.

Summary 2.2.7

Limits of Rational Functions

To evaluate a limit of a rational function,

$$\lim_{x \to a} \frac{p(x)}{q(x)},$$

first evaluate the limits of the numerator and the denominator. If $q(a)$, the limit of the denominator, is not zero, then Theorem 2.2.6 can be used without trouble, and

$$\lim_{x \to a} \frac{p(x)}{q(x)} = \frac{p(a)}{q(a)}.$$

If the limit of the denominator is zero, that is, $q(a) = 0$, and the limit of the numerator is nonzero, that is, $p(a) \neq 0$, then, as in Example 2.1.7, the rational function will have no limit as $x \to a$.

If the numerator and denominator both tend to 0 as $x \to a$, that is, $p(a) = q(a) = 0$, then the Factor Theorem (Theorem 1.3.5) (p. 12) says that both $p(x)$ and $q(x)$ are divisible by $x - a$. Perform the long division (cf. Example 1.3.8, p. 12), and write $p(x) = (x - a)P(x), q(x) = (x - a)Q(x)$. Then, provided $x \neq a$, $\frac{p(x)}{q(x)} = \frac{(x-a)P(x)}{(x-a)Q(x)} = \frac{P(x)}{Q(x)}$, so

$$\lim_{x \to a} \frac{p(x)}{q(x)} = \lim_{x \to a} \frac{P(x)}{Q(x)}.$$

Now we can start again, trying to evaluate this limit. With luck, at least one of the numerator and denominator will have a nonzero value at a, and then we can deal with the limit. If both are zero at a, then we may have to divide again, but this process will eventually stop.

EXAMPLE 6   Find $\lim_{x \to 2} \frac{x^3 - 3x^2 + 4}{x^3 - 6x^2 + 12x - 8}$.

Solution   Both numerator and denominator are zero when $x = 2$. So both can be divided by $(x - 2)$. We find that $x^3 - 3x^2 + 4 = (x - 2)(x^2 - x - 2)$ and that $x^3 - 6x^2 + 12x - 8 = (x - 2)(x^2 - 4x + 4)$. Accordingly,

$$\lim_{x \to 2} \frac{x^3 - 3x^2 + 4}{x^3 - 6x^2 + 12x - 8} = \lim_{x \to 2} \frac{(x - 2)(x^2 - x - 2)}{(x - 2)(x^2 - 4x + 4)} = \lim_{x \to 2} \frac{x^2 - x - 2}{x^2 - 4x + 4}.$$

Again both numerator and denominator are zero at $x = 2$, so we divide again:

$$\lim_{x \to 2} \frac{x^2 - x - 2}{x^2 - 4x + 4} = \lim_{x \to 2} \frac{(x - 2)(x + 1)}{(x - 2)^2} = \lim_{x \to 2} \frac{(x + 1)}{(x - 2)}.$$

Here the numerator tends to 3 as $x \to 2$, and the denominator tends to 0. So we know that the limit of the quotient does not exist, and the limit that we started with does not exist. ◢

Theorem 2.2.5 and our work with polynomials, which have the "expected" limit as x approaches any a, might lead us to think that the limit as $x \to a$ of $f(x)$ will always be related to the value $f(a)$ of $f(x)$ at a. But in the above examples we have discussed limits of functions that are not even defined at $x = a$. In fact, though it is very convenient when a limit does equal the value of the function, the limit itself does not necessarily equal that value, even if it is defined.

EXAMPLE 7   Let $f(x)$ be the function defined as follows:

$$f(x) = \begin{cases} x^2 - 3x + 1, & \text{if } x \neq 1, \\ 5, & \text{if } x = 1. \end{cases}$$

In Figure 2.2.7 we leave an empty circle at the point $(1, -1)$ to indicate that that point on the parabola is not included in the graph of $f(x)$. Then as $x \to 1$, as usual we are concerned only with x "approaching" 1 but not actually equaling 1. So the value of $f(x)$ right at $x = 1$ is irrelevant. As far as the limit process is concerned, the limit as $x \to 1$ of $f(x)$ is exactly the same as the limit of the polynomial $x^2 - 3x + 1$, since they are equal at all points other than $x = 1$. ◢

This example shows that the limit of a function as $x \to a$ may exist and be different from the value of the function at $x = a$. Of course this example was specially made up to demonstrate this phenomenon, and it may seem rather contrived, but the point it illustrates is important.

Just as the limit $\lim_{x \to a} f(x)$ does not depend on $f(a)$, nor even on whether $f(x)$ is defined when $x = a$, the limit also does not depend on the values of $f(x)$ for x far away from a. Using Figure 2.2.8, we see that if there is some interval (c, d) containing a so that $f(x) = g(x)$ for every $x \in (c, d)$, with the possible exception of a itself, then $f(x)$ and $g(x)$ have the same limit as $x \to a$. We record this useful result.

Figure 2.2.7

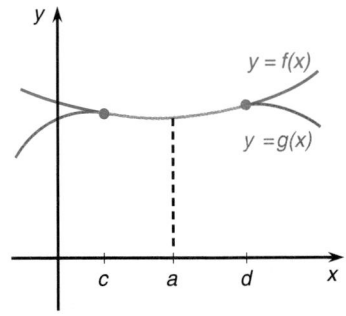

Figure 2.2.8

PROPOSITION 2.2.8

Suppose $f(x)$ and $g(x)$ are functions and a a number. Suppose there is an interval (c, d) containing a (i.e., $c < a < d$) so that for every $x \in (c, d)$ except possibly $x = a$, we have

$$f(x) = g(x).$$

Then

$$\lim_{x \to a} f(x) = \lim_{x \to a} g(x),$$

in the sense that if either of these limits exists, then the other also exists and they are equal.

EXAMPLE 8

(i) Notice that we can use Proposition 2.2.8 to establish the limit discussed in Example 7. After all, with $f(x)$ as defined in that example and $g(x) = x^2 - 3x + 1$, we see that these functions are equal everywhere except at $x = 1$. So the proposition applies, using any interval containing 1, for example, $(0, 2)$. We conclude that $\lim_{x \to 1} f(x) = \lim_{x \to 1} g(x) = \lim_{x \to 1} (x^2 - 3x + 1) = -1$. See Figure 2.2.9.

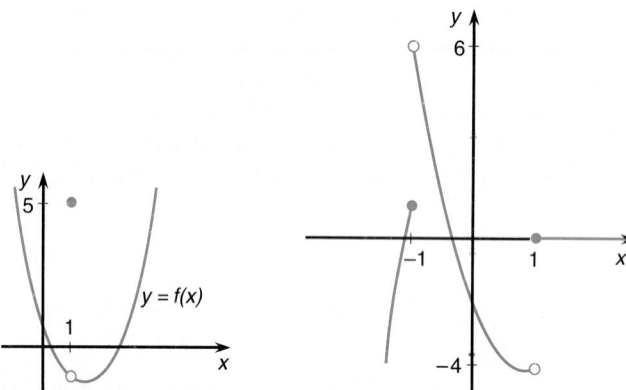

Figure 2.2.9   **Figure 2.2.10**

(ii) Define a function as follows:

$$f(x) = \begin{cases} 3x^2 - 5x - 2, & \text{if } -1 < x < 1, \\ 2x^3 - 3x^2 + 6, & \text{if } x \leq -1, \\ 0, & \text{if } x \geq 1. \end{cases}$$

Then for any a with $-1 < a < 1$, $f(x)$ behaves like $3x^2 - 5x - 2$ as $x \to a$. For example, $\lim_{x \to 0} f(x) = \lim_{x \to 0} (3x^2 - 5x - 2) = -2$. Similarly, $\lim_{x \to 1/2} f(x) = -\frac{15}{4}$. See Figure 2.2.10.

The function $f(x)$ does not have a limit as $x \to 1$ or $x \to -1$.

For any a with $a < -1$, $\lim_{x \to a} f(x)$ will be the same as $\lim_{x \to a} (2x^3 - 3x^2 + 6)$, which equals $2a^3 - 3a^2 + 6$. For any a with $a > 1$, $\lim_{x \to a} f(x)$ will equal $\lim_{x \to a} 0 = 0$.

Another observation we should make is that although in the examples we have done so far we have always used x as the variable, there is really no reason to insist on this choice. Any other letter would do just as well, as long as we are consistent. So for example, we could write

$$\lim_{t \to 7} (t^2 - 3t + 1) = 29, \qquad \lim_{u \to -1} \frac{3u + 7}{u^2 + 1} = \frac{4}{2} = 2,$$

$$\lim_{s \to 0} \frac{s^3 - 3s^2 + 5s}{s} = \lim_{s \to 0} (s^2 - 3s + 5) = 5.$$

Exercises 2.2

Find each of the following limits.

1. $\lim_{x \to 2} (x^2 - 2x + 1)$

2. $\lim_{x \to 4} (x^2 - 10)$

3. $\lim_{x \to 1} (3x^2 - x - 5)$

4. $\lim_{x \to -3} (x^3 + 20)$

5. $\lim_{x \to 0} (x^4 + 17x - 1)$

6. $\lim_{x \to 3} (x^4 - 1)$

7. $\lim_{t \to -1} (t^3 + 3t^2 - 1)$

8. $\lim_{x \to \sqrt{3}} (x^4 + 2x^3 - 4x - 1)$

9. $\lim_{t \to \sqrt{5}} (t^5 - 3t^3 + t - 1)$

10. $\lim_{u \to \sqrt{2}} (u^7 + 2u^6 - 4u^3 + 2u - 1)$

11. $\lim_{x \to 2} \frac{x^2 + x - 1}{x^3 - x^2 + x - 5}$

12. $\lim_{t \to -1} \frac{t^2 + 2t + 2}{t^3 + 3t^2 + 4t - 1}$

13. $\lim_{s \to 0} \frac{s^9 + s^4 + 5}{s^5 - 3s^4 + s^2 - s + 5}$

14. $\lim_{r \to 4} \frac{r^2 - 2r - 6}{r^2 + 2r - 11}$

15. $\lim_{x \to -1} \frac{x^5 + 2x^4 - 1}{x^3 + 2x^2 + 3x - 1}$

16. $\lim_{u \to 0} \frac{u^3 - 2u^2 + u - 1}{u^2 - u + 20}$

17. $\lim_{t \to \sqrt{3}} \frac{t^4 - 6t^2 + 9}{t^2 - 3}$

18. $\lim_{y \to 2^{1/3}} \frac{y^3 - 2}{y^6 - 4}$

19. $\lim_{x \to (1 + \sqrt{2})} \frac{(x - 1)^4 - 4}{(x - 1)^2 - 2}$

20. $\lim_{u \to (1 + \sqrt{5})} \frac{u^2 - 2u - 4}{(u - 1)^2 - 5}$

In each of the following cases, decide whether or not the limit exists. If it does, calculate it. If it does not, explain why.

21. $\lim_{x \to 1} \frac{x^2 - 1}{x - 1}$

22. $\lim_{t \to -1} \frac{t^2 + 2t + 1}{t^2 + 3t + 2}$

23. $\lim_{r \to 3} \frac{r^2 - r - 6}{r - 3}$

24. $\lim_{u \to 0} \frac{2u^3 - 3u^2 + 7u}{u^3 - 3u^2}$

25. $\lim_{y \to -2} \frac{y^2 + 4y + 4}{y^2 - 4}$

26. $\lim_{x \to -1} \frac{2x^2 + 4x + 2}{x^3 + 3x^2 + 3x + 1}$

27. $\lim_{t \to -1} \frac{7t^3 + 14t^2 + 7t}{t^2 + 2t + 1}$

28. $\lim_{s \to 0} \frac{3s^9 - s^5}{s^7 - 2s^6 + 3s^5}$

29. $\lim_{x \to 1} \frac{x^3 - x^2 + x - 1}{x^4 - x^3 + x^2 - x + 1}$

30. $\lim_{u \to 2} \left(\frac{u^2 - 4u + 4}{u^2 - 4u + 3} \right)^3$

31. $\lim_{u \to 3} (u^2 - 5u + 1)$

32. $\lim_{t \to 1} \frac{t^2 - 3t + 1}{2t^3 - t^2 + t + 2}$

33. $\lim_{x \to -2} \frac{x^2 + 2x}{x^2 + 1}$

34. $\lim_{r \to 2} \frac{r^3 + r^2 - 5r - 2}{r^3 - 2r^2 - r + 2}$

35. $\lim_{y \to 0} \frac{2y^3 - 6y^2 + 13y}{y^3 - 2y^2}$

36. $\lim_{x \to 1} \frac{2x^3 + x - 3}{3x^3 - 3x^2 - 3x + 3}$

37. $\lim_{u \to 2} \frac{5u^2 - 4u - 10}{u^3 + u - 6}$

38. $\lim_{r \to -1} \frac{2r^2 - 2}{3r^2 - 3}$

39. $\lim_{x \to 1} \frac{x^2 - 2x + 1}{2(x^2 - 1)^2}$

40. $\lim_{t \to -1} \left(\frac{3t + 3}{t^3 + 1} \right)^4$

For each of the following functions $f(x)$, find all values a so that $f(x)$ does not have a limit as $x \to a$. Find the value of the limit for all other values of a (i.e., your answer will depend on a).

41. $f(x) = \frac{x^2 - 9}{x - 3}$

42. $f(x) = \frac{x - 3}{x^2 - 9}$

43. $f(x) = \frac{(x - 1)(x - 2)(x - 5)}{(x + 1)(x + 2)}$

44. $f(x) = \frac{(x+3)^2(x+1)}{x^2-9}$ **48.** $f(x) = \frac{\sqrt{(x-4)^4}}{(x-4)^{3/2}}$

45. $f(x) = \frac{x(x+1)(x-2)^2}{(x+1)(x-1)(x-2)}$ **49.** $f(x) = \frac{x^2-c^2}{(x-c)^2}$

46. $f(x) = \frac{x^3-2x^2+x}{x^2-x}$ **50.** $f(x) = \frac{(x-c)^2}{(x-c)^3}$

47. $f(x) = \frac{\sqrt{(x-1)^2}}{x-1}$

II

*51. Suppose $f(x)$ and $g(x)$ are polynomials that can be written in the form $f(x) = (x-a)^m F(x)$, $g(x) = (x-a)^n G(x)$, where F and G are polynomials satisfying $F(a) \neq 0$, $G(a) \neq 0$. Evaluate the following limit, when it exists:

$$\lim_{x \to a} \frac{f(x)}{g(x)}.$$

(Your answer will depend on the numbers m and n, and should include a precise statement about when the limit exists.)

III

52. Consider the function $f(x) = \frac{x^3-3x^2+4}{x^3-6x^2+12x-8}$, which we encountered in Example 6. Use a graphics calculator or a computer to sketch the graph. Begin with the graph for $-5 \le x \le 5$ and then zoom in to higher magnification around $x = 2$. Notice in particular how the graph behaves to the left and to the right of $x = 2$.

Find the limit, if it exists, of each of the following functions at the points where the denominator equals 0. Graph each function to confirm your answer.

53. $\frac{x^2+3x+2}{x^2-1}$ **56.** $\frac{x-1}{x+1}$

54. $\frac{x^3-3x^2}{x^2-9}$ **57.** $\frac{x^3+4x^2+3x}{x+3}$

55. $\frac{x+1}{x-1}$ **58.** $\frac{x^4-8x^2+16}{x+2}$

2.3 Working with Limits

POWERS
COMPOSITE FUNCTIONS
THE SQUEEZE THEOREM

We have seen that limits of polynomials and rational functions are very easy to find, but of course it would be desirable to be able to find limits of other, more complicated functions. In this section we learn how to find limits of powers of x with nonintegral exponents (e.g., $x^{2/3}$), and also how to deal with limits of functions like $\sqrt{x+1}$, which are "composite" functions, meaning that one first takes the function $f(x) = x+1$ and then "composes" it with the square root function, to get $\sqrt{f(x)} = \sqrt{x+1}$.

The section ends with a more technical result, the Squeeze Theorem, concerning functions whose graphs lie between the graphs of two other functions whose limits are known. This result is of great theoretical importance and is often used to find new limit rules, but for us its practical applicability will be limited.

Limits of Noninteger Powers of x

EXAMPLE 1   If $a > 0$, find $\lim_{x \to a} \sqrt{x}$.

Solution   The graph of $y = \sqrt{x}$ is shown in Figure 2.3.1. It seems reasonable to guess that $\lim_{x \to a} \sqrt{x} = \sqrt{a}$, which is equivalent to showing that

$$\lim_{x \to a} (\sqrt{x} - \sqrt{a}) = 0.$$

It is this that we shall show. Since $\sqrt{x}$ always means the nonnegative square root, we know that $0 \le \sqrt{x}$.
Write

$$\sqrt{x} - \sqrt{a} = (\sqrt{x} - \sqrt{a})\frac{\sqrt{x} + \sqrt{a}}{\sqrt{x} + \sqrt{a}}.$$

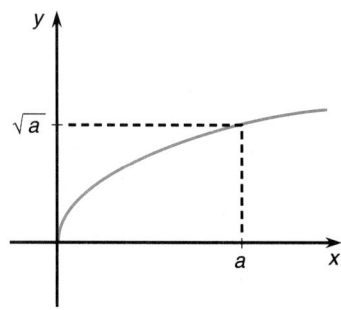

Figure 2.3.1

Multiplying out the difference of squares in the numerator, we find this equals

$$\frac{x-a}{\sqrt{x}+\sqrt{a}}.$$

Now $0 \le \sqrt{x}$, so

$$\frac{1}{\sqrt{x}+\sqrt{a}} \le \frac{1}{0+\sqrt{a}} = \frac{1}{\sqrt{a}}.$$

So when $x - a > 0$,

$$0 < \sqrt{x}-\sqrt{a} = \frac{x-a}{\sqrt{x}+\sqrt{a}} \le \frac{x-a}{\sqrt{a}},$$

and when $x - a < 0$,

$$0 > \sqrt{x}-\sqrt{a} \ge \frac{x-a}{\sqrt{a}}.$$

The function

$$\frac{x-a}{\sqrt{a}}$$

is just $x - a$ times the constant $\frac{1}{\sqrt{a}}$. By Theorem 2.2.3(iii) its limit as $x \to a$ is 0.

We have shown that $\sqrt{x}-\sqrt{a}$ lies between the constant function 0 and $\frac{x-a}{\sqrt{a}}$. These two functions both approach 0 as $x \to a$. The situation is illustrated in Figure 2.3.2.

If two functions both approach 0 as $x \to a$, it certainly seems that anything that lies between them should be forced to approach 0 also. This is a particular instance of what we shall call the Squeeze Theorem (Theorem 2.3.5), which says that something that is "squeezed" between two functions whose limits are known to be the same must itself approach the same limit.

Accepting this reasoning, we have proved what we wanted to know, that

$$\lim_{x \to a} (\sqrt{x}-\sqrt{a}) = 0,$$

or

$$\lim_{x \to a} \sqrt{x} = \sqrt{a}.$$

Strictly speaking, the argument we have given only works when $a > 0$, but something similar is true even when $a = 0$. A comparable result holds not just for square roots, but for roots of any order, or indeed for any power of x, to the extent that it makes sense.

When the denominator gets smaller, the quotient gets larger. . . .

The right-hand inequality comes from multiplying the previous one by $x - a$ . . .

Multiplying by the negative number $x - a$ reverses the inequality. . . .

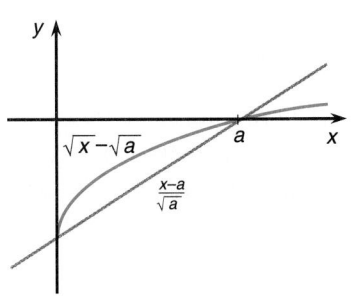

Figure 2.3.2

THEOREM 2.3.1

Let r be any real number. Then

$$\lim_{x \to a} x^r = a^r,$$

provided it makes sense.

Specifically, if $r \le 0$, then it is necessary to require that $a \ne 0$. Moreover, it does not make sense to consider nonpositive values of a except when r is a rational number, $r = \frac{p}{q}$, with q an odd number.

EXAMPLE 2

(i) We know that $x^{1/2} = \sqrt{x}$ makes sense only for $x \geq 0$, and we have just seen that for $a > 0$, $\lim_{x \to a} \sqrt{x} = \sqrt{a}$.

(ii) The cube root of x makes sense for any $x \in \mathbb{R}$ (e.g., $(-8)^{1/3} = -2$). So we can write $\lim_{x \to -8} x^{1/3} = (-8)^{1/3} = -2$, or for any $a \in \mathbb{R}$, $\lim_{x \to a} x^{1/3} = a^{1/3}$.

(iii) The function $x^{-8/5}$, which can be written as

$$(x^{-1/5})^8 = \frac{1}{(x^{1/5})^8},$$

makes sense for all x except $x = 0$ (because in that case the denominator would be zero). So for $a \neq 0$ we know that $\lim_{x \to a} x^{-8/5} = a^{-8/5}$. For example, if $a = -32$, we get $\lim_{x \to -32} x^{-8/5} = (-32)^{-8/5} = \left((-32)^{-1/5}\right)^8 = \left(-\frac{1}{2}\right)^8 = \frac{1}{256}$.

Now that we know how to deal with limits of powers of x and with sums, products, and quotients of limits, we can evaluate a great many complicated limits. The idea is to evaluate the simple parts of a complicated limit and then see how they combine.

EXAMPLE 3

(i) Evaluate

$$\lim_{x \to 3} \frac{x^2 - 9 - \sqrt{x}}{x^{3/2}}.$$

Solution   We can easily find that the limit of the numerator is $3^2 - 9 - \sqrt{3} = 9 - 9 - \sqrt{3} = -\sqrt{3}$ and that the limit of the denominator is $(3)^{3/2} = 3\sqrt{3}$. Applying Theorem 2.2.6(ii), we find that

$$\lim_{x \to 3} \frac{x^2 - 9 - \sqrt{x}}{x^{3/2}} = \frac{-\sqrt{3}}{3\sqrt{3}} = -\frac{1}{3}.$$

(ii) Find

$$\lim_{x \to -1} \frac{x^{1/5} + 3}{x + 1}.$$

Solution   The limit of the numerator is $(-1)^{1/5} + 3 = -1 + 3 = 2$, and the limit of the denominator is 0. This means that as $x \to -1$, the numerator approaches 2 and the denominator approaches 0, so the quotient is something near 2 divided by something very small. This means it will be very large, either very large positive or very large negative, and it will continue to get larger. So the limit does not exist.

(iii) Find

$$\lim_{x \to 0} \frac{x^{1/3} + x^{7/3}}{x^3 - 2x}.$$

Solution   This time we find that the numerator and denominator both tend to 0, so the theorem on quotients does not apply. In a situation like this, something else is needed, and the first thing to try is factoring the numerator and denominator. We can write $x^{1/3} + x^{7/3} = x^{1/3}(1 + x^2)$, so

$$\frac{x^{1/3} + x^{7/3}}{x^3 - 2x} = \frac{x^{1/3}(1 + x^2)}{x(x^2 - 2)} = \frac{1 + x^2}{x^{2/3}(x^2 - 2)}.$$

Now we can try again to evaluate the limit. This time we find that the numerator has 1 as its limit and the denominator tends to 0, so as above the quotient does not have a limit.

> This example illustrates the basic approach to finding limits of complicated expressions. Try evaluating the limit of each simple piece separately, and then put them together. If you encounter a quotient in which the denominator tends to 0 and the numerator tends to a limit other than 0, then the limit of the quotient does not exist. If you encounter a quotient in which both numerator and denominator approach 0, then try to rearrange the expression to make it easier to work with. A good first step is to attempt to factor either of them and cancel any common factors.

EXAMPLE 4   Find

$$\lim_{t \to 0} \frac{|t|^2 + |t|^{6/5}}{|t|^{1/5} + |t|^{1/10}}.$$

Solution   Note that both numerator and denominator tend to 0. The thing to look for is something we can divide out of both the numerator and the denominator, and the most obvious choice is a power of $|t|$.

Dividing by a power of $|t|$ amounts to changing the exponents on all the powers of $|t|$ occurring in both numerator and denominator. After doing that we still need to find the limits of numerator and denominator, so we should avoid having negative powers of $|t|$, which do not tend to any limit as $t \to 0$. The natural thing to do is divide by the lowest power of $|t|$ that occurs, namely, $|t|^{1/10}$. We find that

$$\lim_{t \to 0} \frac{|t|^2 + |t|^{6/5}}{|t|^{1/5} + |t|^{1/10}} = \lim_{t \to 0} \frac{|t|^{19/10} + |t|^{11/10}}{|t|^{1/10} + 1}.$$

Now the numerator still tends to 0, but the denominator tends to 1 as $t \to 0$, so

$$\lim_{t \to 0} \frac{|t|^2 + |t|^{6/5}}{|t|^{1/5} + |t|^{1/10}} = \frac{0}{1} = 0.$$

EXAMPLE 5   Find $\lim_{u \to 8} \frac{u - 8}{u^{1/3} - 2}$.

Solution   Both numerator and denominator tend to 0 as u tends to 8, but this time we cannot divide out by a power of u, since both numerator and denominator contain constant terms. Instead, we should try factoring them both.

On the face of it, the numerator cannot be factored, but suppose we take a hint from the fact that the denominator contains $u^{1/3}$ and write $x = u^{1/3}$. Then the

As $u \to 8$, we have $x = u^{1/3} \to 2$. This suggests evaluating $x^3 - 8$ at $x = 2$; we get 0, so $(x - 2)$ divides $x^3 - 8$.

numerator is $x^3 - 8$, which does factor: $x^3 - 8 = (x - 2)(x^2 + 2x + 4)$. Returning to the variable u, we find the following:

$$\frac{u - 8}{u^{1/3} - 2} = \frac{(u^{1/3} - 2)(u^{2/3} + 2u^{1/3} + 4)}{u^{1/3} - 2} = u^{2/3} + 2u^{1/3} + 4,$$

so

$$\lim_{u \to 8} \frac{u - 8}{u^{1/3} - 2} = \lim_{u \to 8} (u^{2/3} + 2u^{1/3} + 4) = 8^{2/3} + 2(8)^{1/3} + 4$$

$$= 4 + 4 + 4 = 12.$$

Limits of Composite Functions

EXAMPLE 6 Suppose we want to find

$$\lim_{x \to 2} \sqrt{x + 1}.$$

We would like to say that if $x \to 2$, then $x + 1$ approaches 3, and as t approaches 3, we know that $\sqrt{t}$ approaches $\sqrt{3}$. Letting $t = x + 1$, we expect $\sqrt{x + 1}$ to approach $\sqrt{3}$.

This is true, and it is a particular case of the following result.

THEOREM 2.3.2

Suppose f and g are functions, with $\lim_{x \to a} f(x) = L$, $\lim_{t \to L} g(t) = M$, and $g(L) = M$. Then

$$\lim_{x \to a} g(f(x)) = M.$$

This can also be written as

$$\lim_{x \to a} g(f(x)) = \lim_{t \to L} g(t) = M.$$

The condition $g(L) = M$ says that not only should $g(t)$ have a limit as $t \to L$, but it should have the "right value" at $t = L$. In the exercises we shall see the theorem fails without this condition.

In Example 6 we have $f(x) = x + 1$, $g(t) = \sqrt{t}$, $a = 2$, $L = 3$, and $M = \sqrt{3}$. The theorem tells us how to evaluate the limit of $g(f(x))$, which is the function obtained by first applying to x the function f and then taking the answer you get and applying g to that. In the example this means $\sqrt{x + 1}$. In the statement of the theorem we use t as the variable for $g(t)$ in an attempt to avoid confusion, since we intend to substitute $f(x)$ for t later on.

Also notice that the statement of the theorem considers the limit L of $f(x)$ as $x \to a$, and then considers the limit of $g(t)$ as t approaches *this limit L*. Of course this is reasonable, since the idea is to evaluate the limit of $g(f(x))$, and $f(x)$ will be approaching L.

EXAMPLE 7 Find the following limits: (i) $\lim_{x \to 4} \sqrt{x^2 - 1}$; (ii) $\lim_{x \to 1} (x + 2)^5$; (iii) $\lim_{x \to 0} (1 + \sqrt{x + 4})^3$; (iv) $\lim_{x \to 0} \frac{2 - \sqrt{4 - x}}{x}$; (v) Find $\lim_{t \to 0} t^{1/3}$ without using Theorem 2.3.1, *assuming* the limit exists.

Solution To solve any of these questions, the idea is to figure out first what functions f and g we need to use in order to apply the theorem.

(i) We can write $\sqrt{x^2 - 1} = \sqrt{f(x)}$, with $f(x) = x^2 - 1$. Letting $g(t) = \sqrt{t}$, we are ready for the theorem. Since $\lim_{x \to 4} f(x) = \lim_{x \to 4} (x^2 - 1) = 15$ and $\lim_{t \to 15} g(t) = \lim_{t \to 15} \sqrt{t} = \sqrt{15} = g(15)$, the theorem tells us that $\lim_{x \to 4} \sqrt{x^2 - 1} = \sqrt{15}$.

(ii) Here we let $f(x) = x + 2, g(t) = t^5$. So $g(f(x)) = g(x + 2) = (x + 2)^5$. As $x \to 1$, $f(x)$ tends to 3, and $g(f(x))$ tends to $g(3) = 3^5 = 243$. So $\lim_{x \to 1} (x + 2)^5 = 243$.

(iii) For this part we need to evaluate the limit of a function that is the cube of something. As in part (ii), the limit will be the cube of the limit of the something. So we must begin by finding that "inner" limit, that is, $\lim_{x \to 0} (1 + \sqrt{x + 4})$. But as $x \to 0$, we know that $x + 4 \to 4$, so $\sqrt{x + 4} \to \sqrt{4} = 2$.
   This means $\lim_{x \to 0} (1 + \sqrt{x + 4}) = 1 + 2 = 3$, so

$$\lim_{x \to 0} (1 + \sqrt{x + 4})^3 = 3^3 = 27.$$

(iv) The denominator obviously tends to 0. For the numerator, notice that as $x \to 0$, $4 - x$ tends to 4, so $\sqrt{4 - x}$ tends to $\sqrt{4} = 2$, and $2 - \sqrt{4 - x}$ tends to 0.
   Here there is no obvious way to divide out any common factor from the numerator and denominator. Instead, we should remember Example 5 and try to get rid of the square root by multiplying both numerator and denominator by $2 + \sqrt{4 - x}$. We find

$$\lim_{x \to 0} \frac{2 - \sqrt{4 - x}}{x} = \lim_{x \to 0} \frac{2 - \sqrt{4 - x}}{x} \times \frac{2 + \sqrt{4 - x}}{2 + \sqrt{4 - x}} = \lim_{x \to 0} \frac{4 - (4 - x)}{x(2 + \sqrt{4 - x})}$$

$$= \lim_{x \to 0} \frac{x}{x(2 + \sqrt{4 - x})} = \lim_{x \to 0} \frac{1}{2 + \sqrt{4 - x}} = \frac{1}{2 + \sqrt{4}}$$

$$= \frac{1}{4}.$$

(v) We can use the second formula in Theorem 2.3.2. Letting $t = x^3$, we observe that $\lim_{x \to 0} t = \lim_{x \to 0} x^3 = 0$. So with $g(t) = t^{1/3}$, $f(x) = x^3$, we find $\lim_{x \to 0} f(x) = 0$, and

$$\lim_{t \to 0} t^{1/3} = \lim_{t \to 0} g(t) = \lim_{x \to 0} g(x^3) = \lim_{x \to 0} x = 0.$$

Notice the way we used the theorem in part (v) of Example 7. In effect we turned it around: Here it is easy to find the limit of $g(f(x)) = x$, and we use this to find the limit of $g(t)$. This approach cannot be used to show the limit exists, but it does allow us to find the value, *assuming* the limit exists.

Another fact we need to record involves the sign of a limit. If f is always positive, say, then it seems likely that the same should be true of its limits. In fact, this is not quite true.

EXAMPLE 8   Let

$$f(u) = \begin{cases} |u|, & \text{if } u \neq 0 \\ 6, & \text{if } u = 0. \end{cases}$$

Its graph is shown in Figure 2.3.3. By Proposition 2.2.8, $f(u)$ has the same limit as $|u|$, so

$$\lim_{u \to 0} f(u) = \lim_{u \to 0} |u| = 0.$$

Even though the function was always positive, the limit is zero.

On the other hand, it could never be negative, and this is the result we want.

THEOREM 2.3.3

> If $f(x)$ is a function, a is a number so that $\lim_{x \to a} f(x)$ exists, and (c, d) is an interval containing a (i.e., $c < a < d$) so that $f(x) \geq 0$ for all $x \in (c, d)$, except possibly $x = a$, then
>
> $$\lim_{x \to a} f(x) \geq 0.$$

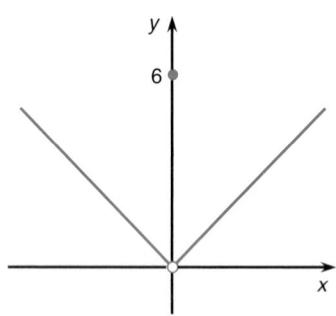

Figure 2.3.3

EXAMPLE 9

(i) Suppose, in the notation of Theorem 2.3.3, that $f(x) \leq 0$ for all $x \in (c, d)$, except possibly $x = a$. Then $-f(x) \geq 0$, and the theorem tells us that $\lim_{x \to a} \left(-f(x) \right) \geq 0$. But by Theorem 2.2.3(iii), $\lim_{x \to a} \left(-f(x) \right) = -\lim_{x \to a} f(x)$, so

$$\lim_{x \to a} f(x) \leq 0.$$

(ii) Suppose K is a constant and $f(x) \geq K$ for all $x \in (c, d)$. Then $f(x) - K \geq 0$, so $\lim_{x \to a} \left(f(x) - K \right) \geq 0$, but this says that $\left(\lim_{x \to a} f(x) \right) - \left(\lim_{x \to a} K \right) \geq 0$, or

$$\lim_{x \to a} f(x) \geq K.$$

A similar result can be found if $f(x) \leq K$.

One final fact we need to record is a complement to Theorem 2.3.3. That theorem states that if $f(x) \geq 0$ on some interval around a, and $\lim_{x \to a} f(x)$ exists, then $\lim_{x \to a} f(x) \geq 0$. The following theorem is more or less the reverse.

THEOREM 2.3.4

> Suppose $f(x)$ is a function and $a \in \mathbb{R}$ so that $\lim_{x \to a} f(x)$ exists.
> If $\lim_{x \to a} f(x) > 0$, then there is some interval (c, d) containing a (i.e., $c < a < d$) so that for every $x \in (c, d)$, except possibly $x = a$ itself,
>
> $$f(x) > 0.$$

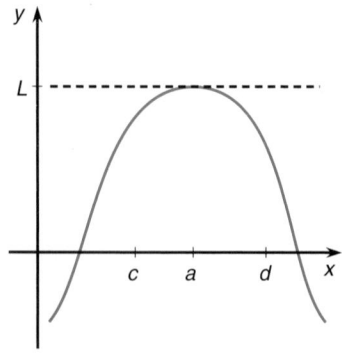

Figure 2.3.4

The idea is very simple. If $f(x)$ approaches some positive limit L as $x \to a$, then close to a, $f(x)$ must get so close to L that it is bigger than 0. This result will be used in the next chapter. See Figure 2.3.4.

The Squeeze Theorem

There are many limits that we can now evaluate, but it would be desirable to extend our techniques for finding limits to other types of functions. One useful way to do this involves comparing a difficult limit with easier ones that are already known (cf. Example 1).

Figure 2.3.5

Figure 2.3.6

Figure 2.3.7

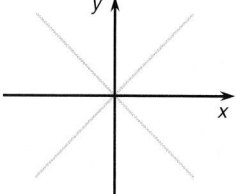

Figure 2.3.8

Specifically, suppose we have two functions that are both known to have a limit as $x \rightarrow a$, and also suppose they have the same limit as $x \rightarrow a$. The situation is illustrated in Figure 2.3.5. The most obvious case would be the one in which both functions are polynomials or rational functions, since we have no trouble evaluating their limits.

Now suppose there is another function, possibly quite complicated, whose graph lies between those two graphs. Then as $x \rightarrow a$, this function is "squeezed" between the other two and is forced to approach the same limit as they do. See Figure 2.3.6.

EXAMPLE 10 Find $\lim_{x \to 0} x \sin\left(\frac{1}{x}\right)$.

Solution The behavior of $\sin\left(\frac{1}{x}\right)$ as $x \rightarrow 0$ is very complicated. After all, if x approaches 0 from the right, then it is small and positive, so $\frac{1}{x}$ is large and positive and gets larger and larger. Now as u gets larger and larger, $\sin(u)$ goes up and down infinitely many times. So as x approaches 0 from the right, $\sin\left(\frac{1}{x}\right)$ goes up and down infinitely often between $y = 1$ and $y = -1$. Its behavior to the left of 0 is similar. In particular, it does not approach any limit. See Figure 2.3.7.

To get an idea of the function $x \sin\left(\frac{1}{x}\right)$, we first draw the two lines $y = x$ and $y = -x$ (see Figure 2.3.8). Then think of the function $x \sin\left(\frac{1}{x}\right)$ as x multiplied by the function $\sin\left(\frac{1}{x}\right)$, which goes back and forth between 1 and -1. So the resulting product function goes back and forth between x and $-x$, and we can visualize it as what you get by squeezing $\sin\left(\frac{1}{x}\right)$ between the two lines $y = x$ and $y = -x$. See Figure 2.3.9(i). Since $\sin\left(\frac{1}{x}\right)$, being a sine, is never bigger than 1 or smaller than -1, we see that when $x > 0$, the product $x \sin\left(\frac{1}{x}\right)$ never gets above the line $y = x$ or below $y = -x$. When $x < 0$, it never lies below $y = x$ or above $y = -x$. See Figure 2.3.9(ii).

Now we are also set up to find the limit. Both the straight lines $y = x$ and $y = -x$ go through the origin or, in terms of limits,

$$\lim_{x \to 0} x = \lim_{x \to 0} (-x) = 0.$$

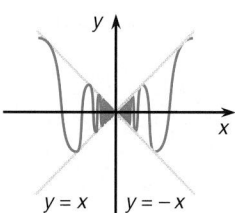

$y = x$ $y = -x$

Figure 2.3.9(i)

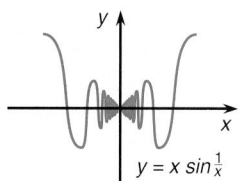

$y = x \sin \frac{1}{x}$

Figure 2.3.9(ii)

The function $x \sin\left(\frac{1}{x}\right)$ is "squeezed" between these two lines, so it must have the same limit:

$$\lim_{x \to 0} \left(x \sin\left(\frac{1}{x}\right) \right) = 0.$$

This example and Example 1 illustrate the use of the following result.

THEOREM 2.3.5

The Squeeze Theorem

Suppose $g(x)$ and $h(x)$ are functions and a is some number so that

$$\lim_{x \to a} g(x) = \lim_{x \to a} h(x) = L.$$

Suppose $f(x)$ is another function and (c, d) is an interval containing a (i.e., $c < a < d$). Suppose that on the interval (c, d), except possibly at $x = a$, the graph of the function $f(x)$ lies between the graphs of $g(x)$ and $h(x)$, that is, either

$$g(x) \leq f(x) \leq h(x)$$

or

$$h(x) \leq f(x) \leq g(x).$$

Then

$$\lim_{x \to a} f(x) = L.$$

See Figure 2.3.10.

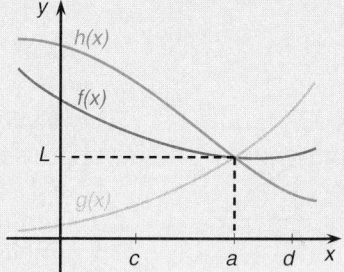

Figure 2.3.10

EXAMPLE 11 Suppose $F = F(x)$ is a function that is defined for every $x \in (0, 2)$. Find $\lim_{x \to 1} \frac{x^2 + (F(x))^2}{1 + (F(x))^2}$.

Solution We begin by writing the function in the form

$$\frac{x^2 + \left(F(x)\right)^2}{1 + \left(F(x)\right)^2} = \frac{x^2 - 1 + 1 + \left(F(x)\right)^2}{1 + \left(F(x)\right)^2} = \frac{x^2 - 1}{1 + \left(F(x)\right)^2} + \frac{1 + \left(F(x)\right)^2}{1 + \left(F(x)\right)^2}$$

$$= \frac{x^2 - 1}{1 + \left(F(x)\right)^2} + 1.$$

Since the constant function 1 tends to the limit 1, we should concentrate on finding $\lim_{x \to 1} \frac{x^2 - 1}{1 + (F(x))^2}$.

Consider the interval $(0, 2)$, which contains 1. On the left half of this interval, $(0, 1)$, the numerator $x^2 - 1$ is always negative, and on the right half, $(1, 2)$, the numerator is always positive. The denominator is always positive, being a sum of squares, so the quotient is also negative on $(0, 1)$ and positive on $(1, 2)$.

We also observe that $1 + \left(F(x)\right)^2 \geq 1$, since the square is nonnegative, and taking reciprocals, we find that $\frac{1}{1 + (F(x))^2} \leq \frac{1}{1} = 1$. (Notice that the inequality is reversed by taking reciprocals.) On the interval $(1, 2)$, where $x^2 - 1$ is positive, we multiply this last inequality by $x^2 - 1$ to find that $\frac{x^2 - 1}{1 + (F(x))^2} \leq \frac{x^2 - 1}{1} = x^2 - 1$.

On the interval $(0, 1)$, where $x^2 - 1$ is negative, we multiply the inequality by $x^2 - 1$, being careful to reverse the inequality, and find that $\frac{x^2 - 1}{1 + (F(x))^2} \geq \frac{x^2 - 1}{1} = x^2 - 1$.

Combining all this information, we see that on $(1, 2)$,

$$0 \leq \frac{x^2 - 1}{1 + \left(F(x)\right)^2} \leq x^2 - 1,$$

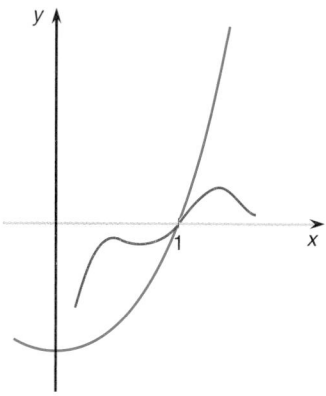

Figure 2.3.11

and on $(0, 1)$,

$$0 \geq \frac{x^2 - 1}{1 + \left(F(x)\right)^2} \geq x^2 - 1.$$

In both cases, the function $\frac{x^2-1}{1+(F(x))^2}$ lies between the functions 0 and $x^2 - 1$ (see Figure 2.3.11). Since $\lim_{x \to 1} (x^2 - 1) = 0 = \lim_{x \to 1} 0$, the Squeeze Theorem applies, and we can conclude that $\lim_{x \to 1} \frac{x^2-1}{1+(F(x))^2} = 0$.

From this we find that the original limit is

$$\lim_{x \to 1} \left(\frac{x^2 - 1}{1 + \left(F(x)\right)^2} + 1 \right) = \lim_{x \to 1} \frac{x^2 - 1}{1 + \left(F(x)\right)^2} + \lim_{x \to 1} 1 = 0 + 1 = 1.$$

Notice that it was possible to find this limit without knowing anything at all about F. Notice too that the function $\frac{x^2-1}{1+(F(x))^2}$ is above 0 and below $x^2 - 1$ when $x > 1$ and the other way around when $x < 1$. What is important is that it is *between* them all the time.

Theorem 2.3.5 requires that $f(x)$ lie between $g(x)$ and $h(x)$ for x in some interval containing a. This is very much in the spirit of Proposition 2.2.8, which says that the behavior of $f(x)$ far away from a is irrelevant to the limit as $x \to a$. It is useful

With a graphics calculator we can try out Example 11 with various functions $F(x)$. For instance, we try $F(x) = \frac{1}{2} \cos x = 0.5 \cos x$.

TI-85	SHARP EL9300
GRAPH F1$(y(x) =)$ (F1(x) x^2 + (.5 COS F1(x))x^2 ÷ (1 + (.5 COS F1(x)) x^2) M2(RANGE) 0.1 ▼ 1.9 ▼ 1 ▼ (−) 1 ▼ 3 ▼ 1 F5(GRAPH)	⫪ (x/θ/T x^2+ (.5 cos x/θ/T)x^2) ÷ (1 + (.5 cos x/θ/T)x^2) RANGE .1 ENTER 1.9 ENTER 1 ENTER −1 ENTER 3 ENTER 1 ENTER ⫪

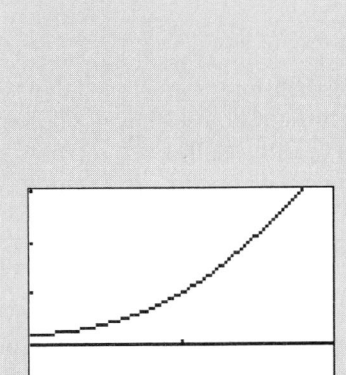

CASIO f_x-7700 GB/f_x-6300 G	HP 48SX
Cls EXE Range .1 EXE 1.9 EXE 1 EXE −1 EXE 3 EXE 1 EXE EXE EXE EXE Graph (X x^2 + (.5 cos X) x^2) ÷ (1 + (.5 cos X) x^2) EXE	↰ PLOT PLOTR ERASE ATTN′ αY = () αX y^x2 + ().5 × COS αX ▶ ▶ y^x 2 ▶ ÷ () 1 + ().5 × COS αX ▶ ▶ y^x 2 ↱ PLOT ↰ DRAW ERASE .1 ENTER 1.9 XRNG 1 +/− ENTER 3 YRNG DRAW

Locate the position of $x = 1$ in the picture. You could also draw the horizontal line $y = 1$ to make it easier to see the limit.

Now you can edit these keystroke sequences and try other functions, such as $F(x) = \sin \pi x$, $F(x) = x^3 - 2$, etc.

Figure 2.3.12

Figure 2.3.13

Figure 2.3.14

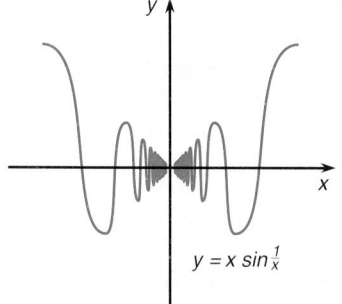

$y = x \sin\frac{1}{x}$

Figure 2.3.15

because it is often much easier to find suitable g and h on such an interval than on the whole line. See Figure 2.3.12.

The hard thing about using Theorem 2.3.5 is finding the functions $g(x)$ and $h(x)$. Usually we are just given $f(x)$, and we have to think of $g(x)$ and $h(x)$ for ourselves. The idea is to replace difficult components of $f(x)$, like $\frac{1}{1+(F(x))^2}$ in the above example, with simpler things that are bigger or smaller (in the example, 0 is smaller than $\frac{1}{1+\left(F(x)\right)^2}$, and 1 is bigger). Similarly, in Example 10 we replaced the messy function $\sin\left(\frac{1}{x}\right)$ with the smaller function -1 and the larger function 1.

EXAMPLE 12   Let $f(t)$ be a function such that

$$|f(t)| \le M, \qquad \text{for all } t,$$

where M is some positive constant (we say that f is "bounded" by M). Show that

$$\lim_{t \to 0} tf(t) = 0.$$

Solution   To say that $|f(t)| \le M$ is to say that $-M \le f(t) \le M$, as shown in Figure 2.3.13, which implies that

$$|tf(t)| \le M|t|,$$

which means that the graph of $tf(t)$ lies between the graphs of $y = Mt$ and $y = -Mt$. See Figure 2.3.14.

Since $\lim_{t \to 0}(-Mt) = \lim_{t \to 0} Mt = 0$, the Squeeze Theorem applies, and we conclude that

$$\lim_{t \to 0} tf(t) = 0.$$

The Squeeze Theorem is extremely useful for establishing some limits that we don't know, but as the examples suggest, it is often difficult to use. The good news is that we will not use it very often, and when we do, it will usually be to see how to evaluate some new type of limit, as we did for the square root function in Example 1.

The Meaning of Limits

We end this section with a brief discussion of the fundamental problem of what a limit is. We have repeatedly used phrases like "approaches" and "tends to," but some of our examples show that these words are not really adequate to describe precisely what we mean.

For instance, the function $f(x) = x\sin\left(\frac{1}{x}\right)$, discussed in Example 10, has limit equal to 0 as x tends to 0, but from Figure 2.3.15 we see that it does not always "approach" 0. On the contrary, it approaches 0 (and actually reaches it) and then moves away again. In fact, it does this infinitely many times, so we cannot justify saying that it keeps approaching 0.

What is true is that it does get close to 0, and while it may move away again, it does not move away too far. By making x very small, we can make sure that $f(x)$ is as close as we like to 0. This idea can be made into a precise definition of limit, but since it is fairly complicated, it will be postponed until Section 2.7.

Exercises 2.3

I

Exercises 1–10 use Theorem 2.3.1, about limits of nonintegral powers of x; Exercises 11–20 use Theorem 2.3.2, about limits of composite functions; Exercises 21–30 use Theorem 2.3.5, the Squeeze Theorem.

Find each of the following limits, explaining your reasoning.

1. $\lim_{x \to 0} \frac{x^{1/3} - 2x^{5/3}}{2x^{1/3} - 3x^2}$

2. $\lim_{t \to 0} \frac{4(-t)^{1/5} + t^{3/5}}{(-t)^{1/7}}$

3. $\lim_{y \to 0} \frac{3y^{2/3} - 5y}{(2y - y^{1/3})^2}$

4. $\lim_{r \to 1} \frac{r^3 - 3r^{5/3} + \sqrt{r}}{r^2 + r^{7/2}}$

5. $\lim_{x \to 9} \frac{3 - \sqrt{x}}{x - 9}$

6. $\lim_{x \to -1} \frac{x+1}{x^{1/3} + 1}$

7. $\lim_{x \to 16} \frac{x - 16}{2 - x^{1/4}}$

8. $\lim_{x \to 0} \frac{|x|^{5/2} + x^2}{x\sqrt{|x|}}$

9. $\lim_{x \to 3} \frac{x - 3}{3^{1/2} - x^{1/2}}$

10. $\lim_{x \to 0} \frac{(x-4)^2 - 16}{\sqrt{|x|}}$

11. $\lim_{x \to 5} \sqrt{x + 4}$

12. $\lim_{t \to 3} \sqrt{t^3 - 2}$

13. $\lim_{y \to 1} (y^2 - 3y + 1)^{1/3}$

14. $\lim_{x \to 2} \frac{(x^2 - 1)^{3/2}}{\sqrt{x+1}}$

15. $\lim_{x \to 1} (3x - \sqrt{x+1})^{5/2}$

16. $\lim_{y \to -1} \frac{(y^2 + 2)^{1/2}}{(10 + y)^{1/4}}$

17. $\lim_{t \to 3} \frac{2 - \sqrt{t+1}}{t - 3}$

18. $\lim_{s \to 0} \frac{s^3 - 2s^2}{3 - \sqrt{9-s}}$

19. $\lim_{r \to -1} \frac{r^{1/3} + 1}{1 - r^{2/5}}$

20. $\lim_{x \to 0} \frac{(8-x)^{1/3} - 2}{x}$

21. $\lim_{x \to 0} x^2 \sin\left(\frac{1}{x}\right)$

22. $\lim_{x \to 0} x^3 \cos\left(\frac{1}{x}\right)$

23. $\lim_{x \to 0} x \sin\left(\frac{1}{x^2 - 2x + 13}\right)$

24. $\lim_{x \to 1} (x - 1) \cos\left(\frac{1}{x^2}\right)$

25. $\lim_{x \to 1} (x^2 - 1) \sin\left(\frac{1}{x-1}\right)$

26. $\lim_{x \to 0} (x^2 - 2x) \cos\left(\frac{x^2 - x + 1}{x^2 + x}\right)$

27. $\lim_{x \to 0} \frac{x}{3 - \cos(\frac{1}{x})}$

28. $\lim_{t \to 0} \frac{3t^2 - 2t^3 \cos(t^{-1})}{t^3 - t^2}$

29. $\lim_{u \to 0} u^2 \sin^2\left(\frac{1}{u^3}\right)$

30. $\lim_{s \to 1} \frac{(s-1)^2 \sin((s-1)^{-1})}{s^2 - 1}$

Find whether each of the following limits exists. Evaluate those that do exist and explain why the others do not.

31. $\lim_{x \to 0} \frac{2x^{1/3} - 4x^2}{x^{2/3} - 7x^3}$

32. $\lim_{t \to 1} \frac{(t-1) - 8(t-1)^{4/5}}{t - 1 + 2(t-1)^{2/3}}$

33. $\lim_{u \to 1} \frac{u^{3/2} - 1}{u - 1}$

34. $\lim_{x \to 0} \frac{\sqrt{9-x} - 2}{x}$

35. $\lim_{x \to 0} \frac{1}{\sin(1/x)}$

36. $\lim_{x \to 0} \left(\cos^2\left(\frac{1}{x}\right) + \sin^2\left(\frac{1}{x}\right) \right)$

37. $\lim_{u \to 9} \frac{u^{1/2} - 3}{u^2 - 81}$

38. $\lim_{x \to 16} \frac{4 - \sqrt{x}}{2 - x^{1/4}}$

39. $\lim_{s \to 0} \frac{3 - \sqrt{9-s}}{s^3 + 2s^2}$

40. $\lim_{t \to 2} \frac{t^2 - 4}{t^{3/2} - 2t^{1/2}}$

II

41. Suppose $r, s, t, u \in \mathbb{R}$ satisfy $r < s$ and $t < u$. Consider the following limit:

$$\lim_{x \to 0} \frac{|x|^r + |x|^s}{|x|^t + |x|^u}.$$

(i) Find conditions on r, s, t, u so that the limit does or does not exist.

(ii) Evaluate the limit when it does exist.

(iii) Do parts (i) and (ii) again with the hypotheses changed to $r \le s$ and $t \le u$.

42. Let c be a positive real number. Evaluate

(i) $\lim_{x \to c^2} \frac{x^{1/2} - c}{x - c^2}$,

(ii) $\lim_{x \to c^{1/2}} \frac{x - c^{1/2}}{x^{1/2} - c^{1/4}}$,

(iii) $\lim_{x \to 0} \frac{x^{1/3} - c}{x}$.

*43. Several times we have used the fact that

$$\lim_{x \to a} |x| = |a|,$$

which is true, but we have never explained why. Using the theorems of this section, give an argument to explain why it is true. (*Hint:* You might want to deal separately with the cases where $a = 0$ or $a \neq 0$.)

*44. The hypotheses of Theorem 2.3.2 include the requirement that $g(L) = M$ in addition to the statement about the limit of $g(t)$. Find an example to show that this is really necessary. In other words, find an example that satisfies all the conditions of the theorem except that $g(L) \neq M$, and for which the conclusion of the theorem is false. (*Hint:* Try letting $f(x) = x$.)

*45. Suppose $f(x)$ has a limit as $x \rightarrow a$ for every $a \in \mathbb{R}$.
  (i) Express $\lim_{x \rightarrow a} f(x + c)$ as a limit of $f(x)$.
  (ii) Express $\lim_{x \rightarrow a} f(c - x)$ as a limit of $f(x)$.
  (iii) Can you say when your results for parts (i) and (ii) remain true if $f(x)$ has a limit as $x \rightarrow a$ only for certain values of a?

*46. (i) Suppose $f(x)$ and $g(x)$ are functions so that $|f(x)| \leq M$ for all x (i.e., $f(x)$ is "bounded" by M) and $\lim_{x \rightarrow a} g(x) = 0$. Show that $\lim_{x \rightarrow a} f(x)g(x) = 0$.
  (ii) Suppose $\lim_{x \rightarrow a} g(x) = 0$ and that $-4 \leq f(x) \leq 3$ for all x. Show that

$$\lim_{x \rightarrow a} f(x)g(x) = 0.$$

  (Be careful; it is possible that $g(x) < 0$.)

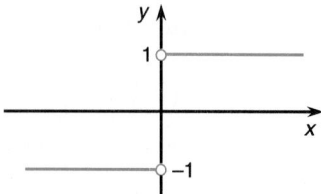

47. In Example 11 (p. 88) we discussed a limit involving a function $F(x)$ without knowing what $F(x)$ is. Use a graphics calculator or computer to sketch the graph of $y = \frac{x^2 + (F(x))^2}{1 + (F(x))^2}$ for $x \in (0, 2)$ if $F(x) = \sin x$. Check that the limit as $x \rightarrow 1$ does equal 1.

48. Use a graphics calculator or computer to sketch the graph of $y = \frac{x^2 + (F(x))^2}{1 + (F(x))^2}$ for $x \in (0, 2)$, if (i) $F(x) = \ln x$, (ii) $F(x) = \tan \frac{\pi x}{4}$. Check that in both cases the limit as $x \rightarrow 1$ does equal 1.

49. Graph the function occurring in the limit of Example 4 for $x \in [-1, 1]$. Then magnify the graph by using $[-0.1, 0.1]$ or $[-0.01, 0.01]$ to convince yourself of the limit.

50. Plot the function $y = \frac{|x|^{3/8} - 2|x|^{1/3}}{|x|^{1/4}(4|x|^{5/3} + 3|x|^{1/12})}$ and, by looking at the graph, try to determine the limit as $x \rightarrow 0$. Evaluate the limit and confirm your answer.

51. We considered the function $y = \sin \frac{1}{x}$ in Example 10 (p. 87). Draw its graph for $x \in [-1, 1]$, $x \in [-\frac{1}{10}, \frac{1}{10}]$, $x \in [-\frac{1}{100}, \frac{1}{100}]$, $x \in [-\frac{1}{1000}, \frac{1}{1000}]$.

52. Draw the graph of the function $y = x \sin \frac{1}{x}$ for $x \in [-1, 1]$, $x \in [-\frac{1}{10}, \frac{1}{10}]$, $x \in [-\frac{1}{100}, \frac{1}{100}]$, $x \in [-\frac{1}{1000}, \frac{1}{1000}]$.

Find the limit, if it exists, of each of the following functions as $x \rightarrow 1$. Graph each function to confirm your answer.

53. $\frac{x-1}{\sqrt{x}-1}$

54. $\frac{x-1}{x^{1/3}-1}$

2.4 One-Sided and Infinite Limits and Limits at Infinity

LIMITS FROM ABOVE/BELOW
INFINITE LIMITS
LIMITS AT INFINITY
LIMITS OF RATIONAL FUNCTIONS

We have already encountered situations in which a function $f(x)$ approaches one value if x approaches a from the right side and a different value if x approaches a from the left (cf. Example 2.1.6, in which the function $f(x) = \frac{x}{|x|}$ was seen to approach 1 if x approaches 0 from the right and -1 if x approaches 0 from the left; see Figure 2.4.1). This phenomenon is sufficiently common that there is special terminology for it.

Figure 2.4.1

DEFINITION 2.4.1

One-Sided Limits

A function $f(x)$ is said to have a limit **from the right** or **from above** if $f(x)$ approaches some value L as x approaches a through values to the right of a, that is, with $x > a$. (Strictly speaking, we need to assume that $f(x)$ is defined for every x in an interval (a, s), for some $s > a$, and possibly at $x = a$.) In this case we write

$$\lim_{x \to a^+} f(x) = L.$$

Similarly, we can speak of limits **from the left** or **from below** (assuming that $f(x)$ is defined on (r, a), for some $r < a$, and possibly at $x = a$) and write

$$\lim_{x \to a^-} f(x) = M.$$

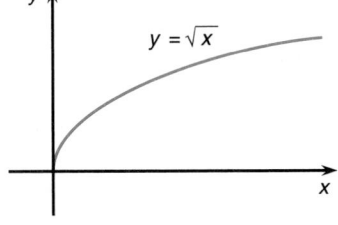

$y = \sqrt{x}$

Figure 2.4.2

One-sided limits sometimes arise with functions that are only defined for x on one side of a. An example of this is

$$\lim_{x \to 0^+} \sqrt{x} = 0,$$

since $\sqrt{x}$ only makes sense when $x \geq 0$. See Figure 2.4.2.

But as we have already seen with $f(x) = \frac{x}{|x|}$, they can also be found with functions that are defined on both sides of the point $x = a$.

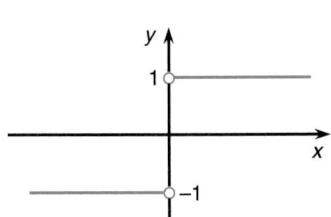

Figure 2.4.3

EXAMPLE 1

(i) The function $f(x) = \frac{x}{|x|}$, graphed in Figure 2.4.3, satisfies

$$\lim_{x \to 0^+} f(x) = 1, \qquad \lim_{x \to 0^-} f(x) = -1.$$

(ii) The function $g(t) = \sqrt{t - 3}$ is only defined when $t \geq 3$, since the square root only makes sense for positive numbers (see Figure 2.4.4). If $t \to 3^+$, that is, $t > 3$ and t approaches 3, then $t - 3$ approaches 0, so its square root should approach 0 too. In other words, $g(t)$ approaches 0, or

$$\lim_{t \to 3^+} \sqrt{t - 3} = 0.$$

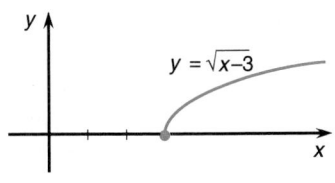

$y = \sqrt{x-3}$

Figure 2.4.4

(iii) Let $h(u) = \frac{\sqrt{(u-1)^2}}{u^2 - u}$. Recall that the square root sign always means the nonnegative square root, so $\sqrt{(u - 1)^2} \geq 0$. If $u > 1$, then $u - 1 > 0$, and $\sqrt{(u - 1)^2} = u - 1$. But if $u < 1$, then $u - 1 < 0$, and the nonnegative square root of $(u - 1)^2$ is not $u - 1$ but rather $-(u - 1) = 1 - u$.

So if $u > 1$, then

$$\frac{\sqrt{(u - 1)^2}}{u^2 - u} = \frac{u - 1}{u(u - 1)} = \frac{1}{u},$$

and

$$\lim_{u \to 1^+} \frac{\sqrt{(u - 1)^2}}{u^2 - u} = \lim_{u \to 1^+} \frac{1}{u} = 1.$$

Similarly, if $u < 1$, then

$$\frac{\sqrt{(u-1)^2}}{u^2 - u} = \frac{1-u}{u(u-1)} = -\frac{1}{u},$$

and

$$\lim_{u \to 1^-} \frac{\sqrt{(u-1)^2}}{u^2 - u} = \lim_{u \to 1^-} -\frac{1}{u} = -1.$$

◢

The nice thing about one-sided limits is that almost all the techniques we have for dealing with limits work just the same way for one-sided limits. Theorems 2.2.3 and 2.2.6 about limits of sums, products, and quotients can be stated for one-sided limits without any changes. The one-sided limits of polynomials and rational functions are both equal to the two-sided limits (cf. Theorems 2.2.5 and 2.2.6 and the discussion after Example 2.2.5).

Proposition 2.2.8, which says that two functions f and g have the same limit as $x \to a$ if $f(x) = g(x)$ for all $x \neq a$ in some interval containing a, has a one-sided analog. It says that the one-sided limits are equal if f and g are equal on some interval on the appropriate side of a.

The Squeeze Theorem (Theorem 2.3.5) works for the one-sided limit of a function that lies between two other functions in an interval on one side of a. See Figure 2.4.5.

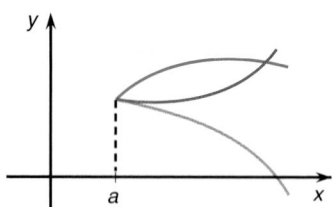

Figure 2.4.5

Theorem 2.3.3, concerning the nonnegativity of the limit of a nonnegative function, carries over to one-sided limits.

The one-sided limits of a power of x with noninteger exponent equal the two-sided limit (Theorem 2.3.1). We can also fill in the gap we left about limits of these powers at $x = 0$. Indeed,

$$\lim_{x \to 0^+} x^r = 0, \qquad \text{whenever } r > 0,$$

$$\lim_{x \to 0^-} x^r = 0, \qquad \text{whenever } r > 0 \text{ and}$$

$r = \frac{a}{b}$ is a rational number with b an odd number (i.e., whenever it makes sense to consider negative x).

The one slight complication arises with Theorem 2.3.2, the result about the limit of a composite function $g\big(f(x)\big)$. To get a one-sided version of it, we need to know that the one-sided limit of $f(x)$ is L and that the *two-sided* limit of $g(t)$ as $t \to L$ equals $g(L)$. It is not enough to know that $g(t)$ has a one-sided limit.

But apart from this minor subtlety, we can work with one-sided limits just as we do with two-sided limits. The main usefulness of one-sided limits comes from their relation to two-sided limits. If $\lim_{x \to a} f(x) = L$, then both one-sided limits exist, and they must also equal L. (If $f(x)$ approaches L as $x \to a$, then it must certainly approach L as $x \to a^+$ or $x \to a^-$.) The converse is also true: If both one-sided limits exist and have the same value, then the two-sided limit exists and equals the one-sided limits. See Figure 2.4.6. We record these facts for later use.

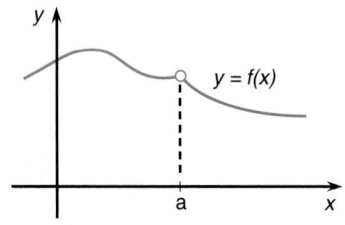

Figure 2.4.6

THEOREM 2.4.2

If

$$\lim_{x \to a} f(x) = L,$$

then $\lim_{x \to a^+} f(x)$ and $\lim_{x \to a^-} f(x)$ both exist, and

$$\lim_{x \to a^+} f(x) = \lim_{x \to a^-} f(x) = L.$$

Conversely, if the one-sided limits exist and are equal, that is,

$$\lim_{x \to a^+} f(x) = \lim_{x \to a^-} f(x) = L,$$

then the two-sided limit exists and equals the common value:

$$\lim_{x \to a} f(x) = L.$$

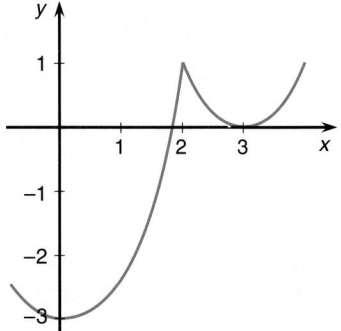

Figure 2.4.7

EXAMPLE 2

(i) Let $f(x)$ be defined as follows:

$$f(x) = \begin{cases} x^2 - 3, & \text{if } x \le 2, \\ x^2 - 6x + 9, & \text{if } x > 2. \end{cases}$$

We consider the one-sided limits of $f(x)$ as x approaches 2 from either side. The function is graphed in Figure 2.4.7.

First, when $x < 2$, $f(x)$ equals the polynomial $x^2 - 3$. By the one-sided version of Proposition 2.2.8, we know that $f(x)$ and $x^2 - 3$ will have the same limit from the left. Theorem 2.2.5(ii) tells us how to find the two-sided limit of the polynomial $x^2 - 3$, which will then also be the one-sided limit, by Theorem 2.4.2. All in all,

$$\lim_{x \to 2^-} f(x) = \lim_{x \to 2^-} (x^2 - 3) = \lim_{x \to 2} (x^2 - 3) = 1.$$

Similarly,

$$\lim_{x \to 2^+} f(x) = \lim_{x \to 2^+} (x^2 - 6x + 9) = \lim_{x \to 2} (x^2 - 6x + 9) = 1.$$

Using the above theorem again, we see that since the one-sided limits exist and are equal, the two-sided limit exists and equals them, or

$$\lim_{x \to 2} f(x) = 1.$$

(ii) If

$$g(x) = \begin{cases} -1, & \text{if } x < -1, \\ x^2, & \text{if } -1 < x < 1, \\ 2x - 1, & \text{if } x > 1, \end{cases}$$

then, reasoning as above we find that

$$\lim_{x \to -1^-} g(x) = \lim_{x \to -1^-} (-1) = -1, \qquad \lim_{x \to -1^+} g(x) = \lim_{x \to -1^+} (x^2) = 1,$$

$$\lim_{x \to 1^-} g(x) = \lim_{x \to 1^-} (x^2) = 1, \qquad \lim_{x \to 1^+} g(x) = \lim_{x \to 1^+} (2x - 1) = 1.$$

Since the one-sided limits at -1 are different, the two-sided limit there does not exist, but at $x = 1$, where the one-sided limits are both the same, the two-sided limit also exists, and

$$\lim_{x \to 1} g(x) = 1.$$

See Figure 2.4.8.

Figure 2.4.8

There are different ways to get a graphics calculator to plot functions that are defined by different formulas on different intervals. Here is one:

Consider the function

$$f(x) = \begin{cases} x^3 + 1, & \text{if } x < 0, \\ x^2, & \text{if } x > 0. \end{cases}$$

We begin by noting that $\frac{|x|}{x}$ equals 1 for $x > 0$ and -1 for $x < 0$. So $1 + \frac{|x|}{x}$ equals 2 and 0, respectively, and $\left(1 + \frac{|x|}{x}\right)/2$ is 1 when $x > 0$ and 0 when $x < 0$. Similarly, $\left(1 - \frac{|x|}{x}\right)/2$ is 0 when $x > 0$ and 1 when $x < 0$.

The function $f(x)$ given above can be expressed as

$$f(x) = (x^3 + 1)\frac{\left(1 - \frac{|x|}{x}\right)}{2} + x^2\frac{\left(1 + \frac{|x|}{x}\right)}{2}.$$

We plot the graph.

TI-85	SHARP EL9300
GRAPH F1$(y(x) =)$ (F1(x)^3 + 1) $\times$ (1− MATH-F1-F5(abs) EXIT (F1(x))÷ F1(x)) ÷ 2+ F1$(x)$$x^2$ $\times$ (1+ MATH-F1-F5(abs) EXIT (F1(x))÷ F1(x)) ÷ 2 M2(RANGE) (−)1.5 ▼ 1.5 ▼ ▼ (−)3 ▼ 2 F5(GRAPH)	⟋⟍ (x/θ/T a^b3 ▶ +1) $\times$ (1− MATH 1 x/θ/T ▶ ÷ x/θ/T) ÷ 2 + x/θ/T x^2 $\times$ (1+ MATH 1 x/θ/T ▶ ÷ x/θ/T) ÷ 2 RANGE −1.5 ENTER 1.5 ENTER 1 ENTER −3 ENTER 2 ENTER 1 ENTER ⟋⟍
CASIO f_x-7700GB/f_x-6300G	HP 48SX
Cls EXE Range −1.5 EXE 1.5 EXE 1 EXE −3 EXE 2 EXE 1 EXE EXE EXE EXE Graph (X x^y 3 + 1) $\times$ (1− Abs (X) ÷ X) ÷ 2 + X x^2 $\times$ (1+ Abs (X) ÷ X) ÷ 2 EXE	↰ PLOT PLOTR ERASE ATTN′ αY = () αX y^x 3 + 1▶ $\times$ () 1− MTH PARTS ABS αX ▶ ÷ αX ▶ ÷ 2 + αX y^x 2 $\times$ () 1+ MTH PARTS ABS αX ▶ ÷ αX ▶ ÷ 2 ↱ PLOT ↰ DRAW ERASE 1.5 +/− ENTER 1.5 XRNG 3 +/− ENTER 2 YRNG DRAW

How would you adapt this approach to plot the function

$$f(x) = \begin{cases} \cos x, & \text{if } x < 1, \\ x^2 - 2, & \text{if } x > 1? \end{cases}$$

Figure 2.4.9

Figure 2.4.10

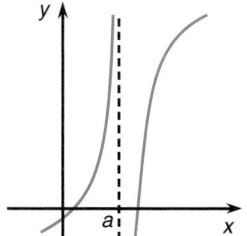

Figure 2.4.11

Infinite Limits

We have considered functions like $h(x) = \frac{1}{(x-1)^2}$, which does not have a limit as x approaches 1. This is because the denominator tends to 0, so the quotient becomes larger and larger, eventually getting bigger than any possible limit L. See Figure 2.4.9.

In a situation like this we say that $h(x)$ becomes infinitely large, or "tends to infinity," and we write $h(x) \to \infty$.

There are several things to be careful about. One is that it is not enough to say merely that a function keeps getting larger. For instance, the function $f(x) = 1 - |x|$, illustrated in Figure 2.4.10, keeps getting larger as x approaches 0 but certainly does not tend to infinity. To tend to infinity, a function must get larger and stay larger than any fixed number L, provided that x is sufficiently close to a (cf. the discussion of ordinary limits at the end of Section 2.3).

Another difficulty is that we must be careful about the sign of the function. If it is getting large in the *negative* direction, then it will tend to *negative infinity*. It frequently happens that a function will tend to infinity as x approaches a from one side and tend to negative infinity as x approaches a from the other side, and then we must talk about one-sided infinite limits. See Figure 2.4.11.

Finally, while we talk about infinite limits and will shortly begin using the usual sort of limit symbols for them, it is very important to bear in mind that infinite limits are *not* limits in the ordinary sense. When we described limits right at the beginning of Section 2.2, we said that there had to be a *number* L so that $f(x)$ approached L. Infinity and negative infinity are *not* numbers. So when we say that $f(x)$ tends to infinity, this means in particular that it does *not* have a limit in the ordinary sense.

This apparent confusion is not as bad as it first appears, and is justified by the usefulness of being able to talk about infinite limits. We just have to bear in mind that infinite limits, though perfectly legitimate, are different from ordinary limits.

DEFINITION 2.4.3

Infinite Limits

If $f(x)$ tends to infinity as $x \to a$, we say the "limit of $f(x)$ as x tends to a is **infinity**," and we write

$$\lim_{x \to a} f(x) = \infty.$$

Similarly we can talk about the limit being **negative infinity**, and we write

$$\lim_{x \to a} f(x) = -\infty.$$

One-sided limits can also be infinite, and the obvious meaning will be conveyed by

$$\lim_{x \to a^+} f(x) = \infty, \quad \lim_{x \to a^-} f(x) = -\infty, \text{ etc.}$$

In each of these situations it is also appropriate to say that $f(x)$ *does not* have a limit as $x \to a$ (or $x \to a^+$, etc.).

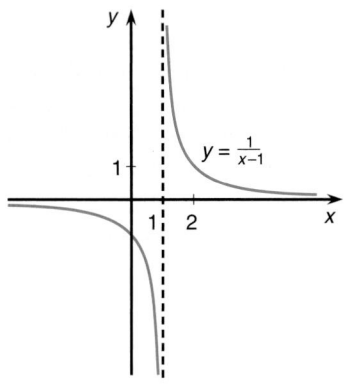

$y = \frac{1}{x-1}$

Figure 2.4.12

EXAMPLE 3

(i) Consider the function $f(x) = \frac{1}{x-1}$, shown in Figure 2.4.12. As $x \to 1$, we know that $(x-1) \to 0$. The complication is that if $x > 1$, then $x - 1$ is positive, so $\frac{1}{x-1}$ is positive and gets very large, but when $x < 1$, the function is negative and gets very large in the negative direction. In other words,

$$\lim_{x \to 1^+} \frac{1}{x-1} = \infty, \qquad \lim_{x \to 1^-} \frac{1}{x-1} = -\infty.$$

None of $\lim_{x \to 1} f(x), \lim_{x \to 1^+} f(x)$, or $\lim_{x \to 1^-} f(x)$ exists (i.e., the *ordinary* limits do not exist, even though some of the corresponding infinite ones do).

(ii) If $g(x) = \frac{x-1}{(x-2)^3(x-3)^2}$, then we know that $g(x)$ has (ordinary) limits as x approaches any number except 2 or 3. What happens at these points?

Suppose x is very close to 2. We may as well assume that it is bigger than 1 and less than 3, so $x - 1 > 0$, $x - 3 < 0$, and $(x-3)^2 > 0$. Now $(x-2)^3$ will be positive when $x > 2$ and negative when $x < 2$, and we can use all this to find the sign of $g(x)$. Indeed,

$$g(x) = \frac{x-1}{(x-2)^3(x-3)^2} = \frac{1}{(x-2)^3}\frac{x-1}{(x-3)^2}.$$

We have just seen that for x close to 2, the last quotient is always positive, and the previous one depends on the sign of $x - 2$. So for x near 2, $g(x)$ is positive when $x > 2$ and negative when $x < 2$.

Since the denominator of $g(x)$ approaches zero as x approaches 2 and the numerator approaches something other than zero (namely, 1), $g(x)$ will tend to infinity or negative infinity depending from which side x approaches 2:

$$\lim_{x \to 2^+} g(x) = \infty, \qquad \lim_{x \to 2^-} g(x) = -\infty.$$

Similarly, if x is near 3 (specifically if $x > 2$), then $\frac{x-1}{(x-2)^3} > 0$, and since $(x-3)^2 \geq 0$ for any x but tends to 0 as $x \to 3$, we write $g(x) = \frac{x-1}{(x-2)^3}\frac{1}{(x-3)^2}$ and see that

$$\lim_{x \to 3^+} g(x) = \infty, \qquad \lim_{x \to 3^-} g(x) = \infty.$$

Since the function tends to infinity as x approaches 3 from either side,

$$\lim_{x \to 3} g(x) = \infty.$$

In the discussion after Example 2.2.5, we saw that a rational function $\frac{p(x)}{q(x)}$ has a limit as $x \to a$ if $q(a) \neq 0$. If $q(a) = 0$, the situation is more complicated. If both $q(a)$ and $p(a)$ are zero, then $q(x)$ and $p(x)$ can both be divided by $x - a$, and the common factor can be canceled.

Repeating this process as many times as necessary, we may rewrite the function so that at most one of $q(a)$ and $p(a)$ is zero. If $q(a) \neq 0$, then the function has a limit, which we already know how to find.

The remaining case occurs if $q(a) = 0$ and $p(a) \neq 0$. Since $q(a) = 0$, we know by the Factor Theorem (Theorem 1.3.5) that $q(x)$ is divisible by $x - a$. We divide and write $q(x) = (x-a)R(x)$ for some polynomial $R(x)$. If $R(a) = 0$, we divide again, and so on until we can write $q(x) = (x-a)^n Q(x)$, with some positive integer n and some polynomial $Q(x)$ that satisfies $Q(a) \neq 0$.

We have divided out the largest possible power of $(x - a)$ from $q(x)$.

Then

$$\frac{p(x)}{q(x)} = \frac{1}{(x-a)^n} \frac{p(x)}{Q(x)}.$$

The last quotient, $\frac{p(x)}{Q(x)}$, is one in which neither numerator nor denominator is zero at $x = a$, so it has a limit, namely, $\frac{p(a)}{Q(a)}$, which is not zero.

The behavior of $\frac{1}{(x-a)^n}$ as x approaches a depends on whether n is even or odd. If n is even, then $(x-a)^n$ is a square and is always nonnegative, so $\frac{1}{(x-a)^n} \to \infty$. If n is odd, then $(x-a)^n$ will have the same sign as $x - a$, so

$$\lim_{x\to a^-} \frac{1}{(x-a)^n} = -\infty, \qquad \lim_{x\to a^+} \frac{1}{(x-a)^n} = \infty.$$

Finally, the original function $\frac{p(x)}{q(x)}$ is the product of $\frac{1}{(x-a)^n}$ with something that tends to the nonzero number $\frac{p(a)}{Q(a)}$. If this number is positive, then $\frac{p(x)}{q(x)}$ will tend to ∞ or $-\infty$ exactly as $\frac{1}{(x-a)^n}$ does. If the number $\frac{p(a)}{Q(a)}$ is negative, then $\frac{p(x)}{q(x)}$ will tend to $-\infty$ when $\frac{1}{(x-a)^n} \to \infty$, and to $+\infty$ when $\frac{1}{(x-a)^n} \to -\infty$.

Summary 2.4.4   Limits of Rational Functions

1. If $q(a) \neq 0$, then $\lim_{x\to a} \frac{p(x)}{q(x)} = \frac{p(a)}{q(a)}$.

2. If $p(a) = q(a) = 0$, then divide $p(x)$ and $q(x)$ by $(x - a)$. Repeat until $p(a) \neq 0$ and/or $q(a) \neq 0$.

3. If $\frac{p(x)}{q(x)}$ is a rational function and $q(a) = 0, p(a) \neq 0$, write

$$q(x) = (x-a)^n Q(x), \qquad \text{with } Q(a) \neq 0.$$

   Suppose $\frac{p(a)}{Q(a)} > 0$.

   (i) If n is even, then $\lim_{x\to a} \frac{p(x)}{q(x)} = \infty$.

   (ii) If n is odd, then $\lim_{x\to a^-} \frac{p(x)}{q(x)} = -\infty, \lim_{x\to a^+} \frac{p(x)}{q(x)} = +\infty$.

   If $\frac{p(a)}{Q(a)} < 0$, then the signs are all reversed, that is, $-\infty$ is changed to ∞ and vice versa.

These rules may seem complicated, but they are really quite easy when you bear in mind the behavior as $x \to a$ of $\frac{1}{(x-a)^n}$. Do not memorize rules like this; it is much better to figure out how they work so that memorizing them becomes unnecessary.

EXAMPLE 4   Discuss the limiting behavior of $\frac{x^3-4x^2+5x+1}{x^3-x^2-x+1}$ as x approaches 1 and -1.

Solution   Note that the numerator is not zero when $x = 1$ or $x = -1$ and that the denominator is zero at both these points.

Dividing the denominator by $(x+1)$, we get $x^3 - x^2 - x + 1 = (x+1)(x^2 - 2x + 1) = (x+1)(x-1)^2$. To discuss the limiting behavior at $x = -1$, we write

$$\frac{x^3 - 4x^2 + 5x + 1}{x^3 - x^2 - x + 1} = \frac{1}{x+1} \frac{x^3 - 4x^2 + 5x + 1}{(x-1)^2}.$$

Evaluating the last quotient at $x = -1$, we find that it equals $\frac{-1-4-5+1}{4} = -\frac{9}{4}$. In particular, it is negative. The exponent n on $(x+1)$ is $n = 1$. As above, we conclude that

$$\lim_{x \to -1^-} \frac{p(x)}{q(x)} = \infty, \qquad \lim_{x \to -1^+} \frac{p(x)}{q(x)} = -\infty.$$

To discuss the behavior at $x = 1$, we write

$$\frac{x^3 - 4x^2 + 5x + 1}{x^3 - x^2 - x + 1} = \frac{1}{(x-1)^2} \frac{x^3 - 4x^2 + 5x + 1}{x+1}.$$

Evaluating the last quotient at $x = 1$, we find that it equals $\frac{1-4+5+1}{1+1} = \frac{3}{2} > 0$. In this case the exponent n is 2, so

$$\lim_{x \to 1} \frac{p(x)}{q(x)} = \infty.$$

Limits at Infinity

We have considered what happens to $f(x)$ as x approaches some value a, and have found that $f(x)$ sometimes approaches infinity. Another possibility is to consider what happens to $f(x)$ as x tends to infinity.

EXAMPLE 5   Discuss the behavior as x tends to infinity of (i) $f(x) = \frac{1}{1+x^2}$; (ii) $g(x) = \frac{2x^3 - 2x + 3}{x^3 + 1}$.

Solution

(i) As $x \to \infty$, $1 + x^2$ gets very large, so $\frac{1}{1+x^2}$ gets very small and approaches the limit 0. We write

$$\lim_{x \to \infty} \frac{1}{1 + x^2} = 0.$$

The same argument shows that

$$\lim_{x \to -\infty} \frac{1}{1 + x^2} = 0.$$

(ii) Notice that as $x \to \infty$, since x gets very large, $\frac{1}{x} \to 0$, so $\frac{1}{x^2} \to 0$, $\frac{1}{x^3} \to 0$, etc. Dividing both numerator and denominator of $g(x)$ by x^3, we get

$$g(x) = \frac{2x^3 - 2x + 3}{x^3 + 1} = \frac{2 - \frac{2}{x^2} + \frac{3}{x^3}}{1 + \frac{1}{x^3}}.$$

By the remarks above, the numerator tends to $2 - 0 + 0 = 2$ and the denominator tends to $1 + 0 = 1$ as $x \to \infty$, so the quotient tends to $\frac{2}{1} = 2$. In other words,

$$\lim_{x\to\infty} g(x) = \lim_{x\to\infty} \frac{2x^3 - 2x + 3}{x^3 + 1} = 2.$$

This example illustrates the technique to use when finding the limit at infinity (or at negative infinity) of any rational function. Factor out the highest power of x that occurs in the denominator. In the simplest case, when this power is the same for both numerator and denominator, we can cancel them out. In general, cancel as much as possible and then evaluate the limit, remembering that $\frac{1}{x}$ or any positive power of $\frac{1}{x}$ must tend to 0 as $x \to \infty$ or $x \to -\infty$.

EXAMPLE 6   Find the following (possibly infinite) limits:

(i) $\lim_{x\to\infty} \frac{x^4 - 3x^3 + 2x - 1}{2x^4 + 2x^2 + 2}$;   (ii) $\lim_{x\to-\infty} \frac{3x^3 - x - 2}{2x^3 + 4}$;   (iii) $\lim_{x\to-\infty} \frac{2x^3 + x^2 - 1}{x^4 + 1}$;

(iv) $\lim_{x\to\infty} \frac{x^3 - x + 2}{x^2 - 1}$; (v) $\lim_{x\to-\infty} (3x^3 + 2x^2 - x + 1)$.

Solution

(i) The highest power of x that occurs in the denominator is x^4, so we divide by it and write

$$\lim_{x\to\infty} \frac{x^4 - 3x^3 + 2x - 1}{2x^4 + 2x^2 + 2} = \lim_{x\to\infty} \frac{1 - \frac{3}{x} + \frac{2}{x^3} - \frac{1}{x^4}}{2 + \frac{2}{x^2} + \frac{2}{x^4}} = \frac{1 - 0 + 0 - 0}{2 + 0 + 0} = \frac{1}{2}.$$

(ii) The highest power is x^3, so

$$\lim_{x\to-\infty} \frac{3x^3 - x - 2}{2x^3 + 4} = \lim_{x\to-\infty} \frac{3 - \frac{1}{x^2} - \frac{2}{x^3}}{2 + \frac{4}{x^3}} = \frac{3}{2}.$$

(iii) The highest power is x^4, and

$$\lim_{x\to-\infty} \frac{2x^3 + x^2 - 1}{x^4 + 1} = \lim_{x\to-\infty} \frac{\frac{2}{x} + \frac{1}{x^2} - \frac{1}{x^4}}{1 + \frac{1}{x^4}} = \frac{0 + 0 - 0}{1 + 0} = 0.$$

(iv) Dividing by x^2, we find

$$\lim_{x\to\infty} \frac{x^3 - x + 2}{x^2 - 1} = \lim_{x\to\infty} \frac{x - \frac{1}{x} + \frac{2}{x^2}}{1 - \frac{1}{x^2}}.$$

The x in the numerator tends to ∞ and $-\frac{1}{x} + \frac{2}{x^2} \to 0$, so the numerator tends to ∞. The denominator tends to 1, so the rational function is something that tends to ∞ over something that tends to 1. Accordingly,

$$\lim_{x\to\infty} \frac{x^3 - x + 2}{x^2 - 1} = \infty.$$

Similar considerations allow us to see that

$$\lim_{x\to-\infty} \frac{x^3 - x + 2}{x^2 - 1} = -\infty.$$

(v) In this case the function $3x^3 + 2x^2 - x + 1$ is a polynomial; there is no denominator (or, if you prefer, the denominator is the constant 1). The highest power of x occurring in the numerator is x^3, so we can write $3x^3 + 2x^2 - x + 1 = x^3\left(3 + \frac{2}{x} - \frac{1}{x^2} + \frac{1}{x^3}\right)$. As $x \to -\infty$, the quantity in parentheses tends to $(3 + 0 - 0 + 0) = 3$; the factor x^3 tends to $-\infty$, so the product tends to $-\infty$.

Summary 2.4.5

Limits of Rational Functions at $\pm\infty$

To find the limits at $\pm\infty$ of the rational function

$$\frac{p(x)}{q(x)} = \frac{a_m x^m + a_{m-1} x^{m-1} + \ldots + a_1 x + a_0}{b_n x^n + b_{n-1} x^{n-1} + \ldots + b_1 x + b_0},$$

we factor out x^n, the highest power of x occurring in the denominator. We can write

$$\frac{p(x)}{q(x)} = \frac{x^m \left(a_m + \frac{a_{m-1}}{x} + \ldots + \frac{a_1}{x^{m-1}} + \frac{a_0}{x^m}\right)}{x^n \left(b_n + \frac{b_{n-1}}{x} + \ldots + \frac{b_1}{x^{n-1}} + \frac{b_0}{x^n}\right)}$$

$$= \frac{x^{m-n} \left(a_m + \frac{a_{m-1}}{x} + \ldots + \frac{a_1}{x^{m-1}} + \frac{a_0}{x^m}\right)}{b_n + \frac{b_{n-1}}{x} + \ldots + \frac{b_1}{x^{n-1}} + \frac{b_0}{x^n}}.$$

First, if $m = n$, then $x^{m-n} = x^0 = 1$ and

$$\lim_{x \to \infty} \frac{p(x)}{q(x)} = \lim_{x \to -\infty} \frac{p(x)}{q(x)} = \frac{a_m}{b_m}.$$

Second, if $m < n$, then $m - n < 0$ and $x^{m-n} \to 0$. Since the rest of the formula tends to $\frac{a_m}{b_n}$ and the original rational function is the product of these two factors,

$$\lim_{x \to \infty} \frac{p(x)}{q(x)} = \lim_{x \to -\infty} \frac{p(x)}{q(x)} = 0.$$

Third, if $m > n$, x^{m-n} will get large and the other factors will still tend to $\frac{a_m}{b_n}$. Specifically, if $m - n$ is even, then $x^{m-n} \to \infty$. If $m - n$ is odd, then $x^{m-n} \to \infty$ as $x \to \infty$, and $x^{m-n} \to -\infty$ as $x \to -\infty$. So if $\frac{a_m}{b_n} > 0$, then

$$\lim_{x \to \infty} \frac{p(x)}{q(x)} = \infty, \qquad \lim_{x \to -\infty} \frac{p(x)}{q(x)} = \begin{cases} \infty, & \text{if } m - n \text{ is even,} \\ -\infty, & \text{if } m - n \text{ is odd.} \end{cases}$$

If $\frac{a_m}{b_n} < 0$, then

$$\lim_{x \to \infty} \frac{p(x)}{q(x)} = -\infty, \qquad \lim_{x \to -\infty} \frac{p(x)}{q(x)} = \begin{cases} -\infty, & \text{if } m - n \text{ is even,} \\ \infty, & \text{if } m - n \text{ is odd.} \end{cases}$$

Once again, these rules are too complicated to memorize. After doing a few exercises you will get the idea and be able to figure out what you need without having to memorize the details.

Limits at infinity behave pretty much like one-sided limits (after all, x approaches ∞ only from one side, namely, from below, and approaches $-\infty$ only from above). The results we have recognized for one-sided limits (see the discussion after Example 1) carry over to limits at infinity without much change.

We have seen how to evaluate limits of rational functions, and in a sense this includes polynomials (rational functions whose denominator happens to be 1). Of course the limit of any nonconstant polynomial as $x \to \infty$ or $x \to -\infty$ is infinite (i.e., either ∞ or $-\infty$).

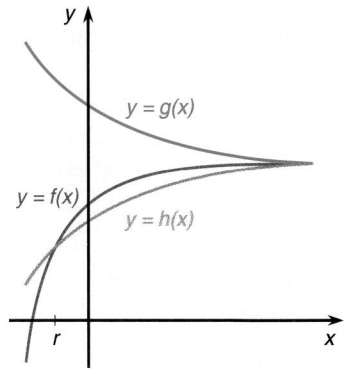

Figure 2.4.13

In keeping with the remark about the one-sided version of Theorem 2.3.2, about composite functions, there is a version in which $x \to a$ is replaced by $x \to \infty$ or $x \to -\infty$, that is, $L = \infty$ or $-\infty$. However, M must be an ordinary finite limit.

Also, to make sense of versions at infinity of Proposition 2.2.8, about two functions being equal on an interval having the same limit, and the Squeeze Theorem, (Theorem 2.3.5), we have to find a way of expressing the idea of an interval on one side of ∞ (or $-\infty$). In fact, this is easy to do: An interval to the left of ∞ will be an interval of the form (r, ∞), meaning $\{x : r < x\}$. An interval to the right of $-\infty$ will be an interval of the form $(-\infty, s)$, meaning $\{x : x < s\}$. Replacing the intervals in Theorems 2.2.8 and 2.3.5 with this kind of "intervals," we obtain versions of them for limits at infinity. See Figure 2.4.13.

The rules for infinite limits, that is, limits in which the value of the limit is ∞ or $-\infty$, are less straightforward. Since they are *not* limits in the ordinary sense, we should not expect them to obey the ordinary rules. In particular, we cannot perform ordinary arithmetic with infinity, so even Theorems 2.2.3 and 2.2.6 about sums, products, and quotients of limits don't necessarily make sense when infinite limits are involved.

EXAMPLE 7   Let

$$f(x) = \frac{2}{(x-1)^2}, \qquad g(x) = \frac{-1}{(x-1)^2},$$

$$h(x) = \frac{-3}{(x-1)^2}, \qquad k(x) = 2x - \frac{2}{(x-1)^2}.$$

We see immediately that

$$\lim_{x \to 1} f(x) = \infty, \qquad \lim_{x \to 1} g(x) = -\infty,$$
$$\lim_{x \to 1} h(x) = -\infty, \qquad \lim_{x \to 1} k(x) = -\infty.$$

Also

$$\lim_{x \to 1} \big(f(x) + g(x)\big) = \lim_{x \to 1} \frac{1}{(x-1)^2} = \infty,$$

$$\lim_{x \to 1} \big(f(x) + h(x)\big) = \lim_{x \to 1} \frac{-1}{(x-1)^2} = -\infty,$$

$$\lim_{x \to 1} \big(f(x) + k(x)\big) = \lim_{x \to 1} 2x = 2.$$

This illustrates that adding functions with infinite limits gives unpredictable results: Adding two functions whose limits are ∞ and $-\infty$ can result in a function whose limit is ∞, $-\infty$ or even a finite limit.

There are some things we can say with assurance, however.

THEOREM 2.4.6

Suppose $f(x)$ and $g(x)$ are two functions. If $\lim_{x \to a} f(x) = \infty$ and $\lim_{x \to a} g(x)$ is either finite or ∞ (but not $-\infty$), then

$$\lim_{x \to a} \big(f(x) + g(x)\big) = \infty.$$

A similar result holds if one limit is $-\infty$ and the other is finite or $-\infty$; the limit of the sum is then $-\infty$.

If one limit is ∞ and the other is $-\infty$, then the limit of the sum is unpredictable and depends on the specific functions.

If $\lim_{x \to a} f(x) = \infty$ or $-\infty$, and $\lim_{x \to a} g(x) \neq 0$, (i.e., the limit of $g(x)$ can be finite or infinite but not zero), then the limit of the product $f(x)g(x)$ is either ∞ or $-\infty$. To decide which, we merely consider the signs of the two limits, as we would for ordinary products:

$$\lim_{x \to a} f(x)g(x) = \begin{cases} \infty, & \text{if the signs of the limits are the same,} \\ -\infty, & \text{if the signs of the limits are different.} \end{cases}$$

(Here we understand that ∞ has a positive sign and $-\infty$ has a negative sign.)

If one of the limits is zero and the other is ∞ or $-\infty$, then the limit of the product is unpredictable: It could be ∞, $-\infty$, 0, or any nonzero finite limit. It also might not exist at all.

All the above remarks also apply to one-sided limits and to limits at infinity (i.e., when $a = \infty$ or $a = -\infty$).

Exercises 2.4

I

Evaluate each of the following finite or infinite limits.

1. $\lim_{x \to 2+} \frac{x^3 + 3}{x - 2}$

2. $\lim_{t \to 2-} \frac{t^3 + 3}{t - 2}$

3. $\lim_{x \to 0+} \frac{x}{x}$

4. $\lim_{x \to 0-} \frac{x}{x}$

5. $\lim_{u \to \infty} \frac{u^3 - u^2 + 3}{2u^3 - u + 1}$

6. $\lim_{x \to -\infty} \frac{x^2 - 1}{x^3 + 1}$

7. $\lim_{x \to 0+} \frac{x^3 - 3x^2 + x}{x^3 - 2x^2}$

8. $\lim_{x \to 2+} \frac{x^2 - 2x + 3}{x^2 - 4}$

9. $\lim_{x \to \infty} (x^3 - 3x^2 + x - 1)$

10. $\lim_{x \to -\infty} \frac{x^3 - 2x + 1}{2x^2 + 2x + 3}$

Decide whether each of the following limits exists, and evaluate those that do, explaining your reasons. For those limits that do not exist, distinguish those that tend to ∞ or $-\infty$ and those that do neither.

11. $\lim_{r \to 3} \frac{r+6}{r^2 - 9}$

12. $\lim_{u \to \infty} \sqrt{u^2 + 1}$

13. $\lim_{x \to 0+} \frac{x^2 + 1}{x}$

14. $\lim_{x \to 0-} \frac{x^2 + 1}{x}$

15. $\lim_{x \to 1+} \frac{x^2 + 3x + 1}{x^2 - 1}$

16. $\lim_{x \to 1-} \frac{x^3 + 1}{x^2 - 1}$

17. $\lim_{t \to 1} \frac{1}{\sqrt{(t-1)^2}}$

18. $\lim_{y \to 0+} y^{-1/3}$

19. $\lim_{x \to 0+} \frac{1}{x^2} \sin(1/x)$

20. $\lim_{x \to 0-} \frac{1}{x} \sin^2(1/x)$

II

*21. Find $\lim_{x \to 0+} \sqrt{x}$.

*22. Some of the functions in Exercises 31–40 of Exercises 2.2 do not have limits at the specified points. Find the one-sided and/or two-sided infinite limits at these points if they exist.

23. Find $\lim_{x \to \infty} (2x^4 - 3x^3 + 7x - 1)$.

24. Find $\lim_{x \to -\infty} (4 - 5x + 2x^2 - 5x^3)$.

*25. Assume that $f(x)$ has one-sided limits at every value of x.
  (i) Evaluate $\lim_{x \to a+} g(x), \lim_{x \to a-} g(x)$, if $g(x) = f(x + c)$.
  (ii) Evaluate $\lim_{x \to a+} h(x), \lim_{x \to a-} h(x)$, if $h(x) = f(c - x)$.

*26. The functions in Exercises 41–50 in Exercises 2.2 do not have limits at certain points. Find the one-sided and/or two-sided infinite limits at these points.

27. Let $f(x)$ be the function defined in Example 2.2.8(ii). Find (i) $\lim_{x \to -1-} f(x)$, (ii) $\lim_{x \to -1+} f(x)$, (iii) $\lim_{x \to 1-} f(x)$, (iv) $\lim_{x \to 1+} f(x)$.

*28. Assume that $p(x)$ and $q(x)$ are polynomials, and define

$$f(x) = \begin{cases} p(x), & \text{if } x < a, \\ q(x), & \text{if } x > a, \\ A, & \text{if } x = a. \end{cases}$$

Evaluate (i) $\lim_{x \to a+} f(x)$, (ii) $\lim_{x \to a-} f(x)$, (iii) $\lim_{x \to b} f(x)$ for any $b \neq a$.

*29. Find examples of functions with the following behavior at some point a.

(i) $\lim_{x \to a^+} f(x) = +\infty, \lim_{x \to a^-} f(x) = -\infty.$

(ii) $\lim_{x \to a} |f(x)| = L$, but $f(x)$ has no limit as $x \to a$.

(iii) $\lim_{x \to a} |f(x)| = \infty$, but $f(x)$ has no limit, including ∞ or $-\infty$, as $x \to a$.

*30. Find examples of functions $f(x)$ and $g(x)$ and a number a so that

$$\lim_{x \to a} f(x) = 0, \qquad \lim_{x \to a} g(x) = \infty,$$

but for which $\lim_{x \to a} f(x)g(x)$ equals (i) 0, (ii) ∞, (iii) $-\infty$, (iv) 3, (v) -2.

For each of the following functions, describe the one-sided limits from both sides at the specified point(s) $x = a$ (including possible infinite one-sided limits). Use a graphics calculator or a computer to sketch the graph of the function and confirm your description.

31. $f(x) = \frac{x}{|x|}, \quad a = 0$

32. $f(x) = \frac{1}{x-1}, \quad a = 1$

33. $f(x) = \frac{x}{(x-1)^2}, \quad a = 1$

34. $f(x) = \frac{1}{x^3(x-1)^2}, \quad a = 1$

35. $f(x) = \frac{x}{|x-1|}, \quad a = 1$

36. $f(x) = \frac{2x-1}{x^2-1}, \quad a = \pm 1$

37. Let $f(x) = \frac{x^3 - 3x + 2}{(x^2 - 1)x^2}$. At each point where $f(x)$ is not defined, predict the one-sided limits (including possible infinite one-sided limits). Use a graphics calculator or a computer to sketch the graph of the function and confirm your results.

38. Let $g(x) = \frac{x^3 + x^2 - 6x}{(x^2 - 4)^2 (x-2)^2}$. At each point where $f(x)$ is not defined, predict the one-sided limits (including possible infinite one-sided limits). Use a graphics calculator or a computer to sketch the graph of the function and confirm your results.

For each of the following two functions it is not easy to solve for the points at which the denominator is zero. (i) Use your computer or graphics calculator to find approximate values for the zeroes of the denominator. Zoom to higher magnification to find the values to three decimal places. Also note the sign of the denominator on either side of each zero. (ii) Predict the one-sided behavior of each function near each of the points you found in (i). Graph the function and confirm your predictions.

39. $F(x) = \frac{x^2 - 2x + 3}{x^3 - 3x^2 + 2x - 1}$

40. $G(x) = \frac{3x - 5}{x^4 - 4x^2 + x + 1}$

2.5 Continuity

POLYNOMIALS

RATIONAL FUNCTIONS

JUMP DISCONTINUITY

REMOVABLE DISCONTINUITY

INFINITE DISCONTINUITY

INTERMEDIATE VALUE THEOREM

CLOSED INTERVALS

In this section we discuss what it means for a function to be "continuous." In terms of the picture, it means that the graph of the function can be drawn "continuously," that is, without lifting the pencil from the page.

A function that "jumps" or has a "hole" in it is not continuous (see Figure 2.5.1(i)).

Unfortunately, this simple notion is not easy to express in a mathematically precise way. It turns out that the best way to do it uses limits. When a function is continuous, it should have a limit as $x \to a$ for any value of a in the domain. Moreover, to avoid "jumps," the value of the function at $x = a$ should equal the limit. (It should be clear from Figure 2.5.1(ii) that if the value $f(a)$ does not equal the limit $\lim_{x \to a} f(x)$, then the function is not continuous at $x = a$.)

Figure 2.5.1(i)

Figure 2.5.1(ii)

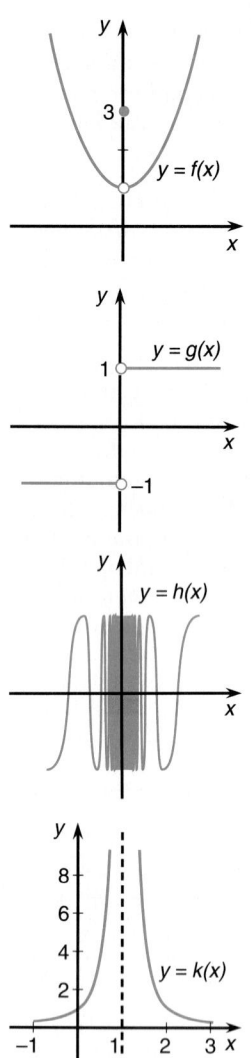

Figure 2.5.2

In Theorem 2.2.5 we saw that polynomials $f(x)$ have the "right" limit as x approaches a, namely, $f(a)$. Rational functions $f(x)$ also have the "right" limit, at least when $f(a)$ makes sense (cf. Theorem 2.2.6). But it is easy to think of functions that do not.

EXAMPLE 1   If

$$f(x) = \begin{cases} x^2 + 1, & \text{whenever } x \neq 0, \\ 3, & \text{when } x = 0, \end{cases}$$

$$g(x) = \frac{x}{|x|},$$

$$h(x) = \sin(1/x),$$

$$k(x) = \frac{1}{(x-1)^2},$$

then we see immediately that

$$\lim_{x \to 0} f(x) = 1 \neq f(0),$$

$$\lim_{x \to 0^+} g(x) = 1 \neq -1$$

$$= \lim_{x \to 0^-} g(x),$$

and $h(x) = \sin(1/x)$ does not have a limit, nor even one-sided limits, as $x \to 0$. The function $k(x)$ does not have a (finite) limit as $x \to 1$, but

$$\lim_{x \to 1} k(x) = \infty.$$

See Figure 2.5.2.

The function $f(x)$ has a limit at $x = 0$, but f was deliberately changed to make it have the "wrong" value. The function $g(x)$ has different one-sided limits and jumps from one to the other, and $h(x)$ has no limit at all. Each of these types of behavior is unlike what we know happens for polynomials and rational functions. The function $k(x)$ is a rational function, but $x = 1$ is a point where its denominator is zero, and $k(x)$ does not have a (finite) limit as $x \to 1$.

DEFINITION 2.5.1

A function $f(x)$ is **continuous at** a if $f(x)$ has a limit as x approaches a and

$$\lim_{x \to a} f(x) = f(a).$$

If $a < b$, we say "$f(x)$ is continuous on the open interval $(a, b) = \{x : a < x < b\}$" to mean that $f(x)$ is continuous at every point of the interval.

If $a < b$, we say "$f(x)$ is continuous on the closed interval $[a, b] = \{x : a \leq x \leq b\}$" to mean that $f(x)$ is continuous at every point of the open interval $(a, b) = \{x : a < x < b\}$ and "continuous from the right" at $x = a$ and "continuous from the left" at $x = b$, that is,

$$f(a) = \lim_{x \to a^+} f(x), \qquad f(b) = \lim_{x \to b^-} f(x).$$

A function $f(x)$ is "continuous" if it is continuous at every $a \in \mathbb{R}$.

EXAMPLE 2

(i) Every polynomial is continuous (by Theorem 2.2.5).

(ii) Every rational function is continuous except at the points where its denominator vanishes (by Theorem 2.2.6).

(iii) $F(x) = |x|$ is continuous.

(iv) The function $\frac{x}{|x|}$ is not continuous at $x = 0$ but is continuous everywhere else.

(v) $\sin(1/x)$ is not continuous at $x = 0$.

DEFINITION 2.5.2

(i) The function $f(x)$ has a **removable discontinuity** at $x = a$ if $\lim_{x \to a} f(x)$ exists and $f(a)$ either does not equal the limit or is not defined. The discontinuity can be "removed" by changing $f(x)$ at $x = a$ so that $f(a) = \lim_{x \to a} f(x)$.

(ii) The function $f(x)$ has a **jump discontinuity** at $x = a$ if both one-sided limits exist but are unequal, that is, $\lim_{x \to a^+} f(x) \neq \lim_{x \to a^-} f(x)$.

(iii) The function $f(x)$ has an **infinite discontinuity** at $x = a$ if at least one of the one-sided limits $\lim_{x \to a^+} f(x)$, $\lim_{x \to a^-} f(x)$ is ∞ or $-\infty$.

Notice that a function can fail to be continuous at $x = a$ without having any of these three types of discontinuity.

EXAMPLE 3 In Example 1 the function f has a removable discontinuity at $x = 0$, the function $g(x) = \frac{x}{|x|}$ has a jump discontinuity there, and the function $k(x) = \frac{1}{(x-1)^2}$ has an infinite discontinuity at $x = 1$. The function $h(x) = \sin\left(\frac{1}{x}\right)$ is not continuous at $x = 0$, but it does not have a removable or a jump or an infinite discontinuity. See Figure 2.5.2.

THEOREM 2.5.3

Suppose $f(x)$ and $g(x)$ are continuous at $x = a$. Then so are

(i) $f(x) + g(x)$,

(ii) $cf(x)$ for any $c \in \mathbb{R}$,

(iii) $f(x)g(x)$,

(iv) $\frac{f(x)}{g(x)}$, provided $g(a) \neq 0$.

PROOF These are immediate consequences of the corresponding facts about limits (Theorems 2.2.3 and 2.2.6).

THEOREM 2.5.4

Suppose $f(x)$ is continuous at $x = a$ and $g(x)$ is continuous at $x = f(a)$. Define the composite function $h(x)$ by

$$h(x) = g\big(f(x)\big).$$

Then $h(x)$ is continuous at $x = a$.

PROOF   This is also an immediate consequence of the corresponding theorem about limits of composite functions, Theorem 2.3.2.   ◢

EXAMPLE 4   Find the points where each of the following functions is continuous: (i) $h(x) = \sqrt{x^2 + x}$; (ii) $k(x) = \left(\frac{x^2-16}{x+2}\right)^{-1/3}$.

Solution

(i)   We apply the theorem with $f(x) = x^2 + x$ and $g(x) = \sqrt{x}$. Since $f(x)$, being a polynomial, is continuous everywhere and $g(x) = \sqrt{x}$ is continuous whenever it is defined, that is, whenever $x \geq 0$, we see that the composite function is continuous whenever it is defined. And it is defined whenever the quantity inside the square root sign is nonnegative, that is, when $x^2 + x \geq 0$.

Drawing a sketch of the parabola $y = x^2 + x$ in Figure 2.5.3, we see that it is negative precisely when $-1 < x < 0$, which means that the composite function $h(x)$ is defined everywhere else. So $h(x) = \sqrt{x^2 + x}$ is defined and continuous when $x \leq -1$ and when $x \geq 0$. Notice that the continuity at $x = -1$ and $x = 0$ is only one-sided; $h(x)$ is continuous "from the left" at $x = -1$ and "from the right" at $x = 0$.

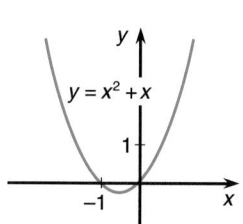

$y = x^2 + x$

Figure 2.5.3

(ii)   Here we write $k(x) = g\big(f(x)\big)$ with $f(x) = \frac{x^2-16}{x+2}$ and $g(x) = x^{-1/3}$. Now $g(x)$ is defined for every x except $x = 0$, and $f(x)$ is defined for every x except $x = -2$. So $k(x) = g\big(f(x)\big)$ will be defined for every value of x except $x = -2$ and any values of x where $f(x) = 0$. It is easy to see that $f(x) = 0$ precisely when $x = 4, -4$, so $k(x)$ is defined except at $x = -2, 4, -4$.

Moreover, $f(x)$ is continuous at every point except $x = -2$, and $g(x)$ is continuous everywhere except $x = 0$, so the composite function will be continuous at every x except $x = -2, 4, -4$.   ◢

Up to now, we have had several results that dealt with open intervals (a, b), that is, the set of all points between a and b but not including a or b. Now we must also consider closed intervals, that is, intervals that include their endpoints. We write

$$[a, b] = \{x : a \leq x \leq b\}.$$

There are three important and profound facts about the behavior of continuous functions on closed intervals. To prove them requires a subtle understanding of the properties of the real numbers. We shall content ourselves instead with stating the results, illustrating them with examples, and trying to explore their significance.

THEOREM 2.5.5

Intermediate Value Theorem

Suppose $a < b$ and $f(x)$ is continuous on the closed interval $[a, b]$. Suppose k is a number that lies between the values of $f(x)$ at $x = a$ and $x = b$ (i.e., $f(a) < k < f(b)$ or $f(a) > k > f(b)$). Then there is a number c between a and b (i.e., $a < c < b$) such that

$$f(c) = k.$$

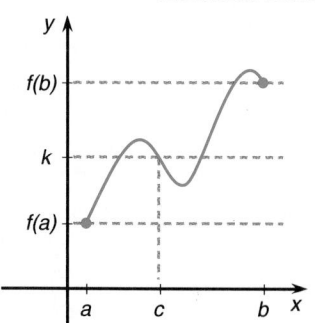

Figure 2.5.4

The theorem is illustrated in Figure 2.5.4. The number k lies between the values of f at a and b (so k is an "intermediate value"). The graph "can't get from $f(a)$ to $f(b)$ without passing through k." The figure also illustrates that there may be more than one possible choice of c.

THEOREM 2.5.6

> Suppose $a < b$ and $f(x)$ is continuous on $[a, b]$. Then $f(x)$ is "bounded" on $[a, b]$, that is, there is a constant M so that $|f(x)| \le M$, for all $x \in [a, b]$.

Figure 2.5.5(i) illustrates the theorem, showing that the graph of the function $f(x)$ between a and b doesn't get bigger than M (nor smaller than $-M$). We also see that the result is false if the interval is not closed: The function $y = \frac{1}{x}$ is continuous on the open interval $(0, 1)$ but is not bounded (as $x \to 0^+, \frac{1}{x} \to \infty$). See Figure 2.5.5(ii). The idea of the theorem is that if $f(x)$ is continuous on $[a, b]$, it is "tied down" at both ends and can't get too big in between.

Figure 2.5.5(i)

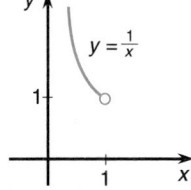

Figure 2.5.5(ii)

THEOREM 2.5.7

> **Extreme Value Theorem**
>
> Suppose $a < b$ and $f(x)$ is continuous on $[a, b]$. Then there is a $c \in [a, b]$ so that $f(c) \ge f(x)$, for all $x \in [a, b]$. There is also a $d \in [a, b]$ so that $f(d) \le f(x)$, for all $x \in [a, b]$.

The theorem says that $f(c)$ is the "maximum value" of $f(x)$ on $[a, b]$, and $f(d)$ is the "minimum." The theorem can be abbreviated by saying "$f(x)$ achieves its maximum (minimum) value at some point of $[a, b]$."

For the function $f(x)$ illustrated in Figure 2.5.6(i), the minimum occurs at $d = a$, and there are two possible choices for c (corresponding to the two "humps" in the graph). Figure 2.5.6(ii) shows that for this theorem too the interval must be closed: The function $y = x^2$ is continuous on $(0, 1)$ but has neither a maximum nor a minimum there (the maximum and minimum are at the "missing" endpoints $x = 1$ and $x = 0$).

Theorems 2.5.5 and 2.5.7 both assert the existence of a value of x at which $f(x)$ has a special value. This can be extremely useful, but the shortcoming of these results is that neither gives us any idea at all about how to find the x in question. In fact, these are both very difficult problems.

Figure 2.5.6(i)

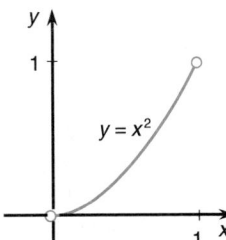

Figure 2.5.6(ii)

EXAMPLE 5

(i) Show that $f(x) = x^4 + 2x^3 - 2x^2 + x + 2$ has a zero between $x = -3$ and $x = -2$.

(ii) Find consecutive integers n and $n + 1$ so that $g(x) = x - \sqrt{x} - 10$ has a zero between n and $n + 1$.

Solution

(i) We evaluate $f(-3) = (-3)^4 + 2(-3)^3 - 2(-3)^2 - 3 + 2 = 81 - 54 - 18 - 3 + 2 = 8$ and $f(-2) = (-2)^4 + 2(-2)^3 - 2(-2)^2 - 2 + 2 = 16 - 16 - 8 - 2 + 2 = -8$. Since $f(x)$ is positive at $x = -3$ and negative at $x = -2$, the Intermediate Value Theorem says that it must have a zero between these points.

(ii) We evaluate $g(x)$ at various integers x: $g(0) = -10, g(1) = 1 - 1 - 10 = -10$, $g(2) = 2 - \sqrt{2} - 10 = -\sqrt{2} - 8$. Each of these numbers is negative.

We can continue in this way, but let us observe that if we write $g(x) = -\sqrt{x} - (10 - x)$, then it will certainly be negative if $10 - x$ is positive, that is, if $x < 10$. Since we are looking for an integer at which g is positive, we may as well start with $x = 10$.

Now $g(10) = -\sqrt{10} + 10 - 10 = -\sqrt{10} < 0$, $g(11) = -\sqrt{11} + 11 - 10 = 1 - \sqrt{11}$. This is also negative, since $\sqrt{11} > 1$. With $x = 12$ we find that $g(12) = 2 - \sqrt{12}$. Since $\sqrt{12} > \sqrt{9} = 3$, we see that $g(12) < 0$; similarly, $g(13) = 3 - \sqrt{13}$, and this is also negative, since $\sqrt{13} > 3$. The next value is $g(14) = 4 - \sqrt{14}$, and this is positive, since $\sqrt{14} < \sqrt{16} = 4$.

By the Intermediate Value Theorem we conclude that there is a zero between $x = 13$ and $x = 14$. ◢

On the facing page, we use a graphics calculator for a similar type of problem.

Techniques for finding the points at which a function achieves its maximum and minimum values are an important part of calculus, and we shall explore them quite fully later on.

For the present we must content ourselves with the observation that it is sometimes not too difficult to find explicit bounds for a function on a closed interval, as promised by Theorem 2.5.6.

EXAMPLE 6   Find a bound for each of the following functions on the specified interval.

(i) $f(x) = x^3 - 2x^2 + 4x - 1$, on $[-3, 2]$,

(ii) $g(x) = \frac{3x^2 - 2}{x^2 + 4}$, on $[-1, 1]$.

Solution

(i) The existence of a bound is guaranteed by Theorem 2.5.6. We note that if $x \in [-3, 2]$, then $|x| \le 3$, and applying the triangle law (cf. Theorem 1.2.2), we find that $|f(x)| = |x^3 - 2x^2 + 4x - 1| \le |x^3| + |2x^2| + |4x| + |1| \le 3^3 + 2(3^2) + 4(3) + 1 = 27 + 18 + 12 + 1 = 58$. So $|f(x)| \le 58$ for $x \in [-3, 2]$. This does not say that we have found the best possible bound for $f(x)$. It might be possible to find smaller numbers that are also bounds for $f(x)$ on $[-3, 2]$,

The polynomial $x^4 - 3x^2 + 2x + 1$ is negative at $x = -1$ and positive at $x = 0$. We know there must be a zero in $[-1, 0]$. We can plot the graph using a graphics calculator and then zoom in or use the equation-solving feature to find the zero. On some calculators it is not even necessary to plot the graph first.

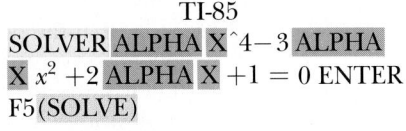

TI-85	SHARP EL9300
SOLVER ALPHA X ^4− 3 ALPHA X x^2 +2 ALPHA X +1 = 0 ENTER F5 (SOLVE)	SOLVER EQTN CL x/θ/T a^b 4 ▶ − 3 x/θ/T x^2 + 2 x/θ/T +1 = 0 ENTER −1 ENTER ENTER ENTER
CASIO f_x-7700 GB/f_x-6300 G	HP 48SX
Cls EXE Range −1 EXE 0 EXE 1 EXE −3 EXE 1 EXE 1 EXE EXE EXE EXE Graph X x^y 4 − 3 X x^2 + 2 X +1 EXE	↰ PLOT PLOTR ERASE ATTN′ $αY$ = $αX$ y^x 4 − 3 × $αX$ y^x 2 + 2 × $αX$ + 1 ↱ PLOT ↰ DRAW ERASE 1 +/− ENTER 0 XRNG 3 +/− ENTER 1 YRNG DRAW FCN ROOT

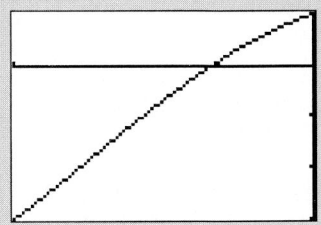

We find that the zero is at approximately $x \approx -0.336532274$ (the exact result may vary on different machines).

but at least we have found one bound. In this case, in fact, we happen to have found the best possible bound, since $f(-3) = -58$.

(ii) This part is trickier. To show that a quotient is smaller than something, we need to show that the numerator is smaller than something and the denominator is *bigger* than something.

On $[-1, 1]$, $|x| \le 1$, so the numerator satisfies $|3x^2 - 2| \le |3x^2| + |2| \le 3 + 2 = 5$. For the denominator, on the other hand, remembering that we want an inequality going the other way, we use the other form of the triangle law (cf. Theorem 1.2.3) (p. 5), and find that $|x^2 + 4| \ge |4| - |x^2| \ge 4 - 1 = 3$.

So on the specified interval,

$$|g(x)| \le \frac{5}{x^2 + 4} \le \frac{5}{3}.$$

In fact, we could also have noticed that since $x^2 \ge 0$, it follows that $x^2 + 4 \ge 4$, so $|g(x)| \le \frac{5}{4}$, which is smaller.

This example illustrates two ideas. One is the use of the triangle law to find simple bounds for polynomials. The other is that if we are asked to find a bound for a function on some interval, it is only necessary to find some bound, and not to worry about finding the smallest possible bound. Of course that might be an interesting question in its own right, but if all we need is some bound, any bound, that tends to be much easier to do.

Exercises 2.5

I

Find all points at which each of the following functions (a) is continuous, (b) has a removable discontinuity, (c) has a jump discontinuity, (d) has an infinite discontinuity, or (e) none of the above.

1. $f(x) = x^7 - 2x^2 + x$

2. $f(x) = \frac{x}{|x|}$

3. $f(x) = \frac{x^3 - 2x + 1}{x(x^2 + 1)}$

4. $f(x) = \frac{x^2 - 9}{x + 3}$

5. $f(x) = \frac{1}{x^2 - 3} \frac{x}{|x|}$

6. $f(x) = \frac{x^2}{x^2 - x}$

7. $f(x) = \frac{x}{|x|}(x^2 + 3)$

8. $f(x) = \frac{x}{|x|}(x^3 - 4x)$

9. $f(x) = (3x + 1)\left(\frac{x}{|x|}\right)^2$

10. $f(x) = \frac{x-1}{|x-1|} + \frac{x-2}{|x-3|}$

Decide which of the following functions are bounded on the stated domain. For those that are, specify a bound M.

11. $f(x) = x^2 - 1$, on the interval $[-1, 3]$

12. $f(x) = x^2 - 1$, on the interval $(-1, 3)$

13. $f(u) = \frac{1}{u-1}$, on the interval $[0, 2]$

14. $f(u) = \frac{1}{u-1}$, on the interval $[-1, 0]$

15. $f(x) = \sin x$, on the entire real axis

16. $f(x) = -x^2 + 3x + 1$, on the entire real axis

17. $f(x) = \sin \frac{1}{x}$, on the interval $(0, 1)$

18. $f(x) = 3x^4 - 2x^3 + 5x^2 - 2x + 4$, on the interval $[-2, 1]$.

19. $f(x) = \frac{2x^2 + x - 1}{x^3 + 3}$, on the interval $[-1, 1]$.

20. $f(x) = \frac{3x^2 - 2x + 4}{x^4 + 2}$, on the interval $[0, 1]$.

Decide which of the following achieve their maximum and/or minimum values on the stated domain. For those that do, specify points at which the maximum and/or minimum is achieved. (If you're stuck, draw a sketch.)

21. $f(x) = x^2 - 2x + 3$, on the interval $[-1, 3]$

22. $f(x) = x^2 - 2x + 3$, on the interval $(-1, 3)$

23. $f(x) = \frac{1}{x} \sin \frac{1}{x}$, on the interval $(0, 1)$

24. $f(x) = \sin x$, on the interval $\left(-\frac{\pi}{4}, 3\pi\right)$

25. $f(t) = \frac{1}{t+2}$, on the interval $[0, 1]$

26. $f(s) = \frac{s^2 - 1}{s - 1}$, on the interval $[0, 1)$

27. $f(u) = |u - 4|$, on the interval $[-2, 5]$

28. $f(x) = 3x + 5$, on the interval $[2, 9]$

29. $f(t) = \frac{1}{t^2 - 4t + 2}$, on the interval $[0, 5]$

30. $f(u) = u^3 - 3u + 2$, on the interval $(-2, 2)$
(*Hint:* Consider $u = -2, -1, 0, 1, 2$.)

II

*31. Show that the square root function $f(x) = \sqrt{x}$ is continuous on $[0, \infty)$:
   (i) Show that $f(x)$ is "continuous from the right" at $x = 0$, that is, $f(0) = \lim_{x \to 0+} f(x)$.
   (ii) Show that $f(x)$ is continuous at every $a > 0$.

*32. Show that the power function $g(x) = x^r$ is continuous wherever it makes sense. You should deal separately with the cases in which r is positive or negative and with the points where $x > 0$, $x = 0$, or $x < 0$.

33. Find all points where each of the following functions is continuous:
   (i) $f(x) = (x + 4)^{1/4}$
   (ii) $g(t) = \frac{t+1}{|t+1|}$
   (iii) $h(u) = \begin{cases} \frac{u(u-1)^2}{|u(u-1)^2|}, & \text{if } u \neq 0, 1 \\ 1, & \text{if } u = 0 \text{ or } 1 \end{cases}$

*34. Show that any polynomial is bounded on any open interval (a, b).

35. Show that $f(x) = 3x^4 - 8x^3 - 18x^2 + 6$ has a zero between $x = 0$ and $x = 1$, and another zero between $x = -1$ and $x = 0$.

36. Find consecutive integers so that $g(x) = x - 20 - x^{1/3}$ has a zero between them.

*37. Give an example to show that the conclusion of Theorem 2.5.4 may not be true if $g(x)$ has a removable discontinuity at $x = f(a)$.

*38. Find examples to show that each of Theorems 2.5.5, 2.5.6, and 2.5.7 fails if we do not require the function to be continuous.

*39. Which (if any) of Theorems 2.5.5, 2.5.6, and 2.5.7 fail and which (if any) remain true for functions that are continuous on $[a, b]$ except for one removable discontinuity at a point in (a, b)? (Give examples of any that fail.)

*40. (i) Show that any polynomial whose degree is odd must have at least one root.
   (ii) Find examples of nonconstant polynomials without any (real) roots.

41. In Example 5(i) (p. 110) we showed that $f(x) = x^4 + 2x^3 - 2x^2 + x + 2$ has a zero between $x = -3$ and $x = -2$. By evaluating $f\left(-\frac{5}{2}\right)$, decide whether the zero is between -3 and $-\frac{5}{2}$ or between $-\frac{5}{2}$ and -2.

42. Give examples of functions with (i) a removable discontinuity at $x = 2$, (ii) removable discontinuities at $x = -2$ and $x = 3$, (iii) a jump discontinuity at $x = -1$ and a removable discontinuity at $x = 2$.

43. In Example 5(i) we saw that $f(x) = x^4 + 2x^3 - 2x^2 + x + 2$ has a zero between $x = -3$ and $x = -2$. Use a graphics calculator or computer to find this point, correct to four decimal places.

44. In Example 5(ii) we saw that $g(x) = x - \sqrt{x} - 10$ has a zero between $x = 13$ and $x = 14$. Use a graphics calculator or computer to find this point, correct to four decimal places.

45. Use a graphics calculator or computer to locate the zeroes of $x^5 - 2x^4 + 1$. Estimate each to four decimal places. How many are there?

46. Use a graphics calculator or computer to locate the zeroes of $x^5 - x^4 + 1$. Estimate each to four decimal places. How many are there?

47. Sketch the graph of $y = x^3 - 2x^2 + 4x - 1$ for $x \in [-3, 2]$. Confirm the bound that we found in Example 6(i).

48. Sketch the graph of $y = \frac{3x^2 - 2}{x^2 + 4}$ for $x \in [-1, 1]$. Confirm the bound that we found in Example 6(ii). Can you use your graph to find a better bound?

POINT TO PONDER

You might be wondering at this point why we bother with the concept of continuity. After all, the most familiar kinds of functions, polynomials and rational functions, are continuous except where rational functions are not defined. In the next section we shall learn, among other things, that the same is true of the trigonometric functions $\sin x$, $\cos x$, etc. Moreover, the examples we have considered of functions that are not continuous seem rather contrived. They have been deliberately concocted to display unusual behavior.

In a sense, it is precisely because the "nice" functions to which we are most accustomed are continuous that this property is useful. It becomes one of the ways in which we can say that some function is "nice," that is, that at least in this respect it is like polynomials and rational functions. See Figure 2.5.7.

The other importance of continuity is that in doing calculus we are from time to time forced to deal with functions that are not continuous. In Example 2.1.1 we encountered the function $\frac{x^2 - x}{x - 1}$, which is not even defined at $x = 1$. Yet it was precisely its behavior at this point that interested us.

To study calculus, we need to confront discontinuous functions, and we need to appreciate the significance of continuity to help understand them.

Figure 2.5.7

2.6 Limits of Trigonometric Functions

CONTINUITY

$$\lim_{x \to 0} \frac{\sin x}{x} = 1$$

$$\lim_{x \to 0} \frac{\cos x - 1}{x} = 0$$

In this brief section we show that the trigonometric functions $\sin x$ and $\cos x$ are continuous, from which it follows that the other trigonometric functions, $\tan x$, $\cot x$, $\sec x$, and $\csc x$, are continuous whenever they are defined.

We also establish two limits involving trigonometric functions that will be needed later on.

It should be emphasized that the angles are always measured in radians. This might appear to be a nuisance, but the work of this section would actually be considerably more difficult if we were to use degrees. If you are not comfortable with radian measure, it would make sense to review it now, before continuing (see Section 1.7).

If we draw the graphs of $\cos x$ and $\sin x$, they appear to be continuous functions (see Figure 2.6.1), but it takes a little work to prove it. We first prove that they are continuous at $x = 0$ and use this to prove continuity everywhere.

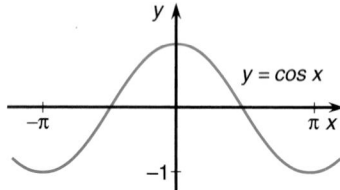

Figure 2.6.1

THEOREM 2.6.1

(i) $\lim_{x \to 0} \sin x = 0$,

(ii) $\lim_{x \to 0} \cos x = 1$,

(iii) $\lim_{x \to a} \sin x = \sin a$, for every $a \in \mathbb{R}$,

(iv) $\lim_{x \to a} \cos x = \cos a$, for every $a \in \mathbb{R}$.

PROOF

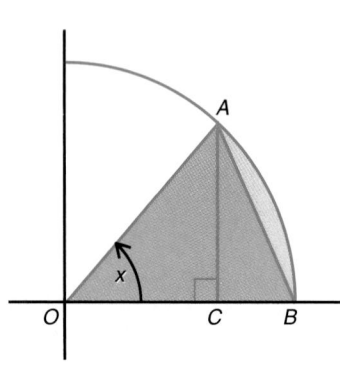

Figure 2.6.2

(i) Suppose that the angle AOC in Figure 2.6.2 has radian measure x, and that OA and OB are each 1 unit long. Then AC has length $\sin x$, OC has length $\cos x$, and by the definition of radian measure the arc of the circle from A to B has length x.

Now ABC is a right triangle, so its hypotenuse AB is longer than either of its other sides. But its hypotenuse, the straight line AB, is definitely shorter than the circular arc from A to B. So: length of $AC <$ length $AB <$ length of arc from A to B, that is,

$$\sin x < \text{ length } AB < x.$$

If we draw the same sort of figure with a negative angle x, then both $\sin x$ and x are negative and $x < \sin x < 0$. In both cases, $|\sin x| < |x|$, that is, $\sin x$ lies between x and $-x$.

Now apply the Squeeze Theorem (Theorem 2.3.5). We have just seen that $\sin x$ lies between the functions x and $-x$, and we know that both of them approach 0 as $x \to 0$. From this, part (i) is immediate.

(ii) This proof is similar, since CB is also shorter than AB. Since the length of OC is $\cos x$ and the length of OB is 1, the length of CB is $1 - \cos x$. The length of CB is less than the length of the hypotenuse AB, which is smaller than x, so $|1 - \cos x| < |x|$, and (ii) follows from the Squeeze Theorem, as above.

(iii), (iv) We notice that instead of $x \to a$, we can write $x = a + h$ and let $h \to 0$. So $\lim_{x \to a} \sin x = \lim_{h \to 0} \sin(a + h) = \lim_{h \to 0} \big(\sin(a) \cos(h) + \cos(a) \sin(h) \big)$, using the addition formula for sin (cf. Theorem 1.7.11). This then equals

$$\lim_{h \to 0} \sin(a) \cos(h) + \lim_{h \to 0} \cos(a) \sin(h)$$
$$= \sin(a) \lim_{h \to 0} \cos(h) + \cos(a) \lim_{h \to 0} \sin(h)$$
$$= \sin(a)(1) + \cos(a)(0) = \sin a,$$

so (iii) follows.

Similarly,

$$\lim_{x \to a} \cos x = \lim_{h \to 0} \cos(a + h) = \lim_{h \to 0} \big(\cos(a) \cos(h) - \sin(a) \sin(h) \big)$$
$$= \lim_{h \to 0} \cos(a) \cos(h) - \lim_{h \to 0} \sin(a) \sin(h)$$
$$= \cos(a) \lim_{h \to 0} \cos(h) - \sin(a) \lim_{h \to 0} \sin(h)$$
$$= \cos(a)(1) - \sin(a)(0) = \cos a,$$

as desired.

COROLLARY 2.6.2

(i) $\sin x$ and $\cos x$ are continuous.

(ii) $\tan x, \cot x, \sec x$, and $\csc x$ are continuous wherever they are defined.

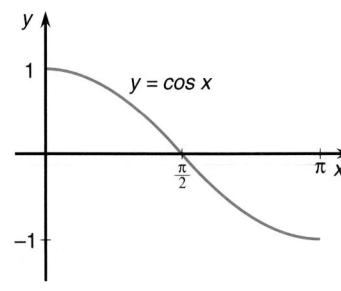

Figure 2.6.3

PROOF   (i) is just a restatement of parts (iii) and (iv) of the theorem while (ii) follows from it by Theorem 2.5.3(iv).

EXAMPLE 1   Evaluate (i) $\lim_{x \to \pi/2} \frac{\sin x - \cos(2x)}{3x}$; (ii) $\lim_{x \to (\frac{\pi}{2})+} \tan x$.

Solution

(i) $\lim_{x \to \pi/2} 3x = \frac{3\pi}{2} \neq 0$, so

$$\lim_{x \to \pi/2} \frac{\sin x - \cos(2x)}{3x} = \frac{\sin(\pi/2) - \cos(\pi)}{3\pi/2} = \frac{1 - (-1)}{3\pi/2} = \frac{4}{3\pi}.$$

(ii) Write $\tan x = \frac{\sin x}{\cos x}$. As $x \to \frac{\pi}{2}^+$, we know that $\sin x \to \sin \frac{\pi}{2} = 1$ and also that $\cos x \to \cos \frac{\pi}{2} = 0$. So the answer is that the limit does not exist.

However, we might well ask about the possibility that there is an infinite limit. Since the numerator tends to 1, the question amounts to asking whether or not the denominator either stays positive (in which case the quotient will approach ∞) or stays negative (in which case the quotient will approach $-\infty$). As $x \to \frac{\pi}{2}^+$, we see from Figure 2.6.3 that $\cos x < 0$, at least for values of x reasonably near $\frac{\pi}{2}$, so the quotient is negative, and we see that

$$\lim_{x \to \frac{\pi}{2}^+} \tan x = -\infty.$$

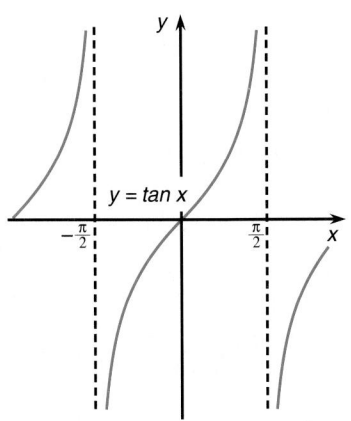

Figure 2.6.4

(If we were to consider the limit as $x \to \frac{\pi}{2}^-$, we would find that it is $+\infty$, for similar reasons, which certainly is consistent with the graph of $y = \tan x$ shown in Figure 2.6.4. This shows, incidentally, that there is no two-sided limit, not even an infinite one.)

THEOREM 2.6.3

(i) $\lim_{x \to 0} \frac{\sin x}{x} = 1$,

(ii) $\lim_{x \to 0} \frac{\cos x - 1}{x} = 0$.

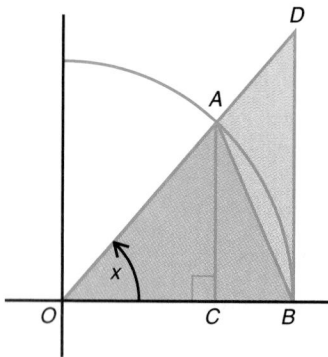

Figure 2.6.5

PROOF We consider first the case in which $x > 0$; refer to Figure 2.6.5. Since the circle has radius $r = 1$, the whole circle has area $\pi r^2 = \pi$. The area of a circular sector like AOB depends on the angle AOB. If the sector was the full circle, the angle would be 2π radians, and the area would be π. If the sector was half a circle, the angle would be π, and the area would be $\frac{\pi}{2}$. In general, the area is proportional to the angle, and we find that the area in square units is one-half the radian measure of the angle. In our case the angle is x, so the area of the circular sector AOB is $\frac{x}{2}$.

From the diagram we see that the triangle AOB is contained in the circular sector AOB, which is itself contained in triangle DOB. The area of triangle AOB is $\frac{1}{2}|OB||AC| = \frac{1}{2}(1)\sin x$.

To find the length $|BD|$, we note that $\frac{|BD|}{|OB|} = \tan x$. Since $|OB| = 1$, we see that $|BD| = \tan x$.

Now the area of triangle DOB is $\frac{1}{2}|OB||DB| = \frac{1}{2}(1)\tan x = \frac{1}{2}\frac{\sin x}{\cos x}$. Comparing these with $\frac{x}{2}$, the area of the circular sector AOB, we get

$$\frac{1}{2}\sin x < \frac{x}{2} < \frac{1}{2}\frac{\sin x}{\cos x}.$$

Dividing by $\frac{1}{2}\sin x$, we get

$$1 < \frac{x}{\sin x} < \frac{1}{\cos x}.$$

Taking reciprocals, we reverse the inequalities and get

$$1 > \frac{\sin x}{x} > \cos x.$$

We showed this under the assumption that $x > 0$. A nearly identical argument can be made if $x < 0$, but it is quicker to notice that $\sin(-x) = -\sin x$ (cf. Theorem 1.7.8, p. 48). Now if $x < 0$, write it as $x = -t$ for some positive t. Then

$$\frac{\sin x}{x} = \frac{\sin(-t)}{-t} = \frac{-\sin t}{-t} = \frac{\sin t}{t}.$$

But we found that this last function lies between 1 and $\cos t = \cos(-x) = \cos x$. Since $\lim_{x \to 0} 1 = \lim_{x \to 0} \cos x = 1$, the Squeeze Theorem applies and proves part (i).

For part (ii), write

$$\lim_{x \to 0} \frac{\cos x - 1}{x} = \lim_{x \to 0} \frac{\cos x - 1}{x} \frac{\cos x + 1}{\cos x + 1}$$

$$= \lim_{x \to 0} \frac{\cos^2 x - 1}{x(\cos x + 1)} = \lim_{x \to 0} \frac{-\sin^2 x}{x(\cos x + 1)}$$

$$= -\lim_{x \to 0} \frac{\sin x}{x}\frac{\sin x}{\cos x + 1} = -\left(\lim_{x \to 0}\frac{\sin x}{x}\right)\left(\lim_{x \to 0}\frac{\sin x}{\cos x + 1}\right)$$

$$= -1\left(\frac{0}{2}\right) = 0,$$

We use the identity $\cos^2 x - 1 = -\sin^2 x$.

Notice how we split off $\frac{\sin x}{x}$, whose limit we know . . .

as required.

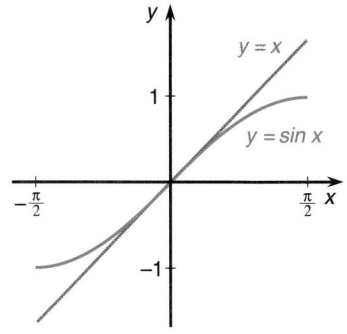

Figure 2.6.6

Part (i) of Theorem 2.6.3 says that as x approaches 0, the value of $\sin x$ gets very close to x. In Figure 2.6.6 the graph of $y = \sin x$ is plotted together with the graph of $y = x$. The idea is that near $x = 0$ the two graphs are very close together, which is certainly apparent from the picture.

EXAMPLE 2   Evaluate $\lim_{x \to 0} \frac{2x + \sin(5x)}{\sin(3x)}$.

Solution   Write

$$\frac{2x + \sin(5x)}{\sin(3x)} = \frac{2x}{\sin(3x)} + \frac{\sin(5x)}{\sin(3x)} = \frac{2}{3}\frac{3x}{\sin(3x)} + \frac{\sin(5x)}{5x}\frac{5x}{3x}\frac{3x}{\sin(3x)}$$

$$= \frac{2}{3}\frac{3x}{\sin(3x)} + \frac{5}{3}\frac{\sin(5x)}{5x}\frac{3x}{\sin(3x)}.$$

Now, as $x \to 0$, we have $5x \to 0$. Writing $u = 5x$, we see that $\lim_{x \to 0} \frac{\sin(5x)}{5x} = \lim_{u \to 0} \frac{\sin u}{u} = 1$.

Similarly, $3x \to 0$ and $\frac{3x}{\sin(3x)} \to 1$. Combining everything, we find that

$$\lim_{x \to 0} \frac{2x + \sin(5x)}{\sin(3x)} = \frac{2}{3} + \frac{5}{3} = \frac{7}{3}.$$

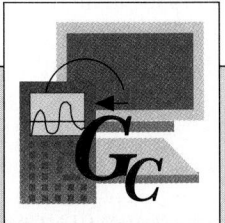

To work this example, we try to find expressions of the form $\frac{\sin u}{u}$ or $\frac{u}{\sin u}$, whose limits we know as $u \to 0$.

We can use the same approach as in Example 2 to evaluate $\lim_{x \to 0} \frac{\sin 5x}{\sin(x/2)}$. This is an interesting function to graph.

TI-85	SHARP EL9300
GRAPH F1 $(y(x) =)$ SIN (5 F1 (x)) $\div$ SIN (F1$(x) \div$ 2) M2(RANGE) $(-)$ π ▼ π ▼ ▼ $(-)$ 3 ▼ 11 F5(GRAPH)	sin (5 x/θ/T) $\div$ sin (x/θ/T $\div$ 2) RANGE $-\pi$ ENTER π ENTER 1 ENTER -3 ENTER 11 ENTER 1 ENTER

CASIO f_x-7700GB/f_x-6300G	HP 48SX
Cls EXE Range $-\pi$ EXE π EXE 1 EXE -3 EXE 11 EXE 1 EXE EXE EXE EXE Graph sin (5 X) $\div$ sin (X $\div$ 2) EXE	↰ PLOT PLOTR ERASE ATTN′ αY = SIN 5 × αX ▶ $\div$ SIN $\alpha X\div$ 2 ↱ PLOT ↰ DRAW ERASE π +/− ENTER π XRNG 3 +/− ENTER 11 YRNG DRAW

Check that the picture agrees with your calculation of the limit as $x \to 0$. Then find $\lim_{x \to 0} \frac{\sin kx}{\sin(x/2)}$ for different values of k and try plotting the corresponding functions.

Exercises 2.6

I

Evaluate the following finite or infinite limits.

1. $\lim_{x\to 0} \frac{3x^2+x-1+\sin(x+\pi/2)}{\cos(2x)}$

2. $\lim_{x\to \pi/2} \frac{x^2+\cos x}{\sin x}$

3. $\lim_{x\to \pi} \frac{\cos^2 x+\sin(x/2)}{\cos x}$

4. $\lim_{x\to \pi/4} \tan x$

5. $\lim_{x\to \pi/2} \csc^3 x$

6. $\lim_{x\to \pi} x + \sec x$

7. $\lim_{x\to 0} \frac{\sin x}{3x}$

8. $\lim_{x\to 0} \frac{\cos^2 x-1}{x}$

9. $\lim_{x\to \pi/2} \tan^2 x$

10. $\lim_{x\to \pi} \left(\cos^2(7x) + \sin^2(7x)\right)$

Decide whether or not each of the following finite or infinite limits exists, and evaluate those that do.

11. $\lim_{x\to 0} \frac{\cos x}{x}$

12. $\lim_{x\to 0} \frac{\sin(7x)}{\sin(4x)}$

13. $\lim_{x\to 0} \frac{\sin x}{\cos x-1}$

14. $\lim_{x\to 0^+} \frac{\sin 2x}{\cos x-1}$

15. $\lim_{x\to 0} \frac{\cos(2x)-2}{x}$

16. $\lim_{x\to 0} \frac{\sin x}{2x^2+x}$

17. $\lim_{x\to 2} \sin(x^2 - 3x + 2)$

18. $\lim_{x\to \frac{1}{2}} \tan(\pi x)$

19. $\lim_{x\to 0^-} \tan x$

20. $\lim_{x\to -1} \csc^2(\pi x)$

Evaluate the following finite or infinite limits.

21. $\lim_{x\to 2} \sin\left(\pi(x^2 - 4)\right)$

22. $\lim_{x\to -1} \cos\left(\pi(1 + x^2)\right)$

23. $\lim_{t\to 3} \frac{\sin(9-t^2)}{9-t^2}$

24. $\lim_{t\to 3} \frac{\sin(9-t^2)}{3-t}$

25. $\lim_{u\to \frac{\pi}{2}} \frac{\sin(u)-1}{u-\frac{\pi}{2}}$

26. $\lim_{r\to -\frac{\pi}{2}} \frac{\cos r}{r+\frac{\pi}{2}}$

27. $\lim_{t\to \frac{1}{2}} \frac{\cos(\pi t)}{\sin(2\pi t)}$

28. $\lim_{x\to 2} \frac{\sin(\pi x)}{\cos(\frac{\pi}{4}x)}$

29. $\lim_{x\to -1} \frac{\tan(\pi x)}{\tan(2\pi x)}$

30. $\lim_{x\to 0} \frac{\cos^2(2x)-1+\sin^2(2x)}{x}$

II

31. (i) Find all values of a at which $\tan x$ does not have a limit, that is, at which $\lim_{x\to a} \tan x$ does not exist.

    (ii) For the values you found in part (i), decide whether or not $\tan x$ has one-sided limits, infinite limits, one-sided infinite limits, or none of the above.

32. Repeat Exercise 31 for the function $\sec x$.

33. Repeat Exercise 31 for the functions $\cot x$ and $\tan^2 x$.

*34. In Example 2.1.4 (p. 66) we expressed the area of a circle of radius $r = 1$ as a limit. Evaluate the limit.

III

35. Use a graphics calculator or computer to graph the function $y = \frac{\cos x-1}{x}$ for $x \in [-1, 1]$, for $x \in [-0.1, 0.1]$, and for $x \in [-0.01, 0.01]$. Confirm the limit in Theorem 2.6.3(ii).

36. Graph the function $y = \frac{\sin 3x}{x}$ for $x \in [-1, 1]$, for $x \in [-0.1, 0.1]$, and for $x \in [-0.01, 0.01]$. Confirm that the limit is as expected.

P ← **POINT TO PONDER**

In this section we have studied various limits involving trigonometric functions. The two fundamental results were Theorem 2.6.1, which just says that $\sin x$ and $\cos x$ are continuous (i.e., that $\lim_{x\to a} \sin x = \sin a$ and $\lim_{x\to a} \cos x = \cos a$), and Theorem 2.6.3, which found that $\lim_{x\to 0} \frac{\sin x}{x} = 1$ and $\lim_{x\to 0} \frac{\cos x-1}{x} = 0$.

The arguments we gave for all these limits were quite geometrical. For Theorem 2.6.1 we had to talk about the length of a piece of the arc of a circle, and

in Theorem 2.6.3 we needed to talk about the area of a sector of a circle.

These arguments are quite plausible, especially if you look at the pictures, and they are correct. However, we have not really defined what is meant by the length of an arc of a circle, and we do not really know how to find the area of a circular sector. So the arguments we have used for these limits rely on concepts that we must take for granted.

In fact, both these concepts will be discussed at considerable length later on. Calculating areas is one of the main uses of the technique of integration, which is the subject of Chapters 5 and 6, and the lengths of curves will be discussed in Chapter 6 and again in Chapters 8 and 15.

None of this should really trouble anybody too much. We have not really given careful proofs of many of our results and have chosen instead to concentrate on developing our skills in working with them. On the other hand, it is interesting to acknowledge our shortcomings.

2.7 The Precise Definition of Limits

Figure 2.7.1

Figure 2.7.2

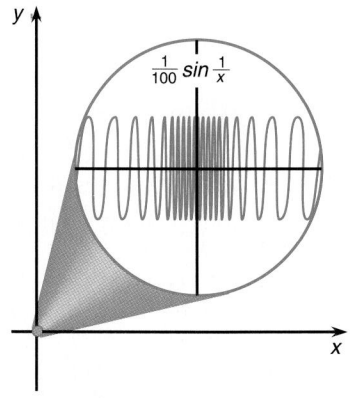

Figure 2.7.3

In this section we discuss how limits can be defined with mathematical precision. We have contented ourselves up to now with a vague description and found that it was enough for our purposes, enabling us to develop a good working knowledge of limits and a feel for what they mean. We have also hinted that the precise definition is difficult to understand and use.

The problem is to state exactly what we mean when we say that $f(x)$ "approaches" L when x "approaches" a. We have already mentioned (at the end of Section 2.3) that it is not simply a matter of saying that $f(x)$ keeps getting closer to L. There we mentioned that $f(x)$ can approach and perhaps even *equal* L and then move away again; what is more important is that it not move away too far. See Figure 2.7.1.

So we should begin by saying that $f(x)$ gets close to L and stays close, but this leaves the question of "how close?" (see Figure 2.7.2). It would not be good enough to say within 0.01 or even within 0.00000001 because it would always be possible to find a function that did stay that close but did not have a limit (after all, think of $\frac{1}{100} \sin \frac{1}{x}$ as $x \to 0 \ldots$). See Figure 2.7.3.

In fact, no specified degree of closeness will work. What is needed is to say that $f(x)$ will eventually get as close as we like to L. In other words, it will eventually get within 0.01 of L, and eventually within 0.0001, and eventually within 0.00000001, and so on. This is not to say that it will satisfy all these conditions simultaneously, but rather that it will eventually satisfy any particular one of them.

So to say that $\lim_{x \to a} f(x) = L$ is to say that if we are given any particular degree of closeness, we can be sure that $f(x)$ will eventually be that close to L. By "eventually" we mean when x is sufficiently close to a.

To specify what we have been calling a "degree of closeness," we should give a number and say "not farther than this much away."

So we need to say that given any positive number $\epsilon > 0$, we can be sure that $f(x)$ will "eventually" be a distance less than ϵ away from L. (The Greek letter ϵ **(epsilon)** is traditionally used in this context.)

By "eventually" we mean that we can be sure it will be true once x is within some small distance of a. To specify this small distance, we can say there must be a positive number $\delta > 0$ so that $f(x)$ will be within ϵ of L whenever x is within a distance δ of a. (The Greek letter δ **(delta)** is traditionally used here.)

Figure 2.7.4

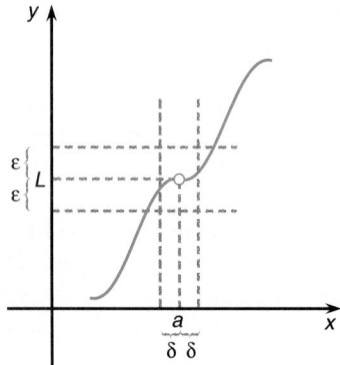

Figure 2.7.5

In Figure 2.7.4 it is possible to see how δ is used as a measure of the horizontal closeness of x to a and ϵ is used as a measure of the (vertical) closeness of $f(x)$ to L. We see that when x is within a distance of δ from a, the value of $f(x)$ is within ϵ of L.

The important thing to bear in mind is that it must work for any choice of $\epsilon > 0$. That is, given any $\epsilon > 0$, there must be a $\delta > 0$ so that $|f(x) - L| < \epsilon$ whenever x is within δ of a. We should expect that δ will depend on ϵ, that is, that for different choices of ϵ it will be necessary to use different choices of δ. See Figure 2.7.5.

The above discussion is an attempt to explain and justify the following definition.

DEFINITION 2.7.1

Suppose the interval (c, d) contains the point a, and suppose the function $f = f(x)$ is defined at every point in (c, d) except possibly the point a. A number L is said to be the **limit** of $f(x)$ as x approaches a, and we write

$$\lim_{x \to a} f(x) = L,$$

provided that for every $\epsilon > 0$ there exists a $\delta > 0$ so that $|f(x) - L| < \epsilon$ whenever $0 < |x - a| < \delta$.

Notice that $f(x)$ has to be defined at every point "near $x = a$" but not necessarily at a itself.

EXAMPLE 1   In Section 2.1 and again in Section 2.2 we agreed that $\lim_{x \to 1} x = 1$. Let us see how the definition applies in this case. Notice that $f(x) = x$, $a = 1$, and $L = 1$.

What is needed is to make x close to 1 or to make $|x - 1|$ small. Suppose we are given $\epsilon > 0$ and want $|x - 1| < \epsilon$. To do this, we are allowed to insist that x be close to 1, that is, that $|x - 1|$ be less than some well-chosen δ.

The answer is almost too easy. To make $|f(x) - L| < \epsilon$, that is, $|x - 1| < \epsilon$, all we need is to require that $|x - 1| < \delta$, with $\delta = \epsilon$.

So for any $\epsilon > 0$ we certainly know how to pick a $\delta > 0$: Just let $\delta = \epsilon$. If $0 < |x - 1| < \delta$, then $|f(x) - L| = |x - 1|$ is indeed less than ϵ. This proves that $\lim_{x \to 1} x = 1$.

On the other hand, it was a little too simple.   ◢

EXAMPLE 2   Let $f(x) = 2x$ and show that $\lim_{x \to -3} f(x) = -6$.

Solution $|f(x) - (-6)| = |2x + 6| = 2|x + 3| = 2|x - (-3)|$. Given any $\epsilon > 0$, we want to make $|f(x) - (-6)| < \epsilon$, that is, $2|x - (-3)| < \epsilon$, which is obviously true if $|x - (-3)| < \epsilon/2$.

Letting $\delta = \epsilon/2$, we see that if $|x - (-3)| < \delta = \epsilon/2$, then $2|x - (-3)| < \epsilon$, as needed. Notice how δ is different for different values of ϵ.   ◢

EXAMPLE 3   Let $f(x) = x^2 - x + 3$. What is the limit of $f(x)$ as x approaches 2?

Solution We know that the limit is $f(2) = 5$. And $|f(x) - 5| = |x^2 - x - 2| = |(x - 2)(x + 1)| = |x - 2||x + 1|$. To make this less than ϵ, we deal with each factor separately.

First, we notice that if x is close to 2, this restricts the size of $|x + 1|$. Indeed, $|x + 1| = |x - 2 + 3| \leq |x - 2| + 3$. We are going to assume that $|x - 2| < \delta$, so $|x + 1| < \delta + 3$. If $\delta \leq 1$, this shows that $|x + 1| < 1 + 3 = 4$. Knowing this, we could

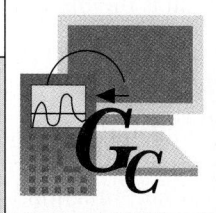

A graphics calculator can be used to visualize how a particular δ works with a particular ϵ. The drawback is that it is usually difficult to do anything that shows anything about *every* ϵ.

We plot the function from Example 3. Suppose we want to consider $\epsilon = 0.01$. The example suggests choosing $\delta = 0.0025$. This means considering $x \in [2 - 0.0025, 2 + 0.0025]$, and letting the vertical range be between $5 - \epsilon$ and $5 + \epsilon$, that is, $4.99 \leq y \leq 5.01$.

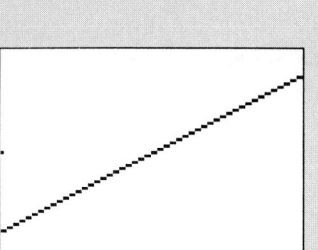

TI-85	SHARP EL9300
GRAPH F1 (y(x) =) F1(x) x^2- F1 (x) + 3 M2(RANGE) 1.9975 ▼ 2.0025 ▼ ▼ 4.99 ▼ 5.01 F5(GRAPH)	⤶ x/θ/T x^2 − x/θ/T + 3 RANGE 1.9975 ENTER 2.0025 ENTER .001 ENTER 4.99 ENTER 5.01 ENTER .01 ENTER ⤶
CASIO f_x-7700GB/f_x-6300G	HP 48SX
Cls EXE Range 1.9975 EXE 2.0025 EXE .001 EXE 4.99 EXE 5.01 EXE .01 EXE EXE EXE EXE Graph X x^2- X + 3 EXE	⇱ PLOT PLOTR ERASE ATTN′ $\alpha Y = \alpha X \ y^x \ 2 - \alpha X + 3$ ↱ PLOT ⇱ DRAW ERASE 1.9975 ENTER 2.0025 XRNG 4.99 ENTER 5.01 YRNG DRAW

For each $x \in [5 - \delta, 5 + \delta]$, the function lies in the specified range $5 - \epsilon \leq y \leq 5 + \epsilon$, that is, $4.99 \leq y \leq 5.01$. This means that the δ we have chosen is sufficiently small. If the graph ran off the top or the bottom of the picture, we would have to choose a smaller δ. In fact, from the picture it appears that we could have chosen a slightly bigger δ . . .

also ask that $|x - 2| < \epsilon/4$, so that $|f(x) - 5| = |x - 2||x + 1| < (\epsilon/4)(4) = \epsilon$, as required. A value of δ that will achieve all this is therefore the minimum of 1 and $\epsilon/4$, that is, $\delta = \min(1, \epsilon/4)$.

If $\delta = \min(1, \epsilon/4)$ and $0 < |x - 2| < \delta$, then $|f(x) - 5| < \epsilon$, and we have found that $\lim_{x\to 2} f(x) = 5$. ◢

This is not the only δ that will work; obviously, any smaller δ will be just as good. Fortunately, all that is required is to find one suitable value.

EXAMPLE 4   Let $f(x) = -3$ be the constant function. What is the limit of $f(x)$ as $x \to 10$?

Solution   We know that the limit is -3. And $|f(x) - (-3)| = |-3 - (-3)| = 0$, which is less than ϵ for any $\epsilon > 0$. So δ could be anything at all; for example, $\delta = 1$ works for any $\epsilon > 0$. If $0 < |x - 10| < 1$, then $|f(x) - (-3)| = |-3 + 3| = 0 < \epsilon$, as needed. Of course, the same argument works for any constant function. ◢

EXAMPLE 5   Show that the function $f(x) = \frac{x}{|x|}$ does not have a limit as $x \to 0$.

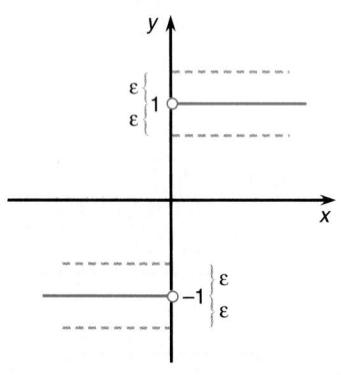

Figure 2.7.6

Solution   Suppose it did have a limit L, and suppose $\epsilon < 1$. Then there would be a $\delta > 0$ so that $|f(x) - L| < \epsilon$ whenever $0 < |x - 0| < \delta$, that is, $0 < |x| < \delta$. If $-\delta < x < 0, f(x) = -1$, so $|-1 - L| < \epsilon < 1$, so $L < 0$. See Figure 2.7.6.

On the other hand, if $0 < x < \delta, f(x) = 1$, and $|1 - L| < \epsilon < 1$, so $L > 0$. But L cannot be greater than 0 *and* less than 0, and this contradiction proves that there is no limit. ◢

> Notice that for a limit to exist, it must be possible to find a δ for every ϵ, but to show that it doesn't exist, all that is needed is one value of ϵ for which no δ can be found.

EXAMPLE 6   Evaluate $\lim_{x\to 2} \frac{1}{3x+1}$.

Solution   We know what the limit should be: $\lim_{x\to 2} \frac{1}{3x+1} = \frac{1}{3(2)+1} = \frac{1}{7}$.

Suppose $\epsilon > 0$ is given. We must consider the difference $\frac{1}{3x+1} - \frac{1}{7}$ and make it small. Rewriting

$$\frac{1}{3x+1} - \frac{1}{7} = \frac{7 - (3x+1)}{7(3x+1)} = \frac{6 - 3x}{21x+7},$$

we have made the difference into a rational function whose numerator is $6 - 3x = -3(x - 2)$. As $x \to 2$, this numerator approaches zero, so to make the quotient small, we only have to make sure the denominator does not also become small.

We are going to consider a condition that $0 < |x - 2| < \delta$. The possible trouble with the denominator would occur when it gets small, which happens when x gets close to $-\frac{1}{3}$. If, for instance, we assume that $\delta \leq 1$, then $|x - 2| < 1$, and this means that $x > 1$. If $x > 1$, then $21x > 21$ and $21x + 7 > 28$.

So, with this understanding that $\delta \leq 1$, we see that

$$\left|\frac{6 - 3x}{21x+7}\right| \leq \frac{|6 - 3x|}{28} = \frac{3}{28}|x - 2|.$$

This is what we have to make smaller than ϵ, and that will certainly be true if $|x - 2| < \frac{28}{3}\epsilon$. In other words, we can choose δ to be less than or equal to $\frac{28}{3}\epsilon$. Bearing in mind our earlier assumption about δ, we let

$$\delta = \min\left(1, \frac{28}{3}\epsilon\right).$$

If $0 < |x - 2| < \delta$, then

$$\left|\frac{1}{3x + 1} - \frac{1}{7}\right| = \left|\frac{3(2 - x)}{21x + 7}\right| \leq \frac{3}{28}|x - 2| < \frac{3}{28}\frac{28}{3}\epsilon = \epsilon.$$

For any ϵ we have shown how to find δ, which proves that

$$\lim_{x \to 2} \frac{1}{3x + 1} = \frac{1}{7},$$

as desired.

The examples that we have just done demonstrate that Definition 2.7.1 is not easy to use. Even relatively simple limits can be quite difficult to prove.

It can help us to understand and appreciate why certain things work the way they do, but as a practical matter, knowing the definition will not even be much use when it comes to evaluating any of the limits we have encountered so far.

This is why we have postponed this section to the end of the chapter. It was not really necessary to learn it to work out the limits we need, and it is undeniably difficult. On the other hand, it can be very interesting, and it is very important for theoretical work with limits.

It is clearly too cumbersome to have to use the definition directly every time we evaluate a limit. The idea is that all the theorems about limits can be proved rigorously by starting with the definition. Then, of course, it is possible to use them, as we have already been doing, without going back to ϵ and δ all the time. We have chosen to bypass the time-consuming proofs of the theorems and concentrate instead on learning to use them.

However, we illustrate the approach by giving the proof of Theorem 2.2.3(i), which says that the limit of the sum of two functions with limits is the sum of their limits.

THEOREM 2.2.3

Suppose $f(x)$ and $g(x)$ are functions and a is a number so that

$$\lim_{x \to a} f(x) = L \quad \text{and} \quad \lim_{x \to a} g(x) = M.$$

(Recall that this means that both limits exist and they have the specified values.)

Then

$$\lim_{x \to a} \big(f(x) + g(x)\big) = L + M.$$

PROOF   Given $\epsilon > 0$, we must show that there is a $\delta > 0$ so that

$$\left| \big(f(x) + g(x)\big) - (L + M) \right| < \epsilon,$$

whenever $0 < |x - a| < \delta$.

Notice that $\left| \big(f(x) + g(x)\big) - (L + M) \right| = \left| \big(f(x) - L\big) + \big(g(x) - M\big) \right| \leq |f(x) - L| + |g(x) - M|$. The last two terms are the things we know can be made small because $f(x) \to L$ and $g(x) \to M$.

To make their sum less than ϵ, it would be enough to make each of them less than $\frac{\epsilon}{2}$. Since

$$\lim_{x \to a} f(x) = L,$$

given the number $\frac{\epsilon}{2}$, we know there exists a $\delta_1 > 0$ so that

$$|f(x) - L| < \frac{\epsilon}{2} \qquad \text{whenever} \qquad 0 < |x - a| < \delta_1.$$

Similarly, there is a $\delta_2 > 0$ so that

$$|g(x) - M| < \frac{\epsilon}{2} \qquad \text{whenever} \qquad 0 < |x - a| < \delta_2.$$

Now let

$$\delta = \min(\delta_1, \delta_2).$$

Whenever $0 < |x - a| < \delta$, then $0 < |x - a| < \delta_1$ and $0 < |x - a| < \delta_2$. So

$$|f(x) - L| < \frac{\epsilon}{2} \qquad \text{and} \qquad |g(x) - M| < \frac{\epsilon}{2}.$$

From this we find that

$$\left| \big(f(x) + g(x)\big) - (L + M) \right| \leq |f(x) - L| + |g(x) - M| < \frac{\epsilon}{2} + \frac{\epsilon}{2} = \epsilon.$$

We have shown that, given any $\epsilon > 0$, it is possible to find a $\delta > 0$ so that $\left| \big(f(x) + g(x)\big) - (L + M) \right| < \epsilon$ whenever $0 < |x - a| < \delta$, and the proof is complete. ◀

Exercises 2.7

In each of the following cases, find $\lim_{x \to a} f(x)$, and then use Definition 2.7.1 to prove that your answer is correct.

1. $f(x) = x - 4, a = -1$
2. $f(x) = 3x + 2, a = \frac{1}{3}$
3. $f(x) = x^2 + 3x - 1, a = 2$
4. $f(x) = 2x^2 - x + 1, a = -2$
5. $f(x) = x^3 + 9, a = -2$
6. $f(x) = x^3 - x^2 + 2x - 1, a = 1$
7. $f(x) = 2x + 1, a = -\frac{1}{2}$
8. $f(x) = x^{10}, a = 1$

9. $f(x) = x(x + 1), a = 2$
10. $f(x) = (x + 1)(x - 1), a = 1$

In each of the following cases, use Definition 2.7.1 to determine whether or not $f(x)$ has a limit as x approaches a, and determine the limit if it exists.

11. $f(x) = \frac{2}{x+1}, a = -1$
12. $f(x) = \frac{x^2-1}{x+1}, a = 1$
13. $f(x) = \frac{x^2+1}{x-1}, a = 1$
14. $f(x) = |x|, a = 0$
15. $f(x) = \frac{x^2}{|x|}, a = 0$
16. $f(x) = \frac{|x|}{x^2}, a = 0$

17. $f(x) = 2 - x^2 \sin(1/x)$, $a = 0$

18. $f(x) = \frac{x^3 + 2x^2 + x - 4}{x - 1}$, $a = 1$

19. $f(x) = \frac{x^2 + x \sin(1/x)}{x}$, $a = 0$

20. $f(x) = \frac{x - x^2 \sin(1/x)}{x}$, $a = 0$

Decide whether or not each of the following functions has a limit at the point $a = 0$ and prove that your answer is correct, using Definition 2.7.1.

21. $f(x) = \cos(1/x)$

22. $g(t) = 1/t$

23. $h(x) = \cos^2(1/x)$

24. $k(x) = \cos^2(1/x) + \sin^2(1/x)$

II

25. Use Definition 2.7.1 to show that
$$\lim_{x \to a} x = a.$$

*26. Use Definition 2.7.1 to show that
  (i) $\lim_{x \to a} (cx + d) = ca + d$, assuming $c \neq 0$.
  (ii) $\lim_{x \to a} (Ax^2 + Bx + C) = Aa^2 + Ba + C$.

*27. (i) Mimic Definition 2.7.1 of a limit to formulate definitions in terms of ϵ and δ for one-sided

limits
$$\lim_{x \to a^+} f(x) \quad \text{and} \quad \lim_{x \to a^-} f(x).$$

  (ii) Use your definition to find the one-sided limits at $x = 0$ of $f(x) = \frac{x}{|x|}$.

  (iii) Using your definitions, prove Theorem 2.4.2, that $\lim_{x \to a} f(x)$ exists and equals L if and only if both $\lim_{x \to a^+} f(x)$ and $\lim_{x \to a^-} f(x)$ exist and both equal L.

*28. Use your definition from Exercise 27 to prove that $\lim_{x \to 0^+} \sqrt{x} = 0$. (This problem was unexpectedly difficult earlier on, but if we use ϵ's and δ's, it is actually not too bad.)

III

*29. Let $f(x) = x^{1/3} + x^{2/5}$. Then $\lim_{x \to 0} f(x) = 0$. Use a graphics calculator or computer to find a value of δ that corresponds to each of the following values of ϵ: $\epsilon = 0.1$, $\epsilon = 0.01$, $\epsilon = 0.001$, $\epsilon = 0.0001$.

*30. With $f(x)$ as in the previous question, we see that $\lim_{x \to 1} f(x) = 2$. Use a graphics calculator or computer to find a value of δ that corresponds to each of the following values of ϵ: $\epsilon = 0.1$, $\epsilon = 0.01$, $\epsilon = 0.001$, $\epsilon = 0.0001$.

POINT TO PONDER

At first glance, many people find the definition of a limit (Definition 2.7.1) confusing and complicated. Try to find a better one, bearing in mind that it should give the same answers as Definition 2.7.1 for each of the examples and exercises we have tried. (This is a problem for which there is probably no solution. If you think about it for a while, you will see that any other definition is just as complicated, or even worse.)

Chapter Summary

§2.1
tangent line
area

§2.2
limit
limits of sums
limits of products
limits of quotients
limits of polynomials

§2.3
limits of powers x^r
limits of composite functions
The Squeeze Theorem

§2.4
one-sided limits
limits from above/below
limits from the right/left
infinite limits

limits of rational functions
infinity
negative infinity
limits at infinity

§2.5
continuous function
continuity of polynomials
continuity of rational functions
removable discontinuity

jump discontinuity
infinite discontinuity
continuity of composite functions
Intermediate Value Theorem
boundedness of continuous functions on $[a, b]$
continuous functions achieve maximum/minimum values on $[a, b]$
Extreme Value Theorem

§2.6
limits of trigonometric functions

continuity of trigonometric functions
$\lim_{x \to 0} \frac{\sin x}{x} = 1$
$\lim_{x \to 0} \frac{\cos x - 1}{x} = 0$

§2.7
limit
precise definition of limits
epsilon (ϵ)
delta (δ)

Review Exercises

I

Calculate each of the following, provided that it exists.

1. $\lim_{x \to 0} (x^3 - 4x^2 + 2x - 1)$
2. $\lim_{x \to -1} (x^2 + 2x + 1)$
3. $\lim_{x \to 2} \frac{x+1}{x+2}$
4. $\lim_{x \to -1} \frac{1 + 3x - x^2}{x^2 + x + 3}$
5. $\lim_{x \to 1} \frac{x^2 - 1}{x - 1}$
6. $\lim_{x \to -1} \frac{x^3 + x^2 + x + 1}{x + 1}$
7. $\lim_{x \to 4} \frac{x^2 - 3x - 4}{x - 3}$
8. $\lim_{x \to -3} \frac{x^2 + 2x - 1}{x^2 - 9}$
9. $\lim_{x \to 0} \frac{2x^3 - 3x^2 + 5x}{x^2 - 2x}$
10. $\lim_{x \to -1} \frac{(x+1)^5 - 4}{(x+1)^2 - x + 2}$
11. $\lim_{x \to 1} \sqrt{x^2 + 3}$
12. $\lim_{x \to -1} (x^2 - 2x + 1)^{-3/2}$
13. $\lim_{x \to \sqrt{5}} \frac{x^2 - 5}{x - \sqrt{5}}$
14. $\lim_{x \to -1} \frac{4 - (x^2 - 6x + 1)^{2/3}}{2 - (x^2 - 6x + 1)^{1/3}}$
15. $\lim_{x \to 8} \frac{x^{1/3} - 2}{x - 8}$
16. $\lim_{x \to -27} \frac{x^{1/3} + 3}{x + 27}$
17. $\lim_{x \to 81} \frac{x^{3/4} - 27}{9 - \sqrt{x}}$
18. $\lim_{x \to 2} \frac{1}{3 - \sqrt{2x^2 + 1}}$
19. $\lim_{x \to -8} \left(\frac{64 - x^2}{2x + 16} \right)^{1/3}$
20. $\lim_{x \to \sqrt{3}} \frac{x^4 - 8}{x^2 - 2}$

21. $\lim_{x \to 1} \sin(\pi x)$
22. $\lim_{x \to -1} \tan(\pi x) + 2$
23. $\lim_{x \to 1/2} \sin(\pi x)$
24. $\lim_{x \to 1} \cot\left(\frac{\pi}{4} x \right)$
25. $\lim_{x \to 0} \frac{1}{x} \sin(\pi x)$
26. $\lim_{x \to 1} \cos\left(\frac{\pi}{2} x \right) \left(\frac{1}{x - 1} \right)$
27. $\lim_{x \to 0} \frac{\sin(\pi x)}{\sin(2x)}$
28. $\lim_{x \to 0} \frac{1 - \cos(\pi x)}{x}$
29. $\lim_{x \to 0} 3x \cot x$
30. $\lim_{x \to 0} \frac{1}{x} (3x - \cos^2 x + \cos x)$

31. $\lim_{x \to 0+} \sqrt{x}$
32. $\lim_{x \to 0-} \frac{|x|^2}{x^2}$
33. $\lim_{x \to 0+} \frac{x}{|1 - x|}$
34. $\lim_{x \to 1-} \frac{\sin x}{|x|}$
35. $\lim_{x \to 0-} \frac{2 - x}{|4x|}$
36. $\lim_{x \to 1+} \frac{\sin(\pi x)}{|x - 1|}$
37. $\lim_{x \to 0-} x |\cot 2x|$
38. $\lim_{x \to 3-} \frac{\tan(\pi x)}{|9 - x^2|}$
39. $\lim_{x \to \pi-} \frac{\sin(2x)}{\sin |x|}$
40. $\lim_{x \to 0-} \frac{\sin^2(\pi x)}{x}$

Evaluate each of the following finite or infinite limits, provided that it exists.

41. $\lim_{x \to \infty} \frac{x^3 - 4x + 2}{2x^3 + 5}$
42. $\lim_{x \to -\infty} \frac{x^3 + 2x^2 - x - 4}{3x - 2}$
43. $\lim_{x \to 0+} \cot(-3x)$
44. $\lim_{x \to -\infty} \frac{\sin x}{3x}$

45. $\lim_{x \to 1+} \frac{1}{x^2 - 1}$

46. $\lim_{x \to -\infty} \frac{x^3 + x + 2}{2x^2 - 3x}$

47. $\lim_{x \to 2} \frac{1}{x^3 - 8}$

48. $\lim_{x \to \infty} \frac{x + 2}{2x\sqrt{x} - 2x + 1}$

49. $\lim_{x \to 2-} \frac{1}{\sin(\pi x)}$

50. $\lim_{x \to 0-} \frac{\cos^2 x + \sin^2 x}{x}$

II

51. Suppose a is a positive number. Evaluate

$$\lim_{x \to a^2} \frac{x - a^2}{a - \sqrt{x}}.$$

52. Suppose a is any real number. Evaluate

$$\lim_{x \to a^3} \frac{x - a^3}{a - x^{1/3}}.$$

53. To evaluate the limit

$$\lim_{x \to 0} \frac{\sin(2x)\cos(2x)}{3x},$$

it is possible to proceed in two different ways. One would be to break the numerator up as a product and deal with each part separately, and the other would be to recognize that the numerator can be expressed in terms of $\sin(4x)$. Evaluate the limit both ways and compare your answers.

*54. Suppose it is known that for every $x \in (1,3)$, $|k(x)| \le \sin^2 x$. Find

$$\lim_{x \to 2} \frac{x^2 - 4}{2 - k(x)}.$$

*55. Find examples of functions $f(x)$ and $g(x)$ and a number a so that

$$\lim_{x \to a} f(x) = 0, \qquad \lim_{x \to a} g(x) = \infty,$$

and for which $\lim_{x \to a^-} f(x)g(x) = -1$ and $\lim_{x \to a^+} f(x)g(x) = 1$.

*56. Find examples of functions $f(x)$ and $g(x)$ and a number a so that

$$\lim_{x \to a} f(x) = 0, \qquad \lim_{x \to a} g(x) = \infty,$$

and for which $\lim_{x \to a^-} f(x)g(x)$ and $\lim_{x \to a^+} f(x)g(x)$ do not exist, even as infinite limits.

*57. At each of the points where the function in Example 2.5.4(ii) is not continuous, describe the type of discontinuity.

*58. Find (i) $\lim_{x \to 0+} \frac{\sin x}{\sqrt{x}}$, (ii) $\lim_{x \to 0+} \frac{\sin(\sqrt{x})}{\sqrt{x}}$, (iii) $\lim_{x \to 0} \frac{\sin x}{x^{2/3}}$.

III

59. Use a graphics calculator or computer to draw the graph of the function $y = x^2 \sin \frac{1}{x}$ for $x \in [-1, 1]$, $x \in [-\frac{1}{10}, \frac{1}{10}]$, $x \in [-\frac{1}{100}, \frac{1}{100}]$, $x \in [-\frac{1}{1000}, \frac{1}{1000}]$.

60. In Section 2.1 we noted that the height of an object dropped from 128 ft at time $t = 0$ is $h(t) = 128 - 16t^2$. (i) Graph $A(t)$, the average speed of the object from 0 to t. (ii) Graph $S(t)$, the average speed of the object between t and $t + 0.1$. Repeat with $T(t)$, the average speed between t and $t + 0.01$.

For each of the following functions it is not easy to solve for the points at which the denominator is zero. (i) Use your computer or graphics calculator to find approximate values for the zeroes of the denominator. Zoom to higher magnification to find the values to three decimal places. Also note the sign of the denominator on either side of each zero. (ii) Predict the one-sided behavior of each function near each of the points you found in (i). Then graph the function and confirm your predictions.

61. $\frac{3x - 2}{x^3 - 4x + 1}$

62. $\frac{x^3 - 3x}{x^5 + 3x^3 + 2}$

63. $\frac{x^3 - 3x^2 + 3x - 1}{x^3 + 2x^2 - x - 2}$

64. $\frac{x^2 - x + 3}{x^4 - 2x^3 + 5x^2 - 4x + 4}$

Derivatives 3

Newton studied gravity, reputedly after watching an apple fall from a tree. Combining this work with his recently developed techniques of calculus, he was able to show how the laws of gravity explained the motions of the planets. This had been a great mystery, and his solution established both Newton's reputation and the importance of calculus.

INTRODUCTION

In this chapter we introduce and study the concept of the derivative. It is the first of two basic ideas in calculus (the other one is the integral, which we will study in Chapter 5). In Section 3.1 we discuss several different examples, which help to show where the idea arises and how it may be useful. In Section 3.2, derivatives are defined, and we learn the basic rules for working with them. In particular, we learn how to differentiate (i.e, find the derivatives of) polynomials and rational functions.

In Section 3.3 we see how derivatives can be used to solve what are known as *maximum/minimum problems*, that is, finding the value of x at which a particular function $f(x)$ has its maximum or minimum value. Solving this type of problem is an extremely important application of calculus. In Section 3.4 we learn how to find the derivatives of trigonometric functions. In Section 3.5 we learn the derivatives of logarithms and exponentials, and we also discuss what are called *hyperbolic functions* and their derivatives.

Section 3.6 is devoted to what is known as the *Chain Rule,* which is a formula for finding the derivative of composite functions. This idea is used in Section 3.7 to develop a technique called "implicit differentiation," which often saves a great deal of work in doing calculations. Derivatives of complicated functions are often found more easily by beginning with their logarithms, using a process known as *logarithmic differentiation.* Section 3.8 studies what are called the *inverse trigonometric functions* and the *inverse hyperbolic functions* and discusses finding their derivatives.

Finally, in Section 3.9 there is a discussion of second and higher-order derivatives. The idea is that if we begin with a function and find its derivative, the derivative will be a function in its own right. The derivative of this function is called the *second derivative* of the original function (since it is obtained by performing twice the operation of finding the derivative).

3.1 Introduction to Differentiation

TANGENT LINE
SLOPE
DIFFERENCE QUOTIENT
DISPLACEMENT
VELOCITY
RATE OF CHANGE

In Section 2.1 we discussed the question of finding the slope of the tangent line to a curve $y = f(x)$ at some specified point on the curve. Ordinarily, the way to find the slope of a line is to start with two points on the line, (x_1, y_1) and (x_2, y_2), say, and to use the formula

$$slope = \frac{rise}{run} = \frac{y_2 - y_1}{x_2 - x_1}.$$

See Figure 3.1.1. The difficulty with applying this approach to tangents is that while we know the point of tangency, call it $\big(a, f(a)\big)$, there is often no apparent way to find a second point on the tangent line. See Figure 3.1.2.

One way around this problem is to consider a second point on the curve, $\big(x, f(x)\big)$, and consider the line through both it and the first point $\big(a, f(a)\big)$. This line is called a secant, and if $\big(x, f(x)\big)$ is chosen to be close to $\big(a, f(a)\big)$, it will likely be close to the tangent line, as Figure 3.1.3 shows. It also seems reasonable to expect

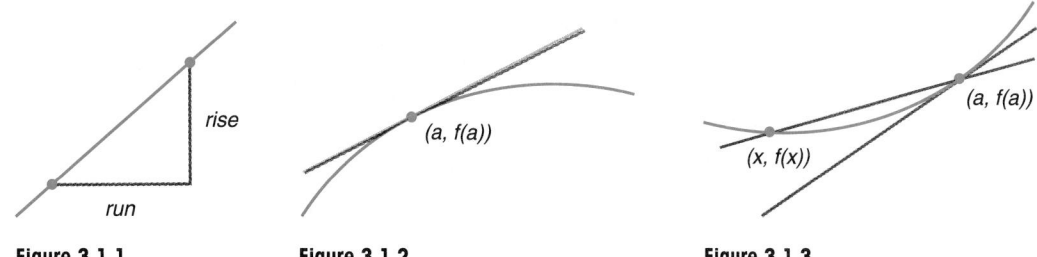

Figure 3.1.1   **Figure 3.1.2**   **Figure 3.1.3**

that its slope will be close to the slope of the tangent line, which is what we want to find. Rather than worry about the awkward question of "how close?," we can let x approach a and find the limit of the slopes of the corresponding secants.

EXAMPLE 1   (i) Find the slope of the line that is tangent to the curve $y = x^3 - 2x + 3$ at the point $(1, 2)$.

If (x, y) is on the curve, then $y = x^3 - 2x + 3$.

Figure 3.1.4

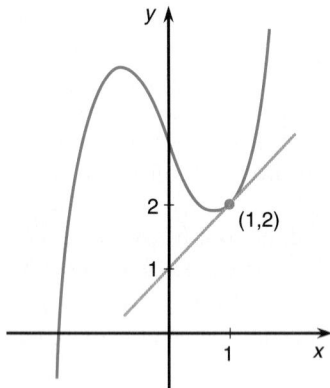

Figure 3.1.5

Solution   First observe that the point $(1, 2)$ is indeed on the curve. Then consider a second point, $(x, x^3 - 2x + 3)$, and the secant through that point and $(1, 2)$. See Figure 3.1.4.

Using the formula we can find the slope of the secant; it is

$$slope = \frac{rise}{run} = \frac{(x^3 - 2x + 3) - 2}{x - 1} = \frac{x^3 - 2x + 1}{x - 1}.$$

The slope of the tangent should be the limit of the slope of the secant as $x \to 1$, so we need to evaluate $\lim_{x \to 1} \frac{x^3 - 2x + 1}{x - 1}$. Applying our techniques for limits of rational functions (see Summary 2.2.7), we notice that both numerator and denominator tend to zero as $x \to 1$. Dividing the numerator by $x - 1$ (see Theorem 1.3.5, the Factor Theorem, p. 12), we find

$$\frac{x^3 - 2x + 1}{x - 1} = \frac{(x - 1)(x^2 + x - 1)}{x - 1} = x^2 + x - 1.$$

Taking the limit is easy now:

$$\lim_{x \to 1} \frac{x^3 - 2x + 1}{x - 1} = \lim_{x \to 1} (x^2 + x - 1) = 1.$$

The slope of the tangent at $(1, 2)$ should be the limit of the slopes of the secants as $x \to 1$, which is 1 (see Figure 3.1.5).

(ii) Find the slope of the tangent to the parabola $y = x^2$ at the point (a, a^2).

Solution   Consider the secant through (a, a^2) and (x, x^2). Its slope is $\frac{x^2 - a^2}{x - a}$. The slope of the tangent should be

$$\lim_{x \to a} \frac{x^2 - a^2}{x - a} = \lim_{x \to a} \frac{(x - a)(x + a)}{x - a} = \lim_{x \to a} (x + a) = 2a.$$

Notice that the slope of the tangent will be different at different points. For large values of a, that is, at points far out along the parabola, the slope will be large, meaning that the tangent rises steeply up to the right. When $a = 0$, that is, at the origin, the slope of the tangent is zero, meaning that it is a horizontal line. When $a < 0$, the slope is negative, meaning that the tangent line slopes down to the right (see Figure 3.1.6).

Figure 3.1.6

With a graphics calculator we can plot the function $y = x^2$ and some of its tangent lines. For instance, by Example 1, the tangent line at $(1, 1)$ has slope $2(1) = 2$. Since it passes through $(1, 1)$, its equation is $\frac{y-1}{x-1} = 2$, that is, $y - 1 = 2(x - 1)$, that is, $y = 2x - 1$. Similarly, the tangent line at $(-1, 1)$ has slope -2 and equation $\frac{y-1}{x+1} = -2$, that is, $y = 1 - 2(x + 1) = -2x - 1$.

TI-85	SHARP EL9300
GRAPH F1 $(y(x) =)$ F1 $(x) x^2$ ▼ 2	x/θ/T x^2 2ndF ▼ 2 x/θ/T $- 1$
F1 $(x) - 1$ ▼ $(-)2$ F1 $(x) - 1$	2ndF ▼ $- 2$ x/θ/T $- 1$ RANGE -2
M2(RANGE) $(-)2$ ▼ 2 ▼ ▼ $(-)2$	ENTER 2 ENTER 1 ENTER -2 ENTER
▼ 4 F5(GRAPH)	4 ENTER 1 ENTER

CASIO f_x-7700GB/f_x-6300G	HP 48SX
Cls EXE Range -2 EXE 2 EXE 1 EXE	PLOT PLOTR ERASE ATTN '
-2 EXE 4 EXE 1 EXE EXE EXE EXE	αY ▬ αX y^x 2 ↱ PLOT ↰ DRAW
Graph X x^2 EXE Graph 2 X $- 1$	ERASE 2 +/- ENTER 2 XRNG 2 +/-
EXE Graph $- 2$ X $- 1$ EXE	ENTER 4 YRNG DRAW ATTN '
	αY ▬ 2 $\times$ $\alpha X - 1$ ↱ PLOT ↰
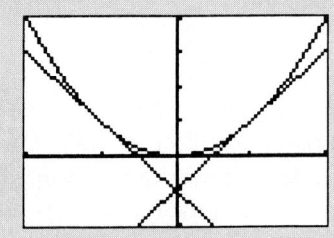	DRAW DRAW ATTN ' αY ▬ 2 +/-
	$\times$ $\alpha X - 1$ ↱ PLOT ↰ DRAW
	DRAW

Upon reflection it is clear that we have used the notion of a tangent line without ever really saying what it meant, and then figured out what its slope ought to be. On the other hand, now that we know what its slope ought to be, we are in a position to say what a tangent line is.

DEFINITION 3.1.1

Consider the function $y = f(x)$, and suppose a is a point in the domain of f, so $(a, f(a))$ is a point on the graph of $y = f(x)$. The **tangent line** to $y = f(x)$ at the point $(a, f(a))$ is the straight line through $(a, f(a))$ whose slope is

$$m = \lim_{x \to a} \frac{f(x) - f(a)}{x - a},$$

provided the limit exists. In particular, if the limit m does exist, the equation of the line is

$$y - f(a) = m(x - a), \qquad \text{or, equivalently,} \qquad y = f(a) + m(x - a).$$

If the limit does not exist, then neither does the tangent line.

Figure 3.1.7

Figure 3.1.8

Figure 3.1.9

Figure 3.1.10

For obvious reasons, the quantity $\frac{f(x)-f(a)}{x-a}$ is called a **difference quotient**. It equals the slope of the secant through $\big(a, f(a)\big)$ and $\big(x, f(x)\big)$.

EXAMPLE 2 (i) Find the equation of the tangent line to the curve $y = f(x)$, where $f(x) = x^2 - 4x - 1$, at the point $(1, -4)$.

Solution The slope of the tangent line is

$$m = \lim_{x \to 1} \frac{f(x) - f(1)}{x - 1} = \lim_{x \to 1} \frac{(x^2 - 4x - 1) - \big(1^2 - 4(1) - 1\big)}{x - 1}$$

$$= \lim_{x \to 1} \frac{x^2 - 4x + 3}{x - 1} = \lim_{x \to 1} \frac{(x - 1)(x - 3)}{x - 1} = \lim_{x \to 1} (x - 3) = -2.$$

Since the line passes through $(1, -4)$, its equation is

$$y - (-4) = -2(x - 1), \qquad \text{or} \qquad y = -2x - 2.$$

See Figure 3.1.7.

(ii) If we consider the graph of $y = f(x)$, where $f(x) = |x|$, and ask about the tangent line at the point $(0, 0)$, we have to evaluate the limit

$$\lim_{x \to 0} \frac{f(x) - f(0)}{x - 0} = \lim_{x \to 0} \frac{|x| - 0}{x} = \lim_{x \to 0} \frac{|x|}{x}.$$

As we saw in Example 2.4.1(i), the function $\frac{|x|}{x}$ approaches different limits from above and below ($+1$ as $x \to 0^+$ and -1 as $x \to 0^-$). Accordingly, the (two-sided) limit does not exist (by Theorem 2.4.2), and therefore the tangent line does not exist (see Figure 3.1.8). The idea is that there is no single line that is an obvious candidate to be the tangent line; if we approach 0 from the left, it appears that the slope of the tangent should be -1, but if we approach from the right, the slope appears to be $+1$.

Another idea that we discussed in Section 2.1 was the concept of instantaneous speed. Suppose an object is moving in a straight line, and suppose some point on the line has been marked as a reference for making measurements (following the usual convention, we label this point O and call it the origin). See Figure 3.1.9.

Suppose we know where the moving object is at any particular moment. In effect this means that we know its position relative to the origin at any time t, or more specifically that we know a function $f(t)$ that tells us its position at time t. See Figure 3.1.10. Once again we follow the convention and measure positions on the line in such a way that when $f(t) > 0$, it represents a point to the right of the origin, and when $f(t) < 0$, it represents a point to the left of the origin. The absolute value $|f(t)|$ indicates the distance from the origin to the object. For example, if $f(3) = -2$, we know that at time $t = 3$ the object is 2 units to the left of the origin. Similarly, if $f(1.5) = \sqrt{2}$, then at time $t = 1.5$ the object is $\sqrt{2} \approx 1.41421$ units to the right of the origin. We say that $f(t)$ is the **displacement** of the object from the origin at time t. Displacement measures the distance, but its sign also tells us the direction from the origin. In contrast, when we speak of distance, we always mean a nonnegative number, and no direction is implied.

The average speed of a moving object is simply the distance it travels divided by the time it takes. To find the average speed of the object between times $t = 2$ and

$t = 4$, for example, we could consider the quotient

$$\frac{f(4) - f(2)}{4 - 2} = \frac{f(4) - f(2)}{2}.$$

There is a catch, however. The numerator is the difference between the *displacements* of the object at the two times $t = 2$ and $t = 4$, which is almost the same as the distance between the corresponding positions, except that it might be negative. The difference quotient we have written down is the change in the displacement divided by the length of time. If the quotient is negative, it represents the average speed of a motion to the left, and if it is positive, it is the average speed of a motion to the right. Strictly speaking, between $t = 2$ and $t = 4$ the object might easily move to the right some of the time and to the left some of the time, but the sign of the quotient tells us the direction of the *net* motion.

To distinguish this from the average speed, which we understand always to be nonnegative, it is called the average **velocity**. The absolute value of the velocity is the speed of a motion, and the sign of the velocity indicates the direction (positive means to the right and negative means to the left).

The difference quotient we considered above, the change in the displacement divided by the time, represents the average velocity of the object between $t = 2$ and $t = 4$. If we want to know its velocity at the "instant" when $t = 2$, we cannot simply take a quotient. Instead, we should take difference quotients corresponding to very brief intervals starting at $t = 2$ and evaluate the limit as the length of the intervals approaches zero.

Specifically, if we consider an interval of time of length h, say, what we mean is the interval from $t = 2$ to $t = 2 + h$. The change in the displacement is $f(2 + h) - f(2)$, and the length of time elapsed is $(2 + h) - 2 = h$. The average velocity is

$$\frac{change\ in\ displacement}{time} = \frac{f(2 + h) - f(2)}{h}.$$

To find the instantaneous velocity, we should let $h \to 0$ and evaluate the limit. In other words, the instantaneous velocity at $t = 2$ is

$$\lim_{h \to 0} \frac{f(2 + h) - f(2)}{h}.$$

This limit is very similar to the one we used in Definition 3.1.1 for the slope of the tangent line. It arises so often that there is a special name and notation for it.

DEFINITION 3.1.2

The **derivative** of $f(x)$ at $x = a$, written $f'(a)$, equals the following limit, provided it exists:

$$f'(a) = \lim_{x \to a} \frac{f(x) - f(a)}{x - a} = \lim_{h \to 0} \frac{f(a + h) - f(a)}{h}.$$

(Notice that the two apparently different limits are really the same, since letting $h \to 0$ makes $(a + h) \to a$, so the first expression is obtained from the second by replacing $a + h$ with x.)

Both $\frac{f(x) - f(a)}{x - a}$ and $\frac{f(a+h) - f(a)}{h}$ are called **difference quotients**.

EXAMPLE 3   (i) Suppose the displacement of a moving object at time t sec is given, in feet, by the function $f(t) = t^2 - 8t - 4$. What is its (instantaneous) velocity at time $t = 3$?

Solution   The instantaneous velocity at $t = 3$ is

$$f'(3) = \lim_{h \to 0} \frac{f(3+h) - f(3)}{h}.$$

Simplifying the difference quotient, we find that

$$\frac{f(3+h) - f(3)}{h} = \frac{\left((3+h)^2 - 8(3+h) - 4\right) - \left(3^2 - 8(3) - 4\right)}{h}$$

$$= \frac{9 + 6h + h^2 - 24 - 8h - 4 - 9 + 24 + 4}{h} = \frac{-2h + h^2}{h}$$

$$= -2 + h \text{ ft/sec.}$$

So the instantaneous velocity at $t = 3$ is

$$f'(3) = \lim_{h \to 0} (-2 + h) = -2 \text{ ft/sec.}$$

This means that at the instant $t = 3$ sec the object is moving to the left at a speed of 2 ft/sec.

The difference quotient $\frac{f(3+h) - f(3)}{h}$ is precisely the average velocity of the moving object between the times $t = 3$ and $t = 3 + h$.

(ii) At what time t will the above moving object be exactly at rest, that is, have velocity equal to zero?

Solution   The velocity at $t = a$ is

$$f'(a) = \lim_{h \to 0} \frac{f(a+h) - f(a)}{h}$$

$$= \lim_{h \to 0} \frac{\left((a+h)^2 - 8(a+h) - 4\right) - (a^2 - 8a - 4)}{h}$$

$$= \lim_{h \to 0} \frac{a^2 + 2ah + h^2 - 8a - 8h - 4 - a^2 + 8a + 4}{h}$$

$$= \lim_{h \to 0} \frac{2ah + h^2 - 8h}{h}$$

$$= \lim_{h \to 0} (2a + h - 8) = 2a - 8.$$

This equals zero precisely when $a = 4$.

Notice that when $a < 4$, the velocity at $t = a$ will be negative, and when $a > 4$, the velocity will be positive. So the object will be moving to the left until $t = 4$, and after $t = 4$ it will be moving to the right. The instant when $t = 4$ is the instant at which the object comes to rest as it "turns around."

Now that we have the notation for derivatives, it is easier to deal with tangent lines.

EXAMPLE 4   (i) Find the slope of the tangent at each point of $y = x^2 - 2x + 2$.

Solution   Let $f(x) = x^2 - 2x + 2$. Any point on the graph can be written in the form $(a, a^2 - 2a + 2)$, and the slope of the tangent at that point will just be

$$f'(a) = \lim_{h \to 0} \frac{f(a + h) - f(a)}{h}$$

$$= \lim_{h \to 0} \frac{(a + h)^2 - 2(a + h) + 2 - (a^2 - 2a + 2)}{h}$$

$$= \lim_{h \to 0} \frac{a^2 + 2ah + h^2 - 2a - 2h + 2 - a^2 + 2a - 2}{h}$$

$$= \lim_{h \to 0} \frac{2ah + h^2 - 2h}{h} = \lim_{h \to 0} (2a + h - 2)$$

$$= 2a - 2.$$

(ii) Find the equation(s) of the tangent lines to the curve $y = x^2 - 2x + 2$ that pass through the point $(2, 0)$.

Solution   Observe that the point $(2, 0)$ does not lie on the curve (it does not satisfy the equation). So what the question asks us is to find lines that are tangent to the curve at some point on the curve and that also happen to pass through $(2, 0)$.

Consider the tangent at the point $(a, a^2 - 2a + 2)$. Its equation is $y = f(a) + f'(a)(x - a)$. (Notice that we have written $f'(a)$ instead of the awkward limit m that appeared in Definition 3.1.1.) The question we must answer is for what values of a does this line pass through $(2, 0)$?

Substituting into this equation of the tangent the fact that $f(a) = a^2 - 2a + 2$ and the result of part (i) that $f'(a) = 2a - 2$, we find that the tangent is

$$y = a^2 - 2a + 2 + (2a - 2)(x - a) = a^2 - 2a + 2 + (2a - 2)x - 2a^2 + 2a,$$

or

$$y = (2a - 2)x - a^2 + 2.$$

For $(2, 0)$ to be on this line means $0 = (2a - 2)(2) - a^2 + 2 = 4a - 4 - a^2 + 2$, or $a^2 - 4a + 2 = 0$.

Solving by using the Quadratic Formula (Theorem 1.3.4), we find

$$a = \frac{4 \pm \sqrt{16 - 8}}{2} = 2 \pm \frac{\sqrt{8}}{2} = 2 \pm \sqrt{2}.$$

We have found that the tangent lines to $y = f(x)$ that pass through $(2, 0)$ are the ones whose points of tangency are at $x = 2 + \sqrt{2}$ and $x = 2 - \sqrt{2}$.

We check this result: When $a = 2 + \sqrt{2} \approx 3.4142$, the tangent line, as we found above, is

$$y = (2a - 2)x - a^2 + 2 = (4 + 2\sqrt{2} - 2)x - (4 + 4\sqrt{2} + 2) + 2,$$

or

$$y = (2 + 2\sqrt{2})x - 4 - 4\sqrt{2}.$$

This line certainly passes through $(2, 0)$.

A similar calculation shows that when $a = 2 - \sqrt{2} \approx 0.5858$, the tangent line is

$$y = (2 - 2\sqrt{2})x - 4 + 4\sqrt{2},$$

which also contains $(2, 0)$. See Figure 3.1.11.

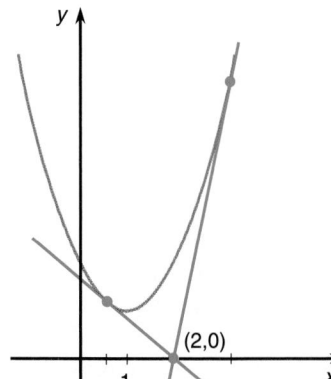

Figure 3.1.11

EXAMPLE 5 **Ideal Gas Law** Suppose a quantity of some kind of gas is confined in a container at a fixed temperature. The **Ideal Gas Law** of physics says that the volume of the gas, V, and the pressure, P, are related by the following equation:

$$V = \frac{k}{P},$$

where k is some constant.

In other words, the volume is *inversely proportional* to the pressure, which is a precise way of stating the obvious fact that the harder you squeeze the gas, the smaller will be the volume it occupies.

If the pressure is P, then the volume will be $V = \frac{k}{P}$. If we change the pressure slightly, the volume will change in response. For instance, if we change the pressure from P to $P + h$, then the volume will change to $V = \frac{k}{P+h}$.

Thinking of V as a function of P, we can find its derivative. For fixed P,

$$V'(P) = \lim_{h \to 0} \frac{V(P+h) - V(P)}{h} = \lim_{h \to 0} \frac{\frac{k}{P+h} - \frac{k}{P}}{h} = \lim_{h \to 0} \frac{\frac{kP - k(P+h)}{(P+h)P}}{h}$$

$$= \lim_{h \to 0} \frac{\frac{-kh}{(P+h)P}}{h} = \lim_{h \to 0} \frac{-k}{(P+h)P}$$

$$= \frac{-k}{P^2}.$$

In doing this calculation we think of P as a *constant*, since it is some *fixed* value of the pressure from which we start. So, for instance, in the evaluation of the last limit we regard it as a rational function in the variable h, and the other two symbols, k and P, are constants.

The difference quotient whose limit we took was the ratio of the change in V to the change in P. We think of the change in volume V as the result of the change in pressure P. If the limit, that is, the derivative, is large, it suggests that a small change in pressure will result in quite a large change in volume; if the derivative is small, a change in the pressure will cause only a small change in the volume. The fact that the derivative is negative reflects the fact that an increase in the pressure results in a *decrease* in the volume and vice versa. If, for example, $V'(P) = -2$, it means that an increase in the pressure by 1 unit, starting from P, should be expected to cause a decrease in the volume of about 2 units.

These considerations lead us to think of the derivative $V'(P)$ as the *rate of change* of the volume with respect to the pressure.

In the earlier discussion of velocity we could also have thought of the velocity as the rate of change of the displacement with respect to time. The bigger the velocity is, the faster the displacement will change as time passes.

> In general, if $y = f(x)$ is some function of the variable x, the derivative $f'(a)$ can be thought of as the **rate of change** of y with respect to x at $x = a$. If we imagine ourselves making a small change in the value of x near a, the derivative tells us what kind of change we should expect in the value of $y = f(x)$. The sign of the derivative tells us whether to expect an increase or a decrease.

Now consider another type of example. Suppose a factory produces a single product, blenders, for instance. Let N be the number of blenders produced in a day, the "level of production." The profit P earned by the factory will depend on N, probably in a very complicated way (it will be best when the factory is operating at its proper capacity; operating too far below or above capacity will likely reduce profits).

Now suppose somebody is able to describe the way profits depend on production level by giving us a function $P = f(N)$. In other words, for any given level of production N, the value $P = f(N)$ will be the resulting profit. In actual situations, such a function is usually impossible to find exactly, though it is not so difficult to make approximations.

If the factory is producing N_0 blenders per day, it would be reasonable to want to know whether this is a good level of production. One way to accomplish that would be to ask what would be the effect on the profit of increasing or decreasing production. An answer to this question would certainly help in deciding how best to run the factory.

To answer the question, we could ask what happens to P if we increase production by some small amount h. The new level of production would be $N_0 + h$, and the corresponding profit would be $P = f(N_0 + h)$. The change in the profit would be $f(N_0 + h) - f(N_0)$, and to some extent the question is answered by knowing this change. Specifically, if it is positive, then increased production increases profit, and if it is negative, then increased production decreases profit. However, this simple observation does not take into account the size of h. It is clearly important to know how much of a change in P will result from a particular change in N.

The difference quotient $\frac{f(N_0+h)-f(N_0)}{h}$ represents the change in profit *per unit change in production*. If this quotient equals \$5, for example, it means that increasing production by h blenders will result in a \$5h increase in profit, or \$5 per additional blender produced.

In keeping with the spirit of the previous examples we can let $h \to 0$ and consider the limit

$$\lim_{h \to 0} \frac{f(N_0 + h) - f(N_0)}{h},$$

which we can think of as the *rate of change of profit per unit change in production*. What it tells us is roughly how much profit will increase if we increase production by one unit. In economics this limit is called the **marginal increase in profit**.

EXAMPLE 6   For a factory producing N blenders per day, suppose the profit, in dollars, is given by $P = P(N) = 6{,}000N - N^2 - 8{,}985{,}000$.

We find that the derivative $P'(N)$ equals

$$\lim_{h \to 0} \frac{6{,}000(N + h) - (N + h)^2 - 8{,}985{,}000 - (6{,}000N - N^2 - 8{,}985{,}000)}{h}$$

$$= \lim_{h \to 0} \frac{6{,}000N + 6{,}000h - N^2 - 2hN - h^2 - 8{,}985{,}000 - 6{,}000N + N^2 + 8{,}985{,}000)}{h}$$

$$= \lim_{h \to 0} \frac{6{,}000h - 2hN - h^2}{h}$$

$$= \lim_{h \to 0} (6{,}000 - 2N - h)$$

$$= 6{,}000 - 2N.$$

The interesting thing about this derivative is that it is easy to see when it is negative and when it is positive. In fact, it is positive when $N < 3{,}000$ and negative when $N > 3{,}000$. This suggests that if $N < 3{,}000$, a small increase in production would result in an increase in profits but that if $N > 3{,}000$, a small increase in production would result in lower profits.

From this it seems reasonable to expect that the most profit will be obtained with $N = 3{,}000$, that is, that the most profitable way to operate the factory is to produce 3,000 blenders per day.

We can check this to some extent by direct calculation. If $N = 3{,}000$, we find that $P = P(3{,}000) = 6{,}000(3{,}000) - 3{,}000^2 - 8{,}985{,}000 = 18{,}000{,}000 - 9{,}000{,}000 - 8{,}985{,}000 = 15{,}000$, or a profit of \$5.00 for each of the 3,000 blenders produced. If, on the other hand, we were to produce one more: $N = 3{,}001$, then $P = 6{,}000(3{,}001) - 3{,}001^2 - 8{,}985{,}000 = 18{,}006{,}000 - 9{,}006{,}001 - 80{,}985{,}000 = 14{,}999$. In other words, total profits would decrease by \$1.00. You can check that if $N = 2{,}999$, the profit would also be less than 15,000.

The above calculation encourages us to think that our reasoning may be correct, but it certainly does not prove conclusively that it is impossible to get higher profits by using some other value of N.

On the other hand, if we take the formula $P = 6{,}000N - N^2 - 8{,}985{,}000$ and complete the square, we find

$$P = -(N^2 - 6{,}000N + 8{,}985{,}000) = -(N^2 - 6{,}000N + 9{,}000{,}000 - 15{,}000)$$
$$= -(N^2 - 6{,}000N + 9{,}000{,}000) + 15{,}000 = 15{,}000 - (N - 3{,}000)^2.$$

This last expression is 15,000 minus a square, and since any square must always be nonnegative, the expression must always be less than or equal to 15,000. Moreover, the only way for it to equal 15,000 is for the square to be zero, which can happen only if $N = 3{,}000$.

So in fact the highest possible profit is 15,000, which will be obtained precisely when $N = 3{,}000$. ◣

It is important to mention once again that while this is fine as far as it goes, we assumed at the beginning that we knew how to express the profit P as a function of production N, but in the real world it is unlikely that we could do that.

Exercises 3.1

I

For each of the following functions $f(x)$, calculate the limit of the difference quotient to find the derivative $f'(a)$ at the specified point a.

1. $f(x) = x^2 + 1, a = 2$

2. $f(x) = x^2 - 3, a = -1$

3. $f(x) = x^2 - x, a = 1$

4. $f(x) = x^2 - 3x, a = 0$

5. $f(x) = 2x^2 - x + 1, a = 2$

6. $f(x) = 3x^2 - 2x + 1, a = -3$

7. $f(x) = -3x^2 + 2x - 2, a = 0$

8. $f(x) = -x^2 + 4x, a = 7$

9. $f(x) = x^3, a = 1$

10. $f(x) = x^3 + 4x - 1, a = -2$

11.– 20. For each of the functions $f(x)$ in Exercises 1 through 10, find the equation of the line that is tangent to the graph of $y = f(x)$ at the point $\big(a, f(a)\big)$.

II

21. Find all values a for which the tangent line to $y = f(x) = x^2 - 2x + 3$ at $\big(a, f(a)\big)$ is horizontal.

22. Find all values a for which the tangent line to $y = f(x) = 2x^2 + 2x - 5$ at $\big(a, f(a)\big)$ is horizontal.

23. (i) Find all tangent lines to the graph of $y = x^2$ that pass through the point $(0, -1)$.

(ii) Find all tangent lines to the graph of $y = x^2$ that pass through the point $(0, 1)$. (Draw a picture.)

24. (i) If the height of a falling object t sec after its release is $H(t) = 128 - 16t^2$ ft, find its velocity $H'(t)$ at time t.

(ii) What is the significance of the sign of the velocity?

25. Suppose a ball is thrown upward in such a way that its height above the ground after t sec in flight is $H(t) = 6 + 32t - 16t^2$ ft.

(i) Find its velocity after t sec in flight.

(ii) When is it rising and when is it falling?

(iii) What is the greatest height it will reach?

26. Suppose a train starts at the origin and moves along a straight track. Suppose after t minutes it has gone $1000t - 50t^2$ ft.

(i) What is its velocity after t min in feet per minute?

(ii) What is the significance of the sign of the velocity?

(iii) When will the train come to rest?

(iv) How far will it have traveled at the moment when it comes to rest?

(v) What will be the greatest speed it reaches before coming to rest?

(vi) Express the answer to part (v) in miles per hour.

(vii) Where will the train be 20 min after it starts?

27. Suppose a balloon contains 0.4 quart of air at a pressure of 15 pounds/sq in (psi).

(i) In the terminology of Example 5, find the constant k for this situation.

(ii) Suppose the pressure is increased to 30 psi. What will be the volume?

(iii) Compare the rate of change of volume with respect to pressure in these two situations (i.e., when $P = 15$ and $P = 30$). In which case will increasing the pressure by 1 psi cause the greater change in the volume?

(iv) Use your answer to part (iii) when $P = 15$ psi to estimate what the volume will be if the pressure is increased by 1 psi to 16 psi.

(v) Calculate the exact volume when $P = 16$ psi, using the formula of Example 5. (The discrepancy arises because the rate of change we found when $P = 15$ psi will not be the rate of change when $P = 16$ psi or 15.5 psi or other points in between. It will be reasonably close, so this approach can be used for estimating, but it will not give exact answers.)

28. Suppose the profit of a factory producing N electric blankets per week is given by $P = 400N - N^2 - 38{,}000$ dollars.

(i) What would you recommend the manager do to increase profits if the level of production was $N = 100$?

(ii) What would you suggest if $N = 500$?

(iii) Can you suggest a strategy that will describe what to do for any value of N?

(iv) What is the greatest possible profit this factory can make under the present circumstances, and how can they make this level of profit?

*29. (i) Use Definition 3.1.2 (p. 133) to find the derivative $f'(1)$ if $f(x) = Ax^2 + Bx + C$, where A, B, C are all constants.

(ii) Use Definition 3.1.2 to find the derivative $f'(a)$, where $f(x)$ is as above and a is some fixed but unspecified number.

*30. (i) Use Definition 3.1.2 to find the derivative at $x = 2$ of $f(x) = x^3$.

(ii) Use Definition 3.1.2 to find the derivative $f'(a)$, where $f(x) = x^3$.

*31. Let $f(x) = |x|$. Specify the points a at which the derivative $f'(a)$ does not exist, and find the derivative at all the points where it does exist.

*32. (i) Show that the derivative of a straight line $y = mx + b$ equals the slope of the line.

(ii) Notice that the straight line $y = mx + b$ is never a vertical line. What goes wrong with part (i) above if we try to do the same thing for a vertical line?

33. In Example 1 (p. 129) we discussed the tangent line to $y = x^3 - 2x + 3$ at the point $(1, 2)$. Use a graphics calculator or computer to draw this curve and the secant lines through (i) $(1, 2)$ and $(2, 7)$, (ii) $(1, 2)$ and $\left(\frac{3}{2}, \frac{27}{8}\right)$, (iii) $(1, 2)$ and $(1.1, 2.131)$.

34. In Example 2 (p. 132) we discussed the tangent line to $y = f(x) = x^2 - 4x - 1$ at the point $(1, -4)$. Graph the curve and the secant through $(1, -4)$ and $\left(1 + h, f(1 + h)\right)$, where (i) $h = 1$, (ii) $h = \frac{1}{2}$, (iii) $h = 0.1$, (iv) $h = 0.01$.

35. Consider the moving object of Example 3 (p. 134). At $t = 3$ its displacement is $f(3) = 3^2 - 8(3) - 4 = -19$. Let $D(h)$ be the net displacement away from this point at time $t = 3 + h$, that is, $D(h) = f(3 + h) - f(3)$.

(i) Graph $D(h)$.

(ii) Let $V(h)$ be the object's average velocity between $t = 3$ and $t = 3 + h$. Graph $V(h)$, and use the picture to confirm the result we found for the instantaneous velocity at $t = 3$.

36. Use your graphics calculator or computer as in Exercise 35 to find the instantaneous velocity of the same moving object at $t = 6$.

37. Graph the profit function $P(N) = 6{,}000N - N^2 - 8{,}985{,}000$ of Example 6. We found that the highest profit is $P = 15{,}000$, obtained when $N = 3{,}000$. Use the picture to find all the values of N for which the profit is at least $10{,}000$.

3.2 Differentiation Rules; Derivatives of Polynomials and Rational Functions

DERIVATIVES OF POLYNOMIALS
PRODUCT RULE
QUOTIENT RULE
DERIVATIVES OF RATIONAL FUNCTIONS
EQUATIONS OF TANGENT LINES

We have defined the derivative of a function $f(x)$ at a point a to be the limit of the difference quotient:

$$f'(a) = \lim_{h \to 0} \frac{f(a+h) - f(a)}{h},$$

provided the limit exists. We saw in Example 3.1.2(ii) that the derivative does not always exist (in that example we saw that $f(x) = |x|$ does not have a derivative at $x = 0$). If the derivative $f'(a)$ exists (i.e., the above limit exists), we say the function $f(x)$ is **differentiable** at $x = a$. To say that $f(x)$ is "differentiable on an interval (c, d)" means that $f(x)$ is differentiable at every $a \in (c, d)$. If $f(x)$ is differentiable at every point in its domain, we simply say that $f(x)$ is differentiable.

Roughly speaking, the derivative $f'(a)$ is the slope of the tangent line to the graph of $y = f(x)$ at the point $(a, f(a))$. The reason why $f(x) = |x|$ is not differentiable at $x = 0$ is that it does not have a tangent line there, and that happens because the graph has a "corner" in it (see Figure 3.2.1).

If the function $f(x)$ has a tangent at $x = a$, the graph of $y = f(x)$ must lie "close" to the graph of the tangent line near the point $(a, f(a))$. This suggests that $f(x)$ ought to be continuous at $x = a$. Certainly, if $f(x)$ had a removable or jump discontinuity at $x = a$, it is hard to imagine that it could have a tangent there. See Figure 3.2.2.

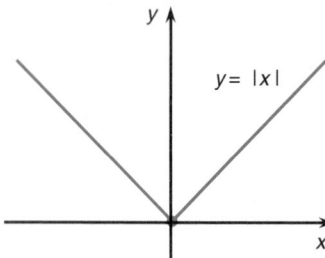

Figure 3.2.1

THEOREM 3.2.1

> Suppose $f(x)$ is differentiable at $x = a$. Then $f(x)$ must be **continuous** at $x = a$.

PROOF   To show that $\lim_{x \to a} f(x) = f(a)$ amounts to the same thing as showing that $\lim_{x \to a} (f(x) - f(a)) = 0$. And

$$\lim_{x \to a} (f(x) - f(a)) = \lim_{x \to a} \frac{f(x) - f(a)}{x - a}(x - a)$$

$$= \left(\lim_{x \to a} \frac{f(x) - f(a)}{x - a} \right) \left(\lim_{x \to a} (x - a) \right)$$

$$= (f'(a))(0) = 0.$$

Note that dividing by $x - a$ is not allowed at $x = a$, but to find the limit, we need to worry only about the points where $x \neq a$ . . .

So $\lim_{x \to a} f(x) = f(a)$, which says that $f(x)$ is continuous at $x = a$, as required. ◀

The theorem says that if a function is differentiable, then it will also be continuous. However, being continuous is not enough to guarantee that a function will be differentiable, as the function $f(x) = |x|$ shows. After all, $f(x)$ is continuous everywhere, but it is not differentiable at $x = 0$.

As with limits, there are some derivatives that are very easy to evaluate.

Figure 3.2.2

Figure 3.2.3

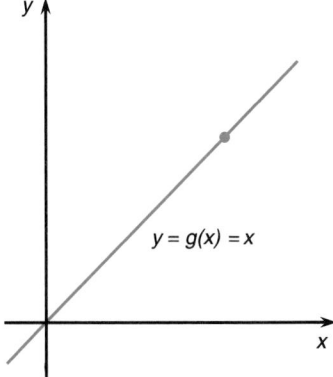

Figure 3.2.4

EXAMPLE 1   Let $f(x) = c$ be a constant function; let $g(x) = x$. Find $f'(x)$ and $g'(x)$ for all values of x.

Solution   We simply write down the difference quotients and evaluate the limits:

$$f'(x) = \lim_{h \to 0} \frac{f(x+h) - f(x)}{h} = \lim_{h \to 0} \frac{c - c}{h} = \lim_{h \to 0} \frac{0}{h} = 0.$$

Similarly,

$$g'(x) = \lim_{h \to 0} \frac{g(x+h) - g(x)}{h} = \lim_{h \to 0} \frac{(x+h) - x}{h} = \lim_{h \to 0} \frac{h}{h} = 1.$$

In other words, the derivative of any constant function is zero, and the derivative of $g(x) = x$ is the constant function 1. In particular, constant functions and the function $g(x) = x$ are differentiable at every point. See Figure 3.2.3, which shows that the tangent line to any constant function is a horizontal line, so its slope is zero, and Figure 3.2.4, which shows that the tangent line to $y = g(x) = x$ is the line $y = x$, whose slope is 1.

There is another notation for derivatives with which we should become familiar. When we defined $f'(a)$, the number a was fixed, but we have already seen that it is often possible to work out a formula for $f'(a)$ that tells what it will be for any choice of a. In this light the derivative itself becomes a function. We write $f'(x)$ for the *function* whose value is $f'(a)$ at each point $x = a$ where $f(x)$ is differentiable. The process of finding the derivative $f'(x)$ from $f(x)$ is called differentiation.

The disadvantage of this notation is that the function usually needs to have a name (like $f(x)$ or $g(x)$, etc.). The alternate notation for the derivative of $f(x)$ is

$$\frac{df}{dx} = \frac{d}{dx}(f).$$

These symbols are read "*d f* by *d x*" and "*d* by *d x* of *f*." The notation includes the idea that $\frac{df}{dx}$ is a function (whose value at x is $\frac{df}{dx}(x)$), but it also allows us to write, for instance,

$$\frac{d}{dx}(x^3 - 3x^2 + 5x - 2)$$

for the function that is the derivative of $x^3 - 3x^2 + 5x - 2$, without having to give a name to that polynomial. This may not seem important, but it can be very useful.

For instance, the results of Example 1 can be written as

$$\frac{d}{dx}(c) = 0, \qquad \frac{d}{dx}(x) = 1.$$

When we write $\frac{df}{dx}$ or $\frac{d}{dx}(f)$, it is not a fraction. The symbol makes sense only as a whole and cannot be broken into its parts. This is emphasized by the way we read it, which is different from the way we would read it if it were a quotient.

If we were considering functions for which the variable was t, say, or u, we would write $\frac{d}{dt}(f)$ or $\frac{d}{du}(u^3 - 2u + 7)$, etc.

Suppose $f(x)$ and $g(x)$ are two functions that are both differentiable at $x = a$. What can we say about their sum, $f(x) + g(x)$? We could attempt to find the derivative of the sum at $x = a$ by evaluating the limit of the difference quotient:

$$\frac{d(f+g)}{dx}(a) = \lim_{h \to 0} \frac{\big(f(a+h) + g(a+h)\big) - \big(f(a) + g(a)\big)}{h}$$

$$= \lim_{h \to 0} \frac{\big(f(a+h) - f(a)\big) + \big(g(a+h) - g(a)\big)}{h}$$

$$= \lim_{h \to 0} \left(\frac{\big(f(a+h) - f(a)\big)}{h} + \frac{\big(g(a+h) - g(a)\big)}{h} \right).$$

Here we have the limit of the sum of two functions. We know that by Theorem 2.2.3(i) it equals the sum of the two limits. Also we know what the two limits are; the first difference quotient approaches $f'(a)$, and the second approaches $g'(a)$. So the derivative of the sum $f + g$ equals the sum of the derivatives of f and g.

A similar but easier calculation shows that the derivative of cf, with c a constant, is just $cf'(a)$, which follows from the corresponding fact about limits (Theorem 2.2.3(iii)).

We have just given the proof of the following extremely important theorem. The statement is written in both derivative notations to help us become accustomed to using them both.

THEOREM 3.2.2

Suppose c is a constant and $f(x)$ and $g(x)$ are two functions that are both differentiable at $x = a$. Then the functions $f + g$ and cf are both differentiable at $x = a$, and their derivatives are given by the following:

$$\frac{d(f + g)}{dx}(a) = \frac{df}{dx}(a) + \frac{dg}{dx}(a),$$

$$\frac{d(cf)}{dx} = c\frac{df}{dx}.$$

Alternatively, these formulas can be written as

$$(f + g)'(a) = f'(a) + g'(a),$$

$$(cf)'(a) = cf'(a),$$

and, if the point a is not important, as

$$\frac{d(f + g)}{dx} = \frac{df}{dx} + \frac{dg}{dx} \quad \text{or} \quad (f + g)' = f' + g',$$

$$\frac{d(cf)}{dx} = c\frac{df}{dx} \quad \text{or} \quad (cf)' = cf'.$$

Of course the result about derivatives of sums applies to sums of three or more functions too.

EXAMPLE 2

(i) Suppose $a, b \in \mathbb{R}$ are constants. Find the derivative of $y = ax + b$.

Solution Differentiating, we have

$$\frac{dy}{dx} = \frac{d}{dx}(ax + b) = \frac{d}{dx}(ax) + \frac{d}{dx}(b)$$

$$= a\frac{d}{dx}(x) + \frac{d}{dx}(b) = a(1) + 0 = a.$$

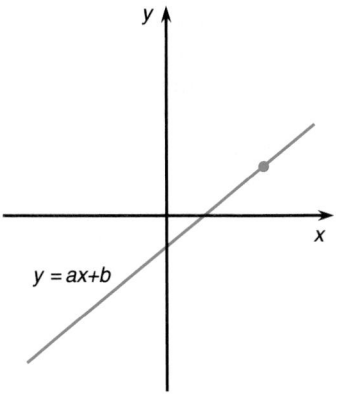

Figure 3.2.5

Here we used Theorem 3.2.2 for the second and third equalities and Example 1 for the fourth.

(ii) Find the tangent line to the graph $y = ax + b$ at any point on the graph.

Solution The result of part (i) says that the derivative of a straight line $y = ax + b$ equals the slope of the line (which is just a). If we combine this observation with the definition of the tangent line (Definition 3.1.1), we find that the tangent line to $y = ax + b$ at any point on the graph is a line with the same slope as $y = ax + b$ and passing through some point on $y = ax + b$; the only way this can happen is for the tangent line to be $y = ax + b$ also. In other words, the tangent line to a straight line is the line itself, which is hardly surprising. See Figure 3.2.5. ◢

Suppose we want to differentiate a polynomial $p(x) = a_m x^m + a_{m-1} x^{m-1} + \ldots + a_1 x + a_0$.

Using Theorem 3.2.2, we find that

$$\frac{dp}{dx} = a_m \frac{d}{dx}(x^m) + a_{m-1} \frac{d}{dx}(x^{m-1}) + \ldots + a_1 \frac{d}{dx}(x) + a_0 \frac{d}{dx}(1),$$

so to find the derivative, all we really need is to be able to find the derivative of x^n for $n = 1, 2, 3, \ldots, m$. Of course we already know the derivative of $x = x^1$; by Example 1 it is the constant function 1.

To find the derivative of x^n, for $n = 2, 3, 4, \ldots$, we must consider the difference quotient $\frac{(x+h)^n - x^n}{h}$. First we recall the Binomial Theorem (Theorem 1.3.10, p. 15), which says that

$$(x + h)^n = x^n + \binom{n}{n-1} x^{n-1} h + \binom{n}{n-2} x^{n-2} h^2$$

$$+ \ldots + \binom{n}{1} x h^{n-1} + \binom{n}{0} h^n.$$

Recalling that $\binom{n}{n-1} = \frac{n!}{(n-1)!1!} = n$, we find that

$$(x + h)^n - x^n = n x^{n-1} h + \binom{n}{n-2} x^{n-2} h^2 + \ldots + \binom{n}{1} x h^{n-1} + \binom{n}{0} h^n.$$

Dividing by h to get the difference quotient, we find

$$\frac{(x + h)^n - x^n}{h} = n x^{n-1} + \binom{n}{n-2} x^{n-2} h + \ldots + \binom{n}{1} x h^{n-2} + \binom{n}{0} h^{n-1}.$$

To find $\frac{d}{dx}(x^n)$, we let $h \to 0$ and take the limit. In the above expression, each term on the right side after the first contains some positive power of h, so as $h \to 0$, each term except the first tends to 0. In other words,

$$\frac{d}{dx}(x^n) = \lim_{h \to 0} \frac{(x + h)^n - x^n}{h}$$

$$= \lim_{h \to 0} \left(n x^{n-1} + \binom{n}{n-2} x^{n-2} h + \ldots + \binom{n}{1} x h^{n-2} + \binom{n}{0} h^{n-1} \right)$$

$$= n x^{n-1}.$$

We have just proved that the derivative of x^n is nx^{n-1}. To find the derivative of a power x^n, we reduce the exponent from n to $n-1$ and multiply by the original exponent n, to get nx^{n-1}. Notice that when $x \neq 0$, this is consistent with our result that $\frac{d}{dx}(x) = 1$, since $x = x^1$, and $1(x^0) = 1(1) = 1$. Moreover, if we think of $1 = x^0$, then the derivative of 1 should be $0(x^{-1})$, which is more or less correct. This last formula equals 0 whenever it is defined, and the derivative of 1 is 0. Unfortunately, $0(x^{-1})$ is not defined when $x = 0$, so we should not really use this formula. Similarly, x^0 is not defined when $x = 0$, so the formula should not be used for the derivative of $x = x^1$ either. On the other hand, it is sometimes helpful as a mnemonic device.

We record these important facts as a theorem.

THEOREM 3.2.3

Let n be any positive integer. Then x^n is differentiable at every point x, and for $n \geq 2$,

$$\frac{d}{dx}(x^n) = nx^{n-1}.$$

When $n = 1$ we know that

$$\frac{d}{dx}(x) = 1.$$

Also the constant function 1 is differentiable at every x, and

$$\frac{d}{dx}(1) = 0.$$

EXAMPLE 3 Differentiate the following functions: (i) x^3; (ii) x^5; (iii) $4x^3 - 2x^2 + 5x - 1$.

Solution   We simply apply Theorem 3.2.3 to find that (i) $\frac{d}{dx}(x^3) = 3x^2$, (ii) $\frac{d}{dx}(x^5) = 5x^4$.

For (iii) we must use Theorem 3.2.2 to break the polynomial into pieces:

$$\frac{d}{dx}(4x^3 - 2x^2 + 5x - 1) = \frac{d}{dx}(4x^3) + \frac{d}{dx}(-2x^2) + \frac{d}{dx}(5x) + \frac{d}{dx}(-1)$$

$$= 4\frac{d}{dx}(x^3) - 2\frac{d}{dx}(x^2) + 5\frac{d}{dx}(x) - 1\frac{d}{dx}(1).$$

Now we can take the derivative of each power of x and find that the above expression equals

$$4(3x^2) - 2(2x^1) + 5(1) - 1(0) = 12x^2 - 4x + 5.$$

The procedure we just used to find the derivative of $4x^3 - 2x^2 + 5x - 1$ will work for any polynomial. It is not necessary to write out all the steps; with a little practice, most of them can be done in your head. Theorem 3.2.2 says that we can work with each term of the polynomial separately. It also says that the derivative of the term $a_n x^n$ will be the product of a_n with the derivative of x^n.

So we work our way through the polynomial. For each term $a_n x^n$ we can write down its derivative: $a_n(nx^{n-1})$. When we have done this for each term, we will have found the derivative of the original polynomial.

EXAMPLE 4   Find the derivative of each of the following polynomials: (i) $3x^3 - 7x^2 - x + 4$; (ii) $2x^{18} - 3x^4 + x + 2$.

Solution

(i) We write down the polynomial and then find its derivative term by term:

$$\frac{d}{dx}(3x^3 - 7x^2 - x + 4) = 9x^2 - 14x - 1.$$

Notice that we did not bother to write down the derivative of the constant term, since it is 0 anyway, and that we did not write the derivative of $-x$ as $-1(1)$ but simply combined them all at one step.

(ii) Again,

$$\frac{d}{dx}(2x^{18} - 3x^4 + x + 2) = 36x^{17} - 12x^3 + 1.$$

We have established the following result.

THEOREM 3.2.4

Consider a polynomial

$$p(x) = a_m x^m + a_{m-1} x^{m-1} + \ldots + a_1 x + a_0.$$

It is differentiable at every point x, and its derivative is given by

$$\frac{dp}{dx} = m a_m x^{m-1} + (m-1) a_{m-1} x^{m-2} + \ldots + a_1.$$

This makes it very much easier to do problems involving derivatives.

EXAMPLE 5   Find the tangent line to the curve $y = 4x^5 - 5x^3 + 3x^2 + 2x - 2$ at the point $(1, 2)$.

Solution $\frac{dy}{dx} = 20x^4 - 15x^2 + 6x + 2$. So the slope of the tangent, which is just the value of the derivative at the point $x = 1$, is $20 - 15 + 6 + 2 = 13$. Accordingly, the equation of the tangent line can be found by using Definition 3.1.1. It is

$$y = 2 + 13(x - 1) \qquad \text{or} \qquad y = 13x - 11.$$

(Notice that this line does indeed go through the point $(1, 2)$. This is an easy way to check that your answer is reasonable.)

EXAMPLE 6   Find all the points at which the tangent line to the curve $y = 2x^3 - 9x^2 + 12x - 6$ is horizontal.

Solution   For the tangent line to be horizontal, its slope must equal 0. The slope of the tangent is just the derivative, which is

$$6x^2 - 18x + 12.$$

To find when this is zero, we factor it:

$$6x^2 - 18x + 12 = 6(x^2 - 3x + 2) = 6(x - 2)(x - 1),$$

which is zero precisely when $x = 1$ or $x = 2$.

So there are exactly two points where the tangent line is horizontal. They are $(1, -1)$ and $(2, -2)$. See Figure 3.2.6.

See the Graphic Calculator box on the next page for a plot of this example.

Figure 3.2.6

With a graphics calculator it is possible to plot both the function and the tangent line from Example 5.

TI-85	SHARP EL9300
GRAPH F1($y(x)$ =) 4 F1(x)^5 − 5 F1(x)^3 + 3 F1(x) x^2 + 2 F1(x) − 2 ▼ M2(RANGE) (−) 2 ▼ 2 ▼ ▼ (−) 10 ▼ 10 F5(GRAPH)	◿ 4 x/θ/T a^b 5 ▶ − 5 x/θ/T a^b 3 ▶ + 3 x/θ/T x^2 + 2 x/θ/T − 2 2ndF ▼ 13 x/θ/T −11 RANGE −2 ENTER 2 ENTER 1 ENTER −10 ENTER 10 ENTER 1 ENTER ◿

CASIO f_x-7700GB/f_x-6300G	HP 48SX
Cls EXE Range −2 EXE 2 EXE 1 EXE −10 EXE 10 EXE 1 EXE EXE EXE EXE Graph 4 X x^y 5 − 5 X x^y 3 + 3 X x^2 + 2 X − 2 EXE Graph 13 X − 11 EXE	◥ PLOT PLOTR ERASE ATTN ′ αY = 4 × αX y^x 5 − 5 × αX y^x 3 + 3 × αX y^x 2 + 2 × αX − 2 ↱ PLOT ◥ DRAW ERASE 2 +/− ENTER 2 XRNG 10 +/− ENTER 10 YRNG DRAW ATTN ′ αY = 13 × αX − 11 ↱ PLOT ◥ DRAW DRAW

EXAMPLE 7 **Weather Balloon** A weather balloon is released at time $t = 0$ and allowed to rise. Suppose its height above the ground at time t sec is given by $H(t) = \frac{1}{3}t^3 − 5t^2 + 23t$ meters.

(i) Find the velocity of the balloon t sec after its release.

(ii) Find when the balloon is rising and when (if ever) it is falling.

(iii) Because of unforeseen problems, the balloon bursts after 11 sec of flight. How fast is it rising at the moment it bursts?

Solution

(iv) The velocity is simply the derivative $H'(t) = 3\left(\frac{1}{3}t^2\right) − 2(5t) + 23 = t^2 − 10t + 23$ m/sec.

(v) The derivative is the rate of change of the height, so when $H'(t) > 0$, it means that the height is increasing, or that the balloon is rising. Similarly, $H'(t) < 0$ means the balloon is falling. We must study the sign of $H'(t)$.

The easiest way to determine the sign of a quadratic expression like $H'(t) = t^2 − 10t + 23$ is to complete the square:

$$t^2 − 10t + 23 = (t^2 − 10t + 25) − 2 = (t − 5)^2 − 2.$$

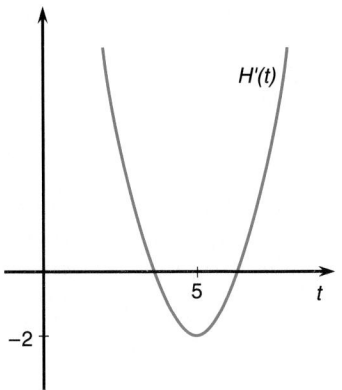

$H'(t)$

5

t

−2

Figure 3.2.7

This is obviously positive when $(t - 5)^2 > 2$ and negative when $(t - 5)^2 < 2$. Now $(t - 5)^2 < 2$ precisely when $|t - 5| < \sqrt{2}$, which means $-\sqrt{2} < t - 5 < \sqrt{2}$, which is the same as $5 - \sqrt{2} < t < 5 + \sqrt{2}$.

So the derivative is negative, which means the balloon is falling, when t is between $5 - \sqrt{2} \approx 3.586$ sec and $5 + \sqrt{2} \approx 6.414$ sec. The derivative is zero, which means the balloon is neither rising nor falling, when $t = 5 \pm \sqrt{2} \approx 3.586$ or 6.414 sec. The rest of the time the derivative is positive, which means the balloon is rising when $t < 5 - \sqrt{2}$ and when $t > 5 + \sqrt{2}$. See the illustration of the graph of $H'(t)$ in Figure 3.2.7.

(vi) To find how fast the balloon is rising 11 sec into its flight, we simply find $H'(11) = (11)^2 - 10(11) + 23 = 121 - 110 + 23 = 34$ m/sec. The balloon is rising at a speed of 34 m/sec.

The solution to part (ii) of Example 7 was not particularly hard, but a little thought should convince you that to do it without knowing about derivatives would be quite difficult.

We used Theorem 2.2.3, which says that limits "preserve" sums and that constants can be "pulled outside" of limits, to prove Theorem 3.2.2, which says that differentiation "preserves" sums and that constants can be "pulled outside" of derivatives.

Since the rules for the limit of a product or a quotient are also quite simple, we might ask about the derivatives of products and quotients. Unfortunately, the rules for them are not quite so simple.

THEOREM 3.2.5

The Product Rule

If $f(x)$ and $g(x)$ are both differentiable at $x = a$, then the product $H(x) = f(x)g(x)$ is also differentiable at $x = a$, and its derivative is given as follows:

$$H'(a) = f(a)g'(a) + f'(a)g(a).$$

This may also be written as

$$\frac{d}{dx}\big(f(x)g(x)\big)(a) = f(a)\frac{dg}{dx}(a) + \frac{df}{dx}(a)g(a).$$

If it is not particularly important to specify the point a, the **Product Rule** can be written more briefly as

$$(fg)' = fg' + f'g \qquad \text{or} \qquad \frac{d}{dx}(fg) = f\frac{dg}{dx} + \frac{df}{dx}g.$$

PROOF   The difference quotient for $H(x)$ is

$$\frac{H(a + h) - H(a)}{h} = \frac{f(a + h)g(a + h) - f(a)g(a)}{h}$$

$$= \frac{f(a + h)g(a + h) - f(a + h)g(a) + f(a + h)g(a) - f(a)g(a)}{h}.$$

We have added and subtracted the term $f(a + h)g(a)$ in the numerator, which of course changes nothing. But when we regroup the terms, the difference quotient becomes easier to work with:

$$\frac{f(a+h)\big(g(a+h)-g(a)\big)+\big(f(a+h)-f(a)\big)g(a)}{h}$$

$$=f(a+h)\frac{g(a+h)-g(a)}{h}+\frac{f(a+h)-f(a)}{h}g(a).$$

The second term is the difference quotient for $f(x)$ multiplied by $g(a)$, so as $h \to 0$, it will tend to $f'(a)g(a)$. The first term is $f(a+h)$ times the difference quotient for $g(x)$. As $h \to 0$, the difference quotient will tend to $g'(a)$, and $f(a+h)$ will tend to $f(a)$, since $f(x)$ is continuous at $x=a$, by Theorem 3.2.1. In symbols,

$$H'(a) = \lim_{h\to 0}\frac{H(a+h)-H(a)}{h}$$

$$= \lim_{h\to 0}\left(f(a+h)\frac{g(a+h)-g(a)}{h}+\frac{f(a+h)-f(a)}{h}g(a)\right)$$

$$= \lim_{h\to 0}\left(f(a+h)\frac{g(a+h)-g(a)}{h}\right)+\lim_{h\to 0}\left(\frac{f(a+h)-f(a)}{h}g(a)\right)$$

$$= \left(\lim_{h\to 0}f(a+h)\right)\left(\lim_{h\to 0}\frac{g(a+h)-g(a)}{h}\right)+\left(\lim_{h\to 0}\frac{f(a+h)-f(a)}{h}\right)g(a)$$

$$= f(a)g'(a)+f'(a)g(a).$$

The idea of this proof was to rearrange the difference quotient into pieces, in each of which only one function (f or g) changed. In each piece we found a difference quotient for f or g times some other factor, so we were able to take the limit. The last ingredient was realizing that $\lim_{h\to 0}f(a+h)=f(a)$, that is, that $f(x)$ is continuous at $x=a$, which is a consequence of Theorem 3.2.1.

Since the Product Rule makes sense wherever both $f(x)$ and $g(x)$ are differentiable, we can also write it in the form

$$(fg)'(x) = f(x)g'(x)+f'(x)g(x),$$

with the understanding that it makes sense wherever the right side of the formula makes sense. Since polynomials, for example, are differentiable everywhere, this form of the Product Rule is particularly convenient for them.

EXAMPLE 8   Differentiate the following products: (i) $(x+3)(x^3-2x-4)$; (ii) $(t^2+3t+4)^2$; (iii) $(x^2+1)(x^3-2)(2x^2+x)$.

Solution

(i) We simply apply the Product Rule, with $f(x)=x+3$ and $g(x)=x^3-2x-4$. We find that

$$\frac{d}{dx}\big((x+3)(x^3-2x-4)\big) = (x+3)(3x^2-2)+(1)(x^3-2x-4)$$

$$= 3x^3+9x^2-2x-6+x^3-2x-4$$

$$= 4x^3+9x^2-4x-10.$$

We could have multiplied out the product $f(x)g(x)$ first and then just differentiated the resulting polynomial. You should try it and check that the answer you get is the same.

(ii) This time we let $f(t) = g(t) = t^2 + 3t + 4$:

$$\frac{d}{dt}\left((t^2 + 3t + 4)^2\right) = (t^2 + 3t + 4)(2t + 3) + (2t + 3)(t^2 + 3t + 4).$$

Noticing that these two terms are really the same, we combine them and find that the derivative is

$$2(t^2 + 3t + 4)(2t + 3) = 2(2t^3 + 9t^2 + 17t + 12) = 4t^3 + 18t^2 + 34t + 24.$$

(iii) We think of the product as

$$(x^2 + 1)(x^3 - 2)(2x^2 + x) = (x^2 + 1)\left((x^3 - 2)(2x^2 + x)\right).$$

We apply the Product Rule with $f(x) = x^2 + 1, g(x) = (x^3 - 2)(2x^2 + x)$. To do this, we need to know $g'(x)$.

To find $g'(x)$, we apply the Product Rule and find

$$g'(x) = (x^3 - 2)(4x + 1) + (3x^2)(2x^2 + x)$$

$$= 4x^4 + x^3 - 8x - 2 + 6x^4 + 3x^3 = 10x^4 + 4x^3 - 8x - 2.$$

Now the original derivative is

$$\frac{d}{dx}\left((x^2 + 1)\left((x^3 - 2)(2x^2 + x)\right)\right) = f(x)g'(x) + f'(x)g(x)$$

$$= (x^2 + 1)(10x^4 + 4x^3 - 8x - 2) + (2x)(x^3 - 2)(2x^2 + x)$$

$$= 10x^6 + 4x^5 + 10x^4 - 4x^3 - 2x^2 - 8x - 2 + (2x^4 - 4x)(2x^2 + x)$$

$$= 10x^6 + 4x^5 + 10x^4 - 4x^3 - 2x^2 - 8x - 2 + 4x^6 + 2x^5 - 8x^3 - 4x^2$$

$$= 14x^6 + 6x^5 + 10x^4 - 12x^3 - 6x^2 - 8x - 2.$$

The diligent student can compare this result with what is obtained by multiplying the three factors first and then differentiating.

Next we should find how to differentiate a quotient $\frac{f(x)}{g(x)}$ at $x = a$, assuming that f and g are both differentiable at $x = a$ and that $g(a) \neq 0$. Notice that if $g(x)$ is differentiable at $x = a$, then Theorem 3.2.1 says it is continuous at $x = a$. If in addition $g(a) \neq 0$, then by Theorem 2.3.4, for x sufficiently near a, $g(x)$ cannot be zero either (close to a, $g(x)$ must be closer to $g(a)$ than 0 is and therefore cannot be 0). The importance of this is that for small values of h we know that $g(a + h) \neq 0$, which we need to know, since $g(a + h)$ will appear in the denominator of the difference quotient.

If we write $Q(x) = \frac{f(x)}{g(x)}$, then the difference quotient is

$$\frac{Q(a + h) - Q(a)}{h} = \frac{1}{h}\left(Q(a + h) - Q(a)\right) = \frac{1}{h}\left(\frac{f(a + h)}{g(a + h)} - \frac{f(a)}{g(a)}\right)$$

$$= \frac{1}{h}\frac{f(a + h)g(a) - f(a)g(a + h)}{g(a + h)g(a)}$$

$$= \frac{1}{g(a + h)g(a)} \times \frac{1}{h} \times \left(f(a + h)g(a) - f(a)g(a + h)\right).$$

Again we can use the trick of adding and subtracting the same quantity in the numerator, to allow breaking the numerator into pieces in each of which only one function changes. The last expression equals

$$\frac{1}{g(a+h)g(a)} \frac{f(a+h)g(a) - f(a)g(a) + f(a)g(a) - f(a)g(a+h)}{h}$$

We added and subtracted
$f(a)g(a)$.

$$= \frac{1}{g(a+h)g(a)} \frac{\big(f(a+h) - f(a)\big)g(a) + f(a)\big(g(a) - g(a+h)\big)}{h}$$

$$= \frac{1}{g(a+h)g(a)} \left(\frac{f(a+h) - f(a)}{h} g(a) - f(a) \frac{g(a+h) - g(a)}{h} \right).$$

Taking the limit as $h \to 0$ is now easy. The first quotient tends to $\frac{1}{g(a)^2}$, the second to $f'(a)$, and the third to $g'(a)$.

We have just shown that the derivative of $\frac{f(x)}{g(x)}$ at $x = a$ is $\frac{f'(a)g(a) - f(a)g'(a)}{g(a)^2}$. This result is known as the Quotient Rule.

THEOREM 3.2.6

The Quotient Rule

Suppose $f(x)$ and $g(x)$ are differentiable at $x = a$ and $g(a) \neq 0$. Then $\frac{f(x)}{g(x)}$ is differentiable at $x = a$, and its derivative there is

$$\left(\frac{f}{g}\right)'(a) = \frac{f'(a)g(a) - f(a)g'(a)}{g(a)^2}.$$

If it is not necessary to specify the point a, this can be written as

$$\left(\frac{f}{g}\right)' = \frac{f'g - fg'}{g^2}.$$

EXAMPLE 9   Differentiate (i) $Q(x) = \frac{x+1}{x-1}$; (ii) $R = \frac{1}{x^2+2}$; (iii) $S = \frac{x^3+3x^2-x-2}{2x-1}$.

Solution

(i) $Q'(x) = \frac{(1)(x-1) - (x+1)(1)}{(x-1)^2} = \frac{-2}{(x-1)^2}.$

(ii) $R'(x) = \frac{(0)(x^2+2) - (1)(2x)}{(x^2+2)^2} = -\frac{2x}{(x^2+2)^2} = -2x(x^2+2)^{-2}.$

(iii) $S'(x) = \frac{(3x^2+6x-1)(2x-1) - (x^3+3x^2-x-2)(2)}{(2x-1)^2} = \frac{(6x^3+9x^2-8x+1) - (2x^3+6x^2-2x-4)}{(2x-1)^2}$

$= \frac{4x^3+3x^2-6x+5}{(2x-1)^2}.$

The use of the product and quotient rules can be complicated and sometimes even tedious, but there is nothing essentially difficult about them. It is, however, absolutely essential to memorize the rules themselves.

Incidentally, since we already know how to differentiate any polynomial, the Quotient Rule tells us how to differentiate any rational function, as the above example illustrates.

THEOREM 3.2.7

If $R(x) = \frac{f(x)}{g(x)}$ is a rational function, then $R(x)$ is differentiable at every point a for which $g(a) \neq 0$.

Moreover, its derivative is given by the Quotient Rule:

$$R'(x) = \frac{f'(x)g(x) - f(x)g'(x)}{g(x)^2},$$

assuming $g(x) \neq 0$.

EXAMPLE 10   Let $f(x) = \frac{x}{1+x^2}$. Find all values of x at which $f'(x) = 0$.

Solution   Notice that the denominator is never zero, so the function is differentiable everywhere. Using the Quotient Rule, we find that its derivative is

$$\frac{(1)(1 + x^2) - x(2x)}{(1 + x^2)^2} = \frac{1 + x^2 - 2x^2}{(1 + x^2)^2} = \frac{1 - x^2}{(1 + x^2)^2}.$$

The question asks when this is zero. Since the denominator is never zero, this is equivalent to asking when the numerator is zero, which is easy: $1 - x^2 = 0$ precisely when $x = \pm 1$.

EXAMPLE 11   Find the derivative of (i) $\frac{1}{x}$, (ii) $\frac{1}{x^n}$, where n is a positive integer.

Solution

(i) Apply the Quotient Rule, with $f(x) = 1$ and $g(x) = x$. We find that

$$\frac{d}{dx}\left(\frac{1}{x}\right) = \frac{0(x) - 1(1)}{x^2} = -\frac{1}{x^2}.$$

(ii) Similarly,

$$\frac{d}{dx}\left(\frac{1}{x^n}\right) = \frac{0(x) - 1(nx^{n-1})}{(x^n)^2} = -\frac{nx^{n-1}}{x^{2n}} = -nx^{-n-1}.$$

Notice that the formula established in Example 11(ii) is consistent with the formula for the derivative of x^n when $n > 0$, as stated in Theorem 3.2.3. We record this result below.

THEOREM 3.2.8

Let n be any integer. If $n > 0$, then x^n is differentiable at every x. If $n < 0$, then x^n is differentiable except at $x = 0$. The constant function 1 is differentiable at every x.

The derivatives of x and 1 are given as follows:

$$\frac{d}{dx}(x) = 1, \qquad \frac{d}{dx}(1) = 0.$$

For $n \neq 0, 1$, whenever x^n is differentiable, its derivative is given as follows:

$$\frac{d}{dx}(x^n) = nx^{n-1}.$$

Exercises 3.2

I

Find the derivative of each of the following functions.

1. $x^3 - 2x^2 + x - 1$
2. $x^4 + 2x^3 - 5x + 7$
3. $2x^3 + 3x^2 + 6x - 4$
4. $3x^{67}$
5. $(3x^2 - 5)(2x + 3)$
6. $(2x + 1)(x - 1)$
7. $(4x + 3)^2$
8. $(x^2 - 2x)^2$
9. $\frac{2x+1}{x^2+1}$
10. $\frac{1}{1+x^4}$

Find the derivatives of the following functions wherever they are differentiable. Specify all values of x at which they are not differentiable.

11. $\frac{2x-1}{x-1}$
12. $\frac{5x+2}{x}$
13. $\frac{x^2}{x^2-4}$
14. $\frac{x}{x^2+3x+2}$
15. $\frac{x^2}{3x^4-2x^3}$
16. $\frac{x-1}{(x-1)^2}$
17. $\frac{3x^2+x-1}{x^2+3x-4}$
18. $\frac{x^3+x^2+x+1}{x+3}$
19. $\frac{x^4}{1+x^3}$
20. $\frac{1+x^3}{x^4}$

II

21. Differentiate the following functions, specifying any points where they are not differentiable:
    (i) $(x+1)(x^2+x)(3x-2)$
    (ii) $(4x^3-1)(3x+3)x^2$
    (iii) $(x+2)\frac{x}{x^2-x-2}$

22. Find the derivative of each of the following:
    (i) $(2x^2+x-1)^2$
    (ii) $(2x^2+x-1)^4$
    (iii) $(2x^2+x-1)^8$
    (iv) $(2x^2+x-1)^{16}$

23. Find the tangent line to each of the following curves at the specified point:
    (i) $y = x^3 - 2x^2 + 2x + 1$, $(1, 2)$
    (ii) $y = x^4 + x^2 + 2x$, $(-1, 0)$

24. Find the tangent line to each of the following curves at the specified point:
    (i) $y = x^5 - 3x^4 + 4x^2 - 5$, $(2, -5)$
    (ii) $y = x^3 - 2x + 4$, $(a, a^3 - 2a + 4)$

25. Find all tangent lines to the curve $y = x^3$ that pass through the point $(0, 16)$.

26. (i) Find all tangent lines to the curve $y = x^4$ that pass through the point $(0, 1)$.
    (ii) Find all tangent lines to the curve $y = x^4$ that pass through the point $(0, -3)$.

27. (i) Calculate the derivative of $H(x) = x^2$ in two ways. First do it by simply applying the formula. Second, write $H(x) = f(x)g(x)$ with $f(x) = g(x) = x$ and apply the Product Rule.
    (ii) With $f(x)$, $g(x)$ as above, calculate the product of the derivatives, $f'(x)g'(x)$, and remark that it is definitely different from the derivative of $H = fg$.

*28. (i) Find a Product Rule for the derivative of the product of three functions: $P(x) = f(x)g(x)h(x)$.
    (ii) A Product Rule for $P = fgh$ can be found by writing $P = fgh = (fg)h$ and then applying the Product Rule for two functions twice. But it is also possible to write $P = f(gh)$ and then apply the Product Rule. Check that these two calculations give the same answer for the derivative P'.

29. If you are familiar with the idea of proof by induction, use it to prove the formula
    $$\frac{d}{dx}(x^n) = nx^{n-1}, \quad \text{for} \quad n \geq 1.$$

*30. (i) Find the derivative of $p(x)^3$ if $p(x)$ is a polynomial. (Express your answer in terms of $p(x)$ and $p'(x)$.)
    (ii) Find the derivative of $f(x)^3$ at $x = a$ if $f(x)$ is a function that is differentiable at $x = a$.
    (iii) Repeat part (ii) with $f(x)^4$ in place of $f(x)^3$.
    (iv) Repeat part (ii) with $f(x)^n$ in place of $f(x)^3$, where n is a positive integer. (*Hint:* Use induction.)

*31. Suppose the height of a helicopter above the ground at time t sec is given by $H(t) = t^3 - 3t^2 + 6$ ft.
    (i) When is the helicopter rising and when is it falling?
    (ii) What is the lowest point in its flight? (Think about when it is rising and falling, according to your answer to part (i).)
    (iii) Will it touch the ground?

*32. (i) If $f(x)$ is differentiable at $x = a$ and $f(a) \neq 0$, show that the derivative of $\frac{1}{f(x)}$ at $x = a$ equals $-\frac{f'(a)}{f(a)^2}$.
    (ii) Find the derivative of $f(x)^{-2}$ at $x = a$.
    (iii) Find the derivative of $f(x)^{-n}$ at $x = a$ for any positive integer n.

*33. Suppose $g(x)$ is a function that is differentiable on an interval (c, d) and such that $g(x) > 0$ for all $x \in (c, d)$. Let $f(x) = \sqrt{g(x)}$.

(i) Express $f'(x)$ in terms of $g(x)$ and $g'(x)$ for $x \in (c, d)$. (*Hint:* Write $g(x) = f(x)f(x)$.)

(ii) Use part (i) to find the derivative of $f(x) = \sqrt{1 + x^2}$.

(iii) Find $\frac{d}{dx}\left((x^2 + 2x + 2)^{1/2}\right)$.

(iv) If $g(x)$ is differentiable on (c, d), figure out the derivative of $f(x) = \left(g(x)\right)^{1/3}$, when it exists, and specify when it does and does not exist.

*34. Find all points x at which $f'(x) = 0$ if (i) $f(x) = \frac{x}{x^2+2}$, (ii) $f(x) = \frac{x^3}{x^2+2}$.

*35. Let $f(x) = x^4 + 2x^3 - x^2 + 3$, and let $g(x) = f'(x)$.

(i) Calculate $g(x)$.

(ii) Calculate $g'(x)$. This function, the derivative of the derivative of $f(x)$, is called the second derivative of $f(x)$.

(iii) What is the second derivative of $g(x)$?

*36. (i) Suppose $f(x)$ is differentiable at $x = a$. Suppose $g(x)$ is a function and suppose there is an interval (c, d) containing a (i.e., $c < a < d$) so that $f(x) = g(x)$ for every $x \in (c, d)$. Show that $g(x)$ is differentiable at $x = a$ and that $g'(a) = f'(a)$.

(ii) Use part (i) to find the derivative of $f(x) = \frac{x-1}{x-1}$.

(iii) Use part (i) to find the derivative of $f(x) = \frac{x^2-9}{x+3}$.

*37. Suppose that neither of the polynomials $p(x)$ and $q(x)$ is the zero polynomial, and suppose that the derivative of their product $p(x)q(x)$ equals the product of their derivatives. Show that $p(x)$ and $q(x)$ must both be constant polynomials.

*38. Use the result of Exercise 32(i) and the Product Rule to find the Quotient Rule.

III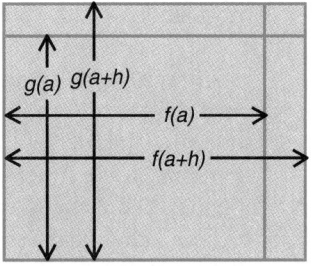

39. Use a graphics calculator or computer to sketch the curve $y = x^3 - 3x^2 + 1$, and on the same picture draw the tangent lines at $(0, 1)$, $(1, -1)$, $(2, -3)$.

40. Sketch the curve $y = \frac{x}{x-1}$, and on the same picture draw the tangent lines at $(0, 0)$, $(2, 2)$, $\left(3, \frac{3}{2}\right)$.

41. Sketch the curve $y = 3x^4 - 2x^3 + 2x^2 + 2$, and on the same picture draw the tangent lines at $(0, 2)$, $(-1, 9)$, $(1, 5)$.

42. We have not yet discussed the derivative of the exponential function $e^x = \exp(x)$, but we can use computer graphics to estimate its derivative. Graph the difference quotient $Q(h) = \frac{\exp(0+h) - \exp(0)}{h}$ and use the graph to estimate the derivative $\exp'(0)$. Using your estimate, draw the tangent line to $y = \exp(x)$ at $(0, 1)$. Graph $y = \exp(x)$ on the same picture, and see whether your line does appear to be a tangent.

43. Use the idea of Exercise 42 to estimate $\exp'(1)$ correct to four decimal places.

44. Use the idea of Exercise 42 to estimate the derivative of $f(x) = \sin x$ at $x = 0$ (with x in *radians*).

45. Use the idea of Exercise 42 to estimate the derivative of $f(x) = \sin x$ at $x = 1$ correct to four decimal places.

P ← **POINT TO PONDER**

Now that we have learned the Product Rule and become familiar with its use, let us remember that our first guess would probably have been that the derivative of a product $f(t)g(t)$ ought to be the product of the derivatives of f and g. We have seen otherwise, and the proof of the Product Rule is not difficult, but on the other hand it is not completely clear where the correct formula comes from.

Suppose $f(t)$ and $g(t)$ are two functions that are both differentiable at $t = a$. Let $A(t) = f(t)g(t)$ be the product, which we can think of as the area of a rectangle whose sides have lengths $f(t)$ and $g(t)$ at time t. We should think of the rectangle changing size as time passes.

The derivative of $A(t)$, if it exists, should be the rate of change of the area of this rectangle. The numerator of the difference quotient, $A(a + h) - A(a) = f(a + h)g(a + h) - f(a)g(a)$, is the difference between the areas of the rectangles at time $t = a + h$ and $t = a$. In Figure 3.2.8 we see the rectangle at time $t = a$ and then the rectangle at time $t = a + h$. The illustration

g(a) g(a+h)

← f(a) →

← f(a+h) →

Figure 3.2.8

shows the rectangle at $t = a + h$ being slightly larger in both dimensions than the one at $t = a$, though there are other possibilities.

The numerator of the difference quotient is the change in the area of the rectangle, which is the area of the extra little strips along two sides of the rectangle. Let us ignore the tiny rectangle in the corner and concentrate on the two long thin strips (see Figure 3.2.9). Their long sides have lengths $f(a)$ and $g(a)$, respectively. The height of the horizontal strip at the top is just the change in $g(t)$ between $t = a$ and $t = a + h$, or in other words, $g(a + h) - g(a)$. The width of the vertical strip at the right side is $f(a + h) - f(a)$, the change in $f(t)$ between $t = a$ and $t = a + h$.

So, ignoring that little rectangle in the corner, the numerator of the difference quotient is $f(a)\big(g(a + h) - g(a)\big) + g(a)\big(f(a + h) - f(a)\big)$. If we form the difference quotient, it is

$$\frac{f(a)\big(g(a + h) - g(a)\big) + g(a)\big(f(a + h) - f(a)\big)}{h}$$
$$= f(a)\frac{g(a + h) - g(a)}{h} + g(a)\frac{f(a + h) - f(a)}{h}.$$

Taking the limit, we know that the first quotient tends to $g'(a)$ and the second to $f'(a)$, which leads us to expect that $A'(t) = f(a)g'(a) + g(a)f'(a)$.

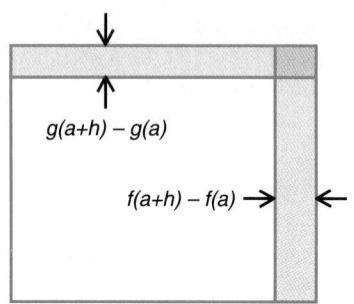

g(a+h) – g(a)

f(a+h) – f(a)

Figure 3.2.9

Roughly speaking, the two terms in this formula come from the areas of those two strips. If we think of the strips having their long side constant and their short side changing, their areas will be the constant length times the changing thickness, and the rate of change of their areas will be the constant length times the rate of change of the thickness, or $f(a)g'(a)$ in one case and $g(a)f'(a)$ in the other.

This explains to some extent why the rate of change of the whole area (i.e., of the product $f(t)g(t)$) ought to be given by the Product Rule $(fg)' = fg' + f'g$.

The argument we have just given is not entirely airtight, since we have simply ignored the area of the little rectangle in the corner. In fact, as we have already seen, that area tends to zero in the limit.

3.3 Maximum/Minimum Problems

CLOSED INTERVAL
ENDPOINTS

In this section we begin to see how useful derivatives can be, by learning how to use them to solve a very important type of problem. There are a great many problems that can be described as follows: For some function $f(x)$, at what point x will $f(x)$ have its maximum (or minimum) value? These are known as **maximum/minimum problems.**

This sort of problem arises in a huge variety of situations. The function $f(x)$ might represent the profit for a factory as a function of production, and the problem would be to find the level of production that would result in the highest profits. On the other hand, the function $f(x)$ could represent the cost of operating a piece of equipment as a function of the speed at which it runs; in this case the problem would be to find the speed that minimizes the cost.

In a chemical reaction we might be able to express the concentration of one of the products as a function $f(t)$ of time, and it could be extremely useful to know when the concentration is highest (if it is a desirable product) or lowest (if it is undesirable).

Figure 3.3.1

Figure 3.3.2

Figure 3.3.3

Figure 3.3.4

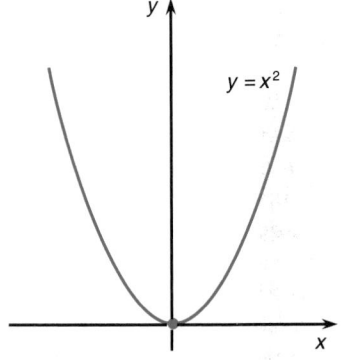

Figure 3.3.5

Foresters might like to know when to cut a stand of trees in order to get the greatest amount of usable timber; farmers might want to know when to harvest crops in order to get the highest yield or the lowest waste; engineers might want to know what shape of steel beam gives the greatest strength; pedestrians might want to know the shortest route from one place to another.

Suppose we are looking for the maximum or minimum of a differentiable function $f(x)$. Recall that the derivative $f'(a)$ is the slope of the tangent at $x = a$. If the slope of the tangent is positive, it means that the tangent slopes upward to the right, as in Figure 3.3.1. In this situation we say that the function $f(x)$ is increasing because for values of x slightly to the right of $x = a$ the value $f(x)$ will be higher than $f(a)$, that is, $f(x)$ gets bigger as x gets bigger. Similarly, if $f'(a) < 0$, the tangent line at $x = a$ slopes down to the right, and we say the function $f(x)$ is decreasing. See Figure 3.3.2.

If $f(x)$ has its maximum value when $x = a$, then as Figure 3.3.3 shows, it should be increasing to the left of $x = a$ and decreasing to the right of $x = a$. This is just another way of saying that the function comes up to its maximum as $x \to a^-$ and then goes down as x moves away on the right side of a. The derivative will be positive to the left of $x = a$ and negative to the right of $x = a$.

This strongly suggests that right at $x = a$ the derivative $f'(a)$ should equal zero. On the graph this means that the tangent line at $x = a$ should be horizontal (i.e., have slope equal to zero), which is also consistent with the idea that the graph has a sort of "hump" at $x = a$. See Figure 3.3.4.

Now working backward, if we do not know where $f(x)$ has its maximum value, we expect it to be at a point where $f'(a) = 0$. So we could start by finding all the points where $f'(a) = 0$. With luck there will not be too many of them, and then we should easily be able to decide at which of them $f(x)$ has the biggest value. Minimum values of $f(x)$ could occur in "valleys," so the same procedure will simultaneously find the minimum value.

There is a complication, however.

EXAMPLE 1   (i) Find the maximum and minimum values of $y = f(x) = x^2$.

Solution   The derivative is $\frac{dy}{dx} = 2x$, which is zero when $x = 0$ and nowhere else. This reflects the familiar fact that the tangent line at $x = 0$ (which is just the x-axis) has slope equal to 0. The function obviously has a minimum at $x = 0$ (after all, it is 0 there and positive everywhere else). Finding where $f'(x) = 0$ helped us to locate where $f(x)$ has its minimum value. See Figure 3.3.5.

But it does not seem to help find the maximum value. The reason is fairly obvious from the picture: There is no maximum value. In fact, $\lim_{x \to -\infty} f(x) = \infty$ and $\lim_{x \to \infty} f(x) = \infty$, which means that $f(x)$ cannot have any maximum because it gets larger than any number we might choose.

So the answer to the problem is that the minimum value of $f(x) = x^2$ is 0, which occurs when $x = 0$, and there is no maximum value.

(ii) Find the maximum and minimum values of $f(x) = x^3$.

Solution   We could begin by noticing that $f'(x) = 3x^2$, which is zero when $x = 0$, but if we remark that $\lim_{x \to -\infty} f(x) = -\infty$ and $\lim_{x \to \infty} f(x) = \infty$, we see that $f(x)$ gets very large as $x \to \infty$—bigger than any possible maximum—and very small as

Figure 3.3.6

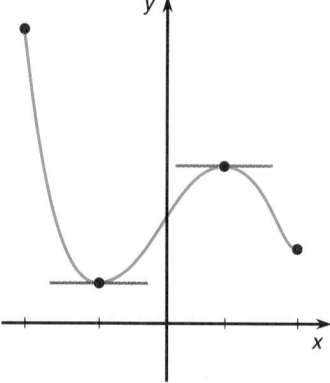

Figure 3.3.7

$x \to -\infty$—smaller than any possible minimum. As Figure 3.3.6 suggests, there is no maximum or minimum.

One way around this difficulty is to appeal to Theorems 2.5.6 and 2.5.7. They say that any continuous function on a closed interval must be **bounded,** that is, it cannot go off to infinity, and that it actually does have a maximum and a minimum value.

So given a differentiable function, such as a polynomial, for instance, on a closed interval, we know it must have a maximum and a minimum there. To find them, we could start by finding all points where the derivative is zero. These will be all the points where the tangent is horizontal, so they will include all the "humps" and "valleys." However, as Figure 3.3.7 shows, there is also the possibility that the maximum or minimum might occur at one of the endpoints of the interval (in the illustration the maximum occurs at the left endpoint). This can happen without the derivative being zero there because the maximum is in a sense artificial. It is there only because the interval comes to an end, not because the function goes up and then comes down again.

So we must revise our approach. First, we should find out where the derivative is zero. Then we should evaluate the function at each of those points *and* at the endpoints of the interval. The biggest value will be the maximum, and the smallest will be the minimum, and we will also know at which point(s) each of them occurs.

EXAMPLE 2   Find the maximum and minimum values of the function $y = f(x) = -3x^4 + 4x^3 + 12x^2 - 2$ on the interval $[-2, 3]$.

Solution   The derivative is $f'(x) = -12x^3 + 12x^2 + 24x$. To find where this equals zero, we should factor it:

$$-12x^3 + 12x^2 + 24x = -12x(x^2 - x - 2) = -12x(x-2)(x+1).$$

So $f'(x) = 0$ when $x = -1, 0, 2$.

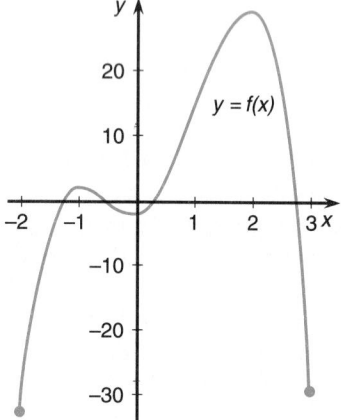

Figure 3.3.8

We can evaluate $f(x)$ at each of these points and at the endpoints $x = -2$ and $x = 3$. We find $f(-2) = -34, f(-1) = 3, f(0) = -2, f(2) = 30$, and $f(3) = -29$. The biggest value is 30, which is the value of $f(x)$ when $x = 2$, and the smallest is -34, which is the value of $f(x)$ when $x = -2$. Figure 3.3.8 is drawn with different scales on the two axes in order to fit the whole graph in.

This technique is extremely useful, so we summarize it for future reference.

Summary 3.3.1

Maximum/Minimum Problems on Closed Intervals

Suppose $f(x)$ is continuous on the closed interval $[c, d]$ and differentiable on the corresponding open interval (c, d), and we want to find the maximum or minimum values of $f(x)$ on $[c, d]$.

1. If it is not stated directly in the problem, identify the function $f(x)$ whose maximum or minimum values we need to find, and the interval $[c, d]$ on which it is defined.

2. Find the derivative $f'(x)$, checking that it is defined on the whole interval (c, d).

3. Solve the equation $f'(x) = 0$ to find all the points $a \in (c, d)$ at which $f'(a) = 0$. The points at which $f(x)$ achieves its maximum and minimum values must be among these points and the endpoints c and d.

4. Evaluate $f(x)$ at $x = c$, $x = d$, and all points where $f'(x) = 0$; the largest value will be the maximum, and the smallest will be the minimum. Note that the maximum and/or the minimum value may occur at more than one point.

Notice that the hypothesis that $f(x)$ be continuous on $[c, d]$ and differentiable on (c, d) is satisfied for any polynomial $f(x)$ by Definition 2.5.2(i) and Theorem 3.2.4. It also holds for a rational function, provided that the denominator is never zero on $[c, d]$, by Definition 2.5.2(ii) and Theorem 3.2.6.

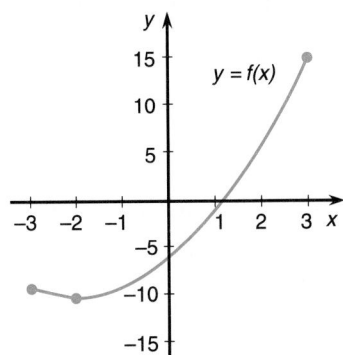

Figure 3.3.9

EXAMPLE 3   Find the maximum and minimum values of each of the following functions on the specified interval: (i) $f(x) = x^2 + 4x - 6$, on $[-3, 3]$; (ii) $g(x) = x^2 + 6x + 2$, on $[-1, 1]$; (iii) $h(x) = x^3 - 3x + 1$, on $[-2, 2]$; (iv) $k(t) = t^4 - 2t^2 + 1$, on $[-1, 2]$.

Solution

(i) We find that $f'(x) = 2x + 4 = 2(x + 2)$, which is zero when $x = -2$. So we have to evaluate $f(x)$ at $x = -2$ and at the endpoints, $x = -3, 3$.

   Now $f(-2) = -10, f(-3) = -9$, and $f(3) = 15$. Clearly the maximum is 15, which occurs at the right endpoint $x = 3$. The minimum is -10, which occurs at $x = -2$. The graph of $y = f(x)$ is of course a parabola; we could have found where the minimum was by completing the square, but finding where $f'(x) = 0$ is actually easier. See Figure 3.3.9.

(ii) Again the graph will be a parabola. The derivative is $g'(x) = 2x + 6$, which is zero when $x = -3$, so that is the minimum point on the graph of the parabola. Unfortunately, the problem asked us to find the maximum and minimum on the interval $[-1, 1]$. The point $x = -3$ where the derivative is zero is outside this interval.

   The procedure we are following is to find all points in the interval at which the derivative is zero and evaluate the function at them and at the endpoints. In this case, since there are no points in the interval at which the derivative is zero, this just means evaluating at the endpoints.

   We find that $g(-1) = -3, g(1) = 9$, so the minimum is -3, which occurs at $x = -1$, and the maximum is 9, which occurs at $x = 1$.

   If we look at Figure 3.3.10, we realize that the entire interval $[-1, 1]$ is to the right of the minimum point and the graph keeps increasing all along the

Figure 3.3.10

Figure 3.3.11

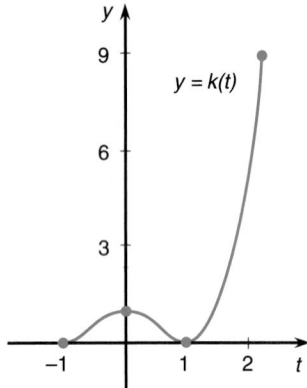

Figure 3.3.12

interval, so it is not surprising that the minimum will be at the left endpoint and the maximum at the right endpoint.

(iii) Here we calculate $h'(x) = 3x^2 - 3 = 3(x^2 - 1)$, which can be factored as $h'(x) = 3(x + 1)(x - 1)$, so we see that $h'(x) = 0$ exactly at $x = \pm 1$.

We evaluate $h(x)$ at $x = -2, -1, 1, 2$: $h(-2) = -1, h(-1) = 3, h(1) = -1$, and $h(2) = 3$. The maximum value is 3, and it occurs at both $x = -1$ and $x = 2$. The minimum value is -1 , and it occurs at both $x = -2$ and $x = 1$. See Figure 3.3.11.

(iv) The derivative is $k'(t) = 4t^3 - 4t = 4t(t^2 - 1) = 4t(t + 1)(t - 1)$, which is zero at $t = 0, -1, 1$. Notice that $t = -1$ is a point at which the derivative is zero and is also one of the endpoints of the interval.

Evaluating $k(-1) = 0, k(0) = 1, k(1) = 0, k(2) = 9$, we see that the maximum is 9, which occurs at $t = 2$. The minimum is 0, which occurs at $t = -1$ and at $t = 1$. See Figure 3.3.12. ◢

All the functions in Example 3 are polynomials, but it has already been remarked that the technique outlined in Summary 3.3.1 for finding maximum and minimum values will work for rational functions too, provided that the denominator is never zero on the interval in question.

EXAMPLE 4   Find the maximum and minimum values of each of the following functions on the specified interval: (i) $R(u) = \frac{1}{u^2 + 1}$ on $[-1, 2]$; (ii) $S(x) = \frac{x+1}{x-1}$ on $[-2, 2]$; (iii) $T(x) = \frac{x+1}{x-1}$ on $[-2, 0]$.

Solution

(i) Applying the quotient rule (Theorem 3.2.6), we can calculate the derivative $R'(u) = \frac{(0)(u^2+1) - 1(2u)}{(u^2+1)^2} = -\frac{2u}{(u^2+1)^2}$. It is easier than it looks to find when this is zero, because we can forget about the denominator and simply ask when the numerator is zero, which is obviously when $u = 0$.

Evaluating at this point and at the endpoints, we find that $R(-1) = \frac{1}{2}, R(0) = 1, R(2) = \frac{1}{5}$. The minimum value is $\frac{1}{5}$, which occurs at $u = 2$, and the maximum is 1, which occurs at $u = 0$.

(ii) We notice that the denominator is zero at $x = 1$, which lies *inside* the interval $[-2, 2]$, so our technique is not applicable. (In fact, there is neither a maximum nor a minimum on this interval, since it is easy to see that $\lim_{x \to 1^-} S(x) = -\infty, \lim_{x \to 1^+} S(x) = \infty. \ldots$)

The important thing is to remember to check that the denominator of a rational function is not zero anywhere in the interval before you try to apply Summary 3.3.1.

(iii) This time the denominator is zero at $x = 1$, but this point is *outside* the interval $[-2, 0]$. We differentiate: $T'(x) = \frac{(1)(x-1) - (x+1)(1)}{(x-1)^2} = -\frac{2}{(x-1)^2}$, which is never zero on the specified interval.

Evaluating at the endpoints, we find $T(-2) = \frac{1}{3}, T(0) = -1$. The minimum, -1 , occurs at the right endpoint $x = 0$, and the maximum, $\frac{1}{3}$, occurs at the left endpoint $x = -2$. ◢

With a graphics calculator it is possible to plot the functions from Example 4. It is best to plot them *after* you have found the maximum and minimum values as a way of checking your work.

TI-85	SHARP EL9300
GRAPH F1$(y(x) =)$1÷(F1$(x) x^2 +$ 1) M2(RANGE) (−) 1 ▼ 2 ▼ ▼ 0 ▼ 1 F5(GRAPH)	◣↲ 1 ÷ (x/θ/T $x^2 + 1$) RANGE − 1 ENTER 2 ENTER 1 ENTER 0 ENTER 1 ENTER 1 ENTER ◣↲
CASIO f_x-7700GB/f_x-6300G	HP 48SX
Cls EXE Range −1 EXE 2 EXE 1 EXE 0 EXE 1 EXE 1 EXE EXE EXE EXE Graph 1 ÷ (◼ $x^2 + 1$) EXE	◼ PLOT PLOTR ERASE ATTN ' αY ◼ 1 ÷ ◼ $\alpha X \, y^x$ 2 + 1 ↵ PLOT ◼ DRAW ERASE 1 +/− ENTER 2 XRNG 0 ENTER 1 YRNG DRAW

Try graphing the other functions from the example, and compare the results of the example with the pictures.

We can use these ideas to solve many interesting problems.

Helicopter bringing hikers to mountain summit in the Alsek Range in southeast Alaska.

EXAMPLE 5   **Helicopter**   Suppose the height above the ground of a helicopter is given by $H(t) = 60t^2 - 6t^3$ ft, where t is the time, in minutes, after the takeoff.

(i) What is the greatest height the helicopter will reach in the first ten minutes of its flight?

(ii) At what moment will it be climbing the fastest?

(iii) Notice that $H(10) = 0$, which means that after flying for 10 minutes the helicopter will land again. If a helicopter touches down while descending at a rate in excess of 5 ft/sec, it will likely sustain damage. Calculate the rate of descent and decide whether or not this helicopter will land safely.

Solution   Part (i) asks us to find the maximum value of $H(t)$ for t in the interval $[0, 10]$, that is, for $0 \le t \le 10$. We find the derivative is $H'(t) = 120t - 18t^2 = 6t(20 - 3t)$. This is zero when $t = 0$ or $t = \frac{20}{3}$. Both these times are in the interval $[0, 10]$, so the maximum could occur at $t = 0, \frac{20}{3}$, or 10.

We evaluate and find that $H(0) = 0, H\left(\frac{20}{3}\right) = \frac{8000}{9} \approx 888.9, H(10) = 0$. The maximum height is approximately 888.9 ft, which is reached $\frac{20}{3}$ minutes after takeoff, or 6 min, 40 sec after takeoff.

Part (ii) asks us to find the maximum value of the rate at which the helicopter is climbing. As we know, the rate at which the helicopter is climbing is exactly the rate of change of the height, which is $H'(t)$. We have just calculated that $H'(t) = 120t - 18t^2$, and to find the maximum of *this* function, we have to find *its* derivative.

The derivative of $H'(t)$, that is, the derivative of the derivative of $H(t)$, is called the second derivative of $H(t)$ and is written $H''(t)$. It is easily calculated:

$$H''(t) = 120 - 36t.$$

This is zero when $t = \frac{120}{36} = \frac{10}{3}$.

Evaluating $H'(0) = 0$, $H'\left(\frac{10}{3}\right) = 200$, $H'(10) = -600$, we see that the helicopter will be climbing fastest at the moment $\frac{10}{3} = 3\frac{1}{3}$ min after takeoff, which means 3 min, 20 sec after takeoff. At this moment the rate of climb will be $H'\left(\frac{10}{3}\right) = 200$ ft/min, or $\frac{200}{60} = 3\frac{1}{3}$ ft/sec.

For part (iii) we evaluate $H'(10) = -600$. The minus sign reflects the fact that the helicopter is descending. Its rate of descent is 600 ft/min, which we must convert to $\frac{600}{60}$ ft/sec $= 10$ ft/sec.

The helicopter will land too fast and will likely be damaged.

In Example 5 the function $H(t)$ was given to us as part of the question, so the solution amounted to finding various maximum values of specified functions. A more interesting but more difficult type of problem arises when we are given a situation and a question to answer but have to figure out the function for ourselves.

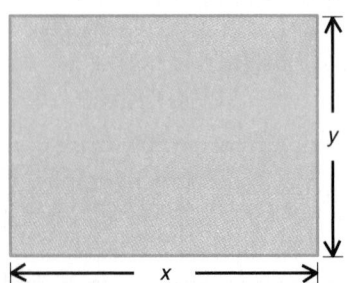

Figure 3.3.13

EXAMPLE 6   Suppose we wish to enclose a rectangular field in a fence. If we are given 600 running feet of fencing material, what is the field of largest area that we can enclose?

Solution   First we draw a diagram. Label the lengths of two adjacent sides of the field by x and y (see Figure 3.3.13). The area of the field is then $A = xy$.

Since the perimeter of the field is $2x + 2y$, this must equal the total length of fencing, that is,

$$2x + 2y = 600.$$

We can solve for y. We know $2y = 600 - 2x$, so $y = 300 - x$. From this we see that $A = xy = x(300 - x) = 300x - x^2$.

We have expressed the area A as a function of x, the length of one side of the field: $A(x) = 300x - x^2$. The problem is to find the maximum possible value of $A(x)$. Since x is a length, it must be positive, and it cannot be bigger than 300; if it were, then the two sides of length x would use up more than the total fencing available. In other words, what we need is the maximum value of $A(x)$ for $x \in [0, 300]$. This is a problem we know how to solve.

The derivative is $A'(x) = 300 - 2x$, which is zero when $x = 150$. Evaluating here and at the endpoints, we find $A(0) = 0$, $A(150) = 22{,}500$, $A(300) = 0$, so the maximum occurs when $x = 150$.

Since $y = 300 - x$, we see that when $x = 150$, the value of y is also 150. In other words, the field of largest possible area is obtained by making a square field with all four sides 150 ft long. Its area is 22,500 sq ft.

Summary 3.3.2

To do Example 6, it was necessary to decide what it was that was to be maximized (or minimized). In this case the area of the field was to be maximized. Then this quantity had to be expressed as a function. The complication is that there may be more than one variable (in this case x and y, the lengths of the two sides). To avoid this difficulty, it is necessary to use the other information in the problem to *eliminate* all but one of them. Typically, this is done by solving for some variables in terms of the others until only one is left (in Example 6 we solved for y and found $y = 300 - x$). Substituting back into the function results in a function with only one variable. Information in the problem or just common sense should also allow us to restrict our attention to values of the variable in some closed interval, and then it should be possible to solve the problem by using the technique outlined in Summary 3.3.1.

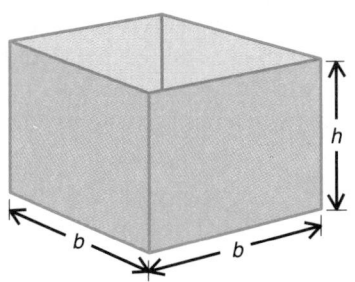

Figure 3.3.14

EXAMPLE 7   **Volume of Box**   An open rectangular box is made so that its bottom is a square and the total surface area (i.e., the outside surface area) is 12 square units. What is the maximum possible volume of the box? (By an open rectangular box, we mean a box without a lid, that is, with a bottom but no top.)

Solution   Let b be the length of each side of the base, and let h be the height of the box (see Figure 3.3.14). The area of the bottom is b^2 and the area of each side is bh, so the total surface area is $b^2 + 4bh$. The volume is $V = b^2 h$.

At this point we have the function representing the volume, which we want to maximize, but there are two variables, b and h. We need to eliminate one of them.

Knowing that $b^2 + 4bh = 12$, we could try to solve for either b or h. If we try to solve for b, it will be messy because we have a quadratic equation in the variable b and the solution will involve square roots. On the other hand, it is easy to solve for h.

From $4bh = 12 - b^2$ we get $h = \frac{12-b^2}{4b}$.

Substituting this back into the formula for V, we find

$$V = b^2 h = b^2 \frac{12 - b^2}{4b} = \frac{1}{4}(12b - b^3).$$

We have to use some initiative to find the appropriate interval ...

This is a function of only one variable, which is what we want. Once again, a length should not be negative, so $b \geq 0$. On the other hand, the total area is 12 and the area of the base is b^2, so we must have $b \leq \sqrt{12}$. The problem is to find the maximum value of $V = \frac{1}{4}(12b - b^3)$ for $b \in [0, \sqrt{12}]$.

The derivative is $V'(b) = \frac{1}{4}(12 - 3b^2)$, which is zero when $12 = 3b^2$, or $b^2 = 4$. This happens when $b = \pm 2$, but of these two possibilities, only $2 \in [0, \sqrt{12}]$. Evaluating at this point and the endpoints, we find $V(0) = 0$, $V(2) = \frac{1}{4}(24 - 8) = 4$, $V(\sqrt{12}) = \frac{1}{4}\left(12\sqrt{12} - (\sqrt{12})^3\right) = \frac{1}{4}\left(12\sqrt{12} - (\sqrt{12})^2\sqrt{12}\right) = 0$. The maximum value is $V = 4$, which occurs when $b = 2$.

The largest possible volume is 4 cubic units. A box of this volume can be made by making the sides of the base 2 units long, in which case the height of the box is $h = \frac{12-b^2}{4b} = \frac{12-4}{8} = 1$ unit.

Notice that we have used letters (b, h, V for "base," "height," "Volume") that help us remember what they stand for.

Example 7 illustrates that the possible choices of which variable to eliminate may not all be equally easy to work with. It is *possible* to solve this problem by eliminating b and working with h, but it is definitely harder and cannot easily be done with our present knowledge of derivatives.

← 30 cm →

Figure 3.3.15

EXAMPLE 8   **Folded Box**   Starting with a square piece of sheet metal whose sides are each 30 cm long, an open rectangular box is made by cutting out four equal squares from the corners and folding up the sides, as shown in Figure 3.3.15. What is the box of largest volume that can be made in this way?

Solution   It is the volume that is to be maximized, but we have to decide what to use as our variable. Let us try letting x be the length of the sides of the little squares cut out of the corners. When the box is folded up, this becomes the height of the box. The base of the box is a square, outlined by dotted lines in Figure 3.3.15.

The length of the side of this "inner" square is $30 - 2x$ (the length of the side of the big square minus two corners). So the area of the base is $(30 - 2x)^2$, and the volume of the box is $V = V(x) = x(30 - 2x)^2 = x(900 - 120x + 4x^2) = 900x - 120x^2 + 4x^3$. Certainly, x cannot be less than 0 nor greater than 15 (since that would mean that the four corners used up all of the big square).

The derivative is $V'(x) = 900 - 240x + 12x^2$. To find when it is zero, we use the Quadratic Formula. The two points are

$$x = \frac{240 \pm \sqrt{(-240)^2 - 4(12)900}}{24}$$

$$= \frac{240 \pm \sqrt{57,600 - 43,200}}{24} = \frac{240 \pm \sqrt{14,400}}{24}$$

$$= \frac{240 \pm 120}{24} = 5, 15.$$

Evaluating at these points and the endpoints gives $V(0) = 0$, $V(5) = 2,000$, and $V(15) = 0$, so the maximum occurs when $x = 5$.

The largest possible box is made by cutting out corners that measure 5 cm on each side. The base will be 20 cm on each side, and the volume will be 2,000 cu cm.

> It is interesting to notice that the derivative is zero at $x = 15$. This corresponds to removing from each corner a square with 15 cm sides, which would leave no bottom at all. It is not uncommon to get some ridiculous possible answers like this. We call them **extraneous solutions,** and they will be eliminated when we evaluate the function.

EXAMPLE 9   **Fruit Tree Yield**   A certain species of fruit tree yields 20 bushels of fruit if it is grown with the trees evenly spaced so that there are 100 trees to the acre. A farmer realizes that putting more trees to the acre will increase the total yield, even though crowding will reduce the yield per tree.

Experience has taught the farmer that each additional tree per acre will reduce the yield per tree by $\frac{1}{10}$ bushel. He also knows that more than 250 trees per acre will result in poor quality fruit. What number of trees per acre will give the highest total yield?

Solution   Let N be the number of trees per acre. First we can calculate the yield per tree. It is 20 when $N = 100$ and reduces by $\frac{1}{10}$ for every tree over $N = 100$. The number of trees over 100 is of course $N - 100$, so the reduction in yield per tree is $\frac{1}{10}(N - 100)$. From this we see that the yield per tree is $20 - \frac{1}{10}(N - 100)$.

The total yield from N trees is therefore

$$Y = N\left(20 - \frac{1}{10}(N - 100)\right) = N\left(20 - \frac{1}{10}N + 10\right) = N\left(30 - \frac{1}{10}N\right)$$

$$= 30N - \frac{1}{10}N^2.$$

We calculate the derivative $\frac{dY}{dN} = 30 - \frac{1}{5}N$. It is zero when $N = 150$. Evaluating the yield at this point and at the endpoints 100, 250 shows that $Y(100) = 2{,}000$, $Y(150) = 2{,}250$, $Y(250) = 1{,}250$.

The highest yield will be obtained by planting 150 trees per acre.

> Example 9 is typical of many problems in which two competing factors affect the function. In this case, increasing the number of trees both reduced the yield per tree and increased the number of trees yielding fruit. It is not easy to guess the best answer with this kind of problem.

EXAMPLE 10 **Closest Point on a Curve** Find the point on the curve $y = x^2 + 2x - 1$ that is closest to the point $(2, -2)$.

Solution The distance from $(2, -2)$ to (x, y) is

$$D = \sqrt{(x - 2)^2 + \left(y - (-2)\right)^2}.$$

If (x, y) is on the curve, we have $y = x^2 + 2x - 1$, so

$$D = \sqrt{(x - 2)^2 + (x^2 + 2x - 1 + 2)^2} = \sqrt{(x - 2)^2 + (x^2 + 2x + 1)^2}$$

$$= \sqrt{x^2 - 4x + 4 + x^4 + 4x^3 + 6x^2 + 4x + 1}$$

$$= \sqrt{x^4 + 4x^3 + 7x^2 + 5}.$$

We want to find the minimum of D, but the square root is awkward. It will be much easier to minimize $D^2 = x^4 + 4x^3 + 7x^2 + 5$, and a point at which D^2 has a minimum value will also be a point at which D has a minimum value and vice versa. (Here we are relying on the fact that $D \geq 0$.)

The derivative of D^2 is

$$(D^2)'(x) = 4x^3 + 12x^2 + 14x = x(4x^2 + 12x + 14).$$

We can find when the quadratic factor $4x^2 + 12x + 14$ is zero by applying the Quadratic Formula. We find that the discriminant is $b^2 - 4ac = (12)^2 - 4(4)14 = 144 - 224 = -80$, so there are no roots. This says that the only place where the derivative is zero is $x = 0$.

The corresponding point on the curve is $(0, -1)$, and the value of D^2 when $x = 0$ is 5. This means that $D = \sqrt{5}$, which means that the point $(0, -1)$ is a distance of $\sqrt{5}$ from $(2, -2)$.

The other ingredient we need is an interval on which to find the minimum. We already have one point, $(0, -1)$, which is $\sqrt{5}$ units away from $(2, -2)$, so there is no need to look at any points that are farther away than that. In particular, the minimum

We found that $D^2 = 5$ when $x = 2$. Also $D^2 = (x-2)^2 + (y+2)^2 \geq (x-2)^2$. Since $(x-2)^2 \geq 3^3 = 9$ outside $[-1,5]$, we see that $D^2 \geq 9$ outside $[-1,5]$, so the minimum must occur *inside* $[-1,5]$.

cannot occur at any point (x, y) for which x is more than $\sqrt{5}$ units from 2, or, to make it easier to work with, when x is more than 3 units from 2. So the minimum cannot occur where $x > 5$ or $x < -1$, so it must occur for some $x \in [-1, 5]$.

We have seen that the derivative of D^2 is zero when $x = 0$, so we evaluate D^2 at this point and at the endpoints: $D^2(-1) = 9, D^2(0) = 5, D^2(5) = 1305$. The minimum is certainly at $x = 0$.

The point on the curve $y = x^2 + 2x - 1$ that is closest to $(2, -2)$ is $(0, -1)$. The distance between these two points is $\sqrt{5}$.

One trick we learned in this example was to minimize (or maximize) a square root by minimizing (or maximizing) its square. Apart from that, the hardest part of the solution was figuring out what interval would be appropriate.

Exercises 3.3

I

Find the maximum and minimum values of each of the following functions on the specified interval (identify *all* the points where the maximum and minimum values occur).

1. $f(x) = x^2 + 4x + 3$, on $[-1, 1]$
2. $g(x) = x^2 + 4x + 3$, on $[-3, -1]$
3. $h(t) = -3t^2 + 3t - 2$, on $[0, 1]$
4. $k(u) = u^3 + 3$, on $[-2, 4]$
5. $F(s) = s^4 - 4s^2 + 1$, on $[-2, 1]$
6. $G(x) = 3x^4 - 8x^3 + 6x^2$, on $[-2, 2]$
7. $H(x) = x^4 - 4x^3 + 4x^2 + 3$, on $[0, 3]$
8. $K(t) = t^5 - 15t^3 + 7$, on $[0, 5]$
9. $L(u) = 2u^7 - 14u + 2$, on $[-2, 2]$
10. $R(x) = x(x - 2)(x - 4)$, on $[0, 2]$
11. $f(x) = \frac{1}{2+x^2}$, on $[-4, 4]$
12. $f(x) = \frac{x+2}{3-x}$, on $[-1, 1]$
13. $f(x) = -\frac{x+2}{3-x}$, on $[-1, 1]$
14. $f(x) = \frac{2}{x}$, on $[1, 4]$
15. $f(x) = \frac{x^2}{x+3}$, on $[-2, 2]$
16. $f(x) = \frac{x+3}{x^2}$, on $[-8, -2]$
17. $f(x) = \frac{x^3}{x-1}$, on $[-1, 0]$
18. $f(x) = \frac{x-1}{x^3}$, on $[1, 2]$
19. $f(x) = \frac{x^2-1}{x+1}$, on $[0, 2]$
20. $f(x) = \frac{x-1}{x^2-1}$, on $[2, 4]$

Practice your differentiation skills by finding the derivative of each of the following functions.

21. $f(x) = 3x^7 - 2x^4 + x - 1$
22. $f(x) = 12x^{12} + 4x^4 + x$
23. $f(x) = \frac{x-1}{x+1}$
24. $f(x) = \frac{1}{x^2+1}$
25. $f(x) = \frac{1}{2x+3}$
26. $f(x) = \frac{x^2+2}{x^2+1}$
27. $f(x) = x^{-3}(2x + 1)$
28. $f(x) = x^{-k}(x^k + 1)$
29. $f(x) = (x^4 + 2x^2 + 4)^{-1}$
30. $f(x) = (x^2 + 1)^{-2}$

II

31. Suppose we wish to enclose a rectangular field of area 2, 500 square meters. What lengths of the sides will permit us to do this with a minimum of fencing material?

32. What is the largest possible volume of a closed rectangular box (i.e., including a lid) whose base is square and whose (outside) surface area is 54 square inches?

33. Suppose an open box is made from a rectangular sheet of metal whose dimensions are 20 cm by 30 cm by cutting equal squares from the corners and folding up the sides. What is the largest possible volume for such a box?

34. A rectangular pasture is to be made beside a river by fencing three sides and letting the straight riverbank be the fourth side (see Figure 3.3.16). How should this be done in order to make the pasture's area as large as possible if 300 yd of fencing material is available?

Figure 3.3.16

35. An open rectangular box must be made with a square base and a volume of 100 cubic inches. Find how to make such a box with a minimum of material (ignoring the thickness of the material and any waste in the cutting).

36. What are the dimensions of the rectangle whose perimeter is P and that has the greatest possible area?

37. (i) What point on the line $y = 2x + 1$ is closest to the point $(0, 0)$?

   (ii) What point on the line $y = 3 - x$ is closest to the point $(2, 2)$?

38. (i) What point on the curve $y = x^2$ is closest to the point $(0, -3)$?

   (ii) What point on the curve $y = x^2 - 4x + 4$ is closest to the point $(-1, 0)$?

*39. A rectangle whose perimeter is 50 cm is rotated around one of its sides to form a right circular cylinder (see Figure 3.3.17). Find the dimensions of the rectangle that results in the cylinder with the largest possible volume.

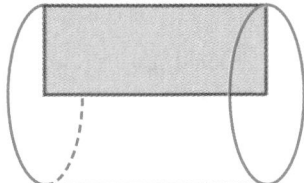

Figure 3.3.17

*40. A farmer has 30 plants in a plot, and each yields an average of 7 kg of vegetables. For each additional 10 plants that are put in the plot, crowding causes the average yield to decrease by 1 kg per plant. How many plants should she have to maximize the total yield?

*41. A piece of wire 1 m long is cut into two pieces. One piece is bent into the shape of a square, the other into a circle. What should the lengths of the pieces be in order that the sum of the areas of the circle and the square should be (i) a minimum, (ii) a maximum?

*42. A piece of wire 16 ft long is cut into two pieces. Each piece is bent into a square. What should the lengths of the pieces be in order that the sum of the areas of the two squares should be (i) a minimum, (ii) a maximum?

*43. Market research has shown that if a certain store sells a kitchen gadget for $10, they will be able to sell 800 units. Moreover, for each additional dollar added to the price, the number the store will sell decreases by 100.

   (i) At what price should the store sell the gadgets in order to maximize the total revenue?

   (ii) If the gadgets cost the store $5 each, what price will maximize the profits?

*44. To set up the equipment for a certain manufacturing process costs $1,000, and once it is set up it costs $10 for each unit produced. If 1,000 units are produced, they will sell for $15 each, and experience has shown that for each additional 1,000 units produced, the unit price will have to be lowered by $1 in order to sell them all. How many units should be produced to maximize profits, and at what price should they sell?

For many functions $f(x)$ it is difficult to find exactly where $f'(x) = 0$, but it is possible to use computer graphics to get good estimates.

For the following two functions $f(x)$, use a graphics calculator or computer to graph $y = f'(x)$ at various magnifications and estimate each zero of $f'(x)$, accurate to three decimal places. Use this information and the graph of $y = f(x)$ to estimate the maximum and minimum values of $f(x)$ on the specified interval.

45. $f(x) = x^5 - x^3 + 2x^2 - x + 1$, for $x \in [-1, 1]$

46. $f(x) = x^4 - 2x^3 + x$, for $x \in [-0.7, 0.7]$

Use a graphics calculator or computer to estimate the maximum and minimum values of each of the following functions on the specified interval, correct to four decimal places.

47. $f(x) = x^4 - 4x^2 + 3x - 2$, $[-2, 2]$

48. $f(x) = x^6 - 2x^4 + x$, $[-1.5, 1]$

49. $f(x) = x \sin x$, $[0, \pi]$

50. $f(x) = x^2 \cos(\pi x)$, $[-1, 1]$

3.4 Derivatives of Trigonometric Functions

$\sin'(x) = \cos x$
$\cos'(x) = -\sin x$

In this section we learn how to differentiate the trigonometric functions. To calculate the derivatives, we must evaluate some limits involving trigonometric functions, so it will be necessary to refer to results about limits established in Chapter 2, in particular Theorem 2.6.3. You may wish to refresh your memory about that result before proceeding. Use will also be made of the addition formulas (Theorem 1.7.11 p. 51) for sin and cos and of various trigonometric identities (e.g., Theorem 1.7.2), so you may wish to review them too.

Throughout this section, angles will be measured in radians unless it is explicitly stated otherwise. By the end of the section it will be reasonably clear that this choice is a good one and that using degrees would make differentiation much more complicated.

To find the derivative of $\sin x$, we must evaluate

$$\lim_{h \to 0} \frac{\sin(x + h) - \sin(x)}{h}.$$

To simplify the numerator, we appeal to the addition formula for sin (Theorem 1.7.11), which says

$$\sin(x + h) = \sin(x)\cos(h) + \cos(x)\sin(h).$$

Substituting this into the limit gives

$$\begin{aligned}
\frac{d}{dx}(\sin x) &= \lim_{h \to 0} \frac{\sin(x + h) - \sin(x)}{h} \\
&= \lim_{h \to 0} \frac{\sin(x)\cos(h) + \cos(x)\sin(h) - \sin(x)}{h} \\
&= \lim_{h \to 0} \frac{\sin(x)\cos(h) - \sin(x) + \cos(x)\sin(h)}{h} \\
&= \lim_{h \to 0} \left(\frac{\sin(x)\big(\cos(h) - 1\big)}{h} + \frac{\cos(x)\sin(h)}{h} \right) \\
&= \lim_{h \to 0} \frac{\sin(x)\big(\cos(h) - 1\big)}{h} + \lim_{h \to 0} \frac{\cos(x)\sin(h)}{h} \\
&= \sin(x) \lim_{h \to 0} \frac{\cos(h) - 1}{h} + \cos(x) \lim_{h \to 0} \frac{\sin(h)}{h}.
\end{aligned}$$

At this point we apply Theorem 2.6.3 to evaluate the two limits. It tells us that

$$\lim_{h \to 0} \frac{\cos h - 1}{h} = 0 \quad \text{and} \quad \lim_{h \to 0} \frac{\sin h}{h} = 1.$$

Substituting these limits into the above formula, we find that

$$\frac{d}{dx}(\sin x) = \sin(x)(0) + \cos(x)(1) = \cos x.$$

In other words, the derivative of $\sin x$ is $\cos x$.

A similar computation can be used to find the derivative of $\cos x$. Before you read the next few lines, which tell how to do it, remind yourself of the addition formula for $\cos x$ (Theorem 1.7.11), and try to work out the derivative for yourself.

The addition formula for $\cos x$ is
$$\cos(x + h) = \cos(x)\cos(h) - \sin(x)\sin(h).$$

Using this, we can calculate

$$\frac{d}{dx}(\cos x) = \lim_{h \to 0} \frac{\cos(x + h) - \cos(x)}{h}$$

$$= \lim_{h \to 0} \frac{\cos(x)\cos(h) - \sin(x)\sin(h) - \cos(x)}{h}$$

$$= \lim_{h \to 0} \frac{\cos(x)\cos(h) - \cos(x) - \sin(x)\sin(h)}{h}$$

$$= \lim_{h \to 0} \frac{\cos(x)\big(\cos(h) - 1\big) - \sin(x)\sin(h)}{h}$$

$$= \lim_{h \to 0} \frac{\cos(x)\big(\cos(h) - 1\big)}{h} - \lim_{h \to 0} \frac{\sin(x)\sin(h)}{h}$$

$$= \cos(x)\lim_{h \to 0} \frac{\cos(h) - 1}{h} - \sin(x)\lim_{h \to 0} \frac{\sin(h)}{h}.$$

Again substituting for these two limits using Theorem 2.6.3, we find that

$$\frac{d}{dx}(\cos x) = \cos(x)(0) - \sin(x)(1) = -\sin x.$$

In other words, the derivative of $\cos x$ is $-\sin x$. We have just proved the following result.

THEOREM 3.4.1

The functions $\sin x$ and $\cos x$ are differentiable at every x, and

$$\frac{d}{dx}(\sin x) = \cos x, \qquad \frac{d}{dx}(\cos x) = -\sin x.$$

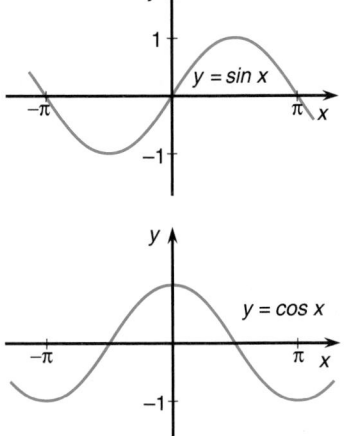

Figure 3.4.1

These two derivatives are extremely important and should be memorized. If you are having trouble remembering which one has the minus sign, consider Figure 3.4.1. It shows the graphs of $\sin x$ and $\cos x$. At $x = 0$ the graph of $\sin x$ passes through the origin. The tangent at that point clearly slopes upward to the right, which means that its slope is positive.

In fact, the slope of its tangent is $\sin'(0)$, which we have just seen is $\cos(0) = 1$. This confirms that the derivative of $\sin x$ should be $\cos x$ and cannot have a minus sign.

It is interesting to look at the two graphs in Figure 3.4.1 and observe how the value of $\cos x$ changes to reflect the slope of the tangent to $y = \sin x$; $\cos x$ is positive when $\sin x$ is increasing and negative when $\sin x$ is decreasing. Similar observations can be made about the value of $-\sin x$, the derivative of $\cos x$.

EXAMPLE 1   Differentiate the following functions: (i) $x^2 \sin x$; (ii) $\sin^2 x$; (iii) $\frac{1}{\cos x}$; (iv) $\tan x$.

Solution

(i) Writing $x^2 \sin x$ as the product of x^2 and $\sin x$ and using the Product Rule and Theorem 3.4.1, we find that its derivative is $2x \sin x + x^2 \cos x$.

(ii) Again we use the Product Rule, thinking of $\sin^2 x$ as the product of $\sin x$ with itself. We find that its derivative is

$$\cos x \sin x + \sin x \cos x = 2 \sin x \cos x.$$

(iii) For this question we use the Quotient Rule, and find that, provided $\cos x \neq 0$,

$$\frac{d}{dx}\left(\frac{1}{\cos x}\right) = \frac{0(\cos x) - 1(-\sin x)}{\cos^2 x} = \frac{\sin x}{\cos^2 x} = \frac{\sin x}{\cos x}\frac{1}{\cos x}$$
$$= \tan x \sec x.$$

Since the original function $\frac{1}{\cos x}$ is the same thing as $\sec x$, we can write this as

$$\frac{d}{dx}(\sec x) = \sec x \tan x,$$

which makes sense whenever $\cos x \neq 0$, which means x cannot be $\pm\frac{\pi}{2}$, $\pm\frac{3\pi}{2}$, etc. In other words, x cannot be an *odd* multiple of $\frac{\pi}{2}$. In symbols this can be written as $x \neq (2n + 1)\frac{\pi}{2}$, for any integer n.

(iv) To use Theorem 3.4.1, we first express $\tan x$ in terms of $\sin x$ and $\cos x$, which is easy, since $\tan x = \frac{\sin x}{\cos x}$. Then we can apply the quotient rule to obtain that, provided $\cos x \neq 0$,

$$\frac{d}{dx}(\tan x) = \frac{d}{dx}\left(\frac{\sin x}{\cos x}\right) = \frac{\cos x \cos x - \sin x(-\sin x)}{\cos^2 x}$$
$$= \frac{\cos^2 x + \sin^2 x}{\cos^2 x} = \frac{1}{\cos^2 x}$$
$$= \sec^2 x.$$

Here we used the identity $\cos^2 x + \sin^2 x = 1$ (cf. Theorem 1.7.2).

This is true whenever $x \neq (2n + 1)\frac{\pi}{2}$, for any integer n. ◢

Example 1 suggests that we can use the quotient rule in combination with Theorem 3.4.1 to find the derivatives of the other trigonometric functions. Parts (iii) and (iv) of Example 1 asked us to find the derivatives of $\sec x$ and $\tan x$, respectively. Similar calculations give the derivatives of $\csc x$ and $\cot x$.

THEOREM 3.4.2

The trigonometric functions $\sec x$, $\tan x$, $\csc x$, and $\cot x$ are differentiable wherever they are defined. Moreover, their derivatives are as follows:

$$\frac{d}{dx}(\sec x) = \sec x \tan x, \qquad \frac{d}{dx}(\csc x) = -\csc x \cot x,$$
$$\frac{d}{dx}(\tan x) = \sec^2 x, \qquad \frac{d}{dx}(\cot x) = -\csc^2 x.$$

You might find it helpful to remember when differentiating the trigonometric functions sin, cos, sec, tan, csc, and cot that three of the six derivatives contain a minus sign, and they are exactly the derivatives of the "co-" functions: cos (cosine), csc (cosecant), and cot (cotangent). If you learn the derivatives of sec and tan, you can get the derivatives of the corresponding "co" functions csc and cot by changing the functions in the derivative to the corresponding "co" functions and putting in a minus sign.

PROOF   Parts (iii) and (iv) of Example 1 were sec x and tan x. The others are very similar:

$$\frac{d}{dx}(\csc x) = \frac{d}{dx}\left(\frac{1}{\sin x}\right) = \frac{0(\sin x) - 1(\cos x)}{\sin^2 x}$$

$$= -\frac{\cos x}{\sin^2 x} = -\frac{\cos x}{\sin x}\frac{1}{\sin x} = -\csc x \cot x.$$

$$\frac{d}{dx}(\cot x) = \frac{d}{dx}\left(\frac{\cos x}{\sin x}\right) = \frac{-\sin x \sin x - \cos x \cos x}{\sin^2 x}$$

$$= -\frac{\sin^2 x + \cos^2 x}{\sin^2 x} = -\frac{1}{\sin^2 x} = -\csc^2 x.$$

These results hold whenever the functions involved are defined. For sec x and tan x this is whenever $\cos x \neq 0$, that is, when $x \neq (2n+1)\frac{\pi}{2}$, for any integer n. For csc x and cot x it is whenever $\sin x \neq 0$, that is, when $x \neq n\pi$, for any integer n. ◢

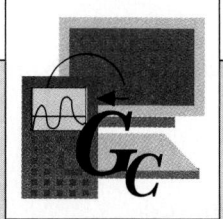

It is interesting to use a graphics calculator to plot a function together with its derivative. For instance, we plot $y = \cot x$ and its derivative $y = -\csc^2 x = -\frac{1}{\sin^2 x}$.

TI-85	SHARP EL9300
GRAPH F1$(y(x) =)$ 1÷ TAN F1(x) ▼ $(-)$ 1 ÷ (SIN F1(x)) x^2 M2(RANGE) 0 ▼ π ▼ ▼ $(-)$ 10 ▼ 10 F5(GRAPH)	$\not{A\!\!\!/}$ 1 ÷ tan x/θ/T 2ndF ▼ -1 ÷ (sin x/θ/T) x^2 RANGE 0 ENTER π ENTER 1 ENTER -10 ENTER 10 ENTER 1 ENTER $\not{A\!\!\!/}$

CASIO f_x-7700GB/f_x-6300G	HP 48SX
Cls EXE Range 0 EXE π EXE 1 EXE -10 EXE 10 EXE 1 EXE EXE EXE EXE Graph 1 ÷ tan ▮ EXE Graph $-$ 1 ÷ (sin ▮) x^2 EXE	▮ ▮▮▮ PLOTR ERASE ATTN ′ αY ▮ 1÷ TAN αX ↱ ▮▮▮ ▮ DRAW ERASE 0 ENTER ▮ XRNG 10 +/− ENTER 10 YRNG DRAW ATTN ′ αY ▮ 1 +/− ÷ ▮ SIN αX ▶ y^x 2 ↱ ▮▮▮ ▮ DRAW DRAW

Identify which graph is which in the picture. Notice how the function $y = \cot x$ decreases more rapidly when its derivative is more negative.

EXAMPLE 2   Find all points at which the derivative of each of the following functions is zero: (i) tan x; (ii) sec x; (iii) $\cos(2x)$.

Solution

(i)   The derivative of tan x is $\sec^2 x = \frac{1}{\cos^2 x}$. It is defined only when $\cos x \neq 0$, and it is never zero (the numerator is always 1).

Figure 3.4.2

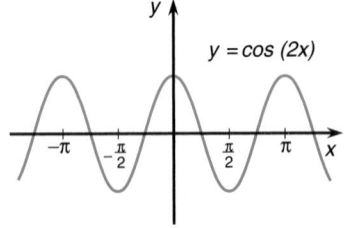

Figure 3.4.3

(ii) The derivative of sec x is $\sec x \tan x = \frac{\sin x}{\cos^2 x}$. This will be zero whenever $\sin x = 0$, which happens whenever x is an integer multiple of π, that is, when $x = n\pi$, for any integer n. We see from Figure 3.4.2 that these are exactly the points at which the tangent line is horizontal.

(iii) To do this, we observe that $\cos(2x) = \cos(x + x)$. By the addition formula (Theorem 1.7.11) this is $\cos(2x) = \cos x \cos x - \sin x \sin x = \cos^2 x - \sin^2 x$. So its derivative is

$$\frac{d}{dx}\big(\cos(2x)\big) = \frac{d}{dx}(\cos^2 x - \sin^2 x) = \frac{d}{dx}(\cos x \cos x - \sin x \sin x)$$
$$= -\sin x \cos x + \cos x(-\sin x) - \cos x \sin x - \sin x \cos x$$
$$= -4\cos x \sin x.$$

This is zero whenever $\cos x$ or $\sin x$ is zero, that is, whenever $x = n\frac{\pi}{2}$, for any integer n. This is also very reasonable when we look at the graph of $y = \cos(2x)$ in Figure 3.4.3.

EXAMPLE 3 Let c be any nonzero real number, and find the derivative of $f(x) = \sin(cx)$.

Solution We know that

$$f'(x) = \lim_{h \to 0} \frac{f(x + h) - f(x)}{h} = \lim_{h \to 0} \frac{\sin\big(c(x + h)\big) - \sin(cx)}{h}$$
$$= \lim_{h \to 0} \frac{\sin(cx + ch) - \sin(cx)}{h}$$
$$= \left(\lim_{h \to 0} \frac{\sin(cx + ch) - \sin(cx)}{ch}\right) c.$$

Now as $h \to 0$, the product ch will also tend to 0, so if we write $u = ch$, the above limit becomes $\lim_{u \to 0} \frac{\sin(cx + u) - \sin(cx)}{u}$, which is just the derivative of sin evaluated at cx, which is $\cos(cx)$.

Including the constant c at the end of the line, we find that

$$\frac{d}{dx}\big(\sin(cx)\big) = \cos(cx)\, c.$$

The argument we just gave is quite similar to the argument used in Example 2.6.2. A few moments' thought should convince you that there is nothing particularly special about the function $\sin x$. We could have started with any differentiable function $g(x)$ and let $f(x) = g(cx)$. Then we would have found that $f(x)$ is differentiable at every x for which g is differentiable at cx. Moreover, the derivative of $f(x)$ is $g'(cx)\, c$.

We record this result for future reference.

THEOREM 3.4.3

Let $g(x)$ be a function, let c be a nonzero real number, and define $f(x) = g(cx)$.
  If $g(x)$ is differentiable at $x = a$, then $f(x)$ is differentiable at $x = \frac{a}{c}$, and, whenever f is differentiable,

$$f'(x) = g'(cx)\, c.$$

EXAMPLE 4   Find the derivative of (i) $\cos(3x)$; (ii) $\tan(-5x)$; (iii) $\sin(2x)$.

Solution

(i) By Theorem 3.4.3,

$$\frac{d}{dx}\big(\cos(3x)\big) = \big(-\sin(3x)\big)(3) = -3\sin(3x).$$

(ii) Similarly,

$$\frac{d}{dx}\big(\tan(-5x)\big) = \sec^2(-5x)(-5) = -5\sec^2(-5x).$$

(iii) The same sort of reasoning tells us the derivative of $\sin(2x)$ is $\cos(2x)(2) = 2\cos(2x)$. However, it is interesting to do this calculation another way. Using the addition formula for $\sin$ (Theorem 1.7.11), we write $\sin(2x) = \sin(x + x) = \cos x \sin x + \sin x \cos x = 2\cos x \sin x$. Now we can use the Product Rule to differentiate, and we find

$$\frac{d}{dx}\big(\sin(2x)\big) = \frac{d}{dx}(2\cos x \sin x) = 2(-\sin x \sin x + \cos x \cos x)$$
$$= 2(\cos^2 x - \sin^2 x).$$

On the face of it we have two different answers for the derivative of $\sin(2x)$. It turns out that they are really the same, because the addition formula (Theorem 1.7.11) for $\cos$ tells us that $\cos(2x) = \cos x \cos x - \sin x \sin x = \cos^2 x - \sin^2 x$.

> The last example illustrates an important fact. Finding a derivative in two different ways sometimes gives two apparently different answers. When this happens, there is always the possibility that we have made an error somewhere, but it is also possible that the two formulas are actually equal and merely appear to be different. This is especially likely to happen with trigonometric functions, because the various trigonometric identities say that apparently different formulas may actually be equal.

EXAMPLE 5   Find the maximum and minimum values of $f(t) = \sin t \cos t$ for $t \in \left[0, \frac{\pi}{2}\right]$.

Solution   The derivative is $f'(t) = \cos t \cos t + \sin t(-\sin t) = \cos^2 t - \sin^2 t$. To find the maximum and minimum values, we must find when the derivative is zero, that is, when $\cos^2 t = \sin^2 t$.

Since $\cos^2 t + \sin^2 t = 1$, we have $\sin^2 t = 1 - \cos^2 t$, and substituting, we see that we are asking when does $\cos^2 t = 1 - \cos^2 t$, or $2\cos^2 t = 1$, or $\cos^2 t = \frac{1}{2}$. This happens when $\cos t = \pm\frac{1}{\sqrt{2}}$, which happens when $t = \pm\frac{\pi}{4}, \pm\frac{3\pi}{4}, \ldots$

The only one of these points that lies in the specified interval $\left[0, \frac{\pi}{2}\right]$ is $\frac{\pi}{4}$. Evaluating there and at the endpoints, we find that $f(0) = 0, f\left(\frac{\pi}{4}\right) = \sin\frac{\pi}{4}\cos\frac{\pi}{4} = \frac{1}{\sqrt{2}}\frac{1}{\sqrt{2}} = \frac{1}{2}, f\left(\frac{\pi}{2}\right) = \sin\frac{\pi}{2}\cos\frac{\pi}{2} = 1(0) = 0.$

The maximum value of $f(t)$ on $\left[0, \frac{\pi}{2}\right]$ is $\frac{1}{2}$, which occurs when $t = \frac{\pi}{4}$. The minimum value is 0, which occurs when $t = 0$ and when $t = \frac{\pi}{2}$.

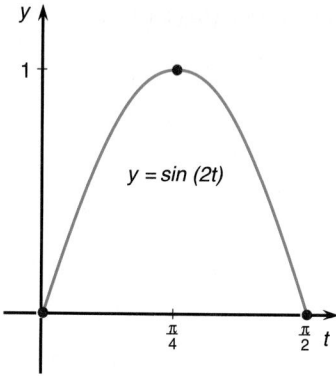

Figure 3.4.4

If we realize that $\sin t \cos t = \frac{1}{2} \sin(2t)$, the answer makes good sense. After all, on the interval $[0, \pi]$ the function $\sin t$ has one "hump" from 0 up to 1. The function $\sin(2t)$ has a corresponding "hump" on the interval $\left[0, \frac{\pi}{2}\right]$, with maximum in the middle at $t = \frac{\pi}{4}$. See Figure 3.4.4.

EXAMPLE 6   Degrees Versus Radians   In this example we explore what happens if we try to measure angles in degrees instead of radians. Recall that an angle of $180°$ is the same thing as π radians. This means that $1°$ must equal $\frac{\pi}{180}$ radians.

For the purposes of this example only, we write sindeg(x) for the function sin, with x measured in *degrees*. So, for example, sindeg$(90) = 1$, sindeg$(180) = 0$, sindeg$(270) = -1$, sindeg$(45) = \frac{1}{\sqrt{2}}$, etc.

As we have just noted, x degrees equals $\frac{\pi x}{180}$ radians. This allows us to write

$$\text{sindeg}(x) = \sin\left(\frac{\pi x}{180}\right),$$

which just expresses the conversion from degrees into radians.

What we want to do is to differentiate sindeg(x). We are able to do it by applying Theorem 3.4.3 to the above conversion formula. After all, the formula says that sindeg(x) equals $\sin(cx)$ with $c = \frac{\pi}{180}$, so Theorem 3.4.3 tells us that

$$\frac{d}{dx}\big(\text{sindeg}(x)\big) = \sin'(cx)(c) = c\cos(cx)$$

$$= \frac{\pi}{180} \cos\left(\frac{\pi x}{180}\right).$$

If we write cosdeg(x) for the cos function with the angle x measured in *degrees*, this can be written

$$\frac{d}{dx}\big(\text{sindeg}(x)\big) = \frac{\pi}{180} \text{cosdeg}(x).$$

What this says is that if we measure angles in degrees, then the derivative of sin is not cos, but cos multiplied by that awkward constant $\frac{\pi}{180}$. When we first use radians, it seems strange to have π turning up all the time, but now we see that by doing that we make the derivative of sin very simple. Moreover, if we refuse to use radians and insist on degrees, then the π turns up in the differentiation formula anyway.

In a sense the use of radians is a choice to have the unpleasant constant in the angles so that everything else will work out well. There is no way to avoid having something awkward somewhere, and this choice reduces the confusion in many calculations.

Exercises 3.4

Find the derivative of each of the following functions, stating where the derivative does and does not exist.

1. $f(x) = \cos^2 x$

2. $g(t) = t \cos t$

3. $h(u) = \frac{1}{2 - \cos u}$

4. $k(r) = \frac{\cos r}{1 + r^2}$

5. $F(w) = (w^2 - 2w - 1)\sin(2w)$

6. $G(x) = \tan(x/2)$

7. $H(t) = \frac{\sin(2t)}{\sin(3t)}$

8. $K(u) = \frac{\sin u}{u}$

9. $L(r) = r^3 \tan r - \sec r$

10. $T(x) = \sin^2(2x) + \cos^2(2x)$

11. $f(x) = \sec^2 x$

12. $f(x) = \csc^2 x$

13. $f(x) = \sin x \csc x$

14. $f(x) = \sin x \sec x$

15. $f(x) = \cos x \csc x$

16. $f(x) = \cot(3x)$

17. $f(x) = \cot x \csc x$

18. $f(x) = \frac{1}{\csc x}$

19. $f(x) = \frac{1}{\cot x}$

20. $f(x) = 1 + \sec^2 x$

Find all points at which the derivative of each of the following functions is zero.

21. $f(x) = \csc x$

22. $g(t) = \cos t \sin t$

23. $h(u) = \sin^2 u$

24. $k(s) = \cot s$

25. $f(v) = \sin(2v)$

26. $g(x) = \sin(3x)$

27. $h(r) = r - \tan r$

28. $k(t) = 2t - \tan(2t)$

29. $f(u) = 3u - \cot(7u)$

30. $g(s) = 7s + \cos s + 2\sin s$

Practice differentiation by finding the derivative of each of the following functions.

31. $f(x) = (1 + 2x^2)\sin x$

32. $f(x) = \frac{x + \sin x}{3x^2 + 4}$

33. $f(x) = \frac{1}{1 + \sin^2 x}$

34. $f(x) = \frac{1 - \cos^2 x}{1 - \cos x}$

35. $f(x) = \frac{\cos x}{x^2 + 2x + 5}$

36. $f(x) = \frac{\sin^2 x}{1 + \cos x}$

37. $f(x) = (x^3 - 8)(\sin x + 2\cos x)$

38. $f(x) = (\sin^2 x + 3)^{-2}$

39. $f(x) = x^2 + 2x\tan x + \tan^2 x$

40. $f(x) = (3x + \sec x)^3$

II

41. Find all points at which each of the following functions has a horizontal tangent line: (i) $g(\theta) = \tan\theta + 2\sec\theta$, (ii) $f(\theta) = \tan\theta - \sec\theta$.

42. Find all points at which the tangent line to $y = \cot x$ has slope equal to -2.

43. Make a rough sketch of the graph of $y = \sin x$ for $x \in [-\pi, \pi]$, and indicate the parts of the interval $[-\pi, \pi]$ where the derivative is positive, the parts where it is negative, and the places where it is zero.

44. Make a rough sketch of the graph of $y = \sec x$ for $x \in [-\pi, \pi]$, and indicate the parts of the interval $[-\pi, \pi]$ where the derivative is positive, the parts where it is negative, and the places where it is zero. Confirm your geometrical impressions by calculating the derivative.

*45. Let $g(x) = x\sin x$.

(i) Evaluate $g'(x)$ and $g'(0)$.

(ii) Show that there must be at least one value of x between $\frac{\pi}{2}$ and $\frac{3\pi}{2}$ at which the derivative g' is zero, that is, $g'(x) = 0$. (*Hint:* Apply the Intermediate Value Theorem (Theorem 2.5.5) to g'.)

*46. (i) Let $f(x) = \sin\left(x + \frac{\pi}{2}\right)$. Show that the derivative is $f'(x) = \cos\left(x + \frac{\pi}{2}\right)$.

(ii) Let c be any fixed real number and define $g(x) = \sin(x + c)$. Find the derivative $g'(x)$.

(iii) What is the derivative of $h(x) = \sin(3x + 5)$?

(iv) If c and d are constants, what is the derivative of $k(x) = \sin(cx + d)$?

*47. Find all points x at which the tangent line to $y = \sec x$ has slope equal to $\sqrt{2}$.

*48. Are there any values of x at which the tangents to $y = \cos x$ and $y = \sin x$ have the same slope?

III

49. Use a graphics calculator or computer to plot both $\sin x$ and its derivative on the same graph. Confirm that the derivative is zero where the function has a horizontal tangent, and that where the derivative is positive the function slopes up to the right.

50. Plot $y = \cos x$ and its derivative on the same graph, checking the behavior of the function where the derivative is zero, positive, or negative.

51. Plot $y = \tan x$ and its derivative on the same graph. Notice how the function rises (or falls) faster when the derivative is large (or is negative with large absolute value).

52. Plot each of $y = \sec x$ and $y = \csc x$ along with its derivative.

*53. Plot the difference quotient $\frac{f(x+h) - f(x)}{h}$ for an appropriate (small) value of h to get an approximation to the graph of $f'(x)$. Use this technique to verify Theorem 3.4.3 (p. 170) for (i) $g(x) = \sin x$, with $c = 2$, $-1, 3, \frac{1}{5}, 0$; (ii) $g(x) = x^2$, $c = 2, -2, 3, -\frac{1}{10}, 0$.

3.5 Derivatives of Logarithms, Exponentials, and Hyperbolic Functions

COMMON LOGARITHMS
NATURAL LOGARITHMS
EXPONENTIAL FUNCTION
HYPERBOLIC FUNCTIONS AND
THEIR DERIVATIVES

In this brief section we learn the derivatives of several important functions. We have not yet learned much about the functions themselves; that will have to wait until Chapter 7. However, the differentiation rules are easily stated, so we can begin to work with them now.

We begin with a brief review of what we do know about logarithms (Section 1.8). Much as trigonometric functions can be defined using either degrees or radians, there are two standard ways of defining logarithms. The familiar way is to say that the logarithm of x is the power to which you have to raise the number 10 to get x, that is,

$$x = 10^{\log x},$$

or

$$\log(10^s) = s. \tag{3.5.1}$$

So, for instance, $\log 10 = 1, \log(1{,}000) = 3, \log(1) = 0, \log\left(\frac{1}{10}\right) = -1$, etc. Strictly speaking, $\log x$ only makes sense if $x > 0$, since 10 raised to any power has to be positive.

Much of the usefulness of logarithms arises from the fact that

$$\log(xy) = \log x + \log y, \tag{3.5.2}$$

which is just another way of saying that $10^{a+b} = 10^a 10^b$, since

$$10^{\log x + \log y} = 10^{\log x} 10^{\log y} = xy.$$

Logarithms of this kind are called **common logarithms**, or *logarithms to the base 10*. However, we could have used some other number instead of 10 when we defined logarithms. It is not unusual to consider logarithms to the base 2, sometimes written $\log_2$. They are defined by the relation

$$x = 2^{\log_2(x)}.$$

So, for example, $\log_2(8) = 3, \log_2\left(\frac{1}{16}\right) = -4, \log_2(1) = 0$, etc.

Just as trigonometric functions were defined using radians, an apparently peculiar choice of angle measurement, there is an unexpected choice that is often made for the base of logarithms. It is a number called e, and it is approximately $2.718\,2818$. Logarithms to the base e are so important that they are given a special name. They are called **natural logarithms**, and for any positive number x we write $\ln x$ for the natural logarithm of x. It satisfies relations analogous to Equation 3.5.1:

$$x = e^{\ln x},$$

or

$$\ln(e^s) = s.$$

So $\ln(e) = 1, \ln(e^2) = 2, \ln(1) = 0, \ln(10) \approx 2.302\,585$, etc. Natural logarithms have the disadvantage that they have complicated values for "nice" numbers like 10.

The reason for using them is that they make derivatives much easier to work with, just as radian measure does for trigonometric functions.

The natural logarithm of a product is the sum of the natural logarithms, just as for common logarithms:

$$\ln(xy) = \ln x + \ln y. \tag{3.5.3}$$

Recall that for any numbers r, s, b with $b > 0$, we know that $(b^r)^s = b^{rs}$. If we write this out with e in place of b and $r = \ln(10)$, we find

$$(e^{\ln(10)})^s = e^{\ln(10)(s)}.$$

Since $e^{\ln(10)} = 10$, this says that

$$10^s = e^{\ln(10)(s)}.$$

Now suppose $s = \log x = \log_{10}(x)$. Then $10^s = 10^{\log(x)} = x$. The above equation becomes

$$x = e^{\ln(10)\log x}.$$

Applying ln to both sides "gets rid" of the e on the right side and gives

$$\ln x = \ln(10)\log x,$$

from which we find

$$\log x = \frac{1}{\ln 10}\ln x. \tag{3.5.4}$$

> It is easy to remember which way this relation goes by letting $x = 10$, so $\log x = 1 \ldots$

The graph of $y = \ln x$ is illustrated in Figure 3.5.1. Notice that it crosses the x-axis at $x = 1$, since $\ln(1) = 0$. Notice too that $\ln x > 0$ when $x > 1$ and $\ln x < 0$ when $0 < x < 1$. The function $\ln x$ is not defined when $x \leq 0$.

The derivative of $\ln x$ is easy to remember, though we will not be able to prove it until Chapter 7.

THEOREM 3.5.5

> The function $\ln x$ is differentiable for all $x > 0$, and
> $$\frac{d}{dx}(\ln x) = \frac{1}{x}.$$

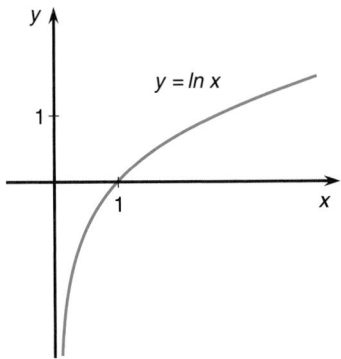

Figure 3.5.1

Refer to the graph of $y = \ln x$ and observe that the derivative is always positive, as the function is always increasing. It is also clear from the picture that the function increases more steeply for small values of x and less steeply for large x. This is reflected in the fact that the derivative $\frac{1}{x}$ is large when x is small, meaning that the slope of the tangent is large then, and smaller when x is larger, meaning that the slope of the tangent is smaller there.

EXAMPLE 1 Differentiate the following functions: (i) $f(x) = x^2 \ln x$; (ii) $g(t) = \ln^2 t$; (iii) $h(u) = \ln(4u^3)$; (iv) $k(s) = \ln\left(\frac{1}{s}\right)$; (v) $F(x) = x\ln x - x$.

Solution

(i) Applying the Product Rule and Theorem 3.5.5, we find $f'(x) = 2x\ln x + x^2\frac{d}{dx}(\ln x) = 2x\ln x + x^2\frac{1}{x} = 2x\ln x + x.$

(ii) Writing $\ln^2 t = \ln t \ln t$ and applying the Product Rule, we find $g'(t) = \frac{1}{t}\ln t + (\ln t)\frac{1}{t} = \frac{2}{t}\ln t.$

(iii) For this part we use the way logarithms work with products (Equation 3.5.3). Realizing that $\ln(4u^3) = \ln 4 + \ln u^3 = \ln 4 + \ln\big((u)(u)(u)\big) = \ln 4 + \ln u + \ln u + \ln u = \ln 4 + 3\ln u$, we can differentiate. Notice that $\ln 4$ is a constant ($\ln 4 \approx 1.38629$), so its derivative is 0.

We have found $h'(u) = 0 + 3\frac{1}{u} = \frac{3}{u}$.

(iv) The rule about logarithms of products can be used to figure out the logarithm of a quotient. Notice that $x = \frac{x}{y}y$, so $\ln x = \ln\left(\frac{x}{y}y\right) = \ln\frac{x}{y} + \ln y$. Rearranging, we find

$$\ln\frac{x}{y} = \ln x - \ln y,$$

so the logarithm of a quotient is the difference of the logarithms of the numerator and the denominator.

Applying this to our problem, we find that $\ln\frac{1}{s} = \ln 1 - \ln s = 0 - \ln s = -\ln s$, so $k'(s) = -\frac{1}{s}$.

(v) $F'(x) = 1(\ln x) + x\frac{1}{x} - 1 = \ln x + 1 - 1 = \ln x.$ ◢

EXAMPLE 2   **Derivative of the Common Logarithm**   Let $f(x) = \log x = \log_{10}(x)$, and find its derivative $f'(x)$.

Solution   We know from Equation 3.5.4 that $f(x) = \log x = \frac{1}{\ln 10}\ln x$, so

$$f'(x) = \frac{1}{\ln 10}\frac{d}{dx}(\ln x) = \frac{1}{\ln 10}\frac{1}{x} \approx 0.4342945\frac{1}{x}.$$ ◢

> Example 2 shows why it is awkward to use common (base 10) logarithms when differentiating (just as it is awkward to measure angles in degrees for trigonometric functions). It is because the derivative of $\ln x$ is so nice that this kind of logarithm is called "natural."

EXAMPLE 3   Find all points where the derivative of $f(x) = x\ln x$ is zero.

Solution   The function $f(x)$ is defined and differentiable whenever $x > 0$. Its derivative is $f'(x) = 1(\ln x) + x\frac{1}{x} = \ln x + 1$. For $f'(x) = 0$ we must have $\ln x = -1$, which means $x = e^{\ln x} = e^{-1} = \frac{1}{e}$. So $f'(x) = 0$ at $x = \frac{1}{e}$ and nowhere else. ◢

EXAMPLE 4   Find the maximum and minimum values of $g(t) = t^2\ln t - \frac{3}{2}t^2$ on the interval $[1, 10]$.

Solution   Notice that $g(t)$ is defined for all $t > 0$, and is differentiable for every such t. Since a differentiable function is continuous, by Theorem 3.2.1, we can apply the technique of Summary 3.3.1 to find its maximum and minimum values.

The derivative is $g'(t) = 2t\ln t + t^2\frac{1}{t} - 3t = 2t\ln t + t - 3t = 2t\ln t - 2t = 2t(\ln t - 1)$. This will be zero only when $\ln t - 1 = 0$, that is, $\ln t = 1$. (Notice that we do not consider $t = 0$, first because it is not in the interval $[1, 10]$ and second because $g(t)$ is not defined at $t = 0$.) If $\ln t = 1$, then $t = e^{\ln t} = e^1 = e$. We evaluate the function $g(t)$ at this point $t = e$ and at the endpoints $t = 1$ and $t = 10$. We find that $g(e) = e^2\ln(e) - \frac{3}{2}e^2 = e^2\left(\ln(e) - \frac{3}{2}\right) = e^2\left(1 - \frac{3}{2}\right) = -\frac{1}{2}e^2 \approx -3.694\,528$. Also $g(1) = 1(\ln 1) - \frac{3}{2}(1)^2 = 1(0) - \frac{3}{2} = -\frac{3}{2}$, and $g(10) = 10^2\ln(10) - \frac{3}{2}(10)^2 \approx 100(2.302\,585 - 1.5) = 80.2585.$

The smallest of these values is $-3.694\,528$, which occurs when $t = e$, and the largest is 80.2585, which occurs when $t = 10$. The minimum value of $g(t)$ on $[1, 10]$ is $-\frac{1}{2}e^2 \approx -3.694\,528$, which occurs when $t = e$. The maximum value is $100\big(\ln(10) - \frac{3}{2}\big)$, which occurs when $t = 10$.   ◢

Log–Log Graph Paper

It is quite common for two quantities x and y to be related by a "power function" like $y = cx^k$. This situation could arise for the length and the circumference of a growing bone, two quantities measured in a chemical reaction, the intensity of a stimulus and the resulting response of a nerve cell, the length and surface area of an amoeba, the dose of some medication and the measured effect, and many other examples.

If a number of measurements can be taken, they can be plotted on a graph and used to estimate the constants c and k. However, it is not easy to guess the exponent k from the shape of the curve.

If we take logarithms of both sides of the equation $y = cx^k$, it becomes

$$\log y = \log c + k \log x.$$

Writing $Y = \log y$, $C = \log c$, and $X = \log x$, we see that the points (X, Y) lie on the straight line $Y = C + kX$. If we plot this line, its slope will be the constant k, and its Y-intercept will be C.

This can be done by finding and plotting the logarithms of all the coordinates, but it is most easily accomplished by using log–log graph paper.

The scales on the paper are "logarithmic," so when we plot (x, y) the horizontal and vertical distances from the origin are $\log x$ and $\log y$. In effect we have plotted (X, Y). In Figure 3.5.2(i) we have plotted a number of points on ordinary graph paper and estimated the shape of a curve joining them. It is difficult to guess the exponent.

However, in Figure 3.5.2(ii) we have plotted the same points on log–log paper, and the slope is easily measured to be about $\frac{3}{4}$. The value of y at the vertical intercept is about 2.1, so the original curve is approximately $y = 2.1x^{3/4}$. (Notice that on log–log paper the vertical intercept occurs where $X = 0$, that is, where $x = 1$.)

Figure 3.5.2(i)

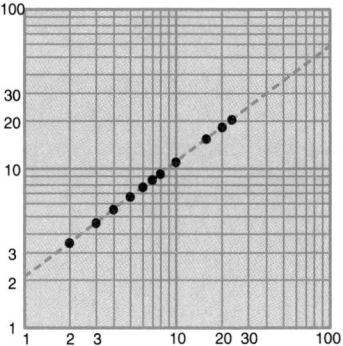

Figure 3.5.2(ii)

The Exponential Function

Next we consider the exponential function e^x, which is sometimes written $\exp(x)$. It is related to the natural logarithm function by the relation

$$\ln(e^x) = x,$$

or

$$\exp(\ln x) = x. \tag{3.5.6}$$

(The first of these equations is true for all x, and the second makes sense only when $x > 0$.) Notice that if x is an integer, then e^x is just the ordinary power of the number e:

$$\exp(n) = e^n \approx (2.7182818)^n.$$

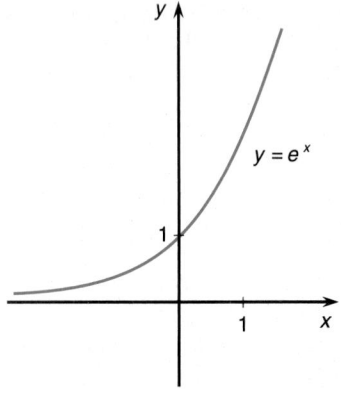

Figure 3.5.3

The graph of $y = e^x$ is illustrated in Figure 3.5.3. Notice that it is defined for all values of x. We see from the graph that $e^0 = 1$, as we should expect. Also notice that $e^x > 1$ when $x > 0$ and $e^x < 1$ when $x < 0$, but that $e^x > 0$ for all values of x. The function e^x increases very fast for large values of x.

Because of the rule about the logarithm of a product (Equation 3.5.3), we observe that

$$\ln(e^{a+b}) = a + b = \ln(e^a) + \ln(e^b) = \ln(e^a e^b).$$

Applying the exponential function to the beginning and end of this equality, we find that

$$\exp\big(\ln(e^{a+b})\big) = \exp\big(\ln(e^a e^b)\big),$$

or, using Equation 3.5.6,

$$e^{a+b} = e^a e^b. \tag{3.5.7}$$

THEOREM 3.5.8

> The function $f(x) = \exp(x) = e^x$ is defined and differentiable for all x. Its derivative is given by
>
> $$\frac{d}{dx}(e^x) = e^x.$$
>
> In words, the exponential function is its own derivative.

It is instructive to reexamine the graph of $y = e^x$ in light of this result. When x is negative, the graph is nearly flat, so its derivative is nearly zero, and the function itself is very small, that is, nearly zero. At $x = 0$ the graph is rising and its derivative is $e^0 = 1$. When $x > 0$, the derivative is bigger than 1 and the function is increasing very fast.

EXAMPLE 5   Find the derivative of each of the following functions: (i) $f(x) = e^{1+x}$; (ii) $g(t) = t^2 e^t$; (iii) $h(u) = e^{-u}$.

Solution

(i) By Equation 3.5.7, $e^{1+x} = e^1 e^x = ee^x$, so $f'(x) = e\frac{d}{dx}(e^x) = ee^x = e^{1+x}$.

(ii) $g'(t) = 2te^t + t^2 e^t = (2t + t^2)e^t$.

(iii) There are two ways to approach this. One is to observe that $e^{-u}e^u = e^{-u+u} = e^0 = 1$, using Equation 3.5.7, so

$$e^{-u} = \frac{1}{e^u}. \tag{3.5.9}$$

From this the derivative can be found by using the Quotient Rule:

$$h'(u) = \frac{0(e^u) - 1(e^u)}{(e^u)^2} = -\frac{e^u}{(e^u)^2} = -\frac{1}{e^u} = -e^{-u}.$$

In other words, $h'(u) = -h(u)$.

Another way to do this question is to appeal to Theorem 3.4.3, letting $c = -1$. So $h(u) = F(-u) = F(cu)$, with $F(x) = e^x$ and $c = -1$. Theorem 3.4.3 then tells us that $h'(u) = F'(cu)c = e^{cu}(-1) = -e^{-u} = -h(u)$.

If we think about ordinary powers with integer exponents, we recall that $b^{mn} = (b^m)^n = (b^n)^m$. We can use this formula to define powers in which the exponent is *not* necessarily an integer. If $b > 0$, then $b = e^{\ln b}$, and we can define $b^r = e^{r \ln b}$.

DEFINITION 3.5.10

> If $b > 0$ and $r \in \mathbb{R}$, define the power b^r by
>
> $$b^r = e^{r \ln b} = \exp(r \ln b).$$

EXAMPLE 6   Let $a > 0$ and define $f(x) = a^x$. Find the derivative $f'(x)$.

Solution   Notice that $f(x) = a^x = \exp(x \ln a)$. Now $\ln a$ is a constant, so we can write $f(x) = e^{cx}$, with $c = \ln a$, and Theorem 3.4.3 tells us that $f'(x) = e^{cx} c = \exp(x \ln a) \ln a = a^x \ln a$.   ◢

Hyperbolic Functions

There are some important functions that are built up from the exponential function. In various ways they are similar or at least analogous to the trigonometric functions, and their names reflect this. They are called hyperbolic sine, hyperbolic cosine, hyperbolic tangent, etc. The symbols used for these functions are the symbols used for trigonometric functions with the letter "h" (for hyperbolic) added at the end.

DEFINITION 3.5.11

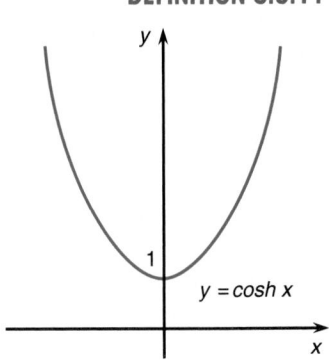

Figure 3.5.4

> The **hyperbolic functions** are defined as follows:
>
> $$\sinh x = \frac{e^x - e^{-x}}{2}, \qquad \cosh x = \frac{e^x + e^{-x}}{2},$$
>
> $$\tanh x = \frac{\sinh x}{\cosh x} = \frac{e^x - e^{-x}}{e^x + e^{-x}}, \qquad \coth x = \frac{\cosh x}{\sinh x} = \frac{e^x + e^{-x}}{e^x - e^{-x}},$$
>
> $$\operatorname{sech} x = \frac{1}{\cosh x} = \frac{2}{e^x + e^{-x}}, \qquad \operatorname{csch} x = \frac{1}{\sinh x} = \frac{2}{e^x - e^{-x}}.$$

They can be read as hyperbolic sine, hyperbolic cos, etc., but often they are read more or less phonetically, with $\cosh x$ pronounced "cawsh x," $\sinh x$ pronounced "sinch x," and $\tanh x$ pronounced "tansh x." There seems to be no consensus about how to pronounce $\operatorname{sech} x$, $\coth x$, or $\operatorname{csch} x$, but fortunately they are not used as frequently.

The graphs of $\cosh x$ and $\sinh x$ are illustrated in Figures 3.5.4 and 3.5.5. Notice that, unlike the trigonometric functions they are not bounded (i.e., they get arbitrarily large), and they are not periodic (i.e., they do not repeat after 2π).

Despite these differences, we have said that in many ways the hyperbolic functions behave like trigonometric functions. For instance,

$$\cosh^2 x - \sinh^2 x = \left(\frac{e^x + e^{-x}}{2}\right)^2 - \left(\frac{e^x - e^{-x}}{2}\right)^2$$

$$= \frac{e^{2x} + 2e^x e^{-x} + e^{-2x}}{4} - \frac{e^{2x} - 2e^x e^{-x} + e^{-2x}}{4}$$

$$= \frac{e^{2x} + 2(1) + e^{-2x} - e^{2x} + 2(1) - e^{-2x}}{4} = \frac{4}{4}$$

$$= 1.$$

Figure 3.5.5

In other words, we have established the hyperbolic version of the fundamental trigonometric identity ($\cos^2 \theta + \sin^2 \theta = 1$), namely,

$$\cosh^2 x - \sinh^2 x = 1. \tag{3.5.12}$$

Notice that there is a minus sign in the hyperbolic identity but that it is remarkably similar to the corresponding trigonometric identity.

EXAMPLE 7   Find the derivatives of the functions $\sinh x$, $\cosh x$, and $\tanh x$.

Solution $\frac{d}{dx}(\sinh x) = \frac{d}{dx}\left(\frac{1}{2}(e^x - e^{-x})\right) = \frac{1}{2}\left(e^x - e^{-x}(-1)\right) = \frac{e^x + e^{-x}}{2} = \cosh x$.
Similarly, $\frac{d}{dx}(\cosh x) = \frac{d}{dx}\left(\frac{1}{2}(e^x + e^{-x})\right) = \frac{1}{2}\left(e^x + e^{-x}(-1)\right) = \frac{e^x - e^{-x}}{2} = \sinh x$. Using the above results and the Quotient Rule, we find that $\frac{d}{dx}(\tanh x) = \frac{d}{dx}\left(\frac{\sinh x}{\cosh x}\right) = \frac{\cosh x \cosh x - \sinh x \sinh x}{\cosh^2 x}$. Applying Equation 3.5.12 to the numerator, we see that this equals $\frac{1}{\cosh^2 x} = \text{sech}^2 x$.

We record the derivatives of all the hyperbolic functions as a theorem. We have actually just worked out the first three of them; the other three, which are similar, will be done as an exercise at the end of the section.

THEOREM 3.5.13

> **Derivatives of Hyperbolic Functions**
>
> The hyperbolic functions are differentiable at every point at which they are defined, and their derivatives are as follows:
>
> $\frac{d}{dx}(\sinh x) = \cosh x,$ $\frac{d}{dx}(\cosh x) = \sinh x,$
>
> $\frac{d}{dx}(\tanh x) = \text{sech}^2 x,$ $\frac{d}{dx}(\coth x) = -\text{csch}^2 x,$
>
> $\frac{d}{dx}(\text{sech } x) = -\text{sech } x \tanh x,$ $\frac{d}{dx}(\text{csch } x) = -\text{csch } x \coth x.$

These formulas are strikingly similar to the formulas for the derivatives of the trigonometric functions (Theorems 3.4.1, 3.4.2). In fact, they are direct translations (adding an "h" to each trigonometric function), *except* for the fact that some of the signs change. For the two basic ones, $\cosh x$ and $\sinh x$, there is no minus sign in their derivatives, which makes them easier to remember than the trigonometric ones.

EXAMPLE 8   Find all points where there is a horizontal tangent line to the graph of (i) the hyperbolic sine, (ii) the hyperbolic cosine.

Solution

(i)   The derivative of $\sinh x$ is $\cosh x$. For the tangent to be horizontal we need the derivative to be zero, that is, $\cosh x = 0$. This amounts to $\frac{1}{2}(e^x + e^{-x}) = 0$, or $e^x + e^{-x} = 0$, which is impossible, since e^x and e^{-x} are both positive. In other words, the tangent to $y = \sinh x$ is never horizontal, which is consistent with the illustration of the graph.

(ii)   The derivative of $\cosh x$ is $\sinh x$, which is zero when $e^x - e^{-x} = 0$, or $e^x = e^{-x}$. Multiplying by e^x, we find that this occurs when $e^x e^x = e^x e^{-x} = 1$, which means $e^x = \pm 1$. But e^x is always positive, so this means $e^x = 1$ and $x = 0$. The only place where the tangent is horizontal is $x = 0$.

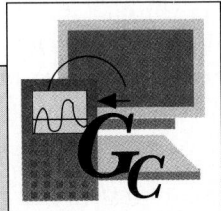

It is interesting to use a graphics calculator to plot each hyperbolic function together with its derivative. For instance, we plot $y = \tanh x$ and its derivative $y = \text{sech}^2 x = \frac{1}{\cosh^2 x}$.

TI-85	SHARP EL9300
GRAPH F1 $(y(x) =)$ MATH F4 F3 (tanh) EXIT F1 (x) ▼ 1 ÷ (MATH F4 F2 (cosh) EXIT F1 (x)) x^2 M2 (RANGE) $(-)$ 3 ▼ 3 ▼ ▼ $(-)$ 1 ▼ 1 F5 (GRAPH)	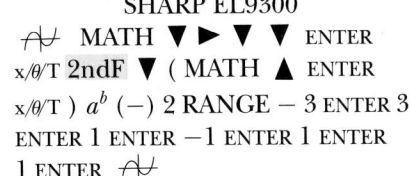 MATH ▼ ► ▼ ▼ ENTER x/θ/T 2ndF ▼ (MATH ▲ ENTER x/θ/T) a^b $(-)$ 2 RANGE $-$ 3 ENTER 3 ENTER 1 ENTER -1 ENTER 1 ENTER 1 ENTER
CASIO f_x-7700 GB/f_x-6300 G	HP 48SX
Cls EXE Range $-$ 3 EXE 3 EXE 1 EXE $-$ 1 EXE 1 EXE 1 EXE EXE EXE EXE Graph hyp tan X EXE Graph (hyp cos X) x^y $-$ 2 EXE (Use MATH HYP on fx7700 GB)	PLOT PLOTR ERASE ATTN ' αY ▬ MTH HYP TANH αX ↱ PLOT ↰ DRAW ERASE 3 $+/-$ ENTER 3 XRNG 1 $+/-$ ENTER 1 YRNG DRAW ATTN ' αY ▬ MTH HYP COSH αX ► y^x 2 $+/-$ ↱ PLOT ↰ DRAW DRAW

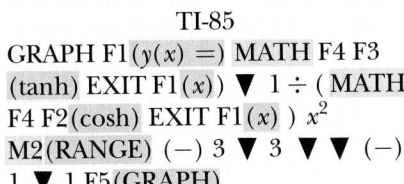

Identify which graph is which in the picture. Notice how the function $y = \tanh x$ increases more rapidly when its derivative is larger. Try plotting the other hyperbolic functions with their derivatives.

EXAMPLE 9   Find the minimum value of (i) $f(t) = t + \frac{1}{t}$ for $t > 0$; (ii) $\cosh x$.

Solution

(i) We cannot apply the technique for finding maximum and minimum values of a function on a closed interval, since the function is defined for all positive values of t. However, we could still ask where $f'(t) = 0$. Since $f'(t) = 1 + \frac{-1}{t^2}$, we see that $f'(t) = 0$ means $\frac{1}{t^2} = 1$, or $t^2 = 1$, or $t = \pm 1$. Since we are assuming $t > 0$, the only place where the derivative is zero is $t = 1$. The value of the function at this point is $f(1) = 2$.

Now an interesting thing about the function f is that $f(t) = t + \frac{1}{t} > t$. So for example, if $t > 2$, then $f(t) > 2$. Also $f(t) = t + \frac{1}{t} > \frac{1}{t}$. So if $t < \frac{1}{2}$, then taking reciprocals, $\frac{1}{t} > 2$, and $f(t) > \frac{1}{t} > 2$. We have found that $f(1) = 2$, and that $f(t) > 2$ when $t > 2$ and when $t < \frac{1}{2}$. This means that in looking for the minimum value there is no need to consider values of t that are greater than 2 or less than $\frac{1}{2}$.

In other words, we need only look for the minimum value of $f(t)$ for t in the closed interval $\left[\frac{1}{2}, 2\right]$, and this we know how to do. In fact we have already found that $t = 1$ is the only point where the derivative is zero, so we need only evaluate $f(t)$ at this point and the endpoints: $f(1) = 2, f\left(\frac{1}{2}\right) = \frac{1}{2} + 2 = \frac{5}{2}, f(2) = 2 + \frac{1}{2} = \frac{5}{2}$. The minimum value is $f(1) = 2$.

(ii) Writing $\cosh x = \frac{1}{2}\left(e^x + e^{-x}\right) = \frac{1}{2}\left(e^x + \frac{1}{e^x}\right) = \frac{1}{2}f(e^x)$, with $f(t) = t + \frac{1}{t}$, as in part (i), we can use the result of part (i) to help solve part (ii).

Indeed, we know that e^x is always positive, so part (i) says that the smallest possible value of $f(e^x)$ will be 2 and that it will occur when $e^x = 1$, that is, when $x = 0$.

The minimum value of $\cosh x$ is $\frac{1}{2}(2) = 1$, which occurs when $x = 0$.

Notice that

$$\tanh x = \frac{e^x - e^{-x}}{e^x + e^{-x}} < \frac{e^x + e^{-x}}{e^x + e^{-x}} = 1,$$

so $\tanh x$ is always less than 1. On the other hand,

$$\tanh x = \frac{e^x - e^{-x}}{e^x + e^{-x}} = \frac{1 - e^{-2x}}{1 + e^{-2x}}.$$

If x is very large, then from the graph of e^x in Figure 3.5.6 it seems reasonable to think that e^x is also very large, so $e^{-2x} = \frac{1}{e^{2x}}$ will be very small and the above quotient will be close to 1. In fact this is true, and

$$\lim_{x \to \infty} \tanh x = 1.$$

A similar argument shows that $\lim_{x \to -\infty} \tanh x = -1$. The graph of $y = \tanh x$ is sketched in Figure 3.5.7. Notice that the function is always increasing but that it is bounded, that is, it never gets bigger than 1 (nor smaller than -1).

Figure 3.5.6

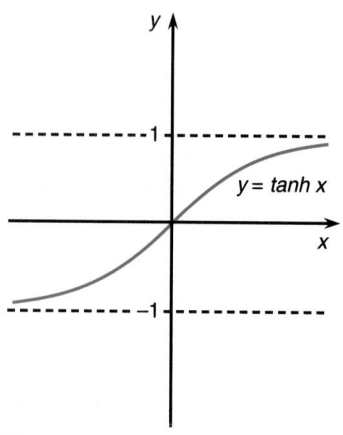

Figure 3.5.7

EXAMPLE 10   Bacterial Growth   Consider a culture of bacteria growing in a glass jar. Suppose the number of bacteria present after t hours have elapsed is $F(t) = 10^5 + 10^6 \tanh\left(\frac{t}{10}\right)$. From the remarks just made about tanh always being less than 1, we see that $F(t)$ will always be less than $10^5 + 10^6 = 1.1 \times 10^6$. On the other hand, as time passes and $t \to \infty$, the number of bacteria $F(t)$ will approach 1.1×10^6.

The biological significance is that the jar cannot support more than 1.1×10^6 bacteria, perhaps because of lack of space, perhaps because of limited food or oxygen, or perhaps because of a number of such considerations. At any rate, the population cannot exceed 1.1×10^6, which is called the carrying capacity of the jar environment.

On the other hand, the bacteria will tend to grow and fill up the jar as much as possible. As they approach the maximum, their growth will slow down (because of lack of space, food, etc.). So the graph of $\tanh t$ has roughly the correct form to describe this kind of growth.

Of course, if the bacteria are left alone, they will eventually die and the population will decrease, but until that happens, a function like $F(t)$ will describe their growth fairly well.

The question we wish to answer is how long does it take for the culture to reach a population of 7×10^5, that is, for what t does $F(t)$ equal $7 \times 10^5 = 700,000$?

Solution What we have to do is solve

$$10^5 + 10^6 \tanh\left(\frac{t}{10}\right) = 7 \times 10^5,$$

or

$$10^6 \tanh\left(\frac{t}{10}\right) = 6 \times 10^5,$$

or

$$\tanh\left(\frac{t}{10}\right) = 0.6.$$

To do this, let us write $u = \exp\left(\frac{t}{10}\right)$ so that $\tanh\left(\frac{t}{10}\right) = \frac{u-u^{-1}}{u+u^{-1}}$. Then we need to solve $\frac{u-u^{-1}}{u+u^{-1}} = 0.6$, or $10(u - u^{-1}) = 6(u + u^{-1})$, or $(10 - 6)u = (10 + 6)u^{-1}$, or $4u = 16u^{-1}$, or $4u^2 = 16$, $u^2 = \frac{16}{4} = 4$.

From this we conclude that $u = \pm 2$. But since u was defined to be an exponential, it is always positive, so $u = 2$, or to return to the original variable,

$$e^{t/10} = 2.$$

Taking natural logarithms of both sides, we find that $\frac{t}{10} = \ln 2 \approx 0.693$, so $t \approx 6.93$.

The answer to our question is that it will take about 6.93 hours for the culture to number 7×10^5.

Exercises 3.5

I

Find the derivative of each of the following functions, specifying any points where it is not defined.

1. $f(t) = t^3 \ln t$

2. $g(u) = \ln^3 u$

3. $h(x) = x \ln x^2$

4. $k(s) = s \ln^2 s$

5. $H(r) = e^r \ln r$

6. $L(z) = (z^3 - 1)e^z$

7. $F(x) = \frac{x}{e^{2x}}$

8. $G(t) = te^{t-6}$

9. $K(u) = e^{\ln u}$

10. $T(w) = \ln(e^w)$

11. $f(x) = \ln(3x)$

12. $g(x) = e^{-7x}$

13. $h(t) = \ln(te^t)$

14. $k(u) = e^{2u + \ln u}$

15. $F(t) = \frac{1}{\ln t}$

16. $G(x) = \frac{1}{e^{\pi x - 1}}$

17. $H(x) = \ln(x^2)$

18. $K(x) = \ln x^n$

19. $R(x) = e^{2x + 4 + 2 \ln x}$

20. $T(x) = e^{\sin^2 x + \cos^2 x}$

21. $f(x) = \cosh x \sinh x$

22. $g(x) = \cosh^2 x$

23. $h(x) = x \tanh 2x$

24. $k(x) = \frac{1}{\coth x}$

25. $U(x) = e^x \sinh x$

26. $V(x) = \sinh(\ln x)$

27. $L(x) = \cosh^2 x - \sinh^2 x$

28. $F(x) = \text{sech}^2 x$

29. $M(x) = \sinh 3x + \cosh 4x$

30. $F(x) = \cosh x \tanh x$

Practice your skill at differentiation by finding the derivatives of the following functions.

31. $f(x) = \frac{\ln x}{2x^3 - 4}$

32. $f(x) = \cos x + 3e^{-x} \tan x$

33. $f(x) = \frac{x^2 + 6}{e^x + \sin x + 2}$

34. $f(x) = \frac{1}{(x-3)^2 + 1} \tan x$

35. $f(x) = (2 + x)e^x \cos x$

36. $f(x) = \frac{e^{-x}}{1 + e^{-2x}}$

37. $f(x) = \frac{x^2 + 3}{\tan^2 x + 2 + \sin x}$

38. $f(x) = \sin^2 x + \sin x + 1$

39. $f(x) = \sin^3 x$

40. $f(x) = \frac{\sin^3 x - 1}{\sin x - 1}$

II

41. Verify the differentiation formulas (Theorem 3.5.13) for sech x, csch x, and coth x.

42. (i) Find an expression for $\log_2 x$ in terms of $\ln x$, in analogy with Equation 3.5.4.

(ii) Find an expression for $\log_2 x$ in terms of $\log_{10} x$.

(iii) If a, b are both greater than 1, say what we should mean by $\log_a x, \log_b x$, and find an expression for $\log_a x$ in terms of $\log_b x$.

43. Find the derivative of the function $f(x) = \log_2 x$.

44. Find all numbers a for which the tangent line to the graph of $y = \ln x$ at $x = a$ passes through the origin.

45. Find all numbers a for which the tangent line to the graph of $y = \exp x$ at $x = a$ passes through the origin.

46. (i) Find the maximum and minimum values of $\cosh x$ on $[-1, 1]$.

(ii) Find the maximum and minimum values of $\operatorname{sech} x$ on $[-1, 1]$.

(iii) Find the maximum and minimum values of $\tanh x$ on $[-2, 3]$.

*47. Let $c > 0$. Find the maximum and minimum values on the interval $[-c, c]$ of each of the following functions: (i) $\cosh x$, (ii) $\operatorname{sech} x$, (iii) $\sinh x$.

*48. Find the addition formula for $\sinh x$, that is, a formula for $\sinh(x + y)$.

*49. (i) Find the addition formula for $\cosh x$, that is, a formula for $\cosh(x + y)$.

(ii) Use the answer to part (i) to give a formula for $\cosh(2x)$.

50. Recall the trigonometric identity $1 + \tan^2 x = \sec^2 x$. Find an analogous identity for hyperbolic functions.

*51. Find a function $f(x)$ whose derivative is $f'(x) = \ln^2 x$. (*Hint:* Look at Example 1(v), p. 176.)

*52. (i) Let $f(x) = e^x$ and pretend that we do not know its derivative except that we know $f'(0) = 1$. Use this and other properties of e^x to show that $f'(x) = f(x)$ for all x.

(ii) Let $g(x) = \ln x$ and assume that we only know that $g'(1) = 1$. Use this and other properties of $\ln x$ to show that $\frac{d}{dx}(\ln x) = \frac{1}{x}$ for all $x > 0$.

*53. Suppose a culture of bacteria is growing in a jar in such a way that the number of bacteria after t minutes is $F(t) = 10^5 \times \left(3 + \tanh\left(\frac{t}{100}\right)\right)$.

(i) What is the carrying capacity of the jar?

(ii) How long will it take until the culture reaches 90% of the carrying capacity?

*54. Suppose bacteria are known to grow so that the number of milligrams of bacteria after t hours is given by $M(t) = 2 + 17 \tanh(kt)$, where k is some constant. Suppose we measure and find that when $t = 2$, the value of M is $M(2) = 10$ (i.e., there are 10 mg after 2 hours). Find the value of the constant k.

55. Use a graphics calculator or computer to draw the graphs of $y = \ln x$ and $y = \ln(2x)$ on the same picture, and confirm that $\ln(2x) = \ln x + \ln 2$.

56. Draw the graphs of $y = e^x$ and $y = 2e^x$ on the same picture, and confirm that $2e^x$ equals $e^{x+\ln 2}$, whose graph is the graph of e^x shifted to the left by a distance of $\ln 2$.

57. Draw the graph of $y = \ln x$. Using successive magnifications, find the value of e to four decimal places by noting where the graph crosses the horizontal line $y = 1$.

58. Use the graph of $y = e^x$ to estimate $\ln 10$ to four decimal places.

59. Sketch the graphs of $y = e^x$, $y = 2^x$, $y = 3^x$, and $y = \left(\frac{1}{2}\right)^x$ on the same graph. How many times do any two of them cross?

60. Compare the graphs of $y = \sinh x$ and the difference quotient $Q(x) = \frac{\cosh(x+h) - \cosh(x)}{h}$, with (i) $h = 1$, (ii) $h = 0.5$, (iii) $h = 0.1$.

61. Use the graph of $y = \sinh x$ to find the value of x for which $\sinh x = 1000$. Can you solve for this value without using the graph?

62. We found that the bacterial culture in Example 10 will number 7×10^5 after 6.93 hours. Graph the function $F(t)$ and use the picture to find (i) how many there will be after another 6.93 hours, (ii) when there will be 3.5×10^5.

POINT TO PONDER

The trigonometric functions $\sin t$ and $\cos t$ were first introduced for dealing with various problems involving angles and circles. They satisfy the fundamental identity $\cos^2 t + \sin^2 t = 1$, which can be interpreted to mean that the point $(\cos t, \sin t)$ is on the circle of radius 1 whose center is the origin (after all, the equation of this circle is $x^2 + y^2 = 1$, which is satisfied by $x = \cos t$, $y = \sin t$).

The fundamental identity for the hyperbolic functions is $\cosh^2 t - \sinh^2 t = 1$, which means that the point $(\cosh t, \sinh t)$ satisfies the equation $x^2 - y^2 = 1$. Of course this is the equation of a hyperbola, and we see that the points $(\cosh t, \sinh t)$ all lie on the hyperbola. In fact, since $\cosh t = \frac{1}{2}(e^t + e^{-t})$ is always positive, these points are all on the right branch of the hyperbola. As t takes on every possible real value, the point $(\cosh t, \sinh t)$ sweeps out the whole right branch of the hyperbola. See Figure 3.5.8.

Of course this is why these functions are called "hyperbolic"; in some old-fashioned books the trigonometric functions are referred to as "circular functions."

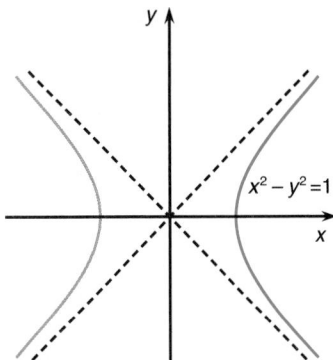

Figure 3.5.8

3.6 The Chain Rule

COMPOSITE FUNCTIONS
DERIVATIVES OF POWERS

In this section we learn to differentiate composite functions. This will allow us to work with a great many new functions built up from the ones we already know about. Before proceeding, you should be familiar and comfortable with the derivatives of polynomials, rational functions, trigonometric functions, logarithms, and exponentials.

In Theorem 2.3.2 we discussed the limit of a composite function $h(t) = g(f(t))$. The *Chain Rule* is a technique for evaluating the derivative of such a composite function. (The name suggests that the functions fit together to form a chain.)

To get some idea of what to expect, let us think of the derivative as a rate of change. If $f'(a) = 2$, for example, it means that for t near a, $f(t)$ changes approximately twice as fast as t does. To put it another way, a small change in t will result in a change of approximately twice as much in $f(t)$. So if t changes from $t = a$ to $t = a + h$, say, then we expect $f(t)$ to change from $f(a)$ by *about* $2h$, to something close to $f(a) + 2h$.

To consider the composite function $h(t) = g(f(t))$, assume $g(u)$ is differentiable at $u = f(a)$. For the sake of our example, suppose $g'(f(a)) = 3$. As above, this means that the value of $g(u)$ changes approximately 3 times as fast as the value of u, provided that u is near $u = f(a)$.

Now suppose we allow t to change slightly near $t = a$ and ask what happens to $h(t) = g(f(t))$. As t changes, we know that $f(t)$ changes by approximately twice as much. Then $g(f(t))$ should change by approximately 3 times as much as that, or $(3)(2) = 6$ times as much as the change in t.

This suggests that the rate of change of $h(t) = g(f(t))$ should be 6, or, to put it another way, that $h'(a) = 6$. Of course, the values we chose for $f'(a)$ and $g'(f(a))$ were not important; what is important is that the derivative of $h(t) = g(f(t))$ is the product of the derivatives $f'(a)$ and $g'(f(a))$. This is what is known as the Chain Rule.

Notice that when we form the composite function $h(t) = g(f(t))$ and consider its behavior near $t = a$, we must of course know what $f(t)$ does for t near $t = a$. But as for $g(u)$, what matters is its behavior near $u = f(a)$. In fact, a moment's reflection will convince you that what $g(u)$ does near $u = a$ is irrelevant. It is possible that $g(u)$ might not even be defined at $u = a$. This is why the derivative of g that enters into the Chain Rule for $h(t)$ is $g'(f(a))$.

EXAMPLE 1 Let $h(t) = (t^2 - 5)^3$. If we want to find its derivative $h'(t)$, we can express $h(t)$ as the composition of two functions: $h(t) = g(f(t))$, with $g(u) = u^3$ and $f(t) = t^2 - 5$.

Of course, we know that $f'(t) = 2t$ and $g'(u) = 3u^2$, so the Chain Rule says that

$$h'(t) = f'(t)g'(f(t)) = (2t)3(f(t))^2 = 6t(t^2 - 5)^2 = 6t(t^4 - 10t^2 + 25)$$
$$= 6t^5 - 60t^3 + 150t.$$

To check this, since after all we do not really know the Chain Rule yet, we could instead just expand $h(t)$ and differentiate. Doing this, we find

$$h(t) = (t^2 - 5)^3 = t^6 - 15t^4 + 75t^2 - 125.$$

Differentiating, we get

$$h'(t) = 6t^5 - 60t^3 + 150t,$$

and we do get the same result both ways.   ◢

Depending on how quickly we could expand $(t^2 - 5)^3$, the two methods are about equally easy in this example, but there are many examples in which the Chain Rule is much quicker, and there are many more in which there is no way to differentiate without the Chain Rule.

EXAMPLE 2 Differentiate $h(x) = \cos(x^2 - 3x + 7)$.

Solution If we write $h(x) = g(f(x))$, with $g(u) = \cos u$ and $f(x) = x^2 - 3x + 7$, then we expect that the derivative of $h(x)$ will be

$$h'(x) = f'(x)g'(f(x)) = (2x - 3)\left(-\sin(x^2 - 3x + 7)\right)$$
$$= (3 - 2x)\sin(x^2 - 3x + 7).$$

Notice that in this case there is no other way to find the derivative.   ◢

Now that we have begun to see how the Chain Rule is used, it is time to state it precisely.

THEOREM 3.6.1

The Chain Rule

Suppose $f(x)$ is differentiable at $x = a$ and $g(u)$ is differentiable at $u = f(a)$. Then $h(x) = g(f(x))$ is differentiable at $x = a$, and its derivative is

$$h'(a) = g'(f(a))f'(a).$$

As a function, the derivative is

$$\frac{d}{dx}(h(x)) = \frac{d}{dx}\left(g(f(x))\right) = g'(f(x))f'(x).$$

PROOF The proof of the Chain Rule is actually quite delicate. You may wish to skip over it the first time through and proceed to see how it is used in the subsequent examples.

To evaluate $h'(x) = \lim_{k \to 0} \frac{h(x+k)-h(x)}{k}$, the natural thing to try first is writing

$$\frac{h(x+k)-h(x)}{k} = \frac{g\big(f(x+k)\big) - g\big(f(x)\big)}{k}$$

$$= \frac{g\big(f(x+k)\big) - g\big(f(x)\big)}{f(x+k) - f(x)} \times \frac{f(x+k) - f(x)}{k},$$

hoping that as $k \to 0$, the first factor will tend to $g'\big(f(a)\big)$ and the second will tend to $f'(a)$. This is more or less the right idea, but the difficulty is that division by $f(x+k) - f(x)$ may not be permissible because $f(x+k) - f(x)$ may be zero even when $k \neq 0$.

Instead we fix the value of x and define a useful function $Q(r)$ for small values of r by

$$Q(r) = \begin{cases} \frac{g\big(f(x)+r\big)-g\big(f(x)\big)}{r}, & \text{if } r \neq 0, \\ g'\big(f(x)\big), & \text{if } r = 0. \end{cases}$$

Since $\lim_{r \to 0} Q(r) = \lim_{r \to 0} \frac{g\big(f(x)+r\big)-g\big(f(x)\big)}{r} = g'\big(f(x)\big) = Q(0)$, we see that $Q(r)$ is continuous at $r = 0$.

Observe that as $k \to 0$, $f(x+k) \to f(x)$, so $f(x+k) - f(x) \to 0$. By Theorem 2.3.2, about the limit of a composite function, we see that as $k \to 0$, we have $Q\big(f(x+k) - f(x)\big) \to Q(0) = g'\big(f(x)\big)$.

Now consider the product

$$Q\big(f(x+k) - f(x)\big)\frac{f(x+k) - f(x)}{k}.$$

If $f(x+k) \neq f(x)$, then the product equals

$$\frac{g\big(f(x) + f(x+k) - f(x)\big) - g\big(f(x)\big)}{f(x+k) - f(x)} \times \frac{f(x+k) - f(x)}{k}$$

$$= \frac{g\big(f(x+k)\big) - g\big(f(x)\big)}{k}.$$

If, on the other hand, for some nonzero k, $f(x+k) = f(x)$, that is, $f(x+k) - f(x) = 0$, then the first factor in the product, $Q\big(f(x+k) - f(x)\big)$, is equal to $Q(0) = g'\big(f(x)\big)$, and the second is obviously zero. So the product is 0, which equals the same difference quotient $\frac{g\big(f(x+k)\big)-g\big(f(x)\big)}{k}$ in this case too. So

$$\lim_{k \to 0} \frac{g\big(f(x+k)\big) - g\big(f(x)\big)}{k} = \lim_{k \to 0} \left(Q\big(f(x+k) - f(x)\big)\frac{f(x+k) - f(x)}{k} \right)$$

$$= \left(\lim_{k \to 0} Q\big(f(x+k) - f(x)\big) \right)\left(\lim_{k \to 0} \frac{f(x+k) - f(x)}{k} \right)$$

$$= Q(0)f'(x) = g'\big(f(x)\big)f'(x),$$

as desired.

The formula

$$\frac{d}{dx}\Big(g\big(f(x)\big)\Big) = g'\big(f(x)\big)f'(x)$$

is called the Chain Rule.

EXAMPLE 3   Find the derivative of each of the following functions: (i) $h(x) = \sin(x^2 + 1)$; (ii) $h(x) = (x^2 - 5x + 2)^7$; (iii) $h(x) = e^{2\tan x}$; (iv) $h(x) = \sin^4(x^3 - 2)$.

Solution   Each of these functions $h(x)$ can be expressed as a composite function $h(x) = g\big(f(x)\big)$.

To find $h(x) = g\big(f(x)\big)$, we first evaluate $f(x)$ and then apply g to the result. In part (i) we get $\sin(x^2 + 1)$ by evaluating $f(x) = x^2 + 1$ and then taking sin of that. In part (ii) we evaluate $f(x) = x^2 - 5x + 2$ and find its seventh power.

(i)   Here we let $g(u) = \sin u$, $f(x) = x^2 + 1$. The Chain Rule tells us that $h'(x) = g'\big(f(x)\big)f'(x) = \cos(x^2 + 1)(2x) = 2x\cos(x^2 + 1)$.

(ii)   Here $g(u) = u^7$ and $f(x) = x^2 - 5x + 2$. So $\frac{d}{dx}\big((x^2 - 5x + 2)^7\big) = 7(x^2 - 5x + 2)^6(2x - 5)$. Notice that in this case to expand the power first and then differentiate would be an unpleasant task.

(iii)   Here $g(u) = e^u$ and $f(x) = 2\tan x$. So $h'(x) = \exp(2\tan x)(2\sec^2 x)$ (since $g'(u) = g(u) = e^u$). This formula only makes sense when $\tan x$ is differentiable, which means when $x \neq (2n + 1)\frac{\pi}{2}$, for any integer n.

(iv)   This function $h(x) = \sin^4(x^3 - 2)$ is the fourth power of the function $\sin(x^3 - 2)$, so if we let $g(u) = u^4$ and $f(x) = \sin(x^3 - 2)$, then we know that

$$h'(x) = 4\big(f(x)\big)^3 f'(x) = 4\sin^3(x^3 - 2)f'(x).$$

However, to find $f'(x)$, we have to use the Chain Rule a second time. Writing $f(x) = \sin(x^3 - 2) = G\big(F(x)\big)$, with $G(u) = \sin u$ and $F(x) = x^3 - 2$, we see that

$$f'(x) = G'\big(F(x)\big)F'(x) = \cos(x^3 - 2)(3x^2).$$

Substituting this into the formula for $h'(x)$, we find that

$$h'(x) = 4\sin^3(x^3 - 2)\cos(x^3 - 2)(3x^2) = 12x^2\sin^3(x^3 - 2)\cos(x^3 - 2).$$

The last part of Example 3 illustrates the possibility of using the Chain Rule repeatedly to differentiate a function that is the composition of three or more functions.

With a little practice, the Chain Rule becomes easy to use, and problems like the ones above can be done mostly in your head without having to write out carefully which functions you are combining to form the composite function. If you are having difficulty figuring out which function is which, consider that in a composite function $h(x) = g\big(f(x)\big)$, what happens first is that x is changed into $f(x)$, and *then* the result is changed into $g\big(f(x)\big)$.

So, for instance, to construct the function $(x^9 - 3x^5 + 2)^{11}$, we first transform x to $x^9 - 3x^5 + 2$ and *then* take the eleventh power. Written as a composite function, $(x^9 - 3x^5 + 2)^{11} = g\big(f(x)\big)$, with $f(x) = x^9 - 3x^5 + 2$ and $g(u) = u^{11}$.

With a graphics calculator it is possible to sketch the graph of the function $h(x) = \sin(x^2 + 1)$ of part (i) of Example 3 together with the graph of its derivative.

TI-85	SHARP EL9300
GRAPH F1$(y(x) =)$ SIN(F1(x) $x^2 +$ 1) ▼ 2 F1(x) COS (F1(x) $x^2 + 1$) M2(RANGE) $(-)$ 3 ▼ 3 ▼ ▼ $(-)$ 6 ▼ 6 F5(GRAPH)	$\curlyvee$ sin (x/θ/T $x^2 + 1$) 2ndF ▼ 2 x/θ/T cos (x/θ/T $x^2 + 1$) RANGE $-$3 ENTER 3 ENTER 1 ENTER $-$ 6 ENTER 6 ENTER 1 ENTER $\curlyvee$

CASIO f_x-7700GB/f_x-6300G	HP 48SX
Cls EXE Range $-$3 EXE 3 EXE 1 EXE $-$6 EXE 6 EXE 1 EXE EXE EXE EXE Graph sin (X $x^2 + 1$) EXE Graph 2 X cos (X $x^2 + 1$) EXE	⬑ PLOT PLOTR ERASE ATTN ' αY = SIN αX y^x 2+1 ↱ PLOT ⬑ DRAW ERASE 3 +/- ENTER 3 XRNG 6 +/- ENTER 6 YRNG DRAW ATTN ' αY = 2 $\times$ αX $\times$ COS αX y^x 2+1 ↱ PLOT ⬑ DRAW DRAW

Be sure to identify which curve is $h(x)$ and which is the derivative $h'(x)$. Notice how the absolute value of the derivative is larger when the function is changing more rapidly. You might like to try plotting these curves on different intervals, or plotting the other functions from the example.

In describing composite functions we have always been careful to write $g(u)$ and $f(x)$, using different variables for the two functions to avoid confusion. In making the composite function we let $u = f(x)$, so $h(x) = g(u) = g\big(f(x)\big)$. The Chain Rule involves two derivatives; the first is $g'(u)$, which can be written $\frac{d}{du}\big(g(u)\big)$ or $\frac{dg}{du}$. The second is $f'(x) = \frac{d}{dx}\big(f(x)\big)$, and with the understanding that $u = f(x)$, this can be written $\frac{du}{dx}$.

Now the Chain Rule says that $h'(x) = g'\big(f(x)\big)f'(x) = g'(u)f'(x)$, or

$$\frac{dh}{dx} = \frac{dg}{du}\frac{du}{dx}. \tag{3.6.2}$$

This version of the Chain Rule can be very useful, but it is important to interpret it correctly. Specifically, the right side of Equation 3.6.2 should not be interpreted as the product of two fractions (cf. the discussion after Example 3.2.1). If the derivative symbols are read as fractions, then there is a great temptation to cancel the du from denominator and numerator. In a way this is helpful for remembering the Chain Rule, but it is not a correct calculation.

The version of the Chain Rule given in Equation 3.6.2 is sometimes helpful for finding derivatives. We return to the example mentioned above.

EXAMPLE 4   Differentiate (i) $(x^9 - 3x^5 + 2)^{11}$; (ii) $\cos(\ln x)$.

Solution

(i) Writing $(x^9 - 3x^5 + 2)^{11} = f(x)^{11}$, to differentiate, we use Equation 3.6.2 and take the derivative of the eleventh power first. The idea is that the derivative of the eleventh power of some function starts, as we expect, with 11 times the tenth power of the function, but the Chain Rule (Equation 3.6.2) says that we have to multiply that by the derivative of the function:

$$\frac{d}{dx}\left((x^9 - 3x^5 + 2)^{11}\right) = 11(x^9 - 3x^5 + 2)^{10}(9x^8 - 15x^4).$$

(ii) To differentiate $\cos(\ln x)$, we notice that the derivative of cos of some function is $-\sin$ of the function multiplied by the derivative of that function. So

$$\frac{d}{dx}\left(\cos(\ln x)\right) = -\sin(\ln x)\frac{1}{x}.$$

◢

EXAMPLE 5   In Theorem 3.4.3 we discussed the derivative of a function of the form $g(cx)$, where c was some nonzero constant. We now can see that this function is the composite function $g\big(f(x)\big)$, with $f(x) = cx$. Since $f'(x) = c$, the result of Theorem 3.4.3 is an immediate consequence of the Chain Rule:

$$\frac{d}{dx}\big(g(cx)\big) = g'(cx)\,c.$$

◢

EXAMPLE 6   Let $r \in \mathbb{R}$ and define $f(x) = x^r$. Find the derivative $f'(x)$ and state where it exists.

Solution   Using Definition 3.5.10, the definition of a power, if $x > 0$ we write

$$f(x) = x^r = e^{r \ln x}.$$

The Chain Rule tells us that

$$f'(x) = e^{r \ln x}\frac{d}{dx}(r \ln x) = x^r\left(r\frac{1}{x}\right) = rx^{r-1}.$$

This will certainly make sense whenever $x > 0$.

However, there are some powers that make sense even for negative values of x, such as $x^{1/3}$. In fact, as we noticed in Theorem 2.3.1, this will happen for x^r when r is a rational number, $r = \frac{p}{q}$, with the denominator q an *odd* number.

We can also use the Chain Rule to evaluate the derivative of this type of function when $x < 0$. We start with $F(x) = x^{1/q}$, where q is an odd integer. Write $g(u) = u^q$, and notice that $h(x) = g\big(F(x)\big) = F(x)^q = (x^{1/q})^q = x$. So of course $h'(x) = 1$, but it is also possible to calculate $h'(x)$ using the Chain Rule.

We find that $h'(x) = g'\big(F(x)\big)F'(x) = qF(x)^{q-1}F'(x)$. But $h'(x) = 1$, so

$$1 = q(x^{1/q})^{q-1}\,F'(x),$$

and

$$F'(x) = \frac{1}{qx^{\frac{q-1}{q}}} = \frac{1}{q}x^{-(1-1/q)} = \frac{1}{q}x^{1/q-1}.$$

Now we consider the function $f(x) = x^{p/q} = (x^{1/q})^p = F(x)^p$, and differentiate using the Chain Rule and the above derivative of $x^{1/q}$. We find that

$$\frac{d}{dx}(x^{p/q}) = \frac{d}{dx}\left((x^{1/q})^p\right) = p(x^{1/q})^{p-1}\left(\frac{1}{q}x^{1/q-1}\right)$$

$$= \frac{p}{q}x^{\frac{p-1}{q}+\frac{1}{q}-1} = \frac{p}{q}x^{p/q-1}.$$

Strictly speaking, to apply the Chain Rule, we have to know that the function $x^{1/q}$ is differentiable. It is, but we cannot really show why, since we do not actually know very much about the power function. Later on it will be possible to discuss this point with greater precision, but for now the result will be very useful. ◢

In summary, then, we have shown that the derivative of $f(x) = x^r$ is $f'(x) = rx^{r-1}$, whenever it makes sense. This is consistent with Theorems 3.2.3 and 3.2.8, which say that $\frac{d}{dx}(x^n) = nx^{n-1}$, for any integer n, provided that the formula makes sense. We record this important result.

THEOREM 3.6.3

The derivative of the rth power function $f(x) = x^r$ is

$$\frac{d}{dx}(x^r) = rx^{r-1},$$

whenever it makes sense.

Specifically, the function x^r is differentiable whenever $x > 0$. If $r = \frac{p}{q}$ is a rational number whose denominator q is odd, then x^r is differentiable at every x if $r \geq 1$ and at every x except $x = 0$ if $r < 1$.

The most important case of Theorem 3.6.3 is $r = \frac{1}{2}$, since $x^{1/2} = \sqrt{x}$.

COROLLARY 3.6.4

The function $f(x) = \sqrt{x} = x^{1/2}$ is differentiable for every $x > 0$, and its derivative is

$$\frac{d}{dx}(x^{1/2}) = \frac{1}{2}x^{-1/2} = \frac{1}{2x^{1/2}},$$

or

$$\frac{d}{dx}(\sqrt{x}) = \frac{1}{2\sqrt{x}}.$$

We often apply the Chain Rule to a power $f(x)^n$ of a function $f(x)$.

COROLLARY 3.6.5

The Generalized Power Rule

$$\frac{d}{dx}\left(f(x)^r\right) = rf(x)^{r-1}f'(x),$$

whenever it makes sense.

EXAMPLE 7   Find the derivative of (i) $h(x) = \sqrt{x^2 + 2x + 2}$; (ii) $\sqrt{1 + \sin\theta}$.

Solution

(i) $h(x) = g\big(f(x)\big)$, with $f(x) = x^2 + 2x + 2$ and $g(u) = \sqrt{u}$. So $g'(u) = \frac{1}{2}u^{-1/2}$, and

$$h'(x) = \frac{1}{2}(x^2 + 2x + 2)^{-1/2}(2x + 2) = \frac{x + 1}{\sqrt{x^2 + 2x + 2}}.$$

For this formula to make sense we have to know that $x^2 + 2x + 2$ is not zero, but completing the square shows that $x^2 + 2x + 2 = x^2 + 2x + 1 + 1 = (x + 1)^2 + 1 > 1$, so it is never zero and there is no difficulty.

(ii) Using Equation 3.6.2, we see that

$$\frac{d}{d\theta}(\sqrt{1 + \sin\theta}) = \frac{1}{2\sqrt{1 + \sin\theta}}\frac{d}{d\theta}(1 + \sin\theta) = \frac{1}{2\sqrt{1 + \sin\theta}}\cos\theta$$

$$= \frac{\cos\theta}{2\sqrt{1 + \sin\theta}}.$$

This will be true whenever $1 + \sin\theta > 0$, which is the same thing as $\sin\theta > -1$. Since $\sin\theta \geq -1$ for every θ, it will be true except when $\sin\theta = -1$. The function will be differentiable and will have the derivative stated above except when $\theta = -\frac{\pi}{2} + 2n\pi$, for some integer n.

Notice that apart from anything else we have to avoid these points because the formula for the derivative will have a zero in the denominator there. ◢

The Chain Rule can also be used to find the derivatives of functions that arise in other problems. For instance, there are many maximum/minimum problems in which the function to be maximized or minimized must be differentiated with the Chain Rule.

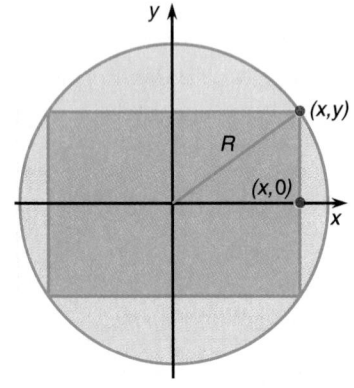

Figure 3.6.1

EXAMPLE 8   **Largest Beam in a Log**   A rectangular beam is cut from a cylindrical log of radius R. What is the largest possible cross-sectional area of the beam?

Solution   Suppose we place the origin at the center of the circular cross-section of the log (see Figure 3.6.1). Since the radius is R, the circle at the outside of the log is $x^2 + y^2 = R^2$. If the point where the edge of the beam crosses the positive x-axis is $(x, 0)$, as shown, then the width of the beam is $2x$. Since the upper right corner of the beam lies on the circle, we can calculate that its coordinates must be (x, y) with $x^2 + y^2 = R^2$, or $y = \sqrt{R^2 - x^2}$. The height of the beam is then $2y = 2\sqrt{R^2 - x^2}$.

The cross-sectional area of the beam is the product of the width and the height, or $A = (2x)(2y) = 4xy = 4x\sqrt{R^2 - x^2}$. We must find the maximum value of this function, and we know that $x \geq 0$ and $x \leq R$, so we must find the maximum on $[0, R]$.

The derivative is

$$A'(x) = \frac{d}{dx}(4x\sqrt{R^2 - x^2}) = 4\sqrt{R^2 - x^2} + 4x\frac{1}{2}(R^2 - x^2)^{-1/2}(-2x)$$

$$= 4\sqrt{R^2 - x^2} - \frac{4x^2}{\sqrt{R^2 - x^2}}.$$

The derivative is zero when

$$4\sqrt{R^2 - x^2} = \frac{4x^2}{\sqrt{R^2 - x^2}}.$$

Multiplying by $\sqrt{R^2 - x^2}$, we see that this is the same as

$$4(R^2 - x^2) = 4x^2,$$

or $4R^2 = 8x^2,$
or $x^2 = \frac{1}{2}R^2,$
or $x = \frac{1}{\sqrt{2}}R.$

Without even evaluating, we realize that this must give the maximum area, since the area is zero when $x = 0$ or $x = R$ (just look at the picture). So the answer is that the maximum area occurs when $x = \frac{1}{\sqrt{2}}R$, in which case $y = \sqrt{R^2 - x^2} = \sqrt{R^2 - \frac{1}{2}R^2} = \sqrt{\frac{1}{2}R^2} = \frac{1}{\sqrt{2}}R$. The beam with the largest cross-sectional area has a square cross-section, and the area is $2R^2$.

Exercises 3.6

I

Find the derivative of each of the following functions.

1. $h(x) = (3x^4 - 2x^2 - 1)^3$

2. $h(x) = (x^5 - 1)^{24}$

3. $h(x) = (x^2 + 4x + 5)^{-2}$

4. $h(x) = \sin^{-2} x$

5. $h(x) = \cos(2x + 1)$

6. $h(x) = \sin(x^2 + 3)$

7. $h(x) = \ln(x^2 + 4)$

8. $h(x) = e^{x^2 - x}$

9. $h(x) = \cosh(\sin x)$

10. $h(x) = \ln\!\big(\exp(x)\big)$

Find the derivative of each of the following functions, stating where it exists.

11. $f(t) = \sqrt{t^2 - 2t + 1}$

12. $k(x) = \sqrt{\sin x}$

13. $g(x) = x^{1/3}$

14. $h(x) = (x^3 + 1)^{2/3}$

15. $F(r) = \ln(1 + \cos r)$

16. $K(x) = \sin(\sqrt{x})$

17. $G(x) = e^{\ln x}$

18. $H(x) = \sin^2(\sqrt{x}) + \cos^2(\sqrt{x})$

19. $f(x) = \cosh(\sinh x)$

20. $g(x) = \sqrt{\tanh x}$

Find the derivative of each of the following functions.

21. $f(x) = \ln^3(x^2 + 7)$

22. $g(x) = \frac{1}{2 + \cos x^2}$

23. $h(x) = \sin\!\big(\ln(x^4 + 2)\big)$

24. $k(x) = \ln\!\big(\cosh(x^2)\big)$

25. $P(x) = 10^{x^2}$

26. $Q(x) = 15^{x+3}$

27. $F(x) = \sqrt{\cosh x}$

28. $G(x) = \frac{1}{\exp(\cos x)}$

29. $K(x) = \cosh^2\!\big(\ln(e^{2x})\big) - \sinh^2\!\big(\ln(e^{2x})\big)$

30. $H(x) = \exp\!\Big(\ln\!\big(5 + \cos(2x)\big)\Big)$

31. $f(x) = \frac{1}{(x^2 + 2)^2}$

32. $g(x) = \frac{1}{x^2 + 3}$

33. $h(x) = \sqrt{4 - \cos x}$

34. $k(x) = (x^2 + 4x + 7)^{7/2}$

35. $W(x) = \left(\frac{x+3}{x^2+1}\right)^4$

36. $P(x) = \exp\!\big(\sin(x^2 + 1)\big)$

37. $F(x) = \sin\!\big(\exp(x^2 + 1)\big)$

38. $G(x) = \frac{1}{\sqrt{1 + \cos^2 x}}$

39. $K(x) = (x^5)^{1/7}$

40. $H(x) = (1 + \sin^2 x)^{1/2}$

II

41. Find a Chain Rule for $F(x) = h\Big(g\big(f(x)\big)\Big)$, assuming that f, g, h are differentiable at the appropriate points.

42. Let $f(t) = t^2$, $g(s) = 2s - 1$, and calculate the derivatives of $h(x) = g(f(x))$ and $k(x) = f(g(x))$. Observe that they are different, which shows that the Chain Rule is not simply a matter of multiplying the two derivatives together.

*43. In Example 3.3.10 we had to maximize a distance that was expressed as a square root $D = \sqrt{(x-2)^2 + (x^2 + 2x + 1)^2}$. In that example we solved the problem by working not with D itself but with its square, D^2. Now that we know how to differentiate a square root, try to work that example again, this time working directly with D. (Of course, you should get the same answer, and you will also see that working with D^2 is actually easier, since the derivative of the square root is more complicated to work with.)

44. Find all points at which the graph of each of the following functions has a horizontal tangent:

   (i) $f(x) = x\sqrt{x^2 + 1}$ (ii) $f(x) = x^2\sqrt{x^2 + 1}$

   (iii) $f(x) = x\exp(-x^2)$ (iv) $f(x) = \exp(\sin x)$

45. Find the maximum and minimum value of each of the following functions on the interval $[-1, 1]$:

   (i) $f(x) = x^2\sqrt{x^2 + 1}$ (ii) $g(x) = x\exp(-x^2)$

46. Suppose a rectangular beam is cut from a cylindrical log of radius R, as in Example 8. For what dimensions of the beam will the perimeter of the rectangular cross-section be a maximum?

*47. The strength of a rectangular beam is proportional to the product of its width and the square of the height of its cross-section. If we want to cut the strongest possible beam from a cylindrical log whose radius is R, what should be the dimensions of the beam?

*48. In a semicircle of radius R, a rectangle is inscribed with one side along the semicircle's diameter and the two opposite corners on the semicircle (see Figure 3.6.2). What is the largest possible area for such a rectangle?

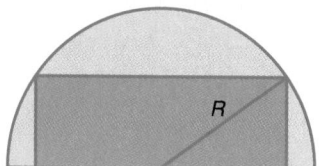

Figure 3.6.2

*49. Suppose $f(t)$ is a function that satisfies $f(\tan x) = x$, for every $x \in \left(-\frac{\pi}{2}, \frac{\pi}{2}\right)$. (So $f(t)$ is a function that "undoes" what $\tan x$ does, much as e^t "undoes" what $\ln x$ does. A function like this is called an *inverse function*; so e^t is the inverse of $\ln x$, and $f(x)$ is the inverse of $\tan x$.)

  (i) Apply the Chain Rule to the equation $f(\tan x) = x$ to find an expression for the derivative $f'(\tan x)$.

  (ii) The formula found in part (i) expresses $f'(\tan x)$ in terms of trigonometric functions of x. Use a trigonometric identity to express $f'(t)$ in terms of t.

*50. Repeat Exercise 49 with $\sin x$ for $x \in \left(-\frac{\pi}{2}, \frac{\pi}{2}\right)$ in place of $\tan x$.

III

51. Use a graphics calculator or computer to sketch the graphs of $y = x^{1/2}$ and its derivative $y' = \frac{1}{2}x^{-1/2}$ on the same picture. Notice how the value of the derivative gives the slope of the original curve.

52. Sketch the graphs of $y = x^{1/3}$ and its derivative $y' = \frac{1}{3}x^{-2/3}$ on the same picture.

53. Sketch the graphs of $y = x^{3/2}$ and its derivative $y' = \frac{3}{2}x^{1/2}$ on the same picture.

54. In Example 8 we found the formula $A = 4x\sqrt{R^2 - x^2}$ for the area of the beam with one side x. Sketch the graph of the area with $R = 2$ and $R = 5$ and confirm our result for the maximum in these cases.

55. Observe that if $y = e^{f(x)}$, then $y' = e^{f(x)}f'(x)$, which is zero precisely when $f'(x) = 0$. So $e^{f(x)}$ has a horizontal tangent where $f(x)$ does. Confirm this by sketching $f(x)$ and $e^{f(x)}$ on the same picture, with (i) $f(x) = x^2$, (ii) $f(x) = -x^3$, (iii) $f(x) = x^3 - x^2$.

In defining the derivative we considered the difference quotient

$$\frac{f(x) - f(a)}{x - a} = \frac{f(a + h) - f(a)}{h}.$$

The numerator is the difference between the values of f at two values of the variable x, and the denominator is the difference between these two values of x.

In many books a difference quotient like this is written as $\frac{\Delta f}{\Delta x}$. The symbol Δ is the capital Greek letter delta; it corresponds to our letter d, and for this reason it reminds us of the word "difference." So $\frac{\Delta f}{\Delta x}$ means "the difference between the values of f divided by the difference between the values of x."

The derivative of f at a is the limit of this quotient as $x \rightarrow a$, which is the same thing as $\Delta x \rightarrow 0$. In symbols,

$$f'(a) = \lim_{x \to a} \frac{f(x) - f(a)}{x - a} = \lim_{h \to 0} \frac{f(a + h) - f(a)}{h}$$
$$= \lim_{\Delta x \to 0} \frac{\Delta f}{\Delta x}.$$

The idea of this notation is that as we take the limit, the Greek letter Δ changes into the letter d, and the derivative is

$$\lim_{\Delta x \to 0} \frac{\Delta f}{\Delta x} = \frac{df}{dx}.$$

We have not used this notation (and will not), but we mention it here because you might encounter it elsewhere. It is really just a way of referring to difference quotients.

3.7 Implicit Differentiation

IMPLICIT FUNCTION

MAXIMUM/MINIMUM PROBLEMS

LOGARITHMIC DIFFERENTIATION

We frequently encounter equations relating x and y that cannot readily be solved explicitly for y as a function of x or for x as a function of y. Such an equation may nonetheless determine y as a function of x or vice versa; such a function is called an **implicit function**.

The equation of the circle $x^2 + y^2 = 1$ can be solved for y in terms of x. In this case we can solve explicitly $y = \sqrt{1 - x^2}$ or $y = -\sqrt{1 - x^2}$ (see the graphs of these two functions in Figure 3.7.1). Notice that there is no single function that will give all the points on the graph of the circle, since each value of x between -1 and 1 corresponds to *two* points on the circle. On the other hand, the graphs of these two functions together give all the points on the circle. However, let us pretend for a moment that we do not know how to find these solutions (either we do not know about square roots or perhaps we simply failed to notice how to simplify).

Figure 3.7.1

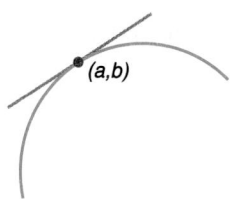

Figure 3.7.2

Suppose we have some point (a, b) that is on the curve (i.e., that satisfies the equation). Then if we change x a little bit near a, it seems reasonable to expect the point to move a little bit along the curve and y to change a little bit near b. We could ask what is $\frac{dy}{dx}$, the rate of change of y with respect to x. This could be interpreted as the slope of the tangent through (a, b), which is exactly what it would be if we knew how to write y as a function of x explicitly. See Figure 3.7.2.

If we simply *assume* that y is a function of x, even though we have no formula for it, we can differentiate the original equation and find a formula for the derivative $y' = \frac{dy}{dx}$.

EXAMPLE 1 Find $\frac{dy}{dx}$ if x and y satisfy $x^2 + y^2 = 1$.

Solution Differentiating $x^2 + y^2 = 1$, we get

$$2x + \frac{d}{dx}(y^2) = 0,$$

If y is a function of x, the Generalized Power Rule (Corollary 3.6.5) gives $\frac{d}{dx}y^n = ny^{n-1}y'$; here we have the case with $n = 2$.

and using the Chain Rule, we have $\frac{d}{dx}(y^2) = 2y\frac{dy}{dx} = 2yy'$. Substituting this into the previous equation, we find that

$$2x + 2yy' = 0,$$

which we can solve: $yy' = -x$, so $y' = -\frac{x}{y}$, provided $y \neq 0$.

We can check that this agrees with ordinary differentiation if $y = -\sqrt{1 - x^2}$, say. After all, in this case,

$$y' = \frac{d}{dx}\left(-(1 - x^2)^{1/2}\right) = -\frac{1}{2}(1 - x^2)^{-1/2}(-2x) = x(1 - x^2)^{-1/2}$$

$$= \frac{-x}{-(1 - x^2)^{-1/2}} = -\frac{x}{y}.$$

A similar calculation works for the other function $y = \sqrt{1 - x^2}$.

We found the derivative without expressing y as a function of x, simply by differentiating the equation $x^2 + y^2 = 1$ and using the Chain Rule. This process is called **implicit differentiation**, since we assume that y is implicitly determined as a function of x, even though we have not written y explicitly as a function. In this particular case it was possible after all to write y as a function of x, and we checked that the derivative of this function was the same as what we found by the process of implicit differentiation.

Several things need to be mentioned here. One is that the result for y' may include y instead of just x in the formula. This is perhaps inconvenient but not disastrous. Second, the derivative may not always exist. In the above example it does not exist when $y = 0$ (i.e., at $(1, 0)$ and $(-1, 0)$, where the tangents to the circle are vertical; see Figure 3.7.3). Third, we could just as easily have assumed that x was a function of y and found $\frac{dx}{dy}$. Fourth, when we assume that y is a function of x or vice versa, we can probably assume that this is true for some values of x, but we should not expect that there will be a function that works for all values of x, as the circle example shows. Finally, although it seems reasonable, our expectation that a small change in x will result in a small change in y is not always valid. For example, if we start at the point $(0, 0)$, which satisfies the equation $x^2 + y^2 = 0$, and change x a

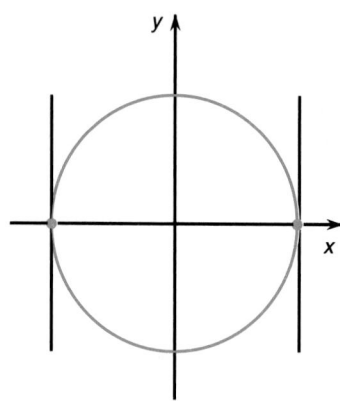

Figure 3.7.3

little bit, then there will be no y that satisfies the equation (since $(0,0)$ is the only solution of $x^2 + y^2 = 0$).

Nonetheless, the assumption that y is a function of x or vice versa, at least near some starting point (a, b), is reasonable and usually justified. We shall not concern ourselves with it, but simply use the process of implicit differentiation to find derivatives.

Summary 3.7.1   Implicit Differentiation

Suppose x and y satisfy some relation, say, $f(x, y) = 0$. Assuming that y is a function of x, at least for certain values of x, differentiate the relation, using the Chain Rule.

The result will be an equation involving x, y, and y', which can be solved for y'. Finding y' in this way is called implicit differentiation.

EXAMPLE 2   Find $y' = \frac{dy}{dx}$ if (x, y) satisfies $xy^3 - 2x^2y + x^3 - 1 = 0$.

Solution   No obvious solution for y in terms of x springs to mind, but using the Product Rule and the Chain Rule, we can differentiate both sides of the equation with respect to x:

$$y^3 + x(3y^2)y' - 4xy - 2x^2y' + 3x^2 = 0,$$

that is,

$$(3xy^2 - 2x^2)y' = 4xy - 3x^2 - y^3.$$

So

$$y' = \frac{4xy - 3x^2 - y^3}{3xy^2 - 2x^2}.$$

This is of limited use unless we know a specific point on the curve, but noting that $(1, 0)$ satisfies the equation, we can substitute these values for x and y and find that the derivative $\frac{dy}{dx}$ at the point $(1, 0)$ has the value $\frac{dy}{dx} = \frac{-3}{-2} = \frac{3}{2}$.   ◢

EXAMPLE 3

(i) Find the equation of the tangent line to the circle $x^2 + y^2 = 2$ at the point $(1, 1)$.

Solution   Differentiating implicitly, we find $2x + 2yy' = 0$, so $yy' = -x$ and $y' = -\frac{x}{y}$. At the point in question, $(1, 1)$, the slope of the tangent is therefore $y' = -1$. By the point-slope form, the equation of the tangent line is $y - 1 = (-1)(x - 1)$, or $x + y = 2$.

Of course, it is possible to solve for y as a function of x in this case and differentiate to find $\frac{dy}{dx}$. However, if you try it, you will see that the calculations are more complicated if you do it that way.

(ii) Find the equation of the tangent line to the curve $x^3 + xy^2 - y^3 + 3 = 0$ at the point $(1, 2)$.

Solution This time we find $3x^2 + y^2 + 2xyy' - 3y^2 y' = 0$, or $(3x^2 + y^2) + (2xy - 3y^2)y' = 0$, so

$$y' = -\frac{3x^2 + y^2}{2xy - 3y^2}.$$

At the specified point, this equals $\frac{7}{8}$, so the slope of the tangent is $\frac{7}{8}$, and the equation of the tangent line is $y - 2 = \frac{7}{8}(x - 1)$, or

$$7x - 8y + 9 = 0.$$

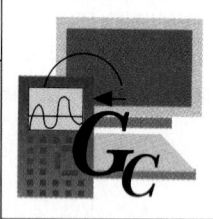

With a graphics calculator it is possible to sketch the circle and the tangent line of part (i) of Example 3. For the circle we write $y^2 = 2 - x^2$, so $y = \pm\sqrt{2 - x^2}$, and we graph both semicircles.

TI-85	SHARP EL9300
GRAPH F1$(y(x) =)\sqrt{(2 - \text{F1}(x)x^2)}$ ▼ $(-)\sqrt{(2 - \text{F1}(x)x^2)}$ ▼ $2 - \text{F1}(x)$ M2(RANGE) $(-)$ 3 ▼ 3 ▼ ▼ $(-)$ 2 ▼ 2 F5(GRAPH)	⟿ $\sqrt{}$ $2 - $ x/θ/T x^2 2ndF ▼ $-$ $\sqrt{}$ $2 - $ x/θ/T x^2 2ndF ▼ $2 - $ x/θ/T RANGE -3 ENTER 3 ENTER 1 ENTER -2 ENTER 2 ENTER 1 ENTER ⟿

CASIO f_x-7700GB/f_x-6300G	HP 48SX
Cls EXE Range -3 EXE 3 EXE 1 EXE -2 EXE 2 EXE 1 EXE EXE EXE EXE Graph $\sqrt{}$ ($2 -$ X x^2) EXE Graph $-\sqrt{}$ ($2 -$ X x^2) EXE Graph $2 -$ X EXE	↰ PLOT PLOTR ERASE ATTN ' $\alpha Y = \sqrt{x}$ () $2 - \alpha X$ y^x 2 ↱ PLOT ↰ DRAW ERASE 3 +/− ENTER 3 XRNG 2 +/− ENTER 2 YRNG DRAW ATTN ' $\alpha Y = -\sqrt{x}$ () $2 - \alpha X$ y^x 2 ↱ PLOT ↰ DRAW DRAW ATTN ' $\alpha Y = 2 - \alpha X$ ↱ PLOT ↰ DRAW DRAW

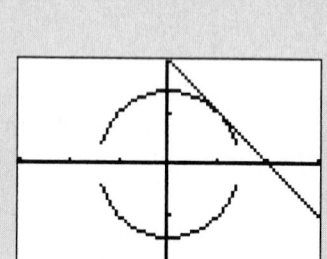

Differences in the horizontal and vertical scales may make the circle appear elliptical.

Implicit Differentiation and Maximum/Minimum Problems

An important use of implicit differentiation arises in maximum/minimum problems. In Example 3.3.6 we considered how to maximize the area A of a rectangular field whose perimeter is 600 ft. We approached the problem by letting the sides of the field be x and y, so $A = xy$ and the condition about the perimeter is $2x + 2y = 600$. It was easy to solve this last equation for y in terms of x and substitute the answer into the formula for A. However, sometimes it is difficult to solve for either variable. In such a case we can "assume" that y is a function of x and use implicit differentiation.

The derivative of the function we wish to maximize or minimize (A' in the above example) will contain y'. But implicit differentiation of the other equation relating x and y will allow us to solve for y', and it may be relatively easy to substitute this into the formula for A' and find where $A' = 0$.

EXAMPLE 4   Find the maximum value of $f(x, y) = x + y$ on the ellipse $2x^2 + 3xy + 3y^2 = 8$.

Solution   Differentiating both equations implicitly, we find that $\frac{df}{dx} = 1 + y'$ and that $4x + 3y + 3xy' + 6yy' = 0$. We need to find where $\frac{df}{dx} = 0$, that is, $1 + y' = 0$, that is, $y' = -1$.

Substituting $y' = -1$ into the other equation gives

$$4x + 3y - 3x - 6y = 0,$$

or $x - 3y = 0,$

or $x = 3y.$

Looking for points on the ellipse that satisfy $x = 3y$, we substitute $x = 3y$ into $2x^2 + 3xy + 3y^2 = 8$ and get

$$18y^2 + 9y^2 + 3y^2 = 8,$$

that is, $30y^2 = 8,$

so $y^2 = \frac{4}{15}, \qquad y = \pm\frac{2}{\sqrt{15}},$

and $x = 3y = \pm\frac{6}{\sqrt{15}} = \pm 2\sqrt{\frac{3}{5}}.$

The two points on the ellipse at which $\frac{df}{dx} = 0$ are $\left(2\sqrt{\frac{3}{5}}, \frac{2}{\sqrt{15}}\right)$ and $\left(-2\sqrt{\frac{3}{5}}, -\frac{2}{\sqrt{15}}\right)$. It is clear that $f(x, y) = x + y$ is larger at the first point and smaller at the second one. If we knew there had to be a maximum and a minimum, they would have to be at these points.

We can see that this is in fact true, as follows. The equation of the ellipse is a quadratic equation. Without actually writing it all out, we know we can solve for y as either of two functions of x, one the "upper half" of the ellipse, and the other the "lower half," corresponding to taking the plus sign or the minus sign in the Quadratic Formula. Each of these functions y is a continuous function on a closed interval, and consequently so is $x + y$. By Theorem 2.5.7, $x + y$ has a maximum and a minimum on each piece of the ellipse and hence on all of it.

The maximum value of $f(x, y) = x + y$ occurs at the point $\left(2\sqrt{\frac{3}{5}}, \frac{2}{\sqrt{15}}\right)$. The value of f at this point is $2\sqrt{\frac{3}{5}} + \frac{2}{\sqrt{15}} = \frac{8}{\sqrt{15}} \approx 2.06559.$

This example illustrates the principal drawback in using implicit differentiation for maximum/minimum problems: While it often makes it easy to find points at which the derivative is zero, it may still be very difficult to decide whether the function has a maximum or minimum value there, and the technique offers no help about whether a maximum or minimum value exists at all.

Logarithmic Differentiation

When we have to differentiate a complicated expression, the work required may sometimes be reduced by differentiating not the expression itself but rather its logarithm. This is especially true with functions involving powers or complicated products or quotients. The idea is that the logarithm of such a function may be simpler to work with than the function itself.

EXAMPLE 5   Differentiate the function $f(x) = \frac{\sqrt{x^4+2}}{(3+\cos x)^{1/5}}$.

Solution   To differentiate directly would require an elaborate application of the Quotient and Chain Rules. Instead, we let $y = f(x)$ and consider

$$\ln y = \ln\big(f(x)\big) = \frac{1}{2}\ln(x^4 + 2) - \frac{1}{5}\ln(3 + \cos x).$$

Its derivative is

$$\frac{d}{dx}\Big(\ln\big(f(x)\big)\Big) = \frac{1}{2} \times \frac{4x^3}{x^4 + 2} - \frac{1}{5} \times \frac{(-\sin x)}{3 + \cos x}.$$

We have just found an expression for $\frac{d}{dx}(\ln y) = \frac{1}{y}\frac{dy}{dx}$, so we can multiply by $y = f(x)$ to find

$$\frac{dy}{dx} = y\frac{d}{dx}(\ln y) = f(x)\frac{d}{dx}\Big(\ln\big(f(x)\big)\Big)$$

$$= \frac{\sqrt{x^4 + 2}}{(3 + \cos x)^{1/5}}\left(\frac{2x^3}{x^4 + 2} + \frac{\sin x}{15 + 5\cos x}\right).$$

The calculation we have just done is probably easier than differentiating the original expression using the Quotient Rule. The advantage of *logarithmic differentiation* is that taking the logarithm "pulls apart" quotients and products and allows us to differentiate each factor separately.

In Example 5 we neglected to use absolute values, but a quick look will convince you that the functions involved there are always positive . . .

Strictly speaking, where $f(x)$ is negative, we should really consider $\ln|f(x)|$, but, interestingly, it turns out that the formula we get will be the same. For instance, suppose $f(x) < 0$ for x in some interval. Then for these values of x, $|f(x)| = -f(x)$ and $|f(x)|' = \frac{d}{dx}\big(|f(x)|\big) = \frac{d}{dx}(-f) = -f'$, so $\frac{d}{dx}\big(\ln|f(x)|\big) = \frac{|f(x)|'}{|f(x)|} = \frac{-f'(x)}{-f(x)} = \frac{f'(x)}{f(x)}$. The formula $\frac{d}{dx}\big(\ln|f(x)|\big) = \frac{f'(x)}{f(x)}$ is not really affected by the absolute value.

This means that we can work without worrying about signs. It is often quite difficult to decide when a function is positive and when it is negative, so it is convenient to avoid this complication.

We now summarize the procedure.

Summary 3.7.2

Logarithmic Differentiation

To differentiate $y = f(x)$, we follow these steps:

1. Write out $\ln|f(x)|$, using the properties of the natural logarithm to "pull apart" products, quotients, and powers. Express $\ln|f(x)|$ as a sum of terms $\ln|g(x)|$ for various functions $g(x)$.

2. Differentiate to find an expression for each term: $\frac{d}{dx}\left(\ln|g(x)|\right) = \frac{d}{dx}\left(\ln|g|\right) = \frac{1}{g}\frac{dg}{dx} = \frac{g'}{g}$. To do this, it is not really necessary to consider absolute values.

3. Multiply by the original function to find the derivative

$$\frac{df}{dx} = \frac{dy}{dx} = y\frac{d}{dx}\left(\ln|y|\right) = f(x)\frac{d}{dx}\left(\ln|f(x)|\right).$$

EXAMPLE 6   Differentiate $f(x) = \sqrt{x(x^2 - 2)(x^3 + 4x + 1)}$.

Solution   Letting $y = f(x)$, we write

$$\ln|y| = \frac{1}{2}\left(\ln|x| + \ln|x^2 - 2| + \ln|x^3 + 4x + 1|\right).$$

Differentiating, we find

$$\frac{d}{dx}(\ln|y|) = \frac{1}{2}\left(\frac{1}{x} + \frac{2x}{x^2 - 2} + \frac{3x^2 + 4}{x^3 + 4x + 1}\right),$$

so

$$\frac{dy}{dx} = y\frac{d}{dx}(\ln|y|)$$

$$= \frac{1}{2}\sqrt{x(x^2 - 2)(x^3 + 4x + 1)}\left(\frac{1}{x} + \frac{2x}{x^2 - 2} + \frac{3x^2 + 4}{x^3 + 4x + 1}\right).$$

Exercises 3.7

I

Find $\frac{dy}{dx}$ on each of the following curves.

1. $x^2 + xy - y^2 = 3$

2. $xy^2 - 3x^2y + y^3 + 1 = 0$

3. $\frac{xy + y^3}{1 + x} = 1$

4. $xy - y^2\sin x = \pi$

5. $3x^2 - 9y^3 + xy + 5 = 0$

6. $xe^{xy} = 2$

7. $x = \sin y$

8. $x^2 = \sin(y^2)$

9. $\cos x \sin y = \frac{1}{2}$

10. $x = e^y$

11.– 20. Find $\frac{dx}{dy}$ for each of the curves in Exercises 1 through 10.

Differentiate each of the following.

21. $f(x) = (x^4 + x^2)(x^3 - 8)(x^2 + x + 1)$

22. $f(x) = \sin x \tan x \sec^2 x$

23. $f(x) = x(x^3 + 4x)^{1/3}$

24. $f(x) = \sqrt{(1 + x^2)(1 + x^4)(1 + x^6)}$

25. $f(x) = \sqrt{(x + 1)(x + 2)(x + 3)}$

26. $f(x) = \left(x(x + 1)(x - 2)\right)^{3/4}$

27. $f(x) = \frac{\sqrt{x^3 + 1}}{(x^2 + 2)^4}$

28. $f(x) = \frac{(x + 1)^{8/9}}{x^{3/4}}$

29. $f(x) = \left(\frac{(x + 1)(x - 2)}{(x + 3)(x^2 - 9)}\right)^{3/7}$

30. $f(x) = \sqrt{x(x + 2)\ln(x)}$

II

31. Use implicit differentiation to do Example 3.3.6 (p. 160) again.

32. Use implicit differentiation to do Example 3.3.7 (p. 161) again.

33. Use implicit differentiation to do Example 3.3.8 (p. 162) again.

*34. A rectangle is drawn inside the ellipse $9x^2 + 4y^2 = 36$ so that two of its sides are horizontal and two are vertical. Find the rectangle with the largest possible area.

35. Find the equation of the tangent line to each of the following curves at the specified point: (i) $x^2 - 3xy + y^2 + 1 = 0$ at $(1, 1)$; (ii) $x^3 + 3xy^2 - xy = 8$ at $(2, 0)$.

36. Find the equation of the tangent line to each of the following curves at the specified point: (i) $x^2 + 4y^2 = 1$ at $(-1, 0)$; (ii) $xy^3 - 2x^2y^2 - 4xy + 8 = 0$ at $(1, 2)$.

37. Find the equation of the tangent line to each of the following curves at the specified point: (i) $x \sin y + y \cos x = \pi$ at $(0, \pi)$; (ii) $xe^y + x^2y - xy^3 = 3$ at $(3, 0)$.

*38. If (x, y) lies on the circle $x^2 + y^2 = 1$, then we have seen that $\frac{dy}{dx}$ does not exist at the points $(-1, 0)$ and $(1, 0)$.
  (i) Find $\frac{dx}{dy}$ at these two points.
  (ii) Use your answer to part (i) to find the equations of the tangent lines at those two points. (Be careful; when you interpret $\frac{dx}{dy}$ as a slope, it is the slope of the tangent to the graph of x as a function of y.)

*39. Assuming that $f(x, y) = 3x - 2y$ has a maximum and a minimum value on the curve $3x^2 + xy + y^2 = 1$, find them.

*40. Assuming that $f(x, y) = 2x - 3y$ has a maximum and a minimum value on the curve $x^2 - xy + y^2 - 2x = 1$, find them.

*41. Assuming that $f(x, y) = xy$ has a maximum and a minimum value on the curve $x^2 + xy + y^2 = 12$, find them.

*42. Show that $f(x, y) = 2y - x$ does not have a maximum value nor a minimum value on the curve $y^2 = x^2 + 2x + 2$, even though implicit differentiation does find two points at which $\frac{dy}{dx} = 0$. Draw a picture.

43. Solve Examples 5 (p. 200) and 6 (p. 201) without us-ing logarithmic differentiation. This should convince you of the value of using it.

*44. Let $f(x) = (x^2 - 3)^5(x^3 - 2x)^{-1/7}$.
  (i) Find the values of x for which $f(x) > 0$ and those for which $f(x) < 0$.
  (ii) Find $\frac{df}{dx}$ using logarithmic differentiation, using part (i) to find a formula for $|f(x)|$ on each interval where the sign stays the same.
  (iii) Find $\frac{df}{dx}$ using logarithmic differentiation but without worrying about signs. Confirm the remarks made before Summary 3.7.2 (p. 200) to the effect that the answers will be the same.

45. Use a graphics calculator or computer and the result of Example 1 (p. 196) to sketch the circle $x^2 + y^2 = 1$ and its tangent line at (i) $\left(\frac{1}{\sqrt{2}}; -\frac{1}{\sqrt{2}}\right)$; (ii) $\left(-\frac{1}{2}, \frac{\sqrt{3}}{2}\right)$; (iii) $\left(\frac{3}{5}, \frac{4}{5}\right)$. (*Hint:* To sketch the circle, work with two semicircles.)

46. Sketch the ellipse $x^2 + 4y^2 = 4$. (Work with two halves.) Find the equation of the tangent at the point $\left(-1, \frac{\sqrt{3}}{2}\right)$, and sketch it on the same picture.

*47. In Exercise 39 we considered the curve $3x^2 + xy + y^2 = 1$. Solve for y (think of the equation as a quadratic equation in y). You will get two solutions; graph both of them and identify the curve.

48. Graph the curve $x^2 - xy + y^2 - 2x = 1$ from Exercise 40.

49. Graph the curve $x^2 + xy + y^2 = 12$ from Exercise 41.

50. Graph the curve $y^2 = x^2 + 2x + 2$ from Exercise 42.

51. Sketch the graph of the function $f(x) = (x^2 - 3)^5(x^3 - 2x)^{-1/7}$ from Exercise 44. Compare the picture with your answer to part (i) of that exercise.

3.8 Derivatives of Inverse Trigonometric and Inverse Hyperbolic Functions

INVERSE FUNCTION
ONE-TO-ONE
arcsin x
arctan x
GRAPHS OF INVERSE FUNCTIONS
INVERSE TRIGONOMETRIC AND INVERSE HYPERBOLIC FUNCTIONS AND THEIR DERIVATIVES

This section presents the difficult concept of inverse functions and applies it in particular to trigonometric and hyperbolic functions. The most important result is Theorem 3.8.5, which lists the derivatives of the inverse trigonometric functions. We will need to know the trigonometric identities (Theorem 1.7.2, Equation 1.7.3, Theorem 1.7.11), the derivatives of trigonometric functions (Theorems 3.4.1 and 3.4.2), and the Chain Rule. The last part of the section deals with inverse hyperbolic functions and their derivatives. These topics are of less importance, but to study them, it is necessary to know the derivatives of hyperbolic functions (Theorem 3.5.13) and the hyperbolic function identity (Equation 3.5.12).

EXAMPLE 1   If $f(x) = 2x + 4$ and $g(u) = \frac{u}{2} - 2$, we notice that their composition is $g\big(f(x)\big) = \frac{f(x)}{2} - 2 = \frac{2x+4}{2} - 2 = x + 2 - 2 = x$. This says that whatever the function f does to x, the function g "undoes" it and takes us back to the original value of x. For example, if $x = 3$, then $f(x) = f(3) = 2(3) + 4 = 10$, and $g(10) = \frac{10}{2} - 2 = 5 - 2 = 3$, which is the value of x with which we started.

In this situation the function g is called the inverse of f and is often written $g(u) = f^{-1}(u)$. It is very important not to confuse this with the reciprocal $\frac{1}{f(u)}$. (In the above example, $\frac{1}{f(u)} = \frac{1}{2u+4}$, which is quite different from the inverse function $f^{-1}(u) = g(u) = \frac{u}{2} - 2$.)

DEFINITION 3.8.1

> If $f(x)$ is a function and $g(u)$ is another function so that
>
> $$g\big(f(x)\big) = x$$
>
> for all x in the domain of f, then $g(u)$ is called the **inverse** of $f(x)$, and we write $g(u) = f^{-1}(u)$.
>
> The domain of the inverse function $f^{-1}(u)$ is the set of all values of the function $f(x)$, that is, the domain of $f^{-1}(u)$ is $\{u : u = f(x), \text{ for some } x \in \text{domain } \big(f(x)\big)\}$.

EXAMPLE 2

(i) The function $\ln u$ is the inverse of e^x, since $\ln(e^x) = x$ for every $x \in \mathbb{R}$.

(ii) The function $\ln x$, defined for all $x > 0$, has as its inverse function $g(u) = e^u$, since

$$e^{\ln x} = x,$$

for all $x > 0$. Notice that it does not make sense to ask what happens for values of x that are not in the domain of f.

EXAMPLE 3   Consider the functions $f(x) = x^2$ and $g(u) = \sqrt{u}$. If $x \geq 0$, then $g\big(f(x)\big) = \sqrt{x^2} = x$. But if $x = -3$, for example, then $g\big(f(-3)\big) = g\big((-3)^2\big) = g(9) = \sqrt{9} = 3 \neq -3$ (recall that $\sqrt{a}$ is always nonnegative). So $g\big(f(x)\big) = x$ if $x \geq 0$ but not if $x < 0$, so we can say that $g = f^{-1}$ if the domain of $f(x)$ is $[0, \infty)$.

What complicates this example is that $f(x) = x^2$ may have the same value at different points x. For instance, $f(3) = f(-3)$, and indeed $f(-a) = f(a)$ for any a. So if we try to make a function $g = f^{-1}$ that "undoes" what $f(x)$ does, we have to make a choice (the square root function is conventionally taken to be nonnegative). We could make a different choice: $h(u) = -\sqrt{u}$ is $f^{-1}(u)$ if the domain of f is $(-\infty, 0]$.

In any case, what is essential is that the domain of $f(x)$ not contain two points at which f takes the same value.

DEFINITION 3.8.2

> A function $f(x)$ is **one-to-one** on its domain D if $f(s) \neq f(t)$ for every pair of distinct points $s, t \in D$.

The terminology suggests that the function takes only *one x* to any *one y*.

To talk about the inverse of a function, we must consider the original function on a domain where it is one-to-one. If $f(x)$ is one-to-one, then f^{-1} is automatically defined (just by "undoing" whatever f does).

Figure 3.8.1

EXAMPLE 4 Consider the function $y = \sin x$, illustrated in Figure 3.8.1. It is certainly not one-to-one. However, if we look at $f(x) = \sin x$ on the interval $\left[-\frac{\pi}{2}, \frac{\pi}{2}\right]$, it *is* one-to-one there (because it is increasing, so it cannot "double back" and take the same value twice). So it has an inverse function, $f^{-1}(u)$, defined for $u \in [-1, 1]$ and taking values in the interval $\left[-\frac{\pi}{2}, \frac{\pi}{2}\right]$. Because we so frequently write trigonometric functions with exponents, the notation $\sin^{-1} u$ is especially prone to misinterpretation. Instead, we call this particular inverse function arcsin u.

It is usually not easy to evaluate inverse functions, but a graphics calculator can be used to get approximate evaluations.

To find arcsin $\frac{1}{3}$, that is, the number x so that $\sin x = \frac{1}{3}$, we plot $y = \sin x$ and $y = \frac{1}{3}$.

TI-85	SHARP EL9300
GRAPH F1 $(y(x) =)$ SIN F1 (x) ▼ 1 ÷ 3 M2 (RANGE) 0 ▼ π ÷ 2 ▼ ▼ 0 ▼ 1 F5 (GRAPH)	⋀⋁ SIN x/θ/T 2ndF ▼ 1 ÷ 3 RANGE 0 ENTER π ÷ 2 ENTER 1 ENTER 0 ENTER 1 ENTER 1 ENTER ⋀⋁
CASIO f_x-7700GB/f_x-6300G	HP 48SX
Cls EXE Range 0 EXE π ÷ 2 EXE 1 EXE 0 EXE 1 EXE 1 EXE EXE EXE EXE Graph sin ▮ EXE Graph 1 ÷ 3 EXE	▮▮▮▮▮PLOTR ERASE ATTN ' αY ▮ SIN αX ↱ ▮▮▮ ▮▮ DRAW ERASE ▮ENTER 2 ÷ 0 ▮▮▮ XRNG 0 ENTER 1 YRNG DRAW ATTN ' αY ▮ 1 ÷ 3 ↱ ▮▮▮ ▮▮▮ DRAW DRAW

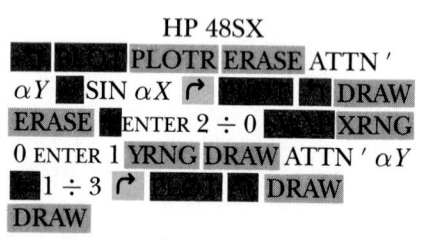

Then by zooming in it is possible to find an approximate value for arcsin $\frac{1}{3}$, the value of x where the curve crosses the line. Depending on which calculator you are using, it is about arcsin $\frac{1}{3} \approx 0.339836909$. It is also possible to use the SOLVE feature of your calculator and find where $\frac{1}{3} - \sin x = 0 \dots$ Most calculators have an inverse tangent key, so in this particular case, that is the easiest of all.

EXAMPLE 5 Suppose we look at the graph of $y = \tan x$, or more specifically the part of the graph with $-\frac{\pi}{2} < x < \frac{\pi}{2}$. It seems from Figure 3.8.2 that $\lim_{x \to -\frac{\pi}{2}^+} \tan x = -\infty$ and $\lim_{x \to \frac{\pi}{2}^-} \tan x = \infty$. By the Intermediate Value Theorem (Theorem 2.5.5) this implies that the function $y = \tan x$ takes on every real value y for some x in the interval $\left(-\frac{\pi}{2}, \frac{\pi}{2}\right)$. Moreover, the function is always increasing, so it cannot take on

Figure 3.8.2

Figure 3.8.3

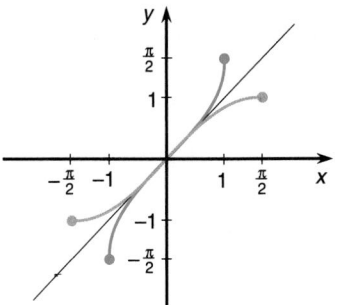

Figure 3.8.4

the same value twice and must be one-to-one. In short, the function $y = \tan x$ on the interval $\left(-\frac{\pi}{2}, \frac{\pi}{2}\right)$ takes on every real value exactly once.

So although it might not be easy to do in practice, it is possible in principle to find what value of x will result in some particular value of y. For instance, if $y = 1$, we can find an x so that $\tan x = 1$, namely, $x = \frac{\pi}{4}$. Of course, that one was deliberately chosen to be easy. It would be harder to specify the value of x for which $\tan x = 3\sqrt{2}$, say. Still, we do know that there must be such an x.

In effect, this means there is an inverse function $g(t)$ so that for every $t \in \mathbb{R}$,

$$\tan\big(g(t)\big) = t.$$

This function g is called arctan t.   ◢

If we graph the function arcsin u, which was defined in Example 4, we notice that $w = \text{arcsin } u$ means $u = \sin w$. Because of this, the graph of $w = \text{arcsin } u$ relative to the u-axis and the w-axis is related to the graph of $u = \sin w$ on the w-axis and the u-axis. Unfortunately, the order of the two axes is reversed, but if you take the graph of $w = \text{arcsin } u$ and turn it $90°$ clockwise, you get the graph of $u = \sin w$ upside-down (i.e., the graph of $u = -\sin w$). Try it with Figure 3.8.3.

Another way of saying this is that the graph of $y = f^{-1}(x)$ is the graph of $y = f(x)$ *reflected* in the line $y = x$, as shown in Figure 3.8.4 (see the box below). The tricky thing about this is that the same letter, x, is used for the variable of $f(x)$ and for the variable of $f^{-1}(x)$, whereas if we start with $y = f(x)$, then f^{-1} should "undo" what f does and take y back to x, suggesting y as the variable for f^{-1}.

In dealing with inverse functions it is very important to keep clearly in mind what is the independent variable for each function. As with composite functions, it often helps to avoid confusion if you use a different letter altogether, for example, $f(x)$ and $f^{-1}(u)$.

Graphs of Inverse Functions

Suppose $f(x)$ is a function that has an inverse $f^{-1}(x)$. If (a, b) is a point on the graph of $y = f(x)$, that is, $b = f(a)$, then $a = f^{-1}(b)$, so (b, a) is on the graph of $y = f^{-1}(x)$. The graph of $y = f^{-1}(x)$ is obtained by reversing the coordinates of the points on the graph of $y = f(x)$, changing (a, b) to (b, a).

On the other hand, what is the effect of reflecting a graph in the line $y = x$? From Figure 3.8.4 we realize that reflecting interchanges the positive x-axis and the positive y-axis. A point (a, b), which is found by moving a units along the x-axis and b units along the y-axis, will be reflected into the point (b, a), which is b units along the x-axis and a units along the y-axis.

This shows that the graph of $y = f^{-1}(x)$ is obtained by reflecting the graph of $y = f(x)$ in the line $y = x$.

To return to Example 4, it is possible to choose other domains on which $\sin x$ is one-to-one (e.g., $\left[\frac{\pi}{2}, \pi\right]$, $\left[\frac{\pi}{2}, \frac{3\pi}{2}\right]$, $\left[0, \frac{\pi}{2}\right] \cup \left(\pi, \frac{3\pi}{2}\right)$, etc.), and $\sin x$ has an inverse on each of them. However, most often we use the interval $\left[\frac{-\pi}{2}, \frac{\pi}{2}\right]$, and then arcsin u is the corresponding inverse function.

It is also easy to verify that $\cos x$ is decreasing on $[0, \pi]$, so it is one-to-one there. Also, $\tan x$ is increasing on $\left(-\frac{\pi}{2}, \frac{\pi}{2}\right)$, $\cot x$ is decreasing on $(0, \pi)$, $\sec x$ is positive and increasing on $\left[0, \frac{\pi}{2}\right)$ and negative and decreasing on $\left[\pi, \frac{3\pi}{2}\right)$, and $\csc x$ is negative and increasing on $\left(-\pi, -\frac{\pi}{2}\right]$ and positive and decreasing on $\left(0, \frac{\pi}{2}\right]$. So each of these functions is one-to-one on the specified domain and has an inverse there (see Figure 3.8.5).

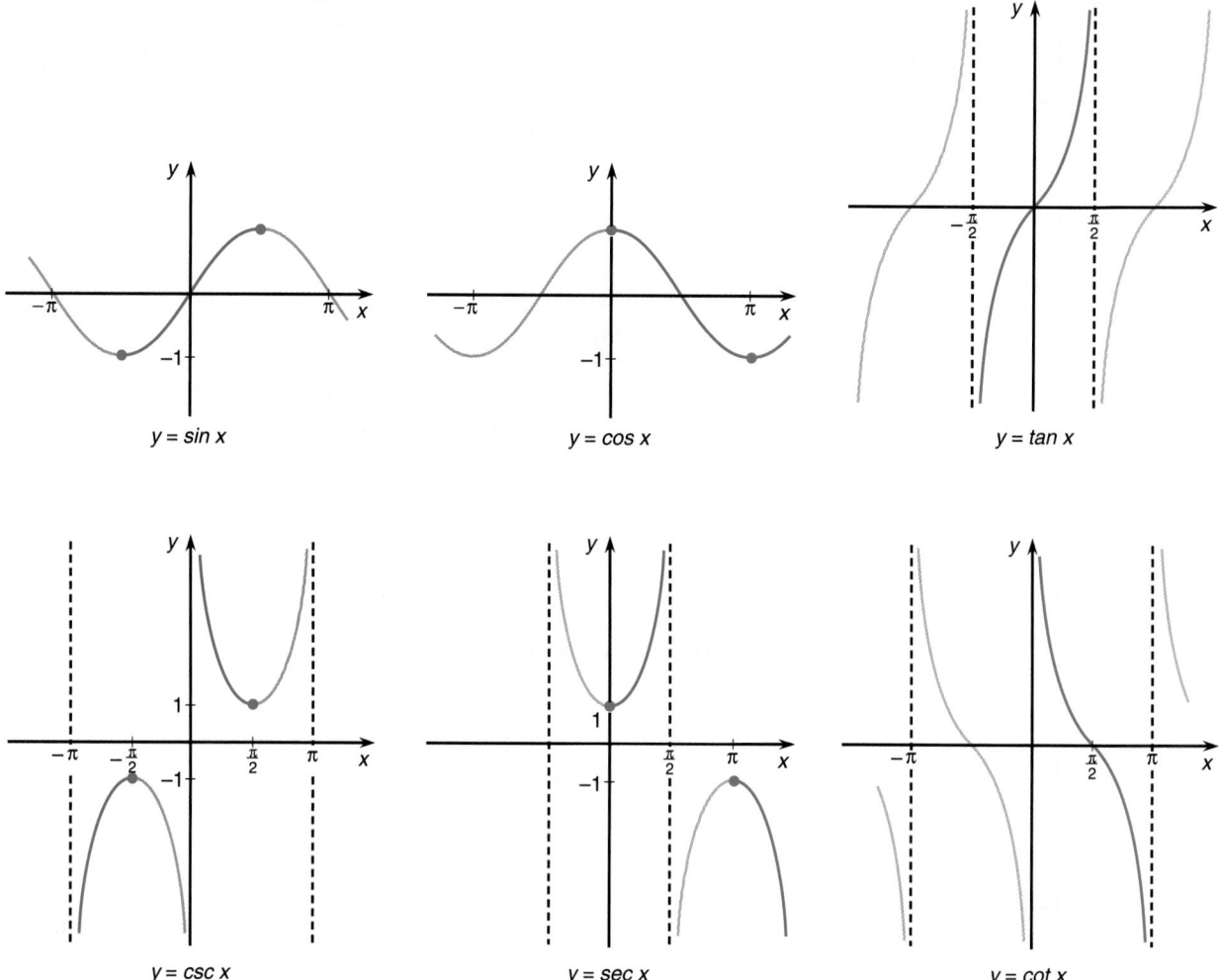

Figure 3.8.5

DEFINITION 3.8.3

The **standard inverse functions of the trigonometric functions** are defined as follows.

For $u \in [-1, 1]$,

$$\arcsin u \in \left[-\frac{\pi}{2}, \frac{\pi}{2}\right] \qquad \text{and} \qquad \arccos u \in [0, \pi].$$

For any $u \in \mathbb{R}$,

$$\arctan u \in \left(-\frac{\pi}{2}, \frac{\pi}{2}\right) \qquad \text{and} \qquad \text{arccot}\, u \in (0, \pi).$$

For u with $|u| \geq 1$,

$$\text{arcsec } u \in \left[0, \frac{\pi}{2}\right) \cup \left[\pi, \frac{3\pi}{2}\right) \quad \text{and} \quad \text{arccsc } u \in \left(-\pi, -\frac{\pi}{2}\right] \cup \left(0, \frac{\pi}{2}\right].$$

Moreover, if $w = \arcsin u$, then $u = \sin w$; if $w = \arccos u$, then $u = \cos w$; if $w = \arctan u$, then $u = \tan w$; if $w = \text{arccot } u$, then $u = \cot w$; if $w = \text{arcsec } u$, then $u = \sec w$; if $w = \text{arccsc } u$, then $u = \csc w$.

EXAMPLE 6   The graphs of $\tan x$, $\arctan u$, $\sec x$, and $\text{arcsec } u$ are sketched in Figure 3.8.6. Note how the graphs of $\tan x$ and $\arctan u$ are each other's reflections in the line $y = x$ and similarly for $\sec x$ and $\text{arcsec } u$.

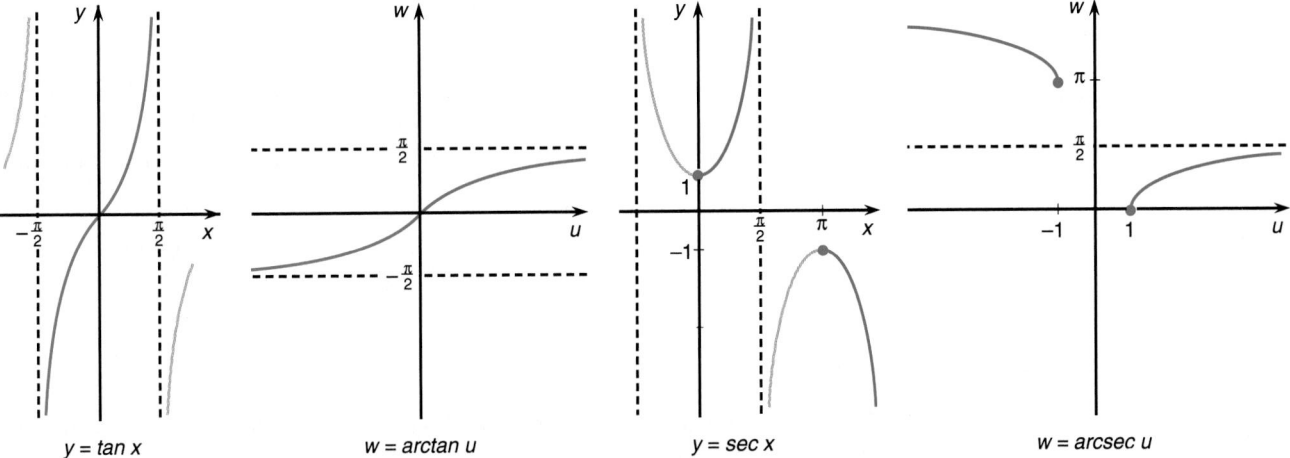

$y = \tan x$ $w = \arctan u$ $y = \sec x$ $w = \text{arcsec } u$

Figure 3.8.6

Next we turn to the question of derivatives of inverse functions. If $f(x)$ is a function that has an inverse, say, $f^{-1}(u)$, then we know that $f^{-1}(f(x)) = x$, for all x in the domain of f. Differentiating this equation, we see that

$$\frac{d}{dx}\left(f^{-1}(f(x))\right) = \frac{dx}{dx} = 1.$$

On the other hand, applying the Chain Rule to the left side suggests that

$$(f^{-1})'(f(x))f'(x) = 1,$$

so

$$(f^{-1})'(f(x)) = \frac{1}{f'(x)},$$

or

$$(f^{-1})'(u) = \frac{1}{f'(f^{-1}(u))}.$$

The difficulty is that to be able to apply the Chain Rule, we have to know that f^{-1} is differentiable, and this is not always true. For instance, the graph of $w = \text{arcsec } u$

above is not even continuous, so it is certainly not always differentiable. Also if $f'(x) = 0$, then the above formula does not make sense, and in fact $f_\bullet^{-1}$ will not be differentiable at the corresponding point $u = f(x)$.

If we think of the graph of f^{-1} as the reflection in the line $y = x$ of the graph of f, then a point at which $f'(x) = 0$ is a point at which the tangent line to $y = f(x)$ is horizontal. This point will be reflected into a point on the graph of f^{-1} at which the tangent line is *vertical*, which is why the derivative does not exist.

On the other hand, this idea also suggests that if $y = f(x)$ has a tangent line that is not horizontal, then at the corresponding point on the graph of $w = f^{-1}(u)$ there will be a tangent line that is not vertical. In particular, f^{-1} should be differentiable there.

EXAMPLE 7

(i) Recall the functions $f(x) = 2x + 4$ and $g(u) = \frac{u}{2} - 2$, which were defined in Example 1. We find that $g'(u) = \frac{1}{2}, f'(x) = 2$, so in this case it certainly is true that $(f^{-1})'(u) = g'(u) = \frac{1}{2} = \frac{1}{f'(f^{-1}(u))}$.

(ii) If $F(x) = e^x$, then from Example 2 we know that $F^{-1}(u) = \ln u$. In this case, $(F^{-1})'(u) = \frac{d}{du}(\ln u) = \frac{1}{u}$. On the other hand, $F'(x) = e^x = F(x)$, so $F'(F^{-1}(u)) = F(F^{-1}(u)) = \exp(\ln u) = u$, and we see that here too $(F^{-1})'(u) = \frac{1}{u} = \frac{1}{F'(F^{-1}(u))}$.

(iii) Considering the functions $f(x) = x^2$ and $g(u) = \sqrt{u}$, as defined in Example 3, we notice that $g = f^{-1}$ and $(f^{-1})'(u) = g'(u) = \frac{1}{2\sqrt{u}}$. On the other hand, $f'(f^{-1}(u)) = 2f^{-1}(u) = 2g(u) = 2\sqrt{u}$. So once again $(f^{-1})'(u) = \frac{1}{2\sqrt{u}} = \frac{1}{f'(f^{-1}(u))}$, provided $u > 0$.

Example 7 illustrates that the formula suggested by the Chain Rule for the derivative of an inverse function does work, although we do have to be careful about the derivative of the original function being nonzero.

THEOREM 3.8.4

Suppose a function $f(x)$, defined on some domain D, has an inverse function. Then $f^{-1}(u)$ is differentiable at $u = a$ if $f(x)$ is differentiable at $x = f^{-1}(a)$ *provided* $f'(f^{-1}(a)) \neq 0$. In this case,

$$(f^{-1})'(a) = \frac{1}{f'(f^{-1}(a))}.$$

In Example 7 we verified that this formula works for various examples of inverse functions. But its usefulness derives from the fact that it can be used even if we do not know much about the inverse function.

EXAMPLE 8   Let $f(x) = x^5 + 2x + 1$. Although it is not obvious, this function is one-to-one on the whole real line, so it has an inverse f^{-1}. Assuming this fact, we can use Theorem 3.8.4 to find the derivative $(f^{-1})'(u)$.

Since $f'(x) = 5x^4 + 2$, we see that $f'(x) \geq 2$, so $f'(x) \neq 0$ for any x, so f^{-1} is differentiable at every point in its domain, and the formula says that

$$(f^{-1})'(u) = \frac{1}{f'(f^{-1}(u))} = \frac{1}{5(f^{-1}(u))^4 + 2}.$$

If we evaluate $f(2) = 32 + 4 + 1 = 37$, for instance, we see that $(f^{-1})'(37) = \frac{1}{f'(f^{-1}(37))} = \frac{1}{f'(2)} = \frac{1}{82}$. We could also write

$$(f^{-1})'(u) = \frac{1}{5x^4 + 2}.$$

Notice that the formula is inevitably awkward. Either the derivative $(f^{-1})'(u)$ is expressed in terms of $f^{-1}(u)$ or else it can be expressed more simply in terms of x, but then it is the value of $(f^{-1})'$ at $u = f(x)$.

EXAMPLE 9   Letting $L(x) = \ln x$, for $x > 0$, we know from Example 2(ii) that $L^{-1}(u) = e^u$. So

$$\frac{d}{du}(e^u) = (L^{-1})'(u) = \frac{1}{L'(L^{-1}(u))} = \frac{1}{L'(e^u)} = \frac{1}{\frac{1}{e^u}} = e^u.$$

This result is not new, but we see that it follows from knowing the derivative of $\ln x$. The interesting thing is that the derivative of e^u is expressed as a function of u itself, rather than as a function of $L^{-1}(u)$.

Notice that we used the fact that $\ln'(w) = \frac{1}{w}$, but that we had to evaluate $\frac{1}{w}$ at $w = L^{-1}(u) = e^u$.

In Example 9 our knowledge of the functions involved allowed us to write a formula for the derivative that was very simple. Most notably, it avoids the difficulty of expressing the derivative of f^{-1} at u in terms of $f^{-1}(u)$. The occurrence of f^{-1} in the formula is concealed.

There are other examples of this phenomenon for familiar functions. One of them we have already seen.

EXAMPLE 10   Let n be a positive integer, and define $R(x) = x^{1/n}$ for $x > 0$. Then $R(x) = f^{-1}(x)$, where $f(x) = x^n$.

We can calculate the derivative of R:

$$R'(x) = \frac{1}{f'(R(x))} = \frac{1}{n(R(x))^{n-1}} = \frac{1}{n(x^{1/n})^{n-1}}$$

$$= \frac{1}{nx^{\frac{n-1}{n}}} = \frac{1}{n}x^{\frac{1-n}{n}} = \frac{1}{n}x^{1/n-1}.$$

Once again the derivative $R'(x)$ is expressed in terms of x. This was possible because we have an explicit formula for $R(x)$.

Next we want to find the derivatives of the inverse trigonometric functions.

EXAMPLE 11   Find the derivative of $\arctan x$.

Solution   By Theorem 3.8.4, we can differentiate

$$\frac{d}{dx}(\arctan x) = \frac{1}{\tan'(\arctan x)} = \frac{1}{\sec^2(\arctan x)}.$$

In a sense this is the answer, but we can do better. Recall the trigonometric identity $\sec^2 t = 1 + \tan^2 t$ (Equation 1.7.3). So the denominator of the derivative above is

> Note that $\tan(\arctan x) = x$, so $\tan^2(\arctan x) = x^2$.

$\sec^2(\arctan x) = 1 + \tan^2(\arctan x) = 1 + x^2$. In other words,

$$\frac{d}{dx}(\arctan x) = \frac{1}{1 + x^2}.$$

The other inverse trigonometric functions can also be differentiated.

THEOREM 3.8.5

Derivatives of Inverse Trigonometric Functions

The standard inverse trigonometric functions (cf. Definition 3.8.3) are differentiable except at the points that are excluded by Theorem 3.8.4 (i.e., f^{-1} is *not* differentiable at u if $f'(f^{-1}(u)) = 0$).

Moreover, their derivatives are as follows:

$$\arcsin'(u) = \frac{1}{\sqrt{1-u^2}}, \qquad \arccos'(u) = -\frac{1}{\sqrt{1-u^2}},$$

$$\arctan'(u) = \frac{1}{1+u^2}, \qquad \operatorname{arccot}'(u) = -\frac{1}{1+u^2},$$

$$\operatorname{arcsec}'(u) = \frac{1}{u\sqrt{u^2-1}}, \qquad \operatorname{arccsc}'(u) = -\frac{1}{u\sqrt{u^2-1}}.$$

PROOF The differentiability of the inverse trigonometric functions as stated above is a direct consequence of Theorem 3.8.4. We have already established the formula for the derivative of arctan u in Example 11.

Now consider the function $\arcsin(u)$. By Theorem 3.8.4 its derivative is

$$\arcsin'(u) = \frac{1}{\cos(\arcsin u)},$$

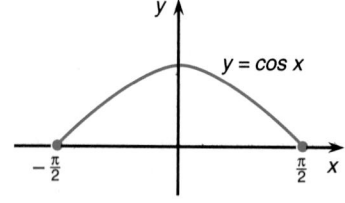

Figure 3.8.7

whenever this formula makes sense. From Definition 3.8.3 we know that the domain of arcsin u is $[-1, 1]$ and that $\arcsin u \in \left[-\frac{\pi}{2}, \frac{\pi}{2}\right]$. Looking at the graph of $\cos x$ for $x \in \left[-\frac{\pi}{2}, \frac{\pi}{2}\right]$, as shown in Figure 3.8.7, we notice that $\cos x \geq 0$ for all x in this interval.

We also know (from Theorem 1.7.2) that $\cos^2 x = 1 - \sin^2 x$, and the non-negativity of $\cos(\arcsin u)$ implies that

$$\cos(\arcsin u) = \sqrt{1 - \sin^2(\arcsin u)} = \sqrt{1 - u^2}.$$

Substituting into the formula for the derivative, we find that

$$\arcsin'(u) = \frac{1}{\sqrt{1 - u^2}},$$

as desired.

Now consider the function arccsc u. Its derivative is

$$\operatorname{arccsc}'(u) = \frac{1}{\csc'(\operatorname{arccsc}(u))}.$$

Now $\csc'(x) = -\csc(x)\cot(x)$, so this time the problem is to express

$$-\csc(\operatorname{arccsc}(u)) \cot(\operatorname{arccsc}(u))$$

in terms of u. The first factor, $-\csc(\operatorname{arccsc}(u))$, is of course just $-u$. For the second, $\cot(\operatorname{arccsc}(u))$, we need to use a trigonometric identity.

If you do not happen to remember any identity involving cot x, the thing to do is write the standard identity $\cos^2 x + \sin^2 x = 1$ and recognize that to get something involving $\cot x = \frac{\cos x}{\sin x}$, we should divide by $\sin^2 x$. What we find is

$$\cot^2 x + 1 = \frac{1}{\sin^2 x} = \csc^2 x,$$

so

$$\cot^2 x = \csc^2 x - 1.$$

Again we consult Definition 3.8.3 and find that $\text{arccsc}(u) \in \left(-\pi, -\frac{\pi}{2}\right] \cup \left(0, \frac{\pi}{2}\right]$. The important thing about these intervals is that $\cot x$ is nonnegative on both of them. In particular, this means that $\cot x = \sqrt{\csc^2 x - 1}$, so

$$\cot\big(\text{arccsc}(u)\big) = \sqrt{\csc^2\big(\text{arccsc}(u)\big) - 1} = \sqrt{u^2 - 1}.$$

Substituting this into the formula for the derivative, we find that

$$\text{arccsc}'(u) = -\frac{1}{u\sqrt{u^2 - 1}},$$

as desired.

The formulas for the derivatives of the remaining inverse trigonometric functions are derived in a similar way, being careful about signs, and are left for the exercises.

The calculation we have just done for the derivative of $\text{arccsc}\, x$ relied on the fact that $\cot x \geq 0$ whenever x is in the range of arccsc. When we defined arccsc and arcsec in Definition 3.8.3, we made what might have seemed to be slightly unexpected choices of intervals, but it was done to make these calculations work out well. It is possible to use other choices, but then the derivatives of arccsc and arcsec will be more complicated.

EXAMPLE 12   Find the derivative of $f(x) = \arctan(2x^2 - x)$.

Solution   Using the Chain Rule, we find that

$$f'(x) = \arctan'(2x^2 - x)(4x - 1) = \frac{1}{1 + (2x^2 - x)^2}(4x - 1) = \frac{4x - 1}{1 + (2x^2 - x)^2}$$

$$= \frac{4x - 1}{4x^4 - 4x^3 + x^2 + 1}.$$

EXAMPLE 13   Find all points at which the tangent line to $y = \arcsin\left(\frac{1}{2+x^2}\right)$ is horizontal.

Solution   The derivative is

$$\frac{dy}{dx} = \arcsin'\left(\frac{1}{2 + x^2}\right)\frac{d}{dx}\left(\frac{1}{2 + x^2}\right) = \frac{1}{\sqrt{1 - \left(\frac{1}{2+x^2}\right)^2}}\frac{0(2 + x^2) - 1(2x)}{(2 + x^2)^2}$$

$$= \frac{1}{\sqrt{1 - \left(\frac{1}{2+x^2}\right)^2}}\frac{-2x}{(2 + x^2)^2}.$$

The first factor is never zero, and the second is zero precisely when $x = 0$. So the answer to the problem is that the only horizontal tangent is at $x = 0$.

EXAMPLE 14 Suppose an observer is watching an aircraft as it approaches. The aircraft is flying directly toward the observer at an altitude of 2 miles. Its speed is 250 mph. As the aircraft approaches, its angle of elevation (the angle between the ground and the observer's line of sight to the aircraft) will change.

(i) How fast is this angle changing when the aircraft is a horizontal distance of 4 miles from the observer?

(ii) Suppose that at the moment when the aircraft is a horizontal distance of 2 miles from the observer, the angle of elevation is increasing at a rate of 1 degree per second. How fast is the aircraft flying (assuming that its altitude is still 2 miles)?

Solution

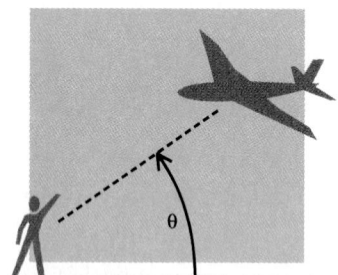

Figure 3.8.8(i)

(i) We begin by drawing a diagram in Figure 3.8.8(i), marking the angle of elevation as θ. If we let x represent the (changing) horizontal distance from the observer to the aircraft, then from Figure 3.8.8(ii) we find that $\tan \theta = \frac{2}{x}$. This amounts to the same thing as saying that $\theta = \arctan \frac{2}{x}$.

What we want to know is the rate of change of θ. It is easily calculated:

$$\frac{d}{dt}(\theta) = \frac{d}{dt}\left(\arctan \frac{2}{x}\right) = \frac{1}{1 + \frac{4}{x^2}}\left(-\frac{2}{x^2}\right)\frac{dx}{dt} = -\frac{2}{x^2 + 4}\frac{dx}{dt}.$$

At the moment in question, $x = 4$, and since the aircraft is *approaching* the observer, $\frac{dx}{dt} = -250$. From this we find that $\frac{d\theta}{dt} = -\frac{2}{4^2+4}(-250) = 25$.

Since the speed was measured in miles per *hour*, this rate of change is measured in radians per hour. It would be more convenient to express it in degrees per second. We find that 25 radians per hour equals $25 \times \frac{180}{\pi} \approx 1432.4$ degrees per hour, which is $25 \times \frac{180}{\pi} \times \frac{1}{3600} \approx 0.3979°/\text{sec}$.

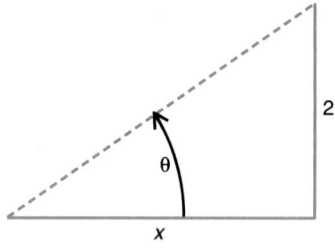

Figure 3.8.8(ii)

(ii) As before, we find that $\frac{d}{dt}(\theta) = -\frac{2}{x^2+4}\frac{dx}{dt}$, with x measured in miles and $\frac{d\theta}{dt}$ measured in radians per hour. Converting this to degrees per second, we see that the rate of change of θ is $-\left(\frac{2}{x^2+4}\frac{dx}{dt} \times \frac{180}{\pi} \times \frac{1}{3600}\right)°/\text{sec}$.

At the specified moment, $x = 2$, and this rate of change equals $1°/\text{sec}$; that is,

$$1 = -\frac{2}{2^2+4}\frac{dx}{dt}\frac{180}{\pi}\frac{1}{3600} = -\frac{360}{8\pi(3600)}\frac{dx}{dt}.$$

From this we find that $\frac{dx}{dt} = (-1)\frac{8\pi(3600)}{360} \approx -251.3$.

The aircraft is approaching the observer at a speed of about 251.3 mph.

Existence of Inverse Functions

One problem that we have largely avoided is deciding whether a function $f(x)$ has an inverse function, that is, whether it is one-to-one. Generally speaking, this is difficult to do, but there is a useful criterion that applies to many of the situations that interest us.

The idea is something we used in Examples 4 and 5. There we noted that if a function was increasing on an interval $[a, b]$, then it could not "double back" and take the same value twice, so it must be one-to-one on $[a, b]$. We are familiar enough with the graphs of the trigonometric functions to be able to see where they are increasing, so we were able to find intervals on which they have inverses. Of course it also works for a function that is always decreasing on some interval; in order *not* to be one-to-one on an interval a function must both increase and decrease there.

To apply this idea to functions that are less familiar, we need a way to decide when they are increasing or decreasing. In the discussion before Example 3.3.1, we noticed that the derivative $f'(a)$, being the slope of the tangent to $y = f(x)$ at $x = a$, gives an indication of the behavior of the function. We expect that when $f'(a) > 0$, the function $f(x)$ should be increasing, at least for values of x very close to $x = a$ (see Figure 3.8.9), and when $f'(a) < 0$, then $f(x)$ should be decreasing near $x = a$ (see Figure 3.8.10). In particular, if $f'(x) > 0$ for all x in some interval, then $f(x)$ should

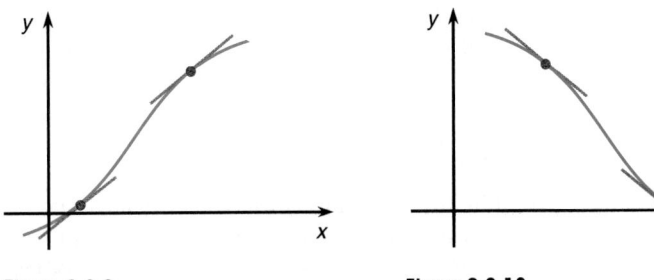

Figure 3.8.9   **Figure 3.8.10**

be increasing on that interval. Similarly, if $f'(x) < 0$ for every x in some interval, then we expect that the function $f(x)$ is decreasing on that interval.

Now suppose that $f'(x) > 0$ for every $x \in (a, b)$. Then $f(x)$ should be increasing on the interval (a, b) and must therefore be one-to-one on (a, b). Accordingly, it has an inverse function f^{-1}.

The value of this criterion is that it gives us something to calculate, namely, the derivative, and if we are able to show the derivative is always positive, then it tells us there must be an inverse function. It does not depend on our knowledge of the function, apart from being able to find and work with its derivative.

Strictly speaking, we have not really yet developed the techniques for discussing when a function is increasing or decreasing (this will be done in the next chapter), but the ideas suggested in the discussion above are in fact correct. We summarize the result as a theorem.

THEOREM 3.8.6

If $f(x)$ is continuous on $[a, b]$, differentiable on (a, b), and $f'(x) > 0$ for every $x \in (a, b)$, then $f(x)$ is one-to-one on $[a, b]$. The inverse function $f^{-1}(u)$ is defined for $u \in \big[f(a), f(b)\big]$, and is differentiable on $\big(f(a), f(b)\big)$, and

$$(f^{-1})'\big(f(x)\big) = \frac{1}{f'(x)},$$

that is,

$$(f^{-1})'(u) = \frac{1}{f'\big(f^{-1}(u)\big)}.$$

The theorem is also true under the condition that $f'(x) < 0$ for every $x \in (a, b)$, except that then $f(a) > f(b)$, so f^{-1} is defined on $\big[f(b), f(a)\big]$ and differentiable on $\big(f(b), f(a)\big)$. The formula for $(f^{-1})'$ remains the same.

There is also a version of the theorem for open intervals, replacing square brackets with round brackets (i.e., with $f(x)$ continuous and differentiable on (a, b) and $f'(x) > 0$ there), except that we cannot then speak of $f(a)$ and $f(b)$. In this case the theorem says simply that there is an inverse function, f^{-1} and that it is defined and differentiable on the set of all values taken by f on (a, b), that is, on $\{y : y = f(x)$ for some $x \in (a, b)\}$.

Comparable remarks apply to half-open intervals (e.g., if $f(x)$ is continuous on $[a, b)$ and differentiable on (a, b) or continuous on $(a, b]$ and differentiable on (a, b) and with $f'(x) > 0$ for all $x \in (a, b)$, then there is an inverse function defined on the set of values taken on by $f(x)$ for $x \in [a, b)$ or $x \in (a, b]$, respectively, and differentiable on $\{y : y = f(x)$ for some $x \in (a, b)\}$).

The theorem is also true with $a = -\infty$ and/or $b = \infty$, though of course then the interval must be open at that end.

Finally, all of these versions still hold true if $f'(x) < 0$ on the interval (a, b), and for all of them the formula for the derivative of f^{-1} is the same.

EXAMPLE 15

(i) The function $\tan x$ has as its derivative $\sec^2 x$, which of course is positive whenever it is defined. In particular, on the interval $\left(-\frac{\pi}{2}, \frac{\pi}{2}\right)$ the derivative is always positive, and the open interval version of the above theorem (Theorem 3.8.6) asserts that the function arctan exists.

(ii) The derivative of $\sin x$ on $\left[-\frac{\pi}{2}, \frac{\pi}{2}\right]$ is $\cos x$, which is positive on the corresponding open interval $\left(-\frac{\pi}{2}, \frac{\pi}{2}\right)$. This implies the existence of the function arcsin.

(iii) The half-open interval versions of Theorem 3.8.6 show that sec is one-to-one on $\left[0, \frac{\pi}{2}\right)$ (since the derivative, $\sec x \tan x$, is positive on $\left(0, \frac{\pi}{2}\right)$) and similarly that it is one-to-one on $\left[\pi, \frac{3\pi}{2}\right)$ (since the derivative is negative there).

To show that it is one-to-one on the union of these two intervals $\left[0, \frac{\pi}{2}\right) \cup \left[\pi, \frac{3\pi}{2}\right)$, we remark that the value of $\sec x$ is positive when $x \in \left[0, \frac{\pi}{2}\right)$ and negative when $x \in \left[\pi, \frac{3\pi}{2}\right)$. Because of this, there is no possibility that $\sec x$ could have the same value for some $x \in \left[0, \frac{\pi}{2}\right)$ as it does for some $x \in \left[\pi, \frac{3\pi}{2}\right)$, so it must be one-to-one on the union of the two intervals.

So Theorem 3.8.6 asserts the existence of the inverse function arcsec u on each of the two intervals, and we can combine them to define arcsec u on the union of the two intervals.

(iv) In Example 8 we discussed the function $f(x) = x^5 + 2x + 1$. Its derivative, $f'(x) = 5x^4 + 2$, is clearly positive on the whole real line, so Theorem 3.8.6 says that $f(x)$ is one-to-one and has an inverse function. When we did Example 8, we had to assume that this was true.

Inverse Hyperbolic Functions

It is also possible to find inverses for the hyperbolic functions (Definition 3.5.11), though, as with the trigonometric functions, it is necessary to restrict the domains of some of them. To begin with, we consider the graphs of the hyperbolic functions in Figure 3.8.11.

From the pictures we see that $\cosh x$ is increasing on $[0, \infty)$ and that $\sinh x$ and $\tanh x$ are increasing on the whole real line. So $\cosh x$ is one-to-one on the positive half-line $[0, \infty)$, and $\sinh x$ and $\tanh x$ are both one-to-one on the entire real line.

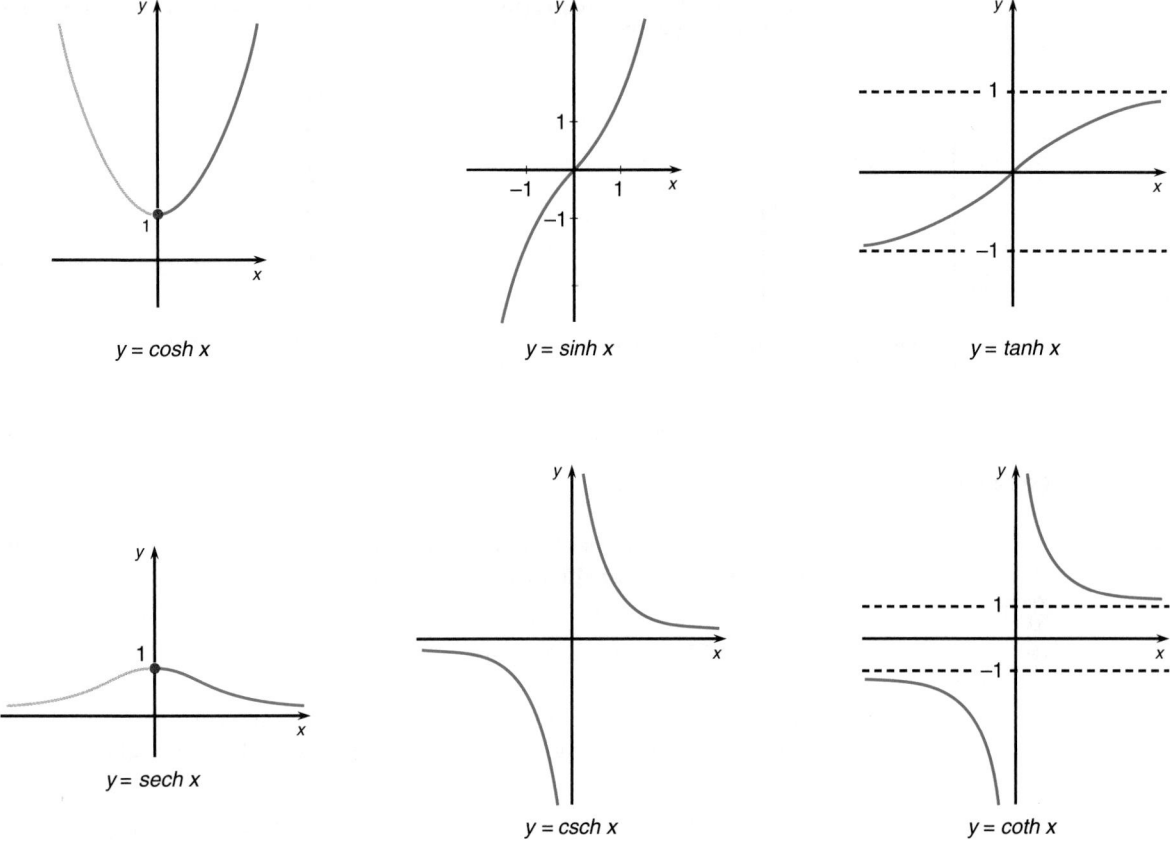

Figure 3.8.11

Also sech x is decreasing on $[0, \infty)$ and one-to-one there, while csch x and coth x are both decreasing on $(-\infty, 0)$ and on $(0, \infty)$ and are one-to-one on the union $(-\infty, 0) \cup (0, \infty)$.

It is possible to check these observations by using the criterion given in Theorem 3.8.6, and this allows us to define the inverse hyperbolic functions.

DEFINITION 3.8.7

The **standard inverse functions of the hyperbolic functions** are defined as follows:

For $u \in [1, \infty)$, arccosh $u \in [0, \infty)$.

For $u \in \mathbb{R}$, arcsinh $u \in \mathbb{R}$, that is, takes all values.

For u with $|u| < 1$, arctanh $u \in \mathbb{R}$, that is, takes all values.

For u with $|u| > 1$, arccoth $u \in (-\infty, 0) \cup (0, \infty)$.

For $u \in (0, 1]$, arcsech $u \in [0, \infty)$.

For $u \in (-\infty, 0) \cup (0, \infty)$, arccsch $u \in (-\infty, 0) \cup (0, \infty)$.

Moreover, if $w = $ arcsinh u, then $u = $ sinh w; if $w = $ arccosh u, then $u = $ cosh w; if $w = $ arctanh u, then $u = $ tanh w; if $w = $ arccoth u, then $u = $ coth w; if $w = $ arcsech u, then $u = $ sech w; if $w = $ arccsch u, then $u = $ csch w.

EXAMPLE 16   The graphs of cosh x, arccosh u, tanh x, and arctanh u are sketched in Figure 3.8.12.

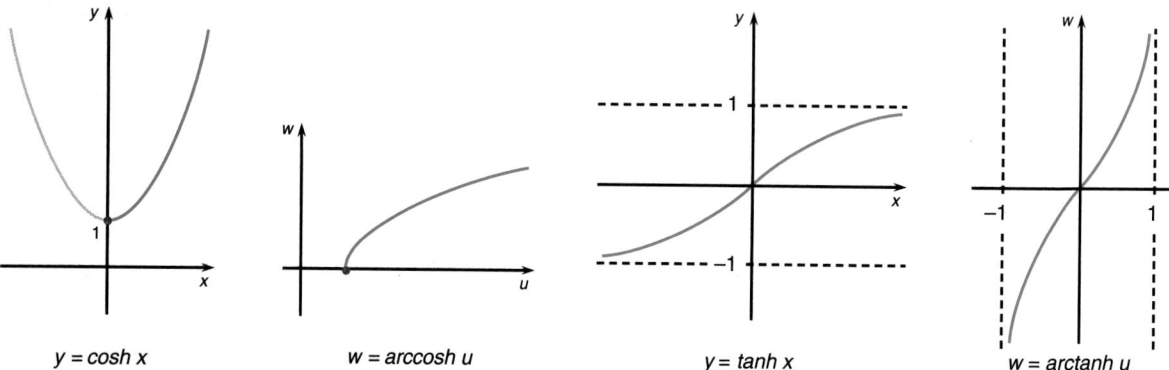

$y = \cosh x$ $w = \text{arccosh } u$ $y = \tanh x$ $w = \text{arctanh } u$

Figure 3.8.12

The derivatives of the inverse hyperbolic functions can be calculated in much the way we did for the inverse trigonometric functions.

EXAMPLE 17   Find the derivative of arctanh u.

Solution   From Theorem 3.5.13 we know that $\tanh'(u) = \text{sech}^2 u$. Noting that $\text{sech } u \neq 0$ for every u, we apply Theorem 3.8.4 to see that $\text{arctanh}'(u) = \frac{1}{\text{sech}^2(\text{arctanh}(u))}$.

Recalling the identity $\cosh^2 t - \sinh^2 t = 1$ from Equation 3.5.12, we divide both sides by $\cosh^2 t$ to get that $1 - \tanh^2 t = \text{sech}^2 t$. So

$$\text{arctanh }'(u) = \frac{1}{\text{sech}^2\big(\text{arctanh }(u)\big)}$$

$$= \frac{1}{1 - \tanh^2\big(\text{arctanh}(u)\big)} = \frac{1}{1 - u^2}.$$

The other inverse hyperbolic functions can also all be differentiated.

THEOREM 3.8.8

Derivatives of Inverse Hyperbolic Functions

The inverse hyperbolic functions defined in Definition 3.8.7 are differentiable except at the points that are excluded by Theorem 3.8.4. In particular, arccosh u is not differentiable at $u = 1$, and arcsech u is not differentiable at $u = 1$. Apart from these exceptions, the inverse hyperbolic functions are differentiable everywhere they are defined.

Moreover, their derivatives are as follows:

$$\text{arcsinh}'(u) = \frac{1}{\sqrt{1 + u^2}},$$

$$\text{arccosh}'(u) = \frac{1}{\sqrt{u^2 - 1}}, \quad u > 1,$$

$$\operatorname{arctanh}'(u) = \frac{1}{1-u^2}, \quad |u| < 1,$$

$$\operatorname{arccoth}'(u) = \frac{1}{1-u^2}, \quad |u| > 1,$$

$$\operatorname{arcsech}'(u) = -\frac{1}{u\sqrt{1-u^2}}, \quad 0 < u < 1,$$

$$\operatorname{arccsch}'(u) = -\frac{1}{|u|\sqrt{1+u^2}}, \quad u \neq 0.$$

PROOF The differentiability follows from Theorem 3.8.4. We have already derived the formula for the derivative of $\operatorname{arctanh} u$ in Example 17. The other formulas are established in a similar way (cf. the proof of Theorem 3.8.5) and are left for the exercises. ◢

In truth the inverse hyperbolic functions and their derivatives are only rarely used. When they are, it is usually cosh, sinh, tanh, and their inverse functions that appear; the other three are almost never seen. The formulas for their derivatives appear here for the sake of completeness and as examples. It is worthwhile to try to work some of them out for yourself (see the exercises), but they are not worth committing to memory.

EXAMPLE 18 Evaluate the derivative of $\operatorname{arcsinh}(t^2 + 1)$.

Solution Using the formula from Theorem 3.8.8 and the Chain Rule, we see that

$$\frac{d}{dt}\Big(\operatorname{arcsinh}(t^2 + 1)\Big) = \frac{1}{\sqrt{1 + (t^2 + 1)^2}}(2t) = \frac{2t}{\sqrt{t^4 + 2t^2 + 2}}.$$ ◢

Exercises 3.8

I

Find the derivative of each of the following functions.

1. $\arctan(2x + 4)$
2. $\arcsin\left(\frac{1}{2}\sin x\right)$
3. $\arccos\left(\frac{1}{1+x^2}\right)$
4. $\arctan(\tan x)$
5. $\operatorname{arcsec}(4 + x^2)$
6. $\operatorname{arccot}(\sin x)$
7. $\operatorname{arccsc}(5 + e^x)$
8. $\ln(2 + \arctan^2 x)$
9. $\exp\big(\operatorname{arccot}(2x)\big)$
10. $\cos(\arctan x)$
11. $\operatorname{arccosh}(\sec^2 x)$
12. $\operatorname{arcsinh}(\tan x)$
13. $\operatorname{arccoth}(x^2 + 2x + 3)$
14. $\operatorname{arctanh}\left(\frac{1}{2}\sin x\right)$
15. $\ln\big(\operatorname{arccosh}(x)\big)$
16. $\operatorname{arctanh}^2\left(\frac{1}{x^2+3}\right)$
17. $\arcsin(\tanh x)$
18. $\operatorname{arcsech}\big(1 - \tanh^2(x)\big)$
19. $\operatorname{arctanh}\big(\operatorname{sech}(x)\big)$
20. $\operatorname{arcsinh}(\tanh x)$

II

For each of the following functions defined on the specified interval, show that the function has an inverse, and calculate the derivative of the inverse function.

21. $g(x) = 5x - 1$, on $[1, 9]$
22. $f(x) = x^2 - 2x + 3$, on $[0, 1]$
23. $h(x) = -2x^2 + 4x - 1$, on $(-\infty, 1]$
24. $F(x) = x^3 - 12x + 1$, on $[-2, 2]$
25. $H(x) = x^3 + 6x^2 + 4$, on $[-2, -1]$
26. $k(x) = \ln(1 + x^2)$, on $[0, \infty)$

Find the formula for the derivative of each of the following functions.

*27. $\operatorname{arccot} u$

*28. (i) $\arccos u$, (ii) $\operatorname{arcsec} u$

*29. (i) arccoth u, (ii) arcsinh u

*30. (i) arccosh u, (ii) arcsech u, (iii) arccsch u

*31. (i) What is the inverse of arcsin x? What is its domain?

(ii) What is the inverse of $\arctan(2x+1)$?

*32. (i) If we know that $\frac{d}{dx}(\ln x) = \frac{1}{x}$ and that e^u is the inverse of $\ln x$, show how to use these facts to conclude that $\frac{d}{du}(e^u) = e^u$.

(ii) If we know that $\frac{d}{dx}(e^x) = e^x$ and that $\ln u$ is the inverse of e^x, show how to use these facts to conclude that $\frac{d}{du}(\ln u) = \frac{1}{u}$.

*33. (i) Suppose $f(x)$ and $g(t)$ are both one-to-one. If the composite function $g\big(f(x)\big)$ is defined, show that it is also one-to-one.

(ii) In the situation of part (i), find a formula for the inverse function of $g\big(f(x)\big)$ in terms of the inverses of f and g.

*34. Find examples of functions $f(x)$ and $g(t)$ for which

(i) $g\big(f(x)\big)$ is one-to-one even though $g(t)$ is not.

(ii) g is one-to-one but $g\big(f(x)\big)$ is not.

(iii) f is one-to-one but $g\big(f(x)\big)$ is not.

*35. Suppose $-\frac{3\pi}{2} < x < -\frac{\pi}{2}$. What is the function $F(x) = \arctan(\tan x)$? (*Hint:* It does not equal the function $f(x) = x$ on $\left(-\frac{3\pi}{2}, -\frac{\pi}{2}\right)$.)

*36. Suppose the aircraft of Example 14 flies past again at an altitude of 1 mile. If the angle of elevation is increasing at 1 degree per second when the aircraft is a horizontal distance of 3 miles from the observer, how fast is the aircraft flying?

37. After the aircraft of Example 14 has gone 3 miles past the observer, its angle of elevation is decreasing at the rate of $\frac{4}{5}°$/sec. How fast is it flying?

*38. Verify that $\csc x$ is one-to-one on $\left[-\frac{\pi}{2}, 0\right) \cup \left(0, \frac{\pi}{2}\right]$. For this exercise *only*, let Arccsc x be the corresponding inverse function. Show that its derivative is $\frac{d}{dx}(\text{Arccsc } x) = -\frac{1}{|x|\sqrt{x^2-1}}$.

39. If we weigh an object in pounds, we might want to convert its weight to kilograms or vice versa. Let $K(x)$ be the function which converts a weight x lb to the corresponding value in kilograms.

Let $P(x)$ be the function that converts a weight x kg to the corresponding value in pounds.

(i) Find formulas for $K(x)$ and $P(x)$. (Note that 1 kg equals 2.2 lb.)

(ii) Show that $P(x)$ and $K(x)$ are inverse functions.

(iii) Find the derivatives $P'(x)$ and $K'(x)$, and confirm that they satisfy the expected relation for derivatives of inverse functions.

40. If we measure a temperature on the Fahrenheit scale, we might want to convert it to the Celsius scale or vice versa. Let $F(t)$ be the function that converts a temperature $t°$C to the corresponding value on the Fahrenheit scale.

Let $C(t)$ be the function which converts a temperature $t°$F to the corresponding value on the Celsius scale.

(i) Find formulas for $F(t)$ and $C(t)$. (Use the facts that the freezing point of water is $0°$C and $32°$F and that the boiling point of water is $100°$C and $212°$F; figure out how many Fahrenheit degrees correspond to one Celsius degree...)

(ii) Show that $F(t)$ and $C(t)$ are inverse functions.

(iii) Find the derivatives $F'(t)$ and $C'(t)$, and confirm that they satisfy the expected relation for derivatives of inverse functions.

41. Use a graphics calculator or computer to graph $y = \arctan x$ and $y = \frac{1}{\tan x} = \tan^{-1} x = \cot x$ on the same picture, and convince yourself once and for all that these are very different functions.

42. Graph $y = \arcsin x$ and $y = \frac{1}{\sin x} = \sin^{-1} x = \csc x$ on the same picture.

43. Graph $y = \ln x$ and $y = \frac{1}{e^x} = e^{-x}$ on the same picture.

44. Graph $y = \tan x$ at various magnifications and find, to four decimal places, the value of arctan 2, that is, the point where $\tan x$ crosses the line $y = 2$.

45. Use the graph of $y = \sin x$ to find $\arcsin \frac{1}{2}$ to four decimal places.

46. Graph the function $y = x^5 + 2x + 1$ of Example 8 (p. 208) and confirm from the picture that it is one-to-one (cf. Example 15(iv)).

3.9 Second and Higher-Order Derivatives

DERIVATIVE OF ORDER k

$f^{(k)}(x)$

ACCELERATION

If we start with a differentiable function $f(x)$, then its derivative is a function $f'(x)$. We might want to differentiate this function; its derivative is conventionally denoted $f''(x)$ and is called the **second derivative** of $f(x)$.

EXAMPLE 1

(i) If $f(x) = x^5 + 3x^4 - 2x^3 + x^2 - 7$, then its derivative is $f'(x) = 5x^4 + 12x^3 - 6x^2 + 2x$. The second derivative is $f''(x) = 20x^3 + 36x^2 - 12x + 2$.

(ii) If $g(t) = \ln(t^2 + 4)$, then

$$g'(t) = \frac{1}{t^2 + 4}(2t) = \frac{2t}{t^2 + 4}.$$

Also

$$g''(t) = \frac{d}{dt}\left(\frac{2t}{t^2 + 4}\right) = \frac{2(t^2 + 4) - (2t)(2t)}{(t^2 + 4)^2} = \frac{8 - 2t^2}{(t^2 + 4)^2}.$$

If we want to use the other notation for derivatives, we can write

$$f''(x) = \frac{d}{dx}\left(\frac{df}{dx}\right),$$

but a more convenient notation is the following:

$$f''(x) = \frac{d^2 f}{dx^2} = \frac{d^2}{dx^2}\big(f(x)\big).$$

The expression in the middle of the above equation is read "d two f by d x squared," *not* "d squared f over d x squared." The expression on the right side is read "d two by d x squared of $f(x)$."

Of course, we might continue taking derivatives. We could write $f'''(x)$ for the third derivative, and so on, but before long this will get confusing. Instead, we usually write $f^{(n)}(x)$ for the nth derivative of $f(x)$, where n could be 3, 4, 5, etc., and occasionally we even use this notation for $n = 2$ or $n = 1$ (i.e., $f^{(2)}(x) = f''(x)$, and $f^{(1)}(x) = f'(x)$). The alternative notation for the **derivative of order n** is

$$f^{(n)}(x) = \frac{d^n f}{dx^n} = \frac{d^n}{dx^n}\big(f(x)\big).$$

EXAMPLE 2

(i) If we return to Example 1(i), where $f(x) = x^5 + 3x^4 - 2x^3 + x^2 - 7$, $f'(x) = 5x^4 + 12x^3 - 6x^2 + 2x$, and $f''(x) = 20x^3 + 36x^2 - 12x + 2$, then we can continue differentiating and find $f^{(3)}(x) = \frac{d}{dx}\big(f''(x)\big) = 60x^2 + 72x - 12$.

Similarly, $f^{(4)}(x) = 120x + 72$, $f^{(5)}(x) = 120$, and $f^{(6)}(x) = 0$.

Subsequent derivatives are all zero, that is, $f^{(n)}(x) = 0$ for all $n \geq 6$.

(ii) Find $G^{(4)}(t)$ if $G(t) = \ln t$.

Solution $G'(t) = \frac{d}{dt}(\ln t) = \frac{1}{t}$. Writing this as t^{-1} (by which we mean the -1 power, *not* the inverse function), we see that its derivative is $G''(t) = -t^{-2}$. Differentiating again, we find that $G^{(3)}(t) = -(-2)t^{-3} = 2t^{-3}$. Finally, $G^{(4)}(t) = 2(-3)t^{-4} = -6t^{-4}$.

Part of the usefulness of second derivatives comes from their physical interpretation. In Sections 2.1 and 3.1 we discussed the idea of velocity. Recall that if an object is moving along a straight line in such a way that its displacement from the origin at time t is given by some function $s(t)$, then the derivative $s'(t)$ is the rate of change of $s(t)$, which is what we call the velocity.

Now consider the second derivative $s''(t)$. It is the rate of change of the velocity $s'(t)$, and this is what is known in physics as the acceleration. After all, if a car is moving along at a constant velocity, we say it is not accelerating, but if the velocity is increasing, we say the car is accelerating.

In the discussion before Example 2.1.3, it was mentioned that if an object held A ft above the ground is released and allowed to fall, its height above the ground t sec after release will be $h(t) = A - 16t^2$. Differentiating, we find that its velocity is $h'(t) = -32t$ ft/sec. This means that the velocity is changing; it starts, when $t = 0$, with the velocity $h'(0) = -32(0) = 0$ (i.e., at the moment of release it is not moving), but after 1 second it is falling at the rate of 32 ft/sec, etc. Notice that the velocity is negative, reflecting the fact that the object is *falling*, that is, its height is decreasing.

Since the velocity is changing, we expect the acceleration to be nonzero. In fact, the acceleration is $h''(t) = \frac{d}{dt}\big(h'(t)\big) = \frac{d}{dt}(-32t) = -32$ ft/sec^2.

The acceleration is constant. This means that each second the velocity changes by the same amount, in this case -32 ft/sec. So from $h'(0) = 0$ ft/sec, it changes after 1 sec to $h'(1) = -32$ ft/sec, and after another second to $h'(2) = -64$ ft/sec, etc. The fact that the acceleration is negative simply means that the velocity is decreasing (becoming more negative), since the acceleration is the rate of change of the velocity.

This represents a fundamental fact in practical physics. Near the surface of the earth a falling object will accelerate due to the force of gravity, and its acceleration will be approximately

$$- 32 \text{ ft/sec}^2$$

(which means 32 ft/sec^2 downward). This is the same thing as

$$- 9.81 \approx -9.8 \text{ m/sec}^2.$$

Of course it is not quite so simple as that. Air resistance will have the effect of slowing down the object's fall, especially for very light objects (a feather does not accelerate nearly as fast as -32 ft/sec^2 ...). Still, it is quite accurate for reasonably heavy objects or, more important, for objects whose surface area is relatively small in comparison with their weight, in which case the air resistance will be insignificant relative to the force of gravity.

Notice that the units in which acceleration is measured are ft/sec^2 or m/sec^2, which can also be written as ft/sec/sec and m/sec/sec, respectively. This is because the acceleration is the rate of change of the velocity, so it tells us, for instance, by how many ft/sec the velocity is changing in one second. To say the acceleration is -32 ft/sec^2 means that with each passing second the velocity changes by -32 ft/sec. The rate of change of the velocity is -32 ft/sec *per second*, or -32 ft/sec/sec, or -32 ft/sec^2.

EXAMPLE 3   Suppose an object is thrown vertically so that it starts at a height A ft above the ground. Suppose moreover that it starts with a velocity B ft/sec. See Figure 3.9.1.

Let $h(t)$ represent the height of the object above the ground at a moment t sec after it is thrown. We know that $h(0) = A$ and that $h'(0) = B$. From the preceding discussion we know that $h''(x) = -32$.

From this we would like to find a formula for $h(t)$. However, rather than try to do that right away, let us ask first for a formula for $h'(t)$. What we need is a function whose derivative is the constant function -32.

One function that obviously satisfies that requirement is $-32t$; its derivative is certainly -32. We want $h'(t)$ to be a function whose derivative is -32 and whose value

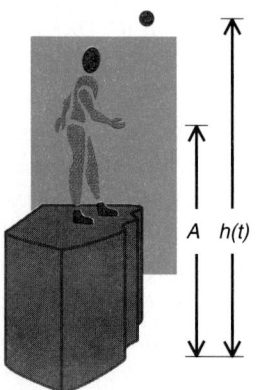

Figure 3.9.1

at $t = 0$ is B. To achieve this, we could let $h'(t) = -32t + B$, which clearly satisfies both conditions. Notice that adding the constant B does not change the derivative.

In fact, as we shall see later, the only functions whose derivative equals -32 are the functions $-32t + c$, where c is some constant. Taking this for granted, the only one of these functions whose value at $t = 0$ is B is

$$h'(t) = -32t + B.$$

The next step is to ask what are the possibilities for $h(t)$? After all, we want $h'(t) = -32t + B$. There is one function that obviously satisfies that condition, namely, $-16t^2 + Bt$. On the other hand, as we have already remarked, adding a constant would not change the derivative, so we should really consider functions of the form $-16t^2 + Bt + C$, where C is any constant.

Now, taking into account the additional condition that $h(0) = A$, we find that if $h(t) = -16t^2 + Bt + C$, then $h(0) = C$, so C must be A. In other words, we have found that the height above the ground at time t must be given by

$$h(t) = -16t^2 + Bt + A.$$

In summary, the height above the ground of an object that has been thrown into the air is

$$h(t) = -16t^2 + V_0 t + H_0,$$

where $V_0 = h'(0)$ is the initial velocity and $H_0 = h(0)$ is the initial height. Notice that if $V_0 > 0$, it means the object was thrown upward; if $V_0 < 0$, it means the object was thrown downward; and if $V_0 = 0$, it means the object started at rest, that is, it was *dropped* rather than thrown.

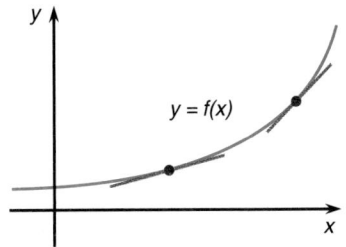

Figure 3.9.2

The other main use of the second derivative arises from its application to drawing graphs. After all, as the rate of change of the derivative of a function $f(x)$, it tells us how the derivative is changing, and specifically it tells us whether the derivative is increasing or decreasing.

For the curve $y = f(x)$, illustrated in Figure 3.9.2, the derivative is always positive, that is, the tangent line has positive slope, so it always slopes upward to the right. But we can say more than this. As we move from left to right along the curve, the tangent line slopes upward more steeply. This means that the slope of the tangent is larger and larger as the point of tangency moves from left to right. In terms of the derivative it means that as x increases, the derivative $f'(x)$ increases.

Now to say that $f'(x)$ is an increasing function suggests that its derivative is positive, that is, that $f''(t) > 0$, or at least that $f''(t) \geq 0$.

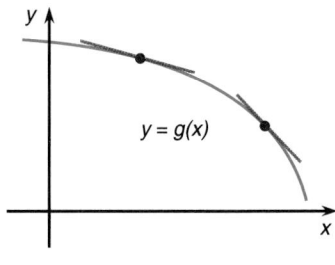

Figure 3.9.3

In Figure 3.9.3 the graph is decreasing, that is, $g'(x) \leq 0$. But as x increases, the graph slopes more steeply downward, that is, $g'(x)$ becomes more negative. This suggests that $g''(x) \leq 0$.

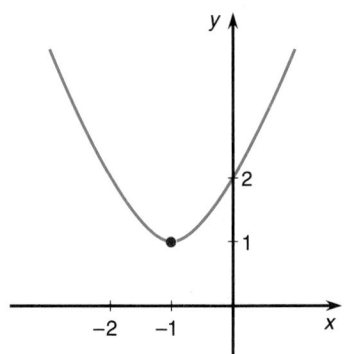

Figure 3.9.4

EXAMPLE 4   Consider the function $y = x^2 + 2x + 2$, whose graph is illustrated in Figure 3.9.4.

Its derivative is $\frac{dy}{dx} = 2x + 2$. We have already discussed in Section 3.3 that to look for maximum or minimum values of y, we should look for places where $y' = 0$. In this case the only such point is $x = -1$. Now consider the second derivative, $\frac{d^2y}{dx^2} = \frac{d}{dx}(2x + 2) = 2$.

The fact that the second derivative is positive means that the first derivative is increasing. Since it is zero when $x = -1$, this suggests that it should be negative to the left of $x = -1$ and positive to the right of $x = -1$. This in turn suggests that the graph is decreasing for points to the left of $x = -1$ and increasing at points to the right of $x = -1$. The only way this can happen is for $f(x)$ to have a *minimum* value at $x = -1$.   ◢

Of course, in this case it is obvious from the picture that this is so, but in fact it is often possible to use this sort of reasoning to decide whether $f(x)$ has a maximum or a minimum value at some particular point where $f'(x) = 0$.

EXAMPLE 5   Let $f(x) = x^3 - 3x + 2$ for $x \in [-2, 3]$. We find that $f'(x) = 3x^2 - 3 = 3(x^2 - 1)$, which is zero when $x = \pm 1$. Calculating that $f''(x) = 6x$, we see that $f''(-1) = -6$ and $f''(1) = 6$.

Near $x = 1$, this suggests that $f'(x)$ is increasing, and since it equals zero at $x = 1$, this means it must be negative to the left of $x = 1$ and positive to the right, so the graph decreases down to $x = 1$ and then increases again. This means that $f(x)$ should have a minimum at $x = 1$.

Similarly, at $x = -1$ the derivative is zero, but the second derivative is negative, so the derivative is decreasing. This suggests that the derivative is positive to the left of $x = -1$ and negative to the right, so the function increases up to its value at $x = -1$ and then decreases again. The function $f(x)$ should have a maximum value at $x = -1$.

The graph of $y = f(x)$ is illustrated in Figure 3.9.5. It is clear that $f(x)$ has a minimum at $x = 1$, but there is also an endpoint minimum at $x = -2$. More distressing is that the maximum value occurs at the endpoint $x = 3$, rather than where we expected it, at $x = -1$. The argument we gave suggested that the value of $f(x)$ at $x = -1$ should be bigger than its values at nearby points, and that is true, but it said nothing about what the function might do somewhere else. In this case it did indeed get larger at the right endpoint.   ◢

The use of second derivatives to find maximum and minimum values is fairly tricky. Specifically, it can only be used to describe what happens for points near the one in question. There is another difficulty in that it might also happen that $f''(x) = 0$, in which case the sort of argument we have been using simply does not apply.

Nonetheless, the second derivative is extremely useful as a tool for finding maximum and minimum values, and in Section 4.4 we will learn in greater detail how to use it. For now, these examples have been presented to give some indication of our purpose in studying second derivatives.

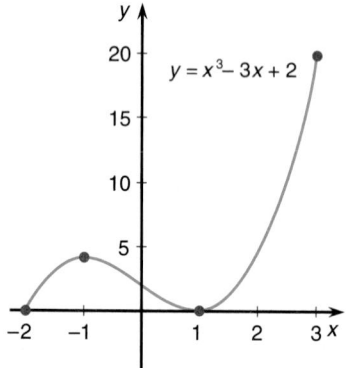

Figure 3.9.5

Exercises 3.9

Evaluate the following.

1. $\frac{d^2}{dx^2}(x^4 - 3x^3 + 2x^2 - x + 3)$

2. $\frac{d^2}{dx^2}(\tan x)$

3. $\frac{d^2}{dx^2}(e^{x^2+1})$

4. $\frac{d^2}{dx^2}(\cosh x)$

5. $\frac{d^3}{dx^3}(e^{7x+1})$

6. $\frac{d^8}{dx^8}(\sinh x)$

7. $\frac{d^3}{dx^3}(\sin x)$

8. $\frac{d^9}{dx^9}(\cos x)$

9. $\frac{d^7}{dx^7}(13x^5 + 7x^4 - 21x^3 + \sqrt{2}x^2 + 9x - 1)$

10. $\frac{d^3}{dx^3}(\sqrt{x})$

11. $\frac{d^2}{dx^2}\left(\frac{1}{x+1}\right)$

12. $\frac{d^2}{dx^2}\left(\frac{1}{x^2+1}\right)$

13. $\frac{d^3}{dt^3}\left(\frac{1}{3t+2}\right)$

14. $\frac{d^4}{dx^4}\left(\cos(2x)\right)$

15. $\frac{d^3}{dx^3}\left(\frac{1}{e^x+1}\right)$

16. $\frac{d^3}{dx^3}\left(\arctan x\right)$

17. $\frac{d^2}{dx^2}\left(\arcsin x\right)$

18. $\frac{d^3}{dx^3}\left(\ln(x^2+4)\right)$

19. $\frac{d^3}{dx^3}\left(\exp(\sin x)\right)$

20. $\frac{d^3}{dx^3}\left(\exp\left(\ln(x^2+4)\right)\right)$

II

21. If n and k are positive integers, find a formula for the kth derivative of x^n. (Your answer will likely distinguish the case in which $k \le n$ from the case in which $k > n$. You might also find it helpful to use the factorial notation (Definition 1.3.8); recall that $m! = m(m-1)(m-2) \times \ldots \times (3)(2)(1)$.)

If n and k are positive integers, find formulas for the following.

22. $\frac{d^k}{dx^k}\left(x^{-n}\right)$

23. $\frac{d^k}{dx^k}\left(\ln x\right)$

24. Show that $f(x) = x^2 - 2x - 6$ cannot have a maximum value, by finding all the points where $f'(x) = 0$ and evaluating the second derivative.

*25. Let $h(x) = 2x^6 - 15x^4 + 24x^2 + 3$.
   (i) Find all the points at which $h'(x) = 0$. (*Hint:* You should be able to factor $h'(x)$ just by looking at it for a while.)
   (ii) Evaluate $h''(x)$ at each of the points you found in part (i). What does this tell you about the behavior of $h(x)$ near each of those points? (It may tell you more for some of them than it does for others.)

*26. Let $g(t) = \sec t$ for $t \in \left(-\frac{\pi}{2}, \frac{\pi}{2}\right)$.
   (i) Find the first and second derivatives of $g(t)$.
   (ii) By considering the first and second derivatives at $t = 0$, show that $g(t)$ has a minimum value at $t = 0$.

27. An object is thrown straight up from ground level at a speed of 96 ft/sec.
   (i) How long after it is thrown will it land again?
   (ii) What is the maximum height it will reach?

28. (i) An object is dropped from a height of 100 ft. How long will it take to reach the ground?
   (ii) If instead the object was thrown down, with a downward speed of 20 ft/sec, how long would it take to fall?
   (iii) Compare the difference between the two times that are the answers to parts (i) and (ii) with the time it would take an object falling at the *constant* speed 20 ft/sec to fall a distance of 100 ft.

29. At what speed must an object be thrown straight up in order to reach a maximum height of 27 m?

*30. On the planet Vruggles an object falling near the planet's surface experiences an acceleration due to gravity of 6 boingles per snoom per snoom. (The boingle is the Vrugglian unit of distance; the snoom is the unit of time.)

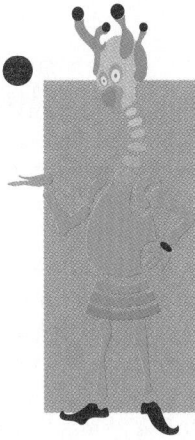

   (i) How many snooms will it take an object to fall 40 boingles if it is dropped from rest?
   (ii) How far will an object dropped from rest fall in 5 snooms?
   (iii) If a ball is thrown straight up from the surface of Vruggles at a speed of 14 boingles per snoom, what is the maximum height it will reach, and how long will it take to fall back to the surface?

*31. Suppose $f(x)$ is a polynomial of degree n.
   (i) Show that $f^{(k)}(x)$ is the zero polynomial if $k > n$.
   (ii) What can you say about $f^{(n)}(x)$?

*32. Find a formula for $\frac{d^n}{dx^n}\left(\sin x\right)$. (*Hint:* n can be written in exactly one of the following forms: $n = 4k+1$, $n = 4k+2$, $n = 4k+3$, or $n = 4k$.)

Find a formula for the following.

*33. $\frac{d^n}{dx^n}\left(\cos x\right)$

*34. $\frac{d^n}{dx^n}\left(e^{2x}\right)$

*35. $\frac{d^n}{dx^n}\left(e^{2x+1}\right)$

If a and b are any constants, find a formula for the following.

*36. $\frac{d^n}{dx^n}\left(e^{ax+b}\right)$

*37. $\frac{d^n}{dx^n}\left(\sin(ax+b)\right)$

*38. $\frac{d^n}{dx^n}\left(\ln(ax+b)\right)$

39. Consider an object thrown upward with a velocity of 50 ft/sec from ground level. Use a graphics calculator or computer to estimate the maximum height it reaches, and compare with the answer you can calculate.

40. Consider an object thrown upward with a velocity of 14 m/sec from 2 m above ground level. Use a graphics calculator or computer to estimate the maximum height it reaches.

41. Sketch the function $y = x^2 + 2x + 2$ and its derivative on the same picture. Compare with the discussion in Example 4 (p. 221).

42. On one picture, graph $f(x) = x^3 - 3x + 2$ and $f'(x)$ and $f''(x)$ for $x \in [-2, 3]$ (cf. Example 5, p. 222).

POINT TO PONDER

Consider an object moving with velocity $v(t)$ and acceleration $a(t)$. Suppose $a(t) > 0$. This means that the velocity $v(t)$ is increasing, but if v happens to be negative, the *speed* $|v(t)|$ will be *decreasing*. This is because the acceleration is in a direction opposite to the motion, so the object slows down.

You can imagine a spaceship flying along; to speed it up, the crew fire the rockets in one direction; to slow it down, they fire the rockets in the opposite direction.

The situation can be summarized by saying that the speed increases if $v(t)$ and $a(t)$ have the same sign, and it decreases if they have opposite signs. Positive acceleration does not necessarily mean that the object moves faster.

Chapter Summary

§3.1
tangent line
slope of tangent line
difference quotient
derivative
displacement
velocity
rate of change
Ideal Gas Law
marginal increase in profit

§3.2
differentiable function
continuity of differentiable functions
$\frac{d}{dx}(constant) = 0$
$\frac{d}{dx}(x^n) = nx^{n-1}$
derivatives of polynomials
Product Rule
finding equations of tangent lines
Quotient Rule
derivatives of rational functions

§3.3
maximum/minimum problems on closed intervals
endpoints

boundedness
extraneous solutions

§3.4
derivatives of trigonometric functions

§3.5
$10^{\log x} = x$
$\log(10^s) = s$
$\log(xy) = \log x + \log y$
common logarithms
$e^{\ln x} = x$
$\ln(e^s) = s$
$\ln(xy) = \ln x + \ln y$
natural logarithms
$\frac{d}{dx}(\ln x) = \frac{1}{x}$, for $x > 0$
$\frac{d}{dx}(e^x) = e^x$
$b^r = e^{r \ln b}$, for $b > 0$
hyperbolic functions
$\cosh^2 x - \sinh^2 x = 1$
derivatives of hyperbolic functions

§3.6
the Chain Rule
the Generalized Power Rule: $\frac{d}{dx}y^r = ry^{r-1}y'$
$\frac{d}{dx}(x^r) = rx^{r-1}$

§**3.7**
implicit function
implicit differentiation
logarithmic differentiation
§**3.8**
one-to-one
inverse function
arcsin
arctan
graphs of inverse functions
standard inverse functions of the trigonometric functions

derivatives of inverse functions
derivatives of inverse trigonometric functions
existence of inverse functions
standard inverse functions of the hyperbolic functions
derivatives of inverse hyperbolic functions
§**3.9**
second derivative
$f^{(n)}(x)$
derivative of order n
$\frac{d^n}{dx^n}\left(f(x)\right)$

Review Exercises

Calculate the derivative of each of the following functions, specifying all points at which it is not differentiable.

1. $f(x) = 4x^5 - 3x^2 + 12$

2. $g(x) = 17 - 2x + 8x^2$

3. $h(x) = 2x^7 - 4x^4 - 3$

4. $k(x) = 1 + x^{3000}$

5. $F(x) = 2x - \sqrt{x}$

6. $G(x) = x^2 - 5x^{1/3}$

7. $H(x) = x^{3/5} + x^{-1/2}$

8. $K(x) = x^3 - 5x^{28/4}$

9. $R(x) = \frac{x^2 - 4}{x^2 + 1}$

10. $S(x) = \frac{x^3 + x^2 + x + 1}{x + 1}$

11. $f(x) = 3\cos x \sin x$

12. $g(x) = \sin(-7x)$

13. $h(x) = (1 + 3x^2)\tan x$

14. $k(x) = \frac{\sin x}{1 + x}$

15. $F(x) = \sqrt{x^4 + 2}$

16. $G(x) = (x^3 - 8)^{1/3}$

17. $H(x) = (x^4 + x^2)^{-1/2}$

18. $K(x) = (1 - x)^{8/6}$

19. $L(x) = e^{2x-5}\cos x$

20. $M(x) = \ln(2 + x^2)\cot x$

Calculate the derivative of each of the following functions.

21. $f(x) = \sinh(x^2 + 3x - 2)$

22. $g(x) = \cosh(2x - 4)$

23. $h(x) = \ln(\cosh x)$

24. $k(x) = (\sinh x)^{-2/3}$

25. $F(x) = \tanh(3x^4 - 2x)$

26. $G(x) = \text{sech}^{-2}(3x^3 - 2)$

27. $H(x) = x\ln x$

28. $K(x) = \frac{x}{\sinh x}$

29. $T(x) = \sin^3 x \cosh^5 x$

30. $U(x) = \cot x \tan x$

Find the maximum and minimum values of each of the following functions on the specified interval.

31. $f(x) = x^2 + 6x - 2$, on $[-5, 0]$

32. $f(x) = x^2 - 2x + 5$, on $[3, 6]$

33. $f(x) = x^3 - 12x + 4$, on $[0, 3]$

34. $f(x) = x^3 - 6x + 1$, on $[-1, 3]$

35. $f(x) = \frac{1}{1 + x^2}$, on $[-2, 3]$

36. $f(x) = \frac{3}{4 - x}$, on $[-2, 2]$

37. $f(x) = \ln x$, on $[1, 8]$

38. $f(x) = \sin x$, on $\left[0, \frac{3}{4}\pi\right]$

39. $f(x) = \tan x$, on $\left[-\frac{\pi}{4}, 0\right]$

40. $f(x) = \sin^2 x + \cos^2 x$, on $\left[-\pi, \frac{4}{5}\pi\right]$

In each of the following situations, find the derivative $\frac{dy}{dx}$.

41. $x^2 + 3xy - y^2 = 3$

42. $xy^3 + x^3 y = 16$

43. $\sin(xy) = \frac{1}{2}$

44. $x^2 e^y + 2y = 4$

45. $\frac{1}{y} = \frac{1}{x}$

46. $\sin x + \cos y = 0$

47. $y\ln x = x^2$

48. $\sqrt{xy^3 + y} = 5$

49. $\frac{1}{x + y} = e^{xy}$

50. $xy^5 - 3x^2 y^2 - 3x = 4y - 2$

Find the derivatives of each the following functions.

51. $\arctan(x + 3)$

52. $\arcsin(1 - x^2)$

53. $\arccos(\sin x)$

54. $\arctan(e^{1 - x^2})$

55. $\text{arccosh}(1 + 2x^2)$

56. $\text{arctanh}(\sin x)$

57. $\exp\left(\arctan(2x)\right)$

58. $\arctan(\ln x)$

59. $\ln(\arcsin x)$

60. $\arctan(2\tan x)$

II

*61. (i) Find a formula for arccosh u as follows. Write out the equation $\cosh x = u$ by expressing $\cosh x$ in terms of e^x and e^{-x}. Multiply through by e^x, and the resulting equation will be a quadratic equation in e^x (i.e., an equation involving e^x and its square). Solve this equation to get an expression for e^x. Finally, use this equation to solve for $x = \text{arccosh } u$.

(ii) Differentiate the formula that you obtained in part (i), and compare the derivative with what we already know.

*62. (i) Find a formula for arcsinh u, following the idea used in Exercise 61.

(ii) Compare the derivative obtained from your answer to part (i) with the derivative as we already know it.

*63. Suppose x and y satisfy $2x\sin^2 y - 3\cos(2x) = 4 - 2x\cos^2 y$. Explain why it is not possible to use this relation to find $\frac{dy}{dx}$.

*64. Verify that $\sec x$ is one-to-one on $\left[0, \frac{\pi}{2}\right) \cup \left(\frac{\pi}{2}, \pi\right]$. For this exercise *only*, let Arcsec x be the corresponding inverse function. Find its derivative $\frac{d}{dx}(\text{Arcsec } x)$.

65. If $f(x)$ and $g(x)$ are differentiable functions that are never zero, verify the Product Rule for $(fg)'$ using logarithmic differentiation.

66. If $f(x)$ and $g(x)$ are differentiable functions that are never zero, verify the Quotient Rule for $\left(\frac{f}{g}\right)'$ using logarithmic differentiation.

*67. (i) The function $|x|$ is not differentiable at $x = 0$, but verify that at every other x, $\frac{d}{dx}(|x|) = \frac{x}{|x|}$.

(ii) Use part (i) and the Chain Rule to find $\frac{d}{dx}(|\sin x|)$.

68. Find (i) $\frac{d}{dx}(|e^x\cos x|)$, (ii) $\frac{d}{d\theta}(|\tan\theta|)$.

III

69. Use a graphics calculator or computer to compare the graphs of $y = \text{arcsinh } x$ and $y = \ln(x + \sqrt{x^2+1})$.

70. Use a graphics calculator or computer to compare the graphs of $y = \text{arcsec } x$ and $y = \text{arccsc } x$.

Suppose $f(x)$ is some differentiable function. If h is small, we expect the difference quotient $Q_h(x) = \frac{f(x+h)-f(x)}{h}$ to be close to $f'(x)$. For each of the following functions $f(x)$ and values of h, sketch $f'(x)$ and the approximation $Q_h(x)$ on the same picture. With sufficient magnification you should be able to see that they are not exactly the same.

71. $f(x) = x^2$, $h = 0.5$
72. $f(x) = x^2$, $h = 0.1$
73. $f(x) = x^3 - 2x^2$, $h = 0.2$
74. $f(x) = \sin x$, $h = 0.1$
75. $f(x) = \frac{1}{x-1}$, $h = 0.01$
76. $f(x) = \ln x$, $h = 0.05$
77. $f(x) = \arctan x$, $h = 0.1$
78. $f(x) = \tanh x$, $h = 0.5$
79. $f(x) = e^{\sin x}$, $h = 0.1$
80. $f(x) = \ln(e^x)$, $h = 1$

*81. Graph $|\tan x|$ and its derivative on the same picture (cf. Exercise 68(ii)).

Applications of Derivatives 4

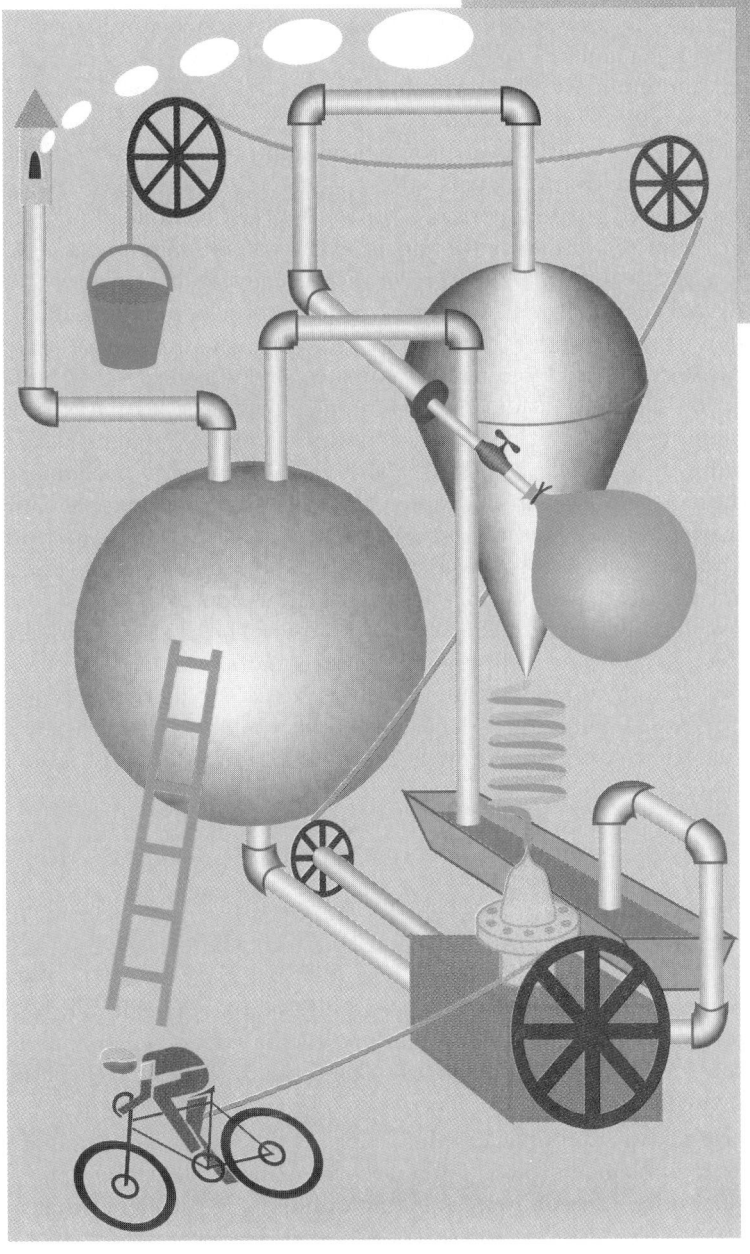

INTRODUCTION

In Chapter 3 we learned about differentiation. We saw how the idea came from looking for the slope of the tangent to a curve and how the same concept arose when we tried to describe the notion of instantaneous velocity or rates of change in general.

We learned how to differentiate many types of functions, and we have become reasonably comfortable with these calculations. In the process we saw that derivatives could be used for various purposes. In addition to finding the slopes of tangent lines and working with velocity and acceleration, we found that they could be used to help find maximum and minimum values of functions.

In this chapter we will learn to use derivatives to solve a great many different types of problems. We will return to maximum/minimum problems and study them more systematically. We will study the use of derivatives and second derivatives in drawing graphs. The interpretation of velocity and acceleration as derivatives will be discussed at greater length. But we will also learn about a number of other types of problems that can successfully be approached using derivatives.

We begin in Section 4.1 with a type of problem known as *related rates;* this refers to a situation in which two different quantities are changing at rates that are somehow related. In Section 4.2 we discuss an important result called the *Mean Value Theorem.* It is the technical tool that allows us to refine the observation we have already made that a function is increasing if its derivative is positive and decreasing if the derivative is negative. We also learn the *First Derivative Test,* which is used to identify maximum and minimum values. In Section 4.3 we introduce the idea of critical points, which is fundamental to the study of maximum/minimum problems, and in Section 4.4 we learn the *Second Derivative Test,* which helps to distinguish maximum points from minimum points.

In Section 4.5 these ideas are applied to *curve sketching,* that is, drawing graphs, and in Section 4.6, two additional refinements are added to our techniques for curve sketching, namely, *concavity* and *inflection points.* In Section 4.7 we discuss maximum/minimum problems at greater length; you are encouraged to do a large number of exercises from this section.

In Section 4.8 we consider the problem of finding an *antiderivative,* which means a function whose derivative equals some specified function. This concept will be considered in much greater depth in Chapters 5, 6, and 8, but it is of immediate use in Section 4.9 for the discussion of velocity and acceleration of moving objects.

Section 4.10 deals with *linear approximation and differentials,* which are techniques that use derivatives to help calculate approximate values of functions that cannot be evaluated exactly. The subject of Section 4.11 is *Newton's Method,* another technique of the same sort; it is used to find approximate values for the roots of equations.

In Section 4.12 we consider *l'Hôpital's Rule,* a technique that uses derivatives to evaluate limits of quotients in situations in which both the numerator and denominator tend to zero. Finally, Section 4.13 discusses *marginal cost, revenue, and profit,* some of the most important applications of the derivative to business and economics.

You may not want to study all these various topics. Depending on your interests, you may choose to skip some of them, but you should probably read Sections 4.1–4.8 at least and add whichever of the remaining sections appeal to you.

Many of the practical problems that we discuss can be broken into two parts. The second part is to apply the techniques of calculus to obtain some information about the behavior of some function. But before it is possible to do that, it is necessary to have the function. Frequently, the first part of the solution is to express the practical question in terms of some function, and very often this is the most difficult aspect of the problem.

4.1 Related Rates

IMPLICIT DIFFERENTIATION
RATE OF CHANGE

For this section it is necessary to be able to differentiate quickly and reliably. It is also important to know about implicit differentiation (cf. Section 3.7), and we need to use both the Pythagorean Theorem (Theorem 1.5.1) and facts about similar triangles (Theorem 1.5.4).

We have already encountered (in Section 3.1 and later) the idea that $\frac{df}{dt}$ represents the rate of change of $f(t)$. There are many situations in which there are two or more quantities, each of them changing with time, in such a way that their rates of change are related in some way. The problem addressed in this section is finding the derivative of one function in terms of the derivative of some other function.

EXAMPLE 1   Expanding Balloon Suppose air is being pumped into a spherical balloon in such a way that its volume is increasing by 40 cu in/sec. How fast is the radius of the balloon increasing at the moment when the radius equals 5 in?

Solution Letting the radius of the balloon be r, we know that its volume, the volume of a sphere of radius r, is $V = \frac{4}{3}\pi r^3$. We are also told that $\frac{dV}{dt} = 40$.

At this point we could take the equation $V = \frac{4}{3}\pi r^3$ and solve it for r. It would then be possible to calculate $\frac{dr}{dt}$, which is what we want. But in fact it is easier to perform an implicit differentiation. We find that

$$\frac{dV}{dt} = \frac{d}{dt}\left(\frac{4}{3}\pi r^3\right) = \frac{4}{3}\pi(3r^2)\frac{dr}{dt} = 4\pi r^2\frac{dr}{dt}.$$

Knowing that the left side of the above equation, $\frac{dV}{dt}$, equals 40, we can now solve for $\frac{dr}{dt}$. We find that

$$\frac{dr}{dt} = \frac{1}{4\pi r^2}\frac{dV}{dt} = \frac{10}{\pi r^2}.$$

When $r = 5$, this tells us that $\frac{dr}{dt} = \frac{10}{25\pi} = \frac{2}{5\pi} \approx 0.1273$. So when the radius of the balloon is 5 in, it is increasing at the rate of $\frac{2}{5\pi}$ in/sec.

As was remarked earlier, this problem could also have been done by solving for r as a function of V and then differentiating. In fact, the calculation is somewhat more complicated that way, and using implicit differentiation actually saves work.

In the course of doing the above calculation we ignored the units, since we expected the answer to come out in inches per second. It is often a good idea to carry the units along through the computation, partly just to be sure of the units for the final answer and partly as a way of checking that the calculation is correct.

EXAMPLE 2   Leaking Balloon Suppose a spherical balloon is leaking air so that its surface area is decreasing at the rate of 1 sq in/sec. At the moment when the radius is 6 in, how fast is the volume decreasing?

Solution We need to calculate $\frac{dV}{dt} = \frac{d}{dt}\left(\frac{4}{3}\pi r^3\right) = 4\pi r^2\frac{dr}{dt}$. What we know is the rate of change of the surface area.

The formula for the surface area of a sphere of radius r is $A = 4\pi r^2$; we will learn how to derive this formula in Chapter 6, but for now we must just accept it, as we do the formula for the volume.

Again we differentiate implicitly and find that

$$\frac{dA}{dt} = \frac{d}{dt}(4\pi r^2) = 8\pi r\frac{dr}{dt}.$$

From this we find that $\frac{dr}{dt} = \frac{1}{8\pi r}\frac{dA}{dt}$. Substituting this into the formula for the derivative of V, we find that

$$\frac{dV}{dt} = 4\pi r^2\frac{dr}{dt} = 4\pi r^2\frac{1}{8\pi r}\frac{dA}{dt} = \frac{r}{2}\frac{dA}{dt}.$$

The question asked us to find $\frac{dV}{dt}$ at the moment when $r = 6$ in, assuming that $\frac{dA}{dt} = -1$ in²/sec. We see that

$$\frac{dV}{dt} = \frac{r}{2}\frac{dA}{dt} = \frac{6\text{ in}}{2}(-1\text{ in}^2/\text{ sec}) = -3\text{ in}^3/\text{sec}.$$

At the specified moment the volume of the balloon is decreasing at the rate of 3 cu in/sec.

There are several things to notice here. One is that the sign of the derivative being negative indicates that the volume is decreasing. This is of course only common sense in this example, but sometimes it is not easy to guess whether a quantity is increasing or decreasing, and close attention to signs will tell us. Another is that after we found a formula relating the two derivatives, we put in numerical values, with units included, and multiplied and divided units just as we would variables. The answer ended up being a number with units. Here again there were no surprises, though occasionally it is not entirely clear what units will be used for the answer. More important is that if we make a mistake in the calculation, there is a reasonable chance that the units will be affected, and this will help us to detect the error. Finally observe that in working with units we wrote cubic inches as in³ and square inches as in² so as to be able to multiply and divide them easily. This is consistent with the way we wrote the units for acceleration as ft/sec² or m/sec², etc.

There are many types of problems in which two distances are related in some geometrical way. It is often possible to find an equation expressing this relationship by using the Pythagorean Theorem (Theorem 1.5.1) or results about similar triangles. Implicit differentiation then gives a relationship between their rates of change.

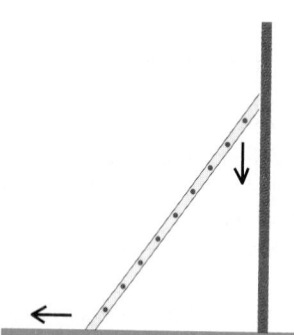

Figure 4.1.1

EXAMPLE 3 **Sliding Ladder** Consider a ladder leaning against a wall. Suppose the foot of the ladder begins to slip away from the wall so that the top of the ladder begins to slide down (see Figure 4.1.1).

If the ladder is 10 ft long and the foot of the ladder is moving away from the wall at a rate of 4 in/sec, how fast is the top descending at the moment when the top is 8 ft above the floor level?

Solution The ladder, the wall, and the floor make a right-angled triangle as shown in Figure 4.1.2. If we let x be the horizontal distance from the wall to the foot of the ladder, in feet, and y the vertical distance from the floor to the top of the ladder, also in feet, then the Pythagorean Theorem says that $x^2 + y^2 = 10^2 = 100$.

Differentiating this equation and recognizing that both x and y are changing quantities, so they should be regarded as functions of t, we find that

$$2x\frac{dx}{dt} + 2y\frac{dy}{dt} = 0,$$

that is,

$$2y\frac{dy}{dt} = -2x\frac{dx}{dt},$$

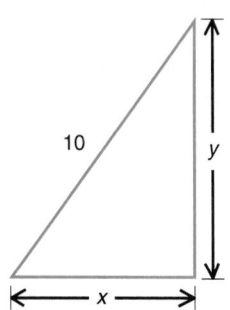

Figure 4.1.2

so

$$\frac{dy}{dt} = -\frac{x}{y}\frac{dx}{dt}.$$

The question asks us to find the rate of descent of the top of the ladder, which is just $\frac{dy}{dt}$. At the moment in question, $y = 8$, and since the Pythagorean Theorem says that $x^2 + y^2 = 100$, we solve and find that $x^2 = 100 - 8^2 = 100 - 64 = 36$. Since x, being a distance, is positive, we must have $x = 6$.

So the rate at which the top of the ladder is descending is

$$\frac{dy}{dt} = -\frac{x}{y}\frac{dx}{dt} = -\frac{6\,\text{ft}}{8\,\text{ft}}(4\,\text{in/sec}) = -\frac{3}{4}(4\,\text{in/sec}) = -3\,\text{in/sec}.$$

At the specified moment the top of the ladder is sliding down the wall at the rate of 3 in/sec.

Notice that the negative sign means that y is decreasing, which confirms our intuition that the top of the ladder is descending. Notice also that although part of the calculation was in feet and part in inches, it all worked out. It would have been possible to do this problem by converting 4 in/sec to $\frac{1}{3}$ ft/sec and proceeding from there, but in fact this was not necessary. Carrying the units along in the computation showed that leaving them as they were was all right; however, to do a calculation using both inches and feet simultaneously without keeping track of units would be unthinkable.

Finally, it is important to notice that there are some quantities in the example that change with time, such as x and y, and others that are constant, such as the length of the ladder. Quantities that change with time must be regarded as functions of t.

Summary 4.1.1

Related Rates

It is difficult to give hard and fast rules that will apply exactly to all related rates problems, but the general idea is roughly the same.

There will be one quantity whose rate of change is known and another whose rate of change we wish to find. If a relation can be found between these two quantities, then implicit differentiation will give an equation that can be solved to express the derivative of one of them in terms of the derivative of the other. The solution may involve various other quantities, so their values at the moment in question will have to be found.

It may be difficult to find a relation between the two quantities but easier to find relations between them and other quantities. In this situation we can express the rate of change of one quantity in terms of the rate of change of a second, that rate of change in terms of a third, and so on until we get to one we know. This situation appeared in Example 2, in which we expressed the rate of change of the volume in terms of the rate of change of the radius and the rate of change of the radius in terms of the rate of change of the area. Together these expressions allowed us to find the rate of change of the volume in terms of the rate of change of the area.

There is no generally applicable method for finding a relation between two quantities. If in doubt, it is usually an excellent first step to draw a diagram. The relation being sought may be a consequence of simple geometry.

A vital step in solving one of these problems is identifying the quantity whose rate of change we know and the one whose rate of change we want. Specifying exactly what they are will often help to uncover the relation between them.

Notice that the question frequently asks for the rate of change of some quantity "at the moment when some other quantity equals some fixed number." It is important to recognize that this other quantity is also varying, so it must be regarded as a function of time. It would be a mistake just to let it equal the given fixed number. Only after differentiating and solving for the required derivative should we substitute in the values of this and other quantities at the specified moment. On the other hand, if a quantity does not vary with time, there is no need to label it with a variable. Its constant value can be substituted before differentiating.

It is always important to keep careful track of units and signs.

Figure 4.1.3

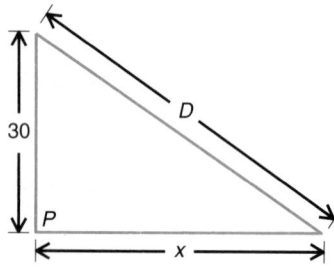

Figure 4.1.4

EXAMPLE 4   A car is traveling along a straight road at a speed of 30 m/sec. An observer is located at a point 30 m off to the side of the road. How fast is the distance between the car and the observer increasing at the moment when the car is 40 m past the point on the road that is closest to the observer?

Solution   First we draw a diagram of the situation, shown in Figure 4.1.3. A second, more idealized diagram shows the distance of 30 m between the observer and P, the point on the road that is closest to the observer (see Figure 4.1.4). Notice that the triangle determined by P, the observer, and the car has a right angle at P.

If we let x be the distance from P to the car and let D be the distance from the observer to the car, then $D^2 = 30^2 + x^2 = 900 + x^2$, and what we want to find is the rate of change of D, that is, $\frac{dD}{dt}$. The other piece of information we know is that x is changing at a rate of 30 m/sec, that is, $\frac{dx}{dt} = 30$.

Implicitly differentiating the equation $D^2 = 900 + x^2$, we get

$$2D\frac{dD}{dt} = 2x\frac{dx}{dt},$$

so

$$\frac{dD}{dt} = \frac{x}{D}\frac{dx}{dt}.$$

At the moment in question, $x = 40$ m and $D = \sqrt{30^2 + 40^2}$ m $= 50$ m, so

$$\frac{dD}{dt} = \frac{40\text{ m}}{50\text{ m}}30\text{ m/sec} = 24\text{ m/sec}.$$

The distance between the observer and the car is increasing at a rate of 24 m/sec.

Notice that while we could have figured out how to write x and D as functions of t, it was not really necessary, since we were able to use implicit differentiation. In many problems it is essential to use this approach, since expressing the various quantities as functions of t may be difficult or impossible.

Also notice that we could have written D as a square root and differentiated that, but as we noticed on an earlier occasion (Example 3.3.10, p. 163), differentiating the square root is an unnecessary complication.

Figure 4.1.5

EXAMPLE 5   **Lamppost and Shadow**   A girl who is 5 ft tall is approaching a post that holds a lamp 15 ft above the ground (see Figure 4.1.5). If she is walking at a speed of 4 ft/sec, how fast is the end of her shadow moving when she is 17 ft away from the base of the lamppost?

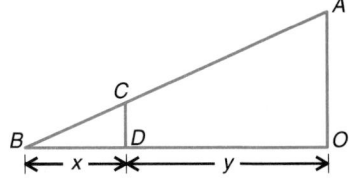

Figure 4.1.6

Note that $x = x(t)$ and $y = y(t)$ are both functions of t, the time.

Figure 4.1.7

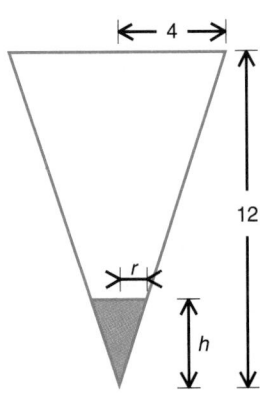

Figure 4.1.8

Solution   We make a diagram, labeling the base of the lamppost O, the light A, the end of the shadow B, the girl's feet D, and her head C (see Figure 4.1.6). Notice that the triangles BCD and BAO are similar (cf. Theorem 1.5.4, p. 24).

If we let x and y be the lengths BD and DO, respectively, then the distance from the end of the shadow to the base of the post is $x + y$, and we need to find $\frac{d}{dt}(x + y)$. By similar triangles we know that

$$\frac{x + y}{15} = \frac{x}{5},$$

so

$$5x + 5y = 15x, \qquad \text{or} \qquad 5y = 10x,$$

so

$$y = 2x, \qquad \text{and} \qquad y' = 2x'.$$

But y' is the velocity at which the girl approaches the lamppost, so, noting that y is *decreasing*, we see that $y' = -4$ ft/sec, $x' = -2$ ft/sec, and $x' + y' = -4 - 2 = -6$ ft/sec.

The end of the shadow is approaching the lamppost at a speed of 6 ft/sec, which is faster than the girl is walking. Incidentally, the length of the shadow is x, and $x' = -2$ ft/sec, so the length of her shadow is decreasing by 2 ft/sec.   ◢

Notice that the distance from the lamppost (17 ft) is irrelevant in this problem.

EXAMPLE 6   **Conical Water Tank**   A water tank has the shape of a cone, with the vertex pointing straight down (see Figure 4.1.7). The tank is 12 ft high, and its radius (at the top) is 4 ft. Water flows out at the bottom. When the water is 3 ft deep at its deepest point, the water is flowing out at a rate of 30 cu in/sec. Find the rate at which the level of the surface of the water is falling at that moment.

Solution   Recall that a cone of radius r and height h has volume $V = \frac{1}{3}\pi r^2 h$ (see Summary 1.5.14, p. 31). Let h be the depth of the water and r be the radius of the surface of the water, as shown in Figure 4.1.8.

Similar triangles tell us that $\frac{12}{4} = \frac{h}{r}$, or $h = 3r$. Consequently,

$$V = \frac{1}{3}\pi r^2 h = \frac{1}{3}\pi r^2 (3r) = \pi r^3.$$

Differentiating with respect to time gives $V' = 3\pi r^2 r'$, so $r' = \frac{V'}{3\pi r^2}$. At the moment in question, $h = 3$ ft, so $r = \frac{h}{3} = 1$ ft $= 12$ in, and $V' = -30$ in^3/sec. Accordingly,

$$r' = \frac{V'}{3\pi r^2} = \frac{-30}{3\pi (12)^2} = -\frac{5}{72\pi} \text{ in/sec.}$$

Since $h = 3r$, we see that $h' = 3r'$, so

$$h' = -3\frac{5}{72\pi} = -\frac{5}{24\pi} \text{ in/sec.}$$

The water level is falling at the rate of $\frac{5}{24\pi} \approx 0.0663$ in/sec.   ◢

Notice that in this question we had to be careful to use consistent units. The distances were measured in feet, but the flow rate was measured in cu in/sec, so we had to convert one or the other (ft to in or in³/sec to ft³/sec). Notice also that there is not enough information to determine what happens at other times (presumably, the flow rate will vary as the depth of the water changes), but that is not necessary for this solution.

Figure 4.1.9

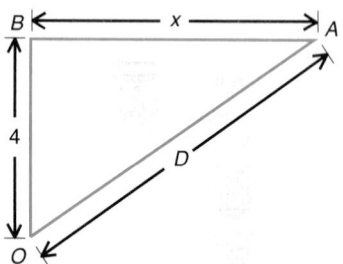

Figure 4.1.10

EXAMPLE 7   **Distance to Airplane**   An airplane is flying horizontally in a straight line at 600 mph. One minute from now, it will be directly over an observer on the ground at an altitude of 4 miles (see Figure 4.1.9). How fast is the distance from the plane to the observer changing at the present moment?

Solution   We begin by drawing a diagram (see Figure 4.1.10). The observer is on the ground at point O, the airplane is at point A, and the flight path of the airplane passes through point B, which is directly over O at an altitude of 4 mi.

The distance AB is the distance the airplane will travel in 1 min $= \frac{1}{60}$ hr, which is $\frac{1}{60}600 = 10$ miles. We let x be the (horizontal) distance from B to the plane, so $\frac{dx}{dt} = -600$ mph. We let D be the distance from the plane to the observer, so what we are asked to find is $\frac{dD}{dt}$.

Since triangle ABO has a right angle at B, the Pythagorean Theorem says that

$$D^2 = 4^2 + x^2,$$

and implicit differentiation gives

$$2D\frac{dD}{dt} = 2x\frac{dx}{dt},$$

so

$$\frac{dD}{dt} = \frac{x}{D}\frac{dx}{dt}.$$

At the specified moment, $x = 10$, so $D = \sqrt{4^2 + 10^2} = \sqrt{116} = 2\sqrt{29}$. Substituting in the formula for $\frac{dD}{dt}$, we find that

$$\frac{dD}{dt} = \frac{x}{D}\frac{dx}{dt} = \frac{10 \text{ mi}}{2\sqrt{29} \text{ mi}}(-600 \text{ mph}) = -\frac{3000}{\sqrt{29}} \text{ mph} \approx -557 \text{ mph}.$$

We conclude that the distance between the plane and the observer is decreasing at the rate of 557 mph.

Exercises 4.1

1. How fast would the volume of a spherical balloon have to be increasing in order for the radius to be increasing at a rate of $\frac{1}{4}$ in/sec at the moment when the radius is 6 in?

2. How fast is the surface area of a spherical balloon increasing if the radius is 4 in and the volume is increasing at the rate of 10 cu in/sec?

3. A 16-ft ladder is leaning against a vertical wall, and the bottom of the ladder begins to slide away from the wall. How fast is the top of the ladder moving when the foot is 10 ft from the wall, sliding at 6 in/sec?

4. A car approaches a man who is standing 100 ft in front of a brick wall so that the car's headlights shine on the wall, leaving a shadow of the man. If the headlights are 2 ft above ground level and the man is 6 ft tall, how fast is the top of the shadow moving when the car is 50 ft from the man and moving at 10 ft/sec?

5. In the situation of Example 6 (p. 233), how fast is the area of the top surface of the water decreasing at the specified moment?

6. Two cars pass a crossroads. One is traveling due north at 60 km/hr. The other passes the crossroads one-half hour after the first, traveling due west at 90 km/hr. How fast is the distance between them changing one hour after the second car passes the crossroads?

7. A weather balloon is rising vertically at the rate of 10 m/sec. A cyclist passes directly underneath it, traveling in a straight line at 3 m/sec, at a moment when the balloon is 40 m above the cyclist. If they both maintain the same speed and direction, how fast will they be moving apart 2 sec later?

*8. Vandals turn the conical tank of Example 6 upside down and insert a hose to fill it up. If water is flowing in at the rate of 30 cu in/sec, how fast is the level rising when the water is 6 ft deep?

*9. For the inverted tank of Exercise 8, how fast must the water be flowing in to make the area of the water surface decrease at the rate of 1 sq ft/min when the water is 6 ft deep?

*10. A water trough is 8 ft long and has as its cross-section an equilateral triangle with a horizontal top. If water is flowing in at the rate of 2 cu ft/min, how fast is the water level rising when the water is 6 in deep?

*11. A family of pigs is steadily drinking water from the water trough described in Exercise 10. When the water in the trough is 3 in deep, the water level is dropping at the rate of $\frac{1}{2}$ in/min. If the pigs continue to drink water at the same rate, how long will it be until the trough is empty? (*Hint:* The answer is not 6 min.)

*12. A motorist who had been charged with speeding on the basis of evidence from a police radar trap claimed that she should be acquitted. She argued as follows. The police car with the radar unit was parked 50 ft off to the side of the road. The radar measured not the speed with which the car was traveling along the road, but the rate of change of the distance from the car to the radar. When the car was 200 ft up the road from the point on the road that is closest to the radar, the radar registered that this speed was 80 mph. Since the law says that it is illegal to travel at 80 mph along the road but says nothing about the rate of change of the distance to a point 50 ft to one side, the motorist argued that the charge ought to be dismissed. The judge had never studied calculus, and let the motorist off with a warning not to do it again. How fast was the motorist in fact traveling along the road? (Assume the road was perfectly straight, and be careful about units.)

*13. A rope 15 m long passes through a pulley 8 m above the ground. On one end of the rope is a weight; the other end is held 1.5 m above the ground by a man who is walking away from the weight at a speed of 2 m/sec. How fast is the weight moving when the man is 7 m from the spot directly under the weight?

*14. A light plane passes over point A at an altitude of 3 miles, traveling due west at 200 mph. One and a half minutes later, an observer 4 miles due south of point A catches sight of the plane. How fast is the distance from the observer to the plane increasing at that moment?

*15. A woman is aiming a camera at a moving train, pivoting so as to follow its motion. The train is traveling due south on a straight track at 60 mph. The woman is standing 100 ft due west of the track, on the same level as the headlight on the engine; she holds the headlight at the center of her viewfinder. See Figure 4.1.11.

Figure 4.1.11

(i) At the moment when the headlight is due east of the woman, that is, when her line of vision points east, how fast is she turning? (Express your answer in radians per second and in revolutions per minute.)

(ii) At the moment when the headlight is exactly northeast of the woman, that is, when her line of vision points northeast, how fast is she turning? (Express your answer in radians per second and in revolutions per minute.)

(iii) At the moment specified in part (ii), how fast is the distance between the woman and the headlight decreasing?

(iv) Suppose a second train passes by, and at the moment when the woman's line of vision to its headlight points northeast she is revolving at 4 rpm (revolutions per minute) to keep aligned with the headlight. How fast is the train traveling?

4.2 Mean Value Theorem, Increasing and Decreasing Functions

LOCAL AND ABSOLUTE
MAXIMUM/MINIMUM

ROLLE'S THEOREM

MEAN VALUE THEOREM

AVERAGE SLOPE

INCREASING/DECREASING
FUNCTION

FIRST DERIVATIVE TEST

When we first explored the concept of the derivative, we thought of it as the slope of the tangent line to a curve. Later on, in Section 3.3 we realized that if the derivative was positive, the tangent sloped upward to the right, so it seemed reasonable to think that the function itself must "slope" upward also, at least near the point of tangency. We described this behavior by saying the function was increasing. Similarly, a negative derivative suggested that the curve was decreasing.

This idea was very useful. It allowed us to look for maximum and minimum values of a function by looking for points at which the derivative is zero. In Section 3.8 it gave us a criterion for a function to have an inverse (Theorem 3.8.6) because a function that is always increasing (or always decreasing) must be one-to-one.

However, although it seems very reasonable, we never really established why a function with a positive derivative must be increasing. This turns out to be quite subtle and difficult, and it is the purpose of this section to develop the techniques that will enable us to do it.

First we need some terminology.

DEFINITION 4.2.1

A function $f(x)$ has its **maximum value** at $x = a$ if $f(a) \geq f(x)$ for all x in the domain of f. In other words, f has its largest value at $x = a$. In this case we sometimes say that $x = a$ is a **maximum** for $f(x)$. The plural of maximum is *maxima*.

A function $f(x)$ has its **minimum value** at $x = a$ if $f(a) \leq f(x)$ for all x in the domain of f. In other words, f has its smallest value at $x = a$. In this case we sometimes say that $x = a$ is a **minimum** for $f(x)$. The plural of minimum is *minima*.

Saying that $x = a$ is a maximum or a minimum for $f(x)$ can seem a little confusing at first, but it is very convenient. A maximum is the point $x = a$ at which $f(x)$ has its largest value. That value, $f(a)$, is referred to as the maximum value. Similar remarks apply to minima and minimum values.

EXAMPLE 1

(i) $f(x) = x^2 + 3$ has its minimum value at $x = 0$ (the minimum value is $f(0) = 3$). It has no maximum value. (See Figure 4.2.1(i).)

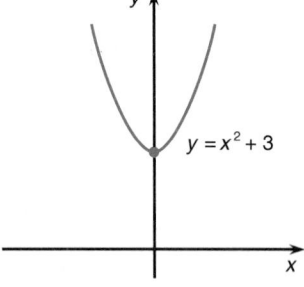

$y = x^2 + 3$

Figure 4.2.1(i)

Figure 4.2.1(ii)

Figure 4.2.1(iii)

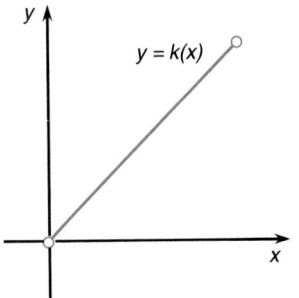

Figure 4.2.1(iv)

(ii) The function $g(x) = \cos x$ has maxima at $x = 0, \pm2\pi, \pm4\pi, \ldots$. It has its minima at $x = \pm\pi, \pm3\pi, \pm5\pi, \ldots$. The maximum value is 1, and the minimum value is -1. (See Figure 4.2.1(ii).) Notice that the maximum and/or minimum value of a function can occur at more than one point.

(iii) The function $h(x) = x$ has neither a maximum nor a minimum value. It gets arbitrarily large and arbitrarily small. (See Figure 4.2.1(iii).)

(iv) The function $k(x) = x$, for $x \in (0, 1)$, is bounded: It never gets bigger than 1 nor smaller than 0. Still it has no maximum and no minimum. The maximum and minimum values should be "at" the missing endpoints 1 and 0. (See Figure 4.2.1(iv).)

(v) The function $p(x) = \arctan x$ is bounded but has no maximum or minimum. (See Figure 4.2.1(v).) The function never quite reaches what might appear to be its maximum and minimum values, namely, $\frac{\pi}{2}$ and $-\frac{\pi}{2}$, respectively.   ◢

EXAMPLE 2   Consider the function $f(x) = x^3 - 3x$, for $x \in [-2, 3]$. We encountered a similar function in Example 3.3.3(iii) (p. 158). The graph is sketched in Figure 4.2.2.

We can calculate $f'(x) = 3x^2 - 3 = 3(x^2 - 1)$, which is zero at $x = \pm1$, negative when $x \in (-1, 1)$, and positive when $x < -1$ or $x > 1$. This leads us to suspect that $f(x)$ will have a minimum at $x = 1$ and a maximum at $x = -1$. The trouble is that $f(3) = 18$, which is greater than $f(-1) = 2$, so there is definitely not a maximum at $x = -1$. It happens that there is a minimum at $x = 1$, but there is also an endpoint minimum at $x = -2$.

On the other hand, from Figure 4.2.2 we see that the behavior of $f(x)$ for x *near* -1 is the sort of thing we expect near a maximum. The function does go up and then come down, so $f(-1)$ is bigger than $f(x)$ for every x near $x = -1$. It turns out that $x = -1$ is not a maximum because of something $f(x)$ does somewhere else, at $x = 3$, which is relatively far away.

Example 2 illustrates a common phenomenon. It frequently happens that a function $f(x)$ has what appears to be a maximum (or a minimum) at $x = a$, provided that we consider only x's that are near a. In this situation we say that $f(x)$ has a local maximum (or local minimum) at $x = a$.   ◢

Figure 4.2.1(v)

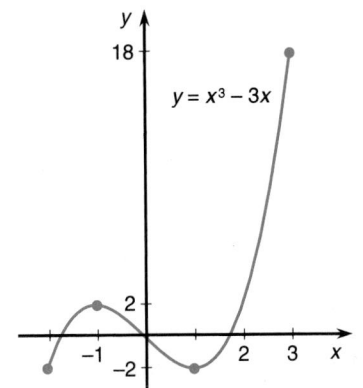

Figure 4.2.2

DEFINITION 4.2.2

A function $f(x)$ has a **local maximum** at the point $x = a$ if there is an interval (c, d) containing a so that $f(a) \geq f(x)$ for every $x \in (c, d)$ that is in the domain of f. We sometimes say that "$x = a$ is a local maximum for $f(x)$."

A function $f(x)$ has a **local minimum** at the point $x = a$ if there is an interval (c, d) containing a so that $f(a) \leq f(x)$ for every $x \in (c, d)$ that is in the domain of f. We sometimes say that "$x = a$ is a local minimum for $f(x)$."

If we want to emphasize that a maximum is not a local maximum, we sometimes refer to it as an **absolute maximum**. Similarly, the term **absolute minimum** means exactly the same thing as "minimum," but is used when it is important to distinguish it from a local minimum. Another term for local maximum or minimum is **relative maximum** or **relative minimum.**

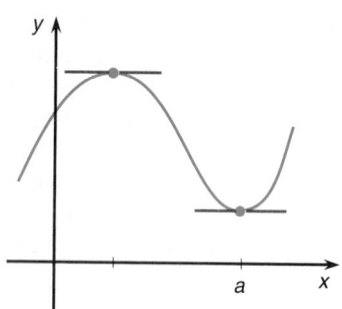

$y = x^4 - 2x^2 + 1$

Figure 4.2.3

EXAMPLE 3

(i) If $x = a$ is a maximum or minimum for $f(x)$, then it is also a local maximum or a local minimum, respectively. The converse is not true, as Example 2 demonstrates. There, as we saw in Figure 4.2.2, the point $x = -1$ is a local maximum but not an absolute maximum.

(ii) For the function $f(x)$ of Example 2, $x = -2$ and $x = 1$ are both absolute minima (and hence also local minima). There is only one absolute maximum, namely, $x = 3$, but there is also a local maximum at $x = -1$.

(iii) Let $h(x) = (x^2 - 1)^2 = x^4 - 2x^2 + 1$. The graph of $y = h(x)$ is sketched in Figure 4.2.3. Since $h(1) = h(-1) = 0$ and $h(x)$, being a square, is always greater than or equal to zero, we see that $x = \pm 1$ are both absolute minima for h. From Figure 4.2.3 it appears that $x = 0$ is a local maximum. We can confirm this as follows. When $x \in (-1, 1)$, $0 \leq x^2 < 1$, so $-1 \leq x^2 - 1 < 0$, and $h(x) = (x^2 - 1)^2 \leq 1$. Since $h(0) = 1$, this shows that $h(0) \geq h(x)$ for all $x \in (-1, 1)$, so $x = 0$ is a local maximum for h. However, it is not an absolute maximum, since, for instance, $h(3) = 64 > h(0)$. ◢

In Section 3.3 we suggested that if $f(x)$ has a maximum or minimum at $x = a$, then we should expect that the graph should have a sort of "hump" or "valley" at $x = a$ (see Figure 4.2.4). It seems reasonable to expect that the tangent line should be horizontal, that is, $f'(a) = 0$. This is true, provided that $f(x)$ is differentiable, and in fact it is true even at *local* maxima and minima.

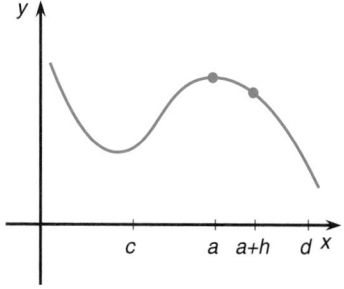

Figure 4.2.4

THEOREM 4.2.3

Suppose the function $f(x)$ is differentiable at $x = a$ and has a local maximum or a local minimum at $x = a$. Then $f'(a) = 0$.

PROOF Remark that because $f(x)$ is differentiable at $x = a$, the domain of f contains an open interval around $x = a$.

First suppose $f(x)$ has a local maximum at $x = a$. So there is an interval (c, d) containing a so that $f(a) \geq f(x)$ for all $x \in (c, d)$. (By the above remark we can assume that (c, d) is contained in the domain of f.) This means that $f(a + h) \leq f(a)$ whenever h is small enough that $a + h \in (c, d)$. See Figure 4.2.5.

Consider the one-sided limit $\lim_{h \to 0^+} \frac{f(a+h) - f(a)}{h}$. Since $f(a + h) \leq f(a)$ if $a + h \in (c, d)$ and $h > 0$, the difference quotient satisfies $\frac{f(a+h) - f(a)}{h} \leq 0$ for all $h > 0$ small enough that $a + h \in (c, d)$.

Figure 4.2.5

The one-sided version of Theorem 2.3.3 implies that

$$\lim_{h \to 0^+} \frac{f(a + h) - f(a)}{h} \le 0.$$

Similarly, if $h < 0$ is small enough that $a + h \in (c, d)$, then $f(a + h) - f(a) \le 0$. Because $h < 0$, the difference quotient satisfies $\frac{f(a+h)-f(a)}{h} \ge 0$. The one-sided version of Theorem 2.3.3 says that

$$\lim_{h \to 0^-} \frac{f(a + h) - f(a)}{h} \ge 0.$$

> Since $f'(a)$ equals both one-sided limits, we have $f'(a) \le 0$ and $f'(a) \ge 0$. We conclude $f'(a) = 0$.

But the differentiability of $f(x)$ at $x = a$ implies that these two one-sided limits are equal and both equal to $f'(a)$ (by Theorem 2.4.2). The only way this can happen is for them both to be zero, that is, $f'(a) = 0$, as desired.

The case in which $f(x)$ has a local minimum at $x = a$ is similar. ◢

EXAMPLE 4

(i) The function $g(x) = x^2$ has a minimum at $x = 0$ and $g'(0) = 0$. See Figure 4.2.6(i).

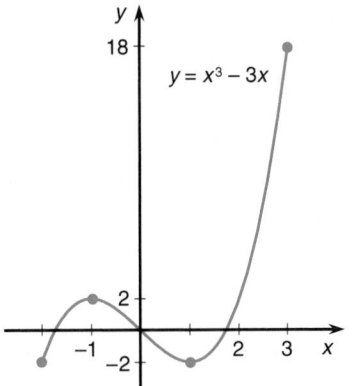

Figure 4.2.6(i)

Figure 4.2.6(ii)

Figure 4.2.6(iii)

(ii) On $[-2, 3]$ the function $f(x) = x^3 - 3x$ of Example 2 has a local maximum at $x = -1$ and $f'(-1) = 0$. It has a minimum at $x = 1$ and $f'(1) = 0$. See Figure 4.2.6(ii).

(iii) On the other hand, notice that the same $f(x)$ has a minimum at $x = -2$ and a maximum at $x = 3$ but $f'(x)$ is not zero at either of these points. This does not contradict the theorem because $f(x)$ is not differentiable at these points. (It is not possible to take the limit *from both sides* at either of these points, since f is not defined on an open interval containing either of them.) This is typical of what happens at endpoint maxima and minima.

(iv) The function $A(t) = |t|$ has its minimum at $t = 0$, but $A(t)$ is not differentiable at $t = 0$, so Theorem 4.2.3 does not apply. See Figure 4.2.6(iii). ◢

> The immediate use of Theorem 4.2.3 is for finding the local maxima and minima of a function $f(x)$. The theorem implies that all local maxima and minima must occur at points at which f' equals zero or f is not differentiable.

Figure 4.2.7(i)

Figure 4.2.7(ii)

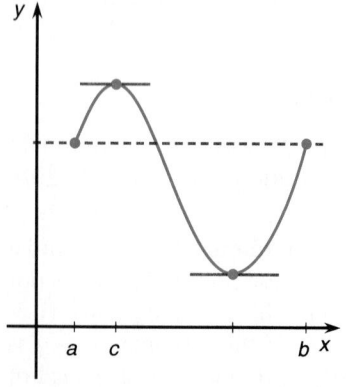

Figure 4.2.8

EXAMPLE 5

(i) Recall the function $h(x) = (x^2 - 1)^2 = x^4 - 2x^2 + 1$ of Example 3(iii). Suppose we want to locate all its local maxima and minima. Theorem 4.2.3 tells us to look where $h'(x) = 0$.

We calculate that $h'(x) = 4x^3 - 4x = 4x(x^2 - 1)$, which is zero at $x = 0$ and at $x = \pm 1$ and nowhere else. Since $h(x)$ is a polynomial, it is differentiable at every x. We know from the discussion in Example 3(iii) that $h(x)$ has absolute minima at $x = \pm 1$ and a local maximum at $x = 0$. See Figure 4.2.7(i).

Since the theorem tells us that all local maxima and minima occur at points where $h(x)$ is not differentiable or at points where $h'(x) = 0$, we know that there cannot be any other local (or absolute) maxima or minima; we have found them all.

(ii) Let $K(s) = s^3$. Then $K'(s) = 3s^2$, which is zero only at $s = 0$. Since $K(s) = s^3$ is positive when $s > 0$ and negative when $s < 0$, the point $s = 0$ cannot be a local maximum or a local minimum. Since $K(x)$ is differentiable everywhere, it cannot have any local maxima or minima. This is consistent with the illustration in Figure 4.2.7(ii). ◢

Example 5 illustrates that the derivative may be zero without the function having a local maximum or local minimum. Still, finding where the derivative is not defined or equals zero does narrow down the search to a relatively small number of points. We are at least assured that there can be no other points that are local maxima or local minima.

This approach to finding local maxima and minima is very useful, and we intend to explore it at greater length in the next section.

Rolle's Theorem and the Mean Value Theorem

Consider a function $f(x)$ on a closed interval $[a, b]$ that takes the same value at both endpoints, that is, $f(a) = f(b)$. If the graph moves up, say, as x moves to the right of a, then it must come down again to get back to the original level at $x = b$. It seems reasonable to expect that there will be a point at which it stops increasing and starts to decrease, and at that point the tangent line should be horizontal. See Figure 4.2.8.

Similar remarks apply if the graph decreases to the right of a, and in both cases we expect that there should be some point at which the derivative is zero. Of course this will only make sense if we assume the function is continuous on $[a, b]$ and differentiable everywhere on (a, b). This important result is named after the French mathematician Michel Rolle (1652–1719).

THEOREM 4.2.4

> **Rolle's Theorem**
>
> Suppose the function $f(x)$ is continuous on the closed interval $[a, b]$ and differentiable on the corresponding open interval (a, b) (i.e., $f'(x)$ exists for every x with $a < x < b$). Suppose $f(a) = f(b)$.
>
> Then there is at least one $c \in (a, b)$ for which $f'(c) = 0$.

Theorem 4.2.4 is illustrated in Figure 4.2.8, which shows two possible points at which the derivative is zero $\left(f'(c) = 0 \right)$.

PROOF If $f(x)$ is a constant function, then we know $f'(x) = 0$ for all x (by Example 3.2.1) and the conclusion holds. Now suppose $f(x)$ is not constant. It is a continuous function on the closed interval $[a, b]$, so by Theorem 2.5.7 it must achieve its maximum and minimum values.

Moreover, since $f(x)$ is not constant, the maximum and minimum cannot both equal $f(a)$, so at least one of them does not occur at $x = a$ or at $x = b$. This means that there is some $c \in (a, b)$ so that $f(x)$ has a maximum or a minimum at $x = c$. Theorem 4.2.3 implies $f'(c) = 0$. ◢

In fact we need another result that is closely related to Rolle's Theorem but slightly more complicated.

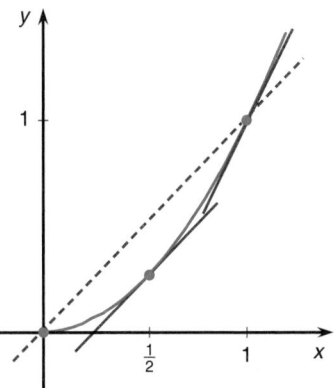

Figure 4.2.9

EXAMPLE 6 Consider the curve $y = x^2$ for x between 0 and 1, which is illustrated in Figure 4.2.9. The dashed line through the endpoints has slope equal to 1. Various tangent lines are illustrated too. The tangent at $x = 0$ has slope $m = 0$, and as x increases, the slope of the tangent increases until at $x = 1$ the tangent has slope $m = 2$. As the slope of the tangent increases from 0 to 2, it seems reasonable that at some point it will have to equal 1, the slope of the line through the endpoints.

We can think of the slope of the dashed line as the **average slope** between the two endpoints, and the above discussion leads us to expect that there should be some point at which the slope of the tangent actually equals the average slope. (In this particular case, $f(x) = x^2$, so $f'(x) = 2x$ and $f'(\frac{1}{2}) = 1$, that is, we see that the slope of the tangent at the point $x = \frac{1}{2}$ is equal to the average slope.) ◢

In this example, what made it work was that there were two points ($x = 0$ and $x = 1$) at which the slopes of the tangent lines were less than and greater than the average slope, respectively, so somewhere in between there should be a point where the tangent has the average slope. This sort of thing should always happen, because it seems unlikely that the slope of the tangent could, for instance, always be greater than the average slope. After all, that would seem to mean that the graph was always rising faster than average. However, this discussion does not constitute a proof.

THEOREM 4.2.5

The Mean Value Theorem

Suppose the function $f(x)$ is continuous on the closed interval $[a, b]$ and differentiable on the corresponding open interval (a, b). Then there is at least one $c \in (a, b)$ at which

$$f'(c) = \frac{f(b) - f(a)}{b - a}.$$

Geometrically, this means that the slope of the tangent at $x = c$ equals the slope of the line through the curve's endpoints, which may be regarded as the average or "mean" slope.

PROOF Let $F(x) = f(x) - \frac{x-a}{b-a}\big(f(b) - f(a)\big)$. Notice that

$$F(x) - f(x) = -\frac{x-a}{b-a}\big(f(b) - f(a)\big) = -\frac{f(b) - f(a)}{b - a}x + \frac{f(b) - f(a)}{b - a}a,$$

which is a polynomial in x, so it is continuous on $[a, b]$ and differentiable on (a, b). Then $F(x)$ is the sum of $f(x)$ and this polynomial, two functions that are both continuous on $[a, b]$ and differentiable on (a, b), so $F(x)$ itself is also continuous on $[a, b]$ and differentiable on (a, b). Moreover, $F(a) = f(a) - 0 = f(a)$, and $F(b) = f(b) - \frac{b-a}{b-a}\left(f(b) - f(a)\right) = f(b) - f(b) + f(a) = f(a) = F(a)$; we see that $F(x)$ has the same value at $x = a$ and $x = b$.

So Rolle's Theorem applies to $F(x)$ and says there is a $c \in (a, b)$ at which $F'(c) = 0$. But

$$F'(x) = f'(x) - \frac{1}{b-a}\left(f(b) - f(a)\right),$$

so $F'(c) = 0$ implies that

$$f'(c) - \frac{f(b) - f(a)}{b - a} = 0,$$

so

$$f'(c) = \frac{f(b) - f(a)}{b - a},$$

as desired.

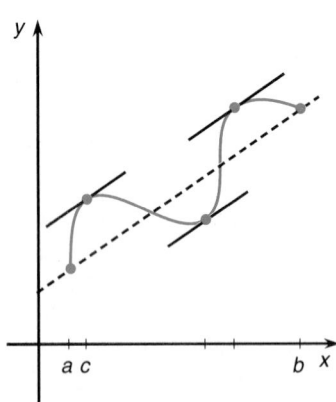

Figure 4.2.10

REMARK 4.2.6

Figure 4.2.10 also illustrates the Mean Value Theorem. Three points are marked at which the slope of the tangent equals the average slope, that is, the slope of the dashed line, which is $m = \frac{f(b)-f(a)}{b-a}$. It should be clear that this is just a "tilted" version of Rolle's Theorem, and indeed the proof just given amounted to "tilting" the graph of $y = f(x)$ until the dotted line was horizontal and then applying Rolle's Theorem.

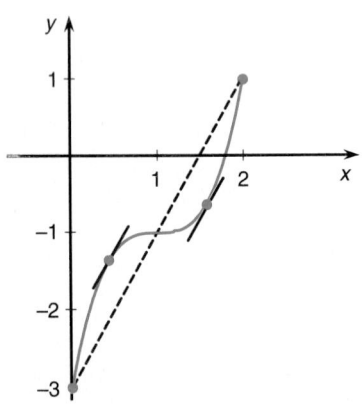

Figure 4.2.11

EXAMPLE 7   Let $f(x) = 2x^3 - 6x^2 + 6x - 3$, for $0 \le x \le 2$. Find the value or values of x at which $f'(x)$ equals the average slope between $x = 0$ and $x = 2$.

Solution $f(0) = -3$, $f(2) = 1$, so the average slope, that is, the slope of the line through $(0, -3)$ and $(2, 1)$, is $\frac{1-(-3)}{2-0} = \frac{4}{2} = 2$. Also $f'(x) = 6x^2 - 12x + 6$, so we need to solve $6x^2 - 12x + 6 = 2$, or $6x^2 - 12x + 4 = 0$, or $3x^2 - 6x + 2 = 0$. By the Quadratic Formula (1.3.4, p. 11), the solutions are

$$x = \frac{6 \pm \sqrt{36 - 24}}{6} = \frac{6 \pm \sqrt{12}}{6} = 1 \pm \frac{\sqrt{3}}{3}.$$

Since $\sqrt{3} < 2$, these solutions both lie between 0 and 2; in fact they are approximately 0.4226 and 1.5774. The graph of f is illustrated in Figure 4.2.11.

The main importance of the Mean Value Theorem is that it allows us to describe increasing and decreasing functions in terms of the signs of their derivatives.

DEFINITION 4.2.7

A function $f(x)$ is said to be **increasing** (resp. **decreasing**) on an interval if whenever u and v are in the interval with $u < v$, we have $f(u) \leq f(v)$ (resp. $f(u) \geq f(v)$).

We say $f(x)$ is increasing (resp. decreasing) if it is increasing (resp. decreasing) on every interval in its domain.

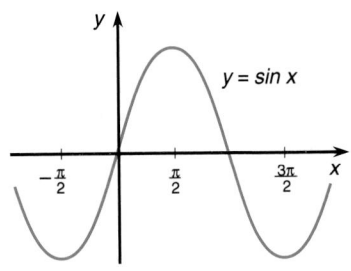

Figure 4.2.12(i)

EXAMPLE 8

(i) The function $f(x) = \sin x$ is increasing on the interval $\left[-\frac{\pi}{2}, \frac{\pi}{2}\right]$ and decreasing on $\left[\frac{\pi}{2}, \frac{3\pi}{2}\right]$. See Figure 4.2.12(i).

(ii) The function $g(x) = x^2$ is decreasing on $(-\infty, 0]$ and increasing on $[0, \infty)$. See Figure 4.2.12(ii).

(iii) The function $h(x) = x$ is increasing. See Figure 4.2.12(iii).

(iv) Any constant function is both increasing and decreasing (!). See Figure 4.2.12(iv).

We are finally able to prove what we have long suspected: that a function with a positive derivative is increasing and one with a negative derivative is decreasing.

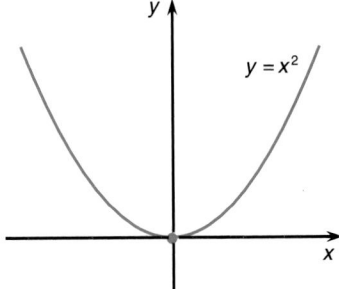

Figure 4.2.12(ii)

THEOREM 4.2.8

If $f(x)$ is differentiable on (a, b) and $f'(x) \geq 0$ for all $x \in (a, b)$, then $f(x)$ is increasing on (a, b). If $f'(x) \leq 0$ for all $x \in (a, b)$, then $f(x)$ is decreasing on (a, b).

PROOF Suppose $f'(x) \geq 0$ for all $x \in (a, b)$ but $f(x)$ is not increasing. Then we can find $u, v \in (a, b)$ with $u < v$ and $f(u) > f(v)$. The Mean Value Theorem (Theorem 4.2.5) applied to $f(x)$ says that there is a c between u and v with $f'(c) = \frac{f(v)-f(u)}{v-u}$. Since the numerator is negative and the denominator positive, this implies $f'(c) < 0$, which is impossible because we assumed $f'(x) \geq 0$ for all $x \in (a, b)$.

The case with $f'(x) \leq 0$ for all $x \in (a, b)$ is similar.

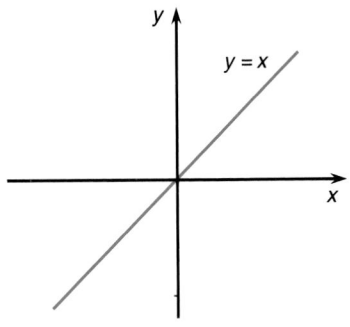

Figure 4.2.12(iii)

This result was stated for open intervals because it is easier to talk about functions being differentiable on open intervals, but a similar result holds for closed intervals.

COROLLARY 4.2.9

Suppose $f(x)$ is continuous on $[a, b]$ and differentiable on (a, b). If $f'(x) \geq 0$ (resp. $f'(x) \leq 0$) for all $x \in (a, b)$, then $f(x)$ is increasing (resp. decreasing) on $[a, b]$.

Similar statements apply if $f(x)$ is continuous on $(a, b]$ or on $[a, b)$ and differentiable on (a, b).

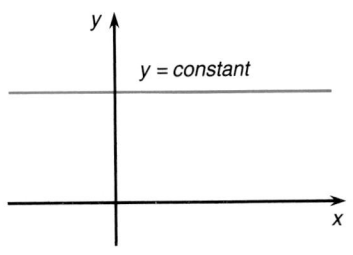

Figure 4.2.12(iv)

PROOF The only questions not covered by the theorem involve the endpoints. Suppose $f'(x) \geq 0$ on (a, b) but $f(b) < f(c)$ for some $c \in (a, b)$. Since $f(x)$ is increasing on (a, b), we know that $f(c) \leq f(x)$ for all x with $c \leq x < b$. By the one-sided version of Example 2.3.9(ii), $\lim_{x \to b^-} f(x) \geq f(c)$. Since the left side of this inequality equals $f(b)$, by the continuity of $f(x)$, we see that $f(b) \geq f(c)$. But we assumed $f(b) < f(c)$; this contradiction shows that $f(x)$ must in fact be increasing on (a, b). A similar proof shows that $f(x)$ is increasing on $[a, b)$ and hence on $[a, b]$. The case in which $f'(x) \leq 0$ can be proved by applying this first part of the result to $-f(x)$.

Knowing when a function is increasing and decreasing allows us to check when it has a maximum or minimum. The following result is obvious but very useful.

THEOREM 4.2.10

The First Derivative Test

Suppose $a \in (c, d)$, and $f(x)$ is a function that is continuous on (c, d) and differentiable at every point in (c, d) except *possibly* $x = a$.

If $f'(x) > 0$ on (c, a) and $f'(x) < 0$ on (a, d), that is, $f(x)$ is increasing to the left of $x = a$ and decreasing to the right, then $f(x)$ has a local maximum at $x = a$.

If $f'(x) < 0$ on (c, a) and $f'(x) > 0$ on (a, d), that is, $f(x)$ is decreasing to the left of $x = a$ and increasing to the right, then $f(x)$ has a local minimum at $x = a$.

If $f'(x)$ has the same sign on (c, a) and (a, d) (either both positive or both negative), then there is no local extremum at $x = a$.

Suppose $f(x)$ is a function whose domain is an interval (possibly infinite) containing a. Suppose $f(x)$ is differentiable at every point in the interval except possibly $x = a$ and the endpoints. If $f'(x) \geq 0$ for every $x < a$ and $f'(x) \leq 0$ for every $x > a$, then there is an absolute maximum at $x = a$. If $f'(x) \leq 0$ for every $x < a$ and $f'(x) \geq 0$ for every $x > a$, then there is an absolute minimum at $x = a$.

EXAMPLE 9   We reconsider Example 8 with the help of Theorem 4.2.8 and Theorem 4.2.10.

(i)   For $f(x) = \sin x$ we know that $f'(x) = \cos x$, so $f'(x) \geq 0$ for $x \in \left[-\frac{\pi}{2}, \frac{\pi}{2}\right]$ and $f'(x) \leq 0$ for $x \in \left[\frac{\pi}{2}, \frac{3\pi}{2}\right]$. By the First Derivative Test there is an (absolute) maximum at $x = \frac{\pi}{2}$.

(ii)   For $g(x) = x^2$ we know $g'(x) = 2x$, so $g'(x) \leq 0$ for $x \in (-\infty, 0]$ and $g'(x) \geq 0$ for $x \in [0, \infty)$. By the First Derivative Test there is an (absolute) minimum at $x = 0$.

(iii)   For $h(x) = x$ the derivative is $h'(x) = 1$, which is always positive. There are no local maxima or minima.

(iv)   The derivative of a constant function is always zero.

The statements we made in Example 8 can now all be proved by applying Theorem 4.2.8 and Corollary 4.2.9. Strictly speaking, although they all seemed perfectly reasonable, some of these assertions would have been quite difficult to prove before. The functions are all graphed in Figure 4.2.13.

Figure 4.2.13

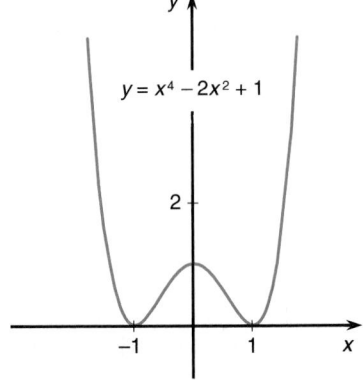

Figure 4.2.14

EXAMPLE 10   Describe the regions where the function $f(x) = x^4 - 2x^2 + 1$ is increasing or decreasing.

Solution $f'(x) = 4x^3 - 4x = 4x(x^2 - 1)$. The sign of $f'(x)$ depends on the sign of the two factors, $4x$ and $x^2 - 1$. The first, $4x$, is negative when x is negative and positive when x is positive. The second, $x^2 - 1$, is positive when $x < -1$ or $x > 1$ and negative when $-1 < x < 1$. Consequently, their product $f'(x)$ is negative on $(-\infty, -1)$ and $(0, 1)$ and positive on $(-1, 0)$ and $(1, \infty)$. The function $f(x)$ is decreasing on $(-\infty, -1]$ and $[0, 1]$ and increasing on $[-1, 0]$ and $[1, \infty)$.

Notice that we can include the endpoints in the intervals on which $f(x)$ is increasing or decreasing. See Figure 4.2.14.

We have observed that the derivative of a constant function is zero. We are about to show that the converse is true too: that a function whose derivative is zero must be constant. This seems plausible, but we need the Mean Value Theorem to prove it. It is a fact that will be very useful.

THEOREM 4.2.11

(i) If $f(x)$ is differentiable on (a, b) and $f'(x) = 0$ for all $x \in (a, b)$, then $f(x)$ is a constant function.

(ii) If $f(x)$ and $g(x)$ are both differentiable on (a, b) and $f'(x) = g'(x)$ for all $x \in (a, b)$, then $f(x) - g(x)$ is a constant, that is, $f(x) = g(x) + C$, for some constant C.

PROOF   Part (ii) follows by applying part (i) to $f(x) - g(x)$. To prove part (i), suppose that $f(x)$ is not constant. Then there are $u, v \in (a, b)$ with $u < v$ and $f(u) \neq f(v)$. The Mean Value Theorem says that there is a c between u and v with $f'(c) = \frac{f(v) - f(u)}{v - u} \neq 0$, contradicting the hypothesis.   ◢

EXAMPLE 11   If $f(x) = \sin^2 x$ and $g(x) = -\cos^2 x$, then by the Product Rule,

$$f'(x) = \sin x \cos x + \cos x \sin x = 2 \sin x \cos x,$$

and

$$g'(x) = -\cos x(-\sin x) - (-\sin x)\cos x = 2\sin x \cos x$$
$$= f'(x).$$

So $f(x) - g(x)$ is a constant; in fact we already know this, because $f(x) - g(x) = \sin^2 x - (-\cos^2 x) = \sin^2 x + \cos^2 x = 1$.

EXAMPLE 12 Suppose $f(x)$ is a differentiable function with $f'(x) = 6x^2 - 4x + 3$ and $f(0) = 7$. What is $f(x)$?

Solution It is not difficult to think of a function whose derivative is $6x^2 - 4x + 3$. After all, we can work with each term separately, and a function whose derivative is 3 is $3x$, a function whose derivative is $-4x$ is $-2x^2$, and a moment's thought shows that a function whose derivative is $6x^2$ is $2x^3$. Adding these together, we find that $g(x) = 2x^3 - 2x^2 + 3x$ has the desired derivative, that is, $g'(x) = 6x^2 - 4x + 3$. By part (ii) of Theorem 4.2.11, $f(x)$ must be $g(x) + C$, for some constant C.

To find C, we notice that $f(0) = 7$, so $7 = g(0) + C = 0 - 0 + 0 + C = C$. The answer is that $f(x) = 2x^3 - 3x^2 + 3x + 7$.

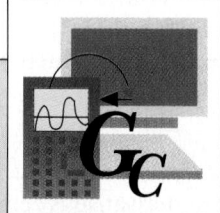

With a graphics calculator we can plot both the function $f(x)$ and the derivative $f'(x)$ from Example 12.

TI-85	SHARP EL9300
GRAPH F1 (y(x) =) 2 F1 (x)^3 − 2 F1 (x)^2 + 3 F1 (x) + 7 ▼ 6 F1 (x)x² − 4 F1 (x) + 3 M2(RANGE) (−)2 ▼ 2 ▼ ▼ (−) 10 ▼ 20 F5 (GRAPH)	⌿ 2 x/θ/T a^b 3 ▶ − 2 x/θ/T x^2 + 3 x/θ/T + 7 2ndF ▼ 6 x/θ/T x^2 − 4 x/θ/T + 3 RANGE −2 ENTER 2 ENTER 1 ENTER −10 ENTER 20 ENTER 5 ENTER ⌿
CASIO f_x-7700GB/f_x-6300G	HP 48SX
Cls EXE Range −2 EXE 2 EXE 1 EXE −10 EXE 20 EXE 1 EXE EXE EXE EXE Graph 2 X x^y 3 − 2 X x^2 + 3 X + 7 EXE Graph 6 X x^2 − 4 X + 3 EXE	⬆ PLOT PLOTR ERASE ATTN ′ αY ▬ 2 × αX y^x 3 − 2 × αX y^x 2 + 3 × αX + 7 ↱ PLOT ⬆ DRAW ERASE 2 +/− ENTER 2 XRNG 10 +/− ENTER 20 YRNG DRAW ATTN ′ αY ▬ 6 × αX y^x 2 − 4 × αX + 3 ↱ PLOT ⬆ DRAW DRAW

Notice that $f(0) = 7$, and consider how $f'(x)$ gives the rate of increase of $f(x)$ (i.e., the slope of the tangent).

EXAMPLE 13   (i) Find all functions whose derivative equals $x^4 - 2x + 1$.

Solution   As above, one function with the specified derivative is easy to find. We can let $g(x) = \frac{1}{5}x^5 - x^2 + x$ and observe that $g'(x) = \frac{1}{5}5x^4 - 2x + 1 = x^4 - 2x + 1$.

By part (ii) of Theorem 4.2.11, any function with the same derivative must be of the form $g(x) + C$, so it must be $\frac{1}{5}x^5 - x^2 + x + C$, for some constant C.

(ii) Find all functions whose derivative equals $2x + \sin x$.

Solution   One function with the specified derivative is $x^2 - \cos x$ (we simply look for something whose derivative is $2x$, and x^2 comes to mind, and for something whose derivative is $\sin x$, and $-\cos x$ comes to mind).

Again by part (ii) of Theorem 4.2.11, any function with the same derivative must be this one plus a constant. The answer is that all functions whose derivative is $2x + \sin x$ are of the form $x^2 - \cos x + C$.

In doing these problems the first step was to guess something with the specified derivative. To do this, we could work with each term of a sum separately, since derivatives work that way. Having found one function with the desired derivative, we applied the theorem to see that every such function is obtained by adding a constant to the one we know.

Given a function $h(x)$, the process of finding a function whose derivative is $h(x)$ is called finding an *antiderivative* of $h(x)$. It is an extremely important question, which we shall consider more fully in Section 4.8 and in Chapters 5 and 8.

Exercises 4.2

Find where each of the following functions has local and/or absolute maxima and/or minima.

1. $f(x) = \sin x$
2. $f(x) = x^6$
3. $f(x) = 2 - \cos x$
4. $f(x) = 1 - x^4$
5. $f(x) = 2 - |x|$
6. $f(x) = |2 - |x||$
7. $f(x) = \sqrt{|x|}$
8. $f(x) = x^3 - 3x$
9. $f(x) = \cosh x$
10. $f(x) = \sec x$

In each of the following situations, find a point c so that $f'(c)$ equals the average slope of $f(x)$ on the specified interval.

11. $f(x) = x^2$, $[-1, 0]$
12. $f(x) = x^2$, $[1, 2]$
13. $f(x) = x^3$, $[0, 1]$
14. $f(x) = x^3 + 1$, $[-1, 1]$
15. $f(x) = x^6$, $[-3, 3]$

16. $f(x) = x^6 - 2x^4 + x^2 - 3$, $[-4, 4]$
17. $f(x) = \sin x$, $[0, 2\pi]$
18. $f(x) = \cos x$, $[-\frac{\pi}{7}, \frac{\pi}{7}]$
19. $f(x) = x^3 - 2x^2 + x + 1$, $[0, 1]$
20. $f(x) = x^4 - x^3$, $[0, 1]$

Find the regions where each of the following functions is increasing or decreasing.

21. $f(x) = 2x + 5$
22. $g(x) = x^2 + 3$
23. $h(x) = x^3 + 6x - 2$
24. $k(t) = t^3 - 2t^2 + t + 3$
25. $F(x) = x^2 - 4x + 2$
26. $G(u) = \frac{|u|}{u}$
27. $H(\theta) = \tan \theta$
28. $K(t) = \arctan t$
29. $R(x) = e^x$
30. $T(x) = x^4 - 4x^3 + 4x^2 + 3$

II

31. Let $f(x) = x^2$, $b > a$, and find $c \in [a, b]$ so that $f'(c)$ equals the average slope of $f(x)$ on the interval $[a, b]$.

32. Let $f(x) = x^2 + 4x + 3$, and find c so that $f'(c)$ equals the average slope of $f(x)$ on the interval $[a, b]$.

33. Let $g(x) = x^8 - 3x^6 + 4x^4 - x^2 - 4$. For any $a > 0$, consider the interval $[-a, a]$ and find a number $c \in [-a, a]$ so that $g'(c) = \frac{g(a) - g(-a)}{2a}$.

*34. Let $P(t)$ be any polynomial and let $f(x) = P(x^2)$. For any $a > 0$, find $c \in [-a, a]$ so that $f'(c) = \frac{f(a) - f(-a)}{2a}$ (cf. Exercise 33).

35. Find a function whose derivative is (i) $4x^3 - 4x - 3$, (ii) $x^4 - 3x^3 + 2x - 1$.

36. Find a function whose derivative is (i) $x^3 + \cos x$, (ii) e^{2x}.

37. In each part, find a function that satisfies the given conditions: (i) $f'(x) = 8x^3 - 6x^2 + 4x - 3$ and $f(1) = -1$, (ii) $f'(x) = \cos x$ and $f(0) = 5$.

38. In each part, find a function that satisfies the given conditions: (i) $f'(x) = 0$ and $f(3) = 4$, (ii) $f'(x) = \frac{1}{1+x^2}$ and $f(0) = 1$.

*39. For $x \geq 0$, let $h(x) = \sqrt{x}$. Let $a > 0$ and find a point between 0 and a at which the slope of the tangent equals the average slope of the graph of $y = h(x)$ between $x = 0$ and $x = a$.

*40. Find an example in which the conclusion of Rolle's Theorem (Theorem 4.2.4, p. 240) fails for a function that is not differentiable at exactly one point.

*41. (i) Suppose $f(x)$ has a local maximum at $x = a$. Show that $g(x) = 3 - f(x)$ has a local minimum at $x = a$.

 (ii) Suppose $f(x)$ has a local minimum at $x = a$. What can you say about $g(x) = -f(x)$?

 (iii) What happens in parts (i) and (ii) if we replace the word "local" with the word "absolute"?

*42. (i) Suppose $f'(x) = 2$, for every x. Show that the graph of $y = f(x)$ is a straight line.

 (ii) What can you say if $f'(x) = C$, for some constant C?

 (iii) What can you say if $f'(x) = 2$ for every $x \neq 3$ but $f(x)$ is not differentiable at $x = 3$?

*43. If $f(x)$ is differentiable at every x and c is some fixed point, decide whether it is always possible to find $a \neq b$ so that

$$f'(c) = \frac{f(b) - f(a)}{b - a}.$$

*44. Suppose we know the following facts about the exponential function: $\frac{d}{dx}(e^x) = e^x$, $e^{a+b} = e^a e^b$, $e = e^1 > 2$, and $e^x > 0$ for all x. Using these facts, prove that $\lim_{x \to \infty} e^x = \infty$ and that $\lim_{x \to -\infty} e^x = 0$. (*Hint:* For $n > 0$, $e^n = e^1 \times e^1 \times \ldots \times e^1 > 2^n$.)

*45. Show that the graph of $y = x^2$ lies *above* any of its tangent lines.

*46. Show that the graph of $y = \ln x$ lies *below* any of its tangent lines.

*47. (i) Show that if $a > 0$, then the tangent line to the graph of $y = x^3$ at the point (a, a^3) passes below the origin (i.e., its y-intercept is negative).

 (ii) What happens if $a < 0$?

*48. Suppose $r < s$ and $f(x)$ is a twice-differentiable function on (r, s), with $f''(x) > 0$ for all $x \in (r, s)$. Suppose $a \in (r, s)$ and show that on the interval (r, s) the graph of $y = f(x)$ lies above the tangent line to $y = f(x)$ at $x = a$.

III

Use a graphics calculator or computer to sketch each of the following functions on the specified interval. Locate the local maxima and minima, and use greater magnification to identify them correct to three decimal places.

49. $x^4 - 3x^3 + 2x$, $[-2, 3]$

50. $\sin(x^2)$, $[-3, 3]$

51. $x^3 \sin x$, $[0, 2\pi]$

52. $2\sin x - x$, $[0, \pi]$

*53. Let $f(x) = x^4 - 3x^3 + x^2 + x$, for $x \in [0, 2]$. Estimate (to four decimal places) the location of two points $c \in [0, 2]$ for which $f'(c)$ equals the average slope.

4.3 Critical Points and Extrema

STATIONARY POINT
LOCAL/ABSOLUTE EXTREMUM

This brief section introduces some terminology and presents some examples.

From Theorem 4.2.3 (p. 238) we know that if a function $f(x)$ is differentiable at $x = a$ and $f(x)$ has a local or absolute maximum or minimum at $x = a$, then $f'(a) = 0$. From this we realized that to find the local maxima and minima, we should first locate all points at which the derivative does not exist or at which it does exist and equals zero.

Such points play such an important role that there is a special terminology for them.

DEFINITION 4.3.1

Suppose $f(x)$ is a function. A point a in the domain of $f(x)$ is called a **critical point** of $f(x)$ if $f'(a) = 0$ *or* if $f(x)$ is not differentiable at $x = a$. Sometimes the critical points at which the derivative does exist (and so equals zero) are called **stationary points**.

Very often we have to find maximum and/or minimum values of a function, but frequently we do not know what to expect. It is irritating to keep having to refer to "maxima or minima," and it is much more convenient to have a single term to cover both possibilities.

DEFINITION 4.3.2

A point $x = a$ is called an **extremum** for $f(x)$ if it is either a maximum or a minimum for $f(x)$. It is a **local extremum** if it is a local maximum or a local minimum, and one can speak of an **absolute extremum** to distinguish it from a local extremum. The plural of extremum is extrema, and an alternate term for extremum is *extreme point*. The value $f(a)$ of the function $f(x)$ at the extremum $x = a$ is an *extreme value*.

The process of finding the extrema of a function $f(x)$ has two steps. The first is to locate the critical points of $f(x)$. The next is to try to decide which of them are local or absolute maxima or minima and which of them are not extrema at all. Sometimes it is possible to do this by relatively simple methods, requiring only a certain amount of ingenuity, but there are occasions when it can be extremely difficult.

EXAMPLE 1   Find all critical points of $f(x) = \frac{1}{1+x^2}$. Decide which of them are maxima or minima.

Solution   By the Quotient Rule, $f'(x) = \frac{-2x}{(1+x^2)^2}$. For this to equal zero, the numerator must equal zero, so $x = 0$. Since the derivative exists at every x, this is the only critical point.

Since $1 + x^2 \geq 1$, we see that for any x, $f(x) = \frac{1}{1+x^2} \leq \frac{1}{1} = 1 = f(0)$, so $x = 0$ is an (absolute) maximum for $f(x)$.

Taking reciprocals reverses inequalities.

EXAMPLE 2   Find all critical points of $f(x) = \frac{x}{1+x^2}$, and find which are maxima or minima.

Solution   The derivative is

$$f'(x) = \frac{1(1+x^2) - x(2x)}{(1+x^2)^2} = \frac{1-x^2}{(1+x^2)^2},$$

so $f'(x) = 0$ precisely when $1 - x^2 = 0$, that is, when $x = \pm 1$. Since the derivative exists at every x, these are the only critical points.

Notice that $f(-1) = -\frac{1}{2}$, $f(1) = \frac{1}{2}$, and $\lim_{x \to \infty} f(x) = \lim_{x \to -\infty} f(x) = 0$. The derivative $f'(x)$ is negative when $x < -1$ or $x > 1$ and positive when $-1 < x < 1$.

Accordingly, $f(x)$ is decreasing on $(-\infty, -1]$ and on $[1, \infty)$ and increasing on $[-1, 1]$. Since $f(x)$ decreases on $[1, \infty)$ but is never negative there, $f(x)$ must have an absolute minimum at $x = -1$ and similarly an absolute maximum at $x = 1$. The graph of $y = f(x)$ is illustrated in Figure 4.3.1. ◢

Figure 4.3.1

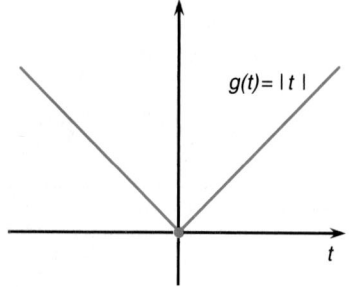

Figure 4.3.2

See the Graphics Calculator Example on the facing page for a plot of this example.

EXAMPLE 3 Find all local extrema of $g(t) = |t|$.

Solution When $t > 0$, $g(t) = t$, so $g'(t) = 1$. (Here we notice that the limit that defines $g'(t)$ only depends on the behavior of the function near t, by Proposition 2.2.8; the derivative of $g(t)$ will be the same as the derivative of t for $t > 0$.) Similarly, for $t < 0$, $g(t) = -t$ and $g'(t) = -1$. As we have already remarked, the derivative does not exist at $t = 0$.

So the only critical point is $t = 0$. Of course, it is an absolute minimum for $g(t)$, since for all t, $g(t) = |t| \geq 0 = g(0)$. See Figure 4.3.2. ◢

EXAMPLE 4 Find and classify the critical points of $f(x) = (x - 1)e^x$.

Solution We find that $f'(x) = e^x + (x - 1)e^x = (1 + x - 1)e^x = xe^x$. Since e^x is never zero, this derivative is zero precisely when $x = 0$. Since the derivative exists for all values of x, the only critical point is $x = 0$.

When the question asks us to *classify* the critical points, it means that we should decide which are local or absolute maxima or minima. In this case we can easily determine the sign of $f'(x)$. After all, e^x is always positive, so the sign of xe^x is the same as the sign of x.

In particular, when $x < 0$, the derivative is negative and $f(x)$ is decreasing; when $x > 0$, the derivative is positive and $f(x)$ is increasing. By the First Derivative Test the critical point $x = 0$ is a minimum. ◢

EXAMPLE 5 Find and classify the critical points of $h(x) = x^5 + 5x^2$.

Solution We differentiate: $h'(x) = 5x^4 + 10x = 5x(x^3 + 2)$. So $h'(x)$ exists for every x and equals zero when $x = 0$ or when $x^3 + 2 = 0$, that is, when $x^3 = -2$, which means $x = -2^{1/3}$. These two points are the only critical points.

We see from the formula for $h'(x)$ that when $x > 0$, we have $h'(x) > 0$. Also notice that the factor $x^3 + 2$ is positive when $x > -2^{1/3}$ and negative when $x < -2^{1/3}$. So when $x < -2^{1/3}$, both factors, $5x$ and $x^3 + 2$, are negative and $h'(x) > 0$. When $-2^{1/3} < x < 0$, then the factor $5x$ is negative and the factor $x^3 + 2$ is positive. Their product, $h'(x)$, must be negative.

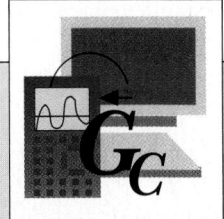

With a graphics calculator we can plot both the function $f(x)$ and the derivative $f'(x)$ from Example 4.

TI-85	SHARP EL9300

CASIO f_x-7700GB/f_x-6300G	HP 48SX

We see the critical point and note that the derivative is zero there. Observe that the calculator picture alone is not really enough to tell us what happens as $x \to \pm\infty$.

Figure 4.3.3

We sketch the real line in Figure 4.3.3, marking the critical points. We also indicate the sign of the derivative $h'(x)$ (where the derivative is positive, we place a $+$, and where it is negative, we place a $-$). Notice that the derivative $h'(x)$ can only change sign at a critical point, in other words that it has the same sign on any interval in the domain that does not contain any critical points.

Using Figure 4.3.3, we find that the function $h(x)$ is increasing for $x < -2^{1/3}$, decreasing between $-2^{1/3}$ and 0, and increasing for $x > 0$. This means that $x = -2^{1/3}$ is a local maximum and $x = 0$ is a local minimum.

Finally, note that $\lim_{x \to \infty} h(x) = \infty$ and $\lim_{x \to -\infty} h(x) = -\infty$, which implies that $h(x)$ has neither an absolute maximum nor an absolute minimum, since it gets arbitrarily large and also arbitrarily large in the negative direction. So neither critical point is an absolute extremum.

EXAMPLE 6   Find and classify the critical points of $R(t) = t^3 e^t$.

Solution   The derivative is $R'(t) = 3t^2 e^t + t^3 e^t = (3t^2 + t^3)e^t = t^2(3 + t)e^t$. It is zero at $t = 0$ and $t = -3$. These are the only critical points.

As above, we sketch the real line in Figure 4.3.4, marking the critical points, and we indicate the sign of the derivative on each part of the domain, that is, on each interval between critical points (or between a critical point and the end of the line). Notice that when $t > 0$, each factor in the derivative is positive, so $R'(t) > 0$. When $t < -3$, we have $t^2 > 0$, $3 + t < 0$ and $e^t > 0$, so their product, $R'(t)$, is negative. When $-3 < t < 0$, the factors are all positive, so $R'(t) > 0$.

Figure 4.3.4

This means that the function is decreasing to the left of $t = -3$ and increasing to the right of $t = -3$, which implies that $t = -3$ is a minimum for $R(t)$. Notice that the critical point $t = 0$ is neither a local maximum nor a local minimum. Also notice that the derivative does not change sign at $t = 0$, even though it is a critical point. ◢

In Example 6 it was possible to figure out the sign of the derivative on each interval by working out the sign of each factor. On the other hand, as we observed in Example 5, the sign of the derivative can only change at a critical point, so to find what the sign is on each interval, we need only evaluate the derivative at one point in each interval.

For instance, in Example 6 the derivative is $R'(t) = 3t^2 e^t + t^3 e^t$, which we can easily evaluate at, say, $x = 1$, $x = -1$, and $x = -4$. We find that $R'(1) = 3e^1 + e^1 = 4e$, $R'(-1) = 3e^{-1} - e^{-1} = 2e^{-1} = \frac{2}{e}$, and $R'(-4) = 3(16e^{-4}) - 64e^{-4} = -16e^{-4}$. This last number is negative and the other two are positive, which shows that the derivative $R'(x)$ is positive at one point in $(0, \infty)$, positive at one point in $(-3, 0)$, and negative at one point in $(-\infty, -3)$. Since the sign of the derivative does not change on each of these intervals, we see that $R'(x)$ is positive for $x > 0$, positive for $-3 < x < 0$, and negative for $x < -3$, which agrees with what we found earlier.

To do this calculation, we had to choose a point in each interval in the domain between critical points. We chose $x = 1, -1, -4$ because they would be relatively easy to work with (better than fractions, for example), but it is possible to do it with any other convenient choice of one point in each interval between critical points.

In this case the two methods of finding the sign of the derivative $R'(x)$ are more or less equally easy, but there are cases in which evaluating at selected points is the only practical way to do it. In many other cases it is the easiest thing to try.

EXAMPLE 7   Find and classify all critical points of the function $k(t) = t^3 (t^2 - 1)^2$, defined for $t \in [-2, 2]$.

Solution   We find that the derivative is $k'(t) = 3t^2(t^2 - 1)^2 + 2t^3(t^2 - 1)(2t) = t^2(t^2 - 1)(3t^2 - 3 + 4t^2) = t^2(t^2 - 1)(7t^2 - 3)$. It is defined for all $t \in (-2, 2)$ but strictly speaking not at $t = \pm 2$. Even though the formula makes sense at $t = \pm 2$, the function $k(t)$ is not differentiable there because it is not defined on an open interval containing $t = -2$ or $t = 2$.

The derivative is zero at $t = 0$, $t = \pm 1$, $t = \pm\sqrt{\frac{3}{7}}$. The critical points of the function $k(t)$ are $t = 0$, $t = \pm 1$, $t = \pm\sqrt{\frac{3}{7}}$, and $t = \pm 2$. (The derivative is zero at $t = 0, \pm 1, \pm\sqrt{\frac{3}{7}}$ and does not exist at $t = \pm 2$.)

As before, we draw the real axis in Figure 4.3.5(i), marking the critical points. We could work out the sign of each factor of $k'(t)$ on each interval, but it is probably easier to evaluate. For a point between 0 and $\sqrt{\frac{3}{7}}$ we could try $t = \sqrt{\frac{2}{7}}$; between $\sqrt{\frac{3}{7}}$ and 1 we could try $t = \sqrt{\frac{4}{7}}$. Between 1 and 2 we can use $t = \frac{3}{2}$. Then

$$k'\left(\sqrt{\frac{2}{7}}\right) = \left(\sqrt{\frac{2}{7}}\right)^2 \left(\left(\sqrt{\frac{2}{7}}\right)^2 - 1\right)\left(7\left(\sqrt{\frac{2}{7}}\right)^2 - 3\right)$$

$$= \frac{2}{7}\left(\frac{2}{7} - 1\right)\left(7\left(\frac{2}{7}\right) - 3\right) = \frac{2}{7}\left(-\frac{5}{7}\right)(-1) = \frac{10}{49} > 0,$$

$$-2 \qquad -1 -\sqrt{\tfrac{3}{7}} \quad 0 \quad \sqrt{\tfrac{3}{7}} \quad 1 \qquad 2$$

$$+ + + - + + + + - + + +$$

Figure 4.3.5(i)

$$k'\left(\sqrt{\frac{4}{7}}\right) = \left(\sqrt{\frac{4}{7}}\right)^2\left(\left(\sqrt{\frac{4}{7}}\right)^2 - 1\right)\left(7\left(\sqrt{\frac{4}{7}}\right)^2 - 3\right)$$

$$= \frac{4}{7}\left(\frac{4}{7} - 1\right)\left(7\left(\frac{4}{7}\right) - 3\right)$$

$$= \frac{4}{7}\left(-\frac{3}{7}\right)(1) = -\frac{12}{49} < 0,$$

$$k'\left(\frac{3}{2}\right) = \left(\frac{3}{2}\right)^2\left(\left(\frac{3}{2}\right)^2 - 1\right)\left(7\left(\frac{3}{2}\right)^2 - 3\right)$$

$$= \frac{9}{4}\left(\frac{9}{4} - 1\right)\left(7\left(\frac{9}{4}\right) - 3\right)$$

$$= \frac{9}{4}\left(\frac{5}{4}\right)\left(\frac{63}{4} - \frac{12}{4}\right) > 0.$$

So $k'(t)$ is positive between $t = 0$ and $t = \sqrt{\frac{3}{7}}$, negative between $t = \sqrt{\frac{3}{7}}$ and $t = 1$, and positive between $t = 1$ and $t = 2$.

We could work it out for negative values of t, but since $k'(-t) = k'(t)$, it is not necessary to do any work. We mark the signs of the derivative on Figure 4.3.5(i).

Using the First Derivative Test, we see that there will be local maxima at $t = \sqrt{\frac{3}{7}}$, $t = 2$, and $t = -1$ and local minima at $t = -\sqrt{\frac{3}{7}}$, $t = -2$, and $t = 1$. There is no local maximum or minimum at $t = 0$.

To determine absolute maxima and minima, we evaluate:

$$k(0) = 0, \qquad k(\pm 1) = 0,$$

$$k\left(\sqrt{\frac{3}{7}}\right) = \left(\frac{3}{7}\right)^{3/2}\left(\frac{3}{7} - 1\right)^2 = \left(\frac{3}{7}\right)^{3/2}\left(\frac{16}{49}\right) \approx 0.091613,$$

$$k\left(-\sqrt{\frac{3}{7}}\right) = -\left(\frac{3}{7}\right)^{3/2}\left(\frac{3}{7} - 1\right)^2 = -\left(\frac{3}{7}\right)^{3/2}\left(\frac{16}{49}\right) \approx -0.091613,$$

$$k(2) = 8(4 - 1)^2 = 72, \qquad k(-2) = 8(4 - 1)^2 = -72.$$

There is an absolute maximum at $t = 2$, an absolute minimum at $t = -2$, local maxima at $t = \sqrt{\frac{3}{7}}$ and $t = -1$, and local minima at $t = -\sqrt{\frac{3}{7}}$ and $t = 1$.

We sketch part of the graph in Figure 4.3.5(ii).

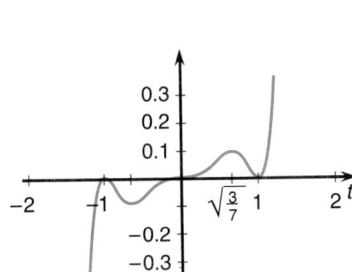

Figure 4.3.5(ii)

Notice that we must include the endpoints in our list of critical points, since $k(t)$ is not defined on open intervals containing them and so cannot possibly be differentiable at either of them. This may seem confusing because the function $f(t) = t^3(t^2 - 1)^2$ *is* differentiable at $t = \pm 2$; it is very important to distinguish this function from $k(t)$, which is only defined for $t \in [-2, 2]$. These two functions are given by the same formula when they are both defined, but they are defined on different domains, so they are *different functions*. In particular they do not have the same critical points.

If we had not made this distinction, we would not have found the absolute maximum at the endpoint $t = 2$.

Exercises 4.3

I

Find all critical points of the following functions.

1. $f(x) = x^3 - 12x$
2. $f(x) = x^4 - 18x^2 + 3$
3. $f(x) = \cos x$
4. $f(x) = \tan x$
5. $f(x) = \exp(2x - 4)$
6. $f(x) = \sin \frac{1}{x}$
7. $f(x) = \exp(\sin x)$
8. $f(x) = \sin(e^x)$
9. $f(x) = |x^2 - 1|$
10. $f(x) = \exp|x|$

Find all critical points of the following functions, and classify which are local or absolute maxima or minima.

11. $f(x) = x^5$
12. $f(x) = 3x^5 - 5x^3 + 2$
13. $f(x) = x^5 + 2x^3$
14. $f(x) = |x^2 - 4x + 3|$
15. $f(x) = (x^2 - 4)^2$
16. $f(x) = (x^2 - 9)^3$
17. $f(x) = \tan(2x)$
18. $f(x) = \csc x$
19. $f(x) = \exp(x^2)$
20. $f(x) = \arctan(x^2)$

For each of the following functions, find and classify all the critical points.

21. $f(x) = x^2 - 2, x \in [-1, 2]$
22. $f(x) = x^2 - 2, x \in (-1, 2)$
23. $f(x) = x^3 + 1, x \in [-3, 3]$
24. $f(x) = -3x^3, x \in (-1, 5)$
25. $f(x) = x^3 - 12x, x \in [-3, 3)$
26. $f(x) = \ln x - x, x > 0$
27. $f(x) = \sin x, x \in \left[0, \frac{3\pi}{4}\right]$
28. $f(x) = e^x \sin x, x \in [0, 2\pi]$
29. $f(x) = \exp(x^2 + 2x), x \in [0, 3]$
30. $f(x) = 1 + \tan^2 x - \sec^2 x, x \in (-1, 1)$

II

31. Suppose $a \neq b$ and let $f(x) = (x - a)(x - b)$. Find and classify the critical points of $f(x)$. (Draw a sketch.)

32. (i) Find and classify the critical points of $F(x) = \exp(x^2 - 2x - 3)$.
    (ii) Find and classify the critical points of $G(x) = \exp(-x^2 + 2x + 3)$.

*33. (i) Suppose $f(x)$ is a function that is defined and differentiable on the whole real line. Let $H(x) = e^{f(x)}$, and show that the critical points of $H(x)$ are exactly the same as the critical points of $f(x)$.
    (ii) Show that the critical points of $H(x)$ are of the same "type" for $H(x)$ as they are for $f(x)$, that is, a local maximum for $f(x)$ is a local maximum for $H(x)$, etc.
    (iii) If $K(x) = e^{-f(x)}$, what statements can you make about the critical points of $K(x)$ in analogy with parts (i) and (ii)?

*34. Suppose $f(x)$ can be written in the form $f(x) = (x - a)^k g(x)$, where $k \geq 2$ is an integer and $g(x)$ is a function that is differentiable at $x = a$ and that satisfies $g(a) \neq 0$.
    (i) Show that $x = a$ is a critical point of $f(x)$.
    (ii) Give conditions (in terms of k and $g(a)$) for this critical point to be a local maximum, a local minimum, or neither.
    (iii) Think of examples of functions $f(x)$ that illustrate each type of critical point discussed in part (ii).

*35. (i) Show that every polynomial of even degree has at least one critical point.
    (ii) Find examples of polynomials without any critical points (find examples with at least two different degrees, more if possible).

*36. Find and classify all critical points of $f(x) = |x - 1|^3$.

III

Use a graphics calculator or computer to graph the following functions.

37. $f(x) = \frac{1}{1 + x^2}$ (Example 1, p. 249)
38. $f(x) = \frac{x}{1 + x^2}$ (Example 2, p. 249)
39. $f(x) = (x - 1)e^x$ (Example 4, p. 250)
40. $h(x) = x^5 + 5x^2$ (Example 5, p. 250)
41. $R(t) = t^3 e^t$ (Example 6, p. 251)
42. $k(t) = (t - 1)\sqrt{t}$ (Example 7, p. 252)

4.4 The Second Derivative Test and Classifying Extrema

LOCAL MAXIMUM
LOCAL MINIMUM

Sections 4.2 and 4.3 showed how derivatives could be used to determine when a differentiable function was increasing or decreasing and then used this idea to see that local extrema occur at critical points. We observed that it is possible to have a critical point that is not a local extremum. This is what happens with $f(x) = x^3$; the

Figure 4.4.1

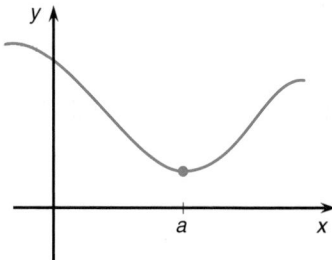

Figure 4.4.2

function has exactly one critical point, at $x = 0$, but it is increasing on the entire real line and has no local extrema at all, let alone absolute extrema. See Figure 4.4.1.

If we want to find the extrema of some function $f(x)$, we can start by finding all the critical points, but then it is necessary to figure out which of them, if any, are in fact extrema and which kind they are (local or absolute maxima or minima). In several examples in Section 4.3 it was possible to do this by drawing a picture and remembering various facts (especially the fact that a square is always nonnegative), but a lot depended on our familiarity with the particular example. In this section we show how the second derivative can frequently be used to identify local maxima and local minima. The method is not always applicable, but when it does work, it is very useful and relatively easy.

Suppose the function $f(x)$ has a local minimum at $x = a$. Then from Figure 4.4.2 we expect that $f(x)$ will be decreasing to the left of $x = a$ and increasing to the right of $x = a$, at least if we stay nearby. If $f(x)$ is differentiable, this means that the derivative $f'(x)$ should be negative for $x < a$ and positive for $x > a$ (and of course zero at the minimum $x = a$).

This in turn leads us to expect that the derivative $f'(x)$ should be increasing near $x = a$, from something negative to something positive. To say that the derivative is increasing suggests that *its* derivative, the second derivative $f''(x)$, should be positive near $x = a$.

This is not foolproof because we know that it is possible for a function to be increasing but to have its derivative equal zero at some point (as with x^3 at $x = 0$), but at the very least we expect that if $x = a$ is a local minimum for $f(x)$, then $f''(a) \geq 0$. Similarly, the second derivative should be less than or equal to zero at a local maximum.

EXAMPLE 1

(i) For $f(x) = x^2$, which has an absolute minimum at $x = 0$, we calculate and find that $f'(x) = 2x$, $f''(x) = 2$, which is certainly greater than 0. See Figure 4.4.3(i).

(ii) For $g(x) = \sin x$, which has an absolute maximum at $x = \frac{\pi}{2}$, we calculate $g'(x) = \cos x$ and $g''(x) = -\sin x$ and observe that $g''\left(\frac{\pi}{2}\right) = -\sin \frac{\pi}{2} = -1$, which is negative. See Figure 4.4.3(ii).

(iii) For $h(t) = t^4$, which has an absolute minimum at $t = 0$, we find that $h'(t) = 4t^3$ and $h''(t) = 12t^2$, so $h''(0) = 0$. This illustrates that it is definitely possible to have a minimum at which the second derivative is not positive but actually equal to zero. See Figure 4.4.3(iii).

Figure 4.4.3(i)

Figure 4.4.3(ii)

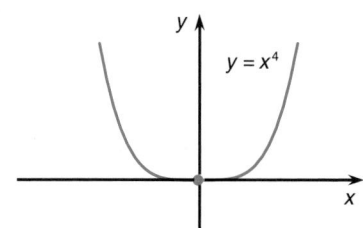

Figure 4.4.3(iii)

The importance of this idea comes from using it the other way around. If we know that $x = a$ is a critical point of $f(x)$ and that $f''(a)$ exists and is positive, then we can conclude that $x = a$ is a local minimum of $f(x)$; if $f''(a) < 0$, then $x = a$ is a local maximum.

THEOREM 4.4.1

> **The Second Derivative Test**
>
> Suppose $f(x)$ is differentiable on an interval containing $x = a$ and $f'(x)$ is itself differentiable at $x = a$. Suppose too that $f'(a) = 0$, that is, that $x = a$ is a critical point.
>
> If $f''(a) < 0$, then $x = a$ is a local maximum.
>
> If $f''(a) > 0$, then $x = a$ is a local minimum.
>
> If $f''(a) = 0$, the test is inconclusive.

PROOF Suppose $f''(a) < 0$. Since f'' is the derivative of f', this says that

$$0 > f''(a) = \lim_{x \to a} \frac{f'(x) - f'(a)}{x - a}$$

$$= \lim_{x \to a} \frac{f'(x)}{x - a}.$$

| Note that $f'(a) = 0$.

Applying Theorem 2.3.4, we see that when x is sufficiently close to a, $\frac{f'(x)}{x-a}$ is negative. Since the denominator $x - a$ is negative when $x < a$ and positive when $x > a$, the numerator $f'(x)$ must be positive when $x < a$ and x is close to a, and it must be negative when $x > a$ and close to a.

Applying Corollary 4.2.9 (p. 243) to $f(x)$, we see that, close to $x = a$, $f(x)$ is increasing for $x \leq a$ and decreasing for $x \geq a$. This implies that $x = a$ is a local maximum.

If $f''(a) > 0$, we may use a similar proof, or simply consider the function $-f(x)$, whose second derivative is $-f''(x)$, which is negative at $x = a$. The argument just completed implies that $-f(x)$ has a local maximum at $x = a$, so $f(x)$ has a local minimum there. ◢

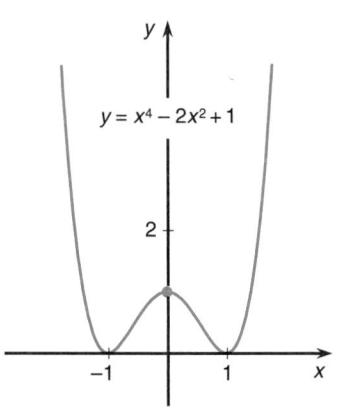

Figure 4.4.4

EXAMPLE 2 Find and classify the local extrema of the function $f(x) = x^4 - 2x^2 + 1$, which we have already considered in Example 4.2.3(iii) (p. 238).

Solution The derivative is $f'(x) = 4x^3 - 4x = 4x(x^2 - 1)$, so the critical points are $x = -1, 0, 1$. Now $f''(x) = 12x^2 - 4$, so $f''(-1) = 12 - 4 = 8$, $f''(0) = -4$, and $f''(1) = 8$, so by the Second Derivative Test (Theorem 4.4.1) the points $x = \pm 1$ are local minima and $x = 0$ is a local maximum. See Figure 4.4.4. ◢

In fact, $f(x)$ has *absolute* minima at $x = \pm 1$. The Second Derivative Test does not distinguish between local and absolute extrema.

Figure 4.4.5(i)

Figure 4.4.5(ii)

Figure 4.4.5(iii)

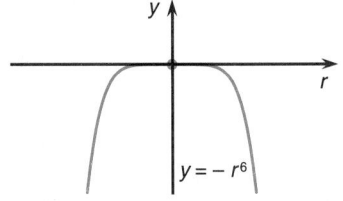

Figure 4.4.5(iv)

EXAMPLE 3   Find and classify the local extrema of (i) $f(x) = x^2$, (ii) $K(s) = s^3$, (iii) $h(t) = t^4$, and (iv) $g(r) = -r^6$. Their graphs are illustrated in Figures 4.4.5(i), (ii), (iii), and (iv).

Solution

(i) $f'(x) = 2x$ and $f''(x) = 2$. The only critical point is $x = 0$, and since $f''(0) = 2 > 0$, the Second Derivative Test tells us that $x = 0$ is a local minimum. In fact we know it is an absolute minimum, since for any $x \neq 0$ we know that $f(x) = x^2 > 0 = f(0)$.

(ii) $K'(s) = 3s^2$ and $K''(s) = 6s$. The only critical point is $s = 0$, and since $K''(0) = 0$, the Second Derivative Test does not apply, and we are unable to use it to decide whether the critical point $s = 0$ is a local extremum. In fact it is not because, as we observed in Example 4.2.5(ii) (p. 240), the function $K(s)$ has no local extrema.

(iii) $h'(t) = 4t^3$ and $h''(t) = 12t^2$. The only critical point is $t = 0$, and $h''(0) = 0$, so the Second Derivative Test is not applicable. In this case we can easily see that $t = 0$ is in fact an absolute minimum, since $h(t) = t^4 \geq 0 = h(0)$, since $t^4 = (t^2)^2$, being a square, is always nonnegative.

(iv) $g'(r) = -6r^5$ and $g''(r) = -30r^4$, so the only critical point is at $r = 0$. Since $g''(0) = 0$, the Second Derivative Test does not apply. In this particular case it is easy to see that $r = 0$ is in fact an absolute maximum, since $g(r) = -r^6 = -(r^3)^2$, being the negative of a square, is never bigger than $0 = g(0)$. ◢

Example 3 illustrates not only how we can use the Second Derivative Test, but also how it cannot be used when the second derivative is zero at a particular critical point. Parts (ii), (iii), and (iv) of Example 3 demonstrate that in this situation it is possible to have maxima or minima or points that are neither.

Despite these shortcomings, the Second Derivative Test is extremely useful. When it can be applied, that is, when the second derivative is nonzero, it is the easiest way to identify local extrema.

The way the Second Derivative Test works is to describe the behavior of a function $f(x)$ for values of x very near to some critical point $x = a$. It can tell us whether the function has a local maximum or local minimum at $x = a$. On the other hand, it is unable to tell us whether or not any of these local extrema are absolute extrema. That depends on the behavior of the function on its entire domain, not just very near $x = a$.

To identify absolute extrema, it is necessary to use our knowledge of the function. There is no simple test that can give that sort of information.

In addition to the Second Derivative Test and using the first derivative to find regions of increase and decrease, it is often helpful to find limits and one-sided limits of $f(x)$ as x approaches $\pm\infty$ or points of discontinuity or points that are not in the domain of $f(x)$.

EXAMPLE 4   (i) Find the extrema of the function $f(x) = x^3 + 3x^2 + 2$ on the interval $[-4, 1]$.

Solution   Since $f'(x) = 3x^2 + 6x = 3x(x + 2)$, the critical points are $x = 0$, $x = -2$, and the endpoints $x = -4$ and $x = 1$. Since $f(x)$ is continuous on a closed interval, it must have extrema, and they must be among these four points.

Figure 4.4.6

Figure 4.4.7

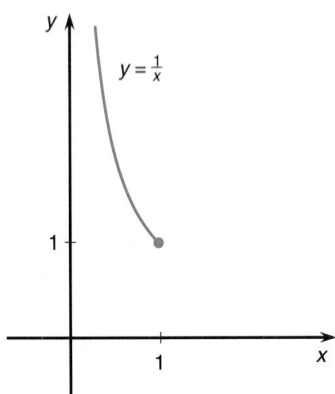

Figure 4.4.8

Evaluating, we find $f(-4) = -14, f(-2) = 6, f(0) = 2$, and $f(1) = 6$. The minimum value is -14, occurring at $x = -4$, and the maximum value is 6, occurring at $x = -2$ and $x = 1$.

The second derivative is $f''(x) = 6x + 6$, so $f''(0) = 6$, which is positive, so the Second Derivative Test says that there is a local minimum at $x = 0$.

The graph of $y = f(x)$ is sketched in Figure 4.4.6.

(ii) Find the extrema of the function $F(x) = x^3 + 3x^2 + 2$ on the interval $(-4, 1)$.

Solution This time the endpoints are not included in the domain. Since by the definition of a critical point (Definition 4.3.1, p. 249), a critical point of $F(x)$ must be in the domain of $F(x)$, the only critical points of $F(x)$ are $x = -2$ and $x = 0$. We know that $F(-2) = 6$ and $F(0) = 2$ and also that

$$\lim_{x \to -4^+} F(x) = -14 \qquad \text{and} \qquad \lim_{x \to 1^-} F(x) = 6.$$

The first limit says that values of $F(x)$ approach -14, which tells us that the value 2 at $x = 0$ is not an absolute minimum (even though the Second Derivative Test tells us it is a local minimum, since $F''(0) = 6 > 0$). So there is no absolute minimum, the absolute maximum occurs at $x = -2$, and there is a local minimum at $x = 0$.

The graph of $y = F(x)$ is illustrated in Figure 4.4.7.

EXAMPLE 5 Find the extrema of $g(x) = \frac{1}{x}$ on the interval $(0, 1]$.

Solution Since $g'(x) = -\frac{1}{x^2}$, which is never zero, the only critical point is the endpoint $x = 1$. Notice that $x = 0$ is not in the domain of $g(x)$.

Since $g'(x) < 0$ for all $x \in (0, 1]$, we can apply Corollary 4.2.9 (p. 243) to see that $g(x)$ is decreasing on $(0, 1]$. This implies that the absolute minimum occurs at the endpoint $x = 1$.

Since $\lim_{x \to 0^+} g(x) = \infty$, there is no maximum. The graph of $y = g(x)$ is sketched in Figure 4.4.8.

EXAMPLE 6 Find the extrema of $h(x) = 3x^4 + 4x^3 - 24x^2 - 48x + 7$.

Solution $h'(x) = 12x^3 + 12x^2 - 48x - 48 = 12(x^3 + x^2 - 4x - 4)$. We want to find when this derivative is zero. In a situation like this, the easiest thing to do is to look for a rational root using the Rational Root Theorem (Theorem 1.3.7, p. 13). We see that $h'(0) = -48 \neq 0$, $h'(1) = -72 \neq 0$, $h'(-1) = 0$, so $x = -1$ is a critical point.

Now since $x = -1$ is a root of $h'(x)$, the Factor Theorem (Theorem 1.3.5, p. 12) tells us that $h'(x)$ can be divided by $x - (-1) = x + 1$. We perform a long division (cf. Example 1.3.8, p. 12), and find that

$$h'(x) = 12(x^3 + x^2 - 4x - 4) = 12(x + 1)(x^2 - 4) = 12(x + 1)(x - 2)(x + 2).$$

The critical points are $x = -1$, $x = 2$, and $x = -2$.

The second derivative is $h''(x) = 36x^2 + 24x - 48$, so we see that $h''(-2) = 48$, $h''(-1) = -36$, and $h''(2) = 144$; $x = -1$ is a local maximum, and $x = \pm 2$ are local minima. Since $h(-2) = 23$, $h(2) = -105$, there is an absolute minimum at $x = 2$ and a local minimum at $x = -2$. Since $\lim_{x \to \pm\infty} h(x) = \infty$, there is no absolute maximum.

It is interesting to use a graphics calculator to plot the function $f(x)$ from Example 4 on the same picture with its derivative $f'(x)$ and its second derivative.

TI-85	SHARP EL9300
GRAPH F1 $(y(x) =)$ F1 (x) ^3 + 3 F1 (x) x^2 + 2 ▼ 3 F1 (x) x^2 + 6 F1 (x) ▼ 6 F1 (x) + 6 M2(RANGE) (−) 4 ▼ 1 ▼ ▼ (−) 20 ▼ 10 F5 (GRAPH)	x/θ/T a^b 3 ▶ + 3 x/θ/T x^2 + 2 2ndF ▼ 3 x/θ/T x^2 + 6 x/θ/T 2ndF ▼ 6 x/θ/T + 6 RANGE − 4 ENTER 1 ENTER 1 ENTER −20 ENTER 10 ENTER 5 ENTER
CASIO f_x-7700GB/f_x-6300G	HP 48SX
Cls EXE Range −4 EXE 1 EXE 1 EXE −20 EXE 10 EXE 5 EXE EXE EXE EXE Graph ▪ x^y 3 + 3 ▪ x^2 + 2 EXE Graph 3 ▪ x^2 + 6 ▪ EXE Graph 6 ▪ + 6 EXE	▪ PLOT PLOTR ERASE ATTN ' αY ▪ αX y^x 3 + 3 × αX y^x 2 + 2 ↱ PLOT ▪ DRAW ERASE 4 +/− ENTER 1 XRNG 20 +/− ENTER 10 YRNG DRAW ATTN ' αY ▪ 3 × αX y^x 2 + 6 × αX ↱ PLOT ▪ DRAW DRAW ATTN ' αY ▪ 6 × αX + 6 ↱ PLOT ▪ DRAW DRAW

Observe the sign of the second derivative at each critical point, and note the corresponding behavior of the first derivative and the original function.

EXAMPLE 7   Find the extrema of

$$f(x) = \frac{x^2 - 1}{x^2 + 2}.$$

Solution   The derivative is

$$f'(x) = \frac{2x(x^2 + 2) - 2x(x^2 - 1)}{(x^2 + 2)^2} = \frac{2x^3 + 4x - 2x^3 + 2x}{(x^2 + 2)^2}$$

$$= \frac{6x}{(x^2 + 2)^2}.$$

So the only critical point is $x = 0$. (Since the denominator is never zero, the function $f(x)$ is always differentiable.)

Notice that $\lim_{x \to \infty} f(x) = \lim_{x \to -\infty} f(x) = 1$ and $f(0) = -\frac{1}{2}$. From the formula for $f'(x)$ we see that $f'(x) < 0$ when $x < 0$ and $f'(x) > 0$ when $x > 0$, so $f(x)$ is decreasing for $x < 0$ and increasing for $x > 0$. By the First Derivative Test we conclude that $f(x)$ has a minimum at $x = 0$ (with less trouble than it would have been to calculate the second derivative) and has no maximum.

In this example we see that analyzing the regions of increase and decrease of $f(x)$ may be easier than applying the Second Derivative Test, especially if the second derivative is complicated to calculate. Of course, there is also the advantage that this approach is always applicable, unlike the Second Derivative Test, which tells us nothing when the second derivative is not defined or equals zero at a critical point.

EXAMPLE 8 Find all local and absolute extrema of $g(t) = \frac{t^3 - 3t}{1+t^2}$.

Solution The derivative is

$$g'(t) = \frac{(3t^2 - 3)(1 + t^2) - 2t(t^3 - 3t)}{(1 + t^2)^2} = \frac{3t^4 - 3 - 2t^4 + 6t^2}{(1 + t^2)^2} = \frac{t^4 + 6t^2 - 3}{(1 + t^2)^2}.$$

To find the critical points, we must solve $t^4 + 6t^2 - 3 = 0$.

To do this, we solve for t^2 first because the equation is quadratic in t^2; that is, letting $u = t^2$, the equation becomes $u^2 + 6u - 3 = 0$, and so

$$u = \frac{-6 \pm \sqrt{36 + 12}}{2} = -3 \pm 2\sqrt{3}.$$

Since $\sqrt{3} > 1.5$, we know $2\sqrt{3} > 3$, so $-3 + 2\sqrt{3} > 0$, and we can let $t = \pm\sqrt{-3 + 2\sqrt{3}} \approx \pm 0.68125$. The other value $u = -3 - 2\sqrt{3}$ is negative and cannot equal t^2 for any real number t.

To work out the second derivative would be straightforward but messy. Instead, it is relatively easy to calculate $g'(-1) = 1$, $g'(0) = -3$, and $g'(1) = 1$, so $g(t)$ is increasing for $t < -\sqrt{-3 + 2\sqrt{3}}$ and for $t > \sqrt{-3 + 2\sqrt{3}}$, and decreasing for $-\sqrt{-3 + 2\sqrt{3}} < t < \sqrt{-3 + 2\sqrt{3}}$. This shows that $-\sqrt{-3 + 2\sqrt{3}}$ is a local maximum and $\sqrt{-3 + 2\sqrt{3}}$ is a local minimum. Since $\lim_{t \to \infty} g(t) = \infty$ and $\lim_{t \to -\infty} g(t) = -\infty$, there are no absolute extrema. ◢

EXAMPLE 9 Consider the function $F(x)$ that is graphed in Figure 4.4.9. At which of the indicated points is $F(x)$ not differentiable, and at which of them is $F'(x) < 0$,

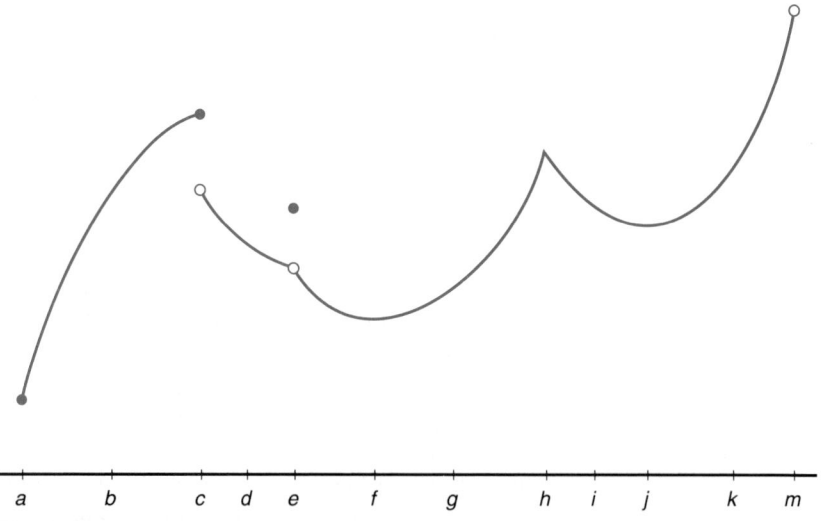

Figure 4.4.9

$F'(x) > 0, F'(x) = 0$? At which of the points does $F(x)$ have a jump discontinuity or a removable discontinuity? Describe all local and absolute extrema of $F(x)$.

Solution $F(x)$ is not differentiable at $x = a$ (the endpoint), $x = c$, $x = e$, $x = h$, and $x = m$. At $x = b$ the tangent line slopes upward to the right, so $F'(b) > 0$; also $F'(d) < 0$, $F'(f) = 0$, $F'(g) > 0$, $F'(i) < 0$, $F'(j) = 0$, and $F'(k) > 0$. There is a jump discontinuity at $x = c$ and a removable discontinuity at $x = e$. This certainly implies that $F(x)$ is not differentiable at those points. It is continuous at $x = h$ but not differentiable there.

There are local minima at $x = a, f$, and j; local maxima at $x = c, e$, and h; an absolute minimum at $x = a$; and no absolute maxima. ◢

EXAMPLE 10 Find the extrema of $h(\theta) = \sec \theta$ for $-\frac{\pi}{2} < \theta < \frac{\pi}{2}$.

Solution The derivative is $h'(\theta) = \sec(\theta)\tan(\theta)$, and $\sec \theta = \frac{1}{\cos \theta}$ is never zero for $\theta \in \left(-\frac{\pi}{2}, \frac{\pi}{2}\right)$. Also the only $\theta \in \left(-\frac{\pi}{2}, \frac{\pi}{2}\right)$ for which $\tan \theta = 0$ is $\theta = 0$, so $\theta = 0$ is the only critical point.

Since $h''(\theta) = \sec \theta \tan \theta \tan \theta + \sec \theta \sec^2 \theta = \sec \theta \tan^2 \theta + \sec^3 \theta$, we find $h''(0) = 0 + 1 = 1$, so $\theta = 0$ is a local minimum. Since $\cos \theta > 0$ for $\theta \in \left(-\frac{\pi}{2}, \frac{\pi}{2}\right)$, we see that $\sec \theta > 0$ also, and $\lim_{\theta \to -\frac{\pi}{2}^+} \sec \theta = \lim_{\theta \to \frac{\pi}{2}^-} \sec \theta = \infty$, so $\theta = 0$ is the absolute minimum, and there are no maxima. ◢

Exercises 4.4

I

Find all local and absolute extrema for each of the following functions. (If the domain is not specified, assume that it is the whole real line.)

1. $f(x) = x^2 - 2x - 2$

2. $f(x) = x^2 + 4x - 6$

3. $f(x) = x^3 - 6x + 3$

4. $f(x) = x^3 - 6x + 3, x \in [-2, 2]$

5. $f(x) = x^3 - 6x + 3, x \in [-1, 1]$

6. $f(x) = x^3 - 6x + 3, x \in [-1, 2]$

7. $f(x) = x^3 - 6x + 3, x \in (-2, 2)$

8. $f(x) = x^3 - 6x + 3, x \in [-2, 1)$

9. $f(x) = x^4 - 8x^2 + 16$

10. $f(x) = 2x^4 + x - 1, x \in [-2, 2]$

11. $g(x) = x^4 - 2x^3 + 3$

12. $h(t) = \frac{t+1}{t^2+1}$

13. $S(\theta) = \sin^2 \theta$

14. $T(\phi) = \tan \phi, \phi \in \left[-\frac{\pi}{4}, \frac{\pi}{4}\right]$

15. $f(\theta) = \cos^2 \theta$

16. $G(t) = \frac{1}{t+1}, t \geq 0$

17. $H(\theta) = \csc \theta, \theta \in (-\pi, 0)$

18. $k(x) = \frac{3x}{x^2+4}$

19. $K(t) = (t - 1)e^t$

20. $F(x) = \cos\left(x - \frac{\pi}{4}\right)$

II

21. (i) Find and classify all critical points of the constant function $K(x) = 3$.

(ii) Find and classify all critical points of the function $A(t) = \frac{|t|}{t}$.

22. Find and classify all critical points of the quadratic function $q(x) = ax^2 + bx + c$. (Assume $a \neq 0$.)

*23. (i) Let $C(t) = t^3 + at^2 + bt + c$. Find conditions on a, b, and c so that $C(t)$ will have no critical points, exactly one critical point, or two critical points.

(ii) Classify the critical points in each of the above cases.

(iii) Find examples of specific numbers a, b, and c that illustrate the various possibilities.

*24. If $Q(x) = x^4 + ax^3 + bx^2 + cx + d$, what are the possible answers to the questions: (i) How many critical points does $Q(x)$ have? (ii) Of what types can the critical points of $Q(x)$ be (local minimum, absolute maximum, etc.)? Give examples.

25. Consider the function $f(x)$ that is graphed in Figure 4.4.10. At which of the indicated points is $f(x)$ not differentiable, and at which of them does it have a removable or jump discontinuity? At which points is the derivative $f'(x)$ positive, negative, or zero? Describe all local and absolute extrema.

Figure 4.4.10

26. Repeat Exercise 25 for the function $g(x)$ graphed in Figure 4.4.11.

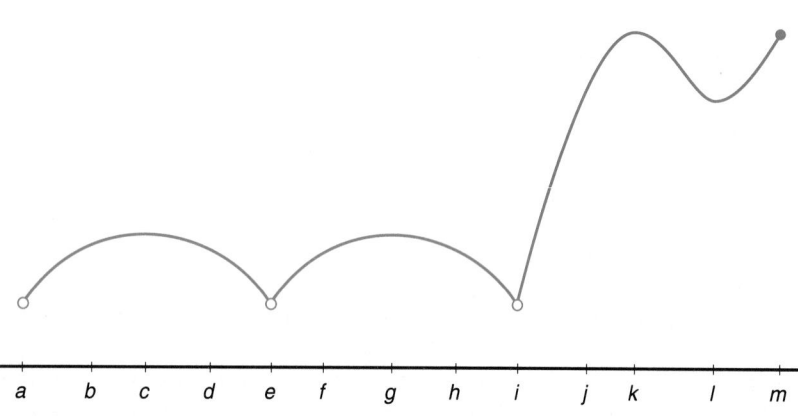

Figure 4.4.11

*27. Let k be a positive integer and define $f(x) = x^k$. Find and classify the critical points of $f(x)$. (Your answer should depend on k.)

*28. (i) Find and classify all critical points of $f(x) = \sinh x$. (You may need to use the fact that $\lim_{x \to \infty} e^x = \infty$.)

   (ii) Find and classify all critical points of $g(x) = \cosh x$.

Use a graphics calculator or computer to graph the following functions.

29. $h(x) = 3x^4 + 4x^3 - 24x^2 - 48x + 7$ (Example 6, p. 258)

30. $f(x) = \frac{x^2 - 1}{x^2 + 2}$ (Example 7, p. 259)

31. $g(t) = \frac{t^3 - 3t}{1 + t^2}$ (Example 8, p. 260)

32. $h(\theta) = \sec\theta$ (Example 10, p. 261)

4.5 Curve Sketching and Asymptotes

EVEN/ODD FUNCTIONS
ASYMPTOTE
VERTICAL ASYMPTOTE
RATIONAL FUNCTIONS

The material of the last few sections gives a great deal of information about the behavior of the graph of a function $y = f(x)$. We know how to find when the graph is increasing or decreasing, and we know how to find local and absolute extrema. All of this can be used to sketch the graph and obtain a reasonably accurate picture of what it looks like. This can be very useful in itself, but the process of sketching the graph will also help us to understand the function $f(x)$.

Computers and graphics calculators can sketch graphs quite quickly and accurately, but it is recommended that you attempt the examples and exercises of this section without these aids and use them only as a means of checking your work afterward, if at all. The finished sketch of a graph may be of considerable use and interest, but learning how to produce it is also an excellent way to exercise your understanding of the concepts we have learned.

In this section we learn how to sketch the graphs of functions, concentrating on polynomials and rational functions. These ideas will be refined and extended to other types of functions in the next section.

EXAMPLE 1   Sketch the graph of $y = f(x) = x^2 + 4x + 2$.

Solution   First we calculate the derivative $f'(x) = 2x + 4$. The only critical point is $x = -2$. Since $f''(x) = 2 > 0$, this critical point is a local minimum.

The derivative $f'(x) = 2x + 4$ is positive when $2x + 4 > 0$, that is, $2x > -4$, that is, when $x > -2$, and $f'(x) < 0$ when $x < -2$. From this we conclude that $f(x)$ is decreasing when $x < -2$ and increasing when $x > -2$ and in particular that $x = -2$ is an absolute minimum. We sketch the real axis in Figure 4.5.1, indicating the sign of the derivative $f'(x)$.

It is easy to see that $\lim_{x \to -\infty} f(x) = \lim_{x \to \infty} f(x) = \infty$, which shows that there is no absolute maximum, and we expect the graph to rise "up to infinity" at either end.

We also get some additional idea of the behavior of the graph by noticing that not only is $f'(x) = 2x + 4 = 2(x - (-2))$ positive when $x > -2$, but it gets bigger and bigger the farther x gets to the right of -2. Similarly, it gets more and more negative the farther x is to the left of -2. This means that the tangents to $y = f(x)$ are steeper and steeper the farther the point of tangency is from $x = -2$. The graph rises more and more steeply as we move farther and farther to the right of $x = -2$ and falls more and more steeply as we move farther and farther to the left of $x = -2$.

Finally, we plot a few points on the graph to give us a rough idea of where it lies. The point on the graph corresponding to the critical point $x = -2$ is $(-2, f(-2)) = (-2, -2)$. We can also plot a few nearby points, such as $(-1, f(-1)) = (-1, -1)$ and $(-3, -1)$, $(0, 2)$, and $(-4, 2)$ (see Figure 4.5.2). While we are at it, we should attempt to find the x-intercepts, the points at which the curve crosses the x-axis. To do this, we simply solve for $y = 0$, that is, for $x^2 + 4x + 2 = 0$. We can use the Quadratic Formula or simply complete the square: $x^2 + 4x + 2 = x^2 + 4x + 4 - 2 = (x + 2)^2 - 2$. This is zero when $x + 2 = \pm\sqrt{2}$, that is, when $x = -2 \pm \sqrt{2}$. So these are the x-intercepts, $x = -2 \pm \sqrt{2} \approx -0.586, -3.414$.

Figure 4.5.1

Figure 4.5.2

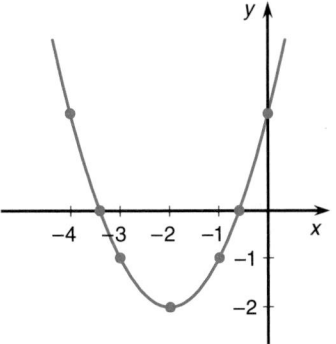

Figure 4.5.3

Finally, we attempt to "fill in" the curve, joining the points we have plotted, bearing in mind what we know about critical points and regions of increase and decrease. In this particular instance we can hardly be surprised to find that the graph is a parabola. See Figure 4.5.3.

> Example 1 illustrates the essential ideas we have at our disposal for sketching graphs. We find the derivative and use it to locate the critical points. The sign of the derivative between critical points tells us where the graph is increasing or decreasing. Using that information and/or the second derivative, we classify the critical points and find local extrema.
>
> Limits of the function at infinity or as we approach points not in the domain are also useful for describing the behavior of the graph of $y = f(x)$ and may in particular be helpful for deciding whether or not there are absolute extrema.
>
> To this largely qualitative information we can add exact information by plotting some points on the graph. The more points we plot, the more detailed and precise will be our information. It makes sense to plot points whose coordinates are easy to calculate. In the above example we tried letting x equal several small integers, which tend to be convenient for computations. Notice that $x = 0$ is likely to be particularly easy to work with, especially if $f(x)$ is a polynomial. It amounts to finding the y-intercept, and if possible we should also find the x-intercepts, that is, the points at which $f(x) = 0$.

EXAMPLE 2   Sketch the graph of $y = f(x)$, where $f(x) = \frac{x}{1+x^2}$.

Solution   We have already considered this function in Example 4.3.2 (p. 249). There we found that its derivative was

$$f'(x) = \frac{1(1 + x^2) - x(2x)}{(1 + x^2)^2} = \frac{1 - x^2}{(1 + x^2)^2},$$

so the critical points are $x = \pm 1$.

Drawing the real line and marking the critical points, we can easily find the sign of the derivative on each of the three intervals. Indeed, $f'(0) = 1, f'(-2) = -\frac{3}{5}$, and $f'(2) = -\frac{3}{5}$, so the derivative is positive between $x = -1$ and $x = 1$ and negative on $(-\infty, -1)$ and $(1, \infty)$. See Figure 4.5.4.

We also find that $\lim_{x \to -\infty} f(x) = \lim_{x \to \infty} f(x) = 0$, and notice that since the denominator of $f(x)$ is always positive, the function $f(x)$ itself is positive when $x > 0$ and negative when $x < 0$. The only x-intercept is at $x = 0$.

It is easy to plot a number of points, such as $(0, 0)$, $\left(1, \frac{1}{2}\right)$, $\left(-1, -\frac{1}{2}\right)$, $\left(2, \frac{2}{5}\right)$, $\left(-2, -\frac{2}{5}\right)$, $\left(3, \frac{3}{10}\right)$, and $\left(-3, -\frac{3}{10}\right)$. From these points we can fill in the rest of the curve, bearing in mind all the qualitative information we have discovered. See Figure 4.5.5.

> Remember that it is enough to evaluate f' at one point in each interval.

Figure 4.5.4

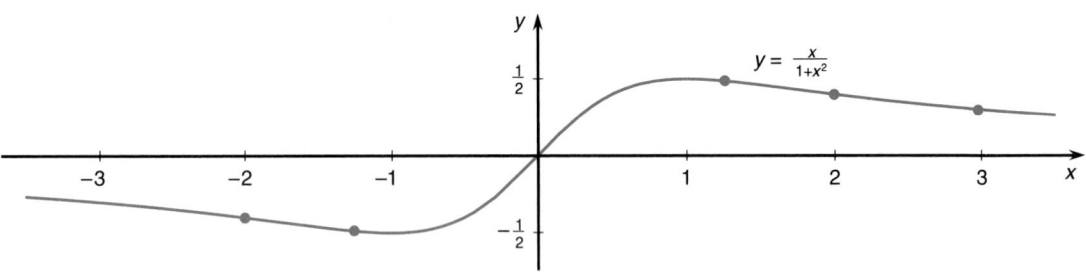

Figure 4.5.5

EXAMPLE 3   Sketch the graph of $y = g(x)$ if $g(x) = x^3 - 3x^2 + 2$.

Solution   We calculate the derivative $g'(x) = 3x^2 - 6x = 3x(x - 2)$, so the critical points are $x = 0$ and $x = 2$. It is very easy to compute $g'(-1) = 9$, $g'(1) = -3$, and $g'(3) = 9$, from which we realize that $g'(x)$ is positive on $(-\infty, 0)$ and $(2, \infty)$ and negative on $(0, 2)$. From this we are able to draw the real line and mark the sign of the derivative on each of the three intervals determined by the critical points. See Figure 4.5.6.

Figure 4.5.6

Without even using the Second Derivative Test, we see that $x = 0$ is a local maximum and $x = 2$ is a local minimum. Since

$$\lim_{x \to -\infty} g(x) = -\infty \qquad \text{and} \qquad \lim_{x \to \infty} g(x) = \infty,$$

we see that there are no absolute extrema.

We plot some points, including the local extrema at the critical points: $(0, 2)$, $(2, -2)$, $(1, 0)$, $(3, 2)$, and $(-1, -2)$ (see Figure 4.5.7). By accident we have found one x-intercept, at $x = 1$. The Factor Theorem (Theorem 1.3.5, p. 12) says that we can divide $g(x)$ by $x - 1$. Carrying out the long division, we find $g(x) = x^3 - 3x^2 + 2 = (x - 1)(x^2 - 2x - 2)$, so to find the other points at which $g(x) = 0$, we need to solve $x^2 - 2x - 2 = 0$. From the Quadratic Formula we find that the other x-intercepts are $1 \pm \sqrt{3} \approx -0.7321, 2.7321$. From this it is possible to fill in the shape of the curve (see Figure 4.5.8).

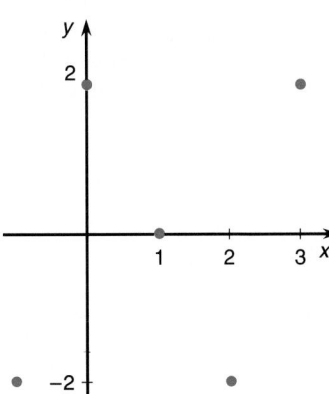

Figure 4.5.7

EXAMPLE 4   Sketch the graph of $y = h(x)$, where $h(x) = \frac{1}{x^2 + 2}$.

Solution   Writing $h(x) = (x^2 + 2)^{-1}$ and applying the Chain Rule, we find that the derivative is

$$h'(x) = -\frac{1}{(x^2 + 2)^2}(2x) = -\frac{2x}{(x^2 + 2)^2}.$$

The only critical point is $x = 0$, and we see that $h'(x) > 0$ when $x < 0$ and $h'(x) < 0$ when $x > 0$, so $x = 0$ is an absolute maximum.

The function $h(x)$ is always positive, and $\lim_{x \to \infty} h(x) = \lim_{x \to -\infty} h(x) = 0$. From this we see that $h(x)$ never equals zero, but it gets arbitrarily close to zero as x gets large, so there is no absolute minimum.

We sketch in a few points, such as $\left(0, \frac{1}{2}\right)$, $\left(1, \frac{1}{3}\right)$, $\left(-1, \frac{1}{3}\right)$, $\left(2, \frac{1}{6}\right)$, and $\left(-2, \frac{1}{6}\right)$, and remark that $h(x)$ is never zero, so there are no x-intercepts. At this stage we notice that $h(-x) = h(x)$ for any x, which can save us trouble when we start to plot points (i.e., when we work out a point, we get another one free). It also tells us that when we plot the curve, it will be "symmetric" about the y-axis, that is, the part of the curve to the left of the axis will be the reflection of the part to the right.

Keeping this in mind and remembering what we know about the regions of increase and decrease, we can plot the curve. See Figure 4.5.9.

Figure 4.5.8

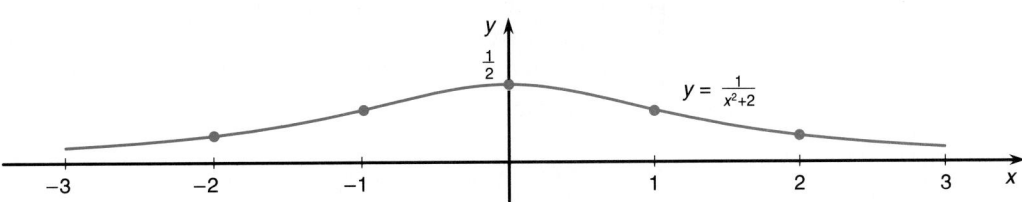

Figure 4.5.9

Using the symmetry of the graph is a very important idea. Not only can it save us a lot of time and effort, but it is an important geometrical property of a curve and should not be missed. The function of Example 4 satisfied $h(x) = h(-x)$. This had the effect of making the graph symmetrical about the y-axis, that is, the part of the graph to the right of the origin is the reflection of the part to the left. Equivalently, it means that reflecting in the y-axis leaves it unchanged.

Another possibility, illustrated by the function $f(x) = \frac{x}{1+x^2}$ of Example 2, is that a function may satisfy $f(-x) = -f(x)$. In terms of the picture, this means that the part of the graph to the right of the origin is obtained by reflecting the part to the left of the origin in the y-axis and then taking its negative. Another way to describe it is that we can draw a line from each point on the graph through the origin and out the same distance on the other side of the origin, and it will end in another point on the graph. Two such points are said to be each other's reflections in the origin, so we can say that reflecting the graph in the origin leaves it unchanged. See Figure 4.5.10.

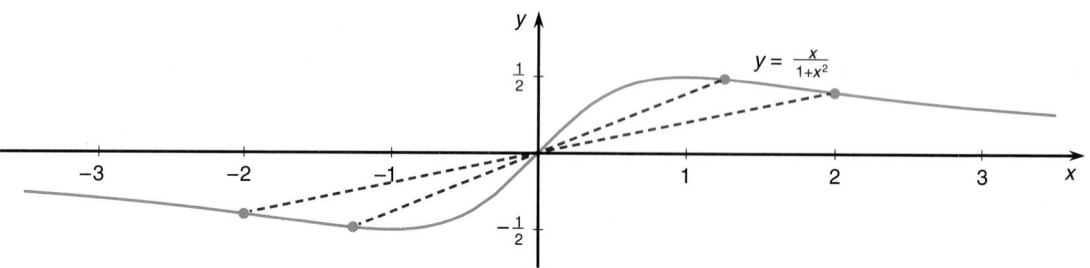

Figure 4.5.10

DEFINITION 4.5.1

> A function $f(x)$ is **even** if $f(-x) = f(x)$ for every x in the domain of f. In particular, if x is in the domain of f, then $-x$ must also be in the domain.
>   A function $f(x)$ is **odd** if $f(-x) = -f(x)$ for every x in the domain of f. In particular, if x is in the domain of f, then $-x$ must also be in the domain.

Notice that some functions are even, some are odd, but many are neither even nor odd.

EXAMPLE 5   The function $f(x) = \frac{x}{1+x^2}$ of Example 2 is odd; the function $h(x) = \frac{1}{x^2+2}$ of Example 4 is even; the functions $x^2 + 4x + 2$ of Example 1 and $x^3 - 3x^2 + 2$ of Example 3 are neither even nor odd.   ◢

> We have just seen that the significance of even and odd functions is that the graph of an even function is symmetric about the y-axis and the graph of an odd function is left unchanged by reflection through the origin.

EXAMPLE 6   Sketch the graph of $F(x) = \frac{|x|}{x}\left(\sqrt{|x|} - |x|\right)$.

Solution   Notice that the function is not defined when $x = 0$ (because of the x in the denominator). In dealing with formulas involving absolute values we have already seen that the thing to do is consider separately what happens when $x > 0$ and when $x < 0$.

First consider the case in which $x > 0$. In this case, $F(x) = \frac{x}{x}(\sqrt{x} - x) = \sqrt{x} - x = x^{1/2} - x$. The derivative is

$$F'(x) = \frac{1}{2}x^{-1/2} - 1 = \frac{1}{2\sqrt{x}} - 1.$$

To find critical points, we see that this will be zero precisely when $2\sqrt{x} = 1$, that is, $\sqrt{x} = \frac{1}{2}$, that is, $x = \frac{1}{4}$.

We notice that when $x > \frac{1}{4}$, then $\sqrt{x} > \frac{1}{2}$, $\frac{1}{2\sqrt{x}} < 1$, so $F'(x) < 0$. Similarly, when $0 < x < \frac{1}{4}$, $F'(x) > 0$. In Figure 4.5.11 we draw the positive real axis and mark on it the critical point and this information about the sign of the derivative. In particular we see that the critical point $x = \frac{1}{4}$ is an absolute maximum (at least on the *positive* axis).

To look for x-intercepts on the positive axis, we must find when $\sqrt{x} - x = 0$, that is, when $\sqrt{x} = x$. The only positive x that satisfies this is $x = 1$, so it is the x-intercept.

Next we should find out what happens as $x \to 0$ and as $x \to \infty$. We see that

$$\lim_{x \to 0^+} F(x) = \left(\lim_{x \to 0^+} \sqrt{x} \right) - \left(\lim_{x \to 0^+} x \right) = 0 - 0 = 0.$$

However, the behavior at ∞ is more difficult to describe.

We expect that as $x \to \infty$, $\sqrt{x}$ will tend to infinity. Indeed, $\sqrt{x}$ gets bigger and bigger. After all, if A is any large number, $\sqrt{x} > A$ whenever $x > A^2$. This says that $\sqrt{x}$ eventually gets bigger than any fixed number A, which is exactly what it means to say $\lim_{x \to \infty} \sqrt{x} = \infty$. As $x \to \infty$, $\sqrt{x}$ and x both tend to infinity, so we do not know what will happen to their difference. It seems reasonable to think that $\sqrt{x}$ is smaller than x (at least for $x > 1$), so $\sqrt{x}$ will go to infinity more slowly than x itself, so perhaps we should expect that $\sqrt{x} - x$ will tend to $-\infty$.

The easiest way to show this would be to show that $F(x)$ is less than some function that we know tends to $-\infty$, at least for large values of x. Suppose $x > 4$. Then $x = \sqrt{x}\sqrt{x} > 2\sqrt{x}$ and $-x < -2\sqrt{x}$. So for $x > 4$, we see that $F(x) = \sqrt{x} - x < \sqrt{x} - 2\sqrt{x} = -\sqrt{x}$. Since we have just seen that $\sqrt{x} \to \infty$, we know that $-\sqrt{x} \to -\infty$, so $\lim_{x \to \infty} F(x) = -\infty$.

Using all the information we have collected, we can draw a sketch of the part of the graph corresponding to positive values of x. See Figure 4.5.12.

The next step is to consider what happens when $x < 0$. It is possible to analyze this case in much the same way as we have just done when $x > 0$, but it is much easier to notice that $F(x)$ is an odd function. (After all, the parts of the formula involving $|x|$ do not change when x is replaced by $-x$, and the quotient $\frac{|x|}{x}$ will change to its negative.)

From this we realize that the part of the graph to the left of the origin is just the reflection through the origin of the part to the right of the origin, which we have already sketched. We can just fill in the remainder of the sketch without further trouble. See Figure 4.5.13.

In drawing the graph we noticed that the function is not defined at $x = 0$, so we have left a little open circle in the picture. On the other hand, both one-sided limits exist there, and in fact they are both equal to 0. This means that there is a removable discontinuity at $x = 0$. We can also remark that there are no absolute extrema, since the function tends to ∞ at one end and to $-\infty$ at the other.

Example 6 demonstrates that recognizing even or odd functions can save a great deal of work.

Figure 4.5.11

Figure 4.5.12

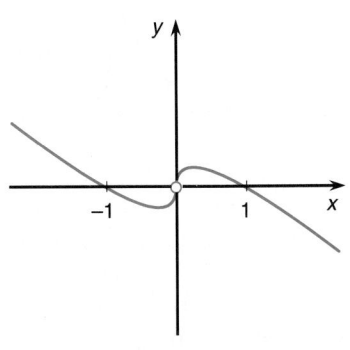

Figure 4.5.13

EXAMPLE 7   Sketch the graph of $G(x) = \frac{x^2}{x^2+2}$.

Solution   We could begin by finding the derivative of $G(x)$ and looking for critical points, but in this case there is an easy way to do the problem. We notice that

$$G(x) = \frac{x^2}{x^2+2} = \frac{x^2+2}{x^2+2} - \frac{2}{x^2+2} = 1 - \frac{2}{x^2+2}.$$

> We added and subtracted $\frac{2}{x^2+2}$.

In Example 4, we sketched the graph of $h(x) = \frac{1}{x^2+2}$. The graph of $y = \frac{2}{x^2+2}$ will be the graph of $h(x)$ made twice as high, that is, stretched by a factor of 2 in the vertical direction. The graph of $y = -\frac{2}{x^2+2}$ will be *that* graph turned upside down. See Figure 4.5.14.

Since $G(x) = 1 - \frac{2}{x^2+2}$, its graph will be the graph of $y = -\frac{2}{x^2+2}$ shifted up by one unit (after all, this is the effect of adding 1 to a function). See Figure 4.5.15.

Figure 4.5.14

Figure 4.5.15

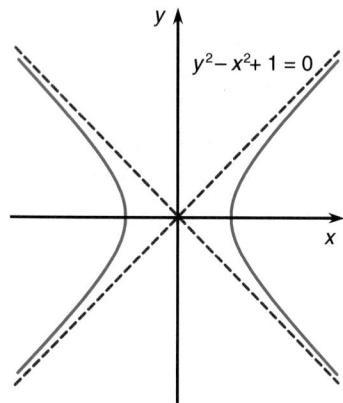

Figure 4.5.16

Asymptotes

One feature of the graph we have just drawn is that the function approaches the limit 1 as $x \to \pm\infty$. In the sketch we see how the curve gets closer and closer to the horizontal line $y = 1$. The functions $f(x) = \frac{x}{1+x^2}$, sketched in Example 2, and $h(x) = \frac{1}{x^2+2}$, sketched in Example 4, also have the property that as $x \to \infty$ and as $x \to -\infty$, the function approaches a certain line (in both these cases it is the line $y = 0$).

Other examples of similar behavior can be found in the graph of a hyperbola (cf. Summary 1.6.4, p. 38). For instance, in Figure 4.5.16 we see the hyperbola $y^2 - x^2 + 1 = 0$. As it stands, this equation does not determine y as a function of x (since there are two values of y for each x). But of course we can choose part of the hyperbola that does define y as a function of x. The most obvious thing to do is to let $y = T(x) = \sqrt{x^2 - 1}$, which defines a function whose graph is the "top half" of

Figure 4.5.17

Figure 4.5.18

Figure 4.5.19(i)

Figure 4.5.19(ii)

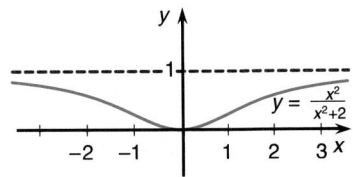

Figure 4.5.19(iii)

the hyperbola. We know that as $x \to \pm\infty$ the hyperbola gets close to one of the two lines we call its asymptotes, getting closer the farther it goes from the origin.

This can be seen from the formula that defines $T(x)$, as follows. Suppose x is a very large number. Then $x^2 - 1$ is close to x^2, and $T(x) = \sqrt{x^2 - 1}$ will be close to $\sqrt{x^2} = x$. This suggests that the graph of $y = T(x)$ will be close to the graph of $y = x$ when x is very large, although it is not clear how close. A similar argument can be used to suggest that when x is a very large negative number, the graph of $y = T(x)$ is close to the graph of the line $y = |x| = -x$. See Figure 4.5.17. The behavior of the bottom half of the hyperbola is similar.

Another similar kind of behavior is found in the function $y = R(x) = \frac{1}{x}$. As $x \to \pm\infty$, the function $R(x)$ tends to zero, which means that the graph of $y = \frac{1}{x}$ gets closer and closer to the line $y = 0$.

But we should also consider what happens near $x = 0$. As x approaches 0 from above, that is, $x \to 0^+$, we know that $R(x) = \frac{1}{x}$ approaches ∞. Similarly, as $x \to 0^-$, x is negative, and $R(x) = \frac{1}{x}$ approaches $-\infty$. In Figure 4.5.18 we see that the graph of $y = R(x)$ gets closer and closer to the y-axis, moving farther and farther away from the origin.

What all of these situations have in common is that the graph of a function approaches some straight line. This situation occurs quite frequently, and there is special terminology for it. Notice that any straight line that is not vertical can be written in the form $y = ax + b$.

DEFINITION 4.5.2

A nonvertical line $y = ax + b$ is an **asymptote** for the function $f(x)$ if the graph of $y = f(x)$ approaches the line as $x \to \infty$ or as $x \to -\infty$. Symbolically, this means

$$\lim_{x \to \infty} \big(f(x) - (ax + b)\big) = 0$$

or

$$\lim_{x \to -\infty} \big(f(x) - (ax + b)\big) = 0.$$

In this situation we also say that $f(x)$ approaches the line $y = ax + b$ **asymptotically** as $x \to \infty$ (or as $x \to -\infty$, as the case may be).

The vertical line $x = a$ is a **vertical asymptote** for the function $f(x)$ if $\lim_{x \to a^+} f(x) = \pm\infty$ or if $\lim_{x \to a^-} f(x) = \pm\infty$. In this situation we also say that $f(x)$ approaches the line $x = a$ asymptotically as $x \to a^+$ (or as $x \to a^-$, as the case may be).

An asymptote that is neither vertical nor horizontal is called an **oblique asymptote**.

EXAMPLE 8

(i) The functions $f(x) = \frac{x}{1+x^2}$ of Example 2 and $h(x) = \frac{1}{x^2+2}$ of Example 4 both approach the line $y = 0$ asymptotically as $x \to \pm\infty$. See Figures 4.5.19(i) and 4.5.19(ii).

(ii) The function $G(x) = \frac{x^2}{x^2+2}$ of Example 7 has the line $y = 1$ as an asymptote as $x \to \pm\infty$. See Figure 4.5.19(iii).

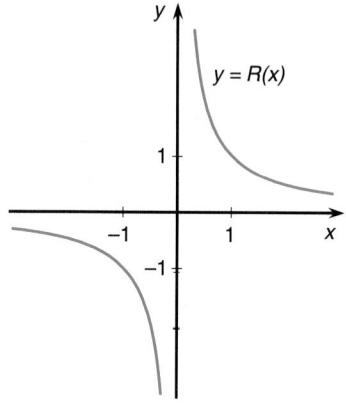

Figure 4.5.19(iv)

(iii) The function $R(x) = \frac{1}{x}$ discussed above has the y-axis $x = 0$ as an asymptote as $x \to 0$. It also approaches the x-axis $y = 0$ asymptotically as $x \to \pm\infty$. See Figure 4.5.19(iv).

(iv) As we have already suggested, the function $T(x) = \sqrt{x^2 - 1}$ approaches the line $y = x$ asymptotically as $x \to \infty$ and approaches the line $y = -x$ asymptotically as $x \to -\infty$. Strictly speaking, we should verify this fact.

What we need to check is that $\lim_{x \to \infty} \left(T(x) - x \right) = 0$. Assuming $x > 0$, we write

$$T(x) - x = \sqrt{x^2 - 1} - x.$$

Now we need a way to deal with the square root; the usual trick in this situation is to multiply and divide by the *sum* of the two terms. What results is

$$T(x) - x = \left(\sqrt{x^2 - 1} - x \right) \frac{\sqrt{x^2 - 1} + x}{\sqrt{x^2 - 1} + x} = \left((x^2 - 1) - x^2 \right) \frac{1}{\sqrt{x^2 - 1} + x}$$

$$= \frac{-1}{\sqrt{x^2 - 1} + x}.$$

Since both terms in the denominator tend to ∞ as $x \to \infty$, this last quotient tends to zero, as desired.

With a graphics calculator we can plot the function $y = T(x)$ of Example 8(iv) on the same picture with the asymptotes $y = \pm x$.

TI-85	SHARP EL9300
GRAPH F1$(y(x) =)$ $\sqrt{\ }$ (F1(x) x^2 $- 1$) ▼ F1(x) ▼ $(-)$ F1(x) M2(RANGE) $(-)$ 4 ▼ 4 ▼ ▼ $(-)$ 4 ▼ 4 F5(GRAPH)	⌐↵ $\sqrt{\ }$ x/θ/T x^2 $- 1$ 2ndF ▼ x/θ/T 2ndF ▼ $-$ x/θ/T RANGE -4 ENTER 4 ENTER 1 ENTER -4 ENTER 4 ENTER 1 ENTER ⌐↵
CASIO f_x-7700GB/f_x-6300G	HP 48SX
Cls EXE Range -4 EXE 4 EXE 1 EXE -4 EXE 4 EXE 1 EXE EXE EXE EXE Graph $\sqrt{\ }$ (X $x^2 - 1$) EXE Graph X EXE Graph $-$ X EXE	⬆ PLOT PLOTR ERASE ATTN ' αY ⬛ $\sqrt{x}$ ◖ αX y^x 2 $- 1$ ↱ PLOT ⬆ DRAW ERASE 4 +/− ENTER 4 XRNG 4 +/− ENTER 4 YRNG DRAW ATTN ' αY ⬛ αX ↱ PLOT ⬆ DRAW DRAW ATTN ' αY ⬛ $- \alpha X$ ↱ PLOT ⬆ DRAW DRAW

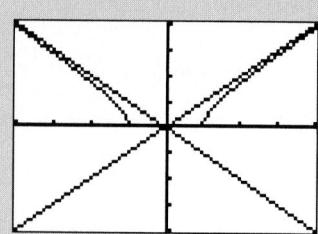

You might like to try plotting different parts of the graph (i.e., for different values of x); you might also like to include the bottom branch of the hyperbola. How close is the hyperbola to its asymptote when $x = 4$? when $x = 10$?

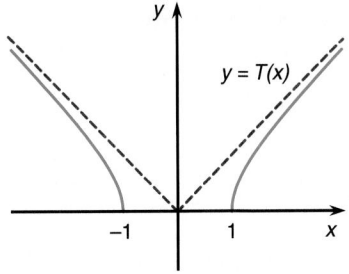

Figure 4.5.19(v)

This shows that $\lim_{x \to \infty} \big(T(x) - x\big) = 0$, which shows that $T(x)$ tends asymptotically to the line $y = x$ as $x \to \infty$. A similar argument shows that $T(x)$ tends asymptotically to the line $y = -x$ as $x \to -\infty$. See Figure 4.5.19(v). ◢

EXAMPLE 9   Sketch the graph of $y = Q(x)$, where $Q(x) = \frac{x+1}{x-1}$.

Solution   The function is not defined at $x = 1$. Elsewhere its derivative is $Q'(x) = \frac{1(x-1)-(x+1)1}{(x-1)^2} = \frac{-2}{(x-1)^2}$. The derivative is never zero, and since the denominator is a square, $Q'(x)$ is always negative. In other words, the function is decreasing on each interval in its domain.

The limits at $\pm\infty$ both exist and are easy to evaluate. In fact, we find that $\lim_{x\to\infty} Q(x) = \lim_{x\to-\infty} Q(x) = 1$. The horizontal line $y = 1$ is an asymptote for $Q(x)$ (at both ∞ and $-\infty$).

As $x \to 1^+$, the numerator is positive and the denominator is positive and tends to zero. Accordingly, the quotient tends to $+\infty$. As $x \to 1^-$, the numerator tends to 2 and the denominator tends to 0 through negative values. Accordingly, the quotient tends to $-\infty$. So $Q(x)$ tends asymptotically to the vertical line $x = 1$ as $x \to 1$ from either side.

The x-intercept is $x = -1$, and it is easy to plot a few points on the graph. For instance, we evaluate and find $(-1, 0)$, $(0, -1)$, $\left(-2, \frac{1}{3}\right)$, $\left(-3, \frac{1}{2}\right)$, $\left(\frac{1}{2}, -3\right)$, $\left(\frac{3}{2}, 5\right)$, $(2, 3)$, and $(3, 2)$.

From all this it is relatively easy to sketch in the graph (see Figure 4.5.20). Notice how we draw the curve to show the asymptotic behavior at $\pm\infty$ and at $x = 1$. ◢

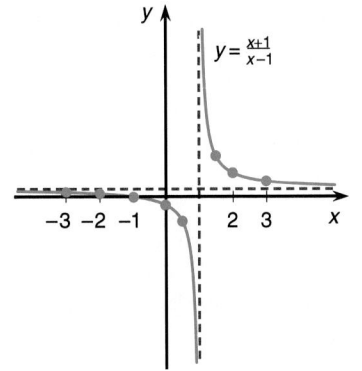

Figure 4.5.20

> Another thing to notice about Example 9 is that the function is decreasing on $(-\infty, 1)$ and on $(1, \infty)$, but it is not decreasing on the whole line (after all, it is bigger to the right of $x = 1$ than it is to the left of $x = 1$). The key point is that since the function is not defined at $x = 1$, its behavior on one side of $x = 1$ is not well related to its behavior on the other side. It decreases on the left side, but it "jumps up" at $x = 1$ before decreasing on the right side. This is why we were careful to say the function is decreasing "on each interval in its domain."

Example 9 also illustrates the way vertical asymptotes can arise with rational functions. If $R(x) = \frac{p(x)}{q(x)}$ is a rational function, suppose $x = a$ is a point for which $q(a) = 0$ but $p(a) \neq 0$. Then depending on the sign of $p(a)$ and on the sign of $q(x)$ as $x \to a^+$ and as $x \to a^-$, the function $R(x)$ will tend to ∞ or to $-\infty$ as $x \to a^+$ and as $x \to a^-$. The vertical line $x = a$ will be an asymptote for the function $R(x)$.

EXAMPLE 10   Sketch the graph of $y = f(x) = \frac{x+1}{x^3 - 2x^2 + x}$.

Solution   First we notice that the denominator factors: $x^3 - 2x^2 + x = x(x^2 - 2x + 1) = x(x-1)^2$, so the function is undefined when $x = 0$ or $x = 1$. Next we calculate the derivative:

$$
\begin{aligned}
f'(x) &= \frac{1(x^3 - 2x^2 + x) - (x+1)(3x^2 - 4x + 1)}{(x^3 - 2x^2 + x)^2} \\
&= \frac{x^3 - 2x^2 + x - 3x^3 + x^2 + 3x - 1}{(x^3 - 2x^2 + x)^2} \\
&= \frac{-2x^3 - x^2 + 4x - 1}{(x^3 - 2x^2 + x)^2}.
\end{aligned}
$$

We try some small values of x in the numerator and find that it is zero when $x = 1$, so the numerator is divisible by $x - 1$. We find $-2x^3 - x^2 + 4x - 1 = (x - 1)(-2x^2 - 3x + 1)$.

So the derivative is

$$f'(x) = \frac{-2x^3 - x^2 + 4x - 1}{(x^3 - 2x^2 + x)^2} = \frac{(x - 1)(-2x^2 - 3x + 1)}{\left(x(x - 1)^2\right)^2} = \frac{-2x^2 - 3x + 1}{x^2(x - 1)^3}.$$

Notice that if we had not factored out the $(x - 1)$ from the numerator and denominator, we might have thought that $f'(x) = 0$ when $x = 1$. This would be incorrect, since the function is not even defined when $x = 1$.

The critical points are the roots of $-2x^2 - 3x + 1$, which are $\frac{3 \pm \sqrt{17}}{-4} \approx -1.781, 0.281$. We draw the real line in Figure 4.5.21, marking these two points and $x = 0, 1$. By evaluating at a point in each interval (e.g., $x = -2$, $x = -1$, $x = \frac{1}{10}$, $x = \frac{1}{2}$, and $x = 2$), we see that $f'(x)$ is positive on $\left(-\infty, -\frac{3 + \sqrt{17}}{4}\right)$ and on $\left(-\frac{3 - \sqrt{17}}{4}, 1\right)$, and negative on the other three intervals.

Figure 4.5.21

As $x \to \pm\infty$, $f(x) \to 0$, so $y = 0$ is an asymptote at both ends of the graph. As $x \to 0^+$, the denominator, which equals $x(x - 1)^2$, is positive and approaches zero. The numerator approaches 1, so the quotient $f(x)$ approaches $+\infty$. As $x \to 0^-$, the denominator is negative, the numerator approaches 1, and $f(x) \to -\infty$. As $x \to 1^+$, the denominator is positive and approaches 0. The numerator approaches 2, so $f(x) \to \infty$. As $x \to 1^-$, both numerator and denominator are positive and $f(x) \to \infty$. We see that $x = 0$ and $x = 1$ are both asymptotes for the graph of $y = f(x)$.

From this information we see that there are no absolute extrema. There is a local maximum at $x = -\frac{3 + \sqrt{17}}{4}$, and a local minimum at $x = \frac{\sqrt{17} - 3}{4}$. The only x-intercept is at $x = -1$. We should plot the local extrema on our graph; for this purpose, approximate values are perfectly adequate. We find $f(-1.781) \approx 0.0567$, $f(0.281) \approx 8.818$.

With the information we now have, it is possible to sketch the graph of $y = f(x)$ reasonably accurately. See Figure 4.5.22.

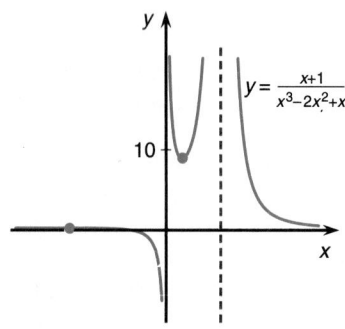

$$y = \frac{x+1}{x^3 - 2x^2 + x}$$

Figure 4.5.22

There is one other type of situation in which asymptotes are readily seen to play a role. Suppose $R(x) = \frac{p(x)}{q(x)}$ is a rational function in which the degree of $p(x)$ is 1 larger than the degree of $q(x)$. In this situation we know that $R(x)$ will tend to ∞ or $-\infty$ as $x \to \infty$ and as $x \to -\infty$.

EXAMPLE 11   Sketch the graph of $y = \frac{x^2 - 1}{2x + 4}$.

Solution   For convenience we write $g(x) = \frac{x^2 - 1}{2x + 4}$. Then $g(x)$ is not defined when $x = -2$. At other points,

$$g'(x) = \frac{2x(2x + 4) - (x^2 - 1)2}{(2x + 4)^2} = \frac{4x^2 + 8x - 2x^2 + 2}{(2x + 4)^2}$$

$$= \frac{2x^2 + 8x + 2}{4(x + 2)^2} = \frac{x^2 + 4x + 1}{2(x + 2)^2}.$$

Figure 4.5.23

The critical points are $x = \frac{-4 \pm \sqrt{12}}{2} = -2 \pm \sqrt{3} \approx -0.268, -3.732$.

By evaluating at $x = -4$, $x = -3$, $x = -1$, and $x = 0$ we find that $g'(x)$ is negative on $(-2 - \sqrt{3}, -2)$ and on $(-2, -2 + \sqrt{3})$ and positive on $(-\infty, -2 - \sqrt{3})$ and on $(-2 + \sqrt{3}, \infty)$ (see Figure 4.5.23).

As $x \to -2^+$, the numerator of $g(x)$ tends to 3 and the denominator, which is obviously positive, tends to zero. We see that $\lim_{x \to -2+} g(x) = \infty$. Similarly, as $x \to -2^-$, the numerator tends to 3 and the denominator is negative and tends to zero, so $\lim_{x \to -2-} g(x) = -\infty$. From this it is apparent that there are no absolute extrema and that there is a local maximum at $x = -2 - \sqrt{3}$ and a local minimum at $x = -2 + \sqrt{3}$.

Finally, we should consider what happens as $x \to \pm\infty$. To do this, we perform a long division and write $g(x) = \frac{x^2-1}{2x+4}$ as a polynomial plus a remainder term. We find

$$g(x) = \frac{x^2 - 1}{2x + 4} = \frac{1}{2}x - 1 + \frac{3}{2x + 4}.$$

Now as $x \to \pm\infty$, the last term $\frac{3}{2x+4}$ tends to zero. This says that as $x \to \pm\infty$, the graph of the function $g(x)$ gets very close to the function $\frac{1}{2}x - 1$. This last function is of course a straight line, and we have just seen that $g(x)$ tends asymptotically to it as $x \to \pm\infty$.

Keeping in mind everything we have learned, we sketch the graph of $y = g(x)$ in Figure 4.5.24. The asymptotes $x = -2$ and $y = \frac{1}{2}x - 1$ are marked as dashed lines. Notice that the remainder term $\frac{3}{2x+4}$, which tells us the difference between $g(x)$ and the straight line, is positive when $x > -2$ and negative when $x < -2$, which says that the graph of $y = g(x)$ is above the asymptote when $x > -2$ and below the asymptote when $x < -2$.

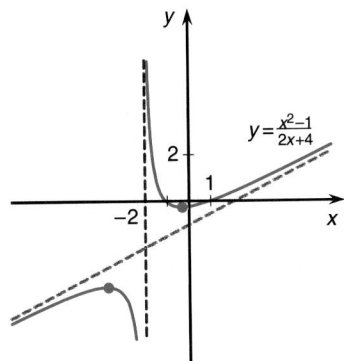

Figure 4.5.24

We began by considering a rational function $R(x) = \frac{p(x)}{q(x)}$ in which the degree of $p(x)$ was 1 larger than the degree of $q(x)$. If we do a long division in this situation, the first term in the answer will be a nonzero constant times x, and the second will be a constant (possibly zero). We get $R(x) = ax + b + \frac{r(x)}{q(x)}$.

In the remainder term the degree of the numerator is smaller than the degree of the denominator, so as $x \to \pm\infty$, $\frac{r(x)}{q(x)}$ will tend to zero.

THEOREM 4.5.3

For a rational function $R(x) = \frac{p(x)}{q(x)}$ in which the degree of $p(x)$ is 1 greater than the degree of $q(x)$, we can write

$$R(x) = ax + b + \frac{r(x)}{q(x)},$$

with $a \neq 0$ and the degree of $r(x)$ less than the degree of $q(x)$. Since $\lim_{x \to \pm\infty} \frac{r(x)}{q(x)} = 0$, we see that $R(x)$ tends asymptotically to the line $y = ax + b$ as $x \to \pm\infty$.

If the degrees of $p(x)$ and $q(x)$ are equal, we can write

$$R(x) = a + \frac{r(x)}{q(x)},$$

with the degree of $r(x)$ less than the degree of $q(x)$ and $a \neq 0$. In this case, $R(x)$ tends asymptotically to the horizontal line $y = a$ as $x \to \pm\infty$.

PROOF The degree of the remainder term $r(x)$ is less than the degree of $q(x)$ because otherwise we could continue the long division.

EXAMPLE 12 Sketch the graph of $y = h(x) = \frac{x^3}{x^2+1}$.

Solution The denominator is never zero, so the function is defined and differentiable everywhere. Its derivative is

$$h'(x) = \frac{3x^2(x^2+1) - x^3(2x)}{(x^2+1)^2} = \frac{x^4 + 3x^2}{(x^2+1)^2} = \frac{x^2(x^2+3)}{(x^2+1)^2}.$$

The only critical point is $x = 0$, and since everything in the derivative is a square, the derivative is positive everywhere except at $x = 0$. It is also clear that $\lim_{x\to-\infty} h(x) = -\infty$ and $\lim_{x\to\infty} h(x) = \infty$, and there are no local maxima or minima. The only x-intercept is $x = 0$. We observe that $h(x)$ is an odd function.

Performing a long division, we find that

$$h(x) = \frac{x^3}{x^2+1} = x - \frac{x}{x^2+1}.$$

As $x \to \pm\infty$, the graph of $y = h(x)$ tends asymptotically to the line $y = x$.

We plot a few points, for example, $(0,0)$, $\left(1, \frac{1}{2}\right)$, $\left(-1, -\frac{1}{2}\right)$, $\left(2, \frac{8}{5}\right)$, and $\left(-2, -\frac{8}{5}\right)$. We are now able to sketch the graph of $y = h(x)$. See Figure 4.5.25.

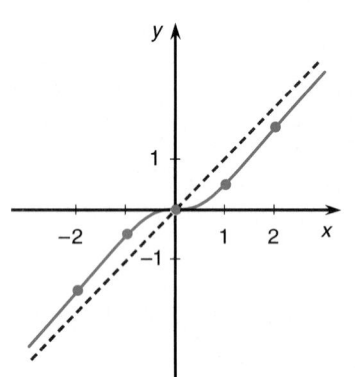

Figure 4.5.25

After we performed the long division in Example 12, we might have observed that since $h(x) = x - \frac{x}{x^2+1}$, we could think of its graph as being obtained from the graph of the straight line $y = x$ by shifting down by the graph of $y = \frac{x}{x^2+1}$. We graphed this last function in Example 2, and if you tilt this page $45°$ clockwise, you can see that the graph of $h(x)$ does appear to be the negative of $y = \frac{x}{x^2+1}$ added to the line $y = x$. (In fact it is stretched out because of the tilting, but the shape is still roughly the same.)

Finally, we summarize the steps we have been using to sketch the graphs of rational functions.

Summary 4.5.4 Sketching Rational Functions

1. Find where the function is undefined, by finding where the denominator is zero. If you cannot factor the denominator, you can use the Rational Root Theorem (Theorem 1.3.7, p. 13) in hopes of finding a small integer value at which the denominator is zero. If the denominator is zero at $x = a$, be sure to check whether the numerator is also zero at $x = a$; if it is, it is possible to factor out $(x - a)$ from both numerator and denominator. Check whether the function happens to be either an even function or an odd function.

2. Find the derivative. Find all critical points. By evaluating at one point in each interval in the domain between critical points, find the sign of the derivative on each interval. Draw a picture of the domain, marking all this information. Find the local extrema.

3. For each point $x = a$ at which the denominator is zero, study the behavior of the rational function as $x \to a^+$ and as $x \to a^-$. This involves the signs of numerator and denominator. Also determine the behavior as $x \to \pm\infty$.

4. If the degree of the numerator is greater than the degree of the denominator by 1, then perform a long division and find the asymptotes at $\pm\infty$. At this stage it should be possible to identify any absolute extrema.

5. Plot a few points, including local and absolute extrema and all x-intercepts that can be found. Sketch the curve.

You might like to review the examples of this section, many of which are rational functions, in light of this summary.

Exercises 4.5

I

Find all asymptotes for each of the following functions.

1. $f(x) = \frac{x^2}{x-1}$

2. $f(x) = \frac{x^2}{x+2}$

3. $f(x) = \frac{x^2+2}{x^2-3}$

4. $f(x) = \frac{x^2+x-1}{x+2}$

5. $f(x) = \frac{x^3}{x^2-4}$

6. $f(x) = \frac{2x^3-x^2+1}{3x^2-1}$

7. $f(x) = \frac{x^3}{x^2+x}$

8. $f(x) = \frac{x^4-x}{2x^2-3x+1}$

9. $f(x) = \frac{x^7}{(x-3)(x+2)x^2}$

10. $f(x) = \frac{x^2-1}{x+1}$

Sketch the graphs of the following functions, showing all asymptotes.

11. $f(x) = \frac{x}{x-1}$

12. $f(x) = \frac{x^2}{x-1}$

13. $f(x) = \frac{x^3}{x^2-1}$

14. $f(x) = \frac{x^3}{x^2-4}$

15. $f(x) = \frac{x^3-3x^2+3x-1}{x^2-2x}$

16. $f(x) = \frac{x^3}{x^2+4}$

17. $f(x) = \frac{1}{x}$

18. $f(x) = \frac{x^3+x^2+4}{x^2+4}$

19. $f(x) = \frac{x^3+x+1}{x^2+1}$

20. $f(x) = \frac{x^4-2x^2-3}{x^2+1}$

II

21. Sketch the graph of $y = f(x) = \frac{x^2-9}{2x+4}$.

22. Sketch the graph of $y = g(x) = \frac{4x^2-1}{3x-6}$.

*23. Consider a rational function $R(x) = \frac{p(x)}{q(x)}$, where $q(x) = (x-a)^k g(x)$ and $g(a) \neq 0$, $p(a) \neq 0$, $k \geq 0$. Under what circumstances will $\lim_{x\to a^+} R(x) = \lim_{x\to a^-} R(x)$?

24. (i) Show that if $f(x)$ is even and $g(x)$ is odd, then $f(x)g(x)$ is odd.

  (ii) Show that if $f(x)$ and $g(x)$ are both even or both odd, then their product is even.

  (iii) Show that the sum of two even functions is even and the sum of two odd functions is odd.

*25. Find conditions on a polynomial $P(x) = a_n x^x + a_{n-1}x^{n-1} + \ldots + a_1 x + a_0$ that guarantee that $P(x)$ is (i) even, (ii) odd, (iii) neither even nor odd.

**26. Let $F(x)$ be a function whose domain is the whole real line. Show that it is possible to write $F(x) = F_{even}(x) + F_{odd}(x)$, where $F_{even}(x)$ is an even function and $F_{odd}(x)$ is an odd function.

*27. (i) Let $f(x) = 3x^4 - 2x^3 - x^2 + 7x - 1$. Write $f(x)$ as the sum of an even function and an odd function.

  (ii) If $P(x) = a_n x^n + a_{n-1}x^{n-1} + \ldots + a_1 x + a_0$, write $P(x)$ as the sum of an even and an odd function.

*28. Let $g(x)$ be both even and odd. What can you say about $g(x)$?

*29. If $H(x) = H_e(x) + H_o(x)$, where H_e and H_o are even and odd, respectively, show that this can be done in only one way, that is, show that if $H(x) = f(x) + g(x)$ with f even and g odd, then $f(x) = H_e(x)$ and $g(x) = H_o(x)$.

*30. Find an example of a function $G(x)$ that cannot be written as the sum of an even function and an odd function. (*Hint:* Think about the domain of $G(x)$.)

*31. (i) Suppose $f(x)$ is an odd function that is differentiable at every x. Show that $f'(x)$ is an even function.

  (ii) Suppose $f(x)$ is an even function that is differentiable at every x. What can you say about $f'(x)$?

  (iii) What happens in parts (i) and (ii) if we remove the assumption that $f(x)$ is differentiable everywhere and just let $f'(x)$ be the derivative when it does exist and be undefined when it does not?

*32. Write each of the following functions as the sum of an even function and an odd function: (i) $\cos\left(x + \frac{\pi}{4}\right)$, (ii) e^x, (iii) $(x-3)^4$.

Use a graphics calculator or computer to sketch the following functions considered in the examples in the text.

33. $f(x)$ (Example 2)   **35.** $h(x)$ (Example 4)
34. $g(x)$ (Example 3)   **36.** $G(x)$ (Example 7)

37. $Q(x)$ (Example 9)   **39.** $g(x)$ (Example 11)
38. $f(x)$ (Example 10)   **40.** $h(x)$ (Example 12)

41. Sketch the function $G(x)$ of Example 7 (p. 268) on the same graph with $y = -\frac{2}{x^2+2}$.

42. Use a graphics calculator or computer to find how large x has to be before $T(x) = \sqrt{x^2 - 1}$ stays within 0.001 of x (cf. Example 8(iv), p. 270).

POINT TO PONDER

In the examples of asymptotes we have considered so far, the graph of a function has usually approached the asymptotic line from above or below (cf. Figures 4.5.26(i) and 4.5.26(ii)). However, it is perfectly possible for the function to cross the asymptote.

As an example, consider the graph of $y = f(x) =$ $\frac{x}{2} + \frac{1}{x}\sin(5x)$, shown in Figure 4.5.26(iii). As $x \to \pm\infty$, we see that $\frac{1}{x}\sin(5x) \to 0$, so $f(x)$ approaches the line $y = \frac{x}{2}$, which is the asymptote. However, the function $f(x)$ crosses and recrosses this line as $x \to \pm\infty$.

Notice that $\lim_{x\to 0} f(x) = \lim_{x\to 0} \frac{\sin(5x)}{x} = 5$.

Figure 4.5.26(i)

Figure 4.5.26(ii)

Figure 4.5.26(iii)

4.6 Concavity and Inflection Points; Applications to Curve Sketching

CONCAVE UP
CONCAVE DOWN
SECOND DERIVATIVE AND CONCAVITY
INFLECTION POINT
CUSP

In this section we introduce the notion of concavity and discuss how to test for it and how to apply it to curve sketching problems. In the process we learn about *inflection points,* which are points where the concavity changes.

In Figure 4.6.1 there are several short curves. The first one is increasing, the second is decreasing, and the third is decreasing at first and then increasing. What they all have in common is the way in which they are bending.

This may be clearer by comparison with Figure 4.6.2, in which there are also three curves. One is increasing, one is decreasing, and one is increasing at first and then decreasing.

Figure 4.6.1

Figure 4.6.2

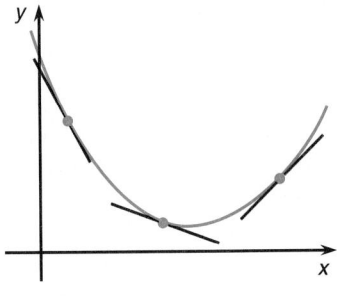

Figure 4.6.3

When a curve bends, we can think of it having a concave side and a convex side. The three curves in Figure 4.6.1 all have their concave side facing up, while the ones in Figure 4.6.2 all have their concave side facing down. Notice that this is independent of whether a curve is increasing or decreasing, in the sense that the concave side can face up for increasing or decreasing curves and can likewise face down for increasing or decreasing curves.

The first thing we need to do is make a precise definition of what we mean by concavity. Once that is accomplished, we should look for ways to find out which side of a particular curve is concave. We will only consider the question at a point where the function is differentiable.

One great advantage of this restriction is that it means that we can draw the tangent line. In Figure 4.6.3 we have drawn a curve that is "concave up," that is, whose concave side faces upward. The tangent lines have been drawn in at several points, and from the picture we see that the curve lies *above* each tangent line.

Figure 4.6.4 shows a function that is "concave down," and the curve lies below all its tangent lines. This certainly suggests a way of defining what we mean by "concave up" and "concave down," but we have to be careful, as the next example shows.

EXAMPLE 1   Consider the function $y = g(x) = x^3 - 3x^2 + 2$, whose graph we sketched in Example 4.5.3 (p. 265) (see Figure 4.6.5). The part of the graph near $x = 0$, which looks like a hump, is concave down. However, the part of the curve near $x = 2$, which looks like a valley, is concave up. In the first place, this suggests that it is perfectly possible for a function to be concave up at some points and concave down at other points.

Now if we try the idea of looking at the tangent lines, we find that the part of the curve that is the hump does in fact lie below the tangent line at $x = 0$ (which of course is horizontal). However, the curve bends up again and crosses this tangent line and eventually rises above it. Similarly, the part of the curve that is the valley does lie above the tangent line at $x = 2$, but part of the curve farther to the left bends down and actually crosses below the tangent.

We have chosen the two horizontal tangents because they are easy to draw, but similar remarks apply to other tangent lines. The curve may well lie above or below the tangent near the point of tangency, but farther away it may loop back and cross the line.

Figure 4.6.4

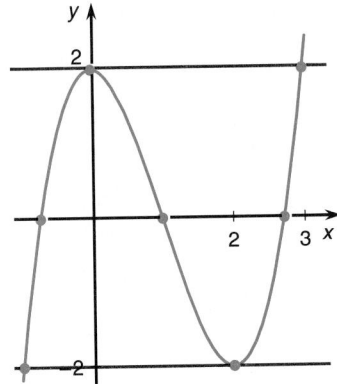

Figure 4.6.5

The answer to this difficulty is to realize that the property of a curve being concave up or concave down at some point $x = a$ is something that has to do only with its behavior near that point. So the definition should be that the curve lies above or below the tangent line *near* $x = a$, that is, on some open interval containing $x = a$.

Recall that the tangent line to $y = f(x)$ at the point $x = a$ has the equation $y = f(a) + f'(a)(x - a)$ (cf. Definition 3.1.1).

DEFINITION 4.6.1

A function $f(x)$ that is differentiable at $x = a$ is **concave up** at $x = a$ if there is an interval (c, d) containing a so that the graph $y = f(x)$ lies above the tangent line to $f(x)$ at $x = a$ at every point in the interval (c, d) except $x = a$. It is **concave down** at $x = a$ if there is an interval (c, d) containing a so that the graph $y = f(x)$ lies below the tangent line to $f(x)$ at $x = a$ at every point in the interval (c, d) except $x = a$.

In symbols this means that $f(x)$ is concave up at $x = a$ if

$$f(x) > f(a) + f'(a)(x - a), \qquad \text{for every } x \in (c, d) \text{ except } x = a$$

and concave down if

$$f(x) < f(a) + f'(a)(x - a) \qquad \text{for every } x \in (c, d) \text{ except } x = a.$$

A function is said to be concave up (respectively concave down) on an interval if it is concave up (respectively concave down) at every point in that interval. It is said to be concave up (respectively concave down) if it has this property at every point in its domain.

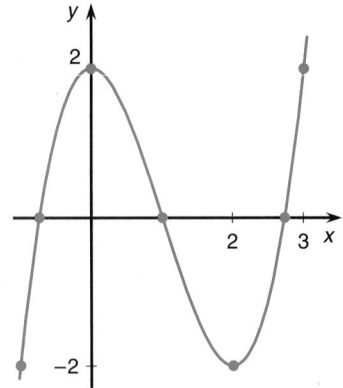

Figure 4.6.6

EXAMPLE 2

(i) The function $g(x) = x^3 - 3x^2 + 2$ discussed in Example 1 is concave down on $(-\infty, 1)$ and concave up on $(1, \infty)$ (see Figure 4.6.6). Notice that although the curve does eventually rise above the horizontal tangent at $x = 0$, there is an interval (e.g., $(-1, 1)$ will do in this case) on which the curve is always below the tangent. It is possible to find such an interval for the tangent at any point $x = a$ with $a < 1$ (although the interval might not continue very far to the right of a if a is very close to 1). If $a > 1$, an interval can be found on which the curve is above the tangent at $x = a$.

(ii) The parabola $y = x^2$ is concave up (i.e., concave up at every point). See Figure 4.6.7.

(iii) The function $y = x^3$, sketched in Figure 4.6.8, is concave down on $(-\infty, 0)$ and concave up on $(0, \infty)$.

Figure 4.6.7

Figure 4.6.8

Figure 4.6.9

Figure 4.6.10

Figure 4.6.11

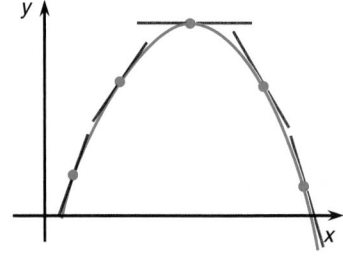

Figure 4.6.12

(iv) The function $\sin x$ is concave down on $(0, \pi)$ and concave up on $(\pi, 2\pi)$. See Figure 4.6.9.

(v) The function $\sqrt{x}$ is concave down on $(0, \infty)$. Notice that it is only defined for $x \geq 0$, and it does not make sense to ask about its concavity at $x = 0$, since it is not differentiable there. See Figure 4.6.10. ◢

Knowing the concavity of $f(x)$ is a very good way to improve the accuracy of our graphs. What we want is to be able to say where the function is concave up and where it is concave down, just as we began our sketching efforts by looking for where it was increasing or decreasing. Regions of increase and decrease were identified by using the derivative $f'(x)$, and it turns out that the concavity can be determined from the second derivative $f''(x)$.

In Figure 4.6.11 we see a curve that is concave up. At the left end, the curve is decreasing, so its derivative is negative. As we move to the right, the tangents slope less steeply, so the derivative becomes closer to zero. At the point $x = a$ the tangent is horizontal and the slope is zero. Farther to the right, the slope becomes positive, and the tangents slope more and more steeply upward, meaning that the slope becomes larger and larger.

In short, the derivative is increasing all the time. If we look at the graph of a function that is concave down, we see in a similar fashion that its derivative is always decreasing. See Figure 4.6.12.

If the derivative is increasing, it means that the tangent lines slope more and more steeply up (or less and less steeply down), which suggests that the curve is bending toward the region above the curve, which in turn suggests that it will be concave up. A decreasing derivative suggests a curve that is concave down. We can detect when the derivative is increasing or decreasing by looking at *its* derivative, the second derivative.

THEOREM 4.6.2

The Second Derivative and Concavity

Suppose $f''(x)$ exists at a point $x = a$.

If $f''(a) > 0$, then $f(x)$ is concave up at $x = a$.

If $f''(a) < 0$, then $f(x)$ is concave down at $x = a$.

The proof of Theorem 4.6.2 is surprisingly tricky. Instead of reading it right away, you might prefer to skip ahead to Example 3 to get some idea of how the theorem is used. Using the result is actually quite easy.

PROOF   Suppose $f''(a) > 0$. In particular, this implies that $f'(x)$ must exist on some open interval containing $x = a$. Moreover, since $f''(a) = \lim_{x \to a} \frac{f'(x) - f'(a)}{x - a}$ is *positive*, we can apply Theorem 2.3.4 to conclude that there is an open interval (r, s) containing a so that

$$\frac{f'(x) - f'(a)}{x - a} > 0, \qquad \text{whenever } x \in (r, s) \qquad (\text{except} \quad x = a).$$

We will show that on this interval (r, s), the curve $y = f(x)$ lies above the tangent line to the curve at $x = a$. The vertical distance between the curve $y = f(x)$ and the tangent line $y = f(a) + f'(a)(x - a)$ at any particular value of x is

$$D = f(x) - \big(f(a) + f'(a)(x - a)\big)$$

Figure 4.6.13

Figure 4.6.14(i)

Figure 4.6.14(ii)

Figure 4.6.14(iii)

Figure 4.6.15

(see Figure 4.6.13). We have to show that $D > 0$ whenever $x \in (r, s)$, except for $x = a$.

Now

$$D = f(x) - \big(f(a) + f'(a)(x - a)\big) = f(x) - f(a) - f'(a)(x - a)$$

$$= \left(\frac{f(x) - f(a)}{x - a} - f'(a)\right)(x - a).$$

The Mean Value Theorem (Theorem 4.2.5) shows that the average slope $\frac{f(x)-f(a)}{x-a}$ equals $f'(c)$ for some c between a and x. In particular, $c \in (r, s)$. So

$$D = \big(f'(c) - f'(a)\big)(x - a)$$

$$= \frac{f'(c) - f'(a)}{c - a}(c - a)(x - a).$$

The first factor is positive, since $c \in (r, s)$. Since c and x are on the same side of a, the factors $(c - a)$ and $(x - a)$ have the same sign, so the product $(c - a)(x - a)$ is also positive. This shows that $D > 0$ whenever $x \in (r, s)$, as desired.

The case in which $f''(a) < 0$ can be proved similarly, or by considering the function $-f(x)$. ◢

EXAMPLE 3

(i) The function $f(x) = x^2$ satisfies $f'(x) = 2x$, $f''(x) = 2$. Since the second derivative is always positive, the function is concave up at every point. See Figure 4.6.14(i).

(ii) The function $g(x) = x^3$ satisfies $g'(x) = 3x^2$, $g''(x) = 6x$. The function $g''(x) = 6x$ is positive when $x > 0$ and negative when $x < 0$, so Theorem 4.6.2 tells us that $g(x)$ is concave up when $x > 0$ and concave down when $x < 0$. See Figure 4.6.14(ii).

(iii) The function $h(x) = e^x$ satisfies $h'(x) = h''(x) = e^x$. Since e^x is always positive, the function $h(x)$ is concave up at every point. See Figure 4.6.14(iii). ◢

Now when we sketch the graph of a function, we can determine not only the regions in which it is increasing or decreasing, but also the regions in which it is concave up or concave down. Just as we drew a picture of the real line and marked on it the sign of the derivative, we can mark the type of concavity. One way to do this is simply to draw little U-shaped curves in the regions where the function is concave up and inverted U's where it is concave down (see Figure 4.6.15). This information can easily be found from the second derivative and will be useful when the sketch of the curve is drawn.

EXAMPLE 4 Recall the function $h(x) = \frac{1}{x^2+2}$, which we considered in Example 4.5.4. There we calculated the derivative

$$h'(x) = -\frac{2x}{(x^2 + 2)^2}.$$

Differentiating this, we find the second derivative:

$$h''(x) = -\frac{2(x^2 + 2)^2 - (2x)2(x^2 + 2)2x}{(x^2 + 2)^4} = -\frac{2x^2 + 4 - 8x^2}{(x^2 + 2)^3}$$

$$= \frac{6x^2 - 4}{(x^2 + 2)^3}.$$

The denominator is always positive. The numerator is positive when $x < -\sqrt{\frac{2}{3}}$ and when $x > \sqrt{\frac{2}{3}}$.

We conclude that the second derivative is positive on $\left(-\infty, -\sqrt{\frac{2}{3}}\right)$ and on $\left(\sqrt{\frac{2}{3}}, \infty\right)$ and similarly that it is negative on $\left(-\sqrt{\frac{2}{3}}, \sqrt{\frac{2}{3}}\right)$.

We draw the real line in Figure 4.6.16, marking these intervals and indicating the concavity of $h(x)$, marking concave up regions (where $h''(x) > 0$) with little U-shaped curves and concave down regions with little inverted U-shaped curves.

When we sketch the graph, we can use this information to give the graph the correct shape. See Figure 4.6.17.

Figure 4.6.16

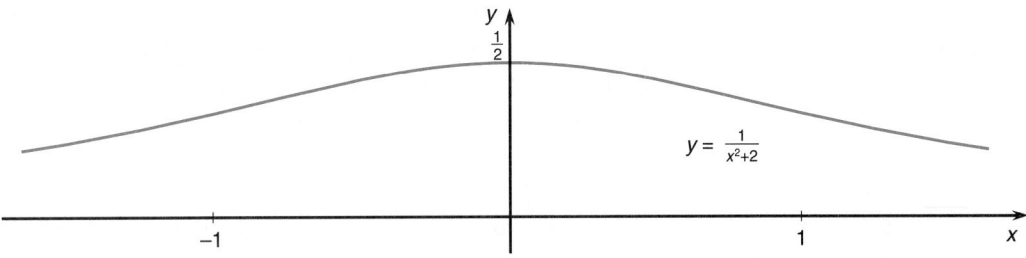

Figure 4.6.17

EXAMPLE 5   Sketch the graph of $y = S(\theta) = \sec\theta$.

Solution   Note that $S(\theta) = \frac{1}{\cos\theta}$ is not defined when $\cos\theta = 0$, that is, when $\theta = \pm\frac{\pi}{2}, \pm\frac{3\pi}{2}, \pm\frac{5\pi}{2}, \ldots$. Recall that $S'(\theta) = \sec\theta\tan\theta = \frac{\sin\theta}{\cos^2\theta}$. So $S'(\theta) = 0$ for $\theta = n\pi$, with $n = 0, \pm 1, \pm 2, \ldots$.

Let us consider the graph between $\theta = 0$ and $\theta = 2\pi$. For $\theta \in \left(0, \frac{\pi}{2}\right)$, $\sin\theta$ is positive (see Figure 4.6.18), and since $\cos^2\theta > 0$, we see that $S'(\theta) > 0$ for $\theta \in \left(0, \frac{\pi}{2}\right)$. Similarly, $\sin\theta > 0$ for $\theta \in \left(\frac{\pi}{2}, \pi\right)$, so $S'(\theta) > 0$ there, and $S'(\theta) < 0$ for $\theta \in \left(\pi, \frac{3\pi}{2}\right)$ and for $\theta \in \left(\frac{3\pi}{2}, 2\pi\right)$.

By the Product Rule, $S''(\theta) = \sec\theta\tan^2\theta + \sec^3\theta = \sec\theta(\tan^2\theta + \sec^2\theta)$. Since the sum of squares is never negative, $S''(\theta)$ is positive for $\theta \in \left(0, \frac{\pi}{2}\right)$ and $\theta \in \left(\frac{3\pi}{2}, 2\pi\right)$. It is negative for $\theta \in \left(\frac{\pi}{2}, \frac{3\pi}{2}\right)$. We sketch the part of the real line from 0 to 2π in Figure 4.6.19 and mark the concavity of $S(\theta)$ in the usual way.

As $\theta \to \frac{\pi}{2}^-$, $\cos\theta$ approaches zero from above, that is, $\cos\theta > 0$, so $S(\theta) = \sec\theta = \frac{1}{\cos\theta} \to \infty$, that is, $\lim_{\theta\to\pi/2^-} S(\theta) = \infty$. As $\theta \to \frac{\pi}{2}^+$, $\cos\theta < 0$ and $\lim_{\theta\to\pi/2^+} S(\theta) = -\infty$.

Similarly,

$$\lim_{\theta\to 3\pi/2^-} S(\theta) = -\infty \qquad \text{and} \qquad \lim_{\theta\to 3\pi/2^+} S(\theta) = \infty.$$

Putting all this together, we can sketch the graph of $S(\theta)$ for $\theta \in [0, 2\pi]$. Since $\cos\theta$ is periodic, that is, repeats itself after 2π, its reciprocal $\sec\theta$ is also periodic, and we can sketch as much of the graph as we wish. See Figure 4.6.20.

Figure 4.6.18

Figure 4.6.19

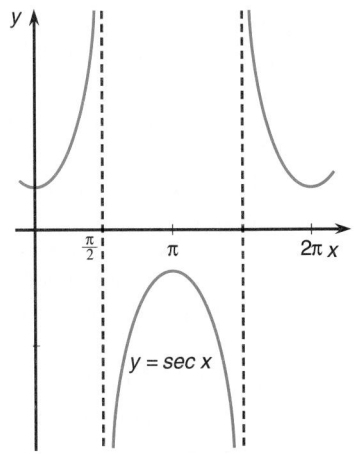

Figure 4.6.20

Inflection Points

When we sketch the graph of $y = x^3$, we find that the second derivative $\frac{d^2y}{dx^2}$ equals $6x$, which is positive when $x > 0$ and negative when $x < 0$. It is interesting to look

Figure 4.6.21

Figure 4.6.22

Figure 4.6.23

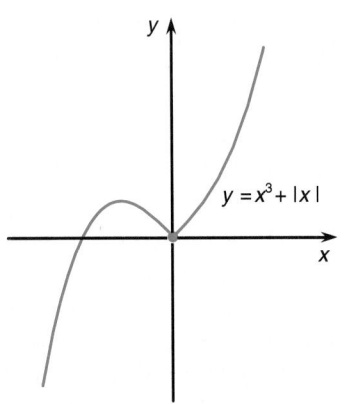

Figure 4.6.24

more closely at the point $(0, 0)$ at which the concavity changes from concave down to concave up (see Figure 4.6.21).

There is a special term for this type of point.

DEFINITION 4.6.3

A point on the graph of f is an **inflection point** for $f(x)$ if the concavity changes there. In other words, $(a, f(a))$ is an inflection point if $f(x)$ is continuous at $x = a$ and is concave up on some interval (c, a) on the left side of a and concave down on some interval (a, d) on the right side of a, or vice versa. We say that "$f(x)$ has an inflection point at $x = a$."

Notice that the function $f(x)$ does not have to be differentiable at $x = a$, but it does have to be defined there and continuous.

EXAMPLE 6

(i) For the function $h(x) = \frac{1}{x^2+2}$, which we graphed in Example 4, we found that the function was concave up for $x < -\sqrt{\frac{2}{3}}$ and for $x > \sqrt{\frac{2}{3}}$ and concave down for $-\sqrt{\frac{2}{3}} < x < \sqrt{\frac{2}{3}}$. It has inflection points at $x = \sqrt{\frac{2}{3}}, -\sqrt{\frac{2}{3}}$. Since the second derivative does not change sign at any other points, they are the only ones. See Figure 4.6.22.

(ii) The function $C(x) = x^3$ has second derivative $6x$, which changes from negative to positive at $x = 0$, so there is an inflection point at $x = 0$, and it is the only one (cf. Figure 4.6.21).

(iii) Let

$$F(x) = \begin{cases} x^2, & \text{when } x \le 0, \\ \sqrt{x}, & \text{when } x \ge 0. \end{cases}$$

The graph of $F(x)$ is easily sketched (after all, it is just the familiar parabola $y = x^2$ to the left of the y-axis and the familiar square root to the right). See Figure 4.6.23.

When $x < 0$, $F'(x) = \frac{d}{dx}(x^2) = 2x$ and $F''(x) = 2 > 0$. When $x > 0$, $F'(x) = \frac{d}{dx}(\sqrt{x}) = \frac{1}{2}x^{-1/2}$ and $F''(x) = -\frac{1}{4}x^{-3/2} < 0$. This means that $F(x)$ is concave up when $x < 0$ and concave down when $x > 0$, so by definition it has an inflection point at $x = 0$. There are no others.

Notice that $F(x)$ is not differentiable at $x = 0$. This is because the limit from the left of the difference quotient is 0, the derivative of x^2 at $x = 0$, and the limit from the right is ∞. This reflects the fact that the tangent line to $y = x^2$ at $x = 0$ is horizontal (i.e., has slope 0) and the tangent lines to $y = \sqrt{x}$ near $x = 0$ approach the vertical. It is possible to have an inflection point even when the function is not differentiable. It frequently happens at a point where the graph has a "corner," as in this example. This inflection point also happens to be a minimum.

(iv) Let $H(x) = x^3 + |x|$. When $x < 0$, $H(x) = x^3 - x$, $H'(x) = 3x^2 - 1$, $H''(x) = 6x$. When $x > 0$, $H(x) = x^3 + x$, $H'(x) = 3x^2 + 1$, $H''(x) = 6x$. The origin is an inflection point; see Figure 4.6.24.

(v) Let $g(x) = x^4 - 6x^2 + 8x - 2$. Then $g'(x) = 4x^3 - 12x + 8$ and $g''(x) = 12x^2 - 12$. Trying a few small values of x, we find that $g'(1) = 0$, so we can divide by $(x - 1)$. We find that $g'(x) = 4x^3 - 12x + 8 = 4(x^3 - 3x + 2) = 4(x - 1)(x^2 + x - 2) = 4(x - 1)(x + 2)(x - 1) = 4(x - 1)^2(x + 2)$. Evaluating at $x = -3$, $x = 0$, and $x = 2$, we find that $g'(x) > 0$ on $(-2, 1)$ and $(1, \infty)$, and $g'(x) < 0$ on $(-\infty, -2)$. At the critical point $x = -2$, the second derivative is $g''(-2) = 36 > 0$, so $x = -2$ is a local minimum. At $x = 1$ the second derivative is zero, but since the function is increasing on both sides of this point, it is neither a local maximum nor a local minimum.

Observing that the second derivative is continuous everywhere, we see that the only way to have an inflection point, a point at which the second derivative changes from negative to positive or vice versa, is to have $g''(a) = 0$. So we look for points at which $g''(x) = 0$ and find that the only possibilities are ± 1. Also $g''(x) = 12x^2 - 12$ is positive on $(-\infty, -1)$ and on $(1, \infty)$ and negative on $(-1, 1)$. This shows that $g(x)$ has inflection points at both $x = 1$ and $x = -1$. We draw the real axis in Figure 4.6.25, marking the critical points and the inflection points and indicating the sign of the first derivative and the type of concavity.

Plotting a few points, for example, $(0, -2)$, $(1, 1)$, $(2, 6)$, $(-2, -26)$, and $(-1, -17)$, we are able to fill in the curve, being careful to draw the change of concavity at the inflection points. See Figure 4.6.26.   ◢

Figure 4.6.25

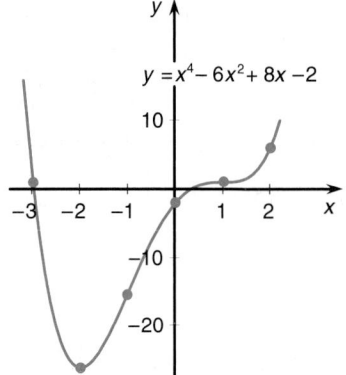

Figure 4.6.26

Example 6 illustrates two important ideas. One is that the function does not have to be differentiable at an inflection point. The other is that if the second derivative exists and is continuous at an inflection point, then the second derivative must be zero there.

So to find the inflection points of a function $f(x)$, we should consider any points at which the second derivative does not exist and any points at which the second derivative equals zero (i.e., at the critical points of $f'(x)$ *rather than* those of $f(x)$). To decide whether or not such a point is an inflection point, it is necessary to decide whether or not the second derivative has different signs on either side.

EXAMPLE 7 Find all the inflection points of (i) $f(x) = x^4 - 24x^2 + 64x + 1$, (ii) $g(x) = x^4 - 3x + 7$, (iii) $h(x) = \sin x$, (iv) $k(x) = \tan x$, (v) $r(x) = x^{1/3}$.

Solution

(i) We differentiate and find that $f'(x) = 4x^3 - 48x + 64, f''(x) = 12x^2 - 48$. This is continuous everywhere and zero at $x = \pm 2$. Since $f''(x) = 12x^2 - 48$ is positive when $x < -2$ and when $x > 2$, and negative when $-2 < x < 2$, there are inflection points at $x = -2$, $x = 2$, and they are the only ones.

(ii) $g'(x) = 4x^3 - 3$, $g''(x) = 12x^2$. The only point at which $g''(x) = 0$ is $x = 0$. Since $g''(x) > 0$ at every other x, there are no inflection points.

(iii) $h'(x) = \cos x, h''(x) = -\sin x$. The second derivative is zero at $x = n\pi$, for every integer n. Since the graph of $h''(x) = -\sin x$ crosses the axis at each of these points, it has different signs on either side of each of them and so has an inflection point at each of them. There are no others. See Figure 4.6.27.

Figure 4.6.27

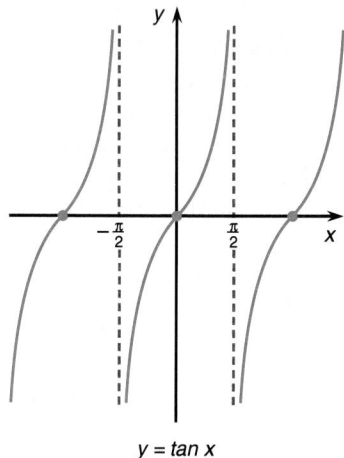

$y = \tan x$

Figure 4.6.28

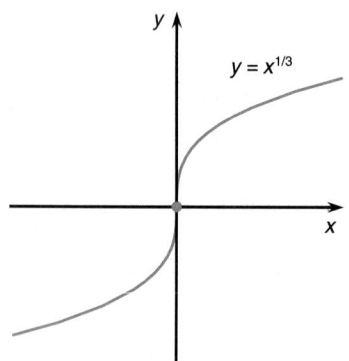

$y = x^{1/3}$

Figure 4.6.29

(iv) $k'(x) = \sec^2 x$, $k''(x) = 2 \sec x \sec x \tan x = 2 \sec^2 x \tan x$. This function is not defined whenever its denominator ($\cos^3 x$) is zero, that is, at $x = \frac{\pi}{2} + n\pi$, for any integer n. However, the original function $k(x) = \tan x$ is not defined there either, so there cannot be an inflection point at any of these points.

The second derivative is zero whenever $\tan x = 0$, which means $x = n\pi$, for any integer n. Since $\tan x$ is periodic and repeats itself after π, that is, $\tan(x + \pi) = \tan x$ for every x, it will be enough to consider the point $x = 0$. We consider the values of $k''(x)$ near $x = 0$.

If x is a small positive number, $\cos x$ is close to 1 and $\sin x$ is near 0 and positive. So $\tan x$ is positive. On the other hand, if x is near zero and negative, then $\cos x$ is near 1 and $\sin x$ is small and negative. In this case, $\tan x < 0$. From these observations we see that the second derivative does change sign at $x = 0$, so there is an inflection point. Because of the periodic nature of $\tan x$, the same is true of $x = n\pi$ for every integer n. See Figure 4.6.28.

(v) $r'(x) = \frac{1}{3} x^{-2/3}$, which exists whenever $x \neq 0$. Differentiating again, we find that $r''(x) = -\frac{2}{9} x^{-5/3}$, which also exists for all $x \neq 0$ and is never zero. The only possible inflection point is at $x = 0$, where the second derivative does not exist.

We observe that $x^{-5/3}$ has the same sign as $x^{1/3}$, which has the same sign as x. So $r''(x) = -\frac{2}{9} x^{-5/3}$ is positive when $x < 0$ and negative when $x > 0$. The graph is concave up for $x < 0$ and concave down for $x > 0$, so there is an inflection point at $x = 0$.

If we realize that $r(x) = x^{1/3}$ is the inverse function of $C(x) = x^3$, then its graph is the reflection in the line $y = x$ of the graph of $y = C(x) = x^3$, so we can draw it easily (see Figure 4.6.29). The tangent line to $y = x^3$ at the origin is horizontal; it is reflected into the tangent line to $y = r(x) = x^{1/3}$, which is vertical. This can also be seen by considering

$$\lim_{x \to 0} r'(x) = \lim_{x \to 0} \left(\frac{1}{3} x^{-2/3} \right) = \infty$$

and recognizing that a slope that tends to infinity corresponds to a vertical tangent.

Summary 4.6.4   Inflection Points

The important things to remember about inflection points are the following:

1. If the second derivative exists at an inflection point, it must equal zero.

2. The second derivative may not exist at an inflection point; even the first derivative may not exist at an inflection point, but the function itself must be defined and continuous there.

3. To find the inflection points of a function $f(x)$, find all points where $f''(x) = 0$ and all points where $f''(x)$ does not exist but $f(x)$ is defined. To check whether each of these candidates is an inflection point, determine whether the sign of the second derivative $f''(x)$ changes there. As we learned when finding the sign of the first derivative of a function, it is enough to evaluate at one convenient point in each interval in the domain of $f''(x)$ between points where $f''(x) = 0$ or where f'' does not exist.

With a graphics calculator we can locate inflection points and discuss the concavity of a curve. We plot a function $y = f(x)$ on the same picture with the second derivative $f''(x)$. For instance, consider $f(x) = x^4 - 6x^2 + 3$ (so $f''(x) = 12x^2 - 12$).

TI-85	SHARP EL9300
GRAPH F1$(y(x) =)$ F1(x)ˆ4 − 6 F1(x) x^2 + 3 ▼ 12 F1(x) x^2 − 12 M2(RANGE) (−) 3 ▼ 3 ▼ ▼ (−) 15 ▼ 10 F5(GRAPH)	x/θ/T a^b 4 ▶ − 6 x/θ/T x^2 + 3 2ndF ▼ 12 x/θ/T x^2 − 12 RANGE −3 ENTER 3 ENTER 1 ENTER −15 ENTER 10 ENTER 5 ENTER ◿◺

CASIO f_x-7700GB/f_x-6300G	HP 48SX
Cls EXE Range − 3 EXE 3 EXE 1 EXE − 15 EXE 10 EXE 5 EXE EXE EXE EXE Graph X x^y 4 − 6 X x^2 + 3 EXE Graph 12 X x^2 − 12 EXE	◤ PLOT PLOTR ERASE ATTN ' αY ▆ αX y^x 4 − 6 × αX y^x 2 + 3 ↱ PLOT ◤ DRAW ERASE 3 +/− ENTER 3 XRNG 15 +/− ENTER 10 YRNG DRAW ATTN ' αY ▆ 12 × αX y^x 2 − 12 ↱ PLOT ◤ DRAW DRAW

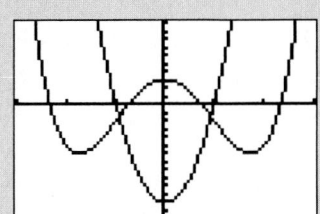

We note how the second derivative changes sign at two points and observe how the function $f(x)$ changes concavity at these points, which are inflection points. You might also like to plot the first derivative on the same picture.

Cusps

Besides the behavior exhibited in Example 7(v) of a function with an inflection point and a vertical tangent line, there is another phenomenon associated with vertical tangents.

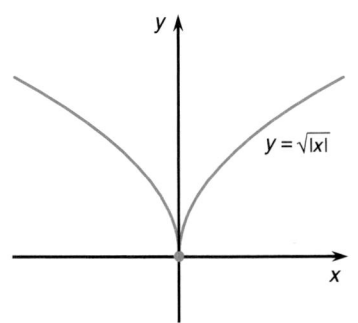

$y = \sqrt{|x|}$

Figure 4.6.30

EXAMPLE 8   Let $f(x) = (x^2)^{1/4}$. When $x > 0$, this is of course the same thing as $x^{2(1/4)} = x^{1/2} = \sqrt{x}$. When $x < 0$, we notice that $x^2 = |x|^2$, so $f(x) = (x^2)^{1/4} = (|x|^2)^{1/4} = (|x|)^{2(1/4)} = |x|^{1/2} = \sqrt{|x|}$.

We can easily sketch the graph of $f(x)$ from this information; it is simply the graph of $y = \sqrt{x}$ for $x > 0$ and the graph of $y = \sqrt{|x|}$ for $x < 0$ (see Figure 4.6.30). We may ask what happens to the derivative at $x = 0$, and find that for $x > 0$,

$$f'(x) = \frac{d}{dx}(\sqrt{x}) = \frac{1}{2}x^{-1/2} = \frac{1}{2\sqrt{x}}.$$

The function is even, so its graph is symmetric about the y-axis. When $x < 0$, $f(x) = \sqrt{|x|} = \sqrt{-x}$ (since $x < 0$, we know that $-x$ is positive . . .). So when $x < 0$,

$$f'(x) = \frac{d}{dx}(\sqrt{-x}) = \frac{1}{2}(-x)^{-1/2}(-1) = -\frac{1}{2\sqrt{-x}}.$$

In each case the denominator is positive and tends to 0 as x approaches 0. This means that

$$\lim_{x \to 0^+} f(x) = \infty \qquad \text{and} \qquad \lim_{x \to 0^-} f(x) = -\infty.$$

The slope of the tangent approaches ∞ from one side and $-\infty$ from the other. The curve appears to come to a point. Notice that the concavity is in the same direction on both sides of the point.

DEFINITION 4.6.5

A point $\big(a, f(a)\big)$ is called a **cusp** for $f(x)$ if $f(x)$ is continuous at $x = a$ and $f'(x)$ tends to ∞ as x approaches a from one side and tends to $-\infty$ as x approaches a from the other side. We say that "$f(x)$ has a cusp at $x = a$."

A cusp is never an inflection point because $f(x)$ will be concave in the same direction on both sides of $x = a$. The function $f(x)$ always has a local maximum or a local minimum at any point where it has a cusp.

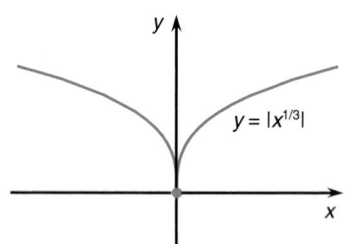

Figure 4.6.31

EXAMPLE 9

(i) Let $g(x) = |x^{1/3}|$. We sketched $x^{1/3}$ in Example 7(v), so it is very easy to sketch $y = g(x)$ (see Figure 4.6.31). If $x > 0$, we find that $g(x) = x^{1/3}$, so $g'(x) = \frac{1}{3}x^{-2/3}$, which tends to ∞ as $x \to 0^+$. On the other hand, if $x < 0$, $g(x) = -x^{1/3}$ and $g'(x) = -\frac{1}{3}x^{-2/3}$, which is negative, and $g'(x) \to -\infty$ as $x \to 0^-$.

We see that $g(x)$ has a cusp at $x = 0$.

(ii) Let $h(x) = \frac{1}{x^2}$. Then $h'(x) = \frac{d}{dx}(x^{-2}) = -2x^{-3}$. Since x^{-3} has the same sign as x, we find that

$$\lim_{x \to 0^+} h'(x) = -\infty \qquad \text{and} \qquad \lim_{x \to 0^-} h'(x) = \infty.$$

However, $h(x)$ does not have a cusp at $x = 0$ because the function $h(x)$ itself is not continuous there (it goes to infinity). See Figure 4.6.32.

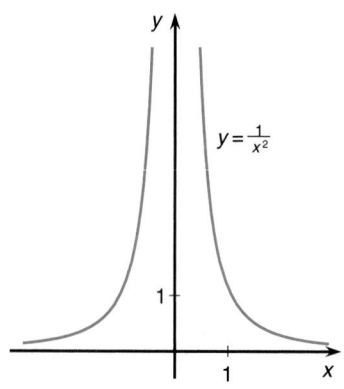

Figure 4.6.32

Inflection points are an important piece of information about a function and should always be marked on the graph. The same is true of cusps, but they are relatively rare.

EXAMPLE 10   In the graph pictured in Figure 4.6.33, mark all the regions in which the function is increasing and the regions in which it is decreasing. Mark the regions

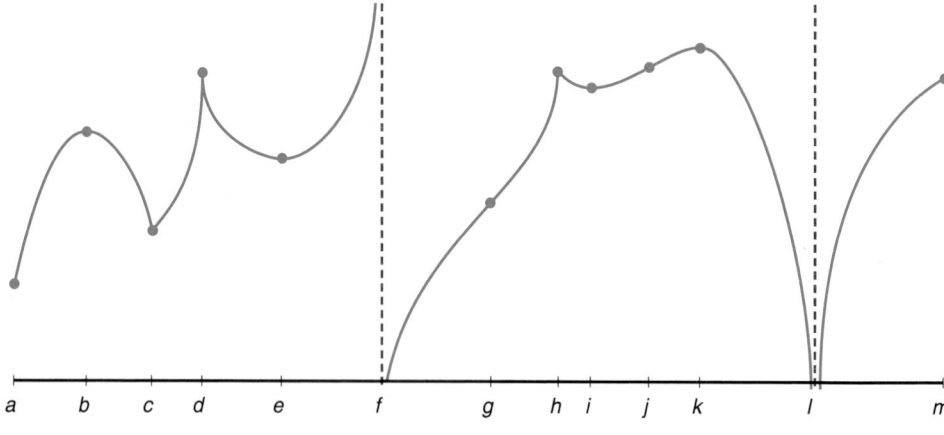

Figure 4.6.33

where it is concave up and the regions where it is concave down. Identify all extrema, inflection points, and cusps. (The dashed lines are asymptotes.)

Solution   The graph is increasing on (a, b), (c, d), (e, f), (f, h), (i, k), and (ℓ, m). It is decreasing on (b, c), (d, e), (h, i), and (k, ℓ). It is concave up on (c, d), (d, f), (g, h), and (h, j) and concave down on (a, c), (f, g), (j, ℓ), and (ℓ, m).

There are no absolute extrema. There are local maxima at b, d, h, k, and m and local minima at a, c, e, and i. There are inflection points at c, g, and j and a cusp at d. Notice that h is neither an inflection point nor a cusp.

We draw the axis and indicate the regions of increase and decrease by marking the sign of the derivative. We indicate the concavity with little U-shaped curves. Notice that the concavity changes at f, but it is not an inflection point, since the function is not defined there. See Figure 4.6.34.

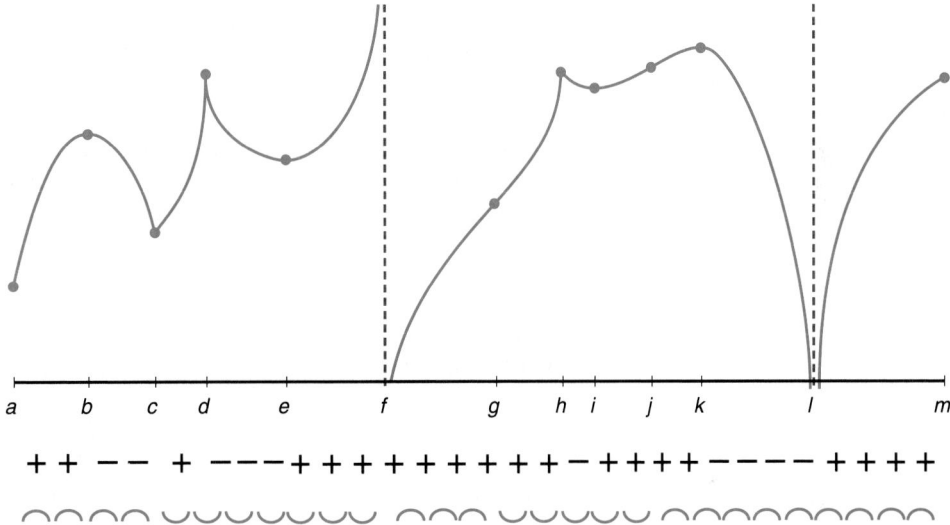

Figure 4.6.34

EXAMPLE 11   Sketch the graph of the function $f(x) = \frac{x}{1+x^2}$, which we have already considered in Example 4.5.2 (p. 264).

Solution   In Example 4.5.2 we found that

$$f'(x) = \frac{1 - x^2}{(1 + x^2)^2},$$

so we can differentiate again to find the second derivative:

$$f''(x) = \frac{-2x(1 + x^2)^2 - (1 - x^2)(2)(1 + x^2)2x}{(1 + x^2)^4}$$

$$= \frac{-2x(1 + x^2) - (1 - x^2)(2)(2x)}{(1 + x^2)^3}$$

$$= \frac{-2x - 2x^3 - 4x + 4x^3}{(1 + x^2)^3} = \frac{2x^3 - 6x}{(1 + x^2)^3}$$

$$= \frac{2x(x^2 - 3)}{(1 + x^2)^3}.$$

Figure 4.6.35

The denominator is always positive, and the numerator is positive on $(-\sqrt{3}, 0)$ and $(\sqrt{3}, \infty)$ and negative on $(-\infty, -\sqrt{3})$ and $(0, \sqrt{3})$. The function is concave up on the first two intervals and concave down on the last two. There are inflection points at $x = -\sqrt{3}, 0$, and $\sqrt{3}$.

We mark all this information on the real line in Figure 4.6.35, and then we sketch the graph in Figure 4.6.36, incorporating all we know about the curve.

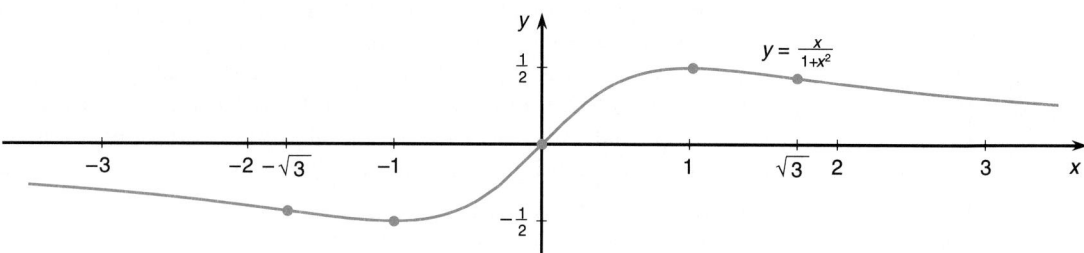

Figure 4.6.36

In the examples we have done so far, the functions were simple enough that it was always easy to find exactly where the second derivative was zero. As the next example shows, this is certainly not always the case.

EXAMPLE 12   Sketch the graph of $G(x) = \frac{e^x}{1+x^2}$.

Solution   The derivative is

$$G'(x) = \frac{e^x(1+x^2) - e^x(2x)}{(1+x^2)^2} = \frac{e^x(x^2 - 2x + 1)}{(1+x^2)^2} = \frac{e^x(x-1)^2}{(1+x^2)^2}.$$

It is positive except at $x = 1$, where it is zero, so the function is always increasing. In Section 4.12 we will learn how to show that $\lim_{x \to -\infty} G(x) = 0$ and $\lim_{x \to \infty} G(x) = \infty$.

The second derivative is found by differentiating $G'(x)$:

$$G''(x) = \frac{\left(e^x(x-1)^2 + e^x(2)(x-1)\right)(1+x^2)^2 - e^x(x-1)^2(2)(1+x^2)(2x)}{(1+x^2)^4}$$

$$= \frac{e^x(1+x^2)(x-1)\left((x-1)(x^2+1) + 2(x^2+1) - 4x(x-1)\right)}{(1+x^2)^4}$$

$$= e^x(x-1)\frac{x^3 - x^2 + x - 1 + 2x^2 + 2 - 4x^2 + 4x}{(1+x^2)^3}$$

$$= e^x(x-1)\frac{x^3 - 3x^2 + 5x + 1}{(1+x^2)^3}.$$

So $G''(x) = 0$ when $x = 1$, but to find any other points at which $G''(x) = 0$, we must solve $A(x) = 0$, where for convenience we write $A(x) = x^3 - 3x^2 + 5x + 1$. We first look for integer roots of $A(x)$, but the Rational Root Theorem (Theorem 1.3.7, p. 13) tells us we only need to check ± 1, and neither of them works. Next we calculate the derivative $A'(x) = 3x^2 - 6x + 5 = 3(x^2 - 2x + 1) + 2 = 3(x-1)^2 + 2$, and having completed the square, we see that $A'(x) > 0$ for every x. This means that $A(x)$ is always increasing, so it has only one zero.

There is a complicated formula for the zero of a cubic equation, but instead we illustrate the possibility of finding the numerical value approximately.

Noticing that $A(-1) = -8 < 0$ and $A(0) = 1 > 0$, we see that the zero must lie between these two points $x = -1$ and $x = 0$ (by the Intermediate Value Theorem (Theorem 2.5.5)). We could narrow down its location more precisely; for example, $A(-\frac{1}{2}) = -2\frac{3}{8}$, and since this is negative, the zero must lie between it and $x = 0$. We could then evaluate at $x = -\frac{1}{4}$ and find whether the zero is to the left or right of it, and then evaluate at $x = -\frac{3}{8}$ or $x = -\frac{1}{8}$, whichever is closer, and so on.

Continuing this process, we could find that the zero is approximately $x \approx -0.18$. Knowing that $A(x)$ is negative to the left of this point and positive to the right of it, we see from the formula for $G''(x)$ that $G''(x)$ is positive to the left of this point and to the right of $x = 1$ and negative between them.

So $G''(x)$ changes sign at $x = 1$ and at the point we have approximated, so there are inflection points at both of them. We can plot the curve now (see Figure 4.6.37). Notice that two significant digits is amply accurate for purposes of plotting the graph.

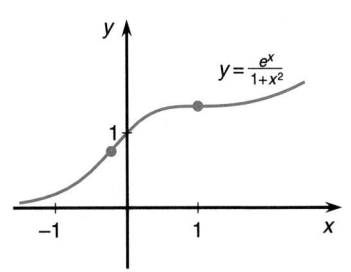

$$y = \frac{e^x}{1+x^2}$$

Figure 4.6.37

There are other ways to calculate approximate solutions of equations, and they will be the subject of Sections 4.10 and 4.11.

Exercises 4.6

I

For each of the following functions, find the regions on which it is concave up and the regions on which it is concave down.

1. $f(x) = x^2 - 3x + 11$

2. $f(x) = x^3 + 1$

3. $f(x) = x^3 + 2x - 3$

4. $f(x) = x^3 - 2x^2 - x + 7$

5. $f(x) = \frac{1}{x^2+4}$

6. $f(x) = \frac{x^2}{x^2+4}$

7. $f(x) = \cos(2x)$

8. $f(x) = xe^x$

9. $f(x) = e^{-x^2}$

10. $f(x) = x \ln x$

Find all the inflection points of each of the following functions.

11. $f(x) = 2x^3 + 7x - 1$

12. $f(x) = x^4 - 2x^3 - 12x^2 + 5x + 7$

13. $f(x) = x^5 - 10x^2 - x + 2$

14. $f(x) = 3x^5 - 10x^3 - 7x + 13$

15. $f(x) = \cos x$

16. $f(x) = \cot(2x)$

17. $f(x) = \cos x \sin x$

18. $f(x) = \exp(-x^2)$

19. $f(x) = \sinh x$

20. $f(x) = \frac{x^3}{|x|}$

II

Sketch the graph of each of the following functions, indicating all extrema and inflection points.

21. $y = f(x) = x^3 - 6x + 7$

22. $y = g(x) = x^3 + 6x^2 + 12x + 1$

23. $y = h(x) = x^3 + 6x - 2$

24. $y = k(x) = (x - 2)^3$

25. $y = F(x) = x^4 - 2x^2 + 1$

26. $y = G(x) = 3x^4 - 8x^3 + 6x^2 - 3$

27. $y = H(x) = \frac{x^2}{x-1}$

28. $y = K(x) = \frac{x}{x^2-1}$

29. $y = L(x) = \frac{x+1}{x^2-1}$

30. $y = M(x) = \frac{x^2+2x+2}{x+2}$

31. $y = N(x) = e^{-x^2}$

32. $y = P(x) = xe^x$

33. $y = Q(x) = \arctan x$

34. $y = R(x) = x \arctan x$

35. $y = S(x) = x \ln x$, for $x > 0$

36. $y = T(x) = \ln(1 + x^2)$

*37. Let $k \geq 3$ be an integer and define $f(x) = x^k$. For which values of k does $f(x)$ have an inflection point at $x = 0$?

*38. Suppose $F(x)$ is a function for which we know the concavity, that is, for which we know all the regions on which it is concave up or concave down and all the inflection points. Let $G(x) = F(x) + 3x - 2$.

  (i) Show that $F(x)$ and $G(x)$ have inflection points and upward and downward concavity at exactly the same places.

  (ii) Show that the same is true of $H(x) = F(x) + ax + b$ for any constants a and b.

39. Determine which of the following functions has an inflection point at $x = 0$ and which of them has a cusp:

  (i) $f(x) = x^{1/5}$, (ii) $g(x) = x^{3/7}$, (iii) $h(x) = x^{4/5}$, (iv) $k(x) = x^{4/3}$.

*40. Let p, q be positive integers with q odd and $p < q$. Let $f(x) = x^{p/q}$. Specify conditions on p and q that determine whether $f(x)$ has an inflection point (with a vertical tangent) or a cusp at $x = 0$.

*41. Carry out the approximation method of Example 12 (p. 288) sufficiently many times to find the third significant digit in the approximation of the location of the inflection point of $G(x)$.

**42. Sketch the graph of $k(x) = \arccos(1 - |x|)$, for $x \in [-1, 1]$.

43. (i) Use a graphics calculator or computer to sketch the function $g(x) = x^3 - 3x^2 + 2$ of Example 2(i) (p. 278). (ii) On the same graph, draw the tangent line at $a = \frac{3}{4}$. (iii) Find an interval around a on which the curve lies below the tangent line. (iv) Repeat parts (ii) and (iii) with $a = 0.99$.

44. Predict the regions of upward and downward concavity for $R(\theta) = \csc \theta$ (cf. Example 5, p. 281). Use a graphics calculator or computer to sketch the graph and verify your answer.

45. Sketch the function $g(x) = x^4 - 6x^2 + 8x - 2$ of Example 6(v) (p. 283), along with $g'(x)$ and $g''(x)$ on the same picture.

46. Sketch the function $f(x) = x^4 - 24x^2 + 64x + 1$ of Example 7(i) (p. 283), along with $f'(x)$ and $f''(x)$ on the same picture.

47. Sketch the function $g(x) = x^4 - 3x + 7$ of Example 7(ii), along with $g'(x)$ and $g''(x)$ on the same picture.

48. Sketch the graphs of $G(x) = \frac{e^x}{1+x^2}$ and $A(x) = x^3 - 3x^2 + 5x + 1$ on the same picture (see Example 12, p. 288). Find the root of $A(x)$ correct to six decimal places.

4.7 Applied Maximum/Minimum Problems

EXTRANEOUS SOLUTIONS
INTEGER SOLUTIONS

Figure 4.7.1

Figure 4.7.2

In Section 3.3 we saw how derivatives could be used to find maximum and minimum values of functions on closed intervals, and we learned how to use this technique to solve a variety of problems. We began to run into difficulties with some problems when the functions we found turned out not to be defined on closed intervals.

Now that our knowledge of extrema is much more detailed and our skill at differentiating more fully developed, we can deal with a wider range of problems.

EXAMPLE 1   Drinking Trough   A drinking trough is made by taking a long piece of sheet metal and folding up two sloping sides (see Figure 4.7.1). Suppose the metal is 75 cm wide and the folds divide it into three equal widths (of 25 cm each). Suppose that the two sides are folded up at the same angle. At what angle should they be folded up in order to make a trough of the largest possible volume?

Solution   The volume of the trough is just the length times the cross-sectional area, so to solve the problem, we must find out how to maximize the cross-sectional area. We draw a diagram of the cross-section in Figure 4.7.2, and we mark the angle θ as shown, the angle through which the sides are bent up.

Figure 4.7.3

Figure 4.7.4

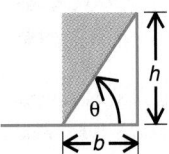

Figure 4.7.5

Next we must express the area of the cross-section as a function of θ. To do this, we divide it into a rectangle and two right-angled triangles, as shown in Figure 4.7.3. The base of the rectangle is 25 cm, and to find its height, we redraw in Figure 4.7.4 the part of the diagram that involves the right side of the trough. In the resulting right-angled triangle, whose height h we want to find, the hypotenuse has length 25 cm, so the height is $h = 25 \sin \theta$. While we are looking at that triangle, we observe that its base is $b = 25 \cos \theta$.

Returning to the rectangle in the middle, we have just found that its height, in centimeters, is $25 \sin \theta$. Since its base is 25, its area is the product $(25)(25 \sin \theta) = 625 \sin \theta$.

Now we turn our attention to the two triangles at the sides of the trough. They both have the same area, so let us concentrate on the one at the right side, shown in Figure 4.7.5. Its height is the same as the height of the rectangle that we just calculated, and its width is the same as the base of the bottom triangle, which as we observed above is $25 \cos \theta$. The area of the triangle is $\frac{1}{2} width \times height = \frac{1}{2} 25 \cos \theta \times 25 \sin \theta = \frac{625}{2} \cos \theta \sin \theta$. The combined area of the triangles at the left and right sides is twice as much, or $625 \cos \theta \sin \theta$.

The total area of the cross-section is $A(\theta) = 625 \sin \theta + 625 \cos \theta \sin \theta$. To maximize it, we calculate its derivative:

$$A'(\theta) = 625 \cos \theta + 625(-\sin^2 \theta + \cos^2 \theta).$$

This will equal zero when $\cos \theta + \cos^2 \theta - \sin^2 \theta = 0$, but this appears difficult to solve.

We begin by writing $\sin^2 \theta = 1 - \cos^2 \theta$, so the equation becomes

$$\cos \theta + \cos^2 \theta - 1 + \cos^2 \theta = 0$$

or

$$2 \cos^2 \theta + \cos \theta - 1 = 0.$$

What we want is to solve this for θ, but it will be easier to start by solving it for $\cos \theta$. In fact it is a quadratic equation in $\cos \theta$.

Letting $u = \cos \theta$, we can write the equation as

$$2u^2 + u - 1 = 0.$$

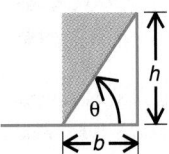

We can use the quadratic formula or just factor $2u^2 + u - 1 = (2u - 1)(u + 1)$, so the solutions are $u = \frac{1}{2}, -1$. The answer we want is $u = \frac{1}{2}$. After all, since $u = \cos \theta$, to say that $u = -1$ is to say that $\cos \theta = -1$ or that $\theta = \pi$. This would correspond to folding the side all the way over onto the bottom of the trough (see Figure 4.7.6), which would give cross-sectional area equal to 0; this point is an endpoint minimum.

To say that $u = \cos \theta = \frac{1}{2}$ is to say that $\theta = \frac{\pi}{3}$. The trough with the maximum volume is made by bending up the sides at an angle of $\frac{\pi}{3}$ radians, that is, 60°.

Example 1 illustrates that it is perfectly possible to solve maximum/minimum problems involving trigonometric functions. One thing to observe is that we had to use a trigonometric identity to convert one stage of the problem into something we could solve. A second is that we solved not for the angle θ directly, but for $\cos \theta$, from which we were able to find θ.

The most important remark, however, is that somebody had to come up with the idea to use the angle θ as the variable. In Section 3.3 we realized that choosing the right quantity to be the variable is a vital and frequently difficult part of solving maximum/minimum problems. If you attempt Example 1 using some other variable, you will find that it is not easy.

With a graphics calculator we can plot the function $y = A(\theta)$ of Example 1. In fact, we plot $y = 625 \sin x + 625 \cos x \sin x$. While we are at it, we can plot the derivative $625 \cos x + 625(\cos^2 x - \sin^2 x)$ too.

TI-85	SHARP EL9300
GRAPH F1$(y(x) =)$ 625 SIN F1(x) + 625 SIN F1(x) COS F1(x) ▼ 625 COS F1(x) + 625 ((COS F1(x)) x^2 − (SIN F1$(x))x^2)$ M2(RANGE) 0 ▼ π ▼ ▼ (−) 800 ▼ 1200 F5(GRAPH)	∿ 625 sin x/θ/T + 625 sin x/θ/T cos x/θ/T 2ndF ▼ 625 cos x/θ/T + 625 ((cos x/θ/T) x^2 − (sin x/θ/T) x^2) RANGE 0 ENTER π ENTER 1 ENTER − 800 ENTER 1200 ENTER 200 ENTER ∿
CASIO f_x-7700GB/f_x-6300G	HP 48SX
Cls EXE Range 0 EXE π EXE 1 EXE −800 EXE 1200 EXE 200 EXE EXE EXE EXE Graph 625 sin X + 625 sin X cos X EXE Graph 625 cos X + 625 ((cos X) x^2 − (sin X) x^2) EXE	⬑ PLOT PLOTR ERASE ATTN ' αY ▭ 625 × SIN αX ▶ + 625 × SIN αX ▶ × COS αX ↱ PLOT ⬑ DRAW ERASE 0 ENTER π XRNG 800 +/− ENTER 1200 YRNG DRAW ATTN ' αY ▭ 625 × COS αX ▶ + 625 × () COS αX ▶ y^x 2 − SIN αX ▶ y^x 2 ↱ PLOT ⬑ DRAW DRAW

We see the maximum value and note the sign of the derivative in various regions.

EXAMPLE 2   **Window**   A window is made so that it is 2 ft wide, in the shape of a rectangle with an isosceles triangle on top (see Figure 4.7.7). A lazy architect has 12 ft of weatherstripping available, and rather than go to the hardware store for more, he decides to design the window to have a perimeter of exactly 12 ft and the largest possible area. What will the dimensions of the window be?

Figure 4.7.7

Solution   This time we choose our variable to be the angle θ between the top edge of the window and the horizontal, as shown in Figure 4.7.8. We know that the width of the window is 2 ft, but we do not know the height h of the rectangular part of the window.

On the other hand, we do know that the perimeter is 12 ft. It is also possible to figure out the perimeter in terms of h and θ. If we draw the little triangle made by half of the top triangle, we find a right-angled triangle whose base is 1 ft long and with an angle θ as shown in Figure 4.7.9. So $\cos\theta$ equals 1 over the length of the hypotenuse, and the length of the hypotenuse is $\frac{1}{\cos\theta} = \sec\theta$. Accordingly, the perimeter of the entire window is $2 + 2h + 2\sec\theta$.

Figure 4.7.8

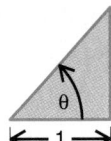

Figure 4.7.9

From the equation $12 = 2 + 2h + 2\sec\theta$, we find that $10 = 2h + 2\sec\theta$, $5 = h + \sec\theta$, so

$$h = 5 - \sec\theta.$$

To find the area of the triangle at the top of the window, we need to observe that the height of this triangle is the same as the height of the little triangle we considered above. That height is $\tan\theta$. The base of the big triangle is 2, so its area is $\frac{1}{2} base \times height = \frac{1}{2}2\tan\theta = \tan\theta$.

> In the little triangle, $\frac{height}{base} = \tan\theta$, and the base has length 1, so $height = \tan\theta$

The total area of the window is

$$A = A(\theta) = 2h + \tan\theta$$
$$= 10 - 2\sec\theta + \tan\theta.$$

> Here we used $h = 5 - \sec\theta$.

Its derivative is

$$A'(\theta) = -2\sec\theta\tan\theta + \sec^2\theta$$
$$= \sec\theta(-2\tan\theta + \sec\theta).$$

For this to equal zero means that

$$\tan\theta = \frac{1}{2}\sec\theta.$$

Multiplying this equation by $\cos\theta$, we get

$$\cos\theta\tan\theta = \frac{1}{2}\cos\theta\sec\theta,$$

or

$$\sin\theta = \frac{1}{2}.$$

Since the angle in an isosceles triangle must be between 0 and $\frac{\pi}{2}$, this means that $\theta = \frac{\pi}{6}$.

What remains is to convince ourselves that this really is the angle that results in the maximum area.

One way to do this is to rewrite the derivative:

$$A'(\theta) = \sec\theta(\sec\theta - 2\tan\theta)$$
$$= \sec^2\theta(1 - 2\sin\theta),$$

(since $\tan\theta = \frac{\sin\theta}{\cos\theta} = \sin\theta\sec\theta$).

In this form we see that the derivative is positive when $\sin\theta < \frac{1}{2}$ and negative when $\sin\theta > \frac{1}{2}$, which is to say $A'(\theta) > 0$ when $\theta < \frac{\pi}{6}$ and $A'(\theta) < 0$ when $\theta > \frac{\pi}{6}$. This means that $A(\theta)$ is increasing on $\left[0, \frac{\pi}{6}\right)$ and decreasing on $\left(\frac{\pi}{6}, \frac{\pi}{2}\right)$, so $\theta = \frac{\pi}{6}$ is indeed a maximum.

The dimensions of the window of greatest possible area are as follows. The width is of course 2 ft. The height of the rectangular part is $h = 5 - \sec\theta = 5 - \sec\frac{\pi}{6} = 5 - \frac{2}{\sqrt{3}} \approx 3.845$ ft. The length of each slanted side at the top is $\sec\theta = \frac{2}{\sqrt{3}} \approx 1.155$ ft.

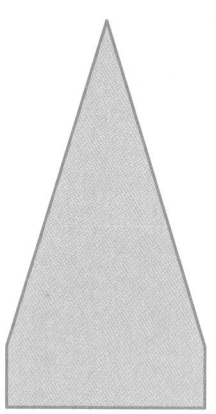

Figure 4.7.10

We have been using our intuition about the behavior of the graph of $A(\theta)$ to decide whether or not the point in question was a maximum. Finding where the derivative was zero was the easy part of the problem. The hard parts were choosing the variable and setting up the function in the first place and deciding whether or not the point we found was really a maximum.

Also observe that once the function $A(\theta)$ was constructed, we thought of it as a function on $\left[0, \frac{\pi}{2}\right)$. In fact this doesn't make sense in terms of the original problem, since if θ is nearly $\frac{\pi}{2}$, that is, nearly a right angle, then the triangle will be so high that the perimeter will have to be greater than 12 (see Figure 4.7.10). If we calculate $A(\theta)$ for values of θ very near $\frac{\pi}{2}$, we will find that it is negative, so it does not make sense to think of it as an area.

Still it does make sense to look for the maximum of this function, and if it happens to be at a point at which the problem does make sense, then we have solved the problem. This is in fact exactly what we did.

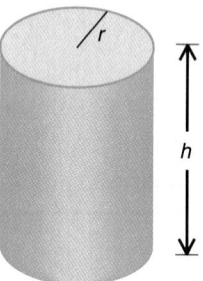

Figure 4.7.11

EXAMPLE 3   Cylindrical Drum   A cylindrical storage drum is to be made with a volume of 16 cu ft. How can it be made so that the surface area is a minimum, that is, so that it will take the least possible amount of material to construct it?

Solution Suppose the radius and height of the cylinder are r and h, respectively (see Figure 4.7.11). We know that the volume is the height times the cross-sectional area, that is,

$$V = \pi r^2 h.$$

Figure 4.7.12

The top and the bottom each have area πr^2. To find the area of the curved sides, cut off the ends, slit the open cylinder, and unroll it into a rectangle as shown in Figures 4.7.12 and 4.7.13. The rectangle will have one dimension equal to h, the height of the cylinder, and the other equal to the circumference, or $2\pi r$. Its area will be $2\pi rh$. The total surface area of the cylinder is $A = 2\pi r^2 + 2\pi rh$.

Knowing that $V = \pi r^2 h = 16$, we can solve and find that

$$h = \frac{16}{\pi r^2}.$$

Substituting this into the area formula, we find that

$$A = 2\pi r^2 + 2\pi rh = 2\pi r^2 + 2\pi r\frac{16}{\pi r^2} = 2\pi r^2 + \frac{32}{r}.$$

Figure 4.7.13

The derivative is

$$A' = 4\pi r - \frac{32}{r^2}.$$

This is zero when $4\pi r = \frac{32}{r^2}$, that is, when $4\pi r^3 = 32$, that is, when $r^3 = \frac{8}{\pi}$, or

$$r = \frac{2}{(\pi)^{1/3}} \approx 1.366.$$

The function $A = A(r) = 2\pi r^2 + \frac{32}{r}$ is defined on $(0, \infty)$. It is easy to find its limiting behavior at both ends of its domain: $\lim_{r \to 0^+} A(r) = \infty$ and $\lim_{r \to \infty} A(r) = \infty$. Since the function is differentiable on the whole interval and has only one critical point, that critical point must be a minimum.

The drum of minimum area has radius approximately equal to 1.366 ft and height approximately equal to 2.731 ft.

EXAMPLE 4   Angle of Reflection   Consider a mirror lying horizontally as shown in Figure 4.7.14. Light travels from a source at point A, is reflected from the mirror at P, and travels on to point B. It is a fundamental idea in optics, known as *Fermat's Principle*, that light travels in such a way as to take the minimum amount of time. Since this amounts to the same thing as traveling the minimum possible distance, we will be able to find the point on the mirror from which the ray of light reflects. Which point of reflection P will require the light to travel the shortest possible total distance?

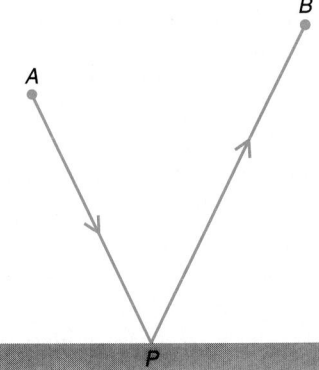

Figure 4.7.14

Solution   The problem does not specify any units of distance, so we must choose our own. We let the height of A above the mirror be h and the height of B above the mirror be k. The horizontal distance between A and B (i.e., measured along the surface of the mirror) we call d. As our variable we try the horizontal distance x from the point directly beneath A to the point of reflection P, as shown in Figure 4.7.15.

It is now easy to find the total distance traveled by the light in going from A to P and from P to B. It is

$$D = \sqrt{h^2 + x^2} + \sqrt{k^2 + (d-x)^2},$$

using the Pythagorean Theorem (Theorem 1.5.1, p. 22).

To find the minimum possible D, we calculate

$$D'(x) = \frac{1}{2}\frac{1}{\sqrt{h^2+x^2}}(2x) + \frac{1}{2}\frac{1}{\sqrt{k^2+(d-x)^2}}\Big(2(d-x)(-1)\Big)$$

$$= \frac{x}{\sqrt{h^2+x^2}} - \frac{d-x}{\sqrt{k^2+(d-x)^2}}.$$

This is zero when

$$\frac{x}{\sqrt{h^2+x^2}} = \frac{d-x}{\sqrt{k^2+(d-x)^2}}.$$

The denominators of the two sides are just the distances traveled by the ray of light between A and P and between P and B, respectively.

If we call these distances d_1 and d_2, respectively, then the condition becomes

$$\frac{x}{d_1} = \frac{d-x}{d_2}.$$

Figure 4.7.15

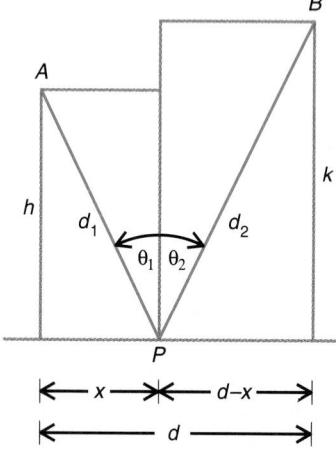

Figure 4.7.16

From Figure 4.7.16 we see that these quotients are the sines of the angles between the two rays and the vertical. If we call these angles θ_1 and θ_2, as shown in the picture, then the condition is that $\sin \theta_1 = \sin \theta_2$.

Since both angles are between 0 and $\frac{\pi}{2}$, this implies that the angles themselves are equal, that is, $\theta_1 = \theta_2$.

The ray of light from A to P, the ray that hits the mirror, is called the *incident ray*. The ray from P to B is called the *reflected ray*, the angle θ_1 is the *angle of incidence*, and the angle θ_2 the *angle of reflection*. What we have shown is that when light is reflected from a mirror, *the angle of incidence equals the angle of reflection*, a law that is well known in physics.

EXAMPLE 5   **Trip from Island**   A woman on an island wishes to go to a small town on the mainland. The island is 1 mile from shore. From the point on the coast closest to the island it is 4 miles along the straight shoreline to the town. See Figure 4.7.17.

If the woman can row her boat at a speed of 2 mph and walk at a speed of 4 mph along the beach, where should she land the boat in order to get to town as quickly as possible?

Solution   We call the point on the shore that is closest to the island P. Let x be the distance from P to the point at which the boat lands. See Figure 4.7.18.

The woman must walk $4 - x$ miles along the beach at 4 mph, so the time it takes her to walk is $\frac{4-x}{4}$ hr. The quickest way to row to the specified point on shore will of course be to row in a straight line, as illustrated. The Pythagorean Theorem (Theorem 1.5.1) tells us that the distance traveled by the boat is $\sqrt{1 + x^2}$ mi, so the time it takes is $\frac{\sqrt{1+x^2}}{2}$ hr.

The total time for the trip is the sum of these two times. The total time, in hours, is

$$T = \frac{4 - x}{4} + \frac{\sqrt{1 + x^2}}{2}.$$

$$time = \frac{distance}{speed}.$$

Figure 4.7.17

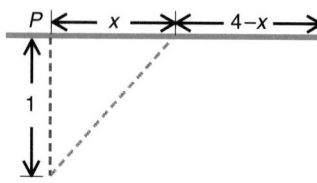

Figure 4.7.18

The problem is to minimize T. To do this, we find the derivative

$$T'(x) = -\frac{1}{4} + \frac{1}{2}\frac{1}{2\sqrt{1+x^2}}(2x) = -\frac{1}{4} + \frac{x}{2\sqrt{1+x^2}}.$$

To find critical points, we ask when this equals zero, which is when

$$\frac{x}{2\sqrt{1+x^2}} = \frac{1}{4}$$

or

$$2x = \sqrt{1 + x^2}$$

or

$$4x^2 = 1 + x^2$$

or

$$3x^2 = 1$$

or

$$x = \pm\frac{1}{\sqrt{3}}.$$

In fact the negative value we have found for x is not a critical point, as is easily seen by evaluating $T'\left(-\frac{1}{\sqrt{3}}\right)$. (It arose because we squared both sides of the second equation above and in doing so lost track of the sign.) The function $T(x)$ is differentiable at every x and has exactly one critical point, at $x = \frac{1}{\sqrt{3}}$.

Notice from the formula for $T'(x)$ that $T'(x) < 0$ whenever $x < 0$. This means that $T(x)$ is decreasing on $(-\infty, 0)$, and since the sign of the derivative can change only at a critical point, $T(x)$ must be decreasing on $\left(-\infty, \frac{1}{\sqrt{3}}\right)$. Next we consider the sign of the derivative to the right of $x = \frac{1}{\sqrt{3}}$. By the usual sort of argument it is enough to calculate $T'(x)$ at a single convenient point x. Since $1 > \frac{1}{\sqrt{3}}$, we evaluate $T'(1) = -\frac{1}{4} + \frac{1}{2\sqrt{2}}$. Since $\sqrt{2} < 2$, $2\sqrt{2} < 4$, and $\frac{1}{2\sqrt{2}} > \frac{1}{4}$, so $T'(1) > 0$, which means that $T'(x) > 0$ to the right of $x = \frac{1}{\sqrt{3}}$ and the function $T(x)$ is increasing there.

Putting these facts together, we find that the minimum of $T(x)$ must occur at the critical point $\frac{1}{\sqrt{3}}$. The woman will minimize the time it takes to get to town by rowing to the point $\frac{1}{\sqrt{3}} \approx 0.577$ mi along the beach from the point closest to the island. Using this route, her trip will take about 1.433 hr. ◢

In doing this example, techniques that were developed for purposes of sketching graphs were used to simplify the problem.

In hindsight, it is certainly ridiculous to expect that the best route would be to land farther from town than the closest point to the island, so we could have deduced that it is only necessary to consider positive values of x. Similarly, landing beyond the town would not be sensible, so we can restrict attention to $x \in [0, 4]$. By these remarks it is possible to reduce the problem to one on a closed interval, which can be approached by the techniques of Summary 3.3.1 (p. 157).

To do the problem this way would require evaluations at the endpoints $x = 0$ and $x = 4$. This approach is not more difficult, but we have shown how graphical considerations can be used to reduce the complexity of a problem. There are instances in which major simplifications result from this idea.

Figure 4.7.19

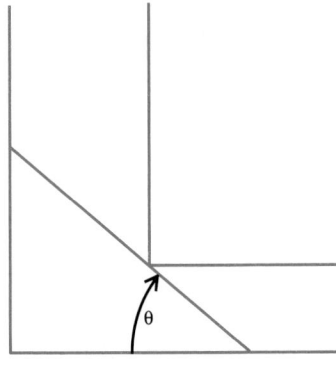

Figure 4.7.20

EXAMPLE 6 **Carrying a Tube Around a Corner** Movers wish to carry a long tube around a corner in a corridor, as shown in Figure 4.7.19. The tube cannot be tilted because it is filled with delicate electronic equipment. If the corridors are 2 m wide and 1 m wide and the tube is 4.2 m long, will the movers be able to get it around the corner?

Solution The question we shall answer is slightly different: What is the longest object that can be carried around the corner?

We consider a long, thin rod placed in the corridor as shown in Figure 4.7.20. Letting θ be the angle marked, we think of the length of the rod as a function L of θ. If we find the **minimum** of this function, we will have found the shortest space available for a rod going around the corner, so we will know the longest rod that can go around the corner.

To find the length of the rod as a function of θ, we consider two small right-angled triangles as shown in Figure 4.7.21. The length of the rod is the sum of the lengths of the hypotenuses of the two triangles. We find the length L_2 of the hypotenuse of the upper left triangle by observing that the base of that triangle has length 2 (the width of the corridor), so $\cos\theta = \frac{2}{L_2}$ and $L_2 = \frac{2}{\cos\theta} = 2\sec\theta$.

To find the length of the hypotenuse of the other triangle, we observe that $\frac{1}{L_1} = \sin\theta$, so $L_1 = \frac{1}{\sin\theta} = \csc\theta$.

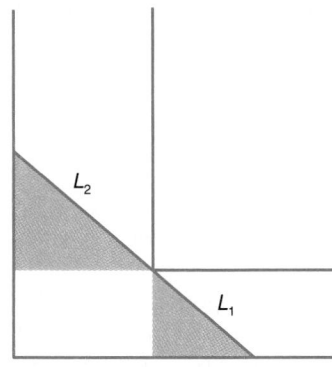

Figure 4.7.21

Adding together the lengths of the hypotenuses, we find the length of the rod is

$$L = L_1 + L_2 = \csc\theta + 2\sec\theta.$$

To minimize L, we find its derivative:

$$L'(\theta) = -\csc\theta\cot\theta + 2\sec\theta\tan\theta = -\frac{\cos\theta}{\sin^2\theta} + 2\frac{\sin\theta}{\cos^2\theta}.$$

To find when the derivative is zero, we cross-multiply and find

$$2\sin^3\theta = \cos^3\theta$$
$$\frac{\sin^3\theta}{\cos^3\theta} = \frac{1}{2}$$
$$\tan\theta = 2^{-1/3}$$
$$\theta = \arctan(2^{-1/3}).$$

We can think of $L(\theta)$ as a function defined for $\theta \in \left(0, \frac{\pi}{2}\right)$ (in terms of the picture of the rod in the corridor it makes no sense for θ to be greater than $\frac{\pi}{2}$). We have found the only place the derivative is zero. Since $\lim_{\theta\to 0+} L(\theta) = \infty$ and $\lim_{\theta\to\pi/2-} L(\theta) = \infty$, the point we have found is a minimum.

We conclude that the longest rod that can be carried around the corner has length $L\big(\arctan(2^{-1/3})\big) = \csc\big(\arctan(2^{-1/3})\big) + 2\sec\big(\arctan(2^{-1/3})\big) \approx 4.1619$ m. In particular the tube whose length is 4.2 m will not fit around the corner.

Notice that we did not consider the thickness of the tube. We showed that an object 4.2 m long will not fit around the corner, no matter how thin it is. If the answer had been that the thin rod of a certain length would fit, then it would have been necessary to consider what thickness it could have and still fit, and this actually turns out to be more difficult.

EXAMPLE 7   An Inventory Problem   A store sells an average of one television set per day, every day of the year. It costs \$70 to store a television set for a year. It costs \$100 to place an order for more sets.

The problem is to decide how often to place an order so as to minimize the overall costs. Ordering often will reduce storage costs but increase ordering charges; ordering infrequently will reduce ordering charges but increase storage costs. What is the best strategy?

Solution   The question assumes that one television is sold each day. In fact this might not be a very good assumption, since it ignores seasonal variations and random fluctuations. However, under this assumption, suppose an order is placed every x days.

Then the number of orders per year will be $\frac{365}{x}$, and the cost of ordering will be $100\left(\frac{365}{x}\right)$ dollars. Next we have to find the storage costs.

If an order is placed every x days, it will have to be for exactly x sets (one to be sold each day until the next order). What we need to know is the average number of sets in storage.

On the day a new order arrives, there will be x sets in storage; on the last day before the next order, there will be just one set left. Figure 4.7.22 shows a bar graph of the number of sets in storage on each day between consecutive orders, and we

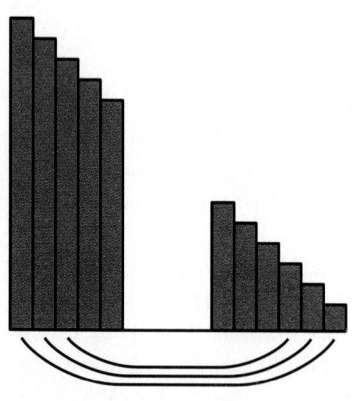

Figure 4.7.22

can make the following observation. The total of the numbers for the first day and the last day is $x + 1$, so the average for these two days is $\frac{x+1}{2}$. Then if we consider the second day and pair it with the second-last day, we see that the second day has one fewer set than the first day but the second-last has one more set than the last day, so the total for these two days remains the same, $x + 1$, and the average is $\frac{x+1}{2}$.

Similarly, the average for the third and third-last days is $\frac{x+1}{2}$, and so on. What we find is that all the days can be grouped in pairs for which the average number of sets in storage is $\frac{x+1}{2}$, *except* that if x is odd, there will be one day right in the middle that will not be paired with any other day. But if x is odd, then the day in the middle is the day on which the number of sets in storage is $\frac{x+1}{2}$.

So in any case the overall average number of sets in storage is $\frac{x+1}{2}$. In terms of Figure 4.7.22 this amounts to "leveling off" the bar graph until all the bars have the same height and observing that this height will then be $\frac{x+1}{2}$ (see Figure 4.7.23). Knowing that it costs \$70 to store a set for a year, we conclude that to store an average of $\frac{x+1}{2}$ sets will cost $\$70\frac{x+1}{2} = \$35(x + 1)$ per year. After all, over the entire year there are, on the average, $\frac{x+1}{2}$ sets in storage.

We want to minimize the sum of the cost of ordering and the cost of storing the sets. This total "inventory cost" is

$$C(x) = \frac{365}{x}(100) + 35(x + 1).$$

Its derivative is

$$C'(x) = -\frac{36{,}500}{x^2} + 35,$$

which is zero when

$$35 = \frac{36{,}500}{x^2}, \qquad \text{or} \qquad x^2 = \frac{36{,}500}{35}$$

or

$$x = \sqrt{\frac{36{,}500}{35}} \approx 32.29.$$

Since $C(x)$ tends to ∞ as $x \to 0^+$ and as $x \to \infty$, this point must be the absolute minimum. The most efficient scheme would be to order every 32.29 days.

Unfortunately, it is not practical to order once every 32.29 days. In a situation like this it is necessary to find the best *integer* solution. A quick look at the formula for $C'(x)$ will convince you that the function $C(x)$ is decreasing to the left of the critical point and increasing to the right of it, so we only have to consider the two nearest integers, 32 and 33. It is easy to evaluate $C(x)$ for $x = 32, 33$, and what we find is that $C(32) \approx \$2295.63$ and $C(33) \approx \$2296.06$. Ordering every 32 days will be slightly better than ordering every 33 days and better than any other scheme.

Problems in textbooks frequently work out to very "nice" answers, just to prevent bogging down in the calculations, but it is very important to recognize that in practice the answer to a problem is unlikely to work out to a nice round number. Sometimes this is unimportant, but very often it is necessary to give some thought, as we did in this example, to finding an answer that is an integer, even though it might not be the absolute minimum.

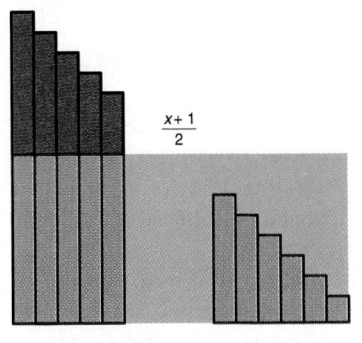

$\frac{x+1}{2}$

Figure 4.7.23

Exercises 4.7

I

1. Suppose a trough was made, as in Example 1 (p. 290), by bending up the sides of a sheet of metal. This time, assume that the metal is 100 cm wide and that two sides are folded up at the same angle, with each side 25 cm wide (so the bottom is 50 cm wide).

   (i) Find the cosine of the angle at which the sides should be bent up in order to make the trough with the maximum possible volume.

   (ii) Use a calculator to find an approximate value for the corresponding angle.

2. A rectangular page is to have margins of 2 in at the top and the bottom and 1 in at each side and is to contain a rectangular printed space with area 50 sq in in the middle. What dimensions for the page will minimize the total amount of paper?

3. Suppose the height above the ground of a helicopter is given by $H(t) = \frac{5,000t^2}{t^2+16}$ ft, where t is the time, in minutes, after the takeoff.

   (i) What is the greatest height it will reach in the first 10 minutes of its flight?

   (ii) At what moment will it be climbing the fastest?

4. Suppose the cylindrical storage drum of Example 3 (p. 294) is to be made without a top.

   (i) If the idea is to make a drum of the same volume (16 cu ft) with a minimum of material, that is, minimum surface area (*not* including the top), how do you expect the best possible dimensions will compare with those found in Example 3 with the top included? Will the radius be smaller and the height bigger or vice versa?

   (ii) Find the exact answer, that is, find the radius and height of the topless drum of volume 16 cu ft for which the total area of sides and bottom is a minimum.

5. Two towns on opposite sides of a river want to be joined by a telephone line. The river is perfectly straight and 1 km wide. One town is exactly 3 km upstream from the other. Suppose laying underwater telephone cable costs $100,000 per kilometer but laying cable on dry land costs only $50,000 per kilometer. What will be the least expensive route for the cable?

6. Suppose a second cable company makes a last-minute bid for the telephone cable joining the two towns of Exercise 5. The second company offers to lay cable on dry land for only $40,000 per kilometer, but expect that laying underwater cable will cost $120,000 per kilometer.

   (i) First try to guess how the best route for the second company to take will differ from the best route for the first company.

   (ii) With each company following the route that is best for it, which company will be able to do the job for less?

   (iii) The town suggests that the companies share the contract, one doing the underwater work and the other doing the dry land part. If the town takes the lower bid for each section of cable, what will be the best route?

   (iv) Will this shared approach cost the town less than either contractor alone?

7. A company wishes to manufacture a tin can, closed both at the top and the bottom, with a volume of 48 cu in. If the material for the top and bottom costs twice as much per square inch as the material for the rounded sides, what are the dimensions of the cheapest can that fills the requirements?

8. (i) If the can in Exercise 7 is replaced with one that has no top, do you expect the radius of the cheapest can to be smaller or larger than the radius found in Exercise 7?

   (ii) Find the cheapest way to make a can with a bottom and no top, whose volume is 48 cu in, assuming that the material for the bottom is twice as expensive as the material for the sides.

9. (i) If Exercise 8 is changed by having the material for the bottom cost half as much as the material for the sides, what effect do you expect this to have on the answer?

   (ii) Find the cheapest way to make an open can (bottom but no top) with volume 48 cu in, if the bottom is made from material that is half as expensive per square inch as the sides.

*10. A cube and a sphere are to be made so that the sum of their surface areas is 12. What are the maximum possible and minimum possible total volume of the two objects? (A sphere of radius r has volume $\frac{4}{3}\pi r^3$ and surface area $4\pi r^2$.)

11. Suppose A is a point 2 units above a horizontal mirror and B is another point 4 units above the mirror and 1 unit to the right of A. (i) Find the point of reflection that results in the shortest total distance for a light ray traveling from A to the mirror and then to B. (ii) Compare the total distance using this point of reflection with the distance obtained by reflecting in the point on the mirror $\frac{1}{2}$ unit to the right of A, that is, halfway between A and B in the horizontal direction.

12. Billiard balls bounce according to the same law that governs the reflection of light: that the angle of incidence equals the angle of reflection. Use this fact to find where a pool shark should aim a ball if he wants to bounce it off the side of the table and hit another ball. Specifically, suppose the first ball is 10 in from the side and the ball at which it is aimed is 25 in from the side. Suppose that the distance between the two balls in the direction parallel to the side of the table is 28 in, as shown in Figure 4.7.24. Find the point on the side of the table at which the first ball should be aimed in order to bounce and hit the second ball.

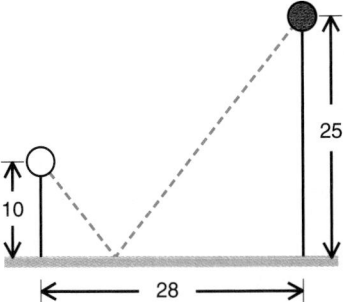

Figure 4.7.24

*13. A more advanced player wants to make a shot in which a ball bounces twice before hitting another ball, as shown in Figure 4.7.25. For the sake of simplicity, let us suppose that the corner of the table is at the origin.

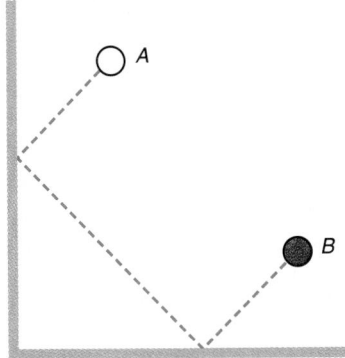

Figure 4.7.25

(i) Consider a ball whose position is point A, whose coordinates are $(3, 9)$ (i.e., it is 3 in to the right and 9 in up from the origin). Suppose the idea is to make this ball bounce off the left wall (the y-axis), then off the bottom wall (the x-axis), and then roll to point B, whose coordinates are $(9, 3)$. Find the angles at which the ball hits the two walls. (*Hint:* Call the angles θ and ϕ, as

shown in Figure 4.7.26; notice that there is a relationship between them. Try solving for $\tan \theta$.)

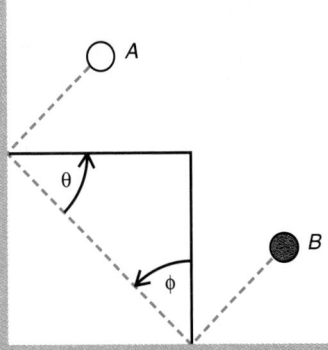

Figure 4.7.26

(ii) If A is the point $(4, 12)$ and B is the point $(8, 4)$, find the angles for the two "reflections." (*Hint:* Again solve for the tangent of one of the angles, and then use your calculator to find the actual angle.)

*14. In finding the law for light reflecting from a mirror, we mentioned that light travels in such a way as to minimize the time of travel and observed that this is the same thing as minimizing the distance. This observation is based on the assumption that the speed of light is always the same, but it is no longer true if the speed can vary. In fact, the speed of light depends on the medium in which it is traveling. The speed of light that is usually quoted in physics texts is the speed in a vacuum, but light travels slightly less fast in air and slower still in glass or water. This fact is behind the everyday observation that an object partially submerged in water appears to bend at the point where it enters the water.

(i) Consider a ray of light traveling from air into water as shown in Figure 4.7.27. If the speed of

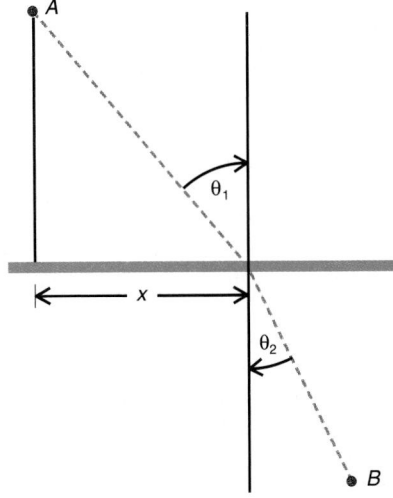

Figure 4.7.27

light in air is v_1 and the speed of light in water is v_2, express the total time for the light to travel from A to B as a function of the variable x (cf. Example 4, p. 295).

(ii) If we label the angles θ_1 and θ_2 as shown, show that the minimum time occurs when x is chosen in such a way that

$$\frac{\sin \theta_1}{\sin \theta_2} = \frac{v_1}{v_2}.$$

The quotient $\frac{v_1}{v_2}$ is called the *index of refraction* of water relative to air and measures how much an object immersed in water appears to bend. The above formula relating the angle of incidence and the angle of refraction to the index of refraction is called **Snell's Law**.

(iii) Assuming that the speed of light in air is known (it is very close to the speed of light in a vacuum, or about 3×10^{10} cm/sec), describe an experiment to determine the speed of light in water.

(iv) Carry out the experiment.

15. Find the maximum length of a thin rod that can be carried without tilting around the corner made by two corridors at right angles that have widths 2 yd and 3 yd. (When the question speaks of a "thin" rod, it means that the thickness is so small that it can be ignored.)

16. What is the maximum length of a thin rod that can be carried without tilting around the corner made by two corridors at right angles that have widths a and b?

*17. Solve the inventory problem of Example 7 (p. 298) again, this time assuming that the prudent store manager wants to keep at least five sets in stock at all times. How does this affect the strategy for when to order and the total cost?

18. (i) If the storage costs remain the same but the cost of ordering new sets rises, what sort of change do you expect there will be to the best strategy for ordering?

(ii) If the cost of ordering rises to $120 and the storage costs remain the same, what would be the best interval between successive orders?

(iii) Some time after the ordering cost rose to $120, news arrived that the storage costs had gone up too. The manager did a calculation and found that the best length of time between orders should be 28 days. What are the new storage charges?

II

The preceding problems were more or less in the same order as the corresponding examples in the text and were sufficiently similar that it was usually clear which sort of technique to apply to each of them. Generally speaking, figuring out which technique to use, and specifically choosing what should be the variable, is one of the most difficult parts of a maximum/minimum problem.

With this in mind, the following problems are in no particular order, and a big part of the question is deciding how to approach them. Of course, in many cases several different approaches are possible, but even then some may be easier to work with than others.

19. A cylinder is placed inside a right circular cone as shown in Figure 4.7.28. Suppose the radius and height of the cone are r and h, respectively. What is the largest possible volume of the cylinder, and what are the dimensions of the cylinder of largest possible volume?

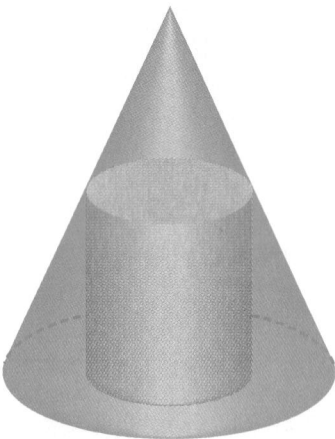

Figure 4.7.28

20. A sphere has radius 1. What is the largest possible volume of a cylinder that can be placed inside the sphere?

21. The intensity of illumination from a small source of light is inversely proportional to the square of the distance from the source, that is, proportional to $\frac{1}{r^2}$, if r is the distance from the source. This is one of many examples in physics of what is known as an *inverse square law*.

Two light bulbs are 100 m apart. One is four times as bright as the other. What point on the straight line between them is darkest, that is, at what point on the line is the total illumination at a minimum?

22. Fences are to be used to enclose a rectangular field and divide it into two pieces by a divider running parallel to two of the sides. If 1200 ft of fencing material are available, what is the maximum possible total area of the enclosure?

23. If a right circular cone has height h and radius r, the distance from the vertex to the circle at the opposite end, measured along the surface of the cone, is $S = \sqrt{h^2 + r^2}$. This quantity is sometimes called the *slant height* of the cone (see Figure 4.7.29). What is the largest possible volume of a cone whose slant height is 40?

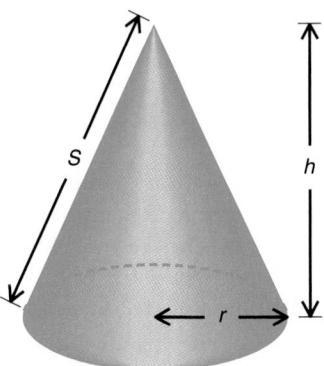

Figure 4.7.29

24. A container is made in the shape of a cylinder with a flat base and a hemispherical cap, as shown in Figure 4.7.30. If the total surface area of the container is to be 80 sq in, what is the maximum possible volume?

Figure 4.7.30

25. A silo is made in the shape of a cylinder with a hemispherical cap, the shape illustrated for Exercise 24. The material available to cover the silo, that is, to cover the cap and the curved sides but not the bottom, will cover 2000 sq ft.

  (i) What dimensions will result in a silo with the maximum possible volume?
  (ii) For many purposes only the cylindrical part of the silo can be used for storage (i.e., the cap remains empty). What dimensions will result in a silo with the maximum possible storage capacity?

26. A tank for compressed gas is made in the shape of a cylinder with a hemispherical cap at each end (see Figure 4.7.31). What is the largest possible volume for such a tank if the surface area is 1 sq m?

Figure 4.7.31

27. A family having a house built must choose the amount of insulation to install. They have been told that they can have as many layers of insulation as they please, and that each layer will add $100 to

their annual mortgage payment. On the other hand, each layer will reduce the heating bill by 20% (so, for example, one layer would reduce heating costs to 80% of the costs for an uninsulated house, and a second layer would reduce costs to 80% of *that*, or 64% of the costs for an uninsulated house . . .).

Thermograph showing the distribution of heat on the external surface of a house. Different levels of heat are indicated by different colors.

  (i) Assuming that the only consideration is to minimize the combined cost of mortgage payments and heating bills, how many layers of insulation would be best if the annual heating costs for an uninsulated house are $1100?
  (ii) What would you recommend if the annual heating costs for an uninsulated house are $1910?
  (iii) If the heating costs are very low for a house in a warm climate, it will not be worthwhile (from the purely financial point of view) to insulate the house. Find the maximum annual heating cost for an uninsulated house for which it is economically unfeasible to insulate at all.

*28. A rectangular painting 5 ft high is hung on a gallery wall with its bottom edge 10 ft above the floor. A person whose eye level is 5 ft above floor level is looking at it. How far from the wall holding the painting must the person stand to maximize the angle between her lines of sight to the bottom and top of the painting?

29. Use a graphics calculator or computer to sketch the function $A(\theta) = 10 - 2\sec\theta + \tan\theta$ of Example 2 (p. 292), for $\theta \in \left[0, \frac{\pi}{2}\right)$. Estimate to four decimal places the value of θ at which $A(\theta)$ crosses the axis to become negative. Can you explain in terms of the original problem why this happens?

30. Sketch the function $L(\theta) = \csc\theta + 2\sec\theta$ of Example 6 (p. 297), and use the graph to verify the answer found in the example.

4.8 Antiderivatives

FUNCTIONS WITH THE SAME DERIVATIVE
ADDING A CONSTANT
COMMON ANTIDERIVATIVES

In Section 4.2 we touched on the idea that two functions with the same derivative must differ by a constant. To put it another way, given a function $f(x)$, if we can guess some function $F(x)$ whose derivative is $f(x)$, then we can find all such functions simply by adding constants to $F(x)$.

It is not terribly difficult to imagine situations in which it is necessary to find a function whose derivative equals some particular function. For instance, if we know that an object moves in such a way that its velocity at time t is $f(t)$, then its position at time t (or more properly its displacement) will be a function $F(t)$ whose derivative is $f(t)$.

More generally, if the rate of change of some quantity is known to be $f(t)$, then the quantity itself will be a function whose derivative is $f(t)$. We often do know the velocity or rate of change of something; if we know the velocity of an airplane, it can be quite important to know where the airplane will be after 3 hours have elapsed, for instance. We might know the rate at which a chemical reaction is proceeding and want to know how much of the product will have been made after a specified period of time. Knowing the rate of population growth, we might like to predict the population ten years from now.

This type of problem is extremely important and will be discussed at greater length in the next four chapters, but in this brief section we make some preliminary remarks that will be sufficient for the purposes of the next section on velocity and acceleration.

We already observed that if $p(x)$ is a polynomial, it is quite easy to find a polynomial whose derivative is $p(x)$.

EXAMPLE 1   Find all functions $F(x)$ defined for all $x \in \mathbb{R}$ and for which $F'(x) = x^3 - 2x^2 + 8x - 3$.

Solution   We begin by looking for just one such function $F(x)$, and to do that, we work with each term of the polynomial separately. A function whose derivative is x^3 is $\frac{1}{4}x^4$, a function whose derivative is $-2x^2$ is $-\frac{2}{3}x^3$, a function whose derivative is $8x$ is $4x^2$, and a function whose derivative is -3 is $-3x$.

Because of the way derivatives add (Theorem 3.2.2), we can put these functions together to find a function whose derivative is the sum of x^3, $-2x^2$, $8x$, and -3. Let $G(x) = \frac{1}{4}x^4 - \frac{2}{3}x^3 + 4x^2 - 3x$. It is obvious that $G'(x) = x^3 - 2x^2 + 8x - 3$.

The question asked for *all* such functions, but this is easier than it sounds. Theorem 4.2.11(ii) says that any such function must differ from $G(x)$ by a constant, which means that it must equal $G(x) + C$, for some constant C. In other words, every function $F(x)$ that satisfies $F'(x) = x^3 - 2x^2 + 8x - 3$ must be of the form

$$F(x) = \frac{1}{4}x^4 - \frac{2}{3}x^3 + 4x^2 - 3x + C,$$

for some constant C. We can find as many particular examples as we like by choosing different values for C (e.g., $C = 0, 1, \sqrt{2}, -\frac{7}{3}, \ldots$).

Finding a function with a specified derivative is so common a problem that it is convenient to have terminology for it. Thinking of this process as the reverse of taking the derivative suggests a name.

DEFINITION 4.8.1

Let $f(x)$ be a function. A function $F(x)$ is called an **antiderivative** of $f(x)$ if

$$F'(x) = f(x),$$

wherever $f(x)$ is defined.

EXAMPLE 2 Find antiderivatives of (i) $f(x) = \cos x$, (ii) $g(x) = x^{-10}$, (iii) $h(x) = e^x$, (iv) $k(x) = \sqrt{x}$, (v) $r(x) = \frac{1}{1+x^2}$.

Solution

(i) By observation, one antiderivative of $f(x) = \cos x$ is $F(x) = \sin x$. Any antiderivative of $f(x)$ is of the form $\sin x + C$, for some constant C.

(ii) Recalling that differentiating a power of x has the effect of lowering the exponent by 1 and multiplying by the original exponent, we see that to have derivative equal to $g(x) = x^{-10}$, a function would have to involve a power of x whose exponent is higher by 1, that is, x^{-9}. But the derivative of x^{-9} is $-9x^{-10}$, so to get rid of the -9 we must let $G(x) = \frac{1}{-9}x^{-9}$. Then $G'(x) = g(x) = x^{-10}$, as required. The antiderivatives of $g(x)$ are of the form $\frac{1}{-9}x^{-9} + C$, for some constant C.

   Notice that $g(x)$ is not defined at $x = 0$ and neither is any of its antiderivatives, so strictly speaking, it is possible to have two different constants C, one when $x < 0$ and another when $x > 0$.

(iii) It is easy to guess an antiderivative for $h(x)$. Any antiderivative for $h(x)$ must be of the form $H(x) = e^x + C$, for some constant C.

(iv) Rewriting $k(x) = \sqrt{x} = x^{1/2}$, we realize that an antiderivative for $k(x)$ is $\frac{2}{3}x^{3/2}$.

   If that was too quick, notice that it should be some constant times x raised to the power $\frac{1}{2} + 1 = \frac{3}{2}$. The constant you can figure out by calculating that $\frac{d}{dx}(x^{3/2}) = \frac{3}{2}x^{1/2}$ and then putting in the constant $\frac{2}{3}$ to cancel the $\frac{3}{2}$.

   Every antiderivative of $k(x)$ is of the form $K(x) = \frac{2}{3}x^{3/2} + C$, for some constant C.

(v) Here we have to remember our formulas for derivatives and recall that $r(x) = \frac{1}{1+x^2}$ is the derivative of $\arctan x$. From this we see that every antiderivative of $r(x)$ must be of the form $R(x) = \arctan x + C$, for some constant C.

In the last example we went to great pains to include the constant C with each antiderivative. It is important to get into the habit of doing this; most of the time it seems to be unnecessary, but there are situations in which it is vital to remember this step. If it is not something you do automatically, then you are liable to forget it in one of these situations, with disastrous consequences.

We record as a theorem what we have already observed about adding constants to antiderivatives.

THEOREM 4.8.2

> If $F(x)$ is an antiderivative of $f(x)$ on an open interval, then so is $G(x) = F(x) + C$, for any constant C. Moreover, every antiderivative of $f(x)$ on the interval is of this form for some constant C.

PROOF   The theorem is really just a restatement of Theorem 4.2.11 (ii). ◢

EXAMPLE 3   Let $f(x) = 3x^2$ for every $x \neq 0$, and let $f(x)$ not be defined for $x = 0$. Find every antiderivative of $f(x)$.

Solution   Of course, we know that x^3 is one antiderivative of $3x^2$ and $x^3 + C$ is an antiderivative for any constant C. But what makes this example more subtle is that the domain of $f(x)$ is not the whole real line $\mathbb{R}$ but is made up of two intervals $(-\infty, 0)$ and $(0, \infty)$.

The theorem says that on each of these intervals, any antiderivative of $f(x)$ must equal $x^3 + C$, for some constant C. What it does *not* say is that it has to be the same constant for both intervals. We can get an antiderivative by letting

$$F(x) = \begin{cases} x^3 + C, & \text{if } x < 0, \\ x^3 + D, & \text{if } x > 0, \end{cases}$$

where C and D are any constants.

In fact these are all the antiderivatives of $f(x)$, since any antiderivative must equal the sum of x^3 and some constant on each of the two intervals in the domain.

◢

> The reason for the above example is to illustrate the importance of the domain for questions involving antiderivatives and to explain the need for the open interval in Theorem 4.8.2.

It is quite easy to find antiderivatives for x^r for any real exponent r, *except $r = -1$*.

EXAMPLE 4   Find the antiderivatives of $f(x) = \frac{1}{x}$.

Solution   When $x > 0$, we know that one antiderivative is $\ln x$, so any antiderivative must be of the form $\ln x + C$.

When $x < 0$, the situation is a bit more complicated. Suppose we let $u = -x$ and consider the function $G(x) = \ln u = \ln(-x)$. By the Chain Rule, $G'(x) = \frac{d}{dx}(\ln u) = \frac{1}{u}\frac{du}{dx} = \frac{1}{-x}(-1) = \frac{1}{x}$. In other words, $G(x) = \ln(-x)$ is an antiderivative of $\frac{1}{x}$ when $x < 0$. (Notice that $G(x)$ is not defined when $x > 0$, since $\ln t$ makes sense only when $t > 0$.)

When $x < 0$, we see that $-x = |x|$, so we can combine the cases in which $x > 0$ and $x < 0$ by letting $H(x) = \ln|x|$. This is an antiderivative of $f(x) = \frac{1}{x}$, and it is defined whenever $x \neq 0$, that is, whenever $f(x)$ is defined.

As in the previous example, any antiderivative of $f(x)$ must differ from $H(x)$ by a constant on each of the two intervals in the domain of $f(x)$, that is, on $(-\infty, 0)$ and $(0, \infty)$. So any antiderivative of $f(x) = \frac{1}{x}$ must be of the form

$$F(x) = \begin{cases} \ln|x| + C, & \text{when } x < 0, \\ \ln|x| + D, & \text{when } x > 0, \end{cases}$$

where C and D are constants.

This is a bit cumbersome, so we often sacrifice some accuracy and say that an antiderivative of $\frac{1}{x}$ is $\ln|x| + C$. This is true, but it neglects the fact that there are other antiderivatives. Similar remarks can be made about antiderivatives of x^r if $r < 0$, since this function is also not defined at $x = 0$ (so its domain is $(-\infty, 0) \cup (0, \infty)$). ◢

Lists of derivatives can now be read the other way around to give lists of antiderivatives. We record a few of the most familiar.

Summary 4.8.3

Common Antiderivatives

For each of the following functions $f(x)$, its antiderivatives are of the form $F(x)$, with $F(x)$ as given in the list below:

$$f(x) = x^r, \quad r \neq 0, -1 \qquad F(x) = \frac{1}{r+1} x^{r+1} + C$$
$$f(x) = 1 \qquad\qquad\qquad F(x) = x + C$$
$$f(x) = \frac{1}{x} \qquad\qquad\qquad F(x) = \ln|x| + C$$
$$f(x) = \sin x \qquad\qquad F(x) = -\cos x + C$$
$$f(x) = \cos x \qquad\qquad F(x) = \sin x + C$$
$$f(x) = e^x \qquad\qquad\quad F(x) = e^x + C$$

For $f(x) = \frac{1}{x}$ and $f(x) = x^r$ with $r < 0$, it is possible to have different constants for $x < 0$ and for $x > 0$ (cf. Example 4).

We can guess antiderivatives for many functions because they are easily recognized as derivatives of known functions. This process is often made easier by breaking up a function as a sum of simpler ones and dealing with each summand separately.

A more complicated problem is to recognize a derivative that is obtained by using the Chain Rule.

EXAMPLE 5 Find the antiderivatives of each of the following functions: (i) $f(x) = \sin 3x$; (ii) $g(x) = (x + 4)^7$; (iii) $h(x) = (2x - 3)^4$; (iv) $k(x) = e^{-5x}$; (v) $r(x) = xe^{x^2}$.

Solution

(i) If we let $u = 3x$, then $\sin 3x = \sin u$, which makes us think of the derivative of $-\cos u$. We could guess that the antiderivative of $\sin 3x$ should be something involving $-\cos 3x$.

   The Chain Rule says that taking the derivative of $-\cos 3x = -\cos u$ results in the product of $\sin 3x$ with the derivative of $u = 3x$. This last derivative is just 3, so we have to "fix up" the function by dividing by 3. The antiderivatives of $f(x) = \sin 3x$ are all of the form $F(x) = -\frac{1}{3}\cos 3x + C$.

(ii) Similarly, $g(x) = (x + 4)^7$ reminds us of the derivative of $\frac{1}{8}u^8$, with $u = x + 4$, except for the factor of $\frac{du}{dx}$. This time we are lucky and $\frac{du}{dx} = 1$, so the antiderivatives of $g(x) = (x + 4)^7$ are $G(x) = \frac{1}{8}(x + 4)^8 + C$.

(iii) If we write $h(x) = (2x - 3)^4 = \frac{1}{2}(2x - 3)^4 \times 2$, we recognize the last two factors $(2x - 3)^4 \times 2$ as the derivative of $\frac{1}{5}(2x - 3)^5$. We conclude that the antiderivatives of $h(x)$ are all of the form $H(x) = \frac{1}{2} \left(\frac{1}{5} \right) (2x - 3)^5 + C = \frac{1}{10} (2x - 3)^5 + C$.

(iv) Noting that $\frac{d}{dx} (e^{-5x}) = e^{-5x}(-5)$, we might write $k(x) = e^{-5x} = -\frac{1}{5} e^{-5x}(-5)$, from which it is easy to see that its antiderivatives are $K(x) = -\frac{1}{5} e^{-5x} + C$.

(v) This part is more complicated. The function $r(x) = xe^{x^2}$ makes us think of e^u with $u = x^2$. If we differentiate $e^u = e^{x^2}$, we obtain $e^u \frac{du}{dx}$. Since $\frac{du}{dx} = 2x$, this derivative is very nearly equal to $r(x)$. We find that the antiderivatives of $r(x) = xe^{x^2}$ are all of the form $R(x) = \frac{1}{2} e^{x^2} + C$.   ◢

Let us examine more closely the last part of Example 5. The reason it worked is that $r(x) = xe^{x^2}$ is the product of two things. One of them, x, is, apart from a constant, the derivative of the function $u = x^2$ occurring in the other, and the other one, e^u, is itself a derivative: $e^u = \frac{d}{du} (e^u)$.

This is exactly what is needed to recognize something as the result of a Chain Rule differentiation. The Chain Rule says that

$$\frac{d}{dx} \Big(G\big(u(x)\big) \Big) = G'(u) \frac{du}{dx},$$

so we must look for some familiar derivative $\big(G'(u)\big)$ evaluated at u, multiplied by the derivative of u. Just to complicate things, we can expect to have to multiply by constants.

EXAMPLE 6   Find antiderivatives for each of the following: (i) $\frac{x}{(x^2+3)^2}$; (ii) $\frac{x}{\sqrt{x^2+1}}$; (iii) $x^2 \cos(x^3 + 4)$.

Solution

(i) We begin by looking for a suitable u, whose derivative appears in the function too. In this case we notice the function $x^2 + 3$, whose derivative $2x$ is essentially the numerator (apart from a factor of 2). We can write

$$\frac{x}{(x^2 + 3)^2} = \frac{1}{2} \frac{1}{(x^2 + 3)^2} (2x),$$

and let $u = x^2 + 3$. We see that

$$\frac{x}{(x^2 + 3)^2} = \frac{1}{2} u^{-2} \frac{du}{dx}.$$

Since u^{-2} is the derivative $\frac{d}{du} (-u^{-1})$ (by Summary 4.8.3), we can see that the antiderivative ought to be $\frac{1}{2} (-u^{-1}) + C = -\frac{1}{2} \frac{1}{x^2+3} + C$.

(ii) Here again we recognize that the x in the numerator is $\frac{1}{2}$ the derivative of the quantity inside the square root. If we let $u = x^2 + 1$, the function becomes $\frac{x}{\sqrt{u}} = xu^{-1/2} = u^{-1/2} \left(\frac{1}{2} \right) \frac{du}{dx}$. Since $\frac{1}{2} u^{-1/2}$ equals $\frac{d}{du} (u^{1/2})$, the antiderivatives of the original function are $u^{1/2} + C = \sqrt{u} + C = \sqrt{x^2 + 1} + C$.

(iii) Here the factor x^2 looks a lot like the derivative of $x^3 + 4$. If we let $u = x^3 + 4$, the function is $x^2 \cos(x^3 + 4) = \frac{1}{3} 3x^2 \cos(x^3 + 4) = \frac{1}{3} \cos u \frac{du}{dx}$. From this we see that the antiderivatives are all of the form $\frac{1}{3} \sin(x^3 + 4) + C$.

Notice that it is always easy to check these answers; just differentiate.   ◢

EXAMPLE 7   Suppose $f(x)$ is differentiable on the whole real line and $f(x) \neq 0$ for any x. Find the antiderivatives of the function $\frac{f'(x)}{f(x)}$.

Solution   The presence of the derivative $f'(x)$ in the numerator suggests that this might be the result of a differentiation using the Chain Rule. If we write it in the slightly different form $\frac{1}{f(x)}f'(x)$, we see immediately that it equals the derivative of $\ln |f(x)|$.

The fact that $f(x)$ is differentiable at every x and never zero implies that $\ln |f(x)|$ is differentiable at every x. The domain is the whole real line, and since $f(x)$ is continuous and never zero, it must be always positive or always negative. (By the Intermediate Value Theorem it cannot change sign.) From this we see that every antiderivative of $\frac{f'(x)}{f(x)}$ must equal $\ln |f(x)| + C$, for some constant C, by Theorem 4.8.2.

Another type of problem we can solve is finding an antiderivative of some function $f(x)$ that also satisfies some additional condition. This has the effect of determining the constant that we usually add to antiderivatives, removing the ambiguity.

EXAMPLE 8   (i) Find the antiderivative $F(x)$ of $f(x) = x^2 - 2x + 1$ that satisfies $F(0) = 3$.

Solution   We know that any antiderivative must be of the form $F(x) = \frac{1}{3}x^3 - x^2 + x + C$, for some constant C. On the other hand, if we substitute $x = 0$ in this formula, we find that $F(0) = \frac{1}{3}0^3 - 0^2 + 0 + C = C$. The constant C must equal 3.

The solution to the problem is the function $F(x) = \frac{1}{3}x^3 - x^2 + x + 3$.

(ii) Find the antiderivative $G(x)$ of $g(x) = e^{2x}$ that satisfies $G(0) = -1$.

Solution   Any antiderivative $G(x)$ must be of the form $G(x) = \frac{1}{2}e^{2x} + C$, for some constant C. The condition says that $-1 = G(0) = \frac{1}{2}e^0 + C = \frac{1}{2} + C$. This implies that $C = -\frac{3}{2}$.

The required function is $G(x) = \frac{1}{2}e^{2x} - \frac{3}{2}$.

(iii) Find an antiderivative $H(x)$ of $h(x) = \frac{1}{x-1}$ that satisfies $H(0) = 4$ and $H(2) = 5$.

Solution   We know from Example 4 that $H(x)$ must be obtained from the function $\ln |x - 1|$ by adding one constant when $x - 1 > 0$ and another constant when $x - 1 < 0$.

When $x = 2$, we see that $\ln |x - 1| = \ln |2 - 1| = \ln 1 = 0$, so the constant that is added to $\ln |x - 1|$ when $x > 1$ must be $H(2) = 5$.

Similarly, when $x = 0$, $\ln |x - 1| = \ln 1 = 0$, so the constant added to $\ln |x - 1|$ when $x < 1$ must be $H(0) = 4$. We see that

$$H(x) = \begin{cases} \ln |x - 1| + 5, & \text{if } x > 1, \\ \ln |x - 1| + 4, & \text{if } x < 1. \end{cases}$$

(iv) Find an antiderivative $K(x)$ of $k(x) = \frac{1}{x}$ that satisfies $K(1) = -2$ and $K(2) = 5$.

Solution   The antiderivative $K(x)$ must be of the form

$$K(x) = \begin{cases} \ln|x| + C, & \text{if } x > 0, \\ \ln|x| + D, & \text{if } x < 0, \end{cases}$$

for some constants C and D.

From this we see that $K(1) = \ln|1| + C = 0 + C = C$. Since $K(1) = -2$, we see that $C = -2$.

Also since $2 > 0$, we see that $K(2) = \ln|2| + C = \ln 2 + (-2) \approx -1.307 \neq 5$.

This says that an antiderivative of $\frac{1}{x}$ that equals -2 at $x = 1$ cannot equal 5 at $x = 2$. It is not possible to find any function satisfying the requirements of the question.

With a graphics calculator we can plot the solution to part (i) of Example 8. We also plot $F(x) = \frac{1}{3}x^3 - x^2 + x + C$, with $C = 0, 4$.

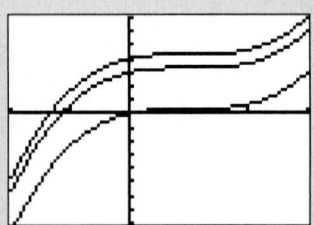

TI-85	SHARP EL9300
GRAPH F1 $(y(x) =)$ F1 (x) ^3 ÷ 3 − F1 (x) x^2 + F1 (x) + 3 ▼ F1 (x) ^3 ÷ 3 − F1 (x) x^2 + F1 (x) + 0 ▼ F1 (x) ^3 ÷ 3 − F1 (x) x^2 + F1 (x) + 4 M2(RANGE) (−) 2 ▼ 3 ▼ ▼ (−) 8 ▼ 7 F5(GRAPH)	↯ x/θ/T a^b 3 ▶ ÷ 3 − x/θ/T x^2 + x/θ/T + 3 2ndF ▼ x/θ/T a^b 3 ▶ ÷ 3 − x/θ/T x^2 + x/θ/T 2ndF ▼ x/θ/T a^b 3 ▶ ÷ 3 − x/θ/T x^2 + x/θ/T + 4 RANGE −2 ENTER 3 ENTER 1 ENTER −8 ENTER 7 ENTER 1 ENTER ↯
CASIO f_x-7700GB/f_x-6300G	HP 48SX
Cls EXE Range −2 EXE 3 EXE 1 EXE −8 EXE 7 EXE 1 EXE EXE EXE EXE Graph X x^y 3 ÷ 3 − X x^2 + X + 3 EXE Graph X x^y 3 ÷ 3 − X x^2 + X EXE Graph X x^y 3 ÷ 3 − X x^2 + X + 4 EXE	↰ PLOT PLOTR ERASE ATTN ' αY ▬ αX y^x 3 ÷ 3 − αX y^x 2 + αX + 3 ↱ PLOT ↰ DRAW ERASE 2 +/− ENTER 3 XRNG 8 +/− ENTER 7 YRNG DRAW ATTN ' αY ▬ αX y^x 3 ÷ 3 − αX y^x 2 + αX ↱ PLOT ↰ DRAW DRAW ATTN ' αY ▬ αX y^x 3 ÷ 3 − αX y^x 2 + αX + 4 ↱ PLOT ↰ DRAW DRAW

Notice that these functions are all the same function shifted up or down by the addition of a constant, so their derivatives are the same. Exactly one of them has the required value 3 at $x = 0$.

EXAMPLE 9   **Bacterial Growth**   Bacteria and other one-celled organisms reproduce by dividing. One cell will divide into two. The rate at which they divide will depend on various factors, such as the temperature, the amount of nutrients available, and so on.

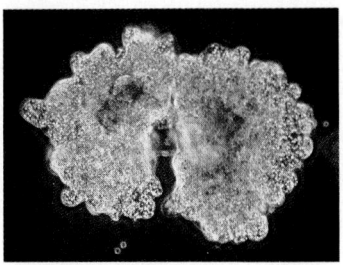

This microscopic view shows the Amoeba proteus *undergoing binary fission, "dividing in two."*

If all these extraneous factors remain the same, then the rate of reproduction (the number of additional cells created per unit of time) will be proportional to the number of cells already present. This simply means that the more cells there are, the more divisions there will be: If one culture contains twice as many cells as another, then the larger one will undergo twice as many divisions as the smaller one, so it will grow twice as fast.

If we write $P(t)$ for the "population," the number of cells in a particular culture at time t, this amounts to saying that

$$\frac{dP}{dt} = kP,$$

where k is some constant, depending on various environmental factors as well as the units in which time t is measured.

Writing $P'(t)$ for $\frac{dP}{dt}$, we find that the above equation is $P'(t) = kP(t)$. Assuming that the population is never zero (a reasonable assumption), we can divide by P and find

$$\frac{P'(t)}{P(t)} = k.$$

By Example 7 we know that an antiderivative of the left side of this equation is $\ln |P(t)|$. This equals $\ln P(t)$, since $P(t)$, being a population, is never negative. An antiderivative of the right side is kt, and these two antiderivatives must differ by a constant. In other words, $\ln P(t) = kt + C$, for some constant C.

To recover a formula for $P(t)$ itself, we take the exponential of each side and find that

$$P(t) = e^{kt+C} = e^{kt}e^{C}$$
$$= C'e^{kt}.$$

Here C' is some constant (it equals e^{C}, the exponential of the other constant C). The notation does *not* mean that it is a derivative; it merely indicates that it is a different constant. Notice that the constant C was the sort of constant that is added to an antiderivative, but since we exponentiated everything, it was converted into a constant that *multiplies* the answer. If we forgot to add the constant C to the antiderivative, then we would not have obtained the constant C', and our answer would be ridiculous. It would say, for instance, that any bacterial culture starts with exactly one organism at the beginning, since $P(0) = e^{k \times 0} = e^{0} = 1$.

Since the constant C came up in the middle of the calculation but it is the final answer that is of the most interest, we often rename the constants and use C to refer to C'. This way we can write

$$P(t) = Ce^{kt}.$$

We say the population of bacteria grows *exponentially*.

This kind of growth is extremely important in a huge variety of contexts. It will arise whenever the rate of growth (or decrease) of some quantity is proportional to that quantity. We will discuss it at greater length in Chapter 7, but mention in passing that it is relevant to such diverse problems as the spread of epidemics, radioactive decay, compound interest, and making yogurt, to name only a few. In each of these situations the rate of change may be proportional to the amount currently present.

EXAMPLE 10   **Bacterial Culture**   Suppose a bacterial culture weighs 10 mg (milligrams). Suppose that 1 hour later it weighs 20 mg. What will it weigh 30 min after that?

Solution   From Example 9 we know that the weight of the culture (which is just a constant times the number of organisms) will be given by a formula of the form

$$W(t) = Ce^{kt},$$

where $W(t)$ is the weight of the culture at time t, in milligrams, and C and k are constants. We may choose to measure time in the most convenient units, which in this case appear to be hours. Next we should find C and k.

If $t = 0$ at the starting time, then the problem tells us that $W(0) = 10$. On the other hand, $W(0) = Ce^{k \times 0} = Ce^0 = C$. This shows that $C = 10$.

The other piece of information we have is that $W(1) = 20$. From the formula we see that $W(1) = Ce^k = 10e^k$, so

$$10e^k = 20,$$
$$e^k = 2,$$

so

$$k = \ln 2.$$

Having calculated both C and k, we know that

$$W(t) = 10e^{t \ln 2}.$$

The question asked us to find the weight of the culture $1\frac{1}{2}$ hours after the starting point. This is $W\left(\frac{3}{2}\right) = 10e^{3/2 \ln 2} = 10(e^{\ln 2})^{3/2} = 10(2^{3/2}) = 10(2\sqrt{2}) = 20\sqrt{2} \approx 28.3$.

After one and one-half hours the culture will weigh about 28.3 mg.

In the course of doing this example we have noticed that in a formula for exponential growth

$$P(t) = Ce^{kt},$$

the constant C is exactly equal to $P(0)$, the *initial population*. The constant k, which depends on the units used, has to do with how fast the population grows. It is often called the *growth constant*.

Exercises 4.8

Find all antiderivatives of each of the following functions.

1. $f(x) = 6x^2 - 4x + 3$
2. $g(x) = x^3 - 4x^2 + 3x - 2$
3. $h(x) = \sin x$
4. $j(x) = \csc^2 x$
5. $r(x) = \dfrac{1}{\sqrt{1-x^2}}$
6. $k(x) = x^2 - 2x + \cos x$
7. $s(x) = \sqrt{x} + x^4$
8. $t(x) = x^{7/4}$
9. $v(x) = \dfrac{3}{\sqrt{x}}$
10. $w(x) = \dfrac{x^2}{1+x^2}$
11. $f(x) = x \sin(2x^2 - 3)$
12. $g(x) = x^2 \cos(x^3 + 2)$
13. $h(x) = (x+1) \exp(x^2 + 2x - 5)$

14. $k(x) = (x^2 - 1)\sin(x^3 - 3x + 7)$

15. $q(x) = \frac{x}{x^2 + 1}$

16. $r(x) = \frac{(x-1)}{x^2 - 2x + 3}$

17. $s(x) = x^3\sqrt{x^4 + 4}$

18. $t(x) = \frac{1}{1 + (x-2)^2}$

19. $v(x) = \frac{1}{e^{-x} + 1}$

20. $w(x) = \cos x \sin^2 x$

II

21. Find an antiderivative $F(x)$ of $f(x) = x^3 - 2x + 3$ that satisfies $F(0) = 5$.

22. Find an antiderivative $G(x)$ of $g(x) = x - \cos x$ that satisfies $G(0) = -2$.

23. Find an antiderivative $H(x)$ of $h(x) = 3x^2 - 4x + 1$ that satisfies $H(1) = 1$.

*24. Is it possible to find an antiderivative $K(x)$ of $k(x) = x^2 + 3x - 2$ that satisfies $K'(3) = 3$? If so, find it.

Find *all* antiderivatives of each of the following functions.

*25. $f(x) = \frac{|x|}{x}$

*26. $g(x) = \frac{3}{x}$

*27. $h(x) = \frac{1}{x-3}$

*28. $k(x) = \frac{1}{(x-2)^3}$

*29. $f(x) = \sec^2 x$

*30. $g(x) = \frac{\cos x}{\sin^2 x}$

*31. $h(x) = \tan x$

*32. $h(x) = \cot x$

33. Suppose a bacterial culture containing 2000 organisms doubles in 40 min.
    (i) How many bacteria will there be 1 hour after the beginning?
    (ii) How long will it take until there are 10,000 organisms?

34. A technician forgot to weigh a bacterial culture when it began growing. He weighed it after 3 hours and found that it weighed 16 mg and weighed it again two hours later, when it weighed 32 mg.
    (i) How much did it weigh at the beginning?
    (ii) What will it weigh 11 hr after the beginning?

*35. Find all antiderivatives $K(x)$ of $k(x) = \sec^2 x$ that satisfy $K\left(-\frac{\pi}{4}\right) = 3$ and $K\left(\frac{\pi}{4}\right) = 5$.

*36. Show that any antiderivative $F(x)$ of $\sec x \tan x$ must satisfy $F\left(\frac{\pi}{6}\right) = F\left(-\frac{\pi}{6}\right)$.

*37. Find as many functions $f(x)$ as you can that satisfy $f''(x) - f'(x) - f(x) = 0$. (*Hint:* Try letting $f(x) = e^{cx}$ and solve for c.)

*38. Find as many functions $f(x)$ as you can that satisfy $f''(x) + 2f'(x) - 3f(x) = 0$. (*Hint:* Try letting $f(x) = e^{cx}$ and solve for c.)

*39. Find as many functions $f(x)$ as you can that satisfy $f''(x) + f'(x) - 6f(x) = e^{4x}$.

**40. Suppose $f(x)$ is a function that is defined on the whole real number line and $F(x)$ is an antiderivative of $f(x)$.
    (i) If $f(x)$ is odd, show that $F(x)$ must be even.
    (ii) Show by example that the corresponding result with "odd" and "even" interchanged is not necessarily true.
    (iii) Show by example that if $f(x)$ is not defined on the whole real line, then even the result of part (i) need not be true.

III

41. (i) Use a graphics calculator or computer to sketch $f(x) = \frac{x}{1 + x^2}$. By noting where $f(x)$ is positive or negative, predict the shape of the graph of an antiderivative $F(x)$ of $f(x)$.
    (ii) Find the antiderivative $F(x)$ that satisfies $F(0) = 0$ and graph it on the same picture, comparing the result with your predictions.

42. (i) Sketch $g(x) = x^3 - 3x^2$ and predict the shape of the graph of an antiderivative $G(x)$.
    (ii) Find the antiderivative $G(x)$ with $G(0) = 4$ and sketch it on the same picture.

43. (i) Sketch the graph of the exponential growth function $f(t) = 7e^{3t}$. Use the picture to find the value of t for which $f(t) = 2f(0)$.
    (ii) Calculate the answer and compare with what you found graphically.

44. (i) Sketch the growth function $g(t) = 2e^{t/4}$. Find to four decimal places how long it takes for $g(t)$ to double starting at $t = 0$.
    (ii) Find how long it takes to double starting at $t = 2$.
    (iii) Find how long it takes to double starting at $t = 0.3$.

4.9 Velocity and Acceleration

CONSTANT ACCELERATION
VERTICAL AND HORIZONTAL
COMPONENTS

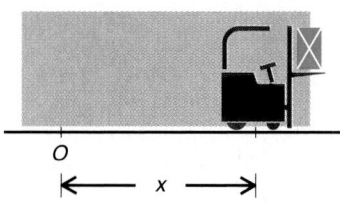

Figure 4.9.1

We recall the discussion in Section 3.1 of the derivative of a distance or a displacement with respect to time and its interpretation as the speed or velocity with which the distance is changing.

Suppose an object is moving along a straight line. Fixing some point on the line, we can call it the origin and measure x, the distance from the object to that point (see Figure 4.9.1). However, if the object moves in both directions, we may need to speak of negative as well as positive values of x, and the word "distance" is usually used only for a positive measurement. To get around this, we use the word "displacement," and the displacement, x, of the moving object from the origin can be any real number. Positive displacements will be in one direction—on one side of the origin—and negative displacements will be in the other direction; it is important to be clear which is which.

The derivative $\frac{dx}{dt}$ of displacement with respect to time is the rate of change of the displacement, which is related to what we call the "speed." But to incorporate the possibility of different signs, we call it the **velocity** of the moving object.

EXAMPLE 1   Suppose an object is moving along the x-axis so that after t seconds it is at the point $x = t^2 - 2t - 1$. Then its velocity is $\frac{dx}{dt} = 2t - 2$. Notice that when $t < 1$, the velocity is negative and the object is moving to the left; when $t > 1$, we have $\frac{dx}{dt} > 0$, and the motion is to the right. At $t = 1$, $\frac{dx}{dt} = 0$, and the object is (momentarily) stationary.

> Frequently the word "speed" is used to refer to the absolute value of the velocity. So at times $t = 0$ and $t = 2$ in the above example the object is moving at velocities of -2 units per second and $+2$ units per second, respectively, but at both times its speed is 2 units per second. This is analogous to the distinction between distance and displacement: At times $t = 1$ and $t = 3$ the object is at displacements of $x = -2$ and $x = +2$, respectively, but at both these times it is at a distance of 2 units from the origin.

EXAMPLE 2   Suppose a moving object has velocity $v(t) = t^3 - t^2 + 3t + 2$ at time t. When $t = 3$, how far will it be from where it was when $t = 0$?

Solution   We know that the velocity, $v = t^3 - t^2 + 3t + 2$, is $\frac{dx}{dt}$, so the displacement x, an antiderivative of this function, must be $\frac{1}{4}t^4 - \frac{1}{3}t^3 + \frac{3}{2}t^2 + 2t + c$, for some constant c. The problem doesn't give us enough information to evaluate c (we would have to know where the object is at some particular time), but fortunately this doesn't matter. When $t = 3$, $x = \frac{1}{4}3^4 - \frac{1}{3}3^3 + \frac{3}{2}3^2 + 2(3) + c = \frac{81}{4} - 9 + \frac{27}{2} + 6 + c = \frac{123}{4} + c$, and when $t = 0$, $x = 0 - 0 + 0 + 0 + c = c$, so the distance between these two points is $\left(\frac{123}{4} + c\right) - c = \frac{123}{4} = 30\frac{3}{4}$.

The derivative of the velocity is called the **acceleration**, and of course it is the second derivative of the displacement. The units of acceleration are of the form $(length)/(time)^2$, for example, ft/sec^2 or cm/sec^2. If, for example, the velocity is decreasing by 3 ft/sec every second, the acceleration is -3 ft/sec/sec, which we write as -3 ft/sec^2.

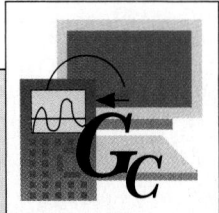

With a graphics calculator we can plot the antiderivative we found in Example 2. It is necessary to use x as the variable. It is also necessary to choose a particular value of c; the simplest choice is $c = 0$.

TI-85	SHARP EL9300
GRAPH F1($y(x)$ =) F1(x)ˆ4 ÷ 4 − F1(x)ˆ3 ÷ 3 + 3 F1(x) cˆ2 ÷ 2 + 2 F1(x) M2(RANGE) 0 ▼ 3 ▼ ▼ 0 ▼ 35 F5(GRAPH)	$↲$ x/θ/T a^b 4 ▶ ÷ 4 − x/θ/T a^b 3 ▶ ÷ 3 + 3 x/θ/T x^2 ÷ 2 + 2 x/θ/T RANGE 0 ENTER 3 ENTER 1 ENTER 0 ENTER 35 ENTER 5 ENTER $↲$
CASIO f_x-7700GB/f_x-6300G	HP 48SX
Cls EXE Range 0 EXE 3 EXE 1 EXE 0 EXE 35 EXE 5 EXE EXE EXE EXE Graph ▣ x^y 4 ÷ 4 − ▣ x^y 3 ÷ 3 + 3 ▣ x^2 ÷ 2 + 2 ▣ EXE	↰ PLOT PLOTR ERASE ATTN ' $αY$▬ $αX$ y^x 4 ÷ 4 − $αX$ y^x 3 ÷ 3 + 3 × $αX$ y^x 2 ÷ 2 + 2 × $αX$ ↱ PLOT ↰ DRAW ERASE 0 ENTER 3 XRNG 0 ENTER 35 YRNG DRAW

From the picture we see quickly that the object moves about 30 units between $x = 0$ and $x = 3$. We could get a more accurate value by zooming in.

A falling object near the earth's surface moves with a constant acceleration, called g (to remind us of gravity). The value of g is about 32.16 ft/sec^2 (often approximated by 32 ft/sec^2) or 9.8 m/sec^2 (or 980 cm/sec^2).

EXAMPLE 3   Ball Thrown Straight Up   A girl throws a ball straight up in the air. It leaves her hand, 5 ft above the ground, traveling at a speed of 50 ft/sec. When will it hit the ground?

Solution   Write $h(t)$ for the height of the ball above the ground, in feet, at time t. We know that $h(0) = 5$, $h'(0) = 50$, and that the acceleration $h''(t)$ is constant, 32 ft/sec^2 *downward*. This means that $h''(t) = -32$.

Now if $h''(t) = -32$, we must have $h'(t) = -32t + c$, for some constant c. But $h'(0) = 50$, so $50 = -32(0) + c = 0 + c = c$, that is, $c = 50$. So $h'(t) = -32t + 50$.

> $h'(t)$ is an antiderivative of $h''(t)$.

But then the antiderivative $h(t)$ must be $-16t^2 + 50t + d$ for some constant d. But $h(0) = 5$, so $5 = h(0) = 16(0^2) + 50(0) + d = d$, so $d = 5$ and $h(t) = -16t^2 + 50t + 5$. To solve the problem, we must find when $h(t) = 0$, which happens when

$$t = \frac{-50 \pm \sqrt{50^2 - 4(-16)5}}{2(-16)} = \frac{50 \pm \sqrt{2820}}{32} \approx \frac{50 \pm 53.1}{32}$$

$$\approx -0.096875 \quad \text{or} \quad 3.22.$$

The first of these is negative, representing a time before the ball was thrown, so it is an extraneous solution. The second is the right answer: $t \approx \frac{103.1}{32} \approx 3.22$ sec. The ball will hit the ground 3.22 seconds after it is released.

REMARK 4.9.1

Motion with Constant Acceleration

Example 3 illustrates a general phenomenon. Suppose an object is moving with constant acceleration a. Writing x for its displacement, we find that $x' = at + c$ and

$$x = \frac{1}{2}at^2 + ct + d.$$

When $t = 0$, $x = d$ and $x' = c$, so c is the *initial velocity* and d is the *initial displacement*. In particular, with $a = -32$ ft/sec^2 for a falling object, the height above the ground is

$$h(t) = -16t^2 + ct + d.$$

EXAMPLE 4   Constant Acceleration   (i) If a car moves with constant acceleration 10 ft/sec^2, beginning with velocity 20 ft/sec, how far will it travel in 10 sec?

Solution   The question tells us that the acceleration (in ft/sec^2) is 10 and that the initial velocity (in ft/sec) is 20. The question does not mention an initial displacement. In effect this means that we can place the origin wherever we please, and a natural place to put it is the starting point, which amounts to saying that the initial displacement is 0.

From Remark 4.9.1 we see that the displacement at time t (time measured in *seconds*) is

$$x(t) = \frac{1}{2}10t^2 + 20t + 0 = 5t^2 + 20t.$$

The displacement of the car after 10 sec will be $x(10) = 5(10)^2 + 20(10) = 700$. After 10 sec, the car will have traveled 700 ft.

(ii) The driver of a car moving at a speed of 40 m/sec slams on the brakes. The car is subjected to a constant deceleration of 5 m/sec^2. How far does it travel before coming to a stop?

Solution   First we must find the time at which the car comes to rest, that is, the time at which the velocity is zero. The fact that the car is decelerating means that the acceleration is negative (the acceleration is the rate of change of the velocity, which is decreasing). So the acceleration is -5, and the initial velocity is 40.

The displacement is $x(t) = \frac{1}{2}(-5)t^2 + 40t + d$, where d is the initial displacement, so the velocity is $x'(t) = -5t + 40$. This is zero when $t = 8$. In other words, the car comes to a standstill after 8 sec.

If we let the origin be the starting point, that is, let $d = 0$, then the displacement at the moment when the car stops will be

$$x(8) = \frac{1}{2}(-5)8^2 + 40(8) + 0 = 160.$$

The car will travel 160 m before stopping.

(iii) A man seated on the edge of a high cliff throws a small heavy object straight up at a speed of 32 ft/sec. It hits the ground at the bottom of the cliff exactly 5 sec later. How high is the cliff?

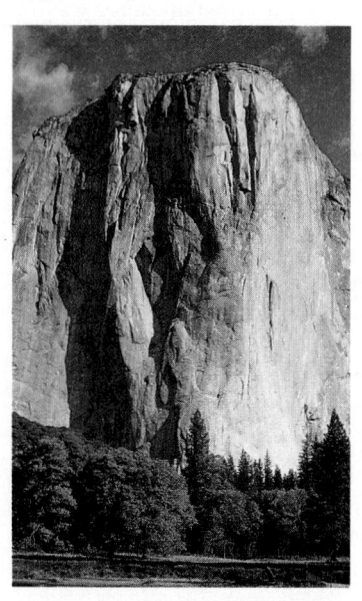

El Capitan, in Yosemite National Park in California, is about 3600 feet high. How long would it take to fall off?

Solution   We could measure height either from the ground level at the bottom of the cliff or from the top of the cliff (or anywhere else, for that matter). Let us measure from the bottom of the cliff and let d be the height of the cliff, in feet.

Then the initial height (displacement) of the object is d (it is d ft above ground level). Since the initial velocity is 32 ft/sec and the acceleration is -32 ft/sec^2, we see that the height at time t is

$$h(t) = -16t^2 + 32t + d.$$

We know that the object hits the ground when $t = 5$, which means that $h(5) = 0$. In terms of the formula, this means that

$$0 = -16(5)^2 + 32(5) + d = -400 + 160 + d = -240 + d.$$

This is only possible if $d = 240$. We have just found that the height of the cliff is 240 ft.

You might like to rework Example 4(iii), making the measurement of height from the top of the cliff. In this case the initial displacement (i.e., initial height) will be zero. Of course, the answer should be the same.   ◢

REMARK 4.9.2

> If somebody throws a rock, it will travel both vertically and horizontally. The surprising thing is that **the vertical and horizontal components of the motion** can be considered independently. That is, the vertical motion (height above the ground) will be described by the formula for the height of a falling object (see Remark 4.9.1), and its horizontal velocity will be constant (ignoring air resistance, which would actually slow it down slightly).

Figure 4.9.2(i)

EXAMPLE 5   A rock is catapulted into the air from ground level so that its path makes an angle of 30° above the ground (see Figure 4.9.2(i)). If its speed is 25 m/sec, how far from its starting point will it fall?

This miniature shows an "engine of war," a catapult from the fifteenth century.

Figure 4.9.2(ii)

Solution   The rock is moving at 25 m/sec in a direction $30°$ above the horizontal. If it continued in a straight line for 1 second, it would travel 25 m along the hypotenuse of the triangle shown in Figure 4.9.2(ii). The horizontal distance it would travel is $25\cos(30°)$ m $= 25\frac{\sqrt{3}}{2}$ m ≈ 21.65 m, and the vertical distance (the vertical side of the triangle) is $25\sin(30°)$ m $= 12.5$ m. This tells us that the initial horizontal velocity is 21.65 m/sec and the initial vertical velocity is 12.5 m/sec.

By Remark 4.9.1, the height is given by $h(t) = \frac{1}{2}at^2 + ct + d$, where the acceleration $a = g = -9.8$ m/sec^2, and the initial height d is zero, since the rock is catapulted from ground level. The initial vertical velocity c is 12.5 m/sec. So $h(t) = -4.9t^2 + 12.5t = t(-4.9t + 12.5)$, and $h(t) = 0$ when $t = 0$ or when $t = \frac{12.5}{4.9} \approx 2.55$ sec. (The solution $t = 0$ tells us what we already know, that the rock starts at ground level initially; the other answer is the one we want.)

Meanwhile, by Remark 4.9.2, the horizontal motion has constant velocity $x' = 21.65$, so $x(t) = 21.65t + c$. If we measure horizontal displacement x from the starting point, $x(0) = 0$, so $c = 0$ and $x(t) = 21.65t$. When the rock hits the ground, $t = 2.55$ sec and $x = 21.65(2.55) \approx 55.2$ m. The rock lands 55.2 m away from its starting point.

Exercises 4.9

I

Evaluate the following.

1. The acceleration at time t of a moving object whose displacement at time t is $x = 5t^3 - 3t^2 + 7t - 12$.

2. The displacement at time t of a moving object whose acceleration at time t is $6t - 2$, whose initial displacement is 6, and whose initial velocity is 5.

3. The displacement at time t of a moving object whose acceleration is given by $12t - 4$ and whose displacement is 2 at $t = 1$ and 12 at $t = 3$.

4. The time or times at which an object is at rest if its acceleration is 12 ft/sec^2 and its initial velocity and displacement are -6 ft/sec and -3 ft, respectively.

5. The maximum height reached by a ball thrown from a point 5 ft above ground level at a speed of 64 ft/sec straight up.

6. The height of a building from which it takes a penny exactly 2 sec to fall to the ground.

7. The time when an object returns to its starting point if its acceleration is -14 m/sec^2 and its initial velocity is 20 m/sec.

8. The greatest distance from its starting point which the object in Exercise 7 reaches before returning to its starting point.

9. The length of time it takes a penny to fall to the ground from a building 144 ft tall.

10. The displacement at $t = 3$ of an object moving with constant acceleration whose displacements at times $t = 0$, $t = 1$, and $t = 2$ are $x = -1$, $x = 2$, and $x = 4$, respectively.

II

11. A ball is projected straight upward from ground level with an initial speed of 60 ft/sec.
    (i) When will it hit the ground?
    (ii) What is the maximum height it will reach?

12. A car traveling at 60 mph drives off the edge of a cliff.
    (i) If the drop is 100 ft to the lake below, how long will it be before the car hits the water?
    (ii) How far from the base of the vertical cliff will the point of impact be?

13. How fast must a ball be thrown upward if it is to reach the top of a building that is 15 m high? (Assume that the ball leaves the thrower's hand at a point 1.5 m above ground level.)

14. Applying a car's brakes causes a constant deceleration of 45 ft/sec^2.
    (i)   How long (in seconds) does it take the car to stop after the brakes are applied with the car moving at 50 mph?
    (ii) How far (in feet) does the car in part (i) travel before stopping?
    (iii) Repeat parts (i) and (ii) with initial speeds of 60 mph and 70 mph.

15. A driver sees a child in the road and slams on her brakes. Suppose she is traveling at 40 mph, her

brakes cause a deceleration of 30 ft/sec^2, and her reaction time (between seeing the child and applying the brakes) is 0.8 sec.

   (i) If she first sees the child at a distance of 105 ft, is there time to stop?

   (ii) How far away must she see the child if her speed is 60 mph?

16. (i) If a daredevil in a barrel takes 3 seconds to descend to the foot of a waterfall, how high is the waterfall?

   (ii) If the waterfall is 160 ft high, how long will the daredevil fall?

*17. A football is thrown at an angle of 45° above the horizontal. It leaves the quarterback's hand at a point 6 ft above the ground, and the highest point in its flight is 30 ft above the ground.

   (i) How far will it fly if it is caught when it is again 6 ft above the ground?

   (ii) How far will it fly if it is allowed to hit the ground?

*18. Repeat Exercise 17 with an angle of 26°.

*19. A projectile is catapulted into the air so that its path makes an angle θ with respect to the ground. Find a formula (involving θ and the initial velocity v_0) for the distance it travels before hitting the ground again. Assume that the ground is perfectly flat and ignore air resistance.

*20. Using the result of Exercise 19, find the angle at which the projectile should be fired in order to maximize the distance traveled. What is the maximum possible distance?

21. A spaceship is at rest in space. It fires rockets for 10 seconds, during which time it undergoes a constant acceleration of 15 m/sec^2. It coasts for 20 seconds, then fires the rockets for an additional 10 seconds. After a further 20 seconds' coasting, how far has the spaceship traveled from its starting position?

22. A small, heavy object is dropped from an airplane.

   (i) If the airplane is traveling at 300 mph and at an altitude of 20,000 ft, how long will the object fall, and what will be the horizontal distance it travels from the drop point to the point of impact?

   (ii) Repeat part (i) for an aircraft traveling at 600 mph.

   (iii) At what angle will the objects in parts (i) and (ii) strike the ground?

23. Experiment has shown that at large distances from the earth, an object will fall toward the earth with an acceleration due to gravity equal to $\frac{k}{R^2}$, where k is some constant and R is the distance from the object to the center of the earth. The earth's radius is 4000 miles, and we know that at the surface of the earth the acceleration due to gravity is 32 ft/sec^2. What is the value of the constant k?

24. A car decelerates at 10 m/sec^2. (i) Find its stopping distance, starting at a speed of 50 km/hr. (ii) Do you expect the stopping distance from 100 km/hr to be exactly twice as far or less or more than twice as far? (iii) Find the stopping distance from 100 km/hr.

25. Suppose an object moves so that its acceleration at time t is $x''(t) = 20t^3 - 24t$ and its initial velocity $x'(0)$ is zero. (i) Find a formula for the displacement $x(t)$. (ii) Use a graphics calculator or computer to find the smallest time $t > 0$ at which the object is 7 units from its initial position.

26. Two objects move in a straight line, both beginning with displacement 0 and velocity 0. One has constant acceleration 1, and the other has acceleration $x''(t) = 2 \cos t$. Use a graphics calculator or computer to find when they collide. How far will they have traveled before colliding?

4.10 Linear Approximation and Differentials

ERROR
ERROR PROPAGATION
PERCENTAGE ERROR

In this section we discuss an application of the derivative to calculating approximate values of functions. It can be extremely useful in situations in which precise calculations are difficult or impossible.

Many functions are actually very difficult to calculate explicitly, except perhaps for a few special values. Even something relatively simple like $\sqrt{63}$ can easily be found by using a calculator, but it is less clear what to do if you want more than eight or ten digits' accuracy. Other functions such as $\sin x$ or $\ln x$ can be *approximated* with a calculator, but there is no way to give an exact evaluation except for a few familiar points.

This is usually not a matter of great concern, since the accuracy that is available on a calculator is generally sufficient for most practical purposes, but a more subtle problem would be to evaluate a function that is described as an inverse function. For instance, the function $g(x) = x^3 + 2x - 5$ has as its derivative $g'(x) = 3x^2 + 2$. This derivative is always positive, so the function $g(x)$ is always increasing, which implies that it has an inverse function, call it $f(x)$. How would we begin to find, say, $f(3)$?

We could try guessing and then try refining our guesses, using the Intermediate Value Theorem to narrow down where the value lies. In this case, for instance, $g(1) = -2$ and $g(2) = 7$, so $g(x)$ must equal 3 for some x between 1 and 2. We followed this approach in Example 4.6.12 (p. 288), realizing that there had to be a zero between two points where a function had opposite signs, but it was fairly cumbersome. There is another approach involving the derivative $f'(x)$.

If $f(x)$ is differentiable at $x = a$, then we can draw the tangent to the graph $y = f(x)$ at the point $\big(a, f(a)\big)$. Near that point, the tangent line lies very close to the graph of $y = f(x)$. For values of x farther away from a, the tangent can easily be quite far from the graph, but for x near a the tangent line is a good approximation to the curve $y = f(x)$. See Figure 4.10.1.

This fact can be very useful. It is very easy to calculate precisely any point on a straight line, so we can find good approximations to $f(x)$ provided that x is close to a.

If $y = f(x)$, we found the derivative $f'(a)$ by taking the limit, as $h \to 0$, of $\frac{f(a+h)-f(a)}{h}$. This idea can be used in reverse: If we already know $f(a)$ and $f'(a)$, then we can estimate nearby values of $f(x)$.

Since $f'(a) = \lim_{h \to 0} \frac{f(a+h)-f(a)}{h}$, we see that for small values of h,

$$\frac{f(a+h) - f(a)}{h} \approx f'(a),$$

that is,

$$f(a+h) - f(a) \approx f'(a)\,h.$$

This can be written in the form

$$f(a+h) \approx f(a) + f'(a)\,h \tag{4.10.1}$$

Figure 4.10.2 shows the tangent line to $y = f(x)$ at $x = a$. Starting from the point $\big(a, f(a)\big)$, we move along the tangent through a horizontal distance of h to find the point $\big(a+h, f(a)+f'(a)h\big)$. The y-coordinate of this point is given by the right side of Formula 4.10.1 and is an approximation to the actual value of $f(a+h)$. The actual value of $f(a+h)$ is the value of y for the point *on the curve* $y = f(x)$; the approximation comes from the corresponding point *on the tangent line*. In Figure 4.10.2 the actual value is slightly lower than the approximation.

This way of approximating $f(a+h)$ is called **linear approximation** because, instead of finding the exact point on the curve, we went along the tangent *line*. Notice that the closer $a+h$ is to a, that is, the smaller h is, the better the approximation will be.

EXAMPLE 1   Use linear approximation to estimate $\sqrt{63}$.

Solution   Let $y = f(x) = \sqrt{x} = x^{1/2}$. We know that $f(64) = 8$, and $f'(x) = \frac{1}{2}x^{-1/2}$, so $f'(64) = \frac{1}{16}$. We want to apply Formula 4.10.1 with $a = 64$, $a+h = 63$, so $h = -1$.

Figure 4.10.1

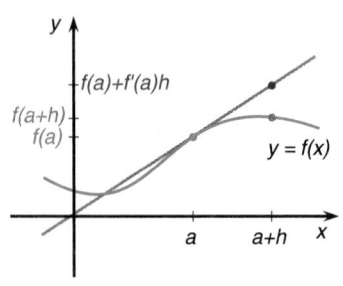

Figure 4.10.2

We find $f(63) = f(64 - 1) \approx f(64) + f'(64)(-1) = 8 + \frac{1}{16}(-1) = 8 - \frac{1}{16} = 7\frac{15}{16} = 7.9375$. In fact, $\sqrt{63} \approx 7.937254\ldots$, so our approximation is reasonably good.

Just how good the approximation is depends on various things, particularly on how small h is. Of course, approximating square roots is not very useful in the age of pocket calculators, but understanding the technique will lead to other applications.

To make this approach work, what is needed is a point $x = a$ for which the values of $f(a)$ and the derivative $f'(a)$ are known. We want an approximate value for f at some nearby point. We write this point as $x = a + h$, so that h is the (small) amount by which it is necessary to move to get from a to $a + h$. We can think of h as the *change* in x from the known point a to the unknown point $a + h$.

EXAMPLE 2 Estimate $\sin(46°)$.

Solution We want to apply Formula 4.10.1 with $f(\theta) = \sin \theta$. It seems reasonable to use $45°$ as our starting point a, so then $h = 1°$. The catch is that we are interpreting $f'(\theta) = \cos \theta$ as the rate of change, and it is the rate of change of $f(\theta) = \sin \theta$ with respect to θ, with θ measured in *radians*.

So we have to convert $45°$ to radians: $a = 45 \times \frac{\pi}{180} = \frac{\pi}{4}$ and $h = 1\left(\frac{\pi}{180}\right)$, so

$$\sin(46°) = \sin\left(\frac{\pi}{4} + \frac{\pi}{180}\right) \approx \sin\left(\frac{\pi}{4}\right) + \cos\left(\frac{\pi}{4}\right)\frac{\pi}{180} = \frac{1}{\sqrt{2}} + \frac{1}{\sqrt{2}}\frac{\pi}{180}$$

$$\approx 0.7194.$$

You can check how close this approximation is by comparing with your calculator.

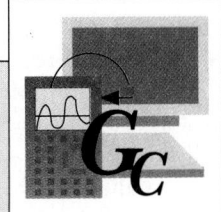

With a graphics calculator we can plot the function $y = \sin x$ and the tangent line to it at $x = \frac{\pi}{4}$. Since $f'\left(\frac{\pi}{4}\right) = \cos\frac{\pi}{4} = \frac{1}{\sqrt{2}}$, the tangent line is $y = \frac{1}{\sqrt{2}} + \frac{1}{\sqrt{2}}\left(x - \frac{\pi}{4}\right)$.

TI-85	SHARP EL9300
GRAPH F1 $(y(x) =)$ SIN F1 (x) ▼ 1 ÷ $\sqrt{2}$ + (F1(x) − π ÷ 4) ÷ $\sqrt{2}$ M2(RANGE) 0 ▼ π÷2 ▼ ▼ 0 ▼ 2 F5(GRAPH)	⟳ sin x/θ/T 2ndF ▼ 1 ÷ $\sqrt{2}$ ▶ + (x/θ/T − π ÷ 4) ÷ $\sqrt{2}$ RANGE 0 ENTER π ÷ 2 ENTER 1 ENTER 0 ENTER 2 ENTER 1 ENTER ⟳
CASIO f_x-7700GB/f_x-6300G	HP 48SX
Cls EXE Range 0 EXE π ÷ 2 EXE 1 EXE 0 EXE 2 EXE 1 EXE EXE EXE EXE Graph sin X EXE Graph 1 ÷ $\sqrt{2}$ + (X − π ÷ 4) ÷ $\sqrt{2}$ EXE	↰ PLOT PLOTR ERASE ATTN ' αY ▬ SIN αX ↱ PLOT ↰ DRAW ERASE 0 ENTER π 2 ÷ XRNG 0 ENTER 2 YRNG DRAW ATTN ' αY ▬ 1 ÷ $\sqrt{x}$ 2 + () αX − π ÷ 4 ▶ ÷ $\sqrt{x}$ 2 ↱ PLOT ↰ DRAW DRAW

Now try graphing the same functions for $x \in \left[\frac{\pi}{4}, \frac{\pi}{4} + \frac{\pi}{180}\right]$, (with $0.7 \leq y \leq 0.72$). The tangent line is extremely close to the curve...

Differentials

If $y = f(x)$ is differentiable, we can take the equation $\frac{dy}{dx} = f'(x)$ and, pretending that dx is a number, just "multiply" by it, to get

$$dy = f'(x)\,dx. \qquad (4.10.2)$$

The symbols dx and dy are called **differentials.** Equation 4.10.2 is just another way of writing $\frac{dy}{dx} = f'(x)$, and it means exactly the same thing, no more and no less.

EXAMPLE 3   Find the differential dy if (i) $y = 3x^3 - \cos x$, (ii) $xy^2 - y^3 + 2x^2y^2 - xy = 1$.

Solution

(i) $\frac{dy}{dx} = 9x^2 + \sin x$, so $dy = (9x^2 + \sin x)\,dx$.

(ii) Here we differentiate implicitly: $y^2 + 2xyy' - 3y^2y' + 4xy^2 + 4x^2yy' - y - xy' = 0$, so $(2xy - 3y^2 + 4x^2y - x)y' = -y^2 - 4xy^2 + y$, and $\frac{dy}{dx} = y' = \frac{-y^2 - 4xy^2 + y}{2xy - 3y^2 + 4x^2y - x}$. Therefore $dy = \frac{-y^2 - 4xy^2 + y}{2xy - 3y^2 + 4x^2y - x}\,dx$.

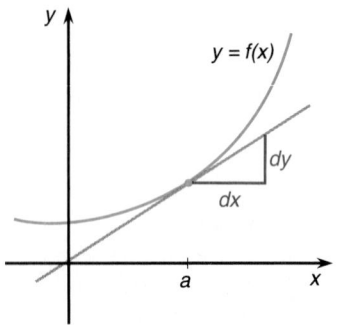

Figure 4.10.3

Although we were careful to observe that $dy = f'(x)\,dx$ is just another way of saying $\frac{df}{dx} = f'(x)$, the formula with differentials can help us to remember Formula 4.10.1 for linear approximation. After all, Formula 4.10.1 says that $f(a + h) - f(a) \approx f'(a)h$. This is quite similar to Equation 4.10.2, the differences being that there is a dx in place of h on the right side and dy in place of $f(a + h) - f(a)$ on the left and that an equals sign has replaced the "approximately equals" sign $\approx$.

If we think of h as the change in x between a and $a + h$, and if we replaced it with dx, we should interpret dx as a small change in x (the letter d might remind us of "difference"). The natural thing would be to let dy represent the corresponding change in y. Now when x changes from a to $a + h$, y will change from $f(a)$ to $f(a + h)$. The change in y will be $f(a + h) - f(a)$, which is exactly the left side of Formula 4.10.1.

So Equation 4.10.2, which says $dy = f'(x)\,dx$, can be interpreted as meaning that the change in y is $f'(x)$ times the change in x. Since $f'(x)$ is the slope of the tangent line, this is the familiar statement that the slope is the rise divided by the run (see Figure 4.10.3). The reason that Equation 4.10.2 is an equality rather than only an approximation is that it refers to changes along the tangent line.

Differentials tell us how a change dx in x is converted into a change dy in y *along the tangent line.* Since the whole idea of linear approximation was to use the tangent line to approximate the curve, differentials tell us *approximately* how the function f changes.

EXAMPLE 4   Notice that the point $(1, 1)$ satisfies the equation in Example 3(ii). Find the approximate value of y for a point (x, y) satisfying the same equation (i.e., "on the curve") with $x = \frac{11}{10}$.

Solution   By Example 3(ii), $dy = \frac{-y^2 - 4xy^2 + y}{2xy - 3y^2 + 4x^2y - x}\,dx$. This suggests that if x changes by dx, the corresponding change in y will be approximately

$$dy = \frac{-y^2 - 4x^2y + y}{2xy - 3y^3 + 4x^2y - x}\,dx.$$

When $x = 1, y = 1$, this equals $\frac{-4}{2} dx = -2 dx$. So if x increases from $x = 1$ to $x = \frac{11}{10}$, that is, $dx = \frac{1}{10}$, then the change in y will be approximately $dy = -2 dx = -\frac{2}{10} = -\frac{1}{5}$. Accordingly, the desired value of y is approximately obtained by adding this change to the original value, which is $y = 1$. The desired approximation is $1 - \frac{1}{5} = \frac{4}{5}$. The point "on the curve" is approximately $\left(\frac{11}{10}, \frac{4}{5} \right)$.

> Notice that there is no obvious way to find approximate solutions to this sort of problem on a calculator, except perhaps by trial and error. The method of linear approximation allows us to find approximate solutions to equations we don't know how to solve exactly, and this can be very useful.

In Example 4 the differentials were used as a convenient way to summarize the calculations needed to do the linear approximation. They have other uses, some of which will be encountered when we discuss functions with several variables in Chapter 16. There is one additional type of example to be considered now.

It concerns what is sometimes called *error propagation*. Most measurements are made subject to possible errors. Here the word "error" is used in the technical sense of "inaccuracy" rather than in the sense of "mistake."

If slightly inaccurate measurements are used to do calculations, the answers that result will themselves be inaccurate. It is very important to know how large an error in the answer can result from a specified error in the data. Frequently, people make the mistake of assuming that the error in the answer will be no worse than the error in the original data, but this is false.

EXAMPLE 5   **Measurement Error**   Suppose somebody measures a square piece of metal and finds that its side is $x = 30$ cm long, with a possible error of ± 3 mm $= \pm 0.3$ cm. This error could come from sloppiness on the part of the person making the measurement, but it might also reflect irregularities in the shape of the square (one side might be slightly longer than another, for example).

Using this measurement, it is easy to see that the area of the square is $A = x^2 = 900$ sq cm. The question is to find the possible error in this value.

Solution   We think of this as an approximation problem. Knowing the value of $A = x^2$ when $x = 30$, we ask by how much A might vary if x varies by up to 0.3.

Letting $dx = 0.3$, we see that $dA = 2x\, dx = 2(30)(0.3) = 18$. This says that the change in A resulting from a change of 0.3 in x is *approximately* 18. It means that a possible error of 3 mm in the measurement of the side of the square will result in a possible error of 18 sq cm in the area.

Of course, at first glance this sounds much bigger than the original error of 0.3 cm, but it is natural for the skeptical to think that although it is a bigger number, it only really makes sense to consider it relative to the number 900. This suggests that we consider the *percentage error*.

EXAMPLE 6   **Percentage Error**   What happens to the percentage error in Example 5?

> If a measurement M is made with an error of E, then the **percentage error** is $\frac{E}{M} \times 100\%$. It expresses the amount of the error as a percentage of the measured value M.

Solution   The percentage error in the original measurement is $\frac{0.3}{30} \times 100\% = 1\%$. The percentage error in the calculated area is $\frac{18}{900} \times 100\% = 2\%$. Calculating the area *doubled* the percentage error.

Doing calculations has the unfortunate effect of piling error upon error. The result is that errors may compound, much the way that compound interest builds interest upon interest. This tendency of errors to progress through the calculation is what is known as error propagation. The effect is often that the errors grow and grow.

In an elaborate calculation, which may involve several measurements, each with its possible errors, the errors may become so large as to make the final answer meaningless. After all, if the possible error is as large as the answer, then the answer is quite unreliable.

This is a serious problem. Errors do sometimes grow alarmingly fast. It can be very important to be able to keep track of just how large they may be.

Exercises 4.10

I

Use linear approximation to estimate each of the following, and compare your answer with what you get with a calculator.

1. $\sqrt{65}$
2. $\sqrt{64.1}$
3. $\sqrt{63.999}$
4. $(126)^{1/3}$
5. $(31)^{3/5}$
6. $(32.01)^{3/5}$
7. $\cos(44°)$
8. $\sin(45.01°)$
9. $\frac{1}{\tan(136°)}$
10. $\cot(-89°)$
11. $\frac{1}{101}$
12. $\frac{1}{(100.01)^3}$

Find the differential dy in each of the following situations.

13. $y = \frac{1}{x+1}$
14. $y = x^3 - 4x + 2$
15. $y = \frac{x}{\sin x}$
16. $y = \tan \theta$
17. $x^2 y - y^2 x + 2 = 0$
18. $6x^2 - 8y^2 = 48$
19. $x + xy \sin x + y^2 \frac{\cos x}{x} = 1$
20. $\cos(x) \sin(y) = \frac{1}{2}$

II

21. (i) Noting that $(1, 2)$ satisfies the equation $x^2 y - xy^2 + 2xy = 2$, find an approximate value of y so that $(1.01, y)$ is also on that curve (i.e., satisfies the same equation).
    (ii) Substitute in the equation to see how nearly the point you found does satisfy it.
22. Repeat Exercise 21, this time approximating a point

on the same curve that has the form $(x, 1.98)$. Check how close it comes to satisfying the equation.

23. A rough block of wood in the shape of a cube is measured and found to have sides 40 in long, with a possible error of 1 in.
    (i) What is the possible error in the volume of the block calculated from this information?
    (ii) Compare the percentage errors in the original measurement and the volume.

24. (i) What will be the possible error in the volume of a sphere calculated from a measurement of the radius that may be wrong by as much as 1%?
    (ii) What will be the possible error in the surface area of the sphere mentioned in part (i)? (Recall that the surface area of a sphere of radius r is $4\pi r^2$.)

25. An artisan makes her own soap, which she sells in cylindrical cakes, each of which has its height exactly equal to its diameter. She feels that, because of irregularities in the shape, it is not possible to measure these dimensions with better than 1% accuracy. She wants to be able to assert with confidence that each cake contains at least 6 cu in of soap. What size (i.e., what height and diameter) cakes should she make?

*26. In studying the effect of calculations upon errors we used differentials, which is another way of saying that we used linear approximation. This is accurate for small errors but becomes less so as the size of errors increases.
    (i) Suppose a quantity x is measured to within an accuracy of 5%. Use differentials to estimate the resulting possible error in the quantity $y = x^{20}$.
    (ii) For the sake of argument, suppose the measured value of x is exactly $x = 1$. If the error is the maximum possible, that is, 5%, it could be that $x = 1.05$. Calculate $y = x^{20}$ and see how much the error actually is.

27. Let $f(x) = x^3 + x^2 + 2x + 5$ and note that $f(-1) = 3$ and $f(-2) = -3$, so $f(x)$ has a zero somewhere in $[-2, -1]$. (i) Use linear approximation to estimate the zero. Do the calculation starting at $a = -2$ and at $a = -1$; compare the values of $f(x)$ at the two answers you find. (ii) Use a graphics calculator or com-

puter to graph $y = f(x)$. Try to decide from the picture which estimate is likely to be better. (It may help to draw the two tangent lines.) (iii) Use the graph to find the zero to four decimal places.

*28. Let $g(x) = x^4 - 20x^2 - x + 97$. Noting that $g(-3) = 1$ and $g'(-3) = 11$, use linear approximation to estimate where $g(x) = 0$. Use a graphics calculator or computer to find to four decimal places where $g(x) = 0$. Use the picture to explain what went wrong.

POINT TO PONDER

Suppose we want to approximate the value of $\sin(\pi e^{0.01})$. We might begin by trying to approximate $t = e^{0.01}$, using the function $f(x) = e^x$, and then try to approximate the value of $\sin(\pi t)$, using the function $g(u) = \sin(\pi u)$. This would involve two approximations. It would also be necessary to evaluate, or at least approximate, the values $g(t)$ and $g'(t)$ in order to perform the second approximation.

On the other hand, we could let $F(x) = \sin(\pi e^x)$

and observe that when $x = 0$, we know that $F(0) = \sin(\pi) = 0$. Since $F'(x) = \cos(\pi e^x)(\pi e^x)$, we see that $F'(0) = \cos(\pi)(\pi) = -\pi$. The linear approximation to F gives the approximation $F(0.01) \approx -\pi(0.01) \approx -0.031\,4159$.

It is easier to do all the approximating at once, rather than doing it step by step. In some cases it can even be much more difficult to do some of the steps than it is to do the single calculation.

4.11 Newton's Method

LINEAR APPROXIMATION
ITERATION

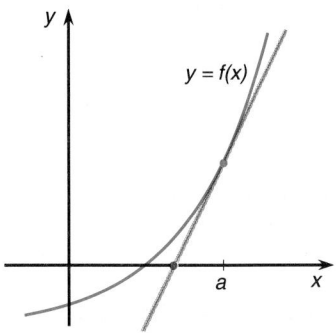

Figure 4.11.1

There are many situations in which we want to know the solution of some equation but are unable to think of a way to solve it. Sometimes it will be sufficient for our purposes to be able to find an approximate value for the solution. Newton's Method is a technique for finding approximate solutions of equations. It uses the idea of linear approximation, but having made one linear approximation, it goes on to make a second, a third, and so on, hoping to improve the approximation each time.

Suppose $y = f(x)$ is a differentiable function, and suppose we want to find a zero of $f(x)$, that is, a value of x for which $f(x) = 0$. Starting at some point $x = a$, consider the tangent line at $(a, f(a))$. With luck, it will cross the x-axis near a point at which the curve $y = f(x)$ crosses the x-axis (as shown in Figure 4.11.1). It is easy to find the point at which the tangent line crosses the x-axis, and we can hope this will give a value of x that is closer to a zero of $f(x)$ than $x = a$ was. What we have done is use the idea of linear approximation from Section 4.10 to find an x that is close to a zero of f. Needless to say, this is likely to work better in cases in which $x = a$ is already close to a zero of $f(x)$.

Since the tangent line to $y = f(x)$ at $(a, f(a))$ is $y = f(a) + f'(a)(x - a)$, to find the point where it crosses the x-axis, we need to let $y = 0$, that is, $0 = f(a) +$

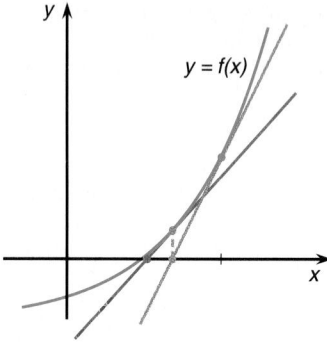

Figure 4.11.2

$f'(a)(x - a)$, so $x - a = -\frac{f(a)}{f'(a)}$, or

$$x = a - \frac{f(a)}{f'(a)}.$$

Having found this x, we hope that we have moved closer to a zero of f. On the other hand, it is unlikely that x will be exactly a zero, and it might well not even be terribly close to one. The idea that is known as Newton's Method is to apply the process again, this time using the tangent line at the new point $(x, f(x))$ (see Figure 4.11.2). With luck, this will move even closer to a zero. After that, we apply it again and again. The result will be a sequence of numbers $x_1, x_2, x_3, \ldots$, which ought to get closer and closer to a zero of f. In particular, if everything works as we hope, by applying the process enough times we should be able to find an approximate value for a zero of f to as many digits' accuracy as we may want.

Summary 4.11.1

Newton's Method

Given a differentiable function $f(x)$, Newton's Method is a procedure for producing a sequence of values of x, labeled $x_1, x_2, x_3, \ldots$, etc. Each successive x_n is obtained from the previous one as follows:

$$x_n = x_{n-1} - \frac{f(x_{n-1})}{f'(x_{n-1})}.$$

The first value, x_1, can be chosen at will but if possible should be chosen near a zero of $f(x)$. Often, though not always, the values $x_1, x_2, x_3, \ldots$, etc. will be better and better approximations of a zero of $f(x)$.

EXAMPLE 1 Use Newton's Method to approximate $\sqrt{2}$.

Solution The square root of 2 is a zero of the polynomial $f(x) = x^2 - 2$. By the Intermediate Value Theorem (Theorem 2.5.5), $f(x)$ will have a root between $x = 1$ and $x = 2$. We try Newton's Method with $x_1 = 2$. Then

$$x_2 = x_1 - \frac{f(x_1)}{f'(x_1)} = 2 - \frac{2^2 - 2}{2(2)} = 2 - \frac{2}{4} = 1.5,$$

$$x_3 = x_2 - \frac{f(x_2)}{f'(x_2)} = 1.5 - \frac{(1.5)^2 - 2}{2(1.5)} \approx 1.416667,$$

$$x_4 = x_3 - \frac{f(x_3)}{f'(x_3)} \approx 1.416667 - \frac{(1.416667)^2 - 2}{2(1.416667)} \approx 1.414216.$$

Similarly, $x_5 \approx 1.414214$, $x_6 \approx 1.414214$. Here only seven significant digits are given, but in fact x_6 will be correct to ten significant digits if you keep track of them all.

Example 1 shows the power of Newton's Method. In a relatively few steps, this method gives surprisingly accurate approximations. The square root key on many pocket calculators uses this method.

EXAMPLE 2   Find an approximate value for a zero of $f(x) = x^5 - 3x^3 + 2x - 1$.

Solution   Noting that $f(0) = -1, f(1) = -1$, and $f(2) = 11$, we see that there must be a zero between $x = 1$ and $x = 2$, by the Intermediate Value Theorem (Theorem 2.5.5). We try $x_1 = 1$. Then $x_2 = x_1 - \frac{f(x_1)}{f'(x_1)} = 1 - \frac{f(1)}{f'(1)} = 1 - \frac{-1}{-2} = 0.5; x_3 = x_2 - \frac{f(x_2)}{f'(x_2)} = 0.5 - \frac{f(0.5)}{f'(0.5)} = 6$. Continuing in this way, we find $x_4 \approx 4.8407\ldots, \quad x_5 \approx 3.9236\ldots, \quad x_6 \approx 3.2029\ldots, \quad x_7 \approx 2.6434\ldots, \quad x_8 \approx 2.2180\ldots, \quad x_9 \approx 1.9084\ldots, \quad x_{10} \approx 1.7036\ldots, \quad x_{11} \approx 1.5967\ldots, \quad x_{12} \approx 1.5656\ldots, x_{13} \approx 1.5631\ldots, x_{14} \approx 1.563\,0959$, where except for the last value we have written down only four decimal places. Continuing the process further does not change the answer on a calculator with eight significant digits, and substituting x_{14} into $f(x)$, we find $f(x_{14}) \approx 0.000\,000\,004$, and x_{14} is indeed very nearly a zero of $f(x)$.

Sir Isaac Newton (1642–1727)

There are several things to notice about Example 2. One is that the first few approximations jumped around: $1, \frac{1}{2}, 6$. Second, though we were looking for a root between $x = 1$ and $x = 2$, the first approximation took us in the wrong direction (from $x_1 = 1$ to $x_2 = \frac{1}{2}$), and the second approximation took us much too far in the opposite direction (to $x_3 = 6$). Nonetheless, the process did eventually give a good approximation to a zero. It might easily happen that the process could find a zero other than where we expect it. (In this case we were looking between $x = 1$ and $x = 2$, but having gone to $x_2 = \frac{1}{2}$, it might have found a root to the left of $x = 1$, etc.)

Another observation is that there does not seem to be any way to find an exact solution. Even with relatively simple polynomials like this, it may be absolutely impossible to find a root exactly. There are very many situations in which finding a root is necessary, and Newton's Method is often the best thing to use.

EXAMPLE 3   Find a value of x for which $x \sin x = \frac{1}{2}$.

Solution   There is no apparent way to solve this exactly, but we can approximate an answer. Letting $f(x) = x \sin x - \frac{1}{2}$, we note that $f(0) = -\frac{1}{2}, f(\pi/2) = \pi/2 - \frac{1}{2} > 0$, so there is a zero between $x = 0$ and $x = \pi/2$, by the Intermediate Value Theorem.

We apply Newton's Method, trying $x_1 = 0$. Then $x_2 = x_1 - \frac{f(x_1)}{f'(x_1)}$. Calculating $f'(x) = \sin x + x \cos x$, we find $f'(0) = 0$, so the formula makes no sense. In this situation the thing to do is to try a different x_1. How about $x_1 = \frac{\pi}{2}$? Then $x_2 = \frac{\pi}{2} - \frac{\pi/2 - 1/2}{1} = \frac{1}{2}$. Continuing, with the aid of a calculator, we find

$$x_3 \approx 0.783\,470\,336,$$
$$x_4 \approx 0.741\,490\,738,$$
$$x_5 \approx 0.740\,841\,123,$$
$$x_6 \approx 0.740\,840\,955,$$
$$x_7 \approx 0.740\,840\,955.$$

Further cycles of the procedure (often called **iterations**) will not change the answer on a calculator that displays nine digits. Be sure to have your calculator evaluate $\sin x$ in radian mode, that is, with x measured in radians.

If you have a programmable calculator or a computer, you can program it to perform one iteration, that is, to find x_{n+1} from x_n. With each cycle you will see a new approximation (some intermediate results may also be displayed); it is entertaining to watch as they approach the answer.

TI-85	SHARP EL9300
PRGM F2(EDIT) NEWT ENTER	✏️ ▼ ▼ (NEW) ENTER ENTER
X − (X SIN X − .5) ÷ (SIN X +	NEWT ENTER X = X − (X sin X
X COS X) ENTER ANS STO X	− .5) ÷ (sin X + X cos X) ENTER
ENTER EXIT π ÷ 2 STO X	COMMAND A 1 (Print) X ENTER
ENTER PRGM F1 F1 (NEWT) ENTER	COMMAND A 6 (End) ENTER 🔳
ENTER ENTER . . .	π ÷ 2 STO X ✏️ (RUN) ENTER
	(NEWT) ENTER ENTER ENTER ENTER
	ENTER ENTER

CASIO f_x-6300G	HP 48SX
Cls EXE MODE 2 EXE Ans → X : X	ATTN CLR ≪≫ → αX ′ αX − ()
− (X sin X − .5) ÷ (sin X + X cos	αX × SIN αX ▶ − .5 ▶ ÷ () SIN
X) : Ans → X : MODE 1 π ÷ 2 EXE	αX ▶ + αX × COS αX ▶ ENTER ′
Prog 0 EXE EXE EXE EXE EXE EXE	NEWTON STO π 2 ÷
(on the f_x-7700GB, : is PRGM F6,	VAR NEWTON
and Prog is PRGM F3.)	VAR NEWTON
	VAR NEWTON
	VAR NEWTON
	VAR NEWTON
	VAR NEWTON

.5
.7834703364
.7414907384
.7408411235
.7408409551
.7408409551
.7408409551

Now edit the program to work for other starting points (e.g., $x = 1$) or for some of the other functions we have discussed.

However, it is important to recognize that Newton's Method is not foolproof.

EXAMPLE 4 Apply Newton's Method to $f(x) = x^3 - 5x$, starting with $x_1 = 1$.

Solution $x_2 = x_1 - \frac{f(x_1)}{f'(x_1)} = 1 - \frac{1-5}{3-5} = -1$, and $x_3 = x_2 - \frac{f(x_2)}{f'(x_2)} = -1 - \frac{-1+5}{3-5} = 1$.

Continued iterations give us alternately 1 and −1, and they never approach a zero. Figure 4.11.3 shows what is happening. We know there is a zero at $x = 0$, but Newton's Method keeps jumping back and forth from one side to the other. ◢

If we chose an x_1 between −1 and 1, Newton's Method would give approximations to the zero at $x = 0$. Values of x_1 outside $[-1, 1]$ would result in approximations to one or other of the other two zeroes. (Factoring the polynomial, $f(x) = x^3 - 5x = x(x^2 - 5)$, we see that the other zeroes are $\pm\sqrt{5}$.)

> The tangent line at $x = 1$ crosses the axis at $x = -1$, and vice versa.

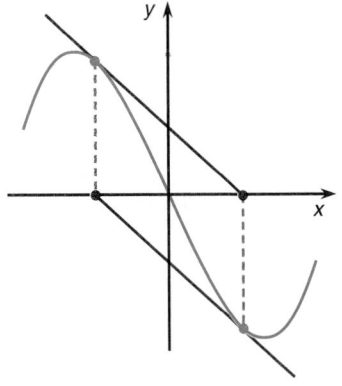

Figure 4.11.3

It is also possible to have examples in which successive values of x_n go to infinity. However, if a particular choice of x_1 does not work, just try a different one; most often there are only a few x_1's that lead to trouble, and the procedure usually works well with most choices of x_1. It is also true that Newton's Method will always approach a root provided that the first choice x_1 is sufficiently close to the root.

Finally, it is worth remarking that in our earlier work on maximum/minimum problems and curve sketching we always worked with polynomials or other functions for which we could find the zeroes. In practice, most polynomials are difficult or impossible to solve, but Newton's Method allows us to find approximate solutions. For a practical maximum/minimum problem or for purposes of sketching a curve, approximate answers are usually perfectly adequate. We leave some examples to the exercises.

Exercises 4.11

In the following exercises, if you have a calculator, try to get as much accuracy as your calculator will permit; if you're doing it by hand, try for three or four significant digits.

I

Find approximate values of the following.

1. $\sqrt{19}$
2. $(103)^{1/3}$
3. $\sqrt{53}$
4. $(71)^{-3/4}$
5. A root of $f(x) = x^3 + 3x^2 + 5x - 2$
6. A root of $g(x) = x^4 - x^3 - 2x^2 + 1$
7. A root of $h(t) = t^7 + 2t - 1$
8. A value of x so that $x^3 - 3x^2 + 4x = 7$
9. A value of t so that $t^4 - 2t^3 + 3t = 4$
10. A value of x so that $x \sin x = 1$

Remembering to use radians, approximate each of the following.

11. $\arcsin(3/4)$
12. $\arccos(0.12)$
13. $\arctan(-1.7)$
14. $\text{arcsec}(31)$
15. A solution of $\cos x + \sin x = 1/3$
16. A solution of $\tan x + 2 \sin x = 3$
17. A value of x so that $3x^2 - \sin x = 5$
18. A value of x satisfying $\frac{\sin x}{x} = \frac{1}{2}$

II

19. Apply Newton's Method to Exercise 4.10.1 (p. 324) to get better approximations.
20. Apply Newton's Method to Exercises 4.10.2 through 4.10.6.
*21. Apply Newton's Method to Exercise 4.10.21.
*22. The function $f(x) = x^2 + 5$ has no zeroes. What happens if you apply Newton's Method to it? Draw a picture.
*23. Work out an exact formula for x_n in terms of x_1 for Newton's Method applied to the function $f(x) = x^2$ (whose only zero is at $x = 0$).
*24. Apply Newton's Method to the function $f(x) = \frac{x}{1+x^2}$, starting with $x_1 = 2$. Perform five iterations, and then sketch the graph and use it to explain your results.
25. Use Newton's Method to find an approximate value for the minimum of $f(x) = x^4 - 2x^3 + x^2 - x + 1$. Assume that we know there is only one critical point.
26. Make an approximate sketch of $g(x) = 2x^4 + 5x^3 - 7x^2 + 8x + 1$.
*27. The function $F(x) = x \sin x$ can be thought of as "$\sin x$ squeezed inside the envelope of $y = x$ and $y = -x$." In particular, it has a local maximum between 0 and π, a local minimum between π and 2π, a local maximum between 2π and 3π, etc.
    (i) Before using Newton's Method to find the approximate locations of these local extrema, try to answer two questions: (a) Will the local maximum between 0 and π be to the left or to the right of $\pi/2$, the local maximum of $\sin x$? (b) Consider the local maximum between $2n\pi$ and $(2n+1)\pi$. As n increases, will the critical point

be shifted more, less, or the same amount away from the midpoint $2n\pi + \pi/2$?

(ii) Use Newton's Method to approximate a few of the critical points, and see whether they confirm your answers to questions (a) and (b) in part (i).

*28. Repeat Exercise 27 for $G(x) = x^2 \sin x$, adding one more preliminary question: (c) Where are the critical points of G in relation to the critical points of $F(x) = x \sin x$—in the same places, slightly to the left, or slightly to the right?

*29. Suppose we want to sketch the graph of $F(x) = x \sin x$ for $x \in [-\pi, \pi]$. The derivative is $F'(x) = \sin x + x \cos x$, so it is easy to see there is a critical point at $x = 0$. The second derivative is $F''(x) = \cos x + \cos x - x \sin x = 2 \cos x - x \sin x$.

  (i) Argue that there is exactly one zero of $F'(x)$ in $(0, \pi)$ and that it is in $\left(\frac{\pi}{2}, \pi\right)$.

 (ii) Use Newton's Method to find approximate values for *all* solution(s) of $F'(x) = 0$ with $x \in [-\pi, \pi]$. (Save some trouble by noting that $F'(x)$ is an odd function.)

(iii) Argue that there is exactly one zero of $F''(x)$ in $(0, \pi)$ and that it is in $\left(0, \frac{\pi}{2}\right)$.

(iv) Use Newton's Method to find approximate values for *all* solution(s) of $F''(x) = 0$ with $x \in [-\pi, \pi]$.

 (v) Sketch the graph of $y = F(x)$ for $x \in [-\pi, \pi]$.

*30. When we use linear approximation to approximate $\sqrt{26}$, we consider the function $f(x) = \sqrt{x} = x^{1/2}$ and work with the tangent line to the graph of $y = f(x)$ at the point $(25, 5)$. If we use Newton's Method for the same purpose, we consider the function $g(x) = x^2 - 26$ and look for a zero of this function. We should try starting with a point that we know is reasonably close to a zero, such as $(5, -1)$.

(i) Perform the linear approximation to $f(x)$, starting at the point $(25, 5)$.

(ii) Perform one single iteration of Newton's Method for the function $g(u)$, starting at the point $x = 5$, that is, let $x_1 = 5$ and find x_2.

(iii) Compare the approximations to $\sqrt{26}$ you get in parts (i) and (ii). Explain the comparison.

31. If you have a programmable calculator or a computer, write a program to find the next iteration of Newton's Method for the function $f(x) = x^5 - 3x^3 + 2x - 1$ of Example 2 (p. 327). Use it to confirm the calculations of that example, and then try it with some different choices for x_1.

32. Modify your program of the previous problem to work for the function $f(x) = x^3 - 5x$ of Example 4 (p. 328), and try some different values of x_1 to confirm the remarks at the end of Example 4. (Check that the answers you get for several choices of $|x_1| > 1$ are approximately $\pm\sqrt{5}$, by squaring them.)

33. Sketch the graph of the function in Exercise 9, and use the picture to decide which starting values x_1 will cause Newton's Method to converge to which zero.

34. In Exercise 17 we found a root of $f(x) = 3x^2 - \sin x - 5$. Note that $f(0) = -5, f(-2) = 7 - \sin(-2) = 7 + \sin(2) \geq 6$. The Intermediate Value Theorem says that there is a zero between $x = 0$ and $x = -2$. Since $f(-1) = 3 - 5 - \sin(1) = -2 - \sin(1) < 0$, it is between $x = -1$ and $x = -2$. Try to find it with Newton's Method.

POINT TO PONDER

It is relatively easy to program a computer or calculator to do an iteration of Newton's Method for some function $f(x)$. Much harder is to decide how many iterations it should perform before stopping. Suppose we are approximating $\sqrt{a}$ but don't know the size of a in advance. We could stop after there is an iteration that does not change the answer, but that is risky: It could happen that because of round-off errors the answer keeps changing back and forth by $+1$ and -1 in the very last digit and never actually stops. Another possibility

would be to stop after some fixed number of iterations, say, ten. This would be fine for moderately sized a's but might cause trouble for very large ones. Another possibility would be to stop if the last change is small relative to the current answer, that is, if $\frac{x_{n+1} - x_n}{x_n}$ is small (say, less than or equal to 10^{-8}).

Another problem is choosing x_1. We could let $x_1 = a$. This is good for small a's but wasteful for big values. We could let $x_1 = a/2$, but this is worse for small values (e.g., if $a \leq 1$). We could always let $x_1 = 1$, no matter what the value of a.

Try out some of these strategies for various values of a (including very large and very small ones), and see whether you can find a good system for choosing x_1 and deciding when to stop.

This sort of question belongs to the beginnings of a subject called Numerical Analysis.

4.12 L'Hôpital's Rule

ONE-SIDED LIMITS
INFINITE LIMITS
INDETERMINATE FORMS
GENERALIZED MEAN VALUE
THEOREM

L'Hôpital's Rule gives a way to evaluate limits of the form $\lim_{x \to a} \frac{f(x)}{g(x)}$ if the numerator and denominator both tend to zero, that is, $\lim_{x \to a} f(x) = \lim_{x \to a} g(x) = 0$. To try to see what to do in this situation, suppose first that both $f(x)$ and $g(x)$ are differentiable at $x = a$. In particular this implies that they are both continuous there, and because we know what the limits are, we see that $f(a) = g(a) = 0$.

This allows us to write

$$\frac{f(x)}{g(x)} = \frac{f(x) - f(a)}{g(x) - g(a)},$$

which makes us begin to think of difference quotients. Writing

$$\frac{f(x)}{g(x)} = \frac{f(x) - f(a)}{g(x) - g(a)} = \frac{f(x) - f(a)}{g(x) - g(a)} \times \frac{x - a}{x - a} = \frac{f(x) - f(a)}{x - a} \times \frac{x - a}{g(x) - g(a)}$$

$$= \frac{f(x) - f(a)}{x - a} \times \frac{1}{\frac{g(x) - g(a)}{x - a}},$$

The Marquis de l'Hôpital (1661–1704)

we can let $x \to a$, and the first factor will tend to $f'(a)$ and the second will tend to $\frac{1}{g'(a)}$. This suggests that

$$\lim_{x \to a} \frac{f(x)}{g(x)} = \frac{f'(a)}{g'(a)}.$$

There are several difficulties with the argument we just gave, among them the possibility that $g'(a)$ or $g(x)$ might equal zero, so various quotients we have written down might not be defined. Nonetheless, it certainly indicates that the limit of $\frac{f}{g}$ has something to do with $\frac{f'}{g'}$. The precise statement is what is known as l'Hôpital's Rule, in honor of the French nobleman the Marquis de l'Hôpital (1661–1704), who was the patron of the Swiss mathematician John Bernoulli, who actually invented the rule.

The rule says that if the numerator and denominator both tend to 0, then the limit of the quotient $\frac{f(x)}{g(x)}$ equals the limit of the quotient of the derivatives $\frac{f'(x)}{g'(x)}$.

EXAMPLE 1   For example, to evaluate $\lim_{x \to 0} \frac{\sin x}{x}$, we note that both numerator and denominator tend to 0, but $\sin' x = \cos x$, $\frac{d}{dx}(x) = 1$, and $\lim_{x \to 0} \frac{\cos x}{1} = 1$, which as we know does equal $\lim_{x \to 0} \frac{\sin x}{x}$.

Johann Bernoulli (1667–1748)

Unfortunately, a proof of l'Hôpital's Rule is not as simple as the rule itself. It hinges on a generalization of the Mean Value Theorem, Theorem 4.12.5. We postpone the proof until the end of the section and concentrate instead on the use of the rule.

THEOREM 4.12.1

> **L'Hôpital's Rule**
>
> Suppose there is an open interval (r, s) containing $x = a$ so that $f(x)$ and $g(x)$ are differentiable at every $x \in (r, s)$ except possibly $x = a$, and that $g'(x) \neq 0$ for every $x \in (r, s)$ except possibly $x = a$. Suppose too that $\lim_{x \to a} f(x) = \lim_{x \to a} g(x) = 0$ and that $\lim_{x \to a} \frac{f'(x)}{g'(x)}$ exists. Then
>
> $$\lim_{x \to a} \frac{f(x)}{g(x)} = \lim_{x \to a} \frac{f'(x)}{g'(x)}.$$
>
> In addition, if $\lim_{x \to a} \frac{f'(x)}{g'(x)}$ is $+\infty$ or $-\infty$, then the equality still holds.

COROLLARY 4.12.2

> (i) **L'Hôpital's Rule for one-sided limits:** Suppose f and g are differentiable on (a, s), $g'(x) \neq 0$ for all $x \in (a, s)$, $\lim_{x \to a^+} f(x) = 0$, $\lim_{x \to a^+} g(x) = 0$, and $\lim_{x \to a^+} \frac{f'(x)}{g'(x)}$ exists. Then
>
> $$\lim_{x \to a^+} \frac{f(x)}{g(x)} = \lim_{x \to a^+} \frac{f'(x)}{g'(x)},$$
>
> where infinite limits are included. A similar statement holds for limits from below, using an interval of the form (r, a).
>
> (ii) **L'Hôpital's Rule for limits at $\pm\infty$:** Suppose f and g are differentiable on (r, ∞), $g'(x) \neq 0$ for all $x \in (r, \infty)$, $\lim_{x \to \infty} f(x) = 0$, $\lim_{x \to \infty} g(x) = 0$, and $\lim_{x \to \infty} \frac{f'(x)}{g'(x)}$ exists. Then
>
> $$\lim_{x \to \infty} \frac{f(x)}{g(x)} = \lim_{x \to \infty} \frac{f'(x)}{g'(x)},$$
>
> where infinite limits are included. A similar statement holds for limits as $x \to -\infty$, using an interval of the form $(-\infty, r)$.

EXAMPLE 2   Evaluate $\lim_{x \to 1} \frac{\ln(x)}{\sin(\pi x)}$.

Solution   Since $\lim_{x \to 1} \ln(x) = \ln(1) = 0$ and $\lim_{x \to 1} \sin(\pi x) = \sin \pi = 0$, we can apply l'Hôpital's Rule. Now $\frac{d}{dx}(\ln x) = \frac{1}{x}$ and $\frac{d}{dx}\big(\sin(\pi x)\big) = \pi \cos(\pi x)$, so $\lim_{x \to 1} \frac{\ln(x)}{\sin(\pi x)} = \lim_{x \to 1} \frac{1/x}{\pi \cos(\pi x)} = \frac{1/1}{\pi \cos \pi} = \frac{1}{-\pi} = -\frac{1}{\pi}$.

EXAMPLE 3   Find $\lim_{x \to \pi/2^-} \frac{\cos x}{1 - \sin x}$.

Solution   First we check that numerator and denominator do both tend to zero as $x \to \frac{\pi}{2}^-$. They are both continuous functions that equal zero at $x = \frac{\pi}{2}$, so their limits are both zero. This means we can apply l'Hôpital's Rule, so $\lim_{x \to \pi/2^-} \frac{\cos x}{1 - \sin x} =$

With a graphics calculator we can plot the original function $y = \frac{\ln x}{\sin(\pi x)}$ from Example 2 and also the quotient of the derivatives $\frac{1/x}{\pi \cos(\pi x)}$.

TI-85	SHARP EL9300
GRAPH F1$(y(x) =)$ LN F1$(x) \div$ SIN $(\pi$ F1$(x))$ ▼ F1(x) $x^{-1} \div (\pi$ COS $(\pi$ F1$(x)))$ M2(RANGE) .5 ▼ 1.5 ▼ ▼ $(-)$ 1 ▼ 0 F5(GRAPH)	↲ ln x/θ/T $\div$ (sin (π x/θ/T)) 2ndF ▼ 1 $\div$ x/θ/T $\div$ (π cos (π x/θ/T)) RANGE .5 ENTER 1.5 ENTER 1 ENTER -1 ENTER 0 ENTER 1 ENTER ↲

CASIO f_x-7700GB/f_x-6300G	HP 48SX
Cls EXE Range .5 EXE 1.5 EXE 1 EXE -1 EXE 0 EXE 1 EXE EXE EXE EXE Graph ln X $\div$ (sin (π X)) EXE Graph 1 $\div$ X $\div$ (π cos (π X)) EXE	↰ PLOT PLOTR ERASE ATTN ′ αY ▬ LN αX ▶ $\div$ ◨ SIN π $\times$ αX ↱ PLOT ↰ DRAW ERASE .5 ENTER 1.5 XRNG 1 +/− ENTER 0 YRNG DRAW ATTN ′ αY ▬ 1 $\div$ αX $\div$ ◨ π $\times$ COS π $\times$ αX ↱ PLOT ↰ DRAW DRAW

The two functions are different, but we see that the limits are the same as $x \to 1$. Explain why the limits are not the same at other points . . .

$\lim_{x \to \pi/2^-} \frac{-\sin x}{-\cos x} = \lim_{x \to \pi/2^-} \tan x$. Now $\lim_{x \to \pi/2^-} \tan x = \infty$ (since $\sin x \to 1$ and $\cos x \to 0$ and both are positive for $x \in (0, \pi/2)$). So

$$\lim_{x \to \pi/2^-} \frac{\cos x}{1 - \sin x} = \lim_{x \to \pi/2^-} \frac{-\sin x}{-\cos x} = \infty.$$ ◢

EXAMPLE 4   Evaluate $\lim_{x \to 0} \frac{\sin(x^3)}{\sin(x^2)}$.

Solution $\lim_{x \to 0} \frac{\sin(x^3)}{\sin(x^2)} = \lim_{x \to 0} \frac{\cos(x^3)(3x^2)}{\cos(x^2)(2x)}$. Here we used l'Hôpital's Rule, only to find that it resulted in another quotient in which both numerator and denominator tend to 0.

But dividing out the common factor of x gives $\lim_{x \to 0} \frac{\cos(x^3)(3x)}{\cos(x^2)(2)}$. As $x \to 0$, the numerator tends to $\cos(0)(3)(0) = (1)(3)(0) = 0$, but the denominator tends to $\cos(0)(2) = 2$. The quotient tends to $\frac{0}{2} = 0$, and we have found that $\lim_{x \to 0} \frac{\sin(x^3)}{\sin(x^2)} = 0$. ◢

It is often necessary to use l'Hôpital's Rule more than once.

EXAMPLE 5 Evaluate $\lim_{x \to \pi} \dfrac{\sin\left((x-\pi)^2\right)}{\sin^2 x}$.

Solution Both numerator and denominator tend to 0 as $x \to \pi$, so we may apply l'Hôpital's Rule and find $\lim_{x \to \pi} \dfrac{\sin\left((x-\pi)^2\right)}{\sin^2 x} = \lim_{x \to \pi} \dfrac{\cos\left((x-\pi)^2\right) 2(x-\pi)}{2\sin(x)\cos(x)}$. The numerator tends to $\cos(0)(2)(0) = 0$, and the denominator tends to $2\sin(\pi)\cos(\pi) = (2)(0)(-1) = 0$. Applying l'Hôpital's Rule again, we get $\lim_{x \to \pi} \dfrac{-\sin\left((x-\pi)^2\right) \times 2(x-\pi) \times 2(x-\pi) + \cos\left((x-\pi)^2\right)(2)}{2\cos^2 x - 2\sin^2 x}$. This time the numerator tends to $\sin(0)(2)(0)(2)(0) + \cos(0)(2) = (0)(0)(0) + (1)(2) = 2$, and the denominator tends to $2\cos^2(\pi) - 2\sin^2(\pi) = 2(1)^2 - 2(0)^2 = 2$, so $\lim_{x \to \pi} \dfrac{\sin\left((x-\pi)^2\right)}{\sin^2 x} = \dfrac{2}{2} = 1$. ◢

> There is a word of warning that must be given here. After you use l'Hôpital's Rule a few times, especially if you use it repeatedly as in Example 5, there is a temptation to be sloppy about verifying the hypotheses. The important one is that both numerator and denominator must tend to zero, and using l'Hôpital's Rule in situations in which this is not true leads to serious errors.
>
> As a simple example, consider $\lim_{x \to 1} \frac{x-1}{x}$. In fact this quotient tends to $\frac{0}{1} = 0$ as $x \to 1$, but if we neglected to notice that the denominator does not tend to zero and mistakenly applied l'Hôpital's Rule, we would obtain the incorrect answer that the limit is 1.

Finally, there is a version of l'Hôpital's Rule for the situation in which $f(x)$ and $g(x)$ both have infinite limits. The proof is quite difficult, so we merely state the result.

THEOREM 4.12.3

L'Hôpital's Rule for Infinite Limits

Suppose $\lim_{x \to a} f(x) = \infty$ or $-\infty$ and $\lim_{x \to a} g(x) = \infty$ or $-\infty$, and $f(x)$ and $g(x)$ are both differentiable on (r, a) and (a, s) for some r, s with $r < a < s$. Suppose that $\lim_{x \to a} \frac{f'(x)}{g'(x)}$ exists. Then

$$\lim_{x \to a} \frac{f(x)}{g(x)} = \lim_{x \to a} \frac{f'(x)}{g'(x)},$$

where infinite limits are included. One-sided versions of this result also hold, as do versions for limits as $x \to \infty$ or $x \to -\infty$.

EXAMPLE 6 Find $\lim_{x \to 1+} \dfrac{\exp\left(\frac{1}{x-1}\right)}{\tan(\pi x/2)}$.

Solution $\lim_{x \to 1+} \tan(\pi x/2) = \lim_{x \to \pi/2+} \tan x = -\infty$ and $\lim_{x \to 1+} \frac{1}{x-1} = +\infty$, so $\lim_{x \to 1+} \exp\left(\frac{1}{x-1}\right) = \infty$ (since $\lim_{t \to \infty} e^t = \infty$). Applying Theorem 4.12.3, we consider

$$\lim_{x \to 1+} \frac{\exp\left(\frac{1}{x-1}\right)\frac{-1}{(x-1)^2}}{\sec^2(\pi x/2)(\pi/2)} = -\frac{2}{\pi} \lim_{x \to 1+} \exp\left(\frac{1}{x-1}\right)\frac{1/(x-1)^2}{1/\cos^2(\pi x/2)}$$

$$= -\frac{2}{\pi} \lim_{x \to 1+} \exp\left(\frac{1}{x-1}\right)\frac{\cos^2(\pi x/2)}{(x-1)^2}.$$

Now $\lim_{x \to 1+} \frac{\cos(\pi x/2)}{x-1} = \lim_{x \to 1+} \frac{-\sin(\pi x/2)(\pi/2)}{1} = -\sin(\pi/2)(\pi/2) = -\frac{\pi}{2}$, so $\lim_{x \to 1+} \frac{\cos^2(\pi x/2)}{(x-1)^2} = \left(-\frac{\pi}{2}\right)^2.$

The limit in question is $-\frac{2}{\pi}\left(\frac{\pi^2}{4}\right)\lim_{x \to 1+} \exp\left(\frac{1}{x-1}\right) = -\infty.$ ◢

There are some other types of limits for which l'Hôpital's Rule can be used.

EXAMPLE 7   Find $\lim_{x \to 0+} x \ln x.$

Solution   Here we have a *product* of two functions, x and $\ln x$, instead of a quotient. One of them, x, goes to 0 and the other, $\ln x$, goes to $-\infty$. To be able to use l'Hôpital's Rule, we need to convert the product into a quotient. One way to accomplish this is to write

$$x \ln x = \frac{\ln x}{\frac{1}{x}}.$$

As $x \to 0^+$, the numerator of this quotient tends to $-\infty$ and the denominator to ∞. We apply l'Hôpital's Rule:

$$\lim_{x \to 0+} x \ln x = \lim_{x \to 0+} \frac{\ln x}{\frac{1}{x}} = \lim_{x \to 0+} \frac{\frac{1}{x}}{-\frac{1}{x^2}} = \lim_{x \to 0+} (-x) = 0.$$ ◢

EXAMPLE 8   Find (i) $\lim_{x \to \infty} (x - \ln x)$; (ii) $\lim_{x \to 0+} (\csc x - \cot x).$

Solution

(i)   Since $x \to \infty$ and $\ln x \to \infty$, the behavior of their difference is unpredictable. However, we can write

$$x - \ln x = x\left(1 - \frac{\ln x}{x}\right).$$

Observing that $\lim_{x \to \infty} \frac{\ln x}{x} = \lim_{x \to \infty} \frac{1/x}{1} = 0$, so $\left(1 - \frac{\ln x}{x}\right) \to 1$, we find that

$$\lim_{x \to \infty} x\left(1 - \frac{\ln x}{x}\right) = \left(\lim_{x \to \infty} x\right)\left(\lim_{x \to \infty} \left(1 - \frac{\ln x}{x}\right)\right) = \left(\lim_{x \to \infty} x\right)(1) = \infty.$$

(ii)   Since $\csc x$ and $\cot x$ both tend to ∞ as $x \to 0^+$, we have to rearrange the function. Writing

$$\csc x - \cot x = \frac{1}{\sin x} - \frac{\cos x}{\sin x} = \frac{1 - \cos x}{\sin x},$$

we now have the quotient of two functions that both tend to 0. L'Hôpital's Rule applies:

$$\lim_{x \to 0^+} (\csc x - \cot x) = \lim_{x \to 0^+} \frac{1 - \cos x}{\sin x} = \lim_{x \to 0^+} \frac{\sin x}{\cos x} = 0.$$

Summary 4.12.4

Indeterminate Forms

1. L'Hôpital's Rule is used to find the limit of a quotient $\frac{f(x)}{g(x)}$, where $f(x)$ and $g(x)$ both tend to 0 or both tend to $\pm\infty$. We refer to these types of limits as indeterminate forms of type $\frac{0}{0}$ and $\frac{\infty}{\infty}$. These symbols indicate the limiting behavior of the numerator and denominator.

2. For a product $f(x)g(x)$ in which $f(x) \to 0$ and $g(x) \to \pm\infty$ it is possible to write

$$f(x)g(x) = \frac{f(x)}{\frac{1}{g(x)}}.$$

In the quotient the numerator and denominator both tend to 0, so l'Hôpital's Rule can be applied. Sometimes it may be easier to write

$$f(x)g(x) = \frac{g(x)}{\frac{1}{f(x)}}.$$

If $\frac{1}{f(x)}$ tends to ∞ or $-\infty$, then l'Hôpital's Rule may be applied (cf. Example 7).

We refer to the limit of $f(x)g(x)$ as an indeterminate form of type $0 \times \infty$.

3. For a difference $f(x) - g(x)$ in which $f(x)$ and $g(x)$ both tend to ∞ (or $-\infty$), it is often possible to perform algebraic manipulations that convert the function into a quotient (cf. Example 8). In particular, if $f(x)$ and $g(x)$ are quotients, it is often helpful to write $f(x) - g(x)$ over a common denominator (cf. Example 8(ii)).

The limit of the function $f(x) - g(x)$ is called an indeterminate form of type $\infty - \infty$.

Proof of l'Hôpital's Rule

Now that we have learned how to use l'Hôpital's Rule, we might want to see how it is proved. The rough argument given at the beginning of this section provides a reasonably good idea of why a result like this ought to be true, but a correct proof is surprisingly delicate and tricky. Understanding the proof will not likely offer much help in using l'Hôpital's Rule, which is probably more important.

The proof of l'Hôpital's Rule rests on the following theorem, which is a generalization of the Mean Value Theorem (Theorem 4.2.5, p. 241).

THEOREM 4.12.5

Suppose $f(x)$ and $g(x)$ are continuous functions on $[a, b]$ that are differentiable on (a, b). Then there is a $c \in (a, b)$ so that

$$\big(f(b) - f(a)\big)g'(c) = \big(g(b) - g(a)\big)f'(c).$$

PROOF   Let $H(x) = \big(f(b) - f(a)\big)g(x) - \big(g(b) - g(a)\big)f(x)$. Then $H(x)$ is also continuous on $[a, b]$ and differentiable on (a, b).

Note that

$$
\begin{aligned}
H(a) &= \big(f(b) - f(a)\big)g(a) - \big(g(b) - g(a)\big)f(a) \\
&= f(b)g(a) - f(a)g(a) - g(b)f(a) + g(a)f(a) \\
&= f(b)g(a) - g(b)f(a).
\end{aligned}
$$

Also

$$
\begin{aligned}
H(b) &= \big(f(b) - f(a)\big)g(b) - \big(g(b) - g(a)\big)f(b) \\
&= f(b)g(b) - f(a)g(b) - g(b)f(b) + g(a)f(b) \\
&= g(a)f(b) - f(a)g(b).
\end{aligned}
$$

In particular, $H(b) = H(a)$. So Rolle's Theorem (Theorem 4.2.4, p. 240) applies to $H(x)$ and says that there is a $c \in (a, b)$ so that $0 = H'(c) = \big(f(b) - f(a)\big)g'(c) - \big(g(b) - g(a)\big)f'(c)$. But this is exactly what we wanted to prove.   ◢

We said that Theorem 4.12.5 is a generalization of the Mean Value Theorem. Notice that if $g(x) = x$, then the statement of Theorem 4.12.5 is exactly the statement of the Mean Value Theorem.

With this theorem available it is now possible to prove l'Hôpital's Rule.

PROOF OF L'HÔPITAL'S RULE   Since none of the hypotheses depends on the values of $f(x)$ and $g(x)$ at $x = a$, we can change them, if necessary, so that $f(a) = 0, g(a) = 0$, making $f(x)$ and $g(x)$ continuous at $x = a$. Then, being differentiable at the other points, they are continuous on all of (r, s).

Fix $x \in (r, s)$ with $x > a$. We consider the interval $[a, x]$ and apply Theorem 4.12.5. So there is a $c \in (a, x)$ so that $\big(f(x) - f(a)\big)g'(c) = \big(g(x) - g(a)\big)f'(c)$.

Notice that if $g(x) = g(a)$, Rolle's Theorem would say that g' has a zero between a and x, contradicting the hypotheses, so $g(x) - g(a) \neq 0$, and we may divide by it and by $g'(c)$ to obtain

$$\frac{f(x) - f(a)}{g(x) - g(a)} = \frac{f'(c)}{g'(c)}.$$

Since $f(a) = g(a) = 0$, this says that $\frac{f(x)}{g(x)} = \frac{f'(c)}{g'(c)}$, with $c \in (a, x)$.

A similar argument shows that if $x < a$, then $\frac{f(x)}{g(x)} = \frac{f'(c)}{g'(c)}$, for some c with $c \in (x, a)$. In particular, since c is always closer to a than x is, c must approach a whenever x approaches a.

Consequently, $\lim_{x\to a}\frac{f(x)}{g(x)} = \lim_{c\to a}\frac{f'(c)}{g'(c)}$, since we are assuming that the right-hand limit exists (or could be ∞ or $-\infty$).

This completes the proof of l'Hôpital's Rule (Theorem 4.12.1).

In fact, since we worked with $x > a$ and then with $x < a$ separately, the above proof is also a proof of the one-sided version of l'Hôpital's Rule, Corollary 4.12.2(i).

Exercises 4.12

I

Evaluate the following limits.

1. $\lim_{x\to 0}\frac{x^2+2x}{\sin x}$

2. $\lim_{x\to 2}\frac{x^2-3x+2}{x^2-4x+3}$

3. $\lim_{x\to 3}\frac{x^4-2x^3-4x^2+2x+3}{x^3-4x^2+4x-3}$

4. $\lim_{x\to -1}\frac{x+1}{x^2-1}$

5. $\lim_{x\to 0}\frac{1-\cos x}{x\sin x}$

6. $\lim_{x\to \pi/2}\frac{\cos^2 x}{1-\sin x}$

7. $\lim_{x\to 0}\frac{\arctan x}{\sin x}$

8. $\lim_{x\to \pi/2}\frac{1-\sin x}{\text{arccot}(x)}$

9. $\lim_{x\to 0}\frac{\sin^2 x}{\sin(x^2)}$

10. $\lim_{x\to 1}\frac{x^2-1}{x^2+1}$

11. $\lim_{x\to 0+}\frac{\sin x}{1-\cos x}$

12. $\lim_{x\to 0-}\frac{\sin x}{1-\cos x}$

13. $\lim_{x\to 1-}\frac{\arccos x}{\ln x}$

14. $\lim_{x\to 0}\frac{x-\sin x}{x^3}$

15. $\lim_{x\to 1+}\frac{(\ln x)^2}{(x-1)^3}$

16. $\lim_{x\to 3-}\frac{\sin \pi x}{x^3-2x^2-3x}$

17. $\lim_{x\to 1+}\frac{(\ln x)^3}{(x-1)^2}$

18. $\lim_{x\to e-}\frac{1-\ln x}{(x-e)^2}$

19. $\lim_{x\to 4+}\frac{\sin \pi x}{x^2-3x-2}$

20. $\lim_{x\to 0}\frac{\sin^7 x}{x^7}$

21. $\lim_{x\to 0+} x^2 \ln x$

22. $\lim_{x\to 0} x\cot(2x)$

23. $\lim_{x\to 0}\left(\frac{1}{x}-\csc x\right)$

24. $\lim_{x\to 0+}\sqrt{x}\ln x$

25. $\lim_{x\to \infty} xe^{-x}$

26. $\lim_{x\to \infty}(x-x^2)$

27. $\lim_{x\to 1+}\ln(x)\tan\frac{\pi x}{2}$

28. $\lim_{x\to 2}\sin(\pi x^2)\cot(\pi x)$

29. $\lim_{x\to 0}\frac{1}{\ln(x)\arctan(x)}$

30. $\lim_{x\to -\infty}\left(x+\sqrt{x^2+1}\right)$

II

*31. (i) Evaluate $\lim_{x\to \pi/2}\frac{\cos^3 x}{(2x-\pi)^3}$. (ii) If k is a positive integer, evaluate $\lim_{x\to 0}\frac{\sin^k x}{x^k}$.

*32. (i) Evaluate $\lim_{x\to 3}\frac{(x-3)^3}{\sin^3(\pi x)}$. (ii) If k is a positive integer, evaluate $\lim_{x\to 0}\frac{x^k}{\sin^k(\pi x)}$. (iii) If k is a positive integer, evaluate $\lim_{x\to 0}\frac{(1-\cos x)^k}{x^{2k}}$.

*33. Let $r > 0$; find $\lim_{x\to 0+} x^r \ln x$.

*34. Evaluate $\lim_{x\to \pi/2-}(2x-\pi)\tan x$.

**35. If $f(x)$ is differentiable and $f'(x)$ is continuous, prove that $\lim_{h\to 0}\frac{f(x+h)-f(x-h)}{2h} = f'(x)$.

**36. Suppose the first and second derivatives of $f(x)$ exist at every x. Evaluate
$$\lim_{h\to 0}\frac{f(x+h)+f(x-h)-2f(x)}{h^2}.$$

**37. Prove the version of l'Hôpital's Rule for limits as $x\to\infty$ with both numerator and denominator tending to zero. (*Hint:* Let $x = \frac{1}{t}$, as $t\to 0^+$.)

**38. Prove a weaker form of the version of l'Hôpital's Rule that applies for limits as $x\to a$ in which $\lim_{x\to a} f(x) = \infty = \lim_{x\to a} g(x)$. Assume that *both* $\lim_{x\to c}\frac{f(x)}{g(x)}$ and $\lim_{x\to a}\frac{f'(x)}{g'(x)}$ exist and are nonzero, and under this assumption prove the formula. (*Hint:* $\frac{f}{g} = \frac{1/g}{1/f}$.)

III

39. Let $f(x) = \sin x$ and $g(x) = x$. Use a graphics calculator or computer to graph $\frac{f(x)}{g(x)}$ and $\frac{f'(x)}{g'(x)}$ on the same picture. Compare their limits as $x\to 0^+$.

40. With $f(x) = \tan^2 x$ and $g(x) = x^2$, sketch $\frac{f}{g}$, $\frac{f'}{g'}$, and $\frac{f''}{g''}$ on the same picture. Compare their limiting behavior near $x = 0$.

4.13 Marginal Cost, Revenue, and Profit

DEMAND FUNCTION
PRICE FUNCTION

In Section 3.1 we saw how various questions coming from the world of business and economics lead quite naturally to the use of derivatives. In this section we will make a more systematic study of these applications.

Suppose a factory produces television sets and the cost of producing x sets is $C(x)$. If the amount for which x sets will sell—the revenue—is $R(x)$, then the profit to be made by producing x sets is $P(x) = R(x) - C(x)$.

Consider the derivatives of these functions. Since $C'(x) = \lim_{h \to 0} \frac{C(x+h) - C(x)}{h}$, if x is large, we will likely get a good approximation to $C'(x)$ by taking $h = 1$ and considering $\frac{C(x+1) - C(x)}{1} = C(x+1) - C(x)$. This is the cost of producing one more set if the factory is already producing x of them. The derivative $C'(x)$ is called the **marginal cost function.**

Similarly, $R'(x)$ is the **marginal revenue function** and $P'(x)$ is the **marginal profit function.** They represent, approximately, the revenue and profit, respectively, to be gained by producing one additional set if current production is x.

EXAMPLE 1   Television Factory   Studies have shown that for a certain television factory, $C(x) = -\frac{1}{10}x^2 + 500x + 30{,}000$ dollars, at least for values of x up to $x = 1000$. Market research indicates that $R(x) = -\frac{1}{2}x^2 + 800x - 10{,}000$ dollars. So $P(x) = R(x) - C(x) = -\frac{2}{5}x^2 + 300x - 40{,}000$ dollars.

The marginal cost is $C'(x) = -\frac{1}{5}x + 500$, the marginal revenue is $R'(x) = -x + 800$, and the marginal profit is $P'(x) = -\frac{4}{5}x + 300$. Let us try to interpret these functions.

The marginal cost $C'(x)$, the approximate cost of producing one additional set, is $500 - \frac{1}{5}x$. This means it costs about \$500 to produce each additional set when production is small, but as volume increases, the unit cost falls, presumably because of increased efficiency and economies of scale. The marginal revenue is $R'(x) = 800 - x$, which means that sets bring in about \$800 each for the first few, but the amount falls as the number increases. This could, for example, mean that as the market becomes saturated, it becomes increasingly difficult, and hence increasingly expensive, to sell additional sets. It could also mean that to sell more, they must offer discounts.

The marginal profit is $P'(x) = -\frac{4}{5}x + 300$. It is about \$300 per set at first but drops off. This says that producing additional sets brings smaller and smaller added profits.

The above example is rather simple, but it illustrates the essential ideas. Presuming that the factory's owners wish to maximize profits, we look for a maximum of $P(x)$.

EXAMPLE 2   In the situation of Example 1, what level of production will give maximum profit?

Solution   We want to maximize $P(x) = -\frac{2}{5}x^2 + 300x - 40{,}000$.

Now $P'(x) = -\frac{4}{5}x + 300$, so the only critical point is $x = 375$. Since $P''(x) = -\frac{4}{5} < 0$, it must be an absolute maximum. The maximum profits will be obtained by producing 375 television sets, and the profit will be $P(375) = \$16{,}250$.

It is interesting to calculate $P(376) = \$16{,}249.60$. This shows that producing one set more than the optimal 375 will actually reduce profits by 40 cents, because costs will be increasing faster than revenues.

Example 2 illustrates a fundamental result in economics.

PRINCIPLE 4.13.1

> The maximum profit will occur at a point where the marginal profit is zero, that is, where marginal cost and marginal revenue are equal.

Strictly speaking, we should assume that $P(x)$ is differentiable, and we should worry about endpoint maxima (e.g., $x = 0$), etc., so this "principle" is a little vague. However, in a situation like this, in which we have simplified and idealized the problem considerably, the general idea is what is important, rather than the technical details.

In a sense, this "principle" is nothing more than Theorem 4.2.3 (p. 238) (that extrema occur at critical points), but it is more than that. What we have described is a simplified special case of a principle with much wider applicability. We should be pleased that our knowledge of calculus makes it seem so natural and easy.

In practice it may be easier to find the marginal cost function $C'(x)$ rather than $C(x)$ itself. Frequently, the assumption is made that $C'(x) = ax + b$ for some constants a and b. Then $C(x)$ must be of the form $C(x) = \frac{a}{2}x^2 + bx + c$ for some constant c. This constant c can be interpreted as the fixed cost, the cost that will be incurred even if no units are produced. It is usually the result of overhead costs or setup costs for a production run.

In practice it might also be easier to determine how many items will sell if they are offered at a given price p, rather than the revenue function $R(x)$. If we assume that the number that will sell is $x = n - up$, then this represents a plausible situation in which higher prices result in fewer sales, or smaller demand. For this reason, when we solve for p, the resulting equation

$$p = p(x) = \frac{n}{u} - \frac{1}{u}x$$

is called the **demand function** or simply the **price function.**

Of course, the total revenue is simply $R(x) = xp(x)$, the unit price times the number of units sold.

EXAMPLE 3   **Kettle Production**   Market studies have shown that if a certain model of kettle is sold for $20, it will be possible to sell 1600 units, but that at $22 only 1200 will sell. Assume that the number x that will sell at price $\$p$ is a linear function, that is, $x = n - up$, for some constants u and n. If the marginal cost is also assumed to be linear, that is, $C'(x) = ax + b$ for constants a and b, and the cost of producing one additional kettle is $12 if 1000 are already being produced and $10 if 1400 are being produced, what production level will maximize profits?

Solution   We are assuming that $x = n - up$, and we know that when $p = 20$, x is 1600 and when $p = 22$, x is 1200. This says that $1600 = n - 20u$ and $1200 = n - 22u$. Subtracting these two equations gives $400 = 2u$, so $u = 200$. Substituting this into either equation gives $n = 5600$. So $x = 5600 - 200p$, and $200p = 5600 - x$, so $p = 28 - (1/200)x$. The total revenue is $R(x) = xp = 28x - \frac{1}{200}x^2$.

Similarly, $C'(x) = ax + b$, and, approximately, $C'(1000) = 12$, $C'(1400) = 10$, that is, $12 = 1000a + b$ and $10 = 1400a + b$. Subtracting, we get $2 = -400a$, $a = -\frac{1}{200}$, and substituting, $b = 17$, so $C'(x) = -\frac{1}{200}x + 17$.

The marginal profit is $P'(x) = R'(x) - C'(x) = 28 - \frac{1}{100}x - (-\frac{1}{200}x + 17) = 11 - \frac{1}{200}x$, which is zero when $x = 2200$. Since $P''(x) = -\frac{1}{200} < 0$, this is a maximum. So producing 2200 kettles will maximize profits. ◢

Notice that to solve this problem, we didn't actually have to know the cost function $C(x)$ but just the marginal cost function $C'(x)$. If in addition we knew that the fixed costs were \$10,000, that is, $C(0) = 10,000$, then from $C'(x) = -\frac{1}{200}x + 17$ we would find that $C(x) = -\frac{1}{400}x^2 + 17x + 10,000$. So at a production level of 2200 kettles the total cost is $C(2200) = \$35,300$. Also $R(2200) = \$37,400$, so the total profit is $P(2200) = R(2200) - C(2200) = \$2100$.

The average cost, or cost per unit, is $\frac{C(x)}{x}$, which, for $x = 2200$, works out to about \$16.05. The average revenue is $\frac{R(x)}{x}$, and for $x = 2200$ this is \$17.00. The average or per unit profit is $\frac{P(x)}{x} \approx \$0.95$.

Exercises 4.13

1. Find the production level x that will maximize profits if the cost function is $C(x) = 300 + 20x - \frac{1}{100}x^2$ and the revenue function is $R(x) = 38x - \frac{1}{10}x^2$.

2. Find the optimal production level for a company whose cost function for producing x units per week is $C(x) = 200 + 30x - \frac{1}{50}x^2$ and whose sales will be $x = 875 - 25p$ if the unit selling price is p.

3. A manufacturer needs to choose one of two products to manufacture. For one, a type of electric motor, the cost function is $C_M(x) = 400 + 15x - \frac{1}{80}x^2$, and the revenue function is $R_M(x) = 20x - \frac{1}{40}x^2$. For the other, a kind of generator, the marginal cost is $C'_G(x) = 18 - \frac{1}{60}x$, and the sales will be $x = 960 - 40p$ if the unit price is p. Supposing that the fixed costs for generators are $C_G(0) = 1000$, which product can yield higher profits?

4. Suppose that for a certain process the revenue function is $R(x) = -\frac{1}{60}x^2 + 13x - 50$ and the marginal cost is $C'(x) = 10 - \frac{1}{50}x$. (i) If $C(0) = 100$, find the production level that maximizes profits. (ii) In the above situation, suppose it is practical to run the process only in batches of 100. How many batches would maximize profits?

5. Assume that the marginal cost $C'(x)$ is a linear function of x. Assume that the number x of units that will sell at price $\$p$ is a linear function of p. Studies indicate that $C'(45) = 24$ and $C'(180) = 21$. Moreover, they show that if $p = 20$, x will be 450, and if $p = 26$, x will be 225.
  (i) Find the optimal production level.
  (ii) If fixed costs are \$500, what will be the profit

per unit at the optimal production level (as found in part (i))?
  (iii) If fixed costs are \$500, what will be the maximum profit per unit?
  (iv) What conclusion should you draw if fixed costs are \$1000?

*6. Show that when the average cost (i.e., unit cost) is a minimum, the marginal cost equals the average cost.

*7. What conclusions would you draw if marginal cost and marginal revenue are both constant? In particular, how would you decide what level of production to recommend?

*8. In a certain jurisdiction a person whose annual taxable income is $\$x$ must pay taxes of $T(x) = \frac{1}{10}x + \frac{1}{1,000,000}x^2$ (for $x \leq 400,000$).
  (i) Explain why the marginal tax $T'(x)$ is (approximately) the amount of additional tax paid by a taxpayer with taxable income $\$x$ if he earns an additional \$1.
  (ii) If a taxpayer's taxable income is \$30,000, how much will he be able to keep of an additional dollar of income after taxes are paid?
  (iii) How much of each dollar earned does the above taxpayer keep?
  (iv) Repeat parts (ii) and (iii) for a taxpayer whose taxable income is \$300,000.
  (v) If the taxpayer in part (ii) and his wife (whose taxable income is \$60,000) have an opportunity to make an investment that will earn an extra dollar, explain why it makes sense for this investment income to be in his name rather than hers. How much difference does it make?

Chapter Summary

§4.1
related rates
implicit differentiation

§4.2
maximum value
minimum value
maximum
minimum
local maximum
local minimum
absolute maximum
absolute minimum
relative maximum
relative minimum
Rolle's Theorem (Theorem 4.2.4, p. 240)
average slope
Mean Value Theorem (Theorem 4.2.5, p. 241)
increasing/decreasing function
First Derivative Test (Theorem 4.2.10, p. 244)
If f' is always zero, then f is constant.

§4.3
critical point
stationary point
extremum
local extremum
absolute extremum

§4.4
Second Derivative Test (Theorem 4.4.1, p. 256)

§4.5
curve sketching
even function
odd function
asymptote
vertical asymptote
oblique asymptote
sketching rational functions

§4.6
concave up
concave down
second derivative and concavity
inflection point
cusp

§4.7
maximum/minimum problems
Snell's Law

§4.8
antiderivatives

§4.9
velocity
acceleration
motion with constant acceleration
vertical and horizontal components of motion

§4.10
linear approximation
differentials
percentage error
error propagation

§4.11
Newton's Method (Summary 4.11.1, p. 326)
$$x_n = x_{n-1} - \frac{f(x_{n-1})}{f'(x_{n-1})}$$

§4.12
l'Hôpital's Rule (Theorem 4.12.1, p. 332)
l'Hôpital's Rule for one-sided limits
l'Hôpital's Rule for limits at infinity
l'Hôpital's Rule for infinite limits
indeterminate forms

§4.13
marginal cost
marginal revenue
marginal profit
demand function
price function

Review Exercises

I

1. If the volume of a spherical balloon is increasing at the rate of 50 cm^3/sec, how fast is its radius increasing at the moment when the radius is 10 cm?

2. If a spherical balloon is deflating, how fast is its surface area decreasing at the moment when its radius is 4 in and its volume is decreasing by 1 in^3/sec?

3. A 20-ft ladder has slipped down a wall, so Fred is pushing it back. How fast is the top of the ladder rising up the wall at the moment when the bottom of the ladder is 12 ft from the base of the wall, moving toward the wall at a speed of 1.5 ft/sec?

4. Diane is standing on a 15-ft ladder when the bottom starts to slide away from the wall at a speed of 1 in/sec. She jumps free when she notices that the

top of the ladder is sliding down at the rate of $\frac{1}{2}$ in/sec. How far is it from the top of the ladder to the ground at this moment?

5. Consider a conical tank with its vertex pointing downward; suppose the radius at the top is 2 m and the tank is 3 m high. If water is leaking out at the rate of 0.002 m^3/hr, how fast is the area of the surface of the water in the tank decreasing when the water is 2 m deep?

6. In the situation of Exercise 5, how fast is the area decreasing when the water is 1 m deep?

7. A ship is approaching a lighthouse from the east at a speed of 10 knots. A sailboat is sailing due south at 7 knots on a course that will take it past the lighthouse at a distance of 4 nautical miles to the east (see figure below). How fast are the vessels approaching each other when the ship is 6 nautical miles due east of the lighthouse and the sailboat is 1 nautical mile north of the imaginary line between the ship and the lighthouse?

8. A mouse runs in a straight line across the floor, at a speed of 1 ft/sec. It does not see the stalking cat and passes the cat at a distance of 8 ft at the closest point. How fast is the mouse moving away from the watchful cat exactly ten seconds later?

9. A retired scientist is hoping to perfect a cylindrical balloon, which has the property that no matter how much it is inflated, it still maintains the same basic shape, that of a cylinder whose height is four times its radius. How fast would the volume of this balloon be increasing at the moment when its radius is 3 in and the radius is increasing at the rate of $\frac{1}{8}$ in/sec?

10. How fast would the surface area of the cylindrical balloon in Exercise 9 be increasing if the volume was increasing at the rate of 5 in^3/sec and the radius was 4 in?

Find the regions where each of the following functions is increasing or decreasing.

11. $f(x) = x^2 - 4x + 9$
12. $f(x) = -x^2 - 7x + 1$
13. $f(x) = x^3 - 9x + 2$
14. $f(x) = -x^3 - 2x - 4$
15. $f(x) = xe^x$
16. $f(x) = \cosh x$

17. $f(x) = \operatorname{sech} x$
18. $f(x) = \sec x$
19. $f(x) = x \ln x$
20. $f(x) = \cos^2 x - \sin^2 x$

Find all critical points of the following functions, and classify which are local or absolute maxima or minima.

21. $f(x) = x^7$
22. $f(x) = x^8$
23. $f(x) = x^4 + 2x^2 + 3$
24. $f(x) = x^2 e^x$
25. $f(x) = |x^2 - 2x - 3|$
26. $f(x) = |x^3 - 1|$
27. $f(x) = \cot \frac{x}{2}$
28. $f(x) = \frac{1}{1+x}$
29. $f(x) = \frac{x^4}{1+x^4}$
30. $f(x) = \operatorname{sech} x$

Find all local and absolute extrema for the following functions. (If the domain is not specified, assume that it is $\mathbb{R}$.)

31. $f(x) = x^3 - x^2 + x + 2$
32. $f(x) = x^3 - x^2 - x + 3$
33. $f(x) = x^3 - 6x,\ x \in [-1, 1]$
34. $f(x) = x^5 - 5x + 3$
35. $f(x) = |x^2 - 3x|,\ x \in [0, 2]$
36. $f(x) = \frac{1}{2x-3},\ x \in [0, 1]$
37. $f(x) = x + |x|,\ x \in [-3, 1]$
38. $f(x) = \cosh x \sinh x,\ x \in [-1, 1]$
39. $f(x) = (x^2 + x)e^{2x}$
40. $f(x) = \ln(x^2 + 1),\ x \in [-2, 3]$

Sketch the graphs of the following functions.

41. $f(x) = \frac{x}{x^2+4}$
42. $f(x) = \frac{x^2+4}{x}$
43. $f(x) = \frac{x^2}{x+3}$
44. $f(x) = x^2 e^{2x}$
45. $f(x) = \cos^2 x$
46. $f(x) = \frac{x}{2-3x}$
47. $f(x) = \frac{x^2-1}{x}$
48. $f(x) = \frac{x}{x^2-2x}$
49. $f(x) = \frac{x}{x^2-9}$
50. $f(x) = \frac{1}{x} + \frac{1}{x^2}$

Find all antiderivatives of each of the following functions.

51. $f(x) = x^4 - 3x^2 + 4x - 1$
52. $f(x) = e^{3x-1}$
53. $f(x) = 3x^{9/4} + 3x - 4$
54. $f(x) = 3x + \sin x$
55. $f(x) = \sqrt{x+7}$
56. $f(x) = \cos(-x)$
57. $f(x) = \sin(5x)$
58. $f(x) = 2e^x - \cos(3x)$
59. $f(x) = x^{-4} + 2x$
60. $f(x) = 6x^{-1/3}$

Find each of the following.

61. The length of time a ball will be in the air if it is thrown straight up from ground level at a speed of 256 ft/sec.

62. The speed with which a ball must be thrown straight up from ground level in order for it to reach a maximum height of 60 m.

63. The height of a building if a brick takes 2.6 sec to fall from its roof.

64. The speed with which a ball must be thrown upward in order for it to take 3 sec to return to the point from which it was launched.

65. The total distance traveled by a car as it decelerates from 50 mph at a rate of 20 ft/sec^2.

66. The acceleration of a car that starts at rest and accelerates at a constant rate, reaching 100 km/hr in 10 sec.

67. The maximum height reached by a ball that is thrown straight up from ground level at 15 m/sec.

68. The constant deceleration necessary for a "soft landing" of a lunar probe that is 100 m above the moon's surface and falling at a rate of 20 m/sec.

69. The constant rate of deceleration needed for a car traveling at 70 mph to stop in a distance of 300 ft.

70. The average speed of the car in Exercise 69 over the period of its deceleration.

Estimate each of the following using linear approximation.

71. $\sqrt{80.5}$

72. $(26.99)^{1/3}$

73. $\frac{1}{40.01}$

74. $(24.99)^{-1/2}$

75. $(63.999)^{5/3}$

76. $\left(1 + (1.01)^2\right)^{-1}$

77. $\sin(30.2°)$

78. $\tan(59.85°)$

79. $\sec(44.97°)$

80. $\csc(-29.95°)$

Use Newton's Method to find approximate values of the following.

81. $\sqrt{41}$

82. $\sqrt{55}$

83. $(40)^{1/3}$

84. $(12)^{-1/5}$

85. $\arctan 5$

86. $\arcsin\left(-\frac{2}{3}\right)$

87. A root of $x^3 - 4x + 2$

88. A root of $x^5 - 3x^3 + x - 1$

89. A root of $x^4 - 4x^2 + 2$

90. A solution of $x \ln x = 6$

Evaluate the following limits, provided that they exist.

91. $\lim_{x \to 0} \frac{x^2 - 3x}{x^2 - 2}$

92. $\lim_{x \to 1} \frac{x^2 - x}{x^2 + x - 2}$

93. $\lim_{x \to 0} \frac{x^2 - 3x}{\sin(4x)}$

94. $\lim_{x \to -2} \frac{x^2 + 2x}{\sin(\pi x)}$

95. $\lim_{x \to \infty} \frac{x^3 - 2x^2}{e^{2x}}$

96. $\lim_{x \to 1+} (x - 1) \ln(x^2 - 1)$

97. $\lim_{x \to -\infty} xe^{-x}$

98. $\lim_{x \to 3} \frac{x^2 - 2x + 1}{\cos(\pi \frac{x}{2})}$

99. $\lim_{x \to 0} (1 - e^x) x^{-1/2}$

100. $\lim_{x \to 3} \frac{\sin x}{\sin(\sqrt{x})}$

II

*101. The light on the roof of a police car rotates once every second, flashing along a fence that is 50 ft from the light at the nearest point.

  (i) How fast is the spot of light moving along the fence as it passes the point that is closest to the car?

  (ii) How fast is the spot of light moving as it passes the point on the fence that is 40 ft away from the point that is closest to the car (assuming that the fence is straight)?

*102. Two planks, one 10 ft long and the other $7\frac{1}{2}$ ft long, are hinged together at one end. A boy is lifting the hinged joint straight up over his head, allowing the other ends of the planks to drag along the ground (see figure below). How fast is the unhinged end of the 10-ft plank moving toward the boy at the moment when the hinge is 6 ft above the ground if the unhinged end of the short plank is moving toward him at the rate of $\frac{1}{2}$ ft/sec?

103. A rectangular sheet of cardboard measuring $8\frac{1}{2} \times 11$ inches is used to make an open box by cutting equal squares from the four corners and folding up the sides. What is the largest possible volume of a box that can be made in this way?

104. Find the point(s) on the parabola $y = 3x^2 + x - 1$ that are closest to $(2, -3)$.

*105. Sketch the graph of $f(x) = \frac{\sin x}{x}$, for $x \in \left(0, \frac{\pi}{2}\right]$.

*106. Let $g(\theta) = |\sin \theta|$.
  (i) Sketch the graph of $g(\theta)$ for $\theta \in [-\pi, \pi]$.
  (ii) Does g have a cusp at $\theta = 0$?

107. Find all antiderivatives of $\frac{3x^2 - 2x}{x}$.

*108. Let $f(x)$ be the function defined for every nonzero real number x by $f(x) = |x|$; $f(x)$ is undefined for $x = 0$. Find all antiderivatives of $f(x)$.

*109. Suppose a building has n stories (ignoring the basement), each of them 10 ft high (including the thickness of the floors). Find a formula (depending on n) for the length of time it will take a small heavy object to fall down a stairwell from the top floor to the ground floor. A lazy person proposes to drop a ball down a stairwell, time its fall, and use the formula instead of going outside and counting the number of stories. Why would it be less reliable to use this approach to find the number of stories in a building with over ten stories than for one with under three stories?

*110. A toy rocket accelerates at the rate of 100 ft/sec^2 for 2 sec until its fuel is burned and then coasts.
  (i) What height will it reach, assuming that it flies straight up?
  (ii) How long will its flight last?

111. Find the differential dy if (i) $y = x \sin x$, (ii) $x^2 y^3 - 4x^3 y = 5$, (iii) $e^x \sin(x + y) = \frac{1}{2}$.

112. Find the differential dy if (i) $x^2 - 2y^4 = 2 - 7x$, (ii) $\sin(xy^2) = x - y$, (iii) $\tan(x + y^3) = 6$.

*113. Use linear approximation to estimate $10^{1.99}$.

114. (i) What is the possible percentage error in the calculation of the volume of a cylinder if its radius is exactly 3 units and its height is measured to within 2%?
  (ii) What is the possible percentage error in the calculation of the volume of the cylinder if its height is exactly 2 units and its radius is measured to within 2%?

115. What production level should be used by a company wishing to maximize its profits if the cost function is $C(x) = 500 + 30x - \frac{1}{80}x^2$ and the revenue function is $R(x) = -1000 + 50x - \frac{1}{40}x^2$ when the production level is x units per day?

116. Find the optimal price and level of production for a gadget if it costs $C(x) = 20{,}000 + 170x - x^2$ dollars to produce x units per month and the manufacturer

can sell $x = 180 - \frac{9}{10}p$ units when the unit price is \$$p$.

*117. (i) Suppose $f(x)$ is an odd function defined on all of $\mathbb{R}$. Show that $f^{(k)}(0) = 0$ for every even integer $k \geq 0$ for which $f^{(k)}(0)$ exists.
  (ii) Suppose $g(x)$ is an even function defined on all of $\mathbb{R}$. Show that $g^{(k)}(0) = 0$ for every odd integer $k \geq 0$ for which $g^{(k)}(0)$ exists.

118. Sketch the graph of the function $f(x) = x\sqrt{1 - x^2}$ on $[-1, 1]$.

*119. Show that if $a \neq 0$, then $f(x) = x^3 + ax^2$ always has a local extremum. What about absolute extrema?

*120. Discuss the extrema of $g(x) = x^4 + ax^3$.

*121. A 15-ft ladder that is leaning against a wall begins to slide down. How fast is the angle between the ground and the ladder changing at the moment when the top of the ladder is 12 ft above the ground and descending at a rate of 6 in/sec?

122. (i) Suppose $r > s$ and $f(x)$ is differentiable on (r, s). Fix $a \in (r, s)$ and let $P_a^0(x)$ be the polynomial of degree 0, that is, a constant polynomial, that has the same value as $f(x)$ at the point $x = a$. Show that if $\frac{f(x) - P_a^0(x)}{x - a} \geq 0$ for every $a \neq x \in (r, s)$, then $f'(a) \geq 0$ for every $a \in (r, s)$.
  (ii) Suppose $f(x)$ is twice differentiable on (r, s). Fix $a, b \in (r, s)$ and let $P_{a,b}^1(x)$ be the polynomial of degree 1 that has the same value as $f(x)$ at the points $x = a$ and $x = b$. Show that if $\frac{f(x) - P_{a,b}^1(x)}{(x-a)(x-b)} > 0$ for every $a, b \in (r, s)$ and every $x \in (r, s)$ except $x = a, b$, then $f''(a) \geq 0$ for every $a \in (r, s)$.
  (iii) Suppose $f(x)$ is three times differentiable on (r, s). Fix $a, b, c \in (r, s)$ and let $P_{a,b,c}^2(x)$ be the polynomial of degree 2 that has the same value as $f(x)$ at the points $x = a$, $x = b$, and $x = c$. Show that if $\frac{f(x) - P_{a,b,c}^2(x)}{(x-a)(x-b)(x-c)} > 0$ for every $a, b, c \in (r, s)$ and every $x \in (r, s)$ except $x = a, b, c$, then $f^{(3)}(a) \geq 0$ for every $a \in (r, s)$.
  (iv) Can you guess a generalization of this result to the fourth derivative, $f^{(4)}(x)$?
  (v) Can you guess a generalization of this result to the nth derivative, $f^{(n)}(x)$?

*123. Suppose we want to sketch the graph of $F(x) = e^{\sin x}$ for $x \in [-\pi, \pi]$. The derivative is $F'(x) = e^{\sin x} \cos x$, so it is easy to find the critical points.
  But if we compute the second derivative, $F''(x) = e^{\sin x} \cos^2 x + e^{\sin x}(-\sin x) = e^{\sin x}(\cos^2 x - \sin x)$, we see that to find the inflection points, we must know where $\cos^2 x - \sin x = 0$, that is, when $\cos^2 x = \sin x$.
  (i) Use Newton's Method to find approximate values for *all* solution(s) of $F''(x) = 0$ with $x \in [-\pi, \pi]$.
  (ii) Sketch the graph of $y = F(x)$ for $x \in [-\pi, \pi]$.

Integration 5

To find the area of a circle, we tried approximating by a number of triangles and adding up their areas. Then we took a limit as the number of triangles goes to infinity.

To find the area of the region enclosed by a curve, we approximate the region by a number of rectangles, and add up their areas. If we use narrower rectangles, the approximation should be more accurate. The exact area is found by taking a limit as the rectangles get narrower and narrower...

INTRODUCTION

The preceding two chapters have developed the theory of derivatives, which is known as differential calculus. This chapter and Chapters 6 and 8 develop the other major branch of the subject, which is known as *integral calculus.*

Integration is a technique, much the way differentiation is. It arises when we try to find the area of a region with curved boundaries. For instance, it will eventually allow us to explain why the area of a circle is πr^2.

The notion of differentiation also arose in answer to a geometrical problem, trying to find the equation of a tangent line. When we learned how to answer that problem, we found that the ideas we had explored could also be used to answer other questions. Something similar happens with integration.

In addition to enabling us to find areas, it can be used for many other things. One of the more unexpected ones is finding antiderivatives. In Section 4.8 we discussed the idea of an antiderivative of $f(x)$, meaning a function $F(x)$ satisfying $F'(x) = f(x)$. We found that this was a useful notion that had applications to questions involving velocity or other rates of change, but we realized that actually finding antiderivatives could sometimes be quite difficult.

It turns out that the technique of integration can be used to find antiderivatives. In fact, we will see that integration more or less amounts to the same thing as finding antiderivatives. It may still be a difficult problem, but integration will help us to approach it systematically.

Sections 5.1 through 5.4 discuss integration in terms of the problem of finding areas. We begin in Section 5.1 by trying to approximate the area under a graph. Section 5.2 is a technical section on summation notation, which is needed for Sections 5.3 and 5.4, which use Riemann sums to define integrals. Section 5.5 establishes the link between the problem of finding areas and the problem of finding antiderivatives, which is known as the Fundamental Theorem of Calculus. Then we begin to learn how to work with integrals, learning to evaluate definite integrals in Section 5.6 and exploring the technique of substitution in Section 5.7.

5.1 The Area Under a Curve

RECTANGLES
APPROXIMATION

Figure 5.1.1

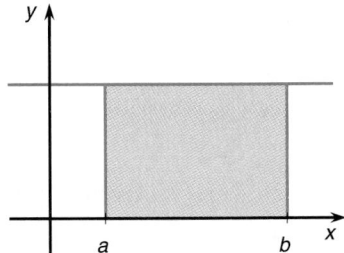

Figure 5.1.2

Suppose $f(x)$ is a continuous function with $f(x) \geq 0$ for all $x \in [a, b]$. The graph of $y = f(x)$ lies above the x-axis, as shown in Figure 5.1.1, and we might ask: What is the area of the shaded region—the region below the graph of $y = f(x)$? Strictly speaking, it is the region whose boundaries are the graph of $y = f(x)$, the x-axis, and the two vertical lines $x = a$ and $x = b$.

There is one situation in which it is easy to find the area—when $y = f(x)$ is a horizontal straight line and the region is a rectangle, as seen in Figure 5.1.2.

EXAMPLE 1   If $f(x) = c$ is a constant function, with $c \geq 0$, then the region under the graph is a rectangle. Its height is c, and its width is $b - a$, the horizontal distance from $x = a$ to $x = b$.

Its area is

$$height \times width = c(b - a).$$

Notice that if $c = 0$, the "rectangle" is really just a line segment, so it is not surprising that the formula says that its area is zero.

For a curve that is not a straight line the problem is more difficult. Rather than calculating the area directly, we can try approximating it by filling the region with rectangles, as shown in Figure 5.1.3. We know how to calculate the total area of the rectangles, and this should be approximately equal to the area under the curve.

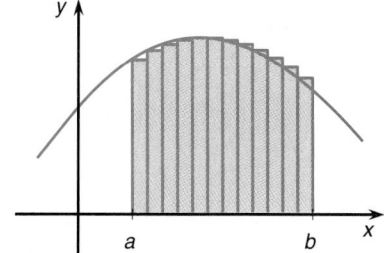

Figure 5.1.3

If we use a larger number of rectangles, as in the second and third figures, the approximation should be better. The total area of the rectangles will be very close to the area of the region under the curve. (The error, that is, the difference between the actual area of the region and the sum of the areas of the rectangles, involves the areas of the little regions between the curve and the tops of the rectangles, which are certainly smaller in the second figure than in the first, and smaller yet in the third).

This suggests trying the approximation with a large number of rectangles, say, n of them, and then taking the limit as $n \to \infty$. This is a good idea but difficult to carry out in practice. To do a simple example, we need to know the following fact.

THEOREM 5.1.1

If n is a positive integer, then

$$1^2 + 2^2 + 3^2 + \ldots + (n-1)^2 + n^2 = \frac{n(n+1)(2n+1)}{6}.$$

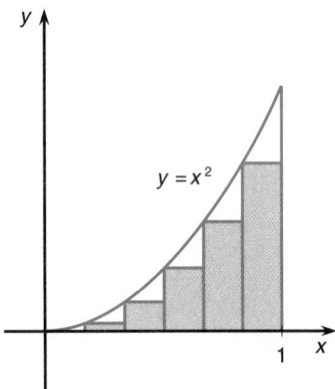

Figure 5.1.4

> If we divide $[0, 1]$ into n equal segments, then the width of each segment is $\frac{1}{n}$.

This formula for the sum of the squares of the integers from 1 to n will be proved in Example 5.2.5 but first let us use it to find an area.

EXAMPLE 2   Find the area under the curve $y = f(x) = x^2$ between $x = 0$ and $x = 1$.

Solution   We divide the interval $[0, 1]$ into n equal segments, and construct a rectangle on each of them, as shown in Figure 5.1.4 (in the illustration, $n = 6$). We let the height of each rectangle be the height of the function at the left endpoint of its segment, so the curve passes through the upper left corner of the rectangle.

The first rectangle lies over the interval $\left[0, \frac{1}{n}\right]$. Its height is the height of $f(x)$ at $x = 0$, so this "rectangle" has area equal to 0.

The second rectangle lies over the interval $\left[\frac{1}{n}, \frac{2}{n}\right]$, so its height is $f\left(\frac{1}{n}\right) = \frac{1}{n^2}$, and its width is $\frac{1}{n}$. Its area is $\frac{1}{n^2} \times \frac{1}{n} = \frac{1}{n^3}$.

The third rectangle lies over the interval $\left[\frac{2}{n}, \frac{3}{n}\right]$, so its height is $f\left(\frac{2}{n}\right) = \frac{2^2}{n^2}$, and its width is $\frac{1}{n}$. Its area is $\frac{2^2}{n^2} \times \frac{1}{n} = \frac{2^2}{n^3}$.

The kth rectangle lies over the interval $\left[\frac{k-1}{n}, \frac{k}{n}\right]$, so its height is $f\left(\frac{k-1}{n}\right) = \left(\frac{k-1}{n}\right)^2$, and its width is $\frac{1}{n}$. Its area is $\left(\frac{k-1}{n}\right)^2 \frac{1}{n} = \frac{(k-1)^2}{n^3}$.

The last rectangle lies over the interval $\left[\frac{n-1}{n}, \frac{n}{n}\right] = \left[\frac{n-1}{n}, 1\right]$, so its height is $f\left(\frac{n-1}{n}\right) = \left(\frac{n-1}{n}\right)^2$, and its area is $\frac{(n-1)^2}{n^3}$.

The total area of all the rectangles is then

$$0 + \frac{1}{n^3} + \frac{2^2}{n^3} + \ldots + \frac{(k-1)^2}{n^3} + \ldots + \frac{(n-1)^2}{n^3}.$$

There is a common denominator of n^3. Factoring it out, we find the simpler expression

$$\frac{1}{n^3}\left(0 + 1 + 2^2 + 3^2 + \ldots + (n-1)^2\right) = \frac{1}{n^3}\left(1^2 + 2^2 + 3^2 + \ldots + (n-1)^2\right).$$

To simplify this, we need Theorem 5.1.1. Using the formula with n replaced by

$n - 1$, we see that the sum is

$$1^2 + 2^2 + \ldots + (n-1)^2 = \frac{(n-1)\big((n-1)+1\big)\big(2(n-1)+1\big)}{6}$$

$$= \frac{(n-1)(n)(2n-1)}{6},$$

so the total area of the rectangles is

$$\frac{1}{n^3}\big(1^2 + 2^2 + \ldots + (n-1)^2\big) = \frac{1}{n^3}\left(\frac{(n-1)n(2n-1)}{6}\right)$$

$$= \frac{(n-1)(2n-1)}{6n^2} = \frac{2n^2 - 3n + 1}{6n^2}.$$

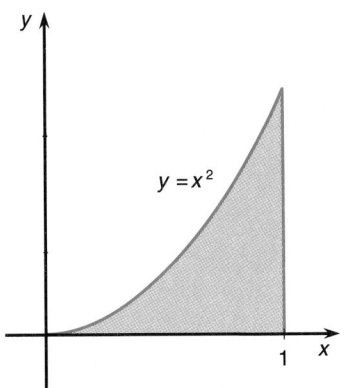

$y = x^2$

Figure 5.1.5

The last step is to take the limit as $n \to \infty$. This amounts to considering more and more rectangles, each of which is narrower and narrower; the total area approaches the area under the curve as shown in Figure 5.1.5.

The area of the region under the curve is

$$\lim_{n \to \infty}\left(\frac{2n^2 - 3n + 1}{6n^2}\right) = \frac{2}{6} = \frac{1}{3}.$$

The idea of approximating the area by large numbers of rectangles and then taking limits does seem to work, but it is awfully complicated in practice. After all, $y = x^2$ is just about the simplest curve we can think of other than a straight line, and still it took a lot of work to find the area. The rest of this chapter is devoted to developing a more practical technique for finding areas.

Despite the difficulties, Example 2 does illustrate the method of approximating areas by filling a region with rectangles.

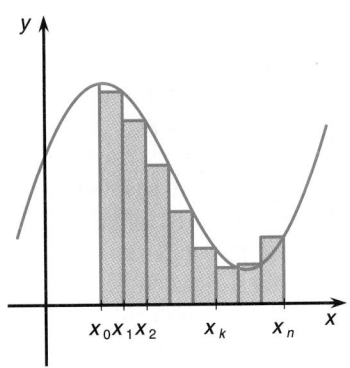

$x_0 x_1 x_2 \quad x_k \quad x_n$

Figure 5.1.6

> The kth point x_k is found by moving k subintervals from $x = a$. Since each subinterval has width $\frac{b-a}{n}$, $x_k = a + k\frac{b-a}{n}$.

EXAMPLE 3 If $f(x) \geq 0$ for all $x \in [a, b]$, approximate the area under the curve $y = f(x)$ between $x = a$ and $x = b$, by dividing the interval into n equal subintervals and using rectangles whose height equals the height of the curve at the *right* endpoint of each of subinterval (see Figure 5.1.6).

Solution The length of the interval $[a, b]$ is $b - a$, so the length of each subinterval will be $\frac{b-a}{n}$. Suppose we label the endpoints of the subintervals as $x_0, \ x_1, \ x_2, \ldots, \ x_n$, so that $x_0 = a$, $x_1 = a + \frac{b-a}{n}$, $x_2 = a + 2\frac{b-a}{n}$, $x_3 = a + 3\frac{b-a}{n}, \ldots, x_k = a + k\frac{b-a}{n}, \ldots, x_{n-1} = a + (n-1)\frac{b-a}{n}$, $x_n = a + n\frac{b-a}{n} = b$.

The kth subinterval will be $[x_{k-1}, x_k] = \left[a + (k-1)\frac{b-a}{n}, a + k\frac{b-a}{n}\right]$. Its right endpoint is $x_k = a + k\frac{b-a}{n}$, so the rectangle over this subinterval has height $f(x_k)$. Its area is $f(x_k)\frac{b-a}{n}$.

The total area of all the rectangles is

$$f(x_1)\frac{b-a}{n} + f(x_2)\frac{b-a}{n} + f(x_3)\frac{b-a}{n} + \ldots + f(x_n)\frac{b-a}{n}$$

$$= \big(f(x_1) + f(x_2) + f(x_3) + \ldots + f(x_n)\big)\frac{b-a}{n}.$$

In Example 3 we chose the height of our rectangles by taking the value of the function at the right endpoint of each subinterval; in Example 2 we used the value

Figure 5.1.7

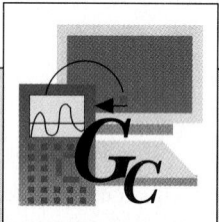

of the function at the left endpoints. We could also have used the midpoints of the subintervals, but the calculations would be slightly more complicated.

Different choices will give different values when we approximate the area under a curve. Our ultimate goal is to take a limit and get the *exact* value of the area, and when we do this, the differences should disappear.

EXAMPLE 4 Use the formula from Example 3 to approximate the area under the curve $y = f(x) = x^3$ between $x = 1$ and $x = 3$, using the values $n = 2, 4, 6$.

Solution With $n = 2$, as shown in Figure 5.1.7, we have $x_0 = 1$, $x_1 = 2$, $x_2 = 3$. The width of each rectangle is $\frac{b-a}{n} = \frac{3-1}{2} = 1$, so the total area of the rectangles is

$$\left(f(x_1) + f(x_2)\right)\frac{b-a}{n} = \left(f(2) + f(3)\right) \times 1$$

$$= f(2) + f(3) = 2^3 + 3^3 = 8 + 27 = 35.$$

A programmable calculator or a computer can be programmed to evaluate the sum in Example 4. The advantage is that the machine can easily do the calculation with larger values of n. The width of each rectangle is $\frac{b-a}{n} = \frac{2}{n}$.

TI-85	SHARP EL9300
PRGM F2 (EDIT) RIEMANN ENTER F3 F1 (Input) N EXIT ENTER 0 STO I ENTER 0 STO S ENTER Lbl B ENTER I+1 STO I ENTER S + (1 + I × 2 ÷ N) ^ 3 STO S ENTER F4 F1 (If) (I TEST F2 (<) N) ENTER Goto B ENTER S × 2 ÷ N ENTER EXIT EXIT EXIT PRGM F1 (NAME) RIEMANN ENTER 6 ENTER ...	▼ ▼ (NEW) ENTER ENTER RIEMANN ENTER COMMAND A 3 (Input) N ENTER I = 0 ENTER S = 0 ENTER COMMAND B 1 (Label) 1 ENTER I = I + 1 ENTER S = S + (1 + I × 2 ÷ N) a^b 3 ENTER COMMAND 3 (If) I COMMAND C 2 (<) N COMMAND B 2 (Goto) 1 ENTER S = S × 2 ÷ N ENTER COMMAND A 1 (Print) S ENTER COMMAND A 6 (End) ENTER (RUN) ENTER (RIEMANN) ENTER 6 ENTER
CASIO f_x-6300G (f_x-7700GB)	HP 48SX
MODE 2 EXE ? → N : 0 → I : 0 → S : Lbl 1 : I + 1 → I : S + (1 + I × 2 ÷ N) x^y 3 → S : I < N ⇒ Goto 1 : S × 2 ÷ N ◢ MODE 1 Prog 0 EXE 6 EXE (on the f_x-7700GB, : is PRGM F6, and Prog is PRGM F3.)	ATTN ↱ CLR ≪≫ "' KEY IN N ▶ SPC PROMPT SPC ' N ▶ STO 0 SPC 1 SPC N SPC FOR I SPC I SPC 2 × N ÷ 1 + 3 y^x + NEXT ▶ 2 × N ÷ ▶ ENTER ' RIEMANN STO VAR RIEMANN 6 ENTER ↰ CONT

Now use the program with other values of n or for some other functions.

?6

ANSWER=
    24.55555556

With $n = 4$ we have $x_0 = 1$, $x_1 = \frac{3}{2}$, $x_2 = 2$, $x_3 = \frac{5}{2}$, $x_4 = 3$. The width of each rectangle is $\frac{b-a}{n} = \frac{3-1}{4} = \frac{1}{2}$, so the total area of the rectangles is

$$\left(f(x_1) + f(x_2) + f(x_3) + f(x_4)\right)\frac{b-a}{n} = \left(f\left(\frac{3}{2}\right) + f(2) + f\left(\frac{5}{2}\right) + f(3)\right) \times \frac{1}{2}$$

$$= \left(\left(\frac{3}{2}\right)^3 + 2^3 + \left(\frac{5}{2}\right)^3 + 3^3\right) \times \frac{1}{2}$$

$$= \left(\frac{27}{8} + 8 + \frac{125}{8} + 27\right) \times \frac{1}{2}$$

$$= 27.$$

With $n = 6$ we have $x_0 = 1$, $x_1 = \frac{4}{3}$, $x_2 = \frac{5}{3}$, $x_3 = 2$, $x_4 = \frac{7}{3}$, $x_5 = \frac{8}{3}$, $x_6 = 3$. The width of each rectangles is $\frac{b-a}{n} = \frac{3-1}{6} = \frac{1}{3}$, so the total area of the rectangles is

$$\left(f(x_1) + f(x_2) + f(x_3) + f(x_4) + f(x_5) + f(x_6)\right)\frac{b-a}{n}$$

$$= \left(\left(\frac{4}{3}\right)^3 + \left(\frac{5}{3}\right)^3 + 2^3 + \left(\frac{7}{3}\right)^3 + \left(\frac{8}{3}\right)^3 + 3^3\right) \times \frac{1}{3}$$

$$= \left(\frac{64}{27} + \frac{125}{27} + 8 + \frac{343}{27} + \frac{512}{27} + 27\right) \times \frac{1}{3}$$

$$= 24\frac{5}{9}.$$

In principle this formula can be used with larger and larger values of n, giving better and better approximations to the area under the curve.

However, this process can only approximate the area, and we want to find a way to calculate it exactly. That is the purpose of the rest of this chapter.

Exercises 5.1

I

Use the formula from Example 3 to approximate the area under each of the following curves $f(x)$ on the given interval, dividing the interval into n equal subintervals with the specified value of n.

1. $f(x) = x^2$ on $[0, 1]$; $n = 3$
2. $f(x) = x^3$ on $[0, 1]$; $n = 4$
3. $f(x) = x^2$ on $[0, 1]$; $n = 5$
4. $f(x) = x^2$ on $[0, 4]$; $n = 4$
5. $f(x) = 3x^2$ on $[0, 1]$; $n = 5$
6. $f(x) = x^2$ on $[-4, 0]$; $n = 4$
7. $f(x) = 2x + 1$ on $[0, 3]$; $n = 6$
8. $f(x) = x^2$ on $[-4, 4]$; $n = 8$
9. $f(x) = x^2 + 3$ on $[0, 1]$; $n = 6$
10. $f(x) = x^3$ on $[0, 3]$; $n = 3$

11. $f(x) = \sin x$ on $[0, \pi]$; $n = 2$
12. $f(x) = \sin x$ on $[0, \pi]$; $n = 4$
13. $f(x) = \cos x$ on $\left[-\frac{\pi}{2}, \frac{\pi}{2}\right]$; $n = 2$
14. $f(x) = \cos x$ on $\left[0, \frac{\pi}{2}\right]$; $n = 2$
15. $f(x) = \sqrt{x}$ on $[0, 4]$; $n = 4$
16. $f(x) = \frac{1}{1+x^2}$ on $[-1, 1]$; $n = 4$
17. $f(x) = \sin^2 x$ on $[0, 4\pi]$; $n = 4$
18. $f(x) = x^4 + 2x + 1$ on $[0, 3]$; $n = 3$
19. $f(x) = \sin^2 x$ on $[0, 4\pi]$; $n = 8$
20. $f(x) = \frac{1}{x}$ on $[1, 6]$; $n = 5$

II

21. Find the area under the curve $y = x^2$ between $x = -1$ and $x = 0$.

22. Find the area under the curve $y = x^2$ between $x = 0$ and $x = 2$.

23. Find the area under the curve $y = x^2$ between $x = 1$ and $x = 2$.

24. Find the area under the curve $y = 5x^2$ between $x = 0$ and $x = 1$.

*25. Consider the straight line $y = f(x) = 2x + 3$, which lies above the x-axis on $[0, 2]$.

   (i) Use the formula from Example 3 to approximate the area under the line between $x = 0$ and $x = 2$, dividing the interval $[0, 2]$ into four subintervals of equal length.

   (ii) If we use rectangles whose height is the value of $f(x)$ at the *left* endpoint of each subinterval, do you expect the approximation to be greater than, less than or the same as the answer to part (i)?

   (iii) Use rectangles whose height is the value of $f(x)$ at the *left* endpoint of each subinterval to approximate the area of the region we considered in part (i).

   (iv) Compare the approximations of parts (i) and (ii) and compare them both with the actual area under the graph.

*26. Consider the constant function $f(x) = -3$ on the interval $[0, 1]$. Suppose we apply the formula from Example 3 with $n = 2$, even though the function $f(x)$ is *negative* on this interval. Compare the number that results from this calculation with the area enclosed

between the graph of $y = f(x)$, the x-axis, and the lines $x = 0$ and $x = 1$. Explain what has happened.

*27. If $g(x) = x^2$, and we use the formula from Example 3 to approximate the area under the graph of $y = g(x)$ between $x = -4$ and $x = -1$, will the approximation be greater than the actual area or less than the actual area? (*Hint:* Draw a picture.)

*28. If $h(x) = \sqrt{x}$ and $0 \le a < b$, suppose we use the formula from Example 3 to approximate the area under the graph of $y = h(x)$ between $x = a$ and $x = b$. Will the approximation be greater than the actual area or less than the actual area? (*Hint:* Draw a picture.)

29. For this exercise you need a personal computer or programmable calculator set to evaluate the formula from Example 3 for $f(x) = x^2$ between $x = 0$ and $x = 2$, using n equal subintervals. Evaluate the approximate area under $y = x^2$ between 0 and 2 with $n = 2, 4, 10, 100$.

30. Use a personal computer or programmable calculator to approximate the area under $y = x^3$ between $x = 1$ and $x = 4$. Use the formula from Example 3 with $n = 4, 16, 64, 256$.

31. Use a personal computer or programmable calculator to approximate the area under $y = \sin x$ between $x = 0$ and $x = \frac{\pi}{2}$. Use the formula from Example 3 with $n = 2, 8, 50, 200$. Can you guess the actual area?

P ← **POINT TO PONDER**

In this section we have talked about the notion of the area of a region in the plane, but we never really specified what it means. This is partly because it is quite difficult to be precise about areas.

There are some regions whose areas we can calculate explicitly. In principle we know how to find the area of any triangle, and from that we can find the area of any region that can be built up from triangles. From Figure 5.1.8 we see how even a complicated polygon (region with a finite number of straight line segments as its boundary) can be broken into triangles, so we could find its area.

In fact these are the only regions we really know how to work with. We know that the area of a circle of radius r is πr^2, but this is a fact we have learned, rather than something we can explain. (For instance, it is not easy to show that the constant must be π; try to convince

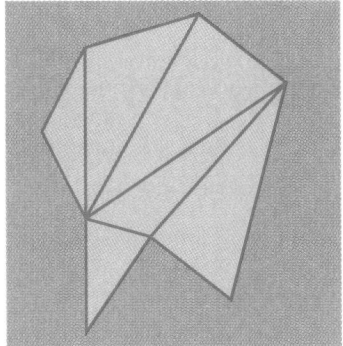

Figure 5.1.8

yourself that it cannot be $\frac{22}{7}$, which is a number close to π . . .).

One thing about area is that we have a good deal of

intuition about what it ought to be. So a possible avenue through which to approach it is to state all the properties we can think of that area should satisfy. For instance, in discussing polygons, we implicitly assumed that areas "add," that is, that the area of a region composed of two nonoverlapping regions should be the sum of their areas. Try to think of other properties that area should satisfy.

There is a surprising fact about areas. Our intuition suggests that it ought to be possible to find the area of any region, but in fact it can be shown that if we specify some reasonable properties for area, then it is impossible to find the area of some regions. To put it another way, there are some regions that are so weird that the idea of area does not make sense for them. However, this is a very difficult subject, and it is extraordinarily hard to find any examples of such a region.

5.2 Summation Notation

INDEX OF SUMMATION
SUM OF INTEGERS
SUM OF SQUARES

In Section 5.1 we developed the idea of approximating the area under a curve by filling the region with rectangles. In Example 5.1.3 we divided the interval into n equal subintervals and on each subinterval constructed the rectangle whose height equals the height of the curve at the right endpoint. Then we calculated the total area of these rectangles.

This provides a formula for calculating approximate areas, but it involves a fairly cumbersome sum with a large number of terms. To deal more efficiently with such sums, we devote this section to introducing a notation for them.

Suppose we want to add up the first n positive integers,

$$1 + 2 + 3 + \ldots + (n-1) + n,$$

or the squares of the first n positive integers,

$$1^2 + 2^2 + 3^2 + \ldots + (n-1)^2 + n^2.$$

Both of these sums can be expressed in the form

$$a_1 + a_2 + a_3 + \ldots + a_{n-1} + a_n,$$

where each a_k is some number determined by k. In the first sum, the sum of the first n positive integers, we can let $a_k = k$, so $a_1 = 1$, $a_2 = 2$, $a_n = n$, etc. For the second we can let $a_k = k^2$, so $a_1 = 1^2 = 1$, $a_2 = 2^2 = 4$, $a_3 = 3^2 = 9$, $a_n = n^2$, etc.

A great many interesting sums can be written as

$$a_1 + a_2 + a_3 + \ldots + a_{n-1} + a_n,$$

for some numbers a_k. This is the sum of the a_k's, as k runs from $k = 1$ to $k = n$. We think of k starting out at $k = 1$ and the sum starting out with a_1. Then k increases to $k = 2$, and we add a_2. Next k increases to $k = 3$, and we add a_3. This continues until k reaches its final value $k = n$, at which point we add a_n and then stop.

There is no particular reason to insist that the sum begin at $k = 1$; we can consider sums beginning and ending with any specified integers:

$$a_m + a_{m+1} + a_{m+2} + \ldots + a_{n-1} + a_n,$$

where m and n are any integers satisfying $m \leq n$.

This process is so common, so useful, and so important that there is special notation for it.

DEFINITION 5.2.1

Suppose m and n are integers with $m \leq n$. Suppose that for each k between m and n (i.e., $m \leq k \leq n$), a_k is a real number.

The **summation notation**

$$\sum_{k=m}^{n} a_k$$

is an abbreviation for the sum of the a_k's as k ranges from $k = m$ to $k = n$. Symbolically,

$$\sum_{k=m}^{n} a_k = a_m + a_{m+1} + a_{m+2} + \ldots + a_{n-1} + a_n.$$

The left side of the above equation is read "the sum as k runs from m to n of a_k" or "the sum of a_k as k runs from m to n."

The summation sign $\sum$ is a stylized capital sigma. Sigma is the letter in the Greek alphabet that corresponds to our letter S, so it reminds us of "sum."

EXAMPLE 1

(i) For the sum of the first n positive integers we can write

$$\sum_{k=1}^{n} k = 1 + 2 + 3 + \ldots + (n - 1) + n.$$

(ii) For the sum of the squares of the first n positive integers we write

$$\sum_{k=1}^{n} k^2 = 1 + 4 + 9 + \ldots + (n - 1)^2 + n^2.$$

By Theorem 5.1.1 we know what this sum equals. It is

$$\sum_{k=1}^{n} k^2 = \frac{n(n + 1)(2n + 1)}{6}.$$

EXAMPLE 2 We reconsider Example 5.1.2 (p. 348), in which we calculated the approximation to the area under the curve $y = f(x) = x^2$ between $x = 0$ and $x = 1$ by n rectangles.

In that example we found that the kth rectangle lies above the interval $\left[\frac{k-1}{n}, \frac{k}{n}\right]$. Its width is $\frac{1}{n}$, and we let its height be the height of the curve at the left endpoint, that is, $f\left(\frac{k-1}{n}\right)$ (see Figure 5.2.1).

The area of the kth rectangle is therefore $f\left(\frac{k-1}{n}\right) \times \frac{1}{n}$, and the total area of all the rectangles is the sum

$$\sum_{k=1}^{n} f\left(\frac{k-1}{n}\right) \times \frac{1}{n} = \sum_{k=1}^{n} \frac{(k-1)^2}{n^2} \times \frac{1}{n} = \sum_{k=1}^{n} \frac{(k-1)^2}{n^3} = \frac{1}{n^3} \sum_{k=1}^{n} (k-1)^2.$$

We observe that while this sum is written in terms of k, it is really $k - 1$ that is most important. While k runs from $k = 1$ to $k = n$, we see that $k - 1$ runs from 0

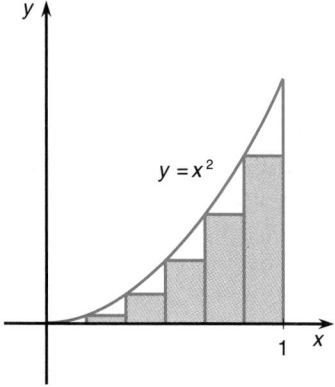

$y = x^2$

Figure 5.2.1

A programmable calculator or a computer can be programmed to evaluate the sum in part (ii) of Example 1. We can use the machine to test the formula for fairly large values of n.

TI-85	SHARP EL9300
PRGM F2(EDIT) SUM ENTER F3 F1(Input) N EXIT ENTER 0 STO K ENTER 0 STO S ENTER Lbl B ENTER K + 1 STO K ENTER S + K^2 STO S ENTER F4 F1(If) (K TEST F2(<) N) ENTER Goto B ENTER S ENTER EXIT EXIT EXIT PRGM F1 (NAME) SUM ENTER 20 ENTER	✎ ▼ ▼ (NEW) ENTER ENTER SUM ENTER COMMAND A 3 (Input) N ENTER K = 0 ENTER S = 0 ENTER COMMAND B 1 (Label) 1 ENTER K = K + 1 ENTER S = S + K x^2 ENTER COMMAND 3 (If) K COMMAND C 2 (<) N COMMAND B 2 (Goto) 1 ENTER N ENTER COMMAND A 1 (Print) S ENTER COMMAND A 6 (End) ENTER ⊞ ✎ (RUN) ENTER (SUM) ENTER 20 ENTER
CASIO f_x-6300G (f_x-7700GB) MODE 2 EXE ? → N : 0 → K : 0 → S : Lbl 1 : K + 1 → K : S + K x^2 → S : K < N ⇒ Goto 1 : S ◢ MODE 1 Prog 0 EXE 20 EXE (on the f_x-7700GB, : is PRGM F6, and Prog is PRGM F3.)	HP 48SX ATTN ↱ CLR ≪ ≫ "" KEY IN N ▶ SPC PROMPT SPC ' N ▶ STO 0 SPC 1 SPC N SPC FOR K SPC K SPC x^2 + NEXT ▼ ENTER ' SUM STO VAR SUM 20 ENTER ↰ CONT

```
?20
ANSWER=
            2870
```

Now use the program with other values of n. Modify your program to find $\sum_{k=1}^{n} k^3$. Can you guess a formula for this sum (in terms of n)?

to $n - 1$. In an attempt to avoid confusion, let us write $\ell = k - 1$, so the last sum is exactly the same thing as

$$\sum_{\ell=0}^{n-1} \ell^2.$$

Moreover, the first term in this sum is $0^2 = 0$, which we can safely ignore. In other words, the sum will be the same if we allow ℓ to run from $\ell = 1$ to $\ell = n - 1$ instead of starting at $\ell = 0$. This means that the sum equals

$$\sum_{\ell=1}^{n-1} \ell^2.$$

This can be evaluated by part (ii) of Example 1.

We find that the total area of the n rectangles is

$$\frac{\sum_{k=1}^{n}(k-1)^2}{n^3} = \frac{1}{n^3}\sum_{\ell=1}^{n-1}\ell^2 = \frac{1}{n^3}\left(\frac{(n-1)\,n(2n-1)}{6}\right)$$

$$= \frac{(n-1)(2n-1)}{6n^2} = \frac{2n^2-3n+1}{6n^2}.$$

This is exactly the same as the calculation we did in Example 5.1.2, except that there we had to write out each sum in detail. The summation notation allows us to write the computation in a neater, more efficient way.

The final observation in Example 2 of Section 5.1 was that we could take the limit as $n \to \infty$ of this approximation by n rectangles, and we find that the area under the graph of $y = f(x) = x^2$ between $x = 0$ and $x = 1$ is

$$\lim_{n\to\infty}\frac{2n^2-3n+1}{6n^2} = \frac{1}{3}.$$

Summation notation can be used to restate the result of Example 5.1.3 more neatly.

EXAMPLE 3   **Approximation by Rectangles**

Suppose $f(x)$ is a continuous function satisfying $f(x) \ge 0$ for all $x \in [a,b]$. Then the area under the graph of the curve $y = f(x)$ between $x = a$ and $x = b$ can be approximated by

$$\frac{b-a}{n}\sum_{k=1}^{n}f(x_k),$$

where $x_k = a + k\frac{b-a}{n}$.

This is the total area of the rectangles constructed by dividing the interval $[a,b]$ into n equal subintervals and constructing on each subinterval a rectangle whose height equals the height of the curve at the right endpoint.

There is essentially no difference between this formula and the one in the original example (Example 5.1.3). The summation notation used in this version makes it more compact and may make some calculations easier to follow, as we saw in Example 2.

When we perform calculations with summation notation, we need to know how it can be manipulated. First is the elementary observation that the distributive law is satisfied.

PROPOSITION 5.2.2

> If C is any real number, m and n are integers with $m \le n$, and a_k is a real number for each k with $m \le k \le n$, then the distributive law holds:
>
> $$\sum_{k=m}^{n}Ca_k = C\sum_{k=m}^{n}a_k.$$

Next is the equally obvious remark that it is permissible to combine sums.

PROPOSITION 5.2.3

Suppose that m and n are integers with $m \le n$ and a_k and b_k are real numbers for each k with $m \le k \le n$. Then

$$\sum_{k=m}^{n} a_k + \sum_{k=m}^{n} b_k = \sum_{k=m}^{n} (a_k + b_k).$$

Next we recall that in the calculation of Example 2 we encountered the sum

$$\sum_{k=1}^{n} (k-1)^2.$$

Recognizing that the terms of the sum were more simply expressed as functions of $k - 1$ rather than k, we let $\ell = k - 1$. As k runs from 1 to n, we see that $\ell = k - 1$ runs from 0 to $n - 1$.

Rather than using k to label the terms in the sum, we can use ℓ, and this means we can rewrite the sum as

$$\sum_{\ell=0}^{n-1} \ell^2.$$

This equals the previous sum, and this technique is very useful. It amounts to changing the label k, sometimes called the **index of summation,** to transform the sum into something simpler or better suited to our purpose.

The important thing to realize here is that when we write a sum

$$\sum_{k=m}^{n} a_k,$$

what matters is the numbers a_k that are being summed, not the indices by which they are labeled. The index k can be shifted without changing the value of the sum.

PROPOSITION 5.2.4

Suppose $m \le n$ are integers and that for each k with $m \le k \le n$, a_k is a real number. Then the sum

$$\sum_{k=m}^{n} a_k$$

is the same as the sum obtained by *shifting* the index k by any fixed integer s. Letting $\ell = k + s$, we have

$$\sum_{k=m}^{n} a_k = \sum_{\ell=m+s}^{n+s} a_{\ell-s}.$$

(Notice how this was done: When $k = m$, we see that $\ell = k + s$ must equal $m + s$, which tells us where the sum over ℓ should begin. Similarly, when $k = n$, we see that $\ell = n + s$, which tells us where the sum should end. Finally, since $\ell = k + s$, we can solve and find $k = \ell - s$, so the terms a_k in the sum are replaced with $a_{\ell-s}$.)

EXAMPLE 4   The sum

$$\sum_{m=5}^{12} (m-4)^2$$

involves terms of the form $(m-4)^2$. It seems reasonable to try writing them using $m-4$ as the index, rather than m.

Letting $h = m-4$, we find that when $m = 5$, we have $h = m-4 = 5-4 = 1$. Also when $m = 12$, we have $h = m-4 = 8$. This allows us to write

$$\sum_{m=5}^{12} (m-4)^2 = \sum_{h=1}^{8} h^2.$$

This last sum is the sum of the squares of the first eight positive integers, which we can evaluate using Theorem 5.1.1, which was repeated in Example 1(ii). The formula says that the sum of the squares of the first n positive integers is $\frac{n(n+1)(2n+1)}{6}$. In this example we are interested in the case in which $n = 8$, so the sum is $\frac{(8)(9)(17)}{6} = 204$.

We have just evaluated the original sum:

$$\sum_{m=5}^{12} (m-4)^2 = 204.$$

Rearranging the sum turned it into another sum that we recognized, which made it easy to evaluate. Without this trick, the only way to evaluate would have been to add up all the terms individually. Of course, in this case that would not have been tremendously difficult, but it is easy to imagine more complicated examples in which adding up all the terms would be extremely tedious.

We conclude this section with a proof of the formula for the sum of the squares of the first n positive integers (cf. Example 1(ii)).

EXAMPLE 5   Prove Theorem 5.1.1:

$$\sum_{k=1}^{n} k^2 = \frac{n(n+1)(2n+1)}{6}.$$

Solution   We begin by finding a formula for the sum

$$S = 1 + 2 + 3 + \ldots + (n-1) + n$$

of the first n positive integers. To do this, we consider the sum

$$T = \sum_{k=0}^{n} \left((k+1)^2 - k^2\right).$$

By Proposition 5.2.3 this sum equals

$$\sum_{k=0}^{n} (k+1)^2 - \sum_{k=0}^{n} k^2.$$

It is possible to adjust the second sum by noticing that when $k = 0$, the term being added is $k^2 = 0^2 = 0$, so the sum will not change if we omit that term. In other words, the second sum can be replaced by $\sum_{k=1}^{n} k^2$.

Now we apply Proposition 5.2.4 to the first sum to shift the index. Adding up $(k+1)^2$ as k runs from 0 to n is exactly the same as adding k^2 with k running from 1 to $n+1$, so the first sum equals $\sum_{k=1}^{n+1} k^2$. By writing the last term (with $k = n+1$) separately we can rewrite this as $(n+1)^2 + \sum_{k=1}^{n} k^2$.

Combining, we find that

$$T = \sum_{k=0}^{n}\left((k+1)^2 - k^2\right) = \sum_{k=0}^{n}(k+1)^2 - \sum_{k=0}^{n}k^2 = \sum_{k=1}^{n+1}k^2 - \sum_{k=1}^{n}k^2$$

$$= (n+1)^2 + \sum_{k=1}^{n}k^2 - \sum_{k=1}^{n}k^2 = (n+1)^2.$$

On the other hand, we can rewrite T using Proposition 5.2.3:

$$T = \sum_{k=0}^{n}\left((k+1)^2 - k^2\right) = \sum_{k=0}^{n}(k^2 + 2k + 1 - k^2) = \sum_{k=0}^{n}(2k+1).$$

By Proposition 5.2.2 we see that

$$T = \sum_{k=0}^{n}2k + \sum_{k=0}^{n}1 = 2\sum_{k=0}^{n}k + \sum_{k=0}^{n}1.$$

The last sum amounts to adding 1 to itself $n+1$ times (once for $k = 0$, once for $k = 1$, etc., up to $k = n$), so it equals $n+1$. The second-last sum is unchanged by omitting the term with $k = 0$, so we find that

$$T = 2\left(\sum_{k=1}^{n}k\right) + (n+1).$$

On the other hand, we already know that $T = (n+1)^2$, so, writing $S = \sum_{k=1}^{n} k$, we have

$$(n+1)^2 = 2S + (n+1)$$

or

$$(n+1)^2 - (n+1) = 2S$$

or

$$(n+1)(n+1-1) = 2S$$

or

$$(n+1)n = 2S.$$

Karl Friedrich Gauss (1777–1855)

One of the great mathematicians of all time was Karl Friedrich Gauss (1777–1855). His elementary school teacher asked the class to add up the numbers from 1 to 100, expecting to keep them busy for a long time. Young Gauss found Formula 5.2.5 and instantly wrote down the correct answer.

From this we find that the sum S of the first n positive integers is given by

$$S = \sum_{k=1}^{n}k = \frac{n(n+1)}{2}. \qquad (5.2.5)$$

To prove the formula for the sum of the squares of the first n positive integers, we proceed in a similar way. Consider the sum

$$U = \sum_{k=0}^{n}\left((k+1)^3 - k^3\right) = \sum_{k=0}^{n}(k+1)^3 - \sum_{k=0}^{n}k^3 = \sum_{k=1}^{n+1}k^3 - \sum_{k=1}^{n}k^3$$

$$= (n+1)^3 + \sum_{k=1}^{n}k^3 - \sum_{k=1}^{n}k^3 = (n+1)^3.$$

Here, just as in the previous argument, we broke the sum into two parts, omitted the irrelevant term with $k = 0$ from the second part, shifted the index of the first part, and split off the term with $k = n + 1$.

On the other hand, we can evaluate U another way:

$$U = \sum_{k=0}^{n} \left((k+1)^3 - k^3 \right) = \sum_{k=0}^{n} \left(k^3 + 3k^2 + 3k + 1 - k^3 \right)$$

$$= \sum_{k=0}^{n} (3k^2 + 3k + 1) = 3\sum_{k=0}^{n} k^2 + 3\sum_{k=0}^{n} k + \sum_{k=0}^{n} 1$$

$$= 3\sum_{k=1}^{n} k^2 + 3\sum_{k=1}^{n} k + (n+1) = 3\sum_{k=1}^{n} k^2 + 3\frac{n(n+1)}{2} + (n+1).$$

Here we used the formula $\sum_{k=1}^{n} k = \frac{n(n+1)}{2}$.

Comparing this with our other evaluation of U, we see that

$$(n+1)^3 = 3\sum_{k=1}^{n} k^2 + 3\frac{n(n+1)}{2} + (n+1),$$

so

$$3\sum_{k=1}^{n} k^2 = (n+1)^3 - 3\frac{n(n+1)}{2} - (n+1)$$

$$= (n+1)\left((n+1)^2 - \frac{3n}{2} - 1 \right)$$

$$= (n+1)\left(n^2 + 2n + 1 - \frac{3n}{2} - 1 \right)$$

$$= (n+1)\left(n^2 + \frac{n}{2} \right)$$

$$= \frac{1}{2}(n+1)(2n^2 + n).$$

Dividing by 3, we find

$$\sum_{k=1}^{n} k^2 = \frac{n(n+1)(2n+1)}{6},$$

which is Theorem 5.1.1, as desired.

A similar procedure can be followed to find the sum of the cubes of the first n positive integers, or the fourth powers, etc. The value of using the summation notation in these arguments should be apparent. They would be much more tedious if each sum were written out in full.

Theorem 5.1.1 can also be proved by using the technique known as mathematical induction, which is discussed in the appendix. However, the proof just given does not require a knowledge of induction, and it also has the advantage that it discovers the answer for us. To use induction, it is necessary to know beforehand what the formula is supposed to be. Since the formula was announced in Theorem 5.1.1, we knew what to expect, but without that it would not have been too easy to guess the answer. If you don't believe that, try to guess a formula for the sum of the *cubes* of the first n positive integers. . . .

Exercises 5.2

I

Evaluate each of the following sums.

1. $\sum_{k=1}^{5} k^3$

2. $\sum_{k=2}^{4} k^4$

3. $\sum_{k=0}^{6} (k^2 - 3)$

4. $\sum_{k=-2}^{2} k^5$

5. $\sum_{k=2}^{4} (4k + 1)$

6. $\sum_{\ell=-1}^{4} (\ell^2 - \ell)$

7. $\sum_{h=1}^{6} (h^2 + 5)$

8. $\sum_{h=-6}^{-1} (h^2 + 5)$

9. $\sum_{i=1}^{12} i^2$

10. $\sum_{j=3}^{3} j^4$

11. $\sum_{j=-3}^{3} (2j + 1)$

12. $\sum_{k=0}^{n} k^2 - \sum_{i=1}^{n-1} i^2$

13. $\sum_{j=1}^{m} (j - 1)^2$

14. $\sum_{r=0}^{n} (r + 1)^2$

15. $\sum_{i=1}^{m} (i + 1)^2$

16. $\sum_{r=0}^{n-1} (r + 1)^2 - \sum_{u=2}^{n+1} (u - 1)^2$

17. $\sum_{j=-m}^{m} (j^2 + 2j + 1)$

18. $\sum_{k=1}^{n} (n - k)$

19. $\sum_{i=2}^{9} 1$

20. $\sum_{k=1}^{n} \left(k - \frac{n+1}{2}\right)$

II

21. Suppose m and n are positive integers satisfying $m \le n$. Find a formula for

$$\sum_{k=m}^{n} k.$$

What is the meaning of your formula when $m = n$?

22. Suppose m and n are positive integers satisfying $m \le n$. Find a formula for

$$\sum_{k=m}^{n} k^2.$$

What is the meaning of your formula when $m = n$?

*23. Suppose m and n are integers, *not necessarily positive,* satisfying $m \le n$. Find a formula for

$$\sum_{k=m}^{n} k.$$

What is the meaning of your formula when $m = n$?

*24. Suppose m and n are integers, *not necessarily positive,* satisfying $m \le n$. Find a formula for

$$\sum_{k=m}^{n} k^2.$$

What is the meaning of your formula when $m = n$?

25. (i) Evaluate $\sum_{i=1}^{5} (-1)^i i$.

    (ii) Evaluate $\sum_{i=1}^{8} (-1)^i i$.

    (iii) If n is a positive integer, find a formula for $\sum_{i=1}^{n} (-1)^i i$. (*Hint:* You may want to distinguish between the cases in which n is even and odd.)

*26. (i) Evaluate $\sum_{j=1}^{5} (-1)^j j^2$.

    (ii) Evaluate $\sum_{j=1}^{8} (-1)^j j^2$.

    (iii) If n is a positive integer, find a formula for $\sum_{j=1}^{n} (-1)^j j^2$. (*Hint:* You may want to distinguish between the cases in which n is even and odd.)

*27. Let n be a positive integer and evaluate $\sum_{k=0}^{n} \cos(k\pi)$.

*28. Let n be a positive integer and evaluate $\sum_{k=0}^{n} \cos\left(k\frac{\pi}{2}\right)$.

29. Use summation notation to write the Binomial Theorem more conveniently. In other words, write the expression given by the Binomial Theorem (Theorem 1.3.10) for $(a + b)^n$ as a summation.

*30. Find a formula for the sum of the cubes of the first n positive integers, that is, for

$$\sum_{\ell=1}^{n} \ell^3.$$

*31. Evaluate (i) $\sum_{k=1}^{5} \left(\sum_{j=1}^{3} jk\right)$, (ii) $\sum_{j=1}^{3} \left(\sum_{k=1}^{5} jk\right)$.

*32. Evaluate

$$\sum_{k=1}^{n} \left(\sum_{j=1}^{k} j\right).$$

III

33. Use a personal computer or programmable calculator to evaluate $\sum_{n=1}^{100} n$; compare with the expected answer.

34. Evaluate $\sum_{k=1}^{50} k^2$ and compare with the expected answer.

35. Evaluate $\sum_{m=0}^{45} 2^{-m}$.

36. Evaluate $\sum_{n=0}^{70} n3^{-n}$.

Use the result of Example 3(p. 356) to approximate the area under the graph of each of the following functions, with the specified n.

37. $x^2 - x^3$, $x \in [0, 1]$, $n = 100$

38. $\cos x$, $x \in \left[0, \frac{\pi}{2}\right]$, $n = 40$

39. $\ln x$, $x \in [1, e]$, $n = 50$

40. e^{-x^2}, $x \in [0, 5]$, $n = 50$

5.3 Riemann Sums

PARTITION
SUBINTERVAL
RIEMANN SUM
INTEGRAL
INTEGRABLE
LIMITS OF INTEGRATION
SIGNED AREA

In Section 5.1 we approximated areas with rectangles (see Figure 5.3.1). There we used subintervals of equal width and rectangles whose height was equal to the height of the curve at the left or right endpoint. To develop ways of calculating areas exactly, we need a different kind of approximation. Ironically, this approach is usually more cumbersome for making approximations, but it is better adapted to the theoretical development which leads to exact evaluations.

As before, we consider the area under a nonnegative function $f(x)$ on an interval $[a, b]$. We will divide the interval $[a, b]$ up into a number of **subintervals**, as shown in Figure 5.3.2. If the endpoints of these subintervals (in increasing order) are $a = x_0, x_1, x_2, \ldots, x_n = b$, then the subintervals are $[x_0, x_1], [x_1, x_2], \ldots, [x_{n-1}, x_n]$; there are n subintervals, and the kth one is $[x_{k-1}, x_k]$.

Notice that the subintervals are not all required to have the same length. We shall write $\Delta x_k = x_k - x_{k-1}$ for the length of the kth subinterval.

DEFINITION 5.3.1

If $a = x_0 < x_1 < x_2 < \ldots < x_n = b$, then the set

$$\mathcal{P} = \left\{x_0, x_1, x_2, \ldots, x_n\right\}$$

is called a **partition** of the interval $[a, b]$. The length of the largest subinterval,

$$\max_{1 \leq k \leq n} \Delta x_k = \max_{1 \leq k \leq n} (x_k - x_{k-1}),$$

is called the **norm** of $\mathcal{P}$ (or the **mesh** of $\mathcal{P}$), and is denoted $\|\mathcal{P}\|$. (The symbol Δ is the capital Greek letter delta. It corresponds to our letter D, which may remind us that Δx_k is the *difference* $x_k - x_{k-1}$ between adjacent points in the partition.)

Figure 5.3.1

Figure 5.3.2

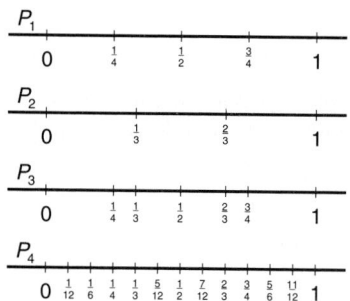

Figure 5.3.3

We shall approximate the area under $y = f(x)$ using partitions $\mathcal{P}$ and then take a sort of limit as $\|\mathcal{P}\| \to 0$, that is, as the subintervals all get smaller and smaller.

EXAMPLE 1 In the interval $[0, 1]$, consider the partitions $\mathcal{P}_1 = \left\{0, \frac{1}{4}, \frac{1}{2}, \frac{3}{4}, 1\right\}$, $\mathcal{P}_2 = \left\{0, \frac{1}{3}, \frac{2}{3}, 1\right\}$. Note that $\|\mathcal{P}_1\| = \frac{1}{4}$ and $\|\mathcal{P}_2\| = \frac{1}{3}$.

Neither of $\mathcal{P}_1$ and $\mathcal{P}_2$ contains the other. One partition that contains both of them is $\mathcal{P}_3 = \left\{0, \frac{1}{4}, \frac{1}{3}, \frac{1}{2}, \frac{2}{3}, \frac{3}{4}, 1\right\}$. Another is $\mathcal{P}_4$, which is obtained by taking all fractions between 0 and 1 that can be written with denominator 12, so

$$\mathcal{P}_4 = \left\{0, \frac{1}{12}, \frac{2}{12}, \frac{3}{12}, \frac{4}{12}, \frac{5}{12}, \frac{6}{12}, \frac{7}{12}, \frac{8}{12}, \frac{9}{12}, \frac{10}{12}, \frac{11}{12}, \frac{12}{12} = 1\right\}$$

$$= \left\{0, \frac{1}{12}, \frac{1}{6}, \frac{1}{4}, \frac{1}{3}, \frac{5}{12}, \frac{1}{2}, \frac{7}{12}, \frac{2}{3}, \frac{3}{4}, \frac{5}{6}, \frac{11}{12}, 1\right\}.$$

Each of the partitions $\mathcal{P}_1, \mathcal{P}_2$, and $\mathcal{P}_4$ has the property that the subintervals it determines are all of equal length; $\mathcal{P}_3$ does not have this property. See Figure 5.3.3.

Suppose we are given a partition $\mathcal{P} = \{x_0, \ldots, x_n\}$ of $[a, b]$, and for each $k = 1, 2, \ldots, n$ a point $x_k^* \in [x_{k-1}, x_k]$. There are of course many ways of choosing the points $\{x_k^*\}$.

Figure 5.3.4

Georg F. B. Riemann (1826–1866)

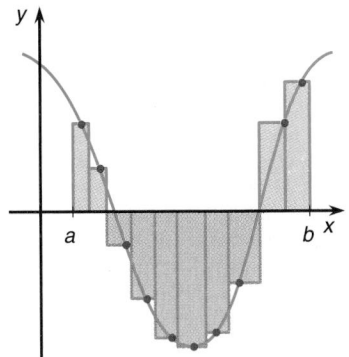

Figure 5.3.5

Note that $x_0 = 1$, $x_1 = 2$, $x_2 = 3$, $x_3 = 4$. Also $\Delta x_1 = x_1 - x_0 = 2 - 1$, $\Delta x_2 = x_2 - x_1 = 3 - 2$, etc.

If $f(x)$ is a continuous nonnegative function on $[a, b]$, we can evaluate $f(x_k^*)$ for each k; it is the height of the graph above the x-axis at the point x_k^*. We could try approximating the area under the graph of $y = f(x)$ by adding up the areas of the rectangles whose heights are $f(x_k^*)$ and whose bases are the intervals $[x_{k-1}, x_k]$. See Figure 5.3.4.

Different choices of $\{x_k^*\}$ will result in different sums, but after all they are only approximations.

DEFINITION 5.3.2

The **Riemann sum** for $f(x)$ associated to the partition

$$\mathcal{P} = \{x_0, \ldots, x_n\} \text{ of } [a, b]$$

and to the choices $x_k^* \in [x_{k-1}, x_k]$ is

$$S(f, \mathcal{P}, \{x_1^*, x_2^*, \ldots, x_n^*\}) = \sum_{k=1}^{n} f(x_k^*)\Delta x_k = \sum_{k=1}^{n} f(x_k^*)(x_k - x_{k-1}).$$

If $f(x) \geq 0$ for all $x \in [a, b]$, this sum is the total area of the rectangles with bases $[x_{k-1}, x_k]$ and heights $f(x_k^*)$ as illustrated in Figure 5.3.4. But the sum makes sense even if $f(x)$ is negative at some values of x. If $f(x_k^*) < 0$, then $f(x_k^*)\Delta x_k$ is the *negative* of the area of the rectangle. In fact this is what we want; we shall call it the **signed area** of the rectangle, meaning the negative of the area if the rectangle is below the x-axis and the actual area if it is above the x-axis. In Figure 5.3.5 the rectangles with positive signed area are shaded blue; those with negative signed area are shaded purple.

The Riemann sum $S(f, \mathcal{P}, \{x_1^*, \ldots, x_n^*\})$ is the sum of the signed areas of the rectangles with bases $[x_{k-1}, x_k]$ and "heights" $f(x_k^*)$.

EXAMPLE 2

(i) With $f(x) = 4x^2 - 1$ on $[1, 4]$ and the partition $\mathcal{P} = \{1, 2, 3, 4\}$ we choose the midpoints $x_1^* = \frac{3}{2} \in [1, 2]$, $x_2^* = \frac{5}{2} \in [2, 3]$, $x_3^* = \frac{7}{2} \in [3, 4]$, and the Riemann sum is

$$S(f, \mathcal{P}, \{x_k^*\}) = \sum_{k=1}^{3} f(x_k^*)\Delta x_k$$

$$= f\left(\frac{3}{2}\right)(2 - 1) + f\left(\frac{5}{2}\right)(3 - 2) + f\left(\frac{7}{2}\right)(4 - 3)$$

$$= \left(4\left(\frac{3}{2}\right)^2 - 1\right) + \left(4\left(\frac{5}{2}\right)^2 - 1\right) + \left(4\left(\frac{7}{2}\right)^2 - 1\right)$$

$$= \left(4\left(\frac{9}{4}\right) - 1\right) + \left(4\left(\frac{25}{4}\right) - 1\right) + \left(4\left(\frac{49}{4}\right) - 1\right)$$

$$= 8 + 24 + 48 = 80.$$

(ii) With $f(x) = \sin x$ on $[0, \pi]$, and the partition $\mathcal{P} = \{0, \frac{\pi}{4}, \frac{\pi}{2}, \frac{3\pi}{4}, \pi\}$, we choose the left endpoints of the subintervals

$$x_1^* = 0, \qquad x_2^* = \frac{\pi}{4}, \qquad x_3^* = \frac{\pi}{2}, \qquad x_4^* = \frac{3\pi}{4}.$$

The corresponding Riemann sum is

$$S(f, \mathcal{P}, \{x_k^*\}) = \sum_{k=1}^{4} f(x_k^*)\Delta x_k$$

$$= f(0)\left(\frac{\pi}{4} - 0\right) + f\left(\frac{\pi}{4}\right)\left(\frac{\pi}{2} - \frac{\pi}{4}\right) + f\left(\frac{\pi}{2}\right)\left(\frac{3\pi}{4} - \frac{\pi}{2}\right)$$

$$+ f\left(\frac{3\pi}{4}\right)\left(\pi - \frac{3\pi}{4}\right)$$

$$= \sin(0)\left(\frac{\pi}{4}\right) + \sin\left(\frac{\pi}{4}\right)\left(\frac{\pi}{4}\right) + \sin\left(\frac{\pi}{2}\right)\left(\frac{\pi}{4}\right)$$

$$+ \sin\left(\frac{3\pi}{4}\right)\left(\frac{\pi}{4}\right)$$

$$= 0 + \frac{1}{\sqrt{2}}\left(\frac{\pi}{4}\right) + (1)\left(\frac{\pi}{4}\right) + \frac{1}{\sqrt{2}}\left(\frac{\pi}{4}\right) = (1 + \sqrt{2})\frac{\pi}{4}.$$

(iii) Suppose $0 \leq a < b$ and let $\mathcal{P}_n$ be the partition that divides the interval $[a, b]$ into n equal subintervals. Let $f(x) = x^m$, with m a fixed positive integer.

If $x_k^* = x_{k-1}$ is the left endpoint of the kth subinterval $[x_{k-1}, x_k]$, let us calculate

$$S(f, \mathcal{P}_n, \{x_1^*, \ldots, x_n^*\}).$$

> The interval $[a, b]$ has length $b - a$, so each little subinterval has length $\frac{b-a}{n}$. The point $x_k = a + k\frac{b-a}{n}$ is k subintervals to the right of $x = a$.

Since x_k is the point that is k subintervals to the right of a, we see that $x_k = a + k\frac{b-a}{n}$. So $\mathcal{P}_n = \left\{a, a + \frac{b-a}{n}, a + 2\frac{b-a}{n}, a + 3\frac{b-a}{n}, \ldots, a + (n-1)\frac{b-a}{n}, b\right\}$. Since $\Delta x_k = \frac{b-a}{n}$, we find that

$$S(f, \mathcal{P}_n, \{x_1^*, \ldots, x_n^*\}) = \sum_{k=1}^{n} f(x_k^*)\Delta x_k = \sum_{k=1}^{n} f(x_{k-1})\frac{b-a}{n}$$

$$= \frac{b-a}{n}\sum_{k=1}^{n} f\left(a + (k-1)\frac{b-a}{n}\right)$$

$$= \frac{b-a}{n}\sum_{k=1}^{n}\left(a + (k-1)\frac{b-a}{n}\right)^m.$$

(iv) With $f(x) = x^m$ and $\mathcal{P}_n$ as in part (iii), we calculate the Riemann sum associated to the choice where $x_k^{**} = x_k$ is the right endpoint of the interval $[x_{k-1}, x_k]$. Here $x_k^{**} = x_k = a + k\frac{b-a}{n}$, so

$$S(f, \mathcal{P}_n, \{x_1^{**}, \ldots, x_n^{**}\}) = \sum_{k=1}^{n} f(x_k^{**})\Delta x_k = \frac{b-a}{n}\sum_{k=1}^{n}\left(a + k\frac{b-a}{n}\right)^m.$$

We expect that the sums in parts (iii) and (iv) of Example 2 should approximate the area under the graph of $y = f(x) = x^m$ between $x = a$ and $x = b$. Consider their difference,

$$S(f, \mathcal{P}_n, \{x_1^{**}, \ldots, x_n^{**}\}) - S(f, \mathcal{P}_n, \{x_1^*, \ldots, x_n^*\})$$

$$= \frac{b-a}{n}\sum_{k=1}^{n}\left(a + k\frac{b-a}{n}\right)^m - \frac{b-a}{n}\sum_{k=1}^{n}\left(a + (k-1)\frac{b-a}{n}\right)^m$$

A programmable calculator or a computer can be programmed to evaluate the sum in part (iv) of Example 2 for specific values of a and b. For example, we try it with $a = 0$ and $b = 2$.

TI-85	SHARP EL9300
PRGM F2(EDIT) RSUM ENTER F3 F1(Input) M ENTER M3 F1(Input) N EXIT ENTER 0 STO K ENTER 0 STO S ENTER Lbl B ENTER K+1 STO K ENTER S+(K × 2 ÷ N)^M STO S ENTER F4 F1(If) (K TEST F2(<) N) ENTER Goto B ENTER S × 2 ÷ N ENTER EXIT EXIT EXIT PRGM F1(NAME) RSUM ENTER 3 ENTER 10 ENTER	▼ ▼ (NEW) ENTER ENTER RSUM ENTER COMMAND A 3 (Input) M ENTER COMMAND A 3 (Input) N ENTER K = 0 ENTER S = 0 ENTER COMMAND B 1 (Label) 1 ENTER K = K + 1 ENTER S = S + (K × 2 ÷ N) a^b M ENTER COMMAND 3 (If) K COMMAND C 2 (<) N COMMAND B 2 (Goto) 1 ENTER S = S × 2 ÷ N ENTER COMMAND A 1 (Print) S ENTER COMMAND A 6 (End) ENTER (RUN) ENTER (RSUM) ENTER 3 ENTER 10 ENTER
CASIO f_x-6300G (f_x-7700GB)	HP 48SX
MODE 2 EXE ? → M : ? → N : 0 → K : 0 → S : Lbl 1 : K + 1 → K : S + (K × 2 ÷ N) x^y M → S : K < N ⇒ Goto 1 : S × 2 ÷ N ◢ MODE 1 Prog 0 EXE 3 EXE 10 EXE (on the f_x-7700GB, : is PRGM F6, and Prog is PRGM F3.)	ATTN ↱ CLR ≪≫ "'" KEY IN M AND N ▶ SPC PROMPT SPC ' N ▶ STO ' M ▶ STO 0 SPC 1 SPC N SPC FOR K SPC K SPC 2 × N ÷ M y^x + NEXT ▶ 2 × N ÷ ▶ ENTER ' RSUM STO VAR RSUM 3 ENTER 10 ENTER ↰ CONT

```
?3
?10
ANSWER=
                4.84
```

Now use the program with various other values of m and n. Can you guess a formula for $\int_0^2 x^m \, dx$?

$$= \frac{b-a}{n}\left(\sum_{k=1}^{n} \left(a + k\frac{b-a}{n} \right)^m - \sum_{k=1}^{n} \left(a + (k-1)\frac{b-a}{n} \right)^m \right).$$

Most of the terms in these last two sums cancel each other. We can see this by changing the second sum, using Proposition 5.2.4 (p. 357). Letting $\ell = k - 1$, we have

$$\sum_{k=1}^{n} \left(a + (k-1)\frac{b-a}{n} \right)^m = \sum_{\ell=0}^{n-1} \left(a + \ell\frac{b-a}{n} \right)^m.$$

With this in mind we can write the previous difference as

$$S\big(f, \mathcal{P}_n, \{ x_1^{**}, \ldots, x_n^{**} \} \big) - S\big(f, \mathcal{P}_n, \{ x_1^{*}, \ldots, x_n^{*} \} \big)$$

$$= \frac{b-a}{n}\left(\sum_{k=1}^{n}\left(a+k\frac{b-a}{n}\right)^{m} - \sum_{\ell=0}^{n-1}\left(a+\ell\frac{b-a}{n}\right)^{m}\right)$$

$$= \frac{b-a}{n}\left(\sum_{k=1}^{n-1}\left(a+k\frac{b-a}{n}\right)^{m} + \left(a+n\frac{b-a}{n}\right)^{m}\right.$$

$$\left. - \sum_{\ell=1}^{n-1}\left(a+\ell\frac{b-a}{n}\right)^{m} - \left(a+0\times\frac{b-a}{n}\right)^{m}\right).$$

We split off the term with $k = n$ from the first sum and the term with $\ell = 0$ from the second.

Since the two summations in this last formula are the same, they cancel, and we are left with

$$S\big(f, \mathcal{P}_n, \{x_1^{**}, \dots, x_n^{**}\}\big) - S\big(f, \mathcal{P}_n, \{x_1^{*}, \dots, x_n^{*}\}\big)$$

$$= \frac{b-a}{n}\left(\left(a+n\frac{b-a}{n}\right)^{m} - \left(a+0\frac{b-a}{n}\right)^{m}\right)$$

$$= \frac{b-a}{n}\big((a+b-a)^{m} - a^{m}\big)$$

$$= \frac{b-a}{n}(b^{m} - a^{m}).$$

In particular, if we let $n \to \infty$, this difference tends to zero, which says that the two different types of sums (left endpoints or right endpoints) get closer and closer together. This is what we expect, since we are hoping that each type of sum will tend to some limit, and both limits will be the same.

We want to say that the value of this common limit is the area under the graph, but the difficulty is that we have considered only two possible choices, $x_k^{*} = x_{k-1}$ and $x_k^{**} = x_k$ (left and right endpoints of each subinterval). For that matter, we have considered only the special partitions $\mathcal{P}_n$ that divide $[a, b]$ into equal subintervals. Different choices of partition or of points in each subinterval could conceivably lead to different limits (or to no limit at all).

What is needed is a way to consider every partition $\mathcal{P}$ and all possible choices $\{x_k^{*}\}$ of points in the intervals determined by $\mathcal{P}$ and to take a limit as the partitions become finer and finer, that is, as $\|\mathcal{P}\| \to 0$. This sort of limit is different from the ones we considered earlier. Roughly speaking, what it means is that for any partition composed of very small subintervals and any choice of points $\{x_k^{*}\}$ in those subintervals, the Riemann sum should be very close to the value of the limit.

As we did with ordinary limits in Chapter 2, we will not stop to discuss the technical details of this type of limit. Instead, we will proceed to see how it is used and present the precise definition at the end of this section.

DEFINITION 5.3.3

Suppose $f(x)$ is a function on the interval $[a, b]$. Consider all possible partitions $\mathcal{P}$ of $[a, b]$, and for each $\mathcal{P}$ consider all possible choices $\{x_k^{*}\}$ of a point in each subinterval determined by $\mathcal{P}$. For such a partition $\mathcal{P}$ and choices $\{x_k^{*}\}$ we form the Riemann sum

$$S\big(f, \mathcal{P}, \{x_k^{*}\}\big).$$

The function $f(x)$ is said to be **integrable** on $[a, b]$ if the following limit exists:

$$\lim_{\|\mathcal{P}\|\to 0} S\big(f, \mathcal{P}, \{x_k^{*}\}\big).$$

In this case the value of the limit is called the **integral** of $f(x)$ from a to b and is written

$$\int_a^b f(x)\,dx.$$

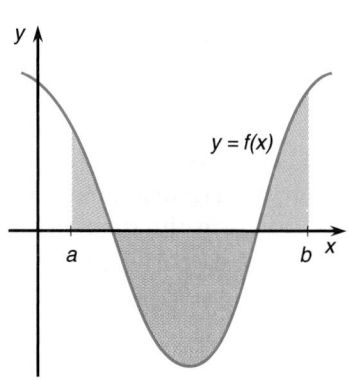

$y = f(x)$

Figure 5.3.6

The way we have defined the integral leads us to expect that $\int_a^b f(x)\,dx$ represents the signed area under the graph of $y = f(x)$ between $x = a$ and $x = b$. See Figure 5.3.6.

In this notation the number a is called the **lower limit of integration** and b is the **upper limit of integration**. They indicate the interval over which the integral is considered, and the word "limit" here means "endpoint" rather than limit in the sense of the previous definition or in the sense in which it was used in Chapter 2.

The integral sign $\int$ is a stylized capital S; the idea is that as the Riemann *sums* $\sum_{k=1}^n f(x_k^*)\Delta x_k$ tend to a limit, the stylized Greek letter $\sum$ (sigma) changes into the corresponding Roman letter S. Similarly, the notation Δx_k changes into dx. The function $f(x)$ is called the **integrand**.

EXAMPLE 3   Suppose $f(x) = c$ is a constant function. Show that $f(x)$ is integrable on any interval $[a, b]$, and find $\int_a^b f(x)\,dx$.

Solution   If $\mathcal{P} = \{x_0, x_1, \ldots, x_n\}$ is any partition and $x_k^* \in [x_{k-1}, x_k]$, then the corresponding Riemann sum is

> What makes this easy is that $f(x_k^*) = c$ for *any* point x_k^* . . .

$$\sum_{k=1}^n f(x_k^*)\Delta x_k = \sum_{k=1}^n c\Delta x_k = c\sum_{k=1}^n \Delta x_k.$$

Since $\sum_{k=1}^n \Delta x_k$ is the sum of the lengths of all the subintervals, it equals the total length of the interval $[a, b]$, or $b - a$.

This shows that every Riemann sum equals $c(b - a)$, so they certainly "tend" to this limit. In other words, $f(x) = c$ is integrable on $[a, b]$, and

$$\int_a^b c\,dx = c(b - a).$$

This is what we expect from our interpretation of the integral as a signed area. (See Figure 5.3.7, which illustrates what happens when $c > 0$ and when $c < 0$. In each case the area of the shaded region is $|c|(b - a)$, and the signed area is $c(b - a)$.)

Figure 5.3.7

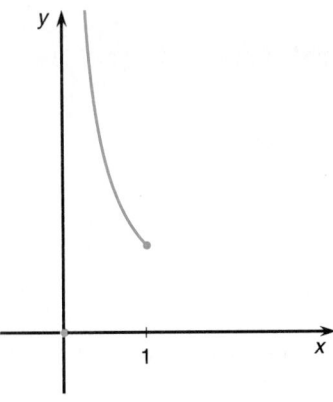

Figure 5.3.8

On the other hand, there are functions that are not integrable.

EXAMPLE 4   Define the function $g(x)$ on the interval $[0, 1]$ by

$$g(x) = \begin{cases} \frac{1}{x}, & \text{if } x > 0, \\ 0, & \text{if } x = 0 \end{cases}$$

(see Figure 5.3.8).

We will show that $g(x)$ is not integrable on $[0, 1]$. For each positive integer n, let $\mathcal{P}_n$ be the partition that divides $[0, 1]$ into n equal subintervals. The first subinterval determined by $\mathcal{P}_n$ is $\left[0, \frac{1}{n}\right]$. Define $\{x_k^*\}$, a choice of points in the subintervals determined by $\mathcal{P}_n$, by $x_1^* = \frac{1}{n^2}$, and for each $k > 1$, $x_k^* = \frac{k}{n}$. For each $k > 1$, x_k^* is the right endpoint of the kth subinterval. Notice that the point $x_1^* = \frac{1}{n^2}$ is in the first subinterval $\left[0, \frac{1}{n}\right]$.

The corresponding Riemann sum is

$$S\big(g, \mathcal{P}_n, \{x_k^*\}\big) = \sum_{k=1}^{n} g(x_k^*)\Delta x_k = g(x_1^*)\Delta x_1 + \sum_{k=2}^{n} g\left(\frac{k}{n}\right)\Delta x_k$$

$$= \frac{1}{\frac{1}{n^2}}\frac{1}{n} + \sum_{k=2}^{n} \frac{n}{k}\frac{1}{n}$$

$$= n + \sum_{k=2}^{n} \frac{1}{k}.$$

Without actually evaluating it, we see that the final sum must be positive. This shows that the Riemann sum is greater than n. Now if $g(x)$ were integrable, we would know that as $n \to \infty$, $\|\mathcal{P}_n\| = \frac{1}{n} \to 0$, so the Riemann sums $S\big(g, \mathcal{P}_n, \{x_k^*\}\big)$ would have to approach a limit (the value of the integral). But since the nth sum is bigger than n, they cannot approach any (finite) limit. This shows that the function $g(x)$ is not integrable on $[0, 1]$. ◢

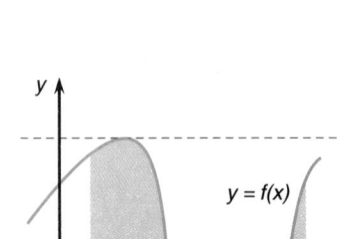

Figure 5.3.9

The essential fact about $g(x)$ that prevented it from being integrable is that it is not bounded on $[0, 1]$. In fact it is also possible to find bounded functions that are not integrable.

Example 4 shows that there are functions that are not integrable. The important thing is that most "reasonable" functions are in fact integrable, which means that it makes sense to speak of the "area" under the graph of the function.

In particular, if $f(x)$ is continuous, then by Theorem 2.5.6 it is *bounded*, and our intuition suggests that the graph encloses a geometrical region (see Figure 5.3.9). The region may be quite complicated, but it seems reasonable to expect that its area is something we should be able to calculate. Calculating the (signed) area can be overwhelmingly difficult in practice, but it is possible in principle, in the sense that a continuous function on a closed interval is always integrable.

THEOREM 5.3.4

If $f(x)$ is a continuous function on a closed interval $[a, b]$, then $f(x)$ is integrable on $[a, b]$.

This theorem is of great importance, but it is extremely difficult to prove. Notice that it does not tell us anything about how to calculate the value of the integral

$$\int_a^b f(x)\, dx.$$

Nonetheless, it does give us enough information that we can do the calculations for a few special cases.

EXAMPLE 5   Suppose $a < b$.

(i) Find $\int_a^b x\, dx$, that is, find the integral from a to b of the function $g(x) = x$.

(ii) Find $\int_a^b x^2 dx$, that is, find the integral from a to b of the function $f(x) = x^2$.

Solution

(i) Theorem 5.3.4 tells us that $g(x) = x$ is integrable over any interval $[a, b]$. What this means is that Riemann sums $S(g, \mathcal{P}, \{x_k^*\})$ tend to a limit provided that $\|\mathcal{P}\| \to 0$, and the limit equals $\int_a^b x\, dx$.

So to calculate the integral, we have only to find this limit for some convenient partitions and choices of points $\{x_k^*\}$. In other words, knowing that the Riemann sums converge to the same limit, no matter what the partitions, we do not have to find the limit for every possible partition. It will be enough to find it for some intelligently chosen partitions.

Not surprisingly, we choose the partitions that divide $[a, b]$ into subintervals of equal length. Let $\mathcal{P}_n$ divide $[a, b]$ into n equal subintervals. For the points in the subintervals we choose the right endpoints, that is,

$$x_k^* = a + k\frac{b-a}{n}.$$

The corresponding Riemann sum is

$$S\big(g, \mathcal{P}_n, \{x_k^*\}\big) = \sum_{k=1}^{n} g(x_k^*)\Delta x_k = \sum_{k=1}^{n}\left(a + k\frac{b-a}{n}\right)\frac{b-a}{n}$$

$$= \frac{b-a}{n}\sum_{k=1}^{n}\left(a + k\frac{b-a}{n}\right) = \frac{b-a}{n}\left(a\sum_{k=1}^{n}1 + \frac{b-a}{n}\sum_{k=1}^{n}k\right)$$

$$= \frac{b-a}{n}\left(an + \frac{b-a}{n}\frac{n(n+1)}{2}\right)$$

$$= (b-a)a + (b-a)^2\frac{n+1}{2n}.$$

> Note: $\sum_{k=1}^{n} 1 = n$ and $\sum_{k=1}^{n} k = \frac{n(n+1)}{2}$.

Now we let $n \to \infty$ in this last formula. Since $\frac{n+1}{n} \to 1$, we see that the Riemann sums tend to the limit

$$(b-a)a + (b-a)^2\frac{1}{2} = ba - a^2 + \frac{1}{2}(b^2 - 2ab + a^2)$$

$$= \frac{1}{2}(b^2 - a^2).$$

What we have just shown is that

$$\int_a^b x\, dx = \frac{1}{2}(b^2 - a^2).$$

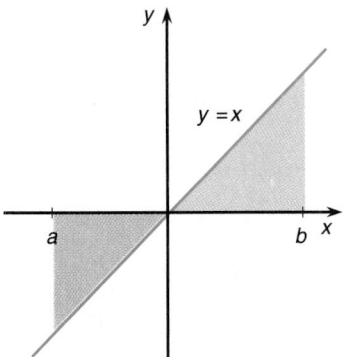

Figure 5.3.10

$\sum_{k=1}^{n} 1$ is obtained by adding n 1's, so it equals n. We have also used Formula 5.2.5 and the result of Example 5.2.5.

You might like to compare this with the signed area under the graph of $y = x$ between $x = a$ and $x = b$ (see Figure 5.3.10).

(ii) Following a similar procedure, using the same partition $\mathcal{P}_n$ and the same points $\{x_k^*\}$, we calculate the Riemann sum

$$S(f, \mathcal{P}_n, \{x_k^*\}) = \sum_{k=1}^{n} f(x_k^*)\Delta x_k = \sum_{k=1}^{n}\left(a + k\frac{b-a}{n}\right)^2 \frac{b-a}{n}$$

$$= \frac{b-a}{n}\sum_{k=1}^{n}\left(a^2 + 2ak\frac{b-a}{n} + k^2\left(\frac{b-a}{n}\right)^2\right)$$

$$= \frac{b-a}{n}\left(a^2\sum_{k=1}^{n}1 + 2a\frac{b-a}{n}\sum_{k=1}^{n}k + \left(\frac{b-a}{n}\right)^2\sum_{k=1}^{n}k^2\right)$$

$$= \frac{b-a}{n}\left(a^2 n + 2a\frac{b-a}{n}\frac{n(n+1)}{2}\right.$$

$$\left. + \left(\frac{b-a}{n}\right)^2\frac{n(n+1)(2n+1)}{6}\right)$$

$$= (b-a)a^2 + a(b-a)^2\frac{n+1}{n}$$

$$+ \frac{(b-a)^3}{6}\frac{(n+1)(2n+1)}{n^2}.$$

Now we let $n \to \infty$ in this last formula. Since $\frac{n+1}{n} \to 1$ and $\frac{(n+1)(2n+1)}{n^2} = \frac{n+1}{n}\frac{2n+1}{n} \to (1)(2) = 2$, we see that the Riemann sums tend to the limit

$$(b-a)a^2 + a(b-a)^2 + \frac{1}{3}(b-a)^3$$

$$= a^2 b - a^3 + ab^2 - 2a^2 b + a^3 + \frac{1}{3}\left(b^3 - 3ab^2 + 3a^2 b - a^3\right)$$

$$= \frac{1}{3}\left(b^3 - a^3\right).$$

What we have just shown is that

$$\int_a^b x^2\,dx = \frac{1}{3}(b^3 - a^3).$$

See Figure 5.3.11.

Unfortunately, this technique is too cumbersome to be very practical. We will have to find better ways to calculate integrals if they are to be of much use to us.

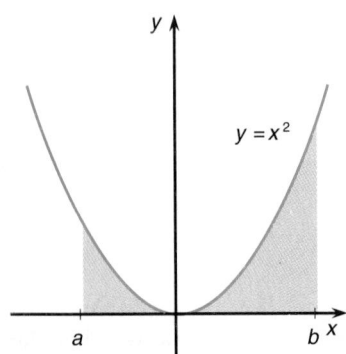

Figure 5.3.11

The Precise Definition of the Integral

Definition 5.3.3 of the integral of $f(x)$ over $[a, b]$ used a special kind of limit,

$$\lim_{\|\mathcal{P}\| \to 0} S(f, \mathcal{P}, \{x_k^*\}).$$

The idea was to consider all possible partitions $\mathcal{P}$ of $[a, b]$ and all possible choices $\{x_k^*\}$ of a point in each of the corresponding subintervals and to take a limit as the mesh $\|\mathcal{P}\|$ of the partitions tends to zero.

This seems a reasonable thing to do, but it is certainly different from the limits we considered previously, and we never specified precisely what it means.

DEFINITION 5.3.5

If $f(x)$ is a function on the interval $[a, b]$, then to say

$$\lim_{\|\mathcal{P}\| \to 0} S(f, \mathcal{P}, \{x_k^*\}) = L$$

means the following:

Given any $\epsilon > 0$, it is possible to find a $\delta > 0$ so that for every partition $\mathcal{P}$ of $[a, b]$ with $\|\mathcal{P}\| < \delta$ and for every choice of $x_k^* \in [x_{k-1}, x_k]$, the kth subinterval corresponding to $\mathcal{P}$, we have

$$\left| S(f, \mathcal{P}, \{x_k^*\}) - L \right| < \epsilon.$$

In this case, $f(k)$ is integrable on $[a, b]$ and $\int_a^b f(x)\,dx = L$.

The idea is that no matter how close we want the Riemann sum to be to the limit L, as measured by ϵ, it is possible to find a measure δ of the mesh of partitions so that any Riemann sum corresponding to a partition whose mesh is smaller than δ must be close enough to L. In other words, it is possible to ensure that Riemann sums are as close as we want to L by requiring that the mesh of the partition is sufficiently small.

There is a close analogy between this definition and the precise definition of ordinary limits (Definition 2.7.1, p. 120). But as with ordinary limits, the precise definition is not needed for us to be able to work with integrals.

One reason for this, as we saw in Example 5, is that once we know the limit exists, it is not necessary to consider all partitions and choices $\{x_k^*\}$ to calculate it. And for continuous functions we already know that the limit exists, by Theorem 5.3.4.

In fact, we will soon learn techniques for calculating integrals that do not require Riemann sums at all.

Exercises 5.3

I

In each of the following situations, let $\mathcal{P}_n$ be the partition that divides the specified interval into n subintervals of equal length, where n is the given number. Let x_k^* be the left endpoint of the kth subinterval determined by $\mathcal{P}_n$, let $f(x)$ be the given function, and calculate the Riemann sum $S(f, \mathcal{P}_n, \{x_k^*\})$.

1. $f(x) = x$ on $[0, 1]$; $n = 4$
2. $f(x) = x$ on $[-1, 0]$; $n = 3$
3. $f(x) = x^2$ on $[0, 1]$; $n = 4$
4. $f(x) = x$ on $[-1, 2]$; $n = 3$
5. $f(x) = x + 3$ on $[0, 4]$; $n = 4$
6. $f(x) = 2x$ on $[-1, 2]$; $n = 3$
7. $f(x) = x^3 - 2$ on $[-2, 3]$; $n = 5$
8. $f(x) = 7$ on $[-5, 9]$; $n = 13$

9. $f(x) = -x$ on $[0, 1]$; $n = 4$
10. $f(x) = \frac{1}{x}$ on $[1, 6]$; $n = 5$

In each of the following questions, let $\mathcal{P}_n$ be the partition that divides the specified interval into n subintervals of equal length, where n is the given number. Let x_k^{**} be the right endpoint of the kth subinterval determined by $\mathcal{P}_n$, let $f(x)$ be the given function, and calculate the Riemann sum $S(f, \mathcal{P}_n, \{x_k^{**}\})$.

11. $f(x) = \sin x$ on $[0, \pi]$; $n = 2$
12. $f(x) = \cos(\pi x)$ on $[-1, 1]$; $n = 4$
13. $f(x) = \frac{1}{1+x^2}$ on $[0, 3]$; $n = 3$
14. $f(x) = \frac{1}{1+x^2}$ on $[-3, 0]$; $n = 3$
15. $f(x) = 0$ on $[-3, 7]$; $n = 5$
16. $f(x) = x^2 - 3x + 1$ on $[-2, 2]$; $n = 4$

17. $f(x) = x + \sin x$ on $[-\pi, \pi]$; $n = 4$

18. $f(x) = \frac{1}{x}$ on $[-2, 2]$; $n = 3$

19. $f(x) = e^x$ on $[-1, 2]$; $n = 3$

20. $f(x) = \sin^2 x$ on $[-\pi, \pi]$; $n = 2$

II

21. Use Theorem 5.3.4 (p. 368) to show that the function $g(x) = \frac{1}{x}$ is integrable on $\left[\frac{1}{2}, 1\right]$.

*22. Show that any rational function $R(x) = \frac{p(x)}{q(x)}$ is integrable on any interval $[a, b]$ on which $q(x)$ is never zero.

23. Draw a picture and use your knowledge of geometry (triangles and rectangles) to calculate the signed area between the straight line $y = x$ and the x-axis between $x = a$ and $x = b$. (Distinguish three cases: $0 \le a \le b$, $a \le 0 \le b$, and $a \le b \le 0$.) Compare your answer with the result of Example 5 (p. 369).

*24. Let $f(x) = cx + d$ be the equation of a straight line; assume $c \ne 0$. Let $a < b$, and consider the interval $[a, b]$.

   (i) If $\mathcal{P}_n$ is the partition that divides $[a, b]$ into n subintervals of equal length, let $x_k^* = a + \frac{k-1}{n}(b - a)$ be the left endpoint of the kth subinterval and x_k^{**} be the right endpoint. Calculate the Riemann sums corresponding to these two choices, namely,

   $$S\left(f, \mathcal{P}_n, \{x_k^*\}\right) \quad \text{and} \quad S\left(f, \mathcal{P}_n, \{x_k^{**}\}\right).$$

   (ii) Find the limit as $n \to \infty$ of each of these two Riemann sums.

   (iii) Suppose $\{x_k^{***}\}$ is *any* choice of points in the subintervals determined by $\mathcal{P}_n$. Show that the corresponding Riemann sum lies between

   $$S\left(f, \mathcal{P}_n, \{x_k^*\}\right) \quad \text{and} \quad S\left(f, \mathcal{P}_n, \{x_k^{**}\}\right).$$

   (*Hint:* First suppose $c > 0$, and notice that this means that $f(x)$ is an increasing function. Draw a picture.)

   (iv) Show that no matter what points $\{x_k^{***}\}$ we choose in the subintervals determined by $\mathcal{P}_n$, the Riemann sums

   $$S\left(f, \mathcal{P}_n, \{x_k^{***}\}\right)$$

   converge to a limit. Calculate the limit.

   (v) Notice that to show that $f(x)$ is integrable on $[a, b]$, we would have to show that the Riemann sums associated to *every* possible partition converge to a limit, not just the ones associated to the special partitions $\mathcal{P}_n$. However, since Theorem 5.3.4 allows us to conclude that $f(x)$, being continuous, is integrable on $[a, b]$, we know that they must converge, and from part (iv) we know

the value of the limit. What is the value of

$$\int_a^b f(x)\,dx,$$

and how does it compare with the signed area under the graph? (*Hint:* To calculate the signed area geometrically, consider separately the parts of the graph above and below the x-axis.)

25. Let $f(x) = x$, $g(x) = x^3 - x$, $h(x) = x^5 - 2x^3$. Each of these functions is an odd function. We wish to work out various Riemann sums for these functions. For $f(x)$ we will consider the interval $[-1, 1]$; for $g(x)$ we will use the interval $[-2, 2]$; and for $h(x)$ we will use the interval $[-4, 4]$. In each case, let $\mathcal{P}_n$ be the partition that divides the specified interval into n equal subintervals, and let $\{x_k^*\}$ be the *midpoint* of the kth subinterval.

   (i) Calculate the Riemann sums for $f(x)$, $g(x)$, and $h(x)$ on $[-1, 1]$, $[-2, 2]$, and $[-4, 4]$, respectively, using the partition $\mathcal{P}_n$ and the points $\{x_k^*\}$, with $n = 2$.

   (ii) Repeat part (i) with $n = 4$.

   (iii) Repeat part (i) with $n = 2$ but using the left endpoints of the subintervals rather than the midpoints as the choices $\{x_k^*\}$.

**26. (i) After doing the previous exercise, show that if $f(x)$ is any odd function that is integrable on the interval $[-1, 1]$, then

   $$\int_{-1}^{1} f(x)\,dx = 0.$$

   (ii) If a is any positive number and $f(x)$ is any odd function that is integrable on the interval $[-a, a]$, show that

   $$\int_{-a}^{a} f(x)\,dx = 0.$$

27. Find $\|\mathcal{P}_3\|$ and $\|\mathcal{P}_4\|$, where $\mathcal{P}_3$ and $\mathcal{P}_4$ are as in Example 1 (p. 362).

**28. Let $f(x)$ be the function that is zero except at $x = 1$, where it equals 1 (see Figure 5.3.12). In symbols,

$$f(x) = \begin{cases} 0, & \text{if } x \ne 1, \\ 1, & \text{if } x = 1. \end{cases}$$

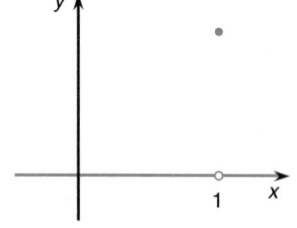

Figure 5.3.12

Show that $f(x)$ is integrable on any interval $[a, b]$. (*Hint:* It is easy if $1 \notin [a, b]$. If $1 \in [a, b]$, notice that in any Riemann sum, all the terms are zero except possibly one or two.)

29. Let $g(x)$ be the function defined as follows:

$$g(x) = \begin{cases} 0, & \text{if } x \le 0, \\ 1, & \text{if } x > 0. \end{cases}$$

Show that $g(x)$ is integrable on any interval $[a, b]$ (see Figure 5.3.13).

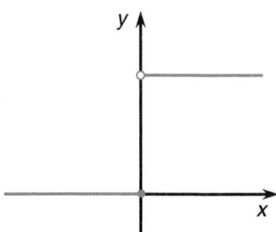

Figure 5.3.13

30. Assuming that the function $g(x)$ of the previous question is integrable, evaluate

$$\int_{-1}^{1} g(x)\,dx.$$

If you have a programmable calculator or can program a computer, make up a program that computes

the Riemann sum for some specified function $f(x)$ on an interval $[a, b]$ relative to the partition $\mathcal{P}_n$ that divides $[a, b]$ into n equal subintervals, using the right endpoints as the chosen points $\{x_k^*\}$. Use your program (or one from a software package) to evaluate the Riemann sum $S(f, \mathcal{P}_n, \{x_k^*\})$ in each of the following cases (x_k^* always refers to right endpoints).

31. $f(x) = x^4 - 3x$ on $[0, 2]$; $n = 20$

32. $f(x) = e^{2x}$ on $[-1, 1]$; $n = 10$

33. $f(x) = \tan x$ on $\left[0, \frac{\pi}{4}\right]$; $n = 15$

34. $f(x) = \frac{1}{x}$ on $[1, 2]$; $n = 25$.

35. If you have suitable software, evaluate some Riemann sums for the functions $f(x) = x^m$ on the interval $[0, 1]$, where m is a positive integer and x_k^* is the right endpoint of the kth subinterval.

(i) Evaluate the Riemann sums $S(f, \mathcal{P}_n, \{x_k^*\})$ for $m = 1, 2, 3, 4, 5$, with $n = 4, 10, 50$.

(ii) Using the information from part (i), try to see a pattern and try to guess the limits. In other words, try to guess the value of

$$\int_{0}^{1} x^m\,dx.$$

(*Hint:* It might be easier to guess the reciprocal,

$$\frac{1}{\int_{0}^{1} x^m\,dx}.)$$

5.4 Properties of Integrals

INTEGRAL MEAN VALUE
THEOREM

JUMP/REMOVABLE/INFINITE
DISCONTINUITIES

We can establish some basic properties of integrals immediately.

THEOREM 5.4.1

Suppose $f(x)$ and $g(x)$ are integrable on $[a, b]$. Then

(i) for any constant K the function $Kf(x)$ is also integrable on $[a, b]$, and

$$\int_{a}^{b} Kf(x)\,dx = K \int_{a}^{b} f(x)\,dx.$$

(ii) $f(x) + g(x)$ is integrable on $[a, b]$, and

$$\int_{a}^{b} \big(f(x) + g(x)\big)\,dx = \int_{a}^{b} f(x)\,dx + \int_{a}^{b} g(x)\,dx.$$

PROOF   For part (i), notice that any Riemann sum $\sum_{k=1}^{n} Kf(x_k^*)\Delta x_k$ for $Kf(x)$ can be rewritten as $K\big(\sum_{k=1}^{n} f(x_k^*)\Delta x_k\big)$, which is K times a Riemann sum for $f(x)$. Since the Riemann sums for $f(x)$ tend to the limit $\int_a^b f(x)\,dx$ as $\|\mathcal{P}\| \to 0$, we see that the Riemann sums for $Kf(x)$ tend to $K \int_a^b f(x)\,dx$. This shows both that $Kf(x)$ is integrable and that its integral has the right value.

For part (ii), note that any Riemann sum for $f(x) + g(x)$ is

$$\sum_{k=1}^n \left(f(x_k^*) + g(x_k^*) \right) \Delta x_k = \sum_{k=1}^n f(x_k^*) \Delta x_k + \sum_{k=1}^n g(x_k^*) \Delta x_k,$$

which is the sum of a Riemann sum for $f(x)$ and one for $g(x)$ (with the same partition). Since Riemann sums for $f(x)$ and for $g(x)$ tend to the limits $\int_a^b f(x)\,dx$ and $\int_a^b g(x)\,dx$, respectively, as $\|\mathcal{P}\| \to 0$, their sum tends to $\int_a^b f(x)\,dx + \int_a^b g(x)\,dx$. So Riemann sums for $f(x) + g(x)$ converge to this limit as $\|\mathcal{P}\| \to 0$, proving (ii). ◢

THEOREM 5.4.2

Suppose $a < b$ and $f(x)$ is a nonnegative function that is integrable on $[a, b]$. Then its integral is also nonnegative:

$$\int_a^b f(x)\,dx \geq 0.$$

PROOF For any partition $\mathcal{P} = \{x_0, x_1, x_2, \ldots, x_n\}$ and any choice $x_k^* \in [x_{k-1}, x_k]$, consider the corresponding Riemann sum

$$\sum_{k=1}^n f(x_k^*) \Delta x_k.$$

Each term in the sum is nonnegative; after all, $f(x) \geq 0$ for every x, so $f(x_k^*) \geq 0$, and Δx_k, the length of the kth subinterval, is positive.

Every Riemann sum is nonnegative, so the integral, which is a limit of Riemann sums, must be nonnegative. ◢

What this theorem says seems geometrically obvious: If $f(x)$ lies above the x-axis on $[a, b]$, then the corresponding signed area is nonnegative (see Figure 5.4.1). Perhaps it is better to say that the theorem confirms that this property of integrals is *consistent* with familiar properties of areas.

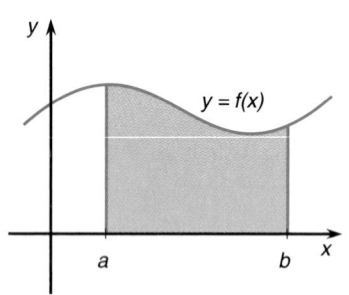

Figure 5.4.1

THEOREM 5.4.3

Suppose $f(x)$ is integrable on $[a, b]$ and M is a number.

(i) Suppose $f(x) \leq M$ for every $x \in [a, b]$. Then

$$\int_a^b f(x)\,dx \leq M(b - a).$$

(ii) Suppose $M \leq f(x)$ for every $x \in [a, b]$. Then

$$M(b - a) \leq \int_a^b f(x)\,dx.$$

(iii) Suppose $|f(x)| \leq M$ for every $x \in [a, b]$. Then

$$\left| \int_a^b f(x)\,dx \right| \leq M(b - a).$$

PROOF   It is possible to prove these results using Riemann sums directly, but it is easier to use the preceding result.

To prove part (i), observe that since $f(x) \leq M$, we can subtract $f(x)$ from both sides to find $0 \leq M - f(x)$, that is, $M - f(x) \geq 0$ for all $x \in [a, b]$. By Theorem 5.4.2,

$$\int_a^b \big(M - f(x)\big) dx \geq 0.$$

But $\int_a^b \big(M - f(x)\big) dx = \int_a^b M dx - \int_a^b f(x) dx$, by Theorem 5.4.1 (ii), and $\int_a^b M dx = M(b - a)$, by Example 5.3.3 (p. 367). So

$$0 \leq \int_a^b \big(M - f(x)\big) dx = \int_a^b M dx - \int_a^b f(x) dx$$

$$= M(b - a) - \int_a^b f(x) dx.$$

Adding $\int_a^b f(x) dx$ to both sides, we find

$$\int_a^b f(x) dx \leq M(b - a),$$

as needed.

Similarly, for part (ii) we know that $M \leq f(x)$, so $f(x) - M \geq 0$ for all $x \in [a, b]$. Accordingly,

$$0 \leq \int_a^b \big(f(x) - M\big) dx = \int_a^b f(x) dx - \int_a^b M dx$$

$$= \int_a^b f(x) dx - M(b - a).$$

Adding $M(b - a)$ to both sides gives

$$M(b - a) \leq \int_a^b f(x) dx.$$

(We could also have proved (ii) by applying part (i) to $-f(x)$ and $-M$.)

For part (iii) we notice that $|f(x)| \leq M$ is exactly the same as $-M \leq f(x) \leq M$. Using (i) and (ii), we conclude that

$$-M(b - a) \leq \int_a^b f(x) dx \leq M(b - a).$$

Then we notice that this is exactly equivalent to the conclusion of the theorem:

$$\left| \int_a^b f(x) dx \right| \leq M(b - a).$$

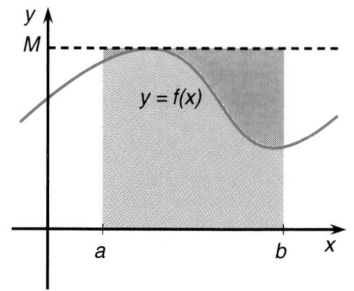

Figure 5.4.2(i)

If $f(x) \geq 0$ for all $x \in [a, b]$, we can think in terms of areas, and part (i) of Theorem 5.4.3 makes the very reasonable statement that the area under the graph of $y = f(x)$ cannot be more than the area of the larger region under the horizontal line $y = M$. After all, the area under this line between $x = a$ and $x = b$ is precisely $M(b - a)$ (see Figure 5.4.2(i)).

In the case in which $f(x)$ may not always be nonnegative, part (iii) says that the signed area cannot be bigger than $M(b - a)$ nor smaller than $-M(b - a)$. If the signed area is positive, it cannot be greater than the area under the horizontal line

Figure 5.4.2(ii)

Figure 5.4.3

Figure 5.4.4

Figure 5.4.5

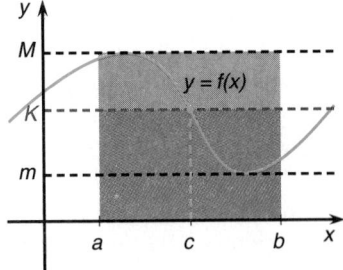

Figure 5.4.6

$y = M$. If it is negative, it cannot be less than the signed area above the line $y = -M$ (which is negative). See Figure 5.4.2(ii).

There will be times when we are unable to evaluate some integral exactly, but knowing a restriction on its magnitude will still be useful. The importance of Theorem 5.4.3 is that it restricts the size of the integral without our having to evaluate it.

Next we are going to prove a Mean Value Theorem for integrals. Suppose $f(x)$ is a continuous function on $[a, b]$. Then by Theorem 2.5.7 (p. 109) it has maximum and minimum values on $[a, b]$. Let M be the maximum value and m the minimum value of $f(x)$ on $[a, b]$, as shown in Figure 5.4.3.

The preceding theorem says that $\int_a^b f(x)\,dx$ lies between $m(b - a)$ and $M(b - a)$, which are just the signed areas determined by the horizontal lines $y = m$ and $y = M$. In the figure we have drawn m and M both positive, but this is not necessary.

The integral of $f(x)$ is some number lying between the signed areas of the lower rectangle and the upper rectangle (see Figure 5.4.4). We can think of it as the signed area of some intermediate rectangle, as shown in Figure 5.4.5. We can let K be the height of this intermediate rectangle and observe that of course K lies between m and M, the maximum and minimum values of $f(x)$.

It makes sense to think of K as the average or mean value of $f(x)$. Because $f(x)$ takes on the values m and M somewhere in $[a, b]$, the Intermediate Value Theorem (Theorem 2.5.5, p. 108) says that there is some $c \in [a, b]$ so that $f(c) = K$. In other words, there is some point in the interval $[a, b]$ at which $f(x)$ takes on its average value (see Figure 5.4.6).

This is what is known as the Integral Mean Value Theorem. The argument we have just given suggests where the idea comes from, but it relied on our geometrical intuition to find the "intermediate rectangle." Below we give a proof that is less subjective, though it is perhaps harder to understand.

THEOREM 5.4.4

The Integral Mean Value Theorem

Suppose $f(x)$ is continuous on $[a, b]$. Then there is some $c \in [a, b]$ so that

$$\int_a^b f(x)\,dx = f(c)(b - a).$$

PROOF Let M and m be the maximum and minimum values, respectively, of $f(x)$ on $[a, b]$. We know that they exist by Theorem 2.5.7. In particular, $m \le f(x) \le M$ for all $x \in [a, b]$, so by Theorem 5.4.3,

$$m(b - a) \le \int_a^b f(x)\,dx \le M(b - a).$$

Dividing by the positive number $b - a$ does not change the inequalities, so we see that

$$m \le \frac{1}{b - a} \int_a^b f(x)\,dx \le M.$$

This shows that the number $K = \frac{1}{b-a} \int_a^b f(x)\,dx$ lies between m and M, that is, $m \le K \le M$. Since there are points in the interval $[a, b]$ at which $f(x)$ equals m and M, the Intermediate Value Theorem (Theorem 2.5.5) tells us that there is a number $c \in [a, b]$ so that $K = f(c)$.

This means that $\frac{1}{b-a} \int_a^b f(x)\,dx = f(c)$, which is the same thing as

$$\int_a^b f(x)\,dx = f(c)(b - a),$$

which is the desired result.

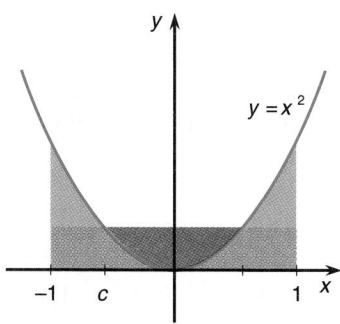

EXAMPLE 1   For the function $f(x) = x^2$ on the interval $[-1, 1]$ we can evaluate the integral using Example 5.3.5 (p. 369). We calculate $\int_{-1}^1 x^2\,dx = \frac{1}{3}\left(1^3 - (-1)^3\right) = \frac{2}{3}$. The length of the interval is 2, and the theorem says that there must be a point c so that $f(c)(b - a) = \int_{-1}^1 x^2\,dx$, that is, so that $f(c)(2) = \frac{2}{3}$, that is, $f(c) = \frac{1}{3}$.

Since $f(x) = x^2$, it is obvious in this case that c can be either $\frac{1}{\sqrt{3}}$ or $-\frac{1}{\sqrt{3}}$. Each of these points satisfies the condition given in Theorem 5.4.4, and both of them are in the interval $[-1, 1]$, so we have verified the theorem in this easy case. See Figure 5.4.7.

Figure 5.4.7

In a sense this example is misleading because it was relatively simple to find the necessary point. The theorem is more important in a situation in which it is not easy to find the point because it assures us that there is such a point, even though it may be extremely difficult to find explicitly.

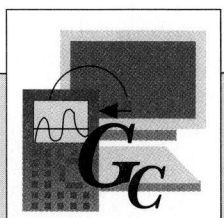

Before long we will be able to find that $\int_{-1}^2 (x^3 - 2x)\,dx = \frac{3}{4}$. The Integral Mean Value Theorem tells us there is a $c \in [-1, 2]$ so that $f(c)(3) = \frac{3}{4}$, that is, $f(c) = \frac{1}{4}$. (Here $f(x)$ is the integrand, $f(x) = x^3 - 2x$.) Using a graphics calculator, we sketch the graphs of $y = f(x)$ and the constant function $y = \frac{1}{4}$.

TI-85	SHARP EL9300
GRAPH F1$(y(x) =)$ F1(x) ^3 $-$ 2 F1(x) ▼ .25 M2(RANGE) $(-)$ 1 ▼ 2 ▼ ▼ $(-)$ 2 ▼ 4 F5(GRAPH)	⌁⌐ x/θ/T a^b 3 ► $-$ 2 x/θ/T 2ndF ▼ .25 RANGE -1 ENTER 2 ENTER 1 ENTER -2 ENTER 4 ENTER 1 ENTER ⌁⌐
CASIO f_x-7700GB/f_x-6300G	HP 48SX
Cls EXE Range -1 EXE 2 EXE 1 EXE -2 EXE 4 EXE 1 EXE EXE EXE EXE Graph ▨ x^y 3 $-$ 2 ▨ EXE Graph .25 EXE	▨ PLOT PLOTR ERASE ATTN ' Y ▨ X y^x 3 $-$ 2 × X ↱ PLOT ▨ DRAW ERASE 1 +/− ENTER 2 XRNG 2 +/− ENTER 4 YRNG DRAW ATTN ' Y ▨ .25 ↱ PLOT ▨ DRAW DRAW

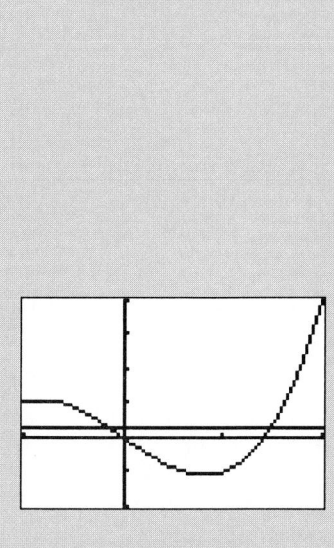

By zooming in (or using the SOLVE feature), we find that $c \approx -0.126\,000\,193$ or $c \approx 1.472\,997\,601$. (Note that there is also a solution to the left of $x = -1$, but it is not in the required interval. . .).

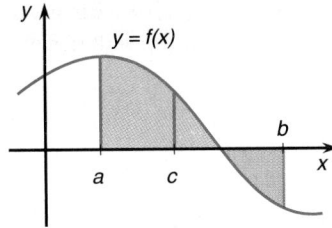

Figure 5.4.8

Suppose $f(x)$ is a function on $[a, b]$ and c is some point between a and b, that is, $a < c < b$. Then we can think of $f(x)$ as a function on the interval $[a, c]$ or on the interval $[c, b]$.

Now suppose $f(x)$ is integrable on $[a, c]$ and on $[c, b]$. It seems reasonable to expect that the total of the signed areas determined by the graph of $y = f(x)$ between $x = a$ and $x = b$ will equal the sum of the signed areas between a and c and the signed areas between c and b (see Figure 5.4.8). This is in fact true. In particular, the function $f(x)$ is integrable on $[a, b]$, and its integral equals the sum of the two smaller integrals.

THEOREM 5.4.5

If $a < c < b$ and $f(x)$ is integrable on $[a, c]$ and on $[c, b]$, then $f(x)$ is integrable on $[a, b]$, and

$$\int_a^b f(x)\, dx = \int_a^c f(x)\, dx + \int_c^b f(x)\, dx.$$

EXAMPLE 2

(i) The continuous function $f(x) = x^2$ is integrable on any closed interval. For instance, it is integrable on both $[-1, 1]$ and $[1, 3]$. Using Example 5.3.5 we can evaluate the corresponding integrals:

$$\int_{-1}^1 x^2\, dx = \frac{1}{3}\left(1^3 - (-1)^3\right) = \frac{2}{3},$$

$$\int_1^3 x^2\, dx = \frac{1}{3}\left(3^3 - 1^3\right) = \frac{26}{3}.$$

On the other hand, we can also evaluate

$$\int_{-1}^3 x^2\, dx = \frac{1}{3}\left(3^3 - (-1)^3\right) = \frac{28}{3}.$$

It is comforting to see that Theorem 5.4.5 is verified in this case:

$$\int_{-1}^1 x^2\, dx + \int_1^3 x^2\, dx = \frac{2}{3} + \frac{26}{3} = \frac{28}{3} = \int_{-1}^3 x^2\, dx.$$

(ii) In fact, if $a < c < b$, we see that

$$\int_a^c x^2\, dx + \int_c^b x^2\, dx = \frac{1}{3}(c^3 - a^3) + \frac{1}{3}(b^3 - c^3)$$

$$= \frac{1}{3}(c^3 - a^3 + b^3 - c^3) = \frac{1}{3}(b^3 - a^3)$$

$$= \int_a^b x^2\, dx.$$

This confirms Theorem 5.4.5 for any $a < c < b$ for the specific function $f(x) = x^2$. We were able to do this because we knew how to evaluate the integral of x^2 by Example 5.3.5.

Figure 5.4.9

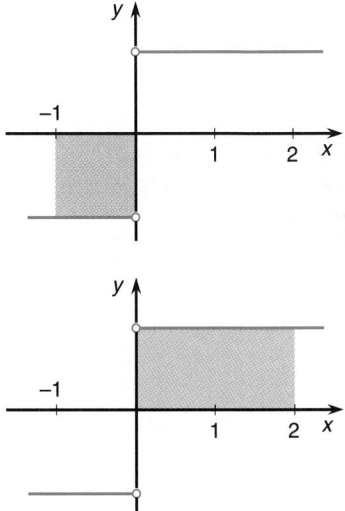

Figure 5.4.10

Integrating Functions with Removable or Jump Discontinuities

EXAMPLE 3   Let $f(x) = \frac{|x|}{x}$, the function whose graph is illustrated in Figure 5.4.9, and calculate $\int_{-1}^{2} f(x)\,dx$.

Solution   The difficulty is that since $f(x)$ is not continuous, we cannot use Theorem 5.3.4 (p. 368) to conclude that it is integrable. On the other hand, Theorem 5.4.5 suggests that the way to evaluate the integral would be to evaluate the integral from $x = -1$ to $x = 0$ and the integral from $x = 0$ to $x = 2$, and add them.

Between $x = -1$ and $x = 0$, we know that $f(x) = -1$, that is, $f(x)$ is a constant function, so it is integrable. Strictly speaking, $f(x)$ is not defined at $x = 0$, so we cannot really use Theorem 5.3.4 to say that it is integrable on $[-1, 0]$. However, $f(x)$ is continuous on $[-1, 0)$ with a removable discontinuity at $x = 0$. In a situation like this we will allow ourselves to remove the discontinuity, making the function integrable on $[-1, 0]$.

Similarly, $f(x)$ is continuous on $(0, 2]$, with a removable discontinuity at $x = 0$. Again we can make the function integrable by removing the discontinuity.

The integral can now be evaluated by doing it in pieces: See Figure 5.4.10.

$$
\begin{aligned}
\int_{-1}^{2} \frac{|x|}{x}\,dx &= \int_{-1}^{0} \frac{|x|}{x}\,dx + \int_{0}^{2} \frac{|x|}{x}\,dx \\
&= \int_{-1}^{0} (-1)\,dx + \int_{0}^{2} 1\,dx \\
&= (-1)\big(0 - (-1)\big) + 1(2 - 0) \\
&= 1.
\end{aligned}
$$

The answer we obtain is certainly consistent with our interpretation of the integral as the total signed area. ◢

In this example we used the idea of Theorem 5.4.5 to integrate $f(x)$ over the interval $[-1, 2]$, even though $f(x)$ is not defined at the point $x = 0$. It has a jump discontinuity at $x = 0$, so there is no way to change $f(x)$ to make it continuous there. On the other hand, if we look at the part of $f(x)$ on the interval $[-1, 0]$, it has a **removable discontinuity** at $x = 0$. Removing it, we can integrate.

Similarly, we can integrate $f(x)$ from $x = 0$ to $x = 2$ after changing the function at $x = 0$. Curiously, the change we make to $f(x)$ in order to integrate it over $[-1, 0]$ is not the same as the change that is necessary to integrate it over $[0, 2]$. (For $[-1, 0]$ we define $f(0) = -1$, and for $[0, 2]$ we define $f(0) = 1$.) Having integrated over both small intervals, we can add the results and call the sum the integral of $f(x)$ over the large interval $[-1, 2]$.

Strictly speaking, the Definition 5.3.3 of integrability was only made for functions that were defined at every point in the interval. On the other hand, the approach we have just used seems very reasonable, and it gives results that fit our geometrical intuition. We want to be able to integrate functions of this type, that is, functions with a jump or removable discontinuity at some point or, for that matter, at several points $d_1,\ d_2, \ldots, d_r$.

The first observation is that if $f(x)$ is defined at every point of $[a, b]$ and is continuous there except for a removable or jump discontinuity at one point $c \in [a, b]$, then it is integrable on $[a, b]$. This result is not very difficult to prove, but we leave the proof to the exercises.

THEOREM 5.4.6

Suppose $f(x)$ is defined at every point in $[a, b]$ and is continuous except for a removable discontinuity or a jump discontinuity at $x = c$. Then $f(x)$ is integrable on $[a, b]$.

This also works if there are removable or jump discontinuities at several points $d_1, d_2, \ldots, d_r$.

Another common situation is that $f(x)$ might not even be defined at certain points $d_1, d_2, \ldots, d_r$. Theorem 5.4.6 suggests that its actual values at these points are not important, so we can try to choose values for $f(d_1), f(d_2), \ldots, f(d_r)$ in such a way that $f(x)$ has at worst a removable or a jump discontinuity at each of them. If this is possible, the resulting "extended" function is integrable, by the theorem, and we will say that the original function is integrable and that its integral equals the integral of the extended function.

DEFINITION 5.4.7

Suppose $f(x)$ is defined on $[a, b]$, except possibly at the points $d_1, d_2, \ldots, d_r$. If $f(x)$ has at worst a removable or a jump discontinuity at each of them, then we say that $f(x)$ is integrable on $[a, b]$, and

$$\int_a^b f(x)\,dx = \int_a^{d_1} f(x)\,dx + \int_{d_1}^{d_2} f(x)\,dx + \ldots + \int_{d_{r-1}}^{d_r} f(x)\,dx$$
$$+ \int_{d_r}^b f(x)\,dx.$$

If $f(x)$ has a removable discontinuity, we evaluate the integral of $f(x)$ by "removing" the discontinuity and integrating the resulting function. If the function has jump discontinuities, it is necessary to evaluate the integral of each continuous piece separately and add the results together, using Theorem 5.4.5, just as we did in Example 3.

To integrate over the subintervals, it might be necessary to remove some removable discontinuities at the endpoints.

EXAMPLE 4   Evaluate (i) $\int_0^4 \frac{x^2-4}{x-2}\,dx$; (ii) $\int_{-2}^2 \frac{|t-1|}{t-1}(t-2)\,dt$.

Solution

(i) The integrand equals $x + 2$ at every point except $x = 2$, where it is not defined. So by Definition 5.4.7 we have

$$\int_0^4 \frac{x^2 - 4}{x - 2}\,dx = \int_0^2 (x+2)\,dx + \int_2^4 (x+2)\,dx$$
$$= \int_0^2 x\,dx + \int_0^2 2\,dx + \int_2^4 x\,dx + \int_2^4 2\,dx$$
$$= \frac{1}{2}(2^2 - 0^2) + 2(2 - 0) + \frac{1}{2}(4^2 - 2^2) + 2(4 - 2)$$
$$= 16.$$

Note that this equals $\int_0^4 (x + 2)\,dx$, the integral we get by removing the discontinuity ...

(ii) The integrand is continuous except at $t = 1$. Accordingly, we proceed to break the integral into two pieces:

$$\int_{-2}^{2} \frac{|t-1|}{t-1}(t-2)\,dt = \int_{-2}^{1} \frac{|t-1|}{t-1}(t-2)\,dt + \int_{1}^{2} \frac{|t-1|}{t-1}(t-2)\,dt$$

$$= \int_{-2}^{1} (-1)(t-2)\,dt + \int_{1}^{2} (1)(t-2)\,dt$$

$$= -\int_{-2}^{1} t\,dt + \int_{-2}^{1} 2\,dt + \int_{1}^{2} t\,dt - \int_{1}^{2} 2\,dt$$

$$= -\frac{1}{2}\left(1^2 - (-2)^2\right) + 2\left(1 - (-2)\right) + \frac{1}{2}(2^2 - 1^2) - 2(2-1)$$

$$= 7.$$

> Note that $\frac{|t-1|}{t-1}$ equals -1 when $t < 1$ and 1 when $t > 1$.

Exercises 5.4

I

For the following exercises, recall that there are several integrals that we know how to calculate. Specifically, we know that $\int_{a}^{b} C\,dx = C(b-a)$ for any constant C, by Example 5.3.3 (p. 367). Also, by Example 5.3.5 (p. 369), we know that $\int_{a}^{b} x\,dx = \frac{1}{2}(b^2 - a^2)$ and that $\int_{a}^{b} x^2\,dx = \frac{1}{3}(b^3 - a^3)$. Using these facts and Theorem 5.4.1 (p. 373), evaluate the following integrals.

1. $\int_{0}^{1} (x+1)\,dx$

2. $\int_{-1}^{2} 2x^2\,dx$

3. $\int_{0}^{3} (x^2 - 1)\,dx$

4. $\int_{-2}^{2} 5x\,dx$

5. $\int_{1}^{4} (x^2 + x - 1)\,dx$

6. $\int_{-3}^{4} (3x^2 + 2)\,dx$

7. $\int_{2}^{3} (x^2 + 2x + 5)\,dx$

8. $\int_{3}^{5} (6x^2 + 4x - 3)\,dx$

9. $\int_{-1}^{1} (6x^2 + 17x + 2)\,dx$

10. $\int_{-\pi}^{\pi} (\sin^2 x + \cos^2 x)\,dx$

Suppose $f(x)$, $g(x)$, and $h(x)$ are continuous functions on $[1, -1]$, and suppose we know the following facts:

$$\int_{-1}^{0} f(x)\,dx = -2, \quad \int_{-1}^{0} g(x)\,dx = 3, \quad \int_{-1}^{1} h(x)\,dx = 5,$$

$$\int_{0}^{1} f(x)\,dx = 1, \quad \int_{-1}^{1} g(x)\,dx = 7, \quad \int_{0}^{1} h(x)\,dx = 11.$$

Use these facts to evaluate the following integrals.

11. $\int_{-1}^{1} f(x)\,dx$

12. $\int_{-1}^{1} \left(f(x) + g(x)\right)dx$

13. $\int_{0}^{1} g(x)\,dx$

14. $\int_{0}^{1} \left(f(x) + 3h(x)\right)dx$

15. $\int_{-1}^{0} h(x)\,dx$

16. $\int_{-1}^{0} \left(2g(x) - f(x)\right)dx$

17. $\int_{0}^{1} \left(4g(x) - 2x^2 + 3\right)dx$

18. $\int_{-1}^{1} \left(2h(x) - f(x) + 4x^2 - 6x\right)dx$

19. $\int_{-1}^{0} \left(g(x) - f(x) + 3h(x) - 3x^2 + 2x\right)dx$

20. $\int_{0}^{1} \left(h(x) + 3g(x) + f(x) - 6x^2 - 2x\right)dx$

II

21. Verify Theorem 5.4.5 (p. 378) for the function $f(x) = x$ on the intervals $[-2, 1]$ and $[1, 2]$.

22. Verify Theorem 5.4.5 for the function $g(x) = x^2 - 2x + 4$ on the intervals $[-3, -1]$ and $[-1, 4]$.

*23. Let $a < c < b$. Verify Theorem 5.4.5 for the function $f(x) = x$ on the intervals $[a, c]$ and $[c, b]$. (*Hint:* Use Example 5.3.5, p. 369).

*24. Let $a < c < b$. Verify Theorem 5.4.5 for the function $f(x) = 3x^2 - 4x + 7$ on the intervals $[a, c]$ and $[c, b]$. (*Hint:* Use Example 5.3.5).

Verify the Integral Mean Value Theorem (Theorem 5.4.4) for each of the following functions.

25. $f(x) = x$ on the interval $[-2, 3]$

*26. $f(x) = x$ on the interval $[a, b]$

27. $f(x) = x^2 - 2x + 3$ on the interval $[0, 4]$

*28. $f(x) = x^2$ on the interval $[a, b]$

29. Find constants c and d so that the function $f(x) = cx + d$ satisfies

$$\int_0^1 f(x)\,dx = 4, \qquad \int_1^3 f(x)\,dx = 14.$$

30. Find constants c and d so that the function $f(x) = cx + d$ satisfies

$$\int_{-1}^2 f(x)\,dx = 3, \qquad \int_2^3 f(x)\,dx = 9.$$

31. Find constants b, c, and d so that the function $g(x) = bx^2 + cx + d$ satisfies

$$\int_{-1}^0 g(x)\,dx = -1, \int_0^1 g(x)\,dx = 1, \int_1^2 g(x)\,dx = 9.$$

32. Find constants b, c, and d so that the function $h(x) = bx^2 + cx + d$ satisfies

$$\int_0^2 h(x)\,dx = 10, \int_{-2}^2 h(x)\,dx = 20, h(1) = 4.$$

*33. Use Theorem 5.4.3 (p. 374) to show that

$$0 \le \int_0^1 x^3\,dx \le 1.$$

*34. Suppose $f(x)$ and $g(x)$ are integrable on $[a, b]$. Show that we cannot expect that $\int_a^b f(x)g(x)\,dx$ will equal the product of the integrals of $f(x)$ and $g(x)$, by finding an example in which they are different.

*35. Suppose $f(x)$ and $g(x)$ are integrable on $[a, b]$, and

$$f(x) \le g(x)$$

for every $x \in [a, b]$. Show that

$$\int_a^b f(x)\,dx \le \int_a^b g(x)\,dx.$$

*36. (i) Show that

$$1 \le \int_1^2 x^4\,dx \le 16.$$

(ii) By writing $\int_1^2 x^4\,dx = \int_1^{3/2} x^4\,dx + \int_{3/2}^2 x^4\,dx$, find better inequalities, that is, show that $C \le \int_1^2 x^4\,dx \le D$ for some constants C and D. These inequalities will be "better" than those in part (i) if $C > 1$ and $D < 16$, because they will then narrow down more closely what the value of the integral might be.

*37. Prove Theorem 5.4.6 (p. 380).

*38. Apply Theorem 5.4.3 to find a number K so that $\left| \int_{-1}^2 (x^2 + 1) \sin(2x)\,dx \right| \le K$.

39. It turns out that $\int_1^4 \sqrt{x}\,dx = \frac{14}{3}$. (i) Use a calculator to find (to four decimal places) the corresponding number $c \in [1, 4]$, which is promised by the Integral Mean Value Theorem (5.4.4). (ii) Use a graphics calculator or computer to sketch $\sqrt{x}$ and the corresponding rectangle on the same graph.

40. Sketch the graphs of $y = x^2$ and $y = \frac{1}{3}$ for $x \in [-1, 1]$ on the same graph. Compare with Example 1 (p. 377).

In each of the following situations the value of an integral is given. Use a graphics calculator or computer to find the number c promised by Theorem 5.4.4, correct to five decimal places.

41. $\int_0^\pi \sin x\,dx = 2$

42. $\int_{-1}^1 x^5\,dx = 0$

43. $\int_1^e \frac{1}{x}\,dx = 1$

44. $\int_0^1 (x^3 - 2x)\,dx = -\frac{3}{4}$

POINT TO PONDER

We have used various facts about limits: In the proof of Theorem 5.4.1 we needed to know that the limit of a sum is the sum of the limits and that multiplying by a constant and taking a limit gives exactly the same result as taking the limit and then multiplying by the constant. For the proof of Theorem 5.4.2 we used the fact that the limit of nonnegative numbers is nonnegative.

Strictly speaking, these are facts we learned in Chapter 2 about conventional limits. Here we have been using analogous facts about the special kind of limits we have been discussing in connection with Riemann sums (limits as the mesh of the partition goes to zero). Convince yourself that the results carry over to this new situation.

5.5 Indefinite Integrals and the Fundamental Theorem of Calculus

DEFINITE/INDEFINITE
INTEGRALS

VARIABLE OF INTEGRATION

DUMMY VARIABLE

ANTIDERIVATIVE

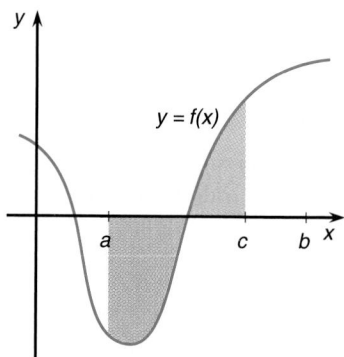

Figure 5.5.1

In this section we explore the ideas behind the Fundamental Theorem of Calculus, which establishes the link between differentiation and integration. We will find that integration is closely related to finding antiderivatives, that is, to "undoing" the process of differentiation.

Suppose $f(x)$ is a continuous function on $[a, b]$. By Theorem 5.3.4 (p. 368), $f(x)$ is integrable on $[a, b]$. For that matter, for any c between a and b, $f(x)$ is continuous on $[a, c]$, so it is integrable on $[a, c]$. The integral

$$\int_a^c f(x)\,dx$$

will depend on c, taking on different values as c changes (see Figure 5.5.1). We can think of c as a variable and of the integral as a function of c.

To emphasize that c is a variable, we can replace it by the letter x and let

$$F(x) = \int_a^x f(t)\,dt, \qquad \text{for any } x \in [a, b].$$

Notice that the variable inside the integral has been changed to t to avoid confusion. It makes no difference what letter is used there as long as it is different from any other symbol being used.

Similarly, there is no particular reason that the lower limit of integration should always be a; for any fixed $d \in [a, b]$ we could consider

$$\int_d^x f(t)\,dt.$$

This leads to trouble in the situation in which $x \leq d$. We avoid the problem by making a definition.

DEFINITION 5.5.1

(i) If $f(x)$ is any function defined at $x = a$, define

$$\int_a^a f(x)\,dx = 0.$$

(ii) If $a < b$ and $f(x)$ is integrable on $[a, b]$, define

$$\int_b^a f(x)\,dx = -\int_a^b f(x)\,dx.$$

REMARK 5.5.2

Part (i) of Definition 5.5.1 has the effect of making the formula in Theorem 5.4.5 (p. 378) remain valid if $c = a$ or $c = b$. It can also be interpreted to say there is zero area between $x = a$ and $x = a$. Part (ii) makes the formula hold even if $c < a$ or $c > b$, provided that the function $f(x)$ is integrable on the larger interval.

DEFINITION 5.5.3

Suppose $f(x)$ is integrable on $[a, b]$ and $d \in [a, b]$. For $x \in [a, b]$ we can define

$$F(x) = \int_d^x f(t)\, dt.$$

The function $F(x)$ is called an **indefinite integral** of $f(x)$. It is a function on $[a, b]$. The name suggests that the integral can have different values as x changes.

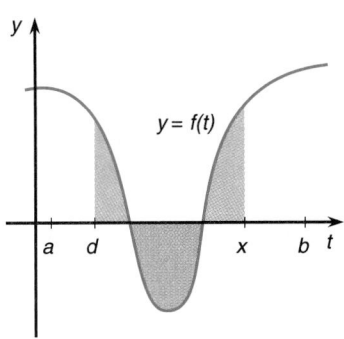

Figure 5.5.2

If $d < x$, the indefinite integral $F(x)$ represents the signed area of the region between $y = f(t)$ and the horizontal axis, between $t = d$ and $t = x$ (see Figure 5.5.2). As x changes, the region changes, and the indefinite integral represents the signed area of different regions.

EXAMPLE 1

(i) If $f(x) = x$, we notice that $f(x)$ is integrable on any interval (since it is always continuous). If d is any fixed number, we can define the indefinite integral

$$F(x) = \int_d^x f(t)\, dt = \int_d^x t\, dt.$$

This last integral can be evaluated, using Example 5.3.5(p. 369). It equals $\frac{1}{2}(x^2 - d^2)$.

So we have found an indefinite integral of $f(x) = x$. It is $\frac{1}{2}(x^2 - d^2)$, where d is any fixed number. See Figure 5.5.3(i).

Notice the way t was used as the variable in the integral. Since x was used as the upper limit of integration, it would have led to confusion if we had used it again inside the integral; this way it is clear that t is a variable that can take on any value between $t = d$ and $t = x$.

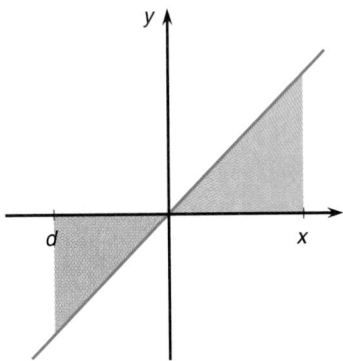

Figure 5.5.3(i)

(ii) Similarly, for $g(x) = c$, where c is a constant, we can define an indefinite integral

$$G(x) = \int_d^x g(t)\, dt = \int_d^x c\, dt.$$

Using Example 5.3.3 (p. 367), we can evaluate this integral; we find that

$$G(x) = c(x - d).$$

See Figure 5.5.3(ii).

Figure 5.5.3(ii)

(iii) For $h(x) = x^2$ we can define the indefinite integral

$$H(x) = \int_d^x h(t)\, dt = \int_d^x t^2\, dt.$$

Using Example 5.3.5, we evaluate and find that

$$H(x) = \frac{1}{3}(x^3 - d^3).$$

See Figure 5.5.3(iii).

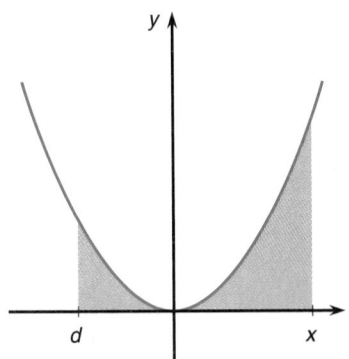

Figure 5.5.3(iii)

EXAMPLE 2 The indefinite integral $F(x)$ of Definition 5.5.3 depends on the choice of $d \in [a, b]$. Consider two possible choices $d_1, d_2 \in [a, b]$ and define

Figure 5.5.4(i)

Figure 5.5.4(ii)

Figure 5.5.4(iii)

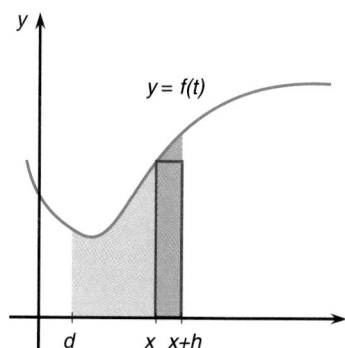

Figure 5.5.5

$$F_1(x) = \int_{d_1}^{x} f(t)\,dt \quad \text{and} \quad F_2(x) = \int_{d_2}^{x} f(t)\,dt.$$

Then

$$F_1(x) - F_2(x) = \int_{d_1}^{x} f(t)\,dt - \int_{d_2}^{x} f(t)\,dt$$

$$= \int_{d_1}^{x} f(t)\,dt + \int_{x}^{d_2} f(t)\,dt,$$

by Definition 5.5.1(ii), and by Remark 5.5.2 this equals

$$\int_{d_1}^{d_2} f(t)\,dt.$$

Notice that the above formula for $F_1(x) - F_2(x)$ is independent of x, that is, it is a constant function. We have just showed that different choices d_1 and d_2 result in indefinite integrals $F_1(x)$ and $F_2(x)$ that differ by a constant. Figures 5.5.4(i), (ii), (iii) show how the difference between $F_1(x)$ and $F_2(x)$ is the signed area between d_1 and d_2. ◢

In particular, since two differentiable functions that differ by a constant have the same derivative, if we knew that $F_1(x)$ and $F_2(x)$ were differentiable, we would know that $F_1'(x) = F_2'(x)$. In fact, not only are they differentiable, but in a sense the derivative is the most important thing about them, as we will soon see.

Derivative of an Indefinite Integral

Suppose $f(x)$ is a continuous function on $[a, b]$. Fix $d \in [a, b]$, and consider the indefinite integral

$$F(x) = \int_{d}^{x} f(t)\,dt.$$

The derivative of $F(x)$ is defined by

$$F'(x) = \lim_{h \to 0} \frac{F(x+h) - F(x)}{h}.$$

To discuss the derivative, we must consider the difference quotient:

$$\frac{F(x+h) - F(x)}{h} = \frac{1}{h}\left(\int_{d}^{x+h} f(t)\,dt - \int_{d}^{x} f(t)\,dt \right)$$

$$= \frac{1}{h}\left(\int_{d}^{x+h} f(t)\,dt + \int_{x}^{d} f(t)\,dt \right).$$

For the last equality we used Definition 5.5.1(ii). Using Theorem 5.4.5 and Remark 5.5.2, we can combine the two integrals in the last line to find that the difference quotient equals

$$\frac{1}{h}\left(\int_{x}^{x+h} f(t)\,dt \right).$$

Suppose for the moment that h is a small positive number. The situation is illustrated in Figure 5.5.5.

The numerator of the difference quotient is the integral

$$\int_x^{x+h} f(t)\, dt,$$

which represents the signed area of the shaded region. This is approximately the same as the area of the tall rectangle shown, whose height is $f(x)$. (Note that x is the point at the left side of the rectangle.) Since the width of the rectangle is h, its (signed) area is $hf(x)$.

The difference quotient is then approximately this area divided by h, or $\frac{hf(x)}{h} = f(x)$. As $h \to 0$, the difference quotient is close to $f(x)$. This suggests the remarkable fact that

$$F'(x) = f(x),$$

which is known as the Fundamental Theorem of Calculus (what makes it fundamental, at least in part, is that it relates derivatives and integrals, the two main ingredients of calculus). In words, what it says is that the derivative of an indefinite integral equals the integrand. Another way to say it is that if we start with a continuous function $f(x)$, construct its indefinite integral $F(x)$, and then take the derivative of that, what we end up with is the original function $f(x)$. So differentiation has the effect of "undoing" the process of forming the indefinite integral.

In making this approximate discussion we assumed that h was positive, and in the illustration we implicitly assumed $f(x) > 0$. You might like to check that it still works if either or both of h and $f(x)$ are negative.

The argument we just gave relied upon noticing that the area of the region under the graph of $y = f(t)$ between $t = x$ and $t = x + h$ was "approximately" equal to the area of a certain rectangle, which does seem reasonable when h is small. It was then necessary to take the limit as $h \to 0$ and to assume that we could ignore the word "approximately." In fact it is possible to show that this all works, but it is not easy. The proof given below uses the Integral Mean Value Theorem (Theorem 5.4.4). Even this proof is not very easy. Since after all it is the result itself that interests us most, you may prefer to jump over the proof and look at the examples that follow it to see how the result can be used.

THEOREM 5.5.4

The Fundamental Theorem of Calculus

Suppose $f(x)$ is continuous on $[a, b]$ and $d \in [a, b]$. Then the indefinite integral

$$F(x) = \int_d^x f(t)\, dt$$

is differentiable on (a, b), and

$$F'(x) = f(x),$$

that is,

$$\frac{d}{dx}\left(\int_d^x f(t)\, dt \right) = f(x).$$

PROOF The difference quotient is

$$\frac{F(x+h) - F(x)}{h} = \frac{1}{h}\left(\int_d^{x+h} f(x)\, dx - \int_d^x f(x)\, dx \right)$$

$$= \frac{1}{h}\left(\int_d^{x+h} f(x)\,dx + \int_x^d f(x)\,dx\right)$$

$$= \frac{1}{h}\left(\int_x^{x+h} f(x)\,dx\right).$$

Assume for the moment that $h > 0$ and apply the Integral Mean Value Theorem (Theorem 5.4.4) to this last integral. We find that it can be written as

$$\int_x^{x+h} f(x)\,dx = f(c)\,(x + h - x) = f(c)\,h,$$

for some $c \in [x, x + h]$. Dividing this by h, we find that the difference quotient equals $f(c)$.

As $h \to 0^+$, the point c will be squeezed between x and $x + h$, which are both approaching x. Accordingly, we see that $c \to x$, and by the continuity of $f(x)$ we know that $f(c) \to f(x)$. Since $f(c)$ equals the difference quotient, we have just shown that the difference quotient approaches $f(x)$ as $h \to 0^+$.

An entirely similar argument applies in the case in which $h < 0$, and we see that

$$\lim_{h \to 0} \frac{F(x + h) - F(x)}{h} = f(x),$$

in the sense that the limit exists and equals $f(x)$. Theorem 5.5.4 is proved.

> Observe that in this proof, x does not change. It plays the role of a constant rather than a variable. Notice that the proof actually works at the endpoints too, showing that $F(x)$ has a one-sided derivative at each endpoint.

It might help to visualize Theorem 5.5.4 in the following way. The derivative $F'(x)$ is the rate at which $F(x)$ changes as x increases. Imagine x moving along the horizontal axis, sweeping out larger and larger regions as it goes. As x moves, the region changes, and the signed area changes too. How can we visualize the rate of change of the signed area?

If $f(x)$ is a positive number, then as x moves to the right, the shaded region is increasing, and the part being added on is above the axis, so it has positive signed area. The bigger $f(x)$ is, the larger will be the area of the region being added on, and the larger will be the rate at which the total signed area is increasing. See Figure 5.5.6(i).

Similarly, if $f(x) < 0$, the region being added on has negative signed area, so the total signed area is decreasing as x moves to the right (see Figure 5.5.6(ii)). Moreover, the rate at which $F(x)$ is decreasing will depend on how far below the axis $f(x)$ is. This is consistent with the fact that the derivative of $F(x)$ equals $f(x)$.

EXAMPLE 3 We can easily check the theorem for the indefinite integrals we found in Example 1. For $f(x) = x$, we found the indefinite integral

$$F(x) = \frac{1}{2}(x^2 - d^2).$$

The derivative of $F(x)$ is $F'(x) = \frac{1}{2}(2x) = x$. Since this equals the original function $f(x)$, we have shown that in this case, $F'(x) = f(x)$, as required.

Figure 5.5.6(i)

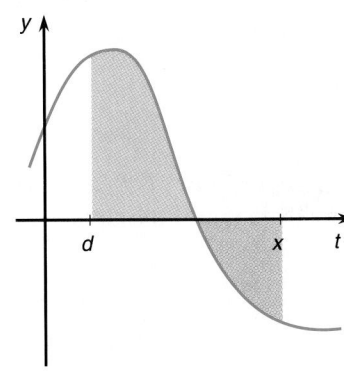

Figure 5.5.6(ii)

For $g(x) = c$ we have the indefinite integral $G(x) = c(x - d)$. Its derivative is $G'(x) = c$, which again equals the original function $g(x)$. Once again Theorem 5.5.4 is verified.

For $h(x) = x^2$ we found the indefinite integral

$$H(x) = \frac{1}{3}(x^3 - d^3),$$

whose derivative is $H'(x) = \frac{1}{3}3x^2 = x^2 = h(x)$. Theorem 5.5.4 is confirmed here too.

With a graphics calculator it is possible to plot both the function $h(x) = x^2$ and the indefinite integral $H(x) = \frac{1}{3}(x^3 - d^3)$. For simplicity we let $d = 0$ and plot $H(x) = \frac{x^3}{3}$.

TI-85	SHARP EL9300
GRAPH F1$(y(x) =)$ F1(x)^2 ▼ F1(x)^3 ÷ 3 M2(RANGE) $(-)$ 2 ▼ 2 ▼ ▼ $(-)$ 4 ▼ 4 F5(GRAPH)	⋀⋁ x/θ/T x^2 2ndF ▼ x/θ/T a^b 3 ▶ ÷ 3 RANGE $-$ 2 ENTER 2 ENTER 1 ENTER -4 ENTER 4 ENTER 1 ENTER ⋀⋁
CASIO f_x-7700GB/f_x-6300G	HP 48SX
Cls EXE Range $-$ 2 EXE 2 EXE 1 EXE $-$ 4 EXE 4 EXE 1 EXE EXE EXE EXE Graph X x^2 EXE Graph X x^y 3 ÷ 3 EXE	↰ PLOT PLOTR ERASE ATTN ' Y = X y^x 2 ↱ PLOT ↰ DRAW ERASE 2 +/- ENTER 2 XRNG 4 +/- ENTER 4 YRNG DRAW ATTN ' Y = X y^x 3 ÷ 3 ↱ PLOT ↰ DRAW DRAW

Identify $h(x)$ and $H(x)$; notice how the indefinite integral increases faster when the graph of $y = h(x)$ is higher above the x-axis.

EXAMPLE 4   Find the critical points of

$$F(x) = \int_1^x (t^2 - 1)\, dt.$$

Solution   The integrand $t^2 - 1$ is continuous on the whole real line, so the Fundamental Theorem tells us that any of its indefinite integrals $F(x)$ is differentiable at every point. Moreover, it tells us that the derivative is

$$F'(x) = x^2 - 1.$$

This function is zero precisely at $x = \pm 1$, and these are the critical points of $F(x)$.

Notice the way the variables x and t were used in Example 4. The variable x is the variable used for the indefinite integral $F(x)$. It serves as the upper limit of integration in

$$F(x) = \int_1^x (t^2 - 1)\, dt.$$

Because x was used this way, we used a different letter, t, as the variable inside the integral.

The tricky thing is that when we find the derivative $F'(x)$, it should of course be a function of x. The Fundamental Theorem says that this derivative equals the integrand, but we have to be careful. In the integral itself the integrand is a function of t. Here, when it equals the derivative of $F(x)$, it must be expressed as a function of the variable x.

The integrand is $t^2 - 1$; the derivative of $F(x)$ is $x^2 - 1$.

It does not really matter what letter we use inside the integral, as long as it is not used anywhere else. The variable inside the integral is sometimes called the **integration variable** or **variable of integration**. It is also sometimes referred to as a **dummy variable** to emphasize the fact that it can be changed without affecting the integral. (This is somewhat analogous to the situation we encountered in Section 5.2, where the index of summation could be changed without affecting the value of a sum.)

The choice of integration variable is not important. In using the Fundamental Theorem it is important to express the *derivative* in terms of the appropriate variable.

EXAMPLE 5 Find the regions where $G(x)$ is increasing or decreasing if

$$G(x) = \int_{-1}^x (u + 1)^3 e^u\, du.$$

Solution $G'(x) = (x + 1)^3 e^x$, and we have to discuss the sign of $G'(x)$. Now $(x + 1)^3$ is negative when $x + 1 < 0$, that is, when $x < -1$. Also $(x + 1)^3 > 0$ when $x + 1 > 0$, that is, when $x > -1$, and $(x + 1)^3 = 0$ when $x = -1$.

Since $e^x > 0$ for all x, the same remarks apply to $G'(x) = (x + 1)^3 e^x$. This shows that $G(x)$ is increasing for $x > -1$, is decreasing for $x < -1$, and has a minimum at $x = -1$.

EXAMPLE 6 Find the derivative of (i) $H(x) = \int_0^{x^2} \sin\theta\, d\theta$; (ii) $K(x) = \int_x^3 \frac{e^t}{t^2+1}\, dt$.

Solution For part (i) we have to use the Chain Rule (Theorem 3.6.1, p. 186). If $F(x) = \int_0^x \sin\theta\, d\theta$, then $H(x) = F(x^2)$, so

$$H'(x) = F'(x^2)(2x) = \sin(x^2)(2x).$$

(Here we have evaluated F' using the Fundamental Theorem.)

For part (ii), write

$$K(x) = \int_x^3 \frac{e^t}{t^2+1}\, dt = -\int_3^x \frac{e^t}{t^2+1}\, dt, \quad \text{so} \quad K'(x) = -\frac{e^x}{x^2+1}.$$

What the Fundamental Theorem of Calculus says is that $\int_d^x f(x)\,dx$ is an **antiderivative** of $f(x)$ (using the terminology of Section 4.8). It will turn out to be very useful to discuss antiderivatives in terms of indefinite integrals.

In fact, we never really had a good notation for antiderivatives, so we can take advantage of the connection with indefinite integrals. We will write $\int f(x)\,dx$, an integral without limits of integration, to mean an antiderivative of $f(x)$. As we mentioned in Section 4.8, it is very important to write antiderivatives with a constant added at the end.

DEFINITION 5.5.5

Suppose $f(x)$ is a continuous function on some interval. Then we write

$$\int f(x)\,dx$$

to mean its general antiderivative. The word "general" means that rather than just one specific antiderivative, we are referring to all possible antiderivatives, which are obtained by adding a constant to any one of them.

EXAMPLE 7   Evaluate the following: (i) $\int x^2\,dx$; (ii) $\int \sin\theta\,d\theta$; (iii) $\int \frac{1}{t^2+1}\,dt$.

Solution

(i) We can easily guess one antiderivative for x^2, namely, $\frac{1}{3}x^3$. The general antiderivative is obtained by adding any constant c to this, so

$$\int x^2\,dx = \frac{1}{3}x^3 + c.$$

(ii) Similarly, an antiderivative of $\sin\theta$ is $-\cos\theta$, so

$$\int \sin\theta\,d\theta = -\cos\theta + c.$$

(iii) We know an antiderivative of $\frac{1}{t^2+1}$ is $\arctan t$, so the general antiderivative is

$$\int \frac{1}{t^2+1}\,dt = \arctan t + c.$$

Exercises 5.5

Find the derivative of each of the following functions.

1. $F(x) = \int_0^x (t^2 - 1)\,dt$
2. $F(x) = \int_a^x e^u\,du$
3. $g(x) = \int_{-2}^x \sin u\,du$
4. $h(x) = \int_1^x \frac{t-1}{t^2+1}\,dt$
5. $G(t) = \int_1^t e^u\,du$
6. $H(u) = \int_a^u \frac{1-x}{e^x+1}\,dx$
7. $K(x) = \int_1^x (e^u - u^2)\,du$
8. $k(x) = \int_2^x e^{-u^2}\,du$
9. $L(x) = \int_{-1}^x 7\,du$
10. $R(x) = \int_0^x 0\,dt$
11. $f(x) = \int_0^{x^2} \cos u\,du$
12. $g(x) = \int_0^{1+x^2} e^t\,dt$

13. $h(x) = \int_{-x}^{2} e^{t+1}\,dt$

14. $k(t) = \int_{t}^{t^2} e^x\,dx$

15. $F(u) = \int_{1-u}^{0} \sin x\,dx$

16. $G(t) = \int_{t}^{t+1} \ln x\,dx$

17. $H(r) = \int_{\cos r}^{\sin r} (t^2 - 1)\,dt$

18. $K(x) = \int_{1}^{\exp(x)} \ln u\,du$

19. $R(t) = \int_{-t}^{t} (u^3 - u)\,du$

20. $T(x) = \int_{-x}^{x} \sin u\,du$

▌▌

21. Find the regions of increase and decrease of $F(x) = \int_{1}^{x} (t^3 - 4t)\,dt$.

22. Find the maximum value of
$$F(x) = \int_{0}^{x} (1 - t^3)\,dt.$$
(Express your answer as an integral, but do not evaluate it.)

23. Find and classify the local extrema of
$$G(x) = \int_{-x}^{x} (t^3 + 4t^2 - t - 1)\,dt.$$

24. Suppose $f(t)$ is continuous on $[-a, a]$, and define
$$H(x) = \int_{-x}^{x} f(t)\,dt,$$
for all $x \in [0, a]$. Find a formula for $H'(x)$.

*25. Find an antiderivative for the continuous function $f(x) = |x|$.

*26. Let $F(x) = \int_{0}^{x} |t|\,dt$. Calculate $F'(x)$ and $F''(x)$, stating at which points $F'(x)$ exists and at which points $F''(x)$ exists.

*27. In discussing the Integral Mean Value Theorem (Theorem 5.4.4, p. 376) we mentioned that the quantity
$$\frac{1}{b-a} \int_{a}^{b} g(x)\,dx$$
could be interpreted as the average value of $g(x)$ on the interval $[a, b]$. Show that if $f(x)$ is differentiable on $[a, b]$ and $f'(x)$ is continuous on $[a, b]$, then the quantity
$$\frac{f(b) - f(a)}{b - a}$$
is in this sense equal to the average value of $f'(x)$ on $[a, b]$.

This says that the average rate of change $\frac{f(b)-f(a)}{b-a}$ equals the average value of the rate of change $f'(x)$. This is not very surprising, but it shows that our notion of the average value of a function is a reasonable one.

*28. Show that every continuous function $f(x)$ on a closed interval $[a, b]$ has an antiderivative. In other words, show that there is a function $F(x)$ on $[a, b]$ so that $F'(x) = f(x)$ for every $x \in (a, b)$ and so that the one-sided derivative of $F(x)$ from above exists at $x = a$ and equals $f(a)$ and the one-sided derivative of $F(x)$ from below exists at $x = b$ and equals $f(b)$.

*29. Suppose $f(x)$ is a continuous function on $[a, b]$ satisfying
$$F(x) = \int_{a}^{x} f(t)\,dt = 0, \qquad \text{for all} \quad x \in [a, b].$$
Show that $f(x) = 0$, for every $x \in [a, b]$.

*30. (i) Suppose $f(x)$ is a continuous function on $[a, b]$ satisfying
$$F(x) = \int_{a}^{x} f(t)\,dt = x - a, \text{ for all } x \in [a, b].$$
Show that $f(x) = 1$, for every $x \in [a, b]$.

(ii) Suppose $f(x)$ and $g(x)$ are continuous functions on $[a, b]$ satisfying
$$\int_{a}^{x} f(t)\,dt = \int_{a}^{x} g(t)\,dt, \text{ for all } x \in [a, b].$$
Show that $f(x) = g(x)$, for every $x \in [a, b]$.

*31. Suppose $f(x)$ is an odd function defined on the whole real line, that is, $f(-x) = -f(x)$ for all x. Let $F(x) = \int_{-x}^{x} f(t)\,dt$. (i) Evaluate $F'(x)$. (ii) Evaluate $F(0)$. (iii) What can you say about the function $F(x)$? (*Hint:* You should be able to say something quite forceful; if you are having trouble, draw a picture and interpret $F(x)$ as a signed area.)

*32. Suppose $F(x)$ is a continuous function whose derivative is not defined at $x = 0$ but so that
$$F'(x) = \begin{cases} 0, & \text{if } x < 0, \\ 2, & \text{if } x > 0. \end{cases}$$
What can you say about $F(x)$?

▐▐▐

33. Use your programmable calculator or personal computer to give numerical confirmation of the Fundamental Theorem of Calculus. Let $F(x) = \int_{1}^{x} e^t\,dt$, and consider $F'(1)$, which should equal $e^1 = e \approx 2.718282$. With $h > 0$, consider the difference quotient $\frac{1}{h}\left(\int_{1}^{1+h} e^t\,dt\right)$. (i) Use the result of Example 5.2.3 (p. 356) with $n = 10$ to estimate this integral for various values of h. Estimate the difference quotient for $h = 1$, $h = 0.1$, $h = 0.01$, and $h = 0.001$. Do they approach the correct value? (ii) Try the same calculations with $n = 20$.

34. Repeat part (i) of Exercise 33 with $h = -1, -0.1, -0.01, -0.001$.

Repeat Exercise 33(i) to estimate the derivative of

each of the following indefinite integrals at the specified value of x.

35. $F(x) = \int_0^x (t^5 - 2t^2) \, dt$; $x = 1$

36. $F(x) = \int_0^x \sin u \, du$; $x = \frac{\pi}{4}$

37. $F(x) = \int_{-1}^x \arctan t \, dt$; $x = 0$

38. $F(x) = \int_1^x \frac{1}{t} \, dt$; $x = 1$

5.6 Definite Integrals

FUNDAMENTAL THEOREM OF CALCULUS (SECOND VERSION)

EVALUATING DEFINITE INTEGRALS

The greatest value of the Fundamental Theorem of Calculus is that it allows us to calculate integrals easily. We have discussed indefinite integrals, by which we mean functions of the form

$$F(x) = \int_d^x f(t) \, dt.$$

As x varies, we are considering the signed area of a changing region, and it is in this sense that the integral is "indefinite."

We also want to be able to calculate integrals of the form $\int_a^b f(t) \, dt$, with specific numbers a and b as the limits of integration. Such an integral is called a **definite integral**.

EXAMPLE 1   Calculate $\int_1^2 (2x + 6x^2) \, dx$.

Solution   Consider the indefinite integral

$$F(x) = \int_1^x (2t + 6t^2) \, dt.$$

The integral we want to evaluate is just $F(2)$. By the Fundamental Theorem of Calculus (Theorem 5.5.4) we know that

$$F'(x) = 2x + 6x^2.$$

So $F(x)$ is an antiderivative of $2x + 6x^2$, so it must be of the form $F(x) = x^2 + 2x^3 + c$, for some constant c. (Note that the derivative of $x^2 + 2x^3$ equals $2x + 6x^2 = F'(x)$, and recall that two functions with the same derivative must differ by a constant, by Theorem 4.2.11 (p. 245). Here this means that $F(x) - (x^2 + 2x^3)$ is a constant, that is, $F(x) = x^2 + 2x^3 + c$ for some $c \in \mathbb{R}$.)

On the other hand, $F(1) = \int_1^1 (2x + 6x^2) \, dx = 0$, that is, $0 = F(1) = 1^2 + 2(1^3) + c = 3 + c$, so $c = -3$. This shows that $F(x) = x^2 + 2x^3 - 3$.

In particular,

$$\int_1^2 (2x + 6x^2) \, dx = F(2) = 2^2 + 2(2^3) - 3 = 17.$$

In Example 1 we could have figured out the integral by using the formulas of Example 5.3.5 for the integrals of x and x^2, but it is easy to think of examples for which we have no such formulas available.

EXAMPLE 2   Evaluate the definite integral

$$\int_{-\pi/2}^{\pi/2} \sin(u)\,du.$$

Solution   If $F(x) = \int_{-\pi/2}^{x} \sin(u)\,du$, then $F'(x) = \sin x$, so $F(x) = -\cos x + c$, for some constant c.

Since $0 = \int_{-\pi/2}^{-\pi/2} \sin(u)\,du = F\left(-\frac{\pi}{2}\right) = -\cos\left(-\frac{\pi}{2}\right) + c = 0 + c$, we find that $c = 0$, so $F(x) = -\cos x$.

In particular,

$$\int_{-\pi/2}^{\pi/2} \sin(u)\,du = F\left(\frac{\pi}{2}\right) = -\cos\left(\frac{\pi}{2}\right) = 0.$$

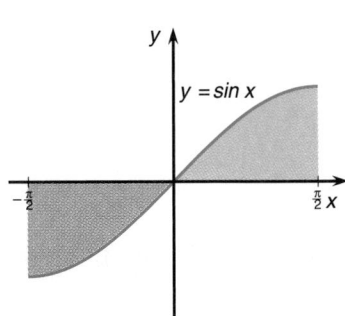

Figure 5.6.1 shows that the integral is the sum of two signed areas. The region between $x = -\frac{\pi}{2}$ and $x = 0$ has negative signed area, and the region between $x = 0$ and $x = \frac{\pi}{2}$ has positive signed area. Since the two regions have the same (unsigned) area, the signed areas cancel, resulting in 0.   ◢

Figure 5.6.1

The preceding examples demonstrate a technique for calculating the definite integral $\int_a^b f(x)\,dx$ that is vastly easier than finding the limit of Riemann sums. However, to make it work, we have to be able to guess an antiderivative $F(x)$ of the integrand $f(x)$. Sometimes this can be difficult or even impossible, but frequently it is reasonably easy, as in the above examples.

Suppose we want to calculate $\int_a^b f(x)\,dx$, and suppose we are able to find an antiderivative $F(x)$ of $f(x)$, that is, a function satisfying $F'(x) = f(x)$.

If $a < d < b$ and we compare $F(x)$ with the indefinite integral $\int_d^x f(t)\,dt$, we see that their derivatives are the same (both equaling $f(x)$), so they must differ by a constant, that is,

$$\int_d^x f(t)\,dt = F(x) + c,$$

for some constant c.

Now

$$\int_a^b f(x)\,dx = \int_a^d f(x)\,dx + \int_d^b f(x)\,dx,$$

by Theorem 5.4.5. The right side of this equation equals

$$-\int_d^a f(x)\,dx + \int_d^b f(x)\,dx,$$

by Definition 5.5.1, and this equals $-\big(F(a) + c\big) + \big(F(b) + c\big) = F(b) - F(a)$. We have just proved the following result.

THEOREM 5.6.1

The Fundamental Theorem of Calculus (Second Version)

If $f(x)$ is continuous on $[a, b]$ and $F'(x) = f(x)$ on $[a, b]$, then

$$\int_a^b f(x)\,dx = F(b) - F(a).$$

If we know an antiderivative $F(x)$ of $f(x)$, Theorem 5.6.1 says that $\int_a^b f(x)\,dx$ equals the difference of the values of $F(x)$ at $x = b$ and $x = a$. Since this will arise so often, we have special notation for it. We write

$$F(x)\Big|_a^b = F(b) - F(a).$$

The symbol on the left side is read "$F(x)$, evaluated from a to b" or "$F(x)$, evaluated between a and b."

The formula in Theorem 5.6.1 can now be written as

$$\int_a^b f(x)\,dx = F(x)\Big|_a^b.$$

EXAMPLE 3   Evaluate the following definite integrals: (i) $\int_0^1 x\,dx$; (ii) $\int_1^3 3x^2\,dx$; (iii) $\int_1^2 x^3\,dx$; (iv) $\int_{-1}^2 (3x^2 - 2x + 5)\,dx$.

Solution   For each integral we need to find an antiderivative. In part (i), for an antiderivative of x we can take $F(x) = \frac{1}{2}x^2$. So

$$\int_0^1 x\,dx = F(x)\Big|_0^1 = \frac{1}{2}x^2\Big|_0^1 = \frac{1}{2}(1)^2 - \frac{1}{2}(0)^2$$

$$= \frac{1}{2}.$$

For part (ii) we can let $F(x) = x^3$ be the antiderivative, and

$$\int_1^3 3x^2\,dx = F(x)\Big|_1^3 = x^3\Big|_1^3 = 3^3 - 1^3$$

$$= 26.$$

For part (iii), take $F(x) = \frac{1}{4}x^4$, and

$$\int_1^2 x^3\,dx = \frac{1}{4}x^4\Big|_1^2 = \frac{1}{4}(2)^4 - \frac{1}{4}(1)^4$$

$$= \frac{15}{4}.$$

For part (iv) we can let $F(x) = x^3 - x^2 + 5x$, so

$$\int_{-1}^2 (3x^2 - 2x + 5)\,dx = (x^3 - x^2 + 5x)\Big|_{-1}^2$$

$$= \left(2^3 - 2^2 + 5(2)\right) - \left((-1)^3 - (-1)^2 + 5(-1)\right)$$

$$= (8 - 4 + 10) - (-1 - 1 - 5)$$

$$= 21.$$

Parts (i), (ii), and (iv) of this example could have been done by using the formulas of Example 5.3.5, but part (iii) could not.

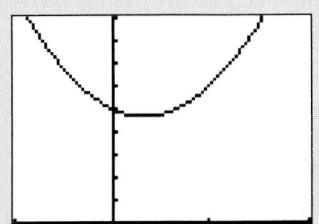

With a graphics calculator we plot the integrand $3x^2 - 2x + 5$ from part (iv) of Example 3, for $x \in [-1, 2]$.

TI-85	SHARP EL9300
GRAPH F1 $(y(x) =)$ 3 F1 (x) ^ 2 − 2 F1 (x) + 5 M2 (RANGE) (−) 1 ▼ 2 ▼ ▼ 0 ▼ 9 F5 (GRAPH)	3 x/θ/T x^2 − 2 x/θ/T + 5 RANGE −1 ENTER 2 ENTER 1 ENTER 0 ENTER 9 ENTER 1 ENTER
CASIO f_x-7700GB/f_x-6300G	HP 48SX
Cls EXE Range −1 EXE 2 EXE 1 EXE 0 EXE 9 EXE 1 EXE EXE EXE EXE EXE Graph 3 X x^2 − 2 X + 5 EXE	⬑ PLOT PLOTR ERASE ATTN ′ $Y \blacksquare 3 \times X \, y^x \, 2 - 2 \times X + 5$ ↱ PLOT ⬑ DRAW ERASE 1 +/− ENTER 2 XRNG 0 ENTER 9 YRNG DRAW

We can try to estimate the area under the graph by looking at the picture. There is a rectangular region 3 units wide and slightly more than 4 units high lying between the x-axis and the lowest point on the graph. This has an area slightly over 12. The remaining part of the region (under the curved ends of the graph) appears to have a total area that is less than the area of this rectangle, but more than half as big.

This suggests the actual area is between $12 + 6 = 18$ and $12 + 12 = 24$, which we see is correct.

EXAMPLE 4   In using this technique for evaluating definite integrals we have simply guessed an antiderivative $F(x)$ of $f(x)$. This may or may not be possible, but even if it can be done, there will be many possible choices for $F(x)$. After all, if $F(x)$ is an antiderivative of $f(x)$, then so is $F(x) + c$ for any constant c. Happily, if we use $F(x) + c$ to evaluate a definite integral of $f(x)$, we get

$$\int_a^b f(x)\,dx = \left(F(x) + c\right)\Big|_a^b = \left(F(b) + c\right) - \left(F(a) + c\right)$$
$$= F(b) + c - F(a) - c = F(b) - F(a)$$
$$= F(x)\Big|_a^b .$$

In other words, the c's cancel, and choosing different antiderivatives does not affect the answer. It will be enough to choose any convenient antiderivative, such as $\frac{1}{3}x^3$ for x^2, $-\cos x$ for $\sin x$, etc.

EXAMPLE 5   Evaluate

$$\text{(i)} \int_0^{\pi/2} \sin\theta\, d\theta; \qquad \text{(ii)} \int_0^{\pi/4} \sec^2 u\, du; \qquad \text{(iii)} \int_a^b x^n\, dx, \qquad (n \neq -1);$$

$$\text{(iv)} \int_0^1 e^{-5x}\, dx; \qquad \text{(v)} \int_{-1}^0 u^2 \cos(u^3 + 4)\, du; \qquad \text{(vi)} \int_0^1 \frac{1}{1 + x^2}\, dx.$$

Solution

(i) $\int_0^{\pi/2} \sin\theta\, d\theta = -\cos\theta \Big|_0^{\pi/2} = -\cos\left(\frac{\pi}{2}\right) - \left(-\cos(0)\right) = -0 - (-1) = 1.$

(ii) For this one we have to remember that $\frac{d}{du}\tan u = \sec^2 u$, and since $\sec^2 u$ is continuous on $\left[0, \frac{\pi}{4}\right]$,

$$\int_0^{\pi/4} \sec^2 u\, du = \tan u \Big|_0^{\pi/4} = \tan\left(\frac{\pi}{4}\right) - \tan(0) = 1 - 0$$
$$= 1.$$

(iii) An antiderivative of x^n is obviously $\frac{1}{n+1}x^{n+1}$, so

$$\int_a^b x^n\, dx = \frac{1}{n+1}x^{n+1}\Big|_a^b = \frac{1}{n+1}b^{n+1} - \frac{1}{n+1}a^{n+1}$$
$$= \frac{1}{n+1}\left(b^{n+1} - a^{n+1}\right).$$

(iv) If we think for a moment, it is possible to guess an antiderivative for e^{-5x}, namely, $F(x) = -\frac{1}{5}e^{-5x}$ (cf. Example 4.8.5(iv), p. 308).

$$\int_0^1 e^{-5x}\, dx = -\frac{1}{5}e^{-5x}\Big|_0^1 = -\frac{1}{5}e^{-5} - \left(-\frac{1}{5}e^0\right)$$
$$= \frac{1}{5}(1 - e^{-5}).$$

(v) To guess an antiderivative for $u^2 \cos(u^3 + 4)$, we should think of the Chain Rule. Writing $u^2 \cos(u^3 + 4) = \left(\frac{1}{3}\right)3u^2 \cos(u^3 + 4)$, we see that an antiderivative is $\frac{1}{3}\sin(u^3 + 4)$ (cf. Example 4.8.6(iii), p. 308). From this we get that

$$\int_{-1}^0 u^2 \cos(u^3 + 4)\, du = \frac{1}{3}\sin(u^3 + 4)\Big|_{-1}^0$$
$$= \frac{1}{3}\sin(4) - \frac{1}{3}\sin(3) = \frac{1}{3}\left(\sin(4) - \sin(3)\right)$$
$$\approx -0.8979.$$

(vi) Recall that $\frac{1}{1+x^2}$ is the derivative of $\arctan x$, so

$$\int_0^1 \frac{1}{1 + x^2}\, dx = \arctan x \Big|_0^1 = \arctan(1) - \arctan(0) = \frac{\pi}{4} - 0$$
$$= \frac{\pi}{4}.$$

In doing these problems we needed to have an antiderivative $\int f(x)\,dx$. Previously, we emphasized that the general antiderivative should always include a constant added at the end. Example 4 tells us that *for this purpose*, the evaluation of definite integrals, that constant is irrelevant. So we allow ourselves to omit it and write, for instance,

$$\int_{-\pi}^{\pi} \cos\theta\,d\theta = \sin\theta \Big|_{-\pi}^{\pi} = \sin(\pi) - \sin(-\pi)$$
$$= 0.$$

When we can guess an antiderivative of the integrand, it is easy to evaluate definite integrals. For some examples, guessing an antiderivative may be *extremely* difficult. Chapter 8 will be devoted to developing techniques for finding indefinite integrals (i.e., antiderivatives) of many different types of functions.

Exercises 5.6

I

Evaluate the following integrals.

1. $\int_0^2 (x^4 - 2x)\,dx$

2. $\int_{-2}^1 x^7\,dx$

3. $\int_1^3 (4x^3 - 6x^2 - 4x)\,dx$

4. $\int_{-2}^2 (x^7 - 3x^3)\,dx$

5. $\int_0^\pi \sin u\,du$

6. $\int_{-\pi}^0 \sin t\,dt$

7. $\int_{-1}^1 \sin x\,dx$

8. $\int_{-1}^1 \cos x\,dx$

9. $\int_0^1 e^x\,dx$

10. $\int_1^e \frac{1}{x}\,dx$

11. $\int_0^1 e^{2x}\,dx$

12. $\int_{-1}^0 e^{-x}\,dx$

13. $\int_0^\pi \sin(2t)\,dt$

14. $\int_0^\pi \sin(5t)\,dt$

15. $\int_0^3 \sin(\pi u)\,du$

16. $\int_0^2 \sec^2\left(\frac{\pi}{8}x\right)dx$

17. $\int_0^4 \sqrt{x}\,dx$

18. $\int_0^1 \cosh t\,dt$

19. $\int_0^1 x(x^2 + 3)^{17}\,dx$

20. $\int_0^1 \frac{x}{x^2+1}\,dx$

21. $\int_0^3 |x - 2|\,dx$

22. $\int_1^4 |x - 2|\,dx$

23. $\int_0^3 |x^2 - 4|\,dx$

24. $\int_{-4}^4 |x^2 - 4|\,dx$

25. $\int_{-4}^4 |x^2 - x - 6|\,dx$

26. $\int_0^3 |t - 1||t - 2|\,dt$

27. $\int_{-2}^3 |x - 2||x^2 - 1|\,dx$

28. $\int_{-3}^2 |x^3 - x|\,dx$

29. $\int_0^2 |x - 1|^2\,dx$

30. $\int_{-3}^3 |x^2 - x - 2|\,dx$

II

Evaluate the following.

31. (i) $\int_0^{\pi/2} \cos(2x)\,dx$; (ii) $\int_{-1}^1 e^{x/2}\,dx$; (iii) $\int_{-1}^0 \sin(\pi x)\,dx$

32. (i) $\int_{-\pi/2}^0 \cos^2 t \sin t\,dt$; (ii) $\int_0^1 \frac{3}{u^2+1}\,du$; (iii) $\int_0^1 \frac{u^2}{u^2+1}\,du$

If you have studied the hyperbolic functions in Section 3.5, evaluate the following definite integrals.

33. (i) $\int_0^1 \cosh t\,dt$; (ii) $\int_{-3}^3 \sinh u\,du$; (iii) $\int_0^2 \frac{1}{\cosh^2 x}\,dx$

34. (i) $\int_0^1 \cosh t\,dt - \int_{-1}^0 \cosh t\,dt$; (ii) $\int_{-1}^0 \sinh 4t\,dt$

The next four questions concern the functions $f(x)$ and $g(x)$ defined below. Let

$$f(x) = \begin{cases} x, & \text{if } x \leq -1, \\ x^2 - 2, & \text{if } -1 \leq x \leq 2, \\ x^3 - 1, & \text{if } 2 < x; \end{cases}$$

$$g(x) = \begin{cases} x^2, & \text{if } x \leq 0, \\ 0, & \text{if } 0 \leq x < 1, \\ 3, & \text{if } 1 \leq x. \end{cases}$$

35. Evaluate (i) $\int_{-2}^3 f(x)\,dx$, (ii) $\int_{-1}^3 f(x)\,dx$, (iii) $\int_{-3}^3 f(x)\,dx$.

36. With $f(x)$ and $g(x)$ as above, evaluate (i) $\int_0^2 g(x)\,dx$, (ii) $\int_{-1}^1 g(x)\,dx$, (iii) $\int_{-1}^2 g(x)\,dx$.

37. With $f(x)$ and $g(x)$ as above, evaluate (i) $\int_0^2 f(x)g(x)\,dx$, (ii) $\int_{-1}^3 \big(f(x) - 2g(x)\big)\,dx$, (iii) $\int_{-2}^2 \sqrt{g(x)}\,dx$.

38. With $f(x)$ and $g(x)$ as above, evaluate (i) $\int_0^1 f(x)g(x)\,dx$, (ii) $\int_{-2}^{-1} \frac{1}{g(x)}\,dx$, (iii) $\int_{-2}^0 f(x)^2 g(x)\,dx$.

*39. (i) Suppose $a > 0$ and evaluate $\int_{-a}^a x\,dx$.

(ii) Suppose $a > 0$ and evaluate $\int_{-a}^a x^3\,dx$.

(iii) Suppose $a > 0$ and evaluate $\int_{-a}^a x^k\,dx$, where k is any odd positive integer.

*40. Suppose $a > 0$ and $f(x)$ is an odd function. What can you say about $\int_{-a}^{a} f(x)\,dx$?

*41. (i) Suppose $a > 0$ and evaluate $\int_{0}^{a} x^2\,dx$, $\int_{-a}^{0} x^2\,dx$.

   (ii) Suppose $a > 0$ and evaluate $\int_{0}^{a} x^4\,dx$, $\int_{-a}^{0} x^4\,dx$.

   (iii) Suppose $a > 0$ and evaluate $\int_{0}^{a} x^k\,dx$, $\int_{-a}^{0} x^k\,dx$, where k is any positive even integer.

*42. Suppose $a > 0$ and $f(x)$ is an even function. What can you say about $\int_{-a}^{0} f(x)\,dx - \int_{0}^{a} f(x)\,dx$?

*43. Evaluate
$$\int_{0}^{1}\left(\int_{1}^{x} 2t\,dt\right)dx.$$

*44. Evaluate
$$\int_{-1}^{2}\left(\int_{1}^{2} xt\,dt\right)dx.$$

*45. The second version of the Fundamental Theorem of Calculus (Theorem 5.6.1) was proved from the first version of the Fundamental Theorem of Calculus (Theorem 5.5.4). Show that it can be done the other way around, that is, that if we assume Theorem 5.6.1, we can use it to prove Theorem 5.5.4, provided that we know that the integrand $f(x)$ has an antiderivative.

*46. The function $f(x) = x^2$ is always nonnegative, so its integral should always be nonnegative, by Theorem 5.4.2. However,
$$\int_{0}^{-1} f(x)\,dx = \frac{1}{3}\left((-1)^3 - 0^3\right) = -\frac{1}{3}.$$

Explain why this is not really a contradiction.

For each of the following integrals, evaluate the Riemann sum approximation from Example 5.2.3 (p. 356) with $n = 100$ and compare with the exact value.

47. $\int_{0}^{4} x^3\,dx$ **49.** $\int_{-1}^{1} \frac{dt}{1+t^2}$

48. $\int_{0}^{\pi} \sin\theta\,d\theta$ **50.** $\int_{0}^{\pi/4} \frac{1}{\cos^2\theta}\,d\theta$

POINT TO PONDER

Now that we have finally developed a convenient way to integrate, it might be interesting to remind ourselves that the chapter began with a question about areas. Try to trace the connection between (signed) areas under curves and expressions of the form $F(b) - F(a)$.

In particular, convince yourself that the two coincide for polygonal regions, that is, regions with straight sides (i.e., integrating functions whose graphs are made up of pieces of straight lines).

For more complicated regions there is no a priori concept of area (cf. the Point to Ponder in Section 5.1), so integrals can provide a way of defining what areas mean. We should at least be sure that they satisfy the properties we expect of areas. If you thought of some of these properties in Section 5.1, try to check that they hold for areas defined by integrals.

5.7 Integration by Substitution

CHAIN RULE
DEFINITE INTEGRALS

In Section 5.6 we learned that evaluating definite integrals is quite easy *provided* that we know an antiderivative of the integrand, that is, an indefinite integral. When we discussed finding antiderivatives in Section 4.8, we found that there were many instances in which it was possible to do it by guessing, but others were more complicated. Now that we have seen the importance of being able to find indefinite integrals, we will have to learn techniques for it. This will be the purpose of all of Chapter 8, but in this section we will begin with the first of these techniques.

When we first encountered antiderivatives in Section 4.8, one kind of function whose antiderivative was less easy to guess was the kind that resulted from applying the Chain Rule to a composite function. We developed an approach to help with this type of problem, and it is the purpose of this section to make that approach more systematic. Being able to speak in terms of indefinite integrals instead of antiderivatives makes it easier to do than when we first attempted the problem.

We will need to use the concept of differentials from Section 4.10.

EXAMPLE 1   Find an antiderivative of xe^{x^2}, that is, find $\int xe^{x^2}\,dx$.

Solution   From Section 4.8 we recall how to approach this sort of question. Writing $xe^{x^2} = \frac{1}{2}2xe^{x^2}$, we recognize the factor $2x$ as the derivative of the function x^2, so $2xe^{x^2}$ is the derivative of $e^{x^2} = e^u$, with $u = x^2$. So an antiderivative of xe^{x^2} is $\frac{1}{2}e^{x^2} + c$.   ◢

This solves the problem, but let us pursue a little further what happens if we try to write it in terms of indefinite integrals.

The approach we just used suggests letting $u = x^2$. So $e^{x^2} = e^u$, and we could try substituting this into the integral:

$$\int xe^{x^2}\,dx = \int xe^u\,dx.$$

The difficulty here is that there are two different variables, x and u, appearing in the same integral. Since $u = x^2$, we could consider trying to write x as a function of u (i.e., solving for x as a function of u). Doing this could help us get rid of the factor x at the beginning of the integrand, but it would leave us with the apparently more difficult problem of what to do about the dx at the end.

Fortunately, the solution is not so difficult. We deal with du and dx by thinking of them as differentials, and since $u = x^2$, we know that

$$du = 2x\,dx.$$

Dividing by 2, we get $\frac{1}{2}du = x\,dx$. This allows us to rewrite the integral:

$$\int xe^{x^2}\,dx = \int e^{x^2}x\,dx = \int e^u\frac{1}{2}du$$

$$= \frac{1}{2}\int e^u\,du.$$

Notice that the factor x in the original integrand combined with the dx to give $\frac{1}{2}du$.

The point of all this is that it is very easy to evaluate the last integral: $\frac{1}{2}\int e^u\,du = \frac{1}{2}e^u + c$. In a sense this is the answer, but the original problem was stated in terms of the variable x, so we ought to convert our answer back into a function of x. Since $u = x^2$, nothing could be easier:

$$\frac{1}{2}e^u + c = \frac{1}{2}e^{x^2} + c.$$

This seems like a lot of work to do something that we already knew how to do with less trouble, but in fact this approach will be extremely useful.

EXAMPLE 2   Find $\int x^2(x^3-1)^{11}dx$.

Solution   We begin by observing that the x^2 factor looks almost like the derivative of the quantity inside parentheses. So we let that quantity equal u, that is, $u = x^3 - 1$.

Then we work with differentials:

$$du = \frac{du}{dx}dx = 3x^2 dx.$$

Since the original integral contained $x^2 dx$, we can divide the differentials by 3 to get $\frac{1}{3}du = x^2 dx$.

Now the original integral can be written as

$$\int x^2(x^3-1)^{11}dx = \int (x^3-1)^{11}x^2 dx = \int u^{11}\frac{1}{3}du$$

$$= \frac{1}{3}\int u^{11}du = \frac{1}{3}\frac{1}{12}u^{12} + c$$

$$= \frac{1}{36}(x^3-1)^{12} + c.$$

There are several things to notice here. One is that because u is defined as a function of x, it is relatively easy to find the differential du, so that is what we did. However, the way we used this formula was really the other way around, allowing us to change dx into something involving du.

In fact it was not dx but rather $x^2 dx$ that we expressed in terms of du. The key to the whole calculation was recognizing that that factor x^2 looked like the derivative of $x^3 - 1$, so if we let $u = x^3 - 1$ then $du = 3x^2 dx$. Because of this we could express the integral in terms of u, and the $x^2 dx$ would turn into a constant times du, leaving the simple integrand u^{11}. In making this recognition we were able to ignore the constant 3, knowing that it would be taken care of in the computations.

The second thing to notice is the very last step, in which the answer was converted back from a function of u to a function of x. This is easy but important.

Finally, as always, we should not forget the "$+c$" at the end of the indefinite integral.

The technique we have illustrated with these two examples is called *integration by substitution* because we substitute the variable u for some function of x and remove the variable x. The idea and the calculations are relatively easy; what can sometimes make it difficult is the absence of a hard and fast rule. All we can do is guess a substitution, try it out, and see if it works. We will agree that it works if the original integral is converted into one that is easily integrated. Otherwise, it is probably not of much use.

It is important to recognize that there is a process of guessing, of trial and error. Nobody expects to see the correct substitution immediately and be certain that it will work. It is necessary to guess and then to try out your guess and, if it does not work, to guess again.

One vital requirement for any substitution to succeed is that the original integral (in the variable x, say) should be converted into one in terms of u, say. There should be no x's in the integral after the substitution. If some occurrences of x do remain, we should try to express them in terms of u. If this is not possible, the attempt must be abandoned and a new substitution found.

Summary 5.7.1

Integration by Substitution

Consider the indefinite integral $\int f(x)\,dx$.

1. Look at the integrand; try to find in it a factor $u'(x)$, so that the remaining part of the integrand can be expressed as a (relatively) simple function of u. At this stage, do not worry too much about constants.

2. Make the substitution. Letting u be the function suggested by the observation just made, calculate the differential $du = u'(x)\,dx$, and rewrite the integral in terms of u. If there are any x's left in the integral, try to express them in terms of u; if this is impossible, go back to Step 1 and look for another substitution.

3. Evaluate the indefinite integral in terms of u, being sure to include the constant c at the end.

4. Substitute back for u in the resulting function to write it as a function of the original variable x.

If this procedure is unsuccessful, something else may be necessary. One trick that is frequently helpful is to express the integral as a sum of two or more integrals and then work with each of them separately. Sometimes different substitutions will work for different parts of the integrand. Far in the back of your mind you should know that there are some integrals that cannot be evaluated by the technique of substitution, but you should not let that knowledge discourage you from trying.

The best thing about indefinite integrals is that it is very easy to check your work. Once you have finished, all that is needed to verify the answer is to differentiate it. The resulting derivative should equal the original integrand.

EXAMPLE 3   Evaluate the following indefinite integrals:

(i) $\int \cos^2 x \sin x\,dx$,   (iv) $\int e^{2-3u}\,du$,

(ii) $\int (x+1)\sqrt{x^2+2x+4}\,dx$,   (v) $\int x(x+1)^6 dx$,

(iii) $\int (3t-1)^3 dt$,   (vi) $\int \tan\theta\,d\theta$.

Solution

(i) Here we notice that, up to a minus sign, $\sin x$ is the derivative of $\cos x$. Letting $u = \cos x$, we find $du = -\sin x\,dx$, so

$$\int \cos^2 x \sin x\,dx = \int u^2(-1)\,du = -\int u^2\,du = -\frac{1}{3}u^3 + c$$

$$= -\frac{1}{3}\cos^3 x + c.$$

(ii) $\int (x+1)\sqrt{x^2+2x+4}\,dx$. Here we can guess that the factor $x+1$ is (approximately) the derivative of the quantity inside the square root sign. Letting u equal that quantity, $u = x^2 + 2x + 4$, we find $du = (2x+2)\,dx = 2(x+1)\,dx$.

So

$$\int (x+1)\sqrt{x^2+2x+4}\,dx = \int \sqrt{x^2+2x+4}(x+1)\,dx = \int \sqrt{u}\frac{1}{2}\,du$$

$$= \frac{1}{2}\int u^{1/2}\,du = \frac{1}{2}\frac{1}{3/2}u^{3/2} + c$$

$$= \frac{1}{3}(x^2+2x+4)^{3/2} + c.$$

This result is easily checked by differentiation.

(iii) $\int (3t-1)^3\,dt$. At first glance there does not seem to be anything that is the derivative of anything else here. However, a nice choice for the function u would be $3t-1$, the function inside parentheses. Its derivative is just 3, and with our usual understanding about constants we should try this substitution.

If $u = 3t-1$, then $du = 3\,dt$, and

$$\int (3t-1)^3\,dt = \int u^3\frac{1}{3}\,du = \frac{1}{3}\int u^3\,du = \frac{1}{3}\times\frac{1}{4}u^4 + c$$

$$= \frac{1}{12}(3t-1)^4 + c.$$

Notice that in this case the original integral was expressed in terms of the dummy variable t instead of x, but that had no effect on the procedure we used.

(iv) $\int e^{2-3u}\,du$. Here again, there is no apparent derivative until we recognize that the exponent $2-3u$ has as its derivative the constant -3. Since in this case the original integral is in terms of the variable u, it would be confusing to use the same letter for the substitution.

Letting $v = 2-3u$, we find $dv = (-3)\,du$, and

$$\int e^{2-3u}\,du = \int e^v\left(-\frac{1}{3}\right)\,dv = -\frac{1}{3}\int e^v\,dv = -\frac{1}{3}e^v + c$$

$$= -\frac{1}{3}e^{2-3u} + c.$$

(v) $\int x(x+1)^6\,dx$. Our first guess here might be to let $u = x+1$. In this case, $du = dx$, which looks fine, but when we substitute, we get $\int x(x+1)^6\,dx = \int xu^6\,du$, and there is still that factor of x in the integral. We could despair, and simply multiply out $(x+1)^6$ and integrate each term. However, first we should try to express that x in terms of u. After all, if $u = x+1$, then $x = u-1$. So we are able to perform the integration:

$$\int x(x+1)^6\,dx = \int xu^6\,du = \int (u-1)u^6\,du$$

$$= \int (u^7 - u^6)\,du = \frac{1}{8}u^8 - \frac{1}{7}u^7 + c$$

$$= \frac{1}{8}(x+1)^8 - \frac{1}{7}(x+1)^7 + c.$$

This was definitely easier than expanding $(x+1)^6$ and integrating. It has the additional advantage that the answer is expressed in a much more compact form.

(vi) $\int \tan\theta \, d\theta$. At first glance there seems to be nothing we can do with $\tan\theta$, but as always with trigonometric functions, the thing to do when you are stumped is to write everything in terms of $\cos x$ and $\sin x$.

$$\int \tan\theta \, d\theta = \int \frac{\sin\theta}{\cos\theta} \, d\theta.$$

Now things look better. Each of $\sin\theta$ and $\cos\theta$ is the other's derivative (up to the usual sign). If we tried $u = \sin\theta$, we would find $du = \cos\theta \, d\theta$. The difficulty with this is that it is not what appears in the integral. There we have $\frac{1}{\cos\theta} \, d\theta$, which is quite different.

Not discouraged, we try the other one. Letting $u = \cos\theta$, we have $du = -\sin\theta \, d\theta$, so

$$\int \tan\theta \, d\theta = \int \frac{\sin\theta}{\cos\theta} \, d\theta = \int \frac{1}{u}(-1)\, du = -\int \frac{1}{u} \, du$$
$$= -\ln|u| + c = -\ln|\cos\theta| + c$$
$$= \ln|\sec\theta| + c. \qquad\blacktriangleleft$$

Part (vi) of Example 3 is interesting in that neither the integrand nor the integral is defined for all real numbers. Depending on the use to which we intend to put it, it will be important to be careful about how it is interpreted.

Notice too that we have allowed ourselves to write

$$\int \frac{1}{u} \, du = \ln|u| + c,$$

instead of the more complicated antiderivative discussed in Example 4.8.4 (p. 306). This is pure laziness, but it will not cause much trouble. It is our intention to use this technique for evaluating definite integrals, and in light of Example 5.6.4 (p. 395), omitting the extra constant there will not matter.

Now that we have begun to get used to the technique of substitution, it is desirable to have a more efficient way of writing down what we are doing. A convenient system is to write down the substitution in a little box beside the integral. Since it is necessary to find the differential of the function being substituted, we write that in the box too.

EXAMPLE 4   Find the following indefinite integrals:

(i) $\int \frac{\sin(\sqrt{t})}{\sqrt{t}} \, dt$,

(ii) $\int \tan^3 x \sec^2 x \, dx$,

(iii) $\int x^3 \sqrt{x^2 + 2} \, dx$,

(iv) $\int \frac{1}{4 + x^2} \, dx$.

Solution

(i) $\int \frac{\sin(\sqrt{t})}{\sqrt{t}} \, dt$. It is often helpful to rewrite square roots as powers with exponent $\frac{1}{2}$. Doing this, we find

$$\int \frac{\sin(\sqrt{t})}{\sqrt{t}} \, dt = \int \frac{\sin(t^{1/2})}{t^{1/2}} \, dt$$
$$= \int \sin(t^{1/2}) t^{-1/2} \, dt.$$

Since, apart from a constant, $t^{-1/2}$ is the derivative of $t^{1/2}$, this suggests what substitution we should try:

$$\int \frac{\sin(\sqrt{t})}{\sqrt{t}}\,dt = \int \sin(t^{1/2})\,t^{-1/2}\,dt$$

$$\boxed{\begin{array}{l} u = t^{1/2} \\[4pt] du = \frac{1}{2}t^{-1/2}\,dt \end{array}} \quad \begin{aligned} &= \int (\sin u)\,2\,du = 2\int \sin u\,du \\ &= -2\cos u + c = -2\cos(t^{1/2}) + c \\ &= -2\cos(\sqrt{t}) + c. \end{aligned}$$

(ii) $\int \tan^3 x \sec^2 x\,dx$. For this one we ought to stare at it for a while, and with luck we will see what to do.

$$\int \tan^3 x \sec^2 x\,dx = \int u^3\,du$$

$$\boxed{\begin{array}{l} u = \tan x \\[4pt] du = \sec^2 x\,dx \end{array}} \quad \begin{aligned} &= \frac{1}{4}u^4 + c \\[6pt] &= \frac{1}{4}\tan^4 x + c. \end{aligned}$$

(iii) $\int x^3\sqrt{x^2 + 2}\,dx$. Here it is not immediately clear what to do. If we try $u = x^3$, we find its derivative disguised in the x^2, but that term is inside the square root, where it does not easily combine with the dx to make the differential du.

A second attempt would be to try the function inside the square root:

$$\int x^3\sqrt{x^2 + 2}\,dx = \int x^2\sqrt{x^2 + 2}\,x\,dx$$

Since $u = x^2 + 2$, we find $x^2 = u - 2$.

$$\boxed{\begin{array}{l} u = x^2 + 2 \\[4pt] du = 2x\,dx \end{array}} \quad \begin{aligned} &= \int x^2\sqrt{u}\,\frac{1}{2}\,du = \frac{1}{2}\int (u - 2)u^{1/2}\,du \\[6pt] &= \frac{1}{2}\int (u^{3/2} - 2u^{1/2})\,du = \frac{1}{2}\left(\frac{2}{5}u^{5/2} - 2\times\frac{2}{3}u^{3/2}\right) + c \\[6pt] &= \frac{1}{5}(x^2 + 2)^{5/2} - \frac{2}{3}(x^2 + 2)^{3/2} + c. \end{aligned}$$

(iv) $\int \frac{1}{4+x^2}\,dx$. Here there does not seem to be any obvious substitution available, but it will help to rewrite the integrand:

$$\frac{1}{4 + x^2} = \frac{1}{4}\left(\frac{1}{1 + \frac{x^2}{4}}\right) = \frac{1}{4}\left(\frac{1}{1 + \left(\frac{x}{2}\right)^2}\right).$$

This suggests the substitution $u = \frac{x}{2}$:

$$\int \frac{1}{4 + x^2}\,dx = \frac{1}{4}\int \frac{1}{1 + \left(\frac{x}{2}\right)^2}\,dx$$

$$\boxed{\begin{array}{l} u = \frac{x}{2} \\[4pt] du = \frac{1}{2}\,dx \end{array}} \quad \begin{aligned} &= \frac{1}{4}\int \frac{1}{1 + u^2}\,(2)\,du = \frac{1}{2}\int \frac{1}{1 + u^2}\,du \\[6pt] &= \frac{1}{2}\arctan u + c \\[6pt] &= \frac{1}{2}\arctan\left(\frac{x}{2}\right) + c. \end{aligned}$$

We include two integrals that will be needed in the next chapter.

EXAMPLE 5   Find the following indefinite integrals:

(i) $\displaystyle\int \cos^2 x\, dx,$ (ii) $\displaystyle\int \sin^2 x\, dx.$

Solution

(i) Since no obvious substitution presents itself, it is necessary to resort to a trick. Recalling the trigonometric identity from Theorem 1.7.12 (p. 51),

$$\cos 2x = \cos^2 x - \sin^2 x,$$

we write $\sin^2 x = 1 - \cos^2 x$ and get

$$\cos 2x = \cos^2 x - (1 - \cos^2 x) = 2\cos^2 x - 1.$$

Adding 1 to both sides gives $\cos(2x) + 1 = 2\cos^2 x$, or $\cos^2 x = \frac{\cos(2x)+1}{2}$. So

$$\int \cos^2 x\, dx = \int \left(\frac{\cos(2x) + 1}{2} \right) dx = \frac{1}{2} \int \cos(2x)\, dx + \frac{1}{2} \int 1\, dx$$

$$\boxed{\begin{array}{l} u = 2x \\ du = 2\, dx \end{array}} \qquad = \frac{1}{2}\left(\frac{1}{2}\right) \int \cos u\, du + \frac{1}{2} \int 1\, dx$$

$$= \frac{1}{4} \sin u + \frac{x}{2} + c$$

$$= \frac{1}{4} \sin(2x) + \frac{x}{2} + c.$$

(ii) We can integrate $\int \sin^2 x\, dx$ by a similar trick, or by using the fact that $\sin^2 x = 1 - \cos^2 x$, so

$$\int \sin^2 x\, dx = \int (1 - \cos^2 x)\, dx = \int 1\, dx - \int \cos^2 x\, dx$$

$$= x - \left(\frac{1}{4} \sin(2x) + \frac{x}{2} \right) + c$$

$$= \frac{x}{2} - \frac{1}{4} \sin(2x) + c.$$

Notice that when we did the integrations in part (ii), we wrote $\frac{1}{4}\sin(2x) + \frac{x}{2}$ for $\int \cos^2 x\, dx$, without the added constant, and added a constant c after all the integrations. This way we end up with the usual "$+c$" at the end. If we included the constant in the integral of $\cos^2 x$, then our final formula would have ended with "$-c$." This complication is unnecessary.

Essentially all that matters is that there must be a "$+c$" with any indefinite integral. As long as it is there when we finish, we do not have to worry about where it came from. In particular, if we have a formula that involves the sum of several integrals, it is not necessary to add a constant to each of them; a single constant is enough. In the above calculation, where we integrated 1 and $\cos^2 x$, all we needed was *one* indefinite integral for each of them. We chose x and $\frac{1}{4}\sin(2x) + \frac{x}{2}$, respectively, and then added the required constant at the end.

Definite Integrals and Substitution

One of the main purposes of finding indefinite integrals is to use them to calculate definite integrals. If we have a definite integral to evaluate, we may find it necessary to use substitution to find the indefinite integral, after which the definite integral can be found in the usual way, using Theorem 5.6.1.

EXAMPLE 6   Evaluate

$$\int_0^{\pi/4} \sec^2 t \sin t \, dt.$$

Solution   First we find the indefinite integral $\int \sec^2 t \sin t \, dt$. The presence of the factor $\sec^2 t$ suggests the $\tan t$ substitution, but that does not seem to get us anywhere. Instead, we write $\sec^2 t = \frac{1}{\cos^2 t}$ and try

$$\int \sec^2 t \sin t \, dt = \int \frac{1}{\cos^2 t} \sin t \, dt$$

$$\boxed{\begin{aligned} u &= \cos t \\ du &= -\sin t \, dt \end{aligned}} \quad = \int u^{-2}(-1)\, du = -\int u^{-2}\, du$$

$$= -\frac{1}{-1} u^{-1} + c = \frac{1}{\cos t} + c$$

$$= \sec t + c.$$

Now it is easy to evaluate the definite integral:

$$\int_0^{\pi/4} \sec^2 t \sin t \, dt = \sec t \Big|_0^{\pi/4} = \sqrt{2} - 1 \approx 0.41421.$$

This is a perfectly reasonable way to evaluate integrals. It is also possible to do it another way, in which the substitution is performed on the definite integral rather than the indefinite integral. The idea is to transform the definite integral in the original variable into a definite integral in the substituted variable. To do this, it is necessary to change the limits of integration in accordance with the substitution.

EXAMPLE 7   Find

$$\int_2^3 \frac{t}{t-1} \, dt.$$

Solution   After some thought, we realize that a good substitution to make would be $u = t - 1$. In this case we find that $du = dt$ and the numerator of the integrand is $t = u + 1$.

It is easy to write down the new form of the integral, but we have to consider what happens to the limits of integration as we make the substitution.

When $t = 2$, we see that $u = t - 1 = 1$, and when $t = 3$, $u = t - 1 = 2$. So we can write the integral as follows:

$$\int_2^3 \frac{t}{t-1} \, dt = \int_1^2 \frac{u+1}{u} \, du = \int_1^2 \left(\frac{u}{u} + \frac{1}{u} \right) du$$

$$\boxed{\begin{aligned} u &= t-1 \\ du &= dt \end{aligned}} = \int_1^2 \left(1+\frac{1}{u}\right) du = (u + \ln|u|)\Big|_1^2$$
$$= (2 + \ln 2) - (1 + \ln 1) = 1 + \ln 2$$
$$\approx 1.693.$$

Either of these methods will work perfectly well, and it is largely a matter of personal taste which one you wish to use. Finding the indefinite integral and then evaluating the definite integral gives the appearance of involving a separate step, but the second method involves figuring out the new limits of integration.

There is one additional subtlety that occasionally causes some trouble. Fortunately, it does not usually matter, but for the second approach to work, it is necessary that the substituted function have a continuous derivative on the interval of integration and that the integrand be continuous and have an antiderivative that is defined at every point of the interval.

EXAMPLE 8 Evaluate

$$\int_{-\pi/2}^{\pi/2} \frac{\cos x}{\sin^2 x}\, dx.$$

Solution This is quite similar to Example 6. We are tempted to try $u = \sin x$, in which case $du = \cos x\, dx$. When $x = -\frac{\pi}{2}$, $u = \sin x = \sin\left(-\frac{\pi}{2}\right) = -1$, and when $x = \frac{\pi}{2}$, $u = \sin\left(\frac{\pi}{2}\right) = 1$.

This seems to suggest the following calculation:

$$\int_{-\pi/2}^{\pi/2} \frac{\cos x}{\sin^2 x}\, dx = \int_{-1}^{1} u^{-2}\, du$$

$$\boxed{\begin{aligned} u &= \sin x \\ du &= \cos x\, dx \end{aligned}} = \frac{1}{-1} u^{-1}\Big|_{-1}^{1} = -\big(1 - (-1)\big)$$
$$= -2.$$

This calculation is *wrong!* For one thing, for x between $-\frac{\pi}{2}$ and $\frac{\pi}{2}$, $\cos x$ is always nonnegative, so the whole integrand is nonnegative. Theorem 5.4.2 says that the integral of a nonnegative integrand should be nonnegative, so it certainly could not equal -2.

The reason it is not permissible to do the calculation suggested above is that the integrand is not defined at $x = 0$, which lies in the interval of integration $\left[-\frac{\pi}{2}, \frac{\pi}{2}\right]$ (see Figure 5.7.1).

In fact the integrand is not integrable over this interval (because it is not bounded there), so any calculation that purports to integrate it will give a meaningless result. If an integrand goes to infinity somewhere in an interval, we cannot integrate it over that interval.

We conclude with a careful statement of the theorem that explains when we can perform a substitution and change the limits of integration.

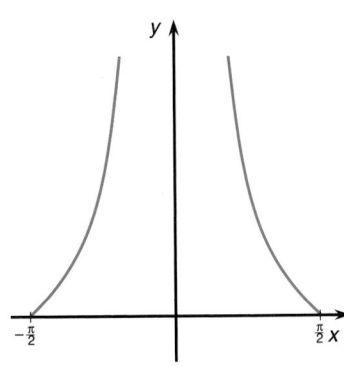

Figure 5.7.1

THEOREM 5.7.2   Suppose $g(x)$ is a function whose derivative $g'(x)$ exists and is continuous on an interval $[a, b]$.

Suppose $f(x)$ is a function that is continuous and has an antiderivative $F(x)$ on an interval that contains all the values of $g(x)$ for $x \in [a, b]$. Then

$$\int_a^b f(g(x))g'(x)\,dx = \int_{g(a)}^{g(b)} f(u)\,du = F(u)\Big|_{g(a)}^{g(b)}.$$

PROOF Consider the function $h(x) = F(g(x))$. It is defined on $[a, b]$, and it is differentiable there. Moreover, by the Chain Rule,

$$h'(x) = F'(g(x))g'(x) = f(g(x))g'(x).$$

In other words, $h(x)$ is an indefinite integral for $f(g(x))g'(x)$. By the second version of the Fundamental Theorem (Theorem 5.6.1) we can evaluate the definite integral using $h(x)$:

$$\int_a^b f(g(x))g'(x)\,dx = h(x)\Big|_a^b = h(b) - h(a)$$
$$= F(g(b)) - F(g(a)).$$

But since $F(x)$ is an indefinite integral of $f(x)$, we know that

$$F(g(b)) - F(g(a)) = \int_{g(a)}^{g(b)} f(u)\,du.$$

This is the desired result. In this last formula, u is just a dummy variable; we could have used x (or t, or anything else), but there is less danger of confusion with u. ◢

This theorem is the precise statement behind the technique of integration by substitution. If we can find a function $u = g(x)$ so that the integrand can be written in the form $f(g(x))g'(x) = f(u)u'$, then the original integral $\int f(g(x))g'(x)\,dx$ can be transformed into $\int f(u)\,du$. It also tells us what happens to the limits of integration.

The technique of substitution that we have learned in this section is extremely useful. It is not easy to master, especially since it involves considerable guesswork. Nothing can take the place of practice and experience, which can help suggest what will and what will not work.

It is very important to do a large number of exercises to consolidate your skill at doing substitutions.

Exercises 5.7

Evaluate the following definite or indefinite integrals.

1. $\int e^{4x-2}\,dx$

2. $\int x\cos(x^2)\,dx$

3. $\int t^2(t^3 - 2)^7\,dt$

4. $\int (x+2)(x^2 + 4x - 3)^5\,dx$

5. $\int_0^3 (t-2)^4\,dt$

6. $\int_1^2 \frac{1}{(x+1)^2}\,dx$

7. $\int_{-1}^2 \sin \pi x\,dx$

8. $\int_0^1 x^2 e^{x^3}\,dx$

9. $\int \tan(\pi x)\sec^3(\pi x)\,dx$

10. $\int x^2\cos(5x^3 + 7)\,dx$

11. $\int_0^{\pi/4} \sin^2 x\,dx$

12. $\int_0^\pi \sin^2 x\,dx$

13. $\int_0^{2\pi} \cos^2 \theta\,d\theta$

14. $\int_{-\pi}^\pi \sin^2 t\,dt$

15. $\int \cos^2(3x)\,dx$

16. $\int \sin^2(-5x)\,dx$

17. $\int_0^1 \cos^2(\pi x)\,dx$

18. $\int_0^1 \frac{1}{1+x}\,dx$

19. $\int_2^3 \frac{x^2}{x^3-1}\,dx$

20. $\int \frac{x^k}{1+x^{k+1}}\,dx,\ k \neq -1$

21. $\int \frac{x}{\sqrt{1+x^2}}\,dx$

22. $\int_0^1 x^2\sqrt{1-x^3}\,dx$

23. $\int \frac{1}{\sqrt{x}(\sqrt{x}+2)^3}\,dx$

24. $\int_0^1 s\sqrt{s+1}\,ds$

25. $\int_1^3 t(3-t)^{1/3}\,dt$

26. $\int e^{ax+b}\,dx$

27. $\int \cos(at)\,dt$

28. $\int_a^b \sin(\pi t)\,dt$

29. $\int (x^2 + x)\sqrt{2x^3 + 3x^2 + 2}\,dx$

30. $\int_0^a \cos\left(\frac{2\pi x}{a}\right)dx$

II

31. Find $\int \frac{1}{9 + t^2}\,dt$.

32. Let $a > 0$. Find $\int \frac{1}{a^2 + t^2}\,dt$.

33. Let $a > 0$ and let $f(x)$ be a function that is integrable on $[0, a]$. Show that $\int_0^a f(a - x)\,dx = \int_0^a f(x)\,dx$. (Draw a picture.)

34. Let $a < b$ and let $f(x)$ be a function that is integrable on $[a, b]$. Show that

$$\int_a^b f(a + b - x)\,dx = \int_a^b f(x)\,dx.$$

(Draw a picture.)

35. For any number a, evaluate

(i) $\int_a^{a+2\pi} \sin x\,dx$,   (ii) $\int_a^{a+2\pi} \sin^2 x\,dx$.

(*Hint:* The answers to parts (i) and (ii) do not depend on a.)

*36. Differentiate $g(x) = x \ln x$, and use your answer to help you guess the indefinite integral

$$\int \ln x\,dx.$$

*37. Suppose k and ℓ are nonnegative integers. Consider

$$\int \cos^k x \sin^{2\ell+1} x\,dx,$$

that is, the integral of the product of any power of $\cos x$ and an odd power of $\sin x$.

(i) Show that this integral can always be transformed into an integral that in principle we

know how to integrate, by making the substitution $u = \cos x$ and writing $\sin^2 x = 1 - \cos^2 x$. (Do not actually integrate the transformed integral.) Evaluate

(ii) $\int \cos^3 x \sin^3 x\,dx$,

(iii) $\int \sin^3 x\,dx$,

(iv) $\int \cos^4 x \sin^3 x\,dx$,

(v) $\int \cos^3 x \sin^2 x\,dx$.

(vi) Show that a similar trick will work for $\int \cos^k x \sin^\ell x\,dx$, provided that at least one of k and ℓ is odd.

*38. (i) Integrate $\int \cos^4 x\,dx$. (*Hint:* Look at Example 5, p. 405).

(ii) Integrate $\int \sin^4 x\,dx$.

*39. If we consider the integral $\int_a^b f(-x)\,dx$ and apply the substitution $u = -x$, we appear to have a minus sign in the resulting integral. But if the function $f(x)$ is always positive, for instance, then so is the function $f(-x)$, and its integral should also be positive. We appear to have found an equation saying that the (positive) integral of $f(-x)$ equals minus a (positive) integral of $f(x)$! Explain why there is really no difficulty here.

*40. Consider the integral $\int_0^1 x^2 (4 - x^3)^6\,dx$. It is clear that the integrand is nonnegative on $[0, 1]$, but after the substitution $u = 4 - x^3$ we obtain an integral of $-\frac{1}{3} u^6$, which is always *nonpositive*. Explain why this is not contradictory.

III

For each of the following integrals, evaluate the Riemann sum approximation from Example 5.2.3 with $n = 100$ and compare with the exact value.

41. $\int_0^1 \sin(\pi x)\,dx$ **43.** $\int_0^1 \frac{1}{4 + t^2}\,dt$

42. $\int_0^1 xe^{x^2}\,dx$ **44.** $\int_0^\pi \cos^2(2x)\,dx$

POINT TO PONDER

Try to see what the process of substitution has to do with signed areas. First consider the integral

$$\int_a^b f(x + 1)\,dx$$

and the substitution $u = x + 1$. Draw a picture.

Second try to figure out what happens in more complicated examples such as

$$\int_1^2 f(6 - 2x)\,dx \quad \text{or} \quad \int_0^1 3x^2 (x^3 + 1)^3\,dx.$$

Chapter Summary

§5.1

area under a curve
rectangle
approximating the area

§5.2

summation notation
index of summation
distributive law
$$\sum_{k=1}^{n} k = \frac{n(n+1)}{2}$$
$$\sum_{k=1}^{n} k^2 = \frac{n(n+1)(2n+1)}{6}$$

§5.3

partition
mesh, norm
subinterval
Riemann sum
signed area
integrable
integral $\int_{a}^{b} f(x)\,dx$
integrand
lower limit of integration
upper limit of integration
limits of integration
precise definition of integral (Definition 5.3.5)
limit as $\|\mathcal{P}\| \to 0$

§5.4

Integral Mean Value Theorem (Theorem 5.4.4)
integrating functions with jump or removable discontinuities

§5.5

indefinite integral
Fundamental Theorem of Calculus (Theorem 5.5.4)
variable of integration
dummy variable
antiderivative
$\int f(x)\,dx$

§5.6

definite integral
Fundamental Theorem of Calculus (second version)
(Theorem 5.6.1)
evaluating definite integrals

§5.7

integration by substitution
substitution and definite integrals

Review Exercises

I

Use the result of Example 5.2.3 (p. 356) to approximate the area under the graph of $y = f(x)$ on the specified interval, dividing the interval into n subintervals of equal length, with the specified value of n.

1. $f(x) = x^2$ on $[0, 2]$; $n = 4$

2. $f(x) = x^2$ on $[0, 2]$; $n = 6$

3. $f(x) = x^2$ on $[-3, 3]$; $n = 4$

4. $f(x) = |x^3|$ on $[-3, 3]$; $n = 6$

5. $f(x) = \frac{1}{x}$ on $[1, 2]$; $n = 3$

6. $f(x) = -\frac{1}{x}$ on $[-2, -1]$; $n = 3$

7. $f(x) = 7$ on $[-3, 4]$; $n = 9$

8. $f(x) = \sin x$ on $\left[0, \frac{\pi}{2}\right]$; $n = 2$

9. $f(x) = -x$ on $[-5, -1]$; $n = 8$

10. $f(x) = \cos x$ on $\left[0, \frac{\pi}{2}\right]$; $n = 2$

Evaluate the following (assume $m, n \geq 1$).

11. $\sum_{k=0}^{5} \ln(k + 1)$

12. $\sum_{k=1}^{n} \ln k$

13. $\sum_{k=0}^{4} 2^{-k}$

14. $\sum_{i=1}^{7} (i - 4)$

15. $\sum_{j=0}^{3} xj^3$

16. $\sum_{k=1}^{4} (k^3 - 2k^2 + k + 2)$

17. $\sum_{k=1}^{m} (k^2 - 2k + 1)$

18. $\sum_{\ell=m-4}^{m} (\ell + 2)$

19. $\sum_{i=-n}^{n} i$

20. $\sum_{j=-m}^{m} j^2$

In each of the following situations, let $\mathcal{P}_n$ be the partition that divides the interval $[0, 4]$ into n subintervals of equal length. Let x_k^ℓ, x_k^r, and x_k^m be the left endpoint, the right endpoint, and the midpoint, respectively, of the kth subinterval. Calculate the following Riemann sums.

21. $S\left(x^2 + 1, \mathcal{P}_4, \{x_k^r\}\right)$

22. $S\left(x + 2, \mathcal{P}_4, \{x_k^m\}\right)$

23. $S\left(3 - 6x, \mathcal{P}_3, \{x_k^\ell\}\right)$

24. $S\left(x^4, \mathcal{P}_4, \{x_k^m\}\right)$

25. $S\left(\sin(\pi x), \mathcal{P}_4, \{x_k^r\}\right)$

26. $S\left(\sin(\pi x), \mathcal{P}_4, \{x_k^m\}\right)$

27. $S\left(\cos(\pi x), \mathcal{P}_4, \{x_k^\ell\}\right)$

28. $S\left(\cos(\pi x), \mathcal{P}_4, \{x_k^m\}\right)$

29. $S\left(\frac{1}{x}, \mathcal{P}_4, \{x_k^r\}\right)$

30. $S\left(\frac{1}{x+1}, \mathcal{P}_4, \{x_k^\ell\}\right)$

Differentiate the following functions of x.

31. $\int_0^x (t^3 - 2t + 1)\, dt$

32. $\int_1^{x^2} \cos u \, du$

33. $\int_{-x}^x (t^7 - 5t^5 + 11t^3 + 6t - 1)\, dt$

34. $\int_{\cos x}^{\sin x} t \, dt$

35. $x \left(\int_1^x (t^2 - 1)\, dt \right)$

36. $\int_1^{\ln x} e^t \, dt$

37. $\int_1^{\exp x} \ln t \, dt$

38. $\left(\int_0^x \sin t \, dt \right) \left(\int_{-x}^0 \cos t \, dt \right)$

39. $\left(\int_1^x t \, dt \right) \left(\int_0^x \frac{2t}{(t^2 - 1)^2} \, dt \right)$

40. $\int_0^3 (x + 3t)\, dt$

Integrate each of the following.

41. $\int_0^5 (x^3 - 3x^2 + 4)\, dx$

42. $\int_1^3 e^t \, dt$

43. $\int_1^2 (4 - 2x^3)\, dx$

44. $\int_0^x e^t \, dt$

45. $\int_0^\pi \sin x \, dx$

46. $\int_1^x \frac{1}{t} \, dt$

47. $\int_0^{\pi/4} \sec^2 x \, dx$

48. $\int_0^1 \sqrt{x} \, dx$

49. $\int_0^8 t^{1/3} (t - 2)\, dt$

50. $\int_1^4 x^{-1/2}\, dx$

51. $\int x\sqrt{1 + 2x^2}\, dx$

52. $\int_0^1 x\sqrt{2 - x^2}\, dx$

53. $\int_1^2 t(t^3 + 2t - 3)\, dt$

54. $\int 5x \sin(x^2 - 1)\, dx$

55. $\int \frac{x-1}{\sqrt{x}+1}\, dx$

56. $\int e^{\ln x}\, dx$

57. $\int \exp(\sqrt{t})\, t^{-1/2}\, dt$

58. $\int_1^8 t^{-2/3} \sin\left(\frac{\pi}{2}(t^{1/3})\right)\, dt$

59. $\int \frac{x-1}{x+1}\, dx$

60. $\int_0^\pi (\sin^2 t + \cos^2 t)\, dt$

II

61. What is the average of the integers from 1 to n?

62. What is the average of the squares of the integers from 1 to n?

*63. Suppose $r \neq 1$. We want to evaluate $S = \sum_{k=0}^n r^k$.
   Consider $A = (1 - r)S$. Expand A by rearranging the sum; you should be able to express it as a simple expression without summations. Then divide by $1 - r$ to find an expression for S. A sum of this sort (consecutive powers of r) is called a *geometric sum*.

*64. Let r be any nonzero number. Find an expression without a summation sign for
$$\sum_{k=m}^n r^k.$$

*65. Evaluate the integral $\int_{-1}^1 \sqrt{1 - x^2}\, dx$ by thinking of it as the area under the graph of a function.

*66. Suppose $f(x)$ is the function whose graph is illustrated below. Sketch the graph of its indefinite integral $F(x) = \int_0^x f(t)\, dt$.

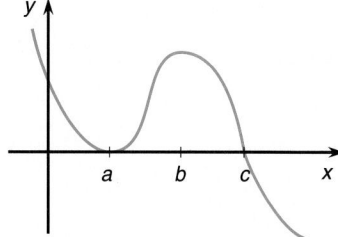

67. Find all local extrema and inflection points of $E(x) = \int_1^x (t^3 + t^2 - 2t)\, dt$.

68. Find all local extrema and inflection points of $G(x) = \int_1^x (t^2 - 4t + 3)\, e^t\, dt$.

*69. Find all local extrema and inflection points of
$$H(x) = \int_{-x}^{1} (t \ln t - 2t)\,dt.$$

*70. Find all local extrema and inflection points of
$$K(x) = \int_{2}^{\exp(x)} (t^2 - 1)\,dt.$$

*71. Suppose $f(x)$ is a continuous function and $F(x) = \int_{d}^{x} f(t)\,dt$. If $F(x)$ has a local maximum at $x = a$, what can we say about the behavior of $f(x)$ at $x = a$?

*72. Suppose $f(x)$ is a differentiable function and $F(x) = \int_{d}^{x} f(t)\,dt$. If $f(x)$ has a local maximum at $x = a$, what can we say about the behavior of $F(x)$ at $x = a$?

*73. Suppose we are told that $\int_{0}^{\pi/2} x \cos x\,dx = \frac{\pi}{2} - 1$.
Assuming this fact, calculate $\int_{0}^{\pi/4} x \cos(2x)\,dx$.

*74. (i) Differentiate $x \sin x$ and guess an antiderivative for $x \cos x$.

  (ii) Guess an antiderivative for $x \sin x$.

75. Evaluate the following: (i) $\int_{-2}^{2} |x^2 - x|\,dx$,
(ii) $\int_{0}^{\pi} \cos t \sin t\,dt$, (iii) $\int_{0}^{4} |(x - 2)(x - 3)|\,dx$.

*76. Evaluate: (i) $\int_{0}^{2} |x(x^2 - 1)^3|\,dx$, (ii) $\int_{0}^{\pi} \cos^3 t \sin t\,dt$.

*77. Find an example of a function $f(x)$ that is differentiable at every $x \in \mathbb{R}$ but so that $f'(x)$ is *not* differentiable at $x = 0$.

*78. Find an example of a function $f(x)$ so that $f(x)$ and $f'(x)$ are differentiable at every $x \in \mathbb{R}$ but so that $f''(x)$ is *not* differentiable at $x = 0$.

*79. If $f(x) = \frac{1}{1+x^2}$, note that $F(x) = \frac{\pi}{3} + \arctan x$ is an antiderivative of $f(x)$. Find a number d so that $F(x) = \int_{d}^{x} f(t)\,dt$.

**80. If $f(x) = \cos x$, then $F(x) = 5 + \sin x$ is an antiderivative. (i) Show that for any x, $4 \le F(x) \le 6$. (ii) Show that the indefinite integral

$$G(x) = \int_{d}^{x} f(t)\,dt = \int_{d}^{x} \cos t\,dt$$

always satisfies $-2 \le G(x) \le 2$, for *any* choice of the starting point d. (*Hint:* Draw a picture.) So although $F(x)$ is an antiderivative it is *never* the signed area under $y = f(x)$. (iii) Explain this.

For each of the following functions $f(x)$, find an antiderivative $F(x)$ and use a graphics calculator or computer to graph $f(x)$ and $F(x)$ on the same picture. Interpret $F(x)$ in terms of signed areas.

81. $f(x) = x$

82. $f(x) = x^2$

83. $f(x) = x^3$

84. $f(x) = x^3 - 3x$

85. $f(x) = e^x$

86. $f(x) = \frac{1}{x}, \; x > 0$

87. $f(x) = \sec^2 x, \; |x| < \frac{\pi}{2}$

88. $f(x) = \cos^2 x$

Applications of Integration

6

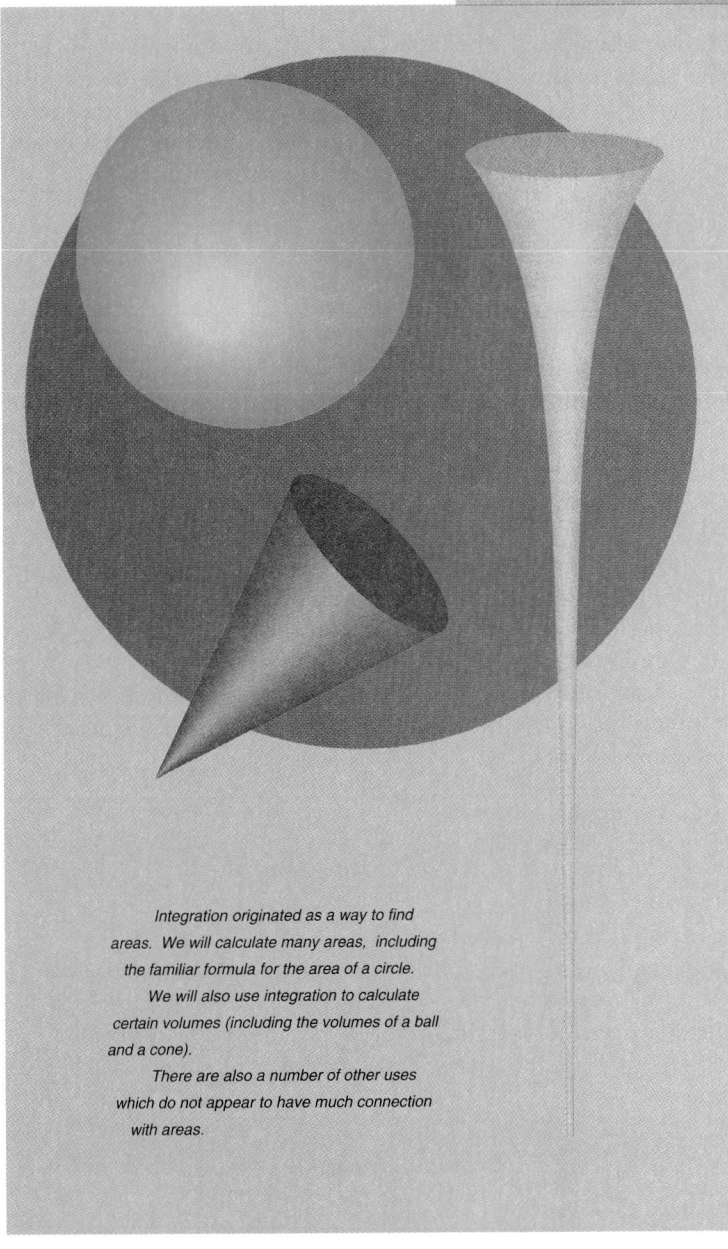

Integration originated as a way to find areas. We will calculate many areas, including the familiar formula for the area of a circle.

We will also use integration to calculate certain volumes (including the volumes of a ball and a cone).

There are also a number of other uses which do not appear to have much connection with areas.

INTRODUCTION

This chapter will explore some of the uses of integration for practical purposes. We have already found that being able to integrate is essentially the same thing as being able to find antiderivatives, and in Section 4.9 we saw the importance of antiderivatives for questions involving velocity and acceleration. If we know, for instance, the velocity of a moving object, then we may want to know its position (i.e., displacement), which is an antiderivative of the velocity. Similarly, the velocity itself is an antiderivative of the acceleration. A similar idea applies in any situation in which the rate of change of some quantity is known; the actual value of that quantity is an indefinite integral of the rate of change.

We have also hinted that integrals have something to do with averages. This arose in the Integral Mean Value Theorem (Theorem 5.4.4), where the number $\frac{1}{b-a}\int_a^b f(x)\,dx$ was regarded as the average value of the function $f(x)$ over the interval $[a, b]$.

Both these ideas will be developed more fully in this chapter. However, we begin with something more basic.

After all, the idea of integration arose in the first place as a means of finding areas. Now that the theory is available to us, it should be possible to use it to find areas of various geometrical regions. Not only will we do this, but we will also be able to use it to find volumes of solid regions, lengths of curves, and areas of curved surfaces.

Section 6.1 concerns areas, and the related Section 6.2 discusses the techniques needed for these calculations. Sections 6.3 and 6.4 concern volume computations, and Section 6.5 discusses the average value of a function. The next four sections are about more specialized topics, and you may wish to omit some of them. Section 6.6 is about work (in the sense in which the term is used in physics), and Section 6.7 discusses the calculation of moments and centroids, a very important topic in physics having to do with various things including rotational motion. Section 6.8, which deals with pressure and forces in a fluid, makes some use of the results of Section 6.7, and Section 6.9 discusses the lengths of curves and the areas of surfaces (such as the surface area of a sphere, etc.). Finally, Section 6.10 is about "numerical integration," which means finding numerical approximations to the value of an integral when none of the methods we know gives us the exact answer. In practice there are many instances in which this is necessary, and numerical integration, especially with the aid of computers or sophisticated calculators, is extremely useful.

6.1 Areas

SIGNED AREA
REGION BETWEEN TWO CURVES

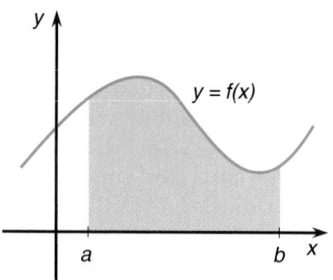

Figure 6.1.1

The integral originally arose in connection with the area under a curve, and it is very useful for calculating areas. Strictly speaking, we do not really know what is meant by the area of a peculiarly shaped region. However, if we consider the region illustrated in Figure 6.1.1, the region under the graph of a nonnegative continuous function $f(x)$ between $x = a$ and $x = b$, it is reasonable to regard $\int_a^b f(x)\, dx$ as the area.

Indeed, the integral was defined as a limit, obtained by approximating the region by collections of rectangles and summing their areas. So if the area of the region is to mean anything at all and be consistent with the areas of rectangles, we can use the integral to define what area means.

If we do not assume that $f(x)$ is nonnegative between $x = a$ and $x = b$, then of course the integral will give the signed area; we can obtain the actual (unsigned) area by integrating $|f(x)|$.

DEFINITION 6.1.1

> If $f(x)$ is a nonnegative continuous function on $[a, b]$, consider the region bounded by the graph $y = f(x)$, the x-axis, and the lines $x = a$ and $x = b$. Then the **area** of this region is defined to be the integral
>
> $$\int_a^b f(x)\, dx.$$
>
> If $f(x)$ is a continuous function on $[a, b]$, the area of the region bounded by the graph $y = f(x)$, the x-axis, and the lines $x = a$ and $x = b$ is defined to be the integral
>
> $$\int_a^b |f(x)|\, dx.$$

Figure 6.1.2

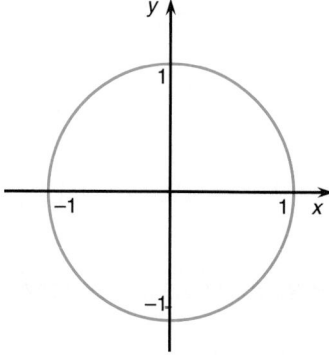

Figure 6.1.3

EXAMPLE 1   If $f(x) = x^n$, then the area under the curve $y = f(x) = x^n$ between $x = 0$ and $x = 1$ is

$$\int_0^1 x^n dx = \frac{1}{n+1} x^{n+1} \bigg|_0^1 = \frac{1}{n+1}.$$

As n increases, the curves $y = x^n$ are lower and lower on the interval $[0, 1]$, so their integrals get smaller and smaller (Figure 6.1.2 shows that $y = x, y = x^2, y = x^3$.)

One area we would like to calculate is the area of a circle (strictly speaking, this means the area of the region inside the circle; see Figure 6.1.3). In fact it is easier to calculate the area of a semicircle. If we consider the circle of radius 1 centered at the origin, its equation is $x^2 + y^2 = 1$, and we know that the top half of the circle is the graph of $y = \sqrt{1 - x^2}$. To find the area, we will have to know how to integrate this function. It can be done by using a technique we will learn in Section 8.3, but for now we simply state the result.

EXAMPLE 2   Check the indefinite integral

$$\int \sqrt{1 - x^2} dx = \frac{1}{2}\left(x\sqrt{1 - x^2} + \arcsin x\right) + c.$$

Solution   All that is needed is to differentiate the right side:

$$\frac{d}{dx}\left(\frac{1}{2}\left(x\sqrt{1 - x^2} + \arcsin x\right) + c\right)$$

$$= \frac{1}{2}\left(\sqrt{1 - x^2} + \frac{1}{2}x\frac{1}{\sqrt{1 - x^2}}(-2x) + \frac{1}{\sqrt{1 - x^2}}\right)$$

$$= \frac{1}{2}\left(\frac{1 - x^2}{\sqrt{1 - x^2}} - \frac{x^2}{\sqrt{1 - x^2}} + \frac{1}{\sqrt{1 - x^2}}\right)$$

$$= \sqrt{1 - x^2}.$$

In the following box we describe a trick for finding this integral.

Consider the indefinite integral $\int \sqrt{1 - x^2} dx$ of Example 2.
   There is no obvious way to integrate by substitution. Instead, we look at the function $x\sqrt{1 - x^2}$. It is not the answer we want, but it may help. Differentiating, we find:

$$\frac{d}{dx}\left(x\sqrt{1 - x^2}\right) = \sqrt{1 - x^2} + x\frac{1}{2}(1 - x^2)^{-1/2}(-2x)$$

$$= \sqrt{1 - x^2} - x^2(1 - x^2)^{-1/2} = \frac{1 - x^2}{\sqrt{1 - x^2}} - \frac{x^2}{\sqrt{1 - x^2}}$$

$$= \frac{1 - 2x^2}{\sqrt{1 - x^2}}.$$

This does not appear to be much good, but suppose for a moment that the 1 in the numerator was a 2. Then we would have

$$\frac{2 - 2x^2}{\sqrt{1 - x^2}} = 2\frac{1 - x^2}{\sqrt{1 - x^2}} = 2\sqrt{1 - x^2},$$

which is practically what we need.

The function $x\sqrt{1 - x^2}$ is not the antiderivative we want, but if we could add to it something that would add $\frac{1}{\sqrt{1-x^2}}$ to its derivative, we would be close. And that is easy if we remember that we know a function whose derivative equals $\frac{1}{\sqrt{1-x^2}}$, namely, arcsin x.

So consider $x\sqrt{1 - x^2} + \arcsin x$, whose derivative is

$$\frac{1 - 2x^2}{\sqrt{1 - x^2}} + \frac{1}{\sqrt{1 - x^2}} = \frac{2 - 2x^2}{\sqrt{1 - x^2}} = 2\sqrt{1 - x^2},$$

which means

$$\int \sqrt{1 - x^2}\,dx = \frac{1}{2}(x\sqrt{1 - x^2} + \arcsin x) + c.$$

EXAMPLE 3   Area of Circle   Find the area of the circle of radius r.

Solution   We may as well let the center of the circle be at the origin, so the equation of the circle is $x^2 + y^2 = r^2$. The upper semicircle is the graph of the function $y = \sqrt{r^2 - x^2}$ illustrated in Figure 6.1.4. It will be easier to calculate the area under this semicircle and then double it to get the area of the circle.

The semicircular region is the region under the function $y = \sqrt{r^2 - x^2}$ between $x = -r$ and $x = r$. So the area of the region under the semicircle is

$$\int_{-r}^{r} \sqrt{r^2 - x^2}\,dx.$$

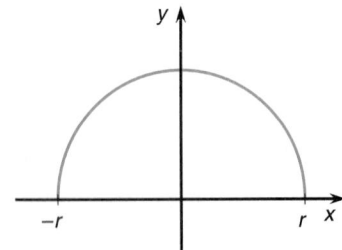

Figure 6.1.4

This is not quite the same as the integral we discussed in the preceding example, because of the r^2 inside the square root. To deal with that, we write the integrand as $\sqrt{r^2 - x^2} = \sqrt{r^2\left(1 - \left(\frac{x}{r}\right)^2\right)} = r\sqrt{1 - \left(\frac{x}{r}\right)^2}.$

This suggests the substitution $u = \frac{x}{r}$. We notice that as x goes from $-r$ to r, the substituted variable $u = \frac{x}{r}$ goes from -1 to 1, so the integral becomes

$$\int_{-r}^{r} \sqrt{r^2 - x^2}\,dx = \int_{-r}^{r} \sqrt{r^2\left(1 - \left(\frac{x}{r}\right)^2\right)}\,dx = \int_{-r}^{r} r\sqrt{1 - \left(\frac{x}{r}\right)^2}\,dx$$

$$\boxed{\begin{array}{c} u = \dfrac{x}{r} \\ du = \dfrac{1}{r}\,dx \end{array}} \qquad = r\int_{-1}^{1} \sqrt{1 - u^2}\,r\,du = r^2\int_{-1}^{1} \sqrt{1 - u^2}\,du$$

$$= r^2\frac{1}{2}\left(u\sqrt{1 - u^2} + \arcsin u\right)\Big|_{-1}^{1}$$

$$= \frac{r^2}{2}\left(1\sqrt{0} + \arcsin(1) - (-1)\sqrt{0} - \arcsin(-1)\right)$$

$$= \frac{r^2}{2}\left(0 + \frac{\pi}{2} - 0 - \left(-\frac{\pi}{2}\right)\right)$$

$$= \frac{1}{2}\pi r^2.$$

This is the area of the semicircle; the area of the circle is twice as much, or πr^2. Of course, we are not surprised, but we should be impressed, at least slightly. After all, this is something we have been told for a long time; now we finally see why it is so.

Example 3 illustrates several interesting points. First, it is important to have a function whose graph encloses the region whose area we want to find. Second, it is frequently convenient to work with only part of the region (in this case the upper semicircle) and determine the remaining areas separately or, as here, by symmetry. Third, it is necessary to find the limits of integration. Sometimes they will be made obvious by drawing a picture, but other times they will have to be calculated.

EXAMPLE 4   Find the area of the region enclosed between the x-axis and the curve $y = x^2 + 6x + 5$.

Solution   We begin with a sketch in Figure 6.1.5 of the curve $y = x^2 + 6x + 5$, in which the specified region is shaded. To find the limits of integration, it is necessary to find the x-intercepts—the points where $x^2 + 6x + 5 = 0$. This is easily done by using the Quadratic Formula (Theorem 1.3.4) or by factoring:

$$0 = x^2 + 6x + 5 = (x + 5)(x + 1),$$

so

$$x = -5, -1.$$

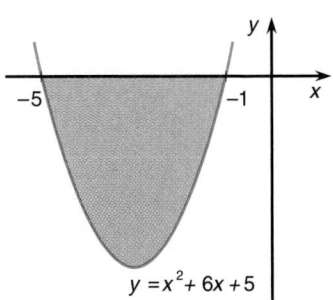

$y = x^2 + 6x + 5$

Figure 6.1.5

The picture suggests that $f(x)$ is negative for $x \in (-5, -1)$. We could confirm this by completing the square: $x^2 + 6x + 5 = (x + 3)^2 - 4$, which is negative when $(x + 3)^2 < 4$, that is, when $|x + 3| < 2$, that is, when $x \in (-5, -1)$. Since $f(x) < 0$, we see that $|f(x)| = -f(x)$. The corresponding integral is easy to evaluate:

$$\int_{-5}^{-1} |f(x)| \, dx = \int_{-5}^{-1} \left(-(x^2 + 6x + 5)\right) dx = -\left(\frac{1}{3}x^3 + 3x^2 + 5x\right)\Bigg|_{-5}^{-1}$$

$$= -\left(\frac{1}{3}(-1) + 3(1) + 5(-1)\right) + \left(\frac{1}{3}(-125) + 3(25) + 5(-5)\right)$$

$$= \frac{32}{3}.$$

The desired area is $\frac{32}{3} = 10\frac{2}{3}$.

If we did not realize that the graph was below the axis, that is, that $f(x)$ was negative, we would have integrated $f(x)$ and obtained the incorrect answer $-\frac{32}{3}$.

EXAMPLE 5   Find the area of the region enclosed between the x-axis and the curve $y = x^3 - 3x$.

To find where $y = x^3 - 3x$ crosses the x-axis, we need to find where $y = 0$, so we solve $x^3 - 3x = 0$.

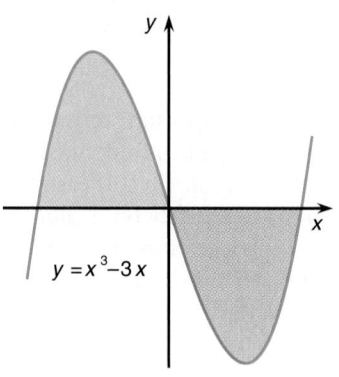

$y = x^3 - 3x$

Figure 6.1.6

Solution   The curve is sketched in Figure 6.1.6. To find where it crosses the x-axis, we solve $0 = x^3 - 3x = x(x^2 - 3) = x(x + \sqrt{3})(x - \sqrt{3})$. The zeroes are at $x = -\sqrt{3}, \ 0, \ \sqrt{3}$. The region in question consists of two parts, the part between $x = -\sqrt{3}$ and $x = 0$, and the part between $x = 0$ and $x = \sqrt{3}$.

The area of the first of these is

$$
\begin{aligned}
\int_{-\sqrt{3}}^{0} (x^3 - 3x)\,dx &= \left(\frac{1}{4}x^4 - \frac{3}{2}x^2 \right) \Big|_{-\sqrt{3}}^{0} \\
&= \left(\frac{1}{4}(0) - \frac{3}{2}(0) \right) - \left(\frac{1}{4}(-\sqrt{3})^4 - \frac{3}{2}(-\sqrt{3})^2 \right) \\
&= 0 - \left(\frac{9}{4} - \frac{9}{2} \right) \\
&= \frac{9}{4}.
\end{aligned}
$$

Since the second part of the region lies below the x-axis, its area is the integral of $|x^3 - 3x| = -(x^3 - 3x)$:

$$
\begin{aligned}
\int_{0}^{\sqrt{3}} \left(-(x^3 - 3x) \right) dx &= -\left(\frac{1}{4}x^4 - \frac{3}{2}x^2 \right) \Big|_{0}^{\sqrt{3}} \\
&= -\left(\frac{1}{4}(\sqrt{3})^4 - \frac{3}{2}(\sqrt{3})^2 \right) + \left(\frac{1}{4}(0) - \frac{3}{2}(0) \right) \\
&= -\left(\frac{9}{4} - \frac{9}{2} \right) + 0 = \frac{9}{4}.
\end{aligned}
$$

The total area is the sum of the areas of the two parts, or $\frac{9}{4} + \frac{9}{4} = \frac{9}{2}$.

It was essential to divide the region into parts according to whether the curve was above or below the axis. If we did not do that but simply integrated all at once, we would evaluate:

$$
\begin{aligned}
\int_{-\sqrt{3}}^{\sqrt{3}} (x^3 - 3x)\,dx &= \left(\frac{1}{4}x^4 - \frac{3}{2}x^2 \right) \Big|_{-\sqrt{3}}^{\sqrt{3}} \\
&= \left(\frac{1}{4}(\sqrt{3})^4 - \frac{3}{2}(\sqrt{3})^2 \right) - \left(\frac{1}{4}(-\sqrt{3})^4 - \frac{3}{2}(-\sqrt{3})^2 \right) \\
&= \left(\frac{9}{4} - \frac{9}{2} \right) - \left(\frac{9}{4} - \frac{9}{2} \right) \\
&= 0.
\end{aligned}
$$

In this case the regions above and below the axis have the same area, so the signed areas cancel. This illustrates the importance of figuring out where the integrand is positive and where it is negative. In practice the integral is usually evaluated by dividing the region into pieces according to the sign of the integrand and integrating each piece separately.

Figure 6.1.7(i)

Figure 6.1.7(ii)

Figure 6.1.8

Figure 6.1.9

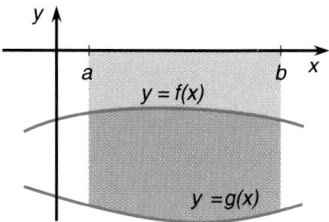

Figure 6.1.10

Another problem that frequently arises involves regions lying between two curves. In Figure 6.1.7 there are two examples. In each case the shaded region lies between $x = a$ and $x = b$ and between $y = f(x)$ and $y = g(x)$. Moreover, in each case $f(x) \geq g(x)$ for all $x \in [a, b]$.

To discuss the area of such a region, we refer to Figure 6.1.8, in which there are two continuous functions $f(x)$ and $g(x)$ with $f(x) \geq g(x)$ for all $x \in [a, b]$. Now $\int_a^b f(x)\,dx$ is the total area of both shaded regions, and $\int_a^b g(x)\,dx$ is the area of the lower shaded region. So the area of the upper shaded region—the region between $y = f(x)$ and $y = g(x)$—is their difference:

$$\int_a^b f(x)\,dx - \int_a^b g(x)\,dx = \int_a^b \big(f(x) - g(x)\big)\,dx.$$

This is still reasonable if either $f(x)$ or $g(x)$ is negative, provided $f(x) \geq g(x)$. In the case illustrated in Figure 6.1.9 we see that $f(x) \geq 0 \geq g(x)$, so the area of the upper shaded region is $\int_a^b f(x)\,dx$, the area of the lower shaded region is $-\int_a^b g(x)\,dx$, and the area of the whole region is their sum, $\int_a^b f(x)\,dx + \big(-\int_a^b g(x)\,dx\big) = \int_a^b \big(f(x) - g(x)\big)\,dx$.

In the case shown in Figure 6.1.10, with $0 \geq f(x) \geq g(x)$, the total area of both parts is $-\int_a^b g(x)\,dx$, the area of the upper part is $-\int_a^b f(x)\,dx$, and the area of the region between the curves is $-\int_a^b g(x)\,dx - \big(-\int_a^b f(x)\,dx\big) = \int_a^b \big(f(x) - g(x)\big)\,dx$.

DEFINITION 6.1.2

Area of the Region Between Two Curves

If $f(x)$ and $g(x)$ are continuous functions satisfying $f(x) \geq g(x)$ for all $x \in [a, b]$, consider the region between $y = f(x)$ and $y = g(x)$, that is, $\{(x, y) : a \leq x \leq b, \; f(x) \geq y \geq g(x)\}$. The area of this region is defined to be

$$\int_a^b \big(f(x) - g(x)\big)\,dx.$$

If $f(x)$ and $g(x)$ are continuous functions on $[a, b]$, the area of the region between $y = f(x)$ and $y = g(x)$ and between $x = a$ and $x = b$ is defined to be

$$\int_a^b |f(x) - g(x)|\,dx.$$

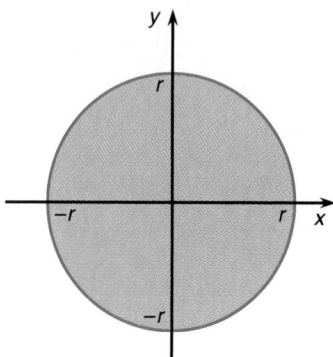

Figure 6.1.11

EXAMPLE 6   Find the area of a circle of radius r (again).

Solution   As in Example 3, we consider the circle $x^2 + y^2 = r^2$ (see Figure 6.1.11). The region in question is the region between the upper semicircle $y = \sqrt{r^2 - x^2}$ and the lower semicircle $y = -\sqrt{r^2 - x^2}$. Its area is

$$\int_{-r}^{r} \left(\sqrt{r^2 - x^2} - (-\sqrt{r^2 - x^2}) \right) dx = 2 \int_{-r}^{r} \sqrt{r^2 - x^2} \, dx.$$

We have already evaluated this integral in Example 3; there we found that the integral equals $\frac{1}{2}\pi r^2$, so the area of the circle is πr^2.

EXAMPLE 7   Find the area of the region enclosed between the curves $y = x^3$ and $y = x$.

Solution   For questions about the region enclosed between two curves, the first step is to find the points where the two curves intersect. In this case, that will happen when $x^3 = x$, that is, when $x^3 - x = 0$, or $0 = x(x^2 - 1) = x(x + 1)(x - 1)$, which means $x = -1, 0, 1$. The points of intersection are $(-1, -1)$, $(0, 0)$, and $(1, 1)$, and we sketch them in Figure 6.1.12.

The part of the region lying between $x = -1$ and $x = 0$ lies below $y = x^3$ and above $y = x$, so its area is

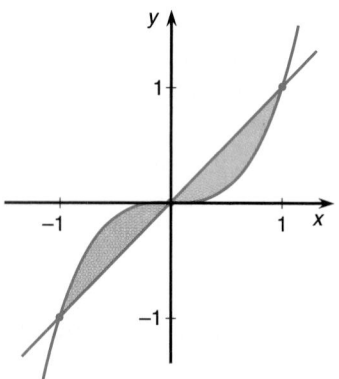

Figure 6.1.12

$$\int_{-1}^{0} (x^3 - x) \, dx = \left(\frac{1}{4}x^4 - \frac{1}{2}x^2 \right) \Big|_{-1}^{0}$$

$$= \left(\frac{1}{4}(0) - \frac{1}{2}(0) \right) - \left(\frac{1}{4}(-1)^4 - \frac{1}{2}(-1)^2 \right)$$

$$= \frac{1}{4}.$$

Between $x = 0$ and $x = 1$, the region lies below $y = x$ and above $y = x^3$, so the area of this part of the region is

$$\int_{0}^{1} (x - x^3) \, dx = \left(\frac{1}{2}x^2 - \frac{1}{4}x^4 \right) \Big|_{0}^{1}$$

$$= \left(\frac{1}{2}(1^2) - \frac{1}{4}(1^4) \right) - \left(\frac{1}{2}(0) - \frac{1}{4}(0) \right)$$

$$= \frac{1}{4}.$$

The total area of the region is the sum of these two parts, or $\frac{1}{4} + \frac{1}{4} = \frac{1}{2}$.

Notice that it was necessary to break the region into parts depending upon which function was greater. This is analogous to what we did earlier in dealing with the region between the x-axis and a single function, where it was necessary to decide where the function was positive (i.e., above the axis) and where it was negative (below the axis).

Also notice that whenever we speak of a region "enclosed" between two curves without mentioning an interval $[a, b]$, we are referring to the part that is *completely* enclosed. For instance in Example 7 we ignore the region between the curves to the right of $x = 1$ and the region between the curves to the left of $x = -1$.

EXAMPLE 8
Find the area of the region enclosed by the parabola $y^2 = x$ and the line $x + y = 2$.

Solution   The region is illustrated in Figure 6.1.13. We find the points of intersection of the line and the parabola by solving for points that satisfy *both* equations. From $x + y = 2$ we get $x = 2 - y$, so $y^2 = x = 2 - y$, or $y^2 + y - 2 = 0$. This we factor as $(y - 1)(y + 2) = 0$ to find $y = 1, -2$. The points of intersection are $(1, 1)$ and $(4, -2)$.

We divide the region into two parts as shown in Figure 6.1.14. The part between $x = 0$ and $x = 1$ is the region between the curves $y = \sqrt{x}$ and $y = -\sqrt{x}$, so its area is

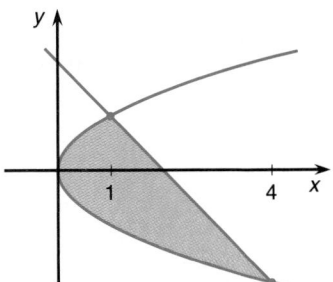

Figure 6.1.13

$$\int_0^1 \left(\sqrt{x} - (-\sqrt{x})\right) dx = 2 \int_0^1 \sqrt{x}\, dx = 2\left(\frac{2}{3} x^{3/2}\right)\Big|_0^1$$

$$= \frac{4}{3}.$$

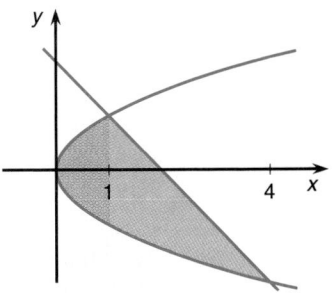

Figure 6.1.14

The part of the region between $x = 1$ and $x = 4$ is the region between $y = 2 - x$ and $y = -\sqrt{x}$, so its area is

$$\int_1^4 \left(2 - x - (-\sqrt{x})\right) dx = \left(2x - \frac{1}{2} x^2 + \frac{2}{3} x^{3/2}\right)\Big|_1^4$$

$$= \left(8 - 8 + \frac{2}{3}(8)\right) - \left(2 - \frac{1}{2} + \frac{2}{3}\right) = \frac{19}{6}.$$

The total area of the original region is $\frac{4}{3} + \frac{19}{6} = \frac{9}{2}$. ◢

Notice that different parts of the region lie between different pairs of curves. To integrate, it is necessary to consider only one such part at a time. Also notice that while the curves were given to us as $y^2 = x$ and $x + y = 2$, it was necessary to express them with y as a function of x in order to be able to integrate. In the case of $y^2 = x$ there were two possible functions of this sort, $y = \sqrt{x}$ and $y = -\sqrt{x}$. Each of them figured in the calculation.

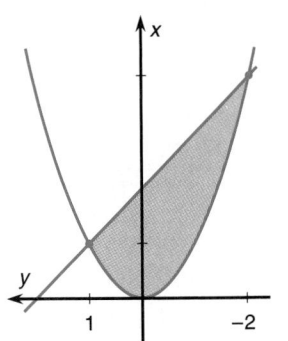

Figure 6.1.15

EXAMPLE 9
Example 8 can also be done another way. Instead of thinking of $y^2 = x$ and $x + y = 2$ as functions $y = \sqrt{x}$, $y = -\sqrt{x}$, and $y = 2 - x$ in which x is the variable, we could just as well think of $x = y^2$ and $x = 2 - y$ as functions of y (see Figure 6.1.15, in which the axes have been rotated). The function $x = 2 - y$ is greater than $x = y^2$ for y between -2 and 1, so the area of the region is

$$\int_{-2}^1 (2 - y - y^2)\, dy = \left(2y - \frac{1}{2} y^2 - \frac{1}{3} y^3\right)\Big|_{-2}^1$$

$$= \left(2 - \frac{1}{2} - \frac{1}{3}\right) - \left(-4 - \frac{1}{2}(-2)^2 - \frac{1}{3}(-2)^3\right) = \frac{9}{2}.$$ ◢

When we rotate the axes, as in Figure 6.1.15, the positive y-axis points to the *left*.

The answer we obtain is exactly the same, but this way the calculation was less work. Specifically, using y it was possible to do the integration for the entire region in one piece instead of having to divide it up. This example illustrates the useful fact that considerable simplifications will sometimes result from using y as the variable of integration for a region, or for some part of it.

EXAMPLE 10   **Metal Tulips**   Decorative tulips are made of painted sheet metal as shown in Figure 6.1.16. Suppose the scale is in inches, and suppose a $1''$ square of sheet metal weighs $\frac{3}{4}$ oz. By finding the weight of each tulip, decide whether or not

A graphics calculator can be used to find the area enclosed between two curves when it is not easy to solve explicitly for the points where they intersect.

For instance, to find the area between $y = x^4 - 3x + 1$ and $y = x$, we plot the graphs, solve for the points a and b where $x^4 - 3x + 1 = x$, i.e., $x^4 - 4x + 1 = 0$, and then evaluate the area $\int_a^b \left(x - (x^4 - 3x + 1) \right) dx = \left(-\frac{x^5}{5} + 2x^2 - x \right) \Big|_a^b$.

TI-85	SHARP EL9300
GRAPH F1$(y(x) =)$ F1(x) ^4 -3 F1(x) $+1$ ▼ F1(x) M2(RANGE) $(-)$ 1 ▼ 2 ▼ ▼ $(-)$ 1 ▼ 3 F5(GRAPH) SOLVER X ^4 $-$ 4 X $+ 1 =$ 0 ENTER 0 F5(SOLVE) EXIT X STO A ENTER SOLVER ENTER 1 F5(SOLVE) EXIT X STO B ENTER $(-)$ B ^5 $\div 5 + 2$ B ^2 $-$ B $+$ A ^5 $\div 5 - 2$ A ^2 $+$ A ENTER	⤶ x/θ/T a^b 4 ▶ $- 3$ x/θ/T $+ 1$ 2ndF ▼ x/θ/T RANGE -1 ENTER 2 ENTER 1 ENTER -1 ENTER 3 ENTER 1 ENTER ⤶ SOLVER x/θ/T a^b 4 ▶ $- 3$ x/θ/T $+ 1 =$ x/θ/T ENTER $(-)$ 1 ENTER ENTER ENTER ENTER ENTER ⊞ X STO A SOLVER ENTER ENTER ENTER 2 ENTER ENTER ENTER ⊞ X STO B $-$ B a^b 5 ▶ $\div 5 + 2$ B x^2 $-$ B $+$ A a^b 5 ▶ $\div 5 - 2$ A $x^2 +$ A ENTER
CASIO f_x-6300G (f_x-7700GB)	HP 48SX
Range -1 EXE 2 EXE 1 EXE -1 EXE 3 EXE 1 EXE EXE EXE EXE Graph X x^y 4 $- 3$ X $+ 1$ EXE Graph X EXE Trace ▶ (move cursor to intersection) X $\rightarrow$ A EXE Graph X EXE Trace ▶ (move cursor to intersection) X $\rightarrow$ B EXE $(-)$ B x^y 5 $\div 5 + 2$ B x^2 $-$ B $+$ A x^y 5 $\div 5 - 2$ A $x^2 +$ A EXE (Use Zoom for more accuracy)	⤶ EQUATION X ⤶ $=$ X y^x 4 ▶ $-$ 3 $\times$ X $+ 1$ ENTER ⤶ SOLVE NEW αE αQ αN ENTER 30 +/– ↪ CST NXT SF ⤶ PLOT PTYPE FUNC PLOTR 1 +/– ENTER 2 XRNG 1 +/– ENTER 3 YRNG DRAW ◀ (Move cursor to intersection) FCN ISECT ATTN ATTN X ENTER ′ A STO DRAW ◀ (Move cursor to intersection) FCN ISECT ATTN ATTN X ENTER ′ B STO B +/– 5 y^x 5 $\div$ B x^2 2 $\times$ + B $-$ A ENTER 5 y^x 5 $\div$ + A ENTER x^2 2 $\times$ $-$ A $+$

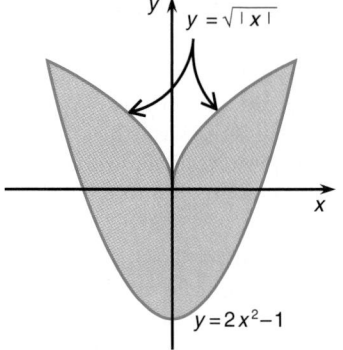

Figure 6.1.16

a "bouquet" of a dozen tulips can be suspended by a nylon thread that can support a load of one pound.

Solution The weight of the tulip will be proportional to its area, and we are told that the metal weighs $\frac{3}{4}$ oz per square inch. To find the area, we divide the tulip into left and right halves; they are symmetrical, so the total area is twice the area of the right half. The upper right corner is the point $(1, 1)$, so the area of the right half is

$$\int_0^1 \left(\sqrt{x} - (2x^2 - 1) \right) dx = \left(\frac{2}{3}x^{3/2} - \frac{2}{3}x^3 + x \right) \Big|_0^1 = 1 \text{ sq in.}$$

The total area of a tulip is 2 sq in, and the weight of the tulip is $2 \times \frac{3}{4} = \frac{3}{2}$ oz.

A dozen tulips weigh 18 oz, so they will break the thread. A stronger support will be needed.

> Example 10 illustrates the useful fact that for a thin object made of uniform material, the weight is proportional to the area. Note that when we found the "total area" of a tulip it was the area of one side, not both the front and the back.

Summary 6.1.3   Areas of Regions in the Plane

1. Find curves that enclose the region whose area we want, expressing y as a function of x.

2. Distinguish intervals on which one curve is always above the other curve. This will involve solving for points of intersection of two curves (or of one curve with the x-axis).

3. Integrate over each interval and add the answers.

4. Consider writing x as a function of y to simplify some or all of the integrals.

Exercises 6.1

In each of the following situations, find the area of the region between the given function $y = f(x)$ and the x-axis over the given interval $[a, b]$, that is, between $x = a$ and $x = b$.

1. $f(x) = x^2$ over $[-1, 1]$
2. $f(x) = x^2$ over $[0, 4]$
3. $f(x) = x^3$ over $[-2, 1]$
4. $f(x) = x^5$ over $[-4, 4]$
5. $f(x) = \sin x$ over $\left[0, \frac{\pi}{2}\right]$
6. $f(x) = \sin x$ over $[0, 2\pi]$
7. $f(x) = \cos x$ over $\left[0, \frac{\pi}{2}\right]$
8. $f(x) = \sin^2 x$ over $\left[0, \frac{\pi}{2}\right]$
9. $f(x) = \frac{1}{x}$ over $[1, 2]$
10. $f(x) = 1/(1 + x^2)$ over $[-1, 1]$

In each of the following situations, find the area of the region enclosed between the given function $y = f(x)$ and the x-axis.

11. $f(x) = 1 - x^2$
12. $f(x) = 4 - x^2$
13. $f(x) = 2 - x^2$
14. $f(x) = x^2 - 2x - 3$
15. $f(x) = x^3 - 4x$

16. $f(x) = x^3 - 6x$

17. $f(x) = x^4 - 16$

18. $f(x) = 1 - \frac{2}{1+x^2}$

19. $f(x) = |x| - 3$

20. $f(x) = x\sqrt{|4 - x^2|}$

In each of the following situations, find the area of the region enclosed between the given functions $y = f(x)$ and $y = g(x)$.

21. $f(x) = x^3$, $g(x) = 4x$

22. $f(x) = 2x$, $g(x) = x^3$

23. $f(x) = x$, $g(x) = x^5$

24. $f(x) = x^5$, $g(x) = 16x$

25. $f(x) = x^2$, $g(x) = 2x^2 - 1$

26. $f(x) = x^2 - 1$, $g(x) = 11 - 2x^2$

27. $f(x) = x^2 - 2x + 1$, $g(x) = x + 1$

28. $f(x) = 2 - x^2$, $g(x) = |x|$

29. $f(x) = 3 - x$, $g(x) = x^2 + 2x + 3$

30. $f(x) = x^4 - 1$, $g(x) = 2x^2 - 2$

II

31. (i) Find the area of the region enclosed between $y = \sin x$ and the x-axis between $x = -\frac{\pi}{2}$ and $x = 0$.

(ii) Find the area of the region enclosed between $y = \sin(2x)$ and the x-axis between $x = -\frac{\pi}{2}$ and $x = 0$.

32. (i) Find the area of the region enclosed between $y = \cos x$ and the x-axis between $x = -\frac{\pi}{2}$ and $x = 0$.

(ii) Find the area of the region enclosed between $y = \cos(3x)$ and the x-axis between $x = -\frac{\pi}{2}$ and $x = 0$.

33. Find the area of the region enclosed between $y = f(x) = x^3 - 3x + 2$ and the x-axis. (*Hint:* $f(1) = 0$.)

34. Find the area of the region enclosed between $y = f(x) = (x^2 - 1)^2$ and the x-axis (cf. Example 4.4.2, p. 256).

*35. Find the area of the region enclosed between $y = |3x|$ and $y = x^2$.

*36. Find the area of the region enclosed between $y = x^2$ and $y = 19 - |5x - 5|$.

*37. Suppose t is some number between -2 and 2, and consider the region between the x-axis and the curve $y = x^3 - 4x$ between $x = -2$ and $x = t$. (i) Find the

area of this region. (ii) For what value(s) of t is the area of the specified region equal to $5\frac{3}{4}$?

*38. Consider the region A enclosed between $y = x - 1$ and $y = 1 - x^2$. Also consider the region B enclosed between the line $3x + 2y = 0$ and the curve $y = x^2 - 1$. (i) Find the area of region A. (ii) Find the area of region B. (iii) Let C be the region that is the intersection of A and B. Find the area of C.

*39. Sheet metal is cut into pieces shaped like the region enclosed between $y = x^2$ and $y = \sqrt{x}$ (with all measurements in centimeters). See Figure 6.1.17. Eight of them are glued to a plastic disk to make a "daisy." If the sheet metal weighs 3 grams per square centimeter, what is the weight of the daisy, ignoring the weight of the disk?

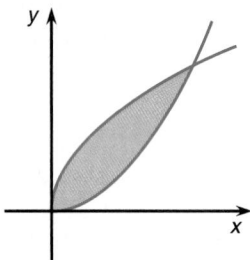

Figure 6.1.17

**40. Consider the region that lies inside the circle $x^2 + y^2 = 1$ and above the line $y = \frac{1}{2}$. Find the area of this region. (You will need to use your calculator to obtain a numerical value.)

*41. Suppose $y = f(x)$ intersects the x-axis at only finitely many points. Show that the region enclosed between the x-axis and the function $y = f(x)$ has the same area as the region enclosed between the x-axis and the function $y = |f(x)|$.

*42. Find an example to show that the region between $f(x)$ and $g(x)$ over an interval $[a, b]$ may not have the same area as the region between $|f(x)|$ and $|g(x)|$ over $[a, b]$.

III

43. Use a graphics calculator or computer to estimate the points x where $\sin x = \frac{x}{2}$ and use this information to find (approximately) the area between the graphs of $y = \frac{x}{2}$ and $y = \sin x$, with $x \geq 0$.

44. Find (approximately) the area lying above the line $y = \frac{x}{3}$ and below $y = \cos x$, with $x \geq 0$.

POINT TO PONDER

As we have already seen in dealing with derivatives of trigonometric functions in Section 3.4, the number π is not accidental. It is a fundamental and profound ingredient of any mathematics involving circles, among other things.

Figure 6.1.18 shows the circle of radius 1. Its area equals $2\int_{-1}^{1} \sqrt{1 - x^2}\,dx$, and this integral does not appear to have anything to do with π. Where does the π in the answer "come from"?

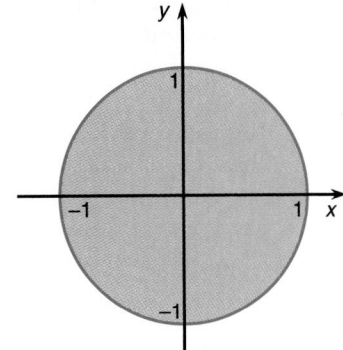

Figure 6.1.18

6.2 Setting Up Integrals; Limits of Integration

STRIP OF WIDTH *dx*
SPECIFIED RATE OF CHANGE

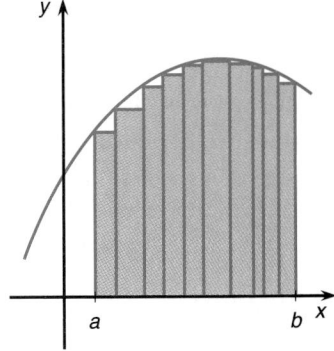

Figure 6.2.1

In Section 6.1 we saw how integration could be used to calculate the areas of geometric regions. The idea was simple enough, but in practice there are two steps that may be difficult. The first was finding the function to integrate, and the second was finding the limits of integration. Usually, the function to be integrated either was given to us explicitly or could be found with relative ease from a picture, but sometimes more effort was required. It frequently took some work to specify the interval over which the function should be integrated.

To some extent these problems must be solved on a case-by-case basis, but in this brief section we discuss some ideas that may make them slightly more manageable.

When we defined integrals, we added the areas of little rectangles of height $f(x_i^*)$ and width Δx_i, as illustrated in Figure 6.2.1, and approximated the integral by a sum

$$\sum_{i=1}^{n} f(x_i^*)\Delta x_i.$$

After taking a limit this became the integral

$$\int_a^b f(x)\,dx.$$

This suggests viewing the integral as a sort of infinite sum of the "areas" of little rectangles of height $f(x)$ and "width" dx. This does not really make sense, but it is often a very helpful way to set up integrals.

EXAMPLE 1   Area of Ellipse   Assuming $a > 0$ and $b > 0$, find the area of the ellipse

$$\frac{x^2}{a^2} + \frac{y^2}{b^2} = 1.$$

Figure 6.2.2

Figure 6.2.3

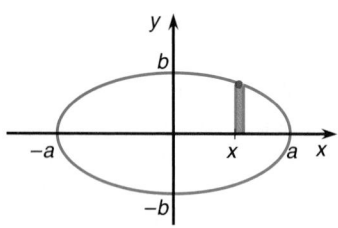

Figure 6.2.4

Solution   The area will be 4 times the area of the part of the region that lies in the first quadrant, the shaded part of the region in Figure 6.2.2.

Suppose we think of this region as being covered with thin little vertical rectangles, as shown in Figure 6.2.3. Each rectangle has width dx, which we regard as a very very small thickness, and we want to figure out the height. Of course, the height is different for different rectangles, so we need to find the height of the rectangle depending on its position. Specifically, if we suppose that the base of the rectangle is at the point x on the x-axis, we should find its height as a function of x. Since the width of the rectangle is so small, we allow ourselves not to worry about the difference between its left side and right side.

Now think about the point at the top of the rectangle (see Figure 6.2.4). We know its x-coordinate is x, and if we let y be its y-coordinate, then we know that the point (x, y) lies on the ellipse, so its coordinates satisfy the equation $\frac{x^2}{a^2} + \frac{y^2}{b^2} = 1$. We can solve this to express y in terms of x. Indeed,

$$\frac{x^2}{a^2} + \frac{y^2}{b^2} = 1$$

so

$$\frac{y^2}{b^2} = 1 - \frac{x^2}{a^2}$$

or

$$y^2 = b^2\left(1 - \frac{x^2}{a^2}\right)$$

and

$$y = b\sqrt{1 - \frac{x^2}{a^2}}.$$

This is the height of the rectangle at x, and its "width" is dx, so its "area" is $b\sqrt{1 - x^2/a^2}\,dx$. To get the area of the whole region, the upper right quarter of the ellipse, we "add up" these rectangles starting with $x = 0$ and going out to the end of the ellipse, which is $x = a$. "Adding up" means integrating, so the area is

$$\int_0^a b\sqrt{1 - \frac{x^2}{a^2}}\,dx = b\int_0^a \sqrt{1 - \frac{x^2}{a^2}}\,dx$$

$$\boxed{\begin{aligned} u &= \frac{x}{a} \\ du &= \frac{1}{a}dx \end{aligned}}$$

$$= b\int_0^1 \sqrt{1 - u^2}\,a\,du = ab\int_0^1 \sqrt{1 - u^2}\,du$$

$$= (ab)\frac{1}{2}\left(u\sqrt{1 - u^2} + \arcsin u\right)\Big|_0^1$$

$$= \frac{ab}{2}\left(0 + \arcsin(1) - 0 - \arcsin(0)\right)$$

$$= \frac{\pi ab}{4}.$$

The area of the whole ellipse is four times as much, or πab.

Notice that if $a = b = r$, then the "ellipse" is just a circle of radius r, whose area is πr^2.

Also notice that this approach using little rectangles of width dx is not ultimately any different than what we did before. It amounts to noticing that the region we wanted is the region lying below the curve $y = b\sqrt{1 - x^2/a^2}$, between $x = 0$ and $x = a$, so its area is the corresponding integral. The advantage is that the rectangles are more geometrical, and that may make it easier to visualize what is being done.

EXAMPLE 2 Find the area of the region enclosed by the curves $y = \frac{1}{2}(x^4 + x)$ and $x = \frac{1}{2}(y^4 + y)$, as shown in Figure 6.2.5.

Solution The first thought is to regard this as the region between two curves $y = f(x)$ and $y = g(x)$. The lower curve is $y = g(x) = \frac{1}{2}(x^4 + x)$, but it is not easy to see how to write the upper curve $x = \frac{1}{2}(y^4 + y)$ in the form $y = f(x)$. On the other hand, it is clearly already written with x as a function of y.

The trick is to divide the region into two parts, separated by the line $y = x$ (see Figure 6.2.6). The region below the line can be filled with little rectangles of width dx whose height is the distance from the line down to the bottom curve, that is, $x - \frac{1}{2}(x^4 + x) = \frac{1}{2}(x - x^4)$. The area of the bottom piece is then the "sum" of these little areas, or

$$\int_0^1 \frac{1}{2}(x - x^4)\,dx = \frac{1}{2}\left(\frac{x^2}{2} - \frac{x^5}{5}\right)\Bigg|_0^1 = \frac{3}{20}.$$

For the top piece we use horizontal rectangles of depth dy and length $y - \frac{1}{2}(y^4 + y) = \frac{1}{2}(y - y^4)$. The area of such a rectangle is $\frac{1}{2}(y - y^4)\,dy$, so the area of the region is

$$\int_0^1 \frac{1}{2}(y - y^4)\,dy = \frac{1}{2}\left(\frac{y^2}{2} - \frac{y^5}{5}\right)\Bigg|_0^1 = \frac{3}{20}.$$

The original region has area $\frac{3}{20} + \frac{3}{20} = \frac{3}{10}$.

To do the integral for the upper piece, we needed to express the upper curve and the line with x as a function of y, so that the length of the little rectangle could be written in terms of y. The (narrow) width of the rectangle is dy, and we integrated in terms of y, that is, using y as the variable of integration. The limits of integration were apparent from the picture.

What makes this example work is that the line $y = x$ can be used either way (as a function of x for the lower piece and as a function of y for the upper region). In hindsight we notice that the upper and lower regions are symmetrical, so we only had to evaluate one integral and then double it.

Finally, we remark again that the little rectangles were not really needed. The lower region was after all the region between two curves, $y = x$ and $y = \frac{1}{2}(x + x^4)$, so we know how to find its area by integrating their difference. Similarly, for the upper region we have to think of y as the variable, but once we have made this choice, the region is again enclosed between two curves, and all that is needed is to integrate their difference. Using rectangles of width dx (or dy) may help us to keep track of what we are doing. It does not change the calculation or give us different answers, but if it makes the situation clearer, then it is useful.

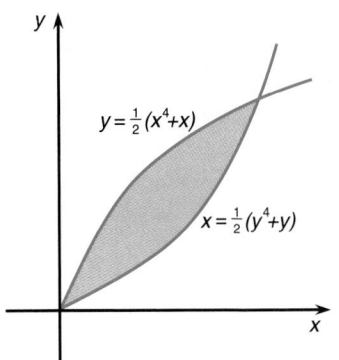

$y = \frac{1}{2}(x^4 + x)$

$x = \frac{1}{2}(y^4 + y)$

Figure 6.2.5

dy

dx

Figure 6.2.6

A graphics calculator can help find the area enclosed between two curves when it is not easy to solve explicitly for the points where they intersect.

For instance, to find the area between $y = \cos x$ and $y = x^2$, we plot the graphs, solve for the points a and b where $\cos x = x^2$, i.e., $\cos x - x^2 = 0$, and then evaluate the area $\int_a^b (\cos x - x^2)\, dx = \left(\sin x - \frac{x^3}{3}\right)\Big|_a^b$. We save some trouble by noting that $a = -b$.

TI-85	SHARP EL9300
GRAPH F1($y(x) =$) COS F1(x) ▼ F1(x) ^ 2 M2(RANGE) $(-)$ 1 ▼ 1 ▼ ▼ $(-)$ 1 ▼ 2 F5(GRAPH) SOLVER COS X $-$ X^ 2 $=$ 0 ENTER 0 F5(SOLVE) EXIT X STO B ENTER $(-)$ X STO A ENTER SIN B $-$ B ^3 $\div$ 3$-$ SIN A $+$ A ^3 $\div$ 3 ENTER	cos x/θ/T 2ndF ▼ x/θ/T x^2 RANGE -1 ENTER 1 ENTER 1 ENTER -1 ENTER 2 ENTER 1 ENTER SOLVER cos x/θ/T $=$ x/θ/T x^2 ENTER $(-)$ 1 ENTER ENTER ENTER ENTER ENTER X STO A $(-)$ X STO B sin B $-$ B a^b 3 ▶ $\div$ 3 $-$ sin A $+$ A a^b 3 ▶ $\div$ 3 ENTER
CASIO f_x-6300G (f_x-7700GB)	HP 48SX
Range -1 EXE 1 EXE 1 EXE -1 EXE 2 EXE 1 EXE EXE EXE EXE Graph cos X EXE Graph X x^2 EXE Trace ▶ (move cursor to intersection) X $\to$ A EXE $(-)$ X $\to$ B EXE sin B $-$ B x^y 3 $\div$ 3 $-$ sin A $+$ A x^y 3 $\div$ 3 EXE (Use Zoom for more accuracy)	EQUATION X y^x 2 ▶ $=$ COS X ENTER SOLVE NEW αE αQ αN ENTER 30 +/− CST NXT SF PLOT PTYPE FUNC PLOTR 1 +/− ENTER 1 XRNG 1 +/− ENTER 2 YRNG DRAW ◀ (Move cursor to intersection) FCN ISECT ATTN ATTN X ENTER ' A STO A +/− ' B STO B SIN B ENTER 3 y^x 3 $\div$ − A SIN − A ENTER 3 y^x 3 $\div$ +

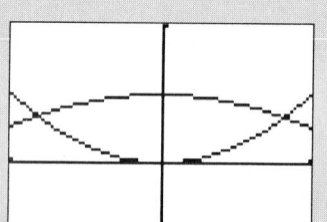

In this example we did not really need the graphs to know which curve is higher and how often they cross. . .

Summary 6.2.1   Setting Up Area Integrals

1. Draw a picture, indicating points where the curves cross each other.
2. Draw little rectangles of width dx, whose height can be expressed as a function of x. It may be necessary to break the region into pieces and find different formulas for the different subregions. It may also be necessary to use y as the variable for some of the subregions (and dy as the "width" of the corresponding rectangles).

In each of the following situations we are given the derivative $F'(t)$ of some function $F(t)$ and one value $F(a)$, for some a. Find the required value $F(b)$.

11. $F'(t) = 3t^2 - 2$, $F(0) = 2$; find $F(3)$.

12. $F'(t) = \sin t$, $F(0) = 0$; find $F(\pi)$.

13. $F'(t) = 1$, $F(2) = -5$; find $F(4)$.

14. $F'(t) = t$, $F(3) = 1$; find $F(\sqrt{2})$.

15. $F'(t) = t^2 - 3$, $F(5) = 2$; find $F(0)$.

16. $F'(t) = |t|$, $F(2) = 4$; find $F(-2)$.

17. $F'(t) = 1/(t^2 + 1)$, $F(-1) = 0$; find $F(1)$.

18. $F'(t) = 1/t$, $F(1) = 6$; find $F(2)$.

19. $F'(t) = e^{2t+1}$, $F(0) = 0$; find $F\left(\frac{1}{2}\right)$.

20. $F'(t) = \cos^2(3t) + \sin^2(3t)$, $F(1) = -2$; find $F(5)$.

‖

21. Find the area of the region enclosed between the curve $x = y^2$ and the right semicircle $x^2 + y^2 = 2$, $x \geq 0$.

22. Find the area of the region enclosed by the graphs of $x = y^3 - 3y$ and $y = x$.

*23. Find the area of the region consisting of all points lying between the graphs of $y = \sin x$ and $y = 1 - \sin x$, with $0 \leq x \leq \pi$.

*24. Find the area of the region enclosed by the graphs of $x = 0$, $y = 3$, and $y = x^3 + 2x$.

**25. Let m and n be positive integers, $m \neq n$.
 (i) Describe the region enclosed between the graphs of $y = x^n$ and $y = x^m$.
 (ii) Find the area of this region.

**26. Let m and n be positive integers, $m \neq n$.
 (i) Describe the region enclosed between the graphs of $y = x^{1/n}$ and $x = y^{1/m}$.
 (ii) Find the area of this region.

Find the areas of the triangles whose vertices are specified below.

27. $(0,0)$, $(3,0)$, and $(1,2)$

28. $(0,0)$, $(1,1)$, and $(1,2)$

*29. $(0,0)$, $(2,1)$, and $(1,2)$

*30. $(-1,0)$, $(2,1)$, and $(1,-2)$

*31. Find the area of the triangle whose vertices are the points $(0,0)$, (a,b), and (c,d), where $0 \leq a \leq c$.

*32. Use integration to verify the formula for the area A of a triangle, $A = \frac{1}{2}bh$, where b is the length of the base and h is the height.

33. (i) In the situation of Example 3 (p. 429), what will be the temperature at 9 A.M.?
 (ii) What will be the temperature at 5 P.M.?

*34. Find a formula for the temperature at time t in the situation of Example 3. (If necessary, simplify so that your formula does not contain an integral sign.)

35. In the situation of Example 4, what will be the volume of the balloon 10 sec after it begins deflating?

*36. (i) In the situation of Example 4, at what time will the volume be a minimum?
 (ii) At the time mentioned in part (i), what will the volume of the balloon be?

37. In the situation of Example 5, how far will the object travel between $t = 1$ and $t = 3$, that is, between the moment 1 sec after starting and the moment 3 sec after starting?

**38. Suppose an object moves so that its velocity at time t is $t^3 - 6t^2 + 5t + 6$ ft/sec.
 (i) Verify that at time $t = 4$ it will have returned to its starting point, that is, the position it was at when $t = 0$.
 (ii) What is the total distance it will have traveled between $t = 0$ and $t = 4$? (*Hint:* The answer is not 0.)

*39. Consider water flowing out of a tank. As the water level falls, the pressure drops and the flow is slower. Suppose the water is flowing out so that at time t the rate of flow is $25e^{-t}$ cu ft/min, with t measured in minutes.
 (i) If there are 25 cu ft of water in the tank when $t = 0$, how much will there be 2 min later?
 (ii) How long will it be until there is only 1 cu ft left?

*40. If money in a bank account grows with *continuous compounding*, it earns interest on its interest and accrues more rapidly than it would with simple interest. If the original amount is A_0 and the rate of interest is $r\%$, then the rate at which the balance is increasing at time t is $A_0 \frac{r}{100} e^{rt/100}$, with t measured in years.
 (i) If $1,000 is placed in an account at an interest rate of 10% with continuous compounding, how much will be there after 1 year?
 (ii) By what percentage has the balance increased in the 1-year period? (This is known as the *effective annual interest rate*, meaning the rate of simple interest that will give the same increase over 1 year as the compound interest does.)
 (iii) What is the effective annual interest rate for continuously compounded interest at a rate of 12%?
 (iv) What is the effective annual interest rate for continuously compounded interest at a rate of 18%?
 (v) How much money would have to be invested at 15%, compounded continuously, so that after 6 *months* there would be $5,000?

*41. Consider the parallelogram shown in Figure 6.2.8, whose base is b units long and whose height is h units. Find its area by integration.

Figure 6.2.8

*42. Notice that the previous question mentioned the height and the length of the base but no other information. Clearly, the area of the parallelogram depends on the height and the length of the base but *not* on the amount by which the parallelogram "tilts." (i) Convince yourself that this is so by a geometrical argument showing that the parallelogram has the same area as the rectangle with base b and height h.

(ii) If the other side has length a and the angle between two adjacent sides is θ, find a formula for the area in terms of a, b, and θ. (iii) What happens if you use the *other* angle between two adjacent sides?

Use a graphics calculator or computer to help you to find the approximate area of the region enclosed between each of the following pairs of curves.

43. $y = x^3 - 2x - 1$, $y = x$

44. $y = \sin x$, $y = x^3$

45. $y = \cos x$, $y = x^4$

46. $y = e^x$, $y = 4 + \ln x$, $x > 0$

47. $y = x^3 + x$, $x = y^2 - 1$

48. $y = \sec^2 x$, $y = 2 - x^2$, $|x| \leq 1$

6.3 Volumes of Solids of Revolution by Slices

SOLID OF REVOLUTION

CYLINDRICAL SLICES

Figure 6.3.1

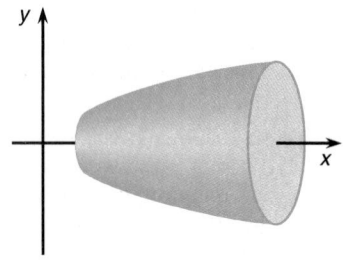

Figure 6.3.2

Integrals can be used to calculate volumes of solid regions in three-dimensional space. We concentrate on what are called *solids of revolution*. If $f(x)$ is a nonnegative function on $[a, b]$, draw its graph $y = f(x)$ as shown in Figure 6.3.1. Then take the xy-plane, containing the graph, and rotate it in three-dimensional space around the x-axis. As the graph revolves, the region under it (shaded in Figure 6.3.1) "sweeps out" a solid three-dimensional region, as shown in Figure 6.3.2. This is a **solid of revolution**.

To find its volume, consider a tiny piece of the interval $[a, b]$ of width dx. Look at the narrow strip of the region under the graph of $y = f(x)$ whose base is this piece of the interval (see Figure 6.3.3). The height of the strip is $f(x)$.

Consider the part of the solid swept out by this strip, as illustrated in Figure 6.3.4. It is a "slice," which is approximately a thin disk. For our purposes a more useful way of seeing it is as a cylinder on its side with radius $f(x)$ and "height" dx. The volume of a cylinder with radius r and height h is $\pi r^2 h$ (cf. Summary 1.5.14(iii)), so the volume of this cylinder is $\pi f(x)^2 dx$.

This suggests that to get the volume of the whole solid, we ought to add up the volumes of all these little slices (cylinders), as shown in Figure 6.3.5, that is,

$$Volume = \int_a^b \pi f(x)^2 dx.$$

Figure 6.3.3

Figure 6.3.4

Figure 6.3.5

It is possible to make this argument more precise, by partitioning $[a, b]$, approximating the volume by the sum of the volumes of cylinders of height Δx_k, and taking a limit. The same integral results, so it seems reasonable to make a definition.

DEFINITION 6.3.1

If $f(x) \geq 0$ for all $x \in [a, b]$, then the volume of the solid of revolution obtained by rotating the graph $y = f(x)$ about the x-axis is

$$Volume = \int_a^b \pi f(x)^2 dx.$$

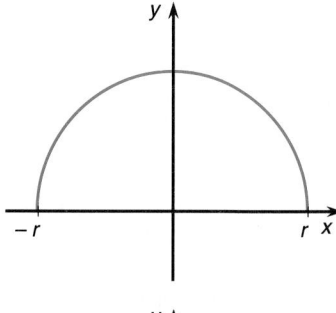

EXAMPLE 1   Volume of a Ball   Find the volume of a ball of radius r.

Solution   We may as well work with the ball whose center is at the origin. It is obtained by revolving the upper semicircle $y = \sqrt{r^2 - x^2}$ around the x-axis, as shown in Figure 6.3.6. So its volume is

$$\int_{-r}^{r} \pi f(x)^2 dx = \pi \int_{-r}^{r} (r^2 - x^2)\, dx = \pi \left(r^2 x - \frac{1}{3} x^3 \right) \Big|_{-r}^{r}$$

$$= \pi \left(r^3 - \frac{1}{3} r^3 - \left(-r^3 + \frac{1}{3} r^3 \right) \right)$$

$$= \frac{4}{3} \pi r^3.$$

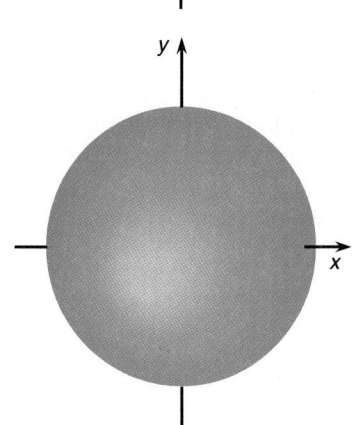

EXAMPLE 2   Volume of a Headlight   Find the volume of the headlight obtained by revolving about the x-axis the parabola $y = \sqrt{x}$ for $0 \leq x \leq 1$. This solid is called a *paraboloid* (see Figure 6.3.7).

Solution   The volume equals

$$\int_0^1 \pi (\sqrt{x})^2 dx = \pi \int_0^1 x\, dx = \frac{\pi}{2} x^2 \Big|_0^1 = \frac{\pi}{2}.$$

Figure 6.3.6

EXAMPLE 3   Volume of Cone   Find the formula for the volume of a right circular cone of radius r and height h (see Figure 6.3.8).

Figure 6.3.7

Figure 6.3.8

Figure 6.3.9

Figure 6.3.10

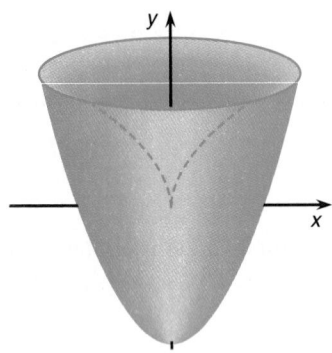

Figure 6.3.11

Since we are integrating with respect to y, the radius is the horizontal distance out from the y-axis. This is why we write $x = f(y)$ and integrate $\pi f(y)^2$.

Figure 6.3.12

Solution We need to express the cone as a solid of revolution. It is obtained by revolving about the x-axis a straight line through the origin. The catch is to find the equation of the line.

To do this, we need two points on the line. One is the origin; what are the coordinates of the point at the upper right corner? As illustrated in Figure 6.3.9, the point is (h, r). This is slightly confusing because the cone is lying on its side, so the "height" is horizontal, but we see from Figure 6.3.8 that it is this dimension that we call the height of the cone.

We find the equation of the line passing through $(0, 0)$ and (h, r). It is $\frac{y-0}{x-0} = \frac{r-0}{h-0}$, or

$$y = \frac{r}{h}x.$$

We are interested in the function $f(x) = \frac{r}{h}x$ with x running from $x = 0$ to $x = h$.

The corresponding volume is

$$\int_0^h \pi \left(\frac{r}{h}x\right)^2 dx = \pi \frac{r^2}{h^2} \int_0^h x^2 dx = \frac{1}{3}\pi \frac{r^2}{h^2} x^3 \bigg|_0^h = \frac{1}{3}\pi r^2 h.$$

EXAMPLE 4 Consider the region in the plane bounded by the y-axis and the curves $y = \sqrt{x}$ and $y = 2x^2 - 1$, shown in Figure 6.3.10. Find the volume of the solid of revolution obtained by revolving this region about the y-axis (see Figure 6.3.11).

Solution Since the region is rotated about the y-axis, we must integrate with respect to y. It is easiest to obtain this solid by taking the paraboloid formed by rotating $y = 2x^2 - 1$ and removing the solid obtained by rotating $y = \sqrt{x}$. The volume we want will be the difference of these volumes. See Figure 6.3.12.

To find the volume of the paraboloid, we must express the parabola $y = 2x^2 - 1$ with x as a function of y, $x = f(y)$. Since $y = 2x^2 - 1$, we see that $y + 1 = 2x^2$, $x^2 = \frac{1}{2}(y + 1)$, $x = \sqrt{\frac{1}{2}(y + 1)} = f(y)$, and the volume of the paraboloid is

$$\int_{-1}^1 \pi \left(f(y)\right)^2 dy = \pi \int_{-1}^1 \left(\sqrt{\frac{1}{2}(y + 1)}\right)^2 dy$$

$$= \pi \int_{-1}^1 \frac{1}{2}(y + 1)\, dy = \frac{\pi}{2}\left(\frac{1}{2}y^2 + y\right)\bigg|_{-1}^1$$

$$= \pi.$$

The other curve $y = \sqrt{x}$ can be written as $x = g(y) = y^2$, and the corresponding solid has volume

$$\int_0^1 \pi \left(g(y)\right)^2 dy = \int_0^1 \pi (y^2)^2 dy = \pi \frac{1}{5}y^5 \bigg|_0^1 = \frac{\pi}{5}.$$

The original solid has volume $\pi - \frac{1}{5}\pi = \frac{4\pi}{5}$.

EXAMPLE 5 Find the volume of the solid obtained by revolving about the x-axis the region between the line $y = \frac{x}{2}$ and the parabola $x = y^2$, between $x = 0$ and $x = 1$ (see Figure 6.3.13).

Figure 6.3.13

Figure 6.3.14

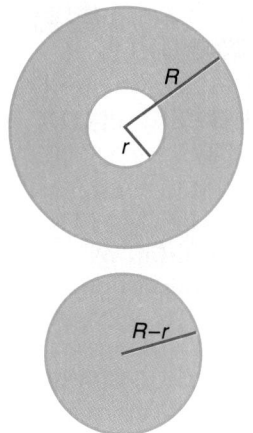

Figure 6.3.15

Solution   Again we find the volumes of two regions and subtract. The volume we want is the volume of the paraboloid minus the volume of the cone (see Figure 6.3.14).

For the paraboloid we are summing up the volumes of slices of thickness dx and radius $y = \sqrt{x}$. The volume is $\int_0^1 \pi (\sqrt{x})^2 dx = \pi \int_0^1 x\, dx = \frac{\pi}{2}$. For the cone we could use the formula from Example 3 or simply calculate $\int_0^1 \pi \left(\frac{x}{2}\right)^2 dx = \frac{\pi}{12}$. The volume of the solid is $\frac{\pi}{2} - \frac{\pi}{12} = \frac{5\pi}{12}$. ◢

There is a temptation to try a shortcut here. In analogy with the procedure for finding the area between two curves, we might try subtracting the lower curve $y = \frac{x}{2}$ from the upper curve $y = \sqrt{x}$ and integrating $\int_0^1 \pi \left(\sqrt{x} - \frac{x}{2}\right)^2 dx$. It is easy to check this gives $\frac{11\pi}{60}$, the *wrong answer*. The reason is that $\left(\sqrt{x} - \frac{x}{2}\right)^2$ does not equal $(\sqrt{x})^2 - \left(\frac{x}{2}\right)^2$.

Geometrically, the reason is that the area between circles of radius R and r is bigger than the area of a circle of radius $R - r$, as shown in Figure 6.3.15. This means that the thin slice between two cylinders has greater volume than the thin cylinder of radius $R - r$. If you are not convinced by the picture, try working it out with $R = 2$ and $r = 1$.

If $f(x) \geq g(x) \geq 0$ on $[a, b]$, then the volume obtained by revolving the region between $y = f(x)$ and $y = g(x)$ about the x-axis is

$$\int_a^b \pi \left(f(x)^2 - g(x)^2\right) dx.$$

EXAMPLE 6   Find the volume of the solid obtained by revolving about the x-axis the region between $f(x) = x^2$ and $g(x) = 1$ for x between $x = -2$ and $x = 2$.

Solution   We begin by drawing a sketch of the situation in Figure 6.3.16. The thing that complicates this problem is that some of the time $f(x)$ is greater than $g(x)$, and

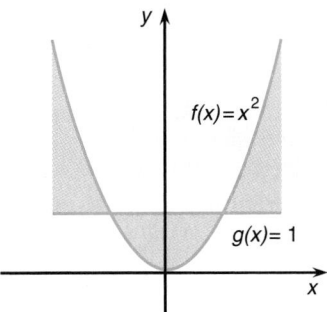

Figure 6.3.16

some of the time it is the other way around. What is required is to find the volume of the solid obtained by revolving the bigger function and subtract from it the volume of the solid obtained from the smaller function. The only way to do this is to work on smaller intervals, on each of which $f(x) \geq g(x)$ all the time or $f(x) \leq g(x)$ all the time.

From the picture we see that $f(x) \leq g(x)$ on $[-1, 1]$ and $f(x) \geq g(x)$ on $[-2, -1]$ and on $[1, 2]$. We work with each of these intervals separately.

If it is not easy to solve explicitly for the points where two curves intersect, a graphics calculator can help find the volume obtained by revolving the region enclosed between the curves.

For instance, to find the volume obtained by revolving the region enclosed between $y = x^4$ and $y = x^3 + 1$, we plot the graphs, solve for the points a and b where $x^4 = x^3 + 1$, i.e., $x^4 - x^3 - 1 = 0$, and then evaluate the volume

$$\pi \int_a^b \left((x^3 + 1)^2 - (x^4)^2 \right) dx = \pi \left(\frac{x^7}{7} + \frac{x^4}{2} + x - \frac{x^9}{9} \right)\Big|_a^b.$$

TI-85

GRAPH F1 $(y(x) =)$ F1 (x) ^ 4
▼ F1 (x) ^ 3 + 1 M2(RANGE) $(-)$ 2
▼ 2 ▼ ▼ $(-)$ 1 ▼ 5 F5(GRAPH)
SOLVER X̂ 4 $-$ X̂ 3 $-$ 1 $=$ 0 ENTER
0 F5(SOLVE) EXIT X STO A ENTER
SOLVER ENTER 1 F5(SOLVE) EXIT
X STO B ENTER π $\times$ (B̂ 7 $\div$ 7 + B̂
4 $\div$ 2 + B $-$ B̂ 9 $\div$ 9 $-$ Â 7 $\div$ 7 $-$
Â 4 $\div$ 2 $-$ A + Â 9 $\div$ 9) ENTER

SHARP EL9300

⫯⫟ x/θ/T a^b 3 ▶ + 1 2ndF ▼ x/θ/T
a^b 4 RANGE -2 ENTER 2 ENTER 1
ENTER -1 ENTER 5 ENTER 1 ENTER
⫯⫟ SOLVER x/θ/T a^b 3 ▶ + 1 $=$
x/θ/T a^b 4 ENTER $(-)$ 1 ENTER ENTER
ENTER ENTER ENTER ⊞ X STO
A SOLVER ENTER 1 ENTER ENTER
ENTER ENTER ENTER ⊞ X STO B π
$\times$ (B a^b 7 ▶ $\div$ 7 + B a^b 4 ▶ $\div$ 2 +
B $-$ B a^b 9 ▶ $\div$ 9 $-$ A a^b 7 ▶ $\div$ 7
$-$ A a^b 4 ▶ $\div$ 2 $-$ A + A a^b 9 ▶ $\div$
9) ENTER

CASIO f_x-6300G (f_x-7700GB)

Range -2 EXE 2 EXE 1 EXE -1 EXE 5
EXE 1 EXE EXE EXE EXE Graph ▦ x^y
3 + 1 EXE Graph ▦ x^y 4 EXE Trace
▶ (move cursor to intersection) ▦
→ ▦ EXE Graph ▦ x^y 4 EXE Trace
▶ (move cursor to intersection) ▦
→ ▦ EXE π $\times$ (▦ x^y 7 $\div$ 7 + ▦ x^y 4
$\div$ 2 + ▦ $-$ ▦ x^y 9 $\div$ 9 $-$ ▦ x^y 7 $\div$ 7
$-$ ▦ x^y 4 $\div$ 2 $-$ ▦ + ▦ x^y 9 $\div$ 9) EXE
(Use Zoom for more accuracy)

HP 48SX

▩ EQUATION X y^x 3 ▶ + 1 ▩ ▩
X y^x 4 ENTER ▩ SOLVE NEW αE
αQ αN ENTER 30 +/− ↱ CST NXT
SF ▩ PLOT PTYPE FUNC PLOTR
2 +/− ENTER 2 XRNG 1 +/− ENTER
5 YRNG DRAW ◀ (Move cursor
to intersection) FCN ISECT ATTN
ATTN X ENTER ' A STO DRAW
◀ (Move cursor to intersection)
FCN ISECT ATTN ATTN X ENTER '
B STO B ENTER 7 y^x 7 $\div$ B ENTER 4
y^x 2 $\div$ + B ENTER + B ENTER 9 y^x 9
$\div$ $-$ A ENTER 7 y^x 7 $\div$ $-$ A ENTER 4
y^x 2 $\div$ $-$ A ENTER $-$ A ENTER 9 y^x 9
$\div$ + ▩ $\times$

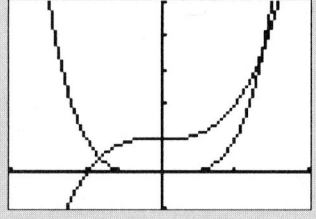

On $[-2, -1]$, the volume is

$$\int_{-2}^{-1} \pi\left(f(x)^2 - g(x)^2\right) dx = \pi \int_{-2}^{-1} \left((x^2)^2 - 1^2\right) dx = \pi \int_{-2}^{-1} \left(x^4 - 1\right) dx$$

$$= \pi\left(\frac{x^5}{5} - x\right)\Bigg|_{-2}^{-1} = \pi\left(\frac{-1}{5} + 1 + \frac{32}{5} - 2\right) = \frac{26}{5}\pi.$$

On $[-1, 1]$, what we need is

$$\int_{-1}^{1} \pi\left(g(x)^2 - f(x)^2\right) dx = \pi \int_{-1}^{1} (1 - x^4) dx = \pi\left(x - \frac{x^5}{5}\right)\Bigg|_{-1}^{1} = \pi\left(2 - \frac{2}{5}\right)$$

$$= \frac{8}{5}\pi.$$

On $[1, 2]$ the calculation is $\pi \int_{1}^{2} (x^4 - 1) dx = \pi\left(\frac{1}{5}x^5 - x\right)\Bigg|_{1}^{2} = \frac{26}{5}\pi.$

The total volume we want is the sum of the volumes for the three intervals, $\frac{26}{5}\pi + \frac{8}{5}\pi + \frac{26}{5}\pi = 12\pi.$

It is very important to recognize that incorrect answers will result from finding the volumes resulting from both functions over the entire interval $[-2, 2]$ and then subtracting. In this example,

$$\int_{-2}^{2} \pi f(x)^2 dx = \pi\frac{1}{5}x^5\Bigg|_{-2}^{2} = \frac{64}{5}\pi \qquad \text{and} \qquad \int_{-2}^{2} \pi g(x)^2 dx = \pi x\Bigg|_{-2}^{2} = 4\pi.$$

The difference of these two is not the correct answer.

This flange is the region contained inside one solid of revolution and outside another solid of revolution. The image was generated by a computer using DesignCAD 3D *software made by* American Small Business Computers. *Note the blue plastic seal and the ease with which the computer can produce different views of the object.*

> We see that solids obtained by revolving the region between two curves are more complicated. It is essential to work on smaller intervals on which one function is always bigger and the other is always smaller.

Summary 6.3.2

Volumes of Revolution by Slices

1. Identify the axis about which the solid is revolved. The corresponding variable will be the variable of integration (i.e., integrate with respect to x for solids of revolution about the x-axis and with respect to y for solids of revolution about the y-axis). We visualize the thin "slice" moving along the axis as x (or y) moves from the lower limit of integration to the upper limit of integration.

2. Express the function f whose graph is being revolved in terms of the variable of integration.

3. Find the limits of integration, drawing a picture if necessary. Set up the integral, $\int_a^b \pi f(x)^2 dx$ or $\int_a^b \pi f(y)^2 dy$ as the case may be.

4. Integrate.

5. For a solid obtained by revolving about an axis the region between two curves, remember to break the interval into smaller intervals so that on each of them

one of the functions is always bigger than the other. If $f(x) \geq g(x) \geq 0$ on $[a, b]$, then the volume obtained by revolving the region between $y = f(x)$ and $y = g(x)$ around the x-axis is

$$\int_a^b \pi\left(f(x)^2 - g(x)^2\right) dx.$$

An analogous formula works for revolution about the y-axis. Finally, combine the results from all the intervals.

Exercises 6.3

I

In each of the following situations, find the volume of the solid obtained by revolving the given function $f(x)$, defined on the specified interval $[a, b]$, about the x-axis. In each case, make a sketch of the solid.

1. $f(x) = 3x$ on $[0, 2]$
2. $f(x) = x^2$ on $[1, 2]$
3. $f(x) = 2 - x$ on $[0, 2]$
4. $f(x) = 2 + x^2$ on $[0, 1]$
5. $f(x) = e^x$ on $[0, 1]$
6. $f(x) = \sin x$ on $[0, \pi]$
7. $f(x) = x^2 - x$ on $[0, 1]$
8. $f(x) = 1 - x^2$ on $[-1, 1]$
9. $f(x) = 2$ on $[2, 5]$
10. $f(x) = x^2$ on $[-1, 1]$

In each of the following situations, find the volume of the solid obtained by revolving about the x-axis the region between the given functions $f(x)$ and $g(x)$ over the specified interval $[a, b]$. In each case, make a sketch of the solid.

11. $f(x) = 2x, g(x) = x$ over $[0, 1]$
12. $f(x) = x^2, g(x) = x$ over $[0, 1]$
13. $f(x) = x, g(x) = \sin x$ over $[0, \pi]$
14. $f(x) = x^3, g(x) = x^2$ over $[1, 2]$
15. $f(x) = 2 - x^2, g(x) = 1$ over $[-1, 1]$
16. $f(x) = |x|, g(x) = x^2$ over $[-2, 2]$
17. $f(x) = 2, g(x) = 1$ over $[0, h]$
18. $f(x) = x, g(x) = 3$ over $[0, 2]$
19. $f(x) = M, g(x) = N$ over $[0, L]$
20. $f(x) = 1 + \sin x, g(x) = 2 + \sin x$ over $\left[0, \frac{\pi}{2}\right]$

In each of the following situations, find the volume of the solid obtained by revolving about the x-axis the region enclosed between the given functions $f(x)$ and $g(x)$. In each case, make a sketch of the solid.

21. $f(x) = x^3, g(x) = x^2$
22. $f(x) = x^2, g(x) = 1 - x^2$
23. $f(x) = x^2, g(x) = 3$
24. $f(x) = x^4, g(x) = 16$
25. $f(x) = x^4, g(x) = x^6$
26. $f(x) = \sqrt{|x|}, g(x) = x^4$
27. $f(x) = 2 - |x - 1|, g(x) = 1$
28. $f(x) = |x^2 - x|, g(x) = 0$
29. $f(x) = 1, g(x) = \sqrt{2 - x^2}$
30. $f(x) = \frac{1}{\sqrt{2}}x^2, g(x) = \left(1 + x^2\right)^{-1/2}$

II

*31. Find the volume of the solid obtained by revolving about the x-axis the region between the graphs of the functions $f(x) = x$ and $g(x) = x^3$ for x between $x = 0$ and $x = 2$. Draw a picture.

*32. Find the volume of the solid obtained by revolving about the x-axis the region between the graphs of the functions $f(x) = x$ and $g(x) = x + \sin x$ for x between $x = 0$ and $x = 2\pi$. Draw a picture.

*33. Consider the functions $f(x) = \sec x$ and $g(x) = 2$ for $x \in \left(-\frac{\pi}{2}, \frac{\pi}{2}\right)$. Rotate the region lying above $f(x)$ and below $g(x)$ about the x-axis, and compute the volume of the resulting solid.

*34. The curves $y = \cos x$ and $y = 1 - \cos x$ cross exactly once for $0 \leq x \leq \frac{\pi}{2}$. Consider the region lying between these curves, to the right of the y-axis and to the left of the point where the curves cross. Find the volume of the solid obtained when this region is rotated about the x-axis.

*35. Consider the circle whose radius is 1 and whose center is at the point $(0, 2)$.

    (i) Describe the solid obtained by revolving the region inside this circle about the x-axis.

    (ii) Find the volume of this solid.

*36. Suppose $R > r$ and consider the circle whose radius is r and whose center is at the point $(0, R)$. Find the volume of the solid obtained by revolving the region inside this circle about the x-axis.

*37. Suppose $a, b > 0$. Find the volume of the "ellipsoid" obtained by revolving about the x-axis the upper half of the ellipse $\frac{x^2}{a^2} + \frac{y^2}{b^2} = 1$. What happens if $a = b$?

*38. Suppose $a, b > 0$. Find the volume of the ellipsoid obtained by revolving the ellipse $\frac{x^2}{a^2} + \frac{y^2}{b^2} = 1$ about the y-axis.

*39. When we talked about solids of revolution and calculating their volumes, we considered a *nonnegative*

function $f(x)$. Discuss what happens if $f(x)$ is allowed to have negative values too. In particular, what is the geometrical meaning of $\int_{-1}^{1} \pi(x)^2 dx$, that is, $f(x) = x$? Make a picture.

*40. Find the volume of the slice of width 2 units cut from the center of a ball of radius 2 units.

In each of the following situations, use a graphics calculator or computer to find the points of intersection of the graphs of $y = f(x)$ and $y = g(x)$. Find the volume of the solid obtained by revolving about the x-axis the region enclosed between the graphs.

41. $f(x) = 3x + 5,$ $g(x) = x^3 + 4$

42. $f(x) = \cos x,$ $g(x) = x^2$

43. $f(x) = 2 \sin x,$ $g(x) = x^3 + x$

44. $f(x) = \frac{1}{4} e^x,$ $g(x) = \sqrt{x}$

6.4 Volumes by Cylindrical Shells

SLICES vs SHELLS

Figure 6.4.1

Figure 6.4.2

There is another way of calculating volumes of solids of revolution that sometimes results in simpler calculations. In the previous section we built up a solid out of thin slices, as shown in Figure 6.4.1. We thought of the thin slice moving along the axis of rotation, "sweeping out" the solid.

    Instead of this, we could build it up out of **cylindrical shells**, as shown in Figure 6.4.2. We can imagine the shells starting as a narrow tube around the axis of rotation and expanding to the outside of the solid, sweeping out the whole volume as they go. "Adding up," that is, integrating, their volumes will give us a way to find the volume of the solid.

EXAMPLE 1   Use cylindrical shells to find the volume of the paraboloid found by rotating $y = \sqrt{x}, 0 \le x \le 4$, about the x-axis.

Solution   As x runs from 0 to 4, $y = \sqrt{x}$ runs from 0 to 2 (see Figure 6.4.3). For each value of y we obtain a cylinder as shown in Figure 6.4.4. Its radius is y and its

Figure 6.4.3

Figure 6.4.4

Figure 6.4.5

Figure 6.4.6

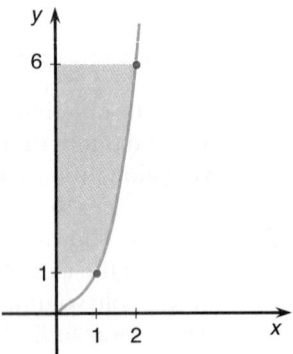

Figure 6.4.7

length is the horizontal distance from the curve to the right end of the paraboloid. The curve is $y = \sqrt{x}$ or $x = y^2$, and the right end of the paraboloid is $x = 4$. The horizontal distance between the curve and this vertical line is $4 - y^2$.

To find the volume of the cylindrical shell, we slit it and flatten it, as shown in Figure 6.4.5. What results is the flat rectangle shown in Figure 6.4.6, whose thickness is dy and with one side $4 - y^2$. The length of the other side used to be the circumference of a circle with radius y, so it is $2\pi y$. The volume of the shell equals the volume of this rectangle, $2\pi y(4 - y^2)\,dy$. We obtain the volume of the whole paraboloid by integrating:

$$Volume = \int_0^2 2\pi y(4 - y^2)\,dy = 2\pi \int_0^2 (4y - y^3)\,dy = 2\pi\left(2y^2 - \frac{1}{4}y^4\right)\Bigg|_0^2$$
$$= 8\pi.$$

The example we have just finished suggests that the volume of a cylindrical shell is $(length) \times (circumference) \times (thickness) = (length) \times 2\pi\,(radius) \times (thickness)$, and this can be used to set up integrals.

In Example 1 the cylindrical shells calculation was no better (perhaps slightly worse) than what we had to do with the method of slices, but there are times when it is clearly easier.

EXAMPLE 2   Consider the curve $y = x^3 - x^2 + x$ between $x = 1$ and $x = 2$, and the region between it and the y-axis, shown in Figure 6.4.7. Find the volume of the solid obtained by rotating this region about the y-axis.

Solution   The derivative $\frac{dy}{dx} = 3x^2 - 2x + 1$ is a quadratic polynomial with no zeroes, so it is always positive, and the graph of $y = x^3 - x^2 + x$ is always increasing.

To apply the method of slices, we would have to find the radius of the slice. This means expressing x as a function of y (see Figure 6.4.8). Since y is a cubic polynomial in x, this will not be easy. Instead, we can use cylindrical shells. This will require integrating with respect to x.

As x goes from 1 to 2, $y = x^3 - x^2 + x$ goes from 1 to 6. We will let x move from 1 to 2 and for each value of x consider the cylindrical shell whose radius is x. The height of the cylindrical shell shown in Figure 6.4.9 is $6 - y = 6 - x^3 + x^2 - x$; after all this is just the difference of the heights at the top (6) and the bottom (y). The thickness of the shell is dx.

Figure 6.4.8   **Figure 6.4.9**

A graphics calculator can be helpful when it is not easy to solve explicitly for the points of intersection.

For instance, to find the volume obtained by revolving about the x-axis the region enclosed between $x = y^4 + 1$ and $x = 5y - y^3$, we plot the graphs and solve for the points a and b where $y^4 + 1 = 5y - y^3$, i.e., $y^4 + y^3 - 5y + 1 = 0$. Since we plot the graphs with "y" as the variable, the positive x-axis points up and the positive y-axis points to the right on the calculator screen. For the actual situation, it is necessary to reflect the picture, as in the second illustration. Then we evaluate the volume $2\pi \int_a^b y\left(5y - y^3 - (y^4 + 1)\right) dy = 2\pi\left(5\frac{y^3}{3} - \frac{y^5}{5} - \frac{y^6}{6} - \frac{y^2}{2}\right)\Big|_a^b$.

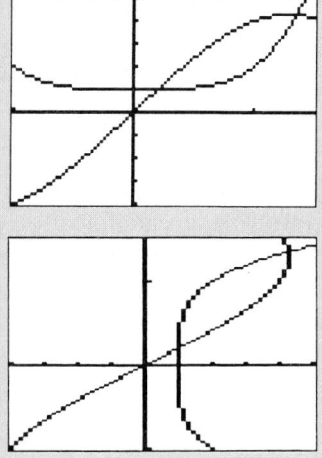

TI-85	SHARP EL9300
GRAPH F1 $(y(x) =)$ 5 F1 (x) − F1 (x) ^3 ▼ F1 (x) ^4 + 1 M2(RANGE) (−) 1 ▼ 1.5 ▼ ▼ (−) 4 ▼ 5 F5(GRAPH) SOLVER X̄ 4 + X̄ 3 − X 5 + 1 = 0 ENTER 0 F5(SOLVE) EXIT X STO A ENTER SOLVER ENTER 1 F5(SOLVE) EXIT X STO B ENTER 2 π × (5 B̄ 3 ÷ 3 − B̄ 5 ÷ 5 − B̄ 6 ÷ 6 − B̄ 2 ÷ 2 − 5 Ā 3 ÷ 3 + Ā 5 ÷ 5 + Ā 6 ÷ 6 + Ā 2 ÷ 2) ENTER	⟳ x/θ/T a^b 4 ▶ + 1 2ndF ▼ 5 x/θ/T − x/θ/T a^b 3 RANGE −1 ENTER 1.5 ENTER 1 ENTER −4 ENTER 5 ENTER 1 ENTER ⟳ SOLVER x/θ/T a^b 4 ▶ + 1 = 5 x/θ/T − x/θ/T a^b 3 ENTER (−) 1 ENTER ENTER ENTER ENTER ENTER ⊞ X STO A SOLVER ENTER 1 ENTER ENTER ENTER ENTER ENTER ⊞ X STO B 2 π × (5 B a^b 3 ▶ ÷ 3 − B a^b 5 ▶ ÷ 5 − B a^b 6 ▶ ÷ 6 − B x^2 ÷ 2 − 5 A a^b 3 ▶ ÷ 3 + A a^b 5 ▶ ÷ 5 + A a^b 6 ▶ ÷ 6 + A x^2 ÷ 2) ENTER
CASIO f_x-6300G (f_x-7700GB)	HP 48SX
Range −1 EXE 1.5 EXE 1 EXE −4 EXE 5 EXE 1 EXE EXE EXE EXE Graph ▆ x^y 4 + 1 EXE Graph 5 ▆ − ▆ x^y 3 EXE Trace ▶ (move cursor to inter-section) ▆ → ▆ EXE Graph 5 ▆ − ▆ x^y 3 EXE Trace ▶ (move cursor to intersection) ▆ → ▆ EXE 2 π × (5 ▆ x^y 3 ÷ 3 − ▆ x^y 5 ÷ 5 − ▆ x^y 6 ÷ 6 − ▆ x^y 2 ÷ 2 − 5 ▆ x^y 3 ÷ 3 + ▆ x^y 5 ÷ 5 + ▆ x^y 6 ÷ 6 + ▆ x^y 2 ÷ 2) EXE (Use Zoom for more accuracy)	▆ EQUATION X y^x 4 ▶ + 1 ▆ ▆ 5 × X − X y^x 3 ENTER ▆ SOLVE NEW αE αQ αN ENTER 30 +/− ↱ CST NXT SF ▆ PLOT PTYPE FUNC PLOTR 1 +/− ENTER 1.5 XRNG 4 +/− ENTER 5 YRNG DRAW ◀ (Move cursor to intersection) FCN ISECT ATTN ATTN X ENTER ′ A STO DRAW ◀ (Move cursor to intersection) FCN ISECT ATTN ATTN X ENTER ′ B STO B ENTER 3 y^x 3 ÷ 5 × B ENTER 5 y^x 5 ÷ − B ENTER 6 y^x 6 ÷ − B ENTER 2 y^x 2 ÷ − A ENTER 3 y^x 3 ÷ 5 × − A ENTER 5 y^x 5 ÷ + A ENTER 6 y^x 6 ÷ + A ENTER 2 y^x 2 ÷ + ▆ × 2 ×

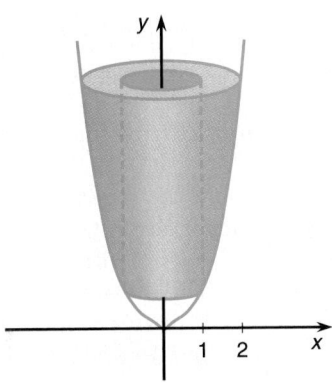

Figure 6.4.10

The volume of the shell is $2\pi x(6-y)\,dx = 2\pi x(6 - x^3 + x^2 - x)\,dx = 2\pi(-x^4 + x^3 - x^2 + 6x)\,dx$. The volume is obtained by integrating:

$$2\pi \int_1^2 (-x^4 + x^3 - x^2 + 6x)\,dx = 2\pi\left(-\frac{1}{5}x^5 + \frac{1}{4}x^4 - \frac{1}{3}x^3 + 3x^2\right)\Bigg|_1^2$$

$$= \frac{253\pi}{30}.$$

This is the volume of the solid obtained by revolving the part of the region between $x = 1$ and $x = 2$; this is the solid without its central "core" (see Figure 6.4.10). The core is a cylinder of radius 1 and height $6 - 1 = 5$, so its volume is 5π. The total volume is $\frac{253\pi}{30} + 5\pi = \frac{403\pi}{30} \approx 42.202$.

If we attempt to write down a formula for the volume of a solid of revolution using the technique of cylindrical shells, either the result will be too restricted to be used in many situations or it will be unnecessarily complicated. Rather than memorizing formulas, it is far better to figure out what is needed in individual cases.

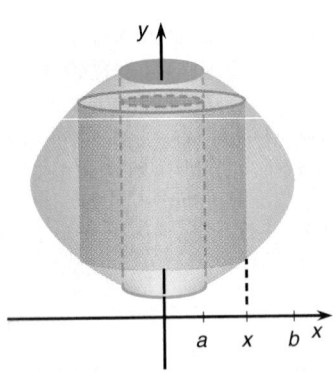

Some people find it helpful to visualize the method of cylindrical shells in the following way. Consider a solid obtained by revolving a region about the y-axis, as shown in the illustration. To find its volume, we integrate with respect to x, with x running from the inner radius a to the outer radius b. At each particular value of x we consider the cylinder whose radius is x and whose height is the vertical cross-section of the solid at radius x. The total volume equals the integral

$$\int_a^b 2\pi x(\textit{height})\,dx.$$

Observe that the integrand $2\pi x(\textit{height})$ is exactly the surface area of the cylinder of radius x. We can think of the integral as the integral as the radius goes from $x = a$ to $x = b$ of the surface area of the cylinder of radius x inside the solid.

Summary 6.4.1

Volumes of Revolution by Cylindrical Shells

1. Identify the axis about which the solid is revolved. The *other* variable will be the variable of integration (i.e., integrate with respect to y for solids of revolution about the x-axis and with respect to x for solids of revolution about the y-axis). We visualize the thin "shell" expanding out from the axis as the radius y (or x) moves from the smallest value (at the inside of the solid) to its largest value (at the outside of the solid).

2. Express the function f whose graph is being revolved in terms of the variable of integration.

Figure 6.4.11

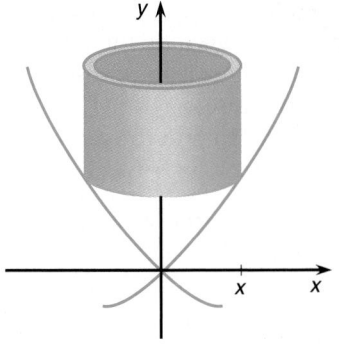

Figure 6.4.12

3. Find the limits of integration, drawing a picture if necessary. The integration will go from the smallest radius of the solid up to the largest radius. Set up the integral, $2\pi \int_a^b x(height)\,dx$ or $2\pi \int_a^b y(height)\,dy$ as the case may be, where "height" refers to the height of the shell in each case. (Note that the "height" may be horizontal.)

4. Integrate.

For a solid obtained by revolving the region between two curves around an axis, it may be necessary to break the interval into smaller intervals so that on each of them one of the functions is always greater than the other. Then find the volumes of the solid constructed from each function and subtract. It may be necessary to use shells for some parts and slices for others. Finally, combine the results from all the intervals.

Slices Vs. Shells

If we rotate a curve $y = f(x)$ about the x-axis, it is usually easy to use the method of slices, because the radius of the slice is $f(x)$, if x is the point where the slice crosses the x-axis (see Figure 6.4.11).

But if we rotate the same curve about the y-axis, then the radius of a slice is x, and we have to express it as a function of y. If $f(x)$ is a quadratic polynomial, this is not too difficult, as we saw in Examples 6.3.4 and 6.3.5, but in general it is complicated and likely to result in a difficult integral. In such a case the method of cylindrical shells is more likely to be suitable. There we would be integrating with respect to x, the radius of the shell would be x, and the length of the shell would be something involving the function $f(x)$. See Figure 6.4.12.

Similar remarks apply to rotating $x = g(y)$ about the y-axis (for which slices are likely best) or about the x-axis (for which shells are likely better).

It is true, however, that there are occasionally instances in which these general guidelines fail and what appears to be the sensible approach gives an integral that is hard to evaluate. In such a case it is sometimes possible to get the answer using the other approach.

So far we have only considered solids formed by revolving about one of the axes, but it is possible to use any straight line.

EXAMPLE 3   Find the volume of the region obtained by revolving the region enclosed between $y = x$ and $y = x^2$ about the line $x = 2$. See Figure 6.4.13.

Solution   We begin by sketching the situation in Figure 6.4.14. If we try the method of slices, we will have to do two calculations, one for the outer region between $y = x$

Figure 6.4.13   **Figure 6.4.14**

Figure 6.4.15

and $x = 2$, with $0 \le y \le 1$, and one for the inner region between $y = x^2$ and $x = 2$, with $0 \le y \le 1$, and then subtract the resulting volumes.

Instead, we try cylindrical shells. From Figure 6.4.15 we see that we should consider values of x between $x = 0$ and $x = 1$. The narrow strip with width dx going from $y = x^2$ to $y = x$ has height $x - x^2$. If we revolve this strip about the line $x = 2$, we obtain a cylindrical shell. Its radius is the distance to the line $x = 2$, or $2 - x$. The volume of the shell is $2\pi \, (radius)\,(height)\, dx = 2\pi(2 - x)(x - x^2)\, dx = 2\pi(x^3 - 3x^2 + 2x)\, dx$.

We integrate to find the volume of the whole solid:

$$2\pi \int_0^1 (x^3 - 3x^2 + 2x)\, dx = 2\pi \left(\frac{1}{4}x^4 - x^3 + x^2 \right) \Big|_0^1 = 2\pi\left(\frac{1}{4}\right)$$

$$= \frac{\pi}{2}.$$

> Rather than trying to memorize formulas for all the possible ways of calculating volumes of revolution about various lines, it is better to realize that it is fairly easy to construct the integral as we have just done. The idea is to rely on the picture and the idea of slices or cylindrical shells, as the case may be. The one new ingredient if the line is not one of the axes is having to find the radius of the slice or the shell, which is just the distance to the line.

Finally, we remark that volumes are related to mass or weight, especially for objects made of material that has uniform density. This simply means that every cubic inch of the material weighs the same amount, which means that the total weight is just that constant times the volume.

EXAMPLE 4   Flashlight Reflector   The glass reflector for a large flashlight is made in the shape of a solid of revolution. It is obtained by revolving about the x-axis the region between the curves $x = \frac{1}{2}y^2$ and $x = \frac{1}{2}y^2 + \frac{1}{8}$, for $0 \le x \le 2$. All measurements are made using inches as units. If the glass used has a density of 1.50 oz/cu in, how much does the reflector weigh?

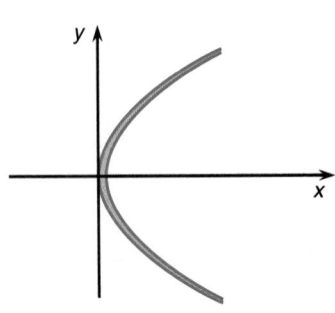

Figure 6.4.16

Solution   We find the volume of the glass reflector. This is the situation of the volume between two solids of revolution. The outer solid is obtained by revolving about the x-axis the curve $x = \frac{1}{2}y^2$ for $0 \le x \le 2$. We use slices and note that $y^2 = 2x$, so its volume is

$$\pi \int_0^2 2x\, dx = \pi x^2 \Big|_0^2 = 4\pi.$$

The inner solid is obtained by rotating the curve $x = \frac{1}{2}y^2 + \frac{1}{8}$ about the x-axis. Notice that for this curve, x is never less than $\frac{1}{8}$ (see Figure 6.4.16). We find that x should run from $\frac{1}{8}$ to $x = 2$, and noting that $y^2 = 2x - \frac{1}{4}$, we find that the volume is

$$\pi \int_{1/8}^2 \left(2x - \frac{1}{4} \right) dx = \pi \left(x^2 - \frac{1}{4}x \right) \Big|_{1/8}^2 = 3\frac{33}{64}\pi.$$

The volume of the reflector is the difference of these two volumes, or $4\pi - 3\frac{33}{64}\pi = \frac{31}{64}\pi \approx 1.52$ cu in.

Knowing that the density of the glass is 1.50 oz/cu in, we see that the reflector weighs $1.52 \times 1.50 \approx 2.28$ oz.

Exercises 6.4

I

In each of the following, revolve the region under the graph of the specified function over the given interval about the x-axis and find the volume of the resulting solid.

1. $x = y^5 + 2y$, for $y \in [0, 1]$
2. $x = y^3 + 3y^2 + 1$, for $y \in [0, 2]$
3. $x = y^3 + 3y$, for $x \in [0, 4]$
4. $x = y^4 + y^2 + 2$, for $x \in [2, 4]$
5. $x = y^2 + 1$, for $y \in [0, 2]$
6. $x = y^4 - 1$, for $x \in [0, 15]$
7. $x = y^3 + y$, for $y \in [0, 1]$
8. $x = y^3 + 2y^2$, for $y \in [0, 1]$
9. $x = y^{3/2}$, for $y \in [1, 4]$
10. $x = y^{4/3}$, for $y \in [1, 2]$
11. $x = y + y^{2/3}$, for $y \in [1, 8]$
12. $x = \frac{\sin y}{y}$, for $y \in \left[\frac{\pi}{4}, \frac{\pi}{2}\right]$
13. $y = \sqrt{1 - x^2}$, for $x \in [-1, 1]$
14. $y = \sqrt{R^2 - x^2}$, for $x \in [0, R]$
15. $y = \sqrt{1 - x^2}$, for $x \in \left[\frac{1}{2}, 1\right]$
16. $y = \sqrt{2 - x^2}$, for $x \in [0, 1]$
17. $x = y^{1/3} + y^{2/5}$, for $y \in [1, 3]$
18. $x = \frac{1}{1+y^2}$, for $y \in [2, 4]$
19. $x^2 = 1 + y^2$, for $y \in [0, 1]$, $x > 0$
20. $x^3 = \frac{1}{1+y^2}$, for $y \in [0, 2]$

In each of the following situations there are two or more curves (and/or lines). Revolve the region bounded by the given curves about the specified line and find the volume of the resulting solid.

21. $y = x^3 + x$, the x-axis, $x = 1$, revolved about the y-axis
22. $y = x^5 + 2x$, the x-axis, $x = 2$, revolved about the y-axis
23. $x = y$, $x + y = 2$, the x-axis, revolved about the x-axis
24. $x = y$, $x + y = 2$, the x-axis, revolved about the y-axis
25. $x = y^3$, $y = x^2$, revolved about the x-axis
26. $y = \sqrt{4 - x^2}$, $-2 \leq x \leq 2$, the x-axis, revolved about the x-axis
27. $x = y^2$, $x = 4$, revolved about the y-axis
28. $x = y^2$, $x = 4$, revolved about the x-axis
29. $y^4 + y = x$, the y-axis, $y = 1$, revolved about the x-axis
30. $x = y^3 + 4y$, the x-axis, $x = 5$, revolved about the x-axis
31. $y = x^3$, $y = x$, $0 \leq x \leq 1$, revolved about the y-axis

32. $y = x^5$, $y = x^2$, revolved about the line $x = 3$
33. $y = x^2$, $x = y^2$, revolved about the line $x = 2$
34. $y = \exp(x^2)$, the y-axis, $y = e$, revolved about the y-axis
35. $y = x^3$, $y = 4x$, between the x-axis and $y = 8$, revolved about the y-axis
36. $x^2 + y^2 = 1$, revolved about the line $x = 1$

II

*37. Find the volume of the region obtained by revolving about the y-axis the region under the graph of $y = 3x^2 - 12x + 14$, for $x \in [0, 4]$.

*38. Find the volume of the region obtained by revolving about the y-axis the region under the graph of $y = \arctan(x^2)$, for $x \in [0, 1]$.

*39. Find the volume of the region obtained by revolving the region enclosed between $y = x$ and $y = x^2$ about the line $x = 4$.

*40. Find the volume of the region obtained by revolving the region enclosed between $y = x$ and $y = x^2$ about the line $y = 1$.

**41. Find the volume of the region obtained by revolving the region enclosed between $y = x$, the x-axis, and $x = 1$ about the line $y = x$.

**42. Find the volume of the region obtained by revolving the region enclosed between $y = x$ and $y = x^2$ about the line $y = x$.

*43. Find the volume that remains if a cylindrical hole of radius 1 in is bored exactly through the center of a sphere of radius 3 in.

*44. Suppose $R > r$ and find the volume that remains if a cylindrical hole of radius r is bored exactly through the center of a sphere of radius R (see Figure 6.4.17).

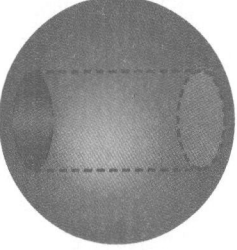

Figure 6.4.17

*45. Let m be a positive integer and consider the solid formed by revolving about the x-axis the region under the curve $y = x^m$, for $0 \leq x \leq a$. Calculate the volume of this solid in two different ways, first using the technique of slices and second using cylindrical shells. Of course, the answers should be the same. (If you are having trouble, try it first with $m = 1$ or $m = 2$.)

**46. A cylindrical hole is bored exactly through the center of a sphere in such a way that the length of the hole in the part that remains is 6 cm, as shown in Figure 6.4.18. Find the volume of the part of the

$\xleftarrow{\qquad}$
6 cm   **Figure 6.4.18**

sphere that is left. (*Hint:* The question does not mention the radius of the original sphere, which appears to be necessary, but in fact it is not needed.)

47. A bowl is made in the shape of the solid of revolution obtained by revolving about the y-axis the region between the curves $4y = 3x^2 + 4$ and $4y = x^4$. Suppose all measurements are made in inches, and suppose the clay from which the bowl is made weighs 2 oz/cu in. How much does the bowl weigh?

48. A piston head is made in the shape of a cylinder with a smaller cylindrical hole bored in its end, as shown in Figure 6.4.19. Suppose the large cylinder has radius 3 cm and height 4 cm and the cylindrical hole has radius 1 cm and height 1 cm. If the material from which the piston head is constructed has a density of $7.6\ \text{g/cm}^3$, will the piston head meet the design specification that it must weigh no more than 850 grams?

Figure 6.4.19

*49. In the situation of Example 4 (p. 444), calculate the volume of the reflector using cylindrical shells. (Be careful near the top.) Of course, the answer should be the same as when it was done with slices.

50. The horn of a musical instrument is made in the shape obtained by revolving the curve $y = e^{x/2}$, for $x \in [-3, 2]$, about the x-axis. If the units are inches, find the volume enclosed in the horn (i.e., if the ends were covered with flat sheets of plastic, how much water would it hold?).

51. A child's top is made by revolving the curve $x = y - 2 + \sqrt{4 - 2y}$, for $0 \le y \le 2$, about the y-axis (see Figure 6.4.20).

  (i) Assuming that the units are inches, find the volume of the top.

  (ii) If the top is made of wood whose density is 0.277 oz/cu in, how much does it weigh?

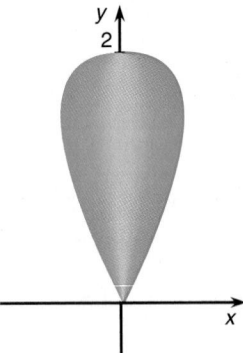

Figure 6.4.20

*52. Consider the curve $x = y^3 + 1$, for $x \in [0, 2]$. Find the volume of the solid obtained by revolving around the x-axis the region between the curve and the x-axis.

Use a graphics calculator or computer to find the volume of the solid obtained by revolving about the x-axis the region enclosed between each of the following pairs of functions.

53. $x = 3y - y^3, \quad x = y^4 + 0.5$

54. $x = \frac{2}{1+y^2}, \quad x = y^2 - 3y + 2$

55. $x = \cos(y^2), \quad x = (y - 1)^2$

56. $x = \frac{1}{2+y^2}, \quad x = \frac{2y^2}{y^4+1}$

6.5 Average Values

Meteorologists frequently speak of average temperatures. If we take the temperature reading at many different times during the course of the day, these values can easily be averaged. Suppose we measure t in hours, running from $t = 0$ at midnight to $t = 24$ the following midnight, and suppose the temperature at time t is $T(t)$. If we take n evenly spaced temperature readings, at times $0 = t_1^*, t_2^*, t_3^*, \ldots, t_n^* = 24$, then the approximate average temperature is $\frac{1}{n}\big(T(t_1^*) + \ldots + T(t_n^*)\big)$.

This looks a lot like a Riemann sum; indeed, it equals

$$\frac{1}{24}\left(T(t_1^*)\frac{24}{n} + \ldots + T(t_n^*)\frac{24}{n}\right),$$

and the sum in the parentheses is a Riemann sum. (Note that partitioning $[0, 24]$ into n equal subintervals gives each subinterval a length of $\frac{24}{n}$.) As n increases, these averages tend to a limit, which is

$$\frac{1}{24}\int_0^{24} T(t)\,dt.$$

The same argument applies to any function $f(x)$ defined on $[a, b]$. Average the values of $f(x)$ at n evenly spaced points (see Figure 6.5.1), and let $n \to \infty$. The averages will tend to the limit $\frac{1}{b-a}\int_a^b f(x)\,dx$. This suggests the following definition.

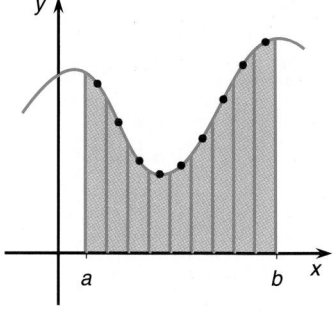

Figure 6.5.1

DEFINITION 6.5.1

> If $f(x)$ is integrable on $[a, b]$, then the **average value** of $f(x)$ on $[a, b]$ is
>
> $$\frac{1}{b-a}\int_a^b f(x)\,dx.$$

Roughly speaking, in a case in which the variable x represents time, say, what this means is to "add up" all the values of $f(x)$, in the sense of integrating them, and then divide by the total time.

EXAMPLE 1   **Average Temperature**   Suppose the outdoor temperature on a particular day is $T(t) = 60 + 16\sin^2\big(\frac{\pi t}{24}\big)\,^\circ$F, with time measured in hours and $t = 0$ at midnight. What is the average temperature for this 24-hour period?

Solution   By the definition above, the average temperature is

$$\frac{1}{24}\int_0^{24}\left(60 + 16\sin^2\left(\frac{\pi t}{24}\right)\right)dt = \frac{1}{24}\left(\int_0^{24} 60\,dt + 16\int_0^{24}\sin^2\left(\frac{\pi t}{24}\right)dt\right)$$

$$\boxed{\begin{array}{l} u = \frac{\pi t}{24} \\ du = \frac{\pi}{24}\,dt \end{array}} \qquad = \frac{60}{24}t\Big|_0^{24} + \frac{16}{24}\int_0^{\pi}\sin^2(u)\frac{24}{\pi}\,du$$

$$= \frac{60}{24}24 + \frac{16}{24}\frac{24}{\pi}\int_0^{\pi}\sin^2(u)\,du$$

$$= 60 + \frac{16}{\pi}\left(\frac{1}{2}u - \frac{1}{4}\sin(2u)\right)\Big|_0^{\pi}$$

$$= 60 + \frac{16}{\pi}\frac{\pi}{2}$$

$$= 68^\circ\text{F}.$$

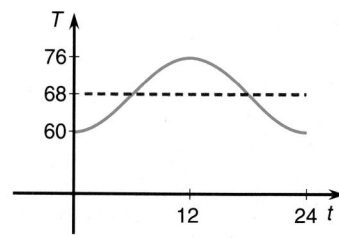

Figure 6.5.2

The graph of $T(t)$ is sketched in Figure 6.5.2, and from the picture it seems very reasonable to expect that the average temperature is 68°F.

In a case like this, in which the function is so symmetrical, it is not difficult to guess the average value, but for more complicated functions that will not be possible.

◢

EXAMPLE 2   Average Velocity   Suppose an object is moving in a straight line so that at time t its velocity is $v(t) = 2t - t^2$.

(i) Find the average velocity between $t = 0$ and $t = 1$.
(ii) Find the average velocity between $t = 0$ and $t = 3$.
(iii) Find the average velocity between $t = 2$ and $t = 4$.

Solution

(i) The average velocity between $t = 0$ and $t = 1$ is $\frac{1}{1-0} \int_0^1 v(t)\, dt = \left(t^2 - \frac{1}{3}t^3\right)\big|_0^1 = \frac{2}{3}$.

(ii) The average velocity between $t = 0$ and $t = 3$ is $\frac{1}{3-0} \int_0^3 v(t)\, dt = \frac{1}{3}\left(t^2 - \frac{1}{3}t^3\right)\big|_0^3 = 0$.
     This reflects the fact that at $t = 3$ the object is back where it started when $t = 0$. It does not mean that it did not move, but that its motion in the positive direction was canceled by its return motion in the negative direction. You might like to sketch the graph of its displacement $s(t) = s(0) + \int_0^t v(u)\, du = s(0) + t^2 - \frac{1}{3}t^3$.

(iii) The average velocity between $t = 2$ and $t = 4$ is $\frac{1}{4-2} \int_2^4 v(t)\, dt = \frac{1}{2}\left(t^2 - \frac{1}{3}t^3\right)\big|_2^4 = \frac{1}{2}\left(16 - \frac{1}{3}64 - 4 + \frac{1}{3}8\right) = -\frac{10}{3}$. The negative sign means that the net displacement and therefore the average velocity are both in the negative direction.

◢

In this particular example we can check our calculations against the familiar notion of average velocity, which after all is just the net displacement divided by the time.

We can find the displacement $s(t)$ by integrating $v(t)$, that is, $s(b) = s(a) + \int_a^b v(t)\, dt$.

As t runs from a to b, say, the object moves a *net* displacement of $s(b) - s(a) = \int_a^b v(t)\, dt$. Its average velocity over this time is

$$\frac{Displacement}{Time} = \frac{s(b) - s(a)}{b - a} = \frac{1}{b - a} \int_a^b v(t)\, dt.$$

This confirms the calculations that we have done and reassures us that the definition of average value (Definition 6.5.1) is a good definition.

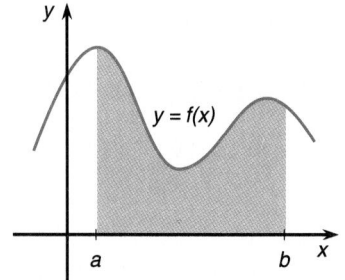

Figure 6.5.3

EXAMPLE 3   Sandbox   Figure 6.5.3 is the graph of a continuous nonnegative function $f(x)$ on $[a, b]$, with the region under the graph shaded. Think of the graph as the top of the sand in a sandbox, and try to "level it off," that is, push the sand from the "peaks" into the "valleys" to get a flat top on the sand, as illustrated by the shaded region in Figure 6.5.4.

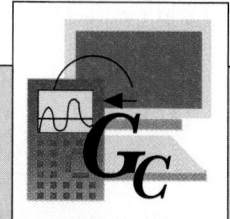

With a graphics calculator we can plot the velocity from Example 2, and plot the various average values, too.

TI-85	SHARP EL9300
GRAPH F1$(y(x) =)$ 2 F1(x) − F1(x) x^2 ▼ 2 ÷ 3 ▼ (−) 10 ÷ 3 M2(RANGE) 0 ▼ 4 ▼ ▼ (−) 8 ▼ 2 F5 (GRAPH)	2 x/θ/T − x/θ/T x^2 2ndF ▼ 2 ÷ 3 2ndF ▼ − 10 ÷ 3 RANGE 0 ENTER 4 ENTER 1 ENTER −8 ENTER 2 ENTER 1 ENTER
CASIO f_x-6300G (f_x-7700GB)	HP 48SX
Range 0 EXE 4 EXE 1 EXE −8 EXE 2 EXE 1 EXE EXE EXE EXE Graph 2 X − X x^2 EXE Graph 2 ÷ 3 EXE Graph − 10 ÷ 3 EXE	↰ PLOT PLOTR ERASE ATTN ′ Y = 2 × X − X y^x 2 ↱ PLOT ↰ DRAW ERASE 0 ENTER 4 XRNG 8 +/− ENTER 2 YRNG DRAW ATTN ′ Y = 2 ÷ 3 ↱ PLOT ↰ DRAW DRAW ATTN ′ Y = 10 ÷ 3 +/− ↱ PLOT ↰ DRAW DRAW

You might like to adapt this example to plot the velocity for $t \in [0, 1]$ with the corresponding average, and similarly for $t \in [0, 3]$ or for $t \in [2, 4]$ with the corresponding averages.

As described in the example, you could also plot the displacement $s(t)$; to do this you have to choose a value for $s(0)$ (one obvious choice is $s(0) = 0 \ldots$).

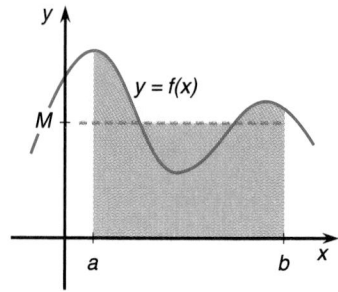

We think of the height M of the sand *after leveling* as the "average height" of the original curve. The area of the original shaded region is $\int_a^b f(x)\,dx$, and the same sand fills up the shaded rectangle whose height is M and whose base is $b - a$. The area of the rectangle is $M(b - a)$, so $M(b - a) = \int_a^b f(x)\,dx$, and we see that

$$M = \frac{1}{b - a} \int_a^b f(x)\,dx.$$

This is consistent with our definition (Definition 6.5.1) of the average.

Figure 6.5.4

We encountered the idea of the average or mean value of a function in Section 5.4, in connection with the Integral Mean Value Theorem (Theorem 5.4.4). The argument given there before the statement of the theorem was based essentially on this picture of the "sandbox."

EXAMPLE 4   **Average Speed of a Car**   Suppose a car moves so that its velocity at time t (in hours) is $v(t) = 15(4t - 3t^2)$ mph. Find the displacement of the car 2 hours after it starts, and find its average speed over those 2 hours.

Solution   We can assume the displacement of the car is $s = 0$ at $t = 0$. Then to find its displacement at $t = 2$, we integrate:

$$s(2) = \int_0^2 s'(t)\,dt = \int_0^2 v(t)\,dt = 15\int_0^2 (4t - 3t^2)\,dt = 15(2t^2 - t^3)\Big|_0^2$$
$$= 0.$$

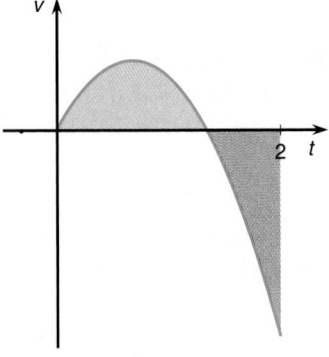

Figure 6.5.5

This means that the car has returned to its starting point after 2 hours. It also means that the average velocity is 0 (cf. Example 2). However, this is not the same as the average speed.

After all, the speed is the absolute value of the velocity. The reason the average velocity is 0 is that some of the time the velocity is positive and some of the time it is negative, and the integral of the positive part and the integral of the negative part cancel (see Figure 6.5.5).

On the other hand, the speed is always positive (being an absolute value), so that sort of cancellation cannot occur. To find the average speed, we have to find the absolute value of $v(t)$, and to do that, we must know where the graph of $v(t)$ crosses the horizontal axis. We find that $v(t) = 0$ means $15(4t - 3t^2) = 0$, or $0 = 4t - 3t^2 = t(4 - 3t)$, which happens exactly when $t = 0$ or when $t = \frac{4}{3}$. From the sketch of the graph (or by evaluating $v(1)$ and $v(2)$, for example) we see that $v(t)$ is positive for $0 < t < \frac{4}{3}$ and negative for $t > \frac{4}{3}$.

So to find the average speed, we must integrate the speed $|v(t)|$, which equals $v(t)$ for $0 < t < \frac{4}{3}$ and equals $-v(t)$ for $t > \frac{4}{3}$ (see Figure 6.5.6). The average speed equals

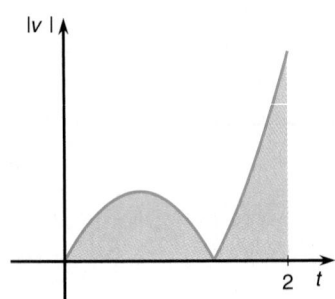

Figure 6.5.6

$$\frac{1}{2 - 0}\int_0^2 |v(t)|\,dt = \frac{1}{2}\left(\int_0^{4/3} v(t)\,dt + \int_{4/3}^2 \big(-v(t)\big)\,dt\right)$$
$$= \frac{15}{2}\left(\int_0^{4/3}(4t - 3t^2)\,dt + \int_{4/3}^2 (3t^2 - 4t)\,dt\right)$$
$$= \frac{15}{2}\left((2t^2 - t^3)\Big|_0^{4/3} + (t^3 - 2t^2)\Big|_{4/3}^2\right)$$
$$= \frac{15}{2}\left(\frac{32}{9} - \frac{64}{27} + 8 - 8 - \frac{64}{27} + \frac{32}{9}\right)$$
$$= 17\frac{7}{9}\quad \text{mph.}$$

Notice that the *instantaneous* speed equals the absolute value of the instantaneous velocity, but this example shows that the *average* speed may not equal the absolute value of the average velocity.

The calculation in this example was not difficult, but the distinction between velocity and speed is important. The other thing to watch in calculating averages is to remember to divide by the length of the interval. There is a risk of getting so wrapped up in doing the integral that we may forget to divide.

Most of the averages considered so far have been averages *with respect to time*. It is also possible to consider averages with respect to other quantities.

EXAMPLE 5   **Average Radius of Cone**   Consider a long conical rod, 2 m long, whose diameter at one end is 2 cm and whose diameter at the other end is 6 cm. As we move from one end to the other, find the average radius and the average cross-sectional area.

Solution

(i) For convenience we place the rod along the x-axis with its narrow end at the origin (see Figure 6.5.7). Measuring in meters, we see that the wider end will be at $x = 2$. The conical shape is obtained by revolving a straight line about the x-axis; since the radius at the small end is 0.01 and the radius at the large end is 0.03, the line passes through $(0, 0.01)$ and $(2, 0.03)$. The equation of the line through these points is easily found. It is

$$y = 0.01 + 0.01x.$$

As x moves from $x = 0$ to $x = 2$, the corresponding radius of the cone is just the height of this line above the axis, which is just $y = 0.01 + 0.01x$. The average radius is easily found. It is

$$\frac{1}{2-0}\int_0^2 y\,dx = \frac{1}{2}\int_0^2 (0.01 + 0.01x)\,dx = \frac{1}{2}\left(0.01x + 0.005x^2\right)\Big|_0^2$$
$$= 0.02 \text{ m} = 2 \text{ cm}.$$

In a sense this is hardly surprising, since the radius goes from 1 cm to 3 cm in a straight line.

(ii) Since the radius is $y = 0.01 + 0.01x$, the cross-sectional area is $\pi r^2 = \pi y^2 = \pi(0.01 + 0.01x)^2 = \frac{\pi}{10,000}(1 + x)^2 = \frac{\pi}{10,000}(1 + 2x + x^2)$. The average cross-sectional area is

$$\frac{1}{2-0}\int_0^2 \frac{\pi}{10,000}(1 + 2x + x^2)\,dx = \frac{\pi}{20,000}\int_0^2 (1 + 2x + x^2)\,dx$$
$$= \frac{\pi}{20,000}\left(x + x^2 + \frac{1}{3}x^3\right)\Big|_0^2 = \frac{13\pi}{30,000} \text{ m}^2$$
$$= \frac{13\pi}{3} \text{ cm}^2 \approx 13.61 \text{ cm}^2.$$

Notice that this is not just halfway between the areas at the two ends. ◢

Figure 6.5.7

This is the line that is the top of the cone.

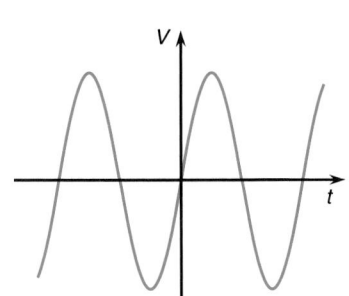

Figure 6.5.8

Figure 6.5.9

EXAMPLE 6   **Electricity and RMS Averages**   Electricity comes from various sources, but in everyday life there are two main types. Ordinary household current is what is called "Alternating Current," or AC; this means that the voltage alternates from positive to negative. The graph of the voltage is a sine curve, as shown in Figure 6.5.8; it changes at the rate of 60 cycles per second, so one complete cycle takes $\frac{1}{60}$ sec. Figure 6.5.9 illustrates "Direct Current" (DC), such as the current that comes from a battery. Its voltage stays constant.

It is easy to measure a DC voltage, but for AC it is less clear what to do. One approach would be to measure the maximum voltage (the height of the peak), but in a way this is misleading. Most of the time the voltage is less than the peak value.

Figure 6.5.10

Figure 6.5.11

Figure 6.5.12

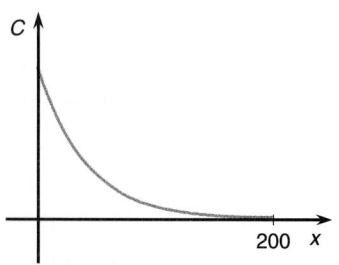

Figure 6.5.13

Another idea would be to find the average voltage. The difficulty here is that since the voltage is negative half the time and positive half the time, the average is zero (see Figure 6.5.10).

We could try averaging the absolute value, but it turns out that it is better to find the average of the *square* of the voltage and then take the square root of that average. This is called the *Root Mean Square* average, and it is extremely important in science and engineering.

For instance, with 60-cycle household current, the voltage can be described as a function of time by $V(t) = A \sin(120\pi t)$. When t increases by $\frac{1}{60}$, we see that $120\pi t$ increases by 2π, so the voltage goes through one complete cycle.

Let us average over one cycle. We find

$$\frac{1}{\frac{1}{60}} \int_0^{1/60} V^2(t)\,dt = 60 \int_0^{1/60} A^2 \sin^2(120\pi t)\,dt$$

$$\boxed{\begin{aligned} u &= 120\pi t \\ du &= 120\pi\,dt \end{aligned}} \qquad = 60A^2 \int_0^{2\pi} \sin^2(u)\,\frac{1}{120\pi}\,du = \frac{A^2}{2\pi}\left(\frac{u}{2} - \frac{1}{4}\sin(2u)\right)\Bigg|_0^{2\pi}$$

$$= \frac{A^2}{2\pi}\pi = \frac{A^2}{2}.$$

This is the average of V^2; its square root is $\frac{A}{\sqrt{2}}$. The root mean square (RMS) average voltage is $\frac{A}{\sqrt{2}}$. When we read that household electricity comes at 115 volts, this means that the RMS average is 115. In particular, $A = 115\sqrt{2}$. Since the maximum value of $V(t) = A\sin(120\pi t)$ equals A, we see that the peak voltage is $115\sqrt{2} \approx 162.6$ volts (see Figure 6.5.11).

EXAMPLE 7   **Drug Concentration**   A certain blood vessel has the approximate shape of a cylinder, 200 mm long and with radius 2 mm. A drug is injected at one end (see Figure 6.5.12) and rapidly spreads along the blood vessel.

Let x measure distance along the cylinder from the point of injection, in millimeters. One second after the injection, the concentration of the drug at a distance x along the blood vessel is

$$C(x) = 80e^{-x/40}\%.$$

Figure 6.5.13 graphs the concentration as a function of x, showing that it is highest at the point of injection and trails off to lower and lower concentrations farther away.

We want to find the average concentration for $x \in [0, 200]$, which is

$$\frac{1}{200}\int_0^{200} C(x)\,dx = \frac{1}{200}\int_0^{200} 80e^{-x/40}\,dx = \frac{2}{5}\int_0^{200} e^{-x/40}\,dx$$

$$\boxed{\begin{aligned} u &= \frac{x}{40} \\ du &= \frac{1}{40}\,dx \end{aligned}} \qquad = \frac{2}{5}\int_0^5 e^{-u}(40)\,du = 16\int_0^5 e^{-u}\,du = -16e^{-u}\Big|_0^5$$

$$= -16(e^{-5} - 1) = 16(1 - e^{-5}) \approx 15.89\%.$$

One way of interpreting this result is that if the drug were evenly distributed along the whole length of the tube, its concentration would be 15.89% (see

Figure 6.5.14

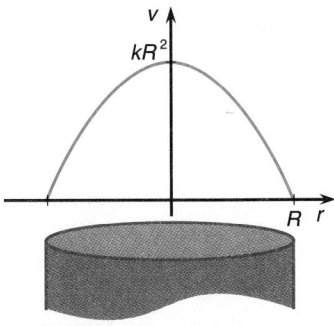

False-color scanning electron micrograph of the interior of an artery in the human liver, showing lymphocytes (white blood cells) adhering to the inner surface. Magnification: 800×.

Figure 6.5.15

Figure 6.5.16

Figure 6.5.14). Since the total volume of the cylinder is $\pi r^2 h = \pi 2^2 (200) = 800\pi \approx 2513.3$ mm^3 = 2.5133 cm^3, we can also find that the total amount of the drug present is 15.89% of this, or

$$0.1589 \times 2.5133 \approx 0.3994 \text{ cm}^3.$$

It appears that the original injection was about 0.4 cc (cubic centimeters) of the drug.

Example 7 ignores irregularities in the shape of the blood vessel and the pumping motion of the blood, which tends to carry the drug along the vessel, but it shows how averages can be used.

EXAMPLE 8   **Blood Flow and Poiseuille's Law**   The walls of a blood vessel exert friction on the blood as it flows past. The effect is that the blood flows more slowly near the walls than it does in the center. Poiseuille's Law says that in a tube of radius R the fluid velocity at a distance r from the center is

$$v(r) = k(R^2 - r^2),$$

where k is some constant (see Figure 6.5.15). When $r = R$ (i.e., right at the walls), the velocity is zero, and the velocity is largest in the center (where $r = 0$ and $v(0) = kR^2$).

In Figure 6.5.16 we imagine the blood in "layers," each a fixed distance from the center. The blood in each thin layer travels at the same velocity, but the inner layers travel faster than the outer layers.

Now consider the average velocity. We could find the average of $v(r)$ for $r \in [0, R]$. It is

$$\frac{1}{R} \int_0^R v(r)\,dr = \frac{k}{R} \int_0^R (R^2 - r^2)\,dr = \frac{k}{R}\left(R^2 r - \frac{r^3}{3}\right)\Bigg|_0^R = \frac{k}{R}\left(R^3 - \frac{R^3}{3}\right)$$

$$= \frac{2}{3}kR^2.$$

However, this average may be misleading. From the picture we see that the slow outer layers are bigger than the fast inner layers. This suggests that most of the blood is moving slowly, so our average, which counts each speed equally, is probably too high.

Another kind of average can be found by asking how much blood passes by in 1 second. To answer this, we consider the blood in a thin layer at a distance r from the center. The velocity of this layer is $v(r)$, so the length that passes a fixed point in 1 sec is $v(r)(1) = v(r)$. Since the circumference is $2\pi r$, the surface area of this part of the layer is $2\pi r v(r)$, and if the thickness of the layer is dr, the volume passing by in this layer is $2\pi r v(r)\,dr$ (see Figure 6.5.17).

Figure 6.5.17

The total flow in 1 sec is the "sum" of the flows in each layer, that is,

$$\int_0^R 2\pi r v(r)\,dr = 2\pi k \int_0^R (rR^2 - r^3)\,dr = 2\pi k \left(R^2 \frac{r^2}{2} - \frac{r^4}{4} \right)\Bigg|_0^R = 2\pi k \frac{R^4}{4}$$

$$= \frac{\pi k R^4}{2}.$$

Figure 6.5.18

To say that the average velocity is v_a should mean that the total flow would be the same if *all* the blood moved at velocity v_a. If the blood all flowed at v_a, then in 1 sec the amount flowing past would be a cylinder of length $v_a(1) = v_a$, as shown in Figure 6.5.18. Since the cross-sectional area is πR^2, the volume passing in 1 sec is $\pi R^2 v_a$.

If this equals the total flow $\frac{\pi k R^4}{2}$, we find that $v_a = \frac{kR^2}{2}$. As expected, this is less than the value $\frac{2kR^2}{3}$ that we found earlier. The smaller average is more meaningful because it reflects more accurately the way the blood flows.

This example shows that deciding what an average actually means can be quite subtle. Care must be used in interpreting averages.

Exercises 6.5

I

In each of the following situations, find the average value of the given function over the specified interval.

1. $f(x) = x^2 + 1$ over $[-1, 1]$
2. $g(x) = x^3 + 3$ over $[-1, 1]$
3. $h(x) = e^x$ over $[0, 1]$
4. $k(x) = \sin x$ over $[0, \pi]$
5. $r(u) = \frac{1}{u}$ over $[1, 5]$
6. $p(t) = \sin t$ over $\left[0, \frac{\pi}{2}\right]$
7. $q(v) = \sin v$ over $[-\pi, \pi]$
8. $s(x) = \sin^2 x$ over $[0, 2\pi]$
9. $w(t) = \sec^2 t$ over $\left[0, \frac{\pi}{4}\right]$
10. $\ell(x) = x\sin(\pi x^2)$ over $[0, 1]$
11. $f(x) = \frac{1}{x^2+1}$ over $[-1, 1]$
12. $f(x) = \frac{x}{x^2+1}$ over $[-1, 1]$
13. $f(x) = \frac{x}{\sqrt{x^2+1}}$ over $[-1, 1]$
14. $f(x) = x\exp(x^2)$ over $[0, 2]$
15. $f(x) = \sin(5\pi x)$ over $[0, 3]$
16. $f(x) = \sin x\cos^3 x$ over $\left[0, \frac{\pi}{2}\right]$
17. $f(x) = x^3\sin(\pi x^4)$ over $[-1, 1]$
18. $f(x) = \sec^2 x\tan^2 x$ over $\left[0, \frac{\pi}{4}\right]$
19. $f(x) = \sin x e^{\cos x}$ over $\left[0, \frac{\pi}{2}\right]$
20. $f(x) = \sin(x)\ln(\cos^2 x + \sin^2 x)$ over $[0, \pi]$

II

21. (i) Find the average value of $f(x) = 3$ over the interval $[1, 2]$.
    (ii) Find the average value of $f(x) = -7$ over the interval $[-1, 4]$.
    (iii) Find the average value of the constant function $f(x) = C$ over the interval $[a, b]$.

22. Define $f(x)$ as follows:
$$f(x) = \begin{cases} 0, & \text{if } x \leq 2, \\ 1, & \text{if } x > 2. \end{cases}$$
    (i) Find the average value of $f(x)$ over the interval $[0, 4]$.
    (ii) Find the average value of $f(x)$ over the interval $[1, 5]$.

*23. (i) Find the average radius of the horn in Exercise 50 of Section 6.4.
     (ii) Find the average cross-sectional area of the horn.

*24. A car travels 50 km at 50 km/hr and then travels the next 50 km at 100 km/hr. What is its average speed? (*Hint:* The answer is not 75 km/hr.)

*25. (i) Find the average radius of the cross-sections of a sphere of radius 1 as the cross-sections move from one side of the sphere to the other.
     (ii) Find the average cross-sectional area of the sphere.

*26. Repeat Exercise 25 for a sphere of radius r.

*27. Consider a bacterial culture in which the weight of the bacteria present at time t is $P(t) = 2e^{t/20}$, where *t* is measured in hours and the weight is measured in grams.

  (i) Find the average weight of the culture for the 24-hour period beginning at $t = 0$.

  (ii) Find the average weight of the culture for the 6-hour period beginning at $t = 4$.

*28. Suppose the population of a city is given by $P(t) = 2{,}000{,}000e^{(t-1980)/15}$, with t being the date, measured in years.

  (i) What was the population in 1990?

  (ii) What was the average population between 1980 and 1990?

  (iii) What was the average population between 1980 and 1985?

*29. (i) If the velocity of a moving object is $v(t) = t^2 - 4t + 3$, what is its average velocity between $t = 0$ and $t = 4$?

  (ii) What is its average speed for the same period?

*30. (i) Suppose the displacement of a moving object, in meters, is $s(t) = t^3 - 6t^2 + 9t + 3$. What is its average velocity between $t = 0$ and $t = 4$?

  (ii) What is its average speed for the same period?

*31. A motorist knows that on a certain freeway his time is monitored to detect speeding. Two tollbooths are 10 miles apart. He sets out from one and travels 5 miles at an average speed of 80 mph. Suddenly, he realizes that he must slow down to avoid a ticket. At what average speed must he travel the second 5 miles in order to have an average speed of 55 mph on the whole 10-mile stretch?

*32. Suppose $f(x)$ is an integrable function on $[a, b]$ and that M and m are numbers for which $m \le f(x) \le M$ for all $x \in [a, b]$. Show that the average value of $f(x)$ on $[a, b]$ lies between m and M.

*33. (i) If $f(x)$ and $g(x)$ are integrable functions on $[a, b]$, show that the average of $f(x) + g(x)$ over $[a, b]$ equals the sum of the averages of $f(x)$ and $g(x)$.

  (ii) Express the average of $3f(x) - 5g(x)$ in terms of the averages of $f(x)$ and $g(x)$.

*34. If $f(x)$ and $g(x)$ are integrable functions on $[a, b]$, show by example that the average of $f(x)g(x)$ over $[a, b]$ does not necessarily equal the product of the averages of $f(x)$ and $g(x)$.

35. (i) Find the average value of x^3 over the interval $[-1, 1]$.

  (ii) Find the average value of x^7 over the interval $[-3, 3]$.

  (iii) Find the average value of $x^5 + 4x^3 - 2x$ over the interval $[-2, 2]$.

*36. (i) Suppose $f(x)$ is an odd function and $a > 0$. Find the average value of $f(x)$ on the interval $[-a, a]$.

  (ii) If $g(x)$ is an even function, compare the average values of $g(x)$ on $[-a, a]$ and $[0, a]$.

*37. For the voltage $V(t) = A\sin(120\pi t)$ of Example 6 (p. 451), find the RMS average over the intervals (i) $\left[0, \frac{1}{120}\right]$, (ii) $\left[\frac{1}{240}, \frac{1}{120}\right]$, (iii) $\left[\frac{1}{160}, \frac{1}{96}\right]$.

*38. If electricity with voltage V flows through a resistance R, such as a toaster or a light bulb, the power consumed is $P = \frac{V^2}{R}$. (i) Consider the AC voltage $V(t)$ of Example 6 passing through a fixed resistance R, and find the average power for one cycle (i.e., between $t = 0$ and $t = \frac{1}{60}$). (ii) Suppose a (constant) DC voltage through the same resistance consumes the same average power. Show that this constant DC voltage equals the RMS average voltage corresponding to the AC voltage $V(t)$. Show that it is different from the average of the absolute value $|V(t)|$.

  This shows why RMS is a good kind of average; it gives the correct value for the electrical power, which is very often the most important quantity to measure.

39. The lower section of a tree trunk is approximately a cylinder of length 10 ft and radius 3 in. Somebody has carelessly spilled a toxic chemical near the roots, and it has entered the tree. If h measures the height above ground, in feet, the concentration of the chemical in the tree at height h is $C(h) = 1.7e^{-h}$ ppm (parts per million). (i) What is the average concentration of the chemical in the lowest 10 ft of the trunk? (ii) What is the average concentration in the lowest 5 ft of the trunk? (iii) Experience has shown that long-term concentrations above 0.5 ppm will kill a tree. Assuming that the chemical disperses evenly through the 10-ft section of trunk and no more enters from the roots, will the tree survive? (iv) How much chemical (in cubic inches) is in this 10-ft section of trunk?

*40. In Example 8 (p. 453) we decided that a sort of weighted average was appropriate. (i) Find the average velocity, in this sense, for the blood in the center of the tube, the part with $0 \le r \le \frac{R}{2}$. (ii) Find the average velocity for the part of the tube with $\frac{R}{2} \le r \le R$. (iii) Without checking, do you expect the average for the whole tube to be exactly halfway between these averages, or smaller, or greater? (iv) Check and explain.

41. (i) Use a personal computer or programmable calculator to approximate the average temperature in Example 1 (p. 447) by averaging the temperature at every hour on the hour from midnight to 11 P.M. (ii) What happens if you take the temperature every half hour? Every 5 minutes?

42. Approximate the average velocity and average speed of the car in Example 4 (p. 449), by averaging the values measured every 5 minutes.

43. Approximate the average value of $\frac{1}{1+x^3}$ over $[0, 4]$.

44. Approximate the average value of $\arcsin x$ over $[0, 1]$.

POINT TO PONDER

In Example 5 (p. 451) we were not surprised to find that the average radius of the cone was halfway between the radii at the two ends. However, a quick check will show that the same is not true of the average cross-sectional area. Another way to say this is that if we take the average radius r_a and compute the corresponding cross-sectional area πr_a^2, what we get will *not* equal the average cross-sectional area.

Explain why this happens and why we should not be surprised.

6.6 Work

FORCE
KINETIC ENERGY
POTENTIAL ENERGY
HOOKE'S LAW
GRAVITATIONAL FORCE AND
WORK

In this section we discuss the notion of work. When a force acts on an object and the object moves, the force performs **work** on the object. This is another way of saying that the force changes the total energy of the object, for example, by making it move faster, which increases its kinetic energy.

There is a simple formula that says that the work done is the product of the force and the distance through which the object moves. This is fine if the force is constant, but many forces vary in intensity, and the simple formula does not apply to them. It turns out that the way to deal with this case is to use integration.

When an object moves a distance s in a straight line under the influence of a force F, the force performs work on the object, and the amount of work done is

$$W = Fs.$$

Performing work means changing the energy of the object. This can happen because of changing its **potential energy** (by changing its position, e.g., lifting an object up increases its potential energy), changing its **kinetic energy** (by changing its speed; the faster an object is moving, the greater is its kinetic energy), or a combination of the two.

Perhaps the most familiar force is gravity.

EXAMPLE 1   Work and the Force of Gravity   Near the earth's surface, an object experiences a force due to gravity, which is called its weight. If an object of weight m at rest at ground level is raised to a height h above the ground, then it travels a distance h. The force required to lift it against gravity is just m, so the work performed is $W = mh$. Since the object starts and ends at rest, there is no kinetic energy involved, and we see that the potential energy has changed by mh.

If instead the object is allowed to fall to the ground from a height h, it moves a distance h under the gravitational force m, so its potential energy decreases by mh. This potential energy is converted into kinetic energy, since the object is moving when it hits the ground. When the object hits the ground, its kinetic energy is converted into various types of energy, including the energy in the sound of the impact, kinetic energy in the object if it bounces, kinetic energy in any little pebbles or dust it may kick up, and so on.

Figure 6.6.1

At first it may seem puzzling that we insist that the object must move in order for work to be done on it. Imagine yourself pushing against a brick wall (see Figure 6.6.1). You may push quite hard and even tire yourself out, but nothing happens to the wall. After you stop, the wall will certainly not be moving, so it has no kinetic energy, and since you have not moved it, you cannot possibly have changed its potential energy. No work has been done to the wall.

The difficulty really arises from the way the word is used. In physics, doing work on an object means changing its energy. In everyday life we feel that if we push hard on something, we have done work, even if it did not move.

Strictly speaking, the force and the motion must point along the same line. If you move an object horizontally, that is, without changing its height, the force of gravity does no work on it. Note, however, that pointing along the same line does allow for the possibility that the force and the motion may point in opposite directions along the same line. Work *is* done by a force that slows a moving object by pushing in the direction opposite to its motion.

A more difficult problem arises when an object is moving under the influence of a force that is not constant. We break the motion up into a succession of small motions, over each of which the force is essentially constant. For each little motion we calculate the work done and then add them up.

Suppose the object moves along the horizontal axis and that its displacement changes from $s = a$ to $s = b$. Suppose that the force, being different at different points along the object's path, can be expressed as a function $F(s)$, that is, a function depending on the displacement s.

We partition the interval $[a, b]$ into n subintervals, each of which is sufficiently short that the value of the force does not change appreciably on the subinterval. If the length of the ith subinterval is Δs_i and s_i^* is any point in the ith subinterval, then the work done as the object moves along the ith subinterval is approximately $F(s_i^*)\Delta s_i$, the product of the force and the displacement. The total work done as the object moves from $s = a$ to $s = b$ is approximately

$$\sum_{i=1}^{n} F\left(s_i^*\right)\Delta s_i.$$

To improve the approximation, we must take finer and finer partitions, and the approximations to the work approach the limit $\int_a^b f(s)\,ds$. This suggests the following.

DEFINITION 6.6.1

The work done by a force $F(s)$ on an object that moves from $s = a$ to $s = b$ along the same line as the direction of the force is

$$W = \int_a^b F(s)\,ds.$$

Just as with area and volume calculations it is possible to think in terms of very small displacements ds to help setting up the integral.

EXAMPLE 2   **Springs and Hooke's Law**   The force exerted by a spring is proportional to the displacement of the end of the spring from its rest position, and in

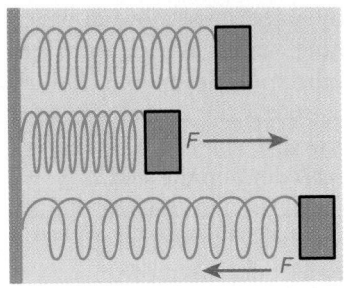

Figure 6.6.2

the direction opposite to the displacement. This is Hooke's Law, named after the English physicist Robert Hooke (1635–1703) and illustrated in Figure 6.6.2.

More informally, what it says is, roughly, that the force of the spring tries to push it back to its rest position (i.e., if you push the spring to the right, it pushes to the left and vice versa) and that the more it is displaced (compressed or extended), the harder it will push back. The top picture shows the spring in its rest position, the second shows it compressed, and the third shows it extended (i.e., stretched).

Hooke's Law says that the force exerted by the spring is proportional to the displacement, that is, when the spring is displaced an amount s from its rest position, the force it exerts is

$$F = -ks \tag{6.6.2}$$

for some positive constant k; the minus sign reflects the fact that the force is in the direction opposite to the displacement. Calculate the work done by the spring when the displacement of the end of the spring is changed from $s = a$ to $s = b$.

Solution   The displacement changes from $s = a$ to $s = b$. At any particular displacement s the force exerted by the spring is $F(s) = -ks$, and the work done in making a small change in displacement ds will be $-ks\,ds$. Adding up the work done in all the little displacements (i.e., integrating), we find that the total work done is

$$W = \int_a^b (-ks)\,ds = -k \int_a^b s\,ds = -k\frac{s^2}{2}\bigg|_a^b$$

$$= -\frac{k}{2}(b^2 - a^2).$$

> Notice that a and b are the displacements of the end of the spring *from the rest position,* not the length of the spring.

At this point we must discuss the units of force and work.

> If force is measured in pounds and displacement in feet, then work is measured in **foot-pounds (ft-lb)**. In the metric system, force is measured in **newtons (N)**, so work is measured in **newton-meters (N-m)**, which are more often called **joules (J)**. An object whose mass is 1 kg weighs 9.80 N at the earth's surface (i.e., gravity exerts a force of 9.80 N on it). An object whose mass is 1 lb weighs 1 lb at the earth's surface. This apparently confusing statement simply means that the units of mass (pounds) correspond to the units of force (pounds) and that unlike what happens in the metric system, there is no conversion factor. It is sometimes necessary to measure mass in a different unit, the "slug," but we will use pounds.

EXAMPLE 3   Find the work done by a person stretching a spring starting from its rest position until it is 1 ft longer than at rest if it exerts a 1 lb force when stretched 1 in.

Solution   If we measure displacement s in feet, we know that when $s = \frac{1}{12}$ (i.e., one inch), $F = -1$ lb, that is, in the direction opposite to the stretching.

Figure 6.6.3

Figure 6.6.4

Figure 6.6.5

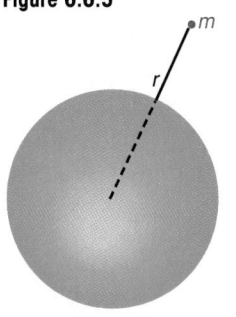

Figure 6.6.6

Since $F = -ks$, we see that $k = 12$. By Example 2, with $a = 0$, $b = 1$, we find that the work done *by the spring* is

$$W = -\frac{k}{2}(b^2 - a^2) = -6 \text{ ft-lb}.$$

The negative sign occurs because the spring pulls in the direction opposite to the motion. The person doing the stretching pulls in the same direction as the motion and so does work of 6 ft-lb.

EXAMPLE 4   Conical Water Tank   Find the work done in pumping all the water out of a conical tank up to a point 2 m above the top of the tank, as shown in Figure 6.6.3. The tank has height 4 m and radius 2 m at the top.

Solution Let h be a variable measuring the height above the base of the tank. Consider a thin horizontal slice of water at height h with thickness dh. It will be lifted through a height of $6 - h$ (see Figure 6.6.4). The volume of the slice is (*Area of circle*) $\times$ (*thickness*) $= \pi r^2 dh$, where r is the radius of the circle at height h. By similar triangles in Figure 6.6.5 we see that $\frac{r}{h} = \frac{2}{4}$, that is, $r = \frac{1}{2}h$.

So the slice has volume $\pi\left(\frac{1}{2}h\right)^2 dh = \frac{1}{4}\pi h^2 dh$, in cubic meters. Since 1 cubic meter contains 1000 liters, and since the mass of 1 liter of water is 1 kg, we know that the mass of 1 m³ of water is 1000 kg. The force of gravity on 1 cubic meter of water is $1000(9.80)$ N $= 9800$ N. Therefore the force of gravity on the slice of water we are considering is $\frac{1}{4}\pi h^2 dh(9800) = 2450\pi h^2 dh$ N, and the work done in pumping it up is $(6 - h)2450\pi h^2 dh$ J.

Adding up the work done for all the slices, that is, for h running from $h = 0$ to $h = 4$, we find that the total work done is

$$W = \int_0^4 (6 - h)2450\pi h^2\, dh = 2450\pi \int_0^4 (6h^2 - h^3)\, dh$$

$$= 2450\pi \left(2h^3 - \frac{1}{4}h^4 \right)\Bigg|_0^4$$

$$\approx 493{,}000 \text{ J}.$$

Gravitational Force and Work

Near the surface of the earth, the force of gravity is related to the mass of an object in a simple way. The force exerted by gravity on a mass of 1 kg is 9.80 N; the force exerted by gravity on a mass of 1 lb is 1 lb. In other words, the force of gravity is a constant times the mass.

For an object that is not necessarily very close to the earth's surface, the force exerted by gravity is $\frac{Km}{r^2}$, where r is the distance from the center of the earth, m is the mass of the object, and K is some constant (see Figure 6.6.6). To find K, we compare this formula with what we know at the earth's surface.

For example, if r is measured in miles and m in pounds, we should have that $\frac{Km}{r^2} = m$ when the object is at the earth's surface (where its weight is m lb). At the earth's surface, r is the earth's radius, or about 4000 miles. So $K \approx 1.6 \times 10^7$ mi².

If r is measured in kilometers and m in kilograms, then at the earth's surface, $r \approx 6400$ and $\frac{Km}{(6400)^2} = 9.8m$ Newtons. So $K \approx 9.8(6400)^2$ N km² kg $\approx 4.0 \times 10^8$ N km²/kg.

EXAMPLE 5   Satellite Launch   Find the energy required to lift a 100-lb satellite to a point 1000 mi above the earth's surface, ignoring the energy needed to lift the rocket and its fuel.

Solution   The satellite rises from the earth's surface (where $r = 4000$ mi) to its destination at $r = 5000$ mi. At any particular r, the force of gravity is $F(r) = \frac{Km}{r^2} = \frac{1.6 \times 10^7 \times 100}{r^2}$ lb, and the work done is

$$\int_{4000}^{5000} F(r)\,dr = \int_{4000}^{5000} 1.6 \times 10^9 r^{-2}\,dr$$

$$= 1.6 \times 10^9 \left(-\frac{1}{r}\right)\Bigg|_{4000}^{5000} = 1.6 \times 10^9 \left(-\frac{1}{5000} + \frac{1}{4000}\right)$$

$$= 8.0 \times 10^4.$$

Since the force was measured in pounds and distances were in miles, the work just calculated is in "mile-pounds." Since 1 mile = 5280 ft, 1 mile-pound = 5280 ft-lb, and the work done is $8.0 \times 10^4 \times 5280 \approx 4.2 \times 10^8 = 420{,}000{,}000$ ft-lb.

Since an actual launch involves lifting the booster rocket and its fuel, which weigh many times as much as the payload, and also usually imparts kinetic energy to the satellite, it actually requires far more energy than what we have calculated to launch a satellite.

Exercises 6.6

I

In each of the following situations, calculate the work done by the force $F(s)$ in moving an object the specified displacement.

1. $F(s) = 3s^2$ lb, moving from $s = 0$ ft to $s = 3$ ft.
2. $F(s) = 12\left(1 - \frac{s}{2}\right)$ lb, moving from $s = 0$ ft to $s = 2$ ft.
3. $F(s) = 3\sin(\pi s)$ lb, moving from $s = 0$ ft to $s = 3$ ft.
4. $F(s) = s\sin(\pi s^2)$ N, moving from $s = 0$ m to $s = 2$ m.
5. $F(s) = \sin(\pi s)\cos(\pi s)$ N, moving from $s = -2$ m to $s = 2$ m.
6. $F(s) = \sin^3 s \cos^4 s$ lb, moving from $s = -4$ ft to $s = 4$ ft.
7. $F(s) = 6$ lb, moving from $s = 1$ ft to $s = 6$ ft.
8. $F(s) = -7$ N, moving from $s = 2$ m to $s = -3$ m.
9. $F(s) = 4s^2$ lb, moving from $s = 2$ ft to $s = -4$ ft.
10. $F(s) = 5s^3$ N, moving from $s = -1$ m to $s = -5$ m.

The following questions concern a spring. The first information tells how far some particular force will stretch the spring from its rest position. Use this to calculate the work done by a person moving the spring between the specified displacements. (All displacements are measured relative to the rest position.)

11. Stretched 1 ft with a force of 2 lb; moved from $s = -1$ ft to $s = 2$ ft.
12. Stretched 1 in with a force of 1 lb; moved from $s = 0$ ft to $s = 3$ ft.
13. Stretched 0.1 m with a force of 1 N; moved from $s = -0.2$ m to $s = 0.3$ m.
14. Stretched 20 cm with a force of 3 N; moved from $s = 0$ cm to $s = 45$ cm.
15. Stretched 1 ft with a force of 30 lb; moved from $s = 1$ ft to $s = -2$ ft.
16. Stretched 1 mm with a force of 0.2 N; moved from $s = -1$ cm to $s = 1$ cm.
17. Stretched 0.1 in with a force of 3 oz; moved from $s = -0.4$ in to $s = -0.5$ in.
18. Stretched 1 in with a force of 30 lb; moved from $s = 0$ ft to $s = 0.1$ ft.
19. Stretched 1 ft with a force of 240 lb; moved from $s = -2$ in to $s = 3$ in.
20. Stretched 1 cm with a force of 2 N; moved from $s = -1$ in to $s = 2$ in.

II

21. A spherical tank of radius 5 ft is resting with its base at ground level. How much work is done in pumping

it full of water if the water comes from a reservoir at ground level? (One cubic foot of water weighs 62.4 lb.)

22. Calculate the work done in pumping all the water out of a full spherical tank of radius 2 m and up to a point 3 m above the top of the tank.

23. A conical tank (with the vertex at the bottom) has height 6 ft, and the radius at the top is 2 ft. The tank has water in it to a depth of 2 ft. Calculate the work done in filling it to a depth of 5 ft, pumping the water up from a river that is 3 ft below the bottom of the tank.

*24. Suppose a spherical tank of radius R ft is full of water. Show that the work done in pumping all the water out of the tank to an outlet valve located right at the top of the tank is exactly the same as the work done in raising the same amount of water (i.e., the amount that would fill the tank) a height of R ft.

*25. Suppose a spring exerts a force of 4 lb when it is extended 1 inch. How far would it be extended if, starting from the rest position, 12 ft-lb of work was done in extending it?

*26. (i) Suppose a spring starts at the rest position, and by doing 1 ft-lb of work it is possible to extend it 1 ft. How much farther would it extend if somebody performed an additional 1 ft-lb of work in extending it?

  (ii) How much work would it take to stretch it from an extension of 1 ft to an extension of 2 ft?

*27. Suppose it takes 8 ft-lb of work to extend a spring 1 ft from its rest position.

  (i) What force will the spring exert when it is extended 1 ft?

  (ii) How much work will it take to extend it an additional foot (i.e., to an extension of 2 ft)?

*28. A spring is compressed 3 in from its rest position. When it is released, it does 6 ft-lb of work pushing against a door until it reaches its rest position.

  (i) How much force will this spring exert when extended 4 in?

  (ii) How much work will it take to move the spring from its first position, compressed 3 in, to the position where it is 4 in extended?

*29. (i) Figure out how much work it would take to lift a 1-lb weight to a height h ft above the surface of the earth, using the formula $F = \frac{Km}{r^2}$ for the force due to gravity.

  (ii) To send a spaceship far off into space, it is necessary to lift it free of the earth's gravitational field. This amounts to giving it enough energy so that it will not eventually fall back to earth. Find out how much work is needed to lift a 1-lb object free of the earth's gravity by letting $h \to \infty$ in your answer to part (i).

*30. Consider a spring that exerts a force of 10 lb when it is extended 1 ft from the rest position.

  (i) Suppose the spring is extended 1 ft from the rest position, and a person moves it until it is *compressed* 1 ft from the rest position. Show that the work done by the spring in the course of this movement equals zero.

  (ii) Explain why no work is done in this situation.

*31. We have two different formulas for the force due to gravity near the earth's surface. On the one hand, the simple approximate version is that an object whose mass is m lb experiences a force of m lb due to gravity. On the other hand, we have seen that the force on an object of mass m is actually $\frac{Km}{r^2}$, if r is the distance from the object to the center of the earth. These two formulas are different, but for objects near the earth's surface they give very similar results.

  (i) If an object weighs 1 lb at the earth's surface, how much does it weigh 1 mile above the earth's surface?

  (ii) If an object weighs 1 lb at the earth's surface, how much does it weigh 1 ft above the earth's surface?

*32. In this question we compare the calculation of the work done in lifting a payload depending on whether we use the actual formula for the force due to gravity or the approximate formula (cf. Exercise 31). (i) Calculate the work done in lifting a load weighing 1 lb at the earth's surface to a height of 1 mile above the earth's surface. Do the calculation using both the correct formula for the force due to gravity and the approximation, and compare the answers.

  Repeat part (i), this time lifting the load to a height of (ii) 10 miles, (iii) 100 miles, (iv) 10,000 miles.

*33. If an object weighing m lb is moving at a speed of v ft/sec, then its kinetic energy *in foot-pounds* is $\frac{1}{64}mv^2$. (If you have studied kinetic energy in physics, you have probably learned a formula for kinetic energy that looks different. The reason is that there is a conversion factor because we have measured mass in pounds and energy in ft-lb, a unit that is well suited to work performed by a force but poorly adapted to kinetic energy.)

  (i) If a force of 3 lb acts on an object weighing 4 lb and pushes it 6 ft, how much work does it perform?

  (ii) If the object in part (i) began at rest and all the work is converted into kinetic energy, how fast will the object be moving when the force stops?

*34. (i) Suppose an object weighing m lb is lifted to a height of h ft above ground level. What is its potential energy (i.e., the work done to it in the process of lifting)? (Use the approximate formula for the force due to gravity.)

  (ii) Suppose the object in part (i) drops to the ground. At the moment of impact, all the potential energy it used to have has been converted into kinetic energy. Use this fact and the information in Exercise 33 to calculate the speed of impact.

(iii) From your answer to part (ii), conclude that Galileo was right: The speed of impact does not depend on the weight of the object, that is, heavy objects fall no faster than light ones.

(iv) Check that your answer to part (ii) agrees with the answer that could be obtained by using Remark 4.9.1 (p. 316).

*35. A rope is 10 feet long and weighs 1 lb. It is lying coiled on the ground, and somebody comes and lifts one end straight up (see Figure 6.6.7). Find the work done in lifting the end of the rope to a height of (i) 5 ft, (ii) 15 ft.

Figure 6.6.7

*36. Water leaks out of a bucket at a constant rate of $\frac{1}{5}$ liter/sec. Starting at ground level, the bucket is lifted at a constant rate of 1 m every 3 sec. The bucket contains 6 liters of water at the beginning, and the

bucket itself weighs 1 kg. Find the work that is performed in lifting the bucket to a height of (i) 5 m, (ii) 10 m, (iii) 15 m.

*37. The bucket from Exercise 36 starts with 5 liters of water and is lifted at a rate of 1 m every 4 sec. The lifting force performs work of 150 J. How far has the bucket risen?

*38. A force pushing a vehicle performs work. Suppose the force is doubled; presumably, the speed will increase. For the sake of example, let us assume that the speed doubles too. Then in 1 second the object will travel twice as far and experience twice as much force, so the work done on it will be *quadrupled*. Since it has gone twice as far, the energy required to make it go a fixed distance has *doubled*. This rough idea is behind the familiar notion that it is less fuel-efficient to drive a car very fast.

If the propelling force doubles, it is actually unlikely that the speed will double too. Suppose that the speed is proportional to the cube root of the force. What will be the effect on the energy required to drive a fixed distance?

39. Water is pumped into an empty spherical tank of radius R from the level of the bottom of the tank.

(i) Find a formula for the work done in pumping in water to a depth d above the bottom.

(ii) Suppose the radius is 2 m and the pumping has used 200,000 J of energy. Use a graphics calculator or computer to find the depth to which the tank has been filled.

6.7 Moments and Centroids

FULCRUM
CENTER OF MASS
UNIFORM DENSITY
SYMMETRY

If you take a thin disk of metal or cardboard, it is possible to balance it on a sharp point held exactly at its center, as jugglers sometimes do with plates (see Figure 6.7.1).

If you take a thin flat object of a more irregular shape, as shown in Figure 6.7.2, it should still be possible to do this, but it will be harder to find the point at which

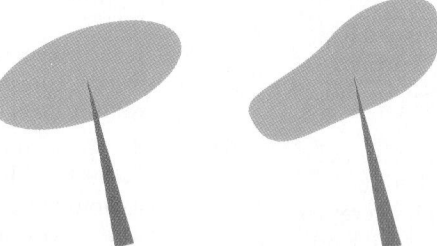

Figure 6.7.1   **Figure 6.7.2**

it will balance, the so-called *centroid* or *center of mass*, or *center of gravity*. It may be possible to find the centroid by trial and error, but the purpose of this section is to learn how to calculate it.

The results we find here will be used in the later parts of the next section, Section 6.8, on fluid pressure.

Perhaps the simplest example of balancing is the seesaw. We are all familiar with the phenomenon of people of different weights being able to balance by sitting at different distances from the support or **fulcrum**.

Let x_1 and x_2 be the displacements of two people from the fulcrum O and write m_1 and m_2 for their respective weights, as in Figure 6.7.3.

A weight on the seesaw will tend to make the seesaw rotate about O. The heavier the weight m or the greater its displacement x from O, the greater will be the tendency to turn. This suggests that the tendency to turn depends on the product mx.

The quantity $m_i x_i$ represents the tendency of the ith person to make the seesaw rotate clockwise about O. (If the quantity is negative, it represents a tendency to rotate counterclockwise.) The total tendency to rotate caused by both persons is $m_1 x_1 + m_2 x_2$.

The seesaw will balance when there is no net tendency to turn, that is, when

$$m_1 x_1 + m_2 x_2 = 0.$$

Notice that since the people are on opposite sides of the fulcrum, one displacement is positive and the other is negative.

This formula reflects the familiar fact that a heavy person should sit closer to the fulcrum than a light person in order to balance. (For example, if the second person is twice as heavy as the first, then $m_2 = 2m_1$, so to balance, $0 = m_1 x_1 + m_2 x_2 = m_1 x_1 + 2m_1 x_2 = m_1(x_1 + 2x_2)$. This means that $x_2 = -\frac{1}{2}x_1$; the heavy person sits half as far from O as the light person.

Figure 6.7.3

Figure 6.7.4(i)

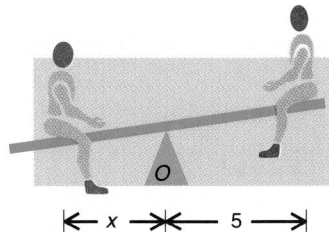

Figure 6.7.4(ii)

> **DEFINITION 6.7.1**
>
> If an object of weight m is at a horizontal displacement x from O, the quantity
>
> mx
>
> is called the **moment** of the object **about the point** O.

Suppose objects weighing $m_1, m_2, \ldots, m_n$ are placed on a seesaw supported at O, and let x_i be the displacement (positive or negative) of the ith object from O (see Figure 6.7.4(i)). The seesaw will balance when the moments cancel, that is, when the total moment is zero (see Figure 6.7.4(ii)):

$$\sum_{i=1}^{n} m_i x_i = 0.$$

EXAMPLE 1   Seesaw   If one person weighing 40 lb sits 5 ft from the fulcrum of a seesaw, where will a person weighing 60 lb have to sit for them to balance?

Solution   The situation is illustrated in Figure 6.7.5, where we have written x for the displacement from the 60-lb person to the fulcrum. The 40-lb person has a displacement of 5 ft from the fulcrum. The total moment is

$$40(5) + 60x = 200 + 60x.$$

Figure 6.7.5

The seesaw will balance when this is zero, that is, when $x = -\frac{200}{60} = -\frac{10}{3}$. The negative sign indicates that this displacement is in the direction opposite to the displacement of the 40-lb person. To balance the seesaw, the 60-lb person must sit $3\frac{1}{3}$ ft from the fulcrum, on the side opposite the other person.

Notice that we ignored the weight of the seesaw itself, presuming that it balances when nobody is on it, so that it contributes nothing to the total moment about the fulcrum.

Now suppose objects with weights $m_1, m_2, \ldots, m_k$ are positioned at displacements $x_1, x_2, \ldots, x_k$ along a thin rod. Let us try to find the point $\bar{x}$ at which the rod can be balanced (see Figure 6.7.6).

If the rod is placed on a pointed support at $\bar{x}$, the support exerts an *upward* force equal to the total weight $M = m_1 + \ldots + m_k$. For the rod to balance, the corresponding moment must cancel the sum of the moments due to the little weights. Since the force of the support is in the direction opposite to the weights, it is $-M$, and we find

$$(-M)\bar{x} + m_1 x_1 + \ldots + m_k x_k = 0,$$

or

$$M\bar{x} = m_1 x_1 + \ldots + m_k x_k,$$

so

$$\bar{x} = \frac{m_1 x_1 + \ldots + m_k x_k}{M} = \frac{m_1 x_1 + \ldots + m_k x_k}{m_1 + \ldots + m_k}.$$

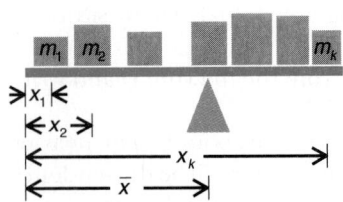

Figure 6.7.6

Now suppose the rod with k weights is replaced by a more realistic thin rod whose weight is distributed continuously along its length (see Figure 6.7.7). We could think of this in terms of letting k, the number of weights, become larger and larger.

If we use the variable x to describe displacement along the rod, with $x \in [a, b]$, then the distribution of weight is conveniently described by a density function, usually denoted $\rho(x)$ (ρ is the Greek letter "rho"). To say that $\rho(x)$ is the density function means that a small piece of the rod centered at x and having length dx has weight $\rho(x)\,dx$ (see Figure 6.7.8). Another way of thinking of it is that the density $\rho(x)$ tells the weight per unit *length* at the point x.

Figure 6.7.7

Figure 6.7.8

> You may be accustomed to using the word "density" in physics to mean the weight per unit *volume*. For a thin rod we measure only one dimension, the length, so we express the density as weight per unit length. In everyday speech we might say that some electrical wire weighs 3 oz per foot or that kite string weights 1 oz per hundred feet...

If the weight of a little piece of length dx is $\rho(x)\,dx$, then the total weight M of the rod is found by "adding up" all these little weights; it is $M = \int_a^b \rho(x)\,dx$. The little piece of length dx centered at x has weight $\rho(x)\,dx$, so its moment about the origin is equal to $x\rho(x)\,dx$. The total moment about the origin of the whole rod is the sum of these moments, or $\int_a^b x\rho(x)\,dx$.

For the rod to balance at a point $\bar{x}$, the moment of the rod must equal the moment obtained by placing the total weight M at $\bar{x}$; that is, $\int_a^b x\rho(x)\,dx = M\bar{x} = \bar{x}\int_a^b \rho(x)\,dx$. Dividing, we find

$$\bar{x} = \frac{1}{M}\int_a^b x\rho(x)\,dx = \frac{\int_a^b x\rho(x)\,dx}{\int_a^b \rho(x)\,dx}. \tag{6.7.2}$$

This formula gives the location of the point at which the object balances, which is called the **center of mass**. By far the most common situation is the one in which the density is constant, that is, the rod has **uniform density**. In this case the center of mass is often called the **centroid**.

EXAMPLE 2   Consider a rod of uniform density whose length is L meters. Find its centroid.

Solution   Let us place the rod with one end at the origin and the rest of it lying along the positive x-axis. The variable x will run from $x = 0$ to $x = L$, and the density function is a constant, $\rho(x) = c$, for some number c. See Figure 6.7.9.

To find the centroid, we need to know the total weight $M = \int_0^L \rho(x)\,dx = \int_0^L c\,dx = c(L - 0) = cL$. The total moment about the origin is $\int_0^L x\rho(x)\,dx = \int_0^L cx\,dx = \frac{c}{2}x^2\big|_0^L = \frac{cL^2}{2}$.

Using the formula above, we find that the centroid is

$$\bar{x} = \frac{\int_0^L x\rho(x)\,dx}{\int_0^L \rho(x)\,dx} = \frac{\frac{cL^2}{2}}{cL} = \frac{L}{2}.$$

The centroid of the rod is $\frac{L}{2}$ meters from the end, which is the same thing as saying the centroid is precisely in the middle, which is hardly surprising.

Notice that the constant c does not appear in the answer. As long as the density is uniform, its exact value is irrelevant to the location of the centroid.

EXAMPLE 3   Consider a rod that is 3 ft long whose density is proportional to the distance from one of the ends. Find its center of mass.

Solution   Again we place the rod on the x-axis with one end at the origin and the other on the positive axis. The variable x ranges from $x = 0$ to $x = 3$.

The density is proportional to the distance from the origin, say, which means that it equals a constant times x. We write $\rho(x) = cx$, for some constant c (whose value we do not know). The total weight is $\int_0^3 \rho(x)\,dx = \int_0^3 cx\,dx = \frac{c}{2}x^2\big|_0^3 = \frac{9c}{2}$.

The total moment about the origin is $\int_0^3 x\rho(x)\,dx = \int_0^3 cx^2\,dx = \frac{c}{3}x^3\big|_0^3 = 9c$. Using the formula for the center of mass, we find that

$$\bar{x} = \frac{\int_0^3 x\rho(x)\,dx}{\int_0^3 \rho(x)\,dx} = \frac{9c}{\frac{9c}{2}} = 2.$$

The center of mass is 2 ft from the end at the origin, that is, the end at which the density is lower.

It is worth noticing that the center of mass is two-thirds of the way from the light end to the heavy end. It makes sense that it ought to be close to where the mass is concentrated, that is, closer to the heavy end. See Figure 6.7.10.

Also remark that the constant c again did not appear in the answer. This is not surprising if we consider what will happen to the quotient

$$\bar{x} = \frac{\int_a^b x\rho(x)\,dx}{\int_a^b \rho(x)\,dx}$$

0   L x

Figure 6.7.9

0                $\bar{x}$ 3 x

Figure 6.7.10

if we multiply the density $\rho(x)$ by some constant k. Both the numerator and the denominator will be multiplied by k, so the quotient will remain the same. In other words, multiplying the density by a constant does not affect the location of the center of mass. In particular, we can assume that the constant c used in either of the preceding examples equals 1 and we will still get the correct answer. This observation is worth recording.

REMARK 6.7.3

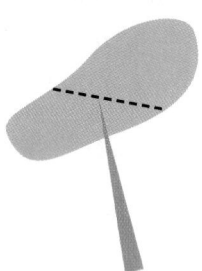

Figure 6.7.11

If we are considering a rod of uniform density, we may as well assume that the density is the constant function $\rho(x) = 1$. In this case the centroid is

$$\bar{x} = \frac{\int_a^b x\,dx}{\int_a^b 1\,dx} = \frac{\int_a^b x\,dx}{b - a} = \frac{1}{b - a} \int_a^b x\,dx.$$

This is exactly the same thing as the *average value* of the function x over the interval $[a, b]$ (cf. Definition 6.5.1, p. 447), so it is not surprising that it turns out to be the value of x at the middle of the rod.

If the density $\rho(x)$ is not a constant, we can still think of the center of mass as a sort of "weighted average" of the function x; we count more heavily the values of x where the rod is heavier and divide by the total weighting. So $\bar{x}$ is a sort of average value of x over the whole rod, taking its density into account.

Next we want to consider the center of mass of a region in the plane. If we attempt to balance a flat object on a point but make a poor choice of point, what will happen is that the object will start to tilt and eventually fall off. When it is tilting, what it is really doing is *rotating* about some line (see Figure 6.7.11). Since moments have to do with an object's tendency to rotate, we should look for a kind of moment having to do with rotation about a line.

Consider the thin flat plate in Figure 6.7.12, which can rotate freely about the y-axis. A small object of weight m is placed on the plate. What will be the tendency of this weight to make the plate rotate about the axis?

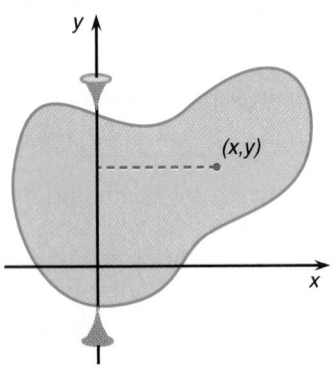

Figure 6.7.12

It seems reasonable to consider the product of the weight m and some displacement. Rather than the displacement from the origin, it makes more sense to consider the displacement from the axis (after all, moving the object up or down parallel to the y-axis should not affect its tendency to rotate the plate). If the object is at the point (x, y) in the plane, then its displacement from the y-axis is simply x. This suggests that the object's moment about the y-axis ought to be mx. Similar considerations apply to moments about the x-axis.

DEFINITION 6.7.4

An object of weight m positioned at the point (x, y) has its moment about the x-axis defined by

$$M_x = my$$

and its moment about the y-axis defined by

$$M_y = mx.$$

Notice that M_x, the moment about the x-axis, involves y, and M_y, the moment about the y-axis, involves x. This is because the moments involve the displacement away from the axis rather than a displacement *along* it.

Figure 6.7.13

Figure 6.7.14

Figure 6.7.15

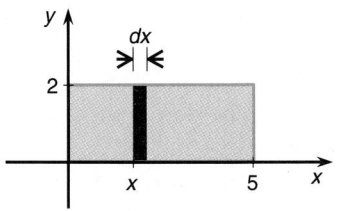

Figure 6.7.16

Now suppose we have a flat object in the shape of some region in the plane, with some specified density. Then, as before, we want to find the total moment of the object about each of the axes and use them to find the center of mass. The most obvious case is that of a flat plate of uniform density.

To find the total moment M_y about the y-axis, we consider the narrow strip in the region of width dx and at a displacement x from the y-axis. Its area is dx times its height, and its weight is $(area) \times (density)$. (Recall that we are assuming that the density equals the constant ρ.) See Figure 6.7.13.

The moment of this strip is $x \times (weight) = x(area)\rho = x(height)\rho \, dx$, and we integrate to get the total moment M_y. The total weight M will be the integral of the weights of all these strips, that is, the integral of $(height)\rho \, dx$.

In particular, if the region is the region lying between two curves $y = f(x)$ and $y = g(x)$, say, for $x \in [a, b]$, and $f(x) \geq g(x)$ for all $x \in [a, b]$, then the height of the strip at a displacement x from the y-axis is $f(x) - g(x)$. See Figure 6.7.14.

Similar remarks apply to the moment about the x-axis. See Figure 6.7.15.

DEFINITION 6.7.5

> Suppose $f(x) \geq g(x)$, for all $x \in [a, b]$. Consider the region lying between the curves $y = f(x)$ and $y = g(x)$ and between the lines $x = a$ and $x = b$, with uniform density ρ.
>
> The moment M_y of this region about the y-axis and the total weight M are given as follows:
>
> $$M_y = \int_a^b x\big(f(x) - g(x)\big)\rho \, dx, \qquad M = \int_a^b \big(f(x) - g(x)\big)\rho \, dx.$$
>
> Similarly, for a region lying between $y = c$ and $y = d$ and between the curves $x = h(y)$ and $x = \ell(y)$ with $h(y) \geq \ell(y)$ for all $y \in [c, d]$, if the density is ρ, then
>
> $$M_x = \int_c^d y\big(h(y) - \ell(y)\big)\rho \, dy, \qquad M = \int_c^d \big(h(y) - \ell(y)\big)\rho \, dy.$$

The centroid will have the property that the total weight concentrated at the centroid will have the same moments about the axes as the original object. If we write $(\bar{x}, \bar{y})$ for the centroid, this means that

$$M_y = M\bar{x}, \qquad M_x = M\bar{y}.$$

These equations can be solved to find the centroid:

$$\bar{x} = \frac{M_y}{M}, \qquad \bar{y} = \frac{M_x}{M}. \tag{6.7.6}$$

EXAMPLE 4   **Centroid of Rectangle**   Find the centroid of the rectangle of uniform density that lies between $(0,0)$, $(0,2)$, $(5,0)$, and $(5,2)$.

Solution   First we observe that by analogy with Remark 6.7.3 we can assume that the density is the constant function 1. The rectangle lies between $y = f(x) = 2$ and $y = g(x) = 0$ and between the vertical lines $x = h(y) = 5$ and $x = \ell(y) = 0$. The height of the vertical strip of width dx that lies at a displacement x from the y-axis is $f(x) - g(x) = 2$ for each $x \in [0, 5]$, and $M_y = \int_0^5 x(height)\rho \, dx = \int_0^5 x(2)(1) \, dx = \int_0^5 2x \, dx = x^2 \big|_0^5 = 25$. See Figure 6.7.16.

Figure 6.7.17

Figure 6.7.18

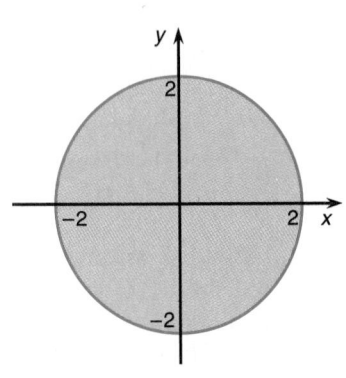

Figure 6.7.19

The total weight is $M = \int_0^5 (height)\rho\, dx = \int_0^5 2\, dx = 10$, so the x-coordinate $\bar{x}$ of the centroid is $\bar{x} = \frac{M_y}{M} = \frac{25}{10} = \frac{5}{2}$.

Similarly, to find the y-coordinate $\bar{y}$ of the centroid, we consider narrow horizontal strips of height dy (see Figure 6.7.17). Each such strip has width 5, and y ranges from $y = 0$ to $y = 2$, so $M_x = \int_0^2 y\big(h(y) - \ell(y)\big)\rho\, dy = \int_0^2 5y\, dy = \frac{5}{2}y^2\big|_0^2 = 10$. Since we have already found that $M = 10$, we find that $\bar{y} = \frac{M_x}{M} = \frac{10}{10} = 1$.

The centroid is $(\bar{x}, \bar{y}) = \left(\frac{5}{2}, 1\right)$, which is the exact center of the rectangle. ◢

■   In fact it is easy to see that the centroid of any rectangle is its center.

For long thin objects the density represented the weight per unit length. Now we are considering thin flat objects, and it makes sense for "density" to refer to weight per unit area. In ordinary situations we might say that some new roofing material weighs 15 kg per square meter or that some sails are being made from special cloth that weighs only 2 lb per hundred square feet.

EXAMPLE 5   Find the centroid of the region with uniform density enclosed between the curves $y = x$ and $y = x^2$, as shown in Figure 6.7.18.

Solution   The region lies between the graphs of the functions $y = f(x) = x$ and $y = g(x) = x^2$, with $f(x) \geq g(x)$ for all $x \in [0, 1]$. Letting $\rho = 1$, we find

$$M_y = \int_0^1 x\big(x - x^2\big)\rho\, dx = \int_0^1 (x^2 - x^3)\, dx = \left(\frac{1}{3}x^3 - \frac{1}{4}x^4\right)\bigg|_0^1$$

$$= \frac{1}{12}.$$

Also the total weight is $M = \int_0^1 (x - x^2)\rho\, dx = \left(\frac{1}{2}x^2 - \frac{1}{3}x^3\right)\big|_0^1 = \frac{1}{6}$.

Looking at the region the other way, with y as the variable, we see that it lies between $x = h(y) = \sqrt{y}$ and $x = \ell(y) = y$, with $h(y) \geq \ell(y)$ for all $y \in [0, 1]$. Consequently,

$$M_x = \int_0^1 y(\sqrt{y} - y)\rho\, dy = \int_0^1 (y^{3/2} - y^2)\, dy = \left(\frac{2}{5}y^{5/2} - \frac{1}{3}y^3\right)\bigg|_0^1$$

$$= \frac{1}{15}.$$

From all this we find $\bar{x} = \frac{M_y}{M} = \frac{1}{12} / \frac{1}{6} = \frac{1}{2}$ and $\bar{y} = \frac{M_x}{M} = \frac{1}{15} / \frac{1}{6} = \frac{2}{5}$. The centroid is the point $(\bar{x}, \bar{y}) = \left(\frac{1}{2}, \frac{2}{5}\right)$. ◢

EXAMPLE 6   **Centroid of Disk**   Find the centroid of a disk of radius 2 with uniform density.

Solution   We may as well assume that the center of the disk is at the origin, so the circle enclosing the disk is $x^2 + y^2 = 4$ (see Figure 6.7.19). The disk is the region lying between $y = f(x) = \sqrt{4 - x^2}$ and $y = g(x) = -\sqrt{4 - x^2}$.

To find M_y, we consider a vertical strip of width dx at a displacement x from the y-axis. The height of the strip is $f(x) - g(x) = 2\sqrt{4 - x^2}$, and if the density is $\rho = 1$,

Figure 6.7.20

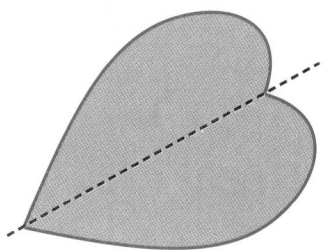

Figure 6.7.21

then

$$M_y = \int_{-2}^{2} x(1)\big(f(x) - g(x)\big)\,dx = \int_{-2}^{2} 2x\sqrt{4 - x^2}\,dx$$

$$\boxed{\begin{array}{l} u = 4 - x^2 \\ du = -2x\,dx \end{array}}$$

$$= \int_{0}^{0} \sqrt{u}(-1)\,du = -\frac{2}{3}u^{3/2}\Big|_{0}^{0}$$

$$= 0.$$

Of course, the calculation for M_x is almost identical, and we find that $M_x = 0$. Without even calculating the total weight M we see that the centroid is $(\bar{x}, \bar{y}) = (0, 0)$. In other words, the centroid of the disk is its center.   ◢

Virtually the same calculation works for a disk of any radius, and we see that the centroid of any disk is its center.

In a way, if we think of the moments M_y and M_x as the *average values* of x and y, respectively, it seems obvious from Figure 6.7.20 that they should both equal zero. This is because the part of the disk to the left of the y-axis and the part to the right are symmetrical, so they contribute the same amount to the average, except that they have opposite signs (i.e., x is positive on the right side and negative on the left side). The two parts cancel, and the total moment is zero.

The same idea can be applied in other situations. If an object is symmetrical about some line, as in Figure 6.7.21, then it has no tendency to rotate one way or the other about that line, so its centroid, the point upon which it balances, must be on the line.

THEOREM 6.7.7

> Suppose a thin plate with uniform density is symmetrical about some line L. Then the centroid lies on the line L.

Figure 6.7.22

EXAMPLE 7   This observation immediately explains and justifies the remarks that the centroid of a rectangle is its center and that the centroid of a circle is its center. For example, a rectangle whose sides are parallel to the axes is symmetrical about the horizontal line through its center and about the vertical line through its center, so the centroid lies on both these lines (see Figure 6.7.22). This means that it must be the point of intersection of these lines, which is the center.

Similarly, a disk is symmetrical about *every* line through its center, so the centroid lies on all these lines. The only point on all these lines is the center (see Figure 6.7.23).

We can also see that the centroid of the region inside an ellipse is the center of the ellipse, since the ellipse is symmetrical about the major axis and the minor axis, which intersect at the center. See Figure 6.7.24.   ◢

Figure 6.7.23

Figure 6.7.24

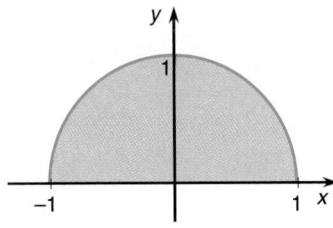

Figure 6.7.25

EXAMPLE 8   Centroid of Semidisk   Find the centroid of the region inside the upper semicircle $y = \sqrt{1 - x^2}$, shown in Figure 6.7.25.

Solution   When we are discussing the *centroid*, we assume that the region has uniform density, so we may as well let $\rho = 1$.

The region is certainly symmetric about the y-axis, so we know that the centroid lies on the y-axis, which means $\bar{x} = 0$.

To find $\bar{y}$, we think of the region as the region between the two sides of the semicircle, that is, the curves $x = h(y) = \sqrt{1 - y^2}$ and $x = \ell(y) = -\sqrt{1 - y^2}$, for $0 \leq y \leq 1$. So

$$M_x = \int_0^1 y\big(h(y) - \ell(y)\big)\rho\, dy = \int_0^1 2y\sqrt{1 - y^2}\, dy$$

$$\boxed{\begin{array}{l} u = 1 - y^2 \\ du = -2y\, dy \end{array}}$$

$$= -\int_1^0 \sqrt{u}\, du = \int_0^1 u^{1/2}\, du = \frac{2}{3} u^{3/2}\Big|_0^1$$

$$= \frac{2}{3}.$$

We also need to evaluate the total weight

$$M = \int_0^1 \big(h(y) - \ell(y)\big)\rho\, dy = \int_0^1 2\sqrt{1 - y^2}\, dy.$$

We could evaluate this integral by using Example 6.1.2, but we could also realize that it is precisely the area of the semicircle, so it equals $\frac{\pi}{2}$.

Now we can find $\bar{y} = \frac{M_x}{M} = \frac{2/3}{\pi/2} = \frac{4}{3\pi} \approx 0.4244$. The centroid of the semicircle is $(\bar{x}, \bar{y}) = \left(0, \frac{4}{3\pi}\right) \approx (0, 0.4244)$. See Figure 6.7.26. ◢

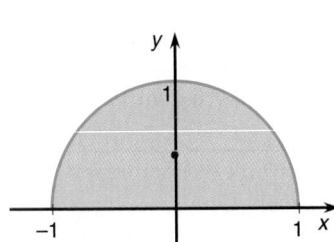

Figure 6.7.26

> Notice that using symmetry saved us some trouble. The other useful thing to notice is that since we let the density be $\rho = 1$, the total weight M just equals the *area* of the region. So whenever we know the area of the region, we can avoid having to do that calculation too.

Having found the centroids of rectangles, circles, semicircles, and ellipses, we should ask about the centroid of a triangle.

Consider the triangle shown in Figure 6.7.27. The dotted line is one of the *medians*, meaning the lines from one of the vertices to the middle of the opposite side. If we break up the triangle into narrow horizontal strips of height dy, as in Figure 6.7.28, then the median cuts each strip in half. The parts of the strip on either side of the median have moments about the median that cancel because they extend the same distance from the line but in opposite directions. This means that the moment of the strip about the median is zero. Each of these strips has no tendency to rotate about the line because it balances.

The total moment of the triangle about the median is the "sum" (i.e., the integral) of the moments of these little strips. Adding up zero many times gives zero, and we see that the total moment about the median is zero. (This argument is not exactly airtight, but it is essentially correct, and it is not terribly difficult to make it precise. One of the exercises suggests how to do it.)

As we observed above, just before Theorem 6.7.7, to say that a region has moment zero about some line means that the centroid must lie on that line. What we have just said indicates that the centroid of a triangle lies on the median. In fact, the same argument works for the other two medians, so the centroid is on each median.

Figure 6.7.27

Figure 6.7.28

Figure 6.7.29

This means that the centroid of the triangle is the point where the medians intersect, shown in Figure 6.7.29.

Happily, this point is extremely easy to find. If the vertices of the triangle are (x_1, y_1), (x_2, y_2), and (x_3, y_3), then the centroid is obtained by "averaging" the coordinates of the vertices:

$$\bar{x} = \frac{x_1 + x_2 + x_3}{3}, \qquad \bar{y} = \frac{y_1 + y_2 + y_3}{3}. \qquad (6.7.8)$$

EXAMPLE 9   **Centroid of Triangle**   Find the centroid of the triangle whose vertices are $(1, 2)$, $(3, 5)$, and $(2, -4)$.

Solution   We simply apply Equations 6.7.8:

$$\bar{x} = \frac{1 + 3 + 2}{3} = 2, \qquad \bar{y} = \frac{2 + 5 - 4}{3} = 1,$$

so the centroid is the point $(2, 1)$.

Frequently, a region is composed of simpler parts, and it is possible to find the centroid of the whole region from knowledge of the centroids and weights of the parts.

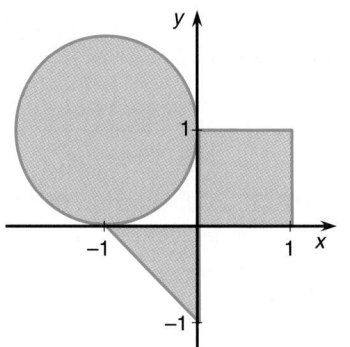

Figure 6.7.30

EXAMPLE 10   Consider a region that is composed of simpler regions. An example is illustrated in Figure 6.7.30. We know the centroids of the disk, the triangle, and the square without doing any work. We want to use this information to find the centroid of the whole shaded region.

Instead of the centroid, it is better to begin with the moments about the axes. First look at the square. Its centroid is, of course, its center, which is $\left(\frac{1}{2}, \frac{1}{2}\right)$. Letting the density be $\rho = 1$, we know that the weight of the square equals its area, which is 1. Since $\bar{x} = \frac{M_y}{M}$, we find that $M_y = M\bar{x} = 1\left(\frac{1}{2}\right) = \frac{1}{2}$. Similarly, the other moment of the square is $M_x = M\bar{y} = \frac{1}{2}$.

It is equally easy to find the moments of the triangle about the axes. By Equations 6.7.8, the centroid is $(\bar{x}, \bar{y}) = \left(-\frac{1}{3}, -\frac{1}{3}\right)$. Since M equals the area, which is $\frac{1}{2}$, we find that $M_x = M\bar{y} = \frac{1}{2}\left(-\frac{1}{3}\right) = -\frac{1}{6}$ and $M_y = M\bar{x} = -\frac{1}{6}$.

Finally, the radius of the circle is 1, so the area is $\pi r^2 = \pi$, so its moments are $M_x = M\bar{y} = \pi(1) = \pi$ and $M_y = M\bar{x} = \pi(-1) = -\pi$.

The moment of the whole region is the sum of the moments of the parts, that is, $M_x = \frac{1}{2} - \frac{1}{6} + \pi = \frac{1}{3} + \pi$ and $M_y = \frac{1}{2} - \frac{1}{6} - \pi = \frac{1}{3} - \pi$. The total weight of the whole region is the sum of the weights of the parts, so $M = 1 + \frac{1}{2} + \pi = \frac{3}{2} + \pi$.

Combining all this, we find the centroid of the whole region.

$$\bar{x} = \frac{M_y}{M} = \frac{\frac{1}{3} - \pi}{\frac{3}{2} + \pi} = \frac{2 - 6\pi}{9 + 6\pi}, \qquad \bar{y} = \frac{M_x}{M} = \frac{\frac{1}{3} + \pi}{\frac{3}{2} + \pi} = \frac{2 + 6\pi}{9 + 6\pi},$$

and the centroid is $\left(\frac{2-6\pi}{9+6\pi}, \frac{2+6\pi}{9+6\pi}\right) \approx (-0.605, 0.749)$.

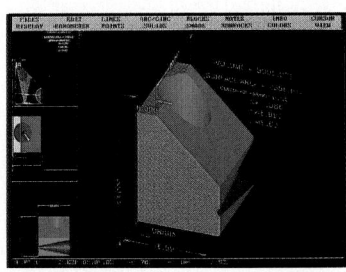

DesignCAD 3D software was used to produce this image of a machine component. A designer can adjust the shape of the object and see the results on the screen. The computer calculates the center of gravity, which may affect the operation of the component.

> Notice that we add up the moments M_x for all the parts of the region to find the corresponding moment for the whole region. It is *not* correct simply to average the coordinates of the centroids of the subregions.

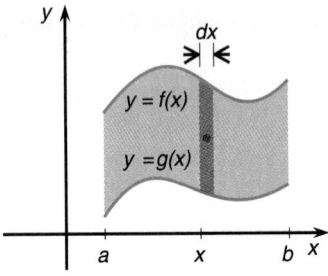

Figure 6.7.31

Returning to the question of calculating moments, we consider again the region between $y = f(x)$ and $y = g(x)$, shown in Figure 6.7.31. The centroid of a thin vertical strip will be at its center, that is, halfway up. Since the top of the strip is at a height $f(x)$ and the bottom at $g(x)$, the midpoint will be at height $\frac{1}{2}\big(f(x) + g(x)\big)$. The moment of the strip about the x-axis will be $\frac{1}{2}\big(f(x) + g(x)\big)(area)\rho$.

The area of the strip is $\big(f(x) - g(x)\big)dx$, so to find the total moment M_x about the x-axis, we "add up" the moments of these strips. In other words, we integrate: $M_x = \int_a^b \frac{1}{2}\big(f(x) + g(x)\big)\big(f(x) - g(x)\big)\rho\,dx = \frac{1}{2}\int_a^b \big(f(x)^2 - g(x)^2\big)\rho\,dx$. We summarize this result below.

THEOREM 6.7.9

> Suppose $f(x) \geq g(x)$, for all $x \in [a, b]$. Consider the region lying between the curves $y = f(x)$ and $y = g(x)$ and between the lines $x = a$ and $x = b$, with uniform density ρ.
>
> The moment M_x of this region about the x-axis is given as follows:
>
> $$M_x = \frac{1}{2}\int_a^b \big(f(x)^2 - g(x)^2\big)\rho\,dx.$$
>
> Similar remarks apply to the moment M_y of a region between two curves $x = h(y)$ and $x = \ell(y)$ (cf. Definition 6.7.5).

See the Graphics Calculator Example on the facing page.

EXAMPLE 11   Centroid of Semidisk Revisited   We can now find the centroid of the upper semicircle more easily (cf. Example 8). The semicircle lies between $f(x) = \sqrt{1 - x^2}$ and $g(x) = 0$, so the formula gives

$$M_x = \frac{1}{2}\int_{-1}^1 (1 - x^2 - 0)\,dx = \frac{1}{2}\left(x - \frac{x^3}{3}\right)\bigg|_{-1}^1 = \frac{2}{3}.$$

This is certainly easier than the calculation we did in Example 8. ◢

Suppose $f(x) \geq g(x) \geq 0$ on $[a, b]$, with $0 \leq a < b$, and let R be the region between their graphs. Consider the solid formed by revolving R around the y-axis, and let us find its volume. Using cylindrical shells, the volume is

$$Volume = \int_a^b 2\pi x\big(f(x) - g(x)\big)dx = 2\pi \int_a^b x\big(f(x) - g(x)\big)dx.$$

This last integral is the moment M_y of the region R about the y-axis, so we can say that the volume is $2\pi M_y$. (In talking about the moment, we are assuming that the density is $\rho = 1$.) Now if the centroid of R is $(\bar{x}, \bar{y})$, then $\bar{x} = \frac{M_y}{M}$, so the volume can be written as

$$Volume = 2\pi M_y = 2\pi M\bar{x}.$$

It is interesting to interpret this result geometrically. Since we have let the density be $\rho = 1$, the mass M is really just the area of R. And $2\pi\bar{x}$ is the circumference of a circle of radius $\bar{x}$. In Figure 6.7.32 we see the region R with its centroid marked, and we see how R is revolved around the y-axis. We also see how the centroid moves around in a circle whose radius is $\bar{x}$.

So $2\pi\bar{x}$ is the distance traveled by the centroid when we revolve R. We have just shown that for a region R lying between two graphs, the volume of the corresponding solid of revolution can be expressed in terms of the area of R and its centroid. This

Figure 6.7.32

With a graphics calculator we can plot the semicircle from Example 11, and plot the centroid

$$(\bar{x}, \bar{y}) = \left(\frac{M_y}{M}, \frac{M_x}{M}\right) = \left(0, \frac{2/3}{\pi/2}\right) = \left(0, \frac{4}{3\pi}\right).$$

TI-85	SHARP EL9300
GRAPH F1 $(y(x) =)$ $\sqrt{}$ (1 − F1 (x) x^2)M2 (RANGE) (−) 1 ▼ 1 ▼ ▼ (−) 0.5 ▼ 1 F5 (GRAPH) PtOff (0 , 4 ÷ 3 ÷ π) ENTER	⤶ $\sqrt{}$ 1 − x/θ/T x^2 RANGE −1 ENTER 1 ENTER 1 ENTER −.5 ENTER 1 ENTER 1 ENTER ⤶ 2ndF PLOT ▼ ▶ ▼ ENTER 0 ENTER 4 ÷ 3 ÷ π ENTER ⤶
CASIO f_x-6300G (f_x-7700GB) Range −1 EXE 1 EXE 1 EXE −.5 EXE 1 EXE 1 EXE EXE EXE EXE Graph $\sqrt{}$ (1 − X x^2) EXE Plot 0 , 4 ÷ 3 ÷ π EXE	HP48SX ⬑ PLOT PLOTR ERASE ATTN ′ Y = $\sqrt{x}$ () 1 − X y^x 2 ↱ PLOT ⬑ DRAW ERASE 1 +/− ENTER 1 XRNG .5 +/− ENTER 1 YRNG ATTN 0 ENTER 4 ENTER 3 ÷ π ÷ PRG OBJ NXT R → C PRG DSPL NXT C → PX PIXOF ⬑ PLOT PLOTR DRAW

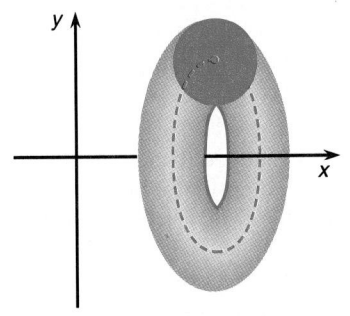

As usual, the semicircle may be somewhat distorted by the unequal horizontal and vertical scales. Can you adjust the scales to remove the distortion?

famous result is named after the Greek mathematician Pappus, who lived in the fourth century A.D. In fact the result is true for more complicated regions too.

THEOREM 6.7.10

Pappus' Theorem

Suppose R is a region in the plane with centroid $(\bar{x}, \bar{y})$. If R is revolved around an axis, the volume of the resulting solid is the product of the area A of R and the distance traveled by its centroid.

In particular, if R is revolved around the y-axis, the resulting solid has volume $2\pi\bar{x}A$, and if R is revolved around the x-axis, the resulting solid has volume $2\pi\bar{y}A$.

Figure 6.7.33

EXAMPLE 12   Volume of a Doughnut   Find the volume of the "doughnut" shape swept out when the disk of radius 1 centered at $(3, 2)$ is revolved around the x-axis (see Figure 6.7.33).

Solution To do this by integration is not extremely difficult, but it is quite messy. But since we know that the centroid of a disk is its center, we know that in this case, $\bar{y} = 2$. The area is $A = \pi$, so by Pappus' Theorem the volume of the doughnut is

$$2\pi\bar{y}A = 2\pi(2)\pi = 4\pi^2.$$

Exercises 6.7

I

Find the center of mass of each of the following objects.

1. A thin rod lying along the interval $[5, 9]$ and having uniform density.

2. A thin rod lying along the x-axis between $x = -2$ and $x = 6$ with density $\rho(x) = x + 2$.

3. A thin rod lying along the x-axis between $x = -3$ and $x = 3$ with density $\rho(x) = 3 + x^2$.

4. A thin rod of uniform density lying along the x-axis between $x = a$ and $x = a + L$.

5. A thin rod lying along the x-axis between $x = 1$ and $x = 2$ with density $\rho(x) = x^2 - 1$.

6. A thin rod lying along the x-axis between $x = 1$ and $x = 4$ with density $\rho(x) = \frac{1}{x}$.

7. A thin rod lying along the x-axis between $x = 0$ and $x = 1$ with density $\rho(x) = \frac{x}{1+x^2}$.

8. A thin rod lying along the x-axis between $x = 0$ and $x = a$ with density $\rho(x) = x^r$, where $a > 0$ and $r > 0$.

9. A thin cone lying along the x-axis between $x = 2$ and $x = 4$, with vertex at $x = 2$ and radius 0.2 at $x = 4$, whose density is proportional to its cross-sectional area.

10. An object composed of two thin rods of the same uniform density, one lying between $x = -4$ and $x = -1$ and the other lying between $x = 1$ and $x = 5$.

Find the centroid of each of the regions described below, assuming that they are all of uniform density.

11. The right semidisk $\{(x, y) : 0 \le x \le \sqrt{1 - y^2}, -1 \le y \le 1\}$.

12. The part of the region inside the circle $x^2 + y^2 = 9$ that lies in the first quadrant (i.e., with $x \ge 0, y \ge 0$).

13. The region inside the bottom half of the ellipse $x^2 + 2y^2 = 4$ (i.e., the intersection of the region inside the ellipse and the region below the x-axis).

14. The intersection of the region inside the ellipse $3x^2 + 4y^2 = 25$ and the third quadrant, that is, the region where $x, y \le 0$.

15. The region inside the triangle with vertices $(-2, 1)$, $(-5, 2)$, and $(1, 3)$.

16. The region inside the triangle with vertices $(2, 3)$, $(-1, 2)$, and $(2, 1)$.

17. The region enclosed between the curve $y = x^2$ and the line $y = 4$.

18. The region enclosed between the curve $y = x^2$ and the line $y = x + 2$.

19. The region enclosed between the curves $y = \sqrt{x}$ and $y = x^2$.

20. The region between the curves $y = x^r$ and $y = x^{1/r}$, for $0 \le x \le 1$, where $r > 0$.

II

21. Suppose the rod of Example 3 (p. 465) is turned around so that its density is proportional to the distance between x and 3 (for $x \in [0, 3]$). Find its center of mass.

22. If a person weighing 40 kg sits on a seesaw 1 meter from the fulcrum and a person weighing 30 kg sits on the other side at a distance of 1.5 meters from the fulcrum, where must a dog weighing 10 kg sit to make the seesaw balance?

23. Find the center of mass of a rod of length L whose density is proportional to the square of the distance from one of the ends.

24. (i) Find the center of mass of a rod of length L whose density is proportional to the cube of the distance from one of the ends.

    (ii) Find the center of mass of a rod of length L whose density is proportional to the sixth power of the distance from one of the ends.

*25. From Example 3 we know that if the density of a thin rod between $x = 0$ and $x = 3$ is $\rho(x) = x$, then the center of mass is $\bar{x} = 2$.

    (i) Suppose the density is changed from x to $\rho(x) = c + x$, where c is a positive constant. Do you expect the center of mass to stay in the same place, to move to the left (i.e., $\bar{x} < 2$), or to move to the right (i.e., $\bar{x} > 2$)?

    (ii) Find $\bar{x}$ when $c = 3$. Is your answer consistent with your answer to part (i)?

    (iii) If the constant c in part (i) is allowed to increase, what effect do you expect it to have on the position of the center of mass?

   (iv) Find $\bar{x}$ when $c = 4$ and when $c = 6$. Is your answer consistent with your answer to part (iii)?

*26. Verify Formula 6.7.8 (p. 471) for the centroid of a triangle. (*Hint:* Place the triangle with one vertex at the origin and its base along the x-axis. Calculate M_x and M_y by integrating.)

*27. Find the centroid of the region illustrated in Figure 6.7.34.

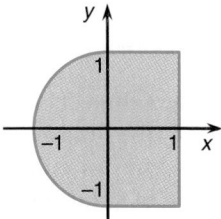

Figure 6.7.34

*28. Find the centroid of the region illustrated in Figure 6.7.35.

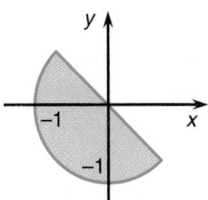

Figure 6.7.35

*29. Find the centroid of the region that consists of all points lying above the x-axis, below the semicircle $y = \sqrt{4 - x^2}$, and above the semicircle $y = \sqrt{1 - x^2}$.

*30. Find the centroid of the region that consists of the region inside the circle of radius 2 with center at the origin *and* the region inside the circle of radius 1 with center at the point $(5, 5)$.

*31. Find the centroid of the region that consists of the union of the region inside the circle of radius 1 with center at the origin, the region inside the circle of radius 2 with center at the point $(0, 8)$, and the rectangular region between the lines $y = 0$, $y = 6$, $x = 1$, and $x = -1$.

 32. (i) Find the centroid of the region enclosed between the curve $y = x^3 - 3x$ and the x-axis.
    (ii) Find the centroid of the region enclosed between the curve $y = x^3 - 3x$ and the x-axis and to the right of the y-axis.

*33. Find the centroid of the region that consists of all points lying inside the circle $x^2 + y^2 = 9$ and outside the square whose corners are $(1, 1)$, $(1, -1)$, $(-1, 1)$, and $(-1, -1)$.

*34. Find the centroid of the region that consists of all points lying inside the circle $x^2 + y^2 = 9$ and outside

the square whose corners are $(1, 1)$, $(1, 0)$, $(0, 1)$, and $(0, 0)$.

*35. Find the centroid of the parallelogram whose vertices are $(0, 0)$, $(3, 0)$, $(1, 2)$, and $(4, 2)$.

*36. Find the centroid of the parallelogram whose vertices are $(0, 0)$, $(a, 0)$, (b, c), and $(a + b, c)$.

 37. Find the centroid of the region enclosed between the curve $y = x^3$ and the line $y = x$.

 38. Find the centroid of the region enclosed between the curves $y = x^2$ and $y = -x^2 + 2x + 4$.

*39. Find the centroid of the region consisting of two disks, one of radius 1 and the other of radius 2, that touch at exactly one point.

*40. Find the centroid of the region consisting of two disks, one of radius 1 and the other of radius r, that touch at exactly one point.

*41. Find the center of mass of the region consisting of two disks touching at a single point, if both disks have the same radius but the density is $\rho = 1$ on one of them and $\rho = 2$ on the other.

*42. Consider a disk of radius 2 with uniform density and place on top of it a disk of radius 1 with the same density so that the smaller disk just touches the outside edge of the larger one at a single point. Find the center of mass of the resulting object. (It might help to visualize two disks made of sheet metal, one welded onto the other.)

*43. Consider the square enclosed between the axes and the lines $x = 2$ and $y = 2$. Let the density function ρ equal r on the smaller square enclosed between the axes and the lines $x = 1$ and $y = 1$, and let ρ equal 1 on the rest of the big square. Find the center of mass of this region if $r = 2$. See Figure 6.7.36.

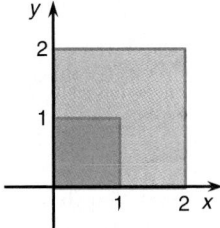

Figure 6.7.36

*44. (i) In the situation of Exercise 43, what would the number r have to be so that the center of mass would be the point $\left(\frac{3}{4}, \frac{3}{4}\right)$?
    (ii) What happens to the center of mass as $r \to \infty$?

 45. Find the centroid of the quadrilateral whose vertices are $(0, 0)$, $(3, 0)$, $(2, 1)$, and $(1, 1)$.

*46. Consider the figure obtained by taking the three points $(-2, 1)$, $(0, 2)$, and $(4, 4)$, joining each pair of these points with a line segment, and looking at the set of all points either lying on one of the segments or enclosed by them.

(i) Find the centroid of the region described above.

(ii) Find the area of the region described above.

47. Find the volume of the solid obtained by revolving the disk of Example 12 around the y-axis.

48. Find the volume of the doughnut obtained by revolving the disk $(x - a)^2 + (y - b)^2 \leq R^2$ around the x-axis. (Assume $0 < R < b$.)

49. Find the volume of the solid obtained by revolving the upper half of the disk in Example 12 around

(i) the y-axis, (ii) the x-axis.

50. Find the volume of the solid obtained by revolving around the x-axis the triangle with vertices $(1, 2)$, $(2, 1)$, and $(3, 3)$.

6.8 Fluid Pressure and Force

DENSITY OF A FLUID

PASCAL'S PRINCIPLE

FLUID FORCE AND CENTROIDS

BUOYANCY

This section is devoted to a discussion of the pressure exerted by a fluid. The fundamental idea is to use integration to "add up" the forces exerted by the fluid at various points on the surface of a submerged object. Later on, it will become clear that many of the calculations we do can be simplified by using the notion of centroid, which was introduced in Section 6.7.

If we imagine an object submerged in a liquid, it seems reasonable to expect that the pressure will depend on the depth; the deeper the object is, the greater will be the force exerted on it by the liquid.

Now suppose we want to find the total force exerted on the object. The tricky thing is that the pressure is not constant. It will be greater at the bottom of the object than at the top, since the bottom is deeper in the liquid. To find the total force, we have to "add up" the force at each point. Since the pressure varies, what is involved here is really an integral rather than a sum.

Figure 6.8.1

Consider a square flat plate whose area is 1 sq ft submerged in water at a depth h ft, lying horizontally, as shown in Figure 6.8.1. What is the force with which the water presses on the top surface of the plate? One reasonable way to approach this question would be to consider the water lying directly above the plate (see Figure 6.8.2). It fills a rectangular solid region of height h and of horizontal cross-sectional area 1 sq ft. The volume of this solid is $h \times 1 \times 1 = h$ ft^3. The density of water is 62.4 lb/ft^3, so the weight of the water over the plate is $62.4h$ lb.

Figure 6.8.2

Whatever the fluid, it is conventional to write ρ, the Greek letter "rho," for its density (i.e., how much the fluid weighs per unit volume). So the weight of the fluid lying directly over the plate is $h\rho$ (h and ρ must be in compatible units, such as ft and lb/ft^3 as above, kg and kg/m^3, in and lb/in^3, etc., and we have assumed that the area of the top of the plate is one square unit). What this says is that the force exerted on the top of a horizontal flat plate of surface area 1 square unit is $h\rho$, where h is its depth and ρ is the density of the fluid.

In fact, there is nothing about this argument that makes it necessary to consider a square plate, or one whose area is 1 square unit. If a flat plate of area A sq ft is lying horizontally in water at a depth of h ft, the same argument shows that the water presses on it with a force of $62.4Ah$ lb (see Figure 6.8.3).

Figure 6.8.3

Another way to say this is that the water presses on the plate with a **pressure** of $62.4h$ lb/sq ft. This is what pressure means, the force per unit area. The advantage of speaking in terms of pressure is that it is not necessary to mention the area of the plate.

Pressures are also frequently measured in *pounds per square inch*, abbreviated *psi*. Since there are $12 \times 12 = 144$ square inches in 1 square foot, we see that the pressure exerted by water on a horizontal flat plate at a depth of h ft is $\frac{62.4h}{144} \approx 0.433h$ psi.

In the metric system we find that the water lying over a horizontal flat plate of area 1 m² at a depth of h m has volume $1 \times h = h$ m³. Since a cubic meter of water weighs 1000 kg, the pressure on a horizontal plate at a depth of h m is $1000h$ kg/m² $= 9800h$ N/m².

> Regardless of the units being used, as long as they are compatible, the pressure on a horizontal plate at depth h in a fluid of density ρ is $h\rho$.

EXAMPLE 1   (i) Find the pressure exerted by water at a depth of 100 ft. (ii) If the top surface area of a large flounder is 2 sq ft, what is the total force exerted on the top of the flounder by the water at a depth of 100 ft?

Solution

(i) In units of pounds per square foot the pressure of the water at a depth of $h = 100$ ft is $62.4h = 6240$ lb/sq ft.

(ii) If the water pressure is 6240 lb/sq ft, then on an area of 2 sq ft the water exerts a force of $2 \times 6240 = 12{,}480$ lb, or a little over 6 tons.

So far we have concentrated on flat horizontal objects (even going so far as to consider a flat fish like the flounder). The remarkable thing about pressure in a fluid is that it is the same in all directions. That is, at a depth of 100 ft in water, as in Example 1, the water exerts downward pressure of 6240 lb/ft² on a horizontal plate. If we ask what the pressure is on the underside of the plate, we realize that it must be *upward* pressure, and it is not clear why the water should be pressing up at all.

However, suppose the pressure on the bottom of the plate was, say, 5000 lb/ft². Then a plate of area 1 ft² would experience downward force of 6240 lb and upward force of 5000 lb. The net force would be 1240 lb downward, which would push the plate downward very hard (and very fast). The fact is that water pressure does not push a plate down or up (or sideways). The only way this can occur is for the pressure to be the same in all directions (up, down, north, south, etc.). See Figure 6.8.4. This fundamental fact is known as Pascal's Principle, after the French mathematician and physicist Blaise Pascal (1623–1662).

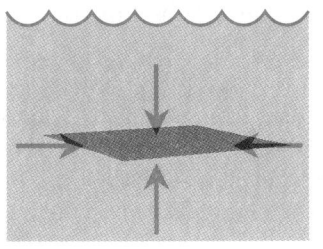

Figure 6.8.4

THEOREM 6.8.1

> **Pascal's Principle**
>
> Fluid pressure is the same in all directions. At a depth of h in a fluid of density ρ the pressure equals $h\rho$.

This will allow us to calculate the force exerted by water pressure on objects that are not necessarily flat and horizontal. The new subtlety that arises here is that since the pressure depends on the depth, the pressure on an object is not uniform but will be different at different places on its surface.

EXAMPLE 2   Consider a square flat plate, each side 2 ft long, suspended vertically in water. If the top edge is horizontal and at a depth of 20 ft, what is the total force exerted on one side of the plate by the water?

Figure 6.8.5

Figure 6.8.6

Figure 6.8.7

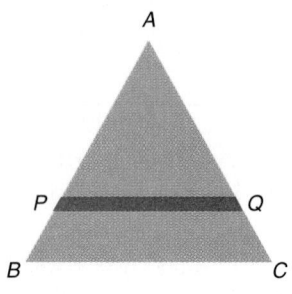

Figure 6.8.8

Solution   We know that the pressure at a depth h is $62.4h$ lb/ft^2. The plate lies between the depths of 20 ft at its top and 22 ft at its lower edge. Consider the depth h as a variable, ranging from $h = 20$ to $h = 22$.

Now at any particular depth h, imagine a narrow strip on the surface of the plate of height dh as shown in Figure 6.8.5. Its width is the width of the plate, or 2 ft. The area of this strip is $2\,dh$, and the pressure on it (at depth h) is $62.4h$ lb/ft^2. Accordingly, the force on it is *(pressure)* $\times$ *(area)* $= 62.4h(2)\,dh = 124.8h\,dh$.

To find the total force, it is necessary to add up the forces on all such little strips as h ranges from $h = 20$ to $h = 22$. As usual, "adding up" means integrating. The total force on the plate is

$$\int_{20}^{22} 124.8h\,dh = 124.8 \int_{20}^{22} h\,dh = 124.8\left(\frac{1}{2}h^2\right)\Bigg|_{20}^{22} = 62.4\left(22^2 - 20^2\right)$$
$$\approx 5240 \text{ lb.}$$

EXAMPLE 3   Consider a triangular plate suspended vertically in water as shown in Figure 6.8.6. The triangle is equilateral with edges 4 ft long. Its bottom edge is horizontal and submerged to a depth of 30 ft. Find the total force exerted on one side of the plate by the water.

Solution   Let h be a variable representing the depth of water. First we should figure out the range of values for h on the plate. To find the height of the plate, that is, the vertical distance from the base to the top, we use the Pythagorean Theorem (Theorem 1.5.1). In one of the little right-angled triangles that make up half the large equilateral triangle, the hypotenuse is 4 ft, the length of one of the sides (see Figure 6.8.7). The horizontal side at the bottom is 2 ft, so the length of the vertical side is $\sqrt{4^2 - 2^2} = \sqrt{12} = 2\sqrt{3}$.

Since the bottom of the triangle is at a depth of 30 ft, the top is at a depth of $30 - 2\sqrt{3}$ ft. The variable h, measuring depth in feet, ranges from $h = 30 - 2\sqrt{3}$ to $h = 30$.

Next consider a small horizontal strip on the triangular plate, of height dh. To find its area, we must find its width, which varies as the strip moves down the height of the triangle.

We draw the triangle in Figure 6.8.8, labeling its vertices A, B, and C and the ends of the strip P and Q. Then the triangle APQ is similar to the triangle ABC. The height of ABC is $2\sqrt{3}$, and if h measures the depth at which the strip PQ lies, then the height of APQ is $h - (30 - 2\sqrt{3})$, since the top of the triangle, A, is at a depth of $30 - 2\sqrt{3}$. Since the length of BC is 4, we can use similar triangles to conclude that the length of PQ is $\frac{(height\ of\ APQ)}{(height\ of\ ABC)} \times 4 = \frac{h-(30-2\sqrt{3})}{2\sqrt{3}}(4) = 2\frac{h-(30-2\sqrt{3})}{\sqrt{3}}$.

The area of the strip is therefore *(length PQ)* $dh = 2\frac{h-(30-2\sqrt{3})}{\sqrt{3}}\,dh$. Since the pressure at a depth of h ft is $h\rho = 62.4h$ lb/ft^2, the force on the strip is *(area)* $\times$ *(pressure)* $= 2\frac{h-(30-2\sqrt{3})}{\sqrt{3}}\,dh \times 62.4h = 124.8h\frac{h-(30-2\sqrt{3})}{\sqrt{3}}\,dh$.

The total force on the triangle is obtained by integrating. It is

$$\int_{30-2\sqrt{3}}^{30} 124.8h\frac{h - (30 - 2\sqrt{3})}{\sqrt{3}}\,dh = \frac{124.8}{\sqrt{3}} \int_{30-2\sqrt{3}}^{30} \left(h^2 - (30 - 2\sqrt{3})h\right)dh$$

$$= \frac{124.8}{\sqrt{3}}\left(\frac{h^3}{3} - (30 - 2\sqrt{3})\frac{h^2}{2}\right)\Bigg|_{30-2\sqrt{3}}^{30}$$

$$\approx 12{,}500 \text{ lb.}$$

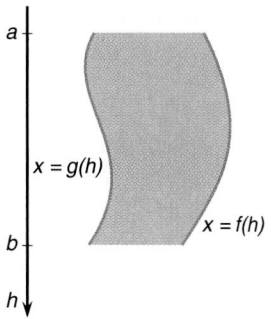

Figure 6.8.9

Before proceeding, we should pause and analyze the calculations we are doing. The variable in the vertical direction is h. It is "upside-down," in the sense that large values of the depth h refer to points far *down* in the fluid. Let us continue to use x for the variable in the horizontal direction.

Consider a plate suspended vertically in the fluid. Suppose it can be described as lying between two curves $x = f(h)$ and $x = g(h)$, for values of h with $a \leq h \leq b$. Also suppose that $f(h) \geq g(h)$ for all $h \in [a, b]$. See Figure 6.8.9.

Then the width of the narrow horizontal strip at depth h is $f(h) - g(h)$, so the area of the strip is $(f(h) - g(h)) \, dh$. The force on the strip is *(pressure)(area)* $= h(f(h) - g(h)) \rho \, dh$, and we integrate to find the total force $\int_a^b h(f(h) - g(h)) \rho \, dh$. See Figure 6.8.10.

This last integral is *exactly* the same as the integral needed to find the moment M_x of the plate about the x-axis, assuming that the plate has uniform density ρ (cf. Definition 6.7.5, p. 467).

On the other hand, the moment equals $M_x = M\bar{h}$, the product of the total "weight" and the vertical coordinate of the centroid. Moreover, the total "weight" is $M = \int_a^b (f(h) - g(h)) \rho \, dh = \rho \int_a^b (f(h) - g(h)) \, dh$. This is the product of ρ and the area of the plate.

Putting all the pieces together, we see that the total force on the plate equals $M_x = M\bar{h} = A\rho\bar{h}$, where A is the area of the plate. Since $\bar{h}\rho$ is just the pressure at depth $\bar{h}$, by Theorem 6.8.1, and multiplying pressure by area gives the total force, what this says is that the total force on the plate is exactly the same as the force $A\bar{h}\rho$ that would be exerted on the plate if it were lying horizontally at a depth of $\bar{h}$.

We have just established the following extremely useful result.

> The force $\int h(f(h) - g(h)) \rho \, dh$ equals $M_x = M\bar{h} = A\bar{h}\rho = A(\bar{h}\rho)$, and $\bar{h}\rho$ is the pressure at $\bar{h}$.

THEOREM 6.8.2

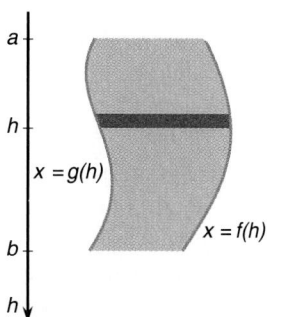

Figure 6.8.10

> The force exerted by a fluid on a flat plate equals the force that would be exerted on the plate if it were suspended horizontally at the depth where its centroid now lies.
>
> If the plate has area A and is suspended with its centroid at depth $\bar{h}$ in a fluid whose density is ρ, then the total force on the plate is $\bar{h}\rho A$.

EXAMPLE 4   Consider a circular plate of radius 6 in suspended vertically in water with its center at a depth of 4 ft. See Figure 6.8.11. What is the force exerted by the water on one side of the plate?

Solution   The centroid of a disk is its center, so the depth of the centroid is $\bar{h} = 4$ ft, the depth of the center of the disk. The radius of the disk is 6 in, but in order to have all the measurements in compatible units we convert this to $r = \frac{1}{2}$ ft.

The area of the disk is $\pi r^2 = \frac{\pi}{4}$. The pressure at a depth of $\bar{h}$ is $\bar{h}\rho = 62.4\bar{h} = 62.4(4) = 249.6$ lb/ft^2, so the total force on the disk equals the force exerted on an area of $\frac{\pi}{4}$ ft^2 by a pressure of 249.6 lb/ft^2. The total force is $249.6\left(\frac{\pi}{4}\right) \approx 196$ lb. ◀

Using Theorem 6.8.2 simplified this example considerably. For purposes of comparison you might like to find the total force by integration, as in Examples 2 and 3.

Figure 6.8.11

EXAMPLE 5   **Underwater Observation Tank**   Find the total force exerted on an underwater observation tank built in the shape of a cube whose sides are all 2 meters long if the bottom of the tank is at a depth of 25 meters. Express the answer in newtons.

Solution   The bottom of the cube is a square of area 4 m² at a depth of 25 m, so the force on it is (*area*)(*pressure*) = 4(25)9800 = 980,000 N. Similarly, the top of the cube is at a depth of 23 m, so the force on it is 4(23)9800 = 901,600 N.

The other sides can be considered separately. Each of them is a square with depth ranging from $h = 23$ to $h = 25$. The centroid, being the center of the square, is at a depth of $\bar{h} = 24$ m. Since the area of the side is 4 m², the total force on it can be found by using Theorem 6.8.2. It is $\bar{h}\rho A = 24(9800)4 = 940,800$ N.

This is the force on each of the four side walls, so the total force on all six walls is $980,000 + 901,600 + 4 \times 940,800 = 5,644,800$ N.

The examples we have considered so far have all involved flat surfaces that are either horizontal or vertical. Naturally, it may be important to deal with curved or sloping surfaces. In fact, curved surfaces will be the subject of the next section, but we can do examples of sloping flat surfaces.

Figure 6.8.12

EXAMPLE 6   **Force in Drainage Ditch**   Consider a drainage trench built so that its cross-section has the shape of a *V*. Suppose the sides of the trench are at an angle of 45° above the horizontal and that they are made of prefabricated concrete slabs, each 10 ft square. See Figure 6.8.12.

Find the force exerted on one of these slabs when the trench is full of water.

Solution   First we find the depth of the water. The surface of the slab forms the hypotenuse of a right-angled triangle, in which the other angles are both 45°. Accordingly, the depth of the water, which is the vertical side of the triangle, is $10\sin(45°) = \frac{10}{\sqrt{2}} \approx 7.07$ ft (see Figure 6.8.13).

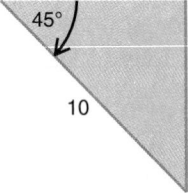

Figure 6.8.13

The centroid, which is just the center of the rectangular slab, is halfway down, so it is at a depth of $\bar{h} = \frac{1}{2}\frac{10}{\sqrt{2}} = \frac{5}{\sqrt{2}} \approx 3.54$ ft. The pressure on the slab is $\bar{h}\rho A = \frac{5}{\sqrt{2}}(62.4)100 \approx 22,060$ lb, or slightly over 11 tons.

It is also possible to solve this problem directly by integrating, as we now demonstrate, partly to show how much easier it is to use centroids and partly to illustrate the approach needed for objects that are tilted.

We let h be a variable representing the depth. It runs from $h = 0$ to $h = \frac{10}{\sqrt{2}}$. Now consider what happens as h changes by a small amount dh. There is a corresponding little strip on the sloping wall of the trench. However, the sloping height of the strip on the wall does not equal dh, but rather $\frac{dh}{\sin(45°)} = \sqrt{2}\,dh$. See Figure 6.8.14.

Figure 6.8.14

Since the slab of concrete is a square, the width of the strip always equals 10 ft, so the area of the strip is $10\sqrt{2}\,dh$ ft². The pressure at depth h is $62.4h$ lb/ft², so the total force on the slab is

$$\int_0^{10/\sqrt{2}} 62.4h \times 10\sqrt{2}\,dh = 624\sqrt{2}\int_0^{10/\sqrt{2}} h\,dh = 624\sqrt{2}\left.\frac{h^2}{2}\right|_0^{10/\sqrt{2}}$$

$$\approx 22,060 \text{ lb.}$$

Buoyancy

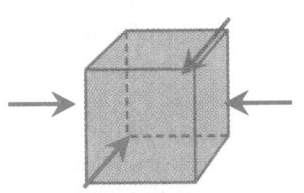

Figure 6.8.15

Consider a cube, whose sides are each k units long, submerged in a fluid. The force on each of the vertical sides is the same, so the forces on pairs of opposite sides cancel, and there is no *net* horizontal force. See Figure 6.8.15.

However, the top and the bottom are at different depths, so the forces on them are not the same. In fact, suppose the top is at a depth h; the bottom is then at depth $h + k$, since each side has length k. The pressure on the top is $h\rho$, and the pressure

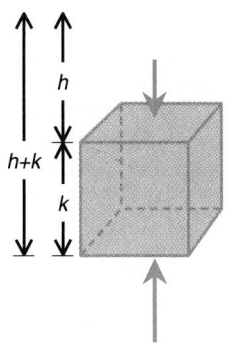

Figure 6.8.16

on the bottom is $(h + k)\rho$. Since each has area k^2, the total force on the top is $hk^2\rho$ *downward*, and the total force on the bottom is $(h + k)k^2\rho$ *upward*. See Figure 6.8.16.

The difference between the force on the bottom and the force on the top is $(h + k)k^2\rho - hk^2\rho = k^3\rho$. Since k^3 is exactly the volume of the cube and ρ is the density of the fluid, the quantity $k^3\rho$ is just the weight of the fluid it would take to fill the cube.

The cube experiences a net upward force equal to the weight of the fluid it would take to fill it up, that is, the amount of fluid "displaced" by the cube.

In fact, the same is true of any submerged object, and the upward force is known as its **buoyancy**. If the object weighs less than the fluid it displaces, then the buoyancy is greater than the weight, and the object floats. If it weighs more than the displaced fluid, it sinks.

This is the famous discovery about which Archimedes is reputed to have shouted "Eureka!" as he jumped out of his bathtub.

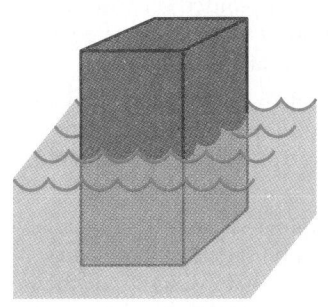

Figure 6.8.17

EXAMPLE 7   **Floating Block of Wood**   Suppose a rectangular block of pine measuring 1 ft $\times$ 1 ft $\times$ 2 ft is floating in water with its long side vertical. If the density of the wood is 30 lb/ft^3, how much of the block will be above the surface?

Solution   If the block is floating, the buoyant force due to the displaced water must exactly equal the weight of the block. Since the volume of the block is 2 ft^3 and its density is 30 lb/ft^3, the block weighs $2 \times 30 = 60$ lb.

The amount of water displaced must weigh 60 lb to balance this weight. Now if the height of the block above the surface is t ft, then the height of the submerged part is $2 - t$, and the volume of the submerged part is $1 \times 1 \times (2 - t) = 2 - t$ ft^3. For this to weigh 60 lb we must have $60 = 62.4 \times (2 - t) = 124.8 - 62.4t$, or $62.4t = 124.8 - 60 = 64.8$, so $t = \frac{64.8}{62.4} \approx 1.04$ ft.

The height to which the block protrudes above the surface is approximately 1.04 ft. See Figure 6.8.17.

Exercises 6.8

I

Each of the flat plates described below is submerged in water. Find the total force exerted by the water on one side of the plate.

1. A square plate, 3 ft on each side, lying horizontally at a depth of 11 ft.

2. A square plate, 2 m on each side, lying horizontally at a depth of 3 m.

3. A square plate, 5 in on each side, lying horizontally at a depth of 6 ft.

4. A square plate, 1 ft on each side, lying horizontally at a depth of 4 m.

5. A square plate, 5 ft on each side, lying vertically with its bottom edge (horizontal) at a depth of 20 ft.

6. A square plate, 3 m on each side, lying vertically with its bottom edge at a depth of 10 m.

7. A square plate, 15 in on each side, lying vertically with its bottom edge at a depth of 8 ft.

8. A square plate, 25 in on each side, lying vertically with its bottom edge at a depth of 1 ft.

9. A round plate, with radius 3 ft, lying vertically with its center at a depth of 10 ft.

10. A round plate, with radius 0.5 m, lying vertically with its center at a depth of 50 m.

Each of the flat plates described below is submerged in water. Find the total force exerted by the water on one side of the plate.

11. A round plate, with radius 30 in, lying vertically with its center at a depth of 20 ft.

12. A round plate, with radius 30 cm, lying vertically with its center at a depth of 4 m.

13. A round plate, with radius 1 ft, lying horizontally at a depth of 26 ft.

14. An elliptical plate, with semiaxes 3 ft and 4 ft, lying with major axis vertical and the lowest point at a depth of 24 ft.

15. An equilateral triangle, with sides 8 ft, lying vertically with its bottom edge horizontal at a depth of 100 ft.

16. An equilateral triangle, with sides 3 ft, lying vertically with its top edge horizontal at a depth of 40 ft.

17. An equilateral triangle, with sides 2 m, lying horizontally at a depth of 14 ft.

18. An equilateral triangle, with sides 20 in, lying vertically with one edge vertical and the opposite vertex at a depth of 40 ft.

19. A right-angled triangle, with base 4 ft long, lying at a depth of 10 ft, and vertical side 3 ft high.

20. A right-angled triangle, with base 6 ft, at a depth of 30 ft and vertical side 6 ft high.

Each of the flat plates described below is submerged in water. Find the total force exerted by the water on one side of the plate.

21. A square plate, 3 ft on each side, lying at a 45° angle with its bottom edge at a depth of 10 ft.

22. A square plate, 2 ft on each side, lying at a 30° angle above the horizontal with its bottom edge at a depth of 10 ft.

23. A round plate, with radius 3 ft, lying at an angle of 45° above the horizontal with its center at a depth of 10 ft.

24. A round plate, with radius 4 m, lying at an angle of 60° above the horizontal with its lowest point at a depth of 30 m.

25. A round plate, with radius 6 ft, lying at an angle of 45° above the horizontal with its center at the surface of the water.

26. A square plate, with sides 4 ft long, lying at an angle of 30° above the horizontal with its bottom edge at a depth of 1 ft.

27. An equilateral triangle, with sides 2 ft, lying at an angle of 45° above the horizontal with its bottom edge at a depth of 100 ft.

28. An equilateral triangle, with sides 5 m, lying at an angle of 30° above the horizontal with its bottom edge at a depth of 10 m.

29. An equilateral triangle, with sides 3 ft, lying at an angle of 60° above the horizontal with its top edge at a depth of 15 ft.

30. A right-angled triangle, with base 4 ft long, lying at a depth of 20 ft, and another side 4 ft long, lying at a 60° angle above the horizontal.

II

31. Find the pressure exerted by the water on one side of a square plate whose sides are all 4 ft long. The plate is lying vertically with each of its sides at an angle of 45° above the horizontal so that the lowest point on the plate is at a depth of 40 ft. See Figure 6.8.18.

40 ft

Figure 6.8.18

32. What is the total force on one side of a square plate of area 1 sq ft if the plate is suspended vertically in a liquid whose density is 50 lb/ft^3, in such a way that its bottom edge lies horizontally at a depth of 3 ft?

*33. Suppose the slab of Example 6 (p. 480) was at an angle of 30° above the horizontal. What would be the total force exerted on it by water that just reaches up to the top of the slab?

*34. Suppose the slab of Example 6 was at an angle of 60° above the horizontal and submerged so that its lower edge was at a depth of 50 ft. What would be the total force exerted on one side of it by the water?

*35. In Example 4 (p. 479) we found the total force by locating the centroid. Do this example directly by evaluating an integral, without using centroids at all.

*36. In Example 5 (p. 479) we found the total force by locating the centroid. Do this example directly by integrating, without using centroids at all.

37. (i) A barge is built in the shape of a rectangular box, 20 ft wide, 40 ft long, and 8 ft high. If the barge weighs 60 tons, how much water will it draw (i.e., to what depth will it sink)?

(ii) If the barge is loaded with a cargo weighing 100 tons, how much water will it draw?

(iii) How heavy a cargo can the barge hold without sinking?

*38. Suppose a block in the shape of a cube is floating in water with its bottom horizontal. If the block is made of material whose density is ρ, figure out a formula for the fraction of the block that is below the surface.

*39. People swimming in seawater find that they float higher in the water, that is, with less of their bodies submerged, than when they swim in fresh water. What does this say about the densities of fresh water and salt water?

People swimming in the Dead Sea float even higher in the water. What does this say about the density of Dead Sea water?

*40. If the density of seawater is 64.3 lb/ft^3 and the density of ice is 57.5 lb/ft^3, what fraction of an iceberg protrudes above the surface of the ocean?

*41. Consider a trough that is 10 ft long, 2 ft high, and 1 ft wide at the base and 3 ft wide at the top, as shown in Figure 6.8.19. The ends of the trough are made of flat sheet metal; they are vertical.

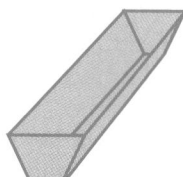

Figure 6.8.19

(i) Suppose the trough is full of water. What is the force exerted by the water on the end of the trough?

(ii) What is the force on the end of the trough if there is water in the trough to a depth of 1 ft?

**42. This problem should be attempted only if you have read Section 6.6 on work.

A hollow cylinder with open ends and radius 3 in stands vertically with its bottom under water and its top above the surface. A piston fits inside the cylinder, moving freely but making a watertight seal (see Figure 6.8.20). How much work is done in pushing the piston down to a depth of 4 ft? (*Hint:* The upward force on the piston comes from the water pressure on the piston head.)

Figure 6.8.20

*43. Consider the cube of Example 5. Show that the magnitude of the net upward buoyant force exactly equals the force due to gravity on the amount of water that it would take to fill the cube.

*44. Find the total force exerted by the water on one side of a round plate, with radius 30 in, lying vertically with its center at a depth of 2 ft.

POINT TO PONDER

When we figured out the pressure at a depth h, we took a flat plate lying horizontally and considered the fluid lying directly over it. This gave us the correct answer, but if we think about it a little more, we may begin to have doubts.

For instance, it is possible that fluid not directly above the plate might still have some effect on it. More troubling is the possibility that there might not be that much fluid lying over the plate, owing to the shape of the container.

As shown in Figure 6.8.21(i), the container could have a narrow neck, in which case there is only a thin column of fluid over the plate. It could even be (as in Figure 6.8.21(ii)) that the plate is lying in a part of the container that is covered so it does not even reach to the

Figure 6.8.21(i)

Figure 6.8.21(ii)

surface level. Could it be that the pressure is different in these situations?

In fact, our original argument does give the correct value, despite these objections. Describe how each of the apparent objections can be answered.

6.9 Arclength; Areas of Surfaces of Revolution

SURFACE OF REVOLUTION
SURFACE AREA

Figure 6.9.1

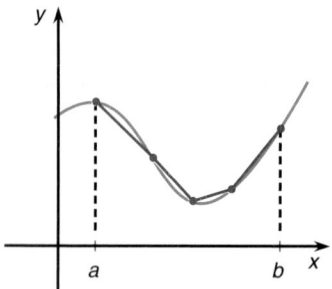

Figure 6.9.2

If we consider a curve drawn in the plane, we might want to find its length. In this section we will discuss this question, beginning by asking what we mean by the length of a curve. We are accustomed to the idea that a circle of radius r has circumference (i.e., length) equal to $2\pi r$; we have used this fact in discussing trigonometric functions in Section 2.6 and for finding volumes of solids of revolution in Section 6.4. However, if we think about it carefully, it becomes less clear what this familiar fact means. In this section we learn how to define and calculate the length of a curve using integrals. Having done this, we will use what we have learned to ask a related question, namely, what do we mean by the surface area of a curved surface, and how do we calculate it?

Consider the graph of a function $y = f(x)$, shown in Figure 6.9.1. We might want to know the length of the part of the curve that lies between $x = a$ and $x = b$. For example, we might like to know the circumference of a circle (strictly speaking, it would be easier to deal with a semicircle $y = \sqrt{r^2 - x^2}$). It could also be that the curve represents the path of a moving object and what we want to know is the total distance it travels.

Based on our experience with similar questions, a reasonable approach to try would be to *approximate* the curve with a number of short straight line segments as in Figure 6.9.2. The sum of their lengths ought to approximate what we should mean by the length of the curve. If we use more and more segments, the sums of their lengths might approach a limit, which should be the length of the curve.

Suppose $\mathcal{P} = \{x_0, \ldots, x_n\}$ is a partition of $[a, b]$. The corresponding points on the curve are $\big(x_0, f(x_0)\big), \big(x_1, f(x_1)\big), \ldots, \big(x_n, f(x_n)\big)$. The kth segment joining consecutive points is the segment from $\big(x_{k-1}, f(x_{k-1})\big)$ to $\big(x_k, f(x_k)\big)$. Its length is

$$\sqrt{(x_k - x_{k-1})^2 + \big(f(x_k) - f(x_{k-1})\big)^2} = \sqrt{(\Delta x_k)^2 + \big(f(x_k) - f(x_{k-1})\big)^2}.$$

The Mean Value Theorem (Theorem 4.2.5, p. 241), applied to the function $f(x)$ on the interval $[x_{k-1}, x_k]$, says that there is some $x_k^* \in [x_{k-1}, x_k]$ so that

$$f(x_k) - f(x_{k-1}) = f'(x_k^*)(x_k - x_{k-1}) = f'(x_k^*)\Delta x_k.$$

Using this, we find that the length of the kth segment is

$$\sqrt{(\Delta x_k)^2 + \big(f'(x_k^*)\Delta x_k\big)^2} = \sqrt{\big(1 + f'(x_k^*)^2\big)\Delta x_k^2} = \sqrt{1 + f'(x_k^*)^2}\Delta x_k.$$

We approximate the length of the curve by the sum of the lengths of all the segments, that is,

$$length \approx \sum_{k=1}^{n} \sqrt{1 + f'(x_k^*)^2}\Delta x_k.$$

As $n \to \infty$, we are considering more and more segments, and if $\|\mathcal{P}\| \to 0$, then it seems reasonable to expect that the segments will all become shorter and shorter and the sum of their lengths will approach the length of the curve. As $n \to \infty$, the sum approaches the integral $\int_a^b \sqrt{1 + f'(x)^2}dx$.

Rather than worrying any more about what is meant by the "length of a curve," we can simply use this integral to define the length, knowing that the argument we have just made makes it seem reasonable.

DEFINITION 6.9.1

Suppose $f(x)$ is a differentiable function on the interval $[a, b]$ and $f'(x)$ is continuous on $[a, b]$. The length of the curve $y = f(x)$ for $x \in [a, b]$, sometimes called the **arclength**, is defined to be

$$s = \int_a^b \sqrt{1 + f'(x)^2}\,dx.$$

It is traditional to use the letter s to represent arclength, and it is also common to encapsulate this definition by writing

$$ds = \sqrt{(dx)^2 + (dy)^2} = \sqrt{1 + \left(\frac{dy}{dx}\right)^2}\,dx.$$

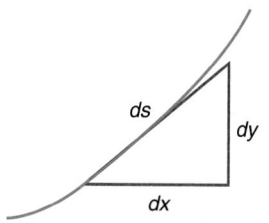

Figure 6.9.3

It may help to remember this last formula if we consider Figure 6.9.3, in which ds, the differential of the arclength, is the hypotenuse of the triangle whose base is dx and whose height is dy.

EXAMPLE 1   Find the length of the curve $y = x^{3/2}$ for x between $x = 1$ and $x = 2$.

Solution   Applying the definition with $f(x) = x^{3/2}$, we find $f'(x) = \frac{3}{2}x^{1/2}$, and the arclength of the curve equals

$$s = \int_1^2 \sqrt{1 + \left(\frac{3}{2}x^{1/2}\right)^2}\,dx = \int_1^2 \sqrt{1 + \frac{9}{4}x}\,dx$$

$$\boxed{\begin{aligned} u &= 1 + \frac{9}{4}x \\ du &= \frac{9}{4}\,dx \end{aligned}} \qquad = \int_{13/4}^{11/2} u^{1/2}\frac{4}{9}\,du$$

$$= \frac{4}{9}\frac{2}{3}u^{3/2}\Big|_{13/4}^{11/2}$$

$$\approx 2.086.$$

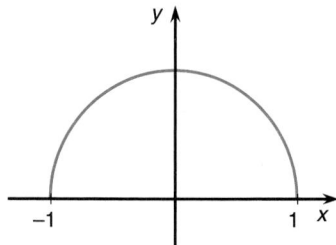

Figure 6.9.4

EXAMPLE 2   **Circumference of a Circle**   Find the circumference of a circle of radius 1.

Solution   It is easier to find the arclength of the upper semicircle and double it. We consider the function $f(x) = \sqrt{1 - x^2}$, for $x \in [-1, 1]$, whose graph is the upper semicircle (see Figure 6.9.4). We find $f'(x) = \frac{1}{2}(1 - x^2)^{-1/2}(-2x) = -\frac{x}{\sqrt{1-x^2}}$.
The length of the semicircle is

$$s = \int_{-1}^1 \sqrt{1 + f'(x)^2}\,dx = \int_{-1}^1 \sqrt{1 + \left(\frac{x^2}{1 - x^2}\right)}\,dx = \int_{-1}^1 \sqrt{\frac{(1 - x^2) + x^2}{1 - x^2}}\,dx$$

$$= \int_{-1}^1 \sqrt{\frac{1}{1 - x^2}}\,dx = \int_{-1}^1 \frac{1}{\sqrt{1 - x^2}}\,dx = \arcsin x\Big|_{-1}^1 = \frac{\pi}{2} - \left(-\frac{\pi}{2}\right) = \pi.$$

The length of the semicircle is π, so the length of the circle of radius 1 is 2π.

With a programmable calculator we could try approximating the arclength by dividing the interval $[-1, 1]$ into n equal subintervals and adding up the distances between adjacent pairs of points. . .

TI-85	SHARP EL9300
PRGM F2(EDIT) ARCL ENTER F3 F1(Input) N EXIT ENTER 0 STO K ENTER 0 STO S ENTER Lbl B ENTER K + 1 STO K ENTER S + $\sqrt{}$ ((2 ÷ N) ^ 2 + ($\sqrt{}$ (1 − ((−) 1 + 2 K ÷ N) x^2) − $\sqrt{}$ (1 − ((−) 1 + 2 (K − 1) ÷ N) x^2)) x^2) STO S ENTER F4 F1(If) (K TEST F2(<) N) ENTER Goto B ENTER S ENTER EXIT EXIT EXIT PRGM F1(NAME) ARCL ENTER 20 ENTER	▼ ▼ (NEW) ENTER ENTER ARCL ENTER COMMAND A 3 (Input) N ENTER K = 0 ENTER S = 0 ENTER COMMAND B 1 (Label) 1 ENTER K = K + 1 ENTER A = $\sqrt{}$ 1 − ((−) 1 + 2 K ÷ N) x^2 ENTER B = $\sqrt{}$ 1 − ((−) 1 + 2 (K − 1) ÷ N) x^2 ENTER S = S + $\sqrt{}$ (2 ÷ N) x^2 + (A − B) x^2) ENTER COMMAND 3 (If) K COMMAND C 2 (<) N COMMAND B 2 (Goto) 1 ENTER COMMAND A 1 (Print) S ENTER COMMAND A 6 (End) ENTER ⊞ ▧ (RUN) ENTER (ARCL) ENTER 20 ENTER
CASIO f_x-6300G (f_x-7700GB)	HP 48SX
MODE 2 EXE ? → N : 0 → K : 0 → S : Lbl 1 : K + 1 → K : S + $\sqrt{}$ ((2 ÷ N) x^2 + ($\sqrt{}$ (1 − ((−) 1 + 2 K ÷ N) x^2) − $\sqrt{}$ (1 − ((−) 1 + 2 (K − 1) ÷ N) x^2)) x^2) → S : K < N ⇒ Goto 1 : S ◢ MODE 1 Prog 0 EXE 20 EXE (on the f_x-7700GB, : is PRGM F6, and Prog is PRGM F3.)	ATTN ↱ CLR ≪≫ "" KEY IN N ▶ SPC PROMPT SPC ′ N ▶ STO 0 SPC 1 SPC N SPC FOR K SPC 2 SPC N ÷ x^2 1 +/− SPC 2 SPC K × N ÷ + x^2 +/− 1 + $\sqrt{x}$ 1 +/− SPC 2 SPC K SPC 1 − × N ÷ + x^2 +/− 1 + $\sqrt{x}$ − x^2 + $\sqrt{x}$ + NEXT ▶ ▶ ENTER ′ ARCL STO VAR ARCL 20 ENTER ↰ CONT

```
?20
ANSWER=
        3.132264619
```

It is interesting to see how close the approximation is to π for various values of n.

The unpleasant thing about calculating arclength is that the integrals that result are often difficult to evaluate. For instance, if we try to find the length of some piece of the parabola $y = x^2$, we have to evaluate the integral $\int_a^b \sqrt{1 + 4x^2}\, dx$, which we do not know how to do.

After we have learned some additional techniques for finding integrals, which is the subject of Chapter 8, we will be able to evaluate the arclengths of more curves.

Figure 6.9.5(i)

Figure 6.9.5(ii)

Figure 6.9.5(iii)

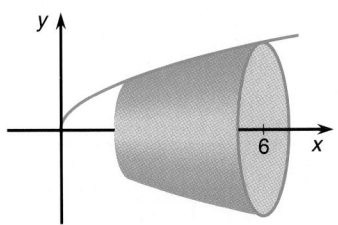

Figure 6.9.6

Surface Areas of Surfaces of Revolution

In many instances it is desirable to find the surface area of some region in space. For example, we might need to know the surface area of a sphere in order to figure out the amount of paint needed to cover a water tank; the surface area of an ellipsoid might be needed to estimate the rate at which a one-celled organism can absorb oxygen through its outer membrane; the surface area of a cone would tell us how much sheet metal was needed to make a funnel. There are many other examples, but it is interesting to notice that the ones mentioned so far are all **surfaces of revolution**, that is, the type of surface obtained by revolving a curve about a line (or, to put it another way, the outside surface of a solid of revolution). It turns out that areas are reasonably easy to find for this type of surface.

EXAMPLE 3   **Area of a Cone**   Consider the cone shown in Figure 6.9.5(i), whose height is h and whose radius at the bottom is r. Find its surface area.

Solution   To find the area of the cone, slit it open and spread it out flat, as shown in Figures 6.9.5(ii) and 6.9.5(iii). What results will be a sector of a disk. The radius of the disk will be the *slant height* of the cone, that is, the distance from top to bottom *along the sloping side*. By the Pythagorean Theorem (Theorem 1.5.1) this is $R = \sqrt{r^2 + h^2}$.

The circumference of the bottom of the cone is $2\pi r$; the circumference of the flattened disk is $2\pi R = 2\pi\sqrt{r^2 + h^2}$. The sector filled by the flattened cone covers a distance of $2\pi r$ at the outside edge, so the fraction of the whole disk that it covers is $\frac{2\pi r}{2\pi R} = \frac{r}{R} = \frac{r}{\sqrt{r^2+h^2}}$.

Since the area of the whole disk of radius R is πR^2, the area of the sector is $\frac{r}{R}\pi R^2 = \pi r R = \pi r\sqrt{r^2 + h^2}$.

Example 3 was done by using a trick and some simple geometry, but we cannot rely on such methods to work for more complicated curves. After all, the cone is obtained as a surface of revolution by revolving the simplest of all curves, a straight line. For more general surfaces it will be necessary to use integration.

EXAMPLE 4   Consider the surface of revolution obtained by revolving the curve $y = f(x) = \sqrt{x}$ about the x-axis. Find the surface area of the part of this surface that lies between $x = 2$ and $x = 6$ (see Figure 6.9.6).

Solution   We consider a short interval of length dx on the x-axis and look at the corresponding short piece of the curve $y = f(x) = \sqrt{x}$. When it is revolved about the axis, it will sweep out a small band in the surface, as shown in Figure 6.9.7(i).

Figure 6.9.7(i)

Figure 6.9.7(ii)

To find the area of this narrow band, slit it and unwind it into a flat strip, as shown in Figure 6.9.7(ii). The width of the strip is just the length of the little piece of curve that we revolved, which is $ds = \sqrt{1 + f'(x)^2}\,dx$. The length of the strip is the circumference of the band, which is 2π times its radius. Since the radius of the band is just the height of the curve above the x-axis, it is $f(x) = \sqrt{x}$, and the circumference is $2\pi\sqrt{x}$.

This suggests that the area of the band is $2\pi f(x)\,ds = 2\pi\sqrt{x}\sqrt{1 + f'(x)^2}\,dx$. The surface area S of the whole surface of revolution is obtained by integrating:

$$S = \int 2\pi f(x)\,ds = 2\pi \int_2^6 \sqrt{x}\sqrt{1 + f'(x)^2}\,dx = 2\pi \int_2^6 \sqrt{x}\sqrt{1 + \left(\frac{1}{2}x^{-1/2}\right)^2}\,dx$$

$$\boxed{\begin{array}{l} u = x + \dfrac{1}{4} \\[2mm] du = dx \end{array}}$$

$$= 2\pi \int_2^6 \sqrt{x}\sqrt{1 + \frac{1}{4x}}\,dx = 2\pi \int_2^6 \sqrt{x + \frac{1}{4}}\,dx = 2\pi \int_{9/4}^{25/4} \sqrt{u}\,du$$

$$= 2\pi \frac{2}{3}u^{3/2}\Big|_{9/4}^{25/4} = \frac{4}{3}\pi \left(\left(\frac{25}{4}\right)^{3/2} - \left(\frac{9}{4}\right)^{3/2}\right) = \frac{4}{3}\pi \left(\left(\frac{5}{2}\right)^3 - \left(\frac{3}{2}\right)^3\right)$$

$$= \frac{49}{3}\pi \approx 51.3.$$

Figure 6.9.8

There are several things to notice here. One is that it is important not to confuse the arclength s with the surface area, which is traditionally denoted S. Second, even though we allowed x to change by the small amount dx, the piece of curve we revolved changed by ds. The idea is that the piece of curve is likely not horizontal, so it will be longer than the horizontal piece dx, and ds is the corresponding length on the slant, as shown in Figure 6.9.8. The effect of this is that the revolved band has width ds and area $2\pi f(x)\,ds$.

To find the surface area, we integrate: $S = \int 2\pi f(x)\,ds$. Notice that at this point it is not possible to put in the limits of integration because the limits are expressed in terms of the variable x and the variable of integration here is s.

The final step is to rewrite the integral in terms of x and then evaluate.

The technique illustrated in the preceding example can be used for any surface of revolution, and in fact the resulting integral can be taken as the definition of surface area.

DEFINITION 6.9.2

Suppose $f(x)$ is a differentiable function for $x \in [a, b]$, with $f(x) \geq 0$ for all $x \in [a, b]$, and $f'(x)$ continuous on $[a, b]$. If the graph of $y = f(x)$ is revolved about the x-axis, the resulting surface of revolution has **surface area** given by the following formula:

$$S = \int 2\pi f(x)\,ds = 2\pi \int_a^b f(x)\sqrt{1 + f'(x)^2}\,dx.$$

If $f(x)$ is not necessarily nonnegative on $[a, b]$, the formula must be modified to

$$S = \int 2\pi |f(x)|\,ds = 2\pi \int_a^b |f(x)|\sqrt{1 + f'(x)^2}\,dx.$$

EXAMPLE 5   **Area of a Sphere**   Find the surface area of a sphere of radius r.

Solution   We think of the sphere as the surface obtained by revolving the curve $y = f(x) = \sqrt{r^2 - x^2}$, for $-r \leq x \leq r$, about the x-axis. The difficulty is that the function $f(x)$ is not differentiable at the endpoints $x = \pm r$. Because of this, we cannot find the surface area of the whole sphere, which would involve integrating from $x = -r$ to $x = r$.

Instead, we consider the part of the sphere between $x = -a$ and $x = a$, for some a with $0 < a < r$ (see Figure 6.9.9). This is the surface obtained by removing equal "caps" from both sides. It is a surface of revolution, and the function $f(x)$ is differentiable on $[-a, a]$, so Definition 6.9.2 applies, and we can find the surface area.

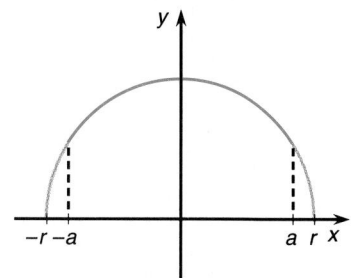

Figure 6.9.9

The formula tells us that the surface area is

$$S = \int 2\pi f(x) \, ds = 2\pi \int_{-a}^{a} f(x) \sqrt{1 + f'(x)^2} \, dx$$

$$= 2\pi \int_{-a}^{a} \sqrt{r^2 - x^2} \sqrt{1 + \left(\frac{-x}{\sqrt{r^2 - x^2}} \right)^2} \, dx$$

$$= 2\pi \int_{-a}^{a} \sqrt{r^2 - x^2} \sqrt{1 + \left(\frac{x^2}{r^2 - x^2} \right)} \, dx$$

$$= 2\pi \int_{-a}^{a} \sqrt{r^2 - x^2} \sqrt{\frac{(r^2 - x^2) + x^2}{r^2 - x^2}} \, dx = 2\pi \int_{-a}^{a} \sqrt{r^2} \, dx$$

$$= 2\pi \sqrt{r^2} \int_{-a}^{a} dx = 2\pi r x \Big|_{-a}^{a}$$

$$= 4\pi r a.$$

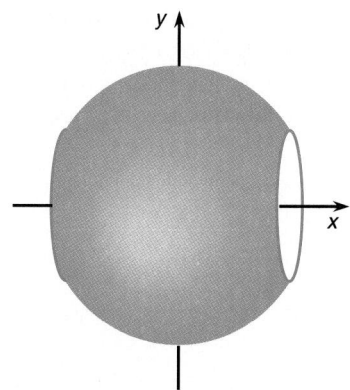

Figure 6.9.10

This is the area of the surface obtained by removing two equal caps from the sphere (see Figure 6.9.10). To find the area of the whole sphere, we let a tend to r and take the limit. This amounts to letting the caps shrink so that the surface that is left approaches the whole sphere. We find that the surface area of the sphere equals

$$S = \lim_{a \to r} 4\pi r a = 4\pi r^2.$$

As a mnemonic device, if you remember that the volume of the sphere is $\frac{4}{3}\pi r^3$, then its surface area $4\pi r^2$ is the derivative of $\frac{4}{3}\pi r^3$.

This formula was used in Section 4.1, even though we did not know how to derive it at that time.

It is also possible to form surfaces by revolving curves about the y-axis.

EXAMPLE 6   Find the area of the surface obtained by revolving about the y-axis the curve $y = x^2$, for $1 \leq x \leq 2$.

Solution   There are two ways to approach this problem. One is to use y as the variable, write $x = \sqrt{y}$, and proceed as we did in Example 4.

On the other hand, we could also let x be the variable. In this case, as x passes through a small interval of width dx, the corresponding part of the curve will sweep out a band in the surface, as shown in Figure 6.9.11. The radius of the band is x, so its circumference is $2\pi x$. Its width is $ds = \sqrt{1 + \left(\frac{dy}{dx} \right)^2} \, dx$, so the area is

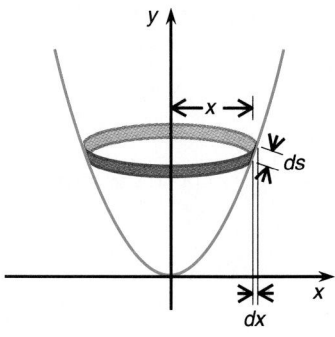

Figure 6.9.11

$2\pi x\sqrt{1+(y')^2}\,dx$. The total area of the surface is

$$S = \int_1^2 2\pi x\sqrt{1+\left(\frac{dy}{dx}\right)^2}\,dx = 2\pi \int_1^2 x\sqrt{1+(2x)^2}\,dx = 2\pi \int_1^2 x\sqrt{1+4x^2}\,dx$$

$$\boxed{\begin{array}{l} u = 1+4x^2 \\ du = 8x\,dx \end{array}} \qquad = 2\pi \int_5^{17} \sqrt{u}\,\frac{1}{8}\,du = \frac{\pi}{4}\frac{2}{3}u^{3/2}\Big|_5^{17}$$

$$= \frac{\pi}{6}\left(17^{3/2} - 5^{3/2}\right) = \frac{\pi}{6}\left(17\sqrt{17} - 5\sqrt{5}\right)$$

$$\approx 30.85.$$

This illustrates a second way of finding surface areas. In the example, y was expressed as a function of x and the graph was revolved about the y-axis, but of course a similar formula can be given if x is expressed as a function of y and the graph is revolved about the x-axis.

All told, there are four possible formulas. Frequently, two of them will apply to the same situation. We list all four below, noting that the first one is the formula of Definition 6.9.2. We have used absolute value signs, but in practice it is often unnecessary, since the functions concerned are usually nonnegative anyway.

As in Definition 6.9.2, we are unable to include the limits of integration on the integrals with respect to s; it is usually only after we express them in terms of x or y that we can find the limits.

THEOREM 6.9.3

Surface Area for Surfaces of Revolution

Suppose $f(x)$ is differentiable and $f'(x)$ is continuous on $[a, b]$. The surface obtained by revolving the graph of $y = f(x)$ about the x-axis has area

$$S = \int 2\pi|f(x)|\,ds = 2\pi \int_a^b |f(x)|\sqrt{1+f'(x)^2}\,dx.$$

The surface obtained by revolving the graph of $y = f(x)$ about the y-axis has area

$$S = \int 2\pi|x|\,ds = 2\pi \int_a^b |x|\sqrt{1+f'(x)^2}\,dx.$$

Suppose $g(y)$ is differentiable and $g'(y)$ is continuous on $[c, d]$. The surface obtained by revolving the graph of $x = g(y)$ about the y-axis has area

$$S = \int 2\pi|g(y)|\,ds = 2\pi \int_c^d |g(y)|\sqrt{1+g'(y)^2}\,dy.$$

The surface obtained by revolving the graph of $x = g(y)$ about the x-axis has area

$$S = \int 2\pi|y|\,ds = 2\pi \int_c^d |y|\sqrt{1+g'(y)^2}\,dy.$$

Since many graphs can be expressed as functions of x or as functions of y, it often happens that two of these formulas are applicable to a particular problem. In this case it is really just a matter of which approach gives the easier integral.

Figure 6.9.12

EXAMPLE 7   **Areas of Cell Membranes**   Consider two one-celled organisms. One is spherical, and the other has the elongated shape of the solid of revolution obtained by revolving the ellipse $x^2 + 2y^2 = 1$ around the x-axis (see Figure 6.9.12). If the two microorganisms have the same volume, it is reasonable to assume that their metabolisms are similar and in particular that they have similar requirements for nutrients.

Nutrients are absorbed through the cell membrane, so the amount that can be absorbed is proportional to the surface area. It has been suggested that the shapes of organisms may be determined by their relative efficiencies. Let us compare the two surface areas.

We begin by finding the volume of the elongated cell. Using the function $y = \frac{1}{\sqrt{2}}\sqrt{1-x^2}$ illustrated in Figure 6.9.13, we find that the volume of the solid of revolution is

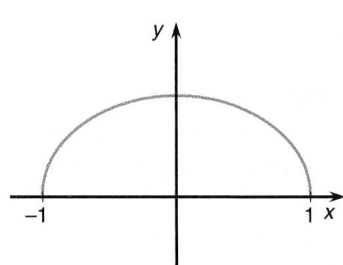

Figure 6.9.13

$$V = \int_{-1}^{1} \pi y^2\, dx = \frac{\pi}{2} \int_{-1}^{1} (1-x^2)\, dx = \frac{\pi}{2} \left(x - \frac{x^3}{3} \right) \Bigg|_{-1}^{1} = \frac{2\pi}{3}.$$

Since the volume of the spherical cell is $\frac{4}{3}\pi r^3$ and it must equal this value $\frac{2\pi}{3}$, we find that the radius satisfies $r^3 = \frac{1}{2}$, so $r = 2^{-1/3}$. By Example 5 the surface area of the spherical cell is $4\pi r^2 = 4\pi 2^{-2/3} = 2^2\pi 2^{-2/3} = 2^{4/3}\pi \approx 7.916$.

Now we find the surface area of the elongated cell. It is

$$S = \int 2\pi y\, ds = \pi\sqrt{2} \int_{-1}^{1} \sqrt{1-x^2}\sqrt{1 + \frac{1}{2}\frac{x^2}{1-x^2}}\, dx$$

$$= \pi\sqrt{2} \int_{-1}^{1} \sqrt{1 - x^2 + \frac{x^2}{2}}\, dx = \pi\sqrt{2} \int_{-1}^{1} \sqrt{1 - \frac{x^2}{2}}\, dx.$$

We know how to integrate $\sqrt{1-x^2}$ from Example 6.1.2; the last integral can be changed by making the substitution $u = \frac{x}{\sqrt{2}}$, $du = \frac{1}{\sqrt{2}}dx$, so

$$S = \pi\sqrt{2} \int_{-1}^{1} \sqrt{1 - \frac{x^2}{2}}\, dx = \pi\sqrt{2} \int_{-1/\sqrt{2}}^{1/\sqrt{2}} \sqrt{1 - u^2}\sqrt{2}\, du$$

$$= 2\pi \int_{-1/\sqrt{2}}^{1/\sqrt{2}} \sqrt{1 - u^2}\, du = \pi\left(u\sqrt{1-u^2} + \arcsin u\right)\Bigg|_{-1/\sqrt{2}}^{1/\sqrt{2}}$$

$$= \pi\left(\frac{1}{\sqrt{2}}\frac{1}{\sqrt{2}} + \frac{\pi}{4} + \frac{1}{\sqrt{2}}\frac{1}{\sqrt{2}} + \frac{\pi}{4} \right) = \pi\left(1 + \frac{\pi}{2} \right)$$

$$= \pi + \frac{\pi^2}{2} \approx 8.076.$$

We find that the area of the elongated cell is only about 2% larger than the area of the spherical cell of the same volume, and this does not seem large enough to be likely to make a decisive difference in the structure of the organisms. ◢

Exercises 6.9

I

Find the arclength of each of the following curves.

1. $y = x^{3/2}$ for $x \in [2, 4]$
2. $y = 2x^{3/2}$ for $x \in [a, b]$
3. $y = x + 2$ for $x \in [-1, 3]$
4. $x = 2y - 3$ for $y \in [2, 3]$
5. $y = x^{3/2} + 7$ for $x \in [2, 4]$
6. $y = 4x^{3/2} - 12$ for $y \in [-8, 20]$
7. $y = (x + 2)^{3/2}$ for $x \in [-1, 2]$
8. $x = (y - 6)^{3/2} + 4$ for $y \in [7, 10]$
9. $x = y^{2/3}$ for $y \in [1, 8]$
10. $y = x^{2/3}$ for $x \in [8, 27]$

Find the surface area of the surface obtained by revolving each of the following curves about the specified axis.

11. $y = 3\sqrt{x}$, $x \in [1, 5]$; x-axis
12. $y = \sqrt{5x}$, $x \in [2, 3]$; x-axis
13. $y = 3x + 1$, $x \in [0, 2]$; x-axis
14. $y = x$, $x \in [-2, 2]$; x-axis
15. $y = 3x + 1$, $x \in [0, 2]$; y-axis
16. $y = |x|$, $x \in [-2, 2]$; x-axis
17. $y = 3$, $x \in [0, 7]$; x-axis
18. $y = x^3$, $x \in [0, 3]$; x-axis
19. $x = r$, $y \in [0, h]$; y-axis
20. $y = x$, $x \in [0, 3]$; y-axis

II

21. Find the arclength of a circle of radius r.
22. Set up (but do not attempt to evaluate) the integral that expresses the length of the parabola $y = x^2$ for x between $x = 0$ and $x = 1$.
23. Verify the formula for the area of a cone found in Example 3 (p. 487) by using Definition 6.9.2 (p. 488) and integrating.
*24. (i) Suppose the function $f(x) = cx + d$ is defined for all $x \in [a, b]$ and is positive there. Find the area of the surface obtained by revolving it about the x-axis.
  (ii) Suppose the function $f(x) = cx + d$ is negative for all $x \in [a, b]$. Find the area of the surface obtained by revolving it about the x-axis.

(iii) What happens if the function $f(x) = cx + d$ is positive for some $x \in [a, b]$ and negative for some $x \in [a, b]$? Find the area of the surface obtained by revolving it about the x-axis.

25. Consider a cone of height h with its vertex pointing down and radius r at the top. Suppose $0 < \ell < h$ and consider the part of the cone that lies above the height ℓ. Find the area of this part of the surface.

*26. Consider Figure 6.9.14. The left-hand curve is $y = \sqrt{x}$, and the right-hand curve is (part of) the circle $x^2 + y^2 = 1$.

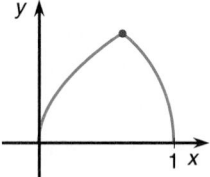

Figure 6.9.14

(i) Find the point of intersection of the two curves.
(ii) Find the area of the surface obtained by revolving the two curves about the x-axis.

*27. Consider the sphere of radius 4 with center at the origin, and find the surface area of the "cap," the part of the sphere that lies above $y = 3$.

*28. Consider the sphere of radius R with center at the origin, and find the surface area of the "cap," the part of the sphere that lies above $y = h$, where $0 < h < R$.

*29. (i) Find the area of the outer surface of the region obtained by drilling a hole of radius $\frac{1}{2}$ precisely through the center of a sphere of radius 3.
  (ii) Find the total surface area of the solid described in part (i), that is, including the area of the surface inside the hole.

*30. (i) Assume that the earth is a perfect sphere of radius 4000 miles, and find a formula for the surface area of the part of the earth that lies north of latitude $\ell°$.
  (ii) What latitude $\ell°$ has the property that the surface to the north of it has area equal to exactly one-quarter of the surface area of the whole planet?

*31. Suppose $0 < a < b$, and find the area of the surface obtained by revolving $y = \sqrt{x}$ for $a \le x \le b$ about the x-axis.

*32. Noticing that $y = \sqrt{x}$ is not differentiable at $x = 0$, find the area of the surface obtained by revolving $y = \sqrt{x}$ for $0 \le x \le b$ about the x-axis.

33. Find the total surface area of an ice cream cone consisting of a cone of height 4 in and radius 1 in at the

top, covered with ice cream in the shape of a hemisphere.

*34. Find the area of the surface obtained by revolving the
curve $y = 3 - \sqrt{x}$ for $1 \leq x \leq 4$ about the line $y = 3$.

*35. Compare the surface areas of the solid obtained by
revolving around the x-axis the curve $x^2 + 4y^2 = 1$
and the sphere of the same volume.

**36. Let $a > 1$ and revolve $x^2 + a^2y^2 = 1$ around the x-
axis. (i) Find the surface area of the resulting solid.
(ii) Compare with the surface area of the sphere with
the same volume when $a = 3, 5, 10$.

37. Use a personal computer or programmable calculator to approximate the arclength of the parabola

$y = x^2$ between $(0,0)$ and $(1,1)$, using the result of
Example 5.2.3 (p. 356) with $n = 10$.

*38. (i) Do you expect the arclength of $y = x^2$ for $x \in [1,2]$ to be greater or less than the arclength of
$y = x^2$ for $x \in [0,1]$, and by approximately what
factor? (You might like to draw the graphs.)

    (ii) Use Example 5.2.3 to check your guess.

Use Example 5.2.3 to estimate the surface area of the
solid obtained by revolving each of the following around
the x-axis.

39. $y = x^3$, $x \in [1,3]$ **42.** $y = \frac{1}{1+x^2}$, $x \in [0,1]$
40. $y = 1 + x^2$, $x \in [-1,1]$ **43.** $y = \sin x$, $x \in [0,\pi]$
41. $y = e^x$, $x \in [0,1]$ **44.** $y = \cos x$, $x \in [0,\pi]$

6.10 Numerical Integration; The Trapezoidal Rule and Simpson's Rule

TRAPEZOIDAL RULE
SIMPSON'S RULE
ERROR ESTIMATES

We have learned how to evaluate many integrals, and in Chapter 8 we will learn
techniques that will permit us to evaluate many more, but there have also been
disconcerting suggestions that there are some integrals that arise in fairly ordinary
ways that we will never be able to do. Not only are there some that are extraordinarily
difficult to find, but there are also some that cannot be done, in the sense that it can
be proved that the indefinite integral cannot be expressed in terms of the functions
we know, that is, rational functions, powers, roots, trigonometric functions, inverse
trigonometric functions, logarithms and exponentials, and combinations of these.

    In any case there are times when it is necessary to find the numerical value of a
definite integral, or at least a good approximation to it, without being able to find
the indefinite integral. In a sense this is how we defined integrals in the first place,
by approximating areas under graphs with collections of rectangles.

    It is possible to evaluate Riemann sums to approximate the value of an integral,
as we did in Example 5.1.3 (p. 349). In this section we also learn two other methods
that usually give better approximations. They are called the Trapezoidal Rule and
Simpson's Rule.

    There are two basic problems. One is that it would be desirable to perform these
calculations efficiently. They are repetitive (adding up many similar terms) and are
well suited to computers and sophisticated calculators, but with all such calculations
efficiency is important. In addition to speed, it is a question of accuracy, since round-
off errors accumulate and become more significant as the calculation becomes
longer. This question is important and interesting, but it will not be discussed here.

    The second problem is that it is all very well to have an accurate approximation,
but its usefulness is very limited unless we know just how accurate it is. To put it
the other way around, we know that the Riemann sums associated to the partitions
dividing an interval into n subintervals of equal length will eventually converge to
the integral. But suppose we need to know the value of some integral *correct to two
decimal places.* How do we choose how big n must be to ensure this level of accuracy?

Suppose we want to estimate $\int_a^b f(x)\,dx$ using Riemann sums. To simplify things, let us decide to use the partition $\mathcal{P}_n$ that divides $[a, b]$ into n equal subintervals and choose x_k^* to be the midpoint of the kth subinterval. The corresponding Riemann sum $R_n = \sum_{k=1}^n f(x_k^*)\Delta x_k = \frac{b-a}{n}\sum_{k=1}^n f(x_k^*)$ is an approximation to the integral.

Assuming that $f(x)$ is differentiable on $[a, b]$, let B_1 be a "bound" for $f'(x)$, that is, a number satisfying $|f'(x)| \le B_1$ for all $x \in [a, b]$. Then the accuracy with which the Riemann sum R_n approximates the integral can be expressed as follows:

$$\left| \int_a^b f(x)\,dx - R_n \right| \le \frac{B_1(b-a)^2}{2n}. \tag{6.10.1}$$

The approximation by a Riemann sum is accurate to at least this amount. In particular, we can use this result to decide how big n needs to be to ensure an approximation of specified accuracy.

EXAMPLE 1   Consider the integral $\int_0^1 x^4\,dx$. How large does n have to be to ensure that the Riemann sum R_n will be accurate to two decimal places?

Solution   To be accurate to two decimal places means to be accurate to within 0.005. In this case the derivative of the integrand is $\frac{d}{dx}(x^4) = 4x^3$, which is always less than or equal to $B_1 = 4$ on $[0, 1]$.

By Inequality 6.10.1 the Riemann sum associated to the midpoints of $\mathcal{P}_n$ is accurate to within $B_1\left(\frac{1^2}{2n}\right) = \frac{B_1}{2n} = \frac{2}{n}$. To make this less than or equal to 0.005, we need $\frac{2}{n} \le 0.005$, or $n \ge 400$. ◢

The difficulty here is that accuracy to two decimal places is not terribly good, and evaluating a Riemann sum with 400 terms in it is quite tedious. We have done this example to illustrate the principle, but it is usually not very practical to use straight Riemann sums to estimate integrals. Fortunately, other methods are available.

The reason that Inequality 6.10.1 involves the size of the derivative is roughly speaking that the error in approximating the integral by the sum arises because the values of $f(x)$ on a particular subinterval will differ from the value of $f(x)$ at the midpoint. The amount by which the values of $f(x)$ can change on a subinterval depends on the derivative $f'(x)$, the rate of change of $f(x)$.

We will prove Inequality 6.10.1 at the end of this section, in Example 5, but for now we will proceed to other approximation techniques.

The Trapezoidal Rule

Suppose $y = f(x)$ is a straight line, as illustrated in Figure 6.10.1. We would like to find the area of the shaded region, which is called a **trapezoid**.

EXAMPLE 2   Find the area of the trapezoid illustrated in Figure 6.10.1.

Solution   In Figure 6.10.2 a dashed line has been drawn on the trapezoid, dividing it into two triangles. A simple way to find the area of the trapezoid is to find the area of each triangle and add them. To find the area of the upper triangle, we think of its vertical right side as the "base," which has length $f(b)$, and the "height" will be

Figure 6.10.1

Figure 6.10.2

Figure 6.10.3

Figure 6.10.4

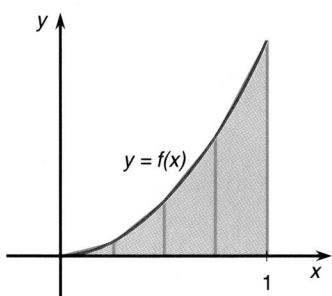

Figure 6.10.5

the distance from this vertical line to the opposite corner, which is just $b - a$ (the length of the interval $[a, b]$). So the area is $\frac{1}{2}\, base \times height = \frac{1}{2}f(b)(b - a)$.

Similarly, the area of the lower triangle is $\frac{1}{2}f(a)(b - a)$, and the total area of the trapezoid is

$$\frac{f(a) + f(b)}{2}(b - a).$$

We can think of $\frac{f(a)+f(b)}{2}$ as the average height of the line $y = f(x)$, and the formula then says the area of the trapezoid is the length of the interval $[a, b]$ multiplied by the average height of the curve. This idea is illustrated in Figure 6.10.3.

To approximate the area of the region under a graph, we can try filling the region with trapezoids, as shown in Figure 6.10.4. We can easily calculate the total area of the trapezoids, which should be approximately equal to the area under the curve.

If we use a larger number of trapezoids, the approximation should be better. This suggests trying the approximation with a large number of trapezoids, say, n of them.

EXAMPLE 3 Estimate the area under the curve $y = f(x) = x^2$ between $x = 0$ and $x = 1$, using n trapezoids of equal width.

Solution We divide the interval $[0, 1]$ into n equal segments and construct a trapezoid on each of them, as shown in Figure 6.10.5 (in the illustration, $n = 4$). The first trapezoid lies over the interval $[0, \frac{1}{n}]$. The heights of its sides (at $x = 0$ and $x = \frac{1}{n}$) are $f(0) = 0^2 = 0$ and $f\left(\frac{1}{n}\right) = \left(\frac{1}{n}\right)^2 = \frac{1}{n^2}$. So as we have seen in Example 1, the area of the trapezoid is $\frac{1}{2}\left(f(0) + f\left(\frac{1}{n}\right)\right)\left(\frac{1}{n} - 0\right) = \frac{1}{2}\left(0 + \frac{1}{n^2}\right)\left(\frac{1}{n}\right)$. Similarly, the area of the second trapezoid, which lies over $[\frac{1}{n}, \frac{2}{n}]$, is $\frac{1}{2}\left(f\left(\frac{1}{n}\right) + f\left(\frac{2}{n}\right)\right)\left(\frac{2}{n} - \frac{1}{n}\right) = \frac{1}{2}\left(\frac{1}{n^2} + \frac{4}{n^2}\right)\left(\frac{1}{n}\right)$.

The kth trapezoid lies over the interval $\left[\frac{k-1}{n}, \frac{k}{n}\right]$, and its area is

$$\frac{1}{2}\left(f\left(\frac{k - 1}{n}\right) + f\left(\frac{k}{n}\right)\right)\left(\frac{k}{n} - \frac{k - 1}{n}\right) = \frac{1}{2}\left(\frac{(k - 1)^2}{n^2} + \frac{k^2}{n^2}\right)\left(\frac{1}{n}\right).$$

The total area of all the trapezoids is then

$$\frac{1}{2}\sum_{k=1}^{n}\left(\frac{(k - 1)^2}{n^2} + \frac{k^2}{n^2}\right)\frac{1}{n} = \frac{1}{2n^3}\sum_{k=1}^{n}\left((k - 1)^2 + k^2\right)$$

$$= \frac{1}{2n^3}\left(\sum_{k=1}^{n}(k - 1)^2 + \sum_{k=1}^{n}k^2\right)$$

$$= \frac{1}{2n^3}\left(\sum_{j=1}^{n-1}j^2 + \sum_{k=1}^{n}k^2\right).$$

> For the second-last sum we shifted to $j = k - 1$ and then dropped off the term with $j = 0$ because it equals zero.

> For the first sum we use Theorem 5.1.1 with n replaced by $n - 1$.

Using Theorem 5.1.1 (p. 348), we see that the total area of the trapezoids is

$$\frac{1}{2n^3}\left(\frac{(n - 1)n(2n - 1)}{6} + \frac{n(n + 1)(2n + 1)}{6}\right)$$

$$= \frac{1}{12n^2}\big((n-1)(2n-1) + (n+1)(2n+1)\big)$$

$$= \frac{1}{12n^2}\big(2n^2 - 3n + 1 + 2n^2 + 3n + 1\big)$$

$$= \frac{1}{12n^2}(4n^2 + 2) = \frac{2}{12n^2}(2n^2 + 1)$$

$$= \frac{1}{6}\left(\frac{2n^2+1}{n^2}\right).$$

As we did in Section 5.1 with rectangles, we could take the limit as $n \to \infty$. This amounts to considering more and more trapezoids, each of which is narrower and narrower; the total area approaches the area under the curve.

The area of the region under the curve is $\lim_{n\to\infty} \frac{1}{6}\left(\frac{2n^2+1}{n^2}\right) = \frac{2}{6} = \frac{1}{3}$.

The same approach can be used to approximate the area under other curves.

EXAMPLE 4 If $f(x) \geq 0$ for all $x \in [a, b]$, approximate the area under the curve $y = f(x)$ between $x = a$ and $x = b$ by dividing the interval into n equal subintervals and using trapezoids.

Solution The length of each subinterval will be $\frac{b-a}{n}$. We label the endpoints of the subintervals as x_0, x_1, x_2, ..., x_n, so $x_0 = a$, $x_1 = a + \frac{b-a}{n}$, $x_2 = a + 2\frac{b-a}{n}$, $x_3 = a + 3\frac{b-a}{n}, \ldots, x_k = a + k\frac{b-a}{n}, \ldots, x_{n-1} = a + (n-1)\frac{b-a}{n}, x_n = a + n\frac{b-a}{n} = b$.

The kth point x_k is found by moving k subintervals from $x = a$. Since each subinterval has width $\frac{b-a}{n}$, $x_k = a + k\frac{b-a}{n}$.

The kth subinterval will be $[x_{k-1}, x_k] = \left[a + (k-1)\frac{b-a}{n}, a + k\frac{b-a}{n}\right]$. The trapezoid over this subinterval has sides whose heights are $f(x_{k-1})$ and $f(x_k)$, so its area is $\frac{f(x_{k-1})+f(x_k)}{2}\frac{(b-a)}{n}$.

The total area of all the trapezoids is

The two summations in the second-last line are equal; we combine them and cancel the 2 in the original denominator $2n$.

$$\sum_{k=1}^{n} \frac{f(x_{k-1}) + f(x_k)}{2}\frac{(b-a)}{n} = \frac{(b-a)}{2n}\sum_{k=1}^{n}\big(f(x_{k-1}) + f(x_k)\big)$$

$$= \frac{(b-a)}{2n}\left(\sum_{k=1}^{n}f(x_{k-1}) + \sum_{k=1}^{n}f(x_k)\right)$$

$$= \frac{(b-a)}{2n}\left(f(x_0) + \left(\sum_{j=1}^{n-1}f(x_j)\right) + \left(\sum_{k=1}^{n-1}f(x_k)\right) + f(x_n)\right)$$

Note that $f(a) = f(x_0)$, $f(b) = f(x_n)$.

$$= \frac{b-a}{2n}\big(f(x_0) + f(x_n)\big) + \frac{b-a}{n}\sum_{k=1}^{n-1}f(x_k).$$

The formula that we obtained in Example 4 can be used even if $f(x)$ is not nonnegative. In this case it approximates the *signed* area. The formula is important enough to have a name.

THEOREM 6.10.2

The Trapezoidal Rule

The signed area enclosed by the graph of $y = f(x)$ between $x = a$ and $x = b$ can be approximated by

$$T_n = \big(f(a) + f(b)\big)\frac{b-a}{2n} + \big(f(x_1) + f(x_2) + \ldots + f(x_{n-1})\big)\frac{b-a}{n}$$

$$= \left(f(x_0) + 2f(x_1) + 2f(x_2) + \ldots + 2f(x_{n-2}) + 2f(x_{n-1}) + f(x_n) \right) \frac{(b-a)}{2n}$$

$$= \frac{b-a}{2n} \left(f(x_0) + f(x_n) \right) + \frac{b-a}{n} \sum_{k=1}^{n-1} f(x_k),$$

where $x_k = a + k\frac{b-a}{n}$, for $k = 0, 1, \ldots, n$.

T_n is the total area of the trapezoids constructed by dividing the interval $[a, b]$ into n equal subintervals.

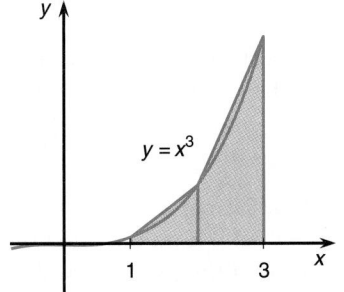

Figure 6.10.6

EXAMPLE 5 Use the Trapezoidal Rule to approximate the area under the curve $y = x^3$ between $x = 1$ and $x = 3$, using the values $n = 2, 4, 6$.

Solution With $n = 2$, as shown in Figure 6.10.6, we have $x_0 = 1$, $x_1 = 2$, $x_2 = 3$, and

$$T_2 = \left(f(1) + f(3) \right)\frac{2}{4} + \left(f(2) \right)\frac{2}{2} = (1 + 27)\frac{1}{2} + 8(1)$$

$$= 14 + 8 = 22.$$

With $n = 4$ we have $x_0 = 1$, $x_1 = \frac{3}{2}$, $x_2 = 2$, $x_3 = \frac{5}{2}$, $x_4 = 3$, and

$$T_4 = \left(f(1) + f(3) \right)\frac{2}{8} + \left(f\left(\frac{3}{2}\right) + f(2) + f\left(\frac{5}{2}\right) \right)\frac{2}{4}$$

$$= (1 + 27)\frac{1}{4} + \left(\frac{27}{8} + 8 + \frac{125}{8} \right)\frac{2}{4} = 7 + \left(\frac{152}{8} + 8 \right)\frac{2}{4} = 7 + (27)\frac{1}{2}$$

$$= 20\frac{1}{2}.$$

With $n = 6$ we have $x_0 = 1$, $x_1 = \frac{4}{3}$, $x_2 = \frac{5}{3}$, $x_3 = 2$, $x_4 = \frac{7}{3}$, $x_5 = \frac{8}{3}$, $x_6 = 3$, and

$$T_6 = \left(f(1) + f(3) \right)\frac{2}{12} + \left(f\left(\frac{4}{3}\right) + f\left(\frac{5}{3}\right) + f(2) + f\left(\frac{7}{3}\right) + f\left(\frac{8}{3}\right) \right)\frac{2}{6}$$

$$= (1 + 27)\frac{1}{6} + \left(\frac{64}{27} + \frac{125}{27} + 8 + \frac{343}{27} + \frac{512}{27} \right)\frac{2}{6}$$

$$= \frac{14}{3} + \left(\frac{1044}{27} + 8 \right)\frac{1}{3} = \frac{14}{3} + \left(38\frac{2}{3} + 8 \right)\frac{1}{3}$$

$$= 20\frac{2}{9}.$$

It is interesting to compare this example with Example 5.1.4 (p. 350), which approximated the same area with Riemann sums. Since we now know that the actual area is $\int_1^3 x^3 \, dx = \frac{x^4}{4} \big|_1^3 = 20$, we find that the Trapezoidal Rule gives much better approximations.

Suppose we want to estimate $\int_a^b f(x) \, dx$ using the Trapezoidal Rule. It turns out that the accuracy we can expect will depend on the size of the second derivative. Assume that $f(x)$ is twice-differentiable on $[a, b]$, and suppose that B_2 is a bound for $f''(x)$ on $[a, b]$, that is, assume $|f''(x)| \leq B_2$ for all $x \in [a, b]$. Let T_n be the Trapezoidal Rule approximation to $\int_a^b f(x)\,dx$ using n subintervals. Then

$$\left| \int_a^b f(x)\,dx - T_n \right| \leq \frac{B_2(b-a)^3}{12n^2}. \tag{6.10.3}$$

For comparison we consider the same integral that we approximated by Riemann sums in Example 1.

EXAMPLE 6   Consider the integral $\int_0^1 x^4\,dx$. How large does n have to be to ensure that the Trapezoidal Rule approximation T_n will be accurate to two decimal places?

Solution   In this case the second derivative of the integrand is $\frac{d^2}{dx^2}(x^4) = 12x^2$, so for a bound we can choose $B_2 = 12$.

By Inequality 6.10.3 the trapezoidal approximation T_n will be accurate to within $B_2\left(1^3/(12n^2)\right) = \frac{1}{n^2}$. To make this less than or equal to 0.005, we need $\frac{1}{n^2} \leq 0.005$, or $n^2 \geq 200$. This amounts to $n \geq \sqrt{200} \approx 14.14$. Since n must be an integer, we should use $n = 15$.

This is a *vast* improvement over the approximation R_n by Riemann sums. Fifteen terms is a sufficiently small sum that it is reasonable to evaluate it on a calculator. Doing this and writing down only five decimal places, we find the points at which we evaluate are $x_0 = 0$, $x_1 = \frac{1}{15}, \dots, x_{15} = 1$, so

$$T_{15} = \frac{b-a}{2n}\left(f(x_0) + f(x_{15}) + 2\sum_{k=1}^{14} f(x_k) \right)$$

$$= \frac{1}{30}\left(0 + 1 + 2\left(\left(\frac{1}{15}\right)^4 + \left(\frac{2}{15}\right)^4 + \dots + \left(\frac{14}{15}\right)^4 \right) \right)$$

$$\approx \frac{1}{30}\big(1 + 2(0.00002 + 0.00032 + 0.0016 + 0.00506 + 0.01235 + 0.0256$$
$$+ 0.04743 + 0.08091 + 0.1296 + 0.19753 + 0.28920 + 0.4096$$
$$+ 0.56417 + 0.75883)\big)$$

$$\approx \frac{1}{30}6.04444 \approx 0.20148.$$

Figure 6.10.7

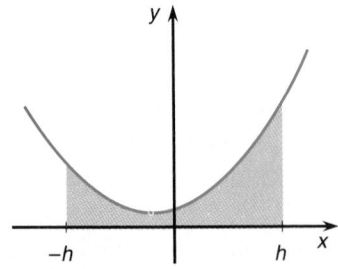

Figure 6.10.8

Simpson's Rule

There is yet another formula for approximating the value of an integral. It is called **Simpson's Rule** after the English mathematician Thomas Simpson (1710–1761), even though he did not invent the rule.

The Riemann sum approximated the area under a graph using rectangles, and the Trapezoidal Rule used trapezoids. To try to get a better approximation, it seems desirable to use some fairly simple region that will fit the shape of the curve more closely. Specifically, it is helpful to use a region that is curved (see Figure 6.10.7).

Probably the simplest of all curved regions is the region under a parabola. We could try to approximate the region under a curve with many pieces, each of which is the region under a suitable parabola. This procedure leads to Simpson's Rule.

To begin with, consider the parabola $y = p(x) = ax^2 + bx + c$. We will look at the part of the parabola lying over the interval $[-h, h]$, shown in Figure 6.10.8, and compute its integral $\int_{-h}^{h} p(x)\, dx = \int_{-h}^{h} (ax^2 + bx + c)\, dx = \left(\frac{a}{3}x^3 + \frac{b}{2}x^2 + cx \right)\big|_{-h}^{h} = \frac{2}{3}ah^3 + 2ch$.

The values of $p(x)$ at the left and right endpoints and at the midpoint of the interval are

$$Y_\ell = p(-h) = ah^2 - bh + c \qquad Y_m = p(0) = c$$
$$Y_r = p(h) = ah^2 + bh + c.$$

With a little algebra it is easy to express the integral we just calculated in terms of these three values. It turns out that

$$\int_{-h}^{h} p(x)\, dx = \frac{2}{3}ah^3 + 2ch = \frac{h}{3}\left(Y_\ell + 4Y_m + Y_r \right). \tag{6.10.4}$$

This formula expresses the integral of the parabola over the interval $[-h, h]$ in terms of the values of the function at the endpoints and the midpoint. It makes no mention of where the interval is. If we consider any interval of length $2h$ and integrate $p(x)$ over it, the integral will be given by the same formula in terms of the values of $p(x)$ at the endpoints and midpoint of the interval. This amounts to shifting the parabola and the interval $[-h, h]$ to the left or to the right, which does not change the integral.

Now we turn to the problem of approximating the integral $\int_a^b f(x)\, dx$. Suppose n is an even integer, and let $\mathcal{P}_n = \{x_0, \ldots, x_n\}$ be the partition that divides the interval $[a, b]$ into n subintervals of equal length. On the first two subintervals we approximate the curve $y = f(x)$ by a parabola. On the next two subintervals we approximate it by another parabola, and so on. On pairs of adjacent subintervals we approximate the integrand $f(x)$ by parabolas. See Figure 6.10.9.

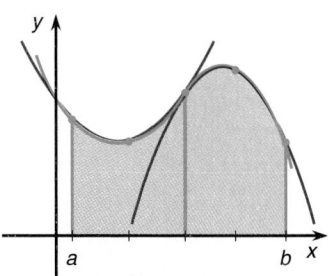

It is not necessary to figure out the equations of the parabolas. We simply choose the parabola that goes through the same points as the graph of $y = f(x)$ at the endpoints and midpoint of the double subintervals. Having done this, we approximate the integral of $f(x)$ with the sum of the integrals of the little parabolas, which we can find by using Equation 6.10.4.

For instance, the width of each little subinterval is $h = \frac{b-a}{n}$, and for the first two subintervals the value of the function at the left endpoint is $Y_\ell = f(x_0)$, the value at the midpoint is $Y_m = f(x_1)$, and the value at the right endpoint is $Y_r = f(x_2)$. The integral of the parabola is found by Equation 6.10.4. It is $\frac{h}{3}\left(Y_\ell + 4Y_m + Y_r \right) = \frac{b-a}{3n}\left(f(x_0) + 4f(x_1) + f(x_2) \right)$.

Similarly, we can find the integral of the parabolic approximation over the third and fourth subintervals. It is $\frac{h}{3}\left(Y_\ell + 4Y_m + Y_r \right) = \frac{b-a}{3n}\left(f(x_2) + 4f(x_3) + f(x_4) \right)$.

We do this for the fifth and sixth subintervals, the seventh and eighth, and so on. Finally, we add up the approximation for each of these pairs of subintervals to get the approximation to the whole integral.

Figure 6.10.9

The values of $f(x)$ at $x = x_0$ and $x = x_n$ are multiplied by 1. The values at the odd-numbered points $x = x_1$, $x = x_3$, ... are multiplied by 4, while the values at the even–numbered points $x = x_2$, $x = x_4$, ... are multiplied by 2.

THEOREM 6.10.5

Simpson's Rule

If n is an even integer, the integral $\int_a^b f(x)\, dx$ can be approximated by

$$S_n = \frac{b-a}{3n}\big(f(x_0) + 4f(x_1) + 2f(x_2) + 4f(x_3) + 2f(x_4) + \ldots$$
$$+ 4f(x_{n-1}) + f(x_n) \big),$$

where $x_k = a + k\frac{b-a}{n}$.

Moreover, if the fourth derivative $f^{(4)}(x)$ exists on $[a, b]$ and B_4 is a number satisfying $|f^{(4)}(x)| \leq B_4$ for all $x \in [a, b]$, then

$$\left| \int_a^b f(x)\, dx - S_n \right| \leq \frac{B_4(b-a)^5}{180 n^4}.$$

EXAMPLE 7   Consider the integral $\int_0^1 x^4\, dx$. Find a Simpson's Rule approximation S_n that is accurate to two decimal places.

Solution   In this case the fourth derivative of the integrand is $\frac{d^4}{dx^4}(x^4) = 24$, so we can take $B_4 = 24$, and we need n to be large enough that $B_4(1^5)/(180)\, n^4 \leq 0.005$, that is, so that $24/(180 n^4) \leq 0.005$, or $n^4 \geq \frac{80}{3} = 26\frac{2}{3}$. This amounts to $n \geq 2.27$, approximately, but since n must be an *even* integer, we must take $n = 4$.

The corresponding approximation by Simpson's Rule is

$$S_4 = \frac{1}{3n}\left(f(0) + 4f\left(\frac{1}{4}\right) + 2f\left(\frac{2}{4}\right) + 4f\left(\frac{3}{4}\right) + f(1) \right)$$

$$\approx \frac{1}{12}\left(0 + 4 \times 0.00391 + 2 \times 0.0625 + 4 \times 0.31641 + 1 \right)$$

$$\approx \frac{1}{12}\left(2.40625 \right)$$

$$\approx 0.20052.$$

The work involved in finding the Simpson's Rule approximation S_n is hardly any more than what is needed for the Trapezoidal Rule *with the same n*, but the accuracy is usually much better. So to get some particular accuracy, Simpson's Rule usually requires a smaller n and consequently is *less* work. Notice that the approximation we obtained in Example 7 is quite good. The integral equals 0.2, and the approximation differs from this by 0.00052, which is much better than the 0.005 we asked for. It is in fact typical of Simpson's Rule that very often the approximation is considerably more accurate than expected. Of course, this is an additional advantage.

The examples we have done so far have all been estimating the integral $\int_0^1 x^4\, dx$, which we can easily evaluate exactly. This was convenient for purposes of comparison, but of course the real interest is in estimating integrals that cannot be evaluated exactly.

EXAMPLE 8   Estimate $\int_0^2 \sqrt{1+x^2}\, dx$ to within an error of less than 0.001.

Solution   Writing $f(x) = \sqrt{1+x^2}$, we find that

$$f'(x) = x(1+x^2)^{-1/2},$$

$$f''(x) = (1+x^2)^{-1/2} - \frac{x}{2}(1+x^2)^{-3/2}(2x) = (1+x^2 - x^2)(1+x^2)^{-3/2}$$

$$= (1+x^2)^{-3/2},$$

$$f^{(3)}(x) = -\frac{3}{2}(1+x^2)^{-5/2}(2x) = -3x(1+x^2)^{-5/2},$$

$$f^{(4)}(x) = -3(1 + x^2)^{-5/2} + \frac{5}{2}(3x)(1 + x^2)^{-7/2}(2x)$$

$$= \left(-3(1 + x^2) + 15x^2\right)(1 + x^2)^{-7/2} = (12x^2 - 3)(1 + x^2)^{-7/2}.$$

We need a bound B_4 for this fourth derivative on $[0, 2]$. There are many possible ways of doing this. First, we can observe that $1 + x^2 \geq 1$, so $(1 + x^2)^{-7/2} \leq 1$. Since $x \in [0, 2]$, we know that $x^2 \leq 4$, so $|12x^2 - 3| \leq |12x^2| + 3 \leq 48 + 3 = 51$. Combining these two observations, we find that

$$|f^{(4)}(x)| \leq \frac{51}{1} = 51,$$

so we can let $B_4 = 51$.

We could also be more sophisticated and recognize that since $0 \leq 12x^2 \leq 48$, we have $-3 \leq 12x^2 - 3 \leq 48 - 3 = 45$, and then let $B_4 = 45$.

So the Simpson's Rule approximation with n subintervals is accurate to within $B_4(b - a)^5/(180n^4) = 51(2 - 0)^5/(180n^4) = 408/(45n^4)$. To make this less than 0.001, we need to have $408/(45n^4) \leq 0.001$, or $n^4 \geq \frac{27,200}{3}$. This amounts to $n \geq$ 9.76, approximately, so we should take $n = 10$, the next higher even integer.

It is frequently useful to summarize a Simpson's Rule calculation in the form of a table, as shown in Table 6.10.1. Each half of each horizontal row corresponds to one of the points of the partition, which is listed in the second column. The next column lists the value of the function $f(x)$ at that point, correct to five decimal places, and the last column shows the number (1, 2, or 4) by which that value is multiplied in the sum in Simpson's Rule.

Table 6.10.1

k	x_k	$f(x_k)$	Multiplier	k	x_k	$f(x_k)$	Multiplier
0	0	1	1	6	1.2	1.56205	2
1	0.2	1.01980	4	7	1.4	1.72047	4
2	0.4	1.07703	2	8	1.6	1.88680	2
3	0.6	1.16619	4	9	1.8	2.05913	4
4	0.8	1.28062	2	10	2	2.23607	1
5	1	1.41421	4				

Now it is reasonably easy to calculate S_n. We multiply each of the values $f(x_k)$ in Table 6.10.1 by the corresponding multiplier (1, 2, or 4) and add the resulting numbers. In this case the total is approximately 44.36827. To find S_n, we multiply this by $\frac{b-a}{3n} = \frac{2}{3(10)} = \frac{1}{15}$. In Table 6.10.1 we wrote down only five decimal places, but if we use a calculator or computer with more accuracy, we find that $S_{10} \approx 2.9578848$.

In Chapter 8 we will learn techniques by which this integral can be evaluated, and its actual value is $\sqrt{5} + \frac{1}{2}\ln|2 + \sqrt{5}| \approx 2.9578857$. Simpson's Rule gives us an approximation that is far more accurate than what we expected.

There is another remark to be made here. The calculations for Simpson's Rule are slightly tedious, but they are not difficult. The hard part was estimating the size of $f^{(4)}(x)$ in order to find B_4. We went to some lengths to find a slightly better estimate (45 instead of 51), but in fact it was not worthwhile. If you work it out using $B_4 = 45$, you will find that what is needed to get the desired accuracy is $n \geq$ 9.46, approximately, so we would still end up using $n = 10$. The point is that slight improvements in the value of B_4 tend to disappear when you take the fourth root.

See the Graphics Calculator Example on the next page.

As a result of this, it is probably not worth the effort to be too careful about finding the smallest possible bound.

At worst, choosing the wrong bound will result in having to evaluate Simpson's Rule with a few more terms in it, which will almost certainly be less trouble than worrying about the bound.

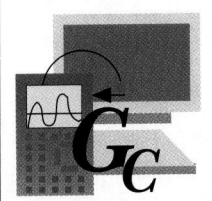

A programmable calculator or a computer can be used to evaluate Simpson's Rule. For instance, we try the integral from Example 8.

TI-85	SHARP EL9300
PRGM F2(EDIT) SIMPSON ENTER F3 F1(Input) N EXIT ENTER 1 STO I ENTER $\sqrt{\ }$ (1 + 0 ^ 2) STO S ENTER S + $\sqrt{\ }$ (1 + 2 ^ 2) + 4 × $\sqrt{\ }$ (1 + (2 ÷ N) ^ 2) STO S ENTER Lbl B ENTER S + 2 × $\sqrt{\ }$ (1 + (I × 4 ÷ N) ^ 2) STO S ENTER S + 4 × $\sqrt{\ }$ (1 + (2 ÷ N + I × 4 ÷ N) ^ 2) STO S ENTER I + 1 STO I ENTER F4 F1(If) (I TEST F2(<) (N ÷ 2)) ENTER Goto B ENTER S × 2 ÷ (3 × N) ENTER EXIT EXIT EXIT PRGM F1(NAME) SIMPSON ENTER 10 ENTER	✏️ ▼ ▼ (NEW) ENTER ENTER SIMP ENTER COMMAND A 3 (Input) N ENTER I = 1 ENTER S = $\sqrt{\ }$ 1 + 0 x^2 ENTER S = S + $\sqrt{\ }$ (1 + 2 x^2 ▶ + 4 × $\sqrt{\ }$ 1 + (2 ÷ N) x^2 ▶ ENTER COMMAND B 1 (Label) 1 ENTER S = S + 2 × $\sqrt{\ }$ (1 + (I × 4 ÷ N) x^2) ENTER S = S + 4 × $\sqrt{\ }$ (1 + (2 ÷ N + I × 4 ÷ N) x^2) ENTER I = I + 1 ENTER COMMAND 3 (If) I COMMAND C 2 (<) N ÷ 2 COMMAND B 2 (Goto) 1 ENTER S = S × 2 ÷ 3 ÷ N ENTER COMMAND A 1 (Print) S ENTER COMMAND A 6 (End) ENTER ⊞ ✏️ (RUN) ENTER (SIMP) ENTER 10 ENTER
CASIO f_x-6300G (f_x-7700GB)	HP 48SX
MODE 2 EXE ? → N : 1 → I : $\sqrt{\ }$ (1 + 0 x^2) → S : S + $\sqrt{\ }$ (1 + 2 x^2) + 4 × $\sqrt{\ }$ (1 + (2 ÷ N) x^2) → S : Lbl 1 : S + 2 × $\sqrt{\ }$ (1 + (I × 4 ÷ N) x^2) → S : S + 4 × $\sqrt{\ }$ (1 + (2 ÷ N + I × 4 ÷ N) x^2) → S : I + 1 → I : I < N ÷ 2 ⇒ Goto 1 : S × 2 ÷ 3 ÷ N ◢ MODE 1 Prog 0 EXE 10 EXE (on the f_x-7700GB, : is PRGM F6, and Prog is PRGM F3.)	ATTN ↵ CLR ≪≫ "" KEY IN N ▶ SPC PROMPT SPC ' N ▶ STO 1 SPC 0 x^2 + $\sqrt{x}$ 1 SPC 2 x^2 + $\sqrt{x}$ + 2 SPC N ÷ x^2 1 + $\sqrt{x}$ 4 × + 1 SPC N SPC 2 ÷ 1 − FOR SPC I SPC I SPC 4 × N ÷ x^2 1 + $\sqrt{x}$ 2 × + I SPC 4 × N ÷ 2 SPC N ÷ + x^2 1 + $\sqrt{x}$ 4 × + NEXT ▶ 2 × N ÷ 3 ÷ ▶ ENTER ' SIMP STO VAR SIMP 10 ENTER ↵ CONT

```
?10
ANSWER=
      2.957884775
```

How large an n do you have to use to get the answer $2.957\,885\,715$ (which is $\sqrt{5} + \frac{1}{2}\ln(2 + \sqrt{5})$ to nine decimal places)?

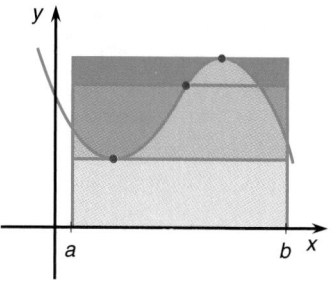

y

a   b   x

Figure 6.10.10

EXAMPLE 9   In this example we prove Inequality 6.10.1 about the accuracy of approximation by Riemann sums.

In Figure 6.10.10 there is one subinterval. The part of the integral corresponding to this subinterval is the shaded area. The corresponding part of the Riemann sum is the area of the middle rectangle, which may be greater or less than the integral or perhaps even equal to it.

Also shown are the big rectangle whose height is the maximum value of $f(x)$ on the subinterval, and the small rectangle whose height is the minimum value of $f(x)$ on the subinterval. We see that the actual integral lies between the areas of these two rectangles and that the area of the rectangle belonging to the Riemann sum with x_k^* equal to the midpoint is also between them.

In particular, the midpoint Riemann sum approximation R_n is no farther from the value of the integral than it is from at least one of the Riemann sums associated to the maximum value or the minimum value on each subinterval. So we can get an inequality describing the accuracy of the Riemann sum approximation by estimating the size of these differences.

Suppose x_k^M and x_k^m are points in the kth subinterval at which $f(x)$ achieves its maximum and minimum values, respectively. We want to estimate $f(x_k^M) - f(x_k^*)$ and $f(x_k^m) - f(x_k^*)$. Assuming that $f(x)$ is differentiable, we know that $f(x_k^M) - f(x_k^*) = \int_{x_k^*}^{x_k^M} f'(x)\,dx$. Assume that B_1 is a positive number chosen so that $|f'(x)| \le B_1$, for all $x \in [a, b]$; we can use Inequality 5.4.3(iii) to conclude that

$$f(x_k^M) - f(x_k^*) = \int_{x_k^*}^{x_k^M} f'(x)\,dx \le B_1 \left| x_k^M - x_k^* \right|.$$

Since x_k^M cannot be farther than $\frac{b-a}{2n}$ from the midpoint x_k^*, we see that $f(x_k^M) - f(x_k^*) \le B_1 \frac{b-a}{2n}$. This is true for each subinterval, so the difference between the "maximum Riemann sum" and the "midpoint Riemann sum" R_n is

$$\sum_{k=1}^{n} f(x_k^M)\,\Delta x_k - \sum_{k=1}^{n} f(x_k^*)\,\Delta x_k = \frac{b-a}{n} \sum_{k=1}^{n} \left(f(x_k^M) - f(x_k^*) \right)$$
$$\le \frac{b-a}{n} \sum_{k=1}^{n} \left(B_1 \frac{b-a}{2n} \right)$$
$$= B_1 \left(\frac{(b-a)^2}{2n^2} \right) \sum_{k=1}^{n} 1 = B_1 \frac{(b-a)^2}{2n}.$$

A similar argument shows that the difference between R_n and the "minimum Riemann sum" is also less than or equal to $B_1(b-a)^2/(2n)$, and from this we conclude that the difference between R_n and the actual integral is also less than or equal to the same number. This is Inequality 6.10.1. ◢

This argument is given here to illustrate the types of techniques used to prove the accuracy estimates. The estimates for the Trapezoidal Rule and for Simpson's Rule are established by more complicated but similar arguments. For our purposes, what is most important is using them.

Exercises 6.10

I

For each of the following integrals, find the Trapezoidal Rule approximation T_n and the Simpson's Rule approximation S_n, for the specified n. Compare these values with the actual value of the integral.

1. $\int_0^4 x \, dx$, $n = 4$

2. $\int_1^9 x^4 \, dx$, $n = 8$

3. $\int_1^4 \sqrt{x} \, dx$, $n = 6$

4. $\int_{-1}^3 (x^3 - 2x) \, dx$, $n = 4$

5. $\int_0^\pi \sin x \, dx$, $n = 4$

6. $\int_0^\pi \sin^2 x \, dx$, $n = 4$

7. $\int_1^5 \frac{1}{x} \, dx$, $n = 8$

8. $\int_1^3 \frac{1}{x}, \, dx$, $n = 6$

9. $\int_0^2 \frac{x}{1+x^2} \, dx$, $n = 4$

10. $\int_{-3}^3 (x^5 - 7x^3 + 2x) \, dx$, $n = 6$

For each of the following integrals, find a value of n that will ensure that the Simpson's Rule approximation S_n will be correct with an error of less than 0.001. Calculate the corresponding S_n and compare it with the exact value of the integral.

11. $\int_0^4 x^4 \, dx$

12. $\int_0^2 x^5 \, dx$

13. $\int_0^{\pi/2} \sin x \, dx$

14. $\int_0^1 e^x \, dx$

15. $\int_{-1}^0 \frac{1}{x-1} \, dx$

16. $\int_0^4 e^{x/4} \, dx$

17. $\int_0^1 x\sqrt{1+x^2} \, dx$

18. $\int_{-1}^1 \cos(\pi x) \, dx$

19. $\int_{-\pi}^\pi \tan(x/4) \, dx$

20. $\int_1^3 \frac{1}{x} \ln(x) \, dx$

For each of the following integrals, find a value of n that will ensure that the Simpson's Rule approximation S_n will be correct with an error of less than 0.001. Calculate the corresponding S_n.

21. $\int_1^2 \ln x \, dx$

22. $\int_1^2 e^x \, dx$

23. $\int_{-1}^0 e^x \, dx$

24. $\int_0^1 \sin(\pi x) \, dx$

25. $\int_1^2 \cos(\pi x) \, dx$

26. $\int_2^4 \cos\left(\frac{\pi}{2} x\right) \, dx$

27. $\int_2^3 x^{5/3} \, dx$

28. $\int_5^7 x^{4/3} \, dx$

29. $\int_0^1 x \sin x \, dx$

30. $\int_{-1}^0 x^2 \sin(\pi x) \, dx$

31. $\int_0^1 \sqrt{1+x^2} \, dx$

32. $\int_0^{1/2} \sqrt{1+x^2} \, dx$

33. $\int_0^1 \frac{1}{\sqrt{1+x^2}} \, dx$

34. $\int_0^{1/2} \frac{1}{\sqrt{1+x^2}} \, dx$

35. $\int_0^1 \sqrt{1+2x^2} \, dx$

36. $\int_0^2 \sqrt{1+x^2/2} \, dx$

37. $\int_0^1 \sqrt{1+x^3} \, dx$

38. $\int_0^1 \sqrt{1+2x^3} \, dx$

39. $\int_0^1 \frac{1}{1+x^3} \, dx$

40. $\int_0^1 \frac{1}{1+x^4} \, dx$

II

*41. (i) Consider the part of the parabola $y = x^2$ between $x = 0$ and $x = 1$. For what value of n will Simpson's Rule give an approximation S_n to the arclength of this curve that is accurate to within 0.001?

(ii) Find an approximation to the arclength of the curve described in part (i), accurate to within 0.001.

*42. (i) Consider the part of the parabola $y = \frac{1}{4}x^2$ between $x = 0$ and $x = 4$. For what value of n will Simpson's Rule give an approximation S_n to the arclength of this curve that is accurate to within 0.001?

(ii) Find an approximation to the arclength of the curve described in part (i), accurate to within 0.001.

**43. With what n can we estimate the arclength of the curve $y = x^3$ between $x = -1$ and $x = 1$ to within 0.001 by Simpson's Rule S_n? (Do not actually evaluate S_n unless you have a computer or programmable calculator.)

**44. With what n can we estimate the arclength of the curve $y = x^4$ between $x = 1$ and $x = 2$ to within 0.001 by Simpson's Rule S_n? (Do not actually evaluate S_n unless you have a computer or programmable calculator.)

*45. Use Simpson's Rule to estimate to within 0.001 the arclength of the curve $y = \sin x$ between $x = 0$ and $x = \pi$.

*46. Use Simpson's Rule to estimate to within 0.001 the arclength of the curve $y = e^x$ between $x = 0$ and $x = 1$.

47. Consider the "multipliers" in the formula of Simpson's Rule, and show that the multipliers in the formula for S_n add up to $3n$. (This means that $\frac{1}{3n}$ times the sum is really a kind of *weighted average*, i.e., the average of $3n$ values of $f(x)$, some of which have been repeated; S_n, which is just this average times $b - a$,

approximates the integral, which equals $b - a$ times the (integral) average value of $f(x)$.)

*48. (i) Show that a parabola of the form $y = ax^2 + bx + c$ is determined by any three points on it (i.e., given three points, there is only one such parabola containing them, i.e., two parabolas containing the three points must actually be the same).

(ii) Can you think of three points that do not lie in such a parabola?

*49. Show that the Simpson's Rule approximation S_n gives an exact evaluation of the integral of any polynomial $p(x)$ whose degree is 3 or less, using *any* positive even integer n.

*50. (i) Suppose that $f(x)$ is an odd function, $a > 0$, and show that for any positive even n, S_n will exactly equal the integral $\int_{-a}^{a} f(x)\, dx$.

(ii) Evaluate the Simpson's Rule approximation S_{20} for $\int_{-6}^{6} (3 + x \cos x)\, dx$.

*51. Use Simpson's Rule to approximate $\ln 2$ to within 0.001. (*Hint:* $\ln 2 = \int_{1}^{2} \frac{1}{x}\, dx$.)

*52. Use Simpson's Rule to approximate $\ln 10$ to within 0.1.

*53. Suppose a and b are two quantities to which we have approximations A and B, respectively. Suppose we know that $|A - a| \le 0.001$ and $|B - b| \le 0.1$ and that $A = 4.7$ and $B = 32.5$.

(i) How good an approximation to the product ab is AB (i.e., estimate the size of the error $|AB - ab|$)?

(ii) How good an approximation to $\frac{a}{b}$ is $\frac{A}{B}$?

*54. (i) Combine the ideas of the previous three problems to estimate the value of $\log_{10}(2)$.

(ii) What degree of accuracy can you guarantee for your estimate?

(iii) Compare your estimate with the actual value.

*55. Recall that $\int_{0}^{1} 1/(1 + x^2)\, dx = \arctan x \big|_{0}^{1} = \pi/4$. We can use Simpson's Rule to approximate $\pi/4$ (and hence π itself).

(i) What value of n is needed to ensure that S_n is within 0.001 of $\pi/4$?

(ii) What value of n is needed to ensure that S_n is within 0.0001 of $\pi/4$?

(iii) What value of n is needed to ensure that S_n is within 0.000001 of $\pi/4$?

(iv) Estimate π to within 0.001, and compare your estimate with the actual value.

**56. Estimate the coordinates of the centroid of the region lying underneath the curve $y = (1 + x^2)^{-1/2}$ between $x = 0$ and $x = 1$. Several integrals are involved here. For each one, first try to evaluate it exactly. Failing that, use Simpson's Rule to estimate it to within 0.001. What accuracy can you guarantee for your answer?

**57. If you have a computer or programmable calculator, program it to evaluate the Simpson's Rule approximation S_n for a specified integrand $f(x)$. Use your program to check your calculations for the exercises in this section.

58. Use Simpson's Rule and a personal computer or programmable calculator to estimate π to within 0.000001 (cf. Exercise 55).

59. Estimate $\ln 2$ to within 0.000001.

*60. Use Simpson's Rule to approximate $\ln 10$ to within 0.001.

Use Simpson's Rule S_n to estimate the surface areas of the surfaces obtained by revolving the following functions around the x-axis. Instead of estimating how large n must be, try it with various values of n until the answer seems to have stopped changing in the first five significant digits.

61. $y = \frac{1}{x^2}$, $x \in [1, 2]$ 64. $y = \ln x$, $x \in [1, e]$

62. $y = x^4 + 1$, $x \in [-2, 2]$ 65. $y = \sec x$, $x \in \left[-\frac{\pi}{4}, \frac{\pi}{4}\right]$

63. $y = \sin^2 x$, $x \in [0, \pi]$ 66. $y = \arctan x$, $x \in [0, 1]$

Chapter Summary

---

§6.1
area
signed area
region between two curves

§6.2
small change dx
strip of width dx
function with specified rate of change

§6.3
solid of revolution
volume by slices (Definition 6.3.1, p. 433)

§6.4
cylindrical shells

§6.5
average value (Definition 6.5.1, p. 447)
velocity/speed

§6.6
work
force
potential energy
kinetic energy
Hooke's Law (Equation 6.6.2, p. 458)
foot-pound (ft-lb)
newton (N)
newton-meter (N-m)
joule (J)

§6.7
centroid
fulcrum
moment about a point
center of mass
uniform density
moment about a line (Definition 6.7.5, p. 467)
symmetry
Pappus' Theorem (Theorem 6.7.10)

§6.8
pressure
density of a fluid
Pascal's Principle (Theorem 6.8.1, p. 477)
fluid force and centroids
buoyancy

§6.9
arclength
$$ds = \sqrt{(dx)^2 + (dy)^2} = \sqrt{1 + (y')^2}\, dx$$
surface of revolution
surface area
$$dS = 2\pi f(x)\, ds$$

§6.10
Trapezoidal Rule (Theorem 6.10.2, p. 496)
Simpson's Rule (Theorem 6.10.5, p. 499)
error estimates

Review Exercises

I

In each of the following situations, find the area of the region enclosed between the functions $f(x)$ and $g(x)$.

1. $f(x) = x^2$, $g(x) = 2 - x^2$
2. $f(x) = x^2$, $g(x) = 9$
3. $f(x) = 4 - x^2$, $g(x) = -2x$
4. $f(x) = x^3 - 3x$, $g(x) = 2x^2$
5. $f(x) = |x^3|$, $g(x) = |x|$
6. $f(x) = x^2 - 4$, $g(x) = |2x - 1|$
7. $f(x) = \sqrt{8 - x^2}$, $g(x) = 2$
8. $f(x) = x^4 - 7x^2 + 10$, $g(x) = x^2 - 6$
9. $f(x) = x^2$, $g(x) = \sqrt{2 - x^2}$
10. $f(x) = x$, $g(x) = \sqrt{x}$

Find the areas of the following regions.

11. Inside $x^2 + y^2 = 1$ and above $y = x$.
12. The intersection of the interiors of $x^2 + y^2 = 1$ and $5x^2 + y^2 = 2$.
13. The intersection of a disk of radius 4 and a square with the same center and sides of length 6.
14. The part of the square in Exercise 13 that is outside the disk.
15. The intersection of two disks of radius 5 with centers 40 units apart.
16. The intersection of two disks of radius r with centers c units apart. (Assume $c < r$.)

17. The intersection of a disk of radius 3 with an ellipse with the same center and semi-minor axis 2 and semi-major axis 4.
18. The intersection of a disk of radius r with an ellipse with the same center and semi-minor axis b and semi-major axis a, with $b < r < a$.
19. The region above the parabola $y = x^2 - 2$ and inside the disk of radius 2 centered at the origin.
20. The region consisting of all points above the parabola $y = x^2$ lying at a distance of 1 unit or less from the point $(1, 0)$.

Find the volume of the solid of revolution formed by revolving the given function $y = f(x)$ on the specified interval about the x-axis. Sketch the solid.

21. $f(x) = \sin x$ on $\left[0, \frac{\pi}{2}\right]$
22. $f(x) = x - 2$ on $[3, 4]$
23. $f(x) = |x|$ on $[-2, 4]$
24. $f(x) = |x^3|$ on $[-1, 1]$
25. $f(x) = \cos^2 x - \sin^2 x$ on $\left[0, \frac{\pi}{4}\right]$
26. $f(x) = \cos x \sin x$ on $\left[0, \frac{\pi}{2}\right]$
27. $f(x) = (1 + x^2)^{-1/2}$ on $[0, 1]$
28. $f(x) = x(x - 1)$ on $[0, 1]$
29. $f(x) = \sec \frac{\pi x}{4}$ on $[-1, 1]$
30. $f(x) = e^x$ on $[-1, 0]$

Find the average value of each of the following functions over the specified interval.

31. $f(x) = x^2 + 3$ on $[0, 2]$

32. $f(x) = x^3$ on $[-2, 2]$

33. $f(x) = \frac{1}{x}$ on $[1, 2]$

34. $f(x) = 3$ on $[-7, 11]$

35. $f(x) = \sin \pi x$ on $[3, 4]$

36. $f(x) = \sin \pi x$ on $[3, 6]$

37. $f(x) = \cos \pi x$ on $[3, 4]$

38. $f(x) = \sin x$ on $\left[0, \frac{\pi}{2}\right]$

39. $f(x) = |x^2 - 1|$ on $[-2, 2]$

40. $f(x) = 1/(1 + x^2)$ on $[0, \sqrt{3}]$

Find the work done by each of the following.

41. Lifting a 3-lb weight through a height of 5 ft.

42. Lifting a 10-kg weight through a height of 3.7 m.

43. Lifting a 30-g weight through a height of 60 cm.

44. Lifting a 3-oz weight through a height of 14 in.

45. Compressing by 1 ft a spring that can be compressed 1 in from rest by a force of 3 lb.

46. A spring relaxing from a stretch of 3 ft to a stretch of 2.5 ft, if it takes 13 lb to compress the spring 1 ft from rest.

47. Gravity on a 10-lb bag of potatoes as it falls 3 ft from the counter to the floor.

48. Gravity on a 5-lb steel ball as it falls 8 ft and then bounces up 7 ft.

49. Pumping all the water from a conical tank of height 10 ft and radius 4 ft at the top and raising it to a point 5 ft above the top of the tank.

50. Pumping the water from a spherical tank of radius 10 ft that is half full and raising it to the outlet valve at the top of the tank.

Find the centroid of each of the following regions.

51. The triangle with vertices $(1, 2)$, $(2, 4)$, $(-3, 0)$.

52. The triangle with vertices $(0, -1)$, $(20, 4)$, $(-3, -1)$.

53. The parallelogram with vertices $(0, 0)$, $(2, 0)$, $(1, 3)$, $(3, 3)$.

54. The parallelogram with vertices $(1, 1)$, $(4, 4)$, $(2, 3)$, $(5, 6)$.

55. The union of the triangle with vertices $(0, 0)$, $(1, 2)$, $(3, 2)$ and the disk of radius 1 centered at $(1, -1)$.

56. The union of the disk of radius 2 centered at $(1, 1)$ and the disk of radius 4 centered at $(10, 0)$.

57. The union of the disk of radius 1 centered at $(0, 0)$ and the disk of radius 1 centered at $(1, 1)$.

58. The union of the disk of radius 1 centered at $(0, 0)$ and the square with vertices $(3, 3)$, $(3, 5)$, $(5, 3)$, $(5, 5)$.

59. The union of the disk of radius 1 centered at $(0, 0)$ and the square with vertices $(0, 0)$, $(1, 0)$, $(0, 1)$, $(1, 1)$.

60. The union of the disk of radius 4 centered at $(0, 0)$ and the square with vertices $(0, 0)$, $(1, 0)$, $(0, 1)$, $(1, 1)$, with density $\rho = 2$ on the square and density $\rho = 1$ on the rest of the disk.

Find the total force on one side of each of the following.

61. A flat plate of area 3 sq ft lying horizontally at a depth of 40 ft in water.

62. A flat plate of area 5 m^2 lying horizontally at a depth of 2 m in water.

63. A square plate with sides 2 ft long, standing vertically in water with its bottom edge at a depth of 20 ft.

64. An equilateral triangle with sides 5 ft long, standing vertically in water with its bottom edge horizontal at a depth of 10 ft.

65. A circular plate of radius 3 m, lying at an angle of $45°$ above the horizontal, with its lowest point at a depth of 6 m.

66. A circular plate of radius 6 ft, lying at an angle of $30°$ above the horizontal with its highest point at a depth of 40 ft.

67. A semicircular plate of radius 3 ft standing vertically in water with its bottom straight edge horizontal at a depth of 10 ft.

68. A semicircular plate of radius 2 m standing vertically in water with its straight edge uppermost at a depth of 5 m.

69. A circular plate of radius 4 ft, standing vertically with its lowest point resting on the bottom of a stream that is 6 ft deep.

70. A semicircular plate of radius 1 ft lying at a $60°$ angle above the horizontal with its straight edge flat on the bottom at a depth of 4 ft.

Find the surface area of the surface obtained by revolving the specified function about the *x*-axis.

71. $f(x) = x$, for $x \in [3, 6]$

72. $f(x) = 3x$, for $x \in [4, 10]$

73. $f(x) = \sqrt{x}$, for $x \in [2, 4]$

74. $f(x) = (3x)^{2/3}$, for $x \in \left[0, \frac{1}{3}\right]$

75. $f(x) = \sqrt{9 - x^2}$, for $x \in [0, 1]$

76. $f(x) = \sqrt{1 - x^2/4}$, for $x \in [-1, 0]$

77. $f(x) = |x - 3|$, for $x \in [0, 6]$

78. $f(x) = x^3$, for $x \in [1, 8]$

79. $f(x) = (x - 1)^3$, for $x \in [2, 9]$

80. $f(x) = \sqrt{2 - x^2}$, for $x \in [0, 1]$

In each of the following cases, find how large n must be so that the Simpson's Rule approximation S_n should be within 0.001 of the given integral. Evaluate S_n for this value of n.

81. $\int_0^1 \left(3x^5 - 2x + 1\right) dx$
86. $\int_0^\pi \sin^2 x \, dx$

82. $\int_1^3 x^6 \, dx$
87. $\int_1^2 \sqrt{2x^2 - 1} \, dx$

83. $\int_0^{\pi/2} \cos x \, dx$
88. $\int_3^4 \sqrt{1 + x^2} \, dx$

84. $\int_0^\pi \sin x \, dx$
89. $\int_{-2}^2 x\sqrt{x^4 + x^2 + 2} \, dx$

85. $\int_{-1}^1 \sin x \, dx$
90. $\int_1^4 \frac{1}{x} \, dx$

II

*91. Find the volume of the intersection of two balls of radius 5 whose centers are 8 units apart.

*92. Find the volume of the intersection of two balls of radius r whose centers are d units apart, with $d < 2r$.

*93. Find the centroid of the region lying between the x-axis and the curve $y = \sin x$, for $0 \le x \le 2\pi$.

*94. Find the centroid of the region lying between the x-axis and the curve $y = \sin x$, for $0 \le x \le \pi$.

Do each of the following examples, using Theorem 6.7.9 (p. 472) to find M_x.

*95. Example 6.7.4 (p. 467)

*96. Example 6.7.5 (p. 468)

*97. Example 6.7.6 (p. 468)

*98. Consider a sphere of radius R and two horizontal planes that are d units apart. Find the surface area of the part of the sphere lying between the two planes. (Assume that both planes intersect the sphere.) Notice that the problem does not say where the planes are; there appears to be insufficient information to solve the problem. In fact, what you will find is that the area of the part of the sphere in question does not depend on where the planes are, but only on how far apart they are. Use this fact to give a snappy answer to Exercise 30(ii) of Section 6.9.

*99. Verify Formula 6.10.4 (p. 499) for the integral of a quadratic function over the interval $[-h, h]$.

*100. Two mathematicians on a fishing trip are having an argument. One says that $\ln 7$ is greater than 2, and the other says that it is less than 2. They are many miles from the nearest calculator or textbook, but there is a nice sandy beach on which they can do calculations with a stick. Since $\ln 7 = \int_1^7 \frac{1}{x} \, dx$, the question can be solved by approximating this integral by Simpson's Rule. The tricky thing is to approximate it with sufficient accuracy.

  (i) Find the Simpson's Rule approximation S_2, and calculate the degree of accuracy we can guarantee for S_2. Based on these two pieces of information, can you tell whether or not $\ln 7 > 2$?

  (ii) Is $\ln 7 > 2$ or not? (No fair using a calculator.)

*101. Suppose the graph of $y = f(x)$ is a straight line. Find the average value of $f(x)$ over the interval $[a, b]$, and show that it equals $f(c)$, where c is the midpoint of the interval.

*102. Find an example of a nonnegative function $f(x)$ and an interval $[a, b]$ so that the average value of $f(x)$ on $[a, b]$ does not equal the RMS average, that is, the square root of the average of $f(x)^2$.

*103. Consider the solid obtained by revolving around the x-axis the square whose vertices are $(1, 1)$, $(1, 2)$, $(2, 2)$, $(2, 1)$. Find its volume using (i) slices, (ii) cylindrical shells, (iii) Pappus' Theorem.

*104. Consider the solid obtained by revolving around the x-axis the square whose vertices are $(1, 1)$, $(2, 0)$, $(2, 2)$, $(3, 1)$. Find its volume using (i) slices, (ii) cylindrical shells, (iii) Pappus' Theorem.

*105. Compare the surface areas of the solid obtained by revolving around the x-axis the curve $y = \sqrt{1 - |x|}$, $-1 \le x \le 1$, and the sphere of the same volume.

**106. Let $a > 0$ and revolve $y = a\sqrt{1 - |x|}$, $-1 \le x \le 1$, around the x-axis. (i) Find the surface area of the resulting solid. (ii) Compare with the surface area of the sphere with the same volume when $a = 2, 5, 10$.

III

*107. How do you expect the arclengths of $y = x^2$, for $x \in [0, 1]$, $y = \frac{x^2}{4}$, for $x \in [0, 2]$, and $y = \frac{x^2}{2}$, for $x \in [0, 2]$ to compare? (Sketch all three graphs.) Use Simpson's Rule to estimate each to five decimal places. What are the ratios of the answers you get? Explain.

Logarithms and
Exponential Functions

7

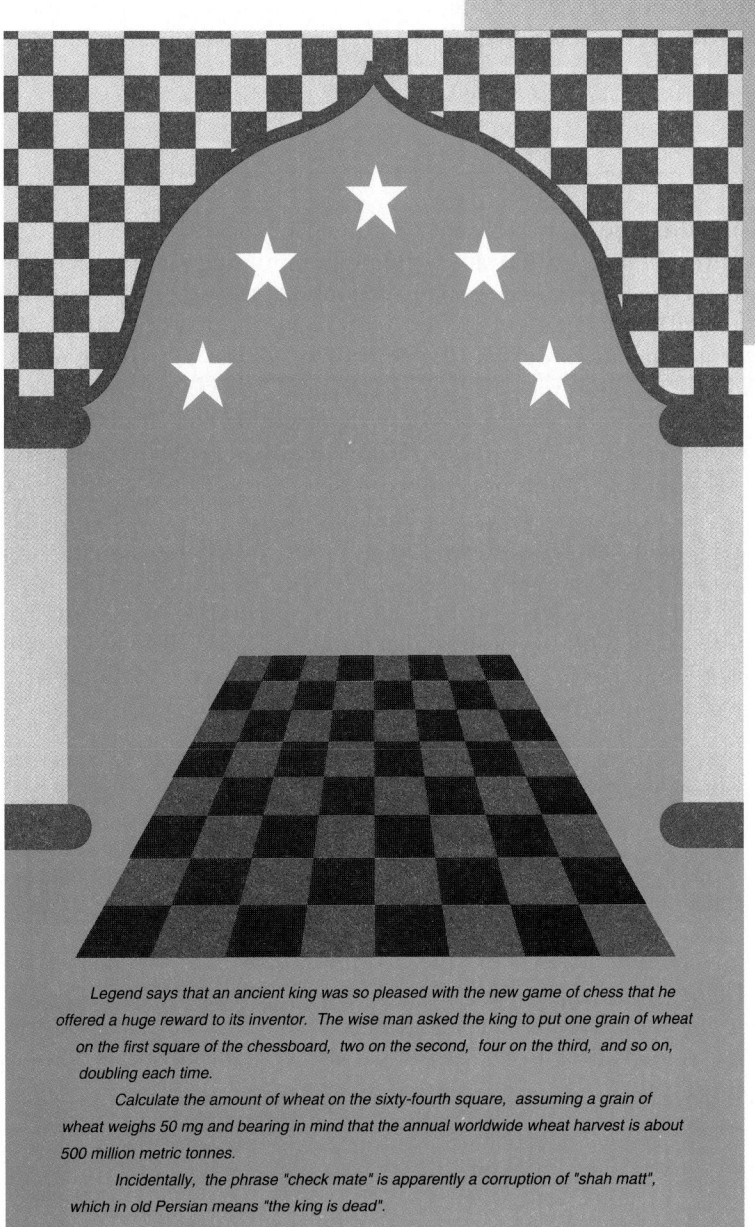

Legend says that an ancient king was so pleased with the new game of chess that he offered a huge reward to its inventor. The wise man asked the king to put one grain of wheat on the first square of the chessboard, two on the second, four on the third, and so on, doubling each time.

Calculate the amount of wheat on the sixty-fourth square, assuming a grain of wheat weighs 50 mg and bearing in mind that the annual worldwide wheat harvest is about 500 million metric tonnes.

Incidentally, the phrase "check mate" is apparently a corruption of "shah matt", which in old Persian means "the king is dead".

INTRODUCTION

We have been working with the exponential and natural logarithm functions for some time, but we have not really studied them very carefully in their own right. In this chapter we learn more about these functions and their practical applications.

The exponential function turns out to be extremely useful. It describes a great many processes of growth or change. Among the examples of exponential growth are population growth in an unrestricted environment (such as a bacterial culture or a population of animals or humans), the concentration of various ingredients or products in many chemical reactions, the amount remaining of an element undergoing radioactive decay, the temperature of an object that is being warmed or cooled by contact with another object at a different temperature, and the amount of money in an account growing under continuously compounded interest.

We begin by studying logarithms in Section 7.1, move from there to exponentials in Section 7.2, and then discuss various applications, including exponential growth and decay in Sections 7.3 and 7.4 and logistic growth in Section 7.5. The applications in Sections 7.3 and 7.4 can be studied any time after you have covered Chapter 6, and Section 7.5 can be covered after Chapter 5.

The final section, Section 7.6, concerns *differential equations*, which are equations involving the *derivatives* of a function. This section is optional; differential equations will be studied later in greater detail. The whole of Chapter 19 is devoted to them.

The relevance of differential equations to this chapter comes from two different observations. The first is that the fundamental fact about all the processes mentioned above is that each is described by a function whose rate of change is related to how big the function is (e.g., the more bacteria there are, the greater the rate at which the population grows). This amounts to a differential equation relating the *derivative* of a function to the *values* of the function. The second thing is that exponential functions play a key role in the solution of several important types of differential equations.

7.1 Logarithms

NATURAL LOGARITHM
$e = 2.71828\ldots$

When we studied the derivatives of trigonometric functions in Section 3.4, we were able to figure out what the derivatives should be by finding some moderately difficult limits and performing some clever calculations. On the other hand, when we got to logarithms and exponentials in Section 3.5, we were able to work with the derivatives, but the actual statement of what the derivatives are was something we just had to learn and take on trust, with no real explanation. The reason for the difference is that we have a definite picture of where the trigonometric functions come from. They have to do with various ratios related to an angle. There is no comparable starting point for logarithms or exponentials.

On the other hand, it is clear in hindsight that we need to have logarithms to be able to evaluate certain common integrals. The function $\ln x$ is an antiderivative of $\frac{1}{x}$.

In light of what we know about indefinite integrals, $\ln x$ is completely determined, as a function on the positive real numbers, by this fact and its value at one point, for example, $\ln 1 = 0$. So if we want to give a precise definition of $\ln x$, for $x > 0$, we can do it this way.

DEFINITION 7.1.1

> The **natural logarithm** $\ln x$ is defined for all $x > 0$ by
>
> $$\ln x = \int_1^x \frac{1}{t}\,dt.$$

We notice immediately from the definition that $\ln(1) = 0$ and by the Fundamental Theorem of Calculus that the derivative is $\frac{d}{dx}\left(\ln x\right) = \frac{1}{x}$.

The other basic thing about logarithms is the way they behave with respect to products. It is interesting to see that this is also a relatively simple consequence of the definition.

THEOREM 7.1.2

> If a and b are positive numbers, then
>
> $$\ln(ab) = \ln a + \ln b.$$

PROOF   We can write

$$\ln(ab) = \int_1^{ab} \frac{1}{t}\,dt = \int_1^a \frac{1}{t}\,dt + \int_a^{ab} \frac{1}{t}\,dt = \ln a + \int_a^{ab} \frac{1}{t}\,dt.$$

Notice that ab might, for instance, be less than 1, in which case the first integral would be negative, and similar remarks apply to the other integrals too.

It remains to show that the last integral equals $\ln b$. To do this, we make the substitution $u = \frac{t}{a}$, $du = \frac{1}{a}\,dt$. The integral becomes

$$\int_a^{ab} \frac{1}{t}\,dt = \int_1^{(ab)/a} \frac{1}{au}\,(a)\,du = \int_1^b \frac{1}{u}\,du = \ln b.$$

This completes the proof.

Repeating this result n times, we see that $\ln(a^n) = n\ln a$, for any positive integer n.

Combining Theorem 7.1.2 with the fact that $\ln(1) = 0$, we see that if $a > 0$, then $0 = \ln 1 = \ln\left(a\frac{1}{a}\right) = \ln a + \ln\frac{1}{a}$, or, after rearranging, $\ln\frac{1}{a} = -\ln a$. From this we also see that if n is a negative integer, $\ln(a^n) = \ln\left(\frac{1}{a^{-n}}\right) = -\ln(a^{-n}) = -(-n)\ln a = n\ln a$, so this formula is valid for any integer n and any $a > 0$.

We make these rearrangements so that we can work with a positive exponent $-n$ ...

Since the integrand $\frac{1}{t}$ is always positive, it is also apparent that $\ln x = \int_1^x \frac{1}{t}\,dt$ is positive if $x > 1$. On the other hand, if $0 < x < 1$, then the limits of integration are "the wrong way around," and $\ln x = \int_1^x \frac{1}{t}\,dt = -\int_x^1 \frac{1}{t}\,dt$, which is negative. In other words, $\ln x > 0$ if $x > 1$ and $\ln x < 0$ if $0 < x < 1$.

Since the derivative of $\ln x$ is $\frac{1}{x}$, the derivative is positive for all $x > 0$, and the function is increasing. The second derivative is $\frac{d}{dx}\left(\frac{1}{x}\right) = -\frac{1}{x^2}$, which is negative, so the graph of $y = \ln x$ is concave down.

Now consider $\ln(2^n) = n\ln 2$. We know that $y = \ln x$ is increasing, so $\ln 2 > \ln 1 = 0$. So $\ln 2$ is a positive number, and as $n \to \infty$, $n\ln 2 \to \infty$ also. This means that $\ln(2^n) \to \infty$ as $n \to \infty$, so $\ln(2^n)$ becomes larger than any fixed number if n is large enough. Since $\ln x$ is increasing, this shows that $\lim_{x\to\infty} \ln x = \infty$.

We could use a similar argument to find $\lim_{x\to 0^+} \ln x$, or we could simply remark that as $x \to 0^+$, we must have $\frac{1}{x} \to \infty$, so

Notice how we substituted $u = \frac{1}{x}$.

$$\lim_{x\to 0^+} \ln x = \lim_{1/x\to\infty} \ln x = \lim_{1/x\to\infty} \left(-\ln\frac{1}{x}\right) = -\lim_{u\to\infty} \ln u = -\infty.$$

We collect all these facts together.

Summary 7.1.3   Properties of ln X

1. For any $x > 0$, $\ln x$ is defined by

$$\ln x = \int_1^x \frac{1}{t}\,dt.$$

2. For any positive numbers a and b and any integer n,

$$\ln(ab) = \ln a + \ln b, \qquad \ln\left(\frac{a}{b}\right) = \ln a - \ln b, \qquad \ln(a^n) = n\ln a.$$

3. $\ln(1) = 0$.

4. $\frac{d}{dx}(\ln x) = \frac{1}{x}$; the function $\ln x$ is increasing and concave downward.

5. If $x > 1$, then $\ln x > 0$; and if $0 < x < 1$, then $\ln x < 0$.

6. $\lim_{x \to \infty} \ln x = \infty$ and $\lim_{x \to 0+} \ln x = -\infty$.

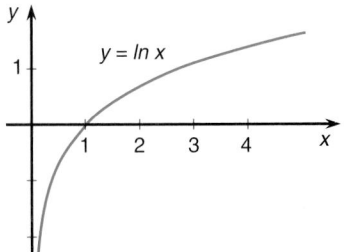

Figure 7.1.1

From Properties 3, 4, 5, and 6 it is possible to sketch the graph of $y = \ln x$.

On the one hand, we know that $\lim_{x \to \infty} \ln x = \infty$, but on the other hand, the derivative is $\frac{1}{x}$, which becomes smaller and smaller, tending to 0 as $x \to \infty$. In terms of the graph, this means that the graph is less and less steep, closer and closer to being horizontal. See Figure 7.1.1.

It seems contradictory that the graph could tend to infinity and yet have tangents that tend to the horizontal, but this is what happens. The idea is that while $\ln x$ does tend to infinity, it does so very *slowly*.

Certainly it goes to infinity more slowly than the straight line $y = x$, as shown in Figure 7.1.2, but the same is true of $y = \sqrt{x}$, for example, the graph of which is shown in Figure 7.1.3. It is not very clear from the pictures whether $\ln x$ or $\sqrt{x}$ goes

Figure 7.1.2

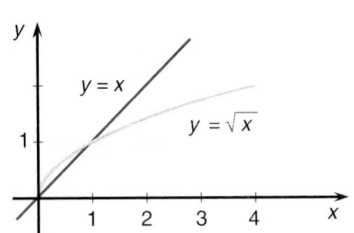

Figure 7.1.3

to infinity more slowly. For that matter we could also consider other powers x^r. If r is any positive number, $\lim_{x \to \infty} x^r = \infty$, but if r is small (i.e., $0 < r < 1$), the graph of $y = x^r$ grows more and more slowly as x increases.

It would be interesting to compare $y = \ln x$ and $y = x^r$ for various positive values of r to see which of them grows most slowly. The answer is contained in the following theorem.

THEOREM 7.1.4

If r is any positive number, then

$$\lim_{x \to \infty} \frac{\ln x}{x^r} = 0.$$

In words, we say that "$\ln x$ grows more slowly than any power x^r."

PROOF We use the version of l'Hôpital's Rule that applies to infinite limits at infinity. It tells us that

$$\lim_{x \to \infty} \frac{\ln x}{x^r} = \lim_{x \to \infty} \frac{\frac{1}{x}}{rx^{r-1}} = \lim_{x \to \infty} \frac{1}{rx^r} = 0.$$

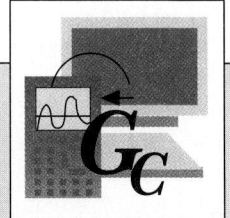

With a graphics calculator we can plot $y = \frac{\ln x}{x^r}$ for various values of r, to see the decrease predicted by the theorem. For instance, with $r = 0.75$,

TI-85	SHARP EL9300
GRAPH F1 $(y(x) =)$ LN F1 (x) ÷ F1 (x) ˆ .75 M2 (RANGE) .5 ▼ 100 ▼ ▼ (−) 1 ▼ .5 F5 (GRAPH)	$\not\!\!A\!\!\downarrow$ ln x/θ/T ÷ (x/θ/T a^b .75 RANGE .5 ENTER 100 ENTER 10 ENTER −1 ENTER .5 ENTER .5 ENTER $\not\!\!A\!\!\downarrow$
CASIO f_x-6300G (f_x-7700GB)	HP 48SX
Range .5 EXE 100 EXE 10 EXE −1 EXE .5 EXE .5 EXE EXE EXE EXE Graph ln X ÷ (X x^y .75) EXE	↰ PLOT PLOTR ERASE ATTN ' Y ≡ LN X ▶ ÷ X y^x .75 ↱ PLOT ↰ DRAW ERASE .5 ENTER 100 XRNG 1 +/− ENTER .5 YRNG DRAW

You could modify these commands to try it with different values of r. You may have to plot the curve for a wider range of x in order to see the decreasing behavior. For instance, try it with $r = 0.1$ or $r = 0.01 \ldots$

In anticipation of the next section we remark that since $\ln x$ equals zero when $x = 1$ and tends to infinity as $x \to \infty$, the graph must at some point cross the horizontal line $y = 1$ (see Figure 7.1.4). In other words, there is some number x whose natural logarithm equals 1. This number is called e, and it is approximately equal to 2.718 281 83. The symbol e, which stands for "exponential," was first introduced by the Swiss mathematician Leonhard Euler (1707–1783).

Leonhard Euler (1707–1783)

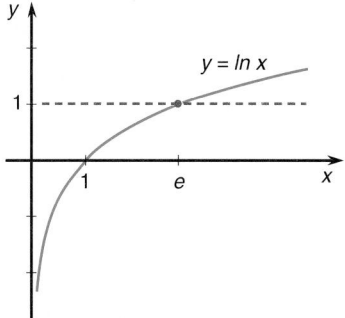

Figure 7.1.4

Exercises 7.1

I

Suppose A and B are positive real numbers, and write $a = \ln A$ and $b = \ln B$. Suppose n and m are integers. Write each of the following quantities in terms of a, b, m, and n.

1. $\ln(AB)$

2. $\ln(A^4 B)$

3. $\ln\left(\frac{A}{B^2}\right)$

4. $\ln\frac{1}{A^2}$

5. $\ln(A^m B^n)$

6. $\ln\big((1)(A)\big)$

7. $\ln(A^{-m})$

8. $\ln(A^m / B^n)$

9. $\ln(\sqrt{A})$

10. $\ln(\sqrt{AB})$

Evaluate the following limits.

11. $\lim_{x\to\infty} \frac{\ln x}{x^3}$

12. $\lim_{x\to\infty} \frac{\ln x}{x+1}$

13. $\lim_{x\to\infty} \frac{\ln(x^5)}{x^{1/3}}$

14. $\lim_{x\to\infty} \frac{x\ln x}{x+4}$

15. $\lim_{x\to 0^+} x\ln x$

16. $\lim_{x\to 0^+} \sqrt{x}\ln x$

17. $\lim_{x\to 2^+} (x-2)^{2/13}\ln(x-2)$

18. $\lim_{x\to 0^+} \sqrt{x}\ln(\sqrt{x})$

19. $\lim_{x\to 1^+} \frac{\ln x}{x-1}$

20. $\lim_{x\to e^+} \frac{x-e}{1-\ln x}$

II

21. Use an argument like the one in Theorem 7.1.2 (p. 510) (making a substitution in an integral) to show directly that for $a > 0$, $\ln\frac{1}{a} = -\ln a$.

22. Suppose $r > 0$ and use an argument like the one in Theorem 7.1.4 (p. 512) to evaluate

$$\lim_{x\to 0^+} (x^r \ln x).$$

23. The formula $\ln(a^n) = n\ln a$ (Summary 7.1.3, Property 2, p. 511) was established in the case in which n is an integer. Give an argument to show that it is still true for $n = \frac{1}{2}$.

*24. (i) Show that for any $a > 0$ and any positive integer n, $\ln(a^{1/n}) = \frac{1}{n}\ln a$. (Notice that $a^{1/n}$ means the positive nth root of a.)

(ii) Show that for any $a > 0$ and any positive rational number r, $\ln(a^r) = r\ln a$.

(iii) Show that for any $a > 0$ and any rational number r, $\ln(a^r) = r\ln a$.

*25. Evaluate $\lim_{x\to\infty} (x - \ln x)$.

26. (i) What is the domain of the function $f(x) = \ln(x^2)$?

(ii) Make a quick sketch of the graphs of $y = f(x) = \ln(x^2)$ and of $y = 2\ln x$.

*27. Imagine the graph of $y = \ln x$, with the units measured in inches. (i) If you go out along the positive x-axis for a distance of 1 mile ($= 5280$ ft) from the origin, what will be the value of $\ln x$? (First find x, recalling that there are 12 inches in one foot.) (ii) What will be the value of $\ln x$ if you travel the distance to the moon (about 240,000 miles)? (iii) What about the distance to the sun (about 93,000,000 miles)?

*28. Now imagine the graph of $y = \ln x$ with units measured in centimeters. How far is it necessary to go from the origin before $\ln x = 100$? (Express your answer in light-years: one light-year is the distance light travels in one year, going at 300,000 km/sec.)

29. In Theorem 7.1.2 (p. 510) we used the integral definition to prove that $\ln(ab) = \ln a + \ln b$. Prove this result another way, by comparing the derivatives of $f(x) = \ln x$ and $g(x) = \ln(ax)$.

30. If n is a positive integer, prove that $\ln(x^n) = n\ln x$, for $x > 0$, by considering the derivative of $h(x) = \ln(x^n)$.

III

31. Use a graphics calculator or computer to sketch the graph of $y = \ln x$, and use the picture to find as many digits as possible in the expression for e. (Recall that $\ln e = 1$.)

32. Use the graph of $y = e^x$ to find as many digits as possible in the expression for $\ln(10)$.

7.2 Exponentials and Powers

EXPONENTIAL FUNCTION

POWERS

LOGARITHM

BASE

COMMON LOGARITHM

In Section 7.1 we studied the natural logarithm. We defined it as the integral $\ln x = \int_1^x \frac{1}{t}\,dt$ and were able to discover most of the familiar properties of the logarithm as simple consequences of the definition.

One of those properties is the fact that $\ln x$ is an increasing function because the derivative $\frac{d}{dx}(\ln x) = \frac{1}{x}$ is positive. We realized in Section 3.8 that an increasing function could not take the same value twice (i.e., at two different values of x); see Figure 7.2.1. In Definition 3.8.2 (p. 203) we called a function with this property *one-to-one* and observed that a function that is one-to-one always has an inverse function.

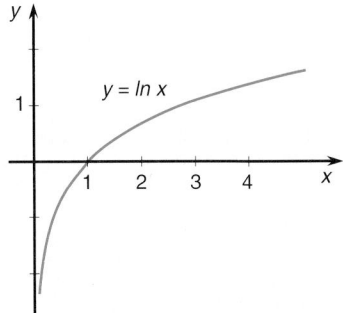

Figure 7.2.1

We define the inverse function of ln x to be the **exponential function** e^x, sometimes written $\exp(x)$. In other words,

$$e^{\ln x} = x, \quad \text{if } x > 0 \qquad \text{and} \qquad \ln(e^x) = x, \quad \text{for any } x. \qquad (7.2.1)$$

Notice that since ln x is only defined for $x > 0$, its inverse, the exponential function, only has positive values, that is, $e^x > 0$, for every x.

Many properties of e^x can easily be deduced from the corresponding properties of ln x, using Equations 7.2.1. For instance,

$$e^{a+b} = \exp(a + b) = \exp\big(\ln(e^a) + \ln(e^b)\big) = \exp\big(\ln(e^a e^b)\big) = e^a e^b.$$

Repeating this several times shows that $e^{nx} = (e^x)^n$, for any positive integer n. (If you know about proofs by induction, then you will recognize this as a result that should be proved that way.)

Since $\ln(1) = 0$, we know from Equations 7.2.1 that $e^0 = 1$, and from this and the above formula about products it follows easily that $e^{-x} = 1/e^x$.

In Theorem 3.8.4 we used the Chain Rule to calculate the derivative of an inverse function. In the present case, if we let $y = e^x$, then $x = \ln(y)$, and the derivative is

$$\frac{dy}{dx} = \frac{d}{dx}(e^x) = \frac{1}{\frac{d}{dy}(\ln y)} = \frac{1}{1/y} = y = e^x.$$

This explains the formula $\frac{d}{dx}(e^x) = e^x$, which we have been using since Chapter 3.

It also shows that the derivative of e^x is always positive, so the function is increasing. In particular, since $e^0 = 1$, we see that $e^1 > 1$, so $e^n = e^{n(1)} = (e^1)^n$; this is the nth power of a number bigger than 1, so it grows larger and larger and tends to infinity. Together with the fact that e^x is increasing, this shows that $\lim_{x \to \infty} e^x = \infty$. An argument like the one used to show $\lim_{x \to 0+} \ln x = -\infty$ shows that $\lim_{x \to -\infty} e^x = 0$.

With these facts and the trivial observation that the second derivative is also positive, it is easy to sketch the graph of $y = e^x$. We summarize.

Summary 7.2.2

Properties of e^x

1. e^x is the inverse function of ln x. This means that

   $$e^{\ln x} = x, \quad \text{if } x > 0 \qquad \text{and} \qquad \ln(e^x) = x, \quad \text{for any } x.$$

2. $e^{a+b} = e^a e^b$, and for any integer n, $e^{na} = (e^a)^n$.

3. $e^0 = 1$.

4. $\frac{d}{dx}(e^x) = e^x$; $y = e^x$ is increasing and concave up.

5. $e^x > 0$ for any x; $0 < e^x < 1$ if $x < 0$, and $e^x > 1$ if $x > 0$.

6. $\lim_{x \to \infty} e^x = \infty$ and $\lim_{x \to -\infty} e^x = 0$.

7. The function $f(x) = e^x$ is uniquely determined by the facts that $\frac{d}{dx}(f) = f$ and $f(0) = 1$; that is, it is the only function defined for every $x \in \mathbb{R}$ and satisfying $f' = f$ and $f(0) = 1$.

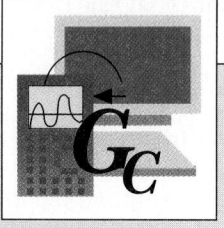

Since $2e^x = e^{x+\ln 2}$, we see that the graph of $y = 2e^x$ is the graph of $y = e^x$ *shifted to the left* by a distance of $\ln 2$.

With a graphics calculator we can plot $y = e^x$, $y = 2e^x$, $y = 4e^x$, and $y = 8e^x$ and see how these curves are all horizontal shifts of each other.

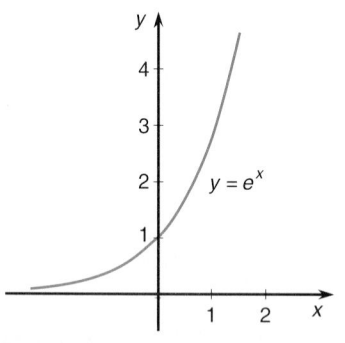

How would you get two more curves, one shifted to the left and one to the right of these, both by the same distance as the distance between $y = e^x$ and $y = 2e^x$?

The graph of $y = e^x$ is shown in Figure 7.2.2. As always happens with inverse functions, we can think of it as shown in Figure 7.2.3 as the result of reflecting the graph of $y = \ln x$ in the line $y = x$ (cf. the discussion after Example 3.8.5, p. 204).

Powers

Now we can use the exponential function to define powers. What makes this a subtle problem is that we already know what we mean by a^n if n is an integer. Even if r is a rational number, a^r has a definite meaning, at least if $a > 0$. So if we want to define a^x for *any* real number x, it will be necessary to check that our definition agrees with the original meaning in case x is a rational number and $a > 0$.

If n is an integer, the power a^n simply refers to the product of n factors of a, if n is positive, and to a product of factors of $\frac{1}{a}$, if $n < 0$. Fractional powers refer to roots ($a^{m/n}$ means the mth power of the nth root of a or, equivalently, the nth root of the mth power of a).

We know that for any $a > 0$ and any integer n, $a^n = (e^{\ln a})^n = e^{n \ln a}$. Next consider the mth root $a^{1/m}$, for any positive integer m. It is the (unique) posi-

Figure 7.2.2

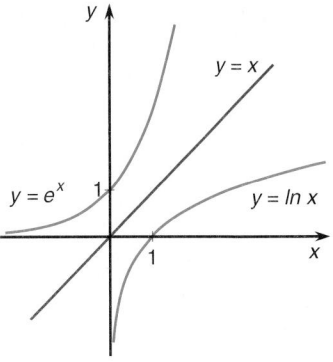

Figure 7.2.3

tive number whose mth power equals a, that is, $a^{1/m}a^{1/m}a^{1/m}\ldots a^{1/m} = a$, where the product on the left side consists of m factors of $a^{1/m}$. Applying the natural logarithm function to both sides of this equation, we find $\ln(a^{1/m}) + \ln(a^{1/m}) + \ln(a^{1/m}) + \ldots + \ln(a^{1/m}) = \ln(a)$, and since there are m terms on the left, this says that $m\ln(a^{1/m}) = \ln a$, or $\ln(a^{1/m}) = \frac{\ln a}{m}$. Exponentiating both sides of the equation, we find $a^{1/m} = \exp\left(\frac{\ln a}{m}\right)$. The left side means the mth root of a, and the right side means the value of $\exp x$ at $x = \frac{\ln a}{m}$.

Now consider the $\frac{n}{m}$th power, for any integers $m \neq 0$ and n. We already know that $\exp\left(\frac{\ln a}{m}\right) = a^{1/m}$, and by the formula about powers in Property 2 of Summary 7.2.2 we can conclude that $\exp\left(n\frac{\ln a}{m}\right) = (a^{1/m})^n$, or $\exp\left(\frac{n}{m}\ln a\right) = a^{n/m}$.

To put it another way, for any $a > 0$ and any rational number r we can express the rational power by using the simple exponential formula

$$a^r = e^{r\ln a}.$$

This formula makes sense for any $r \in \mathbb{R}$, so we can use it to define *any* power.

DEFINITION 7.2.3

> Suppose $a > 0$ and x is any real number. Then the **power** a^x is defined by
> $$a^x = e^{x\ln a}.$$

The remarks preceding the definition show that when x is a rational number, this definition agrees with the usual meaning of the power a^x.

THEOREM 7.2.4

> If $a > 0$, and u, v are any real numbers, then
> (i) $a^u a^v = a^{u+v}$,
> (ii) $(a^u)^v = a^{uv}$.

PROOF

(i) We simply evaluate

$$a^u a^v = e^{u\ln a}e^{v\ln a} = e^{(u\ln a + v\ln a)} = e^{(u+v)\ln a} = a^{(u+v)}.$$

(ii) Similarly,

$$(a^u)^v = \exp\left(v\ln(a^u)\right) = \exp\left(v\ln(e^{u\ln a})\right)$$
$$= \exp\left((v)u\ln a\right) = \exp\left((uv)\ln a\right) = a^{uv}.$$

It is interesting to regard x^r as a function of x, for $x > 0$.

EXAMPLE 1   Let $r \in \mathbb{R}$ be fixed. Find the derivative of the function $y = x^r$, for $x > 0$.

Solution   Since $y = x^r = e^{r\ln x}$, we see that

$$\frac{dy}{dx} = \frac{d}{dx}(e^{r\ln x}) = e^{r\ln x}\frac{d}{dx}(r\ln x) = e^{r\ln x}\left(\frac{r}{x}\right) = x^r\frac{r}{x} = rx^{r-1}.$$

In other words, the familiar formula for the derivative of x^r is valid for any power r. Previously, we knew it only in the case in which r is rational.

The graph of $y = e^x$ rises very fast. Certainly, it goes up faster than the straight line $y = x$, but it is less clear what happens when e^x is compared with other powers x^t. In analogy with the result of Theorem 7.1.4 comparing the growth of x^r with that of $\ln x$, we consider $\lim_{x \to \infty} (x^r/e^x)$.

THEOREM 7.2.5

For any real number r,

$$\lim_{x \to \infty} \left(\frac{x^r}{e^x} \right) = 0.$$

In words, we say that "e^x grows faster than any power of x."

PROOF Consider

$$\ln \left(\frac{x^r}{e^x} \right) = r \ln x - x = \left(r \frac{\ln x}{x} - 1 \right) x.$$

In Theorem 7.1.4 we saw that $\lim_{x \to \infty} (\ln x)/x = 0$. So $\left(\frac{r \ln x}{x} - 1 \right) \to -1$. Since $x \to \infty$, we see that $\ln\left((x^r)/(e^x)\right) = \left(\frac{r \ln x}{x} - 1 \right) x \to -\infty$.

If the logarithm of some quantity tends to $-\infty$, then the quantity itself must tend to 0, since $\lim_{x \to -\infty} e^x = 0$. This completes the proof, showing that $(x^r)/(e^x) \to 0$. ◣

In addition to the rth power function $f(x) = x^r$, it is also possible to consider the function $g(x) = a^x$, where a is any fixed positive number. This is a function that is defined for all values of x, and it is possible to consider its derivative.

EXAMPLE 2 Let $a > 0$ be fixed. Find $\frac{d}{dx}(a^x)$.

Solution We write $a^x = e^{x \ln a}$, so

$$\frac{d}{dx}(a^x) = \frac{d}{dx}(e^{x \ln a}) = e^{x \ln a} \frac{d}{dx}(x \ln a) = a^x \ln a.$$

◣

Note how we wrote $e^{x \ln a}$ as a^x at the end.

Limits of Powers

In evaluating a limit of the form $\lim_{x \to a} \left(f(x)^{g(x)} \right)$, where the base $f(x)$ is a positive function of x, it is frequently helpful to write out the power as

$$f(x)^{g(x)} = e^{g(x) \ln(f(x))}.$$

If we can evaluate the limit of this exponent, say, $\lim_{x \to a} g(x) \ln\left(f(x)\right) = L$, for instance, then we can conclude that

$$\lim_{x \to a} \left(f(x)^{g(x)} \right) = \lim_{x \to a} \left(e^{g(x) \ln(f(x))} \right) = e^L.$$

Frequently, in questions of this kind we find that in the product $g(x) \ln(f(x))$ one factor will tend to ∞ and the other to 0. In such a case it is often helpful to rearrange it into a quotient and use l'Hôpital's Rule. We write

$$g(x) \ln\left(f(x)\right) = \frac{\ln\left(f(x)\right)}{\frac{1}{g(x)}} = \frac{g(x)}{\frac{1}{\ln(f(x))}},$$

and apply l'Hôpital's Rule to either of these quotients. Offhand, it is probably simpler to use the one in which both numerator and denominator tend to zero rather than the other, in which they both tend to ∞.

EXAMPLE 3   Find $\lim_{x\to\infty}\left(1+\frac{1}{x}\right)^x$.

Solution   We write $\left(1+\frac{1}{x}\right)^x = \exp\left(x\ln\left(1+\frac{1}{x}\right)\right)$. As $x\to\infty$, $1+\frac{1}{x}\to 1$, so $\ln\left(1+\frac{1}{x}\right)\to 0$. Following the suggestion made in the discussion above, we write $x\ln\left(1+\frac{1}{x}\right) = \frac{\ln(1+1/x)}{1/x}$ and apply l'Hôpital's Rule. We find

$$\lim_{x\to\infty} x\ln\left(1+\frac{1}{x}\right) = \lim_{x\to\infty} \frac{\ln\left(1+\frac{1}{x}\right)}{\frac{1}{x}} = \lim_{x\to\infty} \frac{\frac{1}{1+\frac{1}{x}}\left(-\frac{1}{x^2}\right)}{-\left(\frac{1}{x^2}\right)}$$

$$= \lim_{x\to\infty} \frac{1}{1+\frac{1}{x}} = 1.$$

This tells us that the natural logarithm of $\left(1+\frac{1}{x}\right)^x$ tends to 1, so $\left(1+\frac{1}{x}\right)^x$ itself tends to $e^1 = e$. We have just found that

$$\lim_{x\to\infty}\left(1+\frac{1}{x}\right)^x = e. \tag{7.2.6}$$

This result is very useful.

EXAMPLE 4   Suppose a is any fixed real number. Find $\lim_{x\to\infty}\left(1+\frac{a}{x}\right)^x$.

Solution   As above, we find the limit of the logarithm $x\ln\left(1+\frac{a}{x}\right)$, using l'Hôpital's Rule:

$$\lim_{x\to\infty}\left(x\ln\left(1+\frac{a}{x}\right)\right) = \lim_{x\to\infty} \frac{\ln\left(1+\frac{a}{x}\right)}{\frac{1}{x}} = \lim_{x\to\infty} \frac{\frac{1}{1+\frac{a}{x}}\left(-\frac{a}{x^2}\right)}{-\frac{1}{x^2}}$$

$$= \lim_{x\to\infty} \frac{1}{1+\frac{a}{x}}\frac{a}{1} = a.$$

Taking exponentials, we see that

$$\lim_{x\to\infty}\left(1+\frac{a}{x}\right)^x = e^a. \tag{7.2.7}$$

Of course, Example 4 includes the previous example, by letting $a = 1$. They will be used in Section 7.4 for the study of compound interest.

Logarithms to Other Bases

The natural logarithm has the property that

$$a = e^{\ln a},$$

for any positive number a.

We could try to replace the number e in this formula with some other positive number and look for an analogous formula. Suppose we fix some positive number

$b \neq 1$ and consider the function $y = b^x$. Given any $a > 0$, we could look for the value of x for which $b^x = a$.

In fact it is easy to find this value of x. After all, if we write $b^x = e^{x \ln b}$, then we see that for this to equal a we must have $x \ln b = \ln a$, or $x = (\ln a)/(\ln b)$.

DEFINITION 7.2.8

Suppose b is a fixed positive number, $b \neq 1$, and consider the function $y = b^x$. We refer to b as the **base** for this function.

Moreover, define the **logarithm to the base b**, written $\log_b$, by

$$a = b^{\log_b(a)},$$

for every $a > 0$.

We saw above that

$$\log_b(a) = \frac{\ln a}{\ln b}.$$

REMARK 7.2.9

From the preceding formula we see that $\log_b$ satisfies many of the same properties as the natural logarithm. Specifically, for $b > 0$, with $b \neq 1$,

(i) $\log_b(xy) = \log_b(x) + \log_b(y)$, for any $x,\ y > 0$.

(ii) $\log_b(1) = 0$.

(iii) $\log_b(x^r) = r \log_b(x)$, for any $x > 0$ and any power r.

(iv) We can easily find $\frac{d}{dx}\big(\log_b(x)\big)$. In fact, using the formula for $\log_b x$, we see that

$$\frac{d}{dx}\big(\log_b(x)\big) = \frac{d}{dx}\left(\frac{\ln(x)}{\ln(b)}\right) = \frac{1}{\ln(b)}\frac{d}{dx}(\ln x) = \frac{1}{\ln(b)}\frac{1}{x}.$$

The most frequently used bases are $b = e$, $b = 10$, and $b = 2$, though in principle it is possible to use any positive number $b \neq 1$ as the base for logarithms. The base $b = 10$ in particular is used so often that logarithms to the base 10 are often called **common logarithms**.

EXAMPLE 5

(i) $\log_{10}(e) = \ln(e)/\ln(10) = \frac{1}{\ln(10)} \approx 0.434294$.

(ii) $\log_2(\sqrt{2}) = \frac{1}{2}$. This we should realize immediately without calculating anything, since after all, $\log_2(\sqrt{2})$ ought to be the power to which we have to raise 2 to get $\sqrt{2}$, so it must equal $\frac{1}{2}$.

(iii) Find $\log_3(81)$. This we might be able to guess, but instead we can calculate $\log_3(81) = \big(\ln(81)\big)/\big(\ln(3)\big)$. This will not help much unless we use a calculator or unless we remember that $81 = (9)(9) = 3^4$. With this in mind we find $\ln(81) = \ln(3^4) = 4\ln(3)$, so $\log_3(81) = \big(\ln(81)\big)/\big(\ln(3)\big) = \big(4\ln(3)\big)/\big(\ln(3)\big) = 4$.

In a sense we ought to have guessed this because all it says is that 81 equals the number 3 raised to the fourth power.

Exercises 7.2

I

Evaluate each of the following. If any limit does not exist, check to see whether it equals $\pm\infty$.

1. $\lim_{x\to\infty} e^{-x}$

2. $\lim_{x\to\infty} xe^{-x}$

3. $\lim_{x\to\infty} \left(\frac{x}{\pi^x}\right)$

4. $\lim_{x\to-\infty} 2^{-x}$

5. $\lim_{x\to-\infty} (\sqrt{2})^x$

6. $\lim_{x\to 0^-} \left(\frac{1}{2}\right)^{1/x}$

7. $\frac{d}{dx}\left(x^{\sqrt{5}}\right)$

8. $\frac{d}{dx}\left((\sqrt{5})^x\right)$

9. $\lim_{x\to 0^+} (x^{-2}e^{-1/x})$

10. $\lim_{x\to\infty} \left(\frac{\pi^x}{x^\pi}\right)$

Evaluate each of the following.

11. $\lim_{x\to\infty} \frac{e^{x/2}}{x^3+3x-2}$

12. $\lim_{x\to-\infty} \left((2x^4+3x+1)e^x\right)$

13. $\int x^{\sqrt{2}}\,dx$

14. $\int x^{2\pi}\,dx$

15. $\int 2^x\,dx$

16. $\int 5^{2x}\,dx$

17. $\int x4^{x^2}\,dx$

18. $\int x^2\pi^{x^3}\,dx$

19. $\int x^{-1/2}10^{\sqrt{x}}\,dx$

20. $\int 2^x e^{7x}\,dx$

Evaluate each of the following.

21. $\log_{10}(100)$

22. $\log_{1/2}(16)$

23. $\log_\pi(\pi^2)$

24. $\log_3(9^{-4/5})$

25. $\frac{d}{dx}\left(\log_{10}(x)\right)$

26. $\frac{d}{dx}\left(\log_5(x^2+3)\right)$

27. $\lim_{x\to\infty} \log_{1/4}(x)$

28. $\lim_{x\to 0^+} \left(\log_{10}(3x)\right)/\left(\log_2(x)\right)$

29. $\int \frac{\log_{10}(x)}{\log_3(x)}\,dx$

30. $\int \frac{\log_2(x)}{x}\,dx$

Evaluate each of the following finite or infinite limits.

31. $\lim_{x\to\infty} \left(1+\frac{3}{x}\right)^x$

32. $\lim_{x\to\infty} \left(1+\frac{1}{2x}\right)^x$

33. $\lim_{x\to\infty} \left(1+\frac{1}{x}\right)^{-x}$

34. $\lim_{x\to-\infty} \left(1+\frac{1}{x}\right)^x$

35. $\lim_{x\to\infty} \left(1-\frac{2}{x}\right)^{2x}$

36. $\lim_{x\to\infty} \left(1-\frac{1}{\pi x}\right)^{x/2}$

37. $\lim_{x\to\infty} \left(1-\frac{1}{x}\right)^x/\left(1+\frac{1}{x}\right)^x$

38. $\lim_{x\to\infty} \left(1+\frac{1}{x^2}\right)^{1/x}$

39. $\lim_{x\to\infty} \left(1+\frac{1}{x^2}\right)^x$

40. $\lim_{x\to\infty} \left(1+\frac{1}{x}\right)^{x^2}$

II

41. Calculate the derivative of $F(x) = (e^2)^x$. Verify that it equals the derivative of $f(x) = e^{2x}$.

42. Suppose $a > 0$ and $b > 0$ and show that for any x, $a^x b^x = (ab)^x$.

43. What is the minimum value of $f(x) = 6^x + 6^{-x}$?

44. What is the minimum value of $f(x) = 2^x + 4^{-x}$?

45. Show that for any $a > 1$ the function $y = a^x$ goes to infinity faster than any power of x in the sense that $\lim_{x\to\infty} (x^r)/(a^x) = 0$.

46. Show that for any $b > 1$ the function $y = \log_b(x)$ goes to infinity more slowly than any positive power of x, in the sense that $\lim_{x\to\infty} \left(\log_b(x)\right)/(x^r) = 0$, for any $r > 0$.

*47. Show that for any $a > 1$ the function $y = a^x$ goes to infinity faster than any polynomial $p(x)$, in the sense that $\lim_{x\to\infty} \left(p(x)\right)/(a^x) = 0$.

*48. Show that for any $b > 1$ the function $y = \log_b(x)$ goes to infinity more slowly than any polynomial $p(x)$, in the sense that $\lim_{x\to\infty} \left(\log_b(x)\right)/\left(p(x)\right) = 0$. What happens if $0 < b < 1$?

49. Find $\lim_{x\to\infty} \left(\frac{1}{6}\right)^x$.

*50. Suppose $a > 0$ and $b > 0$ and let $f(x) = a^x + b^{-x}$. Find the value(s) of x at which $f(x)$ has an absolute minimum. Your answer should contain a condition on a and b, without which there is no absolute minimum.

51. Let $f(x) = 5^x$, defined for every real number x. Find the regions where $f(x)$ is increasing and decreasing.

*52. Suppose $a > 0$ and let $f(x) = a^x$. Find the regions where $f(x)$ is increasing and decreasing. (*Hint:* Your answer will depend on the value of a.)

*53. For $x > 0$, define $f(x) = x^x$. (i) Find the derivative $f'(x)$. (ii) Find $\lim_{x\to 0^+} f(x)$. (iii) Classify the extrema of $f(x)$. (iv) Sketch the graph of $y = f(x)$.

54. If a, b, and c are positive numbers with $a \neq 1$ and $b \neq 1$, show that

$$\left(\log_a(b)\right)\left(\log_b(c)\right) = \log_a(c).$$

*55. Prove the statement made in Property 7 of Summary 7.2.2. (*Hint:* Let $f(x) = e^x$, and suppose $g(x)$ is another function satisfying $g' = g$ and $g(0) = 1$. Since $f(x)$ is never zero, it is possible to consider the function $\frac{g(x)}{f(x)}$; find its derivative ...)

56. Find the derivative of x^{x^x}, for $x > 0$.

57. Use a graphics calculator or computer to sketch the graphs of $y = \ln x$, $y = \log_{10}(x)$, and $y = \frac{\ln(x)}{\ln(10)}$ on the same picture.

58. Use a graphics calculator or computer to sketch the graphs of $y = e^{\sin x}$ and $y = \sin x$ on the same picture.

POINT TO PONDER

In Definition 7.2.3 we defined powers a^r. In Example 1 we found how to differentiate the power x^r; the result was the unsurprising formula $\frac{d}{dx}(x^r) = rx^{r-1}$, for any $r \in \mathbb{R}$.

Using this, we were able to apply l'Hôpital's Rule to find in Theorem 7.2.5 that

$$\lim_{x \to \infty} \frac{x^r}{e^x} = 0,$$

for any $r \in \mathbb{R}$.

In fact we used a similar argument in Theorem 7.1.4

(p. 512) to show that

$$\lim_{x \to \infty} \frac{\ln x}{x^r} = 0,$$

for any $r \in \mathbb{R}$. However, at that time we did not really know what x^r means unless r is a rational number, and we certainly did not know how to differentiate it if r is irrational.

So, strictly speaking, the result of Theorem 7.1.4 was really proved only for rational exponents r, but now, with the benefit of Example 1, we can prove it for every $r \in \mathbb{R}$. The proof is exactly the same.

7.3 Exponential Growth and Decay

GROWTH CONSTANT
DECAY CONSTANT
INITIAL VALUE
HALF-LIFE

There are many practical situations in which the rate of change of some quantity is proportional to the size of the quantity. The first example is population growth, for which the rate of change, that is, the number of new offspring per unit time, is proportional to the population. After all, the larger the population, the more offspring there will be.

This idea applies to many different types of population, including a culture of bacteria, the human population of some region, an animal population, or even some kinds of plant growth. In each of these situations it seems reasonable to assume that the rate of growth is proportional to the population.

In fact it is usually more subtle than that. Imagine a bacterial culture in a jar. The bacteria multiply, and the population grows. It grows faster and faster, but eventually there is a problem because it begins to run out of room. Except in science fiction it is not realistic to think that the culture will keep on growing until it bursts out of the jar. Before the jar is completely full, the growth will slow down.

This will happen for perfectly sensible reasons. For one thing the culture may begin to use up the nutrients in the jar. With less food available the bacteria cannot reproduce as quickly. Even simple overcrowding may reduce the efficiency of the bacterial metabolism.

In any case, the growth will slow down before exceeding the "limits" imposed on its growth by the nature and size of the environment. Similar observations can be made about more complicated populations. Human populations grow at slower rates under the effects of drastic overcrowding. The population of deer in a forest grows dramatically for a time but then falls off as the available food supply becomes insufficient to support all the animals.

Another consideration that we have not even mentioned is that in any natural population there are deaths as well as reproduction. This will of course affect the *net* population growth. In fact in many situations this particular question does not really change the behavior of a population in any qualitative way, so to ignore it may actually not be too serious.

The factors that limit growth and the mechanisms by which they affect the population are numerous and extremely complicated. We do not intend to study them here, other than to remark that such factors do exist and do exert a strong influence upon the growth of various populations. We intend to consider situations in which the rate of growth is proportional to the population. Typically, we can expect this to happen when the population is low in comparison to the levels that the environment can support because in this case the limitations are not likely to have much effect.

Consider a function $f(t)$. We let the variable t represent time, so we may think of $f(t)$ as a changing quantity. Its derivative $\frac{df}{dt}$ is the rate of change of $f(t)$. For convenience, let us assume the values of the function $f(t)$ are always positive, that is, $f(t) > 0$ for all t. An example that you may wish to keep in mind is to let $f(t)$ be the number of bacteria in a jar at time t.

Now suppose that we know that the rate of change of $f(t)$ equals a constant times the function $f(t)$ itself. Symbolically,

$$f'(t) = kf(t),$$

where k is some constant.

From this we see that $\frac{f'(t)}{f(t)} = k$. The left side of this equation is the derivative of $\ln\big(f(t)\big)$, so it says that $\frac{d}{dt}\Big(\ln\big(f(t)\big)\Big) = k$. This means that $\ln\big(f(t)\big)$ is an antiderivative of the constant function k, and that means that it must equal $kt + c'$, for some constant c'.

If $\ln\big(f(t)\big) = kt + c'$, then $f(t) = e^{\ln(f(t))} = e^{kt+c'}$. By the properties of the exponential function this can be rewritten as $e^{kt+c'} = e^{kt}e^{c'} = ce^{kt}$, where $c = e^{c'}$ is a constant. What we have found is that

$$f(t) = ce^{kt}. \tag{7.3.1}$$

> This sort of exponential formula describes the behavior of any function whose derivative equals a constant times the original function. For this reason a process like this is often called **exponential growth**. The constant k is called the **growth constant**. If $k > 0$, then $f(t)$ is an increasing function of t, and if $k < 0$, it is decreasing (in which case it is sometimes called **exponential decay** and k is called the **decay constant**).

> Notice that if we let $t = 0$ in Equation 7.3.1, we find $f(0) = ce^0 = c$. This means that the constant c equals the value of $f(t)$ at $t = 0$; for obvious reasons it is sometimes called the **initial value**. The behavior of an exponential growth process is completely determined once we know the initial value c and the growth constant k.

EXAMPLE 1 Suppose the function $f(t)$ grows exponentially. If $f(0) = 4$ and $f(2) = 20$, find $f(5)$, that is, the value of $f(t)$ when $t = 5$.

Solution We are assuming that $f(t)$ is of the form $f(t) = ce^{kt}$, for some constants c and k. The fact that $f(0) = 4$ tells us that $c = 4$, so $f(t) = 4e^{kt}$.

The other piece of information we have is that $f(2) = 20$. But we know that $f(2) = 4e^{k(2)}$, so $4e^{2k} = 20$ and $e^{2k} = \frac{20}{4} = 5$. Taking logarithms of both sides, we find that $2k = \ln(5)$, or $k = \frac{1}{2}\ln(5)$.

Now we know both constants, so it is possible to write the function precisely: $f(t) = 4e^{t\ln(5)/2}$. In particular, when $t = 5$,

$$f(5) = 4e^{5\ln(5)/2} = 4(e^{\ln(5)})^{5/2} = 4(5)^{5/2} = 100\sqrt{5} \approx 223.6068.$$

The idea was to use the values of $f(t)$ at the two times $t = 0$ and $t = 2$, where the values are known, to calculate the constants c and k. Once these have been found, we know the formula for $f(t)$ at any time t, and calculating it for any particular t is easy.

EXAMPLE 2 Suppose $g(t)$ is a function whose derivative equals a constant times the function itself. If $g(1) = 9$ and $g(4) = 3$, at what time t will $g(t) = 1$?

Solution Once again we write the function in the form of Equation 7.3.1, $g(t) = ce^{kt}$. What we know is that $g(1) = 9$ and $g(4) = 3$. In symbols this means

$$ce^k = 9 \quad \text{and} \quad ce^{4k} = 3.$$

Dividing these two equations, we find that

$$\frac{ce^{4k}}{ce^k} = \frac{3}{9} = \frac{1}{3}, \quad \text{or} \quad e^{3k} = \frac{1}{3}.$$

Taking logarithms, we find that $3k = \ln\frac{1}{3} = -\ln(3)$, so $k = -\frac{1}{3}\ln(3)$.

Next we have to find c. Since we are not told the value of $g(t)$ at $t = 0$, it is necessary to use the information we do have, including the value we have just found for k.

We know that $9 = g(1) = ce^k = ce^{-\ln(3)/3} = c(e^{\ln(3)})^{-1/3} = c(3)^{-1/3}$. From this we find that $c = 9/(3)^{-1/3} = 9(3^{1/3}) = 3^{7/3}$. So $g(t) = ce^{kt} = 3^{7/3}e^{-t\ln(3)/3}$.

To find when this equals 1, we write

$$3^{7/3}e^{-t\ln(3)/3} = 1$$

$$e^{-t\ln(3)/3} = \frac{1}{3^{7/3}}$$

$$-t\ln(3)/3 = -\ln(3^{7/3})$$

So

$$t = 3\frac{\ln(3^{7/3})}{\ln(3)} = 3\frac{\frac{7}{3}\ln(3)}{\ln(3)}$$

$$= 7.$$

The function $g(t)$ equals 1 when $t = 7$.

There are several things to notice about this example. One is that k is negative. This means that the function is "decaying," that is, decreasing; in hindsight this was obvious from the fact that $g(4) < g(1)$.

Second, even though $c = g(0)$ was not given to us in the question, we were able to use the values of $g(t)$ at two times t to find what k is. Knowing this, we were able to find c from either of the known values of $g(t)$.

Finally, it is interesting to observe that in the three units of time between $t = 1$ and $t = 4$ the function decreased by a factor of $\frac{1}{3}$ (from $g(1) = 9$ to $g(4) = 3$). For it to decrease to 1, a further decrease by a factor of $\frac{1}{3}$, it took an additional 3 units of time (from $t = 4$ to $t = 7$). This is in fact characteristic of exponential growth. In a particular length of time, say, I units long (I for "interval"), the function will always change by the same factor, no matter where it starts. In Example 2 the function decreases by a factor of $\frac{1}{3}$ whenever the time increases by 3 units.

The reason is actually very simple. Suppose $f(t) = ce^{kt}$, with $c \neq 0$. If we compare the values of the function at times t and $t + I$, their quotient is

$$\frac{ce^{kt}}{ce^{k(t+I)}} = e^{kt - k(t+I)} = e^{-kI}.$$

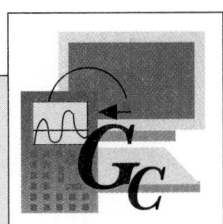

With a graphics calculator we can plot the function $y = g(t)$ from Example 2; we also plot the constant function $y = 1$.

TI-85	SHARP EL9300
3 ^ (7 ÷ 3) STO C ENTER (−) LN (3) ÷ 3 STO K ENTER GRAPH F1$(y(x) =)$ C × e^x (K × F1(x)) ▼ 1 M2(RANGE) 1 ▼ 10 ▼ ▼ (−) 1 ▼ 9 F5(GRAPH)	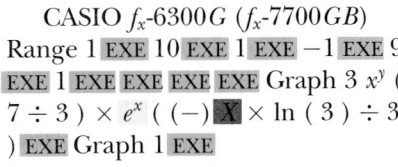 3 a^b 7 ÷ 3 ▶ × e^x (−) x/θ/T × ln 3 ÷ 3 2ndF ▼ 1 RANGE 1 ENTER 10 ENTER 1 ENTER −1 ENTER 9 ENTER 1 ENTER
CASIO f_x-6300G (f_x-7700GB)	HP 48SX
Range 1 EXE 10 EXE 1 EXE −1 EXE 9 EXE 1 EXE EXE EXE EXE Graph 3 x^y (7 ÷ 3) × e^x ((−) X × ln (3) ÷ 3) EXE Graph 1 EXE	← PLOT PLOTR ERASE ATTN ' Y = 3 y^x () 7 ÷ 3 ▶ × e^x X × LN 3 ▶ ÷ 3 +/− ↱ PLOT ← DRAW ERASE 1 ENTER 10 XRNG 1 +/− ENTER 9 YRNG DRAW ATTN ' Y = 1 ↱ PLOT ← DRAW DRAW

Notice where $y = g(t)$ crosses $y = 1$ ...

This number e^{-kI} is independent of t. In other words, the ratio of $f(t)$ to $f(t + I)$ is the same for any t. This is true for any function of exponential growth (or decay).

This remark suggests that it might be interesting to ask how long it will take a particular exponential function to change by some factor R. The most familiar example of this idea occurs in the study of radioactive decay, which we will discuss in Section 7.4.

A substance undergoing radioactive decay is transformed into other substances, and the amount of it that remains at time t is a decaying exponential function $f(t) = ce^{kt}$, with $k < 0$. As usual, the constant k determines how quickly it decays, but a convenient way to describe it is the **half-life**, the length of time it takes for the substance to diminish by one-half. For example, suppose the half-life of a substance was 1 year. Then in one year half of it would decay, and in another year half of what was left would decay, leaving one-quarter of the original amount. In a period of 5 years, a sample of the substance would reduce by half, by half again, by half again, by half again, and by half once more, leaving only $\left(\frac{1}{2}\right)\left(\frac{1}{2}\right)\left(\frac{1}{2}\right)\left(\frac{1}{2}\right)\left(\frac{1}{2}\right) = \frac{1}{32}$ of the original amount.

The half-life is a handy way to describe the rate of decay. It is fairly easy to visualize, so it is used more often than the decay constant k.

Of course, the half-life and the decay constant are closely related.

EXAMPLE 3   Find the half-life of the decay process that is described by the function $f(t) = ce^{-kt}$. (Notice that to emphasize that the process is a decay, we have used the constant $-k$, with $k > 0$.)

Solution   We know that $f(0) = c$; let us find the value of t for which $f(t) = \frac{c}{2}$. This amounts to

$$ce^{-kt} = \frac{c}{2},$$

or

$$e^{-kt} = \frac{1}{2},$$

or

$$-kt = \ln\left(\frac{1}{2}\right) = -\ln(2),$$

or

$$t = \frac{-\ln(2)}{-k} = \frac{\ln(2)}{k}.$$

The half-life is $\ln(2)/k$.

EXAMPLE 4   **Half-Life of a Radioactive Decay**   Suppose some radioactive substance decays in such a way that after 100 years the substance has diminished to 98.79% of its original amount. What is the half-life?

Solution   If we write the amount remaining at time t as $f(t) = ce^{-kt}$, then what we know is that $f(100) = 0.9879 f(0)$. This means that

$$ce^{-100k} = 0.9879c,$$

or

$$e^{-100k} = 0.9879,$$

or

$$-100k = \ln(0.9879),$$

so

$$k = -\frac{\ln(0.9879)}{100}$$

$$\approx 1.2174 \times 10^{-4}.$$

Combining this with the result of Example 3, we find that the half-life is $\ln(2)/k \approx 5700$ years.

This happens to be the approximate half-life of the isotope carbon 14, which is of great importance because of its use for estimating the date of archaeological finds. This is also something we will discuss in Section 7.4.

EXAMPLE 5   Yogurt   Yogurt is made by placing a small amount of yogurt in some warm milk and letting it sit. The bacteria in the yogurt thrive in the warm milk and multiply. In the process, the milk is turned into yogurt.

Suppose we start with 10 grams of yogurt (about 2 teaspoons). Suppose too that the bacteria divide in such a way that the amount of yogurt doubles every 30 minutes. Let us measure time t in hours; the culture doubles every half hour.

If the amount of yogurt present at time t is $Y(t) = ce^{kt}$, measured in grams, then $c = 10$ and $Y\left(\frac{1}{2}\right) = 20$, so $e^{k/2} = 2$ and $\frac{k}{2} = \ln(2)$, $k = 2\ln(2)$. The number of grams of yogurt present at time t is $Y(t) = 10e^{2t\ln(2)}$.

After 3 hours there will be about $Y(3) = 10e^{2(3)\ln(2)} = 10(e^{\ln(2)})^6 = 10(2^6) = 640$ g of yogurt, or a little under 3 cups.

If we consider what will happen in 24 hours, we find that $Y(24) = 10e^{2(24)\ln(2)} = 10(2)^{48} \approx 2.8 \times 10^{15}$ g. To get some idea of the size of this quantity, let us convert it to 2.8×10^{12} kg, which is 2.8×10^9 tonnes. (The "tonne," the metric ton, is 1000 kg, or about 2200 lb, not much different from the ordinary ton, which is 2000 lb.)

Assuming that the density of yogurt is the same as that of water, 1 tonne occupies 1 cubic meter, so after growing for 24 hours, the yogurt will occupy about 2.8×10^9 m^3. If all this yogurt were gathered into a cube-shaped tank, each side of the tank would be $(2.8 \times 10^9)^{1/3} \approx 1400$ m long. This is slightly under one mile.

That is a lot of yogurt.

This 1000× magnification of a smear shows spherical Streptococcus lactis *and rod-shaped* Lactobacillus bulgaris *bacteria in yogurt. The* Lactobacillus bulgaris *bacteria are found in fermenting animal and plant products and produce lactic acid from the fermentation of carbohydrates. They also live in the alimentary tract and vagina of humans and other animals, where they protect against more dangerous bacteria.*

The point of this example is not to tell us anything about yogurt, but to demonstrate the pitfalls that lie in assuming that something will continue indefinitely to grow exponentially. The growth of yogurt will be approximately exponential for a while, but of course it will slow down when it has used up most of the milk available to it. The interesting thing about this example is to see just how fast exponential growth increases. The growth rate suggested here for yogurt is fairly reasonable at the outset, and yet within 24 hours it has become completely ridiculous.

Exercises 7.3

I

In each of the following cases, assume that the function $f(t)$ is of the form of Equation 7.3.1.

1. Find $f(4)$, if $f(0) = 3, f(2) = 6$.
2. Find $f(6)$, if $f(0) = 2, f(2) = 1$.
3. Find $f(3)$, if $f(0) = 1, f(2) = 4$.
4. Find $f(10)$, if $f(0) = 2, f(3) = 20$.
5. Find $f(7)$, if $f(1) = 2, f(3) = 8$.
6. Find $f(4)$, if $f(2) = 5, f(6) = 10$.
7. Find $f(5)$, if $f(2) = 3, f(4) = \frac{1}{3}$.
8. Find $f(0)$, if $f(4) = 8, f(7) = 2$.
9. Find $f(-1)$, if $f(4) = 1, f(6) = 4$.
10. Find $f(3)$, if $f(-3) = 6, f(0) = 18$.

Find the half-life of each of the decay processes described below. Time t is measured in years.

11. $f(0) = 10, f(1) = 5$
12. $f(0) = 2, f(1) = \sqrt{2}$
13. $f(2) = 1, f(5) = 0.75$
14. $f(2) = 6, f(3) = 4$
15. $f(1) = 8, f(2) = 2$
16. $f(-1) = 4, f(1) = 3$
17. $f(6) = 16, f(7) = 1$
18. $f(0) = 1, f(1) = 0.99$
19. $f(-3) = 12, f(0) = 9$
20. $f(-4) = 7, f(-1) = 2$

II

21. If a bacterial culture in an unrestricted environment doubles every half hour, how long will it take to reach ten times its original size?
22. If a bacterial culture in an unrestricted environment increases by 10% each hour, how long will it take to double in size? We can think of this time as a sort of average time for the bacteria to divide in two, in the sense that this is how long it takes for the total number to double.

23. If the half-life of a decay process is 3 years, how long will it take for a sample to reduce to 90% of its original size?
24. If the half-life of a decay process is 8 years, how long will it take for a sample to reduce to 10% of its original size?
25. If it takes 30 years for a sample to reduce to 40% of its original size, what is the half-life of the process?
26. If it takes 30,000 years for a sample to reduce to 1% of its original size, what is the half-life of the process?
27. A bacterial culture is growing exponentially. One hour after it begins growing, it weighs 0.5 g; four hours after *that*, it weighs 3.7 g. How much did it weigh at the beginning?
*28. Grandfather put some money in a bank account 40 years ago. It has been increasing by 6% each year ever since. Assume that the amount grows exponentially (which is correct for continuously compounded interest, as we shall see in the next section). If there is $3085.72 in the account now, how much did Grandfather deposit 40 years ago?
29. Suppose the yogurt of Example 5 (p. 527) were allowed to grow exponentially for 48 hours. Figure out how much yogurt would be produced, and figure out how this compares with the volume of the earth.
*30. Suppose a population of small organisms has a rate of reproduction that is proportional to the population. Suppose the organisms die at a rate that is also proportional to the population. (Roughly speaking, this means that a certain percentage of them die in a given length of time, so the more of them there are, the more will die.) Show that the function $P(t)$, which expresses the total population at time t, is of the form of Equation 7.3.1. What is the significance of the sign of the constant k?

III

31. Use a graphics calculator or computer to sketch the graph of $y = f(t) = e^{t \ln(2)}$. Check the graph at several points to confirm that the value of $f(t)$ doubles whenever t increases by 1. (Be sure to use some values of t that are not integers.)

POINT TO PONDER

When we developed Formula 7.3.1 for a function whose rate of change is proportional to the function itself, we found that $\frac{f'(t)}{f(t)} = k$. Integrating, we saw that $\ln\big(f(t)\big) = kt + c'$.

In this formula the constant c' is just a constant of integration. However, notice the way in which it is changed when we exponentiate:

$$f(t) = e^{\ln(f(t))} = e^{c'} e^{kt}.$$

The constant that was *added* to kt has been transformed into one by which e^{kt} is *multiplied*. Notice the importance of having added the constant of integration. If we had not done so, we would have found that $f(t) = e^{kt}$. This is one of the possible answers, but it is certainly not the only one, and forgetting to add the constant would have caused us to miss all the others.

Notice that we cannot just add a constant wherever we like. Check that if d is a constant, the function $g(t) = ce^{kt} + d$ does *not* have a rate of change that is proportional to the function $g(t)$ (unless $d = 0$).

Finally notice that when we found the function $f(t) = ce^{kt}$, the constant $c = e^{c'}$ is an exponential, so strictly speaking, it is *positive*. There is no reason it should not be positive, but on the other hand a function with a negative constant will still have the basic property that its derivative is a constant times the function itself. Explain why our argument misses the functions with negative constants c.

7.4 Examples of Exponential Growth and Decay

BACTERIAL GROWTH
HUMAN POPULATIONS
RADIOACTIVE DECAY
CARBON DATING
COMPOUND INTEREST
PRESENT VALUE
NEWTON'S LAW OF COOLING
CONCENTRATION OF IMPURITIES

In Section 7.4 we discussed the behavior of a function whose rate of change is proportional to the function itself, that is, $f'(t) = kf(t)$, for some (nonzero) constant k. We found that such a function must be of the form $f(t) = ce^{kt}$, with some constant c, the initial value. The function $f(t)$ is increasing (growing) if $k > 0$ and decreasing (decaying) if $k < 0$.

In this section we discuss a number of practical problems that fit this description. In the back of our minds we ought to remember Example 7.3.5 (p. 527), which showed that applying the mathematical consequences of exponential growth to situations in which they are not justified can easily give ridiculous results. It is particularly dangerous to assume that the trend will continue for very large or very small values of t.

EXAMPLE 1   Bacterial Growth   We have already considered the example of a culture of bacteria. Bacteria are organisms that usually consist of a single cell; the way they reproduce is for each cell to divide into two.

How frequently they do this depends on many factors, including such things as the temperature and the amounts and types of nutrients available. If the environment is reasonably stable, we can assume that all these factors remain unchanged for a period of time. Under these conditions it makes sense to expect that the cells will reproduce at a regular rate.

In other words, if T minutes, say, is the average time between divisions of a particular cell, then every time T minutes elapse, we should expect that, on the

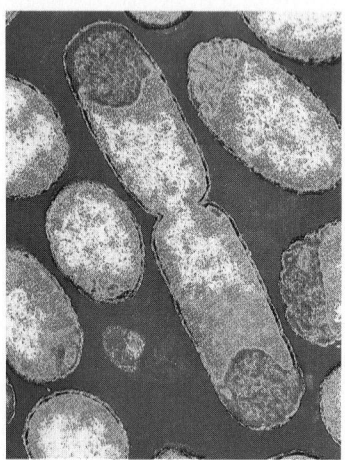

False-color transmission electron micrograph of the bacterium Escherichia coli, *a normal inhabitant of the human intestine. The pair in the center are in the act of separating following binary fission, the process by which a bacterium divides into two. Under certain conditions,* E. coli *bacteria may become harmful and cause infection in the intestinal or urinogenital tracts.* E. coli *is the organism most commonly used in genetic engineering experiments. Magnification: 16,500×.*

average, each cell will have divided once. The effect of this is that the number of cells doubles every T minutes.

For some types of bacteria, under ideal conditions, the average time T between divisions can be less than 15 minutes.

Let $N(t)$ be the number of bacteria present at time t. We should consider what happens in some brief period of length h. Starting at some fixed time t, when the population is $N(t)$, we know that between t and $t + T$ the population will double. In this interval, in effect, each of the original cells will divide exactly once. In a shorter interval, some fraction of them will double.

If we choose an interval of length $\frac{T}{2}$, it seems reasonable to expect half of them to divide. During an interval of length $\frac{T}{6}$, one-sixth of the original cells should divide. During an interval of length h the proportion of the cells that divide ought to be $\frac{h}{T}$.

If h is small, this means that the number of cells that divide will be $\frac{h}{T}N(t)$. The result will be that this many new cells will be created; symbolically, this can be expressed as

$$N(t + h) - N(t) = \frac{h}{T}N(t).$$

To find the derivative of $N(t)$, we form the difference quotient and take the limit:

$$\frac{d}{dt}\big(N(t)\big) = \lim_{h \to 0}\frac{N(t + h) - N(t)}{h} = \lim_{h \to 0}\frac{1}{h}\frac{h}{T}N(t) = \frac{N(t)}{T}.$$

We have just shown that $N'(t) = \frac{1}{T}N(t)$, that is, that $N'(t)$ is proportional to $N(t)$.

From this it follows from the results of the previous section that the population at time t is an exponential function: $N(t) = ce^{kt}$, for some constants c and k.

> We can summarize the above argument as follows. The number of cell divisions in a fixed brief period of time is proportional to the number of cells present. To put it the other way around, the *proportion* of the cells that divide in a given short length of time is a constant, independent of the number of cells present.
>
> This means that the rate of change of the population is proportional to the population, and from this we can conclude the growth is exponential.

The tricky thing about this argument is the assumption that the environment is stable. Eventually, the bacteria themselves will destabilize it if they are allowed to grow unhindered. They will begin to use up the available space or some of the nutrients, and their growth rate will diminish.

Other factors can easily enter the picture too, such as the presence of an organism that eats the bacteria that we are considering.

In spite of all the difficulties, it is still true that bacterial cultures do conform to the rules of exponential growth while the environment remains stable.

EXAMPLE 2   If a bacterial culture increases by 10% in one hour, how long will it take to increase to five times its original size?

Solution   Let $N(t)$ be the population at time t, with t measured in hours. We know that $N(t) = ce^{kt}$, for some constants c and k.

We also know that $N(1) = 1.10 \times N(0)$, that is, $ce^k = 1.10c$. Dividing by c, we find that $e^k = 1.10$, so $k = \ln(1.10) \approx 0.09531$.

The question asks us to find the value of t for which $N(t) = 5N(0)$. This amounts to $ce^{kt} = 5c$, and dividing by c, we find $e^{kt} = 5$, or $kt = \ln(5)$. Knowing the value of k from above, we see that the time we want is

$$t = \frac{\ln(5)}{k} = \frac{\ln(5)}{\ln(1.10)} \approx 16.89 \text{ hr.}$$

One interesting feature of Example 2 is that there is not enough information to determine the value of the constant c. However, it is still possible to solve the problem. Notice that in the calculations the constant c just "divides out."

EXAMPLE 3   **Human Populations**   A population of people grows under the influence of a remarkably complex environment. So many factors affect the growth rate that it is extremely difficult to make accurate predictions.

For one thing, people migrate from region to region, so population growth or decrease due to births and deaths may be masked by immigration and emigration. Diseases and wars may dramatically increase the death rate; medical advances may reduce it. Social and religious factors may have a huge effect on the birth rate; perhaps the most obvious instance is the difference between cultures that do and do not accept contraception.

Another complication is that birth rates depend not so much on the total population as on the population of women of childbearing age.

With all these subtleties to consider, making precise statements about human population growth is hazardous. Even the experts are not terribly reliable, which is hardly surprising, considering the nature of the problem.

Nonetheless, if we assume that the birth and death rates are constant and ignore the effects of migration, the population will grow (or diminish) exponentially, for exactly the same reasons that applied to the bacterial culture discussed in Example 1. We will make these assumptions and see what simple conclusions can be drawn as a result. We will not make any serious attempt to discuss the validity of the assumptions.

EXAMPLE 4   In a country with no immigration or emigration, the population is 10,000,000. One month later, 15,000 births and 6,000 deaths have occurred. Assuming that the birth and death rates are constant, what will the population be 5 years after the starting point?

Solution   From the assumptions we can conclude that the population is described by an exponential function $P(t) = ce^{kt}$. Suppose the time is measured in years, with $t = 0$ at the time when the population is 10,000,000.

We know that the initial value is $c = P(0) = 10,000,000$. We know that one month after that, that is, when $t = \frac{1}{12}$, the population is $10,000,000 + 15,000 - 6,000 = 10,009,000$. This means that $10,000,000e^{k/12} = 10,009,000$, so $\frac{k}{12} = \ln\left(\frac{10,009,000}{10,000,000}\right) = \ln(1.0009)$.

We find that $P(t) = 10,000,000e^{12t\ln(1.0009)}$. What we want to evaluate is $P(5) = 10,000,000e^{12(5)\ln(1.0009)} \approx 10,555,000$.

Of course, the same sort of calculations can be made for populations of animals, subject to the usual assumptions. The situation may be complicated by predators. If the population of rabbits, for instance, begins to grow, then the population of foxes will likely increase. It might well happen that the foxes will grow so fast that, in combination with shortages of food, they will begin to reduce the rabbit population. Before long, the rabbit shortage will devastate the fox population. Because of the resulting reduced predation, the rabbits will rebound.

What happens is a cycle in which the populations grow and fall in a fairly regular pattern. External factors can disrupt the cycle, but there have been wilderness studies that appear to confirm this phenomenon. Folklore almost always seems to suggest that the natural cycles are seven years long.

EXAMPLE 5   **Radioactive Decay**   Radioactive decay is a particularly attractive example because there is no need to make questionable assumptions.

Certain atomic nuclei are unstable. This means that they have a tendency to break apart into pieces, and this "fission" is accompanied by the release of a large amount of energy. Roughly speaking, the energy that is released is what used to be holding the unstable nucleus together.

Consider a sample of some type of unstable nuclei. Let h be a length of time. During any period of length h, the same fixed fraction of the nuclei will decay, no matter how many there were at the beginning of the period.

By an argument similar to the one in Example 1 we conclude that the number of unstable nuclei remaining is an exponentially decaying function of time t. The length of time it takes for half the unstable nuclei of a certain type to decay is called the **half-life** of that type of nuclei. It varies from tiny fractions of a second for some nuclei to billions of years for others, and it is a convenient way to measure how fast the radioactive decay occurs.

The nucleus of an atom is made up of protons and neutrons. The protons carry a positive electric charge, and the neutrons are electrically neutral. Electrons are a type of particle of much lighter weight that carry a negative charge. They are attracted to the positive charge of the nucleus, and it is the combination of the nucleus surrounded by electrons that constitutes an atom.

There is a tendency for the number of electrons to be the same as the number of protons; this has the effect that the positive and negative charges balance, giving the atom no net charge.

Broadly speaking, chemistry is the study of what happens when different atoms interact, which they do essentially through their electrons. One way this happens, for example, is for two nearby atoms to "share" some of their electrons. The resulting pair of atoms, bound together by their electrons, is a simple example of a *molecule*. More complicated molecules may have hundreds or even thousands of atoms. The essential thing is that the chemical behavior is determined by the electrons, and the number of electrons is the same as the number of protons in the nucleus. This number is called the *atomic number* of the nucleus.

For example, the atomic number of carbon is 6, meaning that the carbon nucleus contains 6 protons. But there is also the question of how many neutrons the nucleus contains.

For carbon the answer is usually 6. This means that the nucleus contains 6 protons and 6 neutrons, or a total of 12 particles. Such a nucleus is denoted ^{12}C, with the C standing for carbon and the mass number 12 giving the total number of nuclear particles.

However, it is occasionally possible to form a nucleus with 6 protons and only 5 neutrons. This nucleus, ^{11}C, is unstable. The combination of 6 protons and 5 neutrons does not hold together nearly as well as ^{12}C, and it tends to break apart. The half-life of ^{11}C is about $\frac{1}{3}$ of an hour.

From the chemical point of view, however, ^{11}C is nearly indistinguishable from ^{12}C. It can combine with two oxygen atoms to make carbon dioxide, for example, just as ^{12}C can. We say that ^{11}C is an **isotope** of carbon (meaning that it has atomic number 6).

In addition to the usual stable isotope ^{12}C and the unstable isotope ^{11}C, there is another isotope ^{14}C, with 8 neutrons. It is unstable, but much less so than ^{11}C. Its half-life is about 5730 years.

EXAMPLE 6   The half-life of uranium ^{238}U is 4.5 billion years. If the quantity of this isotope in a rock sample has diminished to 90% of its original amount, how old is the rock sample?

Solution   We know that the amount of ^{238}U remaining t years after the formation of the rock is $Q(t) = ce^{-kt}$, for some constants c and k. Notice that we use a negative sign to help us remember that it is a process of decay.

Since the half-life is 4.5 billion years, we find from Example 7.3.4 that $k = \frac{\ln 2}{4.5 \times 10^9}$.

What we need is the time t for which $Q(t) = 0.90 Q(0)$. This amounts to $e^{-kt} = 0.90$, or $-kt = \ln(0.90)$, which means

$$t = \frac{\ln(0.90)}{-k} = -\frac{\ln(0.90)(4.5 \times 10^9)}{\ln 2} \approx 6.84 \times 10^8.$$

The rock is about 684 million years old.

In practice, the method is a little more complicated than this example appears to indicate. The ^{238}U isotope decays (through several intermediate steps) into the

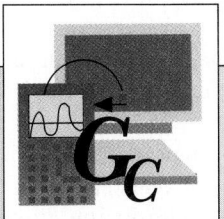

If we want to plot the function of Example 6 we have to fix some value of c. For simplicity we let $c = 1$ and plot $n = Q(t) = e^{-t \ln(2)/(4.5 \times 10^9)}$.

TI-85	SHARP EL9300
GRAPH F1$(y(x) =) e^x$ ((−)	$\not\!\!A\!\!\!\!\downarrow$ e^x (−) x/θ/T × ln 2 ÷
F1(x) × LN (2) ÷ 4500000000	4500000000 RANGE 0 ENTER
) M2(RANGE) 0 ▼ 684000000	684000000 ENTER 100000000 ENTER
▼ ▼ 0 ▼ 1 F5(GRAPH)	0 ENTER 1 ENTER 1 ENTER $\not\!\!A\!\!\!\!\downarrow$
CASIO f_x-6300G (f_x-7700GB)	HP 48SX
Range 0 EXE 684000000 EXE	↰ PLOT PLOTR ERASE ATTN ′
100000000 EXE 0 EXE 1 EXE 1 EXE	$Y\equiv e^x$ X × LN 2 ▶ ÷ 4500000000
EXE EXE EXE Graph e^x ((−) X × ln	+/− ↱ PLOT ↰ DRAW ERASE 0
(2) ÷ 4500000000 EXE	ENTER 684000000 XRNG 0 ENTER 1
	YRNG DRAW

We see how the function decreases to 90% of the original amount. Notice that the part of the curve we have plotted is nearly a straight line...

lead isotope ^{206}Pb. (The Latin name for lead is *plumbum*, which we recognize in our word "plumber," someone who works with lead pipes.) It is possible to measure how much radioactive decay has occurred by comparing the amounts of ^{238}U and ^{206}Pb. The greater the proportion of ^{206}Pb in a rock, the longer it has been in existence. The rock might contain some lead that did not result from radioactive decay, but happily it turns out that this will be a different isotope, ^{204}Pb.

Other combinations of isotopes can also be used to estimate the age of rocks. The first requirement is to find an isotope with a half-life that is roughly comparable to the likely age of the rock. (It is no use trying to date a rock that is millions of years old with an isotope whose half-life is 20 minutes.) Second, the original isotope and/or the products of the decay must be distinguishable from isotopes that may arise in other ways.

The best-known isotope dating is the use of ^{14}C to date archaeological sites.

EXAMPLE 7   **Carbon Dating**   It has already been mentioned that the isotope ^{14}C has a half-life of 5730 years. This is a convenient unit for archaeological materials, whose ages are likely to range from 100 years to perhaps 10,000 or 20,000 years.

Living plants take carbon dioxide from the air and by photosynthesis convert it into products they can use. These processes are *chemical* in nature, so they do not distinguish between carbon dioxide containing ^{12}C and carbon dioxide containing ^{14}C. The effect is that the carbon atoms in any living plant will consist of ^{12}C and ^{14}C in approximately the same proportions as the carbon atoms in the atmosphere.

However, once the plant dies, it no longer absorbs carbon dioxide from the air. Since the ^{14}C isotope undergoes radioactive decay, the amount of it begins to diminish. By measuring the relative amounts of the two isotopes it is possible to measure the length of time the plant has been dead.

The process can be used to date pieces of wood from excavated buildings, for instance, and since animals consume plant material, it can also be used to date animal tissues.

The entire idea relies on the assumption that the proportion of the two isotopes in the atmosphere has remained the same for thousands of years. The ^{14}C isotope is created in nature by the bombardment of nitrogen nuclei in the atmosphere by radiation from space, and the amount of it that is produced appears to have remained unchanged. Comparisons of ^{14}C dates for sites whose age is corroborated by historical data bear out the reliability of the technique. Atmospheric nuclear testing has resulted in elevated ^{14}C levels, so it will not be possible to use carbon dating for organic material deposited in the recent past.

Liquid scintillation counter used to measure the amount of the carbon-14 isotope in a sample being dated.

EXAMPLE 8   If a sample of charred wood from an excavated campsite is found to have 63% of the original amount of ^{14}C, how old is the site?

Solution   As usual, we know that the number of ^{14}C nuclei can be expressed as a function ce^{-kt}. We know that the half-life is 5730 years, which means that $k = \frac{\ln 2}{5730}$.

If t is the age of the site, we know that $e^{-kt} = 0.63$, or $-kt = \ln(0.63)$, so

$$t = -\frac{\ln(0.63)}{k} = \frac{-5730 \ln(0.63)}{\ln 2} \approx 3820.$$

The site is approximately 3820 years old.

Notice that we have been writing exponentially decaying functions in the form ce^{-kt}, with a minus sign in the exponent. This has the effect of making the constant k positive.

It is just as good to omit the minus sign; what matters is to be consistent. Either put it in throughout a particular example or leave it out all the time. It is a question of taste.

EXAMPLE 9   **Compound Interest**   Consider money in a bank account earning 12% annual interest. This means that each year the amount increases by 12%, in effect multiplying the original amount by $\left(1 + \frac{12}{100}\right)$. We say the balance increases by a **factor** of $\left(1 + \frac{12}{100}\right)$; the word "factor" suggests that we *multiply* by this number.

Suppose the interest is compounded semiannually. This means that 6% interest is paid after 6 months (multiplying the original amount by $\left(1 + \frac{6}{100}\right)$), and at the end of the year another 6% interest is paid. This has the effect of multiplying by $\left(1 + \frac{6}{100}\right)$ again, and the overall effect is to multiply the original amount by $\left(1 + \frac{6}{100}\right)^2$.

In one year the amount will increase by a factor of $\left(1 + \frac{6}{100}\right)^2 = 1.1236$. This is slightly more than the factor $1 + \frac{12}{100} = 1.12$ that arises from a single payment of 12% interest. The slight difference comes from paying 6% interest for the second half of the year on the first installment of interest.

The interest could also be compounded quarterly, that is, every three months. The overall effect would be to increase the original amount by a factor

$$\left(1 + \frac{3}{100}\right)\left(1 + \frac{3}{100}\right)\left(1 + \frac{3}{100}\right)\left(1 + \frac{3}{100}\right) = \left(1 + \frac{3}{100}\right)^4 \approx 1.1255.$$

Many banks compound interest monthly. With 12% annual interest, in one year this increases the original amount by $\left(1 + \frac{1}{100}\right)^{12} \approx 1.126825$.

The 12% interest rate is called the **nominal interest rate**. We have just seen that with a nominal rate of 12% and monthly compounding, the annual increase is actually about 12.6825%. This is called the **effective annual interest rate**.

EXAMPLE 10

  (i) What is the effective annual interest rate with quarterly compounding and a nominal rate of 10%?

  (ii) How does this compare with monthly compounding at a nominal rate of 9.9%?

Solution

  (i) Quarterly compounding at a nominal 10% annual rate means paying 2.5% each quarter. Over one year this has the effect of increasing the original deposit by a factor of $\left(1 + \frac{2.5}{100}\right)^4 \approx 1.1038$, so the effective annual interest rate is about 10.38%.

  (ii) Monthly compounding means paying $\frac{1}{12}$ of the nominal rate each month. In this case that means increasing the deposit each month by a factor of $\left(1 + \frac{1}{12}\frac{9.9}{100}\right)$. Over a year, the deposit grows by $\left(1 + \frac{1}{12}\frac{9.9}{100}\right)^{12} \approx 1.1036$, and the effective annual interest rate is about 10.36%, nearly as much as the rate in part (i).

> If the nominal interest rate is $p\%$, let $r = \frac{p}{100}$ so that for simple interest (i.e., not compounded) the effect in one year is to increase the amount by a factor of $(1 + r)$.
>
> Compounding semiannually has the effect of increasing in one year by $\left(1 + \frac{r}{2}\right)^2$; for quarterly compounding, the increase is $\left(1 + \frac{r}{4}\right)^4$. If the interest is compounded n times in one year, the balance increases by the factor
>
> $$\left(1 + \frac{r}{n}\right)^n. \tag{7.4.1}$$
>
> For example, to compound the interest daily, that is, $n = 365$ times a year, we multiply by $\left(1 + \frac{r}{365}\right)^{365}$.

The more often the interest is compounded, the higher will be the effective annual interest. Instead of compounding monthly or daily, we could compound hourly, or every minute, or every second. Compounding more often raises the effective rate, so it is to the advantage of the owner of the account to have interest compounded as often as possible.

If we try compounding every second, say, then in one year the original amount will increase by $\left(1 + \frac{r}{n}\right)^n$, where n is the number of seconds in a year, or about $(365)(24)(60)(60) = 31{,}536{,}000$. We could try compounding every one one-thousandth of a second in hopes of raising the effective rate. This would mean letting $n = 31{,}536{,}000{,}000$. We could try letting n, the number of compoundings, get larger and larger, but $\left(1 + \frac{r}{n}\right)^n$ does not get arbitrarily large. In fact, we saw in Example 7.2.4 (p. 519) that

$$\lim_{n \to \infty} \left(1 + \frac{r}{n}\right)^n = e^r.$$

No matter how often the interest is compounded, the annual increase will never exceed e^r. Rather than take a very large n, corresponding to compounding very many times, we can take this limit and regard it as the effect of **compounding continuously**.

EXAMPLE 11   What is the effective annual rate if the nominal rate is 10% and interest is compounded continuously?

Solution   The factor by which the original amount is increased in one year is $\lim_{n \to \infty} \left(1 + \frac{10}{100n}\right)^n = e^{1/10} \approx 1.10517$. The effective annual interest rate is approximately 10.517%.

It is interesting to compare this with the effective rate resulting from daily compounding. In that situation the amount increases by $\left(1 + \frac{1}{365}\frac{10}{100}\right)^{365} \approx 1.1051558$, and the effective annual rate is approximately 10.51558%.

The difference between continuous compounding and daily compounding is remarkably small.

> It should come as no surprise that in an account with continuously compounded interest, the total amount of money is an exponentially growing function of time t. Essentially, this is because the rate at which the total amount increases is proportional to the amount. (Notice that the *rate* at which the total amount increases is *not* the same thing as the interest *rate;* the interest rate is constant.)

EXAMPLE 12 Mary deposits $1000 in an account with continuously compounded interest at a nominal rate of 8%. One hundred and ten days later, she withdraws $700 from the account. How much will be in the account 40 days after this withdrawal?

Solution Let t represent time, measured in days, with $t = 0$ at the time of the deposit. If $A(t)$ is the amount in the account at time t, in dollars, we know that it grows exponentially, so it must be of the form $A(t) = ce^{kt}$.

The complication is that while it grows according to a function of this form for t between 0 and 110, we know that it changes at $t = 110$ because of the withdrawal. From that time on, it continues to grow exponentially (at least until the next deposit or withdrawal).

First we should find the constant k. If 8% interest is compounded n times annually, the resulting effective increase is $\left(1 + \frac{1}{n}\frac{8}{100}\right)^n$. For continuous compounding we let $n \to \infty$ and take the limit

$$\lim_{n \to \infty} \left(1 + \frac{8}{100n}\right)^n = e^{8/100}.$$

(To evaluate this limit, we used Equation 7.2.7, p. 519).

This means that $A(365) = e^{8/100}A(0)$, or $ce^{365k} = e^{8/100}c$. Dividing by c, we find that $e^{365k} = e^{8/100}$, so $365k = \frac{8}{100}$ and $k = \frac{8}{36,500}$.

First, consider the part of the function between $t = 0$ and $t = 110$. Let us give the name $a(t)$ to this part of the function. Its initial value is 1000, so $a(t) = 1000e^{8t/36500}$, for $0 \le t \le 110$. In particular, just before Mary's withdrawal is made, the amount in the account is $a(110) = 1000e^{880/36500} \approx \$1024.40$.

Now consider separately the amount of money in the account for t between 110 and 150. It will be growing with the same rate of interest, so it will be an exponential function with the *same* growth constant k. Let us use the name $b(t)$ for this function. Its value at $t = 110$ is the amount left after Mary makes her withdrawal, which is $1000e^{880/36500} - 700 \approx 324.40$. So for $110 \le t \le 150$ the amount is $b(t) = c'e^{kt}$, with $k = \frac{8}{36,500}$ as above. To find c', we note that $b(110) = 324.40$, so $c'e^{880/36500} = 324.40$, and $c' = 324.40/e^{880/36500} \approx 316.67$. Using this, we are finally able to evaluate $A(150) = b(150) = 316.67e^{150k} = 316.67e^{1200/36500} \approx 327.25$.

At the specified time the amount in the account is $327.25.

In Figure 7.4.1 we see that the graph of $y = A(t)$ has a discontinuity at $t = 110$, corresponding to Mary's withdrawal. In the above calculation we first considered the initial part of the curve and used it to find the amount in the account after 110 days. Then we found a formula for the part of the function after $t = 110$, and we evaluated it at $t = 110$. This evaluation was made after the withdrawal, so it was $b(110) = 1024.40 - 700 = 324.40$.

In working with $b(t) = c'e^{kt}$ we had to do a small calculation to find c'. This is because the "initial value" we have for $b(t)$ is not at $t = 0$ but rather at $t = 110$. An

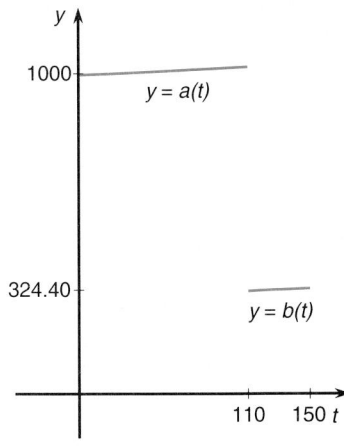

Figure 7.4.1

alternate way to deal with this is to let $b(t) = c'' e^{k(t-110)}$. The advantage of this is that when $t = 110$, the right side of this formula is c'', which we know without any work is $c'' = b(110) = 324.40$. It is easy to verify that this form of the function $b(t)$ is exactly the same as the one used above. (You should verify that you get the same value for $b(150)$.)

Either way of writing $b(t)$ is perfectly acceptable.

In Example 12 we addressed the following problem. If the nominal interest rate is $p\%$, what is the effective rate with continuous compounding? We know that with compounding n times per year the effective rate is $\left(1 + \frac{p}{100n}\right)^n$, and to find the effective rate for continuous compounding, we let $n \to \infty$ and evaluate the limit. Using Formula 7.2.7, we find that in one year an amount of money will increase by the factor

$$\lim_{n \to \infty} \left(1 + \frac{p}{100n}\right)^n = e^{p/100}. \tag{7.4.2}$$

To find the effective annual interest rate *as a percentage*, we take $(e^{p/100} - 1) \times 100\%$.

Incidentally, this reasoning also shows that the growth constant k for the total amount ce^{kt} is $k = \frac{p}{100}$.

EXAMPLE 13   Present Value   Suppose a construction company does some work on Albert's house, and Albert agrees to pay the company $5000 two years from now. He says that he has no money but that he will be able to pay later, and the company was anxious to have the business, since it came during the slow season.

In fact, Albert has plenty of money, but he is just being sneaky. He figures he might as well pay later and earn interest on the money in the interim. However, Albert likes to play the horses, and he is afraid that he might lose all his extra cash, so he decides it would be prudent to put enough money in the bank to cover his obligation.

He finds a bank that will pay him 7% annual interest, compounded continuously, and decides to deposit his money there. Certainly, he does not have to deposit $5000, because the money will earn interest and grow over the two-year period, so he has to figure out exactly how much he must deposit in order to have $5000 in two years.

Let us write D for the amount of the deposit. What we need is to find how much this deposit will be worth after two years. We know that in one year the deposit will grow by a factor of $e^{7/100} \approx 1.0725$. In two years it will do this twice, which means that it will increase by a factor of $\left(e^{7/100}\right)^2 = e^{14/100} \approx 1.1503$.

What we want is that $e^{14/100} D = 5000$, which means $D = 5000 e^{-14/100} \approx 4346.79$. The answer to Albert's question is that he should deposit $4346.79 in the bank. With interest at a nominal 7%, compounded continuously, this will grow in two years to $5000, and he will be able to use it to pay off his debt when it is due.

In a sense, then, an obligation to pay $5000 in two years is an obligation that right now amounts to $4346.79. We say that the **present value** of the obligation is $4346.79.

From the point of view of the construction company, things look slightly different. The company was happy to have the job, but not having been paid, it now has to borrow money. The best deal the company can find is from a bank that will lend it money at 12% interest, compounded semiannually. On the strength of Albert's promise to pay $5000 in two years, the company can borrow the amount of money that will grow to $5000 in two years with 12% interest and semiannual compounding.

An amount L will grow in one year to $L(1.06)^2$ and in two years to $L(1.06)^4 \approx 1.2625L$. For this increased amount to equal 5000 we must have $L = 5000/(1.06)^4 \approx 3960.47$. The amount the bank will agree to lend is $3960.47.

So to the company, the present value of Albert's obligation is $3960.47. The difference arises because of the different interest rates. It is very important to remember that calculations of present value depend on the interest rate used. In practical situations it is often impossible to predict the interest rates over the period of an investment, so calculations of present value are necessarily contingent upon the assumptions being made. Further complications are introduced by inflation, but we will not consider them here.

EXAMPLE 14   **Lottery Ticket**   What is the present value of a "million-dollar" winning lottery ticket that promises to pay $100,000 right away and a further $100,000 each year for a total of ten payments? (Assume 8% interest, compounded quarterly.)

Solution   First let us figure out that in one year an initial amount will increase by a factor of $(1.02)^4 \approx 1.0824$.

Now let us consider each payment separately. The amount paid right now has present value $100,000. The present value of the payment to be made in one year is the amount that will grow to 100,000 in one year, or $100,000/(1.02)^4 \approx 92,384.54$. Similarly, the present value of the payment to be made in two years is $100,000/(1.02)^8 \approx 85,349.04$. We continue like this, finding that the present value of the payment to be made in k years is $100,000/(1.02)^{4k}$. For instance, the last payment, which will be paid in 9 years (not 10), has present value $100,000/(1.02)^{36} \approx 49,022.32$.

The present value of the ticket is the sum of the present values of all the payments, which is

$$100,000 + \frac{100,000}{(1.02)^4} + \frac{100,000}{(1.02)^8} + \cdots + \frac{100,000}{(1.02)^{36}}$$

$$\approx 100,000 + 92,384.54 + 85,349.04 + 78,849.32 + 72,844.58$$
$$+ 67,297.13 + 62,172.15 + 57,437.46 + 53,063.33 + 49,022.32$$
$$= \$718,419.86.$$

The present value of the lottery ticket is only $718,419.86, less than three-quarters of a million dollars.

It is also possible to find the present value of an amount paid some time ago or the present value of a combination of various amounts at various times. As with the lottery ticket in Example 14, the idea is to work with each amount separately and add them up at the end.

EXAMPLE 15   **Christmas Trees**   Five years ago, Alice planted 2000 Christmas tree seedlings at a cost of $2.00 each for materials and labor. One year later and each succeeding year, she had them sprayed, at a cost of $700 each time.

Five and a half years from now, she will harvest the trees. The harvesting and shipping will cost $4.00 per tree, and she will be able to sell them for $22.00 per tree.

Find the present value of her operation, assuming 9% annual interest, compounded continuously.

Solution   Note that with this interest rate an amount of money increases each year by a factor of $e^{0.09} \approx 1.09417$.

It may help to draw a line, representing time in years, and mark the various events on it as in Figure 7.4.2. The dot at the left end represents the planting 5 years ago (i.e., at $t = -5$). The evenly spaced dots (at one-year intervals) represent the sprayings. The large dot represents the present time. The dot at the right end, one-half year after the last spraying, represents the harvesting.

The initial expense was $2000 \times \$2 = \$4000$ for planting. In subsequent years there is an annual expense of $700, and at the time of harvesting there is an expense of $2000 \times \$4 = \$8000$. We draw the line again in Figure 7.4.3, marking these amounts on it.

The next step is to find the present value of each of these amounts. The easy one is the $700 expense at the present time, whose present value is, of course, $700. The present value of the $700 expense one year from now is $700/e^{0.09} = \$639.75$. Similarly, we can find the present value of the expense two years from now, which is $700/(e^{0.09})^2 = \$584.69$, and so on. The present values of the spraying expenses for three, four, and five years from now are $700/(e^{0.09})^3 = \$534.37$, $700/(e^{0.09})^4 = \$488.37$, and $700/(e^{0.09})^5 = \$446.34$.

The present value of the $700 expense one year ago means how much $700 deposited one year ago would be worth today, which is $700 \times (e^{0.09}) = \$765.92$. Similarly, the expense from two years ago would now be worth $700 \times (e^{0.09})^2 = \$838.05$. The expenses from three and four years ago have present values $700 \times (e^{0.09})^3 = \$916.98$ and $700 \times (e^{0.09})^4 = \$1003.33$. The present value of the planting expense is $4000 \times (e^{0.09})^5 = \$6273.25$.

The harvesting will occur in five and a half years, that is, at time $t = 5\frac{1}{2}$. The factor by which an amount will increase in $5\frac{1}{2}$ years is $(e^{0.09})^{5.5} = e^{5.5 \times 0.09} = e^{0.495}$, so the present value of the harvesting cost is $8000/e^{0.495} = \$4876.57$.

We add up all these present values to find the total present value of all the expenses. It comes out to $18,067.62.

The one remaining thing is the money that Alice will earn by selling the trees. It will be $2000 \times 22 = \$44{,}000$ and will be obtained in $5\frac{1}{2}$ years. Its present value is $44{,}000/e^{0.495} = \$26{,}821.12$.

The net present value of the operation is the difference between the present value of the assets ($26,821.12) and the expenses ($18,067.62), or $8,753.50. Because this amount is positive, Alice can expect to make a profit, provided interest rates do not change and there are no unforeseen complications (such as forest fires).

Figure 7.4.2

Figure 7.4.3

EXAMPLE 16   **Newton's Law of Cooling**   If you hold your hand in a refrigerator, it will begin to cool off. If you hold your hand in the freezer, it will cool off more quickly. This is a simple example to illustrate an important principle, which is that an object that is warmer than its surroundings will cool off and moreover that it will do so more quickly the cooler the surroundings are.

Let us consider a reasonably small object at temperature T placed in surroundings where the temperature is A. (We use the letter A to remind us of the **ambient temperature**, which simply means the temperature of the surroundings.) Let us make the further assumption that A is constant. (The idea here is that the warmth from the small object is not enough to have any significant warming effect on the surroundings.) **Newton's Law of Cooling** says that the rate of change of T is proportional to $T - A$; a warm object will cool down at a rate proportional to the difference between its temperature and the ambient temperature. Symbolically,

$$\frac{dT}{dt} = -k(T - A) \tag{7.4.3}$$

for some positive constant k. The negative sign is there to ensure that if $T > A$, then $\frac{dT}{dt}$ is negative, meaning that the small warm object will cool down. Of course, it also works the other way around: A small cold object will warm up, and in either case the rate of change depends on the difference between the two temperatures.

Since A is assumed to be constant, Equation 7.4.3 can be written as $\frac{d}{dt}(T - A) = -k(T - A)$. The rate of change of the function $T - A$ is proportional to the function itself. As usual, this has the consequence that $T - A$ is an exponential function of t. We can write $(T - A)(t) = ce^{-kt}$, that is, $T(t) - A = ce^{-kt}$, or $T(t) = A + ce^{-kt}$, where the constant c is the initial value of $T - A$.   ◢

EXAMPLE 17   A metal spatula is being used to stir-fry some food; its blade is at a temperature of $400°F$. The blade is plunged into a sink full of tepid water whose temperature is $110°F$. After 2 sec in the water the spatula blade has cooled to $270°F$. How long will it take it to cool to $150°F$ so that the cook can safely handle it?

Solution   We assume that there is enough water in the sink so that its temperature will not change appreciably. This means that the ambient temperature is $A = 110$. From Equation 7.4.3 we know that $T(t) = A + ce^{-kt} = 110 + ce^{-kt}$. The constant c is the initial value of $T - A$, which is $400 - 110 = 290$.

To find k, we use the fact that $T(2) = 270$, which means that $110 + 290e^{-2k} = 270$, or $290e^{-2k} = 160$, so $e^{-2k} = \frac{160}{290}$. We take logarithms and see that $-2k = \ln\left(\frac{160}{290}\right)$ and $k = -\frac{1}{2}\ln\left(\frac{16}{29}\right)$.

We need to know when $T(t) = 150$, which means $150 = 110 + 290e^{-kt} = 110 + 290e^{t\ln(16/29)/2}$, or $290e^{t\ln(16/29)/2} = 40$. Dividing by 290 and taking logarithms gives $\frac{t}{2}\ln\left(\frac{16}{29}\right) = \ln\left(\frac{40}{290}\right)$, so $t = 2\ln\left(\frac{40}{290}\right)\frac{1}{\ln(16/29)} \approx 6.66$.

The spatula will be cool enough after 6.66 sec.   ◢

> The thing to remember about questions of this type is that it is not the temperature $T(t)$ that is an exponential function of t, but rather the function $T(t) - A$, the difference between the two temperatures.

EXAMPLE 18   **Concentration of Impurities**   Imagine a polluted lake. Clean water flows into the lake at one end, and dirty water flows out at the other. Suppose that the original source of the pollution has been cleaned up, so the flow of water

Discharge of production waste into a river in Copperhill, Tennessee.

will gradually reduce the concentration of the contaminants in the lake. It is a very important question to estimate the amount of pollution remaining and/or the length of time it will take to flush it out.

Unfortunately, the pollution will never disappear completely, but the concentration will diminish. If we can assume that the fresh water flowing into the lake mixes perfectly with the water already there, which is a fairly reasonable assumption in most cases, then it turns out that the concentration of contaminants will decrease as an exponential function of time.

Consider a lake containing 100,000,000 cubic meters of water. Suppose the rivers flow in and out at the rate of 100,000 m^3 per day. Also suppose that the concentration of the contaminant is 0.1%.

Let t be the time in days, and let $P(t)$ be the number of cubic meters of contaminant at time t. In one day, 100,000 m^3 of polluted water flow out of the lake. This is $\frac{100,000}{100,000,000} = \frac{1}{1000}$ of the water in the lake. So the amount of contaminant that flows out in this water will be $\frac{1}{1000}$ of all the contaminant present, or $\frac{P(t)}{1000}$.

This says that $P(t)$ decreases at the rate of $\frac{P(t)}{1000}$ m^3 per day. In other words, $\frac{dP}{dt} = -\frac{1}{1000}P(t)$.

From this we see that $P(t) = ce^{-t/1000}$. The constant c, the initial value of $P(t)$, is easily calculated to be 0.1% of the volume of the lake, or 100,000. So the amount of contaminant in the lake is $P(t) = 100,000e^{-t/1000}$ m^3, with t measured in days.

For example, after 1 year the amount of contaminant left will be

$$P(365) = 100,000e^{-365/1000} \approx 69,420 \text{ m}^3.$$

The concentration will have diminished from 0.1% to $\frac{69,420}{100,000,000} \times 100\% = 0.06942\%$.

We might ask how long it will take for it to reduce to $\frac{1}{10}$ of its original level, or 0.01%. This means that we need to find t for which $e^{-t/1000} = \frac{1}{10}$, or $-\frac{t}{1000} = \ln\left(\frac{1}{10}\right)$, or $t = -1000\ln(0.1) \approx 2300$. It will take approximately 2300 days, or about 6.3 years for the pollution to reduce to $\frac{1}{10}$ of its original level.

In practice there might be several different rivers flowing into and out of a lake, and some of the ones flowing in might bring in new contaminants. In this kind of situation it is necessary to figure out the *net* inflow and outflow of contaminants, but then the same sort of analysis can be applied. The result will be that the difference between the current level of pollution and the level flowing in will decrease exponentially. We can expect that as $t \to \infty$, the concentration of contaminants in the lake will approach as a limit the concentration of contaminants flowing in. We cannot expect the lake to become cleaner than the water flowing into it.

Exercises 7.4

I

Find the half-life of each of the following decay processes.

1. The original amount reduces to 20% in 3 hr.

2. The original amount reduces to 44% in 9 min.

3. The original amount reduces by 15% in 1 day.

4. The original amount reduces by 89% in 28 sec.

5. After 4 hr the amount is 10 oz; after 6 hr it is 8 oz.

6. After 1 yr the amount is 65 g; after 3 yr it is 30 g.

Find the decay constants for the processes with the following half-lives.

7. 2800 yr

8. 30 days

9. 34 min

10. 2.1 sec

II

11. In one year the population of a small Balkan country grows from 3,200,000 to 3,220,000.

    (i) Assuming that the conditions for exponential growth are satisfied, how long will it be before the population reaches 4,000,000?

    (ii) How long is it since the population was 3,000,000?

12. If the annual birth rate in a certain region is 1.4% and the annual death rate is 0.6%, how long will it take to have a 10% increase in population, assuming the growth is exponential?

13. Suppose a substance decays radioactively with half-life h. What is the decay constant k in the formula $Q(t) = ce^{-kt}$ for the amount remaining at time t?

14. If a substance grows according to the formula $Q(t) = ce^{kt}$, how long will it take to double?

15. Suppose a substance decays radioactively with half-life h. Suppose the initial amount of the substance is I. Find a formula for the amount remaining at time t.

*16. (i) Suppose a radioactive substance decays with half-life h. If the amount of the substance has reduced to r times as much as there was at the beginning of the decay (with $0 < r < 1$), how long has the substance been decaying?

    (ii) Check your answer by applying this formula to Example 8 (p. 534).

*17. (i) The half-life of the isotope polonium 210 is 140 days. Make a quick estimate in your head of how long it will take for a sample to decay to 1% of its original size.

    (ii) Calculate the exact answer to the question posed in part (i). How good was your estimate?

18. If the half-life of a radioactive substance is h, how long will it take to diminish to p% of its original size?

19. Compare the effective annual interest resulting from daily compounding at a nominal annual rate of 9.9% with the effective rate for quarterly compounding at a nominal 10%.

*20. In the text it was stated that the effective interest rate increases as the number of times interest is compounded increases. Prove that this is correct, assuming that the nominal rate is fixed.

21. If the nominal annual interest rate is 10%, what is the effective rate with continuous compounding?

22. What nominal rate of annual interest will give an effective rate of 10% with continuous compounding?

23. If the nominal annual interest rate is p%, what is the effective rate with continuous compounding? Express your answer in terms of $r = \frac{p}{100}$.

24. What nominal rate of annual interest will give an effective rate of p% with continuous compounding? Express your answer in terms of $r = \frac{p}{100}$.

25. A bank account containing $1500 earns interest at the rate of 6%, compounded continuously. No withdrawals or deposits are made for 6 weeks, after which $1000 is deposited. Two weeks after that, $300 is withdrawn. Calculate the amount in the account 12 weeks after this withdrawal.

26. Repeat Exercise 25 with the nominal interest rate of 8%.

*27. Three thousand dollars are deposited in an account bearing interest at the nominal rate of 7.5%, compounded continuously. After 90 days the interest rate rises to 8%. What is the total in the account 1 year after the initial deposit?

28. An account is opened with a deposit of $100, and an additional $100 is deposited each month from then on. If the nominal interest is 9%, compounded continuously, how much money will be in the account after 18 months? (Include the nineteenth deposit.)

29. Compare the effective annual interest rates obtained with a nominal rate of 8% if the compounding is (i) quarterly, (ii) monthly, (iii) daily, (iv) hourly, (v) continuous.

30. Compare the effective annual interest rates obtained with a nominal rate of 11% if the compounding is (i) quarterly, (ii) monthly, (iii) daily, (iv) hourly, (v) continuous.

*31. Work out the effective annual interest rate for daily compounding at 6% nominal interest. Do it two different ways, using a calculator: (i) by multiplying out $\left(1 + \frac{1}{365}\frac{6}{100}\right)^{365}$ (i.e., 364 multiplications) and (ii) using logarithms. Any slight difference in the answers shows the effect of round-off errors.

*32. If we repeat Exercise 31 with 12% nominal interest, do you expect the effective rate to be twice what resulted from 6% nominal interest, or more, or less? Do the calculation and check your intuition.

*33. (i) With interest at a nominal rate of 7% per annum, compare the effective rates obtained by compounding daily in an ordinary year (with 365 days) and in a leap year (with 366 days).

    (ii) How much money would there have to be in an account for the difference in the amounts of annual interest to equal 1 cent?

*34. A car loan is paid off in 36 equal monthly installments (each payment is at the end of the month). If the interest is compounded continuously and the nominal rate is 12% per annum, what will the size of the monthly payment be for an original loan of $8500?

35. What is the present value of an annuity that will pay its owner $20,000 per year for the next 15 years, assuming 8% annual interest, compounded continuously? (Assume that the first payment is today and there are a total of 15 payments.)

36. Uncle Fred makes a deposit of $500 each year for each of his nieces and nephews, beginning on their first birthday and ending on their twentieth.

   (i) Assuming an effective annual interest rate of 8%, what is the present value of the deposit on the tenth birthday of one of the nephews or nieces?

   (ii) What is the present value of the deposit on the twentieth birthday of one of the nephews or nieces?

   (iii) What is the present value of the deposit on the day one of the nephews or nieces is born?

37. For a period of time in the early 1980s the number of reported cases of AIDS was doubling every six months. Ignoring the discrepancies between the number of actual cases and the number of reported cases and supposing that this trend continued, how long after the first case would it take for the entire world population of approximately 5,000,000,000 to be infected?

*38. Discuss the assumptions underlying Exercise 37. How reasonable do you think they are? What significance can be attached to the answer you obtained?

39. A lake contains 500,000,000 ft^3 of water contaminated with chemicals whose concentration is 300 ppm (parts per million). If clean water flows in at the rate of 5,000,000 ft^3 per day, how long will it be until the chemicals are below the minimum standard for drinking water of 10 ppm? (Assume that the clean water mixes perfectly with the lake water and that the outflow from the lake is the same as the inflow.)

40. Suppose a lake just like the one in Exercise 39 (same volume, same water flow) has never been polluted and contains pure water. Suppose that a paper mill opens 3 miles upriver from this lake with the result that the water flowing in is heavily polluted, with the concentration of the contaminants equal to 1.3%.

   (i) How long will it be before the lake water is contaminated at the level of 10 ppm?

   (ii) Show that in the long run the concentration of pollutants in the lake will approach the level in the water flowing in.

41. A lake is round, with a radius of 1 mile and an average depth of 50 ft. A river flows into it; at a point where the river is 40 ft wide and has an average depth of 10 ft, the river flows at a rate of 6 inches per second. This is the only inflow to the lake; a river flows out at the opposite side. The water flowing in is perfectly clean. Suppose the water in the lake is polluted with contaminants at a concentration of 0.07%. Assuming that the water flowing in mixes perfectly with the lake water, how long will it take for the concentration of the pollutant to reach 0.02%?

**42. Two chemists are cleaning flasks after an experiment. Each flask has a capacity of 1 liter and contains 10 g of salt dissolved in 1 liter of water. It is important to remove as much salt as possible because even minute amounts may invalidate the results of the next experiment. Sara empties her flask, leaving 10 milliliters behind. She fills the flask with clean water from the tap, empties it again (except for 10 ml), and so on. In all, she empties the flask six times. Dave, on the other hand, places his flask beneath the tap and lets the water run. Water runs into the flask and mixes with the solution inside, and some of the solution runs over the rim of the flask and down the drain. He leaves it long enough for 5 liters of water to run into the flask and then empties it except for 10 ml. Assume that in both cases the solutions always mix perfectly. The question we want to answer is which method does a better job of cleaning out the salt. Notice that they use almost exactly the same amount of water.

   (i) Try to decide which method leaves less salt behind, using your powers of reasoning but not making any calculations.

   (ii) Calculate how much salt remains in each flask at the end of the cleaning process. By what factor is one method preferable over the other?

43. The rate at which a chemical reaction takes place depends on many things, but if most of them are held fixed (e.g., constant temperature, constant volume, a consistent amount of agitation of the mixture, constant illumination, etc.), the main factor will be the concentration of the reactants. Consider a simple reaction in which a single compound decomposes, in the presence of heat, into two products. Let us assume that the temperature, the volume, and the illumination are all held constant. Let us also assume that the amount of the reactant remaining at any given time is an exponentially decreasing function of time (a fairly reasonable assumption under these conditions).

   (i) Suppose the reaction begins with 1.3 g of the

reactant and that after 25 min there is 0.9 g remaining. How much will there be after 2 hours and 5 minutes?

    (ii) How long will it be until there is 0.5 g remaining?

*44. Suppose a different chemical reaction proceeds in such a way that the rate of change of the concentration of a certain reactant is proportional to the square of the concentration of that reactant (a reaction of this type is said to be "of second order"). Find a formula for the concentration of the reactant in question as a function of time. Interpret any constants in your formula (e.g., as "initial amounts," etc.).

45. Use a programmable calculator or computer to find the effective rate for nominal 6% interest compounded every *hour* by multiplying out the product (cf. Exercise 31).

46. Use a programmable calculator or computer to find the effective rate for nominal 12% interest compounded every *hour* by multiplying out the product (cf. Exercise 32).

POINT TO PONDER

In Example 7.3.5 (p. 527) we saw the effect of continuing calculations of exponential growth for too long a time, beyond the point at which the assumptions are reasonable. In discussing radioactive decay we said that it satisfies the assumptions for exponential decay very well, and it is true that the calculations for this type of problem are extremely accurate. On the other hand, there is a limit to their reliability too.

Consider a sample of radon. It is a chemically inert but radioactive gas, formed as a product of the radioactive decay of radium. Radon decays with a half-life of 3.82 days. A sample of radon weighing 1 gram contains approximately 2.7×10^{21} atoms.

Calculate how long it would take until there are only 1000 atoms left. Then calculate how long it would take until there is only 1 atom left.

At some point the last atom will decay, and there will be no radon left at all. It is not possible to predict exactly when it (or any other single atom) will decay, but it is possible to calculate the probability that it will have decayed by a certain time. After it does decay, the exponential function, which is never zero, will be

(slightly) incorrect in that it will suggest that there is still some radon present.

There is a fundamental difficulty here that arises from the fact that the exponential function is a continuous function, whereas matter is composed of individual atoms. The number of atoms of radon remaining is not a continuous function; it has a jump discontinuity every time an atom decays.

As long as the number of atoms is quite large, the jumps will be relatively small, and the continuous function will be a very accurate approximation to the actual (discontinuous) function. When there are only a handful of atoms, then the randomness in their behavior will become apparent, and the exponential function will be less accurate.

The difficulty is more theoretical than practical because usually for us to be able to measure the number of atoms, they have to be fairly numerous. The situation in which there are few enough for their individual behavior to be noticeable is usually beyond our ability to perceive.

7.5 Logistic Growth

LOGISTIC EQUATION
CARRYING CAPACITY

In Sections 7.3 and 7.4 we discussed exponential growth. We saw how that concept could be applied to the growth of organisms from bacteria to human beings, but we also gained some appreciation of the shortcomings of the approach.

The fundamental difficulty is that while the assumptions from which we can conclude that the growth will be exponential may well hold for a while, they will cease to be valid as the population grows. This can occur in various ways, but usually the growth rate will slow down as the population becomes sufficiently large that the existing population interferes with further unrestricted growth.

In this section we discuss a formula for population growth that attempts to take these ideas into consideration. It is a fairly simple idea, but it turns out to give a surprisingly accurate formula in many situations. Unfortunately, the actual calculations involved are a little less simple than the corresponding ones for exponential growth. They are not tremendously difficult, but they are more complicated.

This section has no bearing on anything appearing in later sections and so may be omitted. However, it should be of special interest to anybody studying life sciences, since the ideas introduced here are widely used in those areas.

The basic assumption underlying the theory of exponential growth is that $\frac{dP}{dt}$, the rate of change of the population, is proportional to the population P. This is equivalent to assuming that the birth rate is *constant*. (The birth rate means the number of births per member of the population per unit of time.) This assumption is frequently quite reasonable for a while, but as the population grows, it will be unable to continue to grow as rapidly. To make a more realistic mathematical model, we will have to incorporate the idea that the birth rate will decrease if the population grows very large.

To simplify the discussion, let us consider a population that does not die, such as bacteria. Instead of assuming that the birth rate is a constant, let us try the simplest decreasing function, a straight line with negative slope, as shown in Figure 7.5.1. This means that we should assume that the birth rate is a function of the form $a - cP$, where a and c are constants, with $c > 0$.

If the birth rate is $a - cP$, then the number of births per unit time will be $(a - cP)P$, and this means that $\frac{dP}{dt} = (a - cP)P$. Letting $L = \frac{a}{c}$, we can rewrite this equation as

$$\frac{dP}{dt} = cP(L - P). \tag{7.5.1}$$

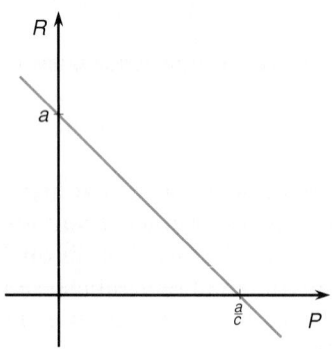

Figure 7.5.1

This is known as the **logistic equation**.

To solve it, we rearrange it in the form

$$\frac{dP}{P(L - P)} = c\, dt,$$

and try to integrate.

To integrate the left side, it will be necessary to use a trick. We notice that $\frac{1}{P(L-P)} = \frac{1}{L}\left(\frac{1}{P} + \frac{1}{L-P}\right)$.

This permits us to integrate the left side of the equation above:

$$\int \frac{dP}{P(L - P)} = \frac{1}{L} \int \left(\frac{1}{P} + \frac{1}{L - P}\right) dP$$

$$= \frac{1}{L}\left(\ln P - \ln(L - P)\right) + d = \frac{1}{L} \ln \frac{P}{L - P} + d.$$

Since the integral of the right side of the equation is ct plus a constant, we get $\frac{1}{L} \ln \frac{P}{L-P} = ct + d'$. Multiplying by L and exponentiating, we see that $\frac{P}{L-P} = Ce^{cLt}$, where the constant C can be determined by evaluating at $t = 0$. If we write P_0 for the population at $t = 0$, then we find that $C = P_0/(L - P_0)$, so

$$\frac{P}{L - P} = \frac{P_0 e^{cLt}}{L - P_0}.$$

Cross-multiplying, we find that $P(L - P_0) = P_0 e^{cLt}(L - P)$.

Collecting the terms containing P, we have $P(L - P_0 + P_0 e^{cLt}) = LP_0 e^{cLt}$, and dividing gives

$$P = P(t) = \frac{LP_0 e^{cLt}}{L - P_0(1 - e^{cLt})} = \frac{P_0 L}{P_0 + (L - P_0)e^{-cLt}}. \tag{7.5.2}$$

In deriving this formula we have implicitly assumed that $L > P$ because we said $\int \frac{dP}{L-P} = -\ln(L - P) + constant$.

Several things are immediately apparent from this formula. It is easy to evaluate the limit of $P(t)$ as $t \to \infty$. Since we know that c and L are positive, we know that $\lim_{t\to\infty} (e^{-cLt}) = 0$. From this we see that

$$\lim_{t \to \infty} P(t) = \frac{P_0 L}{P_0 + 0} = L.$$

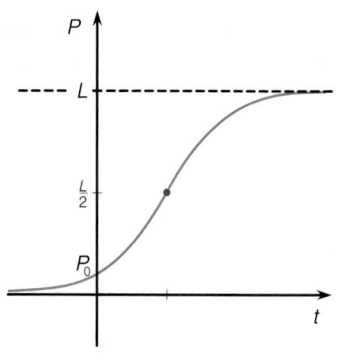

Figure 7.5.2

The population will approach the limit L as t tends to infinity. As $t \to -\infty$, we know that $e^{-cLt} \to \infty$, so the denominator tends to infinity, and $P(t) \to 0$. The shape of the graph of $y = P(t)$ is shown in Figure 7.5.2. It is straightforward but messy to calculate the second derivative $P''(t)$. From this calculation it is possible to see that there is exactly one inflection point. It occurs at the value of t that satisfies $e^{-cLt} = P_0/(L - P_0)$. The value of $P(t)$ at this particular value of t is $\frac{L}{2}$. In terms of the picture, this means that the graph is concave upward until it reaches halfway up from 0 to its limiting value L, and after that it is concave downward.

The limiting population L is often called the **carrying capacity** of the environment. If we think of bacteria in a jar, L represents the maximum number of bacteria that can survive in the jar. It would have a similar interpretation if we were talking about a population of rabbits, deer, or humans.

The logistic formula has proven to be remarkably accurate in many situations. It has had striking successes in predicting human population growth in countries. (It has also failed sometimes but usually because the situation changed in some way, such as a change in the law permitting a large immigration or a medical advance that saved many lives.)

It was mentioned above that we have implicitly assumed that $P < L$, that is, that the population stays below the carrying capacity. If we place a population that is too large in the environment, we should expect it to decrease toward the carrying capacity, and in fact this is reflected in the behavior of any solution of Equation 7.5.1 that begins with a population that is "too large."

Unfortunately, it is more difficult to figure out the constants associated to a logistic curve than it is for exponential curves. In particular, since there are three of them (the initial population P_0, the carrying capacity L, and the constant c in the exponent, which is a sort of growth constant), it is necessary to know the value of $P(t)$ for three values of t.

In terms of the calculation it turns out to be very much easier to work out the constants if we happen to know the values of $P(t)$ for three values of t that are evenly spaced.

EXAMPLE 1   Suppose the population of a small town follows a logistic curve. If the population was 6,000 in 1970, 10,000 in 1980 and 15,000 in 1990, what will it be in the year 2000?

Solution   Let us write the population in the form of Equation 7.5.2, $P(t) = \frac{P_0 L}{P_0 + (L - P_0) \exp(-cLt)}$. It will be more convenient to measure population in thousands, and it will also be convenient to let t be the time in years, with $t = 0$ at the year 1980. We know that P_0 is the *initial population*, meaning the population when $t = 0$, which is $P_0 = 10$ (meaning that in 1980 the population was 10 thousand).

This means that

$$P(t) = \frac{10L}{10 + (L - 10)e^{-cLt}}.$$

We also know that $P(-10) = 6$ and $P(10) = 15$. Writing these out, we have

$$\frac{10L}{10 + (L - 10)e^{10cL}} = 6,$$

$$\frac{10L}{10 + (L - 10)e^{-10cL}} = 15.$$

Cross-multiplying in each of these equations, we obtain

$$10L = 60 + 6(L - 10)e^{10cL},$$
$$10L = 150 + 15(L - 10)e^{-10cL}$$

or

$$10(L - 6) = 6(L - 10)e^{10cL},$$
$$10(L - 15) = 15(L - 10)e^{-10cL}.$$

What makes these equations difficult to solve is the presence of the exponential factors on the right side. But if we multiply the right sides of the last two equations, we see that the two exponential factors cancel. Multiplying the two left sides and the two right sides, we obtain

$$100(L - 6)(L - 15) = 6(15)(L - 10)^2,$$
$$100(L^2 - 21L + 90) = 90(L^2 - 20L + 100),$$
$$100L^2 - 2100L + 9000 = 90L^2 - 1800L + 9000,$$
$$10L^2 - 300L = 0,$$
$$10L(L - 30) = 0.$$

The two possible solutions are $L = 0$ and $L = 30$; the first is an extraneous solution. The carrying capacity is $L = 30$ (i.e., a population of 30,000).

Now it is relatively easy to solve for the constant c. For instance, we have already found that $10(L - 6) = 6(L - 10)e^{10cL}$; putting in the value $L = 30$, we find that

$$10(24) = 6(20)e^{300c},$$
$$240 = 120e^{300c},$$
$$300c = \ln 2,$$
$$c = \frac{\ln 2}{300}.$$

We are now able to write the logistic population formula:

$$P(t) = \frac{300}{10 + 20e^{-t\ln(2)/10}}.$$

From this it is easy to answer the question about the population in the year 2000. Since $t = 0$ in the year 1980, the year 2000 occurs when $t = 20$, so the population then is

$$P(20) = \frac{300}{10 + 20e^{-2\ln(2)}} = \frac{300}{10 + 20(2)^{-2}} = \frac{300}{15} = 20.$$

According to our convention, the population in the year 2000 will be 20,000.

With a graphics calculator we can plot the function of the previous example.

TI-85	SHARP EL9300
GRAPH F1$(y(x) =)$ 300 ÷ (10 + 20 e^x ((−) F1(x) × LN (2) ÷ 10)) M2(RANGE) (−) 60 ▼ 60 ▼ 10 ▼ (−) 10 ▼ 30 ▼ 10 F5(GRAPH)	300 ÷ (10 + 20 e^x (−) x/θ/T × ln 2 ÷ 10 ▶) RANGE −60 ENTER 60 ENTER 10 ENTER − 10 ENTER 30 ENTER 10 ENTER
CASIO f_x-6300G (f_x-7700GB)	HP 48SX
Range −60 EXE 60 EXE 10 EXE − 10 EXE 30 EXE 10 EXE EXE EXE EXE Graph 300 ÷ (10 + 20 e^x ((−) ▓ × ln (2) ÷ 10)) EXE	

We see the characteristic property that the curve approaches a limit as time grows very large.

Incidentally, in the course of solving this problem we have found that the carrying capacity is 30,000, so we expect that the population of the town will level off as it approaches 30,000. This kind of unlimited long-term prediction is notoriously unreliable and should be taken with a large grain of salt in any realistic application.

Notice that it was because we chose times that were evenly spaced that we were able to use the trick of multiplying to get rid of the exponential factors. It is possible to find the constants c, L, and P_0 from the values of the population at any three points, but it is much more difficult. We will not consider the general case.

Exercises 7.5

I

In each of the following situations, find the constants in Formula 7.5.2, assuming that the function is logistic.

1. $P(-1) = 1$, $P(0) = 2$, $P(1) = 3$
2. $P(2) = 5$, $P(4) = 7$, $P(6) = 9$
3. $P(0) = 8$, $P(1) = 10$, $P(2) = 11$
4. $P(4) = 1$, $P(5) = 9$, $P(6) = 12$
5. $P(1950) = 15{,}000$, $P(1960) = 20{,}000$, $P(1970) = 25{,}000$
6. $P(1962) = 20{,}000$, $P(1966) = 22{,}000$, $P(1970) = 23{,}000$
7. $P(10) = 1$, $P(11) = 44$, $P(12) = 45$
8. $P(0) = 3$, $P(1) = 5$, $P(2) = 6$
9. $P(-3) = 4$, $P(-2) = 5$, $P(-1) = 6$
10. $P(-h) = a$, $P(0) = 2a$, $P(h) = 3a$

II

11. A culture of bacteria weighs 2 mg after growing for 1 day, 4 mg after two more days, and 6 mg two days after that.
    (i) What is the carrying capacity of their container, assuming logistic growth?
    (ii) How long will it be after growth begins until the culture reaches 90% of the carrying capacity?

12. A shipwreck releases rabbits on a small island. After 5 years there are 500 rabbits; after 2 years more there are 800 rabbits; 2 years after that there are 1000 rabbits.
    (i) Assuming that the rabbit population follows a logistic curve, what is the carrying capacity of the island for rabbits?
    (ii) How long after the shipwreck will the rabbits have reached half the carrying capacity?

*13. Confirm the assertions made in the text about the location of the inflection point of $y = P(t)$ (i.e., that there is exactly one inflection point, that it occurs when $e^{-cLt} = P_0/(L - P_0)$, and that the value of $P(t)$ at the inflection point is $\frac{L}{2}$). You may find it helpful to use the logistic equation (Equation 7.5.1) instead of calculating the derivative $P'(t)$ directly.

*14. Show that the logistic curve is symmetrical about its inflection point. (This means that if you rotate the graph $180°$ around the inflection point, you get the same graph back again.)

*15. What happens if $L < P_0$? (Give a qualitative description of the type of solution we should expect, and

sketch the graph. Discuss how you would interpret your answer in a real situation, such as a population of rabbits on an island.)

*16. We remarked that it can be quite difficult to find the constants for a logistic curve if we know the values at three unevenly spaced points. Modify the idea demonstrated in Example 1 (p. 548) to deal with the following case: Find the logistic function $P(t)$ that satisfies $P(-1) = 1$, $P(0) = 2$, and $P(2) = 6$. (The carrying capacity L does not work out to a nice round number.)

17. Suppose a bacterial culture is weighed at different times. The first recorded weight was 5 mg; two days later the weight was 7 mg, and two days after that the weight was 9 mg.
    (i) Find the logistic curve that fits these data.
    (ii) How long before the first measurement did the culture weigh 0.1 mg?
    (iii) Suppose we have lost the third measurement, and consider what happens if we use the first two (5 mg and, two days later, 7 mg) and assume that the growth follows an *exponential* curve. Use this approach to answer part (ii), and compare the two answers.
    (iv) Use the logistic curve and the exponential curve of part (iii) to find when the culture will weigh 13 mg. Compare the two answers.

*18. Show that the logistic function (Equation 7.5.2) can also be written in the form

$$P(t) = \frac{L}{1 + e^{-cL(t - t_0)}}$$

if t_0 is the point where $P(t)$ has its inflection point.

*19. Consider the logistic curve

$$Q(t) = \frac{k}{1 + e^{d - at}}.$$

   (i) What is the carrying capacity?
   (ii) At what time t does the inflection point occur?
   (iii) If we write the function $Q(t)$ in the form of Equation 7.5.2, what are the constants P_0 and c?

*20. Show that it is not possible to find a logistic curve for which $P(0) = 9$, $P(1) = 10$, and $P(2) = 12$.

*21. Suppose twin lakes are separated by a barrier. Suppose fish grow in one lake according to the logistic equation

$$P(t) = \frac{L}{1 + e^{d - at}},$$

with t in years. Suppose that at some time t_0 the barrier is removed. If the second lake had the same number of fish, growing according to the same logistic equation, it seems reasonable to expect that the equation for the *total* population of the two joined lakes will now be $T(t) = 2P(t) = \frac{2L}{1+\exp(d-at)}$.

But suppose the barrier is removed at time t_0 and the second lake has no fish in it. What will be the formula for $T(t)$ in this situation? The carrying capacity will have doubled, but what will happen to the constants a and d? (*Hint:* Try redistributing the fish in the obvious way: half in each lake. At this point it may be easier to think about what happens in just one of the lakes and combine them later.)

*22. Suppose that if they are left unmolested, the fish in a lake grow according to the logistic equation

$$P(t) = \frac{L}{1 + e^{d-at}},$$

where t is the time in years. Suppose too that once a year there is a brief fishing season, during which the population is decreased, but that for the rest of the year no fish are caught. The problem is to optimize our fishing in the long run. If we take too many fish this year, the remaining population will be so small that it will grow slowly, and there will not be very many fish next year. If we take too few, in the first place we lose out on our catch, and in the second place the population will be large and the logistic curve will limit their growth; not many will grow to replace the ones we took.

(i) Find the best strategy, that is, the one that gives us the biggest overall catch in the long run. (*Hint:* When we catch some fish at a certain time t_0 and reduce the population at t_0 to A, this amounts to shifting the logistic curve to the right until its value at t_0 is A. In terms of the formula given above, we adjust the constant d but leave a and L unchanged.)

(ii) Discuss *briefly* the best strategy if we want to maximize our catch only for 1, 2, or 3 years.

23. Use a graphics calculator or computer to discuss the effects on a logistic curve of changing the constants P_0, L, and c. For instance, what is the effect of changing P_0 if L and c remain the same? (Sketch several curves.) What is the effect of doubling L and P_0, leaving c the same? What happens if L is doubled but P_0 and c remain unchanged? What is the effect of changing c and leaving L and P_0 the same? What will be the shape of the curve if c is negative?

7.6 Differential Equations

SEPARABLE EQUATION
GENERAL SOLUTION
PARTICULAR SOLUTION
ORDER
CHARACTERISTIC POLYNOMIAL
AUXILIARY EQUATION

In this brief section we begin to explore the subject of differential equations, which arose in the discussion of exponential growth in Section 7.3 and in the discussion of logistic growth in Section 7.5. This topic will be treated in much greater depth in Chapter 19, but it is already possible to indicate some of the important points now.

A differential equation expresses a relation between a function and one or more of its derivatives. We will use the letter t as the variable; this is because in many of the applications the equation deals with the rate of change of some quantity, so the natural variable is *time*. Functions of t will sometimes be written in the usual way as $f(t), g(t)$, and so on, but sometimes we will also use the letter x. The understanding is that x is not a variable but a *function* of t. To emphasize this, we sometimes write $x = x(t)$.

Separable Equations

A differential equation is called **separable** if it can be written in the form

$$F(x)\frac{dx}{dt} = G(t)$$

for some functions F and G. This can be rearranged to

$$F(x)\,dx = G(t)\,dt,$$

with all the x's on the left side and all the t's on the right side. In this form it is not too useful, but we can integrate to obtain

$$\int F(x)\,dx = \int G(t)\,dt + c.$$

This formula expresses a relation between x and t that does not involve derivatives. In particular cases we may be able to solve it to write x as a function of t explicitly.

Strictly speaking, the constant c is understood when we write indefinite integrals. However, this is one instance in which it is vitally important to remember the constant, so we write it out to help remind ourselves.

EXAMPLE 1   Solve the differential equation $\frac{dx}{dt} = xt$.

Solution   To solve a differential equation means to find a function that satisfies the equation. In this case we are looking for the function $x = x(t)$.

As suggested above, we rearrange the equation to $\frac{dx}{x} = t\,dt$. Integrating, we find $\int \frac{1}{x}\,dx = \int t\,dt$, or $\ln|x| = \frac{1}{2}t^2 + c'$.

To find a formula for the function $x = x(t)$, we exponentiate the last equation and find $|x| = e^{c'}e^{t^2/2}$. In rearranging the equation we divided by x, so in effect we have assumed that x is never zero. This means that it cannot change sign. We simplify notation by writing $x = ce^{t^2/2}$, with $c = e^{c'}$ if $x > 0$ and $c = -e^{c'}$ if $x < 0$. We have found the solution

$$x = x(t) = ce^{t^2/2}.$$

The constant $e^{c'}$, being an exponential, is always positive. So $c = \pm e^{c'}$ is never zero. However, the constant function $x(t) = 0$ is also a solution of the original equation; it can be seen as the result of letting $c = 0$ in the solution we have just found. It often happens that we find a solution containing an unspecified constant, with some restrictions on the constant. In this situation we may find that we still get solutions by letting the constant have some of the forbidden values. However, when we proceed this way, it is important to check that the final answer does actually satisfy the original equation.

In solving a separable equation we will encounter a constant of integration. The solution containing this unspecified constant is called the **general solution**. If we know the value of the solution at any particular value of t, then this information can be used to find the constant, and it is then possible to write down the precise solution (i.e., without any undetermined constants). In this case we call it a **particular solution**.

EXAMPLE 2   Find the solution of $\frac{dx}{dt} = 2x - 14$ that satisfies $x(0) = 1$.

Solution   We rearrange the equation in the form $\frac{dx}{x-7} = 2\,dt$. This is in the standard form for a separable equation, and we integrate it to obtain $\ln|x - 7| = 2t + c'$.

To find an expression for the function $x(t)$ itself, we exponentiate: $|x - 7| = e^{c'}e^{2t}$. Letting $c = \pm e^{c'}$, we write $x - 7 = ce^{2t}$, and rearranging, we have

$$x = x(t) = 7 + ce^{2t}.$$

To find the particular solution that satisfies $x(0) = 1$, we must find the correct value for the constant c. Knowing that $x(t)$ is given by the formula above, we see that $1 = x(0) = 7 + ce^{(2)(0)} = 7 + c$, so c must equal -6.

The required particular solution is $x = x(t) = 7 - 6e^{2t}$.

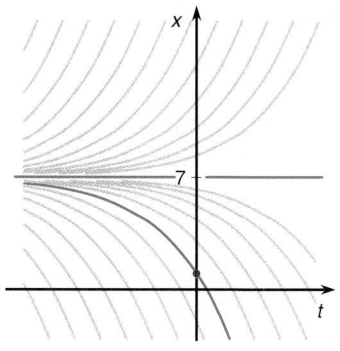

Figure 7.6.1

Notice how the condition $x(0) = 1$ forced the constant c to be negative. Also notice how we used the constant c' when we integrated so that we could use the simpler constant c in the final formula. This is a matter of style and has no other significance.

When we obtain the general solution as above, there is a constant that can have any fixed value. Each different value of the constant gives a function that is a particular solution of the differential equation. There is a whole "family" of solutions. See Figure 7.6.1.

The value of the function $x(t)$ at a particular point is known as an **initial value**. Roughly speaking, the idea is that once we know the value of the function at some "starting time," then the differential equation, which specifies how the function "changes," will determine its values at all other points.

Linear Differential Equations with Constant Coefficients

Consider the differential equation $\frac{df}{dt} = kf$, where k is a constant. We already know its solutions, but it is interesting to realize that it is an example of a separable equation. We write $\frac{df}{f} = k\,dt$, so $\ln|f| = kt + c'$. Exponentiating, we find $f = f(t) = ce^{kt}$, with c a constant.

In fact this is exactly how we solved this type of equation in Section 7.3.

If we write the equation just considered in the form $f' = kf$, or $f' - kf = 0$, we see that it is constructed by multiplying each of f and f' by a constant, adding them up, and setting the resulting formula equal to zero.

Now consider the same sort of thing with f, f', and f''.

EXAMPLE 3   Solve the equation $f'' + f' - 2f = 0$.

Solution   The function $f(t)$ must be one whose derivatives f' and f'' are very closely related to f and each other. Taking our cue from the equation $f' - kf = 0$, we make a guess and try letting $f(t) = e^{kt}$, for some constant k.

This means that $f' = ke^{kt} = kf$ and $f'' = k^2 e^{kt} = k^2 f$. Substituting these derivatives into the original equation, we find

$$0 = f'' + f' - 2f = k^2 f + kf - 2f = (k^2 + k - 2)f.$$

The exponential function $f(t) = e^{kt}$ is never zero, so this equation implies that $k^2 + k - 2 = 0$.

This is just an ordinary quadratic equation for the number k. We solve it by the quadratic formula, or in this case simply by factoring: $0 = k^2 + k - 2 = (k + 2)(k - 1)$, so $k = -2$ or $k = 1$.

We have just found two solutions of the equation: $f_1(t) = e^{-2t}$ and $f_2(t) = e^t$. It is easy to check that they do both satisfy the equation.

It is also easy to check that if a and b are any constants, then $f(t) = af_1(t) + bf_2(t)$ is also a solution. After all, $f' = af_1' + bf_2'$ and $f'' = af_1'' + bf_2''$, so

$$\begin{aligned}
f'' + f' - 2f &= (af_1'' + bf_2'') + (af_1' + bf_2') - 2(af_1 + bf_2) \\
&= a(f_1'' + f_1' - 2f_1) + b(f_2'' + f_2' - 2f_2) \\
&= a(0) + b(0) = 0.
\end{aligned}$$

We group f_1 and its derivatives together, and f_2 and its derivatives together. Each group equals zero because f_1 and f_2 satisfy the original equation.

It is less obvious but also true that this is the general solution, that is, that it gives us all the solutions.

With a graphics calculator we can plot some of the solutions of the previous example. For instance, we try $f(t) = af_1(t) + bf_2(t)$, with $a = 1$, $b = 0$; $a = 0$, $b = 1$; $a = b = 1$; $a = 1$, $b = -2$.

TI-85	SHARP EL9300
GRAPH F1$(y(x) =)$ e^x ($(-)$ 2 F1(x)) ▼ e^x F1(x) ▼ e^x ($(-)$ 2 F1(x)) $+ e^x$ F1(x) ▼ e^x ($(-)$ 2 F1(x)) $- 2$ e^x F1(x) M2(RANGE) $(-)1$ ▼ 1 ▼ ▼ $(-)5$ ▼ 10 F5(GRAPH)	⌐∿ e^x $(-)$ 2 x/θ/T 2ndF ▼ e^x x/θ/T 2ndF ▼ e^x $(-)$ 2 x/θ/T ▶ $+ e^x$ x/θ/T 2ndF ▼ e^x $(-)$ 2 x/θ/T ▶ $- 2$ e^x x/θ/T RANGE -1 ENTER 1 ENTER 1 ENTER -5 ENTER 10 ENTER 5 ENTER ⌐∿
CASIO f_x-6300G (f_x-7700GB)	HP 48SX
Range -1 EXE 1 EXE 1 EXE -5 EXE 10 EXE 5 EXE EXE EXE EXE Graph e^x ($(-)$ 2 **X**) EXE Graph e^x **X** EXE Graph e^x ($(-)$ 2 **X**) $+ e^x$ **X** EXE Graph e^x ($(-)$ 2 **X**) $- 2$ e^x **X** EXE	↰ PLOT PLOTR ERASE ATTN ´ Y ▤ e^x 2 +/− × X ↱ PLOT ↰ DRAW ERASE 1 ENTER +/− 1 XRNG 5 +/− ENTER 10 YRNG DRAW ATTN ´ Y ▤ e^x X ↱ PLOT ↰ DRAW DRAW ATTN ´ Y ▤ e^x 2 +/− × X ▶ $+ e^x$ X ↱ PLOT ↰ DRAW DRAW ATTN ´ Y ▤ e^x 2 +/− × X ▶ $- 2$ × e^x X ↱ PLOT ↰ DRAW DRAW

The solutions behave in quite different ways. Identify which is which.

EXAMPLE 4 Find the general solution of $f''' - f'' - 6f' = 0$.

Solution Again we try letting $f(t) = e^{kt}$ and find that $f' = kf$, $f'' = k^2f$ and $f''' = k^3f$. The differential equation becomes $0 = f''' - f'' - 6f' = (k^3 - k^2 - 6k)f = 0$.

Since $f(t)$ is never zero, this means that $k^3 - k^2 - 6k = 0$, and we can factor: $0 = k(k^2 - k - 6) = k(k - 3)(k + 2)$, so $k = 0$, 3, or -2.

We find three solutions, $f_1(t) = e^{0(t)} = 1$, $f_2(t) = e^{3t}$, and $f_3(t) = e^{-2t}$. As above, *every* solution is of the form $c_1f_1 + c_2f_2 + c_3f_3$, for some constants c_1, c_2, and c_3, and any choice of these constants does give a solution.

A pattern is beginning to emerge from these examples. Consider the differential equation

$$a_nf^{(n)} + a_{n-1}f^{(n-1)} + \ldots + a_2f'' + a_1f' + a_0f = 0.$$

We try letting $f(t) = e^{kt}$, for some constant k, so $f' = kf$, $f'' = k^2f, \ldots, f^{(n)} = k^nf$, and the equation becomes

$$(a_nk^n + a_{n-1}k^{n-1} + \ldots + a_2k^2 + a_1k + a_0)f = 0.$$

Since f is never zero, we have to solve

$$a_n k^n + a_{n-1} k^{n-1} + \ldots + a_2 k^2 + a_1 k + a_0 = 0.$$

This is called the **auxiliary equation**, and its left side is called the **characteristic polynomial** of the original differential equation.

The number n, the order of the highest derivative that occurs in the equation, is called the **order** of the differential equation.

Find the roots $\{k_i\}$ of the characteristic polynomial. For each root k_i the function $f(t) = e^{k_i t}$ is a solution of the differential equation, and if there are n distinct roots, then every solution is of the form

$$c_1 e^{k_1 t} + c_2 e^{k_2 t} + \ldots + c_n e^{k_n t}.$$

This expression is the **general solution** of the differential equation. This approach is exactly what we used to solve the examples above.

There are two possible complications. One is that some roots of the characteristic polynomial may not be real numbers (e.g., the polynomial $k^2 + 1$ has no real roots). This we will not discuss now at all but leave it until Chapter 19.

A second difficulty is that some of the roots may be "repeated."

EXAMPLE 5   Solve $f'' + 4f' + 4f = 0$.

Solution   The characteristic polynomial is $k^2 + 4k + 4$, which we can factor: $k^2 + 4k + 4 = (k + 2)^2$. The only root is $k = -2$, but it is "repeated."

The function $f_1(t) = e^{-2t}$ is a solution of the differential equation, but from the earlier examples we expect to find a second solution. It is not immediately clear why it works, but it is easily checked that $f_2(t) = te^{-2t}$ is also a solution.

It turns out that the general solution is $c_1 f_1 + c_2 f_2 = c_1 e^{-2t} + c_2 te^{-2t}$. This means that every solution is of this form, for some constants c_1 and c_2. ◣

Summary 7.6.1

> The powers of t go up to one less than m, the number of times the root r is repeated.

For each root k_i of the characteristic polynomial we get solutions in this way. If a root r is repeated m times, that is, the characteristic polynomial is divisible by the factor $(k - r)^m$, then $f_1(t) = e^{rt}$, $f_2(t) = te^{rt}$, $f_3(t) = t^2 e^{rt}$, $\ldots$, $f_m(t) = t^{m-1} e^{rt}$ are all solutions of the differential equation. In this way we find solutions corresponding to each root of the characteristic polynomial. We multiply each of them by a constant and add them up to get a more complicated solution, including solutions corresponding to all the roots.

If the total number of roots, *including repetitions*, equals n, the order of the original equation, then this procedure will result in the general solution; that is, it will give all the solutions of the equation.

EXAMPLE 6   Find the general solution of the differential equation $f^{(6)} - f^{(5)} - 2f^{(4)} + 2f^{(3)} + f'' - f' = 0$.

Solution   The characteristic polynomial is $k^6 - k^5 - 2k^4 + 2k^3 + k^2 - k$. We can easily see that it is divisible by k, and it equals $k(k^5 - k^4 - 2k^3 + 2k^2 + k - 1)$. The next problem is to factor this fifth-degree polynomial.

To do this, we try substituting small integer values of k in hopes of finding a root, using the rational root theorem (Theorem 1.3.7) to help choose possible values of k. Our very first attempt, with $k = 1$, gives 0, meaning that the polynomial is divisible by $k - 1$. Dividing, we get $k^5 - k^4 - 2k^3 + 2k^2 + k - 1 = (k - 1)(k^4 - 2k^2 + 1)$.

We could try the same approach again, but if we are lucky, we might notice that the last factor is a square, the square of $(k^2 - 1) = (k - 1)(k + 1)$.

We are now able to factor the characteristic polynomial completely. We have that $k^6 - k^5 - 2k^4 + 2k^3 + k^2 - k = k(k^5 - k^4 - 2k^3 + 2k^2 + k - 1) = k(k - 1)(k^4 - 2k^2 + 1) = k(k - 1)(k^2 - 1)^2 = k(k - 1)(k - 1)^2(k + 1)^2 = k(k - 1)^3(k + 1)^2$.

From the above discussion we see that the root $k = 0$ occurs once; the corresponding solution is $e^{0(t)} = 1$. The root $k = 1$ occurs three times; the corresponding solutions are e^t, te^t, and $t^2 e^t$. The root $k = -1$ occurs twice; the corresponding solutions are e^{-t} and te^{-t}.

Multiplying each of these solutions by a constant and adding, we find the solution

$$c_1 + c_2 e^t + c_3 te^t + c_4 t^2 e^t + c_5 e^{-t} + c_6 te^{-t}.$$

The root $k = 0$ occurs once, the root $k = 1$ occurs three times, and the root $k = -1$ occurs twice. The total number of roots, *counting repetitions*, is 6. Since this equals the order of the original equation, the formula above is the general solution, that is, every solution must be of that form. ◢

Finally, there is the question of finding a particular solution for a differential equation.

For instance, consider the equation $f'' + 4f' + 4f = 0$, which we solved in Example 5. The solution contained two constants, c_1 and c_2. In order for us to be able to solve for them, we will need *two* pieces of information about the solution. For an equation of order n we expect the general solution will have n constants in it, and to determine them all, it will take n pieces of information.

So instead of an initial value, the value of the solution at some point, we need the values of the function at n points, or the values of $f, f', \ldots, f^{(n-1)}$ at some point, or some other combination of n pieces of information.

EXAMPLE 7   Find the solution of $f'' + 4f' + 4f = 0$ that satisfies $f(0) = 1$ and $f'(0) = 4$.

Solution   We know that the solution is of the form $f(t) = c_1 e^{-2t} + c_2 te^{-2t}$, for some constants c_1 and c_2. The value of this function at $t = 0$ is $f(0) = c_1$. The derivative is $f'(t) = -2c_1 e^{-2t} - 2c_2 te^{-2t} + c_2 e^{-2t}$, so $f'(0) = -2c_1 + c_2$.

The condition specified in the question means $c_1 = 1$ and $-2c_1 + c_2 = 4$. Substituting $c_1 = 1$ into the second equation gives $c_2 = 6$.

The required particular solution is $f(t) = e^{-2t} + 6te^{-2t}$. ◢

Exercises 7.6

Solve the following differential equations.

1. $\frac{dx}{dt} = xt^2$

2. $\frac{dx}{dt} = x^2 t$

3. $\frac{dx}{dt} = \frac{t}{x}$

4. $\frac{dx}{dt} = tx^{-2}$

5. $\frac{dx}{dt} = \frac{x}{t}$

6. $\frac{dx}{dt} = x \sin t$

7. $\frac{dx}{dt} = (x - 2)t^2$

8. $\frac{dx}{dt} = x^2 \sin t$

9. $\frac{dx}{dt} = e^{x+t}$

10. $\frac{dx}{dt} = t(x^2 + 4)$

Find the solution of each of the following equations with the given initial value.

11. $\frac{dx}{dt} = \frac{x}{2}$, $x(0) = 2$

12. $\frac{dx}{dt} = 3xt$, $x(0) = 4$

13. $\frac{dx}{dt} = xt^2$, $x(1) = 3$

14. $\frac{dx}{dt} = x^2 \cos t$, $x(\pi) = 1$

15. $\frac{dx}{dt} = tx \ln x$, $x(1) = 1$

16. $\frac{dx}{dt} = x \ln x$, $x(0) = \frac{1}{2}$

17. $\frac{dx}{dt} = \sec x$, $x(0) = \frac{\pi}{4}$

18. $\frac{dx}{dt} = (x^2 + 1)t$, $x(0) = 1$

19. $\frac{dx}{dt} = t\sqrt{x}$, $x(4) = 5$

20. $\frac{dx}{dt} = \cos^2 x \sin^2 t$, $x(0) = \frac{\pi}{4}$

Find the general solution of each of the following differential equations.

21. $f'' - f' - 12f = 0$

22. $f'' + 2f' - 8f = 0$

23. $f''' + 3f'' + 2f' = 0$

24. $f''' - f' = 0$

25. $f'' + 6f' + 9f = 0$

26. $f''' + 3f'' + 3f' + f = 0$

27. $f''' + 2f'' + f' = 0$

28. $f''' + f'' - f' - f = 0$

29. $f^{(4)} - 3f^{(3)} + 3f'' - f' = 0$

30. $f^{(6)} + 6f^{(5)} + 12f^{(4)} + 8f^{(3)} = 0$

Find a solution of each of the following differential equations that satisfies the given initial conditions.

31. $f'' - f = 0$, $f(0) = f'(0) = 1$

32. $f'' - 4f = 0$, $f(0) = f(1) = 4$

33. $f'' + 3f' - 4f = 0$, $f(0) = f'(0) = 0$

34. $f'' - 6f' + 9f = 0$, $f(0) = 0, f'(0) = 1$

35. $f'' = 0$, $f(1) = 0, f(3) = 2$

36. $f''' = 0$, $f(0) = 0 = f'(0), f''(0) = 2$

37. $f''' = 0$, $f(1) = f'(1) = f''(1) = 0$

38. $f''' = 0$, $f(1) = 0 = f'(1), f''(1) = 4$

39. $f''' - 2f'' - 15f' = 0$, $f(0) = f'(0) = 0, f''(0) = 1$

40. $f' = f$, $f(1) = 0 = f'(1), f'(0) = 0$

For each of the following equations, sketch several solutions, that is, find the general solution and then sketch the curve you get by picking several different values of the constant(s). Be careful to label the axes correctly.

41. $\frac{dx}{dt} = x$

42. $\frac{dx}{dt} = x^2$

43. $f'' - f' = 0$

44. $f'' = 0$

45. $f''' = 0$

46. $f'' = 2$

47. Suppose m is a positive integer and consider the differential equation $f^{(m)} = 0$.

   (i) What is its general solution?

   (ii) Find the particular solution that satisfies $f(0) = f'(0) = \ldots = f^{(m-1)}(0) = 0$.

48. Find the particular solution of $f^{(m)} = 0$ that satisfies $f(1) = f'(1) = \ldots = f^{(m-1)}(1) = 0$.

III

49. Find the general solution $f(t) = ae^{ct} + be^{dt}$ of $f'' + 2f' - 3f = 0$, with $c < d$. Use a graphics calculator or computer to sketch on one picture the solutions with $a = b = 1; a = b = 2; a = b = -1; a = 0, b = 1; a = 1, b = 0; a = -1, b = 1; a = 1, b = -1$.

50. The general solution of $f''' - 2f'' - 8f' = 0$ is $f(t) = a + be^{4t} + ce^{-2t}$. Sketch the graph of $2 + 5e^{-2t}$ (i.e., $a = 2, b = 0, c = 5$). Now keeping $a = 2$ and $c = 5$ fixed, sketch the graph of $2 + be^{4t} + 5e^{-2t}$ for several values of b. By trial and error, using the graphs, try to find the value of b for which $f(0) = f(1)$. Can you calculate the correct value of b?

51. Find and sketch the solution of $f''' - f'' - f' + f = 0$ satisfying $f(0) = f(1) = 1, f'(0) = -e$.

52. Find and sketch the solution of $f''' - f'' - f' + f = 0$ satisfying $f(0) = 0, f(1) = 1, f(-1) = 0$.

Find the general solution of each of the following differential equations.

53. $f''' - 4f' + f = 0$

54. $f^{(4)} - 5f^{(2)} + f^{(1)} + f = 0$

55. $f^{(5)} - 5f^{(3)} + 2f^{(2)} = 0$

56. $f^{(4)} - 5f^{(2)} + f^{(1)} = 0$

Chapter Summary

§7.1
natural logarithm
$e \approx 2.71828$

§7.2
exponential function
powers: x^a (Definition 7.2.3, p. 517)
base
logarithm to the base b
$\log_b(x)$
common logarithm

§7.3
exponential growth
growth constant
exponential decay
decay constant
initial value
half-life

§7.4
isotope

factor
compounding
nominal interest rate
effective annual interest rate
continuous compounding
present value
ambient temperature
Newton's Law of Cooling (Example 16, p. 540)

§7.5
logistic equation (Equation 7.5.1, p. 546)
logistic growth
carrying capacity

§7.6
separable equation
general solution
particular solution
auxiliary equation
order
characteristic polynomial

Review Exercises

I

Evaluate each of the following.

1. $\frac{d}{dx}\left(x^{2x}\right)$

2. $\frac{d}{dx}\left((2x)^x\right)$

3. $\frac{d}{dx}\left(x^{3x+5}\right)$

4. $\frac{d}{dx}\left(x^{x^2}\right)$

5. $\frac{d}{dx}\left(x^{\log_2(x)}\right)$

6. $\frac{d}{dx}\left((1+x^2)^{\sin x}\right)$

7. $\lim_{x\to 0+}\left(x^{-\ln x}\right)$

8. $\lim_{x\to\infty}\left(x^{1/\ln x}\right)$

9. $\lim_{x\to 1+}\left(\ln x\right)^{\ln x}$

10. $\lim_{x\to 0+}\left(-\ln x\right)^x$

Find each of the following.

11. The half-life of a substance that diminishes by 5% in 10 minutes.

12. How long it takes a substance to reduce by 20% if its half-life is 3 years.

13. The size after 10 years of an exponentially growing population that begins at 10,000 and grows by 80 in one month.

14. The initial size of an exponentially growing population that is 10,000 after 8 years and 11,000 after 11 years.

15. The temperature after 10 minutes of an object that cools from 80°C to 70°C in 1 min with an ambient temperature of 20°C.

16. The initial temperature of an object that is at 50°F after 3 min and 48°F after 4 min with an ambient temperature of 40°F.

17. The length of time for a bank account to double with 8% interest compounded continuously.

18. The value after 100 months of an initial deposit of $1000 with 6% annual interest compounded continuously.

19. The present value of a contract to make eight payments of $400, one today and one every three months after that, assuming 9.5% interest compounded continuously.

20. The present value of a contract that will pay Josie $200 each month for one year, starting two months from today, in return for her promise to pay back $3000 three years from today, assuming 12% interest, compounded monthly.

Find the general solution of each of the following differential equations.

21. $f'' + f' - f = 0$

22. $f'' - 8f' + 16f = 0$

23. $f''' + 3f'' = 0$

24. $f^{(4)} = 9f^{(2)}$

25. $\frac{dx}{dt} = x^4 \sec^2 t$

26. $\frac{dx}{dt} = -7x$

27. $t^4(t+1)^2 \frac{du}{dt} = t^4 + t^2 + 2t + 1$

28. $\frac{d^2 v}{dt^2} = \frac{dv}{dt} + 6v$

29. $\frac{dy}{dx} = -(x+1)y$

30. $\frac{dP}{dt} = P - P^2$

In each of the following situations, find a solution of the differential equation that satisfies the given initial conditions.

31. $f'' + f' - 6f = 0$, $f(0) = 1$, $f(1) = -1$

32. $2f'' + 5f' + 2f = 0$, $f(1) = f'(1) = 4$

33. $f''' - 16f' = 0$, $f(0) = f(1) = f(2) = 0$

34. $\frac{f}{f'} = 2$, $f(1) = 4$

35. $f'' = 25f$, $f(0) = 1$, $f'(0) = 5$

36. $f'' - f' - 2f = 0$, $f(0) = f'(0)$, $f(1) = 4$

37. $f^{(4)} = f^{(2)}$, $f(0) = 1$, $f'(0) = f''(0) = f'''(0) = 0$

38. $f''' - 2f'' + f' = 0$, $f(0) = f'(0) = f''(0) = 1$

39. $f'' = \pi^2 f$, $f(0) = 1$, $f'(0) = -1$

40. $f''' + 8f = 0$, $f(0) = 1$

II

41. If $f(x)$ is a nonzero differentiable function, verify the formula for $\frac{d}{dx}(f^n)$ using logarithmic differentiation.

42. If $f(x)$ is a differentiable function, verify the formula for $\frac{d}{dx}(e^{f(x)})$ using logarithmic differentiation.

*43. Which is bigger: $(10^{10})^{10}$ or $10^{(10^{10})}$?

*44. (i) If $a > 1$, which of $(a^a)^a$ and $a^{(a^a)}$ is bigger and why?

(ii) What happens if $0 < a < 1$?

*45. A population growing according to a logistic curve begins at exactly half the carrying capacity. One year later it has increased by 2%. How long will it take to reach 75% of the carrying capacity?

*46. A small warm object following Newton's law of cooling begins at $150°$F. One minute later it is at $125°$F, and one minute after that it is at $115°$F. What is the ambient temperature?

*47. (Spread of Rumors) Social scientists study how rumors spread (and similar analyses may apply to the spread of cultural changes and some epidemics).

Consider a population of P people, and let $f(t)$ be the number who have heard the rumor at time t. The more people who have heard the rumor, the more opportunities they will have to pass it on. However, these opportunities also depend on the number of people who have not heard it yet, which is $P - f(t)$. So it seems reasonable that the rate at which new people hear the rumor should be proportional to the product $f(t)(P - f(t))$, that is, $f'(t) = Kf(t)(P - f(t))$, for some constant K.

Consider a small town with population $P = 10,000$. One hundred people hear about a scandal at work. One day later 300 people know about it. How long does our theory predict it will take for half the town to know about it? How long for 90% of the town to know?

Techniques of Integration 8

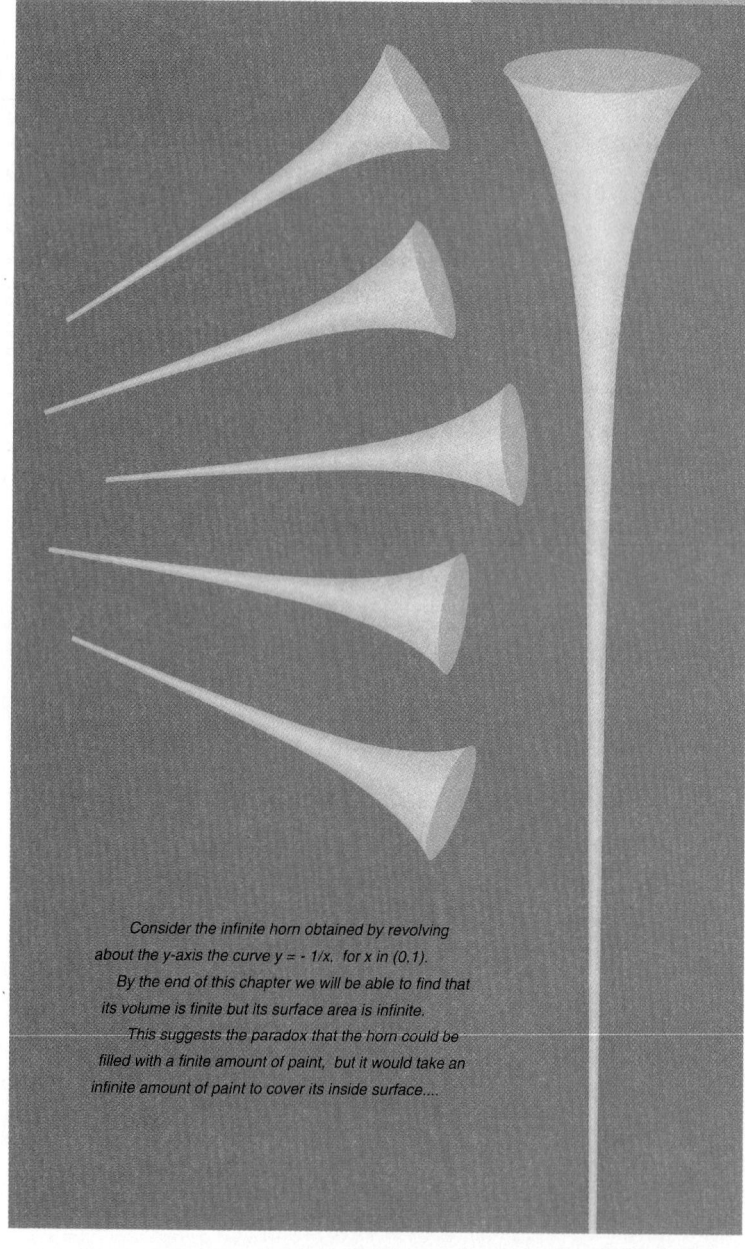

Consider the infinite horn obtained by revolving about the y-axis the curve y = - 1/x, for x in (0,1).

By the end of this chapter we will be able to find that its volume is finite but its surface area is infinite.

This suggests the paradox that the horn could be filled with a finite amount of paint, but it would take an infinite amount of paint to cover its inside surface....

INTRODUCTION

We have seen that the Fundamental Theorem of Calculus (Theorem 5.6.1) gives a good way to evaluate definite integrals:

$$\int_a^b f(x)\,dx = F(x)\Big|_a^b = F(b) - F(a),$$

where $F(x)$ is an antiderivative of the integrand $f(x)$, that is, $F'(x) = f(x)$. In Section 5.7 we discussed the technique of substitution, which allowed us to extend considerably the list of functions we could integrate.

This chapter is devoted to learning a number of techniques that will allow us to integrate many additional types of functions.

It is important to mention that tables of integrals are available. A short list appears inside the covers of this book, and more comprehensive ones can be found in your library. But no table can contain all the integrals you will ever encounter. The idea is to use the table in conjunction with the technique of substitution and the other techniques we shall learn; in this way you will greatly expand your capacity to find indefinite integrals.

The techniques to be learned in this chapter take time and practice to become familiar, and the only way to master them is to do a great many examples and exercises. Probably more so than in any other chapter of this book, it is absolutely *vital* that you work out a large number of problems for yourself. Ask for help, discuss them with your friends, but ultimately you will learn how to integrate by doing problems on your own.

Remember that when you look up an answer in the back of the book or compare with somebody else's work it is possible to have two apparently different answers that are really the same, or the same apart from the addition of a constant. So be careful not to discard your work immediately if it does not match exactly. The good news is that it is always possible to check an integral easily by differentiating.

The exercises in each section are largely ones that use the techniques learned in that section. However, each exercise set ends with a Drill, which is a list of integrals that may require any of the methods learned up to that point. The idea, of course, is that part of the problem of integrating is recognizing which technique to use, and this drill will help you to practice.

The various sections of this chapter are heavily interrelated. Section 8.1, "Integration by Parts," discusses the integral version of the product rule and its use for simplifying integrals. Section 8.2 discusses certain integrals of trigonometric functions and uses several results from Section 8.1. Section 8.3 describes a technique for converting other types of integrals into integrals of trigonometric functions, for which techniques from Section 8.2 are frequently required. The ideas developed in Section 8.4 convert another class of integrals into the kind that are discussed in Section 8.3. Section 8.5 describes how to integrate rational functions. This can always be done, assuming that we know how to factor the denominator of the integrand. In some cases it is necessary to make use of results from Sections 8.3 and 8.4. Section 8.6 describes a trick that converts integrals of trigonometric functions into integrals of rational functions, which can be evaluated as in Section 8.5. The trick is complicated, but it may be necessary in cases in which Section 8.2 does not apply. Section 8.7 discusses some applications, mostly ones we could have done in Chapter 6 except for being unable to evaluate the integrals, something we are now able to do. Finally, Section 8.8 discusses improper integrals, integrals in which the integrand or limits of integration are allowed to become infinite.

Before beginning, we list a number of integrals we already know. Most of them are obtained from familiar rules for differentiation. Some of the formulas contain a constant a; it is often easier to recognize the formula in the case with $a = 1$ (the general case then follows by a simple substitution).

Ignore the last four formulas if you have not studied hyperbolic functions.

$$\int x^r\, dx = \frac{x^{r+1}}{r+1} + c, \text{ if } r \neq -1 \qquad \int \frac{1}{x}\, dx = \ln|x| + c$$

$$\int a\, dx = ax + c \qquad \int e^{ax}\, dx = \frac{1}{a}e^{ax} + c$$

$$\int \sin ax\, dx = -\frac{1}{a}\cos ax + c \qquad \int \cos ax\, dx = \frac{1}{a}\sin ax + c$$

$$\int \sec^2 ax\, dx = \frac{1}{a}\tan ax + c \qquad \int \sec ax \tan ax\, dx = \frac{1}{a}\sec ax + c$$

$$\int \csc^2 ax\, dx = -\frac{1}{a}\cot ax + c \qquad \int \csc ax \cot ax\, dx = -\frac{1}{a}\csc ax + c$$

$$\int \tan ax\, dx = -\frac{1}{a}\ln|\cos ax| + c \qquad \int \frac{1}{a^2+x^2}\, dx = \frac{1}{a}\arctan\frac{x}{a} + c$$

$$= \frac{1}{a}\ln|\sec ax| + c \qquad \int \frac{1}{x\sqrt{x^2-a^2}}\, dx = \frac{1}{a}\operatorname{arcsec}\frac{x}{a} + c$$

$$\int \cot ax\, dx = \frac{1}{a}\ln|\sin ax| + c \qquad \int \frac{1}{\sqrt{a^2-x^2}}\, dx = \arcsin\frac{x}{a} + c$$

$$\int \cos^2 ax\, dx = \frac{x}{2} + \frac{1}{4a}\sin 2ax + c \qquad \int \sin^2 ax\, dx = \frac{x}{2} - \frac{1}{4a}\sin 2ax + c$$

$$\int \sinh ax\, dx = \frac{1}{a}\cosh ax + c \qquad \int \cosh ax\, dx = \frac{1}{a}\sinh ax + c$$

$$\int \operatorname{sech}^2 ax\, dx = \frac{1}{a}\tanh ax + c \qquad \int \operatorname{sech} ax \tanh ax\, dx = -\frac{1}{a}\operatorname{sech} ax + c$$

8.1 Integration by Parts

DEFINITE INTEGRALS BY PARTS
REDUCTION FORMULAS

The technique of substitution (Section 5.7) is in effect the integral version of the Chain Rule. We now consider the integral version of the Product Rule. The integral $\int f(x)g(x)\,dx$ of a product of two functions does not in general equal the product of the integrals of $f(x)$ and $g(x)$ (e.g., try it with $f(x) = x$ and $g(x) = x\ldots$).

The Product Rule (Theorem 3.2.5) says

$$(uv)' = u'v + uv'.$$

Rearranging, we have $uv' = (uv)' - vu'$, and if we integrate, we find

$$\int uv'\,dx = \int (uv)'\,dx - \int vu'\,dx.$$

In this formula we note that the second integral means an antiderivative of the derivative of uv, so we can just replace it with uv. The formula becomes

$$\int uv'\,dx = uv - \int vu'\,dx. \tag{8.1.1}$$

Strictly speaking, we ought to have added a constant c when we did that indefinite integration, but since the formula still contains two other indefinite integrals, we can just absorb the constant into them.

Formula 8.1.1 is the basis of the technique known as *integration by parts*.

EXAMPLE 1 Find $\int x \sin x\,dx$.

Solution We use Formula 8.1.1, letting $u = x$ and $v' = \sin x$. If $v' = \sin x$, then $v = \int \sin x\,dx = -\cos x + c$. Once again we can omit the constant c until the last step, so we take $v = -\cos x$. Then

$$\int x \sin x\,dx = \int uv'\,dx = uv - \int vu'\,dx$$

$$= x(-\cos x) - \int (-\cos x)(1)\,dx = -x\cos x + \int \cos x\,dx$$

$$= -x\cos x + \sin x + c.$$

It is easy to check by differentiating that this is indeed an antiderivative of $x \sin x$. Notice that we did add the constant c at the end of the calculation.

It is possible to write Formula 8.1.1 for integration by parts slightly differently using differentials. Noting that $u'(x)\,dx = du$ and $v'(x)\,dx = dv$, we can write the formula as

$$\int u\,dv = uv - \int v\,du.$$

EXAMPLE 2 Find $\int x^2 e^x\,dx$.

Solution Let $u = x^2$, $dv = e^x dx$. Then $du = 2x\,dx$, and $v = \int dv = \int e^x dx = e^x$

(again we omit constants until the end), and

$$\int x^2 e^x \, dx = \int u \, dv = uv - \int v \, du$$

$$= x^2 e^x - \int e^x (2x) \, dx = x^2 e^x - 2 \int x e^x \, dx.$$

We must apply integration by parts again to evaluate the last integral. Letting $U = x$, $dV = e^x dx$, we find $dU = dx$, $V = \int dV = \int e^x dx = e^x$, so

$$\int x e^x \, dx = UV - \int V \, dU = x e^x - \int e^x (1) \, dx = x e^x - e^x.$$

From this we find that the original integral equals

$$\int x^2 e^x \, dx = x^2 e^x - 2 \int x e^x \, dx = x^2 e^x - 2(x e^x - e^x) + c$$

$$= x^2 e^x - 2x e^x + 2 e^x + c.$$

Notice the constant c in the final answer. In this example it was necessary to use integration by parts twice to get the final answer.

THEOREM 8.1.2

Integration by Parts

$$\int u v' \, dx = uv - \int v u' \, dx$$

$$\int u(x) v'(x) \, dx = u(x) v(x) - \int v(x) u'(x) \, dx$$

$$\int u \, dv = uv - \int v \, du.$$

These formulas hold on any interval where u and v are differentiable with continuous derivatives.

To apply integration by parts to an indefinite integral $\int f(x) \, dx$, it is necessary to break $f(x)$ into two "parts," $f = uv'$. This has to be done in such a way that v' can be integrated, that is, it must be possible to find $v = \int v' dx$. Moreover, the integral $\int v u' \, dx = \int v \, du$ must be easier to integrate than $\int u v' \, dx = \int u \, dv$. As with substitutions, it takes practice and experience to be able to see a suitable u and v. We have also seen that it may be necessary to apply the process more than once.

EXAMPLE 3   (i) $\int \cos^2 x \, dx$.

Solution We have already seen (Example 5.7.5, p. 405) how to integrate using the half-angle formula, but we try integrating by parts, letting $u = \cos x$, $v' = \cos x$. So $v = \sin x$, and $\int \cos^2 dx = \int u \, dv = uv - \int v \, du = \cos x \sin x - \int \sin x (-\sin x) \, dx = \cos x \sin x + \int \sin^2 x \, dx$.

This does not seem to be much of an improvement, but we write $\sin^2 x = 1 - \cos^2 x$, so

$$\int \cos^2 x \, dx = \cos x \sin x + \int (1 - \cos^2 x) \, dx$$

$$= \cos x \sin x + \int 1 \, dx - \int \cos^2 x \, dx$$

$$= \cos x \sin x + x - \int \cos^2 x \, dx.$$

Moving the last integral from the right side to the left, we find

$$2 \int \cos^2 x \, dx = \cos x \sin x + x + c', \quad \text{or} \quad \int \cos^2 x \, dx = \frac{1}{2} \cos x \sin x + \frac{x}{2} + c.$$

(Notice how we added c' when we combined the two integrals and then let $c = \frac{1}{2} c'$.)

(ii) $\int e^x \cos x \, dx$.

Solution   Letting $u = e^x$, $v' = \cos x$, so $u' = e^x$, $v = \sin x$, we find

$$\int e^x \cos x \, dx = uv - \int vu' \, dx = e^x \sin x - \int e^x \sin x \, dx.$$

This does not seem too useful, but let us proceed and do it again, this time with $U = e^x$, $V' = \sin x$, $U' = e^x$, $V = -\cos x$. So

$$\int e^x \sin x \, dx = UV - \int VU' \, dx = -e^x \cos x + \int e^x \cos x \, dx.$$

So

$$\int e^x \cos x \, dx = e^x \sin x - \int e^x \sin x \, dx$$

$$= e^x \sin x - \left(-e^x \cos x + \int e^x \cos x \, dx \right)$$

$$= e^x \sin x + e^x \cos x - \int e^x \cos x \, dx.$$

Moving the last integral to the left side gives

$$2 \int e^x \cos x \, dx = e^x \sin x + e^x \cos x + c',$$

so

$$\int e^x \cos x \, dx = \frac{1}{2} e^x (\sin x + \cos x) + c,$$

being careful, as usual, to add the constant.

This example illustrates an idea that can often be used with integrals involving $\cos x$ or $\sin x$. Performing an integration by parts can change a $\cos x$ into a $\sin x$ or vice versa (perhaps with a minus sign). Doing it twice or applying a trigonometric identity may change it back again. But it often happens that the two integrals do not cancel. (In both parts of the above example they had opposite signs on opposite sides of the equation, so they added up to twice the original integral.)

EXAMPLE 4   Supposing n is a positive integer, find $\int \cos^n x \, dx$.

Solution   Write the integrand as $\cos^n x = (\cos^{n-1} x)(\cos x)$, and integrate by parts, with $u = \cos^{n-1} x$ and $dv = \cos x \, dx$; so $v = \sin x$. We find that

$$\int \cos^n x \, dx = \int \cos^{n-1} x \cos x \, dx = \int u \, dv = uv - \int v \, du$$

$$= \cos^{n-1} x \sin x - \int (\sin x)(n-1)(\cos^{n-2} x)(-\sin x) \, dx$$

$$= \cos^{n-1} x \sin x + (n-1) \int \cos^{n-2} x \sin^2 x \, dx$$

$$= \cos^{n-1} x \sin x + (n-1) \int \cos^{n-2} x (1 - \cos^2 x) \, dx$$

$$= \cos^{n-1} x \sin x + (n-1) \int \cos^{n-2} x \, dx - (n-1) \int \cos^n x \, dx.$$

Moving the last integral to the left side, we find

$$\int \cos^n x \, dx + (n-1) \int \cos^n x \, dx = \cos^{n-1} x \sin x + (n-1) \int \cos^{n-2} x \, dx,$$

or

$$n \int \cos^n x \, dx = \cos^{n-1} x \sin x + (n-1) \int \cos^{n-2} x \, dx,$$

so

$$\int \cos^n x \, dx = \frac{1}{n} \cos^{n-1} x \sin x + \frac{n-1}{n} \int \cos^{n-2} x \, dx.$$

This does not tell us how to integrate $\cos^n x$ because the formula still contains an unevaluated integral. However, that integral is $\int \cos^{n-2} x \, dx$, with the exponent of $\cos x$ reduced by 2. So for instance, if $n = 3$ or $n = 4$, the unevaluated integral would be the integral of $\cos x$ or $\cos^2 x$, respectively, either of which we know how to do.

Whatever n we start with, we can apply the formula to reduce the integral to something involving the integral of $\cos^{n-2} x$. If necessary, we can do it again and reduce to the integral of $\cos^{n-4} x$. Continuing as many times as needed, we eventually get to the integral of $\cos x$ or the integral of $\cos^2 x$, either of which we can do.

So this formula, which *reduces* the integral to a simpler one, provides a means for evaluating the integral of any power $\cos^n x$, even though it does not provide an actual formula for every n. A formula of this type is called a **reduction formula**. An exactly parallel computation establishes a similar formula for the integral of $\sin^n x$.

THEOREM 8.1.3

> **Reduction Formulas**
>
> Suppose n is a positive integer. Then
>
> $$\int \cos^n x \, dx = \frac{1}{n} \cos^{n-1} x \sin x + \frac{n-1}{n} \int \cos^{n-2} x \, dx,$$
>
> $$\int \sin^n x \, dx = -\frac{1}{n} \sin^{n-1} x \cos x + \frac{n-1}{n} \int \sin^{n-2} x \, dx.$$

EXAMPLE 5   (i) Evaluate $\int \cos^5 x \, dx$, (ii) $\int \sin^4 x \, dx$.

Solution

(i) The reduction formula for $\cos^n x$ shows that $\int \cos^5 x \, dx = \frac{1}{5} \cos^4 x \sin x + \frac{4}{5} \int \cos^3 x \, dx$. Now it is necessary to perform another reduction for the integral of $\cos^3 x$.

   We find $\int \cos^3 x \, dx = \frac{1}{3} \cos^2 x \sin x + \frac{2}{3} \int \cos x \, dx = \frac{1}{3} \cos^2 x \sin x + \frac{2}{3} \sin x$. Substituting this into the previous expression, we find that

$$\int \cos^5 x \, dx = \frac{1}{5} \cos^4 x \sin x + \frac{4}{5} \int \cos^3 x \, dx$$

$$= \frac{1}{5} \cos^4 x \sin x + \frac{4}{5} \left(\frac{1}{3} \cos^2 x \sin x + \frac{2}{3} \sin x \right) + c$$

$$= \frac{1}{5} \cos^4 x \sin x + \frac{4}{15} \cos^2 x \sin x + \frac{8}{15} \sin x + c.$$

(ii) Similarly, $\int \sin^4 x \, dx = -\frac{1}{4} \sin^3 x \cos x + \frac{3}{4} \int \sin^2 x \, dx$. We could evaluate this last integral using Example 5.7.5, but we can also just apply the reduction formula again: $\int \sin^2 x \, dx = -\frac{1}{2} \sin x \cos x + \frac{1}{2} \int \sin^0 x \, dx = -\frac{1}{2} \sin x \cos x + \frac{1}{2} \int 1 \, dx = -\frac{1}{2} \sin x \cos x + \frac{x}{2}$.

   Combining, we find

$$\int \sin^4 x \, dx = -\frac{1}{4} \sin^3 x \cos x + \frac{3}{4} \int \sin^2 x \, dx$$

$$= -\frac{1}{4} \sin^3 x \cos x - \frac{3}{8} \sin x \cos x + \frac{3x}{8} + c.$$

It is also possible to use integration by parts for definite integrals. After all, to evaluate $\int_a^b uv' \, dx$, we need an antiderivative of uv', and the formula tells us that we can use $F(x) = uv - \int vu' \, dx$.

THEOREM 8.1.4

Integration by Parts for Definite Integrals

$$\int_a^b u(x) v'(x) \, dx = u(x) v(x) \Big|_a^b - \int_a^b v(x) u'(x) \, dx.$$

This formula holds whenever u and v are differentiable on $[a, b]$ and u' and v' are continuous on $[a, b]$.

EXAMPLE 6   Evaluate $\int_1^4 x \ln x \, dx$.

Solution   A first guess might be to let $u = x$, since du would then be easy to find, but in that case we would have $v' = \ln x$, which we do not know how to integrate to find v. A second guess, which seems a little peculiar, is to let $u = \ln x$ and $v' = x$.

We summarize this information in a box beside the calculation:

$$\int_1^4 x \ln x \, dx = \frac{1}{2} x^2 \ln x \Big|_1^4 - \int_1^4 \frac{1}{2} x^2 \left(\frac{1}{x} \right) dx$$

$$\boxed{\begin{array}{ll} u = \ln x, & dv = x \, dx \\ du = \dfrac{1}{x} dx, & v = \dfrac{1}{2} x^2 \end{array}} \quad = \frac{1}{2}(4^2)\ln(4) - \frac{1}{2}(1^2)\ln(1) - \frac{1}{2}\int_1^4 x \, dx$$

$$= 8\ln(4) - \frac{1}{2}(0) - \frac{1}{4} x^2 \Big|_1^4$$

$$= 8\ln(4) - \left(4 - \frac{1}{4} \right)$$

$$\approx 7.340.$$

In Example 6 we found that it is sometimes necessary to make unexpected choices for u and v' in order to integrate by parts.

EXAMPLE 7 $\int \ln x \, dx$.

Solution   We do not know how to integrate $\ln x$, so we certainly cannot use it for v'. We might try letting it be u, but then what is dv? The answer is almost too easy: Let $dv = dx$. Strangely enough, it works:

$$\int \ln x \, dx = x \ln x - \int x \left(\frac{1}{x} \right) dx$$

$$\boxed{\begin{array}{ll} u = \ln x, & dv = dx \\ du = \dfrac{1}{x} dx, & v = x \end{array}} \quad \begin{array}{l} = x \ln x - \int 1 \, dx \\ = x \ln x - x + c. \end{array}$$

This trick will not always work, but if we are faced with an unknown integrand, we can try letting it be u and letting $dv = dx$. The result will be a formula involving the integral of xu', which may not be an improvement, but there are occasions when it is. It is particularly useful for the integrals of inverse trigonometric functions.

EXAMPLE 8 $\int_0^1 \arctan x \, dx$.

Solution   Following the suggestion made above,

$$\int_0^1 \arctan x \, dx = uv \Big|_0^1 - \int_0^1 vu' \, dx$$

$$\boxed{\begin{array}{ll} u = \arctan x, & v' = 1 \\ u' = \dfrac{1}{1+x^2}, & v = x \end{array}} \quad \begin{array}{l} = x \arctan x \Big|_0^1 - \int_0^1 x \dfrac{1}{1+x^2} \, dx \\[2mm] = 1 \left(\dfrac{\pi}{4} \right) - 0(0) - \int_0^1 \dfrac{x}{1+x^2} \, dx. \end{array}$$

For this last integral we make the substitution $U = 1 + x^2$.

$$\int_0^1 \arctan x \, dx = \frac{\pi}{4} - \int_0^1 \frac{x}{1+x^2} \, dx$$

$$\boxed{\begin{array}{c} U = 1 + x^2 \\ dU = 2x \, dx \end{array}} = \frac{\pi}{4} - \int_1^2 \frac{1}{U} \frac{1}{2} \, dU$$

$$= \frac{\pi}{4} - \frac{1}{2} \ln U \Big|_1^2 = \frac{\pi}{4} - \frac{1}{2} \ln(2)$$

$$\approx 0.439.$$

> Integration by parts can often be used for an integral of the form $\int x f(x) \, dx$, letting $u = x$, $v' = f(x)$. Similarly, $\int x^2 f(x) \, dx$ suggests letting $u = x^2$, $v' = f(x)$, and for $\int x^k f(x) \, dx$ we can try $u = x^k$, $v' = f(x)$. These approaches reduce the power of x but depend on being able to integrate $f(x)$. It is important not to forget that integrals like these sometimes require integration by substitution. We integrate $\int x \cos(x^2) \, dx$ by letting $u = x^2$, $du = 2x \, dx$, so $\int x \cos(x^2) \, dx = \frac{1}{2} \int \cos u \, du = \frac{1}{2} \sin u + c = \frac{1}{2} \sin(x^2) + c$; integration by parts will not work.

Exercises 8.1

I

Evaluate each of the following indefinite integrals.

1. $\int x e^{-x} \, dx$

2. $\int x \sin(2x) \, dx$

3. $\int x^2 e^{3x} \, dx$

4. $\int \cos^4 t \, dt$

5. $\int \sin^5(2u) \, du$

6. $\int x^2 \cos(3x) \, dx$

7. $\int x \sec^2 x \, dx$

8. $\int e^x \sin x \, dx$

9. $\int (x^2 + 3x - 2) e^x \, dx$

10. $\int \ln(1 + 4x) \, dx$

11. $\int \arcsin x \, dx$

12. $\int x \ln x \, dx$

13. $\int 4x/(1 + x^2) \, dx$

14. $\int x \ln(x^4) \, dx$

15. $\int x^2 \ln(x^3) \, dx$

16. $\int e^{ax} \cos(bx) \, dx$

17. $\int x^2 \ln x \, dx$

18. $\int (\ln x)^2 \, dx$

19. $\int e^{2x} \left(3 - \sin(4x)\right) \, dx$

20. $\int 2x \arctan x \, dx$

Evaluate each of the following definite integrals.

21. $\int_{-1}^1 x^2 e^{-2x} \, dx$

22. $\int_1^3 x^3 \ln x \, dx$

23. $\int_0^1 e^{2x} \cos \pi x \, dx$

24. $\int_0^{2\pi} \sin x \cos(2x) \, dx$

25. $\int_{-\pi}^{\pi} \sin(2x) \sin(3x) \, dx$

26. $\int_1^4 \ln x \, dx$

27. $\int_0^2 t^2 \sin\left(\pi(1 - 2t)\right) \, dt$

28. $\int_1^4 (\ln u)^3 \, du$

29. $\int_1^2 \sqrt{x} \arctan(\sqrt{x}) \, dx$

30. $\int_0^2 x^2 \cos^4(\pi x^3) \, dx$

II

31. Establish the reduction formula for $\int \sin^n x \, dx$ given in Theorem 8.1.3 (p. 565).

*32. (i) Let n be any integer except -1. Find $\int x^n \ln x \, dx$.
(ii) What happens if $n = -1$? Evaluate $\int \frac{1}{x} \ln x \, dx$.

*33. Let n be any positive integer. Find a reduction formula for $\int (\ln x)^n \, dx$. Can you use it to get a final formula for $\int (\ln x)^n \, dx$?

*34. Suppose n is an integer with $n \geq 3$. Find a reduction formula for $\int \sec^n x \, dx$. You may wish to use the identity $\tan^2 x = \sec^2 x - 1$ (cf. Equation 1.7.3).

*35. Establish the following reduction formula:

$$\int x^n \cos x \, dx = x^n \sin x - n \int x^{n-1} \sin x \, dx.$$

*36. Integrate $\int x^4 \cos x \, dx$.

*37. Establish a reduction formula for $\int x^n \sin x \, dx$.

*38. Integrate $\int t^5 \sin t \, dt$.

*39. Establish the following reduction formula:

$$\int x^n e^{-x^2} \, dx = -\frac{1}{2} x^{n-1} e^{-x^2} + \frac{n-1}{2} \int x^{n-2} e^{-x^2} \, dx.$$

*40. Establish a reduction formula for $\int x^n e^x \, dx$.

**41. (i) Suppose n and m are positive integers with $m \neq n$. Evaluate
$$\int_0^{2\pi} \cos(m\theta)\cos(n\theta)\,d\theta.$$

  (ii) What happens if we evaluate the integral with $m = -n$?

**42. Suppose n and m are integers. Evaluate
$\int_0^{2\pi} \cos(m\theta)\sin(n\theta)\,d\theta.$

Drill

43. $\int \sin^2 x \cos^2 x\,dx$

44. $\int \sin x \cos x\,dx$

45. $\int x \sin \frac{x}{2}\,dx$

46. $\int (x+3)\,e^{2x}\,dx$

47. $\int x^2 \sin(x^3 + 4)\,dx$

48. $\int xe^{x^2}\,dx$

49. $\int \frac{x}{e^{5x}}\,dx$

50. $\int \frac{x}{1+x}\,dx$

51. $\int \cos x \sin(2x)\,dx$

52. $\int (\cos^2 x - \sin^2 x)\,dx$

8.2 Trigonometric Integrals

INTEGRATING $\cos^m x \sin^n x$
INTEGRATING $\tan^m x \sec^n x$

We have already learned how to evaluate many integrals of trigonometric functions, including powers of $\sin x$ and $\cos x$, certain products of powers of $\sin x$ and $\cos x$, and various other kinds. The first technique we used was substitution, which covers many cases, especially if it is supplemented with clever use of trigonometric identities (cf. Examples 5.7.3(i), (vi) (p. 401); 5.7.4(ii); 5.7.5, and 5.7.6).

In Section 8.1 we saw how integration by parts can be used to establish reduction formulas that enable us to do many more integrals.

In this section we try to organize these results and learn a few additional tricks. Apart from their intrinsic interest, these methods are important to us because in the next section we will learn techniques for converting many other types of integrals into integrals of trigonometric functions. This will be helpful only to the extent that we are then able to evaluate the resulting trigonometric integrals.

Before beginning, we recall the fundamental trigonometric identity $\cos^2 x + \sin^2 x = 1$ (Theorem 1.7.2). Dividing it by $\cos^2 x$, we find $1 + \tan^2 x = \sec^2 x$ (Equation 1.7.3). We record these for convenience:

$$\cos^2 x + \sin^2 x = 1, \tag{8.2.1}$$
$$1 + \tan^2 x = \sec^2 x. \tag{8.2.2}$$

Now suppose m and n are nonnegative integers and consider the integral

$$\int \cos^m x \sin^n x\,dx.$$

If either m or n is odd, it is possible to evaluate using substitution.

EXAMPLE 1 $\int \cos^4 x \sin^3 x\,dx.$

Solution Write the integrand as $\cos^4 x \sin^2 x \sin x$. The idea is to use that last $\sin x$ as the derivative (apart from a sign) of $\cos x$ and perform a substitution. To make this work, it will be necessary to express the $\sin^2 x$ in terms of $\cos x$, but nothing could be easier. We simply apply Equation 8.2.1:

$$\int \cos^4 x \sin^3 x\,dx = \int \cos^4 x (1 - \cos^2 x) \sin x\,dx = \int u^4 (1 - u^2)(-1)\,du$$

$$\boxed{\begin{aligned} u &= \cos x \\ du &= -\sin x\, dx \end{aligned}} \quad = \int (u^6 - u^4)\, du = \frac{1}{7} u^7 - \frac{1}{5} u^5 + c$$

$$= \frac{1}{7} \cos^7 x - \frac{1}{5} \cos^5 x + c.$$

Of course, if m is odd, similar techniques will work, using the substitution $u = \sin x$.

This leaves the more difficult situation in which m and n are both even numbers. If either m or n is zero, then we are integrating a power of $\sin x$ or $\cos x$, and the reduction formulas of the previous section will apply.

If m and n are both even and positive, then two different approaches are possible. One is to use the half-angle identities $\cos^2 x = \frac{1}{2}\big(1 + \cos(2x)\big)$ and $\sin^2 x = \frac{1}{2}\big(1 - \cos(2x)\big)$ (1.7.13). Using them it is possible to express $\int \cos^m x \sin^n x\, dx$ as a combination of integrals involving lower powers of $\cos(2x)$. These can all be done by using the reduction formulas (Theorem 8.1.3, p. 565).

Another possibility is to develop a reduction formula especially designed for this situation.

EXAMPLE 2   We integrate by parts. Making a substitution

$$\boxed{\begin{aligned} u &= \cos^{m-1} x \sin^n x, & dv &= \cos x\, dx, \\ du &= \big(-(m-1)\cos^{m-2} x \sin^{n+1} x + n \cos^m x \sin^{n-1} x\big)\, dx, & v &= \sin x, \end{aligned}}$$

we find

$$\int \cos^m x \sin^n x\, dx = \cos^{m-1} x \sin^{n+1} x + (m-1) \int \cos^{m-2} x \sin^{n+2} x\, dx$$

$$- n \int \cos^m x \sin^n x\, dx$$

$$= \cos^{m-1} x \sin^{n+1} x + (m-1) \int \cos^{m-2} x \sin^n x (1 - \cos^2 x)\, dx$$

$$- n \int \cos^m x \sin^n x\, dx$$

$$= \cos^{m-1} x \sin^{n+1} x + (m-1) \int \cos^{m-2} x \sin^n x\, dx$$

$$- (m-1) \int \cos^m x \sin^n x\, dx - n \int \cos^m x \sin^n x\, dx.$$

> The last two terms are constants times the original integral.

Moving the integrals of $\cos^m x \sin^n x$ to the left side, we find that

$$(m+n) \int \cos^m x \sin^n x\, dx = \cos^{m-1} x \sin^{n+1} x + (m-1) \int \cos^{m-2} x \sin^n x\, dx,$$

so

$$\int \cos^m x \sin^n x\, dx = \frac{1}{m+n} \cos^{m-1} x \sin^{n+1} x + \frac{m-1}{m+n} \int \cos^{m-2} x \sin^n x\, dx.$$

This is the desired reduction formula. A similar calculation finds a formula that reduces the power of $\sin x$ instead of the power of $\cos x$.

THEOREM 8.2.3

> **Reduction Formulas**
>
> (i) $\int \cos^m x \sin^n x\, dx = \frac{1}{m+n} \cos^{m-1} x \sin^{n+1} x + \frac{m-1}{m+n} \int \cos^{m-2} x \sin^n x\, dx,$
>
> (ii) $\int \cos^m x \sin^n x\, dx = -\frac{1}{m+n} \cos^{m+1} x \sin^{n-1} x + \frac{n-1}{m+n} \int \cos^m x \sin^{n-2} x\, dx.$

EXAMPLE 3 $\int \cos^2 x \sin^2 x\, dx.$

Solution   Using Theorem 8.2.3(i), we have

$$\int \cos^2 x \sin^2 x\, dx = \frac{1}{4} \cos x \sin^3 x + \frac{1}{4} \int \sin^2 x\, dx$$

$$= \frac{1}{4} \cos x \sin^3 x + \frac{1}{4} \int \frac{1}{2} \left(1 - \cos(2x)\right) dx$$

$$= \frac{1}{4} \cos x \sin^3 x + \frac{1}{8} \left(x - \frac{1}{2} \sin(2x)\right) + c$$

$$= \frac{1}{4} \cos x \sin^3 x + \frac{x}{8} - \frac{1}{8} \cos x \sin x + c.$$

An alternative approach to this question is to use the identity $\sin 2x = 2 \cos x \sin x$, so $\cos^2 x \sin^2 x = \frac{1}{4} \sin^2 2x$.

We summarize the approach to integrals of products of powers of $\cos x$ and $\sin x$.

Summary 8.2.4

$$\int \cos^m x \sin^n x\, dx$$

Suppose m and n are nonnegative integers, and consider $\int \cos^m x \sin^n x\, dx$.

1. If m or n equals zero, then the integrand is a power of $\sin x$ or $\cos x$, and the reduction formulas (Theorem 8.1.3) from the previous section can be used:

$$\int \cos^m x\, dx = \frac{1}{m} \cos^{m-1} x \sin x + \frac{m-1}{m} \int \cos^{m-2} x\, dx, \qquad (8.1.3(\text{i}))$$

$$\int \sin^n x\, dx = -\frac{1}{n} \sin^{n-1} x \cos x + \frac{n-1}{n} \int \sin^{n-2} x\, dx. \qquad (8.1.3(\text{ii}))$$

2. If m is odd, it is possible to use the substitution $u = \sin x$, $du = \cos x\, dx$. It is necessary to convert the remaining factor of $\cos^{m-1} x$ into an expression involving $\sin x$, using the identity $\cos^2 x = 1 - \sin^2 x$. (Note that $m - 1$ is even.)

   Similarly, if n is odd, it is possible to use the substitution $u = \cos x$, $du = -\sin x\, dx$. It is necessary to convert the remaining factor of $\sin^{n-1} x$ into an expression involving $\cos x$, using the identity $\sin^2 x = 1 - \cos^2 x$. (Note that $n - 1$ is even.)

3. If both m and n are even, several approaches are possible. Since $\cos^m x = (\cos^2 x)^{m/2} = (1 - \sin^2 x)^{m/2}$, the integral can be written as a combination of integrals of powers of $\sin x$. It could also be converted into powers of $\cos x$; in either case it can be integrated as in part 1 above. It is also possible to use the half-angle formulas to transform the integrand into an expression involving

powers of $\cos 2x$, which can then be integrated as in part 1. Finally, if you can remember them, it is possible to use the reduction formulas 8.2.3:

(i) $\displaystyle\int \cos^m x \sin^n x \, dx = \frac{1}{m+n} \cos^{m-1} x \sin^{n+1} x + \frac{m-1}{m+n} \int \cos^{m-2} x \sin^n x \, dx,$

(ii) $\displaystyle\int \cos^m x \sin^n x \, dx = -\frac{1}{m+n} \cos^{m+1} x \sin^{n-1} x + \frac{n-1}{m+n} \int \cos^m x \sin^{n-2} x \, dx.$

$$(8.2.3)$$

At least one of these methods will work for any integral of the form $\int \cos^m x \sin^n x \, dx$, with m and n nonnegative integers. On the other hand, there are instances in which the same integral can be done in different ways (e.g., if m and n are both odd, then it is possible to substitute $u = \sin x$ or $u = \cos x$). In a situation like this it is not unusual to get results that appear to be different, even though they are in fact the same *up to the addition of a constant.*

EXAMPLE 4 $\int \cos^3 x \sin^3 x \, dx.$

Solution   One way, using $u = \sin x$, $du = \cos x \, dx$, we find

$$\int \cos^3 x \sin^3 x \, dx = \int \cos^2 x \sin^3 x \cos x \, dx = \int (1 - \sin^2 x) \sin^3 x \cos x \, dx$$

$$= \int (1 - u^2) u^3 \, du = \frac{1}{4} u^4 - \frac{1}{6} u^6 + c$$

$$= \frac{1}{4} \sin^4 x - \frac{1}{6} \sin^6 x + c.$$

On the other hand, a similar calculation with $u = \cos x$, $du = -\sin x \, dx$ gives us

$$\int \cos^3 x \sin^3 x \, dx = \int \cos^3 x \sin^2 x \sin x \, dx = \int \cos^3 x (1 - \cos^2 x) \sin x \, dx$$

$$= -\int u^3 (1 - u^2) \, du = -\frac{1}{4} u^4 + \frac{1}{6} u^6 + c'$$

$$= -\frac{1}{4} \cos^4 x + \frac{1}{6} \cos^6 x + c'.$$

Notice that we used a different constant c' for this formula.

These two answers do not appear to be the same. It is possible to show that the difference of the two formulas is a constant. In other words, if the constants c and c' are chosen in a suitable way, the formulas are in fact the same.

Indeed, it will suffice to consider the derivative of the difference $\frac{1}{6}(\cos^6 x + \sin^6 x) - \frac{1}{4}(\cos^4 x + \sin^4 x) + c' - c$. The derivative equals

$$\frac{1}{6}\left(6\cos^5 x(-\sin x) + 6\sin^5 x \cos x\right) - \frac{1}{4}\left(4\cos^3 x(-\sin x) + 4\sin^3 x \cos x\right)$$

$$= \cos x \sin x(\sin^4 x - \cos^4 x + \cos^2 x - \sin^2 x)$$

$$= \cos x \sin x\left((\sin^2 x + \cos^2 x)(\sin^2 x - \cos^2 x) + \cos^2 x - \sin^2 x\right)$$

Knowing that the difference is a constant, we can evaluate at any convenient point to find out what it is. Using $x = 0$, we find that the difference is $\frac{1}{6} - \frac{1}{4} = -\frac{1}{12}$.

$$= \cos x \sin x (\sin^2 x - \cos^2 x)(\sin^2 x + \cos^2 x - 1)$$
$$= 0.$$

The derivative is zero, so the two answers differ by a constant. ◢

Integrals Involving Powers of tan x and sec x

Next we consider integrals of the form $\int \tan^m x \sec^n x \, dx$, where m and n are nonnegative integers.

The following example provides an important and useful result. It uses a very clever trick, which is difficult to explain other than by saying that it just happens to work.

EXAMPLE 5 $\int \sec x \, dx.$

Solution   The trick is to multiply the integrand by

$$\frac{\sec x + \tan x}{\sec x + \tan x}.$$

So

$$\int \sec x \, dx = \int \sec x \frac{\sec x + \tan x}{\sec x + \tan x} \, dx = \int \frac{\sec^2 x + \sec x \tan x}{\sec x + \tan x} \, dx.$$

The numerator of this integrand is the derivative of the denominator. So letting $u = \sec x + \tan x$, we find

$$\int \sec x \, dx = \int \frac{\sec^2 x + \sec x \tan x}{\sec x + \tan x} \, dx = \int \frac{du}{u} = \ln |u| + c$$
$$= \ln |\sec x + \tan x| + c.$$ ◢

We already know how to integrate $\tan x$; in Example 5.7.3(vi) we used the substitution $u = \cos x$, $du = -\sin x \, dx$ to find

$$\int \tan x \, dx = \int \frac{\sin x}{\cos x} \, dx = \int \frac{1}{u}(-1) \, du = -\ln |u| + c = \ln |\sec x| + c.$$

These two results can be combined with reduction formulas to integrate $\int \tan^m x \, dx$ or $\int \sec^n x \, dx$. For the mth power of the tangent, with $m \geq 2$, we write

$$\int \tan^m x \, dx = \int \tan^{m-2} x \tan^2 x \, dx = \int \tan^{m-2} x (\sec^2 x - 1) \, dx$$
$$= \int \tan^{m-2} x \sec^2 x \, dx - \int \tan^{m-2} x \, dx.$$

The first of these two integrals is easily evaluated using the substitution $u = \tan x$, $du = \sec^2 x$, and we find $\int \tan^{m-2} x \sec^2 x \, dx = \int u^{m-2} \, du = \frac{1}{m-1} u^{m-1} + c = \frac{1}{m-1} \tan^{m-1} x + c$. Absorbing the constant, we have $\int \tan^m x \, dx = \frac{1}{m-1} \tan^{m-1} x - \int \tan^{m-2} x \, dx$.

For powers of $\sec x$ there is a reduction formula, which can be proved by integration by parts, much as Theorem 8.1.3 was done. We record these two reduction formulas below.

THEOREM 8.2.5

Reduction Formulas

Suppose m is an integer with $m \geq 2$ and n is an integer with $n \geq 2$. Then

(i) $\int \tan^m x \, dx = \frac{1}{m-1} \tan^{m-1} x - \int \tan^{m-2} x \, dx,$

(ii) $\int \sec^n x \, dx = \frac{1}{n-1} \sec^{n-2} x \tan x + \frac{n-2}{n-1} \int \sec^{n-2} x \, dx.$

EXAMPLE 6 (i) $\int \tan^3 x \, dx$, (ii) $\int \sec^4 x \, dx$.

Solution

(i) By Theorem 8.2.5(i), with $m = 3$, we find

$$\int \tan^3 x \, dx = \frac{1}{2} \tan^2 x - \int \tan x \, dx = \frac{1}{2} \tan^2 x - \ln |\sec x| + c.$$

(ii) By Theorem 8.2.5(ii), with $n = 4$, we find

$$\int \sec^4 x \, dx = \frac{1}{3} \sec^2 x \tan x + \frac{2}{3} \int \sec^2 x \, dx$$

$$= \frac{1}{3} \sec^2 x \tan x + \frac{2}{3} \tan x + c.$$

By applying the reduction formulas, repeatedly if necessary, it is possible to reduce the integral of $\tan^m x$ to an integral of $\tan x$, as above, or an integral of $\tan^0 x = 1$. Similarly, the integral of any positive power $\sec^n x$ can be reduced to $\int \sec^2 x \, dx = \tan x + c$, as above, or $\int \sec x \, dx = \ln |\sec x + \tan x| + c$. In each case it is straightforward to evaluate the integral. Of course, if m or n is large, the reduction process will become quite complicated.

Finally, we turn to $\int \tan^m x \sec^n x \, dx$. If m or n is zero, the remarks above tell us what to do. Otherwise, we should look for a substitution. If n is *even*, we can write $\sec^n x = \sec^{n-2} x \sec^2 x$ and use the factor $\sec^2 x \, dx$ as du for the substitution $u = \tan x$. To make this work, it is necessary to convert the remaining $\sec^{n-2} x$ into an expression involving powers of $\tan x$, using $\sec^2 x = 1 + \tan^2 x$ (Equation 8.2.2). (Notice that $n - 2$ is even.)

If m is *odd*, we write $\tan^m x \sec^n x = \tan^{m-1} x \sec^{n-1} x \tan x \sec x$. Since $m - 1$ is even, we can use Equation 8.2.2 to convert $\tan^{m-1} x$ into an expression in powers of $\sec x$ and then evaluate the integral using the substitution $u = \sec x$, $du = \tan x \sec x \, dx$.

These two approaches will deal with every case except one in which m is even and n is odd. In this case, use Equation 8.2.2 to convert $\tan^m x$ into a sum of powers of $\sec x$, so all the resulting integrands will be powers of $\sec x$, which we know how to do.

EXAMPLE 7 (i) $\int \tan^3 x \sec^4 x \, dx$, (ii) $\int \tan^3 x \sec^3 x \, dx$, (iii) $\int \tan^2 x \sec^3 x \, dx$.

Solution

(i) One way to do this is the $u = \tan x$ substitution. In this case, $du = \sec^2 x$, and the remaining factor of $\sec^2 x$ in the integrand must be converted to an expression in tan x, using Equation 8.2.2:

$$\int \tan^3 x \sec^4 x \, dx = \int \tan^3 x \sec^2 x \sec^2 x \, dx$$

$$= \int \tan^3 x (1 + \tan^2 x) \sec^2 x \, dx$$

$$= \int u^3 (1 + u^2) \, du = \frac{1}{4} u^4 + \frac{1}{6} u^6 + c$$

$$= \frac{1}{4} \tan^4 x + \frac{1}{6} \tan^6 x + c.$$

(ii) $\int \tan^3 x \sec^3 x \, dx$. Here we can try the substitution $u = \sec x$, $du = \sec x \tan x$. This time it will be necessary to convert the remaining factor $\tan^2 x$ into an expression involving sec x:

$$\int \tan^3 x \sec^3 x \, dx = \int \tan^2 x \sec^2 x \sec x \tan x \, dx$$

$$= \int (\sec^2 x - 1) \sec^2 x \sec x \tan x \, dx$$

$$= \int (u^2 - 1) u^2 \, du = \frac{1}{5} u^5 - \frac{1}{3} u^3 + c$$

$$= \frac{1}{5} \sec^5 x - \frac{1}{3} \sec^3 x + c.$$

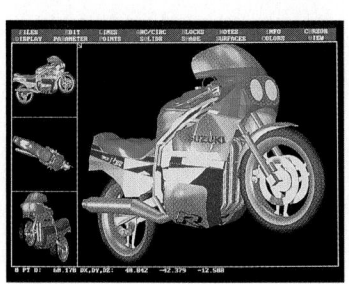

DesignCAD 3D software was used to produce this image of a motorcycle. Computer graphics help designers visualize new products quickly and accurately, and are especially useful for trying out modifications. (Suzuki is a registered trademark of Suzuki Motor Corporation, Japan.)

(iii) $\int \tan^2 x \sec^3 x \, dx$. Here we must change everything into sec x:

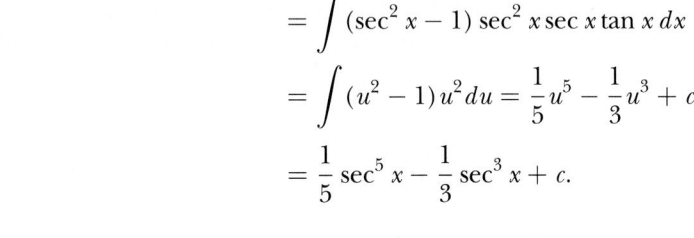

$$\int \tan^2 x \sec^3 x \, dx = \int (\sec^2 x - 1) \sec^3 x \, dx = \int \sec^5 x \, dx - \int \sec^3 x \, dx.$$

These two integrals can be evaluated by using the reduction formula of Theorem 8.2.5(ii). We find that

$$\int \sec^5 x \, dx = \frac{1}{4} \sec^3 x \tan x + \frac{3}{4} \int \sec^3 x \, dx,$$

so

$$\int \tan^2 x \sec^3 x \, dx = \int \sec^5 x \, dx - \int \sec^3 x \, dx$$

$$= \frac{1}{4} \sec^3 x \tan x - \frac{1}{4} \int \sec^3 x \, dx.$$

This last integral can be evaluated by using the reduction formula again, which gives

$$\int \sec^3 x \, dx = \frac{1}{2} \sec x \tan x + \frac{1}{2} \int \sec x \, dx$$

$$= \frac{1}{2} \sec x \tan x + \frac{1}{2} \ln |\sec x + \tan x| + c.$$

Finally, we can combine and find the original integral:

$$\int \tan^2 x \sec^3 x \, dx$$

$$= \frac{1}{4} \sec^3 x \tan x - \frac{1}{4} \left(\frac{1}{2} \sec x \tan x + \frac{1}{2} \ln |\sec x + \tan x| + c \right)$$

$$= \frac{1}{4} \sec^3 x \tan x - \frac{1}{8} \sec x \tan x - \frac{1}{8} \ln |\sec x + \tan x| + c'.$$

Notice that we added the constant c' in the last line. We had the constant c, but it was multiplied by $-\frac{1}{4}$, and it is simpler to write c' than $-\frac{1}{4}c$.   ◢

Summary 8.2.6

$$\int \tan^m x \sec^n x \, dx$$

Let m and n be nonnegative integers and consider $\int \tan^m x \sec^n x \, dx$.

1. If $m = 0$, then we are integrating a power of $\sec x$. If $n = 0$, the integrand is a power of $\tan x$. In either case it is possible to apply a reduction formula.
   Suppose m is an integer with $m \geq 2$ and n is an integer with $n \geq 2$. Then

   (i) $\displaystyle \int \tan^m x \, dx = \frac{1}{m-1} \tan^{m-1} x - \int \tan^{m-2} x \, dx,$

   (ii) $\displaystyle \int \sec^n x \, dx = \frac{1}{n-1} \sec^{n-2} x \tan x + \frac{n-2}{n-1} \int \sec^{n-2} x \, dx.$ (8.2.5)

   By using these it is possible to express the answer in terms of $\int \sec^2 x \, dx = \tan x + c$ or $\int \sec x \, dx = \ln |\sec x + \tan x| + c$ (cf. Example 5) or in terms of $\int \tan x \, dx = \ln |\sec x| + c$ or $\int 1 \, dx = x + c$.

2. If n is nonzero and even, it is possible to use the substitution $u = \tan x$, $du = \sec^2 x \, dx$. It is necessary to convert the remaining factor $\sec^{n-2} x$ into an expression in $\tan x$, using Equation 8.2.2.

3. If m is odd and $n \neq 0$, it is possible to use the substitution $u = \sec x$, $du = \sec x \tan x \, dx$. It is necessary to convert the remaining factor $\tan^{m-1} x$ into an expression in $\sec x$, using Equation 8.2.2.

4. If m is even and n is odd, it is possible to use Equation 8.2.2 to convert $\tan^m x$ into an expression involving $\sec x$. The integral will be a sum of terms, each of which is a constant times the integral of some power of $\sec x$. These can be done as in part 1.

We have seen how to evaluate many types of trigonometric integrals. Of course, there are others that do not fit into any of the categories we have discussed. If you encounter such an integral, you should probably first look to see whether any obvious substitution occurs to you. Failing that, it may help to write everything out in terms of $\sin x$ and $\cos x$. You should also consider integrating by parts.

Exercises 8.2

I

Integrate each of the following.

1. $\int \sin^3 x \, dx$

2. $\int \cos^4 x \, dx$

3. $\int \sin^4 x \, dx$

4. $\int \cos^5 x \, dx$

5. $\int \sec^3 x \, dx$

6. $\int \sec^4 x \, dx$

7. $\int \tan^3 x \, dx$

8. $\int \tan^4 x \, dx$

9. $\int \sin^3 2x \, dx$

10. $\int \cos^3 \frac{x}{3} \, dx$

11. $\int \cos x \sin^3 x \, dx$

12. $\int \cos^3 x \sin x \, dx$

13. $\int \cos^2 x \sin^3 x \, dx$

14. $\int \cos^4 x \sin^2 x \, dx$

15. $\int \tan^3 x \sec^2 x \, dx$

16. $\int \tan^2 x \sec^4 x \, dx$

17. $\int \tan^3 x \sec x \, dx$

18. $\int \tan^2 x \sec x \, dx$

19. $\int \cos^3 2x \sin^3 2x \, dx$

20. $\int \cos^2 x \sin^4 x \, dx$

21. $\int \cos^3 x \sin^7 x \, dx$

22. $\int \cos^4 x \sin^4 x \, dx$

23. $\int \cos^2 3x \sin^2 3x \, dx$

24. $\int (1 - \cos^2 x) \sin^4 x \, dx$

25. $\int \tan^5 (3x) \sec^2 (3x) \, dx$

26. $\int \tan^5 \frac{x}{2} \sec^3 \frac{x}{2} \, dx$

27. $\int \cos x \sin 2x \, dx$

28. $\int \sec^7 x \sin x \, dx$

29. $\int \tan^4 x \sin x \, dx$

30. $\int (\sin^4 x + 2 \cos^2 x \sin^2 x + \cos^4 x) \, dx$

Evaluate each of the following.

31. $\int_0^\pi \sin^2 x \, dx$

32. $\int_0^\pi \sin^2 x \cos x \, dx$

33. $\int_0^{\pi/2} \sin^2 x \cos^3 x \, dx$

34. $\int_0^{\pi/2} \sin^2 x \cos x \, dx$

35. $\int_{-\pi/2}^{\pi/2} \sin^3 x \cos^2 x \, dx$

36. $\int_{-\pi}^{\pi} \sin^5 x \cos^6 x \, dx$

37. $\int_0^1 \sin^2 (\pi x) \cos^2 (\pi x) \, dx$

38. $\int_{-\pi/2}^0 \sin^3 x \cos^3 x \, dx$

39. $\int_0^{2\pi} \sin^2 x \cos^3 x \, dx$

40. $\int_{-\pi}^0 \sin^2 (2x) \cos^2 (2x) \, dx$

II

41. Find the integral in Example 3 (p. 571) using the half-angle formula. Compare the two answers, making sure they are the same up to the addition of a constant.

*42. Do the integral $\int \tan^3 x \, dx$ of Example 6(i) (p. 574) a different way, by writing $\tan x$ out in terms of $\cos x$ and $\sin x$ and making a substitution. Reconcile the answer you get with the answer given for Example 6(i).

*43. Use the trigonometric identity (Equation 8.2.1) to show directly that the two answers to Example 4 (p. 572) agree up to a constant. What is the constant?

**44.   (i) Consider the function $\sin x + \cos x$, and show that it can be written in the form $K \cos(x + s)$ for some numbers K and s. Find K and s. (*Hint:* Think about the formula $\cos(x + s) = \cos x \cos s - \sin x \sin s$.)

   (ii) Integrate $\int \frac{1}{\sin x + \cos x} \, dx$.

   (iii) Assume that a and b are real numbers that are not both zero. Integrate $\int \frac{1}{a \sin x + b \cos x} \, dx$.

*45.   (i) Consider the integral $\int_0^{2\pi} \cos^6 x \sin^3 x \, dx$. Argue that it must equal zero, without actually evaluating it.

   (ii) Find the integral $\int_0^{2\pi} \cos^5 x \sin^8 x \, dx$.

*46.   (i) Suppose m and n are nonnegative integers. Consider the integral

$$\int_{-\pi}^{\pi} \cos^m x \sin^n x \, dx.$$

   Show that it equals zero if either of m and n is odd.

   (ii) Argue that if both m and n are positive even integers, then $\int_0^{2\pi} \cos^m x \sin^n x \, dx > 0$.

*47. Find the arclength of the curve $y = \frac{1}{2} \ln(\cos^2 x)$ for x between $x = 0$ and $x = \frac{\pi}{4}$.

*48. Find $\int \csc^2 x \cot^3 x \, dx$.

∗∗49. (i) Find $\int \csc x \, dx$. (ii) Find a reduction formula for $\int \csc^m x \, dx$.

∗50. Describe how you would set about integrating $\int \cot^m x \csc^n x \, dx$ if m and n are nonnegative integers. (Describe different approaches for different cases.)

∗51. (i) $\int \csc^2 x \cot^2 x \, dx$, (ii) $\int \csc^3 x \cot^3 x \, dx$.

∗52. (i) $\int \csc^4 x \cot^3 x \, dx$, (ii) $\int \csc^3 x \cot^4 x \, dx$.

53. Find $\int \frac{1}{\cos^5 x} \, dx$.

54. Prove the reduction formula of Theorem 8.2.5(ii) (p. 574).

Drill

55. $\int \sin x \cos^3 x \, dx$

56. $\int x \sin 2x \, dx$

57. $\int x \sqrt{x^2 + 1} \, dx$

58. $\int x / \sqrt{x^2 + 1} \, dx$

59. $\int x(x^2 + 1)^{3/2} \, dx$

60. $\int e^{2x} \cos x \, dx$

61. $\int x \cos(x^2 - 3) \, dx$

62. $\int (x + 5) \cos x \sin x \, dx$

63. $\int \frac{\sin^4 x}{\cos^2 x} \, dx$

64. $\int \frac{1}{1 + \sin \theta} \, d\theta$

8.3 Trigonometric Substitutions

TRIGONOMETRIC IDENTITIES
RANGE OF VARIABLES

In Section 8.2 we learned how to do many trigonometric integrals, that is, integrals of trigonometric functions. That is an interesting question in its own right, but the main reason for studying it there is that we will need it for the work of this section. Here we discover ways of changing many different types of integrals into trigonometric integrals. The hope is that the resulting integrals can be done by using the techniques of the last section.

We begin with some examples.

EXAMPLE 1   Find (i) $\int \sqrt{1 - x^2} \, dx$, (ii) $\int_0^3 \sqrt{9 - x^2} \, dx$.

Solution

(i) The problem is that we cannot "get at" the expression inside the square root sign. Things would be better if $1 - x^2$ were equal to the square of something so that the square root could be removed.

At this point, a little bell should be ringing in our memories, reminding us that there is something very familiar about 1 minus the square of something equaling the square of something else. It is the trigonometric identity $\cos^2 \theta + \sin^2 \theta = 1$, which can be rewritten as $1 - \sin^2 \theta = \cos^2 \theta$.

This suggests that we might approach the integral $\int \sqrt{1 - x^2} \, dx$ by letting $x = \sin \theta$. If we do this, the integrand becomes $\sqrt{1 - x^2} = \sqrt{1 - \sin^2 \theta} = \sqrt{\cos^2 \theta} = \cos \theta$. (Actually, we should be more careful; the square root is always nonnegative and $\cos \theta$ may be negative, but let us worry about that later.)

The other thing to do with a substitution is to consider the dx. If $x = \sin \theta$, then $dx = \cos \theta \, d\theta$. We can write the integral as

$$\int \sqrt{1 - x^2} \, dx = \int \sqrt{1 - \sin^2 \theta} \cos \theta \, d\theta = \int \cos^2 \theta \, d\theta$$

$$= \frac{\theta}{2} + \frac{1}{2} \cos \theta \sin \theta + c.$$

Recall that $\int \cos^2 \theta \, d\theta$ $= \frac{1}{2}(\theta + \cos \theta \sin \theta) + c$ (cf. Example 8.1.3, p. 563).

The last step is to return to the original variable x. This is slightly more tricky than what we encountered with other substitutions. We know that $x = \sin \theta$. This means that $\theta = \arcsin x$. With this we are able to convert most of the above formula back into a function of x, but the one remaining difficulty is the function $\cos\theta$. To deal with it, we must write $\cos^2\theta = 1 - \sin^2\theta = 1 - x^2$, so $\cos\theta = \sqrt{1 - x^2}$. Putting all the pieces together, we find the result we mentioned in Example 6.1.2:

$$\int \sqrt{1 - x^2}\, dx = \frac{1}{2}(\arcsin x + x\sqrt{1 - x^2}) + c.$$

It might be instructive to verify this result by differentiating it. The derivative equals

$$\frac{1}{2}\left(\frac{1}{\sqrt{1 - x^2}} + \sqrt{1 - x^2} - \frac{x^2}{\sqrt{1 - x^2}}\right) = \frac{1}{2}\left(\frac{1 - x^2}{\sqrt{1 - x^2}} + \sqrt{1 - x^2}\right)$$

$$= \frac{1}{2}(\sqrt{1 - x^2} + \sqrt{1 - x^2})$$

$$= \sqrt{1 - x^2}.$$

This confirms our integration. Notice that to integrate, we had to use $\sqrt{1 - \sin^2\theta} = \cos\theta$, which means that we have assumed that $\cos\theta \geq 0$.

(ii) Now consider $\int_0^3 \sqrt{9 - x^2}\, dx$. This time it is not 1 minus a square, but 9 minus a square. Since there is no convenient trigonometric identity involving 9, it is necessary to adjust the integrand:

$$\int_0^3 \sqrt{9 - x^2}\, dx = \int_0^3 3\sqrt{1 - \left(\frac{x}{3}\right)^2}\, dx = 3\int_0^3 \sqrt{1 - \left(\frac{x}{3}\right)^2}\, dx.$$

The quantity $1 - \left(\frac{x}{3}\right)^2$ is reminiscent of the left side of the identity $1 - \sin^2\theta = \cos^2\theta$. We take advantage of this observation by letting $\frac{x}{3} = \sin\theta$. As x goes from $x = 0$ to $x = 3$, the quantity $\frac{x}{3}$ goes from 0 to 1. In order for $\sin\theta$ to go from 0 to 1 we let θ go from $\theta = 0$ to $\theta = \frac{\pi}{2}$.

Rewriting the substitution as $x = 3\sin\theta$, we see that $dx = 3\cos\theta\, d\theta$, and $\sqrt{1 - \left(\frac{x}{3}\right)^2} = \sqrt{1 - \sin^2\theta} = \sqrt{\cos^2\theta} = \cos\theta$. Notice that this time, since we know $0 \leq \theta \leq \frac{\pi}{2}$, we know that $\cos\theta \geq 0$, and it was correct to take the square root as we did.

Now we can integrate:

$$\int_0^3 \sqrt{9 - x^2}\, dx = 3\int_0^3 \sqrt{1 - \left(\frac{x}{3}\right)^2}\, dx = 3\int_0^{\pi/2} \sqrt{1 - \sin^2\theta}\, 3\cos\theta\, d\theta$$

$$= 9\int_0^{\pi/2} \cos^2\theta\, d\theta = 9\left(\frac{\theta}{2} + \frac{1}{2}\cos\theta\sin\theta\right)\Big|_0^{\pi/2}$$

$$= \frac{9}{2}\left(\frac{\pi}{2} - 0\right) - \frac{9}{2}(0 - 0) = \frac{9\pi}{4}.$$

These examples illustrate the technique of trigonometric substitution. It is often useful for integrals involving an expression of one of the following forms:

$$\sqrt{a^2 - x^2}, \qquad \sqrt{a^2 + x^2}, \qquad \sqrt{x^2 - a^2}.$$

The quantities inside the square root signs remind us of the left sides of the trigonometric identities

$$1 - \sin^2\theta = \cos^2\theta, \qquad 1 + \tan^2\theta = \sec^2\theta, \qquad \sec^2\theta - 1 = \tan^2\theta,$$

respectively. Accordingly, we try the following substitutions.

Summary 8.3.1   **Trigonometric Substitutions**

Expression in integrand	Substitution	Range
$\sqrt{a^2 - x^2}$, $a > 0$	$x = a\sin\theta$	$-\frac{\pi}{2} \leq \theta \leq \frac{\pi}{2}$
$\sqrt{a^2 + x^2}$, $a > 0$	$x = a\tan\theta$	$-\frac{\pi}{2} < \theta < \frac{\pi}{2}$
$\sqrt{x^2 - a^2}$, $a > 0$	$x = a\sec\theta$	$0 \leq \theta < \frac{\pi}{2}$ or $\pi \leq \theta < \frac{3\pi}{2}$

There are two reasons for making the restrictions on the values of θ. First there is the question of the sign of the square root. For instance, $\sqrt{a^2 - x^2} = \sqrt{a^2(1 - \sin^2\theta)} = a\sqrt{\cos^2\theta}$, and when $-\frac{\pi}{2} \leq \theta \leq \frac{\pi}{2}$, we have $\cos\theta \geq 0$, so $\sqrt{\cos^2\theta} = \cos\theta$, and $\sqrt{a^2 - x^2} = a\cos\theta$.

Similarly, when $-\frac{\pi}{2} < \theta < \frac{\pi}{2}$, we have $\sec\theta > 0$, so $\sqrt{a^2 + a^2\tan^2\theta} = a\sec\theta$, and when $0 \leq \theta < \frac{\pi}{2}$ or $\pi \leq \theta < \frac{3\pi}{2}$, we know that $\tan\theta > 0$, and $\sqrt{a^2\sec^2\theta - a^2} = a\tan\theta$.

Second, the results of evaluating the new integral after one of these substitutions may involve θ. If $x = a\sin\theta$, that is, $\frac{x}{a} = \sin\theta$, then $\theta = \arcsin\left(\frac{x}{a}\right)$. The substitutions $x = a\tan\theta$ and $x = a\sec\theta$ amount to $\theta = \arctan\left(\frac{x}{a}\right)$ and $\theta = \operatorname{arcsec}\left(\frac{x}{a}\right)$, respectively. The restrictions on the range of θ are chosen in a way that is consistent with the standard definitions of the inverse trigonometric functions (cf. Definition 3.8.3, p. 206).

EXAMPLE 2   Find $\int \frac{1}{x\sqrt{x^2-2}}\, dx$.

Solution   Following the above suggestions, we try $x = \sqrt{2}\sec\theta$. So $\sqrt{x^2 - 2} = \sqrt{2\sec^2\theta - 2} = \sqrt{2}\tan\theta$, $dx = \sqrt{2}\sec\theta\tan\theta\, d\theta$, and

$$\int \frac{dx}{x\sqrt{x^2-2}} = \frac{1}{\sqrt{2}}\int \frac{\sqrt{2}\sec\theta\tan\theta}{\sqrt{2}\sec\theta\tan\theta}\, d\theta = \frac{1}{\sqrt{2}}\int 1\, d\theta = \frac{1}{\sqrt{2}}\theta + c.$$

Now $\frac{x}{\sqrt{2}} = \sec\theta$, so $\theta = \operatorname{arcsec}\left(\frac{x}{\sqrt{2}}\right)$, and

$$\int \frac{dx}{x\sqrt{x^2-2}} = \frac{1}{\sqrt{2}}\operatorname{arcsec}\left(\frac{x}{\sqrt{2}}\right) + c.$$

Had we recognized that the integrand $\frac{1}{x\sqrt{x^2-2}}$ is very similar to $\frac{1}{x\sqrt{x^2-1}}$, which is the derivative of arcsec x (cf. Theorem 3.8.5, p. 210), we might have done this example more easily with the substitution $u = \text{arcsec}\left(\frac{x}{\sqrt{2}}\right)$. This remark serves to emphasize the value of knowing the derivatives of as many functions as possible, but meanwhile the example illustrates trigonometric substitution. ◢

Here the answer was simply a constant times θ, which is an inverse trigonometric function of the original variable x. Frequently, the integral will involve trigonometric functions of θ.

EXAMPLE 3 $\int \frac{dx}{x^2\sqrt{a^2-x^2}}$.

Solution Try the substitution $x = a\sin\theta$, so $dx = a\cos\theta\,d\theta$, and

$$\int \frac{dx}{x^2\sqrt{a^2-x^2}} = \int \frac{a\cos\theta}{(a^2\sin^2\theta)(a\cos\theta)}\,d\theta = \frac{1}{a^2}\int \csc^2\theta\,d\theta$$

$$= -\frac{1}{a^2}\cot\theta + c.$$

To express $\cot\theta$ in terms of the original variable x, we draw in Figure 8.3.1(i) a right-angled triangle containing the angle θ. Since our substitution is $\sin\theta = \frac{x}{a}$, we fill in the lengths of the hypotenuse and the vertical side. Finally we use the Pythagorean Theorem (Theorem 1.5.1) to fill in the third side as shown in Figure 8.3.1(ii).

From the figure we can read off that $\cot\theta = \frac{\sqrt{a^2-x^2}}{x}$, so

$$\int \frac{dx}{x^2\sqrt{a^2-x^2}} = -\frac{1}{a^2}\cot\theta + c = -\frac{1}{a^2}\frac{\sqrt{a^2-x^2}}{x} + c.$$ ◢

The same approach can be used for the other trigonometric substitutions.

EXAMPLE 4 $\int \frac{\sqrt{x^2-9}}{x}\,dx$.

Solution $x = 3\sec\theta$, $dx = 3\sec\theta\tan\theta\,d\theta$, so

$$\int \frac{\sqrt{x^2-9}}{x}\,dx = \int \frac{3\sqrt{\tan^2\theta}}{3\sec\theta}3\sec\theta\tan\theta\,d\theta = 3\int \tan^2\theta\,d\theta$$

$$= 3\int (\sec^2\theta - 1)\,d\theta$$

$$= 3\tan\theta - 3\theta + c.$$

In Figure 8.3.2(i) we illustrate the substitution $\frac{x}{3} = \sec\theta$, or $\cos\theta = \frac{3}{x}$. In Figure 8.3.2(ii) the Pythagorean Theorem is used to fill in the length of the third side.

Figure 8.3.1(i)

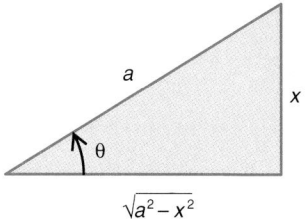

Figure 8.3.1(ii)

Since $\sin\theta$ equals the vertical side over the hypotenuse and $\sin\theta = \frac{x}{a}$, we let the hypotenuse be a and the vertical side be x.

Here we use the identity $\tan^2\theta = \sec^2\theta - 1$.

When we draw these pictures, it is convenient to draw θ as a positive angle and x as a positive number, but once we find the antiderivative, it will work whenever it is defined.

Figure 8.3.2(i)

Figure 8.3.2(ii)

Figure 8.3.3(i)

Figure 8.3.3(ii)

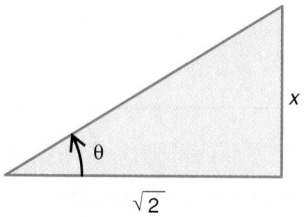

Figure 8.3.4(i)

The $\sqrt{2}$ in the numerator cancels the $\sqrt{2}$ in the denominator.

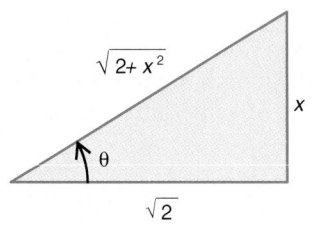

Figure 8.3.4(ii)

We see that $\tan\theta = \frac{\sqrt{x^2-9}}{3}$, so

$$\int \frac{\sqrt{x^2-9}}{x}\,dx = 3\frac{\sqrt{x^2-9}}{3} - 3\arcsec\left(\frac{x}{3}\right) + c$$

$$= \sqrt{x^2-9} - 3\arcsec\left(\frac{x}{3}\right) + c.$$

EXAMPLE 5 $\int \frac{dx}{x^2\sqrt{1+x^2}}$.

Solution   Let $x = \tan\theta$, so $dx = \sec^2\theta\,d\theta$, and

$$\int \frac{dx}{x^2\sqrt{1+x^2}} = \int \frac{\sec^2\theta\,d\theta}{\tan^2\theta\sqrt{1+\tan^2\theta}} = \int \frac{\sec^2\theta\,d\theta}{\tan^2\theta\sec\theta}$$

$$= \int \frac{\sec\theta}{\tan^2\theta}\,d\theta = \int \frac{1}{\cos\theta\frac{\sin^2\theta}{\cos^2\theta}}\,d\theta$$

$$= \int \frac{\cos\theta}{\sin^2\theta}\,d\theta.$$

After the substitution $u = \sin\theta$, $du = \cos\theta\,d\theta$, the integral becomes $\int \frac{du}{u^2} = -\frac{1}{u} + c = -\frac{1}{\sin\theta} + c$.

The triangle illustrating $x = \tan\theta$ is shown in Figure 8.3.3. Filling in the hypotenuse by the Pythagorean Theorem, we see that $\sin\theta = \frac{x}{\sqrt{1+x^2}}$, so

$$\int \frac{dx}{x^2\sqrt{1+x^2}} = -\frac{1}{\sin\theta} + c = -\frac{\sqrt{1+x^2}}{x} + c.$$

EXAMPLE 6 $\int \frac{dx}{\sqrt{2+x^2}}$.

Solution   We want the quantity inside the square root to be $2(1+\tan^2\theta)$, which means $x^2 = 2\tan^2\theta$, so we let $x = \sqrt{2}\tan\theta$, and $dx = \sqrt{2}\sec^2\theta\,d\theta$. With this substitution we integrate:

$$\int \frac{dx}{\sqrt{2+x^2}} = \int \frac{\sec^2\theta}{\sqrt{1+\tan^2\theta}}\,d\theta = \int \frac{\sec^2\theta}{\sec\theta}\,d\theta = \int \sec\theta\,d\theta$$

$$= \ln|\sec\theta + \tan\theta| + c,$$

using Example 8.2.5 (p. 573).

Using Figures 8.3.4(i) and 8.3.4(ii), we find that $\sec\theta = \frac{1}{\cos\theta} = \frac{\sqrt{2+x^2}}{\sqrt{2}}$, so

$$\int \frac{dx}{\sqrt{2+x^2}} = \ln\left|\frac{\sqrt{2+x^2}}{\sqrt{2}} + \frac{x}{\sqrt{2}}\right| + c = \ln\left(\frac{1}{\sqrt{2}}|\sqrt{2+x^2} + x|\right) + c$$

$$= \ln|\sqrt{2+x^2} + x| + \ln\frac{1}{\sqrt{2}} + c.$$

Replacing the constant c with the different constant $c' = \ln\frac{1}{\sqrt{2}} + c$, we have

$$\int \frac{dx}{\sqrt{2+x^2}} = \ln|\sqrt{2+x^2} + x| + c'.$$

In Section 6.9 we noted that we could not evaluate the integrals that arose from trying to find the arclength of a parabolic arc. In Example 6.10.8 (p. 500) we estimated such an integral. Now we can calculate it exactly.

EXAMPLE 7 $\int_0^2 \sqrt{1 + x^2}\, dx$.

Solution   We try $x = \tan\theta$, so $dx = \sec^2\theta\, d\theta$. As x goes from $x = 0$ to $x = 2$, the variable θ will go from $\theta = 0$ to $\theta = \arctan 2$, and

$$\int_0^2 \sqrt{1 + x^2}\, dx = \int_0^{\arctan 2} \sqrt{\sec^2\theta}\,\sec^2\theta\, d\theta = \int_0^{\arctan 2} \sec^3\theta\, d\theta.$$

This is the type of integral that should be done by using the reduction formula of Theorem 8.2.5(ii). We find that $\int \sec^3\theta\, d\theta = \frac{1}{2}\sec\theta\tan\theta + \frac{1}{2}\int\sec\theta\, d\theta = \frac{1}{2}\sec\theta\tan\theta + \frac{1}{2}\ln|\sec\theta + \tan\theta| + c$, so

$$\int_0^2 \sqrt{1 + x^2}\, dx = \int_0^{\arctan 2} \sec^3\theta\, d\theta$$

$$= \left(\frac{1}{2}\sec\theta\tan\theta + \frac{1}{2}\ln|\sec\theta + \tan\theta|\right)\Big|_0^{\arctan 2}$$

$$= \left(\frac{1}{2}\sec(\arctan 2)\tan(\arctan 2) + \frac{1}{2}\ln|\sec(\arctan 2) + \tan(\arctan 2)|\right).$$

Figure 8.3.5(i)

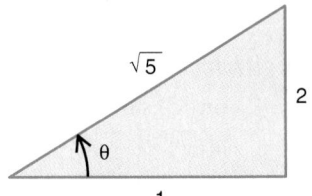

Figure 8.3.5(ii)

From the triangle shown in Figures 8.3.5(i) and (ii) we see that if $\tan\theta = 2$, then $\sec\theta = \frac{1}{\cos\theta} = \frac{1}{1/\sqrt{5}}$, that is, $\sec(\arctan 2) = \sqrt{5}$. Using this, it is now possible to evaluate the integral:

$$\int_0^2 \sqrt{1 + x^2}\, dx$$

$$= \left(\frac{1}{2}\sec(\arctan 2)\tan(\arctan 2) + \frac{1}{2}\ln|\sec(\arctan 2) + \tan(\arctan 2)|\right)$$

$$= \left(\frac{1}{2}\sqrt{5}(2) + \frac{1}{2}\ln|\sqrt{5} + 2|\right) = \sqrt{5} + \frac{1}{2}\ln|2 + \sqrt{5}|.$$

The result of this example was mentioned in Example 6.10.8, in which a numerical approximation to this integral was found.

Exercises 8.3

Evaluate the following integrals.

1. $\int \sqrt{4 - x^2}\, dx$

2. $\int \sqrt{3 - x^2}\, dx$

3. $\int \frac{1}{x\sqrt{x^2 - 1}}\, dx$

4. $\int \frac{1}{x\sqrt{3x^2 - 1}}\, dx$

5. $\int \frac{1}{x^2\sqrt{1 - x^2}}\, dx$

6. $\int \frac{1}{x^2\sqrt{4 - x^2}}\, dx$

7. $\int \frac{1}{x}\sqrt{x^2 - 4}\, dx$

8. $\int \frac{1}{x}\sqrt{x^2 - a^2}\, dx$

9. $\int \frac{1}{x^2\sqrt{x^2 + 4}}\, dx$

10. $\int \frac{1}{\sqrt{x^2 + a^2}}\, dx$

11. $\int (1 - x^2)^{3/2}\, dx$

12. $\int (4 - x^2)^{-1/2}\, dx$

13. $\int (9 - 4x^2)^{1/2}\, dx$

14. $\int (4+9x^2)^{-1/2} dx$

15. $\int (1+5x^2)^{3/2} dx$

16. $\int (4+9x^2)^{-3/2} dx$

17. $\int x(8+x^2)^{3/2} dx$

18. $\int x^2(5-x^2)^{-1/2} dx$

19. $\int x(x^2-1)^{1/2} dx$

20. $\int x^5(4-x^2)^{-1/2} dx$

21. $\int (1+x^2)^{-2} dx$

22. $\int x^3(9-x^2)^{-1/2} dx$

23. $\int (5-x^2)^{-3/2} dx$

24. $\int \frac{1}{9+6x^2+x^4} dx$

25. $\int 2x\sqrt{1-x^4} dx$

26. $\int 2x\sqrt{1+x^4} dx$

27. $\int x^{-3}(1+x^4)^{-1/2} dx$

28. $\int x^3\sqrt{1+x^4} dx$

29. $\int (x+3)\sqrt{4-x^2} dx$

30. $\int x^7\sqrt{1+x^8} dx$

Evaluate the following definite integrals.

31. $\int_0^1 \sqrt{1-x^2} dx$

32. $\int_{-1}^1 \sqrt{2-x^2} dx$

33. $\int_1^2 x^3\sqrt{x^2-1} dx$

34. $\int_2^4 x\sqrt{x^2-1} dx$

35. $\int_0^1 \frac{1}{(2-x^2)^2} dx$

36. $\int_2^4 \frac{\sqrt{x^2-1}}{x} dx$

37. $\int_0^2 \frac{1}{4+x^2} dx$

38. $\int_{-1}^1 x^5\sqrt{4-x^2} dx$

39. $\int_0^3 \frac{x}{1+x^2} dx$

40. $\int_0^1 x(2-x^2)^{-9/2} dx$

II

41. Find the arclength of the parabola $y = x^2$ from $x = 0$ to $x = 1$.

42. Find the arclength of the parabola $y = \frac{1}{2}x^2$ from $x = 1$ to $x = 4$.

43. Find the arclength of the parabola $y = x^2$ from $x = a$ to $x = b$.

*44. Evaluate $\int \frac{1}{1+x^2} dx$ using the substitution $x = \tan\theta$.

45. (i) Evaluate $\int 2x\sqrt{1-x^2} dx$ by the trigonometric substitution $x = \sin\theta$.

   (ii) Compare your answer with what you get by making the more conventional substitution $u = 1 - x^2$.

46. Integrate $\int \sqrt{1-x^2} dx$, using the substitution $x = \cos\theta$, and check that your answer is consistent with what happens using the $\sin\theta$ substitution.

*47. Recall the hyperbolic identity $\cosh^2 t - \sinh^2 t = 1$ from Equation 3.5.12 (p 180). Use it and a suitable substitution to evaluate $\int \frac{1}{\sqrt{1+x^2}} dx$.

**48. Evaluate the integral of the previous question, using a trigonometric substitution, and confirm that the two answers are the same (up to the usual constant).

Drill

49. $\int e^x/(e^{2x}+1) dx$

50. $\int \cos x \sin x\sqrt{1+\sin^2 x} dx$

51. $\int x\sqrt{4+x^2} dx$

52. $\int x^2\sqrt{4-x^2} dx$

53. $\int x^2\sqrt{9+x^2} dx$

54. $\int xe^{x^2+1} dx$

55. $\int \cosh x \sinh x dx$

56. $\int x\ln(1+x^2) dx$

57. $\int x\sin(x^2)\sin(2x^2) dx$

58. $\int x\ln(x^4) dx$

P ← **POINT TO PONDER**

When we first encountered substitutions in Section 5.7, the situation was something like $\int 2x\cos(1-x^2) dx$. The idea there was to make a substitution $u = 1 - x^2$. In this section we encounter an integral like $\int \sqrt{1-x^2} dx$, and we make the substitution $x = \sin\theta$. Notice that the substitution is written quite differently. In the first case we wrote the substituted variable u as a function of the original variable x; in the second it is the other way around, and we have written the original variable x as a function of the new variable θ. In this second situation it is quite straightforward to find dx because we have a formula for x as a function of θ. This is actually easier than what we are used to doing with substitutions of the first type.

Of course, it is possible to write the trigonometric substitution in a form that is consistent with the other kind. If we take $x = \sin\theta$ and write it as $\theta = \arcsin x$, then it looks more like a conventional substitution. However, there is the drawback that $d\theta$ will be expressed with a function of x. In this case, $d\theta = 1/\sqrt{1-x^2} dx$, and in order to be able to integrate we would have to convert that square root back into a function of θ. This would be an unnecessary complication.

Converting the result of our integration back into an expression using the original variable may be more difficult for trigonometric substitutions.

So trigonometric substitutions are in a sense the same kind of thing as the substitutions we already know, but there are some differences.

8.4 Completing the Square in Quadratic Expressions in Integrands

COMPLETING SQUARES
TRIGONOMETRIC SUBSTITUTION

The subject of this section is best described by considering an example.

EXAMPLE 1 $\int \frac{1}{x^2+2x+2}\, dx$.

Solution   It is easy to complete the square (cf. Summary 1.3.3, p. 9): $x^2 + 2x + 2 = x^2 + 2x + 1 + 1 = (x+1)^2 + 1$. So the integral can be written $\int \frac{1}{x^2+2x+2}\, dx = \int \frac{1}{(x+1)^2+1}\, dx$, which suggests the substitution $u = x + 1$, $du = dx$. We find

$$\int \frac{1}{x^2 + 2x + 2}\, dx = \int \frac{1}{(x+1)^2 + 1}\, dx = \int \frac{1}{u^2 + 1}\, du$$

$$= \arctan u + c = \arctan(x + 1) + c.$$

The idea is simple enough. Starting with a quadratic expression $Ax^2 + Bx + C$, we can complete the square and make a substitution to obtain an expression of the form $Au^2 \pm a^2$. The advantage is that this last expression is the kind we may be able to approach using trigonometric substitutions as in Section 8.3. Even if that does not work, it is simpler, and something else might occur to us.

EXAMPLE 2 $\int \sqrt{-2x^2 + 12x - 10}\, dx$.

Solution   We begin by pulling out the common factor of 2 from inside the square root: $\int \sqrt{-2x^2 + 12x - 10}\, dx = \sqrt{2} \int \sqrt{-x^2 + 6x - 5}\, dx$. It is easy to complete the square: $-x^2 + 6x - 5 = -(x^2 - 6x + 5) = -(x^2 - 6x + 9 - 4) = -\left((x-3)^2 - 4\right) = 4 - (x-3)^2$. Making the substitution $u = x - 3$, $du = dx$, we find

$$\int \sqrt{-2x^2 + 12x - 10}\, dx = \sqrt{2} \int \sqrt{4 - (x-3)^2}\, dx = \sqrt{2} \int \sqrt{4 - u^2}\, du.$$

This is the type of integral we know how to evaluate with a trigonometric substitution. Letting $u = 2 \sin \theta$, so $du = 2 \cos \theta\, d\theta$, we find

$$\sqrt{2} \int \sqrt{4 - u^2}\, du = \sqrt{2} \int \sqrt{4 - 4 \sin^2 \theta}\, 2 \cos \theta\, d\theta$$

$$= 2\sqrt{2} \int \sqrt{4(1 - \sin^2 \theta)}\, \cos \theta\, d\theta = 4\sqrt{2} \int \cos^2 \theta\, d\theta$$

$$= 2\sqrt{2}(\theta + \cos \theta \sin \theta) + c.$$

Recall that $\int \cos^2 \theta\, d\theta = \frac{1}{2}(\theta + \cos \theta \sin \theta) + c$ (cf. Example 8.1.3, p. 563).

Considering the triangle shown in Figure 8.4.1, we see that if $u = 2 \sin \theta$, that is, $\frac{u}{2} = \sin \theta$, then $\cos \theta = \frac{1}{2}\sqrt{4 - u^2}$. We find that

$$\int \sqrt{-2x^2 + 12x - 10}\, dx = \sqrt{2} \int \sqrt{4 - u^2}\, du = 2\sqrt{2}(\theta + \cos \theta \sin \theta) + c$$

$$= 2\sqrt{2}\left(\arcsin\left(\frac{u}{2} \right) + \frac{1}{4} u \sqrt{4 - u^2} \right) + c$$

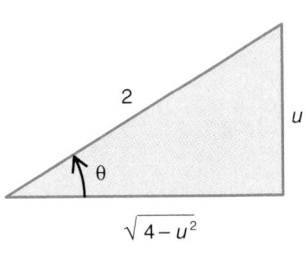

Figure 8.4.1

$$= 2\sqrt{2}\arcsin\left(\frac{x-3}{2}\right) + \frac{1}{\sqrt{2}}(x-3)\sqrt{4-(x-3)^2} + c.$$

◢

EXAMPLE 3 $\displaystyle\int \frac{x^2 - 2x + 3}{x^2 + 4x + 8}\, dx.$

Solution   What is confusing about this integrand is that there are *two* quadratic expressions. We can complete the square in each one without difficulty. The problem is that then we must choose a substitution. The substitution that makes the numerator into something simple ($u^2 + 2$) makes the denominator into something less good ($u^2 + 6u + 13$), and vice versa. The question is which approach will give us a more manageable integral?

The answer is that it is more important to simplify the denominator. We complete the square: $x^2 + 4x + 8 = x^2 + 4x + 4 + 4 = (x + 2)^2 + 4$, and this suggests the substitution $u = x + 2$, $du = dx$. As usual, it is necessary to express the numerator in terms of u.

Since $u = x + 2$, we know that $x = u - 2$, so the numerator is $x^2 - 2x + 3 = (u - 2)^2 - 2(u - 2) + 3 = u^2 - 4u + 4 - 2u + 4 + 3 = u^2 - 6u + 11$. The integral becomes

$$\int \frac{x^2 - 2x + 3}{x^2 + 4x + 8}\, dx = \int \frac{x^2 - 2x + 3}{(x + 2)^2 + 4}\, dx = \int \frac{u^2 - 6u + 11}{u^2 + 4}\, du.$$

Now what? We know how to integrate things like $\frac{u}{u^2+4}$ and things like $\frac{1}{u^2+4}$. The difficulty is that u^2 term in the numerator. The solution is to do a long division. The integrand can be expressed as the sum of a polynomial (i.e., with no denominator) *plus* a rational function in which the numerator has no powers of u higher than the first power.

In this particular case this is easily accomplished without long division just by writing

$$\frac{u^2 - 6u + 11}{u^2 + 4} = \frac{u^2 + 4 - 6u + 7}{u^2 + 4} = \frac{u^2 + 4}{u^2 + 4} + \frac{-6u + 7}{u^2 + 4} = 1 + \frac{-6u + 7}{u^2 + 4}.$$

Now we can integrate:

$$\int \frac{x^2 - 2x + 3}{x^2 + 4x + 8}\, dx = \int \frac{u^2 - 6u + 11}{u^2 + 4}\, du = \int \left(1 + \frac{-6u + 7}{u^2 + 4}\right) du$$

$$= \int 1\, du - \int \frac{6u}{u^2 + 4}\, du + \int \frac{7}{u^2 + 4}\, du.$$

The first integral is trivial, the second can easily be done by substituting $v = u^2 + 4$, so $dv = 2u\, du$, and the third can be done by substituting $w = \frac{u}{2}$, so $dw = \frac{1}{2}\, du$. We find

$$\int \frac{x^2 - 2x + 3}{x^2 + 4x + 8}\, dx = \int 1\, du - \int \frac{6u}{u^2 + 4}\, du + \int \frac{7}{u^2 + 4}\, du$$

$$= u - 3\int \frac{1}{v}\, dv + 7\int \frac{2}{4w^2 + 4}\, dw$$

$$= u - 3\ln|v| + \frac{7}{2}\arctan w + c$$

Since $u^2 + 4 > 0$, we can write $\ln(u^2 + 4)$ without the absolute value.

$$= u - 3\ln(u^2 + 4) + \frac{7}{2}\arctan\left(\frac{u}{2}\right) + c$$

$$= x + 2 - 3\ln\big((x+2)^2 + 4\big) + \frac{7}{2}\arctan\left(\frac{x+2}{2}\right) + c$$

$$= x - 3\ln(x^2 + 4x + 8) + \frac{7}{2}\arctan\left(\frac{x}{2} + 1\right) + c'.$$

Notice how the constants c and $+2$ were combined into one constant c'. Also notice that the integral obtained by completing the square in the denominator and then substituting had to be broken up into three separate pieces, which could be integrated in fairly simple but different ways.

In this case we completed the square in the denominator. Generally speaking, if there are two or more quadratic expressions, it is probably best to complete the square in the one that is *least accessible*. For instance, if one is inside a square root sign, it would be a better choice than one outside the square root. However, there are no overall rules, and it is really a question of trial and error.

EXAMPLE 4 $\int \frac{x}{(x^2 - 2x + 2)^2}\, dx.$

Solution The square is easily completed in the denominator: $x^2 - 2x + 2 = x^2 - 2x + 1 + 1 = (x-1)^2 + 1$. This suggests the substitution $u = x - 1$, $du = dx$, and we find that the numerator becomes $x = u + 1$:

$$\int \frac{x}{(x^2 - 2x + 2)^2}\, dx = \int \frac{u+1}{(u^2+1)^2}\, du = \int \frac{u}{(u^2+1)^2}\, du + \int \frac{1}{(u^2+1)^2}\, du.$$

The first of these integrals can be evaluated using the substitution $v = u^2 + 1$, $dv = 2u\, du$. We find $\int u/(u^2+1)^2\, du = \frac{1}{2}\int v^{-2}\, dv = -\frac{1}{2}\frac{1}{v} + c = -\frac{1}{2}(u^2+1)^{-1} + c = -\frac{1}{2}(x^2 - 2x + 2)^{-1} + c.$

For the second integral the work of the previous section suggests letting $u = \tan\theta$, so $du = \sec^2\theta\, d\theta$. We find

$$\int \frac{1}{(u^2+1)^2}\, du = \int \frac{1}{(\tan^2\theta + 1)^2}\sec^2\theta\, d\theta$$

$$= \int \frac{\sec^2\theta}{(\sec^2\theta)^2}\, d\theta$$

$$= \int \sec^{-2}\theta\, d\theta = \int \cos^2\theta\, d\theta$$

$$= \frac{1}{2}(\theta + \cos\theta\sin\theta) + c'.$$

Since $u = \tan\theta$, we see from Figure 8.4.2 that $\cos\theta = 1/\sqrt{1 + u^2}$. Also $\sin\theta = u/\sqrt{1 + u^2}$ and $\theta = \arctan u$. So

$$\int \frac{1}{(u^2+1)^2}\, du = \frac{1}{2}(\theta + \cos\theta\sin\theta) + c'$$

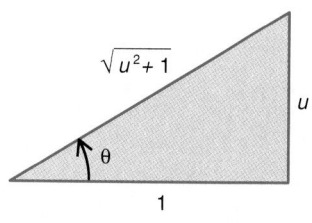

Figure 8.4.2

$$= \frac{1}{2}\left(\arctan u + \frac{1}{\sqrt{1+u^2}}\frac{u}{\sqrt{1+u^2}}\right) + c'$$

$$= \frac{1}{2}\left(\arctan u + \frac{u}{1+u^2}\right) + c'$$

$$= \frac{1}{2}\left(\arctan(x-1) + \frac{x-1}{x^2-2x+2}\right) + c'.$$

Combining, we find

$$\int \frac{x}{(x^2-2x+2)^2}\,dx = \int \frac{u}{(u^2+1)^2}\,du + \int \frac{1}{(u^2+1)^2}\,du$$

$$= -\frac{1}{2}(x^2-2x+2)^{-1}$$

$$+ \frac{1}{2}\left(\arctan(x-1) + \frac{x-1}{x^2-2x+2}\right) + c$$

$$= \frac{1}{2}\left(\arctan(x-1) + \frac{x-2}{x^2-2x+2}\right) + c.$$

REMARK 8.4.1

The technique of Example 4 can be used to evaluate any integral of the form $\int \frac{rx+s}{(Ax^2+Bx+C)^n}\,dx$. The idea is to complete the square in the denominator and make the appropriate substitution so that the denominator will become something of the form $(\pm u^2 \pm a^2)^n$.

Writing the numerator in terms of u, we can break the integral into two parts, one a constant times $\int u/(\pm u^2 \pm a^2)^n\,du$ and the other a constant times $\int 1/(\pm u^2 \pm a^2)^n\,du$. The first of these can be evaluated using the substitution $v = \pm u^2 \pm a^2$, and the second can be done using the appropriate trigonometric substitution (cf. Summary 8.3.1, p. 580).

Exercises 8.4

Evaluate the following integrals.

1. $\int \frac{1}{x^2-4x+5}\,dx$

2. $\int \frac{1}{x^2-2x+5}\,dx$

3. $\int \frac{1}{x^2+6x+14}\,dx$

4. $\int \frac{x}{x^2+2x+2}\,dx$

5. $\int \frac{x+1}{x^2-2x+4}\,dx$

6. $\int \frac{1-3x}{x^2+4x+3}\,dx$

7. $\int \frac{x^2+3}{x^2-2x+2}\,dx$

8. $\int \frac{2x^2}{x^2+4x+4}\,dx$

9. $\int \sqrt{3-2x-x^2}\,dx$

10. $\int \sqrt{7+6x-x^2}\,dx$

11. $\int 1/\sqrt{16-6x-x^2}\,dx$

12. $\int 1/\sqrt{8+2x-x^2}\,dx$

13. $\int \frac{1}{4x-x^2-8}\,dx$

14. $\int \frac{x}{8x-20-x^2}\,dx$

15. $\int \frac{1}{2x^2-4x+4}\,dx$

16. $\int \frac{1}{2x^2+6x+5}\,dx$

17. $\int \frac{x}{\sqrt{x^2+2x+2}}\,dx$

18. $\int (x+2)(x^2-4x+8)^{-1/2}\,dx$

19. $\int (x+1)(x^2-4x+5)^{-3/2}\,dx$

20. $\int x\sqrt{8+2x-x^2}\,dx$

21. $\int (x+1)/\sqrt{3+2x-x^2}\,dx$

22. $\int x(x^2+6x+10)^{-5/2}\,dx$

23. $\int (x+1)(x^2+2x+17)^{-17/2}\,dx$

24. $\int (1-x^2)^{-1/2}\,dx$

25. $\int (5-4x-x^2)^{-3/2}\,dx$

26. $\int (6-2x-3x^2)^{-1/2}\,dx$

27. $\int \frac{x+2}{(x^2+4x+13)^5}\,dx$

28. $\int x^2\sqrt{1-x^2}\,dx$

29. $\int x^3\sqrt{1-x^2}\,dx$

30. $\int x^2\sqrt{2-2x-x^2}\,dx$

What is the degree of $Q(x)$? (*Hint:* Start with long division, resulting in a quotient plus a remainder; consider separately the cases in which the degree of $p(x)$ is or is not less than 6.)

II

*31. Evaluate $\int \frac{x}{(x^2-2x+2)^3}\,dx$.

*32. (i) Let $q(x) = ax^2 + bx + c$ be a quadratic polynomial. Consider a quotient $\frac{p(x)}{q(x)^3}$, where $p(x)$ is a polynomial of degree less than 6. Show that $\frac{p(x)}{q(x)^3}$ can always be written in the form

$$\frac{p(x)}{q(x)^3} = \frac{Ax+B}{q(x)^3} + \frac{Cx+D}{q(x)^2} + \frac{Ex+F}{q(x)}$$

for suitable constants A, B, C, D, E, F. (*Hint:* Write $p(x) = rx^5 + sx^4 + tx^3 + ux^2 + vx + w$. What must E equal?)

(ii) If $p(x)$ is a polynomial of *any* degree, show that $\frac{p(x)}{q(x)^3}$ can be written as the sum of a polynomial $Q(x)$ and an expression like the one in part (i).

Drill

33. $\int \frac{x+1}{x^2+2x+10}\,dx$

34. $\int x\sqrt{x^2+4x+8}\,dx$

35. $\int x\sqrt{x-1}\,dx$

36. $\int \sqrt{2x-x^2}\,dx$

37. $\int (x-2)^3\sqrt{5+4x-x^2}\,dx$

38. $\int e^x\sqrt{e^x-1}\,dx$

39. $\int e^{x/2}\sqrt{e^x-4}\,dx$

40. $\int \frac{\sin t}{\cos^2 t - 2\cos t + 2}\,dt$

41. $\int \frac{1}{x}\sqrt{\ln x - \ln^2 x}\,dx$

42. $\int \frac{\sqrt{2+\tan^2\theta}}{\cos^2\theta}\,d\theta$

8.5 Partial Fractions

RATIONAL FUNCTION

PROPER RATIONAL FUNCTION

IRREDUCIBLE QUADRATIC FACTOR

DISCRIMINANT

In this section we learn a technique for integrating rational functions (functions that are quotients of polynomials). The idea hinges on being able to factor the denominator. Assuming that this is possible, there is a way to express the rational function as a sum of rational functions of several relatively simple types, each of which we can integrate using techniques from the previous sections.

Of course, it is not always possible to see how to factor a polynomial, so in a sense the method is not as complete as it sounds. However, it is extremely useful, and we will always assume that we do know how to factor the denominator.

EXAMPLE 1   If we consider $\frac{1}{1+x} + \frac{1}{1-x}$, we can write it over a common denominator $1 - x^2$, and $\frac{1}{1+x} + \frac{1}{1-x} = \frac{(1-x)+(1+x)}{1-x^2} = \frac{2}{1-x^2}$. Integrating, we find

$$\int \frac{2}{1-x^2}\,dx = \int \frac{1}{1+x}\,dx + \int \frac{1}{1-x}\,dx$$

$$= \ln|1+x| - \ln|1-x| + c = \ln\left|\frac{1+x}{1-x}\right| + c.$$

Here we used substitutions $u = 1 + x$ and $v = 1 - x$ in the two integrals and simplified the answer using properties of the natural logarithm.

This example took a sum of two simple functions, whose integrals we can find very easily, and expressed it as a single rational function over a common denominator. This is really the reverse of what we want; the useful thing would be to take a more complicated rational function and express it as a sum of simple ones. The example suggests the following. Suppose we know how to factor the denominator. Then it might be possible to express the original integrand as a sum of terms, each of which is equal to a constant divided by one of the factors of the denominator. This is not always quite correct, but sometimes it is, and it is a very good idea.

EXAMPLE 2 $\int \frac{2x+6}{x^3+x^2-6x}\, dx.$

Solution   Following the suggestion made above, we factor the denominator. Since there is no constant term, it is divisible by x, so $x^3 + x^2 - 6x = x(x^2 + x - 6) = x(x+3)(x-2)$.

We want to try to write

$$\frac{2x+6}{x^3+x^2-6x} = \frac{A}{x} + \frac{B}{x+3} + \frac{C}{x-2}$$

for suitable constants A, B, C. Clearing denominators, that is, multiplying by $x^3 + x^2 - 6x = x(x+3)(x-2)$, we get

$$2x + 6 = A(x+3)(x-2) + Bx(x-2) + Cx(x+3)$$
$$= A(x^2 + x - 6) + B(x^2 - 2x) + C(x^2 + 3x)$$
$$= (A + B + C)x^2 + (A - 2B + 3C)x - 6A.$$

Here we have collected together all the terms in x^2, all the terms in x, and the constant term. For the above equation to hold, the coefficients of x^2 on the left side and right side must match, that is, $0 = A + B + C$. Similarly, the coefficients of x must be the same: $2 = A - 2B + 3C$, and so must the constant terms: $6 = -6A$. This gives us three equations:

$$A + B + C = 0 \qquad (x^2\text{-terms}),$$
$$A - 2B + 3C = 2 \qquad (x\text{-terms}),$$
$$-6A = 6 \qquad (\text{constant terms}).$$

From the last equation we see that $A = -1$, so the first two equations become

$$B + C = 1,$$
$$-2B + 3C = 3.$$

Adding the last equation to twice the previous one gives $5C = 5$, so $C = 1$ and $B = 1 - C = 0$.

Eliminating B, say, we find $C = 1$, $B = 0$. So

$$\frac{2x+6}{x^3+x^2-6x} = \frac{-1}{x} + \frac{0}{x+3} + \frac{1}{x-2},$$

and $\int \frac{2x+6}{x^3+x^2-6x}\, dx = \int \left(\frac{-1}{x} + \frac{0}{x+3} + \frac{1}{x-2}\right) dx$

$$= -\int \frac{1}{x}\, dx + \int \frac{1}{x-2}\, dx = -\ln|x| + \ln|x-2| + c$$
$$= \ln\left|\frac{x-2}{x}\right| + c.$$

The fact that $B = 0$ means that there is no term with denominator $x + 3$. In hindsight this is because we could have canceled $x + 3$ from the numerator and denominator of the original integrand...

This example illustrates the principle behind the technique of partial fractions, but there are some complications. The first is that we have dealt with rational functions $\frac{p(x)}{q(x)}$ that are in **proper form**, that is, in which the numerator $p(x)$ is of lower degree than the denominator $q(x)$. If we encounter a situation in which $deg\, p(x) \geq deg\, q(x)$, we must first divide (by long division) to write $\frac{p(x)}{q(x)} = f(x) + \frac{r(x)}{q(x)}$, where $f(x)$ is a polynomial and $r(x)$ is the "remainder" in the division. In particular, $deg\, r(x) < deg\, q(x)$. Reducing a rational function to proper form should always be the first step.

EXAMPLE 3 $\int \frac{x^4+1}{x(x+1)(x-1)}$.

Solution   The numerator has degree 4, and the denominator, $x(x+1)(x-1) = x^3 - x$, has degree 3, so the integrand is a rational function that is not proper. Performing a long division, we find $x^4 + 1 = (x^3 - x)x + x^2 + 1$, or $\frac{x^4+1}{x(x+1)(x-1)} = x + \frac{x^2+1}{x(x+1)(x-1)}$. We can concentrate on integrating the remainder term.

The previous discussion suggests trying to write

$$\frac{x^2+1}{x(x+1)(x-1)} = \frac{A}{x} + \frac{B}{x+1} + \frac{C}{x-1}.$$

We multiply by $x(x+1)(x-1)$ to clear the denominators and find

$$x^2 + 1 = A(x+1)(x-1) + Bx(x-1) + Cx(x+1).$$

We want to use this equation to find A, B, and C. We could use the approach of the last example, multiplying everything out, collecting terms according to the power of x, and solving the resulting equations, but there is another way to go about it.

Suppose we let $x = 0$ in the above equation. The last two terms will equal 0, since each of them has a factor of x, so the equation reads $0^2 + 1 = A(1)(-1) + 0 + 0$, or $1 = -A$, so $A = -1$. Similarly, letting $x = 1$ will make $x - 1 = 0$, so the terms containing a factor of $(x-1)$ will equal 0, and the equation will read $1^2 + 1 = 0 + 0 + C(1)(1+1)$, or $2 = 2C$, or $C = 1$. Finally, letting $x = -1$, we find that $(-1)^2 + 1 = 0 + B(-1)(-1-1) + 0$, or $2 = 2B$, or $B = 1$.

We can use this to evaluate the integral:

$$\int \frac{x^4+1}{x(x+1)(x-1)}\, dx = \int \left(x + \frac{x^2+1}{x(x+1)(x-1)} \right) dx$$

$$= \int \left(x - \frac{1}{x} + \frac{1}{x+1} + \frac{1}{x-1} \right) dx$$

$$= \frac{1}{2}x^2 - \ln|x| + \ln|x+1| + \ln|x-1| + c$$

$$= \frac{1}{2}x^2 + \ln\left| \frac{x^2-1}{x} \right| + c.$$

The first new ingredient in this example was dividing to reduce the integrand to proper form. The second was an alternative way to find the constants A, B, and C. The idea was easy: After clearing the denominator we took the resulting equation and substituted cleverly chosen values for x. The values that were chosen were the ones that make one or other of the factors of the denominator equal 0. This had the effect of making each term on the right side except for one equal to 0, so it was easy to read off the value of the constant corresponding to that term. This technique is easier than the earlier method (so much so that you may be able to do it in your head).

The next complication we may encounter is that when we factor the denominator, we may find some factors repeated, that is, occurring with powers greater than 1. Suppose $(x - a)^m$ is the largest power of $(x - a)$ that divides the denominator. Then we have to try terms of the form

$$\frac{A_1}{x - a} + \frac{A_2}{(x - a)^2} + \frac{A_3}{(x - a)^3} + \ldots + \frac{A_m}{(x - a)^m}.$$

EXAMPLE 4 $\int \frac{9x}{(x+2)^2(x-1)}\, dx.$

Solution   We try to write

$$\frac{9x}{(x + 2)^2(x - 1)} = \frac{A}{x + 2} + \frac{B}{(x + 2)^2} + \frac{C}{x - 1}.$$

Clearing denominators, we find

$$9x = A(x + 2)(x - 1) + B(x - 1) + C(x + 2)^2.$$

Letting $x = 1$ gives $9 = 0 + 0 + C(1 + 2)^2 = 9C$, so $C = 1$. Letting $x = -2$ gives $-18 = 0 + B(-2 - 1) + 0 = -3B$, so $B = 6$. There is no obvious substitution we can make for x that will tell us immediately what A is.

We could rewrite the equation using the values we have found for B and C. It becomes

$$9x = A(x^2 + x - 2) + 6(x - 1) + (x + 2)^2,$$
$$A(x^2 + x - 2) = 9x - 6x + 6 - x^2 - 4x - 4$$
$$= -x^2 - x + 2.$$

We find that A must equal -1. So

$$\int \frac{9x}{(x + 2)^2(x - 1)}\, dx = \int \left(\frac{-1}{x + 2} + \frac{6}{(x + 2)^2} + \frac{1}{x - 1} \right) dx$$

$$= -\ln |x + 2| - \frac{6}{x + 2} + \ln |x - 1| + c$$

$$= \ln \left| \frac{x - 1}{x + 2} \right| - \frac{6}{x + 2} + c.$$

Here the trick we developed in Example 3 for finding the values of the constants did not give us a complete answer. It did allow us to find B and C easily, but to find A, we had to fall back on something more like what we did in Example 2.

Of course, the calculations can get quite complicated if the polynomials are a little bigger.

DesignCAD 3D *software was used to produce this image of a house, which permits architects to display their ideas in a form which is readily understood by clients. It is also easy to change the plan and see the effects of the changes.*

EXAMPLE 5 $\int \frac{4x^5 + 2x^4 - 5x^3 + x^2 + 8x - 4}{x^2(x-1)^2(x+2)}\, dx.$

Solution   The denominator is already factored for us; it has degree 5. Since the numerator has degree 5 too, the function is not in proper form, and we must divide. To do this, it is necessary to expand the denominator: $x^2(x - 1)^2(x + 2) = x^5 -$

$3x^3 + 2x^2$. We find

$$\frac{4x^5 + 2x^4 - 5x^3 + x^2 + 8x - 4}{x^2(x-1)^2(x+2)} = \frac{4x^5 + 2x^4 - 5x^3 + x^2 + 8x - 4}{x^5 - 3x^3 + 2x^2}$$

$$= \frac{(4x^5 - 12x^3 + 8x^2) + (12x^3 - 8x^2) + 2x^4 - 5x^3 + x^2 + 8x - 4}{x^5 - 3x^3 + 2x^2}$$

$$= 4 + \frac{2x^4 + 7x^3 - 7x^2 + 8x - 4}{x^5 - 3x^3 + 2x^2}.$$

The next step is to try

$$\frac{2x^4 + 7x^3 - 7x^2 + 8x - 4}{x^5 - 3x^3 + 2x^2} = \frac{A}{x} + \frac{B}{x^2} + \frac{C}{x-1} + \frac{D}{(x-1)^2} + \frac{E}{x+2}.$$

Clearing denominators gives

$$2x^4 + 7x^3 - 7x^2 + 8x - 4 = Ax(x-1)^2(x+2) + B(x-1)^2(x+2)$$
$$+ Cx^2(x-1)(x+2) + Dx^2(x+2) + Ex^2(x-1)^2.$$

The trick of letting x equal certain well-chosen values will not work completely, but it will help. If we let $x = 0$, all the terms on the right side will be zero except for the one with B, and we find $-4 = 2B$, so $B = -2$. Letting $x = 1$ will leave nothing on the right side except $3D$, so $2 + 7 - 7 + 8 - 4 = 3D$, or $6 = 3D$ and $D = 2$. Finally, letting $x = -2$ gives $2(-2)^4 + 7(-2)^3 - 7(-2)^2 + 8(-2) - 4 = E(-2)^2(-2-1)^2$, or $32 - 56 - 28 - 16 - 4 = E(4)(9)$, or $-72 = 36E$ and $E = -2$. The equation becomes

$$2x^4 + 7x^3 - 7x^2 + 8x - 4 = Ax(x-1)^2(x+2) + (-2)(x-1)^2(x+2)$$
$$+ Cx^2(x-1)(x+2) + 2x^2(x+2) - 2x^2(x-1)^2$$
$$= Ax(x-1)^2(x+2) + Cx^2(x-1)(x+2)$$
$$+ (-2)(x-1)^2(x+2) + 2x^2(x+2) - 2x^2(x-1)^2$$
$$= Ax(x-1)^2(x+2) + Cx^2(x-1)(x+2)$$
$$- 2x^4 + 4x^3 + 2x^2 + 6x - 4,$$

so $4x^4 + 3x^3 - 9x^2 + 2x = Ax(x-1)^2(x+2) + Cx^2(x-1)(x+2)$
$$= A(x^4 - 3x^2 + 2x) + C(x^4 + x^3 - 2x^2)$$
$$= (A+C)x^4 + Cx^3 - (3A + 2C)x^2 + 2Ax.$$

Equating the coefficients of each power of x, we find $A + C = 4$, $C = 3$, $3A + 2C = 9$, and $2A = 2$. From the last of these equations we see that $A = 1$. From the second equation we see that $C = 3$.

Now we can integrate:

$$\int \frac{4x^5 + 2x^4 - 5x^3 + x^2 + 8x - 4}{x^2(x-1)^2(x+2)} \, dx$$

$$= \int \left(4 + \frac{1}{x} - \frac{2}{x^2} + \frac{3}{x-1} + \frac{2}{(x-1)^2} - \frac{2}{x+2} \right) dx$$

$$= 4x + \ln|x| + \frac{2}{x} + 3\ln|x-1| - \frac{2}{x-1} - 2\ln|x+2| + c$$

$$= 4x + \frac{2}{x} - \frac{2}{x-1} + \ln\left| \frac{x(x-1)^3}{(x+2)^2} \right| + c.$$

Notice how we combine the ln terms, using $3\ln|x-1| = \ln|(x-1)^3|$ and $-2\ln|x+2| = \ln\frac{1}{|(x+2)^2|}$.

The third complication we may encounter is that it may not be possible to factor the denominator into linear factors. The quadratic polynomial $ax^2 + bx + c$ cannot be factored if $b^2 - 4ac < 0$ and can be factored into the product of two linear factors if $b^2 - 4ac \geq 0$ (cf. Theorem 1.3.4 and Equation 1.3.6). A quadratic polynomial that cannot be factored (i.e., for which $b^2 - 4ac < 0$) is called **irreducible** because it cannot be "reduced" into smaller pieces.

It is possible to prove that any polynomial of degree greater than 2 can be factored. This means that we can factor the denominator and then factor the factors and continue until we have written the denominator as a product of factors, each of which is linear or an irreducible quadratic. We have learned a few tricks to help us factor polynomials (cf. Theorem 1.3.7, p. 13, Example 1.3.10), but in practice it can be very difficult.

If the denominator contains an irreducible quadratic factor $ax^2 + bx + c$, we must look for a term of the form $\frac{Ax+B}{ax^2+bx+c}$ in the expression of the rational function as a sum of simpler terms.

As before, if $b^2 - 4ac < 0$ and $(ax^2 + bx + c)^m$ is the largest power of $ax^2 + bx + c$ that divides the denominator, then we look for an expression of the form

$$\frac{A_1 x + B_1}{ax^2 + bx + c} + \frac{A_2 x + B_2}{(ax^2 + bx + c)^2} + \cdots + \frac{A_m x + B_m}{(ax^2 + bx + c)^m}.$$

> When we factor the denominator, we group together all the copies of an irreducible quadratic factor $ax^2 + bx + c$ as a power $(ax^2 + bx + c)^m$. We have to be careful to include factors that are a constant times $ax^2 + bx + c$. So, for instance, $(2x^2 + 1)(4x^2 + 2)$ should be written as $2(2x^2 + 1)^2$, not considered as two different factors.

EXAMPLE 6 $\int \frac{3x^4 - 9x^3 + 14x^2 - 9x + 2}{(x-1)(x^2 - 2x + 2)^2}\, dx.$

Solution   First note that the rational function is already proper, so division is unnecessary. Next note that the quadratic factor in the denominator is irreducible, since $b^2 - 4ac = (-2)^2 - 4(1)2 = 4 - 8 = -4 < 0$. Try

$$\frac{3x^4 - 9x^3 + 14x^2 - 9x + 2}{(x-1)(x^2 - 2x + 2)^2} = \frac{A}{x-1} + \frac{Bx + C}{x^2 - 2x + 2} + \frac{Dx + E}{(x^2 - 2x + 2)^2}.$$

Clearing denominators, we find

$$3x^4 - 9x^3 + 14x^2 - 9x + 2$$
$$= A(x^2 - 2x + 2)^2 + (Bx + C)(x - 1)(x^2 - 2x + 2) + (Dx + E)(x - 1).$$

Letting $x = 1$, we find $1 = A(1)^2 = A$, so

$$3x^4 - 9x^3 + 14x^2 - 9x + 2 - (x^2 - 2x + 2)^2$$
$$= (Bx + C)(x^3 - 3x^2 + 4x - 2) + Dx^2 + Ex - Dx - E,$$
$$3x^4 - 9x^3 + 14x^2 - 9x + 2 - (x^4 + 4x^2 + 4 - 4x^3 + 4x^2 - 8x)$$
$$= Bx^4 + Cx^3 - 3Bx^3 - 3Cx^2 + 4Bx^2 + 4Cx - 2Bx - 2C + Dx^2 + Ex$$
$$\quad - Dx - E$$
$$2x^4 - 5x^3 + 6x^2 - x - 2$$
$$= Bx^4 + (-3B + C)x^3 + (4B - 3C + D)x^2 + (-2B + 4C + E - D)x$$
$$\quad + (-2C - E).$$

Equating the coefficients of each power of x, we find that we must solve

$$B = 2,$$
$$-3B + C = -5,$$
$$4B - 3C + D = 6,$$

The first of these five equations tells us that $B = 2$; using this the second gives $C = 1$; using this the third gives $D = 1$ and the fifth gives $E = 0$.

$$-2B + 4C - D + E = -1,$$
$$-2C - E = -2.$$

This is easy: $B = 2$, $C = 1$, $D = 1$, and $E = 0$. So

$$\int \frac{3x^4 - 9x^3 + 14x^2 - 9x + 2}{(x-1)(x^2 - 2x + 2)^2}\, dx$$

$$= \int \left(\frac{1}{x-1} + \frac{2x+1}{x^2 - 2x + 2} + \frac{x}{(x^2 - 2x + 2)^2} \right) dx.$$

At this point it helps to complete the square in the quadratic terms in the denominator: $x^2 - 2x + 2 = (x - 1)^2 + 1$. Letting $u = x - 1$, we transform the in-

Some calculators can be used to solve systems of equations like the ones we found in Example 6. In the example there were five equations, but only four variables, B, C, D, E, so we only need to use four of the equations. Sometimes it matters which equations are chosen—certain choices may not determine the solution.

TI-85 SIMULT 4 ENTER 1 ▼ 0 ▼ 0 ▼ 0 ▼ 2 F2(NEXT) (−) 3 ▼ 1 ▼ 0 ▼ 0 ▼ (−) 5 F2(NEXT) 4 ▼ (−) 3 ▼ 1 ▼ 0 ▼ 6 F2(NEXT) (−) 2 ▼ 4 ▼ (−) 1 ▼ 1 ▼ (−) 1 F5(SOLVE)	**SHARP EL9300** MENU 3(MATRIX) MENU B(EDIT) ▶ ENTER 4 ENTER 4 ENTER 1 ENTER −3 ENTER 4 ENTER −2 ENTER 0 ENTER 1 ENTER −3 ENTER 4 ENTER 0 ENTER 0 ENTER 1 ENTER −1 ENTER 0 ENTER 0 ENTER 0 ENTER 1 ENTER QUIT MENU ▼ ENTER 4 ENTER 1 ENTER 2 ENTER −5 ENTER 6 ENTER −1 ENTER QUIT (MAT A ÷ MAT A ÷ MAT A) × MAT B ENTER
CASIO f_x-7700GB MODE 0(MATRIX) F1(A) F6 F1(DIM) 4 EXE 4 EXE 1 EXE 0 EXE 0 EXE 0 EXE −3 EXE 1 EXE 0 EXE 0 EXE 4 EXE −3 EXE 1 EXE 0 EXE −2 EXE 4 EXE −1 EXE 1 EXE PRE F2(B) F6 F1(DIM) 4 EXE 1 EXE 2 EXE −5 EXE 6 EXE −1 EXE PRE F1(A) F4(A^{-1}) F1(C ▶ A) PRE F5(×)	**HP 48SX** ↱ MATRIX 1 ENTER 0 ENTER 0 ENTER 0 ENTER ▼ 3 +/− ENTER 1 ENTER 0 ENTER 0 ENTER ▼ 4 ENTER 3 +/− ENTER 1 ENTER 0 ENTER ▼ 2 +/− ENTER 4 ENTER 1 +/− ENTER 1 ENTER ENTER 1/x ↱ MATRIX 2 ENTER ▼ 5 +/− ENTER ▼ 6 ENTER ▼ 1 +/− ENTER ENTER ×

Notice how we filled in the missing coefficients with 0; for example, the equation $B = 2$ becomes $1 \times B + 0 \times C + 0 \times D + 0 \times E = 2$, and so on.

```
[ 2]
[ 1]
[ 1]
[ 0]
```

tegral into

$$\int \left(\frac{1}{u} + \frac{2u+3}{u^2+1} + \frac{u+1}{(u^2+1)^2} \right) du = \int \frac{1}{u}\,du + \int \frac{2u}{u^2+1}\,du + \int \frac{3}{u^2+1}\,du$$

$$+ \int \frac{u}{(u^2+1)^2}\,du + \int \frac{1}{(u^2+1)^2}\,du.$$

We make the substitutions $v = u^2 + 1$, $dv = 2u\,du$ in the second and fourth integrals in the last expression, and $u = \tan\theta$, $du = \sec^2\theta\,d\theta$ in the fifth, giving

$$\int \frac{1}{u}\,du + \int \frac{1}{v}\,dv + 3\int \frac{1}{1+u^2}\,du + \frac{1}{2}\int \frac{1}{v^2}\,dv + \int \frac{\sec^2\theta}{(\sec^2\theta)^2}\,d\theta$$

$$= \ln|u| + \ln|v| + 3\arctan u - \frac{1}{2}\frac{1}{v} + \int \cos^2\theta\,d\theta$$

$$= \ln|u| + \ln|u^2+1| + 3\arctan u - \frac{1}{2}\frac{1}{1+u^2} + \frac{1}{2}\cos\theta\sin\theta + \frac{\theta}{2} + c.$$

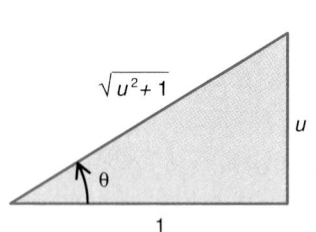

Figure 8.5.1

We use the triangle in Figure 8.5.1, in which $u = \tan\theta$, to find that $\cos\theta\sin\theta = \frac{u}{u^2+1}$. With this and the fact that $\theta = \arctan u$ the above expression becomes

$$\ln|u^3+u| + 3\arctan u - \frac{1}{2}\frac{1}{1+u^2} + \frac{u}{2(u^2+1)} + \frac{1}{2}\arctan u + c$$

$$= \ln|u^3+u| + \frac{7}{2}\arctan u + \frac{1}{2}\frac{u-1}{u^2+1} + c$$

$$= \ln|(x-1)^3 + x - 1| + \frac{7}{2}\arctan(x-1) + \frac{1}{2}\frac{x-2}{(x-1)^2+1} + c.$$

The individual steps in this solution are not too difficult, but the whole procedure is quite complicated, and of course there is the risk that a small error early in the calculation will go undetected and invalidate everything that follows.

Summary 8.5.1

Partial Fractions

This is a procedure for integrating any rational function, that is, a function that is the quotient of two polynomials.

1. If the rational function is not proper, perform a long division and express it as the sum of a polynomial and a proper rational function.

2. Factor the denominator as a product of linear factors and/or irreducible quadratic factors (i.e., factors of the form $q(x) = ax^2 + bx + c$ with $b^2 - 4ac < 0$). If the denominator has integer coefficients, use Theorem 1.3.7 to reduce the work required.

3. Express the proper rational function as a sum of terms of the following two types: If $(x-a)^m$ is the largest power of $(x-a)$ dividing the denominator, include terms of the form

$$\frac{A_1}{x-a} + \frac{A_2}{(x-a)^2} + \ldots + \frac{A_m}{(x-a)^m};$$

if $q(x) = ax^2 + bx + c$, with $b^2 - 4ac < 0$, and $q(x)^m$ is the largest power of $q(x)$

that divides the denominator, include terms of the form

$$\frac{B_1 x + C_1}{q(x)} + \frac{B_2 x + C_2}{q(x)^2} + \ldots + \frac{B_m x + C_m}{q(x)^m}.$$

4. Clear denominators and solve for all the unknown constants. You may want to simplify the work by first substituting for x the root of each linear factor after clearing denominators.

5. Integrate the resulting expression, one term at a time:

$$\int \frac{1}{x - a} \, dx = \ln|x - a| + c$$

$$\int \frac{1}{(x - a)^k} \, dx = -\frac{1}{k - 1} \frac{1}{(x - a)^{k-1}} + c, \qquad \text{if } k > 1.$$

For the quadratic factors of the denominator, complete the square. The resulting integrals can be broken into the following forms:

$$\int \frac{u}{(u^2 + a^2)^k} \, du = \frac{1}{2} \int \frac{dv}{v^k}, \qquad \text{with } v = u^2 + a^2,$$

$$= \begin{cases} -\frac{1}{2(k-1)} \frac{1}{(u^2+a^2)^{k-1}}, & \text{if } k > 1, \\ \frac{1}{2} \ln|u^2 + a^2| + c, & \text{if } k = 1. \end{cases}$$

$$\int \frac{1}{(u^2 + a^2)^k} \, du,$$

which can be integrated by letting $u = a \tan \theta$, transforming the integral into $a^{1-2k} \int \sec^{2-k} \theta \, d\theta$, which can be integrated by the methods of Theorem 8.1.3 (p. 565) or Theorem 8.2.5(ii) (p. 574), depending on the sign of $2 - k$.

It can be shown that it is possible to express every proper rational function in the form outlined in Step 3 above. So in principle the technique of partial fractions will always work. In practice the stumbling block is often Step 2, factoring the denominator. Apart from that the technique is straightforward, though it can easily involve lengthy calculations, as the examples show. Sometimes we will be given the factorization of the denominator, and other times it will be possible to use Theorem 1.3.7 to help us guess an integer root. Long division will then enable us to begin the factorization.

Exercises 8.5

Integrate the following.

1. $\int \frac{1}{(x-1)(x+1)} \, dx$

2. $\int \frac{x}{(x-1)(x+1)} \, dx$

3. $\int \frac{x+1}{(x-1)(x+1)} \, dx$

4. $\int \frac{2x+3}{(x-1)(x+1)} \, dx$

5. $\int \frac{x+2}{(x-2)(x+3)} \, dx$

6. $\int \frac{3x-1}{x(x+4)} \, dx$

7. $\int \frac{x^2-4}{x(x+1)} \, dx$

8. $\int \frac{3x^2+x-1}{(x+2)(x+4)} \, dx$

9. $\int \frac{x^3+3x-1}{x^2(x+1)} \, dx$

10. $\int \frac{x^3+1}{(x+2)^2(x-1)} \, dx$

Evaluate the following integrals.

11. $\int \frac{x^2-x+1}{x(x^2+1)} \, dx$

12. $\int \frac{8-x}{x^3+4x} \, dx$

13. $\int \frac{1}{x^3+x} dx$

14. $\int \frac{x^2-2}{(x-1)(x^2+2x+2)} dx$

15. $\int \frac{x^3}{(x+2)(x^2-4x+8)} dx$

16. $\int \frac{3x^3+x-3}{2x^3+4x^2+4x} dx$

17. $\int \frac{4x^3-x^2+18x-4}{(x^2+4)^2} dx$

18. $\int \frac{6x^2-2x-4}{x^3-4x} dx$

19. $\int \frac{6x^2-4x-1}{(x+1)(2x^2-4x+3)} dx$

20. $\int \frac{x^4-2x^3}{x^5-4x^3} dx$

21. $\int 1/(x^4+2x^2+1) dx$

22. $\int (4x^3+2x^2+2x)/(x^4-1) dx$

23. $\int 1/(x^3-4x) dx$

24. $\int (2x^4+4x^2+2)/(x^2+1)^2 dx$

25. $\int (2x-6)/(x^3-9x) dx$

26. $\int (2x^2-x+1)/(x^3-x^2-x+1) dx$

27. $\int x^4/(x^3-x) dx$

28. $\int 3/(x^3-1) dx$

29. $\int (x^3+3x^2+2x)/(2x^2+6x+4) dx$

30. $\int x/((x-1)^3-4(x-1)) dx$

II

*31. Find $\int \frac{\cos\theta}{\sin^2\theta-2\sin\theta-8} d\theta$.

*32. Find $\int (1+e^t)/(1+e^{2t}) dt$.

**33. Find $\int \frac{1}{x^4+1} dx$. (*Hint:* Factor $x^4+1 = (x^2+rx+1)(x^2+sx+1)$.)

*34. Evaluate $\int \frac{x}{x^8+1} dx$.

*35. Evaluate $\int \frac{x^2+x+1}{(x^2+1)^2} dx$.

*36. Find an example to demonstrate that if there is a repeated factor in the denominator, then it may not be sufficient to consider the simple kind of partial fractions with a sum of constants divided by the linear and irreducible quadratic factors of the denominator.

Drill

37. $\int \frac{1}{x^2+4x+5} dx$

38. $\int \frac{1}{x^2+4x+3} dx$

39. $\int e^x/(e^x+3e^{x/2}+2) dx$

40. $\int \frac{\sin x}{\cos^2 x-1} dx$

41. $\int x\sin^3 x\, dx$

42. $\int \frac{\tan\theta}{2-\cos\theta} d\theta$

43. $\int x\arcsin x\, dx$

44. $\int e^{2x}\cos x\, dx$

45. $\int x^2\arctan x\, dx$

46. $\int x\arctan x\, dx$

47. $\int \ln(x^2+1) dx$

48. $\int \frac{1}{x^2}\arctan x\, dx$

49. $\int x/(1+\exp(x^2)) dx$

50. $\int (x^4+x^3+x^2+x+1)/(x^2+1)^2 dx$

51. $\int 1/(x^4+8x^2-9) dx$

52. $\int \cos(x)\cos(2x)\sin(2x) dx$

53. $\int e^x\sin e^{x/2} dx$

54. $\int \frac{1}{x}\sin(\ln(x^2)) dx$

55. $\int \sqrt{x}\ln x\, dx$

56. $\int \sqrt{x}e^{\sqrt{x}} dx$

III

*57. Consider the integral $\int \frac{1}{x^3-2x^2-x+1} dx$. To apply partial fractions, it is necessary to factor the denominator, but using the Rational Root Theorem (Theorem 1.3.7, p. 13), we see that there are no rational roots.

(i) Use a graphics calculator or computer to find the roots of x^3-2x^2-x+1 with as much accuracy as possible. Use them to factor $x^3-2x^2-x+1 = (x-a)(x-b)(x-c)$, and use partial fractions to integrate.

(ii) Use your answer to evaluate $\int_1^2 \frac{1}{x^3-2x^2-x+1} dx$.

(iii) Estimate this definite integral by Simpson's Rule S_n with $n=10$ and $n=20$.

Using Exercise 57 as a guide, evaluate the following.

58. $\int_{-1}^0 \frac{1}{x^3-4x+2} dx$ **59.** $\int \frac{x+2}{x^3-3x+1} dx$ **60.** $\int \frac{1}{x^3+x+1} dx$

8.6 Rationalizing Substitutions

FRACTIONAL POWERS
tan $(\theta/2)$ SUBSTITUTION

In Section 8.5 we learned how to integrate any rational function, so now it makes sense to try to convert other integrals to integrals of rational functions.

In this short section we learn two techniques for doing this in particular cases. Each of them can be used to convert a particular kind of integral into the integral of a rational function, which in principle we know how to evaluate. In both cases, however, the practical problem is that the rational function is likely to be quite complicated and difficult to integrate. Specifically, the denominator may have a fairly high degree and be difficult to factor.

EXAMPLE 1   Transform $\int \frac{x^{1/3}}{x^{3/4}-1} \, dx$ into the integral of a rational function.

Solution   We need a substitution that will remove the fractional exponents. If $x = u^m$, then $x^{1/3} = u^{m/3}$, $x^{3/4} = u^{3m/4}$, and these powers of u will be integer powers if m is divisible by both 3 and 4. The smallest such m is $m = 12$; letting $x = u^{12}$, $dx = 12u^{11} \, du$, we find

$$\int \frac{x^{1/3}}{x^{3/4}-1} \, dx = 12 \int \frac{u^4}{u^9-1} u^{11} \, du = 12 \int \frac{u^{15}}{u^9-1} \, du.$$

> This sort of technique will always transform integrands that are rational functions of fractional powers of x into rational functions of some new variable u. The idea is to let $x = u^m$, where m is divisible by the denominators of all the exponents in the original integrand.
>
> However, the resulting integral may be difficult to evaluate. (Try to find the factorization of $u^9 - 1$ into linear and quadratic factors. . .).

Another type of integral that can be converted to the integral of a rational function is one whose integrand is a rational function of trigonometric functions, that is, the quotient of two "polynomials" in the *two* "variables" $\sin \theta$ and $\cos \theta$. Such integrals can often be simplified by other methods (substitution, integration by parts, etc.), but if all else fails, there is a substitution that always works. It is to let $x = \tan \frac{\theta}{2}$.

In this case, $\frac{\theta}{2} = \arctan x$, so $\theta = 2 \arctan x$ and $d\theta = \frac{2}{1+x^2} \, dx$. From the triangle shown in Figure 8.6.1 we see that $\sin \frac{\theta}{2} = \frac{x}{\sqrt{1+x^2}}$, $\cos \frac{\theta}{2} = \frac{1}{\sqrt{1+x^2}}$, and from this we find

$$\sin \theta = 2 \cos \frac{\theta}{2} \sin \frac{\theta}{2} = \frac{2x}{1+x^2},$$

$$\cos \theta = \cos^2 \frac{\theta}{2} - \sin^2 \frac{\theta}{2} = \frac{1-x^2}{1+x^2}.$$

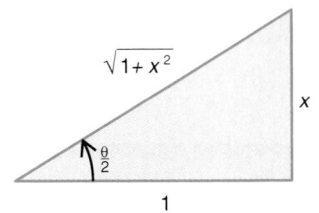

Figure 8.6.1

EXAMPLE 2   Find $\int \frac{d\theta}{\cos \theta + \sin \theta}$.

Solution   Letting $x = \tan \frac{\theta}{2}$, we have $d\theta = \frac{2}{1+x^2} \, dx$, so

$$\int \frac{d\theta}{\cos \theta + \sin \theta} = 2 \int \frac{1}{\frac{1-x^2}{1+x^2} + \frac{2x}{1+x^2}} \frac{1}{1+x^2} \, dx$$

$$= 2 \int \frac{1}{1+2x-x^2} \, dx.$$

To factor $x^2 - 2x - 1$, we find its roots using the Quadratic Formula (Theorem 1.3.4):

$$\frac{2 \pm \sqrt{4+4}}{2} = \frac{2 \pm 2\sqrt{2}}{2} = 1 \pm \sqrt{2},$$

so $x^2 - 2x - 1 = \left(x - (1 + \sqrt{2})\right)\left(x - (1 - \sqrt{2})\right).$

We apply partial fractions:

$$\frac{1}{1 + 2x - x^2} = \frac{A}{x - (1 + \sqrt{2})} + \frac{B}{x - (1 - \sqrt{2})}.$$

Clearing denominators, being careful about signs, we get

$$-1 = Ax - A(1 - \sqrt{2}) + Bx - B(1 + \sqrt{2})$$
$$= (A + B)x - (A + B) + (A - B)\sqrt{2}.$$

So $A + B = 0$ and $(A + B) - (A - B)\sqrt{2} = 1$. We find that $A = -1/2\sqrt{2}$ and $B = 1/2\sqrt{2}$. From this we can integrate:

$$\int \frac{d\theta}{\cos\theta + \sin\theta} = 2 \int \frac{1}{1 + 2x - x^2}\, dx$$

$$= -\frac{1}{\sqrt{2}} \int \frac{1}{x - (1 + \sqrt{2})}\, dx + \frac{1}{\sqrt{2}} \int \frac{1}{x - (1 - \sqrt{2})}\, dx$$

$$= \frac{1}{\sqrt{2}}\left(-\ln|x - (1 + \sqrt{2})| + \ln|x - (1 - \sqrt{2})|\right) + c$$

$$= \frac{1}{\sqrt{2}} \ln\left|\frac{x - 1 + \sqrt{2}}{x - 1 - \sqrt{2}}\right| + c$$

$$= \frac{1}{\sqrt{2}} \ln\left|\frac{\tan\frac{\theta}{2} - 1 + \sqrt{2}}{\tan\frac{\theta}{2} - 1 - \sqrt{2}}\right| + c.$$

Summary 8.6.1   **The $\tan\frac{\theta}{2}$ Substitution**

$$x = \tan\frac{\theta}{2}, \qquad d\theta = \frac{2}{1+x^2}\, dx,$$

$$\sin\theta = \frac{2x}{1+x^2}, \qquad \cos\theta = \frac{1-x^2}{1+x^2}.$$

The substitution $x = \tan\frac{\theta}{2}$ always works in principle, but since the resulting rational function may be difficult to integrate, this approach should not be used before attempting to find an easier technique. It should be regarded as a last resort when everything else has been tried and rejected.

Exercises 8.6

I

Integrate each of the following.

1. $\int x^{1/3}/(x^{1/3}+1)\,dx$

2. $\int x^{1/4}/(x^{1/4}+4)\,dx$

3. $\int x^{1/2}/(x^{1/3}+1)\,dx$

4. $\int x^{1/3}/(x^{1/2}-1)\,dx$

5. $\int \frac{\sqrt{x}}{x+1}\,dx$

6. $\int \frac{\sqrt{x+1}}{x}\,dx$

7. $\int \frac{x-1}{\sqrt{x+1}}\,dx$

8. $\int \frac{1}{x+x^{3/4}}\,dx$

9. $\int x^{1/6}/(x^{2/3}+4x^{1/3})\,dx$

10. $\int \frac{1}{x}(x^2+1)^{1/3}\,dx$

Evaluate the following using the $\tan \frac{\theta}{2}$ substitution.

11. $\int \frac{1}{\sin\theta+1}\,d\theta$

12. $\int \frac{1}{(1-\cos\theta)^2}\,d\theta$

13. $\int \sec\theta\,d\theta$

14. $\int \frac{1}{3\cos\theta-4\sin\theta}\,d\theta$

15. $\int \frac{1}{1+\cos\theta}\,d\theta$

16. $\int \frac{\cos\theta}{1+\cos\theta}\,d\theta$

17. $\int \frac{\cos\theta}{\sin\theta(1+\cos\theta)}\,d\theta$

18. $\int \csc\theta\,d\theta$

19. $\int \frac{1}{1+\sin\theta+\cos\theta}\,d\theta$

20. $\int \frac{1}{4+\sin\theta+\cos\theta}\,d\theta$

II

*21. Suppose a and b are positive numbers with $a > b$. Find $\int \frac{1}{a+b\sin\theta}\,d\theta$.

*22. Suppose a and b are positive numbers. Find $\int \frac{1}{a\cos\theta+b\sin\theta}\,d\theta$.

**23. Find a substitution for integrating rational functions of the hyperbolic functions $\sinh t$ and $\cosh t$. Show that it converts any such integral into the integral of a rational function.

*24. Integrate $\int \frac{1}{1+\sinh t}\,dt$.

Drill

25. $\int \frac{x}{1+\sqrt{x}}\,dx$

26. $\int (x^{2/7}-3x^{-1/9})\,dx$

27. $\int \frac{1}{\sin(2\theta)}\,d\theta$

28. $\int \frac{1}{1+\sin(5\theta)}\,d\theta$

29. $\int \frac{\sin(2\theta)}{\sqrt{1+\sin\theta}}\,d\theta$

30. $\int x(x+8)^{-1/3}\,dx$

31. $\int \cos^2 x\sin x\sqrt{1+2\cos x}\,dx$

32. $\int x\sin(2x-3)\,dx$

33. $\int e^x(4+e^{x/2})^{1/3}\,dx$

34. $\int (1+\sin x)^{1/3}\cot x\,dx$

8.7 Additional Applications

AREA
ARCLENGTH
SURFACE AREA

In this chapter we have learned techniques for evaluating a number of types of integrals. Naturally, this extends the number of practical problems we can approach.

In fact, in Chapter 6 there were several instances in which interesting problems were left unsolved simply because we could not perform the necessary integrations. One place where this was particularly apparent was Section 6.9, where we developed the theory for finding the arclength of a curve but found that even for a simple parabola the resulting integral was beyond our capabilities.

The problem also arose in Section 6.1, where we had to use a complicated string of guesswork to evaluate the integral needed for finding the area of a circle, and in Section 6.7 there were regions whose centroids we could not find because the integrals involved were not ones we knew how to evaluate.

In this brief section we solve some of these problems. There is nothing really new here; we use techniques from the last chapter to solve some applied problem and techniques from this chapter to evaluate the necessary integrals.

EXAMPLE 1 **(cf. Example 6.1.3)** Find the area of the region inside a circle of radius r.

Solution   We did this example in Chapter 6. At the time it was fairly tricky, but now it is easy.

The area we want is twice the area under the upper semicircle $y = \sqrt{r^2 - x^2}$, so it equals $2 \int_{-r}^{r} \sqrt{r^2 - x^2}\, dx$. This integral cries out for a trigonometric substitution. We let $x = r \sin \theta$, so $dx = r \cos \theta\, d\theta$, and

$$2 \int_{-r}^{r} \sqrt{r^2 - x^2}\, dx = 2 \int_{-\pi/2}^{\pi/2} \sqrt{r^2 - r^2 \sin^2 \theta}\ r \cos \theta\, d\theta$$

$$= 2r^2 \int_{-\pi/2}^{\pi/2} \sqrt{\cos^2 \theta}\ \cos \theta\, d\theta = 2r^2 \int_{-\pi/2}^{\pi/2} \cos^2 \theta\, d\theta$$

$$= 2r^2 \frac{1}{2} (\theta + \cos \theta \sin \theta) \Big|_{-\pi/2}^{\pi/2}$$

$$= r^2 \left(\frac{\pi}{2} - \left(-\frac{\pi}{2} \right) + (0)(1) - (0)(-1) \right)$$

$$= \pi r^2.$$

This agrees with the result we found in Example 6.1.3 by guessing the integral in Example 6.1.2. It is comforting to realize that now that we know about trigonometric substitution, there is a clear-cut method for approaching this problem.

EXAMPLE 2   Find the arclength of the parabola $y = x^2$ between $x = 0$ and $x = 1$.

Solution   We know from Definition 6.9.1 that the arclength equals

$$\int ds = \int_0^1 \sqrt{1 + (y')^2}\, dx = \int_0^1 \sqrt{1 + 4x^2}\, dx.$$

Again a trigonometric substitution is called for; letting $2x = \tan \theta$, so $dx = \frac{1}{2} \sec^2 \theta\, d\theta$, we find that the arclength is

$$\int ds = \int_0^1 \sqrt{1 + 4x^2}\, dx = \frac{1}{2} \int_0^{\arctan 2} \sqrt{1 + \tan^2 \theta}\ \sec^2 \theta\, d\theta$$

$$= \frac{1}{2} \int_0^{\arctan 2} \sec^3 \theta\, d\theta.$$

> We found the new upper limit of integration by realizing that it had to be the value of θ for which $\tan \theta = 2$, since $\tan \theta = 2x$ and $2x = 2$ at the original upper limit of integration $x = 1$.

This integral can be evaluated by using the reduction formula of Theorem 8.2.5(ii) for integrals of powers of sec x. We find that

$$\int_0^{\arctan 2} \sec^3 \theta\, d\theta = \frac{1}{2} \sec \theta \tan \theta \Big|_0^{\arctan 2} + \frac{1}{2} \int_0^{\arctan 2} \sec \theta\, d\theta$$

$$= \frac{1}{2} \Big(\sec(\arctan 2)(2) - \sec(0)(0) \Big) + \frac{1}{2} \ln | \sec \theta + \tan \theta| \Big|_0^{\arctan 2}$$

$$= \sec(\arctan 2) + \frac{1}{2} \Big(\ln |\sec(\arctan 2) + 2| - \ln|\sec(0) + \tan(0)| \Big)$$

$$= \sec(\arctan 2) + \frac{1}{2} \Big(\ln|\sec(\arctan 2) + 2| - \ln(1) \Big).$$

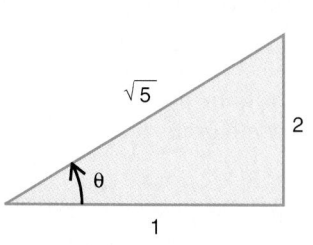

Figure 8.7.1

When $\theta = \arctan 2$, that is, $\tan \theta = 2$, we see from Figure 8.7.1 that $\sec^2 \theta = 1 + \tan^2 \theta = 5$, so $\sec \theta = \sqrt{5}$. The arclength equals

$$\int ds = \frac{1}{2}\left(\sec(\arctan 2) + \frac{1}{2}\ln|\sec(\arctan 2) + 2|\right) = \frac{\sqrt{5}}{2} + \frac{1}{4}\ln|2 + \sqrt{5}|.$$

The numerical value of this expression is approximately 1.479.

EXAMPLE 3   Find the centroid of the region enclosed between the x-axis and the curve $y = \sin x$ between the lines $x = 0$ and $x = \frac{\pi}{2}$, shown in Figure 8.7.2.

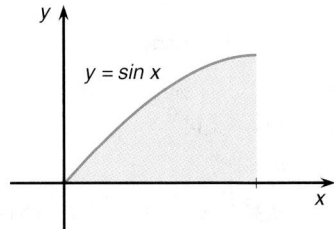

$y = \sin x$

Figure 8.7.2

Solution   Assuming that the density is $\rho = 1$, we find that the total weight is

$$M = \int_0^{\pi/2} \rho \sin x \, dx = -\cos x \Big|_0^{\pi/2} = 0 - (-1) = 1.$$

Using Theorem 6.7.9 (p. 472), we can find the moment about the x-axis:

$$M_x = \frac{1}{2}\int_0^{\pi/2} \rho(\sin^2 x - 0)\,dx = \frac{1}{2}\int_0^{\pi/2} \sin^2 x \, dx$$

$$= \frac{1}{4}(x - \cos x \sin x)\Big|_0^{\pi/2} = \frac{1}{4}\left(\frac{\pi}{2} - 0 - (0)(1) + (1)(0)\right)$$

$$= \frac{\pi}{8}.$$

The moment about the y-axis is

$$M_y = \int_0^{\pi/2} x\rho \sin x \, dx = \int_0^{\pi/2} x \sin x \, dx.$$

To evaluate this, we integrate by parts and find

$$M_y = \int_0^{\pi/2} x \sin x \, dx = -(x\cos x)\Big|_0^{\pi/2} + \int_0^{\pi/2} \cos x \, dx$$

$$\begin{array}{|ll|}\hline u = x, & dv = \sin x\, dx \\ du = dx, & v = -\cos x \\ \hline \end{array}$$

$$= -\left(\frac{\pi}{2}(0) - (0)(1)\right) + \sin x\Big|_0^{\pi/2} = 1 - 0$$

$$= 1.$$

Putting everything together, we find that the centroid is

$$(\bar{x}, \bar{y}) = (M_y/M, M_x/M) = \left(1, \frac{\pi}{8}\right) \approx (1, 0.3927)$$

(see Figure 8.7.3).

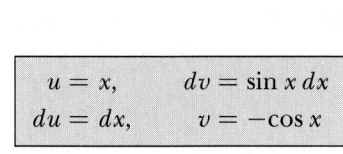

$y = \sin x$

Figure 8.7.3

EXAMPLE 4   Find the arclength of the curve $y = e^x$ between $x = \ln(\sqrt{3})$ and $x = \ln(\sqrt{8})$.

Solution   The arclength equals

$$\int ds = \int_{\ln(\sqrt{3})}^{\ln(\sqrt{8})} \sqrt{1 + e^{2x}} \, dx = \int_{\ln(\sqrt{3})}^{\ln(\sqrt{8})} \sqrt{1 + e^{2x}} \frac{1}{e^{2x}}e^{2x}\,dx$$

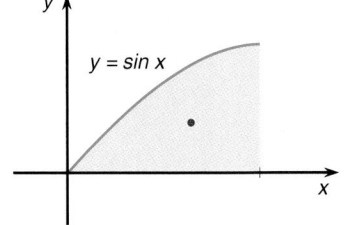

$$\begin{array}{|l|}\hline u = 1 + e^{2x} \\ du = 2e^{2x}\, dx \\ \hline \end{array} \qquad = \frac{1}{2}\int_4^9 \sqrt{u}\,\frac{1}{u-1}\,du$$

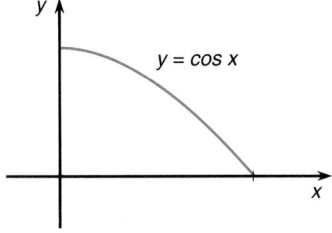

$$= \int_2^3 v \frac{1}{v^2 - 1} v \, dv = \int_2^3 \frac{v^2}{v^2 - 1} \, dv$$

$$= \int_2^3 \left(1 + \frac{1}{v^2 - 1}\right) dv$$

$$= \int_2^3 \left(1 + \frac{1}{2}\left(\frac{1}{v - 1} - \frac{1}{v + 1}\right)\right) dv$$

$$= \left(v + \frac{1}{2}\ln|v - 1| - \frac{1}{2}\ln|v + 1|\right)\Bigg|_2^3$$

$$= 1 + \frac{1}{2}(\ln 2 - \ln 4 - \ln 1 + \ln 3)$$

$\ln 2 - \ln 4 - \ln 1 + \ln 3 =$
$\ln \frac{(2)(3)}{(4)(1)} = \ln \frac{3}{2}$

$$= 1 + \frac{1}{2}\ln\frac{3}{2} \approx 1.2027.$$

In Example 4 we made two successive substitutions and then had to do a (simple) partial fractions calculation. To be able to make the first substitution $u = 1 + e^{2x}$, we multiplied and divided the integrand by e^{2x}, in order that the integrand would contain $e^{2x} dx$. This was needed, along with the constant $\frac{1}{2}$, to form du.

EXAMPLE 5 Find the surface area of the surface obtained by revolving about the x-axis the part of the curve $y = \cos x$ lying between $x = 0$ and $x = \frac{\pi}{2}$. See Figures 8.7.4(i) and (ii).

Solution The surface area is given by

$$S = 2\pi \int \cos x \, ds = 2\pi \int_0^{\pi/2} \cos x \sqrt{1 + (-\sin x)^2} \, dx$$

$$= 2\pi \int_0^{\pi/2} \cos x \sqrt{1 + \sin^2 x} \, dx$$

$$\boxed{\begin{array}{l} u = \sin x \\ du = \cos x \, dx \end{array}} \quad = 2\pi \int_0^1 \sqrt{1 + u^2} \, du$$

$$\boxed{\begin{array}{l} u = \tan\theta \\ du = \sec^2\theta \, d\theta \end{array}} \quad = 2\pi \int_0^{\pi/4} \sqrt{1 + \tan^2\theta} \sec^2\theta \, d\theta$$

$$= 2\pi \int_0^{\pi/4} \sec^3\theta \, d\theta$$

$$= \pi\left(\sec\theta\tan\theta\Big|_0^{\pi/4} + \int_0^{\pi/4} \sec\theta \, d\theta\right)$$

$$= \pi\left(\sqrt{2}(1) - (1)(0) + \ln|\sec\theta + \tan\theta|\Big|_0^{\pi/4}\right)$$

$$= \pi\left(\sqrt{2} + \ln|\sqrt{2} + 1| - \ln|1 + 0|\right)$$

$$= \pi\left(\sqrt{2} + \ln(1 + \sqrt{2})\right)$$

$$\approx 7.2118.$$

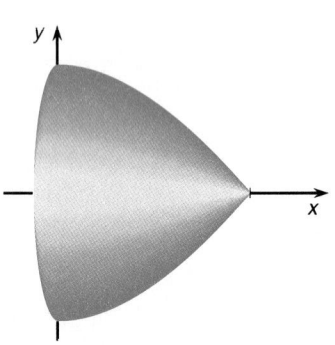

y
y = cos x
x

Figure 8.7.4(i)

y
x

Figure 8.7.4(ii)

Here we needed two substitutions, and then we had to integrate $\sec^3\theta$, which is beginning to be familiar, using the reduction formula of Theorem 8.2.5(ii).

Exercises 8.7

I

Find the area of the region lying between each of the following curves and the x-axis over the specified interval.

1. $y = x \sin x$, $x \in [0, \pi]$

2. $y = x \cos x$, $x \in \left[-\frac{\pi}{4}, \frac{\pi}{4}\right]$

3. $y = \tan^2 x \sec x$, $x \in \left[0, \frac{\pi}{4}\right]$

4. $y = \sec x \sin^2 x$, $x \in \left[-\frac{\pi}{4}, 0\right]$

5. $y = (x + 1)/(x^2 - 4)$, $x \in [-1, 1]$

6. $y = e^x/(e^{2x} - 1)$, $x \in [-2, -1]$

7. $y = xe^x$, $x \in [-1, 1]$

8. $y = \cos^4 x$, $x \in [0, \pi]$

9. $y = \ln x$, $x \in [1, e]$

10. $y = 1/(3x^2 + 6x + 15)$, $x \in [3, 4]$

Find the arclength of each of the following curves.

11. $y = 3x^2$, $1 \le x \le 2$

12. $y = x^2 - 2x$, $0 \le x \le 1$

13. $y = \sqrt{2x}$, $2 \le x \le 8$

14. $y = \sqrt{x + 1}$, $0 \le x \le 8$

15. $y = \sqrt{3x + 1}$, $1 \le x \le 8$

16. $y = \ln(\sin x)$, $\frac{\pi}{6} \le x \le \frac{\pi}{3}$

17. $y = \ln(\cos x)$, $0 \le x \le \frac{\pi}{6}$

18. $y = \frac{1}{5}\ln\left(\sin(5x)\right)$, $\frac{\pi}{20} \le x \le \frac{\pi}{10}$

19. $y = \ln x$, $1 \le x \le 4$

20. $y = 3 + \ln(x^4)$, $1 \le x \le 2$

II

21. Find the arclength of the parabola $y = x^2$ between the points $(0, 0)$ and $(2, 4)$.

22. In Example 6.10.8 (p. 500) we used Simpson's Rule to estimate the integral obtained in figuring out the arclength of the curve $y = \frac{1}{2}x^2$ from $x = 0$ to $x = 2$. Evaluate this integral exactly, and compare with the value given in Example 6.10.8.

*23. Find the arclength of the parabola $y = x^2$ from $x = 0$ to $x = a$, for any $a > 0$.

*24. (i) Find the arclength of the parabola $y = x^2$ from $x = -a$ to $x = 0$, for any $a > 0$.

  (ii) Find the arclength of the parabola $y = x^2$ from $x = a$ to $x = b$, for any $a < b$.

  (iii) Use properties of trigonometric functions and the natural logarithm to show that the answer to part (i) equals the answer to Exercise 23, some-

thing that the geometrical situation shows us must be true.

*25. Find the arclength of the parabola $y = kx^2$ from $x = a$ to $x = b$, for any $a < b$.

*26. Consider the problem of finding the arclength of the parabola $y = 2x^2 + 8x + 10$ between $x = 0$ and $x = 1$.

  (i) Solve this problem by evaluating the appropriate integral.

  (ii) Solve the problem by using Exercise 25 without evaluating any integrals. (*Hint:* Draw a picture.)

*27. Find the arclength of the curve $y = 1 + \frac{1}{2}\ln\left(\cos(2x)\right)$, for x between $x = \frac{\pi}{8}$ and $x = \frac{\pi}{6}$.

*28. Find the arclength of the curve $y = e^{4x}$, for $0 \le x \le \ln 2$.

*29. (i) Find the arclength of the curve $y = \ln x$, for x between $x = 1$ and $x = e$.

  (ii) For the curve described in part (i), find the length of the line segment joining its endpoints, and compare with the arclength of the curve.

*30. (i) Find the arclength of the curve $y = e^{3x}$, for x between $x = 0$ and $x = 1$.

  (ii) Compare the length of the curve with the length of the straight line segment joining its endpoints.

*31. Find the surface area of the surface of revolution obtained by revolving about the x-axis the curve $y = \sin x$, for $x \in \left[0, \frac{\pi}{3}\right]$.

*32. Find the surface area of the surface of revolution obtained by revolving about the x-axis the curve $y = \sin\frac{x}{2}$, for $x \in [0, \pi]$.

*33. A *torus* (doughnut shape) is obtained by revolving about the x-axis the circle of radius 1 with center at $(0, 2)$. See Figure 8.7.5. (The equation of the circle is $x^2 + (y - 2)^2 = 1$.) Find the surface area of the surface. (*Hint:* Work separately with the upper and lower semicircles.)

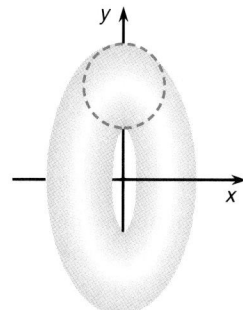

Figure 8.7.5

*34. Suppose $R > r > 0$. A torus is obtained by revolving about the x-axis the circle of radius r with center at $(0, R)$. (The equation of the circle is $x^2 + (y - R)^2 = r^2$.) Find the surface area of the surface.

35. (i) Find the centroid of the region between $y = \sin x$ and the x-axis and between $x = 0$ and $x = \frac{\pi}{4}$.
    (ii) Compare this with the centroid of the triangular

region whose vertices are the "corners" of the region in part (i).

*36. Find the centroid of the region between $y = \cos x$ and the x-axis and between $x = 0$ and $x = \frac{\pi}{3}$.

*37. Find the centroid of the region between $y = x \sin x$ and the x-axis and between $x = 0$ and $x = \frac{\pi}{2}$.

*38. Find the centroid of the region between $y = x^2 \sin x$ and the x-axis and between $x = 0$ and $x = \frac{\pi}{2}$.

8.8 Improper Integrals

DIVERGE
CONVERGE

In this section we consider what happens when we try to integrate a function that is not bounded. The integral of such a function does not exist, but under certain circumstances it is possible to think of it as a limit of integrals that do exist. In a similar way it is possible to think of integrals in which one or both limits of integration are infinite, expressing them as limits of ordinary integrals.

This topic is not really a technique of integration, so in a sense it does not belong in this chapter. However, it is convenient to study it at this point, now that we have become proficient at integrating.

EXAMPLE 1 Suppose we want to integrate $\int_0^\infty e^{-x} dx$. This formula does not make any sense as it stands, but as usual when we see the infinity sign, we should expect it to have something to do with limits.

Although we do not know what it means to have an integral with an infinite limit of integration, we are quite familiar with integrals like $\int_0^t e^{-x} dx$, which can be interpreted as the area under the graph of $y = e^{-x}$ for x between $x = 0$ and $x = t$. If we want to talk about $\int_0^\infty e^{-x} dx$, it ought to be the area under the curve $y = e^{-x}$ along the whole positive x-axis, as shown in Figure 8.8.1. One way to define this area would be to calculate the area under the curve between $x = 0$ and $x = t$, and then take a limit as $t \to \infty$ (see Figure 8.8.2).

This suggests that we try letting

$$\int_0^\infty e^{-x} dx = \lim_{t \to \infty} \int_0^t e^{-x} dx.$$

It is easy to evaluate the integral on the right side: $\int_0^t e^{-x} dx = -e^{-x}\big|_0^t = -e^{-t} - (-e^0) = 1 - e^{-t}$. If we let $t \to \infty$, we see that $e^{-t} = \frac{1}{e^t} \to 0$, so

$$\int_0^\infty e^{-x} dx = \lim_{t \to \infty} \int_0^t e^{-x} dx = \lim_{t \to \infty} (1 - e^{-t})$$
$$= 1.$$

Figure 8.8.1

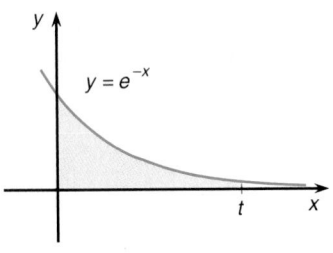

Figure 8.8.2

This example shows that we can define $\int_a^\infty f(x) dx$ to be $\lim_{t \to \infty} \int_a^t f(x) dx$, whenever the limit exists. Similarly, we define $\int_{-\infty}^a f(x) dx = \lim_{t \to -\infty} \int_t^a f(x) dx$, whenever the limit exists.

Finally, we should decide what is meant by $\int_{-\infty}^\infty f(x) dx$. This can be a little more tricky than it seems. A natural first choice might be to consider $\lim_{t \to \infty} \int_{-t}^t f(x) dx$. The difficulty with this is perhaps best explained by an example.

If we let $f(x) = x$, then for any $t > 0$, $\int_{-t}^{t} f(x)\,dx = \int_{-t}^{t} x\,dx = \frac{1}{2}x^2\big|_{-t}^{t} = \frac{1}{2}\left(t^2 - (-t)^2\right) = 0$. From this the limit is easily found: $\lim_{t\to\infty} \int_{-t}^{t} x\,dx = 0$.

The problem is that neither $\int_{0}^{\infty} x\,dx$ nor $\int_{-\infty}^{0} x\,dx$ exists. Indeed, we find $\lim_{t\to\infty} \int_{0}^{t} x\,dx = \lim_{t\to\infty} \frac{1}{2}(t^2 - 0^2) = \lim_{t\to\infty} \frac{t^2}{2}$, and this limit diverges to ∞. Practically the same calculation shows that $\lim_{t\to-\infty} \int_{t}^{0} x\,dx = -\infty$.

Neither of the integrals over half the real axis exists, and the fact that $\int_{-t}^{t} x\,dx$ tends to a limit as $t \to \infty$ is a result of the cancellation of the negative signed area to the left of the origin with the positive signed area to the right of the origin (see Figure 8.8.3).

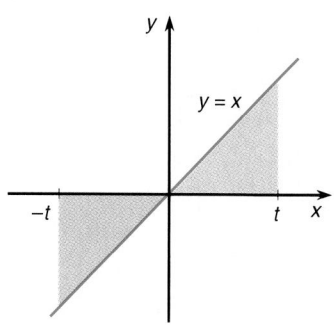

Figure 8.8.3

It turns out that this kind of limit is not sufficient for many purposes, so it is preferable to define $\int_{-\infty}^{\infty} f(x)\,dx = \int_{-\infty}^{a} f(x)\,dx + \int_{a}^{\infty} f(x)\,dx$, *provided* both these integrals exist. It is necessary to take the limit at ∞ and the limit at $-\infty$ *separately* and require that both of them exist. It is not hard to show that the choice of a is not important. If both "one-sided" integrals exist for one choice of a, then the same is true for any other a, and moreover the sum of the two integrals will be the same no matter which a is used.

We summarize.

DEFINITION 8.8.1

Figure 8.8.4(i)

Figure 8.8.4(ii)

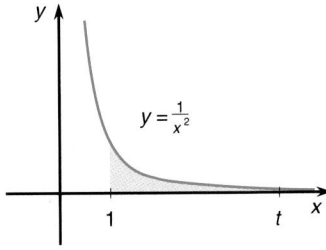

Figure 8.8.4(iii)

If $f(x)$ is a function on $[a, \infty)$, we define the **improper integral**

$$\int_{a}^{\infty} f(x)\,dx = \lim_{t\to\infty} \int_{a}^{t} f(x)\,dx,$$

provided the limit exists.

Similarly, if $f(x)$ is a function on $(-\infty, a]$, we define the **improper integral**

$$\int_{-\infty}^{a} f(x)\,dx = \lim_{t\to-\infty} \int_{t}^{a} f(x)\,dx,$$

provided the limit exists.

Finally, if $f(x)$ is a function on $(-\infty, \infty)$, we define the **improper integral**

$$\int_{-\infty}^{\infty} f(x)\,dx = \lim_{t\to-\infty} \int_{t}^{a} f(x)\,dx + \lim_{t\to\infty} \int_{a}^{t} f(x)\,dx,$$

provided the limits exist. Any choice of a will give the same result.

If any of the limits does not exist, the corresponding improper integral is said to **diverge**. In this case we also say that the integral "does not exist." If the integral does exist, we sometimes say that the integral **converges** or **is convergent**.

EXAMPLE 2

(i) We have already seen that $\int_{0}^{\infty} e^{-x}\,dx = \lim_{t\to\infty} \left(\int_{0}^{t} e^{-x}\,dx\right) = \lim_{t\to\infty} \left(-e^{-x}\right)\big|_{0}^{t}$
$= \lim_{t\to\infty} \left(1 - e^{-t}\right) = 1$. See Figure 8.8.4(i).

(ii) Consider $\int_{-\infty}^{0} x\,dx = \lim_{t\to-\infty} \left(\int_{t}^{0} x\,dx\right) = \lim_{t\to-\infty} \frac{1}{2}(0^2 - t^2) = -\frac{1}{2}$
$\lim_{t\to-\infty} t^2 = -\infty$. Since this limit diverges to $-\infty$, we say that the integral diverges. See Figure 8.8.4(ii).

(iii) Consider $\int_{1}^{\infty} \frac{1}{x^2}\,dx$, illustrated in Figure 8.8.4(iii). By definition it equals
$\lim_{t\to\infty} \left(\int_{1}^{t} \frac{1}{x^2}\,dx\right) = \lim_{t\to\infty} \left(-\frac{1}{x}\right)\big|_{1}^{t} = \lim_{t\to\infty} \left(-\frac{1}{t} + \frac{1}{1}\right) = 1$.

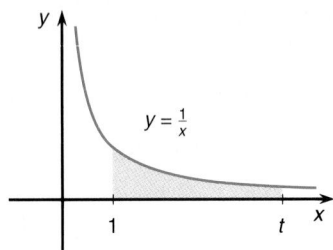

$y = \frac{1}{x}$

Figure 8.8.4(iv)

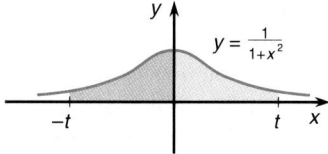

$y = \frac{1}{1+x^2}$

Figure 8.8.4(v)

(iv) Consider $\int_1^\infty \frac{1}{x}\, dx = \lim_{t\to\infty} \left(\int_1^t \frac{1}{x}\, dx\right) = \lim_{t\to\infty} \ln t = \infty$. The limit does not exist, so the integral diverges. See Figure 8.8.4(iv).

(v) The integral $\int_{-\infty}^\infty \frac{1}{1+x^2}\, dx$ must be broken up into two pieces:

$$\int_{-\infty}^\infty \frac{1}{1+x^2}\, dx = \lim_{t\to-\infty} \left(\int_t^0 \frac{1}{1+x^2}\, dx\right) + \lim_{t\to\infty} \left(\int_0^t \frac{1}{1+x^2}\, dx\right)$$

$$= \lim_{t\to-\infty} \left(\arctan x \Big|_t^0\right) + \lim_{t\to\infty} \left(\arctan x \Big|_0^t\right)$$

$$= \lim_{t\to-\infty} (-\arctan t) + \lim_{t\to\infty} (\arctan t).$$

Now as $t \to \infty$, we know that $\arctan t \to \frac{\pi}{2}$, and as $t \to -\infty$, $\arctan t \to -\frac{\pi}{2}$. Both limits exist, so

$$\int_{-\infty}^\infty \frac{1}{1+x^2}\, dx = -\left(-\frac{\pi}{2}\right) + \frac{\pi}{2} = \pi.$$

See Figure 8.8.4(v). ◢

There is another way in which an integral can be "improper." It arises when the integrand tends to $\pm\infty$ at some point in the interval of integration.

EXAMPLE 3   Consider $\int_0^1 \frac{1}{\sqrt{x}}\, dx$. This integral does not make sense because the integrand tends to ∞ as $x \to 0^+$.

However, in the spirit of the previous definition we could let $t > 0$ and look at the integral $\int_t^1 \frac{1}{\sqrt{x}}\, dx$. This integral does exist for any positive t, and we could try taking the limit as $t \to 0^+$:

$$\lim_{t\to0^+} \left(\int_t^1 x^{-1/2}\, dx\right) = \lim_{t\to0^+} \left(\frac{1}{\frac{1}{2}} x^{1/2} \Big|_t^1\right)$$

$$= \lim_{t\to0^+} 2(\sqrt{1} - \sqrt{t}) = 2(1 - 0)$$

$$= 2. \qquad ◢$$

This suggests a way to define the integral in cases in which the integrand tends to $\pm\infty$ at some point.

DEFINITION 8.8.2

Suppose $f(x)$ is a function on the interval $[a, b]$ and that $|f(x)| \to \infty$ as $x \to a^+$. Define the **improper integral** by

$$\int_a^b f(x)\, dx = \lim_{t\to a^+} \int_t^b f(x)\, dx,$$

provided the limit exists.
   Similarly, if $\lim_{x\to b^-} |f(x)| = \infty$, define

$$\int_a^b f(x)\, dx = \lim_{t\to b^-} \int_a^t f(x)\, dx,$$

provided the limit exists.
   In either case if the limit does not exist, the integral is said to **diverge**.

EXAMPLE 4

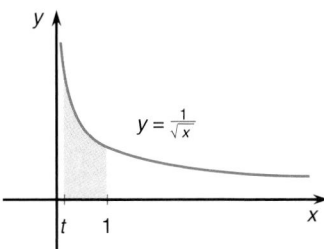

Figure 8.8.5(i)

(i) Example 3 above shows that

$$\int_0^1 \frac{1}{\sqrt{x}}\, dx = \lim_{t \to 0^+} \int_t^1 x^{-1/2} dx = \lim_{t \to 0^+} 2(\sqrt{1} - \sqrt{t}) = 2.$$

See Figure 8.8.5(i).

(ii) Consider $\int_0^2 \ln x\, dx$. The integrand tends to $-\infty$ as $x \to 0^+$, so we should consider the limit

$$\lim_{t \to 0^+} \int_t^2 \ln x\, dx = \lim_{t \to 0^+} (x \ln x - x)\Big|_t^2$$

$$= 2\ln 2 - 2 - \lim_{t \to 0^+} (t \ln t - t).$$

To find $\lim_{t \to 0^+} (t \ln t) = \lim_{t \to 0^+} \left(\frac{\ln t}{1/t}\right)$, we apply l'Hôpital's Rule:

$$\lim_{t \to 0^+} \frac{\ln t}{\frac{1}{t}} = \lim_{t \to 0^+} \frac{\frac{d}{dt}(\ln t)}{\frac{d}{dt}\left(\frac{1}{t}\right)} = \lim_{t \to 0^+} \frac{\frac{1}{t}}{-\frac{1}{t^2}}$$

$$= \lim_{t \to 0^+} (-t) = 0.$$

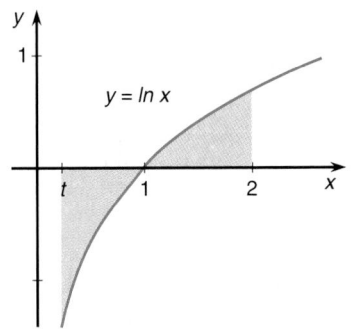

Figure 8.8.5(ii)

From this we find that

$$\int_0^2 \ln x\, dx = \lim_{t \to 0^+} \int_t^2 \ln x\, dx = \lim_{t \to 0^+} (x \ln x - x)\Big|_t^2$$

$$= \lim_{t \to 0^+} (2 \ln 2 - 2 - t \ln t + t) = 2 \ln 2 - 2.$$

The limit exists, so the improper integral exists. See Figure 8.8.5(ii).

(iii) Consider $\int_{-2}^2 \frac{1}{x-1}\, dx$. The difficulty here is that the integrand goes to $\pm\infty$ as x approaches 1 from either side, and 1 is inside the interval $[-2, 2]$ over which we are integrating.

The solution is to break the integral into integrals over $[-2, 1]$ and $[1, 2]$. Each of them is improper. Consider first

$$\int_{-2}^1 \frac{1}{x-1}\, dx = \lim_{t \to 1^-} \int_{-2}^t \frac{1}{x-1}\, dx$$

$$= \lim_{t \to 1^-} \ln|x-1|\Big|_{-2}^t = \lim_{t \to 1^-} (\ln|t-1| - \ln|-2-1|)$$

$$= \lim_{t \to 1^-} (\ln|t-1| - \ln 3).$$

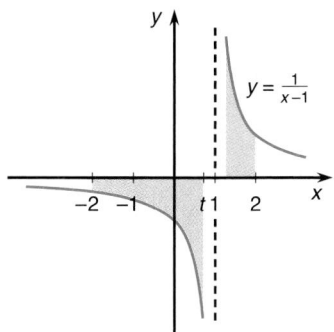

Figure 8.8.5(iii)

Now $\lim_{t \to 1^-} \ln|t-1| = \lim_{x \to 0^-} \ln|x| = \lim_{x \to 0^+} \ln x$, and this limit does not exist. (Recall the graph of $y = \ln x$, shown in Figure 8.8.5(ii), which tends to $-\infty$ as $x \to 0^+$.)

We see that the first integral does not exist, so without even looking at the second we conclude that the original integral $\int_{-2}^2 \frac{1}{x-1}\, dx$ does not exist. (In fact, the second one does not exist either, as can easily be checked.) See Figure 8.8.5(iii).

(iv) The integral $\int_0^\infty \frac{1}{x}\, dx$ is improper in two ways. First, it has an infinite limit of integration, and second, the integrand tends to ∞ as $x \to 0^+$. To deal with such a situation, we should break the integral into pieces, each of which contains only one of the difficulties. We consider $\int_0^1 \frac{1}{x}\, dx$ and $\int_1^\infty \frac{1}{x}\, dx$. See Figure 8.8.5(iv).

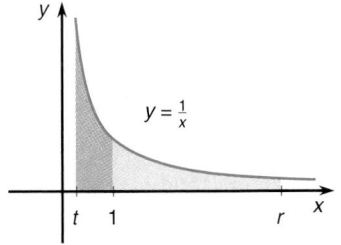

Figure 8.8.5(iv)

The first of these equals

$$\lim_{t \to 0^+} \int_t^1 \frac{1}{x}\, dx = \lim_{t \to 0^+} (\ln 1 - \ln t) = -\lim_{t \to 0^+} \ln t = \infty.$$

In other words, the integral diverges.

The second is

$$\lim_{r \to \infty} \int_1^r \frac{1}{x}\, dx = \lim_{r \to \infty} \ln r = \infty.$$

We see that in fact both parts of the integral diverge. In terms of areas this means that the area under $y = \frac{1}{x}$ between $x = 0$ and $x = 1$ and the area under $y = \frac{1}{x}$ to the right of $x = 1$ are both *infinite*.

Notice that it is possible to have improper integrals that are improper at points *inside* the interval over which we are integrating, as in part (iii) of Example 4. In such a case it is necessary to break the integral into pieces, on each of which the integral is improper at one endpoint. Any convenient breaking point can be used.

It is also possible for an integral to be improper at more than one point, and possibly in different ways. In each case the same strategy is applied: Break the integral into simple pieces, each of which has only one endpoint at which the integral is improper.

In analyzing such an integral we check each piece to see whether it exists. If we find that one of the pieces diverges, then it is not necessary to check the others; we can say immediately that the original integral diverges.

Exercises 8.8

I

Decide whether or not each of the following improper integrals exists, and evaluate those that do.

1. $\int_1^\infty \frac{1}{x^3}\, dx$

2. $\int_{-\infty}^{-1} \frac{1}{x^5}\, dx$

3. $\int_0^1 x^{-1/3}\, dx$

4. $\int_1^\infty \ln x\, dx$

5. $\int_0^\infty \frac{1}{1+x^2}\, dx$

6. $\int_0^\infty \frac{x}{x^2+1}\, dx$

7. $\int_0^\infty \frac{x^2}{x^2+4}\, dx$

8. $\int_1^\infty e^{-3x}\, dx$

9. $\int_0^\infty x e^{-x^2}\, dx$

10. $\int_0^\infty \sin x\, dx$

11. $\int_0^4 \frac{1}{x-2}\, dx$

12. $\int_{-1}^6 \frac{1}{x^2-4}\, dx$

13. $\int_0^{\pi/2} \sec x\, dx$

14. $\int_0^\pi \cot x\, dx$

15. $\int_0^1 \frac{e^x}{e^x-1}\, dx$

16. $\int_{-1}^1 \frac{1}{e^x-e^{-x}}\, dx$

17. $\int_{-\infty}^\infty \frac{1}{x^2+9}\, dx$

18. $\int_0^{\pi/2} x \sec^2 x\, dx$

19. $\int_0^\infty 3\, dx$

20. $\int_{-1}^1 \ln |x|\, dx$

II

21. Let r be a real number and consider $\int_1^\infty x^r\, dx$.
   (i) For which values of r does the improper integral exist?
   (ii) Evaluate the integral when it exists.

22. Let r be a real number and consider $\int_0^1 x^r\, dx$.
   (i)   For which values of r is the integral improper?
   (ii) For which values of r does the improper integral exist?
   (iii) Evaluate the integral when it exists.

*23. Suppose r is a real number and n is a positive integer. If $r \in [a, b]$, show that $\int_a^b \frac{1}{(x-r)^n}\, dx$ does not exist.

*24. Suppose $R(x) = \frac{p(x)}{q(x)}$ is a rational function and r is a real number for which $q(r) = 0$ and $p(r) \neq 0$. Show that if $r \in [a, b]$, then $\int_a^b R(x)\, dx$ does not exist. (*Hint:* See the previous exercise.)

25. In the discussion before Definition 8.8.1 (p. 607) we considered the improper integral $\int_{-\infty}^{\infty} x \, dx$. Show that $\lim_{t \to \infty} \int_{-t}^{t+1} x \, dx$ does not exist.

*26. If $\int_{-\infty}^{\infty} f(x) \, dx$ exists, show that, unlike what happened in the previous exercise,

$$\lim_{t \to \infty} \left(\int_{-t}^{t+1} f(x) \, dx \right)$$

converges *and* equals $\int_{-\infty}^{\infty} f(x) \, dx$.

*27. Give conditions on the real number A that determine whether or not the integral $\int_{a}^{b} \frac{e^x}{e^x - A} \, dx$ exists.

**28. (i) Suppose $f(x)$ and $g(x)$ are continuous functions with $f(x) \geq g(x) \geq 0$ for all $x \in [a, \infty)$, and suppose the improper integral $\int_{a}^{\infty} f(x) \, dx$ ex-

ists. Show that the improper integral $\int_{a}^{\infty} g(x) \, dx$ exists.

(ii) Use the result of part (i) to show that $\int_{1}^{\infty} \frac{1}{x^4 + 3x^3 + 7x^2 + 11x + 5} \, dx$ exists. (*Hint:* Let $f(x) = \frac{1}{x^4}$.)

29. In Example 2(v) (p. 608) we evaluated $\int_{-\infty}^{\infty} \frac{1}{1+x^2} \, dx$ by writing

$$\int_{-\infty}^{\infty} \frac{1}{1 + x^2} \, dx = \int_{-\infty}^{0} \frac{1}{1 + x^2} \, dx + \int_{0}^{\infty} \frac{1}{1 + x^2} \, dx.$$

Confirm that it can also be done with $\int_{-\infty}^{\infty} \frac{1}{1+x^2} \, dx = \int_{-\infty}^{1} \frac{1}{1+x^2} \, dx + \int_{1}^{\infty} \frac{1}{1+x^2} \, dx.$

30. If $\int_{-\infty}^{\infty} f(x) \, dx$ converges, show that the sum $\int_{-\infty}^{a} f(x) \, dx + \int_{a}^{\infty} f(x) \, dx$ does not depend on the choice of a.

POINT TO PONDER

In Example 6.9.5 (p. 489), to find the surface area of a sphere we had to integrate

$$2\pi \int f(x) \sqrt{1 + f'(x)^2} \, dx.$$

Since the function $f(x) = \sqrt{r^2 - x^2}$ is not differentiable at the endpoints $x = \pm r$, we had to integrate from $-a$ to a and then let $a \to r$ (see Figures 8.8.6(i) and (ii)).

This looks similar to what we do for improper integrals, but the reason is slightly different. The integrand does not tend to infinity; the difficulty is that it is not defined at the endpoints. Sometimes this is regarded as another type of improper integral.

Figure 8.8.6(i)

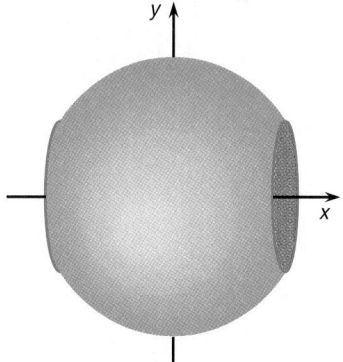

Figure 8.8.6(ii)

Chapter Summary

§8.1
integration by parts
reduction formula
definite integrals by parts
reduction formulas for $\cos^n x$, $\sin^n x$
  (Theorem 8.1.3, p. 565)

§8.2
$\cos^2 x + \sin^2 x = 1$
$1 + \tan^2 x = \sec^2 x$
trigonometric integrals
integrating $\cos^m x \sin^n x$ (Summary 8.2.4, p. 571)
reduction formulas for $\tan^n x$, $\sec^n x$ (Theorem 8.2.5, p. 574)

integrating $\tan^m x \sec^n x$ (Summary 8.2.6, p. 576)

§8.3
trigonometric substitutions

§8.4
completing the square

§8.5
partial fractions
proper rational function
irreducible quadratic polynomial
$b^2 - 4ac < 0$

§8.6
fractional powers
$\tan\left(\frac{\theta}{2}\right)$ substitution (Summary 8.6.1, p. 600)

§8.7
additional applications
arclength

§8.8
improper integral (Definitions 8.8.1, p. 607 and 8.8.2,
    p. 608)
diverge
converge

Review Exercises

I

Evaluate the following integrals.

1. $\int (x + 2) e^{3x} dx$

2. $\int (4e^{2x} + 2e^x)/(1 - e^{2x}) dx$

3. $\int \frac{1}{x^2 - 6x + 13} dx$

4. $\int \sin^3(5x) dx$

5. $\int \sqrt{x} \sin(\sqrt{x}) dx$

6. $\int x^3 \cos(x^2) dx$

7. $\int \frac{2x^4 - 10x^3 + 11x^2 + 4x - 5}{x^3 - 6x^2 + 11x - 6} dx$

8. $\int \frac{x^3 + 2}{x^2 - 1} dx$

9. $\int x \sin^2 x \cos x \, dx$

10. $\int x \cos^3 x \, dx$

11. $\int \sqrt{2x^2 - 12x + 26} \, dx$

12. $\int \frac{1}{12 + \sin x} dx$

13. $\int \frac{1}{x^2 + 6x + 8} dx$

14. $\int \frac{2x - 5}{(x+1)^3} dx$

15. $\int \frac{1}{1 + \sqrt{x}} dx$

16. $\int \tan^2 x \, dx$

17. $\int \cos(\sqrt{x}) dx$

18. $\int e^{\sqrt{x+7}} dx$

19. $\int \ln(1 + x^2) dx$

20. $\int \frac{\cos(2x)}{1 + \cos(2x)} dx$

21. $\int \frac{\sin(2x)}{1 + \sin x} dx$

22. $\int \frac{\cos(2x)}{1 + \cos x} dx$

23. $\int x \cos x \sin x \, dx$

24. $\int x e^{\sqrt{x}} dx$

25. $\int x(x + 8)^{1/3} dx$

26. $\int \frac{1}{x}(x + 1)^{1/3} dx$

27. $\int \sqrt{1 + e^x} \, dx$

28. $\int x \sin x \sec^2 x \, dx$

29. $\int x \arctan x \, dx$

30. $\int x^2 \arctan x \, dx$

31. $\int \csc^3 x \cot^3 x \, dx$

32. $\int x \sec^2(1 + x^2) dx$

33. $\int x \sec^2 x \, dx$

34. $\int \frac{1}{x^2 \sqrt{1 - x^2}} dx$

35. $\int \frac{\sqrt{4 - x^2}}{x^2} dx$

36. $\int \frac{2x^3 + 3x^2 - 8x - 12}{x^4 - 16} dx$

37. $\int \frac{x + 2}{2x^2 - 2x + 1} dx$

38. $\int \frac{3x^{1/3} + 5\sqrt{x}}{x^{2/3} + x} dx$

39. $\int \sqrt{x} \ln(\sqrt{x}) dx$

40. $\int \ln(x^2 - 1) dx$

41. $\int_4^6 \frac{1}{x^3 + 3x^2 + 2x} dx$

42. $\int_{-\pi/4}^{\pi/4} (\sin^5 x \sec^4 x - 2 \tan^3 x \sin^6 x - 3x^5) dx$

43. $\int \frac{\cos^2 \theta}{\sin \theta} d\theta$

44. $\int_0^{\ln 5} \sqrt{4 + e^x} \, dx$

45. $\int \sqrt{x} \ln x \, dx$

46. $\int \frac{x^3 - x^2 - 3x}{x^3 - 1} dx$

47. $\int_0^1 \arcsin\left(\frac{x}{2}\right) dx$

48. $\int_0^{\pi/9} \sec(3x)\,dx$

49. $\int_1^4 \frac{1}{\sqrt{x}} \ln x\,dx$

50. $\int_2^4 \frac{(x+1)^2}{\sqrt{x^2-1}}\,dx$

Decide whether each of the following improper integrals exists. Evaluate those that do.

51. $\int_{-\infty}^{\sqrt{3}} \frac{1}{x^2+1}\,dx$

52. $\int_{-2}^{2} \frac{1}{x^2-1}\,dx$

53. $\int_4^{\infty} e^{3x}\,dx$

54. $\int_{-\infty}^{-1} e^{2x}\,dx$

55. $\int_1^{\infty} \frac{x}{1+x^2}\,dx$

56. $\int_0^{\infty} x^{-1/2}\,dx$

57. $\int_0^{\pi/2} \csc x\,dx$

58. $\int_{-\infty}^{0} \sin(2x)\,dx$

59. $\int_0^1 \frac{x^2-3x}{x^3+x}\,dx$

60. $\int_0^1 \frac{x^2-2x+1}{x^2-1}\,dx$

II

Suppose $a > 0$ and $b > 0$. Evaluate the following indefinite integrals.

61. $\int \frac{x}{\sqrt{ax+b}}\,dx$

62. $\int \frac{x^2}{\sqrt{ax+b}}\,dx$

63. $\int \sqrt{x^2 \pm a^2}\,dx$

64. $\int x^2\sqrt{x^2 \pm a^2}\,dx$

65. $\int \sqrt{a^2 - x^2}\,dx$

66. $\int x^2\sqrt{a^2 - x^2}\,dx$

*67. Find the centroid of the region lying between the x-axis and the curve $y = \sin 3x$ and between the lines $x = 0$ and $x = \frac{\pi}{9}$.

*68. Find the arclength of the curve $y = (x+3)^2$, from $x = 0$ to $x = 2$.

*69. Find the arclength of the curve $y = x^2 - 4x + 1$, from $x = 2$ to $x = 3$.

*70. Consider the integral $\int \frac{1}{x^2-1}\,dx$. Instead of using partial fractions, consider a trigonometric substitution. If $|x| < 1$, try $x = \sin\theta$. If $|x| > 1$, try $x = \sec\theta$. Work out the integral in both these cases, and then see whether you can find a single formula that works in both cases. Finally, reconcile your answer(s) with what you get using partial fractions.

71. (i) Evaluate $\int \frac{5x^3+14x^2-x+6}{x(x+3)(x^2+1)}\,dx$.

    (ii) In writing the integrand in part (i) with partial fractions, $\frac{5x^3+14x^2-x+6}{x(x+3)(x^2+1)} = \frac{A}{x} + \frac{B}{x+3} + \frac{Cx+D}{x^2+1}$, we find that $B = 0$. What is the significance of this fact?

**72. Find the integral $\int (1 + x^2)^{-3/2}\,dx$ using a hyperbolic substitution (cf. Exercise 47 in Section 8.3, p. 584). The answer will involve a hyperbolic function other than the one originally substituted; as with trigonometric substitutions, it is necessary to express this in terms of the function which was substituted.

For trigonometric substitutions this type of calculation was often made easier by using little triangles that showed the relationships among the various trigonometric functions. However, all the triangles really do is to remind us of the fundamental identity $\cos^2\theta + \sin^2\theta = 1$. In the hyperbolic situation we can use the corresponding identity $\cosh^2 t - \sinh^2 t = 1$.

*73. Compare the answers you get for $\int \frac{1}{1+\cos\theta}\,d\theta$ using (i) $x = \tan\frac{\theta}{2}$ or (ii) multiplying numerator and denominator by $1 - \cos\theta$.

Taylor's Formula

9

The straight line $y = x$ is fairly close to $y = \sin x$ near the origin, but $y = x - \frac{x^3}{6}$ is a better fit. The pictures show that $y = x - \frac{x^3}{6} + \frac{x^5}{120}$ is even better and $y = x - \frac{x^3}{6} + \frac{x^5}{120} - \frac{x^7}{5040}$ better yet.

In this chapter we learn how to find polynomials that approximate a complicated function like $y = \sin x$.

INTRODUCTION

In Section 4.10 we discussed the technique of linear approximation. The idea was that if we know the value of some function $f(x)$ at the point $x = a$, then we can estimate its value at a nearby point by assuming that the graph of $y = f(x)$ lies close to the tangent line at $x = a$. This works quite well for values of x near $x = a$ but, of course, is less accurate for points farther away.

The formula for the linear approximation is (cf. Equation 4.10.1)

$$f(a + h) \approx f(a) + f'(a)h.$$

This approximates the function $f(x)$ by its tangent line, which is the line having the same value as $f(x)$ at $x = a$ and whose derivative at $x = a$ is the same as $f'(a)$ (see Figure 9.1.1). If we write $y = \ell(x)$ for the tangent line, this means that $\ell(a) = f(a)$ and $\ell'(a) = f'(a)$.

A natural way to attempt to improve the accuracy of the approximation would be to use a parabola $y = p(x)$ instead of a straight line so that the curve of the parabola could fit more closely the curve of $y = f(x)$, at

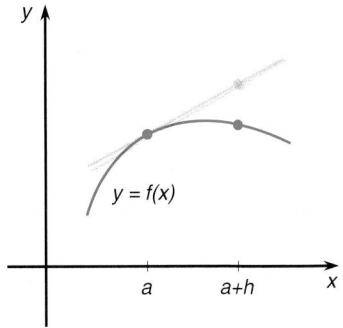

$y = f(x)$

a $a+h$ x

Figure 9.1.1

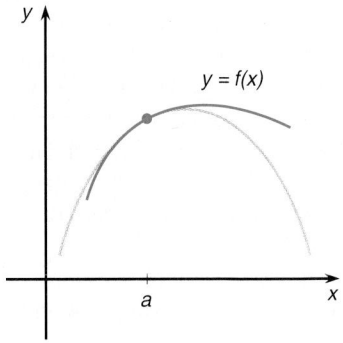

$y = f(x)$

a x

Figure 9.1.2

least near $x = a$. Exactly how to make the parabola fit the curve of $y = f(x)$ is not entirely clear, but a reasonable guess would be to make $p(a) = f(a)$, $p'(a) = f'(a)$, and $p''(a) = f''(a)$. If the parabola is chosen to have the same value, the same derivative and the same second derivative as $f(x)$ at the point $x = a$, then it seems likely that it will be close to $f(x)$, for values of x near a. See Figure 9.1.2.

Notice that it is not possible to do this with a straight line because the second derivative of a straight line function $y = mx + b$ is always zero at every point. There is no way to find a straight line with a specified second derivative unless that second derivative happens to be zero.

Of course, there is no particular reason to stop with the second derivative. If we want to approximate $y = f(x)$ near $x = a$, we could look for a polynomial $q(x)$ of degree n, chosen so that the values of $q(x)$ and its derivatives up to the nth derivative agree with the values of $f(x)$ and its derivatives at $x = a$, that is, $q(a) = f(a)$, $q'(a) = f'(a)$, $q''(a) = f''(a), \ldots$, $q^{(n)}(a) = f^{(n)}(a)$. It seems reasonable to expect that this is the best way to approximate $f(x)$ by a polynomial of degree n. It also seems reasonable to expect that using larger values of n may result in polynomials that give better approximations. (Notice that the $(n + 1)$th and higher derivatives of a polynomial of degree n are always zero, so there is no sense in asking that they agree with the corresponding derivatives of $f(x)$.)

It turns out that this is a very practical way to calculate values of complicated functions. It is relatively easy to evaluate polynomials, so if we want to evaluate a more complicated function, like e^x or $\sin x$, for example, we can approximate it with a polynomial and evaluate that. This is how computers evaluate functions like e^x, $\cos x$, and $\sin x$. Taylor's Formula, the subject of Section 9.1, tells us explicitly what the polynomial should be.

Any approximation is of limited value unless there is some way to assess the accuracy of the answer we calculate. Taylor's Formula with Remainder, discussed in Section 9.2, provides a way to measure the "remainder," the difference between the actual value and the approximation, in effect telling us how close the approximation is to the actual value. It can be used as follows. Suppose we want to know the value of $e^{1.345}$ correct to four decimal places. We can use Taylor's Formula with Remainder to tell us how large n, the degree of the approximating polynomial, will have to be in order to get an approximation that is at least this good.

Once we know that, it is reasonably straightforward to find and evaluate the polynomial. The material of this chapter will be extremely important in Chapter 10.

9.1 Taylor's Formula

TAYLOR POLYNOMIAL
FACTORIAL
MACLAURIN POLYNOMIAL

In this chapter we will write a polynomial $p(x)$ with the constant term first and highest term last:

$$p(x) = a_0 + a_1 x + a_2 x^2 + \ldots + a_{n-1} x^{n-1} + a_n x^n.$$

We are interested in the values of $p(x)$ and its derivatives at some point $x = a$, but for simplicity let us first consider the point $x = 0$. It is not difficult to express the values of $p(x)$ and its derivatives at $x = 0$ in terms of the coefficients a_i of the polynomial.

Certainly the value of $p(x)$ at $x = 0$ is just the constant term, $p(0) = a_0$. If we work out the derivative

$$p'(x) = a_1 + 2a_2 x + \ldots + (n - 1)a_{n-1} x^{n-2} + na_n x^{n-1},$$

we find that its value at $x = 0$ is just *its* constant term, that is, $p'(0) = a_1$. The idea is that when we substitute $x = 0$, any term with a positive power of x in it will become 0; only the constant term a_1 does not contain a positive power of x.

If we work out the second derivative

$$p''(x) = 2a_2 + (3)(2)a_3x + \ldots + (n-1)(n-2)a_{n-1}x^{n-3} + n(n-1)a_nx^{n-2},$$

we find that $p''(0) = 2a_2$. Notice that the terms involving a_0 and a_1 have become zero after two differentiations.

In the expression for the third derivative $p'''(x)$, the constant term is $(3)(2)a_3$, so $p'''(0) = (3)(2)a_3$.

The term a_3x^3 in the original polynomial $p(x)$ changes to $3a_3x^2$ after one differentiation, then to $(3)(2)a_3x$ after another, and finally to $(3)(2)a_3$ after the third. Terms that begin with a power of x lower than the third power will be zero after the third differentiation, and terms that begin with a power of x higher than the third power will still contain a positive power of x and will not contribute to $p'''(0)$.

What happens when we take the kth derivative and evaluate $p^{(k)}(0)$? As above, we have to look for the constant term. Each time we differentiate, each positive power of x is reduced by 1, so terms that begin with a power of x higher than the kth power will still contain a positive power of x after k differentiations. Terms that begin with a power of x lower than the kth power will have become zero, so the only nonzero term in $p^{(k)}(x)$ that does not contain a positive power of x is the term that comes from a_kx^k.

Differentiating a_kx^k gives ka_kx^{k-1}; differentiating again gives $k(k-1)a_kx^{k-2}$. Each differentiation has the effect of multiplying by the exponent and then lowering that exponent by 1. Doing this k times will give

$$\frac{d^k}{dx^k}(a_kx^k) = k(k-1)(k-2)\ldots 2a_k = k!a_k.$$

(Here we have used the abbreviation $k! = k(k-1)(k-2)\ldots(2)(1)$ for the product of the first k positive integers; the symbol $k!$ is read "k **factorial**." See Definition 1.3.8, p. 14.)

Since only the kth term affects $p^{(k)}(0)$, we see that $p^{(k)}(0) = k!a_k$.

It is possible to turn this process around. Suppose we want to construct a polynomial of degree n, and suppose we know in advance the values we want the polynomial and its first n derivatives to have. Suppose $p(0) = b_0$, $p^{(1)}(0) = b_1$, $p^{(2)}(0) = b_2$, $p^{(3)}(0) = b_3, \ldots, p^{(n-1)}(0) = b_{n-1}$, $p^{(n)}(0) = b_n$ are the desired values for the polynomial and its first n derivatives at $x = 0$. Then dividing the above formula by $k!$, we see that the kth coefficient of the polynomial will have to be $a_k = \frac{p^{(k)}(0)}{k!} = \frac{b_k}{k!}$, for $k \geq 1$, and $a_0 = p(0) = b_0$.

If we allow ourselves to write $p^{(0)}(x)$ for $p(x)$ and agree that $0! = 1$, then the formula $a_k = \frac{p^{(k)}(0)}{k!} = \frac{b_k}{k!}$ is correct for $k = 0$ too. It may seem strange at first to let $0! = 1$, but in fact it is the right choice to make; it makes this formula particularly convenient, but there are also other reasons to do it this way. We encountered this convention when we discussed the Binomial Theorem (Theorem 1.3.10). One way to look at it is that we expect that $k! = k(k-1)!$, and for this to be true when $k = 1$ it is necessary that $0! = 1$.

EXAMPLE 1   A Polynomial with Specified Derivatives   Find a polynomial $p(x)$ of degree 3 with $p(0) = 3$, $p^{(1)}(0) = 7$, $p^{(2)}(0) = -2$, and $p^{(3)}(0) = 12$.

Solution If we write $p(x) = a_0 + a_1 x + a_2 x^2 + a_3 x^3$, then the above formula tells us that $a_3 = p^{(3)}(0)/3! = \frac{12}{(3)(2)(1)} = \frac{12}{6} = 2$. Also $a_2 = p^{(2)}(0)/2! = \frac{-2}{(2)(1)} = -1$, and $a_1 = p^{(1)}(0)/1! = \frac{7}{1} = 7$. Finally, $a_0 = p(0)/0! = \frac{3}{1} = 3$.

The polynomial is $p(x) = a_0 + a_1 x + a_2 x^2 + a_3 x^3 = 3 + 7x - x^2 + 2x^3$. It is easy to check that it has the desired derivatives at $x = 0$.

The situation in which we are interested arises when we begin with a function $f(x)$ and want to construct a polynomial $p(x)$ of degree n with the same value and same first n derivatives at $x = 0$. In other words, we want $p(0) = f(0)$, $p^{(1)}(0) = f^{(1)}(0)$, $p^{(2)}(0) = f^{(2)}(0)$, ..., $p^{(n-1)}(0) = f^{(n-1)}(0)$, $p^{(n)}(0) = f^{(n)}(0)$.

The formula we found above tells us that the coefficient a_k of x^k in $p(x)$ must be

$$a_k = \frac{f^{(k)}(0)}{k!}. \qquad (9.1.1)$$

This is the first version of Taylor's Formula.

EXAMPLE 2 Find the polynomial of degree at most 4 that has the same value and same first four derivatives as $f(x) = \sin x$ at $x = 0$.

Solution We calculate the derivatives of $f(x)$, evaluate them at $x = 0$, and then use Equation 9.1.1 to find the coefficients a_i of the polynomial $p(x)$:

$f^{(0)}(x) = f(x) = \sin x,$	$f(0) = \sin 0 = 0,$	$a_0 = \frac{0}{1} = 0;$
$f^{(1)}(x) = f'(x) = \cos x,$	$f^{(1)}(0) = \cos 0 = 1,$	$a_1 = \frac{1}{1!} = 1;$
$f^{(2)}(x) = -\sin x,$	$f^{(2)}(0) = -\sin 0 = 0,$	$a_2 = \frac{0}{2!} = 0;$
$f^{(3)}(x) = -\cos x,$	$f^{(3)}(0) = -\cos 0 = -1,$	$a_3 = \frac{-1}{3!} = -\frac{1}{6};$
$f^{(4)}(x) = \sin x,$	$f^{(4)}(0) = \sin 0 = 0,$	$a_4 = \frac{0}{4!} = 0.$

The polynomial is

$$p(x) = x - \frac{1}{6}x^3.$$

The graph of this polynomial is sketched in Figure 9.1.3; the part of the curve near $x = 0$ does look a lot like the corresponding part of the graph of $y = \sin x$. As we move farther from $x = 0$, the approximation is less good.

EXAMPLE 3 Find the polynomial of degree at most 5 that has the same value and same first five derivatives as $f(x) = \sin x$ at $x = 0$.

Solution We have already done most of the work in Example 2. What is left to do is find $f^{(5)}(x) = \frac{d}{dx}\left(f^{(4)}(x)\right) = \cos x$, so $f^{(5)}(0) = \cos 0 = 1$.

From this we find that the coefficient of x^5 in the polynomial will have to be $a_5 = \frac{1}{5!} = \frac{1}{120}$. The lower coefficients are the same as what we found above, so the polynomial we want is

$$q(x) = x - \frac{1}{6}x^3 + \frac{1}{120}x^5.$$

Figure 9.1.3

$y = p(x)$

$y = \sin x$

Since $a_4 = 0$, the polynomial actually has degree 3, not 4 . . .

The coefficient a_k of x^k is determined by the formula $a_k = \frac{f^{(k)}(0)}{k!}$, so the lower coefficients $a_0, a_1, \ldots, a_4$ are exactly the same as the ones in the previous example . . .

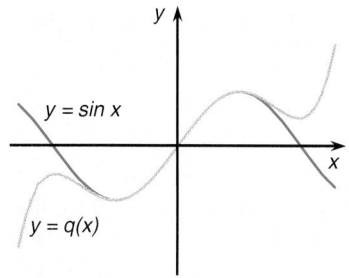

Figure 9.1.4

This polynomial is sketched in Figure 9.1.4, and it is a better approximation to $f(x) = \sin x$ than the polynomial $p(x)$ that we found in the last example, although even this one eventually goes off to infinity and becomes a bad approximation for x sufficiently far from $x = 0$.

The idea is that the higher the degree of the approximating polynomial, the more of its derivatives will agree with those of $f(x)$, so the more closely its behavior will agree with that of $f(x)$ and the better the approximation should be.

Now we would like to do the same thing at any fixed point $x = a$ rather than just the special point $x = 0$. Unfortunately, the derivative of $p(x)$ at $x = a$ is not as simple to evaluate at $x = a$, since it involves all the coefficients instead of only the constant term. The higher derivatives also involve all the coefficients, so it is much more complicated to say what the coefficients have to be in order that $p(x)$ and its derivatives will have specified values at $x = a$.

This difficulty arises because we express the polynomial in terms of powers of x, which are very convenient for evaluating at $x = 0$ but awkward anywhere else. If instead we express the polynomial in terms of powers of $(x - a)$, then the situation will improve considerably. If

$$p(x) = a_0 + a_1(x - a) + a_2(x - a)^2 + \ldots + a_{n-1}(x - a)^{n-1} + a_n(x - a)^n,$$

then $p(a) = a_0$, $p^{(1)}(a) = a_1$, $p^{(2)}(a) = 2a_2$, and so on.

The reason is really the same as it was in the case in which $a = 0$. The kth derivative of $a_k(x - a)^k$ is $k(k - 1) \ldots (2)(1) a_k$; the kth derivative of any term containing a lower power of $(x - a)$ is zero, and the kth derivative of any term containing a power of $(x - a)$ higher than the kth will still contain a positive power of $(x - a)$ and so will be zero at $x = a$.

As in the case with $a = 0$, we see that with $p(x)$ written as above,

$$p^{(k)}(a) = k! a_k.$$

So if we want a polynomial of degree at most n that has the same value and the same first n derivatives at $x = a$ as $f(x)$, then we should use the polynomial $p(x) = a_0 + a_1(x - a) + a_2(x - a)^2 + \ldots + a_{n-1}(x - a)^{n-1} + a_n(x - a)^n$, with

$$a_k = \frac{f^{(k)}(a)}{k!}.$$

It is also convenient to write $p(x)$ using summation notation:

$$p(x) = \sum_{k=0}^{n} \frac{f^{(k)}(a)}{k!}(x - a)^k.$$

If we start with a function $f(x)$ that is n times differentiable, then we have just seen how to construct a polynomial of degree at most n so that its value and the values of its first n derivatives at $x = a$ are the same as those of the function $f(x)$. This polynomial can be used to approximate the original function $f(x)$, and it is extremely important. It is called the Taylor polynomial of degree n, after the English mathematician Brook Taylor (1685–1731). Notice that it depends on the choice of the point $x = a$ and on the degree n.

DEFINITION 9.1.2

Taylor Polynomials

Let $f(x)$ be a function defined on an interval containing $x = a$, and suppose its derivatives $f'(x) = f^{(1)}(x)$, $f^{(2)}(x)$, $\ldots$, $f^{(n)}(x)$ all exist at $x = a$. The corresponding **Taylor polynomial** of degree n at $x = a$ is defined to be the polynomial

$$P_n(x) = a_0 + a_1(x - a) + \ldots + a_{n-1}(x - a)^{n-1} + a_n(x - a)^n,$$

with

$$a_k = \frac{f^{(k)}(a)}{k!}.$$

It can also be written as

$$P_n(x) = \sum_{k=0}^{n} a_k(x - a)^k = \sum_{k=0}^{n} \frac{f^{(k)}(a)}{k!}(x - a)^k.$$

Notice that each $a_k = \frac{f^{(k)}(a)}{k!}$ is a *constant*. Only the powers $(x - a)^k$ depend on x.

Notice too that the polynomial depends on the choice of the starting point a. When we mention $P_n(x)$, it is necessary to make clear what a is being used. In the special case in which $a = 0$, Taylor polynomials are sometimes called **Maclaurin polynomials**, after the Scottish mathematician Colin Maclaurin (1698–1746).

Strictly speaking, the degree of $P_n(x)$ may be less than n (cf. Example 2).

The argument we have given above shows that $P_n(x)$ is determined by the property that it and its first n derivatives equal $f(x)$ and its first n derivatives at $x = a$.

EXAMPLE 4   **Taylor Polynomial of ln x**   Find the Taylor polynomial $P_4(x)$ corresponding to the function $f(x) = \ln x$ at $a = 1$.

Solution   The derivatives are easily calculated and evaluated:

$$f^{(0)}(x) = \ln x, \qquad f^{(0)}(1) = f(1) = \ln(1) = 0;$$
$$f^{(1)}(x) = \frac{1}{x}, \qquad f^{(1)}(1) = \frac{1}{1} = 1;$$
$$f^{(2)}(x) = -\frac{1}{x^2}, \qquad f^{(2)}(1) = -\frac{1}{1} = -1;$$
$$f^{(3)}(x) = \frac{2}{x^3}, \qquad f^{(3)}(1) = \frac{2}{1} = 2;$$
$$f^{(4)}(x) = -\frac{(3)(2)}{x^4}, \qquad f^{(4)}(1) = -\frac{6}{1} = -6.$$

From this we can easily write down the Taylor polynomial:

$$P_4(x) = 0 + \frac{1}{1!}(x - 1) + \frac{-1}{2!}(x - 1)^2 + \frac{2}{3!}(x - 1)^3 + \frac{-6}{4!}(x - 1)^4$$

$$= (x - 1) - \frac{1}{2}(x - 1)^2 + \frac{1}{3}(x - 1)^3 - \frac{1}{4}(x - 1)^4.$$

At first it seems peculiar to write a polynomial in terms of powers of $(x - a)$ instead of powers of x, but in fact it is quite useful. After all, if we want to use $P_n(x)$ to approximate the value of $f(x)$ for some x near a, then we might write $x = a + h$, and $x - a$ is just h. It is actually easier to perform calculations with $x - a$ than it would be to use x.

With a graphics calculator we can plot the function $y = \ln x$ and the Taylor polynomial $P_4(x)$ of Example 4.

TI-85	SHARP EL9300
GRAPH F1$(y(x) =)$ LN F1(x) ▼ F1(x) − 1 − (F1(x) − 1) ^ 2 ÷ 2 + (F1(x) − 1) ^ 3 ÷ 3 − (F1(x) − 1) ^ 4 ÷ 4 M2(RANGE) 0 ▼ 2 ▼ ▼ (−) 2 ▼ 1 F5(GRAPH)	⫽ ln x/θ/T 2ndF ▼ x/θ/T − 1 − (x/θ/T − 1) x² ÷ 2 + (x/θ/ᴛ − 1) a^b 3 ▶ ÷ 3 − (x/θ/T − 1) a^b 4 ▶ ÷ 4 RANGE 0 ENTER 2 ENTER 1 ENTER −2 ENTER 1 ENTER 1 ENTER ⫽
CASIO f_x-6300G (f_x-7700GB)	HP 48SX
Range 0 EXE 2 EXE 1 EXE −2 EXE 1 EXE 1 EXE EXE EXE EXE Graph ln X EXE Graph X − 1 − (X − 1) x² ÷ 2 + (X − 1) x^y 3 ÷ 3 − (X − 1) x^y 4 ÷ 4 EXE	■ PLOT PLOTR ERASE ATTN ' Y ■ LN X ↱ PLOT ■ DRAW ERASE 0 ENTER 2 XRNG 2 +/− ENTER 1 YRNG DRAW ATTN ' Y ■ X − 1 − ▊ X − 1 ▶ y^x 2 ÷ 2 + () X − 1 ▶ y^x 3 ÷ 3 − ▊ X − 1 ▶ y^x 4 ÷ 4 ↱ PLOT ■ DRAW DRAW

Identify which function is which, and notice how the approximation is better for values near $a = 1$.

EXAMPLE 5   Approximating with the Taylor Polynomial   Use the Taylor polynomial $P_4(x)$ from Example 4 to approximate $\ln(1.1)$.

Solution   The Taylor polynomial $P_4(x)$ was found with $a = 1$, so if $x = 1.1 = \frac{11}{10}$, then $x - a = x - 1 = \frac{1}{10} = 0.1$. Since

$$P_4(x) = (x - 1) - \frac{1}{2}(x - 1)^2 + \frac{1}{3}(x - 1)^3 - \frac{1}{4}(x - 1)^4,$$

we find

$$P_4(1.1) = 0.1 - \frac{1}{2}(0.1)^2 + \frac{1}{3}(0.1)^3 - \frac{1}{4}(0.1)^4 \approx 0.095\,3083.$$

This is a pretty good approximation to the actual value, which is $0.095\,310$, correct to six decimal places. Notice that it would actually be slightly more trouble to evaluate a fourth-degree polynomial written in terms of powers of x instead of powers of $(x - 1)$.

EXAMPLE 6   Find the Taylor polynomial $P_4(x)$ for $f(x) = \sin x$, with $a = \frac{\pi}{4}$.

Solution

$$
\begin{array}{ll}
f(x) = \sin x, & f\left(\frac{\pi}{4}\right) = \sin\left(\frac{\pi}{4}\right) = \frac{1}{\sqrt{2}}; \\[2mm]
f^{(1)}(x) = \cos x, & f^{(1)}\left(\frac{\pi}{4}\right) = \cos\left(\frac{\pi}{4}\right) = \frac{1}{\sqrt{2}}; \\[2mm]
f^{(2)}(x) = -\sin x, & f^{(2)}\left(\frac{\pi}{4}\right) = -\sin\left(\frac{\pi}{4}\right) = -\frac{1}{\sqrt{2}}; \\[2mm]
f^{(3)}(x) = -\cos x, & f^{(3)}\left(\frac{\pi}{4}\right) = -\cos\left(\frac{\pi}{4}\right) = -\frac{1}{\sqrt{2}}; \\[2mm]
f^{(4)}(x) = \sin x, & f^{(4)}\left(\frac{\pi}{4}\right) = \sin\left(\frac{\pi}{4}\right) = \frac{1}{\sqrt{2}}.
\end{array}
$$

From this it is possible to write down the Taylor polynomial:

$$
\begin{aligned}
P_4(x) &= \frac{1}{0!\sqrt{2}} + \frac{1}{1!\sqrt{2}}\left(x - \frac{\pi}{4}\right) - \frac{1}{2!\sqrt{2}}\left(x - \frac{\pi}{4}\right)^2 - \frac{1}{3!\sqrt{2}}\left(x - \frac{\pi}{4}\right)^3 \\
&\quad + \frac{1}{4!\sqrt{2}}\left(x - \frac{\pi}{4}\right)^4 \\
&= \frac{1}{\sqrt{2}} + \frac{1}{\sqrt{2}}\left(x - \frac{\pi}{4}\right) - \frac{1}{2\sqrt{2}}\left(x - \frac{\pi}{4}\right)^2 - \frac{1}{6\sqrt{2}}\left(x - \frac{\pi}{4}\right)^3 \\
&\quad + \frac{1}{24\sqrt{2}}\left(x - \frac{\pi}{4}\right)^4.
\end{aligned}
$$

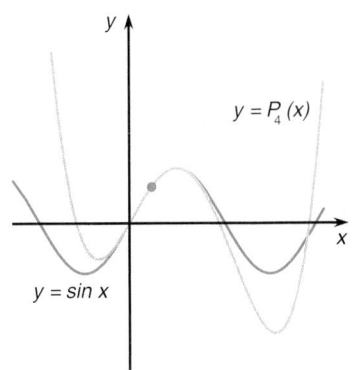

The interesting thing about this example is to compare it with Example 2. There we found the Taylor polynomial of degree 4 for $f(x) = \sin x$ but with $a = 0$. The result was a completely different polynomial.

This is not surprising when we realize that the polynomial in Example 2 approximates the function $f(x) = \sin x$, *starting from the point $x = 0$*, while the one we have just found approximates the same function, *starting from $\frac{\pi}{4}$*. It is graphed in Figure 9.1.5, where we see that it is close to $\sin x$ for x near $\frac{\pi}{4}$.

Figure 9.1.5

Exercises 9.1

I

For each function $f(x)$, find the indicated Taylor polynomial (i.e., $P_n(x)$, with the specified n and the specified value of the starting point a).

1. For $f(x) = \cos x$, $a = 0$, find $P_4(x)$.
2. For $f(x) = \cos x$, $a = \frac{\pi}{2}$, find $P_4(x)$.
3. For $f(x) = e^x$, $a = 0$, find $P_5(x)$.
4. For $f(x) = e^{-x}$, $a = 1$, find $P_4(x)$.
5. For $f(x) = \ln(x + 1)$, $a = 0$, find $P_3(x)$.
6. For $f(x) = \cosh x$, $a = 0$, find $P_4(x)$.
7. For $f(x) = x^3$, $a = 1$, find $P_3(x)$.
8. For $f(x) = x^3$, $a = 2$, find $P_4(x)$.

9. For $f(x) = x^4$, $a = -1$, find $P_{12}(x)$.
10. For $f(x) = e^{\ln x}$, $a = 3$, find $P_6(x)$.

Use the Taylor polynomial of the specified degree and about the specified point a to estimate the following values. In each case, compare your answer with what you get using your calculator.

11. $\sin\left(\frac{1}{100}\right)$, $f(x) = \sin x$, $a = 0$, P_5
12. $\cos\left(\frac{1}{10}\right)$, $f(x) = \cos x$, $a = 0$, P_4
13. $e^{0.1}$, $f(x) = e^x$, $a = 0$, P_4
14. $e^{-0.03}$, $f(x) = e^x$, $a = 0$, P_4
15. $\sin\left(\pi - \frac{1}{5}\right)$, $f(x) = \sin x$, $a = \pi$, P_4

16. $\cos\left(\frac{1}{6} + \pi\right)$, $f(x) = \cos x$, $a = \pi$, P_4

17. $\cosh(0.15)$, $f(x) = \cosh x$, $a = 0$, P_4

18. $\sinh\left(\frac{2}{15}\right)$, $f(x) = \sinh x$, $a = 0$, P_4

19. $\ln(0.9)$, $f(x) = \ln x$, $a = 1$, P_5

20. $f(17)$, with $f(x) = x^2 - 3x + 1$, $a = 2$, P_3

II

21. Find the Taylor polynomial of degree 5 for $f(x) = e^x$ about $a = 0$, and use it to estimate $\sqrt{e} = e^{1/2}$.

22. Find the Taylor polynomial of degree 6 about the point $a = 3\pi$ for the function $f(x) = \cos^2 x + \sin^2 x$.

23. (i) Estimate $\sin\left(\frac{\pi}{4}\right)$ using the Taylor polynomial of degree 4 with $a = 0$.

(ii) Estimate the same value using $a = \frac{\pi}{2}$.

(iii) Which is the better approximation?

24. (i) Estimate $\ln(1.5)$ using the Taylor polynomial of degree 4 about $a = 1$.

(ii) Make the same estimate with the Taylor polynomial of degree 4 about $a = 2$.

(iii) Which is the better approximation?

*25. If $f(x)$ and $g(x)$ are n times differentiable at $x = a$, let $P_n(x)$ and $Q_n(x)$ be their respective Taylor polynomials of degree n about the point $x = a$. Explain why the Taylor polynomial of degree n about $x = a$ of $h(x) = f(x) + g(x)$ is $S_n(x) = P_n(x) + Q_n(x)$.

*26. (i) If $f(x)$ is a polynomial of degree n, what is its Taylor polynomial of degree n about $x = 0$?

(ii) If $f(x)$ is a polynomial of degree n and $m \geq n$, what is its Taylor polynomial $P_m(x)$ about $x = a$?

(iii) Describe $P_m(x)$ if $m < n$.

*27. Suppose $f(x)$ is a function and $p_n(x)$ is its Taylor polynomial of degree n about the point $x = a$. Let r be any nonzero number, let $F(x) = f(rx)$, and verify that the Taylor polynomial of degree n associated to $F(x)$ about $x = \frac{a}{r}$ is $P_n(x) = p_n(rx)$.

*28. Compare the Taylor polynomial $P_4(x)$ for $f(x) = x^4 - 3x^2$ at $a = -1$ with the polynomial you get by writing $x = (x + 1) - 1$ and expanding $x^4 = \left((x + 1) - 1\right)^4$ and $x^2 = \left((x + 1) - 1\right)^2$ by the Binomial Theorem.

*29. (i) Compare the Taylor polynomials of degree 4 about $a = 0$ associated to $f(x) = \cos(2x)$ and $g(x) = \cos x$.

(ii) Compare the Taylor polynomials of degree 4 about $a = 0$ associated to $F(x) = \cos(x^2)$ and $g(x) = \cos x$.

30. Using the Taylor polynomial $P_6(x)$ corresponding to the function $f(x) = \ln(x)$ about $a = 1$, evaluate $P_6(2)$ to estimate $\ln 2$. How good is the estimate?

*31. Sketch the graphs of the Taylor polynomials $P_1(x)$, $P_2(x)$, and $P_4(x)$ associated to the function $f(x) = \cos x$ at $a = 0$.

*32. Sketch the graphs of the Taylor polynomials $P_1(x)$, $P_3(x)$, and $P_5(x)$ associated to the function $f(x) = \sin x$ at $a = 0$.

*33. (i) Let $f(x) = \sin x$ and for each positive integer k evaluate $f^{(k)}(0)$. (*Hint:* Write $k = 4m + i$, with $i = 0, 1, 2,$ or 3.)

(ii) Write an expression for the kth coefficient of the nth Taylor polynomial $P_n(x)$ about $a = 0$ (for $k \leq n$).

*34. Write an expression for the kth coefficient of the nth Taylor polynomial $P_n(x)$ of $g(x) = \cos x$ about $a = 0$ (for $k \leq n$).

III

*35. Use a graphics calculator or computer to plot the graphs of several of the Taylor polynomials $P_1(x), P_2(x), P_3(x), \ldots$ for $f(x) = \sin x$ about the point $a = \frac{\pi}{2}$ (cf. the previous exercises).

*36. (i) With a computer or programmable calculator, write a program to evaluate the Taylor polynomial $P_{10}(x)$ associated to the function $f(x) = e^x$ at $a = 0$.

(ii) Try your program for different values of x, comparing the answers you get with what your calculator tells you to expect. You should find that if x is a reasonably small number, the results are remarkably close; on the other hand, there is clearly some trouble if x is very large (e.g., if $x = 100$). The way your calculator finds e^x likely includes evaluating a Taylor polynomial, just like this, though it probably has a slightly more clever way of dealing with large numbers, but for small numbers this is exactly what it does.

*37. (i) Write a program to evaluate the Taylor polynomial $P_{10}(x)$ for $f(x) = \ln(x)$ about $a = 1$.

(ii) Evaluate $P_{10}(2)$, $P_{10}(3)$, $P_{10}(5)$, and compare with $\ln 2$, $\ln 3$, $\ln 5$.

(iii) Modify your program to evaluate $P_{100}(x)$ and repeat part (ii).

38. Plot $\sin x$ and the corresponding Taylor polynomials about $a = 0$: $P_1(x)$, $P_3(x)$, $P_5(x)$, $P_7(x)$, $P_9(x)$.

39. Plot $\cos x$ and the corresponding Taylor polynomials about $a = 0$: $P_2(x)$, $P_4(x)$, $P_6(x)$, $P_8(x)$, $P_{10}(x)$.

40. Plot $\sin x$ and the corresponding Taylor polynomials about $a = \frac{\pi}{4}$: $P_1(x)$, $P_4(x)$, $P_7(x)$, $P_{10}(x)$.

9.2 Taylor's Formula with Remainder; Approximations

REMAINDER
LAGRANGE FORM
ESTIMATING THE REMAINDER

In Section 9.1 we saw how to construct the Taylor polynomials for a function $f(x)$ about a point $x = a$. The Taylor polynomial of degree n is the polynomial $P_n(x)$ of degree n (or less) that has the same value as $f(x)$ at $x = a$ and that has derivatives with the same values at $x = a$ up to and including the nth derivative. We expect that the polynomial $P_n(x)$ will behave much as $f(x)$ behaves, at least for x's near $x = a$.

We saw how it is possible to use Taylor polynomials to approximate the values of $f(x)$. This can be very useful in situations in which we have no way to evaluate $f(x)$ precisely. The examples that we did suggested that even with fairly small values of n (such as 4 or 5) it is possible to get fairly accurate approximations, especially if we are evaluating at a value x that is near a.

However, if we want to use this technique for calculations, it is essential to have some control over the accuracy of the approximation.

Suppose $f(x)$ is a function that is n times differentiable at $x = a$. Write $P_n(x)$ for the Taylor polynomial of degree n at $x = a$. The problem is to estimate the size of the difference $f(x) - P_n(x)$. This difference is called the **remainder**, and we write

$$R_n(x) = f(x) - P_n(x). \tag{9.2.1}$$

In fact there is a convenient formula that allows us to estimate the size of $R_n(x)$ (and with it the accuracy of the approximation).

THEOREM 9.2.2

Taylor's Formula with Remainder

Suppose the function $f(x)$ is defined on an interval that contains a and x, and suppose its $(n + 1)$th derivative $f^{(n+1)}$ exists on this interval. Then the nth remainder is

$$R_n(x) = \frac{f^{(n+1)}(c)}{(n+1)!}(x - a)^{n+1},$$

for some number c between a and x. The number c depends on x.

This can also be written in the following way:

$$f(x) = P_n(x) + R_n(x)$$

$$= f(a) + \frac{f'(a)}{1!}(x - a) + \ldots + \frac{f^{(n)}(a)}{n!}(x - a)^n$$

$$+ \frac{f^{(n+1)}(c)}{(n+1)!}(x - a)^{n+1}.$$

In this form we see that the formula for the remainder is almost identical to what the next term would be if we continued the Taylor polynomial one more step (i.e., if we found $P_{n+1}(x)$), with the one exception that the $(n + 1)$th derivative $f^{(n+1)}(c)$ is evaluated at c rather than at a. This makes it quite easy to remember.

This formula for the remainder was actually found by Joseph Louis Lagrange (1736–1813), a French-Italian mathematician, and is sometimes called the **Lagrange form of the remainder**. It is proved using Rolle's Theorem (Theorem 4.2.4), but the proof is tricky, so we postpone it until Example 7 at the end of the section. We prefer to consider some examples first in order to see the significance and usefulness of the theorem.

One apparent disadvantage of the Lagrange form of the remainder is that, like Rolle's Theorem itself, it does not give us any way to find the number c at which the $(n+1)$th derivative is evaluated. In practice this is not a terrible handicap; it is usually possible to estimate the size of that derivative without knowing the point exactly. We know that c is between a and x, that is, that $a \leq c \leq x$ or $x \leq c \leq a$, so all that is required is a bound for the size of $f^{(n+1)}(u)$ for all $u \in [a, x]$ or $[x, a]$, as the case may be.

EXAMPLE 1   **Estimating e**   If we find the Taylor polynomial $P_4(x)$ for $f(x) = e^x$ about the point $a = 0$ and use it to estimate the numerical value of e, how good will the approximation be?

Solution   It is very easy to find the derivatives of $f(x) = e^x$; they are all the same: $f^{(k)}(x) = e^x$, so $f^{(k)}(0) = e^0 = 1$. The Taylor polynomial is

$$P_4(x) = 1 + \frac{x}{1!} + \frac{x^2}{2!} + \frac{x^3}{3!} + \frac{x^4}{4!}$$

$$= 1 + x + \frac{x^2}{2} + \frac{x^3}{6} + \frac{x^4}{24}.$$

The approximate value of $e = f(1)$ is $P_4(1) = 1 + 1 + \frac{1}{2} + \frac{1}{6} + \frac{1}{24} = 2\frac{17}{24} \approx 2.708\,333$.

Now Theorem 9.2.2 tells us that the remainder is $R_4(1) = \frac{1}{5!}f^{(5)}(c) = \frac{e^c}{120}$, for some $c \in [0, 1]$. What we need is to know how big this could be, which amounts to knowing how big e^c could be for $c \in [0, 1]$.

That is easy. After all, $f(x) = e^x$ is an increasing function (its derivative, e^x, is always positive), so its largest value on the interval $[0, 1]$ will be at the right end of the interval. We see that $e^c \leq e^1 = e$, so $R_4(1) \leq \frac{e}{120}$.

Knowing that $e < 3$, we see that the remainder is smaller than $\frac{3}{120} = \frac{1}{40} = 0.025$. We conclude that the estimate we found above for the value of e is correct to within an error of at most 0.025.   ◢

In fact, $e \approx 2.718\,282$, so our estimate is off by about 0.009\,949, or less than half the error estimated by the formula for the remainder.

This illustrates an important point. We do not expect to find *exactly* what the remainder is. After all, since $f(x) = P_n(x) + R_n(x)$, if we knew $R_n(x)$ exactly, then we could find $f(x)$ exactly too. That would be nice, but it is too much to expect. So the formula for the remainder allows us to give a bound—a limitation—on how large the remainder is, and that allows us to say with confidence how accurate the approximation is. In practice, the approximation will often be better than what we can guarantee using the estimate for the remainder.

The last example shows how it is possible to use the estimate of Theorem 9.2.2 for the remainder to assess the accuracy of a Taylor polynomial approximation. Most often, however, it is used the other way around. Instead of starting with the polynomial $P_n(x)$ and asking how accurate an approximation it gives for some particular x, we start with the function $f(x)$ and want to approximate its value at some point to within a specified accuracy. The problem we need to answer is: How large must n be in order that $P_n(x)$ has the desired accuracy?

EXAMPLE 2   Calculate e to within 0.000 001.

Solution   We use $a = 0$ as the starting point, and we know that $f^{(k)}(0) = 1$, for every k, so it will be very easy to find the Taylor polynomial $P_n(x)$. The question is which n is the one to use.

The remainder for $P_n(1)$ is $R_n(1) = \frac{1}{(n+1)!} f^{(n+1)}(c)(x - a)^{n+1} = \frac{1}{(n+1)!} e^c (1 - 0)^{n+1} = \frac{1}{(n+1)!} e^c$. Since $0 \le c \le 1$, we see that $e^c \le e^1 = e$, and the remainder is less than or equal to $\frac{e}{(n+1)!}$. Since $e < 3$, we see that $R_n(1) < \frac{3}{(n+1)!}$.

All we need is to find a value of n for which this is less than 0.000 001. We try a few possibilities: $\frac{3}{5!} = \frac{3}{120} = 0.025$. This is much too big. Let us jump to $\frac{3}{8!} = \frac{3}{40,320} \approx 0.000 0744$. This is still too big, but it is much closer. Evaluating $\frac{3}{9!} = \frac{3}{362,880} \approx 0.000 008 27$, we see that it is nearly what we want, but not quite. However, the next one is $\frac{3}{10!} = \frac{3}{3,628,800} \approx 0.000 000 827$, which is definitely smaller than 0.000 001.

We conclude that if $n + 1 = 10$, that is, $n = 9$, the remainder will be small enough, so we approximate $e = f(1)$ by $P_9(1)$:

$$P_9(1) = 1 + \frac{1}{1!} + \frac{1^2}{2!} + \frac{1^3}{3!} + \frac{1^4}{4!} + \frac{1^5}{5!} + \frac{1^6}{6!} + \frac{1^7}{7!} + \frac{1^8}{8!} + \frac{1^9}{9!}$$

$$= 1 + 1 + \frac{1}{2} + \frac{1}{6} + \frac{1}{24} + \frac{1}{120} + \frac{1}{720} + \frac{1}{5040} + \frac{1}{40,320} + \frac{1}{362,880}$$

$$\approx 2.718 281 53.$$

This certainly agrees to within the desired accuracy with $e \approx 2.718 281 83$.

EXAMPLE 3   **Estimating ln 2**   If we want to estimate ln 2 by using a Taylor polynomial $P_n(x)$ associated to the function $f(x) = \ln x$ about the point $a = 1$, how large should n be in order to ensure that the approximation is correct to within 0.0001?

Solution   The remainder associated to $P_n(x)$ is $R_n(x) = \frac{1}{(n+1)!} f^{(n+1)}(c)(x - a)^{n+1}$, for some c between a and x. In this case, $a = 1$ and the x we are interested in is 2.

As we see in the margin, the derivatives are not difficult to find; the pattern is fairly clear. The kth derivative contains the $-k$th power of x. Differentiating it changes the power to the $-(k + 1)$th power and multiplies by $-k$. So the $(k + 1)$th derivative is the product of $x^{-(k+1)}$ and the product of the integers $-1, -2, -3, \ldots,$ $-k$, which is $(-1)^k k! x^{-(k+1)}$.

We see that $R_n(2) = \frac{1}{(n+1)!} \left((-1)^n n! c^{-(n+1)} \right)$, for some $c \in [1, 2]$. Since $x^{-(n+1)}$ is a *decreasing* function, its largest value will occur when $x = 1$ (at the left end of the interval). So

$$|R_n(2)| \le \frac{n!}{(n+1)!} 1^{-(n+1)} = \frac{1}{n+1}.$$

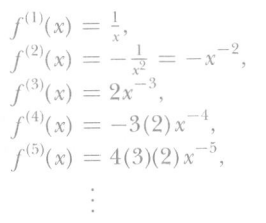
$f^{(1)}(x) = \frac{1}{x},$
$f^{(2)}(x) = -\frac{1}{x^2} = -x^{-2},$
$f^{(3)}(x) = 2x^{-3},$
$f^{(4)}(x) = -3(2)x^{-4},$
$f^{(5)}(x) = 4(3)(2)x^{-5},$
$\vdots$

To ensure the desired accuracy, we need to have $\left|R_n(2)\right| \le 0.0001$, and to do that, we need to have $\frac{1}{n+1} \le 0.0001$, or $n + 1 \ge \frac{1}{0.0001} = 10{,}000$. This means that we need to take $n \ge 9999$ and evaluate $P_{9999}(x)$.

This is frustrating because it is not practical to evaluate so large a polynomial. If nothing else, round-off errors from a calculation this long are likely to make it unreliable anyway. It turns out that for some "nice" functions, like e^x and $\sin x$ and $\cos x$, for instance, the remainder gets quite small after only a moderate number of terms, which makes Taylor polynomials an efficient way of estimating these functions. However, the bad news is that there are other functions for which the remainder stays uncomfortably large unless n is itself extremely large, and as a practical tool Taylor polynomials have very limited use for these functions.

EXAMPLE 4   Find the value of $\sin(3°)$ to within 0.0001.

Solution   As usual, we will have to convert the angle $3°$ into radians; it is $3 \times \frac{\pi}{180} = \frac{\pi}{60}$ radians. If we let $f(x) = \sin x$, what we want to estimate is $f\left(\frac{\pi}{60}\right)$; we can do this by using Taylor polynomials about the point $a = 0$.

Notice that the first derivative is $f'(x) = \cos x$, the second derivative is $-\sin x$, the third derivative is $-\cos x$, and the fourth derivative is $\sin x$. After four derivatives we have come back to the original function $\sin x$. So the fifth, sixth, seventh, and eighth derivatives will be $\cos x$, $-\sin x$, $-\cos x$, and $\sin x$ again, and this pattern will keep repeating again and again. When we want to estimate the size of a remainder $R_n(x)$, we have to estimate the size of the $(n+1)$th derivative $f^{(n+1)}(x)$, and since every derivative is $\pm \sin x$ or $\pm \cos x$, we know that its absolute value is never bigger than 1.

The nth remainder is $R_n(x) = \frac{1}{(n+1)!}f^{(n+1)}(c)(x - a)^{n+1}$, for some $c \in [0, x]$. In our case, $a = 0$ and $x = \frac{\pi}{60}$, so the absolute value of the remainder satisfies $\left|R_n(x)\right| \le \frac{1}{(n+1)!}(1)\left|x - a\right|^{n+1} = \frac{1}{(n+1)!}\left(\frac{\pi}{60}\right)^{n+1}$.

The goal is to make this less than (or equal to) 0.0001, by choosing an appropriate n. Asking that $\frac{1}{(n+1)!}\left(\frac{\pi}{60}\right)^{n+1} \le 0.0001$ is the same as asking that $(n + 1)!\left(\frac{60}{\pi}\right)^{n+1} \ge 10{,}000$. (We just inverted both sides of the inequality, remembering to reverse the direction of the inequality sign.)

With a calculator we find that $\frac{60}{\pi} \ge 19$, so it will be enough to find n so that $(n + 1)!19^{n+1} \ge 10{,}000$ (since $(n + 1)!\left(\frac{60}{\pi}\right)^{n+1}$ is even bigger ...). This we do by trial and error.

First, with $n = 1$ we find that $(n + 1)!19^{n+1} = 2!19^2 = 2 \times 19^2 = 722$, which is not big enough. With $n = 2$ we find that $(n + 1)!19^{n+1} = 3! \times 19^3 = 6 \times 19^3 = 41{,}154$, which is more than sufficient.

In other words, it is only necessary to use the Taylor polynomial $P_2(x)$ in order to have the desired accuracy. Evaluating the derivatives, we find that $P_2(x) = 0 + \frac{1}{1!}(x - 0) + 0 = x$. This Taylor polynomial is extraordinarily simple, but it works.

We find $P_2\left(\frac{\pi}{60}\right) = \frac{\pi}{60} \approx 0.052\,36$. (With a calculator, set to "degrees" for once, we find that $\sin(3°) \approx 0.052\,336$, so our answer does agree to within the specified accuracy of 0.0001.)

Notice that the Taylor polynomial $P_2(x)$ that we used in this example is really just the tangent line at $x = 0$ (see Figure 9.2.1). We have used Lagrange's form of the remainder to check that $P_2(x)$ gives an approximation that is sufficiently accurate,

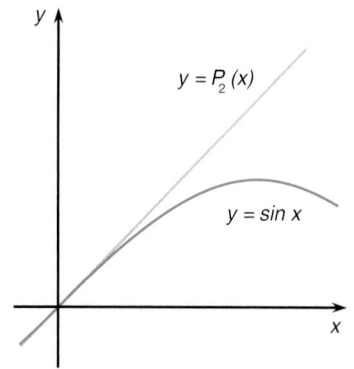

Figure 9.2.1

$f^{(0)}(0) = \sin(0) = 0,$
$f^{(1)}(0) = \cos(0) = 1,$
$f^{(2)}(0) = -\sin(0) = 0.$

for $x = \frac{\pi}{60}$. Then we calculated the value of the Taylor polynomial, which is in effect the same thing as the linear approximation (cf. Section 4.10).

Notice that because the second derivative $f^{(2)}(x)$ is zero at $x = 0$, the Taylor polynomial $P_2(x)$ is actually the same as the Taylor polynomial $P_1(x)$. When we estimated the remainder, the estimate we found for $R_1(x)$ was too big, so we had to use $R_2(x)$, for which we had a smaller estimate. Of course, since we know that $P_1(x)$ and $P_2(x)$ are the same, we see that the remainders are equal too. The point is that the *estimate* we made of the size of the remainder was smaller for $P_2(x)$ than for $P_1(x)$, and it was knowing the size of the estimate that enabled us to use $P_2(x)$ to find a value with confidence.

Since we are only *estimating* the size of the remainder, it may easily happen that the actual remainder is smaller than our estimate. It may also happen that we decide that it is necessary to use some $P_n(x)$ to get the desired accuracy when in fact a Taylor polynomial of some smaller degree would be adequate. This is not serious. At worst it means that we may have to calculate a few more terms of the polynomial than what is strictly necessary.

Another consideration is the choice of the "starting point" a about which we define the Taylor polynomials. So far we have not worried about which a to use, but in practice we are likely to be asked to approximate the value of some function at some particular point, and part of the solution will be to choose the starting point a.

The basic idea is to choose a to be as close as possible to the point at which the function is to be approximated. The catch is that it must be a point at which we know the values of the function and its derivatives so that we can construct the Taylor polynomials.

The formula for the remainder is

$$R_n(x) = \frac{f^{(n+1)}(c)}{(n+1)!}(x-a)^{n+1}.$$

If $|x - a|$ is small, then $(x - a)^{n+1}$ will approach 0 quickly as n increases. This will tend to make the remainder terms small without n having to be too large. If, on the other hand, the point x is far from a, then the remainder will contain increasing powers of the fairly large number $(x - a)$, which will make it less likely to become small for moderate values of n.

EXAMPLE 5   Estimate the value of $\arctan(-0.9)$ to within 0.0005.

Solution   We could try constructing Taylor polynomials about the point $a = 0$, but that is actually fairly far from -0.9. A much better choice would be $a = -1$.

If $f(x) = \arctan x$, then $f^{(1)}(x) = \frac{1}{1+x^2} = (1 + x^2)^{-1}$, and $f^{(2)}(x) = -(1 + x^2)^{-2}(2x) = -2x(1 + x^2)^{-2}$. We then find $f^{(3)}(x) = -2(1 + x^2)^{-2} + (-2)(-2x)(1 + x^2)^{-3}(2x) = -2(1 + x^2)^{-2} + 8x^2(1 + x^2)^{-3}$. Putting this over a common denominator, we find

$$f^{(3)}(x) = \frac{-2(1 + x^2) + 8x^2}{(1 + x^2)^3} = \frac{6x^2 - 2}{(1 + x^2)^3}.$$

Let us use this information to estimate the remainder $R_2(-0.9)$. We know that $R_2(-0.9) = \frac{1}{3!}f^{(3)}(c)\left(-0.9 - (-1)\right)^3 = \frac{(0.1)^3}{6}f^{(3)}(c)$, for some c between -1 and -0.9.

We have to estimate the size of $f^{(3)}(c)$, for $c \in [-1, -0.9]$. It is probably easier to work separately with the numerator and the denominator. The numerator of $f^{(3)}(x)$ is $6x^2 - 2$. This is a function whose derivative $12x$ is always negative on the interval in question, so it is a decreasing function. Its largest value will occur at the left endpoint, where it equals $6(-1)^2 - 2 = 4$. So the numerator is always less than or equal to 4 on the interval.

For the denominator we have to be careful. What we want is to be able to say that $\frac{1}{(1+x^2)^3}$ is less than or equal to some number. Because of the way inequalities reverse when we take reciprocals, this means that we have to be able to say that $(1 + x^2)^3$ is *greater* than or equal to some number; we have to find the *minimum* possible value of the denominator. Certainly, $(1 + x^2)^3$ will have its smallest possible value at $x = -0.9$, so $\frac{1}{(1+x^2)^3} \leq \frac{1}{(1+(-0.9)^2)^3}$.

From this we see that

$$f^{(3)}(c) = \frac{6c^2 - 2}{(1 + c^2)^3} \leq \frac{4}{\left(1 + (-0.9)^2\right)^3} \approx 0.67457 \leq 0.7.$$

Using this it is possible to estimate the remainder:

$$R_2(-0.9) = \frac{f^{(3)}(c)}{3!} \left(-0.9 - (-1)\right)^3 \leq \frac{0.7}{6}(0.1)^3 \approx 0.000\,117.$$

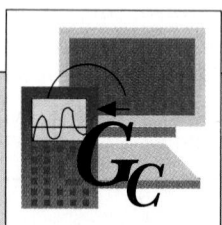

With a graphics calculator we can plot the function $y = \arctan x$ and the Taylor polynomial $P_2(x)$ from Example 5. For comparison we also plot the Taylor polynomial of degree less than or equal to 2 around $a = 0$. Since $f(0) = f''(0) = 0$ and $f'(0) = 1$, we find that it is x (and its degree is actually 1).

TI-85	SHARP EL9300
GRAPH F1$(y(x) =)$ TAN^{-1} F1(x) ▼ $(-) \pi \div 4 + ($ F1$(x) + 1) \div$ 2 $+ ($ F1$(x) + 1)$ ^2 $\div 4$ ▼ F1(x) M2(RANGE) $(-)$ 1 ▼ 0 ▼ ▼ $(-)$ 1 ▼ 0 F5(GRAPH)	⌐⌐ tan^{-1} x/θ/T 2ndF ▼ $(-) \pi \div 4$ $+ ($ x/θ/T $+ 1) \div 2 + ($ x/θ/T $+ 1)$ x^2 $\div 4$ 2ndF ▼ x/θ/T RANGE -1 ENTER 0 ENTER 1 ENTER -1 ENTER 0 ENTER 1 ENTER ⌐⌐
CASIO f_x-6300G (f_x-7700GB)	HP 48SX
Range -1 EXE 0 EXE 1 EXE -1 EXE 0 EXE 1 EXE EXE EXE EXE Graph tan^{-1} ▓ EXE Graph $(-) \pi \div 4 + ($ ▓ $+ 1$ $) \div 2 + ($ ▓ $+ 1)$ $x^2 \div 4$ EXE Graph ▓ EXE	▓ PLOT PLOTR ERASE ATTN ' Y ▓ TAN X ↵ PLOT ▓ DRAW ERASE 1 +/− ENTER 0 XRNG 1 +/− ENTER 0 YRNG DRAW ATTN ' Y ▓ $\pi \div 4$ +/− $+ ($ ▓ $X + 1$ ▶ $\div 2$ $+ ($ ▓ $X + 1$ ▶ $y^x 2 \div 4$ ↵ PLOT ▓ DRAW DRAW ATTN ' Y ▓ X ↵ PLOT ▓ DRAW DRAW

The Taylor polynomial $P_2(x)$ is indistinguishable from arctan x near $x = -1$, but it gets farther away near $x = 0$. The Taylor polynomial around $a = 0$ is indistinguishable from arctan x near $x = 0$, but it gets farther away near $x = -1$.

$$
\begin{aligned}
f(x) &= \arctan x, \\
f(-1) &= \arctan(-1) \\
&= -\tfrac{\pi}{4}; \\
f'(x) &= \tfrac{1}{1+x^2}, \\
f'(-1) &= \tfrac{1}{1+(-1)^2} = \tfrac{1}{2}; \\
f^{(2)}(x) &= \tfrac{-2x}{(1+x^2)^2}, \\
f^{(2)}(-1) &= \tfrac{-2(-1)}{(1+(-1)^2)^2} = \tfrac{1}{2}.
\end{aligned}
$$

Using the Taylor polynomial $P_2(x)$ about the point $x = -1$ will certainly give us sufficient accuracy. To construct $P_2(x)$, we evaluate up to $f^{(2)}(-1)$. We find

$$
\begin{aligned}
P_2(x) &= f(-1) + \frac{1}{1!}f'(-1)\big(x - (-1)\big) + \frac{1}{2!}f^{(2)}(-1)\big(x - (-1)\big)^2 \\
&= -\frac{\pi}{4} + \frac{x+1}{2} + \frac{(x+1)^2}{4}.
\end{aligned}
$$

Evaluating at $x = -0.9$, we find the approximation $P_2(-0.9) = -\frac{\pi}{4} + \frac{0.1}{2} + \frac{(0.1)^2}{4} \approx -0.732\,898$. This is remarkably good; with a calculator we find that $\arctan(-0.9) \approx -0.732\,815$.

The importance of this example becomes clear when we consider what happens if we use $a = 0$ as the starting point. In estimating the size of the remainder $R_2(x)$ in that case we see that the smallest value of the denominator on $[-0.9, 0]$ is at 0, so on this interval $\frac{1}{(1+x^2)^3} \le \frac{1}{1} = 1$. The numerator is more complicated in that it is sometimes positive and sometimes negative; it decreases from $6(-0.9)^2 - 2 = 2.86$ at $x = -0.9$ to $6(0)^2 - 2 = -2$ at $x = 0$. Its absolute value is never bigger than 2.86. The remainder has absolute value

$$
|R_2(c)| = \left| \frac{f^{(3)}(c)}{3!}(-0.9 - 0)^3 \right| \le \left| \frac{1}{6}\frac{2.86}{1}(-0.729) \right| = 0.347\,49.
$$

This is not nearly good enough to give us the accuracy we require. Moreover, to calculate the higher derivatives $f^{(4)}(x)$, $f^{(5)}(x)$, and so on becomes increasingly messy. It would be a major undertaking to find how large n would have to be in order that $P_n(x)$ constructed around $a = 0$ would be sufficiently accurate.

REMARK 9.2.3

1. To measure the accuracy of the approximation by $P_n(x)$, it is necessary to estimate the size of the remainder $R_n(x)$. Generally speaking, this means that we should estimate the *absolute value* of the remainder, which tells how far the estimate is from the actual value. It is sometimes possible to obtain additional information by using the sign of the remainder to say whether the actual value is above or below the approximation.

2. The Lagrange formula for the remainder $R_n(x)$ is the product of $\frac{1}{(n+1)!}$, $f^{(n+1)}(c)$, and $(x - a)^{n+1}$. The first and last are easy to evaluate exactly; it is the derivative that we must estimate, especially because the point c is not determined precisely. What is known is that c is between a and x. If $f^{(n+1)}$ is increasing between a and x (or decreasing), then it is easy to find the maximum possible value or, what is more important, the maximum absolute value. This can then be used in the estimate of the remainder.

3. It is frequently necessary to estimate the size of a quotient $\frac{p(x)}{q(x)}$. It often helps to write $\frac{p(x)}{q(x)} = p(x)\frac{1}{q(x)}$ and work separately with $p(x)$ and $\frac{1}{q(x)}$. To show that $\left|\frac{1}{q(x)}\right| \le B$, for some constant B, it is necessary to show that $|q(x)| \ge \frac{1}{B}$. (It is very important to remember that the inequality is reversed when we take reciprocals.) So the idea is to show that $|p(x)| \le A$, for some constant A, and that $|q(x)| \ge C$, for some constant $C > 0$. From this it is possible to conclude that the original quotient satisfies $\left|\frac{p(x)}{q(x)}\right| \le \frac{A}{C}$.

4. From the formula for $R_n(x)$ we see that the two things that influence the size of the remainder are the size of the $(n + 1)$th derivative of $f(x)$ and the size of $(x - a)$. The size of the various derivatives can vary tremendously. For some functions, such as e^x, $\cos x$, or $\sin x$, we have noticed that on a fixed interval the derivatives are all bounded by some constant that does not depend on n. Since the derivative is divided by $(n + 1)!$, the remainder in such a case will decrease rapidly as n increases. The effect is that approximation by the Taylor polynomial $P_n(x)$ will likely be quite accurate with fairly small values of n. For other functions, such as $\ln x$, the absolute value of the $(n + 1)$th derivative gets larger and larger as n increases, so the remainder does not decrease as rapidly. The practical effect may be that in order to have a sufficiently accurate approximation it will be necessary to use $P_n(x)$ with an unreasonably high n.

The other consideration is the size of $x - a$, that is, how far away from the starting point a is the point at which we wish to estimate $f(x)$. If $x - a$ is very small, then the remainder $R_n(x)$ will likely become small for fairly small values of n even if the derivatives of the function increase in size. In this situation it will still be practical to use Taylor polynomials to approximate the value $f(x)$.

All of the examples we have done so far have estimated values that could be obtained almost instantly on a calculator, with better accuracy, so the importance of Taylor polynomials is not very clear. It is, of course, conceivable that we might want to calculate e to 300 decimal places, and in principle we now know how to set about doing that. There are instances in which calculator accuracy is insufficient, and it is reassuring to know that it is possible to have any desired degree of accuracy by using suitable Taylor polynomials $P_n(x)$. However, these uses are not encountered very often. Without necessarily realizing it, we rely on Taylor polynomials when we use our calculators, since calculators use Taylor polynomials to evaluate many built-in functions.

There are some other situations in which Taylor polynomials are very important and less easily circumvented. It is easy to write down examples of differential equations that we do not know how to solve. In this situation it is often easy to write down the Taylor polynomials of the solution and use them to find approximate values for the solution.

One of the simplest examples of this procedure is estimating an integral, which can be viewed as a differential equation because it asks for an indefinite integral, a function whose derivative equals the integrand.

EXAMPLE 6   Estimate to within 0.01 the following integral, which is very important in statistics:

$$\int_0^{1/2} e^{-x^2} dx.$$

Solution   None of the techniques of integration that we have learned can help us to evaluate this integral. Instead, we consider the indefinite integral $F(x) = \int_0^x e^{-t^2} dt$, realizing that the value we want to estimate is $F\left(\frac{1}{2}\right)$. Let us approximate it by Taylor polynomials about the point $a = 0$. By the Fundamental Theorem of Calculus we know that $F^{(1)}(x) = e^{-x^2}$, and from this we find that

$$F^{(2)}(x) = -2xe^{-x^2}, \qquad\qquad\qquad\qquad F^{(2)}(0) = 0;$$
$$F^{(3)}(x) = -2e^{-x^2} + 4x^2e^{-x^2}, \qquad\qquad\quad F^{(3)}(0) = -2;$$
$$F^{(4)}(x) = 4xe^{-x^2} + 8xe^{-x^2} - 8x^3e^{-x^2} = (12x - 8x^3)e^{-x^2}, \qquad F^{(4)}(0) = 0;$$
$$F^{(5)}(x) = 12e^{-x^2} - 24x^2e^{-x^2} - 24x^2e^{-x^2} + 16x^4e^{-x^2}$$
$$= (12 - 48x^2 + 16x^4)e^{-x^2}, \qquad\qquad F^{(5)}(0) = 12.$$

Let us estimate $|R_3(\frac{1}{2})|$. Since e^{-x^2} is a decreasing function of x for $x \geq 0$, we see that $e^{-x^2} \leq e^0 = 1$, for all $x \in [0, \frac{1}{2}]$. If $c \in [0, \frac{1}{2}]$, then $|c| \leq \frac{1}{2}$, and using the Triangle Inequality we find that

$$|F^{(4)}(c)| = |(12c - 8c^3)e^{-c^2}| \leq \left(12\left(\frac{1}{2}\right) + 8\left(\frac{1}{2}\right)^3\right) \times 1 = 7.$$

From this we find that $|R_3(\frac{1}{2})| \leq \frac{1}{4!} \times 7 \times |(\frac{1}{2} - 0)^4| \approx 0.0182$.

This is not quite good enough, so we should try R_4. We find that $|F^{(5)}(c)| = |(12 - 48c^2 + 16c^4)e^{-c^2}| \leq \left|\left(12 + 48(\frac{1}{2})^2 + 16(\frac{1}{2})^4\right)(1)\right| = (12 + 12 + 1)(1) = 25$. From this we see that $|R_4(\frac{1}{2})| \leq \frac{1}{5!}25(\frac{1}{2} - 0)^5 \approx 0.0065$.

This means that there will be sufficient accuracy if we calculate $P_4(\frac{1}{2})$. To construct the polynomial, we note that $F(0) = 0, F^{(1)}(0) = 1, F^{(2)}(0) = 0, F^{(3)}(0) = -2$, and $F^{(4)}(0) = 0$. The Taylor polynomial is

$$P_4(x) = \frac{0}{0!} + \frac{1}{1!}x + \frac{0}{2!}x^2 + \frac{-2}{3!}x^3 + \frac{0}{4!}x^4$$
$$= x - \frac{1}{3}x^3.$$

Evaluating at $x = \frac{1}{2}$, we find that $P_4(\frac{1}{2}) = \frac{1}{2} - \frac{1}{3}\frac{1}{8} = \frac{11}{24} \approx 0.4583$. From this we conclude that to within an accuracy of 0.01,

$$\int_0^{\frac{1}{2}} e^{-x^2}\, dx \approx 0.46.$$

It should be clear from Example 6 that estimating integrals using Taylor polynomials can be very useful. It is interesting because the answer is not something that can be checked instantly with a calculator. Although we will not deal with them here, similar remarks apply to approximating solutions of differential equations.

Here we present a proof of Taylor's Formula with Remainder.

EXAMPLE 7   Prove Theorem 9.2.2.

Solution   To emphasize that the point at which the remainder is being evaluated is *fixed*, let us call it b (instead of x as in the theorem).

It is necessary to define a special function to which Rolle's Theorem can be applied. We let

$$D(t) = f(b) - \sum_{k=0}^{n} \left(\frac{f^k(t)}{k!} (b-t)^k \right) - R_n(b) \frac{(b-t)^{n+1}}{(b-a)^{n+1}}.$$

When $t = b$, the last quotient equals 0, and since $(b-t) = (b-b) = 0$, each term in the sum is zero except the one with $k = 0$. So $D(b) = f(b) - \frac{f^{(0)}(b)}{0!} = f(b) - \frac{f(b)}{1} = 0$.

When $t = a$ the last quotient equals 1, so

$$D(a) = f(b) - \sum_{k=0}^{n} \left(\frac{f^k(a)}{k!} (b-a)^k \right) - R_n(b) = f(b) - P_n(b) - R_n(b).$$

By the definition of the remainder R_n we see that $D(a) = 0$. Combining these two results, we have that

$$D(a) = D(b) = 0,$$

which means that Rolle's Theorem (Theorem 4.2.4) can be applied to $D(t)$. We conclude that there is a number c between a and b for which $D'(c) = 0$.

What remains is to calculate $D'(c)$.

Let us begin by differentiating the kth term of the sum. By the Product Rule, if $k \neq 0$,

$$\frac{d}{dt} \left(\frac{f^{(k)}(t)}{k!} (b-t)^k \right) = \frac{f^{(k+1)}(t)}{k!} (b-t)^k + \frac{f^{(k)}(t)}{k!} k(b-t)^{k-1}(-1)$$

$$= \frac{f^{(k+1)}(t)}{k!} (b-t)^k - \frac{f^{(k)}(t)}{(k-1)!} (b-t)^{k-1}.$$

Notice that the two terms are very similar. When we add up these derivatives for every k, most of the parts will cancel one another. Using summation notation and applying Proposition 5.2.4 (about shifting the index of summation), we see that

$$\frac{d}{dt} \left(\sum_{k=0}^{n} \frac{f^{(k)}(t)}{k!} (b-t)^k \right) = \sum_{k=0}^{n} \left(\frac{f^{(k+1)}(t)}{k!} (b-t)^k - \frac{f^{(k)}(t)}{k!} k(b-t)^{k-1} \right)$$

$$= \sum_{k=0}^{n} \left(\frac{f^{(k+1)}(t)}{k!} (b-t)^k \right) - \sum_{k=1}^{n} \left(\frac{f^{(k)}(t)}{(k-1)!} (b-t)^{k-1} \right)$$

$$= \sum_{k=0}^{n} \left(\frac{f^{(k+1)}(t)}{k!} (b-t)^k \right) - \sum_{\ell=0}^{n-1} \left(\frac{f^{(\ell+1)}(t)}{\ell!} (b-t)^\ell \right)$$

$$= \frac{f^{(n+1)}(t)}{n!} (b-t)^n.$$

This is the nth term in the first sum; all the other terms cancel.

The last term of $D(t)$ is easily differentiated with respect to t, and we find that

$$D'(t) = -\frac{f^{(n+1)}(t)}{n!} (b-t)^n - R_n(b)(n+1) \frac{(b-t)^n}{(b-a)^{n+1}} (-1).$$

To say that $D'(c) = 0$ means that

$$-\frac{f^{(n+1)}(c)}{n!}(b-c)^n + R_n(b)(n+1)\frac{(b-c)^n}{(b-a)^{n+1}} = 0.$$

Dividing by $(b-c)^n$, we find that

$$R_n(b)\frac{n+1}{(b-a)^{n+1}} = \frac{f^{(n+1)}(c)}{n!},$$

or

$$R_n(b) = \frac{f^{(n+1)}(c)}{(n+1)!}(b-a)^{n+1}.$$

If we return to writing x in place of b, this is the formula of Theorem 9.2.2, which we wanted to prove. ◢

Exercises 9.2

I

In each of the following situations, estimate $f(x)$ to within the specified accuracy, using a Taylor polynomial $P_n(x)$ around the point a. (The first step is to find how large n must be in order to guarantee the desired accuracy.)

1. $f(x) = e^x$, $a = 0$, $x = \frac{1}{2}$, within 0.01

2. $f(x) = e^{2x}$, $a = 0$, $x = \frac{1}{4}$, within 0.01

3. $f(x) = \sin x$, $a = 0$, $x = 1$, within 0.0001

4. $f(x) = \cos x$, $a = 0$, $x = 2$, within 0.001

5. $f(x) = \ln x$, $a = 1$, $x = 1.2$, within 0.2

6. $f(x) = \tan x$, $a = 0$, $x = 0.1$, within 0.01

7. $f(x) = \ln x$, $a = 2$, $x = 1.9$, within 0.1

8. $f(x) = \sin x$, $a = \frac{\pi}{4}$, $x = 1$, within 0.01

9. $f(x) = x^{15}$, $a = 0$, $x = -\frac{1}{10}$, within 0.000 001

10. $f(x) = e^{x^2}$, $a = 0$, $x = \frac{1}{2}$, within 0.01

Estimate each of the following quantities to within the specified accuracy, using an appropriate Taylor polynomial $P_n(x)$ around an appropriate point a. (Specify which function $f(x)$ and which point a you are using, and calculate how large n must be in order to guarantee the desired accuracy.)

11. $\cos 4°$ to within 0.01

12. $\sin 6°$ to within 0.01

13. $\cos 43°$ to within 0.01

14. $\sin 66°$ to within 0.01

15. $\ln(1.2)$ to within 0.1

16. $\ln\left(\frac{5}{6}\right)$ to within 0.1

17. $\tan 3°$ to within 0.01

18. $\cot 46°$ to within 0.01

19. $\sinh(-0.12)$ to within 0.001

20. $\cosh(1.3)$ to within 0.01

II

21. Consider the Maclaurin polynomial $P_5(x)$ for $f(x) = \sin x$. What accuracy can you guarantee for approximations to $f(x)$ by $P_5(x)$ if $-0.5 \leq x \leq 0.5$?

22. What accuracy can you guarantee for the Taylor polynomial $P_6(x)$ associated to $f(x) = e^x$ about the point $a = 1$ if $0 < x < 2$? (You may want to use the inequalities $2.5 < e < 3$.)

*23. Suppose we want to approximate the value of $\sin^2(-2°)$ to within 0.0001. One way to do this would be to use the Taylor polynomials associated to $f(x) = \sin^2 x$ about the point $a = 0$. Unfortunately, the derivatives of $f(x)$ become fairly complicated as n increases. Another approach would be to estimate $\sin(-2°)$ by using Maclaurin polynomials for $g(x) = \sin x$ and squaring the result. This is probably easier, except that slightly more care is needed in determining which polynomial to use.

(i) To within what accuracy must we estimate

$\sin(-2°)$ in order that its square will be accurate to within 0.0001? (*Hint:* Remember that $|\sin x| \le |x|$, and assume that $|P_n(x)| \le |x|$ too.)

(ii) Estimate $\sin^2(-2°)$ to within 0.0001.

Estimate each of the following to within the specified accuracy.

*24. $\cos^2(132°)$ to within 0.001

*25. $\sin^2(1°)\cos(1°)$ to within 0.0005

*26. $\tan(-1°)$ to within 0.001

*27. $e^{\sin(47°)}$ to within 0.0001

*28. $\sec(43°)$ to within 0.001

*29. $\int_0^{1/4} e^{-x^2}\,dx$ to within 0.01

*30. $\int_{-1/8}^{1/8} e^{-x^2}\,dx$ to within 0.01

*31. $\int_0^{1/5} \sin(x^2)\,dx$ to within 0.01

*32. In Example 1 we had to estimate $\frac{e}{120}$, which we did by noting that $e < 3$. Strictly speaking, since the object of the procedure was to approximate e, we should not assume that we already know what e is. Show that $e < 3$ without using your calculator or looking up the value. (One way to proceed is to recall that e is determined by the fact that $\ln e = 1$, i.e., that $\int_1^e \frac{1}{x}\,dx = 1$. Since the derivative of $\ln x$ is positive, the function $\ln x$ is increasing, so it would be enough to show that $\ln 3 > 1$, i.e., that $\int_1^3 \frac{1}{x}\,dx > 1$. We could attempt to show this by noting that the integrand $\frac{1}{x}$ is greater than or equal to $\frac{1}{2}$ when $x \in [1, 2]$ and that it is greater than or equal to $\frac{1}{3}$ when $x \in [2, 3]$, so the integral is greater than or equal to $\int_1^2 \frac{1}{2}\,dx + \int_2^3 \frac{1}{3}\,dx = \frac{1}{2} + \frac{1}{3}$. Unfortunately, this is not greater than 1, so this attempt fails. However, you could use the same idea but divide the interval $[1, 3]$ into more than two pieces . . .).

*33. Consider the differential equation $f' = x^2 f$, and let $f(x)$ be a solution that satisfies $f(0) = 1$. (To say that it is a solution means that $f'(x) = x^2 f(x)$.)

(i) Using the equation, evaluate $f'(0)$. Then find a formula for $f''(x)$; your formula should involve $f'(x)$, but it is possible to substitute for $f'(x)$ using the differential equation and express $f''(x)$ in terms of $f(x)$. Do this and evaluate $f''(0)$.

(ii) Continue in this way and find $f^{(3)}(0)$, $f^{(4)}(0)$, and $f^{(5)}(0)$.

(iii) Write down the Taylor polynomial $P_5(x)$ associated to $f(x)$ at $a = 0$.

(iv) Use $P_5(x)$ to estimate $f(0.1)$.

(v) Recognizing that the differential equation is a separable equation, write $\frac{f'}{f} = x^2$ and integrate to find an explicit formula for $f(x)$. Use the fact that $f(0) = 1$.

(vi) Compare the answer to part (iv) with what your calculator gives you for $f(0.1)$ using the answer to part (v).

*34. Write down the Maclaurin polynomial $P_5(x)$ for the function $f(x)$ that satisfies $f(0) = 4$ and $f'(x) = \cos(x^2)f(x)$.

*35. Estimate to within 0.1 the integral $\int_0^1 e^{-x^2}\,dx$.

*36. Estimate to within 0.01 the integral $\int_0^{1/4} \sin(x^2)\,dx$.

*37. Estimate to within 0.01 the integral
$\int_0^{1/10} \arctan(\sin x)\,dx$.

*38. Estimate to within 0.05 the integral $\int_1^{3/2} \arctan(x^2 + 1)\,dx$.

39. The Maclaurin polynomial $P_6(x)$ for $f(x) = e^x$ gives a good approximation to the function. Use their graphs to estimate the smallest $x > 0$ for which $|f(x) - P_6(x)| \ge 0.0001$ and the smallest $x > 0$ for which $|f(x) - P_6(x)| \ge 0.01$.

40. Use the graphs of $f(x) = \sin x$ and its Maclaurin polynomial $P_5(x)$ to estimate the smallest $x > 0$ for which $|f(x) - P_5(x)| \ge 0.0001$ and the smallest $x > 0$ for which $|f(x) - P_5(x)| \ge 0.01$. (*Hint:* Graph $\sin x - P_5(x)$ and $y = 0.0001$, $y = -0.0001$, or $\sin x - P_5(x) - 0.0001$ and $\sin x - P_5(x) + 0.0001$.)

41. If you evaluate the Maclaurin polynomial $P_n(x)$ for e^x one term at a time, the answer on your calculator or computer will eventually stop changing, that is, additional terms no longer affect the digits you see. How large does n have to be for $P_n(1)$ to stop changing on your calculator? How about $P_n(2)$ or $P_n(10)$?

42. Use the Trapezoidal Rule T_n with $n = 10$ to estimate the integral in Example 6.

*43. The function $f(x) = \cos x$ has a maximum at $x = 0$. The Maclaurin polynomial $P_2(x) = 1 - \frac{x^2}{2}$ is a downward-opening parabola. We expect that $P_4(x) = 1 - \frac{x^2}{2} + \frac{x^4}{4!}$ should be closer to $f(x)$, and since $\frac{x^4}{4!} \ge 0$, $P_2(x)$ lies below $P_4(x)$. This suggests that $P_2(x)$ is likely below the graph of $y = f(x)$, at least near $x = 0$. Sketch $y = f(x)$, $P_2(x)$, $P_4(x)$ on the same picture. This kind of argument is not entirely reliable, but it is a reasonable starting point.

*44. Continuing from Exercise 43, consider the parabola whose vertex is at $(0, 1)$ and that crosses the x-axis at the first points where $y = \cos x$ does, that is, at $\left(\pm\frac{\pi}{2}, 0\right)$. (i) Do you expect this parabola to be above or below $y = \cos x$ near $x = 0$? (ii) Find the equation of the parabola. (iii) Sketch the parabola and $y = \cos x$, $P_2(x)$, and $P_4(x)$ on the same graph.

POINT TO PONDER

In Section 4.11 we discussed the use of Newton's Method to find approximate values of ln a. The idea was that ln a is the value of x for which $f(x) = 0$, where $f(x) = e^x - a$, and Newton's Method is designed to find approximate values for the points at which some function is zero.

To be able to use this technique, it is necessary to be able to calculate the values of $f(x)$ and $f'(x)$ at each of the points $x = x_n$ generated by iterations of Newton's Method. When a computer finds ln a, it uses this approach. When it comes to calculating $f(x_n)$ and $f'(x_n)$, it calculates approximate values, using Taylor polynomials.

The Taylor polynomials give very good approximations to e^x, so they are convenient for calculating $f(x_n)$ and $f'(x_n)$. These values can then be used to find the next approximation x_{n+1} using Newton's Method.

The Taylor polynomials for ln x do not work very efficiently, as we saw in Example 3, so it is not practical to use them to approximate ln a. The combination of Taylor polynomials to approximate e^x and Newton's Method with these approximate values to find ln a is a remarkably efficient method for finding approximate values of natural logarithms.

You might like to try this approach to approximate ln 2. By hand it is pretty tedious, but if you program a calculator or computer to evaluate the nth Taylor polynomial $P_n(x)$ for e^x, then it becomes more manageable. You can try using a fairly small n for the first few steps and then increasing the accuracy of the approximations of e^x as the iterations of Newton's Method bring you closer to the correct answer.

Chapter Summary

§9.1
factorial
$0! = 1$
Taylor polynomial
$P_n(x)$
Maclaurin polynomials

§9.2
remainder
Lagrange form of the remainder
estimating the remainder

Review Exercises

Write down the Taylor polynomial $P_n(x)$ of each of these functions $f(x)$ about the point a.

1. $f(x) = e^{-x}$, $n = 6$, $a = 0$
2. $f(x) = e^{3x}$, $n = 4$, $a = 0$
3. $f(x) = e^x$, $n = 6$, $a = 1$
4. $f(x) = e^{-2x}$, $n = 4$, $a = -1$
5. $f(x) = \sin(\pi x)$, $n = 5$, $a = \frac{1}{2}$
6. $f(x) = \cos\left(\frac{\pi}{2}x\right)$, $n = 4$, $a = -3$
7. $f(x) = \arctan x$, $n = 3$, $a = 1$
8. $f(x) = \cos(x + 5)$, $n = 4$, $a = -5$

9. $f(x) = x^{15}$, $n = 14$, $a = 0$
10. $f(x) = x^3$, $n = 2$, $a = 2$

Estimate each of the following to within the specified accuracy. State clearly which function $f(x)$ you are approximating and which a and n you are using.

11. $\frac{1}{e}$ within 0.0001
12. $\sqrt{e}$ within 0.0001
13. $\sqrt{1.1}$ within 0.005
14. $\sqrt{1.5}$ within 0.01
15. $\sec(1.1°)$ within 0.005
16. $\cos(-2°)$ within 0.005

17. $\sin(1.5°)$ within 0.005   **19.** $\tan(-0.1)$ within 0.005

18. $\sin(10°)$ within 0.01   **20.** $\cot(88°)$ within 0.01

II

*21. Suppose $f(x)$ is an even function (i.e., $f(-x) = f(x)$ for all x). Let $P_n(x)$ be the nth Maclaurin polynomial. Show that $P_n(x)$ is an even polynomial.

*22. Suppose $f(x)$ is an odd function (i.e., $f(-x) = -f(x)$ for all x). Let $P_n(x)$ be the nth Maclaurin polynomial. Show that $P_n(x)$ is an odd polynomial.

*23. Suppose $f(x)$ is n times differentiable at $x = a$. Let $g(x) = f(x + a)$. What is the relationship between the Taylor polynomials of $f(x)$ about the point a and the Maclaurin polynomials of $g(x)$?

Series

10

A hare trying to pass a tortoise must first catch up half the distance between them. Then he must catch up half the remaining distance. Then he must catch up half the remaining distance, and so on.
This argument, known as Zeno's Paradox, suggests that the hare can never overtake the tortoise.

INTRODUCTION

In this chapter we will study *infinite series*, which arise when we try to add up infinitely many numbers. For instance, if we consider

$$
\begin{aligned}
\tfrac{1}{2} &= \tfrac{1}{2} = 0.5, \\
\tfrac{1}{2} + \tfrac{1}{4} &= \tfrac{3}{4} = 0.75, \\
\tfrac{1}{2} + \tfrac{1}{4} + \tfrac{1}{8} &= \tfrac{7}{8} = 0.875, \\
\tfrac{1}{2} + \tfrac{1}{4} + \tfrac{1}{8} + \tfrac{1}{16} &= \tfrac{15}{16} = 0.9375, \\
\tfrac{1}{2} + \tfrac{1}{4} + \tfrac{1}{8} + \tfrac{1}{16} + \tfrac{1}{32} &= \tfrac{31}{32} = 0.96875,
\end{aligned}
$$

it certainly appears that these sums approach the value 1. Each time we add an additional term (equal to one-half the previous term), the total gets closer to 1. It is very tempting to think that if we keep going and add up *infinitely* many terms, the answer ought to be 1. We would like to be able to write

$$
\frac{1}{2} + \frac{1}{4} + \frac{1}{8} + \frac{1}{16} + \frac{1}{32} + \ldots = 1.
$$

It is to make sense of this perfectly reasonable idea that we introduce the concept of series. As the terminology we have just used suggests, the way to define a series will involve a kind of limit.

In Section 10.1 there is a discussion of *sequences*, which provide a way to discuss limits, and in Section 10.2 we consider a special type of series known as *geometric series*; the example just considered was of this type. In Section 10.3 we discuss the precise definition of series in terms of limits and consider the sometimes difficult question of whether the limit exists. A series is said to *converge* if the limit exists, and the next four sections deal with various criteria for deciding whether or not a particular series converges. Section 10.4 presents the *Integral Test*, Section 10.5 presents the *Comparison Test*, and Section 10.6 considers *alternating series* and the concept of *absolute convergence*. In Section 10.7 we discuss the *Ratio Test* and the *Root Test*. In these sections we will need to use l'Hôpital's Rule (p. 332).

In Chapter 9 we considered Taylor polynomials; for each positive n there is a Taylor polynomial $P_n(x)$ of degree n to approximate a function $f(x)$ near $x = a$. The interesting thing is that the terms in $P_n(x)$ are exactly the same as the corresponding terms in $P_{n+1}(x)$. To form $P_{n+1}(x)$, we add one additional term to $P_n(x)$, but the previous terms are left alone.

To get more accuracy, we keep adding more terms to the Taylor polynomial, so it seems reasonable to ask whether it is permissible to add up *all* the infinitely many terms as a series. This is the idea behind *Taylor series*, which are discussed in Section 10.8.

Section 10.9 is devoted to *power series*, a notion that is suggested by Taylor series in the previous section, and Section 10.10 considers how to differentiate and integrate power series.

10.1 Sequences

CONVERGENCE
DIVERGENCE
SQUEEZE THEOREM
SHIFTED SEQUENCE
PRECISE DEFINITION

Before going on to study infinite series in greater depth, we devote some time to the related topic of infinite *sequences*. A sequence just means an infinite list of numbers a_1, a_2, a_3, $a_4, \ldots$ (a series, by contrast, means an infinite *sum* $\sum_{n=1}^{\infty} a_n$). The question we want to study about a sequence is whether or not it converges to a limit; for instance, the sequence $1, \frac{1}{2}, \frac{1}{3}, \frac{1}{4}, \frac{1}{5}, \frac{1}{6}, \ldots$, whose nth term is $\frac{1}{n}$, converges to 0, since $\frac{1}{n}$ approaches 0 as $n \to \infty$.

DEFINITION 10.1.1

> Suppose a_n is a real number, for each $n = 1, 2, 3, \ldots$. The family $\{a_n\}$ of all these numbers is called a **sequence**, and a_n is called its nth **term**.
>
> Another way of describing a sequence is to say that it is a function whose domain is the positive integers. The nth term a_n is the value of the function at the integer n.

Sequences sometimes begin with $n = 0$, $n = -1$, $n = 2$, or some other integer; in this case the sequence is a function whose domain is the set of all integers greater than or equal to 0 or -1 or 2, etc.

We have already observed that the sequence $a_n = \frac{1}{n}$ (i.e., $1, \frac{1}{2}, \frac{1}{3}, \frac{1}{4}, \frac{1}{5}, \ldots$) converges to 0. See Figure 10.1.1(i).

The constant sequence $-2, -2, -2, -2, -2, \ldots$ (i.e., $a_n = -2$, for every n) pretty clearly must converge to -2. See Figure 10.1.1(ii).

The sequence $1, -1, 1, -1, 1, -1, \ldots$ (i.e., $a_n = (-1)^{n+1}$) does not converge to any limit. See Figure 10.1.1(iii).

Figure 10.1.1(i)

Figure 10.1.1(ii)

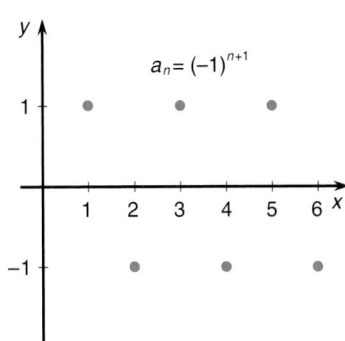

Figure 10.1.1(iii)

The idea of the limit of a sequence is a natural variant of the idea of the limit of a function. We will give the precise definition of this type of limit in Definition 10.1.7, but first we will explore how to use it.

DEFINITION 10.1.2

> The sequence $\{a_n\}$ **converges** to the limit L if its terms approach the number L as a limit as $n \to \infty$. In this case we write
>
> $$\lim_{n \to \infty} a_n = L.$$
>
> We say that the sequence $\{a_n\}$ **converges** (or is **convergent**) if it converges to some limit L and that it **diverges** (or is **divergent**) if it does not converge to any limit.

As with limits of functions, if $\{a_n\}$ converges, then it converges to one and only one limit. (The terms cannot approach two different numbers ...).

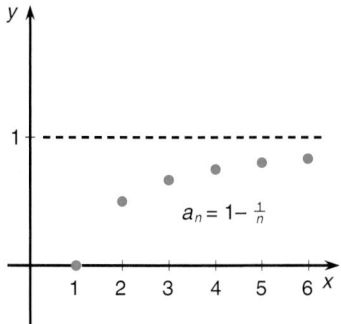

Figure 10.1.2(i)

$b_1 = 3 - 1 = 2$
$b_2 = 3 + \frac{1}{4}$
$b_3 = 3 - \frac{1}{27}$
$b_4 = 3 + \frac{1}{256}$

EXAMPLE 1

(i) Consider the sequence $a_n = 1 - \frac{1}{n}$. Its nth term is slightly less than 1. The amount by which it is less than 1 is $\frac{1}{n}$, which approaches the limit 0, so it seems reasonable to say that $\lim_{n \to \infty} a_n = 1$. See Figure 10.1.2(i).

(ii) Let $b_n = 3 + \left(-\frac{1}{n}\right)^n$. We evaluate a few terms at left. These terms certainly appear to be getting close to 3. The amount by which the nth term differs from 3 is $-\left(-\frac{1}{n}\right)^n$. This is sometimes negative and sometimes positive, which means that some of the terms are greater than 3 and some are less than 3, but the differences do get smaller and smaller. The sequence $\left\{\left(-\frac{1}{n}\right)^n\right\}$ tends to 0 as $n \to \infty$, and $b_n \to 3$ (i.e., $\lim_{n\to\infty} b_n = 3$). See Figure 10.1.2(ii).

(iii) With $c_n = \sin\left(\frac{n\pi}{2}\right)$ we evaluate the first few terms at left to see what happens. After c_4 the sequence "repeats" (since the function $\sin x$ does). The sequence is $1, 0, -1, 0, 1, 0, -1, 0, 1, 0, -1, 0, \ldots$. It does not converge to any limit at all. See Figure 10.1.2(iii).

(iv) If $d_n = 3n^2 - 1$, then the terms get larger and larger and tend to infinity. Because of this they cannot converge to any (finite) limit L, so the sequence diverges. In this situation we sometimes say that the sequence $\{d_n\}$ **diverges to infinity**. See Figure 10.1.2(iv). ◢

Parts (iii) and (iv) of Example 1 illustrate two different ways in which a sequence can diverge. The sequence $\{c_n\}$ in part (iii) diverges because its values "jump around." Some of them approach the limit 1, others -1, and others 0, but there is no one limit that they all approach. In part (iv), on the other hand, the sequence $\{d_n\}$ diverges to infinity because its terms tend to infinity, so they cannot approach any (finite) limit L.

There are other ways in which sequences can diverge, but these are two very common types of divergent sequence.

In discussing all these examples we have relied on the intuition we have about limits from our knowledge of limits of functions. In fact there is an extremely useful result that says that if a sequence $\{a_n\}$ is defined by $a_n = f(n)$ for some function $f(x)$ and if the function $f(x)$ has a limit as $x \to \infty$, then the sequence $\{a_n\}$ will converge to the same limit. Figure 10.1.3 shows the graph of $y = f(x)$, and the sequence $\{a_n\}$ is just its values at the integers.

$c_1 = \sin\frac{\pi}{2} = 1$
$c_2 = \sin\pi = 0$
$c_3 = \sin\frac{3\pi}{2} = -1$
$c_4 = \sin 2\pi = 0.$

Figure 10.1.2(ii)

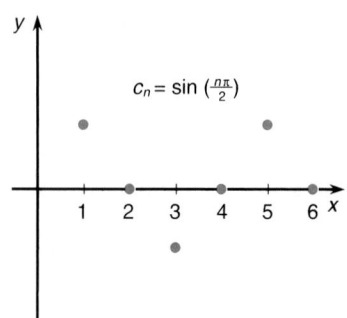

Figure 10.1.2(iii)

> **THEOREM 10.1.3**
> Suppose $f(x)$ is a function for which we know that $\lim_{x\to\infty} f(x) = L$, for some number L. Define a sequence $\{a_n\}$ by $a_n = f(n)$. Then
>
> $$\lim_{n\to\infty} a_n = L.$$
>
> If $\lim_{x\to\infty} |f(x)| = \infty$, then $\{a_n\}$ diverges.

Figure 10.1.2(iv)

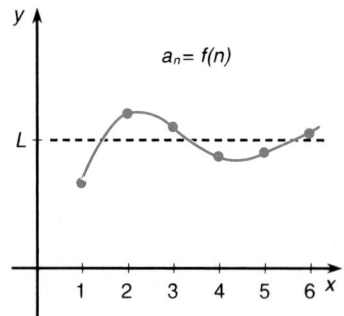

Figure 10.1.3

EXAMPLE 2

(i) We can use Theorem 10.1.3 to show that the sequences in parts (i) and (ii) of Example 1 do have the limits we think they should have. For $a_n = 1 - \frac{1}{n}$ we let $f(x) = 1 - \frac{1}{x}$, which certainly does tend to 1 as $x \to \infty$.

(ii) For $b_n = 3 + \left(-\frac{1}{n}\right)^n$ it is a little more difficult because it does not make sense to consider the function $3 + \left(-\frac{1}{x}\right)^x$. After all, the xth power of something is defined using the natural logarithm, which does not make sense for negative numbers.

It does make sense to consider the function $\left(\frac{1}{x}\right)^x$, but then it is necessary to multiply it by something that equals -1 when x equals an odd integer and is $+1$ when x equals an even integer. This is easier than it sounds. For instance, $\cos(\pi x)$ has that property. So we let $f(x) = 3 + \cos(\pi x)\left(\frac{1}{x}\right)^x$, and then $b_n = f(n)$.

The expected conclusion about b_n will follow if we can show that $\lim_{x \to \infty} f(x) = 3$, and this will follow by the Squeeze Theorem for limits at infinity if we can show that $\lim_{x \to \infty} \left(\frac{1}{x}\right)^x = 0$.

Now $\left(\frac{1}{x}\right)^x = e^{x \ln(1/x)} = e^{-x \ln x}$. For $x > e$, $\ln x > 1$, so $x \ln x > x$, and we conclude that $x \ln x \to \infty$, so

$$\lim_{x \to \infty} \left(\frac{1}{x}\right)^x = \lim_{x \to \infty} e^{-x \ln x} = \lim_{x \to \infty} \frac{1}{e^{x \ln x}} = 0,$$

using the version for limits at infinity of Theorem 2.3.2 about limits of composite functions.

From this we see that $\lim_{x \to \infty} f(x) = 3$, so the theorem applies, and we conclude that $\lim_{n \to \infty} b_n = 3$.

(iii) On the other hand, Theorem 10.1.3 does not help us with part (iii) of Example 1. There we have the sequence $c_n = \sin\left(\frac{n\pi}{2}\right)$ defined as the values $c_n = f(n)$, with $f(x) = \sin\frac{\pi x}{2}$. We also know that this function $f(x)$ does not have a limit as $x \to \infty$.

However, the theorem does not tell us anything in this situation. For instance, if we let $g(x) = \sin(\pi x)$, then $g(x)$ does not have a limit as $x \to \infty$. However, $g(n) = \sin(n\pi) = 0$ for *every* integer n. So if we define a sequence by $d_n = g(n)$, then the sequence $\{d_n\}$ *does* tend to a limit (0) as $n \to \infty$, even though the function $g(x)$ does not tend to any limit as $x \to \infty$.

The point is that even knowing that the values of $g(x)$ at the integers are a nice convergent sequence does not tell us anything about what $g(x)$ might do in between the integers (see Figure 10.1.4). It is because of this that we cannot try to use Theorem 10.1.3 "backwards." If the limit of $f(x)$ exists, then the limit of the corresponding sequence exists and is the same, but if $f(x)$ has no limit, then we cannot conclude that the same is true of the sequence $\{f(n)\}$.

(iv) With $f(x) = 3x^2 - 1$ we find that $\lim_{x \to \infty} f(x) = \infty$, so if $d_n = 3n^2 - 1$, then $\{d_n\}$ diverges.

(v) Let $u > -1$ be a real number and let $k_n = n \ln\left(1 + \frac{u}{n}\right)$. To apply the theorem, we should consider the function $F(x) = x \ln\left(1 + \frac{u}{x}\right)$. We can write $F(x) = \frac{\ln(1+u/x)}{1/x}$ and observe that both numerator and denominator tend to 0 as $x \to \infty$.

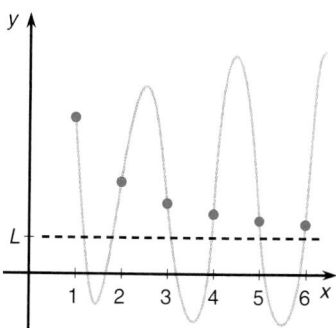

Figure 10.1.4

This suggests applying l'Hôpital's Rule, from which we find that

$$
\lim_{x \to \infty} F(x) = \lim_{x \to \infty} \frac{\ln\left(1 + \frac{u}{x}\right)}{\frac{1}{x}} = \lim_{x \to \infty} \frac{\frac{1}{1 + \frac{u}{x}}\left(-\frac{u}{x^2}\right)}{-\frac{1}{x^2}}
$$

$$
= \lim_{x \to \infty} \frac{1}{1 + \frac{u}{x}} u = u\left(\lim_{x \to \infty} \frac{1}{1 + \frac{u}{x}}\right)
$$

$$
= u.
$$

Now using this result and applying Theorem 10.1.3, we see that the sequence $\{k_n\} = \left\{n \ln\left(1 + \frac{u}{n}\right)\right\}$ converges to the limit u. ◢

In Example 1(i) we considered $a_n = 1 - \frac{1}{n}$ and concluded that $\{a_n\}$ converges to 1 because $\frac{1}{n}$ converges to 0. In a way, what we have done here is to consider the constant sequence 1, 1, 1, 1, 1,..., which converges to 1, and the sequence $\{\frac{1}{n}\}$, which converges to 0, and looked at their difference $a_n = 1 - \frac{1}{n}$. We used this to see that $\{a_n\}$ converges to the difference $1 - 0 = 1$.

This is a familiar property of limits, and it encourages us to record a number of facts about limits of sequences that are very similar to analogous facts about limits of functions.

THEOREM 10.1.4

(i) The constant sequence $\{c\}$, whose nth term is c for every n, converges to c, that is,

$$
\lim_{n \to \infty} c = c.
$$

Now suppose $\{a_n\}$ and $\{b_n\}$ are convergent sequences so that $\lim_{n \to \infty} a_n = L$ and $\lim_{n \to \infty} b_n = M$. Then the sequence formed by multiplying $\{a_n\}$ by a number c and the sequences whose terms are the sum, the product, or the quotient of a_n and b_n all converge to the expected limits:

(ii) $\lim_{n \to \infty} c a_n = cL = c \lim_{n \to \infty} a_n$, for any $c \in \mathbb{R}$,

(iii) $\lim_{n \to \infty} (a_n + b_n) = L + M = \lim_{n \to \infty} a_n + \lim_{n \to \infty} b_n$,

(iv) $\lim_{n \to \infty} (a_n b_n) = LM = (\lim_{n \to \infty} a_n)(\lim_{n \to \infty} b_n)$,

(v) If $b_n \neq 0$ for each n and the limit M of $\{b_n\}$ is not zero, then

$$
\lim_{n \to \infty} \frac{a_n}{b_n} = \frac{L}{M} = \frac{\lim_{n \to \infty} a_n}{\lim_{n \to \infty} b_n}.
$$

(vi) Suppose $h(x)$ is a function that is defined at $x = a_n$ for every n and continuous at $x = L = \lim_{n \to \infty} a_n$, and define $d_n = h(a_n)$. Then the sequence $\{d_n\}$ is convergent, and $\lim_{n \to \infty} d_n = h(\lim_{n \to \infty} a_n) = h(L)$.

(vii) (**The Squeeze Theorem for Sequences**) Suppose $\{a_n\}$, $\{b_n\}$, and $\{c_n\}$ are sequences so that $\{a_n\}$ and $\{c_n\}$ both converge to the same limit L, and so that for each n, b_n lies between a_n and c_n, that is, either $a_n \leq b_n \leq c_n$ or $a_n \geq b_n \geq c_n$. Then the sequence $\{b_n\}$ is convergent, and $\lim_{n \to \infty} b_n = L$. In other words, if the sequence $\{b_n\}$ is "squeezed" between two sequences that both converge to L, then $\{b_n\}$ must also converge to L.

By using Theorems 10.1.3 and 10.1.4(i)–(vii) it is possible with many sequences to decide whether or not they converge and to evaluate the limit if they do.

EXAMPLE 3

See the Graphics Calculator
Example on the next page.

(i) Let $u > -1$ be a real number and let $A_n = \left(1 + \frac{u}{n}\right)^n$. We observe that $\ln A_n = n \ln\left(1 + \frac{u}{n}\right)$, which is the sequence $\{k_n\}$ that we considered in Example 2(v). There we saw that $\lim_{n \to \infty} k_n = u$. Since $A_n = \exp(k_n)$, we can apply part (vi) of Theorem 10.1.4 to see that $\{A_n\}$ converges and $\lim_{n \to \infty} \left(1 + \frac{u}{n}\right)^n = \lim_{n \to \infty} A_n = \exp(\lim_{n \to \infty} k_n) = e^u$.

(ii) Consider the sequence $\{a_n\}$ defined by $a_n = \frac{n}{n+1}$. If we let $f(x) = \frac{x}{x+1}$, then $a_n = f(n)$, and we know that $\lim_{x \to \infty} f(x) = 1$, so Theorem 10.1.3 can be applied to conclude that $\{a_n\}$ converges to 1.

(iii) If $b_n = n + \frac{2}{n}$, we might consider separately the two parts of this sequence. The sequence $\left\{\frac{2}{n}\right\}$ converges to 0 (by Theorem 10.1.3, for instance).

 The other part of the sequence is the sequence $\{n\}$, that is, $1, 2, 3, 4, \ldots$, which certainly does not converge. (It goes to infinity.)

 The sequence $\{b_n\}$ is the sum of two sequences, one of which converges and the other of which diverges. This implies that $\{b_n\}$ itself must also diverge. For suppose it converged to some limit L. Since $\left\{-\frac{2}{n}\right\}$ converges to 0, Theorem 10.1.4(iii) would then tell us that $\left\{b_n + \left(-\frac{2}{n}\right)\right\}$ must converge to $L + 0 = L$. But $b_n + \left(-\frac{2}{n}\right) = n + \frac{2}{n} - \frac{2}{n} = n$, and we have already observed that the sequence $\{n\}$ does *not* converge. This contradiction shows that it was wrong to assume that $\{b_n\}$ converges, and we conclude that $\{b_n\}$ diverges.

(iv) If $c_n = (-1)^n$ and $d_n = (-1)^{n+1}$, that is,

$$\{c_n\} = -1,\ 1,\ -1,\ 1,\ -1,\ 1,\ldots,$$
$$\{d_n\} = 1,\ -1,\ 1,\ -1,\ 1,\ -1,\ldots,$$

then we know that neither of the sequences $\{c_n\}$ or $\{d_n\}$ converges.

 However, their sum is the zero sequence, since $c_n + d_n = (-1)^n + (-1)^{n+1} = (-1)^n\left(1 + (-1)\right) = 0$.

 This example shows that the sum of two divergent sequences may be convergent.

(v) Let $a_n = \frac{1}{2n+(-1)^{n+1}}$. We cannot simply change n to x to make a function $f(x)$ so that $a_n = f(n)$, because $(-1)^x$ makes sense only for certain values of x. However, observing that $(-1)^{n+1} \geq -1$, so $2n + (-1)^{n+1} \geq 2n - 1$ and $a_n = \frac{1}{2n+(-1)^{n+1}} \leq \frac{1}{2n-1}$, we find that $0 \leq a_n \leq \frac{1}{2n-1}$; since $\frac{1}{2n-1} \to 0$, the Squeeze Theorem shows us that $a_n \to 0$.

convergent + convergent = convergent
convergent + divergent = divergent
divergent + divergent = ?

Parts (iii) and (iv) of Example 3 illustrate an interesting phenomenon. The sum of two divergent sequences may be convergent (although it may also be divergent), but the sum of a divergent sequence and a convergent sequence can never be convergent.

Part (iv) of Example 3 also suggests another observation. In a way the sequences $\{c_n\}$ and $\{d_n\}$ defined there can be regarded as essentially the same sequence "shifted over." If $\{a_n\}$ is any sequence, then it is possible to consider the sequence whose nth term is a_{n-1}. This sequence starts with $n = 2$, and it has the same terms as $\{a_n\}$, but they are shifted over by one. For that matter we could consider the sequence $\{a_{n+1}\}$, which starts with $n = 0$, or the sequence that is obtained by shifting the sequence $\{a_n\}$ by k; this is the sequence $\{a_{n-k}\}$. It starts with $n = k + 1$.

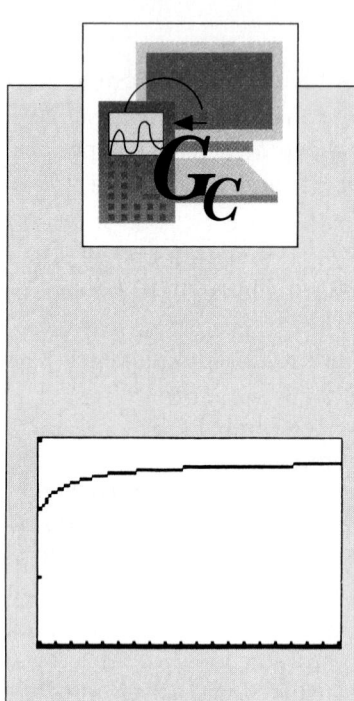

With a graphics calculator we can plot the function $y = f(x) = \left(1 + \frac{1}{x}\right)^x$. This is related to the sequence $\{A_n\}$ of part (i) of Example 3, since $A_n = \left(1 + \frac{u}{n}\right)^n$: If $u = 1$, $A_n = f(n)$.

TI-85	SHARP EL9300
GRAPH F1($y(x) =$) (1 + 1 ÷ F1(x))^F1(x) M2(RANGE) 1 ▼ 20 ▼ ▼ 0 ▼ 3 F5(GRAPH)	⟳ (1 + x/θ/T x^{-1}) a^b x/θ/T RANGE 1 ENTER 20 ENTER 5 ENTER 0 ENTER 3 ENTER 1 ENTER ⟳

CASIO f_x-6300G (f_x-7700GB)	HP 48SX
Range 1 EXE 20 EXE 5 EXE 0 EXE 3 EXE 1 EXE EXE EXE EXE Graph (1 + X x^{-1}) x^y X EXE	⟳ PLOT PLOTR ERASE ATTN ' $Y =$ 1 + 1/x X ▶ ▶ y^x X ↷ PLOT ⟳ DRAW ERASE 1 ENTER 20 XRNG 0 ENTER 3 YRNG DRAW

Notice the limiting behavior as x gets large. You might like to try it with different values of u, for example, $u = 2$, $u = -5$, $u = 20$.

Now if we look at one of these shifted sequences, it will tend to the same limit as the original sequence $\{a_n\}$, provided $\{a_n\}$ is convergent, and will diverge if $\{a_n\}$ diverges. Roughly speaking, the shifted sequence will display the same behavior as the original sequence but will do it "k terms later."

Another related question arises if we consider the sequence $0, 0, 3, \frac{1}{2}, \frac{1}{3}, \frac{1}{4}, \frac{1}{5}, \dots$, whose nth term equals $\frac{1}{n-2}$ when $n \geq 4$ but that has some peculiar terms at the beginning. The important thing about a sequence of this kind is that its limiting behavior is not affected by the first few terms.

Since the sequence eventually equals $\frac{1}{n-2}$, which is the sequence $\frac{1}{n}$ shifted over by 2, and since that sequence converges to 0, we see that the peculiar sequence also converges to 0. We record these two important results as a theorem.

THEOREM 10.1.5

(i) If $\{a_n\}$ is a sequence, let $b_n = a_{n-k}$. The shifted sequence $\{b_n\}$ has the same properties of convergence as the original sequence $\{a_n\}$, that is, either they both diverge or they both converge to the same limit. (Strictly speaking, if the original sequence is defined for n beginning at 1, then $\{b_n\}$ is defined for n beginning at $k + 1$. If $\{a_n\}$ begins with $n = r$, then $\{b_n\}$ begins with $n = k + r$.)

(ii) Suppose $\{c_n\}$ and $\{d_n\}$ are two sequences that are the same after the first few terms. Then they have the same convergence properties, that is, either they both diverge or they both converge to the same limit.

EXAMPLE 4

(i) The first part of Theorem 10.1.5 shows immediately that $\left\{\frac{1}{n+k}\right\}$ converges to 0 for any integer k.

(ii) The second part of Theorem 10.1.5 can be used to show that the sequence $3,\ -2\pi,\ 4,\ -\sqrt{7},\ -e,\ 17,\ 0,\ 0,0,\ 0,\ 0,\ 0,\ 0,\ldots$, which is all zeroes after the first six terms, converges to 0, even though there does not appear to be any reasonable pattern to the first six terms.

(iii) In Example 2(v) we considered the sequence $\{k_n\} = \left\{n\ln\left(1+\frac{u}{n}\right)\right\}$, for $u > -1$. The restriction on u is needed so that $\ln\left(1+\frac{u}{n}\right)$ makes sense for every n. For instance, if we let $u = -4$, then $k_n = n\ln\left(1+\frac{u}{n}\right)$ is not defined for $n \le 4$, but it does make sense for $n > 4$. Part (ii) of Theorem 10.1.5 says that if $u \le -1$, we can ignore the first few missing terms, and the argument given in Example 2(v) will show that $\{k_n\} = \left\{n\ln\left(1+\frac{u}{n}\right)\right\}$ converges to u for *any* real number u.

Similarly, in Example 3(i) we considered $A_n = \left(1+\frac{u}{n}\right)^n$, for $u > -1$. In fact, A_n is defined for any real number u, and by the above remarks and Theorem 10.1.4(vi) we see that $\left\{\left(1+\frac{u}{n}\right)^n\right\}$ converges, and

$$\lim_{n\to\infty}\left(1+\frac{u}{n}\right)^n = \exp\left(\lim_{n\to\infty} n\ln\left(1+\frac{u}{n}\right)\right) = e^u.$$

Part (iii) of Example 4 illustrates an important use of Theorem 10.1.5(ii). Frequently, a sequence is given by a formula $\{f(n)\}$ that does not make sense for a few small values of n. The theorem says that we can safely ignore this problem if $f(n)$ does make sense for all n past some fixed value.

EXAMPLE 5   Powers of c   Suppose c is a positive number and consider the sequence of its powers: $a_n = c^n$. If $c > 1$, the powers c^n grow larger and larger and tend to infinity, so the sequence diverges. If $c = 1$, of course the sequence is a constant sequence and converges to 1.

So suppose $0 < c < 1$. Then $a_n = c^n = e^{n\ln c} = f(n)$, where $f(x) = e^{x\ln c}$. Since $0 < c < 1$, we know that $\ln c < 0$, and $x\ln c \to -\infty$ as $x \to \infty$, so $\lim_{x\to\infty} f(x) = \lim_{x\to\infty} e^{x\ln c} = 0$. By Theorem 10.1.3, $a_n = e^{n\ln c} \to 0$.

Now suppose $-1 < c < 0$ and consider the sequence $c^n = (-|c|)^n = (-1)^n|c|^n$. By the preceding argument we know that $\{|c|^n\}$ converges to 0, and c^n lies between $-|c|^n$ and $|c|^n$ (actually, it alternates between them). Both these sequences converge to 0, so by the Squeeze Theorem, Theorem 10.1.4(vii), $\{c^n\}$ converges to 0.

Finally, if $c = -1$, the sequence $\{(-1)^n\}$ diverges, and if $c < -1$, then $\{c^n\}$ diverges, since the absolute values of the terms tend to infinity.

We summarize:

$$\{c^n\}\quad \begin{cases} \text{diverges}, & \text{if } c > 1 \text{ or if } c \le -1, \\ \text{converges to 1}, & \text{if } c = 1, \\ \text{converges to 0}, & \text{if } -1 < c < 1. \end{cases}$$

EXAMPLE 6   Let c be any positive number and consider the sequence $\{a_n\}$, where

$$a_n = \frac{c^n}{n!}.$$

It is always possible to find an integer N that is greater than c. If $n > N$, we can write $n! = (1)(2)(3)\ldots(N)(N+1)(N+2)\ldots n = N!(N+1)(N+2)\ldots(n-1)n$, so

$$\frac{c^n}{n!} = \frac{c^n}{N!(N+1)(N+2)\ldots(n-1)n} \leq \frac{c^n}{N!(N)(N)\ldots(N)}$$

$$= \frac{c^n}{N!N^{n-N}} = \frac{N^N}{N!}\left(\frac{c}{N}\right)^n.$$

The inequality comes from making the denominator *smaller* by replacing each of $N+1, N+2, \ldots, n$ by N.

Now $0 < \frac{c}{N} < 1$, so by Example 5 we see that $\{\left(\frac{c}{N}\right)^n\}$ converges to 0, and the last line in the computation above also converges to zero as $n \to \infty$ because it is just $\left(\frac{c}{N}\right)^n$ times the constant $\frac{N^N}{N!}$.

Since $\{\frac{c^n}{n!}\}$ lies between this sequence and the zero sequence, the Squeeze Theorem tells us that $\{\frac{c^n}{n!}\}$ converges to 0, that is,

$$\lim_{n\to\infty}\frac{c^n}{n!} = 0.$$

If $c < 0$, then $\frac{c^n}{n!} = (-1)^n\frac{|c|^n}{n!}$, which lies between the sequences $\{\frac{|c|^n}{n!}\}$ and $\{-\frac{|c|^n}{n!}\}$, both of which converge to 0. The Squeeze Theorem implies that $\{\frac{c^n}{n!}\}$ converges to 0.

We have shown that $\{\frac{c^n}{n!}\}$ converges to 0 if c is either positive or negative; if $c = 0$, the sequence is just the zero sequence, so it still converges to 0. So for *every* number c,

$$\lim_{n\to\infty}\frac{c^n}{n!} = 0. \qquad\blacktriangleleft$$

Suppose a sequence $\{a_n\}$ is "increasing," that is, $a_{n+1} \geq a_n$ for every n. Then one possibility is that $\lim_{n\to\infty} a_n = \infty$, that is, the sequence diverges to infinity, and its terms eventually become larger than any number M.

If, on the other hand, there is some number M so that $a_n \leq M$ for every n, then we say the sequence $\{a_n\}$ is "bounded." In this case the terms of the sequence are "trapped" between a_1 and M; they keep increasing but are never greater than M (see Figure 10.1.5). It seems reasonable to expect that in this situation they will have to converge to some limit, and this is in fact true.

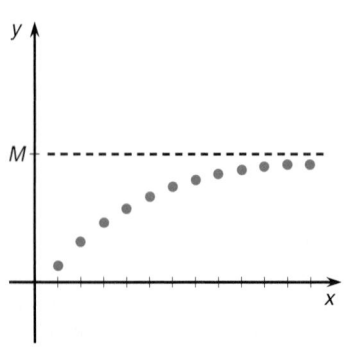

Figure 10.1.5

THEOREM 10.1.6

> Suppose $\{a_n\}$ is an increasing sequence, that is, $a_{n+1} \geq a_n$ for every n. If the sequence is **bounded**, that is, if there is some number M so that $|a_n| \leq M$ for every n, then it must converge. Otherwise, the sequence diverges to infinity, that is, $\lim_{n\to\infty} a_n = \infty$.

Notice that Theorem 10.1.6 does not say what the limit is, simply that there is one. This result is more useful than it sounds; we will use it in Section 10.4.

The Precise Definition of Limits for Sequences

As with limits of functions, we have discussed limits of sequences without giving a precise definition, concentrating instead on learning to work with the limits. Here we give the precise definition and show how it can be used to prove some of the results we have already discussed.

In fact the definition is very similar to the precise definition of limits for functions (Definition 2.7.1, p. 120).

DEFINITION 10.1.7

> The sequence $\{a_n\}$ **converges** to the limit L if and only if the following condition is satisfied.
>
> For every $\epsilon > 0$ it is possible to find an $N > 0$ so that $|a_n - L| < \epsilon$ whenever $n > N$.

In other words, up to any specified "degree of closeness" ϵ, the terms of the sequence must be within a distance of ϵ of the limit L if we go far enough out in the sequence.

EXAMPLE 7 We use this definition to prove the "obvious" fact that the sequence $\{\frac{1}{n}\}$ converges to 0.

Suppose $\epsilon > 0$ is fixed. We want to ensure that $\frac{1}{n} < \epsilon$. This is the same as $n > \frac{1}{\epsilon}$. (Notice the way the inequality is reversed when we take reciprocals.) So if we let N be any integer that is greater than $\frac{1}{\epsilon}$, then whenever $n > N$, we see that $n > N > \frac{1}{\epsilon}$, so $\frac{1}{n} < \epsilon$.

We have shown that given any $\epsilon > 0$, it is possible to find an N (e.g., any integer N that is greater than $\frac{1}{\epsilon}$) so that whenever $n > N$, we have that $\frac{1}{n} < \epsilon$. This is exactly what we need, since the limit is $L = 0$ and $|\frac{1}{n} - L| = |\frac{1}{n} - 0| = \frac{1}{n} < \epsilon$. ◢

In fact it is not very difficult to use Definition 10.1.7 to prove all the Theorems 10.1.3, 10.1.4, 10.1.5 of this section.

Exercises 10.1

I

Say whether or not each of the following sequences $\{a_n\}$ converges, giving reasons, and for those that do converge, find the limit.

1. $a_n = \frac{1}{n} - 1$

2. $a_n = 2 + \frac{3}{n^2}$

3. $a_n = \frac{n-1}{4n}$

4. $a_n = \frac{\cos n\pi}{n}$

5. $a_n = \frac{n^2+n-2}{2n^2-3n} \cos(n\pi)$

6. $a_n = \frac{2n^2-3n+2}{n^3+4n^2-2} \sin \frac{n\pi}{2}$

7. $a_n = \left(\frac{2}{3}\right)^n$

8. $a_n = \left(-\frac{5}{4}\right)^n$

9. $a_n = \left(-\frac{1}{\sqrt{2}}\right)^n$

10. $a_n = \left(\frac{2}{\sqrt{3}}\right)^n$

11. $a_n = \frac{(n-1)!}{n!}$

12. $a_n = \frac{n!}{(n-2)!}$

13. $a_n = \frac{(n!)^2}{(n-1)!(n+1)!}$

14. $a_n = \frac{n!}{4^n}$

15. $a_n = \frac{3^n}{1-3^n}$

16. $a_n = \frac{1+5^n}{6^n}$

17. $a_n = 2\ln(3n)$

18. $a_n = \frac{\cos(n\pi)7^n}{n!}$

19. $a_n = n^{-1/2}$

20. $a_n = \sin(n)\frac{1}{3^n n!}$

II

*21. We know that the function $f(x) = \frac{1}{x}$ approaches a limit as $x \to \infty$, namely, $\lim_{x\to\infty} f(x) = 0$. If we let $a_n = f(n)$, then Theorem 10.1.3 tells us that the sequence $\{a_n\}$ converges to 0. Find another function $g(x)$ so that $a_n = g(n)$ for each $n \geq 1$ (i.e., $g(n) = f(n)$ for each n) but so that $g(x)$ does not have a limit as $x \to \infty$.

*22. Suppose $f(x)$ is a function satisfying $\lim_{x\to\infty} f(x) = L$. Define a sequence $\{a_n\}$ by $a_n = f(n)$. By Theorem 10.1.3, $\lim_{n\to\infty} a_n = L$. Find another function $g(x)$ that does not have a limit as $x \to \infty$ but so that $a_n = g(n)$ for each $n \geq 1$.

23. Show that $\{\frac{1}{n \ln n}\}$ converges and find its limit.

24. Show that $\left\{ \left(\frac{n}{2}\right)^{-n/3} \right\}$ converges and find its limit.

25. Let $a_n = \frac{3^{2n}}{n!}$, for each $n \geq 1$, and decide whether or not $\{a_n\}$ converges, giving reasons and finding the limit if it does converge.

26. Let $b_n = (-1)^n \frac{3^{2n}}{n!}$, for each $n \geq 1$, and decide whether or not $\{b_n\}$ converges, giving reasons and finding the limit if it does converge.

27. Let $c_n = (-1)^{n+1} \frac{2^n(n-1)!}{n!}$, for each $n \geq 1$, and decide whether or not $\{c_n\}$ converges, giving reasons and finding the limit if it does converge.

*28. For what values of c does the sequence $\{n^c\}$ converge, if any? If there are any, find the corresponding limits.

*29. If $\{c_n\}$ is a bounded decreasing sequence (i.e., $c_{n+1} \leq c_n$ for every n), show that it is convergent.

*30. Find two examples of bounded divergent sequences.

10.2 Geometric Sums and Series and Applications

COMMON RATIO
INTEREST
GEOMETRIC SERIES
CONVERGENCE

In the introduction to this chapter we considered the following sums:

$$\begin{aligned}
\tfrac{1}{2} &= \tfrac{1}{2} = 0.5, \\
\tfrac{1}{2} + \tfrac{1}{4} &= \tfrac{3}{4} = 0.75, \\
\tfrac{1}{2} + \tfrac{1}{4} + \tfrac{1}{8} &= \tfrac{7}{8} = 0.875, \\
\tfrac{1}{2} + \tfrac{1}{4} + \tfrac{1}{8} + \tfrac{1}{16} &= \tfrac{15}{16} = 0.9375, \\
\tfrac{1}{2} + \tfrac{1}{4} + \tfrac{1}{8} + \tfrac{1}{16} + \tfrac{1}{32} &= \tfrac{31}{32} = 0.96875.
\end{aligned}$$

In all of these sums, each successive term is equal to one-half the previous term. The first term is $\frac{1}{2}$, the second term is half that, or $\frac{1}{4}$, the third term (if we continue that far) is $\frac{1}{8}$, and so on. The way to find the third term is to start with the first term $\frac{1}{2}$, multiply it by $\frac{1}{2}$ to get the second term, and multiply that by $\frac{1}{2}$ to get the third term, which is $\left(\frac{1}{2}\right)\left(\frac{1}{2}\right)\left(\frac{1}{2}\right) = \left(\frac{1}{2}\right)^3 = \frac{1}{8}$.

Similarly, the fourth term will be $\left(\frac{1}{2}\right)\left(\frac{1}{2}\right)\left(\frac{1}{2}\right)\left(\frac{1}{2}\right) = \left(\frac{1}{2}\right)^4 = \frac{1}{16}$, and the kth term will be the product of k factors of $\frac{1}{2}$, or $\left(\frac{1}{2}\right)^k$.

There is no particular reason for choosing $\frac{1}{2}$. We could just as easily have considered a sum like $\frac{1}{3} + \frac{1}{9} + \frac{1}{27} + \frac{1}{81} + \frac{1}{243} = \frac{1}{3} + \left(\frac{1}{3}\right)^2 + \left(\frac{1}{3}\right)^3 + \left(\frac{1}{3}\right)^4 + \left(\frac{1}{3}\right)^5$, or $-\frac{2}{3} + \left(-\frac{2}{3}\right)^2 + \left(-\frac{2}{3}\right)^3 + \left(-\frac{2}{3}\right)^4 + \left(-\frac{2}{3}\right)^5 + \left(-\frac{2}{3}\right)^6$.

It is also possible to look at $5 + 5^2 + 5^3 + 5^4 + 5^5 + 5^6 + 5^7$, but in this case the terms get larger and larger, so the sum is unlikely to have a (finite) limit as we take more and more terms.

Each of these sums is an example of what is called a *geometric sum*; the corresponding infinite series are called *geometric series*.

DEFINITION 10.2.1

Let a and r be real numbers, and let k be a positive integer. A sum of the form

$$a + ar + ar^2 + ar^3 + \ldots + ar^k = \sum_{n=0}^{k} ar^n$$

$$= a(1 + r + r^2 + r^3 + \ldots + r^k)$$

is called a **geometric sum**. The number r is called the **common ratio**, since it is the ratio of each term to the previous one.

It is actually possible to find a simple formula for evaluating a geometric sum without adding up a lot of terms.

THEOREM 10.2.2

Geometric Sums

Suppose k is a positive integer and a, r are real numbers with $r \neq 1$. Then the geometric sum can be evaluated as follows:

$$a + ar + ar^2 + ar^3 + \ldots + ar^k = \sum_{n=0}^{k} ar^n = a(1 + r + r^2 + r^3 + \ldots + r^k)$$

$$= a\frac{1 - r^{k+1}}{1 - r} = \frac{a - ar^{k+1}}{1 - r}.$$

PROOF   Let us first consider the case in which $a = 1$ and look at the sum $S = 1 + r + r^2 + r^3 + \ldots + r^{k-1} + r^k$. If we multiply the sum S by r we find

$$rS = r(1 + r + r^2 + r^3 + \ldots + r^{k-1} + r^k)$$
$$= r + r^2 + r^3 + r^4 + \ldots + r^k + r^{k+1}.$$

Now what would happen if we were to subtract this from S? Both S and rS are sums of many terms, each of which is a power of r. In fact most of the terms they contain are the same, with the two exceptions that S begins with the term 1, which is not in rS, and rS ends with the term r^{k+1}, which is not in S. So if we subtract them, all the terms except these two will cancel, and we will be left with

$$S - rS = (1 + \ r + r^2 + r^3 + \ldots + r^{k-1} + r^k)$$
$$\qquad - (r + r^2 + r^3 + r^4 + \ldots + r^k + r^{k+1})$$
$$= 1 - r^{k+1},$$

that is,

$$(1 - r)S = 1 - r^{k+1}.$$

Dividing this last equation by $1 - r$ (which is all right, since we assumed that $r \neq 1$) gives the result $S = 1 + r + r^2 + r^3 + r^4 + \ldots + r^{k-1} + r^k = \frac{1-r^{k+1}}{1-r}$.

In the general case, that is, when a is not necessarily equal to 1, the result follows by multiplying both sides of this equation by a.   ◢

In the sum $a(1 + r + r^2 + r^3 + \ldots + r^{k-1} + r^k)$ the number k is the power of r in the last term. Since the first term contains $1 = r^0$, we see that there is a term for each power of r, starting with $r^0 = 1$ and ending with r^k. This means that there are $k + 1$ terms, not just k. We can find r^k by noting that it is the ratio of the last term to the first.

You may find it helpful to think of it this way: The number r is the common ratio of each term to the preceding one. The first term is a, and the last term is ar^k. If we added one more term, it would be ar^{k+1}. So for any particular geometric sum, find the first term a, the ratio r of any two adjacent terms, and the "next term after the last," ar^{k+1}. Then the sum can be evaluated by the formula $\frac{a-ar^{k+1}}{1-r}$.

Observe that if $r = 1$, then the sum is just $a + a + a + a + \ldots + a$, with a total of $k + 1$ terms, so it equals $(k + 1)a$. It was necessary to assume that $r \neq 1$ in order for the formula in the theorem to make sense, but this shows that when $r = 1$, if anything the sum is even easier to evaluate.

EXAMPLE 1   Evaluate each of the following: (i) $1 + \frac{1}{3} + \frac{1}{9}$, (ii) $\frac{1}{2} + \frac{1}{4} + \frac{1}{8} + \frac{1}{16}$, (iii) $\sum_{n=1}^{7} 5^n = 5 + 5^2 + 5^3 + 5^4 + 5^5 + 5^6 + 5^7$.

Solution

(i) The first term is $a = 1$, the common ratio is $\frac{1}{3}$, and the last term is $\frac{1}{9}$, so the "next term" would be $\frac{1}{27}$. (It is possible to check that this equals $\left(\frac{1}{3}\right)^{k+1}$ by noting that the number of terms ought to be $k + 1$, which is 3, which is correct.) The formula tells us that the sum equals

$$\frac{a - ar^{k+1}}{1 - r} = \frac{1 - \frac{1}{27}}{\frac{2}{3}} = \frac{26}{27}\frac{3}{2} = \frac{13}{9}.$$

This is easily checked by adding up the three terms.

ANSWER=
                97655

With a programmable calculator it is not difficult to add up a geometric sum. For instance, we can do the sum $\sum_{n=1}^{7} 5^n$ from part (iii) of Example 1.

TI-85	SHARP EL9300
PRGM F2 (EDIT) GMSUM ENTER 0 STO N ENTER 0 STO S ENTER LbL B ENTER N + 1 STO N ENTER S + 5^N STO S ENTER If (N TEST F2 (<) 7) ENTER Goto B ENTER S ENTER EXIT EXIT EXIT PRGM F1 (NAME) GMSUM ENTER	✎ ▼ ▼ (NEW) ENTER ENTER GSM ENTER N = 0 ENTER S = 0 ENTER COMMAND B 1 (Label) 1 ENTER N = N + 1 ENTER S = S + 5 a^b N ENTER COMMAND 3 (If) N COMMAND C 2 (<) 7 COMMAND B 2 (Goto) 1 ENTER COMMAND A 1 (Print) S ENTER COMMAND A 6 (End) ENTER ▦ ✎ (RUN) ENTER (GSM) ENTER
CASIO f_x-6300G (f_x-7700GB)	HP 48SX
MODE 2 EXE 0 → ▦ : 0 → ▦ : Lbl 1 : ▦ + 1 → ▦ : ▦ + 5 x^y ▦ → ▦ : ▦ < 7 ⇒ Goto 1 : ▦ ◢ MODE 1 Prog 0 EXE (on the f_x-7700GB, : is PRGM F6, and Prog is PRGM F3.)	ATTN ⤴ CLR ▦ 0 SPC 1 SPC 7 SPC FOR N SPC 5 N y^x + NEXT ▼ ENTER ' GSUM STO VAR GSUM ENTER ▦ CONT

You might like to modify this program to evaluate other geometric sums.

(ii) This time we find that $a = \frac{1}{2}$, the common ratio is $r = \frac{1}{2}$, and the last term is $\frac{1}{16}$. The term after that would be $\frac{1}{32}$, so the sum equals

$$\frac{a - ar^{k+1}}{1 - r} = \frac{\frac{1}{2} - \frac{1}{32}}{1 - \frac{1}{2}} = \frac{\frac{16-1}{32}}{\frac{1}{2}} = \frac{15}{16}.$$

This confirms one of the sums evaluated in the introduction to this chapter and could in any event be checked quite easily.

(iii) For this example we find that $a = 5$, $r = 5$, the last term is 5^7, and the following term would be 5^8. The sum equals

$$\frac{a - ar^{k+1}}{1 - r} = \frac{5 - 5^8}{1 - 5} = \frac{5 - 390{,}625}{-4} = \frac{-390{,}620}{-4} = 97655.$$

This was probably a little easier than simply adding up the terms and less prone to errors. ◢

Notice that in part (iii) of Example 1 the index of summation was n, and it ran from $n = 1$ to $n = 7$. The sum will still be a geometric sum no matter where the values of n begin and end, but we need to know the range of n in order to find the first and last terms.

Interest, Mortgages, and Annuities

Consider a bank account that earns interest at the effective rate of 10% per annum. This means that each year the amount in the account is increased by 10%, which is the same thing as multiplying it by 1.10. We say the balance increases by a **factor** of 1.10; the word "factor" reminds us to *multiply* by this constant.

Suppose the owner of the account pays $1000 into it on the first day of each year. At the end of the first year the $1000 in the account will be increased to $1.10 \times \$1000 = \$1100$. Then another $1000 will be deposited, bringing the total to $\$1000 + 1100 = \$2100$ on the first day of the second year. At the end of the second year, interest will bring the total to $1.10 \times (1000 + 1.10 \times 1000)$. Then at the beginning of the third year another $1000 will be added, and the total will be

$$1000 + 1.10 \times (1000 + 1.10 \times 1000) = 1000 + 1.10 \times 1000 + (1.10)^2 \times 1000.$$

A year later, this will have been multiplied by the factor 1.10 as interest was added, and an additional $1000 will have been deposited. The total at the beginning of the fourth year will be

$$1000 + 1.10 \times \left(1000 + 1.10 \times 1000 + (1.10)^2 \times 1000\right)$$
$$= 1000 + 1.10 \times 1000 + (1.10)^2 \times 1000 + (1.10)^3 \times 1000$$
$$= 1000\left(1 + 1.10 + (1.10)^2 + (1.10)^3\right).$$

With each passing year the total is multiplied by 1.10 and an additional $1000 is added. We see that at the beginning of the nth year the balance in the account will be

$$\$1000\left(1 + 1.10 + (1.10)^2 + (1.10)^3 + \ldots + (1.10)^{n-2} + (1.10)^{n-1}\right).$$

At the end of the nth year the amount will have been increased by the factor 1.10, so it will be

$$\$1000\left(1.10 + (1.10)^2 + (1.10)^3 + \ldots + (1.10)^{n-1} + (1.10)^n\right).$$

(Notice that the highest powers occurring in these last two sums are not the same. Each sum consists of n terms, reflecting the fact that in each case there have been n deposits, but in the first the interest has been applied only $n - 1$ times, and in the second the interest has been applied n times.)

Both these amounts are geometric sums, so it is possible to give formulas for evaluating them. For the balance at the beginning of the nth year the first term in the sum is $a = 1000$, the ratio is $r = 1.10$, and the last term is $1000(1.10)^{n-1}$. We see that the total amount is $\left(1000 - 1000(1.10)^n\right)/(1 - 1.10) = \frac{1000}{-0.10}\left(1 - (1.10)^n\right) = 10{,}000\left((1.10)^n - 1\right)$.

It is possible to evaluate the other sum in the same way, but it is easier to notice that it is just the first sum multiplied by 1.10, so it equals $11{,}000\left((1.10)^n - 1\right)$. For instance, at the end of the third year the amount in the account is $11{,}000\left((1.10)^3 - 1\right) = \$3641.00$. This can easily be checked by adding.

However, it would be slightly annoying to have to add up all the terms to find the amount at the end of 20 years. It is easier to calculate $11{,}000\left((1.10)^{20} - 1\right) = \$63{,}002.50$.

EXAMPLE 2   How much money will be in an account after five years if \$100 is deposited on the first of each month and 1% interest is paid at the end of each month?

Solution   This situation is analogous to the discussion above. At the end of the first month, one deposit will have been made and interest will have been paid on it, so the balance will be 1.01×100. At the end of the second month another deposit and another interest payment will have been made; the balance will be $1.01\left(100 + 1.01(100)\right) = 100\left(1.01 + (1.01)^2\right)$.

After n months the balance will be $100\left(1.01 + (1.01)^2 + (1.01)^3 + \ldots + (1.01)^n\right)$. This geometric sum can be evaluated by using the formula. The first term is $100 \times 1.01 = 101$, the common ratio is 1.01, and the last term is $100(1.01)^n$, so the sum equals $\left(101 - 100(1.01)^{n+1}\right)/(1 - 1.01) = \frac{101}{-0.01}\left(1 - (1.01)^n\right) = 10{,}100\left((1.01)^n - 1\right)$.

In particular, in 5 years there are 60 months, so to find the balance after 5 years, we let $n = 60$ and find that the amount is $10{,}100\left((1.01)^{60} - 1\right) = \$8248.64$.   ◢

It is also possible to turn this process around. Consider a twenty-year mortgage for \$100,000. Somebody buying a house might have to borrow this money for the purchase, with the intention of paying it off in equal installments over a *term* of twenty years. For simplicity, let us assume that one payment is made each year (actual mortgages are often paid quarterly or monthly). Let us assume that the interest rate is 10% per year and that the annual payments are all the same size.

We write P for the principal, the original amount borrowed, which in this case is \$100,000. Let r be the factor by which an amount increases in one year due to interest; in this situation, with 10% annual interest, we have $r = 1.1$. Finally, write A for the amount of the annual payment.

At the end of the first year the amount owed will have increased from its initial value P to Pr, and then a payment of A will be made, reducing the balance owed to $Pr - A$. After another year this will have increased to $(Pr - A)r$, due to a year's interest, and the next payment will reduce it to $(Pr - A)r - A = Pr^2 - Ar - A$. After the third year the increase due to interest and the effect of the payment will leave $(Pr^2 - Ar - A)r - A = Pr^3 - Ar^2 - Ar - A$. At the end of the fourth year the balance owed will be $(Pr^3 - Ar^2 - Ar - A)r - A = Pr^4 - Ar^3 - Ar^2 - Ar - A$.

The pattern should be reasonably clear. After k years, that is, at the end of the kth year, the amount still owed will be

$$Pr^k - Ar^{k-1} - Ar^{k-2} - \ldots - Ar^2 - Ar - A$$
$$= Pr^k - A(1 + r + r^2 + r^3 + \ldots + r^{k-2} + r^{k-1}).$$

At the end of the term, that is, after twenty years, the amount owed should be zero, meaning that the mortgage has been paid off. This means that if n is the number of years in the term (in this case $n = 20$), then

$$Pr^n - A(1 + r + r^2 + r^3 + \ldots + r^{n-2} + r^{n-1}) = 0,$$

or

$$Pr^n = A(1 + r + r^2 + r^3 + \ldots + r^{n-2} + r^{n-1}).$$

The right side is a geometric sum that equals $A(1 - r^n)/(1 - r)$. Dividing both sides by the same amount, we can now solve to find a reasonably simple formula for A, namely,

$$A = \frac{Pr^n}{\frac{1-r^n}{1-r}} = \frac{(r - 1)Pr^n}{r^n - 1}. \qquad (10.2.3)$$

Returning to our example, in which $P = 100,000$, $n = 20$, and $r = 1.1$, we see that the annual payment will be $A = (r - 1)Pr^n/(r^n - 1) = (0.1)(100,000)(1.1)^{20} / ((1.1)^{20} - 1) = \$11,745.96$. Notice that just to pay the 10% interest on \$100,000 would take \$10,000, so most of the payment is interest, and only \$1,745.96 of the first payment will go toward reducing the principal. In succeeding years, as the principal begins to diminish, the amount of interest paid will diminish, so a larger part of the payment of \$11,745.96 will go toward principal. See Figure 10.2.1.

Formula 10.2.3 can be used to find the amount of the payment A, assuming that we know the principal P, the number of payments n, and the factor r by which interest increases the balance over the period between two payments. Notice that nothing in our calculation depended on letting the payments be once a year; this formula will work as long as the payments are evenly timed and for the same amount.

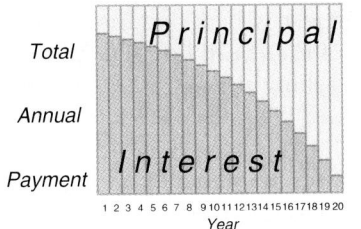

Total

Annual

Payment

1 2 3 4 5 6 7 8 9 10 11 12 13 14 15 16 17 18 19 20
Year

Figure 10.2.1

THEOREM 10.2.4

Mortgage Payments

Suppose equal payments of amount A are made to pay off a debt with principal P, and the interest increases the balance by the factor r between consecutive payments. Then the balance owing after k payments have been made is

$$Pr^k - A\frac{r^k - 1}{r - 1}.$$

If the debt is exactly paid off in n equal payments, then the amount of each payment is

$$A = \frac{(r - 1)Pr^n}{r^n - 1}.$$

EXAMPLE 3   What will be the monthly payment for a car loan of $20,000 paid over 3 years with 15% annual interest, compounded monthly?

Solution   This situation is analogous to the mortgage discussed above. Here the principal is $P = 20,000$, the number of payments is the number of months in 3 years, which is $n = 36$, and the factor by which interest increases a balance in one month (the period between consecutive payments) is $r = 1 + \frac{15}{12}/100 = 1.0125$.

From Theorem 10.2.4 we see that the monthly payment is $A = (r-1)Pr^n/(r^n - 1) = (0.0125)(20,000)(1.0125)^{36}/\left((1.0125)^{36} - 1\right) = \$693.31$.

Notice that since this loan is for a much shorter length of time than the mortgage discussed above, a smaller proportion of each payment will be for interest. In fact, one month's interest on $P = 20,000$ is $\frac{15}{12}\% = 1.25\%$ of P, or $0.0125 \times 20,000 = \$250$, which is much less than half of the monthly payment. See Figure 10.2.2. ◢

A person who is about to retire is likely to be interested in finding a way to provide a secure income for the rest of his or her life. One way of doing this is to purchase what is called an **annuity**. The idea is that the purchaser gives a large amount of money to a financial institution, which promises to pay a regular amount back to the purchaser. There are many possible variations; one type will pay a regular income for the rest of the purchaser's life, while another will last only for a fixed number of years. Some annuities are indexed to allow for inflation, and many other features are available.

However, the basic idea is approximately the same for any annuity.

EXAMPLE 4   Suppose a purchaser pays $50,000 for an annuity that guarantees to pay $7000 a year for 15 years. Assuming that interest is 10% per annum and that interest and disbursements are paid at the end of each year, how much profit will be left at the end of 15 years for the company selling the annuity (ignoring administration costs)?

Solution   This situation is exactly like a mortgage except that the money moves in the opposite direction. If we write $P = 50,000$ for the amount invested, $r = 1.10$ for the factor by which the balance increases in one year due to interest, and $A = 7000$ for the annual payment, then the balance remaining after n years can be found by using Theorem 10.2.4. The amount remaining after n years is $Pr^n - A - Ar - Ar^2 - \ldots - Ar^{n-1} = Pr^n - A(r^n - 1)/(r - 1)$.

In our example, with $n = 15$, we find that the amount left will be $50,000(1.1)^{15} - 7000\left((1.1)^{15} - 1\right)/(1.1 - 1) = -\$13,544.96$. The fact that the balance is negative means that the company had to pay out more money than there was left, which means that this annuity was badly planned and was a losing proposition for the company. ◢

Geometric Series

In Theorem 10.2.2 we found that the geometric sum $a + ar + ar^2 + \ldots + ar^k$ is equal to $a\frac{1-r^{k+1}}{1-r}$. If we let k become larger and larger, which amounts to taking more and more terms in the sum, we can ask what happens to this value of the sum.

For instance, if $a = \frac{1}{2}$ and $r = \frac{1}{2}$, then the sum $\frac{1}{2} + \frac{1}{4} + \frac{1}{8} + \ldots + \frac{1}{2^{k+1}}$ equals $\left(\frac{1}{2} - \left(\frac{1}{2}\right)^{k+2}\right)/\left(1 - \frac{1}{2}\right) = 1 - \left(\frac{1}{2}\right)^{k+1}$. As $k \to \infty$, we know that $\left(\frac{1}{2}\right)^{k+1} \to 0$, so the value of the sum will tend to $1 - 0 = 1$.

We divide 15% by 12 to get the (uncompounded) *monthly* interest.

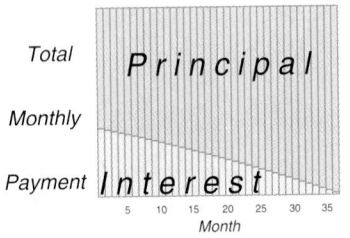

Figure 10.2.2

We can write $\frac{1}{2} + \frac{1}{4} + \ldots + \left(\frac{1}{2}\right)^k + \ldots = 1$, where the "infinite sum" on the left side really means the limit of the finite sums as the number of terms becomes larger and larger.

What made this work was that $\lim_{k \to \infty} \left(\frac{1}{2}\right)^{k+1} = 0$. If we consider the sum $a + ar + ar^2 + \ldots + ar^k = \frac{a - ar^{k+1}}{1-r} = \frac{a}{1-r}(1 - r^{k+1})$, we see that the corresponding question concerns the behavior of r^{k+1} as $k \to \infty$.

If $|r| < 1$, then $\lim_{k \to \infty} r^{k+1} = 0$, by Example 10.1.5, and the value of the sum with k terms tends to $\frac{a}{1-r}(1 - 0) = \frac{a}{1-r}$ as $k \to \infty$. If $|r| > 1$, then this limit does not exist, and if $r = -1$, then r^{k+1} does not tend to a limit as $k \to \infty$ (since $(-1)^{k+1}$ is alternately 1 and -1). If $r = 1$, then Theorem 10.2.2 does not apply and we cannot use this argument, but in this case the sum with k terms is just $a + a + a + \ldots + a = ka$, and unless $a = 0$, this does not tend to a (finite) limit either.

DEFINITION 10.2.5

> Suppose $a \neq 0$. The **geometric series**
>
> $$a + ar + ar^2 + ar^3 + \ldots + ar^k + \ldots = \sum_{n=0}^{\infty} ar^n$$
>
> means the limit as $k \to \infty$, if it exists, of the values of the geometric sum
>
> $$a + ar + ar^2 + ar^3 + \ldots + ar^{k-1} = \sum_{n=0}^{k-1} ar^n = a\frac{1 - r^k}{1 - r}$$
>
> with k terms in it.
> The limit does exist precisely when $|r| < 1$, and in that case
>
> $$a + ar + ar^2 + ar^3 + \ldots + ar^k + \ldots = \sum_{n=0}^{\infty} ar^n = \frac{a}{1 - r}.$$
>
> We say the geometric series **converges** if the limit exists and that it **diverges** if the limit does not exist.

EXAMPLE 5   Decide whether or not each of the following geometric series converges, and evaluate the ones that do: (i) $\sum_{n=0}^{\infty} 3^{-n}$, (ii) $\sum_{n=0}^{\infty} \frac{14}{5^{-n}}$, (iii) $\sum_{k=1}^{\infty} \frac{5}{(-4)^k}$.

Solution

(i) In this case the common ratio is the ratio of the nth term to the $(n-1)$th term, which is $\frac{3^{-n}}{3^{-(n-1)}} = 3^{-n+(n-1)} = 3^{-1} = \frac{1}{3}$. Since this is smaller than 1, the series converges. Its first term is $a = 3^{-0} = 1$, so Definition 10.2.5 tells us that $\sum_{n=0}^{\infty} 3^{-n} = \frac{a}{1-r} = 1/\left(1 - \frac{1}{3}\right) = 1/\left(\frac{2}{3}\right) = \frac{3}{2}$.

(ii) Here the ratio of the nth term to the $(n-1)$th term is $\frac{14}{5^{-n}} / \frac{14}{5^{-(n-1)}} = \frac{14}{5^{-n}} \frac{5^{-(n-1)}}{14} = \frac{5^{-(n-1)}}{5^{-n}} = 5^{-(n-1)-(-n)} = 5^1 = 5 > 1$. Since the ratio is greater than 1, the series diverges. It does not make sense to try to evaluate it.

(iii) The common ratio is $\left(\frac{5}{(-4)^k}\right) / \left(\frac{5}{(-4)^{k-1}}\right) = \frac{5}{(-4)^k} \frac{(-4)^{k-1}}{5} = \frac{(-4)^{k-1}}{(-4)^k} = (-4)^{k-1-k} = (-4)^{-1} = -\frac{1}{4}$. Since the absolute value of this ratio is less than 1, the series converges.

Moreover, its value can easily be found. The first term in the series is the one with $k = 1$, which is $\frac{5}{(-4)^k} = \frac{5}{(-4)} = -\frac{5}{4}$. Since the common ratio is $r = -\frac{1}{4}$, we see that the value of the geometric series is $\frac{-5/4}{1-(-1/4)} = \frac{-5/4}{5/4} = -1$.

> Notice that in part (iii) of Example 5, the sum was indexed by k running from $k = 1$ to $k = \infty$. Where the index runs does not affect the convergence of the series, since that depends only on the ratio of each term to the previous one. Of course, we do have to be careful because the range through which the index runs *does* determine the first term a, which we need to know in order to evaluate the series.

Exercises 10.2

I

Evaluate each of the following geometric sums.

1. $1 + \frac{1}{2} + \frac{1}{4} + \ldots + \frac{1}{64}$

2. $3 + 1 + \frac{1}{3} + \frac{1}{9} + \frac{1}{27}$

3. $1 - \frac{1}{2} + \frac{1}{4} - \ldots - \frac{1}{128}$

4. $\frac{1}{4} - \frac{1}{2} + 1 - 2 + 4 - \ldots + 64$

5. $\sum_{k=0}^{20} 2^{-k}$ **8.** $\sum_{m=2}^{9} 10^{-m}$

6. $\sum_{n=0}^{10} 3^n$ **9.** $\sum_{n=-2}^{6} \frac{2}{(-6)^n}$

7. $\sum_{\ell=1}^{8} 5(4^{-\ell})$ **10.** $\sum_{j=0}^{99} (-1)^j$

In each of the following situations, a loan is to be paid in equal installments. Find the size of the payment.

11. $10,000, 12 annual payments, annual interest 10%.

12. $10,000, 15 annual payments, annual interest 10%.

13. $1000, monthly payments for 2 years, interest rate 12%, compounded monthly.

14. $2000, monthly payments for 4 years, interest 18%, compounded monthly.

15. $40,000, quarterly payments for 16 years, annual interest 10%, compounded quarterly.

16. $60,000, quarterly payments for 20 years, annual interest 12%, compounded quarterly.

17. $100,000, 15 annual payments, annual interest 12%.

18. $100,000, monthly payments for 20 years, interest rate 12%, compounded monthly.

19. $2000, daily payments for 1 year, interest 12%, compounded daily.

20. $2000, weekly payments for 1 year, interest 12%, compounded weekly. (Assume that there are exactly 52 weeks in one year.)

Evaluate each of the following geometric series.

21. $1 + \frac{1}{2} + \frac{1}{4} + \ldots$

22. $49 + 7 + 1 + \frac{1}{7} + \frac{1}{49} + \ldots$

23. $9 - 3 + 1 - \frac{1}{3} + \frac{1}{9} - \ldots$

24. $\frac{3}{4} - \frac{9}{16} + \frac{27}{64} - \ldots$

25. $\sum_{m=0}^{\infty} 2^{-m}$ **28.** $\sum_{m=2}^{\infty} (\sqrt{2})^{-m}$

26. $\sum_{n=0}^{\infty} \left(\frac{3}{5}\right)^n$ **29.** $\sum_{m=-4}^{\infty} \frac{5}{(-3)^m}$

27. $\sum_{k=1}^{\infty} 11(3^{-k})$ **30.** $\sum_{j=0}^{\infty} (-0.9)^j$

II

31. What would be the monthly payment in the situation of Example 3 (p. 654) if interest rates went up to 18%?

32. Suppose a loan for $15,000 is to be paid off in five equal annual installments. If the effective annual interest rate is 9%, what will be the size of the installments?

*33. With what size of annual payment would the company selling the annuity in Example 4 (p. 654) be able to break even?

*34. With what size of annual payment would the company selling the annuity in Example 4 be able to have a surplus of $10,000 when the term of the annuity is over?

*35. (i) If the situation of Example 4 was repeated but interest rates rose to 11%, what effect would this have on the company's position?
(ii) What would happen if the interest rate was 12%?

*36. (i) A certain annuity, which costs $30,000, promises to pay $4402.90 per year for the rest of Grandfather's life. If the effective annual interest rate is 10%, how long will it be before all the money in the plan has been exhausted?

(ii) The best available actuarial predictions say that Grandfather's life expectancy is ten years. If Grandfather lives for exactly ten more years, will the company selling the annuity make or lose money (or break even)?

*37. Suppose m and k are integers with $m < k$. Find a formula for evaluating the sum

$$\sum_{n=m}^{k} ar^n.$$

*38. (i) What happens with the formula you found in the previous question if we allow m and k to be equal? How would you interpret this answer?

(ii) What happens with the formula you found in the previous question if $k = m - 1$? How would you interpret this answer?

*39. Evaluate (i) $\sum_{k=0}^{12} 3^{2k}$, (ii) $\sum_{k=1}^{8} 2^{-3k}$.

*40. Find a formula for evaluating the sum

$$\sum_{n=0}^{k} ar^{2n}.$$

*41. (i) The sum $a + ar + ar^2 + \ldots + ar^n$ is a geometric sum with common ratio r. However, if we turn it around and write it the other way, $ar^n + ar^{n-1} + \ldots + ar^2 + ar + a$, it is a geometric sum with common ratio $\frac{1}{r}$. Find the formulas for evaluating each of these geometric sums.

(ii) Show that the two formulas in part (i) are actually equal.

42. (i) The repeating decimal $0.7777\ldots$ can be written as a geometric series $\frac{7}{10} + \frac{7}{100} + \frac{7}{1000} + \ldots = 7\sum_{n=1}^{\infty} \left(\frac{1}{10}\right)^n$. Express $0.7777\ldots$ as a rational number $\frac{r}{s}$.

(ii) Express the repeating decimal $2.41111\overline{1}\ldots$ as a rational number.

(iii) Express the repeating decimal $1.2121\overline{12}\ldots$ as a rational number.

43. Express the repeating decimal $2.112\,112\,\overline{112}\ldots$ as a rational number.

44. Express the repeating decimal $6.4223\,4223\,\overline{4223}\ldots$ as a rational number.

45. Verify the result of Example 3: With a calculator or computer, start with $P = 20{,}000$, add the monthly interest, and subtract the monthly payment $A = 693.31$. Add the next month's interest, subtract another payment, and so on. What happens after 3 years?

46. The calculation we did before Example 2 (p. 652) showed that $1000\left(1.10 + (1.10)^2 + \ldots + (1.10)^{20}\right) = 63{,}002.50$. Verify this by actually adding up the sum.

POINT TO PONDER

Suppose a rabbit is moving in a straight line, and consider how it moves one unit, that is, from 0 to 1. First it has to move from 0 to $\frac{1}{2}$, then from $\frac{1}{2}$ to $\frac{3}{4}$, then from $\frac{3}{4}$ to $\frac{7}{8}$, and so on. Since the rabbit has to go through all these steps, one after another, we conclude that it never reaches 1 (!).

This argument is known as **Zeno's Paradox**. It suggests that nothing can ever move anywhere (on p. 637 we consider another version, which shows that a fast hare can never overtake a slow tortoise).

Suppose for simplicity that the rabbit's speed is 1 unit per second. Then to get from 0 to $\frac{1}{2}$ takes it $\frac{1}{2}$ sec, to get from $\frac{1}{2}$ to $\frac{3}{4}$ takes it $\frac{1}{4}$ sec, to get from $\frac{3}{4}$ to $\frac{7}{8}$ takes it $\frac{1}{8}$ sec, and so on.

To go through all the infinitely many steps, from 0 to $\frac{1}{2}$, from $\frac{1}{2}$ to $\frac{3}{4}$, from $\frac{3}{4}$ to $\frac{7}{8}$, and so on takes a total time of

$$\frac{1}{2} + \frac{1}{4} + \frac{1}{8} + \ldots.$$

This is a geometric series, which we now know totals 1 sec. In 1 sec the rabbit actually goes through all the steps and does move 1 unit of distance.

The paradox can perhaps be explained by saying that the original argument tacitly assumed that each of the steps from 0 to $\frac{1}{2}$, from $\frac{1}{2}$ to $\frac{3}{4}$, from $\frac{3}{4}$ to $\frac{7}{8}$, and so on takes the same length of time. If the speed is constant, then they do not, and our knowledge of geometric series allows us to find the total time.

10.3 Convergence of Series

PARTIAL SUM
CONVERGENCE
DIVERGENCE
HARMONIC SERIES

In Section 10.2 we discussed geometric series and found that they *converge* when the absolute value of the common ratio is less than 1 and that there is a convenient formula for evaluating the series. In this section we discuss the notion of convergence for other series (i.e., infinite sums). Generally speaking, it can be quite difficult to decide whether or not a particular series converges, and it can be even more difficult to evaluate the series if it does converge.

The question we face is to try to make sense of an infinite sum

$$\sum_{n=1}^{\infty} a_n.$$

To say that this infinite sum equals the number S, say, ought to mean that if we add up enough terms of the sum, the total should approach the value S. In other words, we should consider the sum of the first N terms,

$$S_N = \sum_{n=1}^{N} a_n,$$

and see whether the *sequence* $\{S_N\}$ approaches the limit S.

DEFINITION 10.3.1

Consider the series $\sum_{n=1}^{\infty} a_n$. The sum

$$S_N = \sum_{n=1}^{N} a_n$$

of the first N terms is called the Nth **partial sum** of the series. The series $\sum_{n=1}^{\infty} a_n$ **converges** to a number S if the sequence of partial sums $\{S_N\}$ converges to S.

The series is **convergent** if it converges to some limit S, the *sum*, and **divergent** if it does not converge to any limit.

Let us emphasize again the distinction between sequences and series. A sequence $\{a_n\}$ is a *list* of numbers. A series $\sum_{n=1}^{\infty} b_n$ is like an infinite *sum*. Starting with a series, we can define the partial sums $S_N = \sum_{n=1}^{N} b_n$; each S_N is a number, and together they make a sequence $\{S_N\}$.

If you think back to the way we discussed geometric series, the way we considered convergence or divergence in Definition 10.2.5 was to evaluate the partial sums and see whether or not they tended to a limit. The thing that made that situation special was that there is an explicit formula for the partial sums. For many series no such formula is available.

EXAMPLE 1

(i) The easiest series to evaluate is the one whose terms are all zero, that is, $\sum_{n=1}^{\infty} 0$. If we add up the first N terms to find S_N, we are adding up N zeroes, so $S_N = 0$, for every N. The sequence of partial sums $\{S_N\}$ is the constant sequence $0, 0, 0, 0, \ldots$, which converges to 0. We have the completely unsurprising conclusion that $\sum_{n=1}^{\infty} 0 = 0$.

(ii) Consider the series $\sum_{n=1}^{\infty} a_n$, where the terms a_n are all the same, that is, $a_n = c$, for every n, where c is some constant. But now let us consider the case in which $c \neq 0$.

The partial sum S_N is obtained by adding up N terms, each of which equals c. So $S_N = \sum_{n=1}^{N} c = Nc$. The sequence of partial sums $\{S_N\} = \{Nc\}$ diverges (because the terms Nc tend to ∞ if $c > 0$ and to $-\infty$ if $c < 0$).

Combining parts (i) and (ii), we see that the series $\sum_{n=1}^{\infty} c$ converges to 0 if $c = 0$ and diverges otherwise.

$$S_1 = b_1 = -1$$
$$S_2 = b_1 + b_2 = -1 + 1 = 0$$
$$S_3 = b_1 + b_2 + b_3$$
$$= -1 + 1 + (-1) = -1$$
$$\vdots$$

(iii) Let $b_n = (-1)^n$ and consider $\sum_{n=1}^{\infty} b_n = \sum_{n=1}^{\infty} (-1)^n$. At left we find the first few partial sums. They are alternately -1 and 0; they do not converge to any limit, so the series diverges.

$$S_1 = c_1 = \frac{1}{2}$$
$$S_2 = c_1 + c_2 = \frac{1}{2} + \frac{1}{6} = \frac{2}{3}$$
$$S_3 = c_1 + c_2 + c_3 = S_2 + c_3$$
$$= \frac{2}{3} + \frac{1}{12} = \frac{3}{4}$$
$$S_4 = S_3 + c_4 = \frac{3}{4} + \frac{1}{20} = \frac{4}{5}$$

(iv) Let $c_n = \frac{1}{n^2 + n}$ and consider the series $\sum_{n=1}^{\infty} c_n$. At left we calculate some partial sums. From the ones we have, it is beginning to look as if $S_N = \frac{N}{N+1}$. This is in fact true. The reason is that if $S_N = \frac{N}{N+1}$, then S_{N+1} is obtained from it by adding $c_{N+1} = \frac{1}{(N+1)^2 + (N+1)}$, so

$$S_{N+1} = S_N + \frac{1}{(N+1)^2 + (N+1)} = \frac{N}{N+1} + \frac{1}{(N+1)(N+1+1)}$$

$$= \frac{N(N+2) + 1}{(N+1)(N+2)} = \frac{N^2 + 2N + 1}{(N+1)(N+2)} = \frac{(N+1)^2}{(N+1)(N+2)}$$

$$= \frac{N+1}{N+2}.$$

We know that $S_N = \frac{N}{N+1}$ for $N = 1, 2, 3$, and 4, and we have just seen that it will always be true for the *next* value of N too. From this it is possible to conclude that it is true for *all* values of N. (This is an example of what is called a proof by *induction*; the concept is discussed at greater length in the appendix.)

Since the sequence of partial sums $\{S_N\} = \{\frac{N}{N+1}\}$ converges to 1, we see that the series $\sum_{n=1}^{\infty} c_n$ converges and $\sum_{n=1}^{\infty} c_n = \sum_{n=1}^{\infty} \frac{1}{n^2 + n} = 1$. ◢

In the last part of Example 1 we were able to find an exact formula for the partial sums of the series, which enabled us to find their limit and evaluate the series. Usually, however, this is quite difficult to do.

If the series $\sum_{n=1}^{\infty} a_n$ converges to L, say, then its partial sums must approach L. In particular, the partial sums S_N must eventually get close together (since they will all be close to L).

The difference between two consecutive partial sums is just one of the terms of the series: $S_N - S_{N-1} = a_N$. To say that these partial sums are close together is to say that a_N is small.

So we see that the terms of a convergent series must eventually get small. This is a very important observation.

In Example 1, the series in parts (i) and (iv) both converge. In the first case the terms are all zero, so the terms certainly converge to 0. The series in part (iv) is $\sum_{n=1}^{\infty} \frac{1}{n^2+n}$, and its terms also converge to zero, by Theorem 10.1.3.

The series in parts (ii) and (iii) of Example 1 diverge, and in part (ii) the terms converge to the nonzero number c, while in part (iii) the terms diverge.

THEOREM 10.3.2

The *n*th Term Test

Suppose the series $\sum_{n=1}^{\infty} a_n$ converges. Then its terms converge to 0, that is,

$$\lim_{n \to \infty} a_n = 0.$$

(Notice that this limit refers to the terms of the series as a *sequence*.)

Another way of saying exactly the same thing is this: If $\{a_n\}$ does *not* converge to 0 (i.e., if this sequence diverges or if it converges to something other than 0), then the corresponding series $\sum_{n=1}^{\infty} a_n$ diverges.

PROOF We have already observed that $a_n = S_n - S_{n-1}$. Suppose the series $\sum_{n=1}^{\infty} a_n$ converges to L. Then

$$\lim_{n \to \infty} a_n = \lim_{n \to \infty} (S_n - S_{n-1}) = \left(\lim_{n \to \infty} S_n \right) - \left(\lim_{n \to \infty} S_{n-1} \right) = L - L = 0.$$

(We have used Theorem 10.1.5 to show that the "shifted" sequence $\{S_{n-1}\}$ converges to the same limit as $\{S_n\}$, i.e., to L.)

We have already checked that this result is consistent with Example 1.

The way Theorem 10.3.2 is most often applied is to use the second form of it to prove that a series diverges by noticing that its terms do not converge to 0. It is in this sense that it is a test. We can test whether or not the nth term of a series $\sum_{n=1}^{\infty} a_n$ tends to zero, and if it does not, Theorem 10.3.2 allows us to conclude that the series diverges.

EXAMPLE 2

(i) Theorem 10.3.2 gives an almost instant argument that $\sum_{n=1}^{\infty} (-1)^n$ diverges because the sequence $-1, 1, -1, 1, \ldots$ of its terms does not converge to 0 (or to anything else).

(ii) The series $\sum_{n=1}^{\infty} \frac{n-1}{n}$ diverges because $\lim_{n \to \infty} \frac{n-1}{n} = 1$, by Theorem 10.1.3.

(iii) The series $\sum_{n=1}^{\infty} (3n - 4)$ diverges because $\{(3n - 4)\}$ diverges (it tends to ∞).

We see that Theorem 10.3.2 is extremely useful for proving that certain series diverge. It is important to realize that it does *not* allow us to conclude that if the terms $\{a_n\}$ of a series do tend to zero, then the series must converge. There is a very important example of a series that *diverges* even though its terms converge to zero.

The Harmonic Series

The series $\sum_{n=1}^{\infty} \frac{1}{n}$ whose nth term is $\frac{1}{n}$ is called the **harmonic series**. Its terms tend to 0, but the series diverges.

THEOREM 10.3.3

> The harmonic series diverges.

PROOF Instead of considering all the partial sums, let us concentrate on S_2, S_4, S_8, S_{16}, and so on, that is, on the partial sums S_N for which N is a power of 2.

For instance,

$$S_{16} = 1 + \frac{1}{2} + \frac{1}{3} + \frac{1}{4} + \frac{1}{5} + \frac{1}{6} + \frac{1}{7} + \frac{1}{8} + \frac{1}{9} + \frac{1}{10} + \frac{1}{11} + \frac{1}{12} + \frac{1}{13}$$
$$+ \frac{1}{14} + \frac{1}{15} + \frac{1}{16}$$
$$= (1) + \left(\frac{1}{2}\right) + \left(\frac{1}{3} + \frac{1}{4}\right) + \left(\frac{1}{5} + \frac{1}{6} + \frac{1}{7} + \frac{1}{8}\right)$$
$$+ \left(\frac{1}{9} + \frac{1}{10} + \frac{1}{11} + \frac{1}{12} + \frac{1}{13} + \frac{1}{14} + \frac{1}{15} + \frac{1}{16}\right)$$

Now consider the expressions in the parentheses. Each of them is greater than or equal to $\frac{1}{2}$. The first one is 1; the second is $\frac{1}{2}$. The third is $\frac{1}{3} + \frac{1}{4}$. It consists of two terms, each of which is greater than or equal to $\frac{1}{4}$, so their sum must be greater than or equal to $\frac{1}{2}$. The next pair of parentheses enclose four terms, each greater than or equal to $\frac{1}{8}$, so it must be greater than or equal to $4\left(\frac{1}{8}\right) = \frac{1}{2}$. The last pair of parentheses contains eight terms, each greater than or equal to the last term, which is $\frac{1}{16}$, so the sum is greater than or equal to $8\left(\frac{1}{16}\right) = \frac{1}{2}$.

We could continue this way. If we consider all the terms that come after the 2^nth term up to the 2^{n+1}th term, then there are exactly 2^n such terms, and they are all greater than or equal to the last one, which is $\frac{1}{2^{n+1}} = 2^{-(n+1)}$. The sum of these terms is greater than or equal to $2^n \times 2^{-(n+1)} = \frac{1}{2}$.

So if we consider the partial sum S_{2^n}, we can group the terms in this way in $n+1$ groups (up to 1, from there up to $\frac{1}{2}$, from there up to $\frac{1}{4}$, ..., from $\frac{1}{2^{n-1}}$ up to $\frac{1}{2^n}$, ...). Since the sum of each of these groups is greater than or equal to $\frac{1}{2}$, we see that the partial sum S_{2^n} is greater than or equal to $(n+1)\left(\frac{1}{2}\right) = \frac{n+1}{2}$. (Since the first "group" is 1, we actually have $S_{2^n} \geq \frac{n+2}{2}$ ...).

Since the terms of the sequence $\left\{\frac{n+1}{2}\right\}$ grow to infinity, we see that the partial sums of the harmonic series also diverge, so the harmonic series itself is divergent, as desired. ◢

See the Calculator Example on the next page.

This example is very important for many purposes, but the first thing it does is provide an example of a series whose terms converge to 0 but that is divergent. This shows that while Theorem 10.3.2 can be used to show that various series are divergent, there is no corresponding theorem in the other direction to show that they converge just because their terms approach zero.

We cannot show it yet, but later in the chapter we will learn that if we alter the harmonic series by putting in minus signs on every other term, then the resulting series converges.

With a programmable calculator we can test the idea of the proof of Theorem 10.3.3, for various values of n.

TI-85	SHARP EL9300
PRGM F2(EDIT) HARM ENTER F3 F1(Input) N EXIT ENTER 0 STO K ENTER 0 STO S ENTER Lbl B ENTER K + 1 STO K ENTER S + 1 ÷ K STO S ENTER F4 F1 (If) (K TEST F2(<) 2^N) ENTER Goto B ENTER S ENTER EXIT EXIT EXIT PRGM F1(NAME) HARM ENTER 4 ENTER	[icon] ▼ ▼ (NEW) ENTER ENTER HARM ENTER COMMAND A 3 (Input) N ENTER K = 0 ENTER S = 0 ENTER COMMAND B 1 (Label) 1 ENTER K = K + 1 ENTER S = S + (1 ÷ K) ENTER COMMAND 3 (If) K COMMAND C 2 (<) 2 a^b N COMMAND B 2 (Goto) 1 ENTER COMMAND A 1 (Print) S ENTER COMMAND A 6 (End) ENTER [icon] [icon] (RUN) ENTER (HARM) ENTER 4 ENTER
CASIO f_x-6300G (f_x-7700GB) MODE 2 EXE ? → ■ : 0 → ■ : 0 → ■ : Lbl 1 : ■ + 1 → ■ : ■ + (1 ÷ ■) → ■ : ■ < 2x^y ■ ⇒ Goto 1: S ◢ MODE 1 Prog 0 EXE 4 EXE (on the f_x-7700GB, : is PRGM F6, and Prog is PRGM F3.)	HP 48SX ATTN ↷ CLR [■■] "" KEY IN N ▶ SPC PROMPT SPC ' N ▶ STO 0 SPC 1 SPC 2 SPC N y^x SPC FOR K SPC 1 K ÷ + NEXT ▼ ENTER ' HARM STO VAR HARM 4 ENTER [■■] [■■]

```
?4
ANSWER=
      3.380728993
```

Now use the program with other values of n. With even moderately large values of n it can take quite a while to run. It is interesting to note that the sum up to $k = 2^n$ is always greater than $\frac{n+1}{2}$.

Generally speaking, it is quite delicate and difficult to decide whether or not series converge. The next three sections are devoted to developing some techniques for approaching problems of this sort.

We give one other example to demonstrate the sort of phenomena that can be encountered.

EXAMPLE 3 In Section 10.2 we observed that "shifting" a sequence does not change whether or not it converges, nor its limit if it does converge. The same is clearly true of a series.

For sequences we also observed that changing the first few terms does not affect the limit, but with series this is not true. After all, if we change some terms, the partial sums will change.

For instance, suppose we take a series $\sum_{n=1}^{\infty} a_n$ that converges to L. Define another series $\sum_{n=1}^{\infty} b_n$ by letting $b_1 = a_1 + 1$, $b_2 = a_2 - 2$, $b_3 = a_3 + 5$, and all other b_n's the same as the corresponding a_n's, that is, $b_n = a_n$, for $n > 3$.

Let us write S_N and T_N for the partial sums $S_N = \sum_{n=1}^{N} a_n$ and $T_N = \sum_{n=1}^{N} b_n$. Then

$$T_1 = b_1 = (a_1 + 1) = S_1 + 1,$$
$$T_2 = b_1 + b_2 = (a_1 + 1) + (a_2 - 2) = a_1 + a_2 - 1 = S_2 - 1,$$
$$T_3 = b_1 + b_2 + b_3 = (a_1 + 1) + (a_2 - 2) + (a_3 + 5) = a_1 + a_2 + a_3 + 4$$
$$= S_3 + 4,$$
$$T_4 = b_1 + b_2 + b_3 + b_4 = (a_1 + 1) + (a_2 - 2) + (a_3 + 5) + a_4$$
$$= a_1 + a_2 + a_3 + a_4 + 4 = S_4 + 4.$$

In fact, for every $N > 3$,

$$T_N = \sum_{n=1}^{N} b_n = \sum_{n=1}^{3} b_n + \sum_{n=4}^{N} b_n = \big((a_1 + 1) + (a_2 - 2) + (a_3 + 5)\big) + \sum_{n=4}^{N} a_n$$
$$= 4 + (a_1 + a_2 + a_3) + \sum_{n=4}^{N} a_n = 4 + \sum_{n=1}^{N} a_n$$
$$= 4 + S_N.$$

We see that the first few partial sums change in different ways, but after we reach the stage at which the terms in the two series are the same, the partial sums always differ by 4. Since $\{S_N\}$ converges to L, we see that $\{T_N\}$ converges to $L + 4$, that is, $\sum_{n=1}^{\infty} b_n = \left(\sum_{n=1}^{\infty} a_n\right) + 4$.

Changing the first few terms may change the value of the infinite sum but does not change whether it converges or not.

THEOREM 10.3.4

(i) Consider two series $\sum_{n=1}^{\infty} a_n$ and $\sum_{n=1}^{\infty} b_n$. Suppose that the series are the same after the first few terms, that is, there is an integer K so that $a_n = b_n$ for every $n > K$.

Then either both series converge or else both series diverge. If they converge, then their sums differ in the obvious way owing to the changes in the first K terms:

$$\left(\sum_{n=1}^{\infty} a_n\right) - \left(\sum_{n=1}^{\infty} b_n\right) = \left(\sum_{n=1}^{K} a_n\right) - \left(\sum_{n=1}^{K} b_n\right) = \sum_{n=1}^{K} (a_n - b_n).$$

(ii) Suppose $\sum_{n=1}^{\infty} a_n$ converges. Then the shifted series $\sum_{n=k+1}^{\infty} a_{n-k}$ also converges, and the sums are the same. If either series diverges, then they both do.

(iii) Suppose $\sum_{n=1}^{\infty} a_n$ converges. Then for any $k > 1$ the series $\sum_{n=k}^{\infty} a_n$ that begins with a_k instead of a_1 also converges. If, on the other hand, either of these series diverges, then they both do.

PROOF

(i) Consider the partial sums $S_N = \sum_{n=1}^{N} a_n$ and $T_N = \sum_{n=1}^{N} b_n$. If $N > K$, then

$$S_N - T_N = \left(\sum_{n=1}^{N} a_n\right) - \left(\sum_{n=1}^{N} b_n\right) = \sum_{n=1}^{N} (a_n - b_n)$$

$$= \left(\sum_{n=1}^{K} (a_n - b_n)\right) + \left(\sum_{n=K+1}^{N} (a_n - b_n)\right) = \left(\sum_{n=1}^{K} (a_n - b_n)\right) + 0$$

$$= \left(\sum_{n=1}^{K} a_n\right) - \left(\sum_{n=1}^{K} b_n\right) = S_K - T_K.$$

We see that for $N > K$, $S_N - T_N$ is a constant, independent of N. If either sequence converges, then so must the other, and

$$\left(\sum_{n=1}^{\infty} a_n\right) - \left(\sum_{n=1}^{\infty} b_n\right) = \lim_{N \to \infty} S_N - \lim_{N \to \infty} T_N = \lim_{N \to \infty} (S_N - T_N)$$

$$= \left(\sum_{n=1}^{K} a_n\right) - \left(\sum_{n=1}^{K} b_n\right),$$

as needed.

(ii) Let $S_N = \sum_{n=1}^{N} a_n$, and let $T_M = \sum_{n=k+1}^{M} a_{n-k}$ be the partial sum of the shifted sequence. Then $T_{N+k} = S_N$ for every $N \geq 1$. By Theorem 10.1.5(i) these two sequences S_N and T_N must converge to the same limit (or must both diverge).

(iii) Define a new series by letting $b_n = 0$ if $n < k$ and $b_n = a_n$ if $n \geq k$. Then the series $\sum_{n=1}^{\infty} b_n$ is really the same thing as $\sum_{n=k}^{\infty} a_n$.

On the other hand, the terms a_n and b_n are equal after $n = k$, so by part (i) of this theorem, the two series are either both convergent or both divergent. ◢

EXAMPLE 4

(i) The series $\frac{1}{5} + \frac{1}{6} + \frac{1}{7} + \frac{1}{8} + \frac{1}{9} + \ldots = \sum_{n=5}^{\infty} \frac{1}{n}$ is really just the harmonic series starting with its fifth term. By part (iii) of Theorem 10.3.4 it diverges.

(ii) Let $a_n = \frac{1}{3n+9}$ and consider the series $\sum_{n=1}^{\infty} a_n = \sum_{n=1}^{\infty} \frac{1}{3n+9}$. If we write it as $\sum_{n=1}^{\infty} \frac{1}{3n+9} = \frac{1}{3}\left(\sum_{n=1}^{\infty} \frac{1}{n+3}\right)$, then we see that this last sum is the same as the harmonic series shifted over by 3, starting from the fourth term.

Combining parts (i) and (ii) of Theorem 10.3.4, we see that the series diverges.

(iii) Consider the series with terms $3, 5, -4, 1, \frac{1}{2}, \frac{1}{4}, \frac{1}{8}, \ldots$, which is a geometric series after the first three terms. The first part of the theorem tells us that it converges and even tells us how to find the sum, but it is easier to write

$$3 + 5 - 4 + 1 + \frac{1}{2} + \frac{1}{4} + \frac{1}{8} + \ldots = 3 + 5 - 4 + \sum_{n=0}^{\infty} \left(\frac{1}{2}\right)^n$$

$$= 4 + \frac{1}{1 - \frac{1}{2}} = 4 + 2 = 6.$$

We evaluated this series by breaking off the first few anomalous terms and adding them up separately and then using the formula for geometric series to evaluate the other part of the series. ◢

Exercises 10.3

I

Decide whether or not each of the following series converges, giving reasons, and evaluate the ones that do converge.

1. $\sum_{n=1}^{\infty} 2^{-n}$

2. $\sum_{n=1}^{\infty} 2^{n}$

3. $\sum_{n=1}^{\infty} (2^{-n} - 3)$

4. $\sum_{n=1}^{\infty} \left(n - \frac{1}{n}\right)$

5. $\sum_{n=1}^{\infty} 0$

6. $\sum_{n=1}^{3} (n^2 - n) + \sum_{n=4}^{\infty} 0$

7. $\sum_{n=1}^{7} (e^n + 2) + \sum_{n=8}^{\infty} (-5)$

8. $\sum_{n=4}^{\infty} \frac{2}{n}$

9. $\sum_{n=1}^{\infty} \left(\left(\frac{3}{2}\right)^n - (3^n - 2)/(2^n) \right)$

10. $\sum_{n=4}^{\infty} \left(\sin^n \frac{n\pi}{2}\right)$

Suppose $\sum_{n=1}^{\infty} a_n$ and $\sum_{n=1}^{\infty} b_n$ are convergent series. Suppose $\sum_{n=1}^{\infty} a_n = 3$ and $\sum_{n=1}^{\infty} b_n = -5$. Also suppose $a_1 = 3$, $a_2 = -5$, $a_3 = 0$, $a_4 = 1$, $b_1 = 2$, $b_2 = 2$, $b_3 = 3$, and $b_4 = 2$, and evaluate the following.

11. $\sum_{n=2}^{\infty} a_n$

12. $\sum_{n=3}^{\infty} (2a_n - 3b_n)$

13. $\sum_{n=1}^{\infty} (a_n - b_n)$

14. $\sum_{n=2}^{\infty} (a_n + 2b_n - 5^{-n})$

15. $\sum_{n=1}^{\infty} a_n - \sum_{n=3}^{\infty} a_n$

16. $\sum_{n=5}^{\infty} 4b_n$

17. $\sum_{n=4}^{\infty} b_n - \sum_{n=2}^{\infty} b_n$

18. $\sum_{n=2}^{\infty} b_{n-1}$

19. $\sum_{n=1}^{\infty} a_{n+3} - \sum_{n=4}^{\infty} a_n$

20. $\sum_{n=8}^{\infty} a_{n-7}$

II

*21. Find a series $\sum_{n=1}^{\infty} a_n$ whose Nth partial sum is $S_N = \frac{1}{N}$. Does your series converge, and if so, to what limit?

*22. Find a series $\sum_{n=1}^{\infty} a_n$ whose Nth partial sum is $S_N = \frac{1}{N+1}$. Does your series converge, and if so, to what limit?

*23. Find a series $\sum_{n=1}^{\infty} a_n$ whose Nth partial sum is $S_N = \frac{1}{N^2}$. Does your series converge, and if so, to what limit?

*24. Find a series $\sum_{n=1}^{\infty} a_n$ whose Nth partial sum is $S_N = 2 - \frac{1}{N^3}$. Does your series converge, and if so, to what limit?

*25. Consider the series $1 - 1 + \frac{1}{2} - \frac{1}{2} + \frac{1}{3} - \frac{1}{3} + \frac{1}{4} - \frac{1}{4} + \frac{1}{5} - \frac{1}{5} + \ldots + \frac{1}{n} - \frac{1}{n} + \ldots$, that is, the series whose even-numbered terms are $a_{2n} = -\frac{1}{n}$ and whose odd-numbered terms are $a_{2n-1} = \frac{1}{n}$. Find the partial sums and discuss its convergence.

III

26. Theorem 10.3.3 shows that the harmonic series diverges and in particular that $S_{2^n} = \sum_{k=1}^{2^n} \frac{1}{k} > \frac{n+1}{2}$. Use a programmable calculator or computer to evaluate S_4, S_8, S_{16}, S_{32}, S_{64}, S_{128}, S_{512}, S_{4096}, and verify this result. Although the partial sums S_N tend to ∞, they do it slowly.

P◄ **POINT TO PONDER**

In Example 1(iv) (p. 659) we let $c_n = \frac{1}{n^2+n}$ and considered the series

$$\sum_{n=1}^{\infty} c_n = \sum_{n=1}^{\infty} \frac{1}{n^2 + n}.$$

There is no particular reason to think about partial fractions in connection with series, but if we did, it might occur to us to write $\frac{1}{n^2+n} = \frac{1}{n(n+1)} = \frac{1}{n} - \frac{1}{n+1}$. The Nth

partial sum is

$$S_N = c_1 + c_2 + c_3 + \ldots + c_{N-1} + c_N$$

$$= \left(1 - \frac{1}{2}\right) + \left(\frac{1}{2} - \frac{1}{3}\right) + \left(\frac{1}{3} - \frac{1}{4}\right) + \ldots$$

$$+ \left(\frac{1}{N-1} - \frac{1}{N}\right) + \left(\frac{1}{N} - \frac{1}{N+1}\right).$$

Each pair of parentheses encloses one term c_k. But we observe that the second part of each term cancels the first part of the next (e.g., the $-\frac{1}{2}$ cancels the $\frac{1}{2}$, the $-\frac{1}{3}$ cancels the $\frac{1}{3}$, etc.). In fact, almost everything cancels. The only things left are the 1 from the first term and the $-\frac{1}{N+1}$ from the last.

This is an example of what is called a "telescoping sum"; all the terms in the middle cancel, reminding us of the old-fashioned telescope shown in Figure 10.3.1, whose sections all collapse into one for easy carrying.

We find that $S_N = 1 - \frac{1}{N+1} = \frac{N+1-1}{N+1} = \frac{N}{N+1}$, exactly what we found in the example using a different method. It is usually difficult to find a formula for the partial sums of a series; here it was possible because of the special properties of telescoping sums.

Figure 10.3.1

10.4 The Integral Test

INCREASING SEQUENCE
INTEGRAL TEST
p-SERIES

We have learned about convergence and divergence of series and have learned in detail how to assess convergence and divergence for geometric series. Apart from that we have learned the nth Term Test (Theorem 10.3.2) for the convergence of a series; it is used to show that a series does *not* converge if its terms do not tend to 0, but it is no use for showing that any series *does* converge. We also saw that the harmonic series diverges, but apart from geometric series we have relatively few examples of series that we know are convergent. In this and the next two sections we will learn some additional tests for deciding whether various series converge or diverge. In this section and most of the next we will be considering series whose terms are all positive.

Suppose $\sum_{n=1}^{\infty} a_n$ is a series whose terms are all *positive*, that is, $a_n > 0$ for every n. As we form the partial sums, we see that $S_{N+1} = S_N + a_{N+1} > S_N$. This means that the partial sums are always **increasing**, since each one is obtained from the previous one by adding something positive.

Now from Theorem 10.1.6 we know that there are only two possibilities for an increasing sequence. Either it grows larger and larger and tends to infinity (i.e., "diverges to infinity") *or else* it is a **bounded sequence** and converges up to some limit.

Let us look at the series $\sum_{n=1}^{\infty} \frac{1}{n^2}$, whose terms are smaller than the corresponding terms of the harmonic series. It would be interesting to know whether or not it converges.

One obvious thing about this series is that its terms are found by evaluating the function $f(x) = \frac{1}{x^2}$ at the positive integers. We will find that it is possible to decide whether or not the series $\sum_{n=1}^{\infty} f(n)$ converges by looking at the improper integral $\int_1^{\infty} f(x)\, dx$.

Let us try to think of the series $\sum_{n=1}^{\infty} \frac{1}{n^2}$ in terms of areas. In Figure 10.4.1 we see the graph of $y = f(x) = \frac{1}{x^2}$ with the first few positive integers marked on the

Figure 10.4.1

x-axis. The terms of the series are just the heights of the graph above the axis at the integer points 1, 2, 3, ...

Instead of thinking of them just as heights, we could also think of them as the areas of the rectangles with these heights and each of width equal to 1. These are the little shaded rectangles in Figure 10.4.1, each of which just reaches up to the curve at its right end (where its height is $f(n)$).

Apart from the first one, the rectangles lie below the graph $y = f(x)$ and to the right of the point $x = 1$. So the sum of the series, which equals the total area of the rectangles, should be less than the area of the first rectangle plus the area under the graph, that is, the area of the first rectangle plus the improper integral $\int_1^\infty f(x)\,dx$.

Strictly speaking, we should talk about the first N rectangles rather than all of them at once. What we see from Figure 10.4.1 is that if we could show that the area under the graph of $y = f(x)$ to the right of the point $x = 1$ is less than some number, M, for instance, then the total area of the first N rectangles must always be less than the area of the first rectangle plus M. This just means that the Nth partial sum $\sum_{n=1}^{N} a_n$ is always less than $a_1 + M$, for *every* N.

In particular, the partial sums are bounded, and we remarked at the beginning that an increasing sequence that is bounded must have a limit.

So if the whole area under the curve to the right of $x = 1$ is some (finite) number M, then the sequence of partial sums must be bounded by $a_1 + M$, and so it must converge. If we can show that the improper integral is *finite*, then the series must converge. This is the essence of the Integral Test.

In the particular case we were considering, the improper integral is easy to evaluate:

$$\int_1^\infty f(x)\,dx = \int_1^\infty \frac{dx}{x^2} = \lim_{r \to \infty} \left(\int_1^r \frac{dx}{x^2} \right) = \lim_{r \to \infty} \left(-\frac{1}{x} \right)\Big|_1^r = \lim_{r \to \infty} \left(1 - \frac{1}{r} \right)$$
$$= 1.$$

The integral is finite, so the Integral Test shows that the series converges. Notice that it does not tell us what the value of the series is, just that it is convergent. It is possible to conclude that the series is less than or equal to the sum of the first term and the improper integral, which in this case is 2, so in fact we have some (imprecise) information about the value of the series.

There are a number of subtleties that should be mentioned. One is that it seems awkward to leave out that first rectangle. It might make more sense to consider the integral $\int_0^\infty f(x)\,dx$ so that all the rectangles will lie under its graph.

The difficulty is that this integral may not be finite; in the particular case we have considered, with $f(x) = \frac{1}{x^2}$, the integral is improper at $x = 0$ too, and it diverges there, so the argument would not work.

In a sense, what we are doing is ignoring the first term of the series and showing that $\sum_{n=2}^{\infty} a_n$ is less than $\int_1^\infty f(x)\,dx$. Because of Theorem 10.3.4(iii), the convergence of this series is exactly equivalent to the convergence of the whole series, so this approach is perfectly reasonable, and it saves us from having to worry about the behavior of $f(x)$ near $x = 0$.

Another important point is that we have seen from Figure 10.4.1 that with $f(x) = \frac{1}{x^2}$ the graph of $y = f(x)$ is always above the rectangles. The reason is that if $x \leq n$, then $f(x) \geq f(n) = a_n$; the function $f(x)$ is *decreasing*, so it is always greater than its value at the next integer. To make the Integral Test work, we have to know that we are using a function $f(x)$ that is a decreasing function. In the example we know that $f(x) = \frac{1}{x^2}$ is decreasing because its derivative is $f'(x) = -\frac{2}{x^3}$, which is negative for all $x > 0$.

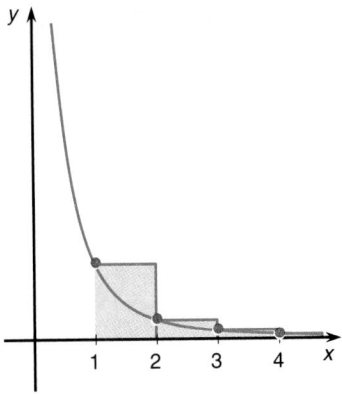

Figure 10.4.2

Finally we should observe that we could have drawn the rectangles a different way, as shown in Figure 10.4.2. Once again the areas of the rectangles equal the terms in the series, but this time the region under the graph of $y = f(x)$ is contained in the rectangles. This means that its area $\int_1^\infty f(x)\,dx$ must be less than the total area of the rectangles, that is, that the integral is less than the series.

These two facts combined tell us that the improper integral lies between the series without its first term and the whole series:

$$\sum_{n=2}^{\infty} a_n = \sum_{n=2}^{\infty} f(n) \leq \int_1^\infty f(x)\,dx \leq \sum_{n=1}^{\infty} a_n = \sum_{n=1}^{\infty} f(n).$$

If the integral is finite, then the series must converge, as we have already seen. If the series converges, then the integral must converge also. In other words, the convergence of the series is exactly equivalent to the convergence of the integral.

We also get some information on the value of the sum if it converges. Since $\sum_{n=2}^{\infty} a_n \leq \int_1^\infty f(x)\,dx$, we add a_1 and find that $\sum_{n=1}^{\infty} a_n \leq \int_1^\infty f(x)\,dx + a_1$, so $\int_1^\infty f(x)\,dx \leq \sum_{n=1}^{\infty} a_n \leq \int_1^\infty f(x)\,dx + a_1$.

THEOREM 10.4.1

The Integral Test

Suppose $f(x)$ is a decreasing function for $x \geq 1$ whose values are always positive, that is, $f(x) > 0$ for all $x \geq 1$. Let $a_n = f(n)$ for each integer $n \geq 1$, and consider the series $\sum_{n=1}^{\infty} a_n$. The series $\sum_{n=1}^{\infty} a_n$ converges if and only if the improper integral $\int_1^\infty f(x)\,dx$ converges, that is, $\int_1^\infty f(x)\,dx < \infty$. Moreover, if the series converges, then

$$\int_1^\infty f(x)\,dx \leq \sum_{n=1}^{\infty} a_n \leq \int_1^\infty f(x)\,dx + a_1.$$

The Integral Test is most often used to show that a series converges by showing that the corresponding integral converges.

EXAMPLE 1

(i) As the calculation done before Theorem 10.4.1 shows, $\int_1^\infty \frac{1}{x^2}\,dx = 1 < \infty$, so the corresponding series $\sum_{n=1}^{\infty} \frac{1}{n^2}$ converges.

(ii) Consider the series $\sum_{n=1}^{\infty} (2n)^{-3}$. Its terms can be found as the values at the positive integers of the function $f(x) = (2x)^{-3}$. This function is easily seen to be decreasing, since its derivative is $-6(2x)^{-4}$, which is negative for all $x \geq 1$. Accordingly, we should consider the integral of $f(x)$:

$$\int_1^\infty f(x)\,dx = \lim_{r \to \infty}\left(\int_1^r \frac{1}{(2x)^3}\,dx\right) = \lim_{r \to \infty}\left(\frac{1}{8}\int_1^r \frac{1}{x^3}\,dx\right)$$

$$= \frac{1}{8}\lim_{r \to \infty}\left(\frac{1}{-2}\frac{1}{x^2}\right)\Big|_1^r = \frac{1}{-16}\lim_{r \to \infty}\left(\frac{1}{r^2} - \frac{1}{1^2}\right) = \frac{1}{16}.$$

The improper integral converges, and the series must be convergent.

(iii) Consider the series $\sum_{n=2}^{\infty} \frac{1}{n \ln(n)}$. If we let $f(x) = \frac{1}{x \ln(x)}$, for $x \geq 2$, then the terms of the series are the values of $f(x)$ at the integers $n \geq 2$.

The derivative of the function f is $f'(x) = -\frac{1}{(x \ln x)^2} \left(\ln x + x \frac{1}{x}\right) = -\frac{1}{(x \ln x)^2}(\ln x + 1)$. For $x \geq 2$ this is always negative, so $f(x)$ is a decreasing function for $x \geq 2$.

To evaluate the integral $\int_2^\infty \frac{1}{x \ln x}\,dx$, we use the substitution $u = \ln x$, $du = \frac{1}{x}\,dx$, and find

$$\int_2^\infty f(x)\,dx = \lim_{r \to \infty} \left(\int_2^r \frac{1}{x \ln x}\,dx\right) = \lim_{r \to \infty}\left(\int_{\ln 2}^{\ln r} \frac{1}{u}\,du\right)$$

$$= \lim_{r \to \infty}(\ln u)\Big|_{\ln 2}^{\ln r} = \lim_{r \to \infty}\big(\ln(\ln r) - \ln(\ln 2)\big) = \infty,$$

because as $r \to \infty$, we know that $\ln r \to \infty$, so the natural logarithm of that must also tend to infinity, that is, $\ln(\ln r) \to \infty$.

We see that the integral diverges, so the series $\sum_{n=2}^\infty \frac{1}{n \ln(n)}$ diverges.

Notice the way we adjusted the Integral Test in part (iii) of Example 1 to allow for the fact that the series started with the $n = 2$ term. We considered the integral from $x = 2$ to ∞. For a series that begins with $n = k$, for instance, we consider the improper integral $\int_k^\infty f(x)\,dx$ and otherwise proceed in the same way.

In Section 10.3 we saw that the harmonic series $\sum_{n=1}^\infty \frac{1}{n}$ diverges, and in this section we have seen that $\sum_{n=1}^\infty \frac{1}{n^2}$ converges. We also considered an example with third powers in the denominator. It is natural to ask about the convergence of the series $\sum_{n=1}^\infty \frac{1}{n^p}$, where p is any positive number. A series of this form is sometimes called a *p*-series.

DesignCAD 3D *software was used to produce this image of a wall unit. The computer can generate pictures of the finished product and also plans for the carpenters who will build it.*

EXAMPLE 2   **Convergence and Divergence of *p*-Series**   Let p be any fixed positive number, and let $f(x) = \frac{1}{x^p}$. Define $a_n = f(n) = \frac{1}{n^p}$, and consider the *p*-series $\sum_{n=1}^\infty \frac{1}{n^p}$.

To apply the Integral Test, we must first show that the function $f(x)$ is decreasing. Its derivative is $f'(x) = -p\frac{1}{x^{p+1}}$, which is negative for all $x > 0$, so $f(x)$ is decreasing and the integral test can be used. Since we have already considered the harmonic series, which is the *p*-series with $p = 1$, let us assume for the moment that $p \neq 1$.

We evaluate the improper integral:

$$\int_1^\infty f(x)\,dx = \lim_{r \to \infty}\left(\int_1^r f(x)\,dx\right) = \lim_{r \to \infty}\left(\int_1^r x^{-p}\,dx\right)$$

$$= \lim_{r \to \infty}\left(\frac{1}{1-p}x^{1-p}\right)\Big|_1^r = \frac{1}{1-p}\lim_{r \to \infty}(r^{1-p} - 1^{1-p}).$$

The important thing is what happens to r^{1-p} as $r \to \infty$, since nothing else depends on r. Of course, as $r \to \infty$, any positive power of it will tend to ∞, and any negative power of it will tend to 0.

We see that the integral converges if $1 - p < 0$, that is, if $p > 1$, and diverges if $1 - p > 0$, that is, if $p < 1$. We also know from Theorem 10.3.3 that the series diverges if $p = 1$.

In summary, the **p-series** $\sum_{n=1}^\infty \frac{1}{n^p}$ *converges if $p > 1$ and diverges if $p \leq 1$.*

EXAMPLE 3   (i) Does the series $\sum_{n=1}^{\infty} \frac{1}{\sqrt{n}}$ converge?

Solution   This is a p-series with $p = \frac{1}{2}$, so it diverges.

(ii) Test the series $\sum_{n=4}^{\infty} (3n+6)^{-3}$ for convergence.

Solution   This does not appear to be a p-series, but it can be rewritten as

$$\sum_{n=4}^{\infty} 3^{-3}(n+2)^{-3} = \frac{1}{27} \sum_{n=4}^{\infty} (n+2)^{-3} = \frac{1}{27} \sum_{m=6}^{\infty} m^{-3}.$$

The series in this last expression is nearly the p-series with $p = 3$; it is that series without its first five terms. The p-series with $p = 3$ converges, and by Theorem 10.3.4 we know that the same will be true of the series without its first five terms. The series converges.

(iii) Test the following series for convergence: $\sum_{n=1}^{\infty} \left(\frac{1}{n^2}\right)^{3/4}$.

Solution   There is a temptation to look at the exponent $\frac{3}{4}$ and conclude that since it is less than 1, the series diverges, but upon closer inspection we realize that the terms of the series are $\left(\frac{1}{n^2}\right)^{3/4} = \left(n^{-2}\right)^{3/4} = n^{-2(3/4)} = n^{-3/2}$.

From this we see that the series is actually the p-series with $p = \frac{3}{2}$, so it converges. ◢

Approximating the Value of a Series

The Integral Test also allows us to approximate the value of a series.

EXAMPLE 4   Corresponding to the p-series $\sum_{n=1}^{\infty} \frac{1}{n^p}$ we have evaluated the integral $\int_1^{\infty} \frac{1}{x^p} dx = \frac{1}{1-p} \lim_{r \to \infty} (r^{1-p} - 1^{1-p})$, and when $p > 1$, this equals $\frac{1}{p-1}$. Since the first term of a p-series is 1, Theorem 10.4.1 tells us that

$$\frac{1}{p-1} \le \sum_{n=1}^{\infty} \frac{1}{n^p} \le \frac{1}{p-1} + 1 = \frac{p}{p-1}.$$

For instance, with $p = \frac{11}{10}$, we see that

$$10 \le \sum_{n=1}^{\infty} \frac{1}{n^{11/10}} \le 11.$$

However, $\frac{1}{n^{1.1}}$ is close to $\frac{1}{n}$, and we know that while $\sum_{n=1}^{\infty} \frac{1}{n}$ diverges, it grows very slowly. From this we expect that $\sum_{n=1}^{\infty} n^{-11/10}$ will grow quite slowly. Using a calculator, we could add up the first few terms, and we find that

$$\sum_{n=1}^{10} n^{-11/10} \approx 2.680\,16, \qquad \sum_{n=1}^{50} n^{-11/10} \approx 3.828\,75, \qquad \sum_{n=1}^{100} n^{-11/10} \approx 4.278\,02.$$ ◢

To get anywhere near the actual value, for example, to get above 10, would take a very large number of terms. The estimate given by the integral is better than what we could find by adding up any reasonable number of terms.

With a programmable calculator we can evaluate $\sum_{n=1}^{m} n^{-11/10}$ for various values of m.

TI-85	SHARP EL9300
PRGM F2(EDIT) PSER ENTER F3 F1(Input) M EXIT ENTER 0 STO N ENTER 0 STO S ENTER Lbl B ENTER N + 1 STO N ENTER S + N^((−) 1.1) STO S ENTER F4 F1(If) (N TEST F2(<) M) ENTER Goto B ENTER S ENTER EXIT EXIT EXIT PRGM F1(NAME) PSER ENTER 10 ENTER	▼ ▼ (NEW) ENTER ENTER PSER ENTER COMMAND A 3 (Input) M ENTER N = 0 ENTER S = 0 ENTER COMMAND B 1 (Label) 1 ENTER N = N + 1 ENTER S = S + N a^b (−) 1.1 ENTER COMMAND 3 (If) N COMMAND C 2 (<) M COMMAND B 2 (Goto) 1 ENTER COMMAND A 1 (Print) S ENTER COMMAND A 6 (End) ENTER (RUN) ENTER (PSER) ENTER 10 ENTER

```
?10
ANSWER=
      2.680155181
```

CASIO f_x-6300G (f_x-7700GB)	HP 48SX
MODE 2 EXE ? → M : 0 → M : 0 → S : Lbl 1 : N + 1 → M : S + N x^y (−) 1.1 → S : N < M ⇒ Goto 1: S ◢ MODE 1 Prog 0 EXE 10 EXE (on the f_x-7700GB, : is PRGM F6, and Prog is PRGM F3.)	ATTN ↱ CLR ⟪⟫ "" KEY IN M ▶ SPC PROMPT SPC ' M ▶ STO 0 SPC 1 SPC M SPC FOR N N SPC 1.1 +/− y^x + NEXT ▼ ENTER ' PSER STO VAR PSER 10 ENTER ↰ CONT

Now use the program with other values of n.

However, the estimate can still be improved considerably.

EXAMPLE 5   We continue with the p-series $\sum_{n=1}^{\infty} n^{-11/10}$. The error in the integral estimate comes from the size of a_1, since the answer lies between $\int_1^{\infty} f(x)\,dx$ and $\int_1^{\infty} f(x)\,dx + a_1$. Having figured out the sum $\sum_{n=1}^{10} n^{-11/10} \approx 2.680\,16$, we have to estimate what is left, that is, $\sum_{n=11}^{\infty} n^{-11/10}$. If we apply the Integral Test here, the error will depend on the first term $a_{11} = 11^{-11/10} \approx 0.071\,53$. We find that

$$\int_{11}^{\infty} x^{-11/10}\,dx \leq \sum_{n=11}^{\infty} n^{-11/10} \leq \int_{11}^{\infty} x^{-11/10}\,dx + a_{11}.$$

Now

$$\int_{11}^{\infty} x^{-11/10}\,dx = \lim_{r\to\infty} \int_{11}^{r} x^{-11/10}\,dx = \lim_{r\to\infty} \left. (-10x^{-1/10}) \right|_{11}^{r}$$

$$= 10(11)^{-1/10} \approx 7.867\,93.$$

So $\sum_{n=11}^{\infty} n^{-11/10}$ lies between $7.867\,93$ and $7.867\,93 + a_{11} \approx 7.939\,46$.

To get the value of the p-series, we add in the first ten terms: $\sum_{n=1}^{\infty} n^{-11/10} = \sum_{n=1}^{10} n^{-11/10} + \sum_{n=11}^{\infty} n^{-11/10}$. Since $\sum_{n=1}^{10} n^{-11/10} \approx 2.680\,16$, we find

$$10.54809 \le \sum_{n=1}^{\infty} n^{-11/10} \le 10.619\,62.$$

EXAMPLE 6 Estimate $\sum_{n=1}^{\infty} \frac{1}{n^2}$ to within 0.01.

Solution We want to add up $\sum_{n=1}^{K} \frac{1}{n^2}$ for some K and estimate the "tail" $\sum_{n=K+1}^{\infty} \frac{1}{n^2}$. The error that we find using the Integral Test depends on the first term in the tail, which is $\frac{1}{(K+1)^2}$. To make this ≤ 0.01, we need $(K+1)^2 \ge 100$, that is, $K \ge 9$. Adding up the first nine terms gives

$$\sum_{n=1}^{9} \frac{1}{n^2} \approx 1.539\,77.$$

The integral corresponding to the tail is

$$\int_{10}^{\infty} \frac{1}{x^2}\,dx = \lim_{r \to \infty} \left(\int_{10}^{r} \frac{1}{x^2}\,dx \right) = \lim_{r \to \infty} \left(-\frac{1}{x} \right)\Big|_{10}^{r} = \frac{1}{10}.$$

The Integral Test says that

$$\frac{1}{10} \le \sum_{n=10}^{\infty} \frac{1}{n^2} \le \frac{1}{10} + a_{10} = \frac{1}{10} + \frac{1}{100} = 0.11.$$

The original p-series $\sum_{n=1}^{\infty} \frac{1}{n^2} = \sum_{n=1}^{9} \frac{1}{n^2} + \sum_{n=10}^{\infty} \frac{1}{n^2}$ lies between $1.539\,77 + \frac{1}{10}$ and $1.539\,77 + 0.11$, that is,

$$1.639\,77 \le \sum_{n=1}^{\infty} \frac{1}{n^2} \le 1.649\,77.$$

In fact, though it is *not* obvious, $\sum_{n=1}^{\infty} \frac{1}{n^2} = \frac{\pi^2}{6}$.

We found that sufficient accuracy would be achieved by summing up to $K = 9$. This does *not* mean that $\sum_{n=1}^{9} \frac{1}{n^2}$ is the correct approximation, but rather that $\sum_{n=1}^{9} \frac{1}{n^2} + \int_{10}^{\infty} \frac{1}{x^2}\,dx$ is the correct approximation. It is accurate to within an error of a_{10}. In general, if $f(x)$ is a nonnegative decreasing function and $\sum_{n=1}^{\infty} a_n$ converges, where $a_n = f(n)$, then the value of the series is estimated by

$$\sum_{n=1}^{K} a_n + \int_{K+1}^{\infty} f(x)\,dx, \tag{10.4.2}$$

accurate to within a_{K+1}.

There is one final remark that should be made about the Integral Test. It is very important to check that the function $f(x)$ is decreasing, but it will sometimes happen that it increases for a little while and *then* decreases after some point $x = K$.

In this situation it is still possible to use the Integral Test because if we ignore the first few terms (the ones for $n < K$), then the rest of the series is obtained as the values at integer points of a decreasing function, and of course the convergence or divergence of a series is unaffected by changing the first few terms. What is important is that *after some point K* the function f must be decreasing.

EXAMPLE 7   Consider the series $\sum_{n=1}^{\infty} ne^{-n/12}$, and test it for convergence.

Solution   If we let $f(x) = xe^{-x/12}$, then its derivative is $f'(x) = e^{-x/12} + xe^{-x/12} \times (-\frac{1}{12}) = e^{-x/12}(1 - \frac{x}{12})$. This derivative is negative for $x > 12$ but positive for $x < 12$.

We conclude that the function $f(x)$ is increasing for $x < 12$ and decreasing for $x > 12$. We can consider the series that begins with the twelfth term; it is constructed from the values at integer points of a decreasing function, so we can apply the Integral Test. The thing to do is to consider the improper integral $\int_{12}^{\infty} f(x)\, dx = \int_{12}^{\infty} xe^{-x/12}\, dx$, which can be integrated by parts. Letting $u = x$ and $dv = e^{-x/12} dx$, we find that $v = \int e^{-x/12}\, dx = -12e^{-x/12}$, so

$$\int_{12}^{\infty} f(x)\, dx = \lim_{r \to \infty} \left(\int_{12}^{r} xe^{-x/12}\, dx \right) = \lim_{r \to \infty} \left(\int u\, dv \right)$$

$$= \lim_{r \to \infty} \left(uv - \int v\, du \right)\Big|_{12}^{r} = \lim_{r \to \infty} \left(-12xe^{-x/12} + 12 \int e^{-x/12}\, dx \right)\Big|_{12}^{r}$$

$$= \lim_{r \to \infty} \left(-12xe^{-x/12} + 12(-12)e^{-x/12} \right)\Big|_{12}^{r}$$

$$= \lim_{r \to \infty} \left(-12re^{-r/12} - 144e^{-r/12} + 12(12)e^{-1} + 144e^{-1} \right).$$

We have to find what happens as $r \to \infty$, and the terms that depend on r are $-12re^{-r/12}$ and $-144e^{-r/12}$. The second of these is easy to deal with because as $r \to \infty$, we know that $-\frac{r}{12} \to -\infty$, so $e^{-r/12} \to 0$.

We write the other term as $-12re^{-r/12} = -12\frac{r}{e^{r/12}}$ and observe that as $r \to \infty$, both the numerator and the denominator of this quotient tend to ∞. This cries out for l'Hôpital's Rule. We find that

$$\lim_{r \to \infty} -12\frac{r}{e^{r/12}} = -12 \lim_{r \to \infty} \frac{1}{\frac{1}{12}e^{r/12}} = -144 \lim_{r \to \infty} e^{-r/12} = 0.$$

We find that the improper integral converges to a finite limit, which means that the series starting at the twelfth term converges. Consequently, the whole series converges.

Exercises 10.4

Decide whether or not each of the following series converges.

1. $\sum_{n=1}^{\infty} \frac{1}{n+5}$

2. $\sum_{n=1}^{\infty} \frac{1}{3n+5}$

3. $\sum_{n=1}^{\infty} \left(\frac{1}{n+7}\right)^4$

4. $\sum_{n=1}^{\infty} \frac{1}{(2n+3)^3}$

5. $\sum_{n=4}^{\infty} \frac{1}{\sqrt{n+7}}$

6. $\sum_{n=0}^{\infty} \frac{1}{\sqrt{(2n+3)^5}}$

7. $\sum_{n=1}^{\infty} \frac{1}{n^2+6n+9}$

8. $\sum_{n=1}^{\infty} \frac{1}{(3n-2)^3}$

9. $\sum_{n=4}^{\infty} ne^{-n^2}$

10. $\sum_{n=1}^{\infty} 4^{-n}$

Test the following series for convergence.

11. $\sum_{n=1}^{\infty} n^2 e^{-n^3}$

12. $\sum_{n=1}^{\infty} \frac{1}{1+n^2}$

13. $\sum_{n=1}^{\infty} (n^2 + 2n + 2)^{-1}$

14. $\sum_{n=1}^{\infty} \frac{1}{n^2+4n+8}$

15. $\sum_{n=2}^{\infty} \ln(n)$

16. $\sum_{n=2}^{\infty} \left(n \ln^2(n) \right)^{-1}$

17. $\sum_{n=1}^{\infty} \frac{\ln(n)}{n^2}$

18. $\sum_{n=2}^{\infty} \frac{1}{n \ln(n^3)}$

19. $\sum_{n=1}^{\infty} \frac{1}{n^2 + 3}$

20. $\sum_{n=2}^{\infty} \left(n \ln^{5/6}(n) \right)^{-1}$

II

*21. Find an example of a bounded sequence that does not converge.

*22. Find an example of an increasing sequence that does not converge.

*23. Test the series $\sum_{n=2}^{\infty} \frac{1}{n^2 - 1}$ for convergence. (*Hint:* After integrating, use properties of the natural logarithm.)

Does each of the following series converge?

*24. $\sum_{n=3}^{\infty} \frac{1}{n^2 - n - 2}$

*25. $\sum_{n=3}^{\infty} \frac{1}{n^2 - 3n + 2}$

*26. $\sum_{n=5}^{\infty} \frac{1}{n^2 - n - 12}$

*27. $\sum_{n=3}^{\infty} \frac{1}{2n^2 - 6n + 4}$

**28. Suppose a, b, and c are any numbers with $a > 0$. Choose an integer K so that $ax^2 + bx + c$ is never zero for $x \geq K$. Show that the following series converges: $\sum_{n=K}^{\infty} \frac{1}{an^2 + bn + c}$.

*29. For what values of q does the following series converge: $\sum_{n=2}^{\infty} \frac{\ln(n)}{n^q}$?

*30. Test the following series for convergence: $\sum_{n=1}^{\infty} \cosh^{-2}(n)$.

*31. Test the following series for convergence: $\sum_{n=1}^{\infty} \cosh(n)$.

32. In Example 4 (p. 670) we mentioned that $\sum_{n=1}^{50} n^{-11/10} \approx 3.82875$. What estimate can you get from this for $\sum_{n=1}^{\infty} n^{-11/10}$?

33. In Example 4 we mentioned that $\sum_{n=1}^{100} n^{-11/10} \approx 4.27802$. What estimate can you get from this for $\sum_{n=1}^{\infty} n^{-11/10}$? Is your answer to this question consistent with what you found for Exercise 32?

34. For the series $\sum_{n=1}^{\infty} ne^{-n/12}$ of Example 7 (p. 673), find the value of K in Equation 10.4.2 for which the approximation is accurate within (i) 0.1, (ii) 0.001. (Do not evaluate.)

35. Estimate $\sum_{n=1}^{\infty} \frac{1}{n^4}$ to within 0.001.

36. How good is the approximation of $\sum_{n=1}^{\infty} \frac{1}{n^6}$ by Equation 10.4.2 with $K = 10$? Evaluate the approximation.

III

37. Use a programmable calculator or computer to confirm the values given in Example 4 for the partial sums $\sum_{n=1}^{N} n^{-11/10}$ with $N = 10, 50, 100$.

38. Estimate $\sum_{n=1}^{\infty} n^{-11/10}$ to within 0.001 (cf. Example 4) and Formula 10.4.2.

39. Estimate $\sum_{n=1}^{\infty} ne^{-n/12}$ to within 0.001 (cf. Example 7).

40. Estimate $\sum_{n=1}^{\infty} n^{-3/2}$ to within 0.001.

POINT TO PONDER

The Integral Test applies to a series $\sum_{n=1}^{\infty} a_n$ that can be expressed by $a_n = f(n)$, for some decreasing function $f(x)$. It is not always easy to find such a function, even apart from deciding whether it is decreasing. For instance, for the series $\sum_{n=1}^{\infty} (-1)^n \frac{1}{n^2}$ we cannot just change the n to x because $(-1)^x$ does not make sense unless x is an integer or a rational number with odd denominator. (For other x's the power involves ln, which is not defined for -1.) On the other hand, the function

$\cos(\pi x)$ equals $(-1)^n$ when $x = n$ is an integer (why?), so we could let $f(x) = \cos(\pi x) \frac{1}{x^2}$. It is not decreasing; in fact the Integral Test cannot possibly apply because the a_n's are not all positive.

For the series $\sum_{n=1}^{\infty} \frac{1}{n!}$ we do not know a function $f(x)$ whose value at $x = n$ is $\frac{1}{n!}$ ($n!$ is only defined when n is an integer). There is no obvious way to use the Integral Test, even though the terms are positive and decrease to zero.

10.5 The Comparison Test

NONNEGATIVE SERIES
COMPARISON TEST
RATIONAL FUNCTIONS

In this section we learn a useful test for convergence of series whose terms are all nonnegative.

THEOREM 10.5.1

The Comparison Test

Suppose $\sum_{n=1}^{\infty} a_n$ and $\sum_{n=1}^{\infty} b_n$ are two series whose terms are all nonnegative. Suppose that $\sum_{n=1}^{\infty} a_n$ converges and that each term b_n is less than or equal to the corresponding term a_n, that is, $0 \leq b_n \leq a_n$. Then $\sum_{n=1}^{\infty} b_n$ converges.

PROOF   To say that the series $\sum_{n=1}^{\infty} a_n$ converges means that its partial sums are bounded. On the other hand, each term b_n is less than or equal to the corresponding a_n, so the partial sums $T_N = \sum_{n=1}^{N} b_n$ must be less than or equal to the corresponding partial sums $S_N = \sum_{n=1}^{N} a_n$, that is, $T_N = \sum_{n=1}^{N} b_n \leq \sum_{n=1}^{N} a_n = S_N$. In particular, if the S_N's are all bounded by some number M, then so are the T_N's.

This means that $\sum_{n=1}^{\infty} b_n$ converges, as required.

EXAMPLE 1

(i) Consider the series $\sum_{n=1}^{\infty} \frac{1}{n^4+1}$. Since $n^4 + 1 > n^4$, we can take reciprocals and find that $\frac{1}{n^4+1} < \frac{1}{n^4}$.

   The series $\sum_{n=1}^{\infty} \frac{1}{n^4}$ is a p-series with $p = 4$, so it is convergent, and the Comparison Test shows that the "smaller" series $\sum_{n=1}^{\infty} \frac{1}{n^4+1}$ is also convergent. This is much easier than trying to apply the Integral Test.

(ii) Consider the series $\sum_{n=1}^{\infty} \frac{1}{e^n+2n+4}$. We note that

$$\frac{1}{e^n + 2n + 4} < \frac{1}{e^n},$$

   for each $n \geq 1$. The series $\sum_{n=1}^{\infty} \frac{1}{e^n}$ is a geometric series with common ratio $\frac{1}{e}$. Since this ratio is less than 1, the geometric series converges, and the Comparison Test then tells us that the series $\sum_{n=1}^{\infty} \frac{1}{e^n+2n+4}$ converges.

(iii) Let us consider the series $\sum_{n=2}^{\infty} \frac{1}{\sqrt{n}-1}$. It appears reasonable to try to compare

$\sqrt{n} - 1 < \sqrt{n}$, and taking reciprocals reverses the inequality.

   its terms with $\frac{1}{\sqrt{n}}$, and we find that $\frac{1}{\sqrt{n}-1} > \frac{1}{\sqrt{n}}$.

   We know that $\sum_{n=2}^{\infty} \frac{1}{\sqrt{n}}$ is a p-series with $p = \frac{1}{2}$, so it diverges. The series with which we began is greater than this divergent series, so it diverges too. After all, if it were *convergent*, then the Comparison Test would tell us that the smaller series $\sum_{n=2}^{\infty} \frac{1}{\sqrt{n}}$ would have to converge, and we know that it does not.

   We conclude that the series $\sum_{n=2}^{\infty} \frac{1}{\sqrt{n}-1}$ diverges.

There are several things to notice in Example 1. The first is that while we might have managed to use the Integral Test for part (i), it would certainly have been more work. In part (ii) it seems that the Integral Test would be very difficult to use. Finally, in part (iii) we used the Comparison Test "backwards." We showed that the terms of the series in question are greater than the terms of a divergent series and concluded that it must also be divergent.

The examples we have considered so far have been constructed deliberately so that they work out well, but of course this will not always happen.

EXAMPLE 2 Suppose we look at the series $\sum_{n=2}^{\infty} \frac{1}{n^4-1}$. It is quite similar to part (i) of Example 1, but its terms are *not* less than or equal to the terms of the convergent series $\sum_{n=2}^{\infty} \frac{1}{n^4}$. They are all slightly *greater* than the corresponding terms.

The key is that word "slightly." The term $\frac{1}{n^4-1}$ is greater than $\frac{1}{n^4}$, but it is less than twice as much. Accepting this for the moment, we can use the Comparison Test because $\sum_{n=2}^{\infty} \frac{2}{n^4}$ is convergent (it is 2 times a convergent p-series without the first term). The terms of the series in which we are interested are less than the terms of this convergent series, so it converges.

What remains is the problem of actually showing that $\frac{1}{n^4-1} < \frac{2}{n^4}$, for $n \geq 2$. To work with an inequality like this, it is often easiest to work with the reciprocals, remembering that it is necessary to reverse the inequality.

So we want to show that $n^4 - 1 > \frac{n^4}{2}$, that is, $2(n^4 - 1) > n^4$. This amounts to showing that $2n^4 - 2 > n^4$, and subtracting n^4 from each side shows that this is the same as $n^4 - 2 > 0$ or $n^4 > 2$. This is certainly true if $n \geq 2$, so we have finished.

With a graphics calculator we can plot the graphs of the functions $\frac{1}{x^4-1}$, $\frac{1}{x^4}$, $\frac{2}{x^4}$, which are related to the terms of the series in Example 2.

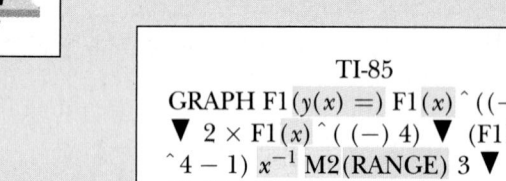

TI-85	SHARP EL9300
GRAPH F1 $(y(x) =)$ F1 (x) ^ $((-)$ 4$)$ ▼ 2 × F1 (x) ^ $((-)$ 4$)$ ▼ $(F1(x)$ ^ 4 − 1$)$ x^{-1} M2(RANGE) 3 ▼ 10 ▼ ▼ 0 ▼ 0.02 F5(GRAPH)	$A\!\!/$ x/θ/T a^b $(-)$ 4 2ndF ▼ 2 × x/θ/T a^b $(-)$ 4 2ndF ▼ $($ x/θ/T a^b 4 ► − 1 $)$ x^{-1} RANGE 3 ENTER 10 ENTER 1 ENTER 0 ENTER .02 ENTER .01 ENTER $A\!\!/$

CASIO f_x-7700GB/f_x-6300G	HP 48SX
Cls EXE Range 3 EXE 10 EXE 1 EXE 0 EXE .02 EXE .01 EXE EXE EXE EXE Graph $\boxed{X}$ x^y $(-)$ 4 EXE Graph 2 × $\boxed{X}$ x^y $(-)$ 4 EXE Graph $($ $\boxed{X}$ x^y 4 − 1 $)$ x^{-1} EXE	↰ PLOT PLOTR ERASE ATTN ′ Y = X y^x 4 +/− ↱ PLOT ↰ DRAW ERASE 3 ENTER 10 XRNG 0 ENTER .02 YRNG DRAW ATTN ′ Y = 2 × X y^x 4 +/− ↱ PLOT ↰ DRAW DRAW ATTN ′ Y = 1/x X y^x 4 − 1 ↱ PLOT ↰ DRAW DRAW

You may have to look closely to identify the three graphs. Check that $\frac{1}{n^4} < \frac{1}{n^4-1} < \frac{2}{n^4}$.

It might be helpful to write down this last argument in the reverse order, now that we have found what to do by working backwards. If $n \geq 2$, then certainly $n^4 > 2$. From this we see that $n^4 - 2 > 0$, and $2n^4 - 2 > n^4$. This can be written as $2(n^4 - 1) > n^4$, or $n^4 - 1 > \frac{n^4}{2}$. Taking reciprocals and reversing the inequality, we find that $\frac{1}{n^4-1} < \frac{2}{n^4}$, when $n \geq 2$.

Combined with the argument we gave above, this shows that $\sum_{n=2}^{\infty} \frac{1}{n^4-1}$ is convergent.

In this example we saw that it is not essential that the terms of the series be less than the terms of a convergent series in order for us to be able to apply the Comparison Test. It is just as good if they are less than some constant times the terms of the convergent series. This is an easy observation, but it is very useful.

A similar observation can be made based on the fact that the convergence of a series is not affected by changing the first few terms. So it will also be possible to use the Comparison Test in situations in which the terms of one series are less than a constant times the terms of another *after the first few terms*. This is also quite obvious, but it will often make the Comparison Test easier to use.

We record these minor modifications to the Comparison Test as a theorem.

THEOREM 10.5.2

> **The Improved Comparison Test**
>
> (i) If $\sum_{n=1}^{\infty} a_n$ is a convergent series and $C > 0$ is a constant so that $0 \leq b_n \leq Ca_n$ for every n greater than some number K, then $\sum_{n=1}^{\infty} b_n$ is convergent.
>
> (ii) If $\sum_{n=1}^{\infty} a_n$ is a divergent series and $C > 0$ is a constant so that $0 \leq Ca_n \leq b_n$ for every n greater than some number K, then $\sum_{n=1}^{\infty} b_n$ is divergent.

> Notice that Theorem 10.5.2 applies to series whose terms are all nonnegative.

EXAMPLE 3   Consider the series $\sum_{n=1}^{\infty} a_n$, where $a_n = \frac{n^4-3n^3+7n^2+1}{2n^5+3n^4-n^2+2n-5}$. We should ask what happens to a_n as $n \to \infty$, and from our work with limits of rational functions we know that the most important things are the terms of highest degree in the numerator and denominator. The idea is that as n becomes very large, the other terms will be so much smaller by comparison that they will not have much effect.

In this example this suggests that as $n \to \infty$, the terms of the series will be close to $\frac{n^4}{2n^5} = \frac{1}{2n}$. The series with these terms is divergent, so it looks as if the series we are considering is probably divergent too.

To make this idea work, we use a trick that we used with rational functions. We write

$$a_n = \frac{n^4 - 3n^3 + 7n^2 + 1}{2n^5 + 3n^4 - n^2 + 2n - 5} = \frac{n^4 \left(1 - \frac{3}{n} + \frac{7}{n^2} + \frac{1}{n^4}\right)}{2n^5 \left(1 + \frac{3}{2n} - \frac{1}{2n^3} + \frac{1}{n^4} - \frac{5}{2n^5}\right)}$$

$$= \frac{n^4}{2n^5} \frac{1 - \frac{3}{n} + \frac{7}{n^2} + \frac{1}{n^4}}{1 + \frac{3}{2n} - \frac{1}{2n^3} + \frac{1}{n^4} - \frac{5}{2n^5}}.$$

The last quotient tends to 1 as $n \to \infty$, and in particular it must eventually be greater than $\frac{1}{2}$. In other words, there is a number K so that that quotient is greater than $\frac{1}{2}$ whenever $n > K$.

This means that whenever $n > K$, $a_n > \frac{1}{2} \frac{1}{2n}$, and part (ii) of the improved version of the Comparison Test tells us that $\sum_{n=1}^{\infty} a_n$ diverges.   ◢

In a way, the idea in Example 3 is to compare the convergence or divergence of $\sum_{n=1}^{\infty} a_n$ with that of $\sum_{n=1}^{\infty} \frac{1}{2n}$, since the terms a_n are eventually close to $\frac{1}{2n}$. We could express this condition by considering the quotient

$$\frac{a_n}{\frac{1}{2n}} = \frac{n^4 - 3n^3 + 7n^2 + 1}{2n^5 + 3n^4 - n^2 + 2n - 5} \frac{2n}{1} = \frac{2n^5 - 6n^4 + 14n^3 + 2n}{2n^5 + 3n^4 - n^2 + 2n - 5}$$

and recognizing that as $n \to \infty$, the quotient tends to 1.

Even if the quotient tended to any other nonzero limit L, we should expect the convergence or divergence of the two series to be related.

THEOREM 10.5.3

> **The Limit Comparison Test**
>
> Suppose $\sum_{n=1}^{\infty} a_n$ and $\sum_{n=1}^{\infty} b_n$ are series with nonnegative terms. Suppose that
>
> $$\lim_{n \to \infty} \frac{a_n}{b_n} = L,$$
>
> for some nonzero number L. Then $\sum_{n=1}^{\infty} a_n$ and $\sum_{n=1}^{\infty} b_n$ *converge or diverge together*, that is, either they both converge or else they both diverge.

PROOF   Suppose $\sum_{n=1}^{\infty} a_n$ converges. Since $\lim_{n \to \infty} \frac{a_n}{b_n} = L$, we know that there is some number K so that $\frac{a_n}{b_n}$ is greater than $\frac{L}{2}$ for every $n > K$. This amounts to $b_n < \frac{2}{L} a_n$, for $n > K$, and Theorem 10.5.2(i) shows that $\sum_{n=1}^{\infty} b_n$ is convergent.

Similarly, suppose $\sum_{n=1}^{\infty} a_n$ diverges. We can find K so that $\frac{a_n}{b_n} < 2L$ for every $n > K$. This says that $b_n > \frac{1}{2L} a_n$ for every $n > K$, and Theorem 10.5.2(ii) says that $\sum_{n=1}^{\infty} b_n$ diverges.   ◢

We can use Theorem 10.5.3 for any series whose terms are "a rational function of n," meaning that there is a rational function $R(x)$ so that the terms of the series are given by $a_n = R(n)$. To decide whether or not such a series converges, we look at the terms of highest degree in the numerator and denominator. Their quotient b_n will be a constant times some power of n, so the corresponding series $\sum_{n=1}^{\infty} b_n$ is a constant times a p-series, and we can immediately decide whether or not it converges.

An argument like the one in Example 3 shows that as $n \to \infty$, the quotient $\frac{a_n}{b_n}$ will tend to 1; the Limit Comparison Test then shows that both series converge or both diverge.

To decide whether or not the simpler series constructed from the highest-order terms is convergent or not, we observe that the p-series $\sum_{n=1}^{\infty} \frac{1}{n^d}$ will converge when $d > 1$. If d is an integer, this means that d must be 2 or greater. So if $R(x) = \frac{p(x)}{q(x)}$ is a rational function and $a_n = R(n)$, then the quotient of the highest-order terms of $p(n)$ and $q(n)$ is a constant times $\frac{1}{n^d}$, where d is the difference of the degrees of the highest-order terms of $q(x)$ and $p(x)$. The series converges when $d = degree\big(q(x)\big) - degree\big(p(x)\big) > 1$, that is, $d \geq 2$.

THEOREM 10.5.4

If $R(x) = \frac{p(x)}{q(x)}$ is a **rational function** and $a_n = R(n)$, then the series $\sum_{n=k}^{\infty} a_n$ converges precisely when $degree\big(q(x)\big) > degree\big(p(x)\big) + 1$, that is, when the degree of the denominator is greater than the degree of the numerator by at least 2.

EXAMPLE 4

(i) The series $\sum_{n=2}^{\infty} \frac{n^2+3n-7}{n^5-2n^3+1}$ converges because the degree of the denominator is 5, which is 3 greater than the degree of the numerator.

(ii) The series $\sum_{n=3}^{\infty} \frac{5n-2}{n^2-4}$ diverges, since the degrees of its denominator and numerator differ by only 1. (Notice that this series starts at $n = 3$ because the denominator is zero when $n = 2$.)

The Integral Test (Theorem 10.4.1) applies to series with decreasing positive terms, and the Comparison Test (Theorems 10.5.1, 10.5.2, 10.5.3) applies to series whose terms are nonnegative.

Exercises 10.5

Testing series for convergence or divergence is a skill that must be learned carefully. As with integration, it takes practice to develop a feel for what might be a suitable approach. It is extremely important to do a large number of exercises.

I

Give reasons why each of the following series is convergent or divergent.

1. $\sum_{n=3}^{\infty} \frac{1}{n+\sin(n\pi)}$

2. $\sum_{n=2}^{\infty} \frac{1}{n^2-2\cos(n\pi)}$

3. $\sum_{n=1}^{\infty} \frac{1}{4n^3+\ln n}$

4. $\sum_{n=1}^{\infty} \frac{1}{n^2\ln^3(n+1)}$

5. $\sum_{n=0}^{\infty} \left(e^n + \frac{n}{2}\right)$

6. $\sum_{n=2}^{\infty} \frac{1}{n^{3/2}-n}$

7. $\sum_{n=1}^{\infty} \frac{3}{2^n+4\ln(n)}$

8. $\sum_{n=-3}^{\infty} \frac{1}{7+\sqrt{n+3}}$

9. $\sum_{n=2}^{\infty} \frac{1}{2n^2\ln(n)}$

10. $\sum_{n=2}^{\infty} \frac{\ln(n)}{n^{3/2}}$

11. $\sum_{n=1}^{\infty} \frac{n^2-3n+4}{n^3-4n^2+n-12}$

12. $\sum_{n=2}^{\infty} \frac{3n^3+n^2-1}{2n^5-2}$

13. $\sum_{n=1}^{\infty} \frac{n^2+5n-2}{n^3\sqrt{n}+3n^3-2n+1}$

14. $\sum_{n=2}^{\infty} \frac{5n^3+2n^2-n+2}{n^5+2n^4-n^3-2}\sqrt{n}$

15. $\sum_{n=1}^{\infty} \frac{n^3-3n^2+n+3}{n^2-n+1}e^{-n/10}$

16. $\sum_{n=1}^{\infty} \frac{3n^3+n^2+4}{2n-3}$

17. $\sum_{n=2}^{\infty} \frac{n^2+2}{n^4-n^2+\cos(n\pi)}$

18. $\sum_{n=1}^{\infty} \frac{2n^3+4n^2+3n}{3n^4+3e^{-n}}$

19. $\sum_{n=1}^{\infty} \frac{4n+1}{n^3-2n^2+3}\big(2-\sin(n\pi)\big)$

20. $\sum_{n=1}^{\infty} \frac{3n^2+6n+5e^n}{4n^5+3n^3+n-1}$

II

*21. Suppose $\sum_{n=1}^{\infty} a_n$ and $\sum_{n=1}^{\infty} b_n$ are series with nonnegative terms and $\sum_{n=1}^{\infty} a_n$ converges. Suppose too that $\lim_{n\to\infty} \frac{b_n}{a_n} = 0$. Show that $\sum_{n=1}^{\infty} b_n$ converges.

*22. Find an example in which $\lim_{n\to\infty} \frac{b_n}{a_n} = 0$, $\sum_{n=1}^{\infty} b_n$ converges and $\sum_{n=1}^{\infty} a_n$ diverges.

*23. Suppose $\sum_{n=1}^{\infty} a_n$ and $\sum_{n=1}^{\infty} b_n$ are series with nonnegative terms and $\sum_{n=1}^{\infty} a_n$ diverges. Suppose too that $\lim_{n\to\infty} \frac{b_n}{a_n} = \infty$. Show that $\sum_{n=1}^{\infty} b_n$ diverges.

*24. Find an example in which $\lim_{n \to \infty} \frac{b_n}{a_n} = \infty$, $\sum_{n=1}^{\infty} b_n$ diverges and $\sum_{n=1}^{\infty} a_n$ converges.

Drill

Give reasons why each of the following series is convergent or divergent.

25. $\sum_{n=1}^{\infty} \frac{1}{2n^2 + 3n^{1/2}}$

26. $\sum_{n=2}^{\infty} \frac{7}{n \ln^3(n)}$

27. $\sum_{n=1}^{\infty} \frac{n}{e^{-3n}}$

28. $\sum_{n=1}^{\infty} \frac{1}{2n + 3e^n}$

29. $\sum_{n=2}^{\infty} \frac{1}{\ln(n) + n \ln(n)}$

30. $\sum_{n=1}^{\infty} \csc(n^{-3})$

31. $\sum_{n=1}^{\infty} \frac{2}{ne^{6n}}$

32. $\sum_{n=1}^{\infty} \frac{5+n}{n^{5/2} \ln(n) + 3}$

33. $\sum_{n=0}^{\infty} \left(\frac{1}{n+1} - \frac{1}{n+2} \right)$

34. $\sum_{n=0}^{\infty} \frac{\cos^2(n+\pi)}{n^2 + e^{-2n}}$

35. (i) Use a programmable calculator or computer to evaluate $\sum_{n=2}^{100} \frac{1}{n^2 \ln(n)}$. (ii) Estimate the accuracy of this approximation to $\sum_{n=2}^{\infty} \frac{1}{n^2 \ln(n)}$ by considering the tail $\sum_{n=101}^{\infty} \frac{1}{n^2 \ln(n)}$. Since $\frac{1}{n^2 \ln(n)} < \frac{1}{n^2}$, argue that $\sum_{n=101}^{\infty} \frac{1}{n^2 \ln(n)} < \sum_{n=101}^{\infty} \frac{1}{n^2}$ and estimate this last series, using the Integral Test.

36. Evaluate $\sum_{n=2}^{250} \frac{1}{n^2 \ln(n)}$. How accurate is this approximation to $\sum_{n=2}^{\infty} \frac{1}{n^2 \ln(n)}$?

Estimate each of the following series to within the specified accuracy.

37. $\sum_{n=1}^{\infty} \frac{1}{n^3 \ln(n+1)}$ to within 0.0001

38. $\sum_{n=1}^{\infty} \frac{1}{n^2 + 5 \ln(n)}$ to within 0.001

39. $\sum_{n=1}^{\infty} \frac{1}{n^2 + e^n}$ to within 0.0001

40. $\sum_{n=1}^{\infty} \frac{1}{n^{5/2} + 3}$ to within 0.0005

10.6 Alternating Series and Absolute Convergence

ALTERNATING SERIES
CONDITIONAL CONVERGENCE

In Sections 10.4 and 10.5 we considered series whose terms were all nonnegative. This was necessary for a result like the Comparison Test, as the following example shows. If we let $a_n = -\frac{1}{n}$, then $\sum_{n=1}^{\infty} a_n$ is the negative of the harmonic series, so it diverges. However, it is "less than" the convergent series $\sum_{n=1}^{\infty} \frac{1}{n^2}$ (in the sense that each term $a_n = -\frac{1}{n}$ is less than the corresponding term $\frac{1}{n^2}$). In fact it is less than *any* series whose terms are nonnegative.

Of course, this is no contradiction, since each a_n is negative and the Comparison Test does not apply. It also demonstrates that it is quite easy to deal with series whose terms are all negative; we simply consider the negative of such a series. The terms of *this* series will be all positive, and the Comparison and Integral Tests can be used for it. In this way we can often find whether or not the original negative series converges.

The situation is more complicated when some of the terms of a series are positive and some are negative. In this section we consider a very special case in which the terms of the series are *alternately* positive and negative.

$$S_1 = -1$$
$$S_2 = -1 + \frac{1}{2} = -\frac{1}{2}$$
$$S_3 = S_2 - \frac{1}{3} = -\frac{5}{6}$$
$$S_4 = S_3 + \frac{1}{4} = -\frac{7}{12}$$
$$S_5 = S_4 - \frac{1}{5} = -\frac{47}{60}$$
$$S_6 = S_5 + \frac{1}{6} = -\frac{37}{60}$$

EXAMPLE 1 Consider the series $\sum_{n=1}^{\infty} (-1)^n \frac{1}{n}$, which is obtained by inserting negative signs on the odd-numbered terms of the harmonic series.

We begin calculating partial sums at left. Since the terms of the series are alternately negative and positive, the partial sums will move alternately up and down. The amounts by which they rise and fall will be $\pm \frac{1}{n}$, and the important thing to notice is that these amounts get smaller and smaller, tending to zero.

The odd-numbered partial sums S_1, S_3, S_5, ... are the "low" values, since the even-numbered sums are obtained from them by adding *positive* numbers. If we think about the odd-numbered sums by themselves for a while, we see that they are increasing. The reason is that to get, say, from S_1 to S_3, we add $a_2 + a_3 = \frac{1}{2} - \frac{1}{3}$, and the negative term $-\frac{1}{3}$ has a smaller absolute value than the positive term $\frac{1}{2}$, so the net effect is to add something positive (in this case $\frac{1}{2} - \frac{1}{3} = \frac{1}{6}$). Similarly, to get from S_3 to S_5, we add $a_4 + a_5 = \frac{1}{4} - \frac{1}{5}$, and once again the positive term $\frac{1}{4}$ outweighs the negative term $-\frac{1}{5}$. The net effect is to add a positive number to S_3.

In exactly the same way, to get from S_{2N+1} to the next odd-numbered partial sum S_{2N+3}, we add $\frac{1}{2N+2} - \frac{1}{2N+3} = \frac{(2N+3)-(2N+2)}{(2N+2)(2N+3)} = \frac{1}{(2N+2)(2N+3)}$, which is positive. This shows that the odd-numbered partial sums are increasing. See Figure 10.6.1.

Almost exactly the same argument shows that the even-numbered partial sums S_{2n} are *decreasing*, since to get from S_{2N} to S_{2N+2}, we add $a_{2N+1} + a_{2N+2} = -\frac{1}{2N+1} + \frac{1}{2N+2} = -\frac{1}{(2n+1)(2N+2)}$, which is negative.

The sequence of odd-numbered partial sums is increasing, and in fact it is *bounded*. To see this, we notice that each odd-numbered partial sum S_{2N+1} is less than the preceding even-numbered sum S_{2N}, since $S_{2N+1} = S_{2N} - \frac{1}{2N+1}$. But the even-numbered sums are decreasing, so S_{2N} must always be less than S_2. Combining these two observations, we see that every odd-numbered partial sum S_{2N+1} is less than S_2.

This shows that the odd-numbered partial sums are a bounded increasing sequence, so they must converge to some limit L, shown as the dotted line in Figure 10.6.1.

Similarly, the even-numbered partial sums are a decreasing sequence. They approach closer and closer to the odd-numbered sums (after all, $S_{2N} = S_{2N-1} + \frac{1}{2N}$, and $\frac{1}{2N} \to 0$), so they must approach the same limit L.

This means that the sequence of all the partial sums $\{S_N\}$ converges to L. We have found the somewhat surprising result that if we put negative signs on alternate terms of the harmonic series, then the resulting series will be *convergent*.

In fact there was nothing so very special about the harmonic series. To make the argument work, all we really needed to know was that the terms were alternately positive and negative and that the absolute values of the terms decreased to zero.

Figure 10.6.1

DEFINITION 10.6.1

A series $\sum_{n=1}^{\infty} a_n$ is an **alternating series** if its terms are alternately positive and negative. Such a series can be written in one of the following two forms:

$$\sum_{n=1}^{\infty} (-1)^n |a_n| \quad \text{or} \quad \sum_{n=1}^{\infty} (-1)^{n+1} |a_n|,$$

depending on whether the first term is negative or positive. These can also be written as

$$\sum_{n=1}^{\infty} (-1)^n b_n \quad \text{or} \quad \sum_{n=1}^{\infty} (-1)^{n+1} b_n \quad \text{with } b_n > 0.$$

THEOREM 10.6.2

Suppose $\sum_{n=1}^{\infty} a_n$ is an alternating series, and suppose the absolute values of its terms *decrease* to zero, that is, $|a_n| \geq |a_{n+1}|$ and $\lim_{n \to \infty} |a_n| = 0$. Then the series is convergent.

> In addition, suppose we attempt to approximate the sum L with a partial sum S_N. Then the absolute value of the error in this approximation will be less than $|a_{N+1}|$ and will have the same sign as a_{N+1}, that is, $|L - S_N| < |a_{N+1}|$ and the sign of $L - S_N$ will be the same as the sign of a_{N+1}.
>
> In other words, if we stop after N terms, the error can be estimated by the first "unused" term. The correct answer will be no farther from the approximation than the next term, and it will be in the same direction.

PROOF The proof is essentially the argument that was given in Example 1. Let us suppose that the first term is negative, so all the odd-numbered terms are negative. Then the odd-numbered partial sums are an *increasing* sequence because to get from S_{2N+1} to S_{2N+3}, we add $a_{2N+2} + a_{2N+3} = |a_{2N+2}| - |a_{2N+3}|$, which is positive, since the absolute values decrease.

On the other hand, the even-numbered partial sums are a decreasing sequence, so in particular they are all less than the first even-numbered sum S_2. Each odd-numbered sum is less than the immediately preceding even-numbered sum, since $S_{2N+1} = S_{2N} + a_{2N+1} = S_{2N} - |a_{2N+1}|$. In particular, each odd-numbered partial sum is less than S_2.

The odd-numbered partial sums are a bounded increasing sequence, so they converge to some limit L. The even-numbered sums approach closer and closer to the odd-numbered sums, since $S_{2N} - S_{2N-1} = a_{2N} \to 0$. This means that the sequence $\{S_N\}$ of all partial sums must converge to L.

The remark about the error is clear when we realize that L must be greater than each odd-numbered partial sum and less than each even-numbered partial sum. The proof is similar in the case in which the first term is positive. ◢

EXAMPLE 2

(i) The series $\sum_{n=1}^{\infty} \frac{(-1)^n}{\ln(n+1)}$ is an alternating series. The absolute value of the nth term is $\frac{1}{\ln(n+1)}$, and since $\ln x$ is an increasing function that tends to ∞ as $x \to \infty$, we see that $\left\{\frac{1}{\ln(n+1)}\right\}$ is a sequence that is decreasing and which tends to 0.

Consequently, we can conclude that the series converges. (Notice that the series without the negative signs, $\sum_{n=1}^{\infty} \frac{1}{\ln(n+1)}$, diverges because for every $n \geq 1$, $n+1 > \ln(n+1)$, so $\frac{1}{\ln(n+1)} > \frac{1}{n+1}$, and the Comparison Test establishes the divergence.)

(ii) The series $\sum_{n=0}^{\infty} \frac{\cos(n\pi)}{\sqrt{n}}$ does not at first glance appear to be an alternating series, but of course $\cos(n\pi)$ equals 1 if n is even and -1 if n is odd, so the series could also be written in the exactly equivalent form $\sum_{n=0}^{\infty} (-1)^n \frac{1}{\sqrt{n}}$. Since $\{\sqrt{n}\}$ is an increasing sequence, we see that $\left\{\frac{1}{\sqrt{n}}\right\}$ is a decreasing sequence, and its limit is certainly zero. We conclude that the series converges.

(iii) Consider the series $\sum_{n=1}^{\infty} (-1)^{n+1} n e^{-n/8}$. It is an alternating series. To check the condition about the terms decreasing, we consider the function $f(x) = xe^{-x/8}$. Its derivative is $f'(x) = e^{-x/8} + xe^{-x/8}\left(-\frac{1}{8}\right) = \left(1 - \frac{x}{8}\right)e^{-x/8}$, which is negative for $x > 8$.

This means that the absolute values of the terms of the sequence decrease after the eighth term, but they increase up to the eighth term. The terms also tend to 0, as can easily be seen by applying l'Hôpital's Rule to the function $f(x) = \frac{x}{e^{x/8}}$.

As we have seen several times by now, the convergence of a series is not affected by the behavior of the first few terms. The fact that after the eighth term this series satisfies the conditions means that the conclusion of Theorem 10.6.2 will still be valid. The series converges.

(iv) Consider the series $\sum_{n=1}^{\infty} \left(1 + 2\cos(n\pi)\right)$. It is an alternating series. However, its terms do not tend to zero. (The even-numbered terms all equal 3, and the odd-numbered terms all equal -1.) A series whose terms do not tend to zero cannot possibly converge, by Theorem 10.3.2. The series diverges, even though it is an alternating series.

With a programmable calculator we can find partial sums and approximate a convergent series. For instance, with the series $\sum_{n=1}^{\infty} (-1)^{n+1} n e^{-n/8}$ of part (iii) of Example 2 we can evaluate the partial sum S_m.

TI-85	SHARP EL9300
PRGM F2(EDIT) PSUM ENTER F3 F1(Input) M EXIT ENTER 0 STO N ENTER 0 STO S ENTER Lbl B ENTER N + 1 STO N ENTER S + N × e^x ((−) N ÷ 8) × ((−) 1) ˆ (N + 1) STO S ENTER F4 F1 (If) (N TEST F2(<) M) ENTER Goto B ENTER S ENTER EXIT EXIT EXIT PRGM F1(NAME) PSUM ENTER 50 ENTER	✏ ▼ ▼ (NEW) ENTER ENTER PSUM ENTER COMMAND A 3 (Input) M ENTER N = 0 ENTER S = 0 ENTER COMMAND B 1 (Label) 1 ENTER N = N + 1 ENTER S = S + N × e^x ((−) N ÷ 8) × ((−) 1) a^b (N + 1) ENTER COMMAND 3 (If) N COMMAND C 2 (<) M COMMAND B 2 (Goto) 1 ENTER COMMAND A 1 (Print) S ENTER COMMAND A 6 (End) ENTER ▦ ✏ (RUN) ENTER (PSUM) ENTER 50 ENTER
CASIO f_x-6300G (f_x-7700GB) MODE 2 EXE ? → M : 0 → ▮ : 0 → S : Lbl 1 : N + 1 → ▮ : ▮ + N × e^x ((−) N ÷ 8) × ((−) 1) x^y (N + 1) → S : N < M ⇒ Goto 1: S ◢ MODE 1 Prog 0 EXE 50 EXE (on the f_x-7700GB, : is PRGM F6, and Prog is PRGM F3.)	HP 48SX ATTN ↱ CLR ▬▬ "" KEY IN M ▶ SPC PROMPT SPC ′ M ▶ STO 0 SPC 1 SPC M SPC FOR N SPC N SPC N 8 +/− ÷ ▮ 1 SPC +/− N SPC 1 + y^x × × + NEXT ▼ ENTER ′ PSUM STO VAR PSUM 50 ENTER ▬ CONT

```
?50
ANSWER=
       .2032963016
```

Since the 51st term of the series is $51 \times e^{-51/8} \approx 0.087$, we find that the approximation is correct to within this accuracy. Now use the program with other values of m. Which m will give accuracy to within 0.01?

Absolute Convergence

The example of the series $\sum_{n=1}^{\infty} (-1)^n \frac{1}{n}$ shows that it is possible for a series to converge even though the series obtained by replacing each term by its absolute value is divergent. (In this example the series with absolute values is the harmonic series.)

If we have a series $\sum_{n=1}^{\infty} a_n$, it may be interesting to consider the related series $\sum_{n=1}^{\infty} |a_n|$ and ask whether or not it converges. If it is convergent, the original series is called *absolutely convergent.*

DEFINITION 10.6.3

A series $\sum_{n=1}^{\infty} a_n$ is said to be **absolutely convergent** (or to **converge absolutely**) if the series of the absolute values of its terms is convergent, that is, if $\sum_{n=1}^{\infty} |a_n|$ is convergent. If $\sum_{n=1}^{\infty} a_n$ is convergent but *not* absolutely convergent, then it is said to be **conditionally convergent** (or to **converge conditionally**).

The importance of the concept of absolute convergence comes from the following theorem.

THEOREM 10.6.4

If $\sum_{n=1}^{\infty} a_n$ is absolutely convergent, then it converges, that is, if $\sum_{n=1}^{\infty} |a_n|$ converges, then $\sum_{n=1}^{\infty} a_n$ converges.

In words: An absolutely convergent series must be convergent.

PROOF   Suppose $\sum_{n=1}^{\infty} |a_n|$ is convergent. Each term a_n satisfies $-|a_n| \le a_n \le |a_n|$, and adding $|a_n|$ to each part of this inequality, we find that

$$0 \le a_n + |a_n| \le 2|a_n|.$$

The series $\sum_{n=1}^{\infty} (a_n + |a_n|)$ has nonnegative terms, and its terms are less than those of the convergent series $\sum_{n=1}^{\infty} 2|a_n|$, so the Comparison Test shows that it converges.

We can write

$$\sum_{n=1}^{\infty} a_n = \sum_{n=1}^{\infty} (a_n + |a_n|) - \sum_{n=1}^{\infty} |a_n|,$$

which shows that the series $\sum_{n=1}^{\infty} a_n$ is the difference of two convergent series. This shows that it is convergent, as desired. ◢

So one way to prove that a series converges is to show that it converges absolutely. Since this involves the convergence of a series with nonnegative terms, it may be possible to use the Integral Test or the Comparison Test (which do not work with series whose terms may be negative).

However, we should always bear in mind that it is possible for a series to be conditionally convergent, that is, it may converge but not converge absolutely.

EXAMPLE 3

(i) The series $\sum_{n=1}^{\infty} (-1)^n \frac{1}{n}$ converges (by the condition for convergence of alternating series), but it is *not* absolutely convergent. It is conditionally convergent.

(ii) The series $\sum_{n=1}^{\infty} (-1)^n \frac{1}{n^{3/2}+2}$ converges absolutely (since $\sum_{n=1}^{\infty} \frac{1}{n^{3/2}+2}$ converges, by the Comparison Test, comparing it to the *p*-series $\sum_{n=1}^{\infty} \frac{1}{n^{3/2}}$).

In particular the original series is convergent. (This part could also be seen by noting that it is an alternating series.) Being absolutely convergent, it is *not* conditionally convergent.

(iii) The series $\sum_{n=2}^{\infty} \left(\cos\left(n\frac{\pi}{3}\right) \frac{2}{n \ln^2(n)} \right)$ is not an alternating series; notice that $\cos\left(2\frac{\pi}{3}\right) = -\frac{1}{2}$, $\cos\left(3\frac{\pi}{3}\right) = -1$, $\cos\left(4\frac{\pi}{3}\right) = -\frac{1}{2}$, $\cos\left(5\frac{\pi}{3}\right) = \frac{1}{2}$, $\cos\left(6\frac{\pi}{3}\right) = 1$, $\cos\left(7\frac{\pi}{3}\right) = \frac{1}{2}$, and after that the "cos" part of the series repeats itself, with three positive terms followed by three negative terms and so on.

If we consider the corresponding positive series

$$\sum_{n=2}^{\infty} \left| \cos\left(n\frac{\pi}{3}\right) \frac{2}{n \ln^2(n)} \right| = \sum_{n=2}^{\infty} \left(\left| \cos\left(n\frac{\pi}{3}\right) \right| \frac{2}{n \ln^2(n)} \right),$$

it looks quite complicated. However, each of its terms is smaller than the corresponding term in the simpler series $\sum_{n=2}^{\infty} \frac{2}{n \ln^2(n)}$, and this looks like a series to which the Integral Test may apply. We consider the improper integral, making the substitution $u = \ln x$, $du = \frac{1}{x} dx$.

$$\int_{2}^{\infty} \frac{2}{x \ln^2(x)} dx = 2 \lim_{r \to \infty} \left(\int_{2}^{r} \frac{1}{x \ln^2(x)} dx \right) = 2 \lim_{r \to \infty} \left(\int_{\ln 2}^{\ln r} \frac{1}{u^2} du \right)$$

$$= 2 \lim_{r \to \infty} \left(-\frac{1}{u} \right) \Big|_{\ln 2}^{\ln r} = 2 \lim_{r \to \infty} \left(-\frac{1}{\ln r} + \frac{1}{\ln 2} \right)$$

$$= \frac{2}{\ln 2}.$$

The improper integral converges, so $\sum_{n=2}^{\infty} \frac{2}{n \ln^2(n)}$ converges, by the Integral Test. By the Comparison Test the series $\sum_{n=2}^{\infty} \left(\left| \cos\left(n\frac{\pi}{3}\right) \right| \frac{2}{n \ln^2(n)} \right)$ converges.

This means that the original series converges absolutely, which in particular means that it converges. ◢

Notice that in part (iii) of Example 3 it was not possible to apply the test for alternating series but that it was possible to show that the series is absolutely convergent (and so convergent). The Integral Test and Comparison Test are only applicable to series with nonnegative terms, but now we see that it may be possible to use them with any series by considering the absolute values of the terms and showing that the series is absolutely convergent.

The concept of absolute convergence is important for another reason. When we are dealing with finite sums, we know that it makes no difference what order we use to add up the terms. With series, however, this is definitely *not* true. In fact, if we take the convergent series $\sum_{n=1}^{\infty} (-1)^n \frac{1}{n}$, it is possible to rearrange the terms so that the rearranged series diverges, and it is also possible to rearrange them so that the rearranged series converges to *any* limit L.

Roughly speaking, the way to find a rearrangement that converges to L is this: Start adding the positive terms $\frac{1}{2} + \frac{1}{4} + \frac{1}{6} + \ldots + \frac{1}{2N}$, and stop at the first N for which this sum is greater than L. Then add to this some negative terms $-1 - \frac{1}{3} - \frac{1}{5} - \ldots$, and stop after the first term with which the whole sum is less than L. Then add additional positive terms (starting with $\frac{1}{2N+2}$) until the whole sum exceeds L, then add additional negative terms until it is less than L, and so on.

Since $\frac{1}{n} \to 0$ as $n \to \infty$, these rearranged partial sums approach closer and closer to L; the rearranged series will converge to L. The "rearrangement" eventually uses all the terms of the original series.

This can be done for *any* value of L; the series can be rearranged so that it will converge to any limit. In fact, the same idea can be applied to any series that is *conditionally* convergent.

Because of this it is extremely important to be careful about rearranging series. On the other hand, for a series that is absolutely convergent there is no problem.

THEOREM 10.6.5

If $\sum_{n=1}^{\infty} a_n$ converges absolutely, then *any* rearrangement of this series also converges absolutely, and moreover the sums are the same. In other words, if $\sum_{n=1}^{\infty} b_n$ is a rearrangement of $\sum_{n=1}^{\infty} a_n$, then it converges absolutely, and

$$\sum_{n=1}^{\infty} b_n = \sum_{n=1}^{\infty} a_n.$$

(To say that $\sum_{n=1}^{\infty} b_n$ is a rearrangement of $\sum_{n=1}^{\infty} a_n$ means that it contains all the same terms but in a different order.)

This theorem is fairly difficult, so we omit the proof. It is another reason for the importance of the concept of absolute convergence and shows that in a sense a series that is absolutely convergent "converges very strongly" to its limit, so much so that even after being rearranged, it will still converge to the same limit.

Let us emphasize once again that "conditionally convergent" means convergent but *not* absolutely convergent.

Exercises 10.6

Decide whether or not each of the following series converges, and give your reasons.

1. $\sum_{n=1}^{\infty} (-1)^n \frac{1}{4n+3}$

2. $\sum_{n=1}^{\infty} (-1)^n \frac{1}{n^{4/3}}$

3. $\sum_{n=1}^{\infty} (-1)^{n+1} \frac{1}{n^{1/3}}$

4. $\sum_{n=1}^{\infty} (-1)^{n-1} \frac{1}{n^{-5/2}}$

5. $\sum_{n=3}^{\infty} (-1)^{n+7} \frac{n^3}{n^4 - 2n^2 + 1}$

6. $\sum_{n=2}^{\infty} (-1)^{n-3} \frac{1}{\ln(n)}$

7. $\sum_{n=2}^{\infty} (-1)^n \frac{\ln(n)}{n+4}$

8. $\sum_{n=1}^{\infty} (-1)^n \frac{\cos^2(n\pi)}{3n+2}$

9. $\sum_{n=2}^{\infty} \cos(n\pi) \frac{\ln(n)}{e^{-n}+1}$

10. $\sum_{n=1}^{\infty} (-1)^{2n} \frac{1}{n^{2/3}}$

11. $\sum_{n=1}^{\infty} \sin\left(n\frac{\pi}{5}\right) \frac{3n}{4n^3 - 2n + 5}$

12. $\sum_{n=2}^{\infty} \sin\left(n\frac{\pi}{7}\right) \frac{1}{n \ln^3(n)}$

13. $\sum_{n=1}^{\infty} (-1)^n \frac{n+3}{3n^2 + 4n + 2}$

14. $\sum_{n=1}^{\infty} \left(\frac{1}{n} - \frac{1}{n+1}\right)$

15. $\sum_{n=7}^{\infty} \cos(n\pi) \frac{\sqrt{n+3}}{4-\sqrt{n+10}}$

16. $\sum_{n=7}^{\infty} \cos(n\pi) \frac{1}{n\ln(n)}$

17. $\sum_{n=5}^{\infty} \sin(n^2-3) \frac{n^{4/3}}{1-n^{15/6}}$

18. $\sum_{n=1}^{\infty} \cos(e^{2n}) \frac{1}{n^2+4}$

19. $\sum_{n=1}^{\infty} (-1)^{n+2} \cos\left(\frac{1}{n}\right)$

20. $\sum_{n=1}^{\infty} \cos(n\pi) \sin\left(\frac{1}{n}\right)$

Decide whether or not each of the following series converges absolutely, conditionally, or neither, and give your reasons.

21. $\sum_{n=1}^{\infty} (-1)^n \frac{1}{5n-3}$

22. $\sum_{n=2}^{\infty} (-1)^n \frac{1}{3n^2-4}$

23. $\sum_{n=1}^{\infty} \cos\left(n\frac{\pi}{5}\right) n^{-7/6}$

24. $\sum_{n=0}^{\infty} \cos^3(n\pi) \frac{1}{\ln(2n^2+4)}$

25. $\sum_{n=1}^{\infty} \cos\left(\pi(n^3-3n^2-n-1)\right)\sqrt{n}$

26. $\sum_{n=1}^{\infty} \sin(n+\pi) \frac{1+\exp(-n)}{1-\exp(n)}$

27. $\sum_{n=1}^{\infty} (-1)^n \tanh(n) n^{-1/7}$

28. $\sum_{n=1}^{\infty} (-1)^{n+1} \frac{\exp(n)}{\sinh(n)}$

29. $\sum_{n=1}^{\infty} \sin(3n^3+2n-\pi) \frac{\exp(n)}{\exp(2n)+1}$

30. $\sum_{n=2}^{\infty} \coth(3n^2-5) \frac{1}{\ln(n)}$

II

Show that each of the following series converges.

*31. $\sum_{n=1}^{\infty} \frac{\sin(n\pi/2)}{n}$

*32. $\sum_{n=1}^{\infty} \frac{\cos(n\pi/2)}{\ln(n+6)}$

33. $\sum_{n=1}^{\infty} \frac{\sin(n)}{n^{3/2}}$

34. $\sum_{n=4}^{\infty} \frac{\cos(n)}{e^{2n}+3n-1}$

*35. Suppose $a_n > 0$ for each n and $\sum_{n=1}^{\infty} (-1)^n a_n$ is conditionally convergent. Show that the series consisting of all the positive terms $\sum_{m=1}^{\infty} a_{2m}$ and the series $\sum_{m=1}^{\infty} (-a_{2m-1})$ consisting of all the negative terms must both be divergent.

*36. Explain how to find a rearrangement of $\sum_{n=1}^{\infty} (-1)^n \frac{1}{n}$ that diverges.

*37. Find the first twenty terms of a rearrangement of

$\sum_{n=1}^{\infty} (-1)^n \frac{1}{n}$ that converges to $\frac{\pi}{2}$, following the suggestion in the text just before Theorem 10.6.5.

*38. Show that the series $\sum_{n=1}^{\infty} \sin\left(n\frac{\pi}{4}\right)\frac{1}{n}$ converges conditionally.

*39. Show that the series $\sum_{n=1}^{\infty} \cos(n^2\pi)\frac{1}{n}$ converges conditionally.

*40. Decide whether or not the series

$$\sum_{n=1}^{\infty} (-1)^{n^2+n} \frac{1}{4n+5}$$

is convergent, and explain your reasons.

Drill

Decide whether or not each of the following series converges, and give your reasons.

41. $\sum_{n=1}^{\infty} \frac{n^2+\sqrt{n}}{2n^4-1}$

42. $\sum_{n=1}^{\infty} \frac{5n}{3n^2+2e^n}$

43. $\sum_{n=2}^{\infty} \frac{3^n}{2^n \ln(n)}$

44. $\sum_{n=1}^{\infty} e^{-2n^2}$

45. $\sum_{n=1}^{\infty} \frac{n+1}{(n^3+1)e^{-n}}$

46. $\sum_{n=1}^{\infty} \frac{ne^{-n}}{2+\cos(n+3)}$

47. $\sum_{n=1}^{\infty} (-1)^n \arctan\left(\frac{1}{n}\right)$

48. $\sum_{n=6}^{\infty} (-1)^n \frac{1}{2+\ln(\ln(n))}$

49. $\sum_{n=2}^{\infty} \frac{(n-2)!}{n!}$

50. $\sum_{n=2}^{\infty} \frac{1}{n^n}$

III

Use a programmable calculator or computer to estimate the following series with the specified accuracy.

51. $\sum_{n=1}^{\infty} (-1)^n \frac{1}{n}$, within 0.005

52. $\sum_{n=2}^{\infty} (-1)^n \frac{\ln(n)}{n}$, within 0.01

53. $\sum_{n=1}^{\infty} \frac{\cos(n)}{n^2}$, within 0.005

54. $\sum_{n=2}^{\infty} \frac{\sin(n^2)}{n^2 \ln(n)}$, within 0.004

10.7 The Ratio Test and the *n*th Root Test

| RATIO TEST |
| nTH ROOT TEST |
| TESTING CONVERGENCE OF SERIES |

In this section we learn two convenient tests for convergence of series. For these tests it is not necessary to assume that the terms of the series are all nonnegative.

When we studied geometric series, we found that the convergence or divergence of a series depended on the *common ratio*, the ratio of each term to the previous one. If the absolute value of this ratio is less than 1, then the series converges; otherwise, it diverges.

For a series other than a geometric series the difficulty is that the ratio of each term to the preceding term will not be the same for different terms, so we cannot use the same idea. However, it sometimes happens that the ratios of the nth term to the preceding term will tend to a limit as $n \to \infty$. Suppose

$$\lim_{n \to \infty} \frac{a_n}{a_{n-1}} = \rho.$$

We use the Greek letter ρ ("rho," pronounced "row" as in "rowboat"), which is the Greek letter corresponding to our letter r, to remind us of the idea of the common ratio but to be sufficiently different that we will remember that it is not the ratio itself, but only a limit.

What this says is that for large values of n, each term is approximately ρ times the previous one. It suggests that the series behaves something like a geometric series whose common ratio is ρ. This is the idea behind the Ratio Test.

THEOREM 10.7.1

The Ratio Test

Consider the series $\sum_{n=1}^{\infty} a_n$, and suppose that $a_n \neq 0$, at least after the first few terms. Suppose the limit

$$\lim_{n \to \infty} \left| \frac{a_n}{a_{n-1}} \right|$$

exists and equals ρ. Then:

If $\rho < 1$, the series converges absolutely,

If $\rho > 1$, the series diverges,

If $\rho = 1$, the test is inconclusive.

PROOF Suppose the limit exists and equals ρ, with $\rho < 1$. Then we can find a number R between ρ and 1, that is, $\rho < R < 1$. Because $\left| \frac{a_n}{a_{n-1}} \right| \to \rho$, we know that there is an integer N so that $\left| \frac{a_n}{a_{n-1}} \right| < R$, for every $n \geq N$, i.e., $|a_n| < R|a_{n-1}|$, for $n \geq N$.

This means that past the Nth term, the absolute value of each term is less than R times the absolute value of the previous one. This means that the series $\sum_{n=N}^{\infty} |a_n|$ has terms that are less than or equal to the geometric series $|a_N|(1 + R + R^2 + R^3 + R^4 + \ldots)$.

This geometric series converges, since $R < 1$, and by the Comparison Test, $\sum_{n=N}^{\infty} |a_n|$ converges. From this we conclude that $\sum_{n=1}^{\infty} |a_n|$ converges, which means that $\sum_{n=1}^{\infty} a_n$ converges absolutely. This proves the first part of the theorem.

Suppose $\rho > 1$. As above, we see that there is an integer N so that $\left| \frac{a_n}{a_{n-1}} \right| > 1$ whenever $n \geq N$. This means that the absolute value of a_n is *greater* than the absolute value of a_{n-1}, for every $n \geq N$, which means that after the Nth term the absolute values of the terms are *increasing*.

In particular, this means that the terms cannot tend to zero, so the series must diverge. This proves the second part of the theorem.

Finally we give two examples to show that it is possible to have $\rho = 1$ for convergent or divergent series, which shows that when $\rho = 1$ it is impossible to draw any conclusions about convergence.

First consider the harmonic series $\sum_{n=1}^{\infty} \frac{1}{n}$. The quotient of its nth term over its $(n-1)$th term is $\left(\frac{1}{n}\right)/\left(\frac{1}{n-1}\right) = \frac{n-1}{n}$. As $n \to \infty$, this tends to 1, that is, $\lim_{n\to\infty} \frac{n-1}{n} = 1$. This is an example of a series that is divergent and for which the limit is $\rho = 1$.

Next consider the series $\sum_{n=1}^{\infty} \frac{1}{n^2}$. The quotient of its nth term over its $(n-1)$th term is $\left(\frac{1}{n^2}\right)/\left(\frac{1}{(n-1)^2}\right) = \frac{(n-1)^2}{n^2}$. As $n \to \infty$, this tends to 1, that is, $\lim_{n\to\infty} \frac{(n-1)^2}{n^2} = 1$. This is an example of a convergent series for which the limit is $\rho = 1$.

The existence of these two examples shows that the Ratio Test tells us nothing when $\rho = 1$. This concludes the proof of the theorem. ◢

Notice that the limit in Theorem 10.7.1 is the limit of the *absolute values* of the ratios of consecutive terms rather than just the limit of the ratios themselves. This actually makes things easier.

Also we should notice that unlike the situation with geometric series, in which a series with common ratio $r = 1$ must diverge, with the Ratio Test we are unable to draw any conclusion if the limit is $\rho = 1$. In this respect the situation is more delicate than it is for geometric series.

EXAMPLE 1

(i) Consider the series $\sum_{n=1}^{\infty} \frac{1}{n!}$. The ratio of its nth term by its $(n-1)$th term is $\left(\frac{1}{n!}\right)/\left(\frac{1}{(n-1)!}\right) = \frac{(n-1)!}{n!} = \frac{1}{n}$, which obviously tends to 0 as $n \to \infty$. This shows that the limit ratio is $\rho = 0$, and since $\rho < 1$, the Ratio Test tells us that the series converges absolutely.

(ii) Now we fix a number c and consider the series $\sum_{n=1}^{\infty} \frac{c^n}{n!}$. The ratio of its nth term over the preceding term is $\left(\frac{c^n}{n!}\right)/\left(\frac{c^{n-1}}{(n-1)!}\right) = \frac{c^n(n-1)!}{c^{n-1}n!} = \frac{c}{n}$. We let $n \to \infty$ and find that $\lim_{n\to\infty} \frac{c}{n} = 0$. The Ratio Test tells us the series converges absolutely (and so it converges).

Notice that this is true for any possible value of c. Also notice that since the terms of a convergent series must tend to 0, we have incidentally proved that $\lim_{n\to\infty} \frac{c^n}{n!} = 0$, something we did with much more work in Example 10.1.6.

(iii) Let $a_n = (-1)^n \frac{n^2}{2^n}$, and consider the series $\sum_{n=1}^{\infty} a_n$. To apply the Ratio Test, we find

$$\rho = \lim_{n\to\infty}\left|\frac{a_n}{a_{n-1}}\right| = \lim_{n\to\infty}\left|\frac{(-1)^n\frac{n^2}{2^n}}{(-1)^{n-1}\frac{(n-1)^2}{2^{n-1}}}\right| = \lim_{n\to\infty}\left|(-1)\frac{n^2}{(n-1)^2}\frac{2^{n-1}}{2^n}\right|$$

$$= \lim_{n\to\infty}\left(\frac{n^2}{(n-1)^2}\frac{1}{2}\right) = \frac{1}{2}\lim_{n\to\infty}\left(\frac{n^2}{n^2-2n+1}\right)$$

$$= \frac{1}{2}.$$

Since the limit ratio $\rho = \frac{1}{2}$ is less than 1, the Ratio Test tells us that the series converges absolutely.

(iv) For what values of c does the series $\sum_{n=1}^{\infty} \left((-1)^n(n^2-2)3^{n-2}c^n\right)$ converge?

Solution   Consider the limit

$$\lim_{n \to \infty} \left| \frac{(-1)^n (n^2 - 2) 3^{n-2} c^n}{(-1)^{n-1} \left((n-1)^2 - 2 \right) 3^{n-3} c^{n-1}} \right| = \lim_{n \to \infty} \left| (-1) \frac{n^2 - 2}{(n-1)^2 - 2} \frac{3^{n-2}}{3^{n-3}} \frac{c^n}{c^{n-1}} \right|$$

$$= \lim_{n \to \infty} \left(\frac{n^2 - 2}{n^2 - 2n - 1} (3) |c| \right)$$

$$= 3|c|.$$

The Ratio Test says that the series will converge absolutely when this ratio is less than 1, that is, when $|c| < \frac{1}{3}$. It will diverge when $|c| > \frac{1}{3}$.

What remains is the more delicate case when $|c| = \frac{1}{3}$, that is, when $c = \pm \frac{1}{3}$. If $c = \frac{1}{3}$, then the nth term of the series is $(-1)^n (n^2 - 2) 3^{n-2} \left(\frac{1}{3} \right)^n = (-1)^n (n^2 - 2) 3^{n-2-n} = (-1)^n (n^2 - 2) 3^{-2}$. As $n \to \infty$, this term does not tend to zero, so the series diverges. When $c = -\frac{1}{3}$, the result is similar; the only thing that changes is the sign of the nth term, which does not affect the fact that it does not tend to 0.

We see that the series converges when $-\frac{1}{3} < c < \frac{1}{3}$ and diverges otherwise. ◢

The Ratio Test tends to be particularly useful for series whose nth term involves factorials or nth powers.

The nth Root Test

The other test we should learn also tends to be applicable to series that involve nth powers in the nth term. The structure of this test is quite similar to that of the Ratio Test.

THEOREM 10.7.2

The nth Root Test

Consider the series $\sum_{n=1}^{\infty} a_n$. Suppose the limit

$$\lim_{n \to \infty} \sqrt[n]{|a_n|}$$

exists and equals ρ. Then:

  If $\rho < 1$, the series converges absolutely,

  If $\rho > 1$, the series diverges,

  If $\rho = 1$, the test is inconclusive.

PROOF   Suppose the limit exists and equals ρ, with $\rho < 1$. Let R be a number between ρ and 1, that is, $\rho < R < 1$. Then there is an integer N so that for every $n \geq N$ we have $\sqrt[n]{|a_n|} < R$. Another way of saying the same thing is that $|a_n| < R^n$, for every $n \geq N$.

Since $R < 1$, $\sum_{n=N}^{\infty} R^n$ is a convergent geometric series. By the Comparison Test, $\sum_{n=N}^{\infty} |a_n|$ is convergent, which means that $\sum_{n=1}^{\infty} |a_n|$ is convergent, which means that $\sum_{n=1}^{\infty} a_n$ is absolutely convergent. This proves the first part of the theorem.

If $\rho > 1$, then there is an integer N so that for every $n > N$ we have $\sqrt[n]{|a_n|} > 1$, which is the same as saying that $|a_n| > 1$ for every $n > N$. Since this means that $\{a_n\}$ cannot tend to 0, we know that the series diverges, and the second part of the theorem is established.

As before, we consider the harmonic series. The nth root of the nth term is $\sqrt[n]{\frac{1}{n}} = \left(\frac{1}{n}\right)^{1/n} = \exp\left(\frac{1}{n}\ln\left(\frac{1}{n}\right)\right)$. We know that as $n \to \infty$, $\frac{1}{n}\ln\left(\frac{1}{n}\right) \to 0$ (using l'Hôpital's Rule, for instance), so $\exp\left(\frac{1}{n}\ln\left(\frac{1}{n}\right)\right)$ tends to $e^0 = 1$.

The harmonic series is an example of a divergent series for which the nth root of the nth term tends to 1.

The nth root of $\frac{1}{n^2}$ is just the square of the nth root of $\frac{1}{n}$, so it must also approach the limit 1 as $n \to \infty$. This means that $\sum_{n=1}^{\infty} \frac{1}{n^2}$ is an example of a series for which the nth root of the nth term tends to 1 but that is convergent.

The existence of these two examples shows that the nth Root Test is inconclusive if $\rho = 1$. This completes the proof.

EXAMPLE 2

(i) Consider the series $\sum_{n=1}^{\infty} \left(\frac{1}{n}\right)^n$. The nth Root Test suggests looking at $\left(\left(\frac{1}{n}\right)^n\right)^{1/n} = \frac{1}{n}$. Since this tends to $\rho = 0$ as $n \to \infty$, the nth Root Test says that the series converges.

(ii) Consider the series $\sum_{n=1}^{\infty} \left(3 - \cos(n\pi)\right)\left(\frac{2}{3}\right)^n$. The nth root of its nth term is

$$\sqrt[n]{\left(3 - \cos(n\pi)\right)\left(\frac{2}{3}\right)^n} = \frac{2}{3}\sqrt[n]{3 - \cos(n\pi)}.$$

See the Calculator Example on the next page.

Now $\cos(n\pi)$ is either 1 or -1, depending on whether n is even or odd, so $3 - \cos(n\pi)$ is either 4 or 2.

Now as $n \to \infty$, $\sqrt[n]{2} = 2^{1/n} = \exp\left(\frac{1}{n}\ln 2\right)$, and since $\frac{1}{n}\ln 2$ tends to 0, we see that $\sqrt[n]{2} \to \exp(0) = 1$. The same argument shows that $\sqrt[n]{4} \to 1$ as $n \to \infty$.

These two observations show that $\lim_{n\to\infty}\sqrt[n]{3 - \cos(n\pi)} = 1$, and from this we see that the nth root of the nth term of the series tends to $\frac{2}{3}\lim_{n\to\infty}\sqrt[n]{3 - \cos(n\pi)} = \frac{2}{3}$. The nth Root Test tells us that the series is convergent.

Notice that in this case the Ratio Test could not be used because when n is even, the ratio of the nth term to the $(n-1)$th term is

$$\frac{\left(3 - \cos(n\pi)\right)\left(\frac{2}{3}\right)^n}{\left(3 - \cos((n-1)\pi)\right)\left(\frac{2}{3}\right)^{n-1}} = \frac{2\left(\frac{2}{3}\right)^n}{4\left(\frac{2}{3}\right)^{n-1}} = \frac{2}{4}\frac{2}{3} = \frac{1}{3},$$

but when n is odd, the ratio is

$$\frac{\left(3 - \cos(n\pi)\right)\left(\frac{2}{3}\right)^n}{\left(3 - \cos((n-1)\pi)\right)\left(\frac{2}{3}\right)^{n-1}} = \frac{4\left(\frac{2}{3}\right)^n}{2\left(\frac{2}{3}\right)^{n-1}} = \frac{4}{2}\frac{2}{3} = \frac{4}{3}.$$

These ratios do not tend to a limit as $n \to \infty$, so it is not possible to use the Ratio Test.

With a programmable calculator we can find partial sums and approximate the series $\sum_{n=1}^{\infty}(3 - \cos n\pi)\left(\frac{2}{3}\right)^n$ of part (ii) of Example 2. Notice that $\cos n\pi = (-1)^n$.

TI-85	SHARP EL9300
PRGM F2(EDIT) PTSUM ENTER F3 F1(Input) M EXIT ENTER 0 STO N ENTER 0 STO S ENTER Lbl B ENTER N +1 STO N ENTER S + (3 − ((−) 1) ^ N) × (2 ÷ 3) ^ N STO S ENTER F4 F1(If) (N TEST F2(<) M) ENTER Goto B ENTER S ENTER EXIT EXIT EXIT PRGM F1(NAME) PTSUM ENTER 20 ENTER	▼ ▼ (NEW) ENTER ENTER PTSM ENTER COMMAND A 3 (Input) M ENTER N = 0 ENTER S = 0 ENTER COMMAND B 1 (Label) 1 ENTER N = N + 1 ENTER S = S + (3 − ((−) 1) a^b N) × (2 ÷ 3) a^b N ENTER COMMAND 3 (If) N COMMAND C 2 (<) M COMMAND B 2 (Goto) 1 ENTER COMMAND A 1 (Print) S ENTER COMMAND A 6 (End) ENTER (RUN) ENTER (PTSM) ENTER 20 ENTER
CASIO f_x-6300G (f_x-7700GB)	**HP 48SX**
MODE 2 EXE ? → M : 0 → N : 0 → S : Lbl 1 : N + 1 → N : S + (3 − ((−) 1) x^y N) × (2 ÷ 3) x^y N → S : N < M ⇒ Goto 1 : S ▶ MODE 1 Prog 0 EXE 20 EXE (on the f_x-7700GB, : is PRGM F6, and Prog is PRGM F3.)	ATTN ↱ CLR ≪≫ "" KEY IN M ▶ SPC PROMPT SPC ' M ▶ STO 0 SPC 1 SPC M SPC FOR N SPC 3 SPC 1 +/− N y^x − 2 SPC 3 ÷ N y^x × + NEXT ▼ ENTER ' PSUM STO VAR PSUM 20 ENTER ↰ CONT

```
?20
ANSWER=
         6.398075337
```

Try to estimate the accuracy of this approximation. (*Hint:* If $n > 20$, then $\left(\frac{2}{3}\right)^n < \left(\frac{2}{3}\right)^{20} \ldots$).

Summary 10.7.3   Testing Convergence of Series

We have learned a number of tests for convergence and divergence of series. Given a series $\sum_{n=1}^{\infty} a_n$, it can be difficult to know which test to apply. In a sense there is no single answer, and we just try each test until we find one that works. However, the following observations may be helpful.

1. It is usually easy to apply the nth Term Test (Theorem 10.3.2), which may immediately show that the series diverges.

2. Next it is probably best to check whether the series is one of the special types we know how to deal with: (i) geometric series, (ii) alternating series (with terms

decreasing in absolute value), (iii) *p*-series (including the harmonic series), (iv) series whose terms are a rational function of *n* (cf. Theorem 10.5.4). It may also be possible to break the series into a sum of simpler series, to factor out constants, or to shift the index of summation to transform the series into one of the familiar types.

3. Next it may be possible to use the Comparison Test (Theorems 10.5.1, 10.5.2) or the Limit Comparison Test (Theorem 10.5.3). Here the difficulty is to find a suitable series with which to compare. Usually, this is done by modifying the original series in some way, often by simplifying or omitting part of the formula for a_n. It is vital to have series with nonnegative terms, but sometimes this means showing absolute convergence by working with $\sum_{n=1}^{\infty} |a_n|$.

4. The integral test may be useful, but only if the series can be expressed in terms of a function that we know how to integrate, and only if the terms are nonnegative. It too can be applied to $\sum_{n=1}^{\infty} |a_n|$.

5. The Ratio Test is most likely to be useful with series whose *n*th term involves factorials or exponentials or *n*th powers. The Root Test will have a better chance of working for series for which the *n*th term is an *n*th power. However, these are only guidelines, and both should be tried.

6. Finally, it may be necessary to combine tests. Often the Comparison Test is used to compare $\sum_{n=1}^{\infty} a_n$ with another series, which is then shown to converge or diverge by using the Integral, Ratio, or Root Test. Another common tactic is to break the series up as a sum and use different tests on each piece.

Some Familiar Convergent and Divergent Series

Geometric series	$a + ar + ar^2 + \dots$ $= \sum_{n=0}^{\infty} ar^n = \frac{a}{1-r}$	Divergent when $|r| \geq 1$, (Absolutely) convergent when $|r| < 1$
Harmonic series	$1 + \frac{1}{2} + \frac{1}{3} + \frac{1}{4} + \dots$ $= \sum_{n=1}^{\infty} \frac{1}{n}$	Divergent
p-series	$1 + \frac{1}{2^p} + \frac{1}{3^p} + \frac{1}{4^p} + \dots$ $= \sum_{n=1}^{\infty} \frac{1}{n^p}$	Convergent when $p > 1$, divergent when $p \leq 1$
Alternating series	$\sum_{n=1}^{\infty} (-1)^n |a_n|$ or $\sum_{n=1}^{\infty} (-1)^{n+1} |a_n|$	Convergent if $|a_n|$ decreases to 0
Rational functions	$\sum_{n=1}^{\infty} \frac{p(n)}{q(n)}$ with $p(x)$, $q(x)$ polynomials	Converges precisely when $deg\ q(x) \geq deg\ p(x) + 2$

You might find it helpful to review the examples of this section in light of this summary. You could try looking at each example and trying to decide which approach to use *before* looking at the solution. Bear in mind that many examples can be done in more than one way.

Exercises 10.7

I

Test the following series for convergence, giving reasons for your conclusions.

1. $\sum_{n=1}^{\infty} \frac{4^n}{(2n)!}$

2. $\sum_{n=1}^{\infty} (-1)^n \frac{6^{n+1}}{(n-1)!}$

3. $\sum_{n=0}^{\infty} (-1)^{n+1} \frac{8^{2n+1}}{(2n+1)!}$

4. $\sum_{n=0}^{\infty} (-1)^n \frac{\pi^{2n}}{(2n)!}$

5. $\sum_{n=1}^{\infty} \frac{2^n}{n^2}$

6. $\sum_{n=1}^{\infty} \left(\frac{2}{3}\right)^n \ln(n+1)$

7. $\sum_{n=1}^{\infty} n^{-2n}$

8. $\sum_{n=1}^{\infty} \frac{n^n}{n!}$

9. $\sum_{n=1}^{\infty} n^{-n} n!$

10. $\sum_{n=1}^{\infty} \left(2 + \sin\left(\frac{n\pi}{2}\right)\right) \left(\frac{4}{5}\right)^n$

11. $\sum_{n=1}^{\infty} \frac{2^n}{\sqrt{n!}}$

12. $\sum_{n=1}^{\infty} \sin^n\left(\frac{1}{n}\right)$

13. $\sum_{n=1}^{\infty} ne^{-n}$

14. $\sum_{n=1}^{\infty} 2^{-1/n}$

15. $\sum_{n=0}^{\infty} 2^{-n} e^{n/2}$

16. $\sum_{n=1}^{\infty} \left(-\frac{1}{2}\right)^n \arctan(n^2)$

17. $\sum_{n=1}^{\infty} (n!)^{-n/3}$

18. $\sum_{n=1}^{\infty} \frac{\sin(n^2+2)}{n!}$

19. $\sum_{n=1}^{\infty} \sin\left(\frac{1}{n!}\right)$

20. $\sum_{n=1}^{\infty} \frac{n!(2n)!}{(3n)!}$

For what values of c is each of the following series convergent?

21. $\sum_{n=0}^{\infty} \frac{c^n}{(2n)!}$

22. $\sum_{n=0}^{\infty} (-1)^n \frac{c^{2n+1}}{(2n+1)!}$

23. $\sum_{n=0}^{\infty} (-1)^n \frac{c^{2n}}{(2n)!}$

24. $\sum_{n=1}^{\infty} \frac{c^n}{n^3}$

25. $\sum_{n=1}^{\infty} (-1)^n \frac{c^n}{n}$

26. $\sum_{n=1}^{\infty} (c-3)^n$

27. $\sum_{n=1}^{\infty} (-5)^n c^{-n}$

28. $\sum_{n=2}^{\infty} \frac{(4c)^n}{\ln(n)}$

29. $\sum_{n=1}^{\infty} (2n)! c^{-n}$

30. $\sum_{n=4}^{\infty} \ln(n)(3c)^{-n}$

II

31. Use the Ratio Test to do Example 2(i) (p. 691).

*32. In Example 2(ii) we showed that $\sum_{n=1}^{\infty} (3 - \cos(n\pi)) \left(\frac{2}{3}\right)^n$ is convergent. To do this, we used the nth root test and remarked that the Ratio Test would not work. Find a way to prove the convergence of this series without the nth Root Test, using the Comparison Theorem and the Ratio Test (or anything else you prefer).

33. For what values of c does the following series converge?

$$1 - \frac{c}{2} + \frac{c^2}{3} - \frac{c^3}{4} + \frac{c^4}{5} - \frac{c^5}{6} + \cdots$$

*34. For what values of c does the following series converge?

$$\left(\frac{1}{2}c\right)^2 - \left(\frac{2}{3}c\right)^3 + \left(\frac{3}{4}c\right)^4 - \left(\frac{4}{5}c\right)^5$$
$$+ \left(\frac{5}{6}c\right)^6 - \left(\frac{6}{7}c\right)^7 + \cdots$$

*35. Suppose $p(x)$ is any polynomial, and consider the series $\sum_{n=1}^{\infty} p(n) r^n$. For what values of r is this series convergent?

*36. Suppose $R(x)$ is any rational function, and consider the series $\sum_{n=N}^{\infty} R(n) r^n$, where N is chosen so that $R(n)$ is defined for every $n \geq N$. For what values of r is this series convergent?

*37. What does the nth Root Test have to say about the convergence of the geometric series $\sum_{n=1}^{\infty} Cr^n$?

*38. Show that $\sum_{n=1}^{\infty} (e^{1/n^2} - 1)$ converges. (*Hint:* Estimate the nth term using the Mean Value Theorem.)

Drill

Test the following series for convergence, giving reasons for your conclusions.

39. $\sum_{n=1}^{\infty} \sin\left(\frac{1}{n^2}\right)$

40. $\sum_{n=1}^{\infty} \sin^2\left(\frac{1}{n}\right)$

41. $\sum_{n=1}^{\infty} \left(\frac{2n-1}{2n}\right)^n$

42. $\sum_{n=1}^{\infty} (-1)^n e^{-\sqrt{n}}$

43. $\sum_{n=1}^{\infty} n(n^2 + 1)^{-5/4}$

44. $\sum_{n=1}^{\infty} ne^{-3n}$

45. $\sum_{n=1}^{\infty} \sin\left(\frac{\pi}{n}\right)$

46. $\sum_{n=1}^{\infty} \cos\left(\frac{\pi}{2} + \frac{2}{n}\right)$

47. $\sum_{n=1}^{\infty} \frac{1}{\sqrt{n}} e^{-\sqrt{n}}$

48. $\sum_{n=1}^{\infty} \exp\left(\frac{n+1}{n^2+1}\right)$

49. $\sum_{n=1}^{\infty} \sqrt{\frac{n}{3^n}}$

50. $\sum_{n=1}^{\infty} \cos\left(\frac{3\ln(n)}{n^4}\right)$

51. $\sum_{n=1}^{\infty} \frac{(2n)!}{n^n}$

52. $\sum_{n=1}^{\infty} \frac{2^n n^{3/2}}{5^n}$

53. $\sum_{n=2}^{\infty} \frac{\arctan(n+1)}{n^2 \ln(n)}$

54. $\sum_{n=1}^{\infty} \left(\frac{n^3+1}{n^6+n^4+2}\right)^{1/5}$

55. $\sum_{n=1}^{\infty} \frac{1}{1+n+\cos(n)}$

56. $\sum_{n=1}^{\infty} \frac{n^n}{(3n)!}$

57. $\sum_{n=1}^{\infty} ne^{-n^3}$

58. $\sum_{n=1}^{\infty} n^{-\sqrt{n}}$

59. (i) Use a programmable calculator or computer to evaluate $\sum_{n=1}^{20} \frac{1}{n!}$. The hard part of this question is: (ii) Estimate how accurately this value approximates $\sum_{n=1}^{\infty} \frac{1}{n!}$.

60. Estimate $\sum_{n=1}^{\infty} \frac{5^n}{n!}$ to within 0.001.

10.8 Taylor Series

TAYLOR SERIES
MACLAURIN SERIES
REMAINDER

In Chapter 9 we studied Taylor polynomials. Given a function $f(x)$ and a point $x = a$, the Taylor polynomial of degree n for f about $x = a$ is a polynomial $P_n(x)$ of degree n (or less) that has the property that its derivatives at $x = a$ up to and including the derivative of order n are equal to the corresponding derivatives of f at $x = a$. The idea was that such a polynomial ought to be "close" to $f(x)$, at least for x near a.

In Definition 9.1.2 we found that

$$P_n(x) = \sum_{k=0}^{n} \frac{f^{(k)}(a)}{k!}(x-a)^k = \sum_{k=0}^{n} a_k(x-a)^k,$$

where $a_k = \frac{f^{(k)}(a)}{k!}$.

In this section we consider the possibility of continuing this finite sum and making it into an infinite series. To do this, we have to know that f can be differentiated infinitely often, that is, that $f^{(k)}(a)$ exists for *every* k. Second, we have to consider the question of whether or not the resulting series converges. Finally, at points at which it does converge, we have to consider the question of whether the limit to which the series converges equals the value of the original function.

Each of these can be a difficult problem, but surprisingly often everything works out very well. For many functions the series does converge to the value of the function, so it can be used to calculate, or at least estimate, the value of the function. It can also be used to study other aspects of the function's behavior.

DEFINITION 10.8.1

Taylor Series

Suppose $f(x)$ is a function that is *infinitely differentiable* at $x = a$, that is, so that $f^{(k)}(a)$ exists for each k.

The **Taylor series** for f about the point $x = a$ is the series

$$\sum_{k=0}^{\infty} \frac{f^{(k)}(a)}{k!}(x-a)^k = \sum_{k=0}^{\infty} a_k(x-a)^k,$$

where $a_k = \frac{f^{(k)}(a)}{k!}$.

EXAMPLE 1

(i) Consider the function $f(x) = e^x$. We know that for each k, $f^{(k)}(x) = \frac{d^k}{dx^k}(e^x) = e^x$, so for $a = 0$, for instance, we find that $a_k = \frac{e^0}{k!} = \frac{1}{k!}$. The Taylor series for e^x is

$$\sum_{k=0}^{\infty} a_k(x-a)^k = \sum_{k=0}^{\infty} \frac{1}{k!}x^k = \sum_{k=0}^{\infty} \frac{x^k}{k!}.$$

In fact we know from the Ratio Test (see Example 10.7.1(ii)), that this series converges for every value of x. Even more, we will soon see that it actually converges to the right value, that is, for every x, $e^x = \sum_{k=0}^{\infty} \frac{1}{k!}x^k$.

(ii) With the same function $f(x) = e^x$ and $a = 1$ we see that the coefficients of the series are now $a_k = \frac{f^{(k)}(1)}{k!} = \frac{e}{k!}$, and the series is

$$\sum_{k=0}^{\infty} a_k(x-a)^k = \sum_{k=0}^{\infty} \frac{e}{k!}(x-1)^k = e\sum_{k=0}^{\infty} \frac{(x-1)^k}{k!}.$$

$$f(x) = \cos x, \qquad f(0) = 1$$
$$f'(x) = -\sin x, \quad f'(0) = 0$$
$$f''(x) = -\cos x, \quad f''(0) = -1$$
$$f^{(3)}(x) = \sin x, \quad f^{(3)}(0) = 0$$
$$f^{(4)}(x) = \cos x, \quad f^{(4)}(0) = 1$$

(iii) Let $f(x) = \cos x$. We calculate its derivatives up to $f^{(4)}(x)$ and observe that after that, they repeat the same cycle of four over and over again. The odd-numbered derivatives are 0 at $x = 0$, and the even-numbered ones alternate between 1 and -1.

To find the coefficients of the Taylor series, we divide the kth derivative by $k!$, so the coefficients, starting with the $k = 0$ coefficient, are $1, 0, -\frac{1}{2!}, 0, \frac{1}{4!}, 0, -\frac{1}{6!}, 0, \frac{1}{8!}, 0, \ldots$. The Taylor series for $f(x) = \cos x$ about $a = 0$ is

$$1 - \frac{x^2}{2!} + \frac{x^4}{4!} - \frac{x^6}{6!} + \frac{x^8}{8!} - \frac{x^{10}}{10!} + \frac{x^{12}}{12!} - \cdots,$$

which can also be written as

$$\sum_{n=0}^{\infty} (-1)^n \frac{x^{2n}}{(2n)!}.$$

(Notice that in this last expression we use $2n$ rather than n to label the powers of x in the series, since we know that the odd powers of x do not appear.)

We can apply the Ratio Test to this sum; the ratio of the term containing x^{2n} over the previous nonzero term (the one containing x^{2n-2}) is

$$\frac{(-1)^n \frac{1}{(2n)!} x^{2n}}{(-1)^{n-1} \frac{1}{(2n-2)!} x^{2n-2}} = (-1)\frac{(2n-2)!}{(2n)!}x^2 = -\frac{x^2}{2n(2n-1)}.$$

The absolute value of this ratio tends to 0 as $n \to \infty$, for any fixed value of x, so by the Ratio Test the series converges absolutely for every x.

(iv) A similar calculation can be made for the function $f(x) = \sin x$. We find that its derivatives evaluated at $a = 0$, starting with $f^{(0)}(0) = f(0)$, are $0, 1, 0, -1, 0, 1, 0, -1, 0, \ldots$, and the Taylor series about $a = 0$ is

$$x - \frac{x^3}{3!} + \frac{x^5}{5!} - \frac{x^7}{7!} + \frac{x^9}{9!} - \cdots = \sum_{n=0}^{\infty} (-1)^n \frac{x^{2n+1}}{(2n+1)!}.$$

Again the Ratio Test can be used to show that the series converges for every value of x. Once again notice that the nth (nonzero) term of the series is not the one containing x^n but rather the one containing x^{2n-1}, because the even powers of x do not appear in the series.

(v) Let $f(x) = \ln x$. This function is defined only for $x > 0$, so let us try $a = 1$. We must find the derivatives of f and evaluate them at $x = 1$. We see that there is a pattern, and

$$f^{(k)}(x) = (-1)^{k+1}(2)(3)(4)(5)\ldots(k-1)\frac{1}{x^k} = (-1)^{k+1}(k-1)!\frac{1}{x^k}.$$

In particular, when we evaluate at $x = 1$, we find that $f^{(0)}(1) = f(1) = \ln(1) = 0$, and for $k \geq 1$, $f^{(k)}(1) = (-1)^{k+1}(k-1)!$. From this we find that for $k \geq 1$ the kth coefficient of the Taylor series is $a_k = \frac{f^{(k)}(1)}{k!} = \frac{(-1)^{k+1}(k-1)!}{k!} = (-1)^{k+1}\frac{1}{k}$.

The Taylor series for $f(x) = \ln x$ about the point $a = 1$ is

$$\sum_{k=1}^{\infty} a_k(x-a)^k = \sum_{k=1}^{\infty} (-1)^{k+1}\frac{1}{k}(x-1)^k = \sum_{k=1}^{\infty} (-1)^{k+1}\frac{(x-1)^k}{k}.$$

(Notice that the series begins with $k = 1$, since $a_0 = 0$.)

For this series it is possible to apply the Ratio Test. The ratio of the kth term to the $(k-1)$th term is $(-1)(x-1)\frac{k-1}{k}$, whose absolute value $|x-1|\frac{k-1}{k}$ tends to $|x-1|$.

From this we conclude that the series converges absolutely when $|x-1| < 1$ and diverges when $|x-1| > 1$. When $x - 1 = 1$, the series is $\sum_{k=1}^{\infty} (-1)^{k+1}\frac{1}{k}$, which converges because it is an alternating series, the absolute values of whose terms decrease to 0.

When $x - 1 = -1$, the series becomes

$$\sum_{k=1}^{\infty} (-1)^{k+1}\frac{(x-1)^k}{k} = \sum_{k=1}^{\infty} (-1)^{k+1}\frac{(-1)^k}{k} = \sum_{k=1}^{\infty} (-1)^{2k+1}\frac{1}{k} = -\sum_{k=1}^{\infty}\frac{1}{k}.$$

In this case we see that the Taylor series is the negative of the harmonic series, so it diverges.

Combining all this information, we see that the Taylor series converges absolutely when $0 < x < 2$, converges conditionally when $x = 2$, and diverges for $x \leq 0$ and for $x > 2$.

In a sense it is not too surprising that the series does not converge when $x \leq 0$, since the original function $f(x) = \ln x$ is not defined for these values of x, but it is disconcerting to find that for $x > 2$, values for which the function $\ln x$ is defined and perfectly reasonable, the series diverges. In fact, this sort of behavior is typical of Taylor series.

Notice that in doing these calculations we had to find the value of the kth derivative of $f(x)$ evaluated at $x = a$. We found the derivative, as a function of x, but we did not evaluate it immediately. The point is that knowing $f^{(2)}(a)$, say, is not enough. We need to find the third derivative, and to do that, it is necessary to differentiate the second derivative, so we need to know the second derivative *as a function of* x. Only after we have found the derivatives $f^{(k)}(x)$ do we substitute $x = a$.

It is extremely important to bear in mind that the Taylor series $\sum_{n=0}^{\infty} a_n(x-a)^n$ is a *function* of x. It has different values for different x's, and as we have just seen, it may converge for some x's and diverge for others.

$$f'(x) = -\frac{1}{(1-x)^2}(-1)$$
$$= \frac{1}{(1-x)^2}$$
$$f''(x) = 2\frac{1}{(1-x)^3}$$
$$f^{(3)}(x) = (2)(3)\frac{1}{(1-x)^4}$$
$$f^{(4)}(x) = (2)(3)(4)\frac{1}{(1-x)^5}$$
$$f^{(5)}(x) = (2)(3)(4)(5)\frac{1}{(1-x)^6}$$
$$\vdots$$

EXAMPLE 2 Consider the function $f(x) = \frac{1}{1-x}$. Let us find its Taylor series about the point $a = 0$.

We find the first few derivatives; the pattern should be clear:

$$f^{(k)}(x) = (2)(3)\ldots(k)\frac{1}{(1-x)^{k+1}} = k!\frac{1}{(1-x)^{k+1}}.$$

If we evaluate at $x = 0$, we find that $f^{(k)}(0) = k!$, and from this we can find that the kth coefficient of the Taylor series is $a_k = \frac{f^{(k)}(0)}{k!} = \frac{k!}{k!} = 1$. The Taylor series for $f(x) = \frac{1}{1-x}$ about the point $a = 0$ is

$$\sum_{k=0}^{\infty} a_k(x-a)^k = \sum_{k=0}^{\infty} x^k = 1 + x + x^2 + x^3 + x^4 + x^5 + \ldots.$$

Having done this, we see that this series is a geometric series with common ratio x, and the formula for evaluating geometric series tells us that if $|x| < 1$, then the series equals $\frac{1}{1-x}$. This is just the function $f(x)$ that we began with.

We see that the Taylor series converges when $|x| < 1$ and that in this case it converges to $f(x)$. The Taylor series for $f(x)$ converges to $f(x)$, provided $|x| < 1$.◀

As with Taylor polynomials, there is a special name for the series about $a = 0$.

DEFINITION 10.8.2

The Taylor series about the point $a = 0$ is sometimes called the **Maclaurin series**. The Maclaurin series of a function $f(x)$ is

$$\sum_{n=0}^{\infty} \frac{f^{(n)}(0)}{n!} x^n.$$

We refer to "Maclaurin series" or "Taylor series about 0" interchangeably.

We have just seen that the Maclaurin series for $f(x) = \frac{1}{1-x}$ is exactly what we would have expected from the formula for geometric series. This observation can save us a lot of trouble when we calculate series.

EXAMPLE 3

(i) Find the Maclaurin series for $F(x) = \frac{1}{1+x}$.

Solution We could try to find a formula for the kth derivative $F^{(k)}$, but it is easier to realize that $F(x) = \frac{1}{1+x} = \frac{1}{1-(-x)} = f(-x)$, where $f(x) = \frac{1}{1-x}$ is the function from Example 2.

This suggests that the series should be obtained by substituting $-x$ for x in the series for $f(x)$. The Maclaurin series for $F(x)$ is

$$\sum_{n=0}^{\infty} (-1)^n x^n = 1 - x + x^2 - x^3 + x^4 - x^5 + x^6 - \ldots.$$

(ii) Find the Maclaurin series for $g(x) = \frac{1}{1+x^2}$.

Solution As above we realize that $g(x) = \frac{1}{1+x^2} = f(-x^2)$, so its series is

$$\sum_{n=0}^{\infty} (-x^2)^n = \sum_{n=0}^{\infty} (-1)^n x^{2n} = 1 - x^2 + x^4 - x^6 + x^8 - x^{10} + \ldots.$$

By the Ratio Test we see that this series converges when $|x^2| < 1$, that is, when $|x| < 1$. From our knowledge of geometric series we know that for these values of x it converges to the "right" thing, $g(x)$.

(iii) Find the Maclaurin series for $h(x) = \arctan x$.

Solution We could begin finding the derivatives of $h(x) = \arctan x$. However, it is very tempting to remember that $\arctan x = \int \frac{1}{1+x^2}\, dx$ and just try integrating the series we have just found for $g(x) = \frac{1}{1+x^2}$.

There are several difficulties with this idea. One is that we will have to deal with the constant C that ought to be added to the indefinite integral; we could try instead to use an intelligently chosen definite integral. Another is that we do not really know anything about what it means to integrate a series. Finally, even if we perform the calculation, we will not know that the result is correct.

Despite all this, however, it does work, and if you find the first few derivatives of $\arctan x$, you will begin to see that that approach is quite messy, so the idea of integrating is quite appealing.

See the Calculator Example on the next page.

We know that $\arctan x = \int_0^x \frac{1}{1+t^2}\, dt$ and that for $|t| < 1$, $\frac{1}{1+t^2} = 1 - t^2 + t^4 - t^6 + \ldots = \sum_{n=0}^{\infty} (-1)^n t^{2n}$, so we can try to write

$$\arctan x = \int_0^x \left(\sum_{n=0}^{\infty} (-1)^n t^{2n} \right) dt = \sum_{n=0}^{\infty} \left(\int_0^x (-1)^n t^{2n}\, dt \right)$$

$$= \sum_{n=0}^{\infty} (-1)^n \frac{1}{2n+1} x^{2n+1}.$$

Since x was the upper limit of integration, we used t as the variable inside the integral, and for this reason we wrote out the series for $\frac{1}{1+t^2}$ in terms of t. This is necessary to avoid confusion.

Although we have no real justification for doing the calculation this way, the result does turn out to be correct. This is the Maclaurin series for $\arctan x$. ◢

The calculation we did in the last part of Example 3 is extremely useful. It was much easier to do it this way than it would have been to find all the derivatives of the function $\arctan x$. We will return in the last section of the chapter to discuss this type of calculation, and there we will see that it really is permissible. *Roughly speaking*, operations like integrating and differentiating Taylor series are permissible when both the original series and the resulting series converge absolutely.

In principle, it is always possible to find Taylor series by finding the kth derivative of the function $f(x)$ at the point $x = a$, but in practice the calculations can be quite complicated. If it is possible to find the series by some trick like the one we have just used, then that may well be easier. There are times when we have no choice, but often we can avoid calculating derivatives.

EXAMPLE 4 (i) Find the Maclaurin series of the function $f(x) = \sin(x^3)$.

Solution We could start calculating derivatives, but instead we could just take the series for $\sin t$, which is

$$t - \frac{t^3}{3!} + \frac{t^5}{5!} - \frac{t^7}{7!} + \frac{t^9}{9!} - \ldots = \sum_{n=0}^{\infty} (-1)^n \frac{t^{2n+1}}{(2n+1)!},$$

and substitute into it $t = x^3$. Doing this gives us the Maclaurin series for $f(x) =$

$\sin(x^3)$; it is

$$x^3 - \frac{x^9}{3!} + \frac{x^{15}}{5!} - \frac{x^{21}}{7!} + \frac{x^{27}}{9!} - \dots = \sum_{n=0}^{\infty} (-1)^n \frac{x^{6n+3}}{(2n+1)!}.$$

(ii) Find the Maclaurin series for the function $g(x) = \ln(1 + x)$.

Solution In Example 2(v) we found that the Taylor series for $\ln(t)$ about $a = 1$ is

$$\sum_{k=1}^{\infty} (-1)^{k+1} \frac{(t-1)^k}{k}.$$

To find the series for $g(x) = \ln(1 + x)$, we try substituting $t = 1 + x$ in this series. We find that $t - 1 = (1 + x) - 1 = x$, and the series equals $\sum_{k=1}^{\infty} (-1)^{k+1} \frac{x^k}{k} = x - \frac{x^2}{2} + \frac{x^3}{3} - \frac{x^4}{4} + \dots$.

With a programmable calculator we can find partial sums and approximate the series for arctan x from part (iii) of Example 3.

TI-85	SHARP EL9300
PRGM F2(EDIT) ARCT ENTER F3 F1(Input) X ENTER F1(Input) M EXIT ENTER (−) 1 STO N ENTER 0 STO S ENTER Lbl B ENTER N + 1 STO N ENTER S + ((−) 1) ^ N × X ^ (2 N + 1) ÷ (2 N + 1) STO S ENTER F4 F1(If) (N TEST F2(<) M) ENTER Goto B ENTER S ENTER EXIT EXIT EXIT PRGM F1(NAME) ARCT ENTER 1 ENTER 20 ENTER	✎ ▼ ▼ (NEW) ENTER ENTER ARCT ENTER COMMAND A 3 (Input) X ENTER COMMAND A 3 (Input) M ENTER N = − 1 ENTER S = 0 ENTER COMMAND B 1 (Label) 1 ENTER N = N + 1 ENTER S = S + ((−) 1) a^b N × X a^b (2 N + 1) ÷ (2 N + 1) ENTER COMMAND 3 (If) N COMMAND C 2 (<) M COMMAND B 2 (Goto) 1 ENTER COMMAND A 1 (Print) S ENTER COMMAND A 6 (End) ENTER ⊞ ✎ (RUN) ENTER (ARCT) ENTER 1 ENTER 20 ENTER

CASIO f_x-6300G (f_x-7700GB)	HP 48SX
MODE 2 EXE ? → X : ? → M : − 1 → N : 0 → S : LbL 1 : N + 1 → N : S + ((−) 1) x^y N × X x^y (2 N + 1) ÷ (2 N + 1) → S : N < M ⇒ Goto 1 : S ◢ MODE 1 Prog 0 EXE 1 EXE 20 EXE (on the f_x-7700GB, : is PRGM F6, and Prog is PRGM F3.)	ATTN ↱ CLR ≪≫ "" KEY IN X AND M ▶ SPC PROMPT SPC ' M ▶ STO SPC ' X ▶ STO SPC 0 SPC 0 SPC M SPC FOR SPC N SPC 1 +/− SPC N y^x X SPC 2 SPC N × 1 + y^x × 2 SPC N × 1 + ÷ + NEXT ▼ ENTER ' ARCT STO VAR ARCT 1 ENTER 20 ENTER ↰ CONT

?1
?20

ANSWER=
.7972961956

The value we get should be close to arctan $1 = \frac{\pi}{4} \dots$

To apply the Ratio Test to this series, we consider the ratio

$$\frac{(-1)^{k+1}\frac{x^k}{k}}{(-1)^k\frac{x^{k-1}}{k-1}} = (-1)(x)\frac{k-1}{k},$$

whose absolute value is $|x|\frac{k-1}{k}$, which tends to $|x|$ as $k \to \infty$. The Ratio Test tells us that the series converges absolutely for $|x| < 1$ and diverges for $|x| > 1$. For $x = 1$ it is an alternating series with terms whose absolute values decrease to 0, so it converges conditionally. If $x = -1$, it is the negative of the harmonic series, so it diverges.

The series converges for $x \in (-1, 1]$ and diverges elsewhere. ◢

Convergence of Taylor Series to the Original Function

Now that we have learned how to find Taylor series and to test them for convergence, we should investigate the question of when they converge to the original function.

Let us begin with the Maclaurin series for $f(x) = e^x$, which is $1 + x + \frac{x^2}{2!} + \frac{x^3}{3!} + \ldots = \sum_{k=0}^{\infty}\frac{x^k}{k!}$. The partial sums of this series are just the Taylor polynomials $P_n(x) = \sum_{k=0}^{n}\frac{x^k}{k!}$.

To check whether or not the Taylor series converges to $f(x) = e^x$, we have to consider the difference $f(x) - P_n(x)$ and show that it tends to 0 as $n \to \infty$.

But this difference is exactly what we called the **remainder** $R_n(x) = f(x) - P_n(x)$ in Equation 9.2.1. In Theorem 9.2.2 we found a formula for the remainder:

$$R_n(x) = \frac{f^{(n+1)}(c)}{(n+1)!}(x-a)^{n+1},$$

for some number c between a and x.

In the case we are considering, $a = 0$ and $f(x) = e^x$, so the derivatives of $f(x)$ are all equal to e^x. Also the function e^x is an increasing function. If $x \geq 0$, then $0 \leq c \leq x$ and $e^c \leq e^x$. If, on the other hand, $x \leq 0$, then $x \leq c \leq 0$ and $e^c \leq e^0 = 1$. In either case, e^c can never be greater than the maximum of e^x and 1.

If we write M for that maximum, we see that

$$|R_n(x)| = \left|\frac{f^{(n+1)}(c)}{(n+1)!}x^{n+1}\right| = \left|\frac{e^c}{(n+1)!}x^{n+1}\right| \leq \frac{M}{(n+1)!}|x|^{n+1} = M\frac{|x|^{n+1}}{(n+1)!}.$$

As $n \to \infty$, the last expression tends to 0 (cf. Example 10.1.6). This means that $\lim_{n\to\infty} R_n(x) = 0$, and since $R_n(x) = f(x) - P_n(x)$, this means that $\lim_{n\to\infty} P_n(x) = f(x)$. Since $P_n(x)$ is the nth partial sum of the Taylor series of $f(x)$ about $a = 0$, we see that the series converges to $f(x)$, for each value of x.

This same approach can be used to show that many Taylor series converge to the appropriate function (for the values for which they converge).

EXAMPLE 5

(i) We have found that the Maclaurin series for $f(x) = \cos x$ is

$$\sum_{n=0}^{\infty}(-1)^n\frac{x^{2n}}{(2n)!}$$

and that it converges for every x. If we let $R_n(x) = f(x) - P_n(x)$ be the remainder for the Taylor polynomial of order n about $a = 0$, then we know that

$$R_n(x) = \frac{f^{(n+1)}(c)}{(n+1)!}(x-a)^{n+1},$$

for some c between $a = 0$ and x.

Now every derivative of $f(x) = \cos x$ is $\pm \sin x$ or $\pm \cos x$, and the absolute value of any one of them is always less than or equal to 1. Because of this,

$$|R_n(x)| = \left| \frac{f^{(n+1)}(c)}{(n+1)!}(x-a)^{n+1} \right| \le \left| \frac{1}{(n+1)!}x^{n+1} \right|$$

$$= \left| \frac{x^{n+1}}{(n+1)!} \right|.$$

We know that this last expression tends to zero as $n \to \infty$, so $\lim_{n\to\infty} |R_n(x)| = 0$. Since $R_n(x) = f(x) - P_n(x)$, this means that $\lim_{n\to\infty} P_n(x) = f(x)$. Since $P_n(x)$ is the nth partial sum of the Taylor series for $f(x)$ about $a = 0$, we see that the Taylor series converges to $f(x)$ for every x.

(ii) In the case of $F(x) = \frac{1}{1-x}$, whose Maclaurin series is $1 + x + x^2 + x^3 + \ldots = \sum_{n=0}^{\infty} x^n$, we saw that the series converges to $F(x)$ because it is a geometric series, and because we knew about geometric series, it was possible to show this convergence without using the formula for the remainder.   ◢

EXAMPLE 6   A Taylor Series That Converges to the Wrong Function   In the examples so far the emphasis has been on finding a Taylor series, after which we have checked that it converges to the "right" value, which it always has. This is frequently what happens, but it is partly because we have chosen "nice" functions. It is possible to find Taylor series that converge, but to the wrong values.

For instance, let

$$f(x) = \begin{cases} e^{-1/x^2}, & \text{if } x \ne 0, \\ 0, & \text{if } x = 0. \end{cases}$$

Then for $x \ne 0$,

$$f'(x) = e^{-1/x^2}\left(\frac{2}{x^3}\right),$$

$$f''(x) = e^{-1/x^2}\left(\frac{2}{x^3}\right)^2 - e^{-1/x^2}\left(\frac{6}{x^4}\right)$$

$$= e^{-1/x^2}\left(\left(\frac{2}{x^3}\right)^2 - \left(\frac{6}{x^4}\right)\right) = e^{-1/x^2}\left(\frac{4 - 6x^2}{x^6}\right).$$

If we continue taking higher derivatives $f^{(n)}(x)$, each of them will be e^{-1/x^2} times a rational function whose denominator is a power of x. By using l'Hôpital's Rule or Theorem 7.2.5 (p. 518) it is easy to see that $\lim_{x\to 0} e^{-1/x^2} x^{-k} = 0$, for any k, and we conclude that $\lim_{x\to 0} f^{(n)}(x) = 0$, for every n.

It takes some work to show that $f^{(n)}(x)$ exists at $x = 0$, for every n, but it does, and the argument just given suggests that $f^{(n)}(0) = 0$, for every $n \geq 0$. The Maclaurin series for $f(x)$ has every coefficient equal to 0, that is, it is $\sum_{n=0}^{\infty} 0(x^n)$. It converges for every x, and it converges to 0.

This gives an example of a function whose Maclaurin series converges for every x to the value 0, even though the function is nonzero for every x except $x = 0$.

Exercises 10.8

I

Find the Taylor series for each of the following functions about the specified point $x = a$.

1. $f(x) = e^{-2x}$ about $a = 0$

2. $f(x) = e^{1-x}$ about $a = 0$

3. $f(x) = \frac{1}{x}$ about $a = 1$

4. $f(x) = \frac{1}{-x}$ about $a = -1$

5. $f(x) = x^3 - 2x^2 - x + 5$ about $a = 0$

6. $f(x) = 2x^3 - 4x^2 + x - 3$ about $a = 2$

7. $f(x) = \cos x$ about $a = \frac{\pi}{2}$

8. $f(x) = \ln(x)$ about $a = 3$

9. $f(x) = \frac{1}{1+2x^2}$ about $a = 0$

10. $f(x) = xe^x$ about $a = 0$

Find the Taylor series for each of the following functions about the specified point $x = a$.

11. $f(x) = 2 - e^{3x}$ about $a = 1$

12. $f(x) = \ln(1 + x^2)$ about $a = 0$

13. $f(x) = \frac{1}{x-2}$ about $a = 0$

14. $f(x) = \arctan(x^2)$ about $a = 0$

15. $f(x) = x^2 + \cos x$ about $a = 0$

16. $f(x) = \cosh x$ about $a = 0$

17. $f(x) = e^{-2x}$ about $a = \frac{1}{2}$

18. $f(x) = \sqrt{x}$ about $a = 1$

19. $f(x) = \cos x + \sin x$ about $a = 0$

20. $f(x) = \cos^2 x + \sin^2 x$ about $a = \frac{\pi}{3}$

II

21. Find the Taylor series of $f(x) = \sqrt{1 + x}$ about the point $a = 0$.

22. (i) Find the Taylor series of the function $f(x) = \sin x$ about the point $a = \frac{\pi}{4}$.
   (ii) For what values of x does this series converge?

23. Find the first six derivatives of $\arctan x$, and evaluate them at $x = 0$. Use these values to find the first seven terms of the Maclaurin series for $\arctan x$, and confirm that they agree with what we found by integrating in Example 3(iii) (p. 699).

24. Find the first four derivatives of $\sin(x^3)$, and evaluate them at $x = 0$. Use these values to find the first five terms of the Maclaurin series for $\sin(x^3)$, and confirm that they agree with what we found in Example 4(i) (p. 699).

25. By differentiating the Taylor series about $a = 0$ for $\ln(1 + x^2)$, find the Taylor series about $a = 0$ for $\frac{x}{1+x^2}$. Confirm that it has the right value when $x = 0$.

26. Find the Maclaurin series for $\ln(1 + x)$ by integrating the geometric series for $\frac{1}{1+x}$.

27. Use the formula for the remainder to show that for every x the Maclaurin series for $f(x) = \sin x$ converges to $f(x)$.

28. Use the formula for the remainder to show that for every x the Taylor series for $f(x) = \cos x$ about the point $a = \frac{\pi}{4}$ converges to $f(x)$.

*29. Suppose $f(x)$ is an even function that is defined for every real number x, that is, $f(-x) = f(x)$. Show that the Maclaurin series for $f(x)$ does not contain any odd powers of x.

*30. Suppose $f(x)$ is an odd function that is defined for every real number x, that is, $f(-x) = -f(x)$. Show that the Maclaurin series for $f(x)$ does not contain any even powers of x.

*31. Use the formula for the remainder to show that the Taylor series for $f(x) = \ln x$ about the point $a = 1$ converges to $f(x)$ for every $x \in [1, 2]$.

*32. Let $p(x)$ be a polynomial, and show that the Maclaurin series for $p(x)$ is exactly equal to $p(x)$.

For each of the following functions $f(x)$, use a programmable calculator or computer to evaluate the partial sum of the corresponding Taylor series about the specified point $x = a$, up to and including the term with $(x - a)^N$, evaluated at the given x.

33. $f(x) = e^x$, $N = 22$, $a = 0$, $x = 3$

34. $f(x) = e^x$, $N = 18$, $a = 1$, $x = -1$

35. $f(x) = \sin(2x)$, $N = 25$, $a = 0$, $x = \pi$

36. $f(x) = \cos(x^2)$, $N = 20$, $a = 0$, $x = \pi$

37. $f(x) = \ln(1 + x^2)$, $N = 300$, $a = 0$, $x = \frac{1}{2}$

38. $f(x) = \sinh(x)$, $N = 39$, $a = 0$, $x = 2$

39. $f(x) = \int_0^x e^{-t^3}\, dt$, $N = 16$, $a = 0$, $x = 3$

40. $f(x) = \int_0^{x^2} e^{-t^2}\, dt$, $N = 22$, $a = 0$, $x = 1$

POINT TO PONDER

It is interesting to consider l'Hôpital's Rule in light of Taylor series. For instance, consider $\lim_{x \to 0} \frac{\sin x}{x}$. The numerator and denominator both tend to 0, so we can evaluate the limit using l'Hôpital's Rule.

If we write out the Maclaurin series for $\sin x$, which is

$$x - \frac{x^3}{3!} + \frac{x^5}{5!} - \frac{x^7}{7!} + \frac{x^9}{9!} - \cdots,$$

and divide by x, we obtain the series for $\frac{\sin x}{x}$,

$$1 - \frac{x^2}{3!} + \frac{x^4}{5!} - \frac{x^6}{7!} + \frac{x^8}{9!} - \cdots.$$

If we evaluate this series at $x = 0$, we find that its value is 1, which is the limit we want.

For another example, consider $\lim_{x \to 0} \frac{x \cos x}{1 - e^x}$. Both numerator and denominator tend to 0. The series for $x \cos x$ can be obtained by multiplying the series for $\cos x$ by x; the result is $x - \frac{x^3}{2!} + \frac{x^5}{4!} - \frac{x^7}{6!} + \frac{x^9}{8!} - \cdots$. The series for $1 - e^x$ is $-x - \frac{x^2}{2!} - \frac{x^3}{3!} - \frac{x^4}{4!} - \frac{x^5}{5!} - \frac{x^6}{6!} - \cdots$.

Each of these series begins with x (i.e., not with a constant term). This reflects the fact that both functions are zero at $x = 0$. We can "factor out" x from each of them: The series for $x \cos x$ is $x\left(1 - \frac{x^2}{2!} + \frac{x^4}{4!} - \frac{x^6}{6!} + \frac{x^8}{8!} - \cdots\right)$. The series for $1 - e^x$ is $x\left(-1 - \frac{x}{2!} - \frac{x^2}{3!} - \frac{x^3}{4!} - \frac{x^4}{5!} - \frac{x^5}{6!} - \cdots\right)$.

The quotient is

$$\frac{x\left(1 - \frac{x^2}{2!} + \frac{x^4}{4!} - \frac{x^6}{6!} + \frac{x^8}{8!} - \cdots\right)}{x\left(-1 - \frac{x}{2!} - \frac{x^2}{3!} - \frac{x^3}{4!} - \frac{x^4}{5!} - \frac{x^5}{6!} - \cdots\right)}$$

$$= \frac{1 - \frac{x^2}{2!} + \frac{x^4}{4!} - \frac{x^6}{6!} + \frac{x^8}{8!} - \cdots}{-1 - \frac{x}{2!} - \frac{x^2}{3!} - \frac{x^3}{4!} - \frac{x^4}{5!} - \frac{x^5}{6!} - \cdots}.$$

As $x \to 0$, the numerator tends to 1 and the denominator tends to -1, so the quotient tends to -1.

In effect we are using l'Hôpital's Rule, because we are really just calculating the quotient of the coefficients of x in the two series, which are equal to the derivatives of the numerator and denominator at $x = 0$.

The argument we have given is not fully justified, because we have not, for instance, defined dividing series or taking limits, but it can be made into a precise argument, and the underlying idea is correct.

10.9 Power Series

RADIUS OF CONVERGENCE

INTERVAL OF CONVERGENCE

CONVERGENCE OF FAMILIAR TAYLOR SERIES

In Section 10.8 we studied Taylor series, which are series of the form

$$\sum_{n=0}^{\infty} a_n (x - a)^n,$$

in which each term contains a *power* of $x - a$. The series obviously depends on the value of x, and it defines a function of x for those x's for which it converges.

In this section we study some properties of this type of series, and in particular we describe, in a fairly simple way, the set of x's for which the series converges.

In the next section we will continue our investigation of these series and consider the possibility of differentiating or integrating them by the simple technique of differentiating or integrating each term separately.

DEFINITION 10.9.1

A series of the form $\sum_{n=0}^{\infty} a_n(x-a)^n$ is called a **power series**.

EXAMPLE 1

(i) The obvious examples of power series are Taylor series. We know a number of examples of series of this type. They include the Maclaurin series for e^x, which is $\sum_{n=0}^{\infty} \frac{x^n}{n!}$. This series converges for every possible x. We also know the series for $\sin x$, which is $\sum_{n=0}^{\infty} (-1)^n \frac{x^{2n+1}}{(2n+1)!}$, and the series for $\cos x$, which is $\sum_{n=0}^{\infty} (-1)^n \frac{x^{2n}}{(2n)!}$. Both these series also converge for every x.

Another familiar series is the series for $\frac{1}{1-x}$, which is the geometric series $\sum_{n=0}^{\infty} x^n$; it converges absolutely when $|x| < 1$ and diverges if $|x| \geq 1$. We also found that the series for $\ln x$ about the point $a = 1$ is $\sum_{n=1}^{\infty} \frac{(-1)^{n+1}}{n}(x-1)^n$, which converges absolutely when $|x-1| < 1$, converges conditionally when $x = 2$, and diverges when $x \leq 0$ or when $x > 2$. Notice that the set of x's for which the series converges is the interval $(0, 2]$.

(ii) It is also possible to "make up" power series without obtaining them as Taylor series. For instance, $\sum_{n=0}^{\infty} 2^{-n}(3n^2 - 2n + 1)(x-3)^n$ is a power series. Using the Ratio Test it is easy to see that it converges absolutely for $|x-3| < 2$ and diverges if $|x-3| > 2$. This means that the set of x's for which it converges must be some sort of interval whose endpoints are 1 and 5; we do not know whether either of the endpoints is included in the interval. In fact, if we substitute $x = 1$ into the series, it becomes $\sum_{n=0}^{\infty} 2^{-n}(3n^2 - 2n + 1)(1-3)^n = \sum_{n=0}^{\infty} 2^{-n}(3n^2 - 2n + 1)(-2)^n = \sum_{n=0}^{\infty} (3n^2 - 2n + 1)(-1)^n$. The terms of this series do not approach 0, so the series diverges. Almost exactly the same thing happens with $x = 5$, and we see that the series converges for $x \in (1, 5)$ and diverges elsewhere.

(iii) Consider the series $\sum_{n=0}^{\infty} n!(x+4)^n$. To apply the Ratio Test, we consider the ratio

$$\frac{n!(x+4)^n}{(n-1)!(x+4)^{n-1}} = n(x+4).$$

If $x + 4 > 0$, this ratio tends to ∞; if $x + 4 < 0$, the ratio tends to $-\infty$; if $x + 4 = 0$, the ratio tends to 0. We see by the Ratio Test that the series diverges unless $x = -4$.

These examples suggest that the Ratio Test is likely to be particularly useful for assessing where a power series converges. Suppose we consider the power series $\sum_{n=0}^{\infty} a_n x^n$. To use the Ratio Test, we have to take the ratio $\left|\frac{a_n x^n}{a_{n-1} x^{n-1}}\right| = \left|\frac{a_n}{a_{n-1}} x\right|$ and find its limit as $n \to \infty$.

Suppose the quotient $\left|\frac{a_n}{a_{n-1}}\right|$ of the absolute values of the nth coefficient and the preceding one tends to some limit ρ, that is,

$$\lim_{n\to\infty}\left|\frac{a_n}{a_{n-1}}\right|=\rho.$$

In this case the ratio we considered for the Ratio Test will tend to the limit

$$\lim_{n\to\infty}\left|\frac{a_n}{a_{n-1}}x\right|=|x|\lim_{n\to\infty}\left|\frac{a_n}{a_{n-1}}\right|=|x|\rho.$$

The Ratio Test tells us that the series converges absolutely if $|x|\rho < 1$ and that it diverges if $|x|\rho > 1$.

If $\rho = 0$, this means that the series will converge for every x. If $\rho = \infty$, it means that the series will only converge when $x = 0$. If ρ is a nonzero number, then the series will converge absolutely when $|x| < \frac{1}{\rho}$ and will diverge when $|x| > \frac{1}{\rho}$.

If we consider a power series $\sum_{n=0}^{\infty} a_n(x-a)^n$ about some point a (instead of just $a = 0$), then an identical argument shows that, provided $\lim_{n\to\infty}\left|\frac{a_n}{a_{n-1}}\right| = \rho$ exists, the series will converge absolutely when $|x-a|\rho < 1$ and diverge when $|x-a|\rho > 1$.

This suggests the following theorem.

THEOREM 10.9.2

> Consider the power series $\sum_{n=0}^{\infty} a_n(x-a)^n$. Then there are three possibilities: Either
>
> (i)   the series converges for every x,
>
> (ii) or the series diverges except for $x = a$,
>
> (iii) or there is a positive number R so that the series converges absolutely when $|x-a| < R$ and diverges when $|x-a| > R$.

The number R is called the **radius of convergence** of the series; in case (i) we say that $R = \infty$, and in case (ii) we say that $R = 0$.

In case (iii) we see that the series converges absolutely when $x \in (a-R, a+R)$ and diverges when $x > a + R$ and when $x < a - R$. It is less clear what it will do for $x = a \pm R$. It may converge or diverge at either of them.

In any event the set of x's at which the series converges is one of the following intervals: $(a-R, a+R)$, $(a-R, a+R]$, $[a-R, a+R)$, or $[a-R, a+R]$. Whichever one it is, this interval is called the **interval of convergence**. In case (i) the interval of convergence is the whole line $(-\infty, \infty)$; in case (ii) the interval of convergence is $[a, a]$, which contains only one point.

Finally, if $\lim_{n\to\infty}\left|\frac{a_n}{a_{n-1}}\right| = \rho$ exists, then it equals the reciprocal of the radius of convergence: $\rho = \frac{1}{R}$. (This is true in a sense even in the case in which $\rho = \infty$, in which $R = 0$, and in the case in which $\rho = 0$, where $R = \infty$.)

In the discussion preceding the statement of the theorem, we have proved it under the assumption that the limit $\lim_{n\to\infty}\left|\frac{a_n}{a_{n-1}}\right|$ exists. Without this assumption the theorem is more difficult to prove.

EXAMPLE 2

See the Calculator Example on the next page.

(i) For the series $\sum_{n=0}^{\infty} \frac{x^n}{n!}$, $\sum_{n=0}^{\infty} (-1)^n \frac{x^{2n}}{(2n)!}$, and $\sum_{n=0}^{\infty} (-1)^n \frac{x^{2n+1}}{(2n+1)!}$, which are the Maclaurin series for e^x, $\cos x$, and $\sin x$, respectively, the radius of convergence is $R = \infty$.

If $a_n = \frac{1}{n!}$ (i.e., the coefficients of the series for e^x), then $\frac{a_n}{a_{n-1}} = \frac{1}{n} \to 0$ as $n \to \infty$. In this case, $\rho = 0$ and $R = \infty$.

For the series for $\cos x$ the odd-numbered coefficients are zero, so the ratio of consecutive coefficients will not make sense; for the series of $\sin x$ the same problem arises because the even-numbered coefficients are zero. (However, in each of these cases the quotient of consecutive *nonzero* coefficients makes sense, and this ratio tends to 0 in each case.)

(ii) Consider the series $\sum_{n=1}^{\infty} (-1)^n \frac{(x-1)^n}{n}$, which is the Taylor series about $a = 1$ of $f(x) = \ln x$. The absolute value of the ratio of consecutive coefficients is

$$\left| \frac{\frac{(-1)^n}{n}}{\frac{(-1)^{n-1}}{n-1}} \right| = \left| (-1) \frac{n-1}{n} \right| = \frac{n-1}{n},$$

which tends to 1 as $n \to \infty$. From the remark at the end of Theorem 10.9.2 we see that the radius of convergence is $R = 1$. From our study of this series earlier we know that the interval of convergence is $(0, 2]$.

(iii) For the series $\sum_{n=1}^{\infty} \frac{(-5)^n}{2n^3+1}(x+7)^{3n+1}$ we observe that the only nonzero coefficients are those corresponding to $(x+7)^4$, $(x+7)^7$, $(x+7)^{10}, \ldots$. The ratio of the coefficients of $(x+7)^m$ and $(x+7)^{m-1}$ is not a useful quantity to consider because at least one of these coefficients is always zero (so the ratio is either 0 or undefined).

However, suppose we rewrite the series as

$$\sum_{n=1}^{\infty} \frac{(-5)^n}{2n^3+1}(x+7)^{3n+1} = (x+7) \sum_{n=1}^{\infty} \frac{(-5)^n}{2n^3+1}(x+7)^{3n}$$

$$= (x+7) \sum_{n=1}^{\infty} \frac{(-5)^n}{2n^3+1}\left((x+7)^3\right)^n$$

$$= (x+7) \sum_{n=1}^{\infty} \frac{(-5)^n}{2n^3+1}u^n,$$

where $u = (x+7)^3$.

We could also consider the ratio of consecutive non-zero coefficients in the original series.

If we look now at this power series in the variable u, we see that it does make sense to consider the absolute value of the ratio of consecutive coefficients. It is

$$\left| \frac{\frac{(-5)^n}{2n^3+1}}{\frac{(-5)^{n-1}}{2(n-1)^3+1}} \right| = \left| (-5) \frac{2n^3 - 6n^2 + 6n - 1}{2n^3 + 1} \right| = 5 \left| \frac{2n^3 - 6n^2 + 6n - 1}{2n^3 + 1} \right|,$$

which tends to 5 as $n \to \infty$.

So the power series in u has radius of convergence equal to $\frac{1}{5}$; this means that it converges absolutely when $|u| < \frac{1}{5}$ and diverges when $|u| > \frac{1}{5}$. Since $u = (x+7)^3$, we see that the series converges absolutely when $|x+7|^3 < \frac{1}{5}$, which is the same thing as saying $|x + 7| < \sqrt[3]{\frac{1}{5}}$. By the same reasoning it diverges when $|x + 7| > \sqrt[3]{\frac{1}{5}} = 5^{-1/3}$.

We see that the radius of convergence of the original series in powers of $x + 7$ is $\sqrt[3]{\frac{1}{5}} = 5^{-1/3} \approx 0.5848$. It converges absolutely on the interval $(-7 - 5^{-1/3}, -7 + 5^{-1/3})$.

With a programmable calculator we can find partial sums for the Maclaurin series for e^x from part (i) of Example 2. We know that for every x the series converges to e^x. It is interesting to see how fast the partial sums converge for various values of x.

In programming with power series it is often more efficient to store the value of the nth term (our variable T) and use it to calculate the next term.

TI-85
PRGM F2(EDIT) EXPML ENTER F3 F1(Input) X ENTER F1(Input) M EXIT ENTER 0 STO N ENTER 1 STO T ENTER 1 STO S ENTER Lbl B ENTER N + 1 STO N ENTER T × X ÷ N STO T ENTER S + T STO S ENTER F4 F1(If) (N TEST F2(<) M) ENTER Goto B ENTER S ENTER EXIT EXIT EXIT PRGM F1(NAME) EXPML ENTER 1 ENTER 10 ENTER

SHARP EL9300
▼ ▼ (NEW) ENTER ENTER EXPML ENTER COMMAND A 3 (Input) X ENTER COMMAND A 3 (Input) M ENTER N = 0 ENTER T = 1 ENTER S = 1 ENTER COMMAND B 1 (Label) 1 ENTER N = N + 1 ENTER T = T × X ÷ N ENTER S = S + T ENTER COMMAND 3 (If) N COMMAND C 2 (<) M COMMAND B 2 (Goto) 1 ENTER COMMAND A 1 (Print) S ENTER COMMAND A 6 (End) ENTER (RUN) ENTER (EXPML) ENTER 1 ENTER 10 ENTER

CASIO f_x-6300G (f_x-7700GB)
MODE 2 EXE ? → X : ? → M : 0 → N : 1 → T : 1 → S : LbL 1 : N + 1 → N : T × X ÷ N → T : S + T → S : N < M ⇒ Goto 1 : S ◢ MODE 1 Prog 0 EXE 1 EXE 10 EXE
(on the f_x-7700GB, : is PRGM F6, and Prog is PRGM F3.)

HP 48SX
ATTN ↻ CLR ≪≫ "" KEY IN X AND M ▶ SPC PROMPT SPC ' M ▶ STO SPC ' X ▶ STO SPC 1 SPC ' T ▶ STO SPC 1 SPC 1 SPC M SPC FOR SPC N SPC T SPC X × N ÷ ' T ▶ STO SPC T + NEXT ▼ ENTER ' EXPL STO VAR EXPL 1 ENTER 10 ENTER ↰ CONT

?1
?10

ANSWER=
    2.718281801

Check how close the answer is to $e = e^1$. Now try the series with $x = 10$; see how close we get to e^{10} using the partial sum with $M = 10 \ldots$

Summary 10.9.3   **Convergence of Familiar Taylor Series**

Function	Series	Radius of convergence	Interval of convergence
e^x	$\sum_{n=0}^{\infty} \frac{x^n}{n!} = 1 + x + \frac{x^2}{2} + \frac{x^3}{6} + \frac{x^4}{24} + \dots$	∞	$(-\infty, \infty)$
$\cos x$	$\sum_{n=0}^{\infty} (-1)^n \frac{x^{2n}}{(2n)!} = 1 - \frac{x^2}{2} + \frac{x^4}{24} - \frac{x^6}{720} + \dots$	∞	$(-\infty, \infty)$
$\sin x$	$\sum_{n=0}^{\infty} (-1)^n \frac{x^{2n+1}}{(2n+1)!} = x - \frac{x^3}{6} + \frac{x^5}{120} - \frac{x^7}{5040} + \dots$	∞	$(-\infty, \infty)$
$\frac{1}{1-x}$	$\sum_{n=0}^{\infty} x^n = 1 + x + x^2 + x^3 + x^4 + \dots$	1	$(-1, 1)$
$\frac{1}{1+x}$	$\sum_{n=0}^{\infty} (-1)^n x^n = 1 - x + x^2 - x^3 + x^4 - \dots$	1	$(-1, 1)$
$\ln x$	$\sum_{n=1}^{\infty} (-1)^{n+1} \frac{(x-1)^n}{n}$ $= (x-1) - \frac{(x-1)^2}{2} + \frac{(x-1)^3}{3} - \frac{(x-1)^4}{4} + \dots$	1	$(0, 2]$
$\ln(1+x)$	$\sum_{n=1}^{\infty} (-1)^{n+1} \frac{x^n}{n} = x - \frac{x^2}{2} + \frac{x^3}{3} - \frac{x^4}{4} + \dots$	1	$(-1, 1]$
$\arctan x$	$\sum_{n=0}^{\infty} (-1)^n \frac{1}{2n+1} x^{2n+1} = x - \frac{x^3}{3} + \frac{x^5}{5} - \frac{x^7}{7} + \dots$	1	$[-1, 1]$
$\cosh x$	$\sum_{n=0}^{\infty} \frac{x^{2n}}{(2n)!} = 1 + \frac{x^2}{2} + \frac{x^4}{24} + \frac{x^6}{720} + \dots$	∞	$(-\infty, \infty)$
$\sinh x$	$\sum_{n=0}^{\infty} \frac{x^{2n+1}}{(2n+1)!} = x + \frac{x^3}{6} + \frac{x^5}{120} + \frac{x^7}{5040} + \dots$	∞	$(-\infty, \infty)$

Exercises 10.9

I

Find the interval of convergence of each of the following power series.

1. $\sum_{n=1}^{\infty} nx^n$

2. $\sum_{n=1}^{\infty} \frac{1}{n}(x+2)^n$

3. $\sum_{n=0}^{\infty} \frac{1}{n^4+3}(x-3)^{n+1}$

4. $\sum_{n=1}^{\infty} \frac{1}{1+\ln(n)}(x-1)^n$

5. $\sum_{n=0}^{\infty} \frac{n!}{(2n)!}(x-\pi)^n$

6. $\sum_{n=1}^{\infty} \frac{n}{n^4+n^2+2}(2x-1)^n$

7. $\sum_{n=3}^{\infty} \frac{4}{(n-1)!}(\sqrt{2}-x)^n$

8. $\sum_{n=2}^{\infty} \left(1-\frac{1}{n}\right)^n (5x+2)^n$

9. $\sum_{n=2}^{\infty} \ln(4n)(x-12)^{n+3}$

10. $\sum_{n=1}^{\infty} 3^{n+2}2^{-n/2}(2x+2)^n$

11. $\sum_{n=1}^{\infty} n^2 x^{2n+1}$

12. $\sum_{n=1}^{\infty} \frac{2n}{n+1}(2x-3)^{5n}$

13. $\sum_{n=2}^{\infty} 2^{-n}\frac{n}{n^3-1}(x-3)^{n+1}$

14. $\sum_{n=2}^{\infty} e^{-2n}(x+3)^n$

15. $\sum_{n=1}^{\infty} \frac{e^n}{n!}(x+5)^n$

16. $\sum_{n=1}^{\infty} \frac{\sqrt{n}}{n}(2-3x)^n$

17. $\sum_{n=1}^{\infty} \frac{n!}{n^5+n}(x-22)^n$

18. $\sum_{n=1}^{\infty} e^{n^2} x^n$

19. $\sum_{n=2}^{\infty} 2^{-3n}\frac{1}{n\ln(n)}(3x)^n$

20. $\sum_{n=1}^{\infty} 5^{2n+1}x^2(3x-1)^n$

II

Find the radius of convergence of each of the following.

21. $\sum_{n=1}^{\infty} 2^{-n^2} x^n$ **23.** $\sum_{n=1}^{\infty} \frac{1}{n}x^{n^2}$

22. $\sum_{n=1}^{\infty} 2^{-n^2}(x-4)^n$

*24. Find the interval of convergence of $\sum_{n=1}^{\infty} 2^n \frac{1}{n} x^{n^2}$.

Find examples of power series whose intervals of convergence are each of the following.

*25. $(3,5)$ ***27.** $[-5,-2)$

*26. $(-1,6]$ ***28.** $[-4,0]$

*29. For the series in Example 2(iii) (p. 707), find the interval of convergence. Specifically, find what happens at the endpoints of the interval.

*30. Suppose $\sum_{n=0}^{\infty} a_n x^n$ is a power series for which $\lim_{n \to \infty} \left| \frac{a_n}{a_{n-1}} \right|$ exists and equals ρ, where ρ could be any nonnegative number *or* ∞. Consider the related series $\sum_{n=0}^{\infty} n a_n x^n$, and show that its radius of convergence is the same as that of the original series.

*31. Suppose $\sum_{n=0}^{\infty} a_n (x + 3)^n$ is a power series for which $\lim_{n \to \infty} \left| \frac{a_n}{a_{n-1}} \right|$ exists and equals A, where A could be any nonnegative number *or* ∞. Consider the related series $\sum_{n=1}^{\infty} \frac{a_n}{n^2} (x + 3)^n$, and show that its radius of convergence is the same as that of the original series.

*32. (i) Find an example of a power series $\sum_{n=1}^{\infty} a_n x^n$ whose interval of convergence is $(-2, 2)$ but so that the interval of convergence of the related series $\sum_{n=1}^{\infty} \frac{a_n}{n^2} x^n$ is $[-2, 2]$.

    (ii) Find an example of a power series $\sum_{n=1}^{\infty} a_n x^n$ whose interval of convergence is $[-4, 4]$ but so that the interval of convergence of the related series $\sum_{n=1}^{\infty} (n^3 + n^2 + 2) a_n x^n$ is $(-4, 4)$.

*33. Suppose the radius of convergence of $\sum_{n=1}^{\infty} a_n (x - a)^n$ is R.

(i) What is the radius of convergence of $\sum_{n=1}^{\infty} 4^n a_n (x - a)^n$?

(ii) If r is any nonzero number, what is the radius of convergence of $\sum_{n=1}^{\infty} r^n a_n (x - a)^n$?

(iii) If $p(x)$ is any polynomial, what is the radius of convergence of $\sum_{n=1}^{\infty} p(n) a_n (x - a)^n$?

*34. (i) If $\sum_{n=0}^{\infty} a_n x^n$ and $\sum_{n=0}^{\infty} b_n x^n$ are power series whose radii of convergence are R and r, respectively, show that the radius of convergence of $\sum_{n=0}^{\infty} (a_n + b_n) x^n$ is at least $\min(R, r)$.

    (ii) Find an example in which the radius of convergence of the "sum" series is exactly equal to the minimum of the radii of convergence of the two original series.

    (iii) Find an example in which the radius of convergence of the "sum" series is greater than the minimum of the radii of convergence of the two original series.

*35. Suppose the series $\sum_{n=0}^{\infty} a_n (x + 2)^n$ diverges with $x = 1$ and converges with $x = -5$. What is its radius of convergence?

POINT TO PONDER

In studying power series it is sometimes necessary to use *complex numbers*. Complex numbers contain the special element i, which satisfies $i^2 = -1$. Every complex number can be written as $z = x + iy$, for some real numbers x and y, and they can be added and multiplied. This allows us to make power series in which the terms are complex numbers. To get a picture, we can associate the complex number $z = x + iy$ to the point (x, y) in the plane; in this way the complex numbers are represented by all the points in the plane. If $y = 0$, then the number $x + iy$ is just the real number x; the complex numbers contain the real numbers, and in the picture of the plane the real numbers correspond to the horizontal axis.

Suppose a is a complex number and a_n is a complex number for each n. Then for any z we can consider the complex power series $\sum_{n=0}^{\infty} a_n (z - a)^n$. This series may or may not converge, but if it does, its limit will be a complex number.

It is not obvious, but it turns out that the set of complex numbers z in the plane for which the series converges is a disk around a. There will be a number R, the *radius of convergence*, so that inside the circle of radius R around a the series converges absolutely and outside the circle it diverges (two examples are shown in Figure 10.9.1). Its behavior on the circle may be complicated.

Now if a and a_n happen to be real numbers, then we can evaluate the series at the real numbers and get an

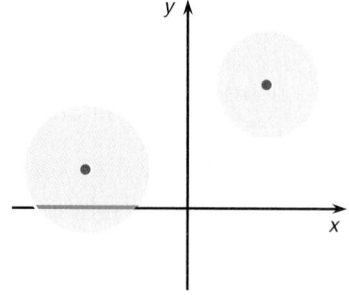

Figure 10.9.1

ordinary real power series. The real numbers at which the series converges will be the points on the real line that also happen to lie inside the disk, as shown in Figure 10.9.2. This set is an interval, and this is why we use the "radius of convergence" terminology for intervals on the line.

Complex numbers will be discussed in greater detail in Chapter 19.

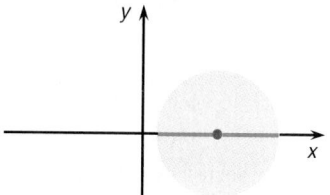

Figure 10.9.2

10.10 Derivatives and Integrals of Power Series

TERM-BY-TERM DIFFERENTIATION AND INTEGRATION

Our study of power series began with Taylor series, and in Section 10.9 we considered power series in their own right. The main result was Theorem 10.9.2, which describes the radius of convergence and the interval of convergence of a power series.

In this section we consider the question of differentiating or integrating a power series. As a preliminary example, consider the Taylor series of the exponential function e^x. It is

$$\sum_{n=0}^{\infty} \frac{x^n}{n!}.$$

Suppose we differentiate each term of this series; the derivative of the nth term is $\frac{d}{dx}\left(\frac{x^n}{n!}\right) = \frac{nx^{n-1}}{n!} = \frac{x^{n-1}}{(n-1)!}$, and we see that this is exactly the $(n-1)$th term. Actually, this is not true for the 0th term 1, whose derivative is 0, but the derivative of each of the other terms equals the preceding term. It is tempting to write

$$\frac{d}{dx}\left(\sum_{n=0}^{\infty} \frac{x^n}{n!}\right) = \sum_{n=0}^{\infty}\left(\frac{d}{dx}\left(\frac{x^n}{n!}\right)\right) = \sum_{n=1}^{\infty} \frac{x^{n-1}}{(n-1)!} = \sum_{m=0}^{\infty} \frac{x^m}{m!}.$$

Notice that after we differentiated, the series began with $n = 1$ (since the derivative of the term with $n = 0$ is zero), but since the nth term then involved the $(n-1)$th power of x, it was simpler to change the index of summation to m, where $m = n - 1$, so m runs from $m = 0$ to ∞.

What makes this calculation interesting is that it gives the right answer. We know from our study of Taylor series that $\sum_{n=0}^{\infty} \frac{x^n}{n!}$ converges for every x to the function e^x. So the derivative of the series ought to be the derivative of e^x, which is e^x, and the series we obtained *is* exactly the Maclaurin series of e^x.

In a similar vein, let us try to find the indefinite integral $\int_0^x \cos t \, dt$ by writing out the Maclaurin series $\sum_{n=0}^{\infty} (-1)^n \frac{t^{2n}}{(2n)!}$ for $\cos t$ and integrating. The calculation appears to be

$$\int_0^x \cos t \, dt = \int_0^x \left(\sum_{n=0}^{\infty} (-1)^n \frac{t^{2n}}{(2n)!}\right) dt = \sum_{n=0}^{\infty} (-1)^n \left(\int_0^x \frac{t^{2n}}{(2n)!} \, dt\right)$$

$$= \sum_{n=0}^{\infty} (-1)^n \left(\frac{1}{2n+1} \frac{t^{2n+1}}{(2n)!}\right)\bigg|_0^x = \sum_{n=0}^{\infty} (-1)^n \frac{x^{2n+1}}{(2n+1)!}.$$

This is the Maclaurin series for sin x, which is exactly what we expect, since

$$\int_0^x \cos t\, dt = \sin t\Big|_0^x = \sin(x) - \sin(0) = \sin x.$$

These examples suggest that it is possible to differentiate and integrate power series by dealing with each term separately. This is called differentiating and integrating *term-by-term*.

Unfortunately, it is not quite as simple as it appears. First there is the usual problem that the series might not converge. Clearly, we should only attempt this approach when the original series converges. Even so, we should then ask whether the series that results from term-by-term differentiation or integration will be convergent.

Now if both the original series and the one that results from the calculation are convergent, there is the question of whether the resulting series converges to the right value, that is, whether it converges to the derivative or the integral we want. We certainly hope it will, but since the series is defined by taking a limit, it is difficult to see what happens when we differentiate or integrate.

Another question arises if we attempt to find an *indefinite integral*. We can find the indefinite integrals of the terms, but we have no way to know what constant to add to each term. If we add peculiar constants to the terms, we can easily make the series divergent.

With all these potential difficulties and complications it is somewhat surprising that it all works.

THEOREM 10.10.1

Suppose $\sum_{n=0}^{\infty} a_n(x-a)^n$ is a power series whose radius of convergence is R, so the series is absolutely convergent on the interval $(a - R, a + R)$ (which means the entire real line if $R = \infty$).

For x in this interval, let f be the function defined by the series

$$f(x) = \sum_{n=0}^{\infty} a_n(x-a)^n.$$

Then

(i) f is differentiable on this interval, and its derivative can be found by differentiating the series **term-by-term**, that is,

$$f'(x) = \sum_{n=1}^{\infty} n a_n(x-a)^{n-1},$$

for all $x \in (a - R, a + R)$.

Moreover, the series for the derivative has the same radius of convergence R as the original series.

(ii) The indefinite integral $\int_a^x f(t)\, dt$ can be found by integrating the series for f **term-by-term**, that is,

$$\int_a^x f(t)\, dt = \sum_{n=0}^{\infty} \frac{1}{n+1} a_n(x-a)^{n+1},$$

for each $x \in (a - R, a + R)$.

Moreover, this series for the indefinite integral has the same radius of convergence R as the original series.

(iii) In particular, if $c, d \in (a - R, a + R)$, then the definite integral can be found by **term-by-term** integration:

$$\int_c^d f(t) \, dt = \sum_{n=0}^{\infty} \left(\frac{a_n}{n+1} \left((d-a)^{n+1} - (c-a)^{n+1} \right) \right).$$

Moreover, this series for the definite integral converges absolutely whenever $c, d \in (a - R, a + R)$.

The proof of this theorem is difficult, and we omit it.

Notice that for the indefinite integral in part (ii) of Theorem 10.1.1 it is extremely important (for once) *not* to include a constant c added on to each term. Doing that might make the series diverge.

Notice too that the differentiated and integrated series have the same radius of convergence as the original series; this means that all three of them converge absolutely on the interval $(a - R, a + R)$ and diverge outside the corresponding closed interval $[a - R, a + R]$. The theorem does not say what happens at the endpoints; as the examples will show, it is possible that the three series may behave differently at the endpoints, but they may also all have the same convergence properties. Also remember that if $R = \infty$, then all three series converge absolutely for every possible value.

EXAMPLE 1

(i) For the series $\sum_{n=0}^{\infty} x^n$, the geometric series that is the Maclaurin series for $f(x) = \frac{1}{1-x}$, we know the radius of convergence is $R = 1$ and the interval of convergence is $(-1, 1)$.

The derivative of this function is $f'(x) = \frac{1}{(1-x)^2} = (1-x)^{-2}$, and Theorem 10.10.1 tells us that for $x \in (-1, 1)$ it equals the series $\sum_{n=0}^{\infty} n x^{n-1} = \sum_{n=1}^{\infty} n x^{n-1}$. Notice the way we started the sum at $n = 1$, since the term with $n = 0$ is zero; to write it starting with $n = 0$ is not wrong, but it is slightly misleading, so it is preferable to make the change.

At the endpoints of the interval of convergence we see that with $x = \pm 1$ the nth term of the series does not tend to 0, so the differentiated series diverges at the endpoints, and its interval of convergence is $(-1, 1)$, the same as the interval of convergence for the original series for f.

(ii) Next we consider the indefinite integral $\int_0^x f(t) \, dt = \int_0^x \frac{1}{1-t} \, dt = -\ln(1 - t)|_0^x = -\ln(1 - x)$. By part (ii) of Theorem 10.10.1, for $x \in (-1, 1)$ it equals $\sum_{n=0}^{\infty} \frac{1}{n+1} x^{n+1} = \sum_{n=1}^{\infty} \frac{1}{n} x^n$. This series converges absolutely for $x \in (-1, 1)$. If $x = 1$, it diverges (it is the harmonic series). If $x = -1$, it is an alternating series whose terms decrease to 0, so it converges. Its interval of convergence is $[-1, 1)$.

(iii) Suppose we integrate again and consider the indefinite integral $\int_0^x \left(-\ln(1 - t) \right) dt = -\int_0^x \ln(1 - t) \, dt$. While it is not terribly difficult, it might take us a moment or two to perform this integration, but writing down the corresponding series is easy. It is $\sum_{n=0}^{\infty} \frac{1}{n+1} \frac{1}{n+2} x^{n+2}$.

Notice that when $x = 1$, this series equals $\sum_{n=0}^{\infty} \frac{1}{(n+1)(n+2)}$, which is convergent. (The nth term is a rational function of n with the degree of the denominator two greater than the degree of the numerator.) The series also converges for $x = -1$, so the interval of convergence is $[-1, 1]$.

(iv) If $f(x) = \cosh x = \frac{1}{2}(e^x + e^{-x})$, then we could calculate its series by finding all its derivatives and evaluating them at $x = 0$, but of course it is much easier to combine the series for e^x and e^{-x}. What we find is

$$\frac{1}{2} \sum_{n=0}^{\infty} \left(\frac{x^n}{n!} + \frac{(-x)^n}{n!} \right) = \frac{1}{2} \sum_{n=0}^{\infty} \left(\frac{1}{n!} + \frac{(-1)^n}{n!} \right) x^n.$$

Since $1 + (-1)^n$ equals 2 when n is even and 0 when n is odd, this series can be written as

$$\frac{1}{2} \sum_{m=0}^{\infty} \frac{2}{(2m)!} x^{2m} = \sum_{m=0}^{\infty} \frac{x^{2m}}{(2m)!}.$$

(Notice the similarity between this series and the series for $\cos x$ ...).

To find the series for the derivative of $\cosh x$, we simply differentiate. We see that the series for $\sinh x$ is

$$\sum_{m=0}^{\infty} \frac{2m x^{2m-1}}{(2m)!} = \sum_{m=1}^{\infty} \frac{x^{2m-1}}{(2m-1)!} = \sum_{n=0}^{\infty} \frac{x^{2n+1}}{(2n+1)!}.$$

Notice the similarity with the series for $\sin x$. Also notice the way we changed the index of summation to run from $m = 1$ to ∞ instead of starting at $m = 0$.

The Ratio Test shows that both the series for $\cosh x$ and the series for $\sinh x$ converge absolutely for every value of x. ◢

In the last part of Example 1 we found the series for $\cosh x$ by combining the Maclaurin series for e^x and e^{-x}. This provides a series that converges for every value of x and that converges to $\cosh x$ for every x (since we know that the series for e^x and e^{-x} converge to these functions). However, we do not really know that what we have found is actually the Maclaurin series of $\cosh x$. On the face of it there is a possibility that there could be two distinct power series that converge to the same function $\cosh x$.

In fact this is not possible, and we can use Theorem 10.10.1 about term-by-term differentiation to see why.

THEOREM 10.10.2

Suppose $f(x)$ is a function. Suppose too that we have a power series $\sum_{n=0}^{\infty} a_n(x-a)^n$ whose radius of convergence is nonzero and that converges to $f(x)$ for each x in some interval about a, that is, for each $x \in (c, d)$, for some $c < a < d$. Then this series is the Taylor series for $f(x)$ about the point $x = a$.

PROOF Theorem 10.10.1 tells us that we can find not only the value of $f(x)$ at $x = a$ from the series, but also the values of all its derivatives $f^{(k)}(a)$. For instance, the derivative of $f(x)$ equals $\sum_{n=1}^{\infty} n a_n(x-a)^{n-1}$, the second derivative equals $\sum_{n=2}^{\infty} n(n-1) a_n(x-a)^{n-2}$, the third derivative equals $\sum_{n=3}^{\infty} n(n-1)$

$(n-2)a_n(x-a)^{n-3}$, and so on. Notice that the series for $f^{(k)}(x)$ starts with the term having $n=k$, since the preceding terms are all zero.

The kth derivative $f^{(k)}(x)$ equals the series

$$\sum_{n=k}^{\infty} n(n-1)\dots(n-k+1)a_n(x-a)^{n-k}.$$

In particular, if we evaluate it at $x=a$, each term except the constant term is zero. The constant term is the term with $n=k$, so $f^{(k)}(a)$ equals the coefficient of this term, that is, $f^{(k)}(a) = k(k-1)\dots(k-k+1)a_k = k(k-1)\dots(1)a_k = k!a_k$.

Turned around, this says that $a_k = \frac{1}{k!}f^{(k)}(a)$, which is exactly the formula for the coefficients of the Taylor series for f. This proves that the series we started with must be the Taylor series, as desired. ◢

This result is extremely useful because it enables us to use various tricks for manipulating power series without having to worry about whether the series that results will in fact be the Taylor series.

EXAMPLE 2

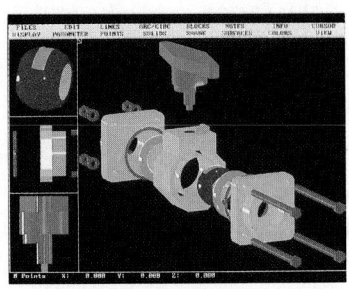

DesignCAD 3D *software was used to produce this exploded picture of valve assembly. The red ball stops fluid flow; the blue component at the top is the control.*

(i) To find the Maclaurin series for $f(x) = e^{-x^2}$, for instance, it is fairly complicated to find a formula for the kth derivative of f, but it is much easier to use the series for e^t and substitute $t = -x^2$. The resulting series is $\sum_{n=0}^{\infty}\frac{1}{n!}t^n = \sum_{n=0}^{\infty}\frac{1}{n!}(-x^2)^n = \sum_{n=0}^{\infty}(-1)^n\frac{1}{n!}x^{2n}$. Since the series for e^t converges to e^t for every value of t, we see that this series converges to e^{-x^2} for every value of x, so by Theorem 10.10.2 it must be the Maclaurin series.

(ii) From our study of geometric series we know that the series $1 - x^2 + x^4 - x^6 + x^8 - \dots = \sum_{n=0}^{\infty}(-1)^n x^{2n}$ converges to $\frac{1}{1+x^2}$ for every $x \in (-1, 1)$. From Theorem 10.10.2 we see that it must be the Maclaurin series of $\frac{1}{1+x^2}$.

By Theorem 10.10.1 we can integrate this series to find the series for arctan x. We find that for $x \in (-1, 1)$,

$$\arctan x = \int_0^x \frac{1}{1+t^2}\,dt = \int_0^x \left(\sum_{n=0}^{\infty}(-1)^n t^{2n}\right)dt$$

$$= \sum_{n=0}^{\infty}\left(\int_0^x (-1)^n t^{2n}\,dt\right) = \sum_{n=0}^{\infty}\left((-1)^n\frac{1}{2n+1}x^{2n+1}\right).$$

Theorem 10.10.1 assures us that for $x \in (-1, 1)$ this series converges to arctan x, and Theorem 10.10.2 then tells us that it must be the Maclaurin series of arctan x.

(iii) Find the tenth and thirteenth derivatives of $f(x) = \arctan x$ at $x = 0$, that is, $f^{(10)}(0)$ and $f^{(13)}(0)$.

Solution Instead of calculating derivatives, we look at the series we just found for arctan x and realize that since it is the Maclaurin series for arctan x, its coefficients are obtained by dividing the derivatives of f by factorials.

Specifically, the coefficient of x^{10} is $\frac{1}{10!}f^{(10)}(0)$, and the coefficient of x^{13} is $\frac{1}{13!}f^{(13)}(0)$. The coefficient of x^{10} is zero, so $f^{(10)}(0) = 10!(0) = 0$. The coefficient of x^{13} (i.e., x^{2n+1}, with $n=6$) is $(-1)^6\frac{1}{13}$. Since this equals $\frac{1}{13!}f^{(13)}(0)$, we see that $f^{(13)}(0) = 13!(-1)^6\frac{1}{13} = 12! = 479{,}001{,}600$. This is easier than calculating directly.

(iv) For another example we consider the function $g(x) = \frac{\sin x}{x}$. We know the Maclaurin series for $\sin x$ is $\sum_{n=0}^{\infty} (-1)^n \frac{1}{(2n+1)!} x^{2n+1}$. To find the series for $g(x)$, the obvious thing to try is just dividing the series for $\sin x$ by x. This suggests the series $\sum_{n=0}^{\infty} (-1)^n \frac{1}{(2n+1)!} x^{2n}$. To be able to use the theorem, we must know that this series converges to $g(x)$ on some interval around $x = 0$.

The difficulty is that we were not very careful when we defined $g(x)$. Specifically, it is not defined when $x = 0$. Of course, we know that $\lim_{x \to 0} g(x) = 1$, so it makes sense to define a slightly different function

$$G(x) = \begin{cases} \frac{\sin x}{x}, & \text{if } x \neq 0, \\ 1, & \text{if } x = 0. \end{cases}$$

The series we found by dividing converges to $G(x)$ when $x = 0$ (since only the first term is nonzero) and also for any nonzero x (since in effect we have divided the series that converges to $\sin x$ by the "constant" x). The series converges to $G(x)$ for every x, so it is the Maclaurin series of $G(x)$.

Strictly speaking, we ought to check that it is possible to differentiate $G(x)$ infinitely often (i.e., so that it has a Maclaurin series in the first place). This is not easy, but in fact it can be derived as a consequence of the fact that there is a power series that converges to $G(x)$ for every point except $x = 0$. In a sense the power series has shown us how to "extend" the function $g(x) = \frac{\sin x}{x}$ (which is not defined at $x = 0$) into the function $G(x)$ (which is defined on the whole real line and can be differentiated infinitely many times).

(v) Evaluate the series $\sum_{n=1}^{\infty} nx^n$.

Solution We recognize that this series is obtained by taking the derivative of the Maclaurin series $\sum_{n=0}^{\infty} x^n$, which is $\sum_{n=1}^{\infty} nx^{n-1}$, and multiplying it by x. Since $\sum_{n=0}^{\infty} x^n$ is the Maclaurin series for $\frac{1}{1-x}$, we see that $\sum_{n=1}^{\infty} nx^n$ converges to $x \frac{d}{dx}\left(\frac{1}{1-x}\right) = x \frac{1}{(1-x)^2} = x(1-x)^{-2}$. The series converges for $x \in (-1, 1)$ to the function $x(1-x)^{-2}$. For instance, if $x = \frac{1}{2}$, we find that $\sum_{n=1}^{\infty} n 2^{-n} = \frac{1}{2}\left(1 - \frac{1}{2}\right)^{-2} = \frac{1}{2}\left(\frac{1}{2}\right)^{-2} = \frac{1}{2}(2^2) = 2$.

We have seen that it is easy to add power series. In fact it's also possible to multiply and divide them. For example, suppose we want to find the series for $\frac{1}{(1+x)(1-2x)}$. One way to proceed would be to multiply the familiar series for $\frac{1}{1-x}$ and $\frac{1}{1-2x}$. We begin by writing down the two series, one above the other:

$$1 + x + x^2 + x^3 + x^4 + x^5 + x^6 + x^7 + x^8 + x^9 + \ldots,$$
$$1 + 2x + (2x)^2 + (2x)^3 + (2x)^4 + (2x)^5 + (2x)^6 + (2x)^7 + (2x)^8 + (2x)^9 + \ldots.$$

Then we multiply, beginning at the left. The idea is to work with one power of x at a time. The zeroth power of x is the constant term, and the only way to get a constant term in the product is to multiply together the constant terms of the two original series.

There are two ways to get terms containing $x = x^1$; we could multiply the constant term of the first series with the x-term of the second series or vice versa. Combining these, we find that the x-term in the product is $(1)(2x) + (x)(1) = 3x$.

For the x^2-term there are three possible combinations: the constant term from the first series times the x^2-term from the second, the product of the x-terms from both series, or the x^2-term from the first series multiplied by the constant term from the second series. Putting them together, we find that the x^2-term in the product series is $(1)(2x)^2 + (x)(2x) + (x^2)(1) = 7x^2$.

Similarly, for the x^3-term we have $(1)(2x)^3 + x(2x)^2 + x^2(2x) + x^3(1) = (8 + 4 + 2 + 1)x^3 = 15x^3$, and for the x^4-term we have $(1)(2x)^4 + x(2x)^3 + x^2(2x)^2 + x^3(2x) + x^4(1) = (16 + 8 + 4 + 2 + 1)x^4 = 31x^4$.

For the x^n-term we have $(1)(2x)^n + x(2x)^{n-1} + x^2(2x)^{n-2} + \ldots + x^{n-1}(2x) + x^n(1) = (2^n + 2^{n-1} + \ldots + 4 + 2 + 1)x^n$. The quantity in parentheses is a geometric sum with common ratio $\frac{1}{2}$, first term 2^n, and last term 1, so it equals $\frac{2^n - \frac{1}{2}}{1 - \frac{1}{2}} = 2^{n+1} - 1$.

This tells us the coefficient of the nth term of the product series. We see that the Maclaurin series for $\frac{1}{(1-x)(1-2x)}$ is

$$\sum_{n=0}^{\infty} (2^{n+1} - 1)x^n.$$

It may help to remember the process by which we multiply as follows. Write down the two series, one above the other. To find the term containing x^n, we must look for all products of one term from the first series with one from the second series whose powers of x add up to n. So for instance, to find the x^5-term, we multiply the $1 = x^0$-term from the first series and the x^5-term from the second, the $x = x^1$-term from the first and the x^4-term from the second, the x^2-term from the first and the x^3-term from the second, the x^3-term from the first and the x^2-term from the second, the x^4-term from the first and the $x = x^1$-term from the second, and finally the x^5-term from the first and the $1 = x^0$-term from the second.

We can illustrate this for the two series we just considered by joining each of these pairs of terms by a line segment, as shown below:

$$1 + x + x^2 + x^3 + x^4 + x^5 + x^6 + x^7 + x^8 + x^9 + \ldots,$$

$$1 + 2x + (2x)^2 + (2x)^3 + (2x)^4 + (2x)^5 + (2x)^6 + (2x)^7 + (2x)^8 + (2x)^9 + \ldots.$$

Of course, it is easy to draw a similar picture for the x^7-term or for any other term in the product.

A related question would be the quotient of two power series. For instance, we might like to find the Maclaurin series for $\tan x$ by writing $\tan x = \frac{\sin x}{\cos x}$ and dividing the series for $\sin x$ and $\cos x$. We could attempt to do this by long division.

$$
\begin{array}{r}
x + \frac{1}{3}x^3 + \frac{2}{15}x^5 + \ldots \\
\left(1 - \frac{x^2}{2} + \frac{x^4}{24} \ldots\right) \overline{\left) x - \frac{x^3}{6} + \frac{x^5}{120} - \ldots \right.} \\
x - \frac{x^3}{2} + \frac{x^5}{24} - \ldots \\
\hline
\frac{1}{3}x^3 - \frac{1}{30}x^5 + \ldots \\
\frac{1}{3}x^3 - \frac{1}{6}x^5 + \ldots \\
\hline
\frac{2}{15}x^5 + \ldots \\
\frac{2}{15}x^5 + \ldots \\
\hline
0 + \ldots
\end{array}
$$

Subtracting the fractions is a little tricky, but otherwise the division is fairly routine. We have found that the Maclaurin series for $\tan x$ begins as $x + \frac{1}{3}x^3 + \frac{2}{15}x^5 + \dots$.

This kind of procedure will work for dividing any series, at least provided the constant term of the denominator is nonzero.

These techniques for multiplying and dividing series are extremely useful. Occasionally, as in the first example, they actually allow us to find a formula for the nth term so that it is possible to write down the entire series. But even when this is too difficult, they provide a technique for finding as many terms of the series as our patience will permit.

EXAMPLE 3 (i) Find the first three nonzero terms of the Maclaurin series for $\cos x \sin x$.

Solution As above, we multiply the familiar series for $\cos x$ and $\sin x$. We begin by writing down the two series, one above the other:

$$1 - \frac{x^2}{2} + \frac{x^4}{24} - \frac{x^6}{720} + \frac{x^8}{40{,}320} - \dots,$$

$$x - \frac{x^3}{6} + \frac{x^5}{120} - \frac{x^7}{5{,}040} + \frac{x^9}{362{,}880} - \dots.$$

Then we multiply, beginning at the left end. Since the second series has no constant term, there can be no constant term in the product. The lowest possible power of x is x itself, and the only way it occurs is by multiplying the 1 from the first series and the x from the second. The product will begin with x. There is no way to multiply a term from the first series and a term from the second to obtain a term containing x^2, but there are two x^3-terms. Adding them up, we get $(1)\frac{-x^3}{6} - \frac{x^2}{2}x = -\left(\frac{1}{6} + \frac{1}{2}\right)x^3 = -\frac{2}{3}x^3$.

There is no x^4-term, but for x^5 we find three: $(1)\frac{x^5}{120} - \frac{x^2}{2}\frac{-x^3}{6} + \frac{x^4}{24}x = \left(\frac{1}{120} + \frac{1}{12} + \frac{1}{24}\right)x^5 = \frac{16}{120}x^5 = \frac{2}{15}x^5$.

In principle it is possible to continue indefinitely in this fashion, but of course it becomes quite tedious. Still, we have found that the series for $\cos x \sin x$ begins with $x - \frac{2}{3}x^3 + \frac{2}{15}x^5 + \dots$.

In fact, in this particular instance there is a better way to do it, which is to notice that $\cos x \sin x = \frac{1}{2}\sin(2x)$, then to take the series for $\sin t$, substitute $t = 2x$, and divide by 2. We find that the Maclaurin series for $\cos x \sin x$ is

$$\frac{1}{2}\sum_{n=0}^{\infty}(-1)^n\frac{2^{2n+1}}{(2n+1)!}x^{2n+1} = \sum_{n=0}^{\infty}(-1)^n\frac{2^{2n}}{(2n+1)!}x^{2n+1},$$

and it is easy to check that the first few terms agree with what we found by multiplying.

(ii) Find the first three nonzero terms in the Maclaurin series for $\sec x$.

Solution Noting that $\sec x = \frac{1}{\cos x}$, we realize that what is needed is to divide the constant function 1 by the function $\cos x$. The Maclaurin series for 1 is just 1 (i.e., a constant term and no other nonzero terms). We perform the division:

$$\left(1 - \frac{x^2}{2} + \frac{x^4}{24} \cdots\right) \overline{\Big)\, 1} \qquad \overset{\displaystyle 1 + \frac{1}{2}x^2 + \frac{5}{24}x^4 + \cdots}{}$$

$$1 - \frac{x^2}{2} + \frac{x^4}{24} - \cdots$$

$$\frac{1}{2}x^2 - \frac{1}{24}x^4 + \cdots$$

$$\frac{1}{2}x^2 - \frac{1}{4}x^4 + \cdots$$

$$\frac{5}{24}x^4 + \cdots$$

$$\frac{5}{24}x^4 + \cdots$$

$$0 \;+ \cdots$$

We see that the series for $\sec x$ begins with $1 + \frac{1}{2}x^2 + \frac{5}{24}x^4 + \cdots$.

(iii) Find $\frac{d^4}{dx^4}(\sec x)\big|_{x=0}$.

Solution Of course, it is possible to do this question by repeated differentiation, but on the other hand we know that the fourth derivative at $x = 0$ equals 4! times the coefficient of x^4 in the Maclaurin series. Having just found that that coefficient is $\frac{5}{24}$, we see that the value at $x = 0$ of the derivative is $4! \times \frac{5}{24} = 5$. ◢

When we found the Taylor series for $\ln(x)$ about $x = 1$, we saw that it is

$$\sum_{n=1}^{\infty} (-1)^{n+1} \frac{1}{n}(x-1)^n.$$

We also realized that the radius of convergence is 1 and that the interval of convergence is $(0, 2]$.

It is very tempting to ask what happens at the endpoint $x = 2$, where the series converges. If we knew that it converges to the "right" value, that is, to $\ln(2)$, we would have the extremely pretty result that $1 - \frac{1}{2} + \frac{1}{3} - \frac{1}{4} + \cdots = \ln(2)$, that is, that the series obtained from the harmonic series by changing the signs of all the even-numbered terms converges to $\ln 2$.

This is in fact true, but it depends on the following theorem, which is very difficult. (This theorem contains the essence of a more technical theorem called Abel's Theorem.)

THEOREM 10.10.3

> Suppose the Taylor series for $f(x)$ about $x = a$ has radius of convergence R. Suppose the series converges to $f(x)$ for all $x \in (a - R, a + R)$. If the series converges at the endpoint $x = a + R$ and if the function f is continuous at $a + R$, then the series converges to $f(a + R)$ when $x = a + R$. The analogous result holds at $x = a - R$.

EXAMPLE 4

(i) As we remarked above, Theorem 10.10.3 implies that

$$\sum_{n=1}^{\infty} (-1)^{n+1} \frac{1}{n} = \ln(2).$$

(ii) In Example 2(ii) we found that the series for $\arctan x$ is $\sum_{n=0}^{\infty} (-1)^n \frac{1}{2n+1} x^{2n+1}$. Its radius of convergence is $R = 1$ (by the Ratio Test), and when $x = 1$, the series is an alternating series with terms whose absolute values decrease to 0. This means that the series converges when $x = 1$, and Theorem 10.10.3 tells us that it converges to the right value, namely, $\arctan 1 = \frac{\pi}{4}$. In symbols,

$$\sum_{n=0}^{\infty} \frac{(-1)^n}{2n+1} = 1 - \frac{1}{3} + \frac{1}{5} - \frac{1}{7} + \frac{1}{9} - \frac{1}{11} + \ldots = \frac{\pi}{4}.$$

This series can be used to estimate the value of π, though unfortunately it takes a large number of terms to achieve much accuracy.

With a programmable calculator we can find partial sums and approximate the series for $\ln(2)$ from part (i) of Example 4.

TI-85	SHARP EL9300
PRGM F2(EDIT) LOG2 ENTER F3 F1(Input) M EXIT ENTER 0 STO N ENTER 0 STO S ENTER Lbl B ENTER N + 1 STO N ENTER S + ((−) 1) ^ (N + 1) ÷ N STO S ENTER F4 F1(If) (N TEST F2(<) M) ENTER Goto B ENTER S ENTER EXIT EXIT EXIT PRGM F1(NAME) LOG2 ENTER 50 ENTER	▼ ▼ (NEW) ENTER ENTER LOG2 ENTER COMMAND A 3 (Input) M ENTER N = 0 ENTER S = 0 ENTER COMMAND B 1 (Label) 1 ENTER N = N + 1 ENTER S = S + ((−) 1) a^b (N + 1) ÷ N ENTER COMMAND 3 (If) N COMMAND C 2 (<) M COMMAND B 2 (Goto) 1 ENTER COMMAND A 1 (Print) S ENTER COMMAND A 6 (End) ENTER (RUN) ENTER (LOG2) ENTER 50 ENTER

?50

ANSWER=
      .6832471606

CASIO f_x-6300G (f_x-7700GB)	HP 48SX
MODE 2 EXE ? → M : 0 → N : 0 → S : Lbl 1 : N + 1 → N : S + ((−) 1) x^y (N + 1) ÷ N → S : N < M ⇒ Goto 1 : S ◢ MODE 1 Prog 0 EXE 50 EXE (on the f_x-7700GB, : is PRGM F6, and Prog is PRGM F3.)	ATTN ↱ CLR ≪≫ "" KEY IN M ▶ SPC PROMPT SPC ' M ▶ STO SPC 0 SPC 1 SPC M SPC FOR SPC N SPC 1 +/− SPC N SPC 1 + y^x N ÷ + NEXT ▼ ENTER ' LOG2 STO VAR LOG2 50 ENTER ↰ CONT

Check how close the answer is to $\ln(2)$. Now try it with different values of M. Then try adapting the program to part (ii) of Example 4.

Exercises 10.10

I

Find the Maclaurin series for each of the following functions.

1. $\int_0^x e^{-t^2}\,dt$

2. $\int_{-x}^x e^{t^4}\,dt$

3. $\frac{d}{dx}\left(\frac{1}{1+x}\right)$

4. $\frac{d}{dx}\left(\frac{1}{1+x^3}\right)$

5. $\ln(2-x)$

6. $\arctan(4x)$

7. $\int_0^x \ln(1+2t)\,dt$

8. $\ln(1-x^2)$

9. $\frac{1}{(1-2x)(1-3x)}$

10. $\sin^2 x$

Find the Maclaurin series for each of the following functions.

11. $x^2 \sin x$

12. $(2x+1)\cos x$

13. $\sin\left(\frac{\pi}{2}-x\right)$

14. $(x^3 - 2x + 3)\ln(1-x)$

15. $\frac{1-\cos x}{x^2}$

16. $\frac{\sin x - x}{x^3}$

17. $\frac{1}{2-x^2}$

18. $\cosh^2 x - \sinh^2 x$

19. $(1-x)^{-4}$

20. $e^{x^2} e^{-x^2}$

Find a formula for the coefficient of x^n in the Maclaurin series of each of the following functions. (For some of them it is likely more convenient to find a formula that is different in different cases, e.g., one formula for odd n and another for even n.)

21. $\frac{1}{1-x} e^x$

22. $\frac{1}{(1-2x)^2}$

23. $\cosh^2 x$

24. $\cos^2 x$

25. $\cos(\pi x)$

26. $\frac{1}{2-(1+x)}$

27. $\frac{x}{x-1}$

28. $\frac{x^2}{2-x}$

29. $\ln\left((1-x)^2\right)$

30. $\sqrt{e^{3x}}$

For each of the following functions, find the first three nonzero terms in its Maclaurin series.

31. $e^x \frac{1}{1-x}$

32. $\cos^3 x$

33. $e^{\sin x}$

34. $e^{\cos x}$

35. $\frac{x}{\sin x}$

36. $\frac{x}{\tan x}$

37. $\tanh x$

38. $x \coth x$

39. $\cot x$

40. $\frac{e^x}{\cos x}$

II

41. Find the indefinite integral in Example 1(iii) (p. 713).

42. Check that the series we found in Example 1(i) by differentiating term-by-term does in fact equal the Maclaurin series of the function $f'(x) = (x-1)^{-2}$, by finding a formula for the derivatives of $f'(x)$ and evaluating them at the point $x = 0$ in the usual way.

*43. Find the Maclaurin series for $\sinh^2 x$. (*Hint:* Square the function *before* you find the series.)

*44. Suppose the power series $\sum_{n=0}^{\infty} a_n x^n$ has radius of convergence $R = 1$. Then the series converges for each $x \in (-1,1)$, and we can use it to define a function on this interval: For each $x \in (-1,1)$, define $f(x) = \sum_{n=0}^{\infty} a_n x^n$. Find an example of such a series that diverges at $x = 1$ but for which $\lim_{x \to 1^-} f(x)$ exists, that is, is a (finite) limit.

Evaluate each of the following series.

*45. $\sum_{n=1}^{\infty} n3^{-n}$

*46. $\sum_{n=0}^{\infty} (n+1)\left(-\frac{1}{2}\right)^n$

*47. $\sum_{n=2}^{\infty} n(n-1)\left(\frac{2}{3}\right)^n$

*48. Evaluate $\sum_{n=1}^{\infty} n^2 x^n$, stating clearly for what values of x the series converges.

*49. (i) Find the ninth derivative of $f(x) = \arctan x$ at $x = 0$, that is, $f^{(9)}(0)$.

   (ii) Find the answer to part (i) in a different way. (If you used series, do it by differentiating, and vice versa.)

50. Find the first five nonzero terms in the Maclaurin series for $\tan x$.

51. Find $\frac{d^6}{dx^6}(\sec x)\Big|_{x=0}$.

*52. Find the Maclaurin series for $\frac{1}{(1-ax)(1-bx)}$.

53. Check the result of Example 3(iii) (p. 719) by differentiating, that is, find $\frac{d^4}{dx^4}(\sec x)\Big|_{x=0}$.

Chapter Summary

§10.1
sequence
term
convergence
divergence
diverges to infinity
bounded
Squeeze Theorem for sequences
bounded shifted sequences
precise definition of limits

§10.2
geometric sum
common ratio
interest
factor
mortgages
annuities
balance
monthly payment
geometric series
convergence
divergence
Zeno's Paradox

§10.3
partial sum
convergence
divergence
nth Term Test
harmonic series

§10.4
increasing sequence
bounded sequence
Integral Test
p-series

§10.5
Comparison Test
Limit Comparison Test
rational function

§10.6
alternating series
absolute convergence
conditional convergence

§10.7
Ratio Test
nth Root Test
testing convergence of series (Summary 10.7.3)

§10.8
Taylor series
Maclaurin series
remainder

§10.9
power series
radius of convergence
interval of convergence
convergence of familiar Taylor series (Summary 10.9.3)

§10.10
term-by-term differentiation
term-by-term integration

Review Exercises

I

Calculate each of the following, provided it exists.

1. $\sum_{n=1}^{\infty} 4(3^{-n})$

2. $\sum_{n=0}^{\infty} (-2)^{-3n}$

3. $\sum_{n=1}^{\infty} \frac{2^{n+1}}{3^n}$

4. $\sum_{n=0}^{\infty} \frac{3^{2n}}{4^n}$

5. The annual payment on a mortgage for $20,000 at 12% compounded annually over a 25-year term.

6. The annual payment on a mortgage for $16,000 at 10% compounded annually over a 20-year term.

7. The balance left owing after 10 years on a $100,000 mortgage with a 20-year term at 12% annual interest.

8. The balance left owing after 15 years on a $45,000 mortgage with a 20-year term at 14% annual interest.

9. The annual payment from an annuity that costs $10,000 and that runs for 15 years with 10% interest.

10. The cost of an annuity that will pay $12,000 per annum for 10 years, assuming interest of 9%.

For each of the following sequences, decide whether or not it converges; if it does, find its limit.

11. $\left\{ \frac{n^2+1}{n^3} \right\}$

12. $\left\{ \sin \frac{1}{n} \right\}$

13. $\left\{ \cos(n\pi) \right\}$

14. $\left\{ \sin(n\pi) \right\}$

15. $\left\{ \frac{\sin(n)}{n} \right\}$

16. $\left\{ \cos^4(n\pi) \right\}$

17. $\left\{ \cos\left(\frac{1}{n}\right) \right\}$

18. $\left\{ \cot\left(\frac{1}{n}\right) \right\}$

19. $\left\{ \sin\left(\frac{\pi}{2} + \frac{1}{n}\right) \right\}$

20. $\left\{ \exp\left(\sin \frac{\pi}{n}\right) \right\}$

Decide whether each of the following series is convergent, and explain your answer.

21. $\sum_{n=1}^{\infty} \frac{3n^4 - n^2 + 4}{10n^3 + 7}$

22. $\sum_{n=1}^{\infty} \frac{2n^3 + 3n^2 + 1}{7n^4 + 2n^3 - 2}$

23. $\sum_{n=2}^{\infty} \frac{n^3 + 5n^2 - 2}{1 - n^{3/2}}$

24. $\sum_{n=0}^{\infty} \frac{\sqrt{n+1}}{n^2 + 4}$

25. $\sum_{n=1}^{\infty} \frac{n^n}{n!}$

26. $\sum_{n=1}^{\infty} (-1)^{n+1} \sin \frac{1}{2n}$

27. $\sum_{n=1}^{\infty} (-1)^n \cos \frac{1}{n^2}$

28. $\sum_{n=1}^{\infty} (-1)^n \left(1 - \cos \frac{1}{n}\right)$

29. $\sum_{n=1}^{\infty} (-1)^n \sqrt{n} \sin \frac{1}{n}$

30. $\sum_{n=1}^{\infty} (-1)^{n+1} \sin\left((-1)^n \frac{1}{n}\right)$

31. $\sum_{n=1}^{\infty} \left(1 - \frac{n-1}{n}\right)$

32. $\sum_{n=1}^{\infty} 3^{n/3} \frac{1}{n!}$

33. $\sum_{n=1}^{\infty} 6^{4n+1} \frac{1}{n!}$

34. $\sum_{n=1}^{\infty} \cos(n^2 - 4) 3^{3-n/5}$

35. $\sum_{n=2}^{\infty} \ln \frac{1}{n}$

36. $\sum_{n=1}^{\infty} \frac{n^3}{n!}$

37. $\sum_{n=2}^{\infty} \ln(n) n^{-4/3}$

38. $\sum_{n=2}^{\infty} \ln(n) 2^{-n/4}$

39. $\sum_{n=1}^{\infty} (-1)^n \frac{1}{1 + \ln(n)}$

40. $\sum_{n=1}^{\infty} n^{3/2} \frac{n!}{(n-3)!}$

For what values of c do each of the following series converge?

41. $\sum_{n=1}^{\infty} c^{-4n}$

42. $\sum_{n=1}^{\infty} nc^{-4n}$

43. $\sum_{n=1}^{\infty} n^{2c}$

44. $\sum_{n=1}^{\infty} \frac{1}{1 + cn^2}$

45. $\sum_{n=1}^{\infty} (\sqrt{n+1})^c$

46. $\sum_{n=1}^{\infty} c^n \frac{n!}{(2n)!}$

47. $\sum_{n=1}^{\infty} (n^3 - 2n + 12)^c$

48. $\sum_{n=1}^{\infty} c^n n^{-n}$

49. $\sum_{n=3}^{\infty} \left(1 - \frac{1}{n}\right)^{cn}$

50. $\sum_{n=1}^{\infty} \frac{n^{1-2c} + 1}{n^c + 2}$

Find the Taylor series of each of the following functions about the specified point.

51. $f(x) = \frac{1}{1 - 3x}$ about $a = 0$

52. $f(x) = \frac{1}{2 - x}$ about $a = 0$

53. $f(x) = \cos x$ about $a = \pi$

54. $f(x) = \frac{1}{x}$ about $a = -1$

55. $f(x) = \ln x$ about $a = 4$

56. $f(x) = \ln(3x)$ about $a = \frac{1}{3}$

57. $f(x) = \ln|x|$ about $a = -1$

58. $f(x) = \ln x^2$ about $a = -2$

59. $f(x) = \arctan 2x$ about $a = 0$

60. $f(x) = x^{1/3}$ about $a = 1$

Find the radius of convergence of each of the following series.

61. $\sum_{n=1}^{\infty} n^3 (2x)^n$

62. $\sum_{n=1}^{\infty} n^{-4} \left(\frac{x}{7}\right)^n$

63. $\sum_{n=1}^{\infty} n! x^{3n}$

64. $\sum_{n=1}^{\infty} n^{n/2} x^n$

65. $\sum_{n=0}^{\infty} (n^2 - 4n + 6)^{1/3} x^{2n}$

66. $\sum_{n=1}^{\infty} \frac{n!}{(3n)!} x^n$

67. $\sum_{n=0}^{\infty} 2^n x^{2n}$

68. $\sum_{n=0}^{\infty} \frac{5^n}{n!} x^n$

69. $\sum_{n=1}^{\infty} \cos\left(\frac{1}{n}\right) x^n$

70. $\sum_{n=0}^{\infty} \sin\left(\frac{1}{n!}\right) x^n$

Find the first three nonzero terms in the Maclaurin series of each of the following functions.

71. $\cos^2 x \sin x$

72. $\cosh x \sinh x$

73. $\frac{1}{1 - \ln(1 - x)}$

74. $(1 - x)^{-3}$

75. $\exp(e^x)$

76. $\frac{x+1}{x-1}$

77. $\cos(e^x)$

78. $\sin(e^x)$

79. $\arctan(\sin x)$

80. $\ln(e^x)$

II

*81. Without actually finding the series, show that the Taylor series of $f(x) = \sin x$ about *any* point $x = a$ converges to $f(x)$ at every x.

*82. (i) Show that the Taylor series of $f(x) = e^x$ about *any* point $x = a$ converges to $f(x)$ at every x.

(ii) Find the Taylor series of $f(x) = e^x$ about the point $x = a$.

83. Suppose the Maclaurin series of $f(x)$ is $\sum_{n=0}^{\infty} a_n x^n$ and the Maclaurin series of $g(x)$ is $\sum_{n=0}^{\infty} b_n x^n$.

(i) What is the Maclaurin series of $f(x) - 3g(x)$?

(ii) Show that if the Maclaurin series for $f(x)$ converges at $x = u$ to $f(u)$, and if the Maclaurin series for $g(x)$ converges at $x = u$ to $g(u)$, then the

Maclaurin series for $f(x) + g(x)$ must converge at $x = u$ to $f(u) + g(u)$.

*84. Find an example of two series $\sum_{n=1}^{\infty} a_n$ and $\sum_{n=1}^{\infty} b_n$ that are both divergent but whose sum $\sum_{n=1}^{\infty} (a_n + b_n)$ converges.

*85. What is $f^{(7)}(0)$ if $f(x) = \frac{1}{1-2x} \frac{1}{1+2x}$?

86. Suppose the Maclaurin series of the function $f(x)$ is $\sum_{n=0}^{\infty} a_n x^n$, and suppose it converges for every x. Assuming that $a_0 \neq 0$, find the first four terms of the Maclaurin series of $\frac{1}{f(x)}$.

87. Graph the function
$$f(x) = \begin{cases} e^{-1/x^2}, & \text{if } x \neq 0, \\ 0, & \text{if } x = 0, \end{cases}$$
whose Maclaurin series we discussed in Example 10.8.6 (p. 702). On the same picture, graph $f'(x)$, $f''(x)$, and as many more derivatives as possible. Do they all tend to 0 as $x \rightarrow 0$?

Conics

11

INTRODUCTION

In Section 1.6 we recalled some basic facts about conics, primarily with the goal of sketching their graphs. In this chapter we make a more systematic study, making use of some of the techniques we have learned.

One way of looking at conics is to take a two-ended cone, as shown at left, and intersect it with a plane. The points of intersection will usually be a conic. (There are some exceptions because it is possible to put the plane in special locations that result in *degenerate conics*, such as a single point or one or two straight lines.) This is the origin of the name "conic."

The illustration shows some of the possibilities.

However, in this chapter we will study conics by beginning with their equations and will not make much mention of the three-dimensional picture with the cone.

In Section 11.1 we recall briefly some of the material of Section 1.6, dealing with identifying and graphing conics from their equations. In Section 11.2 we consider conics from a more geometrical point of view, de-

Ellipse *Hyperbola* *Parabola*

scribing a way of constructing them and discussing some of their properties. Up to this point the conics under consideration all have equations without *cross-terms*, that is, terms involving *xy*. In Section 11.3 we will see that this amounts to insisting that the axes of the conic are parallel to the *x*- and *y*-axes. To study the general case, in which cross-terms are allowed, it is necessary to learn how to rotate the axes. The material of Section 11.3 is not used in later chapters.

11.1 Identifying and Graphing Conics

TRANSLATING AXES
COMPLETING SQUARES
CONICS IN STANDARD FORM

In this brief section we recall material that was covered in Section 1.6. If you are not familiar with conics, you should review that section first.

The equation of a conic is always a quadratic polynomial equation in the two variables *x*, *y*. For the present we will only consider equations without "cross-terms" *xy*, that is, equations of the form $ax^2 + by^2 + cx + dy + q = 0$ with *a*, *b* not both zero.

In Section 1.6 we found that it is often possible to simplify this type of equation by completing squares. Whenever a quadratic polynomial contains both an x^2-term and an *x*-term, we can complete the square and express the polynomial in terms of $(x - h)^2$, say. Letting $x' = x - h$, we rewrite the equation using the variable x'. The resulting polynomial will have an x'^2-term but no x'-term. A similar substitution can be used, if necessary, with *y*.

If the equation does not have an x^2-term, it is possible to combine the *x*-term and the constant term into x', resulting in an equation without a constant term. A similar procedure can be used to remove the constant term from an equation with no y^2-term. The process of replacing *x* with x' and *y* with y' is referred to as *translating coordinates*.

The value of doing this is that it gives a simpler equation that is more easily recognized as one of the standard types of conics. On the other hand, it does mean that we are considering the conic relative to a different set of coordinates (x', y'). The geometrical significance of changing coordinates is that the axes are *translated*, or shifted over; the conic is moved, and the origin changes. Being able to identify the conic is useful only if we keep track of the effect of changing coordinates.

Summary 11.1.1

Figure 11.1.1

Translation of Axes

Suppose *h* and *k* are real numbers; let $O' = (h, k)$ and let $x' = x - h$, $y' = y - k$. Draw the x'-axis and the y'-axis through O'. To graph all the points (x, y) whose coordinates satisfy some equation $F(x, y) = 0$, it is equivalent to express $F(x, y)$ as a function of x' and y', that is,

$$F(x, y) = G(x', y') = G(x - h, y - k),$$

and then graph all points (x', y') that satisfy $G(x', y') = 0$; these points (x', y') must be plotted relative to the x'-axis and the y'-axis. See Figure 11.1.1.

The function *G* can be found by observing that $x = x' + h$ and $y = y' + k$, so

$$G(x', y') = F(x, y) = F(x' + h, y' + k).$$

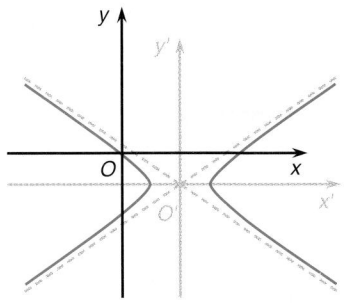

Figure 11.1.2

EXAMPLE 1   Consider the equation $4x^2 - 16x - 9y^2 - 18y + 3 = 0$. Completing the square both in x and y, we find

$$4(x^2 - 4x) - 9(y^2 + 2y) = -3 \ \text{ or } \ 4(x - 2)^2 - 16 - 9(y + 1)^2 + 9 = -3$$

or $\ 4x'^2 - 9y'^2 = 4,$

where $x' = x - 2$ and $y' = y + 1$.

We recognize this as the equation of a hyperbola whose transverse axis is horizontal. It is sketched in Figure 11.1.2, with the translated origin O' at $(x, y) = (2, -1)$ and the x'- and y'-axes shown.

For convenience we recall the standard forms of the various types of conics.

Table 11.1.2   Conics in Standard Form

$y = ax^2, a > 0$	Parabola	Opening upward
$y = ax^2, a < 0$	Parabola	Opening downward
$x = ay^2, a > 0$	Parabola	Opening to the right
$x = ay^2, a < 0$	Parabola	Opening to the left
$\frac{x^2}{a^2} + \frac{y^2}{b^2} = 1, a > b > 0,$	Ellipse	Intercepts $(\pm a, 0), (0, \pm b)$; major axis horizontal
$\frac{x^2}{b^2} + \frac{y^2}{a^2} = 1, a > b > 0,$	Ellipse	Intercepts $(\pm b, 0), (0, \pm a)$; major axis vertical
$\frac{x^2}{a^2} + \frac{y^2}{a^2} = 1, a > 0$	Circle	Radius a, center at $(0, 0)$
$\frac{x^2}{a^2} - \frac{y^2}{b^2} = 1$	Hyperbola	Intercepts $(\pm a, 0)$, transverse axis horizontal, asymptotes $y = \pm \frac{b}{a}x$
$\frac{y^2}{a^2} - \frac{x^2}{b^2} = 1$	Hyperbola	Intercepts $(0, \pm a)$, transverse axis vertical, asymptotes $y = \pm \frac{a}{b}x$

These nine types are illustrated in Figure 11.1.3.

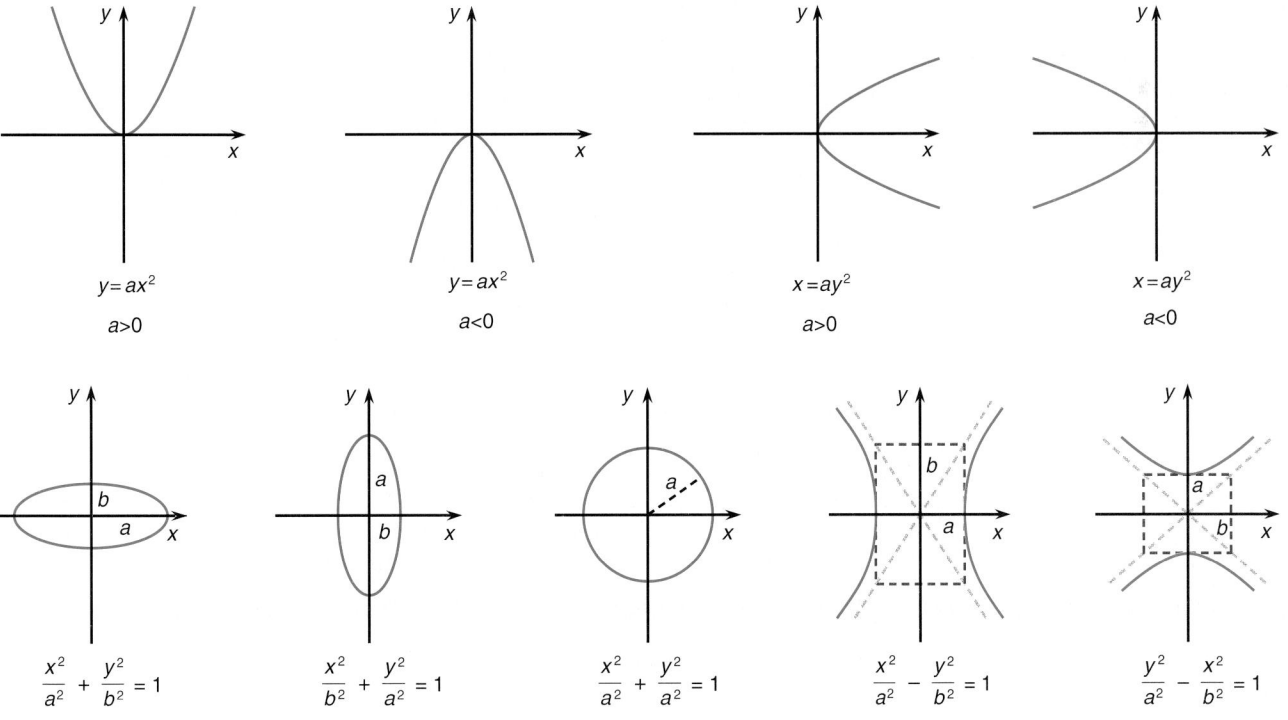

Figure 11.1.3

For the convenience of the reader we repeat the summary from Section 1.6 about how to sketch the graph of a conic.

Summary 11.1.2

Graphing Conics

Consider the quadratic equation $ax^2 + by^2 + cx + dy + q = 0$, which has no xy-term. To draw its graph, follow the following steps.

1. Complete squares if necessary; if the equation contains both an x^2-term and an x-term, that is, if $a \neq 0$ and $c \neq 0$, then complete the square in x. Similarly, if $b \neq 0$ and $d \neq 0$, complete the square in y.

2. Translate axes if necessary. Specifically, if you found it necessary to complete squares in Step 1, and the result of completing the square in x is a term involving $x - h$, then let $x' = x - h$. If completing the square in y gives a term involving $y - k$, let $y' = y - k$.

   If there is no x^2-term, then there must be a y^2-term. In this situation, combine the x-term and the constant term as $c(x - h)$, for some constants c and h, and let $x' = x - h$. Similarly, if there is no y^2-term, combine the y-term and the constant term as $d(y - k)$ and let $y' = y - k$.

3. If the axes have been translated, plot the "new origin" $O' = (h, k)$. (If you changed x to $x' = x - h$ but did not have to translate y, then $O' = (h, 0)$, that is, let $k = 0$. Similarly, if it was not necessary to translate x to x', then let $h = 0$.)

4. Express the equation in terms of x' and y' if translations were necessary. Inspect the equation to see whether it is the equation of a parabola, an ellipse, a circle, or a hyperbola. Find the intercepts.

   For a hyperbola $\frac{x^2}{a^2} - \frac{y^2}{b^2} = 1$ or $\frac{y^2}{a^2} - \frac{x^2}{b^2} = 1$, find the asymptotes and mark them on the picture. You may find it helpful to mark the intercepts (either $(\pm a, 0)$ or $(0, \pm a)$) and also the points on the other axis that are b units from the center, and use them to fill in a rectangle, as shown in Figure 11.1.5. The asymptotes pass through the corners of this box.

   Sketch the graph.

5. Occasionally, we encounter a quadratic equation that is not a parabola or an ellipse or a circle or a hyperbola. This kind of graph is called a **degenerate conic**. There are several possibilities, but we describe just two of them. In an equation like $\frac{x^2}{a^2} + \frac{y^2}{b^2} = -1$, the left side is the sum of two squares, so it is never negative, and we see that no point (x, y) can satisfy this equation. There are *no points* on the graph.

   Another possibility is an equation of the form $\frac{x^2}{a^2} - \frac{y^2}{b^2} = 0$. This is like a hyperbola except that the constant term is zero. Rearranging, we find that $\frac{x^2}{a^2} = \frac{y^2}{b^2}$, or $\frac{x}{a} = \pm \frac{y}{b}$. The graph is the two straight lines $y = \frac{b}{a}x$ and $y = -\frac{b}{a}x$. The name for this kind of graph comes from the observation that this is what results if a hyperbola "degenerates" into its two asymptotes.

In the standard forms for ellipses we always assume $a > b > 0$. In the standard forms for hyperbolas, the right side is always 1 (never -1), and a^2 is the denominator of the *positive* term on the left side. We may have $a > b > 0$ or $b > a > 0$ or $a = b > 0$.

EXAMPLE 2   (i) Sketch the graph of $2x - 4y^2 + 8y - 6 = 0$.

Solution   We begin by completing the square in y, and this gives

$$0 = 2x - 4(y^2 - 2y) - 6 = 2x - 4(y - 1)^2 - 2 = 2(x - 1) - 4(y - 1)^2.$$

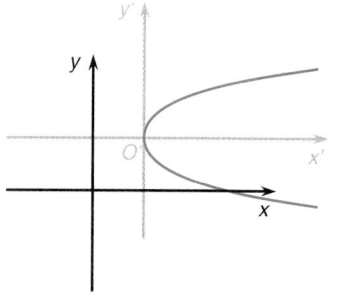

Figure 11.1.4

Letting $x' = x - 1$, $y' = y - 1$, we see that the conic is $2x' = 4y'^2$, which is the equation of a parabola whose vertex is at the "origin" O' and that opens to the right.

We plot the translated origin $O' = (1, 1)$ and the translated axes and then sketch the parabola using them. See Figure 11.1.4.

(ii) Sketch the graph of $9x^2 + 36x - 16y^2 + 180 = 0$.

Solution Completing the square in x, we write the equation as $9(x^2 + 4x) - 16y^2 = -180$, or $9(x + 2)^2 - 36 - 16y^2 = -180$. Letting $x' = x + 2$, we find that this is $9x'^2 - 16y^2 = -144$.

To put this into one of the standard forms of Table 11.1.2, we divide by -144 and find

$$\frac{y^2}{9} - \frac{x'^2}{16} = 1.$$

This is the equation of a hyperbola whose transverse axis is vertical and that, relative to the translated axes has its center at the origin O' and asymptotes $y = \pm\frac{3}{4}x'$.

In Figure 11.1.5 we sketch the translated origin $O' = (-2, 0)$ and the translated axes, the asymptotes, and the hyperbola itself.

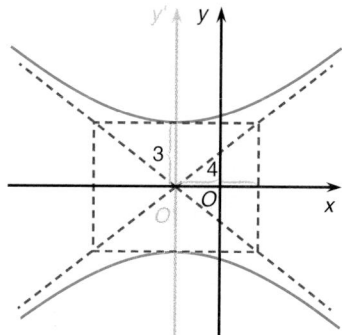

Figure 11.1.5

(iii) Sketch the graph of $7x^2 - 98x + 7y^2 + 28y + 371 = 0$.

Solution Completing both squares, we write the equation as $0 = 7(x^2 - 14x) + 7(y^2 + 4y) + 371 = 7(x - 7)^2 - 343 + 7(y + 2)^2 - 28 + 371$.

This looks like the equation of a circle until we let $x' = x - 7$, $y' = y + 2$ and make the final simplification, writing $7x'^2 + 7y'^2 = 0$, or just $x'^2 + y'^2 = 0$. The only way the left side of this equation can equal zero is for both x' and y' to equal zero, and this means that there is only one point that satisfies the equation. The graph of this equation is the degenerate conic that consists of the single point $(7, -2)$. See Figure 11.1.6.

Figure 11.1.6

With a graphics calculator we can sketch the graphs of the conics in Example 2. We need to express y as a function of x; for the conic $9x^2 + 36x - 16y^2 + 180 = 0$ of part (ii), we write $16y^2 = 9x^2 + 36x + 180$, so $y = \pm\frac{1}{4}\sqrt{9x^2 + 36x + 180}$.

TI-85	**SHARP EL9300**
GRAPH F1$(y(x) =)$ $\sqrt{}$ (9 F1(x) x^2 $+$ 36 F1(x) $+$ 180) $\div$ 4 ▼ $(-)$ $\sqrt{}$ (9 F1(x) x^2 $+$ 36 F1(x) $+$ 180) $\div$ 4 M2(RANGE) $(-)$ 6 ▼ 6 ▼ ▼ $(-)$ 8 ▼ 8 F5(GRAPH)	$\sqrt{}$ $\sqrt{}$ 9 x/θ/T x^2 $+$ 36 x/θ/T $+$ 180 ▶ $\div$ 4 2ndF ▼ $(-)$ $\sqrt{}$ 9 x/θ/T x^2 $+$ 36 x/θ/T $+$ 180 ▶ $\div$ 4 RANGE -6 ENTER 6 ENTER 1 ENTER -8 ENTER 8 ENTER 1 ENTER $\sqrt{}$
CASIO f_x-6300G (f_x-7700GB)	**HP 48SX**
Range -6 EXE 6 EXE 1 EXE -8 EXE 8 EXE 1 EXE EXE EXE EXE Graph $\sqrt{}$ (9 X x^2 $+$ 36 X $+$ 180) $\div$ 4 EXE Graph $(-)$ $\sqrt{}$ (9 X x^2 $+$ 36 X $+$ 180) $\div$ 4 EXE	↰ PLOT PLOTR ERASE ATTN ' Y = $\sqrt{x}$ () 9 × X y^x 2 $+$ 36 × X $+$ 180 ▶ $\div$ 4 ↰ PLOT ↰ DRAW ERASE 6 $+/-$ ENTER 6 XRNG 8 $+/-$ ENTER 8 YRNG DRAW ATTN ' Y = $\sqrt{x}$ () 9 × X y^x 2 $+$ 36 × X $+$ 180 ▶ $\div$ 4 $+/-$ ↰ PLOT ↰ DRAW DRAW

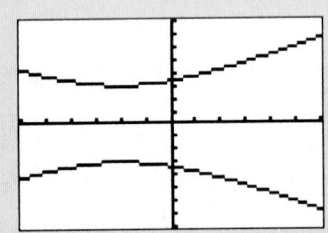

To graph the parabola from part (i) of Example 2, it is necessary to use the quadratic formula to express y in terms of x, and again we find that there are two functions.

Exercises 11.1

Identify and sketch the graph of each of the following.

1. $x^2 - y^2 = -9$
2. $x^2 - 2x + 4y^2 - 16y + 16 = 0$
3. $x^2 + 6x + 2y = 1$
4. $4x^2 + y^2 - y = 0$
5. $2x^2 + y^2 + 4x + 1 = 0$
6. $2x^2 - 4y^2 + 3 = 0$
7. $3x^2 - 12x + y^2 - 4y = 0$
8. $\left(\frac{y}{x}\right)^2 + \frac{4}{x} + 3x^{-2} + 1 = 0$
9. $3y^2 - 12y + x - 7 = 0$
10. $x^4 + 2x^2y^2 + y^4 = 16$

Find the equation of each of the following. (Assume that there are no cross-terms in any of the equations.)

11. The parabola whose vertex is at $(-1, -2)$ that opens upward and contains the point $(-2, 2)$.

12. The ellipse whose center is $(5, -1)$ and whose horizontal and vertical axes have lengths 4 and 9, respectively.

13. The ellipse whose center is $(0, 3)$ and that contains $(0, 4)$ and $(2, 3)$.

14. The ellipse whose center is $(5, 2)$ and that contains $(1, 1)$ and $(3, 0)$.

15. The hyperbola with asymptotes $y = \pm 2(x - 1)$ that contains the point $(2, 0)$.

16. The hyperbola whose center is $(2, 1)$, whose transverse axis is vertical, and that contains $(5, -4)$ and $(2, 5)$.

17. The circle with center $(-3, 4)$ and radius 4.

18. The parabola that is symmetric about the y-axis and contains the points $(2, 0)$ and $(-3, -5)$.

19. The hyperbola that has center $(4, 0)$, that does not intersect the x-axis, and that contains the points $(4, 1)$ and $(3, 2)$.

20. The ellipse whose center is $(1, 2)$, whose minor axis has length 3, and that contains the point $(0, 5)$.

II

As usual, we assume for all the following questions that there are no cross-terms in any of the equations.

21. Sketch the graph of $4x^2 + 8x - 4y^2 - 8y = 0$.

22. There are two ellipses whose center is the origin, whose major axis is twice as long as the minor axis, and that pass through $(2, 1)$. Find the equation of the one whose major axis is longer. Draw a sketch of both ellipses.

23. Find the equations of both parabolas whose vertex is $(3, -1)$ and that pass through $(4, 3)$.

24. Find the equation of the ellipse whose center is $(4, -2)$ and that is tangent to the lines $y = 3$ and $x = 1$.

*25. (i) Find an example of four points that do not lie on any ellipse.

   (ii) Is it possible to find three points that do not lie on any ellipse?

*26. The three points $(0, 0)$, $(1, 1)$, and $(2, 4)$ all lie on the parabola $y = x^2$. Find another parabola on which they all lie.

III

Use a graphics calculator or computer to sketch the graphs of the following conics.

27. $x^2 - 2x - \frac{y^2}{4} = 0$

28. $x^2 + 4x + 9y^2 - 18y + 9 = 0$

29. The ellipse with center $(2, -3)$, containing $(2, 4)$ and $(-1, -3)$.

POINT TO PONDER

When we listed the standard forms of the conics in Table 11.1.2, the idea was to use these forms to identify the type of any particular conic. For this approach to work, we have to know that after translating the axes, every conic can be put into one of these forms.

This is approximately true, but there are some subtleties. First, we have to make our blanket assumption that there are no cross-terms in the equation. Second, it is necessary to allow for the degenerate conics, which are not included in the list.

Some possible degenerate forms are (i) quadratic polynomials depending on only one of the two variables, (ii) equations like the equation of an ellipse or a circle except that the constant on the right side is 0 or -1, (iii) equations like the equation of a hyperbola except that the constant on the right side is 0.

(i) Try to make an argument to explain why every quadratic polynomial equation in x and y with no xy-term can be converted, by completing squares and translating coordinates if necessary into one of the standard forms or into one of these three degenerate forms.

(ii) What are the graphs of each of these types of degenerate conics? Which of them can you form by intersecting a plane with a two-ended cone?

(iii) Beginning with your argument that every (nondegenerate) conic can be expressed as one of the standard types, show that circles and ellipses are the only conics that are *bounded*, that is, contained inside some large circle.

11.2 Geometry of Conics

FOCUS
DIRECTRIX
ECCENTRICITY
REFLECTION PROPERTIES

Figure 11.2.1

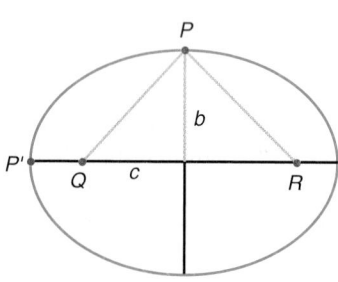

Figure 11.2.2

So far we have concentrated on recognizing the type of a conic and sketching its graph. These are the things most often needed in the study of calculus. Now we discuss briefly some other aspects of conics.

Suppose we take two points in the plane, Q and R, and consider the set of all points P for which the sum of the distances $|PQ|$ and $|PR|$ equals some constant K. (Of course, K will have to be at least as large as the distance between Q and R, or there will be no such points P at all.)

One way to draw the graph of all these points P would be to take a piece of string that is K units long and (with thumbtacks) attach its ends to Q and R. Pulling the string tight with a pencil, we find that the point of the pencil is at a point P on the graph. If we move the pencil, keeping the string tight, it will draw the curve, which turns out to be an ellipse. See Figure 11.2.1.

Each of the points Q and R is called a **focus** of the ellipse. (The plural of "focus" is "foci.")

Now suppose we draw the ellipse and mark on it the major axis (through the foci) and the minor axis (through the midpoint of QR). Consider the point P at one end of the minor axis. Its distance from the major axis is b, the semi-minor axis. Write c for the distance from either focus to the center of the ellipse. We see from Figure 11.2.2 that by the Pythagorean Theorem (Theorem 1.5.1) the distance $|QP|$ from P to a focus is $\sqrt{b^2 + c^2}$. The distance $|RP|$ to the other focus is the same. The constant K is the sum of these two distances, or $2\sqrt{b^2 + c^2}$.

Now suppose P' is the point at one end of the major axis. Since the distance from P' to the center of the ellipse is the semi-major axis, a, we see that its distance from one focus is $|P'Q| = a - c$, and its distance from the other focus is $|P'R| = a + c$. The sum of these distances, which equals K, is $(a - c) + (a + c) = 2a$.

Since we also know that $K = 2\sqrt{b^2 + c^2}$, we see that $a = \sqrt{b^2 + c^2}$, or

$$a^2 = b^2 + c^2. \tag{11.2.1}$$

EXAMPLE 1   **Finding Foci**   Consider the ellipse $\frac{x^2}{4} + \frac{y^2}{9} = 1$. We know that its semi-major axis is $a = \sqrt{9} = 3$ and its semi-minor axis is $b = \sqrt{4} = 2$. From Equation 11.2.1 we see that $c^2 = a^2 - b^2 = 9 - 4 = 5$, so $c = \sqrt{5}$. Since the foci lie on the major axis, which in this case is vertical, we see that the foci are $(0, -\sqrt{5})$ and $(0, \sqrt{5})$. See Figure 11.2.3.

Figure 11.2.3

Figure 11.2.4

Statuary Hall (the Whispering Gallery), U.S. Capitol Building.

There is an interesting property of ellipses having to do with reflections. If an object starts at one focus and travels in a straight line until it hits the outside of the ellipse and then bounces off the edge (i.e., is "reflected"), then it will bounce so that it heads straight for the other focus.

For instance, if an ellipse was made of a reflective surface, the light from a flashbulb fired at one focus would all reflect back to the other focus. Moreover, since each ray of light would travel the same distance in going from one focus to the other (the distance K mentioned above), they would all arrive simultaneously (see Figure 11.2.4).

Some buildings are constructed with a room in the shape of a solid ellipse. Sound emitted at one focus will reflect off the walls and ceiling and bounce to the other focus. A person standing at one focus can hear quite clearly the sound of somebody whispering at the other focus, even though it would be inaudible to somebody standing much closer. Rooms of this design are sometimes called "whispering galleries"; two of the most famous are the one in the United States Capitol and the dome of St. Paul's Cathedral in London, England.

There is another way to describe conics geometrically that applies to parabolas and hyperbolas as well as to ellipses.

Fix a point F and a line d. Now consider all points P whose distance from the line d equals its distance from the point F. If we sketch a number of such points, as in Figure 11.2.5, we begin to suspect that the graph of all of them is a parabola, and this is correct.

We let the point be $F = (0, p)$, on the y-axis, and let the line d be $y = -p$. The advantage of this choice is that the origin lies exactly p units from both F and d, so it is on the graph.

Now consider a point $P = (x, y)$. Its distance from $F = (0, p)$ is $\sqrt{x^2 + (y - p)^2}$. Its distance from the line d is just its vertical distance from d, which is $|y + p|$. See Figure 11.2.6.

For P to lie on the graph, these two distances must be the same, that is, $\sqrt{x^2 + (y - p)^2} = |y + p|$. Squaring both sides, we find that $x^2 + y^2 - 2py + p^2 = y^2 + 2py + p^2$. Subtracting $y^2 - 2py + p^2$ from both sides results in $x^2 = 4py$, or $y = \frac{1}{4p}x^2$, which we recognize as the equation of a parabola with vertex at the origin.

In Figure 11.2.7 the illustration has been drawn with $p > 0$. If $p < 0$, the same calculations work, and in this case the parabola will open downward. If the line d is vertical, then the parabola opens to the left or to the right, depending on the relative positions of the focus and the directrix. Specifically, $y^2 = 4px$ is the equation

Figure 11.2.5

Figure 11.2.6

Figure 11.2.7

Figure 11.2.8

Figure 11.2.9

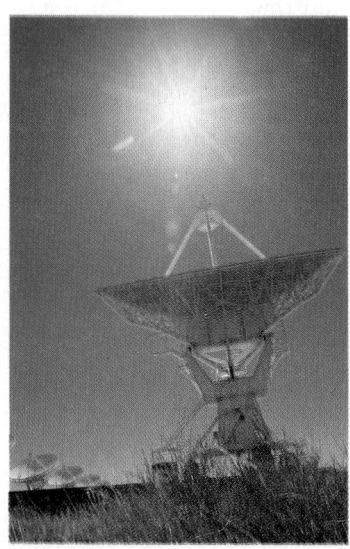

Parabolic antenna for a VLA radio telescope in New Mexico.

of a parabola with vertex at the origin, with $F = (p, 0)$ and with the line d equal to $x = -p$. It opens to the right if $p > 0$ and to the left if $p < 0$.

The point F is called the **focus** of the parabola, and the line d is the **directrix**. Every parabola can be described as the set of points that are equidistant from the focus and the directrix.

There is also a reflection property for parabolas. Suppose an object starts at the focus and moves in a straight line until it bounces off the parabola. Then after reflecting, it will be moving away from the directrix in a direction perpendicular to the directrix, as shown in Figure 11.2.8.

This property is the basic idea behind the parabolic reflector. If a reflecting material is formed in the shape of a parabola and a small light bulb is placed at its focus, then the rays of light moving away from the bulb will all reflect out in parallel lines, forming a narrow beam. Flashlights, automobile headlights, and huge searchlights are all built on this principle. See Figure 11.2.9.

The same idea can also be used in reverse. If parallel rays of light enter a parabolic reflector, they will all be reflected and converge at the focus, as in Figure 11.2.10. This can be used to collect rays of the sun to heat water in a pipe, as in rooftop solar collectors, or even for cooking. It can also be used to collect sound waves by placing a microphone at the focus of a parabolic dish, which is the basis of some directional microphones. It can also be used to collect radio waves, as we see in the design of some radio telescopes and in satellite dishes (which collect radio signals

Figure 11.2.10

from a specific direction, i.e., the direction to the satellite, and concentrate them at the focus of the dish, where they can be picked up and connected to a television set).

EXAMPLE 2   Find the focus and directrix of the parabola $y^2 + 8x = 24$.

Solution   There is no need to complete squares here, but writing the equation as $8(x - 3) = -y^2$ suggests letting $x' = x - 3$ and considering the translated origin $O' = (3, 0)$. The equation now reads $8x' = -y^2$. This is the equation of a parabola whose vertex is O' and that opens to the left (because of the minus sign). If we write it as $x' = -\frac{1}{8}y^2$, that is, $x' = \frac{1}{4p}y^2$, with $p = -2$, the formulas above suggest that the focus F is at the point where $x' = -2$, $y = 0$, and the directrix is the line $x' = -(-2) = 2$.

Converting this, we see that when $x' = -2$, $x = x' + 3 = 1$, and $x' = 2$ is the same as $x = 5$. The focus is $F = (1, 0)$, and the directrix is the line $x = 5$. The parabola is sketched in Figure 11.2.11.

Figure 11.2.11

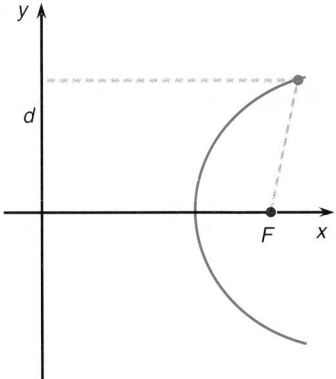

Figure 11.2.12

Now suppose we are given a point F and a line d. If $P = (x, y)$ is any point, we can consider the distance $|PF|$ and the distance from P to the line d. We have just seen that the set of all points P for which these distances are equal is a parabola. Suppose instead we consider the set of all points P for which the distance $|PF|$ is one-half the distance from P to d. Of course, there is no special reason to choose $\frac{1}{2}$; we could try $\frac{1}{3}$ or 17 just as easily. Let us suppose that e is some fixed positive number and consider the set of all $P = (x, y)$ for which the distance $|PF|$ is e times the distance from P to d. It is traditional to use the letter e for this purpose, but here it does *not* refer to the base of the natural logarithm. It is just any positive number.

To be specific, let d be the line $x = 0$ and let $F = (p, 0)$, with $p > 0$, as shown in Figure 11.2.12. The distance from $P = (x, y)$ to the line d is just $|x|$. Since $|PF| = \sqrt{(x - p)^2 + (y - 0)^2} = \sqrt{(x - p)^2 + y^2}$, the condition on P amounts to

$$\sqrt{(x - p)^2 + y^2} = e|x|, \text{ or } (x - p)^2 + y^2 = e^2|x|^2$$
$$\text{or } x^2 - 2px + p^2 + y^2 = e^2 x^2, \text{ or } (1 - e^2)x^2 - 2px + p^2 + y^2 = 0,$$

that is, $(1 - e^2)\left(x^2 - \dfrac{2p}{1 - e^2}x\right) + p^2 + y^2 = 0.$

Completing the square, we have

$$(1 - e^2)\left(x^2 - \frac{2p}{1 - e^2}x + \frac{p^2}{(1 - e^2)^2}\right) - (1 - e^2)\frac{p^2}{(1 - e^2)^2} + p^2 + y^2 = 0$$

$$(1 - e^2)\left(x - \frac{p}{1 - e^2}\right)^2 + p^2 - \frac{p^2}{(1 - e^2)} + y^2 = 0$$

$$(1 - e^2)x'^2 + y^2 = p^2\left(\frac{1}{1 - e^2} - 1\right) = p^2\left(\frac{1 - 1 + e^2}{1 - e^2}\right) = \frac{e^2 p^2}{1 - e^2},$$

or $\dfrac{(1 - e^2)^2}{e^2 p^2}x'^2 + \dfrac{1 - e^2}{e^2 p^2}y^2 = 1.$

We have written $x' = x - \frac{p}{1-e^2}$. The last equation we recognize as the equation of a conic, with center at the point $O' = \left(\frac{p}{1-e^2}, 0\right)$. The line d is called the **directrix**; we chose it to be $x = 0$, that is, $x' = -\frac{p}{1-e^2}$.

This calculation was pretty complicated, but all we have really done is follow the procedures for completing squares and translating axes. Fortunately, the actual calculation is not terribly important for our purposes.

However, let us ask what kind of conic the equation represents. The important things are the signs of the coefficients of x'^2 and y^2. If $e < 1$, then $1 - e^2 > 0$ and both coefficients are positive. In this situation the conic is an ellipse. Since $1 - e^2 < 1$, we see that $(1 - e^2)^2 < 1 - e^2$, so $\frac{1}{(1-e^2)^2} > \frac{1}{1-e^2}$. With $a^2 = \frac{e^2 p^2}{(1-e^2)^2}$ and $b^2 = \frac{e^2 p^2}{1-e^2}$, we see that $a^2 > b^2$ and the major axis of the ellipse is horizontal. Also $(1 - e^2)a^2 = b^2$, so $a^2 - b^2 = a^2(1 - 1 + e^2) = e^2 a^2$ and $e = \frac{\sqrt{a^2 - b^2}}{a}$. Since $a^2 = \frac{e^2 p^2}{(1-e^2)^2}$, $a = \frac{ep}{1-e^2} > 0$, and $p = \frac{a}{e}(1 - e^2)$. The directrix is $x' = -\frac{p}{1-e^2} = -\frac{a}{e}$.

If $e > 1$, the coefficient of y^2 is negative, while the coefficient of x'^2, being a square, is positive. In this situation the conic is a hyperbola, $\frac{x'^2}{a^2} - \frac{y^2}{b^2} = 1$, with its transverse axis horizontal. We also find that $a^2(1 - e^2) = -b^2$, so $a^2 + b^2 = a^2 - a^2(1 - e^2) = e^2 a^2$ and $e = \frac{\sqrt{a^2 + b^2}}{a}$. Since $a^2 = \frac{e^2 p^2}{(1-e^2)^2}$, $a = -\frac{ep}{1-e^2} > 0$, and $p = -\frac{a}{e}(1 - e^2)$. The directrix is $x' = -\frac{p}{1-e^2} = \frac{a}{e}$.

If $e = 0$, we cannot write the last line of the calculation, which involved dividing by e^2, but in the previous line we see the equation becomes $x'^2 + y^2 = 0$, whose graph consists of the single point O', that is, $(x, y) = (p, 0)$.

If $e = 1$, then $1 - e^2 = 0$, so the above calculation, which required dividing by $1 - e^2$, is invalid. However, in that case we have already seen that the conic is a parabola.

Summary 11.2.2

The set of all points $P = (x, y)$ for which $|PF|$ equals e times the distance from P to d is a conic. The point F and the line d are the **focus** and the **directrix** of the conic. The number e is its **eccentricity**. We summarize the above discussion in a table:

Eccentricity	Conic
$e > 1$	Hyperbola
$e = 1$	Parabola
$0 < e < 1$	Ellipse
$e = 0$	Circle

$e = 0.87$

$e = 0.66$

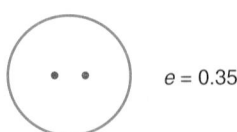

$e = 0.35$

Figure 11.2.13

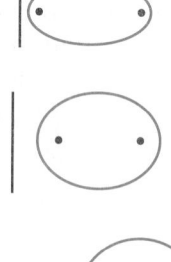

Figure 11.2.14

We have included the circle in this list (its presence gives some explanation of the word "eccentricity": A circle has zero eccentricity; the farther an ellipse is from being round, the greater is its eccentricity, and the more "eccentric" it is.) However, we ought to observe that it is not possible to define a circle by using a focus and directrix, as we did with the other conics. One way to think of it is to consider several ellipses with different eccentricities. The smaller the eccentricity, the closer the ellipse is to being round. See Figure 11.2.13.

The closer the ellipse is to being round, the closer together are its two foci. So if the eccentricity of an ellipse becomes smaller and smaller and approaches 0, then the foci of the ellipse move together. As the eccentricity actually becomes 0, the two foci merge into a single point, which is the center of the circle.

Meanwhile, we could ask what happens to the directrix. As the eccentricity approaches 0, the directrix moves farther and farther away from the ellipse. (After all, suppose the eccentricity e is very small. The distance from a point on the ellipse to a focus is e times the distance to the directrix, so any point on the ellipse must be much closer to the focus than it is to the directrix.) When the ellipse turns into a circle as e becomes 0, the directrix actually "moves off to infinity" (see Figure 11.2.14). This rather vague description is not at all precise, but it is intended to explain why no line will serve as the directrix of a circle.

For a hyperbola of the form $\frac{x^2}{a^2} - \frac{y^2}{b^2} = 1$ the transverse axis is horizontal, and the foci both lie on it. For a hyperbola with a vertical transverse axis the foci lie on the vertical axis. For this reason the transverse axis is sometimes called the **focal axis**.

We should also remark that just as an ellipse is the set of all points whose distances from Q and R add up to some constant K, a hyperbola can be expressed as the set of all points whose distances from the foci Q and R *differ* by some fixed number K, that is, the set of all P for which $|PQ| - |PR| = \pm K$. (We say $\pm K$ because we do not know which of $|PQ|$ and $|PR|$ is larger. In fact, if we just wrote K, we would get one branch of the hyperbola, and $-K$ would give the other branch.)

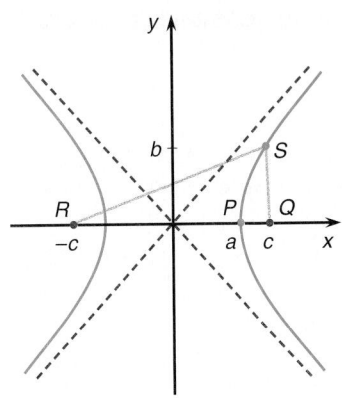

Figure 11.2.15

Finally, in analogy with the ellipse, for which we found that $a^2 = b^2 + c^2$, there are relations between the numbers a and b and the location of the foci of a hyperbola. Consider first the hyperbola $\frac{x^2}{a^2} - \frac{y^2}{b^2} = 1$, as illustrated in Figure 11.2.15. Let us write c for the distance from the origin to either focus, so the foci are the points $Q = (c, 0)$ and $R = (-c, 0)$.

Label the point $P = (a, 0)$ as shown. The distance from P to the left focus is $c + a$, and the distance from P to the right focus is $c - a$. The difference of these distances is $2a$, so the constant K referred to above is $K = 2a$.

Now consider the point S on the hyperbola whose y-coordinate is b and whose x-coordinate is positive. We can solve

$$\frac{x^2}{a^2} = 1 + \frac{y^2}{b^2} = 1 + \frac{b^2}{b^2} = 2,$$

so

$$x^2 = 2a^2$$

and

$$x = \sqrt{2}a.$$

So S is the point $(\sqrt{2}a, b)$. We know that $|SQ| - |SR|$ is $2a$ or $-2a$. Using Formula 1.5.2 for the distance between points, we can express this as

$$\pm 2a = \sqrt{(\sqrt{2}a - c)^2 + b^2} - \sqrt{(\sqrt{2}a + c)^2 + b^2}.$$

We rearrange and square both sides:

$$\pm 2a + \sqrt{(\sqrt{2}a + c)^2 + b^2} = \sqrt{(\sqrt{2}a - c)^2 + b^2}$$

$$4a^2 \pm 4a\sqrt{2a^2 + 2\sqrt{2}ac + c^2 + b^2}$$
$$+ 2a^2 + 2\sqrt{2}ac + c^2 + b^2 = 2a^2 - 2\sqrt{2}ac + c^2 + b^2$$

$$4a^2 \pm 4a\sqrt{2a^2 + 2\sqrt{2}ac + c^2 + b^2} = -4\sqrt{2}ac.$$

We rearrange so that everything is on the right side except the square root, then divide by $4a$ and square again:

$$\pm 4a\sqrt{2a^2 + 2\sqrt{2}ac + c^2 + b^2} = -4a^2 - 4\sqrt{2}ac$$

$$\pm\sqrt{2a^2 + 2\sqrt{2}ac + c^2 + b^2} = -a - \sqrt{2}c$$

$$2a^2 + 2\sqrt{2}ac + c^2 + b^2 = a^2 + 2\sqrt{2}ac + 2c^2$$

$$a^2 + b^2 = c^2.$$

We find that $c = \sqrt{a^2 + b^2}$.

Moreover, all these relations also hold for a hyperbola whose focal axis is vertical.

By applying the description before Summary 11.2.2 we see that $e = \frac{\sqrt{a^2+b^2}}{a}$, so $e = \frac{c}{a}$, and the directrix is $x = \frac{a}{e}$, that is, $x = \frac{a^2}{c}$.

Summary 11.2.3

	Hyperbola	**Hyperbola**	**Ellipse**	**Ellipse**
	$\frac{x^2}{a^2} - \frac{y^2}{b^2} = 1$	$\frac{y^2}{a^2} - \frac{x^2}{b^2} = 1$	$\frac{x^2}{a^2} + \frac{y^2}{b^2} = 1, a > b > 0$	$\frac{y^2}{a^2} + \frac{x^2}{b^2} = 1, a > b > 0$
Foci:	$(\pm c, 0)$, with	$(0, \pm c)$, with	$(\pm c, 0)$, with	$(0, \pm c)$, with
	$c = \sqrt{a^2 + b^2}$	$c = \sqrt{a^2 + b^2}$	$c = \sqrt{a^2 - b^2}$	$c = \sqrt{a^2 - b^2}$
Eccentricity:	$e = \frac{c}{a} = \frac{1}{a}\sqrt{a^2 + b^2}$	$\frac{c}{a} = \frac{1}{a}\sqrt{a^2 + b^2}$	$\frac{c}{a} = \frac{1}{a}\sqrt{a^2 - b^2}$	$\frac{c}{a} = \frac{1}{a}\sqrt{a^2 - b^2}$
Directrix:	$x = \pm\frac{a}{e} = \pm\frac{a^2}{c}$	$y = \pm\frac{a}{e} = \pm\frac{a^2}{c}$	$x = \pm\frac{a}{e} = \pm\frac{a^2}{c}$	$y = \pm\frac{a}{e} = \pm\frac{a^2}{c}$

EXAMPLE 3   Find the equation of the hyperbola with center at the origin, eccentricity $e = 2$, and y-intercepts ± 4.

Solution   Since the hyperbola has y-intercepts, its equation must be of the form $\frac{y^2}{a^2} - \frac{x^2}{b^2} = 1$, and since the intercepts are ± 4, we know that $a = 4$.
We also know that $2 = e = \frac{c}{a} = \frac{1}{4}\sqrt{4^2 + b^2}$. Squaring this equation, we see that $4 = \frac{1}{16}(16 + b^2)$, or $64 = 16 + b^2$, or $b^2 = 48$. The equation is $\frac{y^2}{16} - \frac{x^2}{48} = 1$.

With a graphics calculator we can plot the hyperbola. We rewrite it as $\frac{y^2}{16} = 1 + \frac{x^2}{48}$, or $y = \pm 4\sqrt{1 + \frac{x^2}{48}}$.

TI-85	SHARP EL9300
GRAPH F1$(y(x) =)$ 4 $\sqrt{}$ (1 + F1(x) $x^2 \div 48$) ▼ $(-)$ 4 $\sqrt{}$ (1 + F1(x) $x^2 \div 48$) M2(RANGE) $(-)$ 8 ▼ 8 ▼ ▼ $(-)$ 6 ▼ 6 F5(GRAPH)	◿ 4 $\sqrt{}$ 1 + x/θ/T $x^2 \div 48$ 2ndF ▼ $(-)$ 4 $\sqrt{}$ 1 + x/θ/T $x^2 \div 48$ RANGE -8 ENTER 8 ENTER 1 ENTER -6 ENTER 6 ENTER 1 ENTER ◿
CASIO f_x-6300G (f_x-7700GB)	HP 48SX
Range -8 EXE 8 EXE 1 EXE -6 EXE 6 EXE 1 EXE EXE EXE EXE Graph 4 $\sqrt{}$ (1 + X $x^2 \div 48$) EXE Graph $(-)$ 4 $\sqrt{}$ (1 + X $x^2 \div 48$) EXE	↰ PLOT PLOTR ERASE ATTN ′ Y = 4 × $\sqrt{x}$ () 1 + X y^x 2 ÷ 48 ↱ PLOT ↰ DRAW ERASE 8 +/− ENTER 8 XRNG 6 +/−ENTER 6 YRNG DRAW ATTN ′ Y = 4 +/− × $\sqrt{x}$ () 1 + X y^x 2 ÷ 48 ↱ PLOT ↰ DRAW DRAW

Exercises 11.2

In these exercises, all conics are assumed to have their axes vertical and/or horizontal. In particular, this means that there are no xy-terms in their equations.

I

Find the foci and the directrices of the following conics.

1. $y = x^2 + 2x$
2. $y^2 = 2x + 4y - 8$
3. $\frac{x^2}{4} + y^2 = 1$
4. $2x^2 + 3y^2 = 1$
5. $x^2 - 4y^2 = 1$
6. $x^2 + 2x = y^2$
7. $x^2 + 4y^2 + 4x = 8y$
8. $\frac{x^2}{6} = \frac{2y^2}{3} - 6$
9. $9x^2 = 4y^2 + 18x + 8y + 11$
10. $100x^2 + y^2 = 20y$

Find the equation of each of the following conics.

11. The ellipse with center $(0,0)$, eccentricity $e = \frac{1}{2}$, and foci $(\pm 3, 0)$.
12. The hyperbola with center $(0,0)$ and eccentricity 3 and containing $(5,0)$.
13. The parabola with directrix $x = 4$ and focus $(2,2)$.
14. The hyperbola that opens sideways having transverse axis twice as long as its conjugate axis and center at $(2, -3)$ and that contains $(-2, -2)$.
15. The ellipse whose major axis is three times as long as its minor axis, whose center is $(4, 1)$, and that contains $(3, 3)$ but contains no point to the left of the y-axis.
16. The parabola containing $(1, 2)$, $(2, 4)$, and $(2, 1)$.
17. The hyperbola with center $(2, -1)$, eccentricity $e = 2$, and foci $(2, 1)$ and $(2, -3)$.
18. The conic whose eccentricity is $e = \frac{4}{3}$, whose center is $(0, 2)$, and that contains $(-6, 2)$.
19. The conic whose eccentricity is $e = 1$, that is symmetric about the line $x = 3$, and that contains the points $(3, 4)$ and $(2, 3)$.
20. The conic whose directrix is $x = 1$, that has the point $(4, 2)$ as a focus, and that contains the point $(3, 2)$.

Find each of the following.

21. The eccentricity of the ellipse $2x^2 + 8y^2 = 1$.
22. The eccentricity of $5x^2 - 8y^2 + 4y = 7$.
23. The foci of $3x^2 + 6x + 4y^2 = 12$.
24. The foci of $y^2 - x^2 = 4$.
25. The directrix of $x^2 + 4x - 8y = 12$.
26. The directrices of $x^2 + 4x - y^2 - 6y = 4$.
27. The eccentricity of $5x^2 + 3x - 7y = 12$.
28. The sum of the distances from any point on $4x^2 + 6y^2 = 24$ to the two foci.
29. The distance between the foci of $9x^2 - 4y^2 + 8y = 3$.
30. The possible distances from a focus to a directrix of $4x^2 + 9y^2 - 18y = 27$.

II

31. An architect wishes to fit a rectangular door inside a parabolic arch. If the door is 4 ft wide and 8 ft high and the arch is 6 ft wide at floor level, what is the height above the floor of the highest point of the arch?
32. Suppose a window is to be made in the shape of a horizontal ellipse with eccentricity $e = \frac{1}{2}$. There is to be a rectangular pane of clear glass measuring 1 ft by 2 ft in the center of the ellipse with its four corners just reaching the outer frame. The remaining space of the window will be filled with a design in stained glass. What is the long dimension of the ellipse?
*33. What is the relationship between the eccentricities of the ellipses

$$\frac{x^2}{a^2} + \frac{y^2}{b^2} = 1 \quad \text{and} \quad \frac{x^2}{a^2} + \frac{y^2}{b^2} = 4?$$

*34. What is the relationship between the eccentricities of the hyperbolas $\frac{x^2}{a^2} - \frac{y^2}{b^2} = k$ for different values of k?
*35. Find the equation of the hyperbola with center at the origin, having eccentricity $e = 4$, and containing the point $(2, 8)$.
*36. There are two ellipses with eccentricity $e = \frac{1}{2}$ with center at the origin that contain the point $(2, 1)$. Find their equations.
*37. Suppose the floor of an elliptically shaped room has its long dimension equal to 20 m and its short dimension equal to 15 m. How far out from the walls are the foci?
*38. How many directrices does each type of conic (ellipse, hyperbola, parabola, circle) have?
*39. Consider two circles C, D with different centers so that C is inside D. Consider all circles that lie inside D, tangent to D (at one point), and outside C, tangent to C (at one point). Show that the centers of all

these circles lie on an ellipse. What are the foci of the ellipse?

**40. Consider a line d and a point P that does not lie on d. Consider all circles through P that are tangent to d. Show that the centers of all such circles lie on a parabola.

**41. Let $a > c > 0$, and let $F_1 = (c, 0)$, $F_2 = (-c, 0)$. Find the equation of the set of points $P = (x, y)$ determined by the condition $|PF_1| + |PF_2| = 2a$. You will probably want to let $b^2 = a^2 - c^2$.

**42. The points $F_1 = (1, 1)$ and $F_2 = (-1, -1)$ do not lie on any horizontal or vertical line. Find the equation of the ellipse with F_1, F_2 as foci and semi-major axis of length 2.

**43. How many ellipses are there with eccentricity $e = \frac{1}{2}$ and foci $(0, \pm 2)$?

**44. How many hyperbolas are there with eccentricity $e = 2$ and asymptotes $y = \pm x$?

**45. (Reflection property for parabolas) Consider a point P on the parabola $x = y^2$. In Figure 11.2.16 we draw the tangent line to the parabola at P and the segment PF from P to the focus F. We also show a ray of light coming horizontally from the right. To show that every such ray is reflected to the focus, we need to show that the angle of incidence, α, equals the angle of reflection, β (see Example 4.7.4, p. 295).

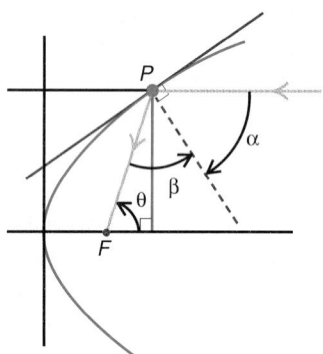

Figure 11.2.16

(i) If $P = (a^2, a)$, find a formula for the slope of the tangent at P; note that this equals $\cot \alpha$.

(ii) Find the coordinates of the focus F. Find a formula for $\tan \theta$ (θ is the angle between PF and the horizontal).

(iii) Use your knowledge of trigonometry to show that $\tan \beta = -\frac{\tan \theta + \tan \alpha}{1 - \tan \theta \tan \alpha}$.

(iv) Use part (iii) to conclude that $\tan \alpha = \tan \beta$, as required.

**46. Prove the reflection property for the parabola $y^2 = 4px$. (Assume $p > 0$.)

*47. Show that the equation of the tangent to the ellipse $\frac{x^2}{a^2} + \frac{y^2}{b^2} = 1$ at the point (u, v) is $\frac{ux}{a^2} + \frac{vy}{b^2} = 1$.

**48. Use Exercise 47 to establish the reflection property for ellipses (following the general idea that we used for parabolas in Exercises 45 and 46).

*49. An airplane starts at the origin and flies in a straight line northeast, that is, up and to the right along the line $y = x$. Transmitters at $(0, 0)$ and $(0, 150)$ (with distances measured in miles) send out synchronized radio signals, and delicate instruments in the plane measure the difference between their arrival times. If the speed of radio signals is known, it is possible to calculate that the plane is 50 miles farther from $(0, 0)$ than from $(0, 150)$. The set of all points at which this is true is one branch of a hyperbola. (i) Find the equation of the hyperbola, and identify the correct branch. (ii) Find the location of the airplane.

*50. In the setting of Exercise 49, suppose winds have blown the aircraft off course, so it is no longer on the line $y = x$. Comparison of radio signals shows that it is 50 miles farther from $(0, 0)$ than from $(0, 150)$. Using signals from a third transmitter, it is possible to see that it is also 100 miles farther from $(0, 0)$ than from $(200, 0)$. Show that the location of the aircraft is determined by the points of intersection of two hyperbolas. (This principle can be used for navigation, but over long distances it is complicated by the fact that the surface of the earth is curved.)

*51. Use a graphics calculator or computer to sketch on one picture several ellipses with different eccentricities. Each should have its center at the origin, major axis horizontal, and semi-major axis equal to 1, and the eccentricities should be $\frac{1}{10}, \frac{1}{4}, \frac{1}{2}, \frac{3}{4}, \frac{9}{10}$. (You might begin by graphing the top half of each ellipse.)

*52. Use a graphics calculator or computer to sketch several hyperbolas on the same picture. Each should have its center at the origin and transverse axis horizontal and should contain the point $(1, 0)$. The hyperbolas should have eccentricities $\frac{11}{10}, \frac{3}{2}, 2, 3, 5$.

11.3 Rotated Coordinates

So far in our study of conics we have always assumed that the axes of the conic are horizontal or vertical, which is the same thing as assuming that the equation of the conic does not contain any cross-term. In this section we learn to deal with conics that do not satisfy this assumption.

Before proceeding, let us think briefly about the technique we learned in Sections 1.6 and 11.1 for *translating the axes*. Roughly speaking, if we consider a conic whose center is not the origin, then its equation will be complicated. It is possible to simplify the equation by considering a new set of axes that are "better placed" relative to the conic; the way to accomplish this is to place the translated origin at the center of the conic (see Figure 11.3.1).

Now suppose we consider an ellipse centered at the origin but placed so that its major and minor axes are not horizontal or vertical, as illustrated in Figure 11.3.2(i). Its equation is likely to be complicated and difficult to recognize, but taking a cue from the discussion above, we might try to find a new set of axes that are placed at an angle relative to the *x*- and *y*-axes, in such a way that they lie along the axes of the ellipse. This is illustrated in Figure 1.3.2(ii).

Relative to these rotated axes, the equation of the ellipse ought to be one of the familiar standard forms, and it should be easy both to identify what it is and also to sketch its graph. This is the idea that we explore in this section.

To make this approach work, it will be necessary to work with rotated axes. Specifically, suppose we know the coordinates (x', y') of a point P relative to a set of axes that are rotated at an angle θ relative to the *x*- and *y*-axes, as shown in Figure 11.3.3(i). Then we need to know how to find its coordinates (x, y) relative to the *x*- and *y*-axes.

It turns out that the most convenient way to approach this question is to use polar coordinates. Suppose the polar coordinates of P, *relative to the x'- and y'-axes*, are (r, ϕ), as shown in Figure 11.3.3(ii). In particular, this means that the rectangular

Figure 11.3.1

Figure 11.3.2(i)

Figure 11.3.2(ii)

Figure 11.3.3(i)

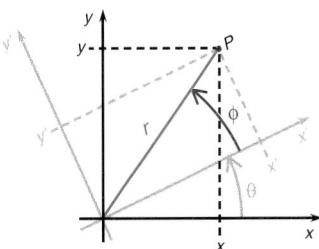

Figure 11.3.3(ii)

coordinates are

$$x' = r \cos \phi, \qquad y' = r \sin \phi$$

(see Equation 1.7.9, p. 50).

Now we consider the same point P relative to the *x*- and *y*-axes. To find its polar coordinates, we observe that the length r is the same, and the angle is $\theta + \phi$. Using

Equation 1.7.9 again, we see that the rectangular coordinates of P relative to the unrotated axes are

$$x = r\cos(\theta + \phi), \qquad y = r\sin(\theta + \phi).$$

Now we simplify, using the addition formulas for sin and cos (cf. Theorem 1.7.11, p. 51):

$$x = r\cos(\theta + \phi) = r\cos\phi\cos\theta - r\sin\phi\sin\theta = x'\cos\theta - y'\sin\theta,$$
$$y = r\sin(\theta + \phi) = r\cos\phi\sin\theta + r\sin\phi\cos\theta = x'\sin\theta + y'\cos\theta.$$

We summarize as a theorem.

THEOREM 11.3.1

Suppose the x'- and y'-axes are rotated through an angle θ relative to the x- and y-axes. If the coordinates relative to the rotated axes of a point P are (x', y'), then its coordinates relative to the (unrotated) x- and y-axes are

$$(x, y) = (x'\cos\theta - y'\sin\theta, x'\sin\theta + y'\cos\theta).$$

Conversely, if the coordinates of P are (x, y) relative to the x- and y-axes, then its coordinates relative to the x'- and y'-axes are

$$(x', y') = (x\cos\theta + y\sin\theta, -x\sin\theta + y\cos\theta).$$

PROOF   The argument given before the statement of Theorem 11.3.1 proves the first formula. It is possible to obtain the second one in the same way, but it is easier to observe that the x- and y-axes are rotated through an angle of $-\theta$ relative to the x'- and y'-axes. Because of this, we can simply use the same formula, replacing θ with $-\theta$. Since $\cos(-\theta) = \cos\theta$ and $\sin(-\theta) = -\sin\theta$, the second formula drops out. ◢

EXAMPLE 1   Consider the point P that lies on the line $y = x$ at a distance of 3 units above and to the right of the origin. It is not difficult to find its coordinates, but let us use Theorem 11.3.1.

We can imagine the x'-axis placed so that it lies along the line $y = x$, as shown in Figure 11.3.4. Then the point P lies on the positive x'-axis at a distance of 3 units from the origin, so its coordinates are $(3, 0)$ relative to the x'- and y'-axes.

Since these axes are rotated through an angle of $45°$ or $\frac{\pi}{4}$ relative to the usual axes, we can use Theorem 11.3.1 with $\theta = \frac{\pi}{4}$ to find that the coordinates of P relative to the usual axes are $x = x'\cos\theta - y'\sin\theta = 3\cos\frac{\pi}{4} - 0 = \frac{3}{\sqrt{2}}$ and $y = x'\sin\theta + y'\cos\theta = 3\sin\frac{\pi}{4} + 0 = \frac{3}{\sqrt{2}}$. The coordinates of P are $(x, y) = \left(\frac{3}{\sqrt{2}}, \frac{3}{\sqrt{2}}\right)$.

This result is not surprising, but it does demonstrate the use of Theorem 11.3.1 and confirm that it works. ◢

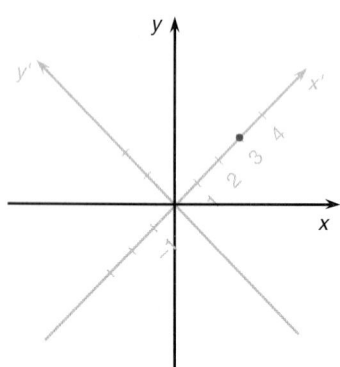

Figure 11.3.4

We really want to use Theorem 11.3.1 to simplify the equation of a conic by removing the cross-term. We begin with an example.

EXAMPLE 2   Consider the conic whose equation is $4x^2 + 2y^2 - 2\sqrt{3}xy = 5$. Notice that it contains a cross-term, that is, a term containing xy.

Let us consider new axes that are rotated through an angle of $\frac{\pi}{3}$ relative to the usual axes. The idea is to express the equation in terms of the rotated coordinates (x', y'). From Theorem 11.3.1 we know that $x = x'\cos\frac{\pi}{3} - y'\sin\frac{\pi}{3} = \frac{x'}{2} - \frac{\sqrt{3}y'}{2}$ and $y = x'\sin\frac{\pi}{3} + y'\cos\frac{\pi}{3} = \frac{\sqrt{3}x'}{2} + \frac{y'}{2}$.

We intend to substitute these formulas into the equation. First we find that

$$x^2 = \frac{1}{4}x'^2 + \frac{3}{4}y'^2 - \frac{\sqrt{3}}{2}x'y',$$

$$y^2 = \frac{3}{4}x'^2 + \frac{1}{4}y'^2 + \frac{\sqrt{3}}{2}x'y',$$

$$xy = \frac{\sqrt{3}}{4}x'^2 - \frac{\sqrt{3}}{4}y'^2 + \left(\frac{1}{4} - \frac{3}{4}\right)x'y' = \frac{\sqrt{3}}{4}x'^2 - \frac{\sqrt{3}}{4}y'^2 - \frac{1}{2}x'y'.$$

Substituting into the original equation, we find that

$$5 = 4x^2 + 2y^2 - 2\sqrt{3}xy$$

$$= (x'^2 + 3y'^2 - 2\sqrt{3}x'y') + \left(\frac{3}{2}x'^2 + \frac{1}{2}y'^2 + \sqrt{3}x'y'\right) - \left(\frac{3}{2}x'^2 - \frac{3}{2}y'^2 - \sqrt{3}x'y'\right),$$

that is,

$$5 = x'^2 + 5y'^2.$$

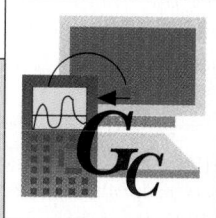

If we want to sketch this ellipse using a graphics calculator, we use the quadratic formula to solve for y. We write $4x^2 + 2y^2 - 2\sqrt{3}xy = 5$ as $2y^2 - (2\sqrt{3}x)y + (4x^2 - 5) = 0$, so $y = \frac{2\sqrt{3}x \pm \sqrt{12x^2 - 8(4x^2 - 5)}}{4} = \frac{2\sqrt{3}x \pm \sqrt{40 - 20x^2}}{4} = \frac{\sqrt{3}x \pm \sqrt{10 - 5x^2}}{2}.$

TI-85	SHARP EL9300
GRAPH F1$(y(x) =)$ ($\sqrt{}$(3) $\times$ F1(x) $+$ $\sqrt{}$ (10 $-$ 5 F1(x) x^2)) $\div$ 2 ▼ ($\sqrt{}$(3) $\times$ F1(x) $-$ $\sqrt{}$ (10 $-$ 5 F1(x) x^2)) $\div$ 2 M2(RANGE) (−) 2 ▼ 2 ▼ ▼ (−) 2 ▼ 2 F5(GRAPH)	♉ ($\sqrt{}$3 ► $\times$ x/θ/T $+$ $\sqrt{}$ 10 $-$ 5 x/θ/T x^2 ►) $\div$ 2 2ndF ▼ ($\sqrt{}$3 ► $\times$ x/θ/T $-$ $\sqrt{}$ 10 $-$ 5 x/θ/T x^2 ►) $\div$ 2 RANGE $-$2 ENTER 2 ENTER 1 ENTER $-$2 ENTER 2 ENTER 1 ENTER ♉
CASIO f_x-6300G (f_x-7700GB)	HP 48SX
Range $-$2 EXE 2 EXE 1 EXE $-$2 EXE 2 EXE 1 EXE EXE EXE EXE Graph ($\sqrt{}$3 $\times$ X $+$ $\sqrt{}$ (10 $-$ 5 X x^2)) $\div$ 2 EXE Graph ($\sqrt{}$3 $\times$ X $-$ $\sqrt{}$ (10 $-$ 5 X x^2)) $\div$ 2 EXE	↰ PLOT PLOTR ERASE ATTN ' $Y =$ () $\sqrt{x}$ 3 $\times$ X $+$ $\sqrt{x}$ () 10 $-$ 5 $\times$ X y^x 2 ► ► $\div$ 2 ↱ PLOT ↰ DRAW ERASE 2 $+/-$ENTER 2 XRNG 2 $+/-$ENTER 2 YRNG DRAW ATTN ' $Y =$ () $\sqrt{x}$ 3 $\times$ X $-$ $\sqrt{x}$ () 10 $-$ 5 $\times$ X y^x 2 ► ► $\div$ 2 ↱ PLOT ↰ DRAW DRAW

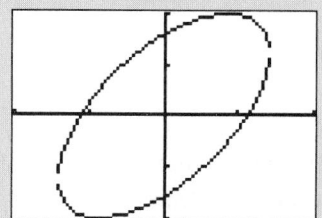

Distortion from unequal scales on the axes may make it difficult to guess the angle through which the conic was rotated relative to the standard position.

Of course, the interesting thing about this new equation is that it contains no cross-term. It is immediately recognized as the equation of an ellipse, and it is quite easy to sketch its graph. We draw in the rotated axes and then sketch the ellipse relative to them. See Figure 11.3.5.

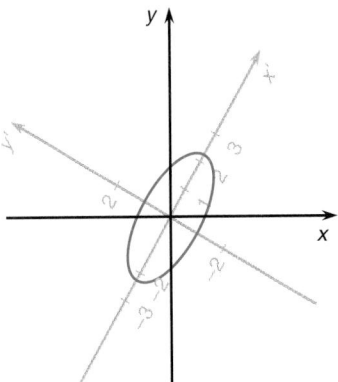

Figure 11.3.5

We were able to use the formulas from Theorem 11.3.1 that relate coordinates relative to the two sets of axes to convert the original equation into an equation in terms of the other set of coordinates. The resulting equation turned out to be easily recognized because it contained no cross-term.

Of course, in this example we were given the angle $\theta = \frac{\pi}{3}$, and in hindsight we see that this gives coordinate axes along the axes of the ellipse. On the other hand, if we simply guessed a value for θ, it is unlikely that it would correspond to the correct angle, and the converted equation would still have a cross-term and would still be difficult to identify and graph. The problem we have to consider now is this: Given an equation that contains a cross-term, how do we find a suitable angle θ through which to rotate the axes in order to remove the cross-term?

Consider the equation $Ax^2 + Bxy + Cy^2 + Dx + Ey = q$. Let us suppose that $B \neq 0$. If we rotate the axes through an angle θ and make the substitutions

$$x = x' \cos\theta - y' \sin\theta, \qquad y = x' \sin\theta + y' \cos\theta,$$

then we could actually do the calculation and work out the transformed equation exactly. But since we are principally interested in removing the cross-term, let us simply find the cross-term in the transformed equation and try to make it zero.

The coefficient of $x'y'$ in Ax^2 is $A(-2\cos\theta\sin\theta) = -2A\cos\theta\sin\theta$; the coefficient of $x'y'$ in Cy^2 is $2C\cos\theta\sin\theta$; and the coefficient of $x'y'$ in Bxy is $B(\cos^2\theta - \sin^2\theta)$. The remaining terms $Dx + Ey$ and k do not contain any $x'y'$-terms. The total $x'y'$-term is

$$-2A\cos\theta\sin\theta + 2C\cos\theta\sin\theta + B(\cos^2\theta - \sin^2\theta)$$
$$= (C - A)\sin 2\theta + B\cos 2\theta.$$

Our goal is to make this equal to zero, that is, $(A - C)\sin 2\theta = B\cos 2\theta$. This is the same as $\frac{A-C}{B} = \cot 2\theta$. So to find the angle θ of rotation that will remove the cross-term, we simply solve $\cot 2\theta = \frac{A-C}{B}$. Notice that we have assumed that $B \neq 0$, so there is no difficulty about dividing.

On the other hand, we are more familiar with tan than cot, so we might like to write the equation as $\frac{B}{A-C} = \tan 2\theta$, or $\theta = \frac{1}{2}\arctan\frac{B}{A-C}$. This might be preferable, especially for people whose calculators have an arctan button and not an arccot button, but it does have the disadvantage of dividing by 0 when $A = C$.

In Example 2 we rotated by $\theta = \frac{\pi}{3}$; if we want $-\frac{\pi}{4} < \theta \le \frac{\pi}{4}$, we could have used $\theta = -\frac{\pi}{6}$.

Let us consider what happens when $A = C$. Then $\cot 2\theta = 0$, which happens, for instance, when $2\theta = \frac{\pi}{2}$, that is, when $\theta = \frac{\pi}{4}$. It is convenient to remember this: When $A = C$, we choose $\theta = \frac{\pi}{4}$; otherwise, we use $\theta = \frac{1}{2} \arctan \frac{B}{A-C}$. As usual we can always assume that $-\frac{\pi}{2} < \arctan x < \frac{\pi}{2}$, so $-\frac{\pi}{4} < \theta \le \frac{\pi}{4}$. We have the following result.

THEOREM 11.3.2

Consider the equation $Ax^2 + Bxy + Cy^2 + Dx + Ey = q$, in which we assume that $B \neq 0$. The cross-term can be removed by rotating the axes through an angle θ, where

$$\theta = \begin{cases} \frac{\pi}{4}, & \text{if } A = C, \\ \frac{1}{2} \arctan \frac{B}{A-C}, & \text{otherwise.} \end{cases}$$

We can assume that $-\frac{\pi}{4} < \theta \le \frac{\pi}{4}$.

EXAMPLE 3   Identify and sketch the graph of $5x^2 + 26xy + 5y^2 = 72$.

Solution   Noticing that the coefficients of x^2 and y^2 are equal, we realize that the angle of rotation should be $\theta = \frac{\pi}{4}$. We consider rotated axes and write

$$x = x' \cos\theta - y' \sin\theta = \frac{x'}{\sqrt{2}} - \frac{y'}{\sqrt{2}},$$

$$y = x' \sin\theta + y' \cos\theta = \frac{x'}{\sqrt{2}} + \frac{y'}{\sqrt{2}}.$$

From this we find that

$$x^2 = \frac{x'^2}{2} + \frac{y'^2}{2} - x'y', \qquad y^2 = \frac{x'^2}{2} + \frac{y'^2}{2} + x'y', \qquad xy = \frac{x'^2}{2} - \frac{y'^2}{2}.$$

Substituting these expressions into the original equation, we find that it becomes

$$72 = 5x^2 + 5y^2 + 26xy$$
$$= 5\left(\frac{x'^2}{2} + \frac{y'^2}{2} - x'y'\right) + 5\left(\frac{x'^2}{2} + \frac{y'^2}{2} + x'y'\right) + 26\left(\frac{x'^2}{2} - \frac{y'^2}{2}\right)$$
$$= 18x'^2 - 8y'^2,$$

or

$$\frac{x'^2}{4} - \frac{y'^2}{9} = 1.$$

This is the equation of a hyperbola, which we sketch in Figure 11.3.6 (after marking the asymptotes).

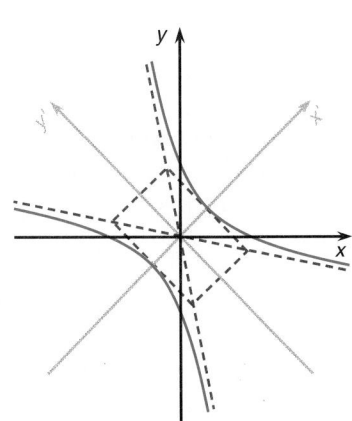

Figure 11.3.6

It is interesting to notice that although the coefficients of x^2 and y^2 in the original equation were both positive, the conic did not turn out to be an ellipse. It is dangerous to try to identify the conic from an equation containing a cross-term.

EXAMPLE 4 Identify and sketch the conic whose equation is $7x^2 + 24xy + 144 = 0$.

Solution From Theorem 11.3.2 we find that the correct angle is $\theta = \frac{1}{2}\arctan\frac{24}{7}$. Writing $2\theta = \arctan\frac{24}{7}$, that is, $\tan 2\theta = \frac{24}{7}$, we find

$$\frac{1}{\cos^2 2\theta} = \sec^2\theta = 1 + \tan^2 2\theta = 1 + \left(\frac{24}{7}\right)^2 = \frac{49 + 576}{49} = \frac{625}{49},$$

so

$$\cos 2\theta = \sqrt{\frac{49}{625}} = \frac{7}{25},$$

and

We use the half-angle formula (Theorem 1.7.13), noting that $\cos\theta > 0 \ldots$

$$\cos\theta = \sqrt{\frac{\cos 2\theta + 1}{2}} = \sqrt{\frac{\frac{7}{25} + 1}{2}} = \sqrt{\frac{\frac{32}{25}}{2}} = \sqrt{\frac{16}{25}} = \frac{4}{5}.$$

Since $\tan 2\theta > 0$, we see that $\theta > 0$ and $\sin\theta > 0$. So $\sin\theta = \sqrt{1 - \cos^2\theta} = \sqrt{1 - \frac{16}{25}} = \sqrt{\frac{9}{25}} = \frac{3}{5}$. We write

$$x = x'\cos\theta - y'\sin\theta = \frac{4}{5}x' - \frac{3}{5}y',$$

$$y = x'\sin\theta + y'\cos\theta = \frac{3}{5}x' + \frac{4}{5}y'.$$

Substituting, we find

$$0 = 7x^2 + 24xy + 144$$
$$= 7\left(\frac{16}{25}x'^2 + \frac{9}{25}y'^2 - \frac{24}{25}x'y'\right) + 24\left(\frac{12}{25}x'^2 - \frac{12}{25}y'^2 + \frac{7}{25}x'y'\right) + 144$$
$$= \frac{400}{25}x'^2 - \frac{225}{25}y'^2 + 144 = 16x'^2 - 9y'^2 + 144$$

or

$$\frac{y'^2}{16} - \frac{x'^2}{9} = 1.$$

This is the equation of a hyperbola, which is easily sketched (see Figure 11.3.7). Observe that one of the asymptotes is the y-axis. Notice that once again it would not have been easy to guess from the original equation which type of conic it was. ◂

EXAMPLE 5 Graph the conic $xy = 1$.

Solution This equation has a nonzero cross-term; in fact, that and the constant term are the only nonzero terms. The coefficients of x^2 and y^2 are both zero; in particular, they are the same, so we rotate the axes by $\theta = \frac{\pi}{4}$. Writing

$$x = \frac{x'}{\sqrt{2}} - \frac{y'}{\sqrt{2}}, \qquad y = \frac{x'}{\sqrt{2}} + \frac{y'}{\sqrt{2}},$$

Figure 11.3.7

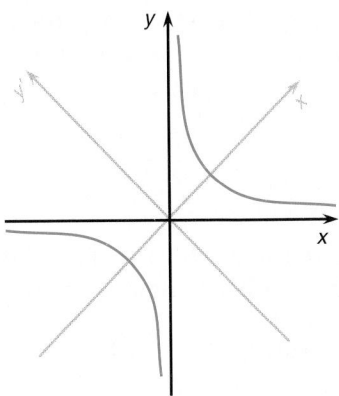

Figure 11.3.8

we find that the equation becomes

$$1 = xy = \frac{x'^2}{2} - \frac{y'^2}{2},$$

which is the equation of a hyperbola.

It is easily sketched, as shown in Figure 11.3.8.

In this particular case we might have rewritten the equation $xy = 1$ as $y = \frac{1}{x}$, and we would have seen how to graph it that way.

None of the examples we have considered so far has had an x-term or a y-term, but the way Theorem 11.3.2 was stated, it applies equally well to conics of that type. The complication is that if we start with an equation containing an x- and/or a y-term, we should expect that after rotating there will be an x'- and/or y'-term. To identify and graph the conic, it will likely be necessary to translate the coordinates *after* rotating them.

EXAMPLE 6 Sketch the graph of $7x^2 + 6\sqrt{3}xy + 13y^2 - (32 - 4\sqrt{3})x - (4 + 32\sqrt{3})y + 32 = 0$.

Solution We begin by finding the angle:

$$\theta = \frac{1}{2}\arctan\frac{B}{A-C} = \frac{1}{2}\arctan\frac{6\sqrt{3}}{-6} = \frac{1}{2}\arctan(-\sqrt{3}) = -\frac{1}{2}\frac{\pi}{3} = -\frac{\pi}{6}.$$

As usual, we write

$$x = x'\cos\theta - y'\sin\theta = \frac{\sqrt{3}}{2}x' + \frac{1}{2}y',$$

$$y = x'\sin\theta + y'\cos\theta = -\frac{1}{2}x' + \frac{\sqrt{3}}{2}y',$$

from which we find that

$$x^2 = \frac{3}{4}x'^2 + \frac{\sqrt{3}}{2}x'y' + \frac{1}{4}y'^2,$$

$$y^2 = \frac{1}{4}x'^2 - \frac{\sqrt{3}}{2}x'y' + \frac{3}{4}y'^2,$$

$$xy = -\frac{\sqrt{3}}{4}x'^2 + \frac{1}{2}x'y' + \frac{\sqrt{3}}{4}y'^2.$$

Substituting in the original equation, we find that it is

$$-32 = 7x^2 + 13y^2 + 6\sqrt{3}xy - (32 - 4\sqrt{3})x - (4 + 32\sqrt{3})y$$

$$= \frac{21}{4}x'^2 + \frac{7\sqrt{3}}{2}x'y' + \frac{7}{4}y'^2 + \frac{13}{4}x'^2 - \frac{13\sqrt{3}}{2}x'y' + \frac{39}{4}y'^2 - \frac{18}{4}x'^2$$

$$+ \frac{6\sqrt{3}}{2}x'y' + \frac{18}{4}y'^2 - (16\sqrt{3} - 6)x' - (16 - 2\sqrt{3})y'$$

$$- (-2 - 16\sqrt{3})x' - (2\sqrt{3} + 48)y'$$

$$= \frac{16}{4}x'^2 + \frac{64}{4}y'^2 + 8x' - 64y' = 4x'^2 + 16y'^2 + 8x' - 64y'.$$

Figure 11.3.9(i)

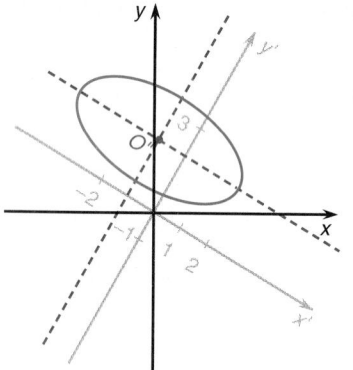

Figure 11.3.9(ii)

The equation has become $4x'^2 + 16y'^2 + 8x' - 64y' = -32$, or $x'^2 + 4y'^2 + 2x' - 16y' = -8$. We can complete the squares to find $(x' + 1)^2 - 1 + 4(y' - 2)^2 - 16 = -8$, or $x''^2 + 4y''^2 = 9$, where $x'' = x' + 1$ and $y'' = y' - 2$. This is

$$\frac{x''^2}{9} + \frac{4y''^2}{9} = 1,$$

which is the equation of an ellipse.

To graph it, we first plot the rotated x'- and y'-axes in Figure 11.3.9(i). Since the conic was translated relative to these axes, we draw the translated origin O'', whose coordinates relative to the x'- and y'-axes are $(x', y') = (-1, 2)$.

In Figure 11.3.9(ii) we draw the ellipse. The idea is to work relative to the rotated axes. The ellipse is drawn with its center at O'', the translated origin, and its axes parallel to the x'- and y'-axes. We could have labeled the axes through O'' as the x''- and y''-axes, but that is probably an unnecessary complication.

In doing this type of question there are two steps. First we rotated the axes to remove the xy-term, and then we translated the coordinates to remove the linear terms (i.e., the x'- and y'-terms). This second step was accomplished by completing the squares.

It is reasonable to ask whether it will work in the opposite order, that is, by completing the squares first. Curiously, it does not. The problem is that if we complete the squares and let $x' = x - k$, $y' = y - \ell$, say, then we can write $x = x' + k$ and $y = y' + \ell$. From this we find that $xy = x'y' + \ell x' + k y' + k\ell$, and translating the coordinates converts the xy-term into something that still contains x'- and y'-terms.

Because of this it makes sense to rotate the axes before trying to translate the axes to remove the linear terms.

Summary 11.3.3   Graphing Quadratic Polynomial Equations

Suppose we want to sketch the graph of the quadratic polynomial equation

$$Ax^2 + Bxy + Cy^2 + Dx + Ey = q.$$

1. If $B = 0$, we proceed according to the description in Summary 11.1.3 for graphing conics.

2. Suppose that $B \neq 0$. Then first it is necessary to rotate the axes to remove the cross-term Bxy. The necessary angle is $\theta = \frac{1}{2}\arctan\frac{B}{A-C}$, with the understanding that $\theta = \frac{\pi}{4}$ if $A = C$ and that in any case, $-\frac{\pi}{4} < \theta \leq \frac{\pi}{4}$.
   Use the rotated coordinates x' and y' defined by
   $$x = x'\cos\theta - y'\sin\theta,$$
   $$y = x'\sin\theta + y'\cos\theta.$$

   Substitute these formulas into the equation; the resulting equation should not have any cross-term.

3. Now the problem is one of graphing an equation that has no cross-term, so we proceed as in Summary 11.1.3. Specifically, we complete the squares and/or translate coordinates as needed to remove the x'-term if there is an x'^2-term and to remove the constant term if there is no x'^2-term (and similarly with the y'-terms).

4. Now it should be possible to express the equation in one of the standard forms and so identify its type.

5. To sketch the graph, the first step is to draw the rotated x'- and y'-axes. If the axes were translated, it is then necessary to plot the translated origin. It is important to plot this new origin *relative to the x'- and y'-axes*. After this is done, it should be straightforward to sketch the graph.

In doing the examples we have considered so far, we have noticed that it is difficult to predict the type of a conic from its equation if the equation contains a cross-term. In fact there is a way to do this, which we now describe.

THEOREM 11.3.4

Consider the equation

$$Ax^2 + Bxy + Cy^2 + Dx + Ey = q.$$

(i) If $B^2 - 4AC < 0$, then the graph is an ellipse or a circle if it is a conic, and if it is degenerate it can be a single point or no points at all.

(ii) If $B^2 - 4AC > 0$, then the graph is a hyperbola if it is a conic, and if it is degenerate it can be two intersecting lines.

(iii) If $B^2 - 4AC = 0$, then the graph is a parabola if it is a conic, and if it is degenerate it can be a line, a pair of parallel lines, or no points at all.

The quantity $B^2 - 4AC$ is called the **discriminant** of the equation $Ax^2 + Bxy + Cy^2 + Dx + Ey = q$. It is not accidental that it is the same formula that we encountered for the discriminant of a quadratic polynomial in one variable (cf. 1.3.4), but we will not pursue the connection here.

If you have trouble remembering which condition in Theorem 11.3.4 corresponds to which type of conic, it is easy to check it on the standard forms.

EXAMPLE 7   Discriminants of Some Conics in Standard Form

(i) For the equation $2x^2 + 7y^2 = 1$ the discriminant is $B^2 - 4AC = 0^2 - 4(2)(7) = -56$. The graph is an ellipse, corresponding to a negative discriminant. For the circle $x^2 + y^2 = 5$ the discriminant is -4.

It is also possible to try examples of degenerate conics. For instance, the equation $x^2 + y^2 = 0$ has only one solution, the point $(0,0)$, and the equation $4x^2 + 9y^2 = -3$ has none at all. Both have negative discriminants.

(ii) The discriminant of $x^2 - y^2 = -1$ is $B^2 - 4AC = 0^2 - 4(1)(-1) = 4$. We see that this hyperbola corresponds to a positive discriminant. The related example $x^2 - y^2 = 0$, which has the same discriminant, can be written as $(x + y)(x - y) = 0$, so its graph is the two lines $x + y = 0$ and $x - y = 0$, which are the asymptotes of the hyperbola we just discussed.

(iii) The discriminant of $y = x^2$, that is, $x^2 - y = 0$, is $0^2 - 4(1)0 = 0$. This parabola has discriminant equal to zero.

It is also possible to find degenerate examples. For instance, the equation $x^2 = 0$ has as its graph the line $x = 0$. The equation $x^2 - x = 0$ can be written as $x(x - 1) = 0$, so its graph is the two (parallel) lines $x = 0$ and $x = 1$. The equation $y^2 = -7$ has no solutions. In each of these cases the discriminant is zero.

The value of the discriminant does not depend on the coefficients of x and y or the constant term. In fact, what makes the theorem work is that when we change an equation by rotating the axes, the discriminant does not change. (You can try this out on the examples we have already done.) Of course, this reflects the observation that rotating the axes does not change the nature of the graph, merely its orientation.

The proof of this fact is moderately messy; the proof of Theorem 11.3.4 is quite complicated. Strangely enough, it is easier to prove both of them by using the techniques of linear algebra. That is a subject outside the province of this book, but if you are familiar with linear algebra, there is a brief discussion in the "Point to Ponder" after the exercises.

The discriminant can be used to identify the type of a conic in a situation in which its exact graph may not be important and the calculations for the rotation are elaborate.

EXAMPLE 8   The equation $7x^2 - 5xy + 2y^2 - 12x + 8y = 13$ has discriminant $B^2 - 4AC = (-5)^2 - 4(7)2 = -31$. We can conclude without doing any more work that it is an ellipse, a circle, a point, or nothing at all. In fact it is an ellipse.   ◢

Exercises 11.3

I

For each of the following questions, the x'- and y'-axes are rotated through the specified angle θ relative to the x- and y-axes. The (x', y') or (x, y) coordinates of a point P are given; find its coordinates relative to the other set of axes.

1. $(x', y') = (1, 2)$, $\theta = \frac{\pi}{4}$
2. $(x', y') = (-5, 5)$, $\theta = \frac{3\pi}{4}$
3. $(x', y') = (0, 3)$, $\theta = \frac{\pi}{3}$
4. $(x', y') = (2, -5)$, $\theta = \pi$
5. $(x, y) = (1, 7)$, $\theta = \frac{\pi}{6}$
6. $(x, y) = (1, -4)$, $\theta = \frac{4\pi}{3}$
7. $(x, y) = (a, 0)$, $\theta = \frac{5\pi}{6}$
8. $(x', y') = (11, -5)$, $\theta = \frac{\pi}{2}$
9. $(x', y') = (0, 0)$, $\theta = \frac{13\pi}{17}$
10. $(x, y) = (1, 0)$, $\theta = \frac{2\pi}{3}$

Identify and sketch each of the following conics.

11. $13x^2 + 10xy + 13y^2 = 2$
12. $x^2 + \sqrt{3}xy = 2$
13. $x^2 - 2\sqrt{3}xy + 3y^2 - 2\sqrt{3}x - 2y = 0$
14. $x^2 + 6\sqrt{3}xy - 5y^2 = 8$
15. $52x^2 + 72xy + 73y^2 = 400$
16. $25x^2 + 120xy + 144y^2 - 156x + 65y = -507$
    (*Hint:* Find $13\cos\theta$.)

17. $x^2 - 8xy + 7y^2 = 4$ (*Hint:* Find $\cos^2\theta$.)
18. $9x^2 - 6xy + 17y^2 = 72$ (*Hint:* Find $\cos^2\theta$.)
19. $(\sqrt{3} - 1)(x^2 + y^2) + 2(\sqrt{3} + 1)xy = 6$
20. $x^2 + xy = 2$

Identify and sketch each of the following conics.

21. $x^2 + 2xy + y^2 + 4x - 2y = 6$
22. $x^2 - 2\sqrt{3}xy - y^2 + x + \sqrt{3}y = -8$
23. $11x^2 + 24xy + 4y^2 - 18x - 16y + 5 = 0$
24. $x^2 - 4xy + 4y^2 + 2\sqrt{5}x - 9\sqrt{5}y = 0$
25. $31x^2 + 10\sqrt{3}xy + 21y^2 + (16 - 36\sqrt{3})x - (36 + 16\sqrt{3})y = 92$
26. $7x^2 - 2xy + 7y^2 + 18\sqrt{2}x - 30\sqrt{2}y + 54 = 0$
27. $x^2 - 6x - y^2 - 8y = 11$
28. $5x^2 + 6xy + 5y^2 - 4x - 12y = 0$
29. $x^2 + 4xy + 4y^2 + 50x = 0$
30. $2xy - 2\sqrt{2}x + 1 = 0$

II

*31. Try sketching the graph of $5x^2 + 26xy + 5y^2 = 72$ by rotating the axes through an angle of $-\frac{\pi}{4}$. Compare your solution with what we did in Example 3 (p. 745) using $+\frac{\pi}{4}$. Is your equation the same? Is your picture the same?

*32. Starting with the equation $x^2 + 3xy + y^2 + 2x - 4y = 12$, try completing the squares before you rotate the axes. Verify that this does not have the effect of removing the linear terms (i.e., the x- and y-terms).

*33. For what values of q is $x^2 - xy + 2x = q$ the equation of a degenerate conic? Sketch the graph for several values of q, including degenerate and non-degenerate cases.

*34. When we rotate a conic, we use the angle θ to know where to draw the rotated axes. On the other hand, for purposes of doing the calculations, all we really need to know is $\cos\theta$ and $\sin\theta$.

If we know that $\phi = \arctan\frac{B}{A-C}$, with $-\frac{\pi}{2} < \phi \le \frac{\pi}{2}$, then using Figure 11.3.10, we see that

$$\cos\phi = \frac{1}{\sqrt{1 + \frac{B^2}{(A-C)^2}}}, \qquad \sin\phi = \frac{\frac{B}{A-C}}{\sqrt{1 + \frac{B^2}{(A-C)^2}}}.$$

The angle of rotation is $\theta = \frac{1}{2}\phi = \frac{1}{2}\arctan\frac{B}{A-C}$.

(i) Use the above formulas and the half-angle formulas (Theorem 1.7.13, p. 52) to find expressions for $\cos\theta$ and $\sin\theta$.

(ii) Check your formula by applying it to Example 4 (p. 746).

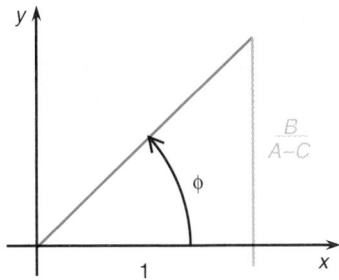

Figure 11.3.10

*35. Prove the statement made in the text that the discriminant does not change if the axes are rotated. (*Hint:* Try it first for an equation with no linear terms, i.e., an equation of the form $Ax^2 + Bxy + Cy^2 = q$.)

*36. Consider the equation $Ax^2 + Bxy + Cy^2 + Dx + Ey = q$, and suppose the coefficients A, B, and C are all integers, with $C \ne A$. Show that it is not possible to rotate the conic by $\theta = \frac{\pi}{6}$ so that the resulting conic has no cross-term.

**37. (i) If $B \ne 0$ and $B^2 - 4AC = 0$, show that $Ax^2 + Bxy + Cy^2$ can be written in the form $Ax^2 + Bxy + Cy^2 = R(ax + by)^2$, for some numbers R, a, and b, with $R \ne 0$.

(ii) Then show that the equation $Ax^2 + Bxy + Cy^2 + Dx + Ey = q$ can be written as $R(ax + by)^2 + S(ax + by) + T(ax - by) = Q$, for some numbers S, T, and Q.

(iii) Use this to show that the graph is either a parabola, one or two straight lines, or the empty set.

*38. (i) Consider the equation $x^2 + y^2 - 8x + 4y = 5$, and rotate it through an angle of $\theta = -\frac{\pi}{6}$. What is the coefficient of the cross-term in the resulting equation?

(ii) What is the coefficient of the cross-term after the original conic is rotated through an angle of $\theta = \frac{\pi}{8}$?

(iii) Find a formula for the coefficient of the cross-term in the equation that results from rotating the original equation through an angle θ.

39. Verify that the discriminant of the conic in Example 4 (p. 746) was not changed by rotating.

40. Verify that the discriminant of the conic in Example 6 (p. 747) was not changed by rotating.

Find the points on each of the following conics that are closest to the origin.

**41. $x^2 + y^2 + 2xy + x - y = 3$

**42. $2x^2 + 2\sqrt{3}xy + 1 = 0$

**43. $15x^2 - 24xy + 8y^2 - 4 = 0$

**44. Find the points on the graph of $x^2 + 2xy + 2y^2 = 0$ that are farthest from the origin.

**45. Find the equation of the ellipse with foci $(2\sqrt{2}, -2\sqrt{2})$, $(-2\sqrt{2}, 2\sqrt{2})$ and semi-major axis 5.

**46. Find the equation of the hyperbola with foci $(-1, 1)$, $(1, -1)$ and the coordinate axes as asymptotes.

|||

47. Use a graphics calculator or computer to sketch the graph of the conic $7x^2 - 5xy + 2y^2 - 12x + 8y = 13$ from Example 8 (p. 750). What is the angle θ through which the axes are rotated?

At the end of this section it was remarked that the results about the discriminant could be proved by using linear algebra. This is the study of matrices and related topics, and it may be unfamiliar to you, but if you do happen to know about it, you may already have observed that the formula given in Theorem 11.3.1 for rotating coordinates can be written as a matrix multiplication:

$$\begin{pmatrix} x \\ y \end{pmatrix} = \begin{pmatrix} \cos(\theta)x' - \sin(\theta)y' \\ \sin(\theta)x' + \cos(\theta)y' \end{pmatrix}$$
$$= \begin{pmatrix} \cos\theta & -\sin\theta \\ \sin\theta & \cos\theta \end{pmatrix} \begin{pmatrix} x' \\ y' \end{pmatrix}.$$

The matrix $\begin{pmatrix} \cos\theta & -\sin\theta \\ \sin\theta & \cos\theta \end{pmatrix}$ in this formula is what is often called a *change of basis matrix*.

Now we have observed that the discriminant $B^2 - 4AC$ does not depend on the linear or constant terms of an equation. If we consider just the quadratic terms, $Ax^2 + Bxy + Cy^2$, we can write them as a matrix product too:

$$Ax^2 + Bxy + Cy^2 = (x \quad y) \begin{pmatrix} A & \frac{B}{2} \\ \frac{B}{2} & C \end{pmatrix} \begin{pmatrix} x \\ y \end{pmatrix}.$$

This matrix product turns out to be a 1×1 matrix, which is just a number, and you can check that it is the right number, that is, $Ax^2 + Bxy + Cy^2$.

Next we notice that for the equation to have no cross-terms is exactly the same thing as for the big matrix in this formula to be diagonal. Removing the cross-terms from the equation is exactly the same problem as "diagonalizing" this matrix. This is one of the major problems that linear algebra enables us to solve.

In fact, if we use the first formula for $\begin{pmatrix} x \\ y \end{pmatrix}$ in terms

of $\begin{pmatrix} x' \\ y' \end{pmatrix}$, this last formula can be written as

$$Ax^2 + Bxy + Cy^2 = (x \quad y) \begin{pmatrix} A & \frac{B}{2} \\ \frac{B}{2} & C \end{pmatrix} \begin{pmatrix} x \\ y \end{pmatrix}$$
$$= (x' \quad y') \begin{pmatrix} \cos\theta & \sin\theta \\ -\sin\theta & \cos\theta \end{pmatrix}$$
$$\begin{pmatrix} A & \frac{B}{2} \\ \frac{B}{2} & C \end{pmatrix} \begin{pmatrix} \cos\theta & -\sin\theta \\ \sin\theta & \cos\theta \end{pmatrix} \begin{pmatrix} x' \\ y' \end{pmatrix}.$$

This is really just the *change of basis formula* for the matrix $\begin{pmatrix} A & B/2 \\ B/2 & C \end{pmatrix}$. In particular, we notice that the discriminant of the conic is just -4 times the determinant of the matrix, and in linear algebra we learn that the determinant is unchanged by changing bases. This explains why rotating does not change the discriminant.

Finally, we should remark that if $B = 0$, so the conic has no cross-term, then the matrix $\begin{pmatrix} A & B/2 \\ B/2 & C \end{pmatrix}$ is diagonal, and the coefficients A and C of x^2 and y^2 are precisely its *eigenvalues*. This was one of the first and is still one of the most important uses of eigenvalues.

If you do not know about linear algebra, then you have probably not managed to read this description, but you will be pleased to know that all the results of this section can be proved without any knowledge of linear algebra. However, some of the proofs are quite complicated; linear algebra actually provides a better framework within which to understand them.

None of this has much bearing on our use of the material in this section. You should *not* go out and try to learn all of linear algebra over the weekend.

Chapter Summary

§11.1
translation of axes
completing the square
conics in standard form (Table 11.1.2)
graphing conics (Summary 11.1.3)

§11.2
focus
reflection property of ellipse
directrix
reflection property of parabola

eccentricity
classification of conics by eccentricity
formulas for foci
focal axis
§11.3
rotated coordinates

$\theta = \frac{1}{2} \arctan \frac{B}{A-C}$
graphing quadratic equations with cross-terms (Summary 11.3.3)
degenerate conics
discriminant

Review Exercises

I

Identify and sketch each of the following conics.

1. $9x^2 + 16y^2 = 144$
2. $9y^2 - 4x^2 = 36$
3. $x - 4y^2 = 2$
4. $x^2 + y^2 = 2$
5. $x^2 + 2y^2 = 4$
6. $2x^2 - 3y^2 + 6 = 0$
7. $x^2 + 4y^2 - 4x = 0$
8. $4x^2 - 9y^2 + 8x + 54y = 113$
9. $x^2 + 4x - 2y = 3$
10. $3x^2 - 12x = 3y^2 + 6y$

Identify and sketch each of the following conics.

11. $41x^2 + 18xy + 41y^2 = 800$
12. $11x^2 - 10\sqrt{3}xy + y^2 = 48$
13. $9x^2 + 24xy + 16y^2 + 40x - 30y = 0$
14. $32x^2 - 52xy - 7y^2 + 180 = 0$
15. $4xy - 6\sqrt{2}x - 2\sqrt{2}y + 8 = 0$
16. $(x+2y)^2 = x$
17. $93x^2 + 48xy + 107y^2 - 180x - 490y = 50$
18. $119x^2 - 240xy - 119y^2 + 312x - 130y = 0$
19. $31x^2 - 24xy - 14y^2 - 8\sqrt{17}x - 32\sqrt{17}y = 238$
20. $(x+y)^2 - (x-y)^2 = 4$

II

*21. For which value(s) of c is the following conic degenerate: $2x^2 + 3y^2 + 4x - 12y = c$? Identify the graph in each of the various cases.

*22. For which value(s) of c is the following conic degenerate: $x^2 - 4y^2 + cx = 4$? Identify the graph in each of the various cases.

*23. Sketch the graph of $x^2 + y^2 + 2x - 2y = c$ for the following values of c: (i) $c = 0$, (ii) $c = 2$, (iii) $c = -1$, (iv) $c = -2$, (v) $c = -3$.

*24. Sketch the graph of $4x^2 - 9y^2 + 16x + c = 0$ for the following values of c: (i) $c = 0$, (ii) $c = 16$, (iii) $c = -20$, (iv) $c = 52$.

*25. Sketch the graph of $x^2 + 2xy + y^2 + c(x-y) = 0$ for the following values of c: (i) $c = 1$, (ii) $c = \sqrt{2}$, (iii) $c = -1$, (iv) $c = 0$.

*26. Sketch the graph of $xy = c$ for the following values of c: (i) $c = 1$, (ii) $c = 2$, (iii) $c = -1$, (iv) $c = 0$.

*27. Consider the ellipse

$$\frac{x^2}{a^2} + \frac{y^2}{b^2} = 1.$$

(i) Show that for any value of t, the point $(x, y) = (a \cos t, b \sin t)$ lies on the ellipse.

(ii) This is analogous to the fact that $(x, y) = (r \cos \theta, r \sin \theta)$ lies on the circle $x^2 + y^2 = r^2$. As θ ranges from 0 to 2π, this point goes once around the circle, counterclockwise, starting and ending at the point $(r, 0)$. Describe what happens to the point $(a \cos t, b \sin t)$ as t ranges from 0 to 2π.

*28. Consider the hyperbola

$$\frac{x^2}{a^2} - \frac{y^2}{b^2} = 1.$$

(i) Show that for any value of t, the point $(x, y) = (a \cosh t, b \sinh t)$ lies on the hyperbola.

(ii) Describe what happens to the point $(a \cosh t, b \sinh t)$ as t ranges from $-\infty$ to ∞.

(iii) Describe a similar construction for points on the other branch of the hyperbola.

*29. Consider a hyperbola centered at the origin with no cross-terms in its equation. If we know that its eccentricity is $e > 1$, what are the possible equations of its asymptotes? (There are usually two possible pairs of asymptotes.)

*30. In Exercise 29 you found the possible asymptotes for a hyperbola with eccentricity e. For what value(s) of e will the two possible pairs of asymptotes actually be the same?

**31. We have considered transforming conics by rotating the axes, but it is possible to think of other ways to move them. One simple type of transformation is the *shear*, in which, for instance, the x-axis remains fixed and the y-axis is pulled sideways, as illustrated in the figure below.

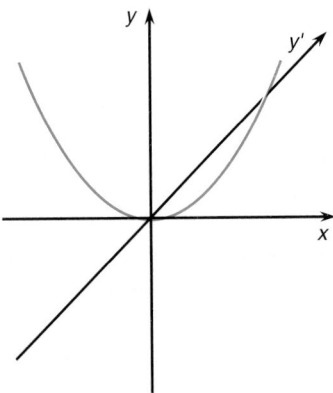

If we label the new axes x' and y', then the coordinates relative to them are

$$x' = x - y, \qquad y' = y.$$

(i) Consider the equation of the parabola $y = x^2$, and express it in terms of the sheared coordinates (x', y').

(ii) Identify the conic whose equation you found in part (i).

(iii) Sketch the graph of $y = x^2$, and then sketch the graph of the equation you found in part (i), but sketch it on a graph in which the x'-axis is horizontal and the y'-axis is vertical (i.e., in which they are not sheared).

It can be shown that the type of the conic does not change even when the axes undergo a shear.

**32. Repeat Exercise 31 for the parabola $x = y^2$.

**33. If the conic $Ax^2 + Bxy + Cy^2 + Dx + Ey = q$ is rotated through an angle θ to remove the cross-term, the slope of the rotated x'-axis is $\tan \theta$. Express this slope in terms of the coefficients A, B, C.

*34. Find the slope of the tangent to the hyperbola $\frac{x^2}{a^2} - \frac{y^2}{b^2} = 1$ at the point (u, v).

**35. Consider the part of the parabola $x = y^2$ with $0 \leq x \leq 1$. (i) Find the volume of the solid obtained by revolving it about the x-axis. (ii) Now consider the solid cone obtained by revolving the part of the line $y = x$ with $0 \leq x \leq 1$, and find its volume. (iii) Notice that the parabolic solid contains the cone and that they have the same cross-sections at $x = 0$, $x = 1$. What is the ratio of the two volumes?

**36. Compare the volumes of the solid obtained by revolving about the x-axis the parabola $x = y^2$, for $0 \leq x \leq a$, and the corresponding cone obtained by revolving the line segment that runs from $(0, 0)$ to $(a, \sqrt{a})$.

*37. (i) Show that the point on a parabola closest to the focus is the vertex.

(ii) A meteorite travels along a parabolic path, with the sun at the focus (see figure below). When

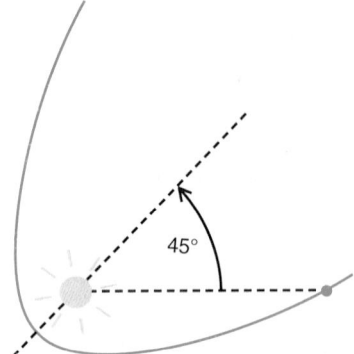

the meteorite is 50,000,000 km from the center of the sun, the line from the meteorite to the sun makes an angle of 45° with the axis of the parabola (the line through the vertex and the focus). How close will the meteorite come to the center of the sun?

Polar Coordinates

12

These curves can all be expressed quite simply in polar coordinates, reflecting the fact that they all have symmetries that involve angles. Some of them are surprisingly difficult to express in rectangular coordinates.

INTRODUCTION

In this chapter we consider polar coordinates and see how to use them with some of the material we have learned.

In Section 12.1 we recall the basic ideas and definitions. In Section 12.2 we discuss drawing graphs of equations that are given in polar coordinates. In Section 12.3 we consider how to use integration to find the area of the region enclosed inside the graph of an equation given in polar coordinates. In Section 12.4 we study conics from the point of view of polar coordinates and discuss the equations of various conics. We will see that the equations of different types of conics are nearly the same in polar coordinates, unlike the situation with rectangular coordinates, for which we had to learn different equations for each type. Finally, in Section 12.5 we discuss *parametric equations*, which are equations that describe the path of a moving point as a function of a parameter t, which usually represents the time. In this section we will usually write these equations in polar coordinates, but in Chapter 13 we will consider them in greater detail, writing them in rectangular coordinates or polar coordinates, depending on the situation.

12.1 Polar Coordinates

RADIUS
POLAR ANGLE
GRAPHING
POLAR/RECTANGULAR
CONVERSIONS

Figure 12.1.1

Figure 12.1.2(i)

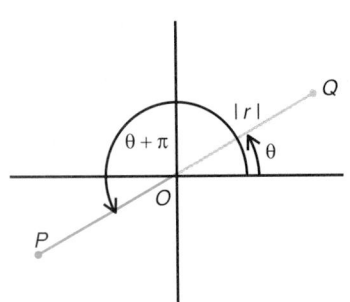

Figure 12.1.2(ii)

In Section 1.7 we discussed **polar coordinates** briefly. The basic idea is quite simple: We identify a point P in the plane by specifying its distance r from the origin O and the angle from the positive horizontal axis to OP. This angle is labeled θ in Figure 12.1.1.

The main complication is that there are many values of θ that all correspond to the same point P. After all, if θ_0 is one such value, then $\theta_0 + 2\pi$ will do just as well, and so will $\theta_0 - 6\pi$ or $\theta_0 + 10\pi$. In fact the angles $\theta_0 + 2k\pi$ all correspond to the same direction in the plane, for any integer k (see Figure 12.1.2(i)).

A second problem is that it is convenient to allow r to be a negative number, in which case (r, θ) refers to a point on the opposite side of the origin. In Figure 12.1.2(ii), where $r < 0$, we see that the point $Q = (-r, \theta) = (|r|, \theta)$ has angle θ, while $P = (r, \theta)$ is at an angle of $\theta + \pi$. Note that $Q = -P$.

The third difficulty is that for the origin itself we have $r = 0$, but it is less clear how to specify the angle θ. We will say that with $r = 0$, *any* angle θ will specify the origin.

EXAMPLE 1 If P is a point in the plane, then its polar coordinates are (r, θ), where $|r|$ is the distance from P to the origin. If $r > 0$, then θ is the angle from the positive horizontal axis to OP, measured in the counterclockwise direction. If $r < 0$, then θ is the angle from the positive horizontal axis to OQ, measured in the counterclockwise direction, where $Q = -P$ is the point opposite P.

If θ_0 is one such angle, then $\theta = \theta_0 + 2k\pi$ is another, for every integer k. If $P = O$ is the origin, then its polar coordinates are $(0, \theta)$, for *any* angle θ. Note that $(-r, \theta + \pi)$ refers to the same point as (r, θ).

The number r is sometimes called the **radius** or the **modulus** of the point P. The angle θ is called its **argument** or its **polar angle** or sometimes just "the angle."

It is easy to convert polar coordinates into rectangular or *Cartesian* coordinates. From Figure 12.1.3 we see that if (r, θ) are the polar coordinates of a point P and

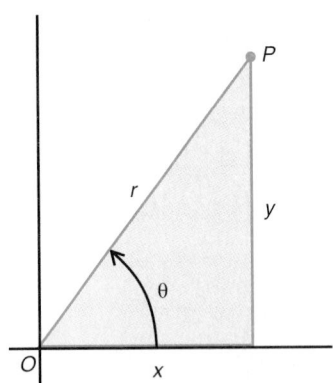

Figure 12.1.3

(x, y) are its rectangular coordinates, then they are related as follows:

$$x = r \cos \theta, \qquad y = r \sin \theta.$$

These formulas still work if $r < 0$; this is why we let θ be the angle of the *opposite* point when $r < 0$.

Going in the opposite direction, we let r be the distance from (x, y) to $(0, 0)$, that is, $r = \sqrt{x^2 + y^2}$. However, it is more complicated to express θ in terms of x and y. From Figure 12.1.3 we see that $\tan \theta = \frac{y}{x}$. The difficulties with this formula are two: First, it does not make sense if $x = 0$, and second, since $\tan(\theta + \pi) = \tan \theta$, the formula does not determine the angle θ completely.

It is necessary to use additional information from the picture to specify the angle. For instance, if we consider the point P whose rectangular coordinates are $(-4, -4)$, as illustrated in Figure 12.1.4, then we see that $r = \sqrt{16 + 16} = 4\sqrt{2}$. The argument θ satisfies $\tan \theta = \frac{y}{x} = \frac{-4}{-4} = 1$. We know that one angle whose tangent equals 1 is $\frac{\pi}{4}$, but as we have just remarked, another possibility is $\frac{\pi}{4} + \pi = \frac{5\pi}{4}$. We use the picture to see that $P = (-4, -4)$ must correspond to the angle $\frac{5\pi}{4}$ and not to $\frac{\pi}{4}$.

It is possible to check this by converting back to rectangular coordinates: $x = r \cos \theta = 4\sqrt{2} \cos \theta$. If $\theta = \frac{\pi}{4}$, then $\cos \theta = \frac{1}{\sqrt{2}}$ and $r \cos \theta = 4$. On the other hand, if $\theta = \frac{5\pi}{4}$, then $r \cos \theta = -4$. Since this is the correct value of x, we see that $\theta = \frac{5\pi}{4}$. Strictly speaking, we should write $\theta = \frac{5\pi}{4} + 2k\pi$, for any integer k.

To figure out the correct value of the argument θ for a point P, we look at Figure 12.1.5. Observe that the argument is 0 for points along the positive x-axis and increases up to $\frac{\pi}{2}$ as P moves to the positive y-axis. In particular we see that we can always choose θ to satisfy $0 < \theta < \frac{\pi}{2}$ if P is above and to the right of the origin. For a point on the negative x-axis the argument is π, so if P moves to the left of the y-axis but stays above the x-axis, its argument will be between $\frac{\pi}{2}$ and π. Similarly, the argument will be between $-\frac{\pi}{2}$ and 0 if P is below and to the right of the origin and will be between $-\pi$ and $-\frac{\pi}{2}$ if P is below and to the left of the origin. These ranges are all indicated in the appropriate regions in Figure 12.1.5. This discussion assumes $r > 0$; if $r < 0$, we add π to the angle θ.

Of course, it is possible to add any integer multiple of 2π to any of these ranges. Notice that you do not have to memorize this picture; it is just a matter of figuring out the argument on each axis and then considering points in between.

It is usually easiest to decide the correct value for θ graphically, as we have just done, but it is also possible to write down a formula. We assume that $\arctan t$ is a number that always lies between $-\frac{\pi}{2}$ and $\frac{\pi}{2}$. Then using Figure 12.1.5, you can check that the value of the argument θ is given by the formula below.

Figure 12.1.4

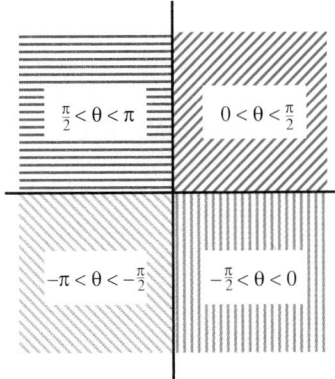

Figure 12.1.5

THEOREM 12.1.1

Polar / Rectangular Conversions

Suppose the point P in the plane has polar coordinates (r, θ). Then its rectangular coordinates are (x, y), where

$$x = r \cos \theta, \qquad y = r \sin \theta.$$

Suppose the point P in the plane has rectangular coordinates (x, y). Then its polar coordinates are (r, θ), where $r = \sqrt{x^2 + y^2}$ and θ is determined by reference to Figure 12.1.5 or as follows.

If $r = 0$, then θ can be *any* angle. Otherwise, with the understanding that $-\frac{\pi}{2} < \arctan t < \frac{\pi}{2}$,

$$\theta = \begin{cases} \arctan \frac{y}{x} + 2k\pi, & \text{if } x > 0, \\ \arctan \frac{y}{x} + \pi + 2k\pi, & \text{if } x < 0, \\ \frac{\pi}{2} + 2k\pi, & \text{if } x = 0 \text{ and } y > 0, \\ -\frac{\pi}{2} + 2k\pi, & \text{if } x = 0 \text{ and } y < 0. \end{cases}$$

(In all these formulas, k is understood to be any integer.)
It is also possible to use $(-r, \theta + \pi)$ in place of (r, θ).

EXAMPLE 2 (i) Find the polar coordinates of the point P whose rectangular coordinates are $(x, y) = (0, -5)$.

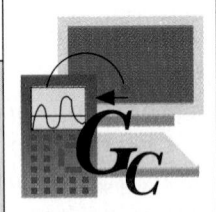

We can use a graphics calculator to plot points in polar coordinates. For instance, we plot the points whose polar coordinates are $(r, \theta) = \left(3, \frac{3\pi}{4}\right), \left(2, \frac{7\pi}{6}\right)$, and $(2, 1)$.

TI-85	SHARP EL9300
MODE ▼ ▼ ▼ ▼ ENTER GRAPH F2(RANGE) (−) 3 ▼ 3 ▼ ▼ (−) 3 ▼ 3 ▼ F5(GRAPH) PtOn(3 COS (3 π ÷ 4) , 3 SIN (3 π ÷ 4)) ENTER PtOn(2 COS (7 π ÷ 6) , 2 SIN (7 π ÷ 6)) ENTER PtOn(2 COS 1 , 2 SIN 1) ENTER	RANGE −3 ENTER 3 ENTER 1 ENTER −3 ENTER 3 ENTER 1 ENTER ⏎ 2ndF PLOT ▼ ▶ ▼ ENTER 3 cos (3 π ÷ 4) ENTER 3 sin (3 π ÷ 4) ENTER ⏎ 2ndF PLOT ▼ ▶ ▼ ENTER 2 cos (7 π ÷ 6) ENTER 2 sin (7 π ÷ 6) ENTER ⏎ 2ndF PLOT ▼ ▶ ▼ ENTER 2 cos 1 ENTER 2 sin 1 ENTER ⏎
CASIO f_x-6300G (f_x-7700GB)	HP 48SX
Range −3 EXE 3 EXE 1 EXE −3 EXE 3 EXE 1 EXE EXE EXE EXE Plot 3 cos (3 π ÷ 4) , 3 sin (3 π ÷ 4) EXE Plot 2 cos (7 π ÷ 6) , 2 sin (7 π ÷ 6) EXE Plot 2 cos 1 , 2 sin 1 EXE	▐ PLOT PLOTR ERASE ATTN 3 ENTER π × 4 ÷ ENTER COS 3 × SWAP SIN 3 × PRG OBJ NXT R → C PRG DSPL NXT C → PX PIXON 7 ENTER π × 6 ÷ ENTER COS 2 × SWAP SIN 3 × PRG OBJ NXT R → C PRG DSPL NXT C → PX PIXON 1 ENTER COS 2 × 1 ENTER SIN 2 × PRG OBJ NXT R → C PRG DSPL NXT C → PX PIXON ↱ PLOT DRAW

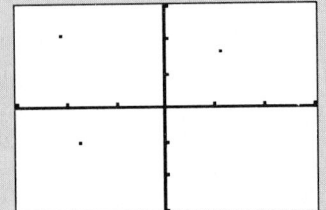

You can try plotting other points. Remember that the picture may be distorted because of the scales. Can you adjust it so that there is no distortion?

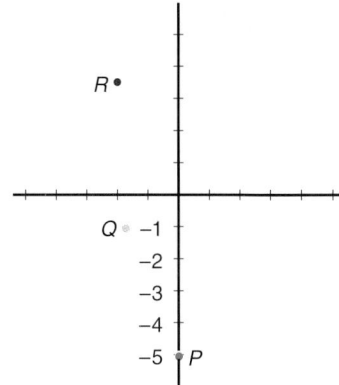

Figure 12.1.6

Solution We find immediately that $r = \sqrt{0^2 + (-5)^2} = 5$. We note that the point lies on the negative y-axis, directly below the origin, so its angle is $\theta = -\frac{\pi}{2}$, or, strictly speaking, $\theta = -\frac{\pi}{2} + 2k\pi$, for any integer k.

The point P is plotted in Figure 12.1.6.

(ii) Find the Cartesian coordinates of the point Q whose polar coordinates are $(r, \theta) = \left(2, \frac{7\pi}{6}\right)$.

Solution We know that $x = r\cos\theta = 2\cos\frac{7\pi}{6} = 2\left(-\frac{\sqrt{3}}{2}\right) = -\sqrt{3}$. Similarly, $y = r\sin\frac{7\pi}{6} = 2\left(-\frac{1}{2}\right) = -1$.

The point is $Q = (-\sqrt{3}, -1)$. It is also plotted in Figure 12.1.6.

(iii) Find the polar coordinates of the point R whose Cartesian coordinates are $R = (-2, 2\sqrt{3})$.

Solution We find that $r = \sqrt{4 + 12} = 4$. Since $\frac{y}{x} = -\sqrt{3}$, we consider $\arctan(-\sqrt{3}) = -\frac{\pi}{3}$. Since $-\frac{\pi}{3}$ corresponds to a point below and to the right of the origin, we realize that the correct angle is $\theta = -\frac{\pi}{3} + \pi + 2k\pi = \frac{2\pi}{3} + 2k\pi$.

The point R is also plotted in Figure 12.1.6.

Exercises 12.1

I

In each of the following questions, a point is specified by either its rectangular coordinates (x, y) or its polar coordinates (r, θ). Find the other kind of coordinates for the point.

1. $(r, \theta) = (2, 0)$
2. $(r, \theta) = \left(1, \frac{3\pi}{4}\right)$
3. $(x, y) = (6, -6)$
4. $(x, y) = \left(0, \frac{3}{2}\right)$
5. $(r, \theta) = \left(0, \frac{11\pi}{17}\right)$
6. $(r, \theta) = \left(\sqrt{3}, \frac{5\pi}{6}\right)$
7. $(x, y) = (-2\sqrt{3}, 2)$
8. $(x, y) = (4, -3)$
9. $(x, y) = (-9, 0)$
10. $(x, y) = (-4, -7)$
11. $(r, \theta) = (-1, 0)$
12. $(r, \theta) = \left(-2, -\frac{\pi}{2}\right)$
13. $(x, y) = (-5, 5)$
14. $(x, y) = (-7, -7)$
15. $(r, \theta) = (7, -23\pi)$
16. $(r, \theta) = \left(2, \frac{9\pi}{8}\right)$
17. $(x, y) = \left(-\frac{4}{\sqrt{3}}, -4\right)$

18. $(x, y) = (-12, 5)$
19. $(x, y) = \left(\frac{1}{\sqrt{2}}, -\frac{1}{\sqrt{2}}\right)$
20. $(x, y) = (0, -\pi)$
21. $(x, y) = (\sqrt{2}, 0)$
22. $(x, y) = (1, -\sqrt{3})$
23. $(r, \theta) = \left(-1, \frac{\pi}{2}\right)$
24. $(r, \theta) = \left(-2, \frac{3\pi}{4}\right)$
25. $(r, \theta) = \left(1, \frac{11\pi}{2}\right)$
26. $(r, \theta) = \left(-2, -\frac{7\pi}{4}\right)$
27. $(x, y) = (5, -12)$
28. $(x, y) = (-24, 7)$
29. $(r, \theta) = (2, \arctan 2)$
30. $(r, \theta) = \left(-4, \arctan \frac{7}{8}\right)$

II

*31. (i) Consider the equation $x^2 + y^2 = 4$. By expressing x and y in terms of the polar coordinates r and θ, write the equation in terms of r and θ.

(ii) Express the equation $x^2 + 2x + y^2 - 4y = 4$ in polar coordinates. What is the graph of this equation?

*32. Express the equation $3x - 4y = 12$ in polar coordinates. What is the graph of this equation?

33. Describe the region $\{(x, y) : x \geq 0, y \geq 0\}$ in terms of polar coordinates (r, θ).

34. Describe the region $\{(x, y) : x \leq y < 0\}$ in terms of polar coordinates (r, θ).

35. Describe the region $\{(x, y) : 1 < x^2 + y^2 < 4, 0 \leq -x \leq y\}$ in terms of polar coordinates (r, θ).

36. Describe the region $\{(x, y) : x^2 + y^2 \leq 9, |x| < y\}$ in terms of polar coordinates (r, θ).

37. Describe the region $\{(r, \theta) : r > 0, -\frac{\pi}{4} < \theta < \frac{\pi}{4}\}$ in terms of rectangular coordinates (x, y).

38. Describe the region $\{(r, \theta) : -1 \leq r \leq 1, \frac{\pi}{4} < \theta < \frac{3\pi}{4}\}$ in terms of rectangular coordinates (x, y).

*39. Describe the region $\{(r, \theta) : r^2 - 2r \leq 0\}$ in terms of rectangular coordinates (x, y).

*40. Describe the region $\{(r, \theta) : r > 0, \cos\theta < \sin\theta\}$ in terms of rectangular coordinates (x, y).

What is the graph of each of the following equations?

*41. $r = 10$

*42. $\theta = \frac{\pi}{6}$

*43. $\cos \theta = \frac{1}{2}$

*44. $\sin \theta = r \cos^2 \theta$

If you have a machine that will plot in polar coordinates, use it to sketch the following curves.

45. $r = \sin(4\theta)$

46. $r = \cos^2 \theta$

47. $r = 5 + \cos(6\theta)$

48. $r = 1 + 2 \sin \theta$

12.2 Graphs in Polar Coordinates

CURVES
SYMMETRY
ROSES
CARDIOID
LEMNISCATE
LIMAÇON
TANGENTS

In this section we consider techniques for graphing equations that are expressed in polar coordinates. Let us begin with some easy examples.

EXAMPLE 1

(i) Consider the equation $r = 3$. The points (r, θ) that satisfy this equation, that is, for which $r = 3$, are the points on the circle of radius 3 around the origin, as sketched in Figure 12.2.1.

(ii) Consider the equation $\theta = \frac{4\pi}{3}$. The points (r, θ) that satisfy this equation, that is, for which $\theta = \frac{4\pi}{3}$, are exactly the points on the line through the origin that makes an angle $\frac{4\pi}{3}$ with the positive horizontal axis. The graph is sketched in Figure 12.2.2. Notice that the origin is included, because the origin can be written as $(r, \theta) = \left(0, \frac{4\pi}{3}\right)$, a point that satisfies the equation. The points below the horizontal axis have $r > 0$, while those above have $r < 0$.

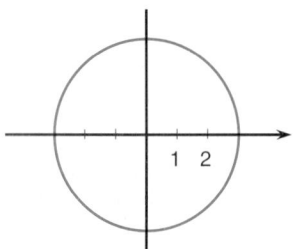

Figure 12.2.1

If we consider a point on the graph of $\theta = \frac{4\pi}{3}$, call it $\left(r, \frac{4\pi}{3}\right)$, for some r, then it is possible to write the same point with different coordinates, such as $\left(r, \frac{4\pi}{3} + 2\pi\right)$ or $\left(r, \frac{4\pi}{3} + 2k\pi\right)$ or $\left(-r, \frac{\pi}{3}\right)$. In any of these other forms, the point does *not* satisfy the equation. So when we draw a graph, we include a point P on it if any one of the possible sets of polar coordinates for P satisfy the equation, even if the others do not.

In this example we included the origin, which can be written as $(r, \theta) = (0, \theta)$ for *any* value of θ. Only one of these values satisfies the equation, but that is enough, and the origin is included.

Next we consider some slightly more complicated examples.

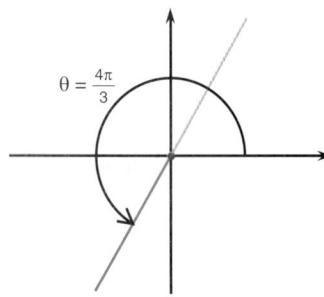

Figure 12.2.2

EXAMPLE 2   **The Spiral of Archimedes**   The equation $r = \theta, \theta \geq 0$ can be visualized as follows. We begin with $\theta = 0$, so $r = 0$, and we plot a point at the origin. Then suppose we let θ increase from 0 to 2π. This means that the points (r, θ) move around the origin counterclockwise. Since $r = \theta$, the distance from the origin also increases from 0 to 2π. The resulting points form a curve that begins at the origin and spirals out, rotating exactly once around the origin. It returns to the positive horizontal axis when $\theta = 2\pi$, so at that point r is also equal to 2π. This curve is illustrated in Figure 12.2.3(i).

However, there is no real reason to stop at $\theta = 2\pi$. We could let θ continue to increase, say, from 2π to 4π. This would correspond to having the curve spiral

Figure 12.2.3(i)

Figure 12.2.3(ii)

Figure 12.2.4

Figure 12.2.5(i)

Figure 12.2.5(ii)

Figure 12.2.5(iii)

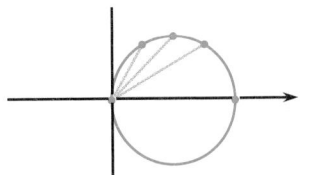

Figure 12.2.6

once more around the origin, this time spreading out from a radius of 2π at the beginning to 4π at the end. After that, θ could continue to increase, and the curve would continue to spiral outward. Each time it winds around the origin, it moves out so that its radius increases by 2π.

The resulting curve is illustrated in Figure 12.2.3(ii), which is drawn with a smaller scale than 12.2.3(i) in order to show more of the curve. This curve is sometimes referred to as the **Spiral of Archimedes**.

In this example we see that different values of θ correspond to different points on the curve. Two values of θ that differ by 2π do *not* correspond to the same point.

EXAMPLE 3   Sketch the graph of $r = \cos\theta$.

Solution   It is helpful to begin by sketching the "ordinary" graph of $y = \cos\theta$, meaning the graph that is drawn by thinking of y and θ as rectangular coordinates. We make the θ-axis horizontal and the y-axis vertical, so the graph is just the familiar trigonometric curve, shown in Figure 12.2.4. It will be useful to refer to this graph when we draw the graph in polar coordinates.

As θ goes from 0 to $\frac{\pi}{2}$, we see that $r = \cos\theta$ starts at 1 and decreases to 0. The corresponding point on the polar graph starts at the point on the horizontal axis 1 unit to the right of the origin. It moves counterclockwise, getting closer and closer to the origin, finally reaching the origin when $\theta = \frac{\pi}{2}$. We expect a graph whose shape is that shown in Figure 12.2.5(i).

Similar remarks apply as θ starts at 0 and decreases to $-\frac{\pi}{2}$; we expect a curve like the one shown in Figure 12.2.5(ii).

The two parts together look a lot like a circle. If we multiply the equation $r = \cos\theta$ by r, it becomes $r^2 = r\cos\theta$. This is easily converted into Cartesian coordinates; it becomes $x^2 + y^2 = x$, or $x^2 - x + y^2 = 0$, or $\left(x - \frac{1}{2}\right)^2 + y^2 = \frac{1}{4}$. This is the equation of a circle of radius $\frac{1}{2}$ centered at $\left(\frac{1}{2}, 0\right)$. The circle is shown in Figure 12.2.5(iii).

Suppose we let θ increase from $\frac{\pi}{2}$ to π. From Figure 12.2.4 we see that $\cos\theta$ starts at zero and decreases to -1. As the angle increases from $\frac{\pi}{2}$ to π, we expect the points on the curve to be above and to the left of the origin. However, because $r = \cos\theta$ is *negative*, the points are actually in the *opposite* direction, below and to the right of the origin.

The part of the curve with $\frac{\pi}{2} \le \theta \le \frac{3\pi}{2}$ retraces the circle we found when $-\frac{\pi}{2} \le \theta \le \frac{\pi}{2}$.

In Example 3 we were able to give a qualitative description of the shape of the curve, and then it was possible to make a lucky guess that it was a circle. Usually, we cannot expect to be so lucky. If the shape of the curve is unfamiliar or difficult to guess, we may have to resort to plotting some points in order to get a rough idea, exactly as we do when plotting in rectangular coordinates.

One way to approach this would be to draw some half-lines lying at some convenient angles, such as $\theta = 0, \frac{\pi}{6}, \frac{\pi}{4}, \frac{\pi}{3}, \frac{\pi}{2}$. For each angle we calculate the corresponding r (in the example it was $r = \cos\theta$). Then we mark a point on the half-line whose distance from the origin is r. This is done in Figure 12.2.6.

Figure 12.2.7

Figure 12.2.8(i)

Figure 12.2.8(ii)

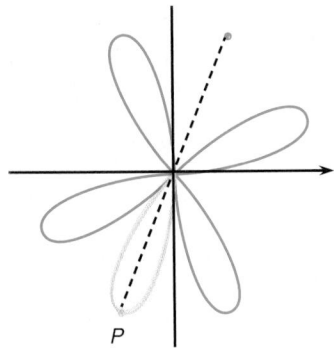

Figure 12.2.8(iii)

For this particular example it is not terribly difficult to proceed in this way, although it is not too easy to measure distances along half-lines unless they are horizontal or vertical. An alternative approach is to consider various points on the curve, such as $(\cos 0, 0)$, $\left(\cos\frac{\pi}{6}, \frac{\pi}{6}\right)$, $\left(\cos\frac{\pi}{4}, \frac{\pi}{4}\right)$, $\left(\cos\frac{\pi}{3}, \frac{\pi}{3}\right)$, $\left(\cos\frac{\pi}{2}, \frac{\pi}{2}\right)$. Instead of trying to plot them directly in polar coordinates, it is possible to convert each of them into rectangular coordinates $(x, y) = (r\cos\theta, r\sin\theta)$ and plot them that way. In many cases this approach is actually easier. You may also be able to obtain polar graph paper, which has angles and distances from the origin marked on it.

All that really matters is that if we cannot guess the shape of the curve, it is possible to plot as many points as necessary until we can see it.

EXAMPLE 4   Plot the graph of $r = \sin(4\theta)$.

Solution   Once again we begin by plotting the ordinary graph $y = \sin(4\theta)$, shown in Figure 12.2.7.

If we start at $\theta = 0$ and let θ increase, we see that $r = \sin(4\theta)$ also starts at zero and increases up to $r = 1$ at $\theta = \frac{\pi}{8}$ and then decreases back down to $r = 0$ at $\theta = \frac{\pi}{4}$. This suggests that the curve will have a loop as shown in Figure 12.2.8(i), lying between the angles $\theta = 0$ and $\theta = \frac{\pi}{4}$ at which $r = 0$.

Instead of letting θ increase further, let us look at the graph of $y = \sin(4\theta)$ in rectangular coordinates in Figure 12.2.7. There we see that y also increases from 0 up to 1 and then decreases back down to 0 when θ increases from $\frac{\pi}{2}$ to $\frac{3\pi}{4}$, and it does it again when θ increases from π to $\frac{5\pi}{4}$ and again between $\frac{3\pi}{2}$ and $\frac{7\pi}{4}$.

We expect three more loops as shown in Figure 12.2.8(ii).

Now let us look at what happens when r is negative. This happens, for instance, when θ is between $\frac{\pi}{4}$ and $\frac{\pi}{2}$. The values of r decrease from 0 to -1 and then increase back up to zero again. Again we expect a loop, but since $r < 0$, we expect to find it on the opposite side of the origin. For instance, when $\theta = \frac{3\pi}{8}$, we have that $r = \sin(4\theta) = -1$. If we try to plot the corresponding point P in rectangular coordinates, we find it is

$$(x, y) = (r\cos\theta, r\sin\theta) = \left(-\cos\frac{3\pi}{8}, -\sin\frac{3\pi}{8}\right).$$

This point P is easily plotted; as we see in Figure 12.2.8(iii), it is exactly opposite the point $(x, y) = \left(\cos\frac{3\pi}{8}, \sin\frac{3\pi}{8}\right)$. We also plot the rest of the loop that occurs as θ increases from $\frac{\pi}{4}$ to $\frac{\pi}{2}$.

Finally, we notice that there will be three more analogous loops when $\theta \in \left[\frac{3\pi}{4}, \pi\right], \left[\frac{5\pi}{4}, \frac{3\pi}{2}\right], \left[\frac{7\pi}{4}, 2\pi\right]$. They are sketched in Figure 12.2.8(iv).

The graph of $r = \sin(4\theta)$, shown in Figure 12.2.8(iv), is an "eight-petaled rose."

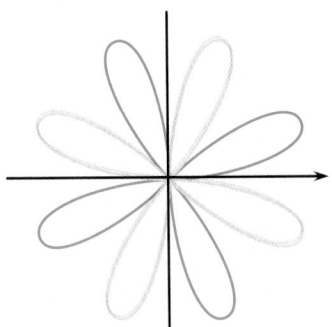

Figure 12.2.8(iv)

In doing this example we saw the usefulness of first sketching the equation in rectangular coordinates. We worked separately with the parts of the curve for which r was positive and negative. It was also helpful to locate the maxima and minima of $r = \sin(4\theta)$; these showed where we should expect loops. Loops begin and end at points where $r = 0$.

We can use a graphics calculator to sketch the graph of $r = \sin n\theta$, for various values of n.

TI-85	SHARP EL9300
MODE ▼ ▼ ▼ ▼ ▶ (Pol) ENTER EXIT 6 STO N ENTER GRAPH F1 $(r(\theta) =)$ SIN (N × F1(θ)) M2(RANGE) 0 ▼ 2π ▼ .01 ▼ (−) 1 ▼ 1 ▼ ▼ (−) 1 ▼ 1 F5(GRAPH)	SET UP E 2 ⤴ sin 6 x/θ/T RANGE −1 ENTER 1 ENTER 1 ENTER −1 ENTER 1 ENTER 1 ENTER 0 ENTER 2 π ENTER .01 ENTER ⤴
CASIO f_x-7700GB	HP 48SX
MODE MODE −(POL) Range −1 EXE 1 EXE 1 EXE −1 EXE 1 EXE 1 EXE 0 EXE 2 π EXE .01 EXE Graph sin 6 θ EXE	↰ RAD ↰ PLOT ′ R ≡ SIN 6 × α ↱ F NEW P O L ENTER PTYPE POLAR PLOTR 1 +/− ENTER 1 XRNG 1 +/− ENTER 1 YRNG ′ α ↱ F INDEP DRAW

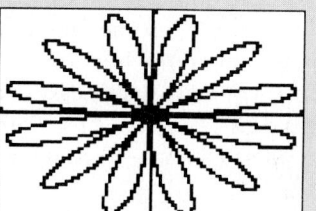

You can try plotting the curve with different values of n.

If we plot the graph of an equation of the form $r = \cos n\theta$ or $r = \sin n\theta$, we will get something like the "rose" we saw in the last example. The number of petals will depend on n. When we graphed $r = \cos\theta$ in Example 3, we found that there was only one petal. When we sketched the graph of $\cos\theta$ in rectangular coordinates, there were two loops as θ went from 0 to 2π, but the points coming from the loop in which r was negative turned out to be the same curve as the other loop, so there was only one petal. On the other hand, in Example 4, when we graphed $r = \sin(4\theta)$, we found that the "negative" loops resulted in new petals and there was a total of eight petals. This turns out to be part of a pattern.

Summary 12.2.1

An equation of the form

$$r = \cos(n\theta) \qquad \text{or} \qquad r = \sin(n\theta)$$

has a graph that is a **"rose."** The number of "petals" is n if n is odd and $2n$ if n is even.

A number of examples are sketched in Figure 12.2.9.

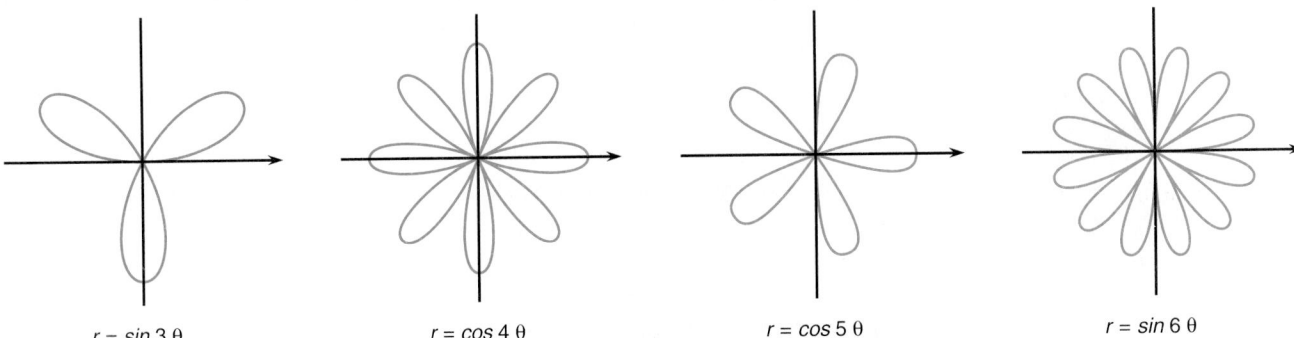

$r = \sin 3\,\theta$ $r = \cos 4\,\theta$ $r = \cos 5\,\theta$ $r = \sin 6\,\theta$

Figure 12.2.9

Before drawing any more graphs, let us remember that when we sketched graphs in rectangular coordinates, it was often possible to save a lot of work by noticing symmetries. For instance, in Example 3 we might have realized that the graph is symmetric about the horizontal axis (see Figure 12.2.5).

In terms of the equation $r = \cos\theta$, this happens because $\cos\theta = \cos(-\theta)$, so the values of r are the same for corresponding points above and below the axis.

If we consider an equation of the form $f(r,\theta) = 0$, where f is some function of two variables, then we see that its graph will be symmetric about the horizontal axis if the equation $f(r,\theta) = 0$ is the same as the equation $f(r,-\theta) = 0$. It is easy to see that it would also be symmetric about the horizontal axis if $f(-r,\pi - \theta) = 0$ gives the same equation as $f(r,\theta) = 0$. This is because the point $(-r,\pi - \theta)$ is obtained by reflecting the point (r,θ) in the horizontal axis (why?).

Similarly, the reflection of (r,θ) in the vertical axis can be written as $(r,\pi - \theta)$ or as $(-r,-\theta)$. From this we see that the graph is symmetric about the vertical axis if the equation is unchanged by substituting either of these points for (r,θ). (The fact that there are different possibilities comes from the different choices of polar coordinates for any particular point.)

Summary 12.2.2   Symmetry

Consider the graph of the equation $f(r,\theta) = 0$.

1. The graph is symmetric about the horizontal axis if the equation $f(r,\theta) = 0$ is unchanged if θ is changed to $-\theta$. It is also symmetric about the horizontal axis if the equation is unchanged when (r,θ) is replaced by $(-r,\pi - \theta)$.

2. The graph is symmetric about the vertical axis if the equation $f(r,\theta) = 0$ is unchanged if θ is changed to $\pi - \theta$. It is also symmetric about the vertical axis if the equation is unchanged when (r,θ) is replaced by $(-r,-\theta)$.

3. Because of the many different ways of writing a point, it is possible for a graph to be symmetric about either axis without the formula satisfying the conditions given here.

Figure 12.2.10

Figure 12.2.11

THEOREM 12.2.3

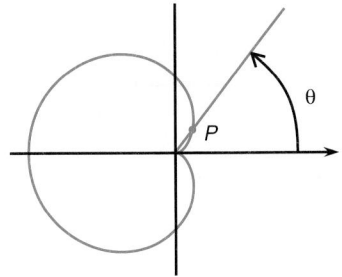

Figure 12.2.12

EXAMPLE 5   Cardioid   Sketch the graph of the curve $r = 1 - \cos\theta$.

Solution   As usual we begin with the plot in rectangular coordinates, shown in Figure 12.2.10. Observing that the equation remains the same if θ is changed to $-\theta$, we conclude that the graph in polar coordinates will be symmetric about the horizontal axis. We also see that the value of r increases from 0 at $\theta = 0$ to 2 at $\theta = \pi$. The graph is sketched in Figure 12.2.11. It is called a **cardioid**, because of its heartlike shape.   ◢

When we sketched the cardioid, we had to guess the exact shape of the curve at the origin. Notice that $r = 1 - \cos\theta$ is not zero except when $\cos\theta = 1$, that is, when $\theta = 2k\pi$. So if θ is near 0 but not equal to 0, we know that $r \neq 0$.

If we write the rectangular coordinates of the corresponding point P, they are $(x, y) = (r\cos\theta, r\sin\theta)$. If we consider the secant OP, as shown in Figure 12.2.12, its slope is $m = \frac{y-0}{x-0} = \frac{y}{x} = \frac{r\sin\theta}{r\cos\theta} = \tan\theta$. To do this calculation, we needed to know that $r \neq 0$.

Now as $\theta \to 0$, we expect that the slope $m = \tan\theta$ will approach $\tan 0 = 0$. This suggests that the tangent line will be horizontal. To put it another way, the cardioid approaches the origin with a horizontal tangent.

In fact, this argument depended on knowing that $r \neq 0$ for θ near 0, and with this assumption it is possible to draw the analogous conclusion for any curve in polar coordinates at points where it approaches the origin.

> Suppose the graph of $r = F(\theta)$ passes through the origin when $\theta = \theta_0$. Suppose that $\frac{dr}{d\theta}$ exists at $\theta = \theta_0$. Then the slope of the tangent line at the origin to the curve near $\theta = \theta_0$ is $m = \tan\theta_0$.

In other words the curve is tangent to the line $\theta = \theta_0$ at the origin.

EXAMPLE 6   Consider the curve $r = 2 - 2\sin\theta$. It passes through the origin when $\theta = \frac{\pi}{2} + 2k\pi$. The theorem says that the tangent line at this point has slope $m = \tan\frac{\pi}{2}$. This does not really make sense, but if we think of $\tan\theta$ as $\theta \to \frac{\pi}{2}^-$, we see that it tends to ∞. This means that secants through the origin and nearby points tend closer and closer to vertical lines, so we expect that the tangent line will be vertical.

The graph is shown in Figure 12.2.13; it is a cardioid. Notice how the tangent line at the origin is vertical. (One way to see immediately that it is a cardioid is to

Figure 12.2.13

Figure 12.2.14(i)

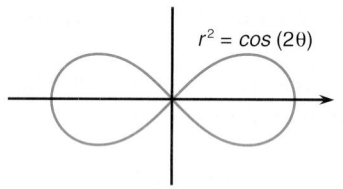

Figure 12.2.14(ii)

observe that $\sin\theta = \cos\left(\theta - \frac{\pi}{2}\right)$, so $r = 1 - \sin\theta$ is just the cardioid $r = 1 - \cos\theta$ rotated through an angle of $\frac{\pi}{2}$.)

EXAMPLE 7   Graph the equation $r^2 = \cos(2\theta)$.

Solution   We begin by sketching the graph of $y = \cos(2\theta)$ in Figure 12.2.14(i). Notice that for it to equal r^2 we must have $\cos(2\theta) \geq 0$, that is, $\theta \in \left[-\frac{\pi}{4}, \frac{\pi}{4}\right]$ or $\theta \in \left[\frac{3\pi}{4}, \frac{5\pi}{4}\right]$ (plus the inevitable $2k\pi$).

The usual sort of analysis suggests that as θ moves from $-\frac{\pi}{4}$ to $\frac{\pi}{4}$, $r^2 = \cos(2\theta)$ must increase from 0 to 1 and then decrease back to 0. So $r = \sqrt{\cos(2\theta)}$ must also increase from 0 to 1 and then decrease back to 0. We expect a loop with tangents along the lines $\theta = \pm\frac{\pi}{4}$.

Similar remarks apply to $\theta \in \left[\frac{3\pi}{4}, \frac{5\pi}{4}\right]$, and we expect a similar loop there. The graph is sketched in Figure 12.2.14(ii). Notice the slopes of the tangents at the origin, and notice the symmetries. (Confirm them by using Summary 12.2.2.)

A curve of this shape is called a **lemniscate**. A similar shape in various positions will result from any of the following equations:

$$r^2 = K\cos(2\theta), \qquad r^2 = K\sin(2\theta).$$

The size of the curve depends on $|K|$; the location depends on whether the equation involves $\cos(2\theta)$ or $\sin(2\theta)$ and also on the sign of K. Some examples are sketched in Figure 12.2.15. You might like to check that it is easy to determine the position of the lemniscate by figuring out where r^2 has its maximum values.

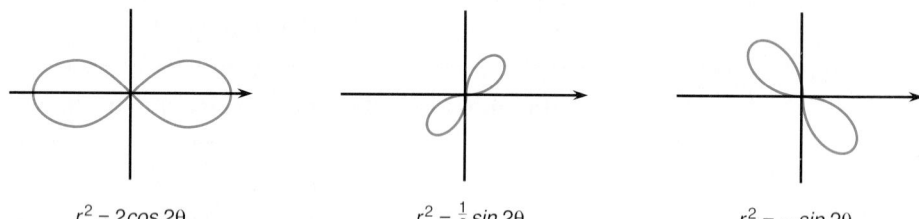

$r^2 = -\cos 2\theta$ $r^2 = 2\cos 2\theta$ $r^2 = \frac{1}{2}\sin 2\theta$ $r^2 = -\sin 2\theta$

Figure 12.2.15

EXAMPLE 8   Find the points of intersection of the lemniscate $r^2 = 2\cos(2\theta)$ and the cardioid $r = 2 - 2\sin\theta$.

Solution   If (r, θ) is on the cardioid, we know that $r^2 = (2 - 2\sin\theta)^2$, so to be on both curves, we need

$$2\cos(2\theta) = (2 - 2\sin\theta)^2 = 4 - 8\sin\theta + 4\sin^2\theta,$$

so

$$2(\cos^2\theta - \sin^2\theta) = 2(2 - 4\sin\theta + 2\sin^2\theta)$$
$$1 - \sin^2\theta - \sin^2\theta = 2 - 4\sin\theta + 2\sin^2\theta$$
$$0 = 1 - 4\sin\theta + 4\sin^2\theta$$
$$0 = (1 - 2\sin\theta)^2.$$

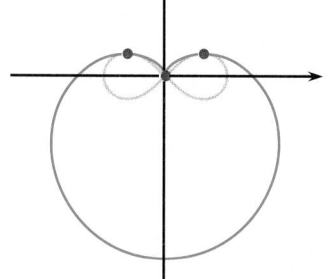

Figure 12.2.16

This means $\sin\theta = \frac{1}{2}$, so $\theta = \frac{\pi}{6}$ or $\frac{5\pi}{6}$. The corresponding points have $r = 2 - 2\sin\theta = 2 - 2\left(\frac{1}{2}\right) = 1$, so they are $(r, \theta) = \left(1, \frac{\pi}{6}\right)$ and $\left(1, \frac{5\pi}{6}\right)$. See Figure 12.2.16.

From the picture we see that the origin is on both curves. It is disconcerting that our calculation did not find it. The reason is that the origin occurs on the cardioid

$r = 2 - 2\sin\theta$ when $\theta = \frac{\pi}{2}$, but on the lemniscate it occurs when $\cos(2\theta) = 0$, that is, when $\theta = \frac{\pi}{4}, \frac{3\pi}{4}, \frac{5\pi}{4}, \frac{7\pi}{4}$. These angles do not coincide, so we do not find them from the formulas.

> In solving for points of intersection of curves in polar coordinates it is important to consider the possibility that a point (r, θ) can also be expressed in different ways: $(r, \theta + 2k\pi)$. The origin is $(0, \theta)$, for *any* θ, so it should be considered separately. Simply check whether the origin is on each curve. A sketch is often helpful.

We recall that the curve $r = 1 - \cos\theta$ is a cardioid. It seems reasonable to consider curves of the form

$$r = a + b\cos\theta \qquad \text{or} \qquad r = a + b\sin\theta.$$

The shape of any of these curves will depend on the signs and relative sizes of a and b. For instance, if $|a| > |b|$, then $|b\cos\theta| \le |b| < |a|$. Because of this, $r = a + b\cos\theta$ is never zero (see Figure 12.2.17(i)). In terms of the picture in polar coordinates this means that the curve will never pass through the origin. Such a curve is illustrated in Figure 12.2.17(ii).

On the other hand, if $|a| < |b|$, then $r = a + b\cos\theta$ will equal 0 for two values of θ, and r will be positive some of the time and negative some of the time. A curve of this type is shown in Figure 12.2.17(iii).

It turns out that there is one more distinction to be made. If $\left|\frac{a}{b}\right| \ge 2$, then the curve has a convex shape, whereas if $1 < \left|\frac{a}{b}\right| < 2$, the curve has a **dimple**; examples of these cases are shown in Figure 12.2.17(iv).

If $|a| = |b|$, the curve is a cardioid. Any of the other possibilities is called a **limaçon**, after the Latin word for snail. It is apparently pure coincidence that the "dimpled limaçon" of Figure 12.2.17(iv) looks like a lima bean, but this may help

Figure 12.2.17(i)

Figure 12.2.17(ii)

Figure 12.2.17(iii)

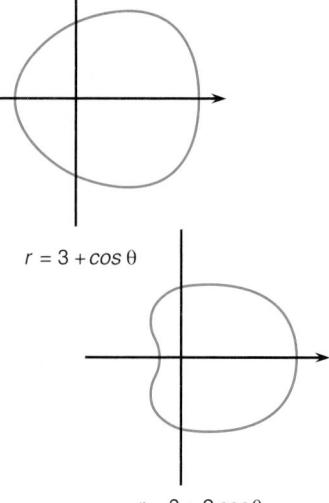

Figure 12.2.17(iv)

you to remember the name. The position of the limaçon is affected by the signs of a and b and the use of $\sin\theta$ or $\cos\theta$ in the equation.

Some of the possibilities for the curve $r = a + b\sin\theta$ are shown in Figure 12.2.18.

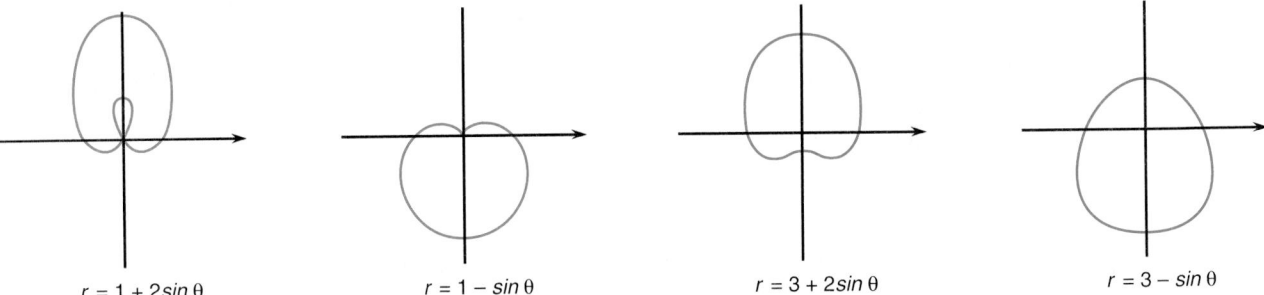

| $r = 1 + 2\sin\theta$ | $r = 1 - \sin\theta$ | $r = 3 + 2\sin\theta$ | $r = 3 - \sin\theta$ |

Figure 12.2.18

Summary 12.2.4

Curves in Polar Coordinates

$r = K$	Circle around origin, radius K
$\theta = K$	Line through origin
$\theta = K, r \geq 0$	Half-line from origin
$r = \theta, \theta \geq 0$	Spiral of Archimedes
$r = K\cos\theta$	Circle with radius $\frac{K}{2}$, center $\left(\frac{K}{2}, 0\right)$
$r = \cos(n\theta)$	Rose with n petals if n is odd, $2n$ petals if n is even
$r = 1 \pm \cos\theta$	Cardioid
$r^2 = \cos(2\theta)$	Lemniscate
$r = a + b\cos\theta$	Cardioid if $|a| = |b|$, otherwise limaçon; two loops if $|a| < |b|$, dimpled if $1 < \left|\frac{a}{b}\right| < 2$, convex if $\left|\frac{a}{b}\right| \geq 2$

Replacing cos by sin rotates any of these curves but does not change its type. Replacing $r = f(\theta)$ by $r = f(\theta + c)$ has the effect of rotating the curve *clockwise* by an angle c.

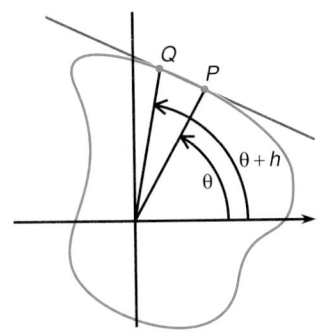

Figure 12.2.19

Slopes of Tangent Lines

Consider a curve $r = f(\theta)$, and let us try to find the slope of the tangent line at point P on the graph. In Figure 12.2.19 we see the point P at $(r, \theta) = \big(f(\theta), \theta\big)$, and draw the nearby point Q at $\big(f(\theta + h), \theta + h\big)$. To find the slope of the secant PQ, we need to know the rectangular coordinates of P and Q.

Since $x = r\cos\theta$, $y = r\sin\theta$, we find that P is

$$(x_P, y_P) = \big(f(\theta)\cos\theta, f(\theta)\sin\theta\big)$$

and Q is

$$(x_Q, y_Q) = \big(f(\theta + h)\cos(\theta + h), f(\theta + h)\sin(\theta + h)\big).$$

The slope of the secant is

$$m = \frac{y_Q - y_P}{x_Q - x_P} = \frac{f(\theta + h)\sin(\theta + h) - f(\theta)\sin(\theta)}{f(\theta + h)\cos(\theta + h) - f(\theta)\cos(\theta)}.$$

To get the slope of the tangent, we should let Q approach P, that is, let $h \to 0$, and take a limit, assuming that the denominator of the difference quotient is not zero for $h \neq 0$.

Assuming that $f(\theta)$ is differentiable, we see that as $h \to 0$, the numerator and denominator both tend to zero. Applying l'Hôpital's Rule, we find that the slope of the tangent is

$$\begin{aligned}
m &= \lim_{h\to 0} \frac{f(\theta + h)\sin(\theta + h) - f(\theta)\sin(\theta)}{f(\theta + h)\cos(\theta + h) - f(\theta)\cos(\theta)} \\
&= \lim_{h\to 0} \frac{f'(\theta + h)\sin(\theta + h) + f(\theta + h)\cos(\theta + h) - 0}{f'(\theta + h)\cos(\theta + h) - f(\theta + h)\sin(\theta + h) - 0} \\
&= \frac{f'(\theta)\sin\theta + f(\theta)\cos\theta}{f'(\theta)\cos\theta - f(\theta)\sin\theta}.
\end{aligned}$$

> We take derivatives with respect to h.

To do this, we have to know that the denominator $x_Q - x_P$ is nonzero for h near 0 but not equal to 0. One way to ensure this is to insist that $\frac{dx}{d\theta} \neq 0$, so the x-coordinate is an increasing (or decreasing) function of θ. Since $x = r\cos\theta = f(\theta)\cos\theta$, we find

$$\frac{dx}{d\theta} = \frac{d}{d\theta}\big(f(\theta)\cos\theta\big) = f'(\theta)\cos\theta - f(\theta)\sin\theta.$$

To say that this is not zero also conveniently ensures that the denominator is nonzero in our slope formula. We have just found the following result.

THEOREM 12.2.5

> Suppose $f(\theta)$ is a differentiable function with $f'(\theta)$ continuous on some interval, and let C be the graph of $r = f(\theta)$. Suppose θ is a point in this interval at which
>
> $$f'(\theta)\cos\theta - f(\theta)\sin\theta \neq 0.$$
>
> Then the slope of the tangent to C at the corresponding point P is
>
> $$m = \frac{f'(\theta)\sin\theta + f(\theta)\cos\theta}{f'(\theta)\cos\theta - f(\theta)\sin\theta}.$$

EXAMPLE 9   For the cardioid $r = 1 + \cos\theta$, the slope of the tangent is

$$\begin{aligned}
m &= \frac{f'(\theta)\sin\theta + f(\theta)\cos\theta}{f'(\theta)\cos\theta - f(\theta)\sin\theta} = \frac{-\sin\theta\sin\theta + (1 + \cos\theta)\cos\theta}{-\sin\theta\cos\theta - (1 + \cos\theta)\sin\theta} \\
&= \frac{\cos^2\theta - \sin^2\theta + \cos\theta}{-2\cos\theta\sin\theta - \sin\theta}.
\end{aligned}$$

In particular, when $\theta = \frac{\pi}{2}$, the slope is $m = \frac{0-1+0}{0-1} = 1$. The tangent at $\theta = \frac{\pi}{2}$ is shown in Figure 12.2.20(i).

If we want to find the highest point on the graph, we find where the slope of the tangent is zero. This means

$$\begin{aligned}
0 &= \cos^2\theta - \sin^2\theta + \cos\theta = \cos^2\theta - (1 - \cos^2\theta) + \cos\theta \\
&= 2\cos^2\theta + \cos\theta - 1.
\end{aligned}$$

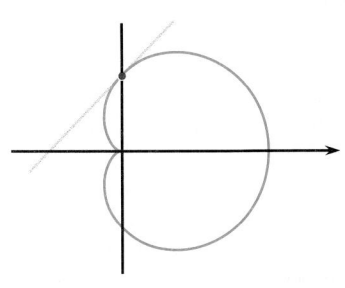

Figure 12.2.20(i)

Writing $u = \cos\theta$ and solving $2u^2 + u - 1 = 0$, we find

$$\cos\theta = u = \frac{-1 \pm \sqrt{1+8}}{4} = \frac{-1 \pm 3}{4} = \frac{1}{2}, -1.$$

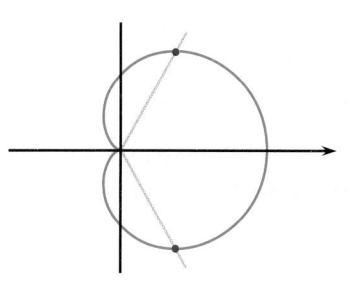

When $\cos\theta = -1$, we have $\sin\theta = 0$, and the denominator in the slope formula is zero, so the formula does not apply. The corresponding point on the graph is the cusp at the origin, where there is no tangent line.

The tangent line is horizontal when $\cos\theta = \frac{1}{2}$, that is, when $\theta = \frac{\pi}{3}$ (at the top point in Figure 12.2.20(ii)) and when $\theta = -\frac{\pi}{3}$ (at the bottom point).

The highest point on the graph occurs when $\theta = \frac{\pi}{3}$, so $r = f\left(\frac{\pi}{3}\right) = 1 + \cos\frac{\pi}{3} = \frac{3}{2}$. Its rectangular coordinates are

Figure 12.2.20(ii)

$$(x, y) = (r\cos\theta, r\sin\theta) = \left(\frac{3}{2}\left(\frac{1}{2}\right), \frac{3}{2}\left(\frac{\sqrt{3}}{2}\right)\right) = \left(\frac{3}{4}, \frac{3\sqrt{3}}{4}\right).$$

Exercises 12.2

I

Sketch the graph of each of the following curves.

1. $r = \sin\theta$
2. $r = 3\cos\theta$
3. $r = -2\cos\theta$
4. $r = 5$
5. $\sin\theta = \frac{1}{\sqrt{2}}, r \geq 0$
6. $r = \cos(5\theta)$
7. $r = 2\sin(2\theta)$
8. $r = \sin\left(3\theta + \frac{\pi}{2}\right)$
9. $r = -\theta, \theta > 0$
10. $r = \cos\theta + \sin\theta$

Sketch and identify the graph of each of the following curves.

11. $r = 1 + \sin\theta$
12. $r = 2 - \cos\theta$
13. $r = 2 - 3\cos\theta$
14. $r = 3 - 4\sin\left(\theta - \frac{\pi}{4}\right)$
15. $r = 1 + 10\cos\theta$
16. $r = \sqrt{\sin(2\theta)}$
17. $r^2 = \cos^2\theta - \sin^2\theta$
18. $r = 3 - 3\cos\theta$
19. $r^2 = \cos^2\theta + \sin^2\theta$
20. $r = 3\sin\theta\cos\theta$

II

Sketch the graph of each of the following curves.

*21. $r = 5\cos\theta - 5\sin\theta$
*22. $r = \cos 2\theta + \sin 2\theta$
*23. $r = 3\cos\theta - 4\sin\theta$
*24. $r = 1 + \cos\theta + \sin\theta$

Find the point(s) of intersection of each of the following pairs of curves. In each case draw a sketch.

*25. The circle $r = \cos\theta$ and the line $r\cos\theta = \frac{1}{2}$.
*26. The circle $r = -2\sin\theta$ and the line $r\sin\theta = -1$.
*27. $r = 3 + 2\cos\theta$ and $r = 6 - 4\cos\theta$
*28. $r = \cos\theta$ and $r^2 = \frac{3}{2}\cos(2\theta)$

*29. $r = \cos(5\theta)$ and $r = \sin(5\theta)$
*30. $r = \sin\theta$ and $r = \frac{2\theta}{\pi}$
**31. $r = 1 + \cos\theta$ and $r = 3 + 5\cos\theta$
**32. $r = \cos\theta - 2$ and $r = 2\cos\theta + 2$
*33. Sketch the graph of $r = \cos^2\left(\frac{\theta}{2}\right)$.
34. In Example 3 (p. 761) it was stated that the part of the curve $r = \cos\theta$ with $\frac{\pi}{2} \leq \theta \leq \frac{3\pi}{2}$ is the same as the circle found when $-\frac{\pi}{2} \leq \theta \leq \frac{\pi}{2}$. Show that this is true by comparing the points whose polar coordinates are $(r, \theta) = (\cos\theta, \theta)$ and $(\cos(\theta + \pi), \theta + \pi)$.
35. Verify that if $r < 0$, then the point whose polar coordinates are (r, θ) is exactly opposite (i.e., the *negative*) of the point whose polar coordinates are $(|r|, \theta)$.
*36. Verify the statement made in Summary 12.2.1 about the number of petals in a rose.

Sketch the graph of each of the following.

*37. $r = |\sin(3\theta)|$
*38. $r = |\cos(5\theta)|$
*39. $r = |\cos\theta|$
*40. $r = |\cos(4\theta)|$
*41. Prove the remark made before Summary 12.2.2 that changing (r, θ) to $(-r, \pi - \theta)$ has the effect of reflecting the point in the horizontal axis.
*42. Prove that replacing the point (r, θ) by $(r, \pi - \theta)$ or by $(-r, -\theta)$ has the effect of reflecting the point in the vertical axis.
*43. Consider the equation $r^2 = \cos\theta$. Show that it is unchanged if we substitute $(-r, -\theta)$ for (r, θ) but that it changes if we substitute $(r, \pi - \theta)$. What can we conclude about its symmetry?

*44. What are the polar coordinates of the points where the Spiral of Archimedes crosses the horizontal axis (cf. Example 2)?

*45. What are the rectangular coordinates of the points where the Spiral of Archimedes crosses the horizontal or vertical axes (cf. Example 2)?

46. When we drew the Spiral of Archimedes in Example 2, we only considered values of θ that are nonnegative. Sketch the graph that results if we let $r = \theta$ and consider *all* possible values of θ.

For each of the following curves, find the rectangular coordinates of the highest and lowest points on the graph.

*47. The right-hand petal of the rose $r = \cos 2\theta$.

*48. The right-hand petal of the rose $r = \cos 3\theta$.

**49. The lemniscate $r^2 = \cos 2\theta$.

*50. The limaçon $r = 1 - 2\cos\theta$.

III

Use a graphics calculator or computer to find the points of intersection of the following pairs of curves. (You may find it easier to begin with the Cartesian graph of $y = f(\theta)$.)

51. $r^2 = 5\sin(2\theta), \quad r = \theta$

52. $r = 1 - 2\sin\theta, \quad r = \frac{\theta}{4}$

*53. $r = \sin\frac{\theta}{2}, \quad r = \sin\theta$

*54. $r^2 = 4\cos(4\theta), \quad r = 3 + 2\cos\theta$

POINT TO PONDER

When we consider the equation $\theta = \frac{4\pi}{3}$, we quite likely think of all points whose polar angle is $\frac{4\pi}{3}$, so it seems reasonable to draw the half-line consisting of all points (r, θ) with $\theta = \frac{4\pi}{3}$ and $r \geq 0$. On the other hand, for many other situations it is much more natural to allow r to be negative. Some books will write the entire line for $\theta = \frac{4\pi}{3}$, others the half-line. The best approach is probably to specify what we mean in any doubtful cases. For instance, for the spiral we drew in Example 2 we wrote $r = \theta$, $\theta \geq 0$. For the half-line we should write $\theta = \frac{4\pi}{3}$, $r \geq 0$.

In working with polar coordinates we often resort to converting to rectangular coordinates in order to draw something. In a way this may seem like cheating; if we are going to have to end up using Cartesian coordinates anyway, why not just use them all along?

In fact, polar coordinates are usually more awkward, especially for drawing graphs. On the other hand, for certain purposes they are very convenient. Obviously, they are suited to situations in which something goes around something else, and they are essential to one of the early successes of calculus, the study of the motion of the planets. The theory of planetary motion can best be expressed in terms of polar coordinates. It is so much more complicated in rectangular coordinates that it might never have been discovered if it had to be done in that form.

You might like to express the equation of a lemniscate in rectangular coordinates just for an example to show that something can be simpler in its polar form.

Polar coordinates are useful because they allow us to express certain problems in a convenient or suggestive form. We then use the techniques we know to try to solve the problems.

12.3 Area in Polar Coordinates

AREA FORMULA
LIMITS OF INTEGRATION
TRACING EXACTLY ONCE

For curves sketched in rectangular coordinates we were able to use integration to find the area contained under the graph of the curve. For curves in polar coordinates it is less clear what we might mean by the area "under" the graph, since the graph tends to wrap around the origin. Instead, it is reasonable to consider the region contained "inside" the curve if this makes sense.

Figure 12.3.1

Figure 12.3.2

Figure 12.3.3

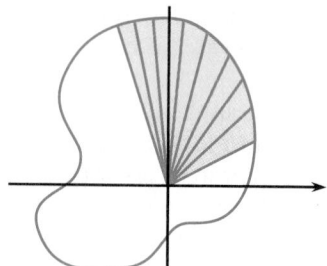

Figure 12.3.4

It certainly would make sense for some curves, like circles and lemniscates and limaçons and cardioids, but it is less clear what it might mean for the spiral shown in Figure 12.3.1.

The spiral does not "contain" anything. On the other hand, we might like to know the area of the shaded region in Figure 12.3.2, which is the region "inside" the spiral $r = \theta$, for $0 \leq \theta \leq \pi$.

For that matter, we might look at the graph of $r = f(\theta)$ and ask about the area of the region inside the graph for $\alpha \leq \theta \leq \beta$, for some angles α and β as shown in Figure 12.3.3. This should include the sort of region that is entirely enclosed inside a curve, but it will also cover other cases.

Our experience with integration suggests that we should break up the region into small sectors, as shown in Figure 12.3.4. A first guess might be to approximate the area of a little sector with the area of a little triangle, but in fact it is better to use a small circular sector.

Consider the small circular sector shown in Figure 12.3.5. Its radius is r and the central angle is $\Delta\theta$. We need to find its area.

The area of the whole circle of radius r is πr^2. The whole circle involves an angle of 2π at the center, so this little sector, with its angle $\Delta\theta$, is $\frac{\Delta\theta}{2\pi}$ of the whole circle, and its area is

$$\frac{\Delta\theta}{2\pi}\pi r^2 = \frac{1}{2}r^2\Delta\theta.$$

As usual, we consider a partition $\mathcal{P} = \{\theta_0, \theta_1, \ldots, \theta_n\}$ of $[\alpha, \beta]$ and break the region inside $r = f(\theta)$ into the little sectors with $\theta_{i-1} \leq \theta \leq \theta_i$. See Figure 12.3.6. The area of the ith sector is approximately equal to the area of the circular sector between θ_{i-1} and θ_i whose radius is $r = f(\theta^*)$ for any $\theta^* \in [\theta_{i-1}, \theta_i]$.

We just found that this area is $\frac{1}{2}r^2\Delta\theta = \frac{1}{2}f(\theta^*)^2(\theta_i - \theta_{i-1})$.

If we choose a point θ_i^* in $[\theta_{i-1}, \theta_i]$ for each i, then we can approximate the area of each little sector in this way and add them up to approximate the area of the whole region. The approximation to the area is

$$\sum_{i=1}^{n}\frac{1}{2}f(\theta_i^*)^2\Delta\theta_i = \sum_{i=1}^{n}\frac{1}{2}f(\theta_i^*)^2(\theta_i - \theta_{i-1}).$$

Of course, this is a Riemann sum.

Now if the mesh of $\mathcal{P}$ tends to 0 as we let $n \to \infty$, we see that the approximations to the area given by these Riemann sums will tend to a limit, which is the area.

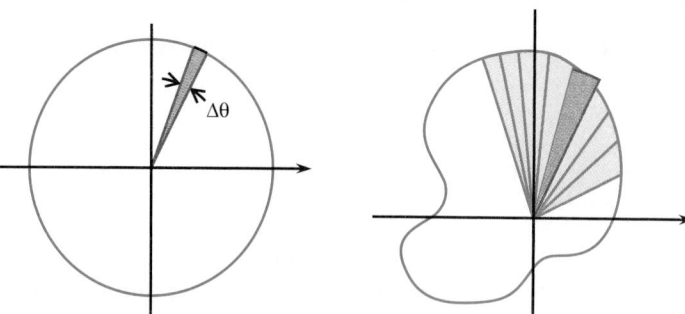

Figure 12.3.5   **Figure 12.3.6**

THEOREM 12.3.1

Consider the curve $r = f(\theta)$. The area of the region enclosed inside this curve as θ runs between the lines $\theta = \alpha$ and $\theta = \beta$ (i.e., for $\alpha \leq \theta \leq \beta$) is

$$Area = \frac{1}{2} \int_{\alpha}^{\beta} f(\theta)^2 \, d\theta.$$

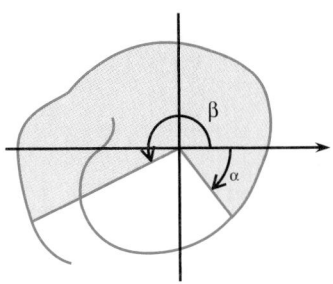

Figure 12.3.7

Such a region is illustrated in Figure 12.3.7.

Notice that the integrand $f(\theta)^2$ is always nonnegative even if the radius $r = f(\theta)$ is negative. So we always have a nonnegative integral and do not have to worry about "signed areas" the way we did for integrals in rectangular coordinates. However, we may have to worry about different parts of the region overlapping.

EXAMPLE 1   Area of Cardioid   Find the area of the region inside the cardioid $r = 1 - \cos\theta$.

Solution   In this case we want the entire region inside the curve, so we should let θ go from 0 to 2π (which takes us once around the origin). From the theorem we see that the area is

$$\frac{1}{2} \int_0^{2\pi} r^2 \, d\theta = \frac{1}{2} \int_0^{2\pi} (1 - \cos\theta)^2 \, d\theta = \frac{1}{2} \int_0^{2\pi} (1 - 2\cos\theta + \cos^2\theta) \, d\theta.$$

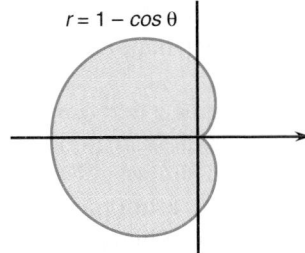

$r = 1 - \cos\theta$

Figure 12.3.8

Using the familiar identity $\cos^2\theta = \frac{1}{2}(1 + \cos 2\theta)$ (cf. Theorem 1.7.13 and Example 5.7.5), we integrate and find that the area equals

$$\frac{1}{2} \int_0^{2\pi} (1 - 2\cos\theta + \cos^2\theta) \, d\theta = \frac{1}{2} \int_0^{2\pi} 1 \, d\theta$$
$$- \int_0^{2\pi} \cos\theta \, d\theta + \frac{1}{4} \int_0^{2\pi} (1 + \cos 2\theta) \, d\theta$$
$$= \pi - (\sin\theta) \Big|_0^{2\pi} + \left(\frac{\theta}{4} + \frac{1}{8} (\sin 2\theta) \right) \Big|_0^{2\pi}$$
$$= \frac{3\pi}{2}.$$

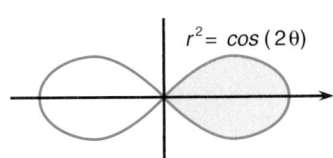

$r^2 = \cos(2\theta)$

Figure 12.3.9(i)

The cardioid is illustrated in Figure 12.3.8. Its area is $\frac{3\pi}{2}$.

We have seen with integrals that finding the limits of integration can be a difficult part of a problem. This is frequently true with integrals in polar coordinates.

EXAMPLE 2   Find the area inside one half of the lemniscate $r^2 = \cos 2\theta$, which is illustrated in Figure 12.3.9(i).

Solution   It may help to consider the graph of $y = \cos 2\theta$ in rectangular coordinates, which is shown in Figure 12.3.9(ii). There we see that $\cos 2\theta$ is positive for $-\frac{\pi}{4} < \theta < \frac{\pi}{4}$, and this is the range that corresponds to the right-hand loop of the lemniscate.

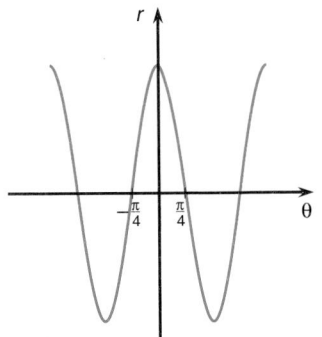

Figure 12.3.9(ii)

The area we want is given by

$$\frac{1}{2} \int_{-\pi/4}^{\pi/4} r^2 \, d\theta = \frac{1}{2} \int_{-\pi/4}^{\pi/4} \cos(2\theta) \, d\theta$$

$$= \frac{1}{4} \left(\sin(2\theta) \right) \Big|_{-\pi/4}^{\pi/4} = \frac{1}{4} \left(\sin\left(\frac{\pi}{2}\right) - \sin\left(-\frac{\pi}{2}\right) \right)$$

$$= \frac{1}{2}.$$

The required area is $\frac{1}{2}$ square unit.

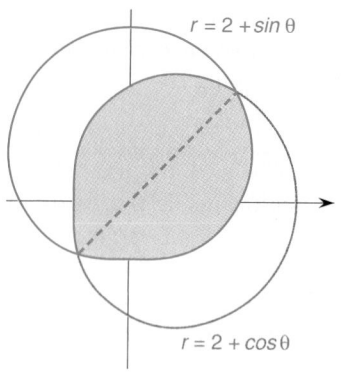

$r = 2 + \sin\theta$

$r = 2 + \cos\theta$

Figure 12.3.10

EXAMPLE 3   Find the area of the region consisting of all points inside the limaçon $r = 2 + \cos\theta$ and also inside the limaçon $r = 2 + \sin\theta$.

Solution   We sketch the curves in Figure 12.3.10. The area we want to find is the shaded region in the picture. We begin by finding the points of intersection of the two curves.

For a point to be on both curves, we need $r = 2 + \cos\theta$ and $r = 2 + \sin\theta$, so $2 + \cos\theta = 2 + \sin\theta$, and $\cos\theta = \sin\theta$. This means $\tan\theta = 1$, which happens when $\theta = \frac{\pi}{4}$ and when $\theta = -\frac{3\pi}{4}$.

In the picture we see that the "upper left" part of the shaded region is determined by lying inside the curve $r = 2 + \cos\theta$, and the "lower right" part of the region lies inside $r = 2 + \sin\theta$. So to find the area of the upper left part, we should integrate $\frac{1}{2}(2 + \cos\theta)^2$. For limits of integration we should begin at $\alpha = \frac{\pi}{4}$. From there we should let the angle sweep counterclockwise until it reaches the other point of intersection of the two limaçons. This means that the angle will increase by π from $\frac{\pi}{4}$, so the upper limit of integration is $\beta = \frac{\pi}{4} + \pi = \frac{5\pi}{4}$. This angle is in the same position as $-\frac{3\pi}{4}$, but it is not the same angle.

To find the area of the lower right region, we integrate $\frac{1}{2}(2 + \sin\theta)^2$ from $\alpha = -\frac{3\pi}{4}$ to $\beta = \frac{\pi}{4}$. The total area is

$$\frac{1}{2} \int_{\pi/4}^{5\pi/4} (4 + 4\cos\theta + \cos^2\theta) \, d\theta + \frac{1}{2} \int_{-3\pi/4}^{\pi/4} (4 + 4\sin\theta + \sin^2\theta) \, d\theta$$

$$= \frac{1}{2} \left(4\theta + 4\sin\theta + \frac{1}{2}\left(\theta + \frac{1}{2}\sin 2\theta\right) \right) \Big|_{\pi/4}^{5\pi/4}$$

$$+ \frac{1}{2} \left(4\theta - 4\cos\theta + \frac{1}{2}\left(\theta - \frac{1}{2}\sin 2\theta\right) \right) \Big|_{-3\pi/4}^{\pi/4}$$

$$= \frac{1}{2} \left(4\pi + 4\left(-\frac{1}{\sqrt{2}} - \frac{1}{\sqrt{2}}\right) + \frac{\pi}{2} + \frac{1}{4} - \frac{1}{4} \right)$$

$$+ \frac{1}{2} \left(4\pi - 4\left(\frac{1}{\sqrt{2}} + \frac{1}{\sqrt{2}}\right) + \frac{\pi}{2} - \frac{1}{4} + \frac{1}{4} \right)$$

$$= \frac{9\pi}{2} - 4\sqrt{2}.$$

By symmetry the two parts of the region have the same area. We could evaluate one integral and double the result.

The area inside the two curves is $\frac{9\pi}{2} - 4\sqrt{2}$.

When we first found the angles θ of the points at which the two curves cross, we found $\frac{\pi}{4}$ and $-\frac{3\pi}{4}$. However, when we needed the limits of integration for the upper left part of the region, we began at $\frac{\pi}{4}$ but we ended at $\frac{5\pi}{4}$. It was important to look at the picture to realize that as θ increases from $\frac{\pi}{4}$, the angle at which it will reach the other intersection point is $\frac{5\pi}{4}$.

EXAMPLE 4   Find the area of the upper half of the inner loop of the limaçon $r = 2\cos\theta - 1$.

Solution   We begin by sketching the graph of $y = 2\cos\theta - 1$ in Figure 12.3.11(i). In particular, it is positive between $\theta = -\frac{\pi}{3}$ and $\theta = \frac{\pi}{3}$, and negative for θ between $\frac{\pi}{3}$ and $\frac{5\pi}{3}$.

From this we can sketch the curve $r = 2\cos\theta - 1$ in Figure 12.3.11(ii). We observe that the inner loop is traced out as θ moves from $-\frac{\pi}{3}$ to $\frac{\pi}{3}$. The upper half of the inner loop will be traced out as θ moves from 0 to $\frac{\pi}{3}$.

Applying Theorem 12.3.1, we find that the area of the upper half of the inner loop is

$$\frac{1}{2}\int_0^{\pi/3}(2\cos\theta - 1)^2\,d\theta = \frac{1}{2}\int_0^{\pi/3}(4\cos^2\theta - 4\cos\theta + 1)\,d\theta$$

$$= \frac{1}{2}(2\theta + \sin 2\theta - 4\sin\theta + \theta)\Big|_0^{\pi/3}$$

$$= \frac{\pi}{3} - 0 + \frac{\sqrt{3}}{4} - 0 - \sqrt{3} + 0 + \frac{\pi}{6} - 0$$

$$= \frac{\pi}{2} - \frac{3\sqrt{3}}{4}.$$

The area of the upper half of the inner loop is $\frac{\pi}{2} - \frac{3\sqrt{3}}{4}$.

EXAMPLE 5   Find the area of the region consisting of all points that lie inside the upper right petal of the three-petaled rose $r = \sin 3\theta$ but outside the circle $r = \sin\theta$.

Solution   We sketch the curves $y = \sin\theta$ and $y = \sin 3\theta$ in rectangular coordinates in Figure 12.3.12(i). Using this sketch, we draw the rose and the circle in Figure 12.3.12(ii), shading the specified region.

Before proceeding, we must know the coordinates of that point where the curves intersect. Since the curves are $r = \sin 3\theta$ and $r = \sin\theta$, we must solve $\sin\theta = \sin 3\theta$. Expanding $\sin 3\theta$ on the right side, we see that this amounts to

$$\sin\theta = \sin\theta\cos 2\theta + \cos\theta\sin 2\theta = \sin\theta\cos 2\theta + 2\cos^2\theta\sin\theta,$$

so $\sin\theta = 0,$ or $1 = \cos 2\theta + 2\cos^2\theta = \cos^2\theta - \sin^2\theta + 2\cos^2\theta,$

that is, $1 = 3\cos^2\theta - (1 - \cos^2\theta) = 4\cos^2\theta - 1.$

From this we see that $2 = 4\cos^2\theta$, so $\cos^2\theta = \frac{1}{2}$.

We consider the point $\theta = \frac{\pi}{4}$, where $\cos\theta = \frac{1}{\sqrt{2}}$, and it is easy to check that $\sin\frac{\pi}{4} = \frac{1}{\sqrt{2}} = \sin\frac{3\pi}{4}$. This is also consistent with Figure 12.3.12(i), where the two curves appear to cross at about $\theta = \frac{\pi}{4}$.

Figure 12.3.11(i)

Figure 12.3.11(ii)

Figure 12.3.12(i)

Figure 12.3.12(ii)

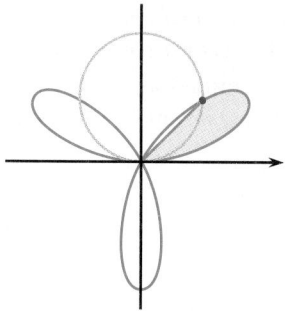

Figure 12.3.12(iii)

To find the desired area, we should find the area inside the rose for θ between 0 and $\frac{\pi}{4}$, which is the large shaded area in Figure 12.3.12(iii), and subtract the area of the part of this region that lies inside the circle. The area we want will be

$$\frac{1}{2}\int_0^{\pi/4}(\sin^2 3\theta - \sin^2\theta)\,d\theta = \frac{1}{4}\int_0^{\pi/4}(1 - \cos 6\theta - 1 + \cos 2\theta)\,d\theta$$

$$= \frac{1}{4}\left(-\frac{1}{6}\sin 6\theta + \frac{1}{2}\sin 2\theta\right)\Bigg|_0^{\pi/4}$$

$$= \frac{1}{4}\left(\frac{1}{6} + 0 + \frac{1}{2} - 0\right) = \frac{1}{6}.$$

The desired area is $\frac{1}{6}$.

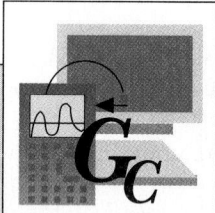

If we consider the similar question of finding the area of the region inside $r = \sin 3\theta$ but outside $r = \cos\theta + \sin\theta$, for $0 \leq \theta \leq \frac{\pi}{2}$, we can use a graphics calculator to sketch the graphs.

TI-85	SHARP EL9300
MODE ▼ ▼ ▼ ▼ ▶ (Pol) ENTER	SET UP E 2 ⤸ 3 sin 3 x/θ/T 2ndF
EXIT GRAPH F1 ($r(\theta) =$) 3 × SIN	▼ cos x/θ/T + sin x/θ/T RANGE 0
(3 × F1 (θ)) ▼ COS F1 (θ) + SIN	ENTER 3 ENTER 1 ENTER 0 ENTER 3
F1 (θ) M2(RANGE) 0 ▼ $\pi \div 2$	ENTER 1 ENTER 0 ENTER $\pi \div 2$ ENTER
▼ ▼ 0 ▼ 3 ▼ ▼ ▼ (−) .5 ▼ 3	.1 ENTER ⤸
F5(GRAPH)	

CASIO f_x-7700GB	HP 48SX
MODE MODE − (POL) Range 0 EXE	↰ RAD ↰ PLOT ′ $R = $ 3 × SIN
3 EXE 1 EXE 0 EXE 3 EXE 1 EXE 0 EXE	3 × α ↱ F NEW P O L ENTER
$\pi \div 2$ EXE .01 EXE Graph 3 sin 3 θ	PTYPE POLAR PLOTR 0 ENTER 3
EXE Graph cos θ + sin θ EXE	XRNG 0 ENTER 3 YRNG ′ α ↱ F
	INDEP DRAW ATTN ′ $R =$ COS α
	↱ F ▶ + SIN α ↱ F ↱ PLOT
	↰ DRAW DRAW

Unlike what happened in Example 5, it is not easy to solve for the intersection of these curves. But we can use the calculator to solve for where $\cos\theta + \sin\theta = 3\sin 3\theta$. We find two solutions $\theta = a, b \in \left[0, \frac{\pi}{2}\right]$, with $a < b$. Then we find that the required area is

$$\frac{1}{2}\int_a^b\left(9\sin^2 3\theta - (\cos\theta + \sin\theta)^2\right)d\theta$$

$$= \left(\frac{9}{4}\left(\theta - \frac{1}{6}\sin 6\theta\right) - \frac{\theta}{2} + \frac{1}{4}\cos 2\theta\right)\Bigg|_a^b.$$

With the help of the calculator we find that this is ≈ 1.603946.

Next we consider the following easy problem.

EXAMPLE 6   **Area of Rose**   Find the area of the five-petaled rose $r = \sin 5\theta$.

Solution   We sketch the function $y = \sin 5\theta$ in rectangular coordinates in Figure 12.3.13(i) and the rose in Figure 12.3.13(ii). It should be easy to find the area; all that is needed is to integrate $\frac{1}{2}r^2$.

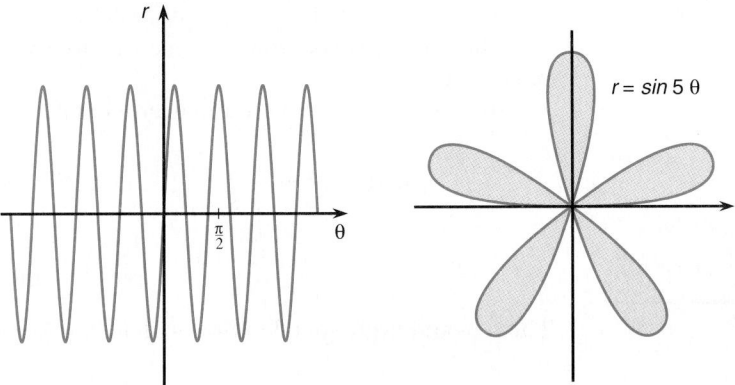

Figure 12.3.13(i)   **Figure 12.3.13(ii)**

Since the rose goes all around the origin, we are tempted to integrate from $\theta = 0$ to $\theta = 2\pi$. This would be a blunder because, as we have already noted about roses with odd numbers of petals, the rose is traced out *twice* as θ goes from 0 to 2π. To go around the rose exactly once, it is necessary to let θ run from 0 to π. The area is

$$\frac{1}{2}\int_0^\pi \sin^2 5\theta\, d\theta = \frac{1}{4}\int_0^\pi (1 - \cos 10\theta)\, d\theta = \frac{1}{4}\left(\theta - \frac{1}{10}\sin 10\theta\right)\Bigg|_0^\pi$$

$$= \frac{\pi}{4} - 0 - \frac{1}{40}(0) + \frac{1}{40}(0) = \frac{\pi}{4}.$$

The area of the rose is $\frac{\pi}{4}$.

> Because we are allowing r to be positive or negative, it is possible that even though θ only runs from 0 to 2π, the moving point may cover part or all of a curve twice. If we want to find the area inside the curve, we have to be careful to choose the limits of integration in a way that traces out the appropriate part of the curve exactly once.

EXAMPLE 7   **Areas of Flowers**   Two botanists are arguing about flowers. The colorful petals attract insects to help with pollination, so it is to the plant's advantage to have as large an area as possible exposed. One botanist argues that a bloom like a daisy with many petals arranged around its center will have a greater area than one with fewer petals of the same length. The other botanist points out that when there are more petals, each one is narrower, and concludes that the area will be the same or maybe even less. Which (if either) is right?

Figure 12.3.14

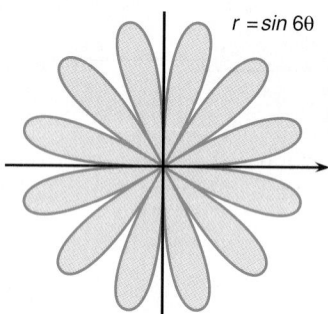

Figure 12.3.15

They decide to compare a hypothetical bloom with ten petals and one with twelve to see whether there is a difference, and they agree that the graphs we have been drawing for "roses" $r = \sin(n\theta)$ seem roughly representative of flower shapes.

In Example 6 we considered $r = \sin(5\theta)$, which has five petals, but we also know that $r = \sin(10\theta)$ has twenty. How do we get ten? The five-petaled rose $r = \sin(5\theta)$ has gaps in its petals: There is a petal with $0 \le \theta \le \frac{\pi}{5}$, but the next has $\frac{2\pi}{5} \le \theta \le \frac{3\pi}{5}$. Strictly speaking, there is a petal for $\frac{\pi}{5} \le \theta \le \frac{2\pi}{5}$, but it is "on the other side" (see Figure 12.3.13(ii)). To get a ten-petaled rose, we should fill in the gaps with another five-petaled rose, as shown in Figure 12.3.14.

The area of this bloom is twice the area we found in Example 6 for the five-petaled rose, so it is $\frac{\pi}{2}$.

In Figure 12.3.15 we see the twelve-petaled rose $r = \sin(6\theta)$. Its area is

$$\frac{1}{2}\int_0^{2\pi} \sin^2(6\theta)\, d\theta = \frac{1}{4}\int_0^{2\pi}\left(1 - \cos(12\theta)\right)d\theta = \left(\frac{\theta}{4} - \frac{1}{48}\sin(12\theta)\right)\Bigg|_0^{2\pi}$$

$$= \frac{2\pi}{4} - 0 - \frac{1}{48}(0) + \frac{1}{48}(0) = \frac{\pi}{2}.$$

The ten- and twelve-petaled flowers have the same area.

Exercises 12.3

I

In each of the following situations, find the area enclosed inside the given curve between the two angles specified.

1. $r = \theta$; $0, \frac{\pi}{2}$
2. $r = \theta$; $\frac{\pi}{2}, \pi$
3. $r = 6 - 2\cos\theta$; $0, 2\pi$
4. $r = 4 - 4\sin\theta$; $0, \pi$
5. $r = 7$; $-\frac{\pi}{4}, \frac{\pi}{4}$
6. $r = 4\sin\theta$; $0, \frac{\pi}{2}$
7. $r^2 = \sin 2\theta$; $\frac{\pi}{4}, \frac{\pi}{2}$
8. $r = \sqrt{2 + \sin\theta}$; $0, \pi$
9. $r = \sin\theta$; $\pi, 2\pi$
10. $r = \cos^2\theta + \sin^2\theta$; $0, 2\pi$

In each of the following situations, find the area enclosed inside the given curve.

11. $r = 2 - 2\sin\theta$
12. $r = 3 - 2\cos\theta$
13. $r = 4\cos\theta$
14. $r = -2\sin\theta$
15. $r = \cos\left(\theta + \frac{\pi}{4}\right)$
16. $r = 3\sin 4\theta$
17. $r^2 = \sin 2\theta$
18. $r^2 = 4\cos 2\theta$
19. $r = 2\sin\left(3\theta - \frac{\pi}{2}\right)$
20. $r^2 = \cos 3\theta$

In each of the following situations, find the area of the region enclosed in both the given curves.

21. $r = \sin\theta$; $r = \cos\theta$
22. $r = 1$; $r = 2\cos\theta$
23. $r = \cos\theta$; $r^2 = \frac{1}{2}\sin 2\theta$
24. $r = 1 + \cos\theta$; $r = 1$
25. $r = 1 + \cos\theta$; $r = 1 - \cos\theta$
26. $r = 1 + \cos\theta$; $r = \cos\theta - 1$
27. $r = 1 + \sin\theta$; $r = 1 + \cos\theta$
28. $r = 1 + \frac{1}{\sqrt{3}}\sin\theta$; $r = 1 + \cos\theta$
29. $r = 2\cos 4\theta$; $r = 1$
30. $r = 4 - \sin 5\theta$; $r = 1 + \sin 2\theta$

II

31. Find the area of the region lying inside the cardioid $r = 1 + \sin\theta$ and below and to the right of the line $y = x$.

32. Find the area of region lying inside the cardioid $r = 2 + 2\cos\theta$ and below and to the left of the line $x + y = 0$.

33. Find the area of the region contained in the "inner loop" of the limaçon $r = 1 - 2\sin\theta$.

**34. (i) Find the area of the region contained in the "inner loop" of the limaçon $r = 1 + 5\sin\theta$.

   (ii) Find the area of the region contained in the "inner loop" of the limaçon $r = 1 + a\sin\theta$, where $a > 1$.

*35. Find the area of the region consisting of all points contained in both $r = \sin 3\theta$ and $r = \cos 3\theta$.

*36. Find the area of the region consisting of all points contained in the circle $r = 2$ and above and to the right of the line $x + y = 2$.

*37. (i) Find the area of the three-petaled rose $r = \cos 3\theta$.

   (ii) Find the area of the seven-petaled rose $r = \cos 7\theta$.

   (iii) Find the area of the n-petaled rose $r = \cos n\theta$ if n is an odd number.

*38. (i) Find the area of the four-petaled rose $r = \cos 2\theta$.

   (ii) Find the area of the eight-petaled rose $r = \cos 4\theta$.

   (iii) Find the area of the $2n$-petaled rose $r = \cos n\theta$ if n is an even number.

*39. Observe that on the rose in Example 6 the petals all have the same area. Calculate the area of one of the petals and multiply by 5 to get the total area.

*40. The botanists from Example 7 realize that in actual flowers, adjacent petals may overlap. In light of this they propose a model that consists of the twelve-petaled rose $r = \sin(6\theta)$ superimposed on the other one $r = \cos(6\theta)$. The total surface area visible to an insect is not the sum of the two areas, since some petals are partly hidden behind others. Find the total visible area. By what percentage is it greater than the area of one twelve-petaled rose?

*41. A leaf has the same shape as the limaçon $r = 3 + 2\cos\theta$, shown in Figure 12.3.16. If it measures 4 cm from the middle of the indentation to the opposite end, what is its surface area?

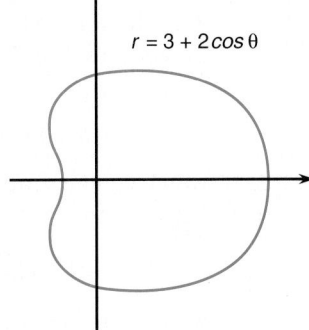

$r = 3 + 2\cos\theta$

Figure 12.3.16

*42. A leaf in the shape of a cardioid measures 2 in from the stem (at the indent) to the opposite end. Find its area.

III

Use a graphics calculator or computer to help you find the area of each of the following regions.

43. $r \leq 3 + \cos\theta,\ r \leq 2 - 2\sin\theta$

44. $r \leq 4 + \cos\theta,\ r \leq 3 - 2\cos\theta$

45. $r \leq 1 - \cos\theta,\ r \leq 2\sin\left(\theta + \frac{\pi}{4}\right)$

46. $r^2 \leq 49\sin 2\theta,\ r \leq 3 + 2\sin\theta$

12.4 Conics in Polar Coordinates

ECCENTRICITY
DIRECTRIX
FOCUS
ROTATIONS
STRAIGHT LINES

In Section 1.6 we reviewed conics from the point of view of sketching their graphs, and we returned to study them in more detail in Chapter 11. In both instances we wrote equations in the usual rectangular coordinates x and y, and we learned to recognize and classify conics from the appearance of their equations. We used various tricks to translate and/or rotate a conic until it is in one of the standard forms. In standard form a parabola has its vertex at the origin, and any other conic has its center at the origin.

Figure 12.4.1

Figure 12.4.2

Figure 12.4.3

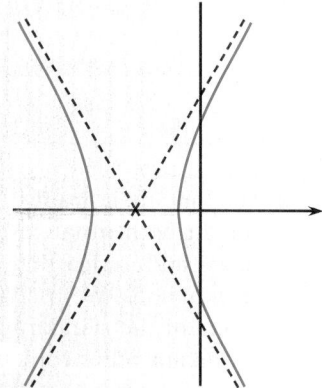

Figure 12.4.4

In this section we consider what happens when we write the equations in polar coordinates r and θ. It turns out that to make the equation simple in polar coordinates, it is better to put a *focus* of the conic at the origin, rather than the center.

Let us consider a conic for which the origin is a focus. Let us also suppose that the directrix is the line $x = -p$, where p is some positive number. The situation is sketched in Figure 12.4.1, where we have marked a point P on the conic with polar coordinates (r, θ). We have also marked the point Q, which is the point on the directrix at the same height as P.

In Section 11.2 we saw that it is possible to describe a conic in terms of the distances from each of its points P to a **focus** and to a **directrix** (see Summary 11.2.2, p. 736). If the **eccentricity** of the conic is e, then the distance from P to the focus equals e times the distance from P to the directrix. In our situation the distance from P to the focus O is just r. The horizontal distance from the point P to the vertical axis is $r \cos \theta$, and the distance from that axis to the directrix $x = -p$ is p.

The equation of the conic is

$$r = e(p + r \cos \theta),$$

which can be simplified to

$$r = \frac{k}{1 - e \cos \theta}, \qquad (12.4.1)$$

where k is some number (actually $k = ep$).

The interesting thing about this equation is that we do not have to know which type of conic we are dealing with. The form of the equation is the same for all types.

Observe that if $e < 1$, then $1 - e \cos \theta$ is never zero. In fact $1 - e \cos \theta \geq 1 - e$, so $r = \frac{k}{1 - e \cos \theta} \leq \frac{k}{1 - e}$. When $e < 1$, the conic is an ellipse, as shown in Figure 12.4.2; the curve is bounded, since r never gets larger than $\frac{k}{1 - e}$.

On the other hand, if $e = 1$, then the denominator $1 - e \cos \theta = 1 - \cos \theta$ tends to 0 as $\theta \to 0$, so $r \to \infty$. This is consistent with what we expect when $e = 1$, since the conic is a parabola (see Figure 12.4.3).

If $e > 1$, then there will be two values of θ at which the denominator is zero, and some of the time r will be positive and some of the time r will be negative. This can be seen in Figure 12.4.4; it is instructive to work out when r is positive and when it is negative, bearing in mind that r is measured from the origin, *not* from the center of the hyperbola. Notice that when the denominator tends to 0 from above or below the quotient $r = \frac{k}{1 - e \cos \theta}$ tends to ∞ or $-\infty$, as we expect for points on a hyperbola.

Finally, if $e = 0$, the equation is $r = k$, which is just the equation of a circle of radius k around the origin.

EXAMPLE 1 **Rotated Ellipse** Consider the ellipse shown in Figure 12.4.5. It is rotated through an angle $\frac{\pi}{4}$ relative to the ellipse given by Equation 12.4.1, with some value of e that satisfies $0 < e < 1$. Find its equation in polar coordinates.

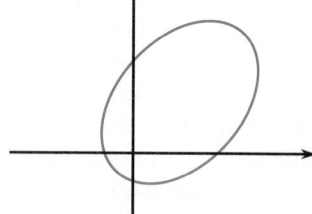

Figure 12.4.5

Solution   Rather than go through the development we used to find Equation 12.4.1, it is much easier to notice that rotating a curve just amounts to "shifting" the angle θ. We should try to write the equation in the form

$$r = \frac{k}{1 - e\cos\theta'},$$

for some θ'.

By analogy with Figure 12.4.1, θ' should be the angle measured from the major axis of the ellipse (see Figure 12.4.6). In other words, θ' should be 0 when $\theta = \frac{\pi}{4}$, so $\theta' = \theta - \frac{\pi}{4}$. The equation of the ellipse is

$$r = \frac{k}{1 - e\cos(\theta - \frac{\pi}{4})}.$$

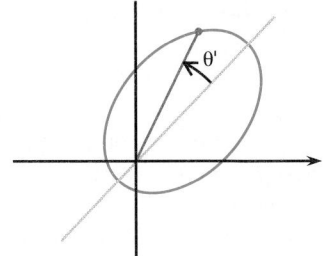

Figure 12.4.6

As Example 1 shows, the equation of a rotated conic is extremely simple, always supposing that the origin is a focus. If you have read Section 11.3, you know that this is in sharp contrast to the situation with rectangular coordinates. There the equations of rotated conics are much more complicated.

We summarize.

Summary 12.4.2

Conics in Polar Coordinates

The equation

$$r = \frac{k}{1 - e\cos\theta}$$

is the equation of a conic with eccentricity e and with the origin as a focus. If $e = 0$, it is a circle of radius k. Otherwise, we let $p = \frac{k}{e}$; the line $x = -p$ is a directrix.

The equation

$$r = \frac{k}{1 - e\cos(\theta - \theta_0)}$$

is the equation of a conic with eccentricity e and with the origin as a focus. This conic is obtained from the previous one by rotating counterclockwise through an angle of θ_0.

Several examples are sketched in Figure 12.4.7.

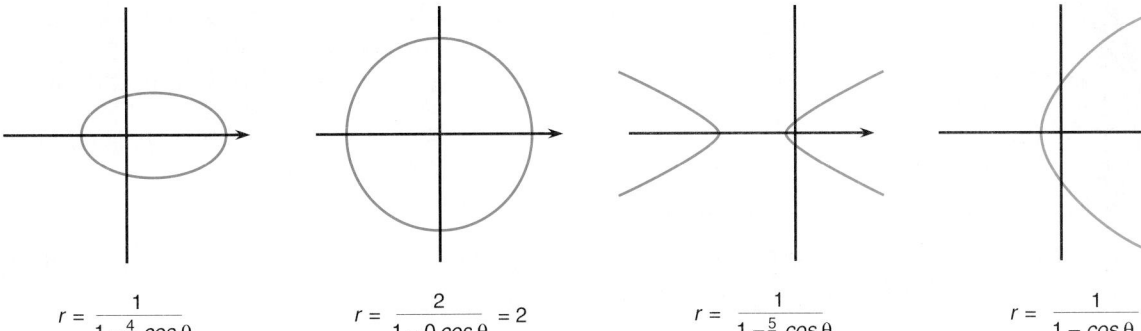

$$r = \frac{1}{1 - \frac{4}{5}\cos\theta} \qquad r = \frac{2}{1 - 0\cos\theta} = 2 \qquad r = \frac{1}{1 - \frac{5}{4}\cos\theta} \qquad r = \frac{1}{1 - \cos\theta}$$

Figure 12.4.7

Notice that if the denominator is never zero, then the conic is either a circle or an ellipse. Circles are easily recognized because the denominator is constant. To distinguish between parabolas and hyperbolas, one approach is to see that for a parabola the denominator is zero at exactly one point, whereas for a hyperbola there are two points at which the denominator is zero (corresponding to the angles of the asymptotes).

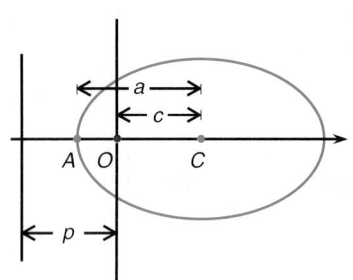

Figure 12.4.8

EXAMPLE 2   Consider the ellipse whose eccentricity is e and whose semi-major axis is a. It is illustrated in Figure 12.4.8, where we have also plotted the focus at the origin and marked the point C which is the center of the ellipse.

We know (see Summary 11.2.3, p. 738) that the distance $c = |OC|$ between the focus and the center satisfies $c = ae$.

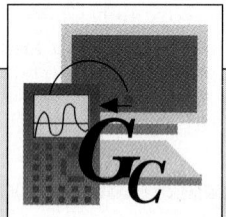

We can use a graphics calculator to sketch the conic $r = \dfrac{k}{1 - e\cos\theta}$ for various values of k and e.

TI-85	SHARP EL9300
MODE ▼ ▼ ▼ ▼ ▶ (Pol) ENTER EXIT GRAPH F1($r(\theta) =$) 1 ÷ (1 − .5 × COS F1(θ)) ▼ 1 ÷ (1 − .8 × COS F1(θ)) ▼ 1 ÷ (1 − .3 × COS F1(θ)) ▼ 3 ÷ (1 − 2 × COS F1(θ)) ▼ 5 ÷ (1 − 3 × COS F1(θ)) M2(RANGE) 0 ▼ 2π ▼ ▼ (−) 5 ▼ 5 ▼ ▼ (−) 5 ▼ 5 F5(GRAPH)	SET UP E 2 ⤸ 1 ÷ (1 − .5 cos x/θ/T) 2ndF ▼ 1 ÷ (1 − .8 cos x/θ/T) 2ndF ▼ 1 ÷ (1 − .3 cos x/θ/T) 2ndF ▼ 3 ÷ (1 − 2 cos x/θ/T) 2ndF ▼ 5 ÷ (1 − 3 cos x/θ/T) RANGE −5 ENTER 5 ENTER 1 ENTER −5 ENTER 5 ENTER 1 ENTER 0 ENTER 2π ENTER .02 ENTER ⤸
CASIO f_x-7700GB	HP 48SX
MODE MODE −(POL) Range −5 EXE 5 EXE 1 EXE −5 EXE 5 EXE 1 EXE 0 EXE 2π EXE .02 EXE Graph 1 ÷ (1 − .5 × cosθ) EXE Graph 1 ÷ (1 − .8 × cosθ) EXE Graph 1 ÷ (1 − .3 × cosθ) EXE Graph 3 ÷ (1 − 2 × cosθ) EXE Graph 5 ÷ (1 − 3 × cosθ) EXE	↰ RAD ↰ PLOT ′ R ▬ 1 ÷ () 1 − .5 × COS α ↱ F NEW P O L ENTER PTYPE POLAR PLOT R 5 +/− ENTER 5 XRNG 5 +/− ENTER 5 YRNG ′ α ↱ F INDEP DRAW ATTN ′ R ▬ 1 ÷ () 1 − .8 × COS α ↱ F ↱ PLOT ↰ DRAW DRAW ATTN ′ R ▬ 1 ÷ () 1 − .3 × COS α ↱ F ↱ PLOT ↰ DRAW DRAW ATTN ′ R ▬ 3 ÷ () 1 − 2 × COS α ↱ F ↱ PLOT ↰ DRAW DRAW ATTN ′ R ▬ 5 ÷ () 1 − 3 × COS α ↱ F ↱ PLOT ↰ DRAW DRAW

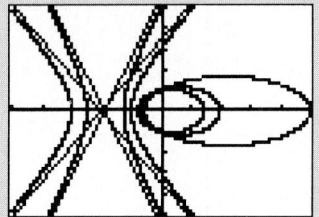

Identify each conic. Then try graphing the rotated conic $r = \dfrac{k}{1 - e\cos(\theta - \theta_0)}$ for various values of k, e, and θ_0.

Let A be the point on the ellipse that is closest to the directrix, as shown in the illustration. From the picture we see that the distance $|AO|$ from A to the focus O is $|AO| = a - c = a - ae = a(1 - e)$. Writing p for the distance from O to the directrix, we use the defining property of conics to see that $|AO|$ is e times the distance from A to the directrix. Since this distance is $p - |AO|$, we have that $|AO| = e(p - |AO|)$, or $|AO|(1 + e) = ep$. Substituting $|AO| = a(1 - e)$ into this, we find that $a(1 - e^2) = ep$, so

$$p = a\frac{1 - e^2}{e}.$$

In writing Equation 12.4.1 we let $k = ep$, so in this case we see that $k = ep = a(1 - e^2)$. The equation of the ellipse is

$$r = \frac{a(1 - e^2)}{1 - e\cos\theta}.$$

It is also possible to use this idea in reverse.

EXAMPLE 3   What are the semi-major and semi-minor axes of the ellipse

$$r = \frac{6}{2 - \cos\theta}?$$

Solution   We begin by dividing numerator and denominator by 2 to write the equation in the form

$$r = \frac{3}{1 - \frac{1}{2}\cos\theta}.$$

From this we see that the eccentricity is $e = \frac{1}{2}$. As we saw in Example 2, $ep = a(1 - e^2)$, and in this case, knowing that $ep = 3$, we find that $a = 3/(1 - e^2) = 3/\left(1 - \frac{1}{4}\right) = 4$.

From Summary 11.2.3 we know that $c = ea = 2$. On the other hand, c also equals $\sqrt{a^2 - b^2}$, so $a^2 - b^2 = 2^2 = 4$. Since $a^2 = 16$, we find that $b^2 = 12$ and $b = 2\sqrt{3}$. The ellipse is illustrated in Figure 12.4.9.

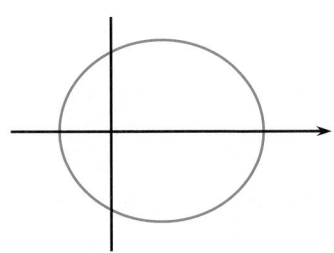

Figure 12.4.9

Much of the importance of describing conics in polar coordinates comes from the fact that the orbits of the planets are ellipses, with the sun at one of the foci. In fact, the orbit of a comet or any other object moving under the influence of the sun's gravity is also a conic, with the sun at a focus. If the orbit is an ellipse, the object is in a closed orbit, and it goes around and around the sun. If the orbit is a parabola or a hyperbola, the object will approach the sun, veer past, and fly off into space, never to return (see Figure 12.4.10).

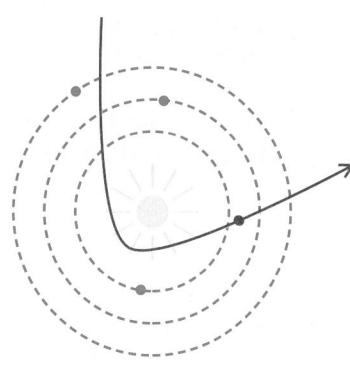

Figure 12.4.10

Circles and Lines

Finally, we consider the equations of lines and of circles whose centers are not necessarily at the origin.

Consider the circle shown in Figure 12.4.11. Its center is the point whose polar coordinates are (ρ, ϕ), and its radius is R. Also shown is a point P on the circle, whose polar coordinates are (r, θ). We apply the Law of Cosines (Theorem 1.7.10,

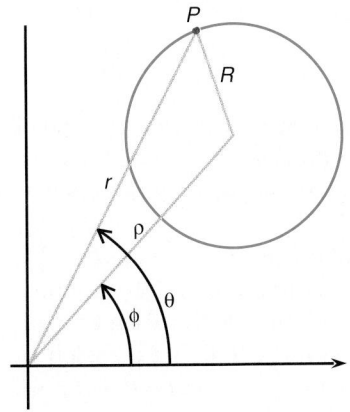

Figure 12.4.11

p. 50) to the triangle made by the origin, the point P, and the center of the circle. Noting that the angle in this triangle at the origin is $\theta - \phi$, we find that $R^2 = r^2 + \rho^2 - 2r\rho\cos(\theta - \phi)$, or

$$r^2 - 2r\rho\cos(\theta - \phi) + \rho^2 - R^2 = 0. \tag{12.4.3}$$

This is the equation of the circle. It is not very convenient to work with, however. For one thing, it does not express r as a function of θ. Instead, it gives a quadratic equation that r must satisfy. From the picture we see that for one value of θ there may be two different points on the circle, and they correspond to the two roots of the polynomial.

In special cases, of course, it will be simpler. For instance, if the center of the circle is at the origin, then $\rho = 0$ and the equation is just $r^2 = R^2$, which is just the ordinary circle $r = R$ of radius r about the origin.

Another special case occurs when $R = \rho$. This means that the origin is one of the points on the circle. The equation simplifies to $r^2 - 2rR\cos(\theta - \phi) = 0$, or

$$r = 2R\cos(\theta - \phi). \tag{12.4.4}$$

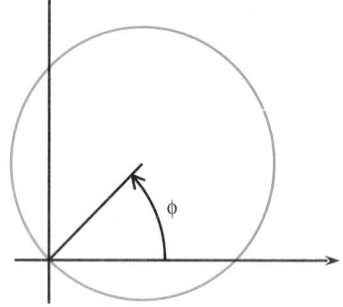

Figure 12.4.12

See Figure 12.4.12.

Next we consider straight lines.

We know that a line through the origin can be described by $\theta = \theta_0$, $r \in (-\infty, \infty)$, for some fixed θ_0.

The equation of the line $x = a$ is just $r\cos\theta = a$, or $r = a\sec\theta$.

Suppose we consider the line shown in Figure 12.4.13. The distance from the line to the origin is d. Suppose we draw the line segment from the origin that meets the line in a right angle, as shown, and suppose the polar coordinates of the point Q where this segment meets the line are (d, ϕ).

We could try to figure out the equation of the line, but it is much easier to notice that it can be obtained by rotating the line $x = d$ through an angle of ϕ. This means that the equation of the line is

$$r = d\sec(\theta - \phi). \tag{12.4.5}$$

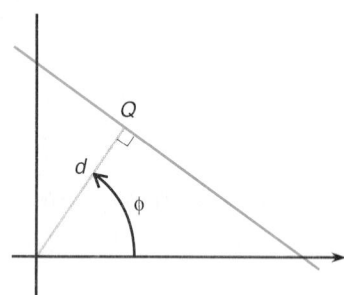

Figure 12.4.13

EXAMPLE 4   Find the equation of the line $x + y = 2$ in polar coordinates.

Solution   One approach is to write $x = r\cos\theta$, $y = r\sin\theta$, so the equation is $r(\cos\theta + \sin\theta) = 2$, that is,

$$r = \frac{2}{\cos\theta + \sin\theta}.$$

It is also possible to use the geometry of the situation. We know that the slope of this line is $m = -1$ (cf. Remark 1.5.10(ii)), so it makes an angle of $\frac{3\pi}{4}$ with the horizontal axis. The angle ϕ between the horizontal axis and the segment perpendicular to the line is $\frac{\pi}{4}$. See Figure 12.4.14.

From Equation 12.4.5 we know that the equation of the line is $r = d\sec\left(\theta - \frac{\pi}{4}\right)$, where d is the distance from the line to the origin. We could try to find the distance from the origin to the line. (Notice that what this means is the *smallest* distance from the origin to the line, so it requires finding the point on the line that is closest to the origin.) On the other hand, knowing that the equation is $r = d\sec\left(\theta - \frac{\pi}{4}\right)$, we could pick some convenient angle and solve.

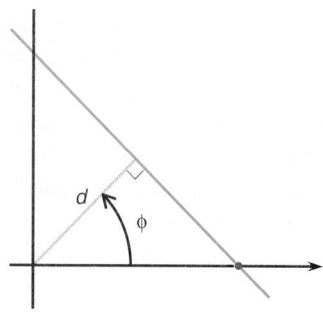

Figure 12.4.14

Suppose we let $\theta = 0$. This means that we are considering points on the horizontal axis, that is, points (x, y) with $y = 0$. The only such point on the line $x + y = 2$ is $(x, y) = (2, 0)$. The polar coordinates of this point are $(r, \theta) = (2, 0)$, and they should satisfy $r = d \sec\left(\theta - \frac{\pi}{4}\right)$, that is, $2 = d \sec\left(-\frac{\pi}{4}\right) = d\sqrt{2}$. We conclude that $d = \sqrt{2}$.

The equation of the line is $r = \sqrt{2} \sec\left(\theta - \frac{\pi}{4}\right)$. It is not immediately obvious that this is the same as the other formula $r = \frac{2}{\cos\theta + \sin\theta} \cdots$

Exercises 12.4

I

Identify the type of each of the following curves and sketch its graph.

1. $r = 1/(1 - \cos\theta)$

2. $r = 1/\left(2 - 2\cos\left(\theta - \frac{\pi}{3}\right)\right)$

3. $r = 1/(2 - \cos\theta)$

4. $r = 4/(3 - \cos\theta)$

5. $r = 2/(2 + \cos\theta)$

6. $r = 5/(3 - \sin\theta)$

7. $r = 1/(1 - 2\cos\theta)$

8. $r = 2/(1 + \sqrt{2}\cos\theta)$

9. $r = 3/\left(1 + \sin\left(\theta + \frac{\pi}{2}\right)\right)$

10. $r = 1/\left(2 - \cos(\theta - \pi)\right)$

11. $r = \cos\theta$

12. $r = 3\sec\left(\theta + \frac{\pi}{3}\right)$

13. $r = 1/(1 - 2\sin\theta)$

14. $r(2 - 3\cos\theta) = 2$

15. $r = 4\csc\theta$

16. $r = -2\sec(\pi - \theta)$

17. $r = 6$

18. $r = 6\sin\left(\theta - \frac{\pi}{6}\right)$

19. $r\sin\left(\theta + \frac{\pi}{4}\right) = 2$

20. $r/\left(3\sin\left(\theta - \frac{\pi}{3}\right)\right) = 5$

II

*21. Find the equation of the conic whose eccentricity is e, that has a focus at the origin, and whose directrix is the line $x = p$, with $p > 0$.

*22. Find the equation of the conic whose eccentricity is e, that has a focus at the origin, and whose directrix is the line $y = -p$, with $p > 0$.

*23. Consider the hyperbola $x^2 - y^2 = 1$. Translate the axes until its right-hand focus is at the origin, and express the equation of the resulting conic in polar coordinates, in the form of Equation 12.4.1.

*24. Consider the hyperbola $y^2 - x^2 = 4$. Translate the axes until its upper focus is at the origin, and express the equation of the resulting conic in polar coordinates, in the form of the second equation in Summary 12.4.2.

*25. What is the (smaller) angle between the asymptotes of the hyperbola
$$r = \frac{1}{1 - 2\cos\theta}?$$

*26. If $e > 1$, what is the angle between the asymptotes of the hyperbola
$$r = \frac{1}{1 - e\cos\theta}?$$

*27. Find the equation of the line $x - y = 3$ in polar coordinates.

*28. Find the equation of the line $x - \sqrt{3}y = 4$ in polar coordinates.

*29. Identify and sketch the conic
$$r = \frac{1}{2 + \sin\theta + \cos\theta}.$$

*30. Consider the hyperbola whose eccentricity is $e > 1$, whose transverse axis is horizontal, and whose right-hand focus is at the origin. If the minimum distance between the two branches is $2a$, what is the equation of the hyperbola in polar coordinates?

*31. Find the *other* focus and the *other* directrix of the ellipse $r = \frac{1}{2 - \cos\theta}$.

**32. When we considered the equation of the line $x = a$, with $a \neq 0$, we wrote it in polar coordinates as $r\cos\theta = a$. Then we rewrote it as $r = a\sec\theta$. This involved dividing by $\cos\theta$, and we cannot do this when $\cos\theta = 0$. Explain why this does not matter, that is, that the set of points that satisfy $r \cos \theta = a$ is exactly the same as the set of points that satisfy $r = a \sec \theta$.

*33. In Example 4 we remarked that it was not necessary to find the point on the line $x + y = 2$ that is closest to the origin in order to find the equation in polar coordinates. Nonetheless, it is possible to proceed this way. Find the point on the line $x + y = 2$ that is closest to the origin, and use it to find the distance from the origin to the line.

34. What point on the conic $r = \frac{2}{2 - \cos(\theta + \pi/2)}$ is closest to the origin?

35. What are the points of intersection of the conics $r = \frac{2}{2 + \cos \theta}$ and $r = \frac{2}{2 + \sin \theta}$? Draw a picture.

36. What are the points of intersection of the conics $r = \frac{2}{2 + 3 \cos \theta}$ and $r = \frac{1}{2 + \cos \theta}$? Draw a picture.

*37. Find the area of the ellipse $r = \frac{2}{2 - \sin \theta}$. (It may be easier not to integrate.)

*38. If $0 < e < 1$, find the area of the ellipse $r = \frac{k}{1 + e \cos \theta}$.

*39. (i) Find the slope of the tangent to $r = \frac{2}{2 - \cos \theta}$ at the point (r, θ).

(ii) Find the rectangular coordinates of the highest point on the ellipse.

*40. (i) If $0 < e < 1$, find the slope of the tangent to $r = \frac{k}{1 + e \cos \theta}$ at the point (r, θ).

(ii) Find the rectangular coordinates of the highest point on the ellipse.

*41. If you have a machine that will plot in polar coordinates, sketch $r = \frac{1}{1 - e \cos \theta}$, with $e = 0, \frac{1}{4}, \frac{1}{2}, \frac{3}{4}, \frac{7}{8}, 1, 1.1,$ 1.5, 2, 4, 10. Plot as many of these conics as possible on the same picture. In each case except $e = 0$ and $e = 1$, locate the second focus.

*42. Sketch the graph of $y = \frac{1}{1 - 2 \cos \theta}$ in rectangular coordinates, and then sketch the corresponding hyperbola $r = \frac{1}{1 - 2 \cos \theta}$ in polar coordinates. Identify the parts of the polar graph where r is positive and negative.

POINT TO PONDER

Strictly speaking, since we used the directrix in deriving Equation 12.4.1, we have in effect assumed that the conic is not a circle, since a circle does not have a directrix (cf. the discussion after Summary 11.2.2). However, if we take Equation 12.4.1 and let $e = 0$, we obtain the equation $r = \frac{k}{1}$, or $r = k$. This is the equation of a circle of radius k around the origin. In Section 11.2 we remarked that it is possible to think of a circle as what happens when an ellipse "degenerates" by letting the two foci run together. In this sense, $r = k$ is the equation of a circle "with focus at the origin."

However, there is a fly in the ointment, which is that in deriving Equation 12.4.1 we let $k = ep$. This seems to suggest that when $e = 0$, we should have $k = 0$, and the equation becomes $r = 0$, the equation of a single point. Fortunately, as we saw in the discussion after Summary 11.2.2, if we think of an ellipse whose foci run together, then its directrix moves out to infinity. So although $e \rightarrow 0$, we also have that $p \rightarrow \infty$.

In Example 2 we saw that $k = ep = a(1 - e^2)$. So if we consider an ellipse with focus at the origin and semi-major axis a and let its eccentricity e tend to 0, then we

$e = \frac{3}{4}$

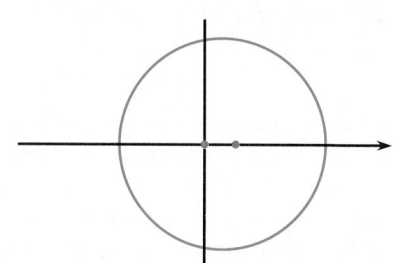

$e = \frac{1}{6}$

Figure 12.4.15

find that $k = a(1 - e^2) \to a$. Also the semi-minor axis b satisfies $a^2 - b^2 = c^2 = e^2 a^2 \to 0$, which means that $b \to a$. From this we see that the two axes of the ellipse approach the same length a, and the ellipse approaches the round shape of a circle.

So as the ellipse degenerates, the graph deforms into a circle of radius a, and its equation $r = a(1 - e^2)/(1 - e\cos\theta)$ approaches the equation $r = \frac{a}{1}$, that is, $r = a$.

This argument shows that it really is legitimate to include the circle as one of the cases of Equation 12.4.1. See Figure 12.4.15.

12.5 Parametric Equations in Rectangular and Polar Coordinates

PARAMETER
MOVING POINT

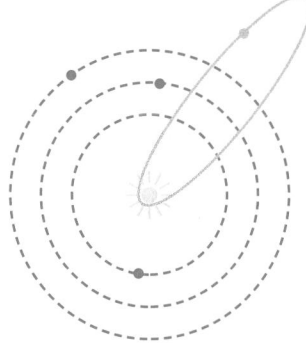

Figure 12.5.1

In Section 12.4 we mentioned the fact that planets and comets move in the solar system along orbits that are conics. Most of the planets move on orbits that are nearly circular (for instance, the eccentricity of the earth's orbit is about 0.017, which is quite close to 0). The only orbits that have eccentricities greater than 0.1 are those of Mercury, whose orbit has eccentricity ≈ 0.206, and Pluto, whose orbit has eccentricity ≈ 0.248. There is a theory that Pluto may have had an orbit that was closer to circular but that it nearly collided with some large object and the gravitational field of the other object distorted its orbit.

Comets often have orbits that are highly eccentric. For instance, the eccentricity of the orbit of Halley's Comet is about 0.97. We can imagine the comet as a point moving around its elliptical orbit (see Figure 12.5.1).

This suggests that its position can be determined at any time t, and we expect to be able to write its coordinates (x, y) as functions $x = x(t)$ and $y = y(t)$ of t. Evaluating these functions at any particular value of t will give the coordinates of the moving point at the time t. In this situation it is customary to refer to the variable t as a **parameter** and to $x = x(t)$ and $y = y(t)$ as **parametric equations** for the curve (or for the moving object).

Of course, in principle it should also be possible to express parametric equations in polar coordinates instead of rectangular coordinates. We should be able to write $\theta = \theta(t)$ and $r = r(t)$.

Naturally, this sort of idea does not apply only to astronomical motions. In Figure 12.5.2 we see a train moving along a straight track; we can certainly describe its position as a function of time.

Figure 12.5.2

Figure 12.5.3

In Figure 12.5.3 there is an illustration of the path of a ball that is thrown upward at an angle of $45°$ above the horizontal. It follows a parabolic arc until it lands, and we suppose that the maximum height it reaches is 16 ft. Here too it is possible to express the position of the moving object as a function of time.

If a fast meteor approaches near the sun, the sun's gravity will pull it from the straight line it would otherwise follow, and its course will bend around the sun. It will fly off in another direction, which depends on how close it passes to the sun, and the

Figure 12.5.4

Figure 12.5.5

path it follows will be a hyperbola, as shown in Figure 12.5.4. The asymptotes of the hyperbola will be the straight line along the direction that the meteor would have followed if the sun had not been there to deflect it and the line along the direction toward which the meteor is deflected. The farther the meteor is from the sun, the less it will be affected by the sun's gravity, and this is why the parts of the hyperbola far away from the focus are very close to straight lines (the asymptotes). Here too we would like to express the position of the meteor as a function of time, giving parametric equations for the hyperbola.

Finally, we can imagine a speck of dust on the rim of a twelve-inch LP as it turns at $33\frac{1}{3}$ rpm. The speck of dust travels along a circle of radius 6, and it ought to be easy to express its position in terms of the time t (see Figure 12.5.5).

In fact it is fairly easy to find parametric equations that describe the position of a point moving along each of the curves mentioned above. The disconcerting thing is that there are many *choices* of formulas. Of course, this reflects the fact that an object may move along the same path in many different ways; even a train moving along a straight track can move at different speeds, can accelerate or decelerate, and for that matter can move in either of two directions. So to make practical use of formulas that express the position of a moving point as a function of t, it is essential to find the right formulas for a particular motion.

However, we begin by trying to find even one possible formula for each of the curves we have discussed.

EXAMPLE 1   **Moving Train**   Suppose a train is moving along a straight track at a constant speed of 30 mph. For convenience let us think of the track as the real axis and let $t = 0$ at the moment when the train passes the origin O. See Figure 12.5.6.

Figure 12.5.6

If we measure the time t in hours, and express the position of the train as $x = x(t)$, since the train moves at a constant speed of 30 mph, we find that $x'(t) = 30$. So $x(t)$ is an antiderivative of the constant function 30, so $x(t) = 30t + c$, for some constant c. Since the train is at the origin when $t = 0$, we find that $0 = 30(0) + c = c$, so $c = 0$ and $x(t) = 30t$.

The way we usually draw our axes, the train will be moving from left to right, and of course there is the other possibility that it moves from right to left. In that case the formula would be $x = x(t) = -30t$.

Notice that because this train is moving along a straight line, it is possible to express its position with a single coordinate x instead of the two needed for positions in the plane (either (x, y) or (r, θ)).

EXAMPLE 2   **Flying Ball**   Next we consider the ball that has been thrown at an angle of $45°$ above the horizontal. Let us measure time t in seconds and assume that $t = 0$ at the moment when the ball is thrown. Let us also assume that its starting point is the origin in the plane.

In Remark 4.9.2 (p. 317) we observed that in this situation the ball will move so that its horizontal velocity is constant and its vertical height in feet will be given by $y = y(t) = -16t^2 + Bt + C$. Here B is the vertical speed with which the ball is thrown and C is the initial height, which we have assumed is 0, so $y(t) = -16t^2 + Bt$.

The maximum height will be reached when $y'(t) = 0$, that is, when $-32t + B = 0$, that is, when $t = \frac{B}{32}$. The height at this moment, which we know should equal 16, is $y\left(\frac{B}{32}\right) = -16\left(\frac{B}{32}\right)^2 + B\left(\frac{B}{32}\right) = -\frac{1}{64}B^2 + \frac{1}{32}B^2 = \frac{1}{64}B^2$. Solving to find when this equals 16, we see that $B = 32$. This tells us that the y-coordinate is given by $y = y(t) = -16t^2 + 32t$. In particular, the vertical velocity is $y'(t) = -32t + 32$, and at the moment when the ball is thrown, $y'(0) = 32$.

For the ball to move at a $45°$ angle above the horizontal, its horizontal and vertical velocities must be equal at $t = 0$, so its horizontal velocity at $t = 0$ must also equal 32 ft/sec. Since the horizontal velocity is constant, we know that it is 32 ft/sec at every time t, so the horizontal position is given by $x = x(t) = 32t$. See Figure 12.5.7.

The position of the ball is given by the parametric equations

$$x = x(t) = 32t \qquad \text{and} \qquad y = y(t) = -16t^2 + 32t.$$

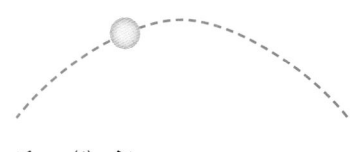

$\longleftarrow x(t) \longrightarrow$

Figure 12.5.7

EXAMPLE 3   Moving Speck on LP   Let us try to find parametric equations for a point moving around the circle $x^2 + y^2 = 36$. It is easy to check that the point $(x, y) = (6\cos u, 6\sin u)$ is on the circle, and this gives us parametric equations for the circle with the parameter u. We know that $(6\cos u, 6\sin u)$ is the point on the circle whose polar angle is $\theta = u$, as shown in Figure 12.5.8.

The complication comes when we want to make sure that the point moves around the circle at $33\frac{1}{3}$ rpm. Suppose we measure time t in minutes. We know that $(6\cos u, 6\sin u)$ goes around the circle exactly once each time u increases by 2π. We want the point to go around the circle $33\frac{1}{3}$ times each minute, so it should go around once in exactly $\frac{1}{33\frac{1}{3}}$ minute. This would be the case if u goes from 0 to 2π when t goes from 0 to $\frac{1}{33\frac{1}{3}}$, and we can arrange for this to happen by letting

$$u = 2\pi\left(33\frac{1}{3}t\right).$$

When $t = 0$, we see that $u = 0$, and when $t = \frac{1}{33\frac{1}{3}}$, we find that $u = 2\pi$.

If we let $x = 6\cos\left(2\pi\left(33\frac{1}{3}t\right)\right)$, $y = 6\sin\left(2\pi\left(33\frac{1}{3}t\right)\right)$, then the point (x, y) will move around the circle of radius 6 at a speed of $33\frac{1}{3}$ rpm.

However, if we look at Figure 12.5.8 and imagine that it is the turntable seen from above, then as the angle u increases, the point moves counterclockwise. In fact, LPs turn clockwise, so we should adjust the parametric equations to make the point move in the opposite direction. The parametric equations we want for the point moving clockwise around the circle of radius 6 at $33\frac{1}{3}$ rpm are

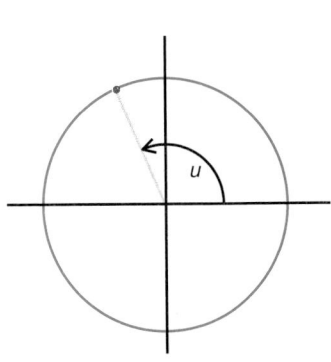

Figure 12.5.8

$$x = 6\cos\left(-66\frac{2}{3}\pi t\right) = 6\cos\left(66\frac{2}{3}\pi t\right)$$

and

$$y = 6\sin\left(-66\frac{2}{3}\pi t\right) = -6\sin\left(66\frac{2}{3}\pi t\right).$$

EXAMPLE 4 **Parametrizing an Ellipse** If we wanted parametric equations for the circle $x^2 + y^2 = 4$ of radius 2 instead, we could let $x = 2\cos t$ and $y = 2\sin t$. This moving point would trace out the circle exactly once as the time t changes from 0 to 2π. It would move in the counterclockwise direction.

Now suppose we want to find parametric equations for a point moving around the ellipse $\frac{x^2}{a^2} + \frac{y^2}{b^2} = 1$. We can check that the point $(a\cos t, b\sin t)$ is on the ellipse. It is a little less clear exactly where it is. Unlike the case of the circle, it is not necessarily true that $(a\cos t, b\sin t)$ has polar angle t. When $t = 0$, for instance, we find that $(a\cos t, b\sin t) = (a, 0)$ does have polar angle 0. When $t = \frac{\pi}{2}$, then the point is $(0, b)$, which also has the expected angle, that is, $\frac{\pi}{2}$.

However, suppose we try $t = \frac{\pi}{4}$ and look at the corresponding point $\left(a\cos\frac{\pi}{4}, b\sin\frac{\pi}{4}\right) = \left(\frac{a}{\sqrt{2}}, \frac{b}{\sqrt{2}}\right)$, as shown in Figure 12.5.9. The polar angle θ cannot equal $\frac{\pi}{4}$ unless $a = b$.

On the other hand, $\tan\theta = \frac{b\sin t}{a\cos t} = \frac{b}{a}\tan t$ has the same sign as $\tan t$, and from this we conclude that as t moves from 0 to $\frac{\pi}{2}$, the angle θ also moves from 0 to $\frac{\pi}{2}$. Similar remarks can be made about the intervals $\left[\frac{\pi}{2}, \pi\right]$, $\left[\pi, \frac{3\pi}{2}\right]$, and $\left[\frac{3\pi}{2}, 2\pi\right]$. This means that the point $(a\cos t, b\sin t)$ moves around the ellipse in the counterclockwise direction, making one complete circuit as t ranges from 0 to 2π.

We find that $(x, y) = (a\cos t, b\sin t)$ are parametric equations for the ellipse $\frac{x^2}{a^2} + \frac{y^2}{b^2} = 1$, but the point does not move at a constant speed. In fact it moves more slowly out near the ends of the major axis than it does near the ends of the minor axis.

Figure 12.5.9

We can use a graphics calculator to sketch the ellipse $\frac{x^2}{9} + \frac{y^2}{4} = 1$, using the parametric equations $x = 3\cos t$ and $y = 2\sin t$. As t runs from 0 to 2π, the point moves around the ellipse counterclockwise.

TI-85	SHARP EL9300
MODE ▼ ▼ ▼ ▼ ► ► (Param)	SET UP E 3 ⌐↵ 3 cos x/θ/T ENTER
ENTER EXIT GRAPH F1 ($E(t) =$)	2 sin x/θ/T ENTER RANGE -3 ENTER
3 COS F1(t) ▼ 2 SIN F1(t)	3 ENTER 1 ENTER -2 ENTER 2 ENTER
M2(RANGE) 0 ▼ 2 π ▼ ▼ ($-$) 3	1 ENTER 0 ENTER 2 π ENTER .02 ENTER
▼ 3 ▼ ▼ ($-$) 2 ▼ 2 F5(GRAPH)	⌐↵
CASIO f_x-7700GB	HP 48SX
MODE MODE × (PARAM) Range	⮐ RAD ' 3 × COS T ► $+ \alpha$
-3 EXE 3 EXE 1 EXE -2 EXE 2 EXE 1	⮐ CST × 2 × SIN T ⮐ PLOT
EXE 0 EXE 2 π EXE .02 EXE Graph 3	NEW P A R ENTER PTYPE PARA
cos T , 2 sin T) EXE	PLOTR ⮐ { } T SPC 0 SPC 6.3
	ENTER INDEP AUTO

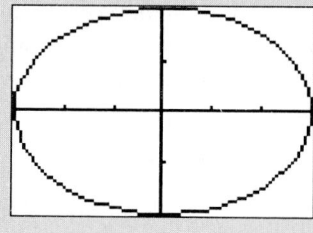

Try graphing $x^2 + 3y^2 = 3$ or $x^2 - y^2 = 1$ using parametric equations.

These parametric equations were relatively easy to write down and work with, but they turn out not to be the correct equations to describe the motion of a comet, for instance.

EXAMPLE 5   **Parametrizing a Hyperbola**   In the same way we might look for parametric equations for a hyperbola. For convenience, consider the hyperbola $x^2 - y^2 = 1$. From looking at this equation we might be inspired to try letting $x = \cosh t$ and $y = \sinh t$. The standard identity $\cosh^2 t - \sinh^2 t = 1$ says that the point (x, y) lies on the hyperbola.

As t ranges from $-\infty$ to ∞, we know that $\sinh t$ ranges from $-\infty$ to ∞, but $\cosh t$ is always greater than or equal to 1. We conclude that as t moves along the real line, the point $(x, y) = (\cosh t, \sinh t)$ moves along the *right branch* of the hyperbola $x^2 - y^2 = 1$ (see Figure 12.5.10). In fact the point moves up from the bottom toward the top of the branch.

Similar considerations show that $(x, y) = (a \cosh t, b \sinh t)$ give parametric equations for the right branch of the hyperbola

$$\frac{x^2}{a^2} - \frac{y^2}{b^2} = 1.$$

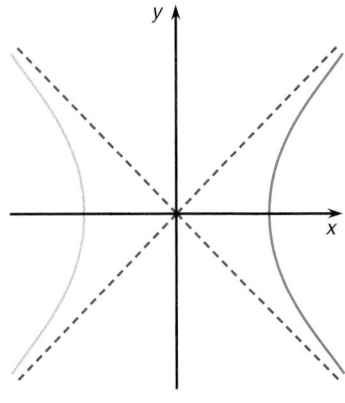

Figure 12.5.10

All the examples up to now have given parametric equations for a moving point by expressing its rectangular coordinates (x, y) as functions of the parameter t. We think of t as the time, so $\big(x(t), y(t)\big)$ represents the position of the point at time t. But it may also be interesting to work with polar coordinates and write parametric equations for r and θ.

EXAMPLE 6   Consider the circle $r = 7$. We already know that the points on this circle are exactly the points whose polar coordinates are $(7, \theta)$, for any value of θ.

Another way of saying this is that as we let θ range over all possible values, the point $(7, \theta)$ will move around the circle. It will go around a full circuit each time θ increases by 2π, and it moves counterclockwise as θ increases.

This is slightly confusing because in a way what we are doing is using θ as the parameter. It might be more clear to write

$$r = 7, \qquad \theta = \theta(t) = t.$$

These are parametric equations; the value of r is the constant function 7, and we just let θ equal the parameter t.

EXAMPLE 7   Consider the ellipse

$$r = \frac{1}{2 - \cos \theta},$$

as illustrated in Figure 12.5.11. In a way this equation can be used to give parametric equations; we let t be the parameter and let $\theta = t$, and then use the equation to write

$$r = \frac{1}{2 - \cos \theta} = \frac{1}{2 - \cos t}.$$

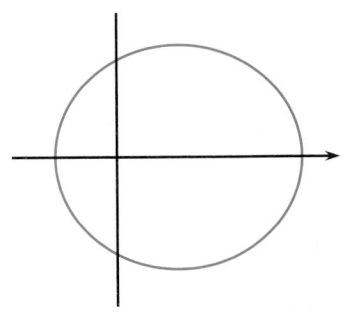

Figure 12.5.11

In this way we have expressed r and θ as functions of the parameter t. As t goes from 0 to 2π, the corresponding point will go around the ellipse exactly once.

The idea from Example 7 can be used for any curve that can be expressed by an equation of the form $r = f(\theta)$. We can use θ as the parameter, or, if we prefer, we can call the parameter t and let $\theta = t$. The equation can then be used to express r as a function of t.

Parametric equations obtained in this way have the property that the moving point "revolves at a constant rate" around the origin, in the sense that the angle θ increases at a constant rate (i.e., $\frac{d\theta}{dt} = \frac{dt}{dt} = 1$). Notice that the parametrization we found in Example 4 for the ellipse $\frac{x^2}{a^2} + \frac{y^2}{b^2} = 1$ does not have this property.

EXAMPLE 8   If we consider the hyperbola

$$r = \frac{1}{1 - 2\cos\theta},$$

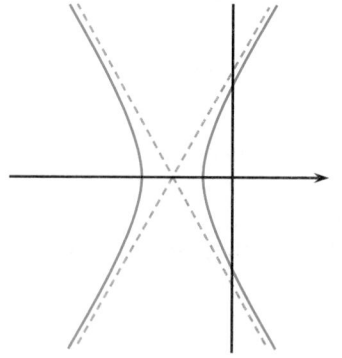

Figure 12.5.12

as shown in Figure 12.5.12, then we can find parametric equations using the idea just discussed. With t as parameter they are

$$\theta = t, \qquad r = \frac{1}{1 - 2\cos t}.$$

The formula makes no sense if $t = \pm\frac{\pi}{3} + 2m\pi$, but otherwise the point (r, θ) always lies on the hyperbola. In fact as t goes from 0 to 2π, missing $\frac{\pi}{3}$ and $\frac{5\pi}{3}$, the point covers the whole hyperbola exactly once. If we want the point to move along just the left branch, we should let t range from $-\frac{\pi}{3}$ to $\frac{\pi}{3}$ or, to be more precise, over the open interval $\left(-\frac{\pi}{3}, \frac{\pi}{3}\right)$. The right branch will be traced out as t moves over $\left(\frac{\pi}{3}, \frac{5\pi}{3}\right)$.

From the picture we see that as t approaches $\pm\frac{\pi}{3}$ from either side, the moving point will move far out on one of the branches, approaching an asymptote. Even a very small change in t will result in a big change in r, and the point will move a large distance. This suggests that using this parametrization will make the point move faster and faster as it approaches the asymptotes. ◢

We have seen that it is not terribly difficult to write down parametrizations for many familiar curves. Often the equation that defines the curve can be used to help express the coordinates in terms of some parameter.

For conics we have seen how to parametrize them in either rectangular or polar coordinates. Unfortunately, none of the parametrizations we have discussed gives the actual motion of a planet, comet, or meteor. One hint can be found in Example 8, in which we found that the speed gets larger and larger way out on the branches of a hyperbola. If a meteor moves on this path, we would expect that as it gets far away from the sun, it will experience very little gravitational force, so it will very nearly be coasting, and it seems logical to expect that its speed will approach a constant. There is certainly no reason to expect it to accelerate.

We will return to this question in Chapter 15, but for now let us remark that in order to solve it we will need to be able to talk about the velocity of an object moving along a curve. Up until now we have been very careful to consider velocities of objects moving in straight lines or ones like the projectile of Example 2 whose motion can be seen as the combination of a motion in the horizontal direction and a motion in the vertical direction, each of which amounts to motion in a straight line. In the next two chapters we will develop the techniques that are needed to make sense of the notion of velocity for curved motion.

Exercises 12.5

I

Find a parametrization of each of the following curves, and specify the range over which the parameter t must run in order to trace out the curve exactly once. Notice that there will be many possible answers.

1. $x^2 + y^2 = 25$
2. $4x^2 + 9y^2 = 36$
3. $r = \cos\theta$
4. $r = \sin 4\theta$
5. $3x - 4y = 7$
6. $y^2 = 2 - x^2$
7. $r = 3\sec\theta$
8. $r = 1 - 2\cos\theta$
9. $r^2 = \cos(2\theta)$
10. $x^2 + 2x + 4y^2 - 8y + 4 = 0$

II

*11. Using the fact that the eccentricity of the earth's orbit is about 0.017, calculate the ratio $\frac{b}{a}$ of its semiminor axis to its semi-major axis.

*12. Using the fact that the eccentricity of Mercury's orbit is about 0.206, calculate the ratio $\frac{b}{a}$ of its semi-minor axis to its semi-major axis.

*13. Using the fact that the eccentricity of Halley's Comet's orbit is about 0.97, calculate the ratio $\frac{b}{a}$ of its semi-minor axis to its semi-major axis.

14. Find parametric equations for the motion of a rock that is thrown upward from ground level at an angle of 45° above the horizontal if the maximum height it reaches is 64 ft.

15. (i) Find parametric equations for a speck of dust moving around the rim of a twelve-inch LP that has mistakenly been put on a turntable revolving at 45 rpm.

   (ii) Find parametric equations for a speck of dust

moving around on a 45 rpm disk at a distance of $1\frac{1}{2}$ in from the center.

16. Find parametric equations for the motion of a speck of dust on the rim of an LP revolving at $33\frac{1}{3}$ rpm so that the position of the speck at time $t = 0$ is the point $(0, 6)$.

17. Find parametric equations for the left-hand branch of the hyperbola $\frac{x^2}{a^2} - \frac{y^2}{b^2} = 1$.

18. Find parametric equations for the upper branch of the hyperbola $\frac{y^2}{a^2} - \frac{x^2}{b^2} = 1$.

*19. Find parametric equations for the upper semicircle of $x^2 + y^2 = 9$ using the parameter t, in such a way that the equation for x is $x = x(t) = t$. Be sure to specify the interval over which t runs.

*20. Find parametric equations for the right branch of $x^2 - y^2 = 4$ using the parameter t, in such a way that the equation for y is $y = y(t) = -3t$. Be sure to specify the interval over which t runs.

What is the equation of the curve that is parametrized by each of the following sets of equations?

*21. $x = t,\ y = t^2 - 1$
*22. $x = t - 3,\ y = t^3 + 2t - 1$
*23. $x = t^5 + 5,\ y = 4t^5 - 9$
*24. $x = t^2,\ y = 2t^2 + 5$

III

If you have a machine that can plot parametric equations, sketch the following curves.

25. $x = 2\sin t,\quad y = 3\cos t$
26. $x = 5\cosh t,\quad y = 4\sinh t$
27. $r = \frac{1}{2 - \sin t},\quad \theta = t$
28. $x = \cos^{1/3} t,\quad y = \sin^{1/3} t$
29. $x = \sin t,\quad y = \cos(3t)$
30. $x = \cos(3t),\quad y = \sin(5t)$

POINT TO PONDER

When we wrote the parametric equations $x = a\cos t$, $y = b\sin t$ for the ellipse $\frac{x^2}{a^2} + \frac{y^2}{b^2} = 1$, we observed that as t goes from 0 to 2π, the point (x, y) moves once around the ellipse, counterclockwise. On the other hand, we observed that t does not always equal the polar angle θ.

In fact it is possible to express θ in terms of t. After all, if $t \in \left[0, \frac{\pi}{2}\right)$, then $\tan\theta = \frac{b\sin t}{a\cos t}$, so $\theta = \arctan\left(\frac{b\sin t}{a\cos t}\right) = \arctan\left(\frac{b}{a}\tan t\right)$. This formula also holds if $t \in \left(-\frac{\pi}{2}, 0\right]$, but it must be modified slightly if $t \in \left(\frac{\pi}{2}, \frac{3\pi}{2}\right)$.

Still, in principle it is not difficult to express the polar angle θ in terms of the parameter t, and it is even easier to write $r = \sqrt{x^2 + y^2} = \sqrt{a^2 \cos^2 t + b^2 \sin^2 t}$.

If $a \neq b$, the polar angle θ is *not* the same as the parameter t. This reflects the fact that the point $(x, y) = (a \cos t, b \sin t)$ does not move in such a way that its angle θ increases at a constant rate.

From our formula for θ we find that

$$\frac{d\theta}{dt} = \frac{1}{1 + \frac{b^2}{a^2} \tan^2 t} \frac{b}{a} \sec^2 t = \frac{ab}{a^2 \cos^2 t + b^2 \sin^2 t}.$$

This formula for the derivative is valid for all values of t. In particular, notice that when $t = 0$, $\frac{d\theta}{dt} = \frac{ab}{a^2} = \frac{b}{a}$, and when $t = \frac{\pi}{2}$, $\frac{d\theta}{dt} = \frac{a}{b}$. This confirms the remark at the end of Example 4 that the point moves more slowly when $t = 0$ than when $t = \frac{\pi}{2}$ (assuming $a > b$).

Chapter Summary

§12.1
polar coordinates
radius
modulus
argument
polar angle
polar/rectangular conversions (Theorem 12.1.1)
graphing points in polar coordinates

§12.2
graphs in polar coordinates
Spiral of Archimedes
roses
symmetry in polar coordinates
cardioid
lemniscate
limaçon

dimple
slope of tangent line

§12.3
areas in polar coordinates
limits of integration
tracing curves exactly once

§12.4
conics in polar coordinates
focus
eccentricity
directrix
$r = \frac{k}{1 - e \cos \theta}$
circles
straight lines

§12.5
parameter
parametric equations

Review Exercises

Sketch each of the following curves.

1. $r = 2 \csc\left(\theta - \frac{\pi}{4}\right)$

2. $r = 3 - 4 \sin \theta$

3. $r = (3 - \sin \theta)^{-1}$

4. $r = (3 \cos \theta - 2)^{-1}$

5. $r^2 = \sin\left(2\theta + \frac{\pi}{4}\right)$

6. $r = 2 \cos \theta \sin \theta$

7. $r = 2 + 2 \sin \theta$

8. $r = \sin\left(5\theta + \frac{\pi}{3}\right)$

9. $2 \sin \theta = 1, r \geq 0$

10. $r + 3\theta = 0$

Find all points of intersection of each of the following pairs of curves. Make a sketch in each case.

11. $r = \theta, \theta \geq 0; \theta = 0, r \geq 0$

12. $r = \frac{1}{2}; r = \cos \theta$

13. $r = \sec\left(\theta + \frac{\pi}{4}\right); r = 1$

14. $r = \sec \theta; r = \sqrt{3} \csc \theta$

15. $r = \frac{1}{2}; r = \cos 2\theta$

16. $r = \csc \theta; r = 2 \sin \theta$

17. $r^2 = \cos 2\theta; r = \frac{1}{\sqrt{2}}$

18. $r^2 = \cos^2 \theta; r = \frac{1}{2}$

19. $r + \sin \theta = 0; \theta = \frac{\pi}{6}, r \geq 0$

20. $r^2 = \sin^2(2\theta) + \cos^2(2\theta); r = \sin(4\theta)$

In each of the following questions, find the area of the region enclosed inside both the curves.

21. $r = 1$; $r = -2\sin\theta$

22. $r = 1$; $r^2 = 2\sin\left(2\theta + \frac{\pi}{3}\right)$

23. $r = \sqrt{3}\cos\theta$; $r = \sin\theta$

24. $r = \cos(4\theta)$; $r = \sin(4\theta)$

25. $r = 2 + \cos\theta$; $r = 2 - \sin\theta$

26. $r = 3 + \sin\theta$; $r = 3 - \cos\theta$

27. $r = \cos\theta$; $r = \sin\left(\theta - \frac{\pi}{2}\right)$

28. $r = 2 - 2\sin\theta$; $r = 2 + 2\cos\theta$

29. $r = 3 + \cos\theta$; $r = 1 - \cos\theta$

30. $r = 1 - \cos\theta$; $r^2 + 2r\cos\theta = 1$

Find a parametrization of each of the following curves, specifying the range over which the parameter must run in order to trace out the curve exactly once.

31. $y = 2x^2 - 3$

32. $2x + 3y^2 = 4$

33. $r = \cos 2\theta$

34. $r^2 = \cos 2\theta$

35. $r = 2\csc\theta$

36. $r = \sin 3\theta$

37. $5x + 4y + 3 = 0$

38. $y = 4$

39. $r + 3 = 3\sin\theta$

40. $r = (2 + 2\sin\theta)^{-1}$

▌▌

Sketch each of the following curves.

41. $r = 2 + \cos\theta$

42. $r = 3 - \sin(4\theta)$

43. $r^2 = \sin 3\theta$

44. $r^2 = \sin^2(3\theta)$

45. $(r - \sec\theta)(r + \csc\theta) = 0$

46. $(r - \sin\theta)\left(r + \sin(2\theta)\right) = 0$

47. $r = \sqrt{\theta}$

48. $r = \theta^2$

49. $r = 2 + |\sin 3\theta|$

50. $r = 3 + \sin\frac{7\theta}{2}$

*51. Consider the point whose polar coordinates are $(1, 0)$. Does it lie on the curve $r = \frac{\theta}{2\pi}$?

*52. Consider the point $(r, \theta) = (1, \pi)$. Does it lie on the curve $r = \cos\theta - 2$?

*53. When we write hyperbolas in rectangular coordinates, we can distinguish those whose transverse axis is horizontal or vertical by recognizing the two different standard forms in Summary 1.6.4 (p. 38):

$$\frac{x^2}{a^2} - \frac{y^2}{b^2} = 1 \quad \text{and} \quad \frac{y^2}{a^2} - \frac{x^2}{b^2} = 1.$$

If we consider hyperbolas with a focus at the origin and write their equations in polar coordinates, what will be the difference between those whose transverse axis is horizontal and those whose transverse axis is vertical?

*54. Sketch the graph of $\theta = \sin r$, for $r \geq 0$.

*55. Find the equations in polar coordinates of two ellipses with one focus at the origin, major axis horizontal, and eccentricity $e = \frac{1}{2}$ and containing the point $(r, \theta) = (1, 0)$.

*56. Find the equations in polar coordinates of two ellipses with one focus at the origin, major axis vertical, and eccentricity $e = \frac{1}{4}$ and containing the point $(r, \theta) = (2, \pi)$.

*57. Sketch the graphs of $r = \sec\theta$ and $r = 1 + \sec\theta$ on the same diagram.

*58. Sketch the graphs of $r = 2\csc\theta$ and $r = 1 + 2\csc\theta$ on the same diagram.

Curves in the Plane

13

The Spiral of Archimedes (top), given by $r = \theta$, expands by the same amount each time it revolves once. It could be used, for instance, to describe the position of a coiled rope.

The second curve is called a Logarithmic Spiral because its equation $r = \theta^k$ is a "straight-line" equation $\ln r = k \ln \theta$ after we take logarithms. It describes the shape of certain snail shells and the positions of leaves around the stems of certain plants, including roses.

INTRODUCTION

In the final section of Chapter 12 we considered parametric equations for various curves. The idea is to think of a point moving along some curve, a circle or a hyperbola, for instance, and to express its position as a function of time. We saw that it is possible to do this for many familiar curves, and it is possible to express the position using either rectangular or polar coordinates.

However, we began to realize that it is possible to find many different parametrizations for the same curve, just as it is possible to move along the curve in many different ways (with different speeds, different directions, different starting points, etc.). If we are to use this idea to study practical problems, it will be necessary to compare the motion described by various parametrizations, and for this we will need to discuss the idea of the velocity of an object moving along a curved path. That is the purpose of this short chapter.

In Section 13.1 we recall briefly the concept of parametric equations and consider some examples. In Section 13.2 we observe that the velocity of a moving point involves not only how fast it is moving (its speed) but also the direction of its motion. Because of this the velocity is best expressed as a *vector*, a quantity that has both a magnitude and a direction. In Section 13.3 we apply some of these ideas to the particular case of the motion of projectiles. Historically, this was one of the early uses of calculus, and it is still very important.

13.1 Parametric Equations

PARAMETER
CURVE
DOMAIN

Suppose an object is moving in the plane, as shown in Figure 13.1.1. At each particular time t it will be at some specific location (x, y), and we can think of the coordinates x and y as functions $x = x(t)$ and $y = y(t)$ of t. If we know these functions, then we know exactly how the object moves, and on the other hand if we know how an object moves, then in principle we ought to be able to find the functions that describe the motion.

It is often necessary to restrict the values that t can take. Usually, t will range over some interval (frequently the whole real line $\mathbb{R}$), but in some cases it will range over a union of intervals.

DEFINITION 13.1.1

Suppose I is an interval (open or closed, possibly infinite), and suppose

$$x = x(t), \qquad y = y(t)$$

are functions defined on I. Assume that they are continuous on I. Then they are called **parametric equations**; the variable t is called the **parameter**. The set of points

$$\{(x, y) = \big(x(t), y(t)\big) : t \in I\}$$

is called the **curve**, and it is sometimes also called the **image** of I. The interval I is called the **domain** of the parametric equations.

Sometimes several curves are combined to make one curve; in this situation it may be necessary to have parametric equations defined on several intervals.

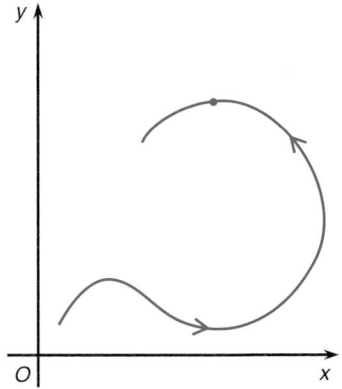

Figure 13.1.1

EXAMPLE 1

(i) Consider the ellipse $\frac{1}{4}x^2 + y^2 = 1$. The equations

$$x = 2\cos t, \qquad y = \sin t$$

are parametric equations. It is easy to check that the points $(x, y) = (2\cos t, \sin t)$ do all lie on the ellipse, and in fact we argued in Example 12.5.4 that as t ranges from 0 to 2π, the point $(x, y) = \big(x(t), y(t)\big)$ moves once around the ellipse counterclockwise, starting and finishing at the point $(x, y) = (2, 0)$. See Figure 13.1.2(i).

We can use the above equations to parametrize the ellipse, but strictly speaking, it is necessary to specify the interval I. One possibility is to let $I = [0, 2\pi)$. This would mean that the moving point would trace out the ellipse exactly once, returning to approach its starting point but never quite reaching it. Another possibility would be to use $I_2 = [0, 2\pi]$. The only difference is that

Figure 13.1.2(i)

Figure 13.1.2(ii)

Figure 13.1.2(iii)

Figure 13.1.3(i)

Figure 13.1.3(ii)

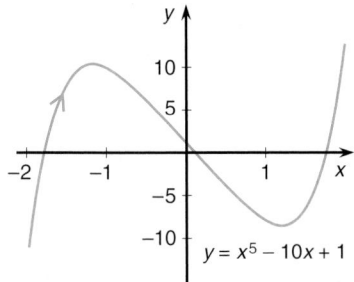

Figure 13.1.4

this way the curve would actually return to its starting point. It would also be possible to use other intervals, such as $[-\pi, \pi)$ or $[-\pi, \pi]$.

Another possibility would be to let $I_3 = \mathbb{R}$. In this case the moving point would go around and around the ellipse infinitely many times. It is also possible to use the same equations but to let the domain be $I_4 = [0, \pi]$. This way the moving point will trace out only the upper half-ellipse. See Figure 13.1.2(ii).

You can also check that the same ellipse can be parametrized by $x = -2 \sin t$, $y = -\cos t$, for $t \in [0, 2\pi]$. In this situation the moving point moves once around the ellipse in the *clockwise* direction, beginning and ending at the point $(0, -1)$. See Figure 13.1.2(iii).

(ii) The straight line $2x - 3y = 2$, shown in Figure 13.1.3(i), can be parametrized in many different ways. To find one, we could try letting $x = t$ and then solving for y. Since $2x - 3y = 2$, we find that $3y = 2x - 2$, so $y = \frac{2x-2}{3}$. If $x = t$, then $y = \frac{2t-2}{3}$. These give parametric equations for the line, with domain $I = \mathbb{R}$.

It is easy to find other parametrizations. We could let $x = 2t$ and $y = \frac{4t-2}{3}$, for $t \in \mathbb{R}$. Another possibility is to let $x = 3t$, $y = 2t - \frac{2}{3}$. This has the slight advantage of involving fewer fractions.

If we let $y = t^2$, we can solve and find that $x = \frac{3}{2}t^2 + 1$. These points are always on the line $2x - 3y = 2$, but since $y = t^2$, we see that y is never negative. For that matter, $x = \frac{3}{2}t^2 + 1$ is never less than 1. We have found a parametrization of the part of the line above and to the right of the point $(1, 0)$; see Figure 13.1.3(ii). Moreover, since t and $-t$ correspond to the same point, the half-line is traced out twice as t runs over $\mathbb{R}$. If we let $I = [0, \infty)$, then the half-line is covered only once.

Finally, we observe that the whole line can also be parametrized by $x = \tan t$, $y = \frac{2}{3} \tan t - \frac{2}{3}$, for $t \in \left(-\frac{\pi}{2}, \frac{\pi}{2}\right)$. This example demonstrates that the domain can be a finite interval even for a curve that is unbounded, like a whole line. (Recall that for $t \in \left(-\frac{\pi}{2}, \frac{\pi}{2}\right)$ the range of the function $\tan t$ is all the real numbers.)

(iii) The graph of $y = x^5 - 10x + 1$ is a curve in the plane (see Figure 13.1.4). We think of a point moving along the graph from left to right. It can be parametrized by letting $x = t$, $t \in (-\infty, \infty)$, and then using the equation to solve for y. The parametric equations are $x = t$, $y = t^5 - 10t + 1$, for $t \in (-\infty, \infty)$.

(iv) Consider the rectangle whose corners are $(0, 0)$, $(3, 0)$, $(3, 2)$, and $(0, 2)$, as illustrated in Figure 13.1.5. If we want to parametrize it, the simplest thing to do would be to parametrize each of the four sides separately. For instance, to parametrize the bottom edge from $(0, 0)$ to $(3, 0)$, we could use the equations $x = t$, $y = 0$, for $t \in [0, 3]$.

Next, to parametrize the right side, we need to let $x = 3$ and then to have y go from 0 to 2. Since we have already let t go from 0 to 3 in parametrizing the bottom edge, it might be confusing to use $[0, 2]$ as the domain for the parametrization of the right side. A better idea would be to let $y = t - 3$, for $t \in [3, 5]$.

Then, to move back along the top edge, we could let $x = 8 - t$ and $y = 2$, for $t \in [5, 8]$. Notice that as t moves from 5 to 8, the value of $x = 8 - t$ will go from 3 back to 0, so the corresponding point will move back from right to left along the top side of the rectangle.

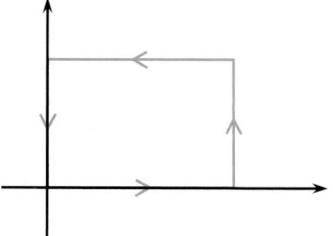

Figure 13.1.5

Finally, to move down the left side of the rectangle, we let $x = 0$, $y = 10 - t$, for $t \in [8, 10]$.

We have parametrized the rectangle by combining parametrizations on four different intervals:

$$(x, y) = \begin{cases} (t, 0), & \text{if } t \in [0, 3], \\ (3, t - 3), & \text{if } t \in [3, 5], \\ (8 - t, 2), & \text{if } t \in [5, 8], \\ (0, 10 - t), & \text{if } t \in [8, 10]. \end{cases}$$

In fact we have fixed up this parametrization so that its domain is the interval $[0, 10]$. Strictly speaking, we began by working on each subinterval separately, but we arranged things so that they "join" at the corners of the rectangle. We could have expressed the parametrization as four separate parametrizations, but it is nice to be able to work with a single interval. There is a slight complication, which is that the functions $x(t)$ and $y(t)$ are not differentiable at $t = 3, 5, 8$, the points that correspond to the corners of the rectangle. To check this, you could begin by sketching the graphs of $x(t)$ and $y(t)$ for $t \in [0, 10]$.

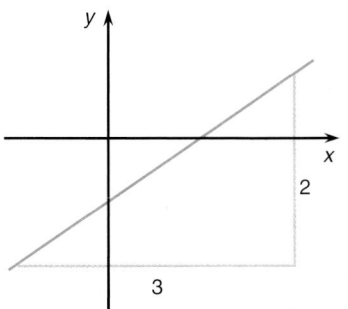

Figure 13.1.6

REMARK 13.1.2

In part (ii) of Example 1 we found the parametrization $x = 3t$, $y = 2t - \frac{2}{3}$ for the line $2x - 3y = 2$. It is interesting to make the following interpretation of these equations. Looking at the formulas $x = 3t$, $y = 2t - \frac{2}{3}$, we see that as t increases, x increases 3 times as fast as t does, and y increases twice as fast as t, which means that y increases $\frac{2}{3}$ as fast as x does.

As we move along the line, this suggests that we move more slowly in the vertical direction than in the horizontal direction, by a factor of $\frac{2}{3}$. This is exactly what we should expect, since the slope of the line is $\frac{2}{3}$; see Figure 13.1.6.

Summary 13.1.3   Parametrizations of Some Familiar Curves

(Many other parametrizations can be found; the ones given here are easy to describe.)

Line	$y = ax + b$;	$x = t$, $y = at + b$, $t \in \mathbb{R}$, if $a \neq 0$, can also use $y = t$, $x = \frac{t-b}{a}$, $t \in \mathbb{R}$,
Graph	$y = f(x)$;	$x = t$, $y = f(t)$, $t \in \text{domain}(f)$,
Circle	$x^2 + y^2 = r^2$;	$x = r \cos t$, $y = r \sin t$, $t \in \mathbb{R}$; counterclockwise, goes around once as t increases by 2π,
		$x = r \cos t$, $y = -r \sin t$, $t \in \mathbb{R}$; clockwise,
Ellipse	$\frac{x^2}{a^2} + \frac{y^2}{b^2} = 1$;	$x = a \cos t$, $y = b \sin t$, $t \in \mathbb{R}$; counterclockwise, goes around once as t increases by 2π,
Hyperbola	$\frac{x^2}{a^2} - \frac{y^2}{b^2} = 1$;	$x = a \cosh t$, $y = b \sinh t$, $t \in \mathbb{R}$; parametrizes right branch,
		$x = -a \cosh t$, $y = b \sinh t$, $t \in \mathbb{R}$; left branch,
Hyperbola	$\frac{y^2}{a^2} - \frac{x^2}{b^2} = 1$;	$x = b \sinh t$, $y = a \cosh t$, $t \in \mathbb{R}$; upper branch,
		$x = b \sinh t$, $y = -a \cosh t$, $t \in \mathbb{R}$; lower branch.

We can use a graphics calculator to sketch the hyperbola $\frac{x^2}{9} - \frac{y^2}{16} = 1$, using the parametric equations $x = 3 \cosh t$, $y = 4 \sinh t$ for the right branch and $x = -3 \cosh t$, $y = 4 \sinh t$ for the left branch.

TI-85	SHARP EL9300
MODE ▼ ▼ ▼ ▼ ► ► (Param) ENTER EXIT GRAPH F1 ($E(t) =$) 3 MATH F4F2 (cosh) M1 (t) ▼ 4 F1 (sinh) M1 (t) ▼ (−) 3 F2 (cosh) M1 (t) ▼ 4 F1 (sinh) M1 (t) EXIT M2 (RANGE) (−) 2 ▼ 2 ▼ ▼ (−) 10 ▼ 10 ▼ ▼ (−) 10 ▼ 10 F5 (GRAPH)	SET UP E 3 〰 3 MATH ▼ ► ▼ ENTER x/θ/T ENTER 4 MATH ▲ ENTER x/θ/T 2ndF ▼ −3 MATH ▼ ENTER x/θ/T ENTER 4 MATH ▲ ENTER x/θ/T ENTER RANGE −10 ENTER 10 ENTER 1 ENTER −10 ENTER 10 ENTER 1 ENTER −2 ENTER 2 ENTER .02 ENTER 〰
CASIO f_x-7700GB	HP 48SX
MODE MODE × (PARAM) Range −10 EXE 10 EXE 1 EXE −10 EXE 10 EXE 1 EXE −2 EXE 2 EXE .02 EXE Graph 3 MATH HYP F2 (csh) T, 4 F1 (snh) T) EXE Graph − 3 F2 (csh) T, 4 F1 (snh) T) EXE	↰ PLOT PLOTR 10 +/− ENTER 10 XRNG 10 +/− ENTER 10 YRNG ATTN ' 3 × MTH HYP COSH T ► + α ↰ CST × 4 × SINH T ↰ PLOT NEW P A R ENTER PTYPE PARA PLOTR ↰ {} T SPC 2 +/− SPC 2 ENTER INDEP DRAW ATTN ' 3 +/− × MTH HYP COSH T ► + α ↰ CST × 4 × SINH T ↱ PLOT ↰ DRAW DRAW

Try finding a different parametrization.

Suppose that we are given some parametric equations $x = x(t)$, $y = y(t)$, for $t \in I$. We know that these equations describe the position of a point moving along some curve in the plane, and it would be interesting to know what that curve is. In particular, we would like to be able to find a single equation that describes the curve. We have seen that sometimes just by looking at the parametric equations it is possible to get some idea of the shape of the curve.

In fact it is often possible to find an equation for the curve. The idea is to try to eliminate the parameter t from the two parametric equations. We illustrate with the following examples.

Figure 13.1.7

Figure 13.1.8

Figure 13.1.9

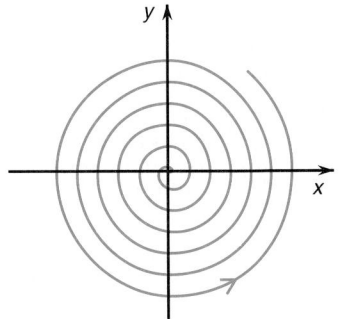

Figure 13.1.10

EXAMPLE 2

(i) Consider the parametric equations

$$x = 2 + 3t, \qquad y = 5t + 1.$$

We can solve for t in each equation and find that $t = \frac{x-2}{3}$, $t = \frac{y-1}{5}$. Equating these two expressions for t, we find that $\frac{x-2}{3} = \frac{y-1}{5}$.

This can be simplified to $5x - 10 = 3y - 3$, or $5x - 3y = 7$. We find that the curve we have parametrized is the straight line $5x - 3y = 7$.

(ii) Consider the parametric equations $x = t^3 - 1$, $y = 3t^6 - 2t^3 + 5$.

From the first of these equations we find that $t^3 = x + 1$, and substituting this into the second equation, we find that

$$y = 3(t^3)^2 - 2(t^3) + 5 = 3(x+1)^2 - 2(x+1) + 5 = 3x^2 + 4x + 6.$$

The curve is the parabola $y = 3x^2 + 4x + 6$, shown in Figure 13.1.7.

(iii) Consider the parametric equations

$$x = t, \qquad y = \sqrt{4 - t^2}.$$

We observe that for the square root to make sense we must have $4 - t^2 \geq 0$, which means $t \in [-2, 2]$.

Squaring the second equation, we find that $y^2 = 4 - t^2$, and since $x = t$, this amounts to $y^2 = 4 - x^2$. The parametric equations parametrize points that lie on the circle $x^2 + y^2 = 4$.

However, because $y = \sqrt{4 - t^2}$ is never negative, we see that they actually parametrize the upper half of the circle. As t runs from -2 to 2, the corresponding point (x, y) moves around the upper semicircle from $(-2, 0)$ to $(2, 0)$, as shown in Figure 13.1.8.

(iv) Consider the parametric equations

$$x = 2t, \qquad y = -\sqrt{1 + t^2}.$$

Here there is no problem about the square root, and t can be any real number.

From the first equation we can solve for t and find $t = \frac{x}{2}$. Squaring the second equation gives $y^2 = 1 + t^2 = 1 + \frac{x^2}{4}$. The point (x, y) is on the curve $y^2 - \frac{1}{4}x^2 = 1$, which is a hyperbola.

Since $y = -\sqrt{1 + t^2}$ is never positive, we see that the parametric equations trace out the bottom branch of the hyperbola, as shown in Figure 13.1.9. ◀

In Example 2 we remarked that certain parametrizations do not trace out the whole curve. In each such case we identified part of the curve on which the moving point lies, but we did not show that it passes through every point. You should convince yourself that the parametrizations actually do cover the stated parts of the curves.

EXAMPLE 3 Parametrize the Spiral of Archimedes, given in polar coordinates by $r = \theta$, for $\theta \geq 0$. See Figure 13.1.10.

Solution We need to describe a point that goes around and around the origin. For instance, the point $(x, y) = (\cos t, \sin t)$ does this. The problem is that it always stays on the circle of radius 1.

> We could also parametrize the spiral *in polar coordinates* by $r = t$, $\theta = t$, $t \in [0, \infty)$, and then find $x = r \cos \theta = t \cos t$, $y = r \sin \theta = t \sin t$.

We want a point that goes around and around but whose distance from the origin equals the angle. One way to achieve this is to let $(x, y) = (t \cos t, t \sin t)$. This point is at the same polar angle relative to the origin as $(\cos t, \sin t)$, so it goes around and around. However, its distance from the origin is $\sqrt{t^2 \cos^2 t + t^2 \sin^2 t} = |t|$. In particular, if $t \geq 0$, the distance is t, and this equals the polar angle. The spiral can be parametrized by the equations $(x, y) = (t \cos t, t \sin t)$, for $t \geq 0$.

Exercises 13.1

I

Find parametrizations for each of the following curves. Specify the domain and say how many times the curve is traced out. Notice that there will be many possible answers to these questions.

1. $x = 4y$
2. $3x + 4y = 5$
3. $x = y^2 + 2$
4. $2x - y^2 + 3y - 6 = 0$
5. $3x^2 + 4y^2 = 12$
6. $x^2 + 2x + y^2 = 3$
7. $y = 2$
8. $x^3 - 4y^2 - 2y + 8 = 0$
9. $y + 2 \sin x = 3$
10. $y = \sec x$

Find an equation for the curve that is traced out by each of the following sets of parametric equations. Notice that there will be various correct answers, but try to put the equation into a standard sort of form from which the type of curve is easily recognized. Identify which portion of the curve is traced out by the moving point, how many times it is covered, and the direction in which the point moves over the curve.

11. $x = \sin t$, $y = \cos t$, $t \in [0, 2\pi]$
12. $x = 4 \cos(2t)$, $y = 9 \sin(2t)$, $t \in [-\pi, \pi]$
13. $x = 3t^4 - 5$, $y = 6 - 4t^4$, $t \in \mathbb{R}$
14. $x = 2t + t^2$, $y = t + 1$, $t \in (-\infty, 0)$
15. $x = 3 + e^t$, $y = 5 - 2e^t$, $t \in [0, \infty)$
16. $x = \frac{1}{1+t}$, $y = \frac{t}{1+t}$, $t \in [0, 1]$
17. $x = t^4$, $y = t^4 - 2t^2$, $t \in \mathbb{R}$
18. $x = 9 \cosh t$, $y = 4 \sinh t$, $t \in \mathbb{R}$
19. $x = \sinh t$, $y = 1 - \cosh t$, $t \geq 0$
20. $x = \sqrt{\arctan t}$, $y = 1 - \arctan t$, $t \in [0, \infty)$

II

21. Find a parametrization for the rectangle of Example 1(iv) that begins and ends at the point $(3, 2)$ and goes around the rectangle in the counterclockwise direction.

22. Find a parametrization for the rectangle of Example 1(iv) that begins and ends at the point $(0, 2)$ and goes around the rectangle in the clockwise direction.

23. Find a parametrization for the rectangle of Example 1(iv) that begins and ends at the point $(2, 0)$ and goes around the rectangle in the counterclockwise direction.

24. Consider the triangle whose vertices are $(1, 1)$, $(3, 1)$, and $(1, 3)$, and find a parametrization for it that begins and ends at $(1, 1)$ and moves counterclockwise.

*25. Consider the curve $y = \arctan x$ (i.e., the set of points in the plane that are the graph of this function). Show that it can be parametrized by either of the following pairs of functions:

$$x = t, \qquad y = \arctan t, \qquad \text{for } t \in \mathbb{R},$$

or

$$x = \tan t, \qquad y = t, \qquad \text{for } t \in \left(-\frac{\pi}{2}, \frac{\pi}{2}\right).$$

*26. At the end of Example 1(iv) it was observed that the functions given in that example for $x(t)$ and $y(t)$ are continuous on $[0, 10]$ but that they are not differentiable at 3, 5, and 8. Sketch the graph of each of these functions and verify these observations.

III

In each of the following exercises, use a graphics calculator or computer to plot the graphs of $x(t)$ and $y(t)$ as functions of t. Look at them and try to predict the approximate shape of the curve $\big(x(t), y(t)\big)$ as t runs over the whole real line $\mathbb{R}$. If possible, graph the parametric curve and confirm your predictions.

27. $x(t) = t^3$, $y(t) = t^6$
28. $x(t) = t^4$, $y(t) = t^8$
29. $x(t) = \cos t$, $y(t) = \sin^2 t$
30. $x(t) = \sin t$, $y(t) = \arctan t$

POINT TO PONDER

In this section we have considered curves that lie in the plane. In practical situations we usually encounter paths lying in three-dimensional space, and it seems that for most purposes it will be necessary to work in that context.

This is partly true, and in Chapter 15 we will learn to deal with curves in space. Most of what we will do with them is analogous to what happens with plane curves but a little more complicated. There are a few things that are genuinely different and have to be considered separately, but on the whole, plane curves are a good introduction to the general case.

It is also true that plane curves are more important in their own right than is apparent at first glance. If we think about the earth orbiting the sun in space, while it is true that it is moving through three-dimensional space, it is also true that the actual path of the earth lies in a flat plane. Since the sun also lies in this plane, everything that matters for studying the earth's motion does in fact lie in a plane.

We can study this motion as if it were just occurring in the ordinary xy-plane. One way to think of it is that if we choose the coordinates correctly, the sun and the

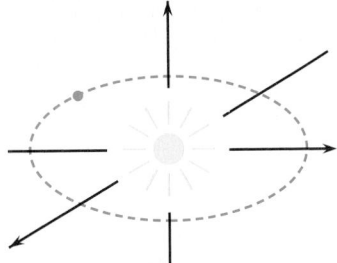

Figure 13.1.11

earth's orbit will all lie in the horizontal plane, and we can simply ignore the third coordinate; see Figure 13.1.11.

So while to some extent studying motions in the plane is an idealization of what actually happens in the real world, it is often quite accurate. We are all accustomed to looking at flat maps that represent the surface of the earth. Provided the region in question is not too large, there is no need to use a round globe even though it would theoretically be more accurate. In just the same way, a great many problems can be described by motions in a well-chosen plane. Sometimes this involves making approximations, and other times it is exact.

13.2 Vectors and Motion

DISPLACEMENT
VELOCITY
SPEED
TANGENT VECTOR
ARCLENGTH

To specify the velocity of an object moving along a straight line as shown in Figure 13.2.1, it is necessary to give its speed and the direction. For an object moving along a curved path, as in Figure 13.2.2, it is less clear what is meant by either of these concepts.

We begin by recalling what we did in the straight-line situation. If the position of the point on the line is given by $x = x(t)$, then to find its velocity at time $t =$

Figure 13.2.1

Figure 13.2.2

a, we considered the difference quotient $\frac{x(a+h)-x(a)}{h}$ and took the limit as $h \to 0$. The difference quotient represents the average velocity of the object over the short interval of time from a to $a + h$. Its numerator $x(a + h) - x(a)$ represents the change in the displacement of the point starting at $t = a$ and ending at $t = a + h$.

When we discussed velocity for a point moving along a straight line, we realized that it was necessary to use the idea of displacement. It is like distance along the line except that it is positive in one direction and negative in the other. On a line, there are only two directions. For motion in the plane it is not possible to specify a direction by a simple sign.

Vectors

What is needed is a way of indicating a direction and a distance. This can be accomplished by using **vectors**. For this section we will think of a vector as an arrow in the plane (see Figure 13.2.3). It begins at some *initial point* and ends with the arrowhead at its *terminal point*. The arrow points in a particular direction, and its length is a nonnegative number that gives a distance. In this way vectors can be used to specify both a direction and a distance.

Frequently, we place the initial point of a vector **v** at the origin. In this case **v** is determined by the coordinates (x, y) of its terminal point. The numbers x and y are called the **components** of **v**; x is the **horizontal component**, and y is the **vertical component**. The arrow points x units horizontally and y units vertically. It is very useful to describe a vector **v** in this way, and we write $\mathbf{v} = (x, y)$.

Recall that a vector has both a length and a direction. The vector $\mathbf{v} = (2, 0)$ points horizontally to the right and has length 2; the vector $\mathbf{u} = (0, -2)$ points straight down and has length 2; the vector $\mathbf{w} = (0, 1)$ points straight up and has length 1; the vector $\mathbf{z} = (1, -1)$ points down and to the right at an angle of $45°$ below the horizontal and has length $\sqrt{2}$. (To do this last one, we noticed that the vector $\mathbf{z} = (1, -1)$ is the arrow from the origin to the point $(1, -1)$ and that the distance from $(0, 0)$ to $(1, -1)$ is $\sqrt{2}$.) See Figure 13.2.4.

We will often use this idea, placing a vector with its initial point at the origin and then representing the vector by its components, the coordinates (a, b) of the terminal point. See Figure 13.2.5(i).

An important property of vectors is that they can be added. For instance, the vectors $\mathbf{u} = (a, b)$ and $\mathbf{v} = (c, d)$ can be added to form the vector $\mathbf{u} + \mathbf{v} = (a + c, b + d)$. Vectors are added by simply adding corresponding components.

To multiply a vector by a number, we just multiply each component by the number. For instance, with $\mathbf{u} = (a, b)$, $2\mathbf{u} = (2a, 2b)$. This is the vector that is twice as long as **u**, pointing in the same direction (see Figure 13.2.5(ii)). If the number by which we multiply is *negative*, then multiplying "turns the vector around," making it point in the opposite direction (see Figure 13.2.5(iii)).

Figure 13.2.3

Figure 13.2.4

Figure 13.2.5(i)

Figure 13.2.5(ii)

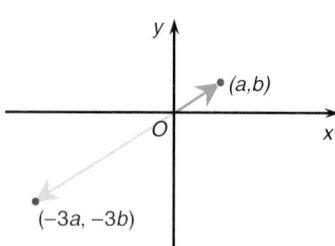

Figure 13.2.5(iii)

It also makes sense to consider the difference of two vectors; with $\mathbf{u} = (a, b)$ and $\mathbf{v} = (c, d)$, we find that $\mathbf{u} - \mathbf{v} = (a - c, b - d)$. For instance, $(3, -5) - (2, -1) = (1, -4)$ (see Figure 13.2.5(iv)).

Notice that the length of the vector $\mathbf{u} - \mathbf{v}$ is $\sqrt{(a - c)^2 + (b - d)^2}$, which is just the distance from the point (a, b) to (c, d). The length of the difference vector is the distance between the terminal points of the two original vectors.

When we specify the position of a point in the plane by giving its coordinates (x, y), we can think of this as a vector that starts at the origin and ends at the point. We refer to it as the **displacement** of the point (meaning its displacement from the origin).

For the problem of a moving point, we want to consider the difference between the displacements of two nearby points. Suppose the position of the point is given by $(x, y) = \big(x(t), y(t)\big)$, so we should consider the difference $\big(x(a + h), y(a + h)\big) - \big(x(a), y(a)\big)$. To make sense of this formula, we think of the two points as *vectors* and write

$$\big(x(a + h), y(a + h)\big) - \big(x(a), y(a)\big) = \big(x(a + h) - x(a), y(a + h) - y(a)\big).$$

In other words, we just subtracted the two x-components and subtracted the two y-components. This is easy to understand graphically if we mark the two points $\big(x(a), y(a)\big)$ and $\big(x(a + h), y(a + h)\big)$ as shown in Figure 13.2.6. Their "difference" $\big(x(a + h), y(a + h)\big) - \big(x(a), y(a)\big) = \big(x(a + h) - x(a), y(a + h) - y(a)\big)$ can be visualized as the vector going from $\big(x(a), y(a)\big)$ to $\big(x(a + h), y(a + h)\big)$: If we start at $\big(x(a), y(a)\big)$ and move $x(a + h) - x(a)$ in the horizontal direction and $y(a + h) - y(a)$ in the vertical direction, then we will end up at the second point $\big(x(a + h), y(a + h)\big)$.

Returning to our moving point, we call the vector $\big(x(a + h) - x(a), y(a + h) - y(a)\big)$ the change in the **displacement** from $\big(x(a), y(a)\big)$ to $\big(x(a + h), y(a + h)\big)$. This change in the displacement includes both the idea of the distance between these points and the direction from one to the other.

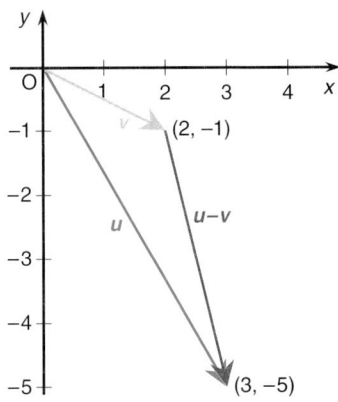

The vectors $\mathbf{u}$ and $\mathbf{v}$ below are drawn with their initial points at the origin; $\mathbf{u} - \mathbf{v}$ is not.

Figure 13.2.5(iv)

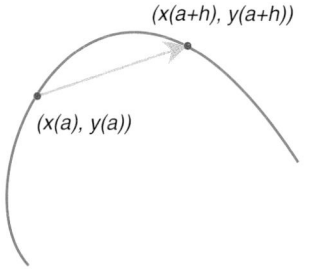

Figure 13.2.6

DEFINITION 13.2.1

If (x, y) and (u, v) are two points in the plane, then the **change in the displacement** from (u, v) to (x, y) is the *vector*

$$(x, y) - (u, v) = (x - u, y - v),$$

which is the difference of the two vectors. We visualize it as an arrow that begins at the point (u, v) and ends at the point (x, y).

The length of the displacement vector is

$$\sqrt{(x - u)^2 + (y - v)^2},$$

which is the distance from (x, y) to (u, v).

EXAMPLE 1

(i) The displacement from the point $(1, -2)$ to the point $(4, 2)$ is the vector $(4, 2) - (1, -2) = (3, 4)$. The distance between these points is the length of the displacement vector, which is $\sqrt{3^2 + 4^2} = 5$. See Figure 13.2.7.

(ii) The displacement from the origin $(0, 0)$ to a point (a, b) is the vector $(a, b) - (0, 0) = (a, b)$. This just means that to get from the origin to (a, b), it is necessary to go along the vector (a, b).

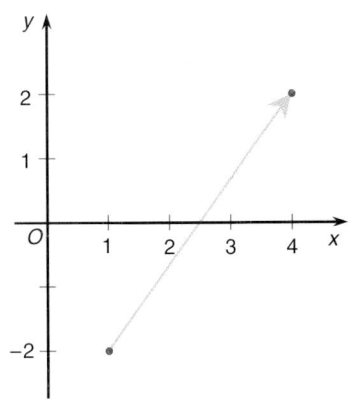

Figure 13.2.7

(iii) It is important to notice that the displacement from P to Q is not the same as the displacement from Q to P; in fact these two displacements are each other's negatives, meaning that they are vectors that have the same length but point in opposite directions.

Of course, this is obvious if we consider that moving from P to Q involves the same distance as moving from Q to P but in the opposite direction. ◢

In part (ii) of Example 1 we wrote the expression (a, b) to mean two very different things. On the one hand, it represents the coordinates of a point in the plane. On the other, it represents a vector by its components. The vector is an arrow, and if we place its initial point at the origin, then its terminal point will be at the point (a, b). See Figure 13.2.8.

Because the vector and the point are closely related in this way, it is very convenient to use the same notation (a, b) for both of them, but in the backs of our minds we should remember that they are different things.

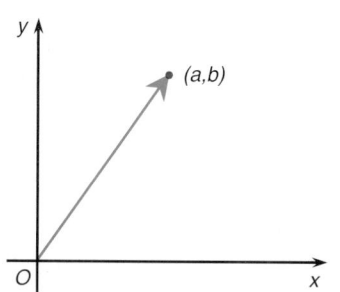

Figure 13.2.8

Velocity

Consider the moving point whose position at time t is the point $\big(x(t), y(t)\big)$. It is sometimes convenient to write

$$\mathbf{r} = \mathbf{r}(t) = \big(x(t), y(t)\big).$$

The letter $\mathbf{r}$ reminds us of "radius"; $\mathbf{r}(t)$ is the vector pointing from the origin out to the position of the moving point at time t.

Now as time t changes from $t = a$ to $t = a + h$, the moving point goes from $\mathbf{r}(a) = \big(x(a), y(a)\big)$ to $\mathbf{r}(a + h) = \big(x(a + h), y(a + h)\big)$. Its displacement changes by $\big(x(a + h) - x(a), y(a + h) - y(a)\big) = \mathbf{r}(a + h) - \mathbf{r}(a)$. By analogy with the straight line case we should try to find the average velocity over this short time. To do this, it is necessary to divide by h or, to put it another way, to multiply by $\frac{1}{h}$. It seems reasonable to expect that the average velocity over the interval from $t = a$ to $t = a + h$ is

To multiply a vector by $\frac{1}{h}$, we simply multiply each of its coordinates by $\frac{1}{h}$.

$$\frac{1}{h}\big(x(a + h) - x(a), y(a + h) - y(a)\big) = \left(\frac{x(a + h) - x(a)}{h}, \frac{y(a + h) - y(a)}{h}\right).$$

Notice that this formula for the average velocity is a *vector*, whose two components are difference quotients. Our original idea was to form the difference quotient for a moving point, allow h to tend to 0, and take the limit. We know that as $h \to 0$, the two difference quotients will both tend to limits: $\frac{x(a+h)-x(a)}{h} \to \frac{dx}{dt}(a) = x'(a)$ and $\frac{y(a+h)-y(a)}{h} \to \frac{dy}{dt}(a) = y'(a)$.

What we have found is that the *vector difference quotient*

$$\frac{1}{h}\big(\mathbf{r}(a + h) - \mathbf{r}(a)\big) = \frac{1}{h}\Big(\big(x(a + h), y(a + h)\big) - \big(x(a), y(a)\big)\Big)$$

$$= \left(\frac{x(a + h) - x(a)}{h}, \frac{y(a + h) - y(a)}{h}\right)$$

tends to the limit $\big(x'(a), y'(a)\big)$ as $h \to 0$. This vector represents the *instantaneous velocity* of the moving point at the time $t = a$; it points in the direction in which the point is moving.

DEFINITION 13.2.2

> Consider a point moving in the plane, and suppose that its position at time t is given by $\mathbf{r}(t) = \big(x(t), y(t)\big)$, where $x(t)$ and $y(t)$ are functions that are differentiable at $t = a$. The **velocity** of the moving point at time $t = a$ is the vector
>
> $$\mathbf{r}'(a) = \big(x'(a), y'(a)\big).$$

We found this formula by following the model of what we did for motion in a straight line, but it turns out to work very well.

EXAMPLE 2   Consider the moving point

$$(x, y) = \mathbf{r}(t) = \big(x(t), y(t)\big) = (\cos t, \sin t).$$

We know that these are parametric equations for the circle of radius 1 around the origin. The point moves around the circle counterclockwise, making one complete revolution each time t increases by 2π.

The velocity vector is given by Definition 13.2.2. It is

$$\mathbf{r}'(t) = (-\sin t, \cos t).$$

Notice that it depends on t, meaning that the velocity changes and will be different at different times.

In Figure 13.2.9(i) we plot the point $(x, y) = \mathbf{r}(t) = (\cos t, \sin t)$. In Figure 13.2.9(ii) we also plot the corresponding velocity vector $\mathbf{r}'(t) = (-\sin t, \cos t)$, placing it so that its initial point is at the moving point $\mathbf{r}(t) = (\cos t, \sin t)$ on the circle. It turns out that the velocity vector is tangent to the circle, and its length is $\sqrt{\sin^2 t + \cos^2 t} = 1$.

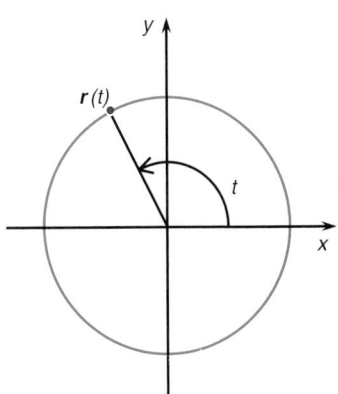

Figure 13.2.9(i)

From Example 2 we can see two very important features of velocity vectors. The first is that the direction in which the vector points is exactly the direction in which the moving point is moving. This direction may be different at different times, but it will be tangent to the curve. The second property is that the length of the velocity vector is the instantaneous *speed* at which the point is moving. In this example the length of the velocity vector is always 1. On the other hand, we know that the point $(\cos t, \sin t)$ goes exactly once around the circle in 2π units of time. Since it travels a distance of 2π, the circumference, we find that its speed is $\frac{2\pi}{2\pi} = 1$, equal to the length of the velocity vector.

We will use these observations to define the concepts of a tangent vector to a curve and of the speed of a moving point.

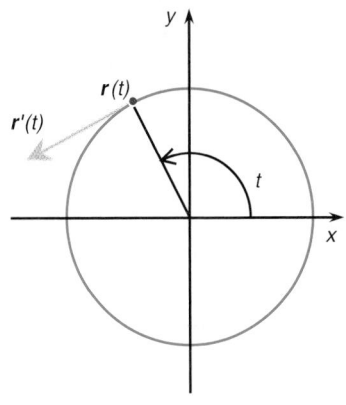

Figure 13.2.9(ii)

DEFINITION 13.2.3

> Consider a curve traced out by the moving point $\mathbf{r}(t) = \big(x(t), y(t)\big)$ for t in some interval I. Suppose $a \in I$ is a point at which both $x(t)$ and $y(t)$ are differentiable, and let $\mathbf{r}'(a) = \big(x'(a), y'(a)\big)$ be the velocity vector. Suppose that $\mathbf{r}'(a)$ is not the zero vector, that is, that $x'(a)$ and $y'(a)$ are not both zero.
>
> Then $\mathbf{r}'(a)$ is a **tangent vector** to the curve at the point $\mathbf{r}(a) = \big(x(a), y(a)\big)$, and its length $\|\mathbf{r}'(a)\| = \sqrt{x'(a)^2 + y'(a)^2}$ is the **speed** of the moving point at $t = a$.

Figure 13.2.10

Figure 13.2.11

Figure 13.2.12

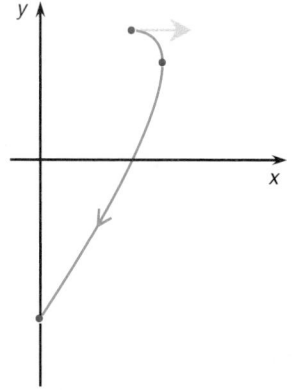

Figure 13.2.13

EXAMPLE 3

(i) Consider the parametric equations $x = 1 + 3t$, $y = 0$. They parametrize the horizontal axis. The velocity vector is $\mathbf{r}'(t) = (3, 0)$. Its horizontal component is 3, and its vertical component is 0.

   Certainly this vector is tangent to the horizontal axis (see Figure 13.2.10). Moreover, the length of the velocity vector is 3, which means that the speed of the moving point is 3. This is confirmed by looking at the original equations. If t increases by 1 unit, then $x = 1 + 3t$ will increase by 3 units; this means that the point moves 3 units of distance in every 1 unit of time, which is exactly what it means to say its speed is 3.

(ii) Consider the parametric equations $x = 2t - 1$, $y = 3t + 2$. Solving for t, we find that $t = \frac{x+1}{2}$ and $t = \frac{y-2}{3}$. Equating these expressions, we find that the curve is the straight line $\frac{x+1}{2} = \frac{y-2}{3}$, or $3x - 2y + 7 = 0$.

   The velocity vector is $\mathbf{v} = \big(x'(t), y'(t)\big) = (2, 3)$. If we consider a time $t = a$ and draw the corresponding point on the line (see Figure 13.2.11), then we should place the velocity vector so that its initial point is at that point on the line. In the picture it seems that the vector is tangent to the line; to prove it, we can check that the slope of the line is $\frac{3}{2}$ (cf. Remark 1.5.10(ii), p. 27). The vector $(2, 3)$ points 2 units horizontally and 3 units vertically, so its slope is also $\frac{3}{2}$, and this is why it is tangent to the line.

   Notice that the length of the velocity vector is $\sqrt{2^2 + 3^2} = \sqrt{13}$. The speed of the moving point is $\sqrt{13}$ units of distance per unit of time.

(iii) The parametric equations $x = t$, $y = t^2$ parametrize the parabola $y = x^2$. The velocity vector $\mathbf{r}'(t) = (1, 2t)$ changes as t varies. This is more complicated than the previous two examples, in both of which the velocity vector is constant.

   Consider the time $t = a$ and the corresponding point (a, a^2) on the parabola, as shown in Figure 13.2.12. We know from our earlier work that the tangent to the parabola at this point has slope $2a$. On the other hand, the velocity vector we have found is $\mathbf{r}'(a) = (1, 2a)$, whose slope is also $\frac{2a}{1} = 2a$. This confirms that the velocity vector is in fact tangent to the curve at each point.

   The speed of the moving point is the length of the velocity vector, which is $\sqrt{1^2 + (2a)^2} = \sqrt{1 + 4a^2}$. This depends on the time $t = a$. Notice that it is larger for large values of a; its smallest value will occur when $a = 0$. In terms of the illustration this means that the moving point moves most slowly at the vertex of the parabola and moves faster the farther it is from the origin.

(iv) Suppose a ship's course is parametrized by the moving point $\mathbf{r}(t) = (3 + 2t - t^2, 4 - t^2)$, for $t \in [0, 3]$. The axes are arranged so that the positive x-axis points due east and the positive y-axis points due north. At what moments is the ship sailing (a) east, (b) north, (c) northwest?

Solution

(a) The velocity vector is $\mathbf{r}'(t) = (2 - 2t, -2t)$. When the ship is sailing due east, the velocity vector points horizontally to the right, so it is $(k, 0)$, for some positive number k. For the vector $\mathbf{r}'(t) = (2 - 2t, -2t)$ to be of this form we must have $t = 0$. At $t = 0$ the velocity is $(2, 0)$, which does point due east. This means that the ship starts out sailing due east, as shown in Figure 13.2.13.

(b) If the ship is sailing due north, the velocity vector points straight up, that is, $\mathbf{r}' = (0, k)$, for some $k > 0$. In particular this means that $2 - 2t$ must equal zero, so $t = 1$. However, if we look at the velocity vector $\mathbf{r}'(1) = (0, -2)$, we realize that it points due *south*. There is no time at which the ship is sailing due north.

(c) If the ship is sailing northwest, its velocity vector must point up and to the left at a $45°$ angle above the horizontal. This means that the vector must be $(-k, k)$, for some $k > 0$. We solve to find when this is possible by setting $-(2 - 2t) = -2t$. Adding $2t$ to each side, we find that $4t - 2 = 0$, so $t = \frac{1}{2}$.

But when $t = \frac{1}{2}$, the velocity is $(1, -1)$, so the ship is sailing *southeast*. There is no time when the ship is sailing northwest.

If $(2 - 2t, -2t) = (-k, k)$, then $k = -2t$ and $-k = 2 - 2t$, so $-(2 - 2t) = -2t$.

Notice that it was possible to answer part (iv) of Example 3 without identifying the curve or finding its equation. The velocity vector is an extremely useful tool for looking at the motion of a point in the plane. It tells us the direction and the instantaneous speed of the motion.

EXAMPLE 4   Recall the Spiral of Archimedes from Example 12.2.2, which we parametrized in Example 13.1.3 by the equations $(x, y) = \mathbf{r}(t) = (t \cos t, t \sin t)$. The velocity vector of the moving point at time t is $\mathbf{r}' = (\cos t - t \sin t, \sin t + t \cos t)$. When t is small, this is almost equal to the vector $(\cos t, \sin t)$, which points in the direction from the origin out to the point $(t \cos t, t \sin t)$. So at the beginning the point is mostly just moving away from the origin.

But when t is large, the corresponding velocity vector is nearly equal to $(-t \sin t, t \cos t)$ (i.e., the other part is much smaller). This is a vector that points in the direction tangent to a circle around the origin. For large t the point moves almost in a circle, with only a relatively small motion away from the origin. See Figure 13.2.14.

Parametric equations can also be used in situations having nothing to do with moving objects.

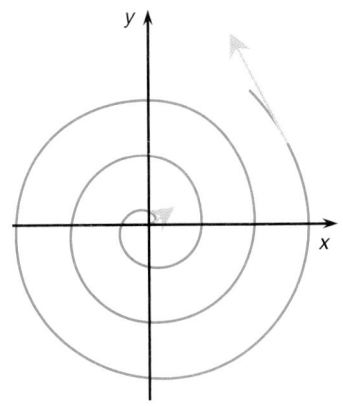

Figure 13.2.14

EXAMPLE 5   **Price Versus Sales**   After a new consumer electronics product is introduced, sales rise quickly and the price gradually comes down. We measure time t in years, with $t = 0$ at the time of the first sales. Suppose the unit price at time t, in hundreds of dollars, is $p(t) = \frac{t^2 + 8}{t^2 + 4}$, and the monthly sales, in units of 100,000, are $S(t) = \frac{t^2 + t + 1}{t^2 + 1}$.

In Figure 13.2.15 we plot the curve $\mathbf{r}(t) = \big(S(t), p(t)\big)$. The dots represent the values at $t = 0, 1, 2, 3, 4$. From the shape of the curve we see that sales rise sharply at first and the price remains almost fixed. But as sales growth slows, competitive pressures force the price down.

A quick calculation shows that $S'(1) = 0$; this is the point at which the tangent vector points straight down. After 1 year, sales have peaked and the price is falling.

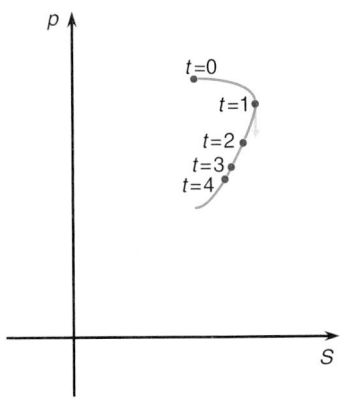

Figure 13.2.15

Arclength

A natural question to ask about a curve $\mathbf{r}(t) = \big(x(t), y(t)\big)$ is how to find its length, that is, its arclength. For instance, we are accustomed to finding that the length of the circle $\mathbf{r}(t) = (R \cos t, R \sin t)$ from $t = 0$ to $t = 2\pi$ is $2\pi R$.

Let us suppose the curve starts at $t = a$, and suppose the length of the curve from $\mathbf{r}(a)$ to $\mathbf{r}(t)$ is called $s = s(t)$. Further let us suppose that s is a differentiable function of t.

Then $s'(t)$ is the rate of change of s, that is, it is the rate at which the length of the curve is increasing at the point $\mathbf{r}(t)$. Another way to put it is that this is the rate of change of the distance traveled by the moving point since $t = a$.

But that is exactly what we call the *speed* of the moving point. We know that the speed is given by the length of the velocity vector $\mathbf{r}'(t) = \big(x'(t), y'(t)\big)$, so we find that

$$s'(t) = \sqrt{x'(t)^2 + y'(t)^2}.$$

Knowing the derivative of $s(t)$, we can find the arclength itself by integrating. The arclength of the curve between the points $\mathbf{r}(a)$ and $\mathbf{r}(b)$ is

$$s(b) = \int_a^b \sqrt{x'(t)^2 + y'(t)^2}\, dt.$$

To find this expression, we assumed that it makes sense to talk of arclength and also assumed that it is a differentiable function of the parameter t. Then we used our intuition about speed to find the formula. In a way this is not very satisfactory because we have never proved that what we call the "speed" of the moving point has anything to do with the rate of change of distance.

In fact these things are all correct, but instead of worrying about it, we will let the above argument explain where the formula for arclength comes from and why it seems reasonable. We will simply use the formula to "define" arclength.

> We could also have approximated the curve by short straight segments; the sum of their lengths gives a Riemann sum. Letting the lengths of the segments tend to zero, we would find that the sum tends to this same integral, reassuring us that we have found the right definition.

DEFINITION 13.2.4

> Suppose $x(t)$ and $y(t)$ are differentiable with continuous derivatives. If $a \leq b$, the **arclength** of the curve $\mathbf{r}(t) = \big(x(t), y(t)\big)$ between the points corresponding to $t = a$ and $t = b$ is defined to be
>
> $$s = \int_a^b \sqrt{x'(t)^2 + y'(t)^2}\, dt.$$

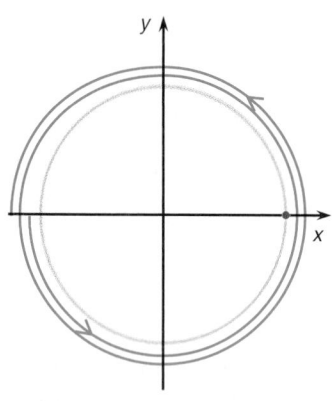

Figure 13.2.16

EXAMPLE 6   **Arclength of a Circle**   Consider the circle $\mathbf{r}(t) = (R \cos t, R \sin t)$. Here the components are $x(t) = R \cos t$ and $y(t) = R \sin t$, so the arclength between $t = a$ and $t = b$ is

$$s = \int_a^b \sqrt{R^2 \sin^2 t + R^2 \cos^2 t}\, dt = \int_a^b R\, dt = R(b - a).$$

So, for instance, between $t = 0$ and $t = 2\pi$, which corresponds to going exactly once around the circle, the arclength is $R(2\pi - 0) = 2\pi R$. Of course, this is the circumference of the circle. See Figure 13.2.16.

Notice that the arclength between $t = -\pi$ and $t = 3\pi$ is $4\pi R$. As t runs from $-\pi$ to 3π, the point $\mathbf{r}(t)$ goes around the circle twice, and the distance it covers is twice the circumference.

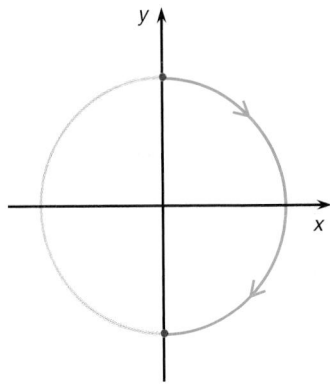

Figure 13.2.17

Of course, it is no great surprise that we can find the arclength of a circle, especially because the parametrization we chose is so convenient. However, the formula can be used for less well-chosen parametrizations.

EXAMPLE 7   Let $\mathbf{r}(t) = \big(\sin(\pi t^2), \cos(\pi t^2)\big)$, for $t \in [0, 1]$.

This point will move along the circle $x^2 + y^2 = 1$, starting at the point $\mathbf{r}(0) = (0, 1)$ and moving along the right half of the circle to $\mathbf{r}(1) = (0, -1)$ (see Figure 13.2.17). Its speed is not constant.

The arclength is given by

$$s = \int_0^1 \sqrt{x'(t)^2 + y'(t)^2}\, dt = \int_0^1 \sqrt{(2\pi t)^2 \cos^2(\pi t^2) + (2\pi t)^2 \sin^2(\pi t^2)}\, dt$$

$$= \int_0^1 2\pi t \sqrt{\cos^2(\pi t^2) + \sin^2(\pi t^2)}\, dt = 2\pi \int_0^1 t\, dt = \pi t^2 \Big|_0^1$$

$$= \pi.$$

It is reassuring to find that even when the point moves at a varying speed, we still get the correct arclength for the semicircle.

EXAMPLE 8   Find the arclength of the curve $\mathbf{r}(t) = (1 + t^2, 3 - 2t^3)$, for $t \in [0, 1]$.

Solution   The arclength is

$$s = \int_0^1 \sqrt{(2t)^2 + (-6t^2)^2}\, dt = \int_0^1 \sqrt{4t^2 + 36t^4}\, dt = \int_0^1 2t\sqrt{1 + 9t^2}\, dt.$$

We substitute $u = 1 + 9t^2$, $du = 18t\, dt$, and get

$$\int_1^{10} \frac{1}{9} u^{1/2}\, du = \frac{1}{9} \frac{2}{3} u^{3/2} \Big|_1^{10} = \frac{2}{27}(10\sqrt{10} - 1).$$

The arclength is $\frac{2}{27}(10\sqrt{10} - 1) \approx 2.268$.

REMARK 13.2.5

> In Definition 6.9.1 (p. 485) we found that the arclength for the curve formed by a graph $y = f(x)$ between $x = a$ and $x = b$ is given by
>
> $$s = \int_a^b \sqrt{1 + f'(x)^2}\, dx.$$
>
> If we want to represent this graph by parametric equations, the easiest thing to do is to write $\mathbf{r}(t) = \big(t, f(t)\big)$, for $t \in [a, b]$, that is, let $x(t) = t$ and $y(t) = f(t)$. Since $x'(t) = \frac{dt}{dt} = 1$, Formula 13.2.4 for arclength using parametric equations gives exactly the same formula as Definition 6.9.1 in this particular case. Of course, Definition 13.2.4 also applies to curves that are not represented as graphs, so it is more general than Definition 6.9.1.

Suppose we consider a curve given in polar coordinates by $r = f(\theta)$. It can be parametrized by $\big(x(\theta), y(\theta)\big) = (r\cos\theta, r\sin\theta) = \big(f(\theta)\cos\theta, f(\theta)\sin\theta\big)$, using θ as

We can use a graphics calculator to sketch the curve from Example 8. To set the ranges, we note that for $t \in [0, 1]$ the value of $x(t) = 1 + t^2$ is in $[1, 2]$, and the value of $y(t) = 3 - 2t^3$ is in $[1, 3]$.

TI-85	SHARP EL9300
MODE ▼ ▼ ▼ ▼ ► ► (Param)	SET UP E 3 ⏎ $1 + $ x/θ/T
ENTER EXIT GRAPH F1 ($E(t) =$)	x^2 ENTER $3 - 2$ x/θ/T a^b 3 ENTER
$1 +$ F1 (t) x^2 ▼ $3 - 2$ F1 (t) ^3	RANGE 0 ENTER 2 ENTER 1 ENTER 0
M2(RANGE) 0 ▼ 1 ▼ ▼ 0 ▼ 2	ENTER 4 ENTER 1 ENTER 0 ENTER 1
▼ ▼ 0 ▼ 4 F5 (GRAPH)	ENTER .01 ENTER ⏎

CASIO f_x-7700GB	HP 48SX
MODE MODE × (PARAM) Range	$' 1 + T\, y^x 2 + \alpha$ ↰ CST × ◼ $3 - 2$
0 EXE 2 EXE 1 EXE 0 EXE 4 EXE 1	× $T\, y^x$ 3 ↰ PLOT NEW P A R
EXE 0 EXE 1 EXE .01 EXE Graph	ENTER PTYPE PARA PLOTR ↰ {}
$1 +$ T x^2, $3 - 2$ T x^y 3 EXE	T SPC 0 SPC 1 ENTER INDEP AUTO

We did not need the picture to find the arclength, but it helps us to see that our answer is reasonable. For instance, how does our answer compare with the distance between the two endpoints of the curve?

the parameter. We find that

$$x'(\theta) = f'(\theta) \cos \theta - f(\theta) \sin \theta, \quad y'(\theta) = f'(\theta) \sin \theta + f(\theta) \cos \theta.$$

Since

$$
\begin{aligned}
x'(\theta)^2 + y'(\theta)^2 &= \big(f'(\theta) \cos \theta - f(\theta) \sin \theta\big)^2 + \big(f'(\theta) \sin \theta + f(\theta) \cos \theta\big)^2 \\
&= f'(\theta)^2 \cos^2 \theta - 2f'(\theta)f(\theta) \cos \theta \sin \theta + f(\theta)^2 \sin^2 \theta \\
&\quad + f'(\theta)^2 \sin^2 \theta + 2f'(\theta)f(\theta) \sin \theta \cos \theta + f(\theta)^2 \cos^2 \theta \\
&= f'(\theta)^2 + f(\theta)^2,
\end{aligned}
$$

we can apply Definition 13.2.4 to find the arclength.

THEOREM 13.2.6

Arclength for Curves in Polar Coordinates

If $f(\theta)$ is differentiable with continuous derivative, then the arclength of the part of the curve $r = f(\theta)$ between $\theta = \alpha$ and $\theta = \beta$ is

$$s = \int_\alpha^\beta \sqrt{f'(\theta)^2 + f(\theta)^2}\, d\theta.$$

EXAMPLE 9   Arclength of a Cardioid   Find the arclength of the top half of the cardioid $r = 1 - \cos \theta$.

Solution   Using the formula from Theorem 13.2.6 with $f(\theta) = 1 - \cos \theta, \theta \in [0, \pi]$, we find $f'(\theta) = \sin \theta$, so

$$f'(\theta)^2 + f(\theta)^2 = \sin^2 \theta + (1 - \cos \theta)^2$$
$$= \sin^2 \theta + 1 - 2\cos \theta + \cos^2 \theta$$
$$= 2 - 2\cos \theta$$
$$= 2(1 - \cos \theta).$$

The arclength is $\sqrt{2} \int_0^\pi \sqrt{1 - \cos \theta} \, d\theta$. We substitute $u = 1 - \cos \theta$, so $\cos \theta = 1 - u$ and $\sin \theta = \sqrt{1 - \cos^2 \theta} = \sqrt{2u - u^2}$. The arclength is

$$\boxed{\begin{aligned} u &= 1 - \cos \theta \\ du &= \sin \theta \, d\theta \end{aligned}}$$

$$\boxed{\begin{aligned} v &= 2 - u \\ dv &= -\, du \end{aligned}}$$

$$s = \sqrt{2} \int_0^\pi \sqrt{1 - \cos \theta} \, d\theta$$
$$= \sqrt{2} \int_0^2 \sqrt{u} \frac{1}{\sqrt{2u - u^2}} \, du = \sqrt{2} \int_0^2 \frac{1}{\sqrt{2 - u}} \, du$$
$$= -\sqrt{2} \int_2^0 v^{-1/2} \, dv = -2\sqrt{2} v^{1/2} \Big|_2^0$$
$$= -2\sqrt{2}(0 - \sqrt{2}) = 4.$$

The arclength of the top half of the cardioid is 4.

Exercises 13.2

|

Find the velocity vector $\mathbf{v}(t)$ for each of the following curves.

1. $x = 4t + 4, y = -3t - 3$
2. $x = t - 5, y = 6$
3. $x = 2t^2 - t + 3, y = 4t^2 - 2t - 6$
4. $x = t^2 - 1, y = 1 - t$
5. $x = \sqrt{1 + t^2}, y = t$
6. $x = 2t, y = \sqrt{1 - t^2}, t \in [-1, 1]$
7. $x = t^2, y = \arctan t$
8. $x = \frac{1}{1+t}, y = e^t, t \ge 0$
9. $x = 2\cos(3t), y = 2\sin(3t)$
10. $x = \sin^2 t + 3 + \cos^2 t, y = -5$

For each of the following curves, find all points at which the point is moving (i) straight up, (ii) horizontally to the left. Express the points by their (x, y)-coordinates and by the corresponding value(s) of the parameter t.

11. $x = \sin t, y = \cos t$
12. $x = t + 1, y = t^3 - 3t + 2$
13. $x = e^t, y = \sin(t^3 - t^2 + 3)$
14. $x = 3, y = t^2 - 4t + 1$
15. $x = 2t^2, y = t^3 - 3t^2 - 9t + 7$
16. $x = 1 - e^t, y = 1 - t^2$
17. $x = \sinh 2t, y = \cosh 2t$
18. $x = \arctan t, y = 1 - t^3$
19. $x = \exp(t^3 - 3t), y = |t|$
20. $x = \exp\left(1 - \arctan(1 - 2t^4) + 3\right),$
    $y = 7 - 2\exp\left(1 - \arctan(1 - 2t^4) + 3\right)$

Find the arclength of the path followed by each of the following moving points for the specified values of the parameter.

21. $\mathbf{r}(t) = (3\sin t, 3\cos t), 0 \le t \le \pi$
22. $\mathbf{r}(t) = (\sin 3t, \cos 3t), 0 \le t \le \pi$
23. $\mathbf{r}(t) = (3t^2, 2t^3), 1 \le t \le 2$

24. $\mathbf{r}(t) = (t^2, t^3), 0 \leq t \leq 1$

25. $\mathbf{r}(t) = \left(4\cos(\pi t^2), -3\cos(\pi t^2)\right), 0 \leq t \leq 1$

26. $\mathbf{r}(t) = (e^{2t}, e^{3t}), -1 \leq t \leq 0$

27. $\mathbf{r}(t) = (\cosh t, t), 0 \leq t \leq 1$

28. $\mathbf{r}(t) = (t, \sqrt{4-t}), -4 \leq t \leq 4$

29. $\mathbf{r}(t) = (3, t^5), 0 \leq t \leq 2$

30. $\mathbf{r}(t) = (\cos t, \cos t), -\frac{\pi}{2} \leq t \leq \frac{\pi}{2}$

II

31. Suppose the course of a ship is given by $x = 5\cos\pi t$, $y = 5\pi t$, where the x-axis points due east, the y-axis points due north, distances are measured in kilometers, and time is measured in hours, with $0 \leq t \leq 1$.

  (i) Find all times at which the ship is sailing northwest.

  (ii) Find the speed of the ship at each time found in part (i).

  (iii) What is the maximum speed of the ship for $0 \leq t \leq 1$?

*32. Consider a ball thrown up into the air so that its path is parametrized by $x = 4t$, $y = 64t - 16t^2$, where t is measured in seconds and x and y are measured in feet.

  (i) Find the velocity vector at time t.

  (ii) When is the velocity vector exactly horizontal or exactly vertical?

  (iii) At what point is the speed of the ball a minimum?

*33. Sketch the curve that is parametrized by $x = \cos t$, $y = \sin 2t$, and find all values of t at which the moving point is going straight up or straight down.

*34. Sketch the curve that is parametrized by $x = \cos t$, $y = \sin kt$, where k is a positive integer, and find all values of t at which the moving point is going straight up or straight down.

*35. Sketch the curve that is parametrized by $x = \cos t$, $y = \sin\frac{t}{3}$, and find all values of t at which the moving point is going straight up or straight down.

*36. Sketch the curve that is parametrized by $x = \cos 2t$, $y = \sin 3t$, and find all values of t at which the moving point is going straight up or straight down.

37. What is the average velocity of the point parametrized by $(x, y) = \mathbf{r}(t) = (t^2 + 1, t^4)$ over the interval from $t = 0$ to $t = 1$? (Remember that velocity is a vector.)

38. What is the average velocity of the point parametrized by $(x, y) = \mathbf{r}(t) = (t^2 + 1, t^4)$ over the interval from $t = -1$ to $t = 1$?

*39. In Example 2 (p. 807) we observed that the velocity vector is always tangent to the circle. If we consider the velocity at time t, the initial point of the velocity vector is at the point $(\cos t, \sin t)$ on the circle, and to show that the velocity vector is tangent, it is enough to show that it is perpendicular to the line segment from the origin to $(\cos t, \sin t)$. Show that this is true.

*40. For the ellipse parametrized by $x = 4\cos t$, $y = 9\sin t$, find all the values of t at which the tangent vector $\mathbf{r}'(t) = \left(x'(t), y'(t)\right)$ is perpendicular to the position vector $\mathbf{r}(t) = \left(x(t), y(t)\right)$, that is, to the segment running from the origin to the point $\mathbf{r}(t) = \left(x(t), y(t)\right)$.

*41. Consider the functions

$$x(t) = \begin{cases} t^2, & \text{if } t < 0, \\ 0, & \text{if } 0 \leq t \leq 1, \\ (t-1)^2, & \text{if } t > 1; \end{cases}$$

$$y(t) = \begin{cases} -t^2, & \text{if } t < 0, \\ 0, & \text{if } 0 \leq t \leq 1, \\ (t-1)^2, & \text{if } t > 1. \end{cases}$$

  (i) Sketch the graphs of each of these two functions of t, and show that they are both differentiable at every t.

  (ii) Sketch the graph of the curve parametrized by the expressions $\mathbf{r}(t) = \left(x(t), y(t)\right)$.

  (iii) Notice that the velocity vector is the zero vector for $t \in [0, 1]$ since both derivatives $x'(t)$ and $y'(t)$ are zero there. What is the tangent vector to the curve at the corresponding point?

*42. Consider the curve parametrized by $x = t^2$, $y = t^3$.

  (i) Find its velocity vector for every t.

  (ii) The vector $\mathbf{r}'(t) = \left(x'(t), y'(t)\right)$ is zero when $t = 0$. In what direction is the point moving when t is slightly greater than zero and in what direction is it moving when t is slightly less than zero? What happens to these directions as t approaches 0 from above and from below?

  (iii) Sketch the curve traced out by the moving point.

Exercises 41 and 42 have shown some of the types of behavior that are possible at points where both derivatives are zero. In particular they explain why it was necessary to assume that at least one derivative is nonzero in Definition 13.2.3.

43. Find the arclength of the circle $r = \sin\theta$.

44. Find the arclength of the right half of the circle $r = \cos\theta$, i.e., the right semicircle.

45. Find the arclength of the cardioid $r = 1 + \sin\theta$.

46. Find the arclength of the part of the cardioid $r = 1 + \cos\theta$ lying to the right of the vertical axis.

Use Simpson's Rule S_n with $n = 10$ to estimate the arclengths of the following curves.

47. $\mathbf{r}(t) = (\cos^2 t, \sin 3t)$, $t \in [0, \pi]$

48. $\mathbf{r}(t) = (1 + t^2, \ln t)$, $t \in [1, 3]$

49. $\mathbf{r}(t) = (t^3 - 2t + 2, t^4 + 2t^2 + 1)$, $t \in [-1, 1]$

50. $\mathbf{r}(t) = (t^2 + t, t + \sin \pi t)$, $t \in [0, 1]$

51. The piece of the hyperbola $x^2 - y^2 = 9$ between $(5, -4)$ and $(5, 4)$.

52. The short piece of the ellipse $4x^2 + 9y^2 = 36$ between $(3, 0)$ and $(0, 2)$.

POINT TO PONDER

Consider the point moving along the real axis whose position is given by $x(t) = t^2$. As t moves from -2 to 0, the point moves from $x = 4$ to $x = 0$. Its average velocity over this interval is $\frac{4-0}{-2-0} = -2$. Its average speed is 2.

As t goes from 0 to 2, the point moves from $x = 0$ to $x = 4$, so over this interval its average velocity is 2 and its average speed is 2.

However, as t goes from -2 to 2, the point starts at $x = 4$ and ends up at $x = 4$, so its net change of displacement is zero. The average velocity over this interval from $t = -2$ to $t = 2$ is 0. But since the total distance it has traveled is $4 + 4 = 8$ and the time it took is $2 + 2 = 4$, its average speed is $\frac{8}{4} = 2$.

So although the instantaneous speed is the absolute value of the velocity, the *average* speed (in this case 2) is not necessarily the absolute value of the *average* velocity (which in this case is 0).

The situation is similar for points moving in the plane. Consider the point $(x, y) = \mathbf{r}(t) = (\cos t, \sin t)$ that moves around the circle of radius 1. We saw that since it goes once around the circle when t increases from 0 to 2π, and the circumference of the circle is 2π, its average speed over the interval from $t = 0$ to $t = 2\pi$ is 1. However, since the point begins at the point $(x, y) = (1, 0)$ when $t = 0$ and returns to the same point when $t = 2\pi$, the net change in its displacement as t runs from 0 to 2π is zero, and its average velocity between $t = 0$ and $t = 2\pi$ is zero, that is, the zero vector $(0, 0)$.

Once again we see that the *average* speed may not equal the length of the *average* velocity vector.

13.3 Projectiles

HORIZONTAL AND VERTICAL COMPONENTS

MUZZLE VELOCITY

ANGLE OF ELEVATION

In this section we consider the particular case of the motion of a **projectile**, an object that has been thrown or "projected." The idea is that the object is made to move by some initial force but after that the only force acting on it is gravity.

This is the situation for a ball or a rock that has been thrown, an egg that has been dropped, or a cannonball that has been fired from a gun. It does not apply to an airplane or a rocket, which have engines that continue to exert a force on them, or to a falling feather, which encounters a considerable force due to air resistance. A hot air balloon, which is held up by the force of buoyancy in the air, or a glider, which experiences lift forces on its wings, does not qualify.

In these examples, we are in effect assuming that the force due to air resistance is so small that it can safely be ignored. This is approximately true for the sorts of examples we will consider.

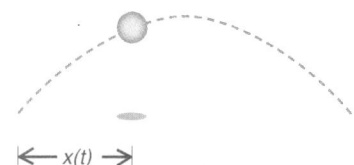

$\overleftarrow{}\, x(t)\, \overrightarrow{}$

Figure 13.3.1

Consider a ball that has been thrown up into the air. Its position will change with time, and we can parametrize its motion by equations

$$x = x(t), \qquad y = y(t).$$

What these equations tell us, in effect, is that the horizontal position of the ball at time t is described by $x(t)$ and its height above the ground at time t is given by $y(t)$. If we imagine the sun shining down from directly overhead, then the ball casts a shadow on the ground, and $x(t)$ just measures the displacement of the shadow at time t. See Figure 13.3.1.

The remarkable thing is that these two "components" of the motion are independent. The vertical motion $y = y(t)$ behaves exactly as if there is no horizontal motion at all, and the horizontal motion is that of an object moving at a constant velocity (cf. Remark 4.9.2).

First we consider the vertical motion, as described by the height $y(t)$. As we saw in Section 4.9, the vertical motion undergoes a constant acceleration due to gravity. It is (approximately) 32 ft/sec^2 or 9.8 m/sec^2. This means that the second derivative is

$$y''(t) = -32,$$

where the minus sign reflects the fact that the acceleration is downward and the units of measurement are feet and seconds. (If we used meters and seconds, the condition would be $y''(t) = -9.8$.)

We conclude that $y'(t)$ must be given by

$$y'(t) = -32t + d,$$

for some constant d, and

$$y(t) = -16t^2 + dt + y_0. \tag{13.3.1}$$

Here y_0 is a constant; substituting $t = 0$ into Equation 13.3.1 shows that $y_0 = y(0)$, that is, that y_0 is the *initial height* or the height of the object at time $t = 0$.

Similarly, substituting $t = 0$ into the equation for y', we find that $d = y'(0)$ is the *initial velocity*, or, strictly speaking, the initial vertical velocity.

For the horizontal motion we are assuming that there is no force acting in a horizontal direction, which means that the horizontal acceleration is zero. So $x''(t) = 0$ and $x'(t) = c$, a constant. From this we find that

$$x(t) = ct + x_0, \tag{13.3.2}$$

where $x_0 = x(0)$ is the initial horizontal displacement and c is the initial horizontal velocity.

Combining Equations 13.3.1 and 13.3.2 we have the following theorem.

THEOREM 13.3.3

For an object moving under the influence of gravity near the earth's surface, experiencing no other forces, its path will be parametrized as follows:

$$x(t) = ct + x_0, \qquad y(t) = -16t^2 + dt + y_0,$$

where x_0 and y_0 are the initial horizontal and vertical displacements and c and d are the initial horizontal and vertical velocities, respectively. Measurements are in feet and seconds. If the units are meters and seconds, the equations become

$$x(t) = ct + x_0, \qquad y(t) = -4.9t^2 + dt + y_0.$$

One of the early applications of calculus was using these techniques to help aim artillery, and they can still be used to solve many practical problems.

EXAMPLE 1 A ball is thrown from ground level so that it reaches a maximum height of 64 ft and lands 128 ft from its starting point. At what speed does it hit the ground?

Solution Placing the origin at the starting point, we have $x_0 = y_0 = 0$, and the path is $x(t) = ct$, $y(t) = -16t^2 + dt$. The maximum value of $y(t)$ is 64, and it will occur at a point where $y'(t) = 0$, that is, $-32t + d = 0$ or $t = \frac{d}{32}$. This means that $64 = y\left(\frac{d}{32}\right) = -16\left(\frac{d}{32}\right)^2 + d\frac{d}{32} = -\frac{d^2}{64} + \frac{d^2}{32} = \frac{d^2}{64}$. We find that $d^2 = 64^2$ and the initial vertical velocity is $d = 64$. Incidentally, the maximum height is reached when $t = \frac{d}{32} = \frac{64}{32} = 2$, that is, 2 sec after starting.

To find when the ball hits the ground, we should solve $y(t) = 0$, that is, $-16t^2 + 64t = 0$, that is, $16t(-t + 4) = 0$, so $t = 0$ or $t = 4$.

We discard $t = 0$, the starting point, and find that the ball lands at $t = 4$. We are also told that the ball lands 128 ft from its starting position, which means that $x(4) = 128$. Since $x(t) = ct$, we find that $c = 32$, and the ball's path is parametrized by

$$x(t) = 32t, \qquad y(t) = -16t^2 + 64t, \qquad 0 \le t \le 4.$$

In particular, the velocity vector is

$$\mathbf{r}'(t) = (32, -32t + 64).$$

At the moment of impact, the velocity is $\mathbf{r}'(4) = (32, -64)$, and the speed equals $\sqrt{32^2 + (-64)^2} = 32\sqrt{5} \approx 71.6$ ft/sec.

Notice that the path of the ball in Example 1, parametrized by $x = 32t$, $y = -16t^2 + 64t$, is the parabola $y = -\frac{1}{64}x^2 + 2x = -\frac{1}{64}(x - 64)^2 + 64$ (see Figure 13.3.2). It is symmetrical about the line $x = 64$ (which goes through the high point of the path). Because of this symmetry the speed of impact also equals the initial speed, that is, the speed at the starting point, which is the length of $\mathbf{r}'(0) = (32, 64)$.

The velocity vector is $\mathbf{r}'(t) = (32, -32t + 64)$. It tells us that the horizontal velocity is always 32 ft/sec and the vertical velocity starts at 64 ft/sec at $t = 0$, when the ball is rising fastest, drops to 0 at $t = 2$, when the ball is at its highest point, and falls to -64 ft/sec at $t = 4$.

The velocity vector is $(32, 0)$ at $t = 2$; this means the ball is moving horizontally at the top of its path at a speed of 32 ft/sec. At any other time the speed is $\sqrt{32^2 + (-32t + 64)^2}$ ft/sec, which is greater than 32 ft/sec.

The observation that the path is a parabola will be true for the motion of any projectile.

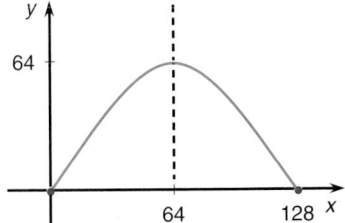

Figure 13.3.2

See the Graphics Calculator Example on the next page.

THEOREM 13.3.4

> If an object moves under the influence of gravity near the earth's surface, experiencing no other forces, then the path it follows will be a parabola or a vertical line segment.

PROOF From Theorem 13.3.3 we know that the path is parametrized by

$$x(t) = ct + x_0, \qquad y(t) = -16t^2 + dt + y_0.$$

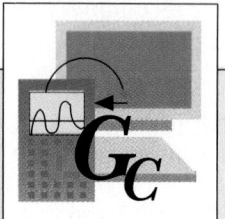

We can use a graphics calculator to sketch the speed of the projectile from Example 1. We plot it as an ordinary function, not using parametric equations.

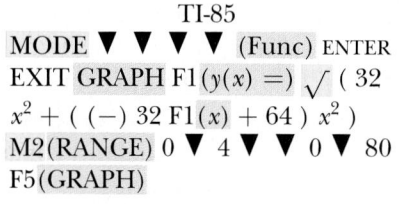

TI-85	SHARP EL9300
MODE ▼ ▼ ▼ ▼ (Func) ENTER EXIT GRAPH F1($y(x) =$) $\sqrt{\ }$ (32 x^2 + ((−) 32 F1(x) + 64) x^2) M2(RANGE) 0 ▼ 4 ▼ ▼ 0 ▼ 80 F5(GRAPH)	⟳ $\sqrt{\ }$ 32 x^2 + (− 32 x/θ/T + 64) x^2 RANGE 0 ENTER 4 ENTER 1 ENTER 0 ENTER 80 ENTER 10 ENTER ⟳
CASIO f_x-7700GB/f_x-6300G	HP 48SX
Cls EXE Range 0 EXE 4 EXE 1 EXE 0 EXE 80 EXE 10 EXE EXE EXE EXE Graph $\sqrt{\ }$ (32 x^2 + (− 32 ◼X◼ + 64) x^2) EXE	↰ PLOT PLOTR ERASE ATTN ' α Y ◼=◼ $\sqrt{x}$ ◼(◼ 32 y^x 2 + ◼(◼ 32 +/− × α X + 64 ▶ y^x 2 ↱ PLOT ↰ DRAW ERASE 0 ENTER 4 XRNG 0 ENTER 80 YRNG DRAW

Notice how the speed has its minimum at the time corresponding to the *highest* point on the trajectory.

If $c = 0$, x is constant and we get a vertical line segment. Otherwise, letting $X = x - x_0 = ct$ and $Y = y - y_0 = -16t^2 + dt$, we see that $t = \frac{X}{c}$, so $Y = -16\left(\frac{X}{c}\right)^2 + d\frac{X}{c} = -\frac{16}{c^2}X^2 + \frac{d}{c}X$, the equation of a parabola, as required.

We can also write it as

$$y - y_0 = -\frac{16}{c^2}(x - x_0)^2 + \frac{d}{c}(x - x_0),$$

or

$$y = -\frac{16}{c^2}x^2 + \left(\frac{d}{c} + 32\frac{x_0}{c^2}\right)x + \left(y_0 - 16\left(\frac{x_0}{c}\right)^2 - \frac{dx_0}{c}\right).$$

EXAMPLE 2   If a ball is thrown at an angle of $\frac{\pi}{6}$ above the horizontal in such a way that it lands 50 ft from its starting point, what is the maximum height it will reach?

Solution   We could do this question by finding the initial velocity, but it is easier to use the fact that the path of the ball is a parabola. Placing the origin at the starting point, we know that the path is a downward-opening parabola, so its equation is of the form $y = -Ax^2 + Bx + C$, and since the origin is on the path, we see that $C = 0$ and the path is $y = -Ax^2 + Bx$, with $A > 0$.

Figure 13.3.3

One of the antique cannons in the Cape Merry Battery in Churchill, Manitoba.

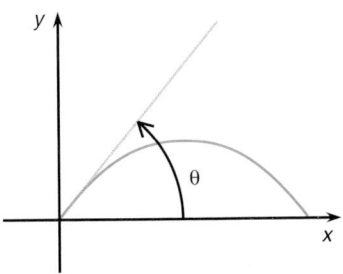

Figure 13.3.4

In Figure 13.3.3 we sketch the parabola. We know that the slope of its tangent at the origin is $\tan\frac{\pi}{6} = \frac{1}{\sqrt{3}}$. But this slope is also the derivative $\frac{dy}{dx}$ at $x = 0$, which is B.

The other thing we know is that the point $(50, 0)$ is on the parabola. Since the equation is $y = -Ax^2 + \frac{1}{\sqrt{3}}x$, we find that $0 = -A(2500) + \frac{1}{\sqrt{3}}(50)$, so $A = \frac{1}{50\sqrt{3}}$. The parabola is $y = -\frac{1}{50\sqrt{3}}x^2 + \frac{1}{\sqrt{3}}x$. It is easy to find that the highest point occurs when $x = 25$ (either by setting the derivative equal to 0 or by looking at the graph and using the symmetry). So the greatest value of y is

$$y = -\frac{1}{50\sqrt{3}}(25)^2 + \frac{1}{\sqrt{3}}(25) = \frac{25}{2\sqrt{3}} \approx 7.2.$$

The maximum height reached by the ball is about 7.2 ft.

EXAMPLE 3   An old cannon shoots a lead ball with a "muzzle velocity" of 200 ft/sec. (i) At what angle of elevation should it be fired in order to reach a distance of 500 ft? (ii) What is its maximum range?

Solution

(i) Notice that the conventional term **muzzle velocity** actually means muzzle *speed*. We sketch the situation in Figure 13.3.4, where the angle of elevation is labeled θ.

As usual, it is convenient to place the origin at the starting point, so that the path of the cannonball is parametrized by $x = ct$, $y = -16t^2 + dt$. In Figure 13.3.5 we draw the initial velocity vector $\mathbf{r}'(0) = (c, d)$ and mark its horizontal and vertical components c and d, respectively. We are told that the initial speed is 200, that is, that $200 = \sqrt{c^2 + d^2}$. Since this is the length of the vector $\mathbf{r}'(0)$, we see from the diagram that $c = 200\cos\theta$ and $d = 200\sin\theta$.

As a result the parametric equations can be written as

$$x = 200\cos(\theta)t, \qquad y = -16t^2 + 200\sin(\theta)t.$$

The cannonball will fall to the ground when $y(t) = 0$, that is, when $0 = -16t^2 + 200\sin(\theta)t = -8t(2t - 25\sin\theta)$, or when $t = \frac{25}{2}\sin\theta$ (we discard the starting point $t = 0$).

At this time the horizontal displacement will be $x\left(\frac{25}{2}\sin\theta\right) = 200\cos(\theta) \times \frac{25}{2}\sin(\theta) = 2500\cos\theta\sin\theta$. We want this to equal 500, that is, for the cannonball to land 500 ft from its starting point. This means that $\cos\theta\sin\theta = \frac{1}{5}$, that is, $\frac{1}{2}\sin(2\theta) = \frac{1}{5}$ or $\sin(2\theta) = \frac{2}{5}$.

Figure 13.3.5

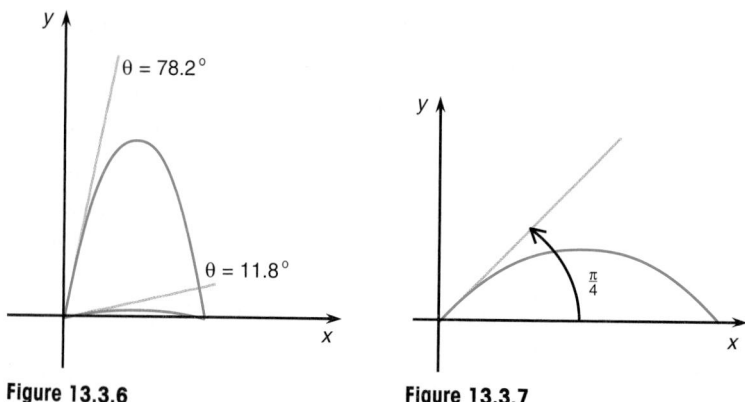

Figure 13.3.6   **Figure 13.3.7**

This means that $\theta = \frac{1}{2}\arcsin\frac{2}{5} \approx 0.206$ radians ($\approx 11.8°$). This is the required **angle of elevation** in order to shoot a distance of 500 ft.

There is another angle that will result in the same range, namely, $\frac{\pi}{2} - \frac{1}{2}\arcsin\frac{2}{5} \approx 1.365$ radians $\approx 78.2°$. See Figure 13.3.6.

(ii) We have already seen that if the cannon is elevated at an angle θ, then its range is $2500\cos\theta\sin\theta = 1250\sin(2\theta)$. Since the maximum value of $\sin 2\theta$ is 1, the maximum range is 1250 ft. Moreover, it occurs when $\sin 2\theta = 1$, which means $\theta = \frac{\pi}{4}$. See Figure 13.3.7.

Exercises 13.3

I

Find each of the following.

1. The speed of impact of a ball thrown so that its maximum height is 16 ft and its horizontal range is 32 ft.

2. The speed of impact of a ball thrown so that its maximum height is 40 ft and its horizontal range is 98 ft.

3. The speed of impact of a ball thrown so that its maximum height is 9 m and its horizontal range is 30 m.

4. The speed of impact of a ball thrown so that its maximum height is 5 m and its horizontal range is 5 m.

5. The horizontal distance covered by a ball thrown from ground level with initial velocity 50 ft/sec so that the maximum height it reaches is 30 ft.

6. The horizontal distance covered by a ball thrown from ground level with initial velocity 30 m/sec so that the maximum height it reaches is 10 m.

7. The initial speed of a ball thrown so that its horizontal range is 80 ft and its maximum height is 24 ft.

8. The maximum height reached by a ball thrown at 25 m/sec that lands after 4 sec.

9. The length of time in the air for a ball thrown at 64 ft/sec that lands 120 ft from its starting point.

10. The maximum height reached by a ball thrown with initial speed 32 ft/sec if it lands after 1 sec.

11. The angle of elevation for a cannonball to have a range of 800 ft if the muzzle velocity is 250 ft/sec.

12. The angle of elevation for a cannonball to have a range of 200 m if the muzzle velocity is 80 m/sec.

13. The range of a cannon with muzzle velocity 150 ft/sec with angle of elevation $\frac{\pi}{3}$.

14. The range of a cannon with muzzle velocity 150 ft/sec with angle of elevation $\frac{\pi}{6}$.

15. The maximum time the cannonball could be in the air if the cannon has muzzle velocity 100 ft/sec.

16. The muzzle velocity of a cannon whose maximum range is 400 m.

17. The maximum range of a cannon with a muzzle velocity of 100 m/sec.

18. The speed with which a ball is thrown if it reaches a maximum height of 40 ft when thrown at a 45° angle above the horizontal.

19. The speed with which a ball must be thrown at an angle of $\frac{\pi}{6}$ above the horizontal in order to land 75 ft from the thrower.

20. The two possible angles of elevation for a cannon with muzzle velocity 250 ft/sec if the range is to be 600 ft.

II

*21. What is the maximum height reached by the cannonball in Example 3 (p. 819) when it is fired so as to land 500 ft from the cannon?

*22. What is the maximum height reached by the cannonball in Example 3 when it is fired so as to have the maximum possible range?

*23. At what elevation should the cannon of Example 3 be fired in order that the cannonball will reach the greatest possible height before falling to the ground? What is the maximum height it can reach?

*24. If a cannon has a muzzle velocity V ft/sec, what is its greatest possible range, and what elevation should be used to achieve this range?

*25. A manufacturer of toy cannons has discovered improvements that double the muzzle velocity. What will this do to the maximum range of the cannons?

*26. A small boy has made a catapult that launches pebbles at a 45° angle above the horizontal; they land 60 feet away.

(i) How fast does his catapult launch the pebbles?

(ii) How fast should he launch a pebble if he wants it to go 80 ft?

*27. A small girl is sitting 8 ft above ground level on a garage roof. She throws a ball at a 45° angle above the horizontal at a speed of 32 ft/sec. How much farther will it go than one thrown by her brother from ground level at the same speed and angle?

*28. Suppose the sister and brother in Exercise 27 agree to throw balls at the same speed (32 ft/sec) but at an inclination of 30° above the horizontal. Before working it out, try to predict whether the difference between their two throws will be greater than, less than, or equal to what it was in Exercise 27 with inclinations of 45°. How much farther will the sister's ball go than her brother's?

III

*29. Use a graphics calculator or computer to plot the trajectories of the cannonball of Example 3 corresponding to angles of elevation of $\theta = \frac{\pi}{10}, \frac{\pi}{8}, \frac{\pi}{6}, \frac{\pi}{5}, \frac{\pi}{4}, \frac{3\pi}{10}, \frac{\pi}{3}, \frac{3\pi}{8}, \frac{2\pi}{5}$. You should be able to do this question even if your equipment cannot plot parametric curves.

POINT TO PONDER

In our study of projectile motion we have assumed that the surface of the earth is perfectly flat. This is a reasonable assumption and quite accurate for motions over fairly short distances.

However, when an object is thrown near the earth's surface, what it actually does is go into an "orbit" around the center of the earth. As with planets and comets, this orbit is an ellipse with the center of the earth as one focus.

Of course, what happens is that the projectile hits the earth before it can follow much of the path, and all we ever see is a tiny piece at the end of a very elongated ellipse (see Figure 13.3.8).

The eccentricity of such a long ellipse is very close to 1, the eccentricity of a parabola. The little piece at the end of the ellipse is almost indistinguishable from a little piece of a parabola, and this is why for practical purposes we can work with parabolas.

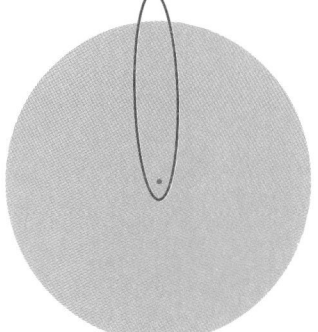

Figure 13.3.8

In effect, assuming that the earth is flat amounts to assuming that the center of the earth is "infinitely far away" (the farther away the center, the flatter the circle...). By the same token, if we take the long elliptical orbit and move the far focus "out to infinity," what happens is that the eccentricity approaches 1 and the ellipse approaches the shape of a parabola.

For motions involving reasonably small distances, the calculations with the parabola will give answers that are extremely close to the more correct ones obtained from the ellipse, so much so that there is no practical advantage in using the ellipse. And there is a practical advantage to using the parabola, that is, to assuming that a small piece of the earth's surface is flat: The calculations tend to be much easier.

Chapter Summary

§13.1
parametric equations
parameter
curve
domain
image
parametrization of lines, circles, ellipses, hyperbolas,
     graphs of functions

§13.2
vector
component
horizontal component
vertical component

displacement
change in the displacement
velocity vector
tangent vector
speed
arclength

§13.3
projectile
horizontal and vertical components
parametric equations for projectile motion
muzzle "velocity"
angle of elevation

Review Exercises

Parametrize each of the following curves.

1. $2x - 7y = 4$

2. $x = 3$

3. $y + 3 = 0$

4. $y = 2x^2 - 4x - 5$

5. $x + 3y^2 - 5y = 8$

6. $x = x^3 - 3y$

7. $2x^2 + 16y^2 = 8$

8. $x^2 + y^2 = 0$

9. $xy = 1$

10. $2xy^2 + 3xy = 12$

11. $x^2 - 2y^2 = 1$, right branch

12. $x^2 - y^2 + 1 = 0$, top branch

13. $5x^2 + 12y^2 = 30$, right half

14. $2x - 5y = 2$, $y \geq 0$

15. $4y^2 - 9x^2 = 36$, bottom branch

16. $y = \ln x$

17. $y = |x|$

18. $x^2 = y^2$, top half

19. $x = \sqrt{y}$

20. $e^x - 2 = \sin y$

Find an equation for each of the following curves, and identify the curve.

21. $x = 3t^2$, $y = t - 4$

22. $x = 3t^3$, $y = 4 - 9t^3$

23. $x = \cos t$, $y = 3 \sin t$

24. $x = \cos(5t)$, $y = \sin(5t)$

25. $x = \cos^2 t - \sin^2 t$, $y = \cos t \sin t$

26. $x = \cos t$, $y = \sin\left(\frac{\pi}{2} - t\right)$

27. $x = -3 \sinh t$, $y = 2 \cosh t$

28. $x = \cos(t^2)$, $y = 4 \sin(t^2)$

29. $x = \frac{1}{1+t^2}$, $y = \frac{t^2}{1+t^2}$

30. $x = 1 + \cos t$, $y = 2 - \sin t$

Find the velocity vector for each of the following parametrizations.

31. $x = 5t + 1$, $y = 2 - t$ **32.** $x = 3t + 2$, $y = 3t - 2$

33. $x = e^{2t}, y = 2t$

34. $x = \sin t, y = \cos(2t)$

35. $x = \sin t, y = \sin(3t)$

36. $x = t + 1, y = \ln t$

37. $x = \sin(t^2 + 2), y = 3 - \cos^2 t$

38. $x = \tan t, y = \sec t$

39. $x = \arctan(3t), y = \arcsin t$

40. $x = 2t + 3, y = x^2 - 1$

Find the length of the path traced out by each of the following moving points for the specified range of the parameter.

41. $\mathbf{r}(t) = (t^3, t^2), 0 \le t \le 1$

42. $\mathbf{r}(t) = (1 + t, 3 - 4t), 2 \le t \le 4$

43. $\mathbf{r}(t) = \left((1 - 2t)^{3/2}, \frac{3}{2}t^2\right), 0 \le t \le \frac{1}{2}$

44. $\mathbf{r}(t) = \left(2t - \frac{t^3}{3}, \sqrt{2t^2}\right), 0 \le t \le 1$

45. $\mathbf{r}(t) = \left(\cos(t + \pi), \sin(t + \pi)\right), 0 \le t \le \pi$

46. $\mathbf{r}(t) = \left(\cos(t + \frac{\pi}{2}), \cos t\right), 0 \le t \le \frac{\pi}{4}$

47. $\mathbf{r}(t) = (2e^{3t/2}, 3e^t), 0 \le t \le 2$

48. $\mathbf{r}(t) = \left(\frac{1}{2}\ln(1 + t^2), \arctan t\right), 0 \le t \le 1$

49. $\mathbf{r}(t) = (t^2, 2t^{5/2}), 0 \le t \le 1$

50. $\mathbf{r}(t) = \left(-4, \sin^2(-4t) + \cos^2(4t)\right), 0 \le t \le 2\pi$

Find each of the following.

51. All points where $x = 3t^3, y = \cos(\pi t)$ is moving horizontally.

52. The points where the tangent to $x = t^3 - 3t + 1, y = e^t$ is vertical.

53. The point where the moving point $x = 5t - 1, y = 2t^2 + 4t - 7$ has minimum speed.

54. The points at which $x = 9t^2, y = 2t^3 + 2$ has minimum speed.

55. The maximum height reached by a ball thrown at a speed of 48 ft/sec at an angle $30°$ above the horizontal.

56. The speed with which a ball must be thrown at an angle of $45°$ above the horizontal in order to reach a maximum height of 10 meters.

57. The length of time a ball is in the air if it reaches a maximum height of 16 ft and goes 84 ft horizontally.

58. The angle of elevation for a projectile that reaches a height of 20 m and covers a horizontal distance of 60 m.

59. The speed of impact of a projectile that was launched from ground level and flew 300 ft in 3 sec.

60. The greatest possible speed of impact for a projectile launched from ground level at a speed of 43 m/sec.

II

*61. Parametrize the circle $x^2 + 2x + y^2 - 4y + 4 = 0$.

*62. Parametrize the ellipse $4x^2 + 8x + 9y^2 = 32$.

*63. Parametrize the upper branch of the hyperbola $y^2 - 4y - 9x^2 - 18x = 14$.

*64. Parametrize the part of the line $2x - y = 0$ that lies above the parabola $y = x^2$.

*65. Parametrize the part of the ellipse $4x^2 + 16y^2 = 16$ that lies below the line $y = x$.

*66. Parametrize the part of the hyperbola $x^2 - y^2 = 1$ that lies between the y-axis and the line $x = 2$.

*67. (i) Consider the curve $\mathbf{r}(t) = \left(x(t), y(t)\right)$, and suppose that $x(t)$ and $y(t)$ are differentiable at $t = a$. Assuming that $x'(a) \ne 0$, show that the slope of the tangent to the curve at the point $\left(x(a), y(a)\right)$ is

$$\frac{y'(a)}{x'(a)}.$$

(*Hint:* Consider the slope of the tangent vector.)

(ii) What happens at a point where $x'(a) = 0$?

*68. Find the slope of the curve $\mathbf{r}(t) = \left(t \cos(t^2), e^{-3t}\right)$ at the point corresponding to $t = 0$.

*69. Two friends each kick a ball from ground level. Both balls reach a maximum height of 12 ft, but one travels 30 ft horizontally and the other travels 72 ft horizontally. Compare the lengths of times the two balls are in the air.

*70. If a ball is thrown horizontally at a speed of 20 ft/sec starting at a point 6 ft above ground level, how far will it travel in the horizontal direction before landing?

III

Use Simpson's Rule S_n with $n = 12$ to approximate the arclengths of the following curves.

71. $\mathbf{r}(t) = \left(\frac{1}{t}, \frac{1}{t^2}\right), t \in [1, 2]$

72. $\mathbf{r}(t) = (t^2, \sin \pi t), t \in [-1, 1]$

73. The top half of the ellipse $2x^2 + y^2 = 1$.

74. The part of the top branch of the hyperbola $4y^2 - x^2 = 4$ between $x = -2$ and $x = 2$.

Vector Geometry

14

These shapes are "regular polyhedra". The faces of each are all the same regular polygon, and the same number of faces meet at each vertex. The "tetrahedron" (pyramid) has four faces which are equilateral triangles. The octahedron has eight faces which are equilateral triangles. The cube has six faces which are squares. The icosahedron has twenty faces which are equilateral triangles. The dodecahedron has twelve faces which are regular pentagons.

These are called the "Platonic solids", and in fact they are the only regular solids (you might like to convince yourself that it is impossible to make one with hexagonal faces....).

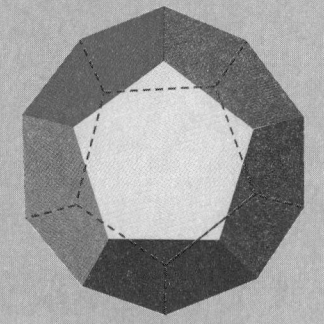

INTRODUCTION

In Chapter 13, in the course of discussing parametric equations for a point moving along a curve, we found that the idea of velocity was most easily described in terms of *vectors*. We were able to discuss velocity vectors and relate them to our everyday intuition about motion.

In this chapter we discuss vectors in a more systematic way. Partly, this will help us understand better some of the steps we have already taken, but it will also lead us in new directions. One of our main purposes will be to lay the groundwork for Chapter 15, in which we will generalize our ideas about motion in a plane to motion in three-dimensional space; to do this, we need to use three-dimensional vectors. Fortunately, they are quite similar to the two-dimensional vectors we have already seen and are not particularly difficult to work with.

The first three sections concern vectors and some of their uses. In Section 14.1 we introduce coordinates in three-dimensional space; in Section 14.2 we discuss vectors, including the useful ideas of dot product and cross product; and in Section 14.3 we discuss various applications of vectors. In Sections 14.4 and 14.5 we use them to discuss lines and planes in space.

14.1 Coordinates in Three-Dimensional Space

RIGHT-HANDED COORDINATES
GRAPHS
DISTANCE

Figure 14.1.1

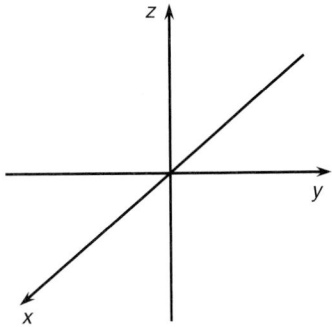

Figure 14.1.2

In the plane the location of a point is specified by two coordinates x and y, which we write as an ordered pair (x, y). To specify the position of a point in space, it is necessary to use three coordinates. This is easy to see if we think in terms of using two coordinates to specify the horizontal position and the third to give the height (see Figure 14.1.1).

We can visualize the ordinary xy-plane lying horizontally in space. A point P has a certain "height" z relative to this plane; if z is positive, it is simply the height of P above the plane; if $z < 0$, then $|z|$ is interpreted as a distance below the plane; and if $z = 0$, the point P lies on the horizontal plane. To locate P, we give its height z and the coordinates (x, y) of the point Q that lies in the horizontal plane directly above or below P. The resulting "ordered triple" (x, y, z) determines P, and we call (x, y, z) the **coordinates** of P.

Three-dimensional space, the set of all such points, is called $\mathbb{R}^3$ (read "$\mathbb{R}$-three"), suggesting that each point corresponds to three real numbers.

Just as for sketching in the plane, it is convenient to draw coordinate axes. We will usually draw them as shown in Figure 14.1.2, with the z-axis vertical, the x-axis coming out from the page pointing slightly down and to the left, and the y-axis horizontal in the plane of the page. The picture is never entirely satisfactory because a two-dimensional diagram cannot do justice to a three-dimensional subject.

Occasionally, it will be necessary to draw the axes in some other position, just as we sometimes translated and rotated axes in the plane to make a graph easier to interpret. Whenever we draw the three axes, they should be perpendicular to one another, but there is one other consideration, sometimes called the "orientation" of the axes. The two sets of axes shown in Figure 14.1.3 can never be rotated into each other because if we move them so that the x- and z-axes are the same, then the two y-axes will point in opposite directions (i.e., the positive direction on one will be the negative direction on the other). There are two *different* types of coordinates, the ones that can be rotated into the first set of axes and the ones that can be rotated into the second set of axes.

To avoid confusion, we will always use the first type of coordinates. If you place your right hand with the thumb along the z-axis and the fingers curling in the

Figure 14.1.3

This image of a guitar was produced by DesignCAD 3D software. It is based on information stored in the computer which represents the coordinates of all the various points on the guitar.

Figure 14.1.4

Figure 14.1.5

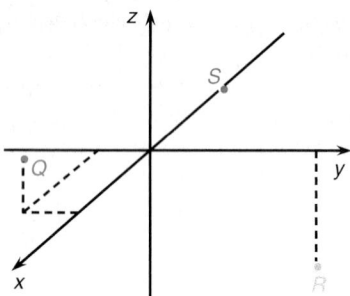

Figure 14.1.6

direction of rotation from the positive x-axis to the positive y-axis, then the thumb will point along the *positive z*-axis. For this reason this is called a **right-handed coordinate system** (see Figure 14.1.4). Another way to say it is that if you place the thumb of your right hand along the positive x-axis and the index finger along the positive y-axis, then the middle finger will point along the positive z-axis. The other kind of coordinates are called left-handed coordinates.

Once we have coordinate axes, it is possible to plot points in $\mathbb{R}^3$. For instance, the point $P = (1, 2, 3)$ can be found by starting at the origin, moving 1 unit in the positive x-direction, then moving 2 units parallel to the direction of the positive y-axis, and then moving 3 units straight up (which is the direction of the positive z-axis). The point P is shown in Figure 14.1.5. It is necessary to show those little line segments on the drawing lying parallel to the axes so that our eyes will see the position of the point.

EXAMPLE 1   Plot the points $Q = (2, -1, 1)$, $R = (0, 3, -2)$, and $S = (-2, 0, 0)$.

Solution   To plot $Q = (2, -1, 1)$, we must move 2 units from the origin along the positive x-axis, -1 unit along the direction of the y-axis, which means 1 unit in the direction of the negative y-axis, and 1 unit up. The point Q is shown in Figure 14.1.6.

To plot $R = (0, 3, -2)$, we move 0 units in the x-direction, that is, not at all, 3 units in the y-direction, and -2 units in the z-direction, that is, 2 units down. The point R is also shown in Figure 14.1.6.

Finally, to plot $S = (-2, 0, 0)$, we move -2 units in the x-direction and not at all in the other directions. This means that the point S is on the x-axis, as shown in Figure 14.1.6.

In doing Example 1 we have observed that points of the form $(x, 0, 0)$, that is, with y and z equal to 0, are on the x-axis. Similarly, points of the form $(0, y, 0)$ are on the y-axis, and points of the form $(0, 0, z)$ are on the z-axis.

A point of the form $(0, y, z)$ lies in the plane that contains the y- and z-axes. This plane is called the yz-plane, and it is illustrated in Figure 14.1.7(i). Similarly, the plane containing the x- and y-axes is the horizontal plane illustrated in Figure 14.1.7(ii); it is called the xy-plane, and it consists of all points that are of the form $(x, y, 0)$. The xz-plane consists of all points $(x, 0, z)$; it is shown in Figure 14.1.7(iii).

Graphs in Three Dimensions

In Section 15.4 we will discuss how to draw the graph of an equation like $x^2 + y^2 = z^2$. When we speak of drawing the graph of an equation, what we mean is drawing the

Figure 14.1.7(i)

Figure 14.1.7(ii)

Figure 14.1.7(iii)

Figure 14.1.8

Figure 14.1.9(i)

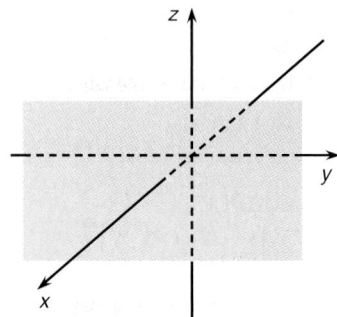

Figure 14.1.9(ii)

set of all points (x, y, z) that satisfy the equation. In Section 15.4 we will learn that the points that satisfy $x^2 + y^2 = z^2$ form a double-ended cone as shown in Figure 14.1.8. However, for now we will draw only a few simple graphs.

EXAMPLE 2   Sketch the graphs of the following equations: (i) $z = 3$, (ii) $x = 2$, (iii) $x = y$, (iv) $y + z = 1$.

Solution

(i) The points (x, y, z) that satisfy $z = 3$ are, of course, all the points of the form $(x, y, 3)$. These are just the points that lie 3 units above the horizontal xy-plane, as illustrated in Figure 14.1.9(i).

(ii) The points satisfying $x = 2$ are just the points that lie 2 units "in front of" the yz-plane. They form the plane that is sketched in Figure 14.1.9(ii).

(iii) The points satisfying $x = y$ can all be written in the form (x, x, z). If we consider the points $(x, x, 0)$, then they are the points in the xy-plane that satisfy $x = y$. From our familiarity with graphs in the plane we know that this is a straight line. It is drawn in the horizontal xy-plane in Figure 14.1.9(iii), where it is the dashed line. (We have to be careful because the x- and y-axes are in unfamiliar positions.)

Now if we consider any point of the form (x, x, z), it lies directly above or below the point $(x, x, 0)$ on that dashed line. This means that the set of all such points consists of all the points lying directly above or below the dashed line (or on it). The graph of $x = y$ is the plane shown in Figure 14.1.9(iii).

(iv) To sketch the graph of $y + z = 1$, we observe that it does not involve the variable x. Taking our cue from part (iii), we look first at points of the form $(0, y, z)$ that satisfy $y + z = 1$. They form a line in the yz-plane, and it is easily drawn. It is sketched as the dashed line in Figure 14.1.9(iv).

Now the set of *all* points (x, y, z) that satisfy $y + z = 1$ consists of all points that can be found by starting with a point $(0, y, z)$ on that dashed line and moving in the x-direction. This set is the plane shown in Figure 14.1.9(iv).

Figure 14.1.9(iii)

Figure 14.1.9(iv)

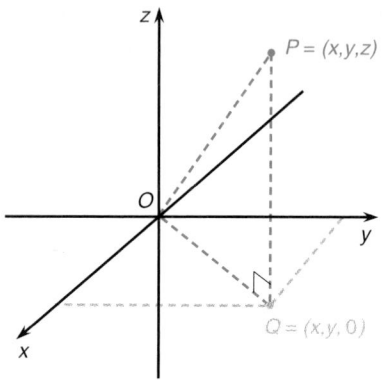

Figure 14.1.10

It comes as no surprise that the formula for distances in $\mathbb{R}^3$ is similar to the corresponding formula in the plane.

We begin by finding the distance from the origin to the point $P = (x, y, z)$. In Figure 14.1.10 we sketch P and the point $Q = (x, y, 0)$ that lies directly below P on the xy-plane. First, working entirely inside the xy-plane, we know that the distance from the origin to $Q = (x, y, 0)$ is $\sqrt{x^2 + y^2}$. This is the length of the dashed segment OQ.

Next we consider the right-angled triangle OPQ. Observing that the length $|PQ|$ is $|z|$, we can apply the Pythagorean Theorem (Theorem 1.5.1) and find that

$$|OP| = \sqrt{|OQ|^2 + |PQ|^2} = \sqrt{x^2 + y^2 + z^2}.$$

For the distance between two points the formula is analogous.

THEOREM 14.1.1

The **distance** between the points (x, y, z) and (a, b, c) is

$$\sqrt{(x - a)^2 + (y - b)^2 + (z - c)^2}.$$

It is not practical to draw three-dimensional pictures with a programmable calculator, but it is possible to do it with a computer. The usual approach is to express z as a function of x and y.

For instance, in Example 4 we will consider the sphere $x^2 + y^2 + z^2 = 4$. We rearrange this to $z = \pm\sqrt{4 - x^2 - y^2}$. We can graph the upper hemisphere $z = \sqrt{4 - x^2 - y^2}$ as follows:

MATHEMATICA	MAPLE
Plot3D [Sqrt[$4 - x\,\hat{}\,2 - y\,\hat{}\,2$], $\{x, -2, 2\}, \{y, -2, 2\}$] ENTER	plot3d(Sqrt($4 - x\,\hat{}\,2 - y\,\hat{}\,2$), $x = -2..2, y = -2..2$); ENTER
DERIVE	MacFUNCTION
O D G ENTER A sqrt ($4 - x\,\hat{}\,2 - y\,\hat{}\,2$) ENTER P O L -2 Tab 2.0 ENTER G 15 Tab 15 ENTER P	sqrt ($4 - x\,\hat{}\,2 - y\,\hat{}\,2$) OK Function Change Domain (*drag to* 2 × 2) OK Options Remove Hidden Surfaces

The computer may have some difficulty near the edges of the graph. The exact keystrokes required may vary slightly on some machines.

Now you can try plotting the lower hemisphere or perhaps the function $z = x^3 + y^2$.

Figure 14.1.11

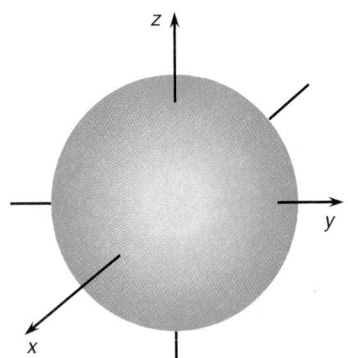

Figure 14.1.12

EXAMPLE 3   (i) Find the distance between the points $(3, -2, 5)$ and $(2, 2, 3)$.

Solution   Applying the formula, we find that the distance is

$$\sqrt{(3-2)^2 + (-2-2)^2 + (5-3)^2} = \sqrt{1 + 16 + 4} = \sqrt{21}.$$

(ii) Find the distance between the points $(5, 3, 0)$ and $(2, 7, 0)$.

Solution   The distance is $\sqrt{(5-2)^2 + (3-7)^2} = \sqrt{3^2 + (-4)^2} = 5$.   ◢

Notice that both points in part (ii) of Example 3 lie in the xy-plane and that the distance between them is just the ordinary distance between two points in the plane. To make sense of this, we have to think of the point $(x, y, 0)$ as the point (x, y) in the plane; it is often very useful to regard the plane consisting of all points (x, y) as the same thing as the xy-plane consisting of all points in $\mathbb{R}^3$ having the form $(x, y, 0)$. See Figure 14.1.11.

EXAMPLE 4   Graph the set of all points (x, y, z) that satisfy $x^2 + y^2 + z^2 = 4$.

Solution   Since $x^2 + y^2 + z^2$ is the square of the distance from (x, y, z) to the origin, the equation just says that (x, y, z) is 2 units from the origin. The set of all such points is the sphere of radius 2 centered at the origin. See Figure 14.1.12.   ◢

Exercises 14.1

I

Sketch each of the following points.

1. $P = (2, 2, 4)$
2. $Q = (2, -1, 1)$
3. $R = (0, 0, 3)$
4. $S = (3, 3, 3)$
5. $T = (1, 1, -1)$
6. $U = (2, 0, -2)$
7. $V = (0, -4, 0)$
8. $W = (-1, 2, 1)$
9. $A = (0, 3, -4)$
10. $B = (\sqrt{2}, \pi, -1)$

Sketch the graph of each of the following equations.

11. $x = -3$
12. $z + 5 = 0$
13. $x = y + 1$
14. $y = z$
15. $3x + 2y = 6$
16. $x - 2z = 0$
17. $x^2 + y^2 + z^2 = 1$
18. $x^2 + y^2 + z^2 = 9$
19. $z^2 = 4$
20. $(x - 1)(z - 2) = 0$

II

Find the (smallest) distance from the point P to the specified plane.

21. $P = (2, 3, 4)$, plane $z = -1$

22. $P = (-1, 2, -1)$, plane $x = 3$
23. $P = (0, 0, 5)$, plane $y = x$
*24. $P = (0, 4, -3)$, plane $y = x$

Sketch the graph of all points (x, y, z) that satisfy the following equations.

*25. Both $x - y = 1$ and $z = 2$
*26. Both $x = y$ and $y = z$
*27. $x^2 + y^2 = 1$
*28. $y^2 + z^2 = 4$
*29. $y = x^2$
*30. If somebody draws x'- and y'-axes in the plane, meeting (at right angles) at some point O', can they always be translated and/or rotated into the usual x- and y-axes so that the positive axes coincide?

III

31. Use a computer to sketch (i) $z = f(x, y) = x^2 + y^2$, (ii) $z = f(x, y) = 4 - x^2 + 2x - 2y^2$, (iii) $z = f(x, y) = x^2 - y^2$. In which of these pictures does z appear to have a maximum or minimum value?

Early in this section it was stated that there are two kinds of coordinate systems, right-handed and left-handed. It certainly seems reasonable that it is impossible to change a left-handed system into a right-handed system by translating and/or rotating the axes, but in effect the statement also says that *any* coordinate system can be changed into one or other of the standard systems illustrated in Figure 14.1.13.

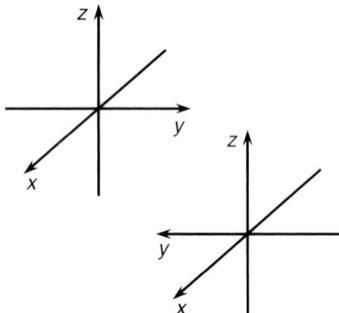

Figure 14.1.13

This is less obvious. Suppose we start with any set of axes. If the origin is not the usual one, we begin by translating them until it is. Then we can rotate the axes until the positive z-axis points up, that is, until it is the same as the ordinary z-axis.

This will mean that the x- and y-axes both lie in the horizontal plane, as shown in Figure 14.1.14. By rotating around the z-axis it is possible to move until the positive x-axis coincides with the ordinary positive

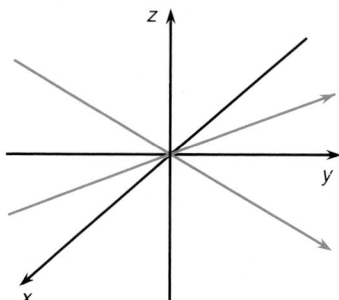

Figure 14.1.14

x-axis in the usual right-handed coordinate system. The z-axis remains unchanged, since we rotated around it. The result is that the positive x- and z-axes coincide with their ordinary right-handed counterparts.

Having done all this, we ask what has become of the y-axis. Since it must be perpendicular to the z-axis, it must lie in the horizontal plane. Since it is also perpendicular to the x-axis, it must lie in the ordinary vertical yz-plane. The only points that lie in both these planes are the points on the ordinary y-axis.

So the y-axis in the new coordinate system has been rotated until it lies along the ordinary y-axis. There are only two possibilities: Either it points in the same direction or it points in the opposite direction. If it is the same, then the system we began with is a right-handed coordinate system; if it is different, then the system must be left-handed.

14.2 Vectors, Dot Product, and Cross Product

COMPONENTS
NORM
TRIANGLE INEQUALITY
STANDARD BASIS VECTORS
ANGLE
ORTHOGONAL
LAGRANGE'S INDENTITY

The basic idea of vectors in three-dimensional space $\mathbb{R}^3$ is the same as in the two-dimensional case. They can be visualized as arrows, so they have both length and direction. They will be used to represent such things as the velocity of a moving point, in which case the direction and length of the velocity vector will be the direction of motion and the speed, respectively.

In Chapter 13 when we used vectors in the plane to discuss velocity, we found that it was necessary to subtract them and multiply by numbers. We also discovered that it was very useful to write a vector in the form $\mathbf{v} = (a, b)$. This meant that $\mathbf{v}$ is

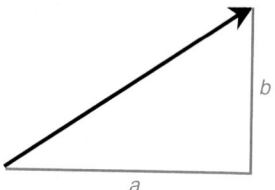

Figure 14.2.1

an arrow whose horizontal displacement is a and whose vertical displacement is b (see Figure 14.2.1).

Of course, the same idea can be applied to vectors in $\mathbb{R}^3$. A vector can be represented as $\mathbf{v} = (a, b, c)$, where the *components a, b,* and *c* are real numbers that represent displacements along the x-, y-, and z-axes, respectively.

We subtracted two vectors in the plane by subtracting their horizontal and vertical components separately: $(a, b) - (u, v) = (a - u, b - v)$. We multiplied a vector by a number $\frac{1}{h}$ by multiplying each component by $\frac{1}{h}$. Both these operations can easily be extended to three-dimensional vectors.

Our original idea of a vector was that it should be an arrow. This is easy to visualize and consistent with the concept of velocity that inspired it. However, it is actually fairly vague. One cause of the vagueness is that it is not very clear where the arrow should start, that is, where its initial point is. In studying velocity we usually choose to place the vector at the moving point, but this is not the only possibility.

Having gained some experience working with vectors in the plane, we have a better idea of how they will be used, and it seems that writing them out in terms of coordinates is very useful. Because of this we will actually define vectors in this way and then observe that they can be represented by arrows.

DEFINITION 14.2.1

(i) A **vector** in the plane is an ordered pair $\mathbf{v} = (a, b)$ of real numbers. A vector in three-dimensional space $\mathbb{R}^3$ is an ordered triple $\mathbf{v} = (a, b, c)$. The entries a, b (or a, b, c) in the vector are called its **components**.

(ii) To add two vectors in the plane or two vectors in space, we simply add the corresponding components:

$$(a, b) + (x, y) = (a + x, b + y),$$
$$(a, b, c) + (x, y, z) = (a + x, b + y, c + z).$$

(iii) To multiply a vector by a real number k, we multiply each component by k:

$$k(a, b) = (ka, kb), \qquad k(a, b, c) = (ka, kb, kc).$$

(iv) Vectors can be represented by arrows; the vector $\mathbf{v} = (a, b)$ is the arrow whose *initial point* or *basepoint* is at the origin and whose *endpoint* or *terminal point* is at the point whose coordinates are (a, b). Similarly, the vector $\mathbf{u} = (a, b, c)$ has initial point at the origin and terminal point at the point (a, b, c).

Sometimes we move the arrow (without rotating) so that its basepoint is somewhere other than the origin, but we still regard it as the *same* vector.

(v) The vector $(0, 0)$ in the plane is called the **zero vector**; the vector $(0, 0, 0)$ in $\mathbb{R}^3$ is called the **zero vector**. We sometimes write 0 to represent the zero vector and rely on the context to distinguish which one is intended.

(vi) If $\mathbf{v}$ is a vector, we write $-\mathbf{v}$ for its **negative**. This is the vector that has the same length and points in the opposite direction. If $\mathbf{v} = (a, b)$, then $-\mathbf{v} = (-a, -b)$; if $\mathbf{u} = (a, b, c)$, then $-\mathbf{u} = (-a, -b, -c)$.

(vii) The length of a vector $\mathbf{v}$ is called its **norm**, and we write it $\|\mathbf{v}\|$. This notation is meant to remind us of the absolute value sign. The norm of a vector is easily found by the Pythagorean Theorem. In fact,

$$\|(a, b)\| = \sqrt{a^2 + b^2}, \qquad \|(a, b, c)\| = \sqrt{a^2 + b^2 + c^2}.$$

Figure 14.2.2(i)

Figure 14.2.2(ii)

Figure 14.2.3

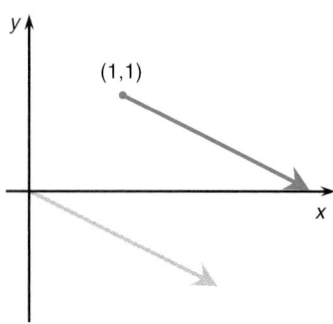

Figure 14.2.4

Let us remark that the rule given in part (ii) of Definition 14.2.1 for the addition of vectors can be given the following graphical interpretation. Place the vector **v** so that its initial point is at the terminal point of **u**. The sum **u** + **v** is the vector whose basepoint is the basepoint of **u** and whose terminal point is the terminal point of **v**. In terms of the picture (see Figure 14.2.2(i)), going along **u** + **v** amounts to going along **u** and then going along **v**. This observation is sometimes called the **Triangle Rule**, for reasons that are clear from the picture. In Figure 14.2.2(ii) we see that **u** + **v** is the diagonal of the parallelogram determined by **u** and **v**. This interpretation, which is essentially the same, is sometimes called the **Parallelogram Rule**.

EXAMPLE 1   If **u** = $(3, -2, 7)$ and **v** = $(2, 5, -4)$, then **u** + **v** = $(5, 3, 3)$. Also $3\mathbf{u} = (9, -6, 21)$ and $-5\mathbf{v} = (-10, -25, 20)$. ◢

EXAMPLE 2   The vector **w** = $(2, -1, 3)$ can be seen as an arrow starting at the origin and ending at the point $(2, -1, 3)$. See Figure 14.2.3. ◢

EXAMPLE 3   If we place the vector **u** = $(2, -1)$ with its initial point at the origin, then its terminal point will be $(2, -1)$. On the other hand, if we place the initial point at $(1, 1)$, then the terminal point will be at $(1, 1) + (2, -1) = (3, 0)$. This is illustrated in Figure 14.2.4, where the vector with basepoint at the origin is lighter and the vector with basepoint at $(1, 1)$ is shaded darker. Notice that the two vectors are parallel and have the same length.

The dark-shaded vector starts at $(1, 1)$ and then goes 2 units horizontally and -1 unit vertically. This is why the endpoint is $(1 + 2, 1 - 1)$. We find it by adding components. ◢

EXAMPLE 4   The vector $(0, 2, 1)$ is shown in Figure 14.2.5 with its basepoint at the origin. The darker vector in that figure is the same vector, but its basepoint is at $(2, 1, 0)$. As in Example 4 we can find the endpoint of the darker vector just by adding: It is $(2, 1, 0) + (0, 2, 1) = (2, 3, 1)$. ◢

EXAMPLE 5   If we place **u** and **v** with their initial points at the origin, as shown in Figure 14.2.6, then consider the vector going from the endpoint of **v** to the endpoint of **u**. In the diagram this is the vector labeled **w**.

Figure 14.2.5

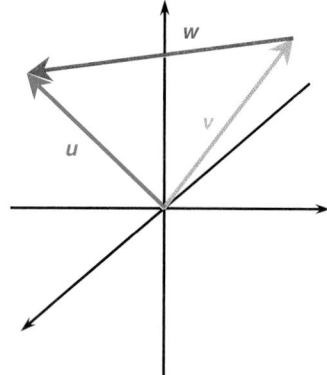

Figure 14.2.6

Using the Triangle Rule, we see that $\mathbf{v} + \mathbf{w} = \mathbf{u}$, and rearranging, we have that $\mathbf{w} = \mathbf{u} - \mathbf{v}$. In effect this is another Triangle Rule, this time for subtraction. ◢

In doing these examples we have learned various properties of vectors. There are a number of other properties that are similar to the corresponding properties for addition and multiplication of real numbers. Most of them are more or less obvious, but we record them as a proposition just for reference.

PROPOSITION 14.2.2

Notice that in part (v) of Proposition 14.2.2 the first 0 is the real number 0 and the other two are vectors. It is important to get used to figuring out from the context what is meant by the symbol 0.

Part (vi) is called the Triangle Inequality because it says that one side of a triangle is shorter than the sum of the lengths of the other two sides (cf. Figure 14.2.2(i)).

Properties of Vectors

The following statements are true for any vectors $\mathbf{u}$ and $\mathbf{v}$ and any real numbers k and ℓ, provided that the vectors $\mathbf{u}$ and $\mathbf{v}$ are both of the same type, that is, either they are both vectors in the plane or else they are both in three-dimensional space $\mathbb{R}^3$.

(i) $\mathbf{u} + \mathbf{v} = \mathbf{v} + \mathbf{u}$ (i.e., vector addition is *commutative*).

(ii) $\mathbf{u} + 0 = \mathbf{u}$.

(iii) $k(\ell \mathbf{u}) = (k\ell)\mathbf{u}$.

(iv) $k(\mathbf{u} + \mathbf{v}) = (k\mathbf{u}) + (k\mathbf{v})$.

(v) $0\mathbf{v} = 0$; $1\mathbf{v} = \mathbf{v}$; $(-1)\mathbf{v} = -\mathbf{v}$; $\mathbf{v} + (-\mathbf{v}) = 0$.

(vi) The "Triangle Inequality": $\|\mathbf{u} + \mathbf{v}\| \le \|\mathbf{u}\| + \|\mathbf{v}\|$ (cf. Theorem 1.5.3, p. 24).

The coordinate axes play an important role in our study of vectors, and it is often useful to distinguish vectors that point along the axes.

DEFINITION 14.2.3

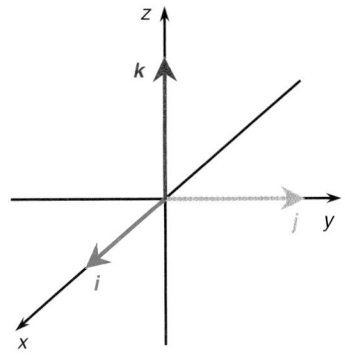

Figure 14.2.7

The **standard basis vectors** are the three vectors in $\mathbb{R}^3$ defined as follows:

$$\mathbf{i} = (1, 0, 0), \qquad \mathbf{j} = (0, 1, 0), \qquad \mathbf{k} = (0, 0, 1).$$

They point along the positive x-axis, the positive y-axis, and the positive z-axis, respectively. See Figure 14.2.7. They also have the property that the length of each of them is 1, that is,

$$\|\mathbf{i}\| = \|\mathbf{j}\| = \|\mathbf{k}\| = 1.$$

Sometimes, in keeping with our understanding that the ordinary plane is the same as the xy-plane in three-dimensional space, we refer to the vectors $\mathbf{i} = (1, 0)$ and $\mathbf{j} = (0, 1)$ in the plane as standard basis vectors.

One reason for the importance of the standard basis vectors is that any vector can be written in terms of them. If $\mathbf{v} = (a, b, c)$, then

$$\mathbf{v} = (a, b, c) = a(1, 0, 0) + b(0, 1, 0) + c(0, 0, 1) = a\mathbf{i} + b\mathbf{j} + c\mathbf{k}. \tag{14.2.4}$$

Dot Product and Cross Product

We have already seen how to multiply a vector by a real number. Real numbers are sometimes referred to as *scalars*, so this kind of multiplication of a vector by a scalar is called **scalar multiplication**.

There are also two different ways of combining two vectors, but they are both slightly unexpected.

DEFINITION 14.2.5

If $\mathbf{u} = (a, b, c)$ and $\mathbf{v} = (x, y, z)$, then the **dot product** of $\mathbf{u}$ and $\mathbf{v}$ is the *real number*

$$\mathbf{u} \cdot \mathbf{v} = ax + by + cz.$$

In words, we multiply corresponding components of the vectors and then add up the products.

If $\mathbf{u} = (a, b)$ and $\mathbf{v} = (x, y)$, then the **dot product** of $\mathbf{u}$ and $\mathbf{v}$ is the *real number*

$$\mathbf{u} \cdot \mathbf{v} = ax + by.$$

EXAMPLE 6

(i) If $\mathbf{u} = (1, -4, 3)$ and $\mathbf{v} = (2, 1, 0)$, then

$$\mathbf{u} \cdot \mathbf{v} = (1)(2) + (-4)(1) + (3)(0) = -2.$$

(ii) $(4, -3) \cdot (2, 4) = (4)(2) + (-3)(4) = -4.$

(iii) For the standard basis vectors $\mathbf{i}, \mathbf{j},$ and $\mathbf{k}$ we have that $\mathbf{i} \cdot \mathbf{i} = \mathbf{j} \cdot \mathbf{j} = \mathbf{k} \cdot \mathbf{k} = 1$ and $\mathbf{i} \cdot \mathbf{j} = \mathbf{i} \cdot \mathbf{k} = \mathbf{j} \cdot \mathbf{k} = 0.$

(iv) Find all vectors $\mathbf{v}$ so that $\mathbf{v} \cdot (2, -3) = 0.$

Solution   We write $\mathbf{v} = (x, y)$, so the desired condition is

$$0 = \mathbf{v} \cdot (2, -3) = (x, y) \cdot (2, -3) = 2x - 3y.$$

In other words, we want $2x = 3y$. If we place the vector $\mathbf{v}$ with its basepoint at the origin, then its terminal point (x, y) lies on the line $2x - 3y = 0$. The vectors $\mathbf{v}$ with $\mathbf{v} \cdot (2, -3) = 0$ are precisely all vectors with terminal point on the line $2x - 3y = 0$. See Figure 14.2.8.

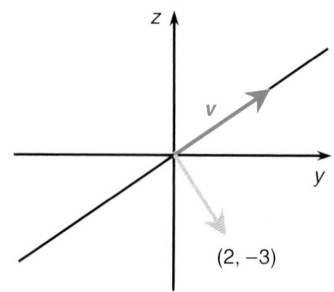

Figure 14.2.8

As with vectors we summarize some obvious properties of dot products as a proposition. These properties will hold for vectors in the plane and for vectors in space, but rather than repeat them all, we write them in a way that makes no mention of which type of vector we are dealing with.

PROPOSITION 14.2.6

Properties of Dot Products

The following statements are true for any vectors $\mathbf{u}, \mathbf{v},$ and $\mathbf{w}$ and any real number k, provided that $\mathbf{u}, \mathbf{v},$ and $\mathbf{w}$ are all vectors of the same type (i.e., all in the plane or all in three-dimensional space $\mathbb{R}^3$).

(i) $\mathbf{u} \cdot \mathbf{v} = \mathbf{v} \cdot \mathbf{u}.$

(ii) $(\mathbf{u} + \mathbf{v}) \cdot \mathbf{w} = \mathbf{u} \cdot \mathbf{w} + \mathbf{v} \cdot \mathbf{w}.$

(iii) $(k\mathbf{u}) \cdot \mathbf{v} = k(\mathbf{u} \cdot \mathbf{v}).$

(iv) $\mathbf{u} \cdot \mathbf{u} \geq 0,$ and $\mathbf{u} \cdot \mathbf{u} = 0$ precisely when $\mathbf{u}$ is the zero vector.

(v) $\mathbf{u} \cdot \mathbf{u} = \|\mathbf{u}\|^2.$

PROOF   These statements are not difficult; we content ourselves with proving the last one. If $\mathbf{u} = (a, b)$, then $\mathbf{u} \cdot \mathbf{u} = (a, b) \cdot (a, b) = a^2 + b^2 = \|\mathbf{u}\|^2$. Similarly, if $\mathbf{u} = (a, b, c)$, then $\mathbf{u} \cdot \mathbf{u} = (a, b, c) \cdot (a, b, c) = a^2 + b^2 + c^2 = \|\mathbf{u}\|^2.$

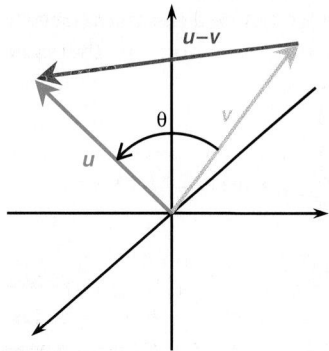

Figure 14.2.9

Let us consider two nonzero vectors **u** and **v**. Place them with their initial points at the origin, as shown in Figure 14.2.9. Then as we saw in Example 5, the third side of the triangle they form is the vector **u** − **v**. All these vectors lie in a plane.

The lengths of the sides of the triangle are $\|\mathbf{u}\|$, $\|\mathbf{v}\|$, and $\|\mathbf{u}-\mathbf{v}\|$. Writing θ for the angle between **u** and **v**, we can apply the Law of Cosines (cf. Theorem 1.7.10, p. 50). It tells us that

$$\|\mathbf{u}-\mathbf{v}\|^2 = \|\mathbf{u}\|^2 + \|\mathbf{v}\|^2 - 2\|\mathbf{u}\|\|\mathbf{v}\|\cos\theta.$$

This can be rearranged to give

$$\cos\theta = \frac{\|\mathbf{u}\|^2 + \|\mathbf{v}\|^2 - \|\mathbf{u}-\mathbf{v}\|^2}{2\|\mathbf{u}\|\|\mathbf{v}\|} = \frac{\mathbf{u}\cdot\mathbf{u} + \mathbf{v}\cdot\mathbf{v} - (\mathbf{u}-\mathbf{v})\cdot(\mathbf{u}-\mathbf{v})}{2\|\mathbf{u}\|\|\mathbf{v}\|}$$

$$= \frac{\mathbf{u}\cdot\mathbf{u} + \mathbf{v}\cdot\mathbf{v} - \mathbf{u}\cdot\mathbf{u} + \mathbf{u}\cdot\mathbf{v} - \mathbf{v}\cdot\mathbf{v} + \mathbf{v}\cdot\mathbf{u}}{2\|\mathbf{u}\|\|\mathbf{v}\|}$$

$$= \frac{\mathbf{u}\cdot\mathbf{v} + \mathbf{v}\cdot\mathbf{u}}{2\|\mathbf{u}\|\|\mathbf{v}\|} = \frac{\mathbf{u}\cdot\mathbf{v}}{\|\mathbf{u}\|\|\mathbf{v}\|}.$$

> Here we needed to know that
> $\mathbf{v}\cdot\mathbf{u} = \mathbf{u}\cdot\mathbf{v}$.

The importance of this calculation is that it allows us to find the angle between two vectors from their dot product. We did it for vectors in $\mathbb{R}^3$, but it also works in the plane.

THEOREM 14.2.7

> The **angle θ between two nonzero vectors u and v** is the angle satisfying $0 \leq \theta \leq \pi$ and
>
> $$\cos\theta = \frac{\mathbf{u}\cdot\mathbf{v}}{\|\mathbf{u}\|\|\mathbf{v}\|}.$$

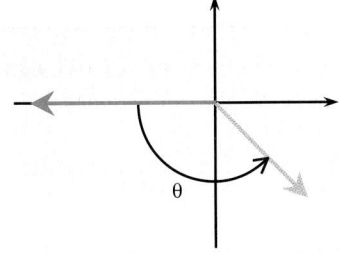

Figure 14.2.10

EXAMPLE 7 Find the angle between each of the following pairs of vectors: (i) **u** = (2, −2), **v** = (−4, 0); (ii) (2, 3), (−2, −1); (iii) (3, −2, −1), (6, −4, −2); (iv) (1, −1, 2), (−2, 2, −4); (v) (3, −2, 7), (6, 2, −2).

Solution

(i) First we calculate that $\|\mathbf{u}\| = \sqrt{2^2 + (-2)^2} = 2\sqrt{2}$ and $\|\mathbf{v}\| = \sqrt{(-4)^2 + 0} = 4$. Since $\mathbf{u}\cdot\mathbf{v} = -8$, we find that $\cos\theta = \frac{-8}{8\sqrt{2}} = -\frac{1}{\sqrt{2}}$. We see that the angle between **u** and **v** is $\frac{3\pi}{4}$, which is confirmed by Figure 14.2.10.

(ii) Here we find that

$$\cos\theta = \frac{(2,3)\cdot(-2,-1)}{\|(2,3)\|\|(-2,-1)\|} = \frac{-7}{\sqrt{13}\sqrt{5}} = -\frac{7}{\sqrt{65}}.$$

With a calculator we find that $\theta = \arccos\left(-\frac{7}{\sqrt{65}}\right) \approx 2.622$ radians $\approx 150.3°$. See Figure 14.2.11.

(iii) The formula tells us that

$$\cos\theta = \frac{(3,-2,-1)\cdot(6,-4,-2)}{\|(3,-2,-1)\|\|(6,-4,-2)\|} = \frac{28}{\sqrt{14}\sqrt{56}} = \frac{28}{\sqrt{784}} = 1.$$

Figure 14.2.11

This means that $\theta = 0$. In hindsight we ought to have guessed this immediately, since the second vector is just twice the first vector. They point in the same direction, so the angle between them is 0.

(iv) Here we find

$$\cos\theta = \frac{(1,-1,2) \cdot (-2,2,-4)}{\|(1,-1,2)\|\|(-2,2,-4)\|} = \frac{-12}{\sqrt{6}\sqrt{24}} = -\frac{12}{\sqrt{144}} = -1.$$

This means that $\theta = \pi$, which just means that the two vectors point in opposite directions. This is obvious once we recognize that the second vector is just (-2) times the first.

(v) Let us begin by finding the dot product $(3,-2,7) \cdot (6,2,-2) = (3)(6) + (-2)(2) + (7)(-2) = 0$. Without even bothering to find the norms of the two vectors we see that $\cos\theta = 0$ and $\theta = \frac{\pi}{2}$. The vectors are perpendicular. ◢

In the last part of Example 7 we saw that if $\mathbf{u} \cdot \mathbf{v} = 0$, then $\mathbf{u}$ and $\mathbf{v}$ are perpendicular. In effect this result is part of Theorem 14.2.7, but it is so important that we will call it a theorem.

THEOREM 14.2.8

> To say that two vectors $\mathbf{u}$ and $\mathbf{v}$ are **perpendicular** (or **orthogonal**) is exactly the same as to say that $\mathbf{u} \cdot \mathbf{v} = 0$.

Notice that according to Theorem 14.2.8, the zero vector is orthogonal to *every* vector.

EXAMPLE 8

(i) The vectors $(5,-2,-3)$ and $(2,2,2)$ are perpendicular.

(ii) Any two of the vectors $\mathbf{i}$, $\mathbf{j}$, and $\mathbf{k}$ are orthogonal to each other.

(iii) Find all vectors that are perpendicular to $(5,-3)$.

Solution Consider the vector $\mathbf{v} = (x, y)$. For it to be orthogonal to $(5,-3)$ we must have that $\mathbf{v} \cdot (5,-3) = 0$, that is, that $5x - 3y = 0$. The vectors that are orthogonal to $(5,-3)$ are the ones whose terminal point is on the line $5x - 3y = 0$ if their initial point is the origin. See Figure 14.2.12.

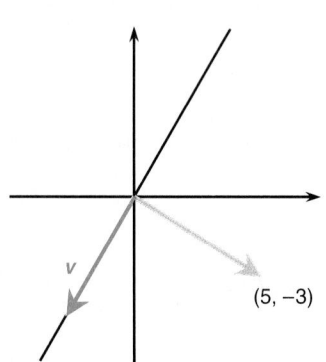

(5, −3)

Another way to describe them is to use parametric equations and say that these are all the vectors $\mathbf{v} = (3t, 5t)$, where t is any real number. ◢

Figure 14.2.12

The dot product takes two vectors and gives a number. It works equally well with plane vectors or vectors in $\mathbb{R}^3$. There is another kind of product, called the *cross product*. It only makes sense for two vectors in three-dimensional space, and the cross product of two vectors is not a number but another *vector*. The formula for it is complicated and must be learned. In the next three sections we will see that it is very useful.

DEFINITION 14.2.9

> If $\mathbf{u} = (a, b, c)$ and $\mathbf{v} = (x, y, z)$, then their **cross product** is the vector
>
> $$\mathbf{u} \times \mathbf{v} = (bz - cy, cx - az, ay - bx).$$

If you know about *determinants*, there is a mnemonic device to help remember this formula. Writing the cross product out as $\mathbf{u} \times \mathbf{v} = (bz - cy, cx - az, ay - bx) = (bz - cy)\mathbf{i} + (cx - az)\mathbf{j} + (ay - bx)\mathbf{k}$, we see that it equals the following determinant:

$$\mathbf{u} \times \mathbf{v} = det \begin{vmatrix} \mathbf{i} & \mathbf{j} & \mathbf{k} \\ a & b & c \\ x & y & z \end{vmatrix} = \begin{vmatrix} b & c \\ y & z \end{vmatrix}\mathbf{i} - \begin{vmatrix} a & c \\ x & z \end{vmatrix}\mathbf{j} + \begin{vmatrix} a & b \\ x & y \end{vmatrix}\mathbf{k}.$$

This is just a trick to help remember the formula, and if you do not know about determinants, it is much easier just to memorize the first formula for cross products.

EXAMPLE 9

See the Calculator Example on the next page.

(i) $(1, 3, -2) \times (2, 0, 1)$
$= \big((3)(1) - (-2)(0), (-2)(2) - (1)(1), (1)(0) - (3)(2)\big) = (3, -5, -6).$

(ii) $\mathbf{i} \times \mathbf{j} = (1, 0, 0) \times (0, 1, 0)$
$= \big((0)(0) - (0)(1), (0)(0) - (1)(0), (1)(1) - (0)(0)\big) = (0, 0, 1) = \mathbf{k}.$

(iii) $(a, b, c) \times (a, b, c) = (bc - cb, ca - ac, ab - ba) = (0, 0, 0).$ This says that the cross product of any vector with itself is the zero vector.

(iv) Suppose $\mathbf{u} = (a, b, c)$ and $\mathbf{v} = (x, y, z)$ as in the definition. Then

$$\mathbf{u} \times \mathbf{v} = (bz - cy, cx - az, ay - bx).$$

and

$$\mathbf{v} \times \mathbf{u} = (cy - bz, az - cx, bx - ay).$$

A quick comparison shows that $\mathbf{v} \times \mathbf{u} = -\mathbf{u} \times \mathbf{v}$. This says that the cross product is *not* commutative.

If we look at part (i) of Example 9, we can easily check that the cross product $(1, 3, -2) \times (2, 0, 1) = (3, -5, -6)$ is orthogonal to both the vectors we started with:

$$(3, -5, -6) \cdot (1, 3, -2) = 0 = (3, -5, -6) \cdot (2, 0, 1).$$

In fact this is no accident. We record the general case of this phenomenon as part (iii) of the following proposition, which lists some of the properties of cross products.

PROPOSITION 14.2.10

Properties of Cross Products

The following statements all hold for any real number k and any vectors $\mathbf{u}$, $\mathbf{v}$, and $\mathbf{w}$ in $\mathbb{R}^3$.

(i) $\mathbf{u} \times \mathbf{v} = -\mathbf{v} \times \mathbf{u}.$

(ii) $\mathbf{u} \times \mathbf{u} = 0.$

(iii) $\mathbf{u} \times \mathbf{v}$ is orthogonal to both $\mathbf{u}$ and $\mathbf{v}$. Symbolically,

$$(\mathbf{u} \times \mathbf{v}) \cdot \mathbf{u} = 0 = (\mathbf{u} \times \mathbf{v}) \cdot \mathbf{v}.$$

(iv) $\mathbf{i} \times \mathbf{j} = \mathbf{k}, \mathbf{j} \times \mathbf{k} = \mathbf{i}, \mathbf{k} \times \mathbf{i} = \mathbf{j}, \mathbf{j} \times \mathbf{i} = -\mathbf{k}, \mathbf{k} \times \mathbf{j} = -\mathbf{i}, \mathbf{i} \times \mathbf{k} = -\mathbf{j}.$

(v) $(k\mathbf{u}) \times \mathbf{v} = k(\mathbf{u} \times \mathbf{v}) = \mathbf{u} \times (k\mathbf{v}).$

(vi) $(\mathbf{u} + \mathbf{v}) \times \mathbf{w} = \mathbf{u} \times \mathbf{w} + \mathbf{v} \times \mathbf{w}.$

With a programmable calculator we can evaluate cross products. We enter the components of each vector in turn and display the components of the cross product.

TI-85	SHARP EL9300
PRGM F2(EDIT) CRSPRD ENTER F3 F1(Input) A ENTER F1(Input) B ENTER F1(Input) C ENTER F1(Input) X ENTER F1(Input) Y ENTER F1(Input) Z ENTER B × Z − C × Y STO U ENTER C × X − A × Z STO V ENTER A × Y − B × X STO W ENTER F3(Disp) U ENTER F3(Disp) V ENTER F3(Disp) W ENTER EXIT EXIT PRGM F1(NAME) CRSPRD 1 ENTER 3 ENTER (−) 2 ENTER , 2 ENTER , 0 ENTER 1 ENTER	▼ ▼ (NEW) ENTER ENTER CROSS ENTER COMMAND A 3 (Input) A ENTER COMMAND A 3 (Input) B ENTER COMMAND A 3 (Input) C ENTER COMMAND A 3 (Input) X ENTER COMMAND A 3 (Input) Y ENTER COMMAND A 3 (Input) Z ENTER $U = B \times Z - C \times Y$ ENTER $V = C \times X - A \times Z$ ENTER $W = A \times Y - B \times X$ ENTER COMMAND A 1 (Print) U ENTER COMMAND A 1 (Print) V ENTER COMMAND A 1 (Print) W ENTER COMMAND A 6 (End) ENTER ▦ (RUN) ENTER (CROSS) ENTER 1 ENTER 3 ENTER −2 ENTER 2 ENTER 0 ENTER 1 ENTER
CASIO f_x-6300G (f_x-7700GB)	HP 48SX
MODE 2 EXE ? → A : ? → B : ? → C : ? → X : ? → Y : ? → Z : B × Z − C × Y → U : C × X − A × Z → V : A × Y − B × X → W : U ◢ V ◢ W ◢ MODE 1 Prog 0 EXE 1 EXE 3 EXE −2 EXE 2 EXE 0 EXE 1 EXE EXE EXE (on the f_x-7700GB, : is PRGM F6, and Prog is PRGM F3.)	ATTN ↱ CLR ≪≫ "" KEY IN VECTORS ▶ SPC PROMPT SPC ' Z ▶ STO SPC ' Y ▶ STO SPC ' X ▶ STO SPC ' C ▶ STO SPC ' B ▶ STO SPC ' A ▶ STO SPC B SPC Z × C SPC Y × − C SPC X × A SPC Z × − A SPC Y × B SPC X × − ENTER ' CROS STO VAR CROS 1 ENTER 3 ENTER 2 +/− ENTER 2 ENTER 0 ENTER 1 ENTER ↰ CONT

```
?1    ?3    ?-2
?2    ?0    ?1
ANSWER =
```

Try out the program with some other examples.

PROOF We have already proved (i) and (ii). We proved one case of (iv), and the others are boringly similar. Both (v) and (vi) are very easy. The only part that remains is (iii).

Letting $\mathbf{u} = (a, b, c)$ and $\mathbf{v} = (x, y, z)$, we have that

$$\mathbf{u} \cdot (\mathbf{u} \times \mathbf{v}) = (a, b, c) \cdot \big((a, b, c) \times (x, y, z)\big)$$

$$= (a, b, c) \cdot (bz - cy, cx - az, ay - bx)$$
$$= a(bz - cy) + b(cx - az) + c(ay - bx)$$
$$= abz - acy + bcx - baz + cay - cbx = 0.$$

This shows that $\mathbf{u}$ is orthogonal to $\mathbf{u} \times \mathbf{v}$. Rather than do it again for $\mathbf{v}$, we remark that $\mathbf{u} \times \mathbf{v} = -\mathbf{v} \times \mathbf{u}$, and the calculation we have just done shows that this is orthogonal to $\mathbf{v}$, as needed.   ◢

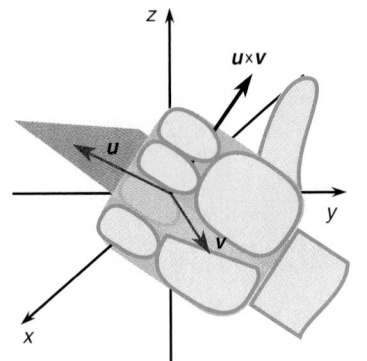

Figure 14.2.13

Part (iv) of Proposition 14.2.10 looks complicated, but there is an easy way to remember it. It says that the cross product of any two of $\mathbf{i}$, $\mathbf{j}$, and $\mathbf{k}$ is either the other one or its negative. To decide on the sign, we draw the circle shown in Figure 14.2.13. The letters $\mathbf{i}$, $\mathbf{j}$, and $\mathbf{k}$ are placed around the circle, and little arrows indicate the direction around the circle in *alphabetical order*, that is, going from $\mathbf{i}$ to $\mathbf{j}$ to $\mathbf{k}$. The sign of the cross product is positive if we take the cross product of two of these vectors in the direction of the arrows and negative if they are in the opposite direction.

For instance, to find $\mathbf{i} \times \mathbf{k}$, we observe that going from $\mathbf{i}$ to $\mathbf{k}$ on the circle is in the direction opposite to the arrows, so this product will have a minus sign; it is $\mathbf{i} \times \mathbf{k} = -\mathbf{j}$. Using the circle again or just because of part (i) of Proposition 14.2.10, we see that $\mathbf{k} \times \mathbf{i} = \mathbf{j}$.

It is much easier to use this picture than to memorize the formulas for cross products of $\mathbf{i}$, $\mathbf{j}$, and $\mathbf{k}$.

Figure 14.2.14

From part (iii) of Proposition 14.2.10 we know that $\mathbf{u} \times \mathbf{v}$ is orthogonal to $\mathbf{u}$ and $\mathbf{v}$. If $\mathbf{u}$ and $\mathbf{v}$ are nonzero and not parallel, they lie in a plane (see Figure 14.2.14) and $\mathbf{u} \times \mathbf{v}$ is perpendicular to that plane.

This determines the direction of $\mathbf{u} \times \mathbf{v}$ *up to a sign*. To find its direction precisely, that is, to determine which of the two perpendicular directions is correct, there is a "right-hand rule." With the fingers of the right hand curling from $\mathbf{u}$ to $\mathbf{v}$, the thumb points along the direction of $\mathbf{u} \times \mathbf{v}$. You might like to confirm this in the situation of Example 9(i). Also notice that this rule is consistent with the fact that $\mathbf{v} \times \mathbf{u} = -\mathbf{u} \times \mathbf{v}$.

EXAMPLE 10   Find a vector that is perpendicular to both $(-1, 3, 2)$ and $(4, -2, 3)$.

Solution   There are many ways to do this question, but one of them is to apply the proposition and realize that the cross product $(-1, 3, 2) \times (4, -2, 3)$ will be orthogonal to both the vectors.

We calculate $(-1, 3, 2) \times (4, -2, 3) = \big((3)(3) - (2)(-2),\ (2)(4) - (-1)(3),$ $(-1)(-2) - (3)(4)\big) = (13, 11, -10)$. It is easy to check that $(13, 11, -10)$ really is perpendicular to both $(-1, 3, 2)$ and $(4, -2, 3)$.   ◢

The final thing we need to do here is find a formula for the norm $\|\mathbf{u} \times \mathbf{v}\|$. Letting $\mathbf{u} = (a, b, c)$ and $\mathbf{v} = (x, y, z)$, we find that

$$\|\mathbf{u} \times \mathbf{v}\|^2 = \|(a, b, c) \times (x, y, z)\|^2 = (bz - cy)^2 + (cx - az)^2 + (ay - bx)^2$$
$$= b^2z^2 - 2bcyz + c^2y^2 + c^2x^2 - 2acxz + a^2z^2 + a^2y^2 - 2abxy + b^2x^2.$$

On the other hand, if we work out the even messier formula

$$\|\mathbf{u}\|^2\|\mathbf{v}\|^2 - (\mathbf{u} \cdot \mathbf{v})^2 = \|(a, b, c)\|^2 \|(x, y, z)\|^2 - \big((a, b, c) \cdot (x, y, z)\big)^2$$

$$= (a^2 + b^2 + c^2)(x^2 + y^2 + z^2) - (ax + by + cz)^2$$

$$= a^2x^2 + a^2y^2 + a^2z^2 + b^2x^2 + b^2y^2 + b^2z^2 + c^2x^2 + c^2y^2$$

$$+ c^2z^2 - a^2x^2 - b^2y^2 - c^2z^2 - 2axby - 2axcz - 2bycz$$

$$= a^2y^2 + a^2z^2 + b^2x^2 + b^2z^2 + c^2x^2 + c^2y^2 - 2abxy$$

$$- 2acxz - 2bcyz,$$

we can compare and see that the two formulas are equal.

We have shown that $\|\mathbf{u} \times \mathbf{v}\|^2 = \|\mathbf{u}\|^2\|\mathbf{v}\|^2 - (\mathbf{u} \cdot \mathbf{v})^2$.

Now, using the fact that $\mathbf{u} \cdot \mathbf{v} = \|\mathbf{u}\|\|\mathbf{v}\|\cos\theta$, where θ is the angle between $\mathbf{u}$ and $\mathbf{v}$, we find that

$$\|\mathbf{u} \times \mathbf{v}\|^2 = \|\mathbf{u}\|^2\|\mathbf{v}\|^2 - (\mathbf{u} \cdot \mathbf{v})^2$$

$$= \|\mathbf{u}\|^2\|\mathbf{v}\|^2 - \|\mathbf{u}\|^2\|\mathbf{v}\|^2\cos^2\theta = \|\mathbf{u}\|^2\|\mathbf{v}\|^2(1 - \cos^2\theta)$$

$$= \|\mathbf{u}\|^2\|\mathbf{v}\|^2\sin^2\theta.$$

Unfortunately, this calculation is not very enlightening, but the result will be useful to us later on, so we record it as a theorem.

THEOREM 14.2.11

> **Lagrange's Identity**
>
> For any vectors $\mathbf{u}$ and $\mathbf{v}$ in $\mathbb{R}^3$ we have
>
> $$\|\mathbf{u} \times \mathbf{v}\|^2 = \|\mathbf{u}\|^2\|\mathbf{v}\|^2 - (\mathbf{u} \cdot \mathbf{v})^2.$$
>
> This can also be written as
>
> $$\|\mathbf{u} \times \mathbf{v}\| = \|\mathbf{u}\|\|\mathbf{v}\|\sin\theta,$$
>
> where θ is the angle between $\mathbf{u}$ and $\mathbf{v}$. (Note that $\sin\theta \geq 0$, since $0 \leq \theta \leq \pi$.)

It is important to remember that cross products only make sense for vectors in three-dimensional space. Dot products are meaningful for vectors in the plane too, but the formula for cross products only makes sense for vectors with three components.

It is true that occasionally we think of the ordinary plane sitting horizontally in three-dimensional space as the xy-plane. In this picture the vector (x, y) in the plane corresponds to the vector $(x, y, 0)$ in $\mathbb{R}^3$, and it is possible to take two such vectors and form their cross product. However, the cross product of two such vectors will be a vertical vector (it has to be perpendicular to the two original horizontal vectors ...), and in particular it is *not* in the plane (see Figure 14.2.15). There is no way to define a product that takes two vectors in the plane and produces another vector in the plane and satisfies the conditions (i), (ii), and (iii) of Proposition 14.2.10 except by letting the product of every pair of vectors be the zero vector. Of course, this is not very interesting.

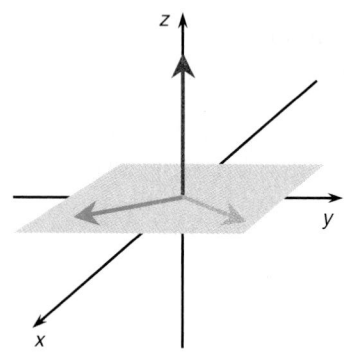

Figure 14.2.15

Exercises 14.2

I

Find each of the following.

1. $(3, -5, -2) + 2(-1, 2, 3)$
2. $3(1, -1, -2) - 4(1, 3, -2)$
3. $(6, -1, 2) \cdot (1, 1, -3)$
4. $(2, 1) \cdot (-2, 4)$
5. $(4, -2, 1) \times (2, -3, 1)$
6. $\|(6, 3, -2)\|$
7. $(7, -11, 16) \times (-7, 11, -16)$
8. $(0, 2, -3) \times (0, 1, -1)$
9. $(\mathbf{j} + \mathbf{k}) \times (\mathbf{k} + \mathbf{j})$
10. $(2\mathbf{i} - 3\mathbf{j} + \mathbf{k}) \times \mathbf{j}$

Find the angle between each of the following pairs of vectors.

11. $(1, 1), (-4, 0)$
12. $(2, 1), (-2, 4)$
13. $(1, 1, 1), (2, -3, 1)$
14. $(1, 2, 2), (1, 0, 1)$
15. $(1, -1, 0), (-1, 2, 1)$
16. $(2, 1, -3), (1, 2, -2)$
17. $(3, -1, 2), (2, -2, 3)$
18. $(3, 4), (2, 1)$
19. $(a, b), (-b, a)$
20. $(1, 3, -2), (7, -5, 1) \times (1, 3, -2)$

II

Find a nonzero vector that is perpendicular to each of the following vectors.

21. $(2, 4)$
22. $(1, -3)$
23. $(2, 1, 7)$
24. $(-4, 2, -1)$
*25. Find two nonzero vectors **u** and **v** that are both orthogonal to $(1, 1, 1)$ but that are not scalar multiples of each other, that is, so that it is not possible to write $\mathbf{u} = k\mathbf{v}$ or $\mathbf{v} = k\mathbf{u}$ with some real number k.

*26. Find two nonzero vectors **u** and **v** that are both orthogonal to $(2, 3, -1)$ but that are not scalar multiples of each other, that is, so that it is not possible to write $\mathbf{u} = k\mathbf{v}$ or $\mathbf{v} = k\mathbf{u}$ with some real number k.

27. Find a nonzero vector that is orthogonal to both $(1, 2, -4)$ and $(-3, 2, 1)$.

28. Find a nonzero vector that is orthogonal to both $(5, -3, 2)$ and $(0, 4, -2)$.

*29. For what value(s) of a is $\mathbf{v} = (a, 2, 3)$ orthogonal to $\mathbf{u} = (-4, 2, -1)$?

*30. For what value(s) of a is $\mathbf{v} = (a, 1, -2)$ orthogonal to $\mathbf{u} = (4 - a, -6, -1)$?

*31. Prove (i) Proposition 14.2.6(i), (ii) Proposition 14.2.6(ii), (iii) Proposition 14.2.6(iii).

*32. Prove Proposition 14.2.6(iv).

*33. (i) What can you say about the cross product of two vectors **u** and **v** that both lie in the vertical yz-plane?
    (ii) Describe the set of all vectors that can be written as $\mathbf{u} \times \mathbf{v}$ with **u** and **v** in the yz-plane.

*34. (i) What can you say about $\mathbf{u} \times \mathbf{k}$ if **u** is any vector in space?
    (ii) Describe the set of all vectors that can be written as $\mathbf{u} \times \mathbf{k}$ with any vector **u**.

35. Show that the cross product is not associative, that is, find vectors **u**, **v**, **w** so that $\mathbf{u} \times (\mathbf{v} \times \mathbf{w}) \neq (\mathbf{u} \times \mathbf{v}) \times \mathbf{w}$.

*36. (i) Find all vectors **v** for which $(1, 3, 2) \times \mathbf{v} = 0$.
    (ii) If $\mathbf{u} = (a, b, c) \neq 0$, show that if $\mathbf{u} \times \mathbf{v} = 0$ then **v** must be a scalar multiple of **u**, that is, $\mathbf{v} = k\mathbf{u}$, for some number k.

III

*37. Plot the points $O = (0, 0)$, $A = (3, 7)$, $B = (7, 3)$, $C = (1, 2)$, $D = (8, 5)$ on your computer screen, using the same scale in the vertical and horizontal directions. Use a protractor to measure the following angles (in degrees): (i) $\angle AOB$, (ii) $\angle AOC$, (iii) $\angle DBA$, (iv) $\angle DCO$. Compare your answers with what you get by calculating.

14.3 Applications of Vectors

DISPLACEMENT
DISTANCE
COMPONENTS
PERPENDICULAR COMPONENT
FORCE AND WORK
PARALLELOGRAM AREA
CYCLOID
CATENARY

In this section we review some of the simple applications of vectors. Most of them involve questions from physics, but it is not necessary to know much physics to understand them.

The importance of these applications is that they explain where the idea of vectors came from and show how convenient vectors can be. They are not specifically needed for the rest of our work with vectors.

Displacements

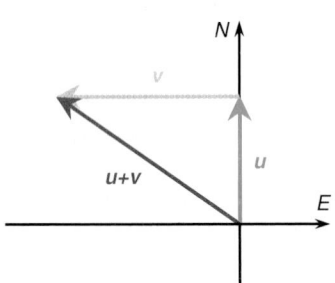

Figure 14.3.1

If a person walks 2 miles due north and then turns and walks 3 miles due west, as shown in Figure 14.3.1, then it is possible to describe these movements in terms of vectors. If the origin is at the person's starting point, with the positive x-axis pointing due east and the positive y-axis pointing due north, then moving 2 miles north means moving to the endpoint of the vector $\mathbf{u} = (0, 2)$. Moving 3 miles west from that point means moving along the vector $\mathbf{v} = (-3, 0)$, placed with its initial point at $(0, 2)$. The final location of the person walking is the point $(-3, 2)$. It can be expressed as the sum of $\mathbf{u}$ and $\mathbf{v}$. After all, $(-3, 2) = (0, 2) + (-3, 0) = \mathbf{u} + \mathbf{v}$.

We regard moving from one point to another as undergoing a *displacement*, and we express the displacement as a vector that points from the initial location to the final location. We have just observed that undergoing one displacement and then undergoing a second displacement amounts to undergoing one single displacement, the one corresponding to the **sum** of the vectors corresponding to the two original displacements.

The distance traveled is equal to the norm of the displacement. For instance, in the example discussed above, the final location is at a distance of $\|\mathbf{u} + \mathbf{v}\| = \|(-3, 2)\| = \sqrt{9 + 4} = \sqrt{13} \approx 3.6$ miles from the starting point.

These observations are equally valid for displacements in three-dimensional space $\mathbb{R}^3$ or in the plane.

REMARK 14.3.1

> **Displacements** in the plane or in three-dimensional space are represented by vectors. If a displacement corresponding to the vector $\mathbf{u}$ is followed by a displacement corresponding to the vector $\mathbf{v}$, then the resulting overall displacement corresponds to the displacement $\mathbf{u} + \mathbf{v}$.
>
> The displacement $\mathbf{v}$ involves moving a **distance** equal to $\|\mathbf{v}\|$.

EXAMPLE 1   A helicopter travels 4 km south, turns and flies 4 km east, and then climbs 2 km straight up. How far is it from its starting position?

Solution   Placing the origin at the starting position with the positive x-axis pointing east, the positive y-axis pointing north, and the positive z-axis pointing straight up, we can represent the first displacement 4 km south by the vector $(0, -4, 0)$, the second displacement of 4 km east by the vector $(4, 0, 0)$, and the 2 km climb by $(0, 0, 2)$.

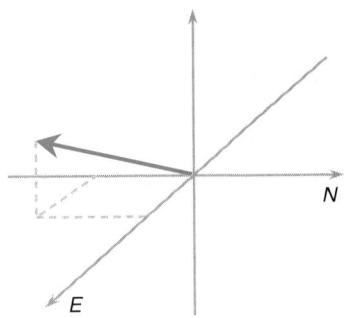

Figure 14.3.2

The overall displacement is the sum of these vectors, or

$$(0, -4, 0) + (4, 0, 0) + (0, 0, 2) = (4, -4, 2).$$

The distance that the helicopter has traveled is $\|(4, -4, 2)\| = \sqrt{16 + 16 + 4} = \sqrt{36} = 6$ km. See Figure 14.3.2.

EXAMPLE 2   **Cycloid**   Consider a wheel of radius 1 rolling along a flat surface at a constant speed. Find parametric equations for the position of a point Q on the rim of the wheel.

Solution   Suppose the moving point Q starts at ground level and let us place the origin at that starting point. Then the center of the wheel, which is 1 unit above ground level, starts at the point $(0, 1)$ (see Figure 14.3.3(i)).

Ignoring for the moment the sideways motion of the wheel, we know that it turns around its center. In Figure 14.3.3(ii) we see the vector $\mathbf{q}(t)$ that points from

Figure 14.3.3(i)

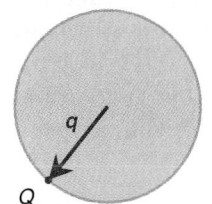

Figure 14.3.3(ii)

the center to the point Q on the rim. The wheel turns clockwise, and since Q begins at the bottom, we see that $\mathbf{q}(t) = (-\sin kt, -\cos kt)$, for some constant k.

The wheel turns around exactly once as t increases from $t = 0$ to $t = \frac{2\pi}{k}$. So in this length of time it rolls a distance equal to its circumference, which is 2π. In a time of $\frac{2\pi}{k}$ it travels a distance 2π, so its speed is $2\pi / \frac{2\pi}{k} = k$.

Since the wheel moves at a constant speed k, the position of the center of the wheel at time t can be described by $\mathbf{p}(t) = (kt, 1)$.

Now the center of the wheel is at position $\mathbf{p}(t)$ at time t, and the vector from the center to Q is $\mathbf{q}(t)$. The position of Q is the sum $\mathbf{r}(t) = \mathbf{p}(t) + \mathbf{q}(t)$. See Figure 14.3.3(iii).

The position of the moving point Q on the rim of the wheel is given by $\mathbf{r}(t) = (kt, 1) + (-\sin kt, -\cos kt) = (kt - \sin kt, 1 - \cos kt)$. The curve it traces out is called a **cycloid**. See Figure 14.3.3(iv).

Figure 14.3.3(iii)

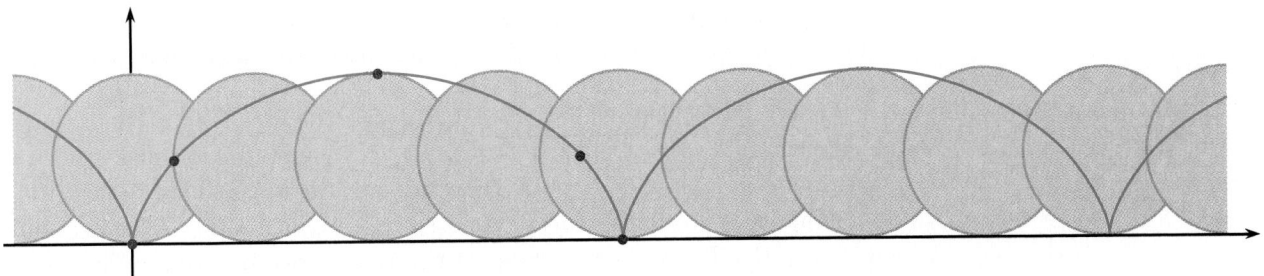

Figure 14.3.3(iv)

Velocity

Another quantity that is conveniently expressed by a vector is the velocity of a moving object. The direction of the vector indicates the direction in which the object is moving, and the length of the velocity vector indicates the speed.

The reason vectors are useful for describing velocities is that the way they add corresponds to the way velocities combine.

Figure 14.3.4(i)

Figure 14.3.4(ii)

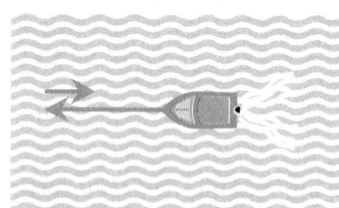

Figure 14.3.4(iii)

EXAMPLE 3

(i) A boat is crossing a river. The river flows at a rate of 3 mph and the speed of the boat in still water is 10 mph. The boat is heading directly across the river, but as we see in Figure 14.3.4(i), it is swept slightly downstream by the current.

Let us place the axes so that the positive x-axis points in the direction of the flow of the current and the positive y-axis points in the direction in which the boat is headed, that is, straight across the river. Then the velocity of the water in the river is the vector $(3, 0)$, and the velocity of the boat *relative to the water* is $(0, 10)$, meaning that it is heading directly across the river at a speed of 10 mph.

The velocity of the boat relative to a fixed observer on the shore will be the sum of these two velocities, that is,

$$\mathbf{v} = (0, 10) + (3, 0) = (3, 10).$$

The speed of the boat will be $\|\mathbf{v}\| = \sqrt{109} \approx 10.44$ mph. Notice that this is slightly faster than the boat will go in still water.

(ii) Now suppose the boat turns and heads downstream. Then its velocity relative to the water is $(10, 0)$ (i.e., 10 mph downstream), and the total velocity is the sum $(10, 0) + (3, 0) = (13, 0)$. We find the unsurprising result that if the boat moves with the current its speed will be 13 mph, the sum of its own speed relative to the water and the speed of the current. See Figure 14.3.4(ii).

(iii) If, on the other hand, the boat turns upstream as shown in Figure 14.3.4(iii), then its velocity relative to the water is $(-10, 0)$, and its velocity relative to the shore is $(-10, 0) + (3, 0) = (-7, 0)$. Its speed going upstream is only 7 mph.

(iv) Finally, we might ask in what direction the boat should head in order that it will actually move straight across the river (see Figure 14.3.5).

We know that the velocity of the water is $(3, 0)$. What we need is to find a vector $\mathbf{v} = (a, b)$ whose length is 10 (the maximum speed of the boat), and so that the overall velocity $\mathbf{v} + (3, 0)$ points straight across the river. This means that $(a, b) + (3, 0) = (a + 3, b)$ should be a vector of the form $(0, k)$, in which the x-component is zero (i.e., there is no "downstream component" to the motion).

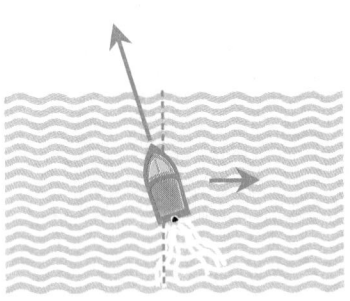

Figure 14.3.5

This will only happen if $a = -3$. In this case the speed of the boat relative to the water is $\|\mathbf{v}\| = \sqrt{(-3)^2 + b^2} = \sqrt{9 + b^2}$. We want this to equal 10, which means $9 + b^2 = 100$, so $b^2 = 91$. There are only two possibilities, $b = \pm\sqrt{91}$.

The two possible velocity vectors for the boat are $(-3, \sqrt{91})$ and $(-3, -\sqrt{91})$. The second of these points back across the river, so the first is the one we want.

The required velocity vector is $\mathbf{v} = (-3, \sqrt{91})$. The boat should steer in the direction of this vector in order to cross the river directly. If we mark the

Figure 14.3.6(i)

Figure 14.3.6(ii)

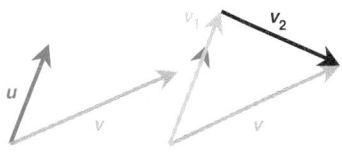

Figure 14.3.7

angle θ as shown in Figure 14.3.6(i), then we see from Figure 14.3.6(ii) that $\tan \theta = \frac{3}{\sqrt{91}}$ so $\theta = \arctan \frac{3}{\sqrt{91}} \approx 17.46°$.

The boat should steer $17.46°$ upstream.

If we write a velocity as a vector $\mathbf{v} = (a, b, c)$, say, then its components a, b, and c represent the "parts" of the velocity that point in the x-, y-, and z-directions, respectively. For instance, with the usual position for the axes, c represents the vertical motion and tells us how the object is rising or falling. It is often called the **vertical component** of the velocity.

If $\mathbf{v} = (a, b, c)$ is a velocity vector, then we can use dot products to find its various components. For instance, the vertical component c is just the dot product of $\mathbf{v}$ with the unit vector $\mathbf{k} = (0, 0, 1)$, that is, $c = \mathbf{v} \cdot \mathbf{k}$. Similarly, the component along the x-axis is $a = \mathbf{v} \cdot \mathbf{i}$.

The same idea will work for other unit vectors. In the first drawing in Figure 14.3.7 we see a vector $\mathbf{v}$ and a unit vector $\mathbf{u}$. In the second we have broken $\mathbf{v}$ up as the sum of a vector $\mathbf{v}_1$ that has the same direction as $\mathbf{u}$ and a vector $\mathbf{v}_2$ whose direction is perpendicular to $\mathbf{u}$. We will call $\mathbf{v}_1$ the **component of v in the direction of u** and $\mathbf{v}_2$ the **component of v perpendicular to u**.

What we have done here is analogous to writing any vector as the sum of a horizontal vector and a vertical vector, but here we use the direction of $\mathbf{u}$ and the direction perpendicular to it instead of the horizontal and vertical directions.

Now since $\mathbf{v}_1$ is parallel to $\mathbf{u}$, we can write $\mathbf{v}_1 = k\mathbf{u}$, for some number k. Let us consider the dot product of $\mathbf{v} = \mathbf{v}_1 + \mathbf{v}_2$ with $\mathbf{u}$. Since $\mathbf{u} \cdot \mathbf{u} = 1$ and $\mathbf{v}_2 \cdot \mathbf{u} = 0$, we find

$$\mathbf{v} \cdot \mathbf{u} = (\mathbf{v}_1 + \mathbf{v}_2) \cdot \mathbf{u} = (k\mathbf{u}) \cdot \mathbf{u} + \mathbf{v}_2 \cdot \mathbf{u} = k\mathbf{u} \cdot \mathbf{u} + 0 = k.$$

We have just proved the following result.

THEOREM 14.3.2

The component $\mathbf{v}_1$ of $\mathbf{v}$ in the direction of a unit vector $\mathbf{u}$ is $k\mathbf{u}$, with $k = \mathbf{v} \cdot \mathbf{u}$, that is,

$$\mathbf{v}_1 = (\mathbf{v} \cdot \mathbf{u})\mathbf{u}.$$

The component $\mathbf{v}_2$ of $\mathbf{v}$ perpendicular to $\mathbf{u}$ is

$$\mathbf{v}_2 = \mathbf{v} - \mathbf{v}_1 = \mathbf{v} - (\mathbf{v} \cdot \mathbf{u})\mathbf{u}.$$

To work with a nonzero vector $\mathbf{w}$ that is not a unit vector, we first form the corresponding unit vector $\mathbf{u} = \frac{1}{\|\mathbf{w}\|}\mathbf{w}$. Then the component of $\mathbf{v}$ in the direction of $\mathbf{w}$ is

$$\mathbf{v}_1 = (\mathbf{v} \cdot \mathbf{u})\mathbf{u} = \frac{1}{\|\mathbf{w}\|^2}(\mathbf{v} \cdot \mathbf{w})\mathbf{w},$$

and the component of $\mathbf{v}$ perpendicular to $\mathbf{w}$ is

$$\mathbf{v}_2 = \mathbf{v} - \mathbf{v}_1 = \mathbf{v} - (\mathbf{v} \cdot \mathbf{u})\mathbf{u} = \mathbf{v} - \frac{1}{\|\mathbf{w}\|^2}(\mathbf{v} \cdot \mathbf{w})\mathbf{w}.$$

These formulas can be used to find the component of a motion in any particular direction.

In Definition 14.2.1 we wrote a vector as (a, b) or (a, b, c) and used the word *component* to refer to each of the numbers a, b, and c. Now we are talking about the *component of a vector in the direction of another vector*, and it is important to recognize that this is a *vector*, not a number. The two uses of the word *component* are related, as we saw in the discussion before Theorem 14.3.2, but they are not the same.

EXAMPLE 4 An airplane takes off flying due north but immediately turns $60°$ toward the west and flies in a straight line at a speed of 400 mph. A radar fix tells the pilot that she is 600 miles from an international border that runs from southwest to northeast. How long will she fly in this direction before reaching the border?

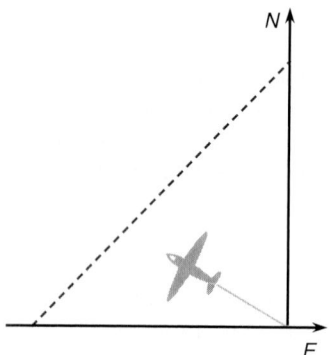

Figure 14.3.8(i)

Solution We begin by drawing a sketch in Figure 14.3.8(i). In Figure 14.3.8(ii) we draw an idealized sketch. The direction of the velocity vector is $60°$ to the west of north, which means $150°$ from the positive x-axis pointing east.

A unit vector pointing in this direction is $(\cos 150°, \sin 150°) = \left(-\frac{\sqrt{3}}{2}, \frac{1}{2}\right)$. Since the airplane's speed is 400 mph, the actual velocity vector is $400\left(-\frac{\sqrt{3}}{2}, \frac{1}{2}\right) = \left(-200\sqrt{3}, 200\right)$.

Next we want to find the component of the velocity that points directly toward the border, that is, in the direction orthogonal to the border; this component is shown as the lighter vector in Figure 14.3.8(ii). Since the border runs from southwest to northeast, the direction toward it is the vector $(-1, 1)$. To find the component, it is necessary to have a unit vector, so we divide by the length to find the unit vector $\mathbf{u} = \frac{1}{\sqrt{2}}(-1, 1) = \left(-\frac{1}{\sqrt{2}}, \frac{1}{\sqrt{2}}\right)$.

The component of the velocity in the direction toward the border, that is, in the direction of $\mathbf{u}$, is

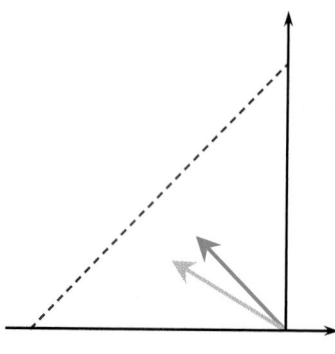

Figure 14.3.8(ii)

$$(\mathbf{v} \cdot \mathbf{u})\mathbf{u} = \left(\left(-200\sqrt{3}, 200\right) \cdot \left(-\frac{1}{\sqrt{2}}, \frac{1}{\sqrt{2}}\right)\right)\left(-\frac{1}{\sqrt{2}}, \frac{1}{\sqrt{2}}\right)$$

$$= \left(100\sqrt{6} + 100\sqrt{2}\right)\left(-\frac{1}{\sqrt{2}}, \frac{1}{\sqrt{2}}\right) = \left(-100\sqrt{3} - 100, 100\sqrt{3} + 100\right).$$

The norm of this component of the velocity is $\|(-100\sqrt{3} - 100, 100\sqrt{3} + 100)\| = (100\sqrt{3} + 100)\|(-1, 1)\| = 100\sqrt{6} + 100\sqrt{2} \approx 386.4$. This means that the airplane is approaching the border at a speed of about 386.4 mph.

In particular, since the border is 600 miles away, it will take $\frac{600}{100\sqrt{6} + 100\sqrt{2}} \approx 1.553$ hr ≈ 1 hr and 33 min to reach the border.

In social sciences especially it is common to measure several different quantities simultaneously, and it is often helpful to write them as the components of a vector.

EXAMPLE 5 In an environmental impact study for a factory we might consider a vector (x, y), in which x represents the amount of pollutant the factory produces and y represents the amount recovered by the waste treatment equipment before it is dumped into the river.

These two values depend on many factors. The amount of pollutant produced, x, depends on the type of operations and the level of productivity, and the amount

Figure 14.3.9(i)

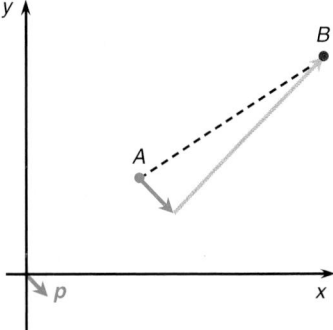

Figure 14.3.9(ii)

recovered, y, depends on the type of treatment equipment, how well it is maintained, and so on. The values of (x, y) for various types of operations can be plotted as points on a graph, as in Figure 14.3.9(i).

The vector $\mathbf{p} = (1, -1)$ points in the direction in which the actual pollution of the environment increases. For instance, suppose that for a certain type of operation the values are $A = (5, 4)$ and that a proposed plant expansion corresponds to $B = (13, 9)$. The vector from A to B is $(8, 5)$; it represents the change in the measurements arising from the expansion. Its component in the direction of $\mathbf{p} = (1, -1)$ is

$$\left((8, 5) \cdot \left(\frac{1}{\sqrt{2}}, -\frac{1}{\sqrt{2}} \right) \right) \left(\frac{1}{\sqrt{2}}, -\frac{1}{\sqrt{2}} \right) = \left(\frac{3}{2}, -\frac{3}{2} \right).$$

In Figure 14.3.9(ii) we see the vector from A to B drawn as the sum of this component and the component orthogonal to $\mathbf{p}$, which equals $(8, 5) - \left(\frac{3}{2}, -\frac{3}{2} \right) = \left(\frac{13}{2}, \frac{13}{2} \right)$. The short piece along the direction of $\mathbf{p}$ corresponds to increased pollution. The longer vector $\left(\frac{13}{2}, \frac{13}{2} \right)$ represents equal increases in pollutant produced and pollutant recovered, so it results in no net increase in pollutant discharged. ◢

In a way Example 5 tells us something we already know. The dot product of (x, y) with $\mathbf{p} = (1, -1)$ equals $(x, y) \cdot (1, -1) = x - y$. If x is the amount of pollutant produced and y is amount recovered, then $x - y$ is the amount discharged. In going from A to B we increase x by 8 and y by 5, so the discharge increases by $8 - 5 = 3$, which does equal the dot product $(8, 5) \cdot \mathbf{p}$.

On the other hand, the picture can sometimes help us get better insight into a situation. In this case we note that the net pollution is the same at two points R and S if $R \cdot \mathbf{p} = S \cdot \mathbf{p}$, that is, if $(R - S) \cdot \mathbf{p} = 0$. This means that the vector from R to S is perpendicular to $\mathbf{p}$. In Figure 14.3.9(iii) we see a number of lines that are perpendicular to $\mathbf{p}$. All the points on any one of these lines have the same net pollution. Points on different lines correspond to different levels of discharge.

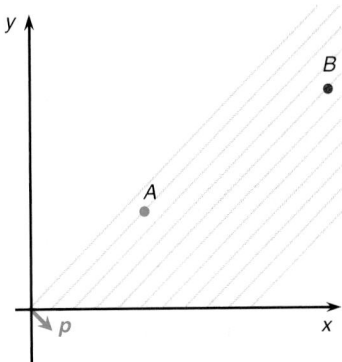

Figure 14.3.9(iii)

The line through the point $A = (5, 4)$ must be moved in the direction of $\mathbf{p}$ to reach the point $B = (13, 9)$, so we can tell at a glance that B gives a higher level of pollution.

Actual studies often measure many quantities, so it may be necessary to use vectors with four or ten or perhaps even hundreds of components. We will study vectors like this in Section 15.6.

Figure 14.3.10(i)

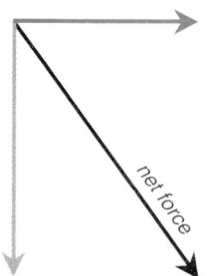

Figure 14.3.10(ii)

Force and Work

Now we observe that forces are another type of physical quantity that can be represented by vectors. The direction of the vector indicates the direction in which the force is acting, and its norm indicates the strength of the force. Once again the way forces combine is consistent with the addition of vectors.

EXAMPLE 6

(i) Two movers are pushing a heavy crate along a floor. One is pushing south with a force of 80 lb, and the other is pushing east with a force of 60 lb. The situation is illustrated in Figure 14.3.10(i).

   In Figure 14.3.10(ii) the forces are represented by the vectors $(0, -80)$ and $(60, 0)$, seen from above. The net force on the crate is the sum $(0, -80) + (60, 0) = (60, -80)$. This force has magnitude equal to $\|(60, -80)\| = 100$, so the net force on the crate is 100 lb. The direction of the net force is the direction of the vector $(60, -80)$, which makes an angle $\arctan \frac{-80}{60} \approx -53°$ with the positive x-axis, that is, with the direction due east. The net force is 100 lb in the direction $53°$ south of east.

(ii) In Figure 14.3.11(i) we see somebody pulling a heavy object with a rope attached to it. The rope makes an angle of $30°$ with the ground, and the force with which it is pulled is 200 newtons.

   In Figure 14.3.11(ii) we see the vector $\mathbf{F}$ that represents the force. Its length is 200, and since a unit vector pointing $30°$ above the horizontal is $(\cos 30°, \sin 30°) = \left(\frac{\sqrt{3}}{2}, \frac{1}{2} \right)$, we multiply by 200 to get

$$\mathbf{F} = (100\sqrt{3}, 100).$$

   From this we see immediately that the horizontal component of the force is $100\sqrt{3}$ N and the vertical component of the force is 100 N. This means that the rope exerts an upward force of 100 N on the object and pulls it with a horizontal force of $100\sqrt{3}$ N.

   Assuming that the object weighs more than 100 N (i.e., 10.2 kg), the upward force will not be sufficient to lift it up. The only visible effect will be that the object will be pulled along the floor with a force of $100\sqrt{3}$ N.

(iii) In Figure 14.3.12(i) we see a wagon on a ramp. The wagon weighs 15 lb, and the angle of inclination of the ramp is θ. The weight of the wagon is a force of 15 lb pointing straight down, so it is the vector $\mathbf{w} = (0, -15)$ shown in Figure 14.3.12(ii).

   The interesting question is the way in which the wagon may move along the ramp. For this it is important to know the component of the force along the direction of the ramp. It is easy to find a unit vector along the downward direction of the ramp: $\mathbf{u} = (\cos \theta, -\sin \theta)$.

Figure 14.3.11(i)

Figure 14.3.11(ii)

Figure 14.3.12(i)

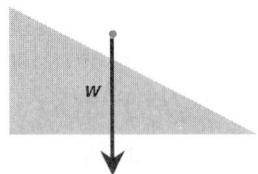

Figure 14.3.12(ii)

From this we find that the component of the force **w** along the direction of the ramp is

$$\mathbf{F} = (\mathbf{w} \cdot \mathbf{u})\mathbf{u} = \big((0, -15) \cdot (\cos\theta, -\sin\theta)\big)\mathbf{u} = 15\sin\theta(\cos\theta, -\sin\theta)$$

$$= (15\sin\theta\cos\theta, -15\sin^2\theta).$$

The magnitude of the component of the force along the ramp is the length of this vector, which is $15\sin\theta$. Notice that when $\theta = 0$, that is, when the ramp is horizontal, the force is zero, exactly as we expect. If the ramp is inclined more steeply, the force increases, which is why the wagon will roll more quickly off a steeply inclined ramp.

The component of the force perpendicular to the surface of the ramp is the vector $\mathbf{p} = \mathbf{w} - \mathbf{F} = (0, -15) - (15\sin\theta\cos\theta, -15\sin^2\theta) = (-15\sin\theta\cos\theta, -15 + 15\sin^2\theta) = (-15\sin\theta\cos\theta, -15\cos^2\theta)$.

This force pushes the wagon into the surface of the ramp (see Figure 14.3.12(iii)). The surface of the ramp exerts a force of the same size but in the opposite direction on the wheels of the wagon. This force exactly balances the force **p**, and this is why the wagon does not move in the direction perpendicular to the ramp surface. ◢

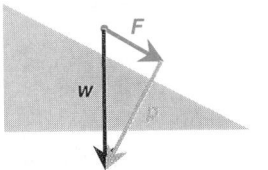

Figure 14.3.12(iii)

Here we discuss a famous example.

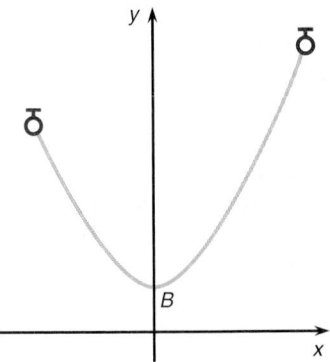

Figure 14.3.13(i)

EXAMPLE 7   The Catenary   Consider a hanging cable, that is, a cable that is freely suspended by its two ends. The question is to find the shape of the stationary curve it forms (see Figure 14.3.13(i)). Let us suppose the curve is the graph of a function $y = f(x)$ and set the axes so that the bottom point B on the curve is at $x = 0$.

Now suppose $P = \big(x, f(x)\big)$ is a point on the curve, and consider the part of the curve between B and P. Since the cable is not moving, there must be no net force acting on it. There are three forces acting on this section of the cable. One is **g**, the downward force of gravity. The other two are the forces $\mathbf{F}_B$ and $\mathbf{F}_P$ acting through the cable at B and P. These forces act along the direction of the cable, that is, in the directions tangent to the curve at those points (see Figure 14.3.13(ii)). The three forces must add up to the zero vector.

Now since B is the bottom point on the curve, the tangent vector $\mathbf{F}_B$ is horizontal; we write $\mathbf{F}_B = (-a, 0)$. The force of gravity points straight down; we write $\mathbf{g} = (0, -w)$, where w is the weight of the section of cable. Since $\mathbf{g} + \mathbf{F}_B + \mathbf{F}_P = 0$, we see that $\mathbf{F}_P = (a, w)$. On the other hand, $\mathbf{F}_P$ is tangent to $y = f(x)$, so the slope is $f'(x) = \frac{w}{a}$.

The weight of a section of cable is proportional to its arclength. If the density of the cable per unit arclength is ρ, then the weight of the section of cable between B and P, that is, between 0 and x, is

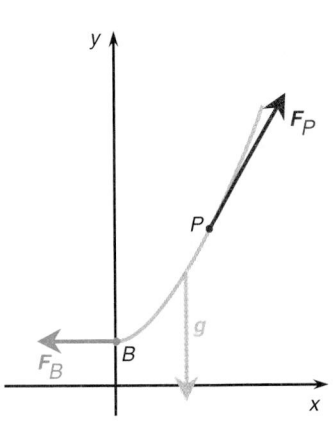

Figure 14.3.13(ii)

$$w = \rho\, arclength = \rho \int_0^x \sqrt{1 + f'(t)^2}\, dt.$$

Combining these equations, we see that $f'(x) = \frac{w}{a} = \frac{\rho}{a}\int_0^x \sqrt{1 + f'(t)^2}\, dt$. Differentiating, we find $f''(x) = \frac{\rho}{a}\sqrt{1 + f'(x)^2}$.

If we write $u(x) = f'(x)$, this equation reads $\frac{du}{dx} = \frac{\rho}{a}\sqrt{1 + u^2}$, or $\frac{1}{\sqrt{1+u^2}}\frac{du}{dx} = \frac{\rho}{a}$. Integrating with respect to x, we find that

$$\int \frac{1}{\sqrt{1 + u^2}}\, du = \int \frac{\rho}{a}\, dx = \frac{\rho}{a}x.$$

Substituting $u = \sinh t$, $du = \cosh t \, dt$, we find the left integral becomes $\int \frac{\cosh t}{\cosh t} \, dt = \int dt = t + c = \text{arcsinh}(u) + c$. The resulting equation $\text{arcsinh}(u) + c = \frac{\rho}{a}x$ can be written as $u = \sinh\left(\frac{\rho}{a}x - c\right)$.

Since $u = f'(x)$ equals 0 when $x = 0$, we see that $c = 0$ and $f'(x) = u = \sinh\left(\frac{\rho}{a}x\right)$. Finally, we integrate to find that $f(x) = \frac{a}{\rho}\cosh\left(\frac{\rho}{a}x\right) + C$.

The shape of the hanging cable is the graph of a hyperbolic cosine with suitable scaling on the axes.

> When we encountered the integral $\int \frac{1}{\sqrt{1+u^2}} \, du$, our immediate reaction might have been to substitute $u = \tan\theta$. This is perfectly reasonable, but if you try it, you will find that in this form the formula for $f'(x)$ is difficult to integrate to find $f(x)$.
>
> The curve $y = K\cosh(kx)$ is called a **catenary**, after the Latin word for a chain.

In Section 6.6 we studied the concept of work. The idea was that a force F acting on an object that moves a displacement s will do an amount of work equal to Fs.

Figure 14.3.14

In fact the situation is more subtle than it first appears. If we consider the wagon on the ramp in Figure 14.3.14, we see that the wagon may roll down the ramp, pulled by the component of the force along the direction of the ramp. The other component of the force, the one perpendicular to the motion, does nothing to the wagon. This suggests that the work performed by a force on an object must depend on the directions of the force and the motion.

If a force acts on an object and the object moves, but not exactly in the direction of the force, then the force will do some work on the object, but just how much will depend on the angle between the force and the motion. If the force and the direction of the motion are perpendicular, then the force does no work on the object. This may remind us of dot products, and in fact the result is the following.

THEOREM 14.3.3

> A constant force **F** acting on an object that moves through a displacement **s** will perform work on the object, and the amount of the work is the dot product $W = \mathbf{F} \cdot \mathbf{s}$.

Figure 14.3.15

EXAMPLE 8 In Example 6(ii) we considered a heavy object being pulled with a force of 200 N by a rope at an angle of $30°$ above the horizontal (see Figure 14.3.15).

We found that this force is $\mathbf{F} = (100\sqrt{3}, 100)$. If we suppose that the object is dragged 12 meters by the force, then it moves through a displacement $\mathbf{s} = (12, 0)$, and the work done on it is

$$W = \mathbf{F} \cdot \mathbf{s} = (100\sqrt{3}, 100) \cdot (12, 0) = 1200\sqrt{3} \approx 2078.$$

The work done on the object is about 2078 joules.

Notice that the horizontal component of the force is the vector $(100\sqrt{3}, 0)$, which has magnitude $100\sqrt{3}$ N. In effect the work done is found by multiplying this by the displacement (since the displacement is horizontal). However, dot products give a more convenient way to express the calculation, and they do not rely on having the displacement in a horizontal direction.

Figure 14.3.16

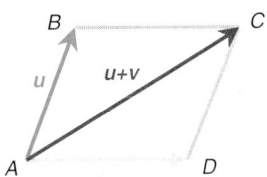

Figure 14.3.17

Cross Products

Finally, we mention a simple use for cross products, though we will discover some others in the next two sections.

Consider the parallelogram illustrated in Figure 14.3.16. One of its vertices is at the origin, and the vectors **u** and **v** form two of its sides. We mark θ, the angle between **u** and **v**, on the diagram, and recall that the area of a parallelogram is *base* $\times$ *height* (cf. Summary 1.5.13). In our diagram the base of the parallelogram is $\|\mathbf{v}\|$, and the height is $h = \|\mathbf{u}\| \sin\theta$.

We see that the area of the parallelogram is $\|\mathbf{v}\|\|\mathbf{u}\| \sin\theta$, and we recognize this as Lagrange's Identity (Theorem 14.2.11) for the norm of the cross product $\mathbf{u} \times \mathbf{v}$. We have just proved the first part of the following proposition.

PROPOSITION 14.3.4

If two vectors **u** and **v** form two adjacent sides of a parallelogram, then the **area of the parallelogram** is

$$Area = \|\mathbf{u} \times \mathbf{v}\|.$$

Moreover, if A, B, C, and D are the vertices of the parallelogram, then the area equals the norm of the cross product of any two of the vectors joining A to B, A to C, or A to D.

PROOF We only need to prove the second observation. In Figure 14.3.17 we see the same parallelogram with the vertices labeled A, B, C, and D. The vector we called **v** runs from A to D and **u** runs from A to B. We know that the norm of their cross product equals the area.

Now consider the vector from A to C. By the Triangle Rule for vectors it equals $\mathbf{u} + \mathbf{v}$, so we consider

$$\mathbf{u} \times (\mathbf{u} + \mathbf{v}) = \mathbf{u} \times \mathbf{u} + \mathbf{u} \times \mathbf{v} = 0 + \mathbf{u} \times \mathbf{v} = \mathbf{u} \times \mathbf{v}.$$

The cross product of **u** with $\mathbf{u} + \mathbf{v}$ equals $\mathbf{u} \times \mathbf{v}$, so its norm equals the area. Virtually the same argument applies to $\mathbf{v} \times (\mathbf{u} + \mathbf{v})$. ◢

EXAMPLE 9 (i) Find the area of the parallelogram whose vertices are $(0, 0, 0)$, $(3, 1, -2)$, $(2, 1, 1)$, and $(5, 2, -1)$.

Solution Let $\mathbf{u} = (3, 1, -2)$ and $\mathbf{v} = (2, 1, 1)$. Then the area of the parallelogram is

$$\|(3, 1, -2) \times (2, 1, 1)\| = \|(3, -7, 1)\| = \sqrt{59}.$$

(ii) Find the area of the parallelogram whose vertices are $(3, 4, 2)$, $(4, 2, 3)$, $(6, 1, 2)$, and $(5, 3, 1)$.

Solution It is necessary to find vectors that form two sides of the parallelogram. Suppose we start at the first point $(3, 4, 2)$. To get to the second point $(4, 2, 3)$, we have to add $\mathbf{u} = (1, -2, 1)$, that is, $(4, 2, 3) = (3, 4, 2) + \mathbf{u}$. Similarly, $(6, 1, 2) = (3, 4, 2) + \mathbf{v}$, with $\mathbf{v} = (3, -3, 0)$.

The area equals $\|\mathbf{u} \times \mathbf{v}\| = \|(1, -2, 1) \times (3, -3, 0)\| = \|(3, 3, 3)\| = 3\sqrt{3}$.

(iii) Find the area of the parallelogram whose vertices are $(4, 2)$, $(3, 3)$, $(2, -1)$, and $(1, 0)$.

Solution To use the proposition, we need to take cross products, which only works for vectors in $\mathbb{R}^3$. One way around this difficulty is to think of the xy-plane as the horizontal plane in three-dimensional space.

The parallelogram has vertices $(4, 2, 0)$, $(3, 3, 0)$, $(2, -1, 0)$, and $(1, 0, 0)$, which are just the points obtained by adding a z-coordinate of 0 to the given points.

Letting $\mathbf{u} = (3, 3, 0) - (4, 2, 0) = (-1, 1, 0)$ and $\mathbf{v} = (2, -1, 0) - (4, 2, 0) = (-2, -3, 0)$, we find that the area of the parallelogram is

$$\|\mathbf{u} \times \mathbf{v}\| = \|(-1, 1, 0) \times (-2, -3, 0)\| = \|(0, 0, 5)\| = 5.$$

The area of the parallelogram is 5 square units.

In using Proposition 14.3.4 the important thing is to use two vectors that start at the *same* vertex of the parallelogram. Pick one vertex as a starting point and find the vectors $\mathbf{u}$ and $\mathbf{v}$ that run from it to any two other vertices. The area is then $\|\mathbf{u} \times \mathbf{v}\|$.

Exercises 14.3

I

Find the distance between the beginning and end points of a journey made up of the following movements:

1. 1 mile north and 3 miles west

2. 5 miles east and 2 miles south

3. 4 miles northwest and 2 miles south

4. 4 miles southeast and 3 miles southwest

5. 2 miles northwest and 2 miles southeast

6. 5 miles southwest and 1 mile straight up

7. 20 feet north and 15 feet down

8. 3 km west and 7 km southeast

9. 7 m up and 5 m horizontally

10. 10 ft down and 7 ft down a 45° ramp

Find the following components.

11. The component along $(1, 0)$ of $(5, -12)$.

12. The component of $(5, -12)$ along the direction of $(1, -1)$.

13. The component of $(3, 4, 2)$ along the direction of $(1, 1, 1)$.

14. The component of $(4, 3, -1)$ along the direction of $(2, -1, -2)$.

15. The component of $(1, 0)$ along the direction of $(1, 1)$.

16. The component of $(1, 0, 0)$ along the direction of $(1, 1, 1)$.

17. The component of $(0, 0, 0)$ along the direction of $(5, -7, 12)$.

18. The vertical component of $(6, -7, -3)$.

19. The component of $(4, 2, 5)$ perpendicular to $(1, -2, -2)$.

20. The component of $(3, -1, 4)$ perpendicular to $(2, 2, -1)$.

In each of the following cases, find the work done.

21. By a 10-lb force pushing north on an object that moves 30 ft north.

22. By a 7-lb force pushing north on an object that moves 8 ft east.

23. By a 6-lb force pushing northwest on an object that moves 12 ft west.

24. By a 40-lb force pushing west on an object that moves 40 ft northeast.

25. By an 80-N force pushing southwest on an object that moves 3 m northwest.

26. By a 60-N force pushing up on an object that moves 10 m up a 45° ramp.

27. By a 10-lb force pushing up on an object that slides 7 ft down a 45° ramp.

28. By a 6-lb force holding up an object that slides 4 ft down a 30° ramp.

29. By gravity on a 12-lb object as it falls 108 ft.

30. By gravity on a 5-kg object as it slides 25 m down a 45° ramp.

II

31. What is the total distance traveled by somebody who climbs a ramp that is inclined at a $45°$ angle if the top of the ramp is 12 feet higher than the bottom?

32. What is the total distance traveled by somebody who climbs a ramp that is inclined at a $30°$ angle above the horizontal if the top of the ramp is 5 m higher than the bottom?

*33. (i) Find parametric equations describing the motion of a point on the rim of a wheel of radius 2 ft that is rolling at a speed of 5 ft/sec.
    (ii) What are the maximum and minimum speeds of the point in part (i)?

*34. (i) Find parametric equations describing the motion of a point halfway between the center and the rim of a wheel of radius 20 inches that is rolling at a speed of 40 in/sec.
    (ii) What are the maximum and minimum speeds of the point in part (i)?

35. Suppose the boat of Example 3 gets a new engine, so its speed in still water is now 15 mph.
    (i) What will be its speed if it steers directly across the river?
    (ii) At what angle should it steer to move directly across the river?

*36. Spring rains increase the current so that a river that is 1 mile wide flows at 6 mph. How long will it take a boat that can go 8 mph in still water to cross the river if it steers so as to go straight across?

*37. At what angle should the boat of Example 3 steer to reach the opposite bank as fast as possible?

*38. A pilot in an airplane that has an airspeed of 300 mph wants to fly due west. Knowing that there is a north wind (i.e., from the north) blowing at 40 mph, where should the pilot steer?

*39. How big a force will it take to hold a 20-lb wagon on a $45°$ incline?

*40. A 50-lb crate is on a ramp inclined $30°$ above the horizontal. The force of friction resisting its sliding down the ramp is 22 lb. Will it slide?

*41. How much work is done by gravity on a 10-lb bag of potatoes as a grocer lifts it 5 ft onto his shoulder, carries it 20 ft to the back of the store, carries it down a ramp that runs across the width of the store for a distance of 50 ft and a drop of 12 ft, and finally lowers it onto a shelf that is 6 inches above the basement floor level?

**42. One way to raise a wagon onto a workbench is simply to lift it straight up. Another way is to roll it up a ramp that slopes up to the table at a $45°$ angle. Suppose the wagon weighs 20 lb and the bench is 3 ft high.
    (i) What force will be needed to push the wagon up the ramp (i.e., to balance the force exerted by the weight of the wagon along the direction of the ramp)?
    (ii) Compare the amounts of work done by using this amount of force to push the wagon up the ramp and by lifting it straight up using a force equal to its weight.

43. Find the area of the parallelogram whose vertices are $(4, 5)$, $(3, 2)$, $(5, 6)$, and $(4, 3)$.

44. Find the area of the parallelogram whose vertices are $(1, 2, 4)$, $(2, -2, 0)$, $(4, 4, 3)$, and $(5, 0, -1)$.

**45. (i) How many parallelograms are there that have $(1, 2)$, $(3, -2)$, and $(2, 4)$ as three of their vertices? (Draw a picture.)
    (ii) Show that all of them have the same area.

 ← POINT TO PONDER

Imagine a satellite in a circular orbit around the earth (see Figure 14.3.18(i)). It is being pulled by the force of gravity in the direction toward the center of the earth. This force is the vector **g** in Figure 14.3.18(ii).

The velocity vector **v** is also shown; it is tangent to the orbit, and in particular it is perpendicular to the vector **g**. The satellite always moves in such a way that its displacement is perpendicular to the direction in which the force of gravity acts.

The result is that gravity does no work on the satellite. Even though the satellite is moving *and* being acted on by this force, no work is done because the force is perpendicular to the displacement.

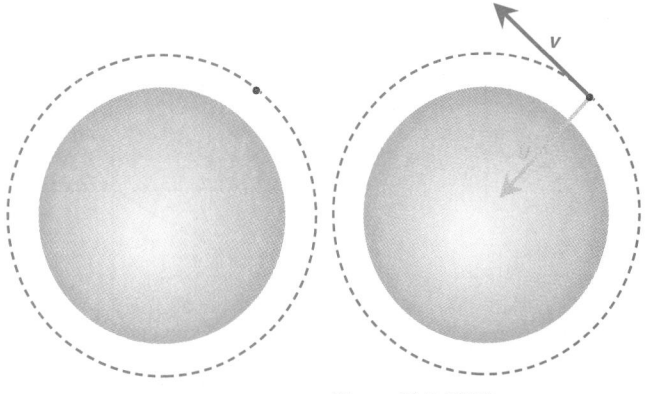

Figure 14.3.18(i)   Figure 14.3.18(ii)

In fact this is why the satellite stays up. If gravity did perform work on it, this would change the total energy of the satellite, and it would move lower and not remain in its circular orbit.

In practice what happens is that collisions with tiny particles in space exert a small force on the satellite.

This force tends to act as a drag on the satellite, that is, in the direction opposite to the displacement, so they do a negative quantity of work and slow the satellite down. Eventually, this effect is sufficiently large that the satellite falls to earth.

14.4 Lines in Three-Dimensional Space

DIRECTION VECTOR
VECTOR EQUATION
SYMMETRIC EQUATIONS
PARALLEL LINES
UNIT VECTOR
DIRECTION COSINES

In this section and the next we use the terminology of vectors to help us describe lines and planes in space.

First we study lines in $\mathbb{R}^3$. For instance, we could consider the z-axis, shown in Figure 14.4.1. One way to describe this line is that it is the set of all points that can be reached by moving in a vertical direction, starting at the origin.

For that matter, it is not really necessary to start at the origin; *any* point on the line will do. If P is such a point, then any other point on the line can be obtained by starting at P and moving some displacement in the vertical direction.

In Figure 14.4.2 the y-axis is shown. Any point on this line can be obtained by starting at the origin, say, and moving in the appropriate direction. In this case the appropriate direction is the direction of the y-axis, which is also the direction of the standard basis vector $\mathbf{j}$.

Now consider the line shown in Figure 14.4.3. It goes through the origin and also through the point $(1, 1, 1)$. Specifying two points on the line is enough to determine it. We would like to describe all the points on the line, and to do this, it is helpful to follow the idea used for the axes. There we described a line by starting at some fixed point on the line and then moving in some specified direction.

In this case it is probably simplest to start at the origin, and then we must give the direction. The required direction is the direction from the origin to the point $(1, 1, 1)$. The line is all the points that can be obtained by starting at the origin and then moving in the direction from the origin to $(1, 1, 1)$.

Now that we are talking about directions, it seems natural to use vectors. The line we have been describing is the set of all points that can be obtained by starting at the origin and moving in the same direction as the vector $\mathbf{v} = (1, 1, 1)$. Actually, this is

Figure 14.4.1

Figure 14.4.2

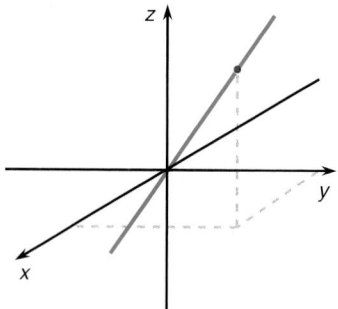

Figure 14.4.3

slightly inaccurate, because we should also be able to move in the opposite direction in order to get the other half of the line. To avoid this difficulty of terminology, we can speak of moving in the direction "parallel to **v**," which includes moving both forward and back. So for instance the line includes the point that is obtained by starting at the origin and moving a displacement of $-2(1,1,1)$; this is the point $(0,0,0) + (-2)(1,1,1) = (-2,-2,-2)$.

One advantage of using vectors is that we are used to placing a vector with its initial point at any chosen point. So if we chose to use a different starting point, our description would still make sense. For instance, the same line can equally well be described as the set of all points that can be obtained by starting at the point $(-2,-2,-2)$ and moving in the direction parallel to the vector $\mathbf{v} = (1,1,1)$.

In Figure 14.4.4 we see a line. Marked on it are a *starting point P* and a direction vector **v**. There are also two other points Q and R. The point Q is obtained by starting at P and moving 4 times the length of **v** in the direction of **v**. It is much easier to say this by using the vector notation and writing $Q = P + 4\mathbf{v}$. Notice that when we write Q and P here, we really mean the *vectors OQ* and *OP*, but not distinguishing points from the corresponding vectors is actually very convenient.

Similarly, the point R is reached from P by moving in the direction opposite to **v**. The distance is $\frac{3}{2}$ the length of **v**, so we write $R = P - \frac{3}{2}\mathbf{v}$.

From these simple examples we see that every point S on the line can be described as $S = P + t\mathbf{v}$, for some real number t. If $t = 0$, we just get P; if $t > 0$, we get points that are on one side of P (i.e., in the direction in which **v** points); and if $t < 0$, we get the points on the other side of P.

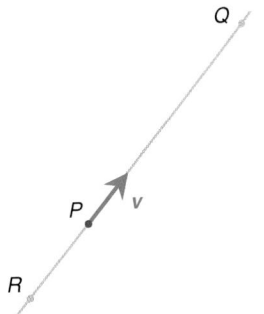

Figure 14.4.4

DEFINITION 14.4.1

Let $P = (X, Y, Z)$ be a point in $\mathbb{R}^3$. Let $\mathbf{v} = (a, b, c)$ be a nonzero vector. The **line** through P with **direction vector v** is the set of all points $Q = P + t\mathbf{v}$, where t is any real number. This is the **vector equation** of the line.

The points on the line can also be written out in terms of their coordinates:

$$Q = (x, y, z) = P + t\mathbf{v} = (X, Y, Z) + t(a, b, c) = (X + ta, Y + tb, Z + tc).$$

These are **parametric equations** for the line.

EXAMPLE 1   Find parametric equations for the line through the points $(2, -1, 3)$ and $(-1, 3, 5)$.

Solution   We can begin by choosing a starting point, and the obvious choice is one or other of the two points we are given. Let us let $P = (2, -1, 3)$ be the starting point.

For the direction vector we can choose the vector from P to the other point, that is, $\mathbf{v} = (-1, 3, 5) - (2, -1, 3) = (-3, 4, 2)$. The equation of the line is $Q = P + t\mathbf{v} = (2, -1, 3) + t(-3, 4, 2) = (2 - 3t, -1 + 4t, 3 + 2t)$.

The parametric equations can also be written as

$$x = 2 - 3t, \qquad y = -1 + 4t, \qquad z = 3 + 2t.$$

Notice that when $t = 0$, we obtain the point $P = (2, -1, 3)$, and when $t = 1$, we have the other point $(-1, 3, 5)$, since we start at P and move along 1 times the vector that goes from P to $(-1, 3, 5)$.

EXAMPLE 2

(i) The same line can be parametrized by choosing the other point as starting point. In this case we would have $(x, y, z) = (-1, 3, 5) + t(-3, 4, 2)$. Here when $t = 0$, the point is $(-1, 3, 5)$, and when $t = 1$, the point is $(-4, 7, 7)$. To get the point P, we have to move *backward* from $(-1, 3, 5)$, so we need to let $t = -1$.

(ii) It is also possible to change the direction vector. After all, $-\mathbf{v}$ or $3\mathbf{v}$ or any nonzero scalar multiple of $\mathbf{v}$ also points in a direction parallel to the direction of $\mathbf{v}$, so any one of them would be equally good as a choice for the direction vector of the line.

For instance, if $\mathbf{u} = -3\mathbf{v} = 3(-3, 4, 2) = (9, -12, -6)$, then $\mathbf{u}$ is also a direction vector for the line. This means that the line can be written as $Q = (-1, 3, 5) + t(9, -12, -6)$.

What value of the parameter in this parametrization will correspond to the point $(-4, 7, 7)$? We begin by considering any one of the three parts of the parametric equations. For instance, we know that $x = -1 + 9t$, so to obtain the point $(-4, 7, 7)$, we need to have that $-1 + 9t = -4$, that is, $9t = -3$ and $t = -\frac{1}{3}$. It is easy to check that with $t = -\frac{1}{3}$ we find the point $Q = (-1, 3, 5) - \frac{1}{3}(9, -12, -6) = (-4, 7, 7)$, as required.

(iii) Is the point $(2, 4, -5)$ on the line? We want to know whether it is possible to find t so that $(2, 4, -5) = (-1, 3, 5) + t(9, -12, -6)$. We proceed just as we did in part (ii) and solve for t from any one of the three equations. For instance, using the first coordinate, we find that $2 = -1 + 9t$, or $3 = 9t$ and $t = \frac{1}{3}$.

Now we can check that $(-1, 3, 5) + \frac{1}{3}(9, -12, -6) = (-1, 3, 5) + (3, -4, -2) = (2, -1, 3)$. This is certainly not the point we wanted. What we have found is that if we look for a point on the line whose first coordinate is 2, then we find $(2, -1, 3)$. From this we conclude that $(2, 4, -5)$ is *not* on the line. ◢

From these examples we see that it is possible to find many different versions of the parametric equations for any particular line. Since it is possible to change either the starting point, or the direction vector, or both, different versions can look quite different, and it may not be easy to tell whether or not they represent the same line.

EXAMPLE 3 (i) Consider the lines $(x, y, z) = (2, 2, -1) + t(4, -6, 2)$ and $(x, y, z) = (4, -1, 1) + t(-2, 3, -1)$. Are they the same?

Solution First we observe that the second direction vector is $-\frac{1}{2}$ times the first, which means that the directions of the two lines are parallel. If they were not parallel, then the lines could not possibly be the same.

Next we might ask whether the starting point of the second line is on the first line, that is, whether $(4, -1, 1)$ can be written as $(4, -1, 1) = (2, 2, -1) + t(4, -6, 2)$. Starting with the first coordinate, we find that this means $4 = 2 + 4t$, so $t = \frac{1}{2}$. With $t = \frac{1}{2}$ we obtain the point $(2, 2, -1) + \frac{1}{2}(4, -6, 2) = (2, 2, -1) + (2, -3, 1) = (4, -1, 0)$. Since this is not the desired point, we conclude that the two lines are different.

In fact the situation is that they are parallel but distinct. See Figure 14.4.5 .

(ii) If we consider the lines $(x, y, z) = \left(5 - 2t, 6 + 3t, -2 + \frac{3}{4}t\right)$ and $(x, y, z) = (2 + t, 5 - 2t, 8 + 4t)$, we see that their direction vectors are $\left(-2, 3, \frac{3}{4}\right)$ and $(1, -2, 4)$, respectively. Since these vectors are not parallel, that is, not scalar multiples of each other, the lines are not even parallel, let alone the same.

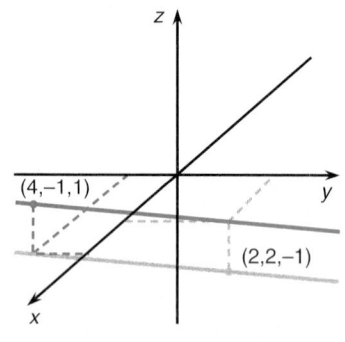

Figure 14.4.5

(iii) Are the following two lines the same: $(x, y, z) = (4, -3, 2) + t(0, 3, -1)$ and $(x, y, z) = (4, 0, 1) + t(0, -3, 1)$?

Solution   Noting that the directions vectors are parallel, being each other's negatives, we ask whether the starting point of the second line lies on the first, that is, whether $(4, 0, 1)$ can be written as $(4, 0, 1) = (4, -3, 2) + t(0, 3, -1)$.

Previously, we began such a calculation by solving with the *x*-coordinate. In this case that gives $4 = 4 + (0)(t)$, which cannot be used to solve for *t*. Instead, we work with the *y*-coordinate: $0 = -3 + 3t$, from which we find that $t = 1$.

Using this value in the parametric equation, we obtain the point $(4, -3, 2) + 1 \times (0, 3, -1) = (4, 0, 1)$. This is the desired point, and we see that the starting point of the second line is in fact on the first line. Since the two direction vectors are parallel, this means that the two lines are the same.   ◢

Another natural type of question involves the intersection of two lines.

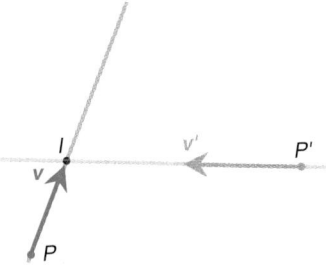

Figure 14.4.6

EXAMPLE 4   Do the two lines $(x, y, z) = (2 - t, 3 + 2t, -1 + t)$ and $(x, y, z) = (-3 + 2t, -5 + 5t, -6 + 3t)$ intersect?

Solution   We begin by drawing a sketch in Figure 14.4.6. On it we show the two lines, each with its starting point and direction vector. Suppose for the sake of the picture that there is some point *I* at which the lines intersect. What we need to do is find how far we have to go along each line from its starting point to reach that intersection point. So if the lines are $(x, y, z) = P + t\mathbf{v}$ and $(x, y, z) = P' + t\mathbf{v}'$, we need to find the value of *t* so that $I = P + t\mathbf{v}$ and the value of *t* so that $I = P' + t\mathbf{v}'$.

What makes this confusing is that the symbol *t* has been used for two different things. In the first place it tells how far we have moved along one line from its starting point, and in the second it tells how far we moved along the other line from its starting point. If the two lines intersect, there is no reason at all to expect that these two things are the same.

Since the two parameters represent two different things, the difficulty arises from using the same letter for both of them. It can be removed by using a different letter, *s*, say, for one of the parameters. We now write the lines as

$$(x, y, z) = (2 - t, 3 + 2t, -1 + t) \quad \text{and} \quad (x, y, z) = (-3 + 2s, -5 + 5s, -6 + 3s).$$

The question becomes: Can we find values of *s* and *t* so that these points are equal?

If we can, then the two *x*-coordinates are equal: $2 - t = -3 + 2s$, and the *y*-coordinates are equal: $3 + 2t = -5 + 5s$. We solve these equations by adding twice the first to the second to eliminate the *t*: $7 = -11 + 9s$. This tells us that $18 = 9s$ and $s = 2$. Substituting this value back into the first equation, we find that $2 - t = -3 + 2s = -3 + 4 = 1$, so $t = 1$.

What we have found is that if it is possible to find values *t* and *s* of the parameters so that the points coincide, then they must be $t = 1$ and $s = 2$. We found them by using the *x*- and *y*-coordinates, so it is still unclear whether the *z*-coordinates will match too.

Substituting $t = 1$ into the equation of the first line, we obtain the point $(x, y, z) = (1, 5, 0)$. Substituting $s = 2$ into the second line, we find $(x, y, z) = (1, 5, 0)$.

DesignCAD 3D *software was used to produce this image. It starts with an architect's floorplan, and "extrudes" it to produce the corresponding three-dimensional picture, making it much easier to visualize the final construction.*

The point $(1, 5, 0)$ is on both lines (on the first when the parameter is $t = 1$ and on the second when the parameter is $s = 2$). In particular we see that the two lines do intersect.

EXAMPLE 5   (i) Do the lines $(x, y, z) = (3 - 2t, 5 + t, 3t - 6)$ and $(x, y, z) = (1 + 2t, 6 - t, -3 - 3t)$ intersect?

Solution   This time we observe that the two direction vectors are scalar multiples of each other, so the lines are parallel. This means that they must either be the same line or else not intersect at all.

To find out which it is, we only need to see whether the starting point $(1, 6, -3)$ of the second line lies on the first. In other words, can we find t so that $(3 - 2t, 5 + t, 3t - 6) = (1, 6, -3)$?

It is easy to see that $t = 1$ works, so we find that the starting point of the second line is on the first, and since the directions are parallel, the two lines must be exactly the same.

(ii) Do the lines $(x, y, z) = (1 - 2t, 3 + t, 2 + 5t)$ and $(x, y, z) = (4 - t, 2 + 3t, 3 + t)$ intersect?

Solution   These two lines are clearly not parallel. As before we begin by using different letters for the parameters of the two lines; let us write the second line as $(x, y, z) = (4 - s, 2 + 3s, 3 + s)$. We set the x-coordinates equal and the y-coordinates equal and solve for t and s. The equations are

$$1 - 2t = 4 - s,$$
$$3 + t = 2 + 3s.$$

Adding three times the first equation to the second to eliminate the s, we find that $6 - 5t = 14$, so $t = -\frac{8}{5}$.

From this and the first equation we find that $1 + \frac{16}{5} = 4 - s$, so $s = -\frac{1}{5}$. The point on the first line corresponding to $t = -\frac{8}{5}$ is $(x, y, z) = \left(\frac{21}{5}, \frac{7}{5}, -6\right)$, and the point on the second line corresponding to $s = -\frac{1}{5}$ is $(x, y, z) = \left(\frac{21}{5}, \frac{7}{5}, \frac{14}{5}\right)$. They are not the same, and we conclude that the lines do not intersect.

There are two things to notice in the second part of this example. One is that the two points we found have the same first two coordinates. This is not surprising, because the way we found them was to set the first two coordinates equal and then solve for the parameters. The real question is whether or not the third coordinates agree. (However, it is probably worthwhile to work out all the coordinates, since it provides a check on the arithmetic.)

The other important observation is that in this example there are two lines that are not parallel but that do not intersect. In the plane we know that any two nonparallel lines must intersect, but in three-dimensional space $\mathbb{R}^3$ there is enough room that they can miss each other entirely. See Figure 14.4.7. Lines that are not parallel are sometimes called **skew**.

When we deal with lines in the plane, we are accustomed to finding an equation for the line that does not involve a parameter. In space it is not possible to find a single equation, but it is possible to describe the line without using parameters.

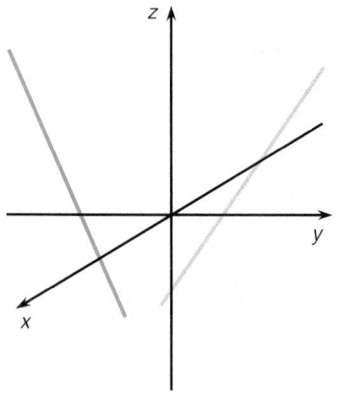

Figure 14.4.7

EXAMPLE 6   Consider the line

$$(x, y, z) = (3 - t, -2 + 3t, 4 + 4t).$$

This can be written as three separate equations for the coordinates, namely, $x = 3 - t$, $y = -2 + 3t$, and $z = 4 + 4t$. It is possible to solve each of these three equations for t; we find

$$t = 3 - x, \qquad t = \frac{y + 2}{3}, \qquad t = \frac{z - 4}{4}.$$

Here we have three expressions for the parameter t; in particular they must all be equal. This means that we can write

$$3 - x = \frac{y + 2}{3} = \frac{z - 4}{4}.$$

These equations determine the line. They are called the symmetric equations of the line. There are two equals signs; there are really three equations (each pair of expressions for t is equal), but any two of them are enough, since the third can always be found from two of them. ◀

It is easy to recover the parametric equations from the symmetric equations. Simply let t equal all three quotients and express x, y, and z in terms of t.

DEFINITION 14.4.2

Equations of the form

$$\frac{x - X}{a} = \frac{y - Y}{b} = \frac{z - Z}{c}$$

are the **symmetric equations** of a line containing the point (X, Y, Z). They determine the line whose parametric equations are

$$x = X + at, \qquad y = Y + bt, \qquad z = Z + ct.$$

Notice that the symmetric equations do not make sense if any of the components a, b, or c of the direction vector is zero.

Summary 14.4.3

Lines in Space

1. There are three ways of writing equations for a line in three-dimensional space $\mathbb{R}^3$. First there is the vector equation $(x, y, z) = P + t\mathbf{v}$, where $\mathbf{v} = (a, b, c)$ is the direction vector and $P = (X, Y, Z)$ is a point on the line.

    Second, these equations can be written slightly differently as the parametric equations, $x = X + at$, $y = Y + bt$, $z = Z + ct$.

    Finally, assuming that a, b, and c are all nonzero, there are the symmetric equations

$$\frac{x - X}{a} = \frac{y - Y}{b} = \frac{z - Z}{c}.$$

2. Each type of equation can be written in many different ways, corresponding to different choices of starting point P and different choices of direction vector. However, any two direction vectors must be parallel, which means that each must be a scalar multiple of the other.

3. To check whether a point lies on a line, find whether or not it satisfies the symmetric equations, or else decide whether or not it is possible to find a value of the parameter t that corresponds to the point.

4. To check whether two lines are the same, check that their direction vectors are parallel, and then check whether any single point on one of them lies on the other.

5. To find whether two lines intersect, try to find values of the parameters for both lines that give the same point. In doing this it is very important to use different symbols for the two parameters.

6. Two lines are **parallel** if their direction vectors are parallel, that is, are scalar multiples of each other. Two lines with parallel direction vectors may be the same line, but otherwise they do not intersect.

7. Two lines that are not parallel may or may not intersect. This is unlike the situation for lines in the plane, where two nonparallel lines must always meet.

One of the things that makes lines awkward to deal with is the fact that there are many possible choices of direction vector. If $\mathbf{v}$ is a direction vector for some line, then so is $k\mathbf{v}$ for any nonzero real number k. It would be convenient if there were a way to specify the direction vector without all this ambiguity.

One attempt to do this is to require that the direction vector be a **unit vector**, that is a vector whose norm is 1. Then there are only two possible choices for direction vector, $\mathbf{u}$ and $-\mathbf{u}$, since any other multiple will have a different norm. Moreover, it is easy to find a direction vector that is a unit vector: If $\mathbf{v}$ is any direction vector, then $\mathbf{u} = \frac{1}{\|\mathbf{v}\|}\mathbf{v}$ is a unit direction vector.

Now suppose $\mathbf{u}$ is a unit vector, and consider the dot product $\mathbf{u} \cdot \mathbf{i}$. If θ is the angle between $\mathbf{u}$ and $\mathbf{i}$, then

$$\mathbf{u} \cdot \mathbf{i} = \|\mathbf{u}\|\|\mathbf{i}\| \cos\theta = \cos\theta.$$

Note that $\|\mathbf{u}\| = \|\mathbf{i}\| = 1$.

This tells us that $\mathbf{u} \cdot \mathbf{i}$ is just the cosine of the angle between $\mathbf{u}$ and the positive x-axis. Similarly, $\mathbf{u} \cdot \mathbf{j}$ and $\mathbf{u} \cdot \mathbf{k}$ are the cosines of the angles between $\mathbf{u}$ and the positive y- and z-axes, respectively. These numbers are called the *direction cosines* of the unit vector $\mathbf{u}$.

DEFINITION 14.4.4

Suppose $\mathbf{u} = (\alpha, \beta, \gamma)$ is a unit vector, that is, $\|\mathbf{u}\| = 1$. Then its components α, β, and γ are the **direction cosines**, the cosines of the angles between $\mathbf{u}$ and the positive x-, y-, and z-axes, respectively.

Exercises 14.4

I

Find each of the following.

1. Parametric equations of the line through the points $(2, -1, 4)$ and $(5, 2, 1)$.

2. Parametric equations of the line through the points $(5, 6, -2)$ and $(3, -2, 4)$.

3. The vector equation of the line through the points $(3, 3, -2)$ and $(4, 4, -1)$.

4. The vector equation of the line through the points $(4, 2, -4)$ and $(3, 5, -2)$.

5. The symmetric equations of the line through the points $(1, 2, -4)$ and $(-3, 5, -2)$.

6. The symmetric equations of the line through the points $(-4, 2, 0)$ and $(2, -2, 4)$.

7. The parametric equations of the line through the point $(2, -3, 5)$ and parallel to the line $(x, y, z) = (t - 1, 2t + 3, -t)$.

8. The parametric equations of the line through the point $(-4, 4, 1)$ and parallel to the line $(x, y, z) = (2t - 1, 3t + 2, t - 5)$.

9. The parametric equations of the line through the point $(-1, -2, 7)$ and parallel to the y-axis.

10. The parametric equations of the vertical line through the point $(5, -4, 9)$.

Answer each of the following.

11. Is the point $(3, 2, 0)$ on the line $(x, y, z) = (t + 2, 3t - 1, 3t + 1)$?

12. Is the point $(2, 4, -2)$ on the line $(x, y, z) = (3t + 2, t - 5, 3 - 2t)$?

13. Is the point $(4, -2, 1)$ on the line $(x, y, z) = (6 - 2t, 1 - 3t, 2)$?

14. Is the point $(2, 2, 2)$ on the line $(x, y, z) = (3t - 1, 3t + 2, 3t + 1)$?

15. Are the lines $(x, y, z) = (t - 2, 2t + 1, 3t)$ and $(x, y, z) = (t, 2t + 5, 3t + 6)$ the same?

16. Are the lines $(x, y, z) = (4t + 2, 1 - 2t, t + 3)$ and $(x, y, z) = (1 - 8t, 4t, 1 - 2t)$ the same?

17. Are the lines $(x, y, z) = (t, t, 1 - t)$ and $(x, y, z) = (3 - 2t, 3 - 2t, 2t + 7)$ the same?

18. Are the lines $(x, y, z) = (2t + 2, 3t - 1, 5t - 4)$ and $(x, y, z) = (4 - 4t, 3 - 6t, 10t - 1)$ the same?

19. Are the lines $(x, y, z) = (2t + 1, t + 2, 4t - 2)$ and $(x, y, z) = (2t - 1, t + 5, 4t + 1)$ parallel?

20. Are the lines $(x, y, z) = (4t, 3t, -2t)$ and $(x, y, z) = (8 - 4t, 6 - 3t, 2t - 4)$ parallel?

Find the points of intersection of the following pairs of lines, if they intersect.

21. $(x, y, z) = (t + 3, 4 - 3t, 2t + 1)$ and $(x, y, z) = (2t - 2, 4 + 6t, 4 - 7t)$.

22. $(x, y, z) = (4, t + 2, 2 - 3t)$ and $(x, y, z) = (t + 2, 5 - 2t, 3t - 1)$.

23. $(x, y, z) = (t, 2t, 5 - 3t)$ and $(x, y, z) = (1 - t, t - 7, 3t + 2)$.

24. $(x, y, z) = (3t - 1, -6, t + 4)$ and $(x, y, z) = (2t + 5, 4t - 2, 3)$.

25. $(x, y, z) = (4t + 1, 2t - 4, 2 - 3t)$ and the line through $(1, 1, 2)$ and $(-2, -5, 7)$.

26. $(x, y, z) = (5t - 3, t - 4, 3 - 2t)$ and the line through $(12, -1, 4)$ and the origin.

27. $(x, y, z) = (2t - 1, 4 - 3t, t + 3)$ and $(x, y, z) = (3 - 2t, 3t - 2, 4 - t)$.

28. The lines through $(4, -3, 5)$ with direction vectors $\mathbf{u} = (3, -7, 12)$ and $\mathbf{v} = (-3, 11, 4)$.

29. The line through $(4, -3, 5)$ with direction vector $\mathbf{u} = (3, 3, -2)$ and the line through the origin with direction vector $\mathbf{v} = (1, 4, -3)$.

30. $(x, y, z) = (3t - 4, 2t + 1, 5 - t)$ and $(x, y, z) = (7t - 4, 1 + 3t, t + 5)$.

II

*31. In Example 2(iii) we solved for the right t to make the first coordinate of the point Q equal to 2. Try instead to find the value of t that makes the second coordinate match the point $(2, 4, -5)$, that is, $y = 4$. What happens if you work with $z = -5$?

*32. Do the two lines $(x, y, z) = (2 - t, 3 + 2t, -1 + t)$ and $(x, y, z) = (-3 + 2t, 13 - 4t, -6 + 3t)$ intersect?

*33. Find the parametric equations of the line that contains the point $(3, 5, -2)$ and that is orthogonal to the lines $(x, y, z) = (3 + t, 2 - 5t, 3t - 1)$ and $(x, y, z) = (2 + 4t, -3 + 2t, 7 - t)$. (Two lines are orthogonal if their direction vectors are orthogonal.)

*34. Find the parametric equations of the line that contains the point $(2, 2, -1)$ and that is orthogonal to the y-axis and to the line $(x, y, z) = (4 - 7t, 3t + 2, 4t - 5)$.

Find the point on each of the following lines that is closest to the point P.

35. $(x, y, z) = (2 + t, t, 1 + t), P = (0, 0, 0)$

36. $(x, y, z) = (3 + 2t, 3, 3 + t), P = (0, 0, 0)$

37. $(x, y, z) = (2 - t, 4 + 2t, t - 3), P = (-1, 10, 0)$

38. $(x, y, z) = (3t, 2 - t, 2t + 1), P = (-6, 4, -3)$

39. $(x, y, z) = (5t - 2, 3 - 2t, 2t + 4), P = (8, -1, 8)$

40. $(x, y, z) = (a + t, b + t, c + t), P = (0, 0, 0)$

14.5 Planes in Three-Dimensional Space

NORMAL VECTOR

POINT-NORMAL FORM

PARALLEL AND ORTHOGONAL PLANES

INTERSECTIONS

Consider the xy-plane lying horizontally in three-dimensional space $\mathbb{R}^3$. One way to describe it is as the set of all points that are orthogonal to the z-axis, that is, orthogonal to the unit vector $\mathbf{k}$. See Figure 14.5.1.

This is slightly misleading because we have allowed ourselves to identify the point (x, y, z) with the vector (x, y, z) in order to be able to talk about it being orthogonal to $\mathbf{k}$.

Now if we consider any plane, it seems reasonable from Figure 14.5.2 that it will be possible to find a vector $\mathbf{n}$ that is orthogonal to the plane. Such a vector is called a **normal vector** to the plane. We have been vague about what is meant by the normal vector being orthogonal to the plane, though in the picture it seems quite clear.

In Figure 14.5.3 we draw a plane and choose two points P and Q lying on it. Now we can consider the vector that runs from P to Q. This vector is shown in the figure, and it lies entirely in the plane. What we want is that all vectors like this will be orthogonal to the normal vector.

Figure 14.5.1

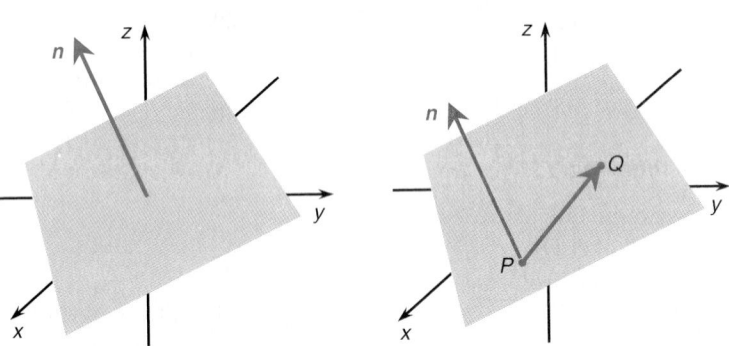

Figure 14.5.2   **Figure 14.5.3**

Now suppose we begin with a normal vector $\mathbf{n} = (a, b, c)$ and a point $P = (X, Y, Z)$ in the plane. How can we specify all the points in the plane?

The idea is to consider a point $Q = (x, y, z)$. For it to be in the plane we require the vector that runs from P to Q to be orthogonal to $\mathbf{n}$. In other words, their dot product should be zero.

DEFINITION 14.5.1

If $\mathbf{n} = (a, b, c)$ is a nonzero vector, and $P = (X, Y, Z)$ is a point, then the **point-normal** form of the equation of the plane containing P and having **normal vector n** is

$$\mathbf{n} \cdot (x - X, y - Y, z - Z) = 0.$$

This equation can also be written out as

$$a(x - X) + b(y - Y) + c(z - Z) = 0,$$

or

$$ax + by + cz = (aX + bY + cZ).$$

> Note that X, Y, Z are fixed, that is, constants, and x, y, z are the variables.

This is the **standard form of the equation of a plane**,

$$ax + by + cz = d,$$

where we let d equal the constant $aX + bY + cZ$.

From this we see that a normal vector for the plane $ax + by + cz = d$ can be found by letting $\mathbf{n} = (a, b, c)$.

EXAMPLE 1 (i) Find the point-normal form of the equation of the plane that contains the point $(3, 0, 2)$ and has normal vector $(2, 5, -4)$.

Solution   Applying the formula from the definition, we find that the equation is

$$(2, 5, -4) \cdot (x - 3, y - 0, z - 2) = 0.$$

Since this can be expanded as $2(x - 3) + 5y - 4(z - 2) = 0$, or $2x + 5y - 4z + 2(-3) - 4(-2) = 0$, the standard form of the equation is

$$2x + 5y - 4z = -2.$$

(ii) Is the point $(5, -1, 3)$ on the plane that contains the point $(4, -1, -2)$ and has normal vector $\mathbf{n} = (1, 1, 1)$?

Solution   The point-normal form of the equation of the plane is $(1, 1, 1) \cdot (x - 4, y + 1, z + 2) = 0$, or $x - 4 + y + 1 + z + 2 = 0$, or $x + y + z = 1$.

To see whether $(5, -1, 3)$ is on this plane, we substitute it into the left side of this equation and find that $5 + (-1) + 3 = 7$, so this point does not satisfy the equation and cannot be on the plane.

(iii) Find the equation of the plane that contains the origin and is perpendicular to the line $(x, y, z) = (3 - 2t, 5 + 3t, -7 + 5t)$.

Solution   To say that the plane is perpendicular to the line means that the direction vector of the line must be a normal vector to the plane. So we can take $\mathbf{n} = (-2, 3, 5)$.

We know immediately that the equation of the plane must be $-2x + 3y + 5z = d$, for some number d. However, since the origin must satisfy this equation, we find that d must equal 0.

The equation of the plane is $-2x + 3y + 5z = 0$.

EXAMPLE 2   Find the equation of the plane containing the points $P = (4, 4, 0)$, $Q = (2, 3, -1)$, and $R = (4, -1, 2)$.

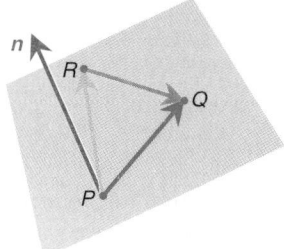

Figure 14.5.4

Solution   Since we know a point on the plane —three in fact—all that is needed is to find a normal vector. In Figure 14.5.4 we see three points P, Q, and R in a plane. From the picture we see that the normal vector is perpendicular to the vectors running between each pair of these points. So an easy way to find a normal vector would be to take two of these vectors and form their cross product. In this case we can take,

Figure 14.5.5

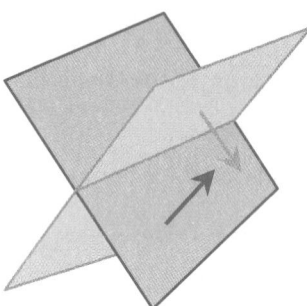

Figure 14.5.6

for instance, the vector that goes from P to Q and the vector that goes from P to R. They are $\mathbf{u} = (2, 3, -1) - (4, 4, 0) = (-2, -1, -1)$ and $\mathbf{v} = (4, -1, 2) - (4, 4, 0) = (0, -5, 2)$.

The cross product of these two vectors is $\mathbf{n} = (-2, -1, -1) \times (0, -5, 2) = (-7, 4, 10)$.

Since this is a normal vector to the plane, we know that the equation of the plane can be written as $-7x + 4y + 10z = d$, for some number d. To find d, we substitute in any of the original three points. For instance, using $Q = (2, 3, -1)$, we find that $-7x + 4y + 10z = (-7)(2) + (4)(3) + (10)(-1) = -12$, so the plane is $-7x + 4y + 10z = -12$. We could also write it as $7x - 4y - 10z = 12$. ◢

> In doing Example 2 we had to choose which pairs of points to use to find vectors in the plane. Since we were going to find their cross product, experience tells us that the work will be easier if some of the components of the vectors are zero. We chose the vector going from P to R because its first component is zero. Any other choice will be just as good, but it is always worthwhile to save trouble and simplify calculations.

In Figure 14.5.5 we see two *parallel* planes. The simplest way to describe this situation is to observe that their normal vectors are parallel. In Figure 14.5.6 there are two planes that are *perpendicular* or *orthogonal*. Here too this is most easily described by observing that their normal vectors are orthogonal. This explains the following definition.

DEFINITION 14.5.2

> Two distinct planes are **parallel** if their normal vectors are parallel, that is, scalar multiples of each other. Two planes are **orthogonal** or **perpendicular** if their normal vectors are perpendicular.

EXAMPLE 3

(i) The planes $2x - 3y + 7z = 2$ and $2x - 3y + 7z = -4$ are parallel.

(ii) The planes $x - y - z = 2$ and $-2x + 2y + 2z + 4 = 0$ have normal vectors $(1, -1, -1)$ and $(-2, 2, 2)$. Since the second is -2 times the first, the normal vectors are parallel. However, the two planes are actually the same, since if we rewrite the second equation as $-2x + 2y + 2z = -4$, we see that it is exactly the same as the first equation multiplied by -2. Any point that satisfies one equation will also satisfy the other.

(iii) The normal vectors of the planes $2x - 3y + z = 7$ and $x + y + z = 2$ are $(2, -3, 1)$ and $(1, 1, 1)$. The dot product of these two vectors is 0, so they are orthogonal, and this means that the planes are orthogonal. ◢

EXAMPLE 4 Find the equation of the plane that is parallel to the plane $3x - 5y + 2z = 1$ and that contains the point $P = (3, -2, -6)$.

Solution Any plane that is parallel to $3x - 5y + 2z = 1$ can be written in the form $3x - 5y + 2z = d$, for some number d. To find the value of d for the plane we want, all that is needed is to observe that the coordinates of P must satisfy the equation.

This means that $(3)(3) - (5)(-2) + (2)(-6) = d$, that is, $7 = d$. The equation of the plane is $3x - 5y + 2z = 7$. See Figure 14.5.7. ◢

Figure 14.5.7

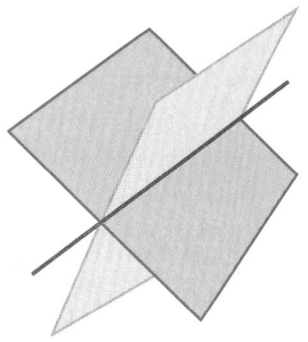

Figure 14.5.8

From the pictures we see that two parallel planes can never intersect. If we write the equations of two parallel planes as $ax + by + cz = d_1$ and $ax + by + cz = d_2$, that is, with the same normal vector (a, b, c) and with $d_1 \neq d_2$, then any point that satisfies one of these equations cannot possibly satisfy the other.

On the other hand, two nonparallel planes must always intersect, and the set of points that forms the intersection is a straight line (see Figure 14.5.8).

EXAMPLE 5   Find equations for the intersection of the planes $x + y + z = 3$ and $2x - 3y + 4z = 6$.

Solution   One way to do this problem is to eliminate one of the variables from these equations. For instance, if we subtract 4 times the first equation from the second, we obtain $-2x - 7y = -6$.

Then we can let one of the remaining variables be the parameter t. For instance, letting $x = t$, we can use the last equation to find that $y = -\frac{1}{7}(2x - 6) = -\frac{1}{7}(2t - 6)$. Substituting this back into the equation of either of the original planes, we can solve for z.

For instance, the first plane $x + y + z = 3$ tells us that $z = 3 - x - y$, so $z = 3 - t + \frac{2}{7}t - \frac{6}{7} = -\frac{5}{7}t + \frac{15}{7}$.

We have just found parametric equations for the line of intersection:

$$x = t, \qquad y = -\frac{2}{7}t + \frac{6}{7}, \qquad z = -\frac{5}{7}t + \frac{15}{7}.$$

Notice that this process could have been done in many different ways. There was a choice of which variable to eliminate, and then there was a choice of which variable to set equal to the parameter t. Making different choices will give different parametrizations of the line.

Here X, Y, Z are constants (the coordinates of a point on the line), and x, y, z are variables.

Now let us think again briefly about the symmetric equations of a line. We can write them as two separate equations, such as $\frac{x-X}{a} = \frac{y-Y}{b}$ and $\frac{x-X}{a} = \frac{z-Z}{c}$. Each of these is the equation of a plane; if we write them as $\frac{1}{a}x - \frac{1}{b}y = \frac{X}{a} - \frac{Y}{b}$ and $\frac{1}{a}x - \frac{1}{c}z = \frac{X}{a} - \frac{Z}{c}$, then we see that their normal vectors are $\left(\frac{1}{a}, -\frac{1}{b}, 0\right)$ and $\left(\frac{1}{a}, 0, -\frac{1}{c}\right)$. The first of these is perpendicular to **k**, so it is horizontal, and the second is perpendicular to **j**. If we had written the other of the three symmetric equations, its normal vector would be perpendicular to **i**.

What this means is that the symmetric equations of a line represent the line as the intersection of two planes, each of which has normal vector that is perpendicular to one of the axes. See Figure 14.5.9; the normal vector of the purple plane is perpendicular to the x-axis, the normal of the blue plane is perpendicular to the y-axis.

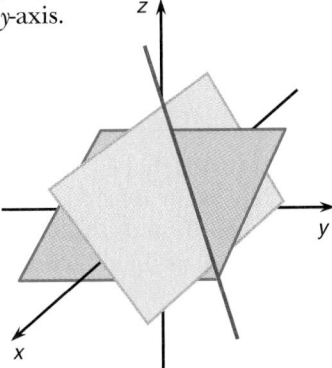

Figure 14.5.9

In Figure 14.5.10 we see a line that is parallel to a plane. Once again it is best to work with the normal vector to the plane, and for the line to be parallel to the plane must mean that its direction vector is perpendicular to the normal.

DEFINITION 14.5.3

Consider a line $(x, y, z) = P + t\mathbf{v}$ and a plane $ax + by + cz = d$. Suppose that the direction vector $\mathbf{v}$ of the line is orthogonal to the normal (a, b, c) to the plane.

Then either the line lies entirely in the plane or else the line is **parallel** to the plane and does not intersect it at all.

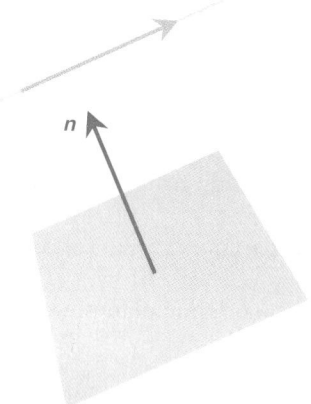

n

Figure 14.5.10

EXAMPLE 6

(i) The line $(x, y, z) = (3, 2, -4) + t(4, 5, -3)$ is parallel to the plane $x + y + 3z = 4$. Notice that the starting point $(3, 2, -4)$ of the line is *not* in the plane, since $3 + 2 + 3(-4) = -7 \neq 4$, so the line is not in the plane.

(ii) The line $(x, y, z) = (3 - t, 4 + 2t, 3 + 3t)$ has direction vector $(-1, 2, 3)$, which is orthogonal to the normal vector to the plane $x + 2y - z = 8$. However, the starting point $(3, 4, 3)$ of the line also lies in the plane, and this means that the whole line lies in the plane.

(iii) Notice that the x-axis and the y-axis are lines that are both parallel to the plane $z = -2$, but they are not parallel to each other. See Figure 14.5.11.   ◢

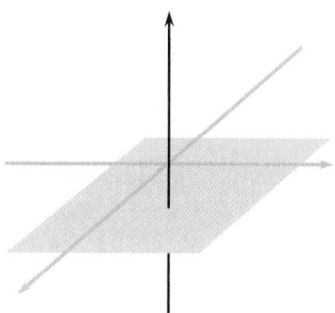

Figure 14.5.11

EXAMPLE 7   Find the point(s) of intersection (if any) of the line $(x, y, z) = (2 + 3t, 1 - t, -3 + t)$ and the plane $2x - y + 3z = 4$.

Solution   We observe that the direction vector of the line is $(3, -1, 1)$, which is *not* orthogonal to the normal $(2, -1, 3)$ to the plane. This means that the line is not parallel to the plane, and we expect them to intersect.

We need to find which points on the line will be on the plane. To do this, we substitute $(x, y, z) = (2 + 3t, 1 - t, -3 + t)$ into the equation of the plane and find $2(2 + 3t) - (1 - t) + 3(-3 + t) = 4$, or $10t = 10$. This will be satisfied when $t = 1$.

The point of intersection is $(x, y, z) = (2 + 3, 1 - 1, -3 + 1) = (5, 0, -2)$.   ◢

EXAMPLE 8   Find the line of intersection of the planes $2x + 5y - 3z = 0$ and $8x - 3y + z = 0$.

Solution   We could proceed as in Example 5, but in this case we have the advantage that it is easy to find one point on the line of intersection. This is because it is clear from the equations that the origin is on both planes.

To find the equation of the line, all that is needed is the direction vector. Since the line lies in each plane, the direction vector must be orthogonal to both normal vectors.

We know how to find a vector that is orthogonal to two given vectors, using the cross product. The direction vector is $(2, 5, -3) \times (8, -3, 1) = (-4, -26, -46)$.

The line is $(x, y, z) = (0, 0, 0) + t(-4, -26, -46)$. Recognizing that $(4, 26, 46)$ or $(2, 13, 23)$ will be just as good as the direction vector, we can write it more simply as

$$(x, y, z) = (2t, 13t, 23t).$$ ◢

Figure 14.5.12(i)

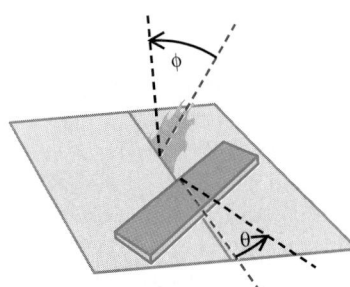

Figure 14.5.12(ii)

EXAMPLE 9   **Compound Miter**   A difficult problem for carpenters is to cut boards which will fit together at an angle. A typical situation occurs at a corner in the moulding at the top of a cupboard.

Suppose the mouldings are at a $45°$ angle above the horizontal. In Figure 14.5.12(i) we see a right-angled corner (actual mouldings are usually rounded or shaped, but that does not change the approach required).

The mouldings can be cut with a table saw, as shown in Figure 14.5.12(ii). The blade can be adjusted in two directions. In the picture the angle marked ϕ is the "tilt," the angle between the blade and the vertical, and the angle marked θ is the angle between the direction of the blade and the line straight across the board. We want to find what angles ϕ and θ are required to cut the mouldings for the corner.

In Figure 14.5.12(iii) we set the axes so that one board is parallel to the x-axis and the other to the y-axis. The flat surfaces of the mouldings lie in two planes through the origin. The plane parallel to the x-axis is $y = z$, and the plane parallel to the y-axis is $x = z$. Their normal vectors are $\mathbf{m} = (0, 1, -1)$ and $\mathbf{n} = (1, 0, -1)$, respectively.

In Figure 14.5.12(iv) we imagine the saw blade lying in the vertical plane $x = y$, whose normal vector is $\mathbf{v} = (1, -1, 0)$.

Figure 14.5.12(iii)

Figure 14.5.12(iv)

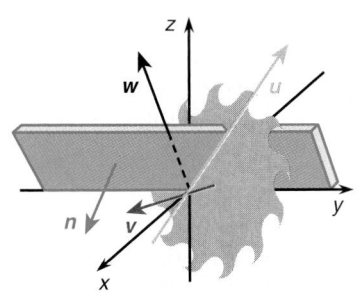

Figure 14.5.12(v)

If $\mathbf{u} = (a, b, c)$ is in the plane $x = y$, we see that $a = b$. If it is in $x = z$, then $a = c$. So $\mathbf{u}$ must be of the form (a, a, a), and $\mathbf{u} = (1, 1, 1)$ is a convenient choice.

Let us concentrate on cutting the board which is in the plane $x = z$. The "tilt" of the blade will be the angle between the blade (in $x = y$) and the normal to the surface of the board ($\mathbf{n}$), that is, $90°$ minus the angle between the normals $\mathbf{n}$ and $\mathbf{v}$. We find it is

$$\phi = 90° - \arccos \frac{\mathbf{n} \cdot \mathbf{v}}{\|\mathbf{n}\|\|\mathbf{v}\|} = 90° - \arccos \frac{(1, 0, -1) \cdot (1, -1, 0)}{\|(1, 0, -1)\|\|(1, -1, 0)\|}$$

$$= 90° - \arccos \frac{1}{\sqrt{2}\sqrt{2}} = 90° - \arccos \frac{1}{2} = 90° - 60° = 30°.$$

In Figure 14.5.12(v) we see the line the blade makes on the surface of the board. It is the intersection of the planes $x = y$ and $x = z$, so its direction vector is $\mathbf{u} = (1, 1, 1)$, which lies in both planes. The line directly across the board (shown dashed in the picture) lies in the plane $x = z$ and is perpendicular to the y-axis. Its direction vector is $\mathbf{w} = (1, 0, 1)$.

The angle θ is the angle between these two lines, so

$$\theta = \arccos \frac{\mathbf{u} \cdot \mathbf{w}}{\|\mathbf{u}\|\|\mathbf{w}\|} = \arccos \frac{(1, 1, 1) \cdot (1, 0, 1)}{\|(1, 1, 1)\|\|(1, 0, 1)\|} = \arccos \frac{2}{\sqrt{3}\sqrt{2}}$$

$$= \arccos \frac{\sqrt{2}}{\sqrt{3}} \approx 35.26°.$$

For the other board the calculations are similar; we find that $\phi = -30°$ and $\theta = -\arccos\frac{\sqrt{2}}{\sqrt{3}} \approx -35.26°$. Instead of resetting the saw, it is easier just to turn the board over...

Exercises 14.5

I

Find the equation of each of the following planes.

1. Containing $(2, 3, 3)$ and with normal vector $(3, -2, 1)$.

2. Containing $(4, -1, 2)$ and with normal vector $(4, 3, -2)$.

3. Containing $(4, -1, 2)$ and orthogonal to the line $(x, y, z) = (3 - 2t, 4t + 5, t - 3)$.

4. Containing $(4, -1, 2)$ and orthogonal to the line containing $(2, 3, 0)$ and $(-1, 4, 5)$.

5. Containing $(4, -1, 2)$, $(3, 3, 1)$ and $(3, 2, 0)$.

6. Containing $(3, 5, -2)$, $(2, 1, -3)$ and $(-1, 2, 3)$.

7. Containing $(4, 3, 1)$ and parallel to $2x - y + 3z = 5$.

8. Containing $(-2, 4, 5)$ and parallel to $x - 2y - 5z = 3$.

9. Containing $(-2, 4, 5)$ and $(2, 2, 1)$ and parallel to the line $(x, y, z) = (t + 3, 3t - 2, 2 - t)$.

10. Containing $(3, -4, 2)$ and not intersecting the plane $5x - 4y + z = 7$.

11. Containing $(5, 3, 7)$ and $(1, 1, 1)$ and perpendicular to $3x - 2y + z = 2$.

12. Containing $(-2, -1, 0)$ and $(3, -3, 2)$ and perpendicular to $x - y = 4$.

13. Containing $(-2, -1, 0)$ and perpendicular to $x - y = 4$ and to $x + 2z = 5$.

14. Containing $(3, 5, -2)$ and perpendicular to the x-axis.

15. Containing $(7, 2, -1)$ and $(-3, -2, 3)$ and perpendicular to the xz-plane.

16. Containing $(5, -5, 3)$ and $(2, 1, 2)$ and parallel to the y-axis.

17. Containing $(4, -3, 2)$ and $(1, 3, -3)$ and lying vertically, that is, perpendicular to the xy-plane.

18. Containing $(7, -12, 14)$ and lying horizontally.

19. Containing $P = (3, 1, -1)$ and the vectors $\mathbf{u} = (3, 3, -2)$ and $\mathbf{v} = (2, -1, 4)$, both with their initial points at P.

20. Containing $(4, -2, 2)$ and the line $(x, y, z) = (2 - 3t, t - 2, 2t + 1)$.

Find parametric equations for each of the following lines; also find symmetric equations if possible. Note that there are many possible answers to these questions.

21. Containing $(3, 2, 1)$ and orthogonal to the plane $3x - 2y + z = 6$.

22. Containing $(5, 6, -2)$ and perpendicular to $3x - y - z = 2$.

23. Containing $(2, 2, -1)$ and parallel to $x + y - z = 3$ and to $2x - 2y + 3z = 7$.

24. Containing $(5, 5, -4)$ and orthogonal to the plane through $(2, 3, -1)$, $(4, 5, 6)$, and $(1, -2, 0)$.

25. The intersection of $3x + 2y - z = 2$ and $4x - y + z = 2$.

26. The intersection of $x + 3y - 4z = 4$ and $5x - 2y - z = -1$.

27. The intersection of $3x - 2z = 6$ and $2x + 2y - 3z = 5$.

28. The intersection of $x + y + z = 4$ and $3x - 2y + 4z = -3$.

29. Through the origin, contained in the plane $x - 2y + z = 0$, and orthogonal to the vector $\mathbf{v} = (3, 4, 2)$.

30. Through the point $(3, 2, -4)$, horizontal, and contained in the plane $x - y + 2z = -7$.

II

*31. Why can a plane and a line not intersect in more than one point unless the line lies entirely in the plane?

*32. If a line is parallel to a plane (but not contained in it), show that they never intersect.

*33. Consider the plane $2x + 2y - z = 3$. Find the equation of the plane that is parallel to it and 1 unit *above* it in the sense that it contains a point that is exactly 1 unit above a point on $2x + 2y - z = 3$.

*34. There are many planes that contain the point $(2, 3, 4)$ and are parallel to the line $(x, y, z) = (3, -1, 5) + t(1, 2, -4)$. If we write such a plane as $x + by + cz = d$, then it is possible to find c and d once we know b.

(i) Find expressions for c and d as functions of b.

(ii) Find the equation of such a plane if it also contains $(2, -1, 3)$.

(iii) Can you find a plane that contains the original point $(2, 3, 4)$ and is parallel to the line but cannot be written in the form $x + by + cz = d$?

**35. Can you guess which point on the plane $2x - y - 2z = 4$ is closest to the origin? (Draw a picture.)

**36. Consider the planes $x + y + z = 1$ and $x + y + z = 0$. How far apart are they? (*Hint:* What point on $x + y + z = 1$ is closest to the origin?)

37. Suppose the mouldings of Example 9 (p. 867) are placed so that their flat sides are $60°$ above the horizontal. What angles are required for setting the saw now?

*38. Suppose the mouldings of Example 9 are placed so that their flat sides are at an angle α above the horizontal. Find a formula for the angles that are required for setting the saw now. (Confirm your answer by checking it against the original example and against the previous exercise.)

Chapter Summary

§14.1
coordinates in $\mathbb{R}^3$
right-handed coordinate system
graphs in three-dimensional space
distance

§14.2
vector
components
zero vector
negative of a vector
norm
Triangle Rule
Parallelogram Rule
properties of vectors (Proposition 14.2.2)
Triangle Inequality
standard basis vectors $\mathbf{i}, \mathbf{j}, \mathbf{k}$
scalar multiplication
dot product
properties of dot products (Properties 14.2.6)
angle θ between two vectors
orthogonal (perpendicular) vectors
cross product
properties of cross products (Proposition 14.2.10)
Lagrange's Identity

§14.3
displacement

addition of displacements
distance
cycloid
vertical component
component of $\mathbf{v}$ in the direction of $\mathbf{u}$
component of $\mathbf{v}$ perpendicular to $\mathbf{u}$
force and work
catenary
area of parallelogram

§14.4
parametric equations of a line
direction vector
vector equation
symmetric equations for a line
parallel lines
skew lines
direction cosines
unit vector

§14.5
normal vector
point-normal form
standard form of the equation of a plane
parallel planes
orthogonal (perpendicular) planes
line parallel to plane
intersection of line and plane

Review Exercises

Find each of the following.

1. $(4, -1, 3) + 2(-1, -2, 0)$

2. $\|(4, 5, -3)\|$

3. $(3, -1, 6) \cdot (4, 2, -1)$

4. $(3, -4) \cdot (1, 1)$

5. $(5, 2, 3) \times (-2, 3, -1)$

6. $(4, 1, -1) \times (3, -2, -1)$

7. The component of $(5, -1)$ along the direction of $(3, -4)$.

8. The component of $(2, 3, 5)$ orthogonal to $(2, -1, -2)$.

9. The angle between $(4, -2)$ and $(1, 2)$.

10. The angle between $(0, -3, 0)$ and $(-2, 2, 0)$.

Find each of the following.

11. The speed of a paper airplane flying at 10 mph across a school bus that is moving forward at 30 mph.

12. The speed of a yoyo that is descending at 3 ft/sec from the hand of somebody who is walking at a speed of 4 ft/sec.

13. The speed relative to the shore of a boat moving at 10 knots heading straight across a river that flows at 4 knots.

14. The speed relative to the water of a boat moving straight across the same river at a speed of 8 knots relative to the shore.

15. The groundspeed of an airplane heading northwest at an airspeed of 400 mph in a 45-mph west wind.

16. The airspeed of an airplane flying southwest at a groundspeed of 350 mph in a 40-mph north wind.

17. The angle at which a boat capable of 15 mph should steer in order to move directly across a 4-mph current.

18. The speed with which a boat capable of going 20 km/hr will travel if it steers at the correct angle so as to move directly across a 7 km/hr current.

19. The heading (i.e., angle relative to due north) of an airplane with an airspeed of 250 mph that is steering west in a 40-mph southeast wind (i.e., the wind is *from* the southeast).

20. The speed of an airplane whose shadow is moving northwest with a speed of 200 mph and that is descending at a rate of 5 ft/sec.

21. The horizontal component of a force of 24 lb transmitted through a rope that is 45° above the horizontal.

22. The vertical component of a force of 35 N transmitted through a rope that is 30° above the horizontal.

23. The force exerted by gravity on a 12-lb cart along the direction of a 30° ramp.

24. The force exerted on the ramp by a 30-lb wagon on a 30° ramp.

25. The work done by an upward force of 10 lb on an object as it moves a distance of 15 ft upward along a stairway 45° above the horizontal.

26. The work done by gravity on a child of 20 kg as she slides down a slide that is 3 m high and 4 m long.

27. The work done on a 20-kg wagon by a man who pushes horizontally with a force of 180 N as the wagon moves 5 m upward along a 30° ramp.

28. The force needed to hold a 30-lb wagon from rolling down a 20° ramp.

29. The angle of a ramp if it takes a 12-lb force to keep a 28-lb cart from rolling down it.

30. The work done on a cart weighing 50 lb by someone who pushes just hard enough to move it 10 ft upward along a 30° ramp.

Find each of the following. For each problem, discuss how many correct answers there are.

31. Parametric equations of the line through $(4, 2, -3)$ and $(1, -3, 0)$.

32. Parametric equations of the line through $(5, 2, 11)$ and $(0, 2, 7)$.

33. Parametric equations of the line through $(3, -4, 5)$ and parallel to $\mathbf{r}(t) = (t, 2 - 3t, 4t + 3)$.

34. Symmetric equations of the line through $(5, -4, 1)$ with direction vector $(3, -2, 4)$.

35. The point of intersection, if any, of the lines $(x, y, z) = (3 - t, 4, 6 + 2t)$ and $(x, y, z) = (3 + t, 6 + 2t, 6 - 2t)$.

36. The point of intersection, if any, of the line $(x, y, z) = (5 + t, 2 - 3t, t - 2)$ and the line through $(2, 3, -1)$ and $(5, -1, -2)$.

37. The equation of the plane containing $(3, 2, -1)$, $(2, 1, 0)$, and $(4, -4, 5)$.

38. The equation of the plane containing $(2, 2, -1)$ and orthogonal to the line $(x, y, z) = (5 - t, 2t + 1, 6t)$.

39. The point of intersection of the line $(x, y, z) = (t, 1 - 3t, 4t + 2)$ and the plane $2x + 4y - z = 7$.

40. The point of intersection of the plane $x + y - 2z = 5$ and the line containing $(3, 4, -1)$ and $(1, 5, -3)$.

Curves and Surfaces in Space

15

The helix or spiral is a curve we recognize from a spiral staircase or a corkscrew. Many springs and telephone wires are helical. The helix also occurs submicroscopically as the fundamental shape of the DNA molecule.

INTRODUCTION

We have already discussed curves in the plane. Now that the necessary facts about vectors have been established, we can extend these ideas to consider curves in three-dimensional space. In Section 15.1 the theory begins with parametric equations just as in the planar case, but in Section 15.2 we discuss velocity and encounter some new ideas that are easier to state using the vector terminology. For plane curves we discussed the velocity but never asked the natural next question: What about acceleration? The acceleration of a moving point is also a vector. Unlike a point moving in a straight line, which can only speed up or slow down, a point moving in a plane or in space can also change its direction by turning. The acceleration vector measures the way the velocity is changing. The more rapidly the direction is changing, the more sharply the curve must be bending, and this leads us in Section 15.3 to discuss its *curvature*. In Section 15.4 we discuss *quadric surfaces*. These are surfaces in three-dimensional space that are analogous to conics in the plane in

871

the sense that they are defined by quadratic polynomial equations. In Section 15.5 we discuss two specialized kinds of coordinates in space, *cylindrical* and *spherical* coordinates. They are three-dimensional analogs of polar coordinates in the plane, and of course they are particularly well suited to studying questions involving rotational motion. The last two sections are more specialized. In Section 15.6 we discuss vectors and coordinates in *n*-dimensional space. This topic is not much different from the corresponding study in two- or three-dimensional space. In Section 15.7 we study *Kepler's laws*, which are the mathematical laws that describe the motion of the planets around the sun. This was one of the early uses of calculus, and it was a very important development in the history of both mathematics and astronomy. ◢

15.1 Curves in Space

PARAMETRIC EQUATIONS

HELIX

VELOCITY VECTOR

TANGENT VECTOR

SPEED

ANGLE BETWEEN CURVES

In this section we observe that as for curves in the plane it is possible to give parametric equations for a curve in space. We can imagine a point moving in space, and the parametric equations give the coordinates of its position at any time t.

The definition is almost identical to the two-dimensional case (cf. Definition 13.1.1, p. 797).

DEFINITION 15.1.1

Suppose I is an interval (open or closed, possibly infinite), and suppose

$$x = x(t), \qquad y = y(t), \qquad z = z(t)$$

are functions defined on I. Assume that they are continuous on I. Then they are called **parametric equations**; the variable t is called the **parameter**. The set of points

$$\{(x, y, z) = (x(t), y(t), z(t)) : t \in I\}$$

is called the **curve**, and it is sometimes also called the **image** of I. The interval I is called the **domain** of the parametric equations.

Sometimes several curves are combined to make one curve; in this situation it may be necessary to have parametric equations defined on several intervals.

To emphasize the idea of the moving point, we sometimes write

$$\mathbf{r} = \mathbf{r}(t) = (x(t), y(t), z(t)).$$

The letter $\mathbf{r}$ reminds us of "radius"; $\mathbf{r}(t)$ is the vector pointing from the origin out to the position of the moving point at time t.

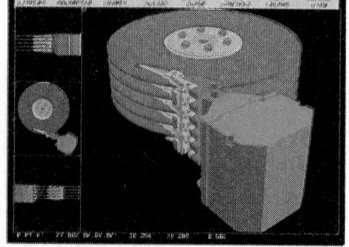

DesignCAD 3D *software was used to produce this image of a computer hard disk. The disks rotate very fast and heads on the moveable arms write and read information to and from the disk. To do this they need to be able to find the exact position of each piece of data on the moving disks.*

EXAMPLE 1   In Section 14.4 we discussed the parametric equations for a straight line. For instance, the equations $(x, y, z) = (-4 + 2t, 2 + 5t, 4 - 3t)$ describe a point moving along a line. They are a very simple example of parametric equations.   ◢

We allow ourselves to use the word "curve" for a straight line even though it does not bend. In Section 15.3 we will discuss the notion of *curvature*, and then we will be able to use it to make such distinctions.

Figure 15.1.1

Figure 15.1.2

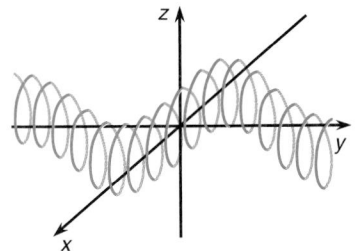

Figure 15.1.3

EXAMPLE 2   Consider the parametric equations

$$x(t) = \cos t, \qquad y(t) = \sin t, \qquad z(t) = t.$$

How does the point $\mathbf{r}(t) = (\cos t, \sin t, t)$ move? As t increases, the z-coordinate, which equals t, also increases. This means that the point is "rising"; its height is increasing.

The point $\big(x(t), y(t), z(t)\big)$ lies directly above or below the point $\big(x(t), y(t), 0\big) = (\cos t, \sin t, 0)$ in the horizontal plane, and from our knowledge of curves in the plane we see that *this* point moves around and around a circle in the plane.

So we have a point that is rising and that lies directly above or below a point tracing out a circle in the horizontal plane. The path it follows is a spiral, as shown in Figure 15.1.1. A curve of this shape is called a (circular) **helix**. ◢

EXAMPLE 3   Sketch the graph of the curve $(x, y, z) = \mathbf{r}(t) = (\cos 12t, t, \sin t + \sin 12t)$.

Solution   First we observe that the y-coordinate is just t, which means that the point is moving steadily to the right if we draw our usual picture. If the z-coordinate was $\sin 12t$, then the x- and z-coordinates $\cos 12t$ and $\sin 12t$ would just trace out a circle, and the graph would be the helix shown in Figure 15.1.2. Because the x- and z-coordinates involve the sin and cos of $12t$, the point goes around the circle more rapidly than in the situation of Example 2. The result is that this helix is "tighter."

However, the z-coordinate is actually $\sin t + \sin 12t$. This means that in addition to going up and down as it goes around the spiral, the point also moves up and down more slowly. The actual curve is the humped spiral shown in Figure 15.1.3.

Notice that between successive "humps" there are 12 little loops. This is because the $\cos 12t$ and $\sin 12t$ trace out a circle 12 times as fast as the $\sin t$ undulates up and down. ◢

Velocity Vectors

Just as for curves in the plane, it is possible to consider the velocity of a moving point.

DEFINITION 15.1.2

Consider the moving point whose position at time t is given by

$$(x, y, z) = \mathbf{r}(t) = \big(x(t), y(t), z(t)\big),$$

where $x(t)$, $y(t)$, and $z(t)$ are differentiable functions. Then its **velocity** at time t is the vector

$$\mathbf{r}'(t) = \big(x'(t), y'(t), z'(t)\big).$$

In Chapter 13 when we discussed velocity for curves in the plane, we observed that the velocity vector can be used to define the idea of a tangent to the curve and that its norm can be used to define the speed of the moving point (cf. Definition 13.2.3, p. 807). These remarks are equally useful for curves in $\mathbb{R}^3$.

DEFINITION 15.1.3

Consider a curve traced out by parametric equations $(x, y, z) = \mathbf{r}(t) = \big(x(t), y(t), z(t)\big)$ for t in some interval I. Suppose $a \in I$ is a point at which $x(t)$, $y(t)$, and $z(t)$ are differentiable, and let $\mathbf{r}'(a) = \big(x'(a), y'(a), z'(a)\big)$ be the velocity vector. Suppose that $\mathbf{r}'(a)$ is not the zero vector, that is, that $x'(a)$, $y'(a)$, and $z'(a)$ are not all zero.

Then $\mathbf{r}'(a)$ is a **tangent vector** to the curve at the point $\big(x(a), y(a), z(a)\big)$ and its length

$$\|\mathbf{r}'(a)\| = \sqrt{x'(a)^2 + y'(a)^2 + z'(a)^2}$$

is the **speed** of the moving point at $t = a$.

Figure 15.1.4

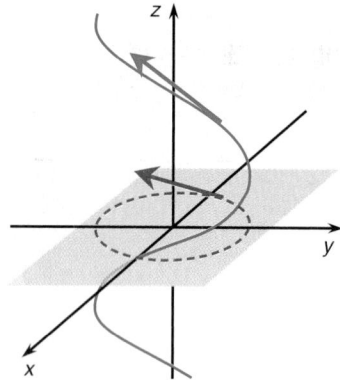

Figure 15.1.5

EXAMPLE 4   Find the velocity vector of each of the following curves:
(i) $(x, y, z) = (-1 + 2t, 4 - t, 7 + 3t)$, (ii) $(x, y, z) = (\cos t, \sin t, t)$, (iii) $(x, y, z) = (\cos t, \sin t, \cos 2t)$.

Solution

(i) This is the parametrization of a straight line; the velocity vector is $\mathbf{r}'(t) = (2, -1, 3)$. It always points along the direction of the line, and its length and direction never change. See Figure 15.1.4.

(ii) This is the helix of Example 2. Its velocity vector is $\mathbf{r}'(t) = (-\sin t, \cos t, 1)$. The fact that its z-component is always 1 reflects the fact that the point is steadily rising (i.e., at a rate of 1 unit of height per unit of time).

The horizontal components of the velocity vector point in the direction that is tangent to the horizontal circle, and the vector itself points in a direction which is tangent to the helix; see Figure 15.1.5.

(iii) Here it is less easy to identify the curve traced out by the point, but it is still quite simple to find the velocity vector. It is $\mathbf{r}'(t) = (-\sin t, \cos t, -2 \sin 2t)$.

Because the horizontal coordinates of the moving point trace out a circle, the curve must lie in the cylinder $x^2 + y^2 = 1$, which is shown in Figure 15.1.6. The vertical component $\cos 2t$ goes up and down twice as fast as the horizontal components go around. The curve itself is shown in Figure 15.1.7, and it is

Figure 15.1.6   **Figure 15.1.7**

We cannot use a graphics calculator to sketch the curve of part (iii) of Example 4, but we can get some idea of its shape by sketching what happens when we project it onto the coordinate planes. We do this by plotting two of the coordinates and just ignoring the third.

Without actually plotting it we see that the projection onto the xy-plane is the circle $(\cos t, \sin t)$. With a graphics calculator we sketch out the projection on the xz-plane, which is $(\cos t, \cos 2t)$.

TI-85	SHARP EL9300
MODE ▼ ▼ ▼ ▼ ► ► (Param)	SET UP E 3 ENTER ⏎ cos
ENTER EXIT GRAPH F1 $(E(t) =)$	x/θ/T ENTER cos (2 x/θ/T) ENTER
COS F1(t) ▼ COS (2F1(t))	RANGE −2 ENTER 2 ENTER 1 ENTER
M2(RANGE) 0 ▼ 4 π ▼ ▼ (−) 2	−2 ENTER 2 ENTER 1 ENTER 0 ENTER
▼ 2 ▼ ▼ (−) 2 ▼ 2 F5(GRAPH)	4 π ENTER .1 ENTER ⏎

CASIO f_x-7700GB	HP 48SX
MODE MODE × (PARAM) Range	⬆ RAD ′ COS T ► + α ⬆ CST ×
−2 EXE 2 EXE 1 EXE −2 EXE 2 EXE 1	COS 2 × T ⬆ PLOT NEW P A R
EXE 0 EXE 4 π EXE .1 EXE Graph cos	ENTER PTYPE PARA PLOTR ⬆
T , cos (2 T) EXE	{ } T SPC 0 SPC 12.6 ENTER INDEP
	AUTO

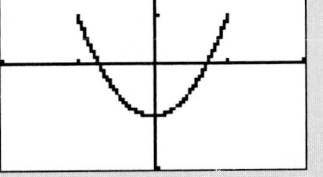

Now plot the projection on the yz-plane, which is $(\sin t, \cos 2t)$. Compare the resulting pictures with the three-dimensional picture.

worthwhile to see how the velocity vector follows the curve. The solid point corresponds to $t = 0$; the velocity vector there is $(0, 1, 0)$. The velocity vector shown at the bottom of the curve corresponds to $t = \frac{\pi}{2}$, and the third one corresponds to $t = \pi$.

EXAMPLE 5   Find all points at which the velocity vector to the curve $(x, y, z) = \mathbf{r}(t) = (3t^6, 4t^2 - 1, t^4 - 8t^2)$ is horizontal.

Solution   Since the velocity vector is $\mathbf{r}'(t) = (18t^5, 8t, 4t^3 - 16t)$, it will be horizontal precisely when its vertical component $4t^3 - 16t$ is zero. This happens when $t = 0, \pm 2$.

The corresponding points $\mathbf{r}(t)$ are $(0, -1, 0)$, $(192, 15, -16)$, and $(192, 15, -16)$. We observe that the points $\mathbf{r}(t)$ corresponding to $t = 2$ and $t = -2$ are the same. This means that as t runs from $t = -2$ to $t = 2$, the moving point $\mathbf{r}(t)$ returns and passes through the same point again.

However, the velocity vectors $\mathbf{r}'(t)$ corresponding to $t = 2$ and $t = -2$ are $(576, 16, 0)$ and $(-576, -16, 0)$, respectively. They are each other's negatives because on the two occasions when the moving point passes through $(192, 15, -16)$ it is going in opposite directions.

In fact, for any value of t the points corresponding to t and $-t$ are the same. This means that the moving point $\mathbf{r}(t)$ traces out a curve as t runs from $-\infty$ to 0 and then turns around and retraces the same curve in the opposite direction as t runs from 0 to ∞. The velocity vector $\mathbf{r}'(t)$ is the negative of the velocity vector $\mathbf{r}'(-t)$, reflecting the fact that the point is moving in opposite directions at these two times.

EXAMPLE 6   Consider the curve parametrized by

$$\mathbf{r}(t) = (t \cos t, t \sin t, t), \qquad \text{for} \qquad t \geq 0.$$

The moving point lies directly above the point $(t \cos t, t \sin t, 0)$; in Example 13.1.3 we saw that this point traces out the Spiral of Archimedes in the plane.

In addition, since the z-component of our moving point equals t, the point is steadily rising. The curve it traces out will be a sort of rising spiral.

The distance from the point $(t \cos t, t \sin t, 0)$ in the horizontal plane to the origin is just t, which is also the height of the point $\mathbf{r}(t) = (t \cos t, t \sin t, t)$ above the horizontal plane, so the curve must lie in the cone consisting of all points whose height above the horizontal plane equals the distance from the z-axis. The curve is a spiral that winds around the cone. See Figure 15.1.8.

The velocity vector at time t is $\mathbf{r}'(t) = (-t \sin t + \cos t, t \cos t + \sin t, 1)$. The fact that the z-component is 1 reflects the fact that the point is rising at a constant rate.

The speed of the point at time t is

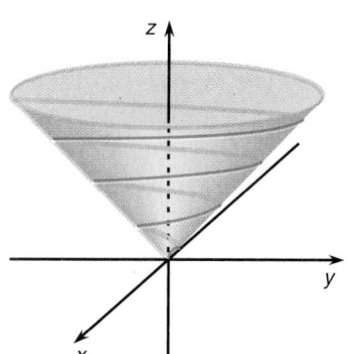

Figure 15.1.8

$$\|\mathbf{r}'(t)\| = \sqrt{(-t \sin t + \cos t)^2 + (t \cos t + \sin t)^2 + 1^2}$$
$$= \sqrt{t^2 \sin^2 t - 2t \sin t \cos t + \cos^2 t + t^2 \cos^2 t + 2t \cos t \sin t + \sin^2 t + 1}$$
$$= \sqrt{t^2 + 2}.$$

We see that as t increases, the speed increases, and the point moves faster and faster.

Imagine a mosquito that takes off and flies in such a way that it follows the spiral path described in Example 6 and illustrated in Figure 15.1.8. We have seen how to calculate its speed at any time t, but another interesting question to ask is how steeply it is climbing. Since the velocity vector $\mathbf{r}'(t)$ is tangent to the curve, this amounts to asking what is the angle between $\mathbf{r}'(t)$ and the horizontal plane.

It turns out that there is a convenient formula for finding the angle between a vector $\mathbf{v}$ and the horizontal plane.

THEOREM 15.1.4

> The angle between the vector $\mathbf{v}$ and the horizontal plane is given by
>
> $$\theta = \arcsin\left(\frac{\mathbf{v} \cdot \mathbf{k}}{\|\mathbf{v}\|}\right).$$
>
> If the angle is negative, it means that the vector points down below the plane.

PROOF   In Figure 15.1.9 we see a vector $\mathbf{v}$, and the angle marked θ is the angle between it and the horizontal plane. We could find this angle by finding the horizontal vector lying directly under $\mathbf{v}$ and using the dot product. However, it is probably eas-

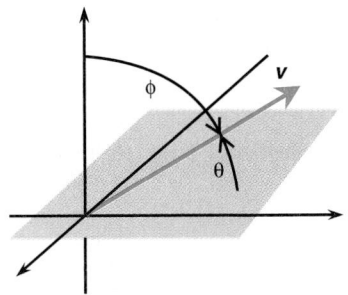

Figure 15.1.9

ier to find the angle marked ϕ in the picture, which is the angle between **v** and the vertical axis. Having found ϕ, we can easily find θ, since their sum is a right angle, that is, $\theta + \phi = \frac{\pi}{2}$.

In fact this relation can be rewritten as $\phi = \frac{\pi}{2} - \theta$, and if we apply the addition formula (Theorem 1.7.11) for cos, we find that $\cos\phi = \cos\frac{\pi}{2}\cos(-\theta) - \sin\frac{\pi}{2}\sin(-\theta) = \sin\theta$.

Now $\cos\phi$ is found by using the dot product:

$$\cos\phi = \frac{\mathbf{v}\cdot\mathbf{k}}{\|\mathbf{v}\|\|\mathbf{k}\|} = \frac{\mathbf{v}\cdot\mathbf{k}}{\|\mathbf{v}\|}.$$

But $\sin\theta = \cos\phi = \frac{\mathbf{v}\cdot\mathbf{k}}{\|\mathbf{v}\|}$, so $\theta = \arcsin\left(\frac{\mathbf{v}\cdot\mathbf{k}}{\|\mathbf{v}\|}\right)$, as required. ◢

EXAMPLE 7 For the spiral curve $\mathbf{r}(t) = (t\cos t, t\sin t, t)$ discussed in Example 6 we know that the velocity vector at time t is

$$\mathbf{r}'(t) = (-t\sin t + \cos t, t\cos t + \sin t, 1).$$

The speed is $\|\mathbf{r}'(t)\| = \sqrt{t^2 + 2}$.

Since $\mathbf{r}'(t)\cdot\mathbf{k} = 1$, the angle θ between $\mathbf{r}'(t)$ and the horizontal plane is

$$\theta = \arcsin\left(\frac{1}{\sqrt{t^2 + 2}}\right).$$

Notice that when $t = 0$, this says that $\theta = \arcsin\frac{1}{\sqrt{2}} = \frac{\pi}{4}$, so the mosquito following the spiral takes off at an angle of 45° above the horizontal.

Later on, as t increases, $\frac{1}{\sqrt{t^2+2}}$ decreases, so $\theta = \arcsin\left(\frac{1}{\sqrt{t^2+2}}\right)$ also decreases. As the mosquito flies, its path levels off.

It is possible to see this in terms of the velocity vector $\mathbf{r}'(t) = (-t\sin t + \cos t, t\cos t + \sin t, 1)$. As t increases, the first two components become larger, and the effect is that the horizontal part of the velocity vector is larger and larger relative to the vertical component. This means that as the point rises on the spiral, its velocity is closer and closer to horizontal. See Figure 15.1.10. ◢

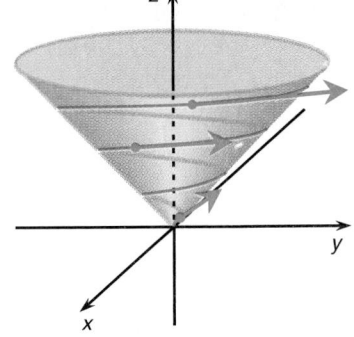

Figure 15.1.10

EXAMPLE 8 Find all points of intersection of the following curves: $\mathbf{r}(t) = (t^3 + t^2 - 3t - 1, t^2 - 1, 3t - 7)$ and $\mathbf{p}(t) = (3t - 1, 2t - 1, t^2 + t - 7)$.

Solution As always it is necessary to use different letters for the two parameters. We need to find values of t and s so that

$$\mathbf{r}(t) = \mathbf{p}(s),$$

that is, so that

$$t^3 + t^2 - 3t - 1 = 3s - 1, \qquad t^2 - 1 = 2s - 1, \qquad 3t - 7 = s^2 + s - 7.$$

Subtracting the second of these equations from the first, we find that $t^3 - 3t = s$. Substituting this into the second equation shows that $2t^3 - t^2 - 6t = 0$, that is, $t(2t^2 - t - 6) = 0$, that is, $t(t - 2)(2t + 3) = 0$. The only possibilities are $t = 0$, $t = 2$, and $t = -\frac{3}{2}$.

If $t = 0$, then $s = t^3 - 3t = 0$. The point on the first curve corresponding to $t = 0$ is $\mathbf{r}(0) = (-1, -1, -7)$; the point on the second corresponding to $s = 0$ is also $(-1, -1, -7)$. This is one point where the two curves intersect.

If $t = 2$, then $s = t^3 - 3t = 2$; the corresponding points are $\mathbf{r}(2) = (5, 3, -1)$ on the first curve and $\mathbf{p}(2) = (5, 3, -1)$. This is a second point where the curves intersect.

If $t = -\frac{3}{2}$, then $s = t^3 - 3t = -\frac{27}{8} + \frac{9}{2} = \frac{9}{8}$. The points are $\mathbf{r}\left(-\frac{3}{2}\right) = \left(\frac{19}{8}, \frac{5}{4}, -\frac{23}{2}\right)$ and $\mathbf{p}\left(\frac{9}{8}\right) = \left(\frac{19}{8}, \frac{5}{4}, -\frac{295}{64}\right)$. Since these points are not the same, there is no third point of intersection.

Notice that in Example 8 we chose two of the components and solved for values of t and s that made them equal. Then it was necessary to check whether the third components were also equal, as we did for intersections of lines.

EXAMPLE 9 The curves $(x, y, z) = (\cos t, \sin t, t)$ and $(x, y, z) = \left(t, \sin(t - 1), t^2 - 1\right)$ meet at the point $(1, 0, 0)$. What is the angle between the two curves at this point (i.e., the angle between their tangent vectors)?

Solution As usual, we use different parameters for the two curves. We notice that the first curve $\mathbf{r}(t) = (\cos t, \sin t, t)$ passes through $(1, 0, 0)$ when $t = 0$. The second curve $\mathbf{p}(s) = \left(s, \sin(s - 1), s^2 - 1\right)$ passes through it when $s = 1$.

The velocity vector to the first curve is $\mathbf{r}'(t) = (-\sin t, \cos t, 1)$, and the velocity vector to the second curve is $\mathbf{p}'(s) = \left(1, \cos(s - 1), 2s\right)$. At the specified point, when $t = 0$, the first of these is $\mathbf{r}'(0) = (0, 1, 1)$, and the second, with $s = 1$, is $\mathbf{p}'(1) = (1, 1, 2)$.

If θ is the angle between these two vectors, then

$$\cos \theta = \frac{(0, 1, 1) \cdot (1, 1, 2)}{\|(0, 1, 1)\| \, \|(1, 1, 2)\|} = \frac{3}{\sqrt{2}\sqrt{6}} = \frac{\sqrt{3}}{2}.$$

The angle between the tangent vectors is $\frac{\pi}{6}$.

Notice how we used the angle between the tangent vectors to define the angle between the curves.

DEFINITION 15.1.5

Suppose two curves

$$\mathbf{r}(t) = \left(f(t), g(t), h(t)\right) \qquad \text{and} \qquad \mathbf{p}(s) = \left(F(s), G(s), H(s)\right)$$

intersect at the point where $t = a$ and $s = b$. Suppose that the functions $f(t)$, $g(t)$, and $h(t)$ are differentiable at $t = a$ and $F(s)$, $G(s)$, and $H(s)$ are differentiable at $s = b$. Also suppose that neither tangent vector at this point is the zero vector. Then the **angle between the curves** at this point of intersection is defined to be the angle between their tangent vectors $\mathbf{r}'(a)$ and $\mathbf{p}'(b)$ at this point.

Exercises 15.1

I

Sketch the graph of each of the following curves.

1. $\mathbf{r}(t) = \left(\cos t, \sin t, \frac{t}{2}\right)$
2. $\mathbf{r}(t) = (\sin t, \cos t, t - \pi)$
3. $\mathbf{r}(t) = (t, \sin 2t, \cos 2t)$
4. $\mathbf{r}(t) = (t, 2t, \sin t)$
5. $\mathbf{r}(t) = (t, t \sin t, t \cos t)$
6. $\mathbf{r}(t) = (t^2 \sin t, t^2 \cos t, t)$
7. $\mathbf{r}(t) = (\cos t, t + \sin t, t)$
8. $\mathbf{r}(t) = (\sin 12t, t + \cos 12t + \cos t, t)$
9. $\mathbf{r}(t) = (\cos t, 1 + 2 \cos t, 1 + \cos t)$
10. $\mathbf{r}(t) = (\sin t, -\cos t, 2 - \sin^2 2t - \cos^2 2t)$

For each of the following curves, find the angle between the velocity vector and the horizontal at the specified time.

11. $\mathbf{r}(t) = (\cos t, \sin t, t)$, $t = \sqrt{3}$
12. $\mathbf{r}(t) = (\cos t, \sin t, -t)$, $t = \frac{1}{\sqrt{3}}$
13. $\mathbf{r}(t) = \left(3 \cos(t^2), 2 \sin(t + 1), \cos(\pi t)\right)$, $t = 1$
14. $\mathbf{r}(t) = \left(4 - \cos(t^3), t - \sin t, \sin t\right)$, $t = 0$
15. $\mathbf{r}(t) = \left(\frac{1}{4} \cos(2t), \frac{1}{2} \sin(2t), 2 - \sin t\right)$, $t = \pi$
16. $\mathbf{r}(t) = (3t^2 - t + 1, t^3 + 4, t^2 - 2t + 7)$, $t = -1$
17. $\mathbf{r}(t) = (t + 3, 2 - \sqrt{2}t, t + 1)$, $t = 17$
18. $\mathbf{r}(t) = (e^t, 2e^t - 1, te^{-t})$, $t = 1$
19. The straight line through $(4, -2, 3)$ and $(7, -1, 3)$ at any point.
20. $\mathbf{r}(t) = (\cos t, \sin t, t^2)$, $t = -\frac{1}{2}$

II

*21. Describe the curve parametrized by the formula given in Example 6 but with the parameter t allowed to run over all real numbers. At what time is the point moving the most slowly?

*22. Find a parametrization for the helix that contains the point $(0, 1, 0)$, winds around the x-axis with a radius of 1, and goes around exactly once each time x increases by 1, winding counterclockwise as we usually draw the picture.

*23. Find a parametrization for the helix that contains the points $(0, 0, 2)$ and $\left(2, \frac{3}{4}, 0\right)$, winds around the y-axis

with a radius of 2, and goes around exactly once each time y increases by 3.

**24. Find a parametrization for the helix that winds around the line $(x, y, z) = (t, t, 0)$ with a radius of 1, winding clockwise from our usual vantage point, and that contains the points $(0, 0, 1)$ and $(2, 2, 1)$ and no points on the segment joining them.

25. Find all points at which the line $(x, y, z) = (t - 1, 2t - 1, 3t - 2)$ intersects the curve $(x, y, z) = (t^3 + 1, 2t^2 - 1, 2 + t)$.

26. Find all points at which the line $\mathbf{r}(t) = (t, 1 - t, t)$ intersects the curve $\mathbf{p}(t) = (t^2 + 1, t + 3, 2t^2 - 2t + 1)$.

*27. Find the angle of the velocity vector above the horizontal plane at time t for the helix $\mathbf{r}(t) = (\sin t, \cos t, t)$.

*28. (i) Find the angle of the velocity vector above the horizontal plane at time t for the curve $\mathbf{r}(t) = (k \cos t, k \sin t, t)$.
   (ii) For what values of k and t will the angle be $\frac{\pi}{3}$?
   (iii) What are the possible angles as k takes on all values in $(0, \infty)$?
   (iv) Describe the curve.

*29. (i) Find a parametrization of the curve that is the intersection of the plane $x + 2y = 0$ with the surface $z = x^2 + y^2$.
   (ii) Find all points on the curve at which the tangent vector is horizontal.

*30. Find the angle between the xz-plane and the tangent vector to the curve $\mathbf{r}(t) = (3 - \cos t, t - 2, 4 + \sin t)$ when $t = 0$.

*31. Find the angle between the yz-plane and the tangent vector to the curve $\mathbf{r}(t) = \left(2t - \sin \frac{\pi}{2}t, t^2, \frac{1}{\sqrt{2}}t^4\right)$ when $t = 1$.

*32. Find the angle between the plane $x + y + z = 2$ and the tangent vector to the curve $\mathbf{r}(t) = (t^2 - 1, t^3 + t, 3 + t^2)$ when $t = -1$.

*33. Suppose $f(t)$ is a function that is differentiable at $t = a$, with $f'(a) \neq 0$. Consider the curve $\mathbf{r}(t) = \left(\frac{1}{\sqrt{2}}f(t), \frac{1}{\sqrt{2}}f(t), f(t)\right)$. What is the angle between this curve and the horizontal when $t = a$?

**34. Suppose $f(t)$ is a function that is differentiable at $t = a$, with $f'(a) \neq 0$. Consider the curve $\mathbf{r}(t) = \left(cf(t), -cf(t), f(t)\right)$. For what value of the constant c is the angle between this curve and the horizontal equal to $\frac{\pi}{6}$ when $t = a$? Describe the curve.

35. Use a graphics calculator or computer to find the maximum and minimum speed of the moving point

$\mathbf{r}(t) = (3t^6, 4t^2 - 1, t^4 - 8t^2)$ of Example 5, for $t \in [-3, 4]$.

***36.** For the humped spiral $\mathbf{r}(t) = (\cos 12t, t, \sin t + \sin 12t)$ of Example 3 (p. 873), try to guess from the figure where the speed is greatest and smallest. Use a graphics calculator or computer to confirm your guess.

15.2 Unit Tangent and Normal Vectors and Arclength

NORMALIZING VECTORS
PARAMETRIZING BY ARCLENGTH

Parametric equations describe the position of a moving point. As the point moves, it traces out a curve, and we have seen how the velocity vector provides a way to define the tangent vector to the curve. See Figure 15.2.1.

We have also seen that it is possible to have many different parametrizations of the same curve. Different parametrizations may have different velocities. The velocity vectors corresponding to different parametrizations are all tangent to the curve, so they must all be scalar multiples of each other. However, they may have different lengths and may point in opposite directions.

To remove this ambiguity, we can consider a *unit tangent vector*, that is, one whose norm is 1. Then there are only two possibilities, and they are each other's negatives.

It is very easy to find a unit tangent vector.

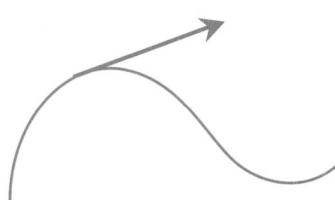

Figure 15.2.1

DEFINITION 15.2.1

If $\mathbf{r}(t)$ is any parametrization of a curve and $\mathbf{r}'(t)$ is not the zero vector, then $\mathbf{r}'(t)$ is a tangent vector, and a **unit tangent vector** can be found by dividing by its length:

$$\mathbf{T} = \mathbf{T}(t) = \frac{1}{\|\mathbf{r}'(t)\|}\mathbf{r}'(t).$$

It is customary to use the letter $\mathbf{T}$ to remind us of "Tangent." The process of dividing a vector by its length to make a unit vector is called **normalizing** the vector.

EXAMPLE 1   A convenient parametrization of the circle of radius R around the origin is

$$\mathbf{r}(t) = (R\cos t, R\sin t).$$

The velocity vector is

$$\mathbf{r}'(t) = (-R\sin t, R\cos t).$$

Several examples of such circles are illustrated in Figure 15.2.2, each with a tangent vector.

The length of the tangent vector $\mathbf{r}'(t)$ is easily found: $\|\mathbf{r}'(t)\| = \sqrt{R^2\sin^2 t + R^2\cos^2 t} = R$. Using the formula from Definition 15.2.1, we find the unit tangent vector

$$\mathbf{T}(t) = \frac{1}{\|\mathbf{r}'(t)\|}\mathbf{r}'(t) = \frac{1}{R}(-R\sin t, R\cos t) = (-\sin t, \cos t).$$

Figure 15.2.2

Figure 15.2.3

Figure 15.2.4

Figure 15.2.5

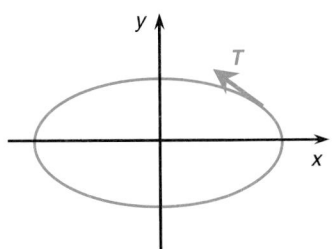

Figure 15.2.6

In Figure 15.2.3 there are circles for several different values of R; notice that the tangent vectors $\mathbf{T}(t)$ marked on them are all of length 1. In Figure 15.2.2, the tangent vectors $\mathbf{r}'(t)$ have different lengths. The reason is that the moving point $\mathbf{r}(t) = (R\cos t, R\sin t)$ goes exactly once around the circle each time t increases by 2π. For a larger circle this means that it must travel faster, and this is why the velocity vectors are longer for the larger circles. ◢

EXAMPLE 2   Consider the circle of radius R whose center is the point with rectangular coordinates $(2, 3)$. It can, for instance, be parametrized by

$$\mathbf{r}(t) = (2 + R\sin t, 3 + R\cos t).$$

The velocity vector is $\mathbf{r}'(t) = (R\cos t, -R\sin t)$, and as in Example 1 we quickly find that $\|\mathbf{r}'(t)\| = R$, so a unit tangent vector is $\mathbf{T}(t) = \frac{1}{R}\mathbf{r}'(t) = (\cos t, -\sin t)$.

Observe that the tangent vector points around the circle in the clockwise direction. This reflects the fact that with this parametrization the point $\mathbf{r}(t)$ moves around the circle *clockwise* (starting at the point $(2, 3+R)$ when $t = 0$). See Figure 15.2.4. ◢

EXAMPLE 3   Find unit tangent vectors for each of the following curves: (i) $y = x^2$, (ii) $\frac{x^2}{4} + y^2 = 1$.

Solution

(i) We begin by parametrizing the curve. In this case a natural choice is $\mathbf{r}(t) = (t, t^2)$. The corresponding velocity vector is $\mathbf{r}'(t) = (1, 2t)$. Its length is $\sqrt{1 + 4t^2}$, so the unit tangent vector is

$$\mathbf{T}(t) = \frac{1}{\sqrt{1 + 4t^2}}(1, 2t).$$

See Figure 15.2.5.

(ii) Here a reasonable parametrization is $\mathbf{r}(t) = (2\cos t, \sin t)$. The velocity vector is $\mathbf{r}'(t) = (-2\sin t, \cos t)$. Its length is $\sqrt{4\sin^2 t + \cos^2 t}$, and the unit tangent vector is

$$\mathbf{T}(t) = \frac{1}{\sqrt{4\sin^2 t + \cos^2 t}}(-2\sin t, \cos t).$$

See Figure 15.2.6. ◢

The difficulty with these unit tangent vectors is that they are still expressed in terms of the parameter t, so a different choice of parameter will give different formulas for the unit tangent vectors. It would be useful to be able to express the tangent vector in terms of the point on the curve rather than in terms of the corresponding value of the parameter. For instance, we should be able to take some point on the ellipse, such as $\left(\sqrt{2}, \frac{1}{\sqrt{2}}\right)$, and have a formula for the unit tangent vector at that point.

EXAMPLE 4   Express the unit tangent vectors of the previous example in terms of the coordinates of the point on the curve.

Solution

(i) We already know that at the point (t, t^2) the unit tangent vector is $\mathbf{T}(t) = \frac{1}{\sqrt{1+4t^2}}(1, 2t)$. If we call the point on the parabola (x, y), then x must be equal to t, so the unit tangent vector is $\frac{1}{\sqrt{1+4x^2}}(1, 2x) = \left(\frac{1}{\sqrt{1+4x^2}}, \frac{2x}{\sqrt{1+4x^2}}\right)$. We might write this as

$$\mathbf{T}(x, y) = \left(\frac{1}{\sqrt{1 + 4x^2}}, \frac{2x}{\sqrt{1 + 4x^2}}\right).$$

(ii) Here we know that at the point $(2\cos t, \sin t)$ the unit tangent vector is $\mathbf{T}(t) = \frac{1}{\sqrt{4\sin^2 t + \cos^2 t}}(-2\sin t, \cos t)$.

This example is slightly more complicated than the last because neither coordinate x nor y is equal to the parameter t. However, what we really need to express the unit tangent vector is $\sin t$ and $\cos t$, and happily, $\sin t$ equals y and $\cos t$ equals $\frac{x}{2}$. The unit tangent vector can be expressed as

$$\mathbf{T}(x, y) = \frac{1}{\sqrt{4y^2 + \frac{x^2}{4}}}\left(-2y, \frac{x}{2}\right).$$

The important thing about these examples is that if we choose a different parametrization, the formulas for the unit tangent vectors *as functions of x and y* should be the same. There is one slight catch, and that is that if the parametrization is chosen in such a way that the moving point traces out the curve in the opposite direction, then the unit tangent vectors will also point in the opposite direction, that is, they will be replaced with their negatives. However, apart from this small ambiguity in the sign, the unit vectors will always be the same.

EXAMPLE 5   Find the unit tangent vector to the helix $\mathbf{r}(t) = (\cos t, \sin t, t)$, and express it in terms of the coordinates (x, y, z) of the point on the curve.

Solution   We find that the tangent vector is $\mathbf{r}'(t) = (-\sin t, \cos t, 1)$. Its length is $\sqrt{\sin^2 t + \cos^2 t + 1} = \sqrt{2}$, so the unit tangent vector is $\mathbf{T}(t) = \frac{1}{\sqrt{2}}(-\sin t, \cos t, 1)$.

Observing that $x = \cos t$, $y = \sin t$, and $z = t$, we find that the unit tangent vector at the point (x, y, z) on the helix is

$$\mathbf{T}(x, y, z) = \left(-\frac{y}{\sqrt{2}}, \frac{x}{\sqrt{2}}, \frac{1}{\sqrt{2}}\right).$$

See Figure 15.2.7.

Normal Vectors

For a curve parametrized by a moving point $\mathbf{r}(t)$ we have seen that the vector $\mathbf{r}'(t)$ is the velocity of the moving point. To find a tangent vector that is independent of

Figure 15.2.7

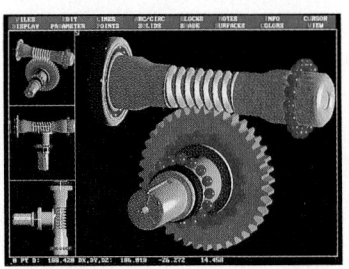

DesignCAD 3D *software was used to produce this image of a screw gear. The screw thread on the grey shaft at the top is approximately a helix.*

the parametrization, we normalize to find the unit tangent vector

$$\mathbf{T}(t) = \frac{1}{\|\mathbf{r}'(t)\|}\mathbf{r}'(t).$$

When we studied motion in a straight line, where the displacement of a moving point is a single function $x(t)$ of time, we found that the derivative $x'(t) = \frac{d}{dt}\big(x(t)\big)$ is the velocity of the point. This idea served as the model when we discussed motion in the plane or in space; now the displacement is a vector $\mathbf{r}(t)$, but we can still form a kind of derivative to get the velocity $\mathbf{r}'(t)$.

For motion in a straight line we also know that the second derivative $x''(t) = \frac{d^2}{dt^2}\big(x(t)\big)$ is the acceleration of the moving point. It is possible to consider the *second derivative* $\mathbf{r}''(t)$ for a moving point in the plane or $\mathbb{R}^3$, and we will do this in the next section. However, in this section we consider the derivative of $\mathbf{T}(t) = \frac{1}{\|\mathbf{r}'(t)\|}\mathbf{r}'(t)$ instead of the derivative $\mathbf{r}''(t)$ of the velocity vector $\mathbf{r}'(t)$. The vector $\mathbf{T}(t)$ is after all a vector that changes with time, and it is perfectly reasonable to consider its derivative. The obvious way to do it would be to write out its components and differentiate each of them.

The importance of the derivative of $\mathbf{T}(t)$ arises from the following easy result.

PROPOSITION 15.2.2

Suppose $\mathbf{u}(t)$ is a unit vector that changes as a function of t. Writing $\mathbf{u}(t) = \big(f(t), g(t), h(t)\big)$, we define its derivative in the obvious way, assuming that $f(t)$, $g(t)$, and $h(t)$ are differentiable functions of t:

$$\mathbf{u}'(t) = \big(f'(t), g'(t), h'(t)\big).$$

Then $\mathbf{u}'(t)$ is orthogonal to $\mathbf{u}(t)$. (If $\mathbf{u}(t)$ is a unit vector in the plane, then we write $\mathbf{u}(t) = \big(f(t), g(t)\big)$ and define $\mathbf{u}'(t) = \big(f'(t), g'(t)\big)$. The conclusion remains the same.)

PROOF   We know that

$$\|\mathbf{u}(t)\|^2 = f(t)^2 + g(t)^2 + h(t)^2 = 1.$$

Differentiating this constant function with respect to t, we find that

$$
\begin{aligned}
0 &= 2f(t)f'(t) + 2g(t)g'(t) + 2h(t)h'(t) \\
&= 2\big(f(t)f'(t) + g(t)g'(t) + h(t)h'(t)\big) \\
&= 2\mathbf{u}(t) \cdot \mathbf{u}'(t).
\end{aligned}
$$

This calculation shows that the dot product $\mathbf{u}(t) \cdot \mathbf{u}'(t)$ equals zero, which means that $\mathbf{u}(t)$ is orthogonal to $\mathbf{u}'(t)$, as claimed. (The calculation is slightly easier for vectors in the plane.) ◢

EXAMPLE 6

(i) For the point $\mathbf{r}(t) = (R\cos t, R\sin t)$ moving around the circle $x^2 + y^2 = R^2$, the unit tangent vector is $\mathbf{T}(t) = (-\sin t, \cos t)$.

Since $-\sin t$ and $\cos t$ are both differentiable functions of t, it is possible to find the derivative

$$\mathbf{T}'(t) = \frac{d}{dt}\big(\mathbf{T}(t)\big) = (-\cos t, -\sin t).$$

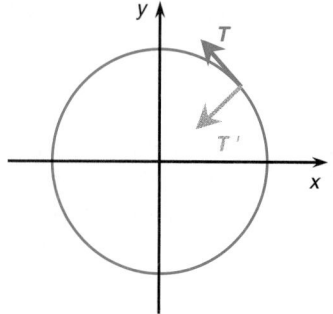

Figure 15.2.8

We check that this is orthogonal to $\mathbf{T}(t)$; after all their dot product is $(-\cos t)(-\sin t) + (-\sin t)(\cos t) = 0$. In fact, $\mathbf{T}'(t)$ is just $-\frac{1}{R}\mathbf{r}(t)$, which means that it points in the direction from $\mathbf{r}(t)$ back to the origin (see Figure 15.2.8).

(ii) For the point $\mathbf{r}(t) = (t, t^2)$ moving along the parabola $y = x^2$ the velocity vector is

$$\mathbf{r}'(t) = (1, 2t),$$

and the unit tangent vector is

$$\mathbf{T}(t) = \left(\frac{1}{\sqrt{1 + 4t^2}}, \frac{2t}{\sqrt{1 + 4t^2}} \right).$$

Its derivative is

$$\mathbf{T}'(t) = \left(\frac{-4t}{(1 + 4t^2)^{3/2}}, \frac{2\sqrt{1 + 4t^2} - (2t)\frac{1}{2}(1 + 4t^2)^{-1/2}(8t)}{1 + 4t^2} \right)$$

$$= \left(\frac{-4t}{(1 + 4t^2)^{3/2}}, \frac{2(1 + 4t^2) - 8t^2}{(1 + 4t^2)^{3/2}} \right)$$

$$= \left(\frac{-4t}{(1 + 4t^2)^{3/2}}, \frac{2}{(1 + 4t^2)^{3/2}} \right).$$

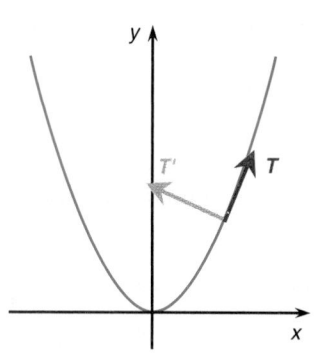

Figure 15.2.9

We check that $\mathbf{T}'(t) \cdot \mathbf{T}(t) = 0$, so $\mathbf{T}'(t)$ is orthogonal to $\mathbf{T}(t)$. See Figure 15.2.9.

Notice that if we do not normalize but simply take the velocity vector $\mathbf{r}'(t) = (1, 2t)$ and differentiate *it*, the resulting vector $\mathbf{r}''(t) = (0, 2)$ is certainly *not* orthogonal to $\mathbf{r}'(t)$. The result depends on the moving vector being a unit vector.

(iii) For the moving point $\mathbf{r}(t) = (\cos t, \sin t, t)$, which we know traces out the helix shown in Figure 15.2.10, the velocity vector is

$$\mathbf{r}'(t) = (-\sin t, \cos t, 1),$$

and the unit tangent vector is

$$\mathbf{T}(t) = \left(-\frac{1}{\sqrt{2}} \sin t, \frac{1}{\sqrt{2}} \cos t, \frac{1}{\sqrt{2}} \right).$$

Figure 15.2.10

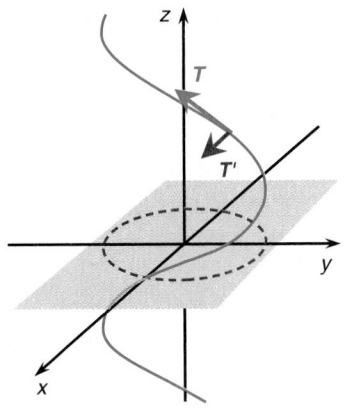

Figure 15.2.11

Its derivative is

$$\mathbf{T}'(t) = \left(-\frac{1}{\sqrt{2}} \cos t, -\frac{1}{\sqrt{2}} \sin t, 0 \right).$$

This vector is orthogonal to $\mathbf{T}(t)$. Since its z-component is 0, it is a horizontal vector. Its first two components are exactly $-\frac{1}{\sqrt{2}}$ times the first two components of $\mathbf{r}(t)$. This means that it points horizontally from the point $\mathbf{r}(t)$ back to the z-axis. See Figure 15.2.11. ◢

Proposition 15.2.2 says that the derivative of a moving *unit* vector is a vector that is perpendicular to the original vector. In studying lines and planes we saw the importance of normal vectors, that is, vectors that are perpendicular to something. What we have found is a way to construct a *normal vector* for a curve.

DEFINITION 15.2.3

> If the moving point $\mathbf{r}(t)$ traces out a curve and
>
> $$\mathbf{T}(t) = \frac{1}{\|\mathbf{r}'(t)\|} \mathbf{r}'(t)$$
>
> is the usual unit tangent vector, then at every t where the components of $\mathbf{T}(t)$ are differentiable functions, the derivative $\mathbf{T}'(t)$ is a **normal vector** to the curve.
> If $\mathbf{T}'(t) \neq 0$, the **unit normal vector**, sometimes called the **principal normal vector**, is obtained by normalizing it:
>
> $$\mathbf{N}(t) = \frac{1}{\|\mathbf{T}'(t)\|} \mathbf{T}'(t).$$

EXAMPLE 7   Find the unit normal vectors for the curves considered in Example 6.

Solution   Since we have already found the normal vector $\mathbf{T}'(t)$ in each case, this is quite easy. We only have to normalize.

(i) For the circle $\mathbf{r}(t) = (R \cos t, R \sin t)$ we found that the unit tangent vector is $\mathbf{T}(t) = (-\sin t, \cos t)$, and its derivative is $\mathbf{T}'(t) = (-\cos t, -\sin t)$. This is already a unit vector, so normalizing does not change it.

The unit normal vector is $\mathbf{N}(t) = (-\cos t, -\sin t)$, which is the vector pointing straight back toward the center of the circle.

(ii) For the parabola $\mathbf{r}(t) = (t, t^2)$ we found that the derivative of the unit tangent vector is

$$\mathbf{T}'(t) = \left(\frac{-4t}{(1 + 4t^2)^{3/2}}, \frac{2}{(1 + 4t^2)^{3/2}} \right).$$

Its length satisfies $\|\mathbf{T}'(t)\|^2 = \frac{16t^2 + 4}{(1 + 4t^2)^3} = \frac{4}{(1 + 4t^2)^2}$. This means that the length is $\|\mathbf{T}'(t)\| = \frac{2}{1 + 4t^2}$, so the unit normal vector is

$$\mathbf{N}(t) = \frac{1}{\frac{2}{1 + 4t^2}} \left(\frac{-4t}{(1 + 4t^2)^{3/2}}, \frac{2}{(1 + 4t^2)^{3/2}} \right) = \left(\frac{-2t}{\sqrt{1 + 4t^2}}, \frac{1}{\sqrt{1 + 4t^2}} \right).$$

(iii) For the helix $\mathbf{r}(t) = (\cos t, \sin t, t)$ we found that the derivative of the unit tangent vector is $\mathbf{T}'(t) = \left(-\frac{1}{\sqrt{2}} \cos t, -\frac{1}{\sqrt{2}} \sin t, 0\right)$. This vector has norm equal to $\frac{1}{\sqrt{2}}$, so the unit normal vector is

$$\mathbf{N}(t) = \frac{1}{\frac{1}{\sqrt{2}}} \left(-\frac{1}{\sqrt{2}} \cos t, -\frac{1}{\sqrt{2}} \sin t, 0\right) = (-\cos t, -\sin t, 0).$$

This vector points in from the curve to the z-axis.   ◀

Figure 15.2.12(i)

As the point $\mathbf{r}(t)$ moves along the curve, the velocity vector is $\mathbf{r}'(t)$. It points in the direction of the motion, and its length indicates the speed (see Figure 15.2.12(i)).

The unit tangent vector $\mathbf{T}(t)$ is obtained by normalizing the velocity; it points in the direction of motion, but its length is always 1. See Figure 15.2.12(ii).

As the point moves, its direction may change; this means that the path "curves." The normal vector $\mathbf{N}(t)$ points in the direction in which the path curves, that is, the direction toward which the tangent vector turns (see Figure 15.2.12(iii)).

Figure 15.2.12(ii)

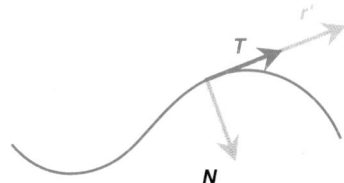

Figure 15.2.12(iii)

Arclength

In Definition 13.2.4 we defined the arclength of a curve $\mathbf{r}(t) = \big(x(t), y(t)\big)$ between $t = a$ and $t = b$ to be

$$s = \int_a^b \sqrt{x'(t)^2 + y'(t)^2} \, dt.$$

The way we arrived at this definition was to recognize that the rate of change of the arclength traced out by a moving point ought to be the speed of the point, and the speed is the norm of the velocity vector. To find the arclength itself, we simply integrate its derivative.

For a curve $\mathbf{r}(t) = \big(x(t), y(t), z(t)\big)$ in three-dimensional space the same idea can be applied. The speed of the moving point in this case is $\|\mathbf{r}'(t)\| = \sqrt{x'(t)^2 + y'(t)^2 + z'(t)^2}$, and we make the following definition.

DEFINITION 15.2.4

Suppose $x(t)$, $y(t)$, and $z(t)$ are differentiable with continuous derivatives. If $a \leq b$, the **arclength** of the curve $\mathbf{r}(t) = \big(x(t), y(t), z(t)\big)$ between the points corresponding to $t = a$ and $t = b$ is defined to be

$$s = \int_a^b \sqrt{x'(t)^2 + y'(t)^2 + z'(t)^2} \, dt.$$

EXAMPLE 8   Find the arclength of one complete loop of the helix $\mathbf{r}(t) = (\cos t, \sin t, t)$, that is, the part of the curve that goes exactly once around the z-axis in the horizontal direction.

The point goes around the axis exactly once each time the parameter increases by 2π. So if we start at $t = a$, we should end at $t = a + 2\pi$. The arclength is

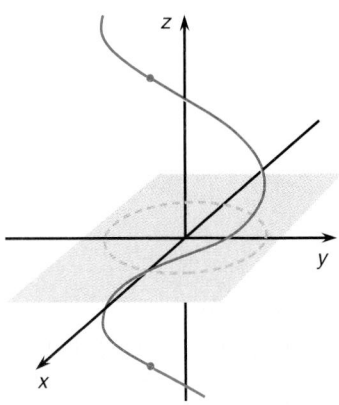

Figure 15.2.13

$$s = \int_a^{a+2\pi} \sqrt{x'(t)^2 + y'(t)^2 + z'(t)^2}\, dt = \int_a^{a+2\pi} \sqrt{\sin^2(t) + \cos^2(t) + 1}\, dt$$

$$= \int_a^{a+2\pi} \sqrt{2}\, dt = \sqrt{2}\big((a + 2\pi) - a\big)$$

$$= 2\sqrt{2}\pi.$$

Notice that the answer does not depend on which a we chose as the starting value of the parameter. In a sense this is because the shape of the helix is the same no matter where we start. See Figure 15.2.13. ◢

Different Parametrizations

Any curve can be parametrized in many different ways. Different parametrizations will have different velocity vectors. For instance, the obvious parametrization of the circle $x^2 + y^2 = 1$ is $\mathbf{p}(t) = (\cos t, \sin t)$. Its velocity vector is $\mathbf{p}'(t) = (-\sin t, \cos t)$, which is already a unit vector.

However, if we choose instead the parametrization $\mathbf{q}(t) = (\cos 2t, \sin 2t)$, it traces out the circle twice as fast. This is reflected in its velocity vector $\mathbf{q}'(t) = (-2\sin 2t, 2\cos 2t)$, which is 2 times $\mathbf{p}'(2t)$.

A more unusual parametrization would be $\mathbf{r}(u) = \big(\cos(1 + u^2), \sin(1 + u^2)\big)$. Its velocity vector is

$$\mathbf{r}'(u) = \big(-2u\sin(1 + u^2), 2u\cos(1 + u^2)\big).$$

Notice that the length of this vector is $|2u|$, which changes as u changes. This means that the speed of the moving point will vary.

Let us consider this last little example in more detail. The parametrization $\mathbf{r}(u) = \big(\cos(1 + u^2), \sin(1 + u^2)\big)$ is related to the first parametrization $\mathbf{p}(t) = (\cos t, \sin t)$. In fact, if we let $w = 1 + u^2$, then $\mathbf{r}(u) = \mathbf{p}(w) = \mathbf{p}(1 + u^2)$. In effect we have made a "change of variables." The curve $\mathbf{r}(u)$ traces out the same curve as $\mathbf{p}(t)$ (or at least part of it). The point $\mathbf{r}(u)$ is the same as the point $\mathbf{p}(t)$, with $t = 1 + u^2$.

Moreover, when we find the derivative (i.e., the tangent vector), we obtain $\mathbf{r}'(u) = 2u\,\mathbf{p}'(1 + u^2) = \frac{dw}{du}\mathbf{p}'(w)$. This looks like the Chain Rule, and in fact it is precisely the Chain Rule, because although we have written it in vector form, if we wrote out the components of the vectors, this formula would just be the Chain Rule for them.

PROPOSITION 15.2.5

Suppose $\mathbf{p}(t)$ is a curve and suppose its components are all differentiable functions of t. If $w(u)$ is a differentiable function and $\mathbf{r}(u) = \mathbf{p}\big(w(u)\big)$ is a different parametrization, then its velocity vector satisfies the following:

$$\mathbf{r}'(u) = w'(u)\mathbf{p}'\big(w(u)\big).$$

PROOF It is simply a matter of writing out the components. We do this for three components, but the case with two is similar. If $\mathbf{p}(t) = \big(x(t), y(t), z(t)\big)$, then $\mathbf{p}'(t) = \big(x'(t), y'(t), z'(t)\big)$. Also

$$\mathbf{r}(u) = \mathbf{p}\big(w(u)\big) = \Big(x\big(w(u)\big), y\big(w(u)\big), z\big(w(u)\big)\Big),$$

so its tangent vector is found by differentiating:

$$\mathbf{r}'(u) = \Big(x'\big(w(u)\big)w'(u), y'\big(w(u)\big)w'(u), z'\big(w(u)\big)w'(u)\Big)$$

$$= w'(u)\Big(x'\big(w(u)\big), y'\big(w(u)\big), z'\big(w(u)\big)\Big)$$

$$= w'(u)\mathbf{p}'\big(w(u)\big),$$

as claimed.   ◢

The formula in Proposition 15.2.5 says that the velocity vector for $\mathbf{r}(u)$ is the velocity vector for $\mathbf{p}\big(w(u)\big)$ *multiplied by the number* $w'(u)$. Using a different parametrization gives a point moving along the same curve, so the direction of the velocity vector must be the same (or at worst the opposite), but the speed may vary, that is, the length of the vector may be different; this is the effect of multiplying by $w'(u)$.

Parametrization by Arclength

When we wanted to find the unit tangent vector $\mathbf{T}(t)$ to a curve, it was necessary to normalize the velocity vector $\mathbf{r}'(t)$, which meant dividing by its length. It seems tempting to wonder whether it might be possible to find a suitable parametrization so that the corresponding velocity vector is always a unit vector.

This means that the speed of the moving point is always 1, meaning that in each unit of time it moves 1 unit of arclength. We can tell how long it has been moving by measuring the arclength it has traveled and vice versa.

In effect we are using the arclength as the parameter; this is exactly what is needed to make the velocity vector a unit vector.

To make this idea work, it is necessary to pick some starting point on the curve and then measure arclength from that point. Actually, it is necessary to use what might be called "signed arclength," which is arclength in one direction from the starting point and the negative of the arclength going in the other direction. Specifying a value s of the signed arclength then completely determines a point on the curve. See Figure 15.2.14, in which the signed arclength is used as a "scale" along a curve.

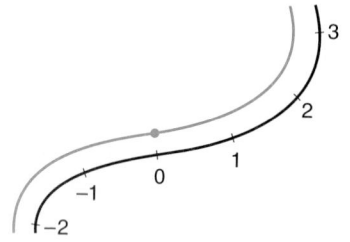

Figure 15.2.14

<table>
<tr><td style="vertical-align:top; text-align:right; padding-right:1em; font-style:italic;">THEOREM 15.2.6</td><td>

Suppose $\mathbf{r}(t)$ is a curve, each of whose components is a differentiable function of t. Also suppose that $\mathbf{r}'(t)$ is never the zero vector. Let $\mathbf{p}(s)$ be the same curve parametrized by arclength, as described above, with $s = 0$ at the point $\mathbf{r}(t_0)$. The point $\mathbf{p}(s)$ is the same as $\mathbf{r}(t)$, where s is the arclength from $\mathbf{r}(t_0)$ to $\mathbf{r}(t)$. In symbols,

$$\mathbf{r}(t) = \mathbf{p}\big(s(t)\big),$$

with

$$s = s(t) = \int_{t_0}^{t} \|\mathbf{r}'(u)\| \, du.$$

Then the velocity vector $\mathbf{p}'(s)$ is always a unit vector.

</td></tr>
</table>

PROOF Applying Proposition 15.2.5 to the formula $\mathbf{r}(t) = \mathbf{p}\big(s(t)\big)$, we find that $\mathbf{r}'(t) = \frac{ds}{dt}\,\mathbf{p}'\big(s(t)\big)$. Since $\frac{ds}{dt} = \|\mathbf{r}'(t)\|$, we see that $\mathbf{r}'(t) = \|\mathbf{r}'(t)\|\,\mathbf{p}'\big(s(t)\big)$,

so

$$\frac{1}{\|\mathbf{r}'(t)\|}\mathbf{r}'(t) = \mathbf{p}'\big(s(t)\big).$$

The vector on the left is a unit vector, so the result follows.

EXAMPLE 9

(i) The usual parametrization

$$\mathbf{r}(t) = (\cos t, \sin t)$$

of the unit circle is already the parametrization in terms of arclength. As t increases by 2π, the point moves exactly once around the circle, traveling a distance equal to its circumference 2π. The distance traveled by the point between $t = \alpha$ and $t = \beta$ is exactly $\beta - \alpha$. (This is essentially the definition of radian measure; see Figure 15.2.15.)

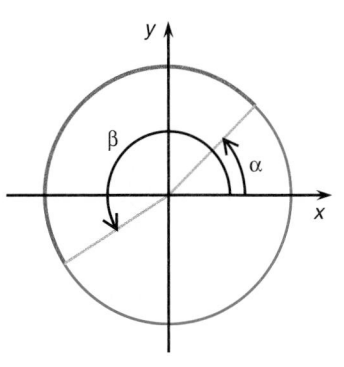

y

β

α

x

Figure 15.2.15

(ii) Now consider the circle of radius R. It is conveniently parametrized by $\mathbf{r}(t) = (R\cos t, R\sin t)$. We find that the velocity vector is $\mathbf{r}'(t) = (-R\sin t, R\cos t)$, and its length is R. This means that the point moves R times as fast as it would if the parameter were the arclength.

We can easily guess the arclength parametrization by letting

$$\mathbf{p}(s) = \left(R\cos\left(\frac{s}{R}\right), R\sin\left(\frac{s}{R}\right)\right).$$

It goes once around the circle each time $\frac{s}{R}$ increases by 2π, that is, as s increases by $2\pi R$, which is the circumference of the circle.

However, let us also work out the calculation in order to become more familiar with it. We choose the starting point to be $(R, 0)$, corresponding to $t = 0$. The arclength is then given by

$$s = s(t) = \int_0^t \|\mathbf{r}'(u)\|\, du = \int_0^t R\, du = Rt.$$

This tells us to let $t = \frac{s}{R}$ and write

$$\mathbf{p}(s) = \mathbf{r}(t) = \mathbf{r}\left(\frac{s}{R}\right) = \left(R\cos\left(\frac{s}{R}\right), R\sin\left(\frac{s}{R}\right)\right).$$

(iii) For the helix $\mathbf{r}(t) = (2\cos t, 2\sin t, t)$ we find the velocity is $\mathbf{r}'(t) = (-2\sin t, 2\cos t, 1)$, so $\|\mathbf{r}'(t)\| = \sqrt{5}$ and $s = \sqrt{5}t$. The arclength parametrization is obtained by letting $t = \frac{s}{\sqrt{5}}$; the helix is

$$\mathbf{p}(s) = \left(2\cos\frac{s}{\sqrt{5}}, 2\sin\frac{s}{\sqrt{5}}, \frac{s}{\sqrt{5}}\right).$$

(iv) Let $\mathbf{r}(t) = (20t^{3/2} + 12t^{5/2}, 15t^2 + 30t)$, for $t \geq 0$. Then

$$\|\mathbf{r}'(t)\| = \left((30t^{1/2} + 30t^{3/2})^2 + (30t + 30)^2\right)^{1/2}$$

$$= \left(900(t + 2t^2 + t^3) + 900(t^2 + 2t + 1)\right)^{1/2}$$

$$= 30(t^3 + 3t^2 + 3t + 1)^{1/2} = 30(t + 1)^{3/2}.$$

$t^3 + 3t^2 + 3t + 1 = (t + 1)^3$

So

$$s = s(t) = \int_0^t \|\mathbf{r}'(u)\| \, du = 30 \int_0^t (u+1)^{3/2} \, du = 30\left(\frac{2}{5}\right)(u+1)^{5/2}\Big|_0^t$$

$$= 12\big((t+1)^{5/2} - 1\big).$$

To parametrize the curve by s, we need to express t in terms of s. We found that $12(t+1)^{5/2} = s + 12$, so $t + 1 = \left(\frac{s+12}{12}\right)^{2/5}$ and $t = \left(\frac{s+12}{12}\right)^{2/5} - 1$. The curve can now be parametrized by s:

$$\mathbf{p}(s) = \mathbf{r}(t) = \mathbf{r}\left(\left(\frac{s+12}{12}\right)^{2/5} - 1\right)$$

$$= \left(20\left(\left(\frac{s+12}{12}\right)^{2/5} - 1\right)^{3/2} + 12\left(\left(\frac{s+12}{12}\right)^{2/5} - 1\right)^{5/2},\right.$$

$$\left. 15\left(\left(\frac{s+12}{12}\right)^{2/5} - 1\right)^2 + 30\left(\left(\frac{s+12}{12}\right)^{2/5} - 1\right)\right).$$

(v) Consider the spiral $\mathbf{r}(t) = (t\cos t, t\sin t)$. Its velocity is $\mathbf{r}'(t) = (\cos t - t\sin t, \sin t + t\cos t)$, so its speed is

$$\|\mathbf{r}'(t)\| = \sqrt{\cos^2 t - 2t\cos t\sin t + t^2\sin^2 t + \sin^2 t + 2t\sin t\cos t + t^2\cos^2 t}$$

$$= \sqrt{1 + t^2}.$$

To find the arclength, we integrate, making the natural choice of $t = 0$ as the starting point. We find

$$s = s(t) = \int_0^t \sqrt{1 + u^2} \, du = \int \sec^3 \theta \, d\theta$$

$$\boxed{\begin{array}{l} u = \tan\theta \\ du = \sec^2\theta \, d\theta \end{array}} \quad = \frac{1}{2}\sec\theta\tan\theta + \frac{1}{2}\int \sec\theta \, d\theta$$

$$= \frac{1}{2}\sec\theta\tan\theta + \frac{1}{2}\ln|\sec\theta + \tan\theta|$$

$$= \frac{1}{2}u\sqrt{1 + u^2} + \frac{1}{2}\ln|u + \sqrt{1 + u^2}|\,\Big|_0^t$$

$$= \frac{1}{2}t\sqrt{1 + t^2} + \frac{1}{2}\ln|t + \sqrt{1 + t^2}|.$$

Note: $1 + u^2 = \sec^2\theta$.

It is not clear how to express the original parameter t in terms of the arclength s. Without being able to do this we are unable to give a formula parametrizing the spiral by its arclength. ◢

Unfortunately, calculating the arclength of a curve tends to be quite difficult except for very simple curves like circles. Even if it is possible to find the arclength, it may be difficult to express the curve in terms of the arclength, as the last part of Example 9 demonstrates. However, the idea that it is possible to parametrize a curve by its arclength and the fact that in this situation the velocity vector is a unit vector are very important. Knowing that it is *possible* is sometimes useful even without actually being able to perform the calculations.

Exercises 15.2

I

Find the unit tangent and unit normal vectors for each of the following curves.

1. $\mathbf{r}(t) = (\sin t, -\cos t)$
2. $\mathbf{r}(t) = (4\cos t, -4\sin t)$
3. $\mathbf{r}(t) = (3\cos 4t, 3\sin 4t)$
4. $\mathbf{r}(t) = \left(2\cos\frac{t}{2}, 2\sin\frac{t}{2}\right)$
5. $\mathbf{r}(t) = (2t, -t, -2t)$
6. $\mathbf{r}(t) = (4t^2 + 1, 6, 5 - 3t^2)$
7. $\mathbf{r}(t) = (\cos t, \sin t, 3t)$
8. $\mathbf{r}(t) = (1 - 2t, \cos 2t, \sin 2t)$
9. $\mathbf{r}(t) = \left(\frac{1}{3}t^3 - t, t^2\right)$
10. $\mathbf{r}(t) = (\cos^2 t - \sin^2 t, -2\cos t \sin t)$

Find the arclength of each of the following curves as a function of t, starting at the point corresponding to $t = 0$ in each case.

11. $\mathbf{r}(t) = (\sin 5t, -\cos 5t)$
12. $\mathbf{r}(t) = (2\sqrt{2}t, \sin t, \cos t)$
13. $\mathbf{r}(t) = (2t, -4t, 4t)$
14. $\mathbf{r}(t) = (3t^2 + 5, -4t^2)$
15. $\mathbf{r}(t) = \left(\frac{1}{3}t^3 - t, t^2\right)$
16. $\mathbf{r}(t) = (6t^2, 12t - t^3)$
17. $\mathbf{r}(t) = \left(\frac{1}{2}\ln(1 + t^2), \arctan t\right)$
18. $\mathbf{r}(t) = (e^{2t} - 2t + 2, 1 + 4e^t)$
19. $\mathbf{r}(t) = (\cos^2 t - \sin^2 t, -2\cos t \sin t)$
20. $\mathbf{r}(t) = (2\sin t \cos t, 1 - 2\sin^2 t, 2t)$

II

21. Express the unit normal vector in Example 7(i) in terms of the coordinates (x, y) of the point on the circle (cf. Example 4).
22. Express the unit normal vector in Example 7(ii) in terms of the coordinates (x, y) of the point on the parabola (cf. Example 4).
23. Express the unit normal vector in Example 7(iii) in terms of the coordinates (x, y, z) of the point on the helix (cf. Example 4).
24. Find a formula for the arclength $s = s(t)$ of the curve $\mathbf{r}(t) = (t^2 \cos t, t^2 \sin t)$, using the point $(0, 0)$ corresponding to $t = 0$ as the starting point.

*25. Show that if $\mathbf{v}(t)$ is a moving vector whose length is always K, for some constant K, then $\mathbf{v}'(t)$ is orthogonal to $\mathbf{v}(t)$. Explain the relevance of this fact to Examples 6(i) and 6(iii).

*26. Prove the following "(dot) product rule" for derivatives of dot products of two vectors: If $\mathbf{r}(t)$ and $\mathbf{v}(t)$ are two moving vectors (i.e., the components of each are functions of t), then

$$\frac{d}{dt}\bigl(\mathbf{r}(t) \cdot \mathbf{v}(t)\bigr) = \mathbf{r}(t) \cdot \mathbf{v}'(t) + \mathbf{r}'(t) \cdot \mathbf{v}(t).$$

(Assume that the components of both vectors are differentiable functions of t.)

*27. Explain why the curves $\mathbf{r}(t) = \bigl(x(t), y(t), z(t)\bigr)$ and $\mathbf{p}(t) = \left(kx\left(\frac{t}{k}\right), ky\left(\frac{t}{k}\right), kz\left(\frac{t}{k}\right)\right)$ have the same speed, in the sense that the speed of the moving point $\mathbf{r}$ at time t equals the speed of the moving point $\mathbf{p}$ at time kt.

*28. Suppose $x(t)$, $y(t)$, and $z(t)$ are functions whose second derivatives are all zero. Show that the "derivative" of the unit tangent vector is the zero vector. This makes it difficult to define the normal vector. Geometrically, the curve is a straight line, and the absence of an obvious normal vector reflects the fact that a line "does not bend."

*29. Find an example of a curve whose components have nonzero second derivatives but for which the "derivative" of the unit tangent vector is always the zero vector.

*30. We are accustomed to expressing any vector in $\mathbb{R}^3$ in terms of the standard basis vectors $\mathbf{i}, \mathbf{j}, \mathbf{k}$. They are three unit vectors, any two of which are orthogonal. Sometimes it is convenient to have vectors with these properties that are adapted to a moving point. If $\mathbf{r}(t)$ is a curve in $\mathbb{R}^3$ whose components are twice-differentiable, let $\mathbf{T}(t)$ be the unit tangent vector and assume $\mathbf{T}'(t) \neq 0$ so we can form the unit normal vector $\mathbf{N}(t)$. Let $\mathbf{B}(t) = \mathbf{T}(t) \times \mathbf{N}(t)$, the "binormal vector." Show that $\mathbf{T}(t), \mathbf{N}(t), \mathbf{B}(t)$ are unit vectors, any two of which are orthogonal. Any vector in $\mathbb{R}^3$ can be expressed in terms of them.

*31. Let $\mathbf{r}(t) = (3\cos t, -5\sin t, 4\cos t)$, with $t \geq 0$, and parametrize the curve by its arclength (i.e., write it as $\mathbf{p}(s)$).

*32. Let $\mathbf{r}(t) = (t^3, t^2)$, with $t \geq 0$, and parametrize the curve by its arclength (i.e., write it as $\mathbf{p}(s)$).

*33. Let $\mathbf{r}(t) = \left(\sqrt{2}\arctan t, 3, \frac{1}{\sqrt{2}}\ln(1 + t^2)\right)$. Find the arclength s as a function $s(t)$.

*34. Let $\mathbf{r}(t) = (\cosh t, t - 1)$. Find the arclength s as a function $s(t)$.

Use a programmable calculator or computer to estimate the arclengths of the following curves using Simpson's Rule S_n with $n = 10$.

35. $\mathbf{r}(t) = (t, t^2, t^3), \quad t \in [-1, 1]$

36. $\mathbf{r}(t) = (2\cos t, 3\sin t), \quad t \in [0, \pi]$

37. $\mathbf{r}(t) = \left(t - e^t, \frac{1}{1+t^2}\right), \quad t \in [0, 1]$

*38. $\mathbf{r}(t) = \left(\frac{1}{t}, \frac{1}{t^2}, \frac{1}{t^3}\right), \quad t \in [1, \infty)$

15.3 Acceleration and Curvature

SECOND DERIVATIVE

CONSTANT CURVATURE

RADIUS/CENTER OF CURVATURE

OSCULATING CIRCLE

We have become accustomed to considering a moving point $\mathbf{r}(t)$ and taking the *derivative* $\mathbf{r}'(t)$ to be its velocity. The obvious next step is to ask about acceleration. For motion in a straight line we know that the acceleration is given by the second derivative of the displacement, so it seems reasonable to try the analogous approach with a curve in the plane or in space.

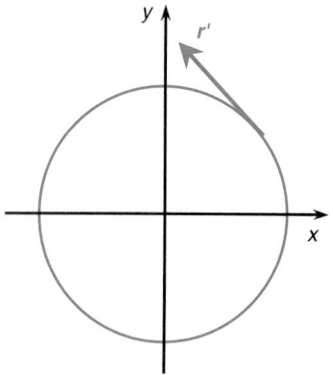

Figure 15.3.1(i)

EXAMPLE 1   Consider the curve $\mathbf{r}(t) = (\cos t, \sin t)$, which traces out a circle in the plane (see Figure 15.3.1(i)). Its velocity vector is $\mathbf{r}'(t) = (-\sin t, \cos t)$, and we have observed that it is always tangent to the circle.

The acceleration ought to have something to do with the way the velocity is changing. In this situation the velocity vectors are hard to compare because they do not start at the same point. When $t = 0$, the velocity vector is $\mathbf{r}'(0) = (0, 1)$. It points straight up, and we usually place it with its initial point at $\mathbf{r}(0) = (1, 0)$.

As t increases, the point begins to move around the circle, and the velocity vector begins to tilt to the left. By the time t reaches $\frac{\pi}{2}$, the velocity vector is the horizontal vector $(-1, 0)$. In Figure 15.3.1(ii) we plot several velocity vectors but place them all with the same initial point. We see that as t increases, the velocity vectors "lean over."

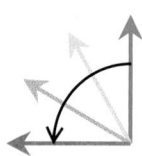

Figure 15.3.1(ii)

If we try taking the *second derivative* of $\mathbf{r}(t)$ in the obvious way, that is, just taking the second derivative of each component, we find that

$$\mathbf{r}''(t) = (-\cos t, -\sin t).$$

In particular, when $t = 0$, we have that $\mathbf{r}''(0) = (-1, 0)$. This vector points to the left, and it suggests that as t increases slightly from $t = 0$, the velocity vector will begin to "tilt" to the left.

If we consider $\mathbf{r}''(t) = (-\cos t, -\sin t)$, we see that it actually equals $-\mathbf{r}(t)$. What this means is that the acceleration vector points from a point on the circle directly back to the origin. Rather than drawing $\mathbf{r}''(t)$ with its initial point at the point $\mathbf{r}(t)$ on the circle, it is convenient to draw it with its initial point at the endpoint of the velocity $\mathbf{r}'(t)$ (see Figure 15.3.1(iii)). Notice that in this case the acceleration $\mathbf{r}''(t)$ is perpendicular to the velocity $\mathbf{r}'(t)$. What the picture suggests is that the acceleration vector points in the direction toward which the velocity vector turns, and this is exactly what happens.

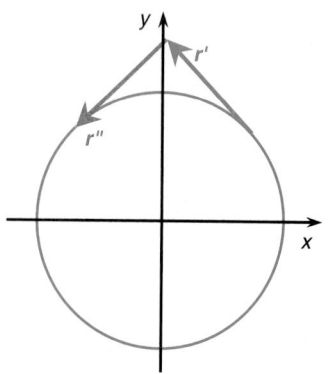

Figure 15.3.1(iii)

The fact that the acceleration equals the negative of the displacement is a special feature of circles, but the idea of discussing acceleration by using second derivatives will work just as well for other curves.

DEFINITION 15.3.1

Suppose $\mathbf{r}(t)$ is a curve whose components are all twice-differentiable functions of t. We form the **second derivative** $\mathbf{r}''(t)$ by taking the second derivative of each component, and call it the **acceleration vector**. It points in the direction toward which the velocity vector is turning.

Figure 15.3.2(i)

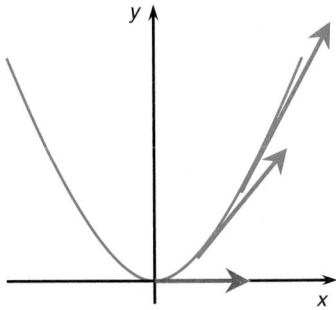

Figure 15.3.2(ii)

EXAMPLE 2   If $\mathbf{r}(t) = (t, t^2)$, then we know that this moving point traces out the parabola $y = x^2$.

Its velocity vector is $\mathbf{r}'(t) = (1, 2t)$, which we plot for several values of t in Figure 15.3.2(i).

The acceleration vector is $\mathbf{r}''(t) = (0, 2)$. It points straight up and does not change with t. This means that the velocity vector is always changing in such a way as to increase its vertical component, leaving its horizontal component unchanged. Of course, this is what happens; the horizontal component of $\mathbf{r}'(t) = (1, 2t)$ is always the same. The vertical component increases.

For instance, at $t = 0$ the velocity vector is horizontal, but as t increases, the velocity vector $\mathbf{r}'(t) = (1, 2t)$ turns so that it points up at an increasingly steep angle. See Figure 15.3.2(ii). Notice that the horizontal component is always 1 unit long.

Also notice that for the parabola, unlike for the circle, the acceleration is not perpendicular to the velocity except at $t = 0$, and it does not equal the negative of the displacement. ◢

EXAMPLE 3   Consider the helix $\mathbf{r}(t) = (\cos t, \sin t, t)$. Its velocity vector is $\mathbf{r}'(t) = (-\sin t, \cos t, 1)$ and the acceleration is $\mathbf{r}''(t) = (-\cos t, -\sin t, 0)$.

The acceleration is a horizontal vector pointing in toward the central axis of the helix (i.e., the z-axis). This reflects the fact that as the moving point winds its way around the helix, its direction keeps turning in toward the center. See Figure 15.3.2(iii).

The vertical component of the acceleration is 0 because the vertical component of the velocity $\mathbf{r}'(t) = (-\sin t, \cos t, 1)$ is constant. We have already observed that the moving point rises at a constant rate. ◢

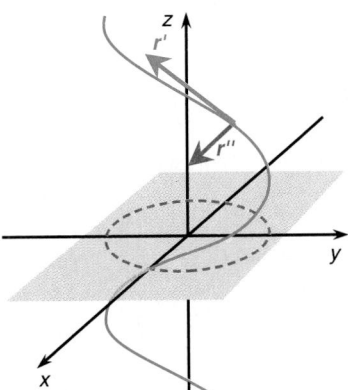

Figure 15.3.2(iii)

Curvature

The length of the velocity vector represents the speed of a moving point, but it is less clear what meaning can be given to the length of an acceleration vector. Roughly speaking, we expect it to have something to do with how fast the velocity vector is changing. The velocity vector changes by changing in length and/or by turning; we want to ask how fast it is turning.

However, this just leaves us with the question: What do we mean by saying "how fast" something is turning? In terms of a picture, this idea seems clear; the first curve in Figure 15.3.3 is turning much faster than the second curve.

If we consider a curve, for instance, the one drawn in Figure 15.3.4(i), and a specific point P on it, we want to ask how fast the curve is turning at P. One way to attempt to answer would be to put a circle through P so that the circle is tangent to the curve at P and has the same curvature there; this is illustrated in Figure 15.3.4(ii). We can use circles as a measure of how fast a curve is "curving." The difficulty is that we have not really said how we can tell what size circle has the same curvature as the curve at the point P.

Another approach would be to begin with the derivative of the velocity, that is, with the second derivative $\mathbf{r}''(t)$ of the displacement, or at least with its norm. However, there is a difficulty here too. If we consider the circle $\mathbf{r}(t) = (\cos t, \sin t)$, its second derivative is $\mathbf{r}''(t) = (-\cos t, -\sin t)$. This is a unit vector. If, on the other hand, we consider another parametrization $\mathbf{p}(t) = (\cos 2t, \sin 2t)$, its second derivative is $\mathbf{p}''(t) = (-4\cos 2t, -4\sin 2t)$. The norm of this vector is 4.

Both these parametrizations describe the same circle $x^2 + y^2 = 1$, shown in Figure 15.3.5. The curvature should not depend on which parametrization we use; after all, the curve is still the same circle. We need a way to define the curvature that does not depend on the particular choice of parametrization.

One way to achieve this would be to use the parametrization by arclength. Then the velocity vector is the unit tangent vector $\mathbf{T}(s)$, and we can consider *its* derivative $\mathbf{T}'(s)$, where this refers to its derivative *with respect to arclength*.

It seems reasonable to expect that $\|\mathbf{T}'(s)\|$ will give a good indication of how fast the curve is turning. Even if we change parametrizations for the curve, its arclength will not change, and this number will be the same. The larger the value of $\|\mathbf{T}'(s)\|$, the more sharply the curve is turning, so it is reasonable to define this number to be the *curvature*.

In this form, however, the definition is not terribly useful, because it is frequently difficult to find the formula for the arclength or to use it to parametrize the curve. Happily, it can be expressed in terms of any parametrization $\mathbf{r}(t)$.

We know that

$$\mathbf{T}(t) = \frac{1}{\|\mathbf{r}'(t)\|}\mathbf{r}'(t),$$

and we also know that $\frac{ds}{dt} = \|\mathbf{r}'\|$, so $\frac{dt}{ds} = \frac{1}{\|\mathbf{r}'\|}$. Using the Chain Rule, we find that

$$\left\|\frac{d}{ds}\big(\mathbf{T}(t)\big)\right\| = \left\|\frac{d}{ds}\left(\frac{1}{\|\mathbf{r}'(t)\|}\mathbf{r}'(t)\right)\right\| = \left\|\frac{d}{dt}\left(\frac{1}{\|\mathbf{r}'(t)\|}\mathbf{r}'(t)\right)\right\|\frac{dt}{ds}$$

$$= \left\|\frac{d}{dt}\big(\mathbf{T}(t)\big)\right\|\left(\frac{dt}{ds}\right) = \frac{1}{\|\mathbf{r}'(t)\|}\left\|\frac{d}{dt}\big(\mathbf{T}(t)\big)\right\|$$

$$= \frac{1}{\|\mathbf{r}'(t)\|}\left\|\frac{d}{dt}\left(\frac{1}{\|\mathbf{r}'(t)\|}\mathbf{r}'(t)\right)\right\|.$$

This is the reason for making the following definition.

Figure 15.3.3

P

Figure 15.3.4(i)

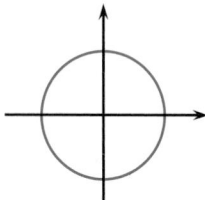

P

Figure 15.3.4(ii)

Figure 15.3.5

DEFINITION 15.3.2

Suppose $\mathbf{r}(t)$ is a curve whose components are all twice-differentiable functions of t. Its curvature is defined at every point at which the velocity vector is not the zero vector. The **curvature** of this curve at such a point $\mathbf{r}(t)$ is

$$\kappa = \|\mathbf{T}'(s)\| = \left\| \frac{d}{ds} \left(\mathbf{T}(s) \right) \right\| = \frac{1}{\|\mathbf{r}'(t)\|} \left\| \frac{d}{dt} \left(\frac{1}{\|\mathbf{r}'(t)\|} \mathbf{r}'(t) \right) \right\|$$

$$= \frac{1}{\|\mathbf{r}'(t)\|} \left\| \frac{d}{dt} \left(\mathbf{T}(t) \right) \right\|.$$

The Greek letter κ (kappa) is used to remind us of "curvature."

EXAMPLE 4

(i) The curve $\mathbf{r}(t) = (\cos t, \sin t)$ is a circle, and it is already parametrized by arclength, that is, the arclength s equals the parameter t.

This means that the curvature can easily be found:

$$\kappa = \left\| \frac{d}{ds} \mathbf{T}(s) \right\| = \left\| \frac{d}{dt} \mathbf{T}(t) \right\| = \left\| \frac{d}{dt} \mathbf{r}'(t) \right\|$$

$$= \left\| \frac{d}{dt} (-\sin t, \cos t) \right\| = \|(-\cos t, -\sin t)\| = 1.$$

We find that the curvature is $\kappa = 1$ at every point on the circle.

(ii) Next we should consider $\mathbf{p}(t) = (\cos 2t, \sin 2t)$, a different parametrization of the same circle. We find that $\mathbf{p}'(t) = (-2 \sin 2t, 2 \cos 2t)$, so $\|\mathbf{p}'(t)\| = 2$. From this we can easily find the curvature:

$$\kappa = \frac{1}{\|\mathbf{p}'(t)\|} \left\| \frac{d}{dt} \left(\frac{1}{\|\mathbf{p}'(t)\|} \mathbf{p}'(t) \right) \right\| = \frac{1}{2} \left\| \frac{d}{dt} \left(\frac{1}{2} (-2 \sin 2t, 2 \cos 2t) \right) \right\|$$

$$= \frac{1}{2} \left\| \frac{d}{dt} \left((-\sin 2t, \cos 2t) \right) \right\| = \frac{1}{2} \|(-2 \cos 2t, -2 \sin 2t)\| = 1.$$

It is reassuring to find that the curvature of the circle is 1 even if we use the second parametrization.

(iii) For the helix $\mathbf{q}(t) = (\cos t, \sin t, 2t)$ we find that $\mathbf{q}'(t) = (-\sin t, \cos t, 2)$, so $\|\mathbf{q}'(t)\| = \sqrt{5}$. The curvature is

$$\kappa = \frac{1}{\|\mathbf{q}'(t)\|} \left\| \frac{d}{dt} \left(\frac{1}{\|\mathbf{q}'(t)\|} \mathbf{q}'(t) \right) \right\| = \frac{1}{\sqrt{5}} \left\| \frac{d}{dt} \left(\frac{1}{\sqrt{5}} (-\sin t, \cos t, 2) \right) \right\|$$

$$= \frac{1}{\sqrt{5}} \left\| \frac{1}{\sqrt{5}} (-\cos t, -\sin t, 0) \right\| = \frac{1}{5}.$$

The curvature of the helix is $\frac{1}{5}$ at every point.

Each of the curves we have discussed so far has **constant curvature**, that is, its curvature is the same at every point. Of course, this is not necessarily true for every curve.

EXAMPLE 5 If $\mathbf{r}(t) = (t^2, 2t)$, then $\mathbf{r}'(t) = (2t, 2)$, and $\|\mathbf{r}'(t)\| = \|(2t, 2)\| = \sqrt{4t^2 + 4} = 2\sqrt{t^2 + 1}$. The curvature is

$$\kappa = \frac{1}{\|\mathbf{r}'(t)\|} \left\| \frac{d}{dt}\left(\frac{1}{\|\mathbf{r}'(t)\|}\mathbf{r}'(t) \right) \right\| = \frac{1}{2\sqrt{t^2+1}} \left\| \frac{d}{dt}\left(\frac{1}{2\sqrt{t^2+1}}(2t, 2) \right) \right\|$$

$$= \frac{1}{2\sqrt{t^2+1}} \left\| \frac{d}{dt}\left(\frac{t}{\sqrt{t^2+1}}, \frac{1}{\sqrt{t^2+1}} \right) \right\|$$

$$= \frac{1}{2\sqrt{t^2+1}} \left\| \left(\frac{(1)\sqrt{t^2+1} - (t)\,t(t^2+1)^{-1/2}}{t^2+1}, \frac{-t}{(t^2+1)^{3/2}} \right) \right\|$$

$$= \frac{1}{2\sqrt{t^2+1}} \left\| \left(\frac{t^2+1-t^2}{(t^2+1)^{3/2}}, \frac{-t}{(t^2+1)^{3/2}} \right) \right\|$$

$$= \frac{1}{2\sqrt{t^2+1}} \left\| \left(\frac{1}{(t^2+1)^{3/2}}, \frac{-t}{(t^2+1)^{3/2}} \right) \right\|$$

$$= \frac{1}{2\sqrt{t^2+1}} \frac{1}{(t^2+1)^{3/2}}\sqrt{1+t^2} = \frac{1}{2(t^2+1)^{3/2}}.$$

We find that the curvature is $\kappa = \frac{1}{2(t^2+1)^{3/2}}$. It is $\kappa = \frac{1}{2}$ when $t = 0$, and for other values of t it is always less than $\frac{1}{2}$. In Figure 15.3.6 we sketch the graph of the curve, which is a parabola. The origin is the point corresponding to $t = 0$, and we see that the curvature is greatest there. At other points the curve turns less sharply. ◢

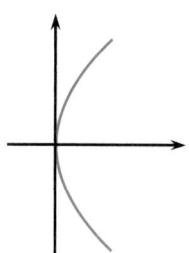

Figure 15.3.6

There is one more formula for curvature that we must learn.

THEOREM 15.3.3

Suppose $\mathbf{r}(t)$ is a curve in $\mathbb{R}^3$ whose components are twice-differentiable functions. Then the curvature can be computed as follows: If $\mathbf{r}'(t)$ is not the zero vector, then the curvature at the point $\mathbf{r}(t)$ is

$$\kappa = \frac{\|\mathbf{r}'(t) \times \mathbf{r}''(t)\|}{\|\mathbf{r}'(t)\|^3} = \frac{\|\mathbf{r}' \times \mathbf{r}''\|}{\|\mathbf{r}'\|^3}.$$

PROOF We begin by noticing that since $\mathbf{T}(t) = \frac{1}{\|\mathbf{r}'(t)\|}\mathbf{r}'(t)$, we can write $\mathbf{r}'(t) = \|\mathbf{r}'(t)\|\mathbf{T}(t)$, or, suppressing the t to simplify the formulas, $\mathbf{r}' = \|\mathbf{r}'\|\mathbf{T}$. Using this, we find

$$\mathbf{r}' \times \mathbf{r}'' = (\|\mathbf{r}'\|\mathbf{T}) \times \left(\frac{d}{dt}(\|\mathbf{r}'\|\mathbf{T}) \right)$$

By the product rule.

$$= (\|\mathbf{r}'\|\mathbf{T}) \times \left(\frac{d}{dt}(\|\mathbf{r}'\|)\mathbf{T} + \|\mathbf{r}'\|\frac{d\mathbf{T}}{dt} \right)$$

$$= (\|\mathbf{r}'\|\mathbf{T}) \times \left(\frac{d}{dt}(\|\mathbf{r}'\|)\mathbf{T} \right) + (\|\mathbf{r}'\|\mathbf{T}) \times \left(\|\mathbf{r}'\|\frac{d\mathbf{T}}{ds}\left(\frac{ds}{dt}\right) \right)$$

Note that $\frac{ds}{dt} = \|\mathbf{r}'\|$.

$$= \left(\|\mathbf{r}'\|\frac{d}{dt}(\|\mathbf{r}'\|) \right)(\mathbf{T} \times \mathbf{T}) + \|\mathbf{r}'\|^3\left(\mathbf{T} \times \frac{d\mathbf{T}}{ds} \right)$$

Note that $\mathbf{T} \times \mathbf{T} = 0$.

$$= 0 + \|\mathbf{r}'\|^3\left(\mathbf{T} \times \frac{d\mathbf{T}}{ds} \right).$$

We know that $\mathbf{T}$ and $\frac{d\mathbf{T}}{ds}$ are orthogonal, by Proposition 15.2.2; in particular the angle between thrm is $\theta = \frac{\pi}{2}$. We also know that $\kappa = \left\|\frac{d\mathbf{T}}{ds}\right\|$. Applying Lagrange's Identity (Theorem 14.2.11, p. 840), we find that

$$\left\|\mathbf{T} \times \frac{d\mathbf{T}}{ds}\right\| = \|\mathbf{T}\| \left\|\frac{d\mathbf{T}}{ds}\right\| \sin\theta = (1)(\kappa)(1) = \kappa.$$

Combined with the previous calculation, this shows that $\|\mathbf{r}' \times \mathbf{r}''\| = \|\mathbf{r}'\|^3 \|\mathbf{T} \times \frac{d\mathbf{T}}{ds}\| = \|\mathbf{r}'\|^3 \kappa$, so

$$\kappa = \frac{\|\mathbf{r}' \times \mathbf{r}''\|}{\|\mathbf{r}'\|^3},$$

as desired.

EXAMPLE 6   Consider the line in three-dimensional space given by the parametric equations

$$x = u + at, \qquad y = v + bt, \qquad z = w + ct.$$

We think of it as a moving point by letting $\mathbf{r}(t) = (u + at, v + bt, w + ct)$. Then $\mathbf{r}'(t) = (a, b, c)$ is a constant vector (which points in the direction of the line), and the second derivative is $\mathbf{r}''(t) = (0, 0, 0)$.

Without doing any more work we see that the formula from Theorem 15.3.3 tells us that the curvature is $\kappa = 0$. It is comforting to find that a straight line has zero curvature.

> Theorem 15.3.3 has been stated for curves in three-dimensional space, but of course it can be used for curves in the plane by the usual trick of placing the xy-plane in $\mathbb{R}^3$.

EXAMPLE 7   Consider the curve $\mathbf{p}(t) = (R\cos t, R\sin t)$, which parametrizes a circle of radius R. To apply the theorem to find the curvature, we write it instead in the form $\mathbf{r}(t) = (R\cos t, R\sin t, 0)$. We find that $\mathbf{r}'(t) = (-R\sin t, R\cos t, 0)$ and $\mathbf{r}''(t) = (-R\cos t, -R\sin t, 0)$. Then

$$\mathbf{r}'(t) \times \mathbf{r}''(t) = (-R\sin t, R\cos t, 0) \times (-R\cos t, -R\sin t, 0)$$
$$= (0, 0, R^2 \sin^2 t + R^2 \cos^2 t) = (0, 0, R^2).$$

Theorem 15.3.3 tells us that the curvature is

$$\kappa = \frac{\|\mathbf{r}' \times \mathbf{r}''\|}{\|\mathbf{r}'\|^3} = \frac{R^2}{R^3} = \frac{1}{R}.$$

We have just found that the curvature of a circle of radius R is $\frac{1}{R}$. This makes sense when we consider a small circle that curves very sharply, that is, has a large curvature, and a large circle that curves more slowly, that is, has a small curvature. See Figure 15.3.7.

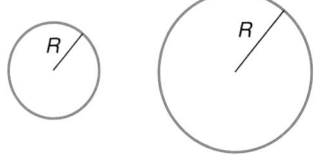

Figure 15.3.7

Radius of Curvature, Osculating Circle

When we began discussing curvature, we thought of the idea of considering the curvature at a point on a curve by finding a circle that would be tangent to the curve

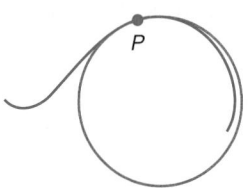

Figure 15.3.8

at that point and also have the same curvature. In light of Example 7 we now know how to do this. If we know that the curvature of the curve at a point P is κ, then we should consider a circle whose radius is $\frac{1}{\kappa}$; its curvature will also be κ. If a circle with this radius is tangent to the curve at P, it will have the same direction and curvature as the curve at P (see Figure 15.3.8).

To arrange this, we have to decide where to put the center of the circle. Clearly, it has to be on the appropriate side of the curve, that is, the side toward which the curve is turning. The way to say this is that the center of the circle is in the direction of the normal vector $\mathbf{N}$ from the point P.

These considerations suggest the following definition.

DEFINITION 15.3.4

> Suppose the curve $\mathbf{r}(t)$ has nonzero curvature κ at the point $P = \mathbf{r}(t)$. Then the **radius of curvature** of the curve at P is
>
> $$R = \frac{1}{\kappa}.$$
>
> The point C that is obtained by moving a distance $R = \frac{1}{\kappa}$ from P in the direction of the normal vector $\mathbf{N}$ is called the **center of curvature**. The circle whose center is C and whose radius is $R = \frac{1}{\kappa}$ is called the **osculating circle**. (The word "osculating" comes from the Latin word for "kiss.")

EXAMPLE 8 Find the osculating circle to the curve $\mathbf{r}(t) = (t^2, 2t - 1)$ at the point corresponding to $t = 1$.

Solution We find that $\mathbf{r}'(t) = (2t, 2)$ and $\mathbf{r}''(t) = (2, 0)$. Writing these vectors as vectors in $\mathbb{R}^3$ and using Theorem 15.3.3, we find that the curvature is

$$\kappa = \frac{\|\mathbf{r}'(t) \times \mathbf{r}''(t)\|}{\|\mathbf{r}'(t)\|^3} = \frac{\|(2t, 2, 0) \times (2, 0, 0)\|}{\|(2t, 2, 0)\|^3} = \frac{\|(0, 0, -4)\|}{(4t^2 + 4)^{3/2}}$$

$$= \frac{1}{2(t^2 + 1)^{3/2}}.$$

The unit tangent vector is $\mathbf{T}(t) = \left(\frac{t}{\sqrt{t^2+1}}, \frac{1}{\sqrt{t^2+1}}, 0 \right)$, so a normal vector is

$$\mathbf{n}(t) = \frac{d\mathbf{T}}{dt} = \left(\frac{\sqrt{t^2+1} - t^2(t^2+1)^{-1/2}}{t^2+1}, \frac{-t}{(t^2+1)^{3/2}}, 0 \right)$$

$$= \left(\frac{(t^2+1) - t^2}{(t^2+1)^{3/2}}, \frac{-t}{(t^2+1)^{3/2}}, 0 \right) = \left(\frac{1}{(t^2+1)^{3/2}}, \frac{-t}{(t^2+1)^{3/2}}, 0 \right).$$

From this it is easy to find that the unit normal vector is

$$\mathbf{N}(t) = \left(\frac{1}{\sqrt{t^2+1}}, \frac{-t}{\sqrt{t^2+1}}, 0 \right).$$

In particular, for $t = 1$ we see that the point is $\mathbf{r}(1) = (1, 1)$, the curvature is $\kappa = \frac{1}{4\sqrt{2}}$, and the unit normal vector is $\mathbf{N}(1) = \left(\frac{1}{\sqrt{2}}, -\frac{1}{\sqrt{2}} \right)$. The center of the osculating circle is found by starting at $(1, 1)$ and going a distance $\frac{1}{\kappa} = 4\sqrt{2}$ in the

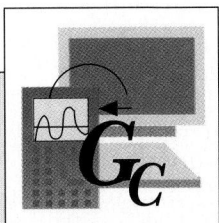

Figure 15.3.9

direction of **N**, so it is

$$C = (1,1) + 4\sqrt{2}\left(\frac{1}{\sqrt{2}}, -\frac{1}{\sqrt{2}}\right) = (5, -3).$$

The osculating circle has radius $4\sqrt{2}$ and center $C = (5, -3)$. Its equation is $(x-5)^2 + (y+3)^2 = 32$, or $x^2 - 10x + y^2 + 6y + 2 = 0$. See Figure 15.3.9.   ◢

Toward the end of this example, notice how we wrote the normal vector $\mathbf{N}(1) = \left(\frac{1}{\sqrt{2}}, -\frac{1}{\sqrt{2}}\right)$ as a vector in the plane rather than in $\mathbb{R}^3$.

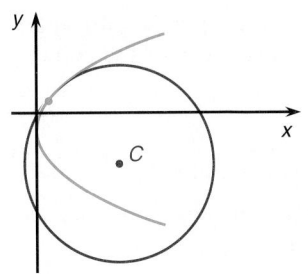

With a graphics calculator we can plot the parabola from Example 8. We can also plot the osculating circle. For instance, when $t = 0$, the curvature is $\kappa = \frac{1}{2}$, so the radius is 2. The normal vector is $(1, 0)$, so the center of the circle is $(2, -1)$.

TI-85	SHARP EL9300
MODE ▼ ▼ ▼ ▼ ► ► (Param) ENTER **EXIT** **GRAPH** F1($E(t) =$) F1(t) x^2 ▼ 2 F1(t) -1 ▼ $2 + 2$ COS ($\pi \times$ F1(t)) ▼ 2 SIN ($\pi \times$ F1(t)) -1 ▼ M2(**RANGE**) (−) 2 ▼ 2 ▼ .01 ▼ (−) 5 ▼ 5 ▼ ▼ (−) 5 ▼ 5 F5(**GRAPH**)	SET UP E 3 ENTER ⊬ x/θ/T x^2 ENTER 2 x/θ/T − 1 ENTER 2ndF ▼ 2 + 2 cos (π x/θ/T) ENTER 2 sin (π x/θ/T) − 1 ENTER RANGE − 5 ENTER 5 ENTER 1 ENTER −5 ENTER 3 ENTER 1 ENTER −2 ENTER 2 ENTER .01 ENTER ⊬
CASIO f_x-7700GB	HP 48SX
MODE MODE × (PARAM) Range −5 **EXE** 5 **EXE** 1 **EXE** −5 **EXE** 3 **EXE** 1 **EXE** −2 **EXE** 2 **EXE** .01 **EXE** Graph **T** x^2 , 2 **T** − 1 **EXE** Graph 2 + 2 cos (π **T**), 2 sin (π **T**) − 1 **EXE**	↰ **RAD** ' T y^x 2 + α ↰ **CST** × **()** 2 × T − 1 ↰ **PLOT** NEW P A R ENTER PTYPE PARA PLOTR ↰ **{}** T SPC 2 +/− SPC 2 ENTER **INDEP AUTO** ATTN ' 2 + 2 × COS π × T ► + α ↰ **CST** × **()** 2 × SIN π × T ► − 1 ↱ **PLOT** ↰ **DRAW** DRAW

Now plot the osculating circles at some other points.

Exercises 15.3

Find the curvature of each of the following curves.

1. $\mathbf{r}(t) = (\cos 3t, \sin 3t)$ **2.** $\mathbf{r}(t) = (\sin 5t, -\cos 5t)$

3. $\mathbf{r}(t) = (3\cos t^2, 3\sin t^2)$

4. $\mathbf{r}(t) = (t, -\sqrt{4 - t^2})$, $t \in (-2, 2)$

5. $\mathbf{r}(t) = (2t, \sin t, \cos t)$

6. $\mathbf{r}(t) = (t, \cos t \sin t, \cos^2 t)$

7. $\mathbf{r}(t) = (3 - 2t, t^2 + t + 1, 4t + 2)$

8. $\mathbf{r}(t) = (t^3, 4t + 2, t^2 - t - 7)$

9. $\mathbf{r}(t) = (3\cos t, 2\cos t)$

10. $\mathbf{r}(t) = (\cos^2 t, \sin^2 t, 2)$

Find the osculating circle for each of the following curves at the specified point.

11. $\mathbf{r}(t) = (3\sin t, 2 - 3\cos t)$, $t = 0$

12. $\mathbf{r}(t) = (2t - 1, 4\cos t, 4\sin t)$, $t = \pi$

13. $\mathbf{r}(t) = (t, \sqrt{9 - t^2}, 3)$, $t = 0$

14. $\mathbf{r}(t) = (5 + \cos t, \sqrt{2}\sin t, 5 - \cos t)$, $t = 0$

15. $\mathbf{r}(t) = (t, t^2)$, $t = 0$

16. $\mathbf{r}(t) = (t^2, t)$, $t = 1$

17. $\mathbf{r}(t) = (t + \cos t, \sin t)$, $t = 0$

18. $\mathbf{r}(t) = (t + \sin t, t + \cos t)$, $t = -\frac{\pi}{2}$

19. $\mathbf{r}(t) = (\sin t + \cos t, \sin t - \cos t)$, $t = \frac{\pi}{8}$

20. $\mathbf{r}(t) = (2t - \cos t, 2t + \cos t, 4t)$, $t = 0$

II

21. Suppose $a \neq 0$ and consider the parabola $\mathbf{r}(t) = (t, at^2)$. What is its curvature at the vertex $(0, 0)$?

22. Find the curvature of the helix $\mathbf{r}(t) = (a\sin t, bt, a\cos t)$.

*23. Show that the normal vector $\mathbf{N}(t)$ can be expressed as $\mathbf{N}(t) = \frac{1}{\kappa}\frac{d\mathbf{T}}{ds}$.

*24. (i) Show that for any curve $\mathbf{r}(t)$ with twice-differentiable components the vector $\mathbf{T}'(t)$ is orthogonal to the velocity vector $\mathbf{r}'(t)$ at every point where $\mathbf{r}'(t)$ is nonzero.

(ii) Find an example of a curve for which $\mathbf{r}''(t)$ is not orthogonal to $\mathbf{r}'(t)$.

*25. Consider the graph of $y = f(x)$ as a curve, and find a formula for its curvature at the point $(x, f(x))$. (You should be able to express κ in terms of $f'(x)$ and $f''(x)$.)

*26. Consider the graphs of two functions $y = f(x)$ and $y = g(x)$. Suppose that at some point $t = a$ the two functions satisfy $f(a) = g(a)$, $f'(a) = g'(a)$, $f''(a) = g''(a)$. Show that the two graphs have the same curvature at the corresponding point $(a, f(a))$.

*27. Consider two curves $\mathbf{r}(t) = (f(t), g(t), h(t))$ and $\mathbf{p}(t) = (F(t), G(t), H(t))$. Suppose the corresponding components of the two curves have the same values and the same first and second derivatives at $t = a$. Is it true that they must have the same curvature at the point $\mathbf{r}(a)$?

*28. Consider the parabola $\mathbf{r}(t) = (t, t^2)$. Find a formula for the center of the osculating circle for the point $\mathbf{r}(t)$ on the parabola.

*29. Consider the curve $\mathbf{r}(t) = \left(\cos(f(t)), \sin(f(t))\right)$. Find the curvature at any point at which $f'(t) \neq 0$.

*30. If $\mathbf{p}(t)$ is a curve and $f(t)$ is a function whose derivative is never zero, let $\mathbf{r}(t) = \mathbf{p}(f(t))$. Show that the curvature of $\mathbf{r}$ at t equals the curvature of $\mathbf{p}$ at $f(t)$.

*31. Consider the parabola $\mathbf{r}(t) = (t, at^2 + bt + c)$, where $a \neq 0$. Find where its curvature is greatest.

*32. (i) Let $\mathbf{r}(t) = (f(t), g(t))$ and let $\mathbf{p}(t) = (g(t), f(t))$. Show that the curvatures of $\mathbf{r}(t)$ and $\mathbf{p}(t)$ are the same whenever $\mathbf{r}'(t)$ is not the zero vector.

(ii) Explain in geometric terms why these two curves should have the same curvature.

33. Find the curvature of the parabola $y = x^2$ at the point (a, a^2).

34. Find the curvature of the curve $y = x^3$ at the point (a, a^3).

35. Suppose the components of $\mathbf{r}(t)$ are all twice-differentiable functions of t. Show that whenever $\mathbf{r}'(t) \neq 0$, the acceleration can be written as

$$\mathbf{r}''(t) = \|\mathbf{r}(t)\|'\mathbf{T}(t) + \kappa\|\mathbf{r}'(t)\|^2\mathbf{N}(t).$$

This expresses $\mathbf{r}''(t)$ in terms of its tangential and normal components, which is sometimes very useful.

36. Apply the result of the previous exercise to the parabola $y = x^2$. For a suitable parametrization $\mathbf{r}(t)$, calculate $\mathbf{r}'(t)$ and $\mathbf{r}''(t)$ and then read off the curvature of the curve at the vertex $(x, y) = (0, 0)$ (where it is easy to guess the normal $\mathbf{N}(t)$).

III

Use a graphics calculator or computer to sketch the following curves and the osculating circle at the specified point.

37. $\mathbf{r}(t) = (t^2, 1 - t)$, $t = 1$

38. $\mathbf{r}(t) = (t^3 - 3t, t^2)$, $t = -1$

39. $y = x^3$, $(2, 8)$

40. $y = x^3 - 3x$, $(1, -2)$

41. $4x^2 + 9y^2 = 36$, $(3, 0)$

42. $4x^2 + 9y^2 = 36$, $(0, 2)$

43. $x^2 - y^2 = 1$, $(-1, 0)$

44. $xy = 1$, $(1, 1)$

15.4 Quadric Surfaces

SPHERE
ELLIPSOID
PARABOLOID
CONE
HYPERBOLOID
SADDLE SURFACE

In Chapter 11 we studied conics, which are the graphs of quadratic polynomials in two variables. In this section we study the graphs of quadratic polynomials in three variables, which are called **quadric surfaces**.

EXAMPLE 1   **Sphere**   In Example 14.1.4 we observed that the set of all points that satisfy $x^2 + y^2 + z^2 = 4$ is the sphere of radius 2 around the origin (see Figure 15.4.1).

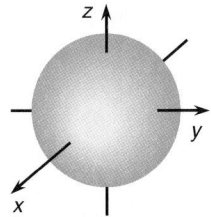

Figure 15.4.1

Let us make it clear once and for all that the word **"sphere"** refers to the surface consisting of all points at a given radius from some center point, and the word **"ball"** refers to the solid region consisting of all these points together with the points *inside*. For instance, the ball of radius 2 centered at the origin is the set of all points satisfying $x^2 + y^2 + z^2 \le 4$, that is, at a distance of 2 or less from the origin.

DesignCAD 3D *software was used to produce this image of cross-sections of an F-18. The computer can use the cross-sections to fill in the shape of the whole aircraft.*

EXAMPLE 2   Consider the equation $36x^2 + 4y^2 + 9z^2 = 36$. It is easy to find its intercepts on the three axes: On the x-axis, that is, for points of the form $(x, 0, 0)$, the only possibilities are $(1, 0, 0)$ and $(-1, 0, 0)$. For points on the y-axis we find $(0, \pm 3, 0)$, and on the z-axis we find $(0, 0, \pm 2)$.

Before we guess what the surface looks like, let us consider the part of it that lies in the yz-plane, that is, the plane $x = 0$. This means that we are looking at points $(0, y, z)$ that satisfy $36x^2 + 4y^2 + 9z^2 = 36$, which means $4y^2 + 9z^2 = 36$. This we recognize as the equation of an ellipse in the yz-plane, and in Figure 15.4.2(i) we sketch it.

For that matter the same sort of calculation shows that the surface goes through the ellipse $36x^2 + 4y^2 = 36$ in the horizontal xy-plane and the ellipse $36x^2 + 9z^2 = 36$ in the vertical xz-plane. These ellipses are sketched in Figure 15.4.2(ii).

Finally, we have to guess the shape of the whole surface $36x^2 + 4y^2 + 9z^2 = 36$, which must go through all these ellipses. It is not surprising that it is the shape illustrated in Figure 15.4.2(iii). This elliptical shape in $\mathbb{R}^3$ is called an *ellipsoid*.

Figure 15.4.2(i)

Figure 15.4.2(ii)

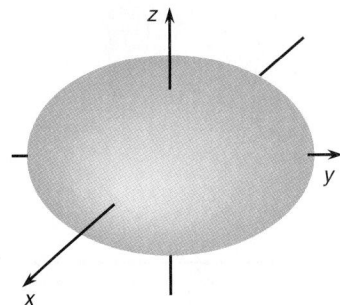

Figure 15.4.2(iii)

DEFINITION 15.4.1

An equation of the form

$$\frac{x^2}{a^2} + \frac{y^2}{b^2} + \frac{z^2}{c^2} = 1,$$

with $a, b, c > 0$, has as its graph a surface called an **ellipsoid** with center at the origin. (If $a = b = c$, it is a sphere of radius a.)

Its intercepts on the coordinate axes are

$$(\pm a, 0, 0), \qquad (0, \pm b, 0), \qquad (0, 0, \pm c).$$

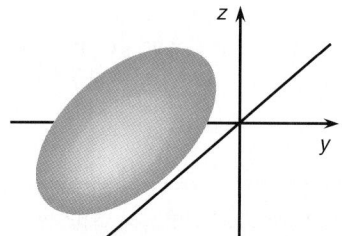

Figure 15.4.3

As with conics, it is possible to translate coordinates.

EXAMPLE 3   Identify the quadric $x^2 - 4x + 4y^2 + 8y + 4z^2 - 8z + 8 = 0$.

Solution   Completing the squares, we find

$$\begin{aligned}
0 &= x^2 - 4x + 4y^2 + 8y + 4z^2 - 8z + 8 \\
&= x^2 - 4x + 4 - 4 + 4y^2 + 8y + 4 - 4 + 4z^2 - 8z + 4 - 4 + 8 \\
&= (x - 2)^2 + 4(y + 1)^2 + 4(z - 1)^2 - 4.
\end{aligned}$$

Letting $x' = x - 2$, $y' = y + 1$ and $z' = z - 1$, we can write the equation in the simpler form

$$x'^2 + 4y'^2 + 4z'^2 = 4.$$

This is an ellipsoid with center at the point $(x, y, z) = (2, -1, 1)$. See Figure 15.4.3.

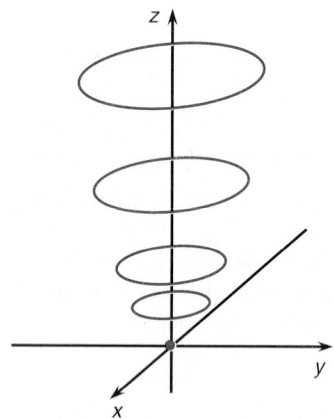

Figure 15.4.4(i)

Notice that the coefficients of y'^2 and z'^2 are equal. This has the effect that the intersection of the ellipsoid with the $y'z'$-plane is a *circle*.

EXAMPLE 4   Consider the equation $z = x^2 + y^2$. Since $x^2 + y^2 \geq 0$, there can be no points satisfying the equation with $z < 0$. This means that the graph lies on or above the horizontal xy-plane.

In fact, its intersection with the xy-plane, that is, with $z = 0$, is the origin $(0, 0, 0)$. If we look at its intersection with the horizontal plane $z = 1$, we must consider points $(x, y, 1)$ satisfying $1 = x^2 + y^2$, which is a circle of radius 1.

Similarly, the intersection of the surface with $z = 2$ is the circle $x^2 + y^2 = 2$ of radius $\sqrt{2}$, and the intersection with the plane $z = A$ is the circle of radius $\sqrt{A}$. Several such circles are shown in Figure 15.4.4(i).

To consider how these circles can fit together into a surface, observe that in the yz-plane, that is, when $x = 0$, we have the parabola $z = y^2$. We are quickly lead to visualize the surface shown in Figure 15.4.4(ii), which is called a **paraboloid**.

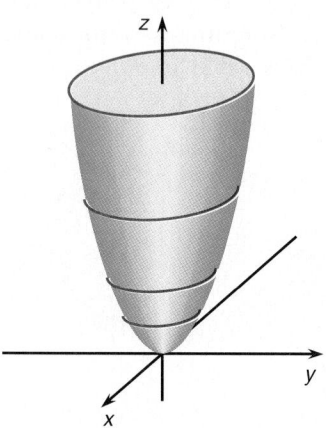

Figure 15.4.4(ii)

One way to understand the shape of the paraboloid is to recognize that $x^2 + y^2$ is the square of the horizontal distance from the point (x, y, z) to the vertical z-axis. The equation $z = x^2 + y^2$ says that the height of the surface above the horizontal plane equals the square of the distance from the z-axis. This is very similar to the familiar parabola $y = x^2$ in the plane, for which the height y equals the square of the horizontal distance from the origin.

Figure 15.4.5

Figure 15.4.6

Figure 15.4.7(i)

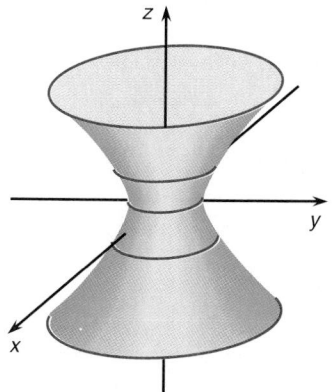

Figure 15.4.7(ii)

EXAMPLE 5   The equation $x^2 - 2y + 4z^2 + 2 = 0$ becomes easier to understand if we write it in the form

$$y = \frac{x^2}{2} + 2z^2 + 1.$$

Since $\frac{x^2}{2} + 2z^2 \geq 0$, we see that y must always be greater than or equal to 1. The intersection with the plane $y = A$ is the curve $\frac{x^2}{2} + 2z^2 = A - 1$. This is an ellipse if $A > 1$, a single point if $A = 1$, and the empty set if $A < 1$. The intersection with the plane $x = A$ is the parabola $y = 2z^2 + 1 + \frac{A^2}{2}$.

The surface is an *elliptical paraboloid*, as shown in Figure 15.4.5. It is like a paraboloid except that its cross-sections are ellipses rather than circles.   ◢

EXAMPLE 6   Consider the quadric $z^2 = x^2 + y^2$. It is tempting to write $z = \sqrt{x^2 + y^2}$. This suggests that the height above the horizontal xy-plane equals the horizontal distance from the z-axis. Strictly speaking, however, it is also possible to have negative values of z, so we should really write $|z| = \sqrt{x^2 + y^2}$.

This makes it fairly clear that the surface is the double-ended **cone** shown in Figure 15.4.6. Note that when $z = 0$, the only possibility is $x = y = 0$; the cone intersects the xy-plane in the single point $(0, 0, 0)$.   ◢

It is worthwhile to think for a minute about the distinction between the cone $z^2 = x^2 + y^2$ and the paraboloid $z = x^2 + y^2$. The difference comes from the power of z. For the paraboloid the equation $z = x^2 + y^2$ says that as a point moves away from the z-axis, its height goes up like the *square* of the horizontal distance from the z-axis, and we expect the familiar parabolic shape. For the cone $z^2 = x^2 + y^2$, writing it as $|z| = \sqrt{x^2 + y^2}$ tells us that the height goes up (or down) like the *first power* of the distance from the z-axis. We expect a surface whose sides have a straight slope.

EXAMPLE 7   Consider the equation $x^2 + y^2 - z^2 = 1$.

Let us resort to the increasingly familiar trick of looking first at the intersection of the surface with the planes $z = A$, for various values of A. For instance, with $A = 0$ the intersection of the quadric with the xy-plane $z = 0$ is the circle $x^2 + y^2 = 1$ of radius 1.

The intersection of the surface with the plane $z = 1$ is the curve $x^2 + y^2 - 1 = 1$, which is the circle $x^2 + y^2 = 2$ of radius $\sqrt{2}$. The same curve is obtained with $z = -1$.

Similarly, the intersection of the surface with the plane $z = A$ is the circle of radius $\sqrt{A^2 + 1}$. This grows larger as $|A|$ increases; several such intersections are shown in Figure 15.4.7(i).

The whole surface is shown in Figure 15.4.7(ii). It is called a **hyperboloid of one sheet**. Notice that its intersection with the yz-plane is the hyperbola $y^2 - z^2 = 1$ and that its intersection with the xz-plane is the hyperbola $x^2 - z^2 = 1$.   ◢

EXAMPLE 8   The equation $z^2 - x^2 - y^2 = 1$ is slightly different. Since $z^2 = 1 + x^2 + y^2$ can never be less than 1, we see that the surface must consist of points lying on or above the plane $z = 1$ or on or below the plane $z = -1$, but not in between.

Figure 15.4.8(i)

Figure 15.4.8(ii)

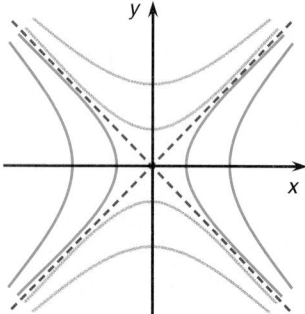

Figure 15.4.9(i)

For $z = \pm 1$ we obtain the two points $(0, 0, \pm 1)$ satisfying the equation. For $z = A$ with $|A| > 1$ we obtain circles $x^2 + y^2 = A^2 - 1$ of radius $\sqrt{A^2 - 1}$. Some of these circles are illustrated in Figure 15.4.8(i).

The whole quadric surface is shown in Figure 15.4.8(ii). It is called a **hyperboloid of two sheets**, a reference to the two parts of the surface.

Notice that the intersection of this quadric surface with the yz-plane is the hyperbola $z^2 - y^2 = 1$ and the intersection of the surface with the xz-plane is the hyperbola $z^2 - x^2 = 1$.

EXAMPLE 9 Consider the equation $z = x^2 - y^2$. If we consider its intersections with various horizontal planes $z = A$, we find that they are hyperbolas $x^2 - y^2 = A$. Depending on the sign of A, they open along the x-axis or the y-axis. In Figure 15.4.9(i) we sketch some hyperbolas for different values of A. The ones with blue lines have $A > 0$, and the ones with purple lines have $A < 0$.

Also in the horizontal plane $z = 0$ we have the equation $x^2 - y^2 = 0$, which factors as $(x + y)(x - y) = 0$. This means that its graph is two straight lines through the origin. They are shown as the dashed lines in Figure 15.4.9(i). As A increases from 0, these lines turn into hyperbolas opening along the x-axis, and if A decreases from 0, they turn into hyperbolas opening along the y-axis.

The overall effect is sketched in Figure 15.4.9(ii). This quadric surface is called a **saddle surface**, for fairly obvious reasons.

We have already observed that the intersection of this surface with the xy-plane is two straight lines. Its intersection with the yz-plane is the downward-opening dashed parabola shown in Figure 15.4.9(iii), and its intersection with the xz-plane is the upward-opening solid-line parabola.

It is interesting to notice that starting at the origin, a point that moves in either direction along the y-axis will go downward and a point that moves in either direction along the x-axis will go up.

The examples we have discussed include the most important kinds of quadric surfaces. There are others, and there are *degenerate quadrics* such as $x^2 + y^2 + z^2 = 0$, whose graph is the single point $(0, 0, 0)$. We will not try to be exhaustive but will summarize the types we have encountered.

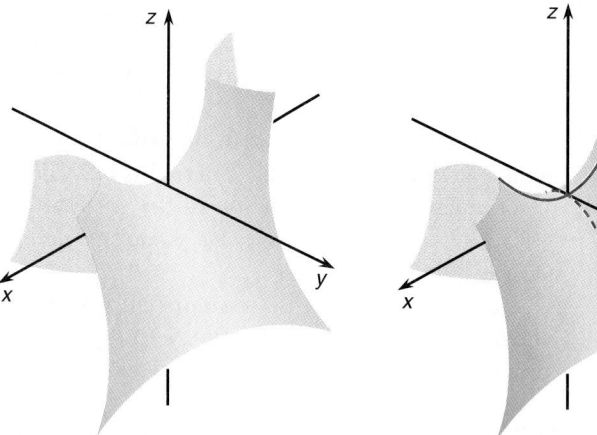

Figure 15.4.9(ii)   **Figure 15.4.9(iii)**

Summary 15.4.2

Quadric Surfaces in Standard Form

In each of the following, assume $a, b, c, R \neq 0$.

Equation	Quadric surface
$x^2 + y^2 + z^2 = R^2$	Sphere of radius R
$\frac{x^2}{a^2} + \frac{y^2}{b^2} + \frac{z^2}{c^2} = 1$	Ellipsoid, intercepts $(\pm a, 0, 0)$, $(0, \pm b, 0)$, $(0, 0, \pm c)$
$z = a(x^2 + y^2)$	Paraboloid
$z^2 = a^2(x^2 + y^2)$	Cone
$\frac{x^2}{a^2} + \frac{y^2}{b^2} - \frac{z^2}{c^2} = 1$	Hyperboloid of one sheet around the z-axis
$\frac{z^2}{a^2} - \frac{x^2}{b^2} - \frac{y^2}{c^2} = 1$	Hyperboloid of two sheets around the z-axis
$z = a^2 x^2 - b^2 y^2$	Saddle surface

A number of the possibilities are illustrated in Figure 15.4.10.

Given a quadratic polynomial in three variables without any cross-terms, we can try to complete squares and see whether it can be made to look like any of these special types of quadrics. If it can, then it should be possible to sketch the graph.

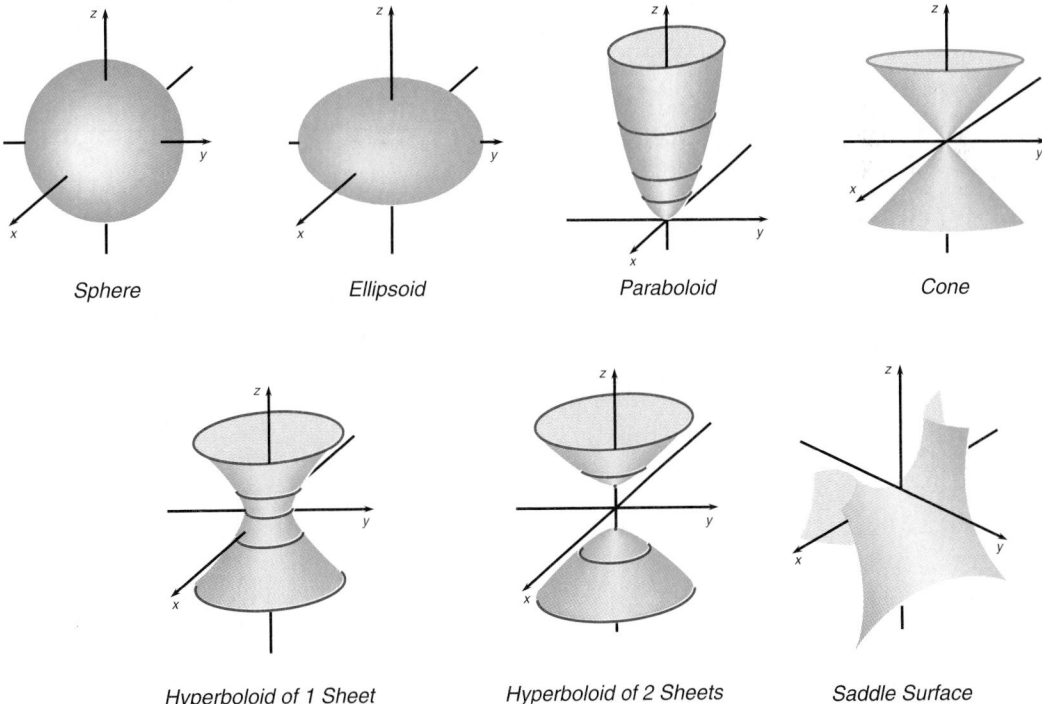

Sphere   Ellipsoid   Paraboloid   Cone

Hyperboloid of 1 Sheet   Hyperboloid of 2 Sheets   Saddle Surface

Figure 15.4.10

If there are cross-terms, it may be more difficult to identify the graph. However, with a computer we can still plot it. For instance, it is not apparent what the graph of $z = x^2 + 4xy + y^2$ will look like, but we can sketch the picture.

Mathematica	Maple
Plot3D[$x\hat{\ }2 + 4\,x * y + y\hat{\ }2$, $\{x, -2, 2\}, \{y, -2, 2\}$] ENTER	plot3d($x\hat{\ }2 + 4 * x * y + y\hat{\ }2$, $x = -2..2$, $y = -2..2$); return

DERIVE	MacFUNCTION
O D G ENTER A $x\hat{\ }2 + 4\,x * y +$ $y\hat{\ }2$ ENTER P O L -2 Tab 2.0 ENTER G 15 Tab 15 ENTER P	$x\hat{\ }2 + 4 * x * y + y\hat{\ }2$ OK Function Change Domain (*drag to* 2 × 2) OK Options Remove Hidden Surfaces

The surface is a saddle surface, but it is rotated relative to the standard position.

Exercises 15.4

I

Identify and sketch each of the following quadric surfaces.

1. $x^2 + y^2 + z^2 = 9$
2. $4x^2 + 9y^2 + z^2 = 36$
3. $\frac{z}{2} = x^2 + y^2$
4. $x^2 + z^2 = -y$
5. $x^2 = y^2 + z^2$
6. $y^2 = 2x^2 + z^2$
7. $x^2 - y^2 + z^2 = 1$
8. $x^2 + y^2 = 1 + z^2$
9. $x^2 - 4y^2 - 4z^2 = 4$
10. $x^2 + y^2 + z^2 + 1 = 0$

Identify and sketch each of the following surfaces.

11. $x^2 + 2x + y^2 + z^2 - 4z = 4$
12. $9x^2 + 4y^2 = 36$

13. $x^2 - y^2 + 2y + z^2 - 2z = 1$
14. $x^2 + 2x + y^2 - 4y - z^2 + 5 = 0$
15. $4x^2 + 8x + 4y^2 - 16y + z^2 - 2z + 17 = 0$
16. $z = x^2 - 2x - y^2 + 4y - 4$
17. $x^2 - 4x + y - z^2 - 4z = 0$
18. $x = 4y^2 + z^2 - 4z$
19. $9x^2 - 18x - 4y^2 + 8y + z^2 = 0$
20. $xy + z^2 = 1$

II

Sketch the following surfaces in $\mathbb{R}^3$:

*21. $x^2 - y^2 = 1$
*22. $y^2 - z^2 = 4$
*23. $x - y = z^2$
*24. $z = (x - y)^3$

*25. Describe the intersection of the saddle surface $z = x^2 - y^2$ with each of the following planes: (i) $y = 0$, (ii) $y = 2$, (iii) $x = -3$, (iv) $z = 1$, (v) $z = -1$.

*26. Find the equation of the ellipsoid whose center is the point $(0, 2, -1)$, which has a semi-axis 4 units long parallel to the x-axis, a semi-axis 2 units long parallel to the y-axis, and a semi-axis 1 unit long parallel to the z-axis.

*27. Find the equation of the saddle surface obtained by taking the saddle surface $x + y^2 - z^2 = 0$ and rotating it around the x-axis through an angle of $\frac{\pi}{4}$ in the counterclockwise direction from the point of view of somebody looking from the positive x-axis toward the origin.

If you have a computer that can sketch three-dimensional graphs, try it out on the following functions.

28. $z = f(x, y) = x^2 + y^2, |x|, |y| \le 1$

29. $z = f(x, y) = 2x^2 + y^2, |x|, |y| \le 2$

30. $z = f(x, y) = \sqrt{x^2 + y^2}, |x|, |y| \le 1$

31. $z = f(x, y) = x^2 - y^2, |x|, |y| \le 1$

POINT TO PONDER

Many of the quadric surfaces we have discussed have a round shape, which suggests they may be surfaces of revolution. For instance, if we take the parabola $z = y^2$ in the yz-plane and revolve it about the z-axis, we expect to get a paraboloid (see Figure 15.4.11(i)).

When we revolve it, let us consider what happens in the horizontal plane $z = A$. The parabola $z = y^2$ contains the points where $y = \pm\sqrt{A}$, $z = A$, each of which is at a horizontal distance of $\sqrt{A}$ from the z-axis. As these points revolve, they stay at a distance of $\sqrt{A}$ from the z-axis and sweep out the circle of points whose horizontal distance from the z-axis is $\sqrt{A}$, that is,

$x^2 + y^2 = A$ (see Figure 15.4.11(ii)). Since this is what happens when $z = A$, we can write $x^2 + y^2 = z$, and this is the equation of a paraboloid.

For the curve $z = y^2$ we think of $|y|$ as the horizontal distance from the z-axis. When we rotate about the z-axis, we should begin to think of horizontal distance in $\mathbb{R}^3$, which is $\sqrt{x^2 + y^2}$. We replace y^2 by $x^2 + y^2$, so rotating the parabola $z = y^2$ gives the paraboloid $z = x^2 + y^2$.

Similarly, rotating the lines $z = \pm y$, that is, $z^2 = y^2$, about the z-axis gives the cone $z^2 = x^2 + y^2$ (see Figure 15.4.11(iii)), and rotating the circle $y^2 + z^2 = 1$ gives the sphere $x^2 + y^2 + z^2 = 1$ (see Figure 15.4.11(iv)).

Figure 15.4.11(i)   **Figure 15.4.11(ii)**

     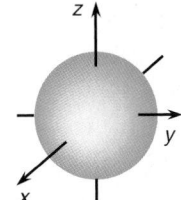

Figure 15.4.11(iii)   **Figure 15.4.11(iv)**

15.5 Cylindrical and Spherical Coordinates

CYLINDRICAL COORDINATES
SPHERICAL COORDINATES
LATITUDE
LONGITUDE

Most often we use rectangular coordinates in the plane, but we have found that there are some purposes for which polar coordinates are better suited. Usually, they are questions that involve rotations or "round" objects.

It is natural to ask what would be comparable for questions in three-dimensional space $\mathbb{R}^3$. In fact there are two answers to this question.

Figure 15.5.1

Figure 15.5.2

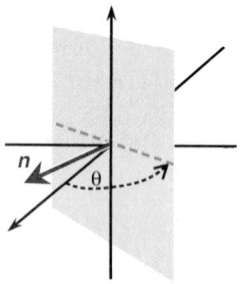

Figure 15.5.3

The first is to place the xy-plane horizontally in space and use polar coordinates in it. The coordinates of a point P in space are (r, θ, z), where as usual z represents the height of the point relative to the horizontal plane and (r, θ) are the polar coordinates of the point Q lying in the plane directly below or above P. See Figure 15.5.1.

Coordinates (r, θ, z) of this type are called **cylindrical coordinates** because they make the equation of a cylinder particularly simple. Specifically, the equation $r = 2$ is satisfied by all points that are 2 units away from the z-axis, which is the cylinder of radius 2 illustrated in Figure 15.5.2. The cylinder of radius a around the z-axis is given by the equation $r = a$. The coordinate r can be interpreted as the distance from the point P to the z-axis.

The points satisfying $\theta = c$ are a vertical plane through the z-axis whose angle relative to the positive x-axis is c. Another way to say this is that a normal to the plane (in rectangular coordinates) is $\mathbf{n} = (\sin c, -\cos c, 0)$. See Figure 15.5.3.

The rectangular coordinates (x, y, z) of P are related to the cylindrical coordinates as follows:

$$x = r \cos \theta, \qquad y = r \sin \theta, \qquad z = z.$$

EXAMPLE 1 Express the helix $\mathbf{r}(t) = (\cos t, \sin t, t)$ shown in Figure 15.5.4 in terms of cylindrical coordinates.

Solution Since the z-coordinate in cylindrical coordinates is the same as the z-coordinate in rectangular coordinates, we know that we can let $z = t$.

The "horizontal" motion of the point is given by $(\cos t, \sin t)$ in the plane, which we know traces out the circle of radius 1 around the origin. What we have to do is express this in terms of polar coordinates.

Fortunately, nothing could be easier. Since the point moves around the circle of radius 1, we know that the first coordinate is $r = 1$. The angle θ is just the argument of the point $(\cos t, \sin t)$ shown in the xy-plane in Figure 15.5.5; practically by definition we see that $\theta = t$.

The curve can be parametrized in cylindrical coordinates by

$$(r, \theta, z) = (1, t, t),$$

which can also be written as

$$r = 1, \qquad \theta = t, \qquad z = t.$$

Figure 15.5.4

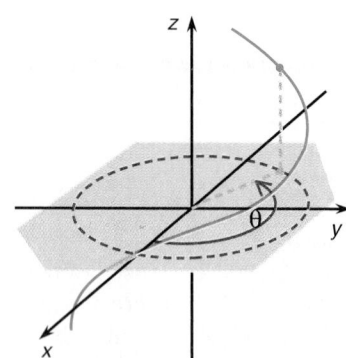

Figure 15.5.5

In words we can read these equations by saying that the radius r is constant and the angle θ and the height z are both equal to t, so they increase with time. We visualize the point winding around the z-axis and rising as it does so.

EXAMPLE 2

(i) Consider the equation $r^2 = z^2$.

Since r is the "radius" for the ordinary polar coordinates in the plane, we know that $r^2 = x^2 + y^2$. So this equation can be written in rectangular coordinates as $x^2 + y^2 = z^2$, which we recognize as the equation of the cone illustrated in Figure 15.5.6.

(ii) The equation $r^2 = z^2 - 1$ can similarly be written in rectangular coordinates as $x^2 + y^2 - z^2 + 1 = 0$, which is the equation of the hyperboloid of two sheets shown in Figure 15.5.7.

Neither of the equations studied in Example 2 involves the variable θ. In terms of the picture this means that they are "the same at every angle." Another useful way to put it is that they are unchanged by any rotation around the z-axis; after all, such a rotation has the effect of changing the variable θ and leaving r and z alone, and this will not change an equation that does not involve θ. Such a surface is a surface of revolution. (See the "Point to Ponder" at the end of Section 15.4.)

Spherical Coordinates

The second way to generalize polar coordinates to points in three-dimensional space is to use what are called **spherical coordinates**. In Figure 15.5.8 we see a point P. We let ρ be the distance from P to the origin O. We let θ be the usual polar angle in the horizontal plane, that is, the angle between the positive x-axis and the segment OQ, where Q lies directly above or below P in the horizontal plane. Finally, we let ϕ be the angle between the positive z-axis and the segment OP. We choose ϕ so that $0 \le \phi \le \pi$; $\phi = 0$ for points on the positive z-axis, and $\phi = \pi$ for points on the negative z-axis.

We want to express the rectangular coordinates (x, y, z) of P in terms of the spherical coordinates (ρ, θ, ϕ). To do this, we begin by finding the coordinates of the point Q lying directly above or below P in the horizontal plane. To find the polar coordinates of Q, we observe from Figure 15.5.9 that the radius is $r = \rho \cos\left(\frac{\pi}{2} - \right.$

Figure 15.5.6

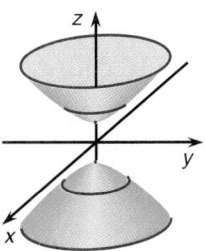

Figure 15.5.7

In spherical coordinates we measure the distance ρ from O to P, while in cylindrical coordinates we measure the horizontal distance r.

Figure 15.5.8

Figure 15.5.9

Figure 15.5.10(i)

Figure 15.5.10(ii)

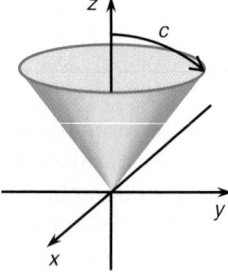

Figure 15.5.10(iii)

$\phi) = \rho \sin \phi$. Since the polar angle is θ, we find that $x = r \cos \theta = \rho \cos \theta \sin \phi$ and $y = r \sin \theta = \rho \sin \theta \sin \phi$.

Using the picture again, we see that $z = \rho \cos \phi$. We have the following:

$$x = \rho \cos \theta \sin \phi, \qquad y = \rho \sin \theta \sin \phi, \qquad z = \rho \cos \phi.$$

Also

$$\rho = \sqrt{x^2 + y^2 + z^2},$$
$$\theta = \arctan \tfrac{y}{x},$$
$$\phi = \arccos \tfrac{z}{\rho} = \arccos \frac{z}{\sqrt{x^2+y^2+z^2}}.$$

(As usual with polar coordinates, the equation $\theta = \arctan \tfrac{y}{x}$ has to be interpreted properly; cf. Theorem 12.1.1 (p. 757). With spherical coordinates it is sometimes useful to allow negative values of ρ, but we will not emphasize this possibility.)

EXAMPLE 3

(i) The equation $\rho = 7$ is the sphere of radius 7, as shown in Figure 15.5.10(i). It is because of this that coordinates of this type are called spherical coordinates. The equation $\rho = c$ is the sphere of radius c around the origin.

(ii) The equation $\theta = c$ is the vertical plane through the z-axis making an angle c with respect to the positive x-axis. See Figure 15.5.10(ii).

(iii) The equation $\phi = c$ is the set of all points P for which OP makes an angle c with the positive z-axis. If $0 < c < \tfrac{\pi}{2}$, then this set is the upper half of a cone (see Figure 15.5.10(iii)); if $\tfrac{\pi}{2} < c < \pi$, then it is the lower half of the cone; the equation $\phi = \tfrac{\pi}{2}$ represents the horizontal plane through the origin (see Figure 15.5.10(iv)).

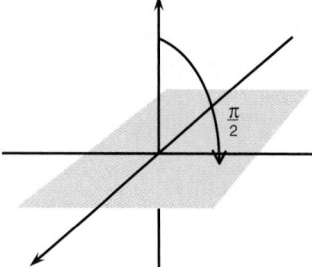

Figure 15.5.10(iv)

EXAMPLE 4   Consider the equation $\rho = -2 \cos \phi$. To try to identify its graph, we write it in rectangular coordinates:

$$\sqrt{x^2 + y^2 + z^2} = -2 \frac{z}{\sqrt{x^2 + y^2 + z^2}}$$
$$x^2 + y^2 + z^2 = -2z$$
$$x^2 + y^2 + z^2 + 2z = 0$$
$$x^2 + y^2 + (z + 1)^2 = 1.$$

Figure 15.5.11

Figure 15.5.12

Figure 15.5.13

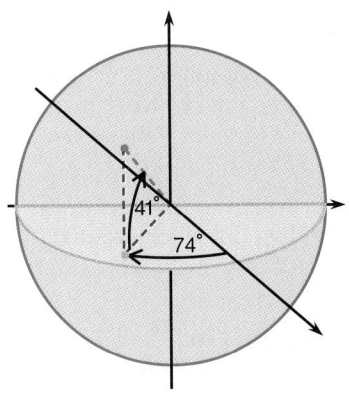

Figure 15.5.14

This is the equation of the sphere of radius 1 whose center is the point with rectangular coordinates $(0, 0, -1)$. See Figure 15.5.11.

EXAMPLE 5   Consider the curve $(\rho, \theta, \phi) = (1, 16t, t)$. The condition $\rho = 1$ says that the moving point is always on the sphere of radius 1 around the origin. The equation $\phi = t$ really makes sense only when $0 \le t \le \pi$, so we consider $t \in [0, \pi]$. As t runs from 0 to π, ϕ also runs from 0 to π, which means that the point moves down from the top of the sphere (i.e., the "north pole") to the bottom (the "south pole").

As it does this, the horizontal polar angle $\theta = 16t$ increases from 0 to 16π. This means that the point winds around the sphere eight times. The curve is the spiral shown in Figure 15.5.12. Notice that the loops are equally spaced on the surface of the sphere.

One important use of spherical coordinates is for locating points on the surface of the earth. Let us assume that the earth is a sphere whose radius is 4000 miles (which is nearly correct). The **longitude** of a point on the surface is measured relative to the *prime meridian*, a north-south semicircle through Greenwich, near London, England. This angle is essentially the same as the polar angle θ; the only catch is to be careful about signs.

The **latitude** of a point is the angle that measures its *elevation* above or below the equator, the horizontal circle around the center of the earth. This angle is 0 at the equator, so it is easily related to the spherical coordinate ϕ. For points in the northern hemisphere the latitude is $90° - \phi$ (with ϕ measured in degrees); in the southern hemisphere the latitude is $\phi - 90°$. See Figure 15.5.13.

EXAMPLE 6   (i) Find the spherical coordinates of New York City, which is located at 41° north latitude and 74° west longitude.

Solution   With miles as the unit of distance we know that $\rho = 4000$. The longitude is 74°, and from Figure 15.5.14 we see that this angle is measured in the *negative* direction in the xy-plane. This means that $\theta = -74°$ or $\theta = 286°$.

Finally, the latitude is 41°, which means that the line from the center of the earth to New York makes an angle of 41° with respect to the horizontal plane. It must make an angle of $90 - 41 = 49°$ with respect to the vertical axis. This means that $\phi = 49°$.

We find that the spherical coordinates of New York City are $(\rho, \theta, \phi) = (4000, 286°, 49°)$. Converting the angles to radians, we find that $(\rho, \theta, \phi) = (4000, 4.99, 0.86)$.

(ii) The city of Sydney, Australia, is located at 151° east longitude and 34° south latitude. Find its spherical coordinates.

Solution   We find that $\theta = 151°$. Since the point is in the southern hemisphere, the angle ϕ must satisfy $90° < \phi < 180°$. In fact, $\phi = 90 + 34 = 124°$. The spherical coordinates are $(\rho, \theta, \phi) = (4000, 151°, 124°) = (4000, 2.64, 2.16)$.

(iii) Describe the Arctic Circle in spherical coordinates.

Solution   The Arctic Circle is the curve that circles the earth at a latitude of $66\frac{1}{2}°$. For a point on the circle we have that $\phi = 90 - 66\frac{1}{2} = 23\frac{1}{2}°$. Also, since it is on the earth's surface, $\rho = 4000$. The longitude can be anything. So the Arctic Circle is the set of all points whose spherical coordinates satisfy $\rho = 4000$, $\phi = 23\frac{1}{2}°$. It can be parametrized as follows: $(\rho, \theta, \phi) = \left(4000, t, 23\frac{1}{2}°\right)$.

Exercises 15.5

I

Find the rectangular coordinates of the points whose cylindrical coordinates (r, θ, z) or spherical coordinates (ρ, θ, ϕ) are given as follows.

1. $(r, \theta, z) = (3, \pi, 2)$
2. $(r, \theta, z) = \left(5, -\frac{\pi}{2}, -3\right)$
3. $(r, \theta, z) = (0, -3\pi, 7)$
4. $(r, \theta, z) = \left(-1, \frac{\pi}{4}, 1\right)$
5. $(\rho, \theta, \phi) = \left(6, -\pi, \frac{\pi}{2}\right)$
6. $(\rho, \theta, \phi) = \left(\frac{1}{2}, \frac{3\pi}{2}, \frac{\pi}{4}\right)$
7. $(\rho, \theta, \phi) = \left(1, \frac{\pi}{2}, \frac{\pi}{2}\right)$
8. $(\rho, \theta, \phi) = \left(2, \frac{\pi}{3}, \frac{5\pi}{6}\right)$
9. $(\rho, \theta, \phi) = \left(4, \frac{5\pi}{7}, \pi\right)$
10. $(\rho, \theta, \phi) = \left(0, \frac{\pi}{5}, \frac{4\pi}{9}\right)$

Perform the following conversions. Express

11. $(x, y, z) = (3, -3, 4)$ in cylindrical coordinates.
12. $(x, y, z) = (-2, 2, 2\sqrt{2})$ in spherical coordinates.
13. $(x, y, z) = (0, 0, -8)$ in spherical coordinates.
14. $(x, y, z) = (0, 4, -3)$ in spherical coordinates.
15. $(r, \theta, z) = \left(3, \frac{\pi}{4}, 4\right)$ in spherical coordinates.
16. $(\rho, \theta, \phi) = \left(5, -\frac{\pi}{2}, \frac{\pi}{6}\right)$ in cylindrical coordinates.
17. $(\rho, \theta, \phi) = \left(9, \frac{3\pi}{8}, \pi\right)$ in cylindrical coordinates.
18. $(r, \theta, z) = \left(6, \frac{2\pi}{3}, 0\right)$ in spherical coordinates.
19. $(r, \theta, z) = (-4, 0, 0)$ in spherical coordinates.
20. $(\rho, \theta, \phi) = \left(5, \frac{\pi}{2}, \frac{\pi}{4}\right)$ in cylindrical coordinates.

Identify the following surfaces, each of which is given in cylindrical or spherical coordinates.

21. $r = 5$
22. $\rho = 3$
23. $\theta = \frac{\pi}{2}$ (cylindrical)
24. $\theta = \frac{\pi}{2}$ (spherical)
25. $\phi = \frac{3\pi}{4}$
26. $\phi = \frac{\pi}{2}$
27. $\rho^2 = \cos 2\phi$
28. $r = \cos \theta$
29. $\rho = 1 - \cos \phi$
30. $r = \cos 3\theta$

II

31. Find the spherical coordinates of Paris, which is at 49° north latitude and 2° east longitude.

32. Find the spherical coordinates of Rio de Janeiro, which is at 23° south latitude and 43° west longitude.

33. Find the spherical coordinates of London, which is at 52° north latitude and 0° longitude.

34. Describe in spherical coordinates the Tropic of Cancer, which is the line around the earth at $23\frac{1}{2}°$ north latitude.

35. Describe in spherical coordinates the Tropic of Capricorn, which is the line around the earth at $23\frac{1}{2}°$ south latitude.

36. Describe in spherical coordinates the Antarctic Circle.

*37. Suppose a surface is described in cylindrical coordinates by an equation that does not involve z, that is, by an equation of the form $F(r, \theta) = 0$. What can you say about the surface? Sketch the surface $r = 1 - \cos \theta$.

*38. Suppose a surface is described in cylindrical coordinates by an equation that does not involve r, that is, by an equation of the form $F(\theta, z) = 0$. What can you say about the surface?

*39. Suppose a surface is described in spherical coordinates by an equation that does not involve ρ, that is, by an equation of the form $F(\theta, \phi) = 0$. What can you say about the surface?

*40. Suppose a surface is described in spherical coordinates by an equation that does not involve θ, that is, by an equation of the form $F(\rho, \phi) = 0$. What can you say about the surface? Sketch $\rho = 1 - \cos \phi$.

*41. Suppose a surface is described in spherical coordinates by an equation that does not involve ϕ, that is, by an equation of the form $F(\rho, \theta) = 0$. What can you say about the surface?

*42. Describe the points of intersection of $r^2 - z^2 = 1$ (in cylindrical coordinates) and $\rho = 4$ (in spherical coordinates).

Use a graphics calculator or computer to sketch a rectangular plot of each of the following equations. Use your graph to describe the surface it gives in cylindrical or spherical coordinates.

43. $r = 3 + \cos 4\theta$

44. $\rho = 4 + \sin 3\phi$

45. $z = \frac{\sin r}{r}$

46. $\rho = \frac{\sin(11\phi)}{\phi}$

15.6 Vectors and Coordinates in n-Dimensional Space

NORM
STANDARD BASIS VECTORS
ANGLES
COORDINATES
AXES
SPHERE
CURVES

In this chapter and the preceding two we have become accustomed to expressing points and vectors in the plane and in three-dimensional space by writing out their components (x, y) or (x, y, z). We have used this approach for many purposes, and it is remarkable how many of the results are the same for both the plane and space.

It is natural to wonder whether we could do the same things with "four-dimensional" vectors (a, b, c, d) or for that matter with "ten-dimensional" vectors $(a, b, c, d, e, f, g, h, i, j)$. We can see how any two vectors of the *same* dimension can be added and how any vector can be multiplied by a scalar. It is also easy to define the dot product. (The cross product is another story, but even for vectors in the plane we could define it only by using a trick, so there is no reason to expect it to extend to other dimensions.) Having the dot product, we can talk about distances and angles, so it should be possible to do some geometry.

For large dimensions it may be awkward to find enough letters to write out several long vectors, so we frequently use subscript notation. We write a vector with five components as $(x_1, x_2, x_3, x_4, x_5)$, or even $(x_1, \ldots, x_5)$. One advantage of this is that we can write a vector with n components $(x_1, x_2, \ldots, x_n)$ without having to specify what n is. Many results about vectors can be proved without specifying n, which means that they will be true for every n.

Another version of this notation is to use a boldface letter, such as **v**, to represent a vector and then label its components with ordinary v's with subscripts. For instance, $\mathbf{v} = (v_1, v_2, \ldots, v_n)$, or $\mathbf{u} = (u_1, u_2, u_3)$, or $\mathbf{x} = (x_1, x_2, \ldots, x_8)$.

DEFINITION 15.6.1

Suppose n is a positive integer. A **vector in n-dimensional space** is an expression of the form

$$\mathbf{a} = (a_1, a_2, \ldots, a_n),$$

where $a_1, a_2, \ldots, a_n$ are real numbers.

Vectors are added by adding corresponding components:

$$(a_1, a_2, \ldots, a_n) + (b_1, b_2, \ldots, b_n) = (a_1 + b_1, a_2 + b_2, \ldots, a_n + b_n).$$

A vector is multiplied by a scalar k by multiplying each component by k:

$$k(a_1, a_2, \ldots, a_n) = (ka_1, ka_2, \ldots, ka_n).$$

The **zero vector** is the vector

$$0 = (0, 0, \ldots, 0).$$

The **negative** of a vector $\mathbf{v} = (v_1, v_2, \ldots, v_n)$ is the vector $-\mathbf{v} = (-v_1, -v_2, \ldots, -v_n)$.

If $\mathbf{u} = (u_1, u_2, \ldots, u_n)$ and $\mathbf{v} = (v_1, v_2, \ldots, v_n)$ are two vectors with the same number of components, then their **dot product** is

$$\mathbf{u} \cdot \mathbf{v} = (u_1, u_2, \ldots, u_n) \cdot (v_1, v_2, \ldots, v_n) = u_1 v_1 + u_2 v_2 + \ldots + u_n v_n.$$

Using this, we define the **norm** of the vector $\mathbf{u}$ to be

$$\|\mathbf{u}\| = \|(u_1, u_2, \ldots, u_n)\| = \sqrt{\mathbf{u} \cdot \mathbf{u}} = \sqrt{u_1^2 + u_2^2 + \ldots + u_n^2}.$$

Two vectors $\mathbf{u}$ and $\mathbf{v}$ are said to be **orthogonal** if $\mathbf{u} \cdot \mathbf{v} = 0$.

It is possible to prove that for any vectors $\mathbf{u}$ and $\mathbf{v}$ their dot product satisfies what is known as the **Cauchy-Schwartz Inequality:**

$$|\mathbf{u} \cdot \mathbf{v}| \leq \|\mathbf{u}\| \|\mathbf{v}\|.$$

This makes it possible to use dot products to define angles. If $\mathbf{u}$ and $\mathbf{v}$ are nonzero vectors, then the **angle** between them is the angle θ defined by

$$\cos \theta = \frac{\mathbf{u} \cdot \mathbf{v}}{\|\mathbf{u}\| \|\mathbf{v}\|},$$

with the understanding that $0 \leq \theta \leq \pi$.

> It is important to emphasize that whenever we add two vectors or find their dot product, the two vectors must have the same number of components, that is, the same n.

Vectors in n-dimensional space behave in many ways like the familiar vectors in the plane or in three-dimensional space. It is not difficult to prove the following result.

PROPOSITION 15.6.2

> Vectors in n-dimensional space have the same properties as vectors in two- and three-dimensional space, as given in Propositions 14.2.2 (p. 833) and 14.2.6 (p. 834).

In three-dimensional space we used the standard basis vectors $\mathbf{i} = (1, 0, 0)$, $\mathbf{j} = (0, 1, 0)$, and $\mathbf{k} = (0, 0, 1)$. In n-dimensional space it is customary to write $\mathbf{e_1} = (1, 0, 0, \ldots, 0)$, $\mathbf{e_2} = (0, 1, 0, 0, \ldots, 0)$, $\ldots$, $\mathbf{e_n} = (0, 0, \ldots, 0, 1)$. For each i from 1 to n, the ith **standard basis vector** $\mathbf{e_i}$ is the vector whose ith component is 1 and whose other components are all 0.

Any vector $\mathbf{v} = (v_1, v_2, \ldots, v_n)$ can be expressed in terms of the standard basis vectors as follows:

$$\mathbf{v} = (v_1, v_2, \ldots, v_n) = v_1 \mathbf{e_1} + v_2 \mathbf{e_2} + \ldots + v_n \mathbf{e_n}.$$

EXAMPLE 1

(i) The sum of $(3, 5, -2, 8, 0)$ and $(4, 5, -2, -1, -1)$ is $(7, 10, -4, 7, -1)$.

(ii) The norm of $(1, -1, 1, 1)$ is $\|(1, -1, 1, 1)\| = \sqrt{4} = 2$. The norm of $(-4, 1, 2, -2)$ is $\|(-4, 1, 2, -2)\| = \sqrt{25} = 5$.

(iii) The angle θ between $\mathbf{u} = (1, -1, 1, 1)$ and $\mathbf{v} = (-4, 1, 2, -2)$ is defined by

$$\cos \theta = \frac{\mathbf{u} \cdot \mathbf{v}}{\|\mathbf{u}\| \|\mathbf{v}\|} = \frac{-5}{10} = -\frac{1}{2}.$$

This means that $\theta = \frac{2\pi}{3}$.

Coordinates in *n*-Dimensional Space

In the plane and in three-dimensional space we began with points and then considered vectors, which were arrows going from one point to another. If the initial point was the origin, we could represent the vector by the coordinates of the second point. Then we learned how to work with vectors written this way.

In *n*-dimensional space we began with vectors $\mathbf{v} = (v_1, v_2, \ldots, v_n)$ and have seen how addition and dot products can be defined for them. However, we can also think of $(x_1, x_2, \ldots, x_n)$ as the **coordinates** of a point P in *n*-dimensional space. The origin is the point $(0, 0, \ldots, 0)$. The distance between two points $P = (x_1, x_2, \ldots, x_n)$ and $Q = (y_1, y_2, \ldots, y_n)$ is

$$\sqrt{(x_1 - y_1)^2 + (x_2 - y_2)^2 + \ldots + (x_n - y_n)^2},$$

which we allow ourselves to write as

$$\|P - Q\| = \sqrt{(P - Q) \cdot (P - Q)}.$$

It is difficult to draw pictures, but otherwise things are quite similar to the familiar cases with $n = 2$ or $n = 3$.

DEFINITION 15.6.3

> Suppose n is a positive integer. The set of all points
>
> $$P = (x_1, x_2, \ldots, x_n),$$
>
> with each $x_i \in \mathbb{R}$, is called **_n_-dimensional space**, or **_n_-dimensional Euclidean space**. It is often symbolized by $\mathbb{R}^n$.

In particular $\mathbb{R}^2$ is the plane, $\mathbb{R}^3$ is three-dimensional space, and $\mathbb{R}^1 = \mathbb{R}$ is the real line.

The **coordinate axes** are the x_1-axis, consisting of all points $(t, 0, \ldots, 0)$, the x_2-axis, consisting of all points $(0, t, 0, \ldots, 0)$, and so on. The x_i-axis consists of all points whose components are all 0 except possibly the *i*th one.

EXAMPLE 2   **Spheres**   Consider the equation $x_1^2 + \ldots + x_n^2 = 4$ in $\mathbb{R}^n$. Since the left side is the square of the distance from $P = (x_1, \ldots, x_n)$ to the origin, this equation just says that P is 2 units from the origin. The set of all such points in $\mathbb{R}^n$ is called the **sphere** of radius 2 around the origin in $\mathbb{R}^n$.

One reason for considering *n*-dimensional space $\mathbb{R}^n$ is that even though we cannot easily draw good pictures, we can often use our geometric intuition to help solve problems.

EXAMPLE 3   Find the point on the unit sphere $x_1^2 + x_2^2 + x_3^2 + x_4^2 + x_5^2 = 1$ that is closest to $P = (1, 2, -1, 1, -3)$.

Figure 15.6.1

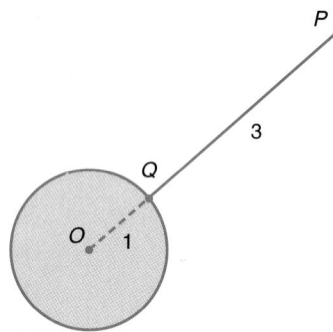

Figure 15.6.2

Solution This sounds difficult, but we can get a hint by thinking about the analogous situation in $\mathbb{R}^3$. In Figure 15.6.1 we see a point P and the unit sphere in $\mathbb{R}^3$, and in that case we know that the point on the sphere that is nearest to P is the point Q where the sphere intersects the segment OP from the origin to P. The way to find Q is to divide the vector from O to P by its length.

Attempting the same thing in $\mathbb{R}^5$, we find that

$$\|(1, 2, -1, 1, -3)\| = \sqrt{1 + 4 + 1 + 1 + 9} = 4,$$

so we try letting $Q = \frac{1}{4}(1, 2, -1, 1, -3) = \left(\frac{1}{4}, \frac{1}{2}, -\frac{1}{4}, \frac{1}{4}, -\frac{3}{4}\right)$. We draw the not-very-accurate picture in Figure 15.6.2, and it suggests that the distance from Q to P is 3. Skeptics can easily check this.

We hope Q will be the point we want, but we still have to prove that it is the closest point. If $U = (x_1, \ldots, x_5)$ is any point on the unit sphere, then the distance from U to P is

$$\sqrt{(U - P) \cdot (U - P)}$$
$$= \sqrt{(x_1 - 1)^2 + (x_2 - 2)^2 + (x_3 + 1)^2 + (x_4 - 1)^2 + (x_5 + 3)^2}.$$

As usual it is easier to work with the square of the distance, and we want to show that $\|U - P\|^2 \geq \|Q - P\|^2 = 3^2 = 9$. But

$$\|U - P\|^2 = (U - P) \cdot (U - P) = U \cdot U - U \cdot P - P \cdot U + P \cdot P$$
$$= \|U\|^2 - 2U \cdot P + \|P\|^2.$$

Now $\|U\|^2 = 1$, since U is a point on the unit sphere, and $\|P\|^2 = 16$. We also know that if θ is the angle between U and P, then

$$|U \cdot P| = \|U\|\|P\||\cos\theta| \leq \|U\|\|P\| = 4.$$

So

$$\|U\|^2 - 2U \cdot P + \|P\|^2 = 17 - 2U \cdot P \geq 17 - (2)(4) = 9.$$

Taking square roots, we find $\|U - P\| \geq 3$, as required.

The point on the unit sphere that is nearest to P is $Q = \left(\frac{1}{4}, \frac{1}{2}, -\frac{1}{4}, \frac{1}{4}, -\frac{3}{4}\right)$. ◢

There are a number of things to notice about Example 3. The main one is the way we used our intuition about $\mathbb{R}^3$ to suggest an answer for $\mathbb{R}^5$. Once we knew where to look, it was still necessary to prove that our guess was correct.

We were also able to use Figure 15.6.2, a vague picture suggested by $\mathbb{R}^2$ or $\mathbb{R}^3$, to help see what to expect.

In the calculations we have allowed ourselves to use the symbol P to refer to a point in $\mathbb{R}^5$ and also to the vector that runs from the origin to that point. This is slightly sloppy, but it is very useful.

Finally, we observe that the proof that Q is the point nearest to P does not really make any mention of $\mathbb{R}^5$. Virtually the same proof will work for similar questions in any $\mathbb{R}^n$.

Curves in $\mathbb{R}^n$

We have discussed curves in $\mathbb{R}^2$ and $\mathbb{R}^3$, and there is no difficulty in doing the same things in $\mathbb{R}^n$.

DEFINITION 15.6.4

Let n be a positive integer. Suppose I is an interval (open or closed, possibly infinite), and suppose

$$x_1 = x_1(t), \qquad x_2 = x_2(t), \dots, \qquad x_n = x_n(t)$$

are functions defined on I. Assume that they are continuous on I. Then they are called **parametric equations**; the variable t is called the **parameter**. The set of points

$$\{(x_1, x_2, \dots, x_n) = \big(x_1(t), x_2(t), \dots, x_n(t)\big) : t \in I\}$$

is called the **curve**. To emphasize the idea of the moving point, we sometimes write

$$\mathbf{r} = \mathbf{r}(t) = \big(x_1(t), x_2(t), \dots, x_n(t)\big).$$

The vector $\mathbf{r}(t)$ points from the origin out to the position of the moving point at time t.

Suppose $t \in I$ is a point at which $x_1(t), x_2(t), \dots, x_n(t)$ are differentiable. Then the **velocity** of the moving point at time t is the vector

$$\mathbf{r}'(t) = \big(x_1'(t), x_2'(t), \dots, x_n'(t)\big).$$

If $\mathbf{r}'(t)$ is not the zero vector, then $\mathbf{r}'(t)$ is a **tangent vector** to the curve at the point $\mathbf{r}(t) = \big(x_1(t), x_2(t), \dots, x_n(t)\big)$, and its length

$$\|\mathbf{r}'(t)\| = \big\| \big(x_1'(t), x_2'(t), \dots, x_n'(t)\big) \big\|$$

is the **speed** of the moving point at time t.

Unit tangent vectors, normal vectors, and arclength are all defined in exactly the same way as in $\mathbb{R}^2$ or $\mathbb{R}^3$.

EXAMPLE 4   (i) Find parametric equations for the straight line in $\mathbb{R}^4$ that contains $P = (3, -1, 0, 2)$ and $(5, -2, -2, 3)$.

Solution   A direction vector of the line can be found by letting $\mathbf{v} = (3, -1, 0, 2) - (5, -2, -2, 3) = (-2, 1, 2, -1)$. The equation of the line is

$$\mathbf{r}(t) = P + t\mathbf{v} = (3, -1, 0, 2) + t(-2, 1, 2, -1) = (3 - 2t, -1 + t, 2t, 2 - t).$$

See Figure 15.6.3.

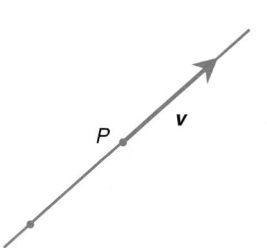

Figure 15.6.3

(ii) To parametrize the same line by its arclength, we begin by noting that the velocity vector is $\mathbf{r}'(t) = \mathbf{v} = (-2, 1, 2, -1)$. The speed is $\|\mathbf{r}'(t)\| = \|\mathbf{v}\| = \sqrt{10}$.

The arclength of the line between $\mathbf{r}(0) = P$ and $\mathbf{r}(t)$ is

$$s = s(t) = \int_0^t \|\mathbf{r}'(u)\| \, du = \int_0^t \sqrt{10} \, du = \sqrt{10}\, t.$$

From this we find that $t = \frac{1}{\sqrt{10}} s$, so

$$\mathbf{r}(s) = (3 - 2t, -1 + t, 2t, 2 - t) = \left(3 - \frac{2}{\sqrt{10}} s, -1 + \frac{1}{\sqrt{10}} s, \frac{2}{\sqrt{10}} s, 2 - \frac{1}{\sqrt{10}} s \right).$$

(iii) Find the unit normal vector to $\mathbf{r}(t) = (3 \cos t, 5 \sin t, 12t, -4 \cos t)$.

Solution The velocity vector is $\mathbf{r}'(t) = (-3\sin t, 5\cos t, 12, 4\sin t)$. Its length is $\|\mathbf{r}'(t)\| = \sqrt{9\sin^2 t + 25\cos^2 t + 144 + 16\sin^2 t} = \sqrt{25 + 144} = 13$. From this we find that the unit tangent vector is

$$\mathbf{T}(t) = \frac{1}{\|\mathbf{r}'(t)\|}\mathbf{r}'(t) = \frac{1}{13}(-3\sin t, 5\cos t, 12, 4\sin t)$$

$$= \left(-\frac{3}{13}\sin t, \frac{5}{13}\cos t, \frac{12}{13}, \frac{4}{13}\sin t\right).$$

Differentiating gives a normal vector

$$\mathbf{n} = \left(-\frac{3}{13}\cos t, -\frac{5}{13}\sin t, 0, \frac{4}{13}\cos t\right).$$

To get a unit normal vector, we divide by the length:

$$\mathbf{N} = \frac{1}{\|\mathbf{n}\|}\mathbf{n} = \frac{1}{\frac{5}{13}}\left(-\frac{3}{13}\cos t, -\frac{5}{13}\sin t, 0, \frac{4}{13}\cos t\right)$$

$$= \left(-\frac{3}{5}\cos t, -\sin t, 0, \frac{4}{5}\cos t\right).$$

Exercises 15.6

I

Find each of the following.

1. $(1, 2, 3, 4, 5) + (9, 8, 7, 6, 5)$

2. $3(2, 0, 4, -1) - 5(3, -1, 2, 1)$

3. $(2, 3, 5, -7) \cdot (4, -3, 1, 2)$

4. $(1, 1, 1, 1, 1) \cdot (3, 5, -5, 2, -4)$

5. $\|(3, -1, 5, 0, 1)\|$

6. $\mathbf{v} \cdot \mathbf{v}$, if $\mathbf{v} = (1, 3, -1, 6, 1, -1)$

7. $\|(4, 2, 1, -3) - (2, 2, 2, -2)\|$

8. $(3\mathbf{e_2} - 4\mathbf{e_4}) \cdot (\mathbf{e_2} + \mathbf{e_4} + \mathbf{e_6})$

9. The angle between $(3, -1, -1, 1)$ and $(-1, 1, 1, -1)$.

10. The angle between $(2, -6, -2, 2)$ and $(3, 3, 0, -3)$.

Find the unit tangent vector for each of the following curves.

11. $\mathbf{r}(t) = (t + 2, 3 - t, 6, t - 4, t)$

12. $\mathbf{r}(t) = (4t - 1, 3 - 2t, 5t, 2t + 7)$

13. $\mathbf{r}(t) = (3t, 4\sin t, -2, 2 - 4\cos t)$

14. $\mathbf{r}(t) = (\sin 3t, 2 - \sin 4t, 4 + \cos 4t, 1 - \cos 3t)$

15. $\mathbf{r}(t) = (t^3 + 1, 4 - 3t^2, -6t, -1)$

16. $\mathbf{r}(t) = (2 - e^t, 3 + e^t, 4 - 3e^t, 5, e^t - 1)$

17. $\mathbf{r}(t) = (\sin t, -\cos t, 3 - t^2, 2t)$

18. $\mathbf{r}(t) = (1 - \sinh t, 3 + \sinh t, \sinh t, -\sinh t)$

19. $\mathbf{r}(t) = (3, \sqrt{1 - t^2}, 4 - t)$

20. $\mathbf{r}(t) = \big(f(t), f(t), f(t), f(t), f(t)\big)$

II

21. Find the equation of the sphere of radius 5 with center at the origin in $\mathbb{R}^6$.

22. Find the equation of the sphere of radius 2 with center at the point $(3, -2, 5, 0, 4)$ in $\mathbb{R}^5$.

*23. Find the point on the sphere of radius 1 with center at the origin in $\mathbb{R}^4$ that is closest to the point $P = (3, -2, 0, 6)$.

*24. Find the point on the sphere of radius 2 with center at the point $C = (4, -2, 3, 1)$ in $\mathbb{R}^4$ that is closest to the point $P = (4, 6, 4, 5)$.

*25. Find the point on the sphere of radius 1 with center at the origin in $\mathbb{R}^5$ that is farthest from the point $P = (2, 0, -4, -2, 1)$.

*26. Find the point on the sphere of radius 4 with center at the point $C = (2, 3, -1, 2)$ in $\mathbb{R}^4$ that is farthest from the point $P = (3, 4, -2, 3)$.

**27. Consider the *ellipsoid*

$$4x_1^2 + x_2^2 + 4x_3^2 + 9x_4^2 = 36.$$

Find the points on it that are nearest the origin and

farthest from the origin. Prove that the points you have found are in fact nearest and farthest.

**28. Consider the *paraboloid* $w = x^2 + y^2 + z^2$ in $\mathbb{R}^4$. Find the point on it that is closest to the point P whose coordinates are $(w, x, y, z) = (-2, 0, 0, 0)$. Prove that your answer is correct.

*29. Find the point on the coordinate *hyperplane* $x_2 = 0$ in $\mathbb{R}^4$ that is closest to the point $P = (3, 2, -1, 5)$.

**30. Find the point on the *hyperplane* $x_1 + x_2 + x_3 + x_4 + x_5 = 0$ in $\mathbb{R}^5$ that is closest to the point $P = (5, -2, 3, 0, 4)$.

 POINT TO PONDER

Before going any farther, we must ask what possible significance four-dimensional vectors can have. It is difficult to draw them on a sketch, since we are used to living in a three-dimensional world, but then again we have been doing not too badly drawing three-dimensional pictures on two-dimensional paper.

Although vectors with four components may not have any obvious geometrical significance, they may still be very useful. For instance, suppose instead of recording just the coordinates of a point (x, y, z) in space we also want to record the time at which we visit it. Then we have four numbers, the three coordinates x, y, and z *and* the time t. It may be convenient to combine them as a vector (x, y, z, t).

This is not merely a convenience; one of the essential steps in developing the theory of relativity is to recognize that it is necessary to consider "space-time," which is precisely the set of all vectors (x, y, z, t). The fact that space and time, though different, are related and cannot be considered separately is fundamental to much of modern physics.

But four-dimensional and higher-dimensional vectors arise in much more down-to-earth places. A social scientist who collects data on a number of subjects may organize them as vectors. For instance, a vector (a, b, c, d, e) for each subject might contain information such as a, the subject's age; b, the subject's weight; c, the subject's height; d, the subject's income; and e, the average length of time the subject sleeps each night. A theory

that tries, for instance, to predict a person's weight from his or her height would attempt to find a function expressing b in terms of c. Another theory might suggest that there is a relation between salary levels and age and sleep patterns and might try to show that all the subjects have vectors that lie on a certain "surface" defined by some equation involving a, d, and e.

In economics it is important to try to predict various quantities, such as the gross national product (GNP), the level of inflation, or the level of unemployment based on a detailed knowledge of various factors such as interest rates and the prices of raw materials. An economist might look for a formula that expresses the level of inflation one year from now as a function of the GNP, the interest rates, and the prices of ten cleverly chosen commodities.

These examples are too oversimplified to be realistic, but the idea is very common. Actual studies may involve dozens of factors. The interesting thing is that using vectors to express the data often makes the data easier to analyze. Much of what we have learned for vectors in the plane and three-dimensional space will work in any dimension, so various familiar techniques are available to help solve problems of this type.

Speaking about higher-dimensional vectors in geometrical terms sometimes makes it easier to understand a problem, even though the original problem may not have much geometric significance.

15.7 Kepler's Laws of Planetary Motion

GRAVITATION CONSTANT
INVERSE SQUARE LAW
RADIAL VECTOR
EQUAL AREAS IN EQUAL TIMES
CONICS

The motion of the planets was a great mystery for many centuries. The Danish astronomer Tycho Brahe (1546–1601) made precise and detailed observations over many years, and Johannes Kepler (1571–1630) was able to use Brahe's data to propose the mathematical laws that govern planetary motion. When Isaac Newton developed calculus, he was able to use it and his knowledge of gravity to prove Kepler's laws. This dramatic success established the importance of calculus beyond any doubt.

One thing that made the problem so difficult is that the earth is not stationary. It is difficult to describe the motion of some other object if the platform from which you are observing it is also moving. A vital first step was to recognize that the earth is not a particularly good reference point.

To put it another way, we choose coordinates so that the sun is at the origin. We have seen many other instances in which a clever choice of coordinates can simplify a mathematical problem, but rarely does it have such philosophical and theological implications as declaring that the earth is not the center of the universe.

Consider two objects, such as the sun and a planet. The gravitational attraction between them depends on their masses M, m, and the distance r between them. The force on the planet is in the direction pointing toward the sun, and its magnitude is $\frac{GMm}{r^2}$, where G is the (universal) **gravitation constant**. If the masses are measured in kilograms and the distance in meters, then

$$G \approx 6.672 \times 10^{-11} \quad \text{N m}^2 \text{ kg}^{-2}.$$

Placing the sun at the origin in $\mathbb{R}^3$, let us write $\mathbf{r}(t)$ for the position of the planet at time t. Then the gravitational force on the planet has magnitude $\frac{GMm}{\|\mathbf{r}(t)\|^2}$, and its direction is the direction from the planet to the sun, which is the direction of $-\mathbf{r}(t)$ (see Figure 15.7.1). Since $\mathbf{u} = \mathbf{u}(t) = \frac{1}{\|\mathbf{r}(t)\|}\mathbf{r}(t)$ is a unit vector pointing in the direction from the sun to the planet, the force can be written as

$$\mathbf{F}(t) = -\frac{GMm}{\|\mathbf{r}(t)\|^2}\frac{1}{\|\mathbf{r}(t)\|}\mathbf{r}(t) = -\frac{GMm}{\|\mathbf{r}(t)\|^3}\mathbf{r}(t). \tag{15.7.1}$$

Newton's second law says that the acceleration $\mathbf{a}$ of an object is related to the force $\mathbf{F}$ acting on it by

$$\mathbf{F} = m\mathbf{a}.$$

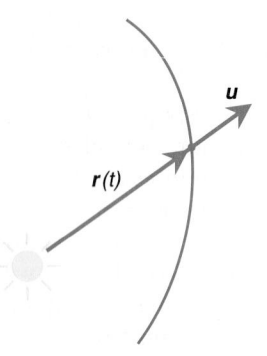

Figure 15.7.1

Roughly speaking, this says that the object accelerates in the direction of the force, which is very reasonable, and that it takes more force to accelerate a heavier object.

The acceleration of the moving planet is $\mathbf{a} = \mathbf{r}''(t)$, so on the one hand the force is $m\mathbf{a} = m\mathbf{r}''(t)$, and on the other it is $-\frac{GMm}{\|\mathbf{r}(t)\|^3}\mathbf{r}(t)$. We find that

$$\mathbf{r}''(t) = -\frac{GM}{\|\mathbf{r}(t)\|^3}\mathbf{r}(t). \tag{15.7.2}$$

In particular, $\mathbf{r}''(t)$ and $\mathbf{r}(t)$ point along the same line (in opposite directions), so (by Proposition 14.2.10(v) and (ii)) their cross product $\mathbf{r}(t) \times \mathbf{r}''(t)$ is zero.

If we consider the vector $\mathbf{c}(t) = \mathbf{r}(t) \times \mathbf{r}'(t)$, we see that its components are made up of sums of products of components of $\mathbf{r}(t)$ and $\mathbf{r}'(t)$, and applying the ordinary product rule to the components we see that the derivative is given by a "product rule for cross products":

$$\mathbf{c}'(t) = \frac{d}{dt}\big(\mathbf{r}(t) \times \mathbf{r}'(t)\big)$$
$$= \mathbf{r}'(t) \times \mathbf{r}'(t) + \mathbf{r}(t) \times \mathbf{r}''(t)$$
$$= 0 + 0 = 0.$$

We have just seen why the last of these cross products is zero, and the previous one is zero because the two vectors are the same.

Figure 15.7.2(i)

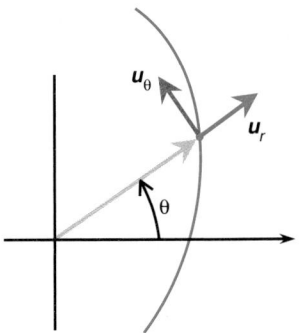

Figure 15.7.2(ii)

The fact that the derivative of $\mathbf{c}(t)$ is the zero vector means that $\mathbf{c}(t) = \mathbf{r}(t) \times \mathbf{r}'(t)$ is a constant vector. If we choose coordinates so that the z-axis points along this constant vector, then it will be orthogonal to both $\mathbf{r}(t)$ and $\mathbf{r}'(t)$. In particular the planet will always move in the xy-plane (see the "Point to Ponder" at the end of Section 13.1), and from now on we will consider motion in the plane with the sun at the origin.

In Figure 15.7.2(i) we see the position vector $\mathbf{r}(t)$ and the polar angle θ. Notice that as the planet moves along its orbit, the angle changes, so we think of θ as a function of t.

It is important to consider the two unit vectors shown in Figure 15.7.2(ii). They are the **unit radial vector** $\mathbf{u}_r$, which points in the direction of the radial vector $\mathbf{r}(t)$, and the unit vector $\mathbf{u}_\theta$, which is orthogonal to $\mathbf{u}_r$, pointing in the direction in which θ increases. Notice that $\mathbf{u}_\theta$ does not necessarily point in the exact direction in which the planet is moving, but it is on the correct side. We can write $\mathbf{u}_r$ in two different ways, either as $\frac{1}{\|\mathbf{r}(t)\|}\mathbf{r}(t)$ or in components as

$$\mathbf{u}_r = (\cos\theta, \sin\theta). \tag{15.7.3}$$

Assuming that the planet is moving counterclockwise, as in the illustration, we then see that

$$\mathbf{u}_\theta = (-\sin\theta, \cos\theta). \tag{15.7.4}$$

The force of gravity points in the direction opposite to $\mathbf{u}_r$, so it will be easy to express the force in terms of $\mathbf{u}_r$ and $\mathbf{u}_\theta$. It is slightly less easy but very helpful to express the velocity and acceleration in terms of $\mathbf{u}_r$ and $\mathbf{u}_\theta$.

Differentiating Equations 15.7.3 and 15.7.4, we find that

$$\frac{d\mathbf{u}_r}{dt} = \mathbf{u}_\theta \frac{d\theta}{dt} \qquad \text{and} \qquad \frac{d\mathbf{u}_\theta}{dt} = -\mathbf{u}_r \frac{d\theta}{dt}. \tag{15.7.5}$$

Let us write $r = \|\mathbf{r}(t)\|$ for the distance from the sun to the planet. It is a function of t. In particular we see that

$$\mathbf{r}(t) = r\mathbf{u}_r. \tag{15.7.6}$$

Differentiating Equation 15.7.6 with respect to t, we find that the velocity is

$$\mathbf{v} = \frac{d}{dt}\big(\mathbf{r}(t)\big) = \frac{dr}{dt}\mathbf{u}_r + r\frac{d\mathbf{u}_r}{dt} = \frac{dr}{dt}\mathbf{u}_r + r\frac{d\theta}{dt}\mathbf{u}_\theta. \tag{15.7.7}$$

Notice how we have broken up the velocity into its radial component and its component in the direction of $\mathbf{u}_\theta$.

Differentiating Equation 15.7.7, we find that the acceleration is

$$\mathbf{a} = \frac{d}{dt}(\mathbf{v}) = \frac{d^2 r}{dt^2}\mathbf{u}_r + \left(\frac{dr}{dt}\right)\frac{d\mathbf{u}_r}{dt} + \left(\frac{dr}{dt}\right)\frac{d\theta}{dt}\mathbf{u}_\theta + r\frac{d^2\theta}{dt^2}\mathbf{u}_\theta + r\left(\frac{d\theta}{dt}\right)\frac{d\mathbf{u}_\theta}{dt}.$$

Substituting into this the expressions from Equations 15.7.5, we find that

$$
\mathbf{a} = \frac{d^2 r}{dt^2}\mathbf{u}_r + \frac{dr}{dt}\mathbf{u}_\theta\frac{d\theta}{dt} + \left(\frac{dr}{dt}\right)\frac{d\theta}{dt}\mathbf{u}_\theta + r\frac{d^2\theta}{dt^2}\mathbf{u}_\theta - r\frac{d\theta}{dt}\mathbf{u}_r\left(\frac{d\theta}{dt}\right)
$$

$$
= \left(\frac{d^2 r}{dt^2} - r\left(\frac{d\theta}{dt}\right)^2\right)\mathbf{u}_r + \left(2\frac{dr}{dt}\left(\frac{d\theta}{dt}\right) + r\frac{d^2\theta}{dt^2}\right)\mathbf{u}_\theta
$$

$$
= \left(\frac{d^2 r}{dt^2} - r\left(\frac{d\theta}{dt}\right)^2\right)\mathbf{u}_r + \left(\frac{1}{r}\frac{d}{dt}\left(r^2\frac{d\theta}{dt}\right)\right)\mathbf{u}_\theta. \tag{15.7.8}
$$

But we know from Equation 15.7.2 that the acceleration points in the direction of $-\mathbf{r}(t)$, which is the opposite of the direction of $\mathbf{u}_r$. The other component in Equation 15.7.8 must be zero, that is,

$$
\frac{1}{r}\frac{d}{dt}\left(r^2\frac{d\theta}{dt}\right) = 0, \text{ so } r^2\frac{d\theta}{dt} \text{ must be a constant.}
$$

In Figure 15.7.3 we have shaded the area that is "swept out" by the position vector $\mathbf{r}(t)$ between time 0 and time t. From Theorem 12.3.1 we know that this area equals

$$
A = \int \frac{1}{2}r^2 \, d\theta = \frac{1}{2}\int_0^t r^2\frac{d\theta}{dt}\, dt.
$$

The Fundamental Theorem of Calculus tells us that

$$
\frac{dA}{dt} = \frac{1}{2}r^2\frac{d\theta}{dt},
$$

and we have just seen that the right side of this equation is a constant.

We have proved **Kepler's Second Law**, which says that the rate at which the position vector sweeps out area is a constant. Another way to put this is that it sweeps out equal areas in equal times.

In Figure 15.7.4 we see the orbit of a planet in which the three shaded regions all have the same area. They must each be swept out in the same length of time, which means that it takes the planet the same length of time to move along the part of its orbit corresponding to each of them. From the picture we see that the planet must move faster when it is closer to the sun and more slowly when it is farther away.

It is interesting to notice how much this argument depended on making the right choice of coordinates. First we had to arrange the orbit in space in such a way that it lies in the xy-plane, which means choosing the z-axis appropriately. Then in the plane it was vital to use the vectors $\mathbf{u}_r$ and $\mathbf{u}_\theta$ and express the velocity and acceleration in terms of them. This amounts to using them as the directions of the coordinates.

Now we choose the coordinates in the plane in such a way that the point on the orbit that is closest to the sun is the point where $\theta = 0$. We can also set $t = 0$ when the planet is at this point. The situation is illustrated in Figure 15.7.5.

When $\theta = 0$, the distance r is a minimum, so $\frac{dr}{dt} = 0$. From Equations 15.7.3 and 15.7.4 we see that $\mathbf{u}_r = (1, 0)$ is a horizontal vector and $\mathbf{u}_\theta = (0, 1)$ points straight up. These facts and Equation 15.7.7 tell us that the velocity at $t = 0$ is

$$
\frac{d\mathbf{r}}{dt}(0) = \left(\frac{dr}{dt}\mathbf{u}_r + r\frac{d\theta}{dt}\mathbf{u}_\theta\right)(0) = 0 + \left(0, r(0)\frac{d\theta}{dt}(0)\right) = (0, v_0),
$$

where we have written $v_0 = \left(r\frac{d\theta}{dt}\right)(0)$ for the speed at $t = 0$.

Figure 15.7.3

Figure 15.7.4

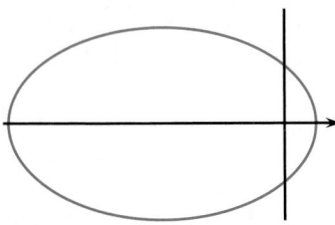

Figure 15.7.5

The last two vectors are written in rectangular coordinates, using the fact that $\mathbf{u}_\theta = (0, 1)$.

By Equation 15.7.2 the acceleration at time t is

$$\frac{d\mathbf{r}'(t)}{dt} = \mathbf{r}''(t) = -\frac{GM}{\|\mathbf{r}(t)\|^3}\mathbf{r}(t) = -\frac{GM}{\|\mathbf{r}(t)\|^2}\mathbf{u}_r.$$

Substituting for $\mathbf{u}_r$, using Equations 15.7.5, we find that

$$\frac{d\mathbf{r}'(t)}{dt} = \frac{GM}{r^2}\left(\frac{1}{\frac{d\theta}{dt}}\right)\frac{d\mathbf{u}_\theta}{dt}.$$

The denominator $r^2\frac{d\theta}{dt}$ is exactly the quantity that we have just recently discovered must be a constant. If we write K for the constant $K = GM/\left(r^2\frac{d\theta}{dt}\right)$, then this equation becomes

$$\frac{d\mathbf{r}'(t)}{dt} = K\frac{d\mathbf{u}_\theta}{dt}.$$

Integrating, we find that

$$\mathbf{r}'(t) = K\mathbf{u}_\theta + \mathbf{C},$$

for some constant vector $\mathbf{C}$. When $t = 0$, we know that $\mathbf{u}_\theta = (0, 1)$ and $\mathbf{r}'(0) = (0, v_0)$, so $\mathbf{C} = (0, v_0 - K)$. This means that

$$\mathbf{r}'(t) = K\mathbf{u}_\theta + \mathbf{C} = K\mathbf{u}_\theta + (0, v_0 - K). \tag{15.7.9}$$

From Equation 15.7.7 we can evaluate the dot product $\mathbf{r}'(t) \cdot \mathbf{u}_\theta = r\frac{d\theta}{dt}$. But from Equation 15.7.9 we find that it also equals $K\mathbf{u}_\theta \cdot \mathbf{u}_\theta + (0, v_0 - K) \cdot \mathbf{u}_\theta = K + (v_0 - K)\cos\theta$, since $\mathbf{u}_\theta = (-\sin\theta, \cos\theta)$. So $r\frac{d\theta}{dt} = K + (v_0 - K)\cos\theta$.

Since $K = GM/\left(r^2\frac{d\theta}{dt}\right)$, we find that $r\frac{d\theta}{dt} = \frac{GM}{Kr}$, and combining with the previous line gives

$$\frac{GM}{Kr} = K + (v_0 - K)\cos\theta.$$

This can be rewritten as

$$r = \frac{GM/K}{K + (v_0 - K)\cos\theta}.$$

Since $v_0 = r(0)\frac{d\theta}{dt}(0)$, we can evaluate at $t = 0$ to find $K = GM/\left(r^2\frac{d\theta}{dt}\right) = \frac{GM}{r(0)v_0}$, and the equation becomes

$$r = \frac{GM/K^2}{1 + (v_0/K - 1)\cos\theta} = \frac{r(0)^2 v_0^2/GM}{1 + \left(r(0)v_0^2/GM - 1\right)\cos\theta}. \tag{15.7.10}$$

This is the equation of a conic. Its eccentricity is

$$e = \frac{r(0)v_0^2}{GM} - 1.$$

This is **Kepler's First Law**, that the orbit of any planet must be a **conic**. It applies not just to planets and comets, whose orbits are ellipses, but to meteors, whose orbits are hyperbolas, and in fact to any object moving under the influence of gravity from a single heavy object (such as the moon orbiting the earth).

Finally, for a planet, or for any object whose orbit is an ellipse, let us write T for the **period** of the orbit, that is, the length of time it takes for one complete revolution. If the semi-major and semi-minor axes of the ellipse are a and b, respectively, then we know that the area of the ellipse is πab (cf. Example 6.2.1, p. 425). But from Theorem 12.3.1 we can also calculate the area by $\int_0^T \frac{1}{2}r^2\frac{d\theta}{dt}\,dt = \frac{1}{2}\int_0^T r(0)v_0\,dt = \frac{1}{2}r(0)v_0 T$.

(Here we used the fact that the constant $r^2 \frac{d\theta}{dt}$ equals its value at $t = 0$, which is $r(0) r(0) \frac{d\theta}{dt}(0) = r(0) v_0$.)

Comparing these two expressions for the area shows that

$$T = \frac{2\pi ab}{r(0) v_0}. \tag{15.7.11}$$

In Example 12.4.2 we found that the equation of an ellipse with eccentricity e and semi-major axis a is

$$r = \frac{a(1 - e^2)}{1 - e\cos\theta}.$$

Combining this with Equation 15.7.10, we find that $r(0)^2 v_0^2 / GM = a(1 - e^2)$, so $r(0)^2 v_0^2 = aGM(1 - e^2)$. But since $e = \frac{c}{a}$, we have that $1 - e^2 = 1 - \frac{c^2}{a^2} = 1 - \frac{a^2 - b^2}{a^2} = \frac{b^2}{a^2}$, so $r(0)^2 v_0^2 = GM\frac{b^2}{a}$.

Squaring Equation 15.7.11 gives

$$T^2 = \frac{(2\pi ab)^2}{r(0)^2 v_0^2} = \frac{(2\pi ab)^2}{GMb^2/a} = \frac{4\pi^2 a^3}{GM}.$$

This is **Kepler's Third Law**, which says that the square of the period of an elliptical orbit is a constant times the cube of the semi-major axis.

Summary 15.7.12

Kepler's Laws

For planetary motion, for which the force is given by Equation 15.7.1, the following three laws hold.

1. The orbit is a conic, whose equation is given by Equation 15.7.10.
2. The rate at which the position vector sweeps out area is a constant, that is, $\frac{dA}{dt} = \frac{1}{2} r^2 \frac{d\theta}{dt}$ is a constant. The planet sweeps out equal areas in equal times.
3. For an elliptical orbit the square of the period T is proportional to the cube of the semi-major axis a. In fact,

$$T^2 = \frac{4\pi^2}{GM} a^3.$$

Kepler's Third Law is perhaps less easy to appreciate than the other two, but roughly speaking, it says that the period of the orbit increases with the semi-major axis (and gives a precise formula describing the increase).

There are two interesting consequences of this. One is that it is slightly surprising that the period does not depend on the semi-minor axis, in the sense that two orbits with the same semi-major axis will have the same period, no matter how different their semi-minor axes. The other is that the ratio $\frac{T^2}{a^3}$ is the same for all the planets. If we know the period of one planet (such as the earth) and the semi-major axes for the earth and any other planet, then Kepler's Third Law can be used to find the period of the other planet.

The same approach will apply for objects orbiting the earth.

EXAMPLE 1 The moon orbits the earth once every $T = 27.3$ days. The semi-major axis of its orbit is about $a = 240{,}000$ miles. The ratio $\frac{T^2}{a^3}$ is approximately $\frac{T^2}{a^3} \approx 5.39 \times 10^{-14}$ day^2/miles^3.

   Now suppose there is a satellite orbiting the earth in a circular orbit of radius 5000 miles. If we let the period of the satellite be T_s, then we know that $\frac{T_s^2}{5000^3} \approx 5.39 \times 10^{-14}$, so $T_s^2 \approx 6.74 \times 10^{-3}$, and $T_s \approx \sqrt{6.74 \times 10^{-3}} \approx 8.21 \times 10^{-2}$. The satellite will orbit the earth in about 8.21×10^{-2} day ≈ 1.97 hours ≈ 118 minutes.

Exercises 15.7

I

Find each of the following.

1. The period of an earth satellite in a circular orbit of radius 6000 miles.

2. The period of an earth satellite in an orbit whose semi-major axis is 8000 miles.

3. The radius of a circular orbit around the earth for which one complete orbit takes 3 hours.

4. The radius of a circular orbit around the earth for which one complete orbit takes half as long as one orbit of the moon.

5. The smallest distance from the surface of the earth to a satellite whose orbit has semi-major axis 16,000 miles and semi-minor axis 12,000 miles. (*Hint*: First measure distances from the center of the earth, which is at a focus of the elliptical orbit.)

6. The greatest distance from the surface of the earth to a satellite whose orbit has semi-major axis 24,000 miles and semi-minor axis 14,000 miles.

7. The greatest distance from the surface of the earth to a satellite whose orbit has semi-minor axis 7000 miles and period 4 hours.

8. The smallest distance from the surface of the earth to a satellite whose orbit has semi-minor axis 8000 miles and period 4 hours.

9. The height above the earth's surface of a satellite in a circular orbit whose period is 100 minutes. (Assume that the earth is a perfect sphere with radius 6400 km.)

10. The period of a satellite in a circular orbit around the earth that is at a height of 500 km above the ground.

11. The period of the planet Mars, which has semi-major axis $\approx 142{,}000{,}000$ miles. (The semi-major axis of the earth's orbit is $\approx 93{,}000{,}000$ miles.)

12. The period of the planet Jupiter, which has semi-major axis $\approx 480{,}000{,}000$ miles.

13. The period of the planet Pluto, which has semi-major axis $\approx 5{,}910{,}000{,}000$ km. (The semi-major axis of the earth's orbit is $\approx 150{,}000{,}000$ km.)

14. The period of the planet Mercury, which has semi-major axis $\approx 58{,}000{,}000$ km.

15. The semi-major axis of the orbit of Venus, which has a period of about 228 days.

16. The semi-major axis of the orbit of Saturn, which has a period of about 29 years.

17. The radius of a circular orbit around the sun if the corresponding period is $\frac{1}{2}$ year.

18. The period of a circular orbit around the sun of radius half that of the earth's orbit.

19. The farthest distance of Mercury from the center of the sun if its orbit has semi-major axis 36 million miles and eccentricity $e \approx 0.2$.

20. The shortest distance between the sun and Pluto if Pluto's orbit has semi-major axis 5.9×10^9 km and eccentricity $e \approx 0.25$.

II

*21. A *geosynchronous* satellite is one that appears to hang motionless in the sky because it orbits the earth exactly once every 24 hours. Such satellites are used for communications purposes because they don't require moving antennas on the ground. What is the radius of the orbit of a geosynchronous satellite in a circular orbit?

*22. Use Kepler's laws to compare the speed of a satellite at its point of nearest approach and its point of greatest distance from the planet about which it is orbiting if the semi-major axis of its orbit is twice as great as its semi-minor axis.

Chapter Summary

§15.1
parametric equations
parameter
curve
domain
image
$\mathbf{r}(t)$
helix
velocity vector
tangent vector
speed
angle between vector and plane
angle between curves

§15.2
unit tangent vector $\mathbf{T}(t)$
normalizing
normal vector
principal (unit) normal vector
arclength
parametrization by arclength

§15.3
second derivative
acceleration vector
curvature
constant curvature
radius of curvature
center of curvature
osculating circle

§15.4
quadric surface
sphere
ball
ellipsoid
paraboloid
cone
hyperboloid of one sheet

hyperboloid of two sheets
saddle surface

§15.5
cylindrical coordinates
spherical coordinates
longitude
latitude

§15.6
vector in n-dimensional space
zero vector
negative of a vector
dot product
norm
orthogonal vector
Cauchy-Schwartz Inequality
angle between two vectors
standard basis vector $\mathbf{e}_i$
coordinates
n-dimensional Euclidean space $\mathbb{R}^n$
coordinate axes
sphere
parametric equations for curve
parameter
curve
velocity
tangent vector
speed

§15.7
gravitation constant
unit radial vector
equal areas in equal times
conics
period
Kepler's laws (Summary 15.7.12)

Review Exercises

Find each of the following.

1. The speed of the moving point $\mathbf{r}(t) = (\cos^2 t, t + 3, \sin^2 t)$ at $t = \frac{\pi}{4}$.

2. The angle of the tangent vector to $\mathbf{r}(t) = (t^3, t^2)$ above the horizontal at $t = 1$.

3. The unit tangent vector to the curve $\mathbf{r}(t) = (t, t^3)$.

4. The principal normal vector to the curve $\mathbf{r}(t) = (3t, t^2, 4t)$.

5. The curvature of the circle $x^2 - 2x + y^2 + 4y + 1 = 0$.

6. The curvature of the parabola $y = ax^2$ at its vertex.

7. The curvature of the ellipse $(x, y) = (4 \cos t, 3 \sin t)$ at $t = 0$.

8. The curvature of the curve $\mathbf{r}(t) = (e^{2t-1}, e^{2t+3}, -e^{2t})$.

9. The maximum curvature for any point on the curve $y = x^4 + 3$.

10. The minimum and maximum curvatures for points on the ellipse $4x^2 + 25y^2 = 100$.

Identify and sketch each of the following quadric surfaces.

11. $x^2 - y^2 - z^2 = 0$

12. $x^2 - y^2 - z^2 = 4$

13. $3x = y^2 + z^2 - 4$

14. $9 - x^2 - y^2 - z^2 = 0$

15. $5x^2 + 3y^2 + 7z^2 = 1$

16. $x^2 - y^2 = z^2 - 1$

17. $x^2 + z^2 = 16$

18. $4x^2 = y + z^2$

19. $x + z^2 = y^2$

20. $x^2 + y^2 + z^2 + 4 = 0$

Find each of the following.

21. The cylindrical coordinates of the point $(x, y, z) = (4, -4, -2)$.

22. The spherical coordinates of the point $(x, y, z) = (-6, -6, 6)$.

23. The spherical coordinates of the point whose cylindrical coordinates are $\left(4, \frac{\pi}{2}, -4\right)$.

24. The cylindrical coordinates of the point whose spherical coordinates are $\left(1, -\frac{\pi}{4}, \frac{\pi}{3}\right)$.

25. The equation in cylindrical coordinates of the plane $x = \sqrt{3}y$.

26. The equation in spherical coordinates of the plane $z = 4$.

27. The equation in cylindrical coordinates of the surface whose equation in spherical coordinates is $\rho = 5$.

28. The equation in spherical coordinates of the surface whose equation in cylindrical coordinates is $r = 3$.

29. Parametric equations in spherical coordinates for the line through $(2, -2, 0)$ and $(2, -2, 5)$.

30. Parametric equations in cylindrical coordinates for the y-axis.

31. $\|(4, 4, -3, 2)\|$

32. $(5, 4, 3, -1, 0) \cdot (3, -2, 1, 4, 2)$

33. The distance between $(1, 1, 1, 1)$ and $(3, 2, 1, -1)$.

34. The angle between $(0, 3, 2, -1)$ and $(4, 5, 0, 1)$.

35. The angle between $(1, 1, 1, 1)$ and $(1, 1, 0, 0)$.

36. The component of $(4, 2, -1, 5, 6)$ along the direction of $(7, -1, 0, -7, -1)$.

37. The component of $(1, 1, 1, 2)$ along the direction of $(3, -5, 4, -1)$.

38. The point on the unit sphere in $\mathbb{R}^5$ that is closest to $(4, -4, 0, -4, -1)$.

39. The point on the sphere of radius 2 around the point $(1, 1, 1, 1)$ that is farthest from the origin.

40. The point on the sphere of radius 3 around the point $(3, -2, 5)$ that is farthest from the point $(1, 0, 4)$.

II

*41. Find parametric equations for the circle of radius R that lies in the plane $x = a$ and whose center is (a, b, c).

*42. Find the speed of the moving point that traces out the cycloid described in Example 14.3.2 (p. 843) at the moments when it touches the ground.

*43. Consider a point P on the circle C whose center is $(2, -3)$ and whose radius is R. Show that the osculating circle is C itself.

*44. Consider the circle C of radius R that lies in the plane $x + y + z = 3$ and whose center is $(1, 1, 1)$. Show that for any point P on the circle C, the osculating circle at P is C itself.

45. Find the spherical coordinates of the summit of Mount Everest, whose height is approximately six miles and which lies at a latitude of $28°$ north and a longitude of $87°$ east. (Assume that the earth has radius 4000 miles.)

*46. Find the arclength of the curve $\mathbf{r}(t) = \left(1 - t^2, \frac{4}{3}t, t^3 + t\right)$ starting at $\mathbf{r}(0)$; that is, express s as a function $s(t)$.

*47. Find the arclength of $\mathbf{r}(t) = \left(2t \sin t, 1 - \frac{t^3}{3}, 2t \cos t\right)$.

*48. Consider the line $\mathbf{r}(t) = P + t\mathbf{v}$. What conditions on P and $\mathbf{v}$ will guarantee that this is the arclength parametrization, that is, that t is the arclength?

Partial Derivatives

16

On a weather map, the isotherms join places where the temperature is the same. Notice how cooler temperatures reach farther south in the mountains on the west side of the continent.

INTRODUCTION

Until now we have been dealing with functions $y = f(x)$, in which a quantity y was expressed as a function of the single variable x. But frequently the quantities we want to consider in the real world depend on many different factors.

In biology, for instance, the rate at which some type of bacteria reproduce is likely to depend on such things as the ambient temperature, the concentrations of various nutrients available in the environment, the "crowding" caused by the number of bacteria already present, and perhaps the level of illumination or the number of competing organisms.

In economics the gross national product (GNP) is extremely complicated and difficult to calculate or even to estimate. It depends on a great many different things, such as the level of unemployment, the rate of inflation, the costs of raw materials and commodities, the size of the workforce, and so on. It is hopelessly unrealistic to think that anybody can find a function that can calculate the value of the GNP if the values of

Figure 16.1.1

Lines of latitude and longitude are clearly marked on this map from the U.S. Department of Defense. Note that distances are distorted because lines of longitude should actually be closer together near the top of the picture due to the curvature of the earth.

all these other factors are given, but at least it is clear that such a function would have to depend on many variables.

A meteorologist might like to express the outdoor temperature at any point on the earth's surface as a function (see Figure 16.1.1). One way to do this would be to use latitude and longitude to determine the location, and this would mean that the temperature would be a function of these two variables. Since the temperature changes, it would be reasonable to expect it to be a function of time too, so the temperature would be a function of three variables: latitude, longitude, and time.

Suppose that we could actually write down the temperature function, some complicated function of these three variables. Then suppose we want to know what happens at one particular place, the place where we live, for instance. This means that we fix the values of the latitude and longitude, specifying this particular location, and having done that, we can now think of the temperature at this place as a function of the single variable, time.

It is quite natural to ask about the rate at which the temperature at this place changes, the derivative of temperature with respect to time. To do this, we need to have a function whose variable is time; we arranged this by fixing the values of the other variables, which amounts to considering what happens at a single location. Then taking the derivative with respect to time is perfectly reasonable.

This is the idea of the *partial derivative*. Starting with a function that depends on several variables, we fix the values of all but one of the variables and think of the function as a function of the one remaining variable. Its derivative with respect to this variable is called the **partial derivative** because, as the name suggests, it concerns only a *part* of the behavior of the function.

In this chapter we begin by studying some properties of functions of several variables in Section 16.1, and in Section 16.2 we discuss the idea of limits and continuity for these functions. In Section 16.3 we introduce the notion of partial derivative and interpret it geometrically, and Section 16.4 is about *differentials*, something we have discussed briefly for functions of one variable but that is more complicated and more interesting for functions of several variables. The Chain Rule is also more complicated for these functions, as we see in Section 16.5. In Section 16.6 we discuss *directional derivatives and gradients*; these are fairly technical ideas that are helpful when we want to deal with partial derivatives and especially for applying them to various geometrical problems. In Section 16.7 we discuss normal vectors and tangent planes to surfaces in $\mathbb{R}^3$, in Section 16.8 we consider maximum/minimum problems for functions of several variables, and in Section 16.9 we discuss the analog of the second derivative test for extrema as it applies to functions of two variables. Section 16.10 is devoted to *Lagrange multipliers*, a clever technique for dealing with a certain type of maximum/minimum problem. In Section 16.11 partial derivatives are applied to a type of approximation problem known as *least squares* or *linear regression*.

16.1 Functions of Several Variables

LEVEL CURVE
LEVEL SURFACE

In this section we will use our knowledge of graphing various curves and surfaces, especially conics, planes, and quadric surfaces (cf. Chapter 11, Chapter 15, and Sections 1.4, 1.5, and 1.6).

We have already encountered examples of functions of several variables. The polynomial $f(x, y) = x^2 + 2x - y^2 - 4$ is a function of the two variables x and y. For any specific values of x and y we can evaluate to get a value of the function. For instance, when $x = 2$ and $y = -3$, we see that $f(2, -3) = 2^2 + 2(2) - (-3)^2 - 4 = -5$.

It is just as easy to make polynomials of three or more variables. For example, $g(x, y, z) = 3x^2yz^3 - 2yz + 7x - y + 5$ is a polynomial in the three variables x, y, and z. For that matter, there is no reason to consider only polynomials. The function $R(x, y, z) = \frac{xyz-3}{xy+2yz-x^2+4}$ is the quotient of two polynomials, and as in the single-variable case such a function is called a rational function. Notice that it will not be defined whenever the denominator is zero (e.g., when $x = 0$, $y = -1$, and $z = 2$, the denominator equals zero, so the function is not defined at this point).

It is also possible to combine other functions we have studied to construct more complicated functions. For example, $h(w, x, y, z) = x \sin(w^2y + 3z + 5) - e^{1-3xy}$ is a function of four variables. If we want to use a large number of variables, it may be difficult to find enough different letters for all of them, and sometimes it is convenient to use subscripts to label variables. For instance, we might let x_1, x_2, x_3, $\ldots$, x_6 be six variables and define a function by

$$k(x_1, x_2, \ldots, x_6) = \frac{x_1 e^{x_2 - 4x_3}}{\ln(x_4^2 + x_5^2 + x_6^2 + 4)}.$$

One of the most important tools we have for understanding functions of one variable is drawing their graphs. For functions of several variables this becomes difficult to do, but for a function of two variables it is still possible. If we consider a function $f(x, y)$ of the two variables x and y, we can try evaluating it at every point (x, y) in the plane. The function may not be defined at some of these points, but in any case the domain will be some subset of the plane. For each point (x, y) that is in the domain of f there will be a value $f(x, y)$. If we think of the xy-plane lying horizontally and as usual think of the value $z = f(x, y)$ as a vertical displacement, then it is possible to "plot" a point lying directly above or below the point (x, y) in the plane, with its displacement above or below the plane determined by the value $f(x, y)$. See Figure 16.1.2.

For instance, if $f(x, y) = \sqrt{x^2 + y^2}$, then the domain of f is the whole plane because $x^2 + y^2$ is always nonnegative. Since the square root is itself always nonnegative, the points we plot will never lie below the horizontal plane.

In fact, the value of the function $f(x, y) = \sqrt{x^2 + y^2}$ is just the distance from the point (x, y) to the origin, so its values will be larger as we consider points (x, y) that are farther from the origin. This means that the points we plot will be higher and higher, the farther they are from the origin. If we plot all the points, that is, one for each pair of values (x, y), the resulting picture will be the top half of the cone $z^2 = x^2 + y^2$ (see Figure 16.1.3). The graph consists of all points (x, y, z) that satisfy $z = \sqrt{x^2 + y^2}$. Such points satisfy $z^2 = x^2 + y^2$, the equation of a cone, so the points on the graph must all lie on the cone. Since the values of $f(x, y)$ are never negative, they actually lie on the top half of the cone. Notice that the point (x, y) is a point in the horizontzal xy-plane; to specify it, we must specify the values of both variables x and y.

Figure 16.1.2

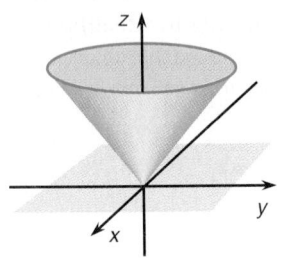

Figure 16.1.3

We visualize the surface lying inside three-dimensional space $\mathbb{R}^3$ with the xy-plane as the horizontal plane and let z be the vertical coordinate. So what we have described as the "graph" of a function $f(x, y)$ can also be described

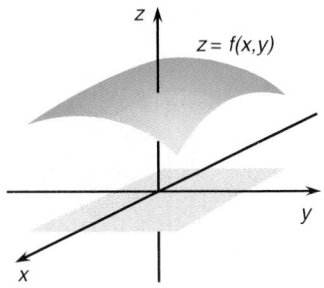

Figure 16.1.4

as the set of all points (x, y, z) in $\mathbb{R}^3$ that satisfy $z = f(x, y)$. In analogy with the single-variable case we will speak of the "graph of $z = f(x, y)$." See Figure 16.1.4.

For another example, let $g(x, y) = y - x + 1$. The graph of $z = g(x, y)$ will consist of points (x, y, z) in space that satisfy $z = g(x, y)$, that is, $z = y - x + 1$. This amounts to $x - y + z = 1$, which is the equation of a plane (see Figure 16.1.5). For each point (x, y) there is exactly one value of z, namely, $z = g(x, y) = y - x + 1$.

Actually sketching the graphs of functions of two variables tends to be quite difficult, or at least tedious. There are computer software packages available that can do it for us, but whether it is done by computer or by hand, there is always the drawback that we are drawing a two-dimensional picture of a surface in three-dimensional space, so it is always a little bit difficult to interpret. It is of course possible to make three-dimensional models, using cardboard, plastic, or modeling clay, but these have other limitations, and it is surprisingly difficult to make them accurately.

In Figure 16.1.6 we see the graphs of a number of functions, most of which are quite familiar.

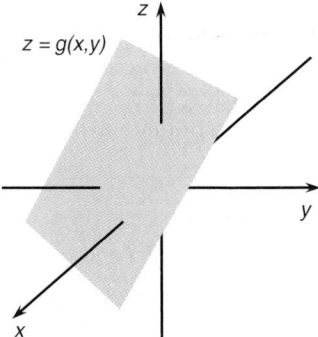

Figure 16.1.5

$z = x^2 + y^2$

$z = 3$

$z = \sqrt{1 - x^2 - y^2}$

$z = \sqrt{x^2 + y^2}$

$z = x^2$

$z = \sqrt{1 + x^2 + y^2}$

Figure 16.1.6

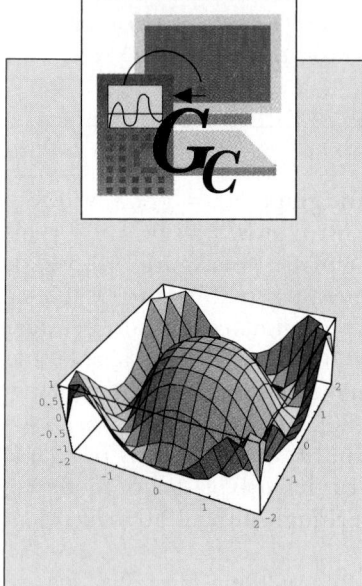

With a computer we can plot the graphs of more complicated functions. For instance, we consider $z = f(x, y) = \cos(x^2 + y^2)$.

MATHEMATICA	MAPLE
Plot3D [Cos [$x^2 + y^2$], $\{x, -2, 2\}$ $\{y, -2, 2\}$] ENTER	plot3d (Cos ($x^2 + y^2$), $x = -2..2$, $y = -2..2$); ENTER

DERIVE	MacFunction
O D G ENTER A COS ($x^2 + y^2$) ENTER P O L 4.0 Tab 4.0 ENTER G 15 Tab 15 ENTER	cos ($x^2 + y^2$) OK Function Change Domain (drag to 2 × 2) OK Options Remove Hidden Surfaces

Now try changing the ranges for x and y. What will the graph of $z = g(x, y) = \sin(x^2 + y^2)$ look like?

Level Curves

There is another technique at our disposal for helping to visualize the graph of a function $f(x, y)$. When meteorologists make a map to show the temperatures around the continent or around the world, they often mark on it what are called *isotherms* (see the illustration on page 928). These are curves that go through all the points that have the same temperature (the prefix "iso" comes from the Greek word "isos," which means "equal"). If isotherms are marked for a number of different temperatures, they give a pretty good idea of how temperatures are distributed over the whole map. We are able to "see" regions where temperatures are particularly warm or particularly cool and how rapidly temperatures change. (If the isotherms are close together, it means that the temperature is changing quickly as one moves past them.)

Meteorologists also make maps to show barometric pressure, and it comes as no surprise that they mark on them what are called *isobars*—curves through points of equal barometric pressure.

A *contour map* is a map on which are marked *contour lines*. These are curves passing through points that are at the same elevation (height above sea level), and with a little practice it is possible to read such a map and obtain a good impression of the nature of the terrain. Specifically, it is easy to recognize high points and low points, ridges and valleys. From the separation between adjacent contour lines it is possible to gauge how quickly the ground is rising or falling. (Again, the closer the lines are to each other, the faster the elevation is changing.) Marine charts often show contour lines for the depth of the ocean floor, enabling sailors to judge the water's depth and to locate dangerous shallows or promising fishing holes.

Figure 16.1.7 shows a contour map of the region around the mouth of a river. On either side of the river we see contour lines rising up from the water level. At first they are fairly far apart, which means that the ground rises slowly. On the right side of the river the terrain rises to a hill (the peak is enclosed inside the two closed loops); on the left side there are two ridges.

Figure 16.1.7

Contour lines on this topographic map show the elevation of the terrain around the Devil's Tower National Monument in Wyoming. The contour lines are close together right around the tower as the ground rises sharply. The lines are much more widely spaced near the meandering Fourche river, where the ground is flatter.

Each of these examples demonstrates a way of visualizing the behavior of some function (the temperature, the atmospheric pressure, the elevation of the earth's surface). If we think of the contour map for a moment, a contour line passes through all the points at a particular height or "level." This suggests the following terminology.

DEFINITION 16.1.1

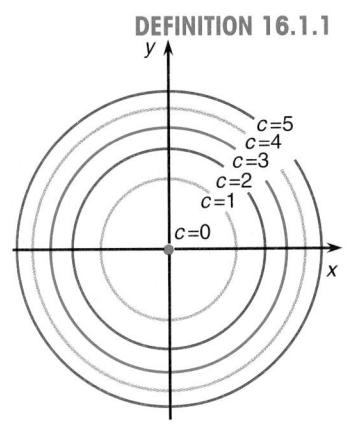

Figure 16.1.8

> Suppose $f(x, y)$ is a function of two variables x and y. Suppose c is some constant, and consider the set of all points (x, y) at which the value of f is c, that is,
>
> $$\{(x, y) : f(x, y) = c\}.$$
>
> This set is called a **level curve** of the function f.

EXAMPLE 1   Discuss the level curves of the function $g(x, y) = x^2 + y^2$.

Solution   The set of points at which $g(x, y) = 1$, for instance, is the set of points (x, y) satisfying $x^2 + y^2 = 1$, which we recognize as the circle of radius 1 around the origin. The level curve associated to the value $c = 2$ is the set of points satisfying $x^2 + y^2 = 2$, which is the circle of radius $\sqrt{2}$ around the origin.

The level curve associated to the value c is the set of all points (x, y) satisfying $x^2 + y^2 = c$, which is the circle around the origin of radius $\sqrt{c}$. Strictly speaking, this only makes sense when $c > 0$. If $c < 0$, there are no points on the level curve, and if $c = 0$, the corresponding level curve consists of exactly one point, the origin $(0, 0)$.

In Figure 16.1.8 we see several level curves. As the values of c increase by regular amounts, 0, 1, 2, 3, and so on, the level curves move away from the origin, but they become less widely spaced the farther they are from the origin. This suggests that the values of the function rise more and more quickly as we move away from the origin.

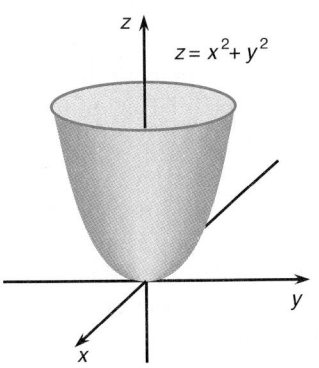

Figure 16.1.9

In Figure 16.1.9 we sketch the graph of $z = g(x, y)$, which is the familiar paraboloid $z = x^2 + y^2$. The level curves suggest that it will have a round shape, and the preceding discussion makes us expect that it will rise faster and faster as we move away from the origin.

The use of level curves may help us to visualize a function or to assess its behavior. It is likely to be easier than sketching the graph.

EXAMPLE 2   Sketch some level curves of the following functions: (i) $h(x, y) = 2x + 3y - 1$, (ii) $k(x, y) = 4x^2 + 8x - 5y^2 + 3$, (iii) $\ell(x, y) = 2x - 3y^2 + 12y$, (iv) $U(x, y) = x + \sin y$.

Solution

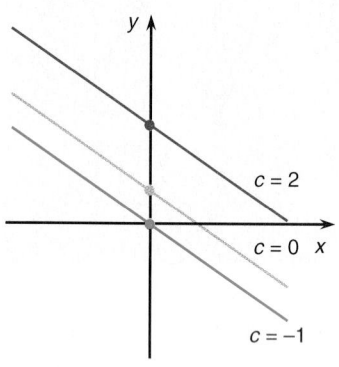

Figure 16.1.10

(i) The level curve for h associated to the value c is the set of all points (x, y) that satisfy $h(x, y) = c$, that is, $2x + 3y - 1 = c$. This amounts to $2x + 3y - (1 + c) = 0$, which is the equation of a straight line.

Notice that for each value c we get a different straight line but that all the lines are parallel. (After all, the slopes are the same, equal to $-\frac{2}{3}$.) The level curves associated to different values of c can never intersect. It is easy enough to sketch these lines; probably the simplest thing to do is to find the y-intercept of the line for each value of c. If c is any fixed number, then the y-intercept of the line $2x + 3y - (1 + c) = 0$ is found by letting $x = 0$ and solving to find that $y = \frac{1+c}{3}$.

So for instance, the y-intercept of the level curve associated to the value $c = 0$ is $\frac{1}{3}$, for $c = 2$ it is 1, for $c = -1$ it is 0, and so on. The lines are easily sketched (see Figure 16.1.10).

(ii) Here it probably helps to complete the square first. We see that $k(x, y) = 4x^2 + 8x - 5y^2 + 3 = 4(x^2 + 2x + 1) - 4 - 5y^2 + 3 = 4x'^2 - 5y^2 - 1$, where $x' = x + 1$.

The set of all points (x, y) at which $k(x, y) = c$ is the set of all points satisfying $4x'^2 - 5y^2 - 1 = c$, which can be written as $4x'^2 - 5y^2 = (c + 1)$. If $c + 1 \neq 0$, this is the equation of a hyperbola. Depending on the sign of $c + 1$, it will have its transverse axis horizontal or vertical. If $c + 1 = 0$, we have the equation of the two asymptotes $4x' = \pm 5y$, or $4(x + 1) = \pm 5y$.

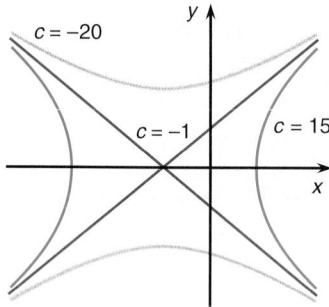

Figure 16.1.11

Each of the hyperbolas described here has its center at the point $O' = (-1, 0)$. When $c + 1 > 0$, that is, when $c > -1$, the transverse axis will be horizontal, and the x'-intercepts will be the points where $y = 0$ and $x' = \pm\frac{1}{2}\sqrt{c + 1}$, that is, $x = x' - 1 = \pm\frac{1}{2}\sqrt{c + 1} - 1$. When $c + 1 < 0$, that is, when $c < -1$, the transverse axis will run vertically between the points $\left(-1, \sqrt{-\frac{c+1}{5}}\right)$ and $\left(-1, -\sqrt{-\frac{c+1}{5}}\right)$.

In Figure 16.1.11 we sketch a number of these hyperbolas, marking each with the corresponding value of c.

(iii) Again we complete the square, writing $\ell(x, y) = 2x - 3y^2 + 12y = 2x - 3(y^2 - 4y) = 2x - 3(y^2 - 4y + 4) + 12 = 2x - 3y'^2 + 12$, with $y' = y - 2$.

For any value of c, $2x - 3y'^2 + 12 = c$ is the equation of a parabola. Writing it as $x = \frac{3}{2}y'^2 - 6 + \frac{c}{2}$, or $x' = \frac{3}{2}y'^2$ with $x' = x + 6 - \frac{c}{2}$, we see that it is a parabola with vertex at the point $O' = \left(\frac{c}{2} - 6, 2\right)$, which opens to the right. We can sketch a number of these parabolas, as in Figure 16.1.12, and we see that they are all just copies of the same parabola, shifted sideways. (In effect, changing c just changes where O', the vertex of the parabola, is.)

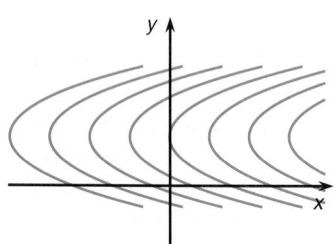

Figure 16.1.12

(iv) Consider what it means for the function $U(x, y) = x + \sin y$ to equal some constant c. This amounts to $x + \sin y = c$, or $x = c - \sin y$. So for instance, if $c = 0$, then $U(x, y) = 0$ means $x = -\sin y$. This is easier to describe than it appears, because it is just the graph of $x = -\sin y$. We usually draw graphs with x as the variable and y as the function, but there is nothing to say that we cannot do it the other way around.

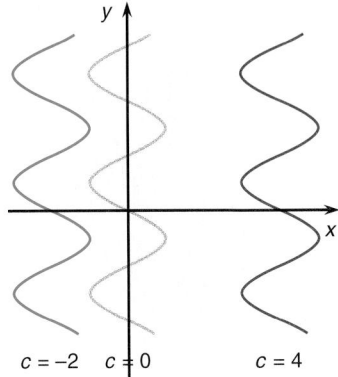

Figure 16.1.13

A convenient way to display a function is to use colors to represent different values. The Total Ozone Mapping Spectrometer aboard the Nimbus-7 satellite measured ozone levels in the northern hemisphere, and the data were used to prepare these images for eight years. The color scale at the bottom shows low total ozone displayed as blue/purple and high total ozone as red/green. The black region in the center of each image is the polar night, where no measurements are made.

Usually there is substantial build-up of total ozone in northern mid- to high-latitudes during late winter and early spring. These images show noticeably less buildup in 1992 than in the earlier years: notice the absence of reds and greens between 50° and 70° north latitude (roughly over Canada) as compared with earlier years.

For that matter, with any c, the graph of $x = c - \sin y$ is the same thing just shifted to the left or to the right. It is easy to draw the level curve; it contains the point $(c, 0)$ and looks like a sine curve drawn vertically. A number of these level curves are sketched in Figure 16.1.13 and labeled with the corresponding values of c.

Level Surfaces

Suppose we want some way to try to visualize the behavior of a function $f(x, y, z)$ of three variables. It is not possible to draw its graph, but it is still possible to consider something analogous to level curves. If c is any constant, we could look at the set of all points (x, y, z) for which $f(x, y, z) = c$.

For instance, if $f(x, y, z) = x^2 + y^2 + z^2$, then the set of all points (x, y, z) at which $f(x, y, z) = 1$ is the set of all points (x, y, z) that satisfy $x^2 + y^2 + z^2 = 1$. This is the sphere of radius 1 with center at the origin.

The set of all points (x, y, z) at which f equals c is the set of points satisfying $x^2 + y^2 + z^2 = c$. Since $x^2 + y^2 + z^2$ is the square of the distance from (x, y, z) to the origin, this is exactly the sphere of radius $\sqrt{c}$ shown in Figure 16.1.14. This sphere is a "surface" (it is the "skin" of the solid ball), and this suggests the following terminology.

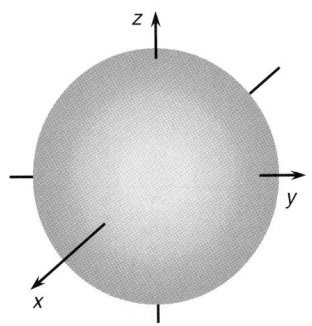

Figure 16.1.14

DEFINITION 16.1.2

Suppose $f(x, y, z)$ is a function of three variables x, y, and z. Suppose c is some constant and consider the set of all points (x, y, z) at which the value of f is c, that is,

$$\{(x, y, z) : f(x, y, z) = c\}.$$

This set is called a **level surface** of the function f.

Figure 16.1.15(i)

Figure 16.1.15(ii)

Figure 16.1.15(iii)

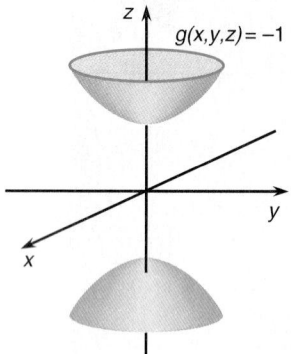

Figure 16.1.15(iv)

EXAMPLE 3   Discuss the level surfaces of the function $g(x, y, z) = x^2 + y^2 - z^2$.

Solution   The set of all points at which g equals 0 is the set of all points satisfying $x^2 + y^2 = z^2$, which is the (double-ended) cone shown in Figure 16.1.15(i).

The level surface associated to $c = 1$ is the set of all points (x, y, z) that satisfy $x^2 + y^2 - z^2 = 1$, or $x^2 + y^2 = 1 + z^2$. Its intersection with any horizontal plane $z = A$ is a circle around the origin. The circles increase in radius the farther the plane is above or below the origin, as seen in Figure 16.1.15(ii). The surface is the hyperboloid of one sheet shown in Figure 16.1.15(iii).

For any positive value of c the corresponding level surface of g will be a hyperboloid of one sheet. The radius of its intersection with the xy-plane is $\sqrt{c}$.

To see what will happen with a negative value of c, consider the special case $c = -1$. The equation of the corresponding level surface is $x^2 + y^2 - z^2 = -1$, or $x^2 + y^2 = z^2 - 1$.

The left side of this equation can never be negative, so the surface lies above the plane $z = 1$ or below the plane $z = -1$ (or perhaps on these planes).

The level surface is the hyperboloid of two sheets shown in Figure 16.1.15(iv). The level surface associated to any negative constant c is also a hyperboloid of two sheets.

The level surfaces of $g(x, y, z) = x^2 + y^2 - z^2$ can be described as follows: The level surface associated to the value $c = 0$ is a cone; the level surface associated to any positive c is a hyperboloid of one sheet, and the level surface corresponding to any negative c is a hyperboloid of two sheets. The main axis of the cone and all the hyperboloids is the z-axis. See Figure 16.1.15(v). ◢

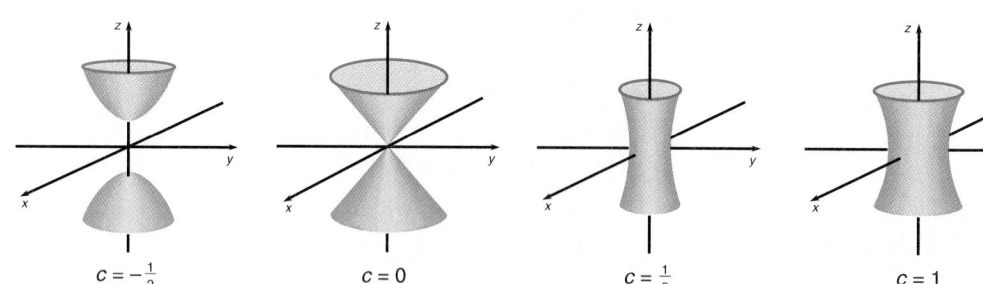

$c = -1$ $c = -\frac{1}{2}$ $c = 0$ $c = \frac{1}{2}$ $c = 1$

Figure 16.1.15(v)

Exercises 16.1

I

Sketch a few level curves of each of the following functions.

1. $f(x, y) = x + y - 4$

2. $f(x, y) = 3x + 2$

3. $f(x, y) = 4y - 7$

4. $f(x, y) = 5$

5. $f(x, y) = x^2 + 2y^2 - 8y + 4$

6. $f(x, y) = 3x^2 - 6x + y + 5$

7. $f(x, y) = \sqrt{x^2 + 2x + y^2 - 4y}$

8. $f(x, y) = x^2 - 2x - 2y^2 + 4y$

9. $f(x, y) = (x^2 + y^2)^5$

10. $f(x, y) = \sin(x + y)$

Sketch a few level surfaces of each of the following functions.

11. $f(x, y, z) = 2x + 2y - 3z$

12. $f(x, y, z) = 2x - 3z$

13. $f(x, y, z) = \sqrt{x^2 + y^2 + z^2}$

14. $f(x, y, z) = y$

15. $f(x, y, z) = x^2 + 2y^2$

16. $f(x, y, z) = 3y^2 + 2z^2$

17. $f(x, y, z) = x^2 - y^2 - z^2$

18. $f(x, y, z) = 2x^2 + y^2 + z^2$

19. $f(x, y, z) = 2x^2 + y^2 - z^2$

20. $f(x, y, z) = x^2 + y^2 + \sin z$

II

21. Sketch the level curves of the function $f(x, y) = y^3 + 1$.

22. Sketch the level curves of the function $f(x, y) = \cos x$.

*23. A function $f(x, y)$ is *independent of x* if its value does not depend on x. This means that $f(x, y) = f(x', y)$ whenever both (x, y) and (x', y) are in the domain of f; that is, as long as the value of y does not change, neither does the value of the function. If $f(x, y)$ is independent of x, what can you say about its level curves?

*24. (i) Define what it means for a function $g(x, y)$ to be independent of y.

(ii) Show that the function $g(x, y) = x^2 - 3x(\cos^2 y + \sin^2 y)$ is independent of y.

(iii) Sketch several level curves of the function $g(x, y)$.

III

Use a computer to sketch the graphs of the following functions.

25. $z = f(x, y) = x^2 + 2y^2$

26. $z = x^2 - y^2$

27. $z = xy$

28. $z = x^2 + 2x + y^2 - 4y$

29. $z = y - x + \sin \pi y$

30. $z = \sin(\sqrt{x^2 + y^2})$

*31. For each of the functions in Exercises 25–30, for which values of c is the level curve associated to c a nonempty set? For which value(s) is it a single point?

P ◂ POINT TO PONDER

In Definition 16.1.1 we defined the level curves of a function $f(x, y)$. This terminology can be inappropriate because that set may not be a curve. To take an extreme example, suppose $f(x, y)$ is the function that equals 0 when $x > 0$, equals -2 when $x < 0$, and equals y when $x = 0$:

$$f(x, y) = \begin{cases} 0, & \text{if } x > 0, \\ -2, & \text{if } x < 0, \\ y, & \text{if } x = 0. \end{cases}$$

The level curve associated to the value $c = 0$ consists of all points (x, y) with $x > 0$, together with the point $(0, 0)$. In other words it is the union of the right half-plane and the set consisting of the origin.

The level curve associated to the value $c = -2$ is the union of the left half-plane and the set consisting of the single point $(0, -2)$. The level curve associated to any other value c consists of exactly one point, namely, $(0, c)$.

By rights, none of these sets should be called a curve. On the other hand, this function was deliberately made up to look terrible. For most "reasonable" functions the level curves *usually* are curves, with the obvious exception that if $f(x, y)$ never equals c, then the corresponding level curve is the empty set.

16.2 Limits and Continuity

PUNCTURED DISK
PROPERTIES OF LIMITS
CONTINUITY
CONTINUITY OF POLYNOMIALS

In Section 16.1 we began to consider functions of several variables, realizing that they arise quite naturally in the study of many different types of problems. In the rest of this chapter (and the next two) we will see how to apply the techniques of calculus to functions of this kind. There are both similarities and differences between this subject and the single-variable calculus that we have already studied, but as with functions of a single variable, one of the very first steps is to discuss the ideas of limits and continuity.

Let us begin with a function $f(x, y)$ of two variables. If (a, b) is some point in the plane, we want to discuss what it will mean to say that $f(x, y)$ approaches the limit L as (x, y) approaches the point (a, b). Roughly speaking, it ought to mean that the value of $f(x, y)$ gets closer and closer to L as (x, y) gets closer and closer to (a, b). When we talk about the value of $f(x, y)$ getting close to L, presumably we mean that $|f(x, y) - L|$ gets small. When we say that (x, y) is close to (a, b), we mean that the distance between these points, $\sqrt{(x - a)^2 + (y - b)^2}$, is small. However, there is a complication here that did not appear in the case of a function of one variable. It has to do with the possibility of approaching a point in the plane from many different directions. If $f(x, y)$ has a limit as (x, y) approaches (a, b), this should mean that it has the same limit no matter from which direction the approach is made.

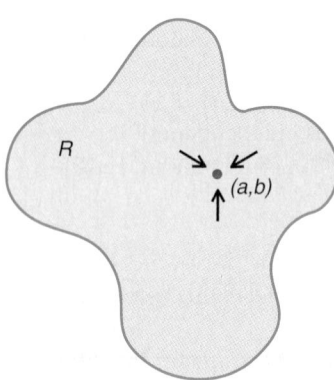

Figure 16.2.1(i)

In particular it should make sense to evaluate $f(x, y)$ as (x, y) approaches (a, b) from any direction. Because of this it is customary to insist that $f(x, y)$ be defined "all around" the point (a, b) if we want to discuss its limit as $(x, y) \to (a, b)$. Suppose the domain of $f(x, y)$ is the region R. If we look at a point (a, b) that is actually *inside* the region (i.e., not on the boundary), then if we approach it from any direction, the point (x, y) will have to be in the domain of f once it gets close enough to (a, b) (see Figure 16.2.1(i)). Even if (a, b) is quite close to the edge, as long as (x, y) is closer than the edge, it will be in the domain of f. We will avoid limits at points on the boundary of R.

Another way of saying this is that it must be possible to draw a small disk around (a, b) so that the whole disk is in the domain of f, except possibly the point (a, b) itself, as shown in Figure 16.2.1(ii). If (a, b) is close to the edge of the region, then it will be necessary to draw a very small disk, but it will still be possible. On the other hand it is clearly impossible to draw a disk around the boundary point P in Figure 16.2.1(ii) so that the disk is entirely contained in the domain R. For that matter it cannot be done for *any* point on the boundary.

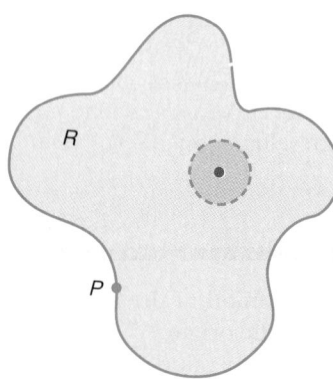

Figure 16.2.1(ii)

With a function of one variable, to consider the limit of $g(x)$ as $x \to a$, say, we did not require that $g(x)$ be defined at $x = a$ itself. With two variables we should not require that the function $f(x, y)$ must be defined at (a, b), but we do insist that it be defined at every other point in some disk around (a, b). A disk with its center removed like this is often called a **punctured disk** (see Figure 16.2.2). So the requirement is that it must be possible to find a punctured disk around (a, b) that is entirely contained in the domain of f. The idea is that as (x, y) approaches (a, b), eventually it will be close enough that it is contained in the punctured disk, and this means that it is in the domain of f, so it makes sense to talk about $f(x, y)$.

We now give the definition of limit. As in Chapter 2 we attempt at first to convey the essential idea without being too complicated, and at the end of the section we will give a precise definition in the same spirit as the precise Definition 2.7.1 of limits with one variable. As in Chapter 2 this precise definition will not be used later.

Figure 16.2.2

DEFINITION 16.2.1

Limits (First Version)

Suppose the function $f(x, y)$ is defined at every point in some punctured disk around (a, b). Then we write

$$\lim_{(x,y) \to (a,b)} f(x, y) = L$$

to mean that the values of $f(x, y)$ approach the limit L as $(x, y) \to (a, b)$, that is, $f(x, y)$ tends to L as (x, y) approaches (a, b) (without actually equaling (a, b)).

Working with limits can be tricky.

EXAMPLE 1   Show that the function

$$f(x, y) = \frac{x^2 - y^2}{x^2 + y^2}$$

does *not* have a limit as $(x, y) \to (0, 0)$.

Figure 16.2.3(i)

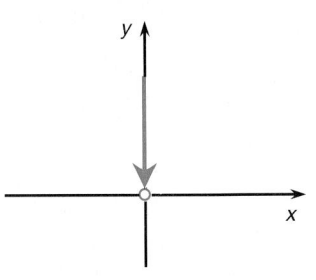

Figure 16.2.3(ii)

Solution   Let us consider what happens as a point approaches the origin along the x-axis, as shown in Figure 16.2.3(i). A point on the x-axis must be of the form $(x, 0)$, and it is easy to evaluate $f(x, 0) = \frac{x^2 - 0}{x^2 + 0} = \frac{x^2}{x^2} = 1$, provided $x \neq 0$. So for any point $(x, 0)$ on the x-axis except the origin $(0, 0)$, the value of $f(x, 0)$ is 1. So certainly the function $f(x, 0)$ "approaches" the value 1 as $(x, 0)$ approaches the origin along the x-axis.

On the other hand we could consider what happens as a point approaches the origin along the y-axis, as shown in Figure 16.2.3(ii). If $(0, y)$ is on the y-axis with $y \neq 0$, then we can easily find $f(0, y) = \frac{0 - y^2}{0 + y^2} = \frac{-y^2}{y^2} = -1$. So as the point $(0, y)$ approaches the origin along the y-axis, the values of $f(0, y)$ "approach" the value -1.

Depending on which axis we choose for approaching the origin, the values of the function $f(x, y)$ will approach 1 or -1. This uncomfortable observation shows that the function $f(x, y)$ does *not* approach any limit as $(x, y) \to (0, 0)$.   ◢

Notice that the function $f(x, y)$ in Example 1 is not defined at the point $(0, 0)$ and that we had to be careful to avoid trying to evaluate the function at $(0, 0)$. (For instance, when evaluating at $(x, 0)$ it was necessary to specify that $x \neq 0$.)

In the exercises we will see that it is actually possible to approach many other values by approaching the origin along other straight lines.

The basic properties of limits that allow us to work with them are virtually identical with the corresponding properties for functions of one variable. Limits behave the way we expect if we add, subtract, multiply, or divide functions. Constant functions have the obvious limits.

In dealing with functions of one variable we needed to know the basic fact that $\lim_{x \to a} x = a$. When there are two variables, x and y, it is slightly less clear what to do. Consider the functions $f(x, y) = x$ and $g(x, y) = y$. Each of them is a function of two variables. (Of course, f does not actually depend on y and g does not depend on x, but they are both functions that are defined on the whole plane.)

Now suppose $(x, y) \to (a, b)$. We can guess what ought to happen to $f(x, y)$ and $g(x, y)$. If the point (x, y) gets closer and closer to (a, b), then its coordinates must approach the coordinates of (a, b). In other words, $x \to a$ and $y \to b$. This suggests $\lim_{(x,y) \to (a,b)} f(x, y) = \lim_{(x,y) \to (a,b)} x = a$, and $\lim_{(x,y) \to (a,b)} g(x, y) = \lim_{(x,y) \to (a,b)} y = b$.

We summarize these and other facts about limits.

THEOREM 16.2.2

Properties of Limits

Limits of functions of two variables satisfy the following properties, which are very similar to the familiar properties of functions of one variable.

Suppose c is a constant, (a, b) is a point, and suppose $f(x, y)$, $g(x, y)$ are functions that satisfy

$$\lim_{(x,y) \to (a,b)} f(x, y) = L, \qquad \lim_{(x,y) \to (a,b)} g(x, y) = M.$$

Then

(i) $\lim_{(x,y) \to (a,b)} \left(f(x, y) + g(x, y) \right) = L + M$
$\lim_{(x,y) \to (a,b)} g(x, y),$ $\qquad = \lim_{(x,y) \to (a,b)} f(x, y) +$

(ii) $\lim_{(x,y) \to (a,b)} \left(c f(x, y) \right) = cL = c\left(\lim_{(x,y) \to (a,b)} f(x, y) \right),$

(iii) $\lim_{(x,y) \to (a,b)} \left(f(x, y) g(x, y) \right) = LM$
$\left(\lim_{(x,y) \to (a,b)} g(x, y) \right),$ $\qquad = \left(\lim_{(x,y) \to (a,b)} f(x, y) \right)$

(iv) $\lim_{(x,y) \to (a,b)} \left(\frac{f(x,y)}{g(x,y)} \right) = \frac{L}{M} = \frac{\lim_{(x,y) \to (a,b)} f(x,y)}{\lim_{(x,y) \to (a,b)} g(x,y)},$ $\qquad$ provided $M \neq 0,$

(v) $\lim_{(x,y) \to (a,b)} c = c,$

(vi) $\lim_{(x,y) \to (a,b)} x = a,$

(vii) $\lim_{(x,y) \to (a,b)} y = b.$

As with functions of one variable these properties make it easy to evaluate limits of polynomials and limits of rational functions at points where the denominator is not zero.

For instance, as $(x, y) \to (1, -3)$, we know that $x \to 1$ and $y \to -3$, by parts (vi) and (vii) of Theorem 16.2.2. From part (iii) we can conclude that $xy \to (1)(-3) = -3$ and $xy^2 \to 1(-3)^2 = 9$. From part (ii) we see that $3xy^2 \to 3(1)(-3)^2 = 27$. Using results like these and part (i) of Theorem 16.2.2 we see that a complicated expression like $3xy^2 + 2x^3 - 6x^7y^2$ will tend to the obvious limit, namely, $3(1)(-3)^2 + 2(1)^3 - 6(1)^7(-3)^2 = 27 + 2 - 54 = -25$. This is the value at the point $(1, -3)$ of the polynomial $3xy^2 + 2x^3 - 6x^7y^2$. The corresponding result holds for rational functions too, except that as usual we have to insist that the denominator be nonzero.

THEOREM 16.2.3

If $P(x, y)$ is any polynomial in the two variables x and y, then for any point (a, b),

$$\lim_{(x,y) \to (a,b)} P(x, y) = P(a, b).$$

If $R(x, y) = \frac{P(x,y)}{Q(x,y)}$ is any rational function and if (a, b) is any point for which $Q(a, b) \neq 0$, then

$$\lim_{(x,y) \to (a,b)} R(x, y) = R(a, b).$$

EXAMPLE 2   Evaluate the following limits, provided they exist:

(i) $\lim_{(x,y)\to(-2,4)} (6x^3 - 3xy + y - 3)$,

(ii) $\lim_{(x,y)\to(-1,-1)} (7xy^4 - x^2y + y^2 - x + 3)$,

(iii) $\lim_{(x,y)\to(1,0)} \left(\frac{xy^2+3x}{xy-1}\right)$,

(iv) $\lim_{(x,y)\to(2,-1)} \left(\frac{x^2-3xy^2+2}{x+2y}\right)$.

Solution

(i) Since $6x^3 - 3xy + y - 3$ is a polynomial, we can find the limit by simply evaluating it at the point $(-2, 4)$. We find that $\lim_{(x,y)\to(-2,4)} (6x^3 - 3xy + y - 3) = 6(-2)^3 - 3(-2)(4) + 4 - 3 = -23$.

(ii) $\lim_{(x,y)\to(-1,-1)} (7xy^4 - x^2y + y^2 - x + 3) = 7(-1)(-1)^4 - (-1)^2(-1) + (-1)^2 - (-1) + 3 = -1$.

(iii) Here we have a rational function. We check that the denominator $xy - 1$ is nonzero at the point $(1, 0)$ (it equals -1 there) and then evaluate: $\lim_{(x,y)\to(1,0)} \left(\frac{xy^2+3x}{xy-1}\right) = \frac{1(0^2)+3(1)}{1(0)-1} = \frac{3}{-1} = -3$.

See the Computer Example on the next page.

(iv) In this rational function the denominator is $x + 2y$. At the point $(2, -1)$ this denominator equals 0, so we cannot use the technique of evaluating the function to find the limit. However, the numerator of the function is also zero at $(2, -1)$. In the case of functions of one variable the Factor Theorem (Theorem 1.3.5) says that it is possible to cancel out something from the numerator and denominator when they are both zero at the same point.

   It is important to realize that this is *not necessarily true* for polynomials of two or more variables. After all, if we could cancel, it would mean in this case that the denominator divides the numerator (since the degree of the denominator is 1, so it cannot be equal to a product). But that is not true. Notice that the denominator is zero at the origin $(0, 0)$, but the numerator is not zero there. But if the denominator divided the numerator, we could write the numerator as $(x + 2y)P(x, y)$, for some polynomial $P(x, y)$, and this would certainly equal 0 at the origin.

   Incidentally, we notice that the denominator is zero on the line $x + 2y = 0$, which passes through the point $(2, -1)$. Because of this the quotient is not defined at any point on this line, and because of that it is impossible to draw a punctured disk around $(2, -1)$ so that the quotient is defined on every point in the punctured disk. Every disk around $(2, -1)$ will contain a piece of the line (see Figure 16.2.4).

   In order for the limit to exist it is necessary that the function be defined at every point in some punctured disk around the limit point, so this limit cannot exist.

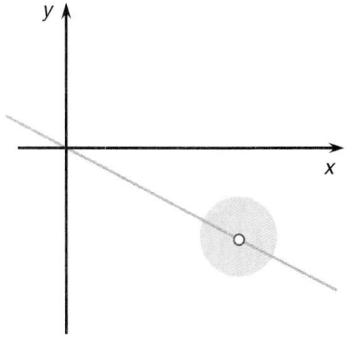

Figure 16.2.4

One other extremely useful result in our study of limits of functions of one variable was Theorem 2.3.2, which concerns limits of composite functions. It says that if $\lim_{x\to a} f(x) = L$, $\lim_{t\to L} g(t) = M$, and $g(L) = M$, then the limit of the composite function is $\lim_{x\to a} g(f(x)) = M$. In other words, if $f(x)$ approaches L, then $g(f(x))$ should approach the limit that $g(t)$ approaches as $t \to L$. We used this fact to help us evaluate limits of functions like $\sin(3x^2 - 4x + 2)$, or $e^{\sin(3x-2)}$.

   It will be very helpful to have a similar result that enables us to evaluate limits of functions like $\sin(x + y)$, or $e^{x \cos y}$.

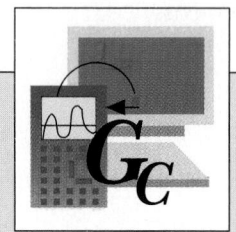

With a computer we can plot the graph of the function from part (iv) of Example 2.

MATHEMATICA	MAPLE
Plot3D [$(x\hat{\ }2 - 3*x*y\hat{\ }2 + 2)/$ $(x + 2y), \{x, 1, 3\}, \{y, -2, 0\}]$ enter	plot3d $((x\hat{\ }2 - 3*x*y\hat{\ }2 + 2)/$ $(x + 2*y), x = 1..3, y = -2..0);$ return
DERIVE	MacFunction
O D G ENTER A ($x\hat{\ }2 - 3x*y\hat{\ }2$ $+ 2$) / ($x + 2y$) ENTER P O L 2.0 Tab 2.0 Tab 20 ENTER C 2.0 Tab -1 ENTER G 15 Tab 15 ENTER	($x\hat{\ }2 - 3*x*y\hat{\ }2 + 2$) / ($x + 2*y$) OK Function Change Domain (drag to 3×3) OK Options Remove Hidden Surfaces Options Autoscaling

Observe how the function behaves erratically as we approach the line $x + 2y = 0$.

THEOREM 16.2.4

Suppose $f(x, y)$ is a function of two variables and $g(t)$ is a function of one variable. Suppose that $\lim_{(x,y)\to(a,b)} f(x, y) = L$, $\lim_{t\to L} g(t) = M$, and $g(L) = M$ (i.e., $g(t)$ is continuous at $t = L$). Then

$$\lim_{(x,y)\to(a,b)} g\big(f(x, y)\big) = M.$$

EXAMPLE 3   Evaluate (i) $\lim_{(x,y)\to(\frac{\pi}{2},0)} \sin(x + y)$, (ii) $\lim_{(x,y)\to(3,-\frac{\pi}{2})} e^{x\cos y}$.

Solution

(i) The function $\sin t$ is continuous at every t, and by Theorem 16.2.3 the limit of the polynomial $f(x, y) = x + y$ is $\frac{\pi}{2} + 0 = \frac{\pi}{2}$. As $(x, y) \to \left(\frac{\pi}{2}, 0\right)$, we see that $x + y$ approaches $\frac{\pi}{2}$ and $\sin(x + y)$ approaches $\sin\left(\frac{\pi}{2}\right)$, that is, $\lim_{(x,y)\to(\frac{\pi}{2},0)} \sin(x + y) = \sin\frac{\pi}{2} = 1$. We could actually have done most of this calculation in our heads.

(ii) With $f(x, y) = x\cos y$ there is a little more work to do. We can easily guess what the limit of f is as $(x, y) \to \left(3, -\frac{\pi}{2}\right)$, but for instance we do not really know that $\cos y$ has a limit, since we are thinking of it as a function of *two* variables.

Happily, the theorem comes to our rescue. We do know that $F(x, y) = y$ does have a limit, by Theorem 16.2.2(vii), and since the function $G(t) = \cos t$ is a continuous function of one variable, the theorem says that $G\big(F(x, y)\big) = \cos(y)$ has a limit and that it equals the value we expect, namely, $\lim_{(x,y)\to(a,b)} \cos\big(F(x, y)\big) = \cos\big(F(a, b)\big) = \cos b$.

Then by Theorem 16.2.2(iii) we see that the product $x \cos y$ also has a limit as (x, y) approaches any point (a, b) and that limit is what it should be, namely, $a \cos b$.

Finally, we can apply Theorem 16.2.4 again, with $f(x, y) = x \cos y$ and $g(t) = e^t$. We find that $\lim_{(x,y)\to(3,-\frac{\pi}{2})} e^{x \cos y} = e^{3\cos(-\pi/2)} = e^{3(0)} = e^0 = 1$. ◢

The argument used in the solution of part (ii) of Example 3 can be used in many other situations. If we have two functions f and g of a *single* variable, then it is easy to construct a function of two variables by considering $H(x, y) = f(x)g(y)$. If $\lim_{x\to a} f(x) = L$ and $\lim_{y\to b} g(y) = K$, then the reasoning used in Example 3 can be used to show that not only does the function $H(x, y)$ have a limit as $(x, y) \to (a, b)$, but the limit is what we expect, LK. We record this useful fact below.

PROPOSITION 16.2.5

Suppose $f(x)$ and $g(y)$ are functions satisfying

$$\lim_{x\to a} f(x) = L = f(a), \qquad \lim_{y\to b} g(y) = K = g(b)$$

(i.e., $f(x)$ is continuous at $x = a$ and $g(y)$ is continuous at $y = b$). Then the function $H(x, y) = f(x)g(y)$ has a limit as $(x, y) \to (a, b)$, and

$$\lim_{(x,y)\to(a,b)} H(x, y) = \lim_{(x,y)\to(a,b)} \big(f(x)g(y)\big) = LK = \Big(\lim_{x\to a} f(x)\Big)\Big(\lim_{y\to b} g(y)\Big).$$

EXAMPLE 4   Find each of the following limits, if it exists:
(i) $\lim_{(x,y)\to(-3,1)} \big(x^3 \tan(\pi y)\big)$, (ii) $\lim_{(x,y)\to(\frac{3}{2},4)} \frac{xy-3}{\cot(\pi x)}$, (iii) $\lim_{(x,y)\to(0,0)} \frac{xy}{2x^2+3y^2}$.

Solution

(i) The function $x^3 \tan(\pi y)$ is the product of a function of x and a function of y. As $x \to -3$, the function x^3 approaches the value $(-3)^3 = -27$. As $y \to 1$, the function $\tan(\pi y) \to \tan(\pi) = 0$.

The product of these functions approaches the product of the limits, that is, $\lim_{(x,y)\to(-3,1)} \big(x^3 \tan(\pi y)\big) = -27(0) = 0$.

(ii) As $x \to \frac{3}{2}$, we know that $\cot(\pi x) \to \cot\big(\frac{3\pi}{2}\big) = 0$. Since the denominator of the function tends to zero, we cannot simply evaluate to find the limit. The numerator $xy - 3$ is a polynomial, so it approaches $\big(\frac{3}{2}\big)(4) - 3 = 3$. When the numerator of a quotient approaches 3, a nonzero number, and the denominator approaches 0, the quotient does not approach any (finite) limit.

> In fact, the function is not defined on any punctured disk around $\big(\frac{3}{2}, 4\big)$, since the denominator is zero on the line $x = \frac{3}{2}$. So the limit does not exist.

(iii) As $(x, y) \to (0, 0)$, both the numerator xy and the denominator $2x^2 + 3y^2$ tend to 0, so it is not clear what will happen.

We can try looking to see what happens as (x, y) approaches the origin along various straight lines. If we look at the point $(x, 0)$ on the x-axis, with $x \neq 0$, then the value of the function at this point is $\frac{x(0)}{2x^2+0} = \frac{0}{2x^2} = 0$. If we look at the point $(0, y)$ on the y-axis, with $y \neq 0$, then the value of the function is $\frac{0(y)}{0+3y^2} = 0$. We see that as $(x, y) \to (0, 0)$ along either axis, the limit is 0.

This is encouraging, but let us try letting (x, y) approach the origin along the line $y = x$. Consider the point (x, x), with $x \neq 0$. The value of the function at this point is $\frac{x(x)}{2x^2+3x^2} = \frac{x^2}{5x^2} = \frac{1}{5}$. So all along the line $y = x$, except right at the origin, the value of the function is $\frac{1}{5}$, which certainly means that the limit as (x, y) approaches $(0, 0)$ along this line must be $\frac{1}{5}$.

Since this is different from the limits we found by approaching along the axes, we have to conclude that the limit does not exist. ◢

In part (ii) of Example 4 we might have asked about the possibility of using infinite limits. The numerator tends to 3, so if (x, y) is close enough to $\left(\frac{3}{2}, 4\right)$, the numerator will be positive (cf. Theorem 2.3.4, p. 86). Since the denominator, $\cot(\pi x)$, is negative as $x \to \frac{3}{2}^+$ and positive as $x \to \frac{3}{2}^-$, the quotient will sometimes be positive and sometimes negative. If we approach $\left(\frac{3}{2}, 4\right)$ "from the left," the function will tend to $+\infty$, and if we approach "from the right," the function will tend to $-\infty$. If we approach along the line $x = \frac{3}{2}$, the function is undefined. It is impossible to say that the limit is either ∞ or $-\infty$.

Notice that there is no simple way of talking about one-sided limits because there are many different directions from which it is possible to approach any point (see Figure 16.2.5). This is more complicated than the situation with functions of a single variable, for which there were essentially only two directions.

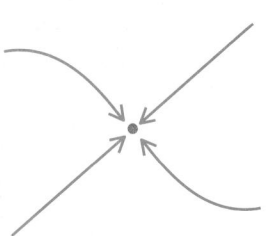

Figure 16.2.5

The other thing to notice in Example 4 was the way we could have been fooled in part (iii). For the first two lines we picked, the limits as (x, y) approaches the origin along these lines (the axes) both exist and both equal zero. It was tempting to jump to the conclusion that the limit exists and equals zero. However, another choice of line led us to a different limit, which shows that there is no limit for the function of two variables.

The situation is this: If we find a line along which there is no limit, or if we find two lines along which there are two different limits, then the function does not have a limit. Testing limits along lines can be used to show that the limit does *not* exist. But even if we check dozens of lines and find the same limit along each of them, we cannot be certain that we have not missed some other line along which the limit is different (or does not exist). And worse, even if we find a way to check the limit along *every* line, it could be that the limit is different if we approach along some other *curve*. Checking limits along lines can never be used to prove that a limit does exist.

> In fact, in one of the exercises we will see an example of a function that has the same limit as we approach the origin along *any* straight line but still does not have a limit. Evaluating limits along lines can only be used to show that a limit does not exist.

Functions of Three or More Variables

We have been working with functions of two variables and have seen that the study of limits is definitely different from what happens with only one variable, even though the general outlines are fairly similar. If we move to functions of three variables, we find that limits behave almost exactly as they do with two variables, and for that matter, there are no real differences with even larger numbers of variables. (The one new complication is that drawing pictures becomes difficult with three variables and impossible with more than three.)

We recall from Theorem 14.1.1 that if $P = (x, y, z)$ and $A = (a, b, c)$ are two points in three-dimensional space, then the distance between them is

$$|AP| = \sqrt{(x - a)^2 + (y - b)^2 + (z - c)^2}.$$

If $f(x, y, z)$ is a function of the three variables x, y, and z, and (a, b, c) is a point, then to say that $\lim_{(x,y,z) \to (a,b,c)} f(x, y, z) = L$ means that $|f(x, y, z) - L|$ becomes small as (x, y, z) gets close to (a, b, c), that is, when the distance $\sqrt{(x - a)^2 + (y - b)^2 + (z - c)^2}$ between (x, y, z) and (a, b, c) becomes small. As with functions of two variables, it is necessary to assume that the function is defined "all around" the limit point (a, b, c). This means that f must be defined at every point (x, y, z) in some punctured ball around the point (a, b, c). A *ball* means the set of points whose distance from the center (a, b, c) is less than some fixed number r, the *radius*, that is, the set

$$\{(x, y, z) : \sqrt{(x - a)^2 + (y - b)^2 + (z - c)^2} < r\}.$$

A *punctured ball* means a ball without its center, that is, the set

$$\{(x, y, z) : 0 < \sqrt{(x - a)^2 + (y - b)^2 + (z - c)^2} < r\}.$$

Suppose $x_1, x_2, \ldots, x_n$ are n variables, $f(x_1, x_2, \ldots, x_n)$ is a function of n variables, and $(a_1, a_2, \ldots, a_n)$ is a point. Then to say that

$$\lim_{(x_1, x_2, \ldots, x_n) \to (a_1, a_2, \ldots, a_n)} f(x_1, x_2, \ldots, x_n) = L$$

means that $|f(x_1, x_2, \ldots, x_n) - L|$ becomes small as the distance

$$\sqrt{(x_1 - a_1)^2 + (x_2 - a_2)^2 + \ldots + (x_n - a_n)^2}$$

between the points $(x_1, x_2, \ldots, x_n)$ and $(a_1, a_2, \ldots, a_n)$ becomes small. Once again it is necessary to assume that the function f is defined on a *punctured ball* around $(a_1, a_2, \ldots, a_n)$; a *ball* means the set of all points whose distance from $(a_1, a_2, \ldots, a_n)$ is less than some positive number r, the *radius*, and a *punctured ball* means a ball without its center.

The properties of limits that we listed in Theorem 16.2.2 are essentially the same for functions of any number of variables. The formulas are nearly identical but more cumbersome to write down, so we summarize without writing all the details.

THEOREM 16.2.6

> **Properties of Limits for Functions of n Variables**
>
> Limits can be added, multiplied, and divided (as long as the denominator is nonzero). The formulas are analogous to Theorem 16.2.2(i)–(v). The result corresponding to 16.2.2(vi) and 16.2.2(vii) is
>
> $$\lim_{(x_1, x_2, \ldots, x_n) \to (a_1, a_2, \ldots, a_n)} x_i = a_i, \qquad i = 1, 2, \ldots, n.$$

THEOREM 16.2.7

> Polynomials of n variables behave the same way as polynomials of two variables (cf. Theorem 16.2.3). To find the limit of a polynomial, we only have to evaluate it at the limit point. The same is true for rational functions at all points where they are defined. In particular the limit exists at every point at which the denominator is not zero.

The other two results we have discussed for limits of functions of two variables also hold for functions of any number of variables, with the obvious modifications.

THEOREM 16.2.8

Suppose $f(x_1, x_2, \ldots, x_n)$ is a function of n variables and $g(t)$ is a function of one variable. Suppose that $\lim_{(x_1, \ldots, x_n) \to (a_1, \ldots, a_n)} f(x_1, \ldots, x_n) = L$, $\lim_{t \to L} g(t) = M$, and $g(L) = M$. Then

$$\lim_{(x_1, \ldots, x_n) \to (a_1, \ldots, a_n)} g(f(x_1, \ldots, x_n)) = M.$$

This result is analogous to Theorem 16.2.4.

THEOREM 16.2.9

Suppose that $f_1(x_1), f_2(x_2), \ldots, f_n(x_n)$ are all functions of one variable and that each f_i is continuous at $x_i = a_i$, with $L_i = f_i(a_i)$. Then the product function $F(x_1, \ldots, x_n) = f_1(x_1) f_2(x_2) \times \ldots \times f_n(x_n)$ has a limit as $(x_1, \ldots, x_n) \to (a_1, \ldots, a_n)$, and

$$\lim_{(x_1, \ldots, x_n) \to (a_1, \ldots, a_n)} F(x_1, \ldots, x_n) = L_1 L_2 \times \ldots \times L_n.$$

This just says that the limit of a product of functions of single variables equals the product of the limits, which is the n-variable version of Proposition 16.2.5.

Using these results it is possible to evaluate many limits of functions of many variables.

EXAMPLE 5 Evaluate the following limits, if they exist:

(i) $\lim_{(x,y,z) \to (2,-1,2)} (xyz^2 - 3xz - 2)$,

(ii) $\lim_{(w,x,y,z) \to (0,3,-2,1)} (y \cos w - 4 \cosh(yz))$,

(iii) $\lim_{(x_1, x_2, \ldots, x_n) \to (2, 2^2, \ldots, 2^n)} \frac{(x_1 x_2 \times \ldots \times x_n)^2}{x_1^{n(n+1)}}$,

(iv) $\lim_{(x,y,z) \to (0,0,0)} \frac{xy + yz - 2xz}{2x^2 + 3y^2 + 5z^2}$.

Solution

(i) The function $xyz^2 - 3xz - 2$ is a polynomial in the three variables x, y, and z. By Theorem 16.2.6 we find its limit by evaluating at the point $(2, -1, 2)$. We find that $\lim_{(x,y,z) \to (2,-1,2)} (xyz^2 - 3xz - 2) = (2)(-1)(2)^2 - 3(2)(2) - 2 = -22$.

(ii) We can deal with each term separately because limits can be added. Consider first the function $y \cos w$. We can think of this as the product of functions of one variable, namely, the function $f_1(w) = \cos w$ and the function $f_3(y) = y$. There are no x's or z's in the function, but we could allow for this by letting $f_2(x) = 1$ and $f_4(z) = 1$ both be the constant function. Then $y \cos w$ equals the product $f_1(w) f_2(x) f_3(y) f_4(z)$. We use Theorem 16.2.9 to find that its limit is the product of the limits. Since $f_1(w) = \cos w \to \cos 0 = 1$, $f_2(x) = 1$, $f_3(y) = y \to -2$, and $f_4(z) = 1$, the limit of $y \cos w$ is $(1)(1)(-2)(1) = -2$, the product of the limits of y and $\cos w$.

Next we consider $\cosh(yz)$. Since yz is a polynomial, its limit is $(-2)(1) = -2$. Since $-4 \cosh(t)$ is a continuous function of one variable, we can apply Theorem 16.2.8 (about composite functions) and find that $\lim_{(w,x,y,z) \to (0,3,-2,1)} (-4 \cosh(yz)) = -4 \cosh(-2) = -2(e^{-2} + e^2)$.

Adding these two results, we find that $\lim_{(w,x,y,z) \to (0,3,-2,1)} (y \cos w - 4 \cosh(yz)) = -2 - 2(e^{-2} + e^2) \approx -17.05$.

(iii) In this case the function is a rational function. Its denominator $x_1^{n(n+1)}$ is certainly not zero at the limit point $(2, 2^2, \ldots, 2^n)$, the point whose ith coordinate is the ith power of 2. So we can use the part of Theorem 16.2.6 about quotients and find the limit by evaluating: $\lim_{(x_1,x_2,\ldots,x_n)\to(2,2^2,\ldots,2^n)} \frac{(x_1 x_2 \times \ldots \times x_n)^2}{x_1^{n(n+1)}} =$ $\frac{\left(2(2^2)(2^3)\times\ldots\times2^n\right)^2}{2^{n(n+1)}} = \frac{(2^{(1+2+3+\ldots+n)})^2}{2^{n(n+1)}}$. This looks pretty awful until we remember Formula 5.2.5 for the sum of the first n positive integers, $1+2+3+\ldots+n = \frac{n(n+1)}{2}$. Using this, we can rewrite the numerator of the limit as $(2^{n(n+1)/2})^2 = 2^{n(n+1)}$. Having done this, we find that the numerator equals the denominator, and the whole limit equals 1.

(iv) This time we have a rational function whose denominator equals zero at the limit point. The numerator also equals zero, so it is difficult to know what to do. The sensible thing to do first is to try to show that the limit does not exist because if that is true, it will likely be reasonably easy to see why.

When we tried to show that limits did not exist for functions of two variables, we did it by looking only at the points in some lines in the plane. Here we might make things easier by looking at points in some particular plane.

For instance, if we let $z = 0$, this means that we are working in the horizontal xy-plane. Since $z = 0$, the function becomes easier to work with. It is $\frac{xy}{2x^2+3y^2}$. This is exactly the function of two variables that we considered in Example 4(iii). There we showed that it does not have a limit as $(x, y) \to (0, 0)$. This shows that our function of three variables does not have a limit either, because if it did, it would have to approach that limit as $(x, y, z) \to (0, 0, 0)$ in every possible direction, including the directions in the xy-plane. We know that it does not do this, so the limit cannot exist.

Continuity

One of the uses of limits in Chapter 2 was the idea of continuity. For functions of several variables we can ask exactly the same questions: Does a function $f(x_1, x_2, \ldots, x_n)$ approach a limit as $(x_1, x_2, \ldots, x_n) \to (a_1, a_2, \ldots, a_n)$, and if so, does the limit equal the value $f(a_1, a_2, \ldots, a_n)$?

DEFINITION 16.2.10

Suppose the function $f(x_1, x_2, \ldots, x_n)$ of n variables is defined at the point $(a_1, a_2, \ldots, a_n)$. We say that the function f is **continuous** at the point $(a_1, a_2, \ldots, a_n)$ provided it has a limit as $(x_1, x_2, \ldots, x_n) \to (a_1, a_2, \ldots, a_n)$ and

$$\lim_{(x_1,x_2,\ldots,x_n)\to(a_1,a_2,\ldots,a_n)} f(x_1, x_2, \ldots, x_n) = f(a_1, a_2, \ldots, a_n).$$

THEOREM 16.2.11

(i) Polynomials of n variables are continuous at every point, and a rational function is continuous at every point where the denominator is nonzero.

(ii) If $f(x_1, x_2, \ldots, x_n)$ is continuous at some point $(a_1, a_2, \ldots, a_n)$ and $g(t)$ is continuous at the point $t = L$, where $L = f(a_1, a_2, \ldots, a_n)$, then the composite function $H(x_1, x_2, \ldots, x_n) = g\big(f(x_1, x_2, \ldots, x_n)\big)$ is continuous at $(a_1, a_2, \ldots, a_n)$.

(iii) If $f_1(x_1), f_2(x_2), \ldots, f_n(x_n)$ are all functions of one variable and each $f_i(x_i)$ is continuous at the point $x_i = a_i$, then their product $F(x_1, x_2, \ldots, x_n) = f_1(x_1)f_2(x_2) \times \ldots \times f_n(x_n)$ is continuous at the point $(a_1, a_2, \ldots, a_n)$.

PROOF   Each of these statements follows from the corresponding fact about limits of these various types of functions. For polynomials and rational functions, Theorem 16.2.7 tells us that the limits have the values we expect, which make these functions continuous. Theorem 16.2.8 says that limits of composite functions also behave nicely, and Theorem 16.2.9 gives the necessary result about limits of products of functions of single variables.   ◢

EXAMPLE 6   Discuss the continuity of each of the following functions:

(i) $\sin(5xy^2z - 3xz + 1)$,

(ii) $\frac{xy + yz - 2xz}{2x^2 + 3y^2 + 5z^2}$,

(iii) $\frac{|xy|}{xy}$,

(iv) $\tan\left(\frac{4}{1 + x^2 + y^2 + z^2}\right)$.

Solution

(i) Since $5xy^2z - 3xz + 1$ is a polynomial, it is continuous everywhere. The function $\sin t$ is a continuous function of the variable t. By Theorem 16.2.8 not only does the composite function have a limit, but the limit is the right one, that is, for any point (a, b, c), $\lim_{(x,y,z)\to(a,b,c)} \sin(5xy^2z - 3xz + 1) = \sin(5ab^2c - 3ac + 1)$. This means that the function is continuous everywhere.

(ii) The rational function $\frac{xy + yz - 2xz}{2x^2 + 3y^2 + 5z^2}$ is continuous except where its denominator is zero. Since the denominator is a sum of squares, each of which is nonnegative, we see that it can only be zero if each of x, y, and z is zero.

In other words, the denominator is nonzero except at the origin. The function is continuous at every point except the origin, where it is not defined. Since we saw in Example 5(iv) that the function does not have a limit at $(0, 0, 0)$, it does not even have a removable discontinuity there.

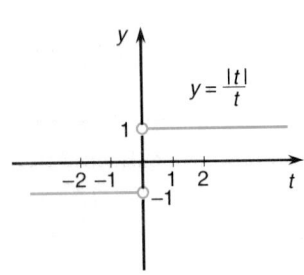

$$y = \frac{|t|}{t}$$

Figure 16.2.6

(iii) If we let $g(t) = \frac{|t|}{t}$, we recall that $g(t)$ is continuous at every point except $t = 0$, where it has a jump discontinuity (it is not even defined at $t = 0$) (see Figure 16.2.6). Since xy is continuous, being a polynomial, we see that $g(xy)$ is continuous except at those points (x, y) where $xy = 0$, that is, where either x or y (or both) equals zero, that is, on the axes.

So $\frac{|xy|}{xy}$ is continuous everywhere except on the axes, and it is not defined on the axes (so it is certainly not continuous there).

(iv) Finally, we note that the rational function $\frac{4}{1 + x^2 + y^2 + z^2}$ is continuous everywhere because its denominator is never zero. In fact, its denominator is never less than 1, so the quotient can never be greater than 4. It is always nonnegative.

Now $\tan t$ is continuous whenever it is defined, but it is not defined at $\pm\frac{\pi}{2}$, $\pm\frac{3\pi}{2}, \pm\frac{5\pi}{2}, \ldots$. We must ask when $\frac{4}{1 + x^2 + y^2 + z^2}$ can equal any of these values.

We have noted that $0 \leq \frac{4}{1 + x^2 + y^2 + z^2} \leq 4$, so the only one of these troublesome values that it could ever possibly equal is $\frac{\pi}{2}$. If it does equal $\frac{\pi}{2}$, that is, $\frac{4}{1 + x^2 + y^2 + z^2} = \frac{\pi}{2}$, then $1 + x^2 + y^2 + z^2 = 4/\frac{\pi}{2} = \frac{8}{\pi}$, and $x^2 + y^2 + z^2 = \frac{8}{\pi} - 1$.

What this means is that if the point (x, y, z) lies on the sphere of radius $\sqrt{\frac{8}{\pi} - 1}$, then the function $\tan\left(\frac{4}{1+x^2+y^2+z^2}\right)$ is not defined. At every other point (x, y, z) it is defined and continuous.

Sometimes in discussing limits and continuity of functions of two variables it is helpful to use polar coordinates. This is especially likely to be true in situations in which the function involves the expression $x^2 + y^2$, which of course is just r^2 in polar coordinates.

EXAMPLE 7   Evaluate the following limits, if they exist: (i) $\lim_{(x,y)\to(0,0)} \frac{3xy^2+2x^3y}{x^2+y^2}$, (ii) $\lim_{(x,y)\to(0,0)} \left((x^2+y^2)\ln(x^2+y^2)\right)$.

Solution

(i) We convert the function to polar coordinates by writing $x = r\cos\theta$ and $y = r\sin\theta$. So the function is $\frac{3xy^2+2x^3y}{x^2+y^2} = \frac{3r\cos\theta(r^2\sin^2\theta)+2r^3\cos^3\theta(r\sin\theta)}{r^2} = 3r\cos\theta\sin^2\theta + 2r^2\cos^3\theta\sin\theta$.

Now as $(x, y) \to (0, 0)$, we see that $r = \sqrt{x^2 + y^2} \to 0^+$ (using Theorem 16.2.8). Since $|\cos\theta| \le 1$ and $|\sin\theta| \le 1$, we can apply the Triangle Inequality and find that

$$0 \le |3r\cos\theta\sin^2\theta + 2r^2\cos^3\theta\sin\theta| \le |3r\cos\theta\sin^2\theta| + |2r^2\cos^3\theta\sin\theta|$$
$$\le (3r + 2r^2).$$

Since the function on the right tends to zero as $r \to 0^+$, it is possible to apply the Squeeze Theorem (Theorem 2.3.5) to conclude that the function tends to zero, that is,

$$\lim_{(x,y)\to(0,0)} \frac{3xy^2 + 2x^3y}{x^2 + y^2} = 0.$$

(ii) Here converting to polar coordinates gives the function $r^2 \ln r^2 = 2r^2 \ln r$. As $r \to 0^+$, r^2 tends to 0 and $\ln r$ tends to $-\infty$. To decide what happens to their product, we can use l'Hôpital's Rule:

$$\lim_{r\to0^+} 2r^2 \ln r = \lim_{r\to0^+} \frac{\ln r}{\frac{1}{2r^2}} = \lim_{r\to0^+} \frac{\frac{1}{r}}{-r^{-3}} = \lim_{r\to0^+} (-r^2) = 0.$$

From this we conclude that $\lim_{(x,y)\to(0,0)} \left((x^2+y^2)\ln(x^2+y^2)\right) = 0$.

Finally, we give the precise definition of limits for a function of n variables. As with the corresponding definition for functions of one variable in Chapter 2, it is really here as a matter of interest, and we will not actually use it.

DEFINITION 16.2.12

Limits (Precise Version)

Suppose the function $f(x_1, x_2, \ldots, x_n)$ is defined at every point in some punctured ball around $(a_1, a_2, \ldots, a_n)$, that is, on the set

$$\{(x_1, x_2, \ldots, x_n) : 0 < \sqrt{(x_1 - a_1)^2 + (x_2 - a_2)^2 + \ldots + (x_n - a_n)^2} < r\},$$

for some $r > 0$. Then we say that the function $f(x_1, x_2, \ldots, x_n)$ approaches the limit L as $(x_1, x_2, \ldots, x_n)$ approaches $(a_1, a_2, \ldots, a_n)$, and write

$$\lim_{(x_1, x_2, \ldots, x_n) \to (a_1, a_2, \ldots, a_n)} f(x_1, x_2, \ldots, x_n) = L,$$

provided that for every $\epsilon > 0$ there exists a $\delta > 0$ so that $|f(x_1, x_2, \ldots, x_n) - L| < \epsilon$ whenever $0 < \sqrt{(x_1 - a_1)^2 + (x_2 - a_2)^2 + \ldots + (x_n - a_n)^2} < \delta$.

Exercises 16.2

I

Evaluate each of the following limits, provided it exists. For those that do not exist, explain why they do not.

1. $\lim_{(x,y) \to (2,-3)} (xy^2 - 3x^2 y)$

2. $\lim_{(x,y) \to (0,2)} \frac{3x + 4y - 5xy}{2xy - x + y - 1}$

3. $\lim_{(x,y) \to (1,1)} \frac{3x^2 y^2 - 4xy}{x - y}$

4. $\lim_{(x,y) \to (2,1)} 3xy^2 \cos(\pi y)$

5. $\lim_{(x,y,z) \to (2,0,-1)} \frac{x + y - 2z^2}{xy + yz + xz}$

6. $\lim_{(x,y,z) \to (3,2,1)} \frac{\cos(\pi xz)}{x - y - z}$

7. $\lim_{(x,y,z) \to (4,4,4)} \left(y^2 \sin \frac{\pi}{x} \cos \frac{\pi}{z} \right)$

8. $\lim_{(w,x,y,z) \to (0,1,4,2)} z \tan\left(w + \pi/(xy) \right)$

9. $\lim_{(x,y) \to (0,0)} \frac{x^2 - y^2}{x + y}$

10. $\lim_{(x,y) \to \left(\frac{1}{2}, \frac{1}{4} \right)} \left(\cos^2(\pi x) + \sin^2(\pi y) \right)$

Discuss the continuity of each of the following functions, stating where it is continuous and where it is not and giving reasons.

11. $f(x, y) = 3xy - 4y^4$

12. $g(x, t) = 5xt^2 - 4xt^{-3}$

13. $h(x, y) = \frac{xy - 3x + 1}{2 + x^2}$

14. $k(x, y) = \frac{\sin x}{3 \cos y}$

15. $R(x, y, z) = \exp\left(\frac{1}{x + y - 2z - 1} \right)$

16. $S(w, x, y, z) = \sqrt{w - x^2 - y^2 - z^2}$

17. $L(x, y) = \ln(2x - 3y + 4)$

18. $F(x, y) = 1/\left(\cosh(3x^3 y^2 - 5x^2 y + 7) \right)$

19. $G(x, y) = \sec\left(\pi \sin(x + y) \right)$

20. $H(x, y) = \arctan\left(\tan(x + 2y) \right)$

Evaluate each of the following limits, provided it exists. For those that do not exist, explain why they do not.

21. $\lim_{(x,y) \to (0,0)} \frac{x^4}{x^2 + y^2}$

22. $\lim_{(x,y) \to (0,0)} \frac{5x^2 y}{x^2 + y^2}$

23. $\lim_{(x,y) \to (0,0)} (x + 2xy) \ln(x^2 + y^2)$

24. $\lim_{(x,y) \to (0,0)} (5x + 3y^2) e^{x^2 + y^2}$

25. $\lim_{(x,y) \to (0,0)} x^3 \cot(x^2 + y^2)$

26. $\lim_{(x,y) \to (0,0)} (x^2 + y^2) \csc(x^2 + y^2)$

27. $\lim_{(x,y) \to (0,0)} x^2 \cot(x^2 + y^2)$

28. $\lim_{(x,y) \to (0,0)} (x + 2y) \csc(x^2 + y^2)$

29. $\lim_{(x,y) \to (0,0)} \frac{x^4 + y^4}{(x^2 + y^2)^2}$

30. $\lim_{(x,y) \to (0,0)} (e^{x^2 + y^2} - 1)/(x^2 + y^2)$

II

*31. Let $f(x, y) = (x^2 - y^2)/(x^2 + y^2)$ be the function defined in Example 1. What happens to the values of $f(x, y)$ as (x, y) approaches the origin along the line $y = x$?

*32. Let $f(x, y) = (x^2 - y^2)/(x^2 + y^2)$ as above, and let a be a fixed real number. What happens to the values of $f(x, y)$ as (x, y) approaches the origin along the line $y = ax$?

*33. Let $f(x, y) = (x^2 - y^2)/(x^2 + y^2)$ as above. What happens to the values of $f(x, y)$ as (x, y) approaches the origin along the curve $y = x^2$?

*34. Let $f(x, y) = (x^2 - y^2)/(x^2 + y^2)$ as above. What happens to the values of $f(x, y)$ as (x, y) approaches the origin along the curve $y = \sqrt{x}$?

*35. Let $f(x, y) = \frac{x^2 y}{x^4 + y^2}$, which is defined for every point (x, y) except $(0, 0)$.
    (i) Show that $f(x, y) \to 0$ as (x, y) approaches $(0, 0)$ along the y-axis.
    (ii) Any straight line through the origin other than the y-axis must be of the form $y = mx$. Show that $f(x, y) \to 0$ as (x, y) approaches $(0, 0)$ along any such line.
    (iii) What happens to $f(x, y)$ as (x, y) approaches the origin along the parabola $y = x^2$?
    (iv) What happens to $f(x, y)$ as (x, y) approaches the origin along the parabola $y = 2x^2$?

(v) Even though $f(x, y) \to 0$ as (x, y) approaches $(0, 0)$ along *every* straight line through the origin, show that $\lim_{(x,y)\to(0,0)} f(x, y)$ does not exist.

*36. Let $g(x, y) = \frac{x^3 y}{x^6 + y^2}$, which is defined for every point (x, y) except $(0, 0)$.

  (i) Show that $g(x, y) \to 0$ as (x, y) approaches $(0, 0)$ along any line through the origin.

  (ii) Show that $g(x, y) \to 0$ as (x, y) approaches $(0, 0)$ along any parabola of the form $y = ax^2 + bx$. (Notice that there is no constant term in the equation of the parabola, so it will go through the origin.)

  (iii) What happens to $g(x, y)$ as (x, y) approaches the origin along the parabola $x = ay^2 + by$?

  (iv) What happens to $g(x, y)$ as (x, y) approaches the origin along the curve $y = x^3$?

  (v) Show that $\lim_{(x,y)\to(0,0)} g(x, y)$ does not exist.

**37. Find an example of a function $h(x, y)$ that tends to 0 as (x, y) approaches the origin along any straight line, along any parabola of the form $y = ax^2 + bx$ or $x = ay^2 + by$, or along any curve of the form $y = ax^3 + bx^2 + cx$ or $x = ay^3 + by^2 + cy$, but that still does not have a limit as $(x, y) \to (0, 0)$. Show that your example does satisfy all the specified conditions.

**38. Let n be a positive integer. Find an example of a function $k(x, y)$ that tends to 0 as (x, y) approaches the origin along any curve of the form $y = p(x)$ or $x = p(y)$, where $p(x)$ is a polynomial in one variable of degree less than n, but so that $k(x, y)$ does not have a limit as $(x, y) \to (0, 0)$.

**39. Find an example of two functions $f(x)$ and $g(y)$ so that $\lim_{x\to0} f(x)$ exists and $\lim_{y\to0} g(y)$ does not exist, but for which the limit of the product does exist, that is, $\lim_{(x,y)\to(0,0)} \big(H(x, y)\big)$ does exist, where $H(x, y) = f(x)g(y)$.

*40. Evaluate

$$\lim_{(x,y)\to(1,-2)} \Big((x - 1)y\ln\big((x - 1)^2 + (y + 2)^2\big)\Big).$$

41. Use a computer to sketch the graph of $z = f(x, y) = \frac{x^2 - y^2}{x^2 + y^2}$, the function discussed in Example 1. Use the picture to discuss its limit as (x, y) approaches the origin along any straight line through the origin. From the picture, try to guess bounds for $f(x, y)$, that is, numbers M and N for which $M \le f(x, y) \le N$ for all $(x, y) \ne (0, 0)$. Can you prove that your guess is correct?

42. Use a computer to sketch the graph of

$$z = f(x, y) = \frac{x^4 - 6x^2y^2 + y^4}{(x^2 + y^2)^2}.$$

Using the picture as inspiration, discuss the limiting behavior of f near the origin, and try to guess bounds, as in Exercise 41.

16.3 Partial Derivatives

| NOTATION FOR PARTIAL DERIVATIVES |
| TANGENT PLANE |
| HIGHER-ORDER PARTIALS |
| MIXED PARTIALS |

In this section we begin to discuss derivatives of functions of two or more variables. As we mentioned in the introduction at the beginning of the chapter, one way to approach this problem is to fix the values of all the variables except one. This has the effect of turning the function into a function of the one remaining variable, and it is possible to consider its derivative. This in a nutshell is what is meant by a partial derivative. The word "partial" refers to the fact that this kind of derivative deals with only "part" of the behavior of the function.

For instance, consider the function $f(x, y) = x^2 e^{3x+4y}$. Because of the properties of exponentials, it can be rewritten as $f(x, y) = x^2 e^{3x} e^{4y}$. Suppose we fix the value of y; then of course this also fixes the value of e^{4y}, and we can think of e^{4y} as a *constant*. So $e^{4y} x^2 e^{3x}$ is a constant times $x^2 e^{3x}$, and it is easy to find its derivative with respect to x, namely, $e^{4y}(2xe^{3x} + 3x^2 e^{3x})$. Notice that we dealt with e^{4y} exactly as we would deal with any other constant, just pulling it out in front of the function as we differentiated.

On the other hand, we could fix the value of x and differentiate with respect to y. This time we would regard $f(x, y) = x^2 e^{3x} e^{4y}$ as a constant times e^{4y}, and its derivative with respect to y would be $4x^2 e^{3x} e^{4y}$.

There are several different notations for partial derivatives. The notation has to make it clear which variable is being used to take the derivative. One way to write the partial derivative is $\frac{\partial f}{\partial x}$ or $\frac{\partial f}{\partial y}$. The "curly" letter ∂ is used to distinguish this from the ordinary derivative $\frac{dF}{dx}$ of a function $F(x)$ of one variable.

Another common notation is to write $f_x(a, b), f_y(a, b)$, respectively, for the partial derivatives of $f(x, y)$ with respect to x and y evaluated at the point (a, b). Let us remind ourselves that to find $f_x(a, b)$, for instance, we should fix the variable y to equal b and then look at the function $f(x, b)$ as a function of the variable x. The derivative of this function of x, evaluated at $x = a$, is what we mean by $f_x(a, b)$.

So, for example, if $g(x, y) = \sin \pi(2x + y)$, then to find $g_x(0, 1)$, we ought to let $y = 1$ and consider $g(x, 1) = \sin \pi(2x + 1)$. The derivative of this function is $\cos\big(\pi(2x + 1)\big)2\pi$, and if we evaluate it at $x = 0$, we get $g_x(0, 1) = 2\pi \cos \pi = -2\pi$.

Similarly, to find $g_y(0, 1)$, we should consider $g(0, y) = \sin \pi(0 + y) = \sin \pi y$, which is easily differentiated to give $\cos(\pi y)(\pi)$. Evaluating at $y = 1$, we find that $g_y(0, 1) = -\pi$.

Another notation for partial derivatives is to write $D_x f$ and $D_y f$ for $f_x = \frac{\partial f}{\partial x}$ and $f_y = \frac{\partial f}{\partial y}$, respectively.

Of course, similar remarks apply to partial derivatives of functions of three or more variables. If $h(x, y, z)$ is a function of three variables, then it makes sense to consider $D_x h = \frac{\partial h}{\partial x}, D_y h = \frac{\partial h}{\partial y}$, and $D_z h = \frac{\partial h}{\partial z}$.

DEFINITION 16.3.1

Suppose $f(x, y)$ is a function of two variables and that there is a disk around the point (a, b) so that $f(x, y)$ is defined for every (x, y) in the disk. Then, provided the limit exists, the **partial derivative** with respect to x is defined as follows:

$$\frac{\partial f}{\partial x}(a, b) = f_x(a, b) = D_x f(a, b) = \lim_{h \to 0} \frac{f(a + h, b) - f(a, b)}{h}.$$

The **partial derivative** with respect to y is defined as follows:

$$\frac{\partial f}{\partial y}(a, b) = f_y(a, b) = D_y f(a, b) = \lim_{h \to 0} \frac{f(a, b + h) - f(a, b)}{h}.$$

For a function $F(x_1, \ldots, x_n)$ of n variables that is defined at every point in a ball around the point $(a_1, \ldots, a_n)$, the partial derivative with respect to x_i is defined by

$$\frac{\partial F}{\partial x_i}(a_1, \ldots, a_n) = F_{x_i}(a_1, \ldots, a_n) = D_{x_i} F(a_1, \ldots, a_n)$$

$$= \lim_{h \to 0} \frac{F(a_1, \ldots, a_i + h, \ldots, a_n) - F(a_1, \ldots, a_n)}{h}.$$

Notice that these limits may not exist, just as ordinary derivatives may not exist at certain points for certain functions.

As with ordinary derivatives of functions of a single variable, the partial derivative $\frac{\partial F}{\partial x_i}(a_1, \ldots, a_n)$ can be evaluated at different points $(a_1, \ldots, a_n)$, and so the partial derivative $\frac{\partial F}{\partial x_i}$ can be regarded as a function $\frac{\partial F}{\partial x_i}(x_1, \ldots, x_n)$ in its own right.

In a way, these definitions make it look complicated. In fact it is not difficult to take partial derivatives. The whole idea is to take the derivative with respect to the appropriate variable in the usual way, remembering to regard all the other variables as constants.

EXAMPLE 1   Find the following: (i) $\frac{\partial}{\partial x}(x^2y - 3xy^2 + 2xy + 5x^4)$, (ii) $D_z(xyz - 3xy^2 - yz^2 + 2)$, (iii) $\frac{\partial}{\partial u}\left(u\cos(3v - 2uv^2)\right)$.

Solution

(i) Here we think of y as a constant, so, for instance, the first term is x^2y, which is just a constant times x^2. So its partial derivative is $2xy$. Similarly, the derivative of $3xy^2$ is $3y^2$ (the derivative of ax, a constant times x, is just a). The derivative of $2xy$ is $2y$, and the derivative of $5x^4$ is $20x^3$.

Adding up all the terms, we find that $\frac{\partial}{\partial x}(x^2y - 3xy^2 + 2xy + 5x^4) = 2xy - 3y^2 + 2y + 20x^3$.

(ii) Here we are dealing with a function $(xyz - 3xy^2 - yz^2 + 2)$ of three variables. To take partial derivative with respect to z, we must regard x and y as constants. The partial derivative with respect to z of xyz is xy. The term $-3xy^2$ is a *constant*, so its partial derivative with respect to z is 0. The partial derivative of $-yz^2$ is $-2yz$, and the partial derivative of the constant 2 is 0.

Adding up these terms, we see that $D_z(xyz - 3xy^2 - yz^2 + 2) = xy - 0 - 2yz = xy - 2yz$.

(iii) The function $\left(u\cos(3v - 2uv^2)\right)$ is a function of the two variables u and v. To find its partial derivative with respect to u, we fix v and think of it as a constant. Then we apply the Product Rule and the Chain Rule and find $\frac{\partial}{\partial u}\left(u\cos(3v - 2uv^2)\right) = \cos(3v - 2uv^2) + u\left(-\sin(3v - 2uv^2)\right)(-2v^2) = \cos(3v - 2uv^2) + 2uv^2\sin(3v - 2uv^2)$.   ◢

EXAMPLE 2   **Effects of Alcohol**   In a study of the effects of alcohol, sober subjects were given a simple test. They were given a second test of comparable difficulty 20 minutes after drinking a measured quantity of alcohol. Their second test scores were expressed as percentages of the first (and were generally lower).

The other factor in the study was the length of time since the subject's last substantial meal. Let A be the amount of alcohol consumed, in ounces, and t be the number of hours since the last meal.

Subjects were tested with a number of combinations of A and t, with $0 \leq A \leq 4$ and $0 \leq t \leq 4$. The data suggested that the second score, as a percentage of the first, could be predicted approximately by the function

$$S(A, t) = 100 - \frac{3}{5}A^2(t^{3/2} + 2).$$

We find that $\dfrac{\partial S}{\partial A} = -\dfrac{6}{5}A(t^{3/2} + 2)$ and $\dfrac{\partial S}{\partial t} = -\dfrac{9}{10}A^2t^{1/2}.$

Both partial derivatives are negative. This suggests that increasing either the amount of alcohol or the length of time after a meal will result in lower test scores. Moreover, these derivatives have larger absolute value as A and/or t increases. The more a person has drunk already, the more seriously his or her score will be affected by additional alcohol. The effects are also greater, the longer it has been since a meal.

If a person in the study group consumed 1 ounce of alcohol 1 hour after a meal, the function predicts a score of $S(1, 1) = 100 - \frac{3}{5}(1^2)(1 + 2) = 98.2$. At this point, $\frac{\partial S}{\partial A} = -\frac{6}{5}(1)(1 + 2) = -\frac{18}{5} = -3.6$, and $\frac{\partial S}{\partial t} = -\frac{9}{10}(1^2)1 = -0.9$. This suggests that small additional amounts of alcohol (fractions of an ounce) will further decrease test scores by about 3.6 percentage points per ounce, and drinking longer after a meal will decrease them by about 0.9 point per hour's wait.

Figure 16.3.1

Computed tomography (CT) is one of several scanning techniques used as diagnostic tools in medicine. Multiple X-ray images from many different angles generate data from which a computer assembles a picture.

In this "false-color" CT brain scan, colors are used to help distinguish different types of tissue. The round purple mass in the center is a large intracranial tumor. Spots inside the tumor are fluid-filled cysts. The outer border of the tumor (pale yellow) has become calcified with age.

This tumor is probably a low-grade glioma, a tumor of supporting glial cells, the specialized connective tissue of the brain and central nervous system. Gliomas account for the majority of intracranial tumors; slow-growing low-grade gliomas may exist for years before rapidly acquiring a malignant character.

These relatively small decreases for a nearly sober subject can be compared with $\frac{\partial S}{\partial A}(3,3) \approx -25.9$ and $\frac{\partial S}{\partial t}(3,3) \approx -14.0$. For somebody who has consumed 3 ounces of alcohol 3 hours after a meal, additional alcohol will drastically affect test performance, and even waiting longer will have a noticeable effect.

In Figure 16.3.1 we have sketched S as a function of A for several values of t. Observe how the slope $\frac{\partial S}{\partial A}$ becomes steeper as A or t increases.

It is reasonably easy to check when partial derivatives exist for many types of functions. At the very least they always exist for polynomials, and they exist for a rational function at every point at which the denominator is not zero.

If we consider a function $f(x, y)$ of two variables, then we can draw the graph of $z = f(x, y)$. Now let us look at the point (a, b) and the corresponding point $\big(a, b, f(a, b)\big)$ on the graph. (This is the point on the graph whose (x, y)-coordinates are (a, b), so it lies directly *above* the point (a, b) in the horizontal xy-plane.)

If we fix the value of y to be $y = b$ and allow x to vary, what this means in terms of the picture is that the point (x, b) moves along a line parallel to the x-axis in the xy-plane. The point in the graph moves along some sort of curve lying directly over this line. See Figure 16.3.2(i).

The partial derivative with respect to x will be the slope of the tangent to this curve. It is the *rate of change* of the function $f(x, b)$ as x varies, which means that it is the rate of change of $f(x, y)$ as we move in the x-direction.

On the other hand, if we fix $x = a$ and allow y to vary, we are letting the point move along a different curve, this one lying directly over a line that is parallel to the y-axis in the xy-plane.

The partial derivative with respect to y is the slope of the tangent to this curve, which is the rate at which $f(x, y)$ changes when we move in the y-direction. See Figure 16.3.2(ii).

Figure 16.3.2(i)

Figure 16.3.2(ii)

Figure 16.3.3(i)

Figure 16.3.3(ii)

Figure 16.3.3(iii)

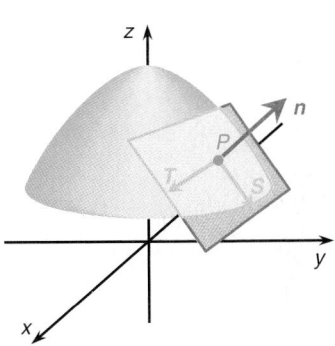

Figure 16.3.3(iv)

Tangent Planes

In Figure 16.3.3(i) we see the graph $z = f(x, y)$ and a point P on the graph. The point directly below P in the xy-plane is (a, b); this means $P = \left(a, b, f(a, b)\right)$.

Also shown are the line through (a, b) that is parallel to the x-axis and the curve lying above it in the surface $z = f(x, y)$. Consider the vector $\mathbf{T} = \left(1, 0, \frac{\partial f}{\partial x}(a, b)\right) = \left(1, 0, f_x(a, b)\right)$. It is perpendicular to the y-axis, and its slope relative to the x- and z-directions is $\frac{f_x(a,b)}{1} = f_x(a, b)$. It is tangent to that curve through P, since as we have just remarked, $f_x(a, b)$ is the slope of the tangent.

In Figure 16.3.3(ii) we see the corresponding picture with the curve lying over a line parallel to the y-axis. In this case we see that the vector $\mathbf{S} = \left(0, 1, f_y(a, b)\right)$ is perpendicular to the x-axis and has slope $\frac{f_y(a,b)}{1} = f_y(a, b)$. It is tangent to the curve lying in the surface.

The vectors $\mathbf{S}$ and $\mathbf{T}$ are tangent to two curves in the graph that pass through P in two different directions. We can use them to find what is called the **tangent plane** to the surface at P, as shown in Figure 16.3.3(iii). It is the plane that contains the tangent vectors at P of all curves through P in the surface.

If $\mathbf{S}$ and $\mathbf{T}$ are both in the tangent plane, then their cross product $\mathbf{S} \times \mathbf{T}$ is perpendicular to it.

Letting $\mathbf{n} = \mathbf{S} \times \mathbf{T}$, we find that $\mathbf{n} = \left(0, 1, f_y(a, b)\right) \times \left(1, 0, f_x(a, b)\right) = \left(f_x(a, b), f_y(a, b), -1\right)$, and $\mathbf{n}$ is a normal vector to the tangent plane. See Figure 16.3.3(iv). The tangent plane contains P, so its equation is $\mathbf{n} \cdot \left((x, y, z) - P\right) = 0$, or $\mathbf{n} \cdot (x, y, z) = \mathbf{n} \cdot P$, that is,

$$f_x(a, b)x + f_y(a, b)y - z = af_x(a, b) + bf_y(a, b) - f(a, b). \qquad (16.3.2)$$

EXAMPLE 3   Find the tangent plane to the paraboloid $z = x^2 + y^2$ at the point $(1, 2, 5)$.

Solution   The paraboloid is the graph $z = f(x, y)$ with $f(x, y) = x^2 + y^2$. We find that $f_x = 2x, f_y = 2y$, so if $(a, b) = (1, 2)$, then $f_x(a, b) = 2, f_y(a, b) = 4$. The normal vector is $\mathbf{n} = \left(f_x(a, b), f_y(a, b), -1\right) = (2, 4, -1)$.

The point of tangency is $P = (1, 2, 5)$, so the equation of the tangent plane is $\mathbf{n} \cdot (x, y, z) = \mathbf{n} \cdot P$, that is,

$$2x + 4y - z = 5.$$

See Figure 16.3.4.

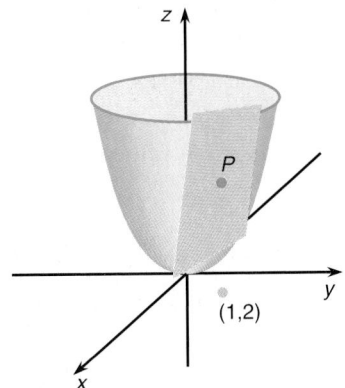

Figure 16.3.4

EXAMPLE 4   Consider the unit sphere $x^2 + y^2 + z^2 = 1$. Find the equation of the tangent plane at each point on the sphere.

Solution   The sphere is not the graph of a function. If we stay on the upper hemisphere, we can write it as $z = \sqrt{1 - x^2 - y^2}$, that is, as the graph of $f(x, y) = \sqrt{1 - x^2 - y^2}$.

We find that $f_x = \dfrac{\partial f}{\partial x} = -\dfrac{x}{\sqrt{1-x^2-y^2}}$ and $f_y = \dfrac{\partial f}{\partial y} = -\dfrac{y}{\sqrt{1-x^2-y^2}}$.

The normal vector at a point $P = (a, b, \sqrt{1 - a^2 - b^2})$ on the hemisphere is $\mathbf{n} = (f_x, f_y, -1) = \left(-\dfrac{a}{\sqrt{1-a^2-b^2}}, -\dfrac{b}{\sqrt{1-a^2-b^2}}, -1\right)$. The tangent plane has this as its normal and passes through P, so its equation is $\mathbf{n} \cdot (x, y, z) = \mathbf{n} \cdot P$, or

$$-\frac{ax}{\sqrt{1 - a^2 - b^2}} - \frac{by}{\sqrt{1 - a^2 - b^2}} - z = -\frac{a^2}{\sqrt{1 - a^2 - b^2}} - \frac{b^2}{\sqrt{1 - a^2 - b^2}}$$
$$-\sqrt{1 - a^2 - b^2}.$$

Multiplying by $-\sqrt{1 - a^2 - b^2}$ gives

$$ax + by + \sqrt{1 - a^2 - b^2}\,z = a^2 + b^2 + 1 - a^2 - b^2 = 1.$$

This equation $ax + by + \sqrt{1 - a^2 - b^2}\,z = 1$ can be simplified by writing $P = (a, b, c)$, that is, by letting $c = \sqrt{1 - a^2 - b^2}$. Then the equation is $ax + by + cz = 1$.

The calculation for points on the lower hemisphere is similar, and the answer is the same: The tangent plane at (a, b, c) is $ax + by + cz = 1$. See Figure 16.3.5. Strictly speaking, these arguments do not work for points on the equator ($z = 0$), which are not inside either hemisphere. The answer is still the same, and it can be found, for instance, by letting y be a function of x and z or x a function of y and z.

We will return to discuss tangent planes in more detail in Section 16.7.

Thinking of partial derivatives as the slopes of curves in the graph suggests another use for them. Suppose $f(x, y)$ is some function and we wish to find its maximum or minimum value. If we knew that f had a maximum, say, at a point (a, b), then in particular the function of x obtained by fixing $y = b$, that is, $f(x, b)$, would have to have a maximum at $x = a$. So its derivative, the partial derivative $\dfrac{\partial f}{\partial x}$, would have to be zero at (a, b). Similarly, $f(a, y)$ would have a maximum at $y = b$, so $\dfrac{\partial f}{\partial y}(a, b) = 0$ too. See Figure 16.3.6.

Just as with functions of one variable, if we want to find extrema, we should look for points at which the partial derivatives are zero (or do not exist). As in the single-variable case, there is no guarantee that these will be extrema, but any extremum will have to be at such a point. This can make it much easier to find them.

EXAMPLE 5   Find the maximum value of the function $f(x, y) = x^2 - 2x + 2y^2 + 4y - 3$.

Solution   We can easily find the partial derivatives: $\dfrac{\partial f}{\partial x} = 2x - 2$ and $\dfrac{\partial f}{\partial y} = 4y + 4$, so $\dfrac{\partial f}{\partial x} = 0$ when $x = 1$ and $\dfrac{\partial f}{\partial y} = 0$ when $y = -1$. The only point at which both partial derivatives are zero is $(1, -1)$.

If there is a maximum value, it must occur at the point $(1, -1)$. We could try experimenting with values of f at several points: $f(1, -1) = 1^2 - 2(1) + 2(-1)^2 +$

n

P

Figure 16.3.5

Figure 16.3.6

$4(-1) - 3 = -6, f(0,0) = -3$. Without finding any more points, we see that $f(x,y)$ cannot possibly have a maximum value. We know that if it did, the maximum would have to occur at $(1, -1)$, and we have just seen that $f(0,0)$ is greater than $f(1, -1)$.

In fact we can complete the squares and write $f(x,y) = x^2 - 2x + 2y^2 + 4y - 3 = (x-1)^2 - 1 + 2(y+1)^2 - 2 - 3 = (x-1)^2 + 2(y+1)^2 - 6$. Since $(x-1)^2 + 2(y+1)^2 \geq 0$ for any choice of x and y, this shows that $f(x,y) \geq -6$ for any value of x and y. This means that $f(x,y)$ has a *minimum* at $(1, -1)$.

In a way we should not be discouraged. Of course, the argument we gave that the partial derivatives must equal zero at a maximum will work just as well for a minimum. So as with functions of one variable, looking for the points where the partials are zero will find both maxima and minima (if any). Our approach did work; it is just that the function did not have the maximum we were looking for, and we found a minimum that we were not looking for. We will return to this topic in Section 16.8.

The *Ideal Gas Law* says that for a fixed quantity of an ideal gas,

$$\frac{PV}{T} = k, \tag{16.3.3}$$

where P is the pressure, V is the volume, T is the temperature, and k is a constant (which depends on the amount of gas and the units in which the other quantities are measured). The temperature must be measured on a scale that begins at *absolute zero*; this is usually done with the Kelvin scale, which has the advantage that a Kelvin degree is the same as a Celsius degree but begins with $0°$K at absolute zero, which is $-273°$C. This means that to convert a temperature in degrees Celsius to degrees Kelvin, we just add 273, so $20°$C (room temperature) is $20 + 273 = 293°$K.

The Ideal Gas Law is not as mysterious as it appears. Suppose the temperature of the gas is held constant. Then $PV = kT$, and the right side is a constant. This says that if the pressure increases, then the volume must decrease and vice versa, which is perfectly reasonable if we think of a balloon, for instance. If we increase the pressure around the balloon, it will shrink.

Similarly, if we hold the pressure constant, then increasing the temperature will make the gas expand, so V increases, and so on.

EXAMPLE 6   Suppose a certain quantity of gas at $27°$C occupies a volume of 1 liter at a pressure of 1 atm (1 atmosphere is the normal atmospheric pressure at sea level, which is about 10,300 kg/sq m or 14.7 lb/sq in).

We can easily calculate the constant k for the Ideal Gas Law for this particular amount of gas. We must first convert the temperature to the Kelvin scale, and we find that $T = 27 + 273 = 300°$K. The Ideal Gas Law says that $\frac{PV}{T} = k$, so $k = \frac{(1 \text{ atm})(1 \text{ liter})}{300°\text{K}} = \frac{1}{300} \frac{\text{atm liter}}{°\text{K}}$.

Now it is possible to express V as a function of the other two quantities: $V = \frac{kT}{P}$. We can think of T and P as variables and V as a function $V(T, P)$ of two variables. It is also easy to find the partial derivatives from this formula:

$$\frac{\partial V}{\partial T}(T, P) = V_T(T, P) = \frac{k}{P}, \qquad \frac{\partial V}{\partial P}(T, P) = V_P(T, P) = -\frac{kT}{P^2}.$$

What this means is that if the pressure is held constant, the rate of change of V with respect to T is $\frac{k}{P}$, and if the temperature is held constant, the rate of change of V with respect to P is $-\frac{kT}{P^2}$. The fact that this last formula is negative reflects the fact that as the pressure P increases, the volume will decrease, and vice versa.

For instance, with the gas as described above, the rate of change of its volume with respect to temperature is $\frac{k}{P} = \frac{1}{300}\frac{\text{atm liter}}{^\circ\text{K}}/(1\text{ atm}) = \frac{1}{300}$ liter/$^\circ$K. This means that a small increase in temperature will result in a small increase in volume; roughly, for each degree by which the temperature increases, the volume will increase by *approximately* $\frac{1}{300}$ liter. As usual, this is an *approximation*, which will be less and less accurate and reliable for larger and larger changes in T.

Similarly, the rate at which volume changes with respect to pressure is $-\frac{kT}{P^2} = -\frac{1}{300}\frac{\text{atm liter}}{^\circ\text{K}}(300^\circ\text{K})/(1\text{ atm}^2) = -1$ liter/atm. A change of 1 atm in the pressure is quite a large change (since $P = 1$ atm already), so we cannot expect the derivative to give a reliable approximation for so large a change, but it seems likely that an increase in pressure of $\frac{1}{100}$ atm, for instance, ought to result in a *decrease* in the volume of *about* $\frac{1}{100}$ liter.

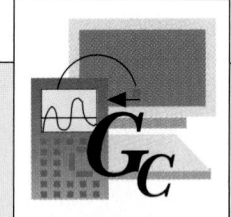

With a graphics calculator we can plot V as a function of P with T held fixed. At 27°C we have $T = 300$, so $V = \frac{kT}{P} = \frac{(1/300)(300)}{P} = \frac{1}{P}$.

We graph $V = \frac{1}{P}$ for $0.9 \le P \le 2.2$ and also graph the tangent line at the point $(P, V) = (1, 1)$. Since its slope is the rate of change, that is, -1, the line is $V = 2 - P$.

TI-85	SHARP EL9300
GRAPH F1 $(y(x) =)$ F1 (x) x^{-1} ▼ 2 − F1 (x) M2 (RANGE) .9 ▼ 2.2 ▼ ▼ 0 ▼ 1.1 F5 (GRAPH)	⟳ x/θ/T x^{-1} 2ndF ▼ 2 − x/θ/T RANGE .9 ENTER 2.2 ENTER 1 ENTER 0 ENTER 1.1 ENTER 1 ENTER ⟳

CASIO f_x-6300G (f_x-7700GB)	HP 48SX
Range .9 EXE 2.2 EXE 1 EXE 0 EXE 1.1 EXE 1 EXE EXE EXE EXE Graph 1 ÷ X EXE Graph 2 − X EXE	↰ PLOT PLOTR ERASE ATTN ' Y = 1/x X ↱ PLOT ↰ DRAW ERASE .9 ENTER 2.2 XRNG 0 ENTER 1.1 YRNG DRAW ATTN ' Y = 2 − X ↱ PLOT ↰ DRAW DRAW

Now try zooming in or redrawing the graphs for $1 \le P \le 1.01$ and see whether the final remarks in Example 6 are sensible.

Higher-Order Partial Derivatives

Just as with ordinary derivatives of functions of a single variable, it is possible to take partial derivatives of a partial derivative or even to repeat the process several times. It can be slightly more complicated, of course, because we could have the partial derivative with respect to x of the partial derivative of f with respect to y, the partial derivative with respect to y of the partial derivative of f with respect to y, and so on. Once again there are several possible notations.

In the notation using ∂ we can write $\frac{\partial}{\partial x}\left(\frac{\partial f}{\partial y}\right)$, and so on, and it is perfectly clear what this means. Another way to write the same thing is $\frac{\partial^2 f}{\partial x \partial y}$. The important thing to notice here is that the ∂x is on the left and the ∂y is on the right in the denominator, and this means that first we take the partial derivative with respect to y and afterward take the partial derivative with respect to x of the result.

It is also possible to have notations like $\frac{\partial^2 f}{\partial x^2}$, $\frac{\partial^3}{\partial x \partial y^2}\left(f(x,y)\right)$, and so on.

In the subscript notation we could write $(f_y)_x$ for $\frac{\partial}{\partial x}\left(\frac{\partial f}{\partial y}\right)$, but this is often abbreviated to f_{yx}. Notice that in this notation, f_{yx} refers to the second-order partial derivative obtained by differentiating *first* with respect to y and *second* with respect to x. This is perfectly reasonable if we realize that it comes from the notation $(f_y)_x$, but it is possible to be confused, since the order is the reverse of what is meant by $\frac{\partial^2 f}{\partial y \partial x}$.

Higher-order partial derivatives are handled in the analogous way, and similar notations can, of course, be used with three or more variables.

DEFINITION 16.3.4

Higher-order partial derivatives are defined as follows:

$$\frac{\partial^2 f}{\partial x \partial y} = f_{yx} = \frac{\partial}{\partial x}\left(\frac{\partial f}{\partial y}\right), \qquad \frac{\partial^2 f}{\partial x^2} = f_{xx} = \frac{\partial}{\partial x}\left(\frac{\partial f}{\partial x}\right).$$

Analogous notation is used for other combinations, for higher-order partial derivatives, and for partial derivatives of functions of three or more variables. Expressions like $\frac{\partial^2 f}{\partial x \partial y}$, h_{yz}, and $\frac{\partial^2 g}{\partial z \partial u}$, are called **mixed partials**.

EXAMPLE 7   Find the following higher-order partial derivatives: (i) $\frac{\partial^2}{\partial x^2}\left(\sin(xy)\right)$, (ii) $\frac{\partial^2}{\partial x \partial y}(3xy^2 - 2x^2y^4)$, (iii) $\frac{\partial^2}{\partial y \partial x}(3xy^2 - 2x^2y^4)$, (iv) $\frac{\partial^3}{\partial x \partial z^2}\left(ze^{xy} + \cosh(xyz)\right)$.

Solution

(i) First we evaluate $\frac{\partial}{\partial x}\left(\sin(xy)\right) = \cos(xy)y$, and then we find $\frac{\partial^2}{\partial x^2}\left(\sin(xy)\right) = \frac{\partial}{\partial x}\left(\cos(xy)y\right) = -\sin(xy)y^2 = -y^2\sin(xy)$.

(ii) $\frac{\partial^2}{\partial x \partial y}(3xy^2 - 2x^2y^4) = \frac{\partial}{\partial x}(6xy - 8x^2y^3) = 6y - 16xy^3$.

(iii) $\frac{\partial^2}{\partial y \partial x}(3xy^2 - 2x^2y^4) = \frac{\partial}{\partial y}(3y^2 - 4xy^4) = 6y - 16xy^3$.

(iv) $\frac{\partial^3}{\partial x \partial z^2}\left(ze^{xy} + \cosh(xyz)\right) = \frac{\partial^2}{\partial x \partial z}\left(e^{xy} + \sinh(xyz)xy\right) = \frac{\partial}{\partial x}\left(0 + \cosh(xyz)(xy)^2\right)$
$= \sinh(xyz)yz(xy)^2 + \cosh(xyz)2xy^2 = x^2y^3z\sinh(xyz) + 2xy^2\cosh(xyz)$.   ◢

Notice that the answers to parts (ii) and (iii) of Example 7 are equal, even though the partial derivatives were done in reverse order.

In fact, this will always happen with mixed partial derivatives of polynomials. After all,

$$\frac{\partial^2}{\partial x \partial y}(x^m y^n) = \frac{\partial}{\partial x}(nx^m y^{n-1}) = nmx^{m-1}y^{n-1}$$

and

$$\frac{\partial^2}{\partial y \partial x}(x^m y^n) = \frac{\partial}{\partial y}(mx^{m-1}y^n) = nmx^{m-1}y^{n-1}.$$

So the mixed partials are the same in either order, and since any polynomial in two variables is made up of terms of this form, the same will be true for any such polynomial.

This can easily be extended to polynomials in three or more variables, for which any mixed partial will be the same no matter what order is used.

It is possible to find examples of functions for which certain mixed partials do change depending on the order of the differentiations, but actually it is not very easy. Most of the time mixed partials do not depend on the order. It is not very easily proved, but the following theorem explains why, since most functions that we can write down will satisfy the conditions required by this theorem.

THEOREM 16.3.5

Suppose there is a disk around (a, b) so that the mixed partials f_{xy} and f_{yx} are both defined at every point in the disk and *continuous* there. Then they are equal at (a, b), that is, $f_{xy}(a, b) = f_{yx}(a, b)$, that is,

$$\frac{\partial^2 f}{\partial y \partial x}(a, b) = \frac{\partial^2 f}{\partial x \partial y}(a, b).$$

Similar results hold for mixed partials of higher order and/or with more than two variables.

Of course, it is easy to find examples of functions $f(x, y)$ for which some higher-order partial derivative is not defined at (a, b), but if it is defined at (a, b), then it will likely be continuous there too. This is not always the case, but it usually is, and this subtlety will not often cause us much difficulty.

EXAMPLE 8   Find $\frac{\partial^5}{\partial z^2 \partial x \partial y \partial x}\left(z e^{x^2 y} \sin(2x - y^2) + xyz^2\right)$.

Solution   We could simply differentiate five times to work this out, but it is possible to save a lot of time and trouble and reduce the chance of making an error by rearranging the order of the partial derivatives.

The first term of the function is $z e^{x^2 y} \sin(2x - y^2)$. If we want to take partial derivatives with respect to z, we think of x and y as constants, which makes this whole expression a constant times z. Its partial derivative with respect to z equals that constant, and its second partial with respect to z is zero (the derivative of the "constant"). So if the differentiations were performed in the opposite order, with the two partial derivatives with respect to z coming first, they would give zero right away, and it would not be necessary to work out the others for this first term.

The other term xyz^2 will also give zero, this time because differentiating twice with respect to x will result in zero. If we rearrange the differentiations so that the two with respect to x are done first, then it becomes clear without doing any work that the whole expression equals 0.

Strictly speaking, it is necessary to check that changing the order of the derivatives is justified. If we use the Chain Rule to differentiate $z e^{x^2 y} \sin(2x - y^2)$ with respect to any combination of the variables x, y, and z, in any order, the result will be a complicated expression made up of terms each of which is the product of $e^{x^2 y}$, $\sin(2x - y^2)$, or $\cos(2x - y^2)$, and a polynomial. Since each of these functions is continuous (by Theorem 16.2.11), their product is continuous.

Any partial derivative of any order of the polynomial xyz^2 is a polynomial, so it is certainly continuous. So any partial derivative of any order must be continuous at every point, and Theorem 16.3.5 tells us that the mixed partials will be the same in any order.

This means that changing the order was legitimate, and the answer 0 we obtained is correct.

Exercises 16.3

I

Evaluate each of the following.

1. $\frac{\partial}{\partial x}(3x^2y - 4xy^3)$

2. $\frac{\partial}{\partial y}(5x^3y + 2x^2y^4 - y^2)$

3. $\frac{\partial}{\partial x}\left(\sin(x^2y^2)\right)$

4. $\frac{\partial}{\partial z}\left(z\cosh(3xy^7 - xy)\right)$

5. $\frac{\partial}{\partial t}\left(\ln\left((1 + t^2)(1 + u^2)(1 + v^2)\right)\right)$

6. $\frac{\partial}{\partial u}\left(\arctan\left(\frac{u}{v}\right)\right)$

7. $\frac{\partial}{\partial x}\left(\frac{xyz}{1 + x^2 + z^2}\right)$

8. $\frac{\partial}{\partial y}\left(\frac{x+y}{y+z}\right)$

9. $\frac{\partial}{\partial x}(x^2\sqrt{4 + x^2y^2})$

10. $\frac{\partial}{\partial t}\left(\coth(xt + 3)\right)$

Find the equation of the tangent plane to each of the following surfaces at the specified point P.

11. $z = x^2 + y^2, \quad P = (2, 2, 8)$

12. $z = 2x^2 + y^2, \quad P = (1, -2, 6)$

13. $z = x^2 - y^2, \quad P = (1, 1, 0)$

14. $z = 3 - x^2 - 2y^2, \quad P = (0, 0, 3)$

15. $z = x^2 - 2x + y^2 + 4y, \quad P = (1, -2, -5)$

16. $z = \sin(x + y), \quad P = (0, \pi, 0)$

17. $z = e^{x-2y}, \quad P = (2, 1, 1)$

18. $x^2 + y^2 + z^2 = 3, \quad P = (1, 1, 1)$

19. $z^2 = x^2 + y^2, \quad P = (12, 5, 13)$

20. $z^2 = x^2 + y^2, \quad P = (3, -4, 5)$

Evaluate each of the following.

21. $\frac{\partial^2}{\partial x \partial y}\left(3xy^4 - 2x^2y^2 + 5y^7\right)$

22. $\frac{\partial^2}{\partial y \partial x}(5x + 7y - 3xy + xy^2)$

23. $\frac{\partial^2}{\partial x^2}\left(\cos(x^2y + 2)\right)$

24. $\frac{\partial^3}{\partial y^2 \partial x}(e^{xy})$

25. $\frac{\partial^2}{\partial x \partial y}\left(\arctan(x + y)\right)$

26. $\frac{\partial^2}{\partial y \partial z}\left(\ln(x + y)\right)$

27. $\frac{\partial^2}{\partial y \partial x}\left(\tan(\pi x^2 y)\right)$

28. $\frac{\partial^2}{\partial t^2}\left(\cos(kx)e^{-kt}\right)$

29. $\frac{\partial^3}{\partial y^2 \partial x}(e^{xy})$

30. $\frac{\partial^2}{\partial x \partial y}\left(\cos^2 2(x + y) + 4\cos^2(x + y)\sin^2(x + y)\right)$

For each of the following functions, check whether or not it satisfies **Laplace's Equation:** $f_{xx} + f_{yy} = 0$. Functions with this property are called **harmonic functions**; they are extremely important for many purposes, particularly in physics.

31. $f(x, y) = \ln(x^2 + y^2)$

32. $f(x, y) = 3x - 12y + 4$

33. $f(x, y) = x^2 - y^2$

34. $f(x, y) = 15x^2y + 2x - 1$

35. $f(x, y) = x^3 - 3xy^2 + y^3$

36. $f(x, y) = xy^3 - x^3y$

37. $f(x, y) = e^x \cosh y$

38. $f(x, y) = (y + 3)/(x^2 + y^2)$

39. $f(x, y) = \sinh x \sinh y$

40. $f(x, y) = \arctan\left(\frac{y}{x}\right)$

II

41. Let $f(x, y) = \cos(x + y)$. Find f_x in two different ways, one by using the Addition Formula (Theorem 1.7.11) and the other by differentiating directly. Compare your answers.

42. Let m and n be positive integers. Find
    (i) $\frac{\partial^{m+n}}{\partial x^m \partial y^n}(e^{xy})$, (ii) $\frac{\partial^{m+n}}{\partial x^m \partial y^n}(e^{2x+3y})$.

43. Assume that $c \neq 0$. Find the equation of the tangent plane to the sphere $x^2 + y^2 + z^2 = 4$ at the point (a, b, c) on the sphere.

*44. Assume that $c \neq 0$. Find the equation of the tangent plane to the sphere $x^2 + y^2 + z^2 = R^2$ at the point (a, b, c) on the sphere.

*45. In Example 4 it was observed that the argument we used does not work for points at which $z = 0$. If $b \neq 0$, consider the point $P = (a, b, 0)$ on the sphere $x^2 + y^2 + z^2 = 1$. Express a hemisphere containing P as the graph of a function $y = g(x, z)$, and find the equation of the tangent plane to the sphere at P. (Distinguish between the two cases $b > 0$ and $b < 0$.)

*46. Find the equation of the tangent planes to the sphere $x^2 + y^2 + z^2 = 1$ at the points $(\pm 1, 0, 0)$. (*Hint:* Use suitable hemispheres.)

47. Evaluate the partial derivative in Example 8 in the order specified, and verify that the answer is the same as what was found in the text.

48. At the end of Example 6 there is an estimate of the effect of increasing the pressure by $\frac{1}{100}$ atm (assuming that the temperature is held constant), using the partial derivative. In fact it is quite easy to use the Ideal

Gas Law (Equation 16.3.3) to calculate exactly what will happen if the pressure is increased this much.

(i) Make this calculation and compare the result with the estimate.

(ii) Calculate the effect of an increase of 1 atm in pressure (at constant temperature), and compare this with the estimate that could be made by using partial derivatives.

49. (i) In the situation of Example 6, use partial derivatives to estimate the effect on the volume of the gas of decreasing the temperature to $15°C$, while holding the pressure constant at 1 atm.

(ii) Calculate the exact answer and compare it with your estimate.

*50. (i) In the situation of Example 6, suppose the temperature is increased to $30°$ and the pressure is decreased by 2%. Estimate the effect on the volume. (*Hint:* Estimate the effect of each change separately.)

(ii) Compare your estimate from part (i) with the actual answer.

*51. The Ideal Gas Law, $PV = kT$ (Equation 16.3.3), allows us to express each of P, V, or T as a function of the other two. This means that it is possible to find the partial derivative of any one of them with respect to either of the others. Show that

$$\frac{\partial P}{\partial V}\frac{\partial V}{\partial T}\frac{\partial T}{\partial P} = -1.$$

*52. Suppose A, B, and C are three quantities that satisfy the equation $A^2B = C^3 + ABC$.

(i) Write B as a function of A and C, and find $\frac{\partial B}{\partial A}$ and $\frac{\partial B}{\partial C}$.

(ii) It is also possible to think of A as a function of B and C. Find $\frac{\partial A}{\partial B}$ and $\frac{\partial A}{\partial C}$.

(iii) It is also possible to think of C as a function of A and B. Find $\frac{\partial C}{\partial A}$ and $\frac{\partial C}{\partial B}$. (*Hint:* This time it is difficult to solve explicitly and write C as some formula involving A and B. Try to differentiate implicitly instead.)

(iv) When you did part (ii), you might have found a formula for A as a function of B and C and differentiated it, or you might have used implicit differentiation, as in part (iii). Whichever method you used, try the other one now, and make sure the answers you get are the same.

*53. Let $f(x, t) = \exp(-k^2 t)\sin(kx)$. Verify that f satisfies the *Heat Equation:* $\frac{\partial f}{\partial t} = \frac{\partial^2 f}{\partial x^2}$. This equation describes the flow of heat in a thin wire if x measures the displacement along the wire and t is the time.

54. In Example 2 we observed that for somebody who had consumed 3 ounces of alcohol 3 hours after a meal, we could estimate from the partial derivative that additional alcohol would lower the test score by 25.9 percentage points per ounce. By looking at Figure 16.3.1 (p. 954), decide whether this estimate will be too high or too low. Confirm your answer by comparing the estimate with the actual value for $A = 3.1$, $t = 3$ and for $A = 4$, $t = 3$.

Find a way to use your computer graphics software to sketch a graph $z = f(x, y)$ *and* the tangent plane at a point $P = (a, b, f(a, b))$. Perhaps you can graph only a small piece of the tangent plane in order not to hide the graph of f.

Sketch the graphs of the following functions, showing the tangent plane at the specified point.

55. $z = x^2 + y^2$, $P = (1, 1, 2)$

56. $z = 1 - x^2 - 3y^2$, $P = (0, 0, 1)$

57. $z = \sin(xy)$, $P = (\pi, \frac{1}{2}, 1)$

58. $z = x^2 - y^2$, $P = (0, 0, 0)$

59. In the situation of Example 2, use a graphics calculator or computer to sketch S as a function of t for $A = 0, 1, 2, 3, 4$. (Plot them all on the same picture if possible.)

16.4 Differentials

APPROXIMATION

(TOTAL) DIFFERENTIAL

DIFFERENTIABILITY

DIFFERENTIABILITY AND CONTINUITY

Now that we know how to work with partial derivatives, it is appropriate to try to use them to do some of the same things we did with ordinary derivatives of functions of one variable. One use we found for derivatives was estimating the values of a function near a point at which we knew the value and the derivative. We described this *linear approximation* in Section 4.10 and found that a convenient way to deal with it was to use *differentials*.

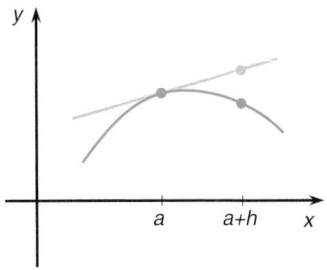

Figure 16.4.1

It turns out that differentials are even better suited to functions of two or more variables, and they are the subject of this section.

If we think about functions of one variable, one way of looking at the derivative is the idea of approximating the function $F(x)$ near $x = a$ by the straight-line approximation $F(a + h) \approx F(a) + F'(a)h$, as shown in Figure 16.4.1 (cf. Equation 4.10.1). We expect this to be a good approximation, especially if h is small.

The definition of the derivative says that the difference quotient $\frac{F(a+h)-F(a)}{h}$ approaches the derivative $F'(a)$. As h becomes small, the difference between the difference quotient and the derivative tends to zero.

Let us write $\epsilon(h)$ for this difference, that is, $\epsilon(h) = \frac{F(a+h)-F(a)}{h} - F'(a)$. With this notation the "error" in the linear approximation $F(a + h) \approx F(a) + F'(a)h$ is $F(a + h) - F(a) - F'(a)h$, which is exactly $\epsilon(h)h$.

Now the definition of the derivative $F'(a)$ says that $\frac{F(a+h)-F(a)}{h}$ approaches $F'(a)$, which is exactly the same as saying that

$$\lim_{h \to 0} \epsilon(h) = 0.$$

In other words, if we try to approximate $F(a + h)$ by $F(a) + F'(a)h$, the *error*, that is, the difference between the actual value $F(a + h)$ and the approximation, is quite small. The error, which is $\epsilon(h)h$, is so small that even if we divide it by h, which tends to 0, the quotient $\epsilon(h)$ will still tend to zero. We could say that the error tends to zero "faster than h does."

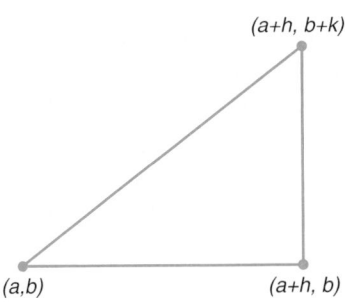

Figure 16.4.2

Now consider a function of two variables, $f(x, y)$. Suppose we know the values of f and its partial derivatives f_x and f_y at the point (a, b), and suppose we want to try to approximate the values of f at nearby points.

It seems natural to write a nearby point as $(a + h, b + k)$, where h and k are small numbers. There are many different ways to move from (a, b) to $(a + h, b + k)$, but an obvious way to do it would be to move from (a, b) along a horizontal line to $(a + h, b)$ and then along a vertical line to $(a + h, b + k)$. See Figure 16.4.2. It should be possible to approximate the corresponding changes in the value of f.

As x changes from $x = a$ to $x = a + h$ and y stays constant at $y = b$, the problem really involves a function of one variable, and we expect that $f(a + h, b) \approx f(a, b) + \frac{\partial f}{\partial x}(a, b)h$, just as in the one-variable case. Next, with x held fixed we let y change from $y = b$ to $y = b + k$, and the same sort of reasoning leads us to expect that the change in the value of the function will be approximately $\frac{\partial f}{\partial y}(a + h, b)k$.

These two changes should approximate the effect on $f(x, y)$ of moving from (a, b) to $(a + h, b)$ and then from there to $(a + h, b + k)$. We expect to have the following approximation:

$$f(a + h, b + k) \approx f(a, b) + \frac{\partial f}{\partial x}(a, b)h + \frac{\partial f}{\partial y}(a + h, b)k.$$

This is perfectly reasonable, but unfortunately it does not quite solve the problem with which we began. The difficulty is that in this formula the partial derivative $\frac{\partial f}{\partial y}$ is evaluated at $(a + h, b)$, rather than at (a, b). We are supposing that the values of f and its partial derivatives are known at (a, b), but we may not find it easy to calculate them at the nearby point $(a + h, b)$. After all, the situation in which approximations are likely to be used is the one in which the precise values of the function are difficult to find, except perhaps at a few special points.

On the other hand, since $(a + h, b)$ is close to (a, b), it seems likely that the values of f_y at these two points will not differ by much. If we knew that this was true, we could replace the awkward quantity $\frac{\partial f}{\partial y}(a + h, b)$ with the much nicer quantity

$\frac{\partial f}{\partial y}(a, b)$. The approximation formula would now read:

$$f(a + h, b + k) \approx f(a, b) + \left(\frac{\partial f}{\partial x}(a, b)\right)h + \left(\frac{\partial f}{\partial y}(a, b)\right)k$$

$$= f(a, b) + (f_x(a, b))h + (f_y(a, b))k. \qquad (16.4.1)$$

This form of the approximation would be much easier to use. The difficulty comes in trying to assess its reliability and accuracy. To make it work, we have to know that $f_y(a + h, b)$ is close to $f_y(a, b)$ if h is small, which will be true if f_y is continuous.

This argument suggests that if $\frac{\partial f}{\partial x}$ and $\frac{\partial f}{\partial x}$ exist and are continuous on a disk around (a, b), then for (x, y) near (a, b) the values of $f(x, y)$ can be approximated by Formula 16.4.1. This approach can be used for approximations, and it is also reminiscent of the situation for functions of one variable.

THEOREM 16.4.2

Suppose there is a disk around the point (a, b) so that f_x and f_y exist at every point in the disk and are continuous functions there. Then it is possible to find functions $\epsilon_1(h, k)$ and $\epsilon_2(h, k)$ so that

$$f(a + h, b + k) = f(a, b) + (f_x(a, b))h + (f_y(a, b))k + (\epsilon_1(h, k))h + (\epsilon_2(h, k))k,$$

and

$$\lim_{(h,k)\to(0,0)} \epsilon_1(h, k) = \lim_{(h,k)\to(0,0)} \epsilon_2(h, k) = 0.$$

For a function $g(x, y, z)$ of three variables the corresponding result is that if there is a ball around the point (a, b, c) on which the partial derivatives f_x, f_y, and f_z are continuous, then it is possible to find three functions ϵ_1, ϵ_2, and ϵ_3 of the variables (h, k, ℓ) so that

$$g(a + h, b + k, c + \ell) = g(a, b, c) + (g_x(a, b, c))h + (g_y(a, b, c))k$$
$$+ (g_z(a, b, c))\ell + (\epsilon_1(h, k, \ell))h + (\epsilon_2(h, k, \ell))k$$
$$+ (\epsilon_3(h, k, \ell))\ell$$

and

$$\lim_{(h,k,\ell)\to(0,0,0)} \epsilon_1(h, k, \ell) = \lim_{(h,k,\ell)\to(0,0,0)} \epsilon_2(h, k, \ell) = \lim_{(h,k,\ell)\to(0,0,0)} \epsilon_3(h, k, \ell) = 0.$$

A similar result can easily be stated for a function of n variables.

PROOF We give the proof for functions of two variables, but for three or more variables it is quite similar.

It is possible to write

$$f(a + h, b + k) = f(a, b) + (f(a + h, b) - f(a, b)) + (f(a + h, b + k) - f(a + h, b))$$

(Notice that most of the terms on the right side cancel.) We begin by considering the first quantity in large parentheses.

If we let $S(x) = f(x, b)$, which is a function of one variable, then $S(x)$ is differentiable on an interval around $x = a$, and its derivative is continuous there (after all, $S'(x) = \frac{\partial f}{\partial x}(x, b)$). The Mean Value Theorem (Theorem 4.2.5) can be applied to S, and it says that there is a number h^* between 0 and h so that $S(a + h) - S(a) = S'(a + h^*)h$. Since $S(x) = f(x, b)$ and $S'(x) = \frac{\partial f}{\partial x}(x, b)$, this says that $f(a + h, b) - f(a, b) = \left(\frac{\partial f}{\partial x}(a + h^*, b)\right)h$.

This can be rewritten as

$$f(a + h, b) - f(a, b) = \left(\frac{\partial f}{\partial x}(a + h^*, b) \right) h$$

$$= \left(\frac{\partial f}{\partial x}(a, b) \right) h + \left(\frac{\partial f}{\partial x}(a + h^*, b) - \frac{\partial f}{\partial x}(a, b) \right) h.$$

This can be written as

$$f(a + h, b) - f(a, b) = \left(\frac{\partial f}{\partial x}(a, b) \right) h + \left(\epsilon_1(h, k) \right) h$$

if we let $\epsilon_1(h, k) = \left(\frac{\partial f}{\partial x}(a + h^*, b) - \frac{\partial f}{\partial x}(a, b) \right)$.

Next we consider the second quantity in large parentheses, $f(a + h, b + k) - f(a + h, b)$. This time we let $T(y) = f(a + h, y)$ and use the Mean Value Theorem to find a k^* between 0 and k for which $T(b + k) - T(b) = T'(b + k^*)k$, that is, $f(a + h, b + k) - f(a + h, b) = \frac{\partial f}{\partial y}(a + h, b + k^*)k$. This can be written as

$$f(a + h, b + k) - f(a + h, b) = \left(\frac{\partial f}{\partial y}(a + h, b + k^*) \right) k$$

$$= \left(\frac{\partial f}{\partial y}(a, b) \right) k + \left(\frac{\partial f}{\partial y}(a + h, b + k^*) - \frac{\partial f}{\partial y}(a, b) \right) k$$

$$= \left(\frac{\partial f}{\partial y}(a, b) \right) k + \left(\epsilon_2(h, k) \right) k,$$

where we have let $\epsilon_2(h, k) = \left(\frac{\partial f}{\partial y}(a + h, b + k^*) - \frac{\partial f}{\partial y}(a, b) \right)$.

We have written

$$f(a + h, b + k) = f(a, b) + \left(\frac{\partial f}{\partial x}(a, b) \right) h + \left(\frac{\partial f}{\partial y}(a, b) \right) k + \left(\epsilon_1(h, k) \right) h$$

$$+ \left(\epsilon_2(h, k) \right) k.$$

All that remains is to show that $\epsilon_1(h, k)$ and $\epsilon_2(h, k)$ both tend to 0 as $(h, k) \to 0$. This follows from the definitions of ϵ_1 and ϵ_2 and the continuity of f_x and f_y, since h^* is between 0 and h and k^* is between 0 and k, so $h^* \to 0$ and $k^* \to 0$ as $(h, k) \to (0, 0)$.

Theorem 16.4.2 says that if the partial derivatives are continuous near (a, b), then the values of $f(x, y)$ near (a, b) can be approximated by $f(a + h, b + k) \approx f(a, b) + f_x(a, b)h + f_y(a, b)k$, or $f(a + h, b + k) - f(a, b) \approx f_x(a, b)h + f_y(a, b)k$. By analogy with the single-variable case this suggests the following definition.

DEFINITION 16.4.3

(Total) Differentials

If $f(x, y)$ is a function of two variables, its **differential** is the expression

$$df = \frac{\partial f}{\partial x} dx + \frac{\partial f}{\partial y} dy$$

or $df(x, y) = f_x(x, y) dx + f_y(x, y) dy$. It is sometimes called the **total differential** of f.

If $f(x, y, z)$ is a function of three variables, its **(total) differential** is the expression

$$df = \frac{\partial f}{\partial x}\, dx + \frac{\partial f}{\partial y}\, dy + \frac{\partial f}{\partial z}\, dz$$

or $df(x, y, z) = f_x(x, y, z)\, dx + f_y(x, y, z)\, dy + f_z(x, y, z)\, dz.$

For a function of n variables the **(total) differential** is

$$df = \frac{\partial f}{\partial x_1}\, dx_1 + \frac{\partial f}{\partial x_2}\, dx_2 + \ldots + \frac{\partial f}{\partial x_n}\, dx_n$$

or $df(x_1, \ldots, x_n) = f_{x_1}(x_1, \ldots, x_n)\, dx_1 + f_{x_2}(x_1, \ldots, x_n)\, dx_2 + \ldots + f_{x_n}$
$(x_1, \ldots, x_n)\, dx_n.$

Notice that the partial derivatives are functions, which have different values at different points. Because of this the differential, which involves the partial derivatives, will be different at different points.

The differential provides an easy way to approach approximation questions. The way it is used is this: Suppose we know $f(a, b)$ and can calculate the partial derivatives of f at (a, b). This means that we can write down the differential at (a, b), $df = f_x(a, b)\, dx + f_y(a, b)\, dy$. The differential is some number times dx plus some number times dy, so it is very easy to evaluate it for any particular values of dx and dy.

The idea is that if x changes by a small amount dx and y changes by a small amount dy, then we expect that the value of $f(x, y)$ will change by an amount which is *approximately* equal to $df = f_x\, dx + f_y\, dy$. We use this to estimate the value of f at the point $(a + dx, b + dy)$ by

$$f(a + dx, b + dy) \approx f(a, b) + df(a, b) = f(a, b) + f_x(a, b)\, dx + f_y(a, b)\, dy.$$

Notice that the partial derivatives are evaluated at (a, b); it is not necessary to know their values at nearby points.

In Figure 16.4.3 the surface is the graph $z = f(x, y)$. Starting at the point P, we change x by dx and y by dy and reach the point Q on the graph. The tangent vectors along the directions of the x- and y-coordinates help us estimate the change in the value of f.

The vertical change from the level of P to the end of each vector represents the estimated change along the x- or y-direction, respectively. These vertical changes are

$$f_x(a, b)\, dx \qquad \text{and} \qquad f_y(a, b)\, dy$$

(since $f_x(a, b)$ and $f_y(a, b)$ are the slopes of the vectors).

The actual changes are the vertical drops to the points on the surface below the endpoints of the vectors. The differential

$$df = f_x(a, b)\, dx + f_y(a, b)\, dy$$

is the sum of the two estimated changes, and it estimates the actual change in f from the level of P to the point Q on the surface.

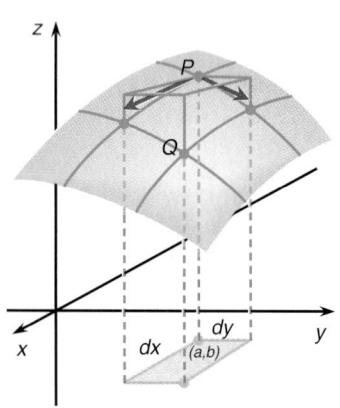

Figure 16.4.3

EXAMPLE 1   (i) Approximate the value $f(0.1, 1.9)$ if $f(x, y) = \sinh(xy^2 + y - 2)$.

Solution   We differentiate and find that $f_x(x, y) = \cosh(xy^2 + y - 2)y^2$ and $f_y(x, y) = \cosh(xy^2 + y - 2)(2xy + 1)$. The differential is

$$df = f_x(x, y)\, dx + f_y(x, y)\, dy$$
$$= y^2 \cosh(xy^2 + y - 2)\, dx + (2xy + 1) \cosh(xy^2 + y - 2)\, dy.$$

The one place where it is especially easy to evaluate $\cosh t$ is at $t = 0$; we notice that the point $(0.1, 1.9)$ is near the point $(0, 2)$, at which the differential can easily be evaluated. We find that $f_x(0, 2) = 2^2 \cosh(0) = 4$ and $f_y(0, 2) = (0 + 1) \cosh(0) = 1$, so the differential at the point $(0, 2)$ is $df = 4\,dx + dy$.

To get from $(0, 2)$ to $(0.1, 1.9)$, we should let $dx = 0.1$ and $dy = -0.1$. The approximate change in the value of f is $df = 4\,dx + dy = 4(0.1) + (-0.1) = 0.3$. Since the value of f at $(0, 2)$ is $f(0, 2) = \sinh(0 + 2 - 2) = \sinh(0) = 0$, we expect that $f(0.1, 1.9) \approx f(0, 2) + df = 0 + 0.3 = 0.3$.

With a calculator it is possible to check that $f(0.1, 1.9) = \sinh\big(0.1(1.9)^2 + 1.9 - 2\big) = \sinh(0.261) = \frac{1}{2}(e^{0.261} - e^{-0.261}) \approx 0.26397$. We see that our estimate of 0.3 was not extremely accurate.

(ii) A rectangular box has dimensions 3 ft $\times$ 2 ft $\times$ 1.5 ft, so its volume is 9 ft^3. If the long dimension is increased by 1% and the other two are both decreased by 1%, estimate the approximate effect on the volume.

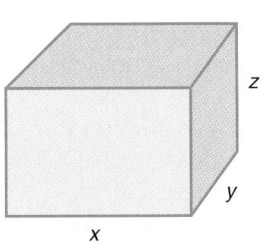

Figure 16.4.4

Solution   If we write x, y, and z for the three dimensions (see Figure 16.4.4), then the volume is $V = xyz$. The differential is $dV = V_x\,dx + V_y\,dy + V_z\,dz = yz\,dx + xz\,dy + xy\,dz$. When $x = 3$, $y = 2$, and $z = 1.5$, the differential is $dV = 3dx + \frac{9}{2}dy + 6dz$.

If the long dimension x increases by 1%, it will increase from $x = 3$ to $x = 3.03$, and the change in x will be $dx = 0.03$. A decrease in y of 1% from $y = 2$ will mean a change of $dy = -0.02$, and a decrease of 1% in z will mean a change of $dz = -0.015$.

We combine these facts to find that the approximate change in the volume is $dV = 3dx + \frac{9}{2}dy + 6dz = 3(0.03) + \frac{9}{2}(-0.02) + 6(-0.015) = -0.09$. The new volume will be approximately $V + dV = 9 - 0.09 = 8.91$ ft^3. If we want to find the percentage change in the volume, we calculate

$$\frac{dV}{V} \times 100\% = \frac{-0.09}{9} \times 100\% = -1\%.$$

The volume *decreases* by about 1%.

Of course, it is easy in this case to evaluate the volume with $x = 3.03$, $y = 1.98$, and $z = 1.485$; we find that $V = 3.03 \times 1.98 \times 1.485 = 8.909109$, and the estimate we made was quite good.

> The examples we have just done are not very convincing because in each case it was easy to evaluate the function directly (or, in the first example, to evaluate approximately with the help of a calculator). On the other hand, the use of differentials can throw some new light even on easy problems like these. For instance, for the volume of the rectangular box we found that $dV = 3dx + \frac{9}{2}dy + 6dz$. By looking at this we see that a small change in z has an effect on the volume that is twice as large as the effect of a change in x of the same amount.
>
> Having made this observation, it is not too difficult in hindsight to realize why it is true. In the formula for the volume $V = xyz$ we see that z is multiplied by xy, y is multiplied by xz, and x is multiplied by yz. In this case the largest of these factors is xy, the factor by which z is multiplied, so a small change in z will have the largest effect on $V = xyz$.

As with functions of one variable, differentials are very useful for estimating the error that may result from calculations with measurements that have been made with some inaccuracy.

EXAMPLE 2   Suppose two sides of a triangular sheet of metal have been measured and found to be 6 ft and 9 ft long, to within $\frac{1}{2}$ in, and the angle between these sides has been measured and found to be $53°$, to within $1°$. If these measurements are used to find the area of one side of the sheet, estimate the largest possible error in the calculated value of the area.

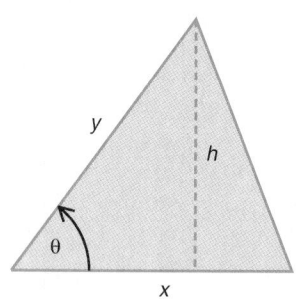

Figure 16.4.5

Solution   We begin by drawing a picture of the triangle in Figure 16.4.5. Recalling that its area is $\frac{1}{2}(base)(height)$, we choose one side to be its base (let us write x for its length), mark the length of the other measured side as y, and mark the angle θ as shown. To find the height of the triangle, we drop a perpendicular line down to the base from the opposite corner and realize that if h is the height, then $\sin \theta = \frac{h}{y}$. This shows that the height is $h = y \sin \theta$. The area is $A = \frac{1}{2}(base)(height) = \frac{1}{2}xy \sin \theta$. (Notice that it does not matter which side we choose to be the base, as this formula is "symmetrical" in x and y.)

Now we consider the function $A(x, y, \theta) = \frac{1}{2}xy \sin \theta$. It is easy to find its value at the given measurements, that is, the calculated area of the triangle is $A(6, 9, 53°) = \frac{1}{2}6(9) \sin(53°)$, which can be evaluated with a calculator to be about 21.6 ft^2.

We will find the approximate amount by which the function A will change if we allow x, y, and θ to change by the largest possible errors in their measurements. Now the possible errors in the measurements of x and y are anything up to $\pm\frac{1}{2}$ in $= \pm\frac{1}{24}$ ft. The possible error in θ is anything up to $\pm1°$. As usual it is necessary to convert the angle to radians, and we find that $53° = \frac{53}{180}\pi \approx 0.925$ radians. The possible error of $\pm1°$ is $\pm\frac{1}{180}\pi \approx \pm0.01745$ radians.

The partial derivatives of A are easily found: $\frac{\partial A}{\partial x} = \frac{1}{2}y \sin \theta$, $\frac{\partial A}{\partial y} = \frac{1}{2}x \sin \theta$, and $\frac{\partial A}{\partial \theta} = \frac{1}{2}xy \cos \theta$. Evaluating these at the point $(6, 9, 0.925)$ with the help of a calculator, we find that $A_x(6, 9, 0.925) \approx 3.594$, $A_y(6, 9, 0.925) \approx 2.396$, and $A_\theta(6, 9, 0.925) \approx 16.249$. From this we find that the differential at this point is approximately $3.594\, dx + 2.396\, dy + 16.249\, d\theta$.

The errors can be anything up to $\pm\frac{1}{24}$ for dx and dy and anything up to ±0.01745 for $d\theta$. The largest possible value of dA will occur when dx, dy, and $d\theta$ are all as large as possible, that is, when $dx = dy = \frac{1}{24}$ and $d\theta = 0.01745$. With these values we find that the error is approximately $dA = 3.594\, dx + 2.396\, dy + 16.249\, d\theta = 3.594\left(\frac{1}{24}\right) + 2.396\left(\frac{1}{24}\right) + 16.249 \times 0.01745 \approx 0.533$.

This suggests that the error in the calculated value of the area could be as much as 0.533 ft^2. This is a percentage error of about $\frac{0.533}{21.6} \times 100\% \approx 2.5\%$.

EXAMPLE 3   A technician weighs a small bacterial culture, finding that it weighs 3 mg and that 1 day later it weighs 5 mg. Assuming that the culture grows exponentially according to the formula $f(t) = Ae^{kt}$, he finds that $A = f(0) = 3$ and $5 = f(1) = Ae^k = 3e^k$, so $k = \ln\frac{5}{3}$. The amount of bacteria after t days is expected to be $f(t) = 3e^{t\ln(5/3)}$ mg.

In particular, after 6 days there should be $f(6) = 3e^{6\ln(5/3)} \approx 64.3$ mg of bacteria.

However, the original measurements were only made to within an accuracy of ±0.1 mg. Estimate the possible percentage error in this conclusion about the six-day growth.

Solution   Let us write A and B for the amounts of bacteria at $t = 0$ and $t = 1$, respectively, so $A = 3 \pm 0.1$ and $B = 5 \pm 0.1$. The amount of bacteria at time t is

$f(t) = Ae^{kt}$, where $B = f(1) = Ae^k$, so $k = \ln \frac{B}{A} = \ln B - \ln A$. The amount of bacteria after 6 days is $f(6) = Ae^{6k} = Ae^{6(\ln B - \ln A)} = A\left(\frac{B}{A}\right)^6 = A^{-5}B^6$.

We think of this expression as a function of A and B, that is,

$$F(A, B) = A^{-5}B^6.$$

The partial derivatives are

$$\frac{\partial F}{\partial A} = -5A^{-6}B^6 \qquad \text{and} \qquad \frac{\partial F}{\partial B} = 6A^{-5}B^5.$$

From this we find the differential

$$dF = \frac{\partial F}{\partial A}\, dA + \frac{\partial F}{\partial B}\, dB = -5A^{-6}B^6\, dA + 6A^{-5}B^5\, dB.$$

The measured values are $A = 3$ and $B = 5$. If we let the errors dA and dB both be 0.1, we find that

$$dF = -5(3^{-6}5^6)(0.1) + 6(3^{-5}5^5)(0.1) = 3^{-6}5^5(-25 + 18)(0.1)$$
$$\approx -3.0.$$

However, we observe that the two terms in this differential have opposite signs, so they partly cancel. We will get a larger error by letting $dA = -0.1$ and $dB = 0.1$. Then

$$dF = -5(3^{-6}5^6)(-0.1) + 6(3^{-5}5^5)(0.1) = 3^{-6}5^5(25 + 18)(0.1)$$
$$\approx 18.4.$$

See the Graphics Calculator Example on the next page.

Since the original calculation gave 64.3 mg, this represents an error of $\frac{18.4}{64.3} \times 100\% \approx 28.6\%$.

When we find an approximate error that is this large, we should suspect that our estimate may not be very accurate. Suppose the error was as bad as possible, which would happen if the initial amount was $A = 3 - 0.1 = 2.9$ and the amount after 1 day was $B = 5 + 0.1 = 5.1$. Then the actual amount after 6 days would be

$$Ae^{6(\ln B - \ln A)} = 2.9e^{6(\ln 5.1 - \ln 2.9)} \approx 85.6.$$

So in the worst possible case the actual error is $85.8 - 64.3 = 21.5$; this is $\frac{21.5}{64.3} \times 100\% \approx 33.4\%$, which is even worse than our estimate. It is especially startling because the errors in the original measurements were only $\frac{0.1}{3} \times 100\% \approx 3.3\%$ and $\frac{0.1}{5} \times 100\% = 2\%$.

Example 3 illustrates that to find the largest possible error, we have to pick dA and dB intelligently. The idea is to do it in such a way that the different terms in the differential do not cancel each other, which is the same thing as saying that they should all have the same sign. The simplest thing to do is to choose the sign for the change in each variable to be the same as the sign of the corresponding partial derivative because when they are multiplied together in the differential, the product will always be positive.

It is also true, just as with differentials of functions with one variable, that differentials can be used to estimate values that cannot easily be calculated, even with the help of a calculator.

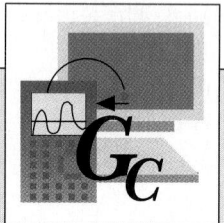

With a graphics calculator we can plot the function F from Example 3, with the measured values $A = 3$ and $B = 5$. Then we plot it with $A = 2.9$ and $B = 5.1$.

TI-85	SHARP EL9300
GRAPH F1 ($y(x) =$) 3 e^x (F1 (x) × LN (5 ÷ 3)) ▼ 2.9 e^x (F1 (x) × LN (5.1 ÷ 2.9)) M2 (RANGE) 0 ▼ 6 ▼ ▼ 0 ▼ 90 F5(GRAPH)	⋀⋁ 3 e^x x/θ/T × ln (5 ÷ 3) 2ndF ▼ 2.9 e^x x/θ/T × ln (5.1 ÷ 2.9) RANGE 0 ENTER 6 ENTER 1 ENTER 0 ENTER 90 ENTER 10 ENTER ⋀⋁

CASIO f_x-6300G (f_x-7700GB)	HP 48SX
Range 0 EXE 6 EXE 1 EXE 0 EXE 90 EXE 10 EXE EXE EXE EXE Graph 3 e^x (X × ln (5 ÷ 3)) EXE Graph 2.9 e^x (X × ln (5.1 ÷ 2.9)) EXE	↰ PLOT PLOTR ERASE ATTN ′ Y = 3 × e^x X × LN 5 ÷ 3 ↱ PLOT ↰ DRAW ERASE 0 ENTER 6 XRNG 0 ENTER 90 YRNG DRAW ATTN ′ Y = 2.9 × e^x X × LN 5.1 ÷ 2.9 ↱ PLOT ↰ DRAW DRAW

Identify which graph is which, and note how a small initial difference can grow. What happens with $A = 3.1$ and $B = 4.9$, or $A = 2.9$ and $B = 4.9$, or $A = 3.1$ and $B = 5.1$?

EXAMPLE 4 Let $f(x, y) = \sinh(xy^2 + y - 2)$ be the function we encountered in Example 1(i). We noted that $f(0, 2) = 0$. Let $g(x, y) = \sin(\pi(y + x^2))$, and note that $g(0, 2) = \sin(2\pi) = 0$. Estimate the coordinates of a point (x, y) near $(0, 2)$ at which $f(x, y) = -0.012$ and $g(x, y) = -0.005$.

Solution We know that $f_x(0, 2) = 4$ and $f_y(0, 2) = 1$, so the differential at the point $(0, 2)$ is $df = 4dx + dy$. Similarly, $g_x = \cos(\pi(y + x^2))2\pi x$ and $g_y = \cos(\pi(y + x^2))\pi$, so $g_x(0, 2) = 0$ and $g_y(0, 2) = \pi$. From this we find that the differential of g at the point $(0, 2)$ is $dg = \pi\, dy$.

What we want is to move from the point $(0, 2)$ by small amounts dx and dy so that the values of f and g change by the appropriate amounts. Now $df = 4\, dx + dy$ and $dg = \pi\, dy$, and what we want is that the value of f should change from 0 to -0.012, that is, that $df = -0.012$ and g should change from 0 to -0.005, that is, that $dg = -0.005$. We try letting

$$4\, dx + dy = -0.012 \quad \text{and} \quad \pi\, dy = -0.005.$$

From the second of these equations we find that $dy = \frac{1}{\pi}(-0.005) \approx -0.00159$, and substituting this value into the first equation gives $4\, dx = -0.012 - dy$, so $dx \approx -0.00260$.

This suggests that the point $(0 + dx, 2 + dy) = (-0.00260, 1.99841)$ ought to be the point we want, at which f will be approximately equal to -0.012 and g is approximately equal to -0.005.

We check our answer with a calculator and find that $f(-0.00260, 1.99841) \approx -0.01197$ and $g(-0.00260, 1.99841) \approx -0.00497$. This shows that our answer is remarkably good.

Notice that there is no obvious way to solve this problem without differentials, even with the help of a calculator, except by trial and error.

Differentiability and Continuity

When we studied functions of one variable, we were able to show that if a function $F(x)$ is differentiable at $x = a$, then it must be continuous there. It seems reasonable to ask whether a function $f(x, y)$ of two variables must be continuous at a point (a, b) if both its partial derivatives $\frac{\partial f}{\partial x}$ and $\frac{\partial f}{\partial y}$ exist at a point (a, b).

Unfortunately, the answer is "no." The reason has to do with the fact that the partial derivatives only involve the values of $f(x, y)$ on two lines, the horizontal and vertical lines through (a, b), namely, $y = b$ and $x = a$ (see Figure 16.4.6). It is not difficult to think of a function that is quite well-behaved on these lines but that is not continuous at (a, b).

EXAMPLE 5   Consider the function

$$f(x, y) = \begin{cases} \frac{xy}{x^2 + y^2}, & \text{if } (x, y) \neq (0, 0), \\ 0, & \text{if } (x, y) = (0, 0). \end{cases}$$

On the x-axis, that is, at the points where $y = 0$, the function is always zero. So $\frac{\partial f}{\partial x}(0, 0) = 0$, since it is the derivative at $x = 0$ of the constant function 0. In the same way, $\frac{\partial f}{\partial y}(0, 0) = 0$ because $f(x, y)$ is also zero at every point on the y-axis.

So both partial derivatives exist at $(0, 0)$ (and in fact they exist at every other point too, since $f(x, y)$ is a rational function whose denominator is nonzero except at $(0, 0)$). On the other hand, if we let (x, y) approach the origin along the line $y = x$ with $x \neq 0$, then the values of the function $f(x, x) = \frac{(x)(x)}{x^2 + x^2} = \frac{x^2}{2x^2} = \frac{1}{2}$ certainly approach the constant $\frac{1}{2}$. Since this does not equal $f(0, 0)$, the function $f(x, y)$ is not continuous at $(0, 0)$.

> This shows that the existence of the partial derivatives at (a, b), or even their existence at *every* point, is not enough to guarantee the continuity of $f(x, y)$. Of course, a similar remark can be made for functions of three or more variables.

With only one variable we found that if a function $F(x)$ is differentiable, then it is possible to make approximations using the derivative. It is not difficult to check that the reverse is also true: If the approximations made this way are accurate, in the sense that $\epsilon(h) \to 0$, where $\epsilon(h)$ is defined by $F(a + h) = F(a) + Ah + \epsilon(h)h$, for some number A, then $F(x)$ must in fact be differentiable, and $F'(a) = A$.

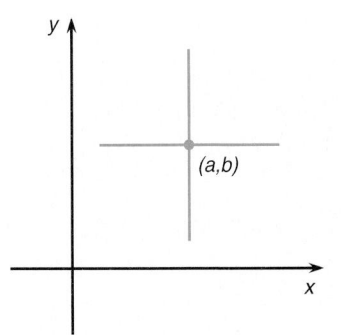

(a,b)

Figure 16.4.6

For functions of two (or more) variables we have seen that the existence of partial derivatives is not enough to guarantee continuity, let alone the stronger approximation condition that is the conclusion of Theorem 16.4.2, so it is not a particularly useful property. The approximation property is much more useful, and in fact it does imply that $f(x, y)$ is continuous at (a, b). It is for these reasons that we use this property to define differentiability for functions of more than one variable. Among other things this avoids the problem that we encountered with partial derivatives, that they deal with the behavior of the function only in two special directions, ignoring what it does in every other direction.

DEFINITION 16.4.4

Differentiability

A function $f(x, y)$ is **differentiable** at a point (a, b) provided that there is a disk around (a, b) on which it is possible to find two functions $\epsilon_1(h, k)$ and $\epsilon_2(h, k)$ so that

$$f(a + h, b + k) = f(a, b) + f_x(a, b)h + f_y(a, b)k + \epsilon_1(h, k)h + \epsilon_2(h, k)k$$

and

$$\lim_{(h,k)\to(0,0)} \epsilon_1(h, k) = \lim_{(h,k)\to(0,0)} \epsilon_2(h, k) = 0.$$

Theorem 16.4.2 gives an easy way to check that a function is differentiable.

For a function of three or more variables the definition is analogous; we omit the details.

Notice that if $f(x, y)$ is differentiable at (a, b), then

$$\lim_{(h,k)\to(0,0)} f(a + h, b + k)$$

$$= \lim_{(h,k)\to(0,0)} \left(f(a, b) + f_x(a, b)h + f_y(a, b)k + \epsilon_1(h, k)h + \epsilon_2(h, k)k \right)$$

$$= f(a, b) + \lim_{(h,k)\to(0,0)} \left(f_x(a, b)h \right) + \lim_{(h,k)\to(0,0)} \left(f_y(a, b)k \right)$$

$$+ \lim_{(h,k)\to(0,0)} \left(\epsilon_1(h, k)h \right) + \lim_{(h,k)\to(0,0)} \left(\epsilon_2(h, k)k \right)$$

$$= f(a, b) + 0 + 0 + 0 + 0 = f(a, b).$$

In other words, $\lim_{(x,y)\to(a,b)} f(x, y) = f(a, b)$, which is exactly what it means to say that $f(x, y)$ is continuous at the point (a, b). So $f(x, y)$ is continuous at every point at which it is differentiable. Of course, a similar remark can be made about functions of three or more variables.

Exercises 16.4

For each of the following functions, find the differential at the specified point.

1. $f(x, y) = 3x^2y + y^3$, $(2, -3)$

2. $f(x, y) = x \sin(2xy + y)$, $(0, \pi)$

3. $f(x, y) = (x^2y^2 + 2)/(3x - 2y)$, $(1, 1)$

4. $f(x, y) = \cos x \cos y - \sin x \sin y$, $\left(\frac{\pi}{2}, -\frac{\pi}{4}\right)$

5. $f(x, y) = e^{x+2xy}$, $(0, 3)$

6. $f(x,y) = \ln(x^2 y^3)$, $(2,1)$

7. $f(x,y,z) = (x^2 + y^2 + z^2)^{-1}$, $(1,0,-2)$

8. $f(x,y,z) = z\sin(x^2 yz)$, $(-1,\pi,3)$

9. $f(x,y,z) = \sqrt{3x^2 - 2y^2 + z}$, $(-2,1,-2)$

10. $f(x,y) = \cos^2 x \cos^2 y + \cos^2 y \sin^2 x + \sin^2 y$, $(2,7)$

In each of the following cases, use the differential to approximate the value of the function at the specified point, using the given starting point.

11. $f(2.1, 1.2)$, with $f(x,y) = e^{x-2y}$, $(a,b) = (2,1)$

12. $f(1.9, 2.01)$, with $f(x,y) = \sqrt{x^3 y}$, $(a,b) = (2,2)$

13. $f(-0.9, 1.99)$, with $f(x,y) = (2xy^2)^{-1/3}$, $(a,b) = (-1,2)$

14. $f\left(\frac{61}{30}, -2.98\right)$, with $f(x,y) = y^2 \cosh(3x+2y)$, $(a,b) = (2,-3)$

15. $f\left(-\frac{13}{24}, 2.02\right)$, with $f(x,y) = xy\sin(\pi x^2 y)$, $(a,b) = \left(-\frac{1}{2}, 2\right)$

16. $f\left(113, 24\frac{3}{4}\right)$, with $f(x,y) = \log\left(\frac{x}{55}\sqrt{y}\right)$, $(a,b) = (110, 25)$

17. $f\left(4\frac{1}{7}, 1.9\right)$, with $f(x,y) = \frac{x}{y} + \frac{y}{x}$, $(a,b) = (4,2)$

18. $f\left(1 + \frac{2}{37}, -3.9\right)$, with $f(x,y) = \exp(xy\ln(2))$, $(a,b) = (1,-4)$

19. $f\left(1 - \frac{1}{12}, 2 + \frac{1}{7}, -0.01\right)$, with $f(x,y,z) = y^2\sqrt{x} + \cos(xyz)$, $(a,b,c) = (1,2,0)$

20. $f(-1.01, 2.01, 0.98)$, with $f(x,y,z) = \tan(\pi xy^2 z^3)$, $(a,b,c) = (-1,2,1)$

Estimate the largest possible percentage error in the following quantities if the measurements are made with the possible errors shown.

21. The area of a rectangle whose sides are 5 ft and 8 ft, each measured to within $\frac{1}{4}$ in.

22. The area of a rectangle whose sides are 3 units and 7 units, each measured to within 1%.

23. The volume of a rectangular box whose sides are 3 m, 2 m, and 4 m, each to within 1 cm.

24. The volume of a rectangular box whose sides are 5 ft, 4 ft, and 4 ft, each to within 0.5%.

25. The area of a triangle, two of whose sides measure 4 m and 5 m, to within 1 cm, with the angle between them measuring 28°, to within 0.5°.

26. The area of a triangle, two of whose sides measure 2 yd and 10 yd, to within 1 in, with the angle between them measuring 42°, to within 0.75°.

27. The height of a tree as found by an observer who measures her distance from the base of the tree to be 100 ft, to within 4 in, and the angle of inclination of her line of sight to the treetop to be 34°, to within 1°.

(Assume that the observation is made from ground level.)

28. The height of a tree whose shadow is 40 m long, to within 10 cm, when the sun is 40° above the horizon, give or take 1°.

29. The distance between $P = (1,2)$ and $Q = (4,-2)$, where the coordinates of both points are measured with a possible error of ± 0.01.

30. The distance between $R = (2,-1,4)$ and $S = (4,1,5)$, where the coordinates of both points are measured to within an accuracy of 1%.

II

31. Consider a rectangular wooden box whose dimensions are 2 ft by 3 ft by 4 ft. Suppose it is to be covered with a plastic laminate that is $\frac{1}{32}$ in thick.
   (i) Approximate the volume of the covered box.
   (ii) Try to decide whether this estimate will be larger or smaller than the actual volume, and explain your reasoning.
   (iii) Find the actual volume, and compare. In particular, is the estimate too large or too small?

32. (i) With the covered wooden box of the previous question, estimate the surface area of the box after the laminate has been applied.
   (ii) Do you expect your estimate to be larger or smaller than the actual area? Explain.
   (iii) Compare your estimate with the exact answer, noting whether it is too large or too small.

*33. From the Ideal Gas Law, $\frac{PV}{T} = k$, where P is the pressure, V the volume, and T the temperature of a fixed quantity of gas and k is a constant, it is possible to express any one of P, V, or T as a function of the other two.
   (i) Express V as a function of P and T and find the differential dV.
   (ii) Using your answer to part (i), express dT in terms of dV and dP.
   (iii) Check that your answer to part (ii) is the same as the differential you would find by writing T as a function of V and P and calculating its partial derivatives.

*34. (i) From the Ideal Gas Law, $\frac{PV}{T} = k$, express P as a function of V and T, and find dP.
   (ii) Check that your answer to part (i) is the same as what you would find by solving from the expression you found for dV in Exercise 33.
   (iii) Check that your answer to part (i) is the same as what you would find by solving from the expression you found for dT in Exercise 33.

35. In the situation of Example 3, calculate the amount of bacteria expected after 8 days' growth, and estimate the corresponding possible percentage error.

36. After Example 3 we did a "worst-case analysis" by evaluating $F(A, B)$ with $A = 2.9$ and $B = 5.1$. Try it "in the other direction," with $A = 3.1$ and $B = 4.9$. Explain your answer.

37. Let $f(x, y) = \sin(x + 2y)$, $g(x, y) = \tan(xy^2 + 2x - 3y)$. Estimate a point (x, y) at which $f(x, y) = 0.002$ and $g(x, y) = -0.008$. (Note that $f(0, 0) = g(0, 0) = 0$.)

38. Let $f(x, y) = e^{\sin(x+y)}$, $g(x, y) = 2x + \cos\left(\frac{\pi}{2}(2x - y)\right)$. Estimate a point (x, y) at which $f(x, y) = 0.99$ and $g(x, y) = 1.02$.

*39. Let $f(x, y, z) = 2\cos\left(\frac{\pi}{3}(xy + z)\right)$, $g(x, y, z) = xz - y$, $h(x, y, z) = yz$. Estimate a point (x, y, z) at which

$f(x, y, z) = -0.95$, $g(x, y, z) = 0.1$, $h(x, y, z) = 1.1$. (*Hint:* Start at $(1, 1, 1)$.)

*40. Let $f(x, y, z) = \frac{xy}{z}$, $g(x, y, z) = \frac{xz}{y}$, $h(x, y, z) = \frac{yz}{x}$. Estimate a point (x, y, z) at which $f(x, y, z) = 1.01$, $g(x, y, z) = 0.99$, $h(x, y, z) = 1.02$.

41. Use a graphics calculator or computer to plot the function $f(t)$ of Example 3 for $0 \le t \le 6$, using the measured values $A = 3$, $B = 5$. On the same picture, plot the functions corresponding to $A = 2.9$, $B = 5.1$; $A = 3.1$, $B = 4.9$; $A = 3.1$, $B = 5.1$; $A = 2.9$, $B = 4.9$.

POINT TO PONDER

In Theorem 16.4.2 we showed that if $f(x, y)$ is a "nice" function defined at the point (a, b), then it is possible to find functions $\epsilon_1(h, k)$ and $\epsilon_2(h, k)$ so that

$$f(a + h, b + k) - f(a, b)$$
$$= f_x(a, b)h + f_y(a, b)k + \epsilon_1(h, k)h + \epsilon_2(h, k)k,$$

and

$$\lim_{(h,k)\to(0,0)} \epsilon_1(h, k) = \lim_{(h,k)\to(0,0)} \epsilon_2(h, k) = 0.$$

It is interesting to observe that it is possible to find many different possible choices of ϵ_1 and ϵ_2.

For example, consider the very simple function $f(x, y) = xy$. We find that $f_x(x, y) = y$ and $f_y(x, y) = x$, so the differential of f at the point (a, b) is $df = b\,dx + a\,dy$. The approximation to $f(a + h, b + k)$ is $f(a, b) + f_x(a, b)h + f_y(a, b)k = ab + bh + ak$.

The *actual value* of $f(a + h, b + k)$ is $(a + h)(b + k) = ab + bh + ak + hk$, so the error between this

and the approximation is hk. We are supposed to be able to express this error as $\epsilon_1(h, k)h + \epsilon_2(h, k)k$, with some functions that tend to 0 as $(h, k) \to (0, 0)$. One particularly easy thing to try would be to let $\epsilon_2(h, k) = 0$ and write the error hk as $\epsilon_1(h, k)h$. This means that $\epsilon_1(h, k) = k$, which certainly does satisfy the requirement that $\epsilon_1(h, k) \to 0$ as $(h, k) \to (0, 0)$.

Of course, we could do it the other way around, letting ϵ_1 be the zero function and letting $\epsilon_2(h, k) = h$. Another possibility would be to let $\epsilon_1(h, k) = ck$, for any constant c, and let $\epsilon_2(h, k) = (1 - c)h$. Then $\epsilon_1(h, k)h + \epsilon_2(h, k)k = ckh + (1 - c)hk = hk$, as required. We have found infinitely many possible choices of ϵ_1 and ϵ_2, one for each constant c.

As an exercise you might like to consider the function $g(x, y) = xy^2$. Find examples of possible functions ϵ_1 and ϵ_2, but show that in this case it is not possible to do it with ϵ_2 equal to the zero function.

16.5 The Chain Rule

INTERMEDIATE VARIABLE

IMPLICIT DIFFERENTIATION

In this section we study the Chain Rule for functions of several variables. One of the main difficulties with this topic is trying to keep track of which quantities are variables and which are functions. This was a problem with functions of one variable, but it is much worse when each function can have several variables.

Let us begin with a simple example. Suppose $f(u, v) = u + v \sin(u + v)$. Now suppose that u and v are both functions of the variable x; suppose, for example, that $u = 3x^2 + 1$ and $v = 2x - 3$. Then u and v depend on x, and since f depends on u and v, it also must depend on x. In particular we can think of f as a function of x, and we can try to differentiate to find $\frac{df}{dx}$. Notice that although f is a function of *two* variables, u and v, if we let u and v be functions of x, then f becomes a function of the *single* variable x. Because of this we write $\frac{df}{dx}$, the *ordinary* derivative, rather than a partial derivative.

Now of course one way to proceed would be to write out f as a function of u and v and then substitute $3x^2 + 1$ for u and $2x - 3$ for v. This would express f as a function of x, and we could differentiate in the ordinary way. However, since we are interested in the Chain Rule, we should attempt to approach the problem through the definition of f as a composite function.

In Figure 16.5.1 we have drawn the function f at the top with its variables u and v below it. Thinking of u and v as functions of the variable x, we have drawn x on the level below u and v. We sometimes refer to u and v as **intermediate variables**. Diagrams like this may help us to keep track of which function is a function of which variables.

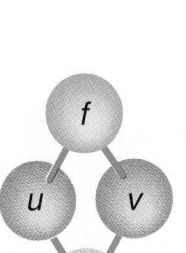

Figure 16.5.1

Now u and v are both functions of the single variable x, so we know how to form their differentials $du = u' \, dx$ and $dv = v' \, dx$. One way of interpreting these differentials is this: Suppose x starts at the value $x = a$ and changes by a small amount dx. Then the value of u will change by approximately $du = u'(a) \, dx$, and the value of v will change by approximately $dv = v'(a) \, dx$.

Now, thinking of $f(u, v)$ as a function of u and v, we know how to find its differential $df = \frac{\partial f}{\partial u} \, du + \frac{\partial f}{\partial v} \, dv$. This can be interpreted to mean that if the variables u and v change by small amounts du and dv, then the value of f will change by approximately $df = \frac{\partial f}{\partial u} \, du + \frac{\partial f}{\partial v} \, dv$.

Combining these observations, we see that a small change dx in the value of x will result in small changes in u and v that are approximately equal to du and dv. These changes in u and v will in turn result in a small change in the value of f, which will be approximately equal to the differential df. We can substitute for du and dv and obtain

$$df = f_u \, du + f_v \, dv = f_u u' \, dx + f_v v' \, dx = (f_u u' + f_v v') \, dx.$$

If we think of f as a function of the single variable x, then we write its differential in the usual way as

$$df = f' \, dx = \frac{df}{dx} \, dx.$$

Comparing these two formulas for df suggests that the derivative $\frac{df}{dx}$ must equal $f_u u' + f_v v'$.

We try out this idea with the example. Since $f(u, v) = u + v \sin(u + v)$, it is easy to find its partial derivatives $f_u = 1 + v \cos(u + v)$ and $f_v = \sin(u + v) + v \cos(u + v)$. Since $u = 3x^2 + 1$, we find that $u' = 6x$, and since $v = 2x - 3$, we see that $v' = 2$. So our formula for the derivative can be written as $f_u u' + f_v v' = \big(1 + v \cos(u + v)\big)(6x) + \big(\sin(u + v) + v \cos(u + v)\big)(2)$. As usual with the Chain Rule we should express everything in terms of the variable x, so we change u to $3x^2 + 1$ and v to $2x - 3$. The result is

$$\frac{df}{dx} = f_u u' + f_v v'$$

$$= \big(1 + v\cos(u+v)\big)(6x) + \big(\sin(u+v) + v\cos(u+v)\big)(2)$$

$$= \big(1 + (2x-3)\cos(3x^2 + 1 + 2x - 3)\big)(6x)$$

$$+ \big(\sin(3x^2 + 1 + 2x - 3) + (2x-3)\cos(3x^2 + 1 + 2x - 3)\big)(2)$$

$$= 6x + (12x^2 - 18x)\cos(3x^2 + 2x - 2) + 2\sin(3x^2 + 2x - 2)$$

$$+ (4x - 6)\cos(3x^2 + 2x - 2)$$

$$= 6x + (12x^2 - 14x - 6)\cos(3x^2 + 2x - 2) + 2\sin(3x^2 + 2x - 2).$$

For the sake of comparison, let us try doing it the other way, that is, writing $f(u,v) = u + v\sin(u+v) = (3x^2 + 1) + (2x-3)\sin(3x^2 + 1 + 2x - 3) = 3x^2 + 1 + (2x-3)\sin(3x^2 + 2x - 2)$. Differentiating this function with respect to x gives

$$\frac{df}{dx} = 6x + (2)\sin(3x^2 + 2x - 2) + (2x-3)\cos(3x^2 + 2x - 2)(6x + 2)$$

$$= 6x + 2\sin(3x^2 + 2x - 2) + (12x^2 - 14x - 6)\cos(3x^2 + 2x - 2).$$

We find that these two formulas are equal.

If F is a function of a single variable u and u is itself a function of x, then the ordinary Chain Rule tells us that $\frac{dF}{dx}$ is the product $\frac{dF}{du}\frac{du}{dx}$ of the derivative of F with respect to u and the derivative of u with respect to x. With a function $f(u,v)$ of two variables the situation is more complicated because f has two partial derivatives. We could consider the product $f_u \frac{du}{dx}$. It tells us the rate at which the value of f changes with respect to x because of the way u changes when x changes. The other product $f_v \frac{dv}{dx}$ tells us the rate at which f changes because of the way v changes when x does. Since both types of changes will affect the value of f, the overall rate of change of f with respect to x will be the sum of these two.

THEOREM 16.5.1

The Chain Rule

If $f(u,v)$ is a function of two variables u and v, and u and v are themselves functions of a variable x, then f can be regarded as a function of the single variable x. If the partial derivatives $\frac{\partial f}{\partial u}$ and $\frac{\partial f}{\partial v}$ exist and are continuous functions and if u and v are differentiable functions of x, then f is a differentiable function of x, and its derivative is

$$\frac{df}{dx} = f_u u' + f_v v' = \frac{\partial f}{\partial u}\frac{du}{dx} + \frac{\partial f}{\partial v}\frac{dv}{dx}.$$

(Strictly speaking, we need a condition like this: Suppose $u(x)$ and $v(x)$ are differentiable at a point $x = c$. Let $a = u(c)$ and $b = v(c)$. Suppose there is a disk around the point (a, b) on which f_u and f_v exist and are continuous. Then f is differentiable at the point $x = c$, and its derivative is given by the above formula evaluated at $x = c$.)

EXAMPLE 1 (i) Suppose $f(u,v) = u^2 v^3 - e^{u+v}$ and $u = x^3$, $v = \sin x$. Find $\frac{df}{dx}$, the derivative of the composite function.

Solution Each of these functions has a continuous derivative (or partial derivatives, as the case may be) at every point, so the Chain Rule tells us that the derivative is

$$\frac{df}{dx} = \frac{\partial f}{\partial u}\frac{du}{dx} + \frac{\partial f}{\partial v}\frac{dv}{dx} = (2uv^3 - e^{u+v})3x^2 + (3u^2v^2 - e^{u+v})\cos x$$

$$= 3x^2(2x^3\sin^3 x - e^{x^3+\sin x}) + \cos x(3x^6\sin^2 x - e^{x^3+\sin x})$$

$$= 6x^5\sin^3 x - 3x^2 e^{x^3+\sin x} + 3x^6\cos x\sin^2 x - \cos x e^{x^3+\sin x}.$$

(ii) If $g(x, y) = \cosh(xy)\cos(xy)$ and $x = 4t$, $y = 3t^4 - 5$, find $\frac{dg}{dt}$.

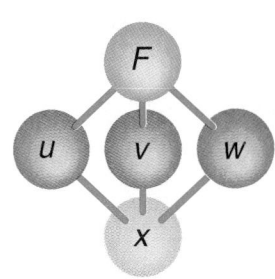

Figure 16.5.2

Solution First we should make sure which are the intermediate variables and what is a function of what. It may help to draw a diagram, as in Figure 16.5.2. At the top we put the function that we want to differentiate, which in this case is g. In the question, g is given to us as a function of the variables x and y, so we put them on the level below g. Each of x and y is a function of t, so we put it on the level immediately below them.

To find the derivative using the Chain Rule, we take the partial derivatives of g with respect to each of the intermediate variables x and y and multiply by their derivatives with respect to t:

$$\frac{dg}{dt} = \frac{\partial g}{\partial x}\frac{dx}{dt} + \frac{\partial g}{\partial y}\frac{dy}{dt}$$

$$= \Big(\sinh(xy)(y)\cos(xy) + \cosh(xy)\big(-\sin(xy)y\big)\Big)(4)$$

$$+ \Big(\sinh(xy)(x)\cos(xy) + \cosh(xy)\big(-\sin(xy)x\big)\Big)(12t^3)$$

$$= \big(4y + (12t^3)x\big)\sinh(xy)\cos(xy) - \big(4y + (12t^3)x\big)\cosh(xy)\sin(xy)$$

$$= (12t^4 - 20 + 48t^4)\big(\sinh(xy)\cos(xy) - \cosh(xy)\sin(xy)\big)$$

$$= (60t^4 - 20)\big(\sinh(12t^5 - 20t)\cos(12t^5 - 20t)$$

$$- \cosh(12t^5 - 20t)\sin(12t^5 - 20t)\big).$$

The calculations themselves are straightforward; we have to decide what is a function of which variable(s) and apply the Chain Rule formula.

It is also possible to consider functions of three or more variables, and for them the Chain Rule is quite similar. For instance, if $f(u, v, w)$ is a function of three variables and each of u, v, and w is a function of x, then f can be regarded as a function of x, and it makes sense to ask about the derivative $\frac{df}{dx}$. Not surprisingly, the **Chain Rule** in this situation is

Figure 16.5.3

$$\frac{df}{dx} = \frac{\partial f}{\partial u}\frac{du}{dx} + \frac{\partial f}{\partial v}\frac{dv}{dx} + \frac{\partial f}{\partial w}\frac{dw}{dx} = f_u u' + f_v v' + f_w w'.$$

f_u, f_v, f_w are given as functions of (u, v, w), but we can express them in terms of x.

Similarly, for a function $F(u_1, u_2, \ldots, u_n)$ of n variables, each of which is itself a function of a single variable x, the Chain Rule says that

$$\frac{dF}{dx} = \frac{\partial F}{\partial u_1}\frac{du_1}{dx} + \frac{\partial F}{\partial u_2}\frac{du_2}{dx} + \ldots + \frac{\partial F}{\partial u_n}\frac{du_n}{dx}.$$

Each $\frac{\partial F}{\partial u_i}$ is a function of $(u_1, u_2, \ldots u_n)$ but can be expressed as a function of x.

EXAMPLE 2 (i) Suppose $F(u, v, w) = u\cos(v + w)$, and $u = 2x$, $v = x^2$, and $w = e^x$. Find $\frac{dF}{dx}$.

Solution We observe that the derivatives and partial derivatives exist and are continuous everywhere. We begin by drawing Figure 16.5.3, which indicates that F is a function of u, v, and w and that each of them is a function of x. The derivative of F

with respect to x could be found by substitution and direct differentiation, but it is also possible to use the Chain Rule to find

$$\frac{dF}{dx} = F_u u' + F_v v' + F_w w'$$

$$= \cos(v+w)(2) + \big(-u\sin(v+w)\big)(2x) + \big(-u\sin(v+w)\big)(e^x)$$

$$= 2\cos(x^2 + e^x) - 2x(2x + e^x)\sin(x^2 + e^x).$$

(ii) Suppose n is an integer greater than 1 and suppose $G(x_1, x_2, \ldots, x_n) = x_1 x_2 \times \ldots \times x_n$ is the product of the n variables $x_1, \ldots, x_n$. If $x_1 = e^t$, $x_2 = e^{2t}$, ..., $x_k = e^{kt}$, ..., $x_n = e^{nt}$, that is, if each x_k equals the corresponding kth power of e^t, then find $\frac{dG}{dt}$.

Solution   In Figure 16.5.4 we show G as a function of the n intermediate variables $x_1, x_2, \ldots, x_n$ and each of them as a function of t. The Chain Rule for a function of n variables tells us that

$$\frac{dG}{dt} = \frac{\partial G}{\partial x_1}\frac{dx_1}{dt} + \ldots + \frac{\partial G}{\partial x_k}\frac{dx_k}{dt} + \ldots + \frac{\partial G}{\partial x_n}\frac{dx_n}{dt}.$$

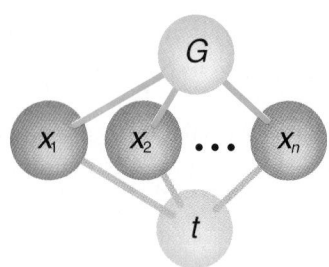

Figure 16.5.4

Let us consider each term on the right side separately. Specifically, to find $\frac{\partial G}{\partial x_k}$, let us write the product $G(x_1, \ldots, x_n) = x_1 \times \ldots \times x_k \times \ldots \times x_n$ in the form Px_k, where P is the product of all the x's *except x_k* itself. When we take the partial derivative with respect to x_k, the product P is regarded as a constant, so $\frac{\partial G}{\partial x_k} = \frac{\partial}{\partial x_k}(Px_k) = P$, which is the product of all the x's except x_k.

This partial derivative must be multiplied by $\frac{dx_k}{dt} = \frac{d}{dt}(e^{kt}) = ke^{kt} = kx_k$. So the kth term in the Chain Rule formula for the derivative of G is $\frac{\partial G}{\partial x_k}\frac{dx_k}{dt} = P \times kx_k$. Since P is the product of all the x's except x_k, we put the x_k "back in" and see that this equals $kx_1 \times \ldots \times x_k \times \ldots \times x_n = kG(x_1, \ldots, x_n)$.

Combining all the terms, we see that

$$\frac{dG}{dt} = \frac{\partial G}{\partial x_1}\frac{dx_1}{dt} + \ldots + \frac{\partial G}{\partial x_k}\frac{dx_k}{dt} + \ldots + \frac{\partial G}{\partial x_n}\frac{dx_n}{dt}$$

$$= (1)G(x_1, \ldots, x_n) + 2G(x_1, \ldots, x_n) + \ldots + nG(x_1, \ldots, x_n)$$

$$= (1 + 2 + \ldots + n)G(x_1, \ldots, x_n)$$

$$= \frac{n(n+1)}{2}e^t e^{2t} \times \ldots \times e^{nt} = \frac{n(n+1)}{2}e^{(1+2+\ldots+n)t}$$

$$= \frac{n(n+1)}{2}e^{n(n+1)t/2}.$$

> We have used Formula 5.2.5 for the sum of the first n positive integers.

In hindsight we might notice that it is easy to express G as a function of t. It is $e^t e^{2t} e^{3t} \times \ldots \times e^{nt} = e^{(1+2+3+\ldots+n)t} = e^{n(n+1)t/2}$. From this it is very easy to find the formula we found. However, we might easily not have realized this, and it is comforting to see how the Chain Rule works.

(iii) Suppose $H(x, y, z) = x^2 y z^3$, and $y = e^x$, $z = \ln x$. Find $\frac{dH}{dx}$.

Solution   The tricky thing about this example is that the variable x plays two different roles. The function H is presented to us as a function of the three intermediate variables x, y, and z. Then we are told that y and z can be expressed as functions of x.

This suggests that we begin our diagram in Figure 16.5.5 by placing H at the top and x, y, and z on the next level. Finally, everything on the second level is expressed

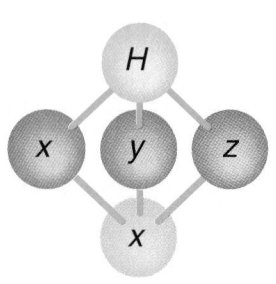

Figure 16.5.5

as a function of x, so we put x on the bottom level. It seems a little silly, but we can think of x as a function of itself: $x = x$.

With this understanding, the derivative is easily found:

$$\frac{dH}{dx} = \frac{\partial H}{\partial x}\frac{dx}{dx} + \frac{\partial H}{\partial y}\frac{dy}{dx} + \frac{\partial H}{\partial z}\frac{dz}{dx}$$

$$= 2xyz^3(1) + x^2z^3(e^x) + 3x^2yz^2\left(\frac{1}{x}\right)$$

$$= 2xe^x\ln^3 x + x^2e^x\ln^3 x + 3xe^x\ln^2 x$$

$$= (2x\ln x + x^2\ln x + 3x)e^x\ln^2 x.$$

Notice that the function (and its derivative) only makes sense when $x > 0$. ◢

Next we want to consider the situation in which a function depends on several variables and each of *them* depends on several variables. For instance, if f is a function of two variables, u and v, we might write $f(u, v)$. If u and v are both functions of two variables x and y, then by substituting we can think of f as a function of x and y. Figure 16.5.6 shows the function f at the top, with its variables u and v on the next level and the variables x and y on the bottom level. This shows that f is a function of intermediate variables u and v, and that they in turn are functions of x and y.

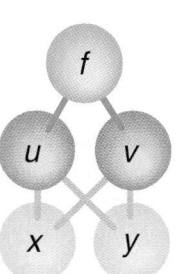

Figure 16.5.6

If we consider the composite function, we can think of f as a function of x and y. One way of writing this would be $f\big(u(x, y), v(x, y)\big)$, which shows how u and v depend on x and y and how f depends on them.

Now with f as a function of x and y, to discuss differentiating it we must consider *partial* derivatives. We need a way to find $\frac{\partial f}{\partial x}$ and $\frac{\partial f}{\partial y}$. To find the partial derivative with respect to x, for example, we fix the value of y and then take the derivative with respect to x.

In Figure 16.5.7 we have shaded the variable y to suggest that it is fixed. The variable x is still allowed to change, and this makes f into a function of x. In fact, if we ignore the presence of y in the diagram, it is really the same as the diagram we used previously when we discussed the situation in which u and v were functions of a single variable x. By fixing the value of y we have in effect returned to that case, and we can use the Chain Rule (Theorem 16.5.1) to differentiate.

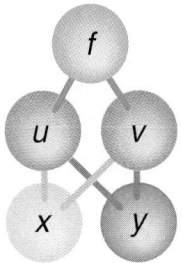

Figure 16.5.7

The only complication is that when we differentiate u or v with respect to x, it is really a partial derivative, since of course u and v are actually functions of two variables x and y. We find that

$$\frac{\partial f}{\partial x} = f_u u_x + f_v v_x = \frac{\partial f}{\partial u}\frac{\partial u}{\partial x} + \frac{\partial f}{\partial v}\frac{\partial v}{\partial x}.$$

Notice the way the ordinary derivatives of u and v in Theorem 16.5.1 have been replaced here with partial derivatives. Of course, the same sort of argument applies almost verbatim to finding the other partial derivative $\frac{\partial f}{\partial y}$.

THEOREM 16.5.2

The Chain Rule

If $f(u, v)$ is a function of the variables u and v, and u and v are functions of the variables x and y, then it is possible to regard f as a function of x and y by writing

$$f\big(u(x, y), v(x, y)\big).$$

Then at points where the functions involved are differentiable, the partial derivatives of f can be found as follows:

$$\frac{\partial f}{\partial x} = f_u u_x + f_v v_x = \frac{\partial f}{\partial u}\frac{\partial u}{\partial x} + \frac{\partial f}{\partial v}\frac{\partial v}{\partial x}.$$

$$\frac{\partial f}{\partial y} = f_u u_y + f_v v_y = \frac{\partial f}{\partial u}\frac{\partial u}{\partial y} + \frac{\partial f}{\partial v}\frac{\partial v}{\partial y}.$$

One familiar situation in which this problem arises is the use of polar coordinates. If $f(x, y)$ is a function on the plane, we can also think of f as a function of the variables r and θ, and it makes sense to consider its partial derivatives with respect to r and θ.

EXAMPLE 3   If $f(x, y) = x^2 y - 3xy$, find $\frac{\partial f}{\partial r}$ and $\frac{\partial f}{\partial \theta}$.

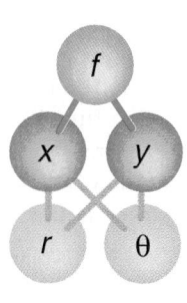

Figure 16.5.8

Solution   We begin with the diagram in Figure 16.5.8, placing f at the top. Since f is given to us in terms of x and y and we wish to find the partial derivatives with respect to r and θ, we place x and y on the middle level and r and θ on the bottom level.

It is probably helpful to remind ourselves of the formulas that express x and y as functions of r and θ, namely,

$$x = r\cos\theta \quad \text{and} \quad y = r\sin\theta.$$

From this we see that

$$\frac{\partial x}{\partial r} = \cos\theta, \qquad \frac{\partial x}{\partial \theta} = -r\sin\theta, \qquad \frac{\partial y}{\partial r} = \sin\theta, \qquad \frac{\partial y}{\partial \theta} = r\cos\theta.$$

The Chain Rule (Theorem 16.5.2) tells us that

$$\begin{aligned}
\frac{\partial f}{\partial r} &= f_x x_r + f_y y_r \\
&= (2xy - 3y)\cos\theta + (x^2 - 3x)\sin\theta \\
&= 2r^2\cos^2\theta\sin\theta - 3r\sin\theta\cos\theta + r^2\cos^2\theta\sin\theta - 3r\cos\theta\sin\theta \\
&= 3r^2\cos^2\theta\sin\theta - 6r\cos\theta\sin\theta.
\end{aligned}$$

Similarly,

$$\begin{aligned}
\frac{\partial f}{\partial \theta} &= f_x x_\theta + f_y y_\theta \\
&= (2xy - 3y)(-r\sin\theta) + (x^2 - 3x)(r\cos\theta) \\
&= -2r^3\cos\theta\sin^2\theta + 3r^2\sin^2\theta + r^3\cos^3\theta - 3r^2\cos^2\theta.
\end{aligned}$$

In each case we have expressed the final answer in terms of r and θ, since they are the variables being used for taking partial derivatives. However, it is certainly also possible to express the answers in terms of x and y.

EXAMPLE 4   **Body Surface Area**   Empirical studies suggest that the surface area of a human body is related to the weight and height. If the weight is w kg and the height is h cm, then the surface area, in square meters, is approximated by

$$S = 0.0072w^{.425}h^{.725}.$$

For instance, a child weighing 17 kg (37.4 lb) and standing 100 cm ($\approx 39.37''$) tall has a surface area of about $S = 0.0072(17^{.425})(100^{.725}) = 0.677$ m$^2 \approx 7.28$ sq ft.

Suppose this child is growing at the rate of 9 cm/year and 2 kg/year; what is the rate at which her surface area is increasing?

Solution   We need to find $\frac{dS}{dt}$. The intermediate variables are w and h, and the Chain Rule gives

$$\frac{dS}{dt} = \frac{\partial S}{\partial w}\frac{dw}{dt} + \frac{\partial S}{\partial h}\frac{dh}{dt}.$$

Since $S = 0.0072w^{.425}h^{.725}$, we find

$$\frac{\partial S}{\partial w} = 0.0072(0.425w^{.425-1})h^{.725} = 0.0072(0.425)w^{-.575}h^{.725},$$

$$\frac{\partial S}{\partial h} = 0.0072w^{.425}(0.725h^{.725-1}) = 0.0072(0.725)w^{.425}h^{-.275}.$$

The rate of change of the surface area is

$$\begin{aligned}
\frac{dS}{dt} &= \frac{\partial S}{\partial w}\frac{dw}{dt} + \frac{\partial S}{\partial h}\frac{dh}{dt}\\
&= 0.0072(0.425)17^{-.575}100^{.725}(2) + 0.0072(0.725)17^{.425}100^{-.275}(9)\\
&\approx 0.078.
\end{aligned}$$

The area is increasing by about 0.078 m^2/yr ≈ 0.84 sq ft/yr.

With a calculator we can calculate the area for a child whose weight is 19 kg and whose height is 109 cm (i.e., the values we get if the growth is constant for a year). Then we can compare this with the original area to find the actual increase.

```
.0072*(19^.425)*
(109^.725)-.0072
*(17^.425)*(100^
.725)
        .0784741314
```

TI-85	SHARP EL9300
.0072 × (19 ^ .425) × (109 ^ .725) − .0072 × (17 ^ .425) × (100 ^ .725) ENTER	.0072 × (19 a^b .425 ▶) × (109 a^b .725 ▶) − .0072 × (17 a^b .425 ▶) × (100 a^b .725 ▶) ENTER
CASIO f_x-6300G (f_x-7700GB)	HP 48SX
.0072 × (19 x^y .425) × (109 x^y .725) − .0072 × (17 x^y .425) × (100 x^y .725) EXE	.0072 ENTER 19 ENTER .425 y^x × 109 ENTER .725 y^x × .0072 ENTER 17 ENTER .425 y^x × 100 ENTER .725 y^x × −

The result is surprisingly close to the growth for one year predicted by the derivative.

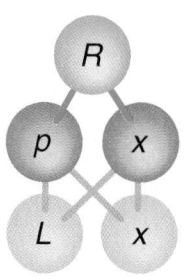

Figure 16.5.9

EXAMPLE 5   **Revenue**   A mechanical device is manufactured and sold at the unit price $\$p$. If x is the number of units sold, in thousands, then the total revenue (in thousands of dollars) is $R = R(p, x) = px$.

The unit price p depends on various things, including L, the hourly cost of labor, and the size of the production run, which we will assume is x. Suppose that with other factors, such as overhead, fixed, the unit price is

$$p = 300 + 20L + \frac{2000}{10 + x^2}.$$

The revenue is $R = R(p, x) = px$, but since p is also a function of L and x, we can think of $R = R(p, x) = R\big(p(L, x), x\big)$ as a function of L and x. The intermediate variables are p and x (see Figure 16.5.9).

We can fix L and use the Chain Rule to find

$$\frac{\partial R}{\partial x} = \frac{\partial R}{\partial p}\frac{\partial p}{\partial x} + \frac{\partial R}{\partial x}\frac{\partial x}{\partial x} = x\left(-\frac{2000}{(10 + x^2)^2}(2x)\right) + (p)(1)$$

$$= -\frac{4000x^2}{(10 + x^2)^2} + p = -\frac{4000x^2}{(10 + x^2)^2} + 300 + 20L + \frac{2000}{10 + x^2}$$

$$= \frac{20{,}000 - 2000x^2}{(10 + x^2)^2} + 300 + 20L.$$

So, for instance, if $L = 12$, $x = 10$ (i.e., labor costs \$12/hr and 10,000 units are being made), then

$$\frac{\partial R}{\partial x} = \frac{20{,}000 - 2000(100)}{(110)^2} + 300 + 240 \approx 525.1.$$

Increases in production will result in increased revenue at the rate of approximately \$525,100 per thousand additional units.

In Example 5, in which we wrote the Chain Rule as

$$\frac{\partial R}{\partial x} = \frac{\partial R}{\partial p}\frac{\partial p}{\partial x} + \frac{\partial R}{\partial x}\frac{\partial x}{\partial x},$$

the symbol $\frac{\partial R}{\partial x}$ appears twice. On the left it refers to the partial derivative of R with respect to x, *thinking of R as a function of L and x*. The second time it refers to R as a function of the intermediate variables p and x, so it means fix p and differentiate with respect to x. It is vitally important to recognize which is which from the context.

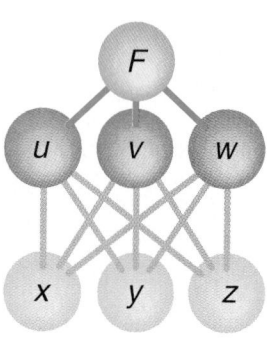

Figure 16.5.10

We have considered the Chain Rule for functions of two variables, each of which is a function of two variables, but of course the same ideas apply to functions of any number of variables. If $F(u, v, w)$ is a function of three variables, and u, v, and w are all functions of x, y, and z, then, as Figure 16.5.10 suggests, it is possible to consider F as a function of x, y, and z. It is not surprising that the Chain Rule in this case tells us that

$$F_x = F_u u_x + F_v v_x + F_w w_x = \frac{\partial F}{\partial u}\frac{\partial u}{\partial x} + \frac{\partial F}{\partial v}\frac{\partial v}{\partial x} + \frac{\partial F}{\partial w}\frac{\partial w}{\partial x},$$

$$F_y = F_u u_y + F_v v_y + F_w w_y = \frac{\partial F}{\partial u}\frac{\partial u}{\partial y} + \frac{\partial F}{\partial v}\frac{\partial v}{\partial y} + \frac{\partial F}{\partial w}\frac{\partial w}{\partial y},$$

$$F_z = F_u u_z + F_v v_z + F_w w_z = \frac{\partial F}{\partial u}\frac{\partial u}{\partial z} + \frac{\partial F}{\partial v}\frac{\partial v}{\partial z} + \frac{\partial F}{\partial w}\frac{\partial w}{\partial z}.$$

It is also possible to consider a function of three variables, each of which is a function of two other variables, or any other possible combination; there is no requirement that the number of variables on the different levels must be the same.

Suppose $G(u_1, \ldots, u_m)$ is a function of m variables and each u_ℓ is a function $u_\ell(x_1, \ldots, x_n)$ of the n variables $x_1, \ldots, x_n$, as shown in Figure 16.5.11. Then the Chain Rule in this case says that for each k the partial derivative of G with respect to x_k is

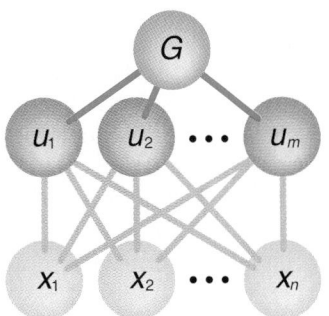

Figure 16.5.11

$$\frac{\partial G}{\partial x_k} = \frac{\partial G}{\partial u_1}\frac{\partial u_1}{\partial x_k} + \frac{\partial G}{\partial u_2}\frac{\partial u_2}{\partial x_k} + \cdots + \frac{\partial G}{\partial u_m}\frac{\partial u_m}{\partial x_k}.$$

Calculations with the Chain Rule can be quite complicated, but the basic idea is fairly simple, especially once we have made it clear what is a function of what.

EXAMPLE 6   Suppose $f(u, v) = uv^2$, and $u = x^2 - y^2$, $v = x + y$. We can make f into a function of the polar coordinates r and θ by expressing x and y in terms of r and θ. Find $\frac{\partial f}{\partial r}$.

Solution   The diagram will be more complicated in this case (see Figure 16.5.12). The function f at the top is a function of u and v, so they are on the second level. Then u and v can be expressed as functions of x and y, so they are on the next level down, and x and y are functions of r and θ, so they are on the bottom level.

We begin by regarding f as a function of u and v, with u and v in turn being functions of r and θ. The Chain Rule in this situation says that

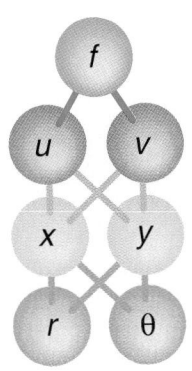

Figure 16.5.12

$$\frac{\partial f}{\partial r} = f_u u_r + f_v v_r = v^2 u_r + 2uv v_r.$$

This leaves us with the problem of having to find u_r and v_r. For this it is necessary to apply the Chain Rule a second time. After all, u and v are functions of x and y, which are functions of r and θ, so

$$u_r = u_x x_r + u_y y_r = 2x\cos\theta - 2y\sin\theta,$$
$$v_r = v_x x_r + v_y y_r = (1)\cos\theta + (1)\sin\theta = \cos\theta + \sin\theta.$$

Combining these results, we can find f_r:

$$
\begin{aligned}
f_r &= f_u u_r + f_v v_r \\
&= v^2(2x\cos\theta - 2y\sin\theta) + 2uv(\cos\theta + \sin\theta) \\
&= (x+y)^2(2r\cos^2\theta - 2r\sin^2\theta) + 2(x^2 - y^2)(x+y)(\cos\theta + \sin\theta) \\
&= (x+y)^2 2r(\cos^2\theta - \sin^2\theta) + 2(x-y)(x+y)^2(\cos\theta + \sin\theta) \\
&= 2(x+y)^2\big(r(\cos^2\theta - \sin^2\theta) + (r\cos\theta - r\sin\theta)(\cos\theta + \sin\theta)\big) \\
&= 2r(x+y)^2\big((\cos^2\theta - \sin^2\theta) + (\cos^2\theta - \sin^2\theta)\big) \\
&= 4r(r\cos\theta + r\sin\theta)^2(\cos^2\theta - \sin^2\theta) \\
&= 4r^3(\cos\theta + \sin\theta)^2(\cos^2\theta - \sin^2\theta).
\end{aligned}
$$

Implicit Differentiation

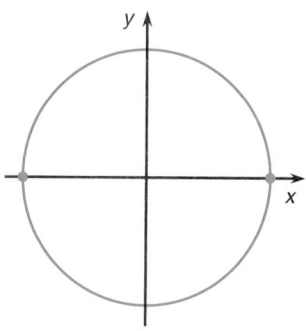

In Section 3.7 we discussed the notion of implicit differentiation. We considered an equation like $x^2 + y^2 = 4$, the equation of a circle, and realized that often it was possible to use such an equation to express either x or y as a function of the other one. In this particular case it is possible to express y as a function of x on either the upper semicircle or the lower semicircle and to express x as a function of y on either the right or left semicircle (see Figure 16.5.13). When it is possible to do this, it is interesting to try to find the derivative of y with respect to x or vice versa. This is especially important in cases in which there is no explicit formula for the functions.

Figure 16.5.13

Suppose for the moment that it is possible to express y as a function of x. Then we can consider the function $f(x, y) = x^2 + y^2$, regarded as a function of x and y. Since y is a function of x, and of course x is a function of x, we can think of both x and y as functions of x and therefore of f as a function of x. We might write $y = y(x)$ and $f = f\big(x, y(x)\big)$.

Now the idea is that for each x in the interval $(-2, 2)$ it is possible to find a value of y so that (x, y) will be a point on the circle, that is, $x^2 + y^2 = 4$. This means that $f(x, y) = x^2 + y^2$ must be the constant function 4.

Now regarding $f = f\big(x, y(x)\big)$ as a function of x in the way we have described, we can apply the Chain Rule to find $\frac{df}{dx}$. Since f is a constant function of x, this derivative must be zero:

$$0 = \frac{df}{dx} = \frac{\partial f}{\partial x}\frac{dx}{dx} + \frac{\partial f}{\partial y}\frac{dy}{dx},$$

so

$$-\frac{\partial f}{\partial x}(1) = \frac{\partial f}{\partial y}\frac{dy}{dx},$$

or

$$\frac{dy}{dx} = -\frac{\frac{\partial f}{\partial x}}{\frac{\partial f}{\partial y}}.$$

Notice that because x and y are functions of the single variable x, we take ordinary derivatives $\frac{dy}{dx}$ and $\frac{dx}{dx} = 1$, but on the other hand f is a function of the two variables x and y, so it is appropriate to consider partial derivatives $\frac{\partial f}{\partial x}$ and $\frac{\partial f}{\partial y}$.

In the particular case we were considering, where $f(x, y) = x^2 + y^2$, it is easy to find $f_x = 2x$ and $f_y = 2y$, so we see that $\frac{dy}{dx} = -(f_x)/(f_y) = -(2x)/(2y) = -\frac{x}{y}$. With this specific example we could have written $y = \sqrt{4 - x^2}$ as an explicit function of x, and it is easy to verify that our formula for the derivative is correct. Frequently, it is difficult or impossible to write the function explicitly, and in such a case this type of calculation is very important.

THEOREM 16.5.3

Suppose c is some constant, and consider the condition $f(x, y) = c$. Suppose that for x in some interval it is possible to express y as a function of x, that is, $y = y(x)$, in such a way that $\big(x, y(x)\big)$ always satisfies the condition $f\big(x, y(x)\big) = c$. Then

$$\frac{dy}{dx} = -\frac{\frac{\partial f}{\partial x}}{\frac{\partial f}{\partial y}},$$

in the sense that the derivative $\frac{dy}{dx}$ exists whenever the partial derivatives exist and the denominator is not zero.

In fact, if $f(a, b) = c$ and $\frac{\partial f}{\partial y} \neq 0$, then it can be shown that there is an open interval around $x = a$ on which y can be expressed as a function of x in such a way that $f\big(x, y(x)\big) = c$. This fact is very difficult to prove and has the practical drawback of not telling us what the interval actually is.

EXAMPLE 7   (i) Suppose y is implicitly defined as a function of x by the condition $x^5 y^2 - x^3 y^6 + 3xy = 3$. Find $\frac{dy}{dx}$. What is the value of this derivative at the point $(1, 1)$?

Solution   We simply apply Theorem 16.5.3, letting $f(x, y) = x^5 y^2 - x^3 y^6 + 3xy$. We find that

$$\frac{\partial f}{\partial x} = 5x^4 y^2 - 3x^2 y^6 + 3y, \qquad \text{and} \qquad \frac{\partial f}{\partial y} = 2x^5 y - 6x^3 y^5 + 3x,$$

so by the formula in Theorem 16.5.3,

$$\frac{dy}{dx} = -\frac{\frac{\partial f}{\partial x}}{\frac{\partial f}{\partial y}} = -\frac{5x^4 y^2 - 3x^2 y^6 + 3y}{2x^5 y - 6x^3 y^5 + 3x}.$$

When $(x, y) = (1, 1)$, we substitute and find that $\frac{dy}{dx} = -\frac{5}{-1} = 5$.

(ii) Suppose x is defined implicitly as a function of y, for values of y in some interval by the condition $2xy \sin(xy) = 1$. Find $\frac{dx}{dy}$.

Solution   We let $f(x, y) = 2xy \sin(xy)$ and calculate the partial derivatives:

$$\frac{\partial f}{\partial x} = 2y \sin(xy) + 2xy^2 \cos(xy), \qquad \frac{\partial f}{\partial y} = 2x \sin(xy) + 2x^2 y \cos(xy).$$

The only thing to watch about this example is that x is a function of y, rather than the other way around. We apply Theorem 16.5.3, being careful to interchange the roles of x and y:

$$\frac{dx}{dy} = -\frac{\frac{\partial f}{\partial y}}{\frac{\partial f}{\partial x}} = -\frac{2x \sin(xy) + 2x^2 y \cos(xy)}{2y \sin(xy) + 2xy^2 \cos(xy)} = -\left(\frac{x}{y}\right) \frac{\sin(xy) + xy \cos(xy)}{\sin(xy) + xy \cos(xy)}$$

$$= -\frac{x}{y}.$$

EXAMPLE 8   **Fuel Consumption**   A speedboat is capable of speeds up to 50 mph. Its speed s, in mph, and its rate of fuel consumption, C, in gallons/hr, have been found to satisfy the equation

$$3C^3 + 10^{-3} Cs^2 - 10^{-6} s^6 = 485.$$

For instance, if $s = 40$ mph, then $C \approx 11.5$ gal/hr.

To make a 100-mile trip at 40 mph takes 2.5 hours, so the total fuel consumed is $(2.5)(11.5) = 28.75$ gal. Estimate the speed at which the 100-mile trip will take 30 gal.

Solution   Assuming that C is a function of the speed s, we need to find $\frac{dC}{ds}$. Letting

$$F(C, s) = 3C^3 + 10^{-3} Cs^2 - 10^{-6} s^6 - 485,$$

we find that

$$\frac{\partial F}{\partial C} = 9C^2 + 10^{-3}s^2, \qquad \frac{\partial F}{\partial s} = 2 \times 10^{-3}Cs - 6 \times 10^{-6}s^5,$$

so $\quad \dfrac{dC}{ds} = -\dfrac{\frac{\partial F}{\partial s}}{\frac{\partial F}{\partial C}} = -\dfrac{2 \times 10^{-3}Cs - 6 \times 10^{-6}s^5}{9C^2 + 10^{-3}s^2}.$

When $s = 40$ and $C = 11.5$, we find $\frac{dC}{ds} \approx 0.515$.

At a speed s the trip takes $\frac{100}{s}$ hours, and the total fuel consumed is $T = C\frac{100}{s}$. We find $\frac{dT}{ds} = \frac{dC}{ds}\frac{100}{s} - \frac{100C}{s^2}$. Substituting our value for $\frac{dC}{ds}$, we find that at $s = 40$,

$$\frac{dT}{ds} \approx (0.515)\frac{100}{40} - (100)\frac{11.5}{1600} \approx 0.57,$$

that is, $dT \approx 0.57\,ds$.

So to make the total T increase from 28.7 to 30, that is, by $dT = 1.25$, s should increase by $ds = \frac{1.25}{0.57} \approx 2.20$. At 42.2 mph the trip will take about 30 gal.

Finally, it is possible to consider a function of three variables $f(x, y, z)$ and consider the set of all points (x, y, z) that satisfy $f(x, y, z) = c$, for some fixed constant c, that is, the level surface for f associated to the value c. It often happens that this condition will in effect define one of the variables as an implicit function of the other two.

For instance, if we think of the function $f(x, y, z) = x^2 + y^2 + z^2$ and consider the level surface $f(x, y, z) = 9$, then of course we know that it is the sphere of radius 3. It is easy to solve for z, say, in terms of x and y: We have $z = \sqrt{9 - x^2 - y^2}$, for instance. This is not the only possible solution, since we could have chosen the negative square root or some combination of the two. Geometrically, what we have done is to find a point on the sphere lying directly over the point (x, y) in the plane. See Figure 16.5.14.

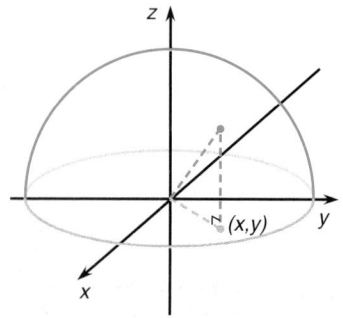

Figure 16.5.14

For each point (x, y) in some region (the region contained in the circle of radius 3 around the origin) we have found a value $z = z(x, y)$ so that the point $(x, y, z) = \big(x, y, z(x, y)\big)$ lies on the sphere, that is, $f\big(x, y, z(x, y)\big) = 9$ for all (x, y) in the region. We have expressed z as a function of x and y (see Figure 16.5.15).

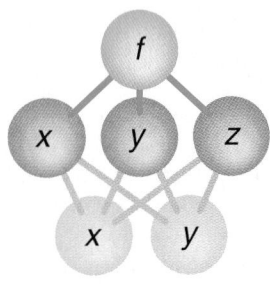

Figure 16.5.15

We can apply the same sort of reasoning we used with a function of two variables and find the partial derivatives of this function z. The idea is that we can think of $f\big(x, y, z(x, y)\big)$ as a function of *two* variables x and y, but for any choice of (x, y) in the disk of radius 3 this function equals the constant function 9, so both its partial derivatives are zero. Then we apply the Chain Rule, thinking of x, y, and z as the intermediate variables. We find that

$$0 = \frac{\partial f}{\partial x} = \frac{\partial f}{\partial x}\frac{\partial x}{\partial x} + \frac{\partial f}{\partial y}\frac{\partial y}{\partial x} + \frac{\partial f}{\partial z}\frac{\partial z}{\partial x}$$

$$= 2x(1) + 2y(0) + 2z\frac{\partial z}{\partial x},$$

so $\quad \dfrac{\partial z}{\partial x} = -\dfrac{2x}{2z} = -\dfrac{x}{z}.$

An almost identical calculation shows that $\frac{\partial z}{\partial y} = -\frac{y}{z}$.

In fact, what we have really shown is that $\frac{\partial z}{\partial x} = -\left(\frac{\partial f}{\partial x}\right)\Big/\left(\frac{\partial f}{\partial z}\right)$ and a corresponding formula for the other partial derivative. We record this result.

THEOREM 16.5.4

Let $f(x, y, z)$ be a function, and suppose z is implicitly defined as a function of x and y by the condition $f(x, y, z) = c$, for some constant c. Then, at points where the functions are differentiable and the denominator is not zero,

$$\frac{\partial z}{\partial x} = -\frac{\frac{\partial f}{\partial x}}{\frac{\partial f}{\partial z}} = -\frac{f_x}{f_z}, \qquad \frac{\partial z}{\partial y} = -\frac{\frac{\partial f}{\partial y}}{\frac{\partial f}{\partial z}} = -\frac{f_y}{f_z}.$$

Of course, the example of the sphere, with $f(x, y, z) = x^2 + y^2 + z^2$, is misleading because there it is possible and even easy to solve for z *explicitly* as a function of x and y. The use of this result usually involves functions for which this is not possible or at least not easy.

EXAMPLE 9   An economist studying the pricing of a certain commodity has discovered that the costs of labor and materials that go into its manufacture and the price that must be charged for it are related. His theory is that if the price is P and the costs of materials and labor are M and L, respectively, then there is a constant K so that

$$P^5 - P^3 - 2M^3L^2 - ML^3 = K.$$

This equation is difficult to solve for P as a function of M and L, but it would be interesting to know $\frac{\partial P}{\partial M}$ and $\frac{\partial P}{\partial L}$.

In particular, the economist has some data that suggest that when $P = 2, M = 2$, and $L = 1$, the partial derivative $\frac{\partial P}{\partial M}$ ought to equal 3.2. How well do the data fit the theory?

Solution   We let $f(L, M, P) = P^5 - P^3 - 2M^3L^2 - ML^3$ and apply Theorem 16.5.4 to find the partial derivatives of P. It says that $\frac{\partial P}{\partial M} = -f_M/f_P$. We begin by finding these partial derivatives: $f_P = \frac{\partial f}{\partial P} = 5P^4 - 3P^2$ and $f_M = \frac{\partial f}{\partial M} = -6M^2L^2 - L^3$. Applying Theorem 16.5.4, we find that

$$\frac{\partial P}{\partial M} = -\frac{f_M}{f_P} = -\frac{-6M^2L^2 - L^3}{5P^4 - 3P^2} = \left(\frac{L^2}{P^2}\right)\frac{6M^2 + L}{5P^2 - 3}.$$

If we substitute $P = 2$, $M = 2$, and $L = 1$, we find that $\frac{\partial P}{\partial M} = \left(\frac{1^2}{2^2}\right)\frac{6(2^2)+1}{5(2^2)-3} = \left(\frac{1}{4}\right)\frac{25}{17} = \frac{25}{68}$. This is so far from equaling the empirical value of 3.2 that we must conclude that the theory does not work too well, at least for values in this range. ◢

This calculation was certainly complicated, but it is interesting to notice that we were able to do it without solving for P as an explicit function of M and L.

Exercises 16.5

Find the required derivative in each of the following cases.

1. $\frac{df}{dx}$ if $f(u, v) = 3u^3v^2 - 7u^2v$, $u = x^2$, $v = 3 - 2x$.

2. $\frac{df}{dx}$ if $f(u, v) = \cos(uv^2 - 3uv)$, $u = 1 + x$, $v = e^x$.

3. $\frac{\partial f}{\partial x}$ if $f(u, v) = e^{uv}$, $u = x^2 + xy$, $v = x - 2y$.

4. $\frac{dg}{du}$ if $g(x, y) = \cosh(xy^2 + 2x^3)$, $x = \sin u$, $y = \cos u$.

5. $\frac{\partial F}{\partial t}$ if $F(x, y) = x^3 + y^3$, $x = \cos t + st$, $y = s^2 - 4t$.

6. $\frac{\partial f}{\partial x}$ if $f(u, v) = \frac{uv}{u^2 + v^2}$, $u = 1 + x^2 + y^2$, $v = xy$.

7. $\frac{dg}{dt}$ if $g(x, y) = xye^{y \sin x}$, $x = 2t^3 + t - 1$, $y = 3t$.

8. $\frac{\partial h}{\partial v}$ if $h(x, y) = (x + 3y)^4$, $x = u^2 v$, $y = 2u - v$.

9. $\frac{\partial F}{\partial y}$ if $F(u, v) = 1/(u + e^v)$, $u = y^4 x^2$, $v = x^2 + 3x - 1$.

10. $\frac{\partial h}{\partial v}$ if $h(x, y) = (x^2 + 4y^2)^4$, $x = 2 \sin u$, $y = \cos v$.

In each of the following situations, assume that the equation given is used to define one variable implicitly as a function of the other variable(s). Find the specified derivative (or partial derivative).

11. Find $\frac{dy}{dx}$ if $x^3 + y^3 = 2$.

12. Find $\frac{dx}{dy}$ if $xy^3 + x^2y^2 + x^3y = 6$.

13. Find $\frac{dy}{dx}$ if $x = \sin(xy)$.

14. Find $\frac{dy}{dx}$ if $x^3 - 3x \ln(xy^2) = 1$.

15. Find $\frac{\partial z}{\partial x}$ if $x^2y + y^2z + 2xz^2 = 4$.

16. Find $\frac{\partial z}{\partial y}$ if $xy \sin(x + z) + e^z = 1$.

17. Find $\frac{\partial w}{\partial u}$ if $ux^2w^4 - xvw + 3u^2vy = 5$.

18. Find $\frac{\partial y}{\partial z}$ if $\sinh(xy) = \cosh(xz)$.

19. Find $\frac{du}{dv}$ if $\cos u = \sin v$.

20. Find $\frac{d\theta}{dr}$ if $r^2 + \theta = \tan \theta$.

‖

21. Verify the calculation of Example 1(i) (p. 976) by substituting for u and v in the formula for $f(u, v)$ and differentiating directly.

22. Verify the calculation of Example 1(ii) (p. 977) by substituting for x and y in the formula for $g(x, y)$ and differentiating directly.

23. Verify the calculation of Example 5 (p. 982) by substituting for p in $R = px$ and differentiating directly.

24. In Example 5, find $\frac{\partial R}{\partial L}$.

*25. Let $G(u, v) = uv$, and suppose that $u = f(x)$ and $v = g(x)$ are both functions of x. Show that the Chain Rule for finding $\frac{dG}{dx}$ amounts to the Product Rule for the derivative of the product $f(x)g(x)$.

*26. Use the Chain Rule to derive the Quotient Rule for the derivative of the quotient $\frac{f(x)}{g(x)}$ at points where $f(x)$ and $g(x)$ are differentiable and $g(x) \neq 0$.

*27. With f, u, and v defined as in Example 6 (p. 983), find $\frac{\partial f}{\partial \theta}$. Express your answer in terms of r and θ.

28. (i) If the condition in Example 7(i) (p. 985) defines x implicitly as a function of y, at least for values of y in some interval, find $\frac{dx}{dy}$.

　　(ii) What is the value of this partial derivative at the point $y = 1$, $x = 1$?

*29. If $f(x, y) = x^2y^3$, regard f as a function of polar coordinates (r, θ), and find $\frac{\partial^2 f}{\partial r^2}$, $\frac{\partial^2 f}{\partial r \partial \theta}$, $\frac{\partial^2 f}{\partial \theta^2}$.

*30. If $g(r, \theta) = r \sec \theta$, find $\frac{\partial g}{\partial x}$, $\frac{\partial g}{\partial y}$.

*31. If $h(r, \theta) = e^{2\theta \sqrt{r}}$, find $\frac{\partial h}{\partial x}$, $\frac{\partial h}{\partial y}$.

*32. Suppose y is defined as an implicit function of x by the equation $x^3y^2 - 3xy = 4$. Find (i) $\frac{dy}{dx}$, (ii) $\frac{d^2y}{dx^2}$.

*33. If the equation $xy^2 - yz^3 + 2x^2yz = 2$ is used to define z as a function of x and y, find (i) $\frac{\partial z}{\partial x}$, (ii) $\frac{\partial^2 z}{\partial x^2}$, (iii) $\frac{\partial^2 z}{\partial x \partial y}$.

**34. If $f(x, y)$ has continuous second partial derivatives, its *Laplacian* $\nabla^2 f$ is defined by $\nabla^2 f = \frac{\partial^2 f}{\partial x^2} + \frac{\partial^2 f}{\partial y^2}$. Show that it can be expressed in polar coordinates by

$$\nabla^2 f = \frac{\partial^2 f}{\partial x^2} + \frac{\partial^2 f}{\partial y^2} = \frac{\partial^2 f}{\partial r^2} + \frac{1}{r}\frac{\partial f}{\partial r} + \frac{1}{r^2}\frac{\partial^2 f}{\partial \theta^2}.$$

35. Use the formula of Example 4 (p. 980) to estimate the surface area of your own body in square meters or square feet.

36. An adolescent weighs 50 kg and stands 150 cm tall. If he is growing at the rate of 4 kg/yr and 8 cm/yr, how fast is his surface area increasing (cf. Example 4)?

‖‖‖

37. In the situation of Example 5 (p. 982), make a picture showing R as a function of x with $L = 8, 10, 12, 20$.

38. Use the "solve" function on your graphics calculator to find a more exact answer to Example 8 (p. 985).

39. Sketch some of the level curves of the surface area function of Example 4 (p. 980).

POINT TO PONDER

If u, v are functions of x and y, and f is a function of u and v, then the Chain Rule says

$$f_x = f_u u_x + f_v v_x,$$
$$f_y = f_u u_y + f_v v_y.$$

If you know about matrices, you might recognize that this can be written as the matrix product

$$(f_x, f_y) = (f_u, f_v) \begin{pmatrix} u_x & u_y \\ v_x & v_y \end{pmatrix}.$$

The matrix $\begin{pmatrix} u_x & u_y \\ v_x & v_y \end{pmatrix}$ of partial derivatives of u and v with respect to x and y is called the *Jacobian matrix*. It will reappear in the "Point to Ponder" in Section 17.9.

You might like to try using matrices to express the Chain Rule with three variables and three intermediate variables or in the more complicated situation of Example 6 (p. 983).

16.6 Directional Derivatives and Gradients

UNIT VECTOR

DIRECTION OF FASTEST INCREASE

GRADIENTS AND LEVEL CURVES

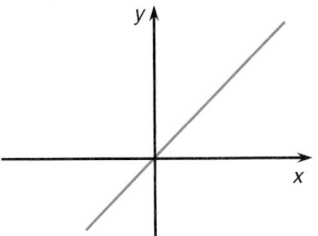

Figure 16.6.1

One of the obvious shortcomings of partial derivatives is that they describe the rates of change of a function along certain chosen directions but say nothing about what happens in all the other possible directions.

For instance, instead of asking about the derivative of $f(x, y)$ in the horizontal direction (i.e., $\frac{\partial f}{\partial x}$), it is just as reasonable to consider the derivative along the direction of the line $y = x$ shown in Figure 16.6.1 or along any other line. This is the idea behind what is called the *directional derivative*, the derivative in some specified direction. In this section we will see that if f is differentiable, then its directional derivative in any direction can be found fairly easily once we know the partial derivatives. This means that it is not really necessary to do the direct calculation of derivatives in every direction because just the two of them along the directions parallel to the axes actually contain all the required information.

The derivative of $f(x, y)$ along a line means the rate at which the value of $f(x, y)$ changes as the point (x, y) moves along the line. Actually, it is necessary to specify in which direction the point moves along the line because if the function increases as the point moves one way, then it must decrease as the point moves the other. So rather than talk about a line, we need to talk about a direction. The easiest way to do this is to give a vector $\mathbf{u} = (u_1, u_2)$ that points in the appropriate direction.

A vector has a direction, but it also has a length; in this situation we are interested only in its direction, so we can avoid being confused by the length by insisting that the vector is a **unit vector**, meaning that its length equals 1, that is, that

$$\|\mathbf{u}\| = \sqrt{u_1^2 + u_2^2} = 1.$$

For instance, the vector $\mathbf{u} = \left(\frac{1}{\sqrt{2}}, \frac{1}{\sqrt{2}}\right)$ is a unit vector whose direction is along the line $y = x$, pointing upward to the right. See Figure 16.6.2.

Suppose we want to consider the derivative at the origin of a function f in the direction of the vector $\mathbf{u}$. The way to do this is to form a difference quotient and take a

Figure 16.6.2

limit. The numerator of the difference quotient ought to be the difference $f(x, y) - f(0, 0)$, where (x, y) is some point near the origin on the line $y = x$. By analogy with the difference quotients we use for ordinary or partial derivatives the denominator should be something to do with the distance from (x, y) to $(0, 0)$. However, it is not quite the distance, since distance is never negative, and the denominators of the other kinds of difference quotients can sometimes be negative.

A simple way to find a point on the line $y = x$ would be to take $h\mathbf{u} = h(u_1, u_2) = (hu_1, hu_2)$, with $\mathbf{u} = (u_1, u_2) = \left(\frac{1}{\sqrt{2}}, \frac{1}{\sqrt{2}}\right)$ as above (see Figure 16.6.3). For any number h this is a point on the line $y = x$. Moreover, it is easy to find that its distance from the origin is

$$\|h\mathbf{u} - (0, 0)\| = \|h\mathbf{u}\| = |h|\|\mathbf{u}\| = |h|.$$

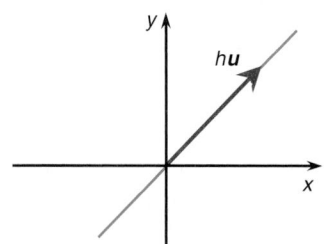

Figure 16.6.3

The distance from $h\mathbf{u}$ to the origin is $|h|$, so the natural thing to do would be to use h as the denominator of the difference quotient. This is very much like what we did with the difference quotient for ordinary derivatives. When $h > 0$, the point $h\mathbf{u}$ is above and to the right of the origin, while for $h < 0$ the point $h\mathbf{u}$ is on the other side of the origin.

The difference quotient is

$$\frac{f(hu_1, hu_2) - f(0, 0)}{h}.$$

The limit of this quotient as $h \to 0$ may not exist, but if it does exist, it is the *directional derivative* in the direction of $\mathbf{u} = (u_1, u_2)$.

All this has been done to find the directional derivative at the origin $(0, 0)$ in the direction of the vector $\mathbf{u} = \left(\frac{1}{\sqrt{2}}, \frac{1}{\sqrt{2}}\right)$. Naturally, it will be necessary to find directional derivatives at any point (a, b) and in the direction of any unit vector $\mathbf{u}$, as shown in Figure 16.6.4. The idea is exactly the same; we should find a point that is near (a, b) *in the direction of* $\mathbf{u}$. The way to do this is to consider the point $(a, b) + h\mathbf{u} = (a, b) + (hu_1, hu_2) = (a + hu_1, b + hu_2)$, where h is a small number. The distance from this point to (a, b) is $\|((a, b) + h\mathbf{u}) - (a, b)\| = \|h\mathbf{u}\| = |h|$. The difference quotient is formed in the analogous way:

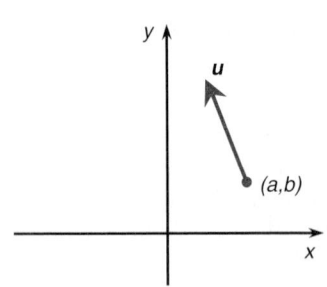

Figure 16.6.4

$$\frac{f(a + hu_1, b + hu_2) - f(a, b)}{h}.$$

The limit of this difference quotient as $h \to 0$, provided it exists, is the directional derivative of f at (a, b) in the direction of the unit vector $\mathbf{u}$.

Curiously enough, it is not very difficult to calculate this limit. Let us write $N(h) = f((a, b) + h\mathbf{u}) = f(a + hu_1, b + hu_2)$. Then the difference quotient is $\frac{f(a+hu_1, b+hu_2) - f(a, b)}{h} = \frac{N(h) - N(0)}{h}$, which is the difference quotient of the function N of the single variable h. If the limit of the difference quotient exists as $h \to 0$, then it must be the derivative at 0 of the function N.

Now it is possible to calculate this derivative N' by using the Chain Rule. We think of f as a function $f(x, y)$, and we let x and y be intermediate variables by letting $x = x(h) = a + hu_1$ and $y = y(h) = b + hu_2$. The Chain Rule tells us that

$$N'(0) = \frac{dN}{dh}(0) = \frac{d}{dh}\big(f(a + hu_1, b + hu_2)\big)\bigg|_{h=0} = \frac{d}{dh}f\big(x(h), y(h)\big)\bigg|_{h=0}$$

$$= \left(\frac{\partial f}{\partial x}\frac{dx}{dh} + \frac{\partial f}{\partial y}\frac{dy}{dh}\right)\bigg|_{h=0} = \left(\frac{\partial f}{\partial x}u_1 + \frac{\partial f}{\partial y}u_2\right)\bigg|_{h=0}$$

$$= \left(\frac{\partial f}{\partial x}(a, b)u_1 + \frac{\partial f}{\partial y}(a, b)u_2\right).$$

Note that
$\frac{\partial x}{\partial h} = \frac{\partial}{\partial h}(a + hu_1) = u_1,$
$\frac{\partial y}{\partial h} = \frac{\partial}{\partial h}(b + hu_2) = u_2.$

Here we used the fact that when $h = 0$, the point is $(x, y) = (a, b)$.

The significance of this calculation is that the directional derivative is simply obtained by multiplying the partial derivatives f_x and f_y by u_1 and u_2 and adding. In particular, if we know the partial derivatives, we can find any directional derivative quite easily.

DEFINITION 16.6.1

Directional Derivatives

If $f(x, y)$ is differentiable at (a, b) and $\mathbf{u} = (u_1, u_2)$ is a unit vector, then the **directional derivative** of f in the direction of $\mathbf{u}$ is defined to be

$$D_{\mathbf{u}}f(a, b) = \lim_{h \to 0} \frac{f\big((a, b) + h\mathbf{u}\big) - f(a, b)}{h} = \lim_{h \to 0} \frac{f(a + hu_1, b + hu_2) - f(a, b)}{h}.$$

It can be calculated as follows:

$$D_{\mathbf{u}}f(a, b) = f_x(a, b)\, u_1 + f_y(a, b)\, u_2 = \frac{\partial f}{\partial x}(a, b)\, u_1 + \frac{\partial f}{\partial y}(a, b)\, u_2.$$

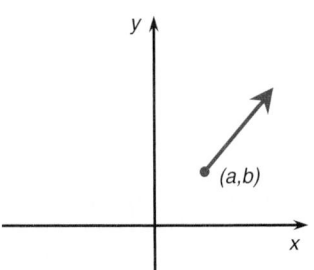

Figure 16.6.5

It is very important to distinguish the roles played by the vector $\mathbf{u} = (u_1, u_2)$ on the one hand and the point (a, b) on the other. The point (a, b) tells us where the directional derivative is being evaluated. It can be any point in the domain of f at which f is differentiable. To differentiate, we need to consider a point nearby that is allowed to move. The vector $\mathbf{u}$ tells us the direction in which we want to let this point move as we measure the rate of change of f. It is very convenient to draw the picture as shown in Figure 16.6.5, with the vector $\mathbf{u}$ placed so that it begins at the point (a, b). This way we see that $\mathbf{u}$ points in a direction in which the second point is allowed to move away from (a, b). If we want to write the vector $\mathbf{u}$ out in terms of its coordinates $\mathbf{u} = (u_1, u_2)$, we are more likely to think of the vector placed so that it begins at the origin, as in Figure 16.6.6.

We will draw the unit vector $\mathbf{u}$ in either of these two ways and rely on the context to let us know which is which.

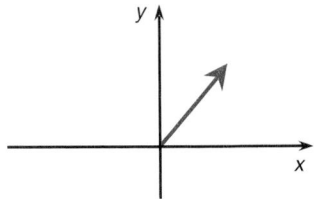

Figure 16.6.6

EXAMPLE 1   (i) Find the directional derivative $D_{\mathbf{u}}f(2, -1)$ if $\mathbf{u} = \left(\frac{1}{\sqrt{2}}, -\frac{1}{\sqrt{2}}\right)$ and $f(x, y) = 3x^2 y - 4y^2$.

Solution To use the formula, we need to know the partial derivatives of f. They are easily found: $f_x = 6xy$ and $f_y = 3x^2 - 8y$. Evaluating at $(2, -1)$, we find $f_x(2, -1) = -12$ and $f_y(2, -1) = 20$. The formula for the directional derivative now gives $D_{\mathbf{u}}f(2, -1) = f_x(2, -1)\, u_1 + f_y(2, -1)\, u_2 = -12\frac{1}{\sqrt{2}} + 20\left(-\frac{1}{\sqrt{2}}\right) = -\frac{32}{\sqrt{2}} = -16\sqrt{2} \approx -22.6$.

(ii) Find the directional derivative of $f(x, y) = e^{xy} + y\sin x$ at the point $(0, -2)$, in the direction that points upward to the right at an angle of $\frac{\pi}{3}$ above the horizontal.

Solution As before, it is easy to find the partial derivatives $f_x = e^{xy}y + y\cos x$ and $f_y = e^{xy}x + \sin x$ and to evaluate them at the specified point: $f_x(0, -2) = e^0(-2) + (-2)\cos 0 = -2 - 2 = -4$ and $f_y(0, -2) = e^0(0) + \sin 0 = 0$.

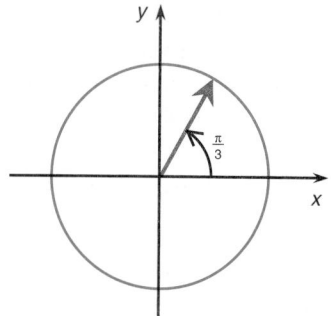

Figure 16.6.7

The difference in this example is that we are not given the unit vector **u**. We need a vector that points upward to the right at an angle of $\frac{\pi}{3}$ above the horizontal. If we sketch the unit circle and draw the angle $\frac{\pi}{3}$ as in Figure 16.6.7, we see that the coordinates of the point at which the line crosses the circle are $\left(\cos\frac{\pi}{3}, \sin\frac{\pi}{3}\right) = \left(\frac{1}{2}, \frac{\sqrt{3}}{2}\right)$. Since this point is exactly 1 unit from the origin, we can use it as our vector **u**, that is, $\mathbf{u} = \left(\frac{1}{2}, \frac{\sqrt{3}}{2}\right)$.

With this information it is now easy to find that the directional derivative is $D_{\mathbf{u}}f(0,-2) = f_x(0,-2)u_1 + f_y(0,-2)u_2 = -4\left(\frac{1}{2}\right) + (0)\frac{\sqrt{3}}{2} = -2$.

(iii) Find the directional derivative of $G(x, y) = \sin(\pi xy)$ at the point $(3, -1)$ in the direction of the vector $\mathbf{v} = (1, -2)$.

Solution The partial derivatives $G_x = \cos(\pi xy)\pi y$ and $G_y = \cos(\pi xy)\pi x$ are easily evaluated at $(3, -1)$; we find $G_x(3, -1) = -\pi\cos(-3\pi) = \pi$ and $G_y(3, -1) = 3\pi\cos(-3\pi) = -3\pi$.

We are given the vector **v** that indicates the direction in which to take the directional derivative, but we must be careful because it is not a *unit vector*. We need to find the unit vector that points in the same direction as **v**, and this is quite easy to do. All that is needed is to divide **v** by its length $\|\mathbf{v}\| = \sqrt{1^2 + (-2)^2} = \sqrt{5}$; we let $\mathbf{u} = \frac{1}{\|\mathbf{v}\|}\mathbf{v} = \frac{1}{\sqrt{5}}(1, -2) = \left(\frac{1}{\sqrt{5}}, \frac{-2}{\sqrt{5}}\right)$. This is a unit vector that points in the same direction as **v**.

Now it is straightforward to find $D_{\mathbf{u}}G(3, -1) = G_x(3, -1)u_1 + G_y(3, -1)u_2 = \frac{\pi}{\sqrt{5}} + \frac{(-3\pi)(-2)}{\sqrt{5}} = \frac{7\pi}{\sqrt{5}}$.

The required directional derivative is $\frac{7\pi}{\sqrt{5}} \approx 9.835$.

(iv) Let $T(x, y) = \arctan\frac{y}{x}$. If $\mathbf{u} = (\cos\phi, \sin\phi)$ is the unit vector whose direction is at an angle ϕ from the positive x-axis, find the directional derivative $D_{\mathbf{u}}T$.

Solution It is easy to calculate

$$T_x = \frac{1}{1 + (y/x)^2}\left(-\frac{y}{x^2}\right) = -\frac{y}{x^2 + y^2},$$

$$T_y = \frac{1}{1 + (y/x)^2}\left(\frac{1}{x}\right) = \frac{x}{x^2 + y^2}.$$

With $\mathbf{u} = (\cos\phi, \sin\phi)$, we see that $D_{\mathbf{u}}T = -\frac{y\cos\phi}{x^2+y^2} + \frac{x\sin\phi}{x^2+y^2} = \frac{-y\cos\phi + x\sin\phi}{x^2+y^2}$.

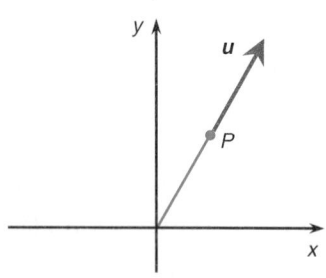

Figure 16.6.8

It is interesting to consider the last part of Example 1 in geometrical terms. The function $T(x, y) = \arctan\frac{y}{x}$ is not defined on the y-axis (i.e., when $x = 0$). On the right half-plane, $T(x, y) = \arctan\frac{y}{x}$ is the angle θ in the polar coordinates of the point (x, y). On the left half-plane it equals $\theta - \pi$, or, in geometrical terms, it is the angle measured from the *negative* x-axis. See Figure 16.6.8.

Now suppose we consider a point $P = (x, y)$ and write its polar coordinates (r, θ), so that $x = r\cos\theta$, $y = r\sin\theta$. One interesting direction in which to take the directional derivative would be the direction OP from the origin out to P. To do this, we should use the unit vector that points in the same direction as OP, which is $\mathbf{u} = (\cos\theta, \sin\theta)$. See Figure 16.6.9.

Now using the formula from part (iv) of Example 1, we find that

$$D_{\mathbf{u}}T = \frac{-y\cos\theta + x\sin\theta}{x^2 + y^2} = \frac{-r\sin\theta\cos\theta + r\cos\theta\sin\theta}{x^2 + y^2} = 0.$$

Figure 16.6.9

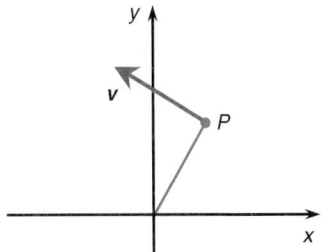

Figure 16.6.10

This result says that if the point (x, y) moves in the direction of $\mathbf{u}$, that is, directly away from the origin, then the rate of change of T is zero. But if we think of T as the angle θ, then this is perfectly reasonable because moving in the direction of $\mathbf{u}$ does not change the angle θ. The value of T stays constant along the line OP, so it is not at all surprising that its directional derivative is 0.

Now suppose instead that we want to move as shown in Figure 16.6.10 in the direction perpendicular to OP. The appropriate unit vector is $\mathbf{v} = (-\sin\theta, \cos\theta)$, and we find that

$$D_{\mathbf{v}} T = \frac{-y(-\sin\theta) + x\cos\theta}{x^2 + y^2} = \frac{r\sin^2\theta + r\cos^2\theta}{x^2 + y^2} = \frac{r}{r^2} = \frac{1}{r}.$$

This says that moving in the direction of the unit vector $\mathbf{v}$ tends to increase the value of T, which is perfectly consistent with our interpretation of T as the angle θ (or $\theta - \pi$ if the point is in the left half-plane).

The directional derivative gives an indication of the behavior of a function in a particular direction. In this example the function T increases if we move in a direction that tends to rotate the point counterclockwise around the origin; it decreases if we move in a direction that tends to rotate the point clockwise; and it remains unchanged if we move along lines through the origin. The directional derivative will be positive for a direction that moves counterclockwise around the origin (and so increases the angle θ); it will be negative for a direction that rotates the point clockwise and zero for a direction that points directly toward or directly away from the origin.

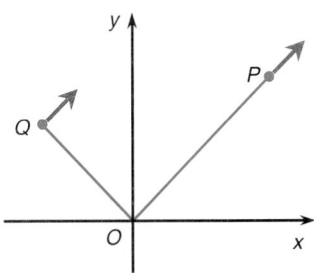

Figure 16.6.11

It is very important to observe that the directional derivative in a direction $\mathbf{u}$ will also depend on the point (x, y) at which it is evaluated. For instance, suppose $\mathbf{u} = \left(\frac{1}{\sqrt{2}}, \frac{1}{\sqrt{2}}\right)$ is the unit vector pointing up and to the right at an angle of $\frac{\pi}{4}$ above the horizontal, and let us evaluate $D_{\mathbf{u}} T$ at several different points.

If $P = (3, 3)$, then $\mathbf{u}$ points in the same direction as OP, and as we have remarked, $D_{\mathbf{u}} T(3, 3) = 0$. If $Q = (-2, 2)$, then moving Q in the direction of $\mathbf{u}$ tends to move it clockwise around the origin, so we expect the value of T to decrease and the directional derivative to be negative (see Figure 16.6.11). Using the formula from the example, we find that

$$D_{\mathbf{u}} T(-2, 2) = \frac{-2\frac{1}{\sqrt{2}} + (-2)\frac{1}{\sqrt{2}}}{(-2)^2 + (2)^2} = \frac{-2\sqrt{2}}{8} = -\frac{1}{2\sqrt{2}}.$$

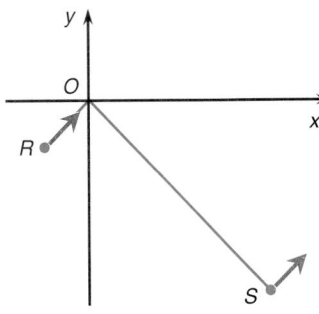

Figure 16.6.12

If $R = (-1, -1)$, then once again $\mathbf{u}$ is in the direction of the line from R through the origin, so $D_{\mathbf{u}} T(-1, -1) = 0$. Finally, if $S = (4, -4)$, we see that moving from S in the direction of $\mathbf{u}$ will tend to increase θ, so we expect $D_{\mathbf{u}} T(4, -4)$ to be positive (see Figure 16.6.12). We find that

$$D_{\mathbf{u}} T(4, -4) = \frac{-(-4)\frac{1}{\sqrt{2}} + 4\frac{1}{\sqrt{2}}}{(4)^2 + (-4)^2} = \frac{4\sqrt{2}}{32} = \frac{1}{4\sqrt{2}}.$$

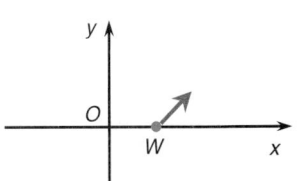

Figure 16.6.13

It is also possible to evaluate at other points, where the situation is less clear-cut. If $W = (1, 0)$, then from Figure 16.6.13 it appears that moving along $\mathbf{u}$ from W will have the effect of moving the point away from the origin in a direction that will also increase the angle θ, so we expect the directional derivative to be positive. A quick calculation shows that $D_{\mathbf{u}} T(1, 0) = \left(-(0)\frac{1}{\sqrt{2}} + (1)\frac{1}{\sqrt{2}}\right)/(1^2 + 0^2) = \frac{1}{\sqrt{2}}.$

EXAMPLE 2   Air Temperature   The boundary between a region of warm air and a region of cooler air is called a *front*. Near it the temperature can change quite rapidly. Let us mark positions on a map by (x, y)-coordinates as usual, with the positive y-axis pointing north and the positive x-axis pointing east. Let us measure distances in kilometers and consider the region where $-4 \leq x, y \leq 4$.

Suppose the air temperature at ground level at (x, y) is given (in degrees Celsius) by the function

$$T(x, y) = 20 - x - y + 0.05x^3 - 0.2y^2.$$

The partial derivatives are

$$\frac{\partial T}{\partial x} = -1 + 0.15x^2 \qquad \text{and} \qquad \frac{\partial T}{\partial y} = -1 - 0.4y.$$

The directional derivative in the direction due north means the directional derivative along the vector $(0, 1)$, so it is

$$(-1 + 0.15x^2)(0) + (-1 - 0.4y)(1) = -1 - 0.4y.$$

The directional derivative in the direction due west is the directional derivative along the vector $(-1, 0)$, so it is $1 - 0.15x^2$. Similarly, in the southeast direction we consider the vector $(1, -1)$; it is not a unit vector, so we divide by its length and get $\left(\frac{1}{\sqrt{2}}, -\frac{1}{\sqrt{2}}\right)$. The direction derivative in the southeast direction is

$$(-1 + 0.15x^2)\frac{1}{\sqrt{2}} + (-1 - 0.4y)\left(-\frac{1}{\sqrt{2}}\right) = \frac{0.15}{\sqrt{2}}x^2 + \frac{0.4}{\sqrt{2}}y.$$

Similarly, the unit vector pointing northeast is $\left(\frac{1}{\sqrt{2}}, \frac{1}{\sqrt{2}}\right)$, and the directional derivative in the northeast direction is

$$(-1 + 0.15x^2)\frac{1}{\sqrt{2}} + (-1 - 0.4y)\frac{1}{\sqrt{2}} = -\frac{2}{\sqrt{2}} + \frac{0.15}{\sqrt{2}}x^2 - \frac{0.4}{\sqrt{2}}y.$$

In particular, at the origin $(0, 0)$ the directional derivative in the northerly direction is $-1 - 0.4y = -1$; in the westerly direction it is $1 - 0.15x^2 = 1$; in the southeasterly direction it is 0; and in the northeasterly direction it is $-\frac{2}{\sqrt{2}} = -\sqrt{2}$.

If somebody starts at the origin and heads north, she will encounter a drop in temperature of about $1°$ for each kilometer she travels (at least for short distances). Going west, the temperature increases by about $1° /$ km; going southeast, it changes very little; and going northeast, it drops by about $\sqrt{2}° /$ km.

In Figure 16.6.14 we see a number of *isotherms*, that is, level curves of T. We see that heading southeast from the origin takes us "along" an isotherm, which explains why the directional derivative is zero in that direction. Temperatures decrease going north and decrease faster going northeast; they increase going west. From the picture we expect that T will increase fastest going southwest from the origin. ◢

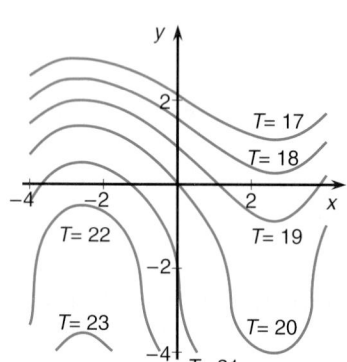

T= 17
T= 18
T= 22
T= 19
T= 23
T= 20
T= 21

Figure 16.6.14

Gradients

If we look at the formula $D_{\mathbf{u}}f(a, b) = f_x(a, b)u_1 + f_y(a, b)u_2$ for the directional derivative of f at (a, b) in the direction of the unit vector $\mathbf{u}$, we realize that it is actually just the dot product of the direction vector $\mathbf{u} = (u_1, u_2)$ with the vector $\left(f_x(a, b), f_y(a, b)\right) = \left(\frac{\partial f}{\partial x}(a, b), \frac{\partial f}{\partial y}(a, b)\right)$.

This vector, whose components are just the partial derivatives of a function evaluated at some point, is extremely important.

DEFINITION 16.6.2

Let $f(x, y)$ be a function. If (a, b) is any point at which the partial derivatives of f exist, then the **gradient** of f at the point (a, b) is the vector

$$\nabla f(a, b) = \big(f_x(a, b), f_y(a, b)\big) = \left(\frac{\partial f}{\partial x}(a, b), \frac{\partial f}{\partial y}(a, b)\right).$$

The symbol ∇ is pronounced "del" (reminding us of the upside-down capital delta), but we often read ∇f as "the gradient of f."

Notice that the gradient ∇f is a *function*, whose value at each point (a, b) is a *vector* $\big(f_x(a, b), f_y(a, b)\big)$. There is another fairly common notation, which is to write grad f for ∇f.

EXAMPLE 3   (i) Find $\nabla f(2, -5)$ if $f(x, y) = x^2 y \cos(3\pi xy)$.

Solution It is easy to find $f_x = 2xy\cos(3\pi xy) - 3\pi x^2 y^2 \sin(3\pi xy)$ and $f_y = x^2 \cos(3\pi xy) - 3\pi x^3 y \sin(3\pi xy)$. It is also easy to evaluate at the point $(2, -5)$ to find that $f_x(2, -5) = 2(2)(-5)\cos(-30\pi) - 3\pi 2^2(-5)^2 \sin(-30\pi) = -20 + 0 = -20$ and $f_y(2, -5) = 2^2 \cos(-30\pi) - 3\pi(2)^3(-5)\sin(-30\pi) = 4 - 0 = 4$.
The gradient is $\nabla f(2, -5) = \big(f_x(2, -5), f_y(2, -5)\big) = (-20, 4)$.

(ii) Find the gradient $\nabla g(a, b)$ if $g(x, y) = xy^2 - 2xy$.

Solution We find that $g_x = y^2 - 2y$ and $g_y = 2xy - 2x$. From this it is easy to calculate the gradient $\nabla g(a, b) = (b^2 - 2b, 2ab - 2a)$.

(iii) Find all points at which $\nabla F = 0$, where $F(x, y) = xe^{xy} - ye^{xy}$.

Since ∇F is a vector, $\nabla F = 0$ means that ∇F is the zero vector.

Solution We find $F_x = e^{xy} + xye^{xy} - y^2 e^{xy} = (1 + xy - y^2)e^{xy}$ and $F_y = x^2 e^{xy} - e^{xy} - xye^{xy} = -(1 + xy - x^2)e^{xy}$. We are looking for points (x, y) at which the vector $\nabla F = (F_x, F_y)$ is zero. For a vector to be zero means that both its components must be zero, so we want $F_x = 0$ and $F_y = 0$.
Since e^{xy}, being an exponential, is never zero, this means that $1 + xy - y^2 = 0$ and $1 + xy - x^2 = 0$. Subtracting these two equations, we find that $y^2 - x^2 = 0$, that is, that $x^2 = y^2$.
One possibility is that $x = y$. In this case, $1 + xy - y^2 = 1 + x^2 - x^2 = 1$. This can never be 0, so the only other possibility is that $y = -x$. In this case, $1 + xy - x^2 = 1 - x^2 - x^2 = 1 - 2x^2$, which is 0 when $x^2 = \frac{1}{2}$, that is, when $x = \pm\frac{1}{\sqrt{2}}$. So the possible points are $\left(\frac{1}{\sqrt{2}}, -\frac{1}{\sqrt{2}}\right)$ and $\left(-\frac{1}{\sqrt{2}}, \frac{1}{\sqrt{2}}\right)$. It is easy to check that F_x and F_y are both zero at these two points, and we know they are the only ones.

The way we first came to think about gradients was by noticing that the directional derivative $D_{\mathbf{u}}f$ of a function f in the direction $\mathbf{u}$ is the dot product of ∇f with $\mathbf{u}$. This makes the formula much simpler to write down:

$$D_{\mathbf{u}}f = \nabla f \cdot \mathbf{u} = (f_x, f_y) \cdot (u_1, u_2) = f_x u_1 + f_y u_2. \qquad (16.6.3)$$

EXAMPLE 4 We return to the situation of Example 2. If

$$T(x, y) = 20 - x - y + 0.05x^3 - 0.2y^2,$$

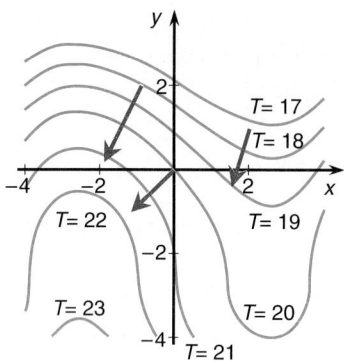

Figure 16.6.15

then

$$\nabla T(x, y) = (-1 + 0.15x^2, -1 - 0.4y).$$

In particular, at the origin, $\nabla T(0, 0) = (-1, -1)$. This vector is shown in Figure 16.6.15, and we see that it points southwest, the direction in which the temperature is increasing most rapidly.

We also find, for example, that $\nabla T(2, 1) = (-0.4, -1.4)$ and $\nabla T(-1, 2) = (-0.85, -1.8)$. These two gradients also appear in the picture. Notice how in each case the gradient points in the direction in which the temperature is increasing.

In Example 2 we calculated various directional derivatives at the origin. They are now more easily found by using Equation 16.6.3. For instance, the directional derivative in the northerly direction starting at (x, y) is

$$\nabla T(x, y) \cdot (0, 1) = (-1 + 0.15x^2, -1 - 0.4y) \cdot (0, 1) = -1 - 0.4y,$$

in the westerly direction we get

$$\nabla T(x, y) \cdot (-1, 0) = (-1 + 0.15x^2, -1 - 0.4y) \cdot (-1, 0) = 1 - 0.15x^2,$$

going southeast gives

$$\nabla T(x, y) \cdot \left(\frac{1}{\sqrt{2}}, -\frac{1}{\sqrt{2}}\right) = (-1 + 0.15x^2, -1 - 0.4y) \cdot \left(\frac{1}{\sqrt{2}}, -\frac{1}{\sqrt{2}}\right)$$

$$= \frac{0.15}{\sqrt{2}}x^2 + \frac{0.4}{\sqrt{2}}y,$$

and going northeast gives

$$\nabla T(x, y) \cdot \left(\frac{1}{\sqrt{2}}, \frac{1}{\sqrt{2}}\right) = (-1 + 0.15x^2, -1 - 0.4y) \cdot \left(\frac{1}{\sqrt{2}}, \frac{1}{\sqrt{2}}\right)$$

$$= -\sqrt{2} + \frac{0.15}{\sqrt{2}}x^2 - \frac{0.4}{\sqrt{2}}y.$$

Three or More Variables

Everything we have been doing with directional derivatives and gradients can be done just as well with functions of three or more variables. The definitions are almost exactly the same.

DEFINITION 16.6.4

Gradients and Directional Derivatives

Suppose $f(x, y, z)$ is a function of three variables. If (a, b, c) is a point at which the partial derivatives of f all exist, then the **gradient** of f at (a, b, c) is the vector

$$\nabla f(a, b, c) = \text{grad } f(a, b, c) = \left(f_x(a, b, c), f_y(a, b, c), f_z(a, b, c)\right)$$

$$= \left(\frac{\partial f}{\partial x}(a, b, c), \frac{\partial f}{\partial y}(a, b, c), \frac{\partial f}{\partial z}(a, b, c)\right).$$

If $\mathbf{u} = (u_1, u_2, u_3)$ is a unit vector (i.e., $\|\mathbf{u}\| = \sqrt{u_1^2 + u_2^2 + u_3^2} = 1$), then the **directional derivative** of f at (a, b, c) in the direction of $\mathbf{u}$ is

$$D_{\mathbf{u}}f(a, b, c) = \nabla f(a, b, c) \cdot \mathbf{u}$$

$$= \frac{\partial f}{\partial x}(a, b, c)u_1 + \frac{\partial f}{\partial y}(a, b, c)u_2 + \frac{\partial f}{\partial z}(a, b, c)u_3.$$

Suppose $F(x_1, \ldots, x_n)$ is a function of n variables. If $(a_1, \ldots, a_n)$ is a point at which the partial derivatives of F all exist, then the **gradient** of F at $(a_1, \ldots, a_n)$ is the vector

$$\nabla F(a_1, \ldots, a_n) = \text{grad } F(a_1, \ldots, a_n)$$
$$= \left(\frac{\partial F}{\partial x_1}(a_1, \ldots, a_n), \frac{\partial F}{\partial x_2}(a_1, \ldots, a_n), \ldots, \frac{\partial F}{\partial x_n}(a_1, \ldots, a_n) \right).$$

If $\mathbf{u} = (u_1, u_2, \ldots, u_n)$ is a unit vector (i.e., $\|\mathbf{u}\| = \sqrt{u_1^2 + u_2^2 + \ldots + u_n^2} = 1$), then the **directional derivative** of F at $(a_1, \ldots, a_n)$ in the direction of $\mathbf{u}$ is

$$D_{\mathbf{u}}F(a_1, \ldots, a_n)$$
$$= \nabla f(a_1, \ldots, a_n) \cdot \mathbf{u}$$
$$= \frac{\partial F}{\partial x_1}(a_1, \ldots, a_n)u_1 + \frac{\partial F}{\partial x_2}(a_1, \ldots, a_n)u_2 + \ldots + \frac{\partial F}{\partial x_n}(a_1, \ldots, a_n)u_n.$$

To find the directional derivative in the direction of a nonzero vector $\mathbf{v}$ that is not a unit vector, we simply find the directional derivative in the direction of the corresponding unit vector $\mathbf{u} = \frac{\mathbf{v}}{\|\mathbf{v}\|}$.

EXAMPLE 5   (i) If $F(x, y, z) = xyz$, find $D_{\mathbf{u}}F(3, -2, 1)$, where $\mathbf{u} = \left(\frac{2}{3}, -\frac{1}{3}, -\frac{2}{3} \right)$.

Solution   We begin by finding the gradient $\nabla F = (yz, xz, xy)$ and then evaluate at the specified point: $\nabla F(3, -2, 1) = (-2, 3, -6)$. From this we find that $D_{\mathbf{u}}F(3, -2, 1) = \nabla F(3, -2, 1) \cdot \mathbf{u} = (-2, 3, -6) \cdot \left(\frac{2}{3}, -\frac{1}{3}, -\frac{2}{3} \right) = -\frac{4}{3} - \frac{3}{3} + \frac{12}{3} = \frac{5}{3}$.

(ii) Find all points (a, b, c) for which $D_{\mathbf{u}}f(a, b, c) = 0$ if $\mathbf{u} = (0, 1, 0)$ and $f(x, y, z) = xz^2 e^{2x-3xz} + y \ln |z - x^2|$.

Solution   We could begin by finding the partial derivatives of f, but we notice that with $\mathbf{u} = (0, 1, 0)$ the directional derivative is $D_{\mathbf{u}}f = \nabla f \cdot \mathbf{u} = (f_x, f_y, f_z) \cdot (0, 1, 0) = f_y$. This means that there is no need to calculate f_x or f_z. We find that $f_y = \ln |z - x^2|$, so $D_{\mathbf{u}}f = \ln |z - x^2|$.

The only way this will be 0 is if $|z - x^2| = 1$, that is, if $z - x^2 = \pm 1$. Another way to say this is that $z = x^2 + 1$ or $z = x^2 - 1$. The points at which the specified directional derivative is zero are exactly the points of the form $(x, y, x^2 + 1)$ or $(x, y, x^2 - 1)$.

(iii) Let $R(x, y, z) = x^2 + y^2 + z^2$. Find ∇R.

Solution   It is easy to see that $R_x = 2x$, $R_y = 2y$, and $R_z = 2z$. This means that the gradient is $\nabla R = (2x, 2y, 2z)$. If we write $P = (x, y, z)$, then the gradient is just twice the vector (x, y, z), which is the vector OP. In other words the gradient is a vector that points directly away from the origin.

(iv) Let $H(x_1, \ldots, x_n) = x_1 x_2 \times \ldots \times x_n$, that is, the product of all the x_k's. Find the gradient of H. Evaluate the gradient at the point $(1, 2, \ldots, n)$, that is, at the point whose kth coordinate equals k.

Solution The gradient is $\nabla H = (x_2 x_3 \times \ldots \times x_n; \; x_1 x_3 \times \ldots \times x_n; \; \ldots; \; x_1 x_2 \times \ldots \times x_{n-1})$; the kth component of the gradient vector is the product of all the x's *except* the kth one x_k.

In particular, at the point $(1, 2, \ldots, n)$ the kth component of the gradient is the product of the integers from 1 to n, *leaving out k*. This can conveniently be described as $\frac{n!}{k}$. So $\nabla H(1, 2, \ldots, n) = \left(n!, \frac{n!}{2}, \frac{n!}{3}, \ldots, \frac{n!}{n-1}, (n-1)! \right) = n! \left(1, \frac{1}{2}, \frac{1}{3}, \ldots, \frac{1}{n} \right).$ ◀

If we think of Formula 16.6.3 for the directional derivative, which says that $D_\mathbf{u} f = \nabla f \cdot \mathbf{u}$, it is obvious that the directional derivative will be 0 when the gradient is perpendicular to $\mathbf{u}$, simply because the dot product of two perpendicular vectors is always 0. The dot product will be positive if $\nabla f \neq 0$ and the angle between ∇f and $\mathbf{u}$ is less than $\frac{\pi}{2}$ and negative when the angle is greater than $\frac{\pi}{2}$. It will have its largest value when ∇f points in the same direction as $\mathbf{u}$ and its smallest (i.e., most negative) value when they point in opposite directions.

For instance, if we let $\mathbf{u} = \frac{1}{\|\nabla f\|} \nabla f$, then the directional derivative in the direction of $\mathbf{u}$, that is, in the direction of the gradient, is $D_\mathbf{u} f = \nabla f \cdot \mathbf{u} = \frac{1}{\|\nabla f\|} \nabla f \cdot \nabla f = \frac{1}{\|\nabla f\|} \|\nabla f\|^2 = \|\nabla f\|$. The directional derivative in the direction of the gradient, which is the largest possible value of the directional derivative, actually equals the length of the gradient vector.

This means that the gradient ∇f can be used to tell us a great deal about the behavior of the function f. The direction in which the values of f change the fastest is the direction along ∇f. The directions perpendicular to the gradient are directions along which the values of f change very little. In particular the gradient vector at a point P points in the direction along which f increases most rapidly as the point moves away from P (see Example 4 and Figure 16.6.15).

If we think of the graph $z = f(x, y)$ of a function of two variables, the value of the function f at the point (x, y) is represented by the height of the graph. To say that the value of f increases as the point moves in a certain direction means that the surface "slopes upward" in that direction. The gradient ∇f tells us the direction in which the graph rises most steeply from the point $(x, y, f(x, y))$. The length of the gradient also tells us just how fast the function increases as the point (x, y) moves in the direction of the gradient. See Figure 16.6.16(i), where the gradient vectors at various places all point roughly toward the point where f has its maximum. The gradients farther from that point are longer, since the surface is steeper there and the values of f rise faster. In Figure 16.6.16(ii) we see the graph of another function and the corresponding gradient vectors in the xy-plane.

This explains the way in which the word "gradient" is used in familiar speech to mean a slope.

Figure 16.6.16(i)

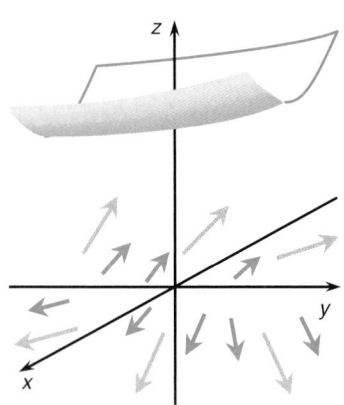

Figure 16.6.16(ii)

EXAMPLE 6 (i) Find the direction in which the function $f(x, y) = e^{2x+3y-1}$ increases most rapidly, starting at the point $(3, 7)$, and calculate its directional derivative in that direction.

Solution We find the gradient $\nabla f = (2e^{2x+3y-1}, 3e^{2x+3y-1}) = e^{2x+3y-1}(2, 3)$. Since $e^{2x+3y-1}$ is always positive, the gradient points in the same direction as $(2, 3)$, and to find a unit vector that points in that direction, we let $\mathbf{u} = \frac{1}{\|(2,3)\|}(2, 3) = \frac{1}{\sqrt{13}}(2, 3) = \left(\frac{2}{\sqrt{13}}, \frac{3}{\sqrt{13}} \right)$. This points in the direction in which f increases most rapidly.

The directional derivative is $D_{\mathbf{u}}f(3,7) = \nabla f(3,7) \cdot \mathbf{u} = e^{2(3)+3(7)-1}(2,3) \cdot$ $\left(\frac{2}{\sqrt{13}}, \frac{3}{\sqrt{13}}\right) = e^{26} \frac{2(2)+3(3)}{\sqrt{13}} = e^{26}\sqrt{13} \approx 7.06 \times 10^{11}$.

Notice that at every point the gradient is some number times the vector $(2,3)$. The gradient always points in the same direction, but its length is different at different points.

(ii) What is the direction in which the function $g(x, y, z) = 3 - x^2 z \cos(y^2 z)$ decreases most rapidly from the point $(3, -2, 0)$, and what is its directional derivative in this direction?

Solution   The gradient is

$$\nabla g = \left(-2xz\cos(y^2 z), 2x^2 yz^2 \sin(y^2 z), -x^2\cos(y^2 z) + x^2 y^2 z \sin(y^2 z)\right).$$

Evaluating at the point $(3, -2, 0)$, we find $\nabla g(3, -2, 0) = (0, 0, -9)$. The unit vector that points in the same direction as this is $\mathbf{u} = (0, 0, -1)$.

The direction in which g increases most rapidly is the direction of its gradient, which, for this function at the specified point, is the direction of $\mathbf{u}$. The question asks for the direction in which g *decreases* most rapidly, which will be the opposite

direction, that is, the direction of $-\mathbf{u} = (0, 0, 1)$. This vector points in the direction of the z-axis, so the function will decrease most rapidly in the z-direction.

The directional derivative is $D_{-\mathbf{u}}g(3, -2, 0) = \nabla g(3, -2, 0) \cdot (0, 0, 1) = (0, 0, -9) \cdot (0, 0, 1) = -9$.

Notice that the directional derivative is negative, corresponding to the fact that g decreases in the direction of $-\mathbf{u}$, and moreover that its absolute value 9 is just the length of the gradient at the point in question.

(iii) **Price, Demand, and Sales**   The sales of a certain product are predicted by the function $S(x, y) = 3x^2 y - 4xy^2$, where x is the demand for the product and y is the price for which it sells. Suppose that at the present time $x = 2$ and $y = 1$ but x is expected to decrease to 1.99. If the company wishes to maintain the current level of sales, what change should it make in the price it charges?

See the Graphics Calculator
Example on the next page.

Solution   The gradient is $\nabla S = (6xy - 4y^2, 3x^2 - 8xy)$, so $\nabla S(2, 1) = (8, -4)$. The idea is to move from the point $(2, 1)$ to a nearby point in such a way that the function S will change as little as possible. A reasonable way to attempt to do this would be to move in the direction in which the directional derivative is zero, which means in a direction perpendicular to $\nabla S(2, 1) = (8, -4)$.

For a vector (v, w) to be perpendicular to $(8, -4)$ means that their dot product should be zero, that is, that $0 = (v, w) \cdot (8, -4) = 8v - 4w$, which means that $4w = 8v$ or that $w = 2v$. We must move from the point $(2, 1)$ to a point (s, t) with $s = 1.99$ and so that the difference between these two points is perpendicular to $(8, -4)$, that is, so that $(s, t) - (2, 1) = (s - 2, t - 1)$ is perpendicular to $(8, -4)$. As we saw above, this means that $t - 1 = 2(s - 2)$, that is, that $t = 2s - 4 + 1 = 2s - 3$. Since s has to be 1.99, we find that $t = 3.98 - 3 = 0.98$.

This suggests that the company ought to lower the price to 0.98 and that this should offset the reduced demand and maintain the level of sales. We can check that $S(1.99, 0.98) \approx 3.998$, which is quite close to the original sales level. For the sake of comparison we note that if the company does nothing, that is, it leaves the price at $y = 1$, then $S(1.99, 1) \approx 3.920$, which represents a drop of 2% in sales.

With a graphics calculator we can plot the level curve of the function in part (iii) of Example 6. Since $S(2, 1) = 4$, we consider $3x^2y - 4xy^2 = 4$. Wanting to solve for y, we write it as a quadratic equation: $4xy^2 - 3x^2y + 4 = 0$, so $y = \frac{3x^2 \pm \sqrt{9x^4 - 64x}}{8x}$.

We also graph the line through the point $(x, y) = (2, 1)$ which is perpendicular to the gradient $(8, -4)$. Since the "slope" of the gradient vector is $-\frac{1}{2}$, the perpendicular line should have slope 2, so it is $y = 2x - 3$.

TI-85	SHARP EL9300
GRAPH F1($y(x)$ =) (3 F1(x) x^2 + $\sqrt{}$ (9 F1(x)^4 − 64 F1(x))) ÷ (8 F1(x)) ▼ (3 F1(x) x^2 − $\sqrt{}$ (9 F1(x)^4 − 64 F1(x))) ÷ (8 F1(x)) ▼ 2 F1(x) −3 M2(RANGE) 1.8 ▼ 3 ▼ ▼ (−) 1 ▼ 2 F5(GRAPH)	◿ (3 x/θ/T x^2 + $\sqrt{}$ 9 x/θ/T a^b 4 ▶ −64 x/θ/T ▶) ÷ (8 x/θ/T) 2ndF ▼ (3 x/θ/T x^2 − $\sqrt{}$ 9 x/θ/T a^b 4 ▶ −64 x/θ/T ▶) ÷ (8 x/θ/T) 2ndF ▼ 2 x/θ/T − 3 RANGE 1.8 ENTER 3 ENTER 1 ENTER −1 ENTER 2 ENTER 1 ENTER ◿
CASIO f_x-6300G (f_x-7700GB)	HP 48SX
Range 1.8 EXE 3 EXE 1 EXE −1 EXE 2 EXE 1 EXE EXE EXE EXE Graph (3 X x^2 + $\sqrt{}$ (9 X x^y 4 − 64 X)) ÷ (8 X) EXE Graph (3 Xx^2 − $\sqrt{}$ (9 X x^y 4 − 64 X)) ÷ (8 X) EXE Graph 2 X − 3 EXE	⬑ PLOT PLOTR ERASE ATTN ' Y ▬ () 3 × X y^x 2 + $\sqrt{x}$ () 9 × X y^x 4 − 64 × X ▶ ▶ ÷ () 8 × X ↱ PLOT ⬑ DRAW ERASE 1.8 ENTER 3 XRNG 1 ENTER +/− 2 YRNG DRAW ATTN ' Y ▬ () 3 × X y^x 2 − $\sqrt{x}$ () 9 × X y^x 4 − 64 × X ▶ ▶ ÷ () 8 × X ↱ PLOT ⬑ DRAW DRAW ATTN ' Y ▬ 2 × X − 3 ↱ PLOT ⬑ DRAW DRAW

Using the direction perpendicular to the gradient amounts to moving along the tangent to the level curve.

Try drawing some other level curves of S.

Gradients and Level Curves

A helpful way to understand gradients is by thinking about the level curves of a function, the curves along which the function is constant. If $f(x, y)$ is a function and c is a constant, consider the level curve $f(x, y) = c$. Suppose that this condition determines y as a function of x. Then we have seen that $\frac{dy}{dx} = -f_x/f_y$, assuming that the partials exist and the denominator is not zero. This derivative equals the slope of the tangent to the level curve.

On the other hand, the gradient vector $\nabla f = (f_x, f_y)$ points along a line whose slope is $\frac{\text{rise}}{\text{run}} = \frac{f_y}{f_x}$. This is the negative of the reciprocal of the slope we just found for the tangent line, and we know that two lines whose slopes are negative reciprocals

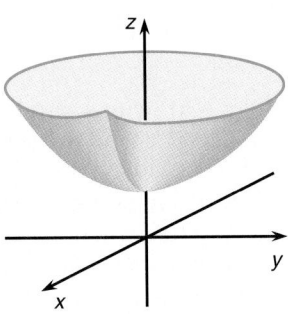

Figure 16.6.17(i)   **Figure 16.6.17(ii)**

must be perpendicular (cf. Theorem 1.5.12). The gradient is perpendicular to the tangent vector to the level curve. Figure 16.6.17(ii) shows several level curves and gradient vectors for the function graphed in Figure 16.6.17(i). The gradient vectors appear perpendicular to the level curves.

In rough terms, what we have just found is that the gradient is always perpendicular to the level curves. We can visualize the function by looking at its level curves, the places at which it takes various different values. The gradient points in the direction in which the function increases most rapidly, that is, moves most quickly to higher level curves. The length of the gradient tells how fast the function increases, which says how far apart the level curves are.

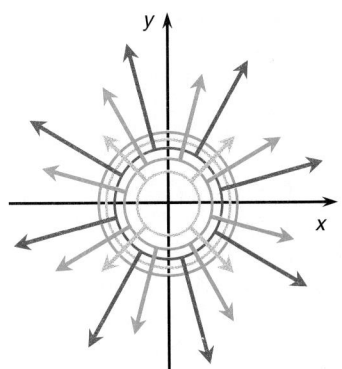

Figure 16.6.18(i)

EXAMPLE 7   In Figure 16.6.18(i) we show the level curves of the function $R(x, y) = x^2 + y^2$. Of course, they are circles. The gradient at each point is $\nabla R = (2x, 2y)$, which is the vector pointing out from the origin, whose length is twice the distance from the origin. Notice that the gradients become longer as we move out from the origin, reflecting the fact that the function R grows faster the farther we are from the origin. The graph of $z = R(x, y)$ is the paraboloid shown in Figure 16.6.18(ii).

For the related function $r(x, y) = \sqrt{x^2 + y^2}$ the level curves are also circles, but they are evenly spaced. The gradient is $\nabla r = \left(\frac{x}{\sqrt{x^2+y^2}}, \frac{y}{\sqrt{x^2+y^2}} \right) = \frac{1}{r}(x, y)$. This is a unit vector (since $\|(x, y)\| = r = \sqrt{x^2 + y^2}$). The gradient at each point is the unit vector pointing away from the origin. In Figure 16.6.19(i) we see that the length of the gradient is the same at each point and that this corresponds to the fact that the level curves are evenly spaced, that is, that the largest possible rate of change at any point is always the same. The graph of $z = r(x, y)$ is the cone shown in Figure 16.6.19(ii).

Figure 16.6.18(ii)

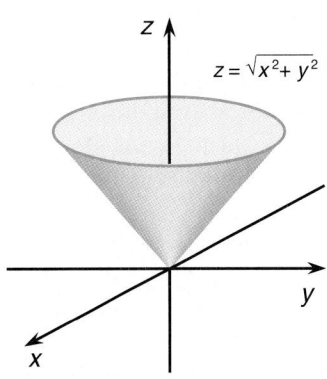

Figure 16.6.19(i)   **Figure 16.6.19(ii)**

Exercises 16.6

I

Find the directional derivative of each of the following functions at the specified point in the direction of the vector **v**.

1. $f(x, y) = 3xy^4$ at $(3, 2)$; $\mathbf{v} = (3, 4)$
2. $f(x, y) = \sin(\pi xy)$ at $(-1, 1)$; $\mathbf{v} = (1, -2)$
3. $f(x, y) = e^{2xy^2}$ at $(0, 3)$; $\mathbf{v} = (3, 4)$
4. $f(x, y) = \ln(xy)$ at $(-1, -1)$; $\mathbf{v} = (3, -3)$
5. $f(x, y) = \frac{x}{y}$ at $(4, 4)$; $\mathbf{v} = (1, 3)$
6. $f(x, y) = x^2 + y^2$ at $\left(\frac{1}{2}, 1\right)$; $\mathbf{v} = (4, -3)$
7. $f(x, y, z) = e^{xyz^2}$ at $(0, 2, -4)$; $\mathbf{v} = (0, 7, -2)$
8. $f(x, y, z) = \frac{1}{x+y+z}$ at $(-1, 0, 0)$; $\mathbf{v} = (1, 1, 1)$
9. $f(x, y, z) = 3x^2 - 2y^2 + z^2$ at $(1, 2, 3)$; $\mathbf{v} = (1, -2, 2)$
10. $f(x, y, z) = \cos^2(xyz) + \sin^2(xyz)$ at $(-\pi, 2, 3)$; $\mathbf{v} = (2, 4, -4)$

Find the gradient of each of the following functions. For those for which a point is specified, evaluate the gradient at that point.

11. $f(x, y) = 3xy^2 - 6x^3y + 2x$
12. $f(x, y) = 3\cos(x + 2y)\sin(xy^2)$
13. $f(x, y) = e^{\sin(xy)}$
14. $f(x, y) = \frac{1}{x+3y}$
15. $f(x, y) = \cosh(xy)$ at $(2, 0)$
16. $f(x, y) = \sec(xy - 6)$ at $(3, 2)$
17. $f(x, y, z) = 3xy^2z^3 - 5x^2yz + 7$
18. $f(x, y, z) = \frac{x+2z}{\sin(\pi y)}$ at $\left(2, \frac{1}{2}, -1\right)$
19. $f(x, y, z) = (xyz)^n$
20. $f(x, y, z) = x^5 + y^5 + z^5$

II

21. Find the direction in which the function $f(x, y) = 2xy - 5y^2$ increases fastest from the point $(2, -3)$.
22. Find the direction in which the function $f(x, y) = \sin(\pi xy^2)$ increases fastest from the point $(-1, 2)$.
23. Find the direction in which the function $f(x, y) = \ln(x^2 + y^2)$ decreases fastest from the point $(6, -8)$.
24. Find the direction in which the function $f(x, y) = 1/(x + e^y)$ decreases fastest from the point $(1, 0)$.
25. (i) Consider the function $F(x, y) = 3x^2y^3 + 5x^2y$. Starting at the point $(4, -2)$, suppose the variable x is changed to 4.01. What change in y can you suggest that will leave the value of F nearly unchanged?

(ii) With a calculator, find the value of F at the point you have suggested. How does it compare with the value of F at the point $(4.01, -2)$?

*26. Consider the function $f(x, y) = x^2y$. Think of f as a function of polar coordinates (r, θ), and find its gradient with respect to these variables.

27. Suppose $f(x, y, z)$ is a differentiable function, and define $E(x, y, z) = e^{f(x,y,z)}$. Show that $\nabla E(x, y, z) = E(x, y, z)\nabla f(x, y, z)$, that is, that $\nabla E = E\nabla f$.

28. Suppose $f(x, y, z)$ is a differentiable function, and $g(t)$ is a differentiable function of a single variable. Define $H(x, y, z) = g\big(f(x, y, z)\big)$. Show that $\nabla H(x, y, z) = g'\big(f(x, y, z)\big)\nabla f(x, y, z)$, that is, that $\nabla H = g'(f)\nabla f$.

29. If $f(x, y, z)$ and $g(x, y, z)$ are two functions, show that $\nabla(f + g) = \nabla f + \nabla g$ (i.e., the sum of two vectors), whenever both ∇f and ∇g are defined.

30. (i) If f and g are two functions of two variables, show that $\nabla(fg) = f\nabla g + g\nabla f$ at every point at which both gradients are defined.

(ii) Repeat part (i) for functions of three variables and then for functions of n variables.

31. Show that if $\nabla f(a, b, c) = (0, 0, 0)$, then for *every* unit vector **u** we must have $D_{\mathbf{u}}f(a, b, c) = 0$.

*32. Sketch the level curves of $f(x, y) = x^2 - y^2$ associated to the constants $c = -2, -1, 0, 1, 2$, and draw the gradient vectors at the following points, each of which lies on one of these level curves: $(1, 0)$, $(0, -1)$, $(\sqrt{2}, 0)$, $(1, 1)$, $(\sqrt{3}, -1)$, $(-\sqrt{2}, 2)$.

*33. A fiberglass mold is made in the shape of the graph of $z = 4x^2 + y^4 + x^2\sin(\pi y^2)$, with $x, y \in [-3, 3]$. If a ball bearing is released at the point $(x, y, z) = (2, -1, 17)$ on the surface, it will roll in the direction in which it descends the fastest. What is this direction?

**34. The shape of a mountain is given by the graph of the function $H(x, y) = 10{,}000 - 10x^4 - 3x^2y^2 - 6xy^3 - 5y^4$, with $|x| \le 4$ and $|y| \le 5$. The horizontal coordinates x, y are measured in kilometers and the height $H(x, y)$ is in meters.

(i) Show that the summit (maximum value of $H(x, y)$) is at $(x, y) = (0, 0)$.

(ii) At the point $(x, y) = (1, 1)$, that is, 1 km east and 1 km north of the summit, in which direction will water flow down the mountain? (Express your answer as $\theta°$ north of due east for some angle θ.)

(iii) In what directions will water flow, starting at points on the line $y = -x$?

(iv) Find all points at which water will flow due west.

(v) In which direction should a climber at $(x, y) = (3, -2)$ move to make the most rapid ascent?

*35. In Example 4 we evaluated the gradient ∇T at the points $(2, 1)$ and $(-1, 2)$. Neither of them lies on any

of the level curves we drew in Figure 16.6.15. Use a graphics calculator or computer to sketch the level curves of T that pass through these two points.

36. Let $F(x, y) = y^3 x + 2x^3 - x^2 + 2$. Sketch the level curves where $F = 2, F = 0, F = 1$. Also show the gradient vector at $(x, y) = (-1, -1)$.

37. Sketch the mold of Exercise 33.

38. Sketch the mountain of Exercise 34.

POINT TO PONDER

In Definition 16.4.4 (p. 967) we discussed the concept of differentiability for functions F of two or more variables. But you may have noticed that we never said what the *derivative* is. In a sense, for a function of two or more variables the gradient ∇F is the analog of the derivative for a function of one variable.

The definition of the derivative $f'(x) = \frac{d}{dx}\big(f(x)\big)$ can be interpreted as saying that $f(a + h)$ is best approximated by $f(a + h) \approx f(a) + f'(a) h$. The definition of differentiability for $F(x, y)$ says that $F(a + h, b + k)$ is

best approximated by

$$F(a + h, b + k) \approx F(a, b) + F_x(a, b) h + F_y(a, b) k$$
$$= F(a, b) + \nabla F(a, b) \cdot (h, k).$$

Here $\nabla F(a, b)$ plays the same role as $f'(a)$ did for a function of one variable. However, we should be careful not to push the analogy too far because there are functions for which ∇F exists at (a, b) but $F(x, y)$ is not differentiable at $(x, y) = (a, b)$.

16.7 Tangent Planes

NORMAL VECTOR
GRADIENT

For a function of one variable the tangent line to a curve was a very important concept. The graph $z = f(x, y)$ of a function of two variables is usually a curved surface in three-dimensional space, and the same is true of a level surface $\{(x, y, z) : F(x, y, z) = C\}$ consisting of all points at which some function $F(x, y, z)$ equals the constant C. In Section 16.3 we saw how partial derivatives could be used to find the *tangent plane* to the surface, which is analogous to the tangent line to a curve.

For example, the level surface $x^2 + y^2 + z^2 = 1$ is a sphere. It is easy to draw a tangent plane to this surface at the "south pole" (i.e., at the point $(0, 0, -1)$); it is just the horizontal plane $z = -1$, on which the sphere "sits" (see Figure 16.7.1). At

Figure 16.7.1

Figure 16.7.2

Figure 16.7.3

Figure 16.7.4

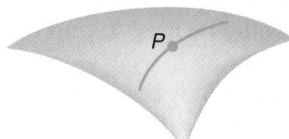

Figure 16.7.5

other points on the sphere it is less easy to write an equation for the tangent plane, but we found how to do it in Example 16.3.4. It is easy to visualize a plane that just touches the sphere at one point (see Figure 16.7.2).

To write the equation of a plane, the most convenient way to begin is to know a **normal vector n**, that is, a vector that is perpendicular to the plane, as shown in Figure 16.7.3. Now consider the level surface defined by $f(x, y, z) = C$, and let us try to find a normal vector to it at some point $P = (a, b, c)$, as in Figure 16.7.4. At the end of Section 16.6 we saw that for a function of two variables the gradient is perpendicular to the tangent line to a level curve; this suggests that the gradient may also be helpful for finding normal vectors to level surfaces.

Let us imagine a curve lying in the surface, passing through the point P, as shown in Figure 16.7.5. Suppose the curve is parametrized by $\mathbf{r}(t) = \big(x(t), y(t), z(t)\big)$, with $\mathbf{r}(0) = P$, that is, that the moving point $\mathbf{r}(t)$ passes through P when $t = 0$. Suppose that the functions $x(t)$, $y(t)$, and $z(t)$ are all differentiable functions of t, and write $\mathbf{r}'(t) = \big(x'(t), y'(t), z'(t)\big)$ for the tangent vector.

In particular, $\mathbf{r}'(0)$ is a tangent vector to the curve at the point P. Since the curve is entirely contained in the level surface of all points at which $f = C$, the tangent vector to the curve must also be a tangent vector to this level surface. See Figure 16.7.6.

Now to say that $\mathbf{r}(t)$ lies in the level surface just means that f equals C all along the curve, that is, $f\big(x(t), y(t), z(t)\big) = C$ for every value of t. In particular, if we think of $f\big(\mathbf{r}(t)\big) = f\big(x(t), y(t), z(t)\big)$ as a function of t, it is a *constant* function.

The Chain Rule shows us how to differentiate this function; it is a composite function with intermediate variables x, y, and z, so

$$\frac{df}{dt} = f_x x'(t) + f_y y'(t) + f_z z'(t) = \nabla f \cdot \mathbf{r}'(t).$$

Since $f\big(x(t), y(t), z(t)\big)$ is a constant function of t, this derivative must be zero. What we have just shown is that the dot product of the gradient and the tangent vector $\mathbf{r}'(t)$ must be zero, which means that these two vectors are perpendicular. Specifically, if we let $t = 0$, we find that the gradient vector at the point $\big(x(0), y(0), z(0)\big) = P = (a, b, c)$ is perpendicular to the tangent vector $\mathbf{r}'(0)$ to the curve at P.

The gradient $\nabla f(a, b, c)$ is perpendicular to the tangent vector to every curve that passes through (a, b, c). It certainly seems reasonable to consider the plane that contains all these tangent vectors and call it the tangent plane to the level surface at the point (a, b, c). We now know that $\nabla f(a, b, c)$ is a normal vector for this plane.

When we began this discussion, we mentioned level surfaces of functions of three variables, but we also mentioned graphs of functions of two variables. Fortunately, this type of graph can also be seen as a level surface. The graph $z = g(x, y)$ is the set of all points (x, y, z) at which $z = g(x, y)$, which is exactly the same thing as the set of points at which $g(x, y) - z = 0$. If we let $f(x, y, z) = g(x, y) - z$, then we see that the level surface of all points at which f is zero is exactly the same thing as the graph of $z = g(x, y)$.

All this suggests the following definition.

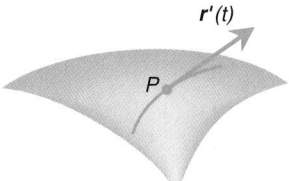

Figure 16.7.6

DEFINITION 16.7.1

Tangent Plane

If $f(x, y, z)$ is a function of three variables and (a, b, c) is a point at which the gradient ∇f exists and is not the zero vector, then consider the level surface of f containing (a, b, c). The gradient $\nabla f(a, b, c)$ is a normal vector to the tangent plane to this surface at the point (a, b, c), so the **tangent plane** is the plane whose equation is

$$\nabla f(a, b, c) \cdot (x - a, y - b, z - c) = 0.$$

Another way to write this is to let $\nabla f(a, b, c) = \mathbf{n} = (n_1, n_2, n_3)$, so the equation becomes

$$n_1 x + n_2 y + n_3 z = K,$$

where $K = n_1 a + n_2 b + n_3 c$.

In the special case in which $g(x, y)$ is a function of two variables and we let $f(x, y, z) = g(x, y) - z$, then $\nabla f = (g_x, g_y, -1)$, and the equation of the tangent plane to the graph of $z = g(x, y)$ is

$$g_x(a, b) x + g_y(a, b) y - z = K,$$

where $K = g_x(a, b) a + g_y(a, b) b - c$ (cf. Equation 16.3.2).

EXAMPLE 1 (i) Find the tangent plane to the sphere $x^2 + y^2 + z^2 = 1$ at the point (a, b, c).

Solution We let $f(x, y, z) = x^2 + y^2 + z^2$ and think of the sphere as the level surface of all points at which $f = 1$. The gradient of f is $\nabla f = (2x, 2y, 2z)$, so the normal vector to the sphere at the point (a, b, c) is $(2a, 2b, 2c)$.

The tangent plane is $2ax + 2by + 2cz = K$, where K can be found by using the formula in the definition: $K = (2a)(a) + (2b)(b) + (2c)(c) = 2(a^2 + b^2 + c^2) = 2f(a, b, c)$. Since (a, b, c) is in the level surface consisting of all points at which $f = 1$, this tells us that $K = (2)(1) = 2$.

The equation of the tangent plane is $2ax + 2by + 2cz = 2$, or, dividing by 2, which still gives the same plane, $ax + by + cz = 1$. See Figure 16.7.7.

Notice that this is slightly easier than the calculation we did in Example 16.3.4.

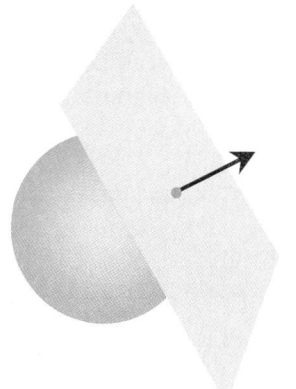

Figure 16.7.7

See the Computer Example on the next page.

(ii) Find the equation of the tangent plane to the graph of the function $g(x, y) = x^3 - 2xy$ at the point $(2, 3, -4)$.

Solution First we consider the function $f(x, y, z) = g(x, y) - z = x^3 - 2xy - z$. Its gradient is $\nabla f = (3x^2 - 2y, -2x, -1)$, and $\nabla f(2, 3, -4) = (6, -4, -1)$. The tangent plane is $6x - 4y - z = K$, but we still have to figure out what the constant K is. Rather than remember the formula, it is often easier to write the equation as we have just done and then notice that since the point $(2, 3, -4)$ must be in the plane, its coordinates must satisfy the equation. We substitute $(2, 3, -4)$ into the left side of the equation and get $6(2) - 4(3) - (-4) = 4$, which says that $K = 4$.

The equation of the tangent plane is $6x - 4y - z = 4$.

(iii) Find the tangent planes to the graph of the function $z = g(x, y) = x^2 + 2y^2 - 1$ at the points of the graph corresponding to $(0, 0)$ and $(2, 1)$.

Solution We let $f(x, y, z) = g(x, y) - z = x^2 + 2y^2 - 1 - z$ and find $\nabla f = (2x, 4y, -1)$.

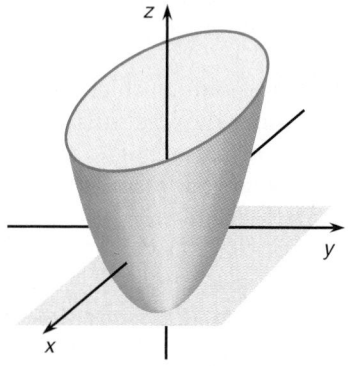

Figure 16.7.8(i)

The points have been specified as points in the xy-plane, since the original function g is a function of x and y. We need to work with the corresponding points on the graph, which are $(0, 0, -1)$ and $(2, 1, 5)$.

Now $\nabla f(0, 0, -1) = (0, 0, -1)$, so the tangent plane is $0(x) + 0(y) - z = K$, that is, $z = -K$, and since the plane contains the point $(0, 0, -1)$, we see that $K = 1$ and the tangent plane is $z = -1$, as shown in Figure 16.7.8(i).

Similarly, $\nabla f(2, 1, 5) = (4, 4, -1)$, so the tangent plane at $(2, 1, 5)$ is $4x + 4y - z = K$, and substituting $(2, 1, 5)$ for (x, y, z), we find that $K = 4(2) + 4(1) - 5 = 7$. The tangent plane is $4x + 4y - z = 7$, as shown in Figure 16.7.8(ii).

Figure 16.7.8(ii)

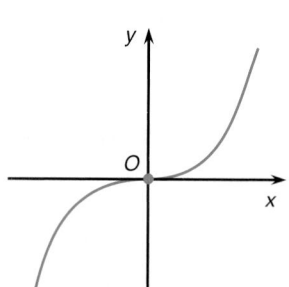

Figure 16.7.9

The familiar function $F(x) = x^3$ gave us an example of a curve that crosses its tangent line. The tangent line at $x = 0$ is easily found; since $F'(0) = 0$, the tangent line is the horizontal line $y = 0$ (i.e., the x-axis). The graph of $y = x^3$ "flattens out" as it approaches the origin, and it crosses the axis there. See Figure 16.7.9.

For functions of two variables the situation with a tangent plane can be much more complicated. Let us consider the function $g(x, y) = x^2 - y^2$. Its graph is the

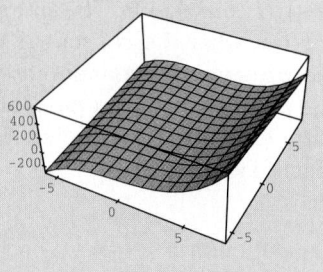

We can use a computer to graph the surface in part (ii) of Example 1.

MATHEMATICA	MAPLE
Plot3D $[x\hat{\,}3 - 2\,x*y, \{x, -6, 8\},$ $\{y, -6, 8\}]$ enter	plot3d $(x\hat{\,}3 - 2*x*y\ x = -6..8,$ $y = -6..8);$ return

DERIVE	MacFunction
O D G ENTER A $x\hat{\,}3 - 2\,x\ y$ ENTER P O L 14 Tab 14 ENTER C 1 Tab 1 ENTER G 15 Tab 15 ENTER	$x\hat{\,}3 - 2*x*y$ OK Function Change Domain (drag to 8×8) OK Options Remove Hidden Surfaces

Notice the vertical scale on the graph.

Figure 16.7.10   **Figure 16.7.11**

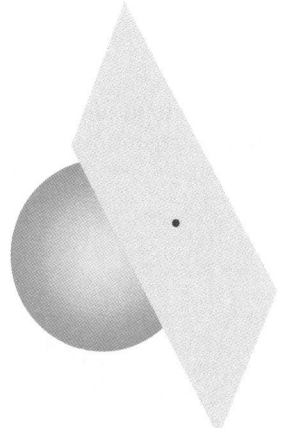

Figure 16.7.12

"saddle surface" shown in Figure 16.7.10. To study this graph, we take the usual first step and define $f(x, y, z) = g(x, y) - z$; the graph of $z = g(x, y)$ is the zero level surface of f. The gradient of f is $\nabla f = (2x, -2y, -1)$, so in particular $\nabla f(0, 0, 0) = (0, 0, -1)$. The tangent plane at the origin is $-z = K$, and since it contains the origin, $K = 0$ and the plane is just the horizontal plane $z = 0$.

From the picture in Figure 16.7.11 we see that the graph is sometimes above the tangent plane and sometimes below it. In fact, since the tangent plane is just $z = 0$, the graph lies in the tangent plane precisely when $g(x, y) = 0$, that is, when $x^2 = y^2$, that is, when $x = \pm y$. So the graph intersects the tangent plane in the two lines $y = x$ and $y = -x$ and in fact crosses the plane at every point in these lines.

This example shows that the behavior of a surface and a tangent plane near their point of tangency can be quite complicated. It can be much more complicated than our very first example of the tangent planes to a sphere, which had the particularly nice property that they intersect the surface in exactly one point (see Figure 16.7.12). In general it is unreasonable to expect that tangent planes will have this or any similar property; for instance, the tangent plane at the origin intersects the saddle surface in two lines that meet at the point of tangency.

Exercises 16.7

I

Find the equation of the tangent plane to the given surface at the specified point.

1. $x^2 + y^2 + z^2 = 4$ at $(2, 0, 0)$

2. $x^2 + 2y^2 - 3z^2 = 6$ at $(1, -2, 1)$

3. $xyz = 12$ at $(3, 4, 1)$

4. $x^2 y - 2xz^2 = 0$ at $(1, 2, 1)$

5. $z = x^2 - y^2$ at $(5, -4, 9)$

6. $x = \sin(\pi yz)$ at $(0, 1, 5)$

7. $xy + yz + xz = 3$ at $(1, 1, 1)$

8. $e^{x+2y-z} = 1$ at $(-1, 2, 3)$

9. $\frac{z}{x+y} = 1$ at $(3, -4, -1)$

10. $xz^3 - 3zy^2 + 8 = 0$ at $(4, -2, 1)$

Find the equation of the tangent plane to the graph of each of the following functions at the given point.

11. $g(x, y) = x^2 + y^2$ at $(2, 3)$

12. $g(x, y) = xy^2 + x$ at $(0, 0)$

13. $g(x, y) = \sin \pi xy$ at $(1, 4)$

14. $g(x, y) = e^{xy}$ at $(-1, -1)$

15. $g(x, y) = \frac{x}{x^2 + y^2}$ at $(2, 3)$

16. $g(x, y) = \arctan \frac{y}{x}$ at $(2, -2)$

17. $g(x, y) = \frac{1}{x+2y}$ at $(2, 1)$

18. $g(x, y) = \cosh 3xy$ at $(3, 0)$

19. $g(x, y) = \ln(x^3 y^4 + xy)$ at $(2, -1)$

20. $g(x, y) = \sin^2(xy) + \cos^2(xy)$ at $(3, -2)$

II

21. Find the equation of the tangent plane to the sphere $x^2 + y^2 + z^2 = 1$ at the following points:
(i) $\left(\frac{1}{\sqrt{3}}, \frac{1}{\sqrt{3}}, \frac{1}{\sqrt{3}}\right)$, (ii) $\left(\frac{3}{5}, 0, -\frac{4}{5}\right)$, (iii) $\left(\frac{2}{3}, -\frac{1}{3}, -\frac{2}{3}\right)$.

22. Find the equation of the tangent plane to the sphere $x^2 + y^2 + z^2 = 16$ at the point (a, b, c) (i.e., assume that this point is on the sphere).

23. Find the equation of the tangent plane to the sphere $(x-1)^2 + (y+2)^2 + (z+1)^2 = 37$ at the point $(2, 4, -1)$.

24. Find the equation of the tangent plane to the ellipsoid $x^2 + 2y^2 + 3z^2 = 6$ at the point $(1, -1, 1)$.

*25. Find the equation of the tangent plane to the sphere $(x-1)^2 + y^2 + z^2 = 4$ at the point (a, b, c). (Assume that this point is on the sphere.)

26. Find the equation of the tangent plane to the ellipsoid $\frac{x^2}{a^2} + \frac{y^2}{b^2} + \frac{z^2}{c^2} = 1$ at the point (u, v, w). (Assume that this point is on the ellipsoid.)

Consider each of the following surfaces. At which points on the surface does the tangent plane not exist?

*27. $x^3 + y^3 + z^3 = 8$

*28. $z^2 = xy$

*29. $z = |x + y|$

*30. $(z-1)^3 = \left((x+2)^2 - (y+1)^3\right)^2$

31. Consider the plane $2x - 3y + z = 5$, and find its tangent plane at the point $(3, 1, 2)$. Confirm that the tangent plane is exactly the same as the original plane.

*32. Show that the tangent plane to any plane $ax + by + cz = d$ at any point in the plane is the same as the original plane, that is, that any plane "is its own tangent plane."

*33. Consider the function $g(x, y) = y^3$. Notice that there is no x in the formula for $g(x, y)$.
(i) Sketch the graph of $z = g(x, y)$.
(ii) Find the tangent plane to the graph at the points $(0, 0)$, $(1, 1)$, and $(3, 1)$.

*34. Where is the tangent plane to the graph of $z = \sin(xy)$ horizontal?

Use a computer to sketch each of the following graphs, and for each observe where the tangent plane is horizontal.

35. $z = \cos(xy)$

36. $z = \sin(x^2 + y^2)$

37. $z = 1 - (2x^2 + y^2 - 1)^2$

38. $z = x^3 - 3x^2 + y^5 - 5y$

39. Sketch $z = x^3 - 3xy^2$, and discuss the intersection of this graph with its tangent plane at the origin.

POINT TO PONDER

We have learned how to find the equation of the tangent plane to a surface, and the procedure has become quite mechanical. However, we should think a little bit about an assumption we have made. Specifically, we have assumed that the gradient vector exists and is *nonzero*.

If the gradient does not exist at a point $P = (a, b, c)$, then our definition of the tangent plane does not apply. It may well be that there is no reasonable way to define a tangent plane. One way this could happen is for the surface to have a *cusp*, meaning that it narrows down to a point at (a, b, c). No plane will be tangent to the surface at such a point. The surface we have sketched in Figure 16.7.13 is $z = -(x^2 + y^2)^{1/4} = -\sqrt{r}$, where r is

Figure 16.7.13

Figure 16.7.14

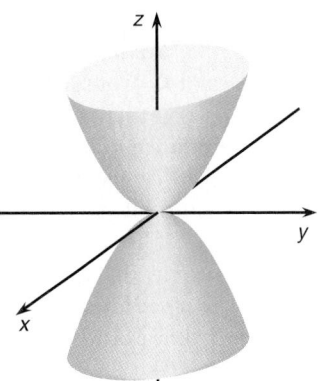

Figure 16.7.15

the radius in polar coordinates. It has no tangent plane at the origin.

Another possibility is that the gradient does exist but equals the zero vector. This does not allow us to use it as the normal vector to define a plane (since *every* vector is perpendicular to the zero vector). An example of this phenomenon is the familiar cone $f(x, y, z) = x^2 + y^2 - z^2 = 0$ at the point $(0, 0, 0)$. The gradient is $\nabla f = (2x, 2y, -2z)$, which exists at every point, but $\nabla f(0, 0, 0) = (0, 0, 0)$. From Figure 16.7.14 it is reasonably clear that there is no tangent plane at the origin. (A first guess might be to try the horizontal *xy*-plane, but it is not really tangent to anything in the

cone; tangent vectors to curves in the cone do not lie in this plane.)

To make up another example, consider the equation $(x^2 + y^2)^2 - z^2 = 0$. This is equivalent to $z = \pm(x^2 + y^2)$, so the graph of this surface is the parabolic bowl $z = x^2 + y^2$ together with its downward-opening reflection $z = -(x^2 + y^2)$. At the origin the gradient is easily seen to be $(0, 0, 0)$, but from the picture in Figure 16.7.15 it seems reasonable to think of the horizontal *xy*-plane as the tangent plane at the origin. Despite this picture, we will use our definition of the tangent plane and say that in this case there is no tangent plane at the origin.

16.8 Maximum/Minimum Problems

ABSOLUTE (LOCAL) MAXIMUM/
MINIMUM
EXTREMUM
CRITICAL POINT
BOUNDED REGION

Figure 16.8.1

Just as with functions of one variable, it is often very important to find maxima and minima of functions of two or more variables. It turns out that the approach is quite similar to what we used for functions of a single variable.

We begin by establishing the terminology, which is almost identical to what we used for functions of one variable. One slight difference arises because it no longer makes sense to talk about an interval around a point. When we were dealing with one variable, an interval around some point was a way of describing all the points that are reasonably close to that point; for functions of two variables the corresponding idea is all the points in a disk around some point (see Figure 16.8.1). With three variables we use a ball around the point, and for *n* variables we use an *n*-dimensional ball.

> **DEFINITION 16.8.1**
> The function $f(x, y)$ has an **absolute maximum** at the point (a, b) if $f(a, b) \geq f(x, y)$ for every point (x, y) in the domain of f. It has an **absolute minimum** at (a, b) if $f(a, b) \leq f(x, y)$ for every point (x, y) in the domain of f.

We say that $f(x, y)$ has a **local maximum** (**local minimum**) at (a, b) if there is a disk around (a, b) so that $f(a, b) \geq f(x, y)$ ($f(a, b) \leq f(x, y)$) for every point (x, y) that is in that disk and in the domain of f. Sometimes they are called a **relative maximum** and **relative minimum**.

We use the word **extremum** to refer to either a maximum or a minimum (and **local extremum** similarly).

The same terminology is used for functions of three variables, except that we say that $F(x, y, z)$ has a **local maximum** (**local minimum**) at (a, b, c) if there is a *ball* around (a, b, c) so that $F(a, b, c) \geq F(x, y, z)$ ($F(a, b, c) \leq F(x, y, z)$) for every point (x, y, z) that is in that ball and in the domain of F.

For a function $G(x_1, \ldots, x_n)$ of n variables the same definition is used, but the word "ball" is now interpreted to mean an n-dimensional ball.

We illustrate these concepts with some pictorial examples.

EXAMPLE 1

(i) The function $f(x, y) = 1 - x^2 - y^2$, graphed in Figure 16.8.2(i), has an absolute maximum at $(x, y) = (0, 0)$ and no other absolute or local extrema. The corresponding point $(0, 0, 1)$ is the highest point on the graph.

(ii) The function $g(x, y) = \cos y$, graphed in Figure 16.8.2(ii), has values between -1 and 1. Its maximum value, 1, occurs at every point on any of the lines $y = 0, \pm 2\pi, \pm 4\pi, \ldots, \pm 2k\pi, \ldots$, and its minimum value -1 occurs on the lines $y = \pm\pi, y = \pm 3\pi, \ldots, \pm(2k + 1)\pi, \ldots$.

There are absolute maxima or minima on these lines. Notice the difference between this and the previous example; here the extrema occur on lines instead of at a single isolated point.

(iii) The function $h(x, y) = f(x, y)^2 = (1 - x^2 - y^2)^2$, the square of the function $f(x, y)$ from part (i), is graphed in Figure 16.8.2(iii). Being a square, it is never negative, so its smallest possible value is 0; there is an absolute minimum at every point on the unit circle $x^2 + y^2 = 1$. From the picture we see that the graph has a "hump" at the origin. The function has a local maximum there, but it is not an absolute maximum, since the function goes to infinity as (x, y) moves very far away from the origin.

(iv) The function $k(x, y) = -x - y$ is illustrated in Figure 16.8.2(iv). Its graph is the plane $z = -x - y$, that is, $x + y + z = 0$. There are no local or absolute extrema at all.

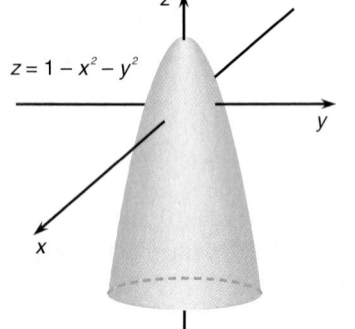

$z = 1 - x^2 - y^2$

Figure 16.8.2(i)

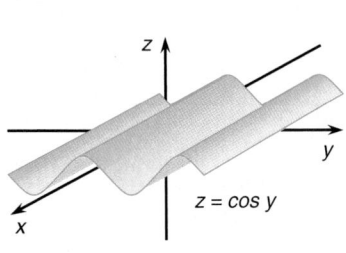

$z = \cos y$

Figure 16.8.2(ii)

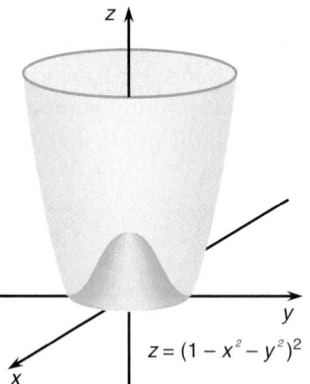

$z = (1 - x^2 - y^2)^2$

Figure 16.8.2(iii)

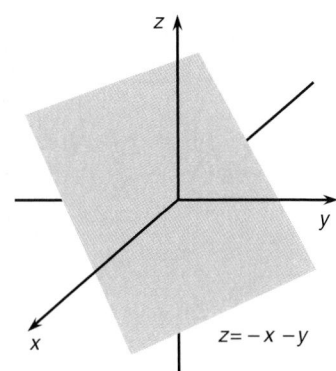

$z = -x - y$

Figure 16.8.2(iv)

Figure 16.8.3

With functions of one variable we made use of Theorems 2.5.3 and 2.5.4, which said that a continuous function on a *closed* interval [*a, b*] must be *bounded* (i.e., never greater than some constant *M* and never less than −*M*) and moreover that it must achieve its maximum and minimum values. It was easy to find examples to show that neither conclusion is necessarily correct for a continuous function on an open interval.

A natural way to attempt to generalize these ideas to functions of two variables would be to consider a closed rectangle in the plane, rather than a closed interval. Another approach would be to use a closed disk (i.e., a disk *including* its boundary circle). See Figure 16.8.3.

Both these types of regions have the property that they do not "go off to infinity." This property is certainly necessary if we hope to have a theorem saying that continuous functions will be bounded because the continuous function $x^2 + y^2$ (the square of the distance from the origin) cannot be bounded on any region that approaches infinity. We will say that a region that does not "go off to infinity" is "bounded," being careful not to confuse this with the other use of the same word when we say that a function is "bounded."

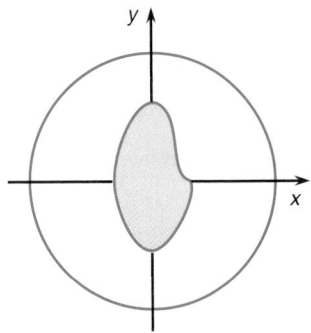

Figure 16.8.4

DEFINITION 16.8.2

A function $f(x, y)$ is **bounded** if there exists a number $M \geq 0$ so that

$$-M \leq f(x, y) \leq M,$$

for every point (x, y) in the domain of f.

A region R in the plane is **bounded** if there is some number K so that every point in R is at a distance of no more than K away from the origin, that is,

$$\sqrt{x^2 + y^2} \leq K,$$

for every point $(x, y) \in R$. Another way to express this is that the region R is enclosed in the circle of radius K around the origin. See Figure 16.8.4.

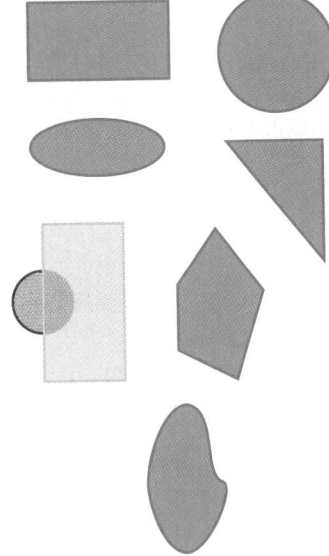

Figure 16.8.5

In fact, results analogous to Theorems 2.5.3 and 2.5.4 do hold for closed rectangles and closed disks and for many other regions in the plane. The important conditions are that the region should be bounded and should include all its boundary points.

Unfortunately, it is surprisingly difficult to specify the second of these conditions precisely. Instead, let us mention some types of regions for which the results are true (see Figure 16.8.5). They are true for the set of all points on or inside a rectangle, a circle, an ellipse, a triangle, or any polygon (a figure with finitely many sides, each of which is a straight line). They are also true for any region that can be written as the intersection of two or more of these regions (e.g., a semidisk is the intersection of the points inside or on the circle and the points inside or on a rectangle, as we see from the illustration). Also allowed are unions of finite numbers of such regions.

The results also hold for any region constructed as follows: Take a curve that makes a closed loop without crossing itself. Consider the region that consists of all points inside or on the curve. The theorems hold for any such region. Disks, semidisks, and all the other examples we have mentioned can be constructed this way.

THEOREM 16.8.3

> Let $f(x, y)$ be a continuous function defined on any of the types of regions discussed above, which are bounded and "contain all their boundary points." Then $f(x, y)$ is **bounded** on this region, that is, there exists a number $M \geq 0$ so that $-M \leq f(x, y) \leq M$ for all points (x, y) in the region.
> Second, the function $f(x, y)$ must achieve its maximum and minimum values in the region, that is, there exists a point (a, b) in the region at which f has an absolute maximum and a point (c, d) at which f has an absolute minimum.

The proof of this theorem is extremely difficult, and we omit it.

Suppose the function $f(x, y)$ has a local maximum at the point (a, b). Consider the function $g(x) = f(x, b)$ (i.e., fix the value of y and regard f as a function g of x). Then g has a local maximum at $x = a$, so it has a critical point at $x = a$, and the derivative of g at $x = a$, if it exists, equals zero. But $g'(a)$ also equals $\frac{\partial f}{\partial x}(a, b)$. So if the partial derivative exists at (a, b), then $\frac{\partial f}{\partial x}(a, b) = g'(a) = 0$.

An almost identical argument shows that $\frac{\partial f}{\partial y}(a, b) = 0$, provided the partial derivative exists. So if f has a local extremum at (a, b), then its partial derivatives must be zero there, if they exist. As with functions of only one variable, this provides a way to look for points at which extrema *may* occur.

EXAMPLE 2   Find the maximum and minimum values of the function $f(x, y) = x^4 - 2x^2 + y^2 + 4y - 2$.

Solution   We find that the partial derivatives are $\frac{\partial f}{\partial x} = 4x^3 - 4x = 4x(x^2 - 1)$ and $\frac{\partial f}{\partial y} = 2y + 4$.

Local extrema can only occur at points at which both partial derivatives are zero. We see that y must equal -2 in order for $\frac{\partial f}{\partial y}$ to equal 0 and that x must equal 0, 1, or -1 in order for $\frac{\partial f}{\partial x}$ to equal 0. The possible points are $(0, -2)$, $(1, -2)$, and $(-1, -2)$.

We evaluate the function at each of these points and find that $f(0, -2) = -6$, $f(1, -2) = -7$, and $f(-1, -2) = -7$. If any of these points is an absolute maximum, it would have to be $(0, -2)$, and if any of them is an absolute minimum, it would have to be $(1, -2)$ and/or $(-1, -2)$.

Now we can resort to the usual trick of completing the squares and write

$$
\begin{aligned}
f(x, y) &= x^4 - 2x^2 + y^2 + 4y - 2 \\
&= (x^4 - 2x^2 + 1) - 1 + (y^2 + 4y + 4) - 4 - 2 \\
&= (x^2 - 1)^2 + (y + 2)^2 - 7.
\end{aligned}
$$

This expresses f as -7 plus the sum of two squares, and we see immediately that $f(x, y)$ can never be less than -7, since squares are never negative. This shows that the points $(1, -2)$ and $(-1, -2)$ are both absolute minima.

The other point $P = (0, -2)$ is more complicated. We note that $f(x, y)$ will tend to ∞ as x and/or y become large, so there is no absolute maximum. There remains the possibility that $P = (0, -2)$ might be a local maximum or a local minimum.

One way to investigate this question is to consider the values of $f(x, y)$ along the horizontal and vertical lines through the point $(0, -2)$. So for instance, fix $y = -2$

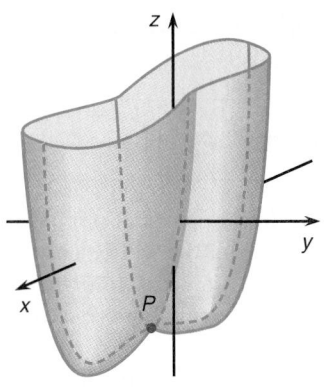

Figure 16.8.6

and let $h(x) = f(x, -2)$. We see that $h'(x) = \frac{\partial f}{\partial x}(x, -2) = 4x^3 - 4x$ and $h''(x) = 12x^2 - 4$. In particular, when $x = 0$, $h'(0) = 0$ and $h''(0) = -4$. From the Second Derivative Test we know that $h(x)$ has a local maximum at $x = 0$.

Similarly, if we fix $x = 0$ and let $k(y) = f(0, y)$, we find that $k'(y) = \frac{\partial f}{\partial y} = 2y + 4$ and $k''(y) = 2$. In particular, $k'(-2) = 0$ and $k''(-2) = 2 > 0$. This tells us that $k(y)$ has a local *minimum* at $y = -2$. See Figure 16.8.6.

This suggests that near the point $(0, -2)$ the function $f(x, y)$ behaves something like the saddle surface shown in Figure 16.8.7; in the x-direction it has a local maximum, and in the y-direction it has a local minimum.

In particular this says that near the point $(0, -2)$ there are points at which $f(x, y)$ is slightly greater than $f(0, -2) = -6$ and other points at which it is slightly less than -6. This means that the point $(0, -2)$ is neither a local maximum nor a local minimum.

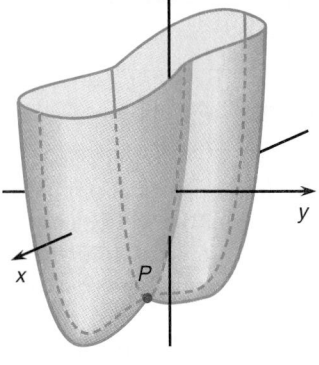

Figure 16.8.7

This example illustrates the essential technique for finding maximum and minimum values. We begin by looking for *critical points*, points at which both partial derivatives are zero (or one or both partials do not exist). If there are any absolute or local extrema, they must be among these critical points.

As with functions of one variable it may be very difficult to decide whether a given critical point is a local or absolute extremum or neither. Notice that in this example we showed that the point $(0, -2)$ was not a local extremum by showing that it was a local maximum along the horizontal line and a local minimum along the vertical line. However, it is very important to recognize that this is one of those techniques that only works in one direction. If a function has a local maximum along one line and a local minimum at the same point along another line, then it cannot have any sort of local extremum at that point. If, on the other hand, we find that it has local maxima, say, along both the horizontal and vertical lines, then it is not correct to conclude that the function has a local maximum. It is not difficult to find examples that have no local extremum at the origin but do have local maxima along both the horizontal and vertical axes.

Since we have already begun to use the terminology, we make the following definition, analogous to the corresponding definition for functions of one variable.

DEFINITION 16.8.4

Critical Points

A point (a, b) in the domain of $f(x, y)$ is a **critical point** of f if both partial derivatives exist at (a, b) and both equal zero, that is, $\frac{\partial f}{\partial x} = \frac{\partial f}{\partial y} = 0$, *or* if at least one of the partial derivatives does not exist at (a, b).

For functions of three or more variables the definition is similar: A point (a, b, c) in the domain of $F(x, y, z)$ is a critical point of F if the partial derivatives all exist at (a, b, c) and all equal zero there *or* if at least one of the partial derivatives does not exist at (a, b, c).

The argument we gave before Example 2 shows that if the partial derivatives all exist at a local extremum, then they must all be zero there, and it must be a critical point. Since it is also a critical point if even one partial derivative does not exist there, we see that local extrema always occur at critical points.

THEOREM 16.8.5

> Any point that is a local extremum of a function f must be a critical point of f.

To find the extrema of a function f, we begin by finding the critical points. If we are lucky, there will be relatively few critical points, and perhaps it will be easy to classify the extrema among them. However, there may be many critical points, and no matter how many there are, it may be very difficult to decide which are maxima or minima.

For functions of one variable the problem of finding extrema was made more complicated by the possibility of *endpoint extrema*, which occur at the endpoints of intervals in the domain of the function. At such points the derivative does not exist, because the function is only defined to one side of the point, but it is easy to overlook them because they do not show up if we only look at points where the derivative is zero.

For functions of two variables the corresponding problem is that extrema can occur at boundary points of the domain of the function. For instance, if we look for the maximum of $f(x, y)$ on a closed rectangle, then looking for critical points will find all possible extrema which occur at *interior points*, that is, points *inside* the rectangle. But there is also the possibility that the maximum might occur on the boundary. It is necessary to consider the boundary points separately.

Figure 16.8.8(i)

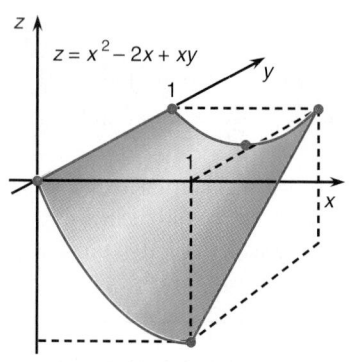

Figure 16.8.8(ii)

EXAMPLE 3   Find the maximum value of the function $F(x, y) = x^2 - 2x + xy$ on the rectangle $0 \leq x, y \leq 1$.

Solution   The partial derivatives are $\frac{\partial F}{\partial x} = 2x - 2 + y$ and $\frac{\partial F}{\partial y} = x$. For $\frac{\partial F}{\partial y}$ to be zero we must have $x = 0$, and from this we find that $\frac{\partial F}{\partial x} = 0$ gives $y = 2$. The point $(0, 2)$ is not in the rectangle; there are no critical points in the rectangle.

The boundary consists of four segments, and we deal with each of them separately.

The bottom boundary of the rectangle is the part of the x-axis with $0 \leq x \leq 1$. Since $y = 0$, there the function is $F(x, 0) = x^2 - 2x = (x - 1)^2 - 1$ for $x \in [0, 1]$; it has a minimum of -1 at $x = 1$ and a maximum of 0 at $x = 0$. These correspond to the points $(1, 0)$ and $(0, 0)$.

The left boundary is $x = 0, 0 \leq y \leq 1$, and the function is $F(0, y) = 0$, the constant function.

The right boundary is $x = 1, 0 \leq y \leq 1$, and the function is $F(1, y) = 1 - 2 + y = y - 1$. Its maximum is 0 at $y = 1$, and its minimum is -1 at $y = 0$. These correspond to the points $(1, 1)$ and $(1, 0)$.

The top boundary is $y = 1, 0 \leq x \leq 1$, and the function there is $F(x, 1) = x^2 - 2x + x = x^2 - x = \left(x - \frac{1}{2}\right)^2 - \frac{1}{4}$. Its minimum is $-\frac{1}{4}$ at $x = \frac{1}{2}$, and its maximum is 0 at $x = 0, 1$. The corresponding points are $\left(\frac{1}{2}, 1\right)$, $(0, 1)$, and $(1, 1)$. We mark all these points on the boundary (see Figure 16.8.8(i)).

The maximum value is 0, which occurs at $(1, 1)$ and all along the left boundary; the minimum is -1, which occurs at $(1, 0)$. The graph of $z = F(x, y)$ is shown in Figure 16.8.8(ii), which has been drawn so that its axes are consistent with Figure 16.8.8(i). ◢

EXAMPLE 4   Find the maximum value of the function $f(x, y) = 9x^2 - 9xy + y^3$ on the closed rectangle whose corners are the points $(1, 1)$, $(1, -1)$, $(-1, 1)$, and $(-1, -1)$.

Solution   We begin by finding that the partial derivatives are $\frac{\partial f}{\partial x} = 18x - 9y$ and $\frac{\partial f}{\partial y} = -9x + 3y^2$. If both partial derivatives are zero, we see that $18x - 9y = 0$, so $y = 2x$. Also $0 = -9x + 3y^2 = -9x + 3(2x)^2 = 12x^2 - 9x = 3x(4x - 3)$, which can only happen if $x = 0$ or if $x = \frac{3}{4}$. So one critical point occurs when $x = 0$ and $y = 2x = 0$, that is, at the origin $(0, 0)$. Another occurs when $x = \frac{3}{4}$ and $y = 2x = \frac{3}{2}$, that is, at the point $\left(\frac{3}{4}, \frac{3}{2}\right)$.

The second of these two points is not in the rectangle, so the only possible local extremum inside the rectangle is at the origin. We evaluate the function at this critical point: $f(0, 0) = 0$.

However, there remains the possibility that the maximum might occur on the boundary. For instance, the top of the rectangle is the part of the line $y = 1$ with $-1 \leq x \leq 1$. The function $f(x, y) = 9x^2 - 9xy + y^3$ becomes $9x^2 - 9x + 1$ when we let $y = 1$; we must consider the function $g(x) = 9x^2 - 9x + 1$ for x in the interval $[-1, 1]$. See Figure 16.8.9(i).

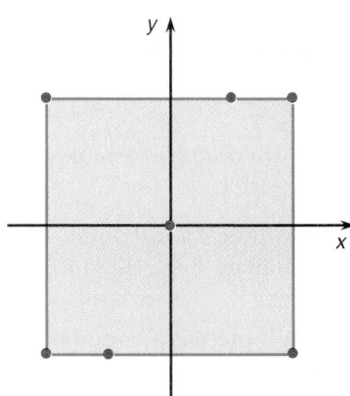

We find that $g'(x) = 18x - 9$, which is zero when $x = \frac{1}{2}$. The possible extrema of g on the given interval are $x = \frac{1}{2}$ and the endpoints $x = \pm 1$. We evaluate and find that $g\left(\frac{1}{2}\right) = \frac{9}{4} - \frac{9}{2} + 1 = -\frac{5}{4}$, $g(-1) = 9 + 9 + 1 = 19$, and $g(1) = 9 - 9 + 1 = 1$. The largest value so far occurs at the left endpoint, which corresponds to the upper left corner of the rectangle.

Figure 16.8.9(i)

Similarly, on the bottom edge of the rectangle we let $y = -1$ and consider the function $G(x) = f(x, -1) = 9x^2 + 9x - 1$. Its derivative is $G'(x) = 18x + 9$, which is zero when $x = -\frac{1}{2}$. We evaluate $G(x)$ at the possible extrema: $G\left(-\frac{1}{2}\right) = \frac{9}{4} - \frac{9}{2} - 1 = -\frac{13}{4}$, $G(-1) = 9 - 9 - 1 = -1$, and $G(1) = 9 + 9 - 1 = 17$. The largest value on the bottom edge is 17, which occurs at the right endpoint, but this is not as large as the value we found at the upper left corner of the rectangle.

Next we must consider the left and right sides of the rectangle. On the left side we let $x = -1$ and consider the function $h(y) = f(-1, y) = 9 + 9y + y^3$. Its derivative is $h'(y) = 3y^2 + 9$, which is never zero. The only possible extrema are at the endpoints; evaluating, we find $h(-1) = 9 - 9 - 1 = -1$ and $h(1) = 9 + 9 + 1 = 19$. This is hardly surprising, since these points correspond to the lower and upper left corners of the rectangle, and we have already found that -1 and 19 are the values of f at those points.

On the right side of the rectangle we let $x = 1$ and consider $H(y) = f(1, y) = 9 - 9y + y^3$. Its derivative is $H'(y) = 3y^2 - 9$, which is zero when $y^2 = 3$, that is, when $y = \pm\sqrt{3}$. The corresponding points are outside the rectangle, so we do not consider them. We could evaluate H at the two endpoints, but this is unnecessary, since we have already evaluated $f(x, y)$ at the two corresponding points, the upper and lower right corners of the rectangle.

The largest value we have found is 19, which is the value of f at the upper left corner of the rectangle.

Incidentally, we have found that $f\left(-\frac{1}{2}, -1\right) = -\frac{13}{4}$ is the minimum value of f on the rectangle. The graph of $z = f(x, y)$ is shown in Figure 16.8.9(ii), with the axes placed as in Figure 16.8.9(i).   ◢

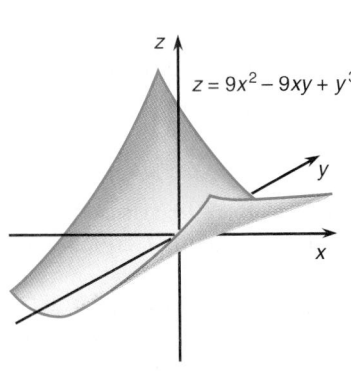

$z = 9x^2 - 9xy + y^3$

Figure 16.8.9(ii)

Notice what we have done in Examples 3 and 4. We found all possible critical points inside the rectangle (in Example 3 there were none; in Example 4 the only one was at the origin). Then, dealing with each side of the rectangle separately, we expressed the values of f on the edge as a function of one variable (x or y, depending on the edge). We found the maximum and minimum values of this function on the interval that corresponds to the side of the rectangle, and finally we compared the maxima obtained from all the edges and from any interior critical points.

Theorem 16.8.3 tells us that there must be a maximum somewhere in or on the rectangle. Theorem 16.8.5 tells us that if it is not on the boundary, then it must occur at a critical point. We found the maximum value on the boundary and the values at any critical points, so the largest of these must be the maximum value.

It was not particularly difficult to find the maximum values on the various pieces of the boundary, since each of them was a vertical or horizontal line segment. For more complicated regions it may be more difficult to find the extrema that lie on the boundary.

Summary 16.8.6

Finding Extrema

To find the local and absolute extrema of a function f of two or more variables, we use the following procedure.

1. Find all critical points of f. Be careful to consider only those critical points that lie in the domain of f, especially if some particular domain is specified in the problem.

2. Find the extrema of f on the boundary of the domain if there is a boundary. It may be necessary to break the boundary into several simple pieces and deal with each one separately. Be careful to remember "endpoint extrema" on portions of the boundary.

3. Compare the values of f at the critical points and the extrema of f on the boundary. If the domain of f is of the type discussed before Theorem 16.8.3, then there must be at least one absolute maximum and at least one absolute minimum, and they must be among the points we have found in the first two steps.

This summary is not quite as useful as it may appear because the second step may be quite difficult, especially if the region is a complicated one.

EXAMPLE 5   Find the maximum value of the function $f(x, y) = x^2 y$ on the closed disk of radius 1 with center at the origin.

Solution   First we find the partial derivatives: $\frac{\partial f}{\partial x} = 2xy$ and $\frac{\partial f}{\partial y} = x^2$. For $\frac{\partial f}{\partial y} = x^2$ to be zero, we must have $x = 0$, in which case $\frac{\partial f}{\partial x} = 2xy$ is zero also. The critical points are exactly the points on the y-axis that are also in the disk. Certainly f equals zero at each of these points.

Next we consider the values of f on the boundary, that is, on the circle $x^2 + y^2 = 1$. It is possible to express y as a function of x on the upper or lower semicircle and using this to express $f(x, y)$ as a function of x and then to apply single-variable techniques.

But it turns out to be easier to parametrize the circle by $\big(x(t), y(t)\big) = (\cos t, \sin t)$ for $t \in [-\pi, \pi]$. The function f is $f(x, y) = f(\cos t, \sin t) = \cos^2 t \sin t$, and we are looking for the maximum of $g(t) = \cos^2 t \sin t$ for $t \in [-\pi, \pi]$.

Now $g'(t) = -2 \cos t \sin^2 t + \cos^3 t = -2 \cos t (1 - \cos^2 t) + \cos^3 t = -2 \cos t + 3 \cos^3 t$. We look for critical points:

$$0 = -2 \cos t + 3 \cos^3 t = \cos t(-2 + 3 \cos^2 t),$$

so $\cos t = 0$ or $\cos^2 t = \frac{2}{3}$. When $\cos t = 0$, that is, $t = \pm\frac{\pi}{2}$, then $\sin t = \pm 1$, and $g(t) = \cos^2 t \sin t = 0$. With $(x, y) = (\cos t, \sin t)$, this tells us that $f(0, 1) = f(0, -1) = 0$.

When $\cos^2 t = \frac{2}{3}$, then $\sin^2 t = 1 - \cos^2 t = \frac{1}{3}$ and $\sin t = \pm\frac{1}{\sqrt{3}}$; the corresponding points are $(x, y) = (\cos t, \sin t) = \left(\sqrt{\frac{2}{3}}, \frac{1}{\sqrt{3}}\right), \left(-\sqrt{\frac{2}{3}}, \frac{1}{\sqrt{3}}\right), \left(\sqrt{\frac{2}{3}}, -\frac{1}{\sqrt{3}}\right),$ $\left(-\sqrt{\frac{2}{3}}, -\frac{1}{\sqrt{3}}\right)$.

We evaluate $f\left(\sqrt{\frac{2}{3}}, \frac{1}{\sqrt{3}}\right) = f\left(-\sqrt{\frac{2}{3}}, \frac{1}{\sqrt{3}}\right) = \frac{2}{3\sqrt{3}}, f\left(\sqrt{\frac{2}{3}}, -\frac{1}{\sqrt{3}}\right) = f\left(-\sqrt{\frac{2}{3}}, -\frac{1}{\sqrt{3}}\right) = -\frac{2}{3\sqrt{3}}$.

The maximum value of $f(x, y) = x^2 y$ is $\frac{2}{3\sqrt{3}}$, and it occurs at the points $\left(\pm\sqrt{\frac{2}{3}}, \frac{1}{\sqrt{3}}\right)$. The graph of $z = x^2 y$ is shown in Figure 16.8.10(i); note the positions of the axes.

Figure 16.8.10(i)

With a graphics calculator we can plot the function $g(t)$ from Example 5.

TI-85	SHARP EL9300
GRAPH F1 ($y(x) =$) (COS F1 (x)) $x^2 \times$ SIN F1 (x) M2 (RANGE) $(-) \pi \blacktriangledown \pi \blacktriangledown \blacktriangledown (-) .4 \blacktriangledown .4$ F5 (GRAPH)	(cos x/θ/T) $x^2 \times$ sin x/θ/T RANGE $-\pi$ ENTER π ENTER 1 ENTER $-.4$ ENTER .4 ENTER 1 ENTER
CASIO f_x-6300G (f_x-7700GB) Range $-\pi$ EXE π EXE 1 EXE $-.4$ EXE .4 EXE 1 EXE EXE EXE EXE Graph (cos X) $x^2 \times$ sin X EXE	HP 48SX ⤴ RAD ⤴ PLOT PLOTR ERASE ATTN ′ $Y =$ COS X ▶ y^x 2 × SIN X ↱ PLOT ⤴ DRAW ERASE 3.2 +/− ENTER 3.2 XRNG .4 ENTER +/− .4 YRNG DRAW

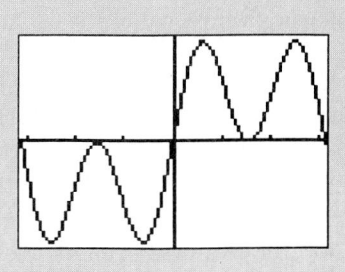

Note the maximum and minimum values.

There are several things to notice about Example 5. One is that we have also found that the minimum value of $f(x, y) = x^2 y$ is $-\frac{2}{3\sqrt{3}}$, and it occurs at the points $\left(\pm\sqrt{\frac{2}{3}}, -\frac{1}{\sqrt{3}}\right)$ on the lower semicircle.

Another observation is that although in this case it was easy to parametrize the boundary, we could also have written y as a function of x on the upper and lower semicircles, but sometimes it is quite difficult to do anything like this. Sometimes it is possible to proceed without expressing y in terms of x or vice versa by using implicit differentiation. (All we really need is to find the derivative with respect to x of the function on each semicircle.)

Finally, it is interesting to think about those critical points along the y-axis. Notice that $f(x, y) = x^2 y$ is nonnegative if $y \geq 0$ and nonpositive if $y \leq 0$. In other words, f is nonnegative above the x-axis and nonpositive below the axis. If $y > 0$, there is a small disk around $(0, y)$ that lies entirely above the x-axis, and since $f(0, y) = 0$, we see that $f(0, y)$ is less than or equal to the value of f at every point in this small disk. This means that f has a local minimum at $(0, y)$ if $y > 0$. See Figure 16.8.10(ii).

Similarly, if $y < 0$, there is a small disk around $(0, y)$ that lies below the axis, so f is nonpositive at every point in this small disk and has a local maximum at $(0, y)$.

At the origin $(0, 0)$, where f is also zero, it has neither a local maximum nor a local minimum, since every disk around the origin will contain points above the axis at which f is positive and points below the axis at which f is negative. See Figure 16.8.10(iii).

By now we see that finding critical points is usually the easier part of the problem, and classifying extrema can be very difficult. With functions of one variable we often were able to use the Second Derivative Test to identify local extrema, but with two or more variables this is not so easy. There is a version of the Second Derivative Test, but it is more complicated and will be postponed until the next section.

Figure 16.8.10(ii)

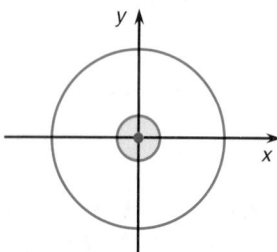

Figure 16.8.10(iii)

Applied Maximum/Minimum Problems

The techniques we have been discussing can be used to solve various types of problems. We will discuss a few of them now, but in fact there are a great many interesting problems that fit into a slightly different framework. These will be the subject of Section 16.10, where we will encounter larger numbers of practical applications.

It is unfortunately true that many of the "applications" that we can make at this point are slightly artificial. They are considered more with the intention of establishing and practicing the techniques than with much serious claim to practical usefulness.

EXAMPLE 6 Find the dimensions of the rectangular box whose volume is 64 and whose surface area is as small as possible.

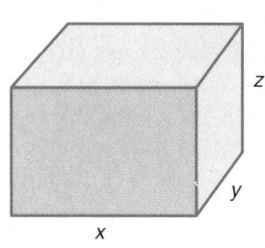

Figure 16.8.11

Solution Letting the dimensions of the box be x, y, and z as shown in Figure 16.8.11, we know that the volume is $V = xyz$ and the surface area is $A = 2xy + 2yz + 2xz$. Using the fact that $V = 64$, we can solve for z as a function of x and y. The equation $64 = xyz$ becomes $z = \frac{64}{xy}$, so $A = 2xy + 2yz + 2xz = 2xy + \frac{128y}{xy} + \frac{128x}{xy} = 2xy + \frac{128}{x} + \frac{128}{y}$.

Thinking of A as a function of x and y, we look for its critical points. Now $A_x = \frac{\partial A}{\partial x} = 2y - \frac{128}{x^2}$, and $A_y = \frac{\partial A}{\partial y} = 2x - \frac{128}{y^2}$.

For these partial derivatives both to equal zero we need to have $2y = \frac{128}{x^2}$, that is, $y = \frac{64}{x^2}$, and $2x = \frac{128}{y^2}$, that is, $x = \frac{64}{y^2}$. Substituting the second of these equations into the first, we find that $y = \frac{64}{x^2} = 64 / \left((64)^2 / y^4\right) = y^4 / 64$.

This amounts to $\frac{1}{64} y^4 - y = 0$, or $\left(\frac{1}{64} y^3 - 1\right) y = 0$, which can happen only if $y = 0$ or if $y^3 = 64$, that is, $y = 4$. Since $V = 0$ if $y = 0$, we only need to consider the other possibility. When $y = 4$, we see that $x = \frac{64}{y^2} = \frac{64}{16} = 4$. Also $z = \frac{64}{xy} = \frac{64}{16} = 4$.

We arrive at the not terribly surprising result that $x = y = z = 4$ are the dimensions at the only possible critical point. This corresponds to the situation in which the rectangular box is a cube. The surface area of the box is $A = 2xy + 2yz + 2xz = 32 + 32 + 32 = 96$.

Strictly speaking, we have not really shown that this point is a minimum for the area. In fact it is possible to show that it is, but it is not very easy. Instead, we just remark that at least it is not a maximum. After all, if we let $x = y = 8$ and $z = 1$, then $V = xyz = 64$, and $A = 2xy + 2yz + 2xz = 128 + 16 + 16 = 160$, which is certainly greater than 96. Our geometrical intuition does suggest to us that there ought to be a minimum somewhere, and we can feel reasonably safe that we have found it.   ◢

EXAMPLE 7   Find the dimensions of the rectangular box whose surface area is 24 and whose volume is as large as possible.

Solution   Letting the dimensions of the box be x, y, and z, we know $2xy + 2yz + 2xz = 24$, that is, $xy + yz + xz = 12$. We could solve for z as a function of x and y, but the calculations become quite difficult.

A better idea is to use implicit differentiation. Knowing that it is *possible* to solve for z as a function of x and y, we can differentiate the equation to find $0 = \frac{\partial}{\partial x}(xy + yz + xz) = y + yz_x + z + xz_x = (y + z) + (x + y)z_x$. From this we find that $z_x = -\frac{y+z}{x+y}$. A similar calculation or Theorem 16.5.4 shows that $z_y = -\frac{x+z}{x+y}$.

Now we regard $V = xyz$ as a function of x and y and look for its critical points. First we find that $V_x = yz + xyz_x$. For this to equal zero we need to have that $xyz_x = -yz$, that is, that $-xy\frac{y+z}{x+y} = -yz$, which amounts to $xy(y + z) = yz(x + y)$. Simplifying, we get $xy^2 + xyz = xyz + y^2z$, or $xy^2 = y^2z$. As before, we can ignore the case in which $y = 0$, so this implies $x = z$.

Almost exactly the same calculation with the partial derivative with respect to y shows that $y = z$. In other words we find once again that the critical point occurs where the three dimensions are equal, that is, when the box is a cube.

In this case the surface area is $6x^2$, and for this to equal 24 we have $x = y = z = 2$. The volume is $V = xyz = 8$.

Once again we have not really shown that this is an absolute maximum, but it is easy to check that it is not an absolute minimum, and our geometric intuition encourages us to expect a maximum. Since the maximum ought to exist, since it must be at a critical point if it does exist, and since we have found the only possible critical point, we will allow ourselves to consider the problem solved. In fact it is true that this is an absolute maximum.   ◢

EXAMPLE 8   Suppose a rectangular box is to be constructed so that its volume is 54 ft^3. Suppose that it costs twice as much per square foot to make the top and bottom as it costs to make the sides. What are the dimensions of the least expensive possible box?

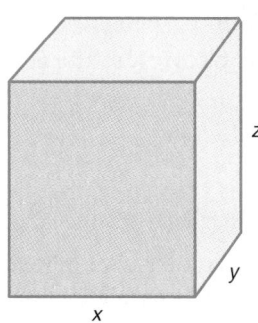

Figure 16.8.12

Solution   As usual we let the dimensions of the box be x, y, and z, with z being the height (see Figure 16.8.12). Then from $54 = V = xyz$ we find that $z = \frac{54}{xy}$. If the cost per square foot of the sides is K, then the cost per square foot of the top and bottom is $2K$. The total cost of the box is $C = (2K)2xy + (K)2(yz + xz)$. We want to minimize C, which is the same thing as minimizing $\frac{C}{K} = 4xy + 2yz + 2xz = 4xy + \frac{108}{x} + \frac{108}{y}$. The partial derivatives are $\frac{\partial}{\partial x}\left(\frac{C}{K}\right) = 4y - \frac{108}{x^2}$ and $\frac{\partial}{\partial y}\left(\frac{C}{K}\right) = 4x - \frac{108}{y^2}$.

For these to be zero we need to have $y = \frac{27}{x^2}$ and $x = \frac{27}{y^2} = 27/(27^2/x^4) = \frac{x^4}{27}$. This can only happen if $x = 0$ or if $x^3 = 27$, that is, if $x = 3$. Since the volume cannot be 54 if $x = 0$, we know that $x = 3$, and then it is easy to calculate $y = \frac{27}{x^2} = \frac{27}{9} = 3$. Since $54 = xyz = 3(3)z$, we see that $z = 6$.

The least expensive box of the required volume is twice as tall as it is wide, which is reasonable, since a box like this uses more of the cheap siding material and less of the expensive material for top and bottom. ◢

EXAMPLE 9 Find the point on the plane $2x - 2y + z = 5$ that is closest to the origin.

Solution We can interpret this question to mean that we should minimize the distance from the point (x, y, z) on the plane to the origin, or, as usual, minimize the square of the distance, which should give the same result.

We want to minimize $F(x, y, z) = x^2 + y^2 + z^2$ for points (x, y, z) on the plane, that is, for which $z = 5 - 2x + 2y$. This means that we can write $F = x^2 + y^2 + z^2 = x^2 + y^2 + (5 - 2x + 2y)^2$, and then it is easy to find the partial derivatives: $\frac{\partial F}{\partial x} = 2x + 2(5 - 2x + 2y)(-2) = 2x - 20 + 8x - 8y = 10x - 8y - 20$ and $\frac{\partial F}{\partial y} = 2y + 2(5 - 2x + 2y)(2) = 2y + 20 - 8x + 8y = -8x + 10y + 20$.

If both partial derivatives are zero, we have that $10x - 8y = 20$ and $8x - 10y = 20$. Multiplying the first of these equations by 10 and the second by 8 and subtracting, we find that $100x - 80y - (64x - 80y) = 200 - 160$, or $36x = 40$, which means $x = \frac{40}{36} = \frac{10}{9}$. Substituting this value in the first of the two equations gives $10\left(\frac{10}{9}\right) - 8y = 20$, so $8y = \frac{100}{9} - 20 = -\frac{80}{9}$, so $y = -\frac{10}{9}$.

Finally, we use the equation of the plane to solve for $z = 5 - 2x + 2y = 5 - \frac{20}{9} - \frac{20}{9} = \frac{5}{9}$.

The critical point is the point $\left(\frac{10}{9}, -\frac{10}{9}, \frac{5}{9}\right)$. Since our geometrical intuition assures us that there must be a point on the plane that is closest to the origin, this must be it. See Figure 16.8.13. ◢

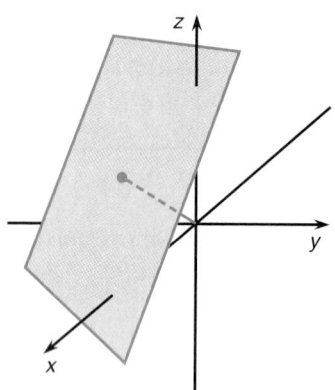

Figure 16.8.13

EXAMPLE 10 What is the maximum possible area of a parallelogram whose perimeter is L?

Solution In Figure 16.8.14 we show the parallelogram; we let x be the length of one of the sides and let θ be the angle between two of the sides.

Since there are two sides of length x and the perimeter is L, we find that the sum of the lengths of the other two sides is $L - 2x$, so each of them has length $\frac{L}{2} - x$. From this we find that the height is $h = \left(\frac{L}{2} - x\right)\sin\theta$ (see the picture), so the area is $A = base \times height = x\left(\frac{L}{2} - x\right)\sin\theta = \left(\frac{L}{2}x - x^2\right)\sin\theta$.

To find critical points, we take the partial derivatives

$$\frac{\partial A}{\partial x} = \left(\frac{L}{2} - 2x\right)\sin\theta, \qquad \frac{\partial A}{\partial \theta} = \left(\frac{L}{2}x - x^2\right)\cos\theta.$$

Suppose they are both zero. First we see from the picture that we can eliminate the possibility that $\sin\theta = 0$, which would result in a completely flat parallelogram. This means that $\frac{L}{2} - 2x$ must be zero, that is, $\frac{L}{2} = 2x$.

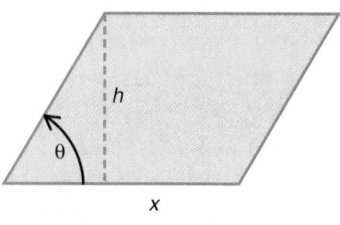

Figure 16.8.14

Substituting this into the partial with respect to θ, we find that $0 = (2x^2 - x^2)\cos\theta = x^2\cos\theta$. Since x also cannot be 0 for a realistic parallelogram, we find that $\cos\theta = 0$, that is, $\theta = \frac{\pi}{2}$.

Along the way we also found that $\frac{L}{2} = 2x$, that is, $x = \frac{L}{4}$, so all the sides have the same length $\frac{L}{4}$. This leads to the plausible conclusion that the biggest parallelogram is a square with sides of length $\frac{L}{4}$.

Exercises 16.8

I

Find all critical points of the following functions.

1. $f(x, y) = 4x^2 - y^2$

2. $f(x, y) = 3xy$

3. $f(x, y) = |x^2 + y^2 - 4|$

4. $f(x, y) = e^{3xy^2 + 2x^2 y}$

5. $f(x, y) = \ln(1 + x^2 + y^2)$

6. $f(x, y) = \ln(1 + |x - y|)$

7. $f(x, y) = e^{-x^2 - y^2}$

8. $f(x, y) = \cos(\pi xy)$

9. $f(x, y) = (x^2 - y^2)/(x^2 + y^2)$

10. $f(x, y) = \sin\left(\frac{1}{1 + x^2 + y^2}\right)$

Find each of the following.

11. The extreme values of $f(x, y) = xy^2 + 2x^2 y$ on the square region whose vertices are $(1, 1)$, $(1, -1)$, $(-1, 1)$, and $(-1, -1)$.

12. The extreme values of $f(x, y) = xy^2 - x^2 y$ on the square region whose vertices are $(0, 0)$, $(0, 1)$, $(1, 0)$, and $(1, 1)$.

13. The extreme values of $f(x, y) = 2xy^2 + 3x^2 y$ on the square region whose vertices are $(1, -1)$, $(3, -1)$, $(1, 1)$, and $(3, 1)$.

14. The extreme values of $f(x, y) = xy^2 - 3x^2 y$ on the rectangular region whose vertices are $(1, 0)$, $(1, 3)$, $(2, 0)$, and $(2, 3)$.

15. The extreme values of $f(x, y) = x^2 + xy^2 - 2$ on the unit disk $x^2 + y^2 \le 1$.

16. The extreme values of $f(x, y) = x + xy$ on the disk $x^2 + y^2 \le 4$.

17. The extreme values of $f(x, y) = 2x^2 + y$ on the disk $x^2 + y^2 \le 1$.

18. The extreme values of $f(x, y) = xy$ on the elliptical region $4x^2 + 9y^2 \le 36$.

19. The extreme values of $f(x, y) = x^2 y$ on the elliptical region $2x^2 + y^2 \le 3$.

20. The extreme values of $f(x, y) = x + y$ on the elliptical region $2x^2 + 3y^2 \le 6$.

II

21. Let $f(x, y) = 9x^2 - 9xy + y^3$ be the function of Example 4 (p. 1015). Find the maximum and minimum values of $\left(f(x, y)\right)^2$ on the closed rectangular region whose corners are $(1, 1)$, $(1, -1)$, $(-1, 1)$, and $(-1, -1)$. Do not attempt to specify all the *points* at which the minimum occurs.

*22. Find and classify the extrema of $f(x, y) = \frac{1 + x^2 + y^2}{1 + x^2}$.

*23. Find and classify the extrema of the function $F(x, y) = (x^2 + y^2 - 1)^2$.

*24. Find and classify the extrema of the function $F(x, y) = (3x^2 + y^2 - 4)^2$.

25. Solve the question of Example 5 (p. 1016) by writing y as a function of x on appropriate parts of the boundary.

*26. Solve the question of Example 5 without writing x and y as functions of t or by writing x as a function of y or y as a function of x but by using implicit differentiation. Be careful about the interval on which the function is defined implicitly.

*27. Find an example of a function that is defined and continuous on an open rectangle (i.e., the set of all the points that are inside a rectangle but not on the boundary) but that is not bounded.

*28. Find an example of a function that is defined and continuous on an open rectangle and that *is* bounded but that does not achieve its maximum value.

*29. Find an example of a function that is defined and continuous on an open disk (i.e., the set of all the points that are inside a circle but not on the circle) but that is not bounded.

*30. Find an example of a function that is defined and continuous on an open disk and that *is* bounded but that does not achieve its maximum value.

*31. Find all critical points of the function $f(x, y) = \ln |x + y|$. Remember that critical points of f must be points that are in the domain of the function.

*32. Find all critical points of the function $f(x, y) = \sqrt{x^2 + y^2 - 1}$.

33. Find the dimensions of a rectangular box whose volume is 24 in^3 and whose surface area is as small as possible.

34. Find the dimensions of a rectangular box whose volume is V and whose surface area is as small as possible.

35. Find the dimensions of a rectangular box whose volume is 32 and whose cost is as small as possible, assuming that its vertical sides cost twice as much per square foot as its top and bottom.

36. Find the dimensions of a rectangular box whose volume is V and whose cost is as small as possible, assuming that its vertical sides cost R times as much per square foot as its top and bottom.

37. Find the dimensions of an open rectangular box whose volume is 40 and whose cost is as small as possible, assuming that its vertical sides cost three times as much per square foot as its bottom.

38. Find the dimensions of a rectangular box whose volume is 250 and whose cost is as small as possible, assuming that its vertical sides cost twice as much per square foot as its top and three times as much as its bottom.

Find the point on each of the following planes that is closest to the specified point.

39. $4x - 2y - 4z = 3$, $(0, 0, 0)$

40. $2x - 3y + z = 4$, $(0, 0, 0)$

41. $x + y + z = 0$, $(2, 3, 4)$

42. $x + 2y - z = 0$, $(1, 1, 4)$

43. $ax + by + cz = d$, $(0, 0, 0)$

44. $ax + by + cz = d$, (u, v, w)

*45. (i) Urban planners have a grid map of a city. There is a hospital at $(0, 0)$, a police station at $(2, 0)$, and a fire station at $(0, 1)$. They propose the function $f(x, y)$, which is the sum of the squares of the distances from (x, y) to these three points, as an indicator of the difficulty of providing emergency services to the location (x, y). Find the points at which $f(x, y)$ has an absolute minimum.

(ii) Another planner suggests the product of the three distances as a better indicator. Where does it have a minimum? (Do not do any work to get this answer.)

**46. Consider three distinct points (a, b), (c, d), (u, v) in the plane, and let $f(x, y)$ be the sum of the squares of the distances from (x, y) to these three points. Find the point(s) at which $f(x, y)$ has an absolute minimum, and interpret the answer geometrically. Interpret the result of Exercise 45 in terms of your answer.

47. Use a computer to sketch the graph of the function $z = x^4 - 2x^2 + y^2 + 4y - 2$ of Example 2 (p. 1012).

48. Sketch the graph of the function $z = 9x^2 - 9xy + y^3$ on $-1 \leq x, y \leq 1$; cf. Example 4 (p. 1015) and Figure 16.8.9(ii).

49. Sketch the graph of the function $z = x^2y$ on $x^2 + y^2 \leq 1$; cf. Example 5 (p. 1016) and Figure 16.8.10(i).

50. Graph the function $A(x, \theta) = \left(\frac{L}{2}x - x^2\right) \sin \theta$ of Example 10 (p. 1020). Explain what various parts of the graph have to do with the original parallelogram.

POINT TO PONDER

In Section 3.3 we did some problems about minimizing the area of a rectangular box with specified volume, maximizing the volume of a box with specified area, or some related questions, which seem very similar to the examples done at the end of this section. The difference is that in Section 3.3 we always had to assume that two of the dimensions of the box are equal, for example, that the base of the box is a square. Then there are only two variables, and by using the condition given by specifying the volume or the area it is possible to solve for one of them as a function of the other. The upshot is that the problem amounts to a problem with only one variable.

In this section we make no assumption about two of the dimensions being equal. This means that we begin with three variables x, y, and z. The condition in the question then makes it possible to solve for one of them as a function of the other two, and we are left with a problem involving two variables. It is interesting to see that for the problems described above, the first thing we discover is that two of the dimensions must be equal after all. Our knowledge of functions of two variables allows us to draw this as a conclusion, whereas in Chapter 3 we had to assume that it was true before we could begin.

In fact this is a very common type of situation. It frequently happens that a problem appears to involve a large number of variables but that there are some conditions that relate them. It may not always be easy to solve for some of the variables as functions of the others, but as we have already seen, it may still be possible to solve the problem.

16.9 The Second Derivative Test

SADDLE POINT
DISCRIMINANT

Figure 16.9.1

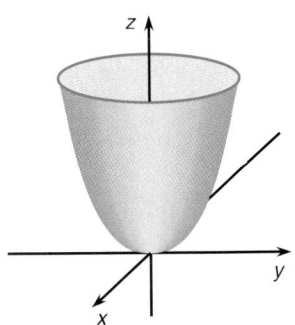

Figure 16.9.2

For functions of a single variable the Second Derivative Test (Theorem 4.4.1, p. 256) is very useful for deciding whether a critical point is a local maximum or minimum. It has its limitations, the main one being that when the second derivative is zero, it tells us nothing.

For functions of two variables the situation is inevitably more complicated. For the function $h(x, y) = x^2 - y^2$, whose graph $z = x^2 - y^2$ is the saddle surface shown in Figure 16.9.1, we find that $\frac{\partial h}{\partial x} = 2x$, $\frac{\partial h}{\partial y} = -2y$, and the only critical point is at the origin. The second derivative in the x-direction is $\frac{\partial^2 h}{\partial x^2} = 2$, which means that as we move along the x-axis $h(x, y)$ has a local minimum at $(0, 0)$.

Similarly, $\frac{\partial^2 h}{\partial y^2} = -2$, and as we move along the y-axis, the function has a local maximum at $(0, 0)$. Of course, this means that the function does not have a local maximum or minimum at the origin, and since $(0, 0)$ is the only critical point, it has no local extrema anywhere.

At a point like this, where in some directions there is a local maximum and in others a local minimum, we say that the function has a **saddle point**.

It is natural and tempting to ask whether a function f must have a local minimum at a critical point where both $\frac{\partial^2 f}{\partial x^2}$ and $\frac{\partial^2 f}{\partial y^2}$ are positive, but unfortunately the answer is "not necessarily."

In the obvious example, $g(x, y) = x^2 + y^2$, the only critical point is $(0, 0)$, and $\frac{\partial^2 g}{\partial x^2} = 2$, $\frac{\partial^2 g}{\partial y^2} = 2$. This function does have a minimum at $0, 0$; its graph is the paraboloid shown in Figure 16.9.2.

However, consider the function $f(x, y) = x^2 - 4xy + y^2$. We find that

$$\frac{\partial f}{\partial x} = 2x - 4y, \qquad \frac{\partial f}{\partial y} = -4x + 2y,$$

so $\frac{\partial f}{\partial x} = 0$ on the line $2x = 4y$, and $\frac{\partial f}{\partial y} = 0$ on the line $4x = 2y$. The only point where $\frac{\partial f}{\partial x}$ and $\frac{\partial f}{\partial y}$ are both zero is the origin.

The second partials are $\frac{\partial^2 f}{\partial x^2} = 2$ and $\frac{\partial^2 f}{\partial y^2} = 2$, so there is a local minimum *along either axis.*

However, consider the values of $f(x, y)$ along the line $y = x$. There we find that $f(x, y) = x^2 - 4xy + y^2 = x^2 - 4x^2 + x^2 = -2x^2$. This function is never positive, and it has a *maximum* when $x = 0$. The graph of $z = f(x, y) = x^2 - 4xy + y^2$ is

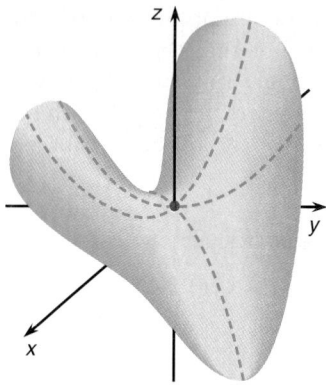

Figure 16.9.3

sketched in Figure 16.9.3. It is nonnegative on both the x- and y-axes but nonpositive on the line $y = x$. The function has a saddle point at the origin.

In particular the function $f(x, y) = x^2 - 4xy + y^2$ demonstrates that the second partials $\dfrac{\partial^2 f}{\partial x^2}$ and $\dfrac{\partial^2 f}{\partial y^2}$ can both be positive at a critical point that is not a local minimum.

As usual the second partials $\dfrac{\partial^2 f}{\partial x^2}$ and $\dfrac{\partial^2 f}{\partial y^2}$ tell about the behavior of $f(x, y)$ along the axes but not elsewhere. This function has minima along both axes but a maximum along the line $y = x$.

So a workable Second Derivative Test will have to be more complicated. It turns out that it must also involve the mixed partial $\dfrac{\partial^2 f}{\partial x \partial y}$.

THEOREM 16.9.1

The Second Derivative Test

Suppose $f(x, y)$ has a **critical point** at (a, b) and the partial derivatives $\dfrac{\partial^2 f}{\partial x^2}$, $\dfrac{\partial^2 f}{\partial y^2}$, $\dfrac{\partial^2 f}{\partial x \partial y}$ are all continuous on some disk around (a, b). Consider the quantity

$$D = \left(\frac{\partial^2 f}{\partial x^2} \frac{\partial^2 f}{\partial y^2} - \left(\frac{\partial^2 f}{\partial x \partial y} \right)^2 \right) (a, b) = (f_{xx} f_{yy} - f_{xy}^2)(a, b)$$

(i.e., evaluate all the partial derivatives at (a, b)).

(i) If $D > 0$ and $\dfrac{\partial^2 f}{\partial x^2} > 0$ or $\dfrac{\partial^2 f}{\partial y^2} > 0$, then $f(x, y)$ has a local minimum at (a, b).

(ii) If $D > 0$, and $\dfrac{\partial^2 f}{\partial x^2} < 0$ or $\dfrac{\partial^2 f}{\partial y^2} < 0$, then $f(x, y)$ has a local maximum at (a, b).

(iii) If $D < 0$, then $f(x, y)$ has a saddle point at (a, b) (and no local extremum there).

(iv) If $D = 0$, then the test is inconclusive.

The quantity D is called the **discriminant** of f at the point (a, b).

EXAMPLE 1

(i) In our earlier discussion of the function $h(x, y) = x^2 - y^2$ we found that the only critical point is $(0, 0)$. The second partial derivatives are $\dfrac{\partial^2 h}{\partial x^2} = 2$, $\dfrac{\partial^2 h}{\partial y^2} = -2$, $\dfrac{\partial^2 h}{\partial x \partial y} = 0$, so the discriminant is $D = h_{xx} h_{yy} - h_{xy}^2 = -4$. Since $D < 0$, the test tells us that h has a saddle point (cf. Figure 16.9.1).

(ii) For the function $g(x, y) = x^2 + y^2$ we also found that the origin is the only critical point. Moreover, $\dfrac{\partial^2 g}{\partial x^2} = 2$, $\dfrac{\partial^2 g}{\partial y^2} = 2$, $\dfrac{\partial^2 g}{\partial x \partial y} = 0$, so the discriminant is $D = g_{xx} g_{yy} - g_{xy}^2 = 4$. Since $D > 0$, we have to look at the sign of $\dfrac{\partial^2 g}{\partial x^2}$ or $\dfrac{\partial^2 g}{\partial y^2}$. Since both are positive, the test tells us that g has a local minimum (cf. Figure 16.9.2).

(iii) The origin is also the only critical point for the function $f(x, y) = x^2 - 4xy + y^2$. Moreover, $\dfrac{\partial^2 f}{\partial x^2} = 2$, $\dfrac{\partial^2 f}{\partial y^2} = 2$, $\dfrac{\partial^2 f}{\partial x \partial y} = -4$, so the discriminant is $D = f_{xx} f_{yy} - f_{xy}^2 = -12$. Since $D < 0$, the Second Derivative Test tells us that f has a saddle point (cf. Figure 16.9.3).

If you find the Second Derivative Test difficult to memorize, it may help to keep these examples in mind. The first two, $h(x, y) = x^2 - y^2$ and $g(x, y) = x^2 + y^2$, have the convenient property that the *mixed* partials $\frac{\partial^2 h}{\partial x \partial y}$ and $\frac{\partial^2 g}{\partial x \partial y}$ are both zero. Because of this the discriminant D is just the product of the two "unmixed" second partials. For the saddle surface $z = x^2 - y^2$ the second partials have opposite signs and their product is negative, while for the paraboloid the second partials have the same sign and their product is positive.

So in the situation in which the mixed partials are zero, D is negative when f_{xx} and f_{yy} have opposite signs, which means that f has a local maximum along one axis and a local minimum along the other. Of course, it cannot have either kind of local extremum.

Similarly, if $f_{xy} = 0$, then D is positive when f_{xx} and f_{yy} have the same sign, so f has the same type of extremum along both axes. The other part of the test, checking the sign of f_{xx} or f_{yy} in this case, determines which type. This is necessary because it is not apparent from D whether f_{xx} and f_{yy} are both positive or both negative.

EXAMPLE 2   Let $f(x, y) = x^2 - 4xy + 6x + y^2$. Find and classify all local extrema.

Solution   We find that

$$f_x = 2x - 4y + 6, \qquad f_y = -4x + 2y.$$

It is easy to find that the only point at which they are both zero is $(x, y) = (1, 2)$. The second partials are

$$f_{xx} = 2, \qquad f_{yy} = 2, \qquad f_{xy} = -4.$$

The discriminant is $D = f_{xx} f_{yy} - f_{xy}^2 = -12$. Because it is negative, there is a saddle point at $(1, 2)$. In particular there are no local extrema. The graph of $z = x^2 - 4xy + 6x + y^2$ is shown in Figure 16.9.4.

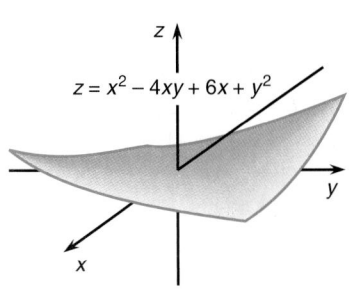

$z = x^2 - 4xy + 6x + y^2$

Figure 16.9.4

Notice that in the original function $f(x, y) = x^2 - 4xy + 6x + y^2$, both x^2 and y^2 occurred with positive coefficients, but even so there is a saddle point. It can be extremely difficult to guess the nature of a critical point just by looking at the equation.

EXAMPLE 3   Let $g(x, y) = x^3 + 5x^2 + 3y^2 - 6xy$. Find and classify all local extrema.

Solution   The partial derivatives are

$$\frac{\partial g}{\partial x} = 3x^2 + 10x - 6y, \qquad \frac{\partial g}{\partial y} = 6y - 6x.$$

At a critical point we must have $0 = \frac{\partial g}{\partial y} = 6y - 6x$, so $x = y$. In that case, $0 = \frac{\partial g}{\partial x} = 3x^2 + 10x - 6y = 3x^2 + 4x = x(3x + 4)$. The only critical points are $(0, 0)$ and $\left(-\frac{4}{3}, -\frac{4}{3}\right)$.

The second partial derivatives are

$$\frac{\partial^2 g}{\partial x^2} = 6x + 10, \qquad \frac{\partial^2 g}{\partial y^2} = 6, \qquad \frac{\partial^2 g}{\partial x \partial y} = -6,$$

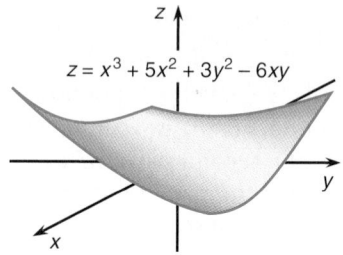

$z = x^3 + 5x^2 + 3y^2 - 6xy$

Figure 16.9.5

Figure 16.9.6(i)

Figure 16.9.6(ii)

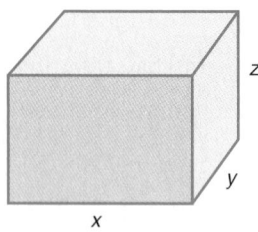

Figure 16.9.7

so the discriminant is $D = (6x + 10)6 - (-6)^2 = 36x + 24$. At $(0, 0)$ we find that $D = 24$, and at $\left(-\frac{4}{3}, -\frac{4}{3}\right)$ we see that $D = -24$.

The Second Derivative Test allows us to conclude that there is a saddle point at $\left(-\frac{4}{3}, -\frac{4}{3}\right)$ but that there is a local extremum at $(0, 0)$. Since both g_{xx} and g_{yy} are positive at $(0, 0)$, we find that there is a local minimum at $(0, 0)$ and no other local extrema. The graph of $z = x^3 + 5x^2 + 3y^2 - 6xy$ is shown in Figure 16.9.5. ◢

Sometimes the Second Derivative Test must be combined with other techniques to find extreme values.

EXAMPLE 4 If $h(x, y) = x^3 - 3xy^2$, find the absolute maximum value of h on the unit disk $x^2 + y^2 \le 1$.

Solution Since $h_x = 3x^2 - 3y^2$ and $h_y = -6xy$, we see that at a critical point $x^2 = y^2$, that is, $x = \pm y$, and $xy = 0$, so $x = y = 0$. The only critical point is the origin.

Also $h_{xx} = 6x$, $h_{yy} = -6x$, and $h_{xy} = -6y$, so the discriminant is $D = -36x^2 - 36y^2$. This equals 0 at the origin, so the Second Derivative Test is not helpful.

On the other hand, if we consider the values of $h(x, y)$ along the x-axis, that is, where $y = 0$, we find that $h(x, 0) = x^3$, which certainly does not have an extremum at $x = 0$.

This means that the extrema must be on the boundary circle. One way to find them is to let $(x, y) = (\cos \theta, \sin \theta)$. As θ ranges from $-\pi$ to π, this point will travel once around the circle. We can assume that $-\pi \le \theta < \pi$, since $\theta = \pi$ and $\theta = -\pi$ correspond to the same point.

The corresponding value of the function is

$$h(x, y) = h(\cos \theta, \sin \theta) = \cos^3 \theta - 3 \cos \theta \sin^2 \theta = \cos \theta (\cos^2 \theta - 3 \sin^2 \theta)$$
$$= \cos \theta (\cos^2 \theta - 3 + 3 \cos^2 \theta) = \cos \theta (4 \cos^2 \theta - 3) = 4 \cos^3 \theta - 3 \cos \theta.$$

Differentiating with respect to θ, we find that

$$\frac{dh(x, y)}{d\theta} = -12 \cos^2 \theta \sin \theta + 3 \sin \theta = \sin \theta (-12 \cos^2 \theta + 3).$$

This is zero when $\sin \theta = 0$ or when $\cos^2 \theta = \frac{1}{4}$, that is, when $\theta = 0$, π, $\frac{\pi}{3}$, $-\frac{\pi}{3}$, $\frac{2\pi}{3}$, $-\frac{2\pi}{3}$.

Thinking of h as a function of θ, we evaluate at each of these points and find $h(0) = 1$, $h(\pi) = -1$, $h\left(\pm\frac{\pi}{3}\right) = \frac{1}{2} - \frac{3}{2} = -1$, $h\left(\pm\frac{2\pi}{3}\right) = -\frac{1}{2} + \frac{3}{2} = 1$.

The maximum value of h is 1, which occurs when $\theta = 0, \pm\frac{2\pi}{3}$. These values of θ correspond to the points $(1, 0)$, $\left(-\frac{1}{2}, \pm\frac{\sqrt{3}}{2}\right)$ on the circle (see Figure 16.9.6(i)).

Incidentally, the minimum value of h is -1, occurring at $(-1, 0)$, $\left(\frac{1}{2}, \pm\frac{\sqrt{3}}{2}\right)$. Figure 16.9.6(ii) shows the graph of $z = x^3 - 3xy^2$. Note that the position of the axes is different from Figure 16.9.6(i). ◢

We now consider a familiar type of example in light of the Second Derivative Test.

EXAMPLE 5 Suppose the dimensions of a rectangular box are x, y, and z, as shown in Figure 16.9.7. Find the largest possible volume of the box if $x + y + z = 1$.

Solution In the past with questions of this sort we have been able to find the right answer but have not been able to show that it is really a maximum.

With a graphics calculator we cannot graph the function $h(x, y)$ from Example 4, but we can plot its values along selected lines. For instance, along the x-axis, we find $h(x, 0) = x^3$.

In Example 4 we found that $\frac{dh}{d\theta} = 0$ when $\theta = 0, \pi, \pm\frac{\pi}{3}, \pm\frac{2\pi}{3}$. We see that $\theta = 0$ and $\theta = \pi$ correspond to points on the x-axis, while $\theta = \frac{\pi}{3}$ and $-\frac{2\pi}{3}$ correspond to points on the line $y = \sqrt{3}x$, and $\theta = -\frac{\pi}{3}$ and $\frac{2\pi}{3}$ correspond to points on the line $y = -\sqrt{3}x$. We graph $h(x, 0) = x^3$ and $h(x, \sqrt{3}x) = x^3 - 3x(3x^2) = x^3 - 9x^3 = -8x^3$, and incidentally $h(x, -\sqrt{3}x) = -8x^3$.

TI-85	SHARP EL9300
GRAPH F1 $(y(x) =)$ F1 (x) ˆ 3 ▼ $(-)$ 8 F1 (x) ˆ 3 M2 (RANGE) $(-)$ 1 ▼ 1 ▼ ▼ $(-)$ 1 ▼ 1 F5 (GRAPH)	⟳ x/θ/T a^b 3 2ndF ▼ -8 x/θ/T a^b 3 RANGE -1 ENTER 1 ENTER 1 ENTER -1 ENTER 1 ENTER 1 ENTER ⟳
CASIO f_x-6300G (f_x-7700GB)	HP 48SX
Range -1 EXE 1 EXE 1 EXE -1 EXE 1 EXE 1 EXE EXE EXE EXE Graph $\boxed{X}$ x^y 3 EXE Graph -8 $\boxed{X}$ x^y 3 EXE	↰ PLOT PLOTR ERASE ATTN ' Y = $\boxed{X}$ y^x 3 ↱ PLOT ↰ DRAW ERASE 1 ENTER +/− 1 XRNG 1 ENTER +/− 1 YRNG DRAW ATTN ' Y = 8 +/− × $\boxed{X}$ y^x 3 ↱ PLOT ↰ DRAW DRAW

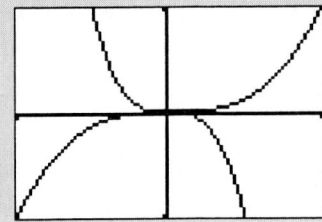

We see where the maximum and minimum values occur. Be sure to relate these graphs to the original problem on the disk. (Notice that the parts of the lines $y = \pm\sqrt{3}x$ that lie inside the unit disk are the parts with $-\frac{1}{2} \le x \le \frac{1}{2}$.)

The volume is $V = xyz$, and from the condition $x + y + z = 1$ we can solve, for instance, to find $z = 1 - x - y$. Using this, we find $V = xyz = xy(1 - x - y) = xy - x^2y - xy^2$.

The partial derivatives are

$$\frac{\partial V}{\partial x} = y - 2xy - y^2, \qquad \frac{\partial V}{\partial y} = x - x^2 - 2xy.$$

Setting them both equal to zero, we have $y - 2xy - y^2 = 0$ and $x - x^2 - 2xy = 0$. Subtracting these equations gives $y - x + x^2 - y^2 = 0$, or $(x - y)(-1 + x + y) = 0$.

We conclude that either $x = y$ or $x + y = 1$. In the second case, where $x + y = 1$, we realize that since $x + y + z = 1$, this means that $z = 0$ and the box has no height. We discard this silly possibility and consider the remaining case in which $x = y$, that is, in which two of the dimensions are equal.

Setting $y = x$ in the equation $\frac{\partial V}{\partial x} = 0$ gives $0 = y - 2xy - y^2 = x - 2x^2 - x^2 = x - 3x^2 = x(1 - 3x)$. This means that $x = 0$ or $3x = 1$. Once again we discard the case $x = 0$, which corresponds to a box with zero volume, and see that $x = \frac{1}{3}$. Since $y = x$, we know that $y = \frac{1}{3}$ and $z = 1 - x - y = 1 - \frac{2}{3} = \frac{1}{3}$. The critical point occurs where $x = y = z = \frac{1}{3}$, that is, when the box is a cube.

Now the second partials are

$$\frac{\partial^2 V}{\partial x^2} = -2y, \qquad \frac{\partial^2 V}{\partial y^2} = -2x, \qquad \frac{\partial^2 V}{\partial x \partial y} = 1 - 2x - 2y,$$

and the discriminant is $D = (-2y)(-2x) - (1 - 2x - 2y)^2$. Rather than multiply this all out, it is easier to evaluate it first by setting $x = y = \frac{1}{3}$. We find that $D = \left(-\frac{2}{3}\right)\left(-\frac{2}{3}\right) - \left(1 - \frac{2}{3} - \frac{2}{3}\right)^2 = \frac{4}{9} - \left(-\frac{1}{3}\right)^2 = \frac{1}{3}$.

Since D is positive, there is a local extremum; since $\frac{\partial^2 V}{\partial x^2} = -\frac{2}{3}$ is negative, the extremum is a local maximum. The biggest volume will be found by making a cube whose sides are all $\frac{1}{3}$. ◢

> Notice how the Second Derivative Test can be used to confirm our intuition about the nature of a critical point. Also observe the way we evaluated the second partials *before* figuring out the discriminant: it is usually easier to work with numbers than with polynomials or other functions.

Exercises 16.9

I

Classify the local extrema of the following functions. (Do not worry about distinguishing local and absolute extrema.)

1. $f(x, y) = 3x^2 + 7y^2$
2. $g(x, y) = 4x^2 - 2y^2$
3. $h(x, y) = x^2 - 4x - 2y^2 + 8y - 1$
4. $k(x, y) = xy$
5. $p(x, y) = \exp(x^2 + 3y^2)$
6. $q(x, y) = 2x^3 + 3x^2 + y^2$
7. $r(x, y) = x^3 + 3x^2 - y^3 + y^2$
8. $s(x, y) = \ln(3 + x^2 + 2xy + 2y^2)$
9. $u(s, t) = 2s^4 + s^2 + 2st + t^2 + s$
10. $w(s, t) = x^2 - \cos(x + y)$

Find the absolute maximum and minimum of each of the following functions on the specified region.

11. $f(x, y) = 2x^2 + 2y^2$ on $x^2 + y^2 \leq 1$
12. $f(x, y) = x^2 + y^2$ on $|x|, |y| \leq 1$
13. $f(x, y) = x^2 + 2x - y^2$ on $x^2 + y^2 \leq 4$
14. $f(x, y) = x^2 + 2xy + y^2$ on $|x|, |y| \leq 2$
15. $f(x, y) = x^2 \arctan(y^2)$ on $|x|, |y| \leq 1$
16. $f(x, y) = x^2 y^2$ on $0 \leq x \leq 1, 0 \leq y \leq 1 - x$

17. $f(x, y) = x^3 - 3x + y^2 + 2y$ on $|x|, |y| \leq 3$
18. $f(x, y) = 2x^2 + 2xy + y^2 + 2x + 2y$ on $|x|, |y| \leq 4$
19. $f(x, y) = y^3 - 3x^2 y$ on $0 \leq x, 0 \leq y; x^2 + y^2 \leq 4$
20. $f(x, y) = x^2 e^{-y^2}$ on $0 \leq x \leq 2, 0 \leq x + y \leq 2$

II

In Section 16.8 there were a number of examples in which we found a critical point but could not really show that it was a local extremum. Show that the critical points in these examples are extrema as indicated.

21. Example 16.8.9, local minimum.
22. Example 16.8.6, local minimum.
23. Example 16.8.8, local minimum.
*24. Example 16.8.7, local maximum.
*25. Let $g(x, y) = \cos^3 x - 3 \cos x + 2 \sin(x + y)$. Find and classify all local extrema.
*26. Let k be a positive integer. Find and classify all local extrema of $f(x, y) = x^k + y^k$.
*27. Let k be a positive integer. Find and classify all local extrema of $f(x, y, z) = x^k + y^k + z^k$.
*28. Suppose k and n are positive integers. Find and classify all local extrema of $f(x_1, x_2, \ldots, x_n) = x_1^k + x_2^k + \ldots + x_n^k$.

*29. Is the local minimum that we found in Example 3 (p. 1025) an absolute minimum or not? (*Hint:* Look along the *x*-axis.)

*30. Show that the discriminant of $-f(x, y)$ is the same as the discriminant of $f(x, y)$. In particular, if f has a local maximum at (a, b), then $-f$ has a local minimum and vice versa. The discriminant by itself cannot distinguish these two cases.

31. Find the (minimum) distance from the point $(0, 0, 0)$ to the plane $x + y + z = 3$. (Use the Second Derivative Test to verify that you have found a local minimum.)

32. Find the (minimum) distance from the point $(1, 2, 3)$ to the plane $x - 2y + z = 5$. (Use the Second Derivative Test to verify that you have found a local minimum.)

*33. Classify all critical points of $f(x, y) = (x^2 + y^2 - 1)^2$.

**34. Find the point(s) on the graph of $z = (x^2 + y^2 - 1)^2$ that are closest to the origin.

*35. A rectangular box is placed in the region $x, y, z \geq 0$ with one corner at the origin, three sides in the coordinate planes, and the corner opposite the origin on the plane $x + y + z = 6$. What is the maximum volume for such a box?

*36. A rectangular box is placed in the region $x, y, z \geq 0$ with one corner at the origin, three sides in the coordinate planes, and the corner opposite the origin on the plane $2x + 3y + 4z = 36$. What is the maximum volume for such a box?

*37. Suppose $a, b, c, d > 0$. A rectangular box is placed in the region $x, y, z \geq 0$ with one corner at the origin, three sides in the coordinate planes, and the corner opposite the origin on the plane $ax + by + cz = d$. What is the maximum volume for such a box?

*38. Suppose $a, b, c, d > 0$. A rectangular box is placed in the region $x, y, z \geq 0$ with one corner at the origin, three sides in the coordinate planes, and the corner opposite the origin on the plane $ax + by + cz = d$. What is the maximum surface area for such a box? Be sure to check whether you have found a maximum or a minimum. What happens to the surface area if the corner opposite the origin is close to one of the coordinate planes?

*39. The U.S. Postal Service will accept a package only if the sum of its length and its girth (the perimeter of a cross-section perpendicular to the length) does not exceed 108 in. What are the dimensions of the rectangular box of largest volume that can be sent?

*40. In the situation of Exercise 39, what are the dimensions of the rectangular package of largest surface area that can be sent?

*41. What are the dimensions of the rectangular box with no top and volume 32 cu ft that takes the smallest amount of material to construct (i.e., has the smallest surface area)?

*42. What are the dimensions of the rectangular box with no top and no front and volume 16 cu ft that takes the smallest amount of material to construct?

*43. What is the rectangular box of maximum volume that can be placed inside a sphere of radius R?

*44. What is the rectangular box of maximum surface area that can be placed inside a sphere of radius R?

*45. A bus shelter is made in the shape of a rectangular box with no front. The material for the roof costs twice as much per square foot as the material for the three sides. What are the dimensions of the least expensive shelter that has volume 108 cu ft?

*46. A factory building is constructed in the shape of a rectangular box with volume 2,500,000 cu ft. The heat loss per square foot is five times as great through the roof as through the walls. What dimensions for the building will minimize the total heat loss? (Before you work it out, do you expect the height of the most efficient building to be greater or smaller than the horizontal dimensions?)

*47. A piece of wire of length L is cut into at most three pieces, and each piece is folded into a square. Which lengths of wire will give the maximum and minimum values for the total area of the squares?

*48. A lump of clay of volume V is divided into at most three lumps, each of which is fashioned into a cube. Which sizes of smaller lumps will give the maximum and minimum total volume?

**49. A strip of metal 3 ft wide is folded to make a trough. A strip of width x along each side is bent up through an angle θ (the width x and the angle θ are the same on both sides). What values of x and θ will give the trough of greatest cross-sectional area (and hence the greatest volume)?

**50. A wire of length L is bent to form two sides of a triangle. (The sides may or may not have the same length.) How do you get the triangle with the maximum possible area?

51. Plot the graph of the function $z = f(x, y) = x^2 - 4xy + y^2$ discussed in Example 1(iii) (p. 1024).

52. Plot the graph of the function $z = g(x, y) = x^3 + 5x^2 + 3y^2 - 6xy$ discussed in Example 3 (p. 1025).

If you know about linear algebra, it may already have occurred to you that the discriminant $D = f_{xx}f_{yy} - f_{xy}^2$ is the determinant of the following matrix:

$$\begin{pmatrix} f_{xx} & f_{xy} \\ f_{xy} & f_{yy} \end{pmatrix}$$

This will make it easier to remember. It may also suggest that a similar formula could be considered for functions of three or more variables. For instance, for the function $g(x, y, z) = x^2 - y^2 - z^2$ the corresponding matrix would be

$$\begin{pmatrix} g_{xx} & g_{xy} & g_{xz} \\ g_{xy} & g_{yy} & g_{yz} \\ g_{xz} & g_{yz} & g_{zz} \end{pmatrix} = \begin{pmatrix} 2 & 0 & 0 \\ 0 & -2 & 0 \\ 0 & 0 & -2 \end{pmatrix}.$$

The determinant is 8, which is exactly the same as the corresponding determinant for $h(x, y, z) = x^2 + y^2 + z^2$. The function h has an absolute minimum at the origin, its only critical point, but g has a saddle point. This shows that looking only at the determinant is not enough; in fact what is needed is to know the signs of the eigenvalues.

Roughly speaking, if there are two positive eigenvalues and one negative eigenvalue, then the graph has the same sort of shape as $x^2 + y^2 - z^2$, while if there are one positive and two negative eigenvalues, then the graph has the same sort of shape as $x^2 - y^2 - z^2$, and so on.

16.10 Lagrange Multipliers

CONSTRAINT

MULTIPLIER

In Section 16.9 we learned how to find maxima and minima of functions of several variables (usually two). In this section we will learn a specialized technique for a related problem, which turns out to be more useful as a practical tool.

Let us consider the question of Example 16.8.5 (p. 1016), which asked for the maximum value of $f(x, y) = x^2 y$ on the closed disk $x^2 + y^2 \leq 1$. After we found all the critical points inside the disk, we had to solve the problem of finding the maximum value of $f(x, y) = x^2 y$ on the circle $x^2 + y^2 = 1$.

This involves a function f of two variables, but there is the condition $x^2 + y^2 = 1$. We are not looking at all points (x, y), but just at those that satisfy the condition, that is, that lie on the circle.

The most obvious way to approach this problem is to solve for y as a function of x and then express the function f as a function of one variable x. In doing this it is important to be very careful about the domain on which y can be expressed as a function of x. (In this example we have to work separately on the upper and lower semicircles and have to be careful about endpoints.)

However, if the condition was not the equation of a circle, it might be difficult or impossible to solve for y as a function of x. In this case it might still be possible to solve the problem using implicit differentiation. This allows us to find the derivative without having an explicit formula for the function itself, and this may be enough to solve the problem (cf. the discussion after Example 16.8.5). However, there are many cases in which this approach will not work.

In this section we will learn another technique for solving this type of problem, called Lagrange multipliers, after Joseph Louis Lagrange (1736–1813). Strictly speaking, it is like finding critical points in the sense that it finds some points that include all possible extrema but does not identify which, if any, are maxima or minima.

The technique works for finding extrema of functions of two or more variables, subject to some condition or conditions. The conditions are often called **constraints**. In the example we have been discussing, we might imagine ourselves looking for the maximum value of the function $f(x, y) = x^2 y$ as the point (x, y) moves around, with the understanding that (x, y) is "constrained" to move on the circle $x^2 + y^2 = 1$.

Lagrange was able to show that to find maximum or minimum values of $f(x, y)$ subject to the constraint $g(x, y) = 0$, we should look for all points (x, y) that satisfy the constraint $g(x, y) = 0$ *and* at which

$$\nabla f = \lambda \nabla g$$

for some number λ (λ is the Greek letter "lambda"). This is the same thing as $\frac{\partial f}{\partial x} = \lambda \frac{\partial g}{\partial x}$ and $\frac{\partial f}{\partial y} = \lambda \frac{\partial g}{\partial y}$.

Any extrema of f must be at points at which $g(x, y) = 0$ and $\nabla f = \lambda \nabla g$, for some number λ. The number λ is called the **Lagrange multiplier**.

This seems like a complicated and peculiar procedure, so let us illustrate with an example. We use the question we discussed in Example 16.8.5.

EXAMPLE 1   Find the maximum value of $f(x, y) = x^2 y$ on the circle $x^2 + y^2 = 1$.

Solution   We express the constraint $x^2 + y^2 = 1$ in the form $g(x, y) = 0$ by letting $g(x, y) = x^2 + y^2 - 1$.

Setting $\nabla f = \lambda \nabla g$ and $g = 0$, we find that

$$2xy = 2x\lambda$$
$$x^2 = 2y\lambda$$
$$x^2 + y^2 = 1.$$

Note:
$\nabla f = (2xy, x^2)$
$\nabla g = (2x, 2y)$

If $y = 0$, the second of these equations implies that $x = 0$, but the point $(0, 0)$ does not satisfy the third equation, so we reject it.

This means we can assume $y \neq 0$. If $x = 0$, the second equation implies that $\lambda = 0$, and the third implies that $y = \pm 1$. These points $(x, y, \lambda) = (0, \pm 1, 0)$ satisfy all three equations.

To look for any other points, we can assume $x \neq 0$, so the first equation tells us that $y = \lambda$. Substituting this into the second equation, we get that $x^2 = 2y^2$. Substituting this into the last equation, we find that $2y^2 + y^2 = 1$, so $y^2 = \frac{1}{3}$, $y = \pm \frac{1}{\sqrt{3}}$. The possible points on the circle are $(x, y) = \left(\pm \sqrt{\frac{2}{3}}, \frac{1}{\sqrt{3}}\right), \left(\pm \sqrt{\frac{2}{3}}, -\frac{1}{\sqrt{3}}\right)$.

So the extrema must be among the points $(0, \pm 1)$, $\left(\pm \sqrt{\frac{2}{3}}, \frac{1}{\sqrt{3}}\right), \left(\pm \sqrt{\frac{2}{3}}, -\frac{1}{\sqrt{3}}\right)$. Evaluating f at each of these points, we find that

$$f(0, \pm 1) = 0, \qquad f\left(\pm \sqrt{\frac{2}{3}}, \frac{1}{\sqrt{3}}\right) = \frac{2}{3\sqrt{3}}, \qquad f\left(\pm \sqrt{\frac{2}{3}}, -\frac{1}{\sqrt{3}}\right) = -\frac{2}{3\sqrt{3}},$$

and we see that the maximum value of $f(x, y)$ on the circle $x^2 + y^2 = 1$ is $\frac{2}{3\sqrt{3}}$, which occurs at the points $\left(\pm \sqrt{\frac{2}{3}}, \frac{1}{\sqrt{3}}\right)$, and the minimum value is $-\frac{2}{3\sqrt{3}}$, which occurs at the points $\left(\pm \sqrt{\frac{2}{3}}, -\frac{1}{\sqrt{3}}\right)$. These points are marked on Figure 16.10.1.

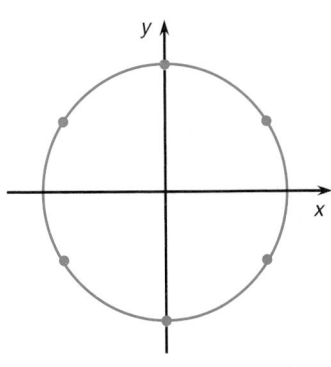

Figure 16.10.1

> Notice the way the multiplier λ was used in the calculation but then disappeared once we found the possible extreme points.

The technique of Lagrange multipliers has two obvious drawbacks. First, as we have already observed, it helps us find some points that must include any extrema, but it gives no information about which are maxima or minima. If we have reasons to expect that there must be a maximum (or a minimum), then it is possible to find it by evaluating the function $f(x, y)$ at all the points and comparing the values. For the most part we will apply this method to problems for which there is a maximum (or minimum), and we will permit ourselves to assume that this is known. Identifying the extrema then becomes a simple matter of evaluating and comparing.

The second difficulty with Lagrange multipliers is that there is no clear-cut way to describe how to find all the points at which the required conditions are satisfied. Roughly speaking, the way to do it is usually to express one variable in terms of the others, to substitute the resulting expression into the equations, to get equations with a smaller number of variables. Then we keep repeating this process until we have an equation with only one variable. With luck it will be possible to solve this equation and then work backward to find all the other variables. Along the way it is quite common to encounter the sort of situation we saw in Example 1, in which it is necessary to consider separately some special cases (some of the variables equaling zero, for instance).

Very much the same approach can be used to find extrema of functions of three variables.

EXAMPLE 2 Find the point on the sphere $x^2 + y^2 + z^2 = 4$ that is closest to the point $(2, -1, -2)$.

Solution We want to find the extrema of the distance between the points (x, y, z) and $(2, -1, -2)$. As we have seen several times before, it is easier to deal with the square of the distance function, whose maxima and minima will occur at the same points as those of the distance itself. Accordingly, we define $f(x, y, z) = (x - 2)^2 + (y + 1)^2 + (z + 2)^2$ and try to find its extrema. The constraint is the condition that the point (x, y, z) should lie on the sphere $x^2 + y^2 + z^2 = 4$, which we express by letting $g(x, y, z) = 0$, where $g(x, y, z) = x^2 + y^2 + z^2 - 4$.

Just as we did with functions of two variables, we ask that $\nabla f = \lambda \nabla g$ and that $g = 0$, that is,

$$2(x - 2) = 2x\lambda$$
$$2(y + 1) = 2y\lambda$$
$$2(z + 2) = 2z\lambda$$
$$x^2 + y^2 + z^2 = 4.$$

If $x = 0$, the first equation says $-4 = 0$, which is impossible, so we can assume $x \neq 0$. Similar arguments show that $y \neq 0$ and $z \neq 0$.

Knowing this, we can solve for λ from each of the first three equations. We find that

$$\lambda = \frac{x - 2}{x} = \frac{y + 1}{y} = \frac{z + 2}{z}.$$

Cross-multiplying in the last equality, we find that $z(y + 1) = y(z + 2)$, or $yz + z = yz + 2y$, so $z = 2y$. From the second equality we find that $y(x - 2) = x(y + 1)$, or $xy - 2y = xy + x$, so $-2y = x$. These facts allow us to express x and z in terms of y; we find that $x = -2y$ and $z = 2y$.

Substituting these formulas into the constraint, we find that $4 = x^2 + y^2 + z^2 = 4y^2 + y^2 + 4y^2 = 9y^2$. From this we get that $y^2 = \frac{4}{9}$, so $y = \pm\frac{2}{3}$. We know how to find x and z from y, and we find two possible points: $(x, y, z) = \left(-\frac{4}{3}, \frac{2}{3}, \frac{4}{3}\right)$ and $(x, y, z) = \left(\frac{4}{3}, -\frac{2}{3}, -\frac{4}{3}\right)$.

It is relatively easy to evaluate f at each of them. We find that $f\left(-\frac{4}{3}, \frac{2}{3}, \frac{4}{3}\right) = \left(-\frac{4}{3} - 2\right)^2 + \left(\frac{2}{3} + 1\right)^2 + \left(\frac{4}{3} + 2\right)^2 = \left(-\frac{10}{3}\right)^2 + \left(\frac{5}{3}\right)^2 + \left(\frac{10}{3}\right)^2 = \frac{100+25+100}{9} = \frac{225}{9}$. Similarly, $f\left(\frac{4}{3}, -\frac{2}{3}, -\frac{4}{3}\right) = \left(\frac{4}{3} - 2\right)^2 + \left(-\frac{2}{3} + 1\right)^2 + \left(-\frac{4}{3} + 2\right)^2 = \left(-\frac{2}{3}\right)^2 + \left(\frac{1}{3}\right)^2 + \left(\frac{2}{3}\right)^2 = \frac{4+1+4}{9} = 1$. The first of these numbers is clearly larger.

We see that the point $\left(-\frac{4}{3}, \frac{2}{3}, \frac{4}{3}\right)$ is the point on the sphere $x^2 + y^2 + z^2 = 4$ that is farthest from the point $(2, -1, -2)$, and the point $\left(\frac{4}{3}, -\frac{2}{3}, -\frac{4}{3}\right)$ is the point on the sphere that is closest to $(2, -1, -2)$. ◢

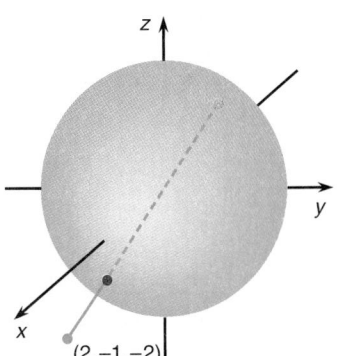

Figure 16.10.2

Notice that the two extreme points we found are on the line through the origin that passes through $(2, -1, -2)$. Of course, this is consistent with our geometrical intuition, which suggests that the point on a sphere that is closest to a point P will be the point on the half-line from the center through P, and the farthest point from P will be the point on the opposite side (on the same line). See Figure 16.10.2, and compare with Example 15.6.3 (p. 915).

Also notice that the procedure was not changed much by the fact that the original function had three variables. The only real difference is that the function f has one more partial derivative.

The technique of Lagrange multipliers depends upon the following theorem, whose proof we omit because it is quite complicated. At the end of the section we will discuss some of the ideas behind the result, but first we concentrate on learning to use it.

THEOREM 16.10.1

> **Lagrange Multipliers**
>
> Suppose f and g are functions of two or more variables with continuous first-order partial derivatives, and consider extrema of f on the set of points at which g equals zero. Then any such extrema must occur at points where $\nabla g = 0$ or where $\nabla f = \lambda \nabla g$, for some number λ.

In the cases we will be considering, ∇g will not be zero, so it is the second possibility that is important. The idea is to look for all points at which $\nabla f = \lambda \nabla g$ and $g = 0$ and then try to decide which of them are extrema.

Of course, it is possible to apply the method of Lagrange multipliers to practical maximum/minimum problems.

EXAMPLE 3   Find the dimensions of the least expensive rectangular box whose volume is 240 cu ft, if the material for the top costs $\$2/\text{ft}^2$, the bottom costs $\$3/\text{ft}^2$, the front and back cost $\$1.50/\text{ft}^2$, and the sides cost $\$1/\text{ft}^2$.

Solution   As usual we let the dimensions be x, y, and z as shown in Figure 16.10.3 (each dimension measured in *feet*). We know that the volume is 240 ft^3, that is, $xyz = 240$. The area of the top is xy, so the cost of the top is $2xy$. The cost of the bottom is $3xy$.

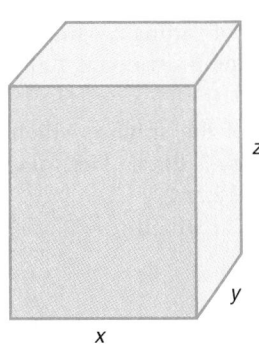

Figure 16.10.3

The combined area of the front and back is $2xz$, so their cost is $1.5 \times 2xz = 3xz$, and the total area of the two sides is $2yz$, so their cost is $1 \times 2yz = 2yz$.

The total cost is $C = 2xy + 3xy + 3xz + 2yz = 5xy + 3xz + 2yz$. What we want is to minimize this function, subject to the constraint $xyz = 240$. Letting $g(x, y, z) = xyz - 240$, we should look for points at which $\nabla C = \lambda \nabla g$, for some λ. In other words,

$$5y + 3z = \lambda yz$$
$$5x + 2z = \lambda xz$$
$$3x + 2y = \lambda xy$$
$$xyz = 240.$$

It is not immediately obvious how to solve these equations, and in fact there are several different ways to do it. One good way is to observe that the right-hand sides of the first three equations are quite similar; each is the product of λ and two of x, y, and z. If we multiply each equation by the remaining variable, we can make the right side of each equation equal $\lambda xyz = 240\lambda$. We find

$$5xy + 3xz = \lambda xyz = 240\lambda,$$
$$5xy + 2yz = \lambda xyz = 240\lambda,$$
$$3xz + 2yz = \lambda xyz = 240\lambda.$$

Subtracting the first two of these equations, we find $3xz - 2yz = 0$, so $(3x - 2y)z = 0$. Since $xyz = 240$, we know that $z \neq 0$, so this can only happen if $3x - 2y = 0$, that is, $3x = 2y$.

Similarly, by subtracting the first and third of these equations we find $0 = 5xy - 2yz = (5x - 2z)y$, and since $y \neq 0$, this means $5x = 2z$.

In other words, $y = \frac{3}{2}x$ and $z = \frac{5}{2}x$; substituting these formulas into the constraint, we find that $240 = xyz = x\left(\frac{3}{2}x\right)\left(\frac{5}{2}x\right) = \frac{15}{4}x^3$. This means $x^3 = \frac{4}{15}(240) = 64$, so $x = 4$. Since $y = \frac{3}{2}x$ and $z = \frac{5}{2}x$, the point we have found is $(4, 6, 10)$. This is the only possible point at which there can be an extremum, so it must be a minimum.

The least expensive box will be 10 ft high, 4 ft wide, and 6 ft deep. ◢

EXAMPLE 4   A tank is to be built in the shape of a cylinder with hemispherical caps on the ends (see Figure 16.10.4). What are the dimensions of the tank with the smallest possible surface area if the volume must be 1000 cu ft?

Figure 16.10.4

Solution   We let the radius of the cylinder be r and the height of the cylinder be h. The volume of the cylindrical part of the tank is then $\pi r^2 h$, and the total volume of the hemispherical caps is the same as the volume of a sphere of radius r, which is $\frac{4}{3}\pi r^3$. The total volume of the tank is $\pi r^2 h + \frac{4}{3}\pi r^3$. This volume must equal 1000, that is, $\pi r^2 h + \frac{4}{3}\pi r^3 = 1000$.

The combined surface area of the two caps is the area of the sphere, which is $4\pi r^2$. The surface area of the cylinder (i.e., without its ends) is $2\pi rh$. The total surface area, which is what we want to minimize, is $A = 4\pi r^2 + 2\pi rh$.

We let $g(r, h) = \pi r^2 h + \frac{4}{3}\pi r^3 - 1000$ and look for points at which $g = 0$ and $\nabla A = \lambda \nabla g$, for some number λ. In other words,

$$8\pi r + 2\pi h = \lambda(2\pi rh + 4\pi r^2)$$
$$2\pi r = \lambda \pi r^2$$
$$\pi r^2 h + \frac{4}{3}\pi r^3 = 1000.$$

Figure 16.10.5

From the last equation we see that $r \neq 0$, and then dividing the second equation by πr^2, we find that $\lambda = \frac{2}{r}$. Substituting this into the first equation, we find that $8\pi r + 2\pi h = \frac{2}{r}(2\pi rh + 4\pi r^2) = 4\pi h + 8\pi r$. Subtracting $8\pi r$ from both sides, we find that $2\pi h = 4\pi h$, or $0 = 2\pi h$. This means that $h = 0$. From the constraint we find that $1000 = 0 + \frac{4}{3}\pi r^3$, so $r^3 = \frac{3}{4\pi} \times 1000$, and $r = \left(\frac{750}{\pi}\right)^{1/3}$.

We find that the dimensions that give the minimum surface area with the specified volume are $h = 0$ and $r = \left(\frac{750}{\pi}\right)^{1/3} \approx 6.2035$. The tank with the minimum area is a sphere with this radius. See Figure 16.10.5.

We could also consider solving Example 4 by using single-variable techniques. We could use the equation $\pi r^2 h + \frac{4}{3}\pi r^3 = 1000$ to solve: $h = \left(1000 - \frac{4}{3}\pi r^3\right)/(\pi r^2)$, and then substitute to find $A = 4\pi r^2 + \frac{2000}{r} - \frac{8}{3}\pi r^2 = \frac{4}{3}\pi r^2 + \frac{2000}{r}$.

With a graphics calculator we can graph this function and see where its minimum occurs.

TI-85	SHARP EL9300
GRAPH F1 $(y(x) =)$ 4 π F1(x) $x^2 \div$ 3 + 2000 $\div$ F1(x) M2(RANGE) 0 ▼ 10 ▼ ▼ 300 ▼ 1200 F5(GRAPH)	⏎ 4 π x/θ/T $x^2 \div$ 3 + 2000 $\div$ x/θ/T RANGE 0 ENTER 10 ENTER 1 ENTER 300 ENTER 1200 ENTER 100 ENTER ⏎

CASIO f_x-6300G (f_x-7700GB)	HP 48SX
Range 0 EXE 10 EXE 1 EXE 300 EXE 1200 EXE 100 EXE EXE EXE EXE Graph 4 π X $x^2 \div$ 3 + 2000 $\div$ X EXE	↰ PLOT PLOTR ERASE ATTN ′ Y = 4 × π × X y^x 2 ÷ 3 + 2000 ÷ X ↱ PLOT ↰ DRAW ERASE 0 ENTER 10 XRNG 300 ENTER 1200 YRNG DRAW

If we wanted to calculate the minimum, we would differentiate and set $A' = 0$, and we would get the same answer.

Summary 16.10.2

Lagrange Multipliers

Suppose we want to find the extreme values of a function f, subject to the constraint $g = 0$. Assume that f and g have continuous partial derivatives.

1. Look for all points at which $g = 0$ and $\nabla f = \lambda \nabla g$, for some number λ. This amounts to saying that each partial derivative of f must equal λ times the corresponding partial derivative of g. When you write down these conditions, remember to write down the condition $g = 0$; otherwise, there is a risk of forgetting to include it.

2. Find all the points described above. Use whatever method you can find, but one useful thing to try is solving for one of the variables (or λ) in terms of the others.

Be careful about special cases; in particular if you have to divide by anything, consider separately the possibility that it might be zero.

3. For each of the points you found, discard the value of λ and evaluate f at the point. By comparing the values of f at the various points, identify the maximum and minimum, if any. If the question asks for a maximum, take this to mean that we can assume that there is a maximum (and likewise for minima). Finding the maximum (or minimum) then becomes a simple matter of evaluating and comparing.

Lagrange Multipliers with Two Constraints

It is also possible to use the technique of Lagrange multipliers in situations in which there are two (or more) constraints. Suppose we want to find the extreme values of the function f subject to the constraints $g = 0$ and $h = 0$. Here the thing to do is to find all possible places where

$$\nabla f = \lambda \nabla g + \mu \nabla h,$$

where λ and μ are numbers (μ is the Greek letter "mu," pronounced "mew"). Apart from this the procedure is the same as what we did with only one constraint.

EXAMPLE 5   Find the points on the intersection of the sphere $(x - 2)^2 + y^2 + z^2 = 4$ and the plane $x + y + z = 0$ that are closest to and farthest from the origin.

Solution   We are asked to find the extrema of the distance from the origin, subject to the constraints $(x - 2)^2 + y^2 + z^2 = 4$ *and* $x + y + z = 0$. As usual it is better to consider the square of the distance, $f(x, y, z) = x^2 + y^2 + z^2$. Letting $g(x, y, z) = (x - 2)^2 + y^2 + z^2 - 4$ and $h(x, y, z) = x + y + z$, we look for points at which $\nabla f = \lambda \nabla g + \mu \nabla h$ and $g = h = 0$, that is,

$$2x = 2(x - 2)\lambda + \mu$$
$$2y = 2y\lambda + \mu$$
$$2z = 2z\lambda + \mu$$
$$(x - 2)^2 + y^2 + z^2 = 4$$
$$x + y + z = 0.$$

Subtracting the second and third of these equations, we find that $2(y - z) = 2(y - z)\lambda$, that is, $y - z - (y - z)\lambda = 0$ or $(y - z)(1 - \lambda) = 0$. This means either $\lambda = 1$ or $y = z$.

If $\lambda = 1$, then the second equation becomes $2y = 2y + \mu$, from which we see that $\mu = 0$. Using this, we can write the first equation as $2x = 2(x - 2) + 0 = 2x - 4$. Subtracting $2x$ from each side, we find the absurd result that $0 = -4$. This shows that the assumption that $\lambda = 1$ is not possible.

So $y = z$, and the last constraint becomes $x + 2y = 0$, or $x = -2y$. Using this, the first constraint becomes $(-2y - 2)^2 + y^2 + y^2 = 4$, or $4y^2 + 8y + 4 + 2y^2 = 4$, or $6y^2 + 8y = 0$, that is, $(6y + 8)y = 0$. There are two possibilities: Either $y = 0$ (and then $z = y = 0$ and $x = -2y = 0$) or $6y + 8 = 0$ and $y = -\frac{4}{3}$. In this case $z = -\frac{4}{3}$ and $x = \frac{8}{3}$.

Figure 16.10.6

Figure 16.10.7(i)

Figure 16.10.7(ii)

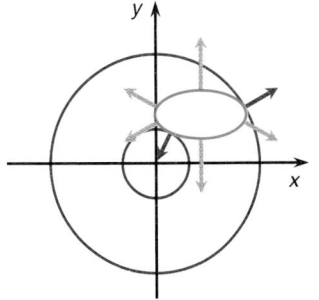

Figure 16.10.8

The points we have found are $(0, 0, 0)$ and $\left(\frac{8}{3}, -\frac{4}{3}, -\frac{4}{3}\right)$. The first is as close to the origin as it is possible to be, and the second must be the point that is farthest from the origin, subject to the constraints.

The situation is illustrated in Figure 16.10.6.

The Idea Behind Lagrange Multipliers

Now that we have learned how to use the technique of Lagrange multipliers to solve various types of problems, we discuss briefly what makes it work.

In the case of functions of two variables the set of points that satisfy a constraint $g(x, y) = 0$ is usually some sort of curve. The problem is to find the point or points on the curve at which $f(x, y)$ has the largest or smallest possible value.

Now suppose we consider the level curves of f, shown in Figure 16.10.7(i). We want to see which of the level curves contain points on the curve $g(x, y) = 0$ and identify the ones on which f has the largest and smallest values. Generally speaking, as a point moves along the curve $g(x, y) = 0$, it will move from one level curve of f to another, that is, it will pass through points at which f has different values.

Suppose we consider a point on the curve $g(x, y) = 0$, and suppose that the gradients of f and g are *not* parallel. Now the gradients are perpendicular to the level curves, or strictly speaking to their tangent vectors. This means that the set of points satisfying $g(x, y) = 0$ (which is the zero level curve of g) will cross the level curves of f at the same angle as the (nonzero) angle between the gradients. In particular the level curves of f are *not* tangent to the curve where $g(x, y) = 0$. See Figure 16.10.7(ii).

So if a point is moving along the curve where $g = 0$, it will cut across the level curves of f. This means that the value of f will be changing. If it is increasing as the point moves in one direction along the curve, then it will decrease if the point moves in the opposite direction. If the value of f can be made to increase by moving one way along the curve and to decrease by moving the other way, then the point in question is certainly not an extreme point.

What this argument shows is that any point at which ∇f and ∇g are not parallel cannot be an extreme point for f on the set of points satisfying $g(x, y) = 0$.

This explains why we find the points at which $\nabla f = \lambda \nabla g$ (i.e., the points at which ∇f and ∇g are parallel). If there are extreme values, they must occur at points with this property.

In Figure 16.10.8 we see illustrated the ellipse $(x - 1)^2 + 4(y - 1)^2 = 1$, and marked on it at several points are the gradients of the corresponding function $g(x, y) = (x - 1)^2 + 4(y - 1)^2 - 1$. Suppose we want to find the point on the ellipse that is farthest from the origin. In the picture it is the point at which the ellipse intersects the large circle.

This problem amounts to finding the maximum value of $R(x, y) = x^2 + y^2$ on the ellipse. We found earlier that the gradients of R always point directly away from the origin, so the technique of Lagrange multipliers tells us to find points on the ellipse at which ∇g points directly away from the origin or directly toward the origin.

From the picture we see that there are two such points (the closest and the farthest point from the origin). At any other point, the ellipse cuts across the level curves of R (which are just circles around the origin). At that maximum point, the ellipse is tangent to the level curve of R (the big circle), and as a point moves along the ellipse there, it does not cross the big circle to reach a higher value of R (i.e., a bigger circle). Similarly, at the point closest to the origin the ellipse is tangent to the corresponding circle, and a point moving along the ellipse does not cross to lower values of R.

In a way this discussion also shows why the points we find by using Lagrange multipliers may not always be points where f has extreme values. It is possible that the curve determined by the condition $g(x, y) = 0$ may be tangent to the level curve of f but also cross it, and at such a point f would not have an extreme value.

For functions of three variables a similar argument can be made, but the difference is that we must consider level surfaces rather than level curves. Still, the gradients are perpendicular to the corresponding level surfaces, so if the gradients are not parallel, then the surface determined by $g(x, y, z) = 0$ (the red surface in Figure 16.10.9) cannot be tangent to the level surface of f (i.e., cannot have the same tangent plane).

This in turn means that it is possible to move along the surface determined by $g(x, y, z) = 0$ in such a way as to move to "higher" or "lower" level surfaces of f (i.e., ones on which the values of f are higher or lower). There cannot be an extreme value of f at a point where this is possible. If there is an extreme value, it must occur at a point where the shaded surface just touches the level surface of f and does not cross it.

Once again we see that any extreme values must occur at points at which the gradients ∇f and ∇g are parallel.

These are the ideas behind Theorem 16.10.1. It is not necessary to understand them in order to apply the technique of Lagrange multipliers, but they may help to make it less mysterious.

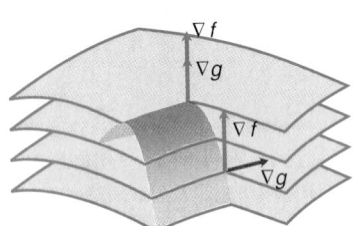

Figure 16.10.9

Exercises 16.10

Solve the following maximum/minimum problems. It is possible to do some of them without using Lagrange multipliers, but you might like to try this approach anyway for the sake of practicing the technique.

1. Find the maximum and minimum values of $f(x, y) = 3x - 4y$ on the circle $x^2 + y^2 = 5$.

2. Find the maximum and minimum values of $f(x, y) = 2x + 3y$ on the circle $(x - 2)^2 + y^2 = 5$.

3. Find the maximum and minimum values of $f(x, y) = 2x + 5y$ on the ellipse $4x^2 + 9y^2 = 36$.

4. Find the maximum and minimum values of $f(x, y) = x + 3y$ on the ellipse $2x^2 + 3y^2 + 6y = 4$.

5. Find the minimum value of the function $f(x, y) = x^2 - y^2$ on the line $x + 2y = 3$.

6. Find the maximum value of the function $f(x, y) = 2x^2 - 3y^2$ on the line $5x - 2y = 3$.

7. Find the maximum value of the function $f(x, y) = 2x - y$ on the parabola $y = x^2$.

8. Find the minimum value of the function $f(x, y) = 2x + 3y$ on the parabola $y = x^2 + 4x + 2$.

9. Find the maximum and minimum values of $f(x, y) = y - \cos x$ on the graph $y = \sin x$.

10. Find the point on the hyperbola $x^2 - (y - 1)^2 = 1$ that is closest to the origin.

11. Find the point on the sphere $x^2 + y^2 + z^2 = 4$ that is closest to the point $(4, 3, -5)$.

12. Find the point on the sphere $x^2 + y^2 + z^2 = 9$ that is farthest from the point $(1, -2, -1)$.

13. Find the point on the sphere $(x - 1)^2 + y^2 + (z + 3)^2 = 2$ that is closest to the origin.

14. Find the point on the sphere $x^2 - 4x + y^2 - 6y + z^2 = -9$ that is closest to the point $(1, 1, -1)$.

15. Find the minimum distance from the origin to the surface $xy - 2z + 2 = 0$.

16. Find the minimum distance from the point $(2, 2, -1)$ to the surface $x^2 + y^2 + z + 4 = 0$.

17. Find the point on the intersection of the planes $x + y + 2z = 3$ and $2x - y + z = 2$ that is closest to the origin.

18. Find the point on the intersection of the planes $z = 2$ and $x - y + z = 3$ that is closest to the point $(2, 1, -1)$.

19. Find the point on the intersection of the sphere $x^2 + y^2 + z^2 = 4$ and the plane $y = -1$ that is closest to the point $(3, 0, 0)$.

20. Find the maximum and minimum distance from the origin to the circle that is the intersection of the sphere $x^2 + y^2 + z^2 = 9$ and the plane $x + y - z = 1$.

II

In each of the following cases, find the maximum and minimum values of the specified function on the given set.

21. $f(x, y) = xy$ on the circle $x^2 + y^2 = 1$

*22. $f(x, y) = xy$ on the circle $x^2 + y^2 = r^2$

*23. $f(x, y) = ax + by$ on $x^2 + y^2 = r^2$ (assume $a \neq 0$, $b \neq 0$)

*24. $f(x, y) = ax + by$ on $(x - c)^2 + (y - d)^2 = r^2$ (assume $a \neq 0$, $b \neq 0$)

*25. $f(x, y) = ax + by$ on the ellipse $cx^2 + dy^2 = 1$ (assume $a \neq 0$, $b \neq 0$)

*26. Argue that the point we found in Example 3 (p. 1033) cannot be a maximum, by considering what happens if we fix $y = 1$ and let $x \to \infty$, subject to the constraint $xyz = 240$.

*27. (i) Explain why the function $f(x, y) = x + 2y$ has neither a maximum nor a minimum on the hyperbola $x^2 - y^2 + 1 = 0$. (Draw a picture.)

　　(ii) What is the significance of the points that are found by using the Lagrange multipliers approach?

*28. Consider the intersection of the plane $x + y + z = 0$ and the sphere $x^2 + (y + 2)^2 + z^2 = 3$. Find the points on this intersection that are closest to the origin and farthest from the origin.

*29. Find the extreme values of $f(x, y, z) = x + 3y$ on the intersection of the ellipsoid $x^2 + 2y^2 + 3z^2 = 6$ and the plane $x - 2y + z = 4$.

**30. Find the points on the sphere $(x - a)^2 + (y - b)^2 + (z - c)^2 = R^2$ that are closest to and farthest from the point (u, v, w).

*31. A storage shed is being built in the shape of a rectangular box. Material for the walls costs \$2 per sq ft, material for the floor costs \$3 per sq ft, and the roof costs \$5 per sq ft. What are the dimensions for a shed of the largest volume that can be constructed for \$384 worth of materials?

*32. A container is made in the shape of a cylinder, open at one end, with a hemispherical cap on the other end. The material for the hemispherical part costs twice as much per square foot as the material for the cylindrical part. What are the dimensions of the cheapest container whose volume is 9π cu ft?

*33. What is the smallest possible value of the sum of the squares of three numbers whose product is 1?

*34. (i) Guess which points (x, y, z), with $x > 0$, $y > 0$, $z > 0$, and $x^2 + y^2 + z^2 = 1$, give a minimum value of $\frac{1}{x} + \frac{1}{y} + \frac{1}{z}$.

　　(ii) Find the minimum value. Is it consistent with your guess?

*35. What are the largest and smallest possible values of the sum of the squares of three nonnegative numbers whose sum is 15? (Why do we need the word "nonnegative" in this question?)

*36. (i) What is the smallest possible value of the sum of the squares of four positive numbers whose sum is 12?

　　(ii) What is the smallest possible value of the sum of the squares of n positive numbers whose sum is K?

*37. A rectangular box is placed inside the ellipsoid $\frac{x^2}{4} + \frac{y^2}{9} + \frac{z^2}{16} = 1$ with sides parallel to the axes. What dimensions will give the box with the maximum possible volume?

*38. A rectangular box is placed inside the ellipsoid $\frac{x^2}{a^2} + \frac{y^2}{b^2} + \frac{z^2}{c^2} = 1$ with sides parallel to the axes. What dimensions will give the box with the maximum possible volume?

*39. Find the point on the sphere $w^2 + x^2 + y^2 + z^2 = 1$ in $\mathbb{R}^4$ that is closest to $(2, -1, 0, -2)$.

*40. Find the point on the sphere $x_1^2 + x_2^2 + \ldots + x_n^2 = 1$ in $\mathbb{R}^n$ that is closest to $(a_1, a_2, \ldots, a_n)$.

*41. A manufacturer must allocate resources between labor and raw material. Experience suggests that with ℓ hours of labor and m units of material the profit will be $P = \ell^{1/2} m^{2/3}$. If labor costs \$15 per hour and each unit of material costs $10, find how to get the maximum profit with a total investment of $35,000 for labor and material. This question is relatively uncomplicated, but it is typical of the way Lagrange multipliers are used in economics.

*42. In the setting of Exercise 41, suppose the profit is $P = \ell^a m^b$, for some constants a, b. How will they get the maximum profit now?

**43. Light sources are placed at the points $(1, 0)$ and $(-1, 0)$; the one at $(-1, 0)$ is twice as bright as the one at $(1, 0)$. Find the points on the circle $x^2 + y^2 = 4$ where the total illumination is greatest and smallest. (The illumination at a distance R from the source has intensity $\frac{K}{R^2}$, where K is some constant.)

**44. In the setting of Exercise 43, make a sketch of the situation, showing the circle $x^2 + y^2 = 4$ and also the circle $x^2 + y^2 = 16$. Plot the points of greatest and smallest total illumination on the smaller circle, and try to *guess* the locations of the points of greatest and smallest illumination on the larger circle. Find the actual points and compare with your guess.

45. In Example 1 (p. 1031) we discussed the function $f(x, y) = x^2 y$. Use a graphics calculator or computer to sketch a number of its level curves, including the ones corresponding to the maximum and minimum values of f on the unit circle.

46. In Example 4 (p. 1034) we discussed the function $A = A(r, h) = 4\pi r^2 + 2\pi rh$. Use a graphics calculator or computer to sketch a number of its level curves, including the one corresponding to the minimum value of A subject to the volume constraint.

16.11 Linear Regression and Least Squares Approximation

BEST FIT
REGRESSION

Figure 16.11.1(i)

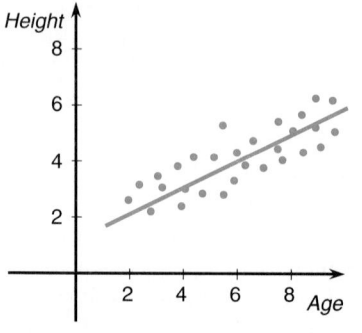

Figure 16.11.1(ii)

In this section we apply partial derivatives and the ideas of Section 16.8 to develop a technique for finding a straight line that best approximates the positions of a number of points.

A scientist often finds data illustrating the relationship between two quantities and then hopes to find a formula that describes this relationship, at least approximately.

For instance, a forester might measure the heights of a number of trees in an attempt to predict the height based on the age of the tree. The results could be plotted as a number of points on a graph, where the coordinates of each point are the age and height of one particular tree. See Figure 16.11.1(i), where age is measured in years and height in feet.

A plot of data like this is a bit discouraging because there is no simple curve that passes through all the points. On the other hand, the picture does seem to suggest that there is a relationship between age and height. Instead of looking for an exact relationship, which is hopelessly unrealistic, we consider instead the possibility of finding an *approximate* relationship. The simplest idea is to look for a straight line that approximately goes through the points.

Such a line is shown in Figure 16.11.1(ii). It seems reasonable to say that the relationship between the ages and heights of these trees is given *approximately* by that line.

In this case the line was drawn by guesswork, but different people would probably guess slightly different lines, and obviously it is desirable to have a standard way of finding the line. So the problem is to start with a number of points in the plane and find the straight line that comes closest to passing through all the points. We speak of the line that gives the **best fit** to the data.

The first difficulty is to decide what we mean by the *best* fit. In Figure 16.11.2(i) we see a plot of four points (x_1, y_1), (x_2, y_2), (x_3, y_3), (x_4, y_4). One natural way to decide how well a line fits these points is to measure the vertical distance from each

Figure 16.11.2(i)

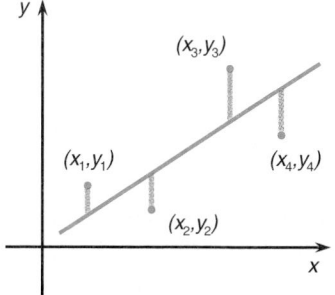

Figure 16.11.2(ii)

point to the line, as shown in Figure 16.11.2(ii). If all these distances are small, then the line is a good fit.

If the line is $y = mx + b$, then the point on the line directly above or below (x_1, y_1) is $(x_1, mx_1 + b)$. The vertical distance from (x_1, y_1) to the point is $|y_1 - (mx_1 + b)|$. Similarly, the vertical distance from the point (x_i, y_i) to the line is $|y_i - (mx_i + b)|$.

The first thing to try would be to add up all these vertical distances and try to make the sum as small as possible. The line with the best fit would be the one for which

$$\sum_{i=1}^{4} |y_i - (mx_i + b)| = \sum_{i=1}^{4} |y_i - mx_i - b|$$

is as small as possible.

Unfortunately, it turns out that this definition is usually difficult to work with. There is another definition that is easier to handle and also gives results that are more useful. Instead of adding up all the vertical distances from the points to the line, we add up the *squares* of those distances. Because the line with the best fit will be the one that makes the sum of the squares as small as possible, it is called the *least squares* approximation. The process of starting with a number of points and finding a line that approximates them is called **linear regression**.

DEFINITION 16.11.1

Linear Regression and Least Squares Approximation

Suppose (x_i, y_i) are points in the plane, for $i = 1, \ldots, n$. For the line $y = mx + b$ we form the sum of the squares of the vertical distances from the points to the line:

$$\sum_{i=1}^{n} (y_i - mx_i - b)^2.$$

The **least squares approximation** to the given points is the line for which this sum of squares is as small as possible.

From the definition it is not even clear that there will always be such a line, but we will ignore this difficulty and use partial derivatives to find it.

We regard the points (x_i, y_i) as fixed; what we want to find is the line $y = mx + b$. This amounts to finding the two numbers m and b. To find them, we need to minimize the sum of the squares, so we let that sum be a function:

$$f(m, b) = \sum_{i=1}^{n} (y_i - mx_i - b)^2.$$

To find the values of m and b at which f has a minimum, we look for critical points. It is easy to evaluate the partial derivatives of each term in the sum:

$$\frac{\partial}{\partial m} \left((y_i - mx_i - b)^2 \right) = -2x_i(y_i - mx_i - b),$$

$$\frac{\partial}{\partial b} \left((y_i - mx_i - b)^2 \right) = -2(y_i - mx_i - b).$$

Since the function $f(m, b)$ is the sum of these terms, its partial derivatives are the sums of the partial derivatives of the terms:

$$\frac{\partial}{\partial m}\big(f(m, b)\big) = \sum_{i=1}^{n} \frac{\partial}{\partial m}\big((y_i - mx_i - b)^2\big) = -2\sum_{i=1}^{n} x_i(y_i - mx_i - b),$$

$$\frac{\partial}{\partial b}\big(f(m, b)\big) = \sum_{i=1}^{n} \frac{\partial}{\partial b}\big((y_i - mx_i - b)^2\big) = -2\sum_{i=1}^{n} (y_i - mx_i - b).$$

These formulas can be rearranged as follows:

$$\frac{\partial}{\partial m}\big(f(m, b)\big) = -2\bigg(\sum_{i=1}^{n} x_iy_i - m\sum_{i=1}^{n} x_i^2 - b\sum_{i=1}^{n} x_i\bigg),$$

$$\frac{\partial}{\partial b}\big(f(m, b)\big) = -2\bigg(\sum_{i=1}^{n} y_i - m\sum_{i=1}^{n} x_i - b\sum_{i=1}^{n} 1\bigg) = -2\bigg(\sum_{i=1}^{n} y_i - m\sum_{i=1}^{n} x_i - bn\bigg).$$

We want to find all critical points, that is, points where both partial derivatives are zero. It is convenient to introduce the following notation: We write $\bar{x} = \frac{1}{n}\sum_{i=1}^{n} x_i$ for the **average** of the x_i's and $\bar{y} = \frac{1}{n}\sum_{i=1}^{n} y_i$ for the **average** of the y_i's. To say that both partial derivatives of $f(m, b)$ are zero is the same as saying that

$$\sum_{i=1}^{n} x_iy_i - m\sum_{i=1}^{n} x_i^2 - b\sum_{i=1}^{n} x_i = 0 \quad \text{and} \quad \sum_{i=1}^{n} y_i - m\sum_{i=1}^{n} x_i - bn = 0.$$

These equations can be written as

$$m\sum_{i=1}^{n} x_i^2 + bn\bar{x} = \sum_{i=1}^{n} x_iy_i \quad \text{and} \quad mn\bar{x} + bn = n\bar{y}.$$

Note that $\sum_{i=1}^{n} x_i = n\bar{x}$, $\sum_{i=1}^{n} y_i = n\bar{y}$.

Multiplying the second of these equations by $\bar{x}$ and subtracting, we find that

$$m = \frac{\sum_{i=1}^{n} x_iy_i - n\bar{x}\bar{y}}{\sum_{i=1}^{n} x_i^2 - n(\bar{x})^2}.$$

Writing the previous equation as $nm\bar{x} + nb = n\bar{y}$, we can divide by n to find that

$$b = \bar{y} - m\bar{x}.$$

It is possible to write a formula for b that does not involve m, but it is complicated. Since we will always want to find m anyway, this formula is easier to use.

Notice that the equation $\bar{y} = m\bar{x} + b$ simply says that the *average point* $(\bar{x}, \bar{y})$ lies on the line. The other observation we should make is that in doing the calculation we divided by $\sum_{i=1}^{n} x_i^2 - n(\bar{x})^2$. In order for this to be legitimate we must assume that this quantity is not zero. Fortunately, it can be shown that it is never zero unless all the x_i's are equal (in which case all the original points lie on a vertical line).

We have the following result.

THEOREM 16.11.2

Suppose the points (x_i, y_i), for $i = 1, \ldots, n$, do not all lie on a vertical line. Then the least squares approximation to these points is the line $y = mx + b$, where

$$m = \frac{\left(\sum_{i=1}^{n} x_i y_i\right) - n\bar{x}\bar{y}}{\left(\sum_{i=1}^{n} x_i^2\right) - n(\bar{x})^2}, \qquad b = \bar{y} - m\bar{x}, \qquad \bar{x} = \frac{1}{n}\sum_{i=1}^{n} x_i, \qquad \bar{y} = \frac{1}{n}\sum_{i=1}^{n} y_i.$$

We have already found that these values of m and b give a critical point for $f(m, b)$, the sum of the squares of the distances from the line $y = mx + b$. To show that it is really a minimum, we can apply the Second Derivative Test (Theorem 16.9.1). However, since the details are slightly complicated, we leave them until the end of the section.

EXAMPLE 1   A forester does a pilot study of tree heights. She measures 6 trees as follows: One is 1 year old and 18 in tall, the next is $1\frac{1}{2}$ years old and 24 in tall, and the others have ages 2, 2, $2\frac{1}{2}$, and 3 years and heights 33, 28, 37, and 40 in, respectively. Find the line that gives the best least squares approximation to these values.

Solution   We are looking for a line that approximates the six points $(1, 18)$, $\left(1\frac{1}{2}, 24\right)$, $(2, 33)$, $(2, 28)$, $\left(2\frac{1}{2}, 37\right)$, $(3, 40)$. We begin by finding the average values $\bar{x} = \frac{1}{6}\left(1 + 1\frac{1}{2} + 2 + 2 + 2\frac{1}{2} + 3\right) = 2$ and $\bar{y} = \frac{1}{6}(18 + 24 + 33 + 28 + 37 + 40) = 30$. Then we need to find $\sum_{i=1}^{6} x_i y_i = 1(18) + \frac{3}{2}(24) + 2(33) + 2(28) + \frac{5}{2}(37) + 3(40) = 388\frac{1}{2}$ and $\sum_{i=1}^{6} x_i^2 = 1^2 + \left(\frac{3}{2}\right)^2 + 2^2 + 2^2 + \left(\frac{5}{2}\right)^2 + 3^2 = 26\frac{1}{2}$. With these we can apply the formula from Theorem 16.11.2 (with $n = 6$) to find

$$m = \frac{\sum_{i=1}^{n} x_i y_i - n\bar{x}\bar{y}}{\sum_{i=1}^{n} x_i^2 - n(\bar{x})^2} = \frac{388\frac{1}{2} - 6(2)30}{26\frac{1}{2} - 6(2^2)} = 11.4,$$

$$b = \bar{y} - m\bar{x} = 30 - 11.4(2) = 7.2.$$

The line we want is $y = mx + b = 11.4x + 7.2$. The original points and the regression line are sketched in Figure 16.11.3.

Since this was only a pilot study, not enough measurements were made to give us any great confidence in the results, but the fact that the slope of the line is $m = 11.4$ suggests that, *roughly*, the trees grow at an average rate of 11.4 in per year.

It is a little less easy to interpret the y-intercept $b = 7.2$, but it suggests that young trees grow a little faster than this average amount. If they grew exactly 11.4 in/year, then the line would pass through the origin; as it is, the line is higher at the left end.

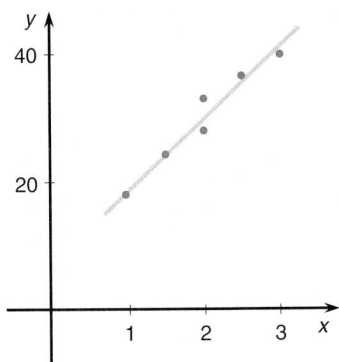

Figure 16.11.3

EXAMPLE 2   An economist wanting to estimate the demand function for a new electronic gadget goes to several small towns of similar size and demographic mix and tries marketing it at different prices. He finds the following data; the first number in each pair is the number sold at a particular price, and the second number is that price (in dollars): $(55, 30)$, $(52, 32)$, $(53, 34)$, $(48, 36)$, $(50, 38)$, $(42, 40)$, $(40, 42)$, $(41, 44)$.

Recall (cf. Section 4.13) that the demand function is $p = K - Lx$, where x is the number sold at price p and K and L are constants. We want to approximate the data with a line.

We will need to find the sum of the x_i's, the sum of the p_i's, the sum of $x_i p_i$, and the sum of x_i^2. It is convenient to organize the data in a table, with one horizontal line for each data point. We put the original values of x and p in the first two columns, xp in the third, and x^2 in the fourth. Then it is easy to add up each column.

	x	p	xp	x^2
	55	30	1,650	3,025
	52	32	1,664	2,704
	53	34	1,802	2,809
	48	36	1,728	2,304
	50	38	1,900	2,500
	42	40	1,680	1,764
	40	42	1,680	1,600
	41	44	1,804	1,681
Total:	381	296	13,908	18,387

We begin by finding the averages of the eight values: $\bar{x} = \frac{1}{8}(381) = 47.625$, $\bar{p} = \frac{1}{8}(296) = 37$.

Then we want to use linear regression to find the best line $p = K - Lx$. We use the formulas from Theorem 16.11.2 (replacing y with p) and find

$$m = -L = \frac{\sum_{i=1}^{n} x_i p_i - n\bar{x}\bar{p}}{\sum_{i=1}^{n} x_i^2 - n(\bar{x})^2} = \frac{13,908 - 8(47.625)37}{18,387 - 8(47.625)^2} \approx -0.78,$$

$$b = K = \bar{p} + L\bar{x} = 37 + (0.781)(47.625) \approx 74.$$

The approximate demand function obtained by applying linear regression to the sales data is $p = 74 - 0.78x$. It suggests that in a town of the same size as the test towns, the demand function will be roughly this. On the other hand, the original points and the least squares line are plotted in Figure 16.11.4, and we see that the fit of the line to the data is not wonderful.

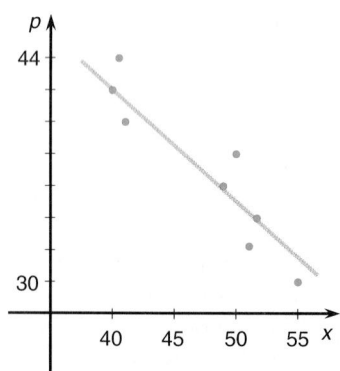

Figure 16.11.4

The regression technique gives us a line, but as we see in Example 2, the line may not always fit the data very well. In a case like this the information we obtain should be used cautiously. It can certainly give us qualitative information, but as a quantitative tool it should be viewed with suspicion.

There always is a least squares line, but unless there is some reason to think that the points do come close to lying on a line, the line may actually be misleading.

Sometimes regression can be used in situations that do not appear to refer to straight lines.

EXAMPLE 3   A chemist has performed a difficult experiment and obtained the following three data points: $(1, 3)$, $(3, 16)$, $(4, 25)$. Theoretical considerations suggest that these points should satisfy an equation of the form $y = cx^k$, for some constants c and k.

We would like to find what the values of c and k are, but this formula does not describe a straight line. However, if we take logarithms, it becomes $\ln y = \ln c + k \ln x$. This says that $(u, v) = (\ln x, \ln y)$ lies on a straight line $v = ku + \ln c$. We have three points and can find the least squares approximation to this line.

We have to take the logarithms of the values of x and y, so the three data points $(u, v) = (\ln x, \ln y)$ become $(\ln 1, \ln 3) \approx (0, 1.1)$, $(\ln 3, \ln 16) \approx (1.1, 2.8)$, $(\ln 4, \ln 25) \approx (1.4, 3.2)$.

Now we can set up a table to help us calculate the necessary sums.

	u	v	uv	u^2
	0	1.1	0	0
	1.1	2.8	3.08	1.21
	1.4	3.2	4.48	1.96
Total:	2.5	7.1	7.56	3.17

From this we find that $\bar{u} = \frac{1}{3}(2.5) \approx 0.833$ and $\bar{v} = \frac{1}{3}(7.1) \approx 2.37$, and the slope of the line is

$$m = \frac{\sum_{i=1}^{3} u_i v_i - 3\bar{u}\bar{v}}{\sum_{i=1}^{3} u_i^2 - 3(\bar{u})^2} \approx \frac{7.56 - 3(0.833)2.37}{3.17 - 3(0.833)^2} \approx 1.5,$$

$$b = \bar{v} - m\bar{u} \approx 2.37 - m(1.5)(0.833) \approx 1.12.$$

This suggests that the *logarithms* $(u, v) = (\ln x, \ln y)$ of x and y lie near the straight line $v = mu + b = 1.5u + 1.12$. Exponentiating gives $y = e^v = e^{1.5u+1.12} = e^{1.12}e^{1.5 \ln x} \approx 3.1x^{1.5}$.

The available data suggest that the constants in the expected formula $y = cx^k$ are $c \approx 3.1$ and $k \approx 1.5$.   ◢

See the Graphics Calculator Example on the next page.

If we have only three data points, it is risky to make any strong conclusions about the validity of a result like this; in particular, notice that we have retained only one decimal place in the constants to emphasize our doubts.

In an actual experimental situation it would be desirable to have more points, but the technique would be the same. For purposes of illustration it is convenient to keep the calculations simple.

For regression problems with large numbers of points the arithmetic can get quite overwhelming. In these situations it is very helpful to use a computer. The calculations are actually quite repetitive and simple, ideal for a computer, and software packages are readily available to perform regressions. For that matter it is not difficult to write your own program in BASIC or almost any other programming language.

Many scientific and financial calculators have built-in regression functions too.

With a graphics calculator we can graph the function from Example 3 and plot the points to see how well they fit.

TI-85	SHARP EL9300
GRAPH F1($y(x)$ =) 3 F1(x) ^ 1.5 M2(RANGE) .5 ▼ 4.5 ▼ ▼ (−) 10 ▼ 35 F5(GRAPH) EXIT PtOff(1,3) ENTER PtOn(3,16) ENTER PtOn(4,25) ENTER	⨏ 3 x/θ/T a^b 1.5 RANGE .5 ENTER 4.5 ENTER 1 ENTER 0 ENTER 35 ENTER 5 ENTER ⨏ 2ndF PLOT ▼ ▶ ▼ ENTER 1 ENTER 3 ENTER ⨏ 2ndF PLOT ▼ ▶ ▼ ENTER 3 ENTER 16 ENTER ⨏ 2ndF PLOT ▼ ▶ ▼ ENTER 4 ENTER 25 ENTER ⨏

CASIO f_x-6300G (f_x-7700GB)	HP 48SX
Range .5 EXE 4.5 EXE 1 EXE 0 EXE 35 EXE 5 EXE EXE EXE EXE Graph 3 X x^y 1.5 EXE Plot 1,3 EXE Plot 3,16 EXE Plot 4,25 EXE	⬑ PLOT PLOTR ERASE ATTN ′ Y = 3 × X y^x 1.5 ↪ PLOT ⬑ DRAW ERASE .5 ENTER 4.5 XRNG 0 ENTER 35 YRNG ATTN 1 ENTER 3 PRG OBJ NXT R → C PRG DSPL NXT C → PX PIXOF ⬑ PLOT PLOTR DRAW ATTN 3 ENTER 16 PRG OBJ NXT R → C PRG DSPL NXT C → PX PIXON ⬑ PLOT PLOTR DRAW ATTN 4 ENTER 25 PRG OBJ NXT R → C PRG DSPL NXT C → PX PIXON ⬑ PLOT PLOTR DRAW

It is also possible to perform the regression on the calculator.

PROOF OF THEOREM 16.11.2   We conclude this section with the application of the Second Derivative Test (Theorem 16.9.1) to show that the critical point we found for $f(m, b)$, the sum of the squares of the vertical distances from the line $y = mx + b$, is really a minimum.

From the formulas

$$\frac{\partial}{\partial m}\big(f(m, b)\big) = \sum_{i=1}^{n} \frac{\partial}{\partial m}\big((y_i - mx_i - b)^2\big) = -2\sum_{i=1}^{n} x_i(y_i - mx_i - b),$$

$$\frac{\partial}{\partial b}\big(f(m, b)\big) = \sum_{i=1}^{n} \frac{\partial}{\partial b}\big((y_i - mx_i - b)^2\big) = -2\sum_{i=1}^{n} (y_i - mx_i - b),$$

we find that

$$\frac{\partial^2}{\partial m^2}\big(f(m,b)\big) = 2\sum_{i=1}^{n} x_i^2,$$

$$\frac{\partial^2}{\partial b^2}\big(f(m,b)\big) = 2\sum_{i=1}^{n} 1 = 2n,$$

$$\frac{\partial^2}{\partial m \partial b}\big(f(m,b)\big) = 2\sum_{i=1}^{n} x_i.$$

The discriminant is

$$D = \frac{\partial^2 f}{\partial m^2}\frac{\partial^2 f}{\partial m^2} - \left(\frac{\partial^2 f}{\partial m \partial b}\right)^2 = \left(2\sum_{i=1}^{n} x_i^2\right)2n - \left(2\sum_{i=1}^{n} x_i\right)^2.$$

In Section 15.6 (p. 913) we mentioned the Cauchy-Schwartz Inequality, which says that for two n-dimensional vectors $\mathbf{u} = (u_1,\ldots,u_n)$ and $\mathbf{v} = (v_1,\ldots,v_n)$ the dot product satisfies $(\mathbf{u}\cdot\mathbf{v})^2 \le \|\mathbf{u}\|^2\|\mathbf{v}\|^2$. Moreover, the inequality is *strict* unless at least one of $\mathbf{u}$ and $\mathbf{v}$ is a scalar multiple of the other. In terms of components this says

$$\left(\sum_{i=1}^{n} u_i v_i\right)^2 < \left(\sum_{i=1}^{n} u_i^2\right)\left(\sum_{i=1}^{n} v_i^2\right).$$

Let us apply this result to the particular vectors $\mathbf{u} = (x_1,\ldots,x_n)$ and $\mathbf{v} = (1,\ldots,1)$, that is, with $u_i = x_i$ and $v_i = 1$ for each i. Since $\|\mathbf{v}\|^2 = n$, the inequality says

$$\left(\sum_{i=1}^{n} x_i\right)^2 < \left(\sum_{i=1}^{n} x_i^2\right)n.$$

Comparing with the formula we found for the discriminant, we see that D is positive, which means that the sum $f(m,b)$ has a local extremum. Since both second partial derivatives are positive, it is a local minimum, as required.

Exercises 16.11

Find the least squares approximation $y = mx + b$ to each of the following sets of points.

1. $(2,3), (5,2), (7,1)$ **3.** $(1,4), (2,2), (3,4)$

2. $(0,0), (2,0), (3,1)$ **4.** $(-1,2), (0,4), (2,6)$

5. $(-3,3), (-1,0), (1,-1), (3,-2)$

6. $(-2,-3), (0,-2), (4,0), (6,1)$

7. $(1,0), (2,2), (3,2), (4,4), (5,3)$

8. $(0,\pi), (1,-\pi), (3,-\pi), (4,\pi)$

9. $(-1,2), (0,2), (1,2), (3,2), (5,2)$

10. $(1,1), (2,2), (4,4), (7,7), (9,9)$

In each of the following situations, approximate the given data by a curve of the specified type.

11. $(1,1), (2,4), (3,9), \quad y = mx + b$

12. $(1,1), (2,4), (3,9), \quad y = cx^k$

13. $(1,5), (3,3), (5,2), \quad y = cx^k$

14. $(1,3), (2,5), (3,6), \quad y = m \ln x + b$

15. $(1,2), (2,3), (3,6), (4,10), (5,20), \quad y = cx^k$

16. $(0,15), (1,15), (2,12), (3,-2), (5,-20),$
$y = mx + b$

II

17. A population of bacteria is believed to be growing exponentially, that is, according to an equation of the form $P = ce^{kt}$, where k and c are constants, P is the population, and t is the time. This can be written as $\ln P = \ln c + kt$, which suggests that the points $(t, \ln P)$ lie on a straight line.

Suppose that the population is measured and it is found that it begins with 0.1 mg, reaches 0.3 mg 2 hours later, and reaches 2.4 mg 2 hours after that. Find the least squares approximation to the line mentioned above, and use your information to estimate the value of the growth constant k.

18. It is expected that the amount of product from a certain chemical reaction increases exponentially with time. In an actual run there is 1 gram after 1 hour, 1.2 g after 2 hours, 1.5 g after 3 hours, and 1.9 g after 4 hours. Estimate the growth constant, and estimate the amount of product that will be present after 6 hours.

19. A store can sell 15 shirts at $20 apiece, 20 of them at $18, or 35 of them at $15. Estimate the demand function $p = K - Lx$, where x is the number sold at price p.

*20. In the situation of Exercise 19, it is also possible to try to find a demand function of the form $p = cx^k$, for some constants c and k.

   (i) Estimate the constants c and k.

   (ii) For the straight line found in Exercise 19 and the curve found in this exercise, evaluate the sum of the squares of the distances from the three data points. Which approach gives the better least squares approximation?

21. A forester measures trees at various ages and finds the following data (each pair represents the age in years and the height in feet of one tree): $(1, 0.9)$, $(1, 1.0), (1.1, 1.1), (1.1, 1.0), (1.2, 1.1), (1.2, 1.2)$, $(1.3, 1.2), (1.3, 1.2), (1.4, 1.3), (1.8, 1.6), (1.9, 1.8)$, $(2, 1.9), (2.1, 2.0), (2.3, 2.1), (2.4, 2.3), (2.8, 2.5)$, $(3, 2.8)$. Find the line that gives the best fit to these data. Use a calculator. Use your answer to estimate the amount of growth per year for these trees.

*22. A botanist studying leaf growth has been asked to test two hypotheses. One idea is that the length and

width of the leaf are proportional to t, the length of time they have been growing, that is, $\ell = ct$ and $w = dt$, for some constants c and d. This would imply that the area of the leaf is proportional to t^2. The other hypothesis is that the weight of plant material is proportional to t.

The easiest way to measure is to weigh the leaves; assuming that the leaves have a uniform thickness, which seems to be roughly true, the weight is proportional to the surface area. So the first hypothesis suggests that the weight should be proportional to t^2, and the second suggests that it should be proportional to t.

Try to decide which hypothesis is more reasonable on the basis of the least squares line approximating the following data (each point represents the growing time and the weight of one leaf): $(4, 20)$, $(5, 25), (5, 27), (6, 40), (6, 37), (7, 55), (8, 65)$.

Notice that this type of analysis does not prove either hypothesis, but it can be used to eliminate at least one of them.

23. If $a \neq c$, consider the two points (a, b), (c, d). Find the least squares line approximating these points and show that it is the line passing through them.

24. If three points lie on a (nonvertical) line, show that the least squares line for them is the line on which they lie.

*25. Suppose (a_i, b_i) lie on a line $y = mx + b$, for $i = 1, \ldots, n$. Explain why the least squares line for these points is $y = mx + b$ itself.

*26. (i) Consider the points $(0,0), (1,0), (2,1)$. Use regression to find the line $y = mx + b$ that gives the best fit.

   (ii) Now think of y as the independent variable and x as the function and approximate by a line $x = ny + c$; that is, apply regression with x and y interchanged.

   (iii) Plot the points and the two lines, all on the same graph.

The significance of this exercise is that the lines are different; the choice of which quantity is the variable affects the answer. Essentially, this happens because the line $y = mx + b$ minimizes in terms of *vertical* distances from the points and $x = ny + c$ minimizes in terms of *horizontal* distances.

*27. (i) Consider the points $(-1,0), (0,1), (1,0)$, and find the line $y = mx + b$ that gives the best fit to these points in the sense that it minimizes the sum of the *fourth* powers of the vertical distances from the points to the line. (*Hint:* If $a^3 = b^3$, then $a = b$.)

   (ii) Compare the answer to part (i) with the least squares line.

*28. (i) Find the line $y = mx + b$ that gives the best fit to points $(-1, 0)$, $(0, 1)$, $(1, 0)$ in the sense that it minimizes the sum of the *sixth* powers of the vertical distances from the points to the line.

(ii) Compare the answer to part (i) with the least squares line.

(iii) Find the line that minimizes the sum of the $2k$th powers of the vertical distances from the points to the line.

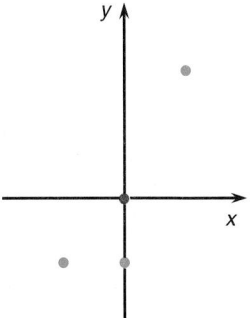

Figure 16.11.5

**29. Consider the points $(-1, -1)$, $(0, -1)$, $(1, 2)$, shown in Figure 16.11.5. The average point is $(\bar{x}, \bar{y}) = (0, 0)$. Let us try to approximate these points by a line, not by using the least squares approach but by minimizing the sum of the vertical distances from the line.

(i) Consider first a line through the average point $(0, 0)$. Such a line must be of the form $y = mx$. Find a formula, in terms of m, for the sum of

the vertical distances from the three points to the line $y = mx$. Your answer should involve absolute values.

(ii) Sketch the graph of the formula you found in part (i) (as a function of m), and find the minimum value of the sum of the vertical distances.

(iii) Now consider the line that passes through the two points $(-1, -1)$ and $(1, 2)$. Show that the sum of the distances from it is smaller than the smallest sum for any line through the origin.

This exercise demonstrates that the average point may not even lie on the line that gives the best approximation if we use the sum of the distances. This is one reason why the least squares approach is preferable. It is also true that absolute values are more difficult to work with than squares.

*30. The last inequality in the proof of Theorem 16.11.2 is a *strict* inequality. Why must this be true?

31. Use a graphics calculator or computer to plot the points and regression lines for the exercises in this section. For each exercise, plot the points first and try to guess the position of the best-fit line before actually plotting it.

POINT TO PONDER

In this section we have considered the idea of approximating a number of points with a straight line. This choice is made for simplicity, but of course it might be better to try to approximate with some slightly more complicated curve. A natural next step would be to try to approximate by a polynomial of degree 2, that is, a parabola, or a polynomial of degree 3 or 4, or so on.

For instance, to approximate a number of points by a parabola $y = ax^2 + bx + c$, we would again form the sum of the squares of the vertical distances from the points to the parabola, as a function of the coefficients (a, b, c), and look for critical points. We would find that there is exactly one, assuming that the given points lie

on at least three vertical lines, and the corresponding curve is the best fit to the points by a parabola, in the least squares sense.

Polynomials of higher degree can be expected to give better approximations to the points, but on the other hand they become somewhat more difficult to work with.

It is also possible to use curves other than polynomials to approximate a collection of points. To obtain meaningful results, it is important to know what type of relationship to expect and to use that type of curve to approximate the data.

Chapter Summary

§16.1
level curve
level surface

§16.2
punctured disk
limits
properties of limits
distance between points in space
continuity
properties of continuity
continuity of polynomials

§16.3
partial derivatives
$\frac{\partial f}{\partial x}, f_x$
tangent plane
higher-order partial derivatives
mixed partials

§16.4
approximation
(total) differential
differentiability
differentiability and continuity

§16.5
intermediate variables
Chain Rule
implicit differentiation

§16.6
unit vector
directional derivative

gradient
∇f
$D_{\mathbf{u}} f = \nabla f \cdot \mathbf{u}$

§16.7
normal vector
tangent plane

§16.8
absolute maximum, minimum
local maximum, minimum
relative maximum, minimum
extremum
local extremum
bounded function
bounded region
critical point

§16.9
saddle point
critical point
discriminant
Second Derivative Test (Theorem 16.9.1)

§16.10
constraint
Lagrange multipliers

§16.11
linear regression
least squares approximation
average
best fit

Review Exercises

I

Calculate each of the following, provided it exists.

1. $\lim_{(x,y) \to (0,0)} (3x^2 y - 4xy + y - 2)$

2. $\lim_{(x,y) \to (0,0)} \sin\left(\pi\left(\frac{x+2y}{2}\right)\right)$

3. $\lim_{(x,y) \to (0,0)} \frac{2x-y+8}{x+3y+4}$

4. $\lim_{(x,y) \to (0,0)} e^{\sin(xy)}$

5. $\lim_{(x,y) \to (0,0)} \frac{x+5y+1}{7x-y}$

6. $\lim_{(x,y) \to (0,0)} \frac{xy}{x}$

7. $\lim_{(x,y) \to (1,1)} e^{\cos(\pi xy)}$

8. $\lim_{(x,y) \to (0,\pi)} \frac{x}{y}$

9. $\lim_{(x,y,z) \to (0,0,0)} \frac{xy+yz-1}{xyz-8}$

10. $\lim_{(x,y) \to (0,-1)} \frac{xy}{xy}$

Find each of the following.

11. $\frac{\partial}{\partial x}(3x^4 y - 7x + 4y - 2)$

12. $\frac{\partial}{\partial y}(e^{3xy+2})$

13. $\frac{\partial}{\partial x}(2xy + 3yz^2 - x^3 z)$

14. $\frac{\partial}{\partial x}(4yz^2 - 14)$

15. $\frac{\partial^2}{\partial x^2}\left(\sin(xy)\right)$

16. $\frac{\partial^2}{\partial x \partial y}(e^{4xy})$

17. $\frac{\partial^3}{\partial x^2 \partial y}(x^3 - 4x^2 y + 3xy^4)$

18. $\frac{\partial^4}{\partial x^2 \partial y^2}(4xy + 3xy^3 - 10^{xy})$

19. $\frac{\partial^3}{\partial x \partial y \partial z}(x^2 y - 7xy^2 + yz^4 - 3xyz^2)$

20. $\frac{\partial^2}{\partial x \partial y}\left(\frac{1}{x+y}\right)$

Evaluate each of the following. (The symbols r and θ that appear in some questions are the polar coordinates of the point (x, y).)

21. $\frac{\partial f}{\partial u}$ if $f(x, y) = 2x + x^2 y$, $x = uv$, $y = u + v$

22. $\frac{\partial f}{\partial v}$ if $f(x, y) = 2x + x^2 y$, $x = uv$, $y = u + v$

23. $\frac{\partial F}{\partial x}$ if $F(u, v) = u \ln(uv)$, $u = 2x - y$, $v = \frac{x}{y}$

24. $\frac{\partial F}{\partial y}$ if $F(u, v) = u \ln(uv)$, $u = 2x - y$, $v = \frac{x}{y}$

25. $\frac{\partial g}{\partial z}$ if $g(u, v) = \sin(3u - 2v)$, $u = x^2 yz$, $v = x^3 yz$

26. $\frac{\partial g}{\partial x}$ if $g(u, v) = \sin(3u - 2v)$, $u = x^2 yz$, $v = x^3 yz$

27. $\frac{\partial G}{\partial r}$ if $G(x, y) = \ln(x^2 + y^2)$

28. $\frac{\partial G}{\partial \theta}$ if $G(x, y) = \ln(x^2 + y^2)$

29. $\frac{\partial h}{\partial \theta}$ if $h(x, y) = (x^2 + y^2)^{5/2}$

30. $\frac{\partial H}{\partial \theta}$ if $H(x, y) = \arctan \frac{y}{x}$

In each of the following cases, find the directional derivative of the function f at the point P in the direction of the vector $\mathbf{v}$.

31. $f(x, y) = x + 3y$, $P = (3, -1)$, $\mathbf{v} = (2, -3)$

32. $f(x, y) = 3x + 2y$, $P = (7, -3)$, $\mathbf{v} = (2, -3)$

33. $f(x, y) = e^{xy - x}$, $P = (3, 5)$, $\mathbf{v} = (-1, 1)$

34. $f(x, y) = \ln xy$, $P = (2, 2)$, $\mathbf{v} = (1, -1)$

35. $f(x, y, z) = \sin xyz$, $P = (\pi, 1, 1)$, $\mathbf{v} = (2, 1, 2)$

36. $f(x, y, z) = \sin xyz$, $P = (\pi, 2, 1)$, $\mathbf{v} = (0, -2, 2)$

37. $f(x, y, z) = x^2 + y^2 + z^2$, $P = (2, 1, -3)$, $\mathbf{v} = (1, -2, 0)$

38. $f(x, y, z) = \frac{1}{x + y + z}$, $P = (4, 3, -2)$, $\mathbf{v} = (1, 1, 1)$

39. $f(x, y, z) = x^2 - yz$, $P = (1, 2, 0)$, $\mathbf{v} = (0, 2, 0)$

40. $f(x, y, z) = \frac{x}{y + z}$, $P = (0, 1, -2)$, $\mathbf{v} = (1, 2, 1)$

Find the following differentials.

41. df if $f(x, y) = 3xy \sin(xy)$

42. df if $f(x, y) = 3x - 4y$

43. df if $f(x, y) = \frac{1}{2x + 3y}$

44. df if $f(x, y) = \arctan\left(\frac{y}{x}\right)$

45. df if $f(x, y) = \cosh^2(xy) - \sinh^2(xy)$

46. df if $f(x, y, z) = \frac{xy}{z}$

47. dz if $x^2 - 2yz = 4$

48. dz if $\sin(xz) - \cos(yz) = \frac{1}{2}$

49. dx if $e^{xy} = 3xyz + 2$

50. dy if $\ln(xy) = \arctan(yz)$

Find the tangent planes to each of the following surfaces at the point P.

51. $2x^2 + 2y^2 + 2z^2 = 3$ at $P = \left(\frac{1}{2}, -\frac{1}{2}, -1\right)$

52. $z - xy = 4$ at $P = (3, 4, 16)$

53. $xyz^2 = 12$ at $P = (-1, -3, -2)$

54. $\sin(2xy) - \cos(3xz) = 1$ at $P = (\pi, 3, -1)$

55. $2x - 3y + 7z = 6$ at $P = (1, 1, 1)$

56. $4x^2 + 2y^2 + 9z^2 = 21$ at $P = (-1, -2, 1)$

57. $xyz = 12$ at $P = (-2, 2, -3)$

58. $2z - \cosh(xy) - 3 = 0$ at $P = (3, 0, 2)$

59. $x^2 + y^2 - z^2 = 0$ at $P = (-3, 4, -5)$

60. $2xy - 3yz + 7xz + 55 = 0$ at $P = (-2, 1, 3)$

Find the absolute maximum and minimum of each of the following functions on the specified region.

61. $f(x, y) = x^2 + 3y^2$ on $|x|, |y| \le 1$

62. $f(x, y) = x^2 - y^2$ on $|x|, |y| \le 1$

63. $f(x, y) = x^2 - xy$ on $-1 \le x, y \le 2$

64. $f(x, y) = \sin^2(\pi x) + y^2$ on $|x|, |y| \le 2$

65. $f(x, y) = x^2 \sin^2\left(\frac{\pi}{2} y\right)$ on $|x|, |y| \le 1$

66. $f(x, y) = x^3 - 3xy + 3x$ on $|x|, |y| \le 4$

67. $f(x, y) = x^3 - 6xy + 3y^2$ on $|x| + |y| \le 4$

68. $f(x, y) = x^2 - x \sin y$ on $|x| \le 1$, $|y| \le \pi$

69. $f(x, y) = (x^2 + y^2) e^x$ on $x^2 + y^2 \le 9$

70. $f(x, y) = y^3 + 3x^2 y - 3y$ on $|x|, |y| \le 2$

Find the extreme values of the following functions subject to the constraints specified.

71. $f(x, y) = x$, subject to $x^2 - 2x + y^2 + 4y = 0$

72. $f(x, y) = x + 2y$, subject to $x^2 + 6x + y^2 - 2y = -9$

73. $f(x, y) = xy$, subject to $x^2 + 2xy + 2y^2 = 2$

74. $f(x, y) = x^2 + 2y$, subject to $x^2 + y^2 = 2$

75. $f(x, y, z) = x + 3y - 2z$, subject to $x^2 + y^2 + z^2 = 4$

76. $f(x, y, z) = 2y - xz$, subject to $x^2 + y^2 + 2y + z^2 = 0$

77. $f(x, y, z) = xyz$, subject to $x^2 + y^2 + z^2 = 0$

78. $f(x, y, z) = xy - z^2$, subject to $x^2 + y^2 + z^2 = 9$

79. $f(x, y, z) = z$, subject to $x^2 + y^2 + 6y + z^2 - 4z = 3$

80. $f(x, y, z) = \sqrt{x^2 + y^2 + z^2 - 4z + 2}$, subject to $x - 3y + 2z = 9$

II

For each of the following functions $f(x, y)$, sketch the level curves corresponding to the specified values.

81. $f(x, y) = 9x^2 - 16y^2 - 2$, $c = 0, -2, 142$

82. $f(x, y) = 5x^2 + 4y^2 - 4$, $c = 0, 16, -5$

83. $f(x, y) = 9x^2 + 9y^2 - 36y$, $c = 0, 13, -20$

84. $f(x, y) = 4xy$, $c = 0, -4, 4, 8$

85. Suppose $f(t)$ is a differentiable function defined for every t. Define $F(x, y) = f(x - y)$. Find the directional derivative of F in the direction of the vector $\mathbf{v} = \langle -4, -4 \rangle$.

*86. We observed that if $y = mx + b$ is the least squares approximation to the points (x_i, y_i), for $i = 1, \ldots, n$, then the average point $(\bar{x}, \bar{y})$ is on the line. Now consider the line that gives the best fit in the sense that the sum of the *fourth* powers of the vertical distances from the points to the line is a minimum. Must the average point be on the line?

87. Consider the function $f(x, y) = x^2 \sin(xy)$, and observe that $f\left(\pi, \frac{1}{2}\right) = \pi^2$. Estimate a value of x for which $f(x, 0.498) = \pi^2 + \frac{1}{100}$, and check how good your estimate is by evaluating the function.

88. Consider the function $g(x, y, z) = ye^{2xy} - yz$, and observe that $g(0, 2, -1) = 4$. Estimate a value of y for which $g(-0.0001, y, -0.998) = 4.001$, and check how good your estimate is by evaluating the function.

89. Find the points in the region $|\bar{x} + y| \le 2$, $|x - y| \le 2$ that are closest to and farthest from the point $(-4, -5)$.

90. Find the points in the region $x^2 + 4y^2 \le 1$ that are closest to and farthest from the point $(3, 2)$.

*91. The U.S. Postal Service will accept a package only if the sum of its length and its girth (the perimeter of a cross-section perpendicular to the length) does not exceed 108 in. What are the dimensions of the cylindrical box of largest volume that can be sent?

*92. In the situation of Exercise 91, what are the dimensions of the cylindrical package of largest surface area that can be sent?

*93. A rectangular box with no top is made from 20 sq ft of material. What are the dimensions of the box that has the greatest volume?

*94. A rectangular box with no top and no front is made from 20 sq ft of material. What are the dimensions of the box that has the greatest volume?

*95. What is the rectangular box of maximum volume that can be placed inside the ellipsoid $4x^2 + y^2 + z^2 = 4$?

*96. What is the rectangular box of maximum surface area that can be placed inside the ellipsoid $x^2 + 9y^2 + z^2 = 36$?

*97. A piece of wire of length L is cut into at most three pieces, and each piece is folded into a circle. Which lengths of wire will give the maximum and minimum values for the total area of the circles?

*98. A lump of clay of volume V is divided into at most three lumps, each of which is fashioned into a sphere. Which sizes of smaller lumps will give the maximum and minimum total volume?

99. Find the least squares line approximating the points $(1, 1)$, $(2, 4)$, $(3, 9)$, $(4, 16)$, $(5, 25)$.

100. Find the least squares line approximating the points $(2, 2)$, $(-2, -2)$, $(2, -2)$, $(-2, 2)$.

*101. An object dropped from rest has displacement $s(t) = h - \frac{1}{2}gt^2$, where h is the initial height and g is the acceleration due to gravity. An experiment is performed in which a ball is dropped and photographed at regular intervals of time by using a strobe light. From the resulting photograph the ball's approximate height is measured at various times. The data are as follows (in each pair the first entry is the time in seconds and the second is the height in feet): $(0.4, 17.4)$, $(0.5, 16)$, $(0.6, 14.2)$, $(0.7, 12.2)$, $(0.8, 9.8)$, $(0.9, 7.0)$, $(1.0, 4.1)$, $(1.1, 0.7)$. Use least squares approximation to estimate the value of the gravitational constant g.

102. If n is a positive integer, let

$$f(x, y) = \frac{\sum_{k=0}^{n} \binom{2n}{2k}(-1)^k x^{2n-2k} y^{2k}}{(x^2 + y^2)^n}.$$

Use a computer to sketch $z = f(x, y)$ for several different values of n, and discuss the continuity of $f(x, y)$.

Multiple Integrals

17

For a positive function of two variables, we will develop the integral to be the volume of the region under the graph. As in the single-variable case, we will find that integration has many additional uses beyond calculating volumes.

The illustration shows a computer-generated picture of the graph of the function $z = 2 + \cos(x^2 + y^2)$. Notice how the graph goes up and down like cos as (x,y) moves away from the origin.

INTRODUCTION

In Chapter 16 we learned how to deal with derivatives of functions of several variables using partial derivatives. In this chapter we discuss the question of integrating functions of several variables.

We begin in Section 17.1 with *double integrals;* this refers to functions of two variables. The basic ideas for functions of three or more variables are not much different from what we see with two variables, and of course it is simpler to deal with only two. An integral of a function of one variable involves an interval in $\mathbb{R}$, specified by the limits of integration. A function of two variables will be integrated over a region in the plane $\mathbb{R}^2$. We begin with rectangular regions but eventually consider other shapes, where what corresponds to *limits of integration* can be quite complicated. This is the topic of Section 17.2. In Section 17.3 we see how integrals of functions of two variables can be used to find the areas of regions in the plane and the volumes of regions in space. If a function in the plane is expressed in terms of polar coordinates rather than rec-

1053

tangular coordinates (x, y), then it may be desirable to integrate using r and θ; we learn how to do this in Section 17.4. In Section 17.5 we see how integration can be used to find the surface areas of many surfaces. In Section 17.6 we consider integrating functions of three variables, so-called *triple integrals*, and Section 17.7 discusses applications to moments and centroids, including *moments of inertia*, which are important for the study of rotational motion. In Section 17.8 we learn to integrate in terms of cylindrical and spherical coordinates in $\mathbb{R}^3$. Finally, in Section 17.9 we discuss *change of variables*, which is the procedure for multiple integrals that corresponds to substitution for integrals with a single variable.

17.1 Double Integrals

ITERATED INTEGRAL
CROSS-SECTION
PARTITION
RIEMANN SUM
DOUBLE INTEGRAL

In Figure 17.1.1 we see the graph of $y = F(x) = x^2$, with shading on the region under the graph and over the interval $[0, 1]$. The area of this region is the integral $\int_0^1 F(x)\,dx$.

In Figure 17.1.2 we see the graph of the function $z = f(x, y) = x^2 + y^2$, with the region lying under it and over the square $0 \leq x \leq 1, 0 \leq y \leq 1$ outlined. We would like to find a way to integrate $f(x, y)$ so that the integral over the square will equal the volume of this region.

For that matter, given any region R in the xy-plane (see Figure 17.1.3(i)) and any continuous nonnegative function $g(x, y)$ defined on R, we want to integrate $g(x, y)$ over the region R and find the volume of the region lying under the graph of $z = g(x, y)$ and over the region R, as shown in Figure 17.1.3(ii).

For functions that may be negative we should be able to consider a sort of signed volume.

Figure 17.1.1

Figure 17.1.2

Figure 17.1.3(i)

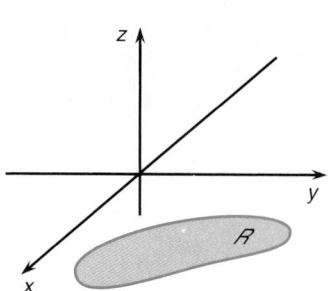

Figure 17.1.3(ii)

EXAMPLE 1   The simplest sort of example would be the integral over a region R in the plane of a *constant* function $f(x, y) = c$. The situation is illustrated in Figure 17.1.4; the graph of $z = c$ is (part of) a horizontal plane. The volume of the region lying under the graph and over the region R is just the product of the height c and the area of R.

Figure 17.1.4

Figure 17.1.5

Figure 17.1.6

Figure 17.1.7

To define the integral for more complicated functions, we take a cue from the way we used partitions and Riemann sums to define integration for functions of a single variable.

To begin with, let us simplify things by assuming that the function $f(x, y)$ is continuous and that the region R is a rectangle in the plane with its sides parallel to the coordinate axes. We divide up the rectangle R into little rectangles as shown in Figure 17.1.5; the sides of the little rectangles are also parallel to the axes. Such a collection of rectangles is called a **partition** of R.

To form a Riemann sum, we should label the rectangles in the partition $\mathcal{P}$ as $R_1, R_2, \ldots, R_n$. The order is of no importance. Suppose that the width and height of R_i are Δx_i and Δy_i, respectively, so the area is $\Delta A_i = \Delta x_i \Delta y_i$. Let us call the **mesh** of $\mathcal{P}$ the maximum of these lengths, that is,

$$mesh(\mathcal{P}) = \|\mathcal{P}\| = max\{\Delta x_i, \Delta y_i : 1 \leq i \leq n\}.$$

For each little rectangle R_i we should pick a point $P_i^* = (x_i^*, y_i^*)$ in R_i and evaluate the integrand at P_i^*.

On the ith little rectangle R_i the function $f(x, y)$ is nearly equal to a constant function with value $f(P_i^*) = f(x_i^*, y_i^*)$. The region under that constant function over R_i is the rectangular box shown in Figure 17.1.6; its volume is $f(x_i^*, y_i^*)\Delta A_i = f(x_i^*, y_i^*)\Delta x_i \Delta y_i$.

We approximate the volume under the graph of $z = f(x, y)$ over the rectangle R by the sum of these approximate volumes for all the little rectangles R_i. This sum is the **Riemann sum**

$$\sum_{i=1}^{n} f(P_i^*)\Delta A_i = \sum_{i=1}^{n} f(P_i^*)\Delta x_i \Delta y_i = \sum_{i=1}^{n} f(x_i^*, y_i^*)\Delta x_i \Delta y_i.$$

See Figure 17.1.7.

As with integrals of functions of one variable we consider finer and finer partitions, that is, with smaller and smaller mesh, and define the integral to be the limit of the Riemann sums (cf. Definition 5.3.5).

DEFINITION 17.1.1

Suppose R is the rectangle in the plane defined by $a \leq x \leq b$ and $c \leq y \leq d$. Suppose $f(x, y)$ is a continuous function defined on R. For each partition $\mathcal{P} = \{R_i\}$ as above and for each choice of a $P_i^* \in R_i$ we form the Riemann sum. The **integral** of $f(x, y)$ over the rectangle R is defined to be the limit of the Riemann sums as the mesh of $\mathcal{P}$ goes to zero. We write

$$\int\int_R f(x, y)\, dA = \int\int_R f(x, y)\, dx\, dy = \lim_{\|\mathcal{P}\| \to 0} \left(\sum_{i=1}^{n} f(P_i^*)\Delta x_i \Delta y_i\right).$$

Notice that the notation involves the double integral sign $\int\int$ and ends with dA or $dx\, dy$. These two notations are used interchangeably. In place of limits of integration we have the subscript R to identify the region over which we are integrating.

EXAMPLE 2 Suppose $f(x, y) = c$ is the constant function. Then for *every* P_i^* in every R_i we know that $f(P_i^*) = c$. For any partition $\mathcal{P} = \{R_i\}$ we find that the Riemann sum is

$$\sum_{i=1}^{n} f(P_i^*)\Delta x_i \Delta y_i = \sum_{i=1}^{n}(c)\,area(R_i) = c\sum_{i=1}^{n} area(R_i) = (c)\,area(R).$$

When we let the mesh of the partition tend to 0 and take a limit, we are taking the limit of a constant, $(c)\,area(R)$, so we find that the integral of a constant function is what it should be:

$$\int\int_R f(x,y)\,dx\,dy = \int\int_R c\,dA = (c)\,area(R).$$

Unfortunately, Definition 17.1.1 does not provide a way to evaluate the integral for anything except constant functions. This is not very different from the situation we encountered with integrals of functions of one variable in Chapter 5; there we had to discover indefinite integrals and the Fundamental Theorem of Calculus in order to evaluate any but the simplest integrals.

EXAMPLE 3 Here we begin by considering the special case of a function that is independent of y, that is, a function $f(x,y)$ that can be written in the form $f(x,y) = F(x)$ for some function $F(x)$. The graph of such a function is shown in Figure 17.1.8. Notice how the function is constant "in the y-direction," meaning that if we let y vary but keep x fixed the value of the function will remain constant.

The integral of $f(x,y)$ over the rectangle R ought to equal the volume under the graph, which is shown in Figure 17.1.9. If we knew the area of the shaded region on the side, then we could easily find the volume by just multiplying that area by the width of the rectangle. But the curve at the top of that shaded area is just the graph of $z = F(x)$, so the shaded area is $\int_a^b F(x)\,dx$. If the rectangle R is defined by $a \le x \le b$ and $c \le y \le d$, then the "width" of the rectangle is $d - c$, so the volume equals $(d-c)\int_a^b F(x)\,dx$, the product of the width of the rectangle and the area of that shaded region. We can write

$$\int\int_R f(x,y)\,dA = \int\int_R F(x)\,dx\,dy = (d-c)\int_a^b F(x)\,dx = (d-c)\int_a^b f(x,y)\,dx.$$

Example 3 suggests a way to evaluate integrals of more general functions. In Figure 17.1.10 we divide the region under the graph of $z = f(x,y)$ into a number of thin slices, much as we did in Section 6.3 when we found the volume of a solid of revolution. If we consider one such slice, the area of its side is just an integral with respect to x of the function whose graph forms the top of the side. That function is $z = f(x,y)$, where x runs from a to b and y is held *fixed*.

The area of the side of the slice is $\int_a^b f(x,y)\,dx$, so if the thickness of the slice is dy, then its volume is approximately $\left(\int_a^b f(x,y)\,dx\right)dy$. To get the volume of the whole region, we "add up" the volumes of all the little slices, and this suggests that the whole volume should equal

$$\int_c^d \left(\int_a^b f(x,y)\,dx\right)dy.$$

Let us consider what this formula means. The outside integral is an integration with respect to y, with y running from c to d. The integrand is the function of y defined by the expression in parentheses. That expression $\int_a^b f(x,y)\,dx$ is itself an

The areas of the little rectangles add up to the area of the big rectangle...

Figure 17.1.8

Figure 17.1.9

Figure 17.1.10

This computer-generated image of a human body begins with cross-sections, shows an outline of the whole surface, and finally fills in the spaces to give a fairly realistic picture.

integral with respect to x. For any particular value of y, $f(x, y)$ is a function of x, so it makes sense to integrate it; from Figure 17.1.10 we see that the inner integral $\int_a^b f(x, y)\,dx$ is the area of a **cross-section** of the solid under the graph of $z = f(x, y)$.

For different y's, $f(x, y)$ is a different function of x, so the integral $\int_a^b f(x, y)\,dx$ will vary, depending on y. In other words it is a *function* of y, and it is this function of y that is the integrand of the outside integral.

To evaluate such a double integral, we perform the *inner* integration first, to express the integrand as a function of y, and *then* perform the *outer* integration.

It is best explained by an example.

EXAMPLE 4   Evaluate

$$\int_0^1 \left(\int_1^2 xy\,dx \right) dy.$$

Solution   We begin by fixing the value of y and performing the inner integration, that is, integrating with respect to x while regarding y as a constant. We find that

$$\int_1^2 xy\,dx = y \int_1^2 x\,dx = y \frac{x^2}{2}\Big|_1^2 = \frac{y}{2}(4 - 1) = \frac{3y}{2}.$$

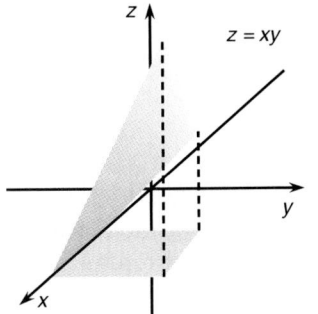

The "constant" y comes outside the integral.

This is the integrand of the outer integral, and we integrate it with respect to y to find

$$\int_0^1 \left(\int_1^2 xy\,dx \right) dy = \int_0^1 \frac{3y}{2}\,dy = \frac{3}{4}y^2\Big|_0^1 = \frac{3}{4}.$$

In Figure 17.1.11 we see the graph of $z = xy$ over the square $1 \le x \le 2, 0 \le y \le 1$. The double integral gives the volume of the region under the graph.

An integral of this form is called an *iterated integral,* meaning that the process of integration is repeated, once for the inner integral and then for the outer integral. In the inner integral $\int_a^b f(x, y)\,dx$ we are integrating with respect to x and holding y constant. An integration of this type is sometimes called a *partial integration.* It gives the (signed) area of the cross-section determined by this value of y.

Figure 17.1.11

DEFINITION 17.1.2

An expression of the form

$$\int_c^d \left(\int_a^b f(x, y)\,dx \right) dy$$

is called an **iterated integral.** The function $f(x, y)$ is called the **integrand,** the expression $\int_a^b f(x, y)\,dx$ is called the **inner integral,** and the other integration is called the **outer integral.**

In the iterated integral

$$\int_a^b \left(\int_c^d f(x, y)\,dy \right) dx$$

the inner integral is $\int_c^d f(x, y)\,dy$, and the outer integral is the integral with respect to x.

To evaluate an iterated integral, it is necessary to perform the inner integration *first* and then the outer integration. Sometimes the expression is written without parentheses, but it is always with this understanding about the order of the integrations. We always integrate *from the inside out*.

However, much of the time it turns out not to matter.

EXAMPLE 5   Perform the iterated integral of Example 4 in the reverse order.

Solution   We want to integrate first with respect to y and then with respect to x. The integral is

$$\int_1^2 \left(\int_0^1 xy \, dy \right) dx = \int_1^2 \left(x \int_0^1 y \, dy \right) dx = \int_1^2 \left(x \frac{y^2}{2} \Big|_0^1 \right) dx$$

$$= \int_1^2 \left(\frac{x}{2} \right) dx = \frac{x^2}{4} \Big|_1^2 = \frac{3}{4}.$$

We find that the iterated integral has the same value as when we did it in the opposite order in Example 4 (cf. Figure 17.1.11).

This is not an accident. The way we developed iterated integrals was by trying to find the volume of the region under the graph of $z = f(x, y)$. We took cross-sections parallel to the x-axis, integrated them and then "added up" the resulting cross-sectional areas by integrating. On the other hand it would be just as reasonable to do it the other way, with cross-sections parallel to the y-axis. Either of these calculations ought to give us the volume, and in particular they ought to be equal. It is possible to find examples in which it does not work, but happily this is usually not a problem for "nice" integrands.

It turns out that for a continuous function on a rectangle the two iterated integrals are always the same. Moreover, the double integral, as defined in Definition 17.1.1 by taking limits of Riemann sums, will always exist and will also equal the two iterated integrals. We omit the difficult proof of this fact, but the result itself is extremely important because it gives us a way to evaluate double integrals quite mechanically.

It is also related to Theorem 5.3.4, which says that the integral of a continuous function on an interval $[a, b]$ always exists.

THEOREM 17.1.3

Suppose R is the rectangle determined by $a \le x \le b$ and $c \le y \le d$. If $f(x, y)$ is a continuous function on R, then the two iterated integrals of $f(x, y)$ over R are equal, that is,

$$\int_c^d \left(\int_a^b f(x, y) \, dx \right) dy = \int_a^b \left(\int_c^d f(x, y) \, dy \right) dx.$$

Moreover, the double integral $\int \int_R f(x, y) \, dx \, dy$ defined by Definition 17.1.1 exists, and it also equals the iterated integrals:

$$\int \int_R f(x, y) \, dA = \int \int_R f(x, y) \, dx \, dy = \int_c^d \left(\int_a^b f(x, y) \, dx \right) dy$$

$$= \int_a^b \left(\int_c^d f(x, y) \, dy \right) dx.$$

Notice that in the second of these four integrals the $dx\, dy$ is equivalent to dA in the double integral. In the third and fourth integrals the dx and dy are separate, each belonging to either the inner or outer integral.

In a way it is because of this theorem that we sometimes use the notation $dx\, dy$ in place of dA. The theorem says that a double integral $\int \int_R f(x, y)\, dA$ can be evaluated by an iterated integral, and the alternative notation $\int \int_R f(x, y)\, dx\, dy$ helps to remind us of this approach.

EXAMPLE 6   Find the volume of the region under the graph of the function $z = f(x, y) = x^2 - y^2 + 4$ over the rectangle $0 \le x \le 3$, $-1 \le y \le 2$.

Solution   Since $-1 \le y \le 2$, we see that $y^2 \le 4$, so $-y^2 + 4 \ge 0$ and $f(x, y) = x^2 - y^2 + 4 \ge x^2 \ge 0$, so there is no difficulty about having to use signed volumes where the function is negative.

Since $f(x, y)$ is a polynomial, we know it is continuous, and we can use the theorem. To evaluate the double integral $\int \int_R f(x, y)\, dx\, dy$ over the given rectangle R, we evaluate the iterated integral.

Evaluate for x between 0 and 3.

$$\int \int_R f(x, y)\, dA = \int_{-1}^{2} \int_{0}^{3} (x^2 - y^2 + 4)\, dx\, dy = \int_{-1}^{2} \left(\left(\frac{x^3}{3} - y^2 x + 4x \right) \Big|_0^3 \right) dy$$

$$= \int_{-1}^{2} (9 - 3y^2 + 12)\, dy = (9y - y^3 + 12y) \Big|_{-1}^{2}$$

$$= (18 - 8 + 24) - (-9 + 1 - 12) = 54.$$

Notice the way we did the partial integration, regarding y as a constant while we did the inner integral with respect to x. Also notice the way the terminology is used. The *double integral* means the integral $\int \int_S f(x, y)\, dA = \int \int_S f(x, y)\, dx\, dy$ defined by limits of Riemann sums. The *iterated integral* refers to what we can actually evaluate by doing the inner integration first and then the outer integration.

In these examples it is possible to perform the iterated integrals in either order, and the calculations, though different, will be similar and of comparable difficulty. There are instances in which an integral is much easier to integrate in one order than in the other.

EXAMPLE 7   Suppose R is the rectangle $1 \le x \le 4$, $|y| \le 2$, and evaluate the double integral $\int \int_R f(x, y)\, dx\, dy$, where

$$f(x, y) = \frac{y}{\sqrt{1 + 3x^4}}.$$

Solution   We could try evaluating

$$\int \int_R f(x, y)\, dx\, dy = \int_{-2}^{2} \int_{1}^{4} \frac{y}{\sqrt{1 + 3x^4}}\, dx\, dy = \int_{-2}^{2} y \left(\int_{1}^{4} \frac{1}{\sqrt{1 + 3x^4}}\, dx \right) dy,$$

but that inner integral is not easily evaluated. On the other hand, in the opposite order,

$$\int\int_R f(x,y)\,dA = \int_1^4 \int_{-2}^2 \frac{y}{\sqrt{1+3x^4}}\,dy\,dx = \int_1^4 \frac{1}{\sqrt{1+3x^4}}\left(\int_{-2}^2 y\,dy\right)dx.$$

Here the inner integral $\int_{-2}^2 y\,dy$ is extremely easy to work out: It equals $\frac{1}{2}y^2\big|_{-2}^2 = \frac{1}{2}(4-4) = 0$. Knowing this, we find that the outer integral is $\int_1^4 0\,dx = 0$. ◢

Integrals of functions of one variable have a number of properties like

$$\int_a^b \big(f(x)+g(x)\big)\,dx = \int_a^b f(x)\,dx + \int_a^b g(x)\,dx,$$

(the integral of a sum equals the sum of the integrals), and knowing how useful they are, we would like to have the same rules for double integrals. If $f(x,y)$ and $g(x,y)$ are continuous functions on a rectangle R in the plane, then consider $\int\int_R \big(f(x,y)+g(x,y)\big)\,dA$.

Knowing that this double integral can be evaluated by iterated integration, we find that if $R = \{(x,y) : a \le x \le b, c \le y \le d\}$, then

$$\int\int_R \big(f(x,y)+g(x,y)\big)\,dA = \int_c^d \left(\int_a^b \big(f(x,y)+g(x,y)\big)\,dx\right)dy.$$

The inner integral is an integral of a function of the *single* variable x, so we know it satisfies the rule for sums. This means that

$$\int_a^b \big(f(x,y)+g(x,y)\big)\,dx = \int_a^b f(x,y)\,dx + \int_a^b g(x,y)\,dx,$$

so

$$\int\int_R \big(f(x,y)+g(x,y)\big)\,dA = \int_c^d \left(\int_a^b \big(f(x,y)+g(x,y)\big)\,dx\right)dy$$

$$= \int_c^d \left(\int_a^b f(x,y)\,dx + \int_a^b g(x,y)\,dx\right)dy$$

$$= \int_c^d \left(\int_a^b f(x,y)\,dx\right)dy + \int_c^d \left(\int_a^b g(x,y)\,dx\right)dy$$

$$= \int\int_R f(x,y)\,dA + \int\int_R g(x,y)\,dA.$$

The fact about integrals of sums of functions of one variable has been used to establish the corresponding fact about double integrals of sums of functions of two variables.

In much the same way it is not difficult to establish the following result about double integrals in the case in which all the functions are continuous (so iterated integration can be used). The result is actually true whenever all the functions involved are integrable, that is, when all the integrals make sense, but in the general case the proof is more complicated.

THEOREM 17.1.4

Suppose R is the rectangle in the plane defined by $a \le x \le b, c \le y \le d$. Suppose $f(x,y)$ and $g(x,y)$ are continuous functions on R and k is a real number. Then

(i) $\int\int_R \big(f(x,y)+g(x,y)\big)\,dx\,dy = \int\int_R f(x,y)\,dx\,dy + \int\int_R g(x,y)\,dx\,dy;$

(ii) $\int \int_R \left(kf(x,y) \right) dx\, dy = k \int \int_R f(x,y)\, dx\, dy$;

(iii) If $f(x,y) \geq 0$ for all $(x,y) \in R$, then $\int \int_R f(x,y)\, dx\, dy \geq 0$;

(iv) If $f(x,y) \leq M$ for all $(x,y) \in R$, then $\int \int_R f(x,y)\, dx\, dy \leq M\ area(R) = M(b-a)(d-c)$;

(v) If $M \leq f(x,y)$ for all $(x,y) \in R$, then $M\ area(R) = M(b-a)(d-c) \leq \int \int_R f(x,y)\, dx\, dy$;

(vi) If $|f(x,y)| \leq M$ for all $(x,y) \in R$, then $|\int \int_R f(x,y)\, dx\, dy| \leq M\ area(R) = M(b-a)(d-c)$.

PROOF We have already proved part (i). Part (ii) is similar. The remaining parts of the theorem are proved in a similar fashion using the corresponding Theorems 5.4.2 and 5.4.3 about integrals of functions of a single variable. ◢

Double integrals can be used to "add up" the values of a function over a region in the plane.

EXAMPLE 8 **Incidence of Disease** A study suggests that a certain illness is more prevalent among older people and that its incidence increases among people who are overweight. In an attempt to explore this phenomenon the authors propose the following *mathematical model* for the population of the city being studied. Let a be a person's age, in years, and w be the number of pounds by which he or she is overweight, relative to the standard tables for weight in terms of height and sex.

Consider people of age a who are overweight by w. Then the number of cases of the illness among people in this group is given by the *density function*

$$D(a,w) = 2 + \frac{aw}{500},$$

at least for $20 \leq a \leq 80$ and $0 \leq w \leq 40$.

Roughly speaking, this function says how many cases there are among people of a given age and weight, but it has to be interpreted properly. Rather than speaking of people who are *exactly* 25 years old and *exactly* 3 lb overweight, we might ask how many there are between 25 and 26 and between 3 and 4 lb overweight, or between 20 and 30 and from 0 to 5 lb overweight, and so on. The answers to these questions are found by integrating the population density $D(a,w)$ over the appropriate ranges of a and w. It is because of this that $D(a,w)$ is called a density function.

So for instance, the number of cases for people between 20 and 30 years of age and between 0 and 5 lb overweight is found by integrating $D(a,w)$ for $a \in [20,30]$ and $w \in [0,5]$. The number is

$$N = \int_0^5 \int_{20}^{30} D(a,w)\, da\, dw = \int_0^5 \int_{20}^{30} \left(2 + \frac{aw}{500} \right) da\, dw$$

$$= \int_0^5 \left(2a + \frac{a^2 w}{1{,}000} \right) \Bigg|_{a=20}^{a=30} dw = \int_0^5 \left(60 - 40 + \frac{900w}{1{,}000} - \frac{400w}{1{,}000} \right) dw$$

$$= \int_0^5 \left(20 + \frac{w}{2} \right) dw = \left(20w + \frac{w^2}{4} \right) \Bigg|_0^5 = 100 + \frac{25}{4} = 106.25.$$

The proposed model leads us to expect about 106 cases of the illness among people in this group. ◢

In this kind of calculation the answer is often not an integer, and we have to round it off to get a result that can be interpreted in terms of population. To use this model intelligently, it would be helpful to know the total number of people in the city in the given ranges of age and weight.

With a graphics calculator we can plot the density $D(a, w)$ as a function of w for several values of a. We try it for $a = 20, 25, 30, 35, 40$.

<table>
<tr><td>

TI-85
GRAPH F1 $(y(x) =)$ 2 + 20 F1 (x)
÷ 500 ▼ 2 + 25 F1 (x) ÷ 500 ▼ 2
+ 30 F1 (x) ÷ 500 ▼ 2 + 35 F1
(x) ÷ 500 ▼ 2 + 40 F1 (x) ÷ 500
▼ M2(RANGE) 0 ▼ 20 ▼ ▼ 0
▼ 4 F5(GRAPH)

</td><td>

SHARP EL9300
◠↩ 2 + 20 x/θ/T ÷ 500 2ndF ▼ 2
+ 25 x/θ/T ÷ 500 2ndF ▼ 2 + 30
x/θ/T ÷ 500 2ndF ▼ 2 + 35 x/θ/T ÷
500 2ndF ▼ 2 + 40 x/θ/T ÷ 500 2ndF
▼ RANGE 0 ENTER 20 ENTER 10
ENTER 0 ENTER 4 ENTER 1 ENTER
◠↩

</td></tr>
<tr><td>

CASIO f_x-6300G (f_x-7700GB)
Range 0 EXE 20 EXE 10 EXE 0 EXE 4
EXE 1 EXE EXE EXE EXE Graph 2 +
20 X ÷ 500 EXE Graph 2 + 25 X ÷
500 EXE Graph 2 + 30 X ÷ 500 EXE
Graph 2 + 35 X ÷ 500 EXE Graph 2
+ 40 X ÷ 500 EXE

</td><td>

HP 48SX
⬑ PLOT PLOTR ERASE ATTN ′
Y ▬ 2 + 20 × X ÷ 500 ↱ PLOT
⬑ DRAW ERASE 0 ENTER 20 XRNG
0 ENTER 4 YRNG DRAW ATTN ′ Y
▬ 2 + 25 × X ÷ 500 ↱ PLOT
⬑ DRAW DRAW ATTN ′ Y ▬ 2 +
30 × X ÷ 500 ↱ PLOT ⬑ DRAW
DRAW ATTN ′ Y ▬ 2 + 35 × X
÷ 500 ↱ PLOT ⬑ DRAW DRAW
ATTN ′ Y ▬ 2 + 40 × X ÷ 500
↱ PLOT ⬑ DRAW DRAW

</td></tr>
</table>

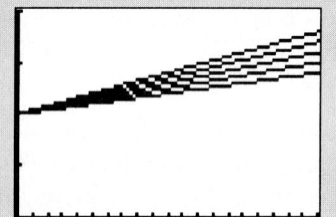

Notice how the incidence of the disease increases with weight. Try plotting the density as a function of a for various fixed values of w.

Exercises 17.1

Evaluate each of the following.

1. $\int_1^2 \left(\int_0^1 xy^2 \, dx \right) dy$

2. $\int_0^1 \left(\int_0^4 (3xy - 2y) \, dx \right) dy$

3. $\int_{-1}^1 \left(\int_0^2 (x + 2y) \, dy \right) dx$

4. $\int_3^4 \left(\int_0^1 7 \, dx \right) dy$

5. $\int_0^1 \left(\int_0^1 y \sin (\pi xy) \, dx \right) dy$

6. $\int_0^1 \left(\int_0^1 (x + \cos \pi y) \, dy \right) dx$

7. $\int_{-1}^1 \left(\int_{-1}^1 e^{x+y} \, dy \right) dx$

8. $\int_1^2 \left(\int_{-\pi}^0 (3 - \cos x) \, dx \right) dy$

9. $\int_a^b \left(\int_c^d 1 \, dx \right) dy$

10. $\int_c^d \left(\int_a^b xy \, dy \right) dx$

In each of the following cases, find the double integral $\int \int_R f(x, y) \, dA$, where $f(x, y)$ is the given function and R is the specified rectangle.

11. $f(x, y) = x^2 y^2$, $\quad R = \{(x, y) : 0 \le x \le 1, 0 \le y \le 1\}$

12. $f(x, y) = xy - y^2$, $R = \{(x, y) : 0 \le x \le 2, 0 \le y \le 3\}$

13. $f(x, y) = (x - y)^2$, $R = \{(x, y) : -1 \le x \le 1, -1 \le y \le 1\}$

14. $f(x, y) = e^{3x-y}$, $R = \{(x, y) : 0 \le x \le 1, -1 \le y \le 0\}$

15. $f(x, y) = \frac{x}{y}$, $R = \{(x, y) : 0 \le x \le 2, 1 \le y \le 2\}$

16. $f(x, y) = \frac{x+1}{y+1}$, $R = \{(x, y) : |x| \le 1, 0 \le y \le 1\}$

17. $f(x, y) = \frac{x}{y^2+1}$, $R = \{(x, y) : 0 \le x \le 1, 0 \le y \le 1\}$

18. $f(x, y) = xy \exp(x^2)$, $R = \{(x, y) : |x| \le 2, 0 \le y \le 4\}$

19. $f(x, y) = 5$, $R = \{(x, y) : |x| \le a, |y| \le c\}$

20. $f(x, y) = \sin^2(x - y) + \cos^2(y - x)$, $R = \{(x, y) : |x| \le 4, |y| \le 3\}$

II

21. Perform the iterated integration of Example 6 (p. 1059) in the opposite order, making sure that you get the same answer.

22. Perform the iterated integration $\int_0^1 \int_0^1 (x^2 y - 2y) \, dx \, dy$ in both orders, making sure that you get the same answer.

*23. Let $f(x, y)$ be the function that equals 1 when (x, y) is in the square $0 \le x \le 1$, $0 \le y \le 1$ and that equals 0 when (x, y) is not in that square.

  (i) Evaluate $\int_0^2 \left(\int_0^2 f(x, y) \, dx \right) dy$.

  (ii) How would you interpret this integral geometrically? (*Hint:* Compare it with the integral $\int_0^1 \int_0^1 f(x, y) \, dx \, dy$.)

*24. Let $f(x, y)$ be the function that equals $x^2 y$ when (x, y) is in the square $0 \le x \le 1$, $0 \le y \le 1$ and that equals 0 when (x, y) is not in that square. Evaluate $\int_0^2 \left(\int_0^2 f(x, y) \, dx \right) dy$.

**25. Let $f(x, y)$ be the function that equals 1 when $x \ge y$ and equals 0 when $x < y$. Evaluate $\int_0^1 \left(\int_0^1 f(x, y) \, dx \right) dy$. Can you interpret the value geometrically?

**26. Let $f(x, y)$ be the function that equals xy when $x \le y$ and equals $2x$ when $x > y$. Evaluate $\int_0^1 \left(\int_0^1 f(x, y) \, dx \right) dy$.

**27. Let $f(x, y)$ be the function that equals $2y^2$ when $x \ge 2y$ and equals xy when $x \le 2y$. Evaluate $\int_{-1}^1 \left(\int_{-1}^1 f(x, y) \, dx \right) dy$.

**28. Let $f(x, y)$ be the function that equals 2 when $x^2 \ge y$ and equals 0 when $x^2 < y$. Evaluate $\int_0^1 \left(\int_{-1}^1 f(x, y) \, dx \right) dy$.

*29. In the situation of Example 8 (p. 1061), how many cases of the illness does the model predict among people between ages 20 and 30 and between 20 and 25 lb overweight?

*30. In the situation of Example 8 (p. 1061), how many cases of the illness does the model predict among people between ages 60 and 70 and between 35 and 40 lb overweight?

*31. Consider $\int \int_R xy \, dA$, where R is the square $0 \le x \le 1$, $0 \le y \le 1$. Let us compute some Riemann sums. For any positive integer n we partition R into n^2 little squares R_i, each with sides $\frac{1}{n}$ long. Within each little square we evaluate the integrand at the point P_i^*, which is the upper right corner. (i) Find a formula for the Riemann sum

$$\sum_{i=1}^{n^2} f(P_i^*) \Delta R_i,$$

where $f(x, y) = xy$. (ii) Evaluate the formula. (iii) By letting $n \to \infty$, find $\int \int_R xy \, dA$. (iv) Check your answer by evaluating the iterated integral.

*32. Repeat Exercise 31 for $\int \int_R x^2 y \, dA$, where R is $-1 \le x \le 1$, $0 \le y \le 1$, and R is partitioned into $2n^2$ little squares with sides $\frac{1}{n}$ long.

III

33. Use a personal computer to sketch the graph of the density function $D(a, w)$ in Example 8 for $20 \le a \le 80$ and $0 \le w \le 40$.

P◄ POINT TO PONDER

In Theorem 17.1.3 we saw that if the integrand is a continuous function, then the iterated integrals in both orders must be equal. This means that we can perform the integration in whichever order we please. Frequently, there is no particular reason to choose one order or the other, but we have seen that sometimes one may be much easier than the other.

This result is so useful and convenient that it is natural to *interchange* the order of the integrals without much thought. However, if the integrand is not continuous, this can be quite tricky, and in fact it is the source of many errors. We will not worry about this question, but we include an example to show how the iterated integrals may not be the same. Let

$$f(x, y) = \begin{cases} \frac{x}{y}, & \text{if } y > 0, \\ 0, & \text{if } y = 0. \end{cases}$$

Then if $y > 0$, $\int_{-1}^{1} f(x, y)\,dx = \int_{-1}^{1} \frac{x}{y}\,dx = \frac{1}{y}\int_{-1}^{1} x\,dx = \frac{1}{2y}x^2\big|_{-1}^{1} = 0$. If $y = 0$, then $\int_{-1}^{1} f(x, 0)\,dx = \int_{-1}^{1} 0\,dx = 0$. We see that in all cases $\int_{-1}^{1} f(x, y)\,dx = 0$. From this we find that the iterated integral is

$$\int_0^1 \int_{-1}^1 \frac{x}{y}\,dx\,dy = \int_0^1 0\,dy = 0.$$

To find the other iterated integral

$$\int_{-1}^1 \int_0^1 \frac{x}{y}\,dy\,dx,$$

it is necessary to evaluate the inner integral $\int_0^1 \frac{x}{y}\,dy = x\int_0^1 \frac{1}{y}\,dy$. Unless x is zero, this improper integral does not exist.

So although one of the iterated integrals exists and equals 0, the other one does not even exist.

17.2 Setting Up Double Integrals

NONRECTANGULAR REGIONS
HORIZONTAL AND VERTICAL
CROSS-SECTIONS

Figure 17.2.1

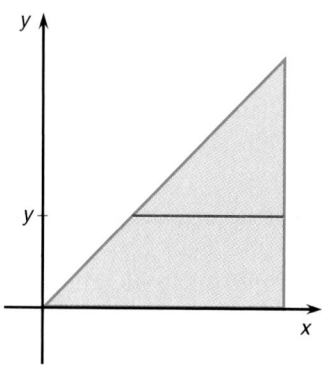

Figure 17.2.2

Our introduction to double integrals has been with integration over rectangular regions R in the plane. Of course, we will want to use other types of regions. The volume of a hemisphere of radius 2 can be calculated by integrating the function $f(x, y) = \sqrt{4 - x^2 - y^2}$; the region over which we should integrate is the disk $x^2 + y^2 \leq 4$ (see Figure 17.2.1).

There are two questions to be considered. One is how we can define the double integral $\int \int_R f(x, y)\,dA = \int \int_R f(x, y)\,dx\,dy$ if the region R is not a rectangle, and the other is how to make sense of iterated integrals for such a region.

Let us begin with an example to illuminate the second question.

EXAMPLE 1 Suppose $f(x, y) = 3x - y$ and R is the triangular region whose vertices are $(0, 0)$, $(1, 0)$, and $(1, 1)$. We fix a value of y and integrate along the corresponding horizontal cross-section (see Figure 17.2.2). To make the iterated integral, we integrate with respect to y to "add up" the volumes of the thin slices of thickness dy.

Suppose we have chosen some value of y; it must be between 0 and 1, or the cross-section will not intersect the triangle at all. We want to set up the inner integral $\int f(x, y)\,dx$, but the question is what the limits of integration are.

The upper limit of integration is 1, since the right side of the triangle is at $x = 1$. The left limit depends on y. For instance, if y is 0, then the corresponding cross-section is the bottom edge of the triangle, so the lower limit of integration is 0. But higher up, the integral begins at the left edge of the triangle. Since that edge is the line $y = x$, we see that for our fixed value of y the x-coordinate of the point at the left end of the cross-section is $x = y$.

The inner integral is $\int_y^1 f(x, y)\,dx$. This is the cross-sectional area. To form the iterated integral, we "add up" the volumes of the slices with this cross-sectional area and thickness dy:

$$\int_0^1 \left(\int_y^1 f(x, y)\,dx \right) dy.$$

Figure 17.2.3

Figure 17.2.4(i)

Figure 17.2.4(ii)

Figure 17.2.4(iii)

Notice that the limits of integration on the outer integral correspond to the values of y at the bottom and top of the triangle.

Remembering that y is held constant in the inner integral, we evaluate it:

$$\int_y^1 f(x, y)\, dx = \int_y^1 (3x - y)\, dx = \left(\frac{3}{2}x^2 - yx \right)\Big|_y^1$$

$$= \left(\frac{3}{2} - y \right) - \left(\frac{3}{2}y^2 - y^2 \right) = \frac{3}{2} - y - \frac{1}{2}y^2.$$

Now it is straightforward to evaluate the outer integral:

$$\int_0^1 \left(\int_y^1 f(x, y)\, dx \right) dy = \int_0^1 \left(\frac{3}{2} - y - \frac{1}{2}y^2 \right) dy = \left(\frac{3}{2}y - \frac{1}{2}y^2 - \frac{1}{6}y^3 \right)\Big|_0^1 = \frac{5}{6}.$$

In Example 1 we figured out what ought to be the iterated integral for integrating over a triangular region. The limits of integration in the inner integral depend on the variable in the outer integral. The reason for wanting iterated integrals as in Example 1 is that they will provide a way to evaluate double integrals $\int \int_R f(x, y)\, dA = \int \int_R f(x, y)\, dx\, dy$ with R a triangle or other nonrectangular region. For this to make any sense, we need to know how to define double integrals for such regions.

What made it easy to use rectangular regions was that they could easily be partitioned into small rectangles, as in Figure 17.2.3. It is less clear what to do for a triangle, let alone a circle. These regions cannot be filled up exactly by little rectangles.

Let us begin with some region R as shown in Figure 17.2.4(i). Let us suppose that it is bounded, which is equivalent to saying that it is contained in some large rectangle, as shown in Figure 17.2.4(ii). Instead of partitioning the irregularly shaped region, we partition this large rectangle as shown in Figure 17.2.4(iii). However, we consider only the little rectangles in this partition that lie entirely inside the original region R. These rectangles are shaded in Figure 17.2.4(iii).

We form a Riemann sum in the usual way, except that we use only those little rectangles in the partition that lie entirely inside R. Labeling these rectangles $\mathcal{P} = \{R_1, \ldots, R_n\}$ in some order, we choose a point $P_i^* = (x_i^*, y_i^*) \in R_i$ for each $i = 1, \ldots, n$. The Riemann sum is

$$\sum_{i=1}^n f(P_i^*) \Delta A_i = \sum_{i=1}^n f(x_i^*, y_i^*) \Delta x_i \Delta y_i,$$

where Δx_i and Δy_i are the width and height of the rectangle R_i.

The mesh of the partition is defined as the maximum length of the sides of the rectangles and is denoted $\|\mathcal{P}\|$.

Using this type of Riemann sum, we define the double integral $\int \int_R f(x, y)\, dA = \int \int_R f(x, y)\, dx\, dy$ in exactly the same way we did for rectangular regions R.

DEFINITION 17.2.1

Suppose R is a bounded region in the plane, and suppose $f(x, y)$ is a continuous function defined on R. For each partition $\mathcal{P} = \{R_i\}$ as above and for each choice of $P_i^* \in R_i$ we form the Riemann sum. The **integral** of $f(x, y)$ over the region R is defined to be the limit of the Riemann sums as the mesh of $\mathcal{P}$ goes to zero, provided the limit exists. We write

$$\int\int_R f(x,y)\,dA = \int\int_R f(x,y)\,dx\,dy = \lim_{\|\mathcal{P}\|\to 0}\left(\sum_{i=1}^{n} f(P_i^*)\Delta x_i\Delta y_i\right).$$

It is possible to show that such integrals satisfy the simple properties of Theorem 17.1.4.

THEOREM 17.2.2

Suppose R is a region in the plane and that $f(x,y)$ and $g(x,y)$ are continuous functions on R. Suppose that $\int\int_R f(x,y)\,dx\,dy$ and $\int\int_R g(x,y)\,dx\,dy$ both exist. Then the conclusions of Theorem 17.1.4 are all true for these integrals, omitting the references to a, b, c, and d in parts (iv), (v), and (vi).

As with rectangular regions this definition does not tell us how to evaluate integrals except in the simplest cases. For that it will be necessary to use iterated integrals.

We have done one example with a triangle, but let us try some regions with curved sides.

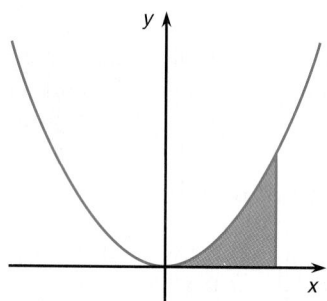

Figure 17.2.5(i)

EXAMPLE 2 Let R be the region that lies under the graph of $y = x^2$ for $0 \le x \le 1$, as shown in Figure 17.2.5(i). Find an iterated integral to evaluate $\int\int_R x\,dA$.

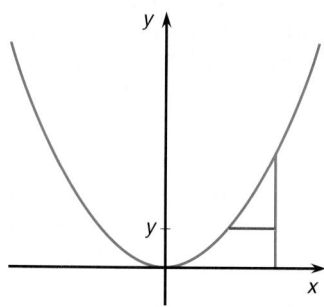

Figure 17.2.5(ii)

Solution Let us fix a value of y between 0 and 1 (the bottom and top of the region) and consider the corresponding horizontal cross-section, as shown in Figure 17.2.5(ii). The left endpoint of that segment is the point on the graph of $y = x^2$ at the height y. This means that its x-coordinate is $x = \sqrt{y}$ (i.e., the point on the parabola is $(\sqrt{y}, y)$).

Since the right end of the segment is at $x = 1$, we see that the inner integral will be $\int_{\sqrt{y}}^1 x\,dx$. This is easy to evaluate, remembering that for the purposes of this calculation y is constant. We find that

$$\int_{\sqrt{y}}^1 x\,dx = \frac{1}{2}x^2\bigg|_{\sqrt{y}}^1 = \frac{1}{2} - \frac{y}{2}.$$

From the picture we see that for the region R the variable y ranges from 0 to 1, so the whole iterated integral is

$$\int_0^1\left(\int_{\sqrt{y}}^1 x\,dx\right)dy = \int_0^1\left(\frac{1}{2} - \frac{y}{2}\right)dy = \left(\frac{y}{2} - \frac{y^2}{4}\right)\bigg|_0^1 = \frac{1}{4}.$$

Just for variety we try answering the same question with an iterated integral performed in the opposite order, that is, with the inner integral in terms of y.

In Figure 17.2.5(iii) we see the vertical cross-section corresponding to a fixed value of x. Its bottom end is at $y = 0$, and its top end is at $y = x^2$. This means that the inner integral is $\int_0^{x^2} x\,dy$. Notice that in this expression the x and the x^2 are *constants*.

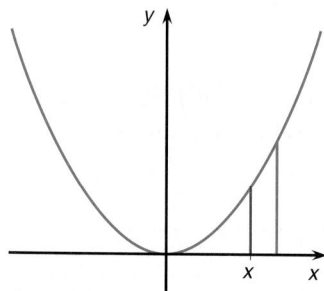

Figure 17.2.5(iii)

It is not difficult to evaluate:

$$\int_0^{x^2} x \, dy = x \int_0^{x^2} 1 \, dy = x \left(y \Big|_0^{x^2} \right) = x^3.$$

The outer integral will be with respect to x, with x running from the left to the right of the region R, that is, from $x = 0$ to $x = 1$. The iterated integral is

$$\int_0^1 \left(\int_0^{x^2} x \, dy \right) dx = \int_0^1 (x^3) \, dx = \frac{1}{4} x^4 \Big|_0^1 = \frac{1}{4}.$$

We see that both iterated integrals equal $\frac{1}{4}$.   ◀

Notice that in this example the calculations, though different for the two iterated integrals, were similar. Perhaps the second one was slightly easier because it was not necessary to solve for $x = \sqrt{y}$. This is not an important distinction in this example, but it is worth noticing what is required. We fix one variable and then try to find the corresponding points on the boundary of the region. If the boundary is given by some equation, then this amounts to solving for the other variable. Obviously, there will be times when one calculation is much easier than the other.

EXAMPLE 3   Find an iterated integral for $\int \int_D (x^2 + y^2) \, dx \, dy$, where D is the disk of radius 1 around the origin.

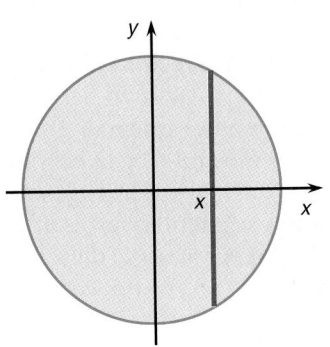

Figure 17.2.6

Solution   Let us try using y as the variable of integration in the inner integral. This means that the inner integral amounts to integrating along a vertical cross-section, as shown in Figure 17.2.6. If we fix the value of x on this segment, then we have to find the values of y at the bottom and top points.

This means solving for y for points on the circle $x^2 + y^2 = 1$. The values of y are $y = \pm\sqrt{1 - x^2}$. For the iterated integral we add up these vertical cross-sections by integrating with respect to x:

$$\int_{-1}^1 \int_{-\sqrt{1-x^2}}^{\sqrt{1-x^2}} (x^2 + y^2) \, dy \, dx.$$

Notice that we have not bothered to put parentheses around the inner integral. Also notice that the limits of integration on the outer integral correspond to the left and right sides of the disk. Finally, observe that the dy comes before the dx, since y is the variable of integration in the inner integral.

Evaluating, we find

$$\int_{-1}^1 \int_{-\sqrt{1-x^2}}^{\sqrt{1-x^2}} (x^2 + y^2) \, dy \, dx = \int_{-1}^1 \left(x^2 y + \frac{1}{3} y^3 \right) \Big|_{y=-\sqrt{1-x^2}}^{y=\sqrt{1-x^2}} dx$$

$$= \int_{-1}^1 \left(2x^2\sqrt{1-x^2} + \frac{2}{3}(1-x^2)\sqrt{1-x^2} \right) dx$$

$$= \int_{-1}^1 \left(\frac{4}{3} x^2 \sqrt{1-x^2} + \frac{2}{3}\sqrt{1-x^2} \right) dx.$$

Notice how we write "$y =$" when we evaluate, to help keep clear how each variable is being used.

Figure 17.2.7

Figure 17.2.8(i)

Figure 17.2.8(ii)

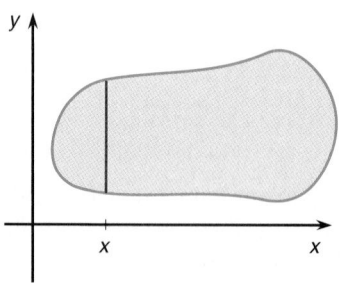

Figure 17.2.8(iii)

To evaluate this last integral, we make the substitution $x = \sin\theta$, $dx = \cos\theta\, d\theta$, with θ running from $-\frac{\pi}{2}$ to $\frac{\pi}{2}$. The integral becomes

$$\int_{-\pi/2}^{\pi/2} \left(\frac{4}{3}\sin^2\theta\cos^2\theta + \frac{2}{3}\cos^2\theta \right) d\theta = \int_{-\pi/2}^{\pi/2} \left(\frac{1}{3}\sin^2(2\theta) + \frac{2}{3}\cos^2\theta \right) d\theta.$$

Next we use the half-angle formulas (Theorem 1.7.13) to see that $\cos^2\theta = \frac{1}{2}\left(1 + \cos(2\theta)\right)$ and $\sin^2(2\theta) = \frac{1}{2}\left(1 - \cos(4\theta)\right)$. The integral equals

$$\int_{-\pi/2}^{\pi/2} \left(\frac{1}{6}\left(1 - \cos(4\theta)\right) + \frac{1}{3}\left(1 + \cos(2\theta)\right) \right) d\theta$$

$$= \left(\frac{\theta}{6} - \frac{1}{24}\sin(4\theta) + \frac{\theta}{3} + \frac{1}{6}\sin(2\theta) \right)\Bigg|_{-\pi/2}^{\pi/2} = \frac{\pi}{2}.$$

If you try setting up the iterated integral in the opposite order, you will find that the calculation is virtually identical, with only the names of x and y interchanged.

We have now worked out some examples of iterated integrals on nonrectangular regions. We have also given the definition of double integrals for such regions, using partitions and taking a limit. The hope is that iterated integrals can be used to evaluate double integrals, and this hope is supported by the picture of iterated integrals as a "sum" of volumes of thin cross-sections (see Figure 17.2.7). However, to prove that there is a connection is quite subtle.

Part of the subtlety arises because we have to know what kind of region is allowed.

In Figure 17.2.8(i) we see a region R in the plane. Suppose we want to perform an iterated integral with x as the integration variable in the inner integral. To do this, we need to find the limits of integration for the inner integral. For fixed y we consider the horizontal cross-section at height y as shown in Figure 17.2.8(ii), and it is necessary to express the left and right endpoints of this segment in terms of y.

If we want to have an iterated integral with y as the inner variable, then it is a question of vertical cross-sections, as shown in Figure 17.2.8(iii). In this case the problem is to express the top and bottom points of the segment as functions of x. The simplest way to do this would be to express the top and bottom boundaries of R as the graphs of functions $y = f(x)$ and $y = g(x)$, as shown in Figure 17.2.8(iv). Then for a given value of x the inner integral would run from $g(x)$ (at the bottom) to $f(x)$ (at the top).

To integrate a function $F(x, y)$ over R, we would try adding up the cross-sections

$$\int_a^b \int_{g(x)}^{f(x)} F(x, y)\, dy\, dx,$$

Figure 17.2.8(iv)

Figure 17.2.8(v)

Figure 17.2.9(i)

Figure 17.2.9(ii)

Figure 17.2.9(iii)

where a and b are the values of x at the left and right ends of R, as shown in the illustration.

If the inner integral is to be with respect to x, then we need to express the left and right boundaries as functions of y, as shown in Figure 17.2.8(v). The left boundary is $x = h(y)$, the right boundary is $x = k(y)$, and the bottom and top of the region occur at $y = c$ and $y = d$, respectively. The iterated integral is

$$\int_c^d \int_{h(y)}^{k(y)} F(x, y)\, dx\, dy.$$

The difficulty with this idea is that not all regions are easily expressed in this form. In Figure 17.2.9(i) we see a region for which some of the horizontal cross-sections do not simply run from one boundary curve to another. Knowing the left and right boundaries is not enough; somehow we have to deal with the piece that is "missing" in the middle.

In this case there are two ways around the problem. One would be to use vertical cross-sections, as shown in Figure 17.2.9(ii). For this particular region the vertical cross-sections are all just single segments. Provided it is possible to express the bottom and top boundaries as functions of x, then it will be possible to use an iterated integral with y as the inner variable.

The second approach is illustrated in Figure 17.2.9(iii), where the region has been divided into two parts. On each of these parts the horizontal cross-sections are single segments, so it is possible to use iterated integrals for them, provided the boundaries can be expressed as functions of y. The integral over the whole original region will be the sum of the integrals over the two parts.

In Figure 17.2.10(i) we see a region for which some of the horizontal cross-sections *and* some of the vertical cross-sections are split. For such a region it will be necessary to divide into smaller regions. One way to do this is shown in Figure 17.2.10(ii); the vertical cross-sections of each of the four pieces are simple segments.

We have used the idea that the integral over a region is the sum of the integrals over smaller regions that make it up. This certainly seems very reasonable, and for the kinds of regions we will use it is always true.

The preceding discussion leads us to consider regions of the following special form. If $f(x)$ and $g(x)$ are continuous functions on $[a, b]$, with $g(x) \leq f(x)$ for all $x \in [a, b]$, consider the set that lies between their graphs, that is,

$$R = \{(x, y) : g(x) \leq y \leq f(x), \quad a \leq x \leq b\}.$$

Figure 17.2.10(i)   **Figure 17.2.10(ii)**

Figure 17.2.11

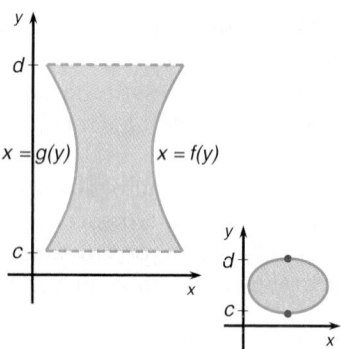

Figure 17.2.12

Two such regions are illustrated in Figure 17.2.11; in the first the two functions $f(x)$ and $g(x)$ are never equal, and in the second they are equal at the endpoints.

The other possibility, illustrated in Figure 17.2.12, is to express x as a function of y on the boundaries and to consider a region of the form

$$S = \{(x, y) : g(y) \le x \le f(y), \quad c \le y \le d\}.$$

For regions of either of these forms, it is easy to write down the appropriate iterated integral, and it turns out that it equals the double integral. We omit the difficult proof of this result.

THEOREM 17.2.3

Suppose $f(x)$ and $g(x)$ are continuous functions on $[a, b]$ with $g(x) \le f(x)$ for all $x \in [a, b]$, and let R be the region between their graphs:

$$R = \{(x, y) : g(x) \le y \le f(x), \quad a \le x \le b\}.$$

If $F(x, y)$ is a continuous function on R, then the double integral $\int \int_R F(x, y)\, dA = \int \int_R F(x, y)\, dx\, dy$ exists, and it equals the iterated integral

$$\int \int_R F(x, y)\, dA = \int \int_R F(x, y)\, dx\, dy = \int_a^b \int_{g(x)}^{f(x)} F(x, y)\, dy\, dx.$$

Similarly, if $f(y)$ and $g(y)$ are continuous functions on $[c, d]$ with $g(y) \le f(y)$ for all $y \in [c, d]$, let S be the region between their graphs:

$$S = \{(x, y) : g(y) \le x \le f(y), \quad c \le y \le d\}.$$

If $F(x, y)$ is a continuous function on S, then the double integral $\int \int_S F(x, y)\, dA = \int \int_S F(x, y)\, dx\, dy$ exists, and it equals the iterated integral

$$\int \int_S F(x, y)\, dA = \int \int_S F(x, y)\, dx\, dy = \int_c^d \int_{g(y)}^{f(y)} F(x, y)\, dx\, dy.$$

Moreover, these results can be used to evaluate integrals over regions that are made up of several regions, each of which is of one of these types. The integrals over the small regions can be evaluated by iterated integrals, and the integral over the whole region is the sum of the integrals over its parts. For this to work, the small regions must not overlap except perhaps on their boundaries.

EXAMPLE 4   Let R be the region lying between the graphs of $y = x^2$ and $x - y + 2 = 0$. Evaluate $\int \int_R (12xy - 15x^2)\, dx\, dy$.

Solution   The region R is illustrated in Figure 17.2.13. We can easily solve and find that the two points where the line intersects the parabola are $(-1, 1)$ and $(2, 4)$.

One way to proceed is to regard R as the region between the graphs of the functions $y = x^2$ and $y = x + 2$. We use y as the variable of integration in the inner integral, and the outer integral will run from $x = -1$ to $x = 2$. We find

$$\int \int_R (12xy - 15x^2)\, dx\, dy = \int_{-1}^2 \int_{x^2}^{x+2} (12xy - 15x^2)\, dy\, dx$$

$$= \int_{-1}^2 (6xy^2 - 15x^2 y) \Big|_{y=x^2}^{y=x+2}\, dx$$

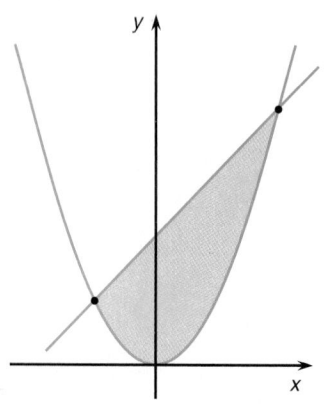

Figure 17.2.13

$$= \int_{-1}^{2} \left(6x(x+2)^2 - 15x^2(x+2) - 6x(x^4) + 15x^2(x^2) \right) dx$$

$$= \int_{-1}^{2} (-6x^5 + 15x^4 - 9x^3 - 6x^2 + 24x) \, dx$$

$$= \left(-x^6 + 3x^5 - \frac{9}{4}x^4 - 2x^3 + 12x^2 \right) \Big|_{-1}^{2}$$

$$= \frac{81}{4}.$$

The iterated integral equals $\frac{81}{4}$. It would have been possible to evaluate the other iterated integral, with x as the inner variable, but it would be more trouble.

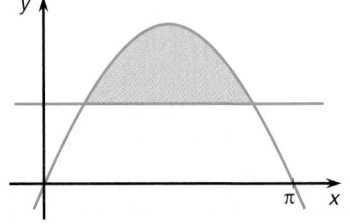

Figure 17.2.14(i)

EXAMPLE 5   Let S be the region lying under the loop of the graph of $y = \sin x$ that lies between $x = 0$ and $x = \pi$, and above the line $y = \frac{1}{2}$, as shown in Figure 17.2.14(i). Integrate $\int \int_S x \, dx \, dy$.

Solution   If we want to use horizontal cross-sections, we need to express the left and right boundaries of S as functions of y. The left boundary is $x = \arcsin y$, and after some thought we see that the right boundary is $x = \pi - \arcsin y$. We could try this approach, but integrals involving arcsin y do not look too promising.

Instead, let us try vertical cross-sections. The left end of the region is at $x = \frac{\pi}{6}$, and the right end is at $x = \frac{5\pi}{6}$, as shown in Figure 17.2.14(ii). Since the top curve is $y = \sin x$ and the bottom is $y = \frac{1}{2}$, the iterated integral is

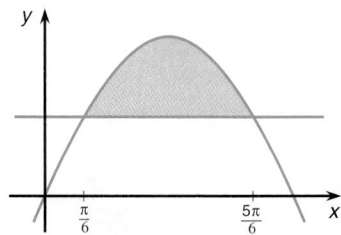

Figure 17.2.14(ii)

$$\int \int_S x \, dx \, dy = \int_{\pi/6}^{5\pi/6} \int_{1/2}^{\sin x} x \, dy \, dx = \int_{\pi/6}^{5\pi/6} xy \Big|_{y=1/2}^{y=\sin x} dx$$

$$= \int_{\pi/6}^{5\pi/6} \left(x \sin x - \frac{x}{2} \right) dx = \left(-x \cos x + \sin x - \frac{x^2}{4} \right) \Big|_{\pi/6}^{5\pi/6}$$

$$= -\frac{5\pi}{6}\left(-\frac{\sqrt{3}}{2} \right) + \frac{1}{2} - \frac{25\pi^2}{144} + \frac{\pi}{6}\frac{\sqrt{3}}{2} - \frac{1}{2} + \frac{\pi^2}{144}$$

$$= \frac{\sqrt{3}}{2}\pi - \frac{\pi^2}{6}.$$

The iterated integral equals $\frac{\sqrt{3}}{2}\pi - \frac{1}{6}\pi^2 \approx 1.076$.

In Example 5 we saw that one iterated integral may be much easier than the other to evaluate. Theorem 17.2.3 tells us how to evaluate double integrals using iterated integrals, but the iterated integrals may be difficult or impossible to evaluate. Not only do we have to decide whether it is possible to express the limits of integration for the inner integral in some reasonable form, but having done that, we still have to work out the integrals involved.

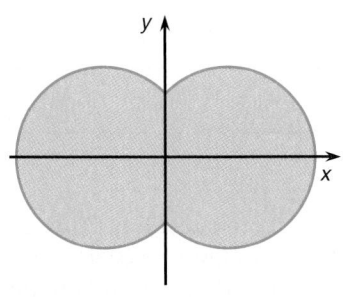

Figure 17.2.15(i)

EXAMPLE 6   In Figure 17.2.15(i) we see the region R consisting of the union of the disks of radius $\sqrt{2}$ whose centers are $(1, 0)$ and $(-1, 0)$. (These disks can be described as $(x-1)^2 + y^2 \leq 2$ and $(x+1)^2 + y^2 \leq 2$.) Find $\int \int_R x \, dA$.

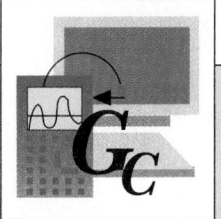

Sometimes it is not easy to see where the graphs intersect. With a graphics calculator we can plot the graphs and solve for the intersection.

For instance, suppose R is the region enclosed between $y = x^4 - 1$ and $y = x^3 + 2$ and evaluate

$$\iint_R x \, dx \, dy = \int_a^b \int_{x^4-1}^{x^3+2} x \, dy \, dx = \int_a^b (-x^4 + x^3 + 3) x \, dx = \left(-\frac{x^6}{6} + \frac{x^5}{5} + \frac{3x^2}{2} \right) \Big|_a^b.$$

TI-85	**SHARP EL9300**
GRAPH F1($y(x) =$) F1(x) ˆ 4 − 1 ▼ F1(x) ˆ 3 + 2 M2(RANGE) (−) 2 ▼ 4 ▼ ▼ (−) 3 ▼ 10 F5(GRAPH) SOLVER X ˆ 4 − X ˆ 3 − 3 ENTER 0 ENTER (−) 2 F5(SOLVE) EXIT X STO A ENTER SOLVER ENTER 0 ENTER 2 F5(SOLVE) EXIT X STO B ENTER (−) B ˆ 6 ÷ 6 + B ˆ 5 ÷ 5 + 3 B x^2 ÷ 2 + A ˆ 6 ÷ 6 − A ˆ 5 ÷ 5 − 3 A x^2 ÷ 2 ENTER	⌐∿ x/θ/T a^b 4 ▶ − 1 2ndF ▼ x/θ/T a^b 3 ▶ + 2 RANGE −2 ENTER 4 ENTER 1 ENTER −3 ENTER 10 ENTER 1 ENTER ⌐∿ SOLVER x/θ/T a^b 4 ▶ − 1 = x/θ/T a^b 3 ▶ + 2 ENTER (−) 2 ENTER ENTER ENTER ENTER ENTER ▦ X STO A ENTER SOLVER ENTER ENTER ENTER 1 ENTER ENTER ENTER ▦ X STO B ENTER (−) B a^b 6 ▶ ÷ 6 + B a^b 5 ▶ ÷ 5 + 3 B x^2 ÷ 2 + A a^b 6 ▶ ÷ 6 − A a^b 5 ▶ ÷ 5 − 3 A x^2 ÷ 2 ENTER
CASIO f_x-6300G (f_x-7700GB)	**HP 48SX**
Range −2 EXE 4 EXE 1 EXE −3 EXE 10 EXE 1 EXE EXE EXE EXE Graph X x^y 4 − 1 EXE Graph X x^y 3 + 2 EXE Trace ▶ (*move cursor to intersection*) X → A EXE Graph X x^y 3 + 2 EXE Trace ▶ (*move cursor to intersection*) X → B EXE (−) B x^y 6 ÷ 6 + B x^y 5 ÷ 5 + 3 B x^2 ÷ 2 + A x^y 6 ÷ 6 − A x^y 5 ÷ 5 − 3 A x^2 ÷ 2 EXE (Use Zoom for more accuracy)	⬑ EQUATION X y^x 4 ▶ − 1 ⬑ = X y^x 3 ▶ + 2 ENTER ⬑ SOLVE NEW E Q N ENTER 2 +/− ↱ CST NXT SF ⬑ PLOT PLOTR 2 +/− ENTER 4 XRNG 3 +/− ENTER 10 YRNG DRAW ◀ (*Move cursor to intersection*) FCN ISECT ATTN ATTN X ENTER ' A STO DRAW ▶ (*Move cursor to intersection*) FCN ISECT ATTN ATTN X ENTER ' B STO B SPC 6 y^x 6 +/− ÷ B SPC 5 y^x 5 ÷ + B x^2 3 × 2 ÷ + A SPC 6 y^x 6 ÷ + A SPC 5 y^x 5 ÷ − A x^2 3 × 2 ÷ −

Solution We could try integrating, either by using vertical cross-sections or by dividing the region along the y-axis and using horizontal cross-sections on each piece.

However, before we do that, let us observe that the integrand is negative on the left half of R and positive on the right half. The integrals over the two halves will

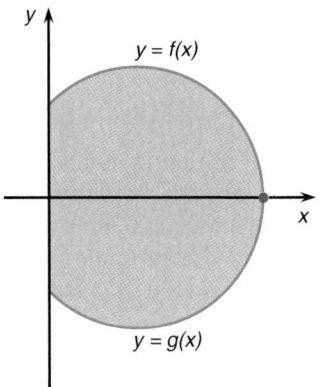

$y = f(x)$

$y = g(x)$

Figure 17.2.15(ii)

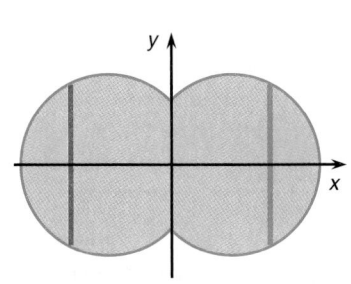

Figure 17.2.15(iii)

have opposite signs. But actually it is much more than that. The two halves have the same size, and although their shapes are not the same, they are each other's reflections.

Looking at the right half of R, let us call the top boundary $y = f(x)$ and the bottom boundary $y = g(x)$, as shown in Figure 17.2.15(ii). So the integral over the right half of R is

$$\int_0^{1+\sqrt{2}} \int_{g(x)}^{f(x)} x \, dy \, dx.$$

On the other hand the top boundary of the left half of R is $y = f(-x)$, and the bottom boundary is $y = g(-x)$. This is just another way of saying that the two halves are symmetrical about the y-axis. So the integral over the left half is

$$\int_{-1-\sqrt{2}}^{0} \int_{g(-x)}^{f(-x)} x \, dy \, dx.$$

Making the substitution $t = -x$, $dt = -dx$ in this integral (strictly speaking in the outer integral), we find that it equals

$$\int_{1+\sqrt{2}}^{0} \int_{g(t)}^{f(t)} (-t) \, dy (-dt) = \int_{1+\sqrt{2}}^{0} \int_{g(t)}^{f(t)} t \, dy \, dt = -\int_0^{1+\sqrt{2}} \int_{g(t)}^{f(t)} t \, dy \, dt.$$

The integral over the left half of R equals the negative of the integral over the right half, so $\int \int_R x \, dA = 0$. ◢

In Figure 17.2.15(iii) we see the two vertical cross-sections of R corresponding to x and $-x$. Because of the symmetry they have the same length. On one the integrand equals the "constant" x and on the other it equals $-x$, so the inner integrals over these two segments will be each other's negatives. This is true for each pair of opposite cross-sections, so when we add them up, we find that the double integrals over the two halves are each other's negatives.

This is a special case of the following result.

THEOREM 17.2.4

Suppose R is a region that is symmetrical about the y-axis, that is, if $(x, y) \in R$, then so is $(-x, y)$. Suppose that $F(x, y)$ is an "odd function in x," that is, $F(-x, y) = -F(x, y)$ for all $(x, y) \in R$. Then, assuming that the double integral exists, it is zero, that is,

$$\int \int_R F(x, y) \, dx \, dy = 0.$$

Similarly, if R is symmetrical about the x-axis (if $(x, y) \in R$, then $(x, -y) \in R$), and if $F(x, y)$ is odd with respect to y (i.e., $F(x, -y) = -F(x, y)$), then the double integral is also zero.

If R is symmetrical about the y-axis and $F(x, y)$ is even with respect to x (i.e., $F(-x, y) = F(x, y)$), then the integrals over the left and right halves of R are equal.

If R is symmetrical about the x-axis and $F(x, y)$ is even with respect to y (i.e., $F(x, -y) = F(x, y)$), then the integrals over the top and bottom halves of R are equal.

In Example 6 we essentially gave the proof of Theorem 17.2.4 assuming that R is the region between two graphs.

In these examples we have seen the basic ideas needed to set up iterated integrals. They are summarized below.

Summary 17.2.5   **Setting Up Iterated Integrals**

1. Begin by sketching the region, identifying the curves that form its boundary.

2. Consider the horizontal and vertical cross-sections. If all the horizontal cross-sections are single segments, then it is possible to use x as the variable of integration in the inner integral. If the vertical cross-sections are single segments, then it is possible to use y. If neither the horizontal nor the vertical cross-sections are single unbroken segments, then it is necessary to divide the region into smaller regions that do satisfy at least one of these conditions.

3. On each region with unbroken cross-sections it is necessary to express the endpoints of the cross-sections as the graphs of two functions, as described in Theorem 17.2.3. This may be difficult, and it may be easier for one variable than for the other. It may also happen that on some regions it is better to use x as the inner variable, while on others y is preferable.

4. On each region that lies between the graphs of two functions, set up the iterated integrals according to Theorem 17.2.3, being careful which curve gives the upper limit and which the lower limit of integration for the inner integral.

5. Evaluate each iterated integral and add up the integrals corresponding to the various pieces that make up the original region. Before actually evaluating integrals, it is worthwhile to check whether there are obvious reasons why any of them equals zero or any two of them may be equal. Such situations are likely to arise when two parts of the region are symmetrical. See Theorem 17.2.4.

   Of course, the actual evaluation of the integrals may turn out to be difficult or impossible. Sometimes when this happens, reversing the order in some of the iterated integrals will make them easier to find.

Exercises 17.2

Evaluate each of the following iterated integrals. In each case describe the corresponding region R in the plane and make a quick sketch of R.

1. $\int_0^1 \int_0^x x^2 y \, dy \, dx$

2. $\int_{-1}^1 \int_x^{2x} xy \, dy \, dx$

3. $\int_0^1 \int_{-y}^y y \, dx \, dy$

4. $\int_1^2 \int_1^{\sqrt{x}} 6xy \, dy \, dx$

5. $\int_1^e \int_1^y \frac{1}{x} dx \, dy$

6. $\int_0^\pi \int_1^y \sin y \, dx \, dy$

7. $\int_0^{\pi/2} \int_0^{\sin y} 1 \, dx \, dy$

8. $\int_0^{\pi/4} \int_0^y 1/(1+x^2) \, dx \, dy$

9. $\int_{-1}^0 \int_0^{1-x} (xy - 2x) \, dy \, dx$

10. $\int_0^{\pi/4} \int_0^y (\sin^2 xy + \cos^2 xy) \, dx \, dy$

11. $\int_2^3 \int_1^{x^2-1} \frac{1}{y^2} dy \, dx$

12. $\int_0^\pi \int_0^{\sin y} 2x \sec y \, dx \, dy$

13. $\int_1^2 \int_1^y \frac{y^2}{x} dx \, dy$

14. $\int_0^1 \int_{-\sqrt{y}}^{\sqrt{y}} \frac{1}{1+y^{3/2}} dx \, dy$

15. $\int_{-1}^1 \int_0^{1-x^2} \frac{x}{y+1} dy \, dx$

16. $\int_{-2}^2 \int_{|x|}^2 xy\sqrt{4-y^2} \, dy \, dx$

17. $\int_1^2 \int_0^{2\ln x} e^{x+y} \, dy \, dx$

18. $\int_0^1 \int_{-y}^y \sinh(y^2) \, dx \, dy$

19. $\int_0^1 \int_{-y^{1/3}}^{y^{1/2}} 3x^2 y \, dx \, dy$

20. $\int_0^{\ln 2} \int_0^x (2 \cosh y \sinh y - \cosh^2 x) \, dy \, dx$

Integrate each of the following functions $f(x, y)$ over the specified region R.

21. $f(x, y) = 2x$, R the region between $y = 0$ and $y = 1 - x^2$.

22. $f(x, y) = x^2 - y$, R the region to the right of $x = y^2$ and to the left of $x + y = 2$.

23. $f(x, y) = 12xy^2$, R the region inside $x^2 + y^2 = 4$ and above the line $x - y + 2 = 0$.

24. $f(x, y) = 6x - 6y$, R the upper semidisk $x^2 + y^2 \leq 1$, $y \geq 0$.

25. $f(x, y) = x - y$, R the region inside the circle of radius 1 with center $(4, -2)$.

26. $f(x, y) = 12x^3$, R the region inside the circle of radius 3 with center $(1, 1)$ and above the x-axis.

27. $f(x, y) = 5x^4 y$, R the region inside the rectangle whose vertices are $(1, 1)$, $(1, -1)$, $(-1, 1)$, $(-1, -1)$.

28. $f(x, y) = 4$, R the region inside the triangle with vertices $(0, 0)$, $(2, 3)$, $(3, 2)$.

29. $f(x, y) = \frac{1}{x}$, R the region between $y = x^2 - 4x + 3$ and the x-axis.

30. $f(x, y) = x^3 \frac{1}{1+y^2}$, R the region inside $x^2 + y^2 = 4$ and outside $x^2 + y^2 = 1$.

31. $f(x, y) = e^{x+y}$, $R = \{(x, y) : 1 \leq x \leq e, 0 \leq y \leq \ln x\}$

32. $f(x, y) = \frac{1}{y}$, $R = \{(x, y) : 1 \leq x \leq 2, 1 \leq y \leq x\}$

33. $f(x, y) = \sin(\pi xy)$, $R = \{(x, y) : |x| \leq 1, |y| \leq 2\}$

34. $f(x, y) = y \sec^2\left(\frac{\pi x}{4}\right)$, $R = \{(x, y) : |x| \leq 1, 1 \leq y \leq 3\}$

35. $f(x, y) = \frac{1}{\sqrt{1-y^2}}$, $R = \{(x, y) : 0 \leq x \leq \frac{\pi}{2}, 0 \leq y \leq \sin x\}$

36. $f(x, y) = \frac{x}{1+y^2}$, $R = \{(x, y) : 0 \leq x \leq 1, |y| \leq x\}$

37. $f(x, y) = \frac{1}{y\sqrt{x}}$, $R = \{(x, y) : y \geq 0, 1 \leq x \leq y^2 \leq 4\}$

38. $f(x, y) = \frac{1}{x^2 y^2}$, $R = \{(x, y) : 1 \leq x \leq y \leq 2x \leq 4\}$

39. $f(x, y) = \frac{1}{x}$, $R = \{(x, y) : 1 \leq x \leq 2 - y, y \geq 0\}$

40. $f(x, y) = \sin(x^2)$, $R = \{(x, y) : 1 \leq x \leq 2y \leq 3x \leq 6\}$

41. $f(x, y) = x^3$, R the disk of radius 2 around $(0, -2)$.

42. $f(x, y) = x^7$, R the region between $y = \sin x$ and the x-axis for $-\pi \leq x \leq \pi$.

43. $f(x, y) = 3 - 4y$, R the disk of radius 5, center $(-2, 0)$.

44. $f(x, y) = -2$, R the two disks of radius 3 around $(1, 2)$ and $(4, -6)$.

45. $f(x, y) = 2x + 3y$, R the two disks of radius 2 around $(-2, 0)$ and $(2, 0)$.

46. $f(x, y) = 5x^3 - y + 4$, R the union of the disks of radius 3 around $(0, 2)$ and $(0, -2)$.

47. $f(x, y) = \sin x$, R the region below $y = |x|$ and above $y = 0$ for $-3 \leq x \leq 3$.

48. $f(x, y) = y \cos y$, R the two squares with horizontal and vertical sides of length 2 and centers $(2, 4)$, $(2, -4)$.

49. $f(x, y) = xy + 2$, R the disk of radius 5 around the origin.

50. $f(x, y) = 0$, R the region enclosed between $y = 0$ and $y = \sin \frac{1}{x}$ for $\frac{1}{20\pi} \leq x \leq \frac{1}{\pi}$.

II

51. (i) Set up the iterated integral for the integral of Example 1 (p. 1064), using y as the variable of integration in the inner integral.

   (ii) Evaluate the iterated integral you have set up in part (i), checking that your answer is the same.

52. (i) Set up the iterated integral for the integral of Example 3 (p. 1067), using x as the variable of integration in the inner integral.

   (ii) Evaluate the iterated integral you have set up in part (i), checking that your answer is the same.

53. Consider $\int \int_R x \cos y \, dx \, dy$, where R is the union of the disks of radius 1 centered at $(4, 4)$ and $(-4, -4)$. Show that it is zero without evaluating it. (*Hint:* Compare the integrals over the two disks.)

54. Evaluate the integrals over both halves of the region in Example 6.

55. (i) Set up the iterated integral for the integral of Example 4 (p. 1070), using x as the variable of integration in the inner integral.

   (ii) Evaluate the iterated integral you have set up in part (i), checking that your answer is the same.

56. Let D be the disk of radius r centered at $(0, 2)$. Evaluate $\int \int_D y \sin(x^3) \, dx \, dy$.

III

57. Let R be the plane region enclosed between the graphs $y = 2x^3 - x$ and $y = x^3 + 2x + 1$. Use a graphics calculator or computer to help find $\int \int_R x \, dA$.

58. Let S be the plane region enclosed between the graphs $y = 2x \sin x$ and $y = x^3$. Use a graphics calculator or computer to help find $\int \int_S x \, dA$.

17.3 Areas and Volumes

AREAS OF REGIONS IN PLANE
VOLUMES OF REGIONS IN SPACE

Figure 17.3.1

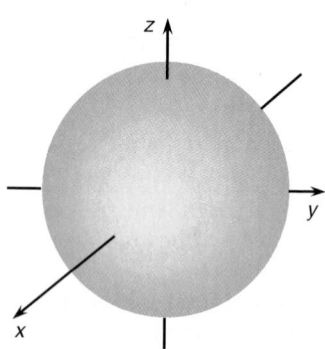

Figure 17.3.2

We defined double integrals by thinking about the volume of the region under the graph of a function (see Figure 17.3.1). We have also observed that the double integral of a constant function is equal to the constant times the area of the region:

$$\int\int_R c \, dA = \int\int_R c \, dx \, dy = (c) \, area(R).$$

In particular, if $c = 1$, we see that $\int\int_R 1 \, dA = area(R)$.

This means that double integrals can be used to find the volumes of many three-dimensional regions and the areas of many two-dimensional ones. It is the purpose of this section to learn how to use them in this way.

EXAMPLE 1 Find the volume of a sphere of radius r.

Solution We consider the region inside the sphere centered at the origin, that is, inside $x^2 + y^2 + z^2 = r^2$, shown in Figure 17.3.2. We can think of this solid region as the set of all points lying between the upper hemisphere $z = \sqrt{r^2 - x^2 - y^2}$ and the lower hemisphere $z = -\sqrt{r^2 - x^2 - y^2}$. Instead of calculating the volume all at once, it is easier to calculate the volume of the upper hemisphere and double it. The volume of the upper hemisphere is just the volume under the graph of the function $z = \sqrt{r^2 - x^2 - y^2}$, so it is just $\int\int_R \sqrt{r^2 - x^2 - y^2} \, dx \, dy$.

The region R over which we must integrate in the xy-plane is the region inside the circle $x^2 + y^2 = r^2$. Let us form cross-sections by letting y run from the lower semicircle $y = -\sqrt{r^2 - x^2}$ to the upper semicircle $y = \sqrt{r^2 - x^2}$, as shown in Figure 17.3.3(i). Adding up the integrals over the corresponding cross-sections (as shown in Figure 17.3.3(ii)), we obtain the iterated integral

$$\int_{-r}^{r} \int_{-\sqrt{r^2-x^2}}^{\sqrt{r^2-x^2}} \sqrt{r^2 - x^2 - y^2} \, dy \, dx.$$

First we must evaluate the inner integral. To simplify the notation, let us write $a = \sqrt{r^2 - x^2}$; the inner integral then becomes

$$\int_{-a}^{a} \sqrt{a^2 - y^2} \, dy.$$

Figure 17.3.3(i)

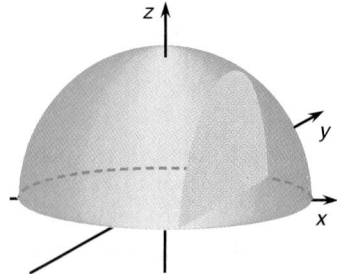

Figure 17.3.3(ii)

Making the substitution $y = a \sin \theta$, $dy = a \cos \theta \, d\theta$, with θ running from $-\frac{\pi}{2}$ to $\frac{\pi}{2}$, we find that this integral equals

$$\int_{-\pi/2}^{\pi/2} (a \cos \theta)(a \cos \theta) \, d\theta = a^2 \int_{-\pi/2}^{\pi/2} \cos^2 \theta \, d\theta = \frac{a^2}{2} \int_{-\pi/2}^{\pi/2} (1 + \cos 2\theta) \, d\theta$$

$$= \frac{a^2}{2} \left(\theta + \frac{1}{2} \sin 2\theta \right) \Bigg|_{-\pi/2}^{\pi/2} = \frac{a^2}{2} \left(\frac{\pi}{2} + 0 - \left(-\frac{\pi}{2} \right) - 0 \right) = \frac{\pi a^2}{2}.$$

Since $a^2 = r^2 - x^2$, this means that the iterated integral is

$$\int_{-r}^{r} \int_{-\sqrt{r^2 - x^2}}^{\sqrt{r^2 - x^2}} \sqrt{r^2 - x^2 - y^2} \, dy \, dx = \int_{-r}^{r} \frac{\pi}{2} (r^2 - x^2) \, dx = \frac{\pi}{2} \left(r^2 x - \frac{x^3}{3} \right) \Bigg|_{-r}^{r}$$

$$= \frac{\pi}{2} \left(2r^3 - \frac{2r^3}{3} \right) = \frac{2\pi r^3}{3}.$$

Doubling, we find that the volume of the whole sphere is $\frac{4}{3}\pi r^3$.

We should probably be slightly embarrassed to realize that it was actually easier to find this volume as the volume of a solid of revolution (cf. Example 6.3.1).

We have an intuitive idea of what is meant by volume, but in fact we cannot really say what it is except for a few simple regions. Double integrals actually give us a way to define what volume means.

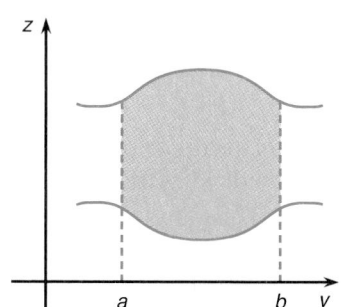

Figure 17.3.4

DEFINITION 17.3.1

Suppose R is a region in the xy-plane and $f(x, y)$ is a continuous nonnegative function defined on R. Let S be the region lying under the graph of $z = f(x, y)$ and above the region R in the xy-plane. Then the **volume** of S is defined to be the following double integral, provided it exists:

$$Volume(S) = \int\int_R f(x, y) \, dA = \int\int_R f(x, y) \, dx \, dy.$$

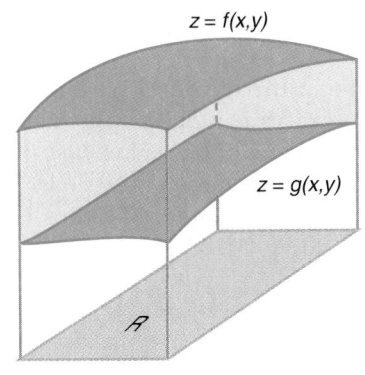

Figure 17.3.5

In doing Example 1 we encountered the region lying between two graphs, rather than the region lying under one graph. This kind of problem is analogous to using integrals with a single variable to find the area of the region lying between two curves $y = f(x)$ and $y = g(x)$, as shown in Figure 17.3.4. There we found that the area is $\int |f(x) - g(x)| \, dx$, or if we know that $f(x) \geq g(x)$, say, then the area is $\int \big(f(x) - g(x)\big) \, dx$.

In Figure 17.3.5 we see the graphs of two functions $z = f(x, y)$ and $z = g(x, y)$. If we integrate $\int\int_R f(x, y) \, dx \, dy$, it will give us the volume under $z = f(x, y)$. The integral $\int\int_R g(x, y) \, dx \, dy$ is the volume of the smaller region under $z = g(x, y)$. The volume between the two graphs will be the difference. This explains the following definition.

DEFINITION 17.3.2

If $f(x, y)$ and $g(x, y)$ are continuous functions defined on a region R in the plane, then the **volume of the region lying between their graphs** is

$$\int\int_R |f(x, y) - g(x, y)| \, dA = \int\int_R |f(x, y) - g(x, y)| \, dx \, dy.$$

> If $f(x, y) \geq g(x, y)$ for all $(x, y) \in R$, then this equals
>
> $$\int\int_R \big(f(x, y) - g(x, y)\big)\,dA = \int\int_R \big(f(x, y) - g(x, y)\big)\,dx\,dy.$$

Note that the functions $f(x, y)$ and $g(x, y)$ do not have to be nonnegative, even though in Figure 17.3.5 they both were. As an exercise you might like to use this idea to find the volume of a sphere directly, rather than by doubling the volume of a hemisphere. Of course, it comes to nearly the same calculation.

EXAMPLE 2   Find the volume of an eraser made in the shape of the region S enclosed between the two paraboloids $z = x^2 + y^2$, $z = 4 - x^2 - y^2$, over the square $|x| \leq 1$, $|y| \leq 1$. (All measurements are in centimeters.)

Solution   We illustrate the region in Figure 17.3.6. Thinking of the upper paraboloid as the graph of $z = f(x, y) = 4 - x^2 - y^2$ and the lower paraboloid as the graph of $z = g(x, y) = x^2 + y^2$, we find that the volume of S is

$$\int_{-1}^{1}\int_{-1}^{1} \big(f(x, y) - g(x, y)\big)\,dx\,dy = \int_{-1}^{1}\int_{-1}^{1} (4 - x^2 - y^2 - x^2 - y^2)\,dx\,dy$$

$$= \int_{-1}^{1}\int_{-1}^{1} (4 - 2x^2 - 2y^2)\,dx\,dy = \int_{-1}^{1}\left(4x - \frac{2}{3}x^3 - 2y^2 x\right)\Bigg|_{x=-1}^{x=1}\,dy$$

$$= \int_{-1}^{1}\left(8 - \frac{4}{3} - 4y^2\right)dy = \left(8y - \frac{4}{3}y - \frac{4}{3}y^3\right)\Bigg|_{-1}^{1} = \frac{32}{3}.$$

The volume of the eraser is $\frac{32}{3}$ cm^3.

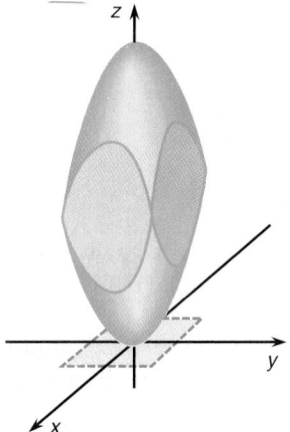

Figure 17.3.6

EXAMPLE 3   Find the volume of the region S specified by the inequalities $x^2 + z^2 \leq 1$ and $y^2 + z^2 \leq 1$.

Solution   First it is necessary to understand the shape of the region S. We begin with the inequality $x^2 + z^2 \leq 1$ and observe that in the xz-plane it defines the disk of radius 1 around the origin. Since it does not involve y at all, the region that it determines in all of $\mathbb{R}^3$ is the cylinder around the y-axis with radius 1, as shown in Figure 17.3.7(i). The other inequality $y^2 + z^2 \leq 1$ gives the cylinder of radius 1 around the x-axis, as shown together with the first cylinder in Figure 17.3.7(ii).

Figure 17.3.7(i)

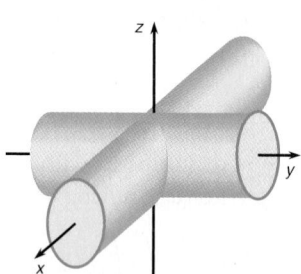

Figure 17.3.7(ii)

The region S is the intersection of these two cylinders, shown on a larger scale in Figure 17.3.7(iii). From the inequality $x^2 + z^2 \leq 1$ we see in particular that x must always lie between -1 and 1, and similarly, from $y^2 + z^2 \leq 1$ we find that $y \in [-1, 1]$. For a fixed value of $y \in [-1, 1]$ we see that z can range between $\pm\sqrt{1 - y^2}$ and still satisfy $y^2 + z^2 \leq 1$. In particular this inequality holds with $z = 0$; with this choice of z the other inequality $x^2 + z^2 \leq 1$ is satisfied with any $x \in [-1, 1]$.

What this means is that the region R lying "under" S in the xy-plane is the square $-1 \leq x \leq 1$, $-1 \leq y \leq 1$ shown in Figure 17.3.7(iv). Looking down on this region R from above, as in Figure 17.3.7(v), we see that the boundary of S is one of the two cylinders over some parts of R and the other cylinder over other parts of R.

For any point $(x, y) \in R$ the top boundary point of S lying above (x, y) will be the point on whichever of the two cylinders is lower at (x, y). Since the tops of the two cylinders are $z = \sqrt{1 - x^2}$ and $z = \sqrt{1 - y^2}$, the first will be lower when $x^2 > y^2$, and the second will be lower when $x^2 < y^2$. The bottom boundary of S is just the negative of the top boundary, $z = -\sqrt{1 - x^2}$ or $z = -\sqrt{1 - y^2}$.

So if $|x| \geq |y|$, then the integrand will be $\sqrt{1 - x^2} - (-\sqrt{1 - x^2}) = 2\sqrt{1 - x^2}$, and where $|x| \leq |y|$, the integrand will be $\sqrt{1 - y^2} - (-\sqrt{1 - y^2}) = 2\sqrt{1 - y^2}$. It will be necessary to integrate separately over the parts of R where $|x| \geq |y|$ and where $|x| \leq |y|$.

Let us integrate first over the part of R where $x \geq 0$ and $-x \leq y \leq x$, as shown in Figure 17.3.7(vi). Here $|x| \geq |y|$, so the integral is

$$\int_0^1 \int_{-x}^x 2\sqrt{1 - x^2}\, dy\, dx = 2\int_0^1 \sqrt{1 - x^2}\left(y\Big|_{-x}^x\right) dx = 2\int_0^1 \sqrt{1 - x^2}\,(2x)\, dx$$

$$= 2\int_1^0 \sqrt{u}\,(-1)\, du = 2\int_0^1 \sqrt{u}\, du = \frac{4}{3}u^{3/2}\Big|_0^1 = \frac{4}{3}.$$

Note the change in the limits of integration when we substituted $u = 1 - x^2$. It is easy to see that the integral over the region where $x \leq 0$ and $x \leq y \leq -x$ is exactly equal to this by symmetry.

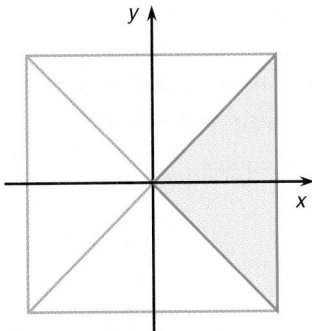

Figure 17.3.7(vi)

Next we consider the integral over the region where $y \geq 0$ and $-y \leq x \leq y$. Here $|x| \leq |y|$, so the integrand is $2\sqrt{1 - y^2}$. The integration is most conveniently carried out with x as the inner variable, and the integral over this part of R is

$$\int_0^1 \int_{-y}^y 2\sqrt{1 - y^2}\, dx\, dy.$$

Figure 17.3.7(iii)

Figure 17.3.7(iv)

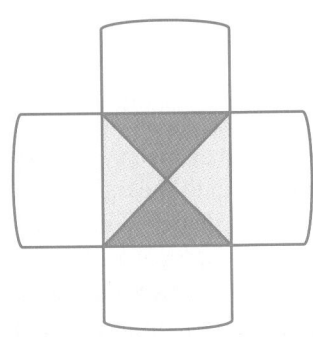

Figure 17.3.7(v)

We could evaluate this integral, but it is easier to see that it is the same as the one we just worked out except that x and y have been interchanged. This does not change the value of the integral, so it also equals $\frac{4}{3}$. An argument about symmetry shows that the integral over the remaining quarter of the square R also equals $\frac{4}{3}$.

We have divided the square R into four regions, over each of which the integral is $\frac{4}{3}$. Adding, we find that the volume of S is $\frac{16}{3}$.

Notice that this region is not a solid of revolution.

EXAMPLE 4   Consider a log of radius 1 ft. A wedge is cut off it by a saw cutting at an angle of 30° from the vertical, as shown in Figure 17.3.8(i). The cut emerges exactly halfway down the circular face of the log. What is the volume of the wedge?

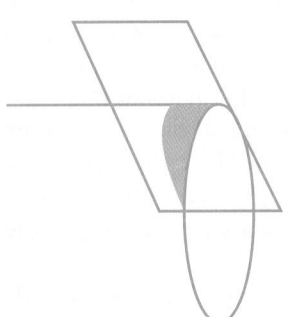

Figure 17.3.8(i)

Solution   For convenience let us place the wedge on the xy-plane with the cut side facing up and so that the semicircular base fits over the right half of the circle $x^2 + y^2 = 1$ around the origin (see Figure 17.3.8(ii)).

The volume we want to find is then the volume lying over that semidisk and under the plane that marks the top of the wedge. We will have to integrate over the semidisk in the plane, but before doing anything else, we need to find the equation of that plane.

Now that the wedge has been moved, the plane makes an angle of 30° with respect to the horizontal xy-plane. It also contains the x-axis. If a plane contains the x-axis, then its normal vector must be of the form $\mathbf{n} = (0, b, c)$. If the plane makes an angle of 30° with respect to the horizontal plane, then we know that their normal vectors make an angle of 30°. This means

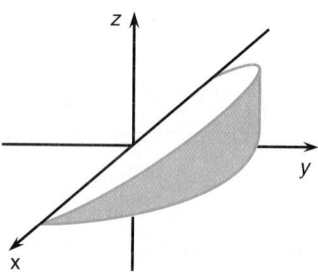

Figure 17.3.8(ii)

$$\frac{(0, 0, 1) \cdot (0, b, c)}{\|(0, 0, 1)\| \|(0, b, c)\|} = \cos 30° = \frac{\sqrt{3}}{2}, \quad \text{or} \quad \frac{c}{\sqrt{b^2 + c^2}} = \frac{\sqrt{3}}{2}.$$

If we let $c = \sqrt{3}$, then we see that $b = \pm 1$. If $b = 1$, it means that the y-component of $\mathbf{n}$ is positive, which means that $\mathbf{n}$ leans to the right, and that means that the plane slopes down to the right. In our picture the plane slopes up to the right, so we choose $b = -1$. The normal vector is $\mathbf{n} = (0, -1, \sqrt{3})$; since the plane goes through the origin, it must be $-y + \sqrt{3}z = 0$.

This can be written as the graph of a function: $z = \frac{y}{\sqrt{3}}$. The volume of the wedge is $\int \int_R \frac{y}{\sqrt{3}} \, dA$, where R is the region contained in the right half of the circle $x^2 + y^2 = 1$. The volume equals

$$\int \int_R \frac{y}{\sqrt{3}} \, dA = \int_0^1 \int_{-\sqrt{1-y^2}}^{\sqrt{1-y^2}} \frac{y}{\sqrt{3}} \, dx \, dy = \frac{1}{\sqrt{3}} \int_0^1 yx \Big|_{-\sqrt{1-y^2}}^{\sqrt{1-y^2}} \, dy$$

$$= \frac{1}{\sqrt{3}} \int_0^1 2y\sqrt{1-y^2} \, dy = \frac{1}{\sqrt{3}} \int_1^0 \sqrt{u}(-1) \, du$$

$$= \frac{1}{\sqrt{3}} \int_0^1 \sqrt{u} \, du = \frac{2}{3\sqrt{3}} u^{3/2} \Big|_0^1 = \frac{2}{3\sqrt{3}}.$$

Again we changed the limits of integration according to the substitution $u = 1 - y^2$. The volume of the wedge is $\frac{2}{3\sqrt{3}} \approx 0.385$ ft³.

It is also possible to use double integrals to find the areas of regions in the plane. This approach is based on something we have already observed, namely, that the integral of a constant equals the product of the constant and the area of the region. As with volumes we know what area means for a few simple regions, and then we have defined areas using integrals of functions of one variable. We can now use double integrals to define areas for other regions, that is, ones that cannot be constructed from the regions under the graphs of functions. Strictly speaking, we ought to check that the two definitions agree for the region lying under the graph of a function, but we omit this step, remarking that it is true.

DEFINITION 17.3.3

Suppose R is a region in the plane that is bounded, that is, contained in some rectangle. If the integral $\int \int_R 1\, dx\, dy$ exists, then we call its value the **area** of R:

$$Area(R) = \int \int_R 1\, dA = \int \int_R 1\, dx\, dy.$$

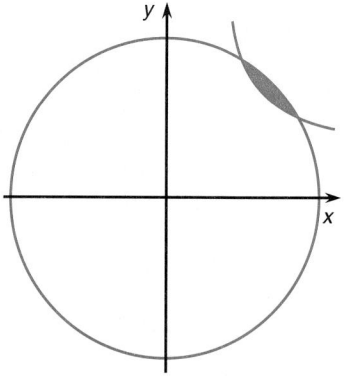

Figure 17.3.9(i)

EXAMPLE 5   Find the area of the region R that is enclosed between the circle $x^2 + y^2 = 4$ and the branch of the hyperbola $xy = \sqrt{3}$ for which $x > 0$.

Solution   The region R is illustrated in Figure 17.3.9(i). It is possible to use either x or y as the inner variable, but let us use x.

First it is necessary to find the points of intersection of the hyperbola and the circle. To do this, we substitute $y = \frac{\sqrt{3}}{x}$ into $x^2 + y^2 = 4$ and find that $x^2 + \frac{3}{x^2} = 4$. Multiplying by x^2, we get the equation $x^4 + 3 = 4x^2$, or $(x^2)^2 - 4x^2 + 3 = 0$. This is a quadratic equation in x^2, and it is easily factored: $(x^2 - 3)(x^2 - 1) = 0$. We see that x^2 must equal 3 or 1; the points of intersection are $(\sqrt{3}, 1)$ and $(1, \sqrt{3})$.

This means that the outer integral ranges from $y = 1$ to $y = \sqrt{3}$. To find the limits of integration for the inner integral, we consult Figure 17.3.9(ii). There we see several horizontal cross-sections, corresponding to different values of y. The left end of each cross-section is on the hyperbola, where $x = \frac{\sqrt{3}}{y}$, and the right end is on the circle, where $x = \sqrt{4 - y^2}$. The area of the region R is the iterated integral

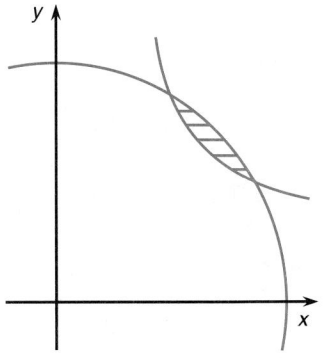

Figure 17.3.9(ii)

$$\int_1^{\sqrt{3}} \int_{\sqrt{3}/y}^{\sqrt{4-y^2}} 1\, dx\, dy = \int_1^{\sqrt{3}} x \Big|_{\sqrt{3}/y}^{\sqrt{4-y^2}} dy = \int_1^{\sqrt{3}} \left(\sqrt{4 - y^2} - \frac{\sqrt{3}}{y} \right) dy.$$

The second of these terms is easily integrated. For the first we let $y = 2\sin\theta$, with θ running from $\frac{\pi}{6}$ to $\frac{\pi}{3}$. The above integral becomes

$$\int_{\pi/6}^{\pi/3} 4\cos^2\theta\, d\theta - \int_1^{\sqrt{3}} \frac{\sqrt{3}}{y}\, dy = 2\int_{\pi/6}^{\pi/3} (1 + \cos 2\theta)\, d\theta - \sqrt{3}\ln y \Big|_1^{\sqrt{3}}$$

$$= 2\left(\theta + \frac{1}{2}\sin 2\theta \right) \Big|_{\pi/6}^{\pi/3} - \sqrt{3}\ln y \Big|_1^{\sqrt{3}}$$

$$= 2\left(\frac{\pi}{3} - \frac{\pi}{6} \right) + \left(\sin\frac{2\pi}{3} - \sin\frac{\pi}{3} \right) - \sqrt{3}\ln\sqrt{3}$$

$$= \frac{\pi}{3} - \frac{\sqrt{3}}{2}\ln 3.$$

The area of R is $\frac{\pi}{3} - \frac{\sqrt{3}}{2}\ln 3 \approx 0.0958$.

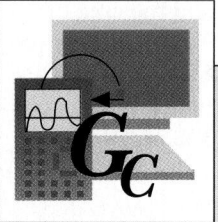

Sometimes it is not easy to see where the graphs intersect, but a graphics calculator can help us solve for the intersection.

For instance, suppose R is the region enclosed between $y = x^4 + x$, $y = 2 - x^2$, $z = 0$ and $z = x + 2$. We solve for the intersections a and b of $y = x^4 + x$ and $y = 2 - x^2$, and then find that the volume of R is

$$\int_a^b \int_{x^4+x}^{2-x^2} \int_0^{x+2} 1 \, dz \, dy \, dx = \int_a^b \int_{x^4+x}^{2-x^2} (x + 2) \, dy \, dx = \int_a^b (x + 2)(-x^4 - x^2 - x + 2) \, dx$$

$$= \int_a^b (-x^5 - 2x^4 - x^3 - 3x^2 + 4) \, dx = \left(-\frac{x^6}{6} - \frac{2x^5}{5} - \frac{x^4}{4} - x^3 + 4x \right) \Bigg|_a^b.$$

TI-85

GRAPH F1($y(x)$ =) F1(x) ^ 4 +
F1(x) ▼ 2 − F1(x) x^2 M2(RANGE)
(−) 2 ▼ 2 ▼ ▼ (−) 2 ▼ 3
F5(GRAPH) SOLVER $\boxed{X}$ ^ 4 + $\boxed{X}$
x^2 + $\boxed{X}$ − 2 ENTER 0 ENTER (−) 2
F5(SOLVE) EXIT $\boxed{X}$ STO A ENTER
SOLVER ENTER 0 ENTER 1
F5(SOLVE) EXIT $\boxed{X}$ STO B ENTER
(−) $\boxed{B}$ ^ 6 ÷ 6 − 2 $\boxed{B}$ ^ 5 ÷ 5 − $\boxed{B}$ ^ 4 ÷
4 − $\boxed{B}$ ^ 3 + 4 $\boxed{B}$ + $\boxed{A}$ ^ 6 ÷ 6 + 2 $\boxed{A}$ ^ 5
÷ 5 + $\boxed{A}$ ^ 4 ÷ 4 + $\boxed{A}$ ^ 3 − 4 $\boxed{A}$ ENTER

SHARP EL9300

⏣ x/θ/T a^b 4 ▶ + x/θ/T 2ndF ▼ 2
− x/θ/T x^2 RANGE −2 ENTER 2
ENTER 1 ENTER −2 ENTER 3 ENTER 1
ENTER ⏣ SOLVER x/θ/T a^b 4
▶ + x/θ/T $=$ 2 − x/θ/T x^2 ENTER (−)
2 ENTER ENTER ENTER ENTER ENTER
$\boxed{\tfrac{+-}{\times\div}}$ $\boxed{X}$ STO A ENTER SOLVER ENTER
ENTER ENTER 0 ENTER ENTER ENTER
$\boxed{\tfrac{+-}{\times\div}}$ $\boxed{X}$ STO B ENTER (−) $\boxed{B}$ a^b 6 ▶ ÷
6 − 2 $\boxed{B}$ a^b 5 ▶ ÷ 5 − $\boxed{B}$ a^b 4 ▶ ÷
4 − $\boxed{B}$ a^b 3 ▶ + 4 $\boxed{B}$ + $\boxed{A}$ a^b 6 ▶ ÷
6 + 2 $\boxed{A}$ a^b 5 ▶ ÷ 5 + $\boxed{A}$ a^b 4 ▶ ÷
4 + $\boxed{A}$ a^b 3 ▶ − 4 $\boxed{A}$ ENTER

CASIO f_x-6300G (f_x-7700GB)

Range −2 EXE 2 EXE 1 EXE −2 EXE
3 EXE 1 EXE EXE EXE EXE Graph $\boxed{X}$
x^y 4 + $\boxed{X}$ EXE Graph 2 − $\boxed{X}$ x^2 EXE
Trace ▶ (*move cursor to intersection*)
$\boxed{X}$ → $\boxed{A}$ EXE Graph 2 − $\boxed{X}$ x^2 EXE
Trace ▶ (*move cursor to intersection*)
$\boxed{X}$ → $\boxed{B}$ EXE (−) $\boxed{B}$ x^y 6 ÷ 6 − 2 $\boxed{B}$ x^y
5 ÷ 5 − $\boxed{B}$ x^y 4 ÷ 4 − $\boxed{B}$ x^y 3 + 4 $\boxed{B}$ +
$\boxed{A}$ x^y 6 ÷ 6 + 2 $\boxed{A}$ x^y 5 ÷ 5 + $\boxed{A}$ x^y 4 ÷
4 + $\boxed{A}$ x^y 3 − 4 $\boxed{A}$ EXE
(Use Zoom for more accuracy)

HP 48SX

↰ EQUATION X y^x 4 ▶ + X ↰
$=$ 2 − X y^x 2 ENTER ↰ SOLVE
NEW E Q N ENTER 2 +/− ↱
CST NXT SF ↰ PLOT PLOTR 2
+/− ENTER 2 XRNG 2 +/− ENTER 3
YRNG DRAW ◀ (*Move cursor to
intersection*) FCN ISECT ATTN
ATTN X ENTER ' A STO DRAW
▶ (*Move cursor to intersection*) FCN
ISECT ATTN ATTN X ENTER ' B
STO B +/− 6 y^x 6 ÷ B SPC 5 y^x 5 ÷
2 × − B SPC 4 y^x 4 ÷ − B SPC 3 y^x
− B SPC 4 × + A SPC 6 y^x 6 ÷ + A
SPC 5 y^x 5 ÷ 2 × + A SPC 4 y^x 4 ÷
+ A SPC 3 y^x + A SPC 4 × −

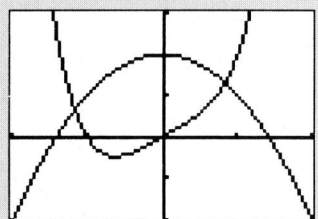

In fact the work we did for Example 5 is hardly any different from what we would have done before we learned about double integrals. Then we would have regarded R as the region between the graphs of two functions and found its area by integrating with respect to a single variable. You might like to work it out that way just to compare the two approaches.

EXAMPLE 6   **Changing the Order in an Iterated Integral**   Consider the integral

$$\int_0^1 \int_{\sqrt{1-y}/2}^{\sqrt{1-y^2}} f(x, y) \, dx \, dy.$$

Express it as an iterated integral in the opposite order.

Solution   To solve this problem, it is essential to understand the region R over which the integral is evaluated. It extends between $y = 0$ and $y = 1$, and horizontally it runs from $x = \frac{1}{2}\sqrt{1-y}$ to $x = \sqrt{1-y^2}$. The first of these amounts to $x^2 = \frac{1}{4}(1-y)$, which is the parabola $y = 1 - 4x^2$. The second can be written as $x^2 = 1 - y^2$, which is the circle of radius 1 with center at the origin. It is easy to find that these curves intersect at $(0, 1)$. The region is shown in Figure 17.3.10(i).

Now we can express the iterated integral in the opposite order, but there is one difficulty. The inner integral will now be over vertical cross-sections, and near the left side of the region the vertical cross-sections will run between the two curves. Near the right side they will run from the x-axis up to the circle (see Figure 17.3.10(ii)). We deal with this problem by dividing the region into two pieces, the first for $0 \le x \le \frac{1}{2}$ and the second for $\frac{1}{2} \le x \le 1$. Then we deal with the two pieces separately.

On the first piece the inner integral runs from the parabola $y = 1 - 4x^2$ up to the circle $y = \sqrt{1 - x^2}$, while on the other it runs from the x-axis $y = 0$ to the circle. The original integral can be written as

$$\int_0^{1/2} \int_{1-4x^2}^{\sqrt{1-x^2}} f(x, y) \, dy \, dx + \int_{1/2}^1 \int_0^{\sqrt{1-x^2}} f(x, y) \, dy \, dx.$$

Figure 17.3.10(i)

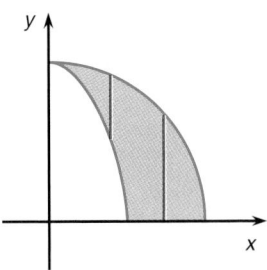

Figure 17.3.10(ii)

Exercises 17.3

Find the volume of each of the following regions.

1. Inside the cylinder $x^2 + y^2 = 1$ and between the planes $z = -2$ and $z = 3$.

2. Inside the cylinder $x^2 + y^2 = 4$ and between the planes $x + z = 0$ and $x + z = 1$.

3. Under the paraboloid $z = x^2 + y^2$ and over the square $0 \le x, y \le 1$.

4. Between the paraboloids $z = x^2 + y^2$ and $z = 3x^2 + 3y^2$ and over the square $0 \le x, y \le 2$.

5. Under the saddle surface $z = 5 + x^2 - y^2$ and over the triangle with vertices $(0, 0)$, $(1, 0)$, $(1, 1)$.

6. Between $z = x^2 + y^2$ and $z = 4 - 2x^2 + y^2$ and over $B = \{(x, y) : 0 \le x \le y \le 1\}$.

7. Inside the cylinder $(x - 1)^2 + y^2 = 1$ and between the planes $y - z = 1$ and $y + z = 5$.

8. Inside the cylinder $(x + 1)^2 + (y - 2)^2 = 1$ and between the xy-plane and the plane $x + y + z = 20$.

9. Inside the cylinder $x^2 + z^2 = 1$ and between the xz-plane and the plane $x + y = 20$.

10. Inside the cylinder $y^2 + z^2 = 9$ and between the yz-plane and the plane $x + y - 2z = 32$.

11. Enclosed by the planes $x = 0$, $y = 0$, $z = 0$, $3x + 2y + z = 6$.

12. Above the triangle with vertices $(0, 0)$, $(1, 0)$, $(1, 2)$ and under the paraboloid $z + x^2 + y^2 = 9$.

13. Inside the sphere $x^2 + y^2 + z^2 = 1$ and above the plane $z = \frac{1}{2}$.

14. Inside the sphere $x^2 + (y-1)^2 + z^2 = 4$ and above the plane $z = \sqrt{3}$.

15. Between the paraboloids $z = x^2 + y^2 - 1$ and $z = 1 - x^2 - y^2$.

16. Inside the sphere $x^2 + y^2 + z^2 = 2$ and above the paraboloid $z = x^2 + y^2$.

17. Between $z = 0$ and $z = x^2 - y^2$ and over the region B enclosed between the graphs of $y = x^2$ and $y = x^3$.

18. Between $z = x + y$ and $z = 2x^2 + 3y^2$ and over the triangle with vertices $(1, 1)$, $(2, 2)$, $(3, 2)$.

19. Inside the cylinder $x^2 + z^2 = 4$ and inside the cylinder $y^2 + z^2 = 4$.

20. Inside both $x^2 + y^2 = 9$ and $y^2 + z^2 = 9$ and with $x \geq 0$, $y \geq 0$, $z \geq 0$.

21. The wedge inside $x^2 + y^2 = 1$, above the xy-plane, and below the plane that contains the y-axis and makes a $45°$ angle with the xy-plane, lying above the positive x-axis.

22. The wedge inside $x^2 + y^2 = 4$, with $z \geq 0$, and below the plane that contains the x-axis and makes a $60°$ angle with the xy-plane, lying above the positive y-axis.

23. Inside $x^2 + y^2 = 1$ and inside the region $|z| \leq 4 - x^2$.

24. Inside $y^2 + z^2 = 12$ and inside the region $|x| \leq 16 - y^2 - z^2$.

25. Inside $x^2 + y^2 = 1$ and inside $x^2 + y^2 + z^2 = 4$.

26. Inside $y^2 + z^2 = 12$ and inside $x^2 + y^2 + z^2 = 16$.

27. The tetrahedron (solid figure with four flat faces) whose faces lie in the planes $x = 0$, $y = 0$, $z = 0$, and $x + y + z = 1$.

28. The tetrahedron (solid figure with four flat faces) whose faces lie in the planes $x = 0$, $z = 0$, $x + y + z = 1$, and $x - y + z = 1$.

29. The tetrahedron whose vertices are $(0, 0, 0)$, $(2, 0, 0)$, $(0, 4, 0)$, and $(0, 0, 1)$.

30. The region enclosed by $x = 0$, $y = 0$, $z = 0$, $x + y = 1$, and $x + y + z = 4$.

Express each of the following as an iterated integral (or integrals) in the opposite order.

31. $\int_0^1 \int_0^y f(x, y)\, dx\, dy$

32. $\int_0^1 \int_0^{x^2} f(x, y)\, dy\, dx$

33. $\int_0^1 \int_{-\sqrt{1-x^2}}^{\sqrt{1-x^2}} f(x, y)\, dy\, dx$

34. $\int_{-2}^2 \int_{|y|-2}^{2-|y|} f(x, y)\, dx\, dy$

35. $\int_{-1}^1 \int_{2x^2}^{x^2+1} f(x, y)\, dy\, dx$

36. $\int_0^1 \int_{2y^2}^{3-y^2} f(x, y)\, dx\, dy$

37. $\int_{-1}^1 \int_{-|x|}^{|x|} f(x, y)\, dy\, dx$

38. $\int_0^1 \int_{x^2}^{2x} f(x, y)\, dy\, dx$

39. $\int_0^1 \int_0^{\sqrt{4-y^2}} f(x, y)\, dx\, dy$

40. $\int_0^1 \int_0^{\exp(y)} f(x, y)\, dx\, dy$

II

41. Find the volume of a sphere of radius r by considering it as the solid region lying between the upper hemisphere and the lower hemisphere, rather than by doubling the volume of a hemisphere as we did in Example 1 (p. 1076).

*42. Find the volume of the region inside the sphere $x^2 + y^2 + z^2 = 16$ but outside the sphere $(x-1)^2 + (y+1)^2 + z^2 = 1$.

*43. Find the volume of the region consisting of all points that lie inside both the cylinder $(x+1)^2 + y^2 = 1$ and the sphere $(x-1)^2 + (y-1)^2 + (z-1)^2 = 1$.

*44. Find the volume of the region consisting of all points that lie inside the cylinder $x^2 + z^2 = a^2$ and inside the cylinder $y^2 + z^2 = a^2$.

*45. Consider two continuous functions defined on the interval $[a, b]$ that satisfy $g(x) \leq f(x)$ for every $x \in [a, b]$. Consider the region R in the plane that lies between the graphs of $y = g(x)$ and $y = f(x)$, for $x \in [a, b]$. Show that the same result is obtained by calculating its area using single-variable techniques or by double integrals.

*46. In Example 4 (p. 1080) we considered a $30°$ wedge cut off a log. Instead, consider a $30°$ wedge cut out of a solid ball (like a segment of an orange). In particular, let R be the region inside the ball $x^2 + y^2 + z^2 \leq 1$, above the xy-plane, and below the plane $z = y/\sqrt{3}$. (i) Set up the integral(s) to find its volume. (ii) Figure out the volume geometrically, without actually integrating.

III

47. With the help of a graphics calculator or computer, find the volume of the region enclosed between $y = x^3 - 2x^2 + 3$, $y = 2x - 1$, $z = 0$, and $z = y$.

48. Find the volume of the region enclosed between $y = x^6 - 24x^2 - x + 50$, $y = x + 20$, $z = 0$, and $z = y + 1$.

POINT TO PONDER

We defined volume in Definition 17.3.1 using double integrals, but in fact we have already learned how to calculate volumes of solids of revolution using integrals of functions of a single variable. For us to feel comfortable about Definition 17.3.1 we ought to check that the new definition gives the same answer as the solid of revolution calculation for solids of this type.

Consider the nonnegative function $F(x)$, for $x \in [a, b]$, and form the solid S by revolving its graph $y = F(x)$ around the x-axis. We found the volume of such a solid by adding up the volumes of little circular slices of thickness dx:

$$V = \int_a^b \pi F(x)^2 dx.$$

See Figure 17.3.11.

On the other hand we can think of the same region in three-dimensional space, as shown in Figure 17.3.12. (Notice that the x-axis points out toward us in the three-dimensional picture.) We want to think of this solid region S as the region between two graphs. First we should ask what is the underlying region R in the plane.

The region R extends for values of x between a and b. For any $x \in [a, b]$ the region R extends between $y = -F(x)$ and $y = F(x)$ (see Figure 17.3.13).

The top of the region S is the graph of a function $f(x, y)$. We could actually figure out that the formula for $f(x, y)$ is $f(x, y) = \sqrt{F(x)^2 - y^2}$, but in fact this is not really needed.

Instead, let us construct the iterated integral

$$\int_a^b \int_{-F(x)}^{F(x)} f(x, y)\, dy\, dx.$$

The inner integral $\int_{-F(x)}^{F(x)} f(x, y)\, dy$ is just the area of the top half of the cross-section of S, as shown in Figure 17.3.14. That cross-section is a semicircle, and its radius is $F(x)$. This means that its area is $\frac{1}{2}\pi F(x)^2$. This is the value of the inner integral.

Using it, we can evaluate the iterated integral for the volume of the top half of S. It is

$$\int_a^b \int_{-F(x)}^{F(x)} f(x, y)\, dy\, dx = \int_a^b \frac{\pi}{2} F(x)^2 dx.$$

The total volume of S is twice the volume of the top half, so it is $\int_a^b \pi F(x)^2\, dx$. This agrees with the volume of revolution formula, and the two approaches are in fact consistent.

Figure 17.3.11

Figure 17.3.13

Figure 17.3.12

Figure 17.3.14

17.4 Double Integrals in Polar Coordinates

POLAR RECTANGLE
$dx\,dy = r\,dr\,d\theta$

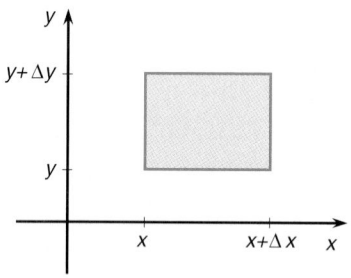

Figure 17.4.1

We have learned how to construct and evaluate double integrals. The integrand $f(x, y)$ is a function on some region R in the plane, and an iterated integral (or perhaps several) is set up to cover the region R. For complicated regions it may be quite difficult to find the appropriate limits of integration. In some cases we found that an iterated integral may be more easily evaluated in one order than in the other.

Some regions that are complicated to express in rectangular coordinates (x, y) may be easier to describe in polar coordinates, and a similar remark can be made about functions: $f(x, y) = (x^2 + y^2)^{-5/2}$ is more easily expressed as $f(r, \theta) = r^{-5}$. It certainly seems possible then that some integrals might be easier to evaluate if they are expressed in polar coordinates. This is most likely to happen if the region R or the function has some sort of "round" symmetry. It is the purpose of this section to learn how to integrate in polar coordinates.

To define double integrals, we used little rectangles whose sides had lengths Δx and Δy. We can visualize the rectangle by letting one variable run from x to $x + \Delta x$ and the other from y to $y + \Delta y$, and of course the rectangle's area is $\Delta A = \Delta x \Delta y$ (see Figure 17.4.1).

In attempting to do the same thing in polar coordinates we could let the radius run from r to $r + \Delta r$ and the polar angle from θ to $\theta + \Delta \theta$. These two angles correspond to half-lines from the origin, and we are considering the region lying between them and between the circles of radius r and $r + \Delta r$. Such a region is shown in Figure 17.4.2. It is called a **polar rectangle** because of the way we constructed it, even though it is not a rectangle in the usual sense: Two of its sides are curved, and the other two are not parallel . . .

Now we want to consider the area of that little polar rectangle. Instead of calculating it exactly, let us approximate it. One side of the rectangle has length Δr. The other is a piece of the circle of radius r corresponding to an angle $\Delta \theta$ at the center of the circle. By the definition of radian measure its arclength is $r \Delta \theta$. If Δr and $\Delta \theta$ are very small, the polar rectangle is very close to an ordinary rectangle with sides Δr and $r \Delta \theta$, so its area is approximately $\Delta A \approx r \Delta r \Delta \theta$.

Notice that the area of the polar rectangle depends on r. This is illustrated in Figure 17.4.3, where we see two polar rectangles. They both have the same $\Delta \theta$ and

Figure 17.4.2

Figure 17.4.3

Figure 17.4.4(i)

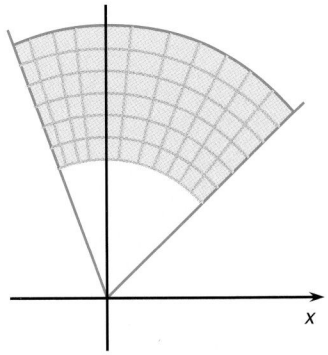

Figure 17.4.4(ii)

the same Δr, but the one that is closer to the origin clearly has a much smaller area. This is why there is the factor r in the formula $\Delta A \approx r \, \Delta r \, \Delta \theta$.

In the "Point to Ponder" at the end of this section we show that the approximations we made in calculating ΔA are actually very reasonable.

Now suppose we want to integrate over a region R that can be expressed as a large polar rectangle, as shown in Figure 17.4.4(i):

$$R = \{(r, \theta) : r_1 \leq r \leq r_2, \theta_1 \leq \theta \leq \theta_2\}.$$

It is easy to divide R into a partition $\mathcal{P}$ consisting of many smaller polar rectangles, as shown in Figure 17.4.4(ii). We call the little rectangles R_i, for $i = 1, \ldots, n$, writing Δr_i and $\Delta \theta_i$ for the changes in r and θ corresponding to R_i. The **mesh** of $\mathcal{P}$ is $\|\mathcal{P}\|$, the maximum length of the diagonals of the R_i's.

Choosing a point $(r_i^*, \theta_i^*) \in R_i$, for each i, we know that the area ΔA_i of R_i is approximately $\Delta A_i \approx r_i^* \Delta r_i \Delta \theta_i$. Using this, we form the polar Riemann sum

$$\sum_{i=1}^{n} f(r_i^*, \theta_i^*) \, \Delta A_i = \sum_{i=1}^{n} f(r_i^*, \theta_i^*) \, r_i^* \, \Delta r_i \, \Delta \theta_i.$$

To integrate $\int \int_R f(r, \theta) \, dA$, we should take a limit as $\|\mathcal{P}\| \to 0$, and this suggests that the integral will be

$$\int \int_R f(r, \theta) \, dA = \int \int_R f(r, \theta) \, r \, dr \, d\theta. \tag{17.4.1}$$

The important thing to notice in this formula is the factor of r before the $dr \, d\theta$.

EXAMPLE 1   (i) Find the area of the circle of radius 1 around the origin.

Solution   The region R is shown in Figure 17.4.5. It consists of all points (r, θ) with $0 \leq \theta \leq 2\pi$ and $0 \leq r \leq 1$. This is a polar rectangle.

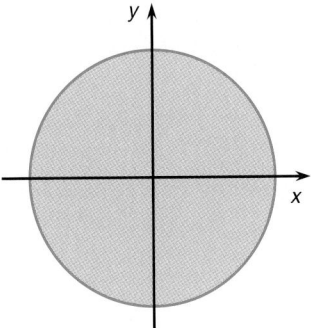

Figure 17.4.5

To find its area, we integrate $\int \int_R 1 \, dA$, which equals $\int \int_R (1) \, r \, dr \, d\theta$. Just as with double integrals using x and y we evaluate this by using an iterated integral. The limits of integration are easy, since r runs from 0 to 1 and θ runs from 0 to 2π. We find that the area is

$$\int \int_R 1 \, dA = \int_0^{2\pi} \int_0^1 r \, dr \, d\theta = \int_0^{2\pi} \left. \frac{r^2}{2} \right|_0^1 d\theta = \int_0^{2\pi} \frac{1}{2} \, d\theta = \pi.$$

Of course, this is the area of the circle.

Figure 17.4.6(i)

Figure 17.4.6(ii)

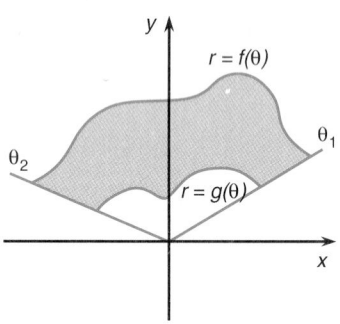

Figure 17.4.7

(ii) Find the volume of the region S enclosed between the paraboloids $z = x^2 + y^2 - 2$ and $z = 6 - x^2 - y^2$.

Solution We illustrate the two paraboloids in Figure 17.4.6(i). An important step in solving the problem will be finding where they intersect. To do this, we consider when the two different expressions for z will be equal:

$$x^2 + y^2 - 2 = 6 - x^2 - y^2.$$

This is the same as $2x^2 + 2y^2 = 8$, or $x^2 + y^2 = 4$. When this is true, $z = x^2 + y^2 - 2 = 4 - 2 = 2$, so we see that the intersection of the two paraboloids lies in the horizontal plane $z = 2$, and it is the circle $x^2 + y^2 = 4$ in that plane.

To integrate, we have to know the region R that lies directly above or below S in the xy-plane, its "shadow" as shown in Figure 17.4.6(ii). In this case R is the disk of radius 2 around the origin, and we will integrate in polar coordinates. The integrand is the difference of the functions $6 - x^2 - y^2$, which forms the top boundary of S, and $x^2 + y^2 - 2$, which is the bottom. We evaluate:

$$\int\int_R (6 - x^2 - y^2 - x^2 - y^2 + 2)\, dy\, dx$$

$$= \int\int_R (8 - 2x^2 - 2y^2)\, dy\, dx = \int_0^{2\pi} \int_0^2 (8 - 2r^2) r\, dr\, d\theta$$

$$= \int_0^{2\pi} \left(4r^2 - \frac{r^4}{2} \right) \Big|_0^2 d\theta = \int_0^{2\pi} (16 - 8)\, d\theta = 8 \int_0^{2\pi} d\theta = 16\pi.$$

The volume of S is 16π.

For double integrals using x and y we began with rectangular regions but then moved on to more complicated regions. In Figure 17.4.7 we see a region R in the plane. It consists of all the points (r, θ) for which θ lies between θ_1 and θ_2 and for which r lies between the graphs of two functions: $g(\theta) \leq r \leq f(\theta)$. This is the sort of region for which it is easy to set up an iterated integral. After all, we let θ run from θ_1 to θ_2, and for each fixed θ we let r run from $g(\theta)$ to $f(\theta)$. We can integrate a function $h(r, \theta)$ over R by the following iterated integral:

$$\int\int_R h(r, \theta)\, dA = \int_{\theta_1}^{\theta_2} \int_{g(\theta)}^{f(\theta)} h(r, \theta) r\, dr\, d\theta.$$

In Figure 17.4.8 we see another region R. It consists of all points (r, θ) for which $r_1 \leq r \leq r_2$ and $g(r) \leq \theta \leq f(r)$. For this R we should let the outer integral be with

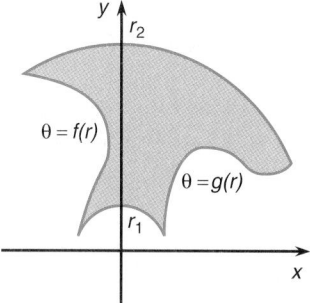

Figure 17.4.8

respect to r. The iterated integral is

$$\int\int_R h(r,\theta)\,dA = \int_{r_1}^{r_2}\int_{g(r)}^{f(r)} h(r,\theta)\,d\theta\,r\,dr.$$

These formulas for iterated integrals in polar coordinates are analogous to the formulas for iterated integrals in rectangular coordinates that we found in Theorem 17.2.3. The way we found them was to copy the definition of double integrals in rectangular coordinates using polar coordinates. The only way they will be very useful is if they give the same answers as rectangular integrals for any region that can be expressed both ways.

We have already seen that finding the area of a circle using these formulas does give the correct answer, and this is only a particular case of the following theorem. The proof is difficult, and we omit it.

> The extra r can be written before or after the $d\theta$, but it must come before the dr.

THEOREM 17.4.2

> In setting up iterated integrals in polar coordinates, be careful to choose the limits of integration so as to cover the region *exactly once*.

Suppose $f(\theta)$ and $g(\theta)$ are continuous functions defined for $\theta \in [\alpha, \beta]$, with $g(\theta) \le f(\theta)$ for all $\theta \in [\alpha, \beta]$. Let R be the region

$$R = \{(r,\theta) : \alpha \le \theta \le \beta, \quad g(\theta) \le r \le f(\theta)\}.$$

Suppose $h(r,\theta)$ is a function that is continuous on R. Then the double integral $\int\int_R h(r,\theta)\,dA$ exists and can be evaluated by an iterated integral as follows:

$$\int\int_R h(r,\theta)\,dA = \int_\alpha^\beta \int_{g(\theta)}^{f(\theta)} h(r,\theta)\,r\,dr\,d\theta.$$

Similarly, if $f(r)$ and $g(r)$ are continuous functions defined for $r \in [a, b]$, with $g(r) \le f(r)$ for all $r \in [a, b]$, let T be the region

$$T = \{(r,\theta) : a \le r \le b, \quad g(r) \le \theta \le f(r)\}.$$

Suppose $h(r,\theta)$ is a function that is continuous on T. Then the double integral $\int\int_T h(r,\theta)\,dA$ exists and can be evaluated by an iterated integral as follows:

$$\int\int_T h(r,\theta)\,dA = \int_a^b \int_{g(r)}^{f(r)} h(r,\theta)\,d\theta\,r\,dr.$$

Figure 17.4.9(i)

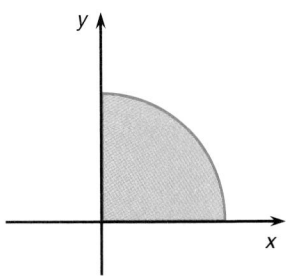

Figure 17.4.9(ii)

EXAMPLE 2   Let R be the region consisting of all points (x, y) inside the circle $x^2 + y^2 = 4$ and satisfying $x \ge 0$, $y \ge 0$. Find the volume of the region lying over R and under the paraboloid $z = x^2 + y^2$, as shown in Figure 17.4.9(i).

Solution   This question does not appear to involve polar coordinates, but if we sketch the region R, as in Figure 17.4.9(ii), we see that it can be described as all points (r, θ) satisfying $0 \le r \le 2$ and $0 \le \theta \le \frac{\pi}{2}$. The volume will be the double integral over R of the function $f(x, y) = x^2 + y^2$, which can be expressed in terms of polar coordinates as $f(r, \theta) = r^2$.

This means that the required volume is

$$\int\int_R f(x,y)\,dA = \int_0^{\pi/2}\int_0^2 (r^2)\,r\,dr\,d\theta = \int_0^{\pi/2}\int_0^2 r^3\,dr\,d\theta$$

$$= \int_0^{\pi/2} \left. \frac{r^4}{4} \right|_0^2 \, d\theta = \int_0^{\pi/2} 4 \, d\theta = \left. 4\theta \right|_0^{\pi/2} = 2\pi.$$

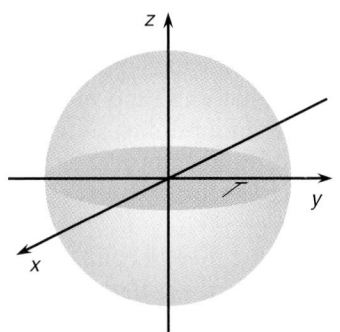

Figure 17.4.10

EXAMPLE 3   Find the volume of the sphere of radius R.

Solution   For convenience we place the sphere with its center at the origin, so its equation is $x^2 + y^2 + z^2 = R^2$. The upper hemisphere is the graph of the function $z = k(x, y) = \sqrt{R^2 - x^2 - y^2}$, and the lower hemisphere is $z = h(x, y) = -\sqrt{R^2 - x^2 - y^2}$. The region T in the xy-plane that lies directly above or below the sphere is the disk $x^2 + y^2 \le R^2$. See Figure 17.4.10.

This means that to find the volume, we should integrate over T the difference $k(x, y) - h(x, y) = 2\sqrt{R^2 - x^2 - y^2}$. Since T is the polar rectangle with $0 \le \theta \le 2\pi$, $0 \le r \le R$, the iterated integral is easily set up:

$$\int_0^{2\pi} \int_0^R 2\sqrt{R^2 - x^2 - y^2} \, r \, dr \, d\theta.$$

Before we can evaluate, it is necessary to express the integrand in polar coordinates. But since $x^2 + y^2 = r^2$, nothing could be easier. The integral is

$$\int_0^{2\pi} \int_0^R 2\sqrt{R^2 - x^2 - y^2} \, r \, dr \, d\theta = \int_0^{2\pi} \int_0^R \sqrt{R^2 - r^2} \, 2r \, dr \, d\theta$$

> We substitute $u = R^2 - r^2$.

$$= -\int_0^{2\pi} \int_{R^2}^0 \sqrt{u} \, du \, d\theta = \int_0^{2\pi} \int_0^{R^2} u^{1/2} \, du \, d\theta = \int_0^{2\pi} \left. \frac{2}{3} u^{3/2} \right|_0^{R^2} d\theta$$

$$= \int_0^{2\pi} \frac{2}{3} R^3 \, d\theta = \left. \frac{2}{3} R^3 \theta \right|_0^{2\pi} = \frac{4}{3} \pi R^3.$$

The volume of the sphere of radius R is $\frac{4}{3} \pi R^3$.

> The integrals we have evaluated so far using polar coordinates are integrals we have done before using rectangular coordinates. The calculation for the volume of the sphere is simpler in polar coordinates, but the rectangular version was not very difficult (cf. Example 17.3.1).
>
> However, there are integrals that are difficult or impossible to evaluate without polar coordinates. One clue that polar coordinates may be helpful is an integrand that involves $x^2 + y^2$, especially inside some other expression, like $\sqrt{x^2 + y^2}$.

EXAMPLE 4   Let R be the interior of the circle $\left(x - \frac{1}{2}\right)^2 + y^2 = \frac{1}{4}$. Evaluate $\int \int_R \sqrt{x^2 + y^2} \, dx \, dy$.

Solution   The region R is sketched in Figure 17.4.11. It is the disk of radius $\frac{1}{2}$ centered at $\left(\frac{1}{2}, 0\right)$. Happily, we already know how to write it in polar coordinates; the circle is $r = \cos\theta$, with $-\frac{\pi}{2} \le \theta \le \frac{\pi}{2}$.

We can think of R as the region lying between the graphs of $r = 0$ and $r = \cos\theta$, with θ between $-\frac{\pi}{2}$ and $\frac{\pi}{2}$. This makes it easy to set up the iterated integral. We find

$$\int \int_R \sqrt{x^2 + y^2} \, dA = \int_{-\pi/2}^{\pi/2} \int_0^{\cos\theta} (r) r \, dr \, d\theta = \int_{-\pi/2}^{\pi/2} \int_0^{\cos\theta} r^2 \, dr \, d\theta$$

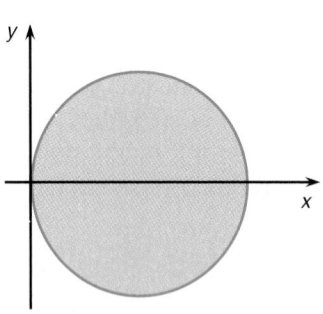

Figure 17.4.11

$$= \int_{-\pi/2}^{\pi/2} \frac{r^3}{3}\Big|_0^{\cos\theta} d\theta = \frac{1}{3}\int_{-\pi/2}^{\pi/2}\cos^3\theta\, d\theta$$

$$= \frac{1}{3}\int_{-\pi/2}^{\pi/2}(1-\sin^2\theta)\cos\theta\, d\theta = \frac{1}{3}\sin\theta\Big|_{-\pi/2}^{\pi/2} - \frac{1}{3}\int_{-1}^{1} u^2\, du$$

$$= \frac{1}{3}\big(1-(-1)\big) - \frac{1}{3}\frac{u^3}{3}\Big|_{-1}^{1} = \frac{2}{3} - \frac{2}{9} = \frac{4}{9}.$$

It is also possible to express the region R from Example 4 as the region with θ lying between the graphs of $\theta = -\arccos r$ and $\theta = \arccos r$, with $0 \le r \le 1$. However, the integral is much more difficult this way. You can also check for yourself that it is very difficult to evaluate the integral in rectangular coordinates.

See the Graphics Calculator
Example on p. 1093.

EXAMPLE 5   Population Density   In many cities the population density is greatest in the center and thins out toward the suburbs. If we place the origin at the center of the city, then $r = \sqrt{x^2+y^2}$ is the distance from the center, in miles, say. Suppose the number of people per square mile, in thousands, is

$$P(x,y) = 100e^{-r/4} = 100e^{-\sqrt{x^2+y^2}/4}.$$

How many people live within 10 miles of the city center?

Solution   We must integrate $P(x,y)$ over the disk D of radius 10 around the origin. The total population, in thousands, is

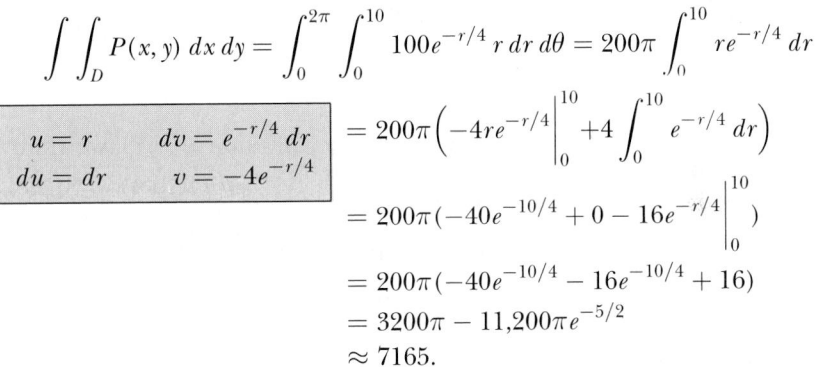

$$\iint_D P(x,y)\,dx\,dy = \int_0^{2\pi}\int_0^{10} 100e^{-r/4}\, r\, dr\, d\theta = 200\pi\int_0^{10} re^{-r/4}\, dr$$

$u = r$	$dv = e^{-r/4}\, dr$
$du = dr$	$v = -4e^{-r/4}$

$$= 200\pi\left(-4re^{-r/4}\Big|_0^{10} + 4\int_0^{10} e^{-r/4}\, dr\right)$$

$$= 200\pi\left(-40e^{-10/4} + 0 - 16e^{-r/4}\Big|_0^{10}\right)$$

$$= 200\pi(-40e^{-10/4} - 16e^{-10/4} + 16)$$

$$= 3200\pi - 11{,}200\pi e^{-5/2}$$

$$\approx 7165.$$

So about 7,165,000 people live within the 10-mile radius.

Figure 17.4.12(i)

Figure 17.4.12(ii)

Figure 17.4.13

Double Integrals Over Unbounded Regions

All our double integrals have been over bounded regions, but it is possible to consider improper integrals involving unbounded regions. For instance, if we wanted to integrate over the infinite region shown in Figure 17.4.12(i), that is, $0 \le y \le 1$, $0 \le x$, the natural thing to try would be integrating over the rectangle $0 \le y \le 1$, $0 \le x \le T$ shown in Figure 17.4.12(ii) and then taking the limit as $T \to \infty$.

Unfortunately, it is more complicated for a region that is unbounded in both directions. Consider an integral over the quadrant $x \ge 0$, $y \ge 0$ shown in Figure 17.4.13. One way to proceed would be to integrate over the rectangle $0 \leq x \leq S$, $0 \leq y \leq T$ and then allow $S \to \infty$ and $T \to \infty$. This often works, but it can be very delicate. It is sometimes possible to get different limits depending on which of S and T goes to infinity first.

Of course, for nonrectangular unbounded regions it is even more complicated. We will not discuss this topic in any detail but will content ourselves with working out one example. This example is very important in the study of statistics and for many other purposes.

EXAMPLE 6   Evaluate $\int_{-\infty}^{\infty} e^{-x^2}\, dx$.

Solution   It is not possible to evaluate this integral directly. There is a famous and clever trick for doing it. It involves considering the square of the integral: $\left(\int_{-\infty}^{\infty} e^{-x^2}\, dx \right) \left(\int_{-\infty}^{\infty} e^{-x^2}\, dx \right)$, which we write as $\left(\int_{-\infty}^{\infty} e^{-x^2}\, dx \right) \left(\int_{-\infty}^{\infty} e^{-y^2}\, dy \right)$ to avoid confusion.

This product can be written as an iterated integral:

$$\left(\int_{-\infty}^{\infty} e^{-x^2}\, dx \right) \left(\int_{-\infty}^{\infty} e^{-y^2}\, dy \right) = \int_{-\infty}^{\infty} \int_{-\infty}^{\infty} e^{-x^2} e^{-y^2}\, dx\, dy$$

$$= \int_{-\infty}^{\infty} \int_{-\infty}^{\infty} e^{-(x^2+y^2)}\, dx\, dy.$$

Letting x and y run from $-\infty$ to ∞ means integrating over the whole plane. Admitting that we have not defined precisely what that means, let us observe that to integrate over the whole plane in polar coordinates, we should let θ run from 0 to 2π and r from 0 to ∞. The integrand can be written as e^{-r^2}. This suggests that the iterated integral above should equal

$$\int_{0}^{2\pi} \int_{0}^{\infty} e^{-r^2} r\, dr\, d\theta.$$

Making the substitution $u = r^2$, we find that the inner integral is $\int_{0}^{\infty} e^{-r^2} r\, dr = \int_{0}^{\infty} e^{-u} \frac{1}{2}\, du = \lim_{T \to \infty} \int_{0}^{T} e^{-u} \frac{1}{2}\, du = \lim_{T \to \infty} \left(-\frac{1}{2} e^{-u} \right) \big|_{0}^{T}$ $= -\frac{1}{2} \lim_{T \to \infty} (e^{-T} - e^{0}) = -\frac{1}{2}(0 - 1) = \frac{1}{2}$. Substituting this value for the inner integral, we evaluate the iterated integral

$$\int_{0}^{2\pi} \int_{0}^{\infty} e^{-r^2} r\, dr\, d\theta = \int_{0}^{2\pi} \frac{1}{2}\, d\theta = \pi.$$

Assuming that all the steps we have taken are really justified, we have shown that $\left(\int_{-\infty}^{\infty} e^{-x^2}\, dx \right)^2 = \pi$, so

$$\int_{-\infty}^{\infty} e^{-x^2}\, dx = \sqrt{\pi}.$$

This famous result is correct, but to be absolutely sure, we would have to know that it was legitimate to convert an iterated integral over the whole plane in rectangular coordinates into the iterated integral in polar coordinates. This is actually not easy to prove.

With a graphics calculator we can plot the population density function from Example 5. We think of it as a function of r, the distance from the city center.

TI-85	SHARP EL9300
GRAPH F1 $(y(x) =)$ 100 e^x $((-)$ F1 (x) ÷ 4) M2(RANGE) 0 ▼ 25 ▼ ▼ 0 ▼ 100 F5 (GRAPH)	100 e^x $(-)$ x/θ/T ÷ 4 RANGE 0 ENTER 25 ENTER 5 ENTER 0 ENTER 100 ENTER 10 ENTER ⏛
CASIO f_x-6300G (f_x-7700GB) Range 0 EXE 25 EXE 5 EXE 0 EXE 100 EXE 10 EXE EXE EXE EXE Graph 100 e^x (− X ÷ 4) EXE	HP 48SX ↰ PLOT PLOTR ERASE ATTN ′ Y = 100 × e^x +/− X ÷ 4 ↱ PLOT ↰ DRAW ERASE 0 ENTER 25 XRNG 0 ENTER 100 YRNG DRAW

We see how the population density falls off as we move out from the city center. A picture like this can be misleading because it seems to suggest that very few people live between, say, 5 and 10 miles from the city center. The *density* is low, but since the 5- to 10-mile ring has a larger area than the disk of radius 5 miles around the city center, it still contains a considerable number of people.

You might like to compare the number of people within 5 miles of the city center with the number between 5 and 10 miles.

Exercises 17.4

▌

Evaluate the following integrals.

1. $\int\int_R (x^2 + y^2 + 2)\,dA$, $R = \{(x, y) : x^2 + y^2 \le 1\}$

2. $\int\int_R (2 - x^2 - y^2)\,dx\,dy$, $R = \{(x, y) : x^2 + y^2 \le 4,\ x \ge 0, y \ge 0\}$

3. $\int\int_R \sqrt{x^2 + y^2}\,dx\,dy$, $R = \{(x, y) : x^2 + y^2 \le 9, y \ge 0\}$

4. $\int\int_R (4 - x^2 - y^2)^{3/2}\,dA$, $R = \{(x, y) : x^2 + y^2 \le 4,\ y \ge |x|\}$

5. $\int\int_R \sqrt{2 - x^2 - y^2}\,dx\,dy$, $R = \{(x, y) : x^2 + y^2 \le 1,\ x \le 0, y \ge 0\}$

6. $\int\int_R \sin\left(\pi(x^2 + y^2)\right)\,dA$, $R = \{(x, y) : x^2 + y^2 \le 1,\ y \le 0\}$

7. $\int\int_R (x^2 + y^2)^{-1/2}\,dx\,dy$, $R = \{(x, y) : 1 \le x^2 + y^2 \le 9\}$

8. $\int\int_R (x^2 + y^2)^{-1}\,dx\,dy$, $R = \{(x, y) : 1 \le x^2 + y^2 \le 4,\ |y| \ge |x|\}$

9. $\int\int_R (x^2 + y^2)^{3/2}\,dA$, $R = \{(x, y) : (x - 1)^2 + y^2 \le 1\}$

10. $\int\int_R (1 + x^2 + y^2)\,dx\,dy$, $R = \{(x, y) : x^2 + (y + 1)^2 \le 1\}$

Find each of the following.

11. The area of the region $\{(r, \theta) : \sin \theta \le r \le \cos \theta,\ 0 \le \theta \le \frac{\pi}{4}\}$.

12. The area of the region inside the circle $x^2 + y^2 = 4$ and above the line $y = 1$.

13. The area inside the cardioid $r = 1 + \cos \theta$.

14. The area inside $r = \cos 2\theta$.

15. The area inside $r = \cos 3\theta$.

16. The area inside $r = \sin 4\theta$.

17. The area inside $r = \sin\left(5\theta + \frac{\pi}{4}\right)$.

18. The volume of the region over the semidisk $x^2 + y^2 \leq 4$, $x \geq 0$, and under the graph of $z = x$.

19. The volume between the xy-plane and the graph of $f(x,y) = \sin(\pi\sqrt{x^2 + y^2})$ over the region $\{(x,y) : x^2 + y^2 \leq 1,\ xy \geq 0\}$.

20. The volume of the region lying over the cardioid $r = 1 + \cos\theta$ and under the graph of $z = r$.

II

21. Find the area of the circle of radius R.

22. Find the volume of the region lying over the semidisk $x^2 + y^2 \leq 9$, $y \geq x$, and under the paraboloid $z = x^2 + y^2$.

23. Find the area inside $r = 3\sin\left(\frac{\pi}{3} - \theta\right)$.

*24. Evaluate the integral in Example 2 (p. 1089) using rectangular coordinates.

*25. Find the area inside $r = \sin(n\theta)$, where n is a positive integer.

*26. Suppose $R \geq a$, and find the area of the region inside the circle of radius R around the origin but to the right of the line $x = a$.

*27. Use the suggestions made before Example 6 (p. 1092) to integrate $f(x,y) = ye^{-x}$ over the "infinite rectangle" $0 \leq y \leq 1$, $0 \leq x$.

*28. Use the suggestions made before Example 6 to integrate $f(x,y) = e^{-xy}$ over the "infinite rectangle" $1 \leq y \leq 2$, $0 \leq x$.

*29. In Section 12.3 we learned how to calculate the area inside the graph of a function $r = f(\theta)$ given in polar coordinates, using integration with respect to the single variable θ. Show that we get the same answer by using a double integral, integrating the constant function 1 over the region inside the graph.

30. Suppose a city's population density in thousands per square mile is $P(x,y) = 100\exp\left(-(x^2 + y^2)/16\right)$. How many people live within a 10-mile radius of the origin?

31. In the situation of Example 5 (p. 1091), compare the number of people who live within 2 miles of the city center with the number of people who live in the suburban ring between 8 and 10 miles from the center.

32. In the situation of Example 5, compare the average population density (in thousands per square mile) in the region within 2 miles of the city center with the average population density (in thousands per square mile) in the suburban ring between 8 and 10 miles from the center.

III

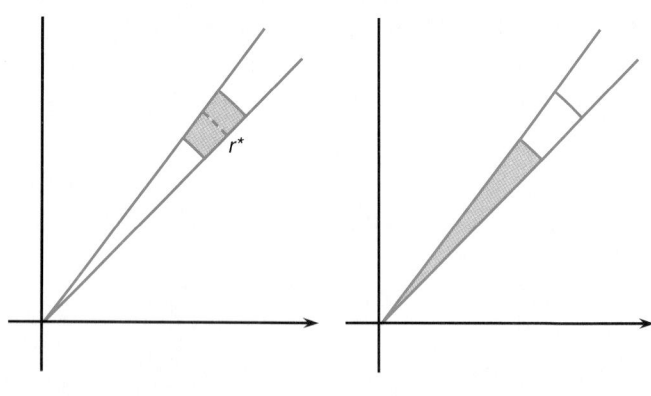

33. Use a graphics calculator or computer to sketch several level curves of the function $P(x,y)$ of Example 5 (p. 1091).

P ← **POINT TO PONDER**

At the beginning of this section we approximated the area of a polar rectangle by pretending that it was nearly an ordinary rectangle. We considered a polar rectangle with radius between r and $r + \Delta r$ and angle between θ and $\theta + \Delta\theta$.

Instead, we could try letting r^* be the radius halfway out in the polar rectangle so that r runs from $r = r^* - \frac{1}{2}\Delta r$ to $r = r^* + \frac{1}{2}\Delta r$ (see Figure 17.4.14(i)).

Now let us find the area of the sector inside the smaller radius $r = r^* - \frac{1}{2}\Delta r$; this sector is shaded in Figure 17.4.14(ii). It is the sector of a circle of radius $r = r^* - \frac{1}{2}\Delta r$ going through an angle of $\Delta\theta$, so its area is $\frac{1}{2}r^2\,\Delta\theta = \frac{1}{2}\left(r^* - \frac{1}{2}\Delta r\right)^2\Delta\theta$.

Figure 17.4.14(i)   **Figure 17.4.14(ii)**

Similarly, the area of the sector inside the larger radius, shown in Figure 17.4.14(iii), is $\frac{1}{2}\left(r^* + \frac{1}{2}\Delta r\right)^2 \Delta\theta$.

The area of the polar rectangle is the difference of the areas of these two sectors, so

$$\Delta A = \frac{1}{2}\left(r^* + \frac{1}{2}\Delta r\right)^2 \Delta\theta - \frac{1}{2}\left(r^* - \frac{1}{2}\Delta r\right)^2 \Delta\theta$$

$$= \frac{1}{2}\left(\left(r^{*2} + r^*\Delta r + \frac{1}{4}\Delta r^2\right)\right.$$

$$\left. - \left(r^{*2} - r^*\Delta r + \frac{1}{4}\Delta r^2\right)\right)\Delta\theta$$

$$= r^* \Delta r \Delta\theta.$$

This shows that our formula $\Delta A = r \Delta r \Delta\theta$ is

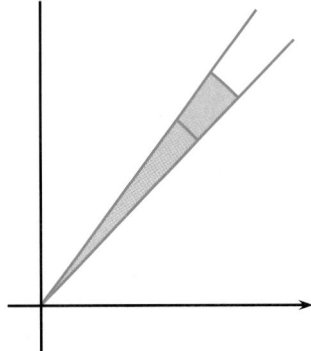

Figure 17.4.14(iii)

correct, assuming that we interpret r to be the radius in the *middle* of the rectangle. Since we use it with very small values of Δr and $\Delta\theta$, this is perfectly reasonable.

17.5 Surface Areas

CURVED SURFACE

$$\int\int_R \sqrt{1 + f_x^2 + f_y^2} \, dx \, dy$$

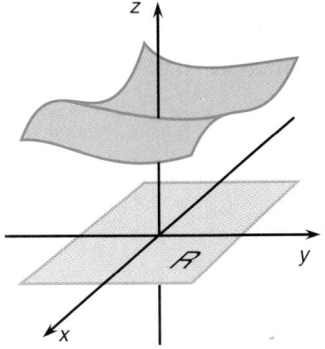

Figure 17.5.1

We have already considered the concept of the surface area of a curved surface in Chapter 6, where we dealt with the special case of a surface of revolution (see Definition 6.9.2, p. 488). Being able to calculate the area of a surface can be important and useful for many purposes. Some examples of problems that involve the area of a surface include finding the amount of paint needed to cover a surface, the quantity of a nutrient that will pass through a membrane, the amount of an impurity that will be adsorbed onto the surface of a filter, the rate at which heat will pass through a heat exchanger, the amount of food a hummingbird needs to sustain the loss of body heat through its skin, or the number of quills on a porcupine if we know the average number of quills per square inch.

We begin by considering a surface that can be written as the graph of a function $z = f(x, y)$ lying over a region R in the xy-plane, as shown in Figure 17.5.1. If we partition R into little rectangles R_i, with $i = 1, \ldots, n$, then we can consider the little piece of the surface that lies above each of the little rectangles, as shown in Figure 17.5.2. If the partition is sufficiently fine, then these little pieces on the surface will all be fairly close to being little parallelograms.

Figure 17.5.2

Figure 17.5.3

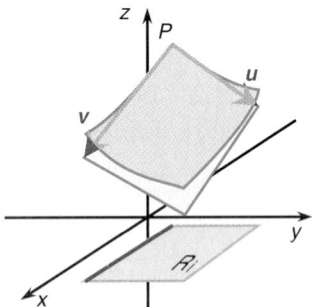

Figure 17.5.4

In Figure 17.5.3 we consider one such little section of the curved surface. It lies over a rectangle R_i in R that has sides of length Δx_i and Δy_i. One corner of the piece of surface is labeled P, and we have drawn the parallelogram that is tangent to the surface at P and lies over the rectangle R_i. Because it lies directly over R_i, two of its sides are parallel to the xz-plane and the other two are parallel to the yz-plane. The idea is to approximate the area of the curved section by the area of the parallelogram.

To find the area of the parallelogram by Lagrange's Identity (Theorem 14.2.11; cf. Proposition 14.3.4, p. 851), we need to know the vectors that are two adjacent sides. Let us consider the sides attached to the point P. The side that is parallel to the xz-plane, the left side in Figure 17.5.4, lies above the left side of the horizontal rectangle R_i. As the point (x, y) moves along the left side of R_i, the side that is drawn with a red line in the illustration, the point directly above it on the parallelogram moves too, and the slope of its path is the partial derivative $\frac{\partial f}{\partial x}$ evaluated at the point under P.

We try to find the vector $\mathbf{v}$ that is the left side of the parallelogram. Since the left side of the rectangle R_i has length Δx_i, we see that the x-component of $\mathbf{v}$ is Δx_i. Since $\mathbf{v}$ is parallel to the xz-plane, its y-component must be 0. In order for its slope to equal the partial derivative f_x we see that $\mathbf{v}$ must be $\mathbf{v} = (\Delta x_i, 0, f_x \Delta x_i)$. (Notice that the slope means the *rise over the run*, and in this situation the *rise* is the z-component, since z is in the vertical direction.)

The vector $\mathbf{u}$ that is the rear side of the parallelogram must be $\mathbf{u} = (0, \Delta y_i, f_y \Delta y_i)$. It is parallel to the yz-plane and has y-component Δy_i and slope f_y. (In this case the *slope* is the z-component over the y-component.)

Using Proposition 14.3.4, we see that the area of the little parallelogram is

$$\|\mathbf{u} \times \mathbf{v}\| = \|(0, \Delta y_i, f_y \Delta y_i) \times (\Delta x_i, 0, f_x \Delta x_i)\|$$

$$= \|(f_x \Delta x_i \Delta y_i, f_y \Delta x_i \Delta y_i, -\Delta x_i \Delta y_i)\| = \sqrt{1 + f_x^2 + f_y^2}\, \Delta x_i \Delta y_i.$$

In this formula the partial derivatives f_x, f_y are both evaluated at the point lying directly under P.

This suggests that if we choose a point $P_i^* \in R_i$ for each i, then the area of the surface lying over the region R is approximated by the sum

$$\sum_{i=1}^{n} \sqrt{1 + f_x(P_i^*)^2 + f_y(P_i^*)^2}\, \Delta x_i \Delta y_i.$$

This is a Riemann sum for the integral $\int \int_R \sqrt{1 + f_x^2(x, y) + f_y^2(x, y)}\, dx\, dy$. This discussion explains why we make the following definition.

DEFINITION 17.5.1

Consider the graph of the differentiable function $z = f(x, y)$ over the region R in the xy-plane. Its **surface area** is defined to be the following integral, provided it exists:

$$\int \int_R \sqrt{1 + f_x^2(x, y) + f_y^2(x, y)}\, dx\, dy.$$

EXAMPLE 1   Find the surface area of a sphere of radius 1.

Solution   As usual we consider the sphere whose center is at the origin, $x^2 + y^2 +$

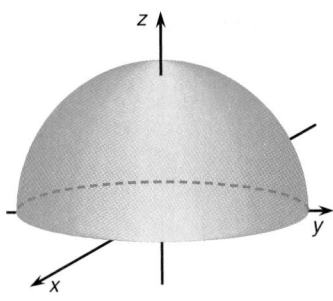

Figure 17.5.5

$z^2 = 1$. To use our formula, we need to express the surface as the graph of a function, so it is more convenient to consider the upper hemisphere (see Figure 17.5.5).

We regard the hemisphere as the graph of the function $z = f(x, y) = \sqrt{1 - x^2 - y^2}$, so we can evaluate the partial derivatives $f_x = \frac{1}{2}(1 - x^2 - y^2)^{-1/2}(-2x) = -x(1 - x^2 - y^2)^{-1/2}$ and $f_y = \frac{1}{2}(1 - x^2 - y^2)^{-1/2}(-2y) = -y(1 - x^2 - y^2)^{-1/2}$. We see that

$$1 + f_x^2 + f_y^2 = 1 + \frac{x^2}{1 - x^2 - y^2} + \frac{y^2}{1 - x^2 - y^2} = \frac{1 - x^2 - y^2 + x^2 + y^2}{1 - x^2 - y^2}$$

$$= \frac{1}{1 - x^2 - y^2}.$$

The surface area of the upper hemisphere is the integral of $\sqrt{1 + f_x^2 + f_y^2}$ over R, the disk of radius 1. It equals

$$\int\int_R \sqrt{1 + f_x^2 + f_y^2}\, dx\, dy = \int\int_R \frac{1}{\sqrt{1 - x^2 - y^2}}\, dx\, dy.$$

This integral cries out for polar coordinates. It becomes

We substitute $u = 1 - r^2$.

$$S = \int_0^{2\pi}\int_0^1 \frac{1}{\sqrt{1 - r^2}}\, r\, dr\, d\theta = \int_0^{2\pi}\int_1^0 u^{-1/2}\left(-\frac{1}{2}\right) du\, d\theta$$

$$= \frac{1}{2}\int_0^{2\pi}\int_0^1 u^{-1/2}\, du\, d\theta = \frac{1}{2}\int_0^{2\pi} 2u^{1/2}\Big|_0^1 d\theta = \frac{1}{2}\int_0^{2\pi} 2\, d\theta = 2\pi.$$

The area of the upper hemisphere is 2π, so the area of the whole sphere of radius 1 is 4π.

A similar calculation for a sphere of radius R confirms the result of Example 6.9.5, in which we found that its surface area is $4\pi R^2$.

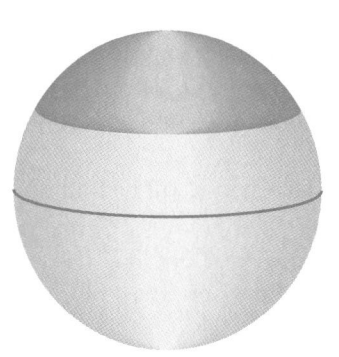

Figure 17.5.6

EXAMPLE 2 Find the surface area of the part of the earth that lies above $30°$ north latitude.

Solution As usual we assume that the earth is a perfect sphere of radius 4000 miles. We have just observed that the total surface area of the entire globe is $4\pi R^2 = 64{,}000{,}000\pi$ sq mi. In Figure 17.5.6 we see the earth with the specified area shaded.

Placing the center of the earth at the origin, we can write the northern hemisphere as $z = f(x, y) = \sqrt{4000^2 - x^2 - y^2}$. We find that $f_x = -x(4000^2 - x^2 - y^2)^{-1/2} = -\frac{x}{f(x,y)}$ and $f_y = -\frac{y}{f(x,y)}$. Then we calculate

$$1 + f_x^2 + f_y^2 = \frac{f(x, y)^2 + x^2 + y^2}{f(x, y)^2} = \frac{4000^2 - x^2 - y^2 + x^2 + y^2}{4000^2 - x^2 - y^2} = \frac{4000^2}{4000^2 - x^2 - y^2}.$$

The integrand is the square root of this quantity.

The question that remains is to find the region R in the xy-plane over which we should integrate. In Figure 17.5.7 we see the upper hemisphere with $30°$ north latitude marked on it. This means that the angle marked in the diagram is $30°$, so the horizontal distance from the z-axis out to this line of latitude is $4000\cos 30° =$

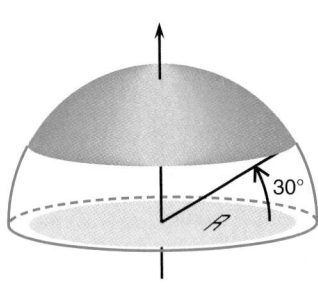

Figure 17.5.7

$4000 \frac{\sqrt{3}}{2} = 2000\sqrt{3}$. The region R is the shaded disk around the origin with radius $2000\sqrt{3}$.

Again using polar coordinates, we find that the surface area is

$$S = \int \int_R \sqrt{1 + f_x^2 + f_y^2}\, dx\, dy = \int \int_R \sqrt{\frac{4000^2}{4000^2 - x^2 - y^2}}\, dx\, dy$$

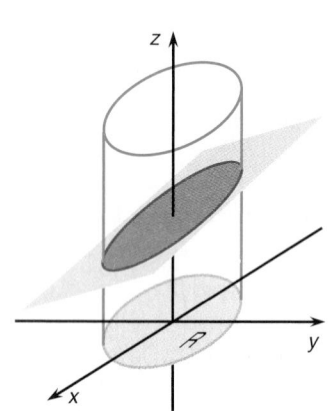

We substitute $u = 4000^2 - r^2$.

$$= \int_0^{2\pi} \int_0^{2000\sqrt{3}} \frac{4000}{\sqrt{4000^2 - r^2}}\, r\, dr\, d\theta = 4000 \int_0^{2\pi} \int_{4000^2}^{2000^2} u^{-1/2} \left(-\frac{1}{2}\right) du\, d\theta$$

$$= 2000 \int_0^{2\pi} \int_{2000^2}^{4000^2} u^{-1/2}\, du\, d\theta = 2000 \int_0^{2\pi} 2u^{1/2} \Big|_{2000^2}^{4000^2}\, d\theta$$

$$= 4000 \int_0^{2\pi} (4000 - 2000)\, d\theta = 8{,}000{,}000(2\pi) = 16{,}000{,}000\pi.$$

The surface area of the part of the earth that lies above $30°$ north latitude is $16{,}000{,}000\pi \approx 50{,}000{,}000$ sq mi. ◢

Notice that the area of the part of the earth above $30°$ north latitude is exactly half the area of the northern hemisphere, or one-quarter the area of the whole globe.

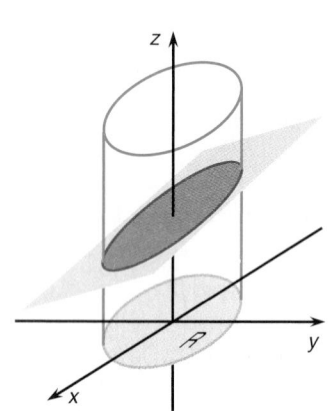

Figure 17.5.8

EXAMPLE 3   Find the area of the piece of the plane $3x - 2y + 4z = 16$ that lies above the ellipse $4x^2 + 9y^2 = 36$ in the xy-plane.

Solution   We illustrate the situation in Figure 17.5.8. To be able to apply the techniques we have learned, we must express the plane as the graph of a function; this is easily done by rewriting the equation as $4z = 16 - 3x + 2y$, or $z = f(x, y) = 4 - \frac{3}{4}x + \frac{1}{2}y$.

From this we find that $f_x = -\frac{3}{4}$ and $f_y = \frac{1}{2}$. The surface area A will be found by integrating over the elliptical region R:

$$A = \int \int_R \sqrt{1 + f_x^2 + f_y^2}\, dA = \int \int_R \sqrt{1 + \frac{9}{16} + \frac{1}{4}}\, dA = \int \int_R \sqrt{\frac{29}{16}}\, dA.$$

This equals $\frac{1}{4}\sqrt{29} \int \int_R 1\, dA = \frac{1}{4}\sqrt{29}\, area(R)$.

Using the fact that the area of an ellipse $\frac{x^2}{a^2} + \frac{y^2}{b^2} = 1$ is πab (cf. Example 6.2.1), we find that the area of the surface in question is $\frac{1}{4}\sqrt{29}\pi(3)2 = \frac{3}{2}\sqrt{29}\pi$. ◢

Notice the way in which we avoided evaluating any integrals in Example 3 by recognizing that $\int \int_R 1\, dA$ is just the area of the region R.

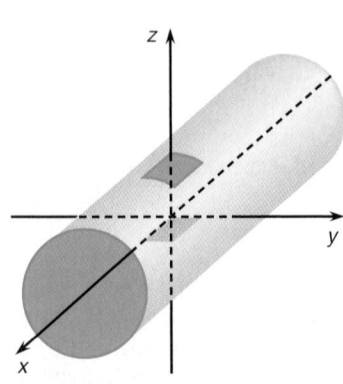

Figure 17.5.9

EXAMPLE 4   Find the area of the part of the cylinder $y^2 + z^2 = 4$ that lies above the rectangle $0 \le x \le 1, 0 \le y \le 1$.

Solution   The surface is illustrated in Figure 17.5.9. It can be expressed as the graph of the function $z = f(x, y) = \sqrt{4 - y^2}$, and we find that $f_x = 0$, $f_y = -\frac{y}{\sqrt{4-y^2}}$. The

area is

$$S = \int_0^1 \int_0^1 \sqrt{1 + f_x^2 + f_y^2}\, dx\, dy = \int_0^1 \int_0^1 \sqrt{1 + \frac{y^2}{4 - y^2}}\, dx\, dy$$

$$= \int_0^1 \int_0^1 \sqrt{\frac{4}{4 - y^2}}\, dx\, dy = \int_0^1 \sqrt{\frac{4}{4 - y^2}}\, x \Big|_0^1 dy = \int_0^1 \frac{2}{\sqrt{4 - y^2}}\, dy$$

$$= 2 \arcsin \frac{y}{2} \Big|_0^1 = 2 \arcsin \frac{1}{2} = 2\frac{\pi}{6} = \frac{\pi}{3}.$$

The area of the surface is $\frac{\pi}{3} \approx 1.047$.

Notice that this area is only slightly bigger than 1, the area of the horizontal square lying under the curved surface. Most of the piece of curved surface is fairly flat, so its area is not very much greater than the flat area beneath it.

With a graphics calculator we cannot plot the surface from Example 4. However, we can plot the cross-section of the surface through the yz-plane, which is just the part of the upper semicircle $y^2 + z^2 = 4$, $z > 0$ with $0 \le y \le 1$.
We also plot the cross-section of the flat square $z = 0$.

TI-85	SHARP EL9300
GRAPH F1($y(x) =$) $\sqrt{}$ $(4 - F1(x)$ x^2) ▼ 0 M2(RANGE) 0 ▼ 1 ▼ ▼ (−) 1 ▼ 3 F5(GRAPH)	$↲$ $\sqrt{}$ $4 - x/\theta/T$ x^2 2ndF ▼ 0 RANGE 0 ENTER 1 ENTER 1 ENTER −1 ENTER 3 ENTER 1 ENTER $↲$

CASIO f_x-6300G (f_x-7700GB)	HP 48SX
Range 0 EXE 1 EXE 1 EXE −1 EXE 3 EXE 1 EXE EXE EXE EXE Graph $\sqrt{}$ (4 − X x^2) EXE Graph 0 EXE	↰ PLOT PLOTR ERASE ATTN ′ Y = $\sqrt{x}$ (4 − X y^x 2 ↱ PLOT ↰ DRAW ERASE 0 ENTER 1 XRNG 1 +/− ENTER 3 YRNG DRAW ATTN ′ Y = 0 ↱ PLOT ↰ DRAW DRAW

Notice how the arclength of the curved cross-section is not much greater than the length of the flat cross-section.

Exercises 17.5

Use integration to find the surface area of each of the following surfaces.

1. The part of the plane $x + y + z = 1$ with $x \ge 0$, $y \ge 0$, $z \ge 0$.

2. The triangle with vertices $(2, 0, 0)$, $(0, 2, 0)$, $(0, 0, 2)$.

3. The part of the graph of $z = x^2$ lying over the rectangle $0 \le x \le 1$, $-1 \le y \le 1$ in the xy-plane.

4. The part of the graph of $z = 2 + x^2$ lying over the triangle $0 \le y \le x$, $0 \le x \le 1$.

5. The part of the graph of $z = 3 - y - x^2$ lying over the region $1 \geq y \geq |x|$.

6. The part of the graph of $z = x + y^2$ lying over the triangle with vertices $(0, 0)$, $(0, 1)$, $(1, 1)$.

7. The part of the paraboloid $x = y^2 + z^2$ lying between $x = 0$ and $x = 16$.

8. The part of the paraboloid $y = x^2 + z^2$ lying between $y = 1$ and $y = 4$.

9. The part of the sphere $x^2 + y^2 + z^2 = 4$ lying between $z = 0$ and $z = 1$.

10. The part of the cylinder $x^2 + z^2 = 16$ lying over the rectangle $-2 \leq x \leq 2$, $0 \leq y \leq 3$.

11. The part of the plane $2x - y - 2z = 4$ lying inside the cylinder $x^2 + y^2 = 1$.

12. The part of the plane $x + 3y - z = 0$ lying inside the cylinder $x^2 + y^2 = 9$.

13. The part of the plane $4x - 4y + 2z = 10$ lying over the region in the xy-plane enclosed between the parabola $y = x^2 - 1$ and the x-axis.

14. The part of the plane $x + 3y - 4z + 15 = 0$ lying over the region in the xy-plane enclosed between $y = x^3$ and $y = x^2$.

15. The part of the paraboloid $z = x^2 + y^2$ lying inside the cylinder $x^2 + y^2 = 4$.

16. The part of the paraboloid $z = 6 - 2x^2 - 2y^2$ lying above $z = -2$.

17. The part of the graph of $z = 15 - x - y^2$ over the triangle with vertices $(0, 0)$, $(2, 0)$, $(2, 3)$.

18. The part of the plane $2x + 3y + 4z = 12$ that lies in the region $x \geq 0$, $y \geq 0$, $z \geq 0$.

19. The part of the graph of $z = 3 + \sqrt{3}y + x^2$ over the square $0 \leq x \leq 2$, $0 \leq y \leq 2$.

20. The part of the graph of $z = 6 + 2x - 2y + \frac{3}{2}y^2$ that lies over the rectangle $-2 \leq x \leq 2$, $-1 \leq y \leq 1$.

II

21. Use the techniques of this section to confirm that the area of a sphere of radius R is $4\pi R^2$.

*22. Use integration to find the surface area of the curved surface of a cylinder of radius R and height h.

*23. Find the area of the portion of the cone $z^2 = x^2 + y^2$ that lies between $z = 0$ and $z = 1$.

*24. Find the area of the portion of the cone $z^2 = 3x^2 + 3y^2$ that lies between $z = 0$ and $z = 12$.

*25. Find the surface area of the part of the cone $y^2 = x^2 + z^2$ that lies between $y = -1$ and $y = 1$.

*26. Find the surface area of the cone whose height is h and whose base has radius R.

*27. Find the area of the part of the sphere $x^2 + y^2 + z^2 = 1$ that lies over the region inside the circle $r = \cos\theta$.

*28. Find the area of the part of the sphere $x^2 + y^2 + z^2 = 4$ that lies over the disk $x^2 + (y - 1)^2 \leq 1$.

*29. (i) Consider the part of the earth that lies above $45°$ north latitude. Do you expect its surface area to be equal to, greater than, or less than one-quarter of the surface area of the whole earth?

   (ii) Find the surface area of the part of the earth specified in part (i).

   (iii) Is the answer to part (ii) equal to, greater than, or less than one-quarter the area of the whole globe?

*30. Find the area of the part of the globe that lies between $30°$ south latitude and $60°$ north latitude.

*31. Find the area of the part of the earth that lies above $\theta°$ north latitude, where $0 \leq \theta \leq 90$.

*32. The area of the earth is about $4\pi R^2 = 4(4000^2)\pi = 64,000,000\pi \approx 200,000,000$ sq mi. Suppose that about 60% of the surface is covered by water, leaving 80,000,000 sq mi of land.

   (i) Of this land area, let us suppose that 30,000,000 sq mi is not fit for human habitation. Supposing that the world population is (roughly) 5,000,000,000, how many people must there be per square mile of habitable land?

   (ii) If everybody was spread out at equal intervals in a rectangular grid pattern, how far would you be from your nearest neighbor?

   (iii) About how many people are there per square mile in the United States, whose area is approximately 4,000,000 sq mi and whose population is about 250,000,000?

   (iv) About how many people are there per square mile in China, whose area is also approximately 4,000,000 sq mi and whose population is about one billion?

P ← **POINT TO PONDER**

There is an interesting question which comes up when we *scale* an object, that is, make it larger or smaller. In Figure 17.5.10 we see two rectangular boxes; the second

Figure 17.5.10

one is twice as big as the first in each dimension. If the length, width, and height of the first box are a, b, and c, then the dimensions of the larger box are $2a$, $2b$, and $2c$.

In particular the volume of the second box is $(2a)(2b)(2c) = 8abc$, which is $8 = 2^3$ times the volume of the first box. On the other hand, the surface area of the second box is $2\big((2a)(2b) + (2a)(2c) + (2b)(2c)\big) = 8(ab + ac + bc)$. This is $4 = 2^2$ times as much as the surface area of the smaller box, which is $2(ab + ac + bc)$.

The same thing happens in general. If we scale an object by a factor of k in each direction, that is, make all its dimensions k times as big, then its volume will change by a factor k^3, and its surface area will change by k^2. Roughly speaking, this is because volume is a three-dimensional quantity and area is a two-dimensional quantity. You should try it out for a sphere.

This observation is fairly simple, but it has a number of very interesting consequences. One arises because the weight of an object is related to its volume. Since all animals have similar density, being made of similar material, this leads us to expect that an elephant, which might be approximately 10 times as large as a dog, will weigh about 1000 times as much. This is not absolutely accurate, because in the first place their densities may be slightly different, but mostly because they are not of exactly the same shape. An elephant does not look like a big dog.

However, even this is worth exploring. Very roughly, an animal's leg has approximately the shape of a cylinder. The strength of a cylinder, the weight it can support, should depend on the area of its cross-section. One way to see this is to think of one thin cylinder, as shown in Figure 17.5.11, and a bundle of 16 of them. It certainly seems reasonable to expect that the bundle of

Figure 17.5.11

16 will support 16 times as much weight. Ignoring the little spaces between the cylinders in the bundle, we see that its cross-sectional area is 16 times the cross-sectional area of a single cylinder.

So again speaking roughly, we see that the strength of a leg is proportional to the cross-sectional area. Now suppose we took a dog and scaled it up by a factor of 10. Its leg would be 10 times as large in each direction, so the cross-sectional area would increase by a factor of $10^2 = 100$. However, the weight it would have to support would increase by $10^3 = 1000$. The scaled-up dog's leg would be 10 times too flimsy to support the oversized dog.

This sounds a little silly until we reflect that proportionally, elephants' legs are much thicker than dogs' legs. And insect legs are proportionally very much thinner. Insects do not need thick legs to support their

weight, because of their size. The giant insects so popular in science fiction movies would collapse under their own weight.

A similar discussion can be made about the loss of body heat through the skin of an animal. Roughly speaking, the amount of heat lost through the skin will be proportional to the surface area of the skin. If we compare a hummingbird and an elephant, we see that the ratio of the surface area to the volume is much

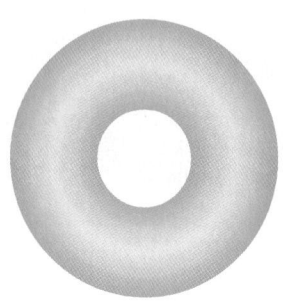

smaller for the elephant than for the hummingbird (since the volume increases as the cube of the scale factor and the area increases only as the square). This means that the heat loss through the skin is smaller for the elephant relative to its weight than it is for the hummingbird.

The significance of this observation is that the heat loss through the skin is a major drain on the hummingbird's metabolism. The hummingbird must consume a great deal of nourishment just to replace the energy it loses this way. A hummingbird consumes several times its own weight in food each day. An elephant eats a great deal, but still its daily meals are only a small fraction of its body weight. Proportionally, the elephant eats much less than the hummingbird.

Of course, many other factors are involved, but in general terms it is true that small animals eat much more relative to their body weight than large animals.

17.6 Triple Integrals

PARTITION
RIEMANN SUM
ITERATED INTEGRAL

We have learned about integration of functions of one variable and functions of two variables, so it is natural to wonder about functions of three variables. In this section we discover that the definitions and properties are very similar to what we did for double integrals. They are defined by considering Riemann sums $\sum_{i=1}^{n} f(P_i^*) \, \Delta V_i$, where ΔV_i is the volume of a little three-dimensional rectangular solid.

The resulting integral

$$\int \int \int_R f(x, y, z) \, dV = \int \int \int_R f(x, y, z) \, dx \, dy \, dz$$

over a three-dimensional region R in $\mathbb{R}^3$ can be used for many purposes (see Figure 17.6.1). Integrating the constant function 1 gives the volume of R: $\int \int \int_R 1 \, dV = volume(R)$. Integration also provides a way to calculate the weight of an unusually shaped object if we know the density of the material of which it is made.

In addition, integrating a function $f(x, y, z)$ over a region R allows us to find the average value of $f(x, y, z)$ on R. In Section 6.7 we have already used this idea for moments, and in the next section we will extend the notion to three-dimensional regions.

Triple integrals are defined in very much the same way as double integrals (cf. Definitions 17.1.1 and 17.2.1). Suppose R is a region in $\mathbb{R}^3$ that is **bounded**, that is, contained in some large rectangular box. Divide up this large rectangular box

Figure 17.6.1

into smaller rectangular boxes whose sides are parallel to the coordinate axes (see Figure 17.6.2). Choose the little rectangular boxes that are entirely contained in R, and call them $R_1, R_2, \ldots, R_n$. This collection of rectangular boxes $\{R_i\}$ is called a **partition** $\mathcal{P}$ of R.

For each box R_i we let its dimensions be $\Delta x_i, \Delta y_i, \Delta z_i$, so its volume is $\Delta V_i = \Delta x_i \, \Delta y_i \, \Delta z_i$. The **mesh** of $\mathcal{P}$ is the maximum length of the sides of the boxes R_i; it is denoted $\|\mathcal{P}\|$.

DEFINITION 17.6.1

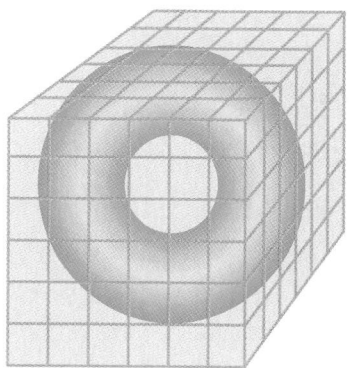

Figure 17.6.2

Suppose R is a bounded region in $\mathbb{R}^3$ as above, $f(x, y, z)$ a continuous function on R, and $\mathcal{P} = \{R_1, \ldots, R_n\}$ a partition of R. For each choice of $P_i^* = (x_i^*, y_i^*, z_i^*) \in R_i$ we form the **Riemann sum**

$$\sum_{i=1}^{n} f(P_i^*) \, \Delta V_i = \sum_{i=1}^{n} f(P_i^*) \, \Delta x_i \, \Delta y_i \, \Delta z_i.$$

The **integral** of $f(x, y, z)$ over R is the limit of the Riemann sums as the mesh of $\mathcal{P}$ goes to zero, provided the limit exists. We write

$$\iiint_R f(x, y, z) \, dV = \iiint_R f(x, y, z) \, dx \, dy \, dz$$

$$= \lim_{\|\mathcal{P}\| \to 0} \left(\sum_{i=1}^{n} f(P_i^*) \Delta x_i \, \Delta y_i \, \Delta z_i \right).$$

Theorem 17.2.3 describes some types of regions in the plane $\mathbb{R}^2$ that are suitable for setting up iterated integrals. It then states that double integrals of continuous functions over such regions exist and can be evaluated by iterated integrals. It seems reasonable to expect a similar theorem for triple integrals, and the corresponding result is true. However, it is difficult to state the theorem precisely because it is cumbersome to describe all the types of regions that can be "covered" by iterated integrals. Certainly it works for three-dimensional rectangles.

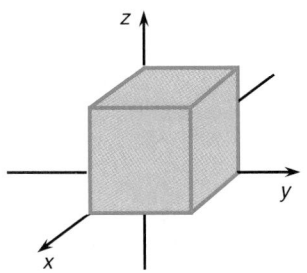

Figure 17.6.3

EXAMPLE 1   Evaluate $\iiint_R (x^2 + yz) \, dV$, where R is the cube $0 \le x \le 1, 0 \le y \le 1, 0 \le z \le 1$.

Solution   This region is the three-dimensional equivalent of a square in the plane (see Figure 17.6.3). The iterated integral could hardly be easier to set up:

$$\iiint_R (x^2 + yz) \, dV = \int_0^1 \int_0^1 \int_0^1 (x^2 + yz) \, dx \, dy \, dz$$

$$= \int_0^1 \int_0^1 \left(\frac{1}{3}x^3 + yzx \right) \Big|_{x=0}^{x=1} dy \, dz$$

$$= \int_0^1 \int_0^1 \left(\frac{1}{3} + yz \right) dy \, dz = \int_0^1 \left(\frac{y}{3} + \frac{1}{2}y^2 z \right) \Big|_{y=0}^{y=1} dz$$

$$= \int_0^1 \left(\frac{1}{3} + \frac{z}{2} \right) dz = \left(\frac{z}{3} + \frac{z^2}{4} \right) \Big|_0^1 = \frac{1}{3} + \frac{1}{4} = \frac{7}{12}.$$

There is one special type of region for which it is relatively easy to set up iterated integrals. Suppose B is a region in the xy-plane, and let R be the region in $\mathbb{R}^3$ that lies between the graphs of two functions $z = g(x, y)$ and $z = f(x, y)$ over B (see Figure 17.6.4).

To integrate over R, we set up an iterated integral for x and y running over B and then let z run from $g(x, y)$ to $f(x, y)$ in the inner integral.

THEOREM 17.6.2

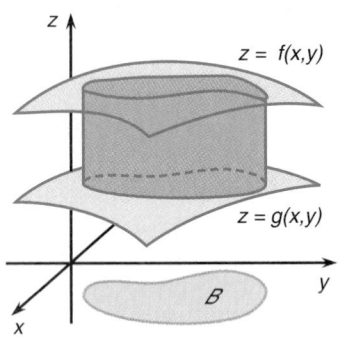

Figure 17.6.4

Suppose B is a region in the xy-plane that is one of the types discussed in Theorem 17.2.3 for which it is possible to set up an iterated integral. Suppose $f(x, y)$ and $g(x, y)$ are continuous functions on B satisfying $g(x, y) \leq f(x, y)$. Let R be the region over B lying between the graphs of $z = g(x, y)$ and $z = f(x, y)$; in symbols,

$$R = \{(x, y, z) : (x, y) \in B, \quad g(x, y) \leq z \leq f(x, y)\}.$$

Suppose $F(x, y, z)$ is a continuous function on R. Then the triple integral

$$\iiint_R F(x, y, z) \, dV = \iiint_R F(x, y, z) \, dx \, dy \, dz$$

exists and equals the iterated integral

$$\iint_B \left(\int_{g(x,y)}^{f(x,y)} F(x, y, z) \, dz \right) dx \, dy.$$

The double integral over B can also be set up as an iterated integral, so the triple integral will be expressed as an iterated integral with three integrations.

EXAMPLE 2 Let R be the upper half of the ball of radius 1 centered at the origin. Evaluate

$$\iiint_R z \, dV = \iiint_R z \, dx \, dy \, dz.$$

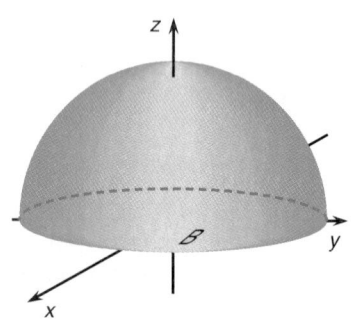

Figure 17.6.5

Solution The region R can be described as $x^2 + y^2 + z^2 \leq 1, z \geq 0$; it is illustrated in Figure 17.6.5. It lies over the unit disk B in the xy-plane, and it lies between the graphs of $z = 0$ and $z = \sqrt{1 - x^2 - y^2}$, that is, the xy-plane and the upper hemisphere.

The iterated integral is easily set up:

$$\iiint_R z \, dx \, dy \, dz = \iint_B \left(\int_0^{\sqrt{1-x^2-y^2}} z \, dz \right) dA = \iint_B \left(\int_0^{\sqrt{1-x^2-y^2}} z \, dz \right) dx \, dy$$

$$= \iint_B \left(\frac{z^2}{2} \Big|_{z=0}^{z=\sqrt{1-x^2-y^2}} \right) dx \, dy = \frac{1}{2} \iint_B (1 - x^2 - y^2) \, dx \, dy.$$

At this point we recognize an integral that should be done in polar coordinates. The unit disk B is given by $0 \leq \theta \leq 2\pi, 0 \leq r \leq 1$, so the integral is

$$\frac{1}{2} \int_0^{2\pi} \int_0^1 (1 - r^2) r \, dr \, d\theta = \frac{1}{2} \int_0^{2\pi} \int_0^1 (r - r^3) \, dr \, d\theta = \frac{1}{2} \int_0^{2\pi} \left(\frac{r^2}{2} - \frac{r^4}{4} \right) \Big|_0^1 dr$$

$$= \frac{1}{2} \int_0^{2\pi} \left(\frac{1}{2} - \frac{1}{4} \right) d\theta = \frac{1}{2} \left(\frac{1}{4} \right) 2\pi = \frac{\pi}{4}.$$

For oddly shaped regions there is the problem of figuring out the limits of integration for the iterated integral. There is also the problem that triple integrals tend to get fairly complicated even for relatively simple functions.

EXAMPLE 3   Express $\int\int\int_R 3yz\,dV$ as an iterated integral, where R is the region consisting of all points lying under the plane $x + 2y + 2z = 4$ and satisfying $x \geq 0$, $y \geq 0$, $z \geq 0$.

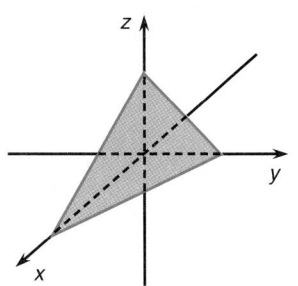

Figure 17.6.6

Solution   The region is shown in Figure 17.6.6. It lies over the triangle in the xy-plane with vertices $(0,0)$, $(4,0)$, $(0,2)$. To integrate over this triangle, we can let x run from 0 to 4 and y from 0 to the line through $(4,0)$ and $(0,2)$. This line is $x + 2y = 4$, which can be written as $y = 2 - \frac{x}{2}$.

The region R lies above the xy-plane and below the plane $x + 2y + 2z = 4$, that is, $z = 2 - y - \frac{x}{2}$, so the iterated integral is

$$\int\int\int_R 3yz\,dV = \int_0^4 \int_0^{2-x/2} \left(\int_0^{2-y-x/2} 3yz\,dz\right) dy\,dx.$$

If you work it out, you will be surprised how much work is needed to integrate this simple polynomial over this comparatively simple region. Incidentally, the answer is $\frac{8}{5}$.

Sometimes an iterated integral will be easier to evaluate if it is expressed as an iterated integral in a different order. However, it can be quite difficult to find the limits of integration for a different order of integration.

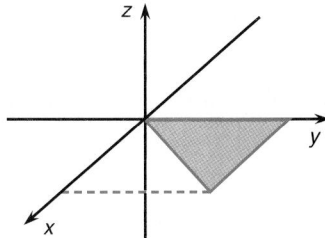

Figure 17.6.7(i)

EXAMPLE 4   Suppose $f(x,y,z)$ is a continuous function. Express the following iterated integral as an iterated integral in which the outer integral is with respect to z and the inner integral is with respect to y:

$$\int_0^1 \int_0^y \int_0^{1-x^2} f(x,y,z)\,dz\,dx\,dy.$$

Solution   In the integral we see that y runs from 0 to 1 and x runs from 0 to y. This means that the "shadow" of the region in $\mathbb{R}^3$ on the xy-plane lies in the triangle $0 \leq x \leq y \leq 1$, shown in Figure 17.6.7(i).

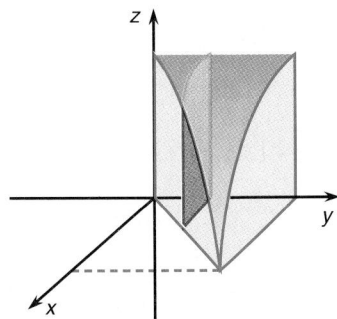

Figure 17.6.7(ii)

The variable z runs from 0 to $1 - x^2$, so the region can be drawn as shown in Figure 17.6.7(ii), where we see the cross-section corresponding to a fixed value of y. The inner double integral in z and x is over such a cross-section. Notice that z ranges from 0 to 1. Since z is always less than or equal to $1 - x^2$, we see that x is never greater than $\sqrt{1-z}$. On the other hand, we know that $x \leq y$.

When we place y in the inner integral, y will range from x to 1.

The rearranged iterated integral is

$$\int_0^1 \int_0^y \int_0^{1-x^2} f(x,y,z)\,dz\,dx\,dy = \int_0^1 \int_0^{\sqrt{1-z}} \int_x^1 f(x,y,z)\,dy\,dx\,dz.$$

In Figure 17.6.7(iii) we see the cross-section for a fixed value of z; on the right side of the equation the inner double integral in y and x is over such a cross-section.

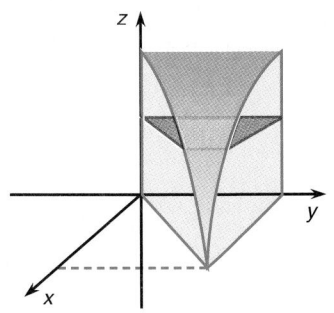

Figure 17.6.7(iii)

One of the uses of triple integrals is to "add up" the values of a function over a region in space.

EXAMPLE 5   Phytoplankton   Phytoplankton are tiny organisms that grow in the ocean, performing photosynthesis using sunlight. They do not live in very deep water where sunlight does not penetrate but also prefer to avoid living right at the surface. Suppose the number of millions of plankton per cubic meter of seawater at a depth of d meters is given by

$$F(d) = 100de^{-d/30}.$$

Find the total number of phytoplankton in 1 square kilometer of ocean, down to a depth of 200 m.

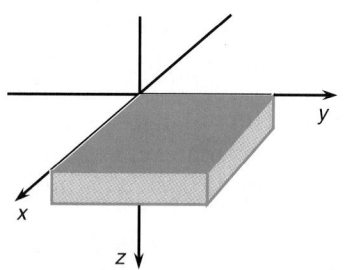

Figure 17.6.8

Solution   We think of this "chunk" of ocean as a rectangular box, measuring horizontal displacements x, y (in meters) from one top corner. It is convenient to place the positive z-axis pointing *down* so that z measures the depth (see Figure 17.6.8). The box is B: $0 \le x, y \le 1000$, $0 \le z \le 200$.

The number of millions of plankton per cubic meter is

$$f(x, y, z) = 100ze^{-z/30}.$$

A tiny box with sides dx, dy, dz has volume $dx\,dy\,dz$, so the number of millions of plankton it contains is $f(x, y, z)\,dx\,dy\,dz$.

The total number of phytoplankton in B is

> We integrate with respect to x and y both at once.

$$\iiint_B f(x, y, z)\,dV = \int_0^{200} \int_0^{1000} \int_0^{1000} 100ze^{-z/30}\,dx\,dy\,dz$$

$$\boxed{\begin{array}{ll} u=z & dv=e^{-z/30}\,dz \\ du=dz & v=-30e^{-z/30} \end{array}} = 1000(1000)100 \int_0^{200} ze^{-z/30}\,dz$$

$$= 10^8 \left(-30ze^{-z/30} \Big|_0^{200} + \int_0^{200} 30e^{-z/30}\,dz \right)$$

$$= 3 \times 10^9 \left(-200e^{-200/30} + 0 - (30e^{-z/30}) \Big|_0^{200} \right)$$

$$= 3 \times 10^9 (-200e^{-200/30} - 30e^{-200/30} + 30e^0)$$

$$\approx 8.9 \times 10^{10}.$$

The total number of phytoplankton is 8.9×10^{10} million $= 8.9 \times 10^{16}$.

Phytoplankton. The colonial blue-green alga Gloeocapsa (here magnified 2000×) consists of two or more cells in a gelatinous sheath.

EXAMPLE 6   Air Pollution   A chimney 100 m tall emits pollutants into the air. They rise and then gradually settle. Let us place the origin at the base of the chimney, so x and y measure horizontal displacements and z measures height above ground level (all in meters). Assume that the pollutant never rises above 1000 m.

Suppose the concentration of pollutant, in grams per cubic meter, at the point (x, y, z) is

$$C(x, y, z) = \frac{100,000 - 100z}{1,000,000 + x^2 + y^2}.$$

Find the total amount of pollutant in the air over the region lying within 10 km of the chimney.

Solution   We must integrate over the region R defined by

$$x^2 + y^2 \le (10,000)^2 = 100,000,000, \qquad 0 \le z \le 1,000.$$

With a graphics calculator we can consider Example 5 and plot the number of phytoplankton per cubic meter as a function of the depth.

TI-85	SHARP EL9300
GRAPH F1$(y(x) =)$ 100 F1(x) e^x ((−) F1(x) ÷ 30) M2(RANGE) 0 ▼ 200 ▼ ▼ (−) 1 ▼ 1500 F5(GRAPH)	⌁ 100 x/θ/T × e^x (−) x/θ/T ÷ 30 RANGE 0 ENTER 200 ENTER 20 ENTER −1 ENTER 1500 ENTER 100 ENTER ⌁
CASIO f_x-6300G (f_x-7700GB) Range 0 EXE 200 EXE 20 EXE −1 EXE 1500 EXE 100 EXE EXE EXE EXE Graph 100 X × e^x (− X ÷ 30) EXE	HP 48SX ⬑ PLOT PLOTR ERASE ATTN ′ Y ◀ 100 × X × e^x +/− X ÷ 30 ↱ PLOT ⬑ DRAW ERASE 0 ENTER 200 XRNG 1 +/− ENTER 1500 YRNG DRAW

Notice how the density is low right near the surface. It grows larger as we descend and then tapers off as we reach depths at which there is not enough light.

The total amount of pollutant is

We use cylindrical coordinates, but it is convenient to put z in the inner integral.

$$P = \int \int \int_R C(x, y, z)\, dV = \int_0^{2\pi} \int_0^{10,000} \int_0^{1000} \frac{100,000 - 100z}{1,000,000 + x^2 + y^2}\, dz\, r\, dr\, d\theta$$

$$= 2\pi \int_0^{10,000} (10^5 z - 50z^2)\Big|_0^{1000} \frac{1}{1,000,000 + r^2}\, r\, dr$$

To get to the second line, we integrated with respect to θ and z in one step.

$$= 2\pi(10^8 - 5 \times 10^7)\frac{1}{2}\ln(10^6 + r^2)\Big|_0^{10,000}$$

$$= 2\pi(5 \times 10^7)\frac{1}{2}\big(\ln(10^6 + 10^8) - \ln(10^6)\big) \approx 7.25 \times 10^8.$$

The total amount of pollutant in the specified region is about 7.25×10^8 g $= 7.25 \times 10^5$ kg $= 725$ tonnes.

Exercises 17.6

I

In each of the following situations, evaluate the integral, where R is the specified region.

1. $\int\int\int_R (x+3y)\,dV$, $0 \le x \le 2, 0 \le y \le 1$, $-1 \le z \le 0$

2. $\int\int\int_R x^2\,dV$, $0 \le x \le 3, -2 \le y \le 2, -1 \le z \le 1$

3. $\int\int\int_R y\sin x\,dx\,dy\,dz$, $0 \le x \le 1, -1 \le y \le 1$, $1 \le z \le 2$

4. $\int\int\int_R \left(\sin^2(xyz) + \cos^2(xyz)\right)dV$, $|x| \le 2, |y| \le 3$, $|z| \le 4$

5. $\int\int\int_R (1+3z^2)\,dx\,dy\,dz$, $x^2+y^2+z^2 \le 4, z \ge 0$

6. $\int\int\int_R x\,dx\,dy\,dz$, $x^2+y^2+z^2 \le 2$

7. $\int\int\int_R (3-x)\,dx\,dy\,dz$, $x^2+y^2 \le 1, 0 \le z \le 2-x$

8. $\int\int\int_R (x^2+y^2)\,dV$, $x^2+y^2 \le 1$, $0 \le z \le 2-x^2-y^2$

9. $\int\int\int_R (x+y)\,dV$, $0 \le y \le x^2, 0 \le x \le 1$, $1 \le z \le 3+x$

10. $\int\int\int_R e^x\,dx\,dy\,dz$, $0 \le x \le 1, 0 \le y \le 1$, $0 \le z \le 2x$

In each of the following situations, evaluate $\int\int\int_R f(x,y,z)\,dV$, where $f(x,y,z)$ is the given function and R is the specified region.

11. $f(x,y,z) = z$, R is $|x| \le 1, |y| \le 1, -x \le z \le 3+y$.

12. $f(x,y,z) = 1+x$, R lies inside $x^2+y^2=4$, between $z = -1, z = x$.

13. $f(x,y,z) = 2-y$, R is $x^2+y^2+z^2 \le 9$.

14. $f(x,y,z) = y^2$, R is $4x^2+4y^2+z^2 \le 4$.

15. $f(x,y,z) = -5$, R is $1 \le x^2+y^2+z^2 \le 4$.

16. $f(x,y,z) = 3$, R is outside $x^2+y^2=1$ and inside $x^2+y^2+z^2=4$.

17. $f(x,y,z) = x$, R is $0 \le x \le z \le 2-y \le 2$.

18. $f(x,y,z) = \frac{1}{z}$, R is $1 \le z \le e^x, 0 \le y \le x \le 1$.

19. $f(x,y,z) = e^{x+y}$, R is $0 \le x \le 1, 0 \le y \le 2, x^2 \le z \le x^2+y+1$.

20. $f(x,y,z) = \sin y$, R is $0 \le x \le y^2, -1 \le y \le 1$, $-1 \le z \le x^2+1$.

II

21. The integral we evaluated in Example 2 (p. 1104) simplified to an integral over the unit disk in the xy-plane. Evaluate the integral from that point on as an iterated integral in the variables x, y.

22. Integrate $f(x,y,z) = x$ over the region R lying inside $y^2+z^2=4$ and between $x+y+z=1$ and $x=4$.

Express each of the following integrals as an iterated integral in which the inner integral is with respect to x and the outer integral is with respect to z.

*23. $\int_0^1\int_0^1\int_0^x f(x,y,z)\,dz\,dx\,dy$

*24. $\int_0^1\int_0^y\int_0^x f(x,y,z)\,dz\,dx\,dy$

*25. $\int_2^4\int_1^x\int_2^{2y} f(x,y,z)\,dz\,dy\,dx$

*26. $\int_0^1\int_0^1\int_0^{x^2} f(x,y,z)\,dz\,dy\,dx$

*27. $\int_0^4\int_0^x\int_0^{y^2} f(x,y,z)\,dz\,dy\,dx$

*28. $\int_0^2\int_0^x\int_0^x f(x,y,z)\,dy\,dz\,dx$

*29. Find the average value of the function z on the upper half of the ball of radius 1 with center at the origin.

*30. Evaluate $\int\int\int_R \sin(xy)\,dV$, where $R = \{(x,y,z) : 0 \le x \le \pi, -1 \le y \le 1, -2 \le z \le 1\}$.

*31. What percentage of the phytoplankton in Example 5 (p. 1106) live (i) below 100 m, (ii) between 0 and 60 m deep?

*32. A different theory of phytoplankton proposes that the number of millions of organisms per cubic meter is $G(d) = 100de^{-d^2/1000}$ at depth d meters. What would this model predict in answer to the problem of Example 5?

*33. Find $\int\int\int_R (x+z)\,dV$, where R is the region outside $x^2+y^2=1$ and inside $x^2+y^2+z^2=4$.

*34. Find $\int\int\int_R y\,dV$, where R is the region outside $x^2+y^2=1$ and inside $x^2+y^2+z^2=2$.

35. In the situation of Example 6 (p. 1106) how much pollutant lies over the region within 5 km of the chimney?

36. In the situation of Example 6, how much pollutant lies over the region between 4 and 8 km of the chimney?

37. Suppose the concentration in Example 6 is given instead by

$$C(x,y,z) = \frac{1000-z}{10,000}\exp\left(-(x^2+y^2)/10,000,000\right).$$

Find the total amount of pollutant over the disk of radius 10 km.

III

38. Use a computer to sketch several level surfaces of the function $C(x,y,z)$ of Example 6.

POINT TO PONDER

In this section we have seen how to integrate functions of three variables, but it should be reasonably clear that a similar definition can be made for functions of n variables. For a continuous function $f(x_1, \ldots, x_n)$ on a bounded region R in $\mathbb{R}^n$ we define the integral

$$\int \int \cdots \int_R f(x_1, \ldots, x_n)\, dx_1 \ldots dx_n.$$

This is done by using limits of Riemann sums as the mesh of the corresponding partitions goes to 0, just as in Definition 17.6.1.

Then it is possible to show that the multiple integral over R can be evaluated by iterated integrals, assuming that R can be expressed in the appropriate form.

We refer to $\int \int \cdots \int_R 1\, dx_1 \ldots dx_n$ as the "n-dimensional volume" of the region R.

As an example we calculate the four-dimensional volume of the four-dimensional ball $w^2 + x^2 + y^2 + z^2 \leq 1$ in $\mathbb{R}^4$. First we have to express it as an iterated integral of the constant function 1 over the ball. The variable z can vary from -1 to 1. Then for fixed z the variable y runs from $-\sqrt{1 - z^2}$ to $\sqrt{1 - z^2}$. Fixing y and z, we find that x runs between $\pm\sqrt{1 - y^2 - z^2}$. Finally, for fixed x, y, z we see that w runs between $\pm\sqrt{1 - x^2 - y^2 - z^2}$. The four-dimensional volume of the ball equals

$$\int_{-1}^{1} \int_{-\sqrt{1-z^2}}^{\sqrt{1-z^2}} \int_{-\sqrt{1-y^2-z^2}}^{\sqrt{1-y^2-z^2}} \int_{-\sqrt{1-x^2-y^2-z^2}}^{\sqrt{1-x^2-y^2-z^2}} 1\, dw\, dx\, dy\, dz.$$

If we evaluate this iterated integral, we find that the four-dimensional volume of the four-dimensional unit ball is $\frac{1}{2}\pi^2$.

17.7 Moments and Centroids

MASS
MOMENT
CENTER OF MASS
SYMMETRY
ANGULAR VELOCITY
MOMENT OF INERTIA

In Section 6.7 we studied the moments of regions in the plane. One of the uses of this idea was to find the centroid $(\bar{x}, \bar{y})$ of a region R. The coordinates $\bar{x}$ and $\bar{y}$ can be viewed as the *average values* of x and y on the region (see Figure 17.7.1).

In Definition 6.7.5 we considered a region R lying between the graphs of $y = f(x)$ and $y = g(x)$, for $x \in [a, b]$, and assumed that the density was given by a function $\rho(x)$. We considered the thin strip between x and $x + dx$. Its height is $f(x) - g(x)$, so its weight is approximately $\rho(x)\big(f(x) - g(x)\big)\, dx$ (see Figure 17.7.2).

Figure 17.7.1

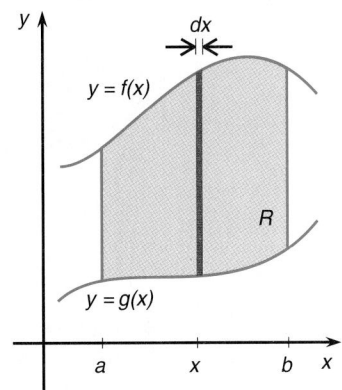

Figure 17.7.2

If we want to find the *weighted average* of x over R, then the contribution from this strip will be $x\rho(x)\big(f(x) - g(x)\big)\,dx$. "Adding up" over all such strips, that is, integrating, gives the moment

$$M_y = \int_a^b x\rho(x)\big(f(x) - g(x)\big)\,dx.$$

We called it M_y, the moment about the y-axis, because we think of x as the displacement from the y-axis.

Now that we know about double integrals, it seems reasonable to find the weighted sum of the values of x on R by integrating $x\rho$ over R. This leads us to consider the double integral $\int\int_R x\rho\,dx\,dy$. Because of the type of region being considered, this is easily expressed as an iterated integral:

$$\int\int_R x\rho\,dA = \int_a^b \int_{g(x)}^{f(x)} x\rho(x)\,dy\,dx.$$

Since the integrand does not involve y, the inner integral is very simple:

$$\int\int_R x\rho\,dA = \int_a^b \int_{g(x)}^{f(x)} x\rho(x)\,dy\,dx = \int_a^b x\rho(x)y\,\Big|_{y=g(x)}^{y=f(x)}\,dx$$

$$= \int_a^b x\rho(x)\big(f(x) - g(x)\big)\,dx.$$

This is exactly our original formula for M_y, which means that we can write $M_y = \int\int_R x\rho\,dA$. One advantage of this formula is that it makes sense even if ρ is not independent of y.

What we have just shown is that the following definition is consistent with Definition 6.7.5 (p. 467).

DEFINITION 17.7.1

If R is a region in the plane and its density is given by the function $\rho(x, y)$, then its **mass** is

$$M = \int\int_R \rho\,dA = \int\int_R \rho(x, y)\,dx\,dy.$$

Its **moments** about the x- and y-axes respectively are

$$M_x = \int\int_R y\rho\,dA = \int\int_R y\rho(x, y)\,dx\,dy,$$

$$M_y = \int\int_R x\rho\,dA = \int\int_R x\rho(x, y)\,dx\,dy.$$

Its **center of mass** is the point $(\bar{x}, \bar{y})$, where

$$\bar{x} = \frac{M_y}{M} = \frac{\int\int_R x\rho\,dx\,dy}{\int\int_R \rho\,dx\,dy}, \qquad \bar{y} = \frac{M_x}{M} = \frac{\int\int_R y\rho\,dx\,dy}{\int\int_R \rho\,dx\,dy}.$$

If the density ρ is *constant*, the **center of mass** is called the **centroid**.

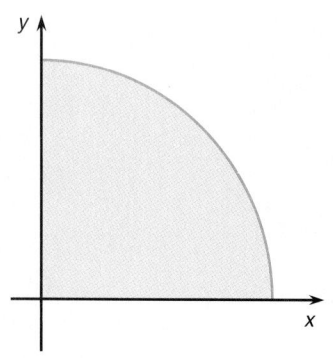

Figure 17.7.3

EXAMPLE 1   Suppose R is the region $x^2 + y^2 \le 1$, $x \ge 0$, $y \ge 0$, as shown in Figure 17.7.3. Find its center of mass if the density function is $\rho(x, y) = \sqrt{x^2 + y^2}$.

Solution We need to evaluate $M = \int\int_R \rho \, dA = \int\int_R \sqrt{x^2 + y^2} \, dx \, dy$, $M_y = \int\int_R x\rho \, dA = \int\int_R x\sqrt{x^2 + y^2} \, dx \, dy$, and $M_x = \int\int_R y\rho \, dA = \int\int_R y\sqrt{x^2 + y^2} \, dx \, dy$. It seems sensible to use polar coordinates, since R is the polar rectangle $0 \le \theta \le \frac{\pi}{2}$, $0 \le r \le 1$. We find

$$M = \int\int_R \sqrt{x^2 + y^2} \, dx \, dy = \int_0^{\pi/2} \int_0^1 (r)\, r \, dr \, d\theta = \int_0^{\pi/2} \left.\frac{r^3}{3}\right|_0^1 d\theta = \frac{\pi}{6},$$

$$M_x = \int\int_R y\sqrt{x^2 + y^2} \, dx \, dy = \int_0^{\pi/2} \int_0^1 r^2 \sin\theta \, r \, dr \, d\theta = \int_0^{\pi/2} \left.\frac{r^4}{4}\right|_0^1 \sin\theta \, d\theta$$

$$= -\frac{1}{4}\cos\theta \Big|_0^{\pi/2} = \frac{1}{4},$$

$$M_y = \int\int_R x\sqrt{x^2 + y^2} \, dx \, dy = \int_0^{\pi/2} \int_0^1 r^2 \cos\theta \, r \, dr \, d\theta = \int_0^{\pi/2} \left.\frac{r^4}{4}\right|_0^1 \cos\theta \, d\theta$$

$$= \frac{1}{4}\sin\theta \Big|_0^{\pi/2} = \frac{1}{4}.$$

Using these, we find $\bar{x} = \frac{M_y}{M} = \frac{1/4}{\pi/6} = \frac{3}{2\pi}$ and $\bar{y} = \frac{M_x}{M} = \frac{1/4}{\pi/6} = \frac{3}{2\pi}$. The centroid is $(\bar{x}, \bar{y}) = \left(\frac{3}{2\pi}, \frac{3}{2\pi}\right) \approx (0.477, 0.477)$.

> Notice that the density in this example depends on both x and y. (It might for instance represent the density of a sector of a wheel that is thicker out near the rim and thin near the center.) Because of this we could not do the calculation using the techniques of Section 6.7.

It is also possible to use double and triple integrals to find moments and centroids of regions in $\mathbb{R}^3$. As with plane regions we suppose that we know the density as a function $\rho = \rho(x, y, z)$. Often $\rho(x, y, z)$ will be a constant.

In the plane we regarded x as the displacement from the y-axis and y as the displacement from the x-axis. In $\mathbb{R}^3$ we think of z, for instance, as the height, which can also be seen as the displacement from the xy-plane. The corresponding moment will be labeled M_{xy} and called the *moment of R about the xy-plane*. Similarly, the moment involving x will be M_{yz}, and the moment involving y will be M_{xz}.

The following definition is analogous to the two-dimensional version.

DEFINITION 17.7.2

If R is a region in $\mathbb{R}^3$ and its density is given by the function $\rho(x, y, z)$, then its **mass** is

$$M = \int\int\int_R \rho \, dV = \int\int\int_R \rho(x, y, z) \, dx \, dy \, dz.$$

Its **moments** about the xy-, xz-, and yz-planes are

$$M_{xy} = \int\int\int_R z\rho \, dV = \int\int\int_R z\rho(x, y, z) \, dx \, dy \, dz,$$

$$M_{xz} = \int\int\int_R y\rho \, dV = \int\int\int_R y\rho(x, y, z) \, dx \, dy \, dz,$$

$$M_{yz} = \int\int\int_R x\rho \, dV = \int\int\int_R x\rho(x, y, z) \, dx \, dy \, dz,$$

respectively. Its **centroid** or **center of mass** is the point $(\bar{x}, \bar{y}, \bar{z})$, where

$$\bar{x} = \frac{M_{yz}}{M} = \frac{\int\int\int_R x\rho\,dx\,dy\,dz}{\int\int\int_R \rho\,dx\,dy\,dz}, \qquad \bar{y} = \frac{M_{xz}}{M} = \frac{\int\int\int_R y\rho\,dx\,dy\,dz}{\int\int\int_R \rho\,dx\,dy\,dz},$$

$$\bar{z} = \frac{M_{xy}}{M} = \frac{\int\int\int_R z\rho\,dx\,dy\,dz}{\int\int\int_R \rho\,dx\,dy\,dz}.$$

EXAMPLE 2   Find the centroid of the upper half of the ball $x^2 + y^2 + z^2 \leq 1$, assuming that the density is uniform.

Solution   To say that the density is uniform means that the density function $\rho(x, y, z)$ is a constant. As we observed in Chapter 6, in this situation the centroid will be the same no matter what the constant, so we may as well assume that $\rho = 1$. The region is illustrated in Figure 17.7.4.

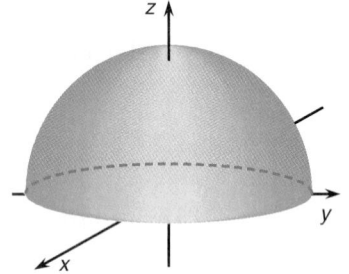

Figure 17.7.4

Its mass is $M = \int\int\int_R \rho\,dV = \int\int\int_R 1\,dV$, and this is just the volume of the hemisphere, which equals $M = \frac{1}{2}\frac{4}{3}\pi r^3 = \frac{2\pi}{3}$.

The moment about the xy-plane is

$$M_{xy} = \int\int\int_R z\,dV = \int_{-1}^{1}\int_{-\sqrt{1-y^2}}^{\sqrt{1-y^2}}\int_0^{\sqrt{1-x^2-y^2}} z\,dz\,dx\,dy$$

$$= \int_{-1}^{1}\int_{-\sqrt{1-y^2}}^{\sqrt{1-y^2}} \frac{z^2}{2}\bigg|_0^{\sqrt{1-x^2-y^2}} dx\,dy = \frac{1}{2}\int_{-1}^{1}\int_{-\sqrt{1-y^2}}^{\sqrt{1-y^2}} (1 - x^2 - y^2)\,dx\,dy.$$

Rather than evaluate this integral in this form, we recognize it as the integral over the unit disk D of the function $1 - r^2$, which is most easily evaluated in polar coordinates:

$$\frac{1}{2}\int\int_D (1 - r^2)\,dA = \frac{1}{2}\int_0^{2\pi}\int_0^1 (1 - r^2)r\,dr\,d\theta = \frac{1}{2}\int_0^{2\pi} \left(\frac{r^2}{2} - \frac{r^4}{4}\right)\bigg|_0^1 d\theta$$

$$= \frac{1}{2}\frac{1}{4}(2\pi) = \frac{\pi}{4}.$$

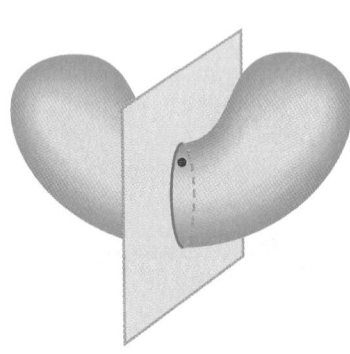

Figure 17.7.5

The moment is $M_{xy} = \frac{\pi}{4}$. We conclude that the vertical coordinate of the centroid is $\bar{z} = \frac{M_{xy}}{M} = \frac{\pi/4}{2\pi/3} = \frac{3}{8}$.

It is possible to find $\bar{x}$ and $\bar{y}$ in exactly the same way, but since the hemisphere is symmetrical about the xz-plane and the yz-plane, we expect the corresponding moments to be zero, in analogy with Theorem 6.7.7 (p. 469). This is easily confirmed, and we see that the centroid is $\left(0, 0, \frac{3}{8}\right)$.

In effect we have used the following result.

THEOREM 17.7.3

Suppose R is a region in $\mathbb{R}^3$ that is symmetrical about some plane. Then its centroid must lie in that plane. In particular, if R is symmetrical about one of the coordinate planes, then its moment about that plane must be 0.

The proof of this result is not very difficult, but it is moderately complicated, so we omit it because the geometric picture is so convincing. See Figure 17.7.5.

Moments of Inertia and Rotational Motion

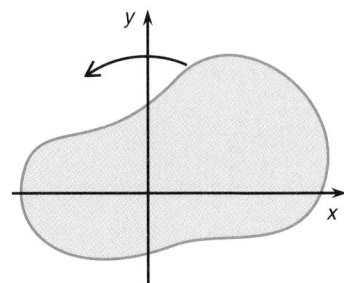

Figure 17.7.6

Now we discuss *moments of inertia*, a concept that is important for studying the motion of rotating objects. Since many important motions are rotational, the value of these ideas cannot be overestimated. However, they do require some familiarity with physics. This topic is not used later on.

First imagine a flat object in the plane, rotating about the origin, as shown in Figure 17.7.6. The point at the origin does not move, points near the origin move quite slowly, and points farther away from the origin move more quickly. This makes it difficult to specify the speed of the rotation.

It is for this reason that we use the **angular velocity**, which measures the speed of rotation in terms of the angle rotated per unit time. For instance, an object turning at a rate of $45°$ per second will rotate $360°$ in 8 seconds. As usual, we will use radian measure for the angles and say instead that this object has an angular velocity of $\frac{\pi}{4}$ radians per second.

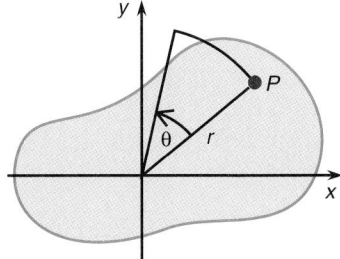

Figure 17.7.7

Suppose an object is rotating at an angular velocity ω (ω is the Greek letter "omega"). Consider a point P on the object at a distance r from the center of rotation, as shown in Figure 17.7.7. As the object moves through an angle θ, this point will move along part of a circle of radius r, and since the circumference of the whole circle is $2\pi r$, it will move through a distance of $\frac{\theta}{2\pi} 2\pi r = \theta r$. In particular, in one second the object will rotate through an angle ω, and the point will move a distance of ωr. We see that a point at a distance r from the center of rotation of an object rotating at ω radians per second will move at a speed of ωr units per second. (The units will be the same as the units of r.)

This is consistent with our earlier observation that points farther from the origin move faster than points near the origin.

From the study of physics, one learns that an object of mass m moving at a speed v has kinetic energy $\frac{1}{2}mv^2$. For this to make sense it is necessary to specify the appropriate units of mass, speed, and energy (cf. Exercise 33 in Section 6.6).

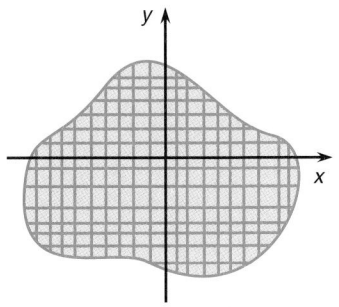

Figure 17.7.8

Suppose we return to the point P at a distance r from the center of rotation of an object rotating at an angular velocity ω. We know that the speed of P is $v = \omega r$, so if it is a small object of mass m, then its kinetic energy is $\frac{1}{2}mv^2 = \frac{1}{2}m(\omega r)^2 = \frac{1}{2}mr^2\omega^2$.

Now consider a more complicated object that covers a region R in the plane and has density ρ. We divide up the region into many little rectangles R_i with $i = 1, \ldots, n$, as shown in Figure 17.7.8. Suppose $P_i^* = (x_i^*, y_i^*)$ is a point in R_i, let $r_i^* = \sqrt{x_i^{*2} + y_i^{*2}}$ be its distance from the origin, and suppose the object rotates about the origin with angular velocity ω.

Then the mass of the part of the object inside R_i is approximately $m = \rho(P_i^*) \Delta A_i$, so its kinetic energy is approximately $\frac{1}{2}m\omega^2 r^2 = \frac{1}{2}\rho(P_i^*)\omega^2 r_i^{*2} \Delta A_i = \frac{1}{2}\rho(x_i^*, y_i^*)\omega^2(x_i^{*2} + y_i^{*2}) \Delta A_i$.

The total kinetic energy of the whole object is approximately the sum

$$\sum_{i=1}^{n} \frac{1}{2}\rho(x_i^*, y_i^*)\omega^2(x_i^{*2} + y_i^{*2}) \Delta A_i = \frac{1}{2}\left(\sum_{i=1}^{n} \rho(x_i^*, y_i^*)(x_i^{*2} + y_i^{*2}) \Delta A_i\right)\omega^2.$$

If we allow the mesh of the partition $\mathcal{P} = \{R_i\}$ to tend to zero, this sum will approach the integral

$$\frac{1}{2}\left(\int\int_R (x^2 + y^2)\rho(x, y)\, dA\right)\omega^2.$$

This suggests the following definition, and the theorem that follows it has essentially been proved by the argument we have just given.

DEFINITION 17.7.4

Suppose R is an object in the plane with density $\rho(x, y)$. Its **moments of inertia** about the x- and y-axes are

$$I_x = \int\int_R y^2 \rho(x, y)\, dA \quad \text{and} \quad I_y = \int\int_R x^2 \rho(x, y)\, dA.$$

Their sum is the **moment of inertia about the origin** (sometimes called the **polar moment of inertia**):

$$I = I_0 = \int\int_R (x^2 + y^2)\rho(x, y)\, dA.$$

THEOREM 17.7.5

If the object described in Definition 17.7.4 rotates about the origin with angular velocity ω, then its kinetic energy is

$$E = \frac{1}{2}I\omega^2 = \frac{1}{2}I_0\omega^2.$$

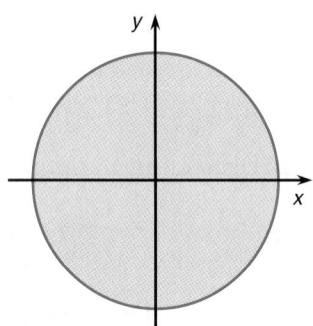

Figure 17.7.9

Notice that the moments of inertia are analogous to the ordinary moments except that they involve the *squares* of y and x inside the integral. They are related to the average values of y^2 and x^2 on R. Since x^2 and y^2 are never negative, there is no cancellation, and the moments of inertia are never zero (unless the mass of R is zero).

EXAMPLE 3   Find the moment of inertia about the origin of D, the unit disk $x^2 + y^2 \leq 1$, if the density is $\rho = 1$. See Figure 17.7.9.

Solution   The polar moment is $I_0 = \int\int_D (x^2 + y^2)\rho\, dA = \int\int_D r^2\, dA$. This is best evaluated by polar coordinates:

$$I_0 = \int\int_D r^2\, dA = \int_0^{2\pi}\int_0^1 (r^2) r\, dr\, d\theta = \int_0^{2\pi} \frac{r^4}{4}\Big|_0^1 d\theta = \frac{1}{4}2\pi = \frac{\pi}{2}.$$

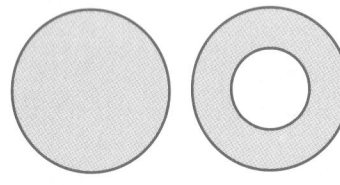

Figure 17.7.10

Notice that this polar moment is not zero even though the region D is symmetrical about the origin.

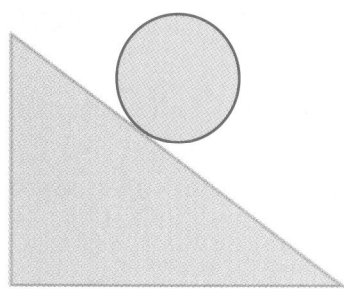

Figure 17.7.11

EXAMPLE 4   **Rolling Wheels**   Consider a wheel cut out of plywood, with diameter 2 feet, and consider a second wheel of the same material and with the same diameter but with a disk of diameter 1 ft cut out from its center, as shown in Figure 17.7.10.

Suppose both wheels are allowed to roll down a ramp inclined at $30°$ as shown in Figure 17.7.11. Which one will reach the bottom faster?

Solution   We think of the solid wheel as D, the disk $x^2 + y^2 \leq 1$ of radius 1. Its density is some constant ρ depending on the type of plywood. We calculate the mass and the polar moment of inertia of each wheel.

For the solid wheel the mass is $M_s = \int \int_D \rho\, dA = \rho\, area(D) = \pi\rho$.

For the punctured wheel the mass is ρ times the area of the disk missing the smaller disk. Since the area of the solid disk is π and the area of the smaller disk of radius $\frac{1}{2}$ is $\pi r^2 = \frac{\pi}{4}$, we see that the mass of the punctured wheel is $M_p = \left(\pi - \frac{\pi}{4}\right)\rho = \frac{3}{4}\pi\rho$. In particular $M_s = \frac{4}{3}M_p$.

To calculate the moments, we use polar coordinates again. For the solid wheel the moment of inertia about the origin is

$$I_s = \int \int_D r^2 \rho\, dA = \rho \int_0^{2\pi} \int_0^1 (r^2)\, r\, dr\, d\theta = \rho \int_0^{2\pi} \left.\frac{r^4}{4}\right|_0^1 d\theta = \rho\frac{1}{4}2\pi = \frac{1}{2}\pi\rho.$$

For the punctured wheel P the moment of inertia about the origin is

$$I_p = \int \int_P r^2 \rho\, dA = \rho \int_0^{2\pi} \int_{1/2}^1 (r^2)\, r\, dr\, d\theta = \rho \int_0^{2\pi} \left.\frac{r^4}{4}\right|_{1/2}^1 d\theta$$

$$= \left(\frac{1}{4} - \frac{1}{64}\right)2\pi\rho = \frac{15}{32}\pi\rho.$$

In particular, $I_s = \frac{16}{15}I_p$.

At this point you might like to pause and try to guess which wheel rolls faster.

In the study of physics it is found that near the surface of the earth the force of gravity has the following effect on a descending object. An object of mass M that descends a vertical distance h will have its kinetic energy increase by Mgh, where g is a constant. In particular the increase in kinetic energy is a constant times the mass.

Now consider our rolling wheels. They descend through some distance h, so the changes in their kinetic energies are

$$E_s = M_s gh = \pi\rho gh \qquad \text{and} \qquad E_p = M_p gh = \frac{3}{4}\pi\rho gh.$$

Without knowing what h is, we see that the increase in kinetic energy will be three-quarters as large for the punctured wheel as for the solid wheel, since this is the ratio of their masses.

All this increase in energy will be converted into the kinetic energy with which the wheels are rolling when they reach the bottom. This kinetic energy is the sum of the rotational energy associated with the spinning of the wheels plus the *translational* kinetic energy arising from the horizontal motion of the wheels.

If we let ω_s and ω_p be the angular velocities of the solid and punctured wheels at the bottom of the ramp, then their rotational energies are $\frac{1}{2}I_s\omega_s^2$ and $\frac{1}{2}I_p\omega_p^2$, respectively.

For a wheel of radius $r = 1$ with angular velocity ω a point on the rim of the wheel moves with a speed $r\omega = \omega$ with respect to the center of the wheel. If we consider the point of contact of the rim with the floor, as shown in Figure 17.7.12, we observe that this point is (momentarily) stationary, so this speed $r\omega = \omega$ is the horizontal speed v of the center of the wheel. The translational kinetic energy of the wheel is $\frac{1}{2}Mv^2$, where M is the mass and v the speed. This means that the translational kinetic energies of the solid wheel and the punctured wheel are $\frac{1}{2}M_s\omega_s^2$ and $\frac{1}{2}M_p\omega_p^2$, respectively.

We combine these results, adding rotational and translational energies, and find that the total kinetic energies of the solid and punctured wheels are, respectively,

$$E_s = \frac{1}{2}I_s\omega_s^2 + \frac{1}{2}M_s\omega_s^2 \qquad \text{and} \qquad E_p = \frac{1}{2}I_p\omega_p^2 + \frac{1}{2}M_p\omega_p^2.$$

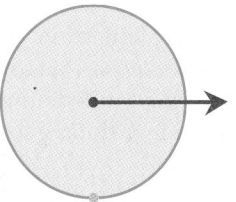

Figure 17.7.12

Using the other expressions we found for the changes in the energy for the two wheels and substituting our formulas for the masses and moments of inertia, we find

$$E_s = \pi \rho g h = \frac{1}{2}\left(\frac{1}{2}\pi\rho\omega_s^2\right) + \frac{1}{2}\pi\rho\omega_s^2 = \frac{1}{2}\pi\rho\left(\frac{3}{2}\omega_s^2\right),$$

$$E_p = \frac{3}{4}\pi \rho g h = \frac{1}{2}\left(\frac{15}{16}\pi\rho\omega_p^2\right) + \frac{1}{2}\left(\frac{3}{4}\pi\rho\omega_p^2\right) = \frac{1}{2}\pi\rho\left(\frac{27}{16}\omega_p^2\right).$$

Multiplying the second of these equations by $\frac{4}{3}$ and subtracting from the first, we find

$$0 = \frac{1}{2}\pi\rho\left(\frac{3}{2}\omega_s^2 - \frac{9}{4}\omega_p^2\right).$$

This means that $\frac{9}{4}\omega_p^2 = \frac{3}{2}\omega_s^2$, so $\frac{3}{2}\omega_p^2 = \omega_s^2$ and $\omega_s = \sqrt{\frac{3}{2}}\omega_p$.

In particular, ω_s is larger than ω_p by a factor of $\sqrt{\frac{3}{2}} \approx 1.22$. The solid wheel will be rolling more quickly at the bottom of the ramp, and accordingly it will roll down the ramp more quickly.

There are several observations to make about this example. One is that the ratio of the speeds of the two wheels does not involve the density ρ. The ratio would be the same for plywood wheels as it would be for wheels made of cast iron; it depends on the shape of the wheels but not on the density.

Second, the ratio does not depend on the height of the ramp. In the original question it was stated that the ramp had an inclination of $30°$, but neither this nor the size of the ramp is needed in the ratio. Of course, they would be used if we wanted to find the exact length of time needed for either of the wheels to roll down the ramp.

Finally, we notice that the moment of inertia is larger relative to the mass for the punctured wheel; this says that the *average value* of $r^2 = x^2 + y^2$ is greater for the punctured wheel, since its mass is concentrated farther out from the center. At a given angular velocity ω the rotational energy $\frac{1}{2}I\omega^2$ will be greater for a punctured wheel than a solid one *of the same mass*. Or to put it the other way around, to have the same energy a punctured wheel will spin at a lower speed than a solid wheel will. So a punctured wheel will roll less quickly to convert its energy into kinetic energy, and this is why the solid wheel rolls down first.

For a rotating three-dimensional object we can ask the same sort of question. In this situation the object must be rotating about some line, the *axis of rotation*, and for convenience we usually let it be one of the coordinate axes. Then we should define the moment of inertia about that axis, and by analogy with the two-dimensional case it should be the integral of the square of the distance from the axis.

DEFINITION 17.7.6

Suppose R is a region in $\mathbb{R}^3$ with density $\rho(x, y, z)$. Its **moments of inertia** about the x-, y-, and z-axes are

$$I_x = \iiint_R (y^2 + z^2)\rho(x, y, z)\, dV, \qquad I_y = \iiint_R (x^2 + z^2)\rho(x, y, z)\, dV,$$

$$I_z = \iiint_R (x^2 + y^2)\rho(x, y, z)\, dV.$$

THEOREM 17.7.7

If the object described in Definition 17.7.6 rotates about the z-axis with angular velocity ω, then its **kinetic energy** is

$$E = \frac{1}{2}I_z\omega^2,$$

and similarly for rotation about the other axes.

EXAMPLE 5   Find the moment of inertia about the z-axis of the ball of radius 2 centered at the origin.

Solution   The moment of inertia is $I_z = \int\int\int_B (x^2 + y^2)\rho\, dV$. When the density is not specified, we assume that it is the constant $\rho = 1$. We can set up an iterated integral as follows:

$$\int\int\int_B (x^2 + y^2)\, dV = \int_{-2}^{2}\int_{-\sqrt{4-x^2}}^{\sqrt{4-x^2}}\int_{-\sqrt{4-x^2-y^2}}^{\sqrt{4-x^2-y^2}} (x^2+y^2)\, dz\, dy\, dx$$

$$= \int_{-2}^{2}\int_{-\sqrt{4-x^2}}^{\sqrt{4-x^2}} \left((x^2+y^2)z\, \Big|_{z=-\sqrt{4-x^2-y^2}}^{z=\sqrt{4-x^2-y^2}} \right) dy\, dx$$

$$= \int_{-2}^{2}\int_{-\sqrt{4-x^2}}^{\sqrt{4-x^2}} 2(x^2+y^2)\sqrt{4-x^2-y^2}\, dy\, dx$$

$$= 2\int_{0}^{2\pi}\int_{0}^{2} r^2\sqrt{4-r^2}\, r\, dr\, d\theta.$$

In the last step we rewrote the integral over the disk in terms of polar coordinates. Making the substitution $u = 4 - r^2$, $du = -2r\, dr$, with u running from 4 to 0, we find that

$$I_z = -\int_{0}^{2\pi}\int_{4}^{0} (4-u)\sqrt{u}\, du\, d\theta = \int_{0}^{2\pi}\int_{0}^{4} (4u^{1/2} - u^{3/2})\, du\, d\theta$$

$$= \int_{0}^{2\pi}\left(\frac{8}{3}u^{3/2} - \frac{2}{5}u^{5/2} \right)\Big|_{0}^{4} d\theta = \int_{0}^{2\pi}\left(\frac{64}{3} - \frac{64}{5} \right) d\theta = \int_{0}^{2\pi} \frac{128}{15}\, d\theta$$

$$= \frac{256}{15}\pi.$$

The moment of inertia is $I_z = \frac{256}{15}\pi \approx 53.6$.

It is interesting to interpret the moment of inertia by thinking about averages. The moment of inertia is $I_z = \int\int\int_B (x^2 + y^2)\rho\, dV$. This is related to the average value of the function $x^2 + y^2$ over the ball; to find the average, we should just divide by the total mass. For the example we have just done we find that the average value of $x^2 + y^2$ on the ball is

$$A = \frac{\int\int\int_B (x^2 + y^2)\rho\, dV}{\int\int\int_B \rho\, dV} = \frac{I_z}{M} = \frac{\frac{256}{15}\pi}{\frac{4}{3}\pi 2^3} = \frac{8}{5} = 1.6.$$

The values of $x^2 + y^2$ on the ball vary from 0 to 4, and their average is 1.6.

Suppose we put all the mass of the ball concentrated at one point and ask that this new object have the same moment of inertia as the ball. How far should the point be from the z-axis?

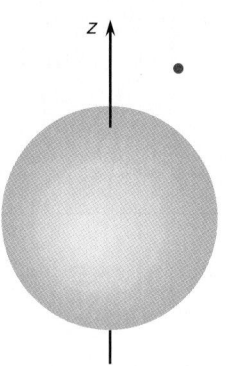

Figure 17.7.13

A body consisting of a mass M at a distance R from the axis has moment of inertia MR^2. If we want this to equal I, we need $MR^2 = I$, so $R^2 = \frac{I}{M}$, and this is what we just decided was the average value of r^2 on B. In the particular case of the ball of radius 2 the answer is that the single point should be at a distance R from the z-axis, where $R^2 = \frac{8}{5}$. See Figure 17.7.13.

This discussion suggests the following definition.

DEFINITION 17.7.8

Consider an object in the plane or in $\mathbb{R}^3$, and suppose it is rotating around the origin or about the z-axis, respectively. Let I be the appropriate moment of inertia (I_0 for a plane object and I_z for an object in space). Let M be the total mass of the object. The **radius of gyration** of the object is

$$R = \sqrt{\frac{I}{M}}.$$

An object whose mass M is the same as that of the original object but that is all concentrated at a distance R from the origin or the z-axis, respectively, will have the

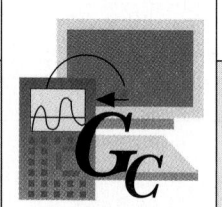

In Example 3 we found that the moment of inertia of the unit disk is $I_0 = \frac{\pi}{2}$. Since its mass is $M = \pi$, we find that its radius of gyration is $R = \sqrt{\frac{I}{M}} = \frac{1}{\sqrt{2}}$.

With a graphics calculator we can plot the unit circle and the line $y = \frac{1}{\sqrt{2}}$.

TI-85	SHARP EL9300
GRAPH F1$(y(x) =)$ $\sqrt{}$ $(1 - \text{F1}(x)$ $x^2)$ ▼ $(-)$ $\sqrt{}$ $(1 - \text{F1}(x)\ x^2)$ ▼ $\sqrt{}$ .5 M2(RANGE) $(-)$ 1.5 ▼ 1.5 ▼ ▼ $(-)$ 1 ▼ 1 F5(GRAPH)	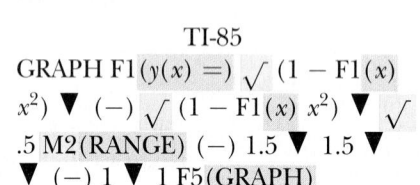 $\sqrt{}$ $1 - x/\theta/T\ x^2$ 2ndF ▼ $-$ $\sqrt{}$ $1 - x/\theta/T\ x^2$ 2ndF ▼ $\sqrt{}$ .5 RANGE -1.5 ENTER 1.5 ENTER 1 ENTER -1 ENTER 1 ENTER 1 ENTER
CASIO f_x-6300G (f_x-7700GB)	HP 48SX
Range -1.5 EXE 1.5 EXE 1 EXE -1 EXE 1 EXE 1 EXE EXE EXE EXE Graph $\sqrt{}$ $(1 - \boxed{X}\ x^2)$ EXE Graph $-$ $\sqrt{}$ $($ $1 - \boxed{X}\ x^2)$ EXE Graph $\sqrt{}$ .5 EXE	↰ PLOT PLOTR ERASE ATTN ′ Y = $\sqrt{x}$ ❪ $1 - X\ y^x\ 2$ ↱ PLOT ↰ DRAW ERASE 1.5 +/− ENTER 1.5 XRNG 1 +/− ENTER 1 YRNG DRAW ATTN ′ Y = +/− $\sqrt{x}$ ❪ $1 - X\ y^x\ 2$ ↱ PLOT ↰ DRAW DRAW ATTN ′ Y = $\sqrt{x}$ .5 ↱ PLOT ↰ DRAW DRAW

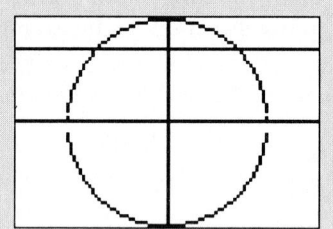

If we imagine the disk rotating about the x-axis, it will behave the same as a point mass lying on this horizontal line.

same moment of inertia. As far as rotations are concerned, it will behave the same way as the original object.

It is possible to find the radius of gyration about any other axis in exactly the same way.

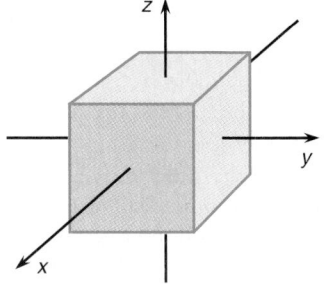

Figure 17.7.14

EXAMPLE 6   Find the radius of gyration of the cube $|x| \leq 1, |y| \leq 1, |z| \leq 1$ about the z-axis.

Solution   We assume that the density is $\rho = 1$. The mass of the cube, seen in Figure 17.7.14, is then just its volume: $M = 8$. The moment of inertia is

$$I_z = \int \int \int_C (x^2 + y^2)\,dV = \int_{-1}^{1} \int_{-1}^{1} \int_{-1}^{1} (x^2 + y^2)\,dz\,dx\,dy$$

$$= \int_{-1}^{1} \int_{-1}^{1} (x^2 + y^2)z \Big|_{z=-1}^{z=1} dx\,dy = 2\int_{-1}^{1} \int_{-1}^{1} (x^2 + y^2)\,dx\,dy$$

$$= 2\int_{-1}^{1} \left(\frac{x^3}{3} + y^2 x\right)\Big|_{x=-1}^{x=1} dy = 2\int_{-1}^{1} \left(\frac{2}{3} + 2y^2\right) dy = 2\left(\frac{2}{3}y + \frac{2}{3}y^3\right)\Big|_{-1}^{1}$$

$$= 2\left(\frac{4}{3} + \frac{4}{3}\right) = \frac{16}{3}.$$

The radius of gyration is $R = \sqrt{\dfrac{I}{M}} = \sqrt{\dfrac{16}{3}\dfrac{1}{8}} = \sqrt{\dfrac{2}{3}} \approx 0.8165.$

Exercises 17.7

I

Find the centroid of each of the following regions, assuming that the density is constant.

1. $x^2 + y^2 \leq 4, y \geq 0$
2. $x^2 + y^2 \leq 9, y \geq 0, x + y \geq 0$
3. $x^2 \leq y \leq 4, x \geq 0$
4. $0 \leq y \leq e^{-x}, -1 \leq x \leq 0$
5. $x^2 \leq y^2 + 1, x \geq 0, 0 \leq y \leq 2$
6. $0 \leq y \leq \sin x, 0 \leq x \leq \frac{\pi}{2}$
7. The triangle with vertices $(2, 3), (5, 3), (3, 2)$.
8. The quadrilateral with vertices $(0, 0), (4, 0), (3, 3), (4, 3)$.
9. $x^2 + y^2 \leq 3, y \geq |x|$
10. $y^2 \leq \cos \pi x, |x| \leq \frac{1}{2}$
11. $x^2 + y^2 \leq 4, 0 \leq z \leq 3$
12. $x^2 + y^2 \leq 9, 0 \leq z \leq x^2 + y^2$
13. $|x| \leq 2, |y| \leq 3, |z| \leq 4$
14. $0 \leq x \leq y \leq 1, 0 \leq z \leq x + y$
15. $2x^2 + 2y^2 \leq z \leq 1 + x^2 + y^2$
16. $x^2 + y^2 + z^2 \leq 4, x \geq 1$

17. The ball of radius 3 around the origin with a ball of radius 1 cut out from its center.
18. The region consisting of the upper half of the unit ball around the origin together with $x^2 + y^2 \leq 1$, $-1 \leq z \leq 0$.
19. The region above $z = x^2 + y^2$ and inside $x^2 + y^2 + z^2 = 2$.
20. The region inside $x^2 + y^2 - z^2 = 1$ and between $z = -1$ and $z = 0$.

Find the radius of gyration for each of the following objects (around the origin for plane objects and around the z-axis for solid objects). Assume that the density is $\rho = 1$.

21. $x^2 + y^2 \leq 1$
22. $x^2 + y^2 \leq 4, y \geq 0$
23. $0 \leq x \leq 1, 0 \leq y \leq 2$
24. $0 \leq x \leq y \leq 1$
25. $x^2 + y^2 + z^2 \leq 9, x \geq 0$
26. $0 \leq x \leq 1, 0 \leq y \leq 1, 0 \leq z \leq 1$
27. $0 \leq z \leq 1 - x^2 - y^2$
28. $(x - 1)^2 + y^2 + z^2 \leq 1$

29. $x^2 + y^2 + z^2 \leq 4, z \leq -1$

30. $x^2 + y^2 + z^2 \leq 3, z^2 \geq 1 + x^2 + y^2$

II

*31. Suppose R is a region in $\mathbb{R}^3$ with uniform density. Show that its centroid is the same no matter what (positive) value is given to the constant ρ.

*32. If an object in the plane has uniform density ρ, show that its radius of gyration (about the origin) does not depend on what the value of ρ is, that is, that R will be the same for any positive constant ρ.

*33. Consider a wheel of radius 1 with a wheel of radius $\frac{3}{4}$ cut out from its center. How will the time it takes for this wheel to roll down a ramp compare with the time for a solid wheel of radius 1 rolling down the same ramp?

*34. Consider a wheel of radius a with a wheel of radius b cut out from its center (with $b < a$). How will

the time it takes for this wheel to roll down a ramp compare with the time for a solid wheel of radius a rolling down the same ramp?

*35. If an object in $\mathbb{R}^3$ has uniform density ρ, show that its radius of gyration (about the z-axis) does not depend on what the value of ρ is, that is, that R will be the same for any positive constant ρ.

*36. Show that the cylinder $x^2 + y^2 \leq 4, 0 \leq z \leq 3$ has the same radius of gyration about the z-axis as the disk $x^2 + y^2 \leq 4$ has about the origin.

*37. Consider a ball of uniform density and radius 1 rolling down a ramp. Consider another ball of the same material and radius but with a hollow center, the cavity being a ball of radius $\frac{1}{2}$ around the center of the large ball. Compare their angular velocities when they reach the bottom of the ramp.

*38. Show by example that the centroid of a region R may not lie inside R.

17.8 Cylindrical and Spherical Coordinates

$dV = r \, dr \, d\theta \, dz$

$dV = \rho^2 \sin \phi \, d\rho \, d\theta \, d\phi$

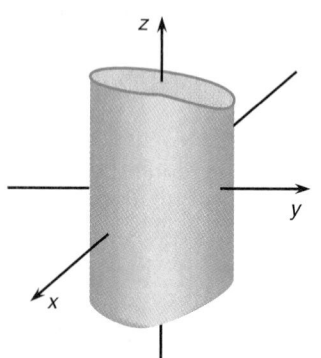

Figure 17.8.1

We have already seen a number of examples in which we start with a triple integral over some region R in $\mathbb{R}^3$, but after performing the inner integration we are left with a double integral over a disk in the xy-plane. Depending on the integrand, it may be that this integral can be evaluated most easily by using polar coordinates.

This situation is most likely to arise when the original region R has a round shape, which often means that R is a type of region that can be described in cylindrical coordinates (cf. Section 15.5 and see Figure 17.8.1). If R can be described in cylindrical coordinates and the integrand can be expressed in terms of cylindrical coordinates as $F(r, \theta, z)$, then it would be convenient to express the integral directly in terms of cylindrical coordinates.

Suppose the integral is an iterated integral of the form

$$\int \int_D \left(\int_{g(x,y)}^{h(x,y)} f(x, y, z) \, dz \right) dx \, dy,$$

where D is some region in the plane and $f(x, y, z)$ is continuous. The outer double integral over D can be expressed in polar coordinates, giving

$$\int \int_D \left(\int_{G(r,\theta)}^{H(r,\theta)} F(r, \theta, z) \, dz \right) r \, dr \, d\theta.$$

Notice that we have expressed the inner limits of integration in polar coordinates by $G(r, \theta) = g(x, y), H(r, \theta) = h(x, y)$ and the integrand as $F(r, \theta, z) = f(x, y, z)$.

An integral of this form is said to be in cylindrical coordinates, and as with other iterated integrals the integrations can be performed in any order (although changing the order will likely change the limits of integration). We have not written explicit limits of integration for the region D in the xy-plane,' preferring to say it is some region that can be described in polar coordinates. We have just shown the following result.

THEOREM 17.8.1

Suppose D is a region in the xy-plane that can be described in polar coordinates, and suppose R is the region in $\mathbb{R}^3$ lying between the graphs of $z = g(x, y)$ and $z = h(x, y)$ over D, that is,

$$R = \{(x, y, z) : g(x, y) \leq z \leq h(x, y), \quad (x, y) \in D\}.$$

If $f(x, y, z)$ is a continuous function on R, we can express f, g, and h as $F(r, \theta, z)$, $G(r, \theta)$, and $H(r, \theta)$, and then the integral of f over R can be evaluated in **cylindrical coordinates** as follows:

$$\iiint_R f(x, y, z)\, dV = \iiint_R f(x, y, z)\, dx\, dy\, dz$$

$$= \iint_D \left(\int_{g(x,y)}^{h(x,y)} f(x, y, z)\, dz \right) dx\, dy$$

$$= \iint_D \left(\int_{G(r,\theta)}^{H(r,\theta)} F(r, \theta, z)\, dz \right) r\, dr\, d\theta.$$

It is also possible to use iterated integrals in other orders, being careful about limits of integration. We sometimes summarize these results by writing

$$\iiint_R f\, dV = \iiint_R f\, dx\, dy\, dz = \iiint_R F\, r\, dr\, d\theta\, dz,$$

or even

$$dV = dx\, dy\, dz = r\, dr\, d\theta\, dz.$$

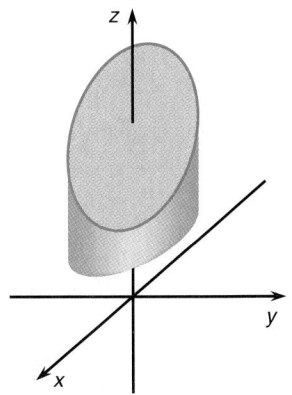

Figure 17.8.2

EXAMPLE 1   Evaluate $\iiint_R \sqrt{x^2 + y^2}\, dx\, dy\, dz$, where R is the region inside the cylinder $x^2 + y^2 = 4$ and between the planes $z = y + 2$ and $z = 6 - x$.

Solution   The region R is illustrated in Figure 17.8.2. Instead of expressing it in terms of the rectangular coordinates, we observe that it can be written as $0 \leq r \leq 2$, $2 + r\sin\theta \leq z \leq 6 - r\cos\theta$. The integral is

$$\iiint_R \sqrt{x^2 + y^2}\, dx\, dy\, dz = \int_0^{2\pi} \int_0^2 \left(\int_{2+r\sin\theta}^{6-r\cos\theta} r\, dz \right) r\, dr\, d\theta$$

$$= \int_0^{2\pi} \int_0^2 \left(rz \Big|_{z=2+r\sin\theta}^{z=6-r\cos\theta} \right) r\, dr\, d\theta$$

$$= \int_0^{2\pi} \int_0^2 r(6 - r\cos\theta - 2 - r\sin\theta)\, r\, dr\, d\theta$$

$$= \int_0^{2\pi} \left(\frac{4}{3}r^3 - (\cos\theta + \sin\theta)\frac{r^4}{4} \right)\Big|_0^2 d\theta$$

$$= \int_0^{2\pi} \left(\frac{32}{3} - 4\cos\theta - 4\sin\theta \right) d\theta = \frac{64}{3}\pi.$$

Of course, we could have begun this example with rectangular coordinates, performed the inner integration, and then evaluated the resulting double integral by polar coordinates, without ever mentioning cylindrical coordinates, but the approach we used is more direct and saves some trouble.

EXAMPLE 2   Find the centroid of the upper half of the ball $x^2 + y^2 + z^2 \leq R^2$.

Solution   Letting the density be $\rho = 1$, we need to find the mass $M = \int\int\int \rho\, dV = \int\int\int dV$ and the moments $M_{xy} = \int\int\int z\, dV$, $M_{xz} = \int\int\int y\, dV$, and $M_{yz} = \int\int\int x\, dV$. Because of symmetry the moments M_{xz} and M_{yz} are both zero, by Theorem 17.7.3. See Figure 17.8.3.

The mass is just the volume of the hemisphere, which is $M = \frac{1}{2}\left(\frac{4}{3}\pi R^3\right) = \frac{2}{3}\pi R^3$. The region can be described in cylindrical coordinates by $0 \leq r \leq R$, $0 \leq z \leq \sqrt{R^2 - r^2}$. We find that

$$M_{xy} = \int\int\int z\, dV = \int_0^{2\pi}\int_0^R\int_0^{\sqrt{R^2-r^2}} z\, dz\, r\, dr\, d\theta$$

$$= \int_0^{2\pi}\int_0^R \left(\frac{z^2}{2}\Big|_{z=0}^{z=\sqrt{R^2-r^2}}\right) r\, dr\, d\theta$$

$$= \frac{1}{2}\int_0^{2\pi}\int_0^R (R^2 - r^2) r\, dr\, d\theta = \frac{1}{2}\int_0^{2\pi}\left(R^2\frac{r^2}{2} - \frac{r^4}{4}\right)\Big|_0^R d\theta$$

$$= \frac{1}{2}\int_0^{2\pi}\left(\frac{R^4}{2} - \frac{R^4}{4}\right) d\theta = \frac{R^4}{4}\pi.$$

From this we find that $\bar{z} = \frac{M_{xy}}{M} = \frac{R^4\pi/4}{2\pi R^3/3} = \frac{3R}{8}$ and the centroid is $\left(0, 0, \frac{3R}{8}\right)$.

You might like to compare this calculation with the one in Example 17.7.2 (p. 1112).

Spherical Coordinates

If we want to integrate a function $F(x, y, z)$ over a region R and both of them can be expressed in terms of spherical coordinates, it seems reasonable to look for a formula for integrating with respect to spherical coordinates.

We imagine a little "spherical box" (shown in Figure 17.8.4) in which the radius varies from ρ to $\rho + \Delta\rho$, the polar angle varies from θ to $\theta + \Delta\theta$, and the third coordinate varies from ϕ to $\phi + \Delta\phi$. We need to approximate the volume of this "box."

If $\Delta\rho$, $\Delta\theta$, and $\Delta\phi$ are all small, it will be very close to an ordinary rectangular box, so we can approximate the volume by finding the product of the lengths of its sides. We consider the three sides a, b, c, which are drawn in red lines in Figure 17.8.5.

The side a along a line through the origin has length $\Delta\rho$. (It runs from ρ to $\rho + \Delta\rho$; see the illustration.) The side labeled c corresponds to an angle $\Delta\phi$, and its distance from the origin is approximately ρ, so its length is approximately $\rho\,\Delta\phi$.

The side labeled b is more complicated. Instead of finding its length directly, let us consider the segment labeled d, which lies in the xy-plane directly underneath b. It corresponds to an angle $\Delta\theta$ at a distance r from the origin, so its length is $r\,\Delta\theta$. To use this, we need to express r in spherical coordinates, and from Figure 17.8.6 we see that $r = \rho\cos\left(\frac{\pi}{2} - \phi\right) = \rho\sin\phi$. So the length of the side labeled b is approximately equal to the length of d, which is approximately $r\,\Delta\theta = \rho\sin\phi\,\Delta\theta$.

The volume of the spherical box is approximately the product of the lengths of a, b, and c, or

$$\Delta V = (\Delta\rho)(\rho\,\Delta\phi)(\rho\sin\phi\,\Delta\theta) = \rho^2\sin\phi\,\Delta\rho\,\Delta\phi\,\Delta\theta.$$

Figure 17.8.3

Figure 17.8.4

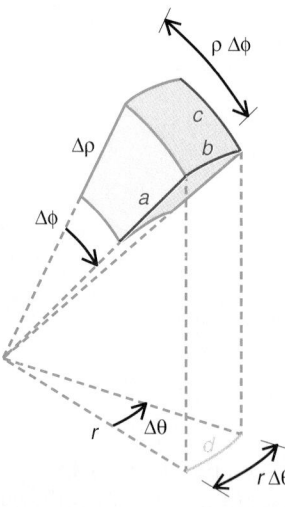

Figure 17.8.5

We can use this to develop a formula for integration in spherical coordinates. We partition a region R into little spherical boxes $R_1, \ldots, R_n$, with volumes $\Delta V_1, \ldots, \Delta V_n$, and let the mesh of the partition be the maximum value of $\max(\Delta \rho_i, \Delta \phi_i, \Delta \theta_i)$, for $1 \leq i \leq n$. Then for any continuous function F on R and points $P_i^* = (\rho_i^*, \theta_i^*, \phi_i^*) \in R_i$ we form the Riemann sum

$$\sum_{i=1}^{n} F(P_i^*) \, \Delta V_i = \sum_{i=1}^{n} F(P_i^*) \, (\rho_i^*)^2 \sin(\phi_i^*) \Delta \rho_i \, \Delta \phi_i \, \Delta \theta_i.$$

Letting the mesh of $\mathcal{P} = \{R_1, \ldots, R_n\}$ tend to zero, we find that the Riemann sum approaches the integral $\int \int \int_R F(\rho, \theta, \phi) \rho^2 \sin \phi \, d\rho \, d\theta \, d\phi$.

DEFINITION 17.8.2

For a region R that can be described in spherical coordinates and a continuous function F on R, we can **integrate in spherical coordinates** as follows:

$$\int \int \int_R F \, dV = \int \int \int_R F(\rho, \theta, \phi) \rho^2 \sin \phi \, d\rho \, d\theta \, d\phi,$$

with appropriate limits on the iterated integral to make it cover the region R. The order of integration may be changed, with the limits of integration changing accordingly.

We sometimes abbreviate by writing

$$dV = dx \, dy \, dz = \rho^2 \sin \phi \, d\rho \, d\theta \, d\phi.$$

Most often this formula is used for integrating over a ball or part of a ball.

EXAMPLE 3   Let B be the unit ball $x^2 + y^2 + z^2 \leq 1$ and U the upper half of B. Find (i) the centroid of U, (ii) the moment of inertia of B about the z-axis, that is, $I_z = \int \int \int_B (x^2 + y^2) \, dV$.

Solution

(i) Assuming that the density of U is 1, the mass is the volume, that is, $M = \frac{2}{3}\pi$. By Theorem 17.7.3 the moments about the xz- and yz-planes are zero; we must find M_{xy}. The upper half-ball is $0 \leq \rho \leq 1, 0 \leq \theta \leq 2\pi, 0 \leq \phi \leq \frac{\pi}{2}$, and $z = \rho \cos \phi$, so

$$M_{xy} = \int \int \int_B z \, dV = \int_0^{\pi/2} \int_0^{2\pi} \int_0^1 (\rho \cos \phi) \rho^2 \sin \phi \, d\rho \, d\theta \, d\phi$$

$$= \int_0^{\pi/2} \int_0^{2\pi} \left(\frac{\rho^4}{4} \Big|_0^1 \right) \cos \phi \sin \phi \, d\theta \, d\phi = \frac{1}{4} \int_0^{\pi/2} \left(\theta \Big|_0^{2\pi} \right) \cos \phi \sin \phi \, d\phi$$

$$= \frac{\pi}{2} \int_0^{\pi/2} \cos \phi \sin \phi \, d\phi = \frac{\pi}{2} \frac{\sin^2 \phi}{2} \Big|_0^{\pi/2} = \frac{\pi}{4}.$$

From this we find that $\bar{z} = \frac{M_{xy}}{M} = \frac{\pi/4}{2\pi/3} = \frac{3}{8}$, and the centroid is $\left(0, 0, \frac{3}{8}\right)$.

(ii) Noting that $x^2 + y^2 = r^2 = \rho^2 \sin^2 \phi$ (as in Figure 17.8.6), we find that

$$I_z = \int \int \int_B (x^2 + y^2) \, dV = \int_0^{\pi} \int_0^{2\pi} \int_0^1 (\rho^2 \sin^2 \phi) \rho^2 \sin \phi \, d\rho \, d\theta \, d\phi$$

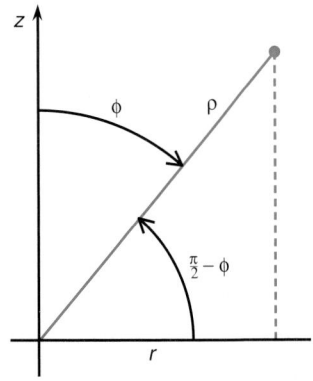

Figure 17.8.6

We substitute $u = \cos\phi$, $du = -\sin\phi\,d\phi$, with u running from 1 to -1.

$$= \int_0^\pi \int_0^{2\pi} \left(\frac{\rho^5}{5}\bigg|_0^1\right)(1 - \cos^2\phi)\sin\phi\,d\theta\,d\phi = \frac{1}{5}\int_1^{-1}\left(\theta\bigg|_0^{2\pi}\right)(1 - u^2)(-1)\,du$$

$$= \frac{2\pi}{5}\int_{-1}^1 (1 - u^2)\,du = \frac{2\pi}{5}\left(u - \frac{u^3}{3}\right)\bigg|_{-1}^1 = \frac{2\pi}{5}\frac{4}{3} = \frac{8\pi}{15}.$$

The calculations of Example 3 can all be done in rectangular or cylindrical coordinates, but they are easier this way.

One word of *caution:* We use the letter ρ for the spherical coordinate that measures the distance from the origin. We are also accustomed to using ρ for the density. In a question involving a density in which spherical coordinates are to be used, it is desirable to use another letter for the density; σ (sigma) and δ (delta) are possible choices.

In Example 3 we found that the centroid of the half-ball U is $\left(0, 0, \frac{3}{8}\right)$. We also found that the moment of inertia of the whole ball is $\frac{8\pi}{15}$, so we see that the moment of inertia of the half-ball is $I = \frac{4\pi}{15}$. Since the mass of the half-ball is $M = \frac{2\pi}{3}$, we find that its radius of gyration about the vertical axis is $R = \sqrt{\frac{I}{M}} = \sqrt{\frac{2}{5}}$.

With a graphics calculator we can plot the upper unit semicircle and the lines $y = \frac{3}{8}$, $x = \sqrt{\frac{2}{5}}$.

TI-85	SHARP EL9300
GRAPH F1$(y(x) =)$ $\surd$ (1 − F1(x) x^2) ▼ 3 ÷ 8 M2(RANGE) (−) 1.5 ▼ 1.5 ▼ ▼ (−) 1 ▼ 1 F5(GRAPH) EXIT Line ($\surd$ (.4), (−) 1, $\surd$ (.4), 1) ENTER	SET UP E 3 ⌐ᴑ cos(π x/θ/T) ENTER sin(π x/θ/T) 2ndF ▼ 3 x/θ/T − 1.5 ENTER 3 ÷ 8 2ndF ▼ $\surd$ .4 ENTER 2 x/θ/T − 1 RANGE 0 ENTER 1 ENTER .01 ENTER ⌐ᴑ −1.5 ENTER 1.5 ENTER 1 ENTER −1 ENTER 1 ENTER 1 ENTER
CASIO f_x-6300G (f_x-7700GB)	HP 48SX
Range −1.5 EXE 1.5 EXE 1 EXE −1 EXE 1 EXE 1 EXE EXE EXE EXE Graph $\surd$ (1 − X x^2) EXE Graph 3 ÷ 8 EXE Plot .6325, −1 EXE Plot .6325 , 1 EXE Line EXE	⬑ PLOT PLOTR ERASE ATTN ′ Y = $\surd$x () 1 − X y^x 2 ↪ PLOT ⬑ DRAW ERASE 1.5 +/− ENTER 1.5 XRNG 1 +/− ENTER 1 YRNG DRAW ATTN ′ Y = 3 ÷ 8 ↪ PLOT ⬑ DRAW DRAW ATTN () .6325, 1 +/− ENTER () .6325 , 1 PRG DSPL LINE ⬑ PLOT PLOTR DRAW

We visualize the picture as the cross-section of the half-ball; the horizontal line gives the height of the centroid. If we imagine the half-ball rotating about the vertical axis, it will behave the same as a point mass lying on the vertical line.

Exercises 17.8

I

Evaluate each of the following integrals, using cylindrical or spherical coordinates.

1. $\iiint_R z \, dV$, R is $x^2 + y^2 \le 1$, $-1 \le z \le 3$.

2. $\iiint_R z^2 \, dV$, R is $x^2 + y^2 \le 1$, $|z| \le 4$.

3. $\iiint_R \sqrt{x^2 + y^2} \, dV$, R is $x^2 + y^2 \le 4$, $x \ge 0$, $x \le z \le 4$.

4. $\iiint_R \exp(x^2 + y^2) \, dV$, R is $x^2 + y^2 \le 9$, $y \le 0$, $0 \le z \le 2$.

5. $\iiint_R \sqrt{1 - x^2 - y^2} \, dV$, R is $x^2 + y^2 \le 1$, $y \le 0$, $x \ge 0$, $0 \le z \le 2$.

6. $\iiint_R (5 - x^2 - y^2)^{-1} \, dV$, R is $x^2 + y^2 \le 4$, $y \ge x$, $0 \le z \le 3 + y$.

7. $\iiint_R z^2 \, dV$, R is $x^2 + y^2 + z^2 \le 4$, $y \ge 0$.

8. $\iiint_R z^4 \, dV$, R is $x^2 + y^2 + z^2 \le 9$.

9. $\iiint_R x \, dV$, R is $x^2 + y^2 + z^2 \le 1$, $x \ge 0$.

10. $\iiint_R (x + y + z) \, dV$, R is $x^2 + y^2 + z^2 \le 4$, $y \le 0$.

Calculate each of the following.

11. $\iiint_R x \, dV$, where R is the cylinder of radius 3 around the z-axis between $z = 1$ and $z = x + 6$.

12. $\iiint_R (x^2 + y^2) \, dV$, where R is the region inside the cylinder of radius 1 around the z-axis between $z = 0$ and $z = 2 - x^2 - y^2$.

13. $\iiint_R \sqrt{x^2 + y^2 + z^2} \, dV$, where R is the ball of radius 4 around the origin.

14. $\iiint_R \sqrt{x^2 + y^2} \, dV$, where R is the ball of radius 4 around the origin.

15. $\iiint_R (3 - z^2) \, dV$, where R is the ball of radius 2 around the origin.

16. $\iiint_R (x^2 + y^2 + z^2)^{3/2} \, dV$, where R is the ball of radius K around the origin.

17. The centroid of the region $x^2 + y^2 + z^2 \le 1$, $x \ge 0$, $y \ge 0$.

18. The centroid of the region $x^2 + y^2 + z^2 \le 9$, $x \ge 0$, $y \ge 0$, $z \ge 0$.

19. The moment of inertia about the z-axis of the region $4 \le x^2 + y^2 + z^2 \le 9$.

20. The moment of inertia about the z-axis of the region $4 \le x^2 + y^2 \le 9$, $x \le z \le x + 1$.

Transform each of the following integrals, expressing the ones given here in rectangular coordinates as integrals in spherical coordinates and vice versa. Some of these questions are tricky; it may help to draw a sketch.

21. $\int_{-2}^{2} \int_{-\sqrt{4-y^2}}^{\sqrt{4-y^2}} \int_{0}^{\sqrt{4-x^2-y^2}} F(x, y, z) \, dz \, dx \, dy$

22. $\int_{0}^{3} \int_{-\sqrt{9-y^2}}^{0} \int_{0}^{\sqrt{9-x^2-y^2}} F(x, y, z) \, dz \, dx \, dy$

23. $\int_{0}^{\pi} \int_{\pi}^{2\pi} \int_{0}^{4} F(\rho, \theta, \phi) \rho^2 \sin\phi \, d\rho \, d\theta \, d\phi$

24. $\int_{\pi/2}^{\pi} \int_{0}^{\pi} \int_{0}^{1} F(\rho, \theta, \phi) \rho^2 \sin\phi \, d\rho \, d\theta \, d\phi$

25. $\int_{-1}^{1} \int_{-\sqrt{1-y^2}}^{0} \int_{0}^{\sqrt{1-x^2-y^2}} F(x, y, z) \, dz \, dx \, dy$

26. $\int_{-1}^{1} \int_{|x|}^{\sqrt{2-x^2}} \int_{-\sqrt{2-x^2-y^2}}^{0} F(x, y, z) \, dz \, dy \, dx$

27. $\int_{0}^{\pi/2} \int_{0}^{\pi/3} \int_{0}^{2} F(\rho, \theta, \phi) \rho^2 \sin\phi \, d\rho \, d\theta \, d\phi$

28. $\int_{0}^{\pi/4} \int_{0}^{2\pi} \int_{0}^{1} F(\rho, \theta, \phi) \rho^2 \sin\phi \, d\rho \, d\theta \, d\phi$

29. $\int_{3\pi/4}^{\pi} \int_{\pi}^{2\pi} \int_{0}^{5} F(\rho, \theta, \phi) \rho^2 \sin\phi \, d\rho \, d\theta \, d\phi$

30. $\int_{0}^{\pi} \int_{0}^{2\pi} \int_{1}^{2} F(\rho, \theta, \phi) \rho^2 \sin\phi \, d\rho \, d\theta \, d\phi$

II

*31. Try evaluating the integral of Example 1 (p. 1121) without ever mentioning r or θ.

32. Find the centroid of the region $x^2 + y^2 + z^2 \le R^2$, $y \ge 0$.

33. Find the centroid of the region $x^2 + y^2 + z^2 \le R^2$, $x \ge 0$, $y \ge 0$.

34. Find the moment of inertia about the z-axis of the ball of radius R about the origin.

*35. Consider the cylinder $x^2 + y^2 \le 1$, $0 \le z \le 3$ and the cylinder $x^2 + y^2 \le 1$, $0 \le z \le 3 + 2x$. Explain why their moments of inertia about the z-axis should be the same. (Make a sketch.) Then calculate and find their moments of inertia about the z-axis.

*36. Find the moment of inertia about the z-axis of the cylinder $x^2 + z^2 \le 4$, $-1 \le y \le x + z$. (*Hint:* Use a "different kind" of cylindrical coordinates.)

Sketch each of the following solids and compute its volume.

*37. $\rho \le \sin\phi$

*38. $\rho \le 1 + \cos\phi$

*39. $0 \le r \le z\cos\theta$, $0 \le z \le 1$

**40. $0 \le r \le |\cos 2\theta|$, $0 \le z \le r$

17.9 Change of Variables

JACOBIAN MATRIX
DETERMINANT
COFACTOR

In this section we discuss the analog for multiple integrals of the process of substitution for integrals with a single variable. Substitution allowed us to take an integral with respect to x, for instance, and "change variables" to get an integral with respect to u or θ or some other variable. The first step was to guess an appropriate new variable, and having done that, it was necessary to figure out the new limits of integration and to express dx in terms of du or $d\theta$ or whatever the appropriate variable might be.

With several variables we have already seen some important examples of a similar technique. We began by writing double integrals as iterated integrals with respect to x and y but later observed that it is possible to use polar coordinates. The formula $dx\,dy = r\,dr\,d\theta$ can be seen as the double integral version of the expressions like $dx = \frac{1}{u'}\,du$ (i.e., $du = u'\,dx$), which we use with single integrals. In a similar way, cylindrical coordinates are an alternative way to evaluate triple integrals, and the relevant formula is $dx\,dy\,dz = r\,dr\,d\theta\,dz$. Another example is spherical coordinates, for which we have $dx\,dy\,dz = \rho^2 \sin\phi\,d\rho\,d\theta\,d\phi$.

What we need is a systematic way of calculating the formula for converting from one set of coordinates to another, something that should include these examples but also work for other cases.

Suppose we are considering an integral over a region R in the plane, and suppose that we have some other coordinates u, v that can be used to describe the region. Let us also suppose that it is possible to express x and y as functions of u and v. One example to keep in the backs of our minds is the case in which u and v are the polar coordinates r, θ. In that case we can write $x = r\cos\theta$, $y = r\sin\theta$.

Suppose we consider a small "rectangle" in which the first coordinate runs from u to $u + \Delta u$ and the second runs from v to $v + \Delta v$. This region may not be an actual rectangle, because its sides may not be perpendicular and may be curved (see Figure 17.9.1(i)), but if Δu and Δv are small, it should be close to being a parallelogram.

The differentials $dx = \frac{\partial x}{\partial u}\,du + \frac{\partial x}{\partial v}\,dv$ and $dy = \frac{\partial y}{\partial u}\,du + \frac{\partial y}{\partial v}\,dv$ can be used to approximate the changes in x and y corresponding to changes Δu and Δv.

Suppose P is the point with coordinates (u, v). It is one of the corners of the "rectangle" we are considering. Changing from u to $u + \Delta u$ with v fixed corresponds to moving along one of the sides adjacent to P, and using differentials, we see that this results in changes of approximately

$$\Delta x = \frac{\partial x}{\partial u}\,\Delta u \qquad \text{and} \qquad \Delta y = \frac{\partial y}{\partial u}\,\Delta u.$$

In other words the tangent vector to the side of the (curved) "rectangle" is $(\Delta x, \Delta y) = \left(\frac{\partial x}{\partial u}\,\Delta u, \frac{\partial y}{\partial u}\,\Delta u\right)$. In a similar way the other side adjacent to P is obtained by fixing u and letting v change by Δv, so its tangent vector is $(\Delta x, \Delta y) = \left(\frac{\partial x}{\partial v}\,\Delta v, \frac{\partial y}{\partial v}\,\Delta v\right)$. See Figure 17.9.1(ii).

In Figure 17.9.1(iii) we see the parallelogram with these two tangent vectors as sides. We will use it to approximate the area of the curved "rectangle." After all, we do not know how to find area in u, v coordinates, but we do know how to find the area of a parallelogram using x, y coordinates. The area of the curved "rectangle" is approximately the area of the parallelogram with sides $\left(\frac{\partial x}{\partial u}\,\Delta u, \frac{\partial y}{\partial u}\,\Delta u\right)$

Figure 17.9.1(i)

Figure 17.9.1(ii)

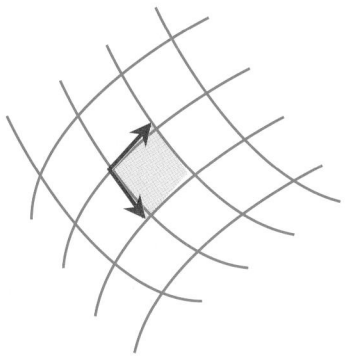

Figure 17.9.1(iii)

and $\left(\frac{\partial x}{\partial v}\Delta v, \frac{\partial y}{\partial v}\Delta v\right)$. By Proposition 14.3.4 (p. 851) this area equals

$$\left\| \left(\frac{\partial x}{\partial u}\Delta u, \frac{\partial y}{\partial u}\Delta u, 0\right) \times \left(\frac{\partial x}{\partial v}\Delta v, \frac{\partial y}{\partial v}\Delta v, 0\right) \right\|$$

$$= \left\| \left(0, 0, \frac{\partial x}{\partial u}\frac{\partial y}{\partial v}\Delta u\,\Delta v - \frac{\partial x}{\partial v}\frac{\partial y}{\partial u}\Delta u\,\Delta v\right) \right\|$$

$$= \left| \frac{\partial x}{\partial u}\frac{\partial y}{\partial v} - \frac{\partial x}{\partial v}\frac{\partial y}{\partial u} \right| \Delta u\,\Delta v.$$

So now suppose we return to the problem of integrating in terms of the coordinates u and v. We might divide the region into little "rectangles" R_i, with sides corresponding to changes Δu_i and Δv_i, for $i = 1, \ldots, n$. If $P_i^* \in R_i$ for each i, what we have just found is that the area of R_i is approximately $\Delta A_i = \left|\frac{\partial x}{\partial u}\frac{\partial y}{\partial v} - \frac{\partial x}{\partial v}\frac{\partial y}{\partial u}\right| \Delta u\,\Delta v$, where the partial derivatives are all evaluated at P_i^*.

To integrate the function $F(u, v)$, we form the Riemann sum

$$\sum_{i=1}^{n} F(P_i^*)\,\Delta A_i = \sum_{i=1}^{n} F(P_i^*) \left|\frac{\partial x}{\partial u}\frac{\partial y}{\partial v} - \frac{\partial x}{\partial v}\frac{\partial y}{\partial u}\right| \Delta u_i\,\Delta v_i,$$

and the corresponding integral will be

$$\int\int_R F(u, v)\,dA = \int\int_R F(u, v) \left|\frac{\partial x}{\partial u}\frac{\partial y}{\partial v} - \frac{\partial x}{\partial v}\frac{\partial y}{\partial u}\right| du\,dv.$$

The expression $\frac{\partial x}{\partial u}\frac{\partial y}{\partial v} - \frac{\partial x}{\partial v}\frac{\partial y}{\partial u}$ that appears in this formula is very important. If you know about linear algebra and in particular about determinants, you will recognize that it equals the determinant of the matrix

$$\begin{pmatrix} \frac{\partial x}{\partial u} & \frac{\partial x}{\partial v} \\ \frac{\partial y}{\partial u} & \frac{\partial y}{\partial v} \end{pmatrix}.$$

This matrix is called the *Jacobian matrix*, after the German mathematician Karl Gustav Jakob Jacobi (1804–1851).

DEFINITION 17.9.1

Suppose u and v are coordinates for some region R in the plane, and suppose x and y can be written as differentiable functions of u and v. The **Jacobian matrix** of x and y with respect to u and v is the matrix

$$J = J\left(\frac{(x, y)}{(u, v)}\right) = \frac{\partial(x, y)}{\partial(u, v)} = \begin{pmatrix} \frac{\partial x}{\partial u} & \frac{\partial x}{\partial v} \\ \frac{\partial y}{\partial u} & \frac{\partial y}{\partial v} \end{pmatrix}.$$

The **Jacobian determinant** is its determinant

$$\det\left(\frac{\partial(x, y)}{\partial(u, v)}\right) = \det J = \det J\left(\frac{(x, y)}{(u, v)}\right) = \frac{\partial x}{\partial u}\frac{\partial y}{\partial v} - \frac{\partial x}{\partial v}\frac{\partial y}{\partial u}.$$

If you do not know about matrices and determinants, the only thing you need to know is this last formula. Notice that $\det J$ is a function on R.

The argument we gave above shows how $\det J$ is used in integrating with respect to the coordinates u and v and demonstrates the following theorem.

THEOREM 17.9.2

Change of Variables for Double Integrals

Suppose u and v are coordinates for a region R in the plane. Suppose x and y can be expressed as differentiable functions of u and v. Then for any continuous function F on R we have

$$\int\int_R F(u, v)\, dA = \int\int_R F(u, v)\left|\det J\left(\frac{(x, y)}{(u, v)}\right)\right| du\, dv$$

$$= \int\int_R F(u, v)\left|\det\frac{\partial(x, y)}{\partial(u, v)}\right| du\, dv.$$

The second and third integrals are to be understood as iterated integrals with the limits of integration chosen so that they "cover" the region R.

This result is best understood by considering examples. The obvious one to try first is polar coordinates.

EXAMPLE 1   Suppose we let the coordinates u and v be the polar coordinates r and θ. Then we know that $x = r\cos\theta$ and $y = r\sin\theta$, so it is easy to evaluate the partial derivatives. We find

$$\frac{\partial x}{\partial u} = \frac{\partial(r\cos\theta)}{\partial r} = \cos\theta, \qquad \frac{\partial x}{\partial v} = \frac{\partial(r\cos\theta)}{\partial \theta} = -r\sin\theta,$$

$$\frac{\partial y}{\partial u} = \frac{\partial(r\sin\theta)}{\partial r} = \sin\theta, \qquad \frac{\partial y}{\partial v} = \frac{\partial(r\sin\theta)}{\partial \theta} = r\cos\theta.$$

The Jacobian determinant is

$$\det J = \frac{\partial x}{\partial u}\frac{\partial y}{\partial v} - \frac{\partial x}{\partial v}\frac{\partial y}{\partial u} = \cos\theta\,(r\cos\theta) - (-r\sin\theta)\sin\theta$$

$$= r(\cos^2\theta + \sin^2\theta) = r.$$

The theorem tells us that for any continuous function $F(x, y)$ on a region R,

$$\int\int_R F(x, y)\, dx\, dy = \int\int_R F(r, \theta)|\det J|\, dr\, d\theta = \int\int_R F(r, \theta)\,r\, dr\, d\theta.$$

This is exactly the formula we have already found for integration in polar coordinates (cf. Equation 17.4.1 and Theorem 17.4.2).

Figure 17.9.2(i)

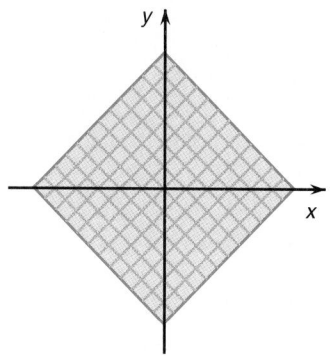

Figure 17.9.2(ii)

EXAMPLE 2   Let R be the square whose corners are $(\pm1, 0)$, $(0, \pm1)$. Integrate $\int\int_R x^2y^2\, dA$.

Solution   The region R is illustrated in Figure 17.9.2(i). We could form the appropriate iterated integrals, but it is tempting to think that it should be possible to use coordinates that are better adapted to this square, that is, coordinates that run parallel to its sides (see Figure 17.9.2(ii)).

To find these coordinates, let us observe that the upper right side of the square is the line through $(1, 0)$ and $(0, 1)$, which is $x + y = 1$. The bottom left side is $x + y = -1$, and the lines $x + y = c$ are all parallel to these sides. If we let c vary from -1 to 1, the line $x + y = c$ will move from the bottom left side of the square up to the top right side.

Similarly, the top left and bottom right sides of the square are $x - y = -1$ and $x - y = 1$, and the lines $x - y = c$ will move between these two sides as c varies from -1 to 1. This suggests that we define coordinates by letting $u = x + y$ and $v = x - y$. The square R is the region $-1 \leq u \leq 1$, $-1 \leq v \leq 1$.

To figure out the Jacobian determinant, we have to express x and y in terms of u and v. Adding the equations $u = x + y$ and $v = x - y$, we find that $u + v = 2x$, so $x = \frac{1}{2}(u + v)$. Similarly, by subtracting we find that $y = \frac{1}{2}(u - v)$.

This makes it extremely easy to find the partial derivatives. We find that

$$\frac{\partial x}{\partial u} = \frac{1}{2}\frac{\partial(u + v)}{\partial u} = \frac{1}{2}, \qquad \frac{\partial x}{\partial v} = \frac{1}{2}\frac{\partial(u + v)}{\partial v} = \frac{1}{2},$$

$$\frac{\partial y}{\partial u} = \frac{1}{2}\frac{\partial(u - v)}{\partial u} = \frac{1}{2}, \qquad \frac{\partial y}{\partial v} = \frac{1}{2}\frac{\partial(u - v)}{\partial v} = -\frac{1}{2}.$$

The Jacobian determinant is $\det J = \frac{\partial x}{\partial u}\frac{\partial y}{\partial v} - \frac{\partial x}{\partial v}\frac{\partial y}{\partial u} = \frac{1}{2}\left(-\frac{1}{2}\right) - \frac{1}{2}\frac{1}{2} = -\frac{1}{4} - \frac{1}{4} = -\frac{1}{2}$. Its absolute value is $|\det J| = \frac{1}{2}$, and we find that

$$\iint_R x^2 y^2 \, dA = \iint_R x^2 y^2 \frac{1}{2} \, du\, dv.$$

Two problems remain. One is to give the limits of integration for this iterated integral. The other is that we have to express the integrand $x^2 y^2$ in terms of the variables of integration u and v. Since $x = \frac{1}{2}(u + v)$ and $y = \frac{1}{2}(u - v)$, we find that $x^2 y^2 = \frac{1}{16}(u + v)^2 (u - v)^2 = \frac{1}{16}(u^2 - v^2)^2 = \frac{1}{16}(u^4 - 2u^2 v^2 + v^4)$.

We have already observed that the square is given by $-1 \leq u \leq 1$, $-1 \leq v \leq 1$, so the integral is

$$\iint_R x^2 y^2 \, dA = \int_{-1}^{1}\int_{-1}^{1} \frac{1}{16}(u^4 - 2u^2 v^2 + v^4)\frac{1}{2} \, du\, dv$$

$$= \frac{1}{32}\int_{-1}^{1}\left(\frac{u^5}{5} - 2\frac{u^3}{3}v^2 + v^4 u \bigg|_{-1}^{1}\right) dv$$

$$= \frac{1}{32}\int_{-1}^{1}\left(\frac{2}{5} - \frac{4}{3}v^2 + 2v^4\right) dv = \frac{1}{32}\left(\frac{2}{5}v - \frac{4}{3}\frac{v^3}{3} + 2\frac{v^5}{5}\bigg|_{-1}^{1}\right)$$

$$= \frac{1}{32}\left(\frac{4}{5} - \frac{8}{9} + \frac{4}{5}\right) = \frac{1}{45}. \qquad \blacktriangleleft$$

Example 2 illustrates the use of Theorem 17.9.2. First we find some new coordinates u and v in which it is relatively easy to express the region R; this is usually mostly guesswork. Then we must express x and y in terms of u and v. From this we can calculate the Jacobian determinant and express the integrand in terms of u and v. Finally, we find the limits of integration needed to express R in terms of u and v, and then we can integrate.

Of course, it can be quite difficult to guess a good substitution. Even if the region is easily described in terms of u and v, there is still the problem that the integral may be difficult to evaluate.

It is worth remarking that the change-of-variables formula can be written, using the alternative notation for Jacobians, as

$$dx\, dy = \left|\det \frac{\partial(x, y)}{\partial(u, v)}\right| du\, dv.$$

This reminds us of the formula for differentials of functions of a single variable, which we could write as

$$dx = \frac{dx}{du}\, du.$$

This is essentially the formula used for substitutions in integrals with a single variable, so we see that the double integral version is actually quite similar.

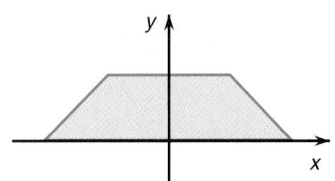

Figure 17.9.3

When we write these partial derivatives, the variables are u and y, so $\frac{\partial y}{\partial u} = 0$.

EXAMPLE 3   Let R be the region $0 \le y \le 1$, $-(2 - y) \le x \le 2 - y$, as illustrated in Figure 17.9.3. Consider the integral $\int\int_R (3 - x^2 + y)\, dx\, dy$.

Of the two conditions that describe R, the first, $0 \le y \le 1$, is very simple. The other is more complicated. If we write $-(2 - y) \le x \le 2 - y$ in the slightly different form $-1 \le \frac{x}{2-y} \le 1$, it looks better. It looks even better if we let $u = \frac{x}{2-y}$, because then the condition is $-1 \le u \le 1$. Let us use u and y as coordinates.

The region is $-1 \le u \le 1$, $0 \le y \le 1$. We have to express the variables x and y in terms of u and y. Clearly, y is already expressed this way, and since $u = \frac{x}{2-y}$, we see that $x = (2 - y)\, u$. From this we see that $\frac{\partial x}{\partial u} = 2 - y$, $\frac{\partial x}{\partial y} = -u$, and $\frac{\partial y}{\partial u} = 0$, $\frac{\partial y}{\partial y} = 1$.

The absolute value of the Jacobian determinant is $|\det J| = \left|\frac{\partial(x,y)}{\partial(u,y)}\right| = \left|\frac{\partial x}{\partial u}\frac{\partial y}{\partial y} - \frac{\partial y}{\partial u}\frac{\partial x}{\partial y}\right| = |(2 - y)(1) - 0(-u)| = 2 - y$.

The integrand $3 - x^2 + y$ can be written as $3 - (2 - y)^2 u^2 + y$, and we already know the limits of integration needed to describe R, so the integral is

$$\int\int_R (3 - x^2 + y)\, dx\, dy = \int_0^1 \int_{-1}^1 \left(3 - (2 - y)^2 u^2 + y\right)|\det J|\, du\, dy$$

$$= \int_0^1 \int_{-1}^1 \left(3 - (2 - y)^2 u^2 + y\right)(2 - y)\, du\, dy.$$

This integral is easily evaluated; it equals $\frac{47}{6}$.

In Example 6.2.1 we found the area of the ellipse $\frac{x^2}{a^2} + \frac{y^2}{b^2} = 1$. There we used an integral with a single variable, but it could be done with the double integral $\int\int_E 1\, dx\, dy$, where E is the region inside the ellipse.

EXAMPLE 4   **Area of Ellipse**   Find the area inside the ellipse $\frac{x^2}{a^2} + \frac{y^2}{b^2} = 1$.

Solution   The area equals $\int\int_E 1\, dx\, dy$, where E is the region $\frac{x^2}{a^2} + \frac{y^2}{b^2} \le 1$. Noticing that x ranges between a and $-a$ on E and y ranges between b and $-b$, we might be tempted to let $u = \frac{x}{a}$ and $v = \frac{y}{b}$. The equation of the ellipse becomes $u^2 + v^2 = 1$, and the region E becomes the disk D, given by $u^2 + v^2 \le 1$.

Since we can also write $x = au$ and $y = bv$, it is easy to find the partial derivatives: $\frac{\partial x}{\partial u} = a$, $\frac{\partial y}{\partial v} = b$, $\frac{\partial x}{\partial v} = \frac{\partial y}{\partial u} = 0$. From this we find that the Jacobian determinant is ab, so the area equals

$$\int\int_E 1\, dx\, dy = \int\int_D |\det J|\, du\, dv = \int\int_D ab\, du\, dv = ab \int\int_D 1\, du\, dv$$

$$= (ab)\, area(D) = \pi ab.$$

In effect what we have done is squeezed and stretched the coordinates until the ellipse is round (see Figure 17.9.4).

> The most interesting thing about Example 4 is the way we reduced the question to finding the area of the disk, which we already know, and so avoided having to do any actual integrating.

Just as it is possible to change variables in a double integral, it is possible to do the same thing for triple integrals. The argument is more complicated, but the result is analogous. To find the factor that relates the two integrals, it is necessary to form a matrix out of the partial derivatives of the coordinates and then take its determinant.

Consider the following **matrix**, which consists of nine numbers in a rectangular array:

$$A = \begin{pmatrix} a & b & c \\ d & e & f \\ g & h & i \end{pmatrix}.$$

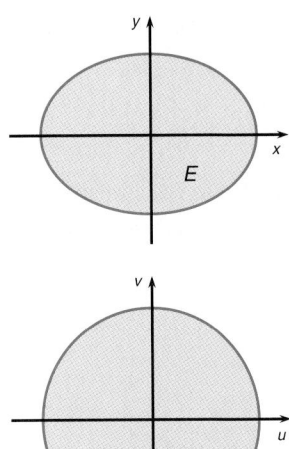

Figure 17.9.4

Figure 17.9.5

Figure 17.9.6

Figure 17.9.7

The numbers $a, b, c, \dots, i$ are called the **entries** of the matrix A. Its **determinant** is a number that is found as follows.

To begin with, the determinant of a 2×2 matrix $\begin{pmatrix} A & B \\ C & D \end{pmatrix}$ is

$$\det \begin{pmatrix} A & C \\ B & D \end{pmatrix} = AD - BC,$$

the product of the diagonal entries A and D, minus the product of the other two, B and C.

Now for each entry in the 3×3 matrix A we can cross out the horizontal row and the vertical column that contain it. In Figure 17.9.5 we have considered the entry b and crossed out the first row and the second column. We think of the numbers that are not crossed out as a little matrix; in this case it is $\begin{pmatrix} d & f \\ g & i \end{pmatrix}$. We find its determinant $di - fg$. If we had begun with the entry c, we would have crossed out the first row and third column, as shown in Figure 17.9.6, and found the determinant $\det\begin{pmatrix} d & e \\ g & h \end{pmatrix} = dh - eg$.

Next we multiply the little determinant by 1 or -1, depending on which entry was crossed out. In Figure 17.9.7 there are alternate plus and minus signs, starting with a plus in the top left corner. We multiply the little determinant by the sign in the position to which it corresponds. The resulting number is called a **cofactor**.

So for instance, the cofactor corresponding to b is $-\det\begin{pmatrix} d & f \\ g & i \end{pmatrix} = fg - di$, and the cofactor corresponding to c is $+\det\begin{pmatrix} d & e \\ g & h \end{pmatrix} = dh - eg$. Similarly, the cofactor corresponding to a is $+\det\begin{pmatrix} e & f \\ h & i \end{pmatrix} = ei - fh$.

To find $\det(A)$, we go along the first row of A, multiplying each entry a, b, c by the corresponding cofactor and adding up the products:

$$\det(A) = (a) \det \begin{pmatrix} e & f \\ h & i \end{pmatrix} + (b)\left(-\det\begin{pmatrix} d & f \\ g & i \end{pmatrix}\right) + (c) \det\begin{pmatrix} d & e \\ g & h \end{pmatrix}$$

$$= a(ei - fh) + b(fg - di) + c(dh - eg)$$

$$= aei - afh + bfg - bdi + cdh - ceg. \tag{17.9.3}$$

It is much better to learn the *procedure* for finding determinants than to memorize this formula.

EXAMPLE 5 Find the following determinant:

$$\det\begin{pmatrix} 2 & 1 & 0 \\ 6 & 2 & -3 \\ -2 & 1 & 2 \end{pmatrix}$$

Solution We need to find the cofactors corresponding to the entries in the first row. The cofactor corresponding to 2 is $\det\begin{pmatrix} 2 & -3 \\ 1 & 2 \end{pmatrix} = 2(2) - (-3)1 = 7$, the one corresponding to 1 is $-\det\begin{pmatrix} 6 & -3 \\ -2 & 2 \end{pmatrix} = -6(2) + (-3)(-2) = -6$, and the one corresponding to 0 is $\det\begin{pmatrix} 6 & 2 \\ -2 & 1 \end{pmatrix} = 6(1) - 2(-2) = 10$.

Multiplying these by 2, 1, 0 and adding gives

$$\det(A) = 2(7) + 1(-6) + 0(10) = 8.$$

In hindsight we didn't really need to find the cofactor corresponding to the 0, since it disappears in the final answer.

We used the first row to calculate the determinant, but in fact we could have used any horizontal row or any vertical column. If some row contains a zero, we might choose that row to reduce the number of cofactors we need to calculate. As an exercise you should calculate the determinant of Example 5 using different rows and columns, being careful about the signs.

Some calculators can evaluate determinants; we try the determinant from Example 5.

TI-85	SHARP EL9300
MATRX F2(EDIT) A ENTER 3 ENTER 3 ENTER 2 ▼ 6 ▼ (−) 2 F2(COL▶) 1 ▲ 2 ▲ 1 F2(COL▶) 0 ▼ (−) 3 ▼ 2 EXIT EXIT MATRX F3(MATH) F1(det) A ENTER	MENU 3(MATRIX) MENU B (EDIT) ▶ ENTER 3 ENTER 3 ENTER 2 ENTER 6 ENTER −2 ENTER 1 ENTER 2 ENTER 1 ENTER 0 ENTER −3 ENTER 2 ENTER QUIT MATH E 6 MAT A ENTER

CASIO f_x-7700GB	HP 48SX
MODE 0(MATRIX) F1(A) F6 F1(DIM) 3 EXE 3 EXE 2 EXE 6 EXE −2 EXE 1 EXE 2 EXE 1 EXE 0 EXE −3 EXE 2 EXE F3(|A|)	↱ MATRIX 2 ENTER 6 ENTER 2 +/− ENTER ▼ 1 ENTER 2 ENTER 1 ENTER ▼ 0 ENTER 3 +/− ENTER 2 ENTER ENTER MTH MATR DET

You can try using the calculator to evaluate other determinants too.

Determinants can also be defined for larger matrices, for example, with 16 numbers in four rows and four columns, or 25 numbers in five rows and five columns, and so on. The idea is to cross out a row and a column, calculate the determinant of what is left, and then multiply by alternate 1's and -1's, starting with $+1$ in the upper left corner. The determinant of the original matrix is found by going along a row or column, multiplying each entry by the corresponding cofactor, and adding up these products.

However, it gets complicated. Even for a 4×4 matrix, each cofactor involves a 3×3 determinant, and it is necessary to find four of them . . .

Using determinants, we can easily state the result about changing variables for triple integrals.

DEFINITION 17.9.4

Suppose u, v, and w are coordinates for some region R in $\mathbb{R}^3$, and suppose x, y, and z can be written as differentiable functions of u, v, and w. The **Jacobian matrix** of x, y, and z with respect to u, v, and w is the matrix

$$J = J\left(\frac{(x, y, z)}{(u, v, w)}\right) = \frac{\partial(x, y, z)}{\partial(u, v, w)} = \begin{pmatrix} \frac{\partial x}{\partial u} & \frac{\partial x}{\partial v} & \frac{\partial x}{\partial w} \\ \frac{\partial y}{\partial u} & \frac{\partial y}{\partial v} & \frac{\partial y}{\partial w} \\ \frac{\partial z}{\partial u} & \frac{\partial z}{\partial v} & \frac{\partial z}{\partial w} \end{pmatrix}.$$

The **Jacobian determinant** is its determinant

$$\det\left(\frac{\partial(x, y, z)}{\partial(u, v, w)}\right) = \det J = \det J\left(\frac{(x, y, z)}{(u, v, w)}\right).$$

Notice that $\det J$ is a function on R.

We use the Jacobian determinant to state the following theorem about integrating with respect to the coordinates u, v, and w.

THEOREM 17.9.5

Change of Variables for Triple Integrals

Suppose u, v, and w are coordinates for a region R in $\mathbb{R}^3$. Suppose x, y, and z can be expressed as differentiable functions of u, v, and w. Then for any continuous function F on R we have

$$\iiint_R F \, dV = \iiint_R F(u, v, w)\left|\det J\left(\frac{(x, y, z)}{(u, v, w)}\right)\right| du \, dv \, dw$$

$$= \iiint_R F(u, v, w)\left|\det \frac{\partial(x, y, z)}{\partial(u, v, w)}\right| du \, dv \, dw.$$

To evaluate the second and third integrals, we would use iterated integrals with the limits of integration chosen so that they "cover" the region R.

EXAMPLE 6   For cylindrical coordinates r, θ, z we can write $x = r\cos\theta$, $y = r\sin\theta$, $z = z$, so the Jacobian matrix is

$$\frac{\partial(x, y, z)}{\partial(r, \theta, z)} = \begin{pmatrix} \frac{\partial(r\cos\theta)}{\partial r} & \frac{\partial(r\cos\theta)}{\partial\theta} & \frac{\partial(r\cos\theta)}{\partial z} \\ \frac{\partial(r\sin\theta)}{\partial r} & \frac{\partial(r\sin\theta)}{\partial\theta} & \frac{\partial(r\sin\theta)}{\partial z} \\ \frac{\partial z}{\partial r} & \frac{\partial z}{\partial\theta} & \frac{\partial z}{\partial z} \end{pmatrix} = \begin{pmatrix} \cos\theta & -r\sin\theta & 0 \\ \sin\theta & r\cos\theta & 0 \\ 0 & 0 & 1 \end{pmatrix}.$$

The determinant is most easily found along the bottom row, where it is only necessary to find one cofactor:

$$\det \begin{pmatrix} \cos\theta & -r\sin\theta & 0 \\ \sin\theta & r\cos\theta & 0 \\ 0 & 0 & 1 \end{pmatrix} = (0)\det \begin{pmatrix} -r\sin\theta & 0 \\ r\cos\theta & 0 \end{pmatrix} + (0)\left(-\det\begin{pmatrix} \cos\theta & 0 \\ \sin\theta & 0 \end{pmatrix}\right)$$

$$+ (1)\det\begin{pmatrix} \cos\theta & -r\sin\theta \\ \sin\theta & r\cos\theta \end{pmatrix}$$

$$= r\cos^2\theta + r\sin^2\theta = r.$$

We find the familiar result that cylindrical coordinates are related to rectangular coordinates by the integration formula

$$dx\, dy\, dz = r\, dr\, d\theta\, dz.$$

EXAMPLE 7   For spherical coordinates ρ, θ, ϕ we know that

$$x = \rho\cos\theta\sin\phi, \qquad y = \rho\sin\theta\sin\phi, \qquad z = \rho\cos\phi.$$

The partial derivatives are easily computed:

$$\frac{\partial x}{\partial \rho} = \cos\theta\sin\phi, \qquad \frac{\partial x}{\partial \theta} = -\rho\sin\theta\sin\phi, \qquad \frac{\partial x}{\partial \phi} = \rho\cos\theta\cos\phi,$$

$$\frac{\partial y}{\partial \rho} = \sin\theta\sin\phi, \qquad \frac{\partial y}{\partial \theta} = \rho\cos\theta\sin\phi, \qquad \frac{\partial y}{\partial \phi} = \rho\sin\theta\cos\phi,$$

$$\frac{\partial z}{\partial \rho} = \cos\phi, \qquad \frac{\partial z}{\partial \theta} = 0, \qquad \frac{\partial z}{\partial \phi} = -\rho\sin\phi.$$

The Jacobian determinant is easily found. We write out the Jacobian matrix and expand the determinant along the bottom row.

$$\det\begin{pmatrix} \cos\theta\sin\phi & -\rho\sin\theta\sin\phi & \rho\cos\theta\cos\phi \\ \sin\theta\sin\phi & \rho\cos\theta\sin\phi & \rho\sin\theta\cos\phi \\ \cos\phi & 0 & -\rho\sin\phi \end{pmatrix}$$

$$= \cos\phi\det\begin{pmatrix} -\rho\sin\theta\sin\phi & \rho\cos\theta\cos\phi \\ \rho\cos\theta\sin\phi & \rho\sin\theta\cos\phi \end{pmatrix}$$

$$+ 0 - \rho\sin\phi\det\begin{pmatrix} \cos\theta\sin\phi & -\rho\sin\theta\sin\phi \\ \sin\theta\sin\phi & \rho\cos\theta\sin\phi \end{pmatrix}$$

$$= \cos\phi(-\rho^2\sin^2\theta\sin\phi\cos\phi - \rho^2\cos^2\theta\sin\phi\cos\phi)$$

$$- \rho\sin\phi(\rho\cos^2\theta\sin^2\phi + \rho\sin^2\theta\sin^2\phi)$$

$$= -\rho^2\big((\sin^2\theta + \cos^2\theta)\sin\phi\cos^2\phi + (\cos^2\theta + \sin^2\theta)\sin^3\phi\big)$$

$$= -\rho^2\sin\phi(\cos^2\phi + \sin^2\phi)$$

$$= -\rho^2\sin\phi.$$

The absolute value of the Jacobian determinant, which is what we need, is $\rho^2\sin\phi$, and we see that

Note: $0 \le \phi \le \pi$, so $\sin\phi \ge 0$.

$$dx\, dy\, dz = \rho^2\sin\phi\, d\rho\, d\theta\, d\phi.$$

This agrees with the formula we have already found for spherical coordinates.

EXAMPLE 8   Let R be the region $x^2 + y^2 \le 1 + z$, $0 \le z \le 3$ in $\mathbb{R}^3$. This is the region lying inside the paraboloid $x^2 + y^2 = 1 + z$ and between $z = 0$ and $z = 3$. Find the integral $I = \int\int\int_R \sin\big(\pi(1 + z)^2\big)\, dx\, dy\, dz$.

Solution   The region can be described in cylindrical coordinates as $r^2 \leq 1 + z, 0 \leq z \leq 3$. The first of these conditions is equivalent to $r \leq \sqrt{1+z}$, so instead of the cylindrical coordinates r, θ, z, let us consider $u = \frac{r}{\sqrt{1+z}}, v = \theta, w = 1 + z$. Then the region R is $0 \leq u \leq 1, 0 \leq \theta \leq 2\pi, 1 \leq w \leq 4$. This is a rectangular box, so it is particularly easy to set up iterated integrals.

Now $x = r \cos\theta = \sqrt{1+z}\, u \cos\theta = \sqrt{w}\, u \cos v, \ y = r \sin\theta = \sqrt{w}\, u \sin v,$ and $z = w - 1$. The partial derivatives are easily evaluated:

$$\frac{\partial x}{\partial u} = \sqrt{w} \cos v, \qquad \frac{\partial x}{\partial v} = -\sqrt{w}u \sin v, \qquad \frac{\partial x}{\partial w} = \frac{1}{2}w^{-1/2} u \cos v,$$

$$\frac{\partial y}{\partial u} = \sqrt{w} \sin v, \qquad \frac{\partial y}{\partial v} = \sqrt{w}u \cos v, \qquad \frac{\partial y}{\partial w} = \frac{1}{2}w^{-1/2} u \sin v,$$

$$\frac{\partial z}{\partial u} = 0, \qquad \frac{\partial z}{\partial v} = 0, \qquad \frac{\partial z}{\partial w} = 1.$$

The Jacobian matrix is

$$\frac{\partial(x, y, z)}{\partial(u, v, w)} = \begin{pmatrix} \sqrt{w} \cos v & -\sqrt{w}u \sin v & \frac{1}{2}w^{-1/2}u \cos v \\ \sqrt{w} \sin v & \sqrt{w}u \cos v & \frac{1}{2}w^{-1/2}u \sin v \\ 0 & 0 & 1 \end{pmatrix}$$

Again it is best to use the bottom row, and the determinant equals

$$0 + 0 + (1) \det \begin{pmatrix} \sqrt{w} \cos v & -\sqrt{w}u \sin v \\ \sqrt{w} \sin v & \sqrt{w}u \cos v \end{pmatrix} = wu \cos^2 v + wu \sin^2 v = wu.$$

The integral is

$$I = \int\!\!\int\!\!\int_R \sin\left(\pi(1+z)^2\right) dx\, dy\, dz = \int\!\!\int\!\!\int_R \sin(\pi w^2)(wu)\, du\, dv\, dw$$

$$= \int_1^4 \int_0^{2\pi} \int_0^1 \sin(\pi w^2)\, wu\, du\, dv\, dw = \frac{1}{2} \int_1^4 \int_0^{2\pi} \sin(\pi w^2)\, wu^2 \Big|_{u=0}^{u=1} dv\, dw$$

$$= \frac{1}{2} \int_1^4 \sin(\pi w^2)\, wv \Big|_{v=0}^{v=2\pi} dw = \pi \int_1^4 \sin(\pi w^2)\, w\, dw$$

$$= \pi \int_1^{16} \sin(\pi U)\, \frac{1}{2}\, dU = -\frac{1}{2} \cos(\pi U) \Big|_1^{16} = -\frac{1}{2}\big((1 - (-1)\big) = -1.$$

(Note how in the last line we substituted $U = w^2, dU = 2w\, dw$.)

Exercises 17.9

|

Find the determinant of each of the following matrices.

1. $\begin{pmatrix} 3 & 4 \\ 5 & 6 \end{pmatrix}$

2. $\begin{pmatrix} 2 & -3 \\ -4 & 6 \end{pmatrix}$

3. $\begin{pmatrix} x & 2y \\ x^2 & 2xy \end{pmatrix}$

4. $\begin{pmatrix} \cos x & \sin x \\ -\sin x & \cos x \end{pmatrix}$

5. $\begin{pmatrix} 2 & 2 & 1 \\ 0 & 3 & 1 \\ 0 & -2 & 3 \end{pmatrix}$

6. $\begin{pmatrix} 1 & 0 & -3 \\ 3 & -2 & 4 \\ 1 & -2 & 1 \end{pmatrix}$

7. $\begin{pmatrix} x^2 & y & 2x \\ 1 & 2y & xy \\ 2y & 2 & 0 \end{pmatrix}$

8. $\begin{pmatrix} 1 & x & x^2 \\ 1 & y & y^2 \\ 1 & z & z^2 \end{pmatrix}$

9. $\begin{pmatrix} 2 & 2 & 2 \\ 2 & 2 & 2 \\ 2 & 2 & 2 \end{pmatrix}$

10. $\begin{pmatrix} 1 & 23 & -7 \\ 0 & 2 & xy \\ 0 & 0 & 4 \end{pmatrix}$

Find the Jacobian determinant $\det\left(\frac{\partial(x,y)}{\partial(u,v)}\right)$ or $\det\left(\frac{\partial(x,y,z)}{\partial(u,v,w)}\right)$ in each of the following cases.

11. $u = 2x + y$, $v = 2y - 3x$

12. $u = 2y - x^5$, $v = x + 7$

13. $u = ax + by$, $v = cx + dy$

14. $x = u^2 - v^2$, $y = 2uv$

15. $u = x + 2y - z$, $v = x + y + z$, $w = x + 3y + z$

16. $u = 3x - y - 2z$, $v = x + y$, $w = x - z$

17. $x = u^3 - 3uv^2$, $y = 3u^2v - v^3$, $z = u + v + w^3$

18. $u = x$, $v = xy$, $w = xyz$

19. $u = x^3$, $v = x + y^3$, $w = x + y + z^3$

20. $x = u \sin v$, $y = w \cos v$, $z = uv^2 w$

Perform each of the following integrations using the substitution indicated.

21. $\int\int_R x^2 \, dA$, where R is the square with vertices $(\pm 1, 0)$, $(0, \pm 1)$; $u = x + y$, $v = x - y$.

22. $\int\int_R (x + 2y^2) \, dA$, where R is the square with vertices $(\pm 2, 0)$, $(0, \pm 2)$; $u = x + y$, $v = x - y$.

23. $\int\int_R x \, dA$, where R is the square with vertices $(0, 0)$, $(3, \pm 3)$, $(6, 0)$; $u = x + y$, $v = x - y$.

24. $\int\int_R 2x^2 \, dA$, where R is the quadrilateral with vertices $(\pm 1, 0)$, $(0, \pm 2)$; $u = 2x + y$, $v = 2x - y$.

25. $\int\int_R x^2 \, dA$, where R is the ellipse $\frac{x^2}{4} + y^2 \le 1$; $u = \frac{x}{2}$, $v = y$.

26. $\int\int_R (x^2 + y^2) \, dA$, where R is the ellipse $\frac{x^2}{9} + \frac{y^2}{4} \le 1$; $u = \frac{x}{3}$, $v = \frac{y}{2}$.

27. $\int\int\int_R x^2 \, dV$, where R is the ellipsoid $\frac{x^2}{16} + \frac{y^2}{9} + \frac{z^2}{9} \le 1$; $u = \frac{x}{4}$, $v = \frac{y}{3}$, $w = \frac{z}{3}$.

28. $\int\int\int_R x \, dV$, where R is the part of the ellipsoid $\frac{x^2}{4} + \frac{y^2}{25} + \frac{z^2}{16} \le 1$ with $x \ge 0$, $y \ge 0$, $z \ge 0$; $u = \frac{x}{2}$, $v = \frac{y}{5}$, $w = \frac{z}{4}$.

29. $\int\int\int_R x^2 \, dV$, where R is the region $4x^2 + 9y^2 \le z + 3 \le 36$; $u = 2x$, $v = 3y$, $w = z + 3$.

30. $\int\int\int_R x \, dV$, where R is the region $z^2 \ge x^2 + y^2$, $x \ge 0$, $1 \le z \le 2$; $u = (x^2 + y^2)/z^2$, $v = \theta$ (the polar angle), $w = z - 1$.

II

Find the area of each of the following ellipses.

31. $\frac{x^2}{9} + \frac{y^2}{4} \le 1$ **32.** $\frac{x^2}{3} + \frac{y^2}{16} \le 1$

Find the volume of each of the following ellipsoidal regions.

*33. $\frac{x^2}{a^2} + \frac{y^2}{b^2} + \frac{z^2}{c^2} \le 1$ ***34.** $\frac{x^2}{4} + \frac{y^2}{9} + \frac{z^2}{16} \le 1$

*35. $\frac{x^2}{2} + \frac{y^2}{8} + \frac{z^2}{50} \le 2$, $x \ge 0$, $y \ge 0$

*36. $\frac{x^2}{2} + \frac{y^2}{4} + \frac{z^2}{25} \le 1$, $x, y \ge 0$, $z \le 0$

*37. Consider the integral $\int\int_D \exp(-x^2 - y^2) \, dx \, dy$, where D is the unit disk.

  (i) Express it in polar coordinates.

  (ii) Express it in terms of the variables $u = r^2 = x^2 + y^2$ and $v = \theta$ (where θ is the argument, the second polar coordinate).

  (iii) Evaluate the integrals you found in parts (i) and (ii).

38. Solve Example 2 (p. 1128) by integrating with respect to x and y.

*39. (i) Find the volume of the region $R = \{(x, y, z) : x, y, z \ge 0; x + y + z \le 1\}$.

  (ii) Find the volume of the region $T = \{(x, y, z) : x, y, z \ge 0; 2x + 3y + 5z \le 1\}$ *without* working out any integrals.

*40. Suppose $a, b, c, d > 0$. Find the volume of the region $R = \{(x, y, z) : x, y, z \ge 0; ax + by + cz \le d\}$ *without* working out any integrals.

**41. Consider the substitution $u = x - 2y$, $v = 4y - 2x$.

  (i) Find $\det\left(\frac{\partial(u,v)}{\partial(x,y)}\right)$.

  (ii) Try to express x and y in terms of u and v. It is not possible to use u and v as coordinates because the values of u and v do not determine the point (x, y) unambiguously. This example illustrates that not every change of variables will work.

**42. Consider the substitution $u = 2x + y$, $v = y + 2z$, $x = x + y + z$.

  (i) Find $\det\left(\frac{\partial(u,v,w)}{\partial(x,y,z)}\right)$.

  (ii) Try to express x, y, and z in terms of u, v, and w.

43. Evaluate the determinant of Example 5 (p. 1132) using the cofactors on (i) the bottom row, (ii) the middle column. Make sure you get the same answer.

44. Evaluate the determinant of Example 7 (p. 1134) using the cofactors along the middle row.

III

If you have a calculator or computer software that evaluates determinants, use it to find the following determinants.

45. Of Example 5.   **49.** Of Exercise 5.

46. Of Example 6, with $\theta = \frac{\pi}{3}$, $r = 4$.   **50.** Of Exercise 6.

47. Of Exercise 1.   **51.** Of Exercise 9.

48. Of Exercise 2.

52. Of Exercise 10, with $x = 17$, $y = -47$.

← POINT TO PONDER

When we developed the change-of-variables formula for double integrals, we made an argument that showed that a "rectangle" with sides Δu and Δv has as its area the product of $\Delta u \, \Delta v$ and the absolute value of the determinant of the Jacobian. It is reasonable to wonder whether a similar argument can be made in the three-dimensional case.

There we should be considering a three-dimensional rectangular "box," whose sides may be curved (see Figure 17.9.8). If it is small enough, it should be nearly the same as a three-dimensional **parallelepiped**, that is, a "solid" parallelogram.

Recall that if the vectors (a, b) and (c, d) are two adjacent sides of a parallelogram in the plane, then the area of the parallelogram is the absolute value of

$$ad - bc = \det \begin{pmatrix} a & b \\ c & d \end{pmatrix}.$$

If the vectors that run along three sides of a parallelepiped are (a, b, c), (d, e, f), and (g, h, i), then

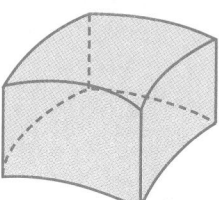

Figure 17.9.8

it turns out that the volume of the parallelepiped is the absolute value of the determinant

$$\det \begin{pmatrix} a & b & c \\ d & e & f \\ g & h & i \end{pmatrix}.$$

It is for this reason that the formula for three-dimensional substitutions involves the absolute value of the Jacobian determinant.

It is also possible to find a formula for substitutions in integrals with n variables, using the Jacobian matrix containing all the partial derivatives and taking the absolute value of its determinant.

Chapter Summary

§17.1
partition
mesh
Riemann sum
double integral
$\int \int_R f(x, y) \, dA$
cross-section
partial integration
iterated integral
integrand
inner integral
outer integral

§17.2
double integrals over nonrectangular regions
horizontal and vertical cross-sections
setting up iterated integrals

§17.3
areas of regions in the plane
volume

§17.4
polar rectangle

$\int \int_R h(r, \theta) \, dA = \int \int_R h(r, \theta) r \, dr \, d\theta$
$\int_{-\infty}^{\infty} e^{-x^2} dx = \sqrt{\pi}$

§17.5
surface area
$S = \int \int_R \sqrt{1 + f_x^2 + f_y^2} \, dx \, dy$

§17.6
partition
mesh
boundedness
triple integrals
Riemann sum
iterated integrals

§17.7
mass
moment about axis
centroid (center of mass)
moment about plane
centroid lies on plane of symmetry
angular velocity
moment of inertia

polar moment of inertia
kinetic energy
radius of gyration

§17.8

cylindrical coordinates: $\int \int \int_R F \, dV = \int \int \int_R F \, r \, dr \, d\theta \, dz$

spherical coordinates: $\int \int \int_R F \, dV =$
$\int \int \int_R F \, \rho^2 \sin \phi \, d\rho \, d\theta \, d\phi$

§17.9

change of variables (coordinates)

Jacobian matrix
Jacobian determinant
double integrals
triple integrals
matrix
entry
determinant of a 3×3 matrix
cofactor
parallelepiped

Review Exercises

Evaluate each of the following.

1. $\int_0^2 \int_{-1}^3 xy^2 \, dx \, dy$

2. $\int_0^1 \int_{-2}^2 x^2 y \, dy \, dx$

3. $\int_0^2 \int_0^x xy \, dy \, dx$

4. $\int_1^2 \int_{-x}^2 x^2 y \, dy \, dx$

5. $\int_0^1 \int_0^x \sin \pi x \, dy \, dx$

6. $\int_1^2 \int_1^y \frac{1}{x^2} \, dx \, dy$

7. $\int_2^3 \int_0^{3y} \sin \pi x \, dx \, dy$

8. $\int_0^1 \int_0^y \frac{1}{1+x^2} \, dx \, dy$

9. $\int_0^1 \int_0^z \int_1^{yz} 6x \, dx \, dy \, dz$

10. $\int_{-1}^1 \int_0^1 \sin(\pi xy) \, dx \, dy$

In each of the following situations, integrate the function F over the region R.

11. $F(x, y) = xy$, R is the square with vertices $(0, 0)$, $(2, 0)$, $(0, 2)$, $(2, 2)$.

12. $F(x, y, z) = x^2 y^2$, R is the triangle with vertices $(-1, 0)$, $(1, 0)$, $(0, 1)$.

13. $F(x, y) = x$, R is the right half of the disk of radius 1 around $(3, 0)$.

14. $F(x, y) = x \sin y$, R is the disk of radius 4 around $(-1, 0)$.

15. $F(x, y, z) = y$, R is the region enclosed between the planes $z = 0$, $z = 1$, $x + y = 1$, $x = -1$, $y = -2$.

16. $F(x, y, z) = x^2 y^4 \sin z$, where $R = \{(x, y, z) : |z| \le |x| + |y| \le 2\}$.

17. $F(x, y) = x$, R is the region $x + y \ge 1$, $x^2 + y^2 \le 1$.

18. $F(x, y, z) = z$, R is the region $3x^2 + 12y^2 \le z \le 12$.

19. $F(x, y) = \sin(x^2 + y^2)$, R is the disk of radius $\sqrt{\pi}$ around the origin.

20. $F(x, y, z) = x^2 + y^2 + z^2$, R is the ball of radius 2 around the origin, with the ball of radius 1 around the origin removed.

Find each of the following.

21. The volume of the region $x^2 + y^2 \le z \le 8 - x^2 - y^2$.

22. The volume of the region $|y| \le 2$, $x^2 + y^2 + z^2 \le 16$.

23. The surface area of the part of the paraboloid $z = x^2 + y^2$ below $z = 25$.

24. The surface area of the part of the paraboloid $z = x^2 + y^2$ below $z = a^2$.

25. The surface area of the solid region enclosed in $x^2 + y^2 + z^2 = 2$ and $z = x^2 + y^2$.

26. The surface area of the region $|z| \le x^2 + y^2 \le 1$.

27. The moment of inertia about the z-axis of the solid rectangular region with vertices $(\pm 1, \pm 2, \pm 4)$.

28. The centroid of the region $x^2 + y^2 \le 4$, $0 \le y \le x$.

29. The centroid of the region $\sqrt{x^2 + y^2} \le z \le 1$.

30. The moment about the xy-plane of the region $x^2 + y^2 \le z^4 \le 16$, $z \le 0$.

31. The moment of inertia about the origin of the region consisting of all points (x, y) satisfying $4 \le x^2 + y^2 \le 9$.

32. The moment of inertia about the origin of the region consisting of all points (x, y) satisfying $9 \le x^2 + y^2 \le 25$.

33. The moment of inertia about the origin of the region consisting of all points (x, y) satisfying $4 \le x^2 + y^2 \le 9$ if the density is $\rho = x^2 + y^2$.

34. The moment of inertia about the origin of the region consisting of all points (x, y) satisfying $9 \le x^2 + y^2 \le 25$ if the density is $\rho = 25 - x^2 - y^2$.

35. The moment of inertia about the z-axis of the region consisting of all points (x, y, z) satisfying $1 \le x^2 + y^2 + z^2 \le 9$.

36. The moment of inertia about the z-axis of the region consisting of all points (x, y, z) satisfying $4 \le x^2 + y^2 + z^2 \le 36$.

37. The moment of inertia about the z-axis of the region consisting of all points (x, y, z) satisfying $1 \le x^2 + y^2 + z^2 \le 9$ if the density is $\sigma = \sqrt{x^2 + y^2 + z^2}$.

38. The moment of inertia about the z-axis of the region consisting of all points (x, y, z) satisfying $4 \le x^2 + y^2 + z^2 \le 16$ if the density is $\sigma = (x^2 + y^2 + z^2)^{-1/2}$.

39. The radius of gyration about the origin of the region consisting of all points (x, y) satisfying $9 \leq x^2 + y^2 \leq 16$.

40. The radius of gyration about the z-axis of the region consisting of all points (x, y, z) satisfying $4 \leq x^2 + y^2 + z^2 \leq 9$.

Express each of the following integrals as an iterated integral with the variables in the opposite order.

41. $\int_0^1 \int_0^{2y} F(x, y) \, dx \, dy$

42. $\int_0^1 \int_{x^2}^1 F(x, y) \, dy \, dx$

43. $\int_{-1}^1 \int_{|x|}^1 F(x, y) \, dy \, dx$

44. $\int_{-4}^4 \int_0^{\sqrt{|y|}} F(x, y) \, dx \, dy$

45. $\int_1^e \int_0^{\ln x} F(x, y) \, dy \, dx$

46. $\int_0^1 \int_{x^3}^{x^2} F(x, y) \, dy \, dx$

47. $\int_0^\pi \int_0^{\sin \theta} F(r, \theta) \, dr \, d\theta$

48. $\int_0^\pi \int_0^{1+\cos \theta} F(r, \theta) \, dr \, d\theta$

49. $\int_0^1 \int_y^{y+1} F(x, y) \, dx \, dy$

50. $\int_0^1 \int_{y-1}^{1-y} F(x, y) \, dx \, dy$

Find limits of integration for an iterated integral in the specified type of coordinates (rectangular: $dx \, dy \, dz$, cylindrical: $r \, dr \, d\theta \, dz$, or spherical: $\rho^2 \sin \phi \, d\rho \, d\theta \, d\phi$) for each of the following regions R.

51. R is the lower half of the ball of radius 3 around the origin; cylindrical coordinates.

52. R is the part of the ball of radius 2 around the origin for which $x \geq 0$; spherical coordinates.

53. R is the ball of radius 2 with center $(x, y, z) = (0, 0, -2)$; spherical coordinates.

54. R is the part of the top half of the solid cone $z^2 \geq x^2 + y^2$ lying inside the ball of radius 5 around the origin; spherical coordinates.

55. R is the part of the cone $z^2 \geq x^2 + y^2$ lying between $z = 0$ and $z = 1$; cylindrical coordinates.

56. R is the region inside the cylinder $x^2 + y^2 = 1$ but outside the cone $z^2 \geq x^2 + y^2$; spherical coordinates.

57. R is the region inside the ball of radius 6 and outside the ball of radius 4 around the origin; spherical coordinates.

58. R is the region outside the cylinder $x^2 + y^2 = 9$ but inside the sphere around the origin of radius 6; spherical coordinates.

59. R is the region $0 \leq \theta \leq \frac{\pi}{2}$, $\frac{\pi}{6} \leq \phi \leq \frac{\pi}{2}$, $0 \leq \rho \leq 1$ in spherical coordinates; rectangular coordinates.

60. R is the region inside the ball of radius 4 with center 4 units directly above the origin and outside the cone $z^2 \geq x^2 + y^2$; spherical coordinates.

II

*61. If $0 \leq a \leq b$, find the moment of inertia about the origin of the set of all (x, y) satisfying $a^2 \leq x^2 + y^2 \leq b^2$.

*62. If $0 \leq a \leq b$, find the moment of inertia about the z-axis of the set of all (x, y, z) satisfying $a^2 \leq x^2 + y^2 + z^2 \leq b^2$.

*63. If $0 \leq a \leq b$, find the radius of gyration about the origin of the set of all (x, y) satisfying $a^2 \leq x^2 + y^2 \leq b^2$.

*64. If $0 \leq a \leq b$, find the radius of gyration about the z-axis of the set of all (x, y, z) satisfying $a^2 \leq x^2 + y^2 + z^2 \leq b^2$.

*65. (i) Consider the unit disk $x^2 + y^2 \leq 1$ with three different densities: $\rho_1 = 1$, $\rho_2 = 3$, $\rho_3 = x^2 + y^2$. How do you expect the corresponding three radii of gyration R_1, R_2, R_3 to compare (i.e., which will be larger)?

  (ii) Find the radius of gyration for each of the three densities.

*66. (i) Consider the unit ball $x^2 + y^2 + z^2 \leq 1$ with three different densities: $\rho_1 = 1$, $\rho_2 = 3$, $\rho_3 = x^2 + y^2 + z^2$. How do you expect the corresponding three radii of gyration R_1, R_2, R_3 to compare (i.e., which will be larger)?

  (ii) Find the radius of gyration for each of the three densities.

**67. (i) Consider the disk of radius R around the origin. Assuming that its density is $\rho = 1$, find its mass and its polar moment of inertia.

  (ii) Find the mass and the polar moment of inertia of the disk of radius $2R$ with density $\rho = 1$.

  (iii) Suppose both disks roll down a ramp whose height is h. Compare the lengths of time it takes each of them to reach the bottom.

**68. (i) Consider a ball of radius R around the origin with density $\rho = 1$. Find its moment of inertia about the x-axis and its mass.

  (ii) Find the mass and the moment of inertia about the x-axis of the ball of radius $2R$ with density $\rho = 1$.

  (iii) Suppose both balls roll down a ramp whose height is h. Compare the lengths of time it takes each of them to reach the bottom.

**69. Consider a solid ball of radius 1 and a wheel of radius 1 (i.e., a disk) rolling down a ramp together. Which will reach the bottom first? (Try to guess before working it out.)

**70. Consider a ball of radius 2 with a ball of radius 1 hollowed out of its center and a wheel of radius 2 (i.e., a disk) with a disk of radius 1 cut out of its center. If they roll down a ramp together, which will reach the bottom first? (Try to guess before working it out.)

*71. Consider the square $0 \leq x \leq 1$, $0 \leq y \leq 1$, and consider the coordinates $u = x + y$, $v = y$.

  (i) Describe the square in the coordinates u and v.

  (ii) Find the Jacobian determinant $\det\left(\frac{\partial(x,y)}{\partial(u,v)}\right)$.

(iii) What does this tell us about the area of the region in the uv-plane described in part (i)?

**72. Use a change-of-variables argument without actually evaluating any integrals to find the area of the parallelogram with vertices $(0,0)$, $(a,0)$, (b,c), $(a+b,c)$.

**73. Use a change-of-variables argument without actually evaluating any integrals to find the volume of the parallelepiped with vertices $(0,0,0)$, $(1,0,0)$, $(1,1,0)$, $(2,1,0)$, $(1,0,1)$, $(2,0,1)$, $(2,1,1)$, $(3,1,1)$.

**74. Use a change-of-variables argument without actually evaluating any integrals to find the volume of the parallelepiped with vertices $(0,0,0)$, $(a,0,0)$, $(b,c,0)$, $(a+b,c,0)$, $(0,0,d)$, $(a,0,d)$, (b,c,d), $(a+b,c,d)$.

**75. If you read the "Point to Ponder" in Section 17.6, find the four-dimensional volume of the four-dimensional ball of radius R.

**76. If you read the "Point to Ponder" in Section 17.6, find the five-dimensional volume of the five-dimensional unit ball.

*77. Define a function f on the square $S = \{(x,y) : 0 < x, y < 1\}$ by $f(x,y) = (xy)^{-1/2}$. As (x,y) approaches either axis, the function $f(x,y)$ approaches ∞. This means that $\int\int_S f(x,y)\,dx\,dy$ must be considered as an improper integral. Figure out how to define this improper integral, decide whether or not it converges, and evaluate it if it does converge.

*78. Let D be the unit disk around the origin in the plane.

(i) Evaluate $\int\int_D (x^2 + y^2)^{s/2}\,dx\,dy$ if $s > 0$.

(ii) Define the integral as an improper integral if $s < 0$.

(iii) Decide for which values of s the improper integral converges, and evaluate it when it does converge.

(iv) If R is the region consisting of all points *outside* D, that is, $R = \{(x,y) : x^2 + y^2 \geq 1\}$, define $\int\int_R (x^2 + y^2)^{s/2}\,dx\,dy$ as an improper integral. Decide for which values of s it converges, and evaluate it when it does converge.

*79. Let B be the unit ball around the origin in $\mathbb{R}^3$.

(i) Evaluate $\int\int\int_B (x^2 + y^2 + z^2)^{s/2}\,dx\,dy\,dz$ if $s > 0$.

(ii) Define the integral as an improper integral if $s < 0$.

(iii) Decide for which values of s the improper integral converges, and evaluate it when it does converge.

(iv) If R is the region consisting of all points *outside* B, that is, $R = \{(x,y,z) : x^2 + y^2 + z^2 \geq 1\}$, define $\int\int\int_R (x^2 + y^2 + z^2)^{s/2}\,dx\,dy\,dz$ as an im-

proper integral. Decide for which values of s it converges, and evaluate it when it does converge.

*80. Exercises 78 and 79 involved improper integrals over a region around the origin in the plane or in three-dimensional space. The corresponding question for one-dimensional space, the line, is the improper integral $\int_{-1}^{1} |x|^s\,dx$.

(i) Evaluate $\int_{-1}^{1} |x|^s\,dx$ if $s > 0$.

(ii) Define the integral as an improper integral if $s < 0$.

(iii) Decide for which values of s the improper integral converges, and evaluate it when it does converge.

(iv) Define $\int_{-\infty}^{-1} |x|^s\,dx + \int_{1}^{\infty} |x|^s\,dx$ as an improper integral. Decide for which values of s it converges, and evaluate it when it does converge.

(v) Compare the answers to this exercise and Exercises 78 and 79.

*81. A marine biologist has developed a theory that predicts that the number of millions of phytoplankton per cubic meter at depth d is $H(d) = Ade^{-d^2/B}$, for some constants A and B. Find A and B if the greatest concentration of phytoplankton occurs at $d = 30$ m and the total number is the same as what we found in Example 17.6.5 (p. 1106).

*82. (i) In the situation of Exercise 81, how do you expect A and B to change if the total number remains the same but the greatest concentration is at $d = 40$ m?

(ii) Find A and B for this case.

*83. In some cities the central core is decayed and not heavily populated. Suppose the population density, in thousands per square mile, in such a city is $P(x,y) = 30(1 + 2r)e^{-r/4}$, with $r = \sqrt{x^2 + y^2}$. Find the total population living within a 10-mile radius of the origin.

*84. In the situation of Exercise 83, compare the average population densities of the region within 2 mi of the city center and the ring between 3 and 6 mi of the center.

*85. Make a change of coordinates, and find the area of the region $\{(x,y) : 0 \leq x + y \leq 2, 0 \leq y \leq 3\}$.

*86. Make a change of coordinates, and find the volume of the region $\{(x,y,z) : |z| \leq 1, |x + 3y| \leq 2, |x - y + z| \leq 2\}$.

*87. Find the volume of the region $\{(x,y,z) : 4y^2 + 9z^2 \leq 36, 0 \leq 5x - y + 2z \leq 10\}$.

Vector Calculus

18

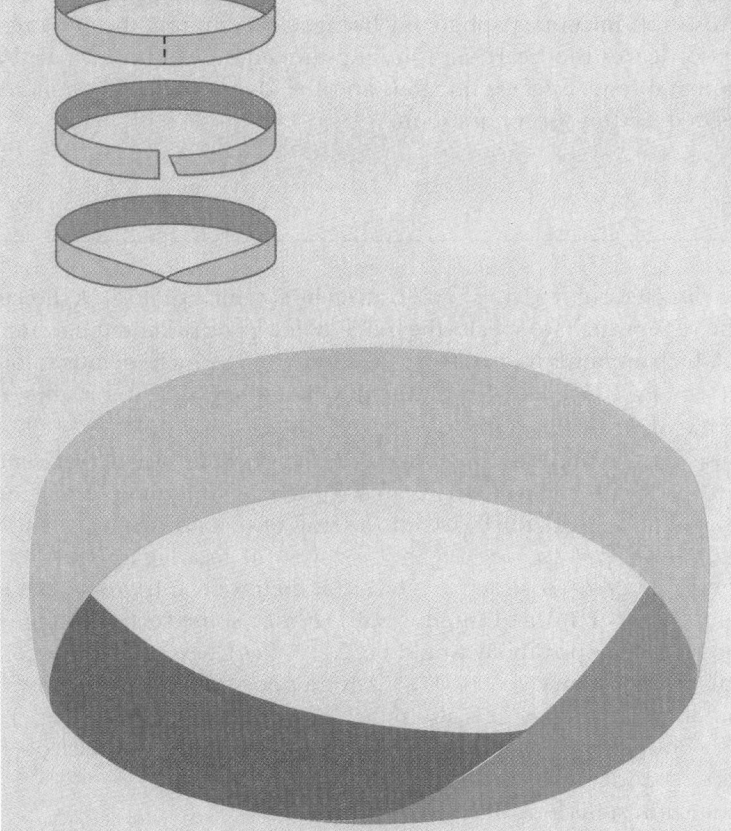

The Möbius strip challenges our intuitive ideas about surfaces. Take a strip of paper, make a single twist in it and tape it into a loop. With a pencil, draw a line down the center of the strip without crossing the edge of the paper. You will find that the pencil line returns on the back of the paper. For this reason people say that the Möbius strip has only one side. If you have performed this experiment, you might like to try another by taking a pair of scissors and cutting along the line down the middle of the strip of paper....

INTRODUCTION

This chapter is devoted to the subject known as *vector calculus.* We have already studied curves in the plane and in space and have seen how the use of vectors can simplify calculations and clarify the ideas involved. An obvious next step would be to consider integrating *along a curve.* This question might arise if we knew the density and wanted to find the total mass of a curved piece of wire. Another instance concerns work; we have seen how a force **F** acting on an object as it moves through a displacement **s** performs work on it equal to **F · s**. In many realistic situations the force is not constant but changes as the object moves along a curve, and in such a case it is necessary to integrate to find the total work done. Examples might include finding the work needed to lift a satellite into its orbit (since the force due to gravity changes as the satellite rises above the surface of the earth) or the work performed by an electric field as an electron moves through it.

We begin in Section 18.1 with what are called *line integrals.* This is the traditional name, but it is actually a misnomer because it refers to integrals along *curves*. In Section 18.2 we discuss the important fact that certain line integrals will give the same value no matter what path we follow to get from some starting point to some terminal point. This property is called *path independence*. Section 18.3 establishes Green's Theorem, a result that can be used to evaluate certain line integrals in a manner reminiscent of the way we use the Fundamental Theorem of Calculus to evaluate ordinary integrals.

Having learned to integrate along curves, in Section 18.4 we consider integrals over surfaces, such as a paraboloid or the surface of a sphere. In Sections 18.5 and 18.6 we consider the Divergence Theorem and Stokes' Theorem, which can be seen as generalizations of Green's Theorem to surface integrals.

In Section 18.7 we apply these ideas to various problems. From what has been said it is clear that many of the problems involve physics. One useful type of application is the study of fluid flows. There it is important to be able to find the amount of fluid flowing past a surface; examples include finding the quantity of a polluting gas that crosses an imaginary sphere enclosing a leak, finding the amount of blood that enters or leaves the heart, and finding the amount of a nutrient that passes through a membrane. Surface integrals and the Divergence Theorem are exactly what is needed for this sort of question.

18.1 Line Integrals

DIRECTED CURVE
INITIAL AND TERMINAL POINTS
DIFFERENTIAL FORM
VECTOR FIELD
WORK

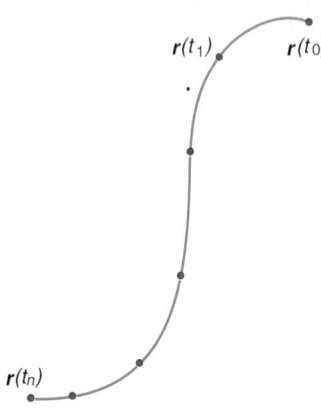

Figure 18.1.1

Consider a directed curve C in $\mathbb{R}^2$ or $\mathbb{R}^3$, given by $\mathbf{r}(t)$ for $t \in [a, b]$. A **directed curve** means a curve for which we specify the initial point $P = \mathbf{r}(a)$ and the terminal point $Q = \mathbf{r}(b)$, which amounts to specifying the direction in which we move along C. Let us write $\mathbf{r}(t) = \big(x(t), y(t), z(t)\big)$, with the understanding that $z(t)$ is always zero for a curve in the plane $\mathbb{R}^2$.

Suppose f is a continuous function on the image of the curve, that is, defined at every point on the curve. We would like to integrate the function f "along the curve." To do this, we might break the curve into a series of short pieces by partitioning the interval $[a, b]$ into $\mathcal{P} = \{a = t_0 < t_1 < \ldots < t_n = b\}$ (see Figure 18.1.1). If t_i^* is in the subinterval $[t_{i-1}, t_i]$, for each $i = 1, \ldots, n$, then we can try to form a Riemann sum by letting $P_i^* = \mathbf{r}(t_i^*)$ and multiplying $f(P_i^*)$ by some sort of change between $\mathbf{r}(t_{i-1})$ and $\mathbf{r}(t_i)$. One possibility would be $\Delta x_i = x(t_i) - x(t_{i-1})$, the x-component of the displacement from $\mathbf{r}(t_{i-1})$ to $\mathbf{r}(t_i)$. Another possibility would be $\Delta y_i = y(t_i) - y(t_{i-1})$, and for a curve in $\mathbb{R}^3$ it is also possible to use $\Delta z_i = z(t_i) - z(t_{i-1})$.

A different idea would be to use Δs_i, the arclength of the part of the curve between $\mathbf{r}(t_{i-1})$ and $\mathbf{r}(t_i)$.

The Riemann sums look like

$$\sum_{i=1}^{n} f(P_i^*)\Delta x_i, \qquad \sum_{i=1}^{n} f(P_i^*)\Delta y_i, \qquad \sum_{i=1}^{n} f(P_i^*)\Delta z_i, \qquad \sum_{i=1}^{n} f(P_i^*)\Delta s_i,$$

and allowing the mesh of the partition $\mathcal{P}$ to tend to zero, we can define the following line integrals as the limits:

$$\int_C f(x, y, z)\, dx, \qquad \int_C f(x, y, z)\, dy, \qquad \int_C f(x, y, z)\, dz, \qquad \int_C f(x, y, z)\, ds.$$

As usual, the definition does not offer much help with evaluating any of these integrals. If we suppose the components $x(t)$, $y(t)$, $z(t)$ of $\mathbf{r}(t)$ are differentiable functions of t, then we know that

$$\Delta x_i \approx \frac{dx}{dt}(t_i^*)\Delta t = x'(t_i^*)\Delta t, \qquad \Delta y_i \approx \frac{dy}{dt}(t_i^*)\Delta t = y'(t_i^*)\Delta t,$$

$$\Delta z_i \approx \frac{dz}{dt}(t_i^*)\Delta t = z'(t_i^*)\Delta t, \qquad \Delta s_i \approx \sqrt{\left(x'(t_i^*)\right)^2 + \left(y'(t_i^*)\right)^2 + \left(z'(t_i^*)\right)^2}\Delta t.$$

So for instance the Riemann sums can be rewritten as

$$\sum_{i=1}^{n} f(P_i^*)\Delta x_i \approx \sum_{i=1}^{n} f(P_i^*)\, x'(t_i^*)\Delta t_i, \qquad \sum_{i=1}^{n} f(P_i^*)\Delta y_i \approx \sum_{i=1}^{n} f(P_i^*)\, y'(t_i^*)\Delta t_i,$$

$$\sum_{i=1}^{n} f(P_i^*)\Delta z_i \approx \sum_{i=1}^{n} f(P_i^*)\, z'(t_i^*)\Delta t_i,$$

$$\sum_{i=1}^{n} f(P_i^*)\Delta s_i \approx \sum_{i=1}^{n} f(P_i^*)\sqrt{\left(x'(t_i^*)\right)^2 + \left(y'(t_i^*)\right)^2 + \left(z'(t_i^*)\right)^2}\Delta t_i.$$

Taking the limit as the mesh of the partition tends to zero suggests the following definition.

DEFINITION 18.1.1

Suppose $\mathbf{r}(t)$, $t \in [a, b]$, is a parametrization of a directed curve C. In component form we write $\mathbf{r}(t) = \big(x(t), y(t)\big)$ or $\mathbf{r}(t) = \big(x(t), y(t), z(t)\big)$ and assume that $x(t)$, $y(t)$, $z(t)$ have continuous derivatives on $[a, b]$. Suppose f is a continuous function defined (at least) at every point on the curve C.

The **line integrals** of f along C are defined as follows:

$$\int_C f(x, y, z)\, dx = \int_a^b f\big(x(t), y(t), z(t)\big) x'(t)\, dt,$$

$$\int_C f(x, y, z)\, dy = \int_a^b f\big(x(t), y(t), z(t)\big) y'(t)\, dt,$$

$$\int_C f(x, y, z)\, dz = \int_a^b f\big(x(t), y(t), z(t)\big) z'(t)\, dt,$$

$$\int_C f(x, y, z)\, ds = \int_a^b f\big(x(t), y(t), z(t)\big) \sqrt{\left(x'(t)\right)^2 + \left(y'(t)\right)^2 + \left(z'(t)\right)^2}\, dt.$$

EXAMPLE 1 Let C be the part of the helix $\mathbf{r}(t) = (\cos t, \sin t, t)$ with $t \in [0, \pi]$. Evaluate (i) $\int_C x^2\, dx$, (ii) $\int_C (x + y + z)\, ds$.

Solution

(i) Since $x(t) = \cos t$, we have that $x'(t) = -\sin t$. Since $f(x, y, z) = x^2$, we substitute $f\big(x(t), y(t), z(t)\big) = x(t)^2 = \cos^2 t$, and

$$\int_c x^2\, dx = \int_0^\pi x(t)^2 x'(t)\, dt = -\int_0^\pi \cos^2 t \sin t\, dt = \left.\frac{1}{3}\cos^3 t\right|_0^\pi = -\frac{2}{3}.$$

(ii) We have $ds = \sqrt{\left(x'(t)\right)^2 + \left(y'(t)\right)^2 + \left(z'(t)\right)^2}\, dt = \sqrt{\sin^2 t + \cos^2 t + 1}\, dt = \sqrt{2}\, dt$, so

$$\int_C (x + y + z)\, ds = \int_0^\pi (\cos t + \sin t + t)\sqrt{2}\, dt = \sqrt{2}\left(\sin t - \cos t + \frac{t^2}{2} \right)\Bigg|_0^\pi$$

$$= \sqrt{2}\left(0 - 0 + 1 + 1 + \frac{\pi^2}{2} \right) = 2\sqrt{2} + \frac{\pi^2}{\sqrt{2}}.$$

Notice the way we substituted the components $x(t)$, $y(t)$, and $z(t)$ of the curve into the integrand $f(x, y, z)$.

EXAMPLE 2 Let C be the upper half of the circle of radius 1 around the origin, running from $(1, 0)$ to $(-1, 0)$. Evaluate $\int_C x\, dy$.

Solution To solve this question, we need a parametrization of C. A natural choice is $\mathbf{r}(t) = (\cos t, \sin t)$, with $t \in [0, \pi]$. We find that $dy = y'(t)\, dt = \cos t\, dt$ and

$$\int_C x\, dy = \int_0^\pi \cos t(\cos t)\, dt = \int_0^\pi \frac{1}{2}(1 + \cos 2t)\, dt = \frac{\pi}{2}.$$

On the other hand, we might just as well have chosen the parametrization $\mathbf{p}(t) = (\cos 2t, \sin 2t)$, with $t \in \left[0, \frac{\pi}{2}\right]$. In this case $dy = y'(t)\, dt = 2\cos 2t\, dt$, and

$$\int_C x\, dy = \int_0^{\pi/2} (\cos 2t)\, 2\cos 2t\, dt = \int_0^{\pi/2} (1 + \cos 4t)\, dt = \frac{\pi}{2}.$$

Strictly speaking, Definition 18.1.1 appears to depend on the parametrization. If we chose two different parametrizations of C, we might get different values for the line integrals. We have just seen in Example 2 that this did not happen for two parametrizations of the semicircle, and in fact it is always true that the line integrals will be the same for any two parametrizations of C, *provided they go in the same direction*, that is, have the same initial point and the same terminal point. This is why we spoke of "directed" curves.

EXAMPLE 3 Consider the integral of Example 2 with the curve in the opposite direction. We sometimes use the notation $-C$ to refer to the curve that is the same as C but in the opposite direction.

One parametrization of $-C$, the upper semicircle beginning at $(-1, 0)$, is $\mathbf{q}(t) = (\cos t, -\sin t)$, with $t \in [\pi, 2\pi]$. We find that $dy = -\cos t\, dt$ and

$$\int_{-C} x\, dy = \int_\pi^{2\pi} \cos t(-\cos t)\, dt = -\frac{1}{2}\int_\pi^{2\pi} (1 + \cos 2t)\, dt = -\frac{\pi}{2}.$$

We find that $\int_{-C} x\, dy = -\int_C x\, dy$. This is a particular case of the following result.

THEOREM 18.1.2

If C is a directed curve, then the line integrals

$$\int_C f\, dx, \qquad \int_C f\, dy, \qquad \int_C f\, dz, \qquad \int_C f\, ds$$

are independent of the parametrization, that is, they are the same for any differentiable parametrization $\mathbf{r}(t)$, $t \in [a, b]$.

If $-C$ is the curve in the opposite direction, then

$$\int_{-C} f\, dx = -\int_C f\, dx, \qquad \int_{-C} f\, dy = -\int_C f\, dy,$$
$$\int_{-C} f\, dz = -\int_C f\, dz, \qquad \int_{-C} f\, ds = \int_C f\, ds.$$

We will not prove this result, but let us draw attention to the fact that reversing the direction changes the sign of the line integrals $\int_C f\,dx$, $\int_C f\,dy$, $\int_C f\,dz$ but leaves $\int_C f\,ds$ *unchanged*. This is because moving in the opposite direction reverses the signs of changes in x, and similarly for y or z, but the arclength is the same in both directions.

REMARK 18.1.3

(i) Frequently, we will want to evaluate integrals like $\int_C f\,dx + \int_C g\,dy + \int_C h\,dz$; in a case like this we economize on integral signs and write $\int_C f\,dx + g\,dy + h\,dz$. The expression $f\,dx + g\,dy + h\,dz$ is called a **differential form**.

(ii) To define line integrals, we assumed that the parametrization was continuously differentiable, but in fact it is possible to use parametrizations that are not differentiable at finitely many points. This is useful for integrating along curves that have "corners" on them.

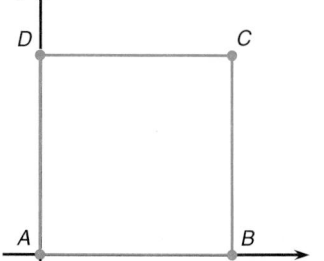

Figure 18.1.2

EXAMPLE 4   Evaluate $\int_Q x\,dx + y\,dy$, where Q goes counterclockwise around the square with corners $(0,0)$, $(1,0)$, $(1,1)$, $(0,1)$.

Solution   The square is shown in Figure 18.1.2, where we have labeled its corners A, B, C, D. We can parametrize the square by running along its sides in order. Let

$$
\mathbf{r}(t) = \begin{cases} (t, 0), & \text{if } t \in [0,1], \\ (1, t-1), & \text{if } t \in [1,2], \\ (3-t, 1), & \text{if } t \in [2,3], \\ (0, 4-t), & \text{if } t \in [3,4]. \end{cases}
$$

It is easy to check that this does parametrize the square. Because the formulas for $x(t)$ and $y(t)$ are different for different parts of the interval $[0,4]$, we have to integrate each piece separately, that is, for $t \in [0,1]$, $t \in [1,2]$, $t \in [2,3]$, and $t \in [3,4]$. For $t \in [0,1]$ we find that $dx = x'(t)\,dt = dt$ and $dy = y'(t)\,dt = 0$. For $t \in [1,2]$ we find that $dx = 0$, $dy = dt$; for $t \in [2,3]$, $dx = -dt$, $dy = 0$; for $t \in [3,4]$, $dx = 0$, $dy = -dt$. Combining, we have

$$
\int_Q x\,dx + y\,dy = \int_0^1 x(t)\,dt + \int_1^2 y(t)\,dt - \int_2^3 x(t)\,dt - \int_3^4 y(t)\,dt
$$

$$
= \int_0^1 t\,dt + \int_1^2 (t-1)\,dt - \int_2^3 (3-t)\,dt - \int_3^4 (4-t)\,dt
$$

$$
= \frac{1}{2} + \left(\frac{3}{2} - 1\right) - \left(3 - \frac{5}{2}\right) - \left(4 - \frac{7}{2}\right) = 0.
$$

The line integral is zero.

For an integral $\int_C f\,dx + g\,dy + h\,dz$ it is tempting to think of the integrand as a dot product $(f, g, h) \cdot (dx, dy, dz)$. Let us define a *vector-valued* function $\mathbf{F} = (f, g, h)$, and let us agree that if $\mathbf{r}(t) = \big(x(t), y(t), z(t)\big)$, then $d\mathbf{r}$ will mean $d\mathbf{r} = (dx, dy, dz) = \big(x'(t)\,dt, y'(t)\,dt, z'(t)\,dt\big)$. Then consider the dot product

$$
\mathbf{F} \cdot d\mathbf{r} = f\big(\mathbf{r}(t)\big) x'(t)\,dt + g\big(\mathbf{r}(t)\big) y'(t)\,dt + h\big(\mathbf{r}(t)\big) z'(t)\,dt.
$$

We see that the original integral can be written as

$$\int_C f\,dx + g\,dy + h\,dz = \int \mathbf{F} \cdot d\mathbf{r}.$$

We can write $d\mathbf{r} = \big(x'(t)\,dt, y'(t)\,dt, z'(t)\,dt\big) = \big(x'(t), y'(t), z'(t)\big)\,dt = \mathbf{r}'(t)\,dt,$ so this formula becomes

$$\int_C f\,dx + g\,dy + h\,dz = \int \mathbf{F} \cdot d\mathbf{r} = \int_a^b \mathbf{F}\big(\mathbf{r}(t)\big) \cdot \mathbf{r}'(t)\,dt.$$

If t is any parameter for the curve C, then we know (cf. Theorem 15.2.6) that $\|\mathbf{r}'(t)\| = \frac{ds}{dt}$, where s is the arclength. From this we see that

$$\mathbf{r}'(t)\,dt = \frac{\|\mathbf{r}'(t)\|}{\|\mathbf{r}'(t)\|}\mathbf{r}'(t)\,dt = \frac{\mathbf{r}'(t)}{\|\mathbf{r}'(t)\|}\,\|\mathbf{r}'(t)\|\,dt = \mathbf{T}(t)\frac{ds}{dt}\,dt = \mathbf{T}\,ds,$$

where $\mathbf{T}$ is the unit tangent vector.

To integrate $\mathbf{F} \cdot \mathbf{r}'(t)\,dt$ along C, we write $\mathbf{r}(t) = \mathbf{p}(s)$ and find the formula

$$\int \mathbf{F}\big(\mathbf{r}(t)\big) \cdot \mathbf{r}'(t)\,dt = \int \mathbf{F}\big(\mathbf{p}(s)\big) \cdot \mathbf{T}(s)\,ds.$$

Some imagination may be needed to plot the curve of Example 4 on a graphics calculator.

TI-85	SHARP EL9300
MODE ▼ ▼ ▼ ▼ ▶ ▶ (PARAM) ENTER EXIT GRAPH F1($E(t)$) F1 (t) ▼ 1 ▼ 1 ▼ 1 − F1(t) M2(RANGE) 0 ▼ 1 ▼ ▼ (−).5 ▼ 1.5 ▼ ▼ (−).5 ▼ 1.5 F5(GRAPH)	SET UP E 3 ⌁ 1 ENTER x/θ/T 2ndF ▼ 1 − x/θ/T ENTER 1 RANGE 0 ENTER 1 ENTER .02 ENTER (−).5 ENTER 1.5 ENTER 1 ENTER (−).5 ENTER 1.5 ENTER 1 ENTER ⌁
CASIO f_x-7700GB	HP 48SX
MODE MODE × (PARAM) Range −.5 EXE 1.5 EXE 1 EXE −.5 EXE 1.5 EXE 1 EXE 0 EXE 1 EXE .02 EXE Graph T, 0) EXE Graph 1 , T) EXE Graph 1 − T , 1) EXE Graph 0 , 1 − T) EXE	↰ PLOT PLOTR ERASE ATTN ' Y = 0 ↱ PLOT ↰ DRAW ERASE .5 +/− ENTER 1.5 XRNG .5 +/− ENTER 1.5 YRNG DRAW ATTN ( ) 0 , 0 ▶ ( ) 1 , 0 PRG DSPL TLINE ( ) 1 , 0 ▶ ( ) 1 , 1 TLINE ( ) 1 , 1 ▶ ( ) 0 , 1 TLINE ( ) 0 , 1 ▶ ( ) 0 , 0 TLINE ↱ PLOT DRAW

You can try using the calculator to plot some of the curves we encounter in other examples.

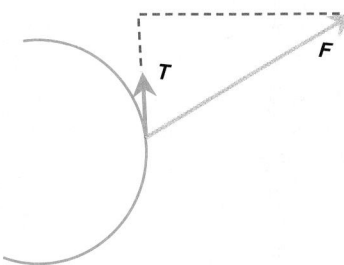

Figure 18.1.3

The quantity $\mathbf{F}\big(\mathbf{p}(s)\big) \cdot \mathbf{T}(s)$ is the *tangential component* of $\mathbf{F}$, the size of the component of $\mathbf{F}$ along the direction of the tangent vector $\mathbf{T}$ (see Figure 18.1.3). The interesting thing about this last formula is that although the integral on the right is with respect to arclength s, it is not actually necessary to know the arclength to evaluate it.

EXAMPLE 5   If C is the curve $y = x^2$, running from $x = 0$ to $x = 1$, and $\mathbf{F}(x, y) = (xy, x + y)$, integrate the tangential component of $\mathbf{F}$ along C.

Solution   The natural parametrization of C is $\mathbf{r}(t) = \big(x(t), y(t)\big) = (t, t^2)$, for $t \in [0, 1]$. Using the formula, we find that

$$\int_C \mathbf{F}(s) \cdot \mathbf{T}(s)\,ds = \int_0^1 \mathbf{F}\big(\mathbf{r}(t)\big) \cdot \mathbf{r}'(t)\,dt = \int_0^1 \big(t(t^2), t + t^2\big) \cdot (1, 2t)\,dt$$

$$= \int_0^1 (t^3 + 2t^2 + 2t^3)\,dt = \left(\frac{3}{4}t^4 + \frac{2}{3}t^3\right)\Bigg|_0^1 = \frac{3}{4} + \frac{2}{3} = \frac{17}{12}.$$

Figure 18.1.4

We have seen that it is not easy to find the arclength of a parabola, but Example 5 was not difficult. The calculation does not really use the arclength at all.

Notice that in Example 5 we worked in the plane using the same formulas we would use in $\mathbb{R}^3$.

DEFINITION 18.1.4

A vector-valued function

$$\mathbf{F}(x, y, z) = \big(u(x, y, z), v(x, y, z), w(x, y, z)\big)$$

is sometimes called a **vector field**. We think of its value at each point (x, y, z) as a vector with initial point (x, y, z) (see Figure 18.1.4). The same terminology is used for functions defined on $\mathbb{R}^2$ with values in $\mathbb{R}^2$.

Physicists often refer to a *force field*, a force acting (as a vector) at each point in a region.

Figure 18.1.5

Line Integrals and Work

Suppose we want to find the work done by a force $\mathbf{F}$ acting on an object as it moves along a curve C parametrized by $\mathbf{r}(t)$, with $t \in [a, b]$ (see Figure 18.1.5). Partitioning $[a, b]$ by $\mathcal{P} = \{t_0, t_1, \ldots, t_n\}$, we can imagine a small displacement $\Delta \mathbf{r}_i = \mathbf{r}(t_i) - \mathbf{r}(t_{i-1})$ (see Figure 18.1.6).

Assuming that the force changes very little during this small movement, we can approximate it by $\mathbf{F}(P_i^*)$, where $P_i^* = \mathbf{r}(t_i^*)$ for some $t_i^* \in [t_{i-1}, t_i]$. The work done as the object moves from $\mathbf{r}(t_{i-1})$ to $\mathbf{r}(t_i)$ is approximately $\mathbf{F}(P_i^*) \cdot \Delta \mathbf{r}_i$, and the total work done is approximated by the Riemann sum

$$\sum_{i=1}^n \mathbf{F}(P_i^*) \cdot \Delta \mathbf{r}_i = \sum_{i=1}^n \mathbf{F}\big(\mathbf{r}(t_i^*)\big) \cdot \Delta \mathbf{r}_i.$$

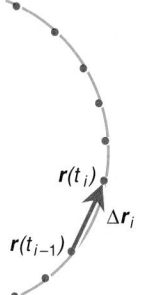

Figure 18.1.6

Taking the limit, we express the total work as the line integral $\int_C \mathbf{F}\big(\mathbf{r}(t)\big) \cdot d\mathbf{r} = \int_a^b \mathbf{F}\big(\mathbf{r}(t)\big) \cdot \mathbf{r}'(t)\,dt$.

We have just derived the following result.

THEOREM 18.1.5

> If an object moves along the curve C given by the continuously differentiable parametrization $\mathbf{r}(t)$, $t \in [a, b]$, under the influence of a force $\mathbf{F}$, then the total work performed on the object by the force is
>
> $$W = \int_C \mathbf{F} \cdot d\mathbf{r} = \int_a^b \mathbf{F}\big(\mathbf{r}(t)\big) \cdot \mathbf{r}'(t)\, dt.$$

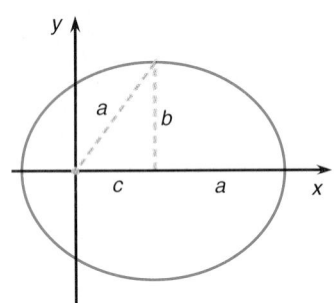

Figure 18.1.7

EXAMPLE 6   **Rolling Ball**   Find the total work done by gravity on a steel ball weighing 4 oz as it rolls down a helical track parametrized by $\mathbf{r}(t) = (\cos \pi t, \sin \pi t, 8 - t)$, with $t \in [0, 8]$ (distances measured in feet). See Figure 18.1.7.

Solution   The force due to gravity is 4 oz $= \frac{1}{4}$ lb pointing straight down, so, using pounds, we can write it as $\mathbf{F} = \left(0, 0, -\frac{1}{4}\right)$. The work done is

$$W = \int_C \mathbf{F} \cdot d\mathbf{r} = \int_0^8 \left(0, 0, -\frac{1}{4}\right) \cdot (-\pi \sin \pi t, \pi \cos \pi t, -1)\, dt = \int_0^8 \frac{1}{4}\, dt = 2.$$

The work done is 2 ft-lb.

This example was relatively simple because the force was constant.

For the next example it will be helpful if you know about the motion of satellites (cf. Section 15.7).

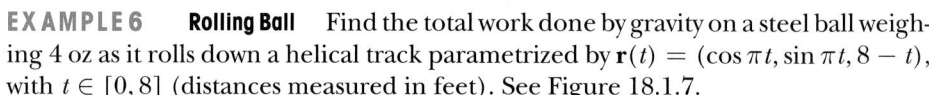

EXAMPLE 7   **Satellite**   Find the work done by gravity on a 100-kg satellite as it moves along its orbit from the highest point (the *apogee*) to the lowest point (the *perigee*) if the orbit is an ellipse with semi-major axis 20,000 km and semi-minor axis 16,000 km.

Solution   We begin by finding a parametrization for the path of the satellite. One approach would be to place the earth at the origin and use polar coordinates to describe the ellipse.

Knowing that the semi-major axis is $a = 20,000$ and the semi-minor axis is $b = 16,000$, we know (cf. Summary 11.2.3, p. 738) that the distance from the center of the ellipse to the focus is $c = \sqrt{a^2 - b^2} = 12,000$. From this we find that the eccentricity is $e = \frac{c}{a} = \frac{12,000}{20,000} = \frac{3}{5}$. See Figure 18.1.8.

This tells us that the equation of the ellipse is

$$r = \frac{K}{1 - e \cos \theta} = \frac{K}{1 - \frac{3}{5} \cos \theta},$$

Figure 18.1.8

(cf. Equation 12.4.1, p. 780), where K is some constant. By considering the point where $\theta = 0$, which is at a distance $a + c = 32,000$ from the origin, we find that $32,000 = \frac{K}{1 - 3/5 \cos 0} = \frac{K}{2/5}$, so $K = \frac{2}{5}(32,000) = \frac{64,000}{5}$. The equation is $r = \frac{64,000/5}{1 - 3/5 \cos \theta} = \frac{64,000}{5 - 3 \cos \theta}$.

We can use this to parametrize the path of the satellite by choosing $t = \theta$ as our parameter and letting it run from 0 to π (see Figure 18.1.9). The position of the

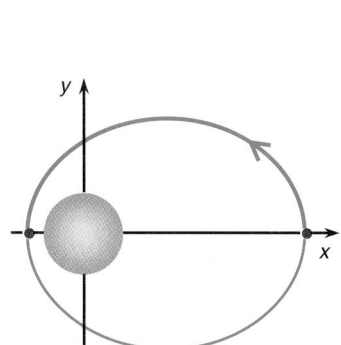

Figure 18.1.9

satellite in rectangular coordinates is

$$\mathbf{r}(t) = \big(x(t), y(t)\big) = (r\cos t, r\sin t) = \left(\frac{64{,}000\cos t}{5 - 3\cos t}, \frac{64{,}000\sin t}{5 - 3\cos t}\right).$$

From this we find that

$$d\mathbf{r}(t) = \big(x'(t)\,dt, y'(t)\,dt\big)$$

$$= 64{,}000\left(\frac{(5 - 3\cos t)(-\sin t) - \cos t(3\sin t)}{(5 - 3\cos t)^2},\right.$$

$$\left.\frac{(5 - 3\cos t)(\cos t) - \sin t(3\sin t)}{(5 - 3\cos t)^2}\right)dt$$

$$= 64{,}000\left(\frac{-5\sin t}{(5 - 3\cos t)^2}, \frac{5\cos t - 3}{(5 - 3\cos t)^2}\right)dt.$$

Next we have to find the force of gravity on the satellite. It points in the direction from the satellite toward the earth, that is, in the same direction as $-\mathbf{r}(t)$. We know that its magnitude is $\frac{C}{\|\mathbf{r}\|^2}$, for some constant C (cf. the discussion before Example 6.6.5 (p. 460) or Section 15.7). The satellite weighs 100 kg at the surface of the earth, so the force exerted on it there would be $100(9.8) = 980$ N. Since the radius of the earth is about 6400 km, we find that $980 = \frac{C}{6400^2}$ so $C = 980(6400^2)$ N km$^2 \approx 4.0 \times 10^{10}$ N km$^2 = 4.0 \times 10^{16}$ N m^2. The force is given by

$$\mathbf{F} = \frac{C}{\|\mathbf{r}\|^2}\left(-\frac{\mathbf{r}}{\|\mathbf{r}\|}\right) = -\frac{C}{\|\mathbf{r}\|^3}\mathbf{r}.$$

> The magnitude of the force is $\frac{C}{\|\mathbf{r}\|^2}$; the unit vector in the direction of the force is $-\frac{\mathbf{r}}{\|\mathbf{r}\|}$.

Noting that $\|\mathbf{r}(t)\| = \left\|\left(\dfrac{64{,}000\cos t}{5 - 3\cos t}, \dfrac{64{,}000\sin t}{5 - 3\cos t}\right)\right\|$

$$= 64{,}000\sqrt{\frac{\cos^2 t + \sin^2 t}{(5 - 3\cos t)^2}} = \frac{64{,}000}{5 - 3\cos t},$$

we find that the total work done is

$$W = \int \mathbf{F} \cdot d\mathbf{r} = 64{,}000\int_0^{\pi} -\left(\frac{C}{\|\mathbf{r}(t)\|^3}\mathbf{r}(t)\right) \cdot \left(\frac{-5\sin t}{(5 - 3\cos t)^2}, \frac{5\cos t - 3}{(5 - 3\cos t)^2}\right)dt$$

$$= 64{,}000\,C\int_0^{\pi} -\frac{1}{\|\mathbf{r}(t)\|^3}\left(\frac{64{,}000\cos t}{5 - 3\cos t}, \frac{64{,}000\sin t}{5 - 3\cos t}\right)$$

$$\cdot \left(\frac{-5\sin t}{(5 - 3\cos t)^2}, \frac{5\cos t - 3}{(5 - 3\cos t)^2}\right)dt$$

$$= 64{,}000^2\,C\int_0^{\pi} \left(\frac{64{,}000}{5 - 3\cos t}\right)^{-3}\frac{5\cos t\sin t - 5\cos t\sin t + 3\sin t}{(5 - 3\cos t)^3}dt$$

$$= \frac{C}{64{,}000}\int_0^{\pi} 3\sin t\,dt = \frac{6C}{64{,}000} = \frac{3}{32{,}000}4.0 \times 10^{16} \approx 3.8 \times 10^{12}\text{J}.$$

The work done by gravity on the satellite as it moves from its apogee to its perigee is about 3.8×10^{12} joules.   ◢

> This work is converted into kinetic energy, making the satellite move faster. This is consistent with the fact we learned in Section 15.7 that the satellite will move faster when it is nearer to the earth.

Exercises 18.1

I

Evaluate each of the following.

1. $\int_C y\,dx - xy\,dy$, where C is the upper semicircle $x^2 + y^2 = 4$, $y \geq 0$, starting at $(2, 0)$.

2. $\int_C x^2\,dx + 2x\,dy$, where C is the lower semicircle $x^2 + y^2 = 1$, $y \leq 0$, starting at $(1, 0)$.

3. $\int_C 2x\,dx + y^2\,dy$, where C is the circle $x^2 + y^2 = 9$, once around clockwise from $(3, 0)$.

4. $\int_C y\,dx$, where C is the circle $x^2 + y^2 = 3$, once around counterclockwise from $(0, \sqrt{3})$.

5. $\int_C x^3\,dy$, where C is the circle $x^2 + y^2 = 4$, twice around counterclockwise from $(0, -2)$.

6. $\int_C y^3\,dx$, where C is the circle $x^2 + y^2 = 8$, three times around clockwise starting at $(2, 2)$.

7. $\int_C xy\,dy$, where C is the square with vertices $(1, 1)$, $(1, -1)$, $(-1, -1)$, $(-1, 1)$, clockwise starting at $(1, 1)$.

8. $\int_C \sin y\,dx$, where C is the rectangle with vertices $(0, 0)$, $(1, 0)$, $(1, \pi)$, $(0, \pi)$, counterclockwise starting at $(0, 0)$.

9. $\int_C \cos x\,dy$, where C is the rectangle with vertices $(0, 0)$, $(0, 2)$, $(2\pi, 2)$, $(2\pi, 0)$, clockwise starting at $(0, 0)$.

10. $\int_C y^2\,dx - x^2\,dy$, where C is the upper semicircle $x^2 + y^2 = 1$, $y \geq 0$, starting at $(1, 0)$, followed by the lower half of the square with corners $(\pm 1, \pm 1)$.

Evaluate $\int_C \mathbf{F} \cdot d\mathbf{r}$ in each of the following cases.

11. $\mathbf{F}(x, y) = (x^2, x)$, C is the upper semicircle $x^2 + y^2 = 1$, $y \geq 0$, starting at $(-1, 0)$.

12. $\mathbf{F}(x, y) = (\cos y, \sin x)$, C is the counterclockwise rectangle with vertices $(0, 0)$, $\left(\frac{\pi}{2}, 0\right)$, $\left(\frac{\pi}{2}, \pi\right)$, $(0, \pi)$, starting at $(0, 0)$.

13. $\mathbf{F}(x, y) = \left(\frac{-y}{x^2 + y^2}, \frac{x}{x^2 + y^2}\right)$, C is the circle $x^2 + y^2 = 1$, counterclockwise, starting at $(0, 1)$.

14. $\mathbf{F}(x, y) = \left(\frac{x^2}{x^2 + y^2}, \frac{y^4}{x^2 + y^2}\right)$, C is the circle $x^2 + y^2 = 4$, clockwise, starting at $(-2, 0)$.

15. $\mathbf{F}(x, y, z) = (x, x, x)$, C is the helix $\mathbf{r}(t) = (\cos t, \sin t, t)$, $t \in [0, 4\pi]$.

16. $\mathbf{F}(x, y, z) = (z, x, y)$, C is the semicircle $x^2 + y^2 = 1$, $y \geq 0$, $z = 1$, starting at $(1, 0, 1)$.

17. $\mathbf{F}(x, y, z) = (x + y, z - 2, z)$, C is the semicircle $y^2 + z^2 = 4$, $z \leq 0$, $x = 2$, starting at $(2, -2, 0)$.

18. $\mathbf{F}(x, y, z) = (x^2, y^2, z^2)$, C is the curve consisting of straight line segments from $(0, 0, 0)$ to $(2, 2, 0)$, from there to $(3, 1, 1)$, and from there to $(5, 4, 3)$.

19. $\mathbf{F}(x, y, z) = (-x, -y, -z)$, C is the curve consisting of straight line segments from $(1, 1, 1)$ to $(4, 0, 3)$, from there to $(3, 2, 0)$, and from there to $(1, 1, 2)$.

20. $\mathbf{F}(x, y, z) = (y - x, 3, 2xz)$, C is the curve consisting of straight line segments from $(3, 2, 1)$ to $(0, 0, 2)$, from there to $(4, 4, 4)$, and from there to $(3, 2, 1)$.

Calculate the work done by the force $\mathbf{F}$ in each of the following situations.

21. $\mathbf{F}(x, y) = (0, -1)$, in pounds, on an object moving along the circle $x^2 + y^2 = 1$ with $y \geq 0$ from $(1, 0)$ to $(0, 1)$, with distances in feet.

22. $\mathbf{F}(x, y) = (y, -x)$, in pounds, on an object moving along the circle $x^2 + y^2 = 4$ with $y \leq 0$ from $(2, 0)$ to $(-2, 0)$, in feet.

23. $\mathbf{F}(x, y) = (2xy, x^2)$, in newtons, on an object moving along the parabola $y = x^2 + 1$ from $(1, 2)$ to $(-2, 5)$, in meters.

24. $\mathbf{F}(x, y) = \left(\frac{-x}{x+3}, \frac{-y}{x+3}\right)$, in pounds, on an object moving along the parabola $y^2 - 2y = x + 1$ from $(-1, 2)$ to $(14, -3)$, in feet.

25. $\mathbf{F}(x, y, z) = (xz, yz, xy)$, in pounds, on an object moving along the hyperbola $x^2 - y^2 = 8$, $z = 3$, from $(-3, 1, 3)$ to $(-3, -1, 3)$, in feet.

26. $\mathbf{F}(x, y, z) = (2, 3, -1)$, in pounds, on an object moving along the portion of the ellipse $x^2 + 4z^2 = 4$, $y = -1$ with $z \leq 0$, from $(2, -1, 0)$ to $(0, -1, -1)$, in feet.

27. $\mathbf{F}(x, y, z) = (y, y, y)$, in newtons, on an object moving along $y = \sin x$, $z = 2$, from $(0, 0, 2)$ to $(\pi, 0, 2)$, measured in meters.

28. $\mathbf{F}$ is the force of gravity on a 10-lb ball being lifted up a ramp that is 4 ft long, inclined at a $30°$ angle above the horizontal.

29. $\mathbf{F}$ is the force of gravity on a 5-lb ball rolling down a spiral staircase of radius 3 ft that makes two complete turns as it descends 10 ft.

30. $\mathbf{F}$ is the force of gravity on a 1-kg ball as it rolls inside a parabolic satellite dish. The dish is horizontal (i.e., it is aimed straight up), with radius 1.5 m and depth 0.5 m. The ball starts at the rim and rolls down to the center of the dish.

II

31. Suppose C is the circle $x^2 + y^2 = 1$, traversed counterclockwise starting at $(1, 0)$, and let n be a positive integer. Find a formula for $\int_C x^n\,dx$.

32. Suppose C is the upper semicircle $x^2 + y^2 = 1$, starting at $(1, 0)$, and let n be a positive integer. Find a formula for $\int_C x^n \, dx$.

33. Suppose C is the square with vertices $(0, 0)$, $(1, 0)$, $(1, 1)$, $(0, 1)$, traversed counterclockwise starting at $(0, 0)$, and let n be a positive integer. Find a formula for (i) $\int_C x^n \, dx$, (ii) $\int_C x^n \, dy$.

34. Suppose C is the square with vertices $(1, 1)$, $(-1, 1)$, $(-1, -1)$, $(1, -1)$, traversed counterclockwise starting at $(1, 1)$, and let n be a positive integer. Find a formula for (i) $\int_C x^n \, dx$, (ii) $\int_C x^n \, dy$.

35. Find the work done by gravity on the satellite of Example 7 (p. 1148) as it moves from its perigee (lowest point) to its apogee (highest point).

36. Find the work done by gravity on the satellite of Example 7 as it moves exactly once around its orbit (i) starting at its apogee, (ii) starting at its perigee.

*37. (i) When the satellite of Example 7 is at its lowest point, how far is it from the center of the earth?

   (ii) When the satellite of Example 7 is at its highest point, how far is it from the center of the earth?

   (iii) Suppose the satellite fell in a straight line toward the center of the earth from the height of its apogee, as found in part (ii), to the height of its perigee, as found in part (i). Calculate the work done on the satellite by the force of gravity during this fall.

   (iv) Compare your answer to part (iii) with the work done by gravity as the satellite moves *along its orbit* from the apogee to the perigee, as found in Example 7. Explain the comparison.

38. Suppose C is a directed **closed curve, that is, one whose terminal point is the same as its initial point. Show that $\int_C \mathbf{F} \cdot d\mathbf{r}$ is independent of the starting point, that is, that the line integral is the same no matter where we start on the curve, as long as we go around the curve once in the specified direction.

**39. Suppose $f(x)$, $g(y)$, and $h(z)$ are continuous functions on $\mathbb{R}$, and let P be the point (a, b, c). If $\mathbf{F} = \big(f(x), g(y), h(z)\big)$ and C is a curve from the origin to P, find a formula for $\int_C \mathbf{F} \cdot d\mathbf{r}$. In particular, conclude that this line integral does not depend on the choice of the curve C joining O to P.

**40. Let $\mathbf{F}(x, y, z) = (xy, z, xyz)$. Show that the line integral $\int_C \mathbf{F} \cdot d\mathbf{r}$ *does* depend on the choice of the curve C running from $(0, 0, 0)$ to $(1, 1, 0)$.

18.2 Path Independence

CONNECTED SET
POTENTIAL FUNCTION
SIMPLY CONNECTED SET
CONSERVATIVE FIELD

Suppose P and Q are two points in $\mathbb{R}^2$ or $\mathbb{R}^3$. There are many different paths that start at P and end at Q (see Figure 18.2.1). When we describe a line integral along a curve C that runs from P to Q, we have to specify the curve, because the integral may depend on which of the many possible curves is used.

On the other hand, there are some important instances in which the line integral will be the same for any path that runs from P to Q.

Figure 18.2.1

EXAMPLE 1

(i) Consider the vector field $\mathbf{F}(x, y) = (1, y)$. If P is the point (r, s) and Q is (u, v), suppose C is a curve from P to Q. Suppose $\mathbf{r}(t) = \big(x(t), y(t)\big)$ is a parametrization of C in which $x(t)$ and $y(t)$ are differentiable functions, with $t \in [a, b]$.
The line integral $\int_C \mathbf{F} \cdot d\mathbf{r}$ can be evaluated as follows:

$$\int_C \mathbf{F} \cdot d\mathbf{r} = \int_a^b \mathbf{F}\big(x(t), y(t)\big) \cdot \big(x'(t), y'(t)\big) \, dt = \int_a^b \big(1, y(t)\big) \cdot \big(x'(t), y'(t)\big) \, dt$$

$$= \int_a^b \Big((1) x'(t) + y(t) y'(t)\Big) \, dt.$$

This last integral is an ordinary integral that is easily evaluated. It equals

$$\int_a^b \big(x'(t) + y(t) y'(t)\big) \, dt = \left(x(t) + \tfrac{1}{2} y(t)^2 \right) \Big|_a^b = x(b) - x(a) + \tfrac{1}{2}\big(y(b)^2 - y(a)^2\big).$$

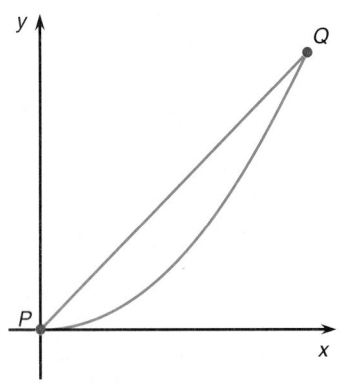

Figure 18.2.2

Since $\mathbf{r}(t) = \big(x(t), y(t)\big)$ is a parametrization of C, we know that $\mathbf{r}(a) = P$, so $\big(x(a), y(a)\big) = P = (r, s)$, and $\mathbf{r}(b) = Q$, so $\big(x(b), y(b)\big) = Q = (u, v)$. Substituting these values into the expression for the integral, we find that it equals $u - r + \frac{1}{2}(v^2 - s^2)$.

The interesting thing about this value is that it makes no mention of the curve C but only depends on the coordinates of the two endpoints of the curve.

(ii) Let $\mathbf{G}(x, y) = (xy, y)$, $P = (0, 0)$ and $Q = (1, 1)$. Consider $\int_C \mathbf{G} \cdot d\mathbf{r}$, where C is a curve running from P to Q.

One such curve is the straight line parametrized by $\mathbf{r}(t) = (t, t)$, for $t \in [0, 1]$. Another is the parabola $\mathbf{p}(t) = (t, t^2)$, for $t \in [0, 1]$ (see Figure 18.2.2).

Evaluating the line integrals corresponding to these two curves, we find

$$\int_0^1 \mathbf{G}\big(x(t), y(t)\big) \cdot d\mathbf{r}(t) = \int_0^1 \mathbf{G}(t, t) \cdot (1, 1)\, dt = \int_0^1 (t^2, t) \cdot (1, 1)\, dt$$

$$= \int_0^1 (t^2 + t)\, dt = \left(\frac{t^3}{3} + \frac{t^2}{2}\right)\bigg|_0^1 = \frac{5}{6},$$

$$\int_0^1 \mathbf{G}\big(x(t), y(t)\big) \cdot d\mathbf{p}(t) = \int_0^1 \mathbf{G}(t, t^2) \cdot (1, 2t)\, dt = \int_0^1 (t^3, t^2) \cdot (1, 2t)\, dt$$

$$= \int_0^1 (t^3 + 2t^3)\, dt = \frac{3t^4}{4}\bigg|_0^1 = \frac{3}{4}.$$

We find that integrating along these two paths gives different values for the line integral. Other paths will result in many different values.

We find that some line integrals depend on the choice of path joining P to Q and others do not. ◢

DEFINITION 18.2.1

> Suppose $\mathbf{F}$ is a vector field defined on some set U in $\mathbb{R}^2$ or $\mathbb{R}^3$, and consider the line integral
>
> $$\int_C \mathbf{F} \cdot d\mathbf{r},$$
>
> where C is any continuously differentiable curve contained in U. We say that this integral is **independent of path** or **path independent** if its value depends only on the initial point P and the terminal point Q of the path C.

In this case we can consider writing $\displaystyle\int_C \mathbf{F} \cdot d\mathbf{r} = \int_P^Q \mathbf{F} \cdot d\mathbf{r}$.

The idea is that if the choice of curve C does not affect the value of the integral there is no need to write it down, and it may be more useful to write down the information that really does matter, that is, the points P and Q.

What we need is a way to decide whether some particular line integral is or is not independent of the path.

In a way the key is hidden in the calculation we did in Example 1(i). When we used a parametrization of the curve to express the line integral as an ordinary integral, we found that the integral could be evaluated using the Fundamental Theorem of Calculus, and this happened because the integrand was a derivative.

So let us consider a vector field $\mathbf{F}(x, y, z) = \big(u(x, y, z), v(x, y, z), w(x, y, z) \big)$ and a curve C parametrized by $\mathbf{r}(t) = \big(x(t), y(t), z(t) \big)$ for $t \in [a, b]$. The line integral is

$$L = \int_C \mathbf{F} \cdot d\mathbf{r} = \int_a^b \mathbf{F}\big(x(t), y(t), z(t) \big) \cdot \big(x'(t), y'(t), z'(t) \big) \, dt$$

$$= \int_a^b \Big(u\big(x(t), y(t), z(t) \big) x'(t) + v\big(x(t), y(t), z(t) \big) y'(t) + w\big(x(t), y(t), z(t) \big) z'(t) \Big) \, dt.$$

What we would like is for the integrand to be a derivative. If $f(x, y, z)$ is a function on $\mathbb{R}^3$ and we let $h(t) = f\big(x(t), y(t), z(t) \big) = f\big(\mathbf{r}(t) \big)$, then the Chain Rule tells us how to differentiate $h(t)$:

$$\frac{dh}{dt} = \frac{\partial f}{\partial x} x'(t) + \frac{\partial f}{\partial y} y'(t) + \frac{\partial f}{\partial z} z'(t).$$

Figure 18.2.3

This suggests that the integrand we found above for the line integral $\int_C \mathbf{F} \cdot d\mathbf{r}$ will be a derivative if $\mathbf{F} = (u, v, w) = \big(\frac{\partial f}{\partial x}, \frac{\partial f}{\partial y}, \frac{\partial f}{\partial z} \big) = \nabla f$. The line integral will be independent of path if the vector field $\mathbf{F}$ is the gradient of some function f. In order for the question to make sense, we have to consider functions that are defined on some **open set**, such as an open disk, a sphere, an open rectangle, or some more complicated set, several examples of which are shown in Figure 18.2.3 (cf. the discussion at the beginning of Section 16.2, p. 938). A calculation like the one in Example 1(i) proves the following result.

THEOREM 18.2.2

Suppose f is a continuously differentiable function on an open set U in $\mathbb{R}^2$ or $\mathbb{R}^3$. Let $\mathbf{F}$ be the vector field defined by $\mathbf{F} = \nabla f$, that is, $\mathbf{F}(x, y, z) = \big(\frac{\partial f}{\partial x}, \frac{\partial f}{\partial y}, \frac{\partial f}{\partial z} \big)$ (or $\mathbf{F}(x, y) = \big(\frac{\partial f}{\partial x}, \frac{\partial f}{\partial y} \big)$ in $\mathbb{R}^2$).

Suppose C is a continuously differentiable curve inside U. Then the line integral $\int_C \mathbf{F} \cdot d\mathbf{r}$ is independent of path, that is, independent of the choice of C.

In fact, if C has initial point P and terminal point Q, then

$$\int_C \mathbf{F} \cdot d\mathbf{r} = \int_C \nabla f \cdot d\mathbf{r} = f\Big|_P^Q = f(Q) - f(P).$$

Of course, if we regard the gradient of f as the analog of the derivative of a function of one variable, then this formula is strongly reminiscent of the Fundamental Theorem of Calculus.

It is not difficult to see that the result is unchanged if the curve C is allowed to have "corners," that is, to be nondifferentiable at finitely many points.

EXAMPLE 2 Evaluate $\int_C \mathbf{F} \cdot d\mathbf{r}$, where $\mathbf{F}(x, y) = (2xy^3, 3x^2y^2)$ and C is a differentiable path running from $(0, 0)$ to $(3, -1)$.

Solution After a moment's thought, we recognize that $\mathbf{F}$ is a gradient, namely, $\mathbf{F} = (2xy^3, 3x^2y^2) = \nabla f$, with $f = x^2y^3$.

Without further ado we can evaluate the integral using Theorem 18.2.2. We find that $\int_C \mathbf{F} \cdot d\mathbf{r} = \int_{(0,0)}^{(3,-1)} \mathbf{F} \cdot d\mathbf{r} = f(3,-1) - f(0,0) = -9$.

If you like physics and have read the material in Section 18.1 about work, you know that one of the uses of line integrals is to calculate the work done on a moving object by a force. The most familiar example is the force of gravity. Somebody lifting a heavy object does work on it, pushing against the object's weight. Near the surface of the earth the force due to gravity points straight down and equals the weight w of the object. This means that the force is the constant vector $\mathbf{F} = (0, 0, -w)$.

Now suppose somebody pushes the object along some path C, parametrized by $\mathbf{r}(t) = \big(x(t), y(t), z(t)\big)$ for $t \in [a, b]$. The work done on the object by gravity during this motion will be

$$\int_C \mathbf{F} \cdot d\mathbf{r} = \int_a^b (0, 0, -w) \cdot \big(x'(t), y'(t), z'(t)\big)\,dt = \int_a^b (-w)z'(t)\,dt$$

$$= -w\int_a^b z'(t)\,dt = -wz(t)\Big|_a^b = -w\big(z(b) - z(a)\big).$$

Since $z(a)$ and $z(b)$ are just the heights of the object at the beginning and end of the path C, the difference $h = z(b) - z(a)$ is just the height through which the object moves.

The work done by gravity on an object rising through a height h is $W = -wh$. The person moving the object has to lift against gravity, that is, in the opposite direction, and so does work equal to $-(-wh) = wh$.

So no matter what path the object follows, to find the work done, it is only necessary to know where it starts and where it ends (and in this case it is actually only necessary to know the two heights). It takes the same amount of work to raise an object 20 ft no matter whether it moves straight up along a vertical line or along a series of ramps or even along a spiral staircase (see Figure 18.2.4).

It is typical of many forces in physics that the corresponding line integrals are independent of path. As an object moves, the force performs work on it, changing its energy, and the energy can be determined from the position of the object. In this situation we speak of the *potential energy* of the object, which refers to the energy it has solely because of its position. (It may also have kinetic energy.)

Now let us work backward. For a force given by $\mathbf{F} = \nabla f$, where f is some function, the work done by $\mathbf{F}$ on an object moving from P to Q is $W = \int_P^Q \mathbf{F} \cdot d\mathbf{r} = \int_P^Q \nabla f \cdot d\mathbf{r} = f(Q) - f(P)$.

We can add a constant to f without changing its gradient, so let us assume that f is zero at the starting point, that is, $f(P) = 0$. This means that the work done by $\mathbf{F}$ equals $f(Q)$. The potential energy of an object at point Q is just the work needed to move it there, which as above is the *negative* of the work done by $\mathbf{F}$: the potential energy of the object when it is at point Q is $-f(Q)$.

Figure 18.2.4

In the world of physics a path-independent line integral $\int_P^Q \nabla f \cdot d\mathbf{r}$ is related to the potential energy $-f$. The minus sign is necessary to make it consistent with our everyday intuition that, for instance, the gravitational potential energy of an object *increases* as it *rises* against the pull of gravity.

Unfortunately, however, the minus sign is annoying, and for many other purposes the function f is called the potential function, *without* the sign. We have to accept the confusion caused by these two different conventions; we will call f the **potential function** and remember that it differs from the potential energy by a sign.

EXAMPLE 3   Gravitational Potential   Consider the force due to gravity for distances that are large relative to the radius of the earth. The force points directly toward the center of the earth, which we place at the origin. Suppose an object of mass m is at the point (x, y, z), and let M be the mass of the earth. The force $\mathbf{F}(x, y, z)$ exerted by the earth on the small object points in the direction of $(-x, -y, -z)$. Its magnitude is $\frac{GMm}{r^2}$, where r is the distance from the center of the earth, that is, $r = \sqrt{x^2 + y^2 + z^2}$, and G is the gravitation constant (cf. Section 6.6, p. 456). From this we find that

$$\mathbf{F}(x, y, z) = -\frac{GMm}{r^2}\frac{1}{r}(x, y, z) = -\frac{GMm}{r^3}(x, y, z)$$

$$= -GMm\left(\frac{x}{(x^2 + y^2 + z^2)^{3/2}}, \frac{y}{(x^2 + y^2 + z^2)^{3/2}}, \frac{z}{(x^2 + y^2 + z^2)^{3/2}}\right).$$

> Note that $-\frac{1}{r}(x, y, z)$ is a unit vector in the direction of the force.

What we would really like is to express this vector field as the gradient of some function, and in fact it is easy to check that $\mathbf{F} = \nabla f$, where

$$f = GMm(x^2 + y^2 + z^2)^{-1/2} = \frac{GMm}{r}.$$

In keeping with the remarks made before Example 3 we find that the potential energy of an object of mass m in the earth's gravitational field is $-f(x, y, z) = -\frac{GMm}{r} = -GMm(x^2 + y^2 + z^2)^{-1/2}$.

> Notice that as the object moves far away from the earth, that is, as $r \to \infty$, the potential energy tends to 0. As the object approaches the earth, its potential energy becomes more negative. We often speak of an object "trapped" in the earth's gravitational field, meaning that it will require some energy to move it free.

We have seen that if $\mathbf{F} = \nabla f$ is the gradient of some function then the line integral $\int_C \mathbf{F} \cdot d\mathbf{r}$ is independent of path. Moreover, the integral can be evaluated using f by a formula that resembles the Fundamental Theorem of Calculus:

$$\int_P^Q \nabla f \cdot d\mathbf{r} = f(Q) - f(P).$$

Now suppose we have a line integral $\int_C \mathbf{F} \cdot d\mathbf{r}$ that we know is path independent. It is reasonable to ask whether or not the vector field $\mathbf{F}$ must be a gradient ∇f and, if so, how we find the function f.

Let us use the preceding formula for inspiration. It suggests that to find the value of f at Q, we should integrate $\int_P^Q \mathbf{F} \cdot d\mathbf{r}$, that is, find a path C running from P to Q and integrate along it (see Figure 18.2.5). Because of the path independence it does not matter which path is used. This should allow us to define a function f, though we still must check whether or not $\mathbf{F}$ is its gradient.

This is a good idea, but there is a complication. We assumed that it is easy to find a path from P to Q, but this may not always be true. For instance, in Figure 18.2.6 we see a set U that consists of two "pieces," and it is clearly impossible to find a continuous path from P to Q that lies *inside* U. This seems like a silly distinction, but it will be important. To make our idea work, it is essential to work with an open set U that is **connected**, meaning that it consists of only one "piece." To make a precise definition of this concept is surprisingly subtle, but for our purposes the idea is perfectly clear from the pictures. In Figure 18.2.7 we see three open sets, of which the first is connected and the others are not.

Figure 18.2.5

Figure 18.2.6

Figure 18.2.7

THEOREM 18.2.3

Suppose the vector field $\mathbf{F}$ is defined on a connected open set U and that the line integral $\int_C \mathbf{F} \cdot d\mathbf{r}$ is independent of path. Define a function f as follows. Pick some point P in U and then for any point Q in U, let

$$f(Q) = \int_P^Q \mathbf{F} \cdot d\mathbf{r} = \int_C \mathbf{F} \cdot d\mathbf{r},$$

where C is a path running from P to Q. Notice that the path independence assures us that it does not matter which path is chosen, and observe that $f(P) = \int_P^P \mathbf{F} \cdot d\mathbf{r} = 0$.
    Then $\mathbf{F} = \nabla f$.

Theorem 18.2.2 says that if $\mathbf{F}$ is the gradient of some function, then $\int_C \mathbf{F} \cdot d\mathbf{r}$ is path independent. Theorem 18.2.3 is the converse of that result; it says that on a connected open set, if $\int_C \mathbf{F} \cdot d\mathbf{r}$ is path independent, then $\mathbf{F}$ must be the gradient of some function f. The function f is called the **potential function** associated to the vector field $\mathbf{F}$.

PROOF   We work with integrals in the plane $\mathbb{R}^2$. If

$$\mathbf{F}(x, y) = \big(u(x, y), v(x, y)\big)$$

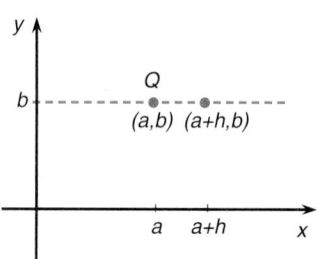

Figure 18.2.8

and $\int_C \mathbf{F} \cdot d\mathbf{r}$ is independent of path, define f as above. Let us calculate $\frac{\partial f}{\partial x}$.
    Fix a point $Q = (a, b)$. To find $\frac{\partial f}{\partial x}(Q) = \lim_{h \to 0} \frac{f(a+h, b) - f(a, b)}{h}$, it is necessary to consider the values $f(a + h, b)$ of f on the horizontal line through $Q = (a, b)$ (see Figure 18.2.8). Actually, what we need is the difference

$$f(a + h, b) - f(a, b) = \int_P^{(a+h, b)} \mathbf{F} \cdot d\mathbf{r} - \int_P^{(a, b)} \mathbf{F} \cdot d\mathbf{r}.$$

These integrals can each be evaluated along any path between their endpoints, but we notice that one way to get from P to $(a + h, b)$ is to take a path from P to $Q = (a, b)$ first and then move horizontally along to $(a + h, b)$. This means that $\int_P^{(a+h, b)} \mathbf{F} \cdot d\mathbf{r} = \int_P^{(a, b)} \mathbf{F} \cdot d\mathbf{r} + \int_{(a,b)}^{(a+h, b)} \mathbf{F} \cdot d\mathbf{r}$. The difference we need is

$$f(a + h, b) - f(a, b) = \int_P^{(a+h, b)} \mathbf{F} \cdot d\mathbf{r} - \int_P^{(a, b)} \mathbf{F} \cdot d\mathbf{r} = \int_{(a,b)}^{(a+h, b)} \mathbf{F} \cdot d\mathbf{r},$$

which is just the integral along the horizontal segment from $Q = (a, b)$ to $(a + h, b)$.
    Parametrizing this segment by $\mathbf{r}(t) = (a + t, b)$, with $t \in [0, h]$, we find that the integral equals

$$\int_{(a,b)}^{(a+h, b)} \mathbf{F} \cdot d\mathbf{r} = \int_0^h \mathbf{F}(a + t, b) \cdot (1, 0)\, dt = \int_0^h u(a + t, b)\, dt.$$

This is just the indefinite integral with respect to t of the function $u(a + t, b)$, so the Fundamental Theorem of Calculus tells us that the limit of the difference quotient is $u(a, b) = u(Q)$.
    We have just shown that $\frac{\partial f}{\partial x}(Q) = u(Q)$, that is, $\frac{\partial f}{\partial x} = u$. An exactly similar calculation shows that $\frac{\partial f}{\partial y} = v$, and together these facts say that $\mathbf{F} = (u, v)$ equals the gradient ∇f.
    The calculation in $\mathbb{R}^3$ is virtually the same.   ◢

Theorem 18.2.3 tells us that path-independent integrals involve gradient vector fields. The difficulty is that it gives us no way to decide whether a particular line integral is or is not path independent.

Suppose $f(x, y)$ is a function on some open set in $\mathbb{R}^2$, and consider its gradient $\nabla f = (u, v) = \left(\frac{\partial f}{\partial x}, \frac{\partial f}{\partial y}\right)$. We know that the mixed partials of f are equal, at least assuming that they are continuous, which means that

$$\frac{\partial^2 f}{\partial x \partial y} = \frac{\partial^2 f}{\partial y \partial x}, \qquad \text{that is,} \qquad \frac{\partial v}{\partial x} = \frac{\partial u}{\partial y}.$$

So for $\mathbf{F} = \big(u(x, y), v(x, y)\big)$ to be the gradient of some function $f(x, y)$ its components u and v must satisfy

$$\frac{\partial v}{\partial x} = \frac{\partial u}{\partial y}. \tag{18.2.4}$$

This gives us a condition that is easily checked; it will tell us that if a vector field does not satisfy this condition, then it is not a gradient, and the corresponding line integral is not independent of path.

For vector fields on open subsets of $\mathbb{R}^3$, all the possible mixed partials must be equal; we find that for $\mathbf{F} = (u, v, w)$ to be a gradient its components must satisfy

$$\frac{\partial v}{\partial x} = \frac{\partial u}{\partial y}, \qquad \frac{\partial v}{\partial z} = \frac{\partial w}{\partial y}, \qquad \frac{\partial w}{\partial x} = \frac{\partial u}{\partial z}. \tag{18.2.5}$$

EXAMPLE 4

(i) The vector field $\mathbf{F}(x, y) = (1, y)$ of Example 1(i) (p. 1151) can be written as $\mathbf{F} = (u, v)$ with $u(x, y) = 1$ and $v(x, y) = y$. We find that $\frac{\partial v}{\partial x} = \frac{\partial y}{\partial x} = 0$ and $\frac{\partial u}{\partial y} = \frac{\partial}{\partial y}(1) = 0$, so Equation 18.2.4 is satisfied with both partials equaling zero.

This encourages us to hope that $\mathbf{F}$ may be a gradient, and with a little thought we realize that $\mathbf{F} = (1, y)$ is the gradient of the potential function $f(x, y) = x + \frac{1}{2}y^2$.

(ii) Consider the vector field $\mathbf{F}(x, y, z) = (z^2 + y \cos x, \sin x, yz)$. Writing it as $\mathbf{F} = (u, v, w)$, we find that $\frac{\partial u}{\partial y} = \cos x$, $\frac{\partial u}{\partial z} = 2z$, $\frac{\partial v}{\partial x} = \cos x$, $\frac{\partial w}{\partial y} = z$, and $\frac{\partial v}{\partial z} = \frac{\partial w}{\partial x} = 0$. We find that the first part of Equations 18.2.5 is true but the other two are not; we conclude that $\mathbf{F}$ is not a gradient and $\int_C \mathbf{F} \cdot d\mathbf{r}$ is not independent of path.

It is tempting to wonder whether or not the condition of Equation 18.2.4 or Equations 18.2.5 is enough to guarantee that a line integral is path independent, that is, that the vector field equals the gradient of some function. Unfortunately, the answer is that it is not quite enough, as the following example illustrates.

EXAMPLE 5 Consider the vector field

$$\mathbf{F}(x, y) = (u, v) = \left(\frac{-y}{x^2 + y^2}, \frac{x}{x^2 + y^2}\right).$$

It is easy to check that $\frac{\partial v}{\partial x} = \frac{(x^2+y^2)-(2x)x}{(x^2+y^2)^2} = \frac{y^2-x^2}{(x^2+y^2)^2}$ and $\frac{\partial u}{\partial y} = \frac{-(x^2+y^2)+(2y)y}{(x^2+y^2)^2} = \frac{y^2-x^2}{(x^2+y^2)^2}$. This shows that $\frac{\partial v}{\partial x} = \frac{\partial u}{\partial y}$, that is, that Equation 18.2.4 is satisfied.

Let us integrate $\int_C \mathbf{F} \cdot d\mathbf{r}$ along the upper semicircle $\mathbf{r}(t) = (\cos t, \sin t)$, for $t \in [0, \pi]$. We find that

$$\int_C \mathbf{F} \cdot d\mathbf{r} = \int_0^\pi \left(\frac{-\sin t}{(\cos^2 t + \sin^2 t)^2}, \frac{\cos t}{(\cos^2 t + \sin^2 t)^2} \right) \cdot (-\sin t, \cos t)\, dt$$

$$= \int_0^\pi (\sin^2 t + \cos^2 t)\, dt = \pi.$$

On the other hand, integrating $\int_C \mathbf{F} \cdot d\mathbf{r}$ along the lower semicircle $\mathbf{r}(t) = (\cos t, -\sin t)$, for $t \in [0, \pi]$ gives

$$\int_C \mathbf{F} \cdot d\mathbf{r} = \int_0^\pi \left(\frac{\sin t}{(\cos^2 t + \sin^2 t)^2}, \frac{\cos t}{(\cos^2 t + \sin^2 t)^2} \right) \cdot (-\sin t, -\cos t)\, dt$$

$$= \int_0^\pi (-\sin^2 t - \cos^2 t)\, dt = -\pi.$$

Both the upper and lower semicircles run from $(1, 0)$ to $(-1, 0)$, but the line integrals along them give different values. The line integral is *not* independent of path, even though it does satisfy Equation 18.2.4.

To try to understand Example 5, we sketch in Figure 18.2.9 the vectors $\mathbf{F}(x, y)$ at a few points (x, y) on the unit circle. They are tangent to the circle and point in

Figure 18.2.9

Tropical storm Javier over the eastern Pacific Ocean. Low level winds are spinning counterclockwise (cyclonic), while cirrus outflow from the thunderstorm in the upper right is moving clockwise (anticyclonic). The circulation shown has a diameter of about 400 nautical miles.

the counterclockwise direction. If we imagine an object moving around the circle, being blown by the force of a swirling hurricane, we notice that the work done will be positive if the object moves with the wind (in this case counterclockwise) and negative if it moves against the wind. This is what the calculation told us in the example.

It is also interesting to notice that if the object went counterclockwise all the way around the circle and back to its starting point, the wind would have done a positive amount of work on it. This is different from the situation for a gradient field, for which no net work would be done.

Example 5 shows that it is possible for a vector field $\mathbf{F}$ to satisfy Equation 18.2.4 or Equations 18.2.5 on some open set U but still not be a gradient, that is, so that the line integral $\int_C \mathbf{F} \cdot d\mathbf{r}$ is not independent of path. In Example 5 we notice that $\mathbf{F}(x, y) = \left(\frac{-y}{x^2+y^2}, \frac{x}{x^2+y^2} \right)$ is not defined at the origin, so the set U on which $\mathbf{F}$ is defined is the *punctured plane* consisting of all points except the origin (see Figure 18.2.10).

If we try to find a potential function f for $\mathbf{F}$ by integrating, as in Theorem 18.2.3, we encounter the following difficulty. Let us start at the point $P = (1, 0)$. To find

Figure 18.2.10

Figure 18.2.11(i)

Figure 18.2.11(ii)

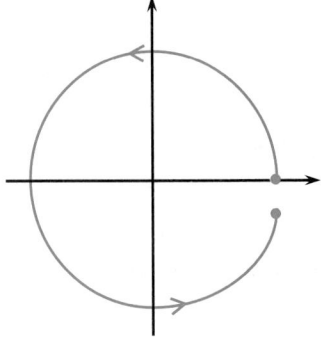

Figure 18.2.11(iii)

$f(Q)$, we want to integrate $\int_C \mathbf{F} \cdot d\mathbf{r}$, where C is some path running from P to Q, as shown in Figure 18.2.11(i).

Now suppose Q is the point $(-1, 0)$. There are many paths from P to Q, but two obvious ones are the upper and lower unit semicircles, as shown in Figure 18.2.11(ii). We have seen that if we integrate over the upper semicircle $\mathbf{r}(t) = (\cos t, \sin t)$, $t \in [0, \pi]$, we obtain $\int_C \mathbf{F} \cdot d\mathbf{r} = \pi$, while the lower semicircle $\mathbf{r}(t) = (\cos t, -\sin t)$, $t \in [0, \pi]$ gives $\int_C \mathbf{F} \cdot d\mathbf{r} = -\pi$.

The line integral has different values depending on which side of the origin the curve goes. There is no single way to choose the path in order to define a potential function. (One suggestion would be to find a path that always goes counterclockwise around the origin, to remove the ambiguity. The trouble is that for points just below P it would be necessary to go almost completely around the origin, as in Figure 18.2.11(iii), so the value of the integral would be nearly 2π. On the other hand, $f(P) = 0$, so the function defined this way is not continuous at P, let alone differentiable ...).

The idea of using line integrals to define a potential function will not necessarily work in situations like this in which there is ambiguity about how to choose a path between P and Q. The problem arises when the vector field $\mathbf{F}$ is defined on a region with *holes* in it. Three examples are shown in Figure 18.2.12; in the first a single point is missing, while in the second and third there are bigger holes. In cases such as these the line integral may be different depending on how the curve goes around the holes.

Figure 18.2.12

Because of this phenomenon it is necessary to consider regions without holes. Disks and rectangles are examples of such regions, and punctured disks are not, but a precise definition is very difficult to make. Fortunately, we can rely on our geometrical intuition to decide whether a region has holes.

DEFINITION 18.2.6

A connected open set U in $\mathbb{R}^2$ is **simply connected** if it has no "holes."

EXAMPLE 6   The first two regions shown in Figure 18.2.13 are not simply connected, but the next three are simply connected.

Note especially that the last region in Figure 18.2.13 is simply connected; there are "peninsulas" cutting into it but no "islands," that is, no holes.

Figure 18.2.13

The importance of simply connected regions comes from the following result, which we do not prove.

> Suppose U is an open set in $\mathbb{R}^2$ that is simply connected, and suppose $\mathbf{F} = (u, v)$ is a continuously differentiable vector field on U satisfying Equation 18.2.4:
>
> $$\frac{\partial v}{\partial x} = \frac{\partial u}{\partial y}.$$
>
> Then $\mathbf{F}$ is a gradient, that is, there is a differentiable potential function f defined on U so that $\mathbf{F} = \nabla f = \operatorname{grad} f$. In particular, $\int_C \mathbf{F} \cdot d\mathbf{r}$ is path independent.

For regions in $\mathbb{R}^3$ the concept of simply connected regions is more subtle, but balls, rectangular solids, ellipsoids, and the whole of $\mathbb{R}^3$ are all simply connected sets. The analog of Theorem 18.2.7 holds for simply connected open sets in $\mathbb{R}^3$.

If we have a vector field $\mathbf{F}$ that satisfies Equation 18.2.4 or Equations 18.2.5 on some simply connected open set U, then we know that $\int_C \mathbf{F} \cdot d\mathbf{r}$ is independent of path and $\mathbf{F} = \nabla f$ for some potential function f, but this leaves the problem of finding f. In practice this is not too difficult.

EXAMPLE 7   **Finding a Potential Function**   Let $\mathbf{F} = (u, v) = (y^2 - 4xy, 2xy - 2x^2 + \cos y)$. It is easy to check that $\frac{\partial v}{\partial x} = 2y - 4x$ and $\frac{\partial u}{\partial y} = 2y - 4x$, so $\frac{\partial v}{\partial x} = \frac{\partial u}{\partial y}$ and $\mathbf{F}$ satisfies Equation 18.2.4. We want to try to find a function f so that $u = \frac{\partial f}{\partial x}$ and $v = \frac{\partial f}{\partial y}$.

To find f satisfying $\frac{\partial f}{\partial x} = u = y^2 - 4xy$, we should perform a *partial integration*. This means holding y fixed, that is, regarding it as a constant, and integrating with respect to x. We find that

$$f = \int (y^2 - 4xy) \, dx = xy^2 - 2x^2 y + C,$$

where C is a "constant." However, in this situation, "constant" means something that does not depend on x, so C may be a function of y. In other words we know that $f = xy^2 - 2x^2 y + h(y)$, for some function $h(y)$.

To find h, we use the other condition, which says that $\frac{\partial f}{\partial y} = v = 2xy - 2x^2 + \cos y$. Knowing that $f = xy^2 - 2x^2 y + h(y)$, we quickly find that $\frac{\partial f}{\partial y} = 2xy - 2x^2 + h'(y)$, so $h'(y)$ must equal $\cos y$, so $h(y) = \sin y + c$, where this time c is a genuine constant.

We have just found that $f(x, y) = xy^2 - 2x^2 y + \sin y + c$, and it is easy to verify that $\nabla f = (y^2 - 4xy, 2xy - 2x^2 + \cos y) = \mathbf{F}$, which is what we wanted.   ◢

EXAMPLE 8   Suppose the force acting on an object moving in space is

$$\mathbf{F}(x, y, z) = (u, v, w) = \left(\frac{x}{\sqrt{x^2 + y^2 + z^2}}, \frac{y}{\sqrt{x^2 + y^2 + z^2}}, \frac{z}{\sqrt{x^2 + y^2 + z^2}} \right).$$

It is possible to evaluate the various partial derivatives needed to check Condition 18.2.5. We find, for instance, that

$$\frac{\partial v}{\partial x} = -\frac{xy}{(x^2 + y^2 + z^2)^{3/2}} = \frac{\partial u}{\partial y}.$$

The other calculations are similar, and Equations 18.2.5 do hold.

This means that $\mathbf{F}$ might be a gradient, but since it is not defined at the origin, we cannot be sure. On the other hand, we could just try to find the potential function f.

For instance, we want $\frac{\partial f}{\partial x}$ to equal $u = \frac{x}{\sqrt{x^2+y^2+z^2}}$. Regarding y and z as constants, we perform a partial integration. Making the substitution $U = x^2 + y^2 + z^2$, $dU = 2x\,dx$, we find that

$$f = \int \frac{x}{\sqrt{x^2 + y^2 + z^2}}\,dx = \frac{1}{2}\int \frac{dU}{\sqrt{U}} = \sqrt{U} + C,$$

where C is a "constant," meaning that it is a function of y and z but not of x.

We can write $f = \sqrt{x^2 + y^2 + z^2} + c(y, z)$. Taking the partial derivative with respect to y and comparing with the function $v = \frac{y}{\sqrt{x^2+y^2+z^2}}$, we find that $v = \frac{\partial f}{\partial y} = \frac{y}{\sqrt{x^2+y^2+z^2}} + \frac{\partial c}{\partial y}$, so $\frac{\partial c}{\partial y} = 0$.

An entirely similar calculation shows that $\frac{\partial c}{\partial z} = 0$, so c is actually a genuine constant and not really a function of y or z.

Of course, it is now easy to see that $\mathbf{F} = \nabla f$, where $f = \sqrt{x^2 + y^2 + z^2}$. (We may as well let $c = 0$.)

The potential energy of an object moving in this force field is $-f = -\sqrt{x^2 + y^2 + z^2}$, which is zero at the origin and more and more negative the farther we go from the origin. If we look at the force $\mathbf{F}$, we see that it points directly away from the origin, so this is an example of a repulsive force. It takes energy to move an object close to the origin, so the potential energy is high near the origin and lower farther away.

We can calculate that the work needed to move the object from the point $(4, -2, -4)$ to the point $(-3, 0, 4)$ is $-f(-3, 0, 4) - (-f)(4, -2, -4) = -\sqrt{9 + 16} + \sqrt{16 + 4 + 16} = -5 + 6 = 1$ unit. Notice that it really depends only on how far each point is from the origin.

This is the work done by somebody opposing $\mathbf{F}$. The work done by $\mathbf{F}$ is -1 unit.

> Example 8 illustrates that if we find a vector field for which Condition 18.2.4 or 18.2.5 holds, then even if we do not know that the region is simply connected, it might be that the vector field is a gradient, and we can attempt to find out by performing the integration.
>
> The startling thing about this example is that the set consisting of $\mathbb{R}^3$ without the origin is in fact a simply connected set. This means that the three-dimensional version of Theorem 18.2.7 does apply after all and tells us that there must be a potential function. However, since we have not really discussed simply connected regions in $\mathbb{R}^3$, we are not in a position to know this. In $\mathbb{R}^3$, "simply connected" does not mean "without holes."

Vector fields that are gradients, that is, that correspond to path-independent line integrals, are sometimes called **conservative vector fields**. The terminology comes from physics, in which motion in a conservative force satisfies the principle of *conservation of energy*. In a nonconservative force, the total energy of an object may change. (For instance, an object starting at rest in a swirling hurricane may be moved by the wind until it returns to its original position. Its potential energy is unchanged because it is at the same position, but its kinetic energy has increased, since it is moving.)

Exercises 18.2

I

Determine whether each of the following line integrals is path independent, and evaluate those that are.

1. $\int_{(0,0)}^{(1,1)} \mathbf{F} \cdot d\mathbf{r}, \mathbf{F} = (x^2, 2y)$

2. $\int_{(0,0)}^{(2,1)} \mathbf{F} \cdot d\mathbf{r}, \mathbf{F} = (2xy, x^2)$

3. $\int_{(0,0)}^{(3,2)} \mathbf{F} \cdot d\mathbf{r}, \mathbf{F} = (xy, xy)$

4. $\int_{(0,0)}^{(2,-2)} \mathbf{F} \cdot d\mathbf{r}, \mathbf{F} = (\sin \pi x, 0)$

5. $\int_{(2,-1)}^{(1,2)} \mathbf{F} \cdot d\mathbf{r}, \mathbf{F} = \left(\frac{1}{x}, 2x - 2y\right)$

6. $\int_{(5,5)}^{(5,5)} \mathbf{F} \cdot d\mathbf{r}, \mathbf{F} = \left(\exp(x^2 + 1), \arctan y^3\right)$

7. $\int_{(0,0,0)}^{(1,3,4)} x^2 \, dx + 6y \, dy - \sin \pi z \, dz$

8. $\int_{(1,3,2)}^{(-1,-1,2)} yz \, dx + xz \, dy + xy \, dz$

9. $\int_{(-2,3,1)}^{(0,2,1)} \left(2x + yze^{xz}\right) dx + e^{xz} \, dy + xye^{xz} \, dz$

10. $\int_{(0,0,0)}^{(a,b,c)} dx + xz \, dy + xy \, dz$

For each of the following vector fields, decide whether or not it has a potential function, and find a potential function if there is one.

11. $\mathbf{F} = (\cos x, e^y)$

12. $\mathbf{F} = (x - 2y, -2x + 2y)$

13. $\mathbf{F} = (2xy^3 - 2y^2 + 6x, 3x^2y^2 - 4xy - 2)$

14. $\mathbf{F} = (-y, x)$

15. $\mathbf{F} = (e^{xy}, e^{xy})$

16. $\mathbf{F} = \left(2x \sin y - (1 + xy)e^{xy}, x^2 \cos y - x^2 e^{xy} + 2\right)$

17. $\mathbf{F} = \left(\frac{y}{1+x^2}, \arctan x + e^y\right)$

18. $\mathbf{F} = (xz + yz \cos xyz, xz \cos xyz, x + xy \cos xyz)$

19. $\mathbf{F} = (3, 3, 3)$

20. $\mathbf{F} = \left(\frac{1}{x}, \frac{1}{y}, \frac{1}{z}\right)$

II

21. With $\mathbf{G}$ as in Example 1(ii) (p. 1151), evaluate $\int_C \mathbf{G} \cdot d\mathbf{r}$, where C is the part of the graph of $y = x^3$ running from $(0,0)$ to $(1,1)$.

22. With $\mathbf{G}$ as in Example 1(ii), evaluate $\int_C \mathbf{G} \cdot d\mathbf{r}$, where C is the part of the graph of $x = y^4$ running from $(0,0)$ to $(1,1)$.

23. Consider the vector field $\mathbf{F}(x, y) = (axy, a^2x^2)$. For what value(s) of a is this vector field equal to the gradient of some function? Find the corresponding function(s).

24. Consider the vector field $\mathbf{F}(x, y) = (ay^2z + bz, cxyz, 2xy^2 + 3x)$. For what value(s) of a, b, and c is this vector field equal to the gradient of some function? Find the corresponding function(s).

25. With the vector field $\mathbf{F}$ of Example 5 (p. 1157), evaluate $\int_C \mathbf{F} \cdot d\mathbf{r}$, where C is the curve that goes twice around the unit circle $x^2 + y^2 = 1$ in the clockwise direction starting at $(0, 1)$.

26. With the vector field $\mathbf{F}$ of Example 5, evaluate $\int_C \mathbf{F} \cdot d\mathbf{r}$, where C is the curve that goes five times around the circle $x^2 + y^2 = 4$ in the counterclockwise direction starting at $(0, 2)$.

*27. Give an example of two connected open sets U and V in $\mathbb{R}^2$ whose intersection $U \cap V$ is not connected. Your answer can consist of a drawing.

*28. Give an example of two connected open sets U and V in $\mathbb{R}^2$, neither of which is simply connected but whose intersection $U \cap V$ is simply connected. Your answer can consist of a drawing.

*29. In Example 5 we considered the vector field $\mathbf{F} = \left(\frac{-y}{x^2+y^2}, \frac{x}{x^2+y^2}\right)$ defined on the punctured plane (without the origin). We found that it satisfies Equation 18.2.4 but the corresponding integral is not path independent. However, if we let U be the part of the plane to the right of $x = 1$, that is, $U = \{(x, y) : x > 1\}$, then U is simply connected. Theorem 18.2.7 says that $\mathbf{F}$ *is* a gradient on U. Find a potential function for $\mathbf{F}$ on U.

**30. Suppose U is a connected open set in $\mathbb{R}^2$, and suppose the vector field $\mathbf{F}$ is defined on U. Suppose that P and Q are two points in U, and suppose that $\int_P^Q \mathbf{F} \cdot d\mathbf{r}$ is independent of path, that is, that its value is the same for any path running from P to Q. Explain why for *any* points A and B in U the line integral $\int_A^B \mathbf{F} \cdot d\mathbf{r}$ must be independent of path. (*Hint:* Try drawing some pictures.) This shows that independence of path can be established just by checking it for paths between two convenient points instead of having to check it for all pairs of points.

**31. Suppose $\mathbf{F}$ is a vector field on an open set U in $\mathbb{R}^2$ having the property that $\int_C \mathbf{F} \cdot d\mathbf{r} = 0$ for every *closed* differentiable path C in U. (A path is closed if its initial and terminal points are the same.) Show that $\int \mathbf{F} \cdot d\mathbf{r}$ is path independent for *any* differentiable path in U.

**32. (i) Consider an object that weighs $m = 1$ lb on the surface of the earth. Assume that the earth's radius is 4000 miles, and calculate the constant GMm mentioned in Example 3 (p. 1155).

(ii) Figure out how much energy it would take to lift the object out of the earth's gravitational field. The idea is to figure out the potential energy at a distance r from the center of the earth, and figure out what the potential energy would be if the object was completely away from the earth's influence, that is, as $r \to \infty$.

If a rocket is launched, it may fall back to the earth or fall into an orbit about the earth; in either case it is "trapped" in the earth's gravity. But if it goes fast enough, it will have enough kinetic energy to break free and keep going off into space. The minimum velocity that is sufficient is called the *escape velocity*, and it is the velocity at which the object's kinetic energy equals the amount needed to break loose. This is exactly the quantity we calculated in part (ii), and you can use your answer to part (ii) to find the escape velocity for an object at the earth's sur-

face. You can also check that it will be the same no matter how much the object weighs.

Use a graphics calculator or computer to sketch the following pairs of curves. In each case, let R be the region enclosed between the curves but not containing any points on the curves. Is R connected?

33. $y = x^4 - 2x^2 + 1, \quad y = x^3 - 2x + 1$

34. $y = x^4 - 2x^2 + 1, \quad y = x^3 - 2x + 2$

35. $y = \sin x, \quad y = \frac{x^2}{10} - \frac{1}{2}$

36. $y = \cos x, \quad y = \frac{x^2}{20} - \cos x$

POINT TO PONDER

The concept of a simply connected region is extremely important, and we have seen that it is related to path independence of line integrals. The geometric picture of a region in the plane without holes is quite easy, but it is surprisingly difficult to make a precise mathematical definition.

It turns out that the way to do it is to take two points P and Q in a region U, and consider paths that run from P to Q, as shown in Figure 18.2.14. We imagine one of

Figure 18.2.14

these paths made out of a flexible rubbery material, with its ends firmly attached at P and Q. What we want to do is to move the rubber curve from its original position until it becomes another curve; we speak of "deforming" the first curve into the second.

In Figure 18.2.15 we see a number of the stages as the straight line segment from $(-1, 0)$ to $(1, 0)$ is deformed into the upper semicircle. Notice how at each stage the endpoints are still the same points $(-1, 0)$ and $(1, 0)$.

Figure 18.2.15

Now suppose we have a set U, shown as the shaded region in Figure 18.2.16, and two points P and Q in it. Now the idea is to deform a path *without letting any part of*

Figure 18.2.16

it move outside U. In the illustration the upper semicircle *cannot* be deformed into the lower semicircle (keeping the ends fixed) without some of it passing outside the region (probably into the hole in the middle).

So a simply connected region can be defined to be one in which any path joining P to Q can be deformed into any other path joining P to Q. We have just seen why the presence of holes in U makes this impossible, and this is why for regions in the plane, simply connected means "without holes."

After Example 8 (p. 1160) we made the unexpected observation that $\mathbb{R}^3$ without the origin is simply connected. Imagining a path joining two points P in Q in $\mathbb{R}^3$ *without going through the origin*, we can see from Figure 18.2.17 that it can be deformed into any other path from P to Q without ever touching the origin. Somehow there is more room to maneuver in $\mathbb{R}^3$ than there is in $\mathbb{R}^2$, and

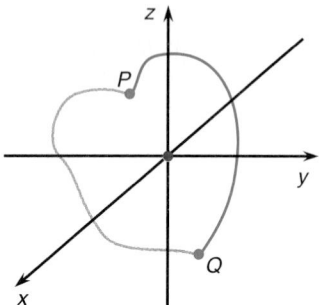

Figure 18.2.17

this is why $\mathbb{R}^3$ without the origin is a simply connected set.

18.3 Green's Theorem

SIMPLE CLOSED PATH
LOOP
OUTER NORMAL
FLUX
    DIVERGENCE
CIRCULATION
CURL

In Section 18.2 we saw that certain line integrals could be evaluated by a formula that looks like the formula we get from the Fundamental Theorem of Calculus for integrals of functions of a single variable. Of course, this result is of limited value because it only applies to line integrals of gradient vector fields.

In this section we discuss another result that allows us to evaluate many line integrals; this result applies to integrals over **closed paths**, that is, curves that start and end at the same point. At the end of the section we will use this theory to study fluid flows.

A closed path is **simple** if it does not cross itself. A simple closed path is often called a **loop** (see Figure 18.3.1). We will usually work with simple closed paths, but observe that this is less restrictive than it sounds because a path that is not simple, such as the path in Figure 18.3.2, can be constructed out of simple closed curves (the two "lobes" of the original curve).

Consider a simple closed curve C that is made up of the graphs of two functions $y = u(x)$ and $y = v(x)$, for $x \in [a, b]$, with $u(x) \leq v(x)$. Let us write C_1 for the bottom graph and C_2 for the top graph (see Figure 18.3.3). We can parametrize

Figure 18.3.1

Figure 18.3.2

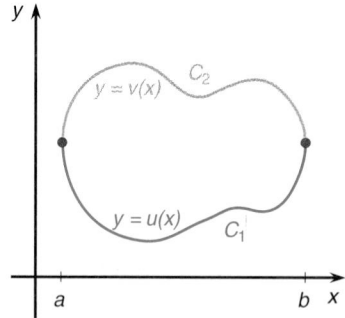

Figure 18.3.3

C_1 by $\mathbf{r}_1(x) = \big(x, u(x)\big)$, with $x \in [a, b]$, and C_2 by $\mathbf{r}_2(x) = \big(x, v(x)\big)$, with $x \in [a, b]$. Notice that these parametrizations use x as the parameter and traverse the curves C_1 and C_2 from left to right.

Now suppose $f(x, y)$ is a differentiable function, and consider the line integral $\int_C f(x, y)\, dx$, where C is *counterclockwise*. To evaluate it, we can integrate first along C_1 and then along the top part of C. This is the curve C_2, *except* that in going around C counterclockwise we move from right to left along the top of the curve, so we integrate along C_1 and then along $-C_2$. We find

$$\int_C f(x, y)\, dx = \int_{C_1} f(x, y)\, dx + \int_{-C_2} f(x, y)\, dx = \int_{C_1} f(x, y)\, dx - \int_{C_2} f(x, y)\, dx$$

$$= \int_a^b f\big(x, u(x)\big)\, dx - \int_a^b f\big(x, v(x)\big)\, dx$$

$$= \int_a^b \Big(f\big(x, u(x)\big) - f\big(x, v(x)\big)\Big)\, dx.$$

Now the integrand $f\big(x, u(x)\big) - f\big(x, v(x)\big)$ can be written as $-f(x, y)\big|_{y=u(x)}^{y=v(x)}$, and this equals $\int_{u(x)}^{v(x)} -\frac{\partial f}{\partial y}(x, y)\, dy$. Substituting into the above expression for the line integral, we find that

$$\int_C f(x, y)\, dx = \int_a^b \left(\int_{u(x)}^{v(x)} -\frac{\partial f}{\partial y}\, dy\right) dx.$$

This is the double integral of $-\frac{\partial f}{\partial y}$ over the region contained between the graphs of $y = u(x)$ and $y = v(x)$. Calling this region R, we have found that

$$\int_C f(x, y)\, dx = \int\int_R \left(-\frac{\partial f}{\partial y}\right) dx\, dy = \int\int_R \left(-\frac{\partial f}{\partial y}\right) dA.$$

This formula relates a line integral over the closed path C to an area integral over the region R contained inside C.

If a curve C is constructed from the graphs of two functions $x = U(y)$ and $x = V(y)$ for $y \in [c, d]$, with $U(y) \le V(y)$, as illustrated in Figure 18.3.4, then it is possible to make a similar calculation for the line integral $\int_C g(x, y)\, dy$. We parametrize the right graph C_1 by $\mathbf{r}_1(y) = \big(V(y), y\big)$ and the left graph C_2 by $\mathbf{r}_2(y) = \big(U(y), y\big)$ and find that

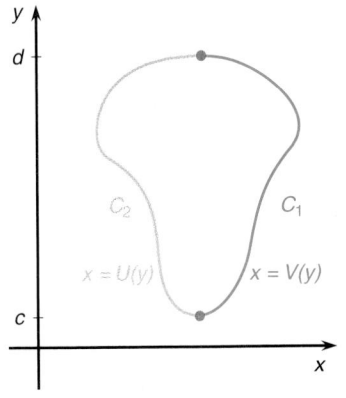

Figure 18.3.4

$$\int_C g(x, y)\, dy = \int_{C_1} g(x, y)\, dy - \int_{C_2} g(x, y)\, dy = \int_c^d g\big(V(y), y\big)\, dy - \int_c^d g\big(U(y), y\big)\, dy$$

$$= \int_c^d \Big(g\big(V(y), y\big) - g\big(U(y), y\big)\Big)\, dy = \int_c^d \left(\int_{x=U(y)}^{x=V(y)} \frac{\partial g}{\partial x}\, dx\right) dy$$

$$= \int\int_R \frac{\partial g}{\partial x}\, dx\, dy,$$

where, as before, R is the region inside C.

If we have a simple closed curve C that can be made up of two graphs $y = u(x)$ and $y = v(x)$ but can also be made up of two graphs in the other direction $x = U(y)$ and $x = V(y)$, then these two results can be combined to give a formula for the line integral $\int_C f(x, y)\, dx + g(x, y)\, dy$. This formula is Green's Theorem, named after the self-taught English physicist and mathematician George Green (1793–1841). It turns out to be true even without the assumption that C can be constructed from two graphs in both the horizontal and vertical directions.

THEOREM 18.3.1

Green's Theorem

Suppose C is a simple closed curve, traversed in the counterclockwise direction and with a parametrization $\mathbf{r}(t)$ that is continuously differentiable at all but (perhaps) a finite number of points. Let R be the region inside C. If $f(x, y)$ and $g(x, y)$ have continuous partial derivatives on some open set containing R, then

$$\int_C f(x, y)\, dx + g(x, y)\, dy = \int \int_R \left(\frac{\partial g}{\partial x} - \frac{\partial f}{\partial y} \right) dA.$$

Figure 18.3.5(i)

PROOF We have already seen how to prove the result when the region R can be expressed as the region between two graphs in both the vertical and horizontal directions.

In Figure 18.3.5(i) we see a region that does not lie between two graphs in the vertical direction, but in Figure 18.3.5(ii) we cut it into two pieces, each of which is the region between two graphs in both directions. This means that Green's Theorem holds for each of these two pieces. Let us write R_1 and R_2 for the two regions, as shown, and C_1 and C_2 for their boundary curves. Each of C_1 and C_2 is traversed counterclockwise; notice in particular the part of each curve that lies along the cut that separates R_1 from R_2. As part of C_1 it is traversed from left to right, but as part of C_2 it is traversed from right to left. We know that

$$\int_{C_1} f(x, y)\, dx + g(x, y)\, dy = \int \int_{R_1} \left(\frac{\partial g}{\partial x} - \frac{\partial f}{\partial y} \right) dx\, dy$$

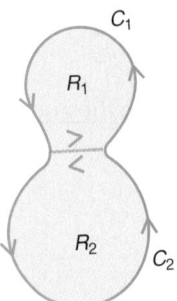

Figure 18.3.5(ii)

and a similar formula with C_1 and R_1 replaced by C_2 and R_2, respectively. Adding these two formulas gives

$$\int_{C_1} f(x, y)\, dx + g(x, y)\, dy + \int_{C_2} f(x, y)\, dx + g(x, y)\, dy$$

$$= \int \int_{R_1} \left(\frac{\partial g}{\partial x} - \frac{\partial f}{\partial y} \right) dx\, dy + \int \int_{R_2} \left(\frac{\partial g}{\partial x} - \frac{\partial f}{\partial y} \right) dx\, dy.$$

The right side of this equation is the sum of integrals over R_1 and R_2, which is just the integral over all of R. The left side integrates along C_1 and C_2. From the picture we see that this includes integrating all around the original curve C, but it also includes two integrals along the little cut that separates R_1 from R_2. However, those two integrals along the cut are in *opposite directions*, so one is the negative of the other, and they cancel.

This means that the left side of the equation equals $\int_C f(x, y)\, dx + g(x, y)\, dy$, and Theorem 18.3.1 is proved for this region.

Similar arguments can be made for more complicated regions. In Figure 18.3.6 we see a region that has been cut in three places to divide it into pieces; each of the pieces lies between two graphs both horizontally and vertically. Green's Theorem holds for each piece, and these results can be combined to get the result for the large region. The key once again is noticing that the line integrals along the cuts are in opposite directions, so they all cancel out. This means that the line integrals over the boundaries of the four pieces add up to the line integral around the boundary of the big piece.

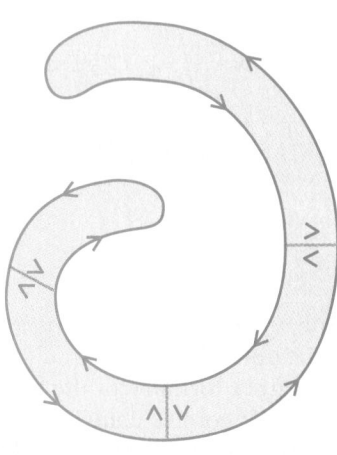

Figure 18.3.6

There is a temptation to try to remember Green's Theorem by abbreviating $\int_C g \, dy = \int \int_R \frac{\partial g}{\partial x} \, dx \, dy$ by writing $g \, dy = \frac{\partial g}{\partial x} \, dx \, dy$, and noticing that this formula looks a lot like simply canceling dx with $\frac{1}{\partial x}$. The trouble is that the other part of Green's Theorem includes a minus sign: $\int_C f \, dx = - \int \int_R \frac{\partial f}{\partial y} \, dx \, dy$, so it is more complicated. In effect what we have to remember is which of the two has the minus sign.

Green's Theorem can be used in many different ways, but of course one way is for evaluating line integrals around closed curves.

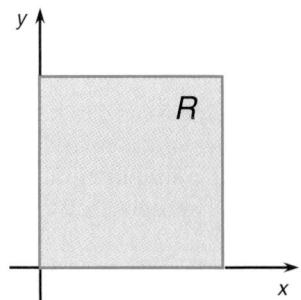

Figure 18.3.7

EXAMPLE 1   Evaluate $\int_C (2x - 4y^2 + e^{x+y}) \, dx + (\arctan y + e^{x+y}) \, dy$, where C is the curve going clockwise around the square with corners $(0,0), (0,1), (1,1), (1,0)$.

Solution   We apply Green's Theorem with $f = 2x - 4y^2 + e^{x+y}$ and $g = \arctan y + e^{x+y}$. The curve C is *clockwise*, so $-C$ is counterclockwise. Since Green's Theorem works for counterclockwise curves, we find that if R is the region enclosed inside the square, as shown in Figure 18.3.7,

$$\int_C (2x - 4y^2 + e^{x+y}) \, dx + (\arctan y + e^{x+y}) \, dy = - \int_{-C} f \, dx + g \, dy$$

$$= - \int \int_R \left(\frac{\partial g}{\partial x} - \frac{\partial f}{\partial y} \right) dx \, dy = - \int \int_R \left((e^{x+y}) - (-8y + e^{x+y}) \right) dx \, dy$$

$$= - \int_0^1 \int_0^1 (8y) \, dx \, dy = - \int_0^1 8y \, dy = -4.$$

By using Green's Theorem we avoided the difficulty of integrating $\arctan y$ and e^{x+y}.

Notice the way we had to change the sign to get a curve going counterclockwise; this step is easy to forget, but the signs in Green's Theorem are only correct for a counterclockwise curve.

EXAMPLE 2   Evaluate $\int_C \pi y \cos \pi x \, dx + \sin \pi x \, dy$, where C is the ellipse $4x^2 + 9y^2 = 72$, traversed counterclockwise.

Solution   Green's Theorem tells us that if R is the region inside the ellipse,

$$\int_C \pi y \cos \pi x \, dx + \sin \pi x \, dy = \int \int_R (\pi \cos \pi x - \pi \cos \pi x) \, dA$$

$$= \int \int_R 0 \, dx \, dy = 0.$$

The value of the line integral around *any* closed curve equals 0.

REMARK 18.3.2

In Example 2 we evaluated a line integral around a closed curve and found that it equals zero. On a simply connected region an integral with this property must be independent of path (cf. Exercise 31 of Section 18.2, p. 1151).

Suppose $p(x, y)$ is some differentiable function on an open set in the plane, and let $\mathbf{F} = \nabla p = \left(\frac{\partial p}{\partial x}, \frac{\partial p}{\partial y} \right)$ be the corresponding gradient field. If C is a closed counterclockwise curve, we can apply Green's Theorem to the line integral $\int_C \mathbf{F} \cdot d\mathbf{r}$ and find that if R is the region inside C:

$$\int_C \mathbf{F} \cdot d\mathbf{r} = \int_C \frac{\partial p}{\partial x}\, dx + \frac{\partial p}{\partial y}\, dy = \int \int_R \left(\frac{\partial^2 p}{\partial x \partial y} - \frac{\partial^2 p}{\partial y \partial x} \right) dx\, dy.$$

If the function $p(x, y)$ has continuous second partial derivatives, then we know (cf. Theorem 16.3.5) that the mixed partials are equal and the expression in the parentheses in the last integral equals zero.

What we find is that when a line integral comes from a gradient vector field $\mathbf{F}$, then Green's Theorem amounts to integrating the difference of the mixed partials. The fact that this difference is zero explains why such a line integral over a closed curve must equal zero, which is consistent with Theorem 18.2.2 about path independence.

The examples we have considered so far use Green's Theorem to evaluate a line integral by converting it into an integral over a region in the plane. Of course, there are times when it can be used the other way around to convert a double integral into a line integral.

One simple instance involves the integral $\int \int_R 1\, dx\, dy$, which evaluates the area of R. Looking at Green's Theorem, we see that if C is the boundary of R in the counterclockwise direction, then

$$Area(R) = \int \int_R 1\, dx\, dy = \int_C x\, dy = \int_C (-y)\, dx.$$

There are other possible formulas obtained by combining these. We have just proved the following result.

THEOREM 18.3.3

If C is a continuously differentiable curve in the plane $\mathbb{R}^2$, counterclockwise, and R is the region enclosed inside C, then

$$Area(R) = \int_C x\, dy = \int_C (-y)\, dx = \frac{1}{2} \int_C x\, dy - y\, dx.$$

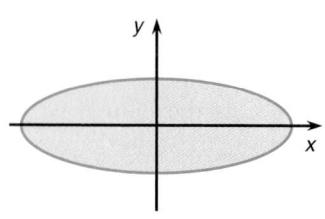

Figure 18.3.8

EXAMPLE 3 Find the area inside the ellipse $\frac{x^2}{a^2} + \frac{y^2}{b^2} = 1$ that is shown in Figure 18.3.8.

Solution Parametrizing the ellipse by $\mathbf{r}(t) = (a \cos t, b \sin t)$, with $t \in [0, 2\pi]$, we find that the area equals

$$\frac{1}{2} \int_C x\, dy - y\, dx = \frac{1}{2} \int_0^{2\pi} \left((a \cos t)(b \cos t) - b \sin t(-a \sin t) \right) dt$$

$$= \frac{1}{2} ab \int_0^{2\pi} (\cos^2 t + \sin^2 t)\, dt = \frac{1}{2} ab(2\pi) = \pi ab.$$

Figure 18.3.9(i)

Figure 18.3.9(ii)

Figure 18.3.10(i)

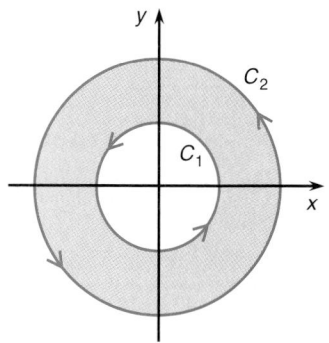

Figure 18.3.10(ii)

We find the familiar result that the area of the ellipse is πab, but this is easier than our earlier methods (cf. Example 6.2.1, p. 425).

It is also possible to use Green's Theorem on some more complicated regions R, including regions with holes in them, as shown in Figure 18.3.9(i).

To make this work, it is necessary to consider the *boundary* of the region, and this will be not just the outer curve that surrounds R but also the inner curves that separate R from the points that are not in R. The trouble is that Green's Theorem tells us that an integral over one of those inner curves is related to a double integral over the region *inside* that curve, and this is part of the region that is *not* in R.

When we think of one of the inner curves shown in Figure 18.3.9(i) as part of the boundary of R, then it must be traversed in the *clockwise* direction, as shown in Figure 18.3.9(ii), where the outer curve is traversed counterclockwise as usual.

A boundary curve with directions fixed in this way is said to be **positively oriented**.

A good way to remember the convention is this: The direction of each piece of the boundary curve is chosen so that as a point moves along it, the region R is on the *left* side. You should check this with the picture.

EXAMPLE 4   Let R be the region consisting of all points (x, y) satisfying

$$1 \le x^2 + y^2 \le 4,$$

that is, all points lying between the circles of radii 1 and 2 around the origin (see Figure 18.3.10(i)). Let C be its boundary curve, with positive orientation, and consider the line integral $\int_C (-x^2 y)\, dx + xy^2\, dy$.

On the one hand we could evaluate it by parametrizing each part of the boundary curve. The outer circle is traversed counterclockwise, so it can be parametrized by $\mathbf{r}_1(t) = (2\cos t, 2\sin t)$, with $t \in [0, 2\pi]$. The inner circle, traversed clockwise, can be parametrized by $\mathbf{r}_2(t) = (\cos t, -\sin t)$, with $t \in [0, 2\pi]$.

The line integral is fairly complicated to evaluate.

On the other hand, Green's Theorem tells us that the line integral equals

$$\int\!\!\int_R (y^2 + x^2)\, dx\, dy = \int_0^{2\pi}\!\!\int_1^2 (r^2)\, r\, dr\, d\theta = \int_0^{2\pi}\!\!\int_1^2 r^3\, dr\, d\theta$$

$$= \int_0^{2\pi} \left(\frac{1}{4} r^4 \Big|_1^2 \right) d\theta = \int_0^{2\pi} \frac{15}{4}\, d\theta = \frac{15\pi}{2}.$$

Let us write C_2 for the outer circle of radius 2 and C_1 for the inner circle of radius 1, both *counterclockwise* as shown in Figure 18.3.10(ii). Then the line integral from Example 4 is $\int_{C_2} (-x^2 y)\, dx + xy^2\, dy - \int_{C_1} (-x^2 y)\, dx + xy^2\, dy$.

Let us consider the two parts of this integral separately. Since C_2 is the boundary of the disk D_2 of radius 2, Green's Theorem applied to C_2 alone tells us that $\int_{C_2} (-x^2 y)\, dx + xy^2\, dy = \int\!\int_{D_2} (x^2 + y^2)\, dx\, dy$. Similarly, $\int_{C_1} (-x^2 y)\, dx + xy^2\, dy = \int\!\int_{D_1} (x^2 + y^2)\, dx\, dy$ (D_1 is the disk of radius 1).

The line integral over the outer circle C_2 equals the integral of $x^2 + y^2$ over the large disk D_2. We want to integrate over R, which is the points in D_2 that are not in D_1, so we should *subtract* the integral over D_1, which equals the line integral over C_1.

The point of this discussion is that the integral over R equals the *difference* of the two line integrals. It is because of this that the inner circle should be in the clockwise direction so that the corresponding integral will have a negative sign.

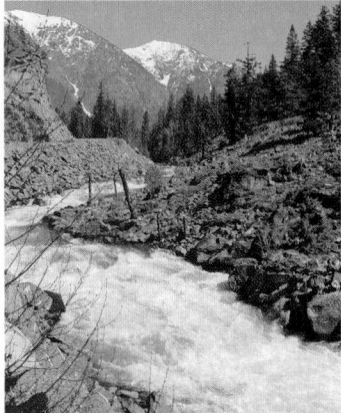

*Rapids in the Icicle River in the
state of Washington.*

Fluid Flows

When we draw a vector field in the plane, as in Figure 18.3.11, we imagine a vector at each point. It might, for instance, represent something that has a different velocity at each point; we can imagine the wind swirling around an obstacle.

Another example would be to imagine a thin layer of some fluid flowing over a flat plane. It might be a puddle of water or oil. If the fluid is in motion, it may move in complicated ways, with different directions and speeds at different points.

It is interesting to interpret Green's Theorem for a vector field $\mathbf{F}$ that represents the velocity of a moving fluid at each point.

In Figure 18.3.12 we see a simple closed curve C in the plane and arrows representing a vector field $\mathbf{F}$ that we will visualize as the velocity of a moving fluid. At different points the fluid flows at different speeds and in different directions.

Now let us parametrize the curve C using the arclength s as the parameter. Suppose it is parametrized in the counterclockwise direction by $\mathbf{r}(s) = \big(x(s), y(s)\big)$, with $s \in [a, b]$. The unit tangent vector is $\mathbf{T}(s) = \big(x'(s), y'(s)\big)$. Thinking of the plane $\mathbb{R}^2$ contained as the horizontal plane in $\mathbb{R}^3$ and letting $\mathbf{k} = (0, 0, 1)$ be the standard basis vector, we see from Figure 18.3.13 that $\mathbf{T} \times \mathbf{k}$ is the unit normal vector $\mathbf{n}$ to the curve. Using the picture and the right-hand rule, we see that $\mathbf{n} = \mathbf{T} \times \mathbf{k}$ is the normal vector that points "out," that is, away from the region R enclosed inside C. This vector is called the **outer normal** to the curve.

It is easy to calculate that

$$\mathbf{n} = \mathbf{T} \times \mathbf{k} = \big(x'(s), y'(s), 0\big) \times (0, 0, 1) = \big(y'(s), -x'(s), 0\big),$$

and thinking of this as a vector in $\mathbb{R}^2$ again, we see that the outer normal vector is $\mathbf{n} = \big(y'(s), -x'(s)\big)$.

Now consider the region R contained inside the curve C, and suppose we want to know how much fluid is flowing into or out of R. In effect this means the amount of fluid flowing across the curve C, which means the amount flowing in the direction of the normal vector (see Figure 18.3.14).

The *normal component* $\mathbf{F} \cdot \mathbf{n}$ of the velocity tells us the rate of flow in the direction of the outer normal. Where it is positive, the fluid has a net outward flow, and where it is negative, the flow is inward. We integrate $\int_C \mathbf{F} \cdot \mathbf{n}\, ds$ to get the total net flow out of the region R. This total net flow is called the **flux** of the fluid across the curve C.

If $\mathbf{F}(x, y) = \big(f(x, y), g(x, y)\big)$, then the flux is

$$\int_C \mathbf{F} \cdot \mathbf{n}\, ds = \int_a^b \Big(f\big(x(s), y(s)\big), g\big(x(s), y(s)\big)\Big) \cdot \big(y'(s), -x'(s)\big)\, ds$$

$$= \int_a^b f\big(x(s), y(s)\big)y'(s) + g\big(x(s), y(s)\big)\big(-x'(s)\big)\, ds = \int_C f\, dy - g\, dx.$$

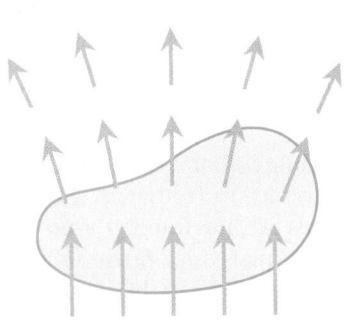

Figure 18.3.13

Figure 18.3.14

We make the following definition.

DEFINITION 18.3.4

Suppose C is a simple closed curve in the plane, with outer unit normal $\mathbf{n}$, and let $\mathbf{F}(x, y) = \big(f(x, y), g(x, y)\big)$ be a vector field. The **flux** of the vector field $\mathbf{F}$ across the curve C is defined to be the line integral

$$\int_C \mathbf{F} \cdot \mathbf{n}\, ds = \int_C (-g)\, dx + f\, dy.$$

Applying Green's Theorem to this expression, we find that the flux of $\mathbf{F}$ across C equals

$$\int_C (-g)\, dx + f\, dy = \int\int_R \left(\frac{\partial f}{\partial x} + \frac{\partial g}{\partial y} \right) dx\, dy.$$

The integrand in this last integral is called the *divergence* of the vector field. The idea is that its value at a point measures the rate at which the fluid is spreading out or *diverging* at that point.

DEFINITION 18.3.5

If $\mathbf{F}(x, y) = \big(f(x, y), g(x, y)\big)$ is a vector field, then its **divergence** is the function

$$\operatorname{div} \mathbf{F} = \frac{\partial f}{\partial x} + \frac{\partial g}{\partial y}.$$

This suggests an interesting notation for the divergence. The divergence is obtained by taking f and g, applying $\frac{\partial}{\partial x}$ and $\frac{\partial}{\partial y}$ to them, and adding the results. If we think of $\left(\frac{\partial}{\partial x}, \frac{\partial}{\partial y}\right)$ as a kind of "vector," then the calculation just described is a kind of "dot product" of $\left(\frac{\partial}{\partial x}, \frac{\partial}{\partial y}\right)$ and $\mathbf{F} = (f, g)$. On the other hand, if we apply this "vector" $\left(\frac{\partial}{\partial x}, \frac{\partial}{\partial y}\right)$ to a function h, what we obtain is $\left(\frac{\partial h}{\partial x}, \frac{\partial h}{\partial y}\right) = \nabla h$.

With this in mind we sometimes write ∇ for the "vector" $\left(\frac{\partial}{\partial x}, \frac{\partial}{\partial y}\right)$, and then the divergence is written

$$\operatorname{div} \mathbf{F} = \frac{\partial f}{\partial x} + \frac{\partial g}{\partial y} = \nabla \cdot \mathbf{F}.$$

This notation for the divergence is very convenient; it is short to write, uses a symbol we already know (∇), and is somewhat self-explanatory. But after all it is just a notation; we only think of ∇ as a vector for convenience of notation.

Green's Theorem tells us that the flux of $\mathbf{F}$ across C equals the integral of the divergence of $\mathbf{F}$ over the region R, that is,

$$\int_C \mathbf{F} \cdot \mathbf{n}\, ds = \int\int_R \operatorname{div} \mathbf{F}\, dx\, dy = \int\int_R \nabla \cdot \mathbf{F}\, dx\, dy.$$

In terms of a fluid flow, this can be interpreted to mean that the net flow into or out from a region R must be the same as the total amount by which the fluid inside the region is running together or spreading out.

In Figure 18.3.15 we see a region containing a *source*, that is, a point from which fluid is flowing out. This means that the velocity field $\mathbf{F}$ will have a positive divergence. The integral of $\nabla \mathbf{F}$ over R will be positive, reflecting the fact that all that fluid will

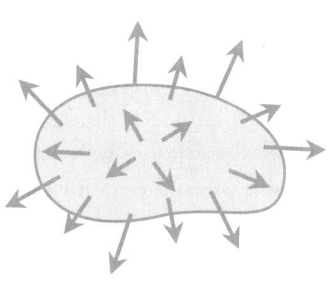

Figure 18.3.15

have to go somewhere, and it will have to do it by having a net flow out of R across the boundary C.

If R contains a *sink*, a point into which fluid flows, then the flux across C will move *into* the region R, so it will be negative.

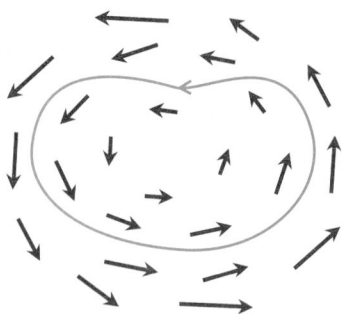

Figure 18.3.16

In Figure 18.3.16 we see a simple closed curve C in the plane and a vector field $\mathbf{F}$ that we visualize as the velocity of a moving fluid. In the situation illustrated here, the fluid is swirling around counterclockwise, as we might expect it to do near a drain hole.

Now suppose the curve C is parametrized in the counterclockwise direction by $\mathbf{r}(s) = \big(x(s), y(s)\big)$, with arclength $s \in [a, b]$. Since the unit tangent vector is $\mathbf{T}(t) = \big(x'(s), y'(s)\big)$, the *tangential component* of $\mathbf{F}$ is $\mathbf{F} \cdot \mathbf{T}$. Roughly speaking, this represents the amount of $\mathbf{F}$ that points along the direction of the tangent to C. In the case illustrated in Figure 18.3.16 it appears that at every point on the curve the vector field will point in approximately the same direction as the tangent vector, so this dot product will always be positive. On the other hand, in Figure 18.3.17 we see the same curve C with a different vector field, and in this case it seems that on the right side of the curve the tangent vector will be in the same general direction as the vector field, but on the left side they will be in directions that are roughly opposite. On the top and bottom the vector field runs across the curve, and it will be nearly perpendicular to the tangent vector.

Now if we integrate this tangential component $\mathbf{F} \cdot \mathbf{T}$ around the curve, what we will be finding is roughly the overall extent to which the fluid is flowing in the direction around the loop. This quantity is called the **circulation** of the vector field $\mathbf{F}$ around C. If it is zero, there is no net circulation of fluid around the loop; this is likely to be nearly the case in the situation illustrated in Figure 18.3.17, in which the flow is sometimes with the direction of the loop and sometimes against it. If the circulation is positive or negative, it means that there is a net flow along (or against) the direction of the curve.

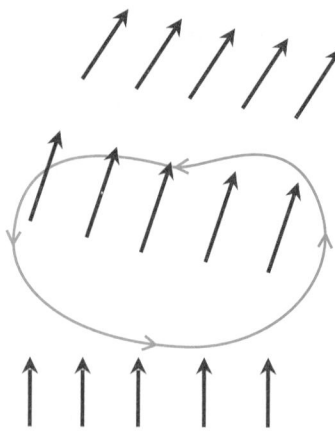

Figure 18.3.17

If we write $\mathbf{F}(x, y) = \big(f(x, y), g(x, y)\big)$, then the circulation is

$$\int_C \mathbf{F} \cdot \mathbf{T}(s)\, ds = \int_a^b \mathbf{F}\big(x(s), y(s)\big) \cdot \big(x'(s), y'(s)\big)\, ds$$

$$= \int_a^b \Big(f\big(x(s), y(s)\big) x'(s) + g\big(x(s), y(s)\big) y'(s) \Big)\, ds$$

$$= \int_C f\, dx + g\, dy.$$

Now Green's Theorem tells us that if R is the region enclosed inside C, this integral for the circulation equals

$$\int\int_R \left(\frac{\partial g}{\partial x} - \frac{\partial f}{\partial y} \right) dx\, dy.$$

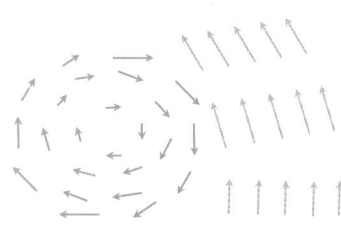

Figure 18.3.18

The integrand in this expression is a quantity that measures the extent to which the vector field $\mathbf{F} = (f, g)$ has a tendency to rotate. Two vector fields are illustrated in Figure 18.3.18; the first appears to rotate clockwise, while the second appears to curl gently to the left. It turns out that $\frac{\partial g}{\partial x} - \frac{\partial f}{\partial y}$ will be positive when the vector field is curling counterclockwise and negative when it is curling clockwise, and the size of its absolute value indicates how sharply it is turning. For this reason this quantity is called the *curl* of the vector field.

DEFINITION 18.3.6

Consider the vector field

$$\mathbf{F}(x, y) = \big(f(x, y), g(x, y)\big),$$

where f and g are differentiable functions. The **curl** of $\mathbf{F}$ is the quantity

$$\text{curl } \mathbf{F} = \frac{\partial g}{\partial x} - \frac{\partial f}{\partial y}.$$

Returning to the situation in which $\mathbf{F}$ represents the velocity of a moving fluid, we see that Green's Theorem says that the circulation of $\mathbf{F}$ around a loop C is equal to the integral of the curl of $\mathbf{F}$ over the region R inside C:

$$Circulation = \int_C \mathbf{F} \cdot \mathbf{T} \, ds = \int_C \mathbf{F} \cdot d\mathbf{r} = \int\int_R \text{curl } \mathbf{F} \, dx \, dy. \qquad (18.3.7)$$

Roughly speaking, we add up the amount by which $\mathbf{F}$ is *curling*, that is, rotating at the points inside the loop, and this will give the overall circulation around the loop.

Both these results about the flux and the circulation give some information about the behavior of the flow on the loop C, by considering its behavior on the region enclosed inside C. They are of great interest to people studying fluid dynamics and have applications in other areas too. For instance, if the vector field $\mathbf{F}$ represents the strength and direction of a magnetic field, then $\int_C \mathbf{F} \cdot \mathbf{n} \, ds$ represents the *magnetic flux* across the loop C.

> We have discussed fluid flows here because they offer a way to visualize vector fields as velocities and make it possible to give a plausible explanation of Green's Theorem in terms of flux and divergence or in terms of circulation and curl. We will return and discuss flows in greater detail in Section 18.7, but for now these results may give us some insight into the way Green's Theorem works. It relates line integrals over loops to double integrals over regions; sometimes one or the other will be easier to evaluate, but it also gives a valuable connection between two apparently different quantities.

Exercises 18.3

I

Evaluate each of the following line integrals using Green's Theorem. (All curves are traversed once in the counterclockwise direction.)

1. $\int_C 2x \, dx - 3x \, dy$, C is $x^2 + y^2 = 1$.

2. $\int_C y \, dx + xy \, dy$, C is $x^2 + y^2 = 4$.

3. $\int_C \cos x^2 \, dx + \sin y^2 \, dy$, C is $x^2 + 4y^2 = 16$.

4. $\int_C y \, dx$, C is $(x-1)^2 + y^2 = 1$.

5. $\int_C -4y \, dx + 3x \, dy$, C is the square with vertices $(\pm 1, \pm 1)$.

6. $\int_C xy^2 \, dx - 3x^3 y^2 \, dy$, C is the square with vertices $(0, 0)$, $(1, 0)$, $(1, 1)$, $(0, 1)$.

7. $\int_C x^2 y^2 \, dx$, C is the triangle with vertices $(0, 0)$, $(3, 0)$, $(3, -2)$.

8. $\int_C x \, dx + (x + 2y) \, dy$, C is the rectangle with vertices $(1, 1)$, $(2, 3)$, $(-3, 3)$, $(-2, 5)$.

9. $\int_C 2y \, dx + x \, dy$, C is the boundary of $0 \leq x \leq 1$, $0 \leq y \leq x^2$.

10. $\int_C -2y \, dx + 3x \, dy$, C is the boundary of $0 \leq x \leq \pi$, $|y| \leq \sin x$.

II

*11. Evaluate the line integral in Example 1 (p. 1167) without using Green's Theorem.

*12. Evaluate the line integral in Example 4 (p. 1169) without using Green's Theorem.

*13. (i) Suppose R is the rectangle $a \le x \le b$, $c \le y \le d$, and let C be its boundary, traversed counterclockwise. Suppose $f(x)$ and $g(y)$ are differentiable functions, and consider $\int_C f(x)g(y)\,dy$. Show that applying Green's Theorem to this line integral just amounts to the Fundamental Theorem of Calculus for f.

(ii) Explain why the sign changes for the analogous calculation for $\int_C f(x)g(y)\,dx$.

14. Suppose C is a simple closed curve in the plane, oriented counterclockwise, and suppose $f(x)$ and $g(y)$ are differentiable functions. Evaluate $\int_C f(x)\,dx + g(y)\,dy$.

*15. If C is the circle $x^2 + y^2 = 9$ in the counterclockwise direction, show that $\int_C x^3\,dy = -\int_C y^3\,dx$. Show that it is easier to evaluate $\int_C (-y^3)\,dx + x^3\,dy$ using Green's Theorem than to evaluate either $\int_C x^3\,dy$ or $\int_C y^3\,dx$ by itself.

*16. (i) If R is the disk $(x-a)^2 + (y-b)^2 \le r^2$, show that
$$\int\int_R (x-a)\,dx\,dy = \int\int_R (y-b)\,dx\,dy = 0.$$

(ii) Use part (i) to evaluate $\int\int_R x\,dx\,dy$ and $\int\int_R y\,dx\,dy$.

(iii) If C is the circle $(x-a)^2 + (y-b)^2 = r^2$, counterclockwise, evaluate $\int_C 2y^2\,dx + 3x^2\,dy$.

*17. Find the area of the region inside the curve parametrized by $x = \cos^3 t$, $y = \sin^3 t$, with $t \in [0, 2\pi]$.

*18. Suppose the simple closed curve C encloses the region R whose area is A. Use Green's Theorem to find the following formulas for the coordinates of the centroid $(\bar{x}, \bar{y})$ of R:
$$\bar{x} = \frac{1}{2A}\int_C x^2\,dy \quad \text{and} \quad \bar{y} = -\frac{1}{2A}\int_C y^2\,dx.$$

*19. Use the result of Exercise 18 to find the centroid of the upper semidisk $x^2 + y^2 \le r^2$, $y \ge 0$.

*20. Use the result of Exercise 18 to find the centroid of the quarter disk $x^2 + y^2 \le r^2$, $x \ge 0$, $y \ge 0$.

*21. Suppose $f(x, y)$ is a twice continuously differentiable function, and let $\mathbf{F} = \nabla f$. Show that the divergence of $\mathbf{F}$ equals
$$\frac{\partial^2 f}{\partial x^2} + \frac{\partial^2 f}{\partial y^2}.$$

Using the ∇ notation for divergence, we can write the divergence of $\mathbf{F} = \nabla f$ as $\nabla \cdot \nabla f$, and stretching the notation a little further, we write
$$\nabla^2 f = \frac{\partial^2 f}{\partial x^2} + \frac{\partial^2 f}{\partial y^2}.$$

This is sometimes called the **Laplacian** of f.

*22. Suppose $f(x, y)$ is a twice continuously differentiable function that satisfies the Laplace Equation
$$\frac{\partial^2 f}{\partial x^2} + \frac{\partial^2 f}{\partial y^2} = 0.$$

(Such a function is called *harmonic*.) Suppose C is a simple closed curve for which Green's Theorem holds (i.e., so that f and its derivatives are defined on an open set containing the region inside C). Show that
$$\int_C \frac{\partial f}{\partial y}\,dx - \frac{\partial f}{\partial x}\,dy = 0.$$

*23. (i) Let C be the straight line segment from (a, b) to (c, d). Show that $\int_C x\,dy - y\,dx = ad - bc$.

(ii) Use part (i) to find the area of the triangle with vertices $(0, 0)$, $(2, 0)$, $(4, 3)$.

*24. Use Exercise 23 to find the area of the triangle with vertices $(-1, 2)$, $(1, 4)$, $(3, 1)$.

*25. Find the area of the quadrilateral with vertices $(0, 0)$, $(2, 5)$, $(4, 3)$, $(3, -2)$. (Be careful about the *order*.)

*26. Find a formula for the area of the polygon whose vertices, listed in counterclockwise order, are $(a_1, b_1), \ldots, (a_n, b_n)$.

**27. Suppose C is a positively oriented differentiable simple closed curve and R is the region inside C. Suppose $f(x, y)$ and $g(x, y)$ are twice continuously differentiable on a region containing R. We will use Green's Theorem to confirm the change of variables formula (Theorem 17.9.2, p. 1128). Suppose u and v are another choice of coordinates, and assume u and v are twice continuously differentiable functions of x and y, and vice versa.

(i) Use Green's Theorem to write $\int\int_R \big(g_x(x, y) - f_y(x, y)\big)\,dx\,dy$ as a line integral around C.

(ii) By expressing the differentials dx and dy in terms of du and dv, write the line integral in the form $\int_C F\,du + G\,dv$, for some functions F and G.

(iii) Use Green's Theorem again to write this line integral as a double integral over R with respect to $du\,dv$.

(iv) Using the Chain Rule, express the partial derivatives f_u, f_v, g_u, g_v in terms of f_x, f_y, g_x, g_y.

(v) Show that the resulting integral agrees with what you get by applying the change-of-variables formula to the original double integral in part (i).

POINT TO PONDER

In Example 5 of Section 18.2 we considered the vector field

$$\mathbf{F}(x, y) = \big(u(x, y), v(x, y)\big) = \left(\frac{-y}{x^2 + y^2}, \frac{x}{x^2 + y^2}\right).$$

It is easy to check that the integral $\int_C \mathbf{F} \cdot d\mathbf{r}$ over the counterclockwise unit circle C is equal to 2π.

On the other hand, we checked that $\frac{\partial v}{\partial x} = \frac{\partial u}{\partial y}$, so $\frac{\partial v}{\partial x} - \frac{\partial u}{\partial y} = 0$, and a quick look at Green's Theorem suggests that the double integral should be zero, which is certainly not equal to the line integral 2π.

The reason this happens is that $\mathbf{F}$ is not defined at the origin (see Figure 18.3.19). The statement of Green's Theorem (Theorem 18.3.1) requires that the functions that appear in the line integral must have continuous partial derivatives on some open set containing R. In this example they are not even defined on all of R itself, let alone some bigger set.

Figure 18.3.19

The hypothesis that the functions be defined on an open set containing R is essential for Green's Theorem. Blindly using the formula without noticing this can lead us to ridiculous results, as in this case, in which we "found" that 2π equals zero.

18.4 Surface Integrals

SURFACE AREA
MOMENTS
CENTROID
MOMENT OF INERTIA

When we studied Green's Theorem in Section 18.3 we found that there is a relation between double integrals over a region R and line integrals over its boundary C. In Sections 18.5 and 18.6 we will study analogous results relating triple integrals over a region R in $\mathbb{R}^3$ and integrals over its boundary. For instance, if R is the ball $x^2 + y^2 + z^2 \leq 4$, then its boundary is the sphere $x^2 + y^2 + z^2 = 4$. If R is the solid cube $-1 \leq x \leq 1$, $-1 \leq y \leq 1$, $-1 \leq z \leq 1$, then its boundary consists of six squares, each of which is part of one of the planes $x = 1$, $x = -1$, $y = 1$, $y = -1$, $z = 1$, $z = -1$ (see Figure 18.4.1).

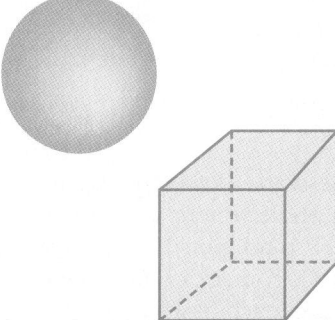

Figure 18.4.1

To be able to discuss these ideas, we first need to be able to integrate over surfaces like pieces of planes or the sphere. To some extent we have already done this in Section 17.5, in which we discussed surface areas. There we found that if S is the surface defined by $z = f(x, y)$, then the **surface area** of the part of S lying over some region T in the xy-plane is given by the double integral

$$A = \int\int_T \sqrt{1 + (f_x)^2 + (f_y)^2} \, dx \, dy.$$

We might abbreviate this by saying that the surface area is given by

$$dS = \sqrt{1 + (f_x)^2 + (f_y)^2} \, dx \, dy = \sqrt{1 + (f_x)^2 + (f_y)^2} \, dA.$$

See Figure 18.4.2.

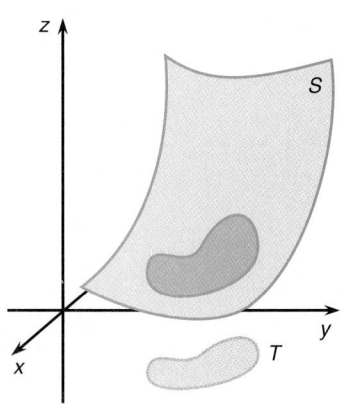

Figure 18.4.2

Finding the surface area of S amounts to integrating the constant function 1 over S, but we may want to integrate other functions. Supposing that S is the part of the surface $z = f(x, y)$ that lies over some region T in the xy-plane as above, then to integrate a function $F(x, y, z)$ over S, we should first observe that on the surface S we have $z = f(x, y)$, so $F(x, y, z) = F\big(x, y, f(x, y)\big)$. Then using the abbreviation just mentioned, we integrate

$$\int\int_S F(x, y, z) \, dS = \int\int_T F\big(x, y, f(x, y)\big) \sqrt{1 + (f_x)^2 + (f_y)^2} \, dx \, dy.$$

It is slightly unfortunate that we have used S to refer to the surface itself and dS in the formula for the integration, but there is little risk of confusion. Also note that the surface integral is written with a double integral sign; this is reasonable because the surface is after all a two-dimensional object (even though it lies in three-dimensional space).

EXAMPLE 1   Let $F(x, y, z) = z$, and integrate F over the upper hemisphere $x^2 + y^2 + z^2 = 1$, $z \geq 0$.

Solution   We can express the hemisphere as the surface S given by $z = f(x, y) = \sqrt{1 - x^2 - y^2}$, for (x, y) in the unit disk D. The formula tells us to integrate as follows:

$$\int\int_S F(x, y, z) \, dS = \int\int_D F\big(x, y, \sqrt{1 - x^2 - y^2}\big) \sqrt{1 + (f_x)^2 + (f_y)^2} \, dx \, dy$$

$$= \int\int_D \sqrt{1 - x^2 - y^2} \sqrt{1 + \left(\frac{-x}{\sqrt{1 - x^2 - y^2}}\right)^2 + \left(\frac{-y}{\sqrt{1 - x^2 - y^2}}\right)^2} \, dx \, dy$$

$$= \int\int_D \sqrt{1 - x^2 - y^2} \sqrt{\frac{1 - x^2 - y^2 + x^2 + y^2}{1 - x^2 - y^2}} \, dx \, dy = \int\int_D 1 \, dx \, dy$$

$$= area(D) = \pi.$$

The integral of $F(x, y, z) = z$ with respect to surface area on the upper hemisphere equals π.

Example 1 suggests an awkward question: Suppose we wanted to integrate some function F over the whole sphere $x^2 + y^2 + z^2 = 1$ instead of just the hemisphere. The difficulty is that it is not possible to express the whole sphere as the graph of a function $z = f(x, y)$. Of course, it is possible to divide the sphere up into the upper and lower hemispheres, each of which *can* be expressed as a graph.

For that matter it is also possible to use different hemispheres; for instance, the "left" hemisphere $x^2 + y^2 + z^2 = 1$, $y \leq 0$ can be expressed as the graph of the function $y = -\sqrt{1 - x^2 - z^2}$ with x and z as variables (see Figure 18.4.3).

It seems reasonably clear what to do in these situations. First we divide the surface into pieces, each of which can be expressed as the graph of some function. For each of them we use a formula for the integral that is analogous to what we used above, making the obvious adjustments depending on which two of x, y, z are the variables.

DEFINITION 18.4.1

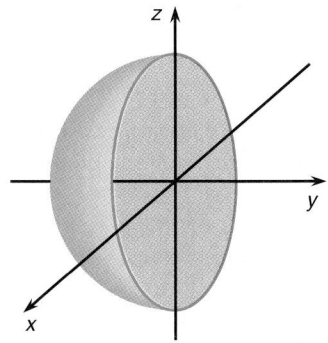

Figure 18.4.3

Consider the surface $S = \{(x, y, z) : z = f(x, y), \text{ for } (x, y) \in T\}$, where f is a differentiable function defined on some region T in the xy-plane. If $F(x, y, z)$ is a continuous function defined on S, then the **surface integral** of F over S is defined to be

$$\int\int_S F(x, y, z)\, dS = \int\int_T F\big(x, y, f(x, y)\big)\sqrt{1 + (f_x)^2 + (f_y)^2}\, dx\, dy.$$

Similarly, if $S = \{(x, y, z) : x = g(y, z), \text{ for } (y, z) \in T\}$, for some differentiable function g defined on a region T in the yz-plane, then

$$\int\int_S F(x, y, z)\, dS = \int\int_T F\big(g(y, z), y, z\big)\sqrt{1 + (g_y)^2 + (g_z)^2}\, dy\, dz.$$

Finally, if $S = \{(x, y, z) : y = h(x, z), \text{ for } (x, z) \in T\}$, for some differentiable function h defined on a region T in the xz-plane, then

$$\int\int_S F(x, y, z)\, dS = \int\int_T F\big(x, h(x, z), z\big)\sqrt{1 + (h_x)^2 + (h_z)^2}\, dx\, dz.$$

EXAMPLE 2   Let S be the sphere $x^2 + y^2 + z^2 = 4$. Integrate $\int\int_S (y^2 + z^2)\, dS$.

Solution   We need to divide the sphere into two hemispheres. Since the integral is expressed in terms of y and z, it probably makes sense to use the corresponding hemispheres $x = \sqrt{4 - y^2 - z^2}$ and $x = -\sqrt{4 - y^2 - z^2}$, that is, $x^2 + y^2 + z^2 = 4$, $x \geq 0$ and $x^2 + y^2 + z^2 = 4$, $x \leq 0$. Calling them S_1 and S_2 and letting D be the disk $y^2 + z^2 \leq 4$, we have

$$\int\int_S (y^2 + z^2)\, dS = \int\int_{S_1} (y^2 + z^2)\, dS + \int\int_{S_2} (y^2 + z^2)\, dS$$

$$= \int\int_D (y^2 + z^2)\sqrt{1 + \frac{(-y)^2}{4 - y^2 - z^2} + \frac{(-z)^2}{4 - y^2 - z^2}}\, dy\, dz$$

$$+ \int\int_D (y^2 + z^2)\sqrt{1 + \frac{y^2}{4 - y^2 - z^2} + \frac{z^2}{4 - y^2 - z^2}}\, dy\, dz$$

$$= 2\int\int_D (y^2 + z^2)\sqrt{\frac{4}{4 - y^2 - z^2}}\, dy\, dz.$$

Using polar coordinates in the yz-plane, with $r^2 = y^2 + z^2$, we see that this is

$$2\int_0^{2\pi}\int_0^2 r^2\frac{2}{\sqrt{4 - r^2}}\, r\, dr\, d\theta = 4\pi\int_0^2 \big((r^2 - 4) + 4\big)\frac{1}{\sqrt{4 - r^2}}\, 2r\, dr$$

We substitute $u = 4 - r^2$, $du = -2r\,dr$.

$$= -4\pi \int_4^0 (4-u)\frac{1}{\sqrt{u}}\,du = 4\pi \int_0^4 (4u^{-1/2} - u^{1/2})\,du$$

$$= 4\pi \left(8u^{1/2} - \frac{2}{3}u^{3/2} \right)\Big|_0^4 = 4\pi \left(16 - \frac{16}{3} \right) = \frac{128\pi}{3}.$$

We see that surface integrals are easily set up; some care may be needed in writing various pieces of a surface as graphs, but the basic idea is not difficult. The resulting integrals are often complicated, but suitable choice of coordinates may simplify the calculations.

Surface Integrals and Moments

One important use of surface integrals is for finding moments. We imagine a surface made of plastic, metal, or some other material and consider the usual problems of finding its total mass and its moments.

Suppose that the density of a surface S is given by a function $\rho(x, y, z)$. (Most often ρ is a constant function.) Then the definitions are almost obvious.

DEFINITION 18.4.2

Let S be a surface, with density $\rho(x, y, z)$. The total **mass** of S is

$$M = \int\int_S \rho(x, y, z)\,dS.$$

The **moments** of S about the xy-plane, the xz-plane, and the yz-plane are

$$M_{xy} = \int\int_S z\rho(x, y, z)\,dS, \qquad M_{xz} = \int\int_S y\rho(x, y, z)\,dS,$$

$$M_{yz} = \int\int_S x\rho(x, y, z)\,dS,$$

respectively.

The **centroid** or **center of mass** of S is the point $(\bar{x}, \bar{y}, \bar{z})$, where

$$\bar{x} = \frac{M_{yz}}{M}, \qquad \bar{y} = \frac{M_{xz}}{M}, \qquad \bar{z} = \frac{M_{xy}}{M}.$$

The **moments of inertia** of S about the x-, y-, and z-axes are

$$I_x = \int\int_S (y^2 + z^2)\,dS, \quad I_y = \int\int_S (x^2 + z^2)\,dS, \quad I_z = \int\int_S (x^2 + y^2)\,dS,$$

respectively.

EXAMPLE 3   Find the centroid of the upper hemisphere $x^2 + y^2 + z^2 = 1$, $z \geq 0$.

Solution   As usual when the density is not specified, we assume it is 1. Expressing the hemisphere as the graph of $z = \sqrt{1 - x^2 - y^2}$, for $(x, y) \in D$, the disk $x^2 + y^2 \leq 1$, we see that the mass is

$$M = \int\int_S \rho\,dS = \int\int_D \sqrt{1 + \frac{x^2}{1 - x^2 - y^2} + \frac{y^2}{1 - x^2 - y^2}}\,dx\,dy$$

$$= \int \int_D \sqrt{\frac{1}{1 - x^2 - y^2}} \, dx \, dy = \int_0^{2\pi} \int_0^1 \frac{1}{\sqrt{1 - r^2}} \, r \, dr \, d\theta$$

$$= 2\pi \int_1^0 \left(-\frac{1}{2}\right) u^{-1/2} \, du$$

$$= (-\pi) 2u^{1/2} \Big|_1^0 = 0 + 2\pi = 2\pi.$$

The moment about the *yz*-plane is

$$M_{yz} = \int \int_S x\rho \, dS = \int \int_D x \frac{1}{\sqrt{1 - x^2 - y^2}} \, dx \, dy = \int_0^{2\pi} \int_0^1 r \cos\theta \frac{1}{\sqrt{1 - r^2}} \, r \, dr \, d\theta.$$

Rearranging the order of integration we find that this equals

$$\int_0^1 \frac{1}{\sqrt{1 - r^2}} r^2 \left(\int_0^{2\pi} \cos\theta \, d\theta\right) dr = \int_0^1 \frac{1}{\sqrt{1 - r^2}} r^2 (0) \, dr = 0.$$

Similarly, $M_{xz} = 0$.

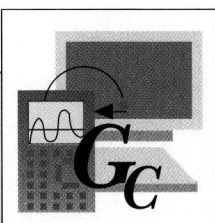

With a graphics calculator we plot the upper unit semicircle and think of it as a cross-section of the upper hemisphere. On the same picture we plot the centroid; it is the point $\left(0, 0, \frac{1}{2}\right)$, so in the cross-section it is $\left(0, \frac{1}{2}\right)$.

On the same picture we also plot the centroid of the upper half ball; it is $\left(0, \frac{3}{8}\right)$ (cf. Example 17.7.2, p. 1066).

TI-85	SHARP EL9300
GRAPH F1 $(y(x) =)$ $\sqrt{}$ (1 − F1 (x) x^2) M2(RANGE) (−) 1 ▼ 1 ▼ ▼ (−) 1 ÷ 3 ▼ 1 F5(GRAPH) PtOff (0 , .375) ENTER PtOff (0 , .5) ENTER	⌂ $\sqrt{}$ 1 − x/θ/T x^2 RANGE (−)1 ENTER 1 ENTER 1 ENTER (−).3 ENTER 1 ENTER 1 ENTER ⌂ 2ndF PLOT ▼ ▶ ENTER 0 ENTER .375 ENTER ⌂ 2ndF PLOT ▼ ▶ ENTER 0 ENTER .5 ENTER

CASIO f_x-6300G (f_x-7700GB)	HP 48SX
Range −1 EXE 1 EXE 1 EXE −.3 EXE 1 EXE 1 EXE EXE EXE EXE Graph $\sqrt{}$ (1 − X x^2) EXE Plot 0 , .375 EXE Plot 0 , .5 EXE	⌐ PLOT PLOTR ERASE ATTN ' Y = $\sqrt{x}$ ◼ 1 − X y^x 2 ↵ PLOT ⌐ DRAW ERASE 1 +/− ENTER 1 XRNG .3 +/− ENTER 1 YRNG ATTN NXT NXT ◼ 1.5 ◼ 1 AXES ATTN ◼ 0 ◼ .5 PRG DSPL NXT PIXON ATTN ◼ 0 ◼ .375 PRG DSPL NXT PIXON ↵ PLOT DRAW

Explain the relative positions of the two centroids.

Finally,

$$M_{xy} = \int\int_S z\rho \, dS = \int\int_D \sqrt{1-x^2-y^2} \frac{1}{\sqrt{1-x^2-y^2}} \, dx \, dy = \int\int_D dA = \pi.$$

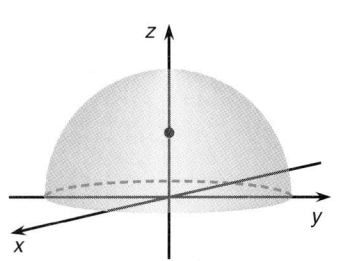

We find that $\bar{x} = \bar{y} = 0$ and $\bar{z} = \frac{M_{xy}}{M} = \frac{\pi}{2\pi} = \frac{1}{2}$. The centroid of the upper hemi-sphere is $(\bar{x}, \bar{y}, \bar{z}) = \left(0, 0, \frac{1}{2}\right)$.

Notice that the centroid of the hemisphere is not *on* the hemisphere at all. Also observe that if we think of $\bar{x}, \bar{y}, \bar{z}$ as the "average values" of x, y, z on S, it is perfectly reasonable to expect that $\bar{x} = \bar{y} = 0$ because of the symmetry (see Figure 18.4.4). Finally, remark that since the density ρ is the constant 1, the "mass" of the surface is just its surface area.

Figure 18.4.4

Exercises 18.4

I

Evaluate $\int\int_S F(x, y, z) \, dS$ in each of the following situations.

1. $F = 1 + 8z$, S is $z = 2x^2 + 2y^2$, $x^2 + y^2 \le 4$
2. $F = z^2$, S is $x^2 + y^2 + z^2 = 9$
3. $F = x$, S is $z^2 = x^2 + y^2$, $x \ge 0$, $y \ge 0$, $x^2 + y^2 \le 1$
4. $F = \sqrt{x^2 + y^2 + z^2}$, S is the sphere of radius 4 around the origin.
5. $F = x^2$, S is $2x + 3y + 4z = 12$, $x \ge 0$, $y \ge 0$, $x + y \le 24$
6. $F = x + y$, S is $x + y + z = 3$, $0 \le x \le 3$, $0 \le y \le 4$
7. $F = y^2(1 + 4z)^{-1/2}$, S is $z = x^2 + y^2$, $x^2 + y^2 \le 9$, $x \ge 0$
8. $F = \sin \pi x$, S is $z^2 = x^2 + y^2$, $0 \le x \le 1$, $-1 \le y \le 0$, $z \ge 0$
9. $F = \sqrt{4 + x^2 + y^2}$, $z = \ln(x^2 + y^2)$, $1 \le x^2 + y^2 \le 4$
10. $F = yz$, S is $x^2 = 2y^2 + 2z^2$, $0 \le z \le y^2 \le 4$, $x \le 0$, $y \ge 0$

Find the centroid of each of the following surfaces.

11. The cylinder $x^2 + y^2 = 4$, $0 \le z \le 2$.
12. The piece of the paraboloid $y = x^2 + z^2$ for which $y \le 25$.
13. The piece of the cone $x^2 = y^2 + z^2$ for which $0 \le x \le 1$.
14. The piece of the cone $y^2 = x^2 + z^2$ for which $-1 \le y \le 2$.
15. The surface $x^2 + y^2 + z^2 = 4$, $z \ge 0$.
16. The surface $x^2 + y^2 + z^2 = 9$, $y \le 0$, $z \ge 0$.

17. The top half (i.e., $z \ge 0$) of the cube with vertices $(\pm 1, \pm 1, \pm 1)$.
18. The closed hemisphere consisting of the hemisphere $x^2 + y^2 + z^2 = 1$, $x \ge 0$, together with its "lid" $y^2 + z^2 \le 1$, $x = 0$.
19. The surface that consists of two parts, the upper hemisphere $x^2 + y^2 + z^2 = 4$, $z \ge 0$, and the part of the lower hemisphere given by $x^2 + y^2 + z^2 = 4$, $z \le -1$.
20. The surface $y^2 + z^2 = 1$, $0 \le x \le z + 2$.

II

21. Solve Exercise 2 using the variables x and y to set up the integration.
*22. Evaluate the double integral of Exercise 3 by expressing the integrand as a partial derivative and using Green's Theorem.
*23. Verify that the center of mass of a surface S is unchanged if the density ρ is replaced by $k\rho$, where k is a constant.
24. Find the centroid of the upper hemisphere of radius R centered at the origin.
*25. Find the moment of inertia about the x-axis of the sphere of radius R centered at the origin.
*26. Find the moment of inertia about the z-axis of the paraboloid S given by $z = x^2 + y^2$, $x^2 + y^2 \le R^2$.
*27. Consider the upper hemisphere $x^2 + y^2 + z^2 = 16$, $z \ge 0$, with either the uniform density $\rho_1(x, y, z) = 1$ or with density $\rho_2(x, y, z) = z$.
   (i) Without finding either of them, what do you expect about the relative positions of the centers of mass found by using these two densities?

(ii) Find the two centers of mass.

*28. (i) For the sphere $x^2 + y^2 + z^2 = 1$, how do you expect that the moments of inertia about the z-axis will compare when the density is $\rho_1(x, y, z) = 1$ or $\rho_2(x, y, z) = |z|$?

    (ii) Calculate the two moments of inertia.

*29. How would you define the radius of gyration about the z-axis of a surface S with density ρ?

*30. Using your definition from Exercise 29, find the radius of gyration about the z-axis of the sphere $x^2 + y^2 + z^2 = 1$, with uniform density. Before working it out, predict how it will compare with the radius of gyration of the solid ball of radius 1.

18.5 The Divergence Theorem

DIVERGENCE
OUTER NORMAL VECTOR
ORIENTABLE SURFACE
FLUX

Toward the end of Section 18.3 we found that Green's Theorem could be used to give some results about vector fields that we interpreted in terms of fluid flows. We began by thinking of a vector field **F** as the velocity of a fluid flow at each point. We compared the flux across a closed curve C with the integral of the "divergence" of the flow, the amount by which it is "spreading out" on the region enclosed inside C, and Green's Theorem enabled us to show that these two quantities are equal. This result reflects the perfectly reasonable idea that if the fluid is spreading out inside C, then it has to go somewhere, and to do that, it has to cross the boundary C. See Figure 18.5.1.

Figure 18.5.1

Figure 18.5.2

In this section we discuss a three-dimensional analog of this result. We consider an open region in space, such as the ball illustrated in Figure 18.5.2, together with its boundary, which in this example is just the sphere, the "skin" of the ball. We will define the *divergence* of a vector field **F** in $\mathbb{R}^3$; it will measure how much the vector field is "spreading out." The main result of this section is the Divergence Theorem (Theorem 18.5.4), which says that the integral of the divergence over a region in $\mathbb{R}^3$ must equal the flux across the boundary of the region.

This says that if a fluid is spreading out on a region R, then some of it must be flowing out of the region by crossing the boundary.

To define the flux, we recall that in the two-dimensional case in Section 18.3 we found the component of **F** along the direction of the outer unit normal vector **n** and integrated that over the boundary. In the three-dimensional case it will first be necessary to define the outer normal vector. In this case the integral over the boundary will be a surface integral (since the boundary is a surface, like the sphere that is the boundary of the ball).

The divergence of a vector field in $\mathbb{R}^3$ is almost the same as the $\mathbb{R}^2$ version.

DEFINITION 18.5.1

Suppose $\mathbf{F}(x, y, z) = \big(f(x, y, z), g(x, y, z), h(x, y, z)\big)$ is a vector field. Its **divergence** is the *function*

$$\operatorname{div} \mathbf{F} = \nabla \cdot \mathbf{F} = \frac{\partial f}{\partial x} + \frac{\partial g}{\partial y} + \frac{\partial h}{\partial z}.$$

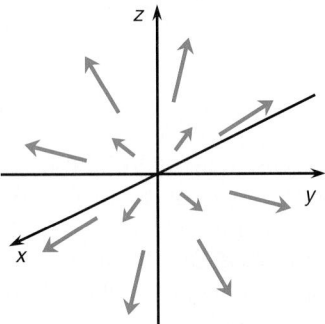

Figure 18.5.3

EXAMPLE 1

(i) Let $\mathbf{F}(x, y, z) = (x, y, z)$. This is the vector field illustrated in Figure 18.5.3, which points out from the origin. Its divergence is

$$\nabla \cdot \mathbf{F} = \frac{\partial f}{\partial x} + \frac{\partial g}{\partial y} + \frac{\partial h}{\partial z} = \frac{\partial x}{\partial x} + \frac{\partial y}{\partial y} + \frac{\partial z}{\partial z} = 1 + 1 + 1 = 3.$$

The divergence is the constant function 3, which we interpret to mean that the flow is diverging at every point.

(ii) In Figure 18.5.4 we see a vector field that represents fluid rotating around the x-axis. The vector field is given by $\mathbf{F}(x, y, z) = (0, -z, y)$, so its divergence is

$$\nabla \cdot \mathbf{F} = \frac{\partial f}{\partial x} + \frac{\partial g}{\partial y} + \frac{\partial h}{\partial z} = \frac{\partial(0)}{\partial x} + \frac{\partial(-z)}{\partial y} + \frac{\partial y}{\partial z} = 0 + 0 + 0 = 0.$$

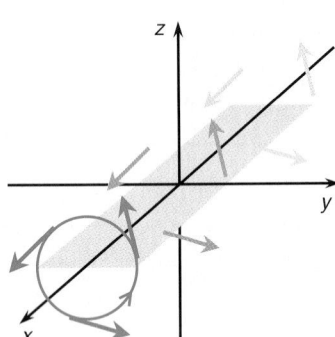

Figure 18.5.4

The divergence is zero, and the fluid is not "spreading out"; after all, rotating does not alter the relative positions of the points, so they stay the same distances from one another.

(iii) In Figure 18.5.5 we see the part of the vector field $\mathbf{F}(x, y, z) = (xz, yz, 1)$ that lies in the xz-plane. Notice that the vertical component of each vector is always 1 and that below the xy-plane the vectors point in toward the origin while above it they point outward.

The divergence is

$$\nabla \cdot \mathbf{F} = \frac{\partial f}{\partial x} + \frac{\partial g}{\partial y} + \frac{\partial h}{\partial z} = \frac{\partial(xz)}{\partial x} + \frac{\partial(yz)}{\partial y} + \frac{\partial(1)}{\partial z} = z + z + 0 = 2z.$$

We find that the divergence $\nabla \cdot \mathbf{F} = 2z$ is zero on the xy-plane where the fluid is flowing straight up, positive above the xy-plane where the fluid is diverging, and negative below the xy-plane where the fluid is being compressed as it flows in toward the origin.

Figure 18.5.5

EXAMPLE 2   **Expanding Universe**   Consider objects in space moving so that an object at the point $P = (x, y, z)$ has velocity

$$\mathbf{v}(x, y, z) = (kx, ky, kz),$$

for some positive constant k. The *relative* velocity between two objects, one at $P = (x, y, z)$ and the other at $Q = (a, b, c)$, is then

$$\mathbf{v}(x, y, z) - \mathbf{v}(a, b, c) = (kx, ky, kz) - (ka, kb, kc)$$
$$= \big(k(x - a), k(y - b), k(z - c)\big).$$

Figure 18.5.6

The Hubble Space Telescope, suspended by the shuttle's manipulator arm prior to the deployment of its solar panels and antennae and its ultimate release. Puerto Rico and the Dominican Republic are barely recognizable at the right side of the frame.

The speed at which the two objects are moving apart is the norm

$$\|\mathbf{v}(x, y, z) - \mathbf{v}(a, b, c)\| = \|\big(k(x - a), k(y - b), k(z - c)\big)\|$$
$$= k\sqrt{(x - a)^2 + (y - b)^2 + (z - c)^2}.$$

This speed is proportional to the distance between P and Q.

Since the motion expands out from the origin, it is a little surprising that the speed at which two points separate is k times the distance between them, *no matter where they are*. (Of course, it is no surprise if one of them, Q, say, is the origin . . .). In a way this is the significance of the fact that

$$\text{div } \mathbf{v} = \frac{\partial}{\partial x}(kx) + \frac{\partial}{\partial y}(ky) + \frac{\partial}{\partial z}(kz) = 3k,$$

that is, that the divergence is constant. The motion is "spreading out" at the same rate *everywhere*.

See the box for a discussion of the astronomical significance of the fact that it is possible to detect this type of expansion from any point, not just from the origin.

The astronomer Edwin P. Hubble discovered the phenomenon of the *red shift*, a change in the wavelengths of light from distant galaxies. From these observations he concluded that the galaxies are moving apart and calculated their relative speeds from the magnitude of the shift. He found that the speed at which a galaxy moves away is proportional to its distance from us.

We have just seen that this is what happens with the velocity field $\mathbf{v}(x, y, z) = (kx, ky, kz)$, which is the velocity field describing motion spreading out from a single point (see Figure 18.5.6). This is the origin of the *Big Bang Theory*, which suggests that the universe expanded from a large explosion.

Normal Vectors

To discuss the flow across the boundary of a three-dimensional region, we need to be able to find the component of the flow in the direction that is normal to the surface. To do this, we just take the dot product with the unit normal vector, or, more precisely, with the *outer* unit normal vector. This will tell us the net rate at which fluid is flowing *out* of the region.

Let us begin by discussing normal vectors. First consider a surface that is the graph of a function $z = f(x, y)$. It can be written as the level surface where the function $z - f(x, y)$ has the value 0. We know that the gradient $\nabla\big(z - f(x, y)\big)$ is a normal vector to the level surface (see Figure 18.5.7). This gradient equals $\left(-\frac{\partial f}{\partial x}, -\frac{\partial f}{\partial y}, 1\right)$. Its z-component is 1, so it always points upward, but its length may vary at different points depending on f.

To get a unit vector, we divide by the norm and get a **unit normal vector**

$$\mathbf{n} = \frac{1}{\sqrt{f_x^2 + f_y^2 + 1}}\left(-f_x, -f_y, 1\right).$$

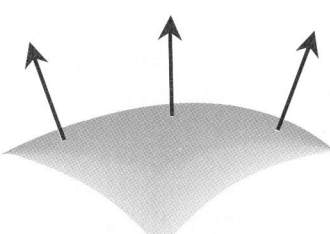

Figure 18.5.7

This was done under the assumption that the surface was a graph $z = f(x, y)$. Generally speaking, this will not be so; even for the familiar sphere we have to break

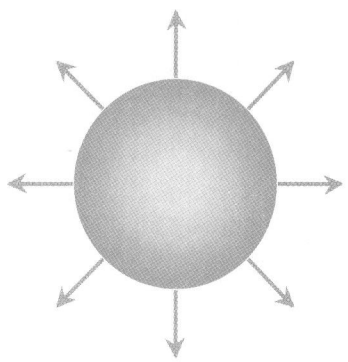

Figure 18.5.8

it into pieces and express, for instance, the upper and lower hemispheres as graphs. Moreover, if we want *outer* normals for the sphere, we must pick vectors that point up on the upper hemisphere and down on the lower hemisphere (see Figure 18.5.8).

In the example of the sphere it is quite clear how to find outer unit normals at every point, and for any region contained between two graphs, $g(x, y) \le z \le f(x, y)$, the normal vectors on the upper boundary $z = f(x, y)$ should point up and the ones on the lower boundary $z = g(x, y)$ should point down. For very complicated regions it can be quite difficult to describe how to choose the outer normal at each point of the boundary. In the "Point to Ponder" at the end of this section we will discuss the fact that there are even some surfaces for which it is impossible to find unit normals at every point without having to "jump" from one side to the other.

To avoid this difficulty, we will assume that for the surfaces we are using, it is always possible to find outer unit normal vectors.

DEFINITION 18.5.2

A surface S is **orientable** if it is possible to find unit normal vectors that do not "jump" from one side of the surface to the other, that is, if there are functions $u(x, y, z)$, $v(x, y, z)$, $w(x, y, z)$ on S so that

$$\mathbf{n}(x, y, z) = \big(u(x, y, z), v(x, y, z), w(x, y, z)\big)$$

is a unit normal vector at each point $(x, y, z) \in S$, with the property that all these unit vectors are on the same "side" of S.

EXAMPLE 3   For the sphere $x^2 + y^2 + z^2 = 1$ we can write the upper hemisphere as $z = \sqrt{1 - x^2 - y^2}$ and find the normal vector $\left(\frac{x}{\sqrt{1 - x^2 - y^2}}, \frac{y}{\sqrt{1 - x^2 - y^2}}, 1\right)$. Dividing by its length, we find the unit normal vector $(x, y, \sqrt{1 - x^2 - y^2})$.

Similarly, on the lower hemisphere $z = -\sqrt{1 - x^2 - y^2}$ we can find the unit normal vector $(-x, -y, \sqrt{1 - x^2 - y^2})$; however, this vector points upward. The *outer* unit normal on the lower hemisphere will be $(x, y, -\sqrt{1 - x^2 - y^2})$.

Now notice that on the upper hemisphere we have $z = \sqrt{1 - x^2 - y^2}$, so the formula for the outer unit normal is $(x, y, \sqrt{1 - x^2 - y^2}) = (x, y, z)$. On the lower hemisphere we have $z = -\sqrt{1 - x^2 - y^2}$, so the outer unit normal is $(x, y, -\sqrt{1 - x^2 - y^2}) = (x, y, z)$.

Combining these two calculations, we find that the outer unit normal can be found *on the whole sphere* by the formula

$$\mathbf{n} = (x, y, z).$$

This is a unit vector since the point (x, y, z) is on the sphere of radius 1.

From now on we will always assume that surfaces are orientable; there are some nonorientable surfaces but it is not very easy to find them.

DEFINITION 18.5.3

Suppose S is some surface in $\mathbb{R}^3$ and that $\mathbf{n} = \mathbf{n}(x, y, z)$ is a unit normal vector at each point $(x, y, z) \in S$. If $\mathbf{F}$ is a vector field, then the **flux** of $\mathbf{F}$ across the surface S (in the direction of $\mathbf{n}$) is the integral over S of the normal component of $\mathbf{F}$, that is,

$$Flux = \int\int_S \mathbf{F} \cdot \mathbf{n}\, dS.$$

In the case in which S is the graph of a function, this is nicer than it appears.

EXAMPLE 4 If S is the surface given by $z = f(x, y)$, for $(x, y) \in D$, for some open set D in $\mathbb{R}^2$, then we have seen that an upward-pointing unit normal is given by

$$\mathbf{n} = \left(\frac{-f_x}{\sqrt{f_x^2 + f_y^2 + 1}}, \frac{-f_y}{\sqrt{f_x^2 + f_y^2 + 1}}, \frac{1}{\sqrt{f_x^2 + f_y^2 + 1}} \right).$$

If the vector field $\mathbf{F}$ is written out as $\mathbf{F} = (r, s, t) = \big(r(x, y, z), s(x, y, z), t(x, y, z)\big)$, then the flux of $\mathbf{F}$ across S is

$$\int\int_S \mathbf{F} \cdot \mathbf{n}\, dS$$

$$= \int\int_D \left(r\frac{-f_x}{\sqrt{f_x^2 + f_y^2 + 1}} + s\frac{-f_y}{\sqrt{f_x^2 + f_y^2 + 1}} + t\frac{1}{\sqrt{f_x^2 + f_y^2 + 1}} \right) dS$$

$$= \int\int_D \left(\frac{-rf_x}{\sqrt{f_x^2 + f_y^2 + 1}} + \frac{-sf_y}{\sqrt{f_x^2 + f_y^2 + 1}} + \frac{t}{\sqrt{f_x^2 + f_y^2 + 1}} \right) \sqrt{1 + f_x^2 + f_y^2}\, dx\, dy$$

$$= \int\int_D (-rf_x - sf_y + t)\, dx\, dy.$$

Recall that $dS = \sqrt{1 + f_x^2 + f_y^2}\, dx\, dy$.

What has happened here is that the square root we needed to make $\mathbf{n}$ a unit vector has canceled out the square root that arises in the formula for surface integrals: $dS = \sqrt{f_x^2 + f_y^2 + 1}\, dx\, dy$. The result is a relatively simple integral in terms of x and y.

Now we are ready to state the Divergence Theorem, which relates the integral of the divergence of a vector field over a three-dimensional region to the flux of the vector field across the boundary of the region.

THEOREM 18.5.4

The Divergence Theorem

Suppose R is a connected open region in $\mathbb{R}^3$, and let S be its boundary surface. Assume that S is orientable. Also suppose $\mathbf{F}$ is a vector field with continuous partial derivatives on an open set containing R and S. Then

$$\int\int\int_R \operatorname{div} \mathbf{F}\, dV = \int\int\int_R \nabla \cdot \mathbf{F}\, dx\, dy\, dz = \int\int_S \mathbf{F} \cdot \mathbf{n}\, dS,$$

where $\mathbf{n}$ is a unit outer normal vector at each point of S.

PROOF This result is difficult to prove in every case, but we will explain the proof for a region that lies between the graphs of two functions.

Rather than prove the whole formula at once, we break it into three simpler pieces, corresponding to the three components of **F**. Writing out

$$\mathbf{F} = (r, s, t) = \big(r(x, y, z), s(x, y, z), t(x, y, z)\big),$$

we know that the divergence is $\nabla \cdot \mathbf{F} = \frac{\partial r}{\partial x} + \frac{\partial s}{\partial y} + \frac{\partial t}{\partial z}$. From this we see that the Divergence Theorem is equivalent to the following formulas:

$$\iiint_R \frac{\partial r}{\partial x}\, dV = \iint_S (r, 0, 0) \cdot \mathbf{n}\, dS,$$

$$\iiint_R \frac{\partial s}{\partial y}\, dV = \iint_S (0, s, 0) \cdot \mathbf{n}\, dS,$$

$$\iiint_R \frac{\partial t}{\partial z}\, dV = \iint_S (0, 0, t) \cdot \mathbf{n}\, dS.$$

We will prove the last of these formulas for a region that lies between the graphs of two functions. Notice that the formula involves not the whole vector field $\mathbf{F} = (r, s, t)$ but just the third component t, so for convenience we will write $\mathbf{G} = (0, 0, t)$.

Let D be an open region in the xy-plane, and let $g(x, y)$ and $f(x, y)$ be continuous functions on D with $g(x, y) \leq f(x, y)$ for every $(x, y) \in D$. Let R be the region that lies between the graphs of g and f, that is,

$$R = \big\{(x, y, z) : g(x, y) \leq z \leq f(x, y), \quad (x, y) \in D\big\}.$$

This region is illustrated in Figure 18.5.9(i).

The first step is to find the outer unit normal vectors, and we do this in three different stages, corresponding to the top of the region, the bottom, and the sides (see Figure 18.5.9(ii)). On the top we use the upward unit normal vector $\mathbf{n}$, and the calculation we did in Example 4 shows that the upward flux of $\mathbf{G}$ across the top boundary equals

$$\iint \mathbf{G} \cdot \mathbf{n}\, dS = \iint_D \big(-0(f_x) - 0(f_y) + t\big)\, dx\, dy = \iint_D t\big(x, y, f(x, y)\big)\, dx\, dy.$$

For the bottom the situation is similar except that the outer normal will be the one pointing *downward*, so the downward flux across the bottom boundary is

$$\iint \mathbf{G} \cdot \mathbf{n}\, dS = \iint_D \big(0(g_x) + 0(g_y) - t\big)\, dx\, dy = -\iint_D t\big(x, y, g(x, y)\big)\, dx\, dy.$$

On the vertical sides of the region the normal vectors will be horizontal. Since the vector field $\mathbf{G} = (0, 0, t)$ has no horizontal components, we find that $\mathbf{G} \cdot \mathbf{n} = 0$ on the vertical sides. This means that the flux, the integral of the normal component $\mathbf{G} \cdot \mathbf{n}$ over S, is really just the sum of the integrals over the top and bottom. So the flux of $\mathbf{G}$ across S is

$$\iint_D \Big(t\big(x, y, f(x, y)\big) - t\big(x, y, g(x, y)\big)\Big)\, dx\, dy.$$

The left side of the equation we want is

$$\iiint_R \frac{\partial t}{\partial z}\, dV = \iint_D \left(\int_{g(x,y)}^{f(x,y)} \frac{\partial t}{\partial z}\, dz\right) dx\, dy = \iint_D \left(t(x, y, z)\Big|_{z=g(x,y)}^{z=f(x,y)}\right) dx\, dy$$

$$= \iint_D \Big(t\big(x, y, f(x, y)\big) - t\big(x, y, g(x, y)\big)\Big)\, dx\, dy,$$

which is exactly the same as the flux we just found.

Figure 18.5.9(i)

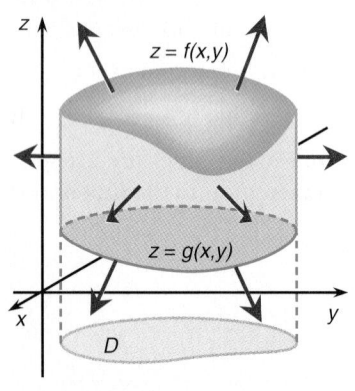

Figure 18.5.9(ii)

This argument establishes the third equation for any region that lies between the graphs of two functions of x and y. A similar argument can be used for the first equation for a region lying between the graphs of two functions of y and z or for the second equation for a region that lies between the graphs of two functions of x and z.

Combining all these arguments gives a proof of the Divergence Theorem for any region that can be expressed as the region between two graphs in each of the three directions; for instance, a ball lies between two hemispheres along each axis, so this argument proves the Divergence Theorem for any ball.

Many regions do not have this convenient property, but on the other hand the regions we will encounter can easily be broken up into pieces, each of which does have this property, and the result can be proved that way.

EXAMPLE 5   Let $\mathbf{F}(x, y, z) = (3x - 2y, x - 5y, x + y + z)$. If R is the ball $x^2 + y^2 + z^2 \leq 1$ and S its boundary $x^2 + y^2 + z^2 = 1$, evaluate $\int \int_S \mathbf{F} \cdot \mathbf{n} \, dS$, where $\mathbf{n}$ is the outer unit normal vector.

Solution   We know that $\mathbf{n}(x, y, z) = (x, y, z)$ for each $(x, y, z) \in S$ (cf. Example 3). We could evaluate the surface integral directly, but it is much easier to use the Divergence Theorem:

$$\int \int_S \mathbf{F} \cdot \mathbf{n} \, dS = \int \int \int_R \nabla \cdot \mathbf{F} \, dV$$

$$= \int \int \int_R \left(\frac{\partial}{\partial x}(3x - 2y) + \frac{\partial}{\partial y}(x - 5y) + \frac{\partial}{\partial z}(x + y + z) \right) dV$$

$$= \int \int \int_R (3 - 5 + 1) \, dV = - \int \int \int_R dV = -\text{vol}(R) = -\frac{4\pi}{3}.$$

The surface integral equals $-\frac{4\pi}{3}$, but it would be considerably more difficult to evaluate it directly.

EXAMPLE 6   Let R be the region enclosed between the coordinate planes and the plane $x + y + z = 3$, as shown in Figure 18.5.10, and let S be its boundary surface. If $\mathbf{n}$ is the outer unit normal vector on S, evaluate $\int \int_S (x - y^2 z, x^2 y - 2y, z - x^2 z) \cdot \mathbf{n} \, dS$.

Solution   By the Divergence Theorem the surface integral equals

$$\int \int \int_R \nabla \cdot (x - y^2 z, x^2 y - 2y, z - x^2 z) \, dV = \int \int \int_R (1 + x^2 - 2 + 1 - x^2) \, dV$$

$$= \int \int \int_R 0 \, dV = 0.$$

Evaluating the surface integral directly would not be too difficult, but it requires finding a unit normal vector and integrating over each of the four sides separately. The integral is clearly zero because the divergence of the vector field is zero.

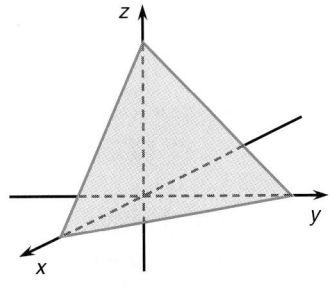

Figure 18.5.10

EXAMPLE 7   Let S be the unit sphere with uniform density $\rho = 1$, shown in Figure 18.5.11. Find its moment of inertia about the z-axis.

Solution   The moment of inertia equals $\int \int_S (x^2 + y^2) \, dS$. The integrand can be interpreted as the dot product $(x, y, 0) \cdot (x, y, z)$, and as we have seen, the outer unit normal vector on the unit sphere is $\mathbf{n} = (x, y, z)$. In other words, the integral equals $\int \int_S (x, y, 0) \cdot \mathbf{n} \, dS$.

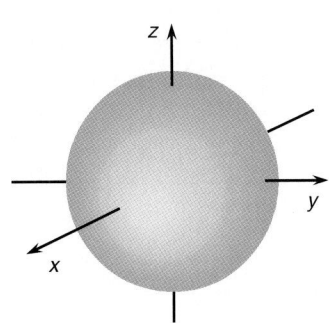

Figure 18.5.11

The Divergence Theorem allows us to integrate much more easily. If the unit ball is R, then

$$\int\int_S (x, y, 0) \cdot \mathbf{n}\, dS = \int\int\int_R \nabla \cdot (x, y, 0)\, dV = \int\int\int_R (1 + 1 + 0)\, dV$$

$$= \int\int\int_R 2\, dV = 2 \times \text{vol}(R) = \frac{8\pi}{3}.$$

This calculation is *much* easier than the calculation needed to find the surface integral directly. We did a similar surface integral in Example 18.4.2 (p. 1177), and it required much more work.

With $\rho = 1$ the mass of the sphere is its area, which is $M = 4\pi$. The radius of gyration of the sphere is $R = \sqrt{\frac{I}{M}} = \sqrt{\frac{2}{3}}$. With a graphics calculator we plot the unit circle and think of it as a cross-section of the sphere. On the same picture we plot the horizontal line $y = \sqrt{\frac{2}{3}}$; we think of it as the location of a small heavy object that has the same properties of rotation about the horizontal axis as the sphere.

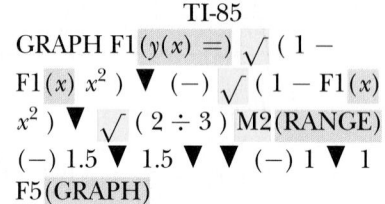

TI-85	SHARP EL9300
GRAPH F1 (y(x) =) √ (1 − F1(x) x²) ▼ (−) √ (1 − F1(x) x²) ▼ √ (2 ÷ 3) M2(RANGE) (−) 1.5 ▼ 1.5 ▼ ▼ (−) 1 ▼ 1 F5(GRAPH)	⟳ √ 1 − x/θ/T x² 2ndF ▼ − √ 1 − x/θ/T x² 2ndF ▼ √ 2 ÷ 3 RANGE (−)1.5 ENTER 1.5 ENTER 1 ENTER (−)1 ENTER 1 ENTER 1 ENTER ⟳
CASIO f_x-6300G (f_x-7700GB)	HP 48SX
Range −1.5 EXE 1.5 EXE 1 EXE −1 EXE 1 EXE 1 EXE EXE EXE EXE Graph √ (1 − X x²) EXE Graph − √ (1 − X x²) EXE Graph √ (2 ÷ 3) EXE	⟳ PLOT PLOTR ERASE ATTN ′ Y = √x () 1 − X yˣ 2 ➤ PLOT ⟳ DRAW ERASE 1.5 +/− ENTER 1.5 XRNG 1 +/− ENTER 1 YRNG DRAW ATTN ′ Y = +/− √x () 1 − X yˣ 2 ➤ PLOT ⟳ DRAW DRAW ATTN ′ Y = √x () 2 ÷ 3 ➤ PLOT ⟳ DRAW DRAW

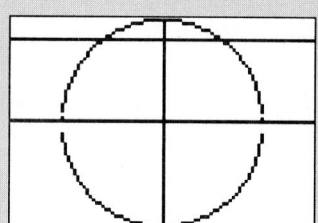

How does the radius of gyration of the unit sphere compare with the radius of gyration of the unit ball?

Exercises 18.5

I

Evaluate each of the following. Here S is a surface, and R is the region enclosed by S. The symbol $\mathbf{n}$ always refers to an outer unit normal vector.

1. $\int\int_S (x, z, y) \cdot \mathbf{n}\, dS$, S is $x^2 + y^2 + z^2 = 1$.

2. $\int\int_S (y - z, z - x, x - y) \cdot \mathbf{n}\, dS$, S is $2x^2 + 3y^2 + 5z^2 = 2$.

3. $\int\int_S (3x + z, z^2 - y, x^3 - 2z) \cdot \mathbf{n}\, dS$, S is $4x^2 + y^2 + 9z^2 = 36$.

4. $\int\int_S (\sin \pi y, \cos \pi x, x + y + z) \cdot \mathbf{n}\, dS$, S is $x^2 + y^2 + z^2 = 25$.

5. $\int\int_S (ax, by, cz) \cdot \mathbf{n}\, dS$, S is $x^2 + y^2 + z^2 = K^2$.

6. $\int\int_S (ax + by + cz)\, dS$, S is $x^2 + y^2 + z^2 = 2$.

7. $\int\int_S (x^2 + z^2)\, dS$, S is $x^2 + y^2 + z^2 = 9$.

8. $\int\int_S (x + 2y - 4z)\, dS$, S is $x^2 + y^2 + z^2 = 4$.

9. $\int\int_S (xy + yz + xz)\, dS$, S is $x^2 + y^2 + z^2 = 16$.

10. $\int\int_S (x + yz, xz, e^{x+y}) \cdot \mathbf{n}\, dS$, R is $x^2 + y^2 \le 1$, $0 \le z \le 2$.

In each of the following situations, verify the Divergence Theorem by evaluating both the surface integral $\int\int_S \mathbf{F} \cdot \mathbf{n}\, dS$ and the integral of $\nabla \cdot \mathbf{F}$ over the corresponding region R in $\mathbb{R}^3$.

11. $\mathbf{F} = (3x, y - z, x + 2y)$, S is $x^2 + y^2 + z^2 = 1$.

12. $\mathbf{F} = (y, x - 2z, x + y + z)$, S is $x^2 + y^2 + z^2 = 4$.

13. $\mathbf{F} = (1 - 3x, 2 + z, y + 3x - z)$, S is $(x - 3)^2 + y^2 + z^2 = 9$.

14. $\mathbf{F} = (y, x - 2z, x + y + z)$, S is $x^2 + y^2 + z^2 = 4$.

15. $\mathbf{F} = (x^2 + y, x + y^3, y + z^3)$, S is $x^2 + y^2 + z^2 = 16$.

16. $\mathbf{F} = (x^2, -2xy, z^2)$, S is $x^2 + y^2 + z^2 = 4$, $z \ge 0$.

17. $\mathbf{F} = (2x, -y^2, z^3)$, R is $|x|, |y|, |z| \le 1$.

18. $\mathbf{F} = (5x + y, y - x, x + 2z)$, R is $0 \le x, y, z \le 4$.

19. $\mathbf{F} = (x, y, z)$, R is $0 \le x, 0 \le y, 0 \le z, x + y \le 1, z \le 1$.

20. $\mathbf{F} = (x + \sin y, -y^2 + z, 2yz - z)$, R is $x^2 + y^2 \le 9$, $|z| \le 2$.

Evaluate each of the following. The symbol $\mathbf{n}$ always refers to an outer unit normal vector.

21. $\int\int_S (2 - x, 4y + z, 3y) \cdot \mathbf{n}\, dS$, S is $x^2 + y^2 + z^2 = 4$, $z \ge 0$.

22. $\int\int_S (x - y, y + z, 2x - z) \cdot \mathbf{n}\, dS$, S is $x^2 + y^2 + z^2 = 4$, $z \ge 1$.

23. $\int\int_S (2x^2 - xy + x, y^2, 3) \cdot \mathbf{n}\, dS$, S is $x^2 + y^2 + z^2 = 1$, $z \le 0$.

24. $\int\int_S (\cos x, y^2, z^2) \cdot \mathbf{n}\, dS$, S is $x^2 + y^2 + z^2 = 1$, $y \ge 0$.

25. $\int\int_S 2x^2\, dS$, S is $x^2 + y^2 + z^2 = 1$.

26. $\int\int_S (2x^2 - 3y^2 + z)\, dS$, S is $x^2 + y^2 + z^2 = 9$.

27. $\int\int_S y^2\, dS$, S is $x^2 + y^2 + z^2 = 4$, $z \ge 0$.

28. $\int\int_S (x^2 + yz)\, dS$, S is $x^2 + y^2 + z^2 = 25$, $z \ge 0$.

29. $\int\int_S (x^2 + y^2)\, dS$, S is $x^2 + y^2 = 1$, $0 \le z \le 1$.

30. $\int\int_S x^2\, dS$, S is $z^2 = x^2 + y^2$, $0 \le z \le 1$.

II

*31. Imagine an explosion at the origin from which objects fly off in different directions and at differing speeds. Suppose that each object maintains a constant speed after the explosion. Consider the objects at time t after the explosion. Show that there is a positive constant k so that an object at (x, y, z) has velocity $\mathbf{v}(x, y, z) = (kx, ky, kz)$ (i.e., show that the constant k is the same at every (x, y, z)).

*32. In Example 2 we found that for points moving with velocity $\mathbf{v} = (kx, ky, kz)$ (with $k > 0$) the relative speed between two points P, Q is k times the distance between P and Q. Think of some other velocity vector fields with this property.

*33. Solve Example 5 (p. 1187) directly (i.e., without the Divergence Theorem).

*34. Solve Example 6 (p. 1187) directly (i.e., without the Divergence Theorem).

*35. Solve Example 7 (p. 1187) directly (i.e., without the Divergence Theorem).

*36. Evaluate the integral of Example 18.4.2 (p. 1177) using the Divergence Theorem.

*37. Find the moment of inertia about the z-axis of the upper hemisphere $x^2 + y^2 + z^2 = 1$, $z \ge 0$, with uniform density $\rho = 1$.

*38. Find the moment of inertia about the y-axis of the upper hemisphere $x^2 + y^2 + z^2 = 1$, $z \ge 0$, with uniform density $\rho = 1$.

*39. Find the moment of inertia about the x-axis of the sphere of radius R with uniform density $\rho = 1$ and center at the origin.

*40. Find the moment of inertia about the z-axis of the surface $x^2 + y^2 + z^2 = R^2$, $0 \le z \le \frac{R}{2}$, with uniform density $\rho = 1$.

*41. Suppose S is an orientable surface and R is the region it encloses. Let $\mathbf{n}$ be an outward-pointing unit normal, and let $\mathbf{F} = (x, y, z)$. Show that

$$Vol(R) = \frac{1}{3} \int \int_S \mathbf{F} \cdot \mathbf{n} \, dS.$$

42. Use Exercise 41 to find the volume of $x^2 + y^2 + z^2 \leq 4$, $z \geq 0$.

43. Find the volume of the cone $z^2 = x^2 + y^2$, $0 \leq z \leq 1$.

44. Find the volume of the cone of radius r and height h.

POINT TO PONDER

The Divergence Theorem involves the outer unit normal vectors on a surface S, and we have always assumed that the surfaces we deal with do have outer unit normal vectors. For the surfaces we are likely to encounter in any practical problem this is a safe assumption, but it is interesting to see that there are surfaces that are not orientable, that is, for which it is not possible to find unit normal vectors without having them "jump" from one side to the other.

The standard example is the Möbius strip. It is constructed by making a loop of paper with a single twist in it.

In Figure 18.5.12 we see what happens when we try to find unit normal vectors on a Möbius strip. We start with the vector at the top, moving counterclockwise around the loop, and find that the normal vector twists around until by the time it reaches the top of the loop again, it is pointing downward. When it returns to the starting point at the top of the loop, it will not be the same as the original starting vector but will be its *negative*.

We have attempted to form a family of unit normal vectors, but they "jump" from one side to the other at that point at the top. If we try starting at other points, the same thing will happen.

In a sense we needed outer unit normal vectors to show us which is the "outside" and which is the inside of a surface. A nonorientable surface like the Möbius strip does not enclose any region R in space; it does not have either an inside or an outside. The statement of the Divergence Theorem does not even make sense for such a surface.

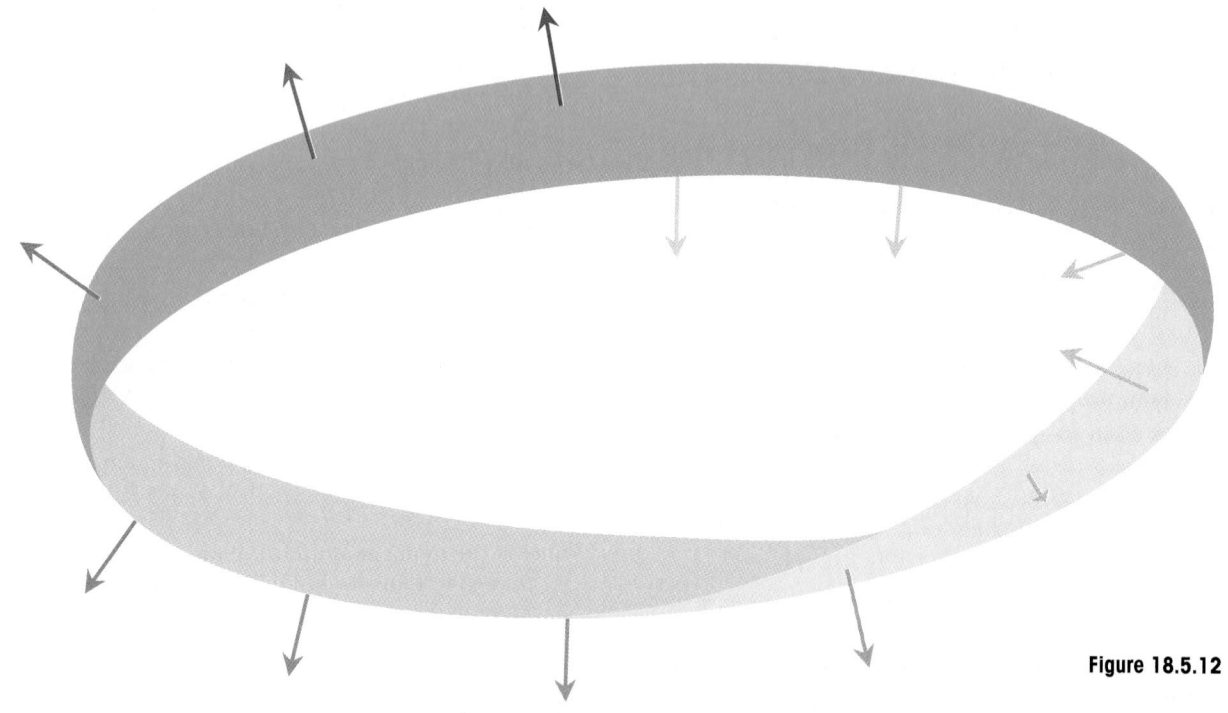

Figure 18.5.12

18.6 STOKES' THEOREM

Figure 18.6.1

Figure 18.6.2

Figure 18.6.3

We have studied Green's Theorem, which relates an integral over a region R in $\mathbb{R}^2$ to a line integral over the curve C that forms the boundary of R. For instance, we might apply Green's Theorem to a disk R whose boundary is a circle C.

There have been a number of instances in which we have found it convenient to think of $\mathbb{R}^2$ as the horizontal plane in $\mathbb{R}^3$, and by doing this we can think of the disk and the circle as subsets of $\mathbb{R}^3$ (see Figure 18.6.1).

On the other hand there are many other ways to include the disk and circle in $\mathbb{R}^3$. We can imagine them lying in any horizontal plane $z = a$, and by changing coordinates it is easy to place them in the xz-plane, the yz-plane, or any plane $x = a$ or $y = a$. For that matter it is easy to visualize a disk in any plane (see Figure 18.6.2).

Now consider the sphere $x^2 + y^2 + z^2 = 9$, and suppose we cut off a small cap from the top of the sphere, as shown in Figure 18.6.3. This surface S is curved, so it is not a disk, but on the other hand it is quite similar to a disk, and its boundary is a circle. It is a slightly "bent" disk. It seems reasonable to look for a formula that relates a surface integral over a surface S to some line integral over the curve C that forms the boundary of S. This idea leads us to what is called Stokes' Theorem, after the Irish mathematician and physicist George Gabriel Stokes (1819–1903).

The statement of Green's Theorem uses partial derivatives with respect to x and y, so it is difficult to see how to express something analogous for a curved surface in $\mathbb{R}^3$. However, near the end of Section 18.3 we found a version of Green's Theorem that makes no apparent reference to partial derivatives. Formula 18.3.7 (p. 1173) uses the *curl* of a vector field, which is supposed to measure how much and in which direction the vector field is rotating. For a vector field in the plane any rotation must be either clockwise or counterclockwise, but to identify a rotation in $\mathbb{R}^3$, it is necessary to specify the axis of the rotation in addition to the amount of rotation. This leads us to expect that in $\mathbb{R}^3$ the curl of a vector field should be a vector.

> **DEFINITION 18.6.1**
>
> If $\mathbf{F} = (f, g, h)$ is a vector field, then its **curl** is the vector field
>
> $$\operatorname{curl}\mathbf{F} = \left(\frac{\partial h}{\partial y} - \frac{\partial g}{\partial z}, \frac{\partial f}{\partial z} - \frac{\partial h}{\partial x}, \frac{\partial g}{\partial x} - \frac{\partial f}{\partial y}\right).$$

Just as we remembered the divergence by using the notation $\operatorname{div}\mathbf{F} = \nabla \cdot \mathbf{F}$, in which ∇ is regarded as some (unspecified) kind of vector, it is possible to abbreviate the formula for curl by thinking of it as the cross product

$$\operatorname{curl}\mathbf{F} = \nabla \times \mathbf{F} = \left(\frac{\partial}{\partial x}, \frac{\partial}{\partial y}, \frac{\partial}{\partial z}\right) \times (f, g, h).$$

Once again this cross product notation does not really mean anything, but it can be helpful as a mnemonic device.

The idea of the curl is that it is a vector that points along the axis around which $\mathbf{F}$ is circulating, and its length gives the speed of the rotation.

EXAMPLE 1

(i) If $\mathbf{F}(x, y, z) = (x \sin y, e^{xyz}, x + 2y + 3z)$, then curl $\mathbf{F} = (2 - xye^{xyz}, 0 - 1, yze^{xyz} - x \cos y) = (2 - xye^{xyz}, -1, yze^{xyz} - x \cos y)$.

(ii) If $\mathbf{F}(x, y, z) = \big(f(x, y), g(x, y), 0\big)$, then $\mathbf{F}$ is a "horizontal" vector field, in the sense that it is a horizontal vector at each point. Since in addition it is independent of the variable z, we find that curl $\mathbf{F} = \big(0, 0, \frac{\partial g}{\partial x} - \frac{\partial f}{\partial y}\big)$. This is a "vertical" vector.

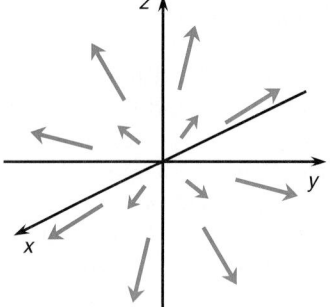

For vector fields in $\mathbb{R}^2$ the axis of rotation must always be in the vertical (z) direction. This explains the definition we made in Definition 18.3.6 (p. 1173) for the curl of a vector field in the xy-plane. We only used the third component $\frac{\partial g}{\partial x} - \frac{\partial f}{\partial y}$, so in that situation the curl was a *function*.

(iii) If $\mathbf{F}(x, y, z) = (x, y, z)$, then curl $\mathbf{F} = (0, 0, 0)$. The vector field $\mathbf{F}$ is illustrated in Figure 18.6.4; it "expands" out from the origin but does not rotate.

(iv) The vector field $\mathbf{F}(x, y, z) = (0, -z, y)$ corresponds to a rotation around the x-axis (see Figure 18.6.5). We evaluate: curl $\mathbf{F} = \big(1 - (-1), 0, 0\big) = (2, 0, 0)$. This vector points in the direction of the positive x-axis, illustrating the way the curl of a vector field is a vector that points along the axis of the rotation.

Figure 18.6.4

To explain Stokes' Theorem let us recall that in Equation 18.3.7 (p. 1173) we interpreted Green's Theorem in the plane to say that the circulation of $\mathbf{F}$ around the boundary curve C equals the integral of curl $\mathbf{F}$ over the interior region S. In terms of a flow, this means that if the fluid is rotating inside S, then there must be some circulation around the boundary. See Figure 18.6.6.

Now suppose we want a similar result for a curved surface in $\mathbb{R}^3$. The circulation around the boundary curve C can be defined in exactly the same way as in the plane; it equals $\int_C \mathbf{F} \cdot \mathbf{T}(s)\, ds = \int_C \mathbf{F} \cdot d\mathbf{r}$.

However, the integral over S is more subtle. Suppose we consider the rotational field $\mathbf{F} = (0, -z, y)$ and take for S a flat horizontal surface, as shown in Figure 18.6.7. There is a lot of rotation, but it will not result in any net circulation around the boundary of S. For that to happen, the vector field would have to be aligned so that there was some rotation *inside* the surface S.

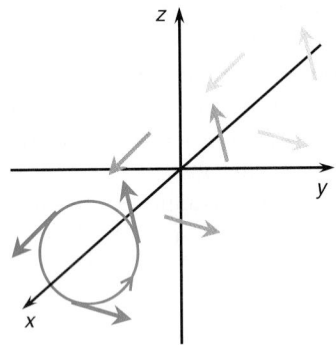

Figure 18.6.5

Another way to say this is that the axis of the rotation would have to point out of the surface. This suggests that what we should integrate over S is the *normal component* of curl $\mathbf{F}$, that is, curl $\mathbf{F} \cdot \mathbf{n}$. This function measures the rotation "in the surface S."

So Stokes' Theorem will say that $\int_C \mathbf{F} \cdot d\mathbf{r} = \int \int_S$ curl $\mathbf{F} \cdot \mathbf{n}\, dS$. The one remaining complication is choosing the direction of the unit normal vectors $\mathbf{n}$ and the direction of the boundary curve C.

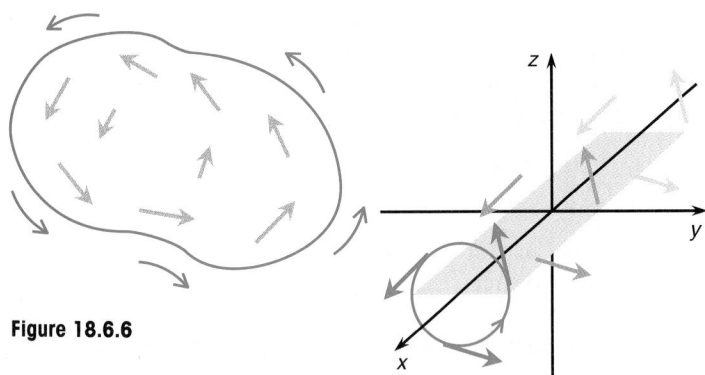

Figure 18.6.6

Figure 18.6.7

The first step is to assume that the surface S is orientable, which means that it is possible to choose the unit normal vectors in a consistent way. Having done this, we must find the right direction for the boundary curve C. The idea is simple: Imagine yourself walking around the boundary curve. Stand so that your head points in the same direction as the normal vectors $\mathbf{n}$; when you walk around the boundary, the region S should be on your left side. If it is, then you are walking in the positive direction around the curve C. See Figure 18.6.8. We say the boundary curve is **positively oriented** relative to the normal vectors $\mathbf{n}$.

Roughly speaking, if you look "down" on the surface, that is, looking back along the normal vectors, then the correct direction for the boundary curve C is "counterclockwise." In effect we have used the normal vectors to give us an idea of which directions are "up" and "down."

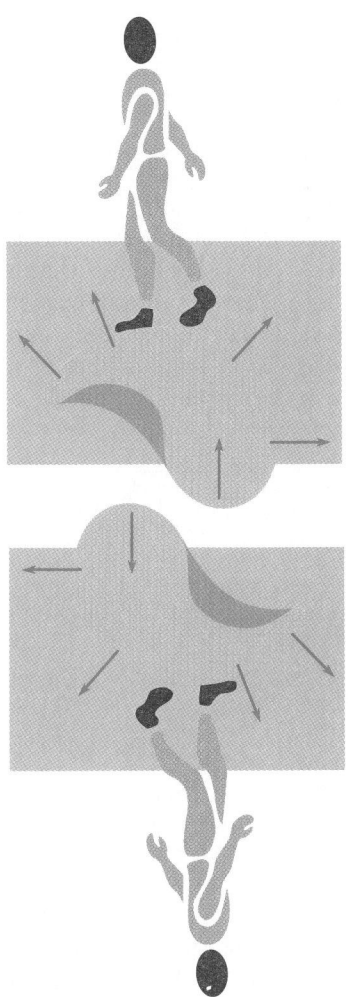

THEOREM 18.6.2

Stokes' Theorem

Suppose S is an orientable surface in $\mathbb{R}^3$ with boundary curve C. Let $\mathbf{n}(x, y, z)$ be a unit normal vector to S at each point $(x, y, z) \in S$, chosen so that $\mathbf{n}$ does not "jump" from one side to the other on S.

Choose the direction so that C is positively oriented (cf. the discussion above).

Suppose $\mathbf{F}$ is a vector field whose components have continuous partial derivatives on an open set that contains S and C. Then

$$\int_C \mathbf{F} \cdot d\mathbf{r} = \int\int_S \operatorname{curl} \mathbf{F} \cdot \mathbf{n} \, dS.$$

The proof of this result is difficult, but at least it is apparent that when S lies in the xy-plane, then it is the same as Green's Theorem.

Figure 18.6.8

EXAMPLE 2   Suppose S is the upper hemisphere $x^2 + y^2 + z^2 = 1$, $z \geq 0$, and $\mathbf{F} = (xy, x^2 z - x, x + y + z)$.

The boundary curve C is the circle $x^2 + y^2 = 1$, $z = 0$ in the xy-plane. If we choose the *outer* unit normals $\mathbf{n}(x, y, z) = (x, y, z)$, then to walk around the boundary with the surface on our left, it is necessary to walk counterclockwise, as shown in Figure 18.6.9.

In this direction the curve C can be parametrized by $\mathbf{r}(t) = (\cos t, \sin t, 0)$, so

$$\int_C \mathbf{F} \cdot d\mathbf{r} = \int_0^{2\pi} \mathbf{F}(\cos t, \sin t, 0) \cdot (-\sin t, \cos t, 0) \, dt$$

$$= \int_0^{2\pi} (\cos t \sin t, 0 - \cos t, \cos t + \sin t) \cdot (-\sin t, \cos t, 0) \, dt$$

$$= \int_0^{2\pi} (-\cos t \sin^2 t - \cos^2 t) \, dt = -\int_0^0 u^2 \, du - \frac{1}{2} \int_0^{2\pi} (1 + \cos 2t) \, dt$$

$$= 0 - \frac{1}{2} 2\pi = -\pi.$$

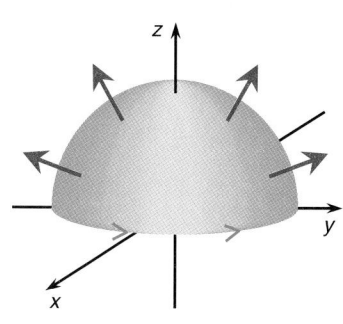

Figure 18.6.9

On the other hand we find that curl $\mathbf{F}$ equals

$$\left(\frac{\partial}{\partial y}(x+y+z) - \frac{\partial}{\partial z}(x^2z - x), \frac{\partial}{\partial z}(xy) - \frac{\partial}{\partial x}(x+y+z), \frac{\partial}{\partial x}(x^2z - x) - \frac{\partial}{\partial y}(xy)\right)$$

$$= (1 - x^2, 0 - 1, 2xz - 1 - x).$$

From this we find that curl $\mathbf{F} \cdot \mathbf{n} = (1 - x^2, -1, 2xz - 1 - x) \cdot (x, y, z) = x - x^3 - y + 2xz^2 - z - xz$. Rather than evaluating the surface integral in detail, let us observe that most of the terms will be zero. In fact, $\int\int_S x \, dS = 0$, because x is an odd function of x, so the integral of x over the part of S where $x \geq 0$ will exactly cancel the integral over the part where $x \leq 0$. The same is true of $-x^3$, $2xz^2$, and $-xz$, and a similar remark can be made about $-y$, since it is an odd function of y. Only one term remains, and letting D be the unit disk in the xy-plane, we have

$$\int\int_S \text{curl } \mathbf{F} \cdot \mathbf{n} \, dS = \int\int_S (-z) \, dS = -\int\int_D \sqrt{1 - x^2 - y^2} \frac{1}{\sqrt{1 - x^2 - y^2}} \, dx \, dy$$

$$= -\int\int_D 1 \, dA = -\text{area}(D) = -\pi.$$

We find that the line integral equals the surface integral, and Stokes' Theorem is verified in this example.

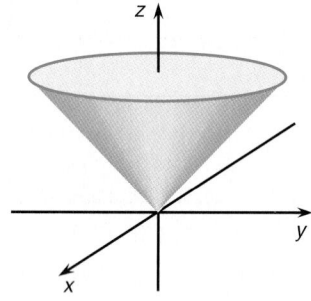

Figure 18.6.10

EXAMPLE 3 Let S be the surface $z^2 = x^2 + y^2$, $0 \leq z \leq 3$, which is the cone illustrated in Figure 18.6.10. Let $\mathbf{F}(x, y, z) = (-y, x, -5)$. Verify Stokes' Theorem in this situation.

Solution First we should identify the boundary curve C. From the picture we see that the boundary is the circle in the plane $z = 3$. The intersection of this plane with $z^2 = x^2 + y^2$ is the circle $x^2 + y^2 = 9$, $z = 3$. The single point at the bottom of the cone is not part of the boundary.

Next we need normal vectors. Writing the cone as the graph of $z = f(x, y) = \sqrt{x^2 + y^2}$, that is, as a level surface of $f(x, y) - z$, we find a normal vector by taking the gradient

$$(f_x, f_y, -1) = \left(\frac{x}{\sqrt{x^2 + y^2}}, \frac{y}{\sqrt{x^2 + y^2}}, -1\right).$$

Observing that the norm of this vector is

$$\sqrt{\frac{x^2}{x^2 + y^2} + \frac{y^2}{x^2 + y^2} + 1} = \sqrt{1 + 1} = \sqrt{2},$$

we divide by $\sqrt{2}$ to find a unit normal vector

$$\mathbf{n} = \left(\frac{x}{\sqrt{2x^2 + 2y^2}}, \frac{y}{\sqrt{2x^2 + 2y^2}}, -\frac{1}{\sqrt{2}}\right).$$

It is easy to find the curl: curl $\mathbf{F} = (0, 0, 2)$. Moreover, for the surface integral we find that $dS = \sqrt{1 + f_x^2 + f_y^2} \, dx \, dy = \sqrt{2} \, dA$, so letting D be the disk $x^2 + y^2 \leq 9$,

we find that

$$\iint_S \text{curl } \mathbf{F} \cdot \mathbf{n} \, dS = \iint_S (0, 0, 2) \cdot \left(\frac{x}{\sqrt{2x^2 + 2y^2}}, \frac{y}{\sqrt{2x^2 + 2y^2}}, -\frac{1}{\sqrt{2}} \right) dS$$

$$= -\iint_S \sqrt{2} \, dS = -\iint_D \sqrt{2} \cdot \sqrt{2} \, dA = -2 \, \text{area}(D)$$

$$= -2\pi r^2 = -18\pi.$$

We notice that the z-component of $\mathbf{n}$ is negative, which means that the normals point downward, as shown in Figure 18.6.11. Imagining ourselves walking around the rim with our heads pointing outward, we see that in order to have the surface on the left side it is necessary to walk in the *clockwise* direction. This means that the boundary curve C can be parametrized by $\mathbf{r}(t) = (3 \cos t, -3 \sin t, 3)$. The circulation around the boundary curve is

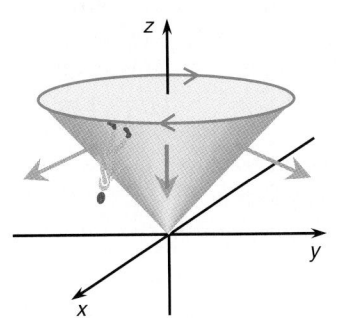

Figure 18.6.11

$$\int_C \mathbf{F} \cdot d\mathbf{r} = \int_0^{2\pi} (-y, x, -5) \cdot (-3 \sin t, -3 \cos t, 0) \, dt$$

$$= \int_0^{2\pi} (3 \sin t, 3 \cos t, -5) \cdot (-3 \sin t, -3 \cos t, 0) \, dt$$

$$= \int_0^{2\pi} (-9 \sin^2 t - 9 \cos^2 t) \, dt = -18\pi.$$

Once again we find that the two integrals are equal, and Stokes' Theorem is verified.

Earlier we said that the correct direction for the boundary curve was "counterclockwise," and in Example 3 we found that it is clockwise. The reason is that the normal vectors we used point *downward*, so as we walk around the rim we should be looking up at the surface *from below*. From this perspective the direction we have chosen is "counterclockwise."

There is another observation to be made about this. We might just as well have chosen the upward unit normal vectors; they are the negatives of the ones we used here. This would change the sign of the surface integral. On the other hand, it would also cause us to walk around the "inside" of the cone, so in order to have it on the left side it would be necessary to walk the opposite direction. This would change the sign of the line integral, and the net result would be that the two integrals would still be equal.

There is no single choice of normals or of boundary direction; the point is to choose the two in a compatible way.

In the examples we have considered, it has been relatively easy to evaluate both the line integral for the circulation and the surface integral of the normal component of the curl, but of course much of the value of Stokes' Theorem comes from situations in which one of these integrals is difficult and the other is easier.

EXAMPLE 4   Let S be the surface $z = x^2 + y^2$, $z \leq 2 - 2x - 2y$, which is part of a paraboloid. If $\mathbf{n}$ is the outer unit normal, that is, the normal that points downward, and $\mathbf{F} = (x - y, 6x + z, 3x + 3y + z)$, evaluate $\iint_S \text{curl } \mathbf{F} \cdot \mathbf{n} \, dS$.

Solution   Setting up the integral over this surface S is moderately complicated, so it is preferable to evaluate the line integral over the boundary.

We begin by observing that the boundary is the intersection of the paraboloid $z = x^2 + y^2$ and the plane $z = 2 - 2x - 2y$. Equating these two expressions, we find that $x^2 + y^2 = 2 - 2x - 2y$, or $x^2 + y^2 + 2x + 2y = 2$. Completing the squares, we find that this is $x^2 + 2x + 1 + y^2 + 2y + 1 = 4$, that is, $(x + 1)^2 + (y + 1)^2 = 4$. The boundary curve is given by $(x + 1)^2 + (y + 1)^2 = 4$, $z = 2 - 2x - 2y$. It is an ellipse in the plane $z = 2 - 2x - 2y$, lying over the circle $(x + 1)^2 + (y + 1)^2 = 4$ in the xy-plane. See Figure 18.6.12.

From the picture we see that C is traversed in the clockwise direction if we look down from the positive z-axis, so it can be parametrized by letting (x, y) traverse the circle $(x + 1)^2 + (y + 1)^2 = 4$ clockwise and then solving for z using $z = 2 - 2x - 2y$. We find that C is

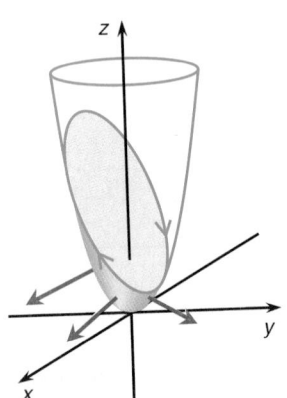

Here we let $x + 1 = 2 \cos t$, $y + 1 = -2 \sin t$ and then solve for z.

$$\mathbf{r}(t) = (-1 + 2 \cos t, -1 - 2 \sin t, 6 - 4 \cos t + 4 \sin t).$$

Stokes' Theorem then tells us that

$$\int \int_S \text{curl } \mathbf{F} \cdot \mathbf{n} \, dS = \int_C \mathbf{F} \cdot d\mathbf{r}$$

$$= \int_0^{2\pi} (2 \cos t + 2 \sin t, 8 \cos t + 4 \sin t, 2 \cos t - 2 \sin t)$$

$$\cdot (-2 \sin t, -2 \cos t, 4 \sin t + 4 \cos t) \, dt$$

$$= \int_0^{2\pi} (-12 \sin^2 t - 8 \cos^2 t - 12 \cos t \sin t) \, dt = -20\pi.$$

The surface integral equals -20π, which we have found by evaluating the line integral and applying Stokes' Theorem. Notice that we never had to find curl $\mathbf{F}$. ◢

Figure 18.6.12

EXAMPLE 5   Let $\mathbf{F} = (x, y, z)$, and let C be the curve that is the intersection of the surface $z = x^4 + y^6$ with the ellipsoid $x^2 + 4y^2 + 9z^2 = 36$. Evaluate $\int_C \mathbf{F} \cdot d\mathbf{r}$, with C parametrized so that (x, y) traces out a counterclockwise curve in the xy-plane.

Solution   It is quite tricky to parametrize the curve C, but if we are lucky, we will remember that curl $\mathbf{F} = (0, 0, 0)$. Because of this, curl $\mathbf{F} \cdot \mathbf{n} = 0$ for *any* vector $\mathbf{n}$, and we see that the surface integral $\int \int_S \text{curl } \mathbf{F} \cdot \mathbf{n} \, dS = \int \int_S 0 \, dS = 0$ for any surface S. Since Stokes' Theorem says that the line integral we want is equal to this surface integral, with S chosen to be the part of the upper half of the ellipsoid that is contained inside C, we find that $\int_C \mathbf{F} \cdot d\mathbf{r} = 0$. ◢

Another way that Stokes' Theorem can be used is illustrated in the following example.

EXAMPLE 6   Let $\mathbf{F} = (4x^2 e^z + x^2 y, -xy^2, x^5 y^3 z^4)$, and let S be the upper hemisphere $x^2 + y^2 + z^2 = 4$, $z \geq 0$. Evaluate $\int \int_S \text{curl } \mathbf{F} \cdot \mathbf{n} \, dS$, where $\mathbf{n}$ is the outer unit normal vector to the hemisphere.

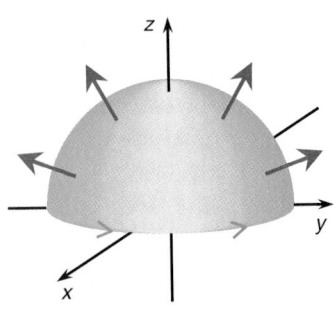

Figure 18.6.13

Solution   Direct evaluation of the surface integral involves a complicated integral. Stokes' Theorem tells us that it equals a line integral around the boundary curve C, which is just the circle $x^2 + y^2 = 4$, $z = 0$ in the xy-plane (see Figure 18.6.13). However, this line integral is not terribly simple to evaluate either.

Now suppose we think of another surface with the same boundary curve C. A particularly simple surface is the disk D in the xy-plane given by $x^2 + y^2 \leq 4$, $z = 0$. Its boundary is the same circle C, so we can apply Stokes' Theorem to it. From the illustration we see that the direction of the curve C is counterclockwise in the horizontal plane; to be consistent with this, we have to choose the unit normals on D that point *up*. So on D we let $\mathbf{n}_D = (0, 0, 1)$ at each point. With this understanding we have that $\int_C \mathbf{F} \cdot d\mathbf{r} = \int\int_D \text{curl } \mathbf{F} \cdot \mathbf{n}_D \, dS$.

To evaluate this last integral, we need to find curl $\mathbf{F}$, but actually, since it is curl $\mathbf{F} \cdot \mathbf{n}_D = \text{curl } \mathbf{F} \cdot (0, 0, 1)$ we need, it is only necessary to find the last component of curl $\mathbf{F}$. This component is $\frac{\partial}{\partial x}(-xy^2) - \frac{\partial}{\partial y}(4x^2 e^z + x^2 y) = -y^2 - x^2$. We find that

$$\int\int_S \text{curl } \mathbf{F} \cdot \mathbf{n} \, dS = \int_C \mathbf{F} \cdot d\mathbf{r} = \int\int_D \text{curl } \mathbf{F} \cdot \mathbf{n}_D \, dS = \int\int_D (-x^2 - y^2) \, dS$$

$$= -\int\int_D r^2 \, dA = -\int_0^{2\pi}\int_0^2 (r^2)\, r \, dr \, d\theta$$

$$= -\int_0^{2\pi}\left(\frac{1}{4}r^4\Big|_0^2\right) d\theta = -\int_0^{2\pi} 4 \, d\theta = -8\pi.$$

> Surface area dS on the disk is just dA.

The original surface integral over the hemisphere S equals the line integral over C, by Stokes' Theorem. Another application of Stokes' Theorem shows that this equals the surface integral over the disk D, which is easily evaluated. It equals -8π. ◢

This example illustrates a simple but useful idea.

With a computer we can try to visualize the intersection of the two surfaces from Example 5. We know what an ellipsoid looks like, and the graph of $z = x^4 + y^6$ must have a sort of bowl shape.

We plot the *maximum* of the bowl function and the function that gives the upper half of the ellipsoid. This gives us a picture of the intersection.

Mathematica	Maple
Plot3D [Max [$x^4 + y^6$], Sqrt [36 − $x^2 - 4y^2$] / 3], {x, −1.2, 1.2}, {y, −1.1, 1.1}] ENTER	plot3d(Max ($x^4 + y^6$, Sqrt (36 − x^2 − $4*y^2$) / 3), x = −1.2..1.2, y = −1.1..1.1); ENTER
DERIVE O D G ENTER A MAX ($x\hat{\ }4 + y\hat{\ }6$, SQRT (36 − $x\hat{\ }2 - 4\,y\hat{\ }2$) / 3) ENTER P O L 2.4 Tab 2.2 ENTER G 15 Tab 15 ENTER P	**MacFunction** max ($x\hat{\ }4 + y\hat{\ }6$, sqrt (36 − $x\hat{\ }2$ − $4*y\hat{\ }2$) / 3) OK Options Remove Hidden Surfaces

Now you can try plotting the *minimum* of the two functions, with $x \in [-3, 3]$ and $y \in [-2, 2]$, or perhaps the part of the first graph near one of the "corners."

THEOREM 18.6.3

> Suppose two surfaces S_1 and S_2 have the same boundary curve C. Suppose unit normal vectors are chosen on S_1 and S_2 in such a way that the corresponding directions on C are the same. If $\mathbf{F}$ is a differentiable vector field on an open set that contains S_1, S_2, and C, then the integrals over the two surfaces are the same:
>
> $$\iint_{S_1} \operatorname{curl} \mathbf{F} \cdot \mathbf{n}\, dS = \iint_{S_2} \operatorname{curl} \mathbf{F} \cdot \mathbf{n}\, dS.$$

Summary 18.6.4

The three main theorems of this chapter have various similarities, but they are distinct.

Green's Theorem:
$$\iint_R \left(\frac{\partial g}{\partial x} - \frac{\partial f}{\partial y} \right) dA = \int_C f\, dx + g\, dy,$$

or
$$\iint_R \operatorname{curl} \mathbf{F}\, dA = \int_C \mathbf{F} \cdot d\mathbf{r} \quad (\mathbf{F} = (f, g)),$$

or
$$\iint_R \nabla \cdot \mathbf{F}\, dA = \int_C \mathbf{F} \cdot \mathbf{n}\, ds;$$

Divergence Theorem:
$$\iiint_R \nabla \cdot \mathbf{F}\, dV = \iint_S \mathbf{F} \cdot \mathbf{n}\, dS;$$

Stokes' Theorem:
$$\iint_S \operatorname{curl} \mathbf{F} \cdot \mathbf{n}\, dS = \int_C \mathbf{F} \cdot d\mathbf{r}.$$

Exercises 18.6

I

Evaluate each of the following. The symbol $\mathbf{n}$ refers to a unit normal vector, pointing upward in the middle of the surface S.

1. $\int_C \mathbf{F} \cdot d\mathbf{r}$, $\mathbf{F} = (2x, -y, 3z)$, C is $x^2 + 4y^2 = 4$, $z = 2$.

2. $\int_C \mathbf{F} \cdot d\mathbf{r}$, $\mathbf{F} = (x, z, -y)$, C is $y^2 + z^2 = 9$, $x = 3$.

3. $\int_C \mathbf{F} \cdot d\mathbf{r}$, $\mathbf{F} = (z, x, 2y)$, C is $x^2 + y^2 = 2$, $z = -1$.

4. $\int_C \mathbf{F} \cdot d\mathbf{r}$, $\mathbf{F} = (y, y, y)$, C is $4x^2 + 9y^2 = 36$, $z = 0$.

5. $\int_C \mathbf{F} \cdot d\mathbf{r}$, $\mathbf{F} = (y^2, 2x, e^{xy})$, C is $9x^2 + 16y^2 = 144$, $z = 4$.

6. $\int_C \mathbf{F} \cdot d\mathbf{r}$, $\mathbf{F} = (3, -5, 13)$, C is $x^2 + 4y^2 = 9$, $x + y + z = 4$.

7. $\int \int_S \operatorname{curl} \mathbf{F} \cdot \mathbf{n}\, dS$, $\mathbf{F} = (yz, xz, xy)$, S is $x^2 + y^2 + z^2 = 9$, $z \geq 0$.

8. $\int \int_S \operatorname{curl} \mathbf{F} \cdot \mathbf{n}\, dS$, $\mathbf{F} = (y, -x, z^2)$, S is $z = x^2 + y^2$, $z \leq 4$.

9. $\int \int_S \operatorname{curl} \mathbf{F} \cdot \mathbf{n}\, dS$, $\mathbf{F} = (yz, -xz, xyz)$, S is $x^2 + y^2 + z^2 = 4$, $z \geq -\sqrt{3}$.

10. $\int \int_S \operatorname{curl} \mathbf{F} \cdot \mathbf{n}\, dS$, $\mathbf{F} = (y, y, 3)$, S is $x^2 + y^2 \leq 4$, $x + y + z = 1$.

Verify Stokes' Theorem in each of the following situations.

11. $\mathbf{F} = (2y, 3x, z)$, S is $x^2 + y^2 + z^2 = 1$, $z \geq 0$.

12. $\mathbf{F} = (x^2 - y, y, z)$, S is $z = x^2 + y^2$, $z \leq 4$.

13. $\mathbf{F} = (x, z, z - y)$, S is $x^2 + y^2 + z^2 = 1$, $x \geq 0$.

14. $\mathbf{F} = (xy, y^2, yz)$, S is $x^2 + y^2 + z^2 = 4$, $y \leq 0$.

15. $\mathbf{F} = (x, 2z, -2y)$, S is $x + y + z = 3$, $x^2 + y^2 \leq 1$.

16. $\mathbf{F} = (y, z, x)$, S is $2x - 3y + 2z = 4$, $y^2 + z^2 \leq 4$.

17. $\mathbf{F} = (z, x, y)$, S is $x^2 + y^2 + z^2 = 1$, $z \geq 0$, $x \geq 0$.

18. $\mathbf{F} = (2y, x, 4z)$, S is $x^2 + y^2 + z^2 = 9$, $x \geq 0$, $y \geq 0$, $z \geq 0$.

19. $\mathbf{F} = (x^3, 3z, -3y)$, S is $x^2 = y^2 + z^2$, $0 \le x \le 5$.

20. $\mathbf{F} = (z, 2y, -2x)$, S is $y = 4$, $0 \le x \le 1$, $0 \le z \le 2$.

In each of the following situations the curve C is parametrized by $\mathbf{r}(t)$. Evaluate the line integral $\int_C \mathbf{F} \cdot d\mathbf{r}$.

21. $\mathbf{r}(t) = (3, \sin t, \cos t)$, $t \in [0, 2\pi]$,
$\mathbf{F} = (x^2 + yz, xy + z, xz)$.

22. $\mathbf{r}(t) = (\sin 2t, -1, \cos 2t)$, $t \in [0, \pi]$,
$\mathbf{F} = \left(3xy - z, \ln(1 + y^4), y^2 z^3 \right)$.

23. $\mathbf{r}(t) = (2 \sin t, 3 \cos t, 4)$, $t \in [0, 2\pi]$,
$\mathbf{F} = (x^2 + xz, yz^2 - 3x, xz)$.

24. $\mathbf{r}(t) = (4 \cos t, 7, 6 \sin t)$, $t \in [0, 2\pi]$,
$\mathbf{F} = (x^2 y, \sin xyz, 2x + 5y^2 z^5)$.

25. $\mathbf{r}(t) = (\cos t, \sin t, \cos t + \sin t)$, $t \in [0, 2\pi]$,
$\mathbf{F} = (x^2 + y, z, z^4 + 2x)$.

26. $\mathbf{r}(t) = (\sin t, \cos t, \cos t + 2 \sin t)$, $t \in [0, 2\pi]$,
$\mathbf{F} = (x - y, x + 2z, y - 3z)$.

27. $\mathbf{r}(t) = (2 \sin t, 3 \cos t, 2 \sin t + 6 \cos t)$, $t \in [0, 2\pi]$,
$\mathbf{F} = (x, x, x)$.

28. $\mathbf{r}(t) = (\cos t, 5 \cos t, \sin t)$, $t \in [0, 2\pi]$,
$\mathbf{F} = (\cos x, z, 2y)$.

29. $\mathbf{r}(t) = (2 \cos 2t, 3 \sin 2t, \cos^2 t - \sin^2 t)$, $t \in [0, \pi]$,
$\mathbf{F} = (x + z, y + 2x - z, 2x - z)$.

30. $\mathbf{r}(t) = (\cos t, 4 \sin t + 2 \cos t, \sin t - \cos t)$,
$t \in [0, 2\pi]$, $\mathbf{F} = (y + z, x + z, x + y)$.

II

*31. Let $\mathbf{F}$ be the vector field whose value $\mathbf{F}(x, y, z)$ at (x, y, z) is a unit vector pointing away from the origin. (i) Find a formula for $\mathbf{F}(x, y, z)$, valid for any point except $(x, y, z) = (0, 0, 0)$. (ii) Find curl $\mathbf{F}$, and compare with Example 1(iii) (p. 1192).

*32. Let S be the sphere $x^2 + y^2 + z^2 = 4$, and let $\mathbf{n}$ be the outer unit normal vector. If $\mathbf{F} = (xyz, 2xy^2 - z, y^2 z + 4x)$, evaluate $\int \int_S$ curl $\mathbf{F} \cdot \mathbf{n} \, dS$.

*33. Suppose $\mathbf{F}$ is a vector field of the form $\mathbf{F}(x, y, z) = \left(f(x), g(y), h(z) \right)$, where f, g, h are continuously differentiable functions. Show that its circulation around any closed loop in $\mathbb{R}^3$ is zero.

*34. Suppose $\mathbf{F}$ is a vector field of the form $\mathbf{F} = \nabla f$, for some twice continuously differentiable function f. Show that the circulation of $\mathbf{F}$ around any closed loop in $\mathbb{R}^3$ is zero.

**35. The surface S given by $x^2 + y^2 = 2z + z^2 + z^3$, $0 \le z \le 2$, is not easily expressed as a graph. If $\mathbf{F} = (y +$

$e^x, z + \cos y, xyz)$ and $\mathbf{n}$ is the outer unit normal to S, evaluate $\int \int_S$ curl $\mathbf{F} \cdot \mathbf{n} \, dS$.

*36. Explain the remark made after the statement of Stokes' Theorem (Theorem 18.6.2) that for a surface S that lies in the xy-plane the result is the same as Green's Theorem.

*37. Let S be the surface $x^2 + y^2 = 1$, $1 \le z \le 5$. (i) What is the boundary of S? (ii) If $\mathbf{n}$ is the *outer* unit normal vector, what is the corresponding direction on the boundary? (iii) Verify Stokes' Theorem for the vector field $\mathbf{F} = (y, -x, z)$. (iv) Verify Stokes' Theorem for the vector field $\mathbf{F} = (yz, -xz, z)$.

*38. Suppose S is the ellipsoid $x^2 + 2y^2 + 4z^2 = 16$, and $\mathbf{F} = \left(x^2 \sin y, \frac{x}{1 + z^2}, \arctan(xyz) \right)$. Evaluate $\int \int_S$ curl $\mathbf{F} \cdot \mathbf{n} \, dS$, where $\mathbf{n}$ is the outer unit normal.

**39. Let C be a continuously differentiable simple closed curve in the xy-plane, in the counterclockwise direction, and let R be the region it encloses. Let S be the part of the graph of $z = xy$ lying over R. Verify Stokes' Theorem for S and its boundary C_S with (i) $\mathbf{F} = (yz, xz, -xy)$, (ii) $\mathbf{F} = (yz, -xz, xy)$, by expressing the integral over the boundary C_S of S as an integral over C and using Green's Theorem.

**40. Let C be a continuously differentiable simple closed curve in the xy-plane, in the counterclockwise direction, and let R be the region it encloses. Let S be the part of the graph of $z = f(x, y)$ lying over R. Use the following steps to prove Stokes' Theorem for S and its boundary curve C_S, with $\mathbf{F} = (r, s, t)$, where r, s, t are continuously differentiable.

(i) By writing dz in terms of dx and dy, express $\int_{C_S} \mathbf{F} \cdot d\mathbf{r}$ as an integral over C.

(ii) Apply Green's Theorem to this line integral over C to get a double integral over R.

(iii) Express $\int \int_S$ curl $\mathbf{F} \cdot \mathbf{n} \, dS$ as a double integral over R.

(iv) Compare these two double integrals. (*Hint:* In part (ii) when you take partial derivatives, remember that r, s, t are functions of $x, y,$ *and* z.)

III

41. Use a computer to sketch the curve C of Example 4 (p. 1195).

42. Use a computer to sketch the surfaces of Example 5 (p. 1196).

18.7 Applications

ROTATION AND CURL
RIGHT-HAND RULE
CONTINUITY EQUATION
INVERSE-SQUARE FORCE
GAUSS' LAW

Now we discuss some applications of the material we have learned in this chapter. We began with line integrals and considered the concept of path independence and the idea of a potential function. They were illustrated by the vector field corresponding to the gravitational force on an object. Path independence amounts to the familiar fact that the work required to lift an object depends on the change in its height but not on the path it follows, and apart from a sign, the potential function amounts to the gravitational potential energy.

Then we moved to Green's Theorem in the plane and the Divergence Theorem and Stokes' Theorem in $\mathbb{R}^3$. They could all be visualized by thinking of fluid flows, either a thin flow on a flat surface for Green's Theorem or a flow in $\mathbb{R}^3$ for the Divergence Theorem or Stokes' Theorem. In each case we let the vector field **F** represent the velocity of the flow at each point.

In this section we discuss some of these visualizations in more detail and see how the results can be used to study various topics. The examples are mostly from physics.

Rotation and Curl

In Section 17.7 we discussed rotational motion and mentioned the idea of angular velocity. Strictly speaking, we were really using the angular speed, because it made no mention of the direction. By analogy with ordinary velocity the angular velocity ought to indicate the direction as well as the rate of rotation. To specify the direction of a rotation with a vector is not so easy, since a vector points in a straight line.

If we want a vector that indicates the direction of a rotation, the best thing to try is letting the vector point along the axis of the rotation, as shown in Figure 18.7.1. The length of the vector can represent the angular speed, that is, the number of radians the object rotates per second.

The only remaining difficulty is to decide which way the vector should point along the axis. One direction should correspond to rotation one way and the other direction to rotation the other way. To do this, we use the **Right-hand Rule**: If you place your right hand so that the fingers curl in the direction of the rotation, then the thumb points in the direction of the angular velocity vector (see Figure 18.7.2).

Consider an object rotating counterclockwise around the z-axis at a rate of s radians/sec, as shown in Figure 18.7.3. Using the Right-hand Rule, we see that the angular velocity vector is $\boldsymbol{\omega} = (0, 0, s)$.

Let us find the velocity **v** of a point (x, y, z) as it rotates; we begin with a point $(x, y, 0)$ in the xy-plane. In Figure 18.7.4 we see the point $P = (x, y)$ in the plane and

Figure 18.7.1

Figure 18.7.2

Figure 18.7.3

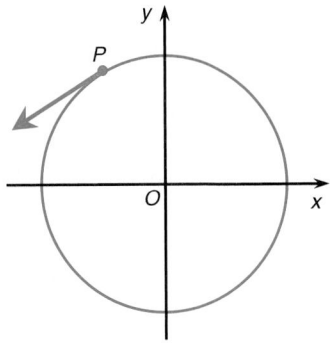

Figure 18.7.4

the circle around which it rotates. The velocity vector is perpendicular to the vector OP, which means that it is some multiple of $(-y, x)$, that is, $\mathbf{v} = k(-y, x)$ for some k. Since the rotation is *counterclockwise*, the number k must be positive.

The point $P = (x, y)$ moves around a circle whose radius equals $r = \sqrt{x^2 + y^2}$, and since its angular speed is s radians per second, its actual speed is $sr = s\sqrt{x^2 + y^2}$ units per second.

But the speed is $\|\mathbf{v}\| = \|k(-y, x)\| = k\|(-y, x)\| = k\sqrt{x^2 + y^2}$, and we see that $k = s$. Finally, we realize that for a point (x, y, z) the velocity does not depend on the height z, so it will be the same as the velocity of $(x, y, 0)$. This means that the velocity vector field is

$$\mathbf{v}(x, y, z) = (-sy, sx, 0).$$

Now write $\mathbf{r} = (x, y, z)$ for the position vector of a point and evaluate the cross product:

$$\boldsymbol{\omega} \times \mathbf{r} = (0, 0, s) \times (x, y, z) = (-sy, sx, 0).$$

We see that this cross product exactly equals the velocity of the rotating point (x, y, z).

In fact, the same is true for a rotation about any axis through the origin. If the angular velocity is $\boldsymbol{\omega}$, then the velocity of the point whose position vector is $\mathbf{r}$ is

$$\mathbf{v} = \boldsymbol{\omega} \times \mathbf{r}.$$

Writing $\boldsymbol{\omega} = (u, v, w)$, we can expand to find that

$$\mathbf{v} = \boldsymbol{\omega} \times \mathbf{r} = (u, v, w) \times (x, y, z) = (vz - wy, wx - uz, uy - vx).$$

Then it is easy to find curl $\mathbf{v}$:

$$\text{curl } \mathbf{v} = \nabla \times \mathbf{v} = \big(u - (-u), v - (-v), w - (-w)\big) = (2u, 2v, 2w) = 2\boldsymbol{\omega}.$$

The point of this calculation is that the curl of a rotational velocity field is, apart from the factor 2, the angular velocity vector. This explains why we think of the curl as a way of detecting the amount of rotation in any vector field.

We record these results as a theorem.

THEOREM 18.7.1

> If an object is rotating with angular velocity $\boldsymbol{\omega}$ about an axis through the origin, then the velocity of the point P whose position vector is $\mathbf{r} = (x, y, z)$ is
>
> $$\mathbf{v} = \boldsymbol{\omega} \times \mathbf{r}.$$
>
> Also
>
> $$\text{curl } \mathbf{v} = \nabla \times \mathbf{v} = 2\boldsymbol{\omega}.$$

Divergence and Flux Density (the Continuity Equation)

In discussing the divergence of a vector field we tried to visualize the velocity field of a fluid flow, and in that case we described the divergence as a measure of how the fluid is "spreading out" at each point. We now describe this idea in greater detail.

Suppose $\mathbf{v}(x, y, z)$ is the velocity at (x, y, z) of a fluid flowing in a region R in $\mathbb{R}^3$, and suppose the density of the fluid at (x, y, z) is given by the function $\rho(x, y, z)$. Let

Figure 18.7.5

Figure 18.7.6

Figure 18.7.7

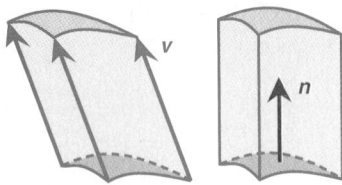

Figure 18.7.8

$\mathbf{F}(x, y, z) = \rho\mathbf{v}$; this vector field points in the same direction as $\mathbf{v}$, which is the direction in which the fluid is flowing. Roughly speaking, it tells us about the movement of the fluid mass.

Now consider the surface integral $\int\int_S \mathbf{F} \cdot \mathbf{n} \, dS$, where the surface S is the boundary of the region R and, as usual, $\mathbf{n}$ is the outer unit normal vector. Roughly speaking, it seems reasonable that this integral describes the mass flowing out of R across the boundary S.

If we consider a small piece of S, with surface area ΔS, we want to find the flow across this small piece of the boundary, as shown in Figure 18.7.5. This would be easy if the flow happened to be perpendicular to the surface, but it is more likely that the flow will be at some other angle relative to the boundary.

Now the velocity tells us that in one second the fluid flows through a displacement approximately equal to the vector $\mathbf{v}(x, y, z)$, so the fluid that flows through ΔS in one second is approximately the shape shown in Figure 18.7.6: it is a solid region whose base is the piece of the surface and whose sides are given by the velocity vector $\mathbf{v}$. The mass of the fluid that flows past in one second will be the product of the density and the volume of this region.

In Figure 18.7.7 we recall that to find the volume of a parallelepiped we take the area of the base and multiply by the height. Another way to say this is that we "straighten up" the parallelepiped until it becomes a rectangular solid.

A similar idea applies to the solid region being discussed here. If we take a very small piece of the surface, then it is virtually flat, and as we see in Figure 18.7.8, the volume of the little solid region is found by straightening it up and multiplying the area of the base by the height. The area of the base is what we are calling ΔS, and the height is just the "vertical" component of $\mathbf{v}$.

Here the word "vertical" means in the direction *normal to the surface*, so the vertical component of $\mathbf{v}$ is $\mathbf{v} \cdot \mathbf{n}$. We find that the volume of fluid that flows across ΔS in one second is approximated by $\mathbf{v} \cdot \mathbf{n} \, \Delta S$. The mass of this fluid is the volume multiplied by the density, or $\rho\mathbf{v} \cdot \mathbf{n} \, \Delta S = \mathbf{F} \cdot \mathbf{n} \, \Delta S$.

Making the usual sort of argument in which the whole surface is divided up into many little pieces, adding up the volumes or masses for all the pieces, and then taking a limit as the size of the pieces tends to zero, we find that the total volume of fluid that flows out of the region R across its boundary in one second is $\int\int_S \mathbf{v} \cdot \mathbf{n} \, dS$. Similarly, the total mass that flows out in one second is $\int\int_S \rho\mathbf{v} \cdot \mathbf{n} \, dS = \int\int_S \mathbf{F} \cdot \mathbf{n} \, dS$.

We summarize by noting that these quantities are just the rates at which the volume and mass of the fluid inside R are decreasing: If V represents the volume of the fluid inside R and M its mass, then

$$\frac{dV}{dt} = -\int\int_S \mathbf{v} \cdot \mathbf{n} \, dS, \qquad \frac{dM}{dt} = -\int\int_S \mathbf{F} \cdot \mathbf{n} \, dS.$$

On the other hand the Divergence Theorem tells us that

$$\int\int_S \mathbf{F} \cdot \mathbf{n} \, dS = \int\int\int_R \operatorname{div} \mathbf{F} \, dx \, dy \, dz = \int\int\int_R \nabla \cdot \mathbf{F} \, dx \, dy \, dz.$$

Let us consider the case in which R is a small ball, and let us divide both sides of this equation by the volume of R. We find that

$$\frac{1}{\operatorname{volume}(R)} \int\int_S \mathbf{F} \cdot \mathbf{n} \, dS = \frac{1}{\operatorname{volume}(R)} \int\int\int_R \nabla \cdot \mathbf{F} \, dx \, dy \, dz.$$

The right side is the average value of the divergence $\nabla \cdot \mathbf{F}$ over the ball R. If we let the radius of the ball tend to zero, then this average will tend to the value of $\nabla \cdot \mathbf{F}$

at the center of the ball. On the other hand the left side of the equation equals the rate of decrease of the mass of fluid *per unit volume*. Since the mass per unit volume is the density, this equals the rate of decrease of the density of the fluid in the ball. As the radius tends to zero, it tends to the rate of decrease of the density *at the center of the ball*.

We have found that at the center of the ball,

$$-\frac{\partial \rho}{\partial t} = \nabla \cdot \mathbf{F}.$$

Since it is possible to place a small ball around any point in the open region R, this equation holds at every point. It is called the **continuity equation** in hydrodynamics:

$$\nabla \cdot \mathbf{F} + \frac{\partial \rho}{\partial t} = 0.$$

It says that the divergence of the vector field $\mathbf{F} = \rho \cdot \mathbf{v}$ equals the rate of decrease of the fluid density at each point. This is consistent with our idea that the divergence represents the extent to which the fluid is "spreading out."

Gauss' Law

We have already noted that various forces in physics obey an **inverse square law**, meaning that the intensity of the force is inversely proportional to the square of the distance from some central point. Examples include gravity and electrostatic forces (cf. the discussion before Example 6.6.5 (p. 460) or in Section 15.7).

If we place the central point at the origin and write $\mathbf{r}$ for the position vector of the point (x, y, z), then this means that the force vector field is of the form

$$\mathbf{F}(x, y, z) = K \frac{1}{\|\mathbf{r}\|^2} \frac{\mathbf{r}}{\|\mathbf{r}\|} = K \frac{1}{\|\mathbf{r}\|^3} \mathbf{r},$$

where K is some constant. If the constant K is positive, it means that the force points away from the origin, as for repulsive electrostatic forces, and if $K < 0$, the force points toward the origin, as it does for attractive electrostatic forces or gravity.

Now consider a sphere S of radius D centered at the origin. The outward unit normal vector at the point (x, y, z) on S is $\mathbf{n} = \frac{1}{\|(x,y,z)\|}(x, y, z) = \frac{1}{D}\mathbf{r}$. We can easily calculate the outward flux across the surface S:

$$\int\int_S \mathbf{F} \cdot \mathbf{n}\, dS = \int\int_S K\frac{1}{\|\mathbf{r}\|^3}\mathbf{r} \cdot \frac{1}{D}\mathbf{r}\, dS = \int\int_S K\frac{1}{D^3}\mathbf{r} \cdot \frac{\mathbf{r}}{D}\, dS$$

$$= \int\int_S K\frac{1}{D^4}D^2\, dS = \frac{K}{D^2}\int\int_S dS = \frac{K}{D^2}4\pi D^2 = 4\pi K.$$

We have used the fact that $\|\mathbf{r}\| = D$ for any point on the sphere, so $\mathbf{r} \cdot \mathbf{r} = D^2$, and also the fact that the surface area of the sphere is $4\pi D^2$ (cf. Example 6.9.5, p. 489).

The interesting thing about this formula is that the outward flux does not depend on the radius D, or to put it another way, that there is the same amount of "flow" out through any sphere around the origin. Of course, when we are talking about gravity or electrostatic forces, it is not entirely clear what is meant by "flow."

Now suppose that instead of a sphere we consider some odd-shaped surface that encloses a region containing the origin (see Figure 18.7.9). If we were actually

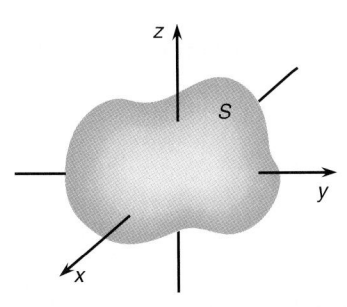

Figure 18.7.9

talking about a fluid flow, it seems reasonable to expect that if the amount flowing out through any sphere centered at the origin is always the same, then the amount flowing out through *any* surface containing the origin should also be the same. After all, this outward flow is just the amount of "fluid" flowing out from the origin.

This "reasonable" fact is Gauss' Law, named after the German mathematician and physicist Karl Friedrich Gauss (1777–1855). It is proved using the Divergence Theorem.

THEOREM 18.7.2

> **Gauss' Law**
>
> For an inverse square force
>
> $$\mathbf{F}(x, y, z) = K \frac{1}{\|\mathbf{r}\|^2} \frac{\mathbf{r}}{\|\mathbf{r}\|} = K \frac{1}{\|\mathbf{r}\|^3} \mathbf{r}$$
>
> the outward flux across any surface S that encloses a region containing the origin is
>
> $$\int\int_S \mathbf{F} \cdot \mathbf{n} \, dS = 4\pi K,$$
>
> where $\mathbf{n}$ is the outer unit normal vector.

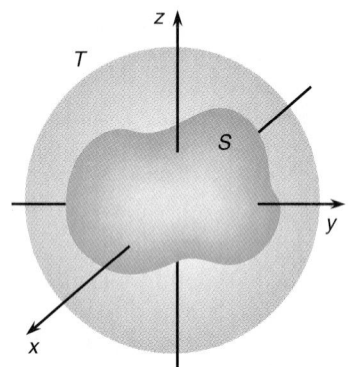

Figure 18.7.10

PROOF   Let us take the surface S and enclose it in a large sphere T, as shown in Figure 18.7.10, and call R the region between the two, that is, the set of all points which are inside T and outside S.

The boundary of R consists of two parts, what we might call the *outside boundary* T and the *inside boundary* S. What is tricky is finding the outer unit normal vectors. On the sphere T the outer unit normal vectors are the familiar vectors $\frac{1}{\|\mathbf{r}\|}\mathbf{r}$ pointing away from the origin. But on the inside boundary S we have to be careful because "outer" unit normal refers to a normal pointing "out" of the region R. Since the region R consists of points *outside* S the "outer" unit normal points out of R into the inside of S. In short the outer unit normal for R is the *negative* of the usual outer unit normal for S.

A routine calculation shows the divergence of $\mathbf{F} = K(x^2 + y^2 + z^2)^{-3/2}(x, y, z)$ equals zero.

The integral over R of the divergence of $\mathbf{F}$ equals zero, and by the Divergence Theorem this equals the surface integral of the outer normal component of $\mathbf{F}$ over the boundary of R. We can write

$$\int\int_T \mathbf{F} \cdot \mathbf{n} \, dS - \int\int_S \mathbf{F} \cdot \mathbf{n} \, dS = 0,$$

where the negative sign reflects the fact that the "outer" normal on the boundary of R equals the negative of the usual outer normal on S.

Put another way, this says that

$$\int\int_S \mathbf{F} \cdot \mathbf{n} \, dS = \int\int_T \mathbf{F} \cdot \mathbf{n} \, dS = 4\pi K,$$

and this is exactly what we wanted to prove.   ◀

This result is of considerable theoretical importance in physics.

Gravitational Force and Spheres

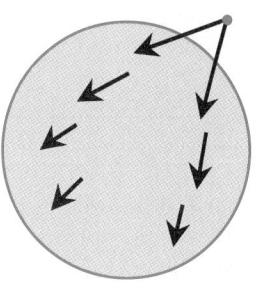

Figure 18.7.11

Finally, we use Gauss' Law to prove another result about gravitational force or indeed any inverse square force. We have mentioned that the force due to gravity is inversely proportional to the square of the distance from the center of the earth; this seems quite reasonable for an object that is very far from the earth, but consider an object close to the surface of the earth, as shown in Figure 18.7.11.

It is attracted by all the matter in the earth, as shown in the illustration, and presumably the parts of the earth that are closest attract it most strongly. It seems surprising that the total attraction should have anything to do with the center of the earth.

It is an important and remarkable result of Newton's that the total gravitational force on the object is exactly the same as the force it would experience if all the mass of the earth were concentrated in a single point at the center of the earth. As far as gravity is concerned, the size of a ball does not affect the force it exerts; all that matters is its mass and the location of the center.

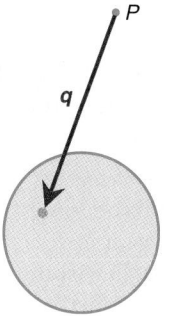

Figure 18.7.12

Consider a point P at a distance D from the center of a ball B of radius b, with $D > b$. If we assume that the density of the material in the ball is $\rho = 1$, then the mass of any part of the ball just equals its volume. Divide the ball up into little pieces of volume ΔV. Suppose the force exerted on P by one of these small pieces of the ball is given by $\frac{K}{\|\mathbf{q}\|^2} \Delta V \frac{1}{\|\mathbf{q}\|} \mathbf{q}$, where $\mathbf{q}$ is the vector from P to the small piece (see Figure 18.7.12). This is exactly what we expect from the inverse square law.

We want to apply Gauss' Law (Theorem 18.7.2) to this inverse square vector field. The constant in the vector field is $K\Delta V$, and the force points *toward* the fragment in the ball, so Gauss' Law says that the flux from this fragment across any surface enclosing the whole ball is $-4\pi K\Delta V$. (The minus sign says that the flux is inward.)

Now the force exerted by the whole ball will be the "sum" of the forces exerted by all the fragments. Of course, "sum" means integral. Accordingly, the flux caused by the force field exerted by the whole ball will be the "sum" of the flux associated to the fragments. Since the ΔV's for all the fragments add up to the total volume of the ball, we find that the total flux is $-4\pi KV$, where V is the volume of the ball.

This is exactly the flux we would get from the inverse square field obtained by placing all the mass of the ball at a single point at its center. However, it is also the same as the flux obtained by placing all the mass at *any* point inside the ball. To show that the force field equals the one with all the mass at the center, it is necessary to use Gauss' Law again.

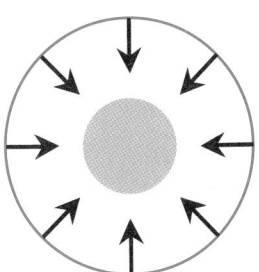

Figure 18.7.13

In Figure 18.7.13 we see the sphere of radius D around the origin. From the symmetry of the ball we see that the force will have the same magnitude at any point on this sphere, and of course it always points directly at the origin. In particular this means that the force vector is normal to the sphere. To find the flux across the sphere, we integrate the normal component of the force, but that is just the negative of the magnitude of the force, which as we have just seen is a constant, $-A$, say. Integrating the constant $-A$ over the sphere gives $-A$ times the surface area, so the flux is $-4\pi AD^2$.

On the other hand we know that the flux is $-4\pi KV$, and we find that the constant A is $A = \frac{KV}{D^2}$. This is the magnitude of the force at any point on the sphere of radius D, which says that the force is an inverse square field. In fact it is exactly the gravitational field resulting from an object of mass V all concentrated at the origin. This is the result we wanted to prove.

THEOREM 18.7.3

For a solid ball of uniform density, consider the gravitational field it exerts on points lying outside the ball. It is the same as the field that would result from concentrating all the mass at the center of the ball.

REMARK 18.7.4

(i) The argument we have given can be roughly summarized as follows. Gauss' Law tells us that the flux must be the same across any surface enclosing the ball. If we consider the surface that is the sphere of radius D around the origin, then its surface area is $4\pi D^2$. The flux is spread out evenly over the surface because of the symmetry, so for the total to remain the same the normal component of the force must be a constant divided by the area, that is, a constant divided by D^2. This means that the force is an inverse square field around the origin, and we evaluate the constant to find that it is the same as what we get by concentrating the mass at the origin.

(ii) In effect we assumed Theorem 18.7.3 several times when we worked with the earth's gravity and assumed that it is the inverse square field around the center of the earth (e.g., Examples 18.1.7 (p. 1148), 18.2.3 (p. 1155)).

(iii) Theorem 18.7.3 refers to the gravitational force field at points *outside* the ball that causes it. It is possible to use a similar argument to find the force field inside the ball too. The answer turns out to be that an object inside the ball feels no net attraction from the part of the ball lying farther from the center than it does. This is because the part of the ball lying farther out surrounds the object and the forces in opposite directions all cancel one another. See Figure 18.7.14.

If an object inside the ball is at a distance D from the center, then the force on it is the same as the force caused by concentrating all the mass of the ball of radius D at the center.

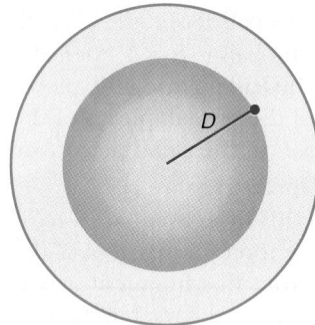

Figure 18.7.14

Exercises 18.7

Find each of the following.

1. The velocity of the point $(1, 3, -2)$ in a rotation about a line through the origin if the angular velocity is $\boldsymbol{\omega} = (4, -2, 3)$ radians/sec.

2. The points that are stationary if $\mathbb{R}^3$ is rotating around a line through the origin with angular velocity $\boldsymbol{\omega} = (2, 1, 5)$ radians/sec.

3. The angular velocity of a rotation about the y-axis if the point $(-3, 2, 4)$ is moving with velocity $\mathbf{v} = (8, 0, 6)$.

4. The angular velocity of a rotation about the line $(x, y, z) = (t, 2t, -2t)$ if the point $(-1, 4, -1)$ is moving with velocity $\mathbf{v} = (4, 2, 4)$.

5. The axis about which the following vector field is rotating: $\mathbf{v}(x, y, z) = (3y + z, -3x - 2z, 2y - x)$.

6. The axis about which the following vector field is rotating: $\mathbf{v}(x, y, z) = (4z - 5y, 5x + z, -4x - y)$.

7. The curl of the velocity field that represents the rotation at 4 radians/sec around the line $(x, y, z) = (2t, 3t, 5t)$ in the direction in which the point $(1, 0, 0)$ moves into a point (x, y, z) with $y > 0$.

8. The velocity field that represents the rotation at π radians/sec around the line $(x, y, z) = (t, t, t)$ in the direction in which the positive z-axis moves into a region where $x < 0$.

9. The magnitude of an inverse square force at a distance of 5 units from its central point if the outward flux across a sphere of radius 3 around the central point is 2 units.

10. The outward flux per unit area across a sphere of radius D around the central point of an inverse square force if the force has magnitude 6 at a distance of 4 units from the central point.

‖

11. (i) Suppose a vector field can be written in the form $\mathbf{F} = \big(f(x,y), g(x,y), h(z)\big)$. Show that curl $\mathbf{F} = \nabla \times \mathbf{F}$ is a "vertical" vector.

(ii) Find an example of a vector field $\mathbf{F}$ that is "horizontal," in the sense that at each point the vector is a horizontal vector, but for which curl $\mathbf{F}$ is *not* "vertical."

*12. We are familiar with the vector field $\mathbf{F} = (x,y,z)$, which always points away from the origin (cf. Example 18.5.1, p. 1182).

(i) Suppose that S is a surface that contains a region R, and show that the flux of $\mathbf{F}$ across S is related to the volume of the region enclosed inside S.

(ii) Show how this result can be used to prove that the unit normal vector $\mathbf{n}(x,y,z)$ on the surface S cannot always be perpendicular to the vector running from the origin to the point (x,y,z). This geometrical fact seems clear from any picture we can draw, but the Divergence Theorem gives a painless way to prove it.

*13. In Theorem 18.7.1 we discussed rotation around an axis *through the origin*. What happens if the axis of the rotation does not contain the origin? For instance, consider rotation around the axis $x = 3$, $z = 0$, which is parallel to the y-axis. (i) If the angular velocity is $\boldsymbol{\omega} = (0, -3, 0)$, find a formula for the velocity $\mathbf{v}$ of the point (x,y,z). (ii) Find the curl of this velocity field.

*14. (i) Find the velocity $\mathbf{v}$ of (x,y,z) for a rotation with angular velocity $\boldsymbol{\omega}$ about an axis containing the point (a,b,c). (ii) Find curl $\mathbf{v} = \nabla \times \mathbf{v}$.

**15. Verify the claims made in Remark 18.7.4(iii).

 ◄— POINT TO PONDER

We have used fluid flows to illustrate a number of the concepts in this chapter. We should actually be a little more careful about the type of fluid we are considering. One property that is important is whether or not the fluid is **incompressible**. This means exactly what it says, that even under high pressure the volume of the fluid does not decrease. Water is a fluid that is incompressible for all practical purposes, but air is an example of a fluid that is very compressible. A diver's tank contains air that has been placed under pressure to compress it to a much smaller volume than usual. In everyday speech we may not regard air as a "fluid," but for technical purposes it is one.

Another way to say that a fluid is incompressible is to say that its density is constant. If an incompressible fluid is flowing in $\mathbb{R}^3$, then its divergence must be zero. The incompressibility means that it cannot be "spreading out," and this is what the divergence measures. The continuity equation $\nabla \cdot \mathbf{F} + \frac{\partial \rho}{\partial t} = 0$ tells us that if ρ is constant, then the divergence $\nabla \cdot \mathbf{F}$ must be zero, where $\mathbf{F} = \rho\mathbf{v}$. Since ρ is constant, this implies that the divergence of the velocity, $\nabla \cdot \mathbf{v}$, is zero too.

On the other hand it is possible to have nonzero divergence by having water pouring into a region from a pipe or out through a drain.

Chapter Summary

§18.1
directed curve
line integrals
initial point, terminal point
differential form
vector field
line integrals and work

§18.2
path independence
open set
potential function
connected set
simply connected set
conservative vector field

§18.3
(simple) closed path
loop
Green's Theorem (Theorem 18.3.1)
$\text{area}(R) = \int_C x\,dy = -\int_C y\,dx$
positively oriented boundary
fluid flow
outer (unit) normal
flux
divergence
circulation
curl

§18.4
surface area
surface integrals
surface integrals and moments
mass
moment
centroid
moments of inertia

§18.5
divergence
(unit) normal vector
positively oriented boundary curve
orientable surface
flux
Divergence Theorem (Theorem 18.5.4, p. 1185)

§18.6
curl **F**
circulation
Stokes' Theorem (Theorem 18.6.2, p. 1193)

§18.7
Right-hand Rule for curl
rotation and curl
Continuity Equation
inverse square force
Gauss' Law (Theorem 18.7.2, p. 1204)
gravity and spheres
incompressible fluid

Review Exercises

Evaluate each of the following.

1. $\int_C x^2\,dy$, where C is the left semicircle $x^2 + y^2 = 4$, $x \le 0$, starting at $(0, 2)$.

2. $\int_C y^3\,dx$, where C is the right semicircle $x^2 + y^2 = 9$, $x \ge 0$, starting at $(0, 3)$.

3. $\int_C x\,dy$, where C is the left semi-ellipse $4x^2 + 9y^2 = 36$, $x \le 0$, starting at $(0, 2)$.

4. $\int_C (x + 2y)\,dy$, where C is the left semi-ellipse $25x^2 + 4y^2 = 100$, $x \le 0$, starting at $(0, -5)$.

5. $\int_C (x^2 - y)\,dx$, where C is the curve $y = x^2 - 2x + 1$, $-1 \le x \le 0$.

6. $\int_C \sin y\,dx$, where C is the curve $y = x^3 + 3x$, $-2 \le x \le 2$.

7. $\int_C x\,dx - y^2\,dy$, where C is the curve $y^2 = x$, $-3 \le y \le 2$.

8. $\int_C y\,dy$, where C is the curve $y^2 = x$, $0 \le y \le 3$.

9. $\int_C (y - 3x)\,dx$, where C is the curve $y^3 = x^5$, $0 \le x \le 1$.

10. $\int_C x\,dy$, where C is the counterclockwise triangle with vertices $(0, 0)$, $(3, 0)$, $(2, 4)$.

Decide whether or not each of the following integrals is path independent; for those that are, find the potential function and evaluate the integral.

11. $\int_{(0,0)}^{(1,1)} \mathbf{F} \cdot d\mathbf{r}, \mathbf{F} = (y, x)$ 12. $\int_{(1,2)}^{(2,1)} \mathbf{F} \cdot d\mathbf{r}, \mathbf{F} = (y^2, x^2)$

13. $\int_{(2,3)}^{(2,4)} \mathbf{F} \cdot d\mathbf{r}, \mathbf{F} = (x^2, y^3)$

14. $\int_{(-1,-1)}^{(1,1)} \mathbf{F} \cdot d\mathbf{r}, \mathbf{F} = (\sin \pi x, \cos \pi y)$

15. $\int_{(5,0,0)}^{(0,5,1)} \mathbf{F} \cdot d\mathbf{r}, \mathbf{F} = (y^2, 2xy, 4z)$

16. $\int_{(3,4)}^{(4,3)} \mathbf{F} \cdot d\mathbf{r}, \mathbf{F} = (ye^{xy}, xe^{xy})$

17. $\int_{(1,3,2)}^{(1,3,2)} \mathbf{F} \cdot d\mathbf{r}, \mathbf{F} = \left(\frac{1}{x^2 yz}, \frac{1}{xy^2 z}, \frac{1}{xyz^2}\right)$

18. $\int_{(0,0)}^{(a,b)} \mathbf{F} \cdot d\mathbf{r}, \mathbf{F} = (x^m, y^n)$

19. $\int_{(1,2)}^{(a,b)} \mathbf{F} \cdot d\mathbf{r}, \mathbf{F} = (mx^{m-1}y^n, nx^m y^{n-1})$

20. $\int_{(1,1,1)}^{(1,1,b)} \mathbf{F} \cdot d\mathbf{r}, \mathbf{F} = \left(z^2, -z^2, 2(x - y)z\right)$

Find each of the following.

21. $\int_C x\,dy$, C is $x^2 + y^2 = 4$, counterclockwise.

22. $\int_C x^2\,dy$, C is $4x^2 + 9y^2 = 4$, counterclockwise.

23. $\int_C y^3\,dx$, C is $(x - 1)^2 + y^2 = 1$, counterclockwise.

24. $\int_C y^5\,dy$, C is $x^2 + 4(y + 3)^2 = 16$, counterclockwise.

25. $\int_C (x + y)\,dy$, C is $x^2 + y^2 = 25$, counterclockwise.

26. $\int_C xy^2\,dx$, C is $9x^2 + (y - 2)^2 = 9$, counterclockwise.

27. $\int_C (y - e^x)\,dx$, C is $x^2 + y^2 = 4$, $y \ge 0$, and $y = 0$, $-2 \le x \le 2$, counterclockwise.

28. $\int_C (x + x^2)\,dy$, C is the triangle with vertices $(-1, 1)$, $(1, 1)$, $(0, 4)$, counterclockwise.

29. $\int_C (3y + e^x)\,dy$, C is the boundary of $0 \le x \le 1$, $-1 \le y \le 0$, counterclockwise.

30. $\int_C (2x - 3y)\,dx$, C is the boundary of $4x^2 + 25y^2 \leq 100$, $x \leq 0$, $y \geq 0$, counterclockwise.

31. $\int\int_S x\,dS$, S is $z^2 = x^2 + y^2$, $x \geq 0$, $0 \leq z \leq 9$.

32. $\int\int_S y^2 z\,dS$, S is $x + 2y + 3z = 4$, $0 \leq x \leq 1$, $0 \leq y \leq x$.

33. $\int\int_S \cos(x^2 + y^2 + z^2)\,dS$, S is $x^2 + y^2 + z^2 = 5\pi$.

34. $\int\int_S x^2\,dS$, S is $x^2 + y^2 + z^2 = 1$, $z \geq 0$, together with $z = 0$, $x^2 + y^2 \leq 1$.

35. The centroid of $x^2 + y^2 + z^2 = 1$, $0 \leq y \leq x$.

36. The centroid of $y^2 = x^2 + z^2$, $0 \leq y \leq 4$.

37. The centroid of $y^2 = 3x^2 + 3z^2$, $1 \leq y \leq 2\sqrt{3}$.

38. The centroid of $z^2 = x^2 + y^2$, $x^2 + y^2 \leq 4$, $x \geq 0$, $y \leq 0$.

39. The centroid of $y^2 = x^2 + z^2$, $x^2 + z^2 \leq 1$, $|x| \leq z$.

40. The centroid of $x^2 = y^2 + z^2$, $|x| \leq 2$.

In each of the following, $\mathbf{n}$ refers to a unit outer normal vector to the surface S. Evaluate.

41. $\int\int_S (2 + x, 2 + 3x, y - z) \cdot \mathbf{n}\,dS$, S is $x^2 + y^2 + z^2 = 4$.

42. $\int\int_S (x - y, y - z, z - x) \cdot \mathbf{n}\,dS$, S is $x^2 + y^2 + z^2 = 9$.

43. $\int\int_S (x + x^2, 2xy, z^3) \cdot \mathbf{n}\,dS$, S is $x^2 + y^2 + z^2 = 1$.

44. $\int\int_S (xy, yz, y^2 z) \cdot \mathbf{n}\,dS$, S is $x^2 + y^2 + z^2 = 4$, $z \geq 0$.

45. $\int\int_S (x + z, y - x, x + 2y) \cdot \mathbf{n}\,dS$, S is the boundary of $0 \leq x, y, z \leq 2$.

46. $\int\int_S (3x + z, y - 2x, 2x + 2y - z) \cdot \mathbf{n}\,dS$, S is the boundary of $0 \leq x \leq 1$, $x \leq y \leq 1$, $0 \leq z \leq x + y$.

47. $\int\int_S (x^3, y, z^2) \cdot \mathbf{n}\,dS$, S is $x^2 + y^2 = 4$, $0 \leq z \leq 5$.

48. $\int\int_S (2x + z, 3y - z, 2y - 6z) \cdot \mathbf{n}\,dS$, S is $x^2 + y^2 = 1$, $0 \leq z \leq x + 3$.

49. $\int\int_S (\sin \pi x, y, 3z) \cdot \mathbf{n}\,dS$, S is $\left(x - \frac{1}{2}\right)^2 + y^2 + z^2 = 36$.

50. $\int\int_S (3x + 2z, y - 7z, 5y - 4z) \cdot \mathbf{n}\,dS$, S is the boundary of $0 \leq x \leq \pi$, $0 \leq y \leq x^2$, $0 \leq z \leq x^2 - y$.

Evaluate $\int_C \mathbf{F} \cdot d\mathbf{r}$, where the curve C is parametrized by $\mathbf{r}$.

51. $\mathbf{F} = (xy, yz, xz)$, $\mathbf{r}(t) = (\cos t, 1, \sin t)$, $t \in [0, 2\pi]$

52. $\mathbf{F} = (yz, xz, xy)$, $\mathbf{r}(t) = (\cos 2t, 1, \sin 2t)$, $t \in [0, \pi]$

53. $\mathbf{F} = (y, x, 4z)$, $\mathbf{r}(t) = (\cos t, \cos 2t, t \sin t)$, $t \in [0, 2\pi]$

54. $\mathbf{F} = (y, y, z + 3)$, $\mathbf{r}(t) = (4, t^2 \sin t, \cos t)$, $t \in [0, 2\pi]$

55. $\mathbf{F} = (x, y, z)$, $\mathbf{r}(t) = (\cos t, \sin t - 3 \cos t, \sin t)$, $t \in [0, 2\pi]$

56. $\mathbf{F} = \left(\frac{1}{x+y+z}, \frac{1}{x+y+z}, \frac{1}{x+y+z}\right)$, $\mathbf{r}(t) = (3 + \sin t, 4 - \sin t, \cos t)$, $t \in [0, \pi]$

57. $\mathbf{F} = (xy, y, z)$, $\mathbf{r}(t) = (\sin t, 3 \sin t + 2 \cos t, \cos t)$, $t \in [0, 2\pi]$

58. $\mathbf{F} = (x^2, y^2, 5z^4)$, $\mathbf{r}(t) = \left(\cos t, t(t - 2\pi), \sin t\right)$, $t \in [0, 2\pi]$

59. $\mathbf{F} = (yz + xy, \cos z, xy + yz)$, $\mathbf{r}(t) = \left(t \sin 2t, 4, \frac{4t}{\pi} + 2 \cos t\right)$, $t \in [0, \pi]$

60. $\mathbf{F} = (2x, y, 3z)$, $\mathbf{r}(t) = \left(t(1 - t), t^2, \sin \pi t\right)$, $t \in [0, 1]$

II

**61. Suppose $\mathbf{F}$ is a vector field on an open set U in $\mathbb{R}^2$ having the property that $\int_C \mathbf{F} \cdot d\mathbf{r}$ is path independent for *any* differentiable path C in U. Show that $\int_C \mathbf{F} \cdot d\mathbf{r} = 0$ for every *closed* differentiable path C in U.

*62. Find an example of a vector field $\mathbf{F}$ on an open set U in $\mathbb{R}^2$ and a closed curve C in U so that $\int_C \mathbf{F} \cdot d\mathbf{r} = 0$ but so that $\int_{C'} \mathbf{F} \cdot d\mathbf{r} \neq 0$ for some other differentiable closed path C' in U.

*63. Suppose R is a region in $\mathbb{R}^3$ that is symmetric about each of the coordinate planes. Suppose $\mathbf{F} = \left(p(x), q(y), r(z)\right)$ is a vector field each of whose coordinates p, q, r is a polynomial containing only even powers. Show that $\int\int_S \mathbf{F} \cdot \mathbf{n}\,dS = 0$, where S is the boundary of R and $\mathbf{n}$ is the outer unit normal vector to S.

*64. Let R be the unbounded region $0 \leq x$, $0 \leq y \leq 1$, and let C be its boundary, which consists of the half-lines $y = 1$, $x \geq 0$, and $y = 0$, $x \geq 0$, and the segment $x = 0$, $0 \leq y \leq 1$. Think of C as a directed curve going from right to left along the top, down the left side, and then from left to right along the bottom.

(i) Verify Green's Theorem for the integral $\int_C ye^{-x}\,dy$, where this line integral and the corresponding area integral are both improper integrals.

(ii) What goes wrong with Green's Theorem for the integral $\int_C x\,dy$? What is it that makes these two examples so different?

*65. In this chapter we have used the following notations: ∇f, $\nabla \cdot \mathbf{F}$, $\mathbf{F} \cdot \mathbf{n}$, $\mathrm{curl}(\mathbf{F})$, $\mathrm{div}(\mathbf{F})$, and $\mathrm{grad}(f)$. In this context, which of f and $\mathbf{F}$ and each of these expressions are functions and which are vectors?

*66. Suppose $\mathbf{F}$ is defined and differentiable on all of $\mathbb{R}^3$. (i) Show that $\int\int_S \mathrm{curl}\,\mathbf{F} \cdot \mathbf{n}\,dS = 0$ if S is $x^2 + y^2 + z^2 = 9$. (ii) What happens if S is $4x^2 + 9y^2 + 16z^2 = 144$?

**67. Suppose B is a ball of radius 2 and S is its boundary sphere. If $\mathbf{F}$ is a vector field and the outward flux is

$$\int\int_S \mathbf{F} \cdot \mathbf{n}\,dS = \frac{16\pi}{3},$$

show that there must be at least one point in B at

which

$$\text{div } \mathbf{F} \geq \frac{1}{2}.$$

Interpret this result in terms of fluid flows.

**68. Is Exercise 67 still correct if we replace $\geq$ with $>$? Or with $\leq$?

*69. Suppose $f(x, y)$ is a twice continuously differentiable function on a simply connected set R and that $f(x, y)$ satisfies Laplace's Equation

$$\frac{\partial^2 f}{\partial x^2} + \frac{\partial^2 f}{\partial y^2} = 0.$$

Show that $\int_C \frac{\partial f}{\partial y} \, dx - \frac{\partial f}{\partial x} \, dy$

is independent of path for any path C in R.

*70. Suppose $f(x, y, z)$ is a twice continuously differentiable function, and that $f(x, y, z)$ is harmonic, that is, satisfies Laplace's Equation

$$\nabla^2 f = \frac{\partial^2 f}{\partial x^2} + \frac{\partial^2 f}{\partial y^2} + \frac{\partial^2 f}{\partial z^2} = 0.$$

Suppose S is an orientable surface and R is the region it encloses. Write $\mathbf{n}$ for the outer unit normal and $D_{\mathbf{n}}f$ for the directional derivative in the direction of $\mathbf{n}$, sometimes called the *normal derivative* of f. Show that

$$\int\int_S D_{\mathbf{n}}f \, dS = 0.$$

*71. Suppose $f(x, y, z)$ is a twice continuously differentiable function. Suppose S is an orientable surface, R

is the region it encloses, and $\mathbf{n}$ is the outer unit normal. Show that

$$\int\int_S \nabla f \cdot \mathbf{n} \, dS = \int\int\int_R \nabla^2 f \, dV.$$

In particular, if f is harmonic, then both integrals equal zero.

*72. Suppose $\mathbf{F}$ has twice-differentiable components, and consider curl (curl $\mathbf{F}$) $= \nabla \times (\nabla \times \mathbf{F})$. Either prove that curl (curl $\mathbf{F}$) $= 0$, for *every* $\mathbf{F}$, or find an example for which it is nonzero.

73. Suppose $\mathbf{F}$ has twice-differentiable components, and consider div (curl $\mathbf{F}$) $= \nabla \cdot (\nabla \times \mathbf{F})$. Either prove that div (curl $\mathbf{F}$) $= 0$ for *every* $\mathbf{F}$, or find an example for which it is nonzero.

*74. Suppose C is a simple closed curve in the plane. Show that $\int_C y \sin(x) \, dx - \big(\cos(x) + 3x\big) \, dy$ does not depend on the shape or location of the curve C, but only on the area of the region it encloses.

**75. Let S be the sphere $x^2 + y^2 + z^2 = M^2$, and evaluate

$$\int\int_S \frac{x^2}{(x^2 + y^2 + z^2)^{3/2}} \, dS.$$

(*Hint:* Argue by symmetry that this integral equals

$$\int\int_S \frac{y^2}{(x^2 + y^2 + z^2)^{3/2}} \, dS$$

or

$$\int\int_S \frac{z^2}{(x^2 + y^2 + z^2)^{3/2}} \, dS;$$

then evaluate the sum of these three integrals without doing much work.)

Differential Equations

19

These graphs represent oscillations, motions that go back and forth as functions of time. Simple examples include the motions of a pendulum or a hanging spring. Oscillations are studied using differential equations.

The last two graphs illustrate damped oscillations, which gradually diminish in intensity. These are typical of the vibrations of a plucked string on a musical instrument. The last graph represents a string which vibrates more rapidly than the previous one; the resulting musical note will have a higher pitch.

INTRODUCTION

In this chapter we study differential equations, which were considered briefly in Section 8.6. A differential equation is an equation involving derivatives. An equation like $x^2 + 3x + 2 = 0$ or $e^x = 5$ is an equation for which the solution, if there is one, is a number, or perhaps several numbers. A differential equation like $f'' + 3x^2 f' + f = \sin x$ is an equation relating a function f and some of its derivatives; the solution, if there is one, is a function, or perhaps a family of functions.

We have already encountered several kinds of differential equations. The most obvious is an equation of the form $f'(x) = 3x^2 - 2x + 7$, which says what the derivative of f must be. To solve it, we integrate, and in this case we find that $f(x) = x^3 - x^2 + 7x + c$, where c is any constant. One example of this type of problem comes up when we know the velocity of an object and want to find its displacement.

When we studied growth problems, we encountered equations like $f'(t) = kf(t)$, where k is some number; it says that the rate of growth of f

is proportional to the value of f, and we found that the solutions are the functions $f(t) = ce^{kt}$, where c is any constant.

Differential equations arise in many situations, in particular whenever we know about the rate of change of some quantity rather than the quantity itself. In addition to the growth problems we have already seen, there are many practical applications of differential equations in all areas of science and social science.

In Section 19.1 we study *homogeneous linear differential equations with constant coefficients*. We considered them briefly in Section 8.6 after learning about exponential functions because exponentials play a key role in solving this type of equation. In Section 19.2 we learn about complex numbers and see how they can be used to help solve some of these equations. In Section 19.3 we learn the method of *undetermined coefficients* for solving nonhomogeneous equations. As an application we study *vibrations* in Section 19.4. They include vibrating strings on a musical instrument, a swinging pendulum, electronic signals, and many other examples. In Section 19.5 we study *first-order differential equations*, a kind that is relatively easy to solve. Included are what are known as *separable equations* and *first-order linear equations*. In Section 19.6 we learn a technique called *variation of parameters* for solving nonhomogeneous equations, and in Section 19.7 we consider the possibility of expressing solutions as infinite series, especially when other approaches fail.

19.1 Homogeneous Linear Differential Equations with Constant Coefficients

ORDER
HOMOGENEOUS EQUATION
CHARACTERISTIC POLYNOMIAL
AUXILIARY EQUATION
GENERAL SOLUTION
MULTIPLICITY OF A ROOT
PARTICULAR SOLUTION
INITIAL CONDITIONS

The material of this section has already been discussed, in Section 8.6. Here we study it more systematically, and in the next section we will go on to consider some additional aspects that we missed the first time around.

A differential equation involves a function and some of its derivatives. To write it, we need to know the function and the variable. We might write $f(x)$, which makes it clear that f is the function and x is the variable. On the other hand, we might let x be the variable and write y for the function, using the notation $\frac{dy}{dx}$ or y' for the derivative.

For questions involving the rate of change of some quantity we often use the variable t to represent time. In this case we might have a function $f(t)$ or a function y, but it is also quite common to use x as the name of a function. (It is reasonable to use x for the horizontal component of the displacement of an object moving in the plane.) It is important to keep clear in our minds what is the variable and what is the function.

For instance, in the equation

$$x^2 y''' - \frac{1}{x^2 + 2} y'' + 2y' - \frac{1}{x} y = e^x$$

the variable is x and y is a function of x. To solve an equation like this means to find all the functions y that satisfy the equation.

The equation above has the property that the derivatives of y occur one at a time, each multiplied by some function of x. This kind of equation tends to be easier to solve than ones in which various derivatives of y are multiplied together.

DEFINITION 19.1.1

A **linear differential equation** is one that can be written in the form

$$a_n(x)y^{(n)} + a_{n-1}(x)y^{(n-1)} + a_{n-2}(x)y^{(n-2)} +$$
$$\ldots + a_2(x)y^{(2)} + a_1(x)y^{(1)} + a_0(x)y = f(x),$$

where $f(x)$ and each $a_i(x)$ are functions of the variable x. The **order** of this equation is n, the order of the highest derivative. (Here we assume that $a_n(x)$ is not the zero function.)

The equation is **homogeneous** if the function $f(x)$ on the right side is zero, that is, if the equation is of the form

$$a_n(x)y^{(n)} + a_{n-1}(x)y^{(n-1)} + a_{n-2}(x)y^{(n-2)} +$$
$$\ldots + a_2(x)y^{(2)} + a_1(x)y^{(1)} + a_0(x)y = 0.$$

The functions $a_i(x)$ are called the **coefficients**; in this section we will consider the special case in which all the coefficient functions are constants. This means that the equation can be written as

$$a_n y^{(n)} + a_{n-1}y^{(n-1)} + a_{n-2}y^{(n-2)} + \ldots + a_2 y^{(2)} + a_1 y^{(1)} + a_0 y = 0. \qquad (19.1.2)$$

EXAMPLE 1   The equation $y' + 3y = 0$ is a homogeneous linear differential equation with constant coefficients. To solve it, we use the trick of guessing that the solution *might* be an exponential function.

If we let $y = e^{rx}$, then $y' = re^{rx} = ry$. Substituting this into the left side of the equation, we have $y' + 3y = ry + 3y = (r + 3)y$. For this to equal zero we must have $r + 3 = 0$, so $r = -3$.

This suggests that we consider letting $y = e^{-3x}$, and we find that this function does satisfy the equation $y' + 3y = 0$. Of course, if we multiply e^{-3x} by any constant, it will still satisfy the equation, so we find a whole family of solutions $y = ce^{-3x}$, where c is any constant.   ◢

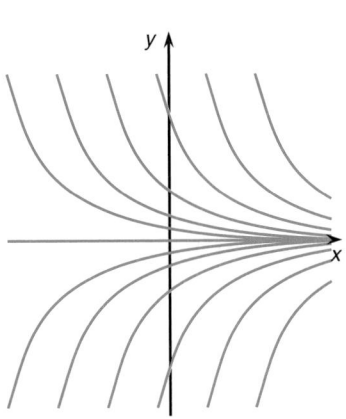

In Figure 19.1.1 we see the graphs of a number of the solutions from Example 1. In a sense the differential equation tells how the function changes, since it says what the derivative is. Once we know the value of the function at one point, then the equation tells us how it will change as we move to any other point. The fact that there are many different solutions corresponds to the fact that there are many possible starting points.

Figure 19.1.1

EXAMPLE 2   Consider the equation $y'' + y' - 2y = 0$. It is a linear homogeneous differential equation with constant coefficients, and once again we try guessing that there may be a solution of the form $y = e^{rx}$, for some number r.

If $y = e^{rx}$, then $y' = re^{rx} = ry$ and $y'' = \frac{d}{dx}(y') = \frac{d}{dx}(ry) = r^2 y$. The equation becomes $r^2 y + ry - 2y = 0$, or $(r^2 + r - 2)y = 0$. We look for values of r that make this expression zero, that is, $r^2 + r - 2 = 0$.

It is possible to factor $r^2 + r - 2 = (r + 2)(r - 1)$, so we should try $r = -2$ and $r = 1$. It is easy to check that both $y_1 = e^{-2x}$ and $y_2 = e^x$ are solutions of the equation.

If a and b are any constants, let us consider $y = ay_1 + by_2$. Substituting this function y into the left side of the differential equation, we find that

$$y'' + y' - 2y = (ay_1 + by_2)'' + (ay_1 + by_2)' - 2(ay_1 + by_2)$$
$$= a(y_1'' + y_1' - 2y_1) + b(y_2'' + y_2' - 2y_2).$$

In the second line, we group all the y_1's together and all the y_2's together.

Since y_1 and y_2 are both solutions of the equation, we see that each of the last two expressions in parentheses equals zero. This means that $y'' + y' - 2y = 0$, so $y = ay_1 + by_2$ is a solution of the original equation.

In other words, for any constants $a, b \in \mathbb{R}$ the function $y = ae^{-2x} + be^x$ is a solution of the original differential equation. In fact it is also true that every solution is of this form, for some a and b. ◢

Example 2 suggests the following idea. If we know some solutions $y_1, y_2, \ldots, y_k$ of a homogeneous linear differential equation, then we can multiply them by constants and add them up, and the resulting function will also be a solution.

THEOREM 19.1.3

Suppose $y_1, y_2, \ldots, y_k$ are all solutions of the homogeneous linear differential equation

$$a_n(x)y^{(n)} + a_{n-1}(x)y^{(n-1)} + a_{n-2}(x)y^{(n-2)} +$$
$$\ldots + a_2(x)y^{(2)} + a_1(x)y^{(1)} + a_0(x)y = 0.$$

Then for any constants $c_1, c_2, \ldots, c_k$, the function $y = c_1y_1 + c_2y_2 + \ldots + c_ky_k$ is also a solution of this equation.

PROOF   We substitute the function $y = c_1y_1 + c_2y_2 + \ldots + c_ky_k$ into the left side of the differential equation. Then we collect together all the terms involving derivatives of y_1, all the terms involving derivatives of y_2, and so on. For instance, the terms involving y_1 and its derivatives will be

$$a_n(x)c_1y_1^{(n)} + a_{n-1}(x)c_1y_1^{(n-1)} + a_{n-2}(x)c_1y_1^{(n-2)} +$$
$$\ldots + a_2(x)c_1y_1^{(2)} + a_1(x)c_1y_1^{(1)} + a_0(x)c_1y_1,$$

and since y_1 is a solution of the original equation, this all equals zero.

Similarly, the terms involving derivatives of y_2 all add up to 0, and the same is true for each y_i, $i = 1, 2, \ldots, k$.

We find that the left side of the equation is zero; in other words, the function y is a solution of the equation, as required. ◢

It is very important to remember that for this theorem to work, the equation must be *homogeneous*.

Consider the following homogeneous linear equation with constant coefficients:

$$a_ny^{(n)} + a_{n-1}y^{(n-1)} + a_{n-2}y^{(n-2)} + \ldots + a_2y^{(2)} + a_1y^{(1)} + a_0y = 0.$$

If we let $y = e^{rx}$, then $y' = ry$, $y'' = r^2y$, $\ldots$, $y^{(n)} = r^ny$. The equation becomes

$$0 = a_ny^{(n)} + a_{n-1}y^{(n-1)} + a_{n-2}y^{(n-2)} + \ldots + a_2y^{(2)} + a_1y^{(1)} + a_0y$$
$$= a_nr^ny + a_{n-1}r^{n-1}y + a_{n-2}r^{n-2}y + \ldots + a_2r^2y + a_1ry + a_0y$$
$$= (a_nr^n + a_{n-1}r^{n-1} + a_{n-2}r^{n-2} + \ldots + a_2r^2 + a_1r + a_0)y.$$

This equation will be satisfied if the expression in parentheses is zero. We look for roots of this polynomial in order to find solutions.

DEFINITION 19.1.4

Consider the following homogeneous linear differential equation with constant coefficients:

$$a_n y^{(n)} + a_{n-1} y^{(n-1)} + a_{n-2} y^{(n-2)} + \ldots + a_2 y^{(2)} + a_1 y^{(1)} + a_0 y = 0.$$

The corresponding polynomial

$$a_n r^n + a_{n-1} r^{n-1} + a_{n-2} r^{n-2} + \ldots + a_2 r^2 + a_1 r + a_0$$

is called the **characteristic polynomial** associated to the equation, and setting it equal to zero gives the **auxiliary equation**:

$$a_n r^n + a_{n-1} r^{n-1} + a_{n-2} r^{n-2} + \ldots + a_2 r^2 + a_1 r + a_0 = 0.$$

THEOREM 19.1.5

For the differential equation given in the definition above, suppose $r_1, r_2, \ldots,$ r_k are the roots of the characteristic polynomial (the solutions of the auxiliary equation). Then for any constants $c_1, c_2, \ldots, c_k$, the function

$$y = c_1 e^{r_1 x} + c_2 e^{r_2 x} + \ldots + c_k e^{r_k x}$$

is a solution of the differential equation. Moreover, if k, the number of distinct roots, equals the order n, then *every* solution is of this form, and we say that we have found the **general solution**.

PROOF   The previous discussion shows that each function $e^{r_i x}$ is a solution, and Theorem 19.1.3 then tells us that adding up constant multiples of them gives a solution too.

The fact that these are all the solutions if there are n distinct roots is quite difficult, and we do not prove it.   ◢

EXAMPLE 3   Find all solutions of $y''' - 5y'' + 6y' = 0$.

Solution   The characteristic polynomial is $r^3 - 5r^2 + 6r$. It is easily factored: $r^3 - 5r^2 + 6r = r(r^2 - 5r + 6) = r(r - 2)(r - 3)$. We find that the roots are $r = 0, 2, 3$. Since there are three of them, and the order of the equation is also 3, the general solution is

$$y = c_1 e^{0x} + c_2 e^{2x} + c_3 e^{3x} = c_1 + c_2 e^{2x} + c_3 e^{3x}.$$ ◢

EXAMPLE 4   Consider the equation $y''' + 2y'' + y' + 2y = 0$. The characteristic polynomial is $r^3 + 2r^2 + r + 2$, which factors as $r^3 + 2r^2 + r + 2 = (r + 2)(r^2 + 1)$.

It has one root $r = -2$, but since the polynomial $r^2 + 1$ is never zero, this is the only root. We find that $y = e^{-2x}$ is a solution, and consequently, so is $y = ce^{-2x}$, for any constant c, but since there is only one root and the order of the equation is 3, we do not know that these are all the solutions.   ◢

In fact, those are not all the solutions of the equation in the example. It is easy to check that $y = \cos x$ and $y = \sin x$ are also solutions. In the next section we will learn how to find solutions like these.

The fact that our trick with exponentials did not find all the solutions in Example 4 is related to the fact that there is only one root for the characteristic polynomial.

Although we did not prove it, Theorem 19.1.5 says that if we find n distinct roots, where n is the degree of the polynomial, then the procedure does give all the solutions.

Example 4 illustrates one way in which this can go wrong: The characteristic polynomial had a factor $r^2 + 1$ that has no real roots. To deal with this, it is necessary to use *complex numbers*, which we will discuss in the next section. Another possible problem arises if the polynomial has *repeated roots*. For instance the polynomial $r^2 - 6r + 9 = (r - 3)^2$ has $r = 3$ as a root, and in a sense it has this same root *twice*, but there are not two *distinct* roots. We need a way to find all the solutions for differential equations whose characteristic polynomials have repeated roots like this.

For instance, if the characteristic polynomial contains $(r - 3)^2$ as a factor, then $r = 3$ is a root, so $y = e^{3x}$ is a solution of the differential equation. It turns out that in this situation $y = xe^{3x}$ is also a solution. As an example, consider the equation $y'' - 6y' + 9y = 0$, whose characteristic polynomial is $r^2 - 6r + 9 = (r - 3)^2$. We can check that $y = xe^{3x}$ is a solution. After all, $y' = e^{3x} + 3xe^{3x}$ and $y'' = 3e^{3x} + 3e^{3x} + 9xe^{3x} = 6e^{3x} + 9xe^{3x}$, so $y'' - 6y' + 9y = 6e^{3x} + 9xe^{3x} - 6e^{3x} - 18xe^{3x} + 9xe^{3x} = 0$.

In general it is not very hard to see that if the characteristic polynomial contains the factor $(r - a)^k$, then the functions

$$e^{ax}, xe^{ax}, x^2 e^{ax}, x^3 e^{ax}, \ldots, x^{k-1} e^{ax}$$

The powers of x go up to $k - 1$, one less than the power of $(r - a)$ in the characteristic polynomial.

are all solutions. Since we can add up constant multiples of solutions and get other solutions, by Theorem 19.1.3, we can find many solutions by finding solutions like this for each root of the characteristic polynomial and then combining them.

THEOREM 19.1.6

For a homogeneous linear differential equation of order n having constant coefficients, it is possible to factor the characteristic polynomial as

$$(r - r_1)^{m_1} (r - r_2)^{m_2} \cdot \ldots \cdot (r - r_k)^{m_k} g(r),$$

where $r_1, r_2, \ldots, r_k$ are the distinct roots and the polynomial $g(r)$ has no real roots. The exponent m_i is called the **multiplicity** of the root r_i; we say that the root r_i *occurs m_i times*. The polynomial $g(r)$ may just be the constant polynomial $g(r) = 1$.

Then $e^{r_i x}$ is a solution for each i, and if the multiplicity m_i is greater than 1, then all the following functions are solutions:

$$e^{r_i x}, xe^{r_i x}, x^2 e^{r_i x}, \ldots, x^{m_i - 1} e^{r_i x}.$$

(The powers of x go up to one less than the multiplicity m_i.)

If the polynomial $g(r)$ is just the constant $g(r) = 1$, then the total number of these solutions is n, the order of the original equation, and in this case every solution of the equation is a sum of multiples of the ones given here for the various roots r_i.

We will not prove this result. It is not very difficult to see that the specified functions actually are solutions, but it is more difficult to show that this procedure gives *all* solutions if the extra factor $g(r)$ is constant.

EXAMPLE 5   Find all solutions of $y^{(5)} + 4y^{(4)} + 4y^{(3)} = 0$.

Solution   The characteristic polynomial is $r^5 + 4r^4 + 4r^3$, which factors as $r^5 +$

$4r^4 + 4r^3 = r^3(r^2 + 4r + 4) = r^3(r + 2)^2$. The roots are $r = 0$, with multiplicity 3, and $r = -2$, with multiplicity 2.

This means that e^{-2x} and xe^{-2x} are solutions, and so are $e^{0x} = 1$, $xe^{0x} = x$, and $x^2 e^{0x} = x^2$. Since the factor $g(r)$ that has no real roots is just 1, we know that all solutions of the equation are of the following form:

$$c_1 e^{-2x} + c_2 xe^{-2x} + c_3 + c_4 x + c_5 x^2,$$

where $c_1, c_2, \ldots, c_5$ are any constants. This is the general solution. ◢

So far we have learned how to find families of solutions to a differential equation. Under certain circumstances we can even find the general solution, meaning a formula that describes all possible solutions. However, for most practical problems the situation will be that we know a differential equation that is satisfied by some quantity, and we want to find a formula for that quantity (without any undetermined constants). To do this, it will be necessary to know some additional information about the solution; in the familiar case of exponential growth we have to know the population at some specified time.

For a homogeneous linear differential equation of order n with constant coefficients, we have seen that in the cases in which we can find the general solution it involves n constants. To determine them, it is necessary to know n facts about the solution.

EXAMPLE 6 Find the function y that satisfies $y'' + y' - 2y = 0$ and $y(0) = 3$, $y(1) = 1$.

Solution The characteristic polynomial is $r^2 + r - 2$, which factors as $r^2 + r - 2 = (r + 2)(r - 1)$, so its roots are $r = -2$ and $r = 1$. As we saw in Example 2, the general solution is $y = ae^{-2x} + be^x$, where a and b are constants.

We also know that $y(0) = 3$ and $y(1) = 1$. From the formula we find that $y(0) = ae^0 + be^0 = a + b$ and $y(1) = ae^{-2} + be^1 = ae^{-2} + be$. This gives us two equations:

$$a + b = 3, \qquad ae^{-2} + be = 1.$$

Subtracting e times the first equation from the second, we find that $ae^{-2} - ae = 1 - 3e$, or $a(e^{-2} - e) = 1 - 3e$, so $a = \frac{1 - 3e}{e^{-2} - e} = \frac{3e - 1}{e - e^{-2}}$. From the first equation we then find that $b = 3 - a = \frac{3e - 3e^{-2}}{e - e^{-2}} - \frac{3e - 1}{e - e^{-2}} = \frac{1 - 3e^{-2}}{e - e^{-2}}$.

The required solution is $y = \dfrac{3e - 1}{e - e^{-2}} e^{-2x} + \dfrac{1 - 3e^{-2}}{e - e^{-2}} e^x$.

It is easy to check that it does actually satisfy the necessary conditions. ◢

See the Graphics Calculator Example on the next page.

We used the letters a and b for the constants because they are easier to write than c_1 and c_2.

Many practical problems involve finding solutions of differential equations with specified values. The specified values are called **initial conditions**, since they describe how the function "begins." However, initial conditions do not always have to be where the variable equals 0.

We take the general solution and solve to find what values of the constants are required to make it satisfy the initial conditions. The resulting function is called a **particular solution**.

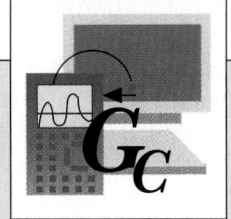

With a graphics calculator we can sketch the graph of the solution to Example 6.

TI-85	SHARP EL9300
e^x 1 STO E ENTER GRAPH F1 $(y(x)$ $=)$ (3**E** $-$ 1) $\div$ (**E** $-$ **E**^$(-)$ 2) $\times$ e^x ($(-)$ 2 F1(x)) $+$ (1 $-$ 3**E**^ $(-)$ 2) $\div$ (**E** $-$ **E**^$(-)$ 2) $\times$ e^x F1(x) M2(RANGE) $(-)$ 1 ▼ 2 ▼ ▼ $(-)$ 1 ▼ 9 F5(GRAPH)	〰 ((3 e^x 1 ▶ $-$ 1) e^x $(-)$ 2 x/θ/T ▶ $+$ (1 $-$ 3 e^x $(-)$ 2 ▶) e^x x/θ/T ▶) $\div$ (e^x 1 ▶ $-$ e^x $(-)$ 2 ▶) RANGE $(-)$1 ENTER 2 ENTER 1 ENTER $(-)$1 ENTER 9 ENTER 1 ENTER 〰
CASIO f_x-6300G (f_x-7700GB)	HP 48SX
Range $-$1 EXE 2 EXE 1 EXE $-$1 EXE 9 EXE 1 EXE EXE EXE EXE Graph ((3 e^x 1 $-$ 1) $\times$ e^x ($-$ 2 X) $+$ (1 $-$ 3 e^x (-2)) $\times$ e^x X) $\div$ (e^x 1 $-$ e^x ($-$ 2)) EXE	↰ PLOT PLOTR ERASE ATTN $'$ Y $=$ () () 3 $\times$ e^x 1 ▶ $-$ 1 ▶ $\times$ e^x 2 +/− $\times$ X ▶ $+$ () 1 $-$ 3 $\times$ e^x 2 +/− ▶ ▶ $\times$ e^x X ▶ ▶ $\div$ () e^x 1 ▶ $-$ e^x 2 +/− ↱ PLOT ↰ DRAW ERASE 1 +/− ENTER 2 XRNG 1 +/− ENTER 9 YRNG DRAW

Note that the solution does satisfy the conditions on $y(0)$ and $y(1)$.

EXAMPLE 7   **Reaction Temperature**   The temperature of a certain chemical reaction is known to satisfy the differential equation

$$2T'' + T' = 0,$$

where $T(t)$ is the temperature in degrees Celsius and the variable t is the time in minutes.

For a particular run of the experiment, the temperature is found to be 30°C at the beginning and 26°C one minute later. What will the temperature be 3 minutes after the first measurement?

Solution   The characteristic polynomial is $2r^2 + r$, which factors as $r(2r + 1)$, so its roots are $r = 0$ and $r = -\frac{1}{2}$. The general solution is

$$T(t) = ae^{0t} + be^{-t/2} = a + be^{-t/2}.$$

We know that $T(0) = 30$ and $T(1) = 26$, and comparing this with the formula gives two equations:

$$a + b = 30,$$
$$a + be^{-1/2} = 26.$$

Subtracting, we find that $b - be^{-1/2} = 4$, that is, $b(1 - e^{-1/2}) = 4$, so $b =$

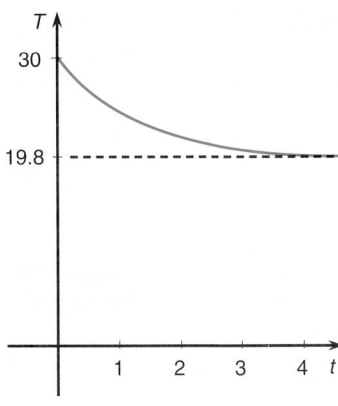

Figure 19.1.2

$\frac{4}{1-e^{-1/2}}$. Since $a+b=30$, we find $a=30-b=30-\frac{4}{1-e^{-1/2}}=\frac{26-30e^{-1/2}}{1-e^{-1/2}}$. The temperature is given by the particular solution

$$T(t) = \frac{26 - 30e^{-1/2}}{1 - e^{-1/2}} + \frac{4}{1 - e^{-1/2}} e^{-t/2} \approx 19.8 + 10.2e^{-t/2}.$$

Three minutes after the first observation, that is, when $t = 3$, we find that $T(3) \approx 19.8 + 10.2e^{-3/2} \approx 22.1°C$.

Notice that as $t \to \infty$, $e^{-t/2} \to 0$, so the temperature $T(t)$ tends to a limit of about 19.8°C. Presumably, the reaction generates heat, but as it slows, the mixture cools down to the ambient temperature of the laboratory. The temperature is graphed in Figure 19.1.2.

> Initial conditions can also involve the derivatives of the solution instead of just the function itself.
>
> There is no reason that the initial conditions have to be at $x = 0$, but the calculations are usually easier if they are.

Exercises 19.1

I

Find the general solution of each of the following differential equations.

1. $y' + 5y = 0$

2. $y'' - 2y' = 0$

3. $y'' + 5y' + 6y = 0$

4. $y'' + 3y' - 4y = 0$

5. $\frac{d^2 y}{dx^2} + \frac{dy}{dx} - 6y = 0$

6. $4\frac{d^2 f}{dt^2} - 5\frac{df}{dt} + f = 0$

7. $y''' - 4y' = 0$

8. $y''' = 6y'' - 11y' + 6y$

9. $y^{(4)} + 2y^{(3)} - y^{(2)} - 2y^{(1)} = 0$

10. $y''' = 3y'$

11. $y'' + 2y' + y = 0$

12. $y'' - 8y' + 16y = 0$

13. $y''' - 2y'' = 0$

14. $y''' - 4y'' + 4y' = 0$

15. $y''' + 3y'' + 3y' + y = 0$

16. $y^{(4)} = 9y^{(2)}$

17. $f^{(4)} = 6f^{(3)}$

18. $g^{(4)} - 10g^{(3)} + 25g^{(2)} = 0$

19. $\frac{d^3 F}{dt^3} = \frac{dF}{dt}$

20. $f^{(10)} = f^{(8)}$

Find the solutions of the following differential equations satisfying the given initial conditions.

21. $y' - 2y = 0$, $y(0) = 3$

22. $y' = 3y$, $y'(0) = 6$

23. $y'' - y = 0$, $y(0) = 1$, $y'(0) = 2$

24. $y'' + 5y' + 6y = 0$, $y(0) = 0$, $y(1) = 4$

25. $y'' - 2y' + y = 0$, $y(0) = 2$, $y(2) = 2$

26. $y' = 0$, $y(7) = 3$

27. $y''' + 4y'' + 4y' = 0$, $y(0) = 1$, $y'(0) = 3$, $y''(0) = -2$

28. $y''' - 6y'' + 9y' = 0$, $y(0) = 2$, $y'(0) = -1$, $y''(0) = 3$

29. $y''' + y'' = 0$, $y(0) = 1$, $y(1) = 3$, $y(2) = 5$

30. $y^{(5)} - 3y^{(4)} + 2y^{(3)} - 11y^{(2)} + y^{(1)} - 7y = 0$, $y(1) = y(2) = y(3) = y(4) = y(5) = 0$

II

*31. Find the general solution of $6y''' + 25y'' - 11y' - 60y = 0$. (*Hint:* $y = e^{-4x}$ is one solution.)

*32. Find the general solution of $20y''' - 9y'' - 14y' + 3y = 0$.

*33. (i) What is the general solution of $y''' = 0$?

  (ii) Find the solution of $y''' = 0$ that satisfies $y(0) = 3$, $y'(0) = -4$, $y''(0) = 8$.

*34. (i) What is the general solution of $y^{(n)} = 0$?

  (ii) Let $a_0, a_1, a_2, \ldots, a_{n-1} \in \mathbb{R}$. Find the solution of

$y^{(n)} = 0$ that satisfies

$$y(0) = y^{(0)}(0) = a_0, \quad y^{(1)}(0) = a_1,$$
$$y^{(2)}(0) = a_2, \quad \ldots, \quad y^{(n-1)}(0) = a_{n-1}.$$

*35. (i) Find the solution of $y''' = 0$ that satisfies $y(1) = 2$, $y'(1) = 5$, $y''(1) = -6$.

    (ii) Express your answer in the form $y = a(x-1)^2 + b(x-1) + c$, for some constants a, b, c. What does this have to do with Taylor polynomials?

*36. Let $a_0, a_1, a_2, \ldots, a_{n-1} \in \mathbb{R}$. Find the solution of $y^{(n)} = 0$ that satisfies

$$y(1) = y^{(0)}(1) = a_0, \quad y^{(1)}(1) = a_1,$$
$$y^{(2)}(1) = a_2, \quad \ldots, \quad y^{(n-1)}(1) = a_{n-1}.$$

(*Hint:* Do Exercise 35.)

*37. After Example 4 (p. 1215) we remarked that $y''' + 2y'' + y' + 2y = 0$ has solutions $y = e^{-2x}$, $y = \cos x$, and $y = \sin x$. Find a solution satisfying $y(0) = 3$, $y'(0) = 4$, $y''(0) = -1$.

*38. (i) Check that the equation $y''' + y'' + 4y' + 4y = 0$ has solutions $y = e^{-x}$, $y = \cos 2x$, $y = \sin 2x$.

    (ii) Find a solution satisfying $y(0) = 3$, $y'(0) = 5$, $y''(0) = -3$.

For each of the following equations, express the general solution as a sum (cf. Theorem 19.1.6, p. 1216). Use a graphics calculator or computer to graph the function in each term in the sum, and then use the picture to *guess* the particular solution satisfying the given conditions. Graph your guess and, if necessary, adjust it.

39. $4y'' + 3y' = 0$, $y(0) = y(3) = 6$

40. $3y'' + 2y' = 0$, $y(0) = 10$, $y'(0) = -1$

41. $y'' - 2y' + y = 0$, $y(0) = y(-1) = 1$

42. $y''' = 0$, $y(0) = 2$, $y'(0) = 0$, $y''(0) = 2$

POINT TO PONDER

In Theorems 19.1.5 and 19.1.6 there are conditions under which we can be sure we have found *all* the solutions of a homogeneous differential equation. In the first case this will be true if the number of distinct roots equals n, the order of the equation. In these circumstances we observed that the n corresponding solutions are enough to construct all solutions. In Theorem 19.1.6 it is similar; if we know that the extra factor $g(r)$ in the characteristic polynomial, the part that has no roots, is just a constant, then we know n solutions and can use them to find all the solutions.

If the equation has order n, then we should expect to find n solutions from which all solutions can be built up. One thing that makes this idea difficult to work with is that just finding n solutions is not enough. For instance, if e^{2x} and e^{5x} are both solutions of a homogeneous equation, then so is $e^{2x} + e^{5x}$, but this does not count as a new solution, since it comes from the first two. We need n solutions that are "independent."

If you have studied linear algebra, you will have encountered the concept of *linear independence*. Functions $f_1, f_2, \ldots, f_n$ are linearly independent if no one of them can be expressed as a sum of multiples of the others. For a homogeneous linear differential equation of order n there are n linearly independent solutions from which all solutions can be constructed.

To determine one solution, we need to specify each of the n constants by which we multiply these n basic solutions. This requires n pieces of information.

The equation $y' = ky$ for exponential growth or decay has order 1, and only one piece of information is needed to pin down a solution. Most often we use the initial population. (Another piece of information is often used to find the growth constant k, but this is part of the differential equation $y' = ky$. Once we know the *equation*, then only one additional piece of information is enough to determine the solution.)

If we consider the equation $y'' + 3y' + 2y = 0$, its characteristic polynomial is $r^2 + 3r + 2 = (r+2)(r+1)$, so the general solution is $y = ce^{-2x} + de^{-x}$. Two pieces of information are required to determine the two constants c and d for a solution.

In Figure 19.1.3(i) we see several solutions that satisfy $y(0) = 1$ but that have different values of $y'(0)$. This means that they all pass through the point $(0, 1)$ but that they have different slopes at that point. Once the value of a solution at $x = 0$ and its slope there are given, the solution is completely specified. In Figure 19.1.3(ii) we see several solutions that satisfy $y(0) = 0$, including one that satisfies $y(1) = 1$.

Figure 19.1.3(i)

Figure 19.1.3(ii)

19.2 Complex Numbers and Linear Homogeneous Equations

REAL AND IMAGINARY PARTS
REAL AND IMAGINARY AXES
PURE IMAGINARY NUMBERS
GENERAL SOLUTION OF
HOMOGENEOUS LINEAR
EQUATIONS WITH CONSTANT
COEFFICIENTS

In Section 19.1 we saw how to find the general solution of a homogeneous linear differential equation, *provided* we could factor the characteristic polynomial and find n roots. On the other hand, if the characteristic polynomial contained a factor like $r^2 + 1$, which has no real roots, we did not know how to proceed. The answer is to use *complex numbers*, which provide roots for every polynomial.

For instance, to find a root for $r^2 + 1$ means to solve $r^2 + 1 = 0$, or $r^2 = -1$. A root of this polynomial will be a number whose square is -1. No real number has this property, but we just invent a number i that satisfies $i^2 = -1$. Then we consider all numbers of the form $a + ib$, where a and b are real numbers. These are the complex numbers, and while it is not obvious, it turns out that every polynomial has a root that is a complex number.

DEFINITION 19.2.1

Let i satisfy $i^2 = -1$. The **complex numbers** are all numbers of the form $a + ib$ with $a, b \in \mathbb{R}$; we denote them by $\mathbb{C}$, so

$$\mathbb{C} = \{a + ib : a, b \in \mathbb{R}\}.$$

If $z = a + ib$ is a complex number, then the real number a is called its **real part**, and the real number b is called its **imaginary part**; we write $a = \text{Re}(z)$, $b = \text{Im}(z)$. The real numbers are contained in the complex numbers; we think of $a + i0$ as the real number a and write $a = a + i0$. The complex numbers of the form ib, that is, with real part equal to 0, are called **pure imaginary** numbers.

It is customary to represent the complex number $a + ib$ as the point in the plane whose coordinates are (a, b) (see Figure 19.2.1). In this picture, the real numbers $a = a + i0$ lie on the horizontal axis, and the pure imaginary numbers lie on the vertical axis, with i corresponding to the point $(0, 1)$. For this reason the horizontal axis is called the **real axis**, and the vertical axis is called the **imaginary axis**. The origin in the plane is the complex number $0 = 0 + i0$.

The distance from $z = a + ib$ to 0 is $\sqrt{a^2 + b^2}$; we call it the **absolute value** of z and write $|z| = |a + ib| = \sqrt{a^2 + b^2}$.

Figure 19.2.1

$a+ib$

b

a

We add and multiply complex numbers in the obvious way, adding and multiplying real numbers as usual and remembering that $i^2 = -1$. For instance,

(i) $(3 - 4i) + (-1 + 6i) = (3 - 1) + (-4 + 6)i = 2 + 2i$,

(ii) $(2 - 5i)(4 - i) = 2(4 - i) - 5i(4 - i) = 8 - 2i - 20i + 5i^2 = 8 - 22i - 5 = 3 - 22i$,

(iii) $(a + ib) + (c + id) = (a + c) + i(b + d)$,

(iv) $(a + ib)(c + id) = ac + iad + ibc + i^2bd = ac + i(ad + bc) - bd = ac - bd + i(ad + bc)$.

EXAMPLE 1   If $z = a + ib$, with $a, b \in \mathbb{R}$, $z \neq 0$, express $\frac{1}{z}$ in the form $c + id$ with $c, d \in \mathbb{R}$.

Solution   We multiply top and bottom by $a - ib$ and find

$$\frac{1}{a + ib} = \frac{1}{a + ib}\frac{a - ib}{a - ib} = \frac{a - ib}{a^2 - (ib)^2} = \frac{a - ib}{a^2 + b^2} = \frac{a - ib}{a^2 + b^2}$$

$$= \frac{a}{a^2 + b^2} - i\frac{b}{a^2 + b^2}.$$

EXAMPLE 2   Consider the polynomial $x^2 + 1$. Writing it as $x^2 - (-1)$, we can think of it as a difference of squares, since -1 is now a square. The formula tells us that

$$x^2 + 1 = x^2 - (-1) = (x + i)(x - i).$$

The fact that $x + i$ and $x - i$ are factors of the polynomial says that $-i$ and i are roots of the polynomial, and of course this is easily verified. For instance, $i^2 + 1 = -1 + 1 = 0$, and we see that i is a root of $x^2 + 1$.

We can work with polynomials using complex numbers just as we did with real numbers. The difference is that polynomials like $x^2 + 1$ that have no real roots *do* have roots when we use complex numbers. We must be careful to specify what kind of roots we mean, saying, for example, that $x^2 + 1$ has no *real* roots but that it has two complex roots $\pm i$.

The remarkable thing is that *every* polynomial has a complex root. We will not prove this extremely important fact.

THEOREM 19.2.2

> **The Fundamental Theorem of Algebra**
>
> Any nonconstant polynomial $p(x)$ with real coefficients has a complex root, and in fact it can be *factored completely* over the complex numbers. This means that it can be written as a product
>
> $$p(x) = a(x - z_1)^{m_1}(x - z_2)^{m_2} \cdot \ldots \cdot (x - z_k)^{m_k}.$$
>
> The complex numbers $z_1, \ldots, z_k$ are the distinct **roots** of $p(x)$, and m_j is the **multiplicity** of z_j.

EXAMPLE 3   Find the roots of $x^2 + x + 1$.

Solution   The Quadratic Formula suggests that the roots are

$$x = \frac{-b \pm \sqrt{b^2 - 4ac}}{2a} = \frac{-1 \pm \sqrt{1 - 4}}{2} = \frac{-1 \pm \sqrt{-3}}{2} = -\frac{1}{2} \pm i\frac{\sqrt{3}}{2}.$$

Note: $\sqrt{-3} = \sqrt{-1}\sqrt{3} = i\sqrt{3}$.

These two points are plotted in Figure 19.2.2. The fact that they have the same real part $\left(-\frac{1}{2}\right)$ means that they lie on the same vertical line. It is easy to check that the absolute value of each of them equals 1. Finally, because of this and the fact that the real part is $-\frac{1}{2}$, we see that the cosine of the upper angle marked in the picture is $-\frac{1}{2}$, so the angle is $\frac{2\pi}{3}$. ◢

EXAMPLE 4   Find the roots of the polynomial $x^3 - 2x^2 + 3x$.

Solution   This polynomial factors as $x^3 - 2x^2 + 3x = x(x^2 - 2x + 3)$. To find the roots of $x^2 - 2x + 3$, we use the Quadratic Formula, which gives

$$x = \frac{2 \pm \sqrt{4 - 12}}{2} = 1 \pm \frac{\sqrt{-8}}{2} = 1 \pm \sqrt{-2} = 1 \pm i\sqrt{2}.$$

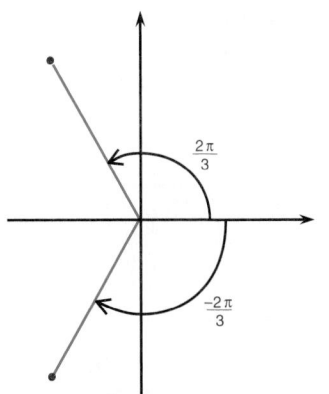
Figure 19.2.2

The roots of the original cubic polynomial $x^3 - 2x^2 + 3x = x(x^2 - 2x + 3)$ are $x = 0$ and $x = 1 \pm i\sqrt{2}$. ◢

Since our work with differential equations has relied so heavily on the exponential function, it seems natural to wonder what happens when we take the exponential of a complex number. One way to consider this question is to use the Maclaurin series

$$e^x = 1 + x + \frac{x^2}{2!} + \frac{x^3}{3!} + \frac{x^4}{4!} + \ldots + \frac{x^n}{n!} + \ldots.$$

We do not really know what it means to consider a series of complex numbers, but it is tempting to substitute and see what we get. To make it easier, we begin with a pure imaginary number it.

EXAMPLE 5   Consider the series for the exponential e^{it}, whatever that means. It is

$$1 + it + \frac{(it)^2}{2!} + \frac{(it)^3}{3!} + \frac{(it)^4}{4!} + \ldots + \frac{(it)^n}{n!} + \ldots$$

$$= 1 + it + \frac{i^2 t^2}{2!} + \frac{i^3 t^3}{3!} + \frac{i^4 t^4}{4!} + \ldots + \frac{i^n t^n}{n!} + \ldots$$

$$= 1 + it + \frac{-t^2}{2!} + \frac{-it^3}{3!} + \frac{t^4}{4!} + \ldots + \frac{i^n t^n}{n!} + \ldots.$$

In simplifying this expression we have to keep track of the powers of i. Observe that $i^2 = -1$, $i^3 = i^2 i = -i$, $i^4 = i^2 i^2 = (-1)(-1) = 1$. After that the powers will repeat in a cycle of 4, that is, $i^5 = i^1 = i$, $i^6 = i^2 = -1$, $i^7 = i^3 = -i$, $i^8 = i^4 = 1$, and so on.

In particular the odd powers of i are imaginary (either i or $-i$), and the even powers are real (either 1 or -1). Collecting together the real terms in the series and the imaginary terms, we find that it equals

$$1 + it + \frac{-t^2}{2!} + \frac{-it^3}{3!} + \frac{t^4}{4!} + \ldots + \frac{i^n t^n}{n!} + \ldots$$

$$= \left(1 - \frac{t^2}{2!} + \frac{t^4}{4!} + \ldots \right) + i \left(t - \frac{t^3}{3!} + \frac{t^5}{5!} + \ldots \right).$$

The two series in parentheses are the series for $\cos t$ and $\sin t$. What this suggests is that

$$e^{it} = \cos t + i \sin t.$$

Strictly speaking, the calculation we have just done uses series of complex numbers, which we have not defined, and involves rearranging a series, which we know can be very tricky at the best of times. In fact it is possible to define complex series and to check that the rearrangement is justified, but for our purposes it is enough to use this calculation to suggest what the definition should be.

DEFINITION 19.2.3

> If t is any real number, define
>
> $$e^{it} = \cos t + i \sin t.$$

EXAMPLE 6   Consider the differential equation $y'' + y = 0$. It is a homogeneous linear differential equation with constant coefficients, and its characteristic polynomial is $r^2 + 1$. The roots of this polynomial are $\pm i$.

If we had found a real root a, we would consider the corresponding exponential function e^{ax}, so let us try the functions e^{ix} and e^{-ix}. They are

$$e^{ix} = \cos x + i \sin x,$$
$$e^{-ix} = e^{i(-x)} = \cos(-x) + i \sin(-x) = \cos x - i \sin x.$$

The sum of these two functions is $(\cos x + i \sin x) + (\cos x - i \sin x) = 2 \cos x$, and their difference is $(\cos x + i \sin x) - (\cos x - i \sin x) = 2i \sin x$. This leads us to expect that $\cos x$ and $\sin x$ ought to be solutions of the equation, and it is easy to check that they are. After all, $\frac{d^2}{dx^2}(\cos x) = \frac{d}{dx}(-\sin x) = -\cos x$, so $(\cos x)'' + \cos x = -\cos x + \cos x = 0$. Almost the same calculation works for $\sin x$.

> We considered the sum of e^{ix} and e^{-ix} because it gives the function $2 \cos x$ that does not involve any complex numbers. Similarly, $\sin x$ is obtained by taking the difference and dividing by $2i$.
>
> Since the original equation had no complex numbers in it, it is comforting to find solutions that also do not involve complex numbers. On the other hand, if we want to allow functions that have complex values, then e^{ix} and e^{-ix} are actually solutions. However, for most practical purposes we are interested in real-valued functions, so it is usually preferable to use $\cos x$ and $\sin x$.

We should also consider exponentials of complex numbers that are not pure imaginary, such as e^{a+ib}. It seems tempting to think that complex exponentials should also obey the rule about sums and products, which can be used as a definition.

DEFINITION 19.2.4

> If $a, b \in \mathbb{R}$, then
>
> $$e^{a+ib} = e^a e^{ib} = e^a (\cos b + i \sin b).$$

THEOREM 19.2.5

For a homogeneous linear differential equation with constant (real) coefficients, if $r = a + ib$ is a root of the characteristic polynomial, then $r = a - ib$ is also a root. In this case

$$e^{ax}\cos(bx) \qquad \text{and} \qquad e^{ax}\sin(bx)$$

are solutions.

Moreover, the roots $a + ib$ and $a - ib$ both occur with the same multiplicity. If the roots $a + ib$ and $a - ib$ occur with multiplicity equal to m and $m > 1$, then the following functions are all solutions:

$$e^{ax}\cos(bx), \quad xe^{ax}\cos(bx), \quad x^2 e^{ax}\cos(bx), \quad \ldots, \quad x^{m-1}e^{ax}\cos(bx),$$

$$e^{ax}\sin(bx), \quad xe^{ax}\sin(bx), \quad x^2 e^{ax}\sin(bx), \quad \ldots, \quad x^{m-1}e^{ax}\sin(bx).$$

If we know all the roots of the characteristic polynomial and their multiplicities, then the **general solution** of the equation is found by combining all the functions of this form for the complex roots *and* the functions suggested by Theorem 19.1.6 for the real roots.

We do not prove this result but remark that the second part is analogous to the corresponding part of Theorem 19.1.6. Notice that the powers of x go up to one less than the multiplicity of the corresponding root.

EXAMPLE 7   Find a function $f(t)$ that satisfies $f'' = 6f' - 13f, f(0) = 3, f'(0) = 2$.

Solution   The differential equation can be rearranged as $f'' - 6f' + 13f = 0$, from which we see that the characteristic polynomial is $r^2 - 6r + 13$. Its roots are $\frac{6 \pm \sqrt{36-52}}{2} = 3 \pm \frac{\sqrt{-16}}{2} = 3 \pm 2i$.

The general solution is $f(t) = ae^{3t}\cos 2t + be^{3t}\sin 2t = e^{3t}(a\cos 2t + b\sin 2t)$. We find that

$$f'(t) = 3e^{3t}(a\cos 2t + b\sin 2t) + e^{3t}(-2a\sin 2t + 2b\cos 2t)$$
$$= e^{3t}\big((3a + 2b)\cos 2t + (3b - 2a)\sin 2t\big).$$

The condition $f(0) = 3$ then becomes $a = 3$, and the condition $f'(0) = 2$ becomes $3a + 2b = 2$, so $2b = 2 - 3a = 2 - 9 = -7$ and $b = -\frac{7}{2}$.

The required solution is $f(t) = e^{3t}\big(3\cos 2t - \frac{7}{2}\sin 2t\big)$.   ◢

See the Graphics Calculator Example on the next page.

EXAMPLE 8   Find the general solution of the equation $y^{(7)} + 8y^{(5)} + 16y^{(3)} = 0$.

Solution   The characteristic polynomial is $r^7 + 8r^5 + 16r^3 = r^3(r^4 + 8r^2 + 16) = r^3(r^2 + 4)^2$. The factor in parentheses can be factored further by writing it as a difference of squares: $r^2 + 4 = r^2 - (-4) = (r + 2i)(r - 2i)$.

This means that the characteristic polynomial is $r^3(r + 2i)^2(r - 2i)^2$ and the roots are $r = 0$, with multiplicity 3, and $r = -2i$, $r = 2i$, each with multiplicity 2.

The general solution is

$$a + bx + cx^2 + d\cos 2x + fx\cos 2x + g\sin 2x + hx\sin 2x.$$ ◢

With a graphics calculator we can sketch the graph of the solution of Example 7.

TI-85	SHARP EL9300
GRAPH F1$(y(x) =) e^x$ (3 F1(x)) × (3 COS (2 F1(x)) −3.5 SIN (2 F1(x))) M2(RANGE) (−) 1 ▼ 1 ▼ ▼ (−) 8 ▼ 4 F5(GRAPH)	∿ e^x 3 x/θ/T ▶ × (3 cos (2 x/θ/T) − 3.5 sin (2 x/θ/T)) RANGE (−)1 ENTER 1 ENTER 1 ENTER (−)8 ENTER 4 ENTER 1 ENTER ∿

CASIO f_x-6300G (f_x-7700GB)	HP 48SX
Range −1 EXE 1 EXE 1 EXE −8 EXE 4 EXE 1 EXE EXE EXE EXE Graph e^x (3 **X**) × (3 cos (2 **X**) − 3.5 sin (2 **X**)) EXE	↰ PLOT PLOTR ERASE ATTN ' Y = e^x 3 × X ▶ × () 3 × COS 2 × X ▶ − 3.5 × SIN 2 × X ↱ PLOT ↰ DRAW ERASE 1 +/− ENTER 1 XRNG 8 +/− ENTER 4 YRNG DRAW

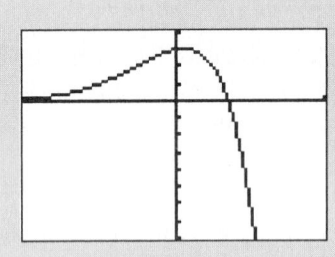

Note that the solution does satisfy the conditions. Observe that the function contains no mention of complex numbers.

You might like to zoom in on the part of the graph near the vertical axis.

We summarize the results of this section and the previous one.

Summary 19.2.6

General Solution of Homogeneous Linear Equations with Constant Coefficients

Consider a homogeneous linear differential equation with constant coefficients

$$a_n y^{(n)} + a_{n-1} y^{(n-1)} + \ldots + a_1 y^{(1)} + a_0 y = 0,$$

where each a_j is a real number.

Suppose we can factor the characteristic polynomial. Let us write the distinct *real* roots of the characteristic polynomial as $r_1, r_2, \ldots, r_k$, and suppose that r_j has multiplicity m_j.

Also suppose that the distinct complex (i.e., nonreal) roots are $a_1 \pm ib_1, a_2 \pm ib_2, \ldots, a_\ell \pm ib_\ell$, and suppose that $a_j + ib_j$ has multiplicity n_j. (In fact, the roots $a_j + ib_j$ and $a_j - ib_j$ will always have the same multiplicity.)

Then each of the following functions is a solution of the original equation:

$$e^{r_j x}, \; xe^{r_j x}, \; x^2 e^{r_j x}, \quad \ldots, \quad x^{m_j-1} e^{r_j x},$$

$$e^{a_j x} \cos(b_j x), \; xe^{a_j x} \cos(b_j x), \; x^2 e^{a_j x} \cos(b_j x), \quad \ldots, \quad x^{n_j-1} e^{a_j x} \cos(b_j x),$$

$$e^{a_j x} \sin(b_j x), \; xe^{a_j x} \sin(b_j x), \; x^2 e^{a_j x} \sin(b_j x), \quad \ldots, \quad x^{n_j-1} e^{a_j x} \sin(b_j x).$$

The powers of x go up to one less than the multiplicity of the corresponding root of the characteristic polynomial.

In particular, the total number of these solutions equals the order n of the original equation. The *general solution* is obtained by multiplying each of these solutions by a constant and adding them up. For each choice of constants this results in a solution, and every solution arises in this way. (Notice that some of the constants may be zero.)

Exercises 19.2

I

Find each of the following.

1. $(3 - 5i)(2 + i)$

2. $(1 + i)(1 - i)$

3. $(1 + i)^3$

4. $(2 - i)^3$

5. $(3 + 4i)^{-1}$

6. $\frac{1}{1+i}$

7. $\frac{2+3i}{12+i}$

8. $\frac{1}{i}$

9. $\text{Im}\left(\frac{3-4i}{5i}\right)$

10. $\text{Im}\left(\frac{2-i}{-4+2i}\right)$

Find the general solution of each of the following differential equations.

11. $y'' + 16y = 0$

12. $y''' + 9y' = 0$

13. $y'' + 2y' + 2y = 0$

14. $y'' - 4y' + 5y = 0$

15. $y''' - 2y'' + 2y' = 0$

16. $y^{(4)} + 2y^{(2)} + y = 0$

17. $y^{(4)} = y$

18. $y''' = y$

19. $\frac{y''}{y} = 25$

20. $\frac{y'}{y'''} = 16$

21. $y'' + 2y' + 5y = 0$

22. $y''' - 2y'' + y' = 0$

23. $y'' + 3y' + 2y = 0$

24. $y''' - 4y'' + 3y' = 0$

25. $y^{(5)} + 2y^{(3)} = 0$

26. $y''' = 8y$

27. $y''' - 3y'' + 3y' - y = 0$

28. $y^{(4)} + 4y^{(3)} + 6y^{(2)} + 4y^{(1)} + y = 0$

29. $y^{(3)} + 4y^{(1)} = 0$

30. $y^{(6)} = y$

In each of the following situations, find the solution of the given differential equation that satisfies the specified conditions.

31. $y'' - 5y' + 4y = 0$, $y(0) = 2$, $y'(0) = -4$

32. $2y'' - 2y' + y = 0$, $y(0) = 0$, $y'(0) = 7$

33. $y'' + y = 0$, $y(0) = y'(0) = 5$

34. $y'' + \pi y = 0$, $y(0) = 2$, $y(1) = 3$

35. $y'' - 6y' + 9y = 0$, $y(1) = 0$, $y'(0) = 2$

36. $y''' + y'' + y' + y = 0$, $y(0) = y'(0) = y''(0) = 1$

37. $y''' + 2y'' + 2y' = 0$, $y(0) = y''(0) = 1$, $y'(0) = 3$

38. $y^{(4)} - 4y^{(2)} + 4y = 0$, $y(0) = 1$, $y'(0) = 3$, $y''(0) = 0$, $y'''(0) = -2$

39. $y'' + 9y = 0$, $y'''(0) = 3$, $y''(0) = 2$

40. $y''' = 0$, $y''(1) = y'(2) = y(3) = 6$

II

41. Use the Binomial Theorem to find $(1 + i)^4$.

42. If z is the complex number $a + ib$, with $a, b \in \mathbb{R}$, then its **complex conjugate** is the complex number $\bar{z} = a - ib$. Verify that $z\bar{z} = |z|^2$.

43. Verify that for any two complex numbers w, z we have $\frac{1}{wz} = \frac{1}{w}\frac{1}{z}$.

44. By multiplying out $(\cos\theta + i\sin\theta)(\cos\phi + i\sin\phi)$, verify that $e^{i\theta}e^{i\phi} = e^{i(\theta+\phi)}$.

*45. Let $p(x) = a_n x^n + \ldots + a_1 x + a_0$, with *real* numbers $a_0, a_1, \ldots, a_n$. If $r = a + ib$ is a root of $p(x)$, show that its complex conjugate $\bar{r} = a - ib$ is *also* a root.

*46. Find an example of a polynomial with *complex* coefficients for which the conclusion of Exercise 45 is false.

III

For each of the following equations, express the general solution as a sum (cf. Summary 19.2.6, p. 1226). Use a graphics calculator or computer to graph each term in the sum, and then use the picture to *guess* the particular solution satisfying the given conditions. Graph your guess and, if necessary, adjust it.

47. $y'' + y = 0$, $y(0) = 2$, $y'(0) = 3$

48. $y'' + 4y = 0$, $y(0) = 1$, $y'(0) = -2$

49. $y'' + 2y' + 2y = 0$, $y(0) = -2$, $y'(0) = 0$

50. $y''' + y' = 0$, $y(0) = y'(0) = y''(0) = 1$

When we studied the hyperbolic functions in Section 3.5 we remarked that in some ways they are analogous to the trigonometric functions. The fundamental hyperbolic identity $\cosh^2 x - \sinh^2 x = 1$ is strongly reminiscent of the trigonometric identity $\cos^2 x + \sin^2 x = 1$.

On the other hand there was no explanation of what the definitions

$$\cosh x = \frac{e^x + e^{-x}}{2} \quad \text{and} \quad \sinh x = \frac{e^x - e^{-x}}{2}$$

have to do with the trigonometric functions.

Now, however, since we know that $e^{ix} = \cos x + i \sin x$ and $e^{-ix} = \cos x - i \sin x$, we can solve and find that

$$\cos x = \frac{e^{ix} + e^{-ix}}{2} \quad \text{and} \quad \sin x = \frac{e^{ix} - e^{-ix}}{2i}.$$

These formulas look a lot like their hyperbolic cousins. They can also be useful for working with trigonometric functions. For instance, suppose we want to derive the addition formula for cosines. We write

$$\cos(a + b)$$
$$= \frac{e^{i(a+b)} + e^{-i(a+b)}}{2}$$
$$= \frac{e^{i(a+b)} + e^{i(a-b)} + e^{i(b-a)} + e^{-i(a+b)}}{4}$$
$$\quad + \frac{e^{i(a+b)} - e^{i(a-b)} - e^{i(b-a)} + e^{-i(a+b)}}{4}$$
$$= \frac{e^{ia} + e^{-ia}}{2} \frac{e^{ib} + e^{-ib}}{2} + \frac{e^{ia} - e^{-ia}}{2} \frac{e^{ib} - e^{-ib}}{2}$$
$$= \frac{e^{ia} + e^{-ia}}{2} \frac{e^{ib} + e^{-ib}}{2} - \frac{e^{ia} - e^{-ia}}{2i} \frac{e^{ib} - e^{-ib}}{2i}$$
$$= \cos a \cos b - \sin a \sin b.$$

19.3 Undetermined Coefficients

COMPLEMENTARY EQUATION
GENERAL SOLUTION: $y_p + y_c$
e^{bx}
$\cos bx$
$\sin bx$
POLYNOMIALS

So far we have been learning how to solve *homogeneous* linear equations, but now we turn our attention to nonhomogeneous linear equations with constant coefficients. For instance, we know how to solve an equation like $y'' + 3y' + 2y = 0$, but how would we find a solution to $y'' + 3y' + 2y = e^{4x}$?

Let us begin by asking ourselves what sort of functions we know whose derivatives involve e^{4x}. The answer is almost too easy: The function e^{4x} and all its derivatives involve e^{4x}; the same is true of ce^{4x}, for any constant c.

So let us consider $y = ce^{4x}$. We find that $y' = 4ce^{4x}$ and $y'' = 16ce^{4x}$. Substituting into the left side of the equation gives $y'' + 3y' + 2y = 16ce^{4x} + 12ce^{4x} + 2ce^{4x} = 30ce^{4x}$. For this to equal the right side e^{4x} means that $c = \frac{1}{30}$, and the desired solution is $y = \frac{1}{30}e^{4x}$.

If we want to solve $a_n y^{(n)} + a_{n-1}y^{(n-1)} + \ldots + a_1 y^{(1)} + a_0 y = e^{bx}$,

we try letting $y = ce^{bx}$ and observe that any derivative of $y = ce^{bx}$ is a constant times e^{bx}, so $a_n y^{(n)} + a_{n-1}y^{(n-1)} + \ldots + a_1 y^{(1)} + a_0 y$ will also be a constant times e^{bx}. All that is needed is to determine the coefficient c.

However, there is a complication because $a_n y^{(n)} + a_{n-1}y^{(n-1)} + \ldots + a_1 y^{(1)} + a_0 y$ might equal 0, and then no choice of the coefficient c would work. After a little experimentation we might try $y = cxe^{bx}$.

It could happen that this is also a solution of the associated homogeneous equation, but then we could try $y = cx^2e^{bx}$. Eventually, this approach will work. We let $y = cx^me^{bx}$, where m is the smallest power so that x^me^{bx} is *not* a solution of the homogeneous equation.

EXAMPLE 1   Find a solution of $y'' - 4y' + 4y = e^{2x}$.

Solution   The characteristic polynomial of the homogeneous equation is $r^2 - 4r + 4 = (r-2)^2$, so $r = 2$ is a root with multiplicity 2. This means (cf. Theorem 19.1.6) that $y = e^{2x}$ and $y = xe^{2x}$ are solutions of the homogeneous equation $y'' - 4y' + 4y = 0$, so of course neither will give the required solution of the nonhomogeneous equation. Instead, we try letting $y = cx^2e^{2x}$, so $y' = 2cxe^{2x} + 2cx^2e^{2x}$ and $y'' = 2ce^{2x} + 4cxe^{2x} + 4cxe^{2x} + 4cx^2e^{2x} = 2ce^{2x} + 8cxe^{2x} + 4cx^2e^{2x}$. Substituting into the left side of the equation, we find that

$$y'' - 4y' + 4y = 2ce^{2x} + 8cxe^{2x} + 4cx^2e^{2x} - 8cxe^{2x} - 8cx^2e^{2x} + 4cx^2e^{2x} = 2ce^{2x}.$$

To get e^{2x}, we let $c = \frac{1}{2}$, and the solution we want is $y = \frac{1}{2}x^2e^{2x}$.   ◢

It is apparent that in solving a nonhomogeneous equation it is necessary to know about the corresponding homogeneous equation.

DEFINITION 19.3.1

> Consider the following nonhomogeneous linear differential equation with constant coefficients:
>
> $$a_ny^{(n)} + a_{n-1}y^{(n-1)} + \ldots + a_1y^{(1)} + a_0y = f(x),$$
>
> where $f(x)$ is some function. The associated homogeneous equation
>
> $$a_ny^{(n)} + a_{n-1}y^{(n-1)} + \ldots + a_1y^{(1)} + a_0y = 0$$
>
> is called the **complementary homogeneous equation**.

The preceding discussion leads to the following technique.

Summary 19.3.2

To find a particular solution of the following nonhomogeneous linear differential equation with constant coefficients:

$$a_ny^{(n)} + a_{n-1}y^{(n-1)} + \ldots + a_1y^{(1)} + a_0y = e^{bx},$$

let $y = cx^me^{bx}$, where m is the smallest power so that x^me^{bx} is *not* a solution of the complementary homogeneous equation

$$a_ny^{(n)} + a_{n-1}y^{(n-1)} + \ldots + a_1y^{(1)} + a_0y = 0.$$

Substitute this function into the nonhomogeneous differential equation, and solve to determine the appropriate coefficient c.

EXAMPLE 2   Find a solution of the equation $y''' + 6y'' + 9y' = 4e^{-3x} + 3$.

Solution   In our problem the right side of the equation is $4e^{-3x} + 3$, which is not a simple exponential function of the form e^{bx}. However, we can deal with each part separately.

First, we try to find a function that will give us $4e^{-3x}$ on the right side. The characteristic polynomial of the complementary equation is $r^3 + 6r^2 + 9r = r(r^2 + 6r + 9) = r(r + 3)^2$. This means that e^{-3x} and xe^{-3x} are solutions of the complementary equation, so we try letting $y = cx^2e^{-3x}$.

Then

$$y' = 2cxe^{-3x} - 3cx^2e^{-3x},$$
$$y'' = 2ce^{-3x} - 6cxe^{-3x} - 6cxe^{-3x} + 9cx^2e^{-3x} = 2ce^{-3x} - 12cxe^{-3x} + 9cx^2e^{-3x},$$
$$y''' = -6ce^{-3x} - 12ce^{-3x} + 36cxe^{-3x} + 18cxe^{-3x} - 27cx^2e^{-3x}$$
$$= -18ce^{-3x} + 54cxe^{-3x} - 27cx^2e^{-3x}.$$

Substituting into the left side of the equation gives

$$y''' + 6y'' + 9y' = -18ce^{-3x} + 54cxe^{-3x} - 27cx^2e^{-3x} + 12ce^{-3x} - 72cxe^{-3x}$$
$$+ 54cx^2e^{-3x} + 18cxe^{-3x} - 27cx^2e^{-3x}$$
$$= -6ce^{-3x}.$$

For this to equal $4e^{-3x}$ we need to have $c = -\frac{2}{3}$. The first part of the solution is $y = -\frac{2}{3}x^2e^{-3x}$.

Second, we look for a function y so that $y''' + 6y'' + 9y' = 3$. If we think of 3 as the "exponential" function $3e^{0x}$, then e^{0x} is a solution of the complementary equation, and we should try $y = cxe^{0x} = cx$. We find that $y' = c$ and $y'' = y''' = 0$, so $y''' + 6y'' + 9y' = 9c$, and we should let $c = \frac{1}{3}$.

Having found functions that satisfy $y''' + 6y'' + 9y' = 4e^{-3x}$ and $y''' + 6y'' + 9y' = 3$, respectively, we only have to add them together to find a function that satisfies $y''' + 6y'' + 9y' = 4e^{-3x} + 3$. The solution we want is

$$y = -\frac{2}{3}x^2e^{-3x} + \frac{x}{3}.$$

In Example 2 we recognized that a constant can be regarded as an exponential function e^{0x}; this apparently silly remark enables us to use the technique of Summary 19.3.2 to solve equations whose right side is a constant.

The other interesting thing about Example 2 is the way we broke up the right side of $y''' + 6y'' + 9y' = 4e^{-3x} + 3$ and dealt with each piece separately. Having found a solution to each part, we just added them to get a solution to the original equation.

We have seen how to find a particular solution of a nonhomogeneous equation with an exponential function on the right side, but it is perfectly natural to ask about the *general solution*, that is, how to find *all* the solutions. In fact this also involves the complementary equation.

THEOREM 19.3.3

The general solution of the equation

$$a_n y^{(n)} + a_{n-1} y^{(n-1)} + \ldots + a_1 y^{(1)} + a_0 y = f(x),$$

where each a_i is a constant and $f(x)$ is a function, is obtained by finding one **particular solution** y_p and adding to it the **general solution** y_c of the complementary equation

$$a_n y^{(n)} + a_{n-1} y^{(n-1)} + \ldots + a_1 y^{(1)} + a_0 y = 0.$$

The general solution of the nonhomogeneous equation is $y_p + y_c$.

PROOF   First we show that any function of the form $y = y_p + y_c$ is a solution:

$$a_n y^{(n)} + a_{n-1} y^{(n-1)} + \ldots + a_1 y^{(1)} + a_0 y$$
$$= a_n (y_p + y_c)^{(n)} + a_{n-1} (y_p + y_c)^{(n-1)} + \ldots + a_1 (y_p + y_c)^{(1)} + a_0 (y_p + y_c)$$
$$= a_n y_p^{(n)} + a_{n-1} y_p^{(n-1)} + \ldots + a_1 y_p^{(1)} + a_0 y_p + a_n y_c^{(n)} + a_{n-1} y_c^{(n-1)} +$$
$$\ldots + a_1 y_c^{(1)} + a_0 y_c$$
$$= f(x) + 0 = f(x).$$

Second, it is necessary to show that any solution must be of this form. Suppose we know one particular solution y_p and y is any other solution. Then both y and y_p satisfy the equation:

$$a_n y^{(n)} + a_{n-1} y^{(n-1)} + \ldots + a_1 y^{(1)} + a_0 y = f(x),$$
$$a_n y_p^{(n)} + a_{n-1} y_p^{(n-1)} + \ldots + a_1 y_p^{(1)} + a_0 y_p = f(x).$$

Subtracting, we find that

$$a_n (y - y_p)^{(n)} + a_{n-1} (y - y_p)^{(n-1)} + \ldots + a_1 (y - y_p)^{(1)} + a_0 (y - y_p) = 0.$$

This says that $y - y_p$ is a solution of the complementary equation, or that y equals y_p plus a solution of the complementary equation, which completes the proof. ◢

EXAMPLE 3   Find the general solution of $y''' - 4y'' + 4y' = -1$.

Solution   The characteristic polynomial of the complementary equation is $r^3 - 4r^2 + 4r = r(r^2 - 4r + 4) = r(r - 2)^2$, so its roots are $r = 0$ with multiplicity 1 and $r = 2$ with multiplicity 2. In particular, constants $k = ke^{0x}$ are solutions of the complementary equation, and we cannot use $y = k$ as a particular solution.

To find a particular solution, we try $y = kx$, so $y' = k$, $y'' = y''' = 0$. The left side of the equation becomes $y''' - 4y'' + 4y' = 4k$, so we take $k = -\frac{1}{4}$ and find a particular solution $y_p = -\frac{1}{4}x$.

On the other hand the general solution of the complementary equation is $y_c = ae^{0x} + be^{2x} + cxe^{2x}$. Adding this to the particular solution gives the general solution of the original equation:

$$y = y_p + y_c = -\frac{1}{4}x + a + be^{2x} + cxe^{2x}.$$

◢

We have seen how to solve equations whose right side is e^{bx}; the approach we used relies upon the fact that we know all the derivatives of e^{bx}. It is only slightly more complicated to consider a function like $\sin bx$ or $\cos bx$.

EXAMPLE 4   Solve $y'' + y' - 6y = \cos 5x$.

Solution   A first guess might be to let $y = y_p = c\cos 5x$. We find that $y' = -5c\sin 5x$ and $y'' = -25c\cos 5x$. From this we see that

$$y'' + y' - 6y = -25c\cos 5x - 5c\sin 5x - 6c\cos 5x = -31c\cos 5x - 5c\sin 5x.$$

Since we want $\cos 5x$ on the right side, this does not work. In a way the problem arises because the various derivatives of $\cos 5x$ are not all constant multiples of $\cos 5x$; they may involve $\sin 5x$. Let us try letting $y = a\cos 5x + b\sin 5x$. Then

$$y' = -5a\sin 5x + 5b\cos 5x,$$
$$y'' = -25a\cos 5x - 25b\sin 5x,$$

and the left side of the equation is $y'' + y' - 6y = -25a\cos 5x - 25b\sin 5x - 5a\sin 5x + 5b\cos 5x - 6a\cos 5x - 6b\sin 5x = (-31a + 5b)\cos 5x + (-31b - 5a)\sin 5x$.

For this to equal $\cos 5x$ we need to have

$$-31a + 5b = 1,$$
$$-5a - 31b = 0.$$

To eliminate the b-terms, we multiply the first equation by 31 and the second by 5 and add; the result is $-31(31)a - 5(5)a = 31$, that is, $-(31^2 + 5^2)a = 31$, so $a = -\frac{31}{986}$. Substituting, we find that $5b = 1 + 31a$, so $b = \frac{1}{5}\left(1 - \frac{961}{986}\right) = \frac{5}{986}$.

We have found a particular solution $y_p = -\frac{31}{986}\cos 5x + \frac{5}{986}\sin 5x$.

Since the characteristic polynomial of the complementary equation is $r^2 + r - 6 = (r + 3)(r - 2)$, the general solution of the complementary equation is $y_c = ae^{-3x} + be^{2x}$. The general solution of the original equation is

$$y = y_p + y_c = -\frac{31}{986}\cos 5x + \frac{5}{986}\sin 5x + ae^{-3x} + be^{2x}.$$

Example 4 suggests that to find a particular solution of an equation whose right side is $\cos bx$ (or $\sin bx$), we should let $y = c\cos bx + d\sin bx$ and solve for c and d.

This will work unless $\cos bx$ or $\sin bx$ is a solution of the complementary equation. In that case we must modify the function by multiplying by a power of x, as before.

It is possible to use a similar approach for solving an equation that has a polynomial on the right side.

EXAMPLE 5   Find a particular solution of $y'' + 4y' + 3y = x^2 - 2$.

Solution   It seems reasonable to expect that it will be possible to find a solution that is a polynomial. Since the right side is a quadratic polynomial, that is, one of degree 2, we first try for a solution that is also quadratic. Let $y = ax^2 + bx + c$.

Then $y' = 2ax + b$ and $y'' = 2a$. The left side of the equation is

$$y'' + 4y' + 3y = 2a + (8ax + 4b) + (3ax^2 + 3bx + 3c)$$
$$= 3ax^2 + (8a + 3b)x + (2a + 4b + 3c).$$

For this to equal the right side $x^2 - 2$ we need to have

$$3a \qquad\qquad = 1,$$
$$8a + 3b \qquad = 0,$$
$$2a + 4b + 3c = -2.$$

The first equation tells us that $a = \frac{1}{3}$, so the second and third equations become

$$3b \qquad = -\frac{8}{3},$$
$$4b + 3c = -\frac{8}{3}.$$

From this we find that $b = -\frac{8}{9}$. Substituting this value into the last equation gives $3c = -\frac{8}{3} - 4b = \frac{8}{9}$ so $c = \frac{8}{27}$.

A particular solution of the equation is $y = \frac{1}{3}x^2 - \frac{8}{9}x + \frac{8}{27}$.

> Notice that although there was no x-term in the right side $x^2 - 2$ of the original equation, we still need an x-term in the solution. What is important is the degree of the polynomial, not whether any specific power occurs in it.

The basic idea is that if the right side of the equation is a polynomial of degree n, then we should try letting $y = c_n x^n + c_{n-1} x^{n-1} + \ldots + c_1 x + c_0$, a polynomial of degree n. In Example 5 the right side is $x^2 - 2$, which is of degree $n = 2$, so we used the polynomial $ax^2 + bx + c$ of degree 2.

As usual there is a complication if some powers of x are solutions of the complementary equation. Roughly speaking, for each power of x that is a solution of the complementary equation (including the constant x^0), we raise by one the powers of x that are included in the particular solution. So if the right side of the equation is a cubic polynomial (i.e., with $n = 3$) and 1 and x are solutions of the complementary equation but x^2 is not, then instead of $c_3 x^3 + c_2 x^2 + c_1 x + c_0$ we raise the powers by 2 and let $y = c_5 x^5 + c_4 x^4 + c_3 x^3 + c_2 x^2$.

Notice that there is no point in including the lower powers $c_1 x + c_0$ in this solution, since they will just give 0 anyway.

Summary 19.3.4   The Method of Undetermined Coefficients (for Nonhomogeneous Equations with Constant Coefficients)

Consider an equation of the form

$$a_n y^{(n)} + a_{n-1} y^{(n-1)} + \ldots + a_1 y^{(1)} + a_0 y = f(x),$$

where each a_j is a real constant and $f(x)$ is some function.

1. The general solution is found by finding a particular solution y_p and the general solution y_c of the complementary equation and adding them: $y = y_p + y_c$.

2. To find a particular solution, we can split up the function $f(x)$. If $f(x) = f_1(x) + f_2(x)$, then we can find particular solutions for

$$a_n y^{(n)} + a_{n-1} y^{(n-1)} + \ldots + a_1 y^{(1)} + a_0 y = f_1(x),$$
$$a_n y^{(n)} + a_{n-1} y^{(n-1)} + \ldots + a_1 y^{(1)} + a_0 y = f_2(x),$$

and add them to get a particular solution for the original equation.

3. For certain types of function $f(x)$ it is possible to find a particular solution y_p by making a guess as shown in the following table and solving for the constants.

Function $f(x)$	Guess for y_p
e^{bx}	$cx^m e^{bx}$, smallest m for which this is not a solution of the complementary equation
$\cos bx$ or $\sin bx$	$cx^m \cos bx + dx^m \sin bx$, smallest m for which this is not a solution of the complementary equation
polynomial of degree n	$c_n x^n + \ldots + c_1 x + c_0$ if 1 is not a solution of the complementary equation; $c_{m+n} x^{m+n} + \ldots + c_m x^m$ if x^{m-1} is a solution of the complementary equation and x^m is not.

In brief, we make a first guess of ce^{bx} if the right side contains e^{bx}, $c \cos bx + d \sin bx$ if it contains $\cos bx$ or $\sin bx$, and a polynomial of degree n if the right side is a polynomial of degree n. In each case this will work unless part of the function in the first guess is a solution of the complementary equation.

In that case we multiply the first guess by the smallest power of x so that no part of the new function is a solution of the complementary equation.

EXAMPLE 6   Parachutist   Let $h(t)$ be the height above ground of a falling object, in feet. If it falls without air resistance, we know $h'' = -32$, in ft/sec^2.

The air resistance on a large object is proportional to the velocity; this means that the object experiences a force that is a constant times h'. By Newton's Second Law $F = ma$, the resulting acceleration a is a constant times the force, so it is also

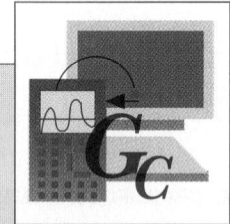

With a graphics calculator we can sketch the graph of the velocity $h'(t) = -32 + 32e^{-t}$ from Example 6.

Noting that the velocity approaches a constant as t increases, what shape do you expect the graph of $h(t)$ will have?

Try plotting the height $y = h(t)$.

a constant times h'. In this situation the height satisfies $h'' = -kh' - 32$, which says that the total acceleration h'' is the upward acceleration $-kh'$ due to the air resistance plus the constant downward acceleration due to gravity.

Writing it as $h'' + kh' = -32$, we find that the general solution of the complementary equation $h'' + kh' = 0$ is $h_c(t) = c_1 + c_2 e^{-kt}$. A particular solution is $h_p(t) = -\frac{32}{k} t$, and the general solution is $h(t) = h_p + h_c = -\frac{32}{k} t + c_1 + c_2 e^{-kt}$.

Consider a parachutist dropping from a stationary helicopter at 5000 ft. This means $h(0) = 5000$ and $h'(0) = 0$. Suppose, for instance, that $k = 2$. From $h(0) = 5000$ we find $c_1 + c_2 = 5000$, and $0 = h'(0) = -\frac{32}{k} - c_2 k$, so $c_2 = -\frac{32}{k^2} = -8$, $c_1 = 5008$.

The height of the parachutist is $h(t) = -16t + 5008 - 8e^{-2t}$ ft. It is interesting to observe that her velocity is $h'(t) = -16 + 16e^{-2t}$. It starts at 0 when she jumps and then decreases. As $t \to \infty$, $h'(t) \to -16$; her velocity approaches the steady downward drift at 16 ft/sec as the air resistance balances the force of gravity. This limiting velocity is called the *terminal velocity*.

Exercises 19.3

I

Find a particular solution of each of the following equations.

1. $y'' - 3y' + y = e^{5x}$

2. $y'' + 4y' + 5y = e^{-x}$

3. $y'' + 4y' + 3y = e^{3x}$

4. $y''' + 3y'' + 2y' = e^x$

5. $y'' - 2y' - 3y = e^{3x}$

6. $y'' - 4y' + 4y = e^{2x}$

7. $y'' + 6y' = 5$

8. $y'' - 5y' + 6y = e^{3x}$

9. $y'' - 2y' = 2 - e^x$

10. $y''' - 4y' = 1 - 5e^{2x}$

11. $y'' + y' + y = \cos 3x$

12. $y'' - y' = \sin 9x$

13. $y''' - 2y'' + y = x^3 - 2x$

14. $y'' + 4y' - y = x^2 + 3x$

15. $y'' + y = \cos x$

16. $y'' - 3y' = x^2 - 2x + 7$

17. $y'' - 6y' + 10y = \cos 2x$

18. $y''' + 9y' = x + \sin 3x$

19. $y''' - y'' + y' - y = \sin x + 3x^2$

20. $y^{(4)} + 2y^{(2)} + y = \sin x$

Find the general solution of each of the following equations.

21. $y'' + y' - 2y = e^{-2x}$ **23.** $y''' = \cos 6x$

22. $y'' + 4y = \sin 2x$ **24.** $y'' - 4y' + 4y = e^{2x}$

25. $y'' + 6y' + 5y = 6x + e^{-x}$

26. $y'' - y' - 2y = e^{2x}$

27. $y'' - 3y' - 4y = 6x^2 + 5x$

28. $y'' - 2y' + y = e^x$

29. $y''' + 16y' = x^3 + 5x + \cos 4x$

30. $y^{(4)} + y^{(2)} = \cos x$

In each of the following situations, find the solution that satisfies the given conditions. (Begin by finding the general solution.)

31. $y'' + y = x$, $y(0) = 1$, $y'(0) = 0$

32. $y'' - 4y' + 4y = e^{2x}$, $y(0) = 0$, $y'(0) = 0$

33. $y'' - 3y' - 4y = e^x$, $y(0) = 1$, $y'(0) = 6$

34. $y'' + 5y' + 6y = x^2$, $y(0) = 6$, $y'(0) = 12$

35. $y'' + 4y = e^x$, $y(0) = 1$, $y'(0) = 4$

36. $y''' = \sin 2x$, $y(0) = 1$, $y'(0) = 6$, $y''(0) = 24$

37. $y''' - 4y' = x$, $y(0) = 2$, $y'(0) = 12$, $y''(0) = 0$

38. $y'' - y = e^{3x}$, $y(0) = 4$, $y'(1) = 3$

39. $y'' + y' + y = 3x^2$, $y(0) = 4$, $y'(0) = -2$

40. $y^{(4)} = 3\cos 2x$, $y(0) = -1$, $y'(0) = 0$, $y''(0) = 4$, $y'''(0) = 0$

II

Find the general solution of each of the following equations.

41. The equation of Example 1 (p. 1229).

42. The equation of Example 2 (p. 1230).

43. The equation of Example 5 (p. 1232).

*44. $y^{(8)} = \cos 3x$

Find a particular solution of each of the following equations.

*45. $y'' + 3y' - y = xe^x$

*46. $y''' - 2y'' + y' + 4y = x \sin x$

*47. $y'' - y' + 5y = a^2 x e^{3x}$

*48. $y'' + 5y' - 3y = x^2 \cos x$

*49. If the parachutist in Example 6 (p. 1234) got a bigger parachute so that the constant k was replaced with $k = 3$, what would the terminal velocity be?

*50. (i) It is not too easy to solve the equation from Example 6 to find when $h(t) = 0$, that is, when the parachutist lands. A way to approximate the answer would be to say $h(t) = -16t + 5008 - 8e^{-2t} \approx -16t + 5008$ and find when this approximation is zero. Find this value of t, and evaluate $h(t)$ to see how close the parachutist is to the ground.

(ii) Use your answer to part (i) to find a better approximation to the time of impact. Again check your result by finding how close $h(t)$ is to zero.

For each of the following equations, express the general solution as a sum (cf. Summary 19.3.4, p. 1233). Use a graphics calculator or computer to graph each term in the sum, and then use the picture to *guess* the particular solution satisfying the given conditions. Graph your guess and, if necessary, adjust it.

51. $y' + y = e^{2x}$, $y(0) = 0$

52. $y'' + y = e^x$, $y(0) = 0$, $y'(0) = 1$

53. $y'' - y' = e^x$, $y(0) = y'(0) = 1$

54. $y'' - y = e^x + x$, $y(0) = y'(0) = 2$

19.4 Vibrations

HOOKE'S LAW
SPRING CONSTANT
SIMPLE HARMONIC MOTION
PERIOD
POUNDS/SLUGS
FREQUENCY
DAMPED VIBRATIONS
DISCRIMINANT

| Remember that it is important to measure the displacement s starting from the rest position.

In this section we consider the type of motion called vibrations. This type of motion can be understood by using differential equations, and it is important for many different practical problems. In addition to the obvious examples like vibrating springs and swaying pendulums, the theory we will develop can also be used to study sound waves, electronic signals, and many other phenomena that have an oscillatory or cyclic character. Even diverse topics such as tidal patterns and animal population cycles may sometimes be considered from this point of view.

In Section 6.6 we discussed Hooke's Law (Equation 6.6.2, p. 458), which describes the relationship between the displacement s of a spring from its rest position and the force F that it exerts. **Hooke's Law** says that

$$F = -ks,$$

where k is a positive constant, called the **spring constant**. The force points in the direction *opposite* to the displacement, tending to push back toward the rest position.

Newton's Second Law says that an object of mass M that is acted on by a force F will experience an acceleration of a and that these quantities are related by the law

$$F = Ma.$$

So consider a heavy object attached to the free end of a spring, as shown in Figure 19.4.1. Suppose the mass of the object is M and it is so much heavier than the spring that we can ignore the mass of the spring. If the spring is at its rest position, there will be no force, and the object will remain stationary.

Figure 19.4.1

The situation is simplified if we assume that the object moves back and forth in a straight line. In this case we can represent the displacement along the line, the velocity, the acceleration, and the force by numbers rather than vectors. Let us agree to measure displacement starting at the rest position, so that $s = 0$ at the rest position and so that s is positive when the spring is extended and negative when the spring is compressed. Let us also understand F to be a force that is positive when it points in the direction in which the spring extends and negative in the direction in which the spring compresses.

Figure 19.4.2

If the spring is stretched out and then released, as shown in Figure 19.4.2, then it will exert a force on the heavy object pulling it back toward the rest position. The force is given by Hooke's Law, so it is $F = -ks$.

On the other hand, Newton's Second Law says that $F = Ma = Ms''$, and equating these two expressions for F gives

$$Ms'' = -ks.$$

Dividing by M, we obtain the differential equation $s'' = -\frac{k}{M}s$, or

$$s'' + \frac{k}{M}s = 0.$$

This is a homogeneous linear differential equation with constant coefficients. Its characteristic polynomial is $r^2 + \frac{k}{M}$, so the roots are $r = \pm i\sqrt{\frac{k}{M}}$. This means that the general solution is

$$s = c\cos\left(\sqrt{\frac{k}{M}}\,t\right) + d\sin\left(\sqrt{\frac{k}{M}}\,t\right),$$

where c and d are constants.

The graph of this function will have the same shape as a sin or cos curve, but it may be shifted sideways. One such curve is illustrated in Figure 19.4.3. Recalling that positive values of the displacement s represent extensions of the spring and negative values of s represent compressions, we see that this graph represents a motion in which the object moves back and forth, compressing and extending the spring.

This is exactly the sort of motion we expect from an object attached to a spring, as shown in Figure 19.4.4. In physics it is called **simple harmonic motion**.

From the formula we see that the function $s = c\cos\left(\sqrt{\frac{k}{M}}\,t\right) + d\sin\left(\sqrt{\frac{k}{M}}\,t\right)$ is *periodic*. It repeats itself after t reaches $\frac{2\pi}{\sqrt{k/M}}$. This is also apparent from the graph of the displacement that we drew in Figure 19.4.3. We say that the **period** of the vibration is $\frac{2\pi}{\sqrt{k/M}}$. It is a fundamental fact in physics that this period does not depend on how far the spring is stretched. The vibration will have the same period for very small displacements as it does for large displacements.

EXAMPLE 1 Suppose a spring exerts a force of 3 newtons when it is extended 10 cm. To be consistent, let us write the displacement 10 cm as 0.1 m. Then the spring constant is $k = \frac{3}{0.1} = 30$ N/m.

If a steel ball weighing 500 g = 0.5 kg is attached to the spring and stretched out and released so that it starts vibrating, then the period of the vibration will be

$$P = \frac{2\pi}{\sqrt{k/M}} = \frac{2\pi}{\sqrt{30/0.5}} = \frac{2\pi}{\sqrt{60}} \approx 0.81.$$

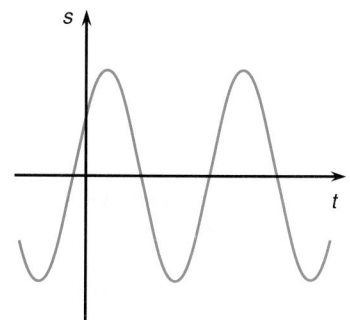

Figure 19.4.3

Figure 19.4.4

We had better check that the units make sense. We recall that 1 newton is the force required to make a mass of 1 kg accelerate at a rate of 1 m/sec^2; this means that 1 N = 1 kg m/sec^2. The units for the spring constant k are N/m, so the units for k/M are N/(m kg). Writing N as kg m/sec^2, we find that the units for k/M are kg m/sec^2/(m kg) = sec^{-2}. This means that the units for $\sqrt{k/M}$ are sec^{-1} and the units for $P = \frac{2\pi}{\sqrt{k/M}}$ are $1/\text{sec}^{-1}$ = sec.

We have just found that the period of the vibration is $\frac{2\pi}{\sqrt{60}} \approx 0.81$ sec.

We used metric units in Example 1 because they are less confusing for this purpose. They have the advantage that mass is measured in kilograms and force is measured in newtons, whereas **pounds** are often used to measure both mass and force.

There is a way around this difficulty, which is to introduce another unit for mass. The new unit of mass is a **slug**. One slug is the amount of mass that will accelerate at a rate of 1 ft/sec^2 if it is acted upon by a force of 1 pound.

To use these units, we have to figure out the mass of an object in terms of its *weight* in pounds. The weight is the force due to gravity, which we know accelerates objects at a rate of 32 ft/sec^2. Suppose the weight (i.e., the gravitational force F) of an object is 1 pound. We know the acceleration is $a = 32$ ft/sec^2, so Newton's Law $F = Ma$ tells us the mass is $M = \frac{F}{a} = \frac{1}{32}$ slug.

An object weighing 1 lb has mass $\frac{1}{32}$ slug. This is potentially confusing, but all we need to remember is this conversion from pounds to slugs:

$$\text{Mass} = \frac{(\text{weight in pounds})}{32} \quad \text{slugs.} \tag{19.4.1}$$

EXAMPLE 2   A spring exerts a force of 6 lb when it is extended 4 inches. What will be the period with which a 1-lb weight will vibrate on the end of the spring?

Solution   Writing 4 in as $\frac{1}{3}$ ft, we see that the spring constant is $k = \frac{6}{1/3} = 18$ lb/ft. The mass of the 1-lb object is $\frac{1}{32}$ slug. From our formula we find that the period of the vibration is

$$P = \frac{2\pi}{\sqrt{k/M}} = \frac{2\pi}{\sqrt{18(32)}} = \frac{2\pi}{24} = \frac{\pi}{12} \approx 0.26.$$

The object will vibrate with a period of $\frac{\pi}{12} \approx 0.26$ sec. This means that it will move back and forth approximately 4 times per second.

EXAMPLE 3   A 5-lb weight on the end of a spring vibrates at a frequency of 30 cycles per minute. What is the spring constant?

Solution   To say that the frequency is 30 cycles per minute means that the weight goes back and forth 30 times each minute, so each cycle takes 2 sec. In other words the period is $P = 2$ sec.

A weight of 5 lb has a mass of $M = \frac{5}{32}$ slug. If the spring constant is k, we know that $P = \frac{2\pi}{\sqrt{k/M}}$, so $\sqrt{k/M} = \frac{2\pi}{P}$ or $\frac{k}{M} = \frac{4\pi^2}{P^2}$ and

$$k = \frac{4\pi^2}{P^2} M = \frac{4\pi^2}{4} \frac{5}{32} = \frac{5\pi^2}{32} \approx 1.54 \text{ lb/ft.}$$

Figure 19.4.5

We have considered the vibration of a mass moving horizontally on the end of a spring, but another common situation is a mass suspended from a hanging spring, as shown in Figure 19.4.5. If the end of the spring is pulled down or up and then released, it will vibrate up and down.

Here we measure displacement vertically from the rest position. We let s be positive when the weight is above the rest position, observing that this is when the spring is compressed. (When we considered horizontal springs, we had $s > 0$ when the spring was *extended*.)

The situation appears more complicated because in addition to the force from the spring there is the force due to gravity. However, the spring stretches the right amount to balance the weight at the "rest position," and then the calculations are the same.

EXAMPLE 4 When a 3-kg weight is suspended from a spring, the spring stretches by 40 cm. Find a formula for the displacement if the weight is pulled down 20 cm below the rest position and then released.

Solution The force of gravity on the weight is $3(9.8) = 29.4$ N. Since it stretches the spring by 40 cm $= 0.40$ m, the spring constant is $k = \frac{29.4}{0.40} = 73.5$ N/m.

The general solution is

$$s(t) = c \cos \sqrt{24.5}t + d \sin \sqrt{24.5}t.$$

We know that the weight was released from 20 cm $= 0.20$ m below the rest position. This means $s(0) = -0.20$. On the other hand, $s(0) = c \cos 0 + d \sin 0 = c$, so $c = -0.20$.

If the weight is held and then released, it begins with no velocity, so $s'(0) = 0$. We calculate $s'(t) = -\sqrt{24.5}c \sin(\sqrt{24.5}t) + \sqrt{24.5}d \cos(\sqrt{24.5}t)$, so $0 = s'(0) = -\sqrt{24.5}c \sin 0 + \sqrt{24.5}d \cos 0 = \sqrt{24.5}d$ and $d = 0$. The solution we want is

$$s(t) = -0.20 \cos \sqrt{24.5}t.$$

We began by imagining a heavy object vibrating on the end of a spring, moving horizontally. In practice there is a complication with this picture.

The object must be resting on something, as shown in Figure 19.4.6. When it moves back and forth, there will be friction between it and the surface.

In our discussion so far we have ignored this effect, which in practical terms amounts to assuming that there is very little friction. We can imagine a very slippery surface or perhaps an object that runs on little wheels.

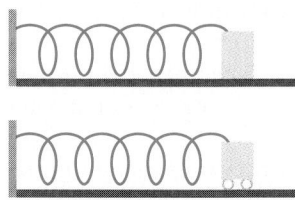

Figure 19.4.6

However, we can also try to take friction into account. If the object is at rest, it experiences no force from friction, but as it starts to move along the surface, friction will tend to slow it down. The faster it moves, the more friction there will be. (The faster you rub your hands together, the more they warm up.)

This suggests that the force due to friction is a constant times the velocity, and since the friction tends to *resist* the motion, the constant is negative. The force due to friction is

$$F_{friction} = -Rv = -Rs',$$

for some positive constant R.

Without considering friction we knew that the force exerted on a weight at the end of a spring is $F_{spring} = -ks$. Now we add the frictional force and find that the total force is

$$F = F_{spring} + F_{friction} = -ks - Rs'.$$

Newton's Second Law says $F = Ma = Ms''$, so $Ms'' = -ks - Rs'$, or

$$s'' + \frac{R}{M} s' + \frac{k}{M} s = 0.$$

This equation describes the displacement s of an object of mass M on a spring with spring constant k if the friction is $F_{friction} = -Rs'$.

This kind of motion is called **damped** motion, and we say that the vibrations are *damped* by friction. A common example of this phenomenon occurs if the object is moving in a fluid, such as water or oil. Instead of friction between the object and surface on which it rests, there is friction between the object and the fluid in which it moves. Happily, the calculations are the same.

The characteristic polynomial is $r^2 + \frac{R}{M} r + \frac{k}{M}$; its discriminant is $D = \frac{R^2}{M^2} - \frac{4k}{M}$. If $D > 0$, there will be two real roots; if $D = 0$, there will be one real root with multiplicity 2; and if $D < 0$, there will be two complex roots.

Let us consider first the case in which $D < 0$. The roots are

$$r = \frac{-\frac{R}{M} \pm \sqrt{\frac{R^2}{M^2} - \frac{4k}{M}}}{2} = -\frac{R}{2M} \pm \frac{\sqrt{R^2 - 4kM}}{2M}.$$

Writing these roots as $r = a \pm ib$, with $a = -\frac{R}{2M}$, $ib = \frac{\sqrt{R^2 - 4kM}}{2M} = i\frac{\sqrt{4kM - R^2}}{2M}$, we observe that $b = \frac{\sqrt{4kM - R^2}}{2M}$ is a real number and $a = -\frac{R}{2M}$ is negative. The general solution of the equation is

$$s(t) = e^{at}(c \cos bt + d \sin bt).$$

The factor $c \cos bt + d \sin bt$ is a curve of the sort we have considered, cycling back and forth as shown in Figure 19.4.7(i). Since $a = -\frac{R}{2M}$ is negative, the factor e^{at} describes an exponential decay. The product $s(t) = e^{at}(c \cos bt + d \sin bt)$ is shown in Figure 19.4.7(ii). It cycles back and forth, but it also decays, moving less and less far from the rest position with each successive cycle.

This is exactly the behavior we expect from an object that is vibrating but slowed down by friction. It vibrates but gradually moves through a shorter and shorter distance.

This kind of motion is not periodic because it does not repeat exactly. On the other hand, it does go regularly back and forth. The length of time it takes to go from the rest position out to one extreme, back through the rest position to the opposite extreme, and then back to the rest position is called the **period**. From the formula we see that the period is the amount of time it takes for bt to increase by 2π, which is $P = \frac{2\pi}{b} = \frac{4\pi M}{\sqrt{4kM - R^2}}$.

Notice that if we ignore the friction, that is, let $R = 0$, then this is just the same as the period $P = \frac{2\pi}{\sqrt{k/M}}$ we found before we considered friction at all. On the other hand, if we do consider friction, then the denominator $\sqrt{4kM - R^2}$ becomes smaller because of the $-R^2$. Making the denominator smaller makes the quotient larger. What this means is that the presence of friction makes the period longer, so the object cycles more slowly, exactly as we would expect.

The *decay constant* $a = -\frac{R}{2M}$ depends on R and M. If R is large, a is a large negative number, and the motion decays rapidly. This corresponds to the physical situation in which a large frictional force will slow the vibration quite rapidly. A small frictional force will still make the motion decay, but more slowly.

Figure 19.4.7(i)

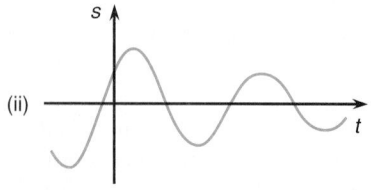

Figure 19.4.7(ii)

Also notice that since the mass M is in the denominator, a large value of M tends to result in a slow decay, consistent with our experience that it is more difficult to slow down a heavy object.

Next we consider the case in which the discriminant is $D = 0$. This means that there is one real root with multiplicity 2, and it is $r = -\frac{R}{2M}$. In particular it is negative. The general solution of the equation is

$$s(t) = ce^{-Rt/2M} + dte^{-Rt/2M}.$$

The shape of the graph varies, depending on the constants c and d, but several possibilities are shown in Figure 19.4.8.

For instance, if $c = 1$ and $d = 0$, the displacement is $s = e^{-Rt/2M}$, which is the exponentially decaying curve shown in Figure 19.4.8(i). This displacement corresponds to the situation in which the weight has been pulled away from the rest position and started with a shove toward the rest position. If $c = 0$ and $d = 1$, the displacement is $s = te^{-Rt/2m}$, whose graph is shown in Figure 19.4.8(ii). This corresponds to the situation in which the weight starts at the rest position $s = 0$ but is given a shove away from the rest position. It moves out for a while, and then the spring drags it back. However, the friction is so great that the weight just slides back to the rest position without going beyond. We can imagine the weight moving in a container of heavy oil.

In Figure 19.4.8(iii) we see the displacement for a situation in which the weight has been pulled away from the rest position and then pushed back so hard that despite the heavy oil it goes past the rest position. Later it rebounds and slides back toward the rest position without going past it again. This type of motion, in which the discriminant of the characteristic polynomial is $D = 0$, is called **critically damped** motion.

(i)

Figure 19.4.8(i)

(ii)

Figure 19.4.8(ii)

(iii)

Figure 19.4.8(iii)

Finally, if the discriminant $D = \frac{R^2}{M^2} - \frac{4k}{M} = \frac{1}{M^2}(R^2 - 4kM)$ is positive, there will be two real roots $r = -\frac{R}{2M} \pm \frac{\sqrt{R^2-4kM}}{2M}$. Because we are assuming $D > 0$, we see that $0 < R^2 - 4kM < R^2$, so $0 < \sqrt{R^2 - 4kM} < R$. This means that both roots $r = -\frac{R}{2M} \pm \frac{\sqrt{R^2-4kM}}{2M}$ will be negative. If we call them r_1 and r_2, then the general solution is

$$s(t) = ce^{r_1 t} + de^{r_2 t}.$$

Since both r_1 and r_2 are negative, this curve decays to 0 as $t \to \infty$. Several possibilities are shown in Figure 19.4.9. They are qualitatively quite similar to the curves in Figure 19.4.8 for critically damped motion.

(i)

Figure 19.4.9(i)

(ii)

Figure 19.4.9(ii)

(iii)

Figure 19.4.9(iii)

When the discriminant D is positive, we speak of **overcritically damped** motion. The various types of damped and undamped vibrations are summarized below, and typical solutions are graphed in Figure 19.4.10.

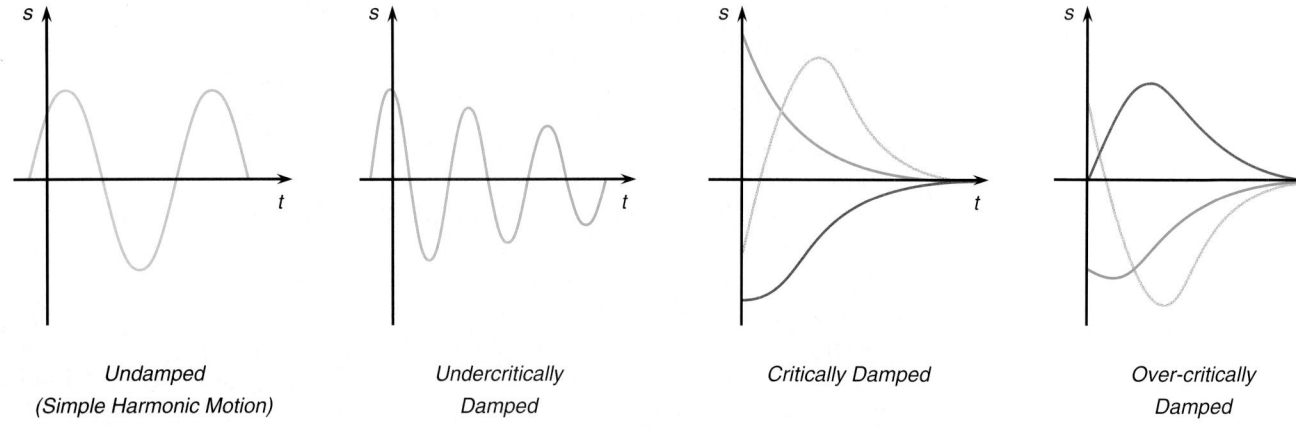

Undamped	*Undercritically*	*Critically Damped*	*Over-critically*
(Simple Harmonic Motion)	*Damped*		*Damped*

Figure 19.4.10

Summary 19.4.2

Undamped vibration (simple harmonic motion)

Equation: $s'' + \frac{k}{M}s = 0$

Roots of characteristic equation: $r = \pm i\sqrt{\frac{k}{M}}$, pure imaginary

General solution: $s(t) = c\cos\left(\sqrt{\frac{k}{M}}\,t\right) + d\sin\left(\sqrt{\frac{k}{M}}\,t\right)$

Period: $T = \frac{2\pi}{\sqrt{\frac{k}{M}}}$

Undercritically damped vibration

Equation: $s'' + \frac{R}{M}s' + \frac{k}{M}s = 0$

Discriminant of characteristic equation: $D = \frac{R^2 - 4kM}{M^2} < 0$

Roots of characteristic equation: $r = a \pm ib = -\frac{R}{2M} \pm i\frac{\sqrt{4kM - R^2}}{2M}$, complex

General solution: $s(t) = e^{at}(c\cos bt + d\sin bt)$

Period: $T = \frac{4\pi M}{\sqrt{4kM - R^2}}$

Critically damped vibration

Equation: $s'' + \frac{R}{M}s' + \frac{k}{M}s = 0$

Discriminant of characteristic equation: $D = \frac{R^2 - 4kM}{M^2} = 0$

Root of characteristic equation: $r = -\frac{R}{2M}$, one negative real root, multiplicity 2

General solution: $s(t) = ce^{-Rt/2M} + dte^{-Rt/2M}$

Overcritically damped vibration

Equation: $s'' + \frac{R}{M}s' + \frac{k}{M}s = 0$

Discriminant of characteristic equation: $D = \frac{R^2 - 4kM}{M^2} > 0$

Roots of characteristic equation: $r = -\frac{R}{2M} \pm \frac{\sqrt{R^2-4kM}}{2M}$, two negative real roots, a, b

General solution: $s(t) = ce^{at} + de^{bt}$.

EXAMPLE 5 Consider a spring that exerts a force of 2 N when it is compressed 20 cm. A mass of 5 kg is attached to the end, and an experiment shows that when the mass is pulled at a speed of 30 cm/sec, it encounters frictional resistance of 1.5 N. If the spring is compressed by 20 cm and released, find a formula for its displacement as a function of time.

Solution For convenience we convert the measurements into meters. From the given information we find that the spring constant is $k = \frac{2}{0.20} = 10$ N/m. The constant R in the expression for the frictional force $F_{friction} = -Rs'$ can be found from the fact that $F_{friction} = -1.5$ N when $s' = 0.30$ m/sec; we find that $R = \frac{1.5}{0.30} = 5$ N/(m/sec) $= 5$ N sec/m.

The differential equation is $s'' + \frac{R}{M}s' + \frac{k}{M}s = 0$, or $s'' + s' + 2s = 0$. Its characteristic polynomial is $r^2 + r + 2$. The discriminant is $D = 1^2 - 8 = -7$, which is negative. There are two complex roots,

$$r = -\frac{1}{2} \pm \frac{i}{2}\sqrt{7}.$$

The vibration is undercritically damped, and the general solution is

$$s(t) = e^{-t/2}\left(c\cos\frac{\sqrt{7}}{2}t + d\sin\frac{\sqrt{7}}{2}t\right),$$

where c and d are constants.

We are told that the spring is compressed 20 cm and released. This suggests that at time $t = 0$ the displacement is $s(0) = -0.20$; notice that the negative sign is used, since the spring is compressed. Also we know that the spring was released, which suggests that it was not given any initial velocity; we interpret this to mean $s'(0) = 0$.

Substituting these facts into the formula for $s(t)$ gives

$$-0.20 = s(0) = e^0(c\cos 0 + d\sin 0) = c,$$

so $c = -0.20 = -\frac{1}{5}$.

Then we differentiate to find

$$s'(t) = -\frac{1}{2}e^{-t/2}\left(c\cos\frac{\sqrt{7}}{2}t + d\sin\frac{\sqrt{7}}{2}t\right)$$

$$+ e^{-t/2}\left(-\frac{\sqrt{7}}{2}c\sin\frac{\sqrt{7}}{2}t + \frac{\sqrt{7}}{2}d\cos\frac{\sqrt{7}}{2}t\right),$$

so $s'(0) = -\frac{1}{2}c + 1\left(\frac{\sqrt{7}}{2}\right)d = 0.$

From this we find that $d = \frac{c}{\sqrt{7}} = -\frac{1}{5\sqrt{7}}$, and the required solution is

See the Graphics Calculator
Example on the next page.

$$s(t) = -e^{-t/2}\left(\frac{1}{5}\cos\frac{\sqrt{7}}{2}t + \frac{1}{5\sqrt{7}}\sin\frac{\sqrt{7}}{2}t\right).$$

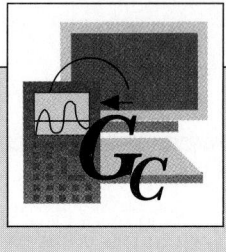

With a graphics calculator we can sketch the graph of the displacement from Example 5.

TI-85	SHARP EL9300
GRAPH F1 $(y(x) =)$ $(-)$ e^x $(\,(-)$ F1$(x) \div 2\,)$ $\times$ (.2 COS (.5 $\sqrt{}$ (7) $\times$ F1(x)) + .2 $\div$ $\sqrt{}$ (7) $\times$ SIN (.5 $\sqrt{}$ (7) $\times$ F1(x))) M2(RANGE) 0 ▼ 10 ▼ ▼ $(-)$ .2 ▼ .1 F5(GRAPH)	$- e^x$ $(-)$ x/θ/T $\div$ 2 ► $\times$ (.2 cos (.5 $\sqrt{}$ 7 ► $\times$ x/θ/T) + .2 $\div$ $\sqrt{}$ 7 ► $\times$ sin (.5 $\sqrt{}$ 7 ► $\times$ x/θ/T)) RANGE 0 ENTER 10 ENTER 1 ENTER $(-)$.2 ENTER .1 ENTER .1 ENTER ↵
CASIO f_x-6300G (f_x-7700GB)	HP 48SX
Range 0 EXE 10 EXE 1 EXE $-$.2 EXE .1 EXE .1 EXE EXE EXE EXE Graph $- e^x$ $(- X \div 2)$ $\times$ (.2 cos (.5 $\sqrt{}$ (7) $\times$ X) + .2 $\div$ $\sqrt{}$ (7) $\times$ sin (.5 $\sqrt{}$ (7) $\times$ X)) EXE	⬆ PLOT PLOTR ERASE ATTN ′ Y = +/− e +/− X $\div$ 2 ► $\times$ (.2 $\times$ COS .5 $\times$ $\sqrt{x}$ 7 $\times$ X ► + .2 $\div$ $\sqrt{x}$ 7 $\times$ SIN .5 $\times$ $\sqrt{x}$ 7 $\times$ X ↱ PLOT ⬆ DRAW ERASE 0 ENTER 10 XRNG .2 +/− ENTER .1 YRNG DRAW

Notice how the vibration is damped.

EXAMPLE 6 For what values of R would the vibration in Example 5 be critically or overcritically damped?

Solution The vibration is critically damped when $D = 0$. In this problem, $D = \frac{R^2 - 4kM}{M^2}$. Since $M = 5$ and $k = 10$, the numerator is $R^2 - 4kM = R^2 - 200$. The discriminant will be zero when $R^2 = 200$, that is, when $R = \sqrt{200} \approx 14.14$.

We conclude that the vibration will be critically damped when $R = \sqrt{200}$ and overcritically damped when $R > \sqrt{200}$. Since we found in Example 5 that $R = 5$, we see that it is not very close to being critically damped.

Exercises 19.4

Find each of the following.

1. The period of undamped vibrations of a 2-kg mass if stretching the spring 10 cm requires 5 N of force.

2. The period of simple harmonic motion of a 3-lb mass if stretching the spring 6 in requires 2 lb of force.

3. The undamped displacement $s(t)$ of a 5-kg mass on a spring that can be stretched 30 cm by a force of 20 N if it is stretched 10 cm from rest and released at time $t = 0$.

4. The undamped displacement $s(t)$ of a 1-lb mass on a spring that can be stretched 1 ft by a force of 5 lb if it is compressed 4 in from rest and released at time $t = 0$.

5. The displacement $s(t)$ of a 4-lb weight on a spring that can be stretched 3 in by a force of 12 lb if it experiences frictional drag of 1 lb at a speed of 1 ft/sec. The spring is compressed 6 in from rest and released.

6. The displacement $s(t)$ of a 1-kg weight on a spring that can be stretched 0.1 m by a force of 9.8 N if it experiences frictional drag of 3 N at a speed of 0.25 m/sec. The spring is extended 10 cm from rest and released.

7. The mass for which critical damping will occur with a spring that can be stretched 2 in by 1 lb of force if the frictional force is 0.3 lb when the weight moves at 1 ft/sec.

8. The mass for which critical damping will occur with a spring that can be stretched 30 cm by 20 N of force, if the frictional force is 3 N when the weight moves at 30 cm/sec.

9. The length of time it takes for the amplitude (i.e., the maximum displacement) of a vibrating 2-kg mass to subside to 10% of its original size, if the spring is compressed 10 cm by 16 N of force and frictional drag is 3 N at a speed of 1 m/sec.

10. The percentage by which the original amplitude has decreased after 1 min with a 100-lb weight on a spring that can be stretched 1 in by 2 lb, if the frictional force is $\frac{1}{10}$ lb at a speed of 1 ft/sec.

II

*11. It takes 5 lb to stretch a spring 3 in. What is the friction constant R if a 1-lb weight vibrates with period $\frac{1}{2}$ sec?

*12. For a mass M on a spring with spring constant k, what is the range of possible periods P for undercritically damped vibrations with different friction constants R?

*13. For a spring with spring constant k and with frictional resistance $F_{friction} = -Rs'$, sketch the graph of the period P of undercritically damped vibration as a function of M. Interpret the shape of the graph qualitatively in terms of vibrations.

*14. For a spring with spring constant k and with frictional resistance $F_{friction} = -Rs'$, consider a mass M in undercritically damped vibration. Sketch the graph of the period P as a function of R. Interpret the shape of the graph qualitatively in terms of vibrations.

*15. A mass M on the end of a spring is vibrating in (undamped) simple harmonic motion with period P.

    (i) Without working it out, imagine that the mass is doubled (i.e, a second weight is added so that the total mass is $2M$). Will this change the period, and if so, will P increase or decrease?

    (ii) Calculate the change in the period.

    (iii) What is the effect on the period P if the mass is multiplied by a constant A, that is, if M is replaced by AM?

*16. (i) What effect do you expect if the mass is increased for a system of undercritically damped motion?

    (ii) Calculate the change in the period for a system in which the spring constant is $k = 3$ N/m, the friction is 1 N at a speed of 1 m/sec, and the mass is doubled from 3 kg to 6 kg.

**17. Imagine a hollow rectangular box whose dimensions are 1 ft by 1 ft by 4 ft, floating upright in water, as shown in Figure 19.4.11. At rest it floats with its bot-

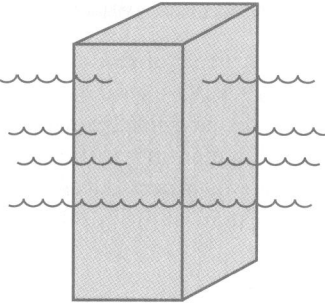

Figure 19.4.11

tom 3 ft below the surface. If somebody pushes the box down and then releases it, the buoyant force of the water will push it up. Let s be the vertical displacement of the box above its rest position.

    (i) How much does the box weigh? (Use the fact that it floats with 3 ft submerged.)

    (ii) Suppose the box is below the rest position by a distance less than 1 ft (so the top is not submerged). This means $-1 < s < 0$. What is the total buoyant force on the box? If we subtract the force needed to balance its weight, what is the net force pushing the box up?

    (iii) If the box is above its rest position, say, $0 < s < 1$, what is the net force pushing it down?

    (iv) Set up a differential equation for the displacement.

    (v) What is the period with which the box bobs up and down?

**18. What happens if the box in Exercise 17 is pushed down so far that its top is submerged?

**19. (i) Find a small rectangular or cylindrical piece of wood, measure and weigh it, and calculate the period with which it will bob in water. Try it out, and compare your calculation with the experimental value. (To measure the period fairly accurately, measure how long it takes for five or ten cycles and divide.)

(ii) If there is a slight damping effect from friction with the water, will this make the actual period longer or shorter than what you calculated?

*20. When we measured the period of a damped vibration, we were careful to measure between times when the object passed through the *rest position* after a com-

plete cycle. This is because the maximum position is not so easy to determine. For example, use a graphics calculator or computer to sketch the graph of $y = e^{-t}\cos(2\pi t)$. Its period is 1. (i) Find the position of the maximum near $t = 0$ and the maximum near $t = 1$ with as much accuracy as possible. (ii) Is the distance between them equal to 1, greater, or less? (iii) Repeat the experiment with two other adjacent peaks; do you expect the difference to be the same, greater, or less?

POINT TO PONDER

One extremely important example of vibrating motion is the pendulum. In Figure 19.4.12 we see a pendulum of length L that has been pulled to one side from its vertical rest position. It has a heavy metal bob at the end, and we can assume that the mass M of this bob is so much greater than the mass of the shaft of the pendulum that the latter can be ignored.

Figure 19.4.12

In Figure 19.4.13 the picture is drawn in more idealized geometrical terms. We consider the horizontal displacement from the rest position, calling it s. If we let θ be the angle shown, between the shaft and the vertical, then $s = L\sin\theta$, and $\sin\theta = \frac{s}{L}$.

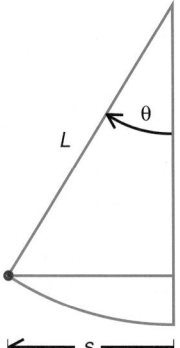

Figure 19.4.13

Gravity exerts a downward force on the bob, which is $F_{gravity} = Mg$, where g is the acceleration due to gravity ($g = 32 \text{ ft/sec}^2 = 9.8 \text{ m/sec}^2$).

This gravitational force points straight downward, as shown in Figure 19.4.14. Part of the weight is supported by the shaft of the pendulum; this is the force represented by the vector on the hypotenuse of the triangle. The horizontal force F satisfies $\frac{F}{Mg} = \tan\theta$. If θ is a reasonably small angle, then $\cos\theta$ is close to 1 and $\tan\theta = \frac{\sin\theta}{\cos\theta} \approx \sin\theta$.

Figure 19.4.14

We find that the horizontal force on the bob is approximately $F = -Mg\sin\theta = -Mg\frac{s}{L}$. Combining with Newton's Second Law, $F = Ma = Ms''$, we get the differential equation $Ms'' = -Mg\frac{s}{L}$, or

$$s'' + \frac{g}{L}s = 0.$$

This is the equation for an undamped vibration; the pendulum moves in simple harmonic motion, with period $P = \frac{2\pi}{\sqrt{g/L}}$. Notice that the period does not depend on the mass M, and as usual it does not depend on the displacement. It is this fact that makes it possible to use a pendulum as the basic element in a clock. The rest of the mechanism of a grandfather clock is basically there to move the hands a certain tiny amount each time the pendulum swings. Since each swing of the pendulum takes the same time, *even as the pendulum runs down,* the clock will keep accurate time.

Notice that the period of the pendulum does depend on the length L of the shaft. This is why grandfather clocks can be adjusted by a screw that lengthens or shortens the pendulum.

19.5 First-Order Differential Equations

ORDER
DEGREE
SEPARABLE
HOMOGENEOUS
INTEGRATING FACTOR

By the **order** of a differential equation we mean the order of the highest derivative occurring in it. A first-order equation includes the first derivative and perhaps the zeroth derivative (the function itself). A second-order equation involves the second derivative and perhaps the first derivative and/or the function itself.

If we write out the equation, it may contain products of various derivatives of the function. To find the degree of a term like $(x^2 + 3)y^4(y'')^3y^{(6)}$, we add up the exponents of y and its derivatives; in this case the sum is $4 + 3 + 1 = 8$. Notice that we ignore completely the powers of the variable x, considering only y and its derivatives. We do this for each term in the equation, and the **degree** of the equation refers to the largest degree among the terms. It is as if we think of the expression as a "polynomial" in y and its derivatives and regard functions involving x as the "constants."

For instance, the equation $3xy''y' - \frac{1}{x}y^2y'y'' + y\cos^2 x = \arctan x$ has order 2 and degree equal to 4.

In particular a **linear equation** means one whose degree equals 1, which means that y and each of its derivatives occurs by itself. In Section 19.1 we saw that such an equation can be written in the form

$$a_n(x)y^{(n)} + a_{n-1}(x)y^{(n-1)} + a_{n-2}(x)y^{(n-2)} +$$
$$\ldots + a_2(x)y^{(2)} + a_1(x)y^{(1)} + a_0(x)y = f(x),$$

where $f(x)$ and each $a_i(x)$ are functions of x. The order of this equation is n, the order of the highest derivative (assuming that $a_n(x)$ is not the zero function).

Later in this section we will consider linear first-order equations, which means equations that can be written in the form

$$a_1(x)y' + a_0(x)y = f(x).$$

Separable Equations

EXAMPLE 1   Consider the first-order differential equation

$$yy' = x^2.$$

We understand it to mean that y is a function of x and that the product of y and its derivative y' equals x^2.

To solve this equation means to find all the functions y that satisfy it. We might begin by using a different notation for the derivative, writing

$$y\frac{dy}{dx} = x^2.$$

Now we have emphasized that the expression $\frac{dy}{dx}$ is not a quotient in the ordinary sense, but suppose we ignore that for the moment and "multiply" by dx to rewrite this equation as

$$y\,dy = x^2\,dx.$$

There seems to be no choice but to integrate; we find that $\int y\,dy = \int x^2\,dx$, which means $\frac{1}{2}y^2 = \frac{1}{3}x^3 + C$, where we have added a single constant C for both integrations. This can be rewritten as $y^2 = \frac{2}{3}x^3 + 2C = \frac{2}{3}x^3 + c$, where we have written c for $2C$ in order to simplify the formula.

This in turn gives $y = \pm\left(\frac{2}{3}x^3 + c\right)^{1/2}$. The steps we took seemed fairly reasonable, but we should check whether or not this function actually satisfies the original equation. Taking the plus sign, we find that $y' = \frac{1}{2}\left(\frac{2}{3}x^3 + c\right)^{-1/2}(2x^2) = \left(\frac{2}{3}x^3 + c\right)^{-1/2}(x^2)$. This means that $yy' = \left(\frac{2}{3}x^3 + c\right)^{1/2}\left(\frac{2}{3}x^3 + c\right)^{-1/2}(x^2) = x^2$.

If we use the minus sign and let $y = -\left(\frac{2}{3}x^3 + c\right)^{1/2}$, then y' will also have a minus sign, and their product will still be x^2.

We have found that for any constant c the function $y = \pm\left(\frac{2}{3}x^3 + c\right)^{1/2}$ is a solution of the equation $yy' = x^2$. ◢

The thing that made this approach work is the fact that it was possible to write the equation with all the y's on the left side and all the x's on the right side, that is, $y\,dy = x^2\,dx$. Then it was easy to integrate both sides, and the resulting equation was relatively easy to solve to express y as a function of x. The important ingredient was being able to "separate" the x's and the y's.

In doing Example 1 we allowed ourselves to break up the derivative notation $\frac{dy}{dx}$. However, now that we have seen what to do, it is possible to take $yy' = x^2$ and just integrate both sides with respect to x. We find $\int yy'\,dx = \int x^2\,dx$. The left side $\int yy'\,dx$ suggests a substitution, since $y'\,dx = dy$. The equation becomes $\int y\,dy = \int x^2\,dx$, and we can proceed as before to integrate. We get exactly the same result.

DEFINITION 19.5.1

A first-order differential equation for a function y of the variable x is **separable** if it can be written in the form

$$\frac{dy}{dx} = \frac{f(x)}{g(y)}.$$

Such an equation can be solved by integrating:

$$\int g(y)\,dy = \int f(x)\,dx$$

and then solving for y in terms of x. It is important not to forget the constant of integration.

In effect we rearrange to put all the y's on the left side and all the x's on the right side.

EXAMPLE 2   Solve $y' = xy$.

Solution   Dividing by y, we can change the equation to $\dfrac{1}{y} y' = x$.

Integrating gives $\int \frac{1}{y} \, dy = \int x \, dx$, or $\ln |y| = \frac{1}{2} x^2 + C$. To get an expression for y, we exponentiate and find

$$|y| = \exp\left(\frac{1}{2} x^2 + C\right) = c e^{x^2/2}.$$

In doing this calculation we let $c = e^C$ to simplify the notation. We know that c, being the exponential of something, must be positive. However, when we remove the absolute value sign from y, we see that $y = \pm c e^{x^2/2}$. The formula will be simpler if we absorb the sign into the constant. This gives $y = c e^{x^2/2}$, the only restriction being that $c \neq 0$. On the other hand, the function $y = c \exp\left(\frac{x^2}{2}\right)$ is a solution of the original equation *even if $c = 0$*.

The solution is $y = c e^{x^2/2}$.   ◢

> The solution to a separable equation will involve a constant, and it is not uncommon that in doing the calculations it is necessary to exclude some values of the constant. In this situation it frequently happens that we still get a solution by using the forbidden values in the final formula. In solving, we divided by y, which only works if $y \neq 0$, but in the end we realize that $y = 0$ is also a solution.

EXAMPLE 3   Solve $xy' - 1 - y^2 = 0$.

Solution   This equation is not in the required form. We leave the y'-term on the left side and move everything else to the right side, obtaining

$$xy' = 1 + y^2.$$

To integrate, we have to move the x to the right side and the y^2 to the left side. We accomplish this by dividing by x and by $1 + y^2$ to obtain

$$\frac{1}{1 + y^2} y' = \frac{1}{x}.$$

Now it is possible to integrate: $\int \frac{1}{1+y^2} \, dy = \int \frac{1}{x} \, dx$, and we find $\arctan y = \ln |x| + c$. Recalling that what we want is an expression for y, we realize that this is the same thing as $y = \tan(\ln |x| + c)$. It is easy to see that this function does satisfy the equation $xy' - 1 - y^2 = 0$ for any value of the constant c.   ◢

> In solving this last equation we were careful to write $\ln |x|$ for the indefinite integral of $\frac{1}{x}$ instead of just $\ln x$. If we had only used $\ln x$, then we would not have found any solutions to the equation that are valid when $x < 0$, but this way we have found many.

Another complication that may arise in the solution of separable equations is that it may not be possible to solve for y as an **explicit function** of x. We may have to be content with an equation that relates y and x; in this case we say that y is an **implicit function** of x.

EXAMPLE 4   Solve $u' = \frac{1}{3tu^2 - 3u^2 + 2t - 2}$.

Solution   At first glance it is not obvious that this equation is separable. However, if we stare at it for a while, we notice that $3u^2$ occurs in two of the terms in the denominator. This suggests writing the denominator as $3tu^2 - 3u^2 + 2t - 2 = 3u^2(t-1) + 2t - 2$, after which we see that it factors as $(3u^2 + 2)(t-1)$.

This means that the equation is $u' = \frac{1}{(3u^2+2)(t-1)}$, and multiplying by $3u^2 + 2$ gives $(3u^2 + 2)\frac{du}{dt} = \frac{1}{t-1}$. Integrating gives

$$\int (3u^2 + 2)\, du = \int \frac{1}{t-1}\, dt,$$

so $u^3 + 2u = \ln|t-1| + c$, for some constant c. Generally speaking, it is difficult to solve cubic equations, and there is no easy way to write down u as a function of t.

Still we will regard this last formula $u^3 + 2u = \ln|t-1| + c$ as the solution of the equation, in the sense that for each value of c it defines the solution u implicitly as a function of t.

The original differential equation relates u and its derivative, so in a sense it also determines solutions implicitly, but the solution we have found is better in the sense that it does not involve any derivatives. In this particular case we could exponentiate and express t as a function of u, so we could for instance sketch the graph. (In effect we would be graphing the inverse function with the axes reversed.) It would not be easy to do this directly from the differential equation.

EXAMPLE 5   Chemical Reaction Rates   Certain chemical reactions involve combining two substances to form a compound called the product. In such cases it often happens that the rate at which the product is produced is proportional to the product of the amounts remaining of the two ingredients.

Let $y = y(t)$ be the amount of product at time t; assume that $y(0) = 0$. Also suppose that the ingredients are measured in appropriate units so that 1 unit of each ingredient combine to make 1 unit of product. Finally, let A be the amount of one ingredient at $t = 0$, and let B be the initial amount of the other.

At time t the amount of product will be $y = y(t)$, and making it will have used up y units of each ingredient. This means that the amounts remaining of the two ingredients will be $A - y$ and $B - y$. Of course, these quantities must be nonnegative.

We are told that the rate of the reaction is proportional to the product of these amounts. This means

$$y' = k(A - y)(B - y),$$

for some constant k. This is a separable equation, which we can write as $\frac{1}{(A-y)(B-y)} y' = k$. Integrating gives $kt = \int \frac{1}{(A-y)(B-y)}\, dy$, and this integral can be done by partial fractions. How to do it depends on whether $A = B$ or $A \neq B$.

First suppose $A \neq B$. Then

$$kt = \int \frac{1}{(A-y)(B-y)}\, dy = \frac{1}{B-A} \int \left(\frac{1}{A-y} - \frac{1}{B-y} \right) dy$$

$$= \frac{1}{B-A}\left(-\ln(A-y) + \ln(B-y) \right) = \frac{1}{B-A}\left(\ln \frac{B-y}{A-y} \right) + c.$$

Multiplying by $B - A$ and exponentiating gives $\dfrac{B-y}{A-y} = Ce^{k(B-A)t}$.

Multiplying by $A - y$ and solving gives $y = \dfrac{B - ACe^{k(B-A)t}}{1 - Ce^{k(B-A)t}}$.

The fact that $y(0) = 0$ means that $B = AC$, so $y = (B - Be^{k(B-A)t})/\left(1 - \frac{B}{A}e^{k(B-A)t}\right)$. Multiplying top and bottom by A gives

$$y = AB \frac{1 - e^{k(B-A)t}}{A - Be^{k(B-A)t}}.$$

On the other hand, if $A = B$, we find that $kt = \int \frac{1}{(A-y)^2}\, dy = \frac{1}{A-y} + c$, so $A - y = \frac{1}{kt-c}$ and $y = A - \frac{1}{kt-c}$.

The fact that $y(0) = 0$ means that $A - \frac{1}{-c} = 0$, so $c = -\frac{1}{A}$. The required solution is $y = A - \frac{1}{kt+1/A} = A - \frac{A}{Akt+1}$, or

$$y = \frac{A^2 kt}{Akt + 1}.$$

Homogeneous First-Order Equations

Some first-order equations that are not separable can be transformed into separable equations by a change of variables. The most important type is what are known as *homogeneous equations*. This terminology can be unfortunate, since we have already discussed homogeneous linear differential equations (with 0 on the right side). What we are discussing now is a different use of the same word, and there is some risk of confusion.

EXAMPLE 6   Consider the equation $y' = \frac{x^2+y^2}{xy}$. It is not separable, but if we rewrite it as $y' = \frac{x^2+y^2}{xy} = \frac{x^2}{xy} + \frac{y^2}{xy} = \frac{x}{y} + \frac{y}{x}$, then it seems natural to let $v = \frac{y}{x}$.

From this we find $y = xv$ and $y' = xv' + v$, so the original equation becomes $xv' + v = \frac{1}{v} + v$. Subtracting v from both sides gives

$$xv' = \frac{1}{v}.$$

This equation is separable, and if we write it as $vv' = \frac{1}{x}$, then integration gives $\frac{1}{2}v^2 = \ln|x| + C$, so $v = \sqrt{2\ln|x| + c}$ or $v = -\sqrt{2\ln|x| + c}$. (Notice the way we wrote $c = 2C$ to simplify the notation.) From this we can find y, since $y = xv$. The solutions are

$$y = x\sqrt{2\ln|x| + c} \qquad \text{or} \qquad y = -x\sqrt{2\ln|x| + c}.$$

The approach used in Example 6 should work for other equations that can be expressed in the form $y' = F\left(\frac{y}{x}\right)$, for some function F.

DEFINITION 19.5.2

> An equation of the form
>
> $$y' = F\left(\frac{y}{x}\right)$$
>
> is called **homogeneous**.

It may be possible to solve it by making the substitution $v = \frac{y}{x}$, so $y = xv$ and $y' = xv' + v$. The equation can be written as $xv' + v = F(v)$, or $x\frac{dv}{dx} = F(v) - v$, which leads to a separable equation for v,

$$v' = \frac{F(v) - v}{x}.$$

It can be solved by integration: $\displaystyle\int \frac{1}{F(v) - v}\, dv = \int \frac{1}{x}\, dx = \ln|x| + c.$

EXAMPLE 7   Find the solution of $y' = \frac{x^3 y + y^4}{x^4}$ that satisfies $y(1) = 2$.

Solution   This equation can be written as $y' = \frac{x^3 y}{x^4} + \frac{y^4}{x^4} = \frac{y}{x} + \left(\frac{y}{x}\right)^4$, or $y' = F\left(\frac{y}{x}\right)$, where F is the function $F(v) = v + v^4$.

The corresponding equation $\frac{dv}{dx} = \frac{F(v) - v}{x}$ is $v' = \frac{v^4}{x}$, which we write as

$$\int \frac{1}{v^4}\, dv = \int \frac{1}{x}\, dx.$$

Integration gives $-\dfrac{1}{3} v^{-3} = \ln|x| + C,$

so $v^{-3} = -3\ln|x| + c$ and $v = (c - 3\ln|x|)^{-1/3}$. Finally, we find that $y = xv = x(c - 3\ln|x|)^{-1/3}$.

We want $y(1) = 2$, which means that $2 = 1(c - 3\ln 1)^{-1/3} = c^{-1/3}$. This means that $c = \frac{1}{8}$, and the desired solution is $y = x\left(\frac{1}{8} - 3\ln|x|\right)^{-1/3}$.

$c = -3C.$

In a homogeneous equation $y' = F\left(\frac{y}{x}\right)$ the function F may be many things. But when the equation is written out in terms of x and y as $y' = R(x, y)$, one common possibility is that the right side $R(x, y)$ is a rational function. In that case it is easy to tell whether or not the equation is homogeneous.

If $F(x, y) = \frac{p(x,y)}{q(x,y)}$ is a rational function, the corresponding equation is homogeneous if all the terms in the numerator $p(x, y)$ and the denominator $q(x, y)$ have the same degree d. For instance, in Example 7, $R(x, y) = \frac{x^3 y + y^4}{x^4}$. There is only one term x^4 in the denominator, and it has degree $d = 4$; the numerator $x^3 y + y^4$ has two terms, but both of them have degree $d = 4$. This is the sort of function that gives a homogeneous equation. To express it in terms of $\frac{y}{x}$, we can divide out any common factors of x or y and then divide both the numerator and the denominator by the dth power of x. (In Example 7, divide by $x^4 \dots$).

This way of identifying homogeneous equations only works when the right side is a *rational function*.

With a graphics calculator we can sketch the solution to Example 7.

TI-85	SHARP EL9300
GRAPH F1$(y(x) =)$ F1$(x) \times$ $(.125 - 3$ LN $($ MATH F1 F5(abs) EXIT F1$(x)))$ ˆ $((-) 1 \div 3)$ M2$($RANGE$)$ $(-)$ 3 ▼ 5 ▼ ▼ $(-)$ 5 ▼ 5 F5$($GRAPH$)$	⏚ x/θ/T $\times$ $(.125 - 1.5$ ln $($ x/θ/T $x^2)$ $)$ a^b $(-)$ 1 $\div$ 3 RANGE $(-)$3 ENTER 5 ENTER 1 ENTER $(-)$5 ENTER 5 ENTER 1 ENTER ⏚
CASIO f_x-6300G (f_x-7700GB)	HP 48SX
Range -3 EXE 5 EXE 1 EXE -5 EXE 5 EXE 1 EXE EXE EXE EXE Graph X $\times$ $(.125 - 1.5$ ln $($ X $x^2))$ x^y $(-1 \div$ 3 $)$ EXE	⬑ PLOT PLOTR ERASE ATTN ' Y = X $\times$ $()$ $()$ $.125 - 1.5 \times$ LN X y^x 2 ▶ ▶ y^x 2 y^x $()$ 1 $\div$ 3 ▶ $\div$ $()$ .125 $- 1.5 \times$ LN X y^x 2 ↱ PLOT ⬑ DRAW ERASE 3 +/– ENTER 5 XRNG 5 +/– ENTER 5 YRNG DRAW

Identify the points where there are vertical asymptotes. What symmetry does the graph have?

Linear First-Order Equations

Consider a linear first-order equation

$$y' + p(x)y = q(x).$$

Now we are not assuming that the coefficients $p(x)$ and $q(x)$ are constants. We can multiply the equation by some function $\rho = \rho(x)$ to obtain

$$\rho(x)y' + \rho(x)p(x)y = \rho(x)q(x).$$

This does not seem very useful, but if we observe that on the left side, one term contains a y' and the other contains a y, we realize that the left side could be the result of applying the Product Rule to something. After all, the derivative of ρy is $\rho y' + \rho' y$, so this suggests trying to choose $\rho(x)$ so that $\rho'(x) = \rho(x)p(x)$.

This is a separable equation $\rho' = \rho p$ for $\rho(x)$. We write $\frac{\rho'}{\rho} = p$, so $\ln \rho = \int p(x)\,dx$ and

$$\rho = e^{\int p(x)\,dx}.$$

Strictly speaking, we should write $\ln |\rho|$ for the integral of $\frac{\rho'}{\rho}$, but since we end up with ρ equaling an exponential, which is always positive, this complication is unnecessary.

So if we choose $\rho(x)$ in this way, we can transform the equation $y' + p(x)y = q(x)$ into the equation

$$\big(\rho(x)y\big)' = \rho(x)q(x).$$

Integrating gives $\rho(x)y = \int \rho(x)q(x)\,dx + c$, so

$$y = \frac{1}{\rho(x)} \int \rho(x)q(x)\,dx + \frac{c}{\rho(x)}.$$

Multiplying by $\rho(x)$ turns the original differential equation into a problem in integration, so $\rho(x) = \exp\big(\int p(x)\,dx\big)$ is called an **integrating factor**.

EXAMPLE 8   Solve $y' = \frac{y}{x} + x$.

Solution   This is $y' + p(x)y = q(x)$, with $p(x) = -\frac{1}{x}$ and $q(x) = x$. The integrating factor is

$$\rho(x) = \exp\left(\int p(x)\,dx\right) = \exp\left(-\int \frac{1}{x}\,dx\right) = \exp(-\ln x) = \frac{1}{x}.$$

At this point it is not necessary to keep the constant of integration, because all we want is any one function ρ with the right derivative. For the same reason we allow ourselves to write $\int \frac{1}{x}\,dx = \ln x$ instead of $\ln |x|$. If $\rho = \frac{1}{|x|}$ satisfies $\rho' = \rho p$, then so does $\rho = \frac{1}{x}$, since the only changes are minus signs on ρ and ρ' when $x < 0$.

The original equation becomes $(\rho y)' = \rho q$, that is,

$$\left(\frac{1}{x}y\right)' = 1,$$

and we integrate to find that $\frac{1}{x}y = x + c$. Multiplying by x, we find that $y = x^2 + cx$. It is easy to verify that for any constant c this function does satisfy the original equation.

Summary 19.5.3   **Linear First-Order Equations**

Consider the equation $y' + p(x)y = q(x)$.

1. Define the **integrating factor** ρ by $\rho(x) = \exp\big(\int p(x)\,dx\big)$. It is not necessary to keep the constant of integration.

2. The equation becomes $(\rho y)' = \rho q$, so integration gives $\rho y = \int \rho(x)q(x)\,dx + c$, and

$$y = \frac{1}{\rho(x)}\left(\int \rho(x)q(x)\,dx + c\right).$$

At *this* stage it is very important to keep the constant c.

EXAMPLE 9   Find the function y that satisfies $y(0) = 3$ and

$$(1 + x^2)y' + 2xy = \cos x.$$

Solution This equation is not in the form $y' + py = q$, but if we divide by $1 + x^2$, we get

$$y' + \frac{2x}{1 + x^2}y = \frac{\cos x}{1 + x^2}.$$

The integrating factor is $\rho(x) = \exp\left(\int \frac{2x}{1+x^2}\,dx\right) = \exp\left(\ln(1 + x^2)\right) = 1 + x^2$. The equation becomes

$$\rho(x)y = \int \rho(x)q(x)\,dx = \int (1 + x^2)\frac{\cos x}{1 + x^2}\,dx = \int \cos x\,dx = \sin x + c.$$

Dividing by $\rho(x) = 1 + x^2$ gives $y = \frac{c + \sin x}{1+x^2}$. The initial condition is that $3 = y(0) = \frac{c + \sin 0}{1} = c$, so the required function is $y = \frac{3 + \sin x}{1+x^2}$.

Exercises 19.5

I

Solve each of the following separable equations, expressing y as an explicit function of x if you can.

1. $y' = 4\frac{x}{y}$

2. $y' = x\cos^2 y$

3. $y' = 3xy$

4. $xy' = -5y$

5. $y' = e^{x+y}$

6. $\frac{dy}{dx} = e^{x-2y}$

7. $\frac{dy}{dx} = y\cos x$

8. $(1 + x^2)y^2y' = 1$

9. $y' = (xy)^{-1/2}$

10. $yy' = (y^2 + 1)x^3$

Solve each of the following homogeneous equations, expressing y as an explicit function of x if you can.

11. $y' = 2\frac{y}{x}$

12. $x^2y' = 2y^2 + xy$

13. $y' = \frac{x+y}{x}$

14. $y' = \frac{x}{x+y}$

15. $y' = \frac{y}{x} + e^{y/x}$

16. $y' = \frac{y}{x} + \cos^2\frac{y}{x}$

17. $xy' = y - 2x$

18. $y' = 1 + \frac{y}{x} + \left(\frac{y}{x}\right)^2$

19. $y' = \frac{y}{x} - \sin\frac{2y-3x}{x}$

20. $\frac{x}{y}y' = 1 + \ln|y| - \ln|x|$

Solve each of the following linear first-order equations, expressing y as an explicit function of x if you can.

21. $y' + 2\frac{y}{x} = x$

22. $y' + 2xy = x$

23. $xy' + y = e^x$

24. $y' + x^2y = x^2$

25. $xy' = y + 1$

26. $xy' = y + x^2\cosh x$

27. $y' = 2xy + e^{x^2}$

28. $\frac{du}{dt} = tu - 3t$

29. $x^2y' + 2xy = \arctan x$

30. $\cos(x)y' = \sin(x)y + \cos(x)$

II

*31. If n is a positive integer, solve $y' + x^ny = x^n$.

*32. If n is a positive integer, solve $x^ny' + y^n - x^{n-1}y = 0$.

*33. Consider the equation $y' + y = \sec^2(e^x)$. It is a non-homogeneous linear differential equation with constant coefficients. We cannot use the method of undetermined coefficients to find a particular solution because we cannot guess what the solution might look like.

  (i) Use an integrating factor to find a particular solution.

  (ii) What is the general solution?

*34. Find the general solution of $y' + 2y = e^{-2x}(1 + x^2)^{-1}$.

*35. In Figure 19.5.1(i) we see the graphs of $y = x^2 + c$ for several values of the constant c. If we consider a point (x, y), then exactly one of these parabolas

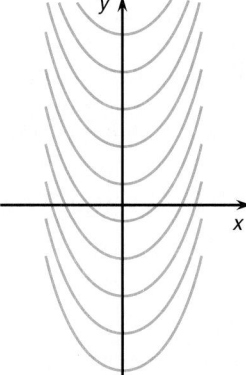

Figure 19.5.1(i)

goes through it, and the slope of the tangent is $\frac{dy}{dx} = 2x$. Now suppose we want to find another family of curves that at each point (x, y) is orthogonal (i.e., perpendicular) to the parabola. This means that the slope of the tangent at (x, y) of one of these curves should be $-\frac{1}{2x}$. (Recall that two lines are orthogonal if their slopes are negative reciprocals.)

(i) What are the curves that satisfy $y' = -\frac{1}{2x}$?

(ii) Find the formula for the particular curve satisfying $y' = -\frac{1}{2x}$ and $y(-1) = 5$. Verify that it is orthogonal to the parabola $y = x^2 + c$ that passes through $(-1, 5)$.

In Figure 19.5.1(ii) we see some of the parabolas sketched along with some of the curves that are orthogonal to them. These curves are called *orthogonal trajectories*.

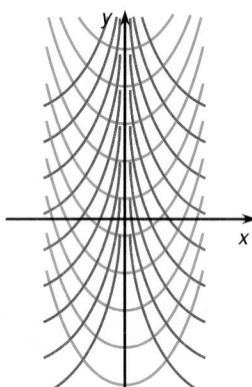

Figure 19.5.1(ii)

*36. (i) Using the terminology of Exercise 35, find the orthogonal trajectories to the family of hyperbolas $x^2 - y^2 = c$. (Here the constant c can be positive or negative.)

(ii) Sketch several hyperbolas (include at least two with $c > 0$ and two with $c < 0$) and several of the orthogonal trajectories.

*37. (i) In the setting of Example 5 (p. 1250), find the limiting value of $y(t)$ as $t \to \infty$ (i.e., $\lim_{t \to \infty} y(t)$). Distinguish the cases $A > B$, $A < B$, and $A = B$.

(ii) Interpret the results of part (i) in terms of the chemical reaction.

*38. (i) In the setting of Example 5, suppose $A > B$ and

that when $t = 1$, we know that $y(1) = \frac{B}{2}$. Find the constant k.

(ii) Repeat part (i) with $A = B$ and $y(2) = \frac{A}{2}$.

*39. A tank contains 40 gallons of water with 10 lb of salt dissolved in it. Brine with 1 lb of salt per gallon pours into the tank at the rate of 5 gal/min, and solution flows out at the same rate. Assume that the solutions are perfectly mixed at all times.

(i) Find a formula for the amount of salt in the tank at time t.

(ii) Can you guess the limiting amount of salt in the tank (i.e., as $t \to \infty$)?

(iii) Using your solution to part (i), find the limiting amount of salt.

**40. A 10-gal tank contains 10 lb of salt. Pure water flows into it at the rate of 1 gal/min and overflows into a 40-gal tank. This tank is full of water, and at $t = 0$ it contains 20 lb of salt. As the water flows into it from the other tank, it also overflows and leaves at the same rate. Assuming that both solutions are perfectly mixed, find a formula for the amount of salt in the large tank at time t. What do you expect the limiting amount to be?

*41. In the situation of Example 8, verify that using $\rho(x) = \frac{1}{|x|}$ does not give any more solutions.

*42. Find a solution of $y' = \frac{y}{x} + 1$ that satisfies $y(-2) = 3$.

43. Graph the solutions of Example 5 (p. 1250) that correspond to (i) $A = 10$, $B = 5$, (ii) $A = B = 10$, (iii) $A = 10$, $B = 20$.

*44. In Example 4 (p. 1250) we found the implicit solution $u^3 + 2u = \ln|t - 1| + c$. By moving the c to the left side and exponentiating, we can express $|t - 1|$ as a function of u. Use a graphics calculator or computer to sketch the solutions corresponding to $c = 0$, ± 1, ± 2 (with t as a function of u).

19.6 Variation of Parameters

NONHOMOGENEOUS LINEAR
EQUATION

In this brief section we learn a technique for finding a particular solution of a nonhomogeneous linear equation, assuming that we know the general solution of the corresponding homogeneous equation. It can be used for equations of any order n, but we will concentrate on equations of order 2. Partly this is because they are easier, but it is also true that they are most often what is needed.

So consider the equation

$$y'' + by' + cy = f(x), \tag{19.6.1}$$

a linear differential equation of order 2 with constant coefficients.

Suppose the general solution of the corresponding homogeneous equation

$$y'' + by' + cy = 0$$

is $y = c_1 y_1 + c_2 y_2$, where c_1 and c_2 are constants. If the right side $f(x)$ is one of the special types of functions we considered in Section 19.3, we can use undetermined coefficients to find a particular solution.

For less convenient functions it may be difficult to guess a solution. However, it turns out to be helpful to guess that there might be a solution of the form

$$y = u y_1 + v y_2,$$

for some functions u and v. Why this should be so is not at all clear, but in fact it is possible to find a solution of this form for any continuous function $f(x)$.

Having made this guess, we need to calculate y' and y''. It is easy to find $y' = u y_1' + v y_2' + u' y_1 + v' y_2$. It is not very difficult to find y'' too, but instead let us remark that it will involve u'' and v'' when we differentiate the last two terms in the expression for y'.

To avoid this and make the formulas simpler, let us ask that those two terms cancel, that is, that $u' y_1 + v' y_2 = 0$. This means that $y' = u y_1' + v y_2'$, so $y'' = u y_1'' + v y_2'' + u' y_1' + v' y_2'$.

The original equation $y'' + by' + cy = f(x)$ becomes

$$\begin{aligned} f(x) &= y'' + by' + cy = u y_1'' + v y_2'' + u' y_1' + v' y_2' + b u y_1' + b v y_2' + c u y_1 + c v y_2 \\ &= u(y_1'' + b y_1' + c y_1) + v(y_2'' + b y_2' + c y_2) + u' y_1' + v' y_2'. \end{aligned}$$

Since y_1 and y_2 are both solutions of the homogeneous equation $y'' + by' + cy = 0$, the expressions in parentheses are both 0, and we have $f(x) = u' y_1' + v' y_2'$.

We now have two equations involving u' and v':

$$u' y_1 + v' y_2 = 0,$$
$$u' y_1' + v' y_2' = f.$$

These equations can be solved by eliminating u', say. If you have studied linear algebra, you know how to solve them using Cramer's Rule; it is easy to check that the result is

$$u' = \frac{-f y_2}{y_1 y_2' - y_1' y_2}, \qquad v' = \frac{f y_1}{y_1 y_2' - y_1' y_2}.$$

In principle we can find u and v by integrating, and then we will have found the particular solution $y = u y_1 + v y_2$.

EXAMPLE 1 Find a particular solution of $y'' + y = x^2$.

Solution Following the idea described above, we begin by solving the complementary homogeneous equation $y'' + y = 0$. The characteristic polynomial is $r^2 + 1$, the roots are $\pm i$, and the general solution is $y = c_1 \cos x + c_2 \sin x$. We let $y_1 = \cos x$ and $y_2 = \sin x$.

With the right side of the equation being $f(x) = x^2$, we use the formulas above and let

$$u' = \frac{-fy_2}{y_1 y_2' - y_1' y_2} = -\frac{x^2 \sin x}{\cos^2 x + \sin^2 x} = -x^2 \sin x,$$

$$v' = \frac{fy_1}{y_1 y_2' - y_1' y_2} = \frac{x^2 \cos x}{\cos^2 x + \sin^2 x} = x^2 \cos x.$$

To find u and v, we integrate. For instance, integrating twice by parts, first with $U = x^2$, $dV = \sin x \, dx$, we find

$$u = -\int x^2 \sin x \, dx = x^2 \cos x - 2 \int x \cos x \, dx = x^2 \cos x - 2\left(x \sin x - \int \sin x \, dx \right)$$

$$= x^2 \cos x - 2x \sin x - 2 \cos x + c.$$

Similarly,

$$v = \int x^2 \cos x \, dx = x^2 \sin x - 2 \int x \sin x \, dx = x^2 \sin x - 2\left(-x \cos x + \int \cos x \, dx \right)$$

$$= x^2 \sin x + 2x \cos x - 2 \sin x + C.$$

The particular solution we are looking for is

$$\begin{aligned} y_p &= u y_1 + v y_2 \\ &= (x^2 \cos x - 2x \sin x - 2 \cos x) \cos x + (x^2 \sin x + 2x \cos x - 2 \sin x) \sin x \\ &= x^2 (\cos^2 x + \sin^2 x) - 2x(\sin x \cos x - \cos x \sin x) - 2(\cos^2 + \sin^2 x) \\ &= x^2 - 2. \end{aligned}$$

It is easy to check that $y_p = x^2 - 2$ is a particular solution of $y'' + y = x^2$.   ◢

Of course, in Example 1 we could have used undetermined coefficients, but it shows how the technique is used.

Notice that when we integrated the formulas for u' and v' to find u and v, we added constants of integration, but when we calculated $y_p = u y_1 + v y_2$, we left them off. If we had included them, we would have found a different solution:

$$(u + c)y_1 + (v + C)y_2 = u y_1 + v y_2 + (c y_1 + C y_2) = y_p + (c y_1 + C y_2).$$

Since the function $c y_1 + C y_2$ is a solution of the complementary equation, this will also be a particular solution of the original equation. On the other hand, all we really needed to find was *one* particular solution, and including the constants just makes the calculation more complicated.

Summary 19.6.2   **Variation of Parameters**

Consider the nonhomogeneous linear differential equation $y'' + by' + cy = f(x)$, and suppose we want to find a particular solution.

1. Find the general solution of the complementary equation $y'' + by' + cy = 0$, and express it in the form $c_1 y_1 + c_2 y_2$.

2. Let

$$u' = \frac{-f y_2}{y_1 y_2' - y_1' y_2}, \qquad v' = \frac{f y_1}{y_1 y_2' - y_1' y_2},$$

and integrate to find the functions u and v. It is not necessary to add the constants of integration.

3. The required particular solution is

$$y_p = u y_1 + v y_2.$$

4. The general solution is

$$y = y_p + y_c = u y_1 + v y_2 + c_1 y_1 + c_2 y_2.$$

Now that we know how to use the technique, we can use it for problems that we cannot solve using undetermined coefficients. Unfortunately, the calculations tend to be quite complicated.

EXAMPLE 2 Find the general solution of $y'' - 4y = e^{2x} \sin x$.

Solution The characteristic polynomial of the complementary equation is $r^2 - 4$, so the general solution is $c_1 e^{2x} + c_2 e^{-2x}$. Letting $y_1 = e^{2x}$ and $y_2 = e^{-2x}$, we let

$$u' = \frac{-f y_2}{y_1 y_2' - y_1' y_2} = -\frac{e^{2x} \sin x e^{-2x}}{-2 e^{2x} e^{-2x} - 2 e^{2x} e^{-2x}} = \frac{1}{4} \sin x,$$

$$v' = \frac{f y_1}{y_1 y_2' - y_1' y_2} = \frac{e^{2x} \sin x e^{2x}}{-2 e^{2x} e^{-2x} - 2 e^{2x} e^{-2x}} = -\frac{1}{4} e^{4x} \sin x.$$

It is easy to integrate to find $u = -\frac{1}{4} \cos x$, omitting the unnecessary constant, but v takes a little more work. Integrating by parts twice, first with $U = \sin x$ and $dV = e^{4x} dx$, we find

$$\int e^{4x} \sin x \, dx = \frac{1}{4} e^{4x} \sin x - \frac{1}{4} \int e^{4x} \cos x \, dx$$

$$= \frac{1}{4} e^{4x} \sin x - \frac{1}{4} \left(\frac{1}{4} e^{4x} \cos x - \frac{1}{4} \int e^{4x} (-\sin x) \, dx \right)$$

$$= \frac{1}{4} e^{4x} \sin x - \frac{1}{16} e^{4x} \cos x - \frac{1}{16} \int e^{4x} \sin x \, dx.$$

Adding $\frac{1}{16} \int e^{4x} \sin x \, dx$ to both sides, we have

$$\frac{17}{16} \int e^{4x} \sin x \, dx = \frac{1}{4} e^{4x} \sin x - \frac{1}{16} e^{4x} \cos x,$$

so $\int e^{4x} \sin x \, dx = \frac{4}{17} e^{4x} \sin x - \frac{1}{17} e^{4x} \cos x.$

Since v is $-\frac{1}{4}$ times this, we have that

$$v = -\frac{1}{4}\int e^{4x}\sin x\,dx = -\frac{1}{17}e^{4x}\sin x + \frac{1}{68}e^{4x}\cos x.$$

The particular solution is

$$y_p = uy_1 + vy_2 = -\frac{1}{4}\cos x e^{2x} + \left(-\frac{1}{17}e^{4x}\sin x + \frac{1}{68}e^{4x}\cos x\right)e^{-2x}$$

$$= -\frac{16}{68}e^{2x}\cos x - \frac{1}{17}e^{2x}\sin x = -\frac{4}{17}e^{2x}\cos x - \frac{1}{17}e^{2x}\sin x.$$

The general solution is

$$y = y_p + y_c = -\frac{4}{17}e^{2x}\cos x - \frac{1}{17}e^{2x}\sin x + c_1 e^{2x} + c_2 e^{-2x}.$$

Exercises 19.6

I

In each of the following questions, use the technique of variation of parameters to find the general solution, and afterward use undetermined coefficients to check your answer.

1. $y'' + y = x$
2. $y'' + y = 4x$
3. $y'' + 4y = \cos 2x$
4. $y'' + 9y = \sin 3x$
5. $y'' + y = e^x$
6. $y'' - 4y = e^x$
7. $y'' - y' = 5x$
8. $y'' + 2y' + y = 2$
9. $y'' - 3y' + 2y = e^x$
10. $y'' - y = \cos x$

Use the technique of variation of parameters to find the general solution of each of the following equations.

11. $y'' + y = \tan x$
12. $y'' + 9y = \cot 3x$
13. $y'' + 4y' + 3y = \cosh x$
14. $y'' - 2y' + y = \sinh x$
15. $y'' + y = \sec^2 x$
16. $y'' + 4y = \csc x$
17. $y'' + y = \sec x$
18. $y'' + 4y = \tan 2x$
19. $y'' + y' - 2y = \cos^2 x$
20. $y'' + y' - 2y = \sin^2 x$

II

21. Find the solution of $y'' + y = \sec x \tan x$ that satisfies $y(0) = 1$, $y'(0) = 0$.

22. Find the solution of $y'' + y = \tan^2 x$ that satisfies $y(0) = 2$, $y'(0) = -2$.

*23. Find the general solution of

$$y'' - 3y' + 2y = \frac{1}{1 + e^{-x}}.$$

*24. Find a particular solution of $y'' + 3y' + 2y = \cos(e^x)$.

*25. Consider the equation $y'' + \frac{1}{x}y' = \frac{1}{x^2}$.

   (i) First solve the complementary equation. (*Hint:* First solve for y', and then integrate.) Express your answer as $c_1 y_1 + c_2 y_2$.

   (ii) Using your answer to part (i) and variation of parameters, find a particular solution of the original nonhomogeneous equation.

   (iii) Find the general solution.

*26. Find the general solution of $y'' + \tan x\, y' = \cos x$.

19.7 Series Solutions

POWER SERIES

TERM-BY-TERM DIFFERENTIATION

We have learned a number of techniques for solving various special kinds of differential equations. Most of them depend on being able to make a good guess, and often one or more integrations are required for the solution.

There are many differential equations that do not fit into any of the special categories we have considered. For some of them it may not be easy to guess a solution. Even the kinds we know how to approach may lead to integrals we cannot do. As we have already noted with integrals, there are relatively simple differential equations for which the solutions can never be expressed in terms of the functions we already know.

Faced with an equation we cannot solve, one approach would be to ask whether there is a solution that can be expressed as a power series

$$y = \sum_{n=0}^{\infty} a_n x^n.$$

Let us begin with two examples we do know how to solve, just to see how the idea works.

EXAMPLE 1   Consider the equation $y' + 2y = 0$. Let us suppose there is a **power series** solution

$$y = \sum_{n=0}^{\infty} a_n x^n,$$

and let us try to find the coefficients a_n.

First we observe that if y is given by this series, then it is possible to differentiate **term by term** to find the derivative

$$y' = \sum_{n=0}^{\infty} a_n(nx^{n-1}) = \sum_{n=1}^{\infty} na_n x^{n-1}.$$

The equation $y' + 2y = 0$ then becomes $\displaystyle\sum_{n=1}^{\infty} na_n x^{n-1} + \sum_{n=0}^{\infty} 2a_n x^n = 0.$

This is hard to use because x appears to the $(n-1)$th power in one series and to the nth in the other. Shifting the index on the first series, we can write

$$0 = \sum_{n=1}^{\infty} na_n x^{n-1} + \sum_{n=0}^{\infty} 2a_n x^n = \sum_{n=0}^{\infty} (n+1)a_{n+1} x^n + \sum_{n=0}^{\infty} 2a_n x^n$$

$$= \sum_{n=0}^{\infty} \big((n+1)a_{n+1} + 2a_n\big) x^n.$$

For this power series to be the zero function, each of its coefficients must be 0. This means that for each n,

$$(n+1)a_{n+1} + 2a_n = 0.$$

A formula like this, which expresses a_{n+1} in terms of previous terms, is called a *recursion relation.*

Another way to put this is that $a_{n+1} = -\frac{2}{n+1} a_n$, which tells us how to find a_{n+1} if we know a_n.

Of course, it is no help for finding a_0. Let us just make $a_0 = 1$ to simplify things. Then $a_1 = -\frac{2}{1} a_0 = -2$, $a_2 = -\frac{2}{2} a_1 = 2$, $a_3 = -\frac{2}{3} a_2 = -\frac{4}{3}$, and so on. We can continue this way as long as we like, but it would be more useful to find a formula for the nth coefficient a_n.

This is not really very difficult. Each time we find the next coefficient, a_{n+1}, say, we multiply by -2 and divide by $n + 1$. So by the time we have reached a_n, we will have multiplied $a_0 = 1$ by $(-2)^n$ and divided by $(1)(2)(3)(4) \ldots (n) = n!$. This means $a_n = 1(-2)^n \frac{1}{n!}$.

The power series is

$$\sum_{n=0}^{\infty} a_n x^n = \sum_{n=0}^{\infty} \frac{(-2)^n}{n!} x^n = \sum_{n=0}^{\infty} \frac{1}{n!}(-2x)^n.$$

Of course, we recognize this as the Taylor series for $y = e^{-2x}$, which is a solution of the original equation.

Notice that we made the simple choice $a_0 = 1$. We could instead have let $a_0 = c$, for any constant c; in that case we would have found $a_n = c\frac{(-2)^n}{n!}$ and $y = \sum_{n=0}^{\infty} a_n x^n = \sum_{n=0}^{\infty} c\frac{(-2)^n}{n!} x^n = c\sum_{n=0}^{\infty} \frac{1}{n!}(-2x)^n = ce^{-2x}$.

EXAMPLE 2   Find power series solutions of $y'' + y = 0$.

Solution   Suppose $y = \sum_{n=0}^{\infty} a_n x^n$. Then the derivative is $y' = \sum_{n=1}^{\infty} na_n x^{n-1}$, and the second derivative is $y'' = \sum_{n=2}^{\infty} n(n-1)a_n x^{n-2}$. The equation $y'' + y = 0$ becomes

$$0 = y'' + y = \sum_{n=2}^{\infty} n(n-1)a_n x^{n-2} + \sum_{n=0}^{\infty} a_n x^n$$

$$= \sum_{n=0}^{\infty} (n+2)(n+1)a_{n+2}x^n + \sum_{n=0}^{\infty} a_n x^n$$

$$= \sum_{n=0}^{\infty} \left((n+2)(n+1)a_{n+2} + a_n\right)x^n.$$

For this series to be the zero function, its coefficients must all be zero, that is, for each $n \geq 0$, $(n+2)(n+1)a_{n+2} + a_n = 0$, which can also be written as $a_{n+2} = -\frac{1}{(n+2)(n+1)} a_n$.

If we know a_0, then this formula tells us how to find a_2, a_4, a_6, and so on but says nothing about the odd-numbered coefficients. On the other hand, if we know a_1, then it allows us to find a_3, a_5, a_7, ...

To find all the coefficients, we need to specify a_0 and a_1. So let $a_0 = c$ and $a_1 = d$. Then

$$a_2 = -\frac{1}{(2)(1)} a_0 = -\frac{1}{2}c, \qquad \text{and} \qquad a_3 = -\frac{1}{(3)(2)} a_1 = -\frac{1}{(3)(2)}d,$$

$$a_4 = -\frac{1}{(4)(3)} a_2 = \frac{1}{(4)(3)(2)} c, \qquad a_5 = -\frac{1}{(5)(4)} a_3 = \frac{1}{(5)(4)(3)(2)}d,$$

$$a_6 = -\frac{1}{(6)(5)} a_4 = -\frac{1}{6!}c, \qquad a_7 = -\frac{1}{(7)(6)} a_5 = -\frac{1}{7!}d,$$

$$\vdots \qquad\qquad\qquad \vdots$$

$$a_{2n} = \frac{(-1)^n}{(2n)!} c \qquad\qquad a_{2n+1} = \frac{(-1)^n}{(2n+1)!} d.$$

We have found that the solution is

$$y = c\sum_{n=0}^{\infty} \frac{(-1)^n}{(2n)!} x^{2n} + d\sum_{n=0}^{\infty} \frac{(-1)^n}{(2n+1)!} x^{2n+1} = c\cos x + d\sin x.$$

Of course, this is not surprising. However, it is interesting to interpret our earlier results about initial conditions in light of this example.

Because this equation has order 2, the formula we get for the coefficients of the power series involves a_{n+2} and a_n. For many second-order equations it would involve a_{n+2}, a_{n+1}, and a_n. In either case it does not allow us to find a_1 from a_0. We have to know both a_0 and a_1 to begin with.

Then the formula tells us how to find the other coefficients. The general solution will involve two constants, which is exactly what we expect for an equation of order 2.

The two examples we have done could very easily be solved by our earlier methods. Let us consider one that cannot.

EXAMPLE 3   Solve $y'' + xy' + y = 0$ in terms of power series.

Solution   Since this equation does not have constant coefficients, we do not know how to solve it. Let us suppose there is a solution of the form $y = \sum_{n=0}^{\infty} a_n x^n$. Since $y' = \sum_{n=1}^{\infty} n a_n x^{n-1}$ and $y'' = \sum_{n=2}^{\infty} n(n-1) a_n x^{n-2}$, the equation becomes

$$0 = y'' + xy' + y = \sum_{n=2}^{\infty} n(n-1) a_n x^{n-2} + x \sum_{n=1}^{\infty} n a_n x^{n-1} + \sum_{n=0}^{\infty} a_n x^n$$

$$= \sum_{n=0}^{\infty} (n+2)(n+1) a_{n+2} x^n + \sum_{n=1}^{\infty} n a_n x^n + \sum_{n=0}^{\infty} a_n x^n$$

$$= \sum_{n=0}^{\infty} \Big((n+2)(n+1) a_{n+2} + n a_n + a_n \Big) x^n$$

$$= \sum_{n=0}^{\infty} \Big((n+2)(n+1) a_{n+2} + (n+1) a_n \Big) x^n.$$

We conclude that $(n+2)(n+1) a_{n+2} + (n+1) a_n = 0$, so $a_{n+2} = -\frac{1}{n+2} a_n$. Again beginning with $a_0 = c$ and $a_1 = d$, we find that

$$a_2 = -\tfrac{1}{2} a_0 = -\tfrac{1}{2} c, \qquad \text{and} \qquad a_3 = -\tfrac{1}{3} a_1 = -\tfrac{1}{3} d,$$

$$a_4 = -\tfrac{1}{4} a_2 = \frac{1}{(4)(2)} c, \qquad\qquad a_5 = -\tfrac{1}{5} a_3 = \frac{1}{(5)(3)} d,$$

$$a_6 = -\tfrac{1}{6} a_4 = -\frac{1}{(6)(4)(2)} c, \qquad\qquad a_7 = -\tfrac{1}{7} a_5 = -\frac{1}{(7)(5)(3)} d,$$

$$\vdots \qquad\qquad\qquad\qquad \vdots$$

$$a_{2n} = (-1)^n \frac{1}{2^n n!} c = \frac{\left(-\frac{1}{2}\right)^n}{n!} c, \qquad a_{2n+1} = (-1)^n \frac{2^n n!}{(2n+1)!} d.$$

The solution is

$$y = c \sum_{n=0}^{\infty} \frac{1}{n!} \left(-\frac{1}{2} \right)^n x^{2n} + d \sum_{n=0}^{\infty} \frac{n!}{(2n+1)!} (-2)^n x^{2n+1}$$

$$= c \sum_{n=0}^{\infty} \frac{1}{n!} \left(-\frac{x^2}{2} \right)^n + d \sum_{n=0}^{\infty} \frac{n!}{(2n+1)!} (-2)^n x^{2n+1}.$$

Note that $(2n) \times \ldots \times (6)(4)(2) = 2^n n!$, $(2n+1) \times \ldots \times (7)(5)(3) = \frac{(2n+1)!}{(2n) \times \ldots \times (6)(4)(2)} = \frac{(2n+1)!}{2^n n!}$.

The first of these two series is the Taylor series for $e^{-x^2/2}$, which is easily seen to be a solution of the original equation. The second series is less easily recognized.

At very least we have used series to find one solution $e^{-x^2/2}$ of this equation. We also have a series expression for another solution, but we do not recognize it. ◢

With a graphics calculator we can find a partial sum for the second solution in Example 3 and sketch its graph.

TI-85	SHARP EL9300
GRAPH F1$(y(x) =)$ F1$(x) - 2$ $\div 3 \times$ F1$(x)\,\hat{}\,3 + 8 \div 120 \times$ F1$(x)\,\hat{}\,5 - 48 \div 5040 \times$ F1$(x)\,\hat{}\,7$ M2(RANGE) $(-)$ 3 ▼ 3 ▼ ▼ $(-)$ 5 ▼ 5 F5(GRAPH)	x/θ/T $- 2 \div 3 \times$ x/θ/T a^b 3 ▶ $+$ $8 \div 120 \times$ x/θ/T a^b 5 ▶ $- 48 \div 5040$ $\times$ x/θ/T a^b 7 RANGE $(-)$3 ENTER 3 ENTER 1 ENTER $(-)$5 ENTER 5 ENTER 1 ENTER
CASIO f_x-6300G (f_x-7700GB)	HP 48SX
Range -3 EXE 3 EXE 1 EXE -5 EXE 5 EXE 1 EXE EXE EXE EXE Graph $\boxed{X} - 2$ $\div 3 \times \boxed{X}\,x^y\,3 + 8 \div 120 \times \boxed{X}\,x^y\,5 -$ $48 \div 5040 \times \boxed{X}\,x^y\,7$ EXE	↰ PLOT PLOTR ERASE ATTN ′ Y ▤ $X - 2 \div 3 \times X\,y^x\,3 + 8 \div 120$ $\times X\,y^x\,5 - 48 \div 5040 \times X\,y^x\,7$ ↱ PLOT ↰ DRAW ERASE 3 +/− ENTER 3 XRNG 5 +/− ENTER 5 YRNG DRAW

You could try including more terms of the series.

In fact, it turns out that knowing one solution of a differential equation often makes it easier to find another.

EXAMPLE 4 Knowing from Example 3 that $y_1 = e^{-x^2/2}$ is a solution of $y'' + xy' + y = 0$, let us look for another solution of the form $y = uy_1$. We find that $y' = u'y_1 + uy_1'$ and $y'' = u''y_1 + 2u'y_1' + uy_1''$, so the equation becomes

$$0 = y'' + xy' + y = u''y_1 + 2u'y_1' + uy_1'' + xu'y_1 + xuy_1' + uy_1$$
$$= u(y_1'' + xy_1' + y_1) + y_1u'' + 2y_1'u' + xy_1u'.$$

The expression in parentheses is zero, so the equation becomes $y_1u'' + (2y_1' + xy_1)u' = 0$, or

$$\frac{u''}{u'} = -\frac{2y_1' + xy_1}{y_1} = -\frac{-2xe^{-x^2/2} + xe^{-x^2/2}}{e^{-x^2/2}} = \frac{xe^{-x^2/2}}{e^{-x^2/2}} = x.$$

Integration gives $\ln u' = \frac{1}{2}x^2$, so $u' = e^{x^2/2}$. The function we want is $u = \int e^{x^2/2}\,dx$, an indefinite integral of $e^{x^2/2}$.

This is an integral that cannot be expressed in terms of the familiar functions. In a sense we know what it is, but there is no simple formula for it.

The additional solution we have found is $uy_1 = e^{-x^2/2} \int e^{x^2/2} \, dx$, and the general solution of $y'' + xy' + y = 0$ is

$$y = c_1 e^{-x^2/2} + c_2 e^{-x^2/2} \int e^{x^2/2} \, dx.$$

Examples 3 and 4 demonstrate some features of the use of series. First, using series allowed us to find a solution $y = e^{-x^2/2}$ for $y'' + xy' + y = 0$, which we did not know how to solve. Second, it gave a power series expression for another solution. Third, having found the first solution using series, we were able to use it to find another solution. However, we found a function that involves an integral we cannot evaluate explicitly.

In a sense this is not unusual. Many differential equations have solutions for which there is no simple formula. Sometimes we have to be content with writing indefinite integrals, but power series can often provide an alternative approach.

For a practical problem we may be interested in finding the value of some particular solution at some value of x, and using power series, we can at least approximate the answer. Especially with a computer to do the work, it is possible to use this idea to get the required information.

In conclusion, however, it is only fair to mention that there can be complications. One is that the power series may not converge at the interesting value of x. Another is that we have been quite casual about differentiating power series. Finally, the whole approach began by assuming that there was a solution that could be written as a power series. This is often true, but not always.

The use of power series for solving differential equations is useful and important, but we have only scratched the surface in this brief introduction. It is discussed in much greater depth in books about differential equations.

Exercises 19.7

I

Find power series solutions $y = \sum_{n=0}^{\infty} a_n x^n$ for each of the following differential equations. Find a formula for the nth coefficient a_n.

1. $y' + y = 0$
2. $3y' - y = 0$
3. $y'' + 4y = 0$
4. $y'' - 4y = 0$
5. $y' - y = x$

6. $y'' + 4y = 3x - 1$
7. $y'' + y' = 0$
8. $y'' - 3y' = 0$
9. $y'' - 3y' + 2y = 0$
10. $y'' - 2y' + y = x$

For each of the following equations, find a series solution $y = \sum_{n=0}^{\infty} a_n x^n$ satisfying the given initial conditions.

11. $y' = 2y, \, y(0) = -2$
12. $y' = -4y, \, y'(0) = 1$
13. $y'' + 4y = 0, \, y(0) = 0, \, y'(0) = 1$

14. $y'' + 9y = 0, \, y(0) = 3, \, y'(0) = -2$
15. $y'' + y = 1, \, y(0) = 2, \, y'(0) = -1$
16. $y'' + y = 2, \, y(0) = 2, \, y'(0) = 0$
17. $y'' + 2y' = x, \, y(0) = 4, \, y'(0) = 0$
18. $xy'' + y = 0, \, y(0) = 0, \, y'(0) = 1$
19. $y''' + y'' = 1, \, y(0) = 12, \, y'(0) = 0, \, y''(0) = 3$
20. $y^{(4)} = y, \, y(0) = 6, \, y'(0) = y''(0) = y'''(0) = 0$

II

*21. (i) Use the results of Example 3 (p. 1263) to find the power series for the function y that satisfies $y'' + xy' + y = 0$ and $y(0) = 1, \, y'(0) = -1$.
(ii) Use the first six terms of the series to approximate $y\left(\frac{1}{2}\right)$.

*22. (i) Find the first eight terms of the Maclaurin series for the function y that satisfies $y'' - 6xy - 9x^4 y = 0$ and $y(0) = 1, \, y'(0) = 0$.

(ii) Use these terms of the series to estimate $y(0.01)$, $y(0.1)$, and $y(1)$.

(iii) Compare your answers with the correct values after checking that the function $y = e^{x^3}$ satisfies the differential equation and the initial conditions.

*23. (i) Find the general solution of $y' + \frac{y}{x} = 2$.

(ii) Try to find a Maclaurin series expression for this general solution, and explain why it is not possible to do so.

*24. Find the solution of $y' + \frac{y}{x} = 2$ that satisfies $y'(0) = 1$. This solution can be expressed as a power series even though the general solution cannot.

*25. Find the general Maclaurin series solution to $y'' + xy = 0$. (*Hint:* What is the coefficient of x^2?)

*26. Find the general Maclaurin series solution to $y'' = 2xy$.

*27. Find the terms up to the x^6-term for the Maclaurin series of the general solution of $y'' + \sin(x)y' + 2y = 0$.

*28. Find the terms up to the x^6-term for the Maclaurin series of the general solution of $y'' + e^x y' - y = 0$.

*29. Find the general Maclaurin series solution to $xy'' + xy' + y = 0$.

*30. Find the general Maclaurin series solution to $xy'' + y' + 3y = 0$.

*31. Are there any polynomial solutions to $2xy''' + 3xy' - 9y = 0$?

*32. Are there any polynomial solutions to $xy'' + xy' - 2y = 2$?

**33. (i) Find the Taylor series of $e^{x^2/2}$ about $x = 0$.

(ii) Integrate term by term to find the Taylor series of $\int e^{x^2/2}\, dx$ (i.e., evaluate $\int_0^x e^{t^2/2}\, dt$).

(iii) Find the terms up to the x^7-term of the product $e^{-x^2/2} \int e^{x^2/2}\, dx$, and compare with the series in Example 3.

*34. (i) Suppose $y = \sum_{n=0}^{\infty} a_n x^n$ is a solution to $y'' + f(x)y' + g(x)y = 0$. If k is a nonzero constant, find a (series) solution to $y'' + kf(x)y' + k^2 g(x)y = 0$.

(ii) Use part (i) and Example 4 (p.1264) to find the general solution of $y'' + 2xy' + 4y = 0$.

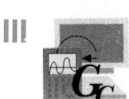

**35. In Example 4 we found the general solution of $y'' + xy' + y = 0$. Find the particular solution satisfying $y(0) = 2$, $y'(0) = 1$. Suppose we want to know the value of this function when $x = 1$. It is difficult to evaluate directly, but there are *two* ways to approximate it. (i) Approximate the value by approximating the integral using Simpson's Rule S_n with $n = 10$. (ii) Approximate the value by adding up the first n terms of the series; experiment with different values of n.

Chapter Summary

§19.1
linear differential equation
order
homogeneous linear differential equation
coefficients
characteristic polynomial
auxiliary equation
general solution
multiplicity of a root
initial conditions
particular solution

§19.2
complex numbers
real and imaginary parts
pure imaginary
real and imaginary axes
absolute value
Fundamental Theorem of Algebra (Theorem 19.2.2)
$e^{it} = \cos t + i \sin t$

general solution of homogeneous linear equations with constant coefficients
complex conjugate

§19.3
undetermined coefficients
complementary equation
general solution $y = y_p + y_c$
undetermined coefficients for e^{bx}, $\cos bx$, $\sin bx$, polynomials

§19.4
Hooke's Law
spring constant
simple harmonic motion
period $\frac{2\pi}{\sqrt{k/M}}$
pounds and slugs
frequency
friction
damped vibrations

critically damped, overcritically damped, undercritically damped

discriminant

§19.5

order

degree

linear differential equation

separable differential equation (Definition 19.5.1)

explicit solution

implicit solution

homogeneous first-order equation, $v = \frac{y}{x}$, $\frac{1}{F(v)-v} v' = \frac{1}{x}$

linear first-order equations

integrating factor (Summary 19.5.3)

§19.6

variation of parameters, $y_p = uy_1 + vy_2$, $u'y_1 + v'y_2 = 0$

$u' = \frac{-fy_2}{y_1 y_2' - y_1' y_2}$, $v' = \frac{fy_1}{y_1 y_2' - y_1' y_2}$

§19.7

power series

term-by-term differentiation

Review Exercises

I

Find the general solution of each of the following equations. These equations are not grouped according to the type of equation; different techniques will be needed to solve them.

1. $y'' + 5y' + 6y = x$
2. $y''' + y' = 2$
3. $y' + y = 3x$
4. $y' + \frac{1}{x}y = 6$
5. $y' = 4xy$
6. $xy'' - y' = 2x$
7. $y'' = 16y + \sin x$
8. $y' = (y^3 + 5x^2 y)/(xy^2 + x^3)$
9. $y''' - y'' + x^2 = 0$
10. $(1 + x^2)y' = 4xy$
11. $y^{(4)} = 16y$
12. $y^{(3)} = y^{(2)}$
13. $y' + y = \arctan x + \frac{1}{1+x^2}$
14. $y' = y^2$
15. $y'' = \frac{x}{y'}$
16. $4\cos^2(x)yy' = 1 + y^2$
17. $y' = 8y^3 x^2$
18. $y^{(4)} + 8y^{(2)} + 16y = x^2$
19. $y''' - 3y'' + 3y' - y = e^x$
20. $y^{(4)} - 4y^{(2)} + 4y = e^{2x}$
21. $y' \sin x + y \cos x = \sin x$
22. $xy' + y = x^7$
23. $xy' - xe^{-2y/x} = y$
24. $y'' + 4y' + 4y = \sinh 2x$
25. $xy' = e^{3x} - y$
26. $(1 + x^2)y' + 2xy = e^{2x}$
27. $xy'' + xy = x^2$

28. $xy'' + y' = 2x$
29. $y'' + 5y' + 6y = x^2 e^{-2x}$
30. $y'' + 3y' - 4y = e^x \cos 2x$

Find each of the functions specified below.

31. $y'' - y = x^2$, $y(0) = y'(0) = 0$
32. $(2 + x^3)y' = 3x^2 y$, $y(0) = 1$
33. $y' - xy = x$, $y(0) = 2$
34. $y''' = 6$, $y(0) = 1$, $y'(0) = 2$, $y''(0) = 3$
35. $y''' - 5y'' + 6y' = e^{2x} + x$, $y(0) = 0$, $y'(0) = 1$, $y''(0) = \frac{1}{6}$
36. $xy' = 2y$, $y(1) = -2$
37. $6y''' + y'' = 5$, $y(0) = 0$, $y'(1) = 1$, $y''(2) = 2$
38. $y' = (1 + 4y^2) \cos x$, $y(0) = 1$
39. $xy' = ny$, $y(1) = 1$
40. $y'' + x^3 y' + 2xy = 0$, $y(0) = y'(0) = 0$

II

41. Find the period of an undamped spring with spring constant 1 lb/in and a 30-lb weight.

42. Find the period of an undamped spring with spring constant 0.3 N/cm and a 5-kg weight.

43. Find the period of a damped spring with spring constant 2 lb/in, frictional resistance $\frac{1}{2}$ lb at a speed of 1 ft/sec, and a 30-lb weight.

*44. What is the weight on a critically damped spring if the spring is compressed 4 in by 10 lb and the frictional resistance is 2 lb at a speed of 1 ft/sec?

*45. Find the general solution of $xy'' + y' = 0$.

*46. Find the general solution of $y'' + y' \tan x = 0$.

**47. (i) Find a complex number z that is a square root of i, that is, satisfying $z^2 = i$. (*Hint:* Write $z = a + ib$.)

(ii) Factor the polynomials $z^2 - i$, $z^2 + i$, $z^4 + 1$.

(iii) Find the general solution of $y^{(8)} = y$.

**48. Find the general solution of $y^{(8)} = 256y$.

**49. For complex numbers $z_1 = r_1 e^{i\theta_1} = r_1 \cos\theta_1 + i r_1 \sin\theta_1$ and $z_2 = r_2 e^{i\theta_2} = r_2 \cos\theta_2 + i r_2 \sin\theta_2$, write their product $z = z_1 z_2$ as $z = r e^{i\theta}$, and find formulas for r and θ in terms of r_1, r_2, θ_1, and θ_2.

**50. Using the result of Exercise 49, find a formula for the square roots of $z = r e^{i\theta}$.

*51. Consider the equation $x^2 y'' + 4xy' + 2y = 0$. To solve it, try letting $y = x^a$, for some constant a. You should be able to find the general solution this way.

*52. Exercise 51 suggests an approach for equations of the form $a_n x^n y^{(n)} + a_{n-1} x^{n-1} y^{(n-1)} + \ldots + a_2 x^2 y^{(2)} + a_1 x^1 y^{(1)} + a_0 y = 0$.

 (i) Find a condition on a so that $y = x^a$ will be a solution of this equation.

 (ii) What happens for "repeated roots," for example, for the equation $x^2 y'' - xy' + y = 0$?

*53. Find the general solution of $x^2 y'' - 5xy' + 3y = 2x$.

*54. Find the general solution of $x^2 y'' + xy' + 4y = 0$. (*Hint:* Write $t^i = e^{i \ln t}$.)

**55. Suppose y_1 is a solution of the equation $y'' + b(x)y' + c(x)y = 0$. To try to find another solution, we consider a function of the form $y = uy_1$.

 (i) By substituting $y = uy_1$ into the original equation, find a differential equation for u.

 (ii) Show that the differential equation for u can be written as an equation for $v = u'$ that is of *first order*.

 In effect this process uses one solution of a second-order differential equation to turn it into a first-order equation for $v = u'$. An integration then gives the required function u and the second solution $y = uy_1$ of the original equation. This procedure is called **reduction of order**. It is illustrated in Exercises 56–58.

*56. Consider the equation $x^2 y'' - 6y = 0$. It is not difficult to see that $y = x^3$ is a solution. Use the technique of reduction of order, as described in Exercise 55, to find another solution, and from that the general solution.

*57. Consider the equation $xy'' - y' + 4x^3 y = 0$. It is not difficult to see that $y = \sin(x^2)$ is a solution. Use the technique of reduction of order to find another solution, and from that the general solution.

*58. It is also possible to use reduction of order to solve nonhomogeneous equations. Consider the equation $y'' + y = \csc x$. Use the fact that $y_1 = \sin x$ is a solution of the complementary equation to solve for a solution $y = uy_1$ of the nonhomogeneous equation.

*59. (i) If a parachutist falls without air resistance, show that he or she will fall 16 ft in the first second.

 (ii) How far will the parachutist of Example 19.3.6 (p. 1234) fall in the first second?

 (iii) In fact, a person falling without a parachute does experience air resistance. Such a person's terminal velocity is approximately 120 mph. Find the resistance constant k, and find how far a person without a parachute will fall in the first second.

*60. (i) Suppose the parachutist of Example 19.3.6 actually drops for 4 seconds before opening the parachute. Making the approximation that she falls without air resistance during this time, find out how far she falls before the parachute opens and how fast she is falling when it opens.

 (ii) Find the formula for the height of the parachutist, starting at $t = 4$ when the parachute opens.

 (iii) Make a quick sketch of the velocity. What is different about the terminal velocity in this case compared with the case discussed in the example, where we assumed that the parachute was open all the time?

*61. (i) Find the general solution of $y' + \frac{2y}{x} = 3x$.

 (ii) Try to find a Maclaurin series expression for the general solution, and explain why it does not work.

*62. Find an expression for the general solution of $(x-1)^2 y'' + (x-1)y' = 0$ as a power series in powers of $(x-1)$.

APPENDIX

Mathematical Induction

Mathematical induction is a technique for proving certain types of theorems. It is most often applied to statements that are asserted for every positive integer n.

Induction proves the result for every n by showing it is true for $n = 1$ and then showing that knowing the result for n allows us to prove it for $n + 1$. If it is known for $n = 1$, then it must be true for the next integer, 2, and then for the next, 3, and so on for every n.

EXAMPLE   **Sum of the first n integers**   If n is a positive integer, show that the sum of the integers from 1 to n equals $\frac{n(n+1)}{2}$.

Solution   First, with $n = 1$, the formula equals $\frac{1(1+1)}{2} = 1$, which *does* equal the sum of the integers from 1 to 1, that is, just 1.

Suppose we know it is true for some n, that is

$$1 + 2 + 3 + \cdots + n = \frac{n(n+1)}{2}.$$

Then to show it is true for $n + 1$ we consider the sum for $n + 1$ and find

$$1 + 2 + 3 + \cdots + n + (n+1) = (1 + 2 + 3 + \cdots + n) + (n+1)$$
$$= \frac{n(n+1)}{2} + (n+1) = (n+1)\left(\frac{n}{2} + 1\right).$$

(Here we used the formula for $1 + 2 + 3 + \cdots + n$ and then took out the common factor $n + 1$.) Now we find this equals

$$(n+1)\left(\frac{n+2}{2}\right) = \frac{(n+1)(n+2)}{2} = \frac{(n+1)((n+1)+1)}{2}.$$

This is exactly the required result (i.e., it is $\frac{n(n+1)}{2}$ with n replaced by $n + 1$.) This completes the proof.

Exercise: Use mathematical induction to prove the following formula for the sum of the squares of the first n positive integers:

$$1^2 + 2^2 + 3^2 + 4^2 + \cdots + n^2 = \frac{n(n+1)(2n+1)}{6}.$$

Answers to Odd-Numbered Exercises

Exercises 1.2 (p. 7)
1. $\frac{1}{84}$ **3.** $\frac{1}{84}$ **5.** $2 - \sqrt{3}$ **7.** $\sqrt{2} + \sqrt{3} - 3$ **9.** $x^2 + 3$
11. $[-3, 5]$ **13.** $(\frac{4}{3}, \frac{8}{3})$ **15.** $(-\infty, 0) \cup (1, \infty)$ **17.** $[-1, 0] \cup [2, 7]$
19. $(-\infty, -\sqrt{3}] \cup [\sqrt{3}, \infty)$

Exercises 1.3 (p. 16)
1. $(x - 3)(x + 4)$ **3.** $(x + 4)(x - 1)$ **5.** $2(x + 1)^2$
7. $(x + 2)(x - 2)$ **9.** $(x - 7 + \sqrt{7})(x - 7 - \sqrt{7})$ **11.** $\frac{-3 + \sqrt{17}}{2}$
13. $2 \pm \sqrt{5}$ **15.** no roots **17.** $0, 6$ **19.** $\frac{4 \pm \sqrt{6}}{10}$ **21.** $(0, 2)$
23. $(-\infty, -3] \cup [-1, \infty)$ **25.** $(-\infty, -4] \cup [4, \infty)$
27. $(-\infty, 2 - \sqrt{2}] \cup [2 + \sqrt{2}, \infty)$ **29.** $(-\infty, \infty)$
31. $a = -1, q(x) = x^2 + 2x + 2$ **33.** $a = 2, q(x) = x^2 - x + 1$
35. $a = 1, q(x) = x^2 + x + 1$ **37.** $a = 3, q(x) = x^2 + 3x + 9$
39. $a = 0, q(x) = x^3 - 5x^2 + 3x - 11$ **41.** $x^3 + 12x^2 + 48x + 64$
43. $x^5 + 5x^4 + 10x^3 + 10x^2 + 5x + 1$ **45.** $x^6 - 6x^4 + 12x^2 - 8$
47. -1 **49.** $1 - 12x + 48x^2 - 64x^3$

Exercises 1.4 (p. 22)
1. $\mathbb{R}$ **3.** $\{x \in \mathbb{R} : x \neq -3\} = (-\infty, -3) \cup (-3, \infty)$ **5.** $[0, \infty)$
7. $(-1, \infty)$ **9.** $[0, 1) \cup (1, \infty)$ **11.** $[5, \infty)$ **13.** $[3, \infty)$ **15.** $[-\frac{1}{12}, \infty)$
17. $[0, \infty)$ **19.** $(0, 1]$

21.

23.

25.

27.

29.
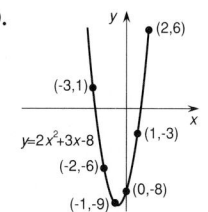

Exercises 1.5 (p. 32)
1. $\sqrt{580} \approx 24.083$ **3.** 5 **5.** 10 units **7.** 4 **9.** 4
11. $y - 2 = x - 1$ or $x - y + 1 = 0$ **13.** $x + 2y = 3$
15. $y + 6 = 2(x - 4)$ or $2x - y = 14$ **17.** $y = -x + 3$
19. $y - \frac{2}{3} = \frac{1}{3}x$ or $x - 3y + 2 = 0$ **21.** $-\frac{3}{2}$ **23.** $x + 2y = 11$
25. $\frac{7}{3}$ **27.** $\frac{7}{2}$ **29.** $(3, 2)$

Exercises 1.6 (p. 42)
1. ellipse, center $(0, 0)$, minor axis between x-intercepts $(\pm 3, 0)$,
major axis between y-intercepts $(0, \pm 4)$ **3.** hyperbola, center
$(0, 0)$, transverse axis between x-intercepts $(\pm 2, 0)$, conjugate axis
between $(0, \pm 1)$, asymptotes $x \pm 2y = 0$ **5.** parabola, vertex
$(2, 0)$, opening to left, x-intercept $(2, 0)$, y-intercepts $(0, \pm 2\sqrt{2})$
7. ellipse, center $(-2, 0)$, major axis between x-intercepts
$(-5, 0)$, $(1, 0)$, minor axis from $(-2, -1)$ to $(-2, 1)$, y-intercepts
$(0, \pm \sqrt{5}/3)$ **9.** parabola, vertex $(-1, -\frac{3}{2})$, opening upward,
x-intercepts $(-1 \pm \sqrt{6}, 0)$, y-intercept $(0, -\frac{5}{4})$ **11.** ellipse, center
$(0, 0)$, major axis between x-intercepts $(\pm \frac{1}{\sqrt{2}}, 0)$, minor axis
between y-intercepts $(0, \pm \frac{1}{2\sqrt{2}})$ **13.** ellipse, center $(0, 0)$, minor
axis between x-intercepts $(\pm \frac{3}{2}, 0)$, major axis between y-intercepts
$(0, \pm 3)$ **15.** parabola, vertex $(\frac{1}{10}, 0)$, opening to the right,
x-intercept $(\frac{1}{10}, 0)$, no y-intercepts **17.** ellipse, center $(2, 0)$,
major axis between x-intercepts $(2 \pm \sqrt{5}, 0)$, minor axis between
$(2, \pm \sqrt{\frac{5}{3}})$, y-intercepts $(0, \pm \frac{1}{\sqrt{3}})$ **19.** parabola, vertex $(\frac{5}{4}, 0)$,
opening to left, x-intercept $(\frac{5}{4}, 0)$, y-intercepts $(0, \pm \frac{2}{\sqrt{5}})$
21. parabola, vertex $(0, \frac{1}{4})$, opening downward, x-intercepts
$(\pm \frac{1}{4}, 0)$, y-intercept $(0, \frac{1}{4})$

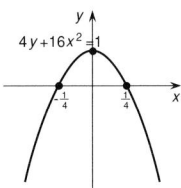

23. hyperbola, center $(0, 0)$, transverse axis between y-intercepts $(0, \pm1)$, no x-intercepts, conjugate axis between $(\pm1, 0)$, asymptotes $y = \pm x$

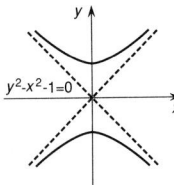

25. hyperbola, center $(0, 0)$, no x-intercepts, transverse axis between y-intercepts $(0, \pm\frac{1}{\sqrt{2}})$, conjugate axis between $(\pm\sqrt{2}, 0)$, asymptotes $2y = \pm x$

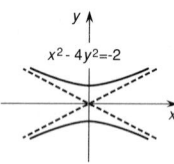

27. ellipse, center $(0, 0)$, major axis between x-intercepts $(\pm\frac{\sqrt{7}}{2}, 0)$, minor axis between y-intercepts $(0, \pm\frac{\sqrt{7}}{3})$

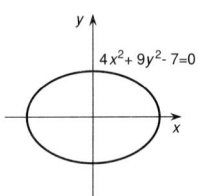

29. circle, center $(-\frac{1}{2}, -\frac{1}{2})$, radius $\sqrt{5}$

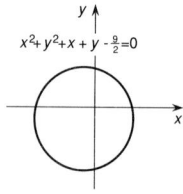

31. $(x + 1)^2 + (y - 3)^2 = 4$ or $x^2 + 2x + y^2 - 6y + 6 = 0$
33. $4x^2 - y^2 = 16$ **35.** $x^2 - 10x + 4y^2 - 16y + 37 = 0$
37. $x^2 + 4x + y^2 = 21$ **39.** $4x^2 - 16x + y^2 + 8y = 0$

Exercises 1.7 (p. 55)
1. -1 **3.** $-\frac{1}{\sqrt{2}}$ **5.** $\sqrt{3}$ **7.** $-\sqrt{2}$ **9.** $-\frac{1}{\sqrt{2}}$
11. $\sqrt{\frac{1}{2} - \frac{1}{2\sqrt{2}}} = \frac{1}{2}\sqrt{2 - \sqrt{2}}$ **13.** $\sqrt{\frac{1}{2} + \sqrt{3}/4} = \frac{1}{2}\sqrt{2 + \sqrt{3}}$
15. $-\frac{1}{2}\sqrt{2 + \sqrt{3}}$ **17.** $\frac{\sqrt{2} + \sqrt{2}}{\sqrt{2} - \sqrt{2}} = 1 + \sqrt{2}$ **19.** $-\frac{\sqrt{2} - \sqrt{2}}{\sqrt{2} + \sqrt{2}} = 1 - \sqrt{2}$
21. $(-3, 0)$ **23.** $(0, 2)$ **25.** $(\frac{1}{2}, -\frac{\sqrt{3}}{2})$ **27.** $(-2\sqrt{2}, 2\sqrt{2})$
29. $(-\sqrt{3}, 1)$ **31.** $(3, 2k\pi)$ **33.** $(2, -\frac{\pi}{2} + 2k\pi)$

35. $(2\sqrt{2}, -\frac{\pi}{4} + 2k\pi)$ **37.** $(4\sqrt{2}, \frac{5\pi}{4} + 2k\pi)$ **39.** $(\frac{1}{2\sqrt{3}}, -\frac{\pi}{6} + 2k\pi)$

Exercises 1.8 (p. 59)
1. $A + B$ **3.** $2A - B$ **5.** $\frac{1}{2}A$ **7.** $\frac{1}{3}A$ **9.** $10^B A$ **11.** 5 **13.** $\frac{1}{2}$
15. $-\frac{1}{3}$ **17.** -3 **19.** $\frac{1}{2}$

Chapter 1 Review Exercises (p. 60)
1. $x \in [-5, -3) \cup (-3, -1]$ **3.** $x \in (-1, -\frac{1}{4}) \cup (\frac{1}{4}, 1)$
5. $(-14, -7) \cup (-5, 2)$ **7.** $\{x: -\frac{1}{2} \le 1 - x^2 \le \frac{1}{2}\}$ or $[-\sqrt{3}/\sqrt{2}, -1/\sqrt{2}] \cup [1/\sqrt{2}, \sqrt{3}/\sqrt{2}]$ **9.** ϕ (the empty set)
11. $(x + 12)(x - 5)$ **13.** $(x + \frac{1}{2} + \frac{\sqrt{5}}{2})(x + \frac{1}{2} - \frac{\sqrt{5}}{2})$
15. $x(x - 7)(x + 5)$ **17.** $(x^2 + 3)(x - \sqrt{5})(x + \sqrt{5})$
19. $x^3(x + 1)(x - 1)$ **21.** 12 **23.** 12 **25.** 0 **27.** 4 **29.** -32
31. $\{x : x \ne \pm\sqrt{3}\}$ **33.** $(-\infty, 1) = \{x : x < 1\}$
35. $\{y : y \ge -7\} = [-7, \infty)$ **37.** $5\sqrt{2} \approx 7.07107$ **39.** 4
41. $\sqrt{58} \approx 7.61577$ **43.** 3 **45.** 4 **47.** $x + y = 7$
49. $3x + 2y + 6 = 0$ **51.** $4x^2 + 9y^2 = 36$
53. $x^2 - 10x + y^2 - 8y + 32 = 0$ **55.** $y^2 - 4x^2 = 9$ **57.** $(-2, 1)$,
$(-\frac{14}{5}, \frac{1}{5})$ **59.** $(-2, 1)$ **61.** parabola, vertex $(0, -2)$, opening upward   **63.** hyperbola, center $(0, 0)$, transverse axis between $(0, \pm2)$, conjugate axis between $(\pm2, 0)$, asymptotes $y = \pm x$
65. ellipse, center $(0, \frac{3}{2})$, major axis between $(\pm\sqrt{17}, \frac{3}{2})$, minor axis between $(0, \frac{1}{2}(3 \pm \sqrt{17}))$ **67.** hyperbola, center $(0, 0)$, transverse axis between $(0, \pm\frac{1}{3})$, conjugate axis between $(\pm\frac{1}{4}, 0)$, asymptotes $3y = \pm4x$ **69.** ellipse, center $(-2, 0)$, major axis between $(-3, 0)$ and $(-1, 0)$, minor axis between $(-2, \pm\frac{1}{\sqrt{3}})$
71. $-\frac{1}{\sqrt{2}}$ **73.** $-\sqrt{\frac{1}{2} - \frac{1}{2\sqrt{2}}} = -\frac{1}{2}\sqrt{2 - \sqrt{2}}$ **75.** $-\sqrt{3}$
77. $(\sqrt{3} - 1)/(2\sqrt{2})$ **79.** $\frac{1}{\sqrt{3}}$ **81.** $(r, \theta) = (4\sqrt{2}, \frac{5\pi}{4} + 2k\pi)$
83. $(r, \theta) = (3\sqrt{5}, -\arctan(2) + 2k\pi) \approx (6.708, -1.10715 + 2k\pi)$
85. $(r, \theta) = (8, -\frac{\pi}{2} + 2k\pi)$ **87.** $(x, y) = (-1, 0)$
89. $(x, y) = (-\frac{7}{\sqrt{2}}, -\frac{7}{\sqrt{2}})$ **91.** $\ln(10) + 2 \ln a + \ln b$
93. $\ln(a)/\ln(10)$ **95.** $\ln a - \ln 2 - \ln b$ **97.** $1/(a^3)$
99. $\ln 2 + \ln a + b \ln 3$

Exercise 2.1 (p. 69)
1. 4 **3.** 2 **5.** 5 **7.** 12 **9.** 0 **11.** $2x - y = 3$
13. $x - y + 4 = 0$ **15.** $19x + y + 20 = 0$ **17.** $11x - y + 14 = 0$
19. $32x + y + 48 = 0$ **21.** $2ax - y = a^2$ **23.** 256 ft
25. $32t$ ft/sec   **27.** 8 ft   **29.** (iii) $t = \frac{v}{32}$, $h = \frac{v^2}{64} + A$ (iv) after $(5 + \sqrt{41})/8$ sec (v) 12 ft/sec   **33.** 25.1327 **37.** (i) $48, \approx 6, \approx 0.6$

Exercises 2.2 (p. 79)
1. 1 **3.** -3 **5.** -1 **7.** 1 **9.** $11\sqrt{5} - 1$ **11.** 5 **13.** 1 **15.** 0
17. 0 **19.** 4 **21.** 2 **23.** 5 **25.** 0 **27.** -7 **29.** 0 **31.** -5
33. 0 **35.** does not exist   **37.** $-\frac{3}{2}$ **39.** $\frac{1}{8}$ **41.** $a + 3$ **43.** no limit for $a = -1, -2$, otherwise $\frac{(a - 1)(a - 2)(a - 5)}{(a + 1)(a + 2)}$.   **45.** no limit for $a = 1$, otherwise $\frac{a(a - 2)}{a - 1}$.   **47.** no limit for $a = 1$, limit is 1 if $a > 1$, -1 if $a < 1$ **49.** if $c \ne 0$, no limit for $a = c$, otherwise $\frac{a + c}{a - c}$; if $c = 0$, 1 for all a **53.** $-\frac{1}{2}$ at -1, does not exist at 1
55. does not exist   **57.** 6 at -3

Exercises 2.3 (p. 91)
1. $\frac{1}{2}$ **3.** 3 **5.** $-\frac{1}{6}$ **7.** -32 **9.** $-2\sqrt{3}$ **11.** 3 **13.** -1
15. $(3 - \sqrt{2})^{5/2} \approx 3.1667$ **17.** $-\frac{1}{4}$ **19.** $\frac{5}{6}$ **21.** 0 **23.** 0 **25.** 0
27. 0 **29.** 0 **31.** does not exist   **33.** $\frac{3}{2}$ **35.** does not exist
37. $\frac{1}{108}$ **39.** does not exist   **45.** (i) $\lim_{x \to a+c} f(x)$ (ii) $\lim_{x \to c-a} f(x)$
53. 2

Exercises 2.4 (p. 104)

1. ∞ **3.** 1 **5.** $\frac{1}{2}$ **7.** $-\infty$ **9.** ∞ **11.** no limit (limit from above is ∞, from below, $-\infty$). **13.** ∞ **15.** ∞ **17.** ∞ **19.** no limit, finite or infinite. **21.** 0 **23.** ∞ **27. (i)** 1 **(ii)** 6 **(iii)** -4 **(iv)** 0 **31.** 1 as $x \to 0^+$, -1 as $x \to 0^-$ **33.** ∞ from both sides **35.** ∞ from both sides **37.** $-\infty$ as $x \to 0$, ∞ as $x \to -1^-$, $-\infty$ as $x \to -1^+$, 0 as $x \to 1$ **39. (i)** $a \approx 2.325$

Exercises 2.5 (p. 112)

1. continuous everywhere **3.** (d) at 0, (a) elsewhere **5.** (c) at 0, (d) at $\pm\sqrt{3}$, (a) elsewhere **7.** (c) at 0, (a) elsewhere **9.** (b) at 0, (a) elsewhere **11.** bounded by $M = 8$ **13.** not bounded **15.** bounded by $M = 1$ **17.** bounded by $M = 1$ **19.** bounded by $M = 2$ **21.** min at $x = 1$, max at $x = 3$, -1 **23.** no max or min **25.** min at $t = 1$, max at $t = 0$ **27.** min at $u = 4$, max at $u = -2$ **29.** no max or min **33. (i)** $x \geq -4$ **(ii)** $\{t : t \neq -1\}$ **(iii)** $\{u : u \neq 0\}$ **43.** -2.7606 **45.** $-0.7748, 1, 1.9276$

Exercises 2.6 (p. 118)

1. 0 **3.** -2 **5.** 1 **7.** $\frac{1}{3}$ **9.** ∞ **11.** does not exist **13.** does not exist **15.** does not exist **17.** 0 **19.** 0 **21.** 0 **23.** 1 **25.** 0 **27.** $\frac{1}{2}$ **29.** $\frac{1}{2}$ **31. (i)** $a = \frac{\pi}{2} + k\pi, k = 0, \pm 1, \pm 2, \ldots$ **33.** $\cot x : a = k\pi, \tan^2 x : a = \frac{\pi}{2} + k\pi, k = 0, \pm 1, \pm 2, \ldots$

Exercises 2.7 (p. 124)

1. -5 **3.** 9 **5.** 1 **7.** 0 **9.** 6 **11.** does not exist **13.** does not exist **15.** 0 **17.** 2 **19.** does not exist **21.** does not exist **23.** does not exist **29.** 0.0001, 0.000 000 3, 0.000 000 000 3, 0.000 000 000 000 3

Chapter 2 Review Exercises (p. 126)

1. -1 **3.** $\frac{3}{4}$ **5.** 2 **7.** 0 **9.** $-\frac{5}{2}$ **11.** 2 **13.** $2\sqrt{5}$ **15.** $\frac{1}{12}$ **17.** $-\frac{9}{2}$ **19.** 2 **21.** 0 **23.** 1 **25.** π **27.** $\frac{\pi}{2}$ **29.** 3 **31.** 0 **33.** 0 **35.** does not exist (∞) **37.** $-\frac{1}{2}$ **39.** -2 **41.** $\frac{1}{2}$ **43.** $-\infty$ **45.** ∞ **47.** does not exist **49.** $-\infty$ **51.** $-2a$ **53.** $\frac{2}{3}$ **57.** removable at $x = -2$, infinite discontinuity at ± 4 **61. (i)** $-2.115, 0.254, 1.861$ **63. (i)** $-2, \pm 1$

Exercises 3.1 (p. 138)

1. 4 **3.** 1 **5.** 7 **7.** 2 **9.** 3 **11.** $4x - y = 3$ **13.** $x - y = 1$ **15.** $7x - y = 7$ **17.** $2x - y = 2$ **19.** $3x - y = 2$ **21.** $a = 1$ **23. (i)** $2x - y = 1, 2x + y = -1$ **(ii)** there are none **25. (i)** $32 - 32t$ ft/sec **(ii)** rising when $t < 1$, falling when $t > 1$ **(iii)** 22 ft **27. (i)** 6 qt-lb/sq in **(ii)** 0.2 qt **(iii)** bigger change when $P = 15$ **(iv)** ≈ 0.37333 **(v)** 0.375 **29. (i)** $2A + B$ **(ii)** $2aA + B$ **31.** $a = 0, f'(x) = 1$ if $x > 0, f'(x) = -1$ if $x < 0$ **37.** $2929 \leq N \leq 3071$, approx.

Exercises 3.2 (p. 152)

1. $3x^2 - 4x + 1$ **3.** $6x^2 + 6x + 6$ **5.** $18x^2 + 18x - 10$ **7.** $32x + 24$ **9.** $(-2x^2 - 2x + 2)/(x^2 + 1)^2$ **11.** $-1/(x - 1)^2$, except at $x = 1$ **13.** $-(8x)/(x^2 - 4)^2$, except at $x = \pm 2$ **15.** $(2 - 6x)/(3x^2 - 2x)^2$, except at $x = 0, \frac{2}{3}$ **17.** $(8x^2 - 22x - 1)/(x^2 + 3x - 4)^2$, except at $x = 1, -4$ **19.** $(x^6 + 4x^3)/(1 + x^3)^2$, except at $x = -1$ **21. (i)** $12x^3 + 12x^2 - 2x - 2$ **(ii)** $72x^5 + 60x^4 - 9x^2 - 6x$ **(iii)** $-(3x^2 + 4x + 4)/(x^2 - x - 2)^2$, except at $x = 2, -1$ **23. (i)** $x - y + 1 = 0$ **(ii)** $4x + y + 4 = 0$

25. $12x - y + 16 = 0$ **31. (i)** falling when $0 < t < 2$, rising when $t > 2$ (or $t < 0$) **(ii)** after 2 sec, height is 2 ft **(iii)** no **33. (i)** $\frac{1}{2}g'(x)/f(x) = \frac{1}{2}g'(x)/\sqrt{g(x)}$ **(ii)** $x/\sqrt{1 + x^2}$ **(iii)** $(x + 1)(x^2 + 2x + 2)^{-1/2}$ **35. (i)** $4x^3 + 6x^2 - 2x$ **(ii)** $12x^2 + 12x - 2$ **(iii)** $24x + 12$ **43.** ≈ 2.7183 **45.** ≈ 0.5403

Exercises 3.3 (p. 164)

1. min 0 at $x = -1$, max 8 at $x = 1$ **3.** min -2 at $t = 0, 1$, max $-\frac{5}{4}$ at $t = \frac{1}{2}$ **5.** min -3 at $s = -\sqrt{2}$, max 1 at $s = 0, -2$ **7.** min 3 at $x = 0, 2$, max 12 at $x = 3$ **9.** min -226 at $u = -2$, max 230 at $u = 2$ **11.** min $\frac{1}{18}$ at $x = \pm 4$, max $\frac{1}{2}$ at $x = 0$ **13.** min $-\frac{3}{2}$ at $x = 1$, max $-\frac{1}{4}$ at $x = -1$ **15.** min 0 at $x = 0$, max 4 at $x = -2$ **17.** min 0 at $x = 0$, max $\frac{1}{2}$ at $x = -1$ **19.** min -1 at $x = 0$, max 1 at $x = 2$ **21.** $21x^6 - 8x^3 + 1$ **23.** $\frac{2}{(x + 1)^2}$ **25.** $-\frac{2}{(2x + 3)^2}$ **27.** $-x^{-4}(4x + 3)$ **29.** $-(4x^3 + 4x)(x^4 + 2x^2 + 4)^{-2}$ **31.** 50 m **33.** ≈ 1056.3 cm^3 **35.** side of base $(200)^{1/3} \approx 5.848$ in, height $\frac{1}{2}(200)^{1/3} = 10(40)^{-1/3} \approx 2.924$ in **37. (i)** $(-\frac{2}{5}, \frac{1}{5})$ **(ii)** $(\frac{3}{2}, \frac{3}{2})$ **39.** $\frac{50}{3}$ cm by $\frac{25}{3}$ cm **41. (i)** square $\frac{4}{\pi + 4}$ m, circle $\frac{\pi}{\pi + 4}$ m **(ii)** all circle, no square **43. (i)** \$9 **(ii)** \$11.50 **45.** max $f(-1) = 4$, min ≈ 0.855 at $x \approx 0.311$ **47.** max $f(2) = 4$, min ≈ -10.4940 at $x \approx -1.5737$ **49.** max ≈ 1.8197 at $x \approx 2.0288$, min $f(0) = f(\pi) = 0$

Exercises 3.4 (p. 172)

1. $-2 \cos x \sin x$ **3.** $-\sin(u)/(2 - \cos u)^2$ **5.** $2(w - 1) \sin(2w) + 2(w^2 - 2w - 1) \cos(2w)$ **7.** $(2 \cos(2t)\sin(3t) - 3 \sin(2t)\cos(3t))/\sin^2(3t)$, not defined for $t = n\frac{\pi}{3}$, n an integer **9.** $3r^2 \tan r + r^3 \sec^2 r - \sec r \tan r$, except $r = \frac{\pi}{2} + n\pi$ **11.** $2 \sec^2 x \tan x$, except for $x = (2n + 1)\frac{\pi}{2}$ **13.** 0, except for $x = n\pi$ **15.** $-\csc^2 x$, except $x = n\pi$ **17.** $-\csc^3 x - \csc x \cot^2 x = -\csc^3 x(1 + \cos^2 x)$, except $x = n\pi$ **19.** $\sec^2 x$, except $x = n\frac{\pi}{2}$ **21.** $x = (2n + 1)\frac{\pi}{2}$ **23.** $u = n\frac{\pi}{2}$ **25.** $v = (2n + 1)\frac{\pi}{4}$ **27.** $r = n\pi$ **29.** never zero **31.** $(1 + 2x^2) \cos x + 4x \sin x$ **33.** $-2 \sin x \cos x/(1 + \sin^2 x)^2$ **35.** $[-(x^2 + 2x + 5) \sin x - (2x + 2) \cos x]/(x^2 + 2x + 5)^2$ **37.** $3x^2(\sin x + 2 \cos x) + (x^3 - 8)(\cos x - 2 \sin x)$ **39.** $2x + 2 \tan x + 2x \sec^2 x + 2 \tan x \sec^2 x$ **41. (i)** $\theta = -\frac{\pi}{6} + 2n\pi$ or $\frac{5\pi}{6} + 2n\pi$ **(ii)** none **43.** positive on $(-\frac{\pi}{2}, \frac{\pi}{2})$, negative on $[-\pi, -\frac{\pi}{2}) \cup (\frac{\pi}{2}, \pi]$, zero at $\pm\frac{\pi}{2}$ **45. (i)** $g'(x) = \sin x + x \cos x, g'(0) = 0$ **47.** $x = \frac{\pi}{4} + 2n\pi$ and $x = \frac{3\pi}{4} + 2n\pi$

Exercises 3.5 (p. 183)

1. $3t^2 \ln t + t^2$, for $t > 0$ **3.** $\ln(x^2) + 2 = 2 \ln |x| + 2$, except $x = 0$ **5.** $e^r \ln r + \frac{1}{r}e^r, r > 0$ **7.** $(1 - 2x)e^{-2x}$ **9.** $1, u > 0$ **11.** $\frac{1}{x}$, for $x > 0$ **13.** $\frac{1}{t} + 1$, for $t > 0$ **15.** $-\frac{1}{t \ln^2 t}$, for $t > 0$ **17.** $\frac{2}{x}$, for $x \neq 0$ **19.** $(2x^2 + 2x)e^{2x+4}$, for $x > 0$ **21.** $\cosh^2 x + \sinh^2 x$ **23.** $\tanh(2x) + 2x \operatorname{sech}^2(2x)$ **25.** $e^x(\sinh x + \cosh x)$ **27.** 0 **29.** $3 \cosh(3x) + 4 \sinh(4x)$ **31.** $((2x^3 - 4)/x - 6x^2 \ln x)/(2x^3 - 4)^2$ **33.** $(2x(e^x + \sin x + 2) - (x^2 + 6)(e^x + \cos x))/(e^x + \sin x + 2)^2$ **35.** $e^x((3 + x)\cos x - (2 + x)\sin x)$ **37.** $(2x(\tan^2 x + 2 + \sin x) - (x^2 + 3)(2 \tan x \sec^2 x + \cos x))/(\tan^2 x + 2 + \sin x)^2$ **39.** $3 \sin^2 x \cos x$ **43.** $\frac{1}{x \ln 2}$ **45.** $a = 1$ **47. (i)** max = $\cosh c = \cosh(-c)$, min = 1 **(ii)** max = 1, min = $\operatorname{sech} c = \operatorname{sech}(-c)$ **(iii)** max = $\sinh c$, min = $\sinh(-c)$ **49. (i)** $\cosh(x + y) = \cosh x \cosh y + \sinh x \sinh y$ **(ii)** $\cosh(2x) = \cosh^2 x + \sinh^2 x$ **51.** $x \ln^2 x - 2x \ln x + 2x$ **53. (i)** 4×10^5 **(ii)** $100 \ln 2$ minutes **57.** $e \approx 2.7183$ **61.** ≈ 7.6009

Exercises 3.6 (p. 193)

1. $3(3x^4 - 2x^2 - 1)^2(12x^3 - 4x)$ **3.** $-2(x^2 + 4x + 5)^{-3}(2x + 4)$
5. $-2\sin(2x + 1)$ **7.** $2x/(x^2 + 4)$ **9.** $\sinh(\sin x)\cos x$
11. $(t - 1)/\sqrt{t^2 - 2t + 1}$, except $t = 1$ **13.** $\frac{1}{3}x^{-2/3}$, except $x = 0$
15. $\frac{-\sin r}{1 + \cos r}$, $r \neq (2n + 1)\pi$ **17.** $1, x > 0$ **19.** $\sinh(\sinh x)\cosh x$
21. $\frac{6x}{x^2 + 7}\ln^2(x^2 + 7)$ **23.** $4x^3\cos(\ln(x^4 + 2))/(x^4 + 2)$
25. $2x\ln(10)10^{x^2}$ **27.** $\frac{1}{2}\frac{\sinh x}{\sqrt{\cosh x}}$ **29.** 0 **31.** $-4x/(x^2 + 2)^3$
33. $\frac{1}{2}\sin x/\sqrt{4 - \cos x}$ **35.** $4(x + 3)^3(-x^2 - 6x + 1)/(x^2 + 1)^5$
37. $2x\cos(\exp(x^2 + 1))\exp(x^2 + 1)$ **39.** $\frac{5}{7}x^{-2/7}$ **45. (i)** max $\sqrt{2}$
at $x = \pm 1$, min 0 at $x = 0$ **(ii)** max $\frac{1}{\sqrt{2}}e^{-1/2}$ at $x = \frac{1}{\sqrt{2}}$, min
$-\frac{1}{\sqrt{2}}e^{-1/2}$ at $x = -\frac{1}{\sqrt{2}}$ **47.** width $= \frac{2}{\sqrt{3}}R$, height $= \frac{2\sqrt{2}}{\sqrt{3}}R$
49. (ii) $\frac{1}{1 + t^2}$

Exercises 3.7 (p. 201)

1. $-\frac{2x + y}{x - 2y}$ **3.** $\frac{1 - y}{x + 3y^2} = \frac{y^3 - y}{(1 + x)(x + 3y^2)}$ **5.** $\frac{6x + y}{27y^2 - x}$ **7.** $\sec y$ **9.** $\tan x$
$\tan y$ **11.** $\frac{2y - x}{2x + y}$ **13.** $\frac{x + 3y^2}{1 - y} = \frac{(x + 3y^2)(1 + x)}{y^3 - y}$ **15.** $\frac{27y^2 - x}{6x + y}$ **17.** $\cos y$
19. $\cot x \cot y$
21. $9x^8 + 8x^7 + 14x^6 - 42x^5 - 35x^4 - 64x^3 - 24x^2 - 16x$
23. $(x^3 + 4x)^{1/3} + \frac{x}{3}(3x^2 + 4)(x^3 + 4x)^{-2/3}$
25. $\frac{1}{2}\left(\sqrt{\frac{(x + 2)(x + 3)}{x + 1}} + \sqrt{\frac{(x + 1)(x + 3)}{x + 2}} + \sqrt{\frac{(x + 1)(x + 2)}{x + 3}}\right)$
27. $\frac{3}{2}\frac{x^2}{(x^2 + 2)^4\sqrt{x^3 + 1}} - \frac{8x\sqrt{x^3 + 1}}{(x^2 + 2)^3}$ **29.** $\left(\frac{(x + 1)(x - 2)}{(x + 3)(x^2 - 9)}\right)^{3/7}$
$\left(\frac{3}{7}\left(\frac{1}{x + 1} + \frac{1}{x - 2} - \frac{1}{x + 3} - \frac{2x}{x^2 - 9}\right)\right)$ **35. (i)** $x + y = 2$ **(ii)** $6x - y = 12$
37. (i) $y = \pi$ **(ii)** $x + 12y = 3$ **39.** max $= \frac{18}{\sqrt{33}}$, min $= -\frac{18}{\sqrt{33}}$
41. max 4, min -12 **47.** ellipse **49.** ellipse

Exercises 3.8 (p. 217)

1. $2/(4x^2 + 16x + 17)$ **3.** $2x/(|x|(1 + x^2)\sqrt{x^2 + 2})$ **5.** $2x/$
$((4 + x^2)\sqrt{x^4 + 8x^2 + 15})$ **7.** $-e^x/((5 + e^x)\sqrt{e^{2x} + 10e^x + 24})$
9. $-2\exp(\text{arccot}(2x))/(1 + 4x^2)$ **11.** $2\sec^2 x\tan x/\sqrt{\sec^4 x - 1}$
13. $-\frac{2x + 2}{(x^2 + 2x + 3)^2 - 1}$ **15.** $1/(\text{arccosh}(x)\sqrt{x^2 - 1})$ **17.** $\text{sech } x$
19. $-\text{csch } x$ **21.** $\frac{1}{5}$ **23.** $\frac{1}{4 - 4h^{-1}(u)} = \frac{1}{\sqrt{8 - 8u}}$ **25.** $(H^{-1})'(u) = 1/$
$(3H^{-1}(u)^2 + 12H^{-1}(u))$ **31. (i)** $\sin x, x \in [-\frac{\pi}{2}, \frac{\pi}{2}]$ **(ii)** $\frac{1}{2}$
$(\tan(u) - 1), u \in (-\frac{\pi}{2}, \frac{\pi}{2})$ **35.** $x + \pi, x \in (-\frac{3\pi}{2}, -\frac{\pi}{2})$ **37.** 104π
mph ≈ 326.7 mph **39. (i)** $K(x) = \frac{x}{2.2}, P(x) = 2.2x$ **45.** ≈ 0.5236

Exercises 3.9 (p. 222)

1. $12x^2 - 18x + 4$ **3.** $(2 + 4x^2)\exp(x^2 + 1)$ **5.** $343e^{7x + 1}$
7. $-\cos x$ **9.** 0 **11.** $2(x + 1)^{-3}$ **13.** $-162(3t + 2)^{-4}$
15. $-(e^{3x} - 4e^{2x} + e^x)(e^x + 1)^{-4}$ **17.** $x(1 - x^2)^{-3/2}$
19. $\exp(\sin x)(\cos^3 x - 3\cos x\sin x - \cos x)$ **21.** $\frac{n!}{(n - k)!}$, if $k \leq n$;
0, if $k > n$ **23.** $(-1)^{k-1}(k - 1)! x^{-k}$ **25. (i)** $0, \pm 1, \pm 2$ **27. (i)** 6
sec **(ii)** 144 ft **29.** $3\sqrt{58.8} \approx 23$ m/sec **33.** $\cos x$, if $n = 4k$;
$-\sin x$, if $n = 4k + 1$; $-\cos x$, if $n = 4k + 2$; $\sin x$, if $n = 4k + 3$
35. $2^n e^{2x + 1}$ **37.** $a^n\sin(ax + b)$, if $n = 4k$; $a^n\cos(ax + b)$, if
$n = 4k + 1$; $-a^n\sin(ax + b)$, if $n = 4k + 2$; $-a^n\cos(ax + b)$, if
$n = 4k + 3$ **39.** $\frac{625}{16}$ ft

Chapter 3 Review Exercises (p. 225)

1. $20x^4 - 6x$ **3.** $14x^6 - 16x^3$ **5.** $2 - \frac{1}{2\sqrt{x}}$, for $x > 0$
7. $\frac{3}{5}x^{-2/5} - \frac{1}{2}x^{-3/2}$, for $x > 0$ **9.** $\frac{10x}{(x^2 + 1)^2}$ **11.** $3\cos^2 x - 3\sin^2 x$
13. $6x\tan x + (1 + 3x^2)\sec^2 x, x \neq (2n + 1)\frac{\pi}{2}$ **15.** $2x^3/\sqrt{x^4 + 2}$
17. $-(2x^3 + x)(x^4 + x^2)^{-3/2}, x \neq 0$ **19.** $e^{2x-5}(2\cos x - \sin x)$
21. $(2x + 3)\cosh(x^2 + 3x - 2)$ **23.** $\tanh x$
25. $(12x^3 - 2)\text{sech}^2(3x^4 - 2x)$ **27.** $\ln x + 1$, for $x > 0$ **29.** $3\sin^2$

$x\cos x\cosh^5 x + 5\sin^3 x\cosh^4 x\sinh x$ **31.** max $f(0) = -2$, min
$f(-3) = -11$ **33.** max $f(0) = 4$, min $f(2) = -12$ **35.** max
$f(0) = 1$, min $f(3) = \frac{1}{10}$ **37.** max $\ln 8$, min 0 **39.** max $f(0) = 0$,
min $f(-\frac{\pi}{4}) = -1$ **41.** $\frac{2x + 3y}{2y - 3x}$ **43.** $-\frac{y}{x}$ **45.** 1 **47.** $(2x - \frac{y}{x})/\ln x$
49. $-\left(1 + (x + y)^2 y e^{xy}\right)/(1 + (x + y)^2 x e^{xy})$ **51.** $\frac{1}{x^2 + 6x + 10}$ **53.** $-\frac{\cos x}{|\cos x|}$
55. $\frac{2x}{|x|\sqrt{x^2 + 1}}$ **57.** $\frac{2}{1 + 4x^2}\exp(\arctan(2x))$ **59.** $1/(\sqrt{1 - x^2}\arcsin x)$
61. (i) $\ln(u + \sqrt{u^2 - 1})$ **67. (ii)** $\frac{|\sin x|}{\sin x}\cos x = \cot x|\sin x|$

Exercises 4.1 (p. 234)

1. $36\pi \approx 113$ in/sec **3.** $\frac{30}{\sqrt{39}} \approx 4.8$ in/sec (downward), or $\frac{5}{2\sqrt{39}}$ ft/
sec **5.** $\frac{5}{3}$ in^2/sec **7.** $\frac{103}{\sqrt{101}} \approx 10.25$ m/sec **9.** 3 ft^3/min **11.** 3
min **13.** $\frac{28}{\sqrt{365}} \approx 1.4656$ m/sec (upward) **15. (i)** 0.88 rad/
sec ≈ 8.4 rpm **(ii)** 0.44 rad/sec ≈ 4.2 rpm **(iii)** $30\sqrt{2} \approx 42.4$ mph
(iv) $\frac{200}{11}\pi \approx 57.1$ mph

Exercises 4.2 (p. 247)

1. abs max at $x = \frac{\pi}{2} + 2n\pi$, abs min at $x = \frac{3\pi}{2} + 2n\pi$ **3.** abs max
at $x = (2n + 1)\pi$, abs min at $x = 2n\pi$ **5.** abs max at $x = 0$
7. abs min at $x = 0$ **9.** abs min at $x = 0$ **11.** $-\frac{1}{2}$ **13.** $\frac{1}{\sqrt{3}}$
15. 0 **17.** $\frac{\pi}{2}$ or $\frac{3\pi}{2}$ **19.** $\frac{1}{3}$ **21.** always increasing **23.** always
increasing **25.** increasing if $x \geq 2$, decreasing if $x \leq 2$
27. increasing on $(-\frac{\pi}{2}, \frac{\pi}{2}), (\frac{\pi}{2}, \frac{3\pi}{2}), (-\frac{3\pi}{2}, -\frac{\pi}{2}), (\frac{3\pi}{2}, \frac{5\pi}{2}), \ldots$
29. always increasing **31.** $\frac{a + b}{2}$ **33.** 0 **35. (i)** $x^4 - 2x^2 - 3x$ **(ii)**
$\frac{1}{5}x^5 - \frac{3}{4}x^4 + x^2 - x$ **37. (i)** $2x^4 - 2x^3 + 2x^2 - 3x$ **(ii)** $\sin(x) + 5$
39. $\frac{a}{4}$ **43.** not always **47. (ii)** above **49.** abs min $x \approx 2.141$,
local min $x \approx -0.432$, local max $x \approx 0.541$ **51.** abs max
$x \approx 2.455$, abs min $x \approx 5.233$ **53.** $\approx 0.7763, 1.8263$

Exercises 4.3 (p. 254)

1. ± 2 **3.** $n\pi, n = 0, \pm 1, \pm 2, \ldots$ **5.** none **7.** $(2n + 1)\frac{\pi}{2}$
9. $\pm 1, 0$ **11.** 0, not any kind of extremum **13.** 0, not any
kind of extremum **15.** 0, local max, ± 2 abs min **17.** none
19. 0 abs min **21.** -1 local max, 0 abs min, 2 abs max
23. -3 abs min, 0 not any kind of extremum, 3 abs max
25. -3 local min, -2 abs max, 2 abs min **27.** 0 abs min, $\frac{\pi}{2}$ abs
max, $\frac{3\pi}{4}$ local min **29.** 0 abs min, 3 abs max **31.** $\frac{a + b}{2}$, abs min
35. (ii) $x, x^3 + 3x$

Exercises 4.4 (p. 261)

1. abs min at 1 **3.** $-\sqrt{2}$ local max, $\sqrt{2}$ local min **5.** -1 abs
max, 1 abs min **7.** $-\sqrt{2}$ abs max, $\sqrt{2}$ abs min **9.** ± 2 abs
min, 0 local max **11.** $\frac{3}{2}$ abs min **13.** $(2n + 1)\frac{\pi}{2}$ abs max, $n\pi$ abs
min **15.** $n\pi$ abs max, $(2n + 1)\frac{\pi}{2}$ abs min **17.** $-\frac{\pi}{2}$ abs min
19. 0 abs min **21. (i)** every point abs max and abs min **(ii)**
$t > 0$ abs max, $t < 0$ abs min **23. (i)** none: $a^2 - 3b < 0$, one:
$a^2 - 3b = 0$, two: $a^2 - 3b > 0$ **25.** not differentiable:
a, c, e, g, i, k, m; jumps: e, i, k; abs max: a, e, i, k; abs min c, g;
local max f; $(k, m]$ local max and local min **27.** $x = 0$ abs min if
k even, no extremum if k odd

Exercises 4.5 (p. 275)

1. $x = 1, y = x + 1$ **3.** $y = 1, x = \pm\sqrt{3}$ **5.** $y = x, x = \pm 2$
7. $y = x - 1, x = -1$ **9.** $x = 3, x = -2$

11.

13.

15.

17.

19.

21.

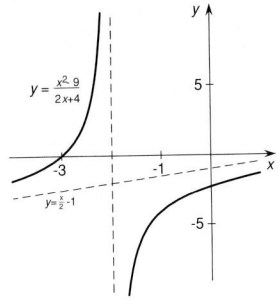

23. if $k = 0$, equal; $k > 0$ even, equal but infinite   **25. (i)** $a_k = 0$, for every *odd* k **(ii)** $a_k = 0$, for every *even* k **27. (i)**
$(3x^4 - x^2 - 1) + (-2x^3 + 7x)$

Exercises 4.6 (p. 289)

1. always concave up   **3.** concave up for $x > 0$, down for $x < 0$
5. concave up $(-\infty, -\frac{2}{\sqrt{3}}) \cup (\frac{2}{\sqrt{3}}, \infty)$, down $(-\frac{2}{\sqrt{3}}, \frac{2}{\sqrt{3}})$ **7.** concave
up $(\frac{\pi}{4} + n\pi, \frac{3\pi}{4} + n\pi)$, down $(-\frac{\pi}{4} + n\pi, \frac{\pi}{4} + n\pi)$ **9.** concave up
$(-\infty, -\frac{1}{\sqrt{2}}) \cup (\frac{1}{\sqrt{2}}, \infty)$, down $(-\frac{1}{\sqrt{2}}, \frac{1}{\sqrt{2}})$ **11.** 0   **13.** 1
15. $(2n + 1)\frac{\pi}{2}$ **17.** $n\frac{\pi}{2}$ **19.** 0   **21.** $-\sqrt{2}$ local max, $\sqrt{2}$ local
min, 0 inflection point

23. 0 infl pt.

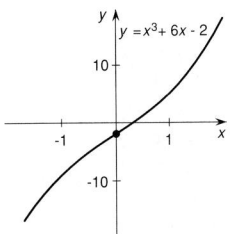

25. ± 1 abs min, 0 local max $\pm\frac{1}{\sqrt{3}}$ infl pts.

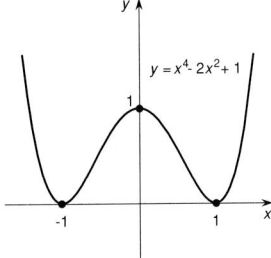

27. 0 local max, 2 local min

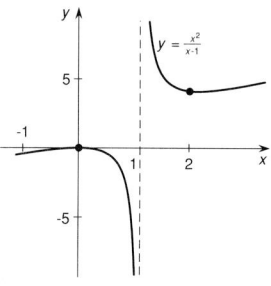

29. no extrema or infl pts.

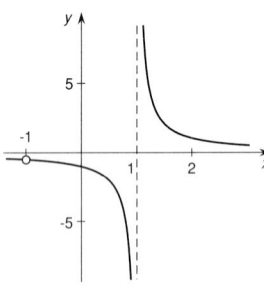

31. 0 max, $\pm\frac{1}{\sqrt{2}}$ infl pts.

33. 0 infl pt.

35. $\frac{1}{e}$ min

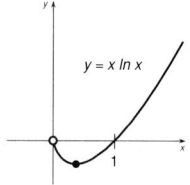

37. $k > 2$ odd   **39. (i)** infl pt. **(ii)** infl pt. **(iii)** cusp **(iv)** neither
41. -0.180 **43. (iii)** $(-\infty, 1.5)$

Exercises 4.7 (p. 300)
1. (i) $\frac{\sqrt{3}-1}{2}$ **(ii)** ≈ 1.196 rad $\approx 68.53°$ **3. (i)** $\frac{500,000}{116} \approx 4310$ ft **(ii)**
$t = \frac{4}{\sqrt{3}} \approx 2.309$ min   **5.** $(3 - \frac{1}{\sqrt{3}})$ km on land   **7.** radius
$(\frac{12}{\pi})^{1/3} \approx 1.563$ in, height $4(\frac{12}{\pi})^{1/3} \approx 6.253$ in   **9. (ii)** radius
$2(\frac{12}{\pi})^{1/3} \approx 3.126$ in, height $(\frac{12}{\pi})^{1/3} \approx 1.563$ in   **11. (i)** one-third of
the way from the point directly below A to the point directly
below B **13. (i)** both $\frac{\pi}{4}$ (45°) **(ii)** $\theta = \arctan \frac{4}{3} \approx 0.927$
rad $\approx 53.1°$, $\phi = \frac{\pi}{2} - \theta \approx 0.644$ rad $\approx 36.9°$
15. $(1 + 3(12^{-1/3}))\sqrt{4 + 12^{2/3}} \approx 7.0235$ yd   **17.** cost increases by
$350   **19.** vol $\frac{4}{27}\pi r^2 h$; radius $\frac{2}{3}r$, height $\frac{1}{3}h$ **21.** 100/
$(1 + 4^{1/3}) \approx 38.65$ m from weaker bulb   **23.** $\frac{128,000\pi}{9\sqrt{3}} \approx 25,796$
25. (i) all hemisphere, no cylinder, radius $10\sqrt{\frac{10}{\pi}} \approx 17.84$ ft **(ii)**
radius $10\sqrt{\frac{10}{3\pi}} \approx 10.30$ ft, height of cylinder $20\sqrt{\frac{10}{3\pi}} \approx 20.60$ ft
27. (i) 4 **(ii)** 7 **(iii)** \$499.99   **29.** $\theta \approx 1.4701$

Exercises 4.8 (p. 312)
1. $2x^3 - 2x^2 + 3x + c$ **3.** $-\cos x + c$ **5.** $\arcsin x + c$
7. $\frac{2}{3}x^{3/2} + \frac{1}{5}x^5 + c$ **9.** $6\sqrt{x} + c$ **11.** $-\frac{1}{4}\cos(2x^2 - 3) + c$
13. $\frac{1}{2}\exp(x^2 + 2x - 5) + c$ **15.** $\frac{1}{2}\ln(x^2 + 1) + c$
17. $\frac{1}{6}(x^4 + 4)^{3/2} + c$ **19.** $\ln(e^x + 1) + c$ **21.** $\frac{1}{4}x^4 - x^2 + 3x + 5$
23. $x^3 - 2x^2 + x + 1$ **25.** $x + c$ for $x > 0$, $-x + d$ for $x < 0$
27. $\ln(x - 3) + c$ for $x > 3$, $\ln(3 - x) + d$ for $x < 3$ **29.** $\tan x$
$+ c_n$, for $x \in (n\pi - \frac{\pi}{2}, n\pi + \frac{\pi}{2})$, i.e., a different constant on each
interval   **31.** $-\ln|\cos x| + c_n$, for $x \in (n\pi - \frac{\pi}{2}, n\pi + \frac{\pi}{2})$, i.e., a
different constant on each interval   **33. (i)** $4000\sqrt{2} \approx 5657$ **(ii)**
$\frac{\ln 5}{\ln 2}(40) \approx 92.88$ min   **35.** $\tan x + c_n$, for $x \in (n\pi - \frac{\pi}{2}, n\pi + \frac{\pi}{2})$,
with $c_0 = 4$ **37.** $c_1 \exp(\frac{1+\sqrt{5}}{2}t) + c_2 \exp(\frac{1-\sqrt{5}}{2}t)$
39. $\frac{1}{14}e^{4x} + c_1 e^{2x} + c_2 e^{-3x}$ **41. (ii)** $\frac{1}{2}\ln(1 + x^2)$
43. $t = \frac{1}{3}\ln 2 \approx 0.231$

Exercises 4.9 (p. 318)
1. $30t - 6$ **3.** $2t^3 - 2t^2 - 13t + 15$ **5.** 69 ft   **7.** $\frac{20}{7}$ sec   **9.** 3
sec   **11. (i)** after $\frac{15}{4}$ sec **(ii)** $\frac{225}{4}$ ft   **13.** $2\sqrt{(13.5)(4.9)} \approx 16.27$
m/sec   **15. (i)** yes **(ii)** $\frac{2992}{15} \approx 199.47$ ft   **17. (i)** 96 ft **(ii)**
$48 + 24\sqrt{5} \approx 101.67$ ft   **19.** $\frac{1}{16}v_0^2 \cos\theta \sin\theta$ **21.** 12,000 m
23. $\approx 1.42737 \times 10^{16}$ ft³/sec² $\approx 1.25673 \times 10^{12}$ mi³/hr²   **25. (i)**
$t^5 - 4t^3$ **(ii)** $t \approx 2.1655$

Exercises 4.10 (p. 324)
1. $8\frac{1}{16}$ **3.** $8 - \frac{1}{16}(0.001) \approx 7.999\,937\,5$ **5.** 7.85
7. $\frac{1}{\sqrt{2}}(1 + \frac{\pi}{180}) \approx 0.7194$ **9.** $-1 - \frac{\pi}{90} \approx -1.0349$ **11.** 0.0099
13. $-\frac{1}{(x + 1)^2}dx$ **15.** $\frac{\sin x - x\cos x}{\sin^2 x}dx$ **17.** $((y^2 - 2xy)/(x^2 - 2xy))\,dx$
19. $((\frac{x\sin x + \cos x}{x^2}y^2 - xy\cos x - y\sin x - 1)/(x\sin x + 2y\frac{\cos x}{x}))\,dx$
21. (i) 2.04   **23. (i)** 4800 in³   **25.** $(\frac{6}{\pi})^{1/3}(1.01) \approx (\frac{6.18}{\pi})^{1/3} \approx 1.253$
in   **27. (i)** -1.7 **(iii)** ≈ -1.6398

Exercises 4.11 (p. 329)
1. 4.3589   **3.** 7.2801   **5.** 0.328 268 856   **7.** 0.496 292 071
9. 1.764 542 338   **11.** 0.848 062 079   **13.** $-1.039\,072\,26$
15. $-0.547\,457\,039$, 2.118 253 365, etc.
17. 1.412 743 866, $-1.166\,290\,826$ **19.** 8.062 257 748
21. 2.038 691 840   **23.** $x_n = \frac{1}{2^{n-1}}x_1$ **25.** $\approx -0.152\,679\,653$

29. (ii) $\pm 2.028\ 757\ 838$ **(iv)** $\pm 1.076\ 873\ 986$
(v)

Exercises 4.12 (p. 338)

1. 2 **3.** $\frac{32}{7}$ **5.** $\frac{1}{2}$ **7.** 1 **9.** 1 **11.** ∞ **13.** $-\infty$ **15.** ∞ **17.** 0
19. 0 **21.** 0 **23.** 0 **25.** 0 **27.** $-\frac{2}{\pi}$ **29.** does not exist
31. (i) $-\frac{1}{8}$ **(ii)** 1 **33.** 0

Exercises 4.13 (p. 341)

1. 100 **3.** motors \$100 profit, generators \$460 loss **5. (i)** 225
(ii) $\approx \$1.28$ **(iii)** $\approx \$1.42$ (at 179 units) **(iv)** at best they will lose
\$212.50

Chapter 4 Review Exercises (p. 342)

1. $\frac{1}{8\pi} \approx 0.0398$ cm/sec **3.** 1.125 ft/sec **5.** 0.002 m^2/hr
7. $\frac{27}{\sqrt 5} \approx 12.075$ knots **9.** $\frac{27}{2}\pi \approx 42.41$ in^3/sec **11.** incrg. $[2, \infty)$,
decrg. $(-\infty, 2]$ **13.** incrg. $(-\infty, -\sqrt 3) \cup [\sqrt 3, \infty)$, decrg.
$[-\sqrt 3, \sqrt 3]$ **15.** decrg. $(-\infty, -1]$, incrg. $[-1, \infty)$ **17.** decrg.
$[0, \infty)$, incrg. $(-\infty, 0]$ **19.** incrg. $[\frac{1}{e}, \infty)$, decrg. $(0, \frac{1}{e}]$ **21.** 0, no
maxima/minima **23.** 0 abs min **25.** 3, -1 abs min, 1 local
max **27.** no critical points **29.** 0 abs min **31.** no extrema
33. -1 abs max, 1 abs min **35.** 0 abs min, $\frac{3}{2}$ abs max, 2 local
min **37.** 1 abs max, abs min at every point in $[-3, 0]$
39. $-1 - \frac{1}{\sqrt 2}$ local max, $-1 + \frac{1}{\sqrt 2}$ abs min **41.** 2 abs max, -2
abs min; 0, $\pm 2\sqrt 3$ infl pts., horizontal asymptote $y = 0$

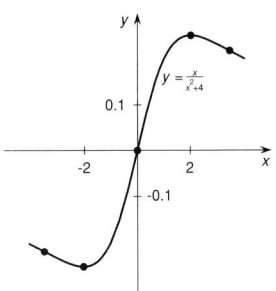

43. 0 local min, -6 local max, vert. asymptote at $x = -3$,
asymptote $y = x - 3$ at $\pm \infty$

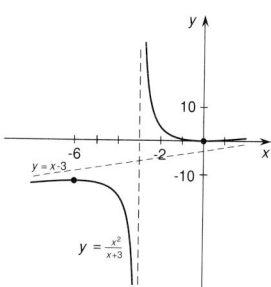

45. $n\pi$ abs max, $(2n + 1)\frac{\pi}{2}$ abs min, $(2n + 1)\frac{\pi}{4}$ infl pt.

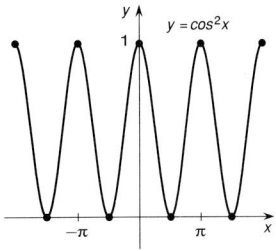

47. vert. asymptote $x = 0$, asymptote $y = x$ at $\pm \infty$

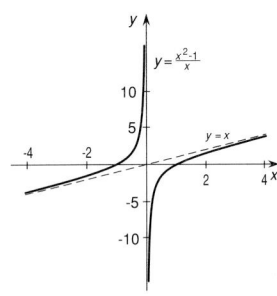

49. vert. asymptotes $x = \pm 3$, horiz. asymptote $y = 0$, infl pt.
$x = 0$

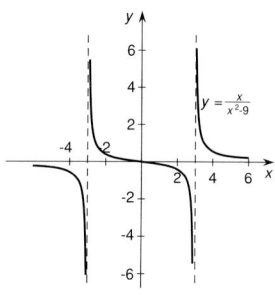

51. $\frac{1}{5}x^5 - x^3 + 2x^2 - x + c$ **53.** $\frac{12}{13}x^{13/4} + \frac{3}{2}x^2 - 4x + c$, for $x > 0$
55. $\frac{2}{3}(x + 7)^{3/2} + c$, for $x > -7$ **57.** $-\frac{1}{5}\cos(5x) + c$
59. $-\frac{1}{3}x^{-3} + x^2 + c$, if $x < 0$, $-\frac{1}{3}x^{-3} + x^2 + d$, if $x > 0$ **61.** 16
sec **63.** $16(2.6)^2 \approx 108$ ft **65.** $\frac{1210}{9} \approx 134$ ft **67.** $\frac{225}{19.6} \approx 11.48$ m
69. $(154)^2/1350 \approx 17.57$ ft/sec^2 **71.** $9 - \frac{1}{36} \approx 8.9722$
73. $0.025 - \frac{1}{160,000} \approx 0.024\ 993\ 75$
75. $1024 - \frac{4}{150} \approx 1023.973\ 333$ **77.** $\frac{1}{2} + \frac{\pi\sqrt 3}{1800} \approx 0.503\ 023$
79. $\sqrt 2 - \frac{\pi\sqrt 2}{6000} \approx 1.413\ 473\ 1$ **81.** $6.403\ 124\ 237$
83. $3.419\ 951\ 893$ **85.** $1.373\ 400\ 767$ **87.** $0.539\ 188\ 873$ or
$1.675\ 130\ 871$ or $-2.214\ 319\ 743$ **89.** $\pm 0.765\ 366\ 865$ or
$\pm 1.847\ 759\ 065$ **91.** 0 **93.** $-\frac{3}{4}$ **95.** 0 **97.** $-\infty$ **99.** 0
101. (i) $100\pi \approx 314.159$ ft/sec **(ii)** $164\pi \approx 515.22$ ft/sec

103. ≈ 66.15 in^3
105.

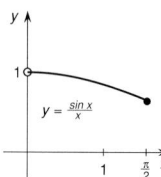

107. $\frac{3}{2}x^2 - 2x + c$ if $x > 0$, $\frac{3}{2}x^2 - 2x + c'$ if $x < 0$ **111. (i)**
$dy = (\sin x + x \cos x)\,dx$ **(ii)** $((2xy^3 - 12x^2y)/(4x^3 - 3x^2y^2))\,dx$ **(iii)**
$(-1 - \tan(x + y))\,dx$ **113.** $100 - \ln 10 \approx 97.697$ **115.** 800
121. $\frac{1}{18}$ rad/sec $= \frac{10}{\pi} \approx 3.183°$/sec ≈ 0.5305 rpm
123.

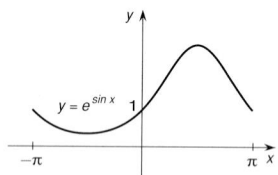

Exercises 5.1 (p. 351)
1. $\frac{14}{27} \approx 0.519$ **3.** $\frac{11}{25} = 0.44$ **5.** $\frac{33}{25} = 1.32$ **7.** $\frac{27}{2}$ **9.** $\frac{739}{216} \approx 3.421$
11. $\frac{\pi}{2}$ **13.** $\frac{\pi}{2}$ **15.** $3 + \sqrt{2} + \sqrt{3} \approx 6.146$ **17.** 0 **19.** 2π
21. $\frac{1}{3}$ **23.** $\frac{7}{3}$ **25. (i)** 11 **(iii)** 9 **27.** less **29.** $5, \frac{15}{4}, \frac{77}{25}, \frac{6767}{2500}$
31. $\approx 1.341, 1.095, 1.016, 1.004$

Exercises 5.2 (p. 361)
1. 225 **3.** 70 **5.** 39 **7.** 121 **9.** 650 **11.** 7 **13.** $\frac{m(m-1)(2m-1)}{6}$
15. $\frac{(m+1)(m+2)(2m+3)}{6} - 1$ **17.** $\frac{1}{3}(2m+1)(m^2 + m + 3)$ **19.** 8
21. $\frac{n^2 + n - m^2 + m}{2}$ **23.** $\frac{n^2 + n - m^2 + m}{2}$ **25. (i)** -3 **(ii)** 4 **27.** 1 if n
even; 0 if n odd **29.** $(a + b)^n = \sum_{k=0}^{n} \binom{n}{k}a^{n-k}b^k$ **31. (i)** 90 **(ii)** 90
37. 0.083 325 **39.** 1.017 120 611

Exercises 5.3 (p. 371)
1. $\frac{3}{8}$ **3.** $\frac{7}{32}$ **5.** 18 **7.** -10 **9.** $-\frac{3}{8}$ **11.** $\frac{\pi}{2}$ **13.** $\frac{4}{5}$ **15.** 0
17. $\frac{x^2}{2}$ **19.** $1 + e + e^2$ **25. (i), (ii)** all zero **(iii)** $-1, -12, -3584$
27. $\|\mathcal{P}_3\| = \frac{1}{4}, \|\mathcal{P}_4\| = \frac{1}{12}$ **31.** ≈ 0.9266 **33.** ≈ 0.373 **35. (ii)**
$\frac{1}{m+1}$

Exercises 5.4 (p. 381)
1. $\frac{3}{2}$ **3.** 6 **5.** $25\frac{1}{2}$ **7.** $16\frac{1}{3}$ **9.** 8 **11.** -1 **13.** 4 **15.** -6
17. $18\frac{1}{3}$ **19.** -15 **29.** $c = 2, d = 3$ **31.** $b = 3, c = 2, d = -1$
39. $\frac{196}{81} \approx 2.4198$ **41.** 0.69011 **43.** 1.71828

Exercises 5.5 (p. 390)
1. $x^2 - 1$ **3.** $\sin x$ **5.** e^t **7.** $e^x - x^2$ **9.** 7 **11.** $\cos(x^2)(2x)$
13. e^{1-x} **15.** $\sin(1 - u)$ **17.** $(\sin^2 r - 1)\cos r + (\cos^2 r - 1)\sin r$
$= -(\cos^3 r + \sin^3 r)$ **19.** 0 **21.** incrg $[-2, 0] \cup [2, \infty)$, decrg
$(-\infty, -2] \cup [0, 2]$ **23.** $-\frac{1}{2}$ local max, $\frac{1}{2}$ local min

Exercises 5.6 (p. 397)
1. $\frac{12}{5}$ **3.** 12 **5.** 2 **7.** 0 **9.** $e - 1$ **11.** $\frac{1}{2}(e^2 - 1)$ **13.** 0 **15.** $\frac{2}{\pi}$
17. $\frac{16}{3}$ **19.** $\frac{1}{36}(4^{18} - 3^{18}) = 1{,}898{,}112{,}673\frac{19}{36}$ **21.** $\frac{5}{2}$ **23.** $\frac{23}{3}$
25. $36\frac{1}{3}$ **27.** $11\frac{1}{12}$ **29.** $\frac{2}{3}$ **31. (i)** 0 **(ii)** $2(e^{1/2} - e^{-1/2})$ **(iii)** $-\frac{2}{\pi}$
33. (i) $\sinh(1)$ **(ii)** 0 **(iii)** $\tanh(2)$ **35. (i)** $10\frac{3}{4}$ **(ii)** $12\frac{1}{4}$ **(iii)** $8\frac{1}{4}$

37. (i) 1 **(ii)** $-\frac{5}{12}$ **(iii)** $2 + \sqrt{3}$ **43.** $-\frac{2}{3}$ **47.** ≈ 65.2864
49. $\approx 1.570\ 762\ 993$

Exercises 5.7 (p. 408)
1. $\frac{1}{4}e^{4x-2} + c$ **3.** $\frac{1}{24}(t^3 - 2)^8 + c$ **5.** $\frac{33}{5}$ **7.** $-\frac{2}{5}$
9. $\frac{1}{3\pi}\sec^3(\pi x) + c$ **11.** $\frac{\pi}{8} - \frac{1}{4}$ **13.** π **15.** $\frac{\pi}{2} + \frac{1}{12}\sin(6x) + c$ **17.** $\frac{1}{2}$
19. $\frac{1}{3}\ln(\frac{26}{7})$ **21.** $\sqrt{1 + x^2} + c$ **23.** $\frac{1}{(\sqrt{x} + 2)^2} + c$ **25.** $\frac{39}{14}2^{1/3}$
27. $\frac{1}{a}\sin(at) + c$ **29.** $\frac{1}{9}(2x^3 + 3x^2 + 2)^{3/2} + c$ **31.** $\frac{1}{3}\arctan(\frac{t}{3}) + c$
35. (i) 0 **(ii)** π **37. (ii)** $\frac{1}{6}\cos^6 x - \frac{1}{4}\cos^4 x + c$ **(iii)** $\frac{1}{3}\cos^3$
$x - \cos x + c$ **(iv)** $\frac{1}{7}\cos^7 x - \frac{1}{5}\cos^5 x + c$ **(v)** $\frac{1}{3}\sin^3 x - \frac{1}{5}\sin^5 x + c$
41. 0.636 567 **43.** 0.231 573

Chapter 5 Review Exercises (p. 410)
1. $\frac{15}{4}$ **3.** $\frac{81}{4}$ **5.** $\frac{37}{60}$ **7.** 49 **9.** 11 **11.** $\ln(6!) = \ln(720) \approx 6.579$
13. $\frac{31}{16}$ **15.** $36x$ **17.** $\frac{m(m-1)(2m-1)}{6}$ **19.** 0 **21.** 34 **23.** -20
25. 0 **27.** 0 **29.** $\frac{25}{12}$ **31.** $x^3 - 2x + 1$ **33.** -2
35. $\frac{4}{3}x^3 - 2x + \frac{2}{3}$ **37.** xe^x **39.** $-x$ **41.** $\frac{205}{4}$ **43.** $-\frac{7}{2}$ **45.** 2
47. 1 **49.** $\frac{216}{7}$ **51.** $\frac{1}{6}(1 + 2x^2)^{3/2} + c$ **53.** $\frac{191}{30}$ **55.** $\frac{2}{3}x^{3/2} - x + c$
57. $2\exp(\sqrt{t}) + c$ **59.** $x - 2\ln|x + 1| + c$ **61.** $\frac{n+1}{2}$ **63.** $\frac{1 - r^{k+1}}{1 - r}$
65. $\frac{\pi}{2}$ **67.** 0 local max, 1 local min, -2 (abs) min; $-\frac{1}{3} \pm \frac{\sqrt{7}}{3}$ infl
pts **69.** $-e^2$ local max; $-e$ infl pt **73.** $\frac{\pi}{8} - \frac{1}{4}$ **75. (i)** $\frac{17}{3}$ **(ii)** 0
(iii) $\frac{17}{3}$ **79.** $-\sqrt{3}$

Exercises 6.1 (p. 423)
1. $\frac{2}{3}$ **3.** $\frac{17}{4}$ **5.** 1 **7.** 1 **9.** $\ln(2) \approx 0.69315$ **11.** $\frac{4}{3}$
13. $\frac{8}{3}\sqrt{2} \approx 3.77124$ **15.** 8 **17.** $\frac{256}{5}$ **19.** 9 **21.** 8 **23.** $\frac{2}{3}$
25. $\frac{4}{3}$ **27.** $\frac{9}{2}$ **29.** $\frac{9}{2}$ **31. (i)** 1 **(ii)** 1 **33.** $\frac{27}{4}$ **35.** 9 **37. (i)**
$\frac{1}{4}t^4 - 2t^2 + 4$, if $t \in [-2, 0]$; $-\frac{1}{4}t^4 + 2t^2 + 4$, if $t \in [0, 2]$ **(ii)** 1
39. 8 g **43.** ≈ 1.895, area $\doteq 0.421$

Exercises 6.2 (p. 430)
1. $\frac{1}{3}$ **3.** $\frac{32}{3}$ **5.** 24 **7.** $\frac{1}{2}(e - 1)$ **9.** $\frac{1}{2}$ **11.** 23 **13.** -3
15. $-\frac{74}{3}$ **17.** $\frac{\pi}{2}$ **19.** $\frac{1}{2}(e^2 - e)$ **21.** $\frac{\pi}{2} + \frac{1}{3}$ **23.** $4\sqrt{3} - 4 - \frac{\pi}{3}$
25. (ii) $2(\frac{1}{n+1} - \frac{1}{m+1})$, if $m > n$ and $m - n$ is even; $(\frac{1}{n+1} - \frac{1}{m+1})$,
if $m > n$ and $m - n$ is odd **27.** 3 **29.** $\frac{3}{2}$ **31.** $\frac{1}{2}|ad - bc|$ **33. (i)**
$15 + \frac{24}{\pi}(1 + \frac{1}{\sqrt{2}}) \approx 28.0°$C **(ii)** $15 + \frac{24}{\pi} - \frac{24}{\pi}\cos(\frac{17\pi}{12}) \approx 24.617°$C
35. 25 in^3 **37.** $23\frac{1}{3}$ m **39. (i)** $25e^{-2} \approx 3.3834$ ft^3 **(ii)**
$\ln 25 \approx 3.2189$ min **41.** bh **43.** ≈ 5.010 **45.** $\approx 1.330\ 78$
47. $\approx 2.114\ 216$

Exercises 6.3 (p. 438)
1. 24π

3. $\frac{8\pi}{3}$

5. $\frac{\pi}{2}(e^2 - 1)$

7. $\frac{\pi}{30}$

9. 12π

11. π

13. $\frac{1}{3}\pi^4 - \frac{1}{2}\pi^2$

15. $\frac{56}{15}\pi$

17. $3xh$

19. $\pi|M^2 - N^2|L$

21. $\frac{2\pi}{35}$

23. $\frac{72\pi}{5}\sqrt{3}$

25. $\frac{8\pi}{117}$

27. $\frac{8\pi}{3}$

29. $\frac{4\pi}{3}$

31. 16π

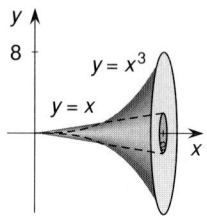

33. $\frac{8}{3}\pi^2 - 2\pi\sqrt{3}$ **35.** **(i)** doughnut ("solid torus") **(ii)** $4\pi^2$
37. $\frac{4}{3}\pi ab^2$
39.

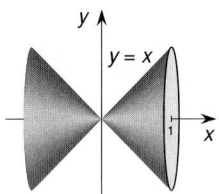

41. vol ≈ 183.511 **43.** 1.769

Exercises 6.4 (p. 445)
1. $\frac{29\pi}{21}$ **3.** $\frac{8\pi}{5}$ **5.** 8π **7.** $\frac{14\pi}{15}$ **9.** $\frac{332}{7}\pi$ **11.** $\frac{2809}{12}\pi$ **13.** $\frac{4\pi}{3}$
15. $\frac{5\pi}{24}$ **17.** $\pi\left(-\frac{13}{42} + \frac{9}{7}(3^{1/3}) + \frac{3}{2}(3^{2/5})\right) \approx 12.166$ **19.** $\frac{\pi}{3}(2 - \sqrt{2})$
21. $\frac{16\pi}{15}$ **23.** $\frac{2\pi}{3}$ **25.** $\frac{2\pi}{5}$ **27.** $\frac{256\pi}{5}$ **29.** π **31.** $\frac{4\pi}{15}$ **33.** $\frac{31\pi}{30}$
35. $\frac{128\pi}{15}$ **37.** 96π **39.** $\frac{7\pi}{6}$ **41.** $\frac{\pi}{3\sqrt{2}}$ **43.** $\frac{64}{3}\sqrt{2}\pi$
45. $\frac{\pi}{2m+1}a^{2m+1}$ **47.** $\frac{28\pi}{3} \approx 29.32$ oz   **51. (i)** $\frac{4\pi}{15}$ in^3 **(ii)** ≈ 0.232
oz   **53.** ≈ 2.598 **55.** ≈ 2.583

Exercises 6.5 (p. 454)
1. $\frac{4}{3}$ **3.** $e - 1$ **5.** $\frac{1}{4}\ln 5$ **7.** 0 **9.** $\frac{4}{\pi}$ **11.** $\frac{\pi}{4}$ **13.** 0 **15.** $\frac{2}{15\pi}$
17. 0 **19.** $\frac{2}{\pi}(e - 1)$ **21. (i)** 3 **(ii)** -7 **(iii)** C **23. (i)**
$\frac{2}{5}(e - e^{-3/2}) \approx 0.998$ in **(ii)** $\frac{\pi}{5}(e^2 - e^{-3}) \approx 4.611$ in^2 **25. (i)** $\frac{\pi}{4}$ **(ii)**
$\frac{2\pi}{3}$ **27. (i)** $\frac{5}{3}(e^{6/5} - 1) \approx 3.867$ g **(ii)** $\frac{20}{3}(e^{1/2} - e^{1/5}) \approx 2.849$ g
29. (i) $\frac{1}{3}$ **(ii)** 1 **31.** $\frac{880}{21} \approx 41.90$ mph   **35.** 0 **37. (i)** $\frac{A}{\sqrt{2}}$ **(ii)** $\frac{A}{\sqrt{2}}$
(iii) $A\sqrt{\frac{1}{2} - \frac{1}{\pi}}$ **39. (i)** $0.17(1 - e^{-10}) \approx 0.16999$ ppm **(ii)**
$0.34(1 - e^{-5}) \approx 0.3377$ ppm **(iv)**
$0.17(1 - e^{-10}) \times 1080\pi \times 10^{-6} \approx 0.000\,577$ in^3 **41.** $68°$
43. ≈ 0.3

Exercises 6.6 (p. 460)
1. 27 ft-lb   **3.** $\frac{6}{\pi} \approx 1.9099$ ft-lb   **5.** 0 **7.** 30 ft-lb   **9.** -96 ft-lb
11. 3 ft-lb   **13.** $\frac{1}{4}$ joule   **15.** 45 ft-lb   **17.** $\frac{9}{1280}$ ft-lb   **19.** $\frac{25}{6}$ ft-lb
21. $52,000\pi$ ft-lb   **23.** 1866.8π ft-lb   **25.** $\frac{1}{\sqrt{2}} \approx 0.707$ ft ≈ 8.485
in   **27. (i)** -16 lb **(ii)** 24 ft-lb   **29. (i)** $h \times 2.112 \times 10^7/$
$(h + 2.112 \times 10^7)$ ft-lb **(ii)** 2.112×10^7 ft-lb   **31. (i)**
$\left(\frac{4000}{4001}\right)^2 \approx 0.9995$ lb **(ii)** $\left(\frac{4000 \times 5280}{1 + 4000 \times 5280}\right)^2 \approx 0.999\,999\,905$ lb   **33. (i)**
18 ft-lb **(ii)** $12\sqrt{2}$ ft/sec   **35. (i)** $\frac{5}{4}$ ft-lb **(ii)** 10 ft-lb
37. $\frac{15}{2} - \frac{1}{14}\sqrt{105^2 - 4(1875)} \approx 3.259$ m
39. (i) $9800\pi\left(\frac{2Rd^3}{3} - \frac{d^4}{4}\right)$J **(ii)** ≈ 1.979 m

Exercises 6.7 (p. 474)

1. 7   **3.** $x = 0$ **5.** $\frac{27}{16}$ **7.** $(2 - \frac{\pi}{2})/\ln(2) \approx 0.6192$ **9.** $\frac{7}{2}$
11. $(\frac{4}{3\pi}, 0)$ **13.** $(0, -\frac{4\sqrt{2}}{3\pi})$ **15.** $(-2, 2)$ **17.** $(0, \frac{12}{5})$ **19.** $(\frac{9}{20}, \frac{9}{20})$
21. 1   **23.** $\frac{3L}{4}$ from the light end, i.e., $\frac{3}{4}$ of the way along the
rod   **25. (ii)** $\frac{5}{3}$ **(iv)** $\frac{18}{11}, \frac{8}{5}$ **27.** $(\frac{2}{12+3\pi}, 0) \approx (0.0933, 0)$
29. $(0, \frac{28}{9\pi})$ **31.** $(0, \frac{212+192\pi}{72+27\pi}) \approx (0, 5.198)$ **33.** $(0,0)$ **35.** $(2,1)$
37. $(0,0)$ **39.** at a distance of $\frac{3}{5}$ from center of larger circle, on
the segment joining the two centers   **41.** $\frac{2}{3}$ of the radius from
the center of the heavy disk, on the segment joining the two
centers   **43.** $(\frac{9}{10}, \frac{9}{10})$ **45.** $(\frac{3}{2}, \frac{5}{12})$ **47.** $6\pi^2$ **49. (i)** $3\pi^2$ **(ii)**
$2\pi^2 + \frac{4\pi}{3}$

Exercises 6.8 (p. 481)

1. $99(62.4) = 6177.6$ lb   **3.** $\frac{25}{24}(62.4) = 65$ lb   **5.** 27,300 lb
7. $\frac{25}{128} \times 62.4 \times 59 \approx 719$ lb   **9.** 5616π lb   **11.** 7800π lb
13. 1622.4π lb   **15.** $16\sqrt{3} \times 62.4 \times (100 - \frac{4}{\sqrt{3}}) \approx 168,934$ lb
17. $18.65 \times 62.4 \times 14 \approx 16,287$ lb   **19.** 3369.6 lb
21. $9(62.4)(10 - \frac{3}{2\sqrt{2}}) \approx 5020$ lb   **23.** 5616π lb
25. $62.4(72\sqrt{2}) \approx 6354$ lb   **27.** $62.4\sqrt{3}(100 - \frac{1}{\sqrt{6}}) \approx 10,764$ lb
29. $\frac{567}{16}(62.4)\sqrt{3} \approx 3830$ lb   **31.** $16(62.4)(40 - 2\sqrt{2}) \approx 37,112$ lb
33. 15,600 lb   **37. (i)** $\frac{1500}{624} \approx 2.4$ ft **(ii)** $\frac{250}{39} \approx 6.41$ ft **(iii)** 199.68
tons   **41. (i)** 208 lb **(ii)** 41.6 lb   **43.** both 78,400 N

Exercises 6.9 (p. 492)

1. $\frac{8}{27}(10^{3/2} - (\frac{11}{2})^{3/2}) \approx 5.548$ **3.** $4\sqrt{2} \approx 5.657$
5. $\frac{8}{27}(10^{3/2} - (\frac{11}{2})^{3/2}) \approx 5.548$ **7.** $\frac{8}{27}(10^{3/2} - (\frac{13}{4})^{3/2}) \approx 7.634$
9. $\frac{8}{27}(10^{3/2} - (\frac{13}{4})^{3/2}) \approx 7.634$ **11.** $4\pi((\frac{29}{4})^{3/2} - (\frac{13}{4})^{3/2}) \approx 171.68$
13. $16\pi\sqrt{10} \approx 158.95$ **15.** $4\pi\sqrt{10} \approx 39.74$ **17.** 42π **19.** $2\pi rh$
21. $2\pi r$ **25.** $\pi r\sqrt{r^2 + h^2}(1 - \ell^2/h^2)$ **27.** 8π **29. (i)** $6\pi\sqrt{35}$ **(ii)**
$7\pi\sqrt{35}$ **31.** $\frac{4\pi}{3}((b + \frac{1}{4})^{3/2} - (a + \frac{1}{4})^{3/2})$ **33.** $\pi(2 + \sqrt{17})$
35. elliptical solid has $\approx 7.7\%$ greater area   **37.** ≈ 1.542
39. actual ≈ 2291.256; with $n = 10, \approx 2773$ **41.** with
$n = 10, \approx 25.039$; with $n = 20, \approx 23.974$ **43.** with
$n = 10, \approx 14.277$; with $n = 20, \approx 14.387$

Exercises 6.10 (p. 504)

1. $T_4 = 8$, $S_4 = 8$, integral $= 8$ **3.** $T_6 \approx 4.661\,488\,4$,
$S_6 \approx 4.666\,563\,1$, int $= 4\frac{2}{3}$ **5.** $T_4 \approx 1.896\,118\,9$,
$S_4 \approx 2.004\,559\,8$, int $= 2$ **7.** $T_8 \approx 1.628\,968$, $S_8 \approx 1.610\,847$,
int $= \ln 5 \approx 1.609\,438$ **9.** $T_4 \approx 0.780\,769\,2$, $S_4 \approx 0.807\,692\,3$,
int $= \frac{1}{2} \ln 5 \approx 0.804\,719\,0$ **11.** $n = 20$, $S_{20} \approx 204.800\,853\,3$,
int $= 204.8$ **13.** $n = 4$, $S_4 \approx 1.000\,134\,6$, int $= 1$ **15.** $n = 4$,
$S_4 \approx -0.693\,25397$, int $= -\ln(2) \approx -0.693\,147\,18$ **17.** $n = 4$,
$S_4 \approx 0.609\,418\,7$, int $= \frac{1}{3}(2\sqrt{2} - 1) \approx 0.609\,475\,7$ (might also
use $n = 6$)   **19.** $n = 12$, $S_{12} = 0$, int $= 0$ **21.** $n = 4$,
$S_4 \approx 0.386\,260$ **23.** $n = 2$, $S_2 \approx 0.632\,334$ **25.** $n = 6$, $S_6 = 0$
27. $n = 2$, $S_2 \approx 4.639\,202$ **29.** $n = 4$, $S_4 \approx 0.301\,101$
31. $n = 4$, $S_4 \approx 1.147\,782\,3$ **33.** $n = 6$, $S_6 \approx 0.881\,374\,6$
35. $S_6 \approx 1.271\,270\,7$ **37.** $S_6 \approx 1.111\,431\,2$
39. $S_{12} \approx 0.835\,650\,9$ **41. (i)** $n = 10$ **(ii)** ≈ 1.479 **43.** $n = 58$
45. $S_{20} \approx 3.820$ **51.** $S_4 \approx 0.693\,25$ **53. (i)** 0.5026 **(ii)**
$\approx 4.77208 \times 10^{-4}$ **55. (i)** $n = 8$ **(ii)** $n = 14$ **(iii)** 40 **(iv)**
$S_{10} \approx 3.141\,592\,614$ **59.** $S_{20} \approx 0.693\,147$ **61.** $\approx 4.456\,65$
63. $\approx 12.001\,51$ **65.** $\approx 13.266\,03$

Chapter 6 Review Exercises (p. 506)

1. $\frac{8}{3}$ **3.** $\frac{20\sqrt{5}}{3}$ **5.** $\frac{1}{2}$ **7.** $2\pi - 4$ **9.** $\frac{\pi}{2} + \frac{1}{3}$ **11.** $\frac{\pi}{2}$
13. $12\sqrt{7} + 32(\arcsin(\frac{3}{4}) - \arcsin(\frac{\sqrt{7}}{4})) \approx 35.7595$ **15.** 0
17. $9\pi + 16 \arcsin\frac{\sqrt{5}}{2\sqrt{3}} - 18 \arcsin\frac{2\sqrt{5}}{3\sqrt{3}} \approx 20.843$ **19.** $3\sqrt{3} + \frac{4\pi}{3}$
21. $\frac{1}{4}\pi^2$

23. 24π

25. $\frac{1}{8}\pi^2$

27. $\frac{1}{4}\pi^2$

29. 8

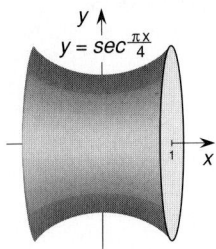

$y = \sec\frac{\pi x}{4}$

31. $\frac{13}{3}$ **33.** $\ln 2$ **35.** $-\frac{2}{\pi}$ **37.** 0 **39.** 1 **41.** 15 ft-lb
43. 0.1764 J **45.** 18 ft-lb **47.** 30 ft-lb **49.** $(62.4)400\pi$ ft-lb
51. $(0, 2)$ **53.** $(\frac{3}{2}, \frac{3}{2})$ **55.** $(\frac{8+3\pi}{6+3\pi}, \frac{8-3\pi}{6+3\pi})$ **57.** $(\frac{1}{2}, \frac{1}{2})$
59. $(\frac{2}{12+9\pi}, \frac{2}{12+9\pi})$ **61.** 7488 lb **63.** 4742.4 lb
65. $(54{,}000 - 13{,}500\sqrt{2})\pi \approx 109{,}667$ kg $\approx 1{,}074{,}737$ N
67. $2808\pi - 1123.2 \approx 7698$ lb
69. $1497.6\sqrt{3} + 1331.2\pi \approx 6776$ lb **71.** $27\sqrt{2}\pi$
73. $\frac{\pi}{6}(17\sqrt{17} - 27)$ **75.** 6π **77.** $18\sqrt{2}\pi$
79. $\frac{\pi}{27}((36{,}865)^{3/2} - 10^{3/2}) \approx 823{,}579.5$ **81.** $S_8 \approx 0.500\,244$
83. $S_4 \approx 1.000\,135$ **85.** $S_4 = 0$ **87.** $S_8 \approx 1.856\,295$ **89.** $S_n = 0$
(all n) **91.** $\frac{28\pi}{3}$ **93.** $(\pi, 0)$ **103.** 3π **105.** sphere:
$4\pi(\frac{3}{4})^{2/3} \approx 10.373$, other: $\frac{\pi}{3}(5^{3/2} - 1) \approx 10.661$
107. $S_{30} \approx 1.47894$, $S_{12} \approx 2.29559$, $S_{36} \approx 2.95789$

Exercises 7.1 (p. 513)
1. $a + b$ **3.** $a - 2b$ **5.** $ma + nb$ **7.** $-ma$ **9.** $\frac{a}{2}$ **11.** 0 **13.** 0
15. 0 **17.** 0 **19.** 1 **25.** ∞ **27.** (i) ≈ 11.06 (ii) ≈ 23.44 (iii)
≈ 29.40 **31.** $\approx 2.718\,281\,828$

Exercises 7.2 (p. 521)
1. 0 **3.** 0 **5.** 0 **7.** $\sqrt{5}x^{\sqrt{5}-1}$ **9.** 0 **11.** ∞
13. $\frac{1}{1+\sqrt{2}}x^{1+\sqrt{2}} + c$ **15.** $\frac{1}{\ln(2)}2^x + c$ **17.** $\frac{1}{2\ln 4}4^{x^2} + c$
19. $\frac{2}{\ln 10}10^{\sqrt{x}} + c$ **21.** 2 **23.** 2 **25.** $\frac{1}{x\ln(10)}$ **27.** $-\infty$
29. $\frac{\ln 3}{\ln 10}x + c$ **31.** e^3 **33.** $\frac{1}{e}$ **35.** e^{-4} **37.** e^{-2} **39.** 1 **41.** $2e^{2x}$
43. 2 **49.** 0 **51.** increasing everywhere **53.** (i) $(1 + \ln x)x^x$ (ii)
1 (iii) abs min at $\frac{1}{e}$
(iv)

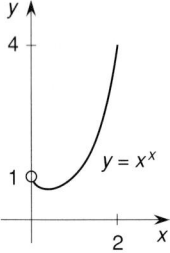

$y = x^x$

Exercises 7.3 (p. 528)
1. 12 **3.** 8 **5.** 128 **7.** $\frac{1}{9}$ **9.** $\frac{1}{32}$ **11.** 1 year
13. $-3\frac{\ln 2}{\ln(0.75)} \approx 7.228$ years **15.** $\frac{1}{2}$ year **17.** $\frac{1}{4}$ year
19. $-3\frac{\ln 2}{\ln(0.75)} \approx 7.228$ years **21.** $\frac{\ln(10)}{2\ln 2} \approx 1.661$ hr
23. $-3\frac{\ln(0.9)}{\ln 2} \approx 0.456$ year **25.** $-30\frac{\ln 2}{\ln(0.4)} \approx 22.694$ years
27. $(0.5)(7.4)^{-1/4} \approx 0.303$ g **29.** $\approx 7.9 \times 10^{29}$ g; 720 times the
volume of earth

Exercises 7.4 (p. 542)
1. $-3\ln(2)/\ln(0.2) = 3\ln(2)/\ln(5) \approx 1.292$ hr **3.** $-\ln(2)/$
$\ln(0.85) \approx 4.265$ days **5.** $-2\ln(2)/\ln(0.8) \approx 6.213$ hr
7. $\frac{\ln 2}{2800} \approx 2.476 \times 10^{-4}$ **9.** $\frac{\ln 2}{34} \approx 0.02039$ **11.** (i) $\ln(\frac{5}{4})/$
$\ln(\frac{161}{160}) \approx 35.8$ years after $3{,}200{,}000$ (ii) $-\ln(\frac{15}{16})/\ln(\frac{161}{160}) \approx 10.36$
years **13.** $\frac{\ln(2)}{h}$ **15.** $Ie^{-t\ln(2)/h}$ **17.** (ii) $-\frac{140}{\ln(2)}\ln(0.01) \approx 930$ days
19. $\approx 10.405\%$, $\approx 10.381\%$ **21.** $(e^{1/10} - 1) \times 100\% \approx 10.517\%$
23. $(e^r - 1) \times 100\%$ **25.** $\approx \$2246.99$ **27.** $\approx \$3245.86$ **29.** (i)
$\approx 8.243\%$ (ii) $\approx 8.300\%$ (iii) $\approx 8.328\%$ (iv) $\approx 8.328\,667\%$
(v) $\approx 8.328\,707\%$ **31.** $\approx 6.183\,131\%$ **33.** (i)
$\approx 7.250\,098\,3\%$, $7.250\,100\,3\%$ (ii) $\$500{,}000$ **35.** $\$181{,}782.73$
37. ≈ 16.1 years **39.** $100\ln(30) \approx 340$ days **41.** $\frac{1}{0.003946}\ln\frac{7}{2}$
≈ 317 days **43.** (i) $1.3e^{5\ln(0.9/1.3)} \approx 0.207$ g (ii) $25\ln(\frac{0.5}{1.3})/$
$\ln(\frac{0.9}{1.3}) \approx 65$ min **45.** $\approx 6.183\,632\,865\%$

Exercises 7.5 (p. 550)
1. $L = 4$, $P_0 = 2$, $c = \frac{1}{4}\ln 3$ **3.** $L = \frac{35}{3}$, $P_0 = 8$, $c = \frac{3}{35}\ln(\frac{11}{4})$
5. $L = 40{,}000$, $P_0 = 40{,}000/(1 + \exp(196\ln(\frac{5}{3})))$, $c = \frac{1}{400{,}000}\ln(\frac{5}{3})$
7. $L = \frac{85{,}096}{1891}$, $P_0 = \frac{85{,}096}{1891}(1 + \frac{83{,}205}{1891}\exp(10\ln(1935)))^{-1}$, $c = \frac{1891}{85{,}096}$
$\ln(1935)$ **9.** $L = 10$, $P_0 = \frac{90}{13}$, $c = \frac{1}{10}\ln(\frac{3}{2})$ **11.** (i) 8 mg (ii) 7
days **17.** (i) $L = 14$, $P_0 = 5$, $c = \frac{1}{28}\ln(\frac{9}{5})$ (ii) $2\ln(\frac{695}{5})/$
$\ln(\frac{9}{5}) \approx 14.8$ days (iii) $2\ln(50)/\ln(\frac{7}{5}) \approx 23.3$ days (iv) logistic: 2
$\ln(\frac{117}{5})/\ln(\frac{9}{5}) \approx 10.73$ days; exponential: $2\ln(\frac{13}{5})/\ln(\frac{7}{5}) \approx 5.68$ days
19. (i) k (ii) $t = \frac{d}{a}$ (iii) $P_0 = k/(1 + e^d)$, $c = \frac{a}{k}$ **21.** same a,
d becomes $d + \ln(2 + e^{at_0 - d})$

Exercises 7.6 (p. 556)
1. $x = c\exp(\frac{t^3}{3})$ **3.** $x = \pm\sqrt{t^2 + c}$ **5.** $x = ct$ **7.** $x = 2 + c\exp(\frac{t^3}{3})$
9. $x = -\ln|c - e^t|$ **11.** $x = 2e^{t/2}$ **13.** $x = 3\exp((t^3 - 1)/3)$
15. $x = 1$ (constant function) **17.** $x = \arcsin(t + \frac{1}{\sqrt{2}})$
19. $x = \frac{1}{16}t^4 + \frac{(-4+\sqrt{5})}{2}t^2 + (-4 + \sqrt{5})^2$ **21.** $c_1e^{4t} + c_2e^{-3t}$
23. $c_1 + c_2e^{-t} + c_3e^{-2t}$ **25.** $c_1e^{-3t} + c_2te^{-3t}$
27. $c_1 + c_2e^{-t} + c_3te^{-t}$ **29.** $c_1 + c_2e^t + c_3te^t + c_4t^2e^t$ **31.** $f(t) = e^t$
33. $f(t) = 0$ **35.** $f(t) = t - 1$ **37.** 0 **39.** $-\frac{1}{15} + \frac{1}{40}e^{5t} + \frac{1}{24}e^{-3t}$

41.

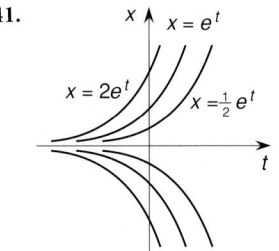

$x = e^t$ $x = 2e^t$ $x = \frac{1}{2}e^t$

43.

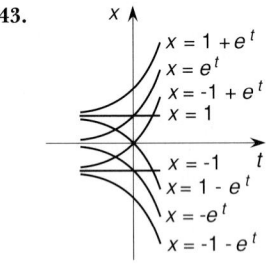

$x = 1 + e^t$
$x = e^t$
$x = -1 + e^t$
$x = 1$
$x = -1$
$x = 1 - e^t$
$x = -e^t$
$x = -1 - e^t$

45.

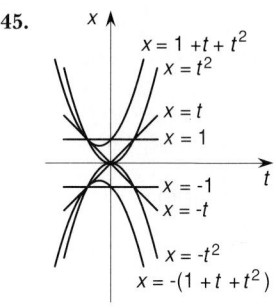

$x = 1 + t + t^2$
$x = t^2$
$x = t$
$x = 1$
$x = -1$
$x = -t$
$x = -t^2$
$x = -(1 + t + t^2)$

47. (ii) $f(t) = 0$ **51.** $(1-e)e^t + (e-1)te^t + ee^{-t}$

Chapter 7 Review Exercises (p. 558)

1. $(2 \ln x + 2)x^{2x}$ **3.** $(3 \ln x + 3 + \frac{5}{x})x^{3x+5}$ **5.** $\frac{2}{x} \log_2(x)x^{\log_2(x)}$
7. 0 **9.** 1 **11.** $10^{\frac{\ln(0.5)}{\ln(0.95)}} \approx 135$ min
13. $10{,}000e^{120 \ln(1.008)} \approx 26{,}017$ **15.** $20 + 60e^{10 \ln(5/6)} \approx 29.7°C$
17. $\frac{1}{0.08} \ln(2) \approx 8.66$ years **19.** \$2949.11 **21.** c_1
$\exp\left(-(\frac{1}{2} - \frac{\sqrt{5}}{2})t\right) + c_2 \exp\left(-(\frac{1}{2} - \frac{\sqrt{5}}{2})t\right)$ **23.** $c_1 + c_2t + c_3e^{-3t}$
25. $-(3 \tan x + c)^{-1/3}$ **27.** $u = -\frac{1}{t+1} - \frac{1}{3t^3} + c$
29. $c \exp(-\frac{1}{2}x^2 - x)$ **31.** $(e^5 - 1)^{-1}(-(e^3 + 1)e^{2t} + (e^3 + e^5)e^{-3t})$
33. 0 **35.** e^{3t} **37.** 1 **39.** $\frac{\pi - 1}{2\pi}e^{\pi t} + \frac{\pi + 1}{2\pi}e^{-\pi t}$ **43.** $10^{(10^{10})}$
45. $\ln(3)/\ln(51/49) \approx 27.46$ years **47.** half: $\frac{\ln 99}{\ln(297/97)} \approx 4.1$ days;
90%: $\frac{\ln 981}{\ln(297/97)} \approx 6.1$ days

Exercises 8.1 (p. 568)

1. $-xe^{-x} - e^{-x} + c$ **3.** $(\frac{1}{3}x^2 - \frac{2}{9}x + \frac{2}{27})e^{3x} + c$ **5.** $-\frac{1}{10} \sin^4(2u)$
$\cos(2u) - \frac{2}{15} \sin^2(2u) \cos(2u) - \frac{4}{15} \cos(2u) + c$ **7.** $x \tan x$
$- \ln|\sec x| + c$ **9.** $(x^2 + x - 3)e^x + c$ **11.** $x \arcsin x$
$+ \sqrt{1 - x^2} + c$ **13.** $2 \ln(1 + x^2) + c$ **15.** $x^3 \ln|x| - \frac{1}{3}x^3 + c$
17. $\frac{1}{3}x^3 \ln|x| - \frac{1}{9}x^3 + c$ **19.** $e^{2x}\left(\frac{3}{2} - \frac{1}{10} \sin(4x) + \frac{1}{5} \cos(4x)\right) + c$
21. $\frac{1}{4}e^2 - \frac{5}{4}e^{-2}$ **23.** $-2(e^2 + 1)/(4 + \pi^2)$ **25.** 0 **27.** $-\frac{2}{\pi}$
29. $\frac{1}{3}\sqrt{2} \arctan(\sqrt{2}) - \frac{1}{3} + \frac{1}{3} \ln(\frac{3}{2}) - \frac{\pi}{6} \approx 1.079\,585$ **41. (i)** 0
(ii) π **43.** $\frac{x}{8} + \frac{1}{8} \cos x \sin^3 x - \frac{1}{8} \cos^3 x \sin x + c$ **45.** $-2x$
$\cos(\frac{x}{2}) + 4 \sin(\frac{x}{2}) + c$ **47.** $-\frac{1}{3} \cos(x^3 + 4) + c$
49. $-\frac{1}{5}xe^{-5x} - \frac{1}{25}e^{-5x} + c$ **51.** $-\frac{2}{3} \cos^3 x + c$

Exercises 8.2 (p. 577)

1. $-\frac{1}{3} \sin^2 x \cos x - \frac{2}{3} \cos x + c$ **3.** $-\frac{1}{4} \sin^3 x \cos x + \frac{3}{8}x - \frac{3}{8} \cos x$
$\sin x + c$ **5.** $\frac{1}{2} \sec x \tan x + \frac{1}{2} \ln|\sec x + \tan x| + c$ **7.** $\frac{1}{2} \tan^2 x$
$+ \ln|\cos x| + c$ **9.** $-\frac{1}{6} \sin^2(2x) \cos(2x) - \frac{1}{3} \cos(2x) + c$ **11.** $\frac{1}{4}$
$\sin^4 x + c$ **13.** $\frac{1}{5} \cos^5 x - \frac{1}{3} \cos^3 x + c$ **15.** $\frac{1}{4} \tan^4 x + c$ **17.** $\frac{1}{3}$
$\sec^3 x - \sec x + c$ **19.** $\frac{1}{8} \sin^4(2x) - \frac{1}{12} \sin^6(2x) + c$ **21.** $\frac{1}{8} \sin^8 x$
$- \frac{1}{10} \sin^{10} x + c$ **23.** $\frac{1}{12} \cos(3x) \sin^3(3x) + \frac{x}{8} - \frac{1}{24} \cos(3x)$
$\sin(3x) + c$ **25.** $\frac{1}{18} \tan^6 3x + c$ **27.** $-\frac{2}{3} \cos^3 x + c$
29. $-\cos x - 2 \sec x + \frac{1}{3} \sec^3 x + c$ **31.** $\frac{\pi}{2}$ **33.** $\frac{2}{15}$ **35.** 0 **37.** $\frac{1}{8}$
39. 0 **43.** $\frac{1}{12}$ **45. (ii)** 0 **47.** $\ln(1 + \sqrt{2})$ **49. (i)** $-\ln|\cot x$
$+ \csc x| + c$ **51. (i)** $-\frac{1}{3} \cot^3 x + c$ **(ii)** $\frac{1}{3} \csc^3 x - \frac{1}{5} \csc^5 x + c$
53. $\frac{1}{4} \sec^3 x \tan x + \frac{3}{8} \sec x \tan x + \frac{3}{8} \ln|\sec x + \tan x| + c$
55. $-\frac{1}{4} \cos^4 x + c$ **57.** $\frac{1}{3}(x^2 + 1)^{3/2} + c$ **59.** $\frac{1}{5}(x^2 + 1)^{5/2} + c$
61. $\frac{1}{2} \sin(x^2 - 3) + c$ **63.** $\tan x - \frac{3}{2}x + \frac{1}{2} \sin x \cos x + c$

Exercises 8.3 (p. 583)

1. $2 \arcsin(\frac{x}{2}) + \frac{x}{2}\sqrt{4 - x^2} + c$ **3.** $\operatorname{arcsec}(x) + c$
5. $-\frac{1}{x}\sqrt{1 - x^2} + c$ **7.** $\sqrt{x^2 - 4} - 2 \operatorname{arcsec}(\frac{x}{2}) + c$

9. $-\frac{1}{4x}\sqrt{4 + x^2} + c$ **11.** $\frac{x}{4}(1 - x^2)^{3/2} + \frac{3}{8}$
$\arcsin x + \frac{3}{8}x\sqrt{1 - x^2} + c$ **13.** $\frac{9}{4} \arcsin(\frac{2x}{3}) + \frac{x}{2}\sqrt{9 - 4x^2} + c$
15. $\frac{x}{4}(1 + 5x^2)^{3/2} + \frac{3x}{8}\sqrt{1 + 5x^2} + \frac{3}{8\sqrt{5}} \ln|\sqrt{1 + 5x^2} + \sqrt{5}x| + c$
17. $\frac{1}{5}(8 + x^2)^{5/2} + c$ **19.** $\frac{1}{3}(x^2 - 1)^{3/2} + c$ **21.** $\frac{1}{2} \arctan x + \frac{1}{2}$
$\frac{x}{1 + x^2} + c$ **23.** $\frac{x}{5}(5 - x^2)^{-1/2} + c$ **25.** $\frac{1}{2}$
$\arcsin(x^2) + \frac{1}{2}x^2\sqrt{1 - x^4} + c$ **27.** $-\frac{1}{2}x^{-2}\sqrt{1 + x^4} + c$
29. $-\frac{1}{3}(4 - x^2)^{3/2} + 6 \arcsin(\frac{x}{2}) + \frac{3}{2}x\sqrt{4 - x^2} + c$ **31.** $\frac{\pi}{2}$
33. $\frac{14\sqrt{3}}{5}$ **35.** $\frac{1}{4} + \frac{1}{4\sqrt{2}} \ln(1 + \sqrt{2})$ **37.** $\frac{\pi}{8}$ **39.** $\frac{1}{2} \ln(10)$
41. $\frac{\sqrt{5}}{2} + \frac{1}{4} \ln(2 + \sqrt{5})$ **43.** $\frac{1}{2}(b\sqrt{1 + 4b^2} - a\sqrt{1 + 4a^2}) + \frac{1}{4}$
$\ln (|2b + \sqrt{1 + 4b^2}|/|2a + \sqrt{1 + 4a^2}|)$ **45.** $-\frac{2}{3}(1 - x^2)^{3/2} + c$
47. $\operatorname{arcsinh} x + c$ **49.** $\arctan(e^x) + c$ **51.** $\frac{1}{3}(4 + x^2)^{3/2} + c$
53. $\frac{x}{4}(9 + x^2)^{3/2} - \frac{9x}{8}\sqrt{9 + x^2} - \frac{81}{8} \ln|x + \sqrt{9 + x^2}| + c$
55. $\frac{1}{2} \cosh^2 x + c$ **57.** $\frac{1}{3} \sin^3(x^2) + c$

Exercises 8.4 (p. 588)

1. $\arctan(x - 2) + c$ **3.** $\frac{1}{\sqrt{5}} \arctan (\frac{x+3}{\sqrt{5}}) + c$ **5.** $\frac{1}{2}$
$\ln(x^2 - 2x + 4) + \frac{2}{\sqrt{3}} \arctan (\frac{x-1}{\sqrt{3}}) + c$ **7.** $x + \ln(x^2 - 2x + 2) + 3$
$\arctan(x - 1) + c$ **9.** $2 \arcsin(\frac{x+1}{2}) + \frac{1}{2}(x + 1)\sqrt{3 - 2x - x^2} + c$
11. $\arcsin(\frac{x+3}{5}) + c$ **13.** $-\frac{1}{2} \arctan(\frac{x-2}{2}) + c$
15. $\frac{1}{2} \arctan(x - 1) + c$
17. $\sqrt{x^2 + 2x + 2} - \ln|x + 1 + \sqrt{x^2 + 2x + 2}| + c$
19. $(3x - 7)(x^2 - 4x + 5)^{-1/2} + c$ **21.** $\sqrt{3 + 2x - x^2} + 2$
$\arcsin(\frac{x-1}{2}) + c$ **23.** $-\frac{1}{15}(x^2 + 2x + 17)^{-15/2} + c$
25. $\frac{1}{9}(x + 2)(5 - 4x - x^2)^{-1/2} + c$ **27.** $-\frac{1}{8}(x^2 + 4x + 13)^{-4} + c$
29. $\frac{1}{5}(1 - x^2)^{5/2} - \frac{1}{3}(1 - x^2)^{3/2} + c$ **31.** $\frac{x-2}{4}(x^2 - 2x + 2)^{-2} + \frac{3}{8}$
$\arctan(x - 1) + \frac{3}{8}(x - 1)(x^2 - 2x + 2)^{-1} + c$
33. $\frac{1}{2} \ln(x^2 + 2x + 10) + c$ **35.** $\frac{2}{5}(x - 1)^{5/2} + \frac{2}{3}(x - 1)^{3/2} + c$
37. $\frac{1}{5}(5 + 4x - x^2)^{5/2} - 3(5 + 4x - x^2)^{3/2} + c$
39. $e^{x/2}\sqrt{e^x - 4} - 4 \ln(e^{x/2} + \sqrt{e^x - 4}) + c$
41. $\frac{1}{8} \arcsin(2 \ln(x) - 1) + \frac{1}{4}(2 \ln(x) - 1)\sqrt{\ln x - \ln^2 x} + c$

Exercises 8.5 (p. 597)

1. $\frac{1}{2} \ln|\frac{x-1}{x+1}| + c$ **3.** $\ln|x - 1| + c$ **5.** $\frac{4}{5} \ln|x - 2| + \frac{1}{5} \ln|x + 3| + c$
7. $x + 3 \ln|x + 1| - 4 \ln|x| + c$ **9.** $x + \frac{1}{x} + 4 \ln|x| - 5 \ln|x + 1| + c$
11. $\ln|x| - \arctan x + c$ **13.** $\ln|x| - \frac{1}{2} \ln(x^2 + 1) + c$ **15.** $x - \frac{2}{5}$
$\ln|x + 2| + \frac{6}{5} \ln(x^2 - 4x + 8) - \frac{4}{5} \arctan(\frac{x-2}{2}) + c$
17. $2 \ln(x^2 + 4) - \frac{1}{2} \arctan(\frac{x}{2}) - (x^2 + 4)^{-1} + c$
19. $\ln|(x + 1)(2x^2 - 4x + 3)| + c$ **21.** $\frac{1}{2} \arctan x + \frac{1}{2}x/(1 + x^2) + c$
23. $\frac{1}{8} \ln|(x^2 - 4)/x^2| + c$ **25.** $\frac{2}{3} \ln|\frac{x}{x+3}| + c$
27. $\frac{1}{2}(x^2 + \ln|x^2 - 1|) + c$ **29.** $\frac{1}{4}x^2 + c$ **31.** $\frac{1}{6} \ln(\frac{4 - \sin \theta}{\sin \theta + 2}) + c$
33. $\frac{1}{4\sqrt{2}} \ln((x^2 + \sqrt{2}x + 1)/(x^2 - \sqrt{2}x + 1)) + \frac{1}{2\sqrt{2}}$
$\arctan(\sqrt{2}x + 1) + \frac{1}{2\sqrt{2}} \arctan(\sqrt{2}x - 1) + c$ **35.** $\arctan x$
$-\frac{1}{2x^2 + 2} + c$ **37.** $\arctan(x + 2) + c$ **39.** $4 \ln(2 + e^{x/2}) - 2$
$\ln(1 + e^{x/2}) + c$ **41.** $-\frac{1}{3}x \sin^2 x \cos x - \frac{2}{3}x \cos x + \frac{1}{9} \sin^3 x + \frac{2}{3}$
$\sin x + c$ **43.** $\frac{1}{2}x^2 \arcsin x - \frac{1}{4} \arcsin x + \frac{1}{4}x\sqrt{1 - x^2} + c$ **45.** $\frac{1}{3}x^3$
$\arctan x - \frac{1}{6}x^2 + \frac{1}{6} \ln(1 + x^2) + c$ **47.** $x \ln(x^2 + 1) - 2x + 2$
$\arctan x + c$ **49.** $-\frac{1}{2} \ln(1 + \exp(-x^2)) + c$ **51.** $\frac{1}{20} \ln|\frac{x-1}{x+1}| - \frac{1}{30}$
$\arctan \frac{x}{3} + c$ **53.** $-2e^{x/2} \cos(e^{x/2}) + 2 \sin(e^{x/2}) + c$ **55.** $\frac{2}{3}x^{3/2} \ln x$
$-\frac{4}{9}x^{3/2} + c$ **57. (ii)** $\approx -0.720\,128\,079$

Exercises 8.6 (p. 601)

1. $x - \frac{3}{2}x^{2/3} + 3x^{1/3} - 3 \ln|1 + x^{1/3}| + c$
3. $\frac{6}{7}x^{7/6} - \frac{6}{5}x^{5/6} + 2x^{1/2} - 6x^{1/6} + 6 \arctan(x^{1/6}) + c$ **5.** $2\sqrt{x} - 2$
$\arctan(\sqrt{x}) + c$ **7.** $\frac{2}{3}x^{3/2} - x + c$ **9.** $2\sqrt{x} - 24x^{1/6} + 48$
$\arctan(\frac{1}{2}x^{1/6}) + c$ **11.** $-2/(1 + \tan(\frac{\theta}{2})) + c$

13. $\ln(|1 + \tan\frac{\theta}{2}|/|1 - \tan\frac{\theta}{2}|) = \ln|\sec\theta + \tan\theta| + c$ **15.** $\tan(\frac{\theta}{2}) + c$
17. $\frac{1}{2}\ln|\tan(\frac{\theta}{2})| - \frac{1}{4}\tan^2(\frac{\theta}{2}) + c$ **19.** $\ln|1 + \tan(\frac{\theta}{2})| + c$
21. $\frac{2}{a}(1 - b^2/a^2)^{-1/2}\arctan([(\tan\frac{\theta}{2}) + b/a](1 - b^2/a^2)^{-1/2}) + c$
25. $\frac{2}{3}x^{3/2} - x + 2\sqrt{x} - 2\ln(1 + \sqrt{x}) + c$ **27.** $\frac{1}{2}\ln|\tan\theta| + c$
29. $\frac{4}{3}(1 + \sin\theta)^{3/2} - 4(1 + \sin\theta)^{1/2} + c$
31. $-\frac{1}{28}(1 + 2\cos x)^{7/2} + \frac{1}{10}(1 + 2\cos x)^{5/2} - \frac{1}{12}(1 + 2\cos x)^{3/2} + c$
33. $\frac{6}{7}(4 + e^{x/2})^{7/3} - 6(4 + e^{x/2})^{4/3} + c$

Exercises 8.7 (p. 605)

1. π **3.** $\frac{1}{\sqrt{2}} - \frac{1}{2}\ln(\sqrt{2} + 1) \approx 0.266$ **5.** $\frac{1}{2}\ln(3) \approx 0.549$
7. $2 - \frac{2}{e} \approx 1.264$ **9.** 1 **11.** $\sqrt{145} - \frac{1}{2}\sqrt{37} + \frac{1}{12}$
$\ln(\frac{12 + \sqrt{145}}{6 + \sqrt{37}}) \approx 9.058$ **13.** $\sqrt{68} - \sqrt{5} + \frac{1}{2}\ln(\frac{4 + \sqrt{17}}{2 + \sqrt{5}})$
15. $\frac{5}{6}\sqrt{109} - \frac{5}{3} + \frac{3}{4}\ln\frac{10 + \sqrt{109}}{9} \approx 7.649$ **17.** $\frac{1}{2}\ln(3)$
19. $\sqrt{17} - \sqrt{2} + \frac{1}{2}\ln((\sqrt{17} - 1)(\sqrt{2} + 1)/[(\sqrt{17} + 1)(\sqrt{2} - 1)])$
21. $\sqrt{17} + \frac{1}{4}\ln(4 + \sqrt{17})$ **23.** $\frac{a}{2}\sqrt{1 + 4a^2} + \frac{1}{4}$
$\ln(2a + \sqrt{1 + 4a^2})$ **25.** $\frac{b}{2}\sqrt{1 + 4k^2b^2} + \frac{1}{4k}$
$\ln(2kb + \sqrt{1 + 4k^2b^2}) - \frac{a}{2}\sqrt{1 + 4k^2a^2} - \frac{1}{4k}\ln(2ka + \sqrt{1 + 4k^2a^2})$
27. $\frac{1}{2}\ln(\frac{2 + \sqrt{3}}{1 + \sqrt{2}})$ **29. (i)** $\sqrt{1 + e^2} - \sqrt{2} + \frac{1}{2}$
$\ln((\sqrt{1 + e^2} - 1)(\sqrt{2} + 1)/(\sqrt{1 + e^2} + 1)(\sqrt{2} - 1)) \approx 2.0035$
(ii) $\sqrt{e^2 - 2e + 2} \approx 1.988$ **31.** $\sqrt{2}\pi - \frac{\pi}{4}\sqrt{5} - \pi\ln(\frac{1 + \sqrt{5}}{2 + 2\sqrt{2}})$
33. $8\pi^2$ **35. (i)** $\bar{x} = \frac{1}{\sqrt{2}}(1 - \frac{\pi}{4})/(1 - \frac{1}{\sqrt{2}}) \approx 0.518\,095, \bar{y} = (\frac{\pi}{16} - \frac{1}{8})/$
$(1 - \frac{1}{\sqrt{2}}) \approx 0.243\,603$ **(ii)** $(\frac{\pi}{6}, \frac{1}{3\sqrt{2}})$ **37.** $\bar{x} = \pi - 2, \bar{y} = \frac{1}{96}\pi^3 + \frac{1}{16}\pi$

Exercises 8.8 (p. 610)

1. $\frac{1}{2}$ **3.** $\frac{3}{2}$ **5.** $\frac{\pi}{2}$ **7.** diverges **9.** $\frac{1}{2}$ **11.** diverges **13.** diverges
15. diverges **17.** $\frac{\pi}{3}$ **19.** diverges **21. (i)** $r < -1$ **(ii)** $-\frac{1}{r + 1}$
27. exists when $\ln A \notin [a, b]$

Chapter 8 Review Exercises (p. 612)

1. $\frac{x}{3}e^{3x} + \frac{5}{9}e^{3x} + c$ **3.** $\frac{1}{2}\arctan(\frac{x - 3}{3}) + c$ **5.** $-2x\cos\sqrt{x} + 4\sqrt{x}$
$\sin\sqrt{x} + 4\cos\sqrt{x} + c$ **7.** $x^2 + 2x + \ln|(x^2 - 3x + 2)/(x - 3)| + c$
9. $\frac{x}{3}\sin^3 x + \frac{1}{9}\sin^2 x\cos x + \frac{2}{9}\cos x + c$
11. $\frac{1}{\sqrt{2}}(x - 3)\sqrt{x^2 - 6x + 13} + 2\sqrt{2}\ln$
$|x - 3 + \sqrt{x^2 - 6x + 13}| + c$ **13.** $\frac{1}{2}\ln\left|\frac{x + 2}{x + 4}\right| + c$ **15.** $2\sqrt{x} - 2$
$\ln(1 + \sqrt{x}) + c$ **17.** $2\sqrt{x}\sin(\sqrt{x}) + 2\cos(\sqrt{x}) + c$ **19.** x
$\ln(1 + x^2) - 2x + 2\arctan x + c$ **21.** $2\sin x - 2\ln(1 + \sin x) + c$
23. $\frac{x}{4}(\sin^2 x - \cos^2 x) + \frac{1}{4}\cos x\sin x + c$
25. $\frac{3}{7}(x + 8)^{7/3} - 6(x + 8)^{4/3} + c$
27. $2\sqrt{1 + e^x} + \ln((\sqrt{1 + e^x} - 1)/(\sqrt{1 + e^x} + 1)) + c$
29. $\frac{1}{2}(x^2 + 1)\arctan x - \frac{x}{2} + c$ **31.** $\frac{1}{3}\csc^3 x - \frac{1}{5}\csc^5 x + c$
33. $x\tan x + \ln|\cos x| + c$ **35.** $-\frac{1}{x}\sqrt{4 - x^2} - \arctan(\frac{x}{2}) + c$
37. $\frac{1}{4}\ln(2x^2 - 2x + 1) + \frac{5}{2}\arctan(2x - 1) + c$ **39.** $\frac{2}{3}x^{3/2}$
$\ln\sqrt{x} - \frac{2}{9}x^{3/2} + c$ **41.** $\ln(\frac{5\sqrt{2}}{7})$ **43.** $\cos\theta + \frac{1}{2}\ln(\frac{1 - \cos\theta}{1 + \cos\theta}) + c$
45. $\frac{2}{3}x^{3/2}\ln x - \frac{4}{9}x^{3/2} + c$ **47.** $\frac{\pi}{6} + \sqrt{3} - 2$ **49.** $8\ln(2) - 4$
51. $\frac{5\pi}{6}$ **53.** diverges **55.** diverges **57.** diverges **59.** $\frac{1}{2}$
$\ln(2) - \frac{3\pi}{4}$ **61.** $\frac{2}{3}a^{-2}(ax + b)^{3/2} - 2ba^{-2}(ax + b)^{1/2} + c$
63. $\frac{x}{2}\sqrt{x^2 \pm a^2} \pm \frac{a^2}{2}\ln|x + \sqrt{x^2 \pm a^2}| + c$ **65.** $\frac{a^2}{2}$
$\arcsin(\frac{x}{a}) + \frac{x}{2}\sqrt{a^2 - x^2} + c$ **67.** $\bar{x} = \frac{1}{9}(3\sqrt{3} - \pi), \bar{y} = \frac{\pi}{6} - \frac{\sqrt{3}}{8}$
69. $\frac{\sqrt{5}}{2} + \frac{1}{4}\ln(2 + \sqrt{5})$ **71. (i)** $2\ln|x| + \frac{3}{2}\ln(x^2 + 1) - $
$\arctan x + c$

Exercises 9.1 (p. 621)

1. $1 - \frac{x^2}{2} + \frac{x^4}{24}$ **3.** $1 + x + \frac{x^2}{2} + \frac{x^3}{6} + \frac{x^4}{24} + \frac{x^5}{120}$ **5.** $x - \frac{x^2}{2} + \frac{x^3}{3}$
7. $1 + 3(x - 1) + 3(x - 1)^2 + (x - 1)^3$
9. $1 - 4(x + 1) + 6(x + 1)^2 - 4(x + 1)^3 + (x + 1)^4$
11. $0.009\,999\,833$ **13.** $1.105\,170\,833$ **15.** $0.198\,666\,667$
17. $1.011\,271\,094$ **19.** $-0.105\,360\,333$
21. $1 + x + \frac{x^2}{2} + \frac{x^3}{6} + \frac{x^4}{24} + \frac{x^5}{120}; e^{1/2} \approx 1.648\,697\,917$
23. (i) $0.704\,652\,651$ **(ii)** $0.707\,429\,207$

31.

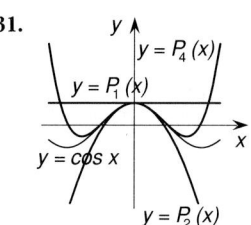

33. (i) $f^{(4m)}(0) = 0, f^{(4m+1)}(0) = 1, f^{(4m+2)}(0) = 0, f^{(4m+3)}(0) = -1$
(ii) $\frac{1}{k!}$, if $k = 4m + 1$; $-\frac{1}{k!}$, if $k = 4m + 3$; 0, if k is even

Exercises 9.2 (p. 633)

1. $P_3(\frac{1}{2}) \approx 1.646$ **3.** $P_7(1) \approx 0.841\,468$ **5.** $P_0(1.2) = 0;$
$P_1(1.2) = 0.2$ **7.** $P_0(1.9) = \ln(2) \approx 0.693; P_1(1.9) \approx 0.643$
9. $P_0(-\frac{1}{10}) = 0$ **11.** $f(x) = \cos x, a = 0, P_1(\frac{4}{180}\pi) = 1$
13. $f(x) = \cos x, a = \frac{\pi}{4}, P_1(a - \frac{2}{180}\pi) \approx 0.7318$
15. $f(x) = \ln(x), a = 1, P_1(a + 0.2) = 0.2$ **17.** $f(x) = \tan x,$
$a = 0, P_1(\frac{3}{180}\pi) = \frac{3}{180}\pi \approx 0.052\,360$
19. $f(x) = \sinh(x), a = 0, P_2(-0.12) = -0.12$
21. $\pm\frac{1}{720}\frac{1}{64} \approx \pm 2.17 \times 10^{-5}$ **23. (i)** ± 0.0014 is sufficient
(ii) 0.001218 (using P_1 for $\sin x$) **25.** using P_0 for $\cos x$ with
$a = 0, P_1$ for $\sin x$ with $a = 0, \sin^2(1°)\cos(1°) \approx 3.046 \times 10^{-4}$
27. $f(x) = e^{\sin x}, a = \frac{\pi}{4}, P_2(a + \frac{2\pi}{180}) \approx 2.077\,919$
29. $f(t) = \int^t \exp(-x^2)\,dx, a = 0, P_2(\frac{1}{4}) = \frac{1}{4}$ **31.** $f(t) = \int_0^t \sin(x^2)\,dx,$
$a = 0, P_1(\frac{1}{5}) = 0$ **33. (iii)** $P_5(x) = 1 + \frac{1}{3}x^3$ **(iv)** $P_5(0.1) = 1 + \frac{1}{3000}$
(v) $\exp(\frac{1}{3}x^3)$ **35.** ≈ 0.743, within 0.1 **37.** 0 or 0.005
39. $\approx 0.892, \approx 1.694$

Chapter 9 Review Exercises (p. 635)

1. $1 - x + \frac{x^2}{2} - \frac{x^3}{6} + \frac{x^4}{24} - \frac{x^5}{120} + \frac{x^6}{720}$
3. $e + e(x - 1) + \frac{e}{2}(x - 1)^2 + \frac{e}{6}(x - 1)^3 + \frac{e}{24}(x - 1)^4 + \frac{e}{120}(x - 1)^5 + \frac{e}{720}(x - 1)^6$ **5.** $1 - \frac{\pi^2}{2}(x - \frac{1}{2})^2 + \frac{\pi^4}{24}(x - \frac{1}{2})^4$
7. $\frac{\pi}{4} + \frac{1}{2}(x - 1) - \frac{1}{4}(x - 1)^2 + \frac{1}{12}(x - 1)^3$ **9.** 0
11. $f(x) = e^x, a = 0, P_7(-1) \approx 0.367\,857$
13. $f(x) = \sqrt{x}, a = 1, P_1(a + 0.1) = 1.05$
15. $f(x) = \sec x, a = 0, P_0(\frac{11\pi}{1800}) = 1$
17. $f(x) = \sin x, a = 0, P_1(1.5\frac{\pi}{180}) = \frac{\pi}{120} \approx 0.026\,180$
19. $f(x) = \tan x, a = 0, P_1(-0.1) = -0.1$ **23.** $P_n(x) = M_n(x - a)$

Exercises 10.1 (p. 647)

1. -1 **3.** $\frac{1}{4}$ **5.** diverges **7.** 0 **9.** 0 **11.** 0 **13.** 1 **15.** -1
17. diverges **19.** 0 **23.** 0 **25.** 0 **27.** diverges

Exercises 10.2 (p. 656)

1. $\frac{127}{64}$ **3.** $\frac{85}{128}$ **5.** $2 - 2^{-20}$ **7.** $\frac{5}{3}(1 - 4^{-8})$ **9.** $\frac{12}{7}(36 + 6^{-7})$
11. $\$1467.63$ **13.** $\$47.07$ **15.** $\$1259.30$ **17.** $\$14,682.42$

19. $\$5.815\,693 \approx \$5.82$ **21.** 2   **23.** $\frac{27}{4}$ **25.** 2   **27.** $\frac{11}{2}$ **29.** $\frac{1215}{4}$
31. $\$723.05$ **33.** $\$6573.69$ **35. (i)** loss of $\$1608.04$ at end
(ii) profit of $\$12,720.29$ at end   **37.** $a(r^m - r^{k+1})/(1 - r)$
39. (i) $\frac{1}{8}(9^{13} - 1)$ **(ii)** $\frac{1}{7}(1 - 8^{-8})$ **43.** $\frac{2110}{999}$

Exercises 10.3 (p. 665)
1. 1   **3.** diverges   **5.** 0   **7.** diverges   **9.** 2   **11.** 0   **13.** 8
15. -2 **17.** -5 **19.** 0   **21.** $a_1 = 1, a_n = -\frac{1}{n(n-1)}$, for $n > 1$
23. $a_1 = 1, a_n = \frac{1-2n}{n^2(n-1)^2}$, for $n > 1$ **25.** converges to 0

Exercises 10.4 (p. 673)
1. diverges   **3.** converges   **5.** diverges   **7.** converges
9. converges   **11.** converges   **13.** converges   **15.** diverges
17. converges   **19.** converges   **23.** converges   **25.** converges
27. converges   **29.** $q > 1$ **31.** diverges   **33.** ≈ 10.58132
35. $1.081\,895$ **39.** $143.903\,745$

Exercises 10.5 (p. 679)
1. diverges   **3.** converges   **5.** diverges   **7.** converges
9. converges   **11.** diverges   **13.** converges   **15.** converges
17. converges   **19.** converges   **25.** converges   **27.** diverges
29. diverges   **31.** converges   **33.** converges
35. (i) $\approx 0.603\,702\,863$ **(ii)** within 0.01   **37.** $\approx 1.604\,61$
39. 0.4147

Exercises 10.6 (p. 686)
1. converges   **3.** converges   **5.** converges   **7.** converges
9. diverges   **11.** converges   **13.** converges   **15.** diverges
17. converges   **19.** diverges   **21.** converges conditionally
23. converges absolutely   **25.** diverges   **27.** converges
conditionally   **29.** converges absolutely   **31.** converges
conditionally   **33.** converges absolutely
37. $\frac{1}{2} + \frac{1}{4} + \cdots + \frac{1}{26} - 1 + \frac{1}{28} + \cdots + \frac{1}{38}$ **41.** converges
43. diverges   **45.** diverges   **47.** converges   **49.** converges
51. $\approx 0.695\,65$ **53.** ≈ 0.3241

Exercises 10.7 (p. 694)
1. converges (Ratio Test)   **3.** converges absolutely (Ratio Test)
5. diverges (Ratio or nth Root Test)   **7.** converges (nth Root
Test)   **9.** converges (Ratio Test)   **11.** converges (Ratio Test)
13. converges (Ratio or Integral Test)   **15.** convergent
geometric series   **17.** converges (Ratio or Limit Comparison
Test)   **19.** converges absolutely (Ratio and Comparison Tests)
21. all c **23.** all c **25.** $\in (-1, 1]$ **27.** $\{c : |c| > 5\}$ **29.** no
nonzero c **33.** $c \in (-1, 1]$ **35.** $|r| < 1$, unless $p(n) = 0$, for all
n, in which case all r **39.** converges (Limit Comparison Test)
41. diverges (nth Term Test)   **43.** converges (Comparison or
Integral Test)   **45.** diverges (Limit Comparison Test)
47. converges (Integral Test)   **49.** converges (Ratio Test)
51. diverges (Ratio or nth Term Test)   **53.** converges
(Comparison Test)   **55.** diverges (Comparison Test)
57. converges (Ratio Test)   **59. (i)** $1.718\,281\,828\,459\,045$
(ii) 8.22×10^{-18}

Exercises 10.8 (p. 703)
1. $\sum_{n=0}^{\infty} \frac{1}{n!}(-2x)^n$ **3.** $\sum_{n=0}^{\infty} (-1)^n(x - 1)^n$ **5.** $5 - x - 2x^2 + x^3$
7. $\sum_{n=0}^{\infty} (-1)^{n+1} \frac{1}{(2n+1)!}(x - \frac{\pi}{2})^{2n+1}$ **9.** $\sum_{n=0}^{\infty} (-1)^n 2^n x^{2n}$
11. $(2 - e^3) - e^3 \sum_{n=1}^{\infty} \frac{1}{n!} 3^n(x - 1)^n$ **13.** $-\frac{1}{2} \sum_{n=0}^{\infty} 2^{-n} x^n$

15. $1 + \frac{x^2}{2} + \sum_{n=2}^{\infty} (-1)^n \frac{1}{(2n)!} x^{2n}$ **17.** $\frac{1}{e} \sum_{n=0}^{\infty} (-1)^n 2^n \frac{1}{n!}(x - \frac{1}{2})^n$
19. $\sum_{n=0}^{\infty} \frac{s(n)}{n!} x^n$, where $s(n) = 1$ if $n = 4k$ or if $n = 4k + 1$, for some
integer k, and $s(n) = -1$ if $n = 4k + 2$ or $n = 4k + 3$ for some
integer k.   **21.** $1 + \frac{x}{2} - \frac{1}{8}x^2 + \cdots + (-1)^{n+1} \frac{1(3)(5) \cdots (2n - 3)}{n! 2^n} x^n + \cdots =$
$1 + \frac{x}{2} + \sum_{n=2}^{\infty} (-1)^{n+1} \frac{1}{n!} \frac{(2n - 3)!}{n! (n - 2)!} 2^{2-2n} x^n$ **25.** $\sum_{n=0}^{\infty} (-1)^n x^{2n+1}$
33. $\approx 20.085\,536\,923$ **35.** $\approx 3.113 \times 10^{-7}$
37. $\approx 0.223\,143\,551$ **39.** $\approx -18,155.343\,286$

Exercises 10.9 (p. 709)
1. $(-1, 1)$ **3.** $[2, 4]$ **5.** $(-\infty, \infty)$ **7.** $(-\infty, \infty)$ **9.** $(11, 13)$
11. $(-1, 1)$ **13.** $[1, 5]$ **15.** $(-\infty, \infty)$ **17.** $[22, 22]$ **19.** $[-\frac{8}{3}, \frac{8}{3})$
21. ∞ **23.** 1   **29.** $[-7 - 5^{-1/3}, -7 + 5^{-1/3}]$ **33. (i)** $\frac{R}{4}$ **(ii)** $\frac{R}{r}$
(iii) R **35.** 3

Exercises 10.10 (p. 721)
1. $\sum_{n=0}^{\infty} \frac{(-1)^n}{n!} \frac{x^{2n+1}}{2n+1}$ **3.** $\sum_{n=1}^{\infty} (-1)^n n x^{n-1}$ **5.** $\ln(2) - \sum_{n=1}^{\infty} \frac{2^{-n}}{n} x^n$
7. $\sum_{n=1}^{\infty} \frac{(-1)^{n+1}}{n(n+1)} 2^n x^{n+1}$ **9.** $\sum_{n=0}^{\infty} (3^{n+1} - 2^{n+1}) x^n$
11. $\sum_{n=0}^{\infty} \frac{(-1)^n}{(2n+1)!} x^{2n+3}$ **13.** $\sum_{n=0}^{\infty} \frac{(-1)^n}{(2n)!} x^{2n}$
15. $-\sum_{n=1}^{\infty} \frac{(-1)^n}{(2n)!} x^{2n-2} = \sum_{m=0}^{\infty} \frac{(-1)^m}{(2m+2)!} x^{2m}$ **17.** $\sum_{n=0}^{\infty} 2^{-n-1} x^{2n}$
19. $\frac{1}{6} \sum_{n=3}^{\infty} n(n - 1)(n - 2) x^{n-3} = \frac{1}{6} \sum_{m=0}^{\infty} (m + 1)(m + 2)(m + 3) x^m$
21. $\frac{1}{0!} + \frac{1}{1!} + \frac{1}{2!} + \cdots + \frac{1}{n!}$
23. $1\frac{1}{(2n)!} + \frac{1}{2!} \frac{1}{(2n-2)!} + \frac{1}{4!} \frac{1}{(2n-4)!} + \cdots + \frac{1}{(2n)!}(1)$ **25.** $\frac{(-1)^n \pi^{2n}}{(2n)!}$
27. -1 for $n \geq 1$, and 0 for $n = 0$ **29.** $-\frac{2}{n}$
31. $1 + 2x + \frac{5}{9}x^2 + \cdots$ **33.** $1 + x + \frac{x^2}{2} + \cdots$
35. $1 + \frac{x^2}{6} + \frac{7}{360}x^4 + \cdots$ **37.** $x - \frac{x^3}{3} + \frac{2}{15}x^5 + \cdots$
39. $\frac{x}{2} - \frac{x}{3} - \frac{x^3}{45} + \cdots$ **41.** $x + (1 - x) \ln(1 - x)$ **43.** $-\frac{1}{2} +$
$\sum_{m=0}^{\infty} \frac{2^{2m}}{(2m)!} x^{2m}$ **45.** $\frac{3}{4}$ **47.** 24   **49. (i)** 40,320   **51.** 61   **53.** 5

Chapter 10 Review Exercises (p. 722)
1. 2   **3.** 4   **5.** $\$2550.00$ **7.** $\$75,644.46$ **9.** $\$1314.74$ **11.** 0
13. diverges   **15.** 0   **17.** 1   **19.** 1   **21.** diverges (nth Term
Test)   **23.** diverges (nth Term Test)   **25.** diverges (nth Term or
Ratio Test)   **27.** diverges (nth Term Test)   **29.** convergent
alternating series   **31.** diverges (harmonic series)   **33.** converges
(Ratio Test)   **35.** diverges (nth Term Test)   **37.** converges
(Integral Test)   **39.** convergent alternating series   **41.** $|c| > 1$
43. $c < -\frac{1}{2}$ **45.** $c < -2$ **47.** $c < -\frac{1}{3}$ **49.** no c **51.** $\sum_{n=0}^{\infty} 3^n x^n$
53. $\sum_{n=0}^{\infty} (-1)^{n+1} \frac{1}{(2n)!}(x - \pi)^{2n}$ **55.** $\ln 4 + \sum_{n=1}^{\infty} (-1)^{n+1} \frac{1}{n 4^n}(x - 4)^n$
57. $-\sum_{n=1}^{\infty} \frac{1}{n}(x + 1)^n$ **59.** $\sum_{n=0}^{\infty} (-1)^n \frac{1}{2n+1} 2^{2n+1} x^{2n+1}$ **61.** $\frac{1}{2}$
63. 0   **65.** 1   **67.** $\frac{1}{\sqrt{2}}$ **69.** 1   **71.** $x - \frac{7}{6}x^3 + \frac{61}{120}x^5 + \cdots$
73. $1 - x + \frac{x^2}{2} + \cdots$ **75.** $e + ex + ex^2 + \cdots$
77. $\cos(1) - \sin(1)x - \frac{1}{2}(\cos(1) + \sin(1))x^2 + \cdots$
79. $x - \frac{x^2}{2} + \frac{3}{8}x^5 + \cdots$ **83. (i)** $\sum_{n=0}^{\infty} (a_n - 3b_n) x^n$ **85.** 0

Exercises 11.1 (p. 730)
1. hyperbola, center $(0, 0)$, asymptotes $y = \pm x$, transverse axis
vertical

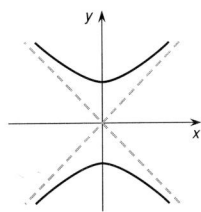

3. parabola, vertex $(-3, 5)$, opening downward

5. ellipse, center $(-1, 0)$, semi-major axis 1, vertical; semi-minor axis $\frac{1}{\sqrt{2}}$

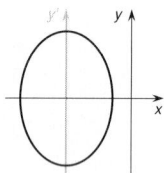

7. ellipse, center $(2, 2)$, semi-major axis 4, vertical; semi-minor axis $\frac{4}{\sqrt{3}}$

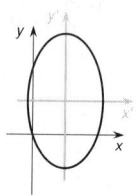

9. parabola, vertex $(19, 2)$, opening left

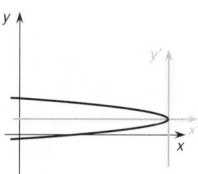

11. $4x^2 + 8x - y + 2 = 0$ **13.** $x^2 + 4y^2 - 24y + 32 = 0$
15. $4x^2 - 8x - y^2 = 0$ **17.** $x^2 + 6x + y^2 - 8y + 9 = 0$
19. $3x^2 - 24x - y^2 + 49 = 0$ **21.** two straight lines, slopes ± 1, through $(-1, -1)$

23. $4x^2 - 24x - y + 35 = 0$, $y^2 + 2y - 16x + 49 = 0$
25. **(i)** $(-1, 0), (1, 0), (0, 1), (0, 2)$, for instance
(ii) $(0, 0), (1, 0), (2, 0)$, for instance **27.** $y = \pm\sqrt{4x^2 - 8x}$
29. $y = -3 \pm 7\sqrt{1 - (x - 2)^2/9}$

Exercises 11.2 (p. 739)

1. focus $(-1, -\frac{3}{4})$, directrix $y = -\frac{5}{4}$ **3.** foci $(\pm\sqrt{3}, 0)$, directrices $x = \pm\frac{4}{\sqrt{3}}$ **5.** foci $(\pm\frac{\sqrt{5}}{2}, 0)$, directrices $x = \pm\frac{2}{\sqrt{5}}$ **7.** foci $(\pm\sqrt{6} - 2, 1)$, directrices $x = -2 \pm \frac{8}{\sqrt{6}}$ **9.** foci $(1 \pm \frac{2\sqrt{13}}{3}, -1)$, directrices $x = 1 \pm \frac{8}{3\sqrt{13}}$ **11.** $\frac{x^2}{36} + \frac{y^2}{27} = 1$ **13.** $x = 3 - \frac{1}{4}(y - 2)^2$, or $y^2 - 4y + 4x - 8 = 0$ **15.** $9(x - 4)^2 + (y - 1)^2 = 13$
17. $(y + 1)^2 - \frac{1}{3}(x - 2)^2 = 1$ **19.** $y = 4 - (x - 3)^2$ **21.** $e = \frac{\sqrt{3}}{2}$
23. $(-1 \pm \frac{\sqrt{5}}{2}, 0)$ **25.** $y = -4$ **27.** $e = 1$ **29.** $2c = \frac{\sqrt{13}}{3}$
31. $\frac{72}{5} = 14.4$ ft **33.** equal **35.** $15y^2 - x^2 = 956$
37. $10 - \frac{5\sqrt{7}}{2} \approx 3.39$ m **39.** centers **41.** $\frac{x^2}{a^2} + \frac{y^2}{b^2} = 1$ **43.** one
49. **(i)** $8(y - 75)^2 - x^2 = 5000$, upper branch
(ii) $(\frac{600 + 200\sqrt{2}}{7}, \frac{600 + 200\sqrt{2}}{7})$ **51.** plot $y = \pm\sqrt{1 - e^2}\sqrt{1 - x^2}$

Exercises 11.3 (p. 750)

1. $(\frac{1}{\sqrt{2}} - \sqrt{2}, \frac{1}{\sqrt{2}} + \sqrt{2})$ **3.** $(-\frac{3\sqrt{3}}{2}, \frac{3}{2})$ **5.** $(\frac{7 + \sqrt{3}}{2}, \frac{7\sqrt{3} - 1}{2})$
7. $(-\frac{a\sqrt{3}}{2}, -\frac{a}{2})$ **9.** $(0, 0)$ **11.** ellipse, center $(0, 0)$, semi-major axis $\frac{1}{2}$, along line $y = -x$; semi-minor axis $\frac{1}{3}$

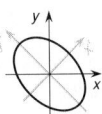

13. parabola, vertex $(0, 0)$, focus $(\frac{\sqrt{3}}{8}, \frac{1}{8})$

15. ellipse, center $(0, 0)$, semi-major axis 4, along line $3x + 4y = 0$; semi-minor axis 2

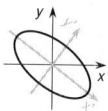

17. hyperbola, center $(0, 0)$, transverse axis along $2x + y = 0$, asymptotes $3y' = \pm x'$

19. hyperbola, center $(0, 0)$, transverse axis along $y = x$, asymptotes $y' = \pm 3^{1/4} x'$

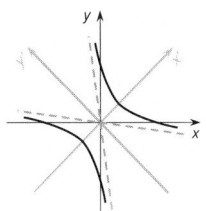

21. parabola, vertex $(\frac{19}{24}, -\frac{31}{24})$, focus $(\frac{5}{12}, -\frac{11}{12})$

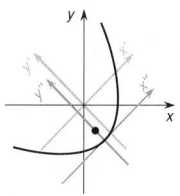

23. hyperbola, center $(\frac{3}{5}, \frac{1}{5})$, transverse axis along $3x - 4y - \frac{13}{5} = 0$

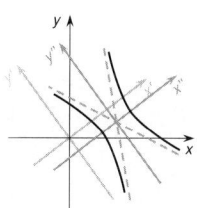

25. ellipse, center $(\frac{\sqrt{3}-1}{2}, \frac{\sqrt{3}+1}{2})$, semi-major axis 3, along $\sqrt{3}x + y = 2$; semi-minor axis 2

27. hyperbola, center $(3, -4)$, transverse axis horizontal, asymptotes $y = x - 7, y + x + 1 = 0$

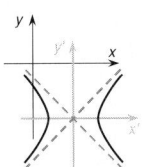

29. parabola, vertex $(-\frac{1}{2}, -\frac{9}{4})$, focus $(-\frac{5}{2}, -\frac{5}{4})$

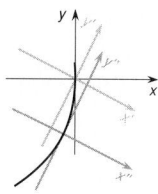

31. $y'^2/4 - x'^2/9 = 1$

33. $q = 0$

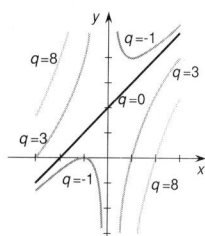

37. (i) $Ax^2 + Bxy + Cy^2 = A(x + \frac{By}{2A})^2$ **41.** $(\frac{3}{2}, -\frac{3}{2})$
43. $\pm(\frac{4}{5\sqrt{6}}, -\frac{3}{5\sqrt{6}})$ **45.** $17x^2 + 17y^2 + 16xy = 225$ **47.** $-\frac{\pi}{8}$

Chapter 11 Review Exercises (p. 753)
1. ellipse, semi-major axis 4, horizontal; semi-minor axis 3

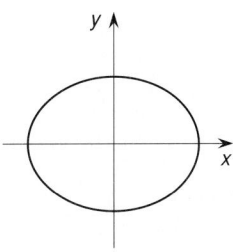

3. parabola, vertex $(2, 0)$, focus $(\frac{33}{16}, 0)$

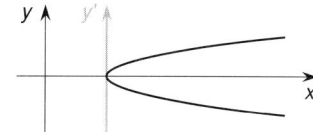

5. ellipse, semi-major axis 2, horizontal; semi-minor axis $\sqrt{2}$

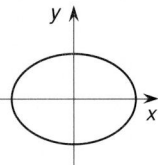

7. ellipse, center $(2, 0)$, semi-major axis 2, horizontal; semi-minor axis 1

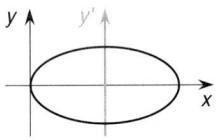

9. parabola, vertex $(-2, -\frac{7}{2})$, focus $(-2, -3)$

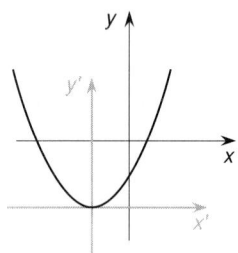

11. $\frac{x'^2}{16} + \frac{y'^2}{25} = 1$, ellipse, center $(0, 0)$, semi-major axis 5, along $y = -x$; semi-minor axis 4

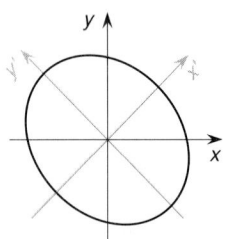

13. $y'^2 + 2x' = 0$, parabola, vertex $(0, 0)$, focus $(-\frac{2}{5}, \frac{3}{10})$

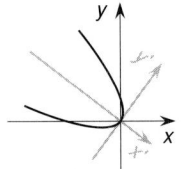

15. $y''^2 - x''^2 = 1$, hyperbola, center $(\frac{1}{\sqrt{2}}, \frac{3}{\sqrt{2}})$, asymptotes $x = \frac{1}{\sqrt{2}}$, $y = \frac{3}{\sqrt{2}}$, transverse axis along $x + y = 2\sqrt{2}$

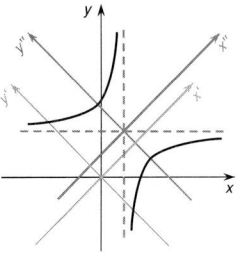

17. $\frac{3}{25}x''^2 + \frac{1}{5}y''^2 = 1$, ellipse, center $(\frac{2}{5}, \frac{11}{5})$, semi-major axis $\frac{5}{\sqrt{3}}$, along $3x + 4y = 10$; semi-minor axis $\sqrt{5}$

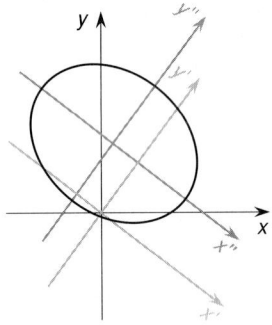

19. $y''^2 - 2x'^2 = 2$, hyperbola, center $(-\frac{4}{\sqrt{17}}, -\frac{16}{\sqrt{17}})$, transverse axis along $y = 4x$

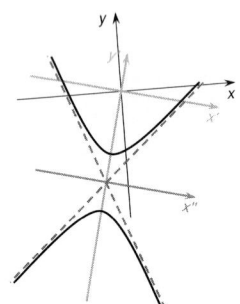

21. ellipse $c > -14$, point $c = -14$, empty set $c < -14$
23. (i) circle, radius $\sqrt{2}$, (ii) circle, radius 2, (iii) circle, radius 1, (iv) point $(-1, 1)$, (v) empty set

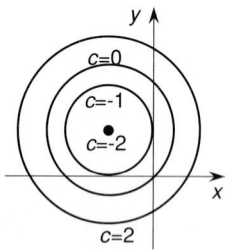

25. **(i), (ii), (iii)** parabolas, **(iv)** line $x + y = 0$

(i)

(ii)

(iii)

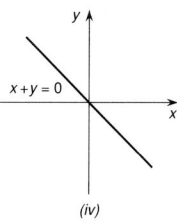

(iv)

27. **(ii)** counterclockwise once around the ellipse, starting at $(a, 0)$ **29.** $y = \pm\sqrt{e^2 - 1}x$, $x = \pm\sqrt{e^2 - 1}y$
31. **(i)** $y'^2 - y' + 2x'y' + x'^2 = 0$ **(ii)** parabola
(iii)

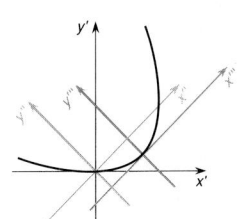

33. if $\frac{B}{A - C} > 0$,

$\sqrt{(\sqrt{(A - C)^2 + B^2} - |A - C|)/(\sqrt{(A - C)^2 + B^2} + |A - C|)}$; the
negative of this if $\frac{B}{A - C} < 0$ **35.** **(i)** $\frac{\pi}{2}$ **(ii)** $\frac{\pi}{3}$ **(iii)** $\frac{3}{2}$
37. **(ii)** $\frac{50,000,000}{\sqrt{14 + 8\sqrt{2}}} \approx 9{,}937{,}843$ km

Exercises 12.1 (p. 759)
1. $(2, 0)$ **3.** $(6\sqrt{2}, -\frac{\pi}{4} + 2k\pi)$ **5.** $(0, 0)$ **7.** $(4, \frac{5\pi}{6} + 2k\pi)$
9. $(9, \pi + 2k\pi)$ **11.** $(-1, 0)$ **13.** $(5\sqrt{2}, \frac{3\pi}{4} + 2k\pi)$ **15.** $(-7, 0)$
17. $(\frac{8}{\sqrt{3}}, -\frac{2\pi}{3} + 2k\pi)$ **19.** $(1, -\frac{\pi}{4} + 2k\pi)$ **21.** $(\sqrt{2}, 2k\pi)$
23. $(0, -1)$ **25.** $(0, -1)$ **27.** $(13, \arctan(-\frac{12}{5}) + 2k\pi)$
29. $(\frac{2}{\sqrt{5}}, \frac{4}{\sqrt{5}})$ **31.** **(i)** $r = 2$ **(ii)** $r^2 + 2r(\cos\theta - 2\sin\theta) = 4$, circle,
radius 3, center $(x, y) = (-1, 2)$ **33.** $r \geq 0$, $0 \leq \theta \leq \frac{\pi}{2}$
35. $1 < r < 2$, $\frac{\pi}{2} \leq \theta \leq \frac{3\pi}{4}$ **37.** $-x < y < x$ **39.** $x^2 + y^2 \leq 4$
41. circle around origin, radius 10   **43.** two half-lines,
$y = \pm\sqrt{3}x$, $x \geq 0$ (i.e., $\theta = \pm\frac{\pi}{3}$); if we allow $r < 0$, get whole lines
$y = \pm\sqrt{3}x$

Exercises 12.2 (p. 770)
1. circle, center $(x, y) = (0, \frac{1}{2})$, radius $\frac{1}{2}$

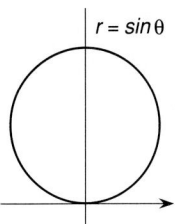

3. circle, center $(x, y) = (-1, 0)$, radius 1

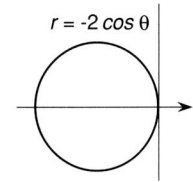

5. two half-lines $y = \pm x$, $y \geq 0$ (i.e., $\theta = \frac{\pi}{4}, \frac{3\pi}{4}$)

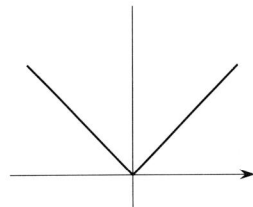

7. four-petaled rose (one petal in each quadrant)

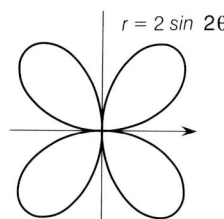

9. counterclockwise spiral, starting horizontally to the left from
the origin

11. cardioid

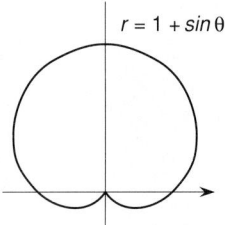

$r = 1 + \sin\theta$

13. limaçon with inner loop

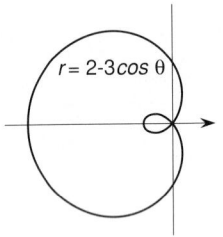

$r = 2 - 3\cos\theta$

15. limaçon with large inner loop

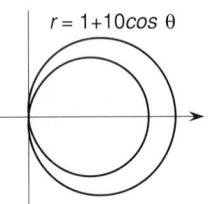

$r = 1 + 10\cos\theta$

17. lemniscate

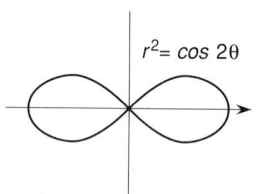

$r^2 = \cos 2\theta$

19. circle

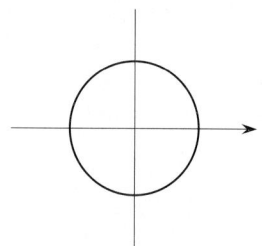

21. circle $r = 5\sqrt{2}\cos(\theta + \frac{\pi}{4})$

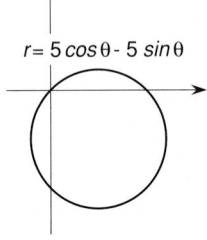

$r = 5\cos\theta - 5\sin\theta$

23. circle, radius $\frac{5}{2}$

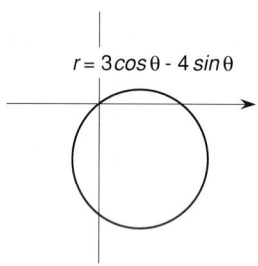

$r = 3\cos\theta - 4\sin\theta$

25. $(r, \theta) = (\frac{1}{\sqrt{2}}, \pm\frac{\pi}{4} + 2k\pi)$, i.e., $(x, y) = (\frac{1}{2}, \pm\frac{1}{2})$

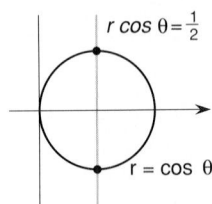

$r\cos\theta = \frac{1}{2}$

$r = \cos\theta$

27. $(r, \theta) = (4, \pm\frac{\pi}{3})$

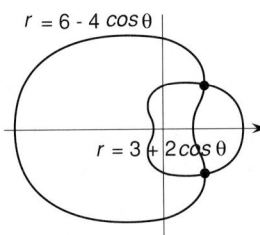

$r = 6 - 4\cos\theta$

$r = 3 + 2\cos\theta$

29. $r = \frac{1}{\sqrt{2}}$, $\theta = \frac{\pi}{20}, \frac{9\pi}{20}, \frac{17\pi}{20}, \frac{25\pi}{20}, \frac{33\pi}{20}$

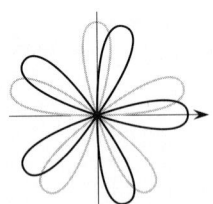

31. $(\frac{1}{2}, \pm\frac{2\pi}{3})$, $(2, 0)$, $(0, \pi)$

33. cardioid

37. six-petaled rose

39. two circles

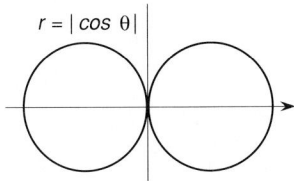

43. symmetric in vertical axis

45. horizontal axis: $((-1)^n n\pi, 0)$, $n = 0, 1, 2, \ldots$; vertical axis: $(0, (-1)^n(\frac{\pi}{2} + n\pi))$, $n = 0, 1, 2, \ldots$ **47.** $(\frac{2}{3}\sqrt{\frac{5}{6}}, \pm\frac{2}{3}\sqrt{\frac{1}{6}})$

49. highest: $(\pm\frac{\sqrt{3}}{2\sqrt{2}}, \frac{1}{2\sqrt{2}})$ lowest: $(\pm\frac{\sqrt{3}}{2\sqrt{2}}, -\frac{1}{2\sqrt{2}})$ **51.** two:
$(r, \theta) = (0, 0)$, $\approx (1.376, 1.376)$ **53.** three:
$(r, \theta) = (0, 0)$, $(\frac{\sqrt{3}}{2}, \frac{2\pi}{3})$, $(\frac{\sqrt{3}}{2}, \frac{\pi}{3})$

Exercises 12.3 (p. 778)
1. $\frac{x^3}{48}$ **3.** 38π **5.** $\frac{49\pi}{4}$ **7.** $\frac{1}{4}$ **9.** $\frac{\pi}{4}$ **11.** 6π **13.** 4π **15.** $\frac{\pi}{4}$
17. 1 **19.** π **21.** $\frac{\pi}{8} - \frac{1}{4}$ **23.** $\frac{\pi}{16}$ **25.** $\frac{3\pi}{2} - 4$ **27.** $\frac{3\pi}{2} - 2\sqrt{2}$
29. $\frac{4\pi}{3} - \sqrt{3}$ **31.** $\frac{3\pi}{4} - \sqrt{2}$ **33.** $\pi - \frac{3\sqrt{3}}{2}$ **35.** $\frac{\pi}{8} - \frac{1}{4}$ **37.** (i) $\frac{\pi}{4}$
41. $\frac{44\pi}{9}$ **43.** ≈ 13.3246 **45.** ≈ 3.5681

Exercises 12.4 (p. 785)
1. parabola, opening right

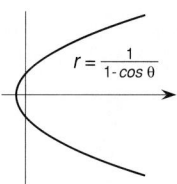

3. ellipse, $e = \frac{1}{2}$, major axis horizontal

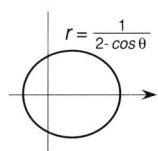

5. ellipse, $e = \frac{1}{2}$, major axis horizontal

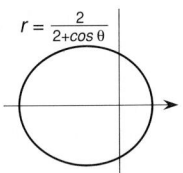

7. hyperbola, $e = 2$, transverse axis horizontal

9. parabola, opening left

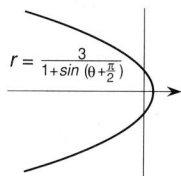

11. circle around $(x, y) = (\frac{1}{2}, 0)$, radius $\frac{1}{2}$

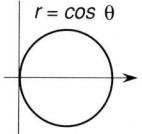

13. hyperbola, $e = 2$, transverse axis vertical

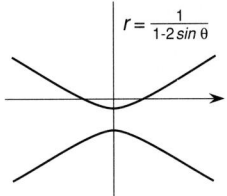

$r = \frac{1}{1 - 2\sin\theta}$

15. horizontal line $y = 4$

$r = 4\csc\theta$

17. circle around origin, radius 6

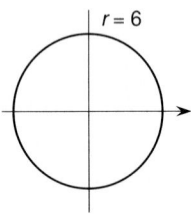

$r = 6$

19. line $x + y = 2\sqrt{2}$

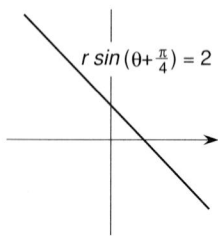

$r\sin\left(\theta + \frac{\pi}{4}\right) = 2$

21. $r = \frac{pe}{1 + e\cos\theta}$ **23.** $r = \frac{1}{1 - \sqrt{2}\cos\theta}$ **25.** $\frac{\pi}{3}$ **27.** $r = \frac{3}{\sqrt{2}}\sec(\theta + \frac{\pi}{4})$
29. ellipse, $e = \frac{1}{\sqrt{2}}$, major axis along $\theta = \frac{5\pi}{4}$

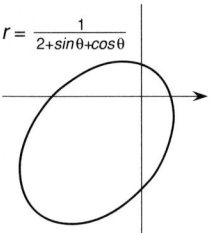

$r = \frac{1}{2 + \sin\theta + \cos\theta}$

31. focus $(x, y) = (\frac{2}{3}, 0)$, directrix $x = \frac{5}{3}$ **33.** $(x, y) = (1, 1)$,
distance $= \sqrt{2}$ **35.** $(r, \theta) = (\frac{2}{1 + 1/\sqrt{2}}, \frac{\pi}{4}), (\frac{2}{2 - 1/\sqrt{2}}, -\frac{3\pi}{4})$

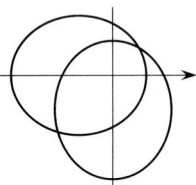

37. $\frac{8\pi}{3\sqrt{3}}$ **39. (i)** $\frac{1 - 2\cos\theta}{2\sin\theta}$ **(ii)** $(x, y) = (\frac{2}{3}, \frac{2\sqrt{3}}{3})$

Exercises 12.5 (p. 793)
1. $(x, y) = (5\cos t, 5\sin t)$; $0 \leq t < 2\pi$ **3.** $(x, y) =$
$(\cos^2 t, \cos t \sin t)$; $-\frac{\pi}{2} \leq t < \frac{\pi}{2}$ **5.** $(x, y) = (4t, 3t - \frac{7}{4})$; $t \in (-\infty, \infty)$
7. $(x, y) = (3, t)$; $t \in (-\infty, \infty)$ **9.** $(r, \theta) = (\sqrt{\cos 2t}, t)$;
$t \in [-\frac{\pi}{4}, \frac{\pi}{4}) \cup [\frac{3\pi}{4}, \frac{5\pi}{4})$, or $(x, y) = (\cos t\sqrt{\cos 2t}, \sin t\sqrt{\cos 2t})$;
$t \in [-\frac{\pi}{4}, \frac{\pi}{4}) \cup [\frac{3\pi}{4}, \frac{5\pi}{4})$ **11.** ≈ 0.99986 **13.** 0.24 **15. (i)** one
answer: $(x, y) = (6\cos(90\pi t), -6\sin(90\pi t))$, **(ii)** $(x, y) =$
$(\frac{3}{2}\cos(90\pi t), -\frac{3}{2}\sin(90\pi t))$ **17.** $(x, y) = (-a\cosh t, b\sinh t)$;
$t \in (-\infty, \infty)$, or $(x, y) = (-a\sqrt{1 + t^2/b^2}, t)$ $t \in (-\infty, \infty)$
19. $(x, y) = (t, \sqrt{9 - t^2})$; $t \in [-3, 3]$ **21.** $y = x^2 - 1$
23. $4x - y = 29$

Chapter 12 Review Exercises (p. 794)
1. straight line, slope $= 1$

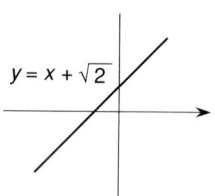

$y = x + \sqrt{2}$

3. ellipse, $e = \frac{1}{3}$

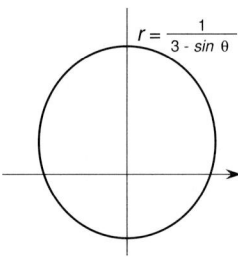

$r = \frac{1}{3 - \sin\theta}$

5. lemniscate, "transverse axis" along $\theta = \frac{\pi}{8}$

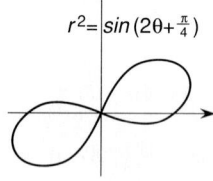

$r^2 = |\sin(2\theta + \frac{\pi}{4})|$

7. cardioid

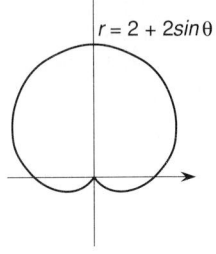

$r = 2 + 2\sin\theta$

9. two half-lines, $\theta = \frac{\pi}{6}, \frac{5\pi}{6}$

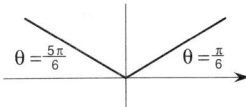

$\theta = \frac{5\pi}{6}$ $\theta = \frac{\pi}{6}$

11. $(r, \theta) = (2k\pi, 2k\pi), k = 0, 1, 2, \ldots$; i.e., $(x, y) = (2k\pi, 0)$

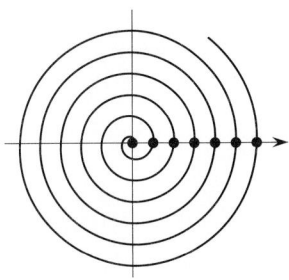

13. $(r, \theta) = (1, -\frac{\pi}{4})$, i.e., $(x, y) = (\frac{1}{\sqrt{2}}, -\frac{1}{\sqrt{2}})$

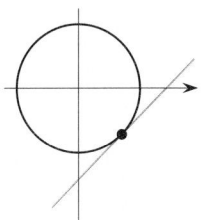

15. $(r, \theta) = (\frac{1}{2}, \theta)$, with $\theta = \pm\frac{\pi}{6}, \pm\frac{5\pi}{6}, \pm\frac{\pi}{3}, \pm\frac{2\pi}{3}$, i.e., $(x, y) = (\pm\frac{\sqrt{3}}{4}, \pm\frac{1}{4}), (\pm\frac{1}{4}, \pm\frac{\sqrt{3}}{4})$

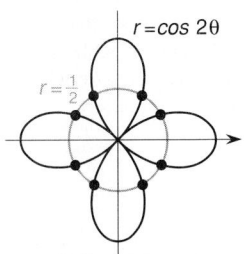

$r = \cos 2\theta$

$r = \frac{1}{2}$

17. $(r, \theta) = (\frac{1}{\sqrt{2}}, \pm\frac{\pi}{6}), (\frac{1}{\sqrt{2}}, \pm\frac{5\pi}{6})$, i.e., $(x, y) = (\pm\frac{\sqrt{3}}{2\sqrt{2}}, \pm\frac{1}{2\sqrt{2}})$

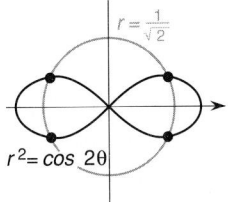

$r = \frac{1}{\sqrt{2}}$

$r^2 = \cos 2\theta$

19. $(r, \theta) = (0, 0)$

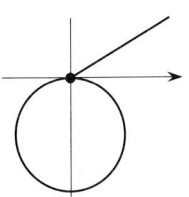

21. $\frac{2\pi}{3} - \frac{\sqrt{3}}{2}$ **23.** $\frac{5\pi}{24} - \frac{\sqrt{3}}{4}$ **25.** $\frac{9\pi}{2} - 4\sqrt{2}$ **27.** 0 **29.** $\frac{3\pi}{2}$
31. $(x, y) = (t, 2t^2 - 3); t \in (-\infty, \infty)$ **33.** $(r, \theta) = (\cos 2t, t);$
$t \in [0, 2\pi)$, or $(x, y) = (\cos(2t) \cos t, \cos(2t) \sin t); t \in [0, 2\pi)$
35. $(x, y) = (t, 2), t \in (-\infty, \infty)$ **37.** $(x, y) = (4t, -5t - \frac{3}{4});$
$t \in (-\infty, \infty)$ **39.** $(r, \theta) = (-3 + 3 \sin t, t); t \in [0, 2\pi)$, or
$(x, y) = (-3 \cos t + 3 \cos t \sin t, -3 \sin t + 3 \sin^2 t); t \in [0, 2\pi)$
41. convex limaçon

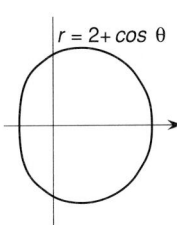

$r = 2 + \cos \theta$

43. three-leaf "clover," six if we allow $r < 0$

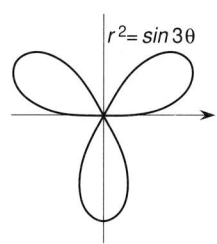

$r^2 = \sin 3\theta$

45. two lines $x = 1, y = -1$

47. tightening spiral

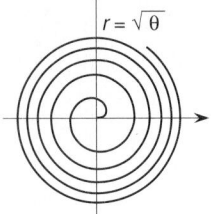

$r = \sqrt{\theta}$

49. solid flower head with six lobes

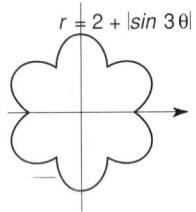

$r = 2 + |\sin 3\theta|$

51. yes   **53.** cos or sin in denominator
55. $r = \frac{1}{2 - \cos\theta}$, $r = \frac{3}{2 + \cos\theta}$
57.

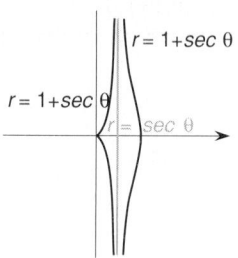

$r = 1 + \sec\theta$

$r = 1 + \sec\theta$ $r = \sec\theta$

Exercises 13.1 (p. 802)

1. $(x, y) = (4t, t)$; $t \in (-\infty, \infty)$, traced once   **3.** $(x, y) = (t^2 + 2, t)$; $t \in (-\infty, \infty)$, once   **5.** $(x, y) = (2 \cos t, \sqrt{3} \sin t)$; $t \in [0, 2\pi)$, once
7. $(t, 2)$; $t \in (-\infty, \infty)$, once   **9.** $(x, y) = (t, 3 - 2 \sin t)$; $t \in (-\infty, \infty)$ once   **11.** $x^2 + y^2 = 1$, circle, once around clockwise
13. $4x + 3y + 2 = 0$, the half-line with $x \geq -5$; point moves in from the right, turns around, and goes out to the right, tracing the half-line twice   **15.** $2x + y = 11$, the half-line with $x \geq 4$, traced once from left to right   **17.** $x^2 - 4x - 2xy + y^2 = 0$, or $4x = (x - y)^2$; parabola, lower part is traced out twice, from right to left and back again   **19.** $(y - 1)^2 - x^2 = 1$; hyperbola, right half of lower branch traced once from left to right   **21.** $(x, y) = (3 - t, 2)$, for $t \in [0, 3)$; $(0, 5 - t)$ for $t \in [3, 5)$; $(t - 5, 0)$ for $t \in [5, 8)$; $(3, t - 8)$ for $t \in [8, 10]$
23. $(x, y) = (2 + t, 0)$, for $t \in [0, 1)$; $(3, t - 1)$ for $t \in [1, 3)$; $(6 - t, 2)$ for $t \in [3, 6)$; $(0, 8 - t)$ for $t \in [6, 8)$; $(t - 8, 0)$ for $t \in [8, 10]$ **27.** parabola   **29.** parabola

Exercises 13.2 (p. 813)

1. $(4, -3)$ **3.** $(4t - 1, 8t - 2)$ **5.** $(t/\sqrt{1 + t^2}, 1)$ **7.** $(2t, \frac{1}{1 + t^2})$

9. $(-6 \sin(3t), 6 \cos(3t))$ **11. (i)** $(x, y) = (-1, 0)$, $t = \frac{3\pi}{2} + 2k\pi$
(ii) $(0, -1)$, $t = \pi + 2k\pi$ **13. (i)** none **(ii)** none   **15. (i)** none
(ii) $(2, 12)$, $t = -1$ **17. (i)** none **(ii)** none   **19. (i)** $(e^{-2}, 1)$, $t = 1$
(ii) none   **21.** 3π **23.** $10\sqrt{5} - 4\sqrt{2}$ **25.** 10   **27.** $\sinh 1 = \frac{1}{2}(e - \frac{1}{e})$ **29.** 32   **31. (i)** $t = \frac{1}{2}$ **(ii)** $5\sqrt{2}\pi$ km/hr **(iii)** $5\sqrt{2}\pi$ km/hr   **33.** "infinity sign"; $t = k\pi$, $k = 0, \pm 1, \pm 2, \ldots$

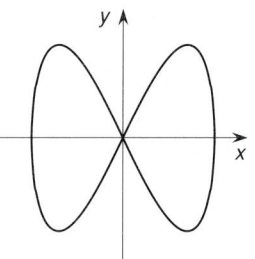

35. $t = k\pi$, $k = 0, \pm 1, \pm 2, \ldots$

37. $(1, 1)$
41.

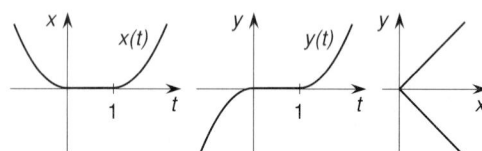

$x(t)$ $y(t)$

41. (iii) does not exist   **43.** π **45.** 8   **47.** 6.5192   **49.** 7.1396
51. 9.1073

Exercises 13.3 (p. 820)

1. $16\sqrt{5} \approx 35.8$ ft/sec   **3.** $\sqrt{298.9} \approx 17.3$ m/sec   **5.** $5\sqrt{174} \approx 66$ ft   **7.** $\sqrt{7808/3} \approx 51.0$ ft/sec   **9.** $\frac{\sqrt{1335}}{16} \approx 2.28$ sec, or $\frac{\sqrt{2761}}{16} \approx 3.28$ sec   **11.** $\approx 12.09° \approx 0.211$ rad, or 77.91°
13. ≈ 609 ft   **15.** 6.25 sec   **17.** ≈ 1020 m   **19.** ≈ 52.6 ft/sec
21. ≈ 26.1 ft if $\theta \approx 11.8°$; ≈ 599 ft if $\theta \approx 78.2°$ (see Figure 13.3.6)   **23.** $\theta = \frac{\pi}{2} = 90°$, i.e., straight up; height = 625 ft
25. quadruple the range   **27.** $16\sqrt{2} - 16 \approx 6.63$ ft farther
29. plot $(x, y) = (200 \cos(\theta)t, -16t^2 + 200 \sin(\theta)t$

Chapter 13 Review Exercises (p. 822)

1. $(x, y) = (7t, 2t - \frac{4}{7})$; $t \in (-\infty, \infty)$ **3.** $(x, y) = (t, -3)$; $t \in (-\infty, \infty)$
5. $(x, y) = (8 + 5t - 3t^2, t)$; $t \in (-\infty, \infty)$ **7.** $(x, y) = (2 \cos t, \frac{1}{\sqrt{2}} \sin t)$; $t \in [0, 2\pi)$ **9.** $(x, y) = (t, \frac{1}{t})$; $t \in (-\infty, 0) \cup (0, \infty)$

11. $(x, y) = (\sqrt{1 + 2t^2}, t)$; $t \in (-\infty, \infty)$, or $(x, y) = (\cosh t, \frac{1}{\sqrt{2}} \sinh t)$; $t \in (-\infty, \infty)$ **13.** $(x, y) = (\sqrt{6} \cos \theta, \sqrt{\frac{5}{2}} \sin \theta)$; $\theta \in [-\frac{\pi}{2}, \frac{\pi}{2}]$
15. $(x, y) = (t, -\frac{3}{2}\sqrt{4 + x^2})$; $t \in (-\infty, \infty)$ **17.** $(x, y) = (t, |t|)$; $t \in (-\infty, \infty)$ **19.** $(x, y) = (\sqrt{t}, t)$; $t \in [0, \infty)$, or (t, t^2); $t \in [0, \infty)$
21. $x = 3(y + 4)^2$, parabola   **23.** $9x^2 + y^2 = 9$, ellipse
25. $x^2 + 4y^2 = 1$; ellipse   **27.** $\frac{y^2}{4} - \frac{x^2}{9} = 1$, hyperbola
29. $x + y = 1$; line segment from $(1, 0)$ to $(0, 1)$, omitting the endpoint $(0, 1)$ **31.** $(5, -1)$ **33.** $(2e^{2t}, 2)$ **35.** $(\cos t, 3 \cos(3t))$
37. $(2t \cos(t^2 + 2), 2 \cos t \sin t)$ **39.** $\left(\frac{3}{1 + 9t^2}, \frac{1}{\sqrt{1 - t^2}}\right)$
41. $\frac{1}{27}(13\sqrt{13} - 8)$ **43.** $\frac{9}{8}$ **45.** π **47.** $2(1 + e^2)^{3/2} - 4\sqrt{2}$
49. $\frac{3886}{9375}\sqrt{29} + \frac{128}{9375}$ **51.** $t = \pm 1, \pm 2, \pm 3, \ldots$ The corresponding points are $(3n^3, \cos \pi n)$, $n = \pm 1, \pm 2, \pm 3, \ldots$
53. $t = -1, (x, y) = (-6, -9)$ **55.** 9 ft   **57.** 2 sec
59. $\sqrt{12304} \approx 111$ ft/sec   **61.** $(x, y) = (-1 + \cos \theta, 2 + \sin \theta)$; $\theta \in [0, 2\pi]$ **63.** $(x, y) = (t, 2 + 3\sqrt{t^2 + 2t + 2})$; $t \in (-\infty, \infty)$, or $(-1 + \sinh t, 2 + 3 \cosh t)$, $t \in (-\infty, \infty)$ **65.** $(x, y) = (2 \cos t, \sin t)$; $t \in (\arctan(2) - \pi, \arctan(2))$ **67. (ii)** if $y'(a) \neq 0$, the tangent is vertical   **69.** equal   **71.** ≈ 0.905 **73.** $\approx 2.701\ 29$

Exercises 14.1 (p. 829)

1.
3.
5. $R = (0,0,3)$

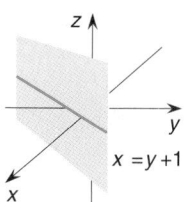

7.
9. $V = (0,-4,0)$

11. vertical plane, parallel to y-axis

13. vertical plane

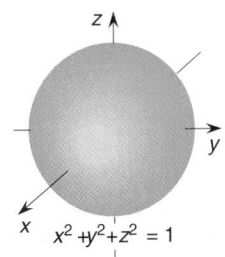

$x = y + 1$

15. vertical plane

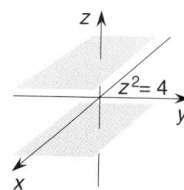

$3x + 2y = 6$

17. sphere around origin, radius = 1

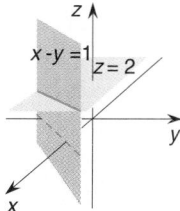

$x^2 + y^2 + z^2 = 1$

19. two horizontal planes

$z^2 = 4$

21. 5   **23.** 0
25. line

$x - y = 1$ $z = 2$

27. cylinder

$$x^2 + y^2 = 1$$

29.

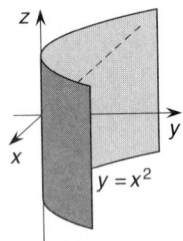

$$y = x^2$$

31. (i) minimum at $(0, 0)$

(ii) maximum at $(1, 0)$

(iii) no max. or min.

Exercises 14.2 (p. 841)

1. $(1, -1, 4)$ **3.** -1 **5.** $(1, -2, -8)$ **7.** $(0, 0, 0)$ **9.** $(0, 0, 0)$

11. $\frac{3\pi}{4}$ **13.** $\frac{\pi}{2}$ **15.** $\frac{5\pi}{6}$ **17.** $\arccos(\sqrt{\frac{14}{17}}) \approx 0.434$ radians $\approx 24.8°$

19. $\frac{\pi}{2}$ **21.** $(2, -1)$, or $t(-2, 1)$ for any $t \neq 0$ **23.** one answer:

$(2, 3, -1)$ **25.** one answer: $(1, -1, 0), (0, 1, -1)$

27. $(1, 2, -4) \times (-3, 2, 1) = (10, 11, 8)$ **29.** $a = \frac{1}{4}$

33. (i) perpendicular to yz-plane **(ii)** all vectors $(a, 0, 0)$, with

$a \in \mathbb{R}$ **35.** one answer $\mathbf{i} \times (\mathbf{j} \times \mathbf{j}) \neq (\mathbf{i} \times \mathbf{j}) \times \mathbf{j}$ **37. (i)** $\approx 43.6°$

(ii) $\approx 3.37°$ **(iii)** $\approx 71.6°$ **(iv)** $\approx 139.8°$

Exercises 14.3 (p. 852)

1. $\sqrt{10}$ miles **3.** $2\sqrt{5} - 2\sqrt{2} \approx 2.95$ miles **5.** 0 **7.** 25 ft

9. $\sqrt{74}$ m **11.** $(5, 0)$ **13.** $(3, 3, 3)$ **15.** $(\frac{1}{2}, \frac{1}{2})$ **17.** $(0, 0, 0)$

19. $(\frac{46}{9}, -\frac{2}{9}, \frac{25}{9})$ **21.** 300 ft-lb **23.** $36\sqrt{2}$ ft-lb **25.** 0

27. $-35\sqrt{2}$ ft-lb **29.** 1296 ft-lb **31.** $12\sqrt{2} \approx 16.97$ ft **33. (i)**

one answer: $(x, y) = (5t - 2 \sin \frac{5t}{2}, 2 - 2 \cos \frac{5t}{2})$ **(ii)** 10 ft/sec,

0 ft/sec **35. (i)** $\sqrt{234} \approx 15.3$ mph **(ii)** $\arctan \frac{3}{\sqrt{216}} \approx 11.54°$

upstream **37.** $0°$, i.e., straight across **39.** $10\sqrt{2} \approx 14.14$ lb

41. 115 ft-lb **43.** 2 **45. (i)** 3 **(ii)** 8

Exercises 14.4 (p. 861)

1. $(x, y, z) = (2 + t, -1 + t, 4 - t)$

3. $(x, y, z) = (3, 3, -2) + t(1, 1, 1)$ **5.** $\frac{x-1}{4} = \frac{y-2}{-3} = \frac{z+4}{-2}$

7. $(x, y, z) = (2 + t, -3 + 2t, 5 - t)$ **9.** $(x, y, z) = (-1, -2 + t, 7)$

11. no **13.** no **15.** yes **17.** no **19.** yes **21.** none

23. $(-2, -4, 11)$ **25.** none **27.** none **29.** none

31. $t = -\frac{1}{12}, t = \frac{5}{3}$ **33.** $(x, y, z) = (3 - t, 5 + 13t, -2 + 22t)$

35. $(1, -1, 0)$ **37.** $(-1, 10, 0)$ (!) **39.** $(8, -1, 8)$

Exercises 14.5 (p. 868)

1. $3x - 2y + z = 3$ **3.** $2x - 4y - z = 10$ **5.** $5x + y - z = 17$

7. $2x - y + 3z = 8$ **9.** $x + z = 3$ **11.** $x + y - z = 1$

13. $2x + 2y - z = -6$ **15.** $2x + 5z = 9$ **17.** $2x + y = 5$

19. $10x - 16y - 9z = 23$ **21.** $(x, y, z) = (3 + 3t, 2 - 2t, 1 + t)$

23. $(x, y, z) = (2 + t, 2 - 5t, -1 - 4t)$

25. $(x, y, z) = (t, 4 - 7t, 6 - 11t)$ **27.** $(x, y, z) = (2 + 4t, \frac{1}{2} + 5t, 6t)$

29. $(x, y, z) = (8t, -t, -10t)$ **33.** $2x + 2y - z = 2$ **35.** $(\frac{8}{9}, -\frac{4}{9}, -\frac{8}{9})$

37. $\theta \approx 26.57°, \phi \approx 37.76°$

Chapter 14 Review Exercises (p. 869)

1. $(2, -5, 3)$ **3.** 4 **5.** $(-11, -1, 19)$ **7.** $(\frac{57}{25}, -\frac{76}{25})$ **9.** $\frac{\pi}{2}$

11. $\sqrt{1000} \approx 31.6$ mph **13.** $\sqrt{116} \approx 10.77$ knots

15. $\sqrt{162,025 - 18,000\sqrt{2}} \approx 369.6$ mph **17.** $\arctan \frac{4}{\sqrt{209}} \approx$

$15.47° \approx 0.2699$ rad **19.** $\approx 84.2°$ west of north

21. $(12\sqrt{2}, 0)$, a horizontal force of ≈ 16.97 lb **23.** 6 lb

25. $75\sqrt{2}$ ft-lb **27.** $450\sqrt{3}$ J **29.** $\arcsin \frac{3}{7} \approx 25.38°$

31. $(x, y, z) = (4 + 3t, 2 + 5t, -3 - 3t)$

33. $(x, y, z) = (3 + t, -4 - 3t, 5 + 4t)$ **35.** $(2, 4, 8)$ **37.** $y + z = 1$

39. $(-\frac{5}{14}, \frac{29}{14}, \frac{4}{7})$

Exercises 15.1 (p. 879)

1. helix around z-axis

3. helix around x-axis

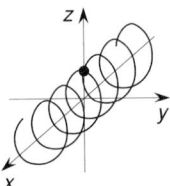

5. "conical spiral" around x-axis

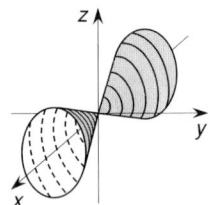

7. "skewed spiral" around the line $(x, y, z) = (0, t, t)$

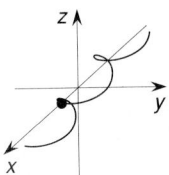

9. line segment from $(-1, -1, 0)$ to $(1, 3, 2)$

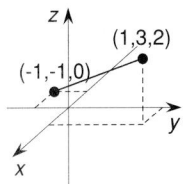

11. $\frac{\pi}{4}$ **13.** 0   **15.** $\frac{\pi}{4}$ **17.** $\frac{\pi}{6}$ **19.** 0   **21.** "two-sided conical spiral"; $t = 0$ slowest   **23.** $(2 \sin \frac{2\pi}{3}t, t, 2 \cos \frac{2\pi}{3}t)$ **25.** $(0, 1, 1)$
27. $\frac{\pi}{4}$ **29.** (i) one answer: $(x, y, z) = (-2t, t, 5t^2)$ (ii) $(0, 0, 0)$
31. $\frac{\pi}{6}$ **33.** $\frac{\pi}{4}$ **35.** min 0, max $\approx 18{,}433.03$

Exercises 15.2 (p. 891)
1. $\mathbf{T} = (\cos t, \sin t)$, $\mathbf{N} = (-\sin t, \cos t)$
3. $\mathbf{T} = (-\sin 4t, \cos 4t)$, $\mathbf{N} = (-\cos 4t, -\sin 4t)$
5. $\mathbf{T} = (\frac{2}{3}, -\frac{1}{3}, -\frac{2}{3})$, $\mathbf{N}$ does not exist   **7.** $\mathbf{T} =$
$(-\frac{1}{\sqrt{10}} \sin t, \frac{1}{\sqrt{10}} \cos t, \frac{3}{\sqrt{10}})$, $\mathbf{N} = (-\cos t, -\sin t, 0)$
9. $\mathbf{T} = (\frac{t^2 - 1}{t^2 + 1}, \frac{2t}{t^2 + 1})$, $\mathbf{N} = (\frac{2t}{t^2 + 1}, \frac{1 - t^2}{t^2 + 1})$ **11.** $s = 5t$ **13.** $6t$
15. $\frac{1}{3}t^3 + t$ **17.** $\ln(t + \sqrt{1 + t^2})$ **19.** $2t$
21. $\mathbf{N} = -\frac{1}{R}(x, y) = (-\frac{x}{R}, -\frac{y}{R})$ **23.** $\mathbf{N} = (-x, -y, 0)$
29. e.g., $(x, y, z) = (t^2, t^2, t^2)$ **31.** $\mathbf{p}(s) = (3 \cos \frac{s}{5}, -5 \sin \frac{s}{5}, 4 \cos \frac{s}{5})$
33. $\sqrt{2} \ln(t + \sqrt{1 + t^2})$ **35.** $\approx 3.726\,042$ **37.** $\approx 0.900\,678$

Exercises 15.3 (p. 899)
1. 1   **3.** $\frac{1}{3}$ **5.** $\frac{1}{5}$ **7.** $4\sqrt{5}(4t^2 + 4t + 21)^{-3/2}$ **9.** 0
11. $x^2 + (y - 2)^2 = 9$ **13.** circle, center $(0, 0, 3)$, radius = 3, lying horizontally; $x^2 + y^2 = 9$, $z = 3$ **15.** $x^2 + (y - \frac{1}{2})^2 = \frac{1}{4}$
17. $(x + 1)^2 + (y - 2)^2 = 8$ **19.** radius = $\sqrt{2}$, center $(0, 0)$
21. $\kappa = 2|a|$ **25.** $\kappa = |f''(x)|(1 + f'(x)^2)^{-3/2}$ **27.** yes   **29.** 1
31. at the vertex $(-\frac{b}{2a}, c - \frac{b^2}{4a})$ **33.** $2(1 + 4a^2)^{-3/2}$ **37.** center $(\frac{7}{5}, 5)$, radius $\frac{5\sqrt{5}}{2}$ **39.** center $(-143, \frac{241}{12})$, radius $\frac{145^{3/2}}{12}$ **41.** center $(\frac{5}{3}, 0)$, radius $\frac{4}{3}$ **43.** center $(-2, 0)$, radius 1

Exercises 15.4 (p. 906)
1. sphere around origin, radius = 3

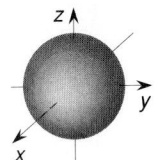

3. paraboloid, vertex $(0, 0, 0)$ opening upward

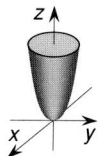

5. two-ended cone around x-axis

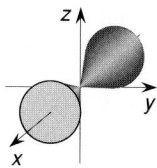

7. hyperboloid of one sheet around y-axis

9. hyperboloid of two sheets around x-axis

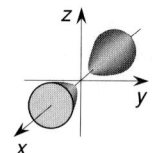

11. sphere, center $(-1, 0, 2)$, radius = 3

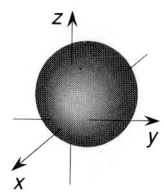

13. hyperboloid of one sheet, center $(0, 1, 1)$, around y'-axis

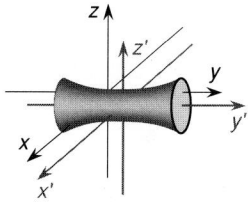

15. ellipsoid, center $(-1, 2, 1)$; semi-axes 1, 1, 2

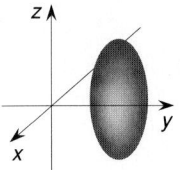

17. saddle surface, center $(2, 0, -2)$

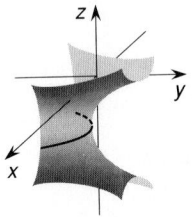

19. hyperboloid of one sheet, center $(1, 1, 0)$, around y'-axis

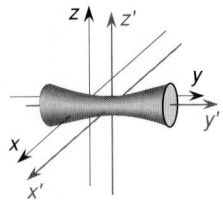

21. hyperbola in xy-plane and all points above and below it

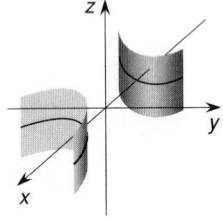

23. "parabolic trough" on its side; vertex along the line $x = y, z = 0$

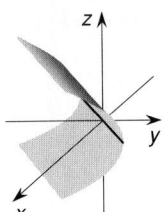

25. (i) parabola opening up **(ii)** parabola opening up
(iii) parabola opening down **(iv)** hyperbola **(v)** hyperbola
27. $x + 2yz = 0$

Exercises 15.5 (p. 912)
1. $(-3, 0, 2)$ **3.** $(0, 0, 7)$ **5.** $(-6, 0, 0)$ **7.** $(0, 1, 0)$
9. $(0, 0, -4)$ **11.** $(3\sqrt{2}, -\frac{\pi}{4}, 4)$ **13.** $(8, \theta, \pi)$, any θ
15. $(5, \frac{\pi}{4}, \arctan\frac{3}{4})$ **17.** $(0, \theta, -9)$, any θ **19.** $(-4, 0, \frac{\pi}{2}) =$
$(4, \pi, \frac{\pi}{2})$ **21.** cylinder around x-axis, radius $= 5$ **23.** half-plane
$x = 0, y \geq 0$ **25.** downward-opening half-cone   **27.** "solid
lemniscate," vertical axis   **29.** surface of revolution obtained by
revolving a cardioid around the z-axis   **31.** $(4000, 2°, 41°)$
33. $(4000, 0°, 38°)$ **35.** $\rho = 4000, \phi = 113\frac{1}{2}°$ **37.** it is all points
lying above or below the points (r, θ) in the horizontal plane
satisfying $F(r, \theta) = 0$

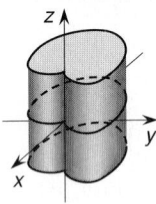

39. it consists of half-lines from the origin, i.e., if a point P is on
the surface, then so is the whole half-line from O through P
41. it consists of vertical semicircles centered at the origin with
endpoints on the z-axis; i.e., if a point P is on the surface, then
so is the whole semicircle through P

Exercises 15.6 (p. 918)
1. $(10, 10, 10, 10, 10)$ **3.** -10 **5.** 6   **7.** $\sqrt{6}$ **9.** $\frac{5\pi}{6}$
11. $(\frac{1}{2}, -\frac{1}{2}, \frac{1}{2}, \frac{1}{2})$ **13.** $(\frac{3}{5}, \frac{4}{5}\cos t, 0, \frac{4}{5}\sin t)$ **15.** $(\frac{t^2}{t^2+2}, \frac{-2t}{t^2+2}, \frac{-2}{t^2+2})$
17. $(\frac{\cos t}{\sqrt{4t^2+5}}, \frac{\sin t}{\sqrt{4t^2+5}}, \frac{-2t}{\sqrt{4t^2+5}}, \frac{2}{\sqrt{4t^2+5}})$ **19.** $(0, -t, -\sqrt{1-t^2})$
21. $x_1^2 + x_2^2 + x_3^2 + x_4^2 + x_5^2 + x_6^2 = 25$ **23.** $(\frac{3}{7}, -\frac{2}{7}, 0, \frac{6}{7})$
25. $(-\frac{2}{5}, 0, \frac{4}{5}, \frac{2}{5}, -\frac{1}{5})$ **27.** $(0, 0, 0, \pm 2), (0, \pm 6, 0, 0)$
29. $(3, 0, -1, 5)$

Exercises 15.7 (p. 925)
1. ≈ 155 min   **3.** 6626 miles   **5.** ≈ 1417 miles from the surface
7. 7900 miles   **9.** 807 km   **11.** 1.89 years   **13.** 247 years
15. $\approx 68,000,000$ miles   **17.** 59,000,000 miles   **19.** 43,200,000
miles   **21.** $\approx 26,500$ miles

Chapter 15 Review Exercises (p. 926)
1. $\sqrt{3}$ **3.** $(\frac{1}{\sqrt{1+9t^4}}, \frac{3t^2}{\sqrt{1+9t^4}})$ **5.** $\frac{1}{2}$ **7.** $\frac{4}{9}$ **9.** $\kappa = \frac{2}{9}7^{7/6}$ at
$x = \pm 56^{-1/6}$ **11.** two-ended cone around x-axis

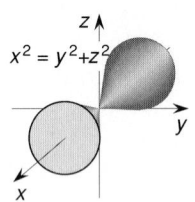

$x^2 = y^2 + z^2$

13. paraboloid, vertex $(-\frac{4}{3}, 0, 0)$, opening around positive x-axis

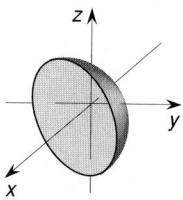

15. ellipsoid, semi-axes $\frac{1}{\sqrt{5}}, \frac{1}{\sqrt{3}}, \frac{1}{\sqrt{7}}$

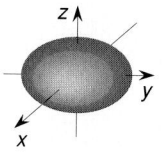

17. cylinder around y-axis, radius 4

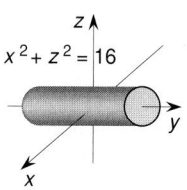

$x^2 + z^2 = 16$

19. saddle surface

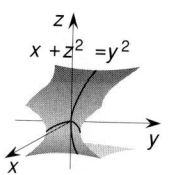

$x + z^2 = y^2$

21. $(4\sqrt{2}, -\frac{\pi}{4}, -2)$ **23.** $(4\sqrt{2}, \frac{\pi}{2}, \frac{3\pi}{4})$ **25.** $\theta = \frac{\pi}{6}, r \in \mathbb{R}$; or
$\theta = \frac{\pi}{6}, \frac{7\pi}{6}, r \geq 0$ **27.** $r^2 + z^2 = 25$ **29.** $(\rho, \theta, \phi) =$
$(2\sqrt{2} \csc t, -\frac{\pi}{4}, t); t \in (0, \pi)$ **31.** $3\sqrt{5}$ **33.** 3   **35.** $\frac{\pi}{4}$
37. $(0, 0, 0, 0)$ **39.** $(2, 2, 2, 2)$ **41.** $(x, y, z) =$
$(a, b + R \cos t, c + R \sin t); t \in [0, 2\pi]$ **45.** $(4006, 87°, 62°)$
47. $\frac{t^3}{3} + 2t$

Exercises 16.1 (p. 937)
1. straight lines, slope $= -1$

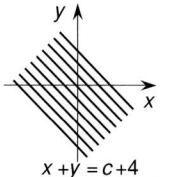

$x + y = c + 4$

3. horizontal lines

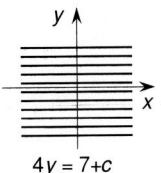

$4y = 7 + c$

5. ellipses

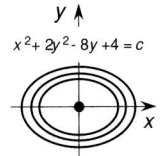

$x^2 + 2y^2 - 8y + 4 = c$

7. circles, radius $\geq \sqrt{5}$

9. circles

11. planes

13. spheres

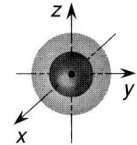

15. cylinders with elliptical cross-sections

$x^2 + 2y^2 = c$

17. hyperboloids of one or two sheets, main axis along x-axis

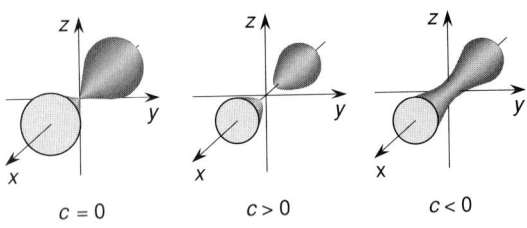

$c = 0$ $c > 0$ $c < 0$

19. hyperboloids of one or two sheets, main axis along z-axis

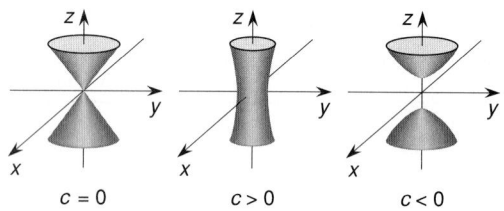

$c = 0$ $c > 0$ $c < 0$

21. horizontal lines

Exercises 16.2 (p. 950)

1. 54   **3.** does not exist   **5.** 0   **7.** 8   **9.** does not exist
11. continuous everywhere   **13.** continuous everywhere
15. continuous except on the plane $x + y - 2z = 1$
17. continuous when $2x - 3y + 4 > 0$, i.e., below the line
$2x - 3y + 4 = 0$ **19.** continuous except on the lines
$x + y = \pm\frac{\pi}{6} + k\pi, k = 0, \pm 1, \pm 2, \ldots$ **21.** 0   **23.** 0   **25.** 0
27. does not exist ($\to$ 1 along x-axis, $\to$ 0 along y-axis)
29. does not exist   **31.** $\to$ 0   **33.** $\to$ 1   **35.** (iii) $\to \frac{1}{2}$
(iv) $\to \frac{2}{5}$ **41.** $M = -1, N = 1$

Exercises 16.3 (p. 961)

1. $6xy - 4y^3$ **3.** $2xy^2 \cos(x^2y^2)$ **5.** $\frac{2t}{1+t^2}$ **7.** $yz(1 - x^2 + z^2)/$
$(1 + x^2 + z^2)^2$ **9.** $2x\sqrt{4 + x^2y^2} + x^3y^2/\sqrt{4 + x^2y^2}$
11. $4x + 4y - z = 8$ **13.** $2x - 2y - z = 0$ **15.** $z = -5$
17. $x - 2y - z = -1$ **19.** $12x + 5y - 13z = 0$ **21.** $12y^3 - 8xy$

23. $-2y \sin(x^2y + 2) - 4x^2y^2 \cos(x^2y + 2)$
25. $-2(x + y)(1 + (x + y)^2)^{-2}$ **27.** $2\pi x \sec^2(\pi x^2 y) + 4\pi^2 x^3 y$
$\sec^2(\pi x^2 y) \tan(\pi x^2 y)$ **29.** $(2x + x^2y)e^{xy}$ **31.** yes   **33.** yes   **35.** no
37. no   **39.** no   **43.** $ax + by + cz = 4$ **45.** $ax + by + cz = 1$
49. (i) decrease by $\frac{1}{25}$ liter to $\frac{24}{25} = 0.96$ liter (ii) 0.96 liter

Exercises 16.4 (p. 972)

1. $-36\,dx + 39\,dy$ **3.** $-7\,dx + 8\,dy$ **5.** $7\,dx$ **7.** $-\frac{2}{25}\,dx + \frac{4}{25}\,dz$
9. $-\frac{3}{\sqrt{2}}\,dx - \frac{1}{\sqrt{2}}\,dy + \frac{1}{4\sqrt{2}}\,dz$ **11.** ≈ 0.7 **13.** $-\frac{311}{600} \approx -0.5183$
15. $-\frac{353}{300} \approx -1.177$ **17.** $\frac{92}{35} \approx 2.6286$ **19.** $5 + \frac{17}{42} \approx 5.405$
21. 0.68%   **23.** 1.08%   **25.** 2.09%   **27.** 4.1%   **29.** 0.56%
31. (i) 24.0677 ft^3 **33.** (i) $dV = -\frac{kT}{p^2}\,dP + \frac{k}{P}\,dT$ (ii) $dT =$
$\frac{P}{k}\,dV + \frac{T}{P}\,dP$ **35.** 39.3%   **37.** $(-0.0014, 0.0017)$
39. $(0.8724, 0.9362, 1.1638)$

Exercises 16.5 (p. 987)

1. $96x^7 - 252x^6 + 162x^5 + 70x^4 - 84x^3$
3. $((x - 2y)(2x + y) + x^2 + xy) \exp((x^2 + xy)(x - 2y))$
5. $3(\cos t + st)^2(s - \sin t) - 12(s^2 - 4t)^2$
7. $(24t^3 + 6t - 3 + 9t^2(6t^2 + 1)(2t^3 + t - 1)$
$\cos(2t^3 + t - 1) + 9t(2t^3 + t - 1) \sin(2t^3 + t - 1))$
$\exp(3t \sin(2t^3 + t - 1))$ **9.** $-4x^2y^3(x^2y^4 + \exp(x^2 + 3x - 1))^{-2}$
11. $-(\frac{x}{y})^2$ **13.** $\frac{1 - y \cos(xy)}{x \cos(xy)}$ **15.** $-(2xy + 2z^2)/(y^2 + 4xz)$
17. $-(x^2w^4 + 6uvy)/(4ux^2w^3 - xv)$ **19.** $-\frac{\cos v}{\sin u}$
27. $2r^4(\cos^4 \theta - 2 \cos \theta \sin \theta - 6 \cos^2 \theta \sin^2 \theta + \sin^4 \theta)$
29. $20r^3 \cos^2 \theta \sin^3 \theta, 5r^4(3 \sin^2 \theta \cos^3 \theta - 2 \cos \theta \sin^4 \theta)$,
$r^5(6 \cos^4 \theta \sin \theta - 17 \cos^2 \theta \sin^3 \theta + 2 \sin^5 \theta)$
31. $(x \arctan(\frac{y}{x}) - 2y)(x^2 + y^2)^{-3/4} \exp(2 \arctan(\frac{y}{x})(x^2 + y^2)^{1/4})$,
$(y \arctan(\frac{y}{x}) + 2x)(x^2 + y^2)^{-3/4} \exp(2 \arctan(\frac{y}{x})(x^2 + y^2)^{1/4})$
33. (i) $(y + 4xz)/(3z^2 - 2x^2)$ (ii) $4z/(3z^2 - 2x^2) + 8x(y + 4xz)/$
$(3z^2 - 2x^2)^2 - 6z(y + 4xz)^2/(3z^2 - 2x^2)^3 =$
$(36z^5 - 48x^2z^3 - 48x^4z - 24xyz^2 - 16x^3y - 6y^2z)/(3z^2 - 2x^2)^3$
(iii) $(15yz^4 + 12xz^5 - 60x^2yz^2 - 64x^3z^3 - 4x^4y + 16x^5z - 12xy^2z)/$
$y(3z^2 - 2x^2)^3 = (y(4x^4y - 16x^5z + 12xy^2z + 48x^2yz^2 -$
$16x^3z^3 - 21yz^4 + 12xz^5))/(2x^2 - 3z^2)$

Exercises 16.6 (p. 1002)

1. $\frac{1296}{5}$ **3.** $\frac{54}{5}$ **5.** $-\frac{1}{2\sqrt{10}}$ **7.** 0   **9.** $\frac{34}{3}$
11. $(3y^2 - 18x^2y + 2, 6xy - 6x^3)$
13. $(y \cos(xy)e^{\sin(xy)}, x \cos(xy)e^{\sin(xy)}) = \cos(xy)e^{\sin(xy)}(y, x)$
15. $(0, 0)$ **17.** $(3y^2z^3 - 10xyz, 6xyz^3 - 5x^2z, 9xy^2z^2 - 5x^2y)$
19. $n(xyz)^{n-1}(yz, xz, xy)$ **21.** $(-6, 34)$ **23.** $(-\frac{3}{25}, \frac{4}{25})$
25. (i) $dy = \frac{17}{4100}, y = -2 + \frac{17}{4100} \approx -1.9959$
33. $(-16, 4 - 8\pi, -272 + 64\pi - 64\pi^2)$

Exercises 16.7 (p. 1007)

1. $x = 2$ **3.** $4x + 3y + 12z = 36$ **5.** $10x + 8y - z = 9$
7. $x + y + z = 3$ **9.** $x + y - z = 0$ **11.** $4x + 6y - z = 13$
13. $4\pi x + \pi y - z = 8\pi$ **15.** $5x - 12y - 169z = -52$
17. $x + 2y + 16z = 8$ **19.** $11x - 30y - 6z = 52 - 6 \ln 6$
21. (i) $x + y + z = \sqrt{3}$ (ii) $3x - 4z = 5$ (iii) $2x - y - 2z = 3$
23. $x + 6y = 26$ **25.** $(a - 1)x + by + cz = 3 + a$ **27.** none

29. $\{(x, -x, 0) : x \in \mathbb{R}\}$
33. (i)

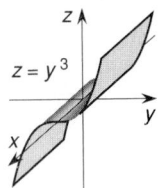

$z = y^3$

33. (ii) $z = 0, 3y - z = 2, 3y - z = 2$

Exercises 16.8 (p. 1021)
1. $(0, 0)$ **3.** $(0, 0)$ and $\{(x, y) : x^2 + y^2 = 4\}$ **5.** $(0, 0)$ **7.** $(0, 0)$
9. the two axes $x = 0, y = 0$, except $(0, 0)$ **11.** max $= 3$ at $(1, 1)$;
min $= -3$ at $(-1, -1)$ **13.** max $f(3, 1) = 33$, min $f(3, -1) = -21$
15. max $f(\pm 1, 0) = -1$, min $f(-\frac{1}{3}, \pm \frac{2\sqrt{2}}{3}) = -\frac{59}{27}$ **17.** max
$f(\pm \frac{\sqrt{15}}{4}, \frac{1}{4}) = \frac{17}{8}$, min $f(0, -1) = -1$ **19.** max $f(\pm 1, 1) = 1$, min
$f(\pm 1, -1) = -1$ **21.** max $f(-1, 1) = 19$, min $f(0, 0) = 0$
23. local max $(0, 0)$; min $\{(x, y) : x^2 + y^2 = 1\}$ **31.** none
33. $x = y = z = (24)^{1/3} = 2(3^{1/3})$ **35.** 4 long, 4 wide, 2 high
37. length $=$ width $= 120^{1/3} \approx 4.932$, height $= (120^{1/3})/3 \approx 1.644$
39. $(\frac{1}{3}, -\frac{1}{6}, -\frac{1}{3})$ **41.** $(-1, 0, 1)$ **43.** $(-\frac{ad}{a^2 + b^2 + c^2}, \frac{bd}{a^2 + b^2 + c^2}, \frac{cd}{a^2 + b^2 + c^2})$
45. (i) $(\frac{2}{3}, \frac{1}{3})$ **(ii)** $(0, 0), (2, 0), (0, 1)$

Exercises 16.9 (p. 1028)
1. min at $(0, 0)$ **3.** saddle point at $(2, 2)$ **5.** (local) min at $(0, 0)$
7. saddle points at $(0, \frac{2}{3}), (-2, 0)$, local max at $(-2, \frac{2}{3})$, local min
at $(0, 0)$ **9.** local min at $(-\frac{1}{2}, \frac{1}{2})$ **11.** max $= 2$ on $x^2 + y^2 = 1$,
min $= 0$ at $(0, 0)$ **13.** max $= 8$ at $(2, 0)$, min $= -\frac{9}{2}$ at $(-\frac{1}{2}, \frac{\sqrt{15}}{2})$
15. max $= \frac{\pi}{4}$ at $(\pm 1, \pm 1)$, min $= 0$ on axes   **17.** max $= 33$ at
$(3, 3)$, min -19 at $(-3, -1)$ **19.** max $= 8$ at $(0, 2)$, min $= -8$ at
$(\sqrt{3}, 1)$ **25.** max $= 4$ at $((2n + 1)\pi, 2m\pi - \frac{\pi}{2})$, $m, n = 0, \pm 1$,
$\pm 2, \pm 3, \ldots$, min $= -4$ at $(2n\pi, 2m\pi - \frac{\pi}{2})$, $m, n = 0, \pm 1, \pm 2$,
$\pm 3, \ldots$ **27.** if k is even, absolute min at $(0, 0, 0)$; if k is odd,
no local extrema   **29.** no   **31.** $\sqrt{3}$ **33.** abs min at every point
on $x^2 + y^2 = 1$, local max $(0, 0)$ **35.** 8   **37.** $d^3/(27abc)$
39. $18 \times 18 \times 36$ inches   **41.** 4 ft long, 4 ft wide, 2 ft high
43. cube with sides $\frac{2R}{\sqrt{3}}$ **45.** 6 ft along the open front, 3 ft
deep, 6 ft high   **47.** min $\frac{L}{3}, \frac{L}{3}, \frac{L}{3}$; max $0, 0, L$ **49.** $x = 1, \theta = \frac{\pi}{3}$

Exercises 16.10 (p. 1038)
1. max $f(\frac{3}{\sqrt{5}}, -\frac{4}{\sqrt{5}}) = 5\sqrt{5}$, min $f(-\frac{3}{\sqrt{5}}, \frac{4}{\sqrt{5}}) = -5\sqrt{5}$ **3.** max
$f(\frac{9}{\sqrt{34}}, \frac{10}{\sqrt{34}}) = 2\sqrt{34}$, min $f(-\frac{9}{\sqrt{34}}, -\frac{10}{\sqrt{34}}) = -2\sqrt{34}$
5. $f(-1, 2) = -3$ **7.** $f(1, 1) = 1$ **9.** min $f(-\frac{\pi}{4} + 2n\pi, -\frac{1}{\sqrt{2}}) =$
$-\sqrt{2}$, max $f(\frac{3\pi}{4} + 2n\pi, \frac{1}{\sqrt{2}}) = \sqrt{2}$ **11.** $(\frac{4\sqrt{2}}{5}, \frac{3\sqrt{2}}{5}, -\sqrt{2})$
13. $(\frac{\sqrt{5}-1}{\sqrt{5}}, 0, \frac{3 - 3\sqrt{5}}{\sqrt{5}})$ **15.** distance $= 1$ at $(0, 0, 1)$ **17.** $(\frac{2}{3}, \frac{1}{3}, 1)$
19. $(\sqrt{3}, -1, 0)$ **21.** max $f(\pm (\frac{1}{\sqrt{2}}, \frac{1}{\sqrt{2}})) = \frac{1}{2}$, min $f(\pm (\frac{1}{\sqrt{2}}, -\frac{1}{\sqrt{2}})) =$
$-\frac{1}{2}$ **23.** max $r\sqrt{a^2 + b^2} = f(\frac{ar}{a^2 + b^2}, \frac{br}{a^2 + b^2})$, min $f(\frac{-ar}{a^2 + b^2}, \frac{-br}{a^2 + b^2}) =$
$-r\sqrt{a^2 + b^2}$ **25.** max $f(\frac{1}{\sqrt{a^2 d + b^2 c}}(a\frac{\sqrt{d}}{\sqrt{c}}, b\frac{\sqrt{c}}{\sqrt{d}})) = \frac{1}{\sqrt{cd}}\sqrt{a^2 d + b^2 c}$, min
$f(\frac{1}{\sqrt{a^2 d + b^2 c}}(-a\frac{\sqrt{d}}{\sqrt{c}}, -b\frac{\sqrt{c}}{\sqrt{d}})) = -\frac{1}{\sqrt{cd}}\sqrt{a^2 d + b^2 c}$ **27. (ii)** max on
lower branch, min on upper branch   **29.** max
$f(\frac{258 + 16\sqrt{129}}{215}, \frac{-258 + 9\sqrt{129}}{215}, \frac{86 + 2\sqrt{129}}{215}) = -\frac{12}{5} + \frac{\sqrt{129}}{5}$, min
$f(\frac{258 - 16\sqrt{129}}{215}, \frac{-258 - 9\sqrt{129}}{215}, \frac{86 - 2\sqrt{129}}{215}) = -\frac{12}{5} - \frac{\sqrt{129}}{5}$

31. $x = y = 4, z = 8$ **33.** 3   **35.** $225, 75$ **37.** $\frac{4}{\sqrt{3}}, 2\sqrt{3}, \frac{8}{\sqrt{3}}$
39. $(\frac{2}{3}, -\frac{1}{3}, 0, -\frac{2}{3})$ **41.** $\ell = 1000, m = 2000$ **43.** max $(-2, 0)$,
min $x = \frac{5(\sqrt{2} - 1)}{2(\sqrt{2} + 1)}, y = \pm \sqrt{4 - x^2}$

Exercises 16.11 (p. 1047)
1. $m \approx -0.3947, b \approx 3.8421$ **3.** $m = 0, b = \frac{10}{3}$ **5.** $m = -\frac{4}{5}, b = 0$
7. $m = \frac{4}{5}, b = -\frac{1}{5}$ **9.** $m = 0, b = 2$ **11.** $m = 4, b = -\frac{10}{3}$
13. $k \approx -0.55, c \approx 5.1$ **15.** $k \approx 1.3860, c \approx 1.5671$
17. $k \approx 0.79451, c \approx 0.084919$ **19.** $L \approx 0.238, K \approx 23.2$
21. $m \approx 0.90991, b \approx 0.0414$; growth ≈ 0.9 ft/year
27. (i) $m = 0, b = (1 + 2^{1/3})^{-1}$ **(ii)** $m = 0, b = \frac{1}{3}$
29. (ii) min $= 4$

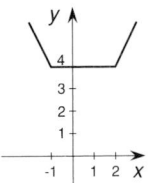

Chapter 16 Review Exercises (p. 1050)
1. -2 **3.** 2   **5.** does not exist   **7.** $\frac{1}{e}$ **9.** $\frac{1}{8}$ **11.** $12x^3 y - 7$
13. $2y - 3x^2 z$ **15.** $-y^2 \sin(xy)$ **17.** -8 **19.** $-6z$
21. $2v + 3u^2 v^2 + 2uv^3$ **23.** $2\ln(\frac{2x}{y} - x) + 2 + \frac{2x - y}{x} = 2$
$\ln(\frac{2x}{y} - x) + 4 - \frac{y}{x}$ **25.** $(3x^2 y - 2x^3 y)\cos((3 - 2x)(x^2 yz))$ **27.** $\frac{2}{r}$
29. 0   **31.** $-\frac{7}{\sqrt{13}}$ **33.** $-\frac{1}{\sqrt{2}}e^{12}$ **35.** $-\pi - \frac{2}{3}$ **37.** 0   **39.** 0
41. $(3y \sin(xy) + 3xy^2 \cos(xy)) dx + (3x \sin(xy) + 3x^2 y \cos(xy)) dy$
43. $-(2x + 3y)^{-2}(2 dx + 3 dy)$ **45.** 0   **47.** $\frac{x}{y} dx - \frac{z}{y} dy$
49. $-\frac{x}{y} dy + 3x/(e^{xy} - 3z) dz$ **51.** $x - y - 2z = 3$
53. $3x + y + 3z = -12$ **55.** $2x - 3y + 7z = 6$
57. $3x - 3y + 2z = -18$ **59.** $3x - 4y - 5z = 0$ **61.** max $= 4$ at
$(\pm 1, \pm 1)$; min $= 0$ at $(0, 0)$ **63.** max $= 6$ at $(2, -1)$; min $= -1$
at $(1, 2)$ **65.** max $= 1$ at $(\pm 1, \pm 1)$; min $= 0$ along both axes
67. max $= 64$ at $(4, 0)$; min $= -64$ at $(-4, 0)$ **69.** max $= 9e^3$ at
$(3, 0)$; min $= 0$ at $(0, 0)$ **71.** max $f(1 + \sqrt{5}, -2) = 1 + \sqrt{5}$, min
$f(1 - \sqrt{5}, -2) = 1 - \sqrt{5}$ **73.** max $f(\pm (\frac{1}{2 - \sqrt{2}}, \frac{1}{1 - \sqrt{2}})) = \frac{1}{4 - 3\sqrt{2}}$,
min $f(\pm (\frac{1}{1 + \sqrt{2}}, \frac{1}{2 + \sqrt{2}})) = \frac{1}{4 + 3\sqrt{2}}$ **75.** max $f(\frac{\sqrt{2}}{\sqrt{7}}, 3\frac{\sqrt{2}}{\sqrt{7}}, -2\frac{\sqrt{2}}{\sqrt{7}}) =$
$2\sqrt{14}$, min $f(-\frac{\sqrt{2}}{\sqrt{7}}, -3\frac{\sqrt{2}}{\sqrt{7}}, 2\frac{\sqrt{2}}{\sqrt{7}}) = -2\sqrt{14}$ **77.** max $=$
min $= f(0, 0, 0) = 0$ **79.** max $f(0, -3, 6) = 6$, min $f(0, -3, -2) = -2$

81.

83.

85. 0   **87.** $x = \pi + \frac{1}{200\pi}, f \approx 9.87946$ **89.** closest $(-\frac{1}{2}, -\frac{3}{2})$; farthest $(0, 2)$ **91.** $h = 36$ in, $r = \frac{36}{\pi}$ in   **93.** $2\sqrt{\frac{5}{3}}$ ft $\times 2\sqrt{\frac{5}{3}}$ ft, $\sqrt{\frac{5}{3}}$ ft high   **95.** $\frac{2}{\sqrt{3}} \times \frac{4}{\sqrt{3}} \times \frac{4}{\sqrt{3}}$ **97.** min $\frac{L}{3}, \frac{L}{3}, \frac{L}{3}$; max $L, 0, 0$ **99.** $y = 6x - 7$ **101.** $g \approx 31.8$ ft/sec^2

Exercises 17.1 (p. 1062)
1. $\frac{7}{6}$ **3.** 8   **5.** $\frac{1}{\pi}$ **7.** $e^2 - 2 + e^{-2}$ **9.** $(b-a)(d-c)$ **11.** $\frac{1}{9}$
13. $\frac{8}{3}$ **15.** $2 \ln 2$ **17.** $\frac{\pi}{8}$ **19.** $20ac$ **23. (i)** 1   **25.** $\frac{1}{2}$ (= the area of the triangle with vertices $(0, 0), (1, 0), (1, 1)$)   **27.** $\frac{4}{3}$
29. 156.25   **31. (ii)** $(n+1)^2/(4n^2)$ **(iii)** $\frac{1}{4}$ **(iv)** $\frac{1}{4}$

Exercises 17.2 (p. 1074)
1. $\frac{1}{10}$

3. $\frac{2}{3}$

5. 1

7. 1

9. $\frac{23}{24}$

11. $1 + \frac{1}{2} \ln \frac{2}{3}$

13. $\frac{8}{3} \ln 2 - \frac{7}{9}$

15. 0

17. e^2

19. $\frac{13}{21}$

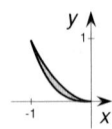

21. 0   **23.** $-\frac{96}{5}$ **25.** 6π **27.** 0   **29.** $4 - 3 \ln(3)$
31. $(e-2)e^e + e$ **33.** 0   **35.** $\frac{\pi^2}{8}$
37. $2 - 2 \ln 2$

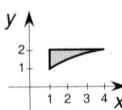

39. $2 \ln(2) - 1$

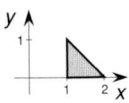

41. 0   **43.** 75π **45.** 0   **47.** 0   **49.** 50π
51. (i) $\int_0^1 \int_0^x (3x - y) \, dy \, dx$
55. (i) $\int_0^1 \int_{-\sqrt{y}}^{\sqrt{y}} (12xy - 15x^2) \, dx \, dy + \int_1^4 \int_{y-2}^{\sqrt{y}} (12xy - 15x^2) \, dx \, dy$

Exercises 17.3 (p. 1083)
1. 5π **3.** $\frac{2}{3}$ **5.** $\frac{8}{3}$ **7.** 6π **9.** 20π **11.** 6   **13.** $\frac{3\sqrt{3}\pi}{8} - \frac{\pi}{3} - \frac{\sqrt{3}}{4}$
15. π **17.** $\frac{32}{105}$ **19.** $\frac{128}{3}$ **21.** $\frac{2}{3}$ **23.** $\frac{15}{2}\pi$ **25.** $\frac{32\pi}{3} - 4\sqrt{3}\pi$
27. $\frac{1}{6}$ **29.** $\frac{4}{3}$ **31.** $\int_0^1 \int_x^1 f(x, y) dy \, dx$ **33.** $\int_{-1}^1 \int_0^{\sqrt{1-y^2}} f(x, y) dx \, dy$

35. $\int_0^1 \int_{-\sqrt{y/2}}^{\sqrt{y/2}} f(x, y)\, dx\, dy + \int_1^2 \int_{-\sqrt{y/2}}^{-\sqrt{y-1}} f(x, y)\, dx\, dy + \int_1^2 \int_{\sqrt{y-1}}^{\sqrt{y/2}} f(x, y)$

$dx\, dy$ **37.** $\int_{-1}^1 \int_{-1}^{-|y|} f(x, y)\, dx\, dy + \int_{-1}^1 \int_{|y|}^1 f(x, y)\, dx\, dy$

39. $\int_0^{\sqrt{3}} \int_0^1 f(x, y)\, dy\, dx + \int_{\sqrt{3}}^2 \int_0^{\sqrt{4-x^2}} f(x, y)\, dy\, dx$ **43.** 0

Exercises 17.4 (p. 1093)

1. $\frac{5\pi}{2}$ **3.** 9π **5.** $\frac{\pi}{3}\sqrt{2} - \frac{\pi}{6}$ **7.** 4π **9.** $\frac{512}{75}$ **11.** $\frac{1}{4}$ **13.** $\frac{3\pi}{2}$
15. $\frac{\pi}{4}$ **17.** $\frac{\pi}{4}$ **19.** 2 **21.** πR^2 **23.** $\frac{9\pi}{4}$ **25.** $\frac{\pi}{4}$ if n is odd, $\frac{\pi}{2}$ if n
is even **27.** $\frac{1}{2}$ **31.** $\approx 907{,}000, \approx 1{,}193{,}000$

Exercises 17.5 (p. 1099)

1. $\frac{\sqrt{3}}{2}$ **3.** $\sqrt{5} + \frac{1}{2}\ln(2 + \sqrt{5})$ **5.** $\frac{\sqrt{2}}{3} + \ln(\sqrt{2} + \sqrt{3})$
7. $\frac{65}{6}\pi\sqrt{65} - \frac{\pi}{6}$ **9.** 4π **11.** $\frac{3\pi}{2}$ **13.** 4 **15.** $\frac{17}{6}\pi\sqrt{17} - \frac{\pi}{6}$
17. $\frac{8}{9}\sqrt{38} + \frac{\sqrt{2}}{9} + \ln(3\sqrt{2} + \sqrt{19})$ **19.** $4\sqrt{5} + 2\ln(2 + \sqrt{5})$
23. $\sqrt{2}\pi$ **25.** $2\sqrt{2}\pi$ **27.** π

29. (ii) $2\pi(1 - \frac{1}{\sqrt{2}})R^2 \approx 2.94 \times 10^7$ sq mi

31. $2\pi(1 - \sin\theta^\circ)R^2 \approx 10^8(1 - \sin\theta^\circ)$ sq mi

Exercises 17.6 (p. 1108)

1. 5 **3.** 0 **5.** $\frac{272}{15}\pi$ **7.** $\frac{25\pi}{4}$ **9.** $\frac{59}{60}$ **11.** 18 **13.** 72π

15. $-\frac{140}{3}\pi$ **17.** $\frac{2}{3}$ **19.** $2(e^3 - e^2)$ **23.** $\int_0^1 \int_0^1 \int_z^1 f(x, y, z)\, dx\, dy\, dz$

25. $\int_2^4 \int_{z/2}^2 \int_2^4 f(x, y, z)\, dx\, dy\, dz + \int_2^4 \int_2^4 \int_z^4 f(x, y, z)\, dx\, dy\, dz$

$+ \int_4^8 \int_{z/2}^4 \int_z^4 f(x, y, z)\, dx\, dy\, dz$ **27.** $\int_0^{16} \int_{\sqrt{z}}^4 \int_y^4 f(x, y, z)\, dx\, dy\, dz$

29. $\frac{3}{8}$ **31. (i)** $\approx 15\%$ **(ii)** $\approx 60\%$ **35.** 512 tonnes **37.** 1570 tonnes

Exercises 17.7 (p. 1119)

1. $(0, \frac{8}{3\pi})$ **3.** $(\frac{3}{4}, \frac{12}{5})$ **5.** $\left(\frac{14}{6\sqrt{5} + 3\ln(2 + \sqrt{5})}, \frac{10\sqrt{5} - 2}{3}\right/$
$(2\sqrt{5} + \ln(2 + \sqrt{5})))$ **7.** $(\frac{10}{3}, \frac{8}{3})$ **9.** $(0, \frac{4\sqrt{2}}{\pi\sqrt{3}})$ **11.** $(0, 0, \frac{3}{2})$
13. $(0, 0, 0)$ **15.** $(0, 0, 1)$ **17.** $(0, 0, 0)$ **19.** $(0, 0, \frac{7}{16\sqrt{2} - 14})$
21. $\frac{1}{\sqrt{2}}$ **23.** $\sqrt{\frac{5}{3}}$ **25.** $3\sqrt{\frac{2}{5}}$ **27.** $\frac{1}{\sqrt{3}}$ **29.** $\sqrt{\frac{53}{50}}$ **33.** slower by a
factor of $\frac{4}{\sqrt{19}} \approx 0.918$ **37.** solid ball faster by a factor of
$(\frac{140 + 33\sqrt{3}}{196})^{1/2} \approx 1.003$

Exercises 17.8 (p. 1125)

1. 4π **3.** $\frac{32\pi}{3} - 8$ **5.** $\frac{\pi}{3}$ **7.** $\frac{64\pi}{15}$ **9.** $\frac{\pi}{4}$ **11.** $\frac{81\pi}{4}$ **13.** 16π

15. $\frac{352}{15}\pi$ **17.** $(\frac{3}{8}, \frac{3}{8}, 0)$ **19.** $\frac{1688}{15}\pi$ **21.** $\int_0^{\pi/2} \int_0^{2\pi} \int_0^2 F(\rho, \theta, \phi)\rho^2$

$\sin\phi\, d\rho\, d\theta\, d\phi$ **23.** $\int_{-4}^4 \int_{-\sqrt{16-x^2}}^0 \int_{-\sqrt{16-x^2-y^2}}^{\sqrt{16-x^2-y^2}} F(x, y, z)\, dz\, dy\, dx$

25. $\int_0^{\pi/2} \int_{\pi/2}^{3\pi/2} \int_0^1 F(\rho, \theta, \phi)\rho^2 \sin\phi\, d\rho\, d\theta\, d\phi$

27. $\int_0^1 \int_0^{\sqrt{3}x} \int_0^{\sqrt{4-x^2-y^2}} F(x, y, z)\, dz\, dy\, dx$

$+ \int_1^2 \int_0^{\sqrt{4-x^2}} \int_0^{\sqrt{4-x^2-y^2}} F(x, y, z)\, dz\, dy\, dx$

29. $\int_{-5/\sqrt{2}}^{5/\sqrt{2}} \int_{-\sqrt{25/2-x^2}}^0 \int_{-\sqrt{25-x^2-y^2}}^{-\sqrt{x^2+y^2}} F(x, y, z)\, dz\, dy\, dx$ **33.** $(\frac{3R}{8}, \frac{3R}{8}, 0)$

35. $\frac{3\pi}{2}$ **37.** $\frac{\pi^2}{4}$

39. $\frac{\pi}{12}$

Exercises 17.9 (p. 1135)

1. -2 **3.** 0 **5.** 22 **7.** $2xy^3 - 2x^3y + 4x - 8xy^2$ **9.** 0 **11.** $\frac{1}{7}$
13. $\frac{1}{ad - bc}$ **15.** $-\frac{1}{4}$ **17.** $27(u^2 + v^2)^2w^2$
19. $\frac{1}{27}(u(v - u^{1/3})(w - u^{1/3} - (v - u^{1/3})^{1/3}))^{-2/3} = \frac{1}{27x^{2}y^{2}z^{2}}$ **21.** $\frac{1}{3}$
23. 54 **25.** 2π **27.** $\frac{576\pi}{15}$ **29.** 162π **31.** 6π **33.** $\frac{4}{3}\pi abc$
35. $\frac{80}{3}\pi$ **37. (iii)** $\pi(1 - \frac{1}{e})$ **39. (i)** $\frac{1}{6}$ **(ii)** $\frac{1}{180}$ **45.** 201.67704

Chapter 17 Review Exercises (p. 1138)

1. $\frac{32}{3}$ **3.** 2 **5.** $\frac{1}{\pi}$ **7.** $\frac{1}{\pi}$ **9.** $-\frac{4}{3}$ **11.** 4 **13.** $\frac{2}{3} + \frac{3\pi}{2}$ **15.** $-\frac{16}{3}$
17. $\frac{1}{6}$ **19.** 2π **21.** 16π **23.** $(101\sqrt{101} - 1)\frac{\pi}{6}$
25. $\frac{23\pi}{6} + \frac{5}{6}\sqrt{5}\pi - 2\sqrt{2}\pi$ **27.** $\frac{320}{3}$ **29.** $(0, 0, \frac{3}{4})$ **31.** $\frac{65\pi}{4}$

33. $\frac{665\pi}{3}$ **35.** $\frac{968\pi}{15}$ **37.** $\frac{1456}{9}\pi$ **39.** $\frac{5}{\sqrt{2}} \approx 3.54$ **41.** $\int_0^2 \int_{x/2}^1 F(x, y)\, dy$

dx **43.** $\int_0^1 \int_{-y}^y F(x, y)\, dx\, dy$ **45.** $\int_0^1 \int_{\exp(y)}^e F(x, y)\, dx\, dy$

47. $\int_0^1 \int_{\arcsin r}^{\pi - \arcsin r} F(r, \theta)\, d\theta\, dr$ **49.** $\int_0^1 \int_{-x}^x F(x, y)\, dy\, dx + \int_1^2 \int_{x-1}^1 F(x, y)$

$dy\, dx$ **51.** $\int_{-3}^0 \int_0^{2\pi} \int_0^{\sqrt{9-z^2}} F(r, \theta, z)\, r\, dr\, d\theta\, dz$

$= \int_0^{2\pi} \int_0^3 \int_{-\sqrt{9-r^2}}^0 F(r, \theta, z)\, dz\, r\, dr\, d\theta$ **53.** $\int_{\pi/2}^\pi \int_0^{2\pi} \int_0^{-4\cos\phi} F(\rho, \theta, \phi)\rho^2$

$\sin\phi\, d\rho\, d\theta\, d\phi$ **55.** $\int_0^1 \int_0^{2\pi} \int_0^z F(r, \theta, z)\, r\, dr\, d\theta\, dz$

57. $\int_0^\pi \int_0^{2\pi} \int_4^6 F(\rho, \theta, \phi)\rho^2 \sin\phi\, d\rho\, d\theta\, d\phi$

59. $\int_0^{\sqrt{3}/2} \int_0^{z/\sqrt{3}} \int_{z2/3 - y^2}^{\sqrt{1-y^2-z^2}} F(x, y, z)\, dx\, dy\, dz$

$+ \int_0^{\sqrt{3}/2} \int_{z/\sqrt{3}}^{\sqrt{1-z^2}} \int_0^{\sqrt{1-y^2-z^2}} F(x, y, z)\, dx\, dy\, dz$ **61.** $\frac{\pi}{2}(b^4 - a^4)$

63. $\sqrt{(a^2 + b^2)/2}$ **65. (ii)** $R_1 = \frac{1}{\sqrt{2}}, R_2 = \frac{1}{\sqrt{2}}, R_3 = \sqrt{\frac{2}{3}}$
67. (i) $M = \pi R^2, I_0 = \frac{\pi}{2}R^4$ **(ii)** $M = 4\pi R^2, I_0 = 8\pi R^4$ **(iii)** equal
69. ball **71. (i)** $0 \le u - v \le 1, 0 \le v \le 1$ **(ii)** 1 **(iii)** area = 1
73. 1 **75.** $\frac{1}{2}\pi^2 R^4$ **77.** 4 **79. (i)** $\frac{4\pi}{3 + s}$ **(iii)** $\frac{4\pi}{3 + s}, s > -3$
(iv) $-\frac{4\pi}{3 + s}, s < -3$ (note: $-\frac{4\pi}{3 + s} > 0$) **81.** $A \approx 98.9, B = 1800$
83. 24,163,000 **85.** 6 **87.** 12π

Exercises 18.1 (p. 1150)

1. $-2\pi - \frac{16}{3}$ **3.** 0 **5.** 24π **7.** 0 **9.** 0 **11.** $\frac{2}{3} - \frac{\pi}{2}$ **13.** 2π
15. 2π **17.** $-8 - 2\pi$ **19.** $-\frac{3}{2}$ **21.** -1 ft-lb **23.** 18 J
25. 0 ft-lb **27.** 2 J **29.** 50 ft-lb **31.** 0 **33. (i)** 0 **(ii)** 1
35. -3.8×10^{12} J **37. (i)** 8000 km **(ii)** 32,000 km
(iii) 3.8×10^{12} J **39.** $\int_0^a f(x)\, dx + \int_0^b g(y)\, dy + \int_0^c h(z)\, dz$

Exercises 18.2 (p. 1162)

1. $\frac{4}{3}$ **3.** no **5.** no **7.** $27\frac{1}{3} = \frac{82}{3}$ **9.** $-2 - 3e^{-2}$
11. $f = \sin x + e^y + c$ **13.** $x^2y^2 - 2xy^2 + 3x^2 - 2y + c$ **15.** no
17. $y \arctan x + e^y + c$ **19.** $3x + 3y + 3z + c$ **21.** $\frac{7}{10}$
23. $a = 0, f = constant; a = \frac{1}{2}, f = \frac{1}{4}x^2y + c$ **25.** -4π **27.** one
answer: $\{(x, y) : 4 < x^2 + y^2 < 9\}$ and $\{(x, y) : -1 < x < 1\}$
29. $\arctan\frac{y}{x}$

Exercises 18.3 (p. 1173)

1. -3π **3.** 0 **5.** 28 **7.** $\frac{108}{5}$ **9.** $-\frac{1}{3}$ **17.** $\frac{3}{8}\pi$ **19.** $(0, \frac{4r}{3\pi})$
23. (ii) 3 **25.** $\frac{31}{2}$

Exercises 18.4 (p. 1180)

1. $\frac{\pi}{40}(4225\sqrt{65} - 1)$ **3.** $\frac{2\sqrt{2}}{3}$ **5.** $6912\sqrt{29}$ **7.** $\frac{81\pi}{8}$ **9.** $\frac{38\pi}{3}$
11. $(0, 0, 1)$ **13.** $(\frac{2}{3}, 0, 0)$ **15.** $(0, 0, 1)$ **17.** $(0, 0, \frac{2}{3})$
19. $(0, 0, \frac{1}{6})$ **25.** $\frac{8}{3}\pi R^4$ **27. (ii)** $(0, 0, 2), (0, 0, \frac{8}{3})$

Exercises 18.5 (p. 1189)

1. $\frac{4\pi}{3}$ **3.** 0 **5.** $\frac{4\pi}{3}(a+b+c)K^3$ **7.** 216π **9.** 0 **11.** $\frac{16\pi}{3}$
13. -144π **15.** $\frac{8192}{5}\pi$ **17.** 24 **19.** $\frac{3}{2}$ **21.** 16π **23.** $\frac{2}{3}\pi$
25. $\frac{8}{3}\pi$ **27.** $\frac{32}{3}\pi$ **29.** 2π **37.** $\frac{4}{3}\pi$ **39.** $\frac{8}{3}\pi R^4$ **43.** $\frac{\pi}{3}$

Exercises 18.6 (p. 1198)

1. 0 **3.** 2π **5.** 24π **7.** 0 **9.** $2\sqrt{3}\pi$ **11.** π **13.** -2π
15. -4π **17.** π **19.** -150π **21.** π **23.** 18π **25.** 2π
27. -18π **29.** 9π **31. (i)** $\frac{1}{\sqrt{x^2+y^2+z^2}}(x,y,z)$ **(ii)** $(0,0,0)$
35. 16π **37. (i)** two circles: $x^2+y^2=1, z=1; x^2+y^2=1, z=5$
(ii) top circle clockwise, bottom counterclockwise **(iii)** both $=0$
(iv) both $=8\pi$ **39. (i)** both $=0$ **(ii)** both $=-4\int\int_R xy\, dA$

Exercises 18.7 (p. 1206)

1. $(-5,11,14)$ **3.** $(0,2,0)$ **5.** $(2t,t,-3t)$ **7.** $\frac{8}{\sqrt{38}}(2,3,5)$
9. $\frac{1}{50\pi}$ **13. (i)** $\omega\times(x-3,y,z)=(-3z,0,3x-9)$ **(ii)** $(0,-6,0)$

Chapter 18 Review Exercises (p. 1208)

1. $-\frac{32}{3}$ **3.** 3π **5.** -2 **7.** $-\frac{265}{6}$ **9.** $-\frac{9}{8}$ **11.** $xy+c, 1$
13. $\frac{1}{3}x^3+\frac{1}{4}y^4+c, \frac{175}{4}$ **15.** $xy^2+2z^2+c, 2$ **17.** $-\frac{1}{xyz}, 0$
19. $x^my^n, a^mb^n-2^n$ **21.** 4π **23.** $-\frac{3\pi}{4}$ **25.** 25π **27.** -2π
29. $e-1$ **31.** $486\sqrt{2}$ **33.** $-20\pi^2$ **35.** $(\frac{1}{\sqrt{2}},1-\frac{1}{\sqrt{2}},0)$
37. $(0,\frac{48\sqrt{3}-2}{33},0)$ **39.** $(0,0,\frac{4\sqrt{2}}{3\pi})$ **41.** 0 **43.** $\frac{32}{15}\pi$ **45.** 16
47. 80π **49.** 1152π **51.** 0 **53.** 0 **55.** 0 **57.** 0 **59.** 0
65. $\nabla f = \text{grad } f$, curl **F** are vectors, ∇ **F** = div **F**, **F** · **n** are
functions **75.** $\frac{4\pi}{3}M$

Exercises 19.1 (p. 1219)

1. ce^{-5x} **3.** $ce^{-2x}+de^{-3x}$ **5.** $ce^{-3x}+de^{2x}$ **7.** $a+be^{2x}+ce^{-2x}$
9. $a+be^x+ce^{-x}+de^{-2x}$ **11.** $ae^{-x}+bxe^{-x}$ **13.** $ae^{2x}+b+cx$
15. $ae^{-x}+bxe^{-x}+cx^2e^{-x}$ **17.** $ae^{6x}+b+cx+dx^2$
19. $a+be^t+ce^{-t}$ **21.** $3e^{2x}$ **23.** $\frac{3}{2}e^x-\frac{1}{2}e^{-x}$
25. $2e^x+(e^{-2}-1)xe^x$ **27.** $\frac{7}{2}-\frac{5}{2}e^{-2x}-2xe^{-2x}$ **29.** $1+2x$
31. $ae^{-4x}+be^{3x/2}+ce^{-5x/3}$ **33. (i)** $a+bx+cx^2$ **(ii)** $3-4x+4x^2$
35. (i) $-3x^2+11x-6$ **(ii)** $-3(x-1)^2+5(x-1)+2$
37. $\frac{2}{5}e^{-2x}+\frac{13}{5}\cos x+\frac{24}{5}\sin x$

Exercises 19.2 (p. 1227)

1. $11-7i$ **3.** $-2+2i$ **5.** $\frac{3}{25}-\frac{4}{25}i$ **7.** $\frac{27}{145}+\frac{34}{145}i$ **9.** $-\frac{3}{5}$
11. $a\cos 4x+b\sin 4x$ **13.** $ae^{-x}\cos x+be^{-x}\sin x$ **15.** $a+be^x$
$\cos x+ce^x\sin x$ **17.** $ae^x+be^{-x}+c\cos x+d\sin x$
19. $ae^{5x}+be^{-5x}$ **21.** $ae^{-x}\cos 2x+be^{-x}\sin 2x$ **23.** $ae^{-x}+be^{-2x}$
25. $a+bx+cx^2+d\cos(\sqrt{2}x)+g\sin(\sqrt{2}x)$
27. $ae^x+bxe^x+cx^2e^x$ **29.** $a+b\cos 2x+c\sin 2x$
31. $-2e^{4x}+4e^x$ **33.** $5\cos x+5\sin x$ **35.** $e^{3x}-xe^{3x}$
37. $\frac{9}{2}-\frac{7}{2}e^{-x}\cos x-\frac{1}{2}e^{-x}\sin x$ **39.** $-\frac{2}{9}\cos 3x-\frac{1}{9}\sin 3x$
41. -4

Exercises 19.3 (p. 1235)

1. $\frac{1}{11}e^{5x}$ **3.** $\frac{1}{24}e^{3x}$ **5.** $\frac{1}{4}xe^{3x}$ **7.** $\frac{5}{6}x$ **9.** e^x-x **11.** $-\frac{8}{73}$
$\cos 3x+\frac{3}{73}\sin 3x$ **13.** $x^3+10x-6$ **15.** $\frac{1}{2}x\sin x$
17. $\frac{1}{30}\cos 2x-\frac{1}{15}\sin 2x$ **19.** $\frac{1}{4}x\cos x-\frac{1}{4}x\sin x-3x^2-6x$
21. $-\frac{1}{3}xe^{-2x}+ae^x+be^{-2x}$ **23.** $-\frac{1}{216}\sin 6x+a+bx+cx^2$
25. $\frac{1}{4}xe^{-x}+\frac{6}{5}x-\frac{36}{25}+ae^{-x}+be^{-5x}$
27. $-\frac{3}{2}x^2+x-\frac{3}{2}+ae^{4x}+be^{-x}$ **29.** $\frac{1}{64}x^4+\frac{37}{256}x^2-\frac{1}{32}x$
$\cos 4x+a+b\cos 4x+c\sin 4x$ **31.** $x+\cos x-\sin x$

33. $-\frac{1}{6}e^x-\frac{3}{10}e^{-x}+\frac{22}{15}e^{4x}$ **35.** $\frac{1}{5}e^x+\frac{4}{5}\cos 2x+\frac{19}{10}\sin 2x$
37. $-\frac{x^2}{8}+\frac{31}{16}+\frac{97}{32}e^{2x}-\frac{95}{32}e^{-2x}$ **39.** $3x^2-6x+4e^{-x/2}$
$\cos(\frac{\sqrt{3}}{2}x)+4\sqrt{3}e^{-x/2}\sin(\frac{\sqrt{3}}{2}x)$ **41.** $-xe^{-2x}+ae^{-x}+be^{-2x}$
43. $\frac{1}{4}x^2-\frac{8}{9}x+\frac{8}{27}+ae^{-x}+be^{-3x}$ **45.** $\frac{1}{3}xe^x-\frac{5}{9}e^x$
47. $\frac{a^2}{11}xe^{3x}-\frac{5a^2}{121}e^{3x}$ **49.** $-\frac{32}{3}$ ft/sec **51.** $\frac{1}{3}e^{2x}-\frac{1}{3}e^{-x}$ **53.** xe^x+1

Exercises 19.4 (p. 1244)

1. $\frac{2\pi}{5}\approx 1.26$ sec **3.** $0.1\cos(\sqrt{\frac{40}{3}}t)$ m $= 10\cos(\sqrt{\frac{40}{3}}t)$ cm
5. $-\frac{1}{2}e^{-4t}\cos(4\sqrt{23}t)-\frac{1}{2\sqrt{23}}e^{-4t}\sin(4\sqrt{23}t)$ ft **7.** $\frac{3}{25}=0.12$ lb
9. $\frac{4}{3}\ln 10\approx 3.07$ sec **11.** $R=\frac{1}{4}\sqrt{40-\pi^2}\approx 1.37$ lb sec/ft
13.

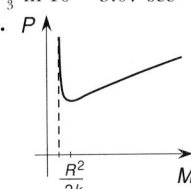

15. (ii) increases by a factor $\sqrt{2}$ **(iii)** increases by $\sqrt{A}$
17. (i) 187.2 lb $\approx$ 5.85 slugs **(ii)** -62.4 s lb **(iii)** -62.4 s lb
(iv) $M s'' = -62.4s$, i.e., $s'' = -\frac{32}{3}s$ **(v)** $\frac{\sqrt{6}\pi}{4}\approx 1.92$ sec
19. (ii) longer

Exercises 19.5 (p. 1255)

1. $y=\pm\sqrt{4x^2+c}$ **3.** $y=ce^{3x^2/2}$ **5.** $y=\ln(1/|e^x+c|)$
7. $y=ce^{\sin x}$ **9.** $y=(3\sqrt{x}+c)^{2/3}$ **11.** $y=cx^2$ **13.** $y=x\ln$
$|x|+cx$ **15.** $y=-x\ln|c-\ln|x||$ **17.** $y=-2x\ln|x|+cx$
19. $x=c\sqrt{|\csc(\frac{2y-3x}{x})+\cot(\frac{2y-3x}{x})|}$ **21.** $y=\frac{1}{4}x^2+cx^{-2}$
23. $y=\frac{1}{x}(e^x+c)$ **25.** $y=cx-1$ **27.** $y=xe^{x^2}+ce^{x^2}$ **29.** $y=\frac{1}{x}$
$\arctan x-\frac{1}{2x^2}\ln(1+x^2)+cx^{-2}$ **31.** $y=1+c\exp(-\frac{x^{n+1}}{n+1})$
33. (ii) $y=e^{-x}\tan(e^x)+ce^{-x}$ **35. (i)** $y=\ln\frac{1}{\sqrt{|x|}}+c$ **(ii)**
$y=5+\ln\frac{1}{\sqrt{|x|}}$ **37. (i)** if $A>B$, $\lim=B$; if $B>A$, $\lim=A$; if
$A=B$, $\lim=A=B$ **39. (i)** $40-30e^{-t/8}$ lb **(iii)** 40 lb

Exercises 19.6 (p. 1260)

1. $x+a\cos x+b\sin x$ **3.** $\frac{1}{8}\cos 2x+\frac{1}{4}x\sin 2x+a\cos 2x+$
$b\sin 2x$ **5.** $\frac{1}{2}e^x+a\cos x+b\sin x$ **7.** $-\frac{5}{2}x^2-5x+a+be^x$
9. $-xe^x+ae^x+be^{2x}$ **11.** $-\cos x\ln|\sec x+\tan x|+a\cos x+$
$b\sin x$ **13.** $\frac{1}{16}e^x+(\frac{x}{4}-\frac{1}{8})e^{-x}+ae^{-x}+be^{-3x}$ **15.** $\sin x\ln$
$|\sec x+\tan x|-1+a\cos x+b\sin x$ **17.** $\cos x\ln|\cos x|+$
$x\sin x+a\cos x+b\sin x$ **19.** $-\frac{1}{4}-\frac{3}{40}\cos 2x+\frac{1}{40}$
$\sin 2x+ae^x+be^{-2x}$ **21.** $x\cos x-\sin x(1+\ln|\cos x|)+\cos x$
23. $(e^{2x}+e^x)(\ln(1+e^x)-x)+ae^x+be^{2x}=$
$(e^{2x}+e^x)\ln(1+e^{-x})+ae^x+be^{2x}$ **25. (iii)** $\frac{1}{2}(\ln|x|)^2+a+b\ln|x|$

Exercises 19.7 (p. 1265)

1. $a_n=(-1)^n\frac{1}{n!}a_0$ **3.** $a_{2n}=(-1)^n\frac{2^{2n}}{(2n)!}a_0, a_{2n+1}=(-1)^n\frac{2^{2n}}{(2n+1)!}a_1$
5. $a_n=\frac{1+a_0}{n!}, n\geq 2; a_1=a_0$ **7.** $a_n=(-1)^{n+1}\frac{1}{n!}a_1, n\geq 1$
9. $a_n=\frac{1}{n!}\big((2-2^n)a_0+(2^n-1)a_1\big)$
11. $\sum_{n=0}^{\infty}-2^{n+1}\frac{1}{n!}x^n$ **13.** $\sum_{m=0}^{\infty}(-1)^m2^{2m}\frac{1}{(2m+1)!}x^{2m+1}$
15. $1+\sum_{m=0}^{\infty}(-1)^m\frac{1}{(2m)!}x^{2m}-\sum_{n=0}^{\infty}(-1)^n\frac{1}{(2n+1)!}x^{2n+1}$
17. $4-\sum_{n=3}^{\infty}(-1)^n2^{n-3}\frac{1}{n!}x^n$ **19.** $12+\frac{1}{2}x^2$
21. (i) $\sum_{n=0}^{\infty}(-1)^n2\frac{1}{n!}x^{2n}-\sum_{m=0}^{\infty}(-1)^m2^m\frac{m!}{(2m+1)!}x^{2m+1}$
(ii) ≈ 0.4224 **23. (i)** $x+\frac{c}{x}$

25. $\sum_{n=0}^{\infty} (-1)^n \left(\frac{c}{(3n)(3n-1) \times (3n-3)(3n-4) \times \ldots \times (3)(2)} x^{3n} + \right.$

$\left. \frac{d}{(3n+1)(3n) \times (3n-2)(3n-3) \times \ldots \times (4)(3)} x^{3n+1} \right) =$

$\sum_{n=0}^{\infty} (-1)^n \left(\frac{c(3n-2)(3n-5)(3n-8) \times \ldots \times 4}{(3n)!} x^{3n} + \right.$

$\left. \frac{d(3n-1)(3n-4)(3n-7) \times \ldots \times 2}{(3n+1)!} x^{3n+1} \right)$

27. $c + dx - cx^2 - \frac{d}{2}x^3 + \frac{c}{3}x^4 + \frac{2d}{15}x^5 - \frac{7c}{90}x^6 + \ldots =$

$c(1 - x^2 + \frac{1}{3}x^4 - \frac{7}{90}x^6 + \ldots) + d(x - \frac{1}{2}x^3 + \frac{2}{15}x^5 + \ldots)$

29. $c\sum_{n=1}^{\infty} (-1)^{n-1} \frac{1}{(n-1)!} x^n = cxe^{-x}$

31. $a(x^3 + 2x)$ **33. (i)** $\sum_{n=0}^{\infty} 2^{-n} \frac{1}{n!} x^{2n}$ **(ii)** $\sum_{n=0}^{\infty} \frac{2^{-n}}{2n+1} \frac{1}{n!} x^{2n+1}$

(iii) $x - \frac{1}{3}x^3 + \frac{1}{15}x^5 - \frac{1}{105}x^7 + \ldots$

Chapter 19 Review Exercises (p. 1267)

1. $\frac{x}{6} - \frac{5}{36} + ae^{-2x} + be^{-3x}$ **3.** $3x - 3 + ae^{-x}$ **5.** ce^{2x^2} **7.** $-\frac{1}{17}$

$\sin x + ae^{4x} + be^{-4x}$ **9.** $\frac{1}{12}x^4 + \frac{1}{3}x^3 + x^2 + a + bx + ce^x$

11. $a \cos 2x + b \sin 2x + ce^{2x} + de^{-2x}$ **13.** $\arctan x + ae^{-x}$

15. $\pm (\frac{x}{2}\sqrt{x^2 + c} + \frac{c}{2} \ln |x + \sqrt{x^2 + c}|) + k$ **17.** $\pm(c - \frac{16}{3}x^3)^{-1/2}$

19. $\frac{1}{6}x^3 e^x + ae^x + bxe^x + cx^2 e^x$ **21.** $-\cot x + a \csc x$

23. $\frac{x}{2} \ln |2 \ln |x|| + c|$ **25.** $\frac{1}{3x}e^{3x} + \frac{c}{x}$ **27.** $x + a \cos x + b \sin x$

29. $\frac{1}{3}x^3 e^{-2x} - x^2 e^{-2x} + 2xe^{-2x} + ae^{-2x} + be^{-3x}$

31. $-x^2 - 2 + e^x + e^{-x}$ **33.** $3e^{x^2/2} - 1$

35. $-\frac{1}{2}xe^{2x} + \frac{x^2}{12} + \frac{5}{36}x + \frac{25}{24}e^{2x} - \frac{13}{54}e^{3x} - \frac{173}{216}$

37. $\frac{5}{2}x^2 + 108e^{1/3} - (4 + 18e^{1/6})x - 108e^{1/3}e^{-x/6}$ **39.** x^n

41. $\frac{\pi\sqrt{15}}{4} \approx 3.04$ sec **43.** $\frac{15\pi}{2\sqrt{359}} \approx 1.244$ sec **45.** $c \ln |x| + d$

47. (i) $\pm(\frac{1}{\sqrt{2}} + \frac{1}{\sqrt{2}}i)$ **(ii)** $z^2 - i = (z - \frac{1}{\sqrt{2}} - \frac{1}{\sqrt{2}}i), (z + \frac{1}{\sqrt{2}} + \frac{1}{\sqrt{2}}i)$

$z^2 + i = (z - \frac{1}{\sqrt{2}} + \frac{1}{\sqrt{2}}i), (z + \frac{1}{\sqrt{2}} - \frac{1}{\sqrt{2}}i), z^4 + 1 = (z - \frac{1}{\sqrt{2}} - \frac{1}{\sqrt{2}}i)$

$(z + \frac{1}{\sqrt{2}} + \frac{1}{\sqrt{2}}i)(z - \frac{1}{\sqrt{2}} + \frac{1}{\sqrt{2}}i)(z + \frac{1}{\sqrt{2}} - \frac{1}{\sqrt{2}}i)$ **(iii)** $ae^x + be^{-x} +$

$c \cos x + d \sin x + Ae^{x/\sqrt{2}} \cos \frac{x}{\sqrt{2}} + Be^{x/\sqrt{2}} \sin \frac{x}{\sqrt{2}} + Ce^{-x/\sqrt{2}}$

$\cos \frac{x}{\sqrt{2}} + De^{-x/\sqrt{2}} \sin \frac{x}{\sqrt{2}}$ **49.** $r = r_1 r_2, \theta = \theta_1 + \theta_2 + 2n\pi, n = 0,$

$\pm 1, \pm 2, \ldots$ **51.** $\frac{a}{x} + \frac{b}{x^2}$ **53.** $-x + ax^{3+\sqrt{6}} + bx^{3-\sqrt{6}}$

55. (i) $y_1 u'' + (2y_1' + b(x)y_1)u' = 0$ **(ii)** $y_1 v' + (2y_1' + b(x)y_1)v = 0$

57. $y_2 = -\frac{1}{2} \cos(x^2), y = a \sin(x^2) + b \cos(x^2)$

59. (ii) $8 + 8e^{-2} \approx 9.08$ ft **(iii)** $k = \frac{2}{11}, 968e^{-2/11} - 792 \approx 15.07$ ft

61. (i) $\frac{3}{4}x^2 + \frac{c}{x^2}$

Index

Boldface page numbers indicate definitions.

above (limits from), **93**

abscissa, 19

absolute convergence, Def 10.6.3, **684,** Thm 10.6.4, 684, 686, 688, 690, 693, 706, 712–3

absolute: maximum/minimum, Def 4.2.2, **238,** 244, 249, Def 16.8.1, **1009**

extremum, Def 4.3.2, **249,** Def 16.8.1, **1010,** 1033

absolute value, **5**–7, Def 19.2.1, **1221**

acceleration, 219–24, 314, 314–9, 892–900

motion with constant, Rmk 4.9.1, 316

acceleration vector, Def 15.3.1, **893,** 900, 920

addition formulas, Thm 1.7.11, 51, 53, 1228

Algebra, Fundamental Theorem of, Thm 19.2.2, 1222

alternating series, Def 10.6.1, 681, 680–7

ambient temperature, 541

angle of elevation, 820

angle of incidence, reflection, 295–6, 301

angle

and dot products, Thm 14.2.7, 835, 914–5

between vectors, Thm 14.2.7, 835, 914–5

between vector and horizontal plane, Thm 15.1.4, 876

between curves, Def 15.1.5, **878**

for compound miter, 867

angular velocity, 1200–1, Thm 18.7.1, 1201, 1207

annual payment, Thm 10.2.4, 653

annuity, 654

antiderivative, 247, Def 4.8.1, **305,** Sum 4.8.3, 307, 304–13, 390, 392–7

apogee, 1148, 1151

apple, 128

Archimedes (287–212 B.C.), 481, 796

Archimedes, spiral of, 761, 801–2, 809

arclength, Def 6.9.1, **485,** 484–93, 504, Def 13.2.4, 810, 811–4, Def 15.2.4, **886,** 886–90

in polar coordinates, Thm 13.2.6, 812

of cardioid, 813

parametrization by, Thm 15.2.6, 888

arcsin, arccos, arctan, arcsec, arccsc, arccot, Def 3.8.3, **206–7,** 202–14, 561

derivatives of, Thm 3.8.5, 210

arcsinh, arccosh, arctanh, arcsech, arccsch, arccoth, Def 3.8.7, **215**

derivatives of, Thm 3.8.8, 216–7

area, 31, 347–53, 363, 374–9, Def 6.1.1, **414,** Def 6.1.2, 419, 414–32, 1056, Def 17.3.3, **1081,** 1076–84, Thm 18.3.3, 1168

of circle, 31, 66–7, 416, 420

of cone, 487

of ellipse, 425–6, 1130

of parallelogram, 30–1, 432, #41–2, Prop 14.3.4, 851

in polar coordinates, Thm 12.3.1, 773, 771–779

of rectangle, 31

of region between two curves, Def 6.1.2, **419**

of rose, 777

of triangle, 30–31, 431, #27–32

of trapezoid, 494

signed, **363**

surface, Def 6.9.2, **488,** Thm 6.9.3, 490, 487–93, Def 17.5.1, **1096,** 1095–1102, 1176

of sphere, 489, 1096

area (*Cont.*)
  of human body, 980
argument, Def 1.7.15, **53, 756**
Aristotle, 62
asymptote, 37–9, Def 4.5.2, **269,** 269–75, 727
  oblique, Def 4.5.2, **269,** Thm 4.5.3, 273
  vertical, Def 4.5.2, **269**
auxiliary equation, 555, Def 19.1.4, **1215**
average slope, 241–2
average, Def 6.5.1, **447,** 446–56
  radius, 451
  speed, 449, 815
  temperature, 447
  value, Def 6.5.1, **447,** 446–56
  velocity, 448, 815
  root mean square (RMS), 451–2
axis, major/minor, 37, 727
  focal, 736
  real, imaginary, Def 19.2.1, **1221**
  semi-major, semi-minor, 37
  translation of axes, 34–5, Sum 11.1.1, 726
  transverse, 38–9, 727

bacterial growth, 310, 522–4, 527, 529–31
balance (for a mortgage), Thm 10.2.4, 653
ball, 433, 901, 1205–6
base, 30–1, Def 7.2.8, **520**
  logarithm to the base *b,* 57, 519–20, Def 7.2.8, **520**
below (limits from), Def 2.4.1, **93,** 92–96
John Bernoulli (1667–1748), 331
Big Bang, 1183
Binomial Theorem, Thm 1.3.10, 15, 143
binormal, 891, #30
blood flow, 453
body, area of, 980
boundary, positively oriented, 1169, 1193
bounded, 109, 156, 368, 646, Def 16.8.2, **1011,** Thm 16.8.3, 1012, 1102
boundedness of continuous functions on [*a,b*], Thm 2.5.6, 109, 156, 154–7, 290
box: volume of, 161
  folded, 162
Tycho Brahe (1546–1601), 920
buoyancy, 480–3

carbon dating, 527, 534
cardioid, 765–70, 773
  arclength of, 886
carrying capacity, 547–51

Cartesian coordinates, 20, 825–6, 907–13
catenary, 849–50
Augustin Louis Cauchy (1789–1857), 914
Cauchy-Schwartz Inequality, 914
cell membrane (area of), 491
center of curvature, Def 15.3.4, **898**
center of mass, 463–76, Eq 6.7.2, 464, Eq 6.7.6, 467, Def 17.7.1, **1110,** Def 17.7.2, **1111–2,** Def 18.4.2, **1178,** 1178–80
centroid, 465, Eq 6.7.2, 464, Eq 6.7.6, **467,** 462–76, Def 17.7.1, **1110,** Def 17.7.2, **1111–2,** Def 18.4.2, **1178,** 1178–80
  of a triangle, 471
  and fluid force, Thm 6.8.2, 479
Chain Rule, Thm 3.6.1, 186, 185–95, Thm 16.5.1, 976, Thm 16.5.2, 979, 974–83
change of variables, Thm 17.9.2, 1128, Thm 17.9.5, 1133, 1126–37
characteristic polynomial, 555, Def 19.1.4, **1215,** 1216, 1220, 1225, Sum 19.2.6, 1226
chemical reaction rate, 1250
chess, 509
circle, arclength of, 810
  area of, 31, 66–7, 416
circulation, 1172
classify (critical points), 250
closed interval, **6**
  max/min problems on, Sum 3.3.1, 157
closed path, 1164
  simple closed path, 1164–6
coefficient, **7,** 1213
  constant coefficient, 553–6, Sum 7.6.1, 555, 1213–6, 1225
cofactor, 1131
common logarithm, 57, 520
common ratio, Def 10.2.1, **648**
Comparison Test, Thm 10.5.1, 675
  Improved, Thm 10.5.2, 677
  Limit, Thm 10.5.3, 678
complementary homogeneous equation, Def 19.3.1, **1229–34**
completing the square, Sum 1.3.3, **9,** 585–9, Rmk 8.4.1, 588
complex number, 710, Def 19.2.1, **1221**
  complex exponential, Def 19.2.3, **1224–7**
complex conjugate, 1227
component, Rmk 4.9.2, 317, Def 14.2.1, 831
  in the direction of **u,** 845
  perpendicular to **u,** 845
composite functions, 84–5, Thm 2.3.2, 84, 107, 185–9, 974–88

and limits, Thm 2.3.2, 84, Thm 16.2.4, 942

chain rule, Thm 16.5.1, 976, Thm 16.5.2, 979

compound interest, 535–40, 651–4, Thm 10.2.4, 653

continuous compounding, 536–7

concave up/down, Def 4.6.1, **278,** 276–90

cone, 31, 725, 903, Sum 15.4.2, 905

area of, 487

volume of, Sum 1.5.14, 31, 433

conics, 32–42, Sum 1.6.5, 39, 725–54

degenerate, 39, 728, 749

graphing, Sum 1.6.5, 39, Sum 11.1.2, 728, Sum 11.3.3, 748

in polar coordinates, 779–87, Eq 12.4.1, 780, Sum 12.4.2, 781

in standard form, 727

conjugate axis, 38

(complex) conjugate, 1227

connected set, 1155

simply connected set, Def 18.2.6, **1159,** 1163–4

conservative vector field, 1161

constant: decay, 523, 1240

gravitational, 920, 1155

growth, 523

spring, 458, 1236

term, 7

constant coefficient, 553–6, Sum 7.6.1, 555, 1213–6, 1225

constraint, 1031, 1036, 1030–8

Continuity Equation, 1203

continuous function, Def 2.5.1, **106,** 105–13, Def 16.2.10, **947,** 972

continuity of polynomials, rational functions, 107, 947

continuity of composite functions, Thm 2.5.4, 107, 948

continuity of trigonometric functions, Cor 2.6.2, 115

continuity of differentiable functions, Thm 3.2.1, 140, 972

integrability of, Thm 5.3.4, 368

continuous compounding, 536–7

converge (improper integrals), Def 8.8.1, **607,** 668

convergence: of a sequence, Def 10.1.2, **639,** Def 10.1.7, **647**

of alternating series, Thm 10.6.2, 681

of geometric series, Def 10.2.5, 655

of series, Def 10.3.1, **658,** 693

absolute, Def 10.6.3, **684**

conditional, Def 10.6.3, **684**

interval of, 706, 709

radius of, 706, 709

tests for convergence of series, Sum 10.7.3, 692

conversion, polar/rectangular coordinates, 54, Thm 12.1.1, 757

Cooling (Newton's Law of), 540–1

coordinate axes, 825–6, 830, 915

coordinates, 19–20, 54, 825–6, 915

cylindrical, 908–9, 1121

spherical, 909–12, 1123

corresponding sides, 24

cosines, Law of, Thm 1.7.10, 50

critical point, Def 4.3.1, **249,** Def 16.8.4, **1013,** Thm 16.8.5, 1014

critically damped motion, Sum 19.4.2, 1241–2

cross-product, Def 14.2.9, **836,** Prop 14.2.10, 837

and **i, j, k,** 837

Lagrange's Identity, Thm 14.2.11, 840

and area of parallelogram, Prop 14.3.4, 851

cross-term, 741–50

cube, 824

culture, bacterial, 312, 522–4, 529–31

curl, Def 18.3.6, **1173,** Def 18.6.1, **1191,** 1200–1

curvature, Def 15.3.2, **895,** 894–6, Thm 15.3.3, 896

radius of, Def 15.3.4, **898**

curve sketching, 263–76, 276–90

curve, Def 13.1.1, **797,** Def 15.1.1, **872,** Def 15.6.4, **917**

area between two curves, Def 6.1.2, **419**

directed, 1142–4

cusp, Def 4.6.5, **286**

cycloid, 843

cylindrical coordinates, 908–9

integration in, Thm 17.8.1, 1121

cylindrical shells, Sum 6.4.1, 442, 439–46

damped motion, 1240–2, Sum 19.4.2, 1242

critically damped motion, 19.4.2, 1242

over-critically damped motion, 19.4.2, 1242

Dead Sea, 483

decay, exponential, 523

decay constant, 523, 1240

decreasing function, Def 4.2.7, **243**

definite integral, 392–7

and integration by parts, Thm 8.1.4, 566

and substitution, 406–8, Thm 5.7.2, 407

definition, precise, of integral, Def 5.3.5, **371**

definition, precise, of limit, Def 2.7.1, **120**

degenerate conic, 39, 728, 749
degree, **7,** 1247
degree/radian conversion, Sum 1.7.5, 46
delta, 119, 362, 371
demand function, 340
density, 464–6, 1109–13
  uniform, Def 6.7.5, 467, 469
  function, 1061
derivative, Def 3.1.2, **133**
  of common logarithm, 176
  of exponential function, Thm 3.5.8,
    178
  of hyperbolic functions, Thm 3.5.13,
    180
  of inverse function, Thm 3.8.4, 208
  of inverse hyperbolic functions, Thm
    3.8.8, 216
  of inverse trigonometric functions, Thm
    3.8.5, 210
  of natural logarithm, Thm 3.5.5, 175
  of powers of x, 143–4, Thm 3.2.3, 144,
    Thm 3.2.8, 151, Thm 3.6.3, 191
  of polynomials, 144–5, Thm 3.2.4, 145
  of powers, Cor 3.6.5, 191
  of products, Thm 3.2.5, 147
  of quotients, Thm 3.2.6, 150
  of rational functions, Thm 3.2.7, 151
  of sum, 142
  of trigonometric functions, Thm 3.4.1,
    167, Thm 3.4.2, 168
  of order n, 219
  First Derivative Test, Thm 4.2.10, 244
  second, 218
  Second Derivative Test, Thm 4.4.1, 256,
    254–62, Thm 16.9.1, 1024
  partial, Def 16.3.1, **952,** 958–60
  directional, Def 16.6.1, **991,** Def 16.6.4,
    996–7, 989–1001
  normal, 1210, #70
René Descartes (1596–1650), 20
determinant, 1127, 1131, 1133
  Jacobian, Def 17.9.1, **1127,** Def 17.9.4,
    1133
difference of squares, Eq 1.3.2, 8
difference quotient, 132–3, 141
differentiable, 140
  function of n variables, Def 16.4.4, **972,**
    971
(total) differential, Def 16.4.3, **965,** 962–71
differentials and linear approximation, Thm
    16.4.2, 964
differential equation, 551–7, 1212, Def
    19.1.1, **1213**
  constant coefficient, 553, Sum 7.6.1,
    555, Def 19.1.4, **1215,** Thm
    19.1.6, 1216, Sum 19.2.6, 1226

  homogeneous, Def 19.1.1, **1213,** Def
    19.1.4, **1215**
  homogeneous first order, Def 19.5.2,
    1252, 1251–3
  linear first order, Sum 19.5.3, 1254,
    1253–5
  separable, 551–3, Def 19.5.1, **1248**
differential form, 1145
differentiation, implicit, Sum 3.7.1, 197,
    195–9, Thm 16.5.3, 984, Thm
    16.5.4, 987
  logarithmic, Sum 3.7.2, 200
  term-by-term, 10.10.1, 1261
n-dimensional space, Def 15.6.1, **913**–9
dimpled limaçon, 767, Sum 12.2.4, 768
directed curve, 1142–4
direction cosines, Def 14.4.4, **860**
direction vector, Def 14.4.1, **855**
directional derivative, Def 16.6.1, **991,** Def
    16.6.4, **996–7,** 989–1001
directrix, 735–6, 780
discontinuity, removable, Def 2.5.2, **107,**
    379
  infinite, Def 2.5.2, **107**
  jump, Def 2.5.2, **107,** 379
discriminant, 11, 749–52, Thm 11.3.4,
    749, 1024
displacement, **132,** 139, 219, 314–319, Def
    13.2.1, **805,** 842, 850
distance between two points, Eq 1.5.2, 23,
    Thm 14.1.1, 828
distributive law, Prop 5.2.2, 356
diverge (improper integrals), Def 8.8.1, **607–
    8,** 668
divergence of sequence, Def 10.1.2, **639,**
    Def 10.1.7, **647**
  of geometric series, Def 10.2.5, 655
  of harmonic series, Thm 10.3.3, 661
  to infinity, 640
  of series, Def 10.3.1, **658,** 693
divergence (*div* **F**), Def 18.3.5, **1171,** Def
    18.5.1, **1182,** 1201–3
Divergence Theorem, Thm 18.5.4, 1185,
    Sum 18.6.4, 1198
dodecahedron, 824
domain, 17–8, Def 13.1.1, **797,** Def
    15.1.1, **872**
  natural, 18
dot product, Def 14.2.5, **834,** 833–6, 914
  and angles, Thm 14.2.7, 835, 914
  and work, Thm 14.3.3, 850, 848–50
double-angle formula, Thm 1.7.12, 51
doughnut, 473–4
down (concave down), Def 4.6.1, **278,** Thm
    4.6.2, 279
dummy variable, 389

$e \approx 2.71828...$, 57, 513
eccentricity, Sum 11.2.2, 736, Sum 11.2.3,
    738, 780, Sum 12.4.2, 781
effective interest rate, 535–8
effects of alcohol, 953
elevation, angle of, 820
ellipse, 36–7, 727, 732–8, Thm 11.3.4,
    749, 780–3, 923
ellipsoid, Def 15.4.1, **902,** Sum 15.4.2, 905
endpoint, **6,** 108–10, 156–7
    endpoint maximum/minimum, 156–7
energy, potential and kinetic, 456, 1149
    kinetic, 456, 1117, 1149
epsilon, 119
equation: of tangent, 63–5, Def 3.1.1, **131,**
    325, 614
    auxiliary, 1215
    differential, 551–7, 1211–68
    separable, 551–3, 1248
error estimates:
    Riemann Sums, Eq 6.10.1, 494, 503
    Simpson's Rule, Thm 6.10.5, 499–500
    Trapezoidal Rule, Eq 6.10.3, 498
error propagation, 323–4
n-dimensional Euclidean space, Def 15.6.1,
    913–9
Euler, Leonhard (1707–1783), 513
evaluating polynomials, 8, 17
evaluating definite integrals, Thm 5.6.1,
    393, 394
even function, Def 4.5.1, **266**
expanding universe, 1182–3
exponent, 56, 516–9
exponential function, 58, 177–8, Eq 7.2.1,
    515, Sum 7.2.2, 515, 1214–6,
    Def 19.2.3, **1224**
exponential growth, 310–2, 523
exponential (complex), Def 19.2.3, **1224,**
    Sum 19.2.6, 1226
extraneous solutions, 162
Extreme Value Theorem, Thm 2.5.7, 109
extremum (absolute, local), Def 4.3.2, **249,**
    Def 16.8.1, **1009–10**

factor, 8–14, 589, 596–7
    integrating, Sum 19.5.3, 1254
    irreducible quadratic, 594, 596–7
Factor Theorem, Thm 1.3.5, 12
factorial, Def 1.3.8, **14**–15, 616
Pierre de Fermat (1601–1665), 295
Fermat's Principle, 295
field, vector, Def 18.1.4, **1147,** 1152–8,
    1170–3, 1181–8
First Derivative Test, Thm 4.2.10, 244

fluid: incompressible, 1207
    density, 476, 1202–3, 1207
    flow, 1170–3
    force and centroids, Thm 6.8.2, 479
    pressure, 476, Thm 6.8.1, 477, 483
flux, 1170, Def 18.3.4, **1171,** Def 18.5.3,
    1184, 1201–3
focal axis, 736
focus, Sum 11.2.2, 736, 732–8, 780
    of ellipse, 735–6, 738
    of hyperbola, 735–6, 738
    of parabola, 734, 738
foot-pound (ft-lb), 458
force, 456–60, 476–9, 920
form, differential, 1145
fractional power, 179, 516–7
fractions, partial, 589–97, Sum 8.5.1, 596
fruit tree yield, 162
fulcrum, 463
function, 17–21
    common logarithm, 57, 520
    composite, 84–5, Thm 2.3.2, 84, 107,
        185–9, 974–88
    continuous, Def 2.5.1, **106,** 105–13, Def
        16.2.10, **947,** 972
    decreasing, Def 4.2.7, **243**
    differentiable, 140
        of n variables, Def 16.4.4, **972,** 971
    domain of a function, 17–8, Def 13.1.1,
        797, Def 15.1.1, **872**
    natural domain, 18
    even, Def 4.5.1, **266**
    exponential, 58, 177–8, Eq 7.2.1, 515,
        Sum 7.2.2, 515, 1214–6, Def
        19.2.3, **1224**
    with specified rate of change, Eq 6.2.2,
        429
    graph of a function, 21
    hyperbolic, Def 3.5.11, 179
    implicit, 195, 1250
    increasing, Def 4.2.7, **243**
    integrable, Def 5.3.3, **366,** Thm 5.3.4,
        368, Def 5.3.5, **371**
    inverse, Def 3.8.1, **203**
    standard inverses of trigonometric
        functions, Def 3.8.3, **206**
    standard inverses of hyperbolic functions,
        Def 3.8.7, 215
    logarithm to the base b, 57, 519–20, Def
        7.2.8, **520**
    natural logarithm, 57, 174, Def 7.1.1,
        510, 510–4
    odd, Def 4.5.1, **266**
    polynomial, 33, 107, 144–5, 726, 901,
        947, 1215
    power function, 81, 179, 516–7

function (*Cont.*)
power series, 704–9, 711–20
range, 18–9
rational, Def 1.3.11, **16**, 589–97, 598–600
of two variables, 939, 949
of *n* variables, 949
of three variables, 944–5
vector-valued, Def 18.1.4, **1147**
Fundamental Theorem of Algebra, Thm 19.2.2, 1222
Fundamental Theorem of Calculus, Thm 5.5.4, 386
second version, Thm 5.6.1, 393

Galileo Galilei (1564–1642), 62
Gas Law, Ideal, 136, Eq 16.3.3, 957
Karl Friedrich Gauss (1777–1855), 359, 1204
Gauss' Law, Thm 18.7.2, 1204
general form of the equation of a line, Eq 1.5.9, 27
general solution of differential equation, 552–5, **1225,** 1226–7, 1231, 1233
of homogeneous linear differential equation with constant coefficients, Thm 19.1.6, 1216, Sum 19.2.6, 1226
Generalized Power Rule, Cor 3.6.5, 191, 196
geometric: sum, Def 10.2.1, **648,** Thm 10.2.2, 649
series, Def 10.2.5, **655**
geosynchronous satellite, 925
gradient, Def 16.6.2, **995,** Def 16.6.4, **996,** 1153–61, Eq 18.2.4–5, 1157, Thm 18.2.7, 1160
and level curves, 1000
grain, 509
graph, 21
gravitation constant, 460, 920, 1155
gravity, 62, 128, 459–60, 920–4, 1203–1206
George Green (1793–1841), 1165
Green's Theorem, Thm 18.3.1, 1166, Sum 18.6.4, 1198
growth, exponential, 310–2, Eq 7.3.1, 523–4
growth constant, 312, 523
gyration, radius of, Def 17.7.8, **1118**

half-angle formulas, Thm 1.7.13, 52
half-life, 526, 528, 534
half-open interval, **6**
hanging cable, 849–50

hare, 637
harmonic function, 961, 1174, #22
harmonic series, Thm 10.3.3, 661
simple harmonic motion, Sum 19.4.2, 1242
headlight, 433
helicopter, 159
helix, 871, 873, 968
homogeneous linear differential equation, Def 19.1.1, 1213, Thm 19.1.5, 1215, Thm 19.2.5, 1225, Sum 19.2.6, 1226
homogeneous first-order differential equation, Def 19.5.2, 1252–3
Robert Hooke (1635–1703), 458
Hooke's Law, 457, Eq 6.6.2, 458, 1236–7
Marquis de l'Hôpital (1661–1704), 331
l'Hôpital's Rule, Thm 4.12.1, 332
one-sided, Cor 4.12.2, 332
at infinity, Cor 4.12.2, 332
for infinite limits, Thm 4.12.3, 334
horizontal component of motion, 317
horn, 560
Hubble, Edwin P., 1183
hyperbola, 38, 37–42, 727–8, 737–8
hyperbolic functions, Def 3.5.11, 179
derivatives of, Thm 3.5.13, 180
inverse, Def 3.8.7, 215
derivatives of, Thm 3.8.8, 216
hyperboloid, 903–4, Sum 15.4.2, 905, 909
hypotenuse, 22

i (complex number), 1221
icosahedron, 824
Ideal Gas Law, 136, Eq 16.3.3, 957
image, Def 13.1.1, **797,** Def 15.1.1, **872**
imaginary axis, Def 19.2.1, **1221**
imaginary part, Def 19.2.1, **1221**
Implicit differentiation, Sum 3.7.1, 197, 195–9, Thm 16.5.3, 984, Thm 16.5.4, 987
implicit function, 195, 1250
improper integral, Def 8.8.1, **607,** Def 8.2.2, **608,** 606–10
Improved Comparison Test, Thm 10.5.2, 677
incompressible fluid, 1207
increasing function, Def 4.2.7, **243**
increasing sequence, 646, Thm 10.1.6, 646, 666
indefinite integral, Def 5.5.3, **384,** Thm 5.5.4, 386, 383–90
independent of path (line integral), Def 18.2.1, **1152,** Thm 18.2.2, 1153, Thm 18.2.3, 1156, Thm 18.2.7, 1160

index of summation, 357, Prop 5.2.4, 357
inequality, 4
  Cauchy-Scwartz, 914
  strict, 4
  triangle, Thm 1.2.2, 1.2.3, 5, Thm
      1.5.3, 24, Prop 14.2.2(vi), 833
infinite discontinuity, Def 2.5.2, **107**
infinite interval, **6**
infinite limits, 97–104, Def 2.4.3, **97**
infinity, 97
  limits at, 100
inflection point, Def 4.6.3, **282**, Sum 4.6.4,
      284
initial conditions, 1217, 1219, 1255
initial point, 804–6, 831
initial value, 524, 553
integrable, Def 5.3.3, **366**, Thm 5.3.4, 368,
      Def 5.3.5, **371**
integral, Def 5.3.3, **366–7**, Def 5.3.5, **371**,
      Def 17.1.1, **1055**, 1058, Def
      17.2.1, **1065**, Def 17.6.1, **1103**
  definite, 392–7, **392**, Thm 5.6.1, 393,
      406–8
  improper, Def 8.8.1, **607**, Def 8.8.2,
      608, 606–10
  indefinite, Def 5.5.3, **384**, Thm 5.5.4,
      386
  independent of path, Def 18.2.1, **1152**,
      Thm 18.2.2, 1153, Thm 18.2.3,
      1156, Thm 18.2.7, 1160
  inner (outer), Def 17.1.2, **1057**
  in polar coordinates, Thm 17.4.2, 1089,
      1128
  in cylindrical coordinates, 1121
  in spherical coordinates, 1123
  iterated, Def 17.1.2, **1057**, Thm 17.1.3,
      1058, Thm 17.2.3, 1070, 1083,
      Thm 17.4.2, 1089, Thm 17.6.2,
      1104
  line, Def 18.1.1, **1143**, 1143–9
  partial, 1057
  precise definition of, Def 5.3.5, **371**
  properties of, 373–8, Thm 17.1.4, 1060,
      Thm 17.2.2, 1066
  surface, Def 18.4.1, **1177**
  trigonometric, Sum 8.2.4, 571, Sum
      8.2.6, 576
  triple, Def 17.6.1, **1103**
  change of variables, Thm 17.9.2, 1128,
      Thm 17.9.5, 1133
Integral Mean Value Theorem, Thm 5.4.4,
      376
Integral Test, Thm 10.4.1, 668
integrand, 367, 1057
integrating factor, Sum 19.5.3, 1254
integration: limits of, 367

  by completing the square, 585–8
  of functions with jump or removable
      discontinuities, Thm 5.4.6, Def
      5.4.7, 380
  by partial fractions, Sum 8.5.1, 596
  by parts, Eq 8.1.1, 562, Thm 8.1.2, 563,
      Thm 8.1.4, 566
  by substitution, Sum 5.7.1, 401, Thm
      5.7.2, 407, 398–408
  by trigonometric substitution, Sum 8.3.1,
      580
  term-by-term, 712
  variable of, 389
x-intercept, y-intercept, 26–7
interest, 651–4, Thm 10.2.4, 653
Intermediate Value Theorem, Thm 2.5.5,
      108
intermediate variable, 975
intersection of sets, **7**
intersection of two lines, 27–8, 857–60
interval, 6, 157
  half—open, **6**
  infinite, **6**
  of convergence, 706, 709
  maximum/minimum on closed, Sum
      3.3.1, 157
  open, closed, **6**
inverse function, Def 3.8.1, **203**
  derivative of, Thm 3.8.4, 208
  existence of, Thm 3.8.6, 213
  graph of, 205
  standard inverses of trigonometric
      functions, Def 3.8.3, **206**
  derivatives of standard inverse
      trigonometric functions, Thm
      3.8.5, 210
  standard inverses of hyperbolic functions,
      Def 3.8.7, 215
  derivatives of standard inverse hyperbolic
      functions, Thm 3.8.8, 216
  and reflection in the line $y = x$, 205
inverse square law, 460, 920, 1203, 1205
irrational numbers, 2–3
irreducible quadratic factor, 594, 596–7
isotherm, 928, 932, 994–6
isotope, 533–4
iterated integral, Def 17.1.2, **1057**, Thm
      17.1.3, 1058, Thm 17.2.3, 1070,
      1083, Thm 17.4.2, 1089, Thm
      17.6.2, 1104
iteration, 327–30

Karl Gustav Jacobi (1804–1851), 1127
Jacobian determinant, Def 17.9.1, **1127**,
      Def 17.9.4, **1133**

Jacobian matrix, 989, Def 17.9.1, **1127,** Def 17.9.4, **1133**
joule (J), 458
jump discontinuity, Def 2.5.2, **107**

Johannes Kepler (1571–1630), 920
Kepler's Laws, Sum 15.7.12, 924
kinetic energy, 456, 1117, 1149

Joseph Louis Lagrange (1736–1813), 624, 840, 1030
Lagrange form of the Remainder, 624, Thm 9.2.2, 623
Lagrange's Identity, Thm 14.2.11, 840
Lagrange Multipliers, 1031, Thm 16.10.1, 1033, Sum 16.10.2, 1035, 1037–8
Marquis Pierre Simon de Laplace (1749–1827), 961, 1174
Laplace's Equation, 961, 1174, #21, 1210, #69
Laplacian, 1174, #21, 1210, #69
latitude, 911–2, 929
leaning tower of Pisa, 62
least squares approximation, Def 16.11.1, **1041**
lemniscate, 766, Sum 12.2.4, 768
length of a curve, Def 6.9.1, **485,** 484–93, 504, Def 13.2.4, **810,** 811–4, Def 15.2.4, **886,** 886–90
level curve, Def 16.1.1, **933,** 1000, 1005
level surface, Def 16.1.2, **935,** 1000, 1037
limaçon, 767, Sum 12.2.4, 768
limit, 62–9, 70–1, Def 2.7.1, **120,** Def 16.2.1, **939,** Def 16.2.12, **949**
  of composite functions, Thm 2.3.2, 84, Thm 16.2.4, 942, Thm 16.2.8, 946
  of functions of two variables, Def 16.2.1, **939,** Def 16.2.12, **949**
  of functions of n variables, Def 16.2.12, **949**
  of functions of three variables, 944–5
  at infinity, 100–4, Sum 2.4.5, 102, Thm 2.4.6, 103
  and polar coordinates, 949
  of polynomials, Thm 2.2.5, 74, Thm 16.2.3, 940, Thm 16.2.7, 945
  of powers x^r, Thm 2.2.5, 74, Thm 2.3.1, 81, 82–4
  of products, Thm 2.2.3, 72, 73, Thm 16.2.2, 940, Thm 16.2.6, 945
  of quotients, Thm 2.2.3, 72, Thm 16.2.2, 940, Thm 16.2.6, 945
  of rational functions, Thm 2.2.6, 75,
    Sum 2.2.7, 76, Sum 2.4.4, 99, Sum 2.4.5, 102, Thm 16.2.4, 940, Thm 16.2.7, 945
  from right/left/above/below, 93
  of sums, Thm 2.2.3, 72, 73, Thm 16.2.2, 940, Thm 16.2.6, 945
  of trigonometric functions, Thm 2.6.1, 114, Cor 2.6.2, 115, Thm 2.6.3, 116
  infinite limits, Def 2.4.3, **97,** 944
  one-sided, Def 2.4.1, 93, Thm 2.4.2, 94
  precise definition of, Def 2.7.1, **120,** Def 16.2.12, **949**
  as $\| \mathcal{P} \| \to 0$, Def 5.3.3, **366,** Def 5.3.5, **371,** Def 17.6.1, **1055,** Def 17.2.1, **1065,** Def 17.6.1, **1103**
Limit Comparison Test, Thm 10.5.3, 678
limit of integration, 367
limit of a sequence, Def 10.1.2, **639,** Def 10.1.7, **647**
line, 25–9, 854–62
  equation of, 63–5, Def 3.1.1, **131,** 325, 614
  general form, Eq 1.5.9, 27
  intersection of two lines, 27–8, 857–8
  in polar coordinates, Eq 12.4.5, 784
  in $\mathbb{R}^3$, Def 14.4.1, **855,** Sum 14.4.3, 859
  line integral, Def 18.1.1, **1143,** 1143–9
  parametric equation of, Def 14.4.1, **855**
  point-normal form, Def 14.5.1, **862**
  point-slope form, Eq 1.5.7, 26
  skew lines, 858
  slope-intercept form, Eq 1.5.8, 26
  symmetric equations of, Def 14.4.2, 859
  tangent, 63–5, Def 3.1.1, **131,** 325, 614
  vector equation of, Def 14.4.1, **855**
line integral, Def 18.1.1, **1143,** 1143–9
  independent of path, Def 18.2.1, **1152,** Thm 18.2.2, 1153, Thm 18.2.3, 1156, Thm 18.2.7, 1160
  and work, Thm 18.1.5, 1148
linear approximation and differentials, Eq 4.10.1, 320, 962–71
linear differential equation, 553–7, Sum 7.6.1, 555, Def 19.1.1, **1213**
  linear first-order differential equation, 1253, Sum 19.5.3, 1254
linear regression, 1040–7, Def 16.11.1, **1041,** Thm 16.11.2, 1043
local extremum, Def 4.3.2, **249,** Def 16.8.1, **1009–10,** Sum 16.8.6, 1016, 1024
local maximum/minimum, Def 4.2.2, **238,** Def 16.8.1, **1009–10,** Sum 16.8.6, 1016, 1024

logarithm, 56, 174, Def 7.1.1, **510,** Sum
　　7.1.3, 511
　common, 57, 520
　derivative of, Thm 3.5.5, 175
　growth of, Thm 7.1.4, 512
　to the base *b,* 57, 519–20, Def 7.2.8,
　　520
　natural, 57, 174, Def 7.1.1, **510,** 510–4
logarithmic differentiation, Sum 3.7.2, 200
log-log graph paper, 177
logistic equation, Eq 7.5.1, 546
logistic growth, Eq 7.5.2, 547
longitude, 929, 911–2
loop, 1164
lower limit of integration, 367
lowest terms, 2

Colin Maclaurin (1698–1746), 619
Maclaurin polynomial, 619
Maclaurin series, Def 10.8.2, **698,** 1261–5
major axis, 37, 727
marginal cost, revenue, profit, 339–41
mass, 458, 467, Def 17.7.1, **1110,** Def
　　17.7.2, **1111–2,** Def 18.4.2, **1178**
　center of, 463–76, Eq 6.7.2, 464, Eq
　　6.7.6, 467, Def 17.7.1, **1110,** Def
　　17.7.2, **1111–2,** Def 18.4.2,
　　1178, 1178–80
matrix, 1131
matrix, Jacobian, 989, Def 17.9.1, **1127,**
　　Def 17.9.4, **1133**
maximum, maxima, 109, 154–65, Def
　　4.2.1, **236,** Def 4.2.2, **238,** 290–
　　303, Def 16.8.1, **1009–10**
　local/absolute/relative, Def 4.2.2, **238**
maximum/minimum on closed interval, Thm
　　2.5.7, 109, Sum 3.3.1, 157
max/min problems, Sum 3.3.1, 157, 290–
　　303, 1013, 1009–23
Mean Value Theorem, Thm 4.2.5, 241
Integral Mean Value Theorem, Thm 5.4.4,
　　376
measurement error, 323
mesh, Def 5.3.1, **362,** 366, 370, 1103
minimum, minima, 109, 154–65, Def
　　4.2.1, **236,** Def 4.2.2, **238,** 290–
　　303, Def 16.8.1, **1009–10**
　local/absolute/relative, Def 4.2.2, **238**
minor axis, 37, 727
mixed partials, Def 16.3.4, **959,** 960
August Möbius (1790–1868), 1141, 1190
Möbius strip, 1141, 1190
modulus, Def 1.7.15, **53,** 756
moment, Def 6.7.1, **463,** Def 6.7.4, **466,**
　　Def 17.7.1, **1110,** Def 17.7.2,
　　1111–2, Def 18.4.2, **1178**

of inertia, Def 18.4.2, **1178**
about a line, Def 6.7.4, **466,** Def 6.7.5,
　　467
monthly payment, Thm 10.2.4, 653
mortgage, 651–4
motion with constant acceleration, Rmk
　　4.9.1, 316
motion, vertical and horizontal components
　　of, Rmk 4.9.2, 317
damped motion, 1240–2, Sum 19.4.2, 1242
　critically damped, 19.4.2, 1242
　over-critically damped, 19.4.2, 1242
multiplicity (of a root), **1216,** 1222, 1225,
　　Sum 19.2.6, 1226
muzzle velocity, 819

n-dimensional space, Def 15.6.1, **913,** Def
　　15.6.3, **915**
n-dimensional Euclidean space, Def 15.6.3,
　　915
n-th Term Test, Thm 10.3.2, 660
n-th Root Test, Thm 10.7.2, 690
natural domain, 18
natural logarithm, 57, 174, Def 7.1.1, **510,**
　　510–4
negative of a vector, Def 14.2.1, **831,** 914
Sir Isaac Newton (1642–1727), 128, 327,
　　920
Newton's Law of Cooling, 540–1
Newton's Method, Sum 4.11.1, 326, 325–
　　31
Newton's Second Law, 1237, 1240
newtons (N), 458
nominal interest rate, 535
non-homogeneous differential equations with
　　constant coefficients, Sum 19.3.2,
　　1229, Thm 19.3.3, 1230, Sum
　　19.3.4, 1233
norm, Def 5.3.1, **362,** Def 14.2.1, **831,**
　　914
normal derivative, 1210, #70
normal vector, Def 14.5.1, **862,** 1181–3,
　　1190–2
　unit, Def 15.2.3, **885,** 1183, 1192
　principal unit normal vector, Def 15.2.3,
　　885
　outer normal, 1170, 1181–5, 1190
　outer unit normal vector, 1184–5
normalizing tangent vectors, Def 15.2.1,
　　880
numerical integration, 493–505
　Simpson's Rule, Thm 6.10.5, 499–500
　error estimate, Thm 6.10.5, 499–500
　Trapezoidal Rule, Thm 6.10.2, 496, 493–
　　505
　error estimates, Eq 6.10.3, 498

oblique asymptote, Def 4.5.2, **269,** Thm
    4.5.3, 273
octahedron, 824
odd function, Def 4.5.1, **266**
one-sided limits, Def 2.4.1, **93,** Thm 2.4.2,
    94
one-to-one, Def 3.8.2, 203
open set, 1153
order (of differential equations), 555, Def
    19.1.1, 1213, 1251, 1257
orientable surface, Def 18.5.2, **1184**
oriented boundary, 1169, 1193
orthogonal planes, Def 14.5.2, **864**
orthogonal trajectories, 1256, #35–36
orthogonal vectors, Thm 14.2.8, 836, 914
oscillation, 1211, 1240–2
    damped, 1240–2, Sum 19.4.2, 1242
osculating circle, Def 15.3.4, **898**
outer normal, 1170, 1181–5, 1190
outer unit normal vector, 1184–5
over-critically damped motion, Sum 19.4.2,
    1241–2

p-series, 669
paint, 560
Pappus' Theorem, Thm 6.7.10, 473
parabola, 33, 39, 727, 733–8
paraboloid, 902, Sum 15.4.2, 905
parachutist, 1236, #49–50, 1268, #59–60
paradox, 560, 637
parallel, 27, 860, 864
    lines, 860
    planes, Def 14.5.2, **864**
    line and plane, Def 14.5.3, **866**
parallelepiped, 1137
parallelogram, 30–1, 432, Prop 14.3.4, 851
Parallelogram Rule, 832
parameter, Def 13.1.1, **797,** Def 15.1.1,
    872, Def 15.6.4, **917**
parametric equation, Def 13.1.1, **797,** Def
    14.4.1, **855,** Def 15.1.1, **872,** Def
    15.6.4, **917**
    of line, Def 14.4.1, **855**
parametrization by arclength, Thm 15.2.6,
    888
partial derivative, Def 16.3.1, **952,** 958–60
    higher order, Def 16.3.4, **959**
    mixed partials, Def 16.3.4, **959,** 960
partial fractions, Sum 8.5.1, 596, 589–98
partial integration, 1057
partial sum, Def 10.3.1, **658**
particular solution, 552, 1217, 1231
partition, Def 5.3.1, **362,** 1055, 1065, 1103
parts, integration by, Eq 8.1.1, 562, Thm

    8.1.2, 563, Thm 8.1.4, 566
    and definite integrals, Thm 8.1.4, 566
Blaise Pascal (1623–1662), 477
Pascal's Principle, Thm 6.8.1, 477
path, closed, 1164
    simple closed path, 1164–6
path independent integral, Def 18.2.1,
    1152, Thm 18.2.2, 1153, Thm
    18.2.3, 1156, Thm 18.2.7, 1160
pendulum, 1246–7
percentage error, 323
perigee, 1148, 1151
period (of orbit), 923
period, 44, 1237, 1240, Sum 19.4.2, 1242
periodic, 44
perpendicular: lines, 28, Thm 1.5.12, 29
    vectors, Thm 14.2.8, 836, 914
    planes, Def 14.5.2, **864**
    component, 845
phytoplankton, 1106
Pisa, leaning tower of, 62
planes in $\mathbb{R}^3$, Def 14.5.1, **862,** 863–7
    parallel planes, Def 14.5.2, **864**
    parallel line and plane, Def 14.5.3, **866**
    perpendicular planes, Def 14.5.2, **864**
    standard form of the equation of, 863
planet, 157, Sum 15.7.12, 924
Platonic solids, 824
Poiseuille's Law, 453
polar angle, 53
polar coordinates, 53–5, Def 1.7.15, **53**
    circles in, 783–4
    conics in, 779–87, Eq 12.4.1, 780, Sum
        12.4.2, 781
    conversion to/from rectangular
        coordinates, 54, Thm 12.1.1, 757
    graphs in, 760–71
    integrals in, Thm 17.4.2, 1089, 1128
    and limits, 949
    lines in, Eq 12.4.5, 784
    polar rectangle, 1086, 1094–5
    and symmetry, Sum 12.2.2, 764
polar moment of inertia, Def 17.7.4, **1114**
polyhedra, regular, 824
polynomial, 33, 107, 144–5, 726, 901,
    947, 1215
    characteristic, 1215, 1226
    continuity of, 107, 947
    derivatives of, 144–5, Thm 3.2.4, 145
    limits of, Thm 2.2.5, 74, Thm 16.2.3,
        940, Thm 16.2.7, 945
    quadratic, 9, 585–9, 726, 901
    Taylor, Eq 9.1.1, 617, Def 9.1.2, **618**
    of two variables, 726
positively oriented boundary, 1193–5
potential energy, 1155

potential function, 1154, Thm 18.2.3, 1156, 1160

gravitational potential, 1155

pound, 458, 1238

Generalized Power Rule, Cor 3.6.5, 191, 196

power series, Def 10.9.1, **705,** 704–9, 711–20

  convergence of, Thm 10.9.2, 706, Sum 10.9.3, 709

  differentiation of, Thm 10.10.1, 712

  solutions to differential equations, 1260–5

powers, fractional, 179, 516–7

  limits of, Thm 2.3.1, 81

precise definition of integral, Def 5.3.5, **371**

precise definition of limits, Def 2.7.1, **120,** Def 16.2.12, **949**

precise definition of limit of sequence, Def 10.1.7, **647**

present value, 538–40

pressure, 476, 476–83

principal normal vector, Def 15.2.3, **885**

products, limits of, Thm 2.2.3, 72, 73, Thm 16.2.2, 940, Thm 16.2.6, 945

Product Rule, Thm 3.2.5, 147

projectile, 317–9, 815–20, Thm 13.3.4, 817

proper rational function, 590, Sum 8.5.1, 596

properties of vectors, Prop 14.2.2, 833, Prop 14.2.6, 834, Prop 15.6.2, 914

punctured ball, 945

punctured disk, 938–9

pure imaginary number, Def 19.2.1, **1221**

Pythagoras, 22

Pythagorean Theorem, 1, Thm 1.5.1, 22

quadratic: polynomial, 9, 585–9, 726, 901

  formula, Thm 1.3.4, 11

  irreducible factor, 594, 596–7

quadric surface, 901, Sum 15.4.2, 905

quotients, limits of, Thm 2.2.3, 72, Thm 16.2.2, 940, Thm 16.2.6, 945

  derivatives of, Thm 3.2.6, 150

Quotient Rule, Thm 3.2.6, 150

$\mathbb{R}^3$, 825–30, Def 14.2.1, **831,** 833–40, 854–60, 862–6, 907

$\mathbb{R}^n$, Def 15.6.3, **915,** 915–9

radian measure, 45, Sum 1.7.5, 46, 755–9

radian/degree conversion, Sum 1.7.5, 46

radius, 31, 66–7, 416, 420, 706, 609, 898, 1118

  of convergence, 706, 709

  of curvature, Def 15.3.4, **898**

  of gyration, Def 17.7.8, **1118**

range, 18–9

rate of change, 136, 227–34

  specified, Eq 6.2.2, 429

chemical reaction rate, 1250

Ratio Test, Thm 10.7.1, 688, Sum 10.7.3, 692–3

rational function, Def 1.3.11, **16,** 75–6, 99–102, 590–6, 940–5

  derivative of, Thm 3.2.7, 151

  limits of, Thm 2.2.6, 75, Sum 2.2.7, 76, Sum 2.4.4, 99, Sum 2.4.5, 102, Thm 16.2.4, 940, Thm 16.2.7, 945

  limits at infinity, Sum 1.4.5, 102

  proper, 590, Sum 8.5.1, 596

  and homogeneous equations, 1253

  terms of series, 679

  sketching, Sum 4.5.4, 274

rational numbers, 2

Rational Root Theorem, Thm 1.3.7, 13

chemical reaction rate, 1250

real axis, Def 19.2.1, **1221**

real numbers ($\mathbb{R}$), 3–6

real part, Def 19.2.1, **1221**

rectangle, area of, 31

  polar rectangle, 1086, 1094–5

reduction formula, 565

  $\cos^n x$, $\sin^n x$, Thm 8.1.3, 565

  $\cos^m x \sin^n x$, Thm 8.2.3, 571

  $\tan^m x$, $\sec^n x$, Thm 8.2.5, 574

reduction of order, 1268

reflections and ellipses, 733, 740, #48

reflections and parabolas, 734, 740, #45–6

reflection in the line $y = x$ and inverse functions, 205

region between two curves, Def 6.1.2, **419**

(linear) regression, 1040–7, Def 16.11.1, **1041,** Thm 16.11.2, 1043

regular polyhedra, 824

related rates, 227–34, Sum 4.1.1, 231

relative maximum/minimum/extremum, Def 4.3.2, **249,** Def 16.8.1, **1009–10,** Sum 16.8.6, 1016, 1024

remainder, Eq 9.2.1, 623, Thm 9.2.2, 623, 701

  Taylor's Formula with, Thm 9.2.2, 623, 701

removable discontinuities, Def 2.5.2, **107**

  integrating functions with, Thm 5.4.6, Def 5.4.7, **380**

revenue, marginal, 339–41
revolution, solid of, **432,** Def 6.3.1, 433,
        Sum 6.3.2, 437, Sum 6.4.1, 442,
        1085
    surface of, **487,** Def 6.9.2, 488, Thm
        6.9.3, 490
Georg Friedrich Bernhard Riemann (1826–
        1866), 363
Riemann sum, Def 5.3.2, **363,** 366, 369–
        71, 1055, 1065, 1087, 1103
right-handed coordinate system, 826
right-hand rule, 839, 1200
RMS average, 451–2
Michel Rolle (1652–1719), 240
Rolle's Theorem, Thm 4.2.4, 240
root, 12
    of characteristic polynomial, 553–6, 1215–
        6, 1222, 1225, Sum 19.2.6, 1226
root mean square (RMS) average, 451–2
n-th Root Test, Thm 10.7.2, 690
rose, 762–4, 768
rotation of axes, Thm 11.3.1, 742, Thm
        11.3.2, 745

saddle surface, 904, Sum 15.4.2, 905
saddle point, 1023, Thm 16.9.1, 1024–5
salt, 1256, #39–40, 482–3, #39–40
sandbox, 448
Hermann Amandus Schwartz (1843–1921),
        914
scientific notation, 3–4
secant, 43, 114, 129–32, 769
second derivative, 218, Def 16.3.4, **959**
    and concavity, Thm 4.6.2, 279
    test, Thm 4.4.1, 256, 254–62, Thm
        16.9.1, 1024
semi-major, semi-minor axis, 37
separable differential equation, 551, Def
        19.5.1, 1248
sequence, Def 10.1.1, **638,** Def 10.1.2,
        639, 640–7, Def 10.1.7, **647,** 658
    shifted, Thm 10.1.5, 644
series, 655, 658–64
    convergence of, Def 10.3.1, **658**
    geometric, Def 10.2.5, **655**
    power, Def 10.9.1, **705,** 704–9, 711–20
        convergence of, Thm 10.9.2, 706, Sum
            10.9.3, 709
        differentiation of, Thm 10.10.1, 712
        solutions to differential equations, 1260–5
    Taylor's, Def 10.8.1, **695,** 696–703,
        1261–5
set, open, 1153
signed area, **363**
significant digits, 3

similar triangles, 24
simple closed path, 1164
simple harmonic motion, Sum 19.4.2, 1242
simply connected set, Def 18.2.6, **1159,**
        1163–4
Thomas Simpson (1710–1761), 498
Simpson's Rule, Thm 6.10.5, 499–500
    error estimate, Thm 6.10.5, 499–500
    sketching: curves, 263–76, 276–90
    conics, Sum 11.1.2. 728, Sum 11.3.3, 748
    rational functions, Sum 4.5.4, 274
skew lines, 858
slices, 437, 443
slope, Def 1.5.5, **26,** Thm 1.5.12, 29, 129,
        Def 3.1.1, **131,** 241, 769
    average, 241–2, Thm 4.2.5, 241
    point-slope form, Eq 1.5.7, 26
    -intercept form, Eq 1.5.8, 26
    slopes of perpendicular lines, Thm
        1.5.12, 29
    of tangent line, Def 3.1.1, **131**
    of tangent line in polar coordinates, Thm
        12.2.5, 769
slug, 458, 1238
Snell's Law, 295–6, 301
solid of revolution, **432,** Def 6.3.1, 433,
        Sum 6.3.2, 437, Sum 6.4.1, 442,
        1085
solids, Platonic, 824
solution: particular, 552, 1217, 1231
    general, 552–5, **1225,** 1226–7, 1231,
        1233, Thm 19.1.6, 1216, Sum
        19.2.6, 1226
speed, 65–6, 314, Def 13.2.3, **807,** Def
        15.1.3, **874,** Def 15.6.4, **917**
sphere, 31, 433, 489, 901, 905, 915, 1096,
        1205–6
    area of, 489, 1096
    gravity and spheres, 1205–6
    volume of, Sum 1.5.14, 31, 433
spherical coordinates, 909–12
    integration in, Def 17.8.2, **1123**
spiral
    of Archimedes, 761, 796, 801–2, 809
    logarithmic, 796
    staircase, 871
square, completing the, Sum 1.3.3, **9,** 585–8
square root, 10
Squeeze Theorem, Thm 2.3.5, 88, 88–90
    for sequences, Thm 10.1.4(vii), 642
standard basis vectors, Def 14.2.3, **833,** 914
standard form:
    of equation of plane, 863
    conics in, 727
standard inverses of trigonometric functions,
        Def 3.8.3, **206**

derivatives of, Thm 3.8.5, 210

standard inverses of hyperbolic functions, Def 3.8.7, 215

derivatives of, Thm 3.8.8, 216

stationary point, Def 4.3.1, **249**

George Gabriel Stokes (1819–1903), 1191

Stokes' Theorem, Thm 18.6.2, 1193, Sum 18.6.4, 1198

straight line, 25–9, 854–62

general form, Eq 1.5.9, 27

point-normal form, Def 14.5.1, **862**

point-slope form, Eq 1.5.7, 26

slope-intercept form, Eq 1.5.8, 26

in polar coordinates, Eq 12.4.5, 784

strict inequality, 4

strip of width dx, 425–7

subinterval, 362

subscript, 7

substitution, Sum 5.7.1, 401, Thm 5.7.2, 407, 398–408

and definite integrals, Thm 5.7.2, 407, 398–408

sum: limit of, Thm 2.2.3, 72, 73, Thm 16.2.2, 940, Thm 16.2.6, 945

of integers from 1 to n, 353–4, Eq 5.2.5, 359

of squares of integers from 1 to n, Thm 5.1.1, 348, 353–4, 358–60

derivative of, 142

Riemann, Def 5.3.2, **363**, 366, 369–71, 1055, 1065, 1087, 1103

telescoping, 666

summation, index of, 357

summation notation, Def 5.2.1, **354**

surface integral, Def 18.4.1, 1177

surface of revolution, 487–90

area of, Def 6.9.2, **488**, Thm 6.9.3, 490

symmetry, 266, Thm 6.7.7, 469

and cylindrical coordinates, 909, 912, #37–8

and polar coordinates, Sum 12.2.2, 764

and spherical coordinates, 912, #39–41

tangent line, 63–5, Def 3.1.1, **131**, 325, 614, 769

equation of, 63–5, Def 3.1.1, **131**, 325, 614

slope of, Def 3.1.1, **131**

slope of, in polar coordinates, Thm 12.2.5, 769

tangent vector, Def 13.2.3, **807**, Def 15.1.3, 874, 878, 880, Def 15.6.4, 917

unit, Def 15.2.1, **880**, 883

tangent plane, 955–6, 1003–9, Def 16.7.1, **1105**

$\tan\frac{\theta}{2}$-substitution, Sum 8.6.1, 600

Brook Taylor (1685–1731), 618

Taylor Polynomials, Eq 9.1.1, 617, Def 9.1.2, **618**

Taylor's Formula with Remainder, Thm 9.2.2, 623, 701

Taylor's series, Def 10.8.1, **695**, 696–703, 1261–5

telescoping sum, 666

temperature, ambient, 541

term-by-term: differentiation, Thm 10.10.1, 712, 1261

integration, Thm 10.10.1, 712–3

terminal point, 804–6, Def 14.2.1, **831**

terminal velocity, 1235, 1268

tetrahedron, 824

tortoise, 637

tower, leaning, of Pisa, 62

trajectory, Thm 13.3.4, 817

orthogonal, 1256, #35–6

translation of axes, Sum 1.6.1, 35, Sum 11.1.1, 726

transverse axis, 38–9, 727

trapezoid, **494**

Trapezoidal Rule, Thm 6.10.2, 496, 493–505

error estimates, Eq 6.10.3, 498

triangle, 22–31, 431

area of, 30–1, 431, #27–32

Triangle Inequality, Thm 1.2.2, 5, Thm 1.2.3, 5, Thm 1.5.3, 24, Prop 14.2.2(vi), 833

Triangle Rule, 832

trigonometric: functions, Def 1.7.1, **43**, 114–6, 167–8, 571, 576, 580

continuity of, Cor 2.6.2, 115

derivatives of, Thm 3.4.1, 167, Thm 3.4.2, 168

limits of, Thm 2.6.1, 114, Cor 2.6.2, 115, Thm 2.6.3, 116

integrals, Sum 8.2.4, 571, Sum 8.2.6, 576

substitutions, Sum 8.3.1, 580

undetermined coefficients, 1228–36, Sum 19.3.2, 1229, Thm 19.3.3, 1230, Sum 19.3.4, 1233

uniform density, 465–6, Def 6.7.5, **467**, Thm 6.7.7, 469

union, 7

unit normal vector, Def 15.2.3, **885**, 1183, 1192

outer, 1184–5

unit radial vector, 921
unit tangent vector, Def 15.2.1, **880,** Def
      15.6.4, 917
unit vector, 833, 845, 880, Prop 15.2.2,
      883, 885, 888, 917, 921, 1183–5,
      1192
up (concave up), Def 4.6.1, **278,** Thm
      4.6.2, 279
upper limit of integration, 367

value: average, Def 6.5.1, **447,** 446–54
   absolute, 5
   initial, 524, 553
   present, 538–40
   of a function, 17
   of a polynomial, 8
Integral Mean Value Theorem, Thm 5.4.4,
      376
Mean Value Theorem, Thm 4.2.5, 241
variable, 8
   dummy, 389
   of integration, 389
variation of parameters, 1256–60, Sum
      19.6.2, 1258
vector, 804–5, Def 14.2.1, **831,** Prop
      14.2.2, 833, Def 15.6.1, **913**
   acceleration, Def 15.3.1, **893**
   equation of line, Def 14.4.1, **855**
   negative of a, 914
   tangent, Def 13.2.3, 807, Def 15.1.3,
      874, Def 15.6.4, **917**
   unit normal, Def 15.2.3, **885,** 1183, 1192
   outer unit normal, 1184–5
   vectors and velocity, Def 15.1.2, **873,**
      Def 15.6.4, **917**
   zero vector, Def 14.2.1, **831,** 914
vector field, Def 18.1.4, **1147**
   conservative, 1161
vector-valued function, Def 18.1.4, **1147**
velocity, 133–4, 139–40, 219–24, 314–9,
      Def 15.1.2, **873,** Def 15.6.4, **917**
   angular, 1200–1, Thm 18.7.1, 1201,
      1207
   muzzle, 819

and vectors, Def 15.1.2, **873,** Def
      15.6.4, **917**
   terminal, 1235, 1268
vertex of parabola, 33–4, 733–4
vertical asymptote, Def 4.5.2, **269**
vertical component of motion, Rmk 4.9.2,
      317
vibration, 1236–47, Sum 19.4.2, 1242
volume, 31, 432, Def 17.3.2, **1077,** 1076–
      85
volume of solids of revolution, 432, 1085
   by slices, Def 6.3.1, **433,** Sum 6.3.2,
      437, 432–9
   by cylindrical shells, Sum 6.4.1, 442,
      439–46
   slices vs shells, 443
   volume of ball, 31, 433, 1076
      of cone, 433
      of headlight, 433

wheat, 509
work, 456–62, Def 6.6.1, **457,** 1147, Thm
      18.1.5, 1148
   and dot products, 848–50, Thm 14.3.3,
      850

x-axis, 19, 825–6, 830
x-coordinate, 19–21, 825–6
x-intercept, 27
xy-, xz-planes, 826–7, 901–2

y-axis, 19, 825–6, 830
y-coordinate, 19–21, 825–6
y-intercept, 26
yogurt, 527
yz-plane, 826, 902

z-axis, 825–6, 830, 905
Zeno's Paradox, 637
zero vector, Def 14.2.1, **831,** 914

Photo Credits

Derivatives

$$\frac{d}{dx}(x^n) = nx^{x-1}$$

$$\frac{d}{dx}(constant) = 0$$

$$\frac{d}{dx}(fg) = fg' + f'g$$

$$\frac{d}{dx}\left(\frac{f}{g}\right) = \frac{f'g - g'f}{g^2}$$

$$\frac{d}{dx}(\sin x) = \cos x$$

$$\frac{d}{dx}(\cos x) = -\sin x$$

$$\frac{d}{dx}(\tan x) = \sec^2 x$$

$$\frac{d}{dx}(\cot x) = -\csc^2 x$$

$$\frac{d}{dx}(\sec x) = \sec x \tan x$$

$$\frac{d}{dx}(\csc x) = -\csc x \cot x$$

$$\frac{d}{dx}(\arctan x) = \frac{1}{1 + x^2}$$

$$\frac{d}{dx}(\arcsin x) = \frac{1}{\sqrt{1 - x^2}}$$

$$\frac{d}{dx}(\operatorname{arcsec} x) = \frac{1}{x\sqrt{1 - x^2}}$$

$$\frac{d}{dx}(\ln x) = \frac{1}{x}$$

$$\frac{d}{dx}(e^x) = e^x$$

$$\frac{d}{dx}(\cosh x) = \sinh x$$

$$\frac{d}{dx}(\sinh x) = \cosh x$$

$$\frac{d}{dx}(\tanh x) = \operatorname{sech}^2 x$$

Integrals

$$\int x^n \, dx = \frac{1}{n + 1}x^{n+1} + c, \quad n \neq -1$$

$$\int \frac{1}{x} \, dx = \ln|x| + c$$

$$\int \frac{1}{a^2 + x^2} \, dx = \frac{1}{a}\arctan\left(\frac{x}{a}\right) + c$$

$$\int \frac{1}{\sqrt{a^2 - x^2}} \, dx = \arcsin\left(\frac{x}{|a|}\right) + c$$

$$\int \frac{1}{(a^2 + x^2)^2} \, dx = \frac{x}{2a^2(a^2 + x^2)} + \frac{1}{2a^3}\arctan\left(\frac{x}{a}\right) + c$$

$$\int \sqrt{a^2 - x^2} \, dx = \frac{x}{2}\sqrt{a^2 - x^2} + \frac{a^2}{2}\arcsin\left(\frac{x}{|a|}\right) + c$$

$$\int \frac{x}{a + bx} \, dx = \frac{x}{b} - \frac{a}{b^2}\ln|a + bx| + c$$

$$\int \frac{dx}{\sqrt{x^2 \pm a^2}} = \ln|x + \sqrt{x^2 \pm a^2}| + c$$

$$\int \frac{1}{a^2 - x^2} \, dx = \frac{1}{2a}\ln\left|\frac{x + a}{x - a}\right| + c$$

$$\int \frac{1}{x\sqrt{x^2 - a^2}} \, dx = \frac{1}{|a|}\operatorname{arcsec}\left|\frac{x}{a}\right| + c$$

$$\int \ln x \, dx = x \ln x - x + c$$

$$\int \sqrt{x^2 \pm a^2} \, dx = \frac{x}{2}\sqrt{x^2 \pm a^2} \pm \frac{a^2}{2}\ln|x + \sqrt{x^2 \pm a^2}| + c$$

If $b^2 - 4ac < 0$:
$$\int \frac{dx}{ax^2 + bx + c} = \frac{2}{\sqrt{4ac - b^2}}\arctan\left(\frac{2ax + b}{\sqrt{4ac - b^2}}\right) + c$$

$$\int \sin x \, dx = -\cos x + c$$

$$\int \cos x \, dx = \sin x + c$$

$$\int \tan x \, dx = \ln|\sec x| + c$$

$$\int \sec^2 x \, dx = \tan x + c$$

$$\int \cos^2 x \, dx = \frac{x}{2} + \frac{1}{2}\cos x \sin x + c$$

$$\int \sin^2 x \, dx = \frac{x}{2} - \frac{1}{2}\cos x \sin x + c$$

$$\int \cos^3 x \, dx = \frac{1}{3}\cos^2 x \sin x + \frac{2}{3}\sin x + c$$

$$\int \sec x \, dx = \ln|\sec x + \tan x| + c$$